AA002458

2018 IEEE Computer Society Annual Symposium on VLSI (ISVLSI 2018)

Hong Kong
8 – 11 July 2018

IEEE Catalog Number: CFP18179-POD
ISBN: 978-1-5386-7100-9

**Copyright © 2018 by the Institute of Electrical and Electronics Engineers, Inc.
All Rights Reserved**

Copyright and Reprint Permissions: Abstracting is permitted with credit to the source. Libraries are permitted to photocopy beyond the limit of U.S. copyright law for private use of patrons those articles in this volume that carry a code at the bottom of the first page, provided the per-copy fee indicated in the code is paid through Copyright Clearance Center, 222 Rosewood Drive, Danvers, MA 01923.

For other copying, reprint or republication permission, write to IEEE Copyrights Manager, IEEE Service Center, 445 Hoes Lane, Piscataway, NJ 08854. All rights reserved.

*** *This is a print representation of what appears in the IEEE Digital Library. Some format issues inherent in the e-media version may also appear in this print version.*

IEEE Catalog Number:	CFP18179-POD
ISBN (Print-On-Demand):	978-1-5386-7100-9
ISBN (Online):	978-1-5386-7099-6
ISSN:	2159-3469

Additional Copies of This Publication Are Available From:

Curran Associates, Inc
57 Morehouse Lane
Red Hook, NY 12571 USA
Phone: (845) 758-0400
Fax: (845) 758-2633
E-mail: curran@proceedings.com
Web: www.proceedings.com

2018 IEEE Computer Society Annual Symposium on VLSI (ISVLSI 2018)

Hong Kong
8 – 11 July 2018

IEEE Catalog Number: CFP18179-POD
ISBN: 978-1-5386-7100-9

Proceedings

2018 IEEE Computer Society Annual Symposium on VLSI

ISVLSI 2018

Proceedings

2018 IEEE Computer Society Annual Symposium on VLSI

9–11 July 2018
Hong Kong, Hong Kong

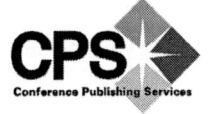

Los Alamitos, California

Washington • Tokyo

2018 IEEE Computer Society
Annual Symposium on VLSI
ISVLSI 2018

Table of Contents

Message from the General Chairs xxii
Message from the Technical Program Chairs xxiii
Conference Committees xxv
Technical Program Committee xxvii
Keynotes xxx
Plenary Talks xxxix

Session 01: Analog and Mixed Signal I

Gyrator-C Based Bandpass Filter with Improved Dynamic Range for Fully Integrated RF Front-End 1
    Lakshmi N S (National Institute of Technology Tiruchirappalli) and
    Bhaskar M (National Institute of Technology Tiruchirappalli)

Replica-Based Low Drop-Out Voltage Regulator with Assistant Power Transistors for Digital VLSI
Systems 6
    Yang Nan (Southern University of Science and Technology, Shenzhen,
    China), Chenchang Zhan (Southern University of Science and Technology,
    Shenzhen, China), Guanhua Wang (Southern University of Science and
    Technology, Shenzhen, China), Linjun He (Southern University of
    Science and Technology, Shenzhen, China), and Han Li (Southern
    University of Science and Technology, Shenzhen, China)

Area Efficient NMOS Based Positive and Negative Voltage Multiplier 10
    Vikas Rana (STMicroelectronics Pvt Ltd, Greater Noida, India)

Session 02: Digital Circuits and FPGA based Design I

Achieving Low Power Classification with Classifier Ensemble 16
    Fanglei Hu (Peking University Shenzhen Graduate School), Min Zhang
    (Peking University Shenzhen Graduate School), and Hailong Jiao (Peking
    University Shenzhen Graduate School)

A Power-Efficient Hybrid Architecture Design for Image Recognition Using CNNs 22
    Jinhang Choi (Pennsylvania State University), Srivatsa Srinivasa
    (Pennsylvania State University), Yasuki Tanabe (Toshiba Corporation),
    Jack Sampson (Pennsylvania State University), and Vijaykrishnan
    Narayanan (Pennsylvania State Universtiy)

Towards Budget-Driven Hardware Optimization for Deep Convolutional Neural Networks Using Stochastic Computing 28

Zhe Li (Syracuse University), Ji Li (University of Southern California), Ao Ren (Syracuse University), Caiwen Ding (Syracuse University), Jeffrey Draper (University of Southern California), Qinru Qiu (Syracuse University), Bo Yuan (City University of New York, City College), and Yanzhi Wang (Syracuse University)

Session 03: Testing, Reliability, and Fault-Tolerance I

Fast Heuristics for Near-Optimal Signal Restoration in Post-Silicon Validation 34

Xiaobang Liu (University of Cincinnati) and Ranga Vemuri (University of Cincinnati)

PGIREM: Reliability-Constrained IR Drop Minimization and Electromigration Assessment of VLSI Power Grid Networks Using Cooperative Coevolution 40

Sukanta Dey (Indian Institute of Technology Guwahati), Satyabrata Dash (Indian Institute of Technology Guwahati), Sukumar Nandi (Indian Institute of Technology Guwahati), and Gaurav Trivedi (Indian Institute of Technology Guwahati)

Silicon Debug with Maximally Expanded Internal Observability Using Nearest Neighbor Algorithm 46

Ankit Jindal (IIT Bombay), Binod Kumar (IIT Bombay), Nitish Jindal (IBM Systems - ISDL, Bengaluru), Masahiro Fujita (University of Tokyo), and Virendra Singh (IIT Bombay)

Session 04: Computer Aided Design and Verification I

Application Specific Networks-on-Chip Synthesis: An Energy Efficient Approach 52

Somayeh Kashi (Iran University of Science and Technology University), Ahmad Patooghy (Boston University, Boston, USA), Dara Rahmatiy (Institute for Research in Fundamental Sciences, Tehran, Iran), Mahdi Fazeli (Iran University of Science and Technology University), and Michel A. Kinsy (Boston University, Boston, USA)

Accurate Models for Optimizing Tapered Microchannel Heat Sinks in 3D ICs 58

Leslie Hwang (University of Illinois at Urbana-Champaign), Beomjin Kwon (Arizona State University), and Martin Wong (University of Illinois at Urbana-Champaign)

Designing and Benchmarking of Double-Row Height Standard Cells 64

Yu-Xiang Chiang (Yuan Ze University), Cheng-Wei Tai (Yuan Ze University), Shang-Rong Fang (Yuan Ze University), Kai-Chun Peng (Yuan Ze University), Yuan-Dar Chung (Yuan Ze University), Jin-Kai Yang (Yuan Ze University), and Rung-Bin Lin (Yuan Ze University)

Session 05: Emerging and Post-CMOS Technologies I

A Dual-Threshold Scheme Along with Security Reinforcement for Energy Efficient Nonvolatile
Processors 70
> Dongqin Zhou (Capital Normal University), Keni Qiu (Capital Normal
> University), Yuanchao Xu (Capital Normal University), Xin Shi
> (Tsinghua University), and Yongpan Liu (Tsinghua University)

A Comprehensive Electro-Optical Model for Silicon Photonic Switches 76
> Xuanqi Chen (Hong Kong University of Science and Technology), Zhifei
> Wang (Hong Kong University of Science and Technology), Yi-Shing Chang
> (Silicon Photonics Product Division, CG/DCG, Intel Corp., USA), Jiang
> Xu (Hong Kong University of Science and Technology), Peng Yang (Hong
> Kong University of Science and Technology), Zhehui Wang (Hong Kong
> University of Science and Technology), and Luan H.K. Duong (Hong Kong
> University of Science and Technology)

CMOS Gates with Second Function 82
> Jan Nevoral (Brno University of Technology, Czech Republic), Richard
> Ruzicka (Brno University of Technology, Czech Republic), and Vaclav
> Simek (Brno University of Technology, Czech Republic)

Session 06: System Design and Security I

TDC: Tagless DRAM Cache 88
> S R Swamy Saranam (Indian Institute of Technology Madras) and Madhu
> Mutyam (Indian Institute of Technology Madras)

CT-Cache: Compressed Tag-Driven Cache Architecture 94
> Haeyoon Cho (Kyungpook National University), Joonho Kong (Kyungpook
> National University), Arslan Munir (Kansas State University), and
> Naresh Kumar Giri (Kansas State University)

High Bandwidth Off-Chip Memory Access Through Hybrid Switching and Inter-Chip Wireless Links 100
> Sri Harsha Gade (Indraprastha Institute of Information Technology
> Delhi), Hemanta Kumar Mondal (CNRS Lab-STICC, University of Southern
> Brittany), and Sujay Deb (Indraprastha Institute of Information
> Technology Delhi)

SPECIAL Session 01: Shall We Jointly Address VLSI Reliability and Security?

Investigating Reliability and Security of Integrated Circuits and Systems 106
> Qiaoyan Yu (University of New Hampshire), Zhiming Zhang (University of
> New Hampshire), and Jaya Dofe (University of New Hampshire)

Reliability and Security in Non-volatile Processors, Two Sides of the Same Coin 112
> Patrick Cronin (University of Delaware), Chengmo Yang (University of
> Delaware), and Yongpan Liu (Tsinghua University)

A Short Survey at the Intersection of Reliability and Security in Processor Architecture Designs 118
> Lake Bu (Boston University, Boston, USA), Miguel Mark (Boston
> University, Boston, USA), and Michel Kinsy (Boston University, Boston,
> USA)

Can Soft Errors be Handled Securely? 124
Senwen Kan (Cypress Semiconductor) and Jennifer Dworak (Southern Methodist University)

Session 07: System Design and Security II

BD-NET: A Multiplication-Less DNN with Binarized Depthwise Separable Convolution 130
Zhezhi He (University of Central Florida), Shaahin Angizi (University of Central Florida), Adnan Siraj Rakin (University of Central Florida), and Deliang Fan (University of Central Florida)

TaiJiNet: Towards Partial Binarized Convolutional Neural Network for Embedded Systems 136
Yingjian Ling (Chongqing University), Kan Zhong (Chongqing University), Yunsong Wu (Chongqing University), Duo Liu (Chongqing University), Jinting Ren (Chongqing University), Renping Liu (Chongqing University), Moming Duan (Chongqing University), Weichen Liu (Nanyang Technological University), and Liang Liang (Chongqing University)

An ECC-Free MLC STT-RAM Based Approximate Memory Design for Multimedia Applications 142
Zihao Liu (Flordia International University), Tao Liu (Florida International University), Jie Guo (Hewlett-Packard Co.), Nansong Wu (Arkansas Tech University), and Wujie Wen (Florida International University)

Robust Timing Attack Countermeasure on Virtual Hardware 148
Kai Yang (University of Florida), Jungmin Park (University of Florida), Mark Tehranipoor (University of Florida), and Swarup Bhunia (University of Florida)

SPECIAL Session 02: Emerging Computing and Memory Technologies at Post-CMOS Era

Towards Theoretical Cost Limit of Stochastic Number Generators for Stochastic Computing 154
Meng Yang (University of Michigan-Shanghai Jiao Tong University Joint Institute), Bingzhe Li (Shanghai Jiao Tong University, China), David J. Lilja (University of Minnesota), Bo Yuan (Shanghai Jiao Tong University, China), and Weikang Qian (University of Michigan-Shanghai Jiao Tong University Joint Institute)

Poster Session

Fully-on-Chip Digitally Assisted LDO Regulator with Improved Regulation and Transient Responses 160
Han Li (Southern University of Science and Technology, Shenzhen, China), Chenchang Zhan (Southern University of Science and Technology, Shenzhen, China), and Ning Zhang (Southern University of Science and Technology, Shenzhen, China)

An Asynchronous Analog to Digital Converter for Surveillance Camera Applications 164
Siddharth R. K. (National Institute of Technology Goa), Sunil R.
(National Institute of Technology Goa), Nithin Kumar Y. B. (National
Institute of Technology Goa), Vasantha M. H. (National Institute of
Technology Goa), and Edoardo Bonizzoni (University of Pavia)

An Integrated MaxFit Genetic Algorithm-SPICE Framework for 2-Stage Op-Amp Design Automation 170
Harsha M. V. (Bangalore University, Bangalore, India) and B. P. Harish
(Bangalore University, Bangalore, India)

Mismatch Resilient 3.5-Bit MDAC with MCS-CFCS 175
Satyajit Mohapatra (IIT Gandhinagar), Hari Shanker Gupta (Space
Applications Centre, Jodhpur Tekra, ISRO, Ahmedabad), and Nihar Ranjan
Mohapatra (IIT Gandhinagar)

Design of Low Power SAR ADC Using Clock Retiming 181
Jalaja S (Bangalore Institute of Technology) and Vijaya Prakash A M
(Bangalore Institute of Technology)

A 375 nA Input Off Current Schmitt Triger LDO for Energy Harvesting IoT Sensors 187
Koichiro Ishibashi (The University of Electro-Communications) and
Shiho Takahashi (The University of Electro-Communications)

Precise Duty Cycle Variation Detection and Self-Calibration System for High-Speed Data Links 191
Karen Khachikyan (Synopsys Armenia CJSC), Abraham Balabanyan (Synopsys
Armenia CJSC), and Hrachya Gumroyan (Synopsys Armenia CJSC)

Parametric Circuit Optimization with Reinforcement Learning 197
Changcheng Tang (Tsinghua University, Institute of Microelectronics),
Zuochang Ye (Tsinghua University, Institute of Microelectronics), and
Yan Wang (Tsinghua University, Institute of Microelectronics)

End-to-End Industrial Study of Retiming 203
Cunxi Yu (EPFL), Chau-Chin Huang (National Taiwan University), Gi-Joon
Nam (IBM T.J. Watson Research Center), Mihir Choudhury (IBM T.J.
Watson Research Center), Victor N. Kravets (IBM T.J. Watson Research
Center), Andrew Sullivan (IBM T.J. Watson Research Center), Maciej
Ciesielski (University of Massachusetts Amherst), and Giovanni De
Micheli (EPFL)

Communication-Aware Module Placement for Power Reduction in Large FPGA Designs 209
Kalindu Herath (Nanyang Technological University, Singapore), Alok
Prakash (Nanyang Technological University, Singapore), Udaree Kanewala
(Nanyang Technological University, Singapore), and Thambipillai
Srikanthan (Nanyang Technological University, Singapore)

A Novel Mixed-Size Fixed-Outline Floorplacement Algorithm 215
Qian Chen (Tsinghua University, Beijing, China) and Sheqin Dong
(Tsinghua University, Beijing, China)

ARCHVerifyr: An Embedded Software-Driven Approach for Architecture Verification 220
Tomás Grimm (Ruhr-University Bochum), Djones Lettnin (Universidade
Federal de Santa Catarina), and Michael Hübner (Ruhr-University
Bochum)

High-Average and Guaranteed Performance for Wireless Networks-on-Chip Architectures 226
Mohammad Baharloo (The University of Tehran, Iran), Ahmad Khonsari (The University of Tehran, Iran), Pouya Shiri (The University of Tehran, Iran), Iman Namdari (The University of Tehran, Iran), and Dara Rahmati (Institute for Research in Fundamental Sciences (IPM), Tehran, Iran)

Hardware Implementation of Reconfigurable Separable Convolution 232
Lei Rao (Xi'an Jiaotong University, Xi'an, China), Bin Zhang (Xi'an Jiaotong University, Xi'an, China), and Jizhong Zhao (Xi'an Jiaotong University, Xi'an, China)

Low Overhead Online Checkpoint for Intermittently Powered Non-volatile FPGAs 238
Xinyi Zhang (University of Pittsburgh), Clay Patterson (Oklahoma State University), Yongpan Liu (Tsinghua University), Chengmo Yang (University of Delaware), Chun Jason Xue (City University of Hong Kong), and Jingtiong Hu (University of Pittsburgh)

Pixel-Parallel Architecture for Neuromorphic Smart Image Sensor with Visual Attention 245
Md Jubaer Hossain Pantho (University of Arkansas), Pankaj Bhowmik (University of Arkansas), and Christophe Bobda (University of Arkansas)

Lightweight ASIC Implementation of AEGIS-128 251
Anubhab Baksi (Nanayang Technological University), Vikramkumar Pudi (Nanayang Technological University), Swagata Mandal (Nanayang Technological University), and Anupam Chattopadhyay (Nanayang Technological University)

Architecture Exploration and Delay Minimization Synthesis for SET-Based Programmable Gate Arrays 257
Chia-Cheng Wu (National Tsing Hua University), Kung-Han Ho (National Chiao Tung University), Juinn-Dar Huang (National Chiao Tung University), and Chun-Yao Wang (National Tsing Hua University)

MRAM-on-FDSOI Integration: A Bit-Cell Perspective 263
Hao Cai (Southeast University, Nanjing, China), You Wang (Beihang University, Beijing, China), Wang Kang (Beihang University, Beijing, China), Lirida Naviner (Télécom-ParisTech, Paris, France), Xinning Liu (Southeast University, Nanjing, China), Jun Yang (Southeast University, Nanjing, China), and Weisheng Zhao (Beihang University, Beijing, China)

High Performance Ternary Multiplier Using CNTFET 269
Subhendu Kumar Sahoo (Birla Institute of Technology & Science, Pilani Hyderabad Campus, India), Krishna Dhoot (Birla Institute of Technology & Science, Pilani Hyderabad Campus, India), and Rasmita Sahoo (Nalla Narasimha Reddy School of Engineering, India)

A Robust Dual Reference Computing-in-Memory Implementation and Design Space Exploration Within STT-MRAM 275
Liuyang Zhang (Beihang University, Beijing, China), Wang Kang (Beihang University, Beijing, China), Hao Cai (Telecom Paristech, University of Paris-Saclay, France), Peng Ouyang (Beihang University, Beijing, China), Lionel Torres (LIRMM/University of Montpellier, France), Youguang Zhang (Beihang University, Beijing, China), Aida Todri-Sanial (LIRMM/University of Montpellier, France), and Weisheng Zhao (Beihang University, Beijing, China)

Biosensing Performance Optimization of DMFET for Fully Filled and Partially Filled Cavity 281
Ankita Porwal (Malaviya National Institute of Technology) and
Chitrakant Sahu (Malaviya National Institute of Technology Jaipur)

A Dynamic Resource Allocation Strategy for NoC Based Multicore Systems with Permanent Faults 287
Suraj Paul (IIEST, Shibpur), Navonil Chatterjee (IIEST, Shibpur), and
Prasun Ghosal (IIEST, Shibpur)

Floorplanning in Graphene Nanoribbon (GNR) Based Circuits 293
Subrata Das (Academy of Technology, Hooghly, India) and Debesh Kumar
Das (Jadavpur University, India)

Generating Safety Guidance for Medical Injection with Three-Compartment Pharmacokinetics Model 299
Cunxi Yu (EPFL), Heinz Riener (EPFL), Francesca Stradolini (EPFL), and
Giovanni De Micheli (EPFL)

A Novel Approach for Nearest Neighbor Realization of 2D Quantum Circuits 305
Anirban Bhattacharjee (Indian Institute of Engineering Science and
Technology Shibpur, India), Chandan Bandyopadhyay (Indian Institute of
Engineering Science and Technology (IIEST) Shibpur, India), Robert
Wille (Johannes Kepler University Linz, Austria), Rolf Drechsler
(University of Bremen & Cyber-Physical Systems, Germany), and Hafizur
Rahaman (Indian Institute of Engineering Science and Technology
Shibpur, India)

A Hardware-Efficient Implementation of CLOC for On-chip Authenticated Encryption 311
Mahmoud A. Elmohr (University of Waterloo, Ontario), Sachin Kumar
(Nanyang Technological University, Singapore), Mustafa Khairallah
(School of Physical and Mathematical Sciences, NTU), and Anupam
Chattopadhyay (Nanyang Technology University, Singapore)

0.9 to 2.5 GHz Sub-Sampling Receiver Architecture for Dynamically Reconfigurable SDR 316
Ajinkya Kale (International Institute of Information Technology
Hyderabad, India and Carinthia University of Applied Sciences,
Villach, Austria), Johannes Sturm (International Institute of
Information Technology Hyderabad, India and Carinthia University of
Applied Sciences, Villach, Austria), and Vijaya Sankara Rao
Pasupureddi (University of Hyderabad, India)

Hardware Obfuscation Using Strong PUFs 321
Soroush Khaleghi (University of Illinois at Chicago) and Wenjing Rao
(University of Illinois at Chicago)

Multi-block APUF with 2-Level Voltage Supply 327
Yunxi Guo (Iowa State University), Timothy Dee (Iowa State
University), and Akhilesh Tyagi (Iowa State University)

Write Energy Optimization for STT-MRAM Cache with Data Pattern Characterization 333
Bi Wu (Beihang University, Beijing, China), Xiaolong Zhang (Beihang
University, Beijing, China), Yuanqing Cheng (Beihang University,
Beijing, China), Zhaohao Wang (Beihang University, Beijing, China),
Dijun Liu (China Academy of Telecommunication Technology (CATT)),
Youguang Zhang (Beihang University, Beijing, China), and Weisheng Zhao
(Beihang University, Beijing, China)

Time Stamp Based Scheduling for Energy Harvesting Systems with Hybrid Nonvolatile Hardware Support 339
  Xin Shi (Tsinghua University), Tongda Wu (Tsinghua University), Keni
  Qiu (Capital Normal University), Huazhong Yang (Tsinghua University),
  and Yongpan Liu (Tsinghua University)

EETD: An Energy Efficient Design for Runtime Hardware Trojan Detection in Untrusted Network-on-Chip ... 345
  Mubashir Hussain (University of New South Wales, Australia), Amin
  Malekpour (University of New South Wales, Australia), Hui Guo
  (University of New South Wales, Australia), and Sri Parameswaran
  (University of New South Wales, Australia)

Combining Symbolic Computer Algebra and Boolean Satisfiability for Automatic Debugging and Fixing of
Complex Multipliers 351
  Alireza Mahzoon (University of Bremen), Daniel Große (University of
  Bremen/DFKI), and Rolf Drechsler (University of Bremen/DFKI)

Enhancing Lifetime of PCM-Based Main Memory with Efficient Recovery of Stuck-at Faults 357
  Marjan Asadinia (University of Arkansas) and Christophe Bobda
  (University of Arkansas)

Student Research Forum

Guessing Your PIN Right: Unlocking Smartphones with Publicly Available Sensor Data 363
  David Berend (Nanyang Technological University, Singapore, University
  of Applied Sciences Wiesbaden, Germany), Bernhard Jungk (Cyber and
  Information Security Group, Fraunhofer Singapore), and Shivam Bhasin
  (Nanyang Technological University, Singapore)

Synthesis, Technology Mapping and Optimization for Emerging Technologies 369
  Debjyoti Bhattacharjee (Nanyang Technological Univeristy) and Anupam
  Chattopadhyay (Nanyang Technological University)

Logic Synthesis for In-memory Computing Using Resistive Memories 375
  Saeideh Shirinzadeh (University of Bremen) and Rolf Drechsler
  (University of Bremen; Cyber-Physical Systems, DFKI GmbH)

Minimalistic Perspective to Public Key Implementations on FPGA 381
  Debapriya Basu Roy (Indian Institute of Technology Kharagpur) and
  Debdeep Mukhopadhyay (Indian Institute of Technology Kharagpur)

Development of High-Stability, Low-Leakage 6Tr-SRAM with Single Data Line and Single Power Supply
Using SOTB Process 387
  Shin Miyamoto (Nihon University) and Nobuaki Kobayashi (Nihon
  University)

Exploiting Principle of Moving Target Defense to Secure FPGA Systems 393
  Zhiming Zhang (University of New Hampshire) and Qiaoyan Yu (University
  of New Hampshire)

Session 08: System Design and Security III

A Highly Flexible Lightweight and High Speed True Random Number Generator on FPGA 399
  Faqiang Mei (Nanjing University of Aeronautics and Astronautics,), Lei Zhang (Nanjing University of Aeronautics and Astronautics,), Chongyan Gu (Queen's University Belfast), Yuan Cao (Hohai University), Chenghua Wang (Nanjing University of Aeronautics and Astronautics,), and Weiqiang Liu (Nanjing University of Aeronautics and Astronautics,)

LUT-Lock: A Novel LUT-Based Logic Obfuscation for FPGA-Bitstream and ASIC-Hardware Protection 405
  Hadi Mardani Kamali (George Mason University), Kimia Zamiri Azar (George Mason University), Kris Gaj (George Mason University), Houman Homayoun (George Mason University), and Avesta Sasan (George Mason University)

ArtiFact: Architecture and CAD Flow for Efficient Formal Verification of SoC Security Policies 411
  Atul Prasad Deb Nath (University of Florida), Swarup Bhunia (University of Florida), and Sandip Ray (University of Florida)

Session 09: Computer Aided Design and Verification II

Identifying Lithography Weak-Points of Standard Cells with Partial Pattern Matching 417
  Yongfu Li (GLOBALFOUNDRIES Singapore Pte. Ltd.), I-Lun Tseng (GLOBALFOUNDRIES Singapore Pte. Ltd.), Zhao Chuan Lee (GLOBALFOUNDRIES Singapore Pte. Ltd.), Valerio Perez (GLOBALFOUNDRIES Singapore Pte. Ltd.), Vikas Tripathi (GLOBALFOUNDRIES Singapore Pte. Ltd.), and Jonathan Yoong Seang Ong (GLOBALFOUNDRIES Singapore Pte. Ltd.)

Feature Based Coverage Analysis of AMS Circuits 423
  Antara Ain (Indian Institute of Technology Kharagpur), Akshay Mambakam (Indian Institute of Technology Kharagpur), and Pallab Dasgupta (Indian Institute of Technology Kharagpur)

SAT Encoding-Based Verification of Sneak Path Problem in Via-Switch FPGA 429
  Ryutaro Doi (Osaka University) and Masanori Hashimoto (Osaka University)

Session 10: Emerging and Post-CMOS Technologies II

RRAM Based Buffer Design for Energy Efficient CNN Accelerator 435
  Kaiyuan Guo (Tsinghua University, China), Jincheng Yu (Tsinghua University, China), Xuefei Ning (Tsinghua University, China), Yiming Hu (Tsinghua University, China), Yu Wang (Tsinghua University, China), and Huazhong Yang (Tsinghua University, China)

A Mixed-Mode Neuron with On-chip Tunability for Generic Use in Memristive Neuromorphic Systems 441
  Sagarvarma Sayyaparaju (University of Tennessee, Knoxville), Ryan Weiss (University of Tennessee, Knoxville), and Garrett S. Rose (University of Tennessee, Knoxville)

Harnessing Emerging Technology for Compute-in-Memory Support 447

Nicholas Jao (The Pennsylvania State University), Akshay Krishna Ramanathan (The Pennsylvania State University), Srivatsa Srinivasa (The Pennsylvania State University), Sumitha George (The Pennsylvania State University), John Sampson (The Pennsylvania State University), and Vijaykrishnan Narayanan (The Pennsylvania State University)

Session 11: Analog and Mixed Signal II

91dB Dynamic Range 9.5nW Low Pass Filter for Bio-Medical Applications 453

Jayaram Reddy M K (National Institute of Technology Karnataka, Surathkal), Sreenivasulu Polineni (National Institute of Technology Karnataka, Surathkal), and Laxminidhi Tonse (National Institute of Technology Karnataka, Surathkal)

A Low Power, High Gain Semi-Floating Gate Amplifier for Resonating Sensors Front-End 458

Luca Marchetti (University College of Southeast Norway), Yngvar Berg (University College of Southeast Norway), and Mehdi Azadmehr (University College of Southeast Norway)

A High-Efficient Current-Mode PWM DC-DC Buck Converter Using Dynamic Frequency Scaling 464

Ankit Rehani (IIIT Delhi), Sujay Deb (IIIT Delhi), Pydi Ganga Bahubalindruni (IIIT Delhi), Bhavin Odedara (Western Digital), and Srikanth Bojja (Western Digital)

Session 12: System Design and Security IV

ReRise: An Adversarial Example Restoration System for Neuromorphic Computing Security 470

Chenchen Liu (Clarkson University), Qide Dong (Exacloud Inc.), Fuxun Yu (George Mason University), and Xiang Chen (George Mason University)

MAT: A Multi-strength Adversarial Training Method to Mitigate Adversarial Attacks 476

Chang Song (Duke University), Hsin-Pai Cheng (Duke University), Huanrui Yang (Duke University), Sicheng Li (Hewlett Packard Labs), Chunpeng Wu (Duke University), Qing Wu (Air Force Research Lab), Yiran Chen (Duke University), and Hai Li (Duke University)

Hu-Fu: Hardware and Software Collaborative Attack Framework Against Neural Networks 482

Wenshuo Li (Tsinghua University; Beijing National Research Center for Information Science and Technology), Jincheng Yu (Tsinghua University; Beijing National Research Center for Information Science and Technology), Xuefei Ning (Tsinghua University; Beijing National Research Center for Information Science and Technology), Pengjun Wang (Tsinghua University; Beijing National Research Center for Information Science and Technology), Qi Wei (Tsinghua University; Beijing National Research Center for Information Science and Technology), Yu Wang (Tsinghua University; Beijing National Research Center for Information Science and Technology), and Huazhong Yang (Tsinghua University Beijing National Research Center for Information Science and Technology)

SPECIAL Session 03: Essential Keys to Manufacturability: Layout Features and Lithography Technologies

Sparse VLSI Layout Feature Extraction: A Dictionary Learning Approach 488
Hao Geng (The Chinese University of Hong Kong), Haoyu Yang (The Chinese University of Hong Kong), Bei Yu (The Chinese University of Hong Kong), Xingquan Li (Fuzhou University), and Xuan Zeng (Fudan University, China)

Pattern Similarity Metrics for Layout Pattern Classification and Their Validity Analysis by Lithographic Responses 494
Atsushi Takahashi (Tokyo Institute of Technology), Shimpei Sato (Tokyo Institute of Technology), Hiroki Ogura (Tokyo Institute of Technology), Yu-Min Sung (National Tsing Hua University), and Ting-Chi Wang (National Tsing Hua University)

Recent Research and Challenges in Multiple Patterning Layout Decomposition 498
Iris Hui-Ru Jiang (National Taiwan University) and Hua-Yu Chang (Synopsys, Inc.)

Guiding Template-Induced Design Challenges in DSA-MP Lithography 500
Shao-Yun Fang (National Taiwan University of Science and Technology) and Kuo-Hao Wu (National Taiwan University of Science and Technology)

Session 13: Digital Circuits and FPGA Based Designs II

FPAP: A Folded Architecture for Efficient Computing of Convolutional Neural Networks 503
Yizhi Wang (Nanjing University, P.R. China), Jun Lin (Nanjing University, P.R. China), and Zhongfeng Wang (Nanjing University, P.R. China)

Hyperdrive: A Systolically Scalable Binary-Weight CNN Inference Engine for mW IoT End-Nodes 509
Renzo Andri (Integrated Systems Laboratory, ETH Zurich, Zurich, Switzerland), Lukas Cavigelli (Integrated Systems Laboratory, ETH Zurich, Zurich, Switzerland), Davide Rossi (DEI, University of Bologna, Bologna, Italy), and Luca Benini (Integrated Systems Laboratory, ETH Zurich, Zurich, Switzerland)

An Optimized Architecture For Decomposed Convolutional Neural Networks 516
Fangxuan Sun (Nanjing University), Jun Lin (Nanjing University), and Zhongfeng Wang (Nanjing University)

Interconnect Delay Analysis for RRAM Crossbar Based FPGA 522
Masanori Hashimoto (Osaka University), Yuki Nakazawa (Osaka University), Ryutaro Doi (Osaka University), and Jaehoon Yu (Osaka University)

SPECIAL Session 04: Emerging Trends in Energy Efficient and Secure Neural Network Acceleration

Enhancing the Robustness of Deep Neural Networks from "Smart" Compression 528
Tao Liu (Florida International University), Zihao Liu (Florida International University), Qi Liu (Florida International University), and Wujie Wen (Florida International University)

Accelerating Low Bit-Width Deep Convolution Neural Network in MRAM 533
Zhezhi He (University of Central Florida), Shaahin Angizi (University
of Central Florida), and Deliang Fan (University of Central Florida)

Emerging Neuromorphic Computing Paradigms Exploring Magnetic Skyrmions 539
Sai Li (Beihang University, Beijing, China), Wang Kang (Beihang
University, Beijing, China), Xing Chen (Beihang University, Beijing,
China), Jinyu Bai (Beihang University, Beijing, China), Biao Pan
(Beihang University, Beijing, China), Youguang Zhang (Beihang
University, Beijing, China), and Weisheng Zhao (Beihang University,
Beijing, China)

Session 14: System Design and Security V

Security-Driven Task Scheduling for Multiprocessor System-on-Chips with Performance Constraints 545
Nan Wang (East China University of Science and Technology), Manting
Yao (East China University of Science and Technology), Dongxu Jiang
(East China University of Science and Technology), Song Chen
(University of Science and Technology of China), and Yu Zhu (East
China University of Science and Technology)

A Hardware Monitor to Protect Linux System Calls 551
George Provelengios (University of Massachusetts Amherst), Arman
Pouraghily (University of Massachusetts Amherst), Russell Tessier
(University of Massachusetts Amherst), and Tilman Wolf (University of
Massachusetts Amherst)

Towards Dynamic Execution Environment for System Security Protection Against Hardware Flaws 557
Kenneth Schmitz (DFKI GmbH), Oliver Keszocze (University of Bremen,
DFKI GmbH), Jurij Schmidt (University of Bremen, DFKI GmbH), Daniel
Große (University of Bremen, DFKI GmbH), and Rolf Drechsler
(University of Bremen, DFKI GmbH)

Session 15: Digital Circuits and FPGA Based Designs III

A Fast and Effective Memristor-Based Method for Finding Approximate Eigenvalues and Eigenvectors of
Non-negative Matrices 563
Chenghong Wang (Syracuse University - Syracuse, NY), Zeinab S. Jalali
(Syracuse University - Syracuse, NY), Caiwen Ding (Syracuse University
- Syracuse, NY), Yanzhi Wang (Syracuse University - Syracuse, NY), and
Sucheta Soundarajan (Syracuse University - Syracuse, NY)

A Low-Power and Small-Area Multiplier for Accuracy-Scalable Approximate Computing 569
Hiroyuki Baba (Fukuoka University), Tongxin Yang (Fukuoka University),
Masahiro Inoue (Fukuoka University), Kaori Tajima (Fukuoka
University), Tomoaki Ukezono (Fukuoka University), and Toshinori Sato
(Fukuoka University)

A Hardware/Software Co-design Method for Approximate Semi-Supervised K-Means Clustering 575
Pengfei Huang Huang (Nanjing University of Aeronautics and
Astronautics), Chenghua Wang (Nanjing University of Aeronautics and
Astronautics), Ruizhe Ma (Nanjing University of Aeronautics and
Astronautics), Weiqiang Liu (Nanjing University of Aeronautics and
Astronautics), and Fabrizio Lombardi (Northeastern University)

xvi

SPECIAL Session 05: Intelligent Methods & Techniques for Reliable and Adaptive Multicore/Manycore System

Robustness for Smart Cyber Physical Systems and Internet-of-Things: From Adaptive Robustness Methods to Reliability and Security for Machine Learning 581

Florian Kriebel (Vienna University of Technology (TU Wien)), Semeen Rehman (Vienna University of Technology (TU Wien)), Muhammad Abdullah Hanif (Vienna University of Technology (TU Wien)), Faiq Khalid (Vienna University of Technology (TU Wien)), and Muhammad Shafique (Vienna University of Technology (TU Wien))

On How to Efficiently Implement Deep Learning Algorithms on PYNQ Platform 587

Luca Stornaiuolo (Politecnico di Milano), Marco Santambrogio (Politecnico di Milano), and Donatella Sciuto (Politecnico di Milano)

Enabling Reliable High Throughput On-chip Wireless Communication for Many Core Architectures 591

Sri Harsha Gade (Indraprastha Institute of Information Technology Delhi), Mitali Sinha (Indraprastha Institute of Information Technology Delhi), Sidhartha Sankar Rout (Indraprastha Institute of Information Technology Delhi), and Sujay Deb (Indraprastha Institute of Information Technology Delhi)

Session 16: Testing, Reliability, and Fault-Tolerance II

Predicting the Tolerance of Extreme Electromagnetic Interference on MOSFETs 597

Nishchay H. Sule (University of New Mexico), Troy Powell (University of New Mexico), Sameer Hemmady (University of New Mexico), and Payman Zarkesh-Ha (University of New Mexico)

Enhancing Observability for Post-Silicon Debug with On-chip Communication Monitors 602

Yuting Cao (University of South Florida), Hernan Palombo (University of South Florida), Sandip Ray (University of Florida), and Hao Zheng (University of South Florida)

Performance Enhancement of Split Length Compensated Operational Amplifiers 608

Donel Anto (National Institute of Technology Goa, India), Abhijeet D. Taralkar (National Institute of Technology Goa, India), Nithin Kumar Y B (National Institute of Technology Goa, India), and Vasantha M.H. (National Institute of Technology Goa, India)

Session 17: System Design and Security VI

Design-Based Fingerprinting Using Side-Channel Power Analysis for Protection Against IC Piracy 614

James Shey (USN Naval Academy/University of Maryland, Baltimore County), Naghmeh Karimi (University of Maryland, Baltimore County), Ryan Robucci (University of Maryland, Baltimore County), and Chintan Patel (University of Maryland, Baltimore County)

PPAP and iPPAP: PLL-Based Protection Against Physical Attacks 620

Prasanna Ravi (TEMASEK LABS, NTU), Shivam Bhasin (TEMASEK LABS, NTU), Jakub Breier (TEMASEK LABS, NTU), and Anupam Chattopadhyay (School of Computer Science and Engineering, Nanyang Technological University, Singapore)

Mystic: Mystifying IP Cores Using an Always-ON FSM Obfuscation Method 626

Ahmad Patooghy (Boston University), Ehsan Aerabi (Boston University), Hamidreza Rezaei (Boston University), Miguel Mark (Boston University), Mahdi Fazeli (Boston University), and Michel A. Kinsy (Boston University)

Session 18: Digital Circuits and FPGA Based Designs IV

FPGA-Based Controllers for Compact Low Power Refreshable Braille Display 632

Suman Muralikrishnan Adhepalli (IIT Delhi), Pulkit Sapra (IIT Delhi), Saurabh Agrawal (IIT Delhi), Piyush Chanana (IIT Delhi), M. Balakrishnan (IIT Delhi), and P.V.M. Rao (IIT Delhi)

Very Large-Scale and Node-Heavy Graph Analytics with Heterogeneous FPGA+CPU Computing Platform 638

Yu Zou (University of Central Florida) and Mingjie Lin (University of Central Florida)

On-chip Data Security Against Untrustworthy Software and Hardware IPs in Embedded Systems 644

SreeCharan Gundabolu (Villanova University) and Xiaofang Wang (Villanova University)

SPECIAL Session 06: Large Scale Integration (mVLSI): Recent Developments and Upcoming Challenges

Design Automation and Test for Flow-Based Biochips: Past Successes and Future Challenges 650

Tsung-Yi Ho (National Tsing Hua University)

Multi-target Many-Reactant Sample Preparation for Reactant Minimization on Microfluidic Biochips 655

Yung-Chun Lei (National Chiao Tung University), Tien-Kuo Lin (National Chiao Tung University), and Juinn-Dar Huang (National Chiao Tung University)

More Effective Randomly-Designed Microfluidics 660

Weiqing Ji (Tsinghua University), Tsung-Yi Ho (National Tsing Hua University), and Hailong Yao (Tsinghua University)

Accelerating Simulation of Particle Trajectories in Microfluidic Devices by Constructing a Cloud Database 666

Junchao Wang (Hangzhou Dianzi University, China), Lingxuan Fu (Hangzhou Dianzi University, China), Liyang Yu (Hangzhou Dianzi University, China), Xiwei Huang (Hangzhou Dianzi University, China), Philip Brisk (University of California Riverside, USA), and William H. Grover (University of California Riverside, USA)

SPECIAL Session 07: Secure Hardware Design for Distributed Agents

PUF-Based Secure Test Wrapper for SoC Testing 672
  Sudeendra Kumar K (National Institute of Technology, Rourkela),
  Saurabh Seth (National Institute of Technology, Rourkela), Sauvagya
  Sahoo (National Institute of Technology, Rourkela), Abhishek Mahapatra
  (National Institute of Technology, Rourkela), Ayas Kanta Swain
  (National Institute of Technology, Rourkela), and Kamalakanta
  Mahapatra (National Institute of Technology, Rourkela)

Detection of Sequential Trojans in Embedded System Designs Without Scan Chains 678
  Pranav Dharmadhikari (University of Cincinnati), Akhilesh Raju (Xilinx
  Inc.), and Ranga Vemuri (University of Cincinnati)

Designing for Security Within and Between IoT Devices 684
  Mike Borowczak (University of Wyoming), Rafer Cooley (University of
  Wyoming), and Shaya Wolf (University of Wyoming)

A Two-Tiered Heterogeneous and Reconfigurable Application Processor for Future Internet of Things 690
  Prasanna Kansakar (Kansas State University) and Arslan Munir (Kansas
  State University)

SPECIAL Session 08: Embedded Multi-Core in Automotive and I4.0

SPECIAL Session 09: Energy Efficient and Hardware Secured Architectures for Smart Electronics I

Solar Cell Based Physically Unclonable Function for Cybersecurity in IoT Devices 697
  S. Dinesh Kumar (University of Kentucky), Carson Labrado (University
  of Kentucky, Lexington, KY, USA), Riasad Badhan (University of
  Kentucky, Lexington, KY, USA), Himanshu Thapliyal (University of
  Kentucky, Lexington, KY, USA), and Vijay Singh (University of
  Kentucky, Lexington, KY, USA)

Designing Scalable Hybrid Wireless NoC for GPGPUs 703
  Hui Zhao (University of North Texas), Xianwei Cheng (University of
  North Texas), Saraju P. Mohanty (University of North Texas), and Juan
  Fang (Beijing University of Technology)

Functional Obfuscation of DSP Cores Using Robust Logic Locking and Encryption 709
  Anirban Sengupta (Indian Institute of Technology Indore) and Saraju P
  Mohanty (University of North Texas, USA)

SPECIAL Session 10: Timing in the Nanometer Era

Timing Macro Modeling for Efficient Hierarchical Timing Analysis 714
  Iris Hui-Ru Jiang (National Taiwan University) and Pei-Yu Lee
  (National Chiao Tung University)

Timing with Virtual Signal Synchronization for Circuit Performance and Netlist Security 715
  Grace Li Zhang (Technical University of Munich (TUM)), Bing Li
  (Technical University of Munich (TUM)), and Ulf Schlichtmann
  (Technical University of Munich (TUM))

SPECIAL Session 11: Attacking Dynamic Optimizations in the Era of Complex Heterogeneous Multi-core Computing I

Realizing Closed-Loop, Online Tuning and Control for Configurable-Cache Embedded Systems: Progress and Challenges 719
Islam Badreldin (University of Florida), Ann Gordon-Ross (University of Florida), Tosiron Adegbija (University of Arizona), and Mohamad Hammam Alsafrjalaniz (University of Miami)

An FPGA-Based Brain Computer Interfacing Using Compressive Sensing and Machine Learning 726
Ritu Ranjan Shrivastwa (Nanyang Technological University), Vikramkumar Pudi (Nanyang Technological University), and Anupam Chattopadhyay (Nanyang Technological University)

SPECIAL Session 12: Energy Efficient and Hardware Secured Architectures for Smart Electronics II

Obfuscation of Fault Secured DSP Design Through Hybrid Transformation 732
Anirban Sengupta (Indian Institute of Technology Indore), Shubha Neema (Indian Institute of Technology Indore), Pallabi Sarkar (Jadavpur University), Sri Harsha P (Indian Institute of Technology Indore), Saraju P Mohanty (University of North Texas), and Mrinal Kanti Naskar (Jadavpur University)

Run Time Mitigation of Performance Degradation Hardware Trojan Attacks in Network on Chip 738
Manoj Kumar JYV (National Institute of Technology, Rourkela, India), Ayas Kanta Swain (National Institute of Technology, Rourkela, India), Sudeendra Kumar K (National Institute of Technology, Rourkela, India), Sauvagya Ranjan Sahoo (National Institute of Technology, Rourkela, India), and Kamalakanta Mahapatra (National Institute of Technology, Rourkela, India)

Exploration on Routing Configuration of HNoC with Reasonable Energy Consumption 744
Juan Fang (Beijing University of Technology), Zeqing Chang (Beijing University of Technology), Yanjin Cheng (Beijing University of Technology), and Hui Zhao (University of North Texas)

SPECIAL Session 13: Design Using Emerging Devices

Nonvolatile Memory and Computing Using Emerging Ferroelectric Transistors 750
Xueqing Li (Tsinghua University) and Longqiang Lai (Tsinghua University)

SPECIAL Session 14: Attacking Dynamic Optimizations in the Era of Complex Heterogeneous Multi-core Computing II

Software Support for Heterogeneous Computing 756
Siqi Wang (National University of Singapore), Alok Prakash (Nanyang Technological University, Singapore), and Tulika Mitra (National University of Singapore, Singapore)

Predictive Modeling for CPU, GPU, and FPGA Performance and Power Consumption: A Survey 763
Kenneth O'Neal (University of California, Riverside) and Philip Brisk
(University of California, Riverside)

Author Index **769**

Message from the General Chairs

It is our great pleasure to welcome all the participants to the 2018 IEEE Computer Society Annual Symposium on VLSI (ISVLSI, http://www.isvlsi.org/) held at Hong Kong, China. The main goal of ISVLSI is to explore emerging trends and novel ideas and concepts in the area of VLSI and provide a platform for both academic and industrial researchers to interact under one roof for research and development which may lead to realization of efficient, robust, and secure VLSI circuits and systems. ISVLSI has been initiated as a sponsored meeting of Technical Committee on VLSI (TCVLSI, http://www.ieee-tcvlsi.org/), of IEEE Computer Society (IEEE-CS) and we are sure that from the active co-operation of all volunteers ISVLSI got an excellent start. It may be noted that TCVLSI is one among 26 technical committees/councils of IEEE-CS which endorse different meetings in the scope of IEEE-CS. TCVLSI endorses a league of successful meetings such as ARITH, ASAP, iNIS, IWLS, and SLIP. ISVLSI 2018 is sponsored by IEEE-CS through TCVLSI and technically co-sponsored by IEEE-CAS as well as IEEE-CEDA. ISVLSI 2018 proceedings is published by IEEE-CS conference publication services (CPS). ISVLSI has attracted attendees from all over the globe. We hope that ISVLSI will continue attracting renowned people from various parts of the globe and continue serving the community in its future years.

ISVLSI 2018 will be hosted at the Hong Kong Polytechnic University. Hong Kong, located at the mouth of the Pearl River Delta in Southern China and bordered by the Shenzhen Special Economic Zone to the north, is a major financial and services center featuring state-of-the-art infrastructure and highly efficient business services. Its stock market is Asia's third-largest one in terms of market capitalization, only behind the Tokyo Stock Exchange and Shanghai Stock Exchange, and the sixth largest in the world. Frequently described as a place where "East meets West", Hong Kong is a cosmopolitan metropolis where traditional Chinese culture blends perfectly with Western culture. Hong Kong is also one global hub for higher education, with one of the world's most impressive concentrations of internationally renowned institutions. We hope that you will spend several days and enjoy the historical and modern monuments of the city and its surrounding regions.

The general chairs would like to thank the steering committee for supporting ISVLSI 2018. We would like to thank the Hong Kong Polytechnic University and other sponsors in helping ISVLSI 2018. ISVLSI 2018 will have 6 keynotes, 5 plenary talks and 32 sessions from high quality researchers around the globe. We would like to specifically thank the keynote and speakers for their support and exciting talks to the ISVLSI 2018 attendee. We would like to specifically thank the program chairs, Hai (Helen) Li, Yu Wang, and Wujie Wen for their excellent job in selecting quality papers for presentations. We would sincerely thank the special session chairs, publication chair, finance chairs, web chair, publicity chairs, local arrangement chair, student symposium chair, registration chairs, and all other active volunteers, for their fantastic job in running the symposium. We would like to thank the sponsors for supporting ISVLSI 2018.

General Co-Chair, **Wei Zhang**
*Hong Kong University of
Science and Technology
Hong Kong*
wei.zhang@ust.hk

General Co-Chair, **Chun Jason Xue**
*City University of Hong Kong
Hong Kong*
jasonxue@cityu.edu.hk

General Co-Chair, **Zili Shao**
*Hong Kong Polytechnic
University, Hong Kong*
zili.shao@polyu.edu.hk

Message from the Technical Program Chairs

It is with great pleasure that we welcome you to the 17th IEEE Computer-Society Symposium on VLSI (ISVLSI 2018, http://www.eng.ucy.ac.cy/theocharides/isvlsi18/) in the city, Hong Kong SAR, China. ISVLSI is a sister conference of a league of successful meetings such as ARITH, ASAP, iNIS, IWLS, and SLIP which are sponsored by the Technical Committee on VLSI (TCVLSI, http://www.ieee-tcvlsi.org/), of IEEE Computer Society (IEEE-CS). ISVLSI explores emerging trends and novel ideas and concepts in the area of VLSI. Over more than a decade the ISVLSI has been a unique forum promoting multidisciplinary research and new visionary approaches in the area of VLSI. The ISVLSI brings together leading scientists and researchers from academia and industry. The papers from this symposium have been published as the special issue to top archival journals and for the ISVLSI 2018 selected papers will be invited for consideration in special issues in the IEEE Transactions on Nanotechnology (TNANO) and IEEE Consumer Electronics Magazine. This is one of the facts that indicate a very high quality of the ISVLSI papers, and we are determined to keep a strong emphasis on this critical aspect of any conference. The symposium proceedings are published by IEEE-CS conference publication services (CPS).

ISVLSI 2018 covers a range of topics: from VLSI circuits, systems and design methods to system level design and system-on-chip (SoC) issues, to bringing VLSI experience to new areas, architectures, and technologies. Future design methodologies as well as new Electronic Design Automation (EDA) tools to support them will also be the key topics. ISVLSI 2018 papers are in the following 6 tracks:
1) Analog and Mixed-Signal Circuits (AMS)
2) Computer-Aided Design and Verification (CAD)
3) Digital Circuits and FPGA based Designs (DCF)
4) Emerging and Post-CMOS Technologies (EPT)
5) System Design and Security (SDS)
6) Testing, Reliability, and Fault-Tolerance (TRF)

ISVLSI 2018 papers can be divided into two separate categories: regular papers and special session papers. The regular session papers have been selected after a rigorous review process. The special session papers have been by invitation from the established researchers from different areas of VLSI, which have also been reviewed by special session chairs and individual special session proposers. Due to the time constraints the papers are either selected oral presentation or poster presentations. However, both oral and poster presentation papers got 6-page budget in the proceedings. This is one of the unique aspects that the accepted papers get similar importance in the proceedings. In addition, ISVLSI 2018 also has a Student Research Forum (SRF) in which the papers are presented as posters and appear in proceedings as 6-page papers. The SRF papers have gone through a different review process by the SRF chairs and other volunteers.

The ISVLSI 2018 has received a very good response for the manuscript submission. The submissions were received from all parts of the globe, with major submissions from United States of America (USA), India, China, Singapore, Japan, Taiwan and Switzerland. Due to time constraints in the 3-day event, ISVLSI 2018 could only accept limited number of papers of high quality. ISVLSI 2018 received 192 submissions, out of these **54** high quality papers are accepted for oral presentation and the proceedings. These are divided into **18** oral sessions. Thus making an acceptance ratio of 29%. This shows the quality and competitiveness of the conference. All submitted papers had undergone double-blind-review process by a strong team of leading experts from around the globe in respective fields and the program committee members and additional reviewers. There are **50** special session oral presentations from prominent authors divided into **14** sessions. In addition, a poster session contains **35** posters from general pool and **6** posters Student Research Forum will enrich the technical discussions.

The success of this magnitude is not possible without since help from various volunteers. We wish to express our sincere appreciation to the hardworking track chairs, dedicated members of the Technical Program Committee and additional reviewers. We also thank the authors and invited speakers for their contributions to an outstanding technical program. Our special thanks go to the Steering Committee and Organizing Committee members of ISVLSI 2018 for their support and cooperation. We acknowledge the high quality editing work of production editor of the IEEE Computer Society Conference Publishing Services for the high-quality and timely production of the ISVLSI 2018 Proceedings.

We wish you a very productive ISVLSI 2018 and hope you will find papers presented at the ISVLSI 2018 to be a valuable source of reference for your current and future research. We hope your stay during the ISVLSI 2018 at Hong Kong, will be an enjoyable experience and help in professional networking.

We also look forward to your participation in ISVLSI 2019 next year to be held at Miami, Florida, USA.

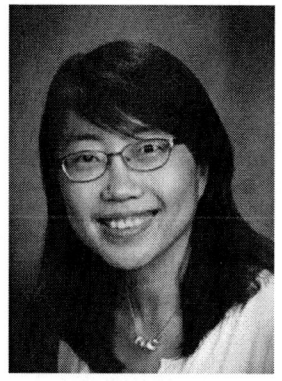  

Program Chair
Hai (Helen) Li
Duke University, USA
hai.li@duke.edu

Program Chair
Yu Wang
Tsinghua University, China
yu-wang@tsinghua.edu.cn

Program Chair
Wujie Wen. Florida
International University, USA
wwen@fiu.edu

ISVLSI 2018 Organizing Committee

General Chairs
Wei Zhang, *Hong Kong Univ. of Science and Technology*
Jason Xue, *City University of Hong Kong*
Zili Shao, *Hong Kong Polytechnic University*

TPC Chairs
Hai Li, *Duke University, USA*
Yu Wang, *Tsinghua University, China*
Wujie Wen, *Florida International Univ., USA*

Special Session Chairs
Bei Yu, *The Chinese University of Hong Kong*
Yuan-Hao Chang, *Academia Sinica, Taiwan*

Web Chair
Theocharis Theocharides, *University of Cyprus, Cyprus*

Student Research Forum Chairs
Anupam Chattopadhyay, *Nanyang Technological University, Singapore*
Zheng Wang, *Shenzhen Institutes of Advanced Technology (SIAT) of the Chinese Academy of Science (CAS)*

Publication Chairs
Mahdi Nikdast, *Colorado State University, USA*
Chenchen Liu, *Clarkson University, USA*

Publicity Chairs
Guangyu Sun, *Peking University, China*
Muhammad Shafique, *Vienna University of Technology, Austria*
Jingtong Hu, *University of Pittsburgh, USA*
Masaaki Kondo, *The University of Tokyo, Japan*
Chun-Yi Lee, *National Tsinghua University, Taiwan*
Dhruva Ghai, *Oriental University, India*

Financial Chair
Duo Liu, *Chongqing University, China*

Registration Chair
Weichen Liu, *Nanyang Technological University, Singapore*

Local Arrangement Chairs
Nan Guan, *Hong Kong Polytechnic University*
Ray Chak-Chung Cheung, *City University of Hong Kong*

Industrial Liaison Chairs
Wei Zhang, *Hong Kong University of Science and Technology, China*
Jürgen Becker, *Karlsruhe Institute of Technology, Germany*
Zhihong Wu, *Tongji University, China*

Steering Committee

Chair
Jürgen Becker, *Karlsruhe Institute of Technology, Germany*

Vice-Chair
Saraju P. Mohanty, *University of North Texas, USA*

Hai (Helen) Li, *Duke University, USA*
Lionel Torres, *University of Montpellier, France*
Michael Hübner, *Ruhr-University of Bochum, Germany*
Nikos Voros, *Technological Educational Institute of Western Greece*
Ricardo Reis, *Universidade Federal do Rio Grande do Sul, Brazil*
Sandip Kundu, *University of Massachusetts, Amherst, USA*
Sanjukta Bhanja, *University of South Florida, USA*
Susmita Sur-Kolay, *Indian Statistical Institute, Kolkata, India*
Vijaykrishnan Narayanan, *Pennsylvanian State University, USA*

Technical Program Committee

Digital Circuits and FPGA based Design Track

Dimitrios Soudris	*National Technical University of Athens*
Xinying Wang	*Intel Programmable Solution Group*
Yu Bai	*California State University Fullerton*
Rakesh Kumar	*zGlue Inc*
Hailong Jiao	*Peking University*
Christophe Bobda	*University of Arkansas*
Liang Men	*University of Arkansas*
Li Zhang	*Technical University of Munich*
Deliang Fan	*University of Central Florida*
Jinhui Wang	*North Dakota State University*
Na Gong	*North Dakota State University*
Zheng Wang	*Shenzhen Institutes of Advanced Technology*
Ronald Demara	*University of Central Florida*
Yingjie Lao	*Clemson University*
Eric Liang	*Peking University*
Miaoqing Huang	*University of Arkansas*
Chien-Wei Lo	*University of Arkansas*
David Bol	*Université catholique de Louvain*
Ann Gordon-Ross	*University of Florida*
Chun-Yi Lee	*National Tsinghua University*
Weiqiang Liu	*Nanjing University of Aeronautics and Astronautics*
Hao Zheng	*University of South Florida*
Balatsoukas-Stimming Alexios	*Ecole Polytechnique Fédérale de Lausanne*
Christophe Jego	*IMS Laboratory*

Computer-Aided Design and Verification Track

Kunal Ganeshpure	*Mentor Graphics Corporation*
Vivek Chaturvedi	*Nanyang Technological University*
Weize Yu	*Old Dominion University*
Xiaowei Xu	*University of Notre Dame*
Shouyi Yin	*Tsinghua University*
Theocharis Theocharides	*University of Cyprus*
Nagi Naganathan	*Avago Technologies*
Miroslav Velev	*Aries Design Automation*

Emerging and Post-CMOS Technologies Track

Mahdi Nikdast	*Colorado State University*
Prasun Ghosal	*Indian Institute of Engineering Science and Technology, Shibpur*
Himanshu Thapliyal	*University of Kentucky*

Sharad Sinha	*Nanyang Technological University*
Yuanqing Cheng	*Beihang University*
Xiaoming Chen	*Institute of Computing Technology, Chinese Academy of Sciences*
Garrett Rose	*University of Tennessee*
Lionel Torres	*LIRMM*
Chenchen Liu	*Clarkson University*
Nezih Pala	*Florida International University*
Jiang Xu	*Hong Kong University of Science and Technology*
Thomas Mikolajick	*NaMLab / TU Dresden*
Saraju Mohanty	*University of North Texas*

System Design and Security Track

Upasna Vishnoi	*Marvell Semiconductor Inc.*
Luciano Ost	*Loughborough University*
Naghmeh Karimi	*University of Maryland, Baltimore County (UMBC)*
Chen Liu	*Intel*
Madhu Mutyam	*Indian Institute of Technology, Madras.*
Keni Qiu	*Capital Normal University*
Arslan Munir	*Kansas State University*
Rance Rodrigues	*University of Massachusetts at Amherst*
Arun Kanuparthi	*Intel*
Fabio Campi	*Silicon Biosystems*
Tanguy Risset	*Citi, INSA-Lyon*
Michail Maniatakos	*New York University*
Marco Wehrmeister	*Federal University of Technology - Parana*
Fernando Moraes	*Pontifical Catholic University of Rio Grande do Sul*
Edoardo Fusella	*UNINA*
Apostolos Fournaris	*Technological Educational Institute of Western Greece*
Ricardo Chaves	*IST / INESC-ID*
Sicheng Li	*Hewlett-Packard (HP)*
Yarui Peng	*University of Arkansas*
Ramesh Karri	*polytechnic institute of NYU*
Mateus Rutzig	*Federal University of Santa Maria*
Nicolas Sklavos	*University of Patras*
Sheng Wei	*University of Nebraska Lincoln*
Guy Gogniat	*Université de Bretagne Sud - UEB*
Nele Mentens	*Katholieke Universiteit Leuven*
David Hely	*Grenoble INP*
Fei Wu	*Huazhong University of Science and Technology*
Yier Jin	*University of Florida*

Testing, Reliability and Fault-Tolerance Track

Mihalis Psarakis	*University of Piraeus*
Matteo Sonza Reorda	*Politecnico di Torino*
Ernesto Sanchez	*Politecnico di Torino*
Xiaoqing Wen	*Kyushu Institute of Technology*
Alessandro Savino	*Politecnico di Torino*
Giorgio Di Natale	*LIRMM*
Eduardo Bezerra	*UFSC*
Elena Ioana Vatajelu	*INP - TIMA Laboratory*
Luigi Dilillo	*LIRMM*
Stefano Di Carlo	*Politecnico di Torino*
Leticia Bolzani Poehls	*Catholic University of Rio Grande do Sul (PUCRS)*
Zebo Peng	*Linkoping University*
Yongfu Li	*GLOBALFOUNDRIES, Singapore*
Michele Portolan	*TIMA*
Dong Xiang	*Tsinghua University*

Analog and Mixed-Signal Circuits Track

Bo Jiang	*OmniVision Technologies. Inc*
Jose Pineda De Gyvez	*NXP Semiconductors*
Munem Hossain	*University of Missouri – Kansas City*
Elias Kougianos	*University of North Texas*
Chi Zhang	*GlobalFoundries*
Aisha Alhammadi	*University of Sharjah*
Manish Goswami	*Indian Institute of Information Technology*
Chuang Zhang	*Broadcom*
Maryam Shojaei Baghini	*India Institute of Technology, Bombay*
Steffen Paul	*University Bremen*

Student Research Forum Track

Anupam Chattopadhyay	*Nanyang Technological University Singapore*
Zheng Wang	*Shenzhen Institutes of Advanced Technology*

KEYNOTE 1

8:30am ~ 9:30am, Monday July 9, 2018

N003

Future Hybrid Circuits for Functionality, Performance and Energy Efficiency

K.-T. Tim Cheng
Professor, Dean of School of Engineering
Department of Electronic and Computer Engineering
Hong Kong University of Science and Technology

Summary

Advances in photonics, flexible electronics, emerging memories, etc. and Si electronics' integration with these devices have enabled new classes of integrated circuits and systems with enhanced functionality, higher performance, or lower power consumption. Driving greater integration of such heterogeneous hybrid chips/systems can facilitate the continued proliferation of low-cost micro-/nano-systems for a wide range of applications. However, achieving their large-scale integration will require design ecosystem and design automation tools/methodologies much like those that enabled electronic integration in previous decades.

In this talk, I will briefly introduce two recent Manufacturing Innovation Institutes, on Integrated Photonics and on Flexible Hybrid Electronics respectively, and a research center on developing 3D Hybrid CMOS-memristor circuits, which bring together academia, industry, and government partners to increase design and manufacturing competitiveness in these areas. I will then describe some of our recent results and highlight the needs, challenges and opportunities in these areas.

About K.-T. Tim Cheng

K.-T. Tim Cheng received his Ph.D. in EECS from the University of California, Berkeley in 1988. He has been serving as Dean of Engineering and Chair Professor of ECE and CSE at Hong Kong University of Science and Technology (HKUST) since May 2016. He worked at Bell Laboratories from 1988 to 1993 and joined the faculty at Univ. of California, Santa Barbara in 1993 where he was the founding director of UCSB's Computer Engineering Program (1999-2002), Chair of the ECE Department (2005-2008) and Associate Vice Chancellor for Research (2013-2016). His current research interests include design automation for photonics IC and flexible hybrid circuits, memristive memories, mobile embedded systems, and mobile computer vision. He has co-authored five books, supervised 50 PhD theses, held 12 U.S. Patents, and published extensively in these areas. He served as Director for US Department of Defense MURI Center for 3D hybrid circuits which aims at integrating CMOS with high-density memristors.

Cheng, an IEEE fellow, received 10+ Best Paper Awards from various IEEE and ACM conferences and journals. He has also received UCSB College of Engineering Outstanding Teaching Faculty Award. He served as Editor-in-Chief

xxx

of *IEEE Design and Test of Computers* and was a board member of IEEE Council of Electronic Design Automation's Board of Governors and IEEE Computer Society's Publication Board.

KEYNOTE 2
1:20pm ~ 2:20pm, Monday July 9, 2018
N003

Low Power High Performance Multicore Hardware and Software Co-Design

Hironori Kasahara
President of IEEE Computer Society
Professor, Department of Computer Science and Engineering
Waseda University

Summary

Multicores have been attracting much attention to improve performance and reduce power consumption of computing systems, from embedded to supercomputing systems. To obtain high performance and low power on multicores, co-design of hardware and software is essential. Especially architecture supports for parallelizing and power reducing compiler are very important. This talk first introduces a parallelizing, memory usage optimizing and power reducing compiler and its performance on various multicores from Intel, IBM, arm, Fujitsu, Infineon, and Renesas for various applications including multimedia, automobile, cancer treatment, and earthquake simulation. It next explains architecture supports for the compiler, such as, global address space, data and group barrier synchronization, vector accelerators, data transfer controllers, and power control using DVFS and Clock and Power Gating. The hardware and software co-design allows us not only high performance and low power but also short development period and low development cost of parallel software.

About Hironori Kasahara

Hironori Kasahara is an IEEE Computer Society President 2018 and a professor in the Department of Computer Science and Engineering at Waseda University. He is an IEEE Fellow, an IPSJ Fellow, a Golden Core Member of the IEEE Computer Society, a professional member of the IEEE Eta Kappa Nu, a member of the Engineering Academy of Japan and the Science Council of Japan. He received a PhD in 1985 from Waseda University, Tokyo, joined its faculty in 1986, and has been a professor of computer science since 1997 and a director of the Advanced Multicore Research Institute since 2004. He was a visiting scholar at University of California, Berkeley, and the University of Illinois at Urbana–Champaign's Center for Supercomputing R&D.

He has served as a chair or member of 250 society and government committees, including a member of the CS Board of Governors and Executive Committee; chair of CS Planning Committee, Constitution & Bylaws Committee, Multicore STC and CS Japan chapter; associate editor of *IEEE Transactions on Computers*; vice PC chair of the 1996 ENIAC 50th Anniversary International Conference on Supercomputing; general chair of LCPC; PC member of

SC, PACT, and ASPLOS; board member of IEEE Tokyo section; and member of the Earth Simulator and K supercomputer committees. Kasahara received the CS Golden Core Member Award, IFAC World Congress Young Author Prize, Sakai Special Research Award, and the Japanese Minister's Science and Technology Prize. He led Japanese national projects on parallelizing compilers and embedded multicores, and has presented 215 papers, 155 invited talks, and 30 patents. His research has appeared in 557 newspaper and Web articles.

KEYNOTE 3
8:30am ~ 9:30am, Tuesday July 10, 2018
N003

Machine Learning Further Improve Physical Design PPA at Advanced Node

Weibin Ding
Cadence Software Engineering Group Director
Head of Cadence Global AI Center of Digital and Signoff Group

Summary

Place and route at advanced process node is becoming much more complicated than ever, to meet both timing and DRC closure need multiple core engines to co-work seamlessly well. Traditionally it is very hard problem to decide a solution finally works or not along the flow, the machine learning method opens a door to give better correlated result in the flow. We have constantly seen the improvement trend in our product development.

About Xianfeng (Sean) Ding

Weibin is based in Shanghai – the APAC Headquarter of Cadence and currently leading the Global AI Center of DSG which is a new strategic function in Cadence since January 2018. Prior to that, he led 100+ software engineers responsible for Innovus product development. He initiated the Machine Learning direction for Innovus in early 2015 and responsible for making it production. He has been working at Cadence for 12 years and held 3 US patents.

Weibin received a B.S. and M.S. degree in Computer Science from Shanghai Jiao Tong University.

KEYNOTE 4
1:20pm ~ 2:20pm, Tuesday July 10, 2018
N003

Power Density and Circuit Aging – System-Level Means for Mitigation

Jörg Henkel
Chair of Embedded System
Professor of Computer Science
Karlsruhe Institute of Technology (KIT), Germany

Summary

Power density will stay a major challenge for the foreseeable future. Despite orders-of-magnitude-improved efficiency, power consumption per area is rising, mainly due to the limits of voltage scaling. To investigate the physical implications of high power densities, we must distinguish between peak and average temperatures and temporal and spatial thermal gradients because they trigger circuit-aging mechanisms and eventually jeopardize the reliability of an on-chip system.

The talk starts by presenting some basic interdependencies in the triangle of power density, circuit aging and reliability and continues with solutions to mitigate the problem via, among others, power density-aware resource management, thermal save power (TSP), efficient power budgeting as well as "Aging Aware Boosting.

About Jörg Henkel

Jörg Henkel (M'95-SM'01-F'15) received the master's and PhD (Summa cum laude) degrees from the Technical University of Braunschweig, Germany. He is with the Karlsruhe Institute of Technology (KIT), Germany. Before he worked at the NEC Laboratories, Princeton, NJ. His current research interests include design and architectures for embedded systems with focus on low power and reliability. He has received various research awards, among them the 2008 DATE Best Paper Award, the 2009 IEEE/ACM William J. Mc Calla ICCAD Best Paper Award, the CODES+ISSS 2011, 2014 and 2015 Best Paper Awards. He was the general chair of major CAD events incl. ICCAD and ESWeek. He is the chairman of the IEEE Computer Society, Germany Section, and was the editor-in-chief of the *ACM Transactions on Embedded Computing Systems* for two terms. He is currently the editor-in-chief of the *IEEE Design and Test* Magazine. He is also an Initiator and Spokesperson of the national priority program on Dependable Embedded Systems of the German Science Foundation and the site coordinator (Karlsruhe site) of the three-university collaborative research center on invasive computing. He is a Fellow of the IEEE and holds ten US patents.

KEYNOTE 5

8:30am ~ 9:30am, Wednesday July 11, 2018

N003

Self-Awareness for Heterogeneous MPSoCs:
A Case Study using Adaptive, Reflective Middleware

Nikil Dutt

Center for Embedded and CyberPhysical Systems (CECS)
Center for Cognitive Neuroscience and Engineering (CENCE)
Chancellor's Professor, Department of Computer Science
University of California, Irvine

Summary

Self-awareness has a long history in biology, psychology, medicine, engineering and (more recently) computing. In the past decade this has inspired new self-aware strategies for emerging computing substrates (e.g., complex heterogeneous MPSoCs) that must cope with the (often conflicting) challenges of resiliency, energy, heat, cost, performance, security, etc. in the face of highly dynamic operational behaviors and environmental conditions. Earlier we had championed the concept of *CyberPhysical-Systems-on-Chip (CPSoC)*, a new class of sensor-actuator rich many-core computing platforms that intrinsically couples on-chip and cross-layer sensing and actuation to enable self-awareness. Unlike traditional MPSoCs, CPSoC is distinguished by an intelligent co-design of the control, communication, and computing (C3) system that interacts with the physical environment in real-time in order to modify the system's behavior so as to adaptively achieve desired objectives and Quality-of-Service (QoS). The CPSoC design paradigm enables self-awareness (i.e., the ability of the system to observe its own internal and external behaviors such that it is capable of making judicious decision) and (opportunistic) adaptation using the concept of cross-layer physical and virtual sensing and actuations applied across different layers of the hardware/software system stack. The closed loop control used for adaptation to dynamic variation -- commonly known as the observe-decide-act (ODA) loop -- is implemented using an adaptive, reflective middleware layer.

In this talk I will present a case study of this adaptive, reflective middleware layer using a holistic approach for performing resource allocation decisions and power management by leveraging concepts from reflective software. Reflection enables dynamic adaptation based on both external feedback and introspection (i.e., self-assessment). In our context, this translates into performing resource management actuation considering both sensing information (e.g., readings from performance counters, power sensors, etc.) to assess the current system state, as well as models to predict the behavior of other system components before performing an action. I will summarize results leveraging our adaptive-reflective middleware toolchain to i) perform energy-efficient task mapping on heterogeneous architectures, ii) explore the design space of novel HMP architectures, and iii) extend the lifetime of mobile devices.

About Nikil Dutt

Nikil Dutt is a Chancellor's Professor of CS, Cognitive Sciences, and EECS at the University of California, Irvine. He received a PhD from the University of Illinois at Urbana-Champaign (1989). His research interests are in embedded systems, EDA, computer architecture and compilers, distributed systems, and brain-inspired architectures and computing. He has received numerous best paper awards and is coauthor of 7 books. Professor Dutt has served as EiC of *ACM TODAES* and AE for *ACM TECS* and *IEEE TVLSI*. He is on the steering, organizing, and program committees of several premier EDA and Embedded System Design conferences and workshops, and has also been on the advisory boards of ACM SIGBED, ACM SIGDA, ACM TECS and IEEE ESL. He is an ACM Fellow, IEEE Fellow, and recipient of the IFIP Silver Core Award.

KEYNOTE 6

1:20pm ~ 2:20pm, Wednesday July 11, 2018

N003

Cogntive Vision Systems: Energy Efficiency Influences from Algorithms to Architectures

Vijaykrishnan Narayanan
Distinguished Professor, Computer Science and Engineering and Electrical Engineering
The Pennsylvania State University

Summary

Shopping is widely considered as a relaxing leisure activity. However, grocery shopping can be a frustrating experience for those with visual impairment. While getting to a grocery shop itself is not as much of a challenge for them, locating and picking the items in the grocery shelf becomes a task as challenging as picking a needle from the haystack. Imagine picking up five items for your dinner recipe from a typical grocery store in the US that carries around 35,000 unique items and can have more than 30 aisles spanning 45,000 square meters. This talk will showcase synergistic advances in algorithms, architectures and interface design for assisting those with visual impairment to do shopping. The talk will focus on multiple energy-efficient solutions that consider the battery life time of the vision system.

About Vijaykrishnan Narayanan

Vijay Narayanan is a Distinguished Professor of Computer Science and Engineering and Electrical Engineering at The Pennsylvania State University. He is the director of the NSF Expeditions-in-Computing Program on Visual Cortex on Silicon and a thrust leader for the JUMP Center on Brain-Inspired Computing. He has published more than 400 papers and won several awards in recognition of his research in power-aware systems, embedded systems and computer architecture. He is a fellow of IEEE and ACM.

Plenary Talk 1

2:30pm ~ 3:00pm, Monday July 9, 2018

Achieving 19 TFLOPS in 10W for deep learning using network pruning on Xilinx Zynq Ultrascale+ FPGA

Yi Shan
CTO and Partner of DeePhi Tech

Summary

Deep learning algorithms such as Convolution Neural Network (CNN) is fast becoming the critical part of image perceptions in embedded vision applications in the automotive, drones, surveillance and industrial vision markets. Applications include multi-object detection, semantic segmentation and image classification. However, when scaling these networks to modern resolutions like HD and 4K, the computational requirements for real-time system could easily go over 10 TFLOPS consuming hundreds of watts of power, which is simply unacceptable for most edge applications. In this talk, we will describe a network/weight pruning methodology that achieves over 10 times performance gain on Zynq Ultrascale+ with very small accuracy loss. The network inference running on Zynq Ultrascale+ has achieved 19 TFLOPS-equivalent of the original SSD network in less than 10W.

About Yi Shan

Yi Shan graduated from Tsinghua University with a Ph.D. degree in Electronic Science and Technology. He was joint educated at Imperial College London. He has been engaged in research work at Microsoft Research, IBM Research and Baidu Research. He was selected as an outstanding talents program in Beijing.

Dr. Shan is the CTO and partner of DeePhi Tech and fully responsible for the technical development and the management of engineering and production teams. In the development and application of the original innovation technology in frontier industries such as artificial intelligence, he has achieved a number of scientific and technological achievements. He led the team to develop DPUs (Deep Processing Units) and its hardware and software systems for AI applications. Based on this, they formed two key directions for deep learning accelerated on FPGA modules and ASIC. Now DeePhi's products are widely used in smart surveillance, data center, ADAS (advanced driver assistance systems) and other fields.

Plenary Talk 2

3:00pm ~ 3:30pm, Monday July 9, 2018

Overcoming Challenges of Accelerating Deep Neural Network Computations

Deming Chen
Donald Biggar Willett Scholar
Professor, Electrical and Computer Engineering
University of Illinois Urbana-Champaign

Summary

Deep Neural Networks (DNNs) are computation intensive. Without efficient hardware implementations of DNNs, many promising AI applications will not be practically realizable. In this talk, we will analyze several challenges facing the AI community for mapping DNNs to hardware accelerators. Especially, we will evaluate FPGA's potential role for accelerating DNNs for both the cloud and edge devices. Although FPGAs can provide desirable customized hardware solutions, they are difficult to program and optimize. We will present a series of effective design techniques for implementing DNNs on FPGAs with high performance and energy efficiency. These include automated hardware/software co-design, the use of configurable DNN IPs, resource allocation across DNN layers, smart pipeline scheduling, Winograd and FFT techniques, and DNN reduction and re-training. We showcase several design solutions including Long-term Recurrent Convolution Network (LRCN) for video captioning, Inception module (GoogleNet) for face recognition, as well as Long Short-Term Memory (LSTM) for sound recognition. We will also present some of our recent work on developing new DNN models and data structures for achieving higher accuracy for several interesting applications such as crowd counting, genomics, and music synthesis.

About Deming Chen

Dr. Deming Chen obtained his BS in computer science from University of Pittsburgh, Pennsylvania in 1995, and his MS and PhD in computer science from University of California at Los Angeles in 2001 and 2005 respectively. He joined the ECE department of University of Illinois at Urbana-Champaign (UIUC) in 2005 and has been a full professor in the same department since 2015. His current research interests include system-level and high-level synthesis, machine learning, computational genomics, GPU and reconfigurable computing, and hardware security. He has given about 100 invited talks sharing these research results worldwide. Dr. Chen is a technical committee member for a series of top conferences and symposia on EDA, FPGA, low-power design, and VLSI systems design. He is an associated editor for several leading IEEE and ACM journals. He received the NSF CAREER Award in 2008, the ACM SIGDA Outstanding New Faculty Award in 2010, and IBM Faculty Award in 2014 and 2015. He also received seven Best Paper Awards and the First Place Winner Award of DAC International Hardware Contest on IoT in 2017. He is included in the List of Teachers Ranked as Excellent in 2008 and 2017. He was involved in two startup companies previously, which were both acquired. In 2016, he co-founded a new startup, Inspirit IoT, Inc., for design and synthesis for machine learning applications targeting the IoT industry. He is the Donald Biggar Willett Faculty Scholar of College of Engineering of UIUC.

Plenary Talk 3

2:30pm ~ 3:00pm, Tuesday July 10, 2018

The Advanced Technologies in the New Generation of Phytium's Processors -- From the Architecture to Implementation

Zhuo Ma
Deputy General Manager
SoC R&D Center of Phytium

Summary

In the past decades, the computer technology comes to a rapid increasing new ear, and the silicon-based high performance processors give the main impetus. To be a member of the first group of CPU design houses, Phytium Technology Co., Ltd. is a fast-growing Chinese IC design company, and is dedicated to design, manufacture high performance and low power CPU chips as well as services around the products.

About Zhuo Ma

Dr. Ma is the deputy general manager of SoC R&D center of Phytium. In the past two decades, Dr. MA focused on the technologies of the implementation of high performance processors. He is leading a group of gifted engineers to build the most high performanced ARM-based CPU chips. In the presentation, He will introduce the advanced design technologies of Phytium's CPUs, and share a lot of successful stories.

Plenary Talk 4

3:00pm ~ 3:30pm, Tuesday July 10, 2018

Security of the Internet of Things: Can Hardware Change the Game?

Swarup Bhunia
Professor, Electrical and Computer Engineering
University of Florida, Gainesville, Florida

Summary

Security has become a critical design challenge for modern electronic hardware. With the emergence of the Internet of Things (IoT) regime that promises exciting new applications from smart cities to connected autonomous vehicles, security has come to the forefront of the system design process. Recent discoveries and reports on numerous security attacks on microchips and circuits violate the well-regarded concept of hardware trust anchors. It has prompted system designers to develop wide array of design-for-security and test/validation solutions to achieve high security assurance for electronic hardware, which supports the software stack. At the same time, emerging security issues and countermeasures have also led to interesting interplay between security, verification, and interoperability. Verification of hardware for security and trust at different levels of abstraction is rapidly becoming an integral part of the system design flow. The global economic trend that promotes outsourcing of design and fabrication process to untrusted facilities coupled with the prevalent practice of system on chip design using untrusted 3rd party intellectual property blocks (IPs), has given rise to the critical need of trust verification of IPs, system-on-chip design, and fabricated chips. The talk will also cover spectrum of security challenges for IoTs and describe emerging solutions in creating secure trustworthy hardware that can enable IoT security for the mass.

About Swarup Bhunia

Swarup Bhunia is a preeminence professor of cybersecurity and Steven Yatauro endowed faculty fellow of Computer Engineering at University of Florida, FL, USA. Earlier he was appointed as the T. and A. Schroeder associate professor of Electrical Engineering and Computer Science at Case Western Reserve University, Cleveland, OH, USA. He has over twenty years of research and development experience with 250+ publications in peer-reviewed journals and premier conferences and six authored/edited books. His research interests include hardware security and trust, adaptive nanocomputing and novel test methodologies. Dr. Bhunia received IBM Faculty Award (2013), National Science Foundation career development award (2011), Semiconductor Research Corporation Inventor Recognition Award (2009), and SRC technical excellence award (2005) as a team member, and several best paper awards/nominations. He is co-founding editor-in-chief of a Springer journal on hardware and systems security. He has been serving as an associate editor of *IEEE Transactions on CAD, IEEE Transactions on Multi-Scale Computing Systems, ACM Journal of Emerging Technologies*, and *Journal of Low Power Electronics*; served as guest editor of *IEEE Design & Test of Computers* (2010, 2013) and *IEEE Journal on Emerging and Selected Topics in Circuits and Systems* (2014). He has served as co-program chair of IEEE IMS3TW 2011, IEEE NANOARCH 2013, IEEE VDAT 2014, and IEEE HOST 2015, and in the program committee of several IEEE/ACM conferences. Dr. Bhunia received his PhD from Purdue University on energy-efficient and robust electronics. He is a senior member of IEEE.

Plenary Talk 5

3:30pm ~ 4:00pm, Tuesday July 10, 2018

Designing Safety-critical Systems: Rapidly and Accurately!

Ulf Schlichtmann
Professor, Electrical and Computer Engineering
Technical University of Munich

Summary

Thanks to still ever increasing advances in manufacturing technology, we can implement ever more complex systems. System-level design is essential to deal with the enourmous complexity of today's advanced systems. At the same time, fast time to market is as essential as ever. And robustness and reliability are increasing in importance, as the focus of the semiconductor industry shifts to applications with high safety requirements such as automotive. High-level system models, known as Virtual Prototypes, are essential to meet all these challenges. Fault injection is the most common technique to evaluate system robustness.

I will outline some recent progress in Virtual Prototypes for safety evaluations. I will especially discuss how both high performance and high accuracy in fault injection can be achieved at the same time. Robustness evaluation has to be extended to Firmware in addition to Hardware. Future challenges for system design will conclude my presentation.

About Ulf Schlichtmann

Ulf Schlichtmann holds a Dipl.-Ing. degree (MSc equivalent) and a doctorate in electrical engineering from TUM, as well as a technology business degree. He spent about 10 years in the semiconductor industry (Siemens, Infineon) in various engineering, management and executive positions, working on design automation, design libraries, IP reuse, and product development.

In 2003, he joined TUM as professor and head of the Chair of Electronic Design Automation. From 2007-2013 he served as Dean and Vice Dean of TUM's Department of Electrical and Computer Engineering (ECE). Since 2016, Ulf is an elected member of TUM's Academic Senate as well as the TUM Board of Trustees. Since 2017, he is a member of the Advisory Council of TUM's Institute for Advanced Studies. Also, since 2017, Ulf serves as Vice Dean of TUM's ECE Department again. Ulf is currently a visiting Professor at Nanyang Technological University (NTU), Singapore as well as Honorary Chair Professor at National Taiwan University of Science and Technology (NTUST), Taipei.

Ulf's current research interests include computer-aided design of electronic circuits and systems, with an emphasis on designing reliable and robust systems. His research increasingly focuses on emerging technologies, such as microfluidic biochips and photonic interconnects.

2018 IEEE Computer Society Annual Symposium on VLSI

Gyrator-C based Bandpass Filter with Improved Dynamic Range for Fully Integrated RF Front-end

Lakshmi N S
Department of Electronics and Communication Engineering
National Institute of Technology Tiruchirappalli
Tamil Nadu, India
Email:lachuns123@gmail.com

Bhaskar M
Department of Electronics and Communication Engineering
National Institute of Technology Tiruchirappalli
Tamil Nadu,India
Email:bhaskar@nitt.edu

Abstract— In this paper, a gyrator-C based bandpass filter with independent tuning of center frequency and quality factor with a focus on improving the noise performance and hence the dynamic rage is presented. The proposed second order filter provides a narrow bandwidth of 80 MHz at a tuned center frequency of 2.45 GHz. RF filter is designed using UMC 180 nm RF CMOS technology library and the post layout simulations are carried out in Cadence design suite at 1.8 V. It is observed from the simulations that the noise figure of the proposed differential active inductor filter is 7.5 dB, which is 10 dB better than the filters with active components reported in literature. The filter occupies an active area of 385 x 285 μm².

Keywords— Regulated cascode, active inductor, gyrator-C, dynamic range

I. INTRODUCTION

Fully integrated transceiver is of great interest in telecommunication industry. RF bandpass filter with narrow band selection is a vital block in the receiver front-end. On-chip implementation of the bandpass filter meeting the system requirements like linearity, noise performance and selectivity is a tedious task [1]. Highly selective preselect filters required at receiver front-end are implemented using off-chip SAW filters [2]. On-chip spiral inductors [3] were used due to the desirable noise characteristics and low power consumption. However, it is not a good choice to use on-chip passive components at high frequency due to substrate losses that leads to poor quality factor (Q) and large area requirement of the spiral inductor. Q-enhanced LC filters [4], [5] were used for fully integrated receiver front-end. Negative resistors are utilized for Q enhancement of spiral inductors at the cost of increased power consumption and poor noise characteristics that offsets the merits of passive implementation. Active inductors provide attractive features like tunable center frequency and quality factor with less silicon area. Tunable feature of active inductors make them a good choice for multiband fully integrated receiver architecture [6].

Several active inductor topologies were reported in literature. Inductor emulated using two transistors with current reuse technique was introduced in [7], [8]. Resistive losses associated with the active inductor was reduced using cascode, regulated and multi-regulated cascode structures to enhance quality factor [9]. Regulated cascode active inductor topology allows tuning of quality factor independent of center frequency tuning which is a difficult task in designing active inductor filters at reduced power consumption. Cascode grounded active inductor with feedback resistor was proposed to improve quality factor and inductance in [10]. Q and center frequency was

independently tuned by MOS varactor and DC biasing in [11], [12]. Class AB active inductor for enhancing the signal handling capability of active inductor was introduced in [13].

This paper proposes the design and implementation of an active inductor based bandpass filter with a center frequency of 2.45 GHz and a bandwidth of 80 MHz (2.4 GHz to 2.48 GHz) which is the frequency allotted for IEEE 802.11.b and Bluetooth. Noise and linearity are two major issues with active inductors. Dynamic range of active inductor is limited by poor noise performance and non-linearity. In this paper, a regulated cascode active inductor is used. Dynamic range is improved by reducing the noise level by suppressing noise of cascode device by feedback device in regulated cascode topology. The organization of the paper is as follows: In Section–II, the design of individual stages of the filter is described. Post layout simulation results and comparison with other RF band pass filter design proposed in literature are given in Section-III. Section IV provides the conclusion.

II. CIRCUIT IMPLEMENTATION

A. Active Inductor based Filter Realization

Gyrator-C topology consisting of two transconductors connected in feedback loop given in [14], [15] is shown in Fig.1. G_{o1} and G_{o2} are the output conductance of transconductors G_{m1} and G_{m2} respectively. C_1 and C_2 are intrinsic capacitance at ports 1 and 2 respectively. This topology emulates an inductor which can be modelled as a parallel RLC tank circuit. The input admittance of this two port network is given by (1)

$$Y_{in} = sC_2 + G_{o2} + \frac{1}{s\frac{C_1}{G_{m1}G_{m2}} + \frac{G_{o1}}{G_{m1}G_{m2}}} \qquad (1)$$

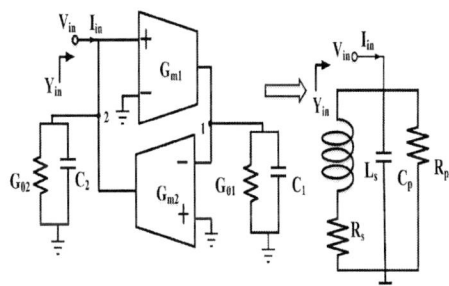

Fig. 1. Gyrator-C topology and equivalent RLC network[15]

978-1-5386-7100-9/18 $31.00 © 2018 IEEE

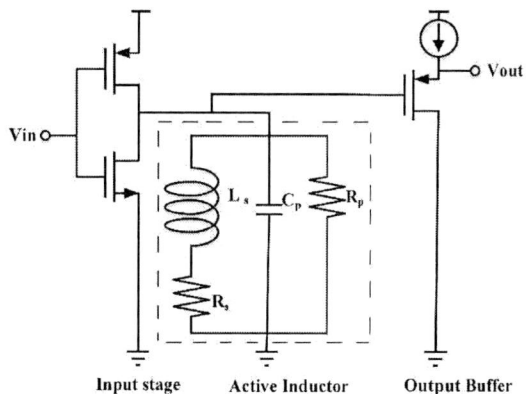

Fig. 2. Active inductor based bandpass filter structure

The equivalent RLC network elements are $R_p = 1/G_{02}$, $L_S = C_1/G_{m1}G_{m2}$, $R_s = G_{01}/G_{m1}G_{m2}$, $C_p = C_2$.

Active inductor based bandpass filter in [12] is realized by input transconductor (common source stage) that converts input voltage to current for active inductor and an output buffer that provides impedance matching to load. Bandpass filter structure proposed in this paper is shown in Fig.2.

B. Input Transconductor

The input transconductor stage is implemented as CMOS inverter as shown in Fig 2. Transconductance of this stage is sum of transconductance of NMOS and PMOS $g_{mN} + g_{mP}$. By means of complementary derivative superposition technique [16], [17], by adjusting bias point and width of transistors, a linear input transconductor is realized.

C. Active Inductor

Regulated cascode topology [18] is used for active inductor implementation [9] as shown in Fig.3. Corresponding to two port gyrator network shown in Fig.1, for regulated cascode-common drain topology, G_{01} is output conductance of regulated cascode amplifier, G_{02} is output conductance of common drain amplifier. Port capacitance C_1 is intrinsic capacitance C_{gs2}, port capacitance C_2 is intrinsic capacitance C_{gs1}. Accordingly, equivalent RLC circuit elements are given below [15].

$$C_p = C_{gs1} \tag{2}$$

$$R_p = \frac{1}{g_{02}} \tag{3}$$

$$L_s = \frac{C_{gs2}}{g_{m1}g_{m2}} \tag{4}$$

$$R_s = \frac{g_{o1}g_{o3}g_{o4}}{g_{m1}g_{m2}g_{m3}g_{m4}} \tag{5}$$

$$\omega_o^2 = \frac{1}{LC_p} = \frac{g_{m1}g_{m2}}{C_{gs1}C_{gs2}} \tag{6}$$

$$Q(\omega_o) = \frac{\omega_o L}{R_s} = \sqrt{\frac{C_{gs2}g_{m1}g_{m2}g_{m3}^2 g_{m4}^2}{C_{gs1}g_{o1}^2 g_{o3}^2 g_{o4}^2}} \tag{7}$$

Transconductance of M1, M2, M3 and M4 are given by g_{m1}, g_{m2} g_{m3}, g_{m4} respectively. Output conductance of M1, M2, M3 and M4 are given by g_{o1}, g_{o2}, g_{o3} and g_{o4} respectively. Center frequency and quality factor for this active inductor is given by (6) and (7) respectively.

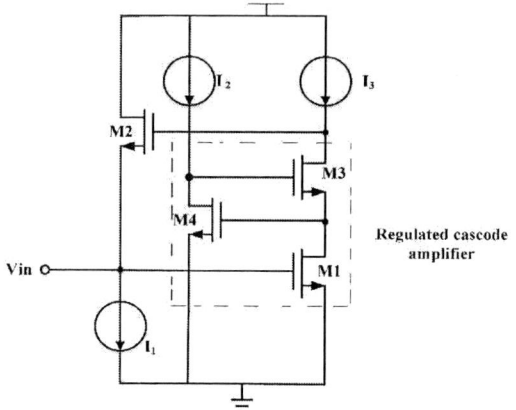

Fig. 3. Regulated cascode active inductor [9]

D. Buffer

Buffer is implemented as common drain stage characterized by high input impedance and low output impedance for providing impedance match to resistive load.

E. Proposed bandpass filter

Fig.4 shows half circuit of proposed differential band pass filter. M_{in1}, M_{in2} constitute CMOS inverter. M_{out} is output buffer with bias current I_5. M1-M4 constitutes active inductor. Bias current I_3 is used to vary transconductance of both M3 and M1. Transconductance of M1which determines center frequency can be independently controlled by varying bias current I_4. Transconductance of M4 which determines the quality factor can be independently controlled by varying bias current I_2. Filter is tuned to desired center frequency by varying bias currents I_1 and I_4. Coarse tuning of quality factor is done by varying I_3. This shifts the center frequency since current through M1 changes and hence its transconductance. Current through M1 should be kept constant for a particular center frequency. This is done by fine tuning bias current I_4. Once the desired center frequency is attained, quality factor is fine tuned by varying bias current I_2. Current sources are implemented using current mirrors.

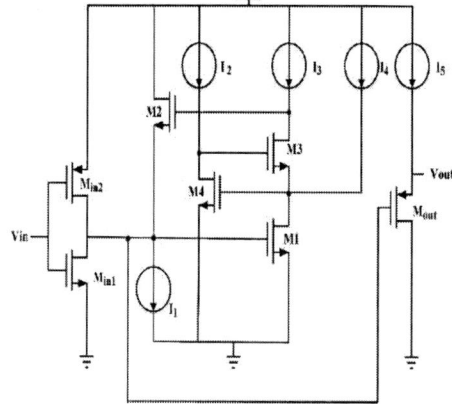

Fig. 4. Half circuit of proposed bandpassfilter

978-1-5386-7100-9/18 $31.00 © 2018 IEEE

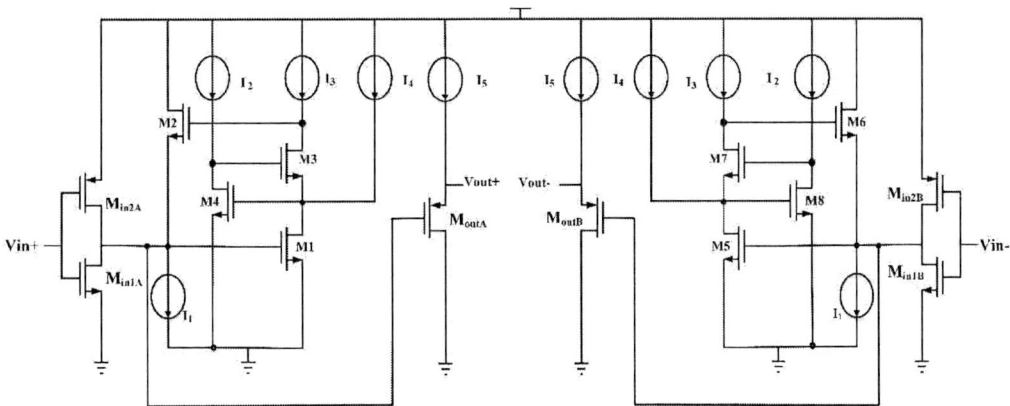

Fig. 5. Proposed bandpass filter

One of the main drawbacks of active inductor based bandpass filter is thermal noise associated with transistor. Compared to cascode topology, regulated cascode topology suppresses the noise of cascode device M3 strongly [18]. Input referred noise of M3 transistor is given in (8)

$$\overline{v_{n3}^2} = \frac{8KT\Delta f}{3g_{m3}}\left(\frac{g_{m3}g_{04}}{g_{m1}g_{m4}}\right)^2 \quad (8)$$

There will be noise due to M4 which is negligible compared to main noise contributor M1. Input referred noise of M1 and M4 are (10) and (9) respectively

$$\overline{v_{n4}^2} = \frac{8KT\Delta f}{3g_{m4}}\left(\frac{g_{01}}{g_{m1}}\right)^2 \quad (9)$$

$$\overline{v_{n1}^2} = \frac{8KT\Delta f}{3g_{m1}} \quad (10)$$

where K-Boltzmann constant, T-Absolute Temperature in Kelvin, Δf-Noise bandwidth in Hertz

The larger the transconductance of M1, the lesser the total noise. To reduce noise without reducing resonant frequency, the method adopted in [19] uses an additional current source rather than increasing width of M1. In this filter implementation, in order to reduce the noise of inherently noisy active inductor, this technique is adopted along with suppression of noise of M3 by increasing g_{m4} using current source I2.

The differential implementation is shown in Fig. 5. M_{in1A}, $M_{in\,2A}$ and M_{in1B}, M_{in2B} constitute the input stages.M1-M4 and M5-M8 constitute active inductors. M_{outA} and M_{outB} constitute output buffers. Layout of the proposed filter is shown in Fig.6.

III. SIMULATION RESULTS

Simulation of differential second order bandpass filter is done in Cadence 0.18 μm RF-CMOS process with 1.8 V supply voltage. The layout of the filter is designed in Cadence Virtuoso Layout XL. Physical verification checks are performed. Filter occupies an active area of 385 μm x 285 μm. At 2.45 GHz resonant frequency, 3 dB bandwidth of proposed bandpass filter is 80 MHz and Q is 30.6.The observed noise figure is 7.5dB and 1-dB compression point (P_{1dB}) is -18 dBm. Magnitude response of bandpass filter tuned to 2.45 GHz is shown in Fig.7. Noise figure at 2.45 GHz is shown in Fig.8.

Fig.9 shows 1-dB compression point characterizes linearity of filter at 2.45 GHz.

Tuning of quality factor from 10 to 30 by varying bias current I2 from 0.3mA to 0.43mA is shown in Fig. 10. Resonant frequency is tuned from 1.9 GHz to 2.45 GHz by varying I4 from 0.12 mA to 0.2 mA in Fig.11.

In Table-I, dynamic range (DR) expressed in dB-Hz is used for performance comparison [20]. Limited dynamic range of bandpass filter affects dynamic range of receiver. Comparison is done using the definition of dynamic range as the ratio of 1-dB compression point (P_{1dB}) to input referred noise level in 1Hz bandwidth (P_n) [20],[21] is given below

$$DR = \frac{P_{1dB}}{P_n} = \frac{P_{1dB}}{kTF} \quad (11)$$

where kT- Thermal noise floor = -174dBm/Hz, F-Noise factor

Figure of Merit (FOM) expressed in dB-Hz/mW used for comparing active bandpass filters is given below as in [1],[20]

$$FOM = \frac{P_{1dB}}{P_n}\frac{1}{P_{DC}} \quad (12)$$

where P_{DC} - DC power consumption in mW.

The proposed filter attains a dynamic range of 149dB-Hz and FOM is 138 dB-Hz/mW which is higher than that reported in literature.

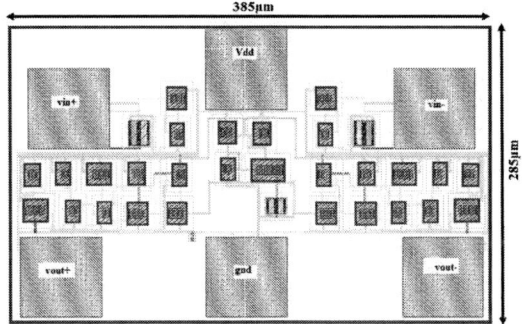

Fig. 6. Layout of proposed filter

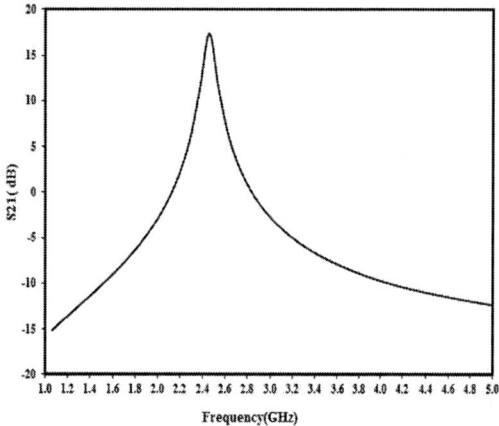

Fig. 7. Magnitude Response of filter tuned at 2.45GHz.

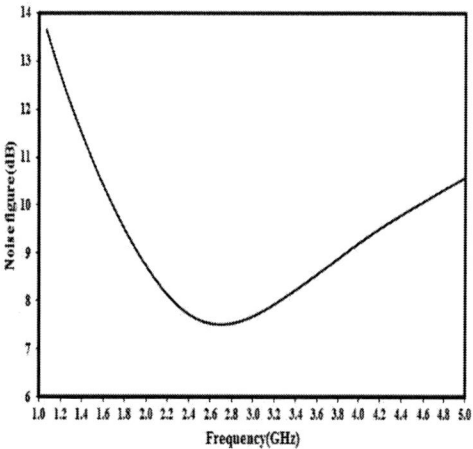

Fig. 8. Noise figure of proposed bandpass filter at 2.45GHz

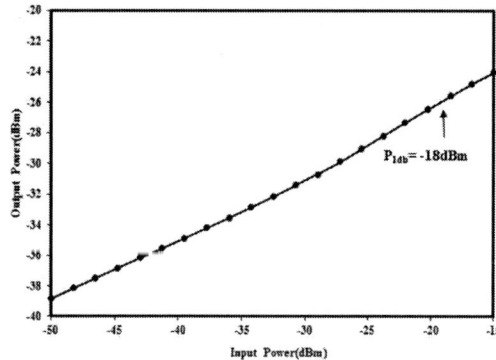

Fig. 9. 1-dB compression

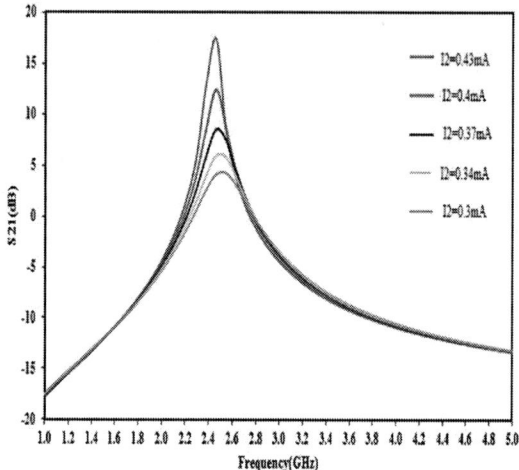

Fig. 10. Quality factor tuning

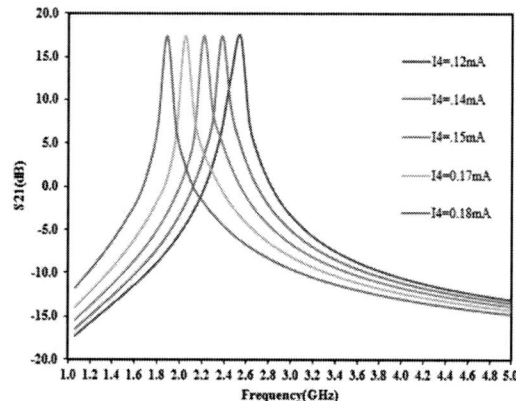

Fig. 11. Center frequency tuning

IV. CONCLUSION

A second order active RF bandpass filter with tunable center frequency and quality factor and low noise figure is implemented.In this paper dynamic range is enhanced by improving the noise performance. Compared to on-chip RF filters using spiral inductors, this active filter consumes less area. This filter can be used for multiband receiver front-end that replaces a bank of SAW filters for tuning to desired frequency band.

978-1-5386-7100-9/18 $31.00 © 2018 IEEE

TABLE I.   PERFORMANCE COMPARISON OF ACTIVE BANDPASS FILTERS

Reference (year)		[5]	[22]	[23]	[6]	*[20]	*This
Parameter	Unit	(2002)	(2003)	(2006)	(2008)	(2017)	work
Technology	μm	0.25	0.35	0.18	0.18	.045	0.18
Order		6	2	3	2	2	2
Center frequency	GHz	2.14	2.19	2.368	2.44	2.511	2.45
3dB bandwidth	MHz	60	53.8	60	60	36.21	80
Mid-band gain	dB	0	-13	-1.8	6	72.76	17.4
Supply voltage	V	2.5	1.3	1.5	1.8	±1	1.8
P_{DC}	mW	17.5	5.2	8.8	10.8	0.168	13.8
Noise figure	dB	19	26.8	18	18	29.62	7.5
P_{1dB}	dBm	-13.4	-30	-20	-15	-1.5	-18
Dynamic Range	dB-Hz	142	117	135	141	125.84	149
FOM	dB-Hz/mW	129	110	125.6	131	133.6	138

*Postlayout Simulation result

REFERENCES

[1] W. B. Kuhn, D. Nobbe, D. Kelly and A. W. Orsborn, "Dynamic range performance of on-chip RF bandpass filters," IEEE Trans. Circuits Syst.II: vol. 50, pp.685-694, October 2003.

[2] R. Pastore, J. A. Kosinski, and H. L. Cui," An Improved Tunable Filter Topology for HF Preselection," Proc. IEEE Int. Freq. Cont. Symp, pp. 575-579,May1998.

[3] C. P. Yue and S. S. Wong, "On-chip spiral inductors with patterned ground shields for Si-based RF ICs," IEEE J. Solid-State Circuits, vol. 33, no. 5, pp. 743–752, May 1998.

[4] W. B. Kuhn. N. K. Yanduru, A. S. Wyszynski, "Q-Enhanced LC bandpass filters for integrated wireless applications," IEEE Trans. Microw. Theory Techn., vol. 46, no. 12, pp.2577- 2586. December 1998.

[5] T. Soorapanth and S. S. Wong, "A 0-dB IL 2140 ± 30 MHz bandpass filter utilizing Q-enhanced spiral inductors in standard CMOS," IEEE J. Solid-State Circuits, vol. 37, no. 5, pp. 579–586, May 2002.

[6] Z. Gao, J. Ma, M. Yu, and Y. Ye, "A fully integrated CMOS active bandpass filter for multiband RF front-ends," IEEE Trans. Circuits Syst. II, vol. 55, no. 8, pp. 718–722,August 2008.

[7] Y. Wu, M. Ismail, and H. Olsson, "A novel CMOS fully differential inductorless RF bandpass filter," in Proc. IEEE Int. Symp. Circuits Syst., vol. 4, pp. 149–152, May 2000.

[8] Y. Wu, X. Ding, M. Ismail, and H. Olsson, "RF bandpass filter design based on CMOS active inductors," IEEE Trans.Circuits Syst. II: Analog and Digital Signal Processing, vol. 50, no. 12, pp. 942–949,December 2003.

[9] A.Thanachayanont and A.Payne "VHF CMOS integrated active inductor", IEEE Electron. Lett, vol 32, pp.999- 1000,May 1996 .

[10] C.Hsaio, C.Kuo,C.Ho and Y.Chan"Improved Quality-Factor of 0.18-um CMOS Active Inductor by a Feedback Resistance Design", IEEE Microw. Wireless Compon. Lett, vol. 12, no. 12, pp. 467–469, December. 2002.

[11] A. Karsilayan and R. Schaumann. "A high-frequency high-Q CMOS active inductor with DC bias control,"IEEE Midwest Symp.Circuits Syst.,vol.1,pp.486-489, August 2000.

[12] Haiqiao Xiao and Rolf Schaumann, "A radio-frequency CMOS active inductor and its application in designing high-Q filters,"in Proc. IEEE Int. Symp. Circuits Syst., vol. 4, pp. 197-200,May 2004.

[13] A. Thanachayanont and S. Sae Ngow, "Class AB VHF CMOS Active Inductor," in Proc. IEEE Midwest Symp. Circuits Syst., vol. 1, pp. 64-67, August 2002.

[14] A. Thanachayanont, "CMOS transistor-only active inductor for IF/RF applications", " in Proc. IEEE Int. Conf. Ind. Technol., Bangkok, Thailand, vol. 2, pp. 1209–1212, December 2002.

[15] Fei Yuan, CMOS Active Inductors and Transformers:Principle, Implementation, and Applications,Springer 2008 edition.

[16] H. K. Subramaniyan, E. A. M. Klumperink, V. Srinivasan, A. Kiaei, and B. Nauta, "RF transconductor linearization robust to process, voltage and temperature variations," IEEE J. Solid-State Circuits, vol. 50, no. 11, pp. 2591–2602, November. 2015.

[17] H. Zhang and E. Sanchez-Sinencio, "Linearization techniques for CMOS low noise amplifiers: A tutorial," IEEE Trans. Circuits Syst. I,vol. 58, no. 1, pp. 22–36, January 2011.

[18] E. Sackinger and W. Guggenbuhl, "A high-swing high-impedance MOS cascode circuit," IEEE J. Solid-State Circuits, vol. 25,no.1, pp. 289–298, February 1990.

[19] R.M. Weng and R. C Kuo, "An ω_0 - Q tunable CMOS active inductor for RF band pass filters," Signals, Systems and Electronics Symp., pp. 571-574, August 2007.

[20] V.Kumar ,R Mehra and A Islam " A 2.5 GHz low power, high-Q, reliable design of active bandpass filter," IEEE Trans. Device Mater Reliab. vol.17, no.1, pp.:229–244 ,March 2017.

[21] X. He and W. B. Kuhn, "A 2.5 - GHz low-power, high dynamic range, self-tuned Q-enhanced LC filter in SOI," IEEE J. Solid-State Circuits,vol. 40, no. 8, pp. 1618–1628, Aug. 2005

[22] F. Dulger, E. Sanchez-Sinencio, and J. Silva-Martinez, "A 1.3-V 5-mW fully integrated tunable bandpass filter at 2.1 GHz in 0.35-μm CMOS," IEEE J. Solid-State Circuits, vol. 38, no. 6, pp. 918–928, June. 2003

[23] J. Kulyk and J. Haslett, "A monolithic CMOS 2368±30 MHz transformer based Q-enhanced series-C coupled resonator bandpass filter," IEEE J. Solid-State Circuits, vol. 41, no. 2, pp. 362-374, February 2006.

Replica-Based Low Drop-Out Voltage Regulator with Assistant Power Transistors for Digital VLSI Systems

Yang Nan, Chenchang Zhan, Guanhua Wang, Linjun He and Han Li

Department of Electrical and Electronic Engineering, Southern University of Science and Technology, Shenzhen, China

Email:nany@mail.sustc.edu.cn; zhancc@sustc.edu.cn; {11649167, helj, 11649124}@mail.sustc.edu.cn

Abstract— An advanced replica-based low drop-out (LDO) voltage regulator with assistant power transistors and bulk modulation digital VLSI systems is proposed in this paper. It utilizes seven current comparators with different aspect sizes to control seven assistant power transistors to compensate for the load current. As a result, the load regulation of the replica LDO is significantly improved, while the wide-load-range-stability which is critical for digital systems is not compromised. The proposed LDO is designed in a standard 0.18-μm CMOS process. Extensive simulation results show that, with a power supply of 1.2 V and output voltage of 1.0 V, the proposed LDO has a drop-out voltage of 200mV when delivering 1~100mA to the load. The load regulation is 0.148mV/mA, and the supported load capacitance range is from 0 to 100nF.

Keywords—low drop-out (LDO) voltage regulator; assistant power transistor; load regulation; current comparator; current sense; bulk modulation

I. INTRODUCTION

With the growing demand of multi-functional system-on-chip (SoC) design, the power management system should be integrated into a single-chip solution as well. A low drop-out (LDO) regulator is widely used in power management IC on due to its simple structure, fast response and low noise characteristic [1-3]. Wherever, for traditional LDO, there is a μF level off-chip capacitor which limits the ability for fully integrated applications [2-5]. Therefore, LDOs without using external capacitors have become popular and attracted extensive research efforts [6]. However, the on-chip loading parasitic capacitance has dramatically wide range, especially for the digital VLSI systems. Consequently, the LDOs need to achieve good stability over wide capacitive range, in addition to large load current range.

To solve this problem, the replica LDO (for NMOS LDO) concept is developed in [7], which could achieve faster response and better stability over wide capacitive and resistive loading range, as the loading is effectively unregulated. However, the conventional replica-based LDO has poor load regulation due to the unregulated loading. With technology scaling, the supply voltage of digital VLSI systems becomes lower and lower, and its tolerance on the supply variation also becomes narrower and narrower.

This work was in part supported by NSFC under grant 61604067 and SZSTI under grant JCYJ201060530191008447.

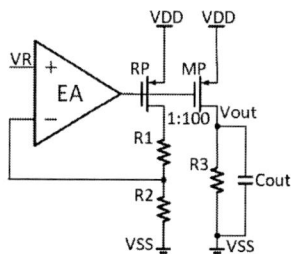

Fig. 1. Traditional replica LDO structure.

Therefore, the poor regulation accuracy in conventional replica LDO is unacceptable. A regulated-replica which could achieve a better load regulation is introduced in [8]. However, it needs a slow op-amp in the replica-regulation loop in order not to affect the main loop's stability, which causes large chip area (500kohm resistor and 5pF capacitor as in [8]), and potentially poor transient responses and stability when the load changes.

In this paper, we propose a load current compensation structure to improve the load regulation and load transient without compromising the stability or using large RC networks. In Section III, the proposed structure will be described. Simulation results will be given in Section III, followed by the concluding remarks in Section IV.

II. LDO REGULATOR WITH ASSISTANT POWER TRANSISTORS

Fig. 1 shows the simplified structure of the traditional replica LDO [7-8]. In this illustration, it is a PMOS LDO and, compared to the NMOS LDO, it achieves low drop-out voltage without additional power supply. It consists of two P-type power transistors (RP and MP) and the two transistors' size ratio is, e.g., 1:100. This means the ratio of currents of the two power transistors is about 1:100 if their drain voltages are also the same as each other, since they share the source and gate terminals. So, RP is a replica of MP and, by regulating the drain voltage of RP, the LDO output voltage would be also the same as the drain voltage of RP if the loading current is, e.g., 100 times of RP's current. The advantage of this structure is that the load part is isolated from the feedback loop, so the stability is good regardless of the loading capacitance. However, it is noted that there is only a regulation loop for RP and there is no feedback from the LDO output. In other words, there is no regulation for MP, hence the load regulation of the whole replica LDO is poor.

978-1-5386-7100-9/18 $31.00 © 2018 IEEE

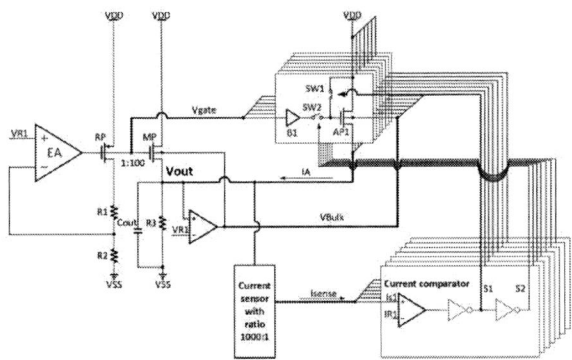

Fig. 2. Structure of the proposed replica LDO with assistant power transistors.

Fig. 2 shows the structure of the proposed replica LDO with assistant power transistors. It consists of a current sensor, current comparators, a buck modulation loop [9], voltage buffers, CMOS transmission gates and the assistant power transistors. Seven assistant power transistors are used in this design. For the traditional replica LDO, the output voltage is the same as the replica loop regulated voltage at the drain of the replica transistor only when the load current is at the right ratio of the replica current. When the load current increases to beyond this point, the power transistor cannot provide enough current to maintain the required output voltage and the output decreases, since the power transistor gate is fixed. On the contrary, in the proposed structure, the seven assistant power transistors are added to aid with the main power transistor to provide sufficient current to maintain the output voltage at the expected value after the replica point. The current sensor, with a sensing ratio of 1000:1, senses the load current I_{Load}. Then, the sensed current I_{sense} is used to compare with seven different reference currents, IR1~IR7. The output of each current comparator is used by two inverters to generate two voltage signals, e.g., S1 and S2. The two voltage signals determine the states of the two switches, e.g., SW1 and SW2, which control the corresponding assistant power transistor. For instance, when I_{sense} is larger than IR1, S1 and S2 will invert and SW1 is off while SW2 is on and the assistant power transistor AP1 is turned on. Note that the states of the two switches SW1 and SW2 are inverted all the time. When SW1 is on and SW2 is off, the first assistant power transistor AP1 is off, such that AP1 cannot provide any current to the load. When SW1 is off and SW2 is on, AP1 is on and its gate voltage is about the same as the gate voltage of the main power transistor due to the buffer B1, and AP1 will provide current certain times of the main power transistor to the load. As the load current increases, the seven assistant power transistors open one by one to achieve improved load regulation over a large load current range.

Due to the digital control, the provided current by the assistant power transistors is discrete and the load regulation is not smooth enough. Hence, in the proposed structure, a bulk modulation part is added to modulate the load current of the assistant power transistors and the main power transistor [9].

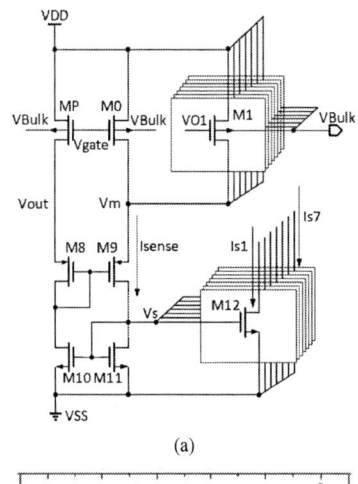

(a)

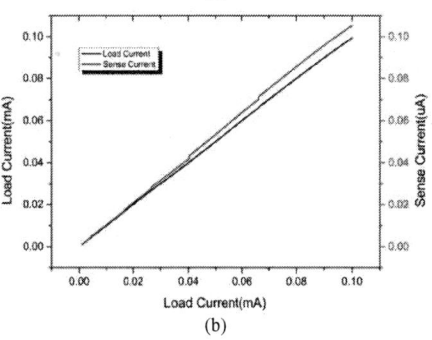

(b)

Fig. 3. Current sensor: (a) schematic and (b) accuracy illustration.

The output of the bulk modulation amplifier is connected to the bodies of those transistors, utilizing body effect to regulate the transistors' output current. Since the bulk modulation loop is isolated from the replica loop, it does not affect the replica loop's stability. The bulk modulation loop does not need to provide high accuracy, so its own stability can be readily achieved across a wide load capacitive and resistive range.

Fig. 3 (a) shows the schematic of the current sensor which uses a common-gate amplifier to improve sensing accuracy [10]. M0 is utilized to sense the load current of MP. However, the goal is to sense the whole load current which consists of the current of MP and AP1~AP7. So, M1~M7 are used to sense the currents of AP1~AP7 and they add up to I_{sense}. M8~M11 make up the common-gate amplifier, which ensures the voltage of Vm and Vout be the same. Since the gate-source voltage of MP and AP1~AP7 is the same as that of M0~M8, with their drain-source voltage the same as each other as well, the sensed current I_{sense} is very close to the predetermined ratio of the whole load current. The sensed current is mirrored over to M12~M18 to perform the current comparison. Fig. 3 (b) illustrates the load current (mA) and the sensed current (μA). The sensed current is very precise at light load, and only shows small deviations at heavy load due to the decreased amplifier gain at larger currents.

Fig. 4 (a) shows the schematic of the current comparator, which consists of a current source, several current mirrors and inverters. M19~M25 with different aspect ratios are used to mirror the reference current IR to generate different biasing

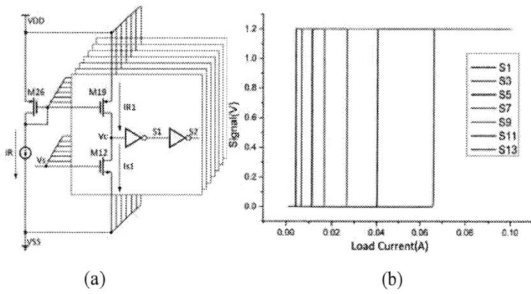

(a)   (b)

Fig. 4. Current comparator: (a) schematic and (b) simulation result.

currents, IR1~IR7. These seven reference currents are used to compare with those sensed currents. When the first sensed current (Is1) is lower than the first reference current (IR1), M19 will charge up Vc to high voltage, so S1 is low and S2 is high to turn off AP1 (explained next). If the load current increases and Is1 increases to be larger than IR1, Vc will be discharged to low voltage and AP1 will be turned on to provide current to support the loading demand. Note that a comparison hysteresis can be created by using Schmitt trigger in the inverter to avoid small oscillations. With further increased load current, the assistant power transistors are turned on one by one. The switching thresholds can be decided based on the load regulation requirement. Fig. 4 (b) shows the switch signals vs. the load current in this design. Different load current thresholds are used for the seven assistant power transistors.

Fig. 5 (a) shows the schematic of the assistant power transistors and their related control elements, including the voltage buffer, transmission gates (TG), and MOS-FET switches. The voltage buffer is used to drive the assistant power transistors without affecting V_{gate} of the main power transistor. TG and SW1 are used as switches to control VO1~VO7. When S1 is low and S2 is high, SW1 is on and SW2 is off, and VO1 is connected with VDD, hence AP1 will be off. On the other hand, when S1 is high and S2 is low, AP1 will be on. The other assistant power transistors are controlled in similar ways. Fig. 5 (b) shows VO1~VO7 vs. load current, illustrating the assistant power transistors' turning sequence when load current increases.

III. SIMULATION RESULTS

The proposed replica LDO with assistant power transistors is designed in a standard 0.18-μm CMOS process. Extensive simulations have been done to verify the performances. Some of the key results are presented next.

The quiescent current of the LDO regulator is 48 μA at light load, and 429.4 μA at full load. Fig. 6 (a) compares the load regulations of three replica LDOs: the traditional replica LDO with and without bulk modulation, and the proposed replica LDO. Obviously, the proposed structure achieves the best result in load regulation. Due to the unregulated load, the traditional LDO without bulk modulation shows the worst regulation. Although it is improved at light load with bulk modulation, the enhancement does not work when the load increases to heavy one. Fig. 6 (b) details the load regulation of the proposed LDO. With the help of the assistant power

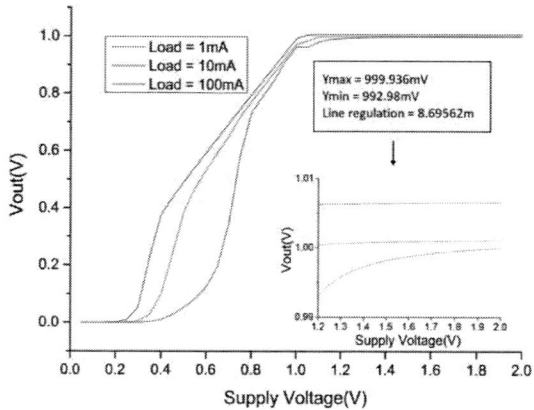

Fig. 7. Simulated V_{out} vs. V_{in}.

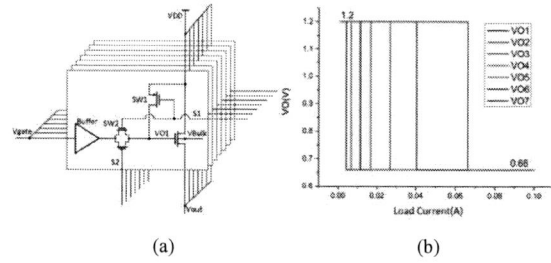

(a)   (b)

Fig. 5. Assistant power transistors: (a) schematic and (b) simulation result.

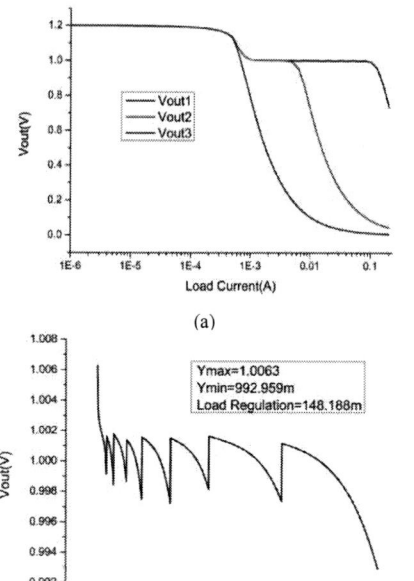

(a)

(b)

Fig. 6. Simulated load regulation: (a) comparison of the load regulation of three replica LDO structures: traditional replica LDO (Vout3), traditional replica LDO with bulk modulation (Vout2) and proposed structure (Vout1), and (b) detailed result of the proposed LDO structure.

transistors, the load regulation is significantly improved to 0.148mV/mA for 1 to 100 mA load.

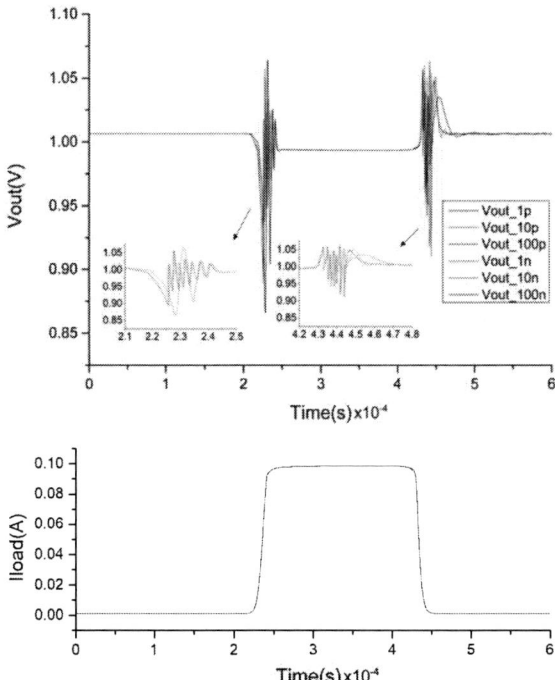

Fig. 8 Simulated load transient responses when I_{Load} jumps between 1mA and 100mA, with capacitive load of 1pF to 100nF.

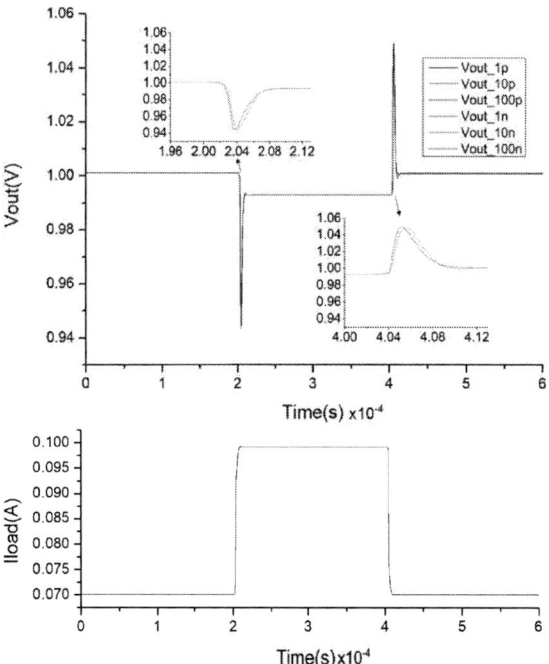

Fig. 9 Simulated load transient responses when I_{Load} jumps between 70mA and 100mA, with capacitive load of 1pF to 100nF.

Fig. 7 shows the simulated V_{out} vs. V_{in} of the proposed LDO from 0 to 2 V, under different load currents. The line regulation for V_{in} from 1.2 to 2 V is as good as 8.69562 mV/V.

Fig. 8 shows the simulated load transient responses when the load current is switched between 1mA and 100 mA with 1pF~100nF load capacitance. The output recovery time is 30μs when I_{Load} changes from 1mA to 100mA and it is 35μs when changes from 100mA to 1mA. The oscillations are due to the seven assistant power transistors which are turned on and off one by one. The load capacitance has little impact on the performance of the load transient, confirming that the stability of the replica LDO is not compromised. Fig. 9 shows the load transient when I_{Load} changes between 70mA and 100mA. In this case, no assistant power transistors are turned on and off. No oscillations happen and the recovery time is shortened to less than 10 μs.

IV. CONCLUSION

A replica-based LDO regulator with assistant power transistors and bulk modulation has been presented. Due to the assistant power transistors, sufficient current can be provided to support a large loading range, and the unregulated loading in a traditional replica LDO is greatly mitigated. Furthermore, the bulk modulation technique helps achieve fine-grained regulation. Simulated results in a standard CMOS design have confirmed that the proposed replica LDO achieves much improved load regulation without compromising the wide-load-range stability. The proposed design is hence a good candidate for digital VLSI systems' power management with large range of load capacitance and currents.

REFERENCES

[1] G. Rincon-Mora and P. E. Allen, "A low-voltage, low quiescent current, low drop-out regulator," *IEEE J. Solid-State Circuits, vol. 33, no.1, pp. 36–44, Jan. 1998.*

[2] M. Al-Shyoukh, H. Lee, and R. Perez, "A transient-enhanced low-quiescent current low-dropout regulator with buffer impedance attenuation," *IEEE J. Solid-State Circuits, vol. 42, no. 8, pp. 1732–1742, Aug. 2007.*

[3] Y. H. Lam and W. H. Ki, "A 0.9 V 0.35 m adaptively biased CMOS LDO regulator with fast transient response," *in IEEE ISSCC Dig. Tech. Papers, 2008, pp. 442–626.*

[4] K. N. Leung and Y. S. Ng, "A CMOS low-dropout regulator with a momentarily current-boosting voltage buffer," *IEEE Trans. Circuits Syst. I, Reg. Papers, vol. 57, no. 9, pp. 2312–2319, Sep. 2010.*

[5] M. Ho, K. N. Leung, and K. L. Mak, "A low-power fast-transient 90 nm low-dropout regulator with multiple small-gain stages," *IEEE J. Solid-State Circuits, vol. 45, no. 11, pp. 2466–2475, Nov. 2010.*

[6] C. Zhan and W. H. Ki, "An output-capacitor-free adaptively biased low-dropout regulator with subthreshold undershoot-reduction for SoC," *IEEE Tran. Circ. Syst. I: Reg. Papers, vol. 59, no. 5, pp. 1119-1131, May 2012.*

[7] G. W. Besten and B. Nauta, "Embedded 5 V-to-3.3 V Voltage Regulator for Supplying Digital IC's in 3.3 V CMOS Technology," *IEEE JSSC, pp. 956-962, Jul. 1998.*

[8] T. Toifl, etc., "A 1.25–5 GHz Clock Generator With High-Bandwidth Supply-Rejection Using a Regulated-Replica Regulator in 45-nm CMOS," *IEEE JSSC, pp. 2901-2910, Nov. 2009.*

[9] Kamyar Keikhosravy and Shahriar Mirabbasi "A 0.13- CMOS Low-Power Capacitor-Less LDO Regulator Using Bulk-Modulation Technique "*IEEE Trans, Regular Papers, VOL. 61, NO. 11, Nov. 2014.*

[10] Y. Liu, C. Zhan, L. Cheng and W. H. Ki, "A chip-area-efficient CMOS low-dropout regulator using wide-swing voltage buffer with parabolic adaptive biasing for portable applications," *IEEE Asian Solid-State Circ. Conf.,* Kobe, Japan, pp. 233-236, Nov. 2012.

Area efficient NMOS based positive and negative voltage multiplier

Vikas Rana
STMicroelectronics Pvt. Ltd
Greater Noida, India
Vikas18feb@gmail.com

Abstract— **This paper presents NMOS based voltage multiplier circuit that can be used to generate both high positive and negative voltages from single charge-pump circuit. Basic, voltage multiplier unit consists of two phase clock signals, charge transfer NMOS transistors and bootstrapped configuration to boost the gate drive of NMOS transistors. Due to use of only NMOS transistors, output resistance of circuit is lower than conventional PMOS based circuits thus able to drive high load current. Electrical conditions of all devices used in the circuit is managed in such a way that there is no electrical stress across any device used in design. Circuit is design and implemented in BCD-110nm technology using conventional (No DMOS) transistors.**

Keywords—*positive charge-pump, negative charge-pump, voltage multiplier, voltage doubler, EEPROM, Flash memory, DC-DC convertor, high voltages*

I.   INTRODUCTION

There is continues increase in demand of new features (like voice access, email and web access, music, video, gaming etc.) in portable devices [1-2]. As the number of features and functions embedded into portable devices are increasing, so does the usage of embedded Non Volatile Memory (eNVM). This eNVM require high positive and negative voltages to perform memory write and erase operations. Generally, these voltages are not provided from external pins, so needs to be generated in-situ with positive and negative charge-pump circuits. A generic approach is to use separate charge-pumps for positive and negative voltages, which require huge on-chip capacitor (separately for positive and negative charge-pumps). In addition, we need to use separate voltage detector and regulators (to regulate output voltage at desired voltages) for charge-pump that again implies using big on-chip resistors. In order to resolve this issue, a new voltage multiplier circuit and associated charge-pump circuit is proposed that can be used to generate both high positive and negative voltages. With this architecture, a single charge-pump can generate both positive and negative voltages thus on-chip capacitor can be reduced to almost half. Similarly, a single voltage regulator system can be used for charge-pump, thus we can reduce the charge-pump area to almost half.

Figure 1a, shows the block diagram of a typical charge-pump circuit. The core of charge-pump consist of switched capacitor based voltage multiplier circuit that can be used in cascaded stages to generate high voltage. A regulation system that consist of a resistive ladder (to detect the voltage level) and an operational amplifier (to compare the detected voltage with a

known reference voltage) is used to regulate the output of charge-pump. Output of operational-amplifier (used as comparator) controls the clock (generated through on-chip oscillator) going to charge-pump and eventually regulates the output to desired voltage level.

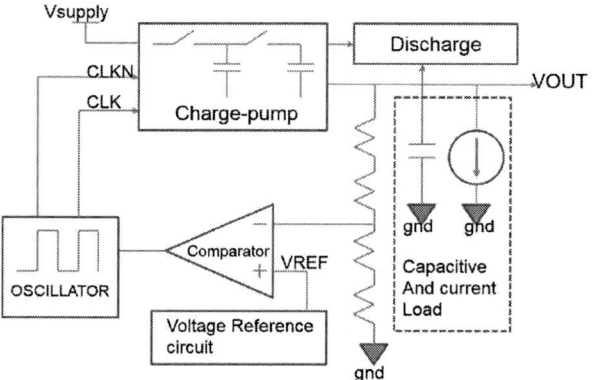

Fig. 1a: Block diagram of conventional charge-pump

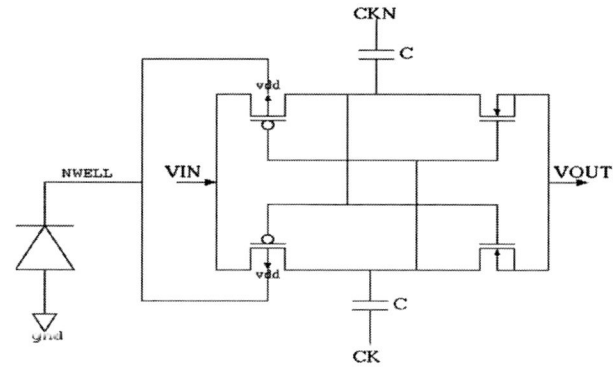

Fig. 1b: Latch based charge-pump [3]

A. Related work:

There have been several important researches done on different charge-pump architectures and few relevant researches are briefed below:

The most common and widely used architecture is "Latch based charge-pump [3]. As shown in figure 1b, this circuit consist of a CMOS latch (using two NMOS and two PMOS transistors) and boost capacitors. Circuit works fine for positive

978-1-5386-7100-9/18 $31.00 © 2018 IEEE

voltage generation but not suitable for negative voltage generation. In negative voltage generation case, bulk of PMOS transistors cannot be connected to source because source can go to negative and in that situation, NWELL to P-SUBSTRATE junction will be forward biased causing functionality failure. If we keep NWELL biased to supply voltage then Bulk to Source/Drain junctions will be electrically stressed for higher negative voltage case. Another problem will be that PMOS will be working in body effect, causing the increase in threshold voltage and thus reducing the efficiency of circuit in negative voltage case.

The work in reference [4] describes a four-phase clock based charge-pump scheme where gate and body of charge transfer switches are controlled to increase the efficiency of charge-pump. This strategy is usable for both positive and negative voltage generation. However, author has not discussed anything about generating positive and negative voltages from a single circuit.

Another architecture is "Dickson charge-pump with Gate Drive Enhancement" [5] where unidirectional transistor's gate drive is enhanced using bootstrapped structure. This structure uses PMOS transistors as switch, which is good for positive voltage generation but not very well suited for negative voltage generation.

In next sections, this paper is organized as follows. Section II explain the architecture: basic voltage multiplier cell, operating principle of basic unit cell, positive voltage charge-pump, negative voltage charge-pump, design considerations for charge-pump and switches. Section III explain the post-layout spice simulation results of design. Finally, section IV draws the conclusion.

II. ARCHITECTURE OF PROPOSED CHARGE PUMP

A. Basic voltage multilier unit

A charge-pump [6] is a switched capacitor based circuit that uses capacitors as energy storage device and Charge transfer switches (CTS) transfer the charge from one stage to other with a timed sequence, to obtain desired voltage conversion. Conventional CMOS technology does not support implementation of these charge transfer switches that can be used for both positive and negative voltages. It is because of following reasons:

1) As shown in figure 2a, NMOS is fabricated in PWELL which is physically connected to P-Substrate. So we can not bias bulk of NMOS to any voltage other than GND. But in negative voltage case, bulk can not be kept at GND otherwise BULK to SOURCE and BULK to DRAIN junction will be forward biased.

2) PMOS can not be used for negative voltages case.if we use it then source can not be connected to bulk (NWELL here), as sourec can go negative and if sourec and bulk are connected then NWELL to P-substrate junction diode (diode D1 in figure 2a) can be forward biased (as P-substrate is always connected to GND).

In order to implement these CTS, we need triple well technology that can support both positive and negative voltages. As shown

in figure 2b, a deep NWELL structure is doped that isolates the PWELL from P-substrate. NMOS is fabricated in this isolated PWELL so that bulk of NMOS is isolated from P-substrate and it can be biased to any voltage different from GND. PMOS can be fabricated within this deep NWELL. As shown in figure 2b, there are two parasitic diodes in this structure: D1 between isolated PWELL and deep NWELL and D2 between deep NWELL to P-Substrate. During designing any circuit in triple well technology, both of these diodes needs to be reverse biased.

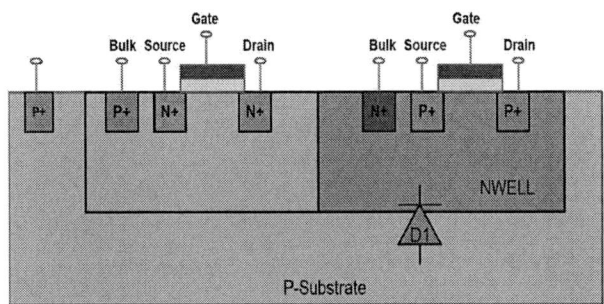

Fig. 2a: NMOS and PMOS in CMOS technology

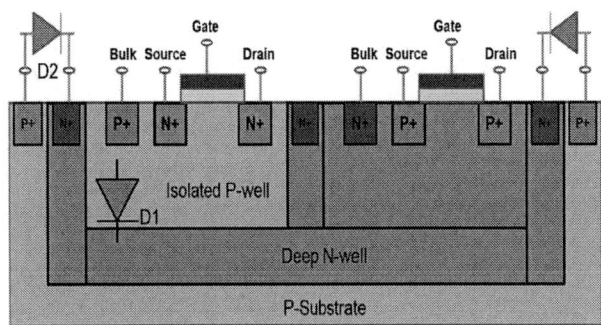

Fig. 2b: NMOS and PMOS in Triple-Well technology

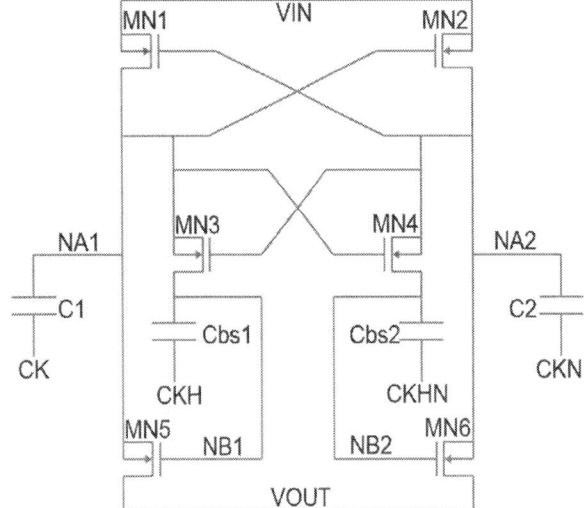

Fig. 3a: Basic Voltage Multiplier unit cell

978-1-5386-7100-9/18 $31.00 © 2018 IEEE

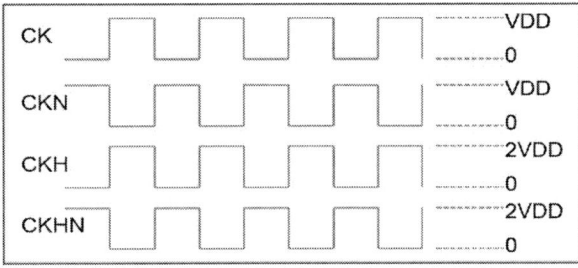

Fig. 3b: Two phase clock used in proposed architecture

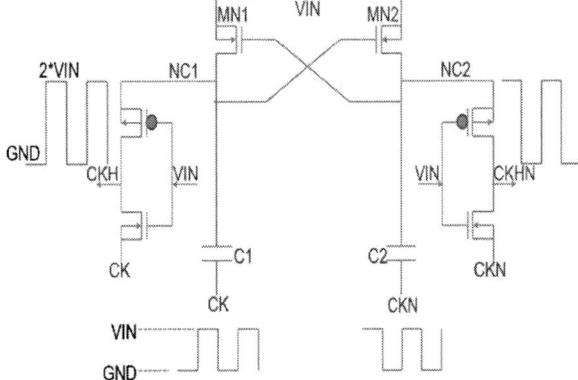

Fig. 3c: Boosted clock generator

Figure 3a, shows the circuit diagram of proposed NMOS based voltage multiplier circuit [7]. NMOS MN1 and MN2 are used in cross-coupled configuration. Capacitors C1 and C2 are boost capacitors driven by CK and CKN clock signals respectively. Switch MN3 combined with capacitor Cbs1 makes a bootstrapped configuration to bias gate of transistor MN5. Similarly, switch MN4 and Cbs2 makes bootstrapped configuration to bias gate of MN6. Other terminal of bootstrapped capacitors Cbs1 and Cbs2 are connected to CKH and CKHN clock signals. As shown in figure 3b, CKH/CKHN signals are in phase with CK/CKN signals but amplitude of CKH/CKHN signals is more than CK/CKN signals. As shown in figure 3c, CKH and CKHN signals are generated using a capacitor based clock level shifter circuit.

B. Operating principle of basic unit cell

This circuit (shown in figure 3a) has capability to generate both positive and negative voltages. When circuit is configure as positive voltage generation, then supply voltage VDD is applied to "VIN" node and with no clock condition, nodes NA1 and NA2 will charge to "VDD – Vtn" voltage levels. As soon as clock is enabled, assuming CK is "VDD", CKN is "0", CKH is "2*VDD" and CKHN is "0". Due to coupling effect across capacitors C1 and C2 (in Figure 3a), node NA1 will charge to "2VDD-Vtn" and NA2 will charge to "VDD". Due to cross-coupled configuration of MN3 and MN4, NB1 and NB2 will charge to "3VDD" and "VDD" respectively. As NB1 is at "3VDD" and NA1 is at "2VDD" so NMOS MN5 has sufficient Vgs to pass "2VDD" voltage from node NA1 to VOUT. In this

way, a positive voltage (higher than input supply voltage) is generated and pass on to output node to drive capacitive and current load. During different clock cycles, node NA1 and NA2 switches between "VDD" and "2VDD". Similarly, NB1 and NB2 toggles between "VDD" and "3VDD".

When the same circuit operates as negative voltage generator, then clock configuration remains same, but input is applied at node "VOUT" and output is taken from terminal "VIN". During negative voltage configuration, "VOUT" node is connected to "GND" and in no clock condition, NA1 and NA2 will discharge to "Vtn" voltage level. When CKH goes to "2VDD" (at this time CK is "VDD") it switch ON the NMOS MN5 and charges node NA1 to "0". During next clock cycle, when CKH switch from "2VDD" to "0" and CK changes state from "VDD" to "0" then node NA1 move from "0" to "–VDD". In addition, node NB1 discharge to "–VDD" via transistor MN3 and switch-OFF the transistor MN5. In this way, node NA1 reaches to "–VDD" voltage level. Due to effect of CKN and CKHN, node NA2 gets charge to "0" via MN6. As NA2 is at "0" and NA1 is at "–VDD", this configuration switch-ON the transistor "MN1" and pass "–VDD" voltage to "VIN" node. In this way, a negative voltage is generated and pass on to "VIN" node. During negative voltage configuration, node NA1 and NA2 toggle between "0/–VDD" and vice versa. Similarly, nodes NB1 and NB2 toggle between "VDD/–VDD" and vice versa.

As explained, a single multiplier circuit can be configure to generate both positive and negative voltages. In order to generate high positive or negative voltages, more number of stages can be cascaded in series that is explained in subsequent sections.

C. High voltage positive charge-pump

Figure 4, shows the complete circuit diagram of four-stage high voltage charge-pump [8] that can be used for both positive and negative voltages. Core of the circuit is designed using series connected stages of basic voltage-multiplier unit cell (discussed in last section). Here four stages are cascaded to generate high voltages (but this is not the limitation of circuit and one can use different number of stages according to specification). An internal oscillator generates clock signal required for charge-pump. Clock signals are adequately buffered before applying to charge-pump. Same clock (CK/CKN) is used to generate boosted clock (CKH/CKHN) using circuit shown in figure 3c. As shown in Figure 4, there are two resistive ladders used in charge-pump: one for positive voltage detection (right one) and another for negative voltage detection (left one). Output of two ladders is multiplexed using analog multiplexer and applied to one terminal of operational amplifier, where it is compared with a known reference voltage to control the clock signals provided to charge-pump thus achieving desired regulated output voltage.

In positive voltage configuration, signal "En_pos_n" is "GND" to switch-ON PMOS transistor "MP1". Similarly, "Vcasc_Neg" is used to enable "Vsupply" path to input of charge-pump. At the same time signal, "En_neg" is disable to cut-off the ground path from node "Vpos". Once oscillator is enable, all clock signals start driving the capacitors and whole

978-1-5386-7100-9/18 $31.00 © 2018 IEEE

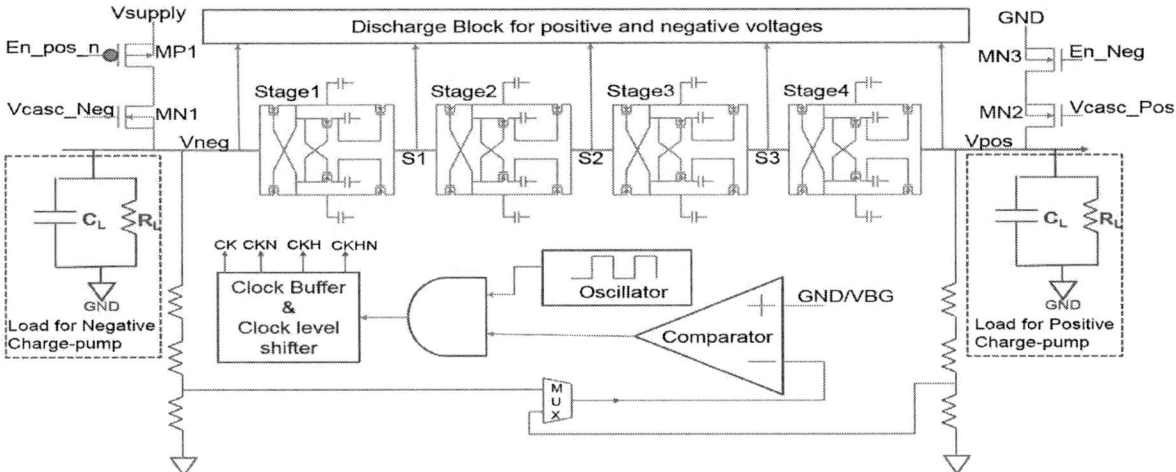

Fig. 4: High voltage positive and negative charge-pump

system operates in positive voltage configuration. Due to the voltage multiplier effect of basic unit cell, we get "2*Vsupply" voltage at node "S1", "3*Vsupply" voltage at node S2, "4*Vsupply voltage at node S3 and finally "5*Vsupply" voltage at node Vpos.

In the positive voltage configuration, output of charge-pump is given by equation 1.

$$Vpos = (VIN + N*V_{CK}) - N*V_{PAR} - N* I_{OUT}*R_{OUT} \qquad (1)$$

For four stages, N = 4, and here we assume that circuit is working at single supply "VDD", so VIN = V_{CK} = VDD, V_{PAR} is the voltage drop across parasitic components at different nodes, I_{OUT} is the current load at output and R_{OUT} is the output impedance of circuit in positive voltage configuration.

$$\text{Thus} \qquad Vpos = 5*VDD - 4*V_{PAR} - 4* I_{OUT}*R_{OUT} \qquad (2)$$

During the positive charge-pump configuration, charge-flow from input to output is as follow:

$$Vsupply \rightarrow Vneg \rightarrow S1 \rightarrow S2 \rightarrow S3 \rightarrow Vpos.$$

D. High voltage negative charge-pump

In the negative voltage configuration, A DC input is applied at "Vpos" node and output is taken from "Vneg" node. Therefore, "Vsupply" path is disable by switching-OFF transistors MP1 and MN1. In addition, transistors MN2 and MN3 are switch-ON to apply "GND" path to "Vpos" node. Once clock signals are enable, circuit starts operating in negative voltage configuration. Due to negative voltage generation feature of voltage multiplier circuit, we get "-Vsupply" at node "S3", "-2*Vsupply" at node "S2", "-3*Vsupply" at node "S1" and eventually "-4*Vsupply" voltage at "Vneg" node.

In negative voltage configuration, output voltage is given by equation 3:

$$Vneg = (Vpos - N*V_{CK}) + N*V_{PAR} + N* I_{OUT}*R_{OUT} \qquad (3)$$

Here V_{PAR} is the voltage drop due to parasitic components at different nodes in circuit, R_{OUT} is the output resistance offered circuit in negative voltage case and IOUT is the load current in circuit. As input is applied at "Vpos" node so Vpos = "0", for four stages N = 4, and V_{CK} = VDD. So equation 3 can be written as:

$$Vneg = -4*VDD + 4*V_{PAR} + 4* I_{OUT}*R_{OUT} \qquad (4)$$

During the negative charge-pump configuration, charge-flow from input to output is as follow:

$$GND \rightarrow Vpos \rightarrow S3 \rightarrow S2 \rightarrow S1 \rightarrow Vneg$$

In this way, through a proper combination of switches and a novel basic voltage multiplier cell we can achieve both high positive and negative voltages from a single circuit.

E. Design considerations for charge-pump and switches

Charge transfer switches (CTS) plays a very important role in determining the efficiency of charge-pump. These switches carry the charge developed across the boost capacitor to next stage and helps in building the high voltage (either positive or negative) at final stage. However, in output equation of charge-pump (referring to equation 2 and 4) there are two main components "$I_{OUT}*R_{OUT}$" and "V_{PAR}" that determines the efficiency of charge-pump. Here "R_{OUT}" is given as:

$$R_{OUT} = 1/f_{clk}*C + R_{ON} \qquad (5)$$

Here C1 = C2 = C is stage capacitance

"f_{clk}" is operating clock frequency

"R_{ON}" is the ON resistance of switch (MN1, MN2, MN5 and MN6 in figure 2a).

For a NMOS transistor, ON resistance can be written as:

$$R_{ON} = L/\mu C_{OX}W (V_{GS}-V_{TH}) \qquad (6)$$

978-1-5386-7100-9/18 $31.00 © 2018 IEEE

From equation 6, it is evident that if we want to have small switch resistance then we need to keep bigger width of charge transfer switch (CTS) in design. Alternatively, in order to reduce the output resistance offered by circuit, we can increase the operating frequency of clock signals (referring to equation 5).

Therefore, if charge-pump has to drive very high current then switch resistance should be very low so that very small voltage drops across switch.

Another important parameter to control is "V_{PAR}". Here "V_{PAR}" is the voltage drop (or charge loss) across the parasitic capacitances seen on different nodes in charge-pump. These parasitic capacitances are due to following contributors:

1) *Capacitances due to routing metals*
2) *Capacitance due to charge transfer switches*

Due to inherit property of NMOS (used as CTS), it offers finite capacitance, which can be written as:

$$C_{SW} = C_{OX}*W*L_{eff} \qquad (7)$$

Here "C_{OX}" is the capacitance due to gate oxide
"W" is width of transistor
"L_{eff}" is the effective length of transistor

A bigger switch area offers more capacitance. Therefore, in order to have less capacitance contribution by switches its width "W" should be small.
As seen from equation 6 and 7, size of switch is a trade-off in charge-pump design. If load current is very high then it is always better to keep bigger switch size but in other case, if load current is very small then it's better to keep small switch sizes so that it's parasitic capacitance contribution is also small.
One figure of merit of charge-pump circuit is power-efficiency that is the measure of ratio of output power to input power.

$$P_{eff} = (P_{out}/P_{in})* 100\% \qquad (8)$$

$$P_{eff} = \{(V_{OUT}*I_{LOAD})/ (VDD*I(VDD))\}*100\% \qquad (9)$$

Here "V_{OUT}" is the output voltage of charge pump (either positive or negative).

In order to take correct numbers, mean value of "V_{OUT}" and "I(VDD) are taken.

For the input power, I(VDD) is the main contributor. Clock buffers used in each stage of charge-pump consume the maximum current. Therefore, clock drivers and charge transfer switches should be sized appropriately so that charge-lose (or voltage drop) is minimum at each stage and maximum power can be transfer from one stage to next stage.

III. SIMULATION RESULTS

This charge-pump circuit is design in 110nm-BCD technology using standard NMOS and PMOS transistors (not using DMOS). MOM capacitors are used in each stage for clock boost and bootstrapped configuration. Total area occupied by charge-pump is 0.0014mm2.

To compare the performance of proposed charge-pump circuit with respect to conventional circuits, both circuits (proposed and latch based charge-pump [3]) are design and simulated in BCD-110nm technology. Post-layout simulations are performed results are compared for both positive and negative voltage configurations. Results are compared considering following figure of merits of charge-pump:

1) Variation in Output voltage with respect to input voltage (considering no current load).

2) Variation in Output voltage (at fixed supply voltage) with respect to output current load.

3) Power effeciency of circuit with different current load conditions.

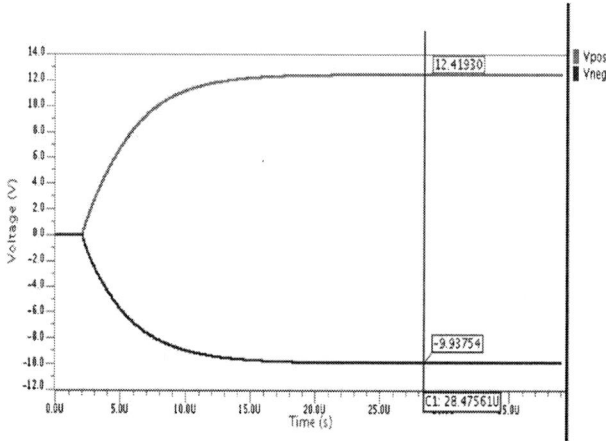

Fig. 5a: Simulation result of proposed circuit in positive and negative configuration

Figure 5a, shows the transient result for proposed circuit for four-stage charge-pump operated in positive and negative configuration. In this simulation, supply voltage is 2.5V, capacitive load is 10pF and operating clock frequency is 50MHz. In positive charge-pump case, we got 12.42V; hence, 80mV of voltage drop (V_{PAR}) occurs due to parasitic elements. For negative charge-pump, we reached up to -9.93V hence 70mV of voltage drop occur due to parasitic elements.

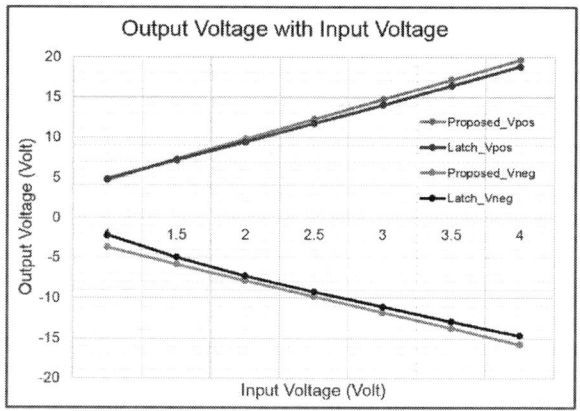

Fig. 5b: Output voltage variation with input supply for both positive and negative configuration

978-1-5386-7100-9/18 $31.00 © 2018 IEEE

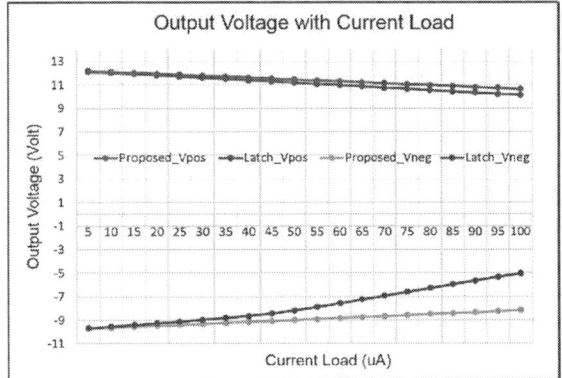

Fig. 5c: Output voltage variation with output current load for both positive and negative configuration

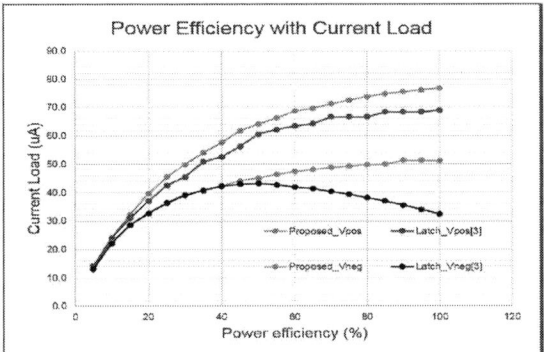

Fig. 5d: Power efficiency with output current load for both positive and negative configuration

Figure 5b, shows the variation in output voltage with input supply voltage. Here input supply voltage is varied from 1V to 4V and accordingly output voltage is checked for both positive and negative voltage configurations. It is quite evident from the results that proposed circuit is able to work properly down to 1V. Results are obtained for both proposed circuit and Latch based circuit in positive and negative voltage configurations and compared. It is evident from results that proposed circuit have better voltage efficiency and generates more positive and negative voltages than conventional circuit.

Figure 5c, shows the variation in output voltage with increase in output load current. Output current is varied from 5uA to 100uA and accordingly output voltage is observed. There is a drop in output voltage due to finite output impedance shown by circuit in both positive and negative charge-pump configuration. As compare to conventional circuit, proposed circuit have less output impedance. It is due to use of NMOS transistors current driving capability is better with proposed architecture thus less voltage drops at output terminal. More specifically for negative voltage generation case, where PMOS starts coming in body effect and thus output impedance start increasing for high negative voltages thus increasing the output impedance.

Figure 5d, shows the power efficiency of proposed circuit in positive and negative configuration and its comparison with

Latch based circuit. Here power efficiency is shown with respect to different output current load. On the X-axis, power efficiency of circuit if plotted and on Y-axis, output current load is plotted. For positive voltage, case maximum efficiency achieved is up to 75% and for negative voltage case, it is up to 51.4%. As compare to conventional architecture, positive charge-pump case have almost 5% better efficiency for high current. For negative voltage case, proposed architecture is much better than Latch based architecture. It is due to use of PMOS transistors in Latch based architecture, PMOS transistor starts operating in high body effect and thus reduces the power efficiency significantly. However, in proposed architecture, we have used NMOS transistors for negative voltage generation and it does not suffer from body effect issue.

IV. CONCLUSION

A novel NMOS based voltage multiplier scheme is discussed that can be used to generate high positive and negative voltages. Due to use of only NMOS transistors, it solve the issue of bulk management faced in latch based negative charge-pump circuits. Proposed circuit is better in terms of voltage efficiency, output impedance and power efficiency with respect to conventional charge-pump. This feature of generating both positive and negative voltages from a single circuit decreases the area of charge-pump by 40% as compared to conventional architectures because now we do not have to replicate pumping capacitors separately for positive and negative voltage circuits. Thus, a single charge-pump can generate positive and negative voltages.

V. REFERENCES

[1] Kangho Lee, Jimmy J. kan, Seung H. Kang, "Unified embedded non-volatile memory for emerging mobile markets" 2014 IEEE/ACM International Symposium on Low Power Electronics and Design (ISLPED), pp. 131–136.

[2] Seung-Hwan Song, Ki Chul-chun, Chris H. Kim, "A Logic compatible Embedded Flash Memory for Zero stand-by power System-on-Chips Featuring a multi-story high voltage switch and a selective refresh scheme" IEEE Journal of Solid State Circuits, Year – 2013, Vol. 48, pp. 1302–1314.

[3] R. Pelliconi; D. Iezzi; A. Baroni; M. Pasotti; P. L. Rolandi, 'Power efficient charge pump in deep submicron standard CMOS technology', IEEE Journal of Solid-State Circuits, 2003, Volume 38, pp. 1068-1071, doi: 10.1109/JSSC.2003.811991

[4] Takanori Yamazoe, Hisanobu Ishida, Yasutaka Nihongi, "A charge pump that generates positive and negative high voltages with low power-supply and low power consumption for Non-Volatile memories" IEEE International Symposium on Circuits and Systems, ISCAS'2009, Year – 2009, pp. 988 – 991.

[5] Hesheng Lin, Wing Chun Chan, Wai Kwong Lee, Zhirong Chen, and Min Zhang: 'Dickson Charge Pump with Gate Drive Enhancement and Area Saving' Journal of Power electronics, May 2016, Volume 16, No. 3, pp. 1209-1217, doi: 10.6113/JPE.2016.16.3.1209.

[6] Janusz A. Starzyk: 'A DC-DC charge pump design based on voltage doubler', IEEE Transaction on Circuits and System-I, March 2001, VOL.48, No.3, pp. 350–360, doi: 10.1109/81.915390

[7] V.Rana, M. Pasotti, F. desantis, "Voltage doubling circuit and charge pump applications for the voltage doubling circuit" US patent 9634562

[8] R.R. Anantha, A. Srivastava, P. K. Ajmera, "Charge pump CMOS circuit based on internal clock voltage boosting for bio-medical applications", proceedings of SPIE, vol. 5763, pp. 11-13, 2005, doi: 10.1117/12.600546

Achieving Low Power Classification with Classifier Ensemble

[1]Fanglei Hu, [2]Min Zhang, and [1]Hailong Jiao

[1]VLSI Lab   [2]Nanoelectronics Lab

School of Electronic and Computer Engineering, Peking University Shenzhen Graduate School
Li-Shui Road 2199, University Town, Nanshan District, Shenzhen, China

Email: jiaohl@pkusz.edu.cn

Abstract—Machine learning algorithms, such as Support Vector Machine (SVM) and Artificial Neural Networks, have been widely applied in many aspects of daily life. Low power/energy integrated circuit implementation of machine learning algorithms with high accuracy is however still a great challenge, which is critical for portable and wearable devices. In this paper, classifier ensemble is investigated for achieving low power classification. The classifier ensemble algorithm Adaboost is employed to combine multiple SVM classifiers with linear kernel to achieve high classification accuracy while reducing the hardware complexity. The proposed classifier ensemble is evaluated on the MNIST dataset by using a 45-nm CMOS technology. Compared to the traditional SVM classifier with second-order polynomial kernel, the proposed classifier ensemble achieves up to 45.7%, 20.3%, and 20.3% savings in total energy consumption, leakage power consumption, and area, respectively, while providing similar classification accuracy.

Keywords—Machine learning, Support Vector Machine, kernel, Adaboost, energy efficiency.

I. INTRODUCTION

Machine learning is extensively used in applications such as facial detection [1], voice recognition [2], driver assistance system [3], and disease detection [4]. Many of the applications run on portable or wearable devices which have stringent power/energy requirement. Efficient implementations of machine learning algorithms on integrated circuits (ICs) are therefore highly desirable.

Support Vector Machine (SVM) and Artificial Neural Networks (ANN) are among the most popular machine learning algorithms due to their excellent performance in various classification and/or detection problems. A variety of low power implementations of SVM and ANN are presented in the literature. In [5], an efficient hyperbolic tangent function is proposed to replace the energy-hungry activate function in ANN. In [6], approximate multipliers are employed to replace the accurate multipliers in ANN to lower the energy consumption. In [7], a specialized processing dataflow "row stationary" is proposed to minimize the data movement between the processing unit and the memory circuits in a convolutional neural network (CNN) accelerator, thereby enhancing the energy efficiency. An SVM implementation is proposed in [4] for epileptic seizure detection. A linear kernel is employed to reduce the number of calculations, thereby minimizing the energy consumption and area. Such an SVM requires a complicated patient-specific training to achieve reasonable classification accuracy due to the utilization of the linear kernel. In [8], a Mercer's kernel (which is an alternative to the Gaussian kernel) is used to simplify the hardware implementation of SVM. The advantage of the Mercer's kernel is that no multipliers are required. However, more tuning of the kernel is required compared to the conventional Gaussian kernel. An SVM implementation is proposed in [9] for biomedical applications. Three different kernels and inline shifters are used to prevent overflow. The energy consumption of this implementation is however not optimized since the energy-intensive Gaussian kernel is used. In [10], similar to [6], approximate multipliers are used in the implementation of SVM to lower the energy consumption while maintaining the classification accuracy.

Different from SVM or ANN, classifier ensemble (CE) is an alternative machine learning algorithm to achieve high classification accuracy. In [11], the hardware implementation of a classifier ensemble is presented for enhanced performance of color classification, which uses decision tree as the base classifier. In [12], error-adaptive classifier boosting (EACB) is used to overcome computational errors that are caused by the process variations of semiconductor devices. In [13], the same idea as [12] is employed in a large-area image sensing and detection system that integrates, on glass, sensors and thin-film transistor (TFT) circuits for classifying images from sensor data. In [14], error-resilient machine learning architecture by employing random forest (RF) is presented to tackle the circuit variations in subthreshold voltage region. In [15], multiple SVM classifiers are assembled to predict the timing failure caused by memory blocks during the floorplan stage in physical design, thereby reducing the iterations in the backend design of ICs. However, potential of using classifier ensemble to achieve low power classification has not been explored in those previous works.

In this paper, classifier ensemble is employed to implement on-chip low-power highly-accurate classifiers. The classifier ensemble algorithm, Adaboost, is used to assemble multiple SVM classifiers with linear kernel to achieve similar classification accuracy as the SVM classifier with second-order polynomial kernel. The computation complexity of the classifier ensemble algorithm is reduced compared to the SVM classifier with polynomial kernel, thereby simplifying

978-1-5386-7100-9/18 $31.00 © 2018 IEEE

the hardware implementation. Significant power, energy, and area savings are therefore achieved by the proposed classifier ensemble.

The paper is organized as follows. The concept of SVM is introduced in Section II. The proposed classifier ensemble algorithm based on Adaboost is presented in Section III. The hardware implementation of the proposed classifier ensemble is described in Section IV. The evaluation and comparison of the proposed classifier ensemble and the traditional SVM classifier are evaluated and compared at both the software and hardware levels in Section V. The paper is summarized in Section VI.

II. SUPPORT VECTOR MACHINE

Support Vector Machine [16] is one of the most popular supervised learning algorithms, which is widely used in classification and regression. SVM consists of two steps: training and testing. During training, a labeled dataset with a number of vectors $\{\vec{x}, y\}$, where $\vec{x}$ is a data sample vector with d dimensions and y (± 1) is a data label, is used to position a hyper plane with maximum margin which can separate this set of d-dimensional data. The samples, which are on the boundary of a class and thereby define the margin of the hyper plane, are the support vectors.

The training yields an SVM-model that consists of the following parameters: support vectors ($\overrightarrow{sv}$), support vector labels (y), Lagrange multipliers (α), and a bias (b). This SVM model is further used in the classification step by following the decision rule:

$$y_u = sign(b + \vec{w} \cdot \vec{u}^T), \qquad (1)$$

where $\vec{w}$ is the weight vector. $\vec{u}$ is the sample vector with an unknown class (the length of $\vec{u}$ is the number of features). y_u is the predicted class of $\vec{u}$. The weight vector is perpendicular to the hyper plane and sets the position of the plane. The bias defines an offset of the plane. $\vec{w}$ can also be described as the weighted sum of support vectors ($\overrightarrow{sv}$),

$$\vec{w} = \sum_{i=0}^{N} y_i \alpha_i \overrightarrow{sv_i}, \qquad (2)$$

where N is the number of support vectors. It is typically not possible to separate two classes by a linear hyper plane for all datasets. The space of a non-linear problem has to be translated with a function ϕ such that the problem becomes linear. The kernel that is used by SVM is

$$K(x_1, x_2) = \phi(x_1) \cdot \phi(x_2)^T. \qquad (3)$$

The three most commonly used SVM kernels are listed in Table I. By combining (1), (2), and (3),

$$y_k = sign(\sum_{i=1}^{N} y_i \alpha_i \phi(\overrightarrow{sv_i}) \phi(\overrightarrow{u_k}) + b). \qquad (4)$$

TABLE I
THE MOST COMMONLY USED SVM KERNELS

Linear	$K(\vec{x}_1, \vec{x}_2) = \vec{x}_1 \cdot \vec{x}_2^T$	
n-order Polynomial	$K(\vec{x}_1, \vec{x}_2) = (1 + \vec{x}_1 \cdot \vec{x}_2^T)^n,$	$P > 0$
Gaussian	$K(\vec{x}_1, \vec{x}_2) = \exp(-\|x_1 - x_2\|^2/(2 * \sigma^2))$	

III. CLASSIFIER ENSEMBLE

Ensemble algorithm is a classical machine learning algorithm to enhance the learning capability [17]. The key of the ensemble algorithm is aggregating the result of multiple individual classifiers which are named base classifiers. Furthermore, the property of each base classifier should be different from each other, so that they can compensate for the weakness of each other to achieve better performance when assembled. In this work, instead of targeting only better classification performance, the potential of power savings with classifier ensemble is explored.

Bagging and boosting are the two most typical ensemble algorithms. Bagging uses the same model for each base classifier. However, the parameters for the model of each base classifier are different. The different sets of parameters are obtained by training each base classifier with different random subsets from the training set. The ensemble makes a prediction by aggregating the predictions of all base classifiers. The aggregation function is typically a statistical model. Different from bagging, boosting uses the whole training set to train each base classifier. The base classifiers are trained sequentially. The base classifiers that are trained later try to avoid the mistakes that are made by the base classifiers that are trained earlier. Adaboost (adaptive boosting) is the most popular boosting algorithm. During training, a new base classifier corrects its predecessor by allocating larger weights to the training instances that are under-fitted. This means that the new classifiers focus more on the difficult cases in the training set. As a natural consequence, different weights are assigned to each base classifier to achieve diversity. In terms of hardware implementation, bagging requires more memory storage compared to boosting since different parameters are required for different base classifiers. Boosting is therefore preferable to bagging for saving power consumption and layout area.

In this paper, Adaboost is used to combine the base classifiers. Assuming M base classifiers are combined, the classification result of Adaboost is

$$sign(f(x)) = sign(\sum_{m=1}^{M} \beta_m G(x)_m), \qquad (5)$$

where $G(x)_m$ is the result of each base classifier. β_m is the weight of each base classifier. SVM with linear kernel is employed as the base classifier in this work. The principle of the proposed algorithm (classifier ensemble of SVM with linear kernel, LSE) is illustrated in Fig. 1.

By putting the linear kernel in Table I to (4),

$$K(x_1, x_2) = \vec{x}_1 \cdot \vec{x}_2^T. \qquad (6)$$

Therefore, the final decision of the classification after ensemble is

$$y_k = sign(\sum_{m=1}^{M} \beta_m sign(\sum_{i=1}^{N} y_i \alpha_i \overrightarrow{sv_i} \overrightarrow{u_k} + b)), \quad (7)$$

where y_k is the classifier result. $\overrightarrow{u_k}$ is the test data. Depending on the sign (positive or negative), the test case can be classified as one class or another.

978-1-5386-7100-9/18 $31.00 © 2018 IEEE

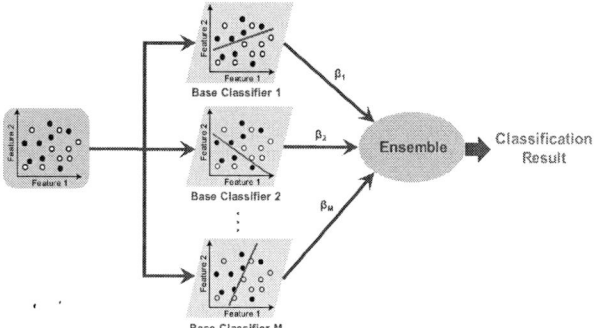

Fig.1. Principle of the proposed algorithm (classifier ensemble of SVM with linear kernel, LSE).

IV. HARDWARE IMPLEMENTATION OF THE CLASSIFIER

The hardware architecture of the proposed classifier ensemble LSE is described in this section. The hardware complexity of classifier ensemble is also compared with the SVM classifier with polynomial kernel (poly-SVM).

A. Hardware Architecture of Classifier Ensemble

In this work, SVM classifier with linear kernel is chosen as the base classifier for the proposed classifier ensemble. According to (2), the weights of support vectors can be pre-computed. The pre-computed result is stored in

$$\vec{w} = \sum_{i=1}^{N} y_i \alpha_i \overrightarrow{sv_i} . \qquad (8)$$

Then in the first on-chip computation step, the test data $\overrightarrow{u_k} = [u_{k1}, u_{k2}, \dots u_{kd}]$ is multiplied with the weight vector $\vec{w}$.

$$q_k = [u_{k1}, u_{k2}, \dots u_{kd}] \begin{bmatrix} w_1 \\ w_2 \\ \vdots \\ w_d \end{bmatrix} . \qquad (9)$$

In the second step, the bias is added to q_k.

$$r_k = q_k + b_k . \qquad (10)$$

In the third step, the base classifier weight β is multiplied with the sign of r_k. Depending on the sign of r_k, either the original value or the two's complementary value of β is used as the result of the third step. All the base classifiers perform their calculations in parallel. Finally, the results of all the base classifiers are aggregated. The pseudo code of the LSE algorithm is shown in Algorithm I.

The data-path of a single SVM classifier with linear kernel is shown in Fig. 2a. The overall architecture of the LSE classifier is shown in Fig. 2b. Two memories are used to store the parameters which are pre-extracted from the training part. The output of the LSE classifier is the combination of every base classifier. The inputs of the finite state machine include a "start" signal, write enable of the memory for storing w, write enable of the memory for storing β, the address, clock, reset, and the "DinW" which is used to write the parameters into the memory. The test data with unknown class and bias are delivered as data flow into the calculation. The outputs of the architecture include a one-bit result (1 or 0) and a one-bit ready signal.

Algorithm I: Classifier ensemble of SVM with linear kernel (LSE)

[ready, result] = **FUNCTION** LSE(state, start, addr, weW, DinW, din, clk, rst)

```
CASE state OF
    reset:
        i, j, k = addr               > forward mem. address
        weW' = weW
        DinW' = DinW
        IF start == True THEN
            weW' = False
            DinW' = False
            q = 0, result = 0, temp = 0
            i = 0, j = 0, k = 0, l = 0
            state = step1
        ENDIF
    step1:
        q = M_i • U_k + q
        i ++, k ++
        IF i == features THEN       > column finish → step2
            k = biasPosition          > select element for next
            state = step2             > select bias pos. in mem.
        ENDIF
    step2:
        temp = (q + U_k)              > add bias
        l++                           > point to β address
        state = step3;
    step3:
        r = r + sign(temp) * β        > multiply β
        k = 0
        q = 0
        state = step1
        IF i == address.end THEN
            state = finish
    finish:
        result = sign(r)
        ready = True
        state = wait
    wait:                            > wait for result to be read
        IF start == False THEN
            ready = False
            state = reset
        ENDIF
    default:
        state = reset
ENDCASE
```

ENDFUNCTION

B. Complexity Comparison with Second-Order Poly-SVM

The SVM classifier with polynomial kernel (poly-SVM) can achieve similar classification accuracy to the SVM classifier with Gaussian kernel in many applications, yet has lower computation complexity. Poly-SVM is therefore widely used [6], [10]. In this sub-section, the computation complexity of the proposed LSE classifier and a second-order poly-SVM is compared. In order to make the comparison, the hardware implementation scheme of poly-SVM is introduced.

By putting the polynomial kernel in Table I to (4), the decision function of second-order poly-SVM is

$$y_k = sign(\sum_{i=1}^{N} y_i \alpha_i (1 + \overrightarrow{sv_i u_k})^2 + b). \qquad (11)$$

A reorganized form of the equation is

$$y_k = sign(U_k \sum_{i=1}^{N} y_i \alpha_i [1, \overrightarrow{sv_i}^T]^T [1, \overrightarrow{sv_i}^T] U_k^T + b), \qquad (12)$$

where $U_k = [1, u_k^T]$. The matrix $\sum_{i=1}^{N} y_i \alpha_i \overrightarrow{sv_i}^T \overrightarrow{sv_i}$ can be pre-computed and put in a matrix A. Therefore, the decision function becomes

$$y_k = sign \left(U_k \left(\begin{bmatrix} a_{11} & \cdots & a_{1D} \\ \vdots & \ddots & \vdots \\ a_{D1} & \cdots & a_{DD} \end{bmatrix} \right) U_k^T + b \right) . \qquad (13)$$

978-1-5386-7100-9/18 $31.00 © 2018 IEEE

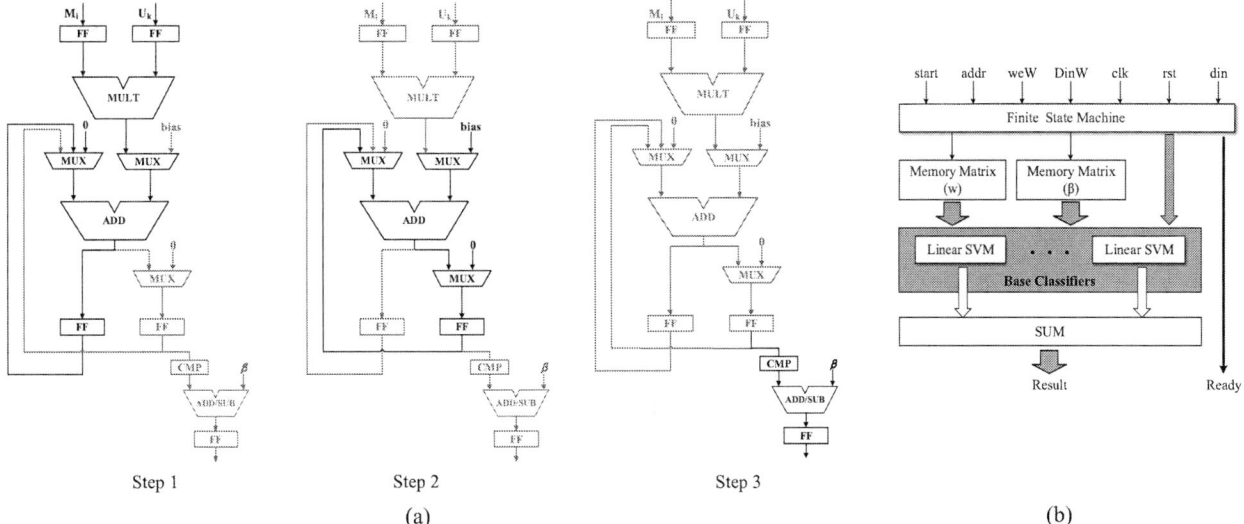

Fig. 2. The data-path and architecture of the proposed LSE classifier. (a) The data-path of a single SVM classifier with linear SVM that is used for ensemble. During one classification, the same architecture is used in the three steps of computation. The blocks that are not used in a certain step are grayed out. (b) The overall architecture of the proposed LSE classifier.

The data-path of the poly-SVM also can be divided into three steps. In the first step, the test data $u_k = [1, u_1, u_2, ... u_d]$ is multiplied with the jth column of matrix A.

$$q_j = \begin{bmatrix} 1 & u_k \end{bmatrix} \begin{bmatrix} a_{1j} \\ \vdots \\ a_{(d+1)j} \end{bmatrix}. \qquad (14)$$

In the second step, q_j is multiplied with the jth value of u_k, resulting in m_j. Then all the m_j's are added together.

$$r = \sum_{\forall j} m_j = \sum_{\forall j} q_j u_j. \qquad (15)$$

Finally, in the third step, the bias is added into (15). The test data with unknown class is classified according to the sign of the final result. The data-path and overall architecture of the second-order poly-SVM is shown in Fig. 3.

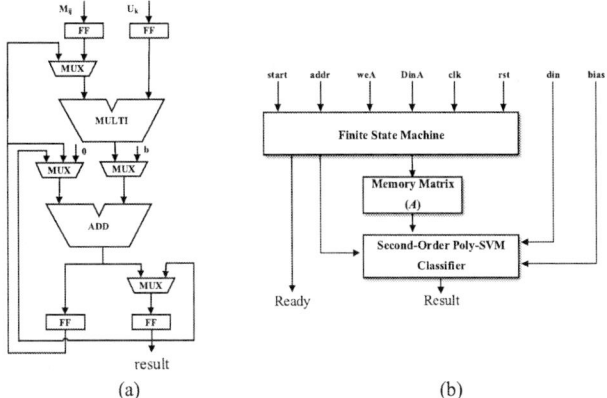

Fig.3.(a) The data-path of the SVM classifier. (b) The overall architecture of the SVM-based platform.

The comparison of the computation complexity between the proposed LSE classifier and the second-order poly-SVM classifier is listed in Table II. Assume the number of features is d. The number of base classifiers for LSE is M. As listed in Table II, LSE has a linear relationship with both d and M. Alternatively, the second-order poly-SVM has a quadratic relationship with d. d is typically significantly larger than M. Therefore, LSE can reduce the number of multiplication, addition, as well as memory access significantly compared to poly-SVM assuming that the same number of features is used for both classifiers, thereby achieving energy savings at the hardware level.

TABLE II
COMPARISON OF THE COMPUTATION COMPLEXITY BETWEEN LSE AND SECOND-ORDER POLY-SVM

	Multiplication	Addition	Memory Access
LSE	$M \times (d+1)$	$M \times (d+2) - 1$	$M \times (d+1)$
Poly-SVM	$(d+1)^2$	$(d+1)^2 + 1$	$(d+1)^2$

V. EVALUATION OF THE CLASSIFIERS

Evaluation of the proposed classifier ensemble and comparison with the poly-SVM classifier in both software and hardware are presented in this section. The MNIST database of handwritten digits [18] is used for the evaluation in this work. All the original digit images in the database have 28×28 pixels. The images are down-sampled to a size of 14×14 pixels. The pixels in each image are then converted in one 196-dimensional vector, where black is represented as a "0" and white as a "1". Without losing generality, the classifiers target to classify the samples in two classes: "digit-4" or "non-digit-4". The training data contains 5000 digit-4's and 5000 non-digit-4's. The test data contains 1000 randomly selected samples from the test dataset of MNIST database.

A. Software (Matlab) Simulation

To reduce the number of features (for lower computation complexity), an efficient feature selection method "F-score" [19] is used. The classification accuracy of different classifiers based on different number of features is listed in Table III. For the proposed classifier ensemble (LSE), three SVM classifiers with linear kernel are aggregated. For the compared individual SVM classifier, three types of kernels are used. When there is no feature selection, the classification accuracy is 97% for the proposed LSE and ~94% for the three individual SVM classifiers. When the feature number is reduced to 10, the classification accuracy of the poly-SVM and Gaussian-SVM decreases slightly to ~92%, while the proposed LSE decreases to 84.8%. To maintain similar (slightly higher) classification accuracy to the individual SVM classifiers, the number of features is increased to 30 for the proposed LSE. The accuracy is thereby revived to 93.4%. As listed in Table III, poly-SVM exhibits competitive performance compared to Gaussian-SVM for this application. Poly-SVM is therefore chosen for further evaluation in this work. A more detailed analysis of the classification with the proposed LSE and poly-SVM is listed in Table IV. The proposed LSE tends to predict the images as "non-digit-4", while poly-SVM displays relatively balanced classification results.

TABLE III
CLASSIFICATION ACCURACY COMPARISON OF DIFFERENT CLASSIFIERS

	Number of features	Classification accuracy
SVM with linear kernel (Linear-SVM)	196	94.1%
	10	80.4%
	30	86.8%
SVM with polynomial kernel (Poly-SVM)	196	94.0%
	10	92.1%
SVM with Gaussian kernel (Gaussian-SVM)	196	94.0%
	10	92.2%
The proposed classifier ensemble with three linear-SVM	196	97.0%
	10	84.8%
	30	93.4%

TABLE IV
DETAILED CLASSIFICATION ANALYSIS OF LSE AND POLY-SVM CLASSIFIERS

		Predict as non-digit-4	Predict as digit-4	Total	Re-score
Poly-SVM	Non-digit-4	821	69	890	0.92
	Digit-4	10	100	110	0.91
	Total	831	169	1000	/
	Pre-Score	0.98	0.59	/	/
LSE	Non-digit-4	853	37	890	0.95
	Digit-4	28	82	110	0.75
	Total	881	119	1000	/
	Pre-Score	0.97	0.69	/	/

To reduce the complexity of hardware, all the on-chip computations are integer-based. However, the offline training is performed with floating point data. The model parameters obtained from the training need to be converted into integers. For poly-SVM, two parameters need conversion: the pre-computed matrix and the bias. Similarly, three parameters need conversion for the proposed LSE: the pre-computed matrix, bias, and β. To minimize the quantization error, every element in pre-computed matrix and bias are multiplied by a factor, which has no effect on the classification result. The conversion is

$$A_{ij} = \left\lceil a_{ij} \frac{2^n}{max\{A,bias\}} \right\rceil, \text{ for } \forall i,j . \quad (16)$$

$$b = \left\lceil bias \frac{2^n}{max\{A,bias\}} \right\rceil . \quad (17)$$

For β,

$$\beta_i' = \lceil \beta_i * 2^n \rceil, \text{ for } \forall i. \quad (18)$$

n is the data width. The variation of the classification accuracy with the data width is shown in Fig. 4. When the data width is larger than eight, the accuracy of both LSE and poly-SVM becomes steady. Therefore, the data width is chosen to be eight in the hardware design.

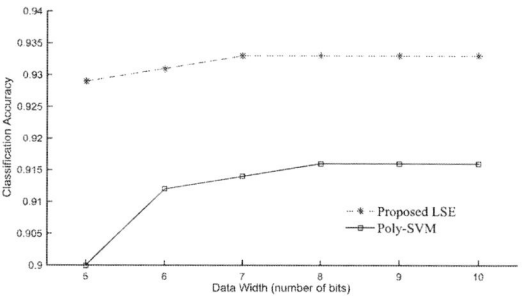

Fig. 4. The variation of the classification accuracy with the data width.

B. Hardware Evaluation

Hardware evaluation of the proposed LSE and second-order poly-SVM is discussed in this sub-section. The RTL code of both classifiers are synthesized using Cadence RTL Compiler with the Cadence open-source 45-nm CMOS technology in the slow NMOS slow PMOS process corner at 125 °C. The power supply voltage is 1.2 V. The synthesis frequency varies from 1 MHz to 400 MHz. The energy consumption per classification, area, and leakage power consumption of the two classifiers are evaluated.

The area of the proposed LSE and poly-SVM classifiers at different synthesis frequencies is shown in Fig 5. When the synthesis frequency varies from 1 MHz to 100 MHz, the area of both classifiers changes slightly since those minimum sized standard cells can satisfy the timing requirements. When the frequency is higher than 100 MHz, higher frequencies lead to larger area. The area of the LSE classifier is smaller than the poly-SVM classifier at all synthesis frequencies. The LSE classifier reduces the area by up to 20.3% and on average 20.1% compared to the poly-SVM classifier.

The comparison of the energy consumption per classification between the proposed LSE classifier and the poly-SVM classifier is shown in Fig. 6. In order to evaluate the energy consumption of the classifiers, Cadence NC-Sim is used to obtain the realistic switching activity of the hardware, in the form of TCF (toggle count format) files. The TCF files are then used for energy evaluation with Cadence RTL compiler. Though the proposed LSE classifier uses three times of feature than the poly-SVM classifier, the number of multiplications and additions involved in the computation is much smaller than the poly-SVM classifier. The LSE classifier thereby achieves significant energy savings at all synthesis frequencies compared to the poly-SVM classifier. The LSE classifier reduces the energy consumption per classification by up to 45.7% and on average 42.3% compared to the poly-SVM classifier, as shown in Fig. 6.

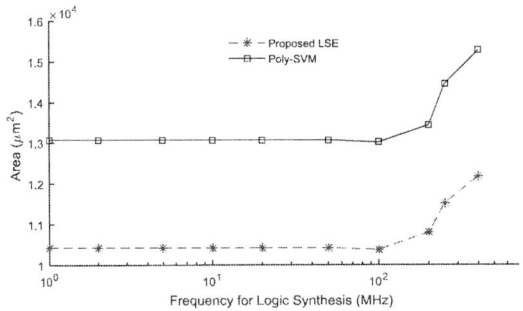

Fig. 5. Area of the proposed LSE and poly-SVM classifiers at different synthesis frequencies.

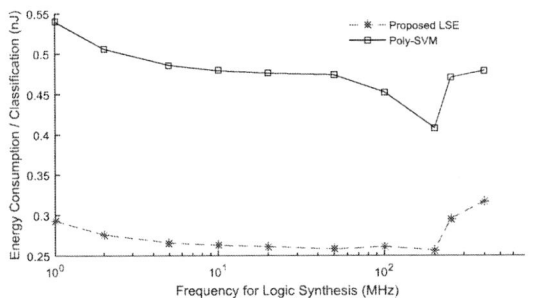

Fig. 6. Energy consumption per classification with the proposed LSE and poly-SVM classifiers at different synthesis frequencies.

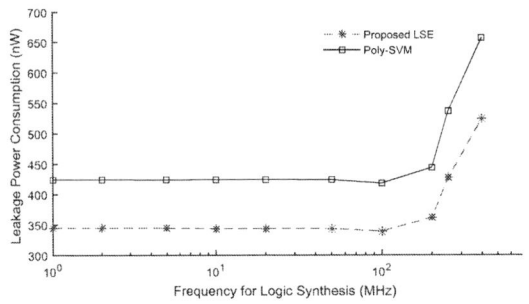

Fig. 7. The leakage power consumption of the LSE and poly-SVM classifiers.

From Fig. 6, it is noticeable that the minimum energy point is not at the lowest synthesis frequency. When the clock frequency is low (e.g. 1 MHz), the leakage currents contribute significantly to the total energy consumption. The leakage power consumption of the LSE and poly-SVM classifiers is shown in Fig. 7. Due to the simplified hardware architecture, the proposed LSE classifier achieves up to 20.3% and on average 19.5% leakage savings compared to the poly-SVM classifier.

VI. CONCLUSION

Classifier ensemble is explored in this paper for on-chip hardware energy savings in classification applications. Classifier ensemble of SVM with linear kernel based on

Adaboost is proposed. While providing 93.4% classification accuracy, the proposed classifier ensemble achieves up to 45.7%, 20.3%, and 20.3% savings in total energy consumption, leakage power consumption, and area, respectively, for an image classification application by using a 45-nm CMOS technology.

ACKNOWLEDGEMENT

This work was supported by the Shenzhen Competitive Fundamental Research Grant JCYJ20170306164939111 and JCYJ20160330100025255.

REFERENCES

[1] K. Bong, S. Choi, C. Kim, D. Han, and H.-J. Yoo, "A low-power convolutional neural network face recognition processor and a CIS integrated with always-on face detector," *IEEE Journal of Solid-States Circuits*, vol. 53, no. 1, pp. 115–123, January 2018.

[2] M. Price *et al.*, "A low-power speech recognizer and voice activity detector using deep neural networks," *IEEE Journal of Solid-States Circuits*, vol. 53, no. 1, pp. 66–75, January 2018.

[3] D. Gerónimo *et al.*, "Survey of pedestrian detection for advanced driver assistance systems," *IEEE Transactions on Pattern Analysis and Machine Intelligence*, vol. 32, no. 7, pp. 1239–1258, July 2010.

[4] J. Yoo *et al.*, "An 8-Channel scalable EEG acquisition SoC with patient-specific seizure classification and recording processor," *IEEE Journal of Solid-States Circuits*, vol. 48, no. 1, pp. 214–228, January 2013.

[5] B. Zamanlooy and M. Mirhassani, "Efficient VLSI implementation of neural networks with hyperbolic tangent activation function," *IEEE Transactions on Very Large Scale Integration (VLSI) Systems*, vol. 22, no. 1, pp. 39–48, January 2014.

[6] V. Mrazek, S. S. Sarwar, L. Sekanina, Z. Vasicek, and K. Roy, "Design of power-efficient approximate multipliers for approximate artificial neural networks," *Proceedings of the IEEE/ACM International Conference on Computer-Aided Design*, pp. 1–7, November 2016.

[7] Y.-H. Chen *et al.*, "Eyeriss: An energy-efficient reconfigurable accelerator for deep convolutional neural networks," *IEEE Journal of Solid-States Circuits*, vol. 52, no. 1, pp. 115–123, January 2017.

[8] D. Anguita, S. Pischiutta, S. Ridella, and D. Sterpi, "Feed-forward support vector machine without multipliers," *IEEE Transactions on Neural Networks*, vol. 17, no. 5, pp. 1328–1331, September 2006.

[9] K. H. Lee and N. Verma, "A low-power processor with configurable embedded machine-learning accelerators for high-order and adaptive analysis of medical-sensor signals," *IEEE Journal of Solid-States Circuits*, vol. 48, no. 7, pp. 1625–1637, July 2013.

[10] M. van Leussen, J. Huisken, L. Wang, H. Jiao, and J. Pineda de Gyvez, "Reconfigurable support vector machine classifier with approximate computing," *Proceedings of the IEEE Computer Society Annual Symposium on VLSI (ISVLSI)*, pp. 13-18, July 2017.

[11] J. Cho *et al.*, "Increased performance of FPGA-based color classification system," *Proceedings of the IEEE International Symposium on Field-Programmable Custom Computing Machines*, pp. 29–32, May 2010.

[12] Z. Wang, R. E. Schapire, and N. Verma, "Error adaptive classifier boosting (EACB): Leveraging data-driven training towards hardware resilience for signal inference," *IEEE Transactions on Circuits and Systems I*, vol. 62, no. 4, pp. 1136–1145, April 2015.

[13] W. Rieutort-Louis *et al.*, "A large-area image sensing and detection system based on embedded thin-film classifiers," *IEEE Journal of Solid-States Circuits*, vol. 51, no. 1, pp. 281–290, January 2016.

[14] S. Zhang and N. R. Shanbhag, "Error-resilient machine learning in near threshold voltage via classifier ensemble," arxiv:1607.00667, 2016.

[15] W.-T. Chan, K. Y. Chung, A. B. Kahng, N. D. MacDonald, and S. Nath, "Learning-based prediction of embedded memory timing failures during initial floorplan design," *Proceedings of the Asia and South Pacific Design Automation Conference*, pp. 178-185, January 2016.

[16] C. Cortes and V. Vapnik, "Support-vector networks," *Machine Learning*, vol. 20, no. 3, pp. 273–297, 1995.

[17] C. M. Bishop, *Pattern Recognition and Machine Learning*, Springer, 2006, ISBN 0-387-31073-8.

[18] "MNIST handwritten digit database." [Online]. Available: http://yann.lecun.com/exdb/mnist/.

[19] Y.-W. Chen and C.-J. Lin, "Combining SVMs with Various Feature Selection Strategies," *Studies in Fuzziness & Soft Computing*, Springer, pp. 315-324, 2006.

2018 IEEE Computer Society Annual Symposium on VLSI

A Power-efficient Hybrid Architecture Design for Image Recognition using CNNs

Jinhang Choi*, Srivatsa Rangachar Srinivasa*, Yasuki Tanabe[†], Jack Sampson*, and Vijaykrishnan Narayanan*

*School of Electrical Engineering and Computer Science, Pennsylvania State University, University Park, PA 16802 USA

[†]Toshiba Corporation, Japan

*{jpc5731, sxr5403, sampson, vijay}@cse.psu.edu, [†]yasuki.tanabe@toshiba.co.jp

Abstract—Convolutional Neural Networks (CNNs) are proving to be highly effective in vision recognition systems. However, it is a challenge to use them in real-time embedded systems because of their requirements for computation-intensive operations and high memory bandwidth. This paper proposes a power-efficient CNN architecture that has a pipelined streaming accelerator coupled to 4,096 SIMD Processing Elements. We reduce memory bandwidth via hierarchical intermediate data buffering and batch processing on the chip. As a result, we achieve high power-efficiency: Our proposed design processes 2,175 regions/second when operating at 500MHz with a power budget less than 7.5 Watts.

Index Terms—Parallel architectures, Neural network hardware, Convolutional neural networks

I. INTRODUCTION

With the growth of high performance systems and the availability of vast training sets of labeled images, Convolutional Neural Networks (CNNs) have changed the landscape of computer vision. Since AlexNet demonstrated large-scale object recognition [1], Deep Convolutional Neural Networks (DCNNs) have been introduced for new innovative applications in object recognition [2], [3]. Given the generalized capability of detecting typical objects, DCNN-based recognition systems are being considered as viable solutions in the automotive industry for Advanced Driver Assistance Systems (ADAS)[1] to reinforce detection of expected objects (e.g. road and lane divisions, other vehicles) as well as non-typical objects such as people, fallen rocks, wood, animals, and so on. However, DCNNs usually require computation-intensive arithmetic operations and large memory footprints. For instance, AlexNet requires 1.2 billion Multiply-and-Accumulate (MAC) operations and 90 million trained parameter accesses for every image recognition, which limits real-time embedded vision systems in employing large-scale DCNNs. For high-performance GPUs, such as the NVIDIA Titan X, running AlexNet takes less than $40msecs$ to process 128 images at 256×256 resolution in a power budget of 227 Watts [5]. Even though commercial GPUs realize the low latencies and high throughput requirements of ADAS, employing a Titan X or equivalent GPU as part of an embedded system is often infeasible due to power budget. Therefore, it is imperative to develop CNN-inspired low-power systems that can still provide high throughput.

In this paper, we target object recognition on region proposals as described in [2]. With fast object localization schemes like BING [6], we can retrieve more than 1,000 region proposals a second in an image. To satisfy ADAS requirements, we propose using 4,096 Single Instruction, Multiple Data (SIMD) Processing Elements (PEs) on a single chip to efficiently accelerate the essential operations of CNNs. Specifically, we design a data reuse approach of exchanging intermediate

This work was supported in part by NSF Expeditions in Computing Program: Visual Cortex on Silicon CCF 1317560.

[1]ADAS typically detects potential obstructions within a few milliseconds to alert the driver about objects such as pedestrians, vehicles, and traffic signs [4].

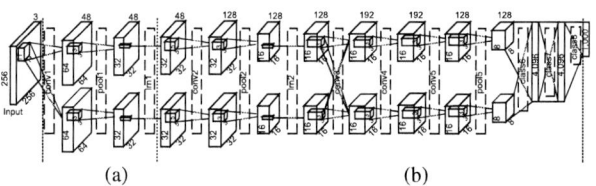

Fig. 1: Modified AlexNet [12] considering memory access alignment: (a) the first convolution layer, (b) the rest of layers.

values through PE interconnection controlled by Very Long Instruction Word (VLIW) microcodes. Due to layer-wise data locality in CNN models, we can mitigate excessive off-chip memory accesses by propagating data stored in adjacent PEs.

Along similar lines to our work, Chen, *et al.* [7] introduce DaDianNao, a multi-chip design consisting of 64 architecture nodes, where each node has a dedicated logic and eDRAMs in an area of $68mm^2$ at $28nm$ consuming around 16 Watts. To support flexible programmability for optimizing diverse neural networks, a domain-specific Instruction Set Architecture (ISA), Cambricon is proposed [8]. A Cambricon-based prototype accelerator presents high performance comparable to DaDianNao in an area of $56mm^2$ at $65nm$ consuming less than 2 Watts. However, neither of them address excessive off-chip memory accesses in end-to-end DCNN models. Due to data sparsity in DCNN models, EIE [9] and Sparse CNN accelerator [10] achieve significant improvement by reducing either data transfer or zero-value operand MAC operations in terms of performance, power, and energy. Instead of focusing on data sparsity, we increase data-level parallelism and decrease off-chip memory access based on a custom VLIW ISA. Alwani, *et al.* [11] propose a cache mechanism for keeping intermediate feature values to preserve inter-layer data locality. Similar to their approach, we focus on inherent inter-layer and intra-layer data locality. However, we do not only keep intermediate values in cache, but also in registers.

The key contributions of this paper include: *a)* A hierarchical intermediate data buffering mechanism that enables high PE utilization. *b)* A SIMD architecture that achieves a high throughput of 2175 regions/second for AlexNet while consuming only 7.5 Watts.

II. CHARACTERISTICS OF CNNs

A prototypical CNN is composed of convolutional layers (*conv*), pooling layers (*pool*), local response normalization layers (*lrn*) and classifier layers (*class*) as shown in Fig. 1. For efficient hardware implementation, we align memory by extending the size of feature map space to a power of two. Therefore, in the case of AlexNet, the input size is changed from 227 to 256, and the subsequent layers are also adjusted accordingly.

978-1-5386-7100-9/18 $31.00 © 2018 IEEE

- *conv* layer computes Z output feature maps such that

$$conv(x, y, z) = bias_z +$$
$$\sum_{k=0}^{K_z} \sum_{i=\lceil -K_x/2 \rceil}^{\lfloor K_x/2 \rfloor} \sum_{j=\lceil -K_y/2 \rceil}^{\lfloor K_y/2 \rfloor} w_z(i, j, k) \cdot f(x+i, y+j, k),$$
$$(1)$$

where x, y, z refer to the position of output feature with respect to height, width, and depth axis. $bias_z$ represents a trained bias of the z-th output feature map. w_z represents a trained weight parameter at (i, j, k) in the z-th convolution kernel. The height and width of a kernel, K_x and K_y, would be different depending on the pattern of convolution. Meanwhile, the depth of a kernel is identical to the input feature map depth divided by the number of convolution groups ($K_z = f_z/g$).

- *pool* layer retrieves the maximum value in a subset of input feature map such that

$$pool(x, y, z) = \max_{\substack{\lceil -K_x/2 \rceil \leq i \leq \lfloor K_x/2 \rfloor \\ \lceil -K_y/2 \rceil \leq j \leq \lfloor K_y/2 \rfloor}} f(sx+i, sy+j, z), \quad (2)$$

where s is the stride of sampling.

- *lrn* layer computes normalized response among neuron activations such that

$$lrn(x, y, z) = \frac{f(x, y, z)}{\left(\alpha \sum_{k=-n/2}^{n/2} f(x, y, z+k)^2 + \beta \right)^\gamma}, \quad (3)$$

where n neighboring input feature maps are used.

- *class* layer computes inner product for every object d by reshaping all input features, f, as a vector:

$$class(d) = \sum_{i=0}^{n} w(d, i) \times f(i) \quad (4)$$

Since every neuron contributes to feature map, *class* layer is also known as fully-connected layer.

- A neuron's activation function fires a signal to the output feature map. Following non-linearity robustness, Rectified Linear Unit function (ReLU) suppresses negative values to zero. ReLU is typically located after other layers except the last *class* layers followed by softmax regression.

Trained network parameters do not only present sparsity in numerical distribution [9], but also in numerical precision [13]. Therefore, architects can practically use fixed-point representation in the existing DCNN's weights and biases for embedded systems [14]. Using the approach proposed by Gysel, *et al.* [13] for evaluating the impact of fixed-point arithmetic operations on Caffe [15], we find, consistent with prior work, that at least 8-bit fractions are necessary to represent a DCNN's trained parameters within negligible precision loss as depicted in Table I. For brevity, we strategically use static precision for the network parameters, and dynamic precision for intermediate feature maps in 16-bit width fixed-point representation.

Meanwhile, a layer can share trained parameters for *batch processing*, where individual input feature maps use the same kernel to complete arithmetic operations for every batch size. In addition, while processing atomic operations in a layer, a kernel only focuses on the neighboring local area in input feature maps. Such CNN's inter-layer and intra-layer data-level parallelisms correspond to matrix-vector and matrix-matrix multiplication, respectively. Most accelerator designs mitigating off-chip memory traffic rely on an optimization of

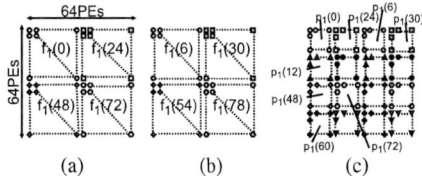

Fig. 2: Output feature allocation to PEs for four 32×32 output feature maps in (a) and (b), and sixteen 16×16 maps in (c).

dense or sparse matrix multiplication. Our proposed approach addresses dense matrix multiplication, where data reuse plays an important role in reducing off-chip memory accesses. Specifically, we focus on data locality-based vectorization by referencing on-chip memory.

III. ARCHITECTURE OVERVIEW

A. Routing Registers in SIMD PEs

The proposed design employs $4,096 (= 64 \times 64)$ PEs in a single-chip SIMD processor. In the proposed architecture, each PE has a dedicated MAC unit and holds feature values at corresponding positions as shown in Fig. 2. For instance, if the size of output feature map is 32×32, 4 output feature maps will be allocated in 64×64 PE array by blocking four 32×32 PE groups as shown in Fig. 2(a) and 2(b). Therefore, $PE_{(33,15)}$, denoting a PE at position $v = 33, h = 15$, calculates a feature value at $v = 1, h = 15$ position in the 48th feature map output, $f_1(48)$. For a 16×16 map, 16 output maps will be allocated as depicted in Fig. 2(c). In other words, each PE in a group should send data to other PEs in the same group to feed input features to a CNN layer's kernel. However, data delivery must not interrupt MAC operations in order to realize high throughput. As exhaustive PE interconnection causes a long routing distance to limit the maximum operating frequency, we implement two hierarchical registers, *VHReg*, and connect them with neighboring PEs. Fig. 3(a) depicts interconnection between 4 PEs. We use the first register to move a feature map vertically and the second register to move it horizontally. Fig. 3(b) shows how we exchange data. I_x denotes an instruction ID in each pipeline stage. For a 3×3 kernel, data is continuously fed to a MAC unit in the following sequence:

Clk0) The 1st register receives input feature $P_{(0,0)}$.
Clk1) The 2nd register receives $P_{(0,-1)}$ from the 1st register in $PE_{(0,-1)}$.
Clk2) The 2nd register receives $P_{(0,0)}$ held by the 1st register in $PE_{(0,0)}$.
Clk3) The 1st register receives input feature $P_{(1,0)}$ from the 1st register in $PE_{(1,0)}$.

In this manner, we can use *VHReg* to exchange data in a two-dimensional access pattern. Consequently, the MAC unit can receive data for a 3×3 convolution kernel from $P_{(-1,-1)}$ to $P_{(1,1)}$ for every clock cycle.

TABLE I: Accuracy loss with respect to fraction bits in fixed-point format of AlexNet, SqueezeNet, and GoogLeNet.

CNN Model	Infrence	Accuracy loss (vs. float32)				Test Dataset
		frac-12	frac-10	frac-8	frac-6	
AlexNet [12]	Top-1	0.1%p	0.1%p	1.4%p	54.2%p	ILSVRC 2012 [16]
	Top-5	0%p	0.2%p	1.4%p	76.9%p	
SqueezeNet [17]	Top-1	0.1%p	0.1%p	0.7%p	17.9%p	
	Top-5	0.1%p	0.1%p	0.5%p	72.7%p	
GoogLeNet [3]	Top-1	0.2%p	0.3%p	0.7%p	15.0%p	
	Top-5	0.1%p	0%p	0.3%p	10.8%p	

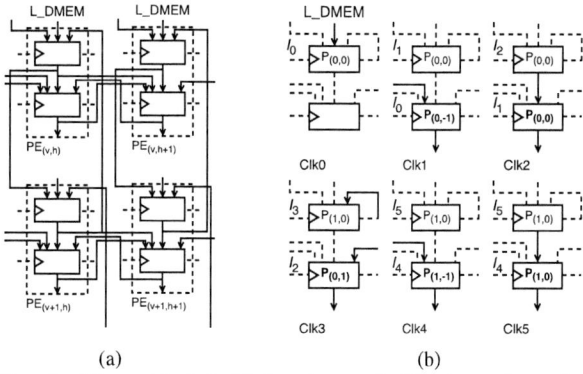

(a)   **(b)**

Fig. 3: Hierarchical buffering in vertical/horizontal routing registers: (a) PE interconnection (b) data exchange example.

B. VLIW Instruction Set Architecture

As described in Section III-A, while PEs perform a CNN layer's MAC operation, *VHReg* enables us to reuse data that is controlled by VLIW words. We propose a 64-bit wide VLIW ISA that has 59-bit width for signaling functional units, and 5-bit width zero padding for instruction word alignment. To provide programmability, we implemented a custom assembler supporting VLIW microcode. Fig. 4 describes an example of assembly codes for computing the first row of 5×5 convolution kernel. A VLIW word starts with setting PATCH field to select $32 \times 32 (= 1,024)$ PE groups. A VLIW word ends with end-of-instruction (EOI ISSUE) by issuing the 64-bit wide instruction word. The order of microcode in a VLIW word does not affect the result of operation. To support loop iteration, a VLIW word can end with repeat (RPT ISSUE) and end-of-repeat (ERTP ISSUE), indicating the starting and ending VLIW word of a loop block, respectively. In this example, three channels (IN_CHANNEL) are iterated to feed RGB color space to the convolution layer. At first, local data memory is accessed from region 1, and four input feature maps will be loaded on four 32×32 PE groups (line 17). Second, L_DMEM propagates input data to V_REG in all PEs (line 19). Third, each PE tries to deliver input data to its right PE according to horizontal shift microcode, MV_HM1 (line 20). For $n \times n$ convolution, each row of kernel requires n VLIW words (line 16–54). As a result, 75 VLIW instruction words will be executed to get an output feature map in Fig. 4.

Similar to *conv* layer, *pool*, *lrn*, and *class* layers are supported by VLIW microcodes. However, AlexNet-like DCNNs shown in Fig. 1(a) present a stride of 4 pixels in a large input layer, which is the unique data access pattern only found in the first convolution layer. It is not efficient either to implement extra interconnections nor to spend extra instruction cycles to support such striding access patterns on VLIW/SIMD processor. Therefore, our proposed design has two accelerators: a dedicated accelerator for the first *conv* and *pool* layers (Conv1HW), and VLIW/SIMD processor that handles the rest of the network.

IV. A PROTOTYPE IMPLEMENTATION

A. Accelerator Design for the First Convolution Layer

In *conv1* layer, each input image of 256×256 resolution requires 142 million multiplications to generate 96 output feature maps of 64×64 resolution. Including *pool1* layer, the amount of processing time has to be balanced between Conv1HW and the VLIW/SIMD processor to increase throughput. Fig. 5 depicts the Conv1HW accelerator designed for the first convolution kernels. Conv1HW consists of three logical functionali-

```
1   .set PATCH4, 32          ; Define the size of PE group
2   .set CO_IN, 1
3   .set CO_OUT, 2
4   .set IN_CHANNEL, 3
5   ;;;;;;;;;;;;;;;;;;;;;;;;;;;
6   ;;; load bias and set it to accumulator.
7   ;;;;;;;;;;;;;;;;;;;;;;;;;;;
8   PATCH `PATCH4            ; START of a VLIW word
9   LD CONV_PMEM            ; load bias
10  SET P_REG CONV_PMEM     ; set bias to parameter reg
11  SET A_REG P_REG         ; set param reg to accumulator
12  EOI ISSUE               ; END of a VLIW word
13  ;;;;;;;;;;;;;;;;;;;;;;;;;;;
14  ;;; iteration start
15  ;;;;;;;;;;;;;;;;;;;;;;;;;; f: input, w: weight parameter
16  PATCH `PATCH4
17  LD L_DMEM (`CO_IN)      ; load f(v,h,k) to PE(v,h)
18  LD CONV_PMEM            ; load w(0,-1)
19  SET V_REG L_DMEM        ; set each PE's f to routing vreg
20  MV_HM1 H_REG V_REG      ; horizontal shift f to PE(v,h+1)
21                          ; i.e. f(v,h-i-1) to PE(v,h-i)
22  SET P_REG CONV_PMEM     ; set w(0,-1) to parameter reg
23  MACC A_REG H_REG P_REG  ; compute bias+f(v,h-1) * w(0,-1)
24  RPT ISSUE `IN_CHANNEL   ; END of a VLIW word
25                          ; start of loop iterations
26  PATCH `PATCH4
27  LD CONV_PMEM            ; load w(0,-2)
28  NOP V_REG               ; keep f(v,h) at routing vreg
29  MV_HM1 H_REG H_REG      ; shift PE's hreg to PE(v,h-1)
30                          ; i.e. f(v,h-i-2) to PE(v,h-i)
31  SET P_REG CONV_PMEM     ; set w(0,-2) to parameter reg
32  MACC A_REG H_REG P_REG  ; accumulate f(v,h-2) * w(0,-2)
33  EOI ISSUE
34  PATCH `PATCH4
35  LD CONV_PMEM            ; load w(0,0)
36  NOP V_REG               ; keep f(v,h) at routing vreg
37  SET H_REG V_REG         ; set f to PE(v,h)
38  SET P_REG CONV_PMEM     ; set w(0,0) to parameter reg
39  MACC A_REG H_REG P_REG  ; accumulate f(v,h) * w(0,0)
40  EOI ISSUE
41  PATCH `PATCH4
42  LD CONV_PMEM            ; load w(0,+1)
43  MV_HP1 H_REG H_REG      ; shift PE's hreg to PE(v,h+1)
44                          ; i.e. f(v,h-i+1) to PE(v,h-i)
45  SET P_REG CONV_PMEM     ; set w(0,+1) to parameter reg
46  MACC A_REG H_REG P_REG  ; accumulate f(y,h+1) * w(0,+1)
47  EOI ISSUE
48  PATCH `PATCH4
49  LD CONV_PMEM            ; load w(0,+2)
50  MV_HP1 H_REG H_REG      ; shift PE's hreg to PE(v,h+1)
51                          ; i.e. f(v,h-i+2) to PE(v,h-i)
52  SET P_REG CONV_PMEM     ; set w(0,+2) to parameter reg
53  MACC A_REG H_REG P_REG  ; accumulate f(v,h+2) * w(0,+2)
54  EOI ISSUE
55  ...
```

Fig. 4: An assembly code of the first row operations in 5×5 convolution: $\sum_{v=0, -2 \leq h \leq 2} w(v, h, k) \times f(x+v, y+h, k)$.

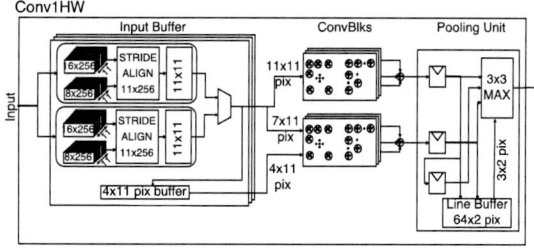

Fig. 5: Conv1HW accelerator design.

ties: input channel buffers, convolution blocks (ConvBlks), and a max value pooling unit. An input channel buffer consists of two ping-pong memory modules to hide the off-chip memory latency. Each ConvBlk needs different 11×11-pixel windows from three input buffers concurrently with a stride pattern of 4. To maximize input data availability, we arrange SRAM blocks in 11 banks. Input channel rows are interleaved among rows of the memory banks to facilitate reading 11 rows of an image in one cycle. For instance, the first row of the image channel is stored at Bank 0, the second row at Bank 1, and so on. Therefore, 11×256 rows are read from the SRAM and stored in a stride-aligned buffer, which is only updated when a new vertical stride happens. For strides, three 4×11 registers preserve 4×11 pixels from the previous read window. Then,

978-1-5386-7100-9/18 $31.00 © 2018 IEEE

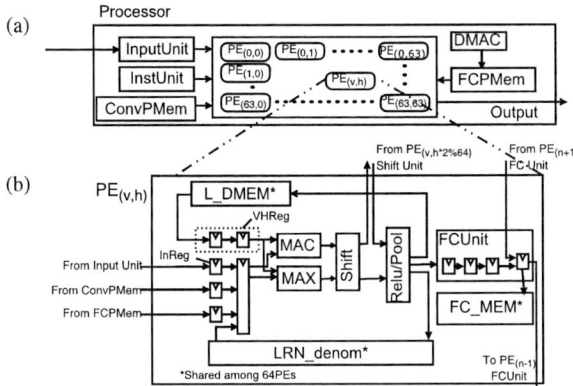

Fig. 6: SIMD design with Processing Elements (PEs).

7×11 pixels from the newly read window are merged with the preserved 4×11 pixels to supply two 11×11-pixel windows to ConvBlks every clock cycle. On every vertical stride, there is a stall on ConvBlks since the 4×11 buffers need to be updated before the second window is ready. However, the stall occurs only once on the vertical stride and never on the horizontal stride.

A ConvBlk receives 11×11 pixels, and multiplies them with preloaded kernel parameters in three 11×11 MAC units for RGB channels. Using adder trees and inter-channel adders, two ConvBlk generate output feature values.

The pooling block has a 64×2 line buffer to store two rows of the incoming feature map data from ConvBlks. When a new feature map arrives, the pooling block performs max-value sampling for the 3×3 overlapping *pool1* layer.

B. VLIW/SIMD Processor Design

Fig. 6(a) depicts the component integration in the SIMD architecture. The VLIW/SIMD processor consists of an input data delivery unit (InputUnit), an instruction buffer and issue unit (InstUnit), a convolution parameter memory (ConvP-Mem), a parameter buffer for *class* layers (FCPMem), and 64×64 PE array.

InputUnit delivers a feature value to a PE's *InReg* every clock cycle in a pipeline-based round-robin fashion whenever it receives the input data from Conv1HW.

InstUnit queues instructions for *conv*, *pool*, and *lrn* layers' atomic operations, as well as a set of special instructions. Special instructions include instructions for moving delivered input data to local data memory in each PE, and instructions for *class* layers. As the stall control of deeply-pipelined wide SIMD processors tends to be a critical path, we implement our proposed processor without stall. Instead, program code should be written to avoid data hazard for convolutions, which is not difficult due to independence between loop iterations in output feature maps. However, special instructions are issued only when either required data or target memory area is available. For example, after InputUnit receives and delivers 4,096 inputs to each PE, InstUnit interrupts queuing atomic operation sets, and queues the special instruction set. InstUnit then issues an instruction to move input data stored in each PE's dedicated register to local data memory, L_DMEM.

ConvPMem provides weights and biases used in *conv* layer operations. Again, the loaded parameters are delivered to PEs using a pipeline for achieving a high clock frequency. Since we process convolutions of a feature map at a time, each trained parameter of a convolution kernel is required just once. To avoid idle PEs, we implement ConvPMem with a simple hardware address manager using read pointer. Therefore, 4/8/16/64 output maps can be simultaneously processed in PE arrays with different kernel parameters as explained in Section III-A.

FCPMem is a 32 KiB SRAM block that stores 64×256 parameters accessed by special instructions for *class* layer. Like ConvPMem, a hardware address manager using a pointer provides the FCPMem memory address.

C. Processing Elements (PEs)

Fig. 6(b) depicts the structure of a single PE, where we denote PE at v, h position of the array as $PE_{(v,h)}$ or PE_n such that $n = v \times 64 + h$.

InREG is a register holding an input data delivered from Conv1HW by InputUnit. If we directly store the input value in L_DMEM, we need to handle write collisions with general processing, which results in pipeline stalls decreasing PE utilization rate. Therefore, InReg temporarily holds the data, and moves it to each PE's L_DMEM by using a special instruction after all PEs receive data.

L_DMEM is a local data memory with an address manager, logically implemented in each PE. In reality, L_DMEM is physically implemented as a block with $128 \times 16 \times 64$ bit SRAM, and shared by 64 PEs. In total, we have 1 MiB local storage on VLIW/SIMD processor. L_DMEM stores intermediate data while processing convolutional layers. Once a layer's atomic operation is finished, memory area for the input feature map can be released. A shallow 128-entry memory space is sufficient to keep all intermediate data and input data. As a result, we do not have to transfer intermediate feature data to/from off-chip. Since all memory access patterns in DCNN are sequential, we implemented a dedicated address manager that controls load and store using 16 pointers with a 16-entry memory region configuration table. Each load/store atomic operation is issued with a pointer identifier and a field to control the pointer. Therefore, we do not need additional hardware or instructions to calculate relative addresses.

MAC unit supports multiply, multiply and accumulate, square, and bypass operations. In contrast to the 16-bit width of other paths, accumulation and its output have 43-bit width to avoid overflow.

MAX unit provides sampling the maximum feature value. VHReg is responsible of controlling kernel reference patterns for an input feature map.

Shift unit performs bit shift operation to convert an output of MAC unit to a user-defined fixed-point precision. Even though supporting variable-bit shift pattern in a cycle will increase hardware cost, the bit shift is required only once for every output feature map. Thus, we provide a feedback loop to support variable-bit shift for flexible precision in each layer.

Relu/Pool unit provides negative value suppression and pooling. In the pooling operation, the input patch needs to be sampled in either horizontal or vertical direction. To support horizontal feature map sampling, we implement special connections such that $PE_{(v,h)}$ will have incoming data from Shift unit of $PE_{(v,(h\times2) \mod 64)}$. As we restrict the special interconnection within a group of 64 PEs, it does not limit the maximum operating frequency. Unfortunately, in case of vertical sampling, the target feature map could be placed in a distant place. Therefore, we pay extra VLIW instruction cycles: horizontally sampled data is evicted to L_DMEM temporarily, and VHReg moves data to the destination PE.

LRN_denom unit supports *lrn* layer operation. Exponentiation and division are too expensive to implement on hardware in terms of logic size. Instead, we use a custom lookup-table that takes a $\sum_{i=-N/2}^{N/2} f(x, y, z+i)^2$ value as an input,

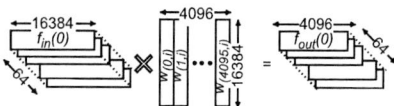

Fig. 7: Computation in *class* layer is matrix multiplication between input feature map f_{in} and weights $W(d, i)$. In batch processing, loaded parameters $W(d, i)$ are shared among multiple input feature maps from $f_{in}(0)$ to $f_{in}(63)$, to generate $f_{out}(0, d)$ to $f_{out}(63, d)$.

and returns a reciprocal of the denominator. The lookup table output will be multiplied by $f(x, y, z)$ using MAC unit.

FCUnit & FC_MEM manage intermediate feature data for *class6* layer. We apply a batch processing approach to reduce the bandwidth requirement as illustrated in Fig. 7. Feature maps allocated to each PE array can be considered as one-dimensional data. If the maps were scattered across PEs to multiply with *class* layer parameters, we should accumulate the output of each PE's MAC unit horizontally and vertically, which results in wide bit-width interconnections. Instead, we arrange and store the intermediate data in FC_MEM. Before AlexNet's *class6* layer, for instance, 256 output feature maps of 8×8 resolution in *pool5* layer have to generate a one-dimensional input feature map. When iterating *pool5* operations, *1)* each PE holds a value of the feature map at FCUnit's registers; *2)* FCUnit delivers the feature map to FC_MEM; *3)* A FC_MEM submodule shares the feature map with 64 PEs. Since 64 PEs associated with a FC_MEM submodule receive 64 individual feature values at a time, it takes 256 cycles for an input feature map to interact with an inner product kernel's parameters in FCPMem. Then, every 64 PEs horizontally collect and accumulate the result of MAC operation in FCUnit. Lastly, the output feature map is generated and it will be stored in FC_MEM for the next *class* layer. Each FC_MEM submodule has a 72 KiB buffer that keeps both input feature map and and intermediate data between *class* layers. In total, assuming that the batch size is set to 64, we need 4.5 MiB memory area for FC_MEM. As LRN_denom and FC_MEM are also shared by 64 PEs like L_DMEM, we have 64 different LRN_denom and FC_MEM physical submodules in the processor.

D. Design synthesis and functional verification

We implement Conv1HW and VLIW/SIMD processor design including ConvPMEM and LRN_denom in Verilog. The designs are synthesized with Synopsys Design Compiler targeting 500 MHz clock frequency, and placed and routed with Synopsys IC Compiler using the SAED $32nm$ generic standard cell library from 0.85V for logic design, and 1.05V for memory. Without loss of generality, functional verification in an FPGA is conducted at 100 MHz on the PESet design representing 512 PEs arranged in 16×32 fashion. Fig. 8(a) and 8(b) shows the actual physical layout of Conv1HW, VLIW/SIMD processor design with 8 PESets, shared L_DMEM and LRN_denom submodules respectively. The clock grid runs throughout the layout in metal7 and metal8 layers. In addition, the clock grid sinks down to the registers in the form of an H-tree structure to reduce the overall skew so that the design can run at 500 MHz. Submodules in every PESet are clustered together, and SRAM banks of ConvPMEM surround the entire layout to avoid the routing and placement congestion. Table II presents power consumption of our design using the activity factors derived from Conv1HW emulation. The power consumption of the memory is compared with McPAT [18]. The entire design consumes very close to 5

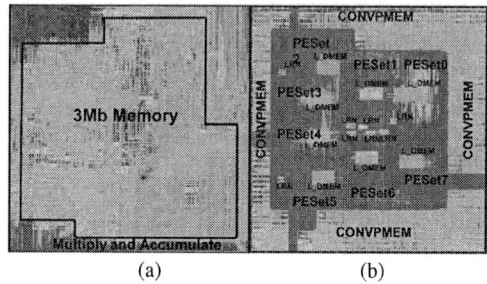

(a)   (b)

Fig. 8: Layout snapshot: (a) the first layer accelerator (Conv1HW) with the placement of input buffer and MAC units, and (b) VLIW/SIMD processor design with a modular placement of 8 PESets surrounded by ConvPMem.

Watts and has a total cell (logic + memory) area of $109mm^2$, and occupies $188.3mm^2$ with actual placement and routing. The layout increase is due to constraints in floorplan, clock mesh/tree routing, and standard cell replacement.

V. EVALUATION

We simulate our proposed design running AlexNet in a cycle accurate model based on FPGA implementation. Table III describes configuration of our target DCNN, AlexNet, and its required clock cycles. f_x, f_y, f_z correspond to the width, height and channel depth of each layer's output feature maps, respectively. Patch size denotes the number of available feature maps in PE array. For example, we can process 16 output maps of *conv4*'s $K_x \times K_y$ convolution kernel at the same time. Therefore, $384/16$ iterations take 41,472 cycles. *pool* layers take account of $K_x \times K_y$ comparisons and f_y iterations of vertical shift using VHReg for every output channel. Similarly, *lrn* layers require accumulations of n powered feature values and an additional multiplication with a LRN lookup result for every output channel. Each *class* layer operates $(f_x^{prev} \times f_y^{prev} \times f_z^{prev}) \times f_z^{current}$ multiplications, which are distributed and mitigated for every 64 PEs in batch processing. Meanwhile, *conv2* layer needs extra cycles to generate dummy output feature maps in *lrn2* layer. For Conv1HW, the elapsed cycles can be hidden by the SIMD processor. Since *1)* VLIW instruction execution regulates provision of input image pixels in a fixed rate, *2)* kernel parameters are shared among PEs in ConvPMEM, *3)* and intermediate layer

TABLE II: Power and Area breakdown.

FPGA Synthesis	Power (Watts)	Cell Area (mm^2)	Layout Area (mm^2)
Conv1HW Accelerator	2.062	42	64
VLIW/SIMD Processor	2.790	67	124.3
Total	4.86	109	188.3

TABLE III: AlexNet specification & target performance.

Layer	f_x	f_y	f_z	g	$K_{x,y}$	patch size	cycles
input	256	256	3	-	-	-	-
conv1	64	64	96	2	11	-	(196,608)
pool1	32	32	96	2	3		
lrn1	32	32	96	2	-	4	144
conv2	32	32	256	2	5	4	76,800
pool2	16	16	256	2	3	16	4,672
lrn2	16	16	256	1	-	16	96
conv3	16	16	384	2	3	16	55,296
conv4	16	16	384	2	3	16	41,472
conv5	16	16	256	2	3	64	27,648
pool5	8	8	256	2	3	-	2,192
class6	1	1	4096	1	-	-	16,384
class7	1	1	4096	1	-	-	4,096
class8	1	1	1000	1	-	-	1,000
Total							229,800

978-1-5386-7100-9/18 $31.00 © 2018 IEEE

TABLE IV: Performance comparison.

Network: AlexNet	Proposed Design	Tesla K20m [20]		Tegra X1 [5]		Titan X [5]	
Design Type	Synthesized	GPU		mGPU		GPU	
Technology	32nm	28nm		28nm		28nm	
Clock (MHz)	500	706		852		1,075	
Batch Size	64	128	1	128	1	128	1
Throughput (Regions/sec)	2,175	649	171	258	67	3,216	405
Energy efficiency (Regions/J)	291.6	4.3	2.0	45.0	13.1	14.2	2.5
Power (Watts)	7.5	150.0	85.0	5.7	5.1	227.0	164.0

data is stored in L_DMEM, the actual utilization of off-chip memory is restricted by inner product parameters of *class* layers. In our proposed design with LPDDR4 [19], 4.5 GiB/s peak bandwidth will be required to load the parameters, which consumes an additional 2.46 Watts. In the meantime, the SIMD PEs run *conv* layer operations for the next batch while it is waiting for the parameters to load. Hence, we hide the DDR latency. In short, a total of 229,800 clock cycles are required for every input image region, which leads to processing over 2,000 regions/second at 500 MHz in 7.5 Watts.

In Table IV, we compare our processor to a 3.52 TFLOPS Tesla K20m GPU [20] running the same DCNN model, AlexNet on Caffe with cuDNN library [21]. Additionally, we benchmark our design against commercial embedded and high-performance GPUs in terms of throughput. A high-performance GPU takes less than $198msec$ per 128 input regions (at best $39.8msec$ in NVIDIA Titan X). In contrast, our processor needs $29.4msec$ per 64 input regions, achieving over $5\times$ better throughput while power consumption is comparable to embedded GPU (Tegra X1 consumes around 5 Watts [5]).

DaDianNao [7] proposed an energy-efficient CNN accelerator which utilizes only on-chip embedded eDRAM. To support the first convolutional layer's computational capability, it is required to deploy a 4-chip DaDianNao in 64 Watts, reporting an $63.35\times$ improvement in performance over a Tesla K20m. However, if the two configurations we examine in Table IV are indicative, a 4-chip DaDianNao capable of running AlexNet has a region/J somewhere between 170–650 depending on which baseline more closely corresponds to the one in [7]. Note that we are able to achieve a region/J rate of 291 utilizing only LPDDR4. We believe that switching to stacked or embedded eDRAM more comparable to DaDianNao would amplify our efficiency by a further order of magnitude (same performance at substantially lower power).

Despite reasonable resource scalability, our proposed design only fits in multiple FPGAs due to its memory requirement as well as the dedicated Conv1HW accelerator. As a point of comparison for FPGA implementation, relative to the 137 GigaOP/second design of Qiu, *et al.* [14] using 2 large, complex PEs, our 4,096 PE design utilizes $7.2\times$ the LUTs, $10.3\times$ the FFs, and $6.6\times$ the DSPs in order to optimize for data reuse. Our hybrid architecture provides an end-to-end pipeline, unlike layer-specific accelerators targeting memory-centric [9] or computation-centric operations [10], [11].

Our current approach does impose some limitations due to fixed-size datapath widths and built-in assumptions about early-stage convolution kernels. Many state-of-the-art DCNNs exploit only small convolution kernels [22], where Conv1HW may not be useful, and direct use of the VLIW/SIMD processor, bypassing Conv1HW, degrades off-chip latency hiding. In addition, network compression for low-bit width quantization [9], [13] indicates the potential benefits of extremely-narrow bit-width SIMD processor design. Our future work aims at designing tunable architectures for CNN parametrization.

VI. CONCLUSION

We present an architecture targeted for DCNN-based image processing. The proposed architecture has been synthesized, routed, and evaluated against GPUs and DaDian-Nao. The resulting design achieves 2 TeraOP/second ($=$ 4096 MACs/cycle $\times$ 500 MHz) and can process 2,175 image regions/second while consuming less than 7.5 Watts due to intensive hierarchical buffering and batch processing in SRAM. This modest power consumption envelope makes our design suitable for many embedded system domains and form factors that would not otherwise see this level of performance for acceptable power expenditure. By providing support for the primitive operations of DCNN layers, this approach can flexibly support a large class of DCNNs with high accuracy.

REFERENCES

[1] A. Krizhevsky, I. Sutskever, and G. E. Hinton, "ImageNet Classification with Deep Convolutional Neural Networks," in *Advances in Neural Information Processing Systems 25 (NIPS)*, 2012, pp. 1097–1105.

[2] R. Girshick *et al.*, "Rich feature hierarchies for accurate object detection and semantic segmentation," in *Proceedings of IEEE Conference on Computer Vision and Pattern Recognition (CVPR)*, 2014.

[3] C. Szegedy *et al.*, "Going Deeper With Convolutions," in *Proceedings of IEEE Conference on Computer Vision and Pattern Recognition (CVPR)*, 2015, pp. 1–9.

[4] Y. Tanabe *et al.*, "A 464GOPS 620GOPS/W heterogeneous multi-core SoC for image-recognition applications," in *Proceedings of IEEE International Solid-State Circuits Conference (ISSCC)*, Feb. 2012, pp. 222–223.

[5] NVIDIA, "GPU-Based Deep Learning Inference: A Performance and Power Analysis," Tech. Rep., Nov. 2015.

[6] M. M. Cheng *et al.*, "BING: Binarized Normed Gradients for Objectness Estimation at 300fps," in *Proceedings of IEEE Conference on Computer Vision and Pattern Recognition (CVPR)*, Jun. 2014, pp. 3286–3293.

[7] Y. Chen *et al.*, "DaDianNao: A Machine-Learning Supercomputer," in *Proceedings of the 47th IEEE/ACM International Symposium on Microarchitecture (MICRO)*, Dec. 2014, pp. 609–622.

[8] S. Liu *et al.*, "Cambricon: An Instruction Set Architecture for Neural Networks," in *Proceedings of ACM/IEEE 43rd International Symposium on Computer Architecture (ISCA)*, Jun. 2016, pp. 393–405.

[9] S. Han *et al.*, "EIE: Efficient Inference Engine on Compressed Deep Neural Network," in *Proceedings of the 43rd ACM International Symposium on Computer Architecture (ISCA)*, 2016, pp. 243–254.

[10] A. Parashar *et al.*, "SCNN: An Accelerator for Compressed-sparse Convolutional Neural Networks," in *Proceedings of the 44th International Symposium on Computer Architecture (ISCA)*, 2017, pp. 27–40.

[11] M. Alwani *et al.*, "Fused-layer CNN accelerators," in *Proceedings of the 49th IEEE/ACM International Symposium on Microarchitecture (MICRO)*, Oct. 2016, pp. 1–12.

[12] E. Shelhamer, "Berkeley AI Research / Berkeley Vision and Learning Center CaffeNet Model," 2014, https://github.com/BVLC/caffe/tree/master/models/bvlc_reference_caffenet.

[13] P. Gysel *et al.*, "Ristretto: A Framework for Empirical Study of Resource-Efficient Inference in Convolutional Neural Networks," *IEEE Transactions on Neural Networks and Learning Systems*. 2018.

[14] J. Qiu *et al.*, "Going Deeper with Embedded FPGA Platform for Convolutional Neural Network," in *Proceedings of the 2016 ACM/SIGDA International Symposium on Field-Programmable Gate Arrays (FPGA)*, 2016, pp. 26–35.

[15] Y. Jia *et al.*, "Caffe: Convolutional architecture for fast feature embedding," in *Proceedings of the 22nd ACM International Conference on Multimedia (MM)*, 2014.

[16] O. Russakovsky *et al.*, "ImageNet Large Scale Visual Recognition Challenge," *International Journal of Computer Vision (IJCV)*, vol. 115, no. 3, pp. 211–252, 2015.

[17] F. N. Iandola *et al.*, "SqueezeNet: AlexNet-level accuracy with 50x fewer parameters and <0.5MB Model size," *arXiv.org*, Feb. 2016.

[18] S. Li *et al.*, "McPAT: An integrated power, area, and timing modeling framework for multicore and manycore architectures," in *Proceedings of the 42nd IEEE/ACM International Symposium on Microarchitecture (MICRO)*, Dec. 2009, pp. 469–480.

[19] JEDEC Solid State Technology Association, "JEDEC Standard: Low Power Double Data Rate 4 (LPDDR4)," 2014, http://www.jedec.org/sites/default/files/docs/JESD209-4.pdf.

[20] NVIDIA, "Tesla K20 GPU Accelerator Board Specification," https://www.nvidia.com/content/PDF/kepler/Tesla-K20-Passive-BD-06455-001-v05.pdf.

[21] NVIDIA, "CUDA® Deep Neural Network library (cuDNN)," 2017. https://developer.nvidia.com/cudnn.

[22] C. Szegedy *et al.*, "Rethinking the Inception Architecture for Computer Vision," in *Proceedings of IEEE Conference on Computer Vision and Pattern Recognition (CVPR)*, 2016, pp. 2818–2826.

2018 IEEE Computer Society Annual Symposium on VLSI

Towards Budget-Driven Hardware Optimization for Deep Convolutional Neural Networks using Stochastic Computing

Zhe Li*, Ji Li[†], Ao Ren*, Caiwen Ding*, Jeffrey Draper[‡†], Qinru Qiu*, Bo Yuan[§], Yanzhi Wang*

*Department of Electrical Engineering and Computer Science, Syracuse University, Syracuse, NY 13244, USA
[†]Department of Electrical Engineering, University of Southern California, Los Angeles, CA 90089, USA
[‡] Information Sciences Institute, University of Southern California, Marina del Rey, CA 90292, USA
[§] Department of Electrical Engineering, City University of New York, City College, New York, NY 10031, USA
{zli89, aren, cading, qiqiu, ywang393}@syr.edu, jli724@usc.edu, draper@isi.edu, byuan@ccny.cuny.edu

Abstract—Recently, Deep Convolutional Neural Network (DCNN) has achieved tremendous success in many machine learning applications. Nevertheless, the deep structure has brought significant increases in computation complexity. Large-scale deep learning systems mainly operate in high-performance server clusters, thus restricting the application extensions to personal or mobile devices. Previous works on GPU and/or FPGA acceleration for DCNNs show increasing speedup, but ignore other constraints, such as area, power, and energy. Stochastic Computing (SC), as a unique data representation and processing technique, has the potential to enable the design of fully parallel and scalable hardware implementations of large-scale deep learning systems. This paper proposed an automatic design allocation algorithm driven by budget requirement considering overall accuracy performance. This systematic method enables the automatic design of a DCNN where all design parameters are jointly optimized. Experimental results demonstrate that proposed algorithm can achieve a joint optimization of all design parameters given the comprehensive budget of a DCNN.

I. INTRODUCTION

As the important branches of deep learning, *Deep Neural Networks (DNNs)* and Recurrent Neural Networks (RNNs), outperforms traditional machine learning techniques in several real-world problems [1–3]. Recently, *Deep Convolutional Neural Network (DCNN)*, which is one of most widely used types of DNNs, has achieved tremendous success in many machine learning applications, such as speech recognition [4], image classification [5], [6], and video classification [7]. DCNN is now the dominant approach for almost all recognition and detection tasks, and approaches human performance on some tasks [8]. Nevertheless, the introduction of deep structure in deep learning brought significant increases in computation complexity. Large-scale deep learning systems mainly operate in high-performance server clusters, thus restricting the application extensions to personal or mobile devices.

General-Purpose Graphics Processing Units (GPGPUs) are widely used for current deep learning research to accelerate DCNNs [9]. GPGPU's major competitor is *Field-Programmable Gate Arrays (FPGAs)*. Considering energy-efficiency, FPGAs are more suitable for portable and embedded DCNN applications. Previous works on FPGA acceleration for DCNNs [10][11][12] show increasing speedup. These implementations, however, focus on improving the throughput of an embedded network, which ignores other constraints to run a network, such as area, power, and energy. A notable trend is that machine learning is running locally on mobile/wearable devices and *Internet-of-Things (IoT)* entities instead of relying on a remote server. In order to bring the success of DCNNs to these resource constrained systems, designers must overcome the challenges of implementing resource-hungry DCNNs in embedded systems with limited area and power budget.

Stochastic Computing (SC) [13], as a unique data representation and processing technique, has the potential to enable the design of fully parallel and scalable hardware implementations

of large-scale deep learning systems. Many complex arithmetic operations can be implemented with very simple hardware logic in stochastic computing framework [13–17], which offers an immense design space for (i) neuron integrations due to the significantly reduced area per neuron, and (ii) performance optimizations with respect to the *budget* of area, error resiliency, power/energy, or speed. We use the term budget to describe the design constraint when implementing a hardware based DCNN using SC. For example, given the area budget of an embedded design of DCNN, SC enables a comprehensive optimization including other design parameters mentioned above.

A node, referred as to *neuron*, in a DCNN can be implemented by different stochastic computing based designs. Given these designs, how to arrange them to structure a complete DCNN achieving preferred design parameter(s) such as constrained energy, promised accuracy, and restricted area leaves blanks to researchers. This paper deals with the problem of deriving the optimal structure of hardware deep learning systems given a design budget and proposes an automatic algorithm driven by budget requirement with the comprehensive design parameters of a network taken into consideration. More specifically, the proposed automatic design allocation algorithm greedily decides implementation for each layer of a DCNN, and then optimizes the complete DCNN jointly to achieve the better objective design parameter(s) by re-allocating different implementations. This systematic method enables the automatic design of a DCNN where design parameters are jointly optimized given the budget.

The contributions of this paper are summarized as follows, *1)* SC paradigm is finely applied to DCNN, i.e., major computing tasks are performed in SC domain. With SC components, the hardware footprints can be significantly reduced for wearable/mobile devices. *2)* We explored different implementations for neurons of different layers in a DCNN. *Accumulative/Approximate Parallel Counter (APC)* [18], [19] and *Multiplexer* based neurons with distinct implementation details are investigated. *3)* An automatic design allocation algorithm is proposed to optimize the complete DCNN hardware design using SC. This algorithm can also evaluate a general budget-driven hardware optimization for DCNNs.

Experimental results have been demonstrated on the problem of classifying handwritten digits in MNIST database using LeNet-5 [20] with the SC based DCNN designs. It reveals that proposed automatic design allocation algorithm can achieve a joint optimization of all design parameters given the comprehensive budget of a DCNN.

II. RELATED WORKS

Experiments in [9][21] showed a high speedup of DCNN implemented on GPGPU. However, the widespread deployment of DCNNs has been hindered by their high complexities and power consumptions of server-based GPGPU-like devices,

978-1-5386-7100-9/18 $31.00 © 2018 IEEE

28

particularly in resource-constrained offline wearable devices and embedded systems. Farabet *et al.* in [10] proposed an FPGA-based processor specific for convolutional networks. The proposed processor was well structured. Nevertheless, the paper lacked evaluation of power and/or energy consumption. Similarly, Cadambi *et al.* of [12] demonstrated a general accelerator working for five typical tasks including DCNN. The paper, however, explored architectural design space instead of considering optimization of on-chip design parameters of a hardware-based DCNN. In [11], authors contributed to developing a co-processor to improve the speed of DCNN, but they didn't mention how to optimize the network performance given hardware budget constraints.

Predecessors have been investigating SC as a candidate for implementing hardware-based DCNN. Inspired by [18], authors in [22] introduced several basic hardware implementations for matrix operations including inner-product calculation which is crucial in DCNN inference process. Based on lots of similar works exploring hardware implementation for stochastic computing, a recent work [19] presented a neuron cell design using SC components, where the progressive precision characteristics of SC was exploited. The aforementioned papers focused on the analysis of performance in SC implementations, but optimization was not accomplished from a higher network-wise view. Even though the synthesis results were listed, but still, there lacked re-design of the implementation for DCNN with constrained hardware resources. The aforementioned researches proved SC is a promising technique for embedded DCNN system. However, there is no existing work designing DCNN using SC with a structural optimization given hardware budget.

III. STOCHASTIC COMPUTING BASED DCNN

A. Deep Convolutional Neural Network Architecture

In this paper, we consider a general DCNN architecture, which consists of a stack of convolutional layers, pooling layers, and fully connected layers. By arranging the topology of the above layers, powerful architectures, such as LeNet [23] and AlexNet [24], can be built for specific applications. Without the loss of generality, we conduct the investigation on the fifth generation of LeNet architecture using SC, which is comprised of two pairs of convolutional and pooling layers, and one fully connected hidden layer with the output layer. Note that the proposed methodology can accommodate other DCNN architectures as well.

A convolutional layer is associated with a set of learnable filters (or kernels), which are activated when specific types of features are found at some spatial positions in the inputs. After obtaining features using convolution, a subsampling step can be applied to aggregate statistics of these features to reduce the dimensions of data and mitigate over-fitting issues. This subsampling operation is realized by a pooling layer in hardware-based DCNNs, where different non-linear functions can be applied, such as max pooling, average pooling, and L2-norm pooling. The activation functions in neurons are non-linear transformation functions, such as Rectified Linear Units (ReLU) $f(x) = max(0, x)$, hyperbolic tangent (tanh) $f(x) = tanh(x)$ or $f(x) = |tanh(x)|$, and sigmoid function $f(x) = \frac{1}{1+e^{-x}}$. We adopt the tanh activation function since it can be implemented efficiently as a finite state machine (FSM) in SC using a stochastic approximation method. The fully connected layer is a normal neural network layer with

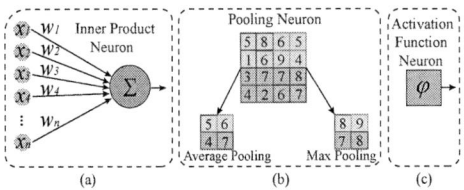

Fig. 1. Neurons in DCNN. (a) Inner Product, (b) pooling, and (c) activation

its inputs fully connected with its previous layer. The loss function of DCNN specifies how the network training penalizes the deviation between the predicted and true labels, and typical loss functions are softmax loss, sigmoid cross-entropy loss or Euclidean loss.

In general, we can define three kinds of basic neurons (nodes) in hardware-based DCNN based on their corresponding operations as Fig.1 shows. Neurons in convolutional layers and fully connection layers calculate the inner product shown in Fig.1 (a) of inputs and weights based on its incoming connection with the previous layer. And the products are subsampled through a pooling neuron shown in Fig.1 (b). We use average pooling here as a case study due to its simple hardware implementation. The subsampled outputs are transformed by an activation function shown in Fig.1 (c) to ensure the inputs of next layer are within $[-1, 1]$.

B. Stochastic Computing Based Neuron

In stochastic computing, the value of a number which lies in $[0, 1]$ is represented by the occurrence probability of 1s in a random bit stream. A 4-bit sequence $X = 0010$, for example, represents $x = P(X = 1) = \frac{1}{4} = 0.25$. An m-bit sequence can only represent numbers in the set $\{\frac{0}{m}, \frac{1}{m}, \frac{2}{m}, \cdots, \frac{m}{m}\}$; Only a small subset of the real numbers in the interval $[0, 1]$ can be expressed exactly in SC. Clearly, the precision and accuracy of SC depend on the length of the stream.

The two most popular representations for stochastic numbers are unipolar and bipolar formats, which interpret values in the intervals $[0, 1]$ and $[-1, 1]$, respectively. Unipolar coding is commonly used in unsigned arithmetic operations, whereas bipolar format is used in signed arithmetic calculations. More specifically, in unipolar coding, the number x carried in a stochastic stream of bits X is $x = P(X = 1) = P(X)$, whereas in the bipolar format, $x = 2P(X = 1) - 1 = 2P(X) - 1$.

In this section, we first conduct a detailed investigation of the energy-accuracy trade-off among two hardware neuron designs using SC, i.e., APC-based neuron and MUX-based neuron, as shown in Fig.2 and Fig.3, respectively. Hardware-based pooling is provided afterward, and finally, we present the structure optimization method for the overall DCNN architecture.

1) APC-Based Neuron: Fig.2 illustrates the APC-based hardware neuron design, where the inner product is calculated using XNOR gates (for multiplication in bipolar coding) [25] and an APC (for addition). To be more specific, we denote the number of bipolar inputs and stochastic stream length by n and m, respectively. Accordingly, n XNOR gates are used to generate n products of inputs ($x_i's$) and weights ($w_i's$), and then the APC accumulates the sum of 1s in each column of the products shown in Fig.2 (a). Instead of an FSM, a saturated up/down counter is used to perform the scaled hyperbolic tangent activation function $Btanh(\cdot)$ for binary inputs as Fig.2 (b) shows. Details and optimization of the $Btanh(\cdot)$ activation

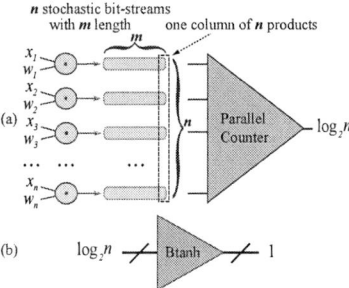

Fig. 2. APC-based neurons. (a) Inner product and (b) activation

function using a saturated up/down counter for binary inputs are demonstrated in [19].

2) MUX-Based Neuron: As shown in Fig.3, the MUX-based neuron is composed of XNOR gates, a MUX, and a K-state FSM. XNOR gates compute the products of bipolar inputs ($x_i's$) and weights ($w_i's$); the n−to−1 MUX sums up all stochastic products; and the hyperbolic tangent activation function is achieved by a K-state FSM, respectively. As the inner product calculated by a MUX is a stochastic number, the K-state FSM design mentioned in [25] can be used here to implement the activation function denoted as $Stanh(K, x) = tanh(\frac{K \cdot x}{2})$ where x is the input.

Nevertheless, two problems challenge the implementation: (i) the inner product calculated by an n input MUX is scaled to $\frac{z}{n}$, where the correct inner-product result is z, and (ii) with the input $\frac{z}{n}$, the K-state FSM calculates $tanh(\frac{K \cdot z}{2 \cdot n})$ instead of the desired value $tanh(z)$. Thus, in order to recover the correct activation, we need to re-scale up the results of MUX by n times and multiply the stream by $\frac{2}{K}$ (or multiply by $\frac{2 \cdot n}{K}$ directly). As opposed to the relatively simple and efficient data conversions on a software platform, such conversions in a hardware-based neuron incur significant hardware overhead, because the linear gain transformation needs one more FSM [25], and the scaling multiplication requires one XNOR gate as well as another bipolar stochastic stream generated.

In this paper, considering an n inputs neuron with inner product denoted by z, we select the state number K such that $\frac{2 \cdot n}{K} = 1$, and the final output of the FSM is calculated as

$$Stanh(K, \frac{z}{n}) = tanh(\frac{K \cdot z}{2 \cdot n}) = tanh(z) \quad (1)$$

In this way, we achieve the desired activation result with no additional bit stream conversion (i.e., no hardware overhead).

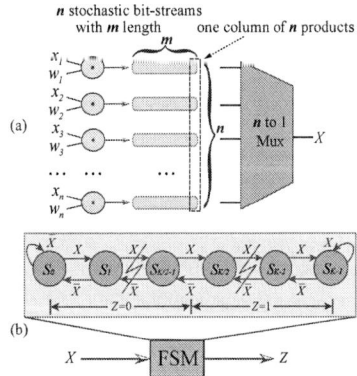

Fig. 3. MUX-based neurons. (a) Inner product and (b) activation

3) Pooling Neuron: In a DCNN, down-sampling steps are performed by the neurons in pooling layers, which reduce the dimensions of neurons for following convolutional layers or fully connection layers. Pooling operation achieves the invariance to input data (i.e., image, video, etc.) transformations and better robustness to noise and clutter. Moreover, the inter-layer connections can be significantly reduced for a hardware DCNN by using pooling layers. In this paper, we adopt the *average pooling* which simply outputs a mean of k inputs. In SC-based hardware, we implement it using a MUX and each input x_i has the same probability to be selected as output. For example, the stochastic arithmetic mean over 4 inputs (2×2 region) is provided by using three 2-to-1 MUXs connected hierarchically, where selection signal for each MUX is a stochastic bit-stream for 0.5.

IV. BUDGET-DRIVEN DESIGN ALLOCATION

Obviously, since neurons in each DCNN layer are homogeneous, we consider neurons in the same layer are implemented using the same design. Design allocation which represents how to arrange different neuron designs for each DCNN's layer is a challenging task. Accuracy, power, area, and energy constraints have to be satisfied. Inspired by the *stable marriage problem* [26], we present a design allocation algorithm which optimizes the overall network design given a preferred parameter. The first step is to find a *minimum feasible solution (MFS)* so that each layer in the DCNN is implemented by a valid neuron design. The second step is to optimize the MFS by re-allocating implementation for neurons in each layer to achieve the desired network-wise objective. The objective of the overall DCNN optimization is evaluated in terms of a comprehensive design score defined as follows

$$Score = \frac{\prod_j C_j^{\omega_j}}{1 - Err} \quad (2)$$

where ω_j is the integer weight of each design parameter, C_j is the overall network cost in terms of one metric (e.g., area or power), and Err is the overall network error rate. As long as any one of the design parameter is the optimization target, corresponding ω_j increases. A small score suggests an optimized design considering both design parameters and network performance.

A. Minimum Feasible Solution

The stable marriage algorithm cannot be applied directly to find a MFS. In the stable marriage problem, an element can be matched to another element freely. However, in our problem, a valid hardware design can implement multiple DCNN layers; several constraints result in limitation to apply a specific hardware design for neurons in a certain DCNN layer. For example, in the case, where using APC-based neuron in a convolutional layer leads to the budget violation, the convolutional layer should not be implemented with the APC-based neuron.

In the minimum feasible design allocation algorithm shown in Algorithm.1, the cost, denoted as c_{lid}, of an implementation i on neurons in a layer l with a specific constraint d is defined as

$$c_{lid} = \psi \cdot u_{lid} \quad (3)$$

where ψ is the number of neurons in the current layer and u_{lid} is the unit cost of constraint d for a neuron in layer l by

Algorithm 1: Minimum Feasible Design Allocation

Data: list of feasible hardware implementations $\mathcal{I}$, list of layers in DCNN $\mathcal{L}$, optimization objective constraint t
Result: minimum feasible design allocation solution $\mathcal{S}$
Initialize all $l \in \mathcal{L}$ to be free
$\mathcal{S} \leftarrow \emptyset$
while $\exists$ *free layer* $l \in \mathcal{L}$ *has no implementation* **do**
 for *each* $i \in \mathcal{I}$ **do**
  $IsValid \leftarrow True$
  if *cost* $c_{lit} > Budget\ b_{lt}$ **then**
   $IsValid \leftarrow False$
  if *IsValid* **then**
   l's implementation $\leftarrow i$
   Append pair (i, l) to $\mathcal{S}$
   break
 if l *is free* **then**
  $\mathcal{S} \leftarrow \emptyset$
  break
Output $\mathcal{S}$

Algorithm 2: Design Allocation Optimization

Data: minimum feasible design allocation solution $\mathcal{S}$, list of feasible hardware implementations $\mathcal{I}$, optimization objective constraint t
Result: optimized design allocation solution $\mathcal{S}'$
$lastScore \leftarrow Scores_{\mathcal{S}}$
$currScore \leftarrow Scores_{\mathcal{S}}$
$iterCount \leftarrow 0$
do
 $lastScore \leftarrow currScore$
 $lastCost \leftarrow 0$
 $currCost \leftarrow 0$
 do
  $\mathcal{S}' \leftarrow \emptyset$
  $lastCost \leftarrow currCost$
  $currCost \leftarrow 0$
  for *each pair* (i, l) *in* $\mathcal{S}$ **do**
   for *each* $i' \in \mathcal{I}$ **do**
    if *cost* $c_{i't} < c_{it}$ **then**
     l's implementation $\leftarrow i'$
   $i^* \leftarrow l$'s implementation
   $currCost \leftarrow currCost + dc_{i^*}$
   Append pair (i^*, l) to $\mathcal{S}'$
 while $lastCost - currCost \leq threshold\ \theta$
 $currScore \leftarrow Score_{\mathcal{S}'}$
 $iterCount \leftarrow iterCount + 1$
while $lastScore \leq currScore$ *or* $iterCount < \tau$
Output $\mathcal{S}'$

---

implementation i. Similarly, the budget for layer l with respect to constraint d is denoted by b_{ld}. Layers in a DCNN are firstly initialized to be free. The MFS algorithm tries to find a valid design for each layer at first. An implementation is *valid* for a specific layer when the cost in every aspect of constraint metric satisfies the requirement. The first valid implementation is used for current layer. However, if all implementations are not valid, the solution is void. And an impossible solution is reported. This greedy process is repeated for each layer of a DCNN to obtain a MFS.

B. Joint Design Allocation Optimization

The MFS guarantees that each DCNN layer has a valid implementation design, but the complete DCNN is not optimized, i.e. the accuracy of DCNN is not optimized for the given constraints. A joint optimization algorithm is proposed in Algorithm.2. The cost in the algorithm is defined in Eqn.(3), where we only calculate the cost for the constraint t. Taking the MFS and the given design budget as the inputs, the algorithm firstly seeks an implementation with lower cost for each DCNN layer. The cost difference between the old implementation and the new one has to be as big as θ to output the new one. After a lower cost design is ensured, the algorithm evaluates the refined implementation. If the accuracy increases, which means the error rate of the network with new implementation is smaller than the previous one, the algorithm will output the new implementation scheme for the DCNN. When the process reaches an iteration limit τ, the current solution is recognized as the optimized one and algorithm terminates.

V. EXPERIMENTAL RESULTS

A. Comparison between APC-based and MUX-based neuron

We use Synopsys Design Compiler to synthesize the neurons with the 45nm Nangate Open Cell Library [27]. For an APC-based neuron, the area, path delay, energy, power, and absolute error with respect to the input size are shown in Fig.4 (a), (b), (c), (d), and (e), respectively. Energy and absolute error differ due to bit-stream length change, while other measures remain constant. To be more specific, as illustrated in Fig.4 (a), (c), and (d), the APC-based neuron shows an exponential increase in area, energy, and power including dynamic and leakage power as the input size of a neuron increases exponentially. This means that the area, energy, and power are linearly proportional to input size. However, path

delay shown in Fig.4 (b) reflects a saturated pattern when the input size of a neuron increases to a certain point, which means that a large input size will not lead to extreme long path delay.

The reason results from: With the efficient implementation of $Btanh(\cdot)$ function, the hardware of $Btanh(\cdot)$ increases logarithmically as the input increases, since the input width of $Btanh(\cdot)$ is $log_2 n$. On the other hand, the number of XNOR gates and the size of the APC grow linearly as the input size increases. Hence, the inner product calculation part, i.e., XNOR array and APC, is dominant in an APC-based neuron, and the area, power, and energy of the entire APC-based neuron cell also increase at the same rate as the inner product part when the input size increases.

We observe that the error is normally distributed with a mean value of 0. In this paper, we denote absolute error as the absolute standard deviation of error's distribution. For different bit-stream lengths, the result, as Fig.4 (e) shows, agrees on the intuitive observation that a longer bit stream can reduce the absolute error of calculation. Moreover, more inputs lead to larger absolute error. The absolute error increases logarithmically with respect to input size. The improvement due to the increase of bit-stream length is non-linear but independent of the input size. Longer bit-stream length helps to reduce the error, but this improvement decreases and converges when bit-stream length gets longer than 1024. Designers should consider the latency and energy overhead caused by long bit streams; the convergence of improvement trend helps designers achieve the desired trade-off between accuracy and overhead.

Similarly, we investigate the performance of the MUX-based neuron with respect to its input size. Fig.5 (a), (b), (c), and (d) show the results of the number of inputs versus the area, path delay, energy, power, and absolute error with respect to the input size. Based on observation on the absolute error the MUX-based neuron can achieve, as shown in Fig.5 (e), when the input size exceeds 64, the absolute errors with different bit-stream length approach 1, which is about 100% error compared to correct value range. Then we plotted the chart for input size from 8 to 64 only. With input size increase linearly, the absolute error increase in a linear pattern. The reason is that MUX addition selects only one bit at a time

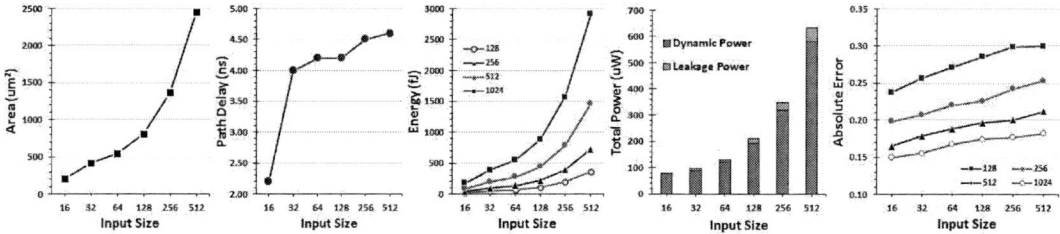

Fig. 4. Input size versus (a) area, (b) path delay, (c) energy, (d) power, and (e) absolute error with different bit-stream lengths for APC-based neuron

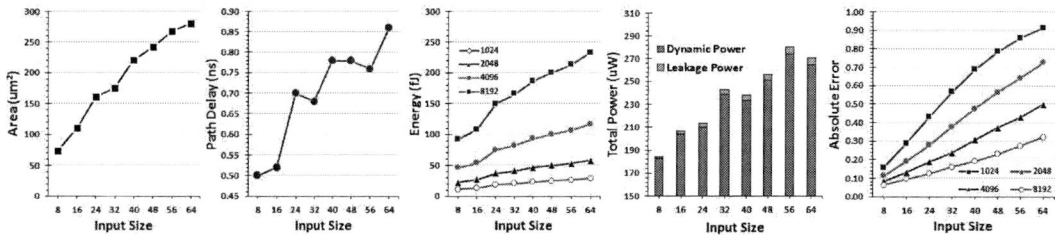

Fig. 5. Input size versus (a) area, (b) path delay, (c) energy, (d) power, and (e) absolute error with different bit-stream lengths for MUX-based neuron

TABLE I
COMPARISON BETWEEN APC-BASED NEURON AND MUX-BASED
NEURON USING 1024 BIT STREAM

	APC-based neuron			MUX-based neuron		
Input size	16	32	64	16	32	64
Absolute error	0.15	0.16	0.17	0.29	0.56	0.91
Area (μm^2)	209.9	417.6	543.2	110.7	175.3	279.8
Path delay (ns)	2.20	4.00	4.20	0.52	0.70	0.68
Power (μW)	80.7	95.9	130.5	206.5	242.9	271.2
Energy (fJ)	177.4	383.7	548.1	110.0	169.1	238.9

and ignores the rest of the bits, leading high absolute errors when input size is large. Furthermore, as APC-based neuron performs, longer bit streams can reduce absolute error. The improvements are independent of input size; the improvements with respect to bit-stream length are logarithmic. When the bit-stream length is increased large enough, the absolute error will not be reduced significantly. In general, MUX-based neuron gains larger absolute error compared to APC-based neuron, which suggests MUX-based should be applied to more error-tolerant arithmetic operations. In addition, one can observe from Fig.5 (a), (c), and (d) that as the number of inputs increases, area, power, and energy of the MUX-based neuron all tend to increase. The synthesis result also shows the path delay for a neuron increases approximate linearly when we enlarge the input size with the same stride. These are because the MUX-based neuron with more inputs requires more XNOR gates and MUXes for inner product calculation, and more states in the FSM ($K = 2 \cdot n$) to compute the activation. Hence, the increased hardware components result in more area, power, path delay, and energy in the neuron cell.

we compare the performance between APC-based neuron and MUX-based neuron using a fixed bit stream length equal to 1024 under different input sizes, as shown in Table.I. Clearly, APC-based neuron is more accurate but occupies more area than MUX-based neuron. Besides, as APC is slower than MUX, the latency of APC-based neuron is larger than MUX-based neuron, which causes APC-based neuron to consume more energy than MUX-based neuron for one calculation. As for the power performance, an APC-based neuron has less switching (due to the long latency) and larger area than the

MUX-based neuron, resulting in less dynamic power, more leakage power, and less overall power.

B. Evaluation of design allocation algorithm

We use LeNet-5 DCNN as a case study in this experiment to evaluate our budget-driven algorithm to optimize a stochastic computing based DCNN. Neurons in LeNet-5 DCNN layers are configured as $784(28 \times 28) - 11520(20 \times 24 \times 24) - 2880(20 \times 24 \times 24) - 3200(50 \times 8 \times 8) - 800(50 \times 4 \times 4) - 500 - 10$. The MNIST handwritten digit image dataset [28] consisting of 60,000 training data and 10,000 testing data with 28x28 grayscale image and 10 classes is used in the experiments. The synthesis results are gathered as mentioned in Section III-B1 using Synopsys Design Complier with the 45nm Nangate Open Cell Library [27].

We first listed several different configurations to implement each DCNN layer as Table.II shows. We fed these configurations into our algorithm given the budget constraint and evaluated the design score, defined as Eqn.(2), for the optimized configuration of the DCNN. To validate our algorithm, we optimized the DCNN with three different optimization targets as an example, i.e., area, power, and energy, given corresponding network-wise constraints. The energy is approximately positively proportional to power; we set the weight for energy as 0 and use power to represent energy approximately. In the experiment, three example cases were studied, where case 1 emphasized area and the rest emphasized power.

- Case1. $Score1 = \frac{Area^2 \cdot Power}{1 - Err}$ and $Area \leq 5mm^2$
- Case2. $Score2 = \frac{Area \cdot Power^2}{1 - Err}$ and $Power \leq 2W$
- Case3. $Score3 = \frac{Area \cdot Power^2}{1 - Err}$ and $Energy \leq 4\mu J$

As shown in Table.III, the proposed joint optimization algorithm picked configuration 4, 7, and 14 for case 1, 2, and 3, respectively. Under certain given constraints, some configurations are not valid, which is filtered out by the algorithm. For comparison, we calculated the scores for those configurations which are not optimized when they are valid (configuration 1, 2, 9, 12). We have conducted a separate exhaustive search, and verified the selected configurations by the proposed algorithm are the best. One can observe from

978-1-5386-7100-9/18 $31.00 © 2018 IEEE

TABLE II
EXAMPLE CONFIGURATIONS

Configuration	Bitstream Length	Layer 1	Layer 2	Layer 3
1	1024	MUX	MUX	MUX
2	1024	MUX	MUX	APC
3	1024	MUX	APC	MUX
4	1024	MUX	APC	APC
5	1024	APC	MUX	MUX
6	1024	APC	MUX	APC
7	1024	APC	APC	MUX
8	1024	APC	APC	APC
9	512	APC	MUX	APC
10	512	APC	APC	MUX
11	512	APC	APC	APC
12	512	APC	APC	APC
...				
13	256	APC	APC	MUX
14	256	APC	APC	APC
...				

TABLE III
EXAMPLES OF CONFIGURATIONS EXPLORED AND GENERATED BY THE ALGORITHM

Configuration	Error (%)	Area(mm^2)	Power (W)	Energy (uJ)	Score 1	Score 2	score 3
1	21.7	3.18	3.08	2.85	38.56	–	38.56
2	11.9	3.69	3.03	4.21	38.50	–	–
4	8.7	4.56	2.75	5.44	37.70	–	–
7	4.3	7.20	1.77	7.63	–	95.80	–
9	4.7	6.83	2.01	3.96	–	–	28.84
12	9.4	6.83	2.01	1.98	–	–	30.34
14	2.0	7.70	1.72	2.36	–	104.24	23.29
...							

the Table.III that the picked configurations have the lowest score whereas other configurations are invalid or have larger scores. Also being observed from Table.III, our algorithm gives the best trade-off between network accuracy and design parameters (with the highest scores in all the cases).

VI. CONCLUSION

This paper introduced hardware implementation for Deep Convolutional Neural Network using stochastic computing. Each distinct stochastic computing based neuron in DCNN is analyzed. A two-step joint optimization algorithm is proposed that given design budgets, re-structuring the SC-based DCNN can achieve optimized hardware footprint with a relative high network accuracy performance. Experimental results showed with restricted design requirements, the optimized SC-based implementation for DCNN achieved the lower error rate with the least design resources requested.

REFERENCES

[1] Caiwen Ding, Siyu Liao, Yanzhi Wang, Zhe Li, Ning Liu, Youwei Zhuo, Chao Wang, Xuehai Qian, Yu Bai, Geng Yuan, et al. Circnn: accelerating and compressing deep neural networks using block-circulant weight matrices. In *Proceedings of the 50th Annual IEEE/ACM International Symposium on Microarchitecture*, pages 395–408. ACM, 2017.
[2] Shuo Wang, Zhe Li, Caiwen Ding, Bo Yuan, Qinru Qiu, Yanzhi Wang, and Yun Liang. C-lstm: Enabling efficient lstm using structured compressiontechniques on fpgas. In *2018 ACM/SIGDA International Symposium on Field-Programmable Gate Arrays*. ACM, 2018.
[3] Sheng Lin, Ning Liu, Mahdi Nazemi, Hongjia Li, Caiwen Ding, Yanzhi Wang, and Massoud Pedram. Fft-based deep learning deployment in embedded systems. In *2018 Design, Automation & Test in Europe Conference & Exhibition, DATE 2018, Dresden, Germany, March 19-23, 2018*, pages 1045–1050, 2018.
[4] Tara N Sainath, Abdel-rahman Mohamed, Brian Kingsbury, and Bhuvana Ramabhadran. Deep convolutional neural networks for lvcsr. In *2013 IEEE International Conference on Acoustics, Speech and Signal Processing*, pages 8614–8618. IEEE, 2013.
[5] Karen Simonyan and Andrew Zisserman. Very deep convolutional networks for large-scale image recognition. *arXiv preprint arXiv:1409.1556*, 2014.

[6] Yanzhi Wang, Caiwen Ding, Zhe Li, Geng Yuan, Siyu Liao, Xiaolong Ma, Bo Yuan, Xuehai Qian, Jian Tang, Qinru Qiu, and Xue Lin. Towards ultra-high performance and energy efficiency of deep learning systems: An algorithm-hardware co-optimization framework. In *Proceedings of the Thirty-Second AAAI Conference on Artificial Intelligence, New Orleans, Louisiana, USA, February 2-7, 2018*, 2018.
[7] Andrej Karpathy, George Toderici, Sanketh Shetty, Thomas Leung, Rahul Sukthankar, and Li Fei-Fei. Large-scale video classification with convolutional neural networks. In *Proceedings of the IEEE conference on Computer Vision and Pattern Recognition*, pages 1725–1732, 2014.
[8] Yann LeCun, Yoshua Bengio, and Geoffrey Hinton. Deep learning. *Nature*, 521(7553):436–444, 2015.
[9] Dan C Ciresan, Ueli Meier, Jonathan Masci, Luca Maria Gambardella, and Jürgen Schmidhuber. Flexible, high performance convolutional neural networks for image classification. In *IJCAI Proceedings-International Joint Conference on Artificial Intelligence*, volume 22, page 1237, 2011.
[10] Clément Farabet, Cyril Poulet, Jefferson Y Han, and Yann LeCun. Cnp: An fpga-based processor for convolutional networks. In *2009 International Conference on Field Programmable Logic and Applications*, pages 32–37. IEEE, 2009.
[11] Srimat Chakradhar, Murugan Sankaradas, Venkata Jakkula, and Srihari Cadambi. A dynamically configurable coprocessor for convolutional neural networks. In *ACM SIGARCH Computer Architecture News*, volume 38, pages 247–257. ACM, 2010.
[12] Srihari Cadambi, Abhinandan Majumdar, Michela Becchi, Srimat Chakradhar, and Hans Peter Graf. A programmable parallel accelerator for learning and classification. In *Proceedings of the 19th international conference on Parallel architectures and compilation techniques*, pages 273–284. ACM, 2010.
[13] Brian R Gaines. Stochastic computing systems. In *Advances in information systems science*, pages 37–172. Springer, 1969.
[14] Ji Li, Ao Ren, Zhe Li, Caiwen Ding, Bo Yuan, Qinru Qiu, and Yanzhi Wang. Towards acceleration of deep convolutional neural networks using stochastic computing. In *22nd Asia and South Pacific Design Automation Conference (ASP-DAC)*, pages 115–120. IEEE, 2017.
[15] Ji Li, Zihao Yuan, Zhe Li, Caiwen Ding, Ao Ren, Qinru Qiu, Jeffrey Draper, and Yanzhi Wang. Hardware-driven nonlinear activation for stochastic computing based deep convolutional neural networks. *arXiv preprint arXiv:1703.04135*, 2017.
[16] Zihao Yuan, Ji Li, Zhe Li, Caiwen Ding, Ao Ren, Bo Yuan, Qinru Qiu, Jeffrey Draper, and Yanzhi Wang. Softmax regression design for stochastic computing based deep convolutional neural networks. In *Proceedings of the Great Lakes Symposium on VLSI*, pages 467–470. ACM, 2017.
[17] Ji Li, Zihao Yuan, Zhe Li, Ao Ren, Caiwen Ding, Jeffrey Draper, Shahin Nazarian, Qinru Qiu, Bo Yuan, and Yanzhi Wang. Normalization and dropout for stochastic computing-based deep convolutional neural networks. *Integration, the VLSI Journal*, 2017.
[18] Behraoz Parhami and Chi-Hsiang Yeh. Accumulative parallel counters. In *Conference Record of The Twenty-Ninth Asilomar Conference on Signals, Systems and Computers*, volume 2, pages 966–970. IEEE, 1995.
[19] Kyounghoon Kim, Jungki Kim, Joonsang Yu, Jungwoo Seo, Jongeun Lee, and Kiyoung Choi. Dynamic energy-accuracy trade-off using stochastic computing in deep neural networks. In *Proceedings of the 53rd Annual Design Automation Conference*, page 124. ACM, 2016.
[20] Yann LeCun, LD Jackel, Leon Bottou, A Brunot, Corinna Cortes, JS Denker, Harris Drucker, I Guyon, UA Muller, et al. Comparison of learning algorithms for handwritten digit recognition. In *International conference on artificial neural networks*, volume 60, pages 53–60, 1995.
[21] Daniel Strigl, Klaus Kofler, and Stefan Podlipnig. Performance and scalability of gpu-based convolutional neural networks. In *PDP*, pages 317–324, 2010.
[22] Pai-Shun Ting and John Patrick Hayes. Stochastic logic realization of matrix operations. In *Digital System Design (DSD), 2014 17th Euromicro Conference on*, pages 356–364. IEEE, 2014.
[23] Yann LeCun, Léon Bottou, Yoshua Bengio, and Patrick Haffner. Gradient-based learning applied to document recognition. *Proceedings of the IEEE*, 86(11):2278–2324, 1998.
[24] Alex Krizhevsky, Ilya Sutskever, and Geoffrey E Hinton. Imagenet classification with deep convolutional neural networks. In *Advances in neural information processing systems*, pages 1097–1105, 2012.
[25] Bradley D Brown and Howard C Card. Stochastic neural computation. i. computational elements. *IEEE Transactions on computers*, 50(9):891–905, 2001.
[26] David Gale and Lloyd S Shapley. College admissions and the stability of marriage. *The American Mathematical Monthly*, 120(5):386–391, 2013.
[27] Nangate 45nm Open Library, Nangate Inc., 2009.
[28] Li Deng. The mnist database of handwritten digit images for machine learning research. *IEEE Signal Processing Magazine*, 29(6):141–142, 2012.

978-1-5386-7100-9/18 $31.00 © 2018 IEEE

Fast Heuristics for Near-optimal Signal Restoration in Post-Silicon Validation

Xiaobang Liu and Ranga Vemuri
Department of Electrical Engineering and Computer Science
University of Cincinnati, Cincinnati, OH, USA
liu2xg@mail.uc.edu, ranga.vemuri@uc.edu

Abstract—In post-silicon validation based on exploiting trace buffer based techniques, signal observability is largely restricted by the size of trace buffer. Moreover, the forward propagation and backward justification (FB) methods which are commonly used for signal restoration can potentially miss a large number of restorable signals yielding low restoration ratios. Optimal restoration based on satisfiability (SAT) methods is usually accompanied by the extremely long runtime for large size benchmark circuits. Therefore, near-optimal yet efficient signal restoration is highly desired. In this paper, we propose efficient heuristic methods for near-optimal signal restoration by combining three key ideas: (1) dividing the restoration window into frames, (2) selectively using SAT and FB methods within and across frames, and (3) employing a prioritization heuristic based on restoration in the preceding time frames. Experimental results show that the proposed method is efficient (up to 87% improvement in run-time compared to non-prioritization) and can achieve significant restoration ratio improvement (up to 16x more when compared to FB based methods and up to 99% of optimal restoration).

I. INTRODUCTION

Post-silicon validation or post-silicon debugging refers to error detection and localization in a manufactured IC design while running applications at the target clock speed. As the physical chip executes "at speed", it could reveal errors not possible to encounter in pre-silicon simulations. However, controllability and observability are considerably weakened when compared to simulations where the internal signals can be easily observed. *Trace buffer* based run-time techniques have been employed to enhance observability by tracking a small number of pre-selected signals across a predefined number of clock cycles [1]. Run-time data of the traced signals is transferred off-chip for analysis. Observability can be improved by computing the values of the untracked states through *state restoration*. Errors are detected if the restored state values differ from the golden values. The more the signals restored and the faster the restoration, the better the quality and efficiency of debug.

The size of trace buffer is defined as the product of its width w, which is the number of trace signals, and depth d, which is the number of clock cycles over which these signal values are recorded. *State Restoration Ratio* (SRR) is commonly used to represent the quality of state restoration based on the traced data and circuit under debug. Suppose N_t represents the total number of signal instances traced over t time steps and that N_r is the total number of signal instances restored by a restoration

algorithm over the same time steps. SRR is then determined as $(N_t + N_r)/N_t$. For a concrete set of signals selected for tracing, the SRR is not only affected by the corresponding data recorded in trace buffer, but also by the state restoration algorithm. Various *trace signal selection algorithms* proposed [2]–[5] favor to store signals which yield better SRR. While some researchers argue that SRR is not necessarily the best metric to evaluate the trace signal quality [6] [7], we use this metric in this paper since (1) As long as the trace buffer technique to augment signal visibility is incorporated in circuit design, better quality of state restoration, given a set of trace signals, does facilitate better chance of error detection and (2) No other figure of merit to compare trace signal quality has been universally agreed upon.

There are two categories of commonly applied state restoration algorithms, namely the *forward propagation and backward justification* (FB) based algorithms [3] [5] and the *satisfiability* (SAT) based methods [8] [9]. FB methods are highly efficient but unable to infer indirect implications, while SAT based methods can achieve optimal restoration of much larger number of states but take much longer time for execution. The lower efficiency of SAT based methods is due to the unrolling of sequential circuits into multiple copies over multiple clock cycles thereby increasing the problem size for SAT solvers. For both trace signal selection and fault localization algorithms which will perform state restoration many times, faster yet high quality of signal restoration is desirable and critical. To practically apply the SAT-based methods, it is necessary to compromise the optimal SRR to a small extent while substantially improve efficiency.

[8] mentions a technique which evenly divides unrolled copies of sequential circuits into *time frames*. A time frame is a series of consecutive clock cycles. Their framing method applies SAT based restoration in each frame locally and the FB method globally. We observe in our preliminary experiments that this frame based algorithm shows very high quality of restoration with much less run-time compared with the optimal restoration. It is worth noting that for a concrete set of signals with their corresponding traced values obtained from trace buffer, the size of the frame determines the SRR in the framing method. In this paper, we propose a faster signal prioritization heuristic for near-optimal state restoration based on combining the framing algorithm and the FB method. In addition, we introduce a much more efficient algorithm with very little

978-1-5386-7100-9/18 $31.00 © 2018 IEEE

restoration ratio drop compared with the first heuristic.

The rest of this paper is organized as follows. Section II discusses the background and related work in signal restoration. Section III describes our state restoration algorithms in detail. The experimental results are reported in Section IV and finally the concluding remarks are provided in Section V.

II. BACKGROUND AND RELATED WORK

State restoration is the process of determining the values (if restorable) of the unknown state elements based on the available signal values over a predefined number of time steps.

A. FB methods

The process of inferring the output value of a gate given some or all the input values is called forward propagation. Conversely, backward justification refers to determining the values of one or more input values of a gate based on the output value and some input values if available. These two processes can be applied to a sequential circuit iteratively over a certain number of clock cycles until no more inference can be made. Figure 1 illustrates the FB method with a simple example. Assume flip-flop G is selected for tracing over 6

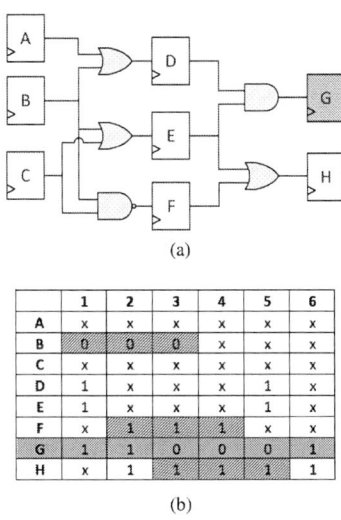

(a)

	1	2	3	4	5	6
A	x	x	x	x	x	x
B	0	0	0	x	x	x
C	x	x	x	x	x	x
D	1	x	x	x	1	x
E	1	x	x	x	1	x
F	x	1	1	1	x	x
G	1	1	0	0	0	1
H	x	1	1	1	1	1

(b)

Fig. 1. Example to illustrate FB method

clock cycles as shown in the figure. Then we can see the values of D and E are inferred as '1' in cycles 1 and 5. H is restored to be value '1' for cycle 2 and 6 as well. This concludes the FB restoration. We know from the figure that $N_r = 6$ and $N_t = 6$ therefore $SRR = 12/6 = 2$.

Because of its high efficiency, FB method is frequently used as part of the existing trace signal selection algorithms. Recent papers on automated trace signal selection like [5] [3] [10] [4] [11] all utilize it for signal restoration. Given that in [6] the authors strongly argue that SRR as a metric for signal selection can overlook useful hints for fault localization, [10] describes a trace signal selection algorithm which operates at functional RTL level with an error detection aware feature and keeps using FB method for restoration. Simulation based

approaches to signal selection such as [12] and [2] employ an efficient bit-parallel algorithm to accelerate state restoration but it limits the maximum number of clock cycles to 64. [5] proposed a faster event-driven state restoration algorithm by removing unnecessary forward and backward operations. All of these FB based approaches can potentially miss a large number of restorable states.

B. SAT methods

SAT solvers are repeatedly invoked in this method for visibility enhancement. Each gate is represented as a Boolean formula in the conjunctive normal form (CNF). The CNF of a combinational circuit is the conjunction of the CNFs of all the constituent gates. A sequential circuit is unrolled into a combinational iterative network across a predefined number of clock cycles. In between the unrolled copies, the input of a flip-flop in each clock cycle is connected to the output of the corresponding flip-flop in the next cycle. The unrolled *combinational* iterative network is then conveniently expressed as CNF. SAT solvers accept a Boolean formula in CNF as input, and yield a single assignment of binary values to all the variables if *satisfiable* (SAT), or simply print UNSAT if *unsatisfiable*. Signal restoration using a SAT solver involves establishing the set of signals with constant value in all the possible satisfiable value assignments. Multiple SAT calls for distinctive solutions are essential in the procedure, hence the low efficiency compared with the FB method.

Reexamining the circuit in figure 1, the entries in the table filled with diagonal stripe are additionally restored by a SAT solver within the 6 clock cycles. N_r in this case is 15 and the new SRR equals to $(15 + 6)/6 = 3.5$, representing a 75% improvement compared to the FB method. Several existing SAT-based state restoration approaches are reviewed in the following paragraphs.

1) Iterative SAT Testing: This algorithm explained in [13] is substantially more efficient compared to a naïve method which identifies variables with consistent values by enumerating all possible assignments. Each variable of a given formula is tested at most once yielding worst case $(n + 1)$ SAT calls, where n is the number of variables. The *probing and recording* approach introduced in [9] is a similar idea.

2) Core-based Algorithm: Similar but unlike iterative SAT testing, this method introduced in [13] centers on testing multiple variables (termed as a *chunk*) at the same time in one SAT call. It makes use of a set of clauses responsible for unsatisfiability known as the *unsat core*. This algorithm can be faster than iterative SAT testing if the chunk size is set properly. Because of its high performance, we adopt it as part of our state restoration algorithms in this paper.

3) Other Related Research: Recently a circuit structure aware SAT-based state restoration algorithm was proposed in [8] by probing the reconvergent fanout signals with priority, utilizing the core-based algorithm mentioned above. It also presents an effective signal ordering heuristic aiming to generating more diverse SAT solutions thus reducing runtime. However this method can't handle large number of clock cycles for large benchmark circuits since it attempts to achieve *optimal* signal restoration. [14] employed SAT restoration but it's not scalable as it aims at maximal restoration too. [9] studied several SAT-based techniques and proposed a novel decision-based All-SAT and Learning algorithm, but the probing algorithm performs significantly better in runtime.

III. PROPOSED ALGORITHMS

As SAT-based optimal state restoration in post-silicon debugging is not feasible for large size circuits, we present two efficient heuristics to significantly improve the run-time by concentrating on near-optimal restoration in this paper. The contribution of our approaches is as follows: (1) We show that the frame based approach to signal restoration can achieve near-optimal (occasionally optimal) results in relatively small amount of time; (2) Execution time can be considerably reduced by prioritizing the state signals restored in *all* the preceding time frames in an incremental fashion due to the structural similarity of the frames to each other; (3) By slightly compromising the SRR of the SAT framing algorithm for a specified frame size, we demonstrate that efficiency can be further elevated substantially.

Let S denote the set of signals in a sequential logic circuit. Let $T \subseteq S$ be the set of state signals being traced. Let the set of states signals be numbered $1, 2, \cdots, M$. Let N be the number of clock cycles over which signal tracing and restoration are performed. Assume V is a $(M \times N)$ matrix storing state signal values at different time steps. The value of state signal i at cycle t is given by $V[i, t]$ and is denoted by $V_i(t)$ where $1 \leq i \leq M$ and $1 \leq t \leq N$. Let F be the time frame size. The number of frames is determined by $f = N/F$. The value of signal i at the tth clock cycle within time frame n $(1 \leq n \leq f; 1 \leq t \leq F)$ is denoted by $V_i^{(n)}(t)$. Therefore $V^{(n)}$ is $(M \times F)$ submatrix of V. Entries in $V^{(n)}$ can be one of the following: {*0, 1, non-unique, open*}. Entries '*0*' and '*1*' designate uniquely determined binary values (traced or restored). '*open*' implies that that signal instance is still open for further inspection. '*non-unique*' entry means within the active local frame that signal is identified as unrestorable, however, it doesn't necessarily imply that its value cannot be inferred and restored globally (across all the N clock cycles). Let $O^{(n)}$ be the set of *open* signal instances in frame n. $O^{(n)} = \{(i, t) | V_i^{(n)}(t) = open; 1 \leq i \leq M; 1 \leq t \leq F\}$. At the end, the number of known instances of signal values is given by $N_k = |\{(i, t) | (V_i(t) = 0 \text{ or } V_i(t) = 1); 1 \leq i \leq M; 1 \leq t \leq N\}|$. Finally, $N_r - N_k$ N_t and $SRR = N_k/N_t$.

A. Prioritization within Time Frames

We combine a learning based signal prioritization with the time frame based SAT method. Our approach is based on the reasoning that a signal instance has a high likelihood of being restored in a time-frame if the corresponding instances of the same signal have been restored in the previous frames. Our method is shown in Algorithm 1 and explained below.

Given a circuit C, an initialized state value matrix V (copying data in trace buffer) and the overall number of clock

cycles N, the first step applies the FB method globally for initial restoration. The next two steps involve unrolling the sequential circuit for N cycles and generating the time frames. What distinguishes our algorithm from the the time frame method in [8] is the heuristic assigning priority to state signals which is motivated by the following observation.

Before loading the values from the trace buffer, the circuit structure in each frame is identical. After performing the first FB restoration, the circuit is simplified based on the restored values. Though the simplification results in dissimilar circuit structures in various frames in general, large portions remain the same. Hence constant value state signals (i.e. restored signals) discovered in the preceding frames exhibit good probability to be restored in the subsequent frames.

Let P be the set of these priority signals to be progressively identified. Let $P_o^{(n)}$ represent all the *open* priority signal instances at time frame n. That is, $P_o^{(n)} = \{(i, t) | V_i^{(n)}(t) = open; i \in P; 1 \leq t \leq F\}$.

Algorithm 1 FRAMES-FB-SAT-PRIO(C, V, N)

1: V ← FB-RESTORE(C, V, N)
2: C_N ← UNROLL(C, N)
3: $C_F[1...f]$ ← GENERATE-FRAMES(C_N, N, F)
4: P = ∅
5: **for** $r = 1$ to f **do**
6:    **if** P ≠ ∅ **then**
7:       $(V, V^{(r)})$ ← PROBE-PRIO-SIG($C_F[r]$, $V^{(r)}$, $P_o^{(r)}$)
8:       V ← FB-RESTORE(C, V, N)
9:    **end if**
10:   $(V, V^{(r)})$ ← SAT-RESTORE($C_F[r]$, $V^{(r)}$, $O^{(r)}$)
11:   P ← UPDATE-PRIO-SIG($V^{(r)}$, P)
12:   V ← FB-RESTORE(C, V, N)
13: **end for**
14: **return** V

We are now at step 4 in Algorithm 1 where P is initialized to be empty. As long as P is empty, we scan the frames one by one. In each frame, we apply a SAT-based restoration procedure SAT-RESTORE which attempts to restore the open signal instances in that frame using the core-based satisfiability method [13]. It returns the updated global restoration matrix V as well as the local restoration matrix for that frame. Any newly restored signals in the current frame now added to the priority list (step 11). Following this the FB method is applied across all cycles (step 12) to restore any additional instances that can be inferred as a consequence of the newly restored signal instances due to the preceding SAT application.

Once a set of priority signals are identified (step 6), for each subsequent frame, those priority signals are probed first if their instances are still open in that frame (step 7). PROB-PRIO-SIG is a SAT-based procedure similar to SAT-RESTORE and attempts to restore the open instances of priority signals in that frame using core-based method. Following this, again FB method is globally applied to infer any new restorations consequent to the recent SAT-based restorations (step 8). The algorithm continues to attempt to restore additional open instances in the current frame using SAT, followed by priority signal set updating and global FB restoration (steps 10-12).

The algorithm terminates when all the frames are inspected. The size of P reaches its maximal value at the last frame. Note that in Steps 7 and 10, although it is possible that some signals being explored are not restorable (i.e. inferred to be *non-unique*), the satisfiable assignments determined in the process will assist in pruning the search space as discussed in *Iterative SAT Testing* in [13].

Algorithm 1 is optimal within each frame since SAT-based restoration is used to restore all open instances in step 10. However, the algorithm is not globally optimal since SAT is never globally applied due to the prohibitive run-time.

To further illustrate this approach, we revisit the example in Figure 1. The frame size is set to $F = 2$ in the experiment of this example. After the execution of this algorithm, the updated V after finishing the first frame is presented in left half of Table I, and the corresponding V after the first 2 frames are processed is shown in the right half of the table. The underlined entries are values restored in addition to initial FB-RESTORE (step 1). The algorithm starts with the first frame ($r = 1$). Since P is empty, it moves to step 10 directly and proceeds. (B,1) and (F,2) are restored by SAT-RESTORE, while (H,3) is restored by FB-RESTORE at step 12. Therefore P consists of signals B and F at this point. For $r = 2$ at step 7, (B,3) and (F,4) are *open* thus prioritized for restoration. Further, SAT-RESTORE shows they are restorable in frame 2. As a result, (H,5) is restored by step 8. Steps 10-12 again are executed but for all the *open* signal instances remained in frame 2 this time. P won't be updated because all the signals in $O^{(2)}$ are *non-unique*. The updated V is presented in the right part of Table I. The procedure for time frame 3 is the same but no more signal instances can be recovered. It concludes the entire process when all the 3 frames are inspected. Note that (B,2), (F,3) and (H,4) are restorable according to Figure 1(b), but they are missed due to the lack of global SAT-RESTORE.

TABLE I
UPDATED V FOR THE FIRST AND SECOND FRAME

	1	2	3	4	5	6	1	2	3	4	5	6
A	x	x	x	x	x	x	x	x	x	x	x	x
B	$\underline{0}$	x	x	x	x	x	$\underline{0}$	x	$\underline{0}$	x	x	x
C	x	x	x	x	x	x	x	x	x	x	x	x
D	1	x	x	x	1	x	1	x	x	x	1	x
E	1	x	x	x	1	x	1	x	x	x	1	x
F	x	$\underline{1}$	x	x	x	x	x	$\underline{1}$	x	$\underline{1}$	x	x
G	1	$\underline{1}$	0	0	0	1	1	1	0	0	0	1
H	x	1	$\underline{1}$	x	x	1	x	1	$\underline{1}$	x	$\underline{1}$	1

B. Trading Frame-Wise Optimality for Efficiency

We propose a heuristic to further promote efficiency with little compromise in SRR. This approach is inspired by the observation that the application of SAT on all *open* signal instances in steps 10 and 12 in Algorithm 1 takes more time than the SAT application on *open* priority signal instances in steps 7 and 8 because the number of priority signal instances is usually much smaller than the other *open* signal instances.

However, many of these other open instances are actually not restorable. Hence, we reason that we need to apply SAT on all open signal instance only in the first few frames (say, the first K frames, for some $1 \le K \le f$) with little impact on overall SRR. This implies that we accumulate the priority signals only based on the restorations during the first K frames and attempt to restore the corresponding instances in the remaining frames. The modified method is shown as Algorithm 2.

Algorithm 2 FRAMES-PRIORITY-SUBOPTIMAL(C, V, N)

1: IDENTICAL TO ALGORITHM 1 FROM LINE 1 to 9
2: **if** $r \le K$ **then**
3: $\quad (V, V^{(r)}) \leftarrow$ SAT-RESTORE($C_F[r]$, $V^{(r)}$, $O^{(r)}$)
4: $\quad$ P $\leftarrow$ UPDATE-PRIO-SIG($V^{(r)}$, P)
5: $\quad$ V $\leftarrow$ FB-RESTORE(C, V, N)
6: **end if**
7: IDENTICAL TO ALGORITHM 1 FROM LINE 13 to 14

The superior efficiency of this algorithm relies on the fact that it enforces optimal restoration only within the first K frames and executes suboptimal restoration for the rest of the frames. Its efficacy depends on the estimation quality of the restorable signal instances in the first few (K) time frames.

IV. EXPERIMENTAL RESULTS

A. Experimental Setup

In this section we show the performance of our algorithms against the recent suboptimal time frame method in [8] with the signal ordering heuristic of its algorithm 2 applied inside each frame. The SRRs and run-time of five commonly used ISCAS-89 benchmark circuits are shown for comparison . The SAT solver we adapted in all experiments is Minisat2.2 which is also used in the core-base algorithm proposed in [13]. Our algorithms are implemented in C++ and the Minisat2.2 APIs are incorporated to avoid costly CNF file read/write operations. The incremental solving feature of Minisat2.2 is also extensively exploited to accelerate the process of solving the same problem under different assumptions. Our experiments are run on an Intel quad-core processor @2.7GHz with 16GB memory. The circuit statistics are shown in Table II.

TABLE II
CIRCUIT STATISTICS

Circuits	# of Flip-Flops	# of Gates	Control Signal [8]
s5378	179	2779	No control signal
s9234	211	5597	No control signal
s15850	534	9772	No control signal
s35932	1728	16065	RESET
s38584	1426	19253	g35

To set up experimentation, Verilog simulations by applying random input vectors to each benchmark circuit are performed for the purpose of loading trace signal data as well as validating the restored data of our algorithms. Control signals in benchmark circuits are handled properly for normal operations. We present results for trace buffer width $w = 8$, 16 and 32 and depth $d = 256$, 512 and 1024. The frame size F is decided to be

TABLE III
PERFORMANCE ANALYSIS OF PROPOSED ALGORITHMS

BM	d	w	SRR_{fb}	SRR_{frm}	SRR_{opt}	T_{frm}	T_{A1}	T_{A2}	T_{A1} vs T_{frm}	T_{A2} vs T_{frm}	A2 SRR loss
s5378	256	8	11.92	12.07	12.07	4.09	3.82	2.9	6.77%	29.25%	0.10%
		16	5.84	6.41	6.43	4.3	3.77	3.3	12.33%	23.29%	0.94%
		32	2.51	3.75	3.75	3.3	2.72	2.4	17.40%	27.12%	0.82%
	512	8	10.79	11.25	11.26	11.58	10.56	6.88	8.83%	40.59%	0.46%
		16	6.08	6.45	6.46	8.31	5.71	4.65	31.25%	43.98%	0.77%
		32	3.86	3.97	3.98	6.82	3.51	2.89	48.60%	57.64%	0.08%
	1024	8	4.22	10.6	10.6	25.02	19.06	11.09	23.82%	55.68%	0.16%
		16	4.29	6.56	6.57	24.8	14.32	9.94	42.24%	59.93%	0.75%
		32	4.14	4.37	4.38	17.42	6.03	5.38	65.36%	69.09%	1.12%
s9234	256	8	6.54	9.01	9.28	11.49	7.73	6.97	32.77%	39.35%	0.74%
		16	2.01	3.29	4.52	13.72	9.62	9	29.87%	34.40%	20.86%
		32	3.54	4.42	4.89	7.15	2.86	2.69	60.04%	62.45%	2.98%
	512	8	2.1	2.78	4.84	47.4	30.74	21.83	35.15%	53.95%	2.40%
		16	2.01	4.84	4.95	30.5	15.23	13.07	50.05%	57.14%	0.94%
		32	3	4.19	4.57	22.38	5.79	5.32	74.10%	76.24%	0.72%
	1024	8	5.85	8.88	8.99	92.17	32.4	27.31	64.85%	70.38%	1.03%
		16	3.99	4.65	5.6	81.56	23.69	19.88	70.95%	75.62%	0.77%
		32	2.27	4.38	4.83	78.61	18.59	17.23	76.35%	78.08%	1.25%
s15850	256	8	1.93	19.46	22.12	59.21	41.73	35.48	29.53%	40.07%	0.85%
		16	3.55	16.23	19.26	47.22	28.62	26.17	39.40%	44.58%	3.68%
		32	2.53	10.65	11.15	37.97	20.1	17.81	47.05%	53.10%	1.24%
	512	8	10.71	23.29	24.71	128.25	48.02	34.41	62.56%	73.17%	1.59%
		16	3.36	15.09	17.11	143.85	64.14	51.35	55.41%	64.31%	3.31%
		32	7.66	12.17	12.64	106.67	28.94	23.23	72.87%	78.23%	0.79%
	1024	8	**1.46**	**25.06**	28.51	459.27	137.43	103.97	70.08%	77.36%	3.54%
		16	2.49	20.04	22.4	414.4	105.88	82.14	74.45%	80.18%	1.82%
		32	7.51	10.63	11.37	380.25	63.27	46.07	**83.36%**	**87.88%**	1.65%
s35932	256	8	7.48	7.48	-	68.33	68.95	50.8	-0.91%	25.67%	0.00%
		16	11.93	11.93	-	62.38	64.5	46.38	-3.40%	25.66%	0.00%
		32	6.07	15.77	-	131.11	125.8	112.7	4.05%	14.04%	**19.34%**
	512	8	9.62	9.62	-	138.53	125.27	82.11	9.57%	40.72%	0.03%
		16	7.06	8.86	-	146.11	135.65	113.84	7.16%	22.09%	17.11%
		32	5.18	11.21	-	818.39	714.48	575.5	12.70%	29.68%	**27.41%**
	1024	8	14.5	14.56	-	287.98	249.9	170.68	13.22%	40.73%	0.29%
		16	5.31	5.54	-	306.77	262.62	193.15	14.39%	37.04%	3.11%
		32	5.58	18.81	-	760.35	680.93	539.75	10.44%	29.01%	17.50%
s38584	256	8	1.25	4.14	-	487.51	354.25	226.2	27.33%	53.60%	1.82%
		16	1.06	4.97	-	353.1	222.1	150.47	37.10%	57.39%	2.41%
		32	1.31	5.12	-	269.59	191.83	140.86	28.85%	47.75%	1.21%
	512	8	1.37	4.47	-	948.83	748.62	427.06	21.10%	54.99%	3.27%
		16	1.77	5.41	-	728.34	507.44	238.73	30.33%	67.22%	1.34%
		32	1.22	6.08	-	537.16	375.79	224.07	30.04%	58.29%	1.39%
	1024	8	1.53	7.25	-	1631.15	1191.22	609.26	26.97%	62.65%	3.58%
		16	1.19	6.24	-	1454.89	974.07	436.34	33.05%	70.01%	1.30%
		32	1.14	6.81	-	1112.09	682.39	352.4	38.64%	68.31%	7.29%

32 in all our experiments and K is set to 4 in Algorithm 2. We will discuss how we determine these values later. Trace signals are selected randomly for each buffer width. Identical set of selected signals is used to analyze different algorithms. We have performed experiments with several randomly selected trace signal sets and several simulation vector sequences and observed similar trends as reported here.

B. Results and Discussion

Table III shows the comparison of run-time and state restoration ratio between our proposed algorithms and the state-of-the-art. The second method presented in [8] to the best of our knowledge is the most efficient hence it's implemented within each frame. The resulting time frame method (referred to as 'FRM' in the rest of the paper) is thus serving as the state-of-the-art method whose runtime and SRR are represented as T_{frm} and SRR_{frm} respectively. Correspondingly, SRR_{fb} and SRR_{opt} stand for the SRR of FB method and optimal SRR (obtained from global SAT restoration) respectively; T_{A1} means the run-time of our proposed Algorithm 1 and T_{A2} is the run-time of Algorithm 2. The optimal SRRs for large circuits s35932 and s38584 are not attained because of the prohibitively long execution time and are shown as '-' in Table III. 'FRM' and 'A1' yield the same SRR whereas that of 'A2'

978-1-5386-7100-9/18 $31.00 © 2018 IEEE

is smaller or equal. Note in some cases SRR_{frm} has the same value as SRR_{opt} in the table but it doesn't necessarily mean the restorations are identical due to rounding. The last column "A2 SRR loss" displays the percentage of restoration lost in A2 compared with A1.

It is evident that 'FRM' and 'A1' achieve near-optimal SRR, considerably better than that of FB in many cases. However, in rare cases, FB method can achieve optimal SRR as well ($d = 256$ and $w = 8$ for s35932). Generally 'A1' and 'A2' show significant runtime reduction. 'A1' can achieve up to 83.36% improvement in runtime compared with 'FRM' while 'A2' can reach even higher 87.88% with 1.65% compromise in SRR. The effect of 'A2' becomes significant in cases like s35932 ($d = 512$ and $w = 8$). By losing 0.03% SRR 'A2' shows runtime is ameliorated to 40.72% less compared with 'FRM', when 'A1' doesn't exhibit impressive progression. However it is also shown in some cases that the better efficiency of 'A2' leads to relatively high penalty in SRR. For example, we notice the scenario of $d = 512$ and $w = 32$ for s35932, reducing runtime by 29.68% from 12.7% yields 27.41% SRR loss, which underlines the need for further research.

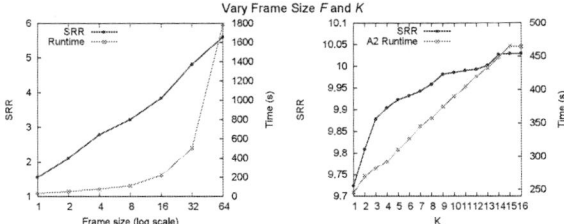

Fig. 2. The effect of frame size and K on SRR and run-time

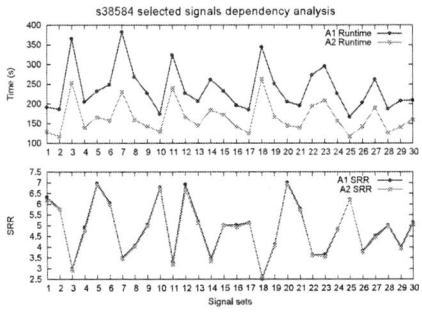

Fig. 3. Signal selection dependency analysis

Figure 2 shows how frame size and K impact the restoration quality and run-time using the circuit s38584. For the left part, in each execution the selected signals and traced values are kept identical except the frame size. We can see from the figure that with the increase in frame size F, the SRR increases in nearly linear fashion whereas the execution time increase resembles the exponential trend. This is expected as larger F means more information is added for restoration yet larger problem instance for SAT solver. We set $F = 32$ because the runtime for $F = 64$ rises drastically. To establish the value of K, we experiment using the same circuit with $d = 512$ and $w = 32$ ($1 \leq K \leq 16$) for a same set of trace signals and vary K. From the right part of the figure, we observe $K = 4$ balances SRR and runtime quite well.

In Figure 3, randomly 30 sets of signals are selected for s38584 to see if the improvement in run-time of 'A1' and 'A2' varies considerably when different sets of signals are traced. As shown in the figure, there is no strong dependency and the SRR loss of 'A2' is negligible.

V. CONCLUSION

In this work, we have presented two algorithms combining the FB method and a SAT solver to effectively reduce the run-time of time frame based method for fast near-maximal restoration in post-silicon debugging utilizing trace buffers. Our Algorithm 1 concentrates on priority signal identification by exploring the spatial resemblance between time frames and shows substantial execution time reduction. Our Algorithm 2 further improves efficiency based on Algorithm 1 with insignificant SRR loss in most situations. Both algorithms can handle large size benchmarks over large number of clock cycles successfully to enable practical applications in post-silicon validation. Further research in intelligent selection of time frames could be conducted for better performance.

REFERENCES

[1] P. Mishra, R. Morad, A. Ziv, and S. Ray, "Post-silicon validation in the soc era: A tutorial introduction," *IEEE Design Test*, vol. 34, no. 3, pp. 68–92, June 2017.

[2] D. Chatterjee, C. McCarter, and V. Bertacco, "Simulation-based signal selection for state restoration in silicon debug," in *2011 IEEE/ACM ICCAD*, Nov 2011, pp. 595–601.

[3] K. Rahmani, P. Mishra, and S. Ray, "Efficient trace signal selection using augmentation and ilp techniques," in *Fifteenth International Symposium on Quality Electronic Design*, March 2014, pp. 148–155.

[4] M. Li and A. Davoodi, "A hybrid approach for fast and accurate trace signal selection for post-silicon debug," *IEEE Transactions on Computer-Aided Design of Integrated Circuits and Systems*, vol. 33, no. 7, pp. 1081–1094, July 2014.

[5] S. BeigMohammadi and B. Alizadeh, "Combinational trace signal selection with improved state restoration for post-silicon debug," in *2016 DATE*, March 2016, pp. 1369–1374.

[6] S. Ma, D. Pal, R. Jiang, S. Ray, and S. Vasudevan, "Can't see the forest for the trees: State restoration's limitations in post-silicon trace signal selection," in *2015 IEEE/ACM ICCAD*, Nov 2015, pp. 1–8.

[7] B. Kumar, A. Jindal, V. Singh, and M. Fujita, "A methodology for trace signal selection to improve error detection in post-silicon validation," in *2017 30th International Conference on VLSID*, Jan 2017, pp. 147–152.

[8] X. Liu and R. Vemuri, "Effective signal restoration in post-silicon validation," in *2017 IEEE ICCD*, Nov 2017, pp. 169–176.

[9] C. S. Zhu, G. Weissenbacher, D. Sethi, and S. Malik, "Sat-based techniques for determining backbones for post-silicon fault localisation," in *2011 IEEE International HLDVT Workshop*, Nov 2011, pp. 84–91.

[10] B. Kumar, K. Basu, M. Fujita, and V. Singh, "Rtl level trace signal selection and coverage estimation during post-silicon validation," in *2017 IEEE International HLDVT Workshop*, Oct 2017, pp. 59–66.

[11] K. Basu, P. Mishra, and P. Patra, "Efficient combination of trace and scan signals for post silicon validation and debug," in *2011 IEEE International Test Conference*, Sept 2011, pp. 1–8.

[12] P. Komari and R. Vemuri, "A novel simulation based approach for trace signal selection in silicon debug," in *2016 IEEE 34th ICCD*, Oct 2016, pp. 193–200.

[13] M. Janota, I. Lynce, and J. Marques-Silva, "Algorithms for computing backbones of propositional formulae," *AI Communications*, vol. 28, no. 2, pp. 161–177, 2015.

[14] X. Liu and R. Vemuri, "Combined inference and satisfiability based methods for complete signal restoration in post-silicon validation," in *2018 31st IEEE VLSI Design (VLSID)*, Jan 2018, pp. 416–421.

2018 IEEE Computer Society Annual Symposium on VLSI

PGIREM: Reliability-Constrained IR Drop Minimization and Electromigration Assessment of VLSI Power Grid Networks using Cooperative Coevolution

Sukanta Dey[†], Satyabrata Dash[‡], Sukumar Nandi[†], Gaurav Trivedi[‡]
[†]*Department of Computer Science and Engineering*
[‡]*Department of Electronics and Electrical Engineering*
Indian Institute of Technology Guwahati, Assam, India
Email: {*sukanta.dey, satyabrata, sukumar, trivedi*}*@iitg.ac.in*

Abstract—Due to the resistance of metal wires in power grid network, voltage drop noise occurs in the form of IR drop which may change the output logic of underlying circuits and may affect the reliability performance of a chip. Further, it is necessary to handle different reliability constraints while designing a robust power grid network for a chip. Any violation of such constraints may increase the occurrences of IR drop. Therefore, there is a need to minimize the IR drop without violating the reliability constraints. In this paper, the IR drop minimization problem is formulated as a single objective large-scale variable minimization problem subjected to different reliability constraints, such as IR drop constraints, electromigration constraints, minimum width constraints, metal area constraints. At first, the large-scale minimization problem is divided into several subproblems using a divide and conquer based decomposition strategy, called Cooperative Coevolution. Secondly, each subproblem is solved using self-adaptive differential evolution with neighborhood search. Finally, electromigration (EM) assessment is done for the power grid networks using Black's equation to demonstrate the optimism in the predicted time-to-failure (TTF) after minimization of the IR drop.

Keywords-Cooperative Coevolution, Electromigration, IR drop noise, Large Scale Optimization, Power Grid Network, Reliability-Constraints, VLSI.

I. INTRODUCTION

Power grid networks of VLSI chip are becoming larger than ever with the scaling of technology node, which introduces the added design challenges and makes the design phase time-consuming and iterative. Generally, IR drop noise occurs in the power grid due to the metal resistance of the grid, which is one of the major concern while designing the power grid [1]. As the IR drop noise can change the voltage level of a logic block, hence it is essential to ensure the IR drop noise below a threshold level. The underlying logic block can malfunction if the IR drop noise exceeds a certain threshold level. Furthermore, EM-induced increase in metal resistances can also change the IR drop level [2]. Therefore, it is necessary to locate the IR drop noises and minimizing the IR drop noises occur in the power grid network.

The IR drop noises are located by doing power grid analysis of the whole network which is a process by which currents and voltages of the equivalent model of the power grid network are determined. In general, voltage drop noise violations are occurred by IR drop (due to the resistances of the metal lines) and Ldi/dt voltage drop (due to inductances of the metal lines and C4 bumps) noises. In this paper, we limit this work only for IR drop minimization based on metal width reduction. IR drop generally depends on the current flowing through the metal lines and the resistances of the metal lines. The amount of current flowing through the metal line is determined by the current drawn by the underlying functional blocks. Therefore, it is necessary to limit the flow of currents through the metal lines, to prevent any occurrences of EM. On the other hand, resistances of the metal lines depend on the width and length of the metal layer. If the length is kept constant, then by increasing the width of the metal layers the resistance of the metal lines can be reduced which will force the IR drop to decrease further.

In industry, there are many tools for IR drop analysis, such as $RedHawk^{TM}$, $PrimeRail^{TM}$, and $Totem^{TM}$. Therefore, using these tools power grid analysis is done and IR drop noises are located. Layout designers try to minimize the IR drop in the located area manually, by increasing the metal width (but confining within the design rules) of the power and ground lines, to decrease the resistances which reduce IR drop. After varying the widths of the metal lines, designers have to perform power grid analysis again to know whether the hotspots created by IR drop is below a certain threshold. And this process continues iteratively until the IR drop comes below the threshold level. As this is a time-consuming process, the manual design of power grid network after analysis becomes cumbersome. Also, no automated tools in the industry are available which can minimize IR drop in power grid network. Therefore, in this paper, we have constructed the IR drop minimization problem as a large-scale optimization problem. We also tried to propose a framework using Cooperative coevolution for minimizing the IR drop by varying the metal layer dimensions of power grid network which would help in automatically projecting metal widths within safer IR drop noise level.

To the best of our knowledge, this paper is the first to study the minimization of IR drop for power grid networks by changing the metal widths considering different reliability

978-1-5386-7100-9/18 $31.00 © 2018 IEEE

constraints. The major contribution of this paper includes:

- The IR drop minimization problem has been formulated as the large-scale optimization problem for a simplified steady-state model of the power grid network.
- Cooperative coevolution based method is employed for the minimization of the IR drop of the power grid network.
- The proposed minimization approach is able to minimize the IR drop without changing the topology of the grid by only changing the metal widths.
- EM assessment of the power grid network is done for the optimized power grid with minimized IR drop which demonstrated optimism in the life time prediction of the chip.

The rest of the paper is arranged as follows. In Section II, the related work on power grid optimization is described. Power grid network model used in the paper is explained in Section III. Problem formulation and all the reliability constraints of the power grid network are described in Section IV. Section V describes the Cooperative Coevolution based approach and how it is implemented for doing IR drop minimization. The experimental results on different power grid benchmark circuits are mentioned in Section VI.

II. RELATED WORK

There are several works on the power grid analysis and verification, such recent works are [3, 4]. The basic objective of the works done in power grid optimization so far is to minimize the area of the metal wires (considering IR drop as a constraint) of the power grid network by constructing two-phase optimization problem, then iteratively solve the two-phase optimization problems using different algorithms. Tan et al. [5] solved the problem using a sequence of linear programmings. Wang et al. [6] solved the same problem with sequential network simplex algorithm. Similarly, there are more works on the metal area minimization. Zeng et al. [7] have done power grid wire sizing optimization problem using locality driven partitioning based two-step optimization algorithm. Chang et al. [8] proposed routing friendly multilayer power grid network by allocating each layer metal width considering the impact of AP layer. However, there is hardly any work so far, which tried to minimize the IR drop by considering different power grid network constraints. Moreover, recent research developments are more concentrated on developing new physics-based EM models for power grids to achieve optimism in mean-time-to-failure (MTTF) [2, 9]. Minimizing IR drop can be one of the alternatives to obtain optimism in MTTF. In this paper, we tried to propose an approach to minimize the total IR drop of the whole power grid network by changing the metal widths of each of the metal segments (branches), which are manually done iteratively by the layout designers in the industry, to minimize the IR drop. Layout designers also use decoupling capacitances at few critical nodes to reduce

the IR drop and Ldi/dt voltage noises. But in this paper, we are only considering IR drop minimization by varying metal widths. And here Cooperative Coevolution approach for solving large-scale optimization problem is employed.

III. POWER GRID NETWORK MODEL

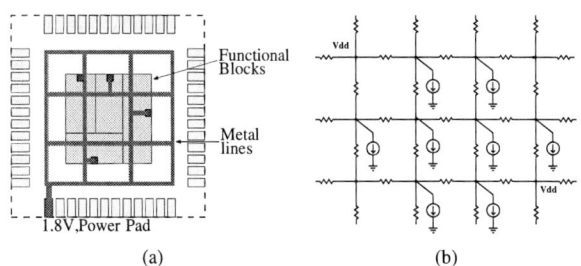

Figure 1. (a) An example of floor-plan and its power grid network(metal lines) with functional blocks (b) Modeling power lines to resistive network.

An illustration of floor-plan and its power grid network with functional blocks is shown in Figure 1(a). For minimization of IR drop voltage noise, steady-state model of the power grid network is considered in this paper, which is shown in Figure 1(b). In this model, only the resistance of the metal lines is considered. The current drawn by the underlying circuit is modeled as the current sinks connected to ground as shown in Figure 1(b). The vias of this model are considered having zero resistance, as vias have very low resistance. Inductances associated with the C4 bumps, to connect V_{dd} and ground connections are not considered here. Also, any other parasitic effects due to inductances and capacitances are not considered here. The steady-state model of the power grid network can be represented as a linear system of equations i.e., $\mathbf{GV} = \mathbf{I}$, where conductances of the metal lines make the $\mathbf{G}$ matrix, current sinks connected to ground contributes to form the $\mathbf{I}$ vector and node voltages of all the nodes generate the $\mathbf{V}$ vector. By using direct solvers, such as KLU solver [10], we can determine $\mathbf{V}$ i.e., voltages of all the nodes. Similarly, from node voltages of all the nodes, we can even find the branch currents.

IV. PROBLEM FORMULATION

A. Objective Function for IR Drop Minimization

Let's consider $G = \{V, E\}$ be a graph corresponding to a power grid network, where $V = \{1, 2, \cdots, n\}$ is the set of all the n nodes of the power grid network and $E = \{1, 2, \cdots, b\}$ is the set of all the b branches of the graph corresponding to the steady state model of power grid network.

If I is the current passing through a metal segment (branch of the graph) of the Power Grid network having resistance R, then the voltage drop occurred across the metal segment can be represented by v :

$$v = IR \qquad (1)$$

With sheet resistance $\rho \ \Omega/\square$ which is constant for a metal layer, having metal segment length and breadth of l and w respectively, the voltage drop can be denoted by the following:

$$v = I\frac{\rho l}{w}, \tag{2}$$

where $R = \frac{\rho l}{w}$ represents the total resistance of the metal segment of the power line. Similarly, the voltage drop of the whole Power Grid network with b number of metal wire segments (or branches) can be written as follows:

$$\begin{aligned} v &= \sum_{i=1}^{b} |I_i| \, R_i \\ &= \sum_{i=1}^{b} |I_i| \frac{\rho l_i}{w_i}, \end{aligned} \tag{3}$$

where $\mathbf{W}, \mathbf{I}, \mathbf{l}$ are set of vectors of metal widths, branch currents and metal lengths respectively i.e., $\mathbf{W} = (w_1, \cdots, w_b)$, $\mathbf{I} = (I_1, \cdots, I_b)$, $\mathbf{l} = (l_1, \cdots, l_b)$. For large value of b, equation (3) can be treated as the large scale optimization problem. In equation (3), I_i and w_i are the variables for the ith metal segment which are non-separable in nature. Non-separable variables are those for which objective function depends on the interacting variables [11]. l_i has been taken as constant for the objective function throughout this work which will be imported from the circuit netlist. Therefore, for the whole power grid network, vectors $\mathbf{I}$ and $\mathbf{W}$ are sets of non-separable variables. Hence, the objective function can be formulated as a large-scale optimization problem with b non-separable variables as follows:

$$v(I_i, w_i) = \sum_{i=1}^{b} |I_i| \frac{\rho l_i}{w_i} \tag{4}$$

Hence the IR drop minimization problem with unequal branch currents I_i is given as follows:

$$\mathscr{P} : \underset{\substack{w_i \in \mathbf{W} \ I_i \in \mathbf{I}}}{minimize} \ v, \tag{5}$$

subject to different reliability constraints which are described in section IV-B.

Theorem 1: Minimization of total IR drop v of (4) reduces the worst case (maximum) IR drop.

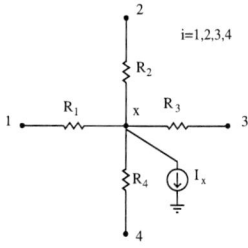

Figure 2. Single node of the power grid network model

Proof: The worst case maximum IR drop $v_{\text{IR max}}$ can be expressed as

$$v_{\text{IR max}} = Max(V_{DD} - V_x) \ \forall x \in V, \tag{6}$$

where V_x is the node voltage of the xth node which depends on the voltage of the neighboring nodes and also depends on the current of the neighboring edges. Using KCL for a node V_i (see Figure 2), carrying current I_i from x to i, the expression of V_x can be written as follows:

$$\begin{aligned} I_i &= \frac{V_x - V_i}{R_i} \ \forall i \in K_s \\ \Rightarrow V_x &= V_i + I_i R_i \\ \Rightarrow V_x &= V_i + I_i \frac{\rho l_i}{w_i}, \end{aligned} \tag{7}$$

where K_s is the set of neighboring nodes of x. Therefore, V_x of (7) depends on the neighboring node voltages, neighboring branch currents and resistances. Also, v of (3) depends on neighboring branch currents and resistances. Therefore, minimizing v of (3) under the constraint $\mathcal{C}_1 : |I_{i \in E}| \, R_{i \in E} \le \xi \ \forall i \in E$ by varying current and resistance will reduce the worst case maximum IR drop $v_{\text{IR max}}$ of (6). ∎

B. Reliability Constraints of Power Grid Network

1) IR Drop Constraints: It can be defined by the following relation:

$$\mathcal{C}_1 : |I_{i \in E}| \, R_{i \in E} \le \xi \tag{8}$$

The above relation should be maintained for all the ith branches of the power grid network. ξ is the maximum tolerance level of voltage drop noise allowed between two consecutive nodes of the power grid network.

2) Metal Area Constraints: The metal area of the power grid network should be restricted to $\mathcal{A}_{max}$:

$$\mathcal{C}_2 : \sum_{i=1}^{b} l_i w_i \le \mathcal{A}_{max} \tag{9}$$

3) Electromigration Constraints: To prevent the current carrying metal lines from electromigration, the current density of the metal lines should be less than I_m

$$\mathcal{C}_3 : \frac{I_{i \in E}}{w_{i \in E}} \le I_m \tag{10}$$

4) Minimum Width Constraints: The minimum width of the metal lines w_{min} is limited by the technology on which the power grid network lies. Therefore, the constraint can be expressed as:

$$\mathcal{C}_4 : w_{i \in E} \ge w_{min} \tag{11}$$

5) KCL Constraints: KCL should be followed at all the n nodes of the power grid network.

$$\mathcal{C}_5 : \sum_{i=1}^{K} I_{j_i} + I_x = 0 \ \forall j \in V \tag{12}$$

where K is the number of neighboring nodes of node j and I_x is the sink current of the model connected to ground.

V. MINIMIZATION USING COOPERATIVE COEVOLUTION

A. Cooperative Coevolution(CC)

Cooperative Coevolution is a divide and conquer based approach to solve large scale variable optimization problems. It decomposes a large scale problem into several simple sub-problems. So the basic phenomenon of CC is that it decomposes an n-dimensional decision vector into n subcomponents and then optimizes each of the subcomponents using standard evolutionary optimization algorithm in a round robin fashion. The basic principle of the evolutionary optimization algorithm is to mimic the biological evolution process in generating good candidate solutions for a given objective function. Generally, candidate solutions of an optimization problem play the role of individuals in a population, and these individuals go under reproduction, mutation and recombination and selection to find the optimum solutions for a given objective function. Cooperative coevolution algorithm is stated in Algorithm 1.

Algorithm 1: Cooperative Coevolution Algorithm

Input: f, x_{min}, x_{max}, n

Output: Optimized value of f and corresponding variables $x_1, x_2, \cdots, x_n$ values

/*grouping based variable decomposition*/;
groups ← grouping(f, x_{min}, x_{max}, n);
/*Optimization stage using evolutionary algorithm*/;
population ← rand(population_size, n);;
for $j \leftarrow 1$ to size(groups) **do**
    group_num ← groups[j];
    subpop ← population[:,group_num];
    subpop ← optimizer(best,subpop,FE);
    population[:,group_num] ← subpop;
    (best,best_val)←min(population);

Theorem 2: CC based algorithm converges to global minimum for large-scale optimization problems if the main optimizer converges to the global minimum.

Proof: Let $f(x_1, x_2, \cdots, x_n)$ be an objective function with n decision variables. Now if n variables have been decomposed by random grouping with each group containing s decision variables, then $t = n/s$ number of subcomponents will be there. In other words, t instances of the objective function with each containing s number of decision variables and $(n - s)$ number of constants will be there. Now each of the t objective functions will be minimized independently using a standard optimization algorithm which will provide us with t local minimum values from t subcomponents. And then random grouping based strategy is applied to co-adapt these t minimum values. Hence, the global minimum will be obtained for the objective function f if the main optimization algorithm converges to the global minimum of each of the t instances of the objective function f. ∎

CC is introduced into Genetic Algorithm for optimization of function by Potter et al. [12]. Liu et al. [13] used CC in large-scale optimization problem by using Fast Evolu-

tionary Programming with Cooperative Coevolution. CC is introduced into PSO by Bergh et al. [14]. CC has also been adapted into Differential Evolution(DE) in [15], [11]. An improved version of DE is Self-Adaptive Differential Evolution with Neighborhood Search(SaNSDE)[16] which self-adapts its scaling factor F, crossover rate CR, and mutation strategy. It is proved that SaNSDE performs quite well compared to the other similar DE algorithms[16]. Yang et al. [11] showed that SaNSDE under CC framework (CC-SaNSDE) for large-scale variable optimization works very well. To deal with the non-separable nature of the problem, random grouping based decomposition strategy of the decision variables is used. Generally, in large-scale problems, only a proportion of variables interact with each other, therefore, the random grouping of variables increase the probability of grouping two interacting variables in the same subcomponent [17]. So CC-SaNSDE has been adapted in this paper to solve total IR drop minimization of power grid network.

B. IR drop minimization using CC-SaNSDE

Algorithm 2: IR drop minimization using CC-SaNSDE

Input: Both branch width & current ranges (for problem $\mathscr{P}$) are given as input from a power grid circuit netlist with b branches and n nodes. Branch lengths from the netlist are also taken as input.

Output: Optimum power grid netlist having minimized IR drop along-with the corresponding optimum resistance budget and metal width budget for the b branches.

1 Search space S is constructed in such a way to incorporate the reliability constraints $\mathcal{C}_1$, $\mathcal{C}_2$, $\mathcal{C}_3$, $\mathcal{C}_4$, and $\mathcal{C}_5$ mentioned in section IV-B.;
2 **while** *inside search space S* **do**
3   Initialize the initial parameters for CC-SaNSDE with random grouping.;
4   Random grouping is employed to decompose the b variables in t subcomponents.;
5   SaNSDE optimization algorithm is used to optimize each of the subcomponents.;
6   Random grouping strategy is also used for the co-adaptation of all the subcomponents.;
7 Optimum metal widths corresponding to minimized IR drop is found and model parameters are updated.;
8 KLU solver is used to find the optimum IR drop.;

The IR drop minimization algorithm using CC-SaNSDE is given in Algorithm 2. For the problem $\mathscr{P}$, number of variables are decomposed to form subcomponents and then each of the subcomponents is minimized using SaNSDE. The subcomponents are formed based on random grouping of variables and each of the subcomponents are minimized using SaNSDE. And finally, random grouping based strategy is used for co-adaptation of the subcomponents. Also, SaNSDE has been modified to incorporate the reliability constraints as mentioned in section IV-B, to keep the search space within the region of validation. Branch widths are calculated corresponding to the resistances of the branches and ranges of branch width is given as input for the problem

$\mathscr{P}$. Apart from the branch width ranges, power grid analysis is done using KLU solver[10] to find the branch current ranges of the power grid network and given as input.

C. EM Assessment

EM assessment is done to predict the lifetime of a chip. EM occurs with two phases naming, nucleation phase, and the growth phase. In the nucleation phase, the voids started to form over a long period of time until the voids are nucleated. In the growth phase, hillocks are started to form and the interconnect metal resistance changes to a point where the resistance exceeds a threshold and failure occurs. The failure rates of the interconnect metal lines can be taken as a measure to check for EM-induced reliability. In extreme case, MTTF of the weakest metal segment can be treated as the life-time of the whole chip. Here, EM assessment of the power grid is done by considering Black's equation [18]. MTTF from Black's equation statistics can be written as follows:

$$MTTF = AJ^{-N}exp\{E_a/kT\}, \qquad (13)$$

where A is constant which depends on the material properties of the metal. Here, k is the Boltzmann's constant and T is the temperature. Value of N is found to be 2, which depends on residual stress and current density J. E_a is the activation energy which also depends on current density J. The Black's equation of MTTF has been controversial and a better physics-based EM model is proposed in [2]. This proposed method [2] of predicting MTTF has also been used here to do the full chip life-time assessment of the power grid networks. In view of this, EM also causes the IR drop to increase as the resistance of the interconnect metal increases chronologically over a long period in the growth phase of EM. Therefore, minimizing the IR drop of the power grid network will surely demonstrate an optimism in the predicted MTTF.

VI. Experimental Results

Table I
Power Grid Benchmark Circuits Data [19]

PG circuits	#Nodes(n)	#Branches(b)	Branch resistance ranges (in Ω)
ibmpg2	127238	208325	(0,1.17]
ibmpg3	851584	1401572	(0,9.36]
ibmpg4	953583	1560645	(0,2.34]
ibmpg5	1079310	1076848	(0,1.51]
ibmpg6	1670494	1649002	(0,17.16]
ibmpgnew1	1461036	2352355	(0,21.6]
ibmpgnew2	1461039	1422830	(0,21.6]

The CC-SaNSDE algorithm with random grouping is implemented in MATLAB and the experiments are performed on a machine with Intel Xeon E5-2650 processor having 32 GB memory and validated by IBM power grid benchmark

Table II
Comparision of IR drop for different Power Grid circuits before and after minimization

PG_circuits	IR drop (in volts)					
	Before minimization			After minimization		
	Max.	Avg.	Min.	Max.	Avg.	Min.
ibmpg2	0.0631	0.0315	0.0129	0.0567	0.0283	0.0116
ibmpg3	0.0310	0.0252	0.0049	0.0279	0.0226	0.0044
ibmpg4	0.0386	0.0220	0.0060	0.0347	0.0198	0.0055
ibmpg5	0.0690	0.0373	0.0146	0.0621	0.0335	0.0134
ibmpg6	0.0598	0.0297	0.0111	0.0538	0.0267	0.0998
ibmpgnew1	0.0353	0.0204	0.0055	0.0317	0.0183	0.0051
ibmpgnew2	0.0516	0.0271	0.0102	0.0464	0.0243	0.0091

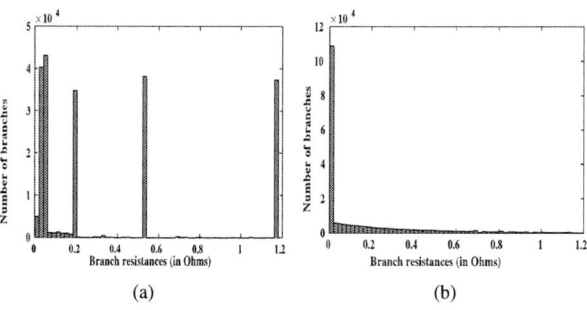

Figure 3. Resistance budget for *ibmpg2* circuit (a) before minimization, (b) after minimization.

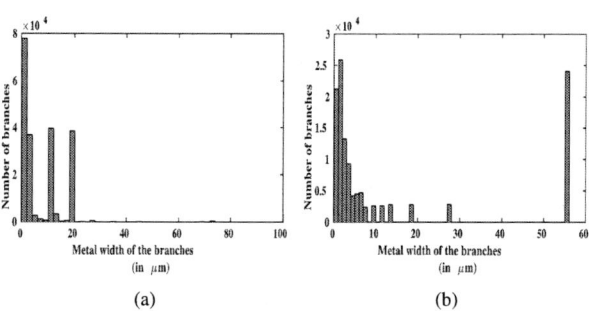

Figure 4. Metal width budget for *ibmpg2* circuit (a) before minimization, (b) after minimization.

Table III
Comparison of MTTF for different Power Grid circuits using Balck's equation and using proposed method of [2] before and after minimization of IR drop.

PG_circuits	Mean time-to-failure (yrs)			
	Before minimization		After minimization	
	Black's	[2]	Black's	[2]
ibmpg2	7.81	15.65	11.21	19.35
ibmpg3	15.75	27.60	19.80	30.67
ibmpg4	12.55	29.25	16.33	32.42
ibmpg5	6.31	23.06	10.52	26.72
ibmpg6	9.49	17.75	13.34	21.10
ibmpgnew1	13.62	22.45	17.50	25.48
ibmpgnew2	12.41	20.10	16.22	23.37

circuits [19]. IBM power grid benchmark data for 7 circuits are listed in the Table I. Experiments are performed for these 7 circuits *i.e.*, *ibmpg2* to *ibmpgnew2*. Current sink values of these circuits are modified so that the initial IR drop is below a threshold voltage level. Although metal width and length information are not available in the IBM power grid circuits, appropriate values of lengths are used and the corresponding width of metal layers are determined by considering sheet resistance of the metal $\rho = 0.02 \ \Omega/\square$ (assuming copper interconnect materials) and from branch resistances using equation (2). Algorithm 2 is tested using these power grid benchmarks and the obtained minimized IR drop data is listed in Table II. For the problem $\mathscr{P}$ power grid analysis is done using KLU solver [10] and all the branch currents are determined, from there branch current ranges have been found to be in the range of $0.1mA$ to $10mA$ for all the benchmark circuits. Subsequently, Algorithm 2 is used to get an optimum budget of metal widths and resistances corresponding to the minimum total IR drop of the power grid network. Figure 3 and Figure 4 show the resistance budget and metal width budget before and after minimization respectively for *ibmpg2* circuit. Fitness Evaluation(FE) used for the experiments is 10^6 as for this value of FE the convergence of the Algorithm 2 is found to be the best.

For the EM assessment, we assume that the power grid will fail when the worst case IR drop exceeds $10\%V_{DD}$. In Black's equation based series model, the circuit is assumed to be failed as soon as any branch fails. Parameters used for calculation of MTTF of different power grid circuits in this paper are same as stated in [2]. Comparison of MTTF values for different power grid benchmarks before and after IR drop minimization for the Black's series model and the physics-based EM-model of [2] are listed in Table III. It is observed from the Table III that the MTTF after minimization of IR drop has increased significantly. That is because MTTF is inversely proportional to the current density raised to the power N (J^N). Since we have minimized IR drop by increasing the metal widths which decreases the J. As a result, we have got an optimistic prediction of the life-time of the power grid network with an increased value of MTTF.

VII. CONCLUSION

This paper presents a method to minimize the IR drop for power grid networks. The IR drop minimization problem for a power grid network is formulated as a large-scale optimization problem and the minimization of the IR drop is done using CC-SaNSDE. IR drop is minimized at the cost of the metal area of the chip. Results show minimization of IR drop for different power grid benchmarks. Further, EM assessment of the optimized power grid network is done to demonstrate the optimism in the life-time prediction of the power grid network.

REFERENCES

[1] F. N. Najm, "Physical design challenges in the chip power distribution network," in *Proceedings of the 2015 Symposium on International Symposium on Physical Design*. ACM, 2015, pp. 101–101.

[2] X. Huang, Y. Tan, V. Sukharev, and S. X.-D. Tan, "Physics-based electromigration assessment for power grid networks," in *Design Automation Conference (DAC), 2014 51st ACM/EDAC/IEEE*. IEEE, 2014, pp. 1–6.

[3] S. Dey, S. Dash, S. Nandi, and G. Trivedi, "Markov chain model using lévy flight for VLSI power grid analysis," in *VLSI Design and 2017 16th International Conference on Embedded Systems (VLSID), 2017 30th International Conference on*. IEEE, 2017, pp. 107–112.

[4] M. Fawaz and F. N. Najm, "Parallel simulation-based verification of RC power grids," in *VLSI (ISVLSI), 2017 IEEE Computer Society Annual Symposium on*. IEEE, 2017, pp. 445–452.

[5] S. X.-D. Tan, C.-J. R. Shi, and J.-C. Lee, "Reliability-constrained area optimization of VLSI power/ground networks via sequence of linear programmings," *IEEE Transactions on Computer-Aided Design of Integrated Circuits and Systems*, vol. 22, no. 12, pp. 1678–1684, 2003.

[6] T.-Y. Wang and C.-P. Chen, "Optimization of the power/ground network wire-sizing and spacing based on sequential network simplex algorithm," in *Quality Electronic Design, 2002. Proceedings. International Symposium on*. IEEE, 2002, pp. 157–162.

[7] Z. Zeng and P. Li, "Locality-driven parallel power grid optimization," *IEEE Transactions on Computer-Aided Design of Integrated Circuits and Systems*, vol. 28, no. 8, pp. 1190–1200, 2009.

[8] W.-H. Chang, M. C.-T. Chao, and S.-H. Chen, "Practical routability-driven design flow for multilayer power networks using aluminum-pad layer," *IEEE Transactions on Very Large Scale Integration (VLSI) Systems*, vol. 22, no. 5, pp. 1069–1081, 2014.

[9] S. Chatterjee, V. Sukharev, and F. N. Najm, "Power grid electromigration checking using physics-based models," *IEEE Transactions on Computer-Aided Design of Integrated Circuits and Systems*, 2017.

[10] T. A. Davis and E. Palamadai Natarajan, "Algorithm 907: KLU, a direct sparse solver for circuit simulation problems," *ACM Transactions on Mathematical Software (TOMS)*, vol. 37, no. 3, p. 36, 2010.

[11] Z. Yang, K. Tang, and X. Yao, "Large scale evolutionary optimization using cooperative coevolution," *Information Sciences*, vol. 178, no. 15, pp. 2985–2999, 2008.

[12] M. A. Potter and K. A. De Jong, "A cooperative coevolutionary approach to function optimization," in *International Conference on Parallel Problem Solving from Nature*. Springer, 1994, pp. 249–257.

[13] Y. Liu, X. Yao, Q. Zhao, and T. Higuchi, "Scaling up fast evolutionary programming with cooperative coevolution," in *Evolutionary Computation, 2001. Proceedings of the 2001 Congress on*, vol. 2. IEEE, 2001, pp. 1101–1108.

[14] F. Van den Bergh and A. P. Engelbrecht, "A cooperative approach to particle swarm optimization," *IEEE transactions on evolutionary computation*, vol. 8, no. 3, pp. 225–239, 2004.

[15] Y.-j. Shi, H.-f. Teng, and Z.-q. Li, "Cooperative co-evolutionary differential evolution for function optimization," *Advances in natural computation*, pp. 428–428, 2005.

[16] Z. Yang, K. Tang, and X. Yao, "Self-adaptive differential evolution with neighborhood search," in *Evolutionary Computation, 2008. CEC 2008.(IEEE World Congress on Computational Intelligence). IEEE Congress on*. IEEE, 2008, pp. 1110–1116.

[17] M. N. Omidvar, X. Li, Z. Yang, and X. Yao, "Cooperative co-evolution for large scale optimization through more frequent random grouping," in *Evolutionary Computation (CEC), 2010 IEEE Congress on*. IEEE, 2010, pp. 1–8.

[18] J. R. Black, "Electromigrationa brief survey and some recent results," *IEEE Transactions on Electron Devices*, vol. 16, no. 4, pp. 338–347, 1969.

[19] S. R. Nassif, "Power grid analysis benchmarks," in *Design Automation Conference, 2008. ASPDAC 2008. Asia and South Pacific*. IEEE, 2008, pp. 376–381.

2018 IEEE Computer Society Annual Symposium on VLSI

Silicon Debug with Maximally Expanded Internal Observability using Nearest Neighbor Algorithm

Ankit Jindal[1], Binod Kumar[1], Nitish Jindal[2], Masahiro Fujita[3] and Virendra Singh[1]

Indian Institute of Technology Bombay, Mumbai[1], IBM Systems - ISDL, Bengaluru[2], University of Tokyo, Japan[3]

Email: {ankitjindal, binodkumar, viren}@ee.iitb.ac.in[1], nitish.jindal92@gmail.com[2], fujita@ee.t.u-tokyo.ac.jp[3]

Abstract—One of the most difficult challenges during the process of silicon debug is overcoming the bottleneck of limited visibility of internal states. Although the application of state restoration technique enhances the limited debug data available through on-chip trace buffers, yet the number of restored signal states are not significant. This paper proposes an approach which addresses the limited observability problem through a machine learning perspective. Based on training with pre-silicon buggy signatures on a relatively smaller design, a model is developed which identifies a set of neighbors for every flip-flop of the design. The application of nearest neighbors principle eliminates the obstacle of unknown signal values despite restoration because these values are obtained from the neighbors. Experimental results on benchmark circuits depict that the proposed approach is able to correctly discover 93% of the total signal values. The methodology is verified with the help of cross-validation of the debug data on designs injected with gate-level error models.

Index Terms—Post-silicon validation, Machine learning, Trace signal selection, Error localization, Nearest Neighbors.

I. Introduction

SoC (System-on-Chip) design complexity is continuously increasing due to demands of rich functionalities. This makes the verification of complete design difficult as well as extremely necessary. To alleviate the obstacle of slow RTL simulation speed, the first silicon can be utilized for validation and debug purposes. This process commonly referred to as *post-silicon validation* suffers from the hindrance of limited observability of design states. On-chip trace buffers can store few selected signals for a fixed number of clock cycles. However, owing to area constraints, only 1-2% of the total signals are generally traced [1]. The application of *state/signal restoration* [2] is useful for observability expansion to some extent as the untraced states are discovered(reconstructed) from the traced states by *forward propagation* and *backward justification* rules. The problem of restricted visibility of the design still remains challenging because only some of the internal signals (flip-flops) values can be known even if a set of highly effective signals are chosen for tracing or an improved signal state restoration technique is used [3].

This paper presents an alternative methodology for enhancement of the observability of internal signals at the post-silicon stage through the usage of a learning algorithm. Based on mock simulation of small designs injected with design bugs, a nearest neighbor model is developed which is independent of the characteristics of the particular design. This model can then be utilized for finding out neighbors of flip-flops

of any arbitrary design given the traced and restored data corresponding to a post-silicon test execution. This ensures full scalability of the proposed technique to industrial-sized large circuits. The basic premise of the proposed methodology lies in the fact that the value of untraced signals can be decided based on the signal state of its neighbors. To the best of our knowledge, this is the first approach which targets the discovery of complete set of signal states and achieves the expanded observability with reasonable accuracy. Specifically, this paper makes following contributions:

- two techniques based on nearest neighbor (NN) algorithm are proposed which identify closely related neighbors for each flip-flop of the design. First, a NN-based simple classification technique is proposed. Second, a learning model based on derivations of structural features is developed to obtain nearest neighbors (flip-flops).
- a debug methodology is proposed which utilizes the expanded states (discovered with help of nearest neighbors) to localize the error to a smaller region of the netlist.

The remainder of the paper is organized as follows. Section II presents preliminaries of the proposed methodology and summarizes related work on application of machine learning in error localization. Section III describes our learning based methodology for post-silicon observability expansion in detail. Section IV explains the proposed gate-level error localization methodology based on the expanded signal states. Section V presents the results of observability expansion and error localization by using the proposed techniques. Section VI finally concludes the paper with directions on future research.

II. Background and Previous Work

A. Post-silicon observability expansion

The objective of design observability expansion can be regarded as a data mining problem. Previously proposed techniques of state restoration attempt to solve this through analysis of the design netlist and decides values of untraced signal states with the help of structural dependencies [2]. However, a very important aspect of this process is the choice of trace signals as the technique of backward justification or forward propagation can reveal the untraced signal values only if appropriate hints (in the form of values of structurally related signals) are supplied. Therefore, attempts have been made to filter out best possible set of trace signal candidates through

978-1-5386-7100-9/18 $31.00 © 2018 IEEE   46

heuristics based on design/netlist analysis [2]. However, attempting only state restoration for expanding the observability is not sufficient. Clustering algorithms like nearest neighbors can assist in deciding the values of unknown signals (which are either untraced or not restored) as they capture the correlation between data values. We exploit this fact to achieve *maximal* (i.e., full with reasonable accuracy) internal visibility.

B. Post-silicon error localization

Park et al. [4] proposed a technique named as *Instruction Footprint Recording and Analysis (IFRA)* for error detection in processor systems using low cost on-chip recorders as an observability mechanism. They perform program analysis after constructing bug localization graphs for error localization to a block-level granularity. However, typically an architectural block in a processor may contain thousands of gates. This makes it very difficult to debug at the gate-level. In the recent years, machine learning techniques have been explored for the purpose of bug triaging or error localization at both the pre-silicon and post-silicon stage. Since during post-silicon validation, a large number of tests can be applied, the amount of logs collected can become very large. Therefore, machine learning techniques like clustering, regression analysis etc., can be suitably deployed to extract hints for bug/error localization from the obtained test response logs. DeOrio et al. [5] proposed a post-silicon bug diagnosis methodology based on data collection from failing tests and then applying a clustering technique to form different signal groups.

III. PROPOSED METHODOLOGY

A. Methodology illustration

As outlined in Section-I, limited observability is one of the main obstacles in the process of post-silicon error localization. This becomes more challenging at the gate-level or at the granularity of flip-flops since the usage of observability enhancement mechanisms provides the states of very few flip-flops. This large mismatch between the number of internal signal values available to us during simulation phase and post-silicon phase makes the debug process difficult.

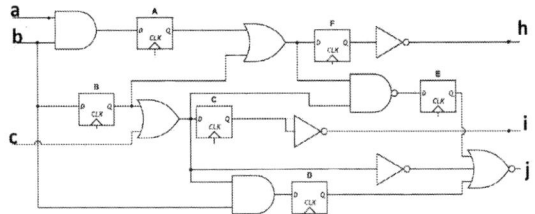

Fig. 1. Example circuit for illustrating methodology

To cope up with this wide disparity of internal signal values, method of *matrix completion* (completing a partially-filled matrix) is used. We formulate this as a machine learning based clustering (particularly, *nearest neighbors*) problem. This assists in obtaining signal states similar to the known (i.e., traced or restored) flip-flop values. Once the neighbors of a particular signal are identified, the state of that particular flip-flop can

be obtained. The state (either 0/1) of a particular flip-flop in any clock cycle is assigned the value most common among states (which are known either through restoration/tracing) of its k-nearest neighbors. Continuing in this manner, the state of all the flip-flops for all clock cycles can be obtained, leading to full internal observability. The *NN* algorithm decides the closely related neighbors (in this case flip-flops) based on some information pertaining to the design. It can be either related to the structure of the design or simulation values. For the sake of illustration, we consider the circuit shown in Fig. 1 (in which a synthetic design bug has been injected). Note that simulation values of bug injected netlist are not a necessity for the success of this method. However, signal values of a buggy netlist provide more useful behavior compared to the original netlist as the logic inconsistencies are pin-pointed. We performed a simulation of this modified netlist for 10 cycles. Table-I shows states of six flip-flops for these ten clock-cycles. Flip-flop A & C were traced and their states are known for all ten cycles. Through the usage of state restoration technique [2], some states of other flip-flops can be computed. The state of many flip-flops can not be discovered. Those states are depicted as X. Note that it is a common assumption in silicon debug to not trace the inputs because of excessive storage requirements due to run-time ranging from hours to days.

TABLE I
RESTORED AND TRACED STATES FOR ILLUSTRATION

FF→	A	B	C	D	E	F
c1	0	X	X	0	X	X
c2	0	X	X	0	X	X
c3	1	1	X	1	1	1
c4	0	X	X	0	X	1
c5	0	X	X	0	1	1
c6	0	1	1	1	1	1
c7	0	X	X	0	1	1
c8	0	X	X	0	1	1
c9	0	1	1	1	1	1
c10	0	X	X	0	1	1

Nearest neighbor algorithm works on discovering some kind of metric (distance) between the targets. One such metric is the Euclidean distance which is the measure of dissimilarity between different data points. The Euclidean distance between all of these flip-flops is shown in Fig. 2. With nearest neighbor algorithm, 4 neighbors obtained for each flip-flop are shown in Table II. If the flip-flops of last column are removed from the above set, we obtain the set of 3 neighbors. Depending on the state of these neighbors, the X values can be determined. This is essentially done by a majority vote of the known values of neighbors at the particular instant. For instance- if 2 neighbors of a flip-flop have "1" and the third has value "0", the target flip-flop is assigned a value of "1". In cases of tie-up between neighbor values, we assume "1" as the default value. Note that the signal state of neighbors of a target flip-flop is dynamically updated as the discovery of values progresses. The difference in the distances of the neighbors justifies the choice of nearest neighbors as obtained in Table II.

Note that it is difficult to fully showcase the merit of this method through a miniature example because here many

TABLE II
OBTAINED NEAREST NEIGHBORS FOR EXAMPLE CIRCUIT

FF→	nbr1	nbr2	nbr3	nbr4
A	D	B	C	E
B	D	A	C	E
C	E	F	D	B
D	B	A	C	E
E	C	F	D	B
F	C	E	D	B

signal states have been reconstructed through state restoration. However, we observed in our experiments, almost (or more than) 60-70% of the internal signal values are always unknown for large circuits even though a highly effective signal selection technique is utilized. This necessitates observability expansion after the application of state restoration.

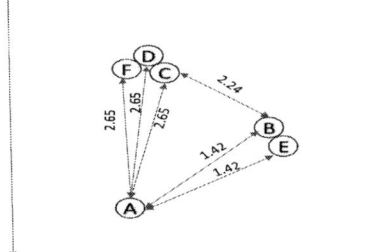

Fig. 2. Euclidean distance between Nearest neighbors for example circuit

The completed expanded (fully known) states are shown in Table III. However, as expected, the derived states are different from the actual ones in some cases. These cases vary with the choice of number of neighbors (4 or 3). The states different from actual ones and those derived with 4-neighbors are depicted by β. Similarly, signal states different from actual ones and those derived with 3-neighbors are depicted by α. For instance, in 2nd clock cycle, state of FF-C is different from the actual signal state when the expansion is done either with the help of 3-neighbors or 3-neighbors. We obtained "0" in this case whereas the actual value is "1". In 5th clock cycle, for FF-C, signal state comes out to be opposite to that of the actual state only in case of discovery with 4-neighbors (depicted by β). As is evident from Table III, the no. of dissimilar state values are fewer in case of less neighbors.

TABLE III
COMPLETELY EXPANDED INTERNAL SIGNAL STATES

FF→	A	B	C	D	E	F
c1	**0**	**0**	0	**0**	0	0
c2	**0**	**0**	$1(0^{\alpha\beta})$	**0**	$1(0^{\alpha\beta})$	$1(0^{\alpha\beta})$
c3	**1**	**1**	1	**1**	1	1
c4	**0**	**0**	0	**0**	0	1
c5	**0**	**0**	$1(0^{\beta})$	**0**	1	1
c6	**0**	**1**	1	**1**	1	1
c7	**0**	**0**	$1(0^{\beta})$	**0**	1	1
c8	**0**	**0**	$1(0^{\beta})$	**0**	1	1
c9	**0**	**1**	1	**1**	1	1
c10	**0**	**0**	$1(0^{\beta})$	**0**	1	1

B. Proposed Methodology

The terminology utilized for formally explaining the proposed methodology is presented in Table IV. In the proposed methodology, we develop a model, M by two training

methods. For the first method, we utilize the error signature obtained from the chip (netlist). In the second method, we employ structural relationships to obtain nearest neighbors. Observability expansion can be achieved by either of these methods. Based on the traced signal states (which assists in obtaining restored signal values) and the obtained neighbors, states of all the flip-flops can be discovered. An very important consideration is to figure out the optimum number of nearest neighbors. Therefore, we iterate the neighbor finding exercise for certain number of times taking the minimum no. of neighbors as 5 and maximum as 500. The procedures shown in $Training1$ and $Training2$ expand the observability of internal states and measure the difference between the expanded signature and the reference complete signature.

TABLE IV
TERMS AND THEIR MEANING

Term	Meaning
M	learning model
TBw	width of trace buffer
D_{tb}	depth of trace buffer
L_i	i^{th} Nearest Neighbor technique
K	no. of iterations of learning
ϕ_j	Numbers of nearest neighbors in M
Cy	No. of cycles in each iteration of learning
n	number of flip-flops in design(netlist)
FF_e	Error signature of design
ξ_i	Features of the design(netlist)
FF_t	Flip-flop signature used for testing
GFF_t	Reference/Golden flip-flop signature

There can be many methodologies to derive nearest neighbors based on the specific implementations. For the purpose of finding nearest neighbors, we utilize *Scikit-learn* [6] library. We have implemented 3 types of nearest neighbor algorithms (*K-D Tree*, *Ball Tree* and *Brute Force*). By normalizing values of specific parameters in these algorithms their variants are obtained. These are denoted by L_i in both these procedures. For each of the NN algorithms, we perform training on 7 configuration of no. of neighbors (denoted by ϕ_j) as 5, 10, 25, 50, 100, 250 and 500. With training on smaller circuits, we can identify the best combination of the no. of neighbors (ϕ_j) and the learning method (L_i). Since the training is done on smaller circuits only, we can perform K (typically between 100-200) iterations in each of the procedures. We take Cy as 1024 which is the typical trace-buffer depth [1].

1: **procedure** $Training1$(In:FF_e,K; Out:M)
2:     $M \leftarrow \emptyset$
3:     $I \leftarrow$ incomplete states (restored and traced ones) for Cy cycles
4:     $FF_e \leftarrow$ state of n flip-flops for Cy cycles
5:     $score_{\phi_j,L_i} \leftarrow 0$
6:     **for** each L_i **do**
7:         **for** z= NN (ϕ_j) **do**
8:             **for** w=1 to K **do**
9:                 $nbrs \leftarrow$ NNmodel(FF_e,L_i)
10:                 $fullval \leftarrow$ fillvalues($nbrs,I$)
11:                 $score \leftarrow fullval$ - FF_e
12:             **end for**
13:             $score_{\phi_j,L_i} \leftarrow \sum score$

978-1-5386-7100-9/18 $31.00 © 2018 IEEE

14:       **end for**
15:   **end for**
16:       $nbrscore = \text{minimum}(score_{\phi_j, L_i})$
17:       $M \leftarrow NN(\phi_j, L_i)$
18: **end procedure**

As is seen in the procedure $Training1$, finding out the nearest neighbors depends on the complete error signature of chip, FF_e (derived from a reference simulation) and the particular learning method (L_i). After particular neighbors (corresponding to specific (ϕ_j)) are identified, the incomplete (restored + traced) signature can be completed leading to a maximal expansion of internal observability. The best configuration is the combination of ϕ_j and L_i for which the lowest difference between the expanded signature and the reference complete erroneous signature (FF_e). $nbrscore$ is the sum of variation between actual and discovered states for this particular choice of ϕ_j and L_i.

We propose a second procedure to develop the model (M) utilizing only the netlist and thus, independent of error signatures. Essentially, only static features of the design description are accounted for in this case. To capture the structural characteristics of any flip-flop ff_i, we define following features on the lines of [7]. For each flip-flop ff_i, a set of traits which are called as features is given in Table V.

TABLE V
FEATURES AND THEIR MEANING

ξ_1(fan-in)	number of different flip-flops connected to ff_i in its fan-in
ξ_2(fan-out)	number of different flip-flops connected to ff_i in its fan-out
ξ_3(gate count)	number of gates in connection to ff_i in its fan-in and fan-out cone
ξ_4(2nd level connectivity)	average of number of different flip-flops in second level fan-in and second level fan-out
ξ_5(ed score)	$\sum_{p=0}^{p=P}(\prod_{g=0}^{g=G} (score\ of\ each\ gate))$

While majority of the features listed above are fairly straightforward and can be easily computed, the last feature is taken from a previous work by Kumar et al. [1] which approximately calculates the error propagation score of each flip-flop. An estimation of the error transmission through each gate can be given by $1/N_{inputs}$ where N_{inputs} is the number of inputs of the gate. Once the score of each gate is calculated, by counting all the incoming paths to different flip-flops, we can approximately find $ed\ score$. Further, we need to consider the score of each gate in a multiplicative manner (no. of gates generalized by G) and sum these scores along all the incoming paths (no. of paths generalized by P) to get the rough estimate of error transmission ability of each flip-flop. As is seen in $Training2$, finding out the nearest neighbors depends on the computed features (depicted by $Feature\ space$) from the netlist and the particular learning method (L_i). Thus, unlike $Training1$ procedure, closely related neighbors are discovered statically without resorting to the analysis of data (i.e., error signatures) from simulation.

1: **procedure** $Training2$(In:$Netlist$,K,n; Out:M)
2:       $M \leftarrow \emptyset$

3:       $Feature\ space \leftarrow \emptyset$
4:       $I \leftarrow$ incomplete states (restored and traced ones) for Cy cycles
5:       $FF_e \leftarrow$ state of n flip-flops for Cy cycles
6:       $score_{\phi_j, L_i} \leftarrow 0$
7:       **for** each flip-flop 1 to n **do**
8:           Calculate all features ($\xi_1,\xi_2,\xi_3,\xi_4,\xi_5$)
9:           $Feature\ space \leftarrow$ features of flip-flop
10:      **end for**
11:      **for** each L_i **do**
12:          **for** z= NN (ϕ_j) **do**
13:              **for** w=1 to K **do**
14:                  nbrs $\leftarrow$ NNmodel($Feature\ space$,L_i)
15:                  $fullval \leftarrow$ fillvalues($nbrs$,I)
16:                  score $\leftarrow fullval$ - FF_e
17:              **end for**
18:              $score_{\phi_j, L_i} \leftarrow \sum$ score
19:          **end for**
20:      **end for**
21:      $nbrscore = \text{minimum}(score_{\phi_j, L_i})$
22:      $M \leftarrow NN(\phi_j, L_i)$
23: **end procedure**

The resultant trained model, M (either through $Training1$ or $Training2$) is used to find the neighbors of any large circuit. This allows us to reduce the requirements of computation time and memory space. With the help of neighbors, we can assign the values to the missing values of internal signals. In our experiments, we reached 100% expansion however with considerably accuracy. This allows us to have better visibility of internal signals and thus better debug. Assuming that we have a "golden" reference model constructed out of a completely verified higher level abstraction of the design-under-validation, we can exactly localize the design bug/error.

IV. PROPOSED ERROR LOCALIZATION METHODOLOGY

Localizing at gate level is difficult due to the restricted visibility of internal states of the chip. Therefore, with the expanded observability by the proposed method, we aim to achieve error localization at the granularity of flip-flops. The debug methodology is formally expressed in $Debug$ procedure shown below. The basic principle of the debug methodology is to catch the difference between the reference signature and the expanded signature. Since, the unknown signal values (either untraced or non-restored) have been resolved, the differences can be more clearly pin-pointed now. For localizing to smaller portion of the design, we divide the complete error signature (consisting of n flip-flops and D_{tb} cycles) into certain number of blocks (which are decided by a factor termed as $bfactor$). Based on the dissimilarity between chunks of flip-flop values (denoted by $data\ block$), we can localize to a smaller region of the netlist. This comparison leads to a ranking of all the $ffblocks$ out of which we choose the top ranked $ffblock$.

1: **procedure** $Debug$(In:FF_t,GFF_t,$bfactor$; Out:$ffblock$)
2:       $FF_t \leftarrow$ flip-flop signature used for testing
3:       $FF_t' \leftarrow$ completed FF_t by $Training1$/$Training2$
4:       $N_t \leftarrow$ length of FF_t

5:    $nffblocks \leftarrow N_t/bfactor$
6:    $\{block\ data1, block\ data2\} \leftarrow \emptyset$
7:    $GFF_t \leftarrow$ state of n flip-flops for Cy cycles
8:    **for do** $ffblock$=1 to $nffblocks$
9:        $block\ data1 \leftarrow$ portion of FF_t
10:       $block\ data2 \leftarrow$ portion of FF_t'
11:      $score_{ffblock}$=diff($block\ data1, block\ data2$)
12:      $\{block\ data1, block\ data2\} \leftarrow \emptyset$
13:    **end for**
14:    Rank $ffblocks$ in decreasing order of $score_{ffblock}$
15:    $Error\ blocks \leftarrow$ Top ranked $ffblock$
16: **end procedure**

Note that post-silicon design fault/bug modeling has not been possible to a large extent. As such, the design bugs can be mapped to their manifestations at the netlist level [1]. Such gate-level design error models consist of random gate replacement, unintended wire exchange, extra inverter insertion etc. We consider an example scenario shown in Fig. 3. Suppose, the wires "p" and "q" are exchanged because of a design error and this slips into the fabricated silicon. As is seen from this figure, the error can appear in 3 different flip-flops ($F2, F3$ and $F4$) depending on the allowed propagation through the combinational gates. An efficient localization approach must not include $F1$ in the suspect list and choose of either/all of the other 3 flip-flops so that their logical cone can be traced and errors like wire-exchange, gate-replacement, shorted inputs, extra insertion or similar ones can be localized.

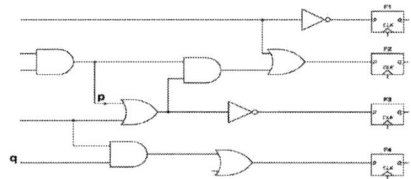

Fig. 3. Illustration of gate-level design error 1

We consider another example of design error in which an unintended inverter (marked as f) is added to the netlist [1]. This is shown below in Fig. 4. Here, the suspect list includes only two flip-flops ($F2$ and $F3$).

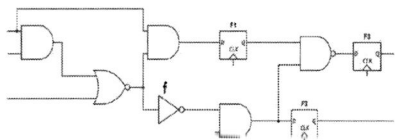

Fig. 4. Illustration of gate-level design error 2

V. EXPERIMENTAL FORMULATION AND RESULT

A. Experimental Setup

As mentioned earlier, *Scikit-learn* [6] library has been used for implementing the proposed methodology. We chose circuits from ISCAS'89, ITC'99 and *Opencore* benchmark suite. The training (i.e., building the model) was done only on design error 1 (wire exchange). We performed 100 iterations of learning on a smaller circuit to select the appropriate no.

of neighbors(ϕ_j) and type of learning(L_i). We chose our configuration as TBw as 32 and D_{tb} as 1024. Characteristic of the benchmarks are noted in Table VI. They serve as representatives of digital blocks inside complex SoCs.

TABLE VI
CHARACTERISTICS FOR BENCHMARK CIRCUITS

Sl.No.	Name	No. of FFs	No. of Gates
1	b17	864	13988
2	b21	429	13098
3	p16c5x	739	4107
4	soft_navre	317	4718
5	usb	1761	10650
6	s13207	638	7951
7	s38417	1636	22179
8	s38584	1452	19253

B. Observability expansion results

After achieving observability expansion, we validate the obtained signal states by comparing with the reference complete signature in each iteration of the learning experiment. We report the dissimilarity in the actual and expanded signal states obtained by the method $Training1$ in Fig. 5. The maximum accuracy occurs for a particular number of neighbors. When we choose very less number of neighbors like 5 or 10 in M, accuracy is very poor. Similarly, if very large number of neighbors like 500 is chosen, the accuracy falls. With successive increase in the number of neighbors, we compute the factor of increase/decrease in the accuracy and the time needed for observability expansion. As expected, there is a continual increase in the time (measured in seconds) spent in unknown signal state discovery as the number of neighbors increases. As is observed from Fig. 5, there is a trend of variation (measured by factor shown on Y-axis) as number of neighbors increases for $b17$ circuit. We achieved the maximum accuracy (at 8%) when 250 neighbors are selected in the learning model. The graph is drawn considering the time spent and the accuracy achieved with $\phi_j = 10$ as the baseline.

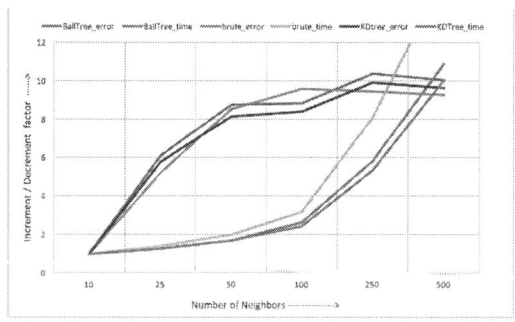

Fig. 5. Observability expansion by $Training1$ for $b17$ circuit

For the second method ($Training2$), we obtained similar trend of variation with the number of neighbors for different learning methods. However, the amount of variation is higher than that of $Training1$. We achieved the maximum accuracy (at 15%) when 50 neighbors are selected in the learning model, M. This justifies that the simulation based observability expansion method works better in terms of exact signal state

978-1-5386-7100-9/18 $31.00 © 2018 IEEE

reconstruction. As is observed from Fig. 6, the rate of increase/decrease in the accuracy is significantly lower compared to that of model developed by $Training1$. The graph in Fig. 6 is drawn considering the time spent in discovery and the accuracy achieved with $\phi_j = 5$ as the baseline.

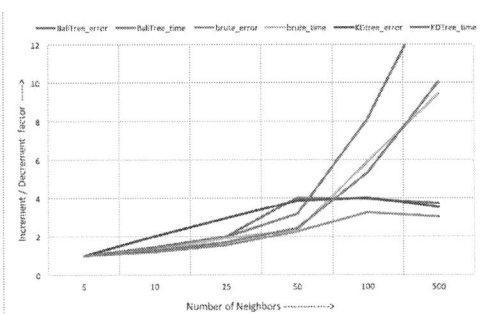

Fig. 6. Observability expansion by $Training2$ for $b17$ circuit

C. Defining error localization metric

We perform a topological connection based analysis of the injected error location. The errors injected at granularity level of nets/gates in the netlist can be localized if the error is detected by flip-flops in the immediate vicinity. So, it is meaningful to analyze the efficacy in following terms:

- The infected flip-flops obtained ($f_{obtained}$) are among the suspect candidates (f_{actual}).
- The proposed method catches least but the important suspect candidates. This reduces the debug effort.

Note that flip-flops in $f_{obtained}$ are derived from $Error~blocks$ obtained from the $Debug$ procedure. We define a localization function (g_{loc}) in Table VII to quantify the efficacy of gate-level error localization with the proposed debug methodology. If the error injection and subsequent localization experiment is carried out K times, the function g_{loc} has the maximum value of K. We report g_{loc} values for the two training methods ($Training1$ and $Training2$ depicted by $Tr1$ and $Tr2$ respectively) and different error injection scenarios ($design~error1$ and $design~error2$ are depicted by $e1$ and $e2$ respectively) in Table VIII.

TABLE VII
LOCALIZATION METRIC DEFINITION

$fn.$	Value	Condition
g_{loc}	1	$f_{obtained} = f_{actual}$
g_{loc}	1	$f_{obtained} \subset f_{actual}$
g_{loc}	0	$f_{obtained} \not\subset f_{actual}$

D. Error localization results

The error injection experiments were iterated for 50 times for both $e1$ and $e2$ cases. Table VIII shows the values of the localization function, g_{loc} obtained in the experiments. Both the training methods achieve comparable results with $Tr2$ achieving better results. The notation e1-$Tr1$ in third column refers to g_{loc} values for case of $e1$ when the observability expansion is achieved by $Tr1$ method.

In cases when f_{actual} is greater than one, $Training2$ gives larger $f_{obtained}$ as compared to $Training1$ for some circuits.

TABLE VIII
LOCALIZATION RESULTS FOR $e1$ AND $e2$ SCENARIO

Sl.No.	Name	e1-$Tr1$	e2-$Tr1$	e1-$Tr2$	e2-$Tr2$
1	$b17$	41	42	43	39
2	$b21$	43	41	46	37
3	$p16c5x$	42	35	40	31
4	$soft\_navre$	50	50	49	50
5	usb	37	30	48	32
6	$s13207$	39	39	37	34
7	$s38417$	44	43	47	41
8	$s38584$	45	45	44	43

Lesser flip-flops in $f_{obtained}$ is expected to assist in quick debug as the intersection of logic-cones of each flip-flop provides the exchanged nets/wires. We obtained maximum value of g_{loc} for one circuit. There is minor variation in the localization for different error models ($e1$ and $e2$). Similarly, we observed minor differences in g_{loc} values when the number of neighbors or the type of learning method in the built nearest neighbor model, M are changed.

VI. CONCLUSION

This paper proposed a methodology of gate-level error localization based on the observability expansion of the limited debug data available through on-chip trace buffers. By discovering nearest neighbors, unknown signal states can be discovered with the help of decision making based on the traced and restored signal states. The proposed methodology reconstructs signal states with reasonable accuracy when compared to the actual signal states. The first methodology involves neighbor finding by distance calculations from simulation values. Structure based analysis method assists in discovering a set of features which discovers neighbors, independent of simulation of designs. The proposed methodology can be extended to localize errors of electrical nature such as bit-flip or stuck-at 0/1 on some nets of the design. Finding the number of neighbors (in the learning model) required to achieve the minimum accuracy between the discovered and actual signal states is also a promising direction of further investigation.

REFERENCES

[1] B. Kumar, A. Jindal, M. Fujita, and V. Singh, "Combining restorability and error detection ability for effective trace signal selection," in *Proceedings of the Great Lakes Symposium on VLSI*, ser. GLSVLSI '17. Banff, Alberta, Canada: ACM, May 2017.

[2] H. F. Ko and N. Nicolici, "Automated trace signals selection using the rtl descriptions," in *IEEE International Test Conference*, 2010, pp. 1–10.

[3] X. Liu and R. Vemuri, "Effective signal restoration in post-silicon validation," in *2017 IEEE International Conference on Computer Design (ICCD)*, Nov 2017, pp. 169–176.

[4] S. B. Park and S. Mitra, "Ifra: Instruction footprint recording and analysis for post-silicon bug localization in processors," in *Design Automation Conference, 2008. DAC 2008. 45th ACM/IEEE*, June 2008, pp. 373–378.

[5] A. DeOrio, Q. Li, M. Burgess, and V. Bertacco, "Machine learning-based anomaly detection for post-silicon bug diagnosis," in *Design, Automation and Test in Europe, DATE 13, Grenoble, France, March 18-22, 2013*, 2013, pp. 491–496.

[6] F. Pedregosa, G. Varoquaux, A. Gramfort, V. Michel, B. Thirion, O. Grisel, M. Blondel, P. Prettenhofer, R. Weiss, V. Dubourg, J. Vanderplas, A. Passos, D. Cournapeau, M. Brucher, M. Perrot, and E. Duchesnay, "Scikit-learn: Machine learning in Python," *Journal of Machine Learning Research*, vol. 12, pp. 2825–2830, 2011.

[7] K. Rahmani and P. Mishra, "Feature-based signal selection for post-silicon debug using machine learning," *IEEE Transactions on Emerging Topics in Computing*, vol. PP, no. 99, pp. 1–1, 2017.

Application Specific Networks-on-Chip Synthesis: An Energy Efficient Approach

Somayeh Kashi*, Ahmad Patooghy‡, Dara Rahmati†, Mahdi Fazeli*, Michel A. Kinsy‡

*Department of Computer Engineering, Iran University of Science and Technology University, Tehran, Iran
†Institute for Research in Fundamental Sciences, Tehran, Iran
‡Department of Electrical and Computer Engineering, Boston University, Boston, USA

Abstract—**Multiple Voltage Supply (MSV) chip fabrication is considered as a viable technique to address the power and thermal challenges of modern many-core systems. Efficiency of this technique has been demonstrated in application specific Network-on-Chips (NoCs) which have lots of cores and various operating voltages/frequencies. In this paper, a four-phase synthesis toolchain is proposed and evaluated for the design of multi-voltage application specific NoCs. The proposed synthesis toolchain performs i) core to router allocation, ii) voltage islanding to match voltages of cores connected to the same router, iii) hierarchical floorplanning to reduce the complexity of power delivery network, and iv) path allocation to connect routers based on the application requirements. The distinguishing feature of the proposed toolchain is that, for the first time, it performs the router allocation phase prior to voltage islanding. This offers more flexibility and more efficiency in the multi-voltage NoC synthesis. Experimental results on real world benchmarks show that the proposed toolchain provides 63 % less energy consumption as well as more than double the design alternatives satisfying the benchmarks requirements compared to existing approaches.**

Index Terms—**application-specific chip, custom NoC synthesis, partitioning, islanding, floorplanning.**

Introduction With the recent advances in VLSI fabrication technology, complex Multi-Processor System on Chips (MP-SoCs) are now widely commercialized in the market. MPSoCs contain several general processing, DSP, I/O controller, and memory cores working with different voltages and frequencies. Such heterogeneous multi-core systems need an efficient on-chip communication architecture to ease data transmission between the cores. In such chips, general purpose Network-on-Chip (NoC) which has been widely used in homogeneous chips [1], [5] is not a viable solution because MPSoCs 1) have cores with different working voltages/frequencies, 2) mainly use multiple voltage islands [4], 3) may have information about the target application at the design time [2], [3], and 4) commonly use the router sharing technique, in which multiple cores are connected to a single router [7]–[9]. As a result, application specific NoC design/synthesis for multi-voltage MPSoCs has been widely addressed by researchers [7]–[9]. Application specific NoC design and synthesis directly affects objectives of modern MPSoCs including performance (both delay and bandwidth), area, power consumption, and temporal and spatial temperature of the chip.

The main steps of a custom NoC synthesis are core to router allocation, router to router connection and path allocation for the communication flows [7]. Seiculescu et al. proposed a synthesis flow for voltage-driven custom NoC design [7]. In their work, cores are assigned to voltage islands based on their nominal operating voltages at the first step. Then, cores are allocated to the routers, using a min-cut based partitioning algorithm in each island. Finally, the Dijkstra algorithm finds the shortest deadlock free path for each flow when establishing the router to router connections and physical links. Todorov et al. proposed a synthesis methodology which uses spectral partitioning and spanning trees to construct a custom NoC [8]. The partitioning step is done according to the communication graph, the floorplan information, and the voltage islands. In [9], Wang et al. have presented voltage-driven frameworks to do voltage island generation, voltage-driven floorplanning and post-floorplan processing. The objective of the voltage assignment is to select the least possible voltage value for cores to minimize the overhead of voltage converters under performance critical constraints. After this step, cores are assigned to the routers by a min-cut based partitioning algorithm in each island. A simulated-annealing based floorplanning is used simultaneously to determine the position of the cores and the routers during voltage-driven floorplanning.

In almost all of the previously proposed methods, the core to router assignment step is done after the voltage islanding. As a result, cores with high communication demand operating on different voltages have to be placed in different voltage islands. This increases the number of hop counts between the communicating cores and imposes a voltage conversion overhead between the islands. The former degrades the chip performance and the latter results in more power consumption and heat generation. To address the problem, our proposal is to do the partitioning step prior to voltage islanding in order to have more flexibility in the partitioning step. To prevent potential voltage fragmentations, we consider the voltage related issues during all steps of the synthesis, including partitioning, islanding and floorplanning. To the best of our knowledge, this is the first work to consider the voltage values in all the steps of the NoC synthesis flow with the concurrent goals of reducing the power consumption, delay and the complexity of the power delivery network.

The rest of the paper is organized as follows. Section II defines the NoC synthesis problem. Section III introduces the key observations motivating our proposal. Our multi-voltage synthesis toolchain is described in detail in Section IV. Experimental results are presented in Section V and finally Section VI concludes the paper.

I. SYNTHESIS PROBLEM DEFINITION

The custom NoC synthesis process starts with a given communication core graph and tries to find a design satisfying application requirements. The core graph $Gcomm(C, E)$ is a directed and weighted graph where each vertex $c_i \in C$ is a core and each directed edge $e_{ij} = (c_i, c_j)$ denotes needed communication flow from core c_i to core c_j. The required bandwidth of the communication flow from core c_i to core c_j is represented by bw_{ij}; the latency constraint for the flow is represented by lat_{ij}. Each core c_i has its own minimum operating voltage v_i subject to performance constraints. The voltage difference between corresponding cores c_i and c_j is represented by $diff_{volt_{ij}} = |v_i - v_j|$.

Weight function: The weight of each edge e_{ij} ($Wcomm_{ij}$) is a combination of bw_{ij}, lat_{ij}, and $diff_{volt_{ij}}$ according to the proposed cost function of equation (1). It is used during the synthesis process.

$$Wcomm_{ij} = \begin{cases} \alpha \cdot \frac{bw_{ij}}{max_{bw}} + \gamma \cdot (1 - \frac{diff_{volt_{ij}}}{max_{diffvolt}}) \\ \quad + \beta \cdot \frac{min_{lat}}{lat_{ij}} & lat_{ij} \neq 0 \\ \alpha \cdot \frac{bw_{ij}}{max_{bw}} + \gamma \cdot (1 - \frac{diff_{volt_{ij}}}{max_{diffvolt}}) & lat_{ij} = 0. \end{cases}$$
(1)

Parameters max_{bw}, min_{lat}, and $max_{diffvolt}$ are the maximum required bandwidth, the worst latency constraint, and the maximum voltage difference between cores respectively. Parameters α, β, and γ, which are the weight coefficients, should satisfy the $\alpha + \beta + \gamma = 10$. According to this cost function, the communication flow between two cores with high bandwidth requirement, low latency constraint and low voltage difference would have a large $Wcomm$ value.

Partitioning Problem: Partition the $Gcomm(C, E)$ into l graph-partitions $p_1, p_2, ..., p_l$ such that 1) $p_1 \bigcup p_2 \bigcup ... \bigcup p_l = Gcomm(C, E)$, 2) $p_i \bigcap p_j = \phi, 1 \leqslant i, j \leqslant l, i \neq j$, and 3) for each partition $p_n, 1 \leqslant n \leqslant l$ sum of inter partition weights be lower than from sum of intra partition weights i.e., equation (2) be satisfied.

$$\sum_{p_m, m \neq n} \sum_{c_i \in p_n, c_j \in p_m} Wcomm_{ij} < \sum_{c_i \& c_j \in p_n} Wcomm_{ij}. \quad (2)$$

We assume that the minimum operating voltage of a core is determined based on the performance constraints of the core. Then, we build an island graph $Gvolt(V, F)$ which is a directed and weighted graph. In the island graph each vertex $v_i \in V$ is a voltage island in which the operating voltage of all cores is v_i. Therefore, the directed edge $f_{ij} = (v_i, v_j)$ connects a lower voltage island (v_i) to a higher voltage island (v_j). Each edge has a weight of $Wvolt_{ij} = (v_j - v_i) \times n_i$ (n_i is the number of cores in island v_i) and denotes the cost of merging islands v_i and v_j with respect to their power consumption. It is clear that in the case of merging two voltage islands, the lower voltage island should work with the voltage of the other island to meet the performance constraints of both islands.

As described, extending the number of on-chip voltage converters increases the power consumption and degrades the performance of the chip. To limit the number of on-chip voltage converters, we need a way to merge some the adjacent voltage islands. This problem is defined as follows.

Voltage Island Merging Problem: Find the smallest set U in the $Gvolt(V, F)$ graph, $U \subset V$, such that 1) $\sum_{i\&j \in U} Wvolt_{ij}$ is minimized, and 2) by merging the islands of set U to their higher voltage islands, the power constraint of the application is satisfied. When such a subset is found, the merging continues till at most two islands remain in the set U.

II. OBSERVATIONS & MOTIVATIONS

Considering the existence of several heterogeneous cores on a MPSoC, multiple-supply voltage (MSV) technique [7] can be effectively used to reduce the power consumption of MPSoCs. In an MSV enabled chip, the performance-critical cores normally have to work with higher supply voltages to meet their performance constraints, while other cores have the chance to work with lower supply voltages. To make data communication feasible between the cores working with different voltages, the use of voltage converters have been proposed [9]. However, even if there are no physical limitations, voltage converters imply significant overheads on both communication power and packet delivery delay. These overheads are obviously in contrast with the main objective of MSV enabled MPSoC design.

A communication core graph with 26 IP cores, D-26-media, [6] is shown in Figure 1. The color of each vertex here represents the working voltage of the core and the numbers next to edges represent the required bandwidth between communicating cores. Some pairs of the cores with different voltages, e.g., (DMA, MEM5), (MEM3, PE1), (ARM, SDRAM1), demand a high communication bandwidth. However, if the voltage islanding takes place prior to the partitioning, cores with different operating voltages and at the same time high communication demands will be placed in different voltage islands. To support this claim, we synthesized 3 real communication core graphs of D-35-bot, D-36-4, and D-38-VOPD under the two policies of 1) single voltage MPSoC, and 2) MSV enabled MPSoC with no limitation on the number of voltage converters. Results of power delay product which simultaneously shows power consumption and delay are depicted in figure Fig. 2. One can conclude from this figure - across all the synthesized benchmarks - that the MSV enabled chips have significantly more power delay product than the single voltage chips. This confirms that in application specific NoC synthesis it is needed to 1) limit the number of voltage converters and 2) place communicating cores as close as possible regardless of their working voltages. Driven by these requirements, our proposed toolchain performs the core partitioning phase prior to the voltage islanding. The proposed synthesis toolchain is described in the Section IV.

III. PROPOSED SYNTHESIS TOOLCHAIN

In this section, we describe the proposed toolchain for application specific NoC synthesis. The proposed chain follows four steps to reach designs with the satisfied power/performance constraints of the given application. Unlike other methods, we

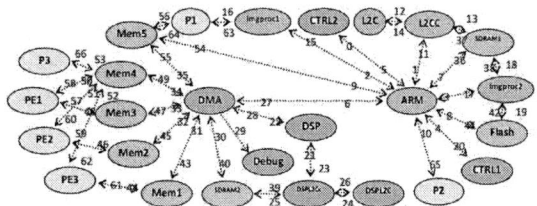

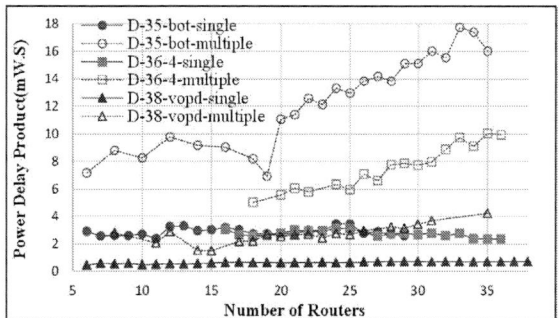

Fig. 1: A sample application core graph with multiple voltages.

Fig. 2: PDP comparison between single and multiple voltage synthesis without limitation on the number of voltage converters.

start with core partitioning, without any assumption on the floorplan of the cores.

A. Core to Router Allocation

The first step of the proposed toolchain assigns cores to on-chip routers based on the bandwidth requirement of cores. When designing an NoC based system for a given class of multi-core applications, the number of routers could theoretically vary from a single global router to one per core. While use of several-port routers will consume more power due to the complex switching structure, using many small routers in turn increases the hop count as well as the switching activity, which inevitably increases the power needed to transfer a packet. Therefore, for a given class of applications, it is necessary to elaborate different scenarios based on the constraints to find the best switching structure. For this purpose, the core to router allocation step tries to assign the cores with high communication bandwidth and/or tight latency demands to the same partition, using a unique router for each partition [7].

Equation (1) calculates $Wcomm_{ij}$, the weight of each communication flow e_{ij}. The parameters α, β, and γ are either set by the designer based on the application characteristics or needed iterations over a range of values. In this case, the objective is to meet the specified constraints.

The number of routers (e.g. i) iterates from one to the number of cores. For every value of i, the input communication graph $Gcomm(C, E)$ is partitioned into the number of i min-cut partitions, namely $Gcomm(C_i, E_i), 1 \leq i \leq n$. To describe intuitively, the partitioning is done in such a way that, the edges that are cut among the partitions will have lower weights than the remaining edges. The number of vertices assigned to each partition is the same as defined in Equation (2). As a result, the communication flows with higher

---

```
Algorithm 1 Merging Island
 1: function MERGEISLANDS()
 2:     for i ← 0 to |partitions| do
 3:         Islanding Partition(i);
 4:         IslandNum ← MakeIslandGraph();
 5:         while IslandNum < 2 do
 6:             integrate vertices belong to the least weighted edge
 7:             IslandNum ← UpdateIslandGraph();
 8:         end while
 9:         if EdgeWeight > Threshold then
10:             break to two partitions
11:         else
12:             integrate to one partition
13:         end if
14:     end for
15: end function
```

Fig. 3: Pseudo-code of merging voltage islands.

bandwidth requirements or tighter latencies or closer voltages are connected to the same router in the same partition. This assignment reduces the hop count and also dissipated power. We have done the partitioning using an efficient partitioning tool called Chaco [10].

B. Merging Voltage Islands

The proposed method performs core partitioning according to $Wcomm$ weight values and without taking into account the core voltages at the first step. This approach may lead to having cores with different voltages in the same partition i.e., the island fragmentation. The fragmentation imposes a large number of voltage converters and a complicated power delivery network. To alleviate this problem, we propose an island merging algorithm to re-adjust the voltages of cores in every partition.

A simplified pseudo code of the island merging algorithm is shown in Fig. 3. For this purpose, cores in each partition are grouped in voltage islands according to their operating voltages (**Islanding()** function in Fig. 3). New edges are added to the graph and appropriate weights ($Wvolt$) compute the merging cost for them. The weight value is defined as $Wvolt_{ij} = (v_j - v_i) \times n_i$, in which v_i and v_j are the supply voltages of the islands. In order to minimize the merging cost, $Wvolt$ values are listed in ascending order. The top of the list denotes the minimum cost and corresponding islands are merged. During this process, cores with supply voltage vi are readjusted to operate on v_j ($v_i < v_j$). As a result, $Gvolt(V, F)$ is updated by removing the lower voltage island and changing the weight values (**UpdateIslandGraph()**).

The island merging (within each partition) continues until a single island emerges. The weight of the edges between the islands is compared with an input threshold (*Integration-threshold*). In case of a lower value the islands are separated into two partitions. Otherwise, all cores of the island are set to work under the same voltage. This process is applied to all the partitions, and thus, the supply voltages of all the cores inside a partition will be the same.

Lemma 1) The proposed method merges the islands with minimal cost.

Proof) Let us consider m islands $V = \{v_0, v_1, ..., v_m\}$ as well as the number of cores in each island to be defined as $n_0, n_1, ..., n_m$. We define the current abstract computation

978-1-5386-7100-9/18 $31.00 © 2018 IEEE

power as $cur = n_0 \times v_0 + ... + n_i \times v_i + .. + n_j \times v_j + .. + n_m \times v_m$. Assume that the voltage of an island v_i is lower than that of an island v_j and these islands are merged. The next abstract power after merging is defined as $next = n_0 \times v_0 + ... + n_i \times v_j + .. + n_j \times v_j + .. + n_m \times v_m$. The difference between the current abstract power and next one is $n_i \times (v_j - v_i)$, which is equal to $Wvolt_{ij}$ on $Gvolt(V, F)$ graph. Therefore, as the order of $Wvolt$ is ascending, the lower values are selected first. As a result, the difference between the current power and next power i.e., cost of merging, is minimized. ■

C. Hierarchical Voltage-Aware Floorplanning

The aim of the partitioning step is to group tightly connected cores (in terms of low delay and high bandwidth constraints) at the same partition to make them connected through the same router. This effectively reduces the communication costs; otherwise, if such cores are placed far away without provisioning the communication requirements, the link power consumption and latency may grow drastically [7], [9]. Although, we have merged the voltage islands within the same partition to solve the voltage fragmentation problem and to reduce the complexity of the power delivery network, islands with similar supply voltages may still remain far away among different partitions. This will potentially increase the complexity of the power delivery network. To tackle this issue, we introduce a hierarchical floorplanning method, which includes three steps; i) core-floorplanning, ii) partition floorplanning and iii) core placement update. The first step places the cores within each partition close to each other to meet their communication requirements. In this step the cores on each partition are floorplanned separately. In the second step, each partition is considered as a fixed block and the blocks are floorplanned to determine the position of the partitions on the chip. This step is done with the aim of placing the blocks with the same voltage and also communication requirements close to each other.

For our hierarchical floorplanning, we use the Parquet floorplanner tool [11] in a two-step approach. The method does *close-proximity-aware* placements of (1) cores within the same partition and (2) islands with the same voltage and high communication constraints. As a result we will have shorter links and a simpler power delivery network.

D. Routing and Path Allocation

This phase tries to establish the required physical links between each pair of routers. A set of links are used to establish the best routing path between the source and destination of a communication flow. We have used the algorithm proposed in [7], [12], [13] to find a deadlock free routing path for each communication data flow. The process starts with sorting the data flows based on their required bandwidth. The flow with the highest requirement is processed first. A fully connected graph is generated where the vertices represent routers. The graph is pre-processed to find the prohibited turns and to break any cyclic dependencies between flows and thus avoid deadlocks [13], [14]. The next step is to assign a weight (cost) to each edge of graph based on its corresponding physical link. Once the weights are assigned, choosing the shortest path is equivalent to finding a path with the least cost to route a data flow. This is done by applying Dijkstra's shortest path algorithm [15]. It should be noted that as the voltage converters between the islands induce overheads in terms of dissipated power and delay, the routing path allocation algorithm finds a path for each data flow such that the number of inter-island links are reduced.

IV. EXPERIMENTAL RESULTS

Our proposed toolchain utilizes Chaco [10] and Parquet floorplanner [11] tools to do the partitioning and floorplanning phases respectively. We also implemented a version of the well-known Sunfloor synthesis flow [7], [12] that we call *Sunfloor inspired* (SFI) and use it to make power/performance comparisons with our design flow. The SFI toolchain needs a given floorplan and assigned voltages of cores to start its synthesis process. This toolchain groups the cores into different voltage islands according to their assigned voltages, and then cores within each voltage island are partitioned to do router allocation; finally, routers are connected to find routing paths for communication flows.

In the experiments, standard core graphs of D-36-4, D-35-bot, D-65-pipe, D-38-tvopd [12] are synthesized under 45nm technology size, and are compared in terms of power delay product. We used Orion2.0/Orion3.0 [16], [17] power models to estimate static/dynamic power consumption of routers and links respectively for our proposed method and our SFI implementation.

In our proposed synthesis toolchain, we treated the lowest allowed voltage as an input parameter i.e., we repeated the synthesis with the lowest allowed voltages of 0.8, 0.9, 1, 1.2 volts. To have a better insight, we implemented the proposed synthesis toolchain in two scenarios. In the first scenario the synthesis is done assuming a given initial core floorplanning, while in the second scenario there is no initial floorplanning given, i.e., the toolchain does a voltage-aware floorplanning.

A. Power, Delay and PDP Results

For each of the mentioned core-graph benchmarks, we let the tools find their synthesis solutions under the specified constraint of the core-graph. The experiments are repeated for different upper bounds on the number of on-chip routers from one to the number of cores in the core-graphs. In cases that tools found a synthesis solution, the power delay product is represented and shown in Fig. 4. As an example, in Fig. 4.a, for the maximum allowed on-chip routers of 10, the proposed toolchain *with* initial floorplanning provides the best synthesis solution in terms of power delay product. The SFI toolchain has not been able to find any synthesis solution that satisfies the application constraints. The key outcome highlighted in Fig. 4 is that in almost all the cases the proposed toolchain offers better designs against the SFI tool in terms of PDP. Our average PDP improvement over the SFI design flow is 59% and 63% for *with* and *without* initial floorplanning scenarios

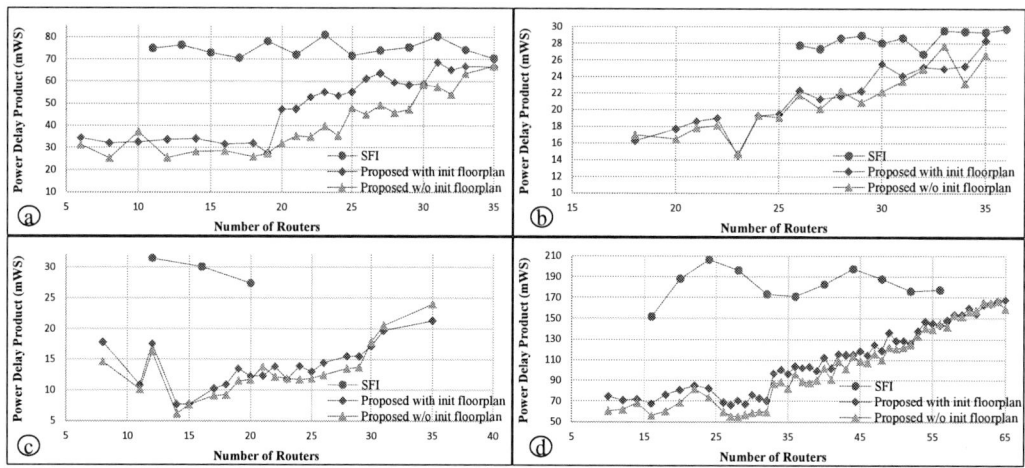

Fig. 4: PDP Comparison of four benchmarks (a) D-35-bot, (b) D-36-4, (c) D-38-tvopd, and (d) D-65-pipe.

respectively. It is clear that when the maximum number of allowed on-chip routers changes, both of our proposed and SFI toolchains have to redo the synthesis flow from start. Due to this reason, solutions for the maximum router limit of K are independent of those for $K+1$ and/or $K-1$.

The general pattern that can be seen in all four core graphs of Fig. 4 is that increasing the maximum allowed number of on-chip routers increases the PDP of solutions found by the proposed synthesis toolchain under both scenarios. The other important point is that, based on the found solutions, our proposed toolchain finds better synthesis solutions when it is not forced to obey an initial floorplanning. In fact, the average PDP improvement of 9% can be seen in all 109 different synthesis cases when we let the proposed toolchain do the floorplanning by itself. The best solutions found by our proposed toolchain and the SFI tool, in terms of power consumption and average network delay, are shown in figures Fig. 5 and Fig. 6. The proposed tool outperforms the SFI design flow in all benchmarks.

Another important observation is that the SFI tool is able to find synthesis solutions satisfying the application constraints for only 37% of the design candidates. Whereas, our proposed synthesis toolchain shows a great efficiency with 73%. In fact, our toolchain gives designers a greater flexibility by providing twice as much valid design alternatives. Using PDP as the metric, Fig. 7 shows the comparative evaluations of four core-graphs. Based on these results, to use of the SFI tool leads to at least 84% - and on average 155% - worse PDP than our proposed toolchain. The improvements seen with our design flow is even larger under the *without* initial floorplanning.

To further validate the PDP reductions seen in the proposed methodology, we count the number of used voltage converters in the output designs for the two toolchains. The results for the examined four benchmarks are presented in Figure Fig. 8. The proposed synthesis toolchain needs at least 40% fewer voltage converters. This positive outcome is due to the fact that our algorithm places communicating cores as close as possible. A higher number of voltage converters leads to more power

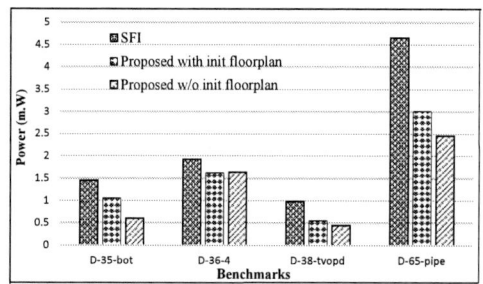

Fig. 5: Design solutions of the proposed and SFI tools based on best-power criteria.

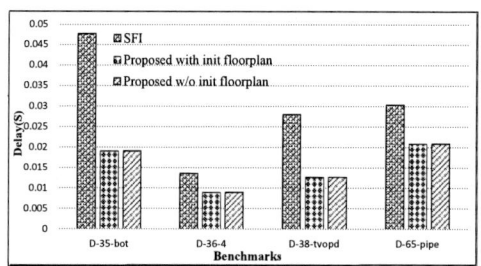

Fig. 6: Design solutions of the proposed and SFI tools based on best-delay criteria.

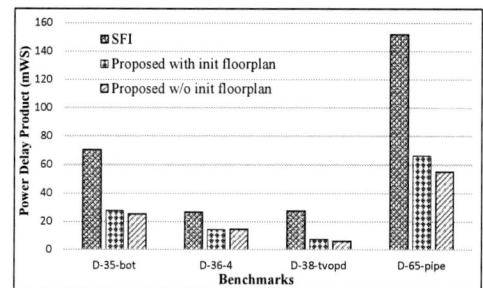

Fig. 7: Design solutions selected based on best-PDP criteria.

consumption and delay.

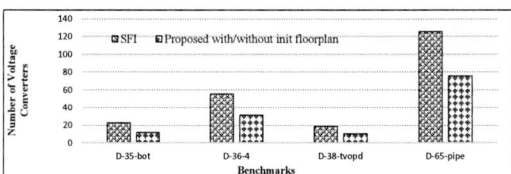

Fig. 8: Average number of required voltage converters.

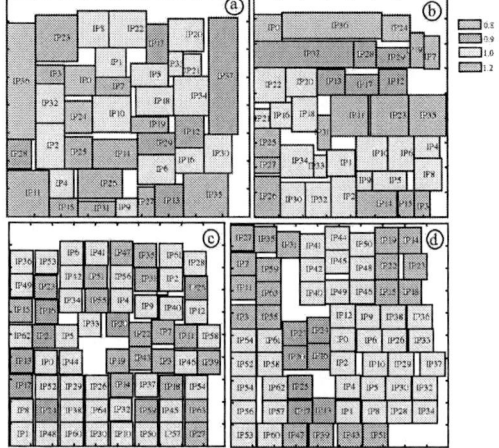

Fig. 9: Final floorplanning of the proposed synthesis toolchain. (a) and (c) benchmarks D-38-tvopd and D-65-pipe *with* initial floorplanning; (b) and (d) the same benchmarks *without* initial floorplanning.

B. Complexity of the Power Delivery Network

As we discussed earlier in the paper, the main objective of voltage-aware floorplanning is to reduce the cost of the power delivery network. In this regard, we compared the final floorplanning of our proposed toolchain under *with* and *without* initial floorplanning scenarios in Fig. 9. Comparing the generated floorplans for D-38-tvopd and D-65-pipe (Fig. 9), one can gather that the hierarchical floorplanning algorithm of the proposed toolchain gives a better core layout when no initial placement is given. Most of the cores with the same voltage are automatically placed in the same groups under the *without* initial floorplanning scheme. This means that synthesized *without* initial floorplans have fewer voltage converters and lower PDN costs. Although our main focus is power and performance, the output designs of the proposed toolchain also show on average 3.56% and 4.64% lower on-chip areas - for the four examined benchmarks - under the *with* and *without* initial floorplanning scenarios, respectively.

The time complexity of the proposed algorithm is $O(|C|^3 ln(|C|)|E|)$, where $|C|$ and $|E|$ are the number of vertices and edges in the core graph $Gcomm(C, E)$ respectively. These two values respectively show the number of on-chip cores and communication flows between cores. The number of routers is varied from 1 to $|C|$ and the tool tries to build a topology under each router count. Also, for each topology, $|C|^2 ln(|C|)$ corresponds to Dijkstra algorithm which is used to find the shortest path for flows. The time complexity of the proposed method is similar to time complexity of the SFI method [7].

V. CONCLUSIONS

In order to benefit from the multi-voltage NoC design, core to router allocation, core to voltage island assignment, voltage-aware floorplanning, and routing paths should be done judiciously. In this paper, we propose heuristic methods to reduce communication power consumption and complexity of the power delivery network. By performing partitioning before islanding, the power consumption and latency are significantly reduced. In addition, with voltage island merging and voltage-aware floorplanning methods, voltage fragmentation and power delivery network complexity can be further minimized. The multi-voltage custom NoCs synthesized from our design flow show 63% *power and delay product* improvement (on average) over the ones synthesized with the SFI tool.

REFERENCES

[1] G. Chen, F. Li, and S.W. Son, M. Kandemir, *Application Mapping for Chip Multiprocessors*, Design Automation Conference (DAC), 2008.

[2] S. Tosun Y. Ar and S. Ozdemir, *Application-specific topology generation algorithms for network-on-chip design*, IET Computers & Digital Techniques,Vol.6, Iss.5,pp.318-333, 2012.

[3] K.S.-M. Li, *CusNoC: Fast Full-Chip Custom NoC Generation*, IEEE Transactions on VLSI Systems, vol. 21, no. 4, pp. 692-705, 2013.

[4] D.E. Lackey, P.S. Zuchowski, T.R. Bednar, D.W. Stout, S.W. Gould, and J.M. Cohn, *Managing power and performance for System-on-Chip designs using Voltage Islands*, IEEE/ACM International Conference on Computer-Aided Design (ICCAD), 2002.

[5] N. Kapadia and S. Pasricha, *A System-Level Cosynthesis Framework for Power Delivery and On-Chip Data Networks in Application-Specific 3-D ICs*, IEEE Transactions on VLSI Systems, 2016.

[6] D. Rahmati, S. Murali, L. Benini, F. Angiolini, G. De Micheli and H. Sarbazi-Azad, *A Method for Calculating Hard QoS Guarantees for Networks-on-Chip*, ICCAD, 2009.

[7] C. Seiculescu, S. Murali, L. Benini, and G.De. Micheli, *NoC Topology Synthesis for Supporting Shutdown of Voltage Islands in SoCs*, Design Automation Conference (DAC), 2009.

[8] V. Todorov, D. Mueller-Gritschneder, H. Reinig, and U. Schlichtmann, *Deterministic Synthesis of Hybrid Application-Specific Network-on-Chip Topologies*, IEEE Transactions on Computer-Aided Design of Integrated Circuits and Systems (TCAD), Vol. 33, No. 10, 2014.

[9] K. Wang, S. Dong, and F. Jiao, *TSF3D: MSV-driven Power Optimization for Application-Specific 3D Network-on-Chip*, IEEE Transactions on Computer-Aided Design of Integrated Circuits and Systems (TCAD), Vol. 36, No. 7, 2017.

[10] B. Hendrickson, R. Leland, *The Chaco User's Guide: Version 2.0*, Sandia Tech Report SAND942692, 1994.

[11] S.N. Adya and I.L. Markov, *Fixed-outline Floorplanning: Enabling Hierarchical Design*, IEEE Transactions on VLSI Systems, Vol. 11, Iss. 6, 2003.

[12] C. Seiculescu, S. Murali, L. Benini, and G. De Micheli, *SunFloor 3D: A Tool for Networks on Chip Topology Synthesis for 3-D Systems on Chips*, IEEE Transactions on Computer-Aided Design of Integrated Circuits and Systems (TCAD), Vol. 29, No. 12, 2010.

[13] M. A. Kinsy, M. H. Cho, T. Wen, E. Suh, M. van Dijk, and S. Devadas, *Application-aware deadlock-free oblivious routing*, ACM International Symposium on Computer architecture, ISCA '09, pages 208–219, 2009.

[14] D. Starobinski, M. Karpovsky, and L.A. Zakrevski, *Application of network calculus to general topologies using turn-prohibition*, IEEE/ACM Transactions on Networking (TON), 2003.

[15] M. A. Kinsy, M. H. Cho, K. S. Shim, M. Lis, G. E. Suh, and S. Devadas, *Optimal and heuristic application-aware oblivious routing*, IEEE Transactions on Computers, 62(1):59–73, 2013.

[16] A. Kahng, B. Li, L.-S. Peh, and K. Samadi, *Orion 2.0: A fast and accurate noc power and area model for early-stage design space exploration*, in Design, Automation Test in Europe Conference Exhibition (DATE), 2009, pp. 423 428.

[17] A.B. Kahng, B. Lin, and S. Nath, *ORION3.0: A Comprehensive NoC Router Estimation Tool*, IEEE Embedded Systems Letters, Vol. 7, No. 2, 2015.

2018 IEEE Computer Society Annual Symposium on VLSI

Accurate Models for Optimizing Tapered Microchannel Heat Sinks in 3D ICs

Leslie K. Hwang*, Beomjin Kwon[†], Martin D. F. Wong*

*Department of Electrical and Computer Engineering, University of Illinois at Urbana-Champaign

{lkhwang, mdfwong}@illinois.edu

[†]School for Engineering of Matter, Transport & Energy, Arizona State University

kwon@asu.edu

Abstract—**High-performance computing systems, especially 3D ICs, are yet facing thermal exacerbation. Inter-tier liquid cooling microchannel layers have been introduced into 3D ICs as an integrated cooling mechanism to tackle thermal degradation. Many research works optimize microchannel designs based on runtime-expensive numerical simulations or inaccurate thermo-fluid models. In this work, we propose accurate closed-form models on tapered microchannel to capture the relationship between channel geometry and heat transfer performance. To improve the accuracy, our correlation is based on developing flow model and derived from numerical simulation using a subset of multiple channel parameters. Our models reduce error by 57 % in Nusselt number and 45 % in pressure drop for channels with inlet width 100-400 μm compared to commonly used fully developed flow based models in optimization. Obtained correlations show potential as solid foundation to achieve close to optimal design through runtime-efficient microchannel design optimization.**

Index Terms—**Microchannel, inter-tier liquid cooling, tapered channel, developing flow model, thermal optimization, 3D IC**

I. INTRODUCTION

More than a decade ago, three-dimensional integrated circuit (3D IC) has been introduced as a promising breakthrough to overcome the physical bottleneck of denser packaging achieved from transistor size miniaturization. Not only it can achieve higher transistor densities and shorter interconnect lengths on a given footprint, but it is also a new paradigm for integrating digital and analog components on a single chip [1]. Recently, 3D memory chip was officially announced and mass production is imminent [2]. However, high-performance multi-processors and heterogeneous device integration have been facing thermal bottlenecks. Heat dissipation has been the most critical barrier to tackle in advancing to core-to-memory or core-to-core stacking, which has higher power density compared to memory-to-memory stacking.

Thermal-aware designs from the high-level system to the physical and manufacturing level implementations have been studied to alleviate the aggravated thermal issues by introducing the third dimension to the chip design [3], [4]. Majority of existing works are based on conductive heat transfer resolved in solids within a chip. Conductive thermal management designs only in combination with conventional air-cooling based heat sinks are often insufficient to keep the chip in operational and reliable temperature range for high performance applications [5]. Upper limit of the heat flux with air-cooling methods for most applications is around

100 W/cm². 3D multi-chip modules that dissipate more than 300 W/cm² at the die are beyond the capability of most conventional air-cooling solutions [6]. Additional to the fact that heat sink is attached to one of the device layers, major heat transfer blockages in 3D IC, that create localized, trapped heat, or hotspots, are resulted mainly from the bonding layer between the device stacks. Bonding layer is composed with very low thermally conductive inter-layer dielectric to isolate the unnecessary electric connections between device tiers. Hence, it is crucial to introduce additional cooling mechanism for thermally isolated device layers, that are distant to the heat sinks.

Engineers have been exploring more effective ways of cooling by pumping liquid coolants directly onto the chips, rather than circulating air around them. The use of liquid coolant, commonly water, has become an attractive option due to higher convective heat transfer coefficient compared to air-cooling. Heat transfer coefficient of water $h_{water} = $ 100-1200 W/m²-K is comparably higher than air $h_{air} = $ 5-37 W/m²-K for natural convection, and becomes even more effective in forced convection, $h_{water} = $ 500-10 000 W/m²-K versus $h_{air} = $ 10-200 W/m²-K. A single phase loop in liquid cooling system consists of a miniaturized pump and heat exchanger (e.g. cold plate or microchannels). Conventional board level liquid-cooling heat exchangers are not suitable for chip level implementations due to bulky modules. There have been interests on compact microchannel heat exchangers that could directly be etched on the back of the silicon dies [7], [8] which also have the benefit to be placed between the die tiers. Hence, the inter-tier liquid-cooling microchannel layer has been gaining attention as an integrated cooling mechanism to tackle thermal degradation in 3D ICs [9], [10].

Majority of the research works on microchannel design optimizations are based on runtime-expensive numerical simulations or oversimplified thermo-fluidic models with incorrect assumptions. Numerical models are typically unsuitable for optimization considering intensive computing. Approximate models or correlations with improper assumptions pose a fundamental limitation to accurately derive the relationships between the channel parameters and the thermal performance. Applying fully developed flow model to analyze short channel that presents developing flow characteristics will cause large discrepancy. In microchannel optimization, inaccuracy on the

978-1-5386-7100-9/18 $31.00 © 2018 IEEE

base channel model will significantly affect the optimality and quality of the resulting design.

In this work, we derive an accurate thermo-fluid correlations of the microchannel to capture the relationship between the channel physical parameters and the thermo-fluid performance. The thermal correlation is based on developing flow model to properly establish the parametric study at the entrance region of the channel. In microchannel design optimization, the flow behavior at the entrance region will be prolonged when microchannel dimensions vary across the flow direction. Therefore, correlation based on developing flow model will serve as a solid foundation for finding the optimal microchannel design. To verify the accuracy of our models, we have compared with commonly used fully developed flow based correlations and reliable numerical simulation.

II. RELATED WORKS

The concept of liquid cooling microchannel integrated to electronics was introduced by Tuckerman and Pease [11]. Since then, many researchers have studied and experimented various channel designs. Kishimoto and Ohsaki [12] numerically simulated non-monotonic relationship of thermal resistance and channel width. They also fabricated board-level via-hole channels and tested the cooling performance. Working with popular 28-nanometer FPGA devices made by Altera Corp., Sarvey et al. have demonstrated a monolithically-cooled chip using microfluidic passages that could operate at temperatures more than 60 percent below those of similar air-cooled chips with heat sink or cooling fans [13].

One of the challenges in single-phase liquid cooling is the increased thermal gradient which leads to performance degradation. To mitigate non-uniform heat distribution, Sabry et al. [14] proposed to modulate the channel width along the traveling direction of the fluid. This work assumes piece-wise constant channel width and is based on rectangular channel approximate model for fully developed flow. In fact, all channels at the inlet starts with the developing flow before it reaches steady-state flow and there is a distinct difference in velocity and thermal profiles between developing and fully developed regions. Thus, applying fully developed flow model to analyze developing region is fundamentally incorrect. Especially for non-uniform channels, flow is more perturbed and fluctuated than uniform channels, thus, likely to remain developing both hydrodynamically and thermally. Although the resulting tapered channel geometry is reasonable in terms of thermal gradient reduction, it is uncertain to consider the design optimal when the microchannel design is optimized based on fully developed flow model.

Hung and Yan [15] studied similar tapered channel but using finite-volume based numerical analysis. The results are insightful to understand the performance with channel parameter variation. However, numerical approach is very compute-intensive and inappropriate for further optimization. In addition, this model is based on single-sided heat source, which is inapplicable for inter-tier cooling stacked between device layers in 3D IC. Thus, it is crucial to establish a

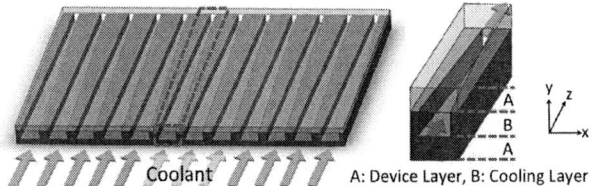

Fig. 1: 3D stacked device layers with inter-tier liquid cooling microchannel layer. Tapered channels are etched at the bottom of the device layer. Top device layer is presented transparent for clear cooling layer illustration. Note that figure is not to scale.

compact model for fast microchannel design optimization that incorporates 3D IC applications.

In this paper, we propose compact and accurate closed-form models of tapered channel for microchannel design optimization. The presented models are of fundamental importance in channel design study. They improve the accuracy with proper flow assumption and computation speed compared to the numerical approach in optimizing various microchannel designs.

III. FUNDAMENTALS OF HEAT TRANSFER

Three dimensional chips are composed of multiple vertically stacked device layers. To effectively dissipate heat from the device layers to outside the chip, liquid cooling microchannel layers are fabricated in between the device tiers. Fig. 1 depicts 3D view of two device tiers with single microchannel layer in between.

In this section, we start by the introducing the parameters of the tapered microchannel dimensions, followed by the metric used to determine the thermal efficiency of liquid cooling heat exchangers. Then, appropriate heat transfer models for tapered microchannel design and analysis are proposed which are further improved with parameter fitting.

A. Microchannel Geometry

Increased thermal gradient across a chip has been the rising challenge in liquid cooling designs. reduce the temperature gradient [14]. While large inlet provides abundant incoming coolant with fixed incoming coolant temperature, small outlet with increased temperature from absorbed heat will benefit from increased coolant velocity. Top view of the tapered rectangular channel is portrayed in Fig. 2.

Given the channel inlet width w_{in}, outlet width w_{out} and length L, the channel angle θ will be expressed as Eq. (1). Opposed to tapered channel, channel with uniform cross-section from inlet to outlet is denoted as uniform channel in the remaining of the paper.

$$\theta = arctan\left(\frac{w_{in} - w_{out}}{2L}\right) \qquad (1)$$

B. Thermal Resistance

Within the microchannel cooling system, conduction and convection are primary modes of heat transfer. To evaluate

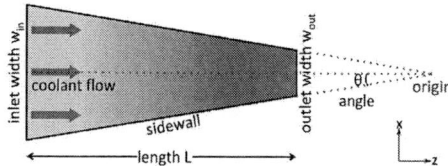

Fig. 2: Top view of the tapered channel with inlet width w_{in}, outlet width w_{out}, length L, and half angle θ.

the thermal effectiveness of the design, thermal resistance is considered. Objectives for most of the thermal optimizations are based on the temperature. However, there hardly exists analytical closed-form temperature model for given design. To replace the metric, we focus on the temperature difference instead of the temperature. Overall temperature drop $\triangle T$ is the product of the heat rate q and the thermal resistance R_{th}, $\triangle T = q * R_{th}$. Total thermal resistance R_{th} is the summation of resistances from conductive $R_{th,cond}$ to convective $R_{th,conv}$ heat transfer as described by Eq. (2). Each term is dependent on the material property, physical dimensions as well as the hydrodynamic conditions in fluid region.

$$R_{th} = R_{th,cond} + R_{th,conv} = \frac{d}{kA_c} + \frac{1}{hA} \quad (2)$$

In IC design, we formulate the conductive resistance only in solid: silicon device layers and walls surrounding the microchannels where majority of the conduction occurs. Conductive thermal resistance $R_{th,cond}$ is defined by distance d, thermal conductivity k and the cross-sectional area in solid A_c. Convective thermal resistance is defined in liquid region: microchannel layer. Convective thermal resistance $R_{th,conv}$ is the inversely proportional to the product of heat transfer coefficient h and wetted area A. Total convective resistance can be considered as parallel connections of each local cross-sectional convective thermal resistance $R_{th,conv,i}$ at location i as in Eq. (3). In tapered channel, both h and A vary across the channel length and local values are represented as h_i and A_i respectively. For each section, A_i is computed with local perimeter P_i and the length segment dz.

$$R_{th,conv} = \left(\int_i \frac{1}{R_{th,conv,i}} \right)^{-1} = \frac{1}{\sum\limits_i h_i A_i} = \frac{1}{\sum\limits_i h_i P_i dz} \quad (3)$$

C. Convective Heat Transfer

Heat dissipated in liquid cooling microchannel is based on forced convection. Heat transfer coefficient h, one of the factors to determine $R_{th,conv}$, depends on numerous parameters, such as coolant properties, fluid velocity, and channel dimensions, and yet there does not exist any closed-form analytical model. Thus, we rely on the empirical or numerical correlations. Similar to the overall thermal resistance, temperature difference is computed by the product of the heat rate, q and the convective thermal resistance, $R_{th,conv}$. Using this relationship, we can derive h as in Eq. (4).

$$h = \frac{1}{R_{th,conv} * A} = \frac{q}{(T_w - T_b) * A} = \frac{q''}{T_w - T_b} \quad (4)$$

where q'' is the heat flux, T_w is the channel wall temperature and T_b is the bulk fluid temperature.

The efficiency ratio of the convection to conduction in fluid is defined as Nusselt number, $Nu = hD_h/k$, where k is the thermal conductivity and D_h is the hydraulic diameter. Hydraulic diameter is computed from area A and wetted perimeter P_{wet}, using equation $D_h = 4A/P_{wet}$. There exist numerous Nusselt number correlations for various cases and designs. Hence, it is commonly used value to be compared with numerical or empirical results to verify the accuracy. Then, h is derived from Nusselt number correlation.

Shah and London [16] have established Nusselt number models on various channel geometries. Among the models, many recent works on microchannel design and optimization have adopted rectangular channel approximate model for fully developed flow [17], [18]. However, depending on the flow assumption and the channel geometry, the model cannot fully capture the phenomenon and becomes inapplicable for accurate optimizations. Fully developed region is defined where flow velocity and normalized temperature gradient profiles remain unchanged along the flow direction. In uniform channel, fully developed hydrodynamic flow is reached after length $L_{fd,h} \approx 0.05Re * D_h$ and fully developed thermal flow is reached after length $L_{fd,t} \approx 0.05Re * Pr * D_h$ for laminar flow ($Re < 2300$), where Re is Reynolds number and Pr is Prandtl number. However, in tapered channel, normalized velocity profile continuously changes based on mass conservation law and therefore, fully developed state will hardly be reached. In addition, piece-wise channel optimization will vary the channel dimensions at each segmented section. Non-constant geometry contributes to alter the flow and extend the length to reach the fully developed flow. Therefore, fluid dynamics and heat transfer in channel with varying cross-sectional shape differ from those in the uniform channel as the flow cannot reach the developed condition.

Local Nusselt number, Nu_z, correlation for developing flow in entrance region for circular and non-circular channel developed by Shah and London [19] is shown in Eq. (5).

$$Nu_z \approx 0.517 * (f * Re)^{1/3} * \left(\frac{z}{D_h * Re * Pr} \right)^{-1/3} \quad (5)$$

where f is Fanning friction factor f and z is position variable. In laminar flow, f is $16/Re$ which leads the term $f * Re$ constant as 16. Pr is also a fixed constant for water at room temperature. In result, the only remaining variables are D_h, Re and z.

IV. CLOSED-FORM THERMO-FLUID MODEL FOR TAPERED CHANNEL

A. Simulator

For numerical simulations, we used commercial CFD solver, ANSYS Fluent v.18. The software uses finite-volume method

TABLE I: Microchannel Parameters

Definitions	Param.	Min	Max
Height	H	100 μm	
Length	L	0.5 mm	
Inlet width	w_{in}	100 μm	400 μm
Outlet width	w_{out}	10 μm	400 μm
Angle	θ	0°	19.8°

TABLE II: Thermal and Fluid Properties

Definitions	Param.	Values
Silicon thermal conductivity	k_{si}	130 W/m-K
Water thermal conductivity at 300 K	k_{water}	0.613 W/m-K
Water kinematic viscosity	ν	1.004×10^{-6} m²/s
Coolant inlet flow velocity	v	1 m/s
Coolant inlet temperature	T_{inlet}	300 K
Prandtl number for water at 300 K	Pr	5.83
Reynolds number	Re	99-159

(FVM) with the support of the semi-implicit-method for pressure-linked-equations method. We applied rectilinear mesh with less than 0.1 maximum skewness, 5 μm mesh element size and double precision simulation setting. The solutions converged within a hundred iterations.

B. Parameters

Microchannel dimensions used in the experiments are shown in Table I. Height and length of the channel are fixed to single values as they are mainly defined by the manufacturing technology and the chip footprint. Channel inlet and outlet widths are varied between the minimum and maximum range listed. The range is selected based on commonly used manufacturing dimensions listed in previous works [11], [17], [20], [21]. To ensure the converging tapered shape, channel outlet width is set less than or equal to the channel inlet width for each design. Tapering angle is computed using Eq. (1). Table II lists the thermo-fluid conditions and the material properties used in the simulations. Water at room temperature is used as coolant and inlet flow velocity is set to 1 m/s. Adding fluid velocity into the optimization variable can be the future extension of this work.

C. Fitted Developing Flow Model for Tapered Channel

First, we have compared the models of different flow conditions and analysis methods. Fig. 3a compares four models: 1) Shah-London fully developed model for rectangular channel [16], 2) Sparrow-Starr fully developed model for tapered cylindrical channel [22], 3) Shah-London developing flow model for general channel [19] compared with 4) FVM numerical simulation. Model comparison was experimented on a single tapered channel with inlet width of 400 μm, outlet width of 200 μm, length of 500 μm with inlet flow velocity of 1 m/s and heat flux of 100 W/cm². Regardless of two different geometry approaches, rectangular channel versus tapered channel, both fully developed models showed large gaps on the local Nusselt number as well as the Nusselt number gradient along the channel to the FVM result. On the contrary, developing flow model for general channel has reasonably well-matching curve to the FVM.

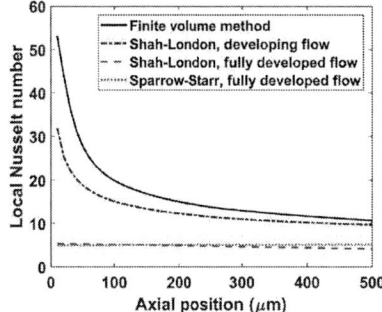

(a) Numerical simulation, fully developed and developing flow correlations comparison

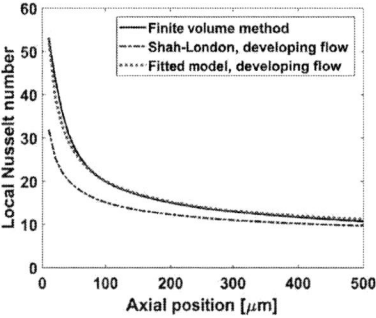

(b) Numerical simulation, general channel developing flow and fitted developing flow model comparison

Fig. 3: Nusselt number comparisons on tapered channel with inlet width 400 μm, outlet width 200 μm, length 500 μm, inlet flow velocity 1 m/s, heat flux 100 W/cm²

1) Nusselt Number: It is a well-known approach to derive closely-fitting Nusselt number or heat transfer coefficient correlation from empirical data [23]. In this work, the model is built on numerical simulation data instead of the experimental results. Based on the Shah-London developing flow model with fixed constants for coefficient and exponents, we have further improved the model by fitting two parameters: 1) coefficient, denoted as α, and 2) exponent, denoted as β, in Eq. (6) to the numerical results. Exponent of $f * Re$ remains unchanged as the term is constant in laminar flow.

$$Nu_{z,fitted} \approx \alpha * (f * Re)^{1/3} * (z^*)^{\beta} \qquad (6)$$

We have simulated 15 channel sizes: combination of three inlet widths and five different angles. Flow velocity and heat flux are fixed to a single value in our model for in-depth study of the channel geometry dimensions. These values can also be varied to induce more comprehensive correlations but it is beyond the scope of this work. Each (α, β) value pair is determined by the least squares method for all data sets. Fig. 3b exhibits the improvement of our fitted correlation from Shah-London general channel developing flow model, matching closer to the FVM result on the same tapered channel used in Fig. 3a.

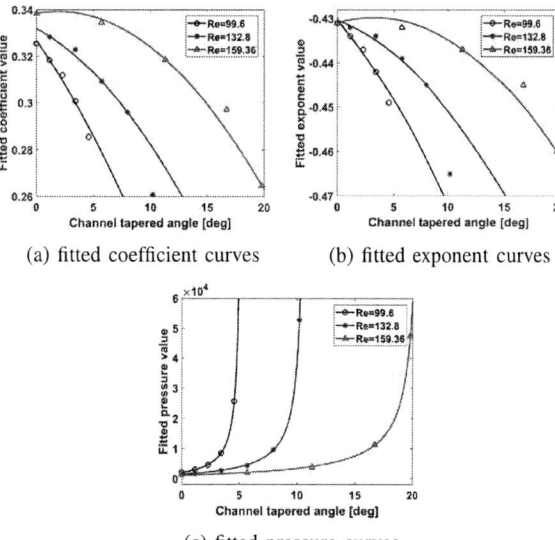

(a) fitted coefficient curves   (b) fitted exponent curves

(c) fitted pressure curves

Fig. 4: 3D surface fitted parameters of developing flow model for tapered channel on various tapering angles and Reynolds numbers.

Next, fitted (α, β) pair values for selected channel geometries are extended to derive the 3D correlations of the parameters as function of two variables: 1) tapered angle θ and 2) Reynolds number Re, using polynomial least squares surface fit. Reynolds number is a representative indicator of the flow condition, defined by flow velocity u, wetted area A, hydraulic diameter D_h and kinematic viscosity ν, $Re = (u * A * D_h)/\nu$. We fixed flow velocity and channel height, thus inlet width is the only variable to determine the Reynolds number. Therefore, fitting parameters can be expressed as equivalent function of channel tapering angle θ and the inlet width w_{in}. 3D correlations were within 95 % accuracy from the individually fitted parameters. The 3D fitted curves of coefficient and exponent values are illustrated in the Fig. 4a and 4b respectively. The derived correlations for α and β are shown in equations (7) and (8). These fitted functions are combined to the Eq. (6) to compute the heat transfer coefficient of developing flow in tapered rectangular channel.

$$\begin{aligned} \alpha(\theta, Re) = &\ 0.3317 - 0.020290\theta - 0.0001991Re \\ &- 0.0002337\theta^2 + 0.000133\theta Re + 1.513 * 10^{-6}Re^2 \end{aligned} \quad (7)$$

$$\begin{aligned} \beta(\theta, Re) = &\ -0.435 - 0.0092960\theta + 8.235 * 10^{-5}Re \\ &- 0.0001101\theta^2 + 6.282 * 10^{-5}\theta Re - 3.586 * 10^{-7}Re^2 \end{aligned} \quad (8)$$

For qualitative and quantitative analysis between the models, mean error values on local Nusselt number for simulated set of channels are compared in Fig. 5. Each model is compared with FVM. Average error percentage of Shah-London and Sparrow-Starr fully developed correlations are 62 % and 72 % respectively and Shah-London developing flow model tremendously drops down to 14 %. Our fitted model further improves up to 4 % error. Deviation is greater on channels

with larger tapered angle, but yet the difference is 1-3.5 %.

Furthermore, Sparrow-Starr model is based on tapered cylindrical model, which incorporates the effect of the slanted channel wall. However, this model is inapplicable for uniform channels with angle 0°. Rightmost bars from Fig. 5a and 5c show uniform channels with same inlet and outlet widths. Both fully developed flow models were comparable between tapered channels, but the error rose significantly for Sparrow-Starr model on the uniform channel and exhibited 97 % error. On the other hand, our model is reliably accurate throughout all angles in tapered channel.

Looking back onto Fig. 3, the discrepancy of analytical models to FVM is larger near the entrance region of the channel. The observation indicates that the fully developed assumption will lead to more unreliable result for channels which are significantly shorter than the entry length. Piecewise channel optimization dissects the channel into small sections and apply the heat transfer model to each channel section. In the optimization based on fully developed flow, every section will carry large error that accumulate throughout the entire channel and exacerbate the effect. Moreover, channel dimensions can alter between the sections and cause the flow unstable. In result, fully developed flow in optimized channel will never be reached in the extreme case.

2) Pressure Drop: Cooling efficiency in microchannel comes at a price which can be mainly quantified by the pressure drop. Given microchannel design and the coolant flow rate, pressure drop is determined. Darcy-Weisbach analytical pressure model for general channel in laminar flow is $\triangle P_{dw} = 0.5 * f_D * L * \rho * u_{mean}^2/D_h$, where f_D is Darcy friction factor, ρ is the density of the fluid, and u_{mean} is the mean flow velocity. Although it is a commonly used model in rectangular channels, it does not preserve the accuracy on tapered channel designs and accurate pressure model for varying cross-sectional channels has not been heavily studied.

Similar to the parameter fitting in Nusselt number correlation, pressure model is derived based on numerical result. The model is expressed in Eq. (9) and fitted curves are demonstrated in Fig. 4c. Tapered channel pressure model is valid for the range where tapering angle does not exceed to fully close the channel outlet. As tapering angle reaches the maximum, pressure drop will be extremely high, reaching infinite value. Valid pressure drop cannot be computed at this angle and becomes singularity point in our fitted model. Beyond this angle, outlet width becomes below 0. Fitted pressure curves demonstrate 1.58 % error to FVM result while Darcy-Weisbach model shows 46.92 % error in average on the same set of channels used in Fig. 4c.

$$\triangle P(\theta, Re) = \frac{795.335 + 38.9285\theta - 11.8544Re + 0.0685Re^2}{1 - 0.0278\theta - 0.0185Re + 0.0001Re^2} \quad (9)$$

V. CONCLUSIONS

We have proposed accurate closed-form correlations for tapered channel based on the developing flow model. Tapered

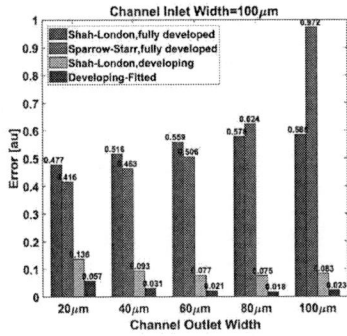
(a) tapered channels with inlet=100 μm

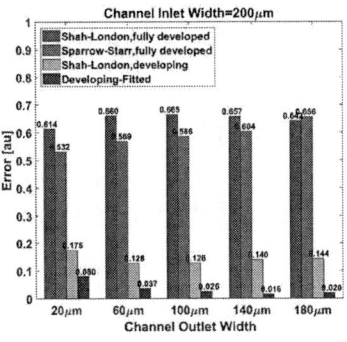
(b) tapered channels with inlet=200 μm

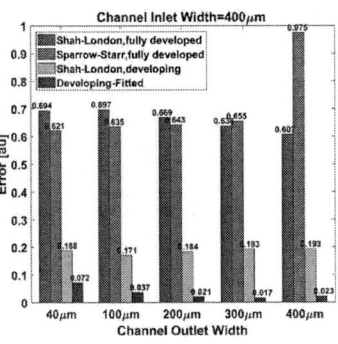
(c) tapered channels with inlet=400 μm

Fig. 5: Average error comparison on various Nusselt number models: Shah-London (fully-developed flow), Sparrow-Starr (fully-developed flow), Shah-London (developing flow) and developing flow-fitted.

channel designs will fluctuate the coolant flow, extend the developing flow region and possibly never reach fully developed state. Nevertheless, widely used correlations were based on the fully developed flow models. Compared to the fully developed flow based thermo-fluid models, our derived correlations have reduced error by 57 % in Nusselt number, 45 % in pressure drop to FVM simulation values for channels with inlet width 100-400 μm and tapering angle 0-19.8°. As the model captures the relationship of fitting parameters and pressure drop in terms of the channel geometric dimensions, it can be applied for the accurate microchannel design optimizations.

Acknowledgment

This work was partially supported by NSF grants CCF-1421563 and CCF-171883.

References

[1] K. Banerjee, S. J. Souri, P. Kapur, and K. C. Saraswat, "3-d ics: A novel chip design for improving deep-submicrometer interconnect performance and systems-on-chip integration," *Proceedings of the IEEE*, vol. 89, no. 5, pp. 602–633, May 2001.

[2] M. Tyson. (2016, Jan) 3d xpoint memory chip samples "just around the corner". [Online]. Available: http://hexus.net/tech/news/industry/89780-3d-xpoint-memory-chip-samples-just-around-corner/

[3] K. Stavrou and P. Trancoso, "Thermal-aware scheduling: A solution for future chip multiprocessors thermal problems," September 2006.

[4] B. Goplen and S. Sapatnekar, "Thermal via placement in 3d ics," in *Proceedings of the International Symposium on Physical Design*, April 2005, pp. 167–174.

[5] J. F. Tullius, R. Vajtai, and Y. Bayazitoglu, "A review of cooling in microchannels," *Heat Transfer Engineering*, vol. 32, no. 7-8, pp. 527–541, November 2011.

[6] (2008, March) Liquid cooling is coming to chips and boards. [Online]. Available: http://www.powerelectronics.com/thermal-management/liquid-cooling-coming-chips-and-boards

[7] R. W. Tjerkstra, M. de Boer, E. Berenschot, J. Gardeniers, A. van der Berg, and M. Elwenspoek, "Etching technology for microchannels," in *Proceedings of the 1997 10th Annual International Workshop on Micro Electro Mechanical Systems, MEMS*, January 1997, pp. 147–152.

[8] E. G. Colgan, B. Furman, M. Gaynes, W. S. Graham, N. C. LaBianca, J. H. Magerlein, R. J. Polastre, M. B. Rothwell, R. J. Bezama, R. Choudhary, K. C. Marston, H. Toy, J. Wakil, J. A. Zitz, and R. R. Schmidt, "A practical implementation of silicon microchannel coolers for high power chips," *IEEE Transactions on Components and Packaging Technologies*, vol. 30, no. 2, pp. 218–225, June 2007.

[9] D. Atienza, "Thermal-aware design of 3d ics with inter-tier liquid cooling," in *Electron Devices Meeting*, December 2010.

[10] A. Sridhar, M. M. Sabry, and D. Atienza, "System-level thermal-aware design of 3d microprocessors with inter-tier liquid cooling," in *17th International Workshop on Thermal Investigations of ICs and Systems (THERMINIC)*, September 2011.

[11] D. Tuckerman and R. Pease, "High-performance heat sinking for vlsi," *IEEE Electron Device Letters*, vol. EDL-2, no. 5, pp. 126–129, May 1981.

[12] T. Kishimoto and T. Ohsaki, "Vlsi packaging technique using liquid-cooled channels," *IEEE Transaction on Components, Hybrids, and Manufacturing Technology*, vol. CHMT-9, no. 4, pp. 328–335, December 1986.

[13] T. E. Sarvey, Y. Zhang, C. Zheng, P. Thadesar, R. Gutala, C. Cheung, A. Rahman, and M. S. Bakir, "Embedded cooling technologies for densely integrated electronic systems," in *Proceedings of the IEEE Custom Integrated Circuits Conference*, September 2015.

[14] M. M. Sabry, A. Sridhar, J. Meng, A. K. Coskun, and D. Atienza, "Greencool: An energy-efficient liquid cooling design technique for 3-d mpsocs via channel width modulation," *IEEE Transactions on Computer-Aided Design of Integrated Circuits and Systems*, vol. 32, no. 4, pp. 524–537, April 2013.

[15] T.-C. Hung and W.-M. Yan, "Effects of tapered-channel design on thermal performance of microchannel heat sink," *International Communications in Heat and Mass Transfer*, vol. 39, no. 9, pp. 1342–1347, November 2012.

[16] R. K. Shah and A. L. London, *Advances in Heat Transfer: Laminar Flow Forced Convection in Ducts*. New York: Academic Press, 1978.

[17] M. M. Sabry, A. Sridhar, and D. Atienza, "Thermal balancing of liquid-cooled 3d-mpsocs using channel modulation," in *Date Automation and Test in Europe*, 2012.

[18] G. Chen, J. Kuang, Z. Zeng, H. Zhang, E. F. Y. Young, and B. Yu, "Minimizing thermal gradient and pumping power in 3d ic liquid cooling network design," in *Proceedings of Design Automation Conference*, June 2017.

[19] R. K. Shah and A. L. London, *Fundamentals of Heat Exchanger Design*. New Jersey: John Wiley and Sons, 2003.

[20] A. K. Coskun, D. Atienza, T. S. Rosing, T. Brunschwiler, and B. Michel, "Energy-efficient variable-flow liquid cooling in 3d stacked architectures," in *Date Automation and Test in Europe*, March 2010.

[21] R. Singhal and M. Z. Ansari, "Flow and pressure drop characteristics of equal section divergent-convergent microchannels," *Procedia Technology*, vol. 23, pp. 447–453, 2016.

[22] E. M. Sparrow and J. B. Starr, "Heat transfer to laminar flow in tapered passages," *ASME Journal of Applied Mechanics*, vol. 32, no. 3, pp. 684–689, September 1965.

[23] J. Fernández-Seara, F. J. Uhía, J. Sieres, and A. Campo, "A general review of the wilson plot method and its modifications to determine convection coefficients in heat exchange devices," *Applied Thermal Engineering*, vol. 27, no. 17-18, pp. 2745–2757, December 2007.

978-1-5386-7100-9/18 $31.00 © 2018 IEEE

2018 IEEE Computer Society Annual Symposium on VLSI

Designing and Benchmarking of Double-Row Height Standard Cells

Yu-Xiang Chiang, Cheng-Wei Tai, Shang-Rong Fang, Kai-Chun Peng, Yuan-Dar Chung, Jin-Kai Yang, Rung-Bin Lin
Computer Science and Engineering
Yuan Ze University
Taoyuan, Taiwan

Abstract—This article presents our experience of designing double-row height standard cell libraries and their use for chip designs. Seven cell libraries are designed based on the 15nm process technology stipulated in FreePDK15. A single-row height of 7.5 M2 tracks is used as a basis for designing double-row height cells. Two minimum-sized transistors, one having two fins and the other having four fins, are employed to design 1X drive-strength cells. Among the seven libraries, two libraries consist of only single-row height cells. The other five libraries each consist of partly single-row height cells and double-row height cells. Our experiments show that a double-row height library can achieve on average -2% to 21% area saving and -23% to 19% smaller power-delay-area product. Our results also show that using a large minimum-sized transistor for designing a double-row height library is not viable if extensive transistor folding is required.

Keywords—mixed-height; multi-row height; double-row height, standard cell; library; FinFET

I. INTRODUCTION

The Intel's 22nm process technology ushers semiconductor fabrication into the era of FinFET [1]. Being able to continually shrink fin pitch and increase fin height is key to Moore's law [2]. A larger fin height leads to a higher current drive per fin. This enables us to use a smaller area to achieve the same current drive as that provided by traditional planar transistors. As a consequence, more and more standard cell libraries are designed with a small cell height, e.g., a cell height of 7 tracks or even 6 tracks [3-6]. However, a small cell height makes signal routing inside a cell more difficult due to fewer horizontal tracks and makes pin access more problematic due to fewer access points per pin [7,8]. This leads to use of more metal layers, e.g., M2 or even M3 layers, for routing signals inside a cell. Furthermore, the width of a high-drive cell or a complex cell becomes exceedingly large and thus may create long horizontal wires inside a cell. This also makes the placement of cells more difficult. One solution to these issues is to employ mixed-height or multi-row height standard cells for chip designs. Such a trend is evidenced by increasing research momentum on the related subjects carried out in academia and industries [9-17]. The related investigation mainly focuses on placement problem, especially on cell legalization. The placement problem for mixed-cell height cells and multi-row height cells are different. For a mixed-height design, several different cell heights are employed by the cells and the cell heights may not have an integral-multiple relationship. Hence, all the cells using the same cell height should be put into a module and placed in a region defined based on the underlying cell height. This sort of placement problem [13] is similar to the placement of modules that use different supply voltages, i.e., placement of voltage islands. On the other hand, for a multi-row height design, all the cells even with different cell heights can be dispersed in the placement region defined for single-row height cells. Although the placement problem of mixed-height cells and multi-row height cells are widely investigated, designing of this sort of cells and their use for chip designs are not much addressed in the literature. To our knowledge, only the work in [16] specifically presents a methodology for designing multi-row height cells and exercises the cells for chip designs. It reports that a double-row height library consisting of 8-track cells and 16-track cells could achieve 14% area saving at 100MHz frequency when compared to a library consisting of only 10-tarck cells. Although, significant advancement has been made in [16], there remains many non-disclosed details and leaves many issues un-investigated. For instance, it is not clear to what extent logic functions are implemented as double-row height cells and to what extent they are used in a design. It is also not clear to what extent the impact on placement and routability double-row height cells will bring about. Hence, in this article we take a deeper look into some issues of designing multi-row height cells and study the influence of their use on chip designs. We design seven standard cell libraries based on the 15nm FinFET process technology stipulated in FreePDK15 [18,19]. The first three cell libraries use a minimum-sized transistor of two fins. The first cell library is designed with single-row height of 7.5 tracks. The second library consists of some cells designed with double-row height. The third library consists of both single-row and double-row height cells for the same logic function. For example, AND2X1 could have both a single-row height version and a double-row height version. We further design three cell libraries using a minimum-sized transistor of four fins to see whether extensive use of double-row height cells will behave differently. The last library comprises cells compiled based on a minimum-sized transistor of two fins and four fins as well. With a single-row height of 7.5 tracks, a minimum-sized transistor of four fins need be folded into two fingers. Extensive experiments have been conducted. Below are our major findings.

- A double-row height library can achieve on average -2% to 21% area saving and -23% to 19% smaller power-delay-area product.

- A library comprising double-row height cells when designed with a minimum-sized transistor (of 4 fins) that need be

978-1-5386-7100-9/18 $31.00 © 2018 IEEE

64

folded is not viable. At the same timing performance level, the library results in using much more area and power when compared to a double-row height library employing a minimum-sized transistor of 2 fins.

- The best timing performance achieved by a double-row height library employing a minimum-sized transistor of 4 fins is not better than that achieved by a double-row height library employing a minimum-sized transistor of 2 fins. That is, it does not have any advantage of using a larger minimum-sized transistor for designing a double-row height library.

The rest of this article is organized as follows. Section II briefly introduces FreePDK15, presents some basics about multi-row height layout, and describes the problem addressed in this work. Section III presents our research approach. Section IV shows some experimental results. Last section draws a conclusion.

II. PROBLEM STATEMENTS

A. FreePDK15

FreePDK15 is a process design kit designed by NCSU in collaboration with Mentor Graphics based on some data published in the open literature [18,19]. FreePDK15 specifies three middle-of-line layers, AIL1, AIL2, and GIL. The local interconnect layer AIL1 is employed to connect fins of the same transistor together, AIL2 is used to connect AIL1 sources/drains to M1 by V0 layer, and GIL is employed to connect poly gates to M1 by V0 layer. Because GIL cannot run over active regions, this severely constrains the placement of M1 pins inside a cell. M1 wires are bidirectional and fabricated using double patterning. Poly gate layer is unidirectional and also adopts double patterning. In this work, M2 wires are routed horizontally and M3 wires are routed vertically. Although the PDK released in 2017 [19] provides LVS and RC rule files, we still use our own rule files for this work because ours are obtained by compiling an MIPT file containing *multigate* (i.e., FinFET) descriptions [20].

B. Multi-row Height Cells

As mentioned in Section I, small cell height is now widely used to design standard cells for minimizing chip area. However, there are several adversary side effects of using small cell height. First, large-drive cells or complex cells tend to be exceedingly wide. Second, to achieve similar rise and fall transition times, the size of transistors on a serial path is usually k times the size of transistors on its dual parallel path. This causes lots of area wasted in the region for the transistors on the dual parallel path. For example, the size of a P-type transistor in NOR4X1 on the pull-up serial path should be four times the size of an N-type transistor on the pull-down parallel path. As shown in Fig. 1, lots of area is wasted in the region for N-type transistors. To address the above two issues, a cell can be designed with multi-row height. In this work, double-row height is used without loss of generality. There are two types of double-row height cells as shown in Fig. 2. One is called *VSS-abut type* and the other is called *VDD-abut type*. Each type has three different N-well topologies. One basic requirement for layout design is that when a double-row height cell abuts any cell, its N-well should join the N-well of its neighbor seamlessly and must not overlap the P-substrate of its neighbor. Note that our cells do not have any

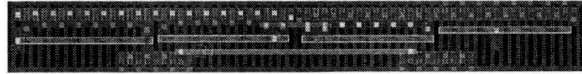

Fig. 1. The layout of NOR4X1 with single-row height. The bottom-half region for N-type tarnsistors are sparsely populated.

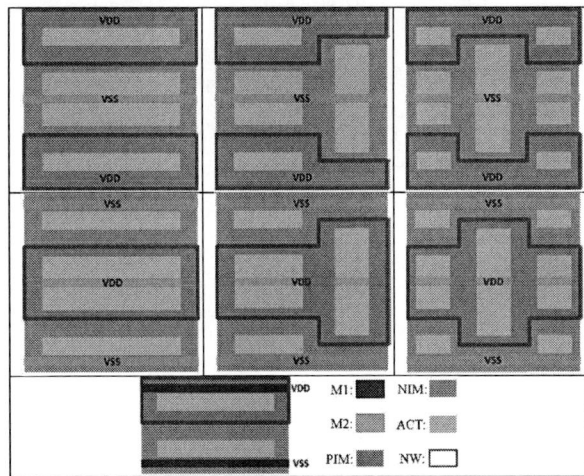

Fig. 2. N-Well topologies: VSS-abut (top row), VDD-abut (middle row), and single-row height.

well or substrate contacts. These contacts should be made through tap cells deployed on cell rows during placement. Note that Fig. 2 also shows the N-well topology for single-row height cells. Obviously, a double-row height cell can be obtained by simply folding a single-row height cell at the middle of the cell. This will reduce cell width approximately to a half, but area wasted in the single-row height cell remains wasted in the corresponding double-row height cell. To tackle this problem, one can make the area underneath the center power/ground (P/G) rail into an active area for implementing large-sized transistors. Cell width can be thus reduced by taking advantage of this active area. Although double-row height enables smaller cell areas, they have their own problems. One problem is that chip routability may be compromised due to difficult placement of double-row height cells.

C. Problem Investigated

The problem investigated in this work is to see how much area saving can be obtained, whether chip routability will be worsened, and whether timing performance will be reduced by employing double-row height cells for chip designs.

III. METHODOLOGY

To address the problem presented in Section II, we first design three new cell libraries based on FreePDK15 with a minimum-sized transistor of two fins. These three libraries are named Mix_2F, DR_2F, and SR_2F, respectively. They are referred here as 2F libraries. TABLE I gives a detailed description of cell height employed by each cell library. In order to study whether double-row height cells play a more important role in a library that uses a larger minimum-sized transistor, we further design three libraries called Mix_4F, DR_4F, and SR_4F. These three libraries use a minimum-sized transistor of four fins. Hence, they are referred here as 4F libraries. Clearly,

TABLE I. CELL HEIGHTS EMPLOYED BY EACH LIBRARY.

Library	Description
SR_2F	single-row height only.
DR_2F	If double-row height could lead to a smaller cell area, use double-row height, otherwise use single-row height.
Mix_2F	Same as DR_2F, but If the difference between single-row height cell area and double-row height cell area is within 10% of single-row height cell area, both the single-row height and double-row height cells will coexist in the library.

a cell in a 4F library will have twice the current drive of its counterpart in a 2F library. Because 4F libraries use larger sized transistors, we expect that DR_4F and Mix_4F will contain more double-row height cells.

Basically, a cell in a 4F library is about twice faster than that its counterpart in a 2F library under the same capacitance load. However, it will also present larger pin capacitance load to a driving cell. This may cancel out the gain of higher current drive and increase power dissipation significantly. To study the larger pin capacitance problem, we further create a library by combining all the cells in DR_2F and DR_4F into one cell library called DR_4/2F. A circuit synthesized using DR_4/2F is supposed to use more high-drive cells on its critical paths and more low-drive cells with less input capacitance on the side paths. It is expected that DR_4/2F will help create a design with even higher timing performance.

Each of the libraries includes AND, NAND, NOR, OR, XNOR, XOR, each of which has two, three, and four inputs. It also includes two flip-flops DFFR(with asynchronous reset), DFFS(asynchronous set). Each of the above logic functions is implemented with a drive strength of 1X, 2X, and 4X. Here, 1X drive strength corresponds to the drive current provided by an inverter made by minimum-sized transistors, i.e., two fins for 2F libraries and four fins for 4F libraries. To provide more choices of drive strength, each library also contains buffers and inverters with drive strengths of 1X, 2X, 4X, 8X, 16X, and 32X. TABLE II shows the numbers of double-row height cells in each library. There are 60 cells in each library except Mix_2F and Mix_4F. In Mix_2F, a logic function such as XOR2X1 can have two different layout designs, one in single-row height and the other in double-row height, if their area difference is within 10% of the single-row height cell's area. These two layouts are counted as two cells even though they have the same drive strength. The above measure is also applied to Mix_4F. One can see that 10 such logic functions in Mix_2F have two layout designs coexistent in the library. There are 11 such logic functions for Mix_4F. So, with regard to the set of cells, DR_2F is a subset of Mix_2F and DR_4F is a subset of Mix_4F. One can also observe that average cell area of SR_4F is much larger than that of DR_4F. Hence, using double-row height cells should be able to effectively reduce circuit area. Note that DR_2F has more than 50% of the cells designed with double-row height. DR_4F has even more. We find that most of the single-row height cells in SR_2F are those with one or two inputs and small drive strength.

The template used for designing single-row height cell is shown on the left of Fig. 3. Its cell height is 7.5 M2 tracks. The M1 P/G rails each take one track. A single poly can provide two

fins for forming a P-type transistor and two fins for forming an N-type transistor. In fact, a single poly could provide up to three fins for each type of transistors. However, so-doing will create pin access and internal signal routing difficulty. The difficulty arises from not allowing a GIL (usually used to bring input pins to M1) to locate on active regions and from having a small number of horizontal tracks. As a consequence, a transistor taking more than 2 fins should be folded. This will certainly increase cell width. Hence, the width of a cell in a 4F library is considerably larger than that of a cell in a 2F library.

As mentioned in II.B, there are two types (i.e., VDD-abut and VSS-abut) of double-row height cells depending on how P/G rails are laid out. In this work, whichever type results in a smaller area is used. Fig. 3 also shows a VDD-abut template for

TABLE II. NUMBER OF DOUBLE-ROW HEIGHT CELLS.

	MIX_2F	DR_2F	SR_2F	Mix_4F	DR_4F	SR_4F
# of single-row cells	34	27	60	15	11	60
# of double-row cells	36	33	0	56	49	0
% of double-row cells	51.4	55	0	78.9	81.7	0
# of VSS-abut cells	25	20	0	38	32	0
Area per cell (um²)	0.538	0.493	0.622	0.708	0.719	1.109

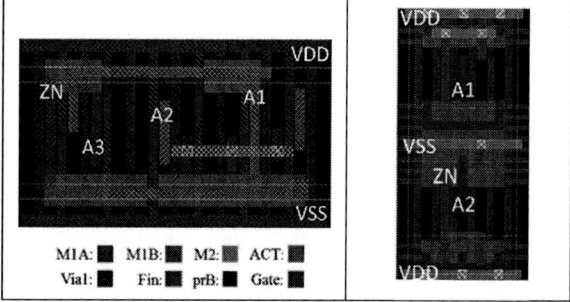

Fig. 3. A template for single-row height cells (left) and a template for VSS-abut double-row height cells (right).

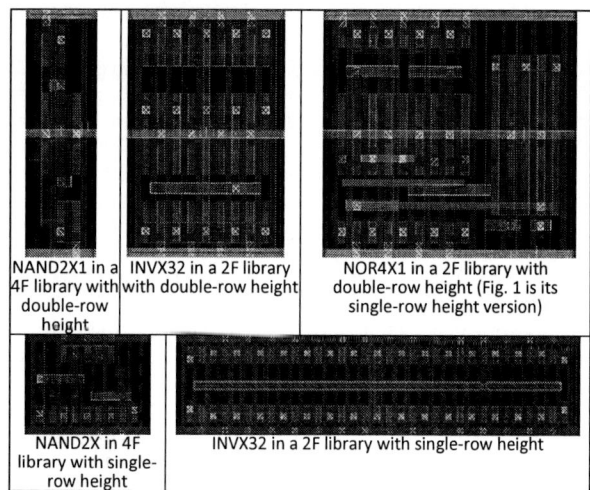

Fig. 4. Single-row height versus double-row height implementations (row height not scaled due to limited column width)

designing double-row height cells. This template can use the region underneath the P/G rail at the center to make a larger transistor up to eight fins. Cell area may thus be reduced significantly. Note that a double-row height cell uses M2 P/G rails rather than M1 P/G rails. The reason for this is that the preferred direction of M2 is horizontal. If the central P/G rail uses M1, the internal signal routing between a terminal on the top row and a terminal on the bottom row need run a vertical M2 wire across the M1 P/G rail. This is against the preferred direction of M2 and will render M2 almost useless for chip-level routing. Fig. 4 shows pairs of cells with the same logic function but implemented respectively with single-row height and double-row height. As one can see here, double-row height effectively addresses the transistor sparsely populated problem for NAND2X1 in a 4F library and NOR4X1 in a 2F library. It also addresses exceedingly large width problem for INVX32 and NOR4X1 in a 2F library.

IV. EXPERIMENTAL RESULTS

In this section, we present our experimental setup that exercises the libraries designed in Section III for chip designs to evaluate the merits of double-row height cells. Design Compiler from Synopsys and SoC Encounter from Cadence are employed to perform logic synthesis and physical design, respectively. Because we do not have a sub-20nm SoC Encounter license, we cannot perform chip routing. To circumvent the license problem, we increase wire width from 28nm to 42nm for a given pitch of 64nm for M1 to M4. So the minimum metal spacing is reduced to 22nm. Accordingly, spacing rules and line-end rules need be also adjusted. The capacitance per unit area, fringing capacitance per unit length, and resistance per square are adjusted too. To consider the surface scattering effect on resistance of a narrow wire, we calculate the resistivity of a narrow wire in the way as it was done in [21]. Five ITC99 [22] circuits and 128/192 AES [23] from OpenCores are used for benchmarking. Logic synthesis is performed for each circuit with the libraries designed in Section III. A circuit is synthesized first without timing constraint so that a minimum-area circuit can be obtained. The circuit is then placed and routed. clock latency is set equal to the targeted clock period for placement and routing. The slew rate at clock sinks and clock skew is 10% of targeted clock period, respectively. After routing, post-layout timing analysis is performed to obtain the longest path delay of the chip. This delay called X will be used later as a basis for specifying tighter clock period constraints for logic synthesis. Power dissipation is calculated by assuming input switching activity of 0.2. As a byproduct, a *wireload* model is created

based on the underlying routed circuit. The so-generated *wireload* model will be used for logic synthesis with tighter timing constraints later.

The first experiment is performed as follows. First, each circuit is synthesized with different targeted clock periods of 1X, 0.5X, and 0.35X. Here, X of a circuit is obtained with Mix_4F in the way described above. It is used as a basis for defining targeted clock periods regardless of which 4F library is used. For example, 0.5X means a targeted clock period is set to 0.5 times X. Similarly, we need obtain a different X for 2F libraries. In this experiment, core utilization is set to 90% for placement and power stripes that take 1% of M3 layers are deployed. M1 through M4 are used for routing. This experiment serves the following purposes:

- To see how many double-row height cells are used in each circuit.
- To evaluate to what extent double-row height cells influence placement and routing.
- To see which cell library can achieve best area, power, and delay products.

As shown in TABLE III, circuits synthesized with a 4F library use much more double-row height cells than they do with a 2F library. The reason for this is that most frequently used cells of a design with a 2F library are still implemented in single-row height. However, a circuit with a 2F library would typically employ more double-row height cells for meeting tighter constraints.

TABLE IV shows total wire length (TWL), number of vias (#via), power consumption, total cell area, longest path delay, and power-delay-area product (PDA) of each circuit with different targeted clock periods. The data of Mix_4F (Mix_2F) and DR_4F (DR_2F) are normalized with respect to that of SR_4F (SR_2F). TABLE V summarizes the results in TABLE IV by averaging their ratios (i.e., averaging normalized data). With regard to total cell area, on average about 20% reduction is observed for 4F libraries whereas on average -2% to 2% (individually up to 6%) reduction is observed for 2F libraries. Inspecting TWL, we find that a circuit designed with Mix_4F/DR_4F is more difficult to route than it is designed with Mix_2F/DR_2F. One reason for this is that the placer tries to optimize placement of a circuit with double-row height cells at the expense of incurring longer wires. This phenomenon is more striking on a design with Mix_4F/DR_4F where the design contains about 80% of double-row height cells. Another reason is that the power/ground rails of a design with double-row height cells are on M2 whereas those of a design with single-row height cells are on M1. With respect to power-delay-area product (PDA) for the designs with 4F libraries, Mix_4F and DR_4F often give considerably better results than SR_4F, on average up to 19% better. However, this does not occur for the designs with 2F libraries except for some cases with B18 and B19.

To further investigate the impact of double-row height cells on chip routability, we increase M3 power stripes from 1% of M3 layer to 2% and repeat the above experiment, i.e., obtaining another 128 layout designs. TABLE VI shows the numbers of DRC errors and shorts in the layouts. As one can see, a circuit designed with Mix_4F/DR_4F under tighter timing constraints has many DRC errors and shorts. However, this phenomenon does not happen to a circuit with Mix_2F/DR_2F (on average

TABLE III. PERCENTAGE OF DOUBLE-ROW HEIGHT CELLS USED IN EACH DESIGN WITH DIFFERENT TARGETED CLOCK PERIODS.

	Mix_4F			DR_4F			Mix_2F			DR_2F		
	1X	0.5X	0.35X	1X	0.5X	0.35X	1X	0.5X	0.35X	1X	0.5X	0.35X
B17	88	79	80	84	78	78	5	8	12	6	7	10
B18	84	79	80	81	78	79	5	5	14	5	6	14
B19	86	79	81	82	77	80	5	6	14	6	6	13
B21	83	77	79	79	76	77	4	9	9	4	10	8
B22	85	74	79	80	77	78	3	11	9	4	9	8
AES	93	92	83	93	92	84	5	5	11	5	5	10
Average	87	80	80	83	80	79	4	8	12	5	7	11

using less than 11% of double-row height cells). The above investigations evidence that a circuit has double-row height cells is less routable than a circuit does not have. Moreover, a circuit with more double-row height cells is also less routable.

We further design an experiment to study area and timing performance tradeoff brought about by DR_2F, DR_4F, and DR_4/2F. Note that DR_4/2F is formed by including all the cells in DR_2F and DR_4F. We synthesize B19, B22, and AES with

TABLE IV. TOTAL WIRE LENGTH, NUMBER OF VIAS, TOTAL CELL AREA, POWER, LONGEST PATH DELAY, AND POWER-DELAY-AREA PRODUCTS.

		Mix_4F			DR_4F			SR_4F			Mix_2F			DR_2F			SR_2F		
		1X	0.5X	0.35X	1X	0.5X	0.35X	1X	0.5X	0.35X	1X	0.5X	0.35X	1X	0.5X	0.35X	1X	0.5X	0.35X
B17	TWL(um)	0.98	1.41	1.55	1.02	1.12	1.32	94669	108690	121843	0.98	0.95	1.22	0.93	0.96	1.01	79516	88667	99023
	#vias	0.80	0.80	0.79	0.83	0.80	0.77	8255	8732	9200	1.04	1.00	0.96	0.98	0.97	0.96	5033	5363	5629
	Area (um²)	0.79	0.80	0.79	0.83	0.80	0.76	8362	9169	9724	1.02	0.97	0.96	0.98	0.97	0.96	5054	5637	5937
	Power (mW)	0.88	0.94	1.09	0.88	0.91	0.95	20.8	40.4	58.1	1.00	1.09	1.11	0.97	1.03	1.11	11.5	21.3	31.9
	Delay (ns)	0.99	1.11	1.73	0.97	1.02	1.16	0.954	0.441	0.320	0.96	1.05	1.04	0.94	0.95	1.13	0.987	0.448	0.335
	PDA	0.69	0.84	1.50	0.71	0.74	0.83	165805	163373	180631	0.98	1.11	1.12	0.90	0.95	1.21	57291	53752	63413
B18	TWL(um)	1.16	1.09	1.45	1.26	1.13	1.07	241651	260016	347222	1.06	1.01	1.07	1.02	1.02	1.06	194583	204667	261562
	#vias	0.87	0.87	1.01	0.91	0.89	0.88	328179	347015	412110	1.06	1.01	1.03	1.02	1.01	1.03	323146	335393	399829
	Area (um²)	0.80	0.81	0.80	0.83	0.82	0.79	22219	23804	27872	1.01	0.98	0.95	0.97	0.96	0.94	13501	14414	17242
	Power (mW)	0.94	0.93	1.02	0.97	0.92	0.91	41.3	82.6	281.0	1.00	0.99	1.06	0.95	1.03	1.01	22.2	42.7	147.6
	Delay (ns)	1.02	0.97	1.40	0.98	1.02	1.04	1.350	0.698	0.456	1.07	0.95	0.92	1.05	0.99	0.90	1.401	0.800	0.514
	PDA	0.77	0.73	1.14	0.79	0.77	0.75	1238477	1372040	3571206	1.08	0.92	0.92	0.97	0.99	0.86	419892	492713	1307616
B19	TWL(um)	1.11	1.51	1.70	1.12	1.38	1.69	459489	480557	619297	1.01	1.07	1.08	0.99	1.02	1.03	377015	397485	490633
	#vias	0.84	0.95	1.09	0.88	0.91	1.08	627127	672565	791296	1.06	1.03	1.05	1.02	1.02	1.02	612148	649316	755196
	Area (um²)	0.79	0.81	0.80	0.83	0.81	0.80	42129	45588	53459	1.01	0.97	0.97	0.97	0.97	0.96	25668	27635	32212
	Power (mW)	0.90	0.97	1.12	0.93	0.95	1.16	82.9	166.9	561.9	1.01	1.07	0.99	0.99	1.01	0.96	43.4	82.6	309.0
	Delay (ns)	1.01	1.48	2.36	0.98	1.09	1.58	1.228	0.652	0.501	1.01	1.00	0.97	0.99	0.97	0.94	1.335	0.709	0.511
	PDA	0.72	1.16	2.10	0.76	0.84	1.47	4288172	4960669	15048283	1.03	1.05	0.94	0.95	0.95	0.87	1486263	1618260	5085685
B21	TWL(um)	1.11	1.41	1.54	1.10	1.52	1.39	29872	43023	50478	1.10	0.78	1.11	1.08	1.12	1.04	25260	39926	35807
	#vias	0.89	0.97	1.03	0.94	0.97	0.97	43574	56718	63795	1.05	0.98	1.09	1.03	1.06	1.05	44723	53806	57058
	Area (um²)	0.83	0.79	0.77	0.88	0.77	0.79	2968	3682	3950	1.01	1.04	1.08	0.99	0.98	0.99	1923	2063	2172
	Power (mW)	1.38	1.17	0.97	1.94	1.10	1.04	8.2	17.8	28.3	1.49	1.64	1.07	1.64	1.34	1.46	3.9	8.9	12.8
	Delay (ns)	0.99	1.05	1.36	0.99	1.04	1.21	1.147	0.576	0.375	0.98	0.96	1.09	0.98	1.05	1.27	1.344	0.669	0.423
	PDA	1.14	0.97	1.02	1.69	0.89	1.00	27879	37656	41867	1.48	1.63	1.26	1.58	1.38	1.83	10080	12327	11793
B22	TWL(um)	1.08	1.35	1.19	1.17	1.32	1.08	45311	57264	87888	1.08	0.87	1.11	1.01	1.05	1.07	37007	50677	56621
	#vias	0.87	0.94	0.93	0.94	0.96	0.93	65006	78363	94876	1.07	0.97	1.07	1.01	1.00	1.04	66090	79965	88861
	Area (um²)	0.80	0.80	0.80	0.83	0.81	0.79	4530	5243	5740	1.00	0.96	0.98	0.96	0.96	0.96	2863	3298	3548
	Power (mW)	0.81	1.03	0.92	0.87	1.11	1.03	13.8	24.4	40.4	0.97	1.08	1.02	1.00	0.78	1.00	7.1	13.7	20.3
	Delay (ns)	0.93	1.02	1.27	0.98	0.97	1.10	1.298	0.619	0.440	1.02	1.03	1.11	1.01	1.04	1.11	1.294	0.626	0.419
	PDA	0.60	0.84	0.94	0.71	0.88	0.90	81381	79106	101916	0.99	1.07	1.11	0.96	0.77	1.07	26151	28250	30123
AES	TWL(um)	0.90	0.89	1.10	0.91	0.91	0.96	64599	67655	85462	1.05	1.05	1.18	1.06	1.05	1.04	48442	49058	60544
	#vias	0.78	0.77	0.95	0.78	0.78	0.89	63518	66201	74533	1.10	1.10	1.25	1.11	1.10	1.14	56852	57120	64743
	Area (um²)	0.77	0.77	0.82	0.78	0.78	0.80	4121	4233	5004	1.05	1.06	1.11	1.05	1.05	1.06	2316	2325	2602
	Power (mW)	1.62	1.27	0.99	1.69	1.01	1.05	11.4	25.9	82.0	0.56	0.96	1.16	0.82	1.03	1.05	10.8	11.6	27.0
	Delay (ns)	0.95	0.93	0.82	0.90	0.95	0.73	0.919	0.509	0.499	1.12	0.95	0.99	1.09	1.02	1.01	1.429	0.783	0.351
	PDA	1.20	0.90	0.66	1.18	0.75	0.62	43178	55903	204838	0.66	0.97	1.29	0.94	1.11	1.13	35588	21144	24647

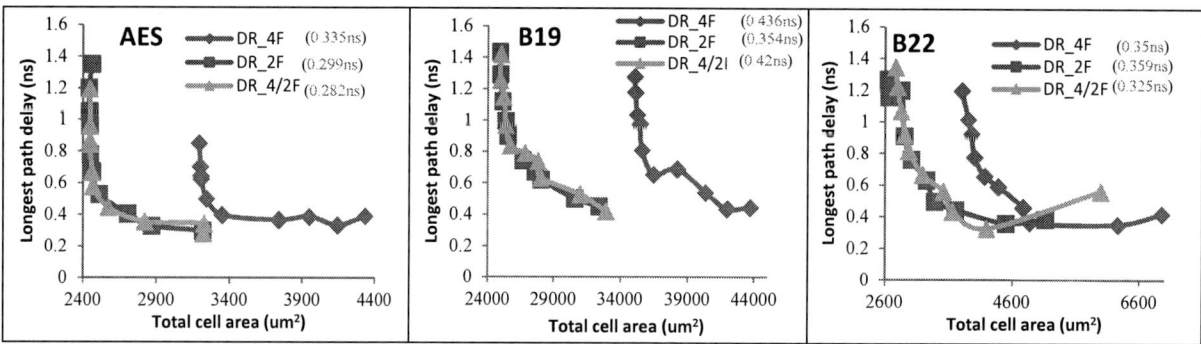

Fig. 5. Delay and area tradeoff for AES, B19, and B22.

TABLE V. AVERAGE RATIOS OF WIRE LENGTH, NUMBER OF VIAS, CELL AREA, POWER-DELAY-AREA PRODUCTS.

	Mix_4F			DR_4F			Mix_2F			DR_2F		
	1X	0.5X	0.35X	1X	0.5X	0.35X	1X	0.5X	0.35X	1X	0.5X	0.35X
TWL	1.06	1.27	1.42	1.10	1.23	1.25	1.05	0.96	1.13	1.02	1.04	1.04
#vias	0.84	0.88	0.97	0.88	0.89	0.92	1.06	1.01	1.07	1.03	1.03	1.04
Area	0.80	0.80	0.80	0.83	0.80	0.79	1.02	1.00	1.01	0.99	0.98	0.98
Power	1.09	1.05	1.02	1.21	1.00	1.02	1.01	1.14	1.07	1.06	1.04	1.10
Delay	0.98	1.09	1.49	0.97	1.02	1.14	1.03	0.99	1.02	1.01	1.00	1.06
PDA	0.85	0.91	1.23	0.97	0.81	0.93	1.04	1.12	1.11	1.05	1.03	1.16

TABLE VI. SHORTS AND DRC ERRORS WITH 2% POWER STRIPES.

		Mix_4F			DR_4F			SR_4F		
		1X	0.5X	0.35X	1X	0.5X	0.35X	1X	0.5X	0.35X
B17	DRC	0	0	41	0	52	42	0	0	0
	Short	0	0	9147	0	7220	7851	0	0	0
B18	DRC	0	1355	298	0	1	88	0	0	0
	Short	0	80816	22403	0	302	27992	0	0	0
B19	DRC	0	10667	14720	0	28483	105	0	0	0
	Short	0	750757	869249	0	1900370	41949	0	0	0
B21	DRC	0	0	18	0	0	0	0	0	0
	Short	0	0	4083	0	2	12	0	0	0
B22	DRC	0	0	25	0	0	0	0	0	0
	Short	0	0	5647	0	0	0	0	0	0
AES	DRC	0	0	14	0	0	144	0	0	0
	Short	0	0	4459	0	0	6634	0	0	0

targeted clock periods of 1X, 0.9X, 0.8X, …, 0.2X, and 0.1X respectively with DR_2F, DR_4F, and DR_4/2F. Note that the values of X are different for DR_2F, DR_4F, and DR_4/2F. Here, we use a core utilization of 85% and 1% of M3 layer for power stripes. We do not set a high core utilization on purpose because we would like to see how much the longest path delay of a circuit can be squeezed at the expense of chip area if designs do not have routability problem. Fig. 5 shows the area and delay tradeoff curves for B19, B22, and AES. The numbers in the parentheses in the legend for each graph gives the smallest longest delays achievable by each library. To our surprise, a pure 4F library DR_4F does not result in best achievable timing performance even though the same circuit designed with it uses a much larger area. One possible reason for this is that the merit of higher driving strength which the cells in DR_4F can provide is offset by larger wire capacitance and higher input capacitance. Another surprise is that the hybrid library DR_4/2F achieves only as good timing performance as DR_2F does. The reason for this is that a circuit synthesized with DR_4/2F has more than 90% of the cells picked from DR_2F. Clearly, the circuit does not employ large drive cells from DR_4F for timing optimization.

V. CONCLUSIONS

This work has presented a method for designing double-row height standard cell libraries. Seven cell libraries are designed. Five of them contain double-row height cells. Extensive experiments are designed to benchmark these libraries. Our experiments show that a double-row height library can achieve on average up to 21% area saving and up to 19% smaller power-delay-area product. The results also show that a double-row height library should not be designed with a large minimum-sized transistor (of 4 fins) which need be folded into two fingers with a single-row height of 7.5 tracks.

REFERENCES

[1] C. Auth, C. Allen, A. Blattner, D. Bergstrom, M. Brazier, M. Bost, et al., "A 22nm high performance and low-power CMOS technology featuring fully-depleted tri-gate transistors, self-aligned contacts and high density MIM capacitors," Symposium on VLSI Technology, pp. 131-132, 2012.

[2] C. G. Auth, A. Aliyarukunju, M. Asoro, D. E. Bergstrom, V. R Bhagwat, J. Birdsall, et al., "A 10nm high performance and low-power CMOS technology featuring 3rd generation FinFET transistors, Self-Aligned Quad Patterning, contact over active gate and cobalt local interconnects," IEDM 2017, pp. 29.1.1-29.1.4.

[3] X. Xu, N. Shah, A. Evans, S. Sinha, B. Cline, and G. Yeric, "Standard cell library design and optimization methodology for ASAP7 PDK," ICCAD, pp. 985-990, 2017.

[4] V. Vashishtha, M. Vangala, and L. T. Clark, "ASAP7 predictive design kit development and cell design technology co-optimization," ICCAD, pp. 978-984, 2017.

[5] H. H. Nguyen, L. L. Chau, T. Pham, and T. D. T. Nguyen, "7-tracks standard cell library," USA Patent, US6938226 B2, 2005.

[6] S. M. Y. Sherazi, et al., "Low track height standard cell design in iN7 using scaling boosters," Proc. SPIE 10148, Design-Process-Technology Co-optimization for Manufacturability XI, 101480Y (4 April 2017).

[7] X. Xu, B. Cline, G. Yeric, B. Yu, and D. Z, Pan, "Self-aligned double patterning aware pin access and standard cell layout co-optimization," IEEE Transactions on CAD, Vol. 34, No. 5, pp. 699-712, May 2015.

[8] J. Seo, J. Jung, S. Kim, and Y. Shin, "Pin accessibility-driven cell layout redesign and placement optimization," DAC, 2017.

[9] Y. C. Chou, J. H. Chen, M. Yang, and S. C. Ein, "A Multiple-row transistor placement system for full custom design," IEEE VLSI-TSA, pp. 136-139, Apr. 2005.

[10] T. R. Gheewala, M. J. Colwell, H. H. Yang, and D. G. Breid, "Dual-height cell with variable width power rail architecture," USA Patent, US7129562 B1, 2006.

[11] C. H. Wang, Y. Y. Wu, J. Chen, Y. W Chang, S. Y. Kuo, W. Zhu, and G. Fan, "An effective legalization algorithm for mixed-cell-height standard cells," ASPDAC, pp. 450-455, 2017.

[12] J. Chen, Z. Zhu, W. Zhu, and Y. W. Chang, "Toward optimal legalization for mixed-cell-height circuit designs," DAC, 2017.

[13] S. Dobre, A. B. Kahng, and J. Li, "Mixed cell-height implementation for improved design quality in advanced nodes," ICCAD, pp. 854-860, 2015.

[14] Y. Y. Chen and Y. W. Chang, "Mixed-cell-height detailed placement considering complex minimum-implant-area constraints," ICCAD, pp. 65-72, 2017.

[15] C. Y. Hung, P. Y. Chou, and W. K. Mak, "Mixed-cell-height standard cell placement legalization," GLSVLSI, pp. 149-154, 2017.

[16] P. Penzes and K. Lampaert, "Mixed-height high speed reduced area cell library," US Patent Application, US2010/0162187 A1.

[17] Y. Lin, B. Yu, X. Xu, J. R. Gao, N. Viswanathan, W. H. Liu, Z. Li, C. J. Alpert, and D. Z. Pan, "MrDP: Multiple-row detailed placement of heterogeneous-sized cells for advanced nodes," TCAD, DOI 10.1109/TCAD.2017.2748025, 2017.

[18] K. Bhanushali, *Design Rule Development for FreePDK15: An Open Source Predictive Process Design Kit for 15nm FinFET Devices*, Master Thesis, Electrical Engineering, North Carolina State University, 2014.

[19] https://www.eda.ncsu.edu/wiki/FreePDK15:Contents

[20] *xCalibrate™ Batch User's Manual*, Version 2015.3, Mentor Graphics.

[21] C. H. Lin, C. S. Chen, Y. H. Chang, Y. T. Zhang, S. R. Fang, S. Santra, and R. B. Lin, "Design space exploration of FinFETs with double fin heights for standard cell library," ISVLSI, pp. 673-678, July 2016.

[22] http://www.cerc.utexas.edu/itc99-benchmarks/bench.html

[23] "128/192 AES," https://opencores.org/project,systemcaes

2018 IEEE Computer Society Annual Symposium on VLSI

A Dual-Threshold Scheme Along with Security Reinforcement for Energy Efficient Nonvolatile Processors

Dongqin Zhou, Keni Qiu, Yuanchao Xu
Information Engineering College
Capital Normal University
Beijing, China

Xin Shi, Yongpan Liu
Department of Electronic Engineering
Tsinghua University
Beijing, China

Abstract—With the increasing scale and decreasing size of the Internet of Things (IoTs) devices, energy harvesting systems have been proposed to power the systems instead of batteries. Addressing the problem that harvested energy is unstable, nonvolatile processors (NVPs) have been proposed to hold intermediate data and avoid frequent program restarting from the beginning. However, NVPs often suffer frequent backup and recovery operations, wasting a lot of energy and system resources. To further improve the performance of NVPs, this paper proposes a dual-threshold method to maximize execution progress by enabling a system to hibernate to wait for power resumption instead of backing up data directly upon power interruptions. In particular, the appropriate retention and backup thresholds are discussed in details in order to achieve the goal of minimizing power failures and maximizing computation progress. In the meantime, the possible attacks to NVPs with dual-threshold and solutions combating these threats are discussed to guarantee NVP's security. The evaluation results show an average of up to 82.3% reduction on power failures and 1.5x speedup on forward progress by the proposed dual-threshold method compared to the conventional single threshold scheme.

Index Terms—NVPs, Dual-threshold, Retention State, Security

I. INTRODUCTION

Recent years have witnessed the fast development of Internet of Things (IoTs) [1]. Implantable and wearable technology enables the devices to keep close contact with users to respond to users' requirements [2]. For these small devices, supplying power is a major hurdle. And battery is no longer a suitable power source due to its large size, large weight and maintenance difficulty. Therefore, energy harvesting systems have been proposed to power the systems instead of batteries due to their potential ultra-long operation time without maintenance [3][4]. Such devices harness energy from the environment or other energy sources and convert it to electrical energy.

Unlike conventional battery powered scenarios, energy harvesting faces a big challenge that ambient energy sources are unstable[5]. The instability of ambient energy makes the traditional systems suffer from frequent data loss and rollbacks, which lowers system efficiency [6]. Addressing this problem, nonvolatile processors (NVPs) have been proposed [7][8][9]. Upon a power failure, the volatile data in NVPs are copied to nonvolatile memory (NVM). Once the power supply resumes, the data stored in NVM are copied back and the system continues normal execution [10]. In this way, the frequent data loss and rollbacks of volatile processors can be avoided. In the

traditional NVPs, the direct backup scheme is established to determine when the volatile data are stored into NVM using a single threshold, namely *backup threshold*. Only when the voltage of capacitor powering the system is lower than the backup threshold, can the system perform backup.

The harvested energy is supposed to be used in forwarding system execution as much as possible instead of backup and recovery. *However, the conventional direct backup strategy has a serious drawback that the frequent data backups and recoveries waste a lot of energy and system resources, slowing the program execution.* On the other hand, we have an interesting observation that the ambient energy voltage often resumes shortly after it drops. This phenomenon hints us that if a system can be retained to wait for power resumption once power interrupts, the program can make more execution progress.

To the end of maximizing execution progress, this paper proposes a more complete *dual-threshold method* to architect NVPs efficiently. In this method, a high threshold, namely *retention threshold*, is set to make the system enter hibernation state. When the power interruption occurs, the capacitor in the system will serves as energy source. If the capacitor voltage drops to the high threshold, the system goes into hibernation state and retains volatile data to wait for power recovery. If the system is restored during hibernation, it can avoid unnecessary backups to make the system run continuously. Furthermore, once the voltage of capacitor reaches *the backup threshold*, the system enters the backup state and stores the volatile data into NVM, which is the same as the conventional systems. In all, the goal of this dual-threshold method is to forward program execution as much as possible by exploiting the retention state. The proposed method is based on the traditional on-board design structure to support the retention state, without need of specific hardware modifications.

In addition, we also consider the security problems of NVPs with dual-threshold scheme. This paper lists two security challenges: avoidance failures of the unnecessary backups due to unreasonable setting and data inconsistency due to incomplete backup. In order to eliminate these threats, we propose strategies in the forms of the following two methods: threshold voltage record and secure checkpointing in this work. In summary, the paper makes the following contributions.

- Addressing the drawback of frequent data backups and recoveries waste in the conventional NVP systems, combin-

978-1-5386-7100-9/18 $31.00 © 2018 IEEE

70

ing with the power trace observation, this paper proposes a dual-threshold method to avoid this waste and maximize execution progress with no hardware modifications.

- The power interruption pattern is analyzed as the reference of hibernation time calculation. The optimal high and low thresholds are determined to reduce power failures and maximize computation progress.
- The evaluations show that the proposed dual-threshold method can achieve 82.3% reduction on power failures and 1.5x on progress forwarding compared to the conventional single threshold scheme.
- The possible attacks to destroy the dual-threshold scheme and induce data inconsistency are discussed for security consideration, the corresponding solutions of threshold voltage record and secure checkpointing are presented.

The rest of the paper is organized as follows. Section II introduces the background and the motivation of this work. Section III presents the proposed dual-threshold method to achieve high computation efficiency. In Section IV, the observations of security threats and the solutions are proposed. Section V shows the experimental results. Finally, conclusions are drawn in Section VI.

II. BACKGROUND AND MOTIVATION

In this section, the details of conventional NVPs are first introduced. Motivated by the observation, the idea of the paper is derived.

A. Background of NVPs with Single Threshold

During the normal execution, NVPs often adopt the *instant backup* strategy. When a power interruption is detected, the capacitor in system is discharging. Then the voltage declines continuously owing to the consumed energy of NVPs exceed the harvested energy. Once the voltage in capacitor is lower than *the backup threshold* voltage, data from volatile memories will be saved to NVM. To guarantee the correctness of backup files, the amount of power reserved in the capacitor should be sufficient to back up the maximum data stack in the program [8]. The system status transitions upon power interruption and resumption of our concerned NVPs can be depicted by Fig. 1.

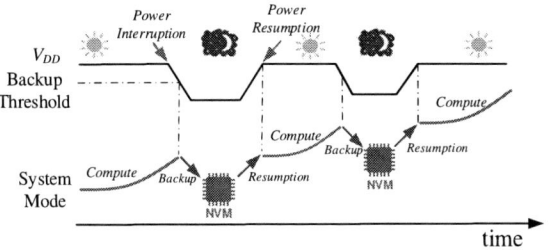

Fig. 1. System status transition of NVPs with single threshold.

B. Retention State Application

In further optimization, retention state has been introduced to NVPs [11]. By exploiting retention state, we can retain the volatile data to wait for power supply instead of backing up data immediately. In this way, there is no need to save volatile data, and computations can be continuous. The retention state

can be very useful for energy harvesting sensors. For example, a tiny wireless sensor called "*nerve dust*" is reported to be implanted into the body to track neural activity, thereby monitoring people's health [12]. It works as a kind of internal and located in deep tissue wearable devices, collecting a large number of data which are difficult to obtain for us. This sensor need not electric wire or battery. It is because that its interior contains piezoelectric crystals which can convert ultrasonic to electricity to power tiny transistors touched neural cell.

These transistors record neural activities and transfer them to receiver outside body using ultrasonic signals. Since the body data are critical for every moment, the continuous working state is highly demanded. Therefore, when the ultrasonic signals are interrupted, the system should retain volatile data to wait for power resumption through the retention state, avoiding abnormal situations that cannot be detected in time during the backup and pause of the system.

However, the retention state in [11] is only supported by specific on-chip hardware design. This paper aims to propose a design to realize the retention function by only imposing very small software modifications on traditional NVPs.

C. Power Trace Observation

IoT devices can be powered by various ambient power [13]. We observed that the energy traces are different in terms of signal strength and intermittency frequency as shown in Fig. 2. They all exhibit a similar phenomenon that the energy voltage often resumes shortly after dropping. The traditional NVPs still choose to continue using system capacitor power when the system is underpowered. In this case, the capacitor voltage will drop rapidly and enter the backup state in a short time, which will likely cause the system to miss the opportunity that could be continued.

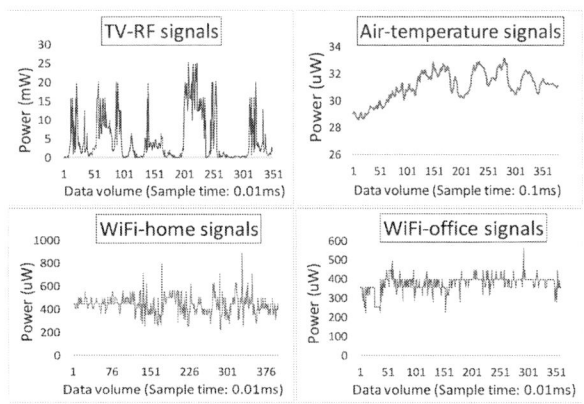

Fig. 2. Power traces of different harvested energy.

D. Motivation

Motivated by the idea of retention state and the observation on power traces, this paper considers a software-based energy management on traditional structure of NVPs. In this paper, we set a high voltage threshold to enable a retention state for the conventional NVPs. Therefore, a dual-threshold method is proposed to assign the working mode of NVPs system. The high threshold and low threshold value, referred as

978-1-5386-7100-9/18 $31.00 © 2018 IEEE

retention threshold and *backup threshold* respectively, control the system's hibernation state and store state. If the capacitance voltage in the system is lower than the *retention threshold*, the system will stop executing and enter hibernation state instead of directly executing to the *backup threshold* for backup. In this way, the limited available energy can be saved to extend the waiting time of power recovery. **If the system can be resumed during hibernation, unnecessary backups and recoveries can be avoided, delivering the benefits from retention state.** Anyway, if the capacitance voltage of the system drops to the *backup threshold*, the system will still enter the backup state to store the volatile data so as to guarantee the next correct recovery.

For the retention state can function normally, the reliability and security of the NVPs also have become a top priority. Therefore, this paper considers some faults appearing by malicious attacks. Such as, the retention threshold and backup threshold are set either too low or too high causing hibernation state cannot avoid unnecessary backups. In addition, the data copied back from NVM also have inconsistency problem if backup operation has been incomplete on checkpoint. Consequently, this paper proposes corresponding solutions.

III. METHODOLOGY

The goal of the dual-threshold method is to maximize program execution progress by efficiently exploiting the retention state. As shown in Fig. 3, the dual thresholds are set in the NVPs to make the system achieve the retention state successfully to avoid data backup, which associated with energy recovery time. If the retention time is greater than the energy recovery time, the system can obtain the energy supply resumption and the program can continue. If the retention time is less than the energy recovery time, the system has to perform a backup operation to store the volatile data into non-volatile memory. Therefore, the retention time calculation is a critical issue in the dual-threshold scheme.

Specifically, we first analyze the power trace to know the energy resumption time and derive the retention time. Then, the high and low voltage thresholds can be determined based on the retention time. These two steps are presented in detail as follows.

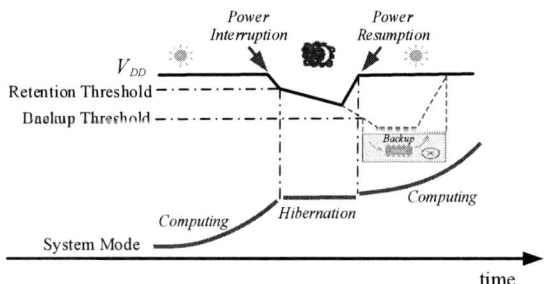

Fig. 3.   System status transition of NVPs with dual thresholds.

A. Power trace analysis

We collected 400,000 WiFi signals at home, 140,000 WiFi signals in office and 1800 TV-RF signals with sampling rate of 0.1s. For air temperature sources, we obtained 9500 signals with ten days of collection. And its sample time is one minute. It is assumed that each signal sends a data package of 1,000 bytes data, such that the sampling rate of every bytes of TV-RF signals is 0.01ms. The pulse sequence of the TV-RF signals can be obtained as shown in Fig. 4. Filtered by the voltage detecting circuit with a specified threshold, the detailed wave profile is indicated by the reading glass. And the threshold is a indication whether the ambient energy can meet the normal execution of the system. It is difficult to tell how soon the power recovers after power interruptions. But the interval of power recover can help us set appropriate sleep time. And the system can wait power resumption during hibernation to avoid actual power failures which will lead to data backups and recoveries.

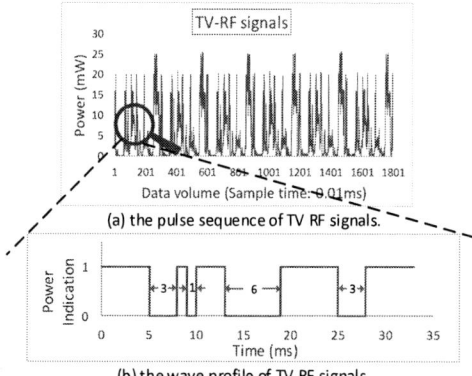

(a) the pulse sequence of TV RF signals.

(b) the wave profile of TV RF signals.

Fig. 4.   the wave profile of TV-RF signals.

B. High and low thresholds setting

Since the high and low voltage thresholds are set as trigger levels for a system to enter retention state and backup state respectively, it is a critical issue how to set these two threshold values.

(1) Low threshold setting There are different sizes of backup stacks at different backup points during program execution. In order to guarantee the success of data backup at any execution point, the system will reserve the maximum backup consumption energy. The energy of data backup E_{backup} is calculated in Equation 1.

$$E_{backup} = S_{backup} * P_{backup} \qquad (1)$$

In Equation 1, S_{backup} denotes the largest data size of all positions which can be obtained using the method in [14] and P_{backup} represents the energy consumption for backing up every byte.

Based on the linear dependence between voltage and energy of the capacitor, we can set the low voltage to make the system enter the store mode. The *backup threshold* voltage V_{low} can be obtained by Equation 2[1] where the parameter C represents capacitance of the capacitor.

---

[1]The amount of energy stored in a capacitor has nothing to do with the length of time, only with its own capacity and the actual voltage between its two pins. The formula is $E = \frac{1}{2} * C * V^2$.

$$V_{low} = \sqrt{\frac{2 * E_{backup}}{C}} \qquad (2)$$

(2) High threshold setting The dual-threshold method is to set the retention threshold before the backup threshold, so that the system can enter hibernation to wait for energy recovery instead of instant backup. Therefore, the high voltage threshold determines how long a system sleeps till entering data backup state. It is a critical parameter to achieve energy failure avoidance. It is assumed that the high threshold voltage is V_{high}. When the capacitance voltage drops to the high threshold, the energy left in the capacitor E_{high} can be obtained by Equation 3.

$$E_{high} = \frac{1}{2} * C * V_{high}^2 \qquad (3)$$

Because the energy for data backup in a capacitor needs to be reserved, the consumption energy during hibernation can be obtained by $E_{sleep}=E_{high}$-E_{backup}. And the time of hibernation T_{sleep} can be calculated by Equation 4 where P_{sleep} is the power consumption when the system sleeps.

$$T_{sleep} = E_{sleep}/P_{sleep} \qquad (4)$$

We can obtain the retention threshold if the hibernation time is known. Since the backup avoidance is decided by comparing the hibernation time and the energy resumption time, the sleeping energy assignment is critical. Here, we can choose different energy resumption time as the system sleep time candidates, such as maximum, minimum or average of the intervals of energy recoveries. Based on these different T_{sleep}, the corresponding retention threshold voltage can be calculated. And the optimal retention threshold will be picked out through evaluating the number of reduction on power failures and the energy consumption. However, there exists a selection range for the sleep time, because the energy left in the capacitor except backup energy reservation is limited. The maximum hibernation time can be calculated by Equation 5.

$$T_{max} = (E_{avail} - E_{backup})/P_{sleep} \qquad (5)$$

Based on the power trace analysis, we pick up the pulse width from a profiling segment to denote the power resumption time. For example the energy resumption pulse widths are 3, 1, 6 and 3 in the signal profiling as shown in Fig. 4(b). Therefore, the maximum 6, the minimum 1 or the uppermost 3 can be as the sleep time candidates to calculate the retention threshold.

IV. Security of NVPs with Dual-Threshold

With the increasing use of NVPs, more researchers focus on non-volatile processor security problems. For example Fig. 5 presents that if one power failure appears after the command 'len++;', this command 'len++;' will be executed once again when energy resuming. Consequently, the value of 'len' has been fault, and the parameter 'x' also be stored in incorrect position. P. Cronin et al. listed several attack vectors and threat models of NVPs suffering likely in their work [15]. 1) If data stored in NVM are not encrypted at the checkpoint, the NVM is likely to be attacked by the attacker, resulting in data leakage. 2) An attacker, aware of checkpoint mechanism of NVPs, could interrupt the power supply in a

precise manner in-between batches, leaving only a portion of the checkpoint completed. At result, this will make the data in NVM inconsistent, leading to potentially major control flow or data faults in the NVPs.

In this work, we also care the security issues of NVPs with the proposed dual-threshold method. In order to ensure NVPs security during execution, we briefly analyze the possible threat models for NVPs with dual-threshold, and put forward corresponding solutions against these attacks.

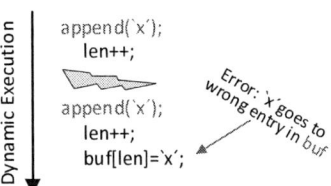

Fig. 5.   Error on data inconsistency.

A. Threat models

(1) Retention state failure As discussed above, when the voltage of capacitor in system drops to the retention threshold, the system goes into hibernation state until the capacitance voltage reaches the backup threshold. In other words, how long the system can sleep to wait for energy recovery depends on the value of the retention threshold and the backup threshold. If the value of retention threshold is too close to backup threshold due to external influences, the hibernation time will be reduced which reduces the probability of energy supply recovery during hibernation. For example, the system may be likely to be attacked by the square wave as shown in Fig. 6 by estimating the ambient energy recovery time. That is, the system is forced to quickly drop/restore the power by controlling the energy supply. Therefore, the energy collected by the system not only increases the frequency of power interruption, but also reduces the recovery time interval compared to the normal one. The time of hibernation selected from energy recovery time will be shorten, which will likely cause the system to miss the opportunity that could be continued. Actually, the retention state fails in this case.

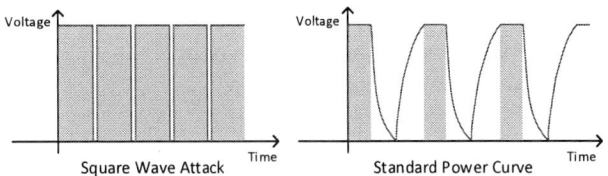

Fig. 6.   Square wave attacks against the NVPs.

(2) Data inconsistency To ensure the completeness of the data backup, we reserve the energy required for the maximum stack size. However, when the circuit is short-circuited or the capacitor is suddenly boosted, the capacitance will be affected to generate instantaneous current and consume a lot of energy in a short time. If the amount of backup data is large at this point, the residual energy in the capacitor will likely be insufficient to complete the backup of all volatile data. However some of the data backed up to non-volatile memory has been already updated. When the system gains energy recovery, these data are copied back from non-volatile memory to continue execution. Consequently, these data are inconsistent with the results of the actual execution of the program, resulting in incorrect execution results.

B. Countermeasures

(1) Dynamic profiling and recording Addressing the retention state failure problem, this paper presents the following two measures to reinforce the security of NVPs. First, we adopt dynamic profiling to estimate the energy resumption time and set the retention threshold, which can prevent the system from getting incorrect sampling due to the square wave-like attack on the initial profiling. Second, the way of maintaining a history record of dynamic retention thresholds is used to detect malicious power supply. If a malicious third party can provide and cut power to the processor at will, the system can discriminate it by comparing it to the threshold record and stop the threshold updating immediately. As a consequence, retention state failures can be avoided because the ill-suited setting for retention threshold can be prevented by this proposed method.

(2) Secure checkpointing Although we have reserved sufficient backup energy to complete the data backup at any backup point, we still use the method of marking system state to ensure the integrity of data backup. We set a flag with the values of 0 and 1 to indicate the system backup begins or finishes respectively. If the capacitance is attacked and the data backup is incomplete, the value of flag is not set to 1. Therefore, we first check the flag value after the system recovery, and then decide whether to read back the data to continue forward. When flag = 1, the successful backup of the system can be continued. When flag = 0 indicating the system's backup fails, and the system needs to rollback to the checkpoint where the data is successfully backed up.

V. EXPERIMENTS

In the experiments, the NVP architecture in [16] is applied. The detailed parameters are shown in Table I. We assume a minimum capacitance for the energy storage C_{bulk}, which must hold enough energy to back up all 128-byte volatile data when receiving energy warning. The benchmarks are from powerstone suite [17], whose program sizes are suitable for application-specific embedded systems. We evaluate power failure, forwarding progress, as well as computation efficiency by comparing our method considering retention state (referred as *dual-th*) to direct backup (referred as *single-th*).

TABLE I
ARCHITECTURAL PARAMETERS.

Parameter	Description
Processor	NVP THU1010N [16]
System frequency	217KHz
Microcontroller	Intel MCS-51
Memory	Registers: Nonvolatile flip-flops On-chip memory: SRAM On-chip NVM: FeRAM

A. Evaluation on power failures

Fig. 7 presents the reduction on power failures with four different high thresholds. It is observed that the number of power failures of *dual-th* can be greatly reduced for all four cases. And the rate of power failure reduction rises as the high threshold increases. Especially for the case where *Th* is 1.33V, the number of power failures of *dual-th* are reduced by 85.2% on average. It is because the system can have more opportunities to resume from hibernation and thus avoiding real power failures.

However, the energy failure reductions for the benchmarks *adpcm* and *blit* do not change with the high threshold of 1.28V. This may be due to the fact that the program reserves more energy for backup, and the sleep time exceeds the maximum hibernation time when *Th* is 1.33V. Actually there is no sufficient time to wait for energy resumption, such that some energy failures cannot be avoided. As a result, we choose 1.28V as the retention threshold, and the energy failure reduction is 82.3% on average.

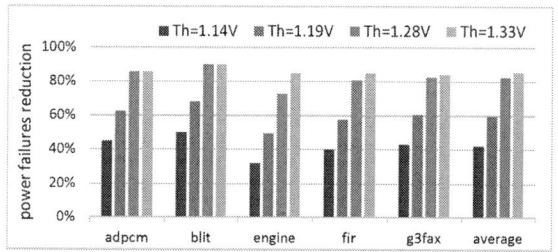

Fig. 7. Reduction on power failures with different high thresholds.

B. Evaluation on forward progress

Hibernation state enables the system directly recovers from power interruption. The program can execute continuously if backup and restore operations are successfully avoided. And this potentially leads to more progress of program execution when compared with the traditional *single-th* strategy [14].

We define the number of instruction steps that are saved by avoiding the backup as the forwarding progress. Since energy consumption during hibernation is very small, we ignore it in the forwarding progress evaluation. The method in [14] is adopted to calculate the average forward execution of each benchmark. This method has known all possible paths of each position by analyzing program, and calculate the number of instructions that continue to execute of the position if an energy warning happens here with the support of energy assigned for instruction execution. Fig. 8 presents the forwarding progress for each test bench, showing an average forwarding steps of 169, indicating that the program can execute 776 more instructions before energy outage with the proposed dual-threshold method while only 507 steps for *single-th*. It achieves a 1.5x improvement in the forwarding progress. This is because the energy consumption is much lower during hibernation state than normal execution so that the system has more time to wait for power recovery. It also avoids unnecessary energy wasting for backup and recovery.

978-1-5386-7100-9/18 $31.00 © 2018 IEEE

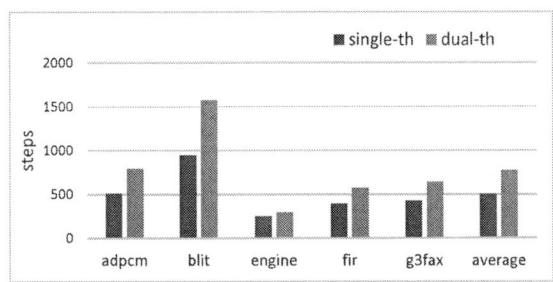

Fig. 8.   Forwarding progress for each benchmark.

C. Evaluation on computation efficiency

When energy is restored during hibernation, data backup and recovery operations can be successfully avoided. This not only saves energy to support program execution, but also increases the overall computation efficiency of the program by avoiding unnecessary time waste. Here the computation efficiency is defined as *the radio of the time of program execution and the time of overall program consumption.*

Fig. 9 presents the evaluation on computation efficiency by comparing *dual-th* with *single-th*. The results show that the computation efficiency of *dual-th* over that of *single-th* is increased by 27% on average, respectively. This is because our method can make system enter hibernation state to wait for power supply, which avoids unnecessary waste for backup and recovery.

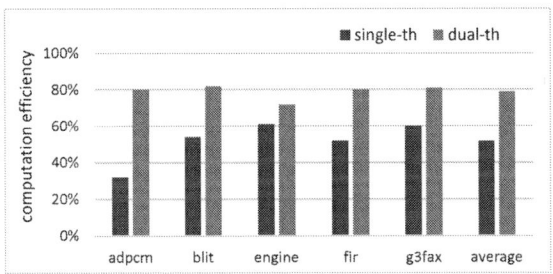

Fig. 9.   Improvement on computation efficiency of each benchmark.

Among all the benchmarks, it is observed that the improvement of *engine* is the smallest. The reason may be that the frequency of power interruption of *engine* is lower than other benchmarks. Therefore, the possibility of avoiding data backup becomes comparatively small. In addition, the computation efficiency improvement of benchmark *blit* is not the highest while its power failures are avoided the most. One reason may account for this observation. That is, the data size of the selected backup position may be larger than other benchmarks, therefore more time is used for data backup and recovery and pause, which lowers computation efficiency improvement.

VI. CONCLUSION

Addressing the drawback of frequent data backups and recoveries waste in the conventional NVP systems, and combining the observation of power traces where the energy voltage can often resume quickly, we propose a dual-threshold method, with which, the system can enter the retention state to wait for power resumption when voltage is below the high threshold. To achieve the goal of maximizing program execution progress, the dual-threshold values are derived in details. In addition, the security problems of NVP with dual-threshold are discussed. Two threat models and solutions against these attacks are proposed. The proposed method delivers significant improvement on program forwarding progress and computation efficiency.

VII. ACKNOWLEDGMENTS

This work is the extended version based on the "Dual-Threshold Directed Execution Progress Maximization for Non-volatile Processors" which was published in ACM International Conference on Computer Frontiers 2018 (CF'18) [18]. It is supported by the grants from Beijing Innovation Center for Future Chip, Beijing Advanced Innovation Center for Imaging Technology, National Natural Science Foundation of China [Project No. 61502321] and the Project of Beijing Municipal Education Commission [Project No. KM201710028016]. The corresponding author is Keni Qiu: qiukn@cnu.edu.cn.

REFERENCES

[1] I. Lee and O. Sokolsky, "Medical cyber physical systems," in *DAC'10*, 2010, pp. 743–748.

[2] M. C. Herrera and J. M. O. et al., "Implantable hemodynamic monitors: Can be conductance catheter system successfully implemented?" in *EMBS'10*, 2010, pp. 3549–3552.

[3] X. Jiang, J. Polastre, and D. Culler, "Perpetual environmentally powered sensor networks," in *IPSN'05*, 2005, pp. 463–468.

[4] S. Sudevalayam and P. Kulkarni, "Energy harvesting sensor nodes: Survey and implications," *IEEE Communications Surveys Tutorials*, vol. 13, no. 3, pp. 443–461, 2011.

[5] A. Kansal and J. H. et al., "Power management in energy harvesting sensor networks," *TECS*, vol. 6, no. 4, 2007.

[6] C. Moser and L. T. et al., "Adaptive power management in energy harvesting systems," in *DATE'07*, 2007, pp. 773–778.

[7] K. Ma and Y. Z. et al., "Architecture exploration for ambient energy harvesting nonvolatile processors," in *HPCA'15*, 2015, pp. 526–537.

[8] F. Su and Y. L. et al., "A ferroelectric nonvolatile processor with 46 μ s system-level wake-up time and 14 μ s sleep time for energy harvesting applications," *TCAS-I'17*, vol. 64, no. 3, pp. 596–607, 2017.

[9] X. Sheng and Y. W. et al., "SPaC: A segment-based parallel compression for backup acceleration in nonvolatile processors," in *DATE'13*, 2013, pp. 865–868.

[10] M. Zwerg and A. B. et al., "An 82ua/mhz microcontroller with embedded FeRAM for energy-harvesting applications," in *ISSCC'11*, 2011, pp. 334–336.

[11] Y. Liu and Z. W. et al., "4.7 a 65nm ReRAM-enabled nonvolatile processor with 6x reduction in restore time and 4x higher clock frequency using adaptive data retention and self-write-termination nonvolatile logic," in *ISSCC'16*, 2016, pp. 84–86.

[12] T. D. Telegraph, "Scientists develop 'nerve dust': monitoring your health in your body," http://www.tech.163.com/16/0909/02/C0G621U300097U81.html, 2016.

[13] X. Li and U. D. H. et al., "RF-powered systems using steep-slope devices," in *NEWCAS'14*, 2014, pp. 73–76.

[14] M. Zhao and C. F. et al., "Stack-size sensitive on-chip memory backup for self-powered nonvolatile processors," *TCAD'17*, vol. 36, no. 11, pp. 1804–1816, 2017.

[15] P. Cronin and C. Y. et al., "'the danger of sleeping', an exploration of security in non-volatile processors," in *AsianHOST'17*, 2017, pp. 121–126.

[16] Y. Wang and Y. L. et al., "A 3us wake-up time nonvolatile processor based on ferroelectric flip-flops," in *ESSCIRC'12*, 2012, pp. 149–152.

[17] J. Scott and L. H. L. et al., "Designing the low-power M*CORE architecture," in *PDMW'98*, 1998, pp. 145–150.

[18] D. Zhou and K. Q. et al., "Dual-threshold directed execution progress maximization for nonvolatile processors," in *CF'18*, 2018.

978-1-5386-7100-9/18 $31.00 © 2018 IEEE

2018 IEEE Computer Society Annual Symposium on VLSI

A Comprehensive Electro-Optical Model for Silicon Photonic Switches

Xuanqi Chen[†], Zhifei Wang[†], Yi-Shing Chang[‡], Jiang Xu[†], Peng Yang[†], Zhehui Wang[†], Luan H.K. Duong[†]

[†]The Hong Kong University of Science and Technology

[‡]Silicon Photonics Product Division, CG/DCG, Intel Corp., USA

Abstract—**Optical networks are revolutionizing computing systems by improving the energy efficiency, bandwidth, and latency of data movements. Silicon photonic switches, such as microresonator and Mach-Zehnder Interferometer (MZI), are the basic building blocks of optical networks. This work proposes a SPICE-compatible electro-optical co-simulation model, BOSIM, to systematically study silicon photonic switches using PN, PIN, and MIS capacitor device technologies. BOSIM holistically models both transient and steady-state properties such as switching speed, power, transmission spectrum, area and carrier distribution. BOSIM is validated by the measured data from eight research groups and companies. Compared to microresonator, BOSIM shows that MZI can provide 1.24X performance for a 128-core multiprocessor using photonic network-on-chip but cost 2.20X energy and 7.3X area.**

I. INTRODUCTION

Optical networks are accepted as a promising candidate in data transmission which possesses high bandwidth, low signal loss and energy consumption [1]. Silicon photonic switches are their basic building blocks and crucial to the network performance. It is essential to study on the network-level impacts of those silicon photonic switches. However, network-level comparisons among various advanced device technologies are difficult and most previous studies ignore the network disparities introduced by silicon photonic switches. Difficulties arise two aspects: i) complexities of the big design space, quick verification and exploration; and ii) the absence of a precise and detailed integration model.

In regard to the first aspect, device-level research stays at the experimental stage and explores silicon photonic switches, such as microresonator (MR) and Mach-Zehnder Interferometer (MZI), of high speed [2–6], high contrast [7], low loss and energy [8], low V_{pp} [9, 10], compacted size [11, 12] and tunable wavelength [13]. These state-of-the-art silicon photonic switches provide guidelines for device optimization and experimental results for system-level reference. The design complexity comes from the multiple layers, cross-disciplines, and large combination possibilities [14, 15]. With sufficient early-stage experiments, silicon photonic switch modeling is necessary for next-stage switch optimization and large-scale network design. Similar to the IC design, silicon photonic switch should specify structure size and doping concentration. The difference is that the features of optical elements are highly related with their geometric parameters. This kind of structure sensitivity brings even worse modeling situations and requires higher simulation accuracy in quick verification and exploration.

In regard to the second aspect, previous modeling work mainly focuses on one type of an optical modulator. Shih *et al.* [16] studied MIS-based MZI modulated by carrier accumulation and clarified the different modeling considerations of optics and electronics. Xu *et al.* [17] targeted an MZI based on a specific interleaved PN diode and validated the applicability of matrix and spectrum analysis on optical modulators. Wu *et al.* [18]

proposed a small-signal model for carrier-injection MRs and highlighted the importance of SPICE compatibility. Their following work [19] validated the correctness of a large–signal model by eye diagram compared to their small–signal model.

To address these problems, we propose BOSIM (Basic Optical Switch Integration Model) for systematic and effective modeling and analysis of silicon photonic switches. In this paper, we mainly have the following contributions:

- A holistic and SPICE-compatible electro-optical co-simulation model, BOSIM, is proposed. Its differences with previous models are summarized in Table I.
- We validate BOSIM with the published data from eight research groups and companies for both transient and steady state.
- Based on BOSIM, we study and compare four typical MZIs and MRs in the scenarios of modulators, filters and switches given for silicon photonic networks.

The rest of the paper is organized as follows. Section II details the modeling process of BOSIM. Its validation is in section III. Section IV compares four oNoC cases based on different silicon photonic switches and section V concludes the work. In this paper, silicon photonic switch is a general concept including the scenarios of modulator, switch and filter.

Table I: Comparison with Related Works

Coverages	[16]	[17]	[18]	BOSIM
Properties				
1: Material				✓
1.1: electrical coefficient				✓
1.2: optical coefficient				✓
2: Electrical Elements		✓		✓
2.1: horizontal PN				✓
2.2: vertical PN				✓
2.3: PIN			✓	✓
2.4: MIS Capacitor	✓			✓
3: Optical Elements				✓
3.1: phase shifter	*	*		✓
3.2: bus waveguide				✓
3.3: directional coupler				✓
3.4: MMI				✓
4: Optical Modulator				✓
4.1: MZI	✓	✓		✓
4.2: MR			✓	✓
5: Dispersion				✓
5.1: chromatic dispersion				✓
5.1.1: material dispersion				✓
5.1.2: waveguide dispersion				✓
5.2: free-carrier plasma dispersion	✓	✓	✓	✓
6: SPICE compatibility			✓	✓
Evaluation Metrics				
1: Latency				✓
2: Power				✓
3: SNR				✓
4: Transmission Spectrum	✓	✓	✓	✓
5: Eye diagram		✓		✓
6: Carrier Distribution	✓			✓
7: Area Consumption				✓
Comparative Studies				
1: MZI vs. MR	✓			✓
2: PN vs. PIN vs. MIS				✓
3: PN-V vs. PN-H				✓

* discussed but not modeled

This work is supported by INTEL11EG01 and PCF.004.16/17.

978-1-5386-7100-9/18 $31.00 © 2018 IEEE

II. BOSIM MODEL

This section introduces BOSIM, including the model overview, the material, the optical elements, the electrical elements and the photonic-electronic circuit.

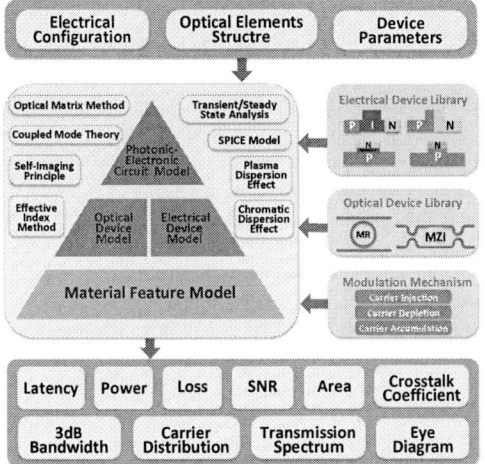

Figure 1: Block diagram of BOSIM.

A. Overview

A holistic optical switch integration model, named BOSIM, is proposed. It promises high accuracy, multiple functions and multiple levels. A system overview is shown in Fig. 1, where the BOSIM analyzer inputs design details, selects related libraries and outputs performance results. The design details include electrical configurations, optical element structures, and device parameters. The input of the electrical configuration involves the driven voltage and circuit parameters, while optical element structure includes the geometry size, film height, etching depth, cladding height and PN interface offset. The device parameters include the doping concentration of the electrical element, its SPICE parameters, and the optical parameters like refractive index and absorption rate.

The BOSIM library contains all the common objects. The electrical objects are a PN diode, PIN diode, and MIS capacitor, while the optical objects are a phase shifter, bus waveguide, directional coupler, and multimode interferometer. These objects can form almost all optical modulators based on the free-carrier plasma dispersion effect. As for the outputs, BOSIM can evaluate latency, power, loss, signal to noise ratio (SNR), transmission spectrum, eye diagram, area consumption and carrier distribution. These aspects are validated by real devices in Section III.

The BOSIM analyzer has three levels: material, device and circuit. In the material level, BOSIM precisely depicts critical physical quantities, such as refractive index and mobility. The device level adopts the classical model for electrical and optical elements with plenty of setup variables, and the circuit level integrates all the components and analyzes the optical modulators with peripheral electrical and optical circuits.

B. Material

Material feature modeling is essential, and we focus on two critical physical quantities of semiconductor materials: electrical mobility and the optical index of refraction.

Electrical mobility defines the moving ability of carriers in semiconductors, which is controlled by scattering and can be affected by acoustic phonons, ionized impurities, and intravalley scattering. The mobility from an acoustic phonon (μ_l) can be formularized from [20], which shows μ_l decreases in temperature, while mobility from ionized impurities (μ_i) can be formularized from [21], which illustrates μ_i decreases with ionized impurity density. Though the total mobility can be combined from μ_l and μ_i using the Matthiessen rule, theoretical defects still exist, like unclear mechanisms of intravalley scattering. Thus, our model uses the experiment fitting curves from [22] to model the mobility of holes and electrons, μ_n and μ_p.

The optical index of refraction is a complex number in the loss medium, affected by two dispersions: free-carrier plasma dispersion and chromatic dispersion. Its real part is the refractive index n, and its image part is the absorption rate α. n and α affect each other based on Kramers-Kronig relations. Free-carrier dispersion refers to the dependence of the index of refraction on the free-carrier concentration. Eq. 1 depicts the changes in the refractive index and absorption caused by the presence of electron concentration N_e and hole concentration N_h [23]. m_{ch} and m_{ce} are the effective mass for holes and electrons. When the concentrations of holes and electrons increase, $\Delta\alpha$ becomes positive, which will cause more propagation loss and worsen data transmission.

$$\begin{aligned} \Delta n &= -(q^2\lambda^2/8\pi c^2\epsilon_0 n)[\Delta N_e/m_{ce}^* + \Delta N_h/m_{ch}^*] \\ \Delta\alpha &= (q^3\lambda^2/4\pi c^3\epsilon_0 n)[\Delta N_e/m_{ce}^*\mu_e + \Delta N_h/m_{ch}^*\mu_h] \end{aligned} \quad (1)$$

Chromatic dispersion represents the dependence of the refractive index on the wavelength, composed of waveguide dispersion and material dispersion. The waveguide dispersion is caused by frequency dependence of the propagation constant of a specific mode. It will be analyzed using the effective index method (EIM), discussed in Section II-D1. The material dispersion can be modeled by the Sellmeier equation in Eq. 2, where coefficients A_i and λ_i for crystalline silicon and silica are measured in experiments. n has a strong negative correlation with wavelength λ, which will influence optical element properties.

$$n^2(\lambda) = \sum_i \frac{A_i}{1-(\lambda_i/\lambda)^2} \quad (2)$$

C. Electrical Elements

1) PN Diode: Carrier depletion is implemented by a reverse-biased PN diode [24] with heavy doping. The carrier distribution of a one-dimensional abrupt junction model is depicted as Eq. 3, where L is the diffusion length; p_n and n_p are the hole and electron concentration in the N region and P region respectively; p_{n0} and n_{p0} are the initial concentrations which are equal to n_i^2/N_D and n_i^2/N_A in the thermal equilibrium; N_A and N_D are the doping density of the acceptors in the P region and donors in N region respectively; W_{Dp} and W_{Dn} are the depletion region width of the P region and N region respectively, related to the driven voltage and doping density; and q is the unit electric charge; k is the Boltzmann constant; and x, V and T are the distance, voltage and temperature. As shown in Eq. 3, the minorities, p_n and n_p, decrease exponentially with the negative V and distance x, and increase exponentially with positive V. Hence, the minorities only impacts the forwarding state of the PN diode with positive V.

$$\begin{aligned} p_n(x,V) &= p_{n0}[exp(\frac{qV}{kT})-1]exp(-\frac{x-W_{Dn}}{L_p}) + p_{n0} \\ n_p(x,V) &= n_{p0}[exp(\frac{qV}{kT})-1]exp(-\frac{x-W_{Dp}}{L_n}) + n_{p0} \end{aligned} \quad (3)$$

We assume majority concentrations remain constant till depletion area and no carriers are distributed in that depletion area. When the driven voltage decreases, depletion area expands.

2) PIN Diode: Carrier injection is implemented by a PIN diode [25] with an intrinsic layer (I-region) sandwiched between a heavily doped P layer and N layer. When the driven voltage increases, the concentration of the minority carriers also increases and ultimately exceeds the doping concentration in the I region, which is defined as high-level injection. Due to the charge neutrality in I region, the concentrations of holes and electrons are equal and can be given by

$$n(x) = p(x) = \frac{\tau_{HL} J_T}{2qL_a} \left[\frac{cosh(x/L_a)}{sinh(d/L_a)} - \frac{sinh(x/L_a)}{2cosh(d/L_a)} \right] \quad (4)$$

where $L_a = \sqrt{D_a \tau_{HL}}$ and the ambipolar diffusion coefficient $D_a = \frac{2D_n D_p}{D_n + D_p}$. τ_{HL} is the minority lifetime of high injection and d is the distance of the I region. The relationship between total current density (J_T) and the total voltage drop (V) for the PIN didoe is given by

$$J_T = \frac{2qD_a n_i}{d} F(\frac{d}{L_a}) e^{qV/2kT} \quad (5)$$

where $F(\frac{d}{L_a}) = \frac{(d/L_a)tanh(d/L_a)}{\sqrt{1 - 0.25tanh^4(d/L_a)}} e^{-qV_M/2kT}$. V_M is the voltage drop across the I region.

On high-level injection, the carrier-carrier scattering effect and the Auger recombination effect are notable. The scattering effect produces a reduction of the mobility for both electrons and holes, while the Auger recombination process reduces the minority lifetime. Both effects are considered in the material feature model. In the I region, the concentrations of holes and electrons are equal and increase exponentially with the driven voltage, also given by Eq. 5. The carrier distribution remains approximately constant with a certain voltage, which is essential for the optical model of a PIN-based phase shifter. Besides this, as P and N regions are heavily doped, the increasing of minority concentrations in these two regions is not comparable to that in the I region.

3) MIS Capacitor: Carrier accumulation is achieved by MIS capacitor [24]. In CMOS manufacturing, polysilicon replaces the metal layer by introducing a self-aligning technology. For silicon photonic switches, the reason for not using metal gate is different because metal has a high absorption rate and will cause more beam dissipation. Thus, we only consider a polysilicon-insulator-silicon capacitor. The solution of the carrier distribution is to get the roots, ψ_{ps} and ψ_{ns}, from Eq. 6 and Eq. 7 with an initial voltage state. ψ_{ps} and ψ_{ns} are the surface potentials of the two layers between the insulator layer, and ε is the electric field intensity. Eq. 6 represents the equal surface charges of the capacitor's two grid plates, and Eq. 7 is the Kirchhoff voltage formula of the MIS capacitor.

$$\epsilon_{poly} \varepsilon(\psi_{ps}) = \epsilon_{silion} \varepsilon(\psi_{ns}) \quad (6)$$
$$V = \psi_{ps} + V_i + \psi_{ns} = \psi_{ps} + d \cdot \varepsilon(\psi_{ns}) + \psi_{ns} \quad (7)$$

where d is the distance between the two grid plates, and $\varepsilon(\psi_{ns})$ is given by

$$\varepsilon^2 = \frac{2q}{\epsilon_s \beta} \{ p_{n0}[exp(-\beta\psi_n) + \beta\psi_n - 1] + n_{n0}[exp(\beta\psi_n) - \beta\psi_n - 1] \} \quad (8)$$

where $\beta = q/kT$, and ϵ_s is the dielectric constant. As Eq. 8 is not a typical nonlinear differential equation, the analytical solution of Eq. 6 and Eq. 7 does not exist. Thus, we use numerical approaches

(specifically, MINPACK's hybrd and hybrj algorithms for solving differential equations). The solution of $\varepsilon(\psi_{ps})$ is similar. Then at a certain driven voltage, ψ_{ps} and ψ_{ns} are determined. ψ_{ps} and ψ_{ns} are the inital condition of the differential equation, Eq. 8 as $\varepsilon = \frac{d\psi}{dx}$. After solving the potential distribution and possion equation, the carrier distributions in the MIS capacitaor can be obtained.

D. Optical Elements

1) Bus Waveguide and Phase shifter: A bus waveguide (channel waveguide) is the primary optical element for beam transmission and confinement. Most waveguides used in a silicon photonic switch are rib waveguides or ridge waveguides. The phase shifter is a particular kind of bus waveguide, which provides a controllable phase shift of the traveling wave and serves as an active region in the optical modulator. This active region is driven by electrical signals and can have different refractive index distributions depending on various modulation mechanisms and driven voltages.

As the phase shifter is the only electro-optic bridge, whose accuracy has a tremendous impact on the modeling, our model details the difference using METRIC [26], an effective refractive index analysis tool. According to different modulation mechanisms, the phase shifter models are divided into three groups in BOSIM: carrier-depletion phase shifter, carrier-injection phase shifter and carrier-accumulation phase shifter.

2) Directional Coupler: A directional coupler consists of two parallel waveguides and its analytical model is based on coupled mode theory. For simplicity, we focus on one standard coupler, a two-channel symmetric directional coupler which is forwarding propagating and works in TE mode. The coupling coefficient κ_c is given by [27]:

$$\tilde{\kappa}_{aa} = \frac{1}{\beta w_E} \cdot \frac{\zeta^2}{2\gamma} (1 - e^{-2\gamma w}) e^{-2\gamma d}, \tilde{\kappa}_{ab} = \frac{2}{\beta w_E} \cdot \frac{\zeta^2 \gamma}{\zeta^2 + \gamma^2} e^{-\gamma d}$$
$$c = \frac{2}{\beta w_E} \frac{\zeta^2}{\zeta^2 + \gamma^2} (d + \frac{e^{-\gamma w}}{\gamma}) e^{-\gamma d}, \kappa_c = \frac{\tilde{\kappa}_{ab} - c^* \tilde{\kappa}_{aa}}{1 - |c|^2} \quad (9)$$

where $\tilde{\kappa}_{aa}$ is the self mode coefficient, $\tilde{\kappa}_{ab}$ is the mode a and mode b coupling coefficient and c is the overlap coefficient. These coefficients can be derived as structure-related formulas: w_E is the effective waveguide width and d is the gap distance between two waveguides. ζ and γ are given by $\zeta = \sqrt{k_1^2 - \beta^2}$ and $\gamma = \sqrt{\beta^2 - k_2^2}$, where β is the wave propogation constant, k_1 is the propogation constant of the transmission layer and k_2 is the propogation constant of the adjacent layers.

As shown in Eq. 9, when the gap distance d increases, all these coefficients, $\tilde{\kappa}_{aa}$, $\tilde{\kappa}_{ab}$ and c, decrease exponentially, which means that all the self mode and inter-mode coupling and overlap effects become weak. The total coupling coefficient κ_c is also reduced by the increasing gap distance. After obtaining κ_c and introducing the third dimention (that is the coupler length L_c), the overall coupling and transmission coefficient are settled by $\kappa = i \cdot sin(\kappa_c L_c) e^{i\beta L_c}$ and $t = cos(\kappa_c L_c) e^{i\beta L_c}$.

3) Multimode Interferometer (MMI): Compared to the directional coupler, MMI is based on self-imaging principle and has a large tolerance to polarization, chromatic dispersion, and fabrication variations. The central structure of a MMI is a multimode waveguide which supports a large number of modes. At its beginning and end, one or two single-mode waveguides are used to launch light into and recover light from the multimode waveguide.

978-1-5386-7100-9/18 $31.00 © 2018 IEEE

As for silicon photonic switches, they are generally 1×2, 2×1 or 2×2 MMI. The MMI modeling uses guided mode propagation analysis [28], which is precise and numerical. Both directional coupler and MMI can serve as the splitter or the combiner for silicon photonic switches.

E. Photonic-Electronic Circuit

The circuit-level model can be divided into an electronic circuit and photonic circuit. In the electronic circuit, we adopt SPICE model which is mature, effective and integrable for large-scale circuit design. The phase shifter is in the $\sim \mu m$ scale, suitable for the classical SPICE model [29]. BOSIM also provides the interface with the Cadence Spectre tool [30]. In the photonic circuit, the ray transfer matrix method is utilized and beam path tracing can be performed by multiplying matrices of optical elements. Two typical silicon photonic switches are discussed further.

An MZI is formed by two arms, one splitter, and one combiner. The splitter and combiner can be a directional coupler or MMI. At least one of two arms should be a phase shifter. Due to the working principle of optical interference, an MZI processes the signal by controlling the phase difference between the two arms. When the output phase difference between two arms is an odd multiple of π or an even multiple of π, destructive interference and constructive interference are generated respectively. The detailed in-out relationship is described as Eq. 10, where M_s and M_c are the matrix of the splitter and combiner, and p_1 and p_2 are the transmission coefficients of the two arms.

$$E_{out} = M_s \begin{bmatrix} p_1 & 0 \\ 0 & p_2 \end{bmatrix} M_c E_{in} \qquad (10)$$

As one typical MR, a microring resonator consists of a looped phase shifter and a directional coupler to access the loop. Due to the working principle of optical resonance, MR processes the signal by switching the resonant state of ring cavity. When the light waves in the loop build up a round trip phase shift that equals the integer times 2π, the waves interfere with itself constructively and the cavity is in resonance. Eq. 11 depicts the relationship of input E_{in} and output E_{out}, where $E_2 = pE_1$ and $L = 2\pi R$ for the modulator scenario.

$$\begin{bmatrix} E_{out} \\ E_2 \end{bmatrix} = \begin{bmatrix} t & \kappa \\ -\kappa^* & t^* \end{bmatrix} \begin{bmatrix} E_{in} \\ E_1 \end{bmatrix} \qquad (11)$$

III. MODEL VALIDATION

The validations cover the transient and steady state of various kinds of MRs and MZIs implemented by different electrical elements and optical elements. As the simulation tools in [16–18] are not publicly available, validations of BOSIM are made with published data from state-of-the-art silicon photonic switches. The simulation profiles come from related publications and some abbreviated values are sourced from [24, 25, 27].

A. Steady State

In the steady state, most of the critical properties like the loss, bandwidth, wavelength shift and crosstalk noise can be obtained and discussed by the transmission spectrum with a significant amount of published data. So in this section, our validation will directly compare the published figures with BOSIM's results. Some representative validations of MZIs and MRs will be introduced.

Fig. 2 is the result of MZI validations. The first verification [4] is MZI with $3 \ mm$ arm length. Compared to the left-side original

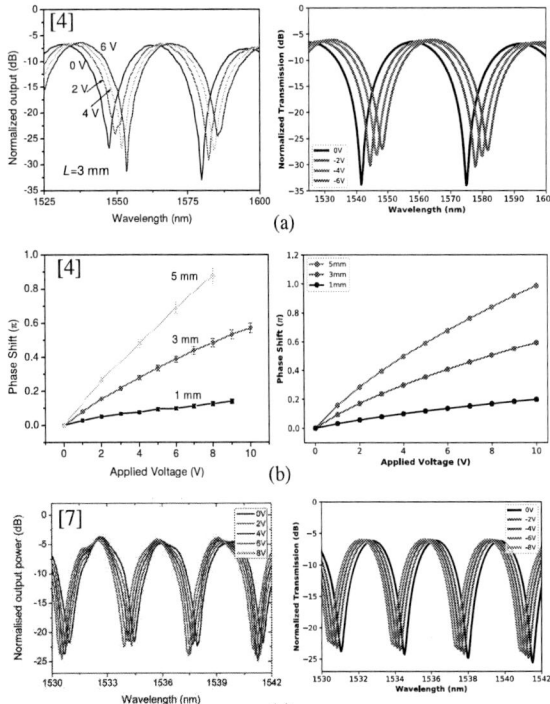

Figure 2: Validation results on MZI transmission spectrum and phase shift vs. driven voltage: (a,b) Intel [4], (c) University of Surrey [7].

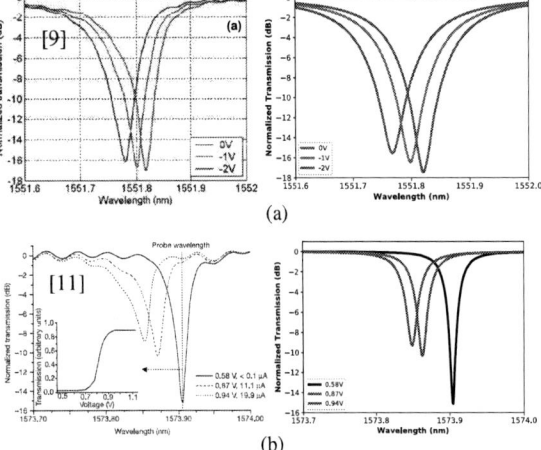

Figure 3: Validation results on MR transmission spectrum: (a) Bell Labs [9], (b) Cornell [11].

figure, the BOSIM spectrum has been validated in passing loss and stopping loss, 3dB bandwidth and the corresponding wavelength at various voltages. Fig. 2(b) shows the relationship of phase shift and applied voltage. The simulation result matches experimental results well. Moreover, the phase shift is proportional to the square root of voltage, which is observed by both experiment and simulation. The second verification [7] is carrier depletion MZI with 1 mm arm length. This detail shows the accuracy of BOSIM.

Fig. 3 is the result of MR validations. The first verification [31] is carrier-depletion MR with 15 μm radius. The second verification

[11] is carrier-injection MR with $6\mu m$ radius. The simulation results also show highly consistent in loss, resonant wavelength and bandwidth with the experiment results. Moreover, the spectrum changing tendencies at different voltages are depicted nicely from the simulation.

B. Transient State

The transient state analysis uses eye diagram and from the eye diagram, the rise time and fall time can be obtained. Based on these measured data from published papers, the transient state validation is shown in Fig. 4. The validation covers various silicon photonic switches in modulator scenario of large bit rate range from different research groups. The simulation results are all within experimental errors of actual value. Besides, BOSIM simulaions satisfy MZI experiments well in a wide arm length range.

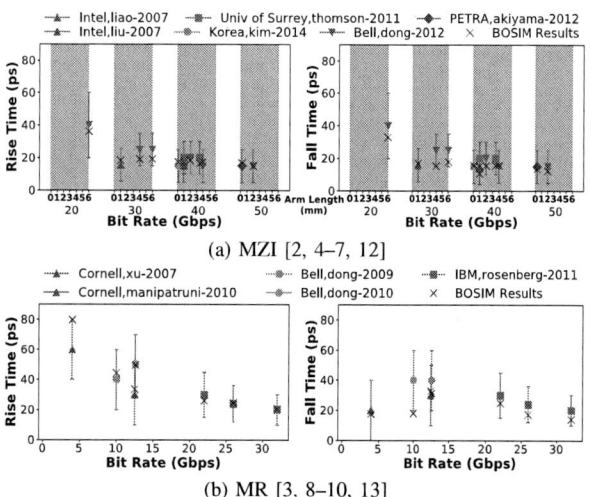

(a) MZI [2, 4–7, 12]

(b) MR [3, 8–10, 13]

Figure 4: Validation results on the transient state.

IV. CASE STUDY

Silicon photonic switch model is crucial for device optimizations and cost-effective network designs. The previous studies on device modelings are limited and lack insights into device impacts on system level. The past network researches are stopped digging into device level and short of silicon photonic switch discussions. In this section, we use BOSIM to analyze four typical silicon photonic switches and evaluate their impacts on a photonic network-on-chip [14] in terms of network performance, energy consumption and area.

A. Simulation Setup

The prototypes and profile parameters of selected silicon photonic switches come from state-of-the-art works [5, 6, 8, 9] and we convert them into scenarios of modulator, switch and filter simulated using BOSIM to minimize passing loss (PL) and insertion loss (IL). The thermal adjustment power is assumed to be $2.4\,mW/nm$. Other profiles remains the same with the prototypes.

The system evaluation is conducted by a cycle accurate simulator JADE [32]. In the oNoC, processor clusters are interconnected by optical waveguides. Each cluster contains four processing cores (32 KB private L1 I/D cache included) and an 128 KB shared L2 cache slice. We assume four wavelengths are used for one waveguide and four parallel waveguides to conduct a communication transaction for fair comparison among the four types of optical devices.

The laser power efficiency is 0.3 and photodetector sensitivity is assumed to be -20 dBm for the energy consumption evaluations.

B. Device-level Evaluation

The device-level simulation results are presented in Fig. 5. Each property is normalized by the largest or the second largest value among four switches. MR-PNH-1 is a compact horizontal PN based MR pursuing low power and compact size. MR-PNH-2 is also formed by horizontal PN diode but targets at high-speed application with 2.3 times longer phase shifter than MR-PNH-1. MZI-PNV is a vertical PN based MZI which promises high extinction ratio in modulator scenario, large 3dB bandwidth in filter scenario and low power but has large footprint. MZI-PIN is formed by PIN diode and sacrifice energy indicators to achieve high speed and compact size.

In the regard to speed, MZI-PIN provides 1.67X the bit rate of MZI-PNH and MR-PNH-2, 2.5X that of MR-PNH-1 in the modulator scenario. Its switch time is also reduced by 57.3% in the switch scenario, compared to MR-PNH-1. The reason is that optical interference based MZI has longer phase shifter, higher extinction ratio, and larger variation tolerance, compared to optical resonance based MR.

The loss and area situations of MZI is worse than those of MR, due to MZI's structure and long length. The PL and IL of MZI in the switch and filter scenario is at least 3.26X PL and 1.86X IL of MR because longer phase shifter brings more propagation loss and MZI has two arms, compared to one ring phase shifter in MR's stucture. Moreover, such ring based structure helps MR to save more footprint than MZI.

Generally, compared to MR, MZI can achieve high speed, large extinction ratio, and large 3dB bandwidth but pay in loss and area viewed in Fig. 5. These tradeoffs express differently in upper level.

C. Network-level Evaluation

The normalized system performance of MR-PNH-1, MR-PNH-2, MZI-PNV and MZI-PIN based networks are as shown in Fig. 6(a). Overall, MZI-based networks can achive higher system performance than MR-based networks. Specifically, MZI-PIN based network can achieve 24.3% and 23.8% performance improvement over MR-PNH-2 based networks in 64-core and 128-core systems, respectively. It is natural to anticipate such performance difference since these networks can provide different bandwidth under similar configurations as evaluated in section IV.B.

In Fig. 6(b), MZI-based networks consume more energy than MR-based networks. The tendencies of modulator, switch and filter power is consistent with their device properties. However, there are hierachical disparities on the power of laser and thermal adjustment. The network of low-power-targeted MZI-PNV costs more laser power while MR-PNH-1 based network consumes more thermal adjustment power. For laser power, this is because MZI-PNV have longest phase shifter (that is, 10X normalized length) so that it has the the worst passing loss and insertion loss in switch and filter scenarios viewed by Fig. 5. Compared to MZI-PNV, MZI-PIN choose PIN diode to increase voltage efficiency and reduce 91.7% phase shifter length, which decreases PL/IL and saves laser power. The thermal adjustment power in MR-based network is huge because 3dB-bandwidth of MR is at least 57.3% less than MZI and the scale of MR is at least 94% less than MZI, which makes it more sensitve to process variation. MR-PNH-1 has

978-1-5386-7100-9/18 $31.00 © 2018 IEEE

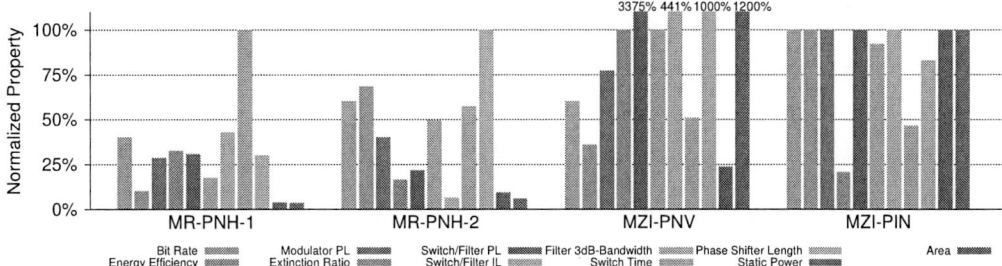

Figure 5: Normalized transient and steady-state properties in the scenarios of modulator, switch and filter.

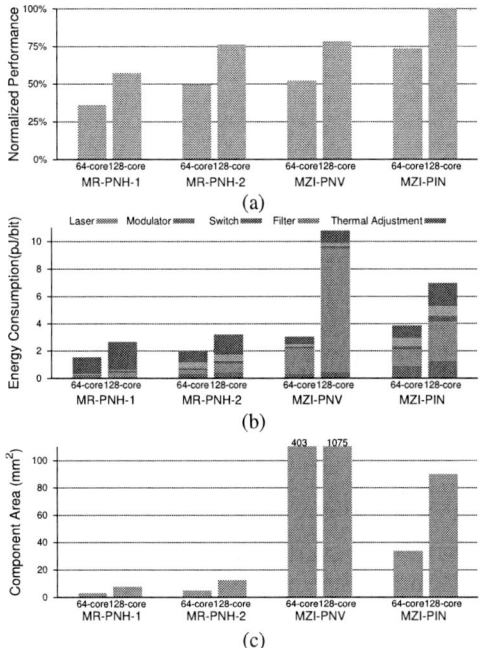

Figure 6: Comparisons among energy consumption, normalized performance and area consumption when the network scale is 64-core and 128-core.

the smallest footprint and narrow 3dB band causing the largeset thermal adjustment power. Totally, MZI-PIN based network costs 1.89X and 2.20X the energy consumption of MR-PNH-2 based network in 64-core and 128-core systems.

The area issue of MZI is also severe. In Fig. 6(c), the compact MZI-PIN based network will consume 7.3 times the component area of high-speed-targeted MR-PNH-2 based network and 12.8 times that of compact MR-PNH-1 based network. The situation for MZI-PNV based network is even worse, which consumes 12X compoenet area of MZI-PNV based network and 87.3X that of MR-PNH-2 based network. Due to the intolerant area consumption, MZI-PNV based network is not suitable for oNoC.

V. CONCLUSIONS

In this work, we propose BOSIM, a basic silicon photonic switch integration model for the optical network simulation. Validations show that BOSIM is accurate in transient and steady state among various types of state-of-the-art silicon photonic switches. BOSIM shows MZI network can provide 1.24X performance improvement to MR network but cost 1.89X and 2.20X energy in 64-core and 128-core systems. MZI consumes at least 7.3X area and area issue

will be critical in PN-based MZI.

REFERENCES

[1] P. Yang et al., "Inter/intra-chip optical interconnection network: Opportunities, challenges, and implementations," in *NOCS*, 2016.
[2] P. Dong et al., "High-speed low-voltage single-drive push-pull silicon Mach-Zehnder modulators," *OE*, 2012.
[3] Q. Xu et al., "12.5 Gbit/s carrier-injection-based silicon micro-ring silicon modulators," *OE*, 2007.
[4] A. Liu et al., "High-speed optical modulation based on carrier depletion in a silicon waveguide," *OE*, 2007.
[5] L. Liao et al., "40 Gbit/s silicon optical modulator for highspeed applications," *EL*, 2007.
[6] S. Akiyama et al., "50-Gb/s silicon modulator using 250-μm-Long phase shifter based-on forward-biased pin diodes," in *GFP*, 2012.
[7] D. J. Thomson et al., "High contrast 40Gbit/s optical modulation in silicon," *OE*, 2011.
[8] J. C. Rosenberg et al., "Low-Power 30 Gbps Silicon Microring Modulator," in *CLEO*, 2011.
[9] P. Dong et al., "Low V_{pp}, ultralow-energy, compact, high-speed silicon electro-optic modulator," *OE*, 2009.
[10] S. Manipatruni et al., "Ultra-low voltage, ultra-small mode volume silicon microring modulator," *OE*, 2010.
[11] Q. Xu et al., "Micrometre-scale silicon electro-optic modulator," *Nature*, 2005.
[12] G. Kim et al., "Compact-sized high-modulation-efficiency silicon Mach–Zehnder modulator based on a vertically dipped depletion junction phase shifter for chip-level integration," *OL*, 2014.
[13] P. Dong et al., "Wavelength-tunable silicon microring modulator," *OE*, 2010.
[14] X. Wu et al., "SUOR: Sectioned Undirectional Optical Ring for Chip Multiprocessor," *JETC*, 2014.
[15] Z. Wang et al., "High-Radix Nonblocking Integrated Optical Switching Fabric for Data Center," *JLT*, 2017.
[16] C. T. Shih et al., "Design and Analysis of Metal-Oxide-Semiconductor-Capacitor Microring Optical Modulator With Solid-Phase-Crystallization Poly-Silicon Gate," *JLT*, 2009.
[17] H. Yu et al., "Demonstration and Characterization of High-Speed Silicon Depletion-Mode Mach-Zehnder Modulators," *JSTQE*, 2014.
[18] R. Wu et al., "Compact models for carrier-injection silicon microring modulators," *OE*, 2015.
[19] R. Wu et al., "Large-Signal Model for Small-Size High-Speed Carrier-Injection Silicon Microring Modulator," in *IPR*, 2016.
[20] J. Bardeen et al., "Deformation Potentials and Mobilities in Non-Polar Crystals," *PR*, 1950.
[21] E. Conwell et al., "Theory of Impurity Scattering in Semiconductors," *PR*, 1950.
[22] C. Bulucea, "Recalculation of Irvin's resistivity curves for diffused layers in silicon using updated bulk resistivity data," *SSE*, 1993.
[23] R. A. Soref et al., "Electrooptical effects in silicon," *QEJ*, 1987.
[24] S. M. Sze et al., *Physics of Semiconductor Devices*. John wiley & sons, 2006.
[25] B. J. Baliga, *Fundamentals of Power Semiconductor Devices*. Springer Science & Business Media, 2010.
[26] A. Ivanova et al., "A variational mode solver for optical waveguides based on quasi-analytical vectorial slab mode expansion," *Arxiv physics*, 2013.
[27] J. Liu, *Photonic Devices*. Cambridge University Press, 2009.
[28] L. B. Soldano et al., "Optical multi-mode interference devices based on self-imaging: Principles and applications," *JLT*, 1995.
[29] J. M. Rabaey et al., *Digital Integrated Circuits*. Prentice hall Englewood Cliffs, 2002.
[30] Cadence Spectre, *Product Version 14.1*, 2014.
[31] P. Dong et al., "High speed silicon microring modulator based on carrier depletion," in *OFC/NFOEC*, 2010.
[32] R. K. V. Maeda et al., "JADE: A Heterogeneous Multiprocessor System Simulation Platform Using Recorded and Statistical Application Models," in *AISTECS*, 2016.

CMOS Gates with Second Function

Jan Nevoral
Faculty of Information Technology
Brno University of Technology
Czech Republic
Email: inevoral@fit.vutbr.cz

Richard Ruzicka
Faculty of Information Technology
Brno University of Technology
Czech Republic
Email: ruzicka@fit.vutbr.cz

Vaclav Simek
Faculty of Information Technology
Brno University of Technology
Czech Republic
Email: simekv@fit.vutbr.cz

Abstract—In this paper, a new approach to design of multi-functional digital circuits is presented. It is based on adoption of polymorphic electronics paradigm which permits digital circuits to exhibit more than one function while preserving the same structure. In that case only components of the circuit (gates) have to be multifunctional. Individual gates have typically built-in sensitivity to the occurrence of some phenomena invoking the function change (e.g. power supply level etc.), which means that no dedicated net is required for that purpose. One of the key advantages of such circuits is the efficiency in terms of size. In this paper, MOS transistors are exploited in an unconventional manner where the circuit function selection depends just on the condition of power supply voltage rails, which is otherwise typical for polymorphic circuits utilizing ambipolar transistors. Furthermore, a first complete set of successfully simulated two-input polymorphic gates was obtained. These gates show the best parameters of all the previously published polymorphic gates – high input impedance and low output impedance, short time of signal propagation, low power consumption and low transistor count being used. Wide range of proposed polymorphic gates (function combinations) may help to obtain more efficient results during synthesis.

Index Terms—Polymorphic electronics, MOSFET, polymorphic gate, digital circuit, gate set.

I. INTRODUCTION

Nowadays, significant majority of logic circuits are typically realised by means of using widespread CMOS technology. Although the first physical implementation of CMOS-based logic circuit has been successfully demonstrated more than 55 years ago [1] and a plenty of research activities have been carried out in the meantime, this technology still prevails as the major choice even in case of the cutting-edge electronic devices. In fact, last several decades have brought only a minor advancement regarding the structural and implementation aspects of basic Boolean functions (logic gates) in CMOS technology. When it comes to the target implementation size of logic circuitry, CMOS gates seem to be very optimal (only few transistors are needed to implement a Boolean function), thus, even very complex behaviour can be realised on a small silicon die. In order to accomplish higher functional density (more features within a similar chip area or amount of transistors), the reconfiguration scheme is sometimes used.

Another idea, how to increase functional density of a digital circuit, was proposed by the team of Adrian Stoica from NASA JPL [2]. This concept is called Polymorphic Electronics and its foundation is based on changing the function of circuit elements while their structure (and mutual interconnection) are preserved. This is in contrary to the principles of classic reconfiguration, where the change of circuit structure is the key for invoking another function. Ability of circuit function changing always incurs additional overhead. But in general, if the overhead is smaller than a cost of two (or more) circuits representing the intended functions, the resulting functional density is higher.

In case of classic reconfiguration, the associated overhead becomes evident in both space and time domain [3]. Polymorphic circuits typically have no time overhead (the function change is immediate) [4] and, moreover, overhead in the size domain is mostly smaller than for a classic reconfiguration. This is given by the fact that structure comprising a lot of switches or multiplexers is typically needed for reconfiguration purposes, as well as corresponding memory resources for storing the configurations. Neither additional memory nor dedicated reconfigurable structure pattern is needed for polymorphic electronics. If the synthesis of a polymorphic circuit is performed cleverly, the cost paid for an additional function could be very low [5].

Utilization of the polymorphic electronics concept as an approach how to design circuits with increased functional density, in comparison to conventional CMOS-oriented techniques, strongly depends on satisfying two principal conditions. First is given by the existence of suitable synthesis methods which are able to found one highly optimized structure for all intended functions. There have been recently proposed several approaches [6] – [7] how to surmount this obstacle.

Second condition is given by the availability of an efficient implementation of polymorphic gates, which should be able to change their functions in accordance to some well-defined and well-controllable global circumstances. However, the change of function for all gates in the circuit should require only minimal overhead when it comes to the propagation of such stimulus. Ideally, no dedicated global net is used for distributing the information about function change to the individual gates. It could be fulfilled for example by means of using power rail as a channel for function-change information distribution. In this case, polymorphic gates with the ability to change their function in accordance to the state of power supply voltage, should be employed [8]. Typically, polymorphic gates exhibit different functions for different levels of supply voltage.

One of the main problems connected with the practical

978-1-5386-7100-9/18 $31.00 © 2018 IEEE

adoption of this approach is given by the analogue nature of the change itself. Such evidence subsequently leads to rather analogue design concepts projected into the polymorphic gates rather than staying in purely digital domain. The result is that some transistors work in a linear mode or short power supply. In this paper, another approach to the design of polymorphic gates controlled by the condition of power supply rail is proposed. This novel approach simply avoids the previously outlined pitfalls of analogue behaviour embedded into formerly proposed polymorphic gates because all transistors are either closed or in saturation and the nature of gates, as well as the indication of function change sent along the power supply rail, is handled in purely discrete way.

The paper is organized as follows: Section II introduces the field of polymorphic electronics and various aspects related to the design of polymorphic gates. New approach to multifunctional digital circuits is proposed in Section III together with a brief description of an evolutionary method used for design of new gates. In Section IV, set of designed polymorfic gates is presented. Finally, Section V provides a conclusion of this paper.

II. POLYMORPHIC ELECTRONICS

Polymorphic circuit represents specific type of a digital circuit that exhibits the natural ability to operate in two or more intentionally specified modes, when different (logic) function is performed in each of them. The structure (interconnections between individual components in such circuit) remains unchanged for all the permissible modes/functions. It is important to emphasize at this point that the multifunctionality itself is not achieved by means of exploiting the reconfiguration scheme for the actual circuit structure, but instead of that the change of behaviour (logic function performed) takes place for some components.

The swap of the function may be arranged in various ways. It could be advantageous (and somewhat typical scenario for polymorphic electronics) if each component demonstrates its sensitivity to a specific phenomenon responsible for the function change. It turns out to be very useful if such phenomenon is inherently present in the circuit (like supply voltage level, temperature etc.). Then, the circuit changes its function in accordance with those circumstances and no special signal (distributed by a global interconnection network) is needed to control this process. Such arrangement helps to keep the overhead paid for "extra" functions low.

The mechanism described above, responsible for the change of polymorphic circuit function, implies the existence of some application domains in which such kind of behaviour may help to get significant advantages. In general, these circuits could be used e.g. for:

- adaptation of the circuit behaviour to variable conditions with the target environment,
- reduction of power consumption or heat dissipation in order to preserve at least the essential circuit functionality, when energy goes low or temperature goes high,

- storing or hiding "extra" functions used as a watermark or other special features [2],
- embedded diagnostics purposes.

It is possible to recognize one important aspect shared across the various application areas of polymorphic electronics suggested above: There typically exists so called primary function, which needs to be implemented first, and another "additional" or "extra" function that might become desirable in some situations or under specific circumstances only. Although the concept of polymorphic electronics generally enables the possibility to have several equally important functions implemented in one circuit, just one main and one or more auxiliary, temporary, supplementary or emergency function are implemented due to practical reasons. In those cases, little bit more complicated or unusual manner of function change doesn't bring any major restrictions.

A. Polymorphic gates

Structure of a polymorphic circuit is typically studied on the gate level of abstraction. Just the gates themselves make the obvious difference between conventional logic and polymorphic circuits. Two main problems of the polymorphic electronics are:

1) The problem of design (synthesis) methods for polymorphic circuits, i.e. how to map a description in the behavioural domain to a description in the structural domain. A lot of polymorphic circuits were designed using evolutionary-based methods, especially those involving Cartesian Genetic Programming [9]. But also non-evolutionary (conventional) design methods were proposed [7].
2) Search for suitable polymorphic components (gates). Implementation of the gate should be efficient (in terms of occupied chip area or transistor count).

The emergence of new gates suitable for polymorphic electronics have been reported very rarely during last years. Most of them are based on unconventional devices or technologies like organic ambipolar transistors [10], nanoelectronic devices [11] etc. These devices allow to realise the changes in behaviour of the circuit at low levels within the design hierarchy – mostly at physical level.

The reason, why multifunctional or polymorphic gates based on CMOS were overlooked, lies in the fact that conventional MOS transistors, used as switches, are very stable devices. Thus, an attempt to develop a multifunctional gate using silicon MOS transistors mostly leads to a complicated structure with inferior parameters. On the contrary, new ambipolar devices hold a significant promise of developing new efficient and highly-refined gates with potentially very good electrical parameters [11]. The research of polymorphic gates utilising ambipolarity of transistors as the way how to switch between two function exhibited by the gate enables also development of new gates based on conventional silicon MOS transistors as described further in this paper.

Rich set of findings and observations gathered throughout numerous experiments with published polymorphic gates, as well as with in-house design of several novel polymorphic gates and practical experiments with applications of digital polymorphic electronics, have allowed to formulate properties and criteria which should be fulfilled by polymorphic gate:

1) high input resistance (the gate must avoid the excessive overloading of previous logic net generated by its inputs),
2) low output resistance (the aim is to achieve unambiguously defined logic levels at the gate and let them closely follow, if possible, the potential of supply source irrespective of the actual output load caused by subsequent logic net, whereby maintaining reasonable circuit noise immunity),
3) short time of signal propagation – definition based on measurement of transfer path delay between input and output ports (t_{pd}),
4) low power consumption,
5) small dimensions.

III. APPROACH TO THE CIRCUIT DESIGN

Polymorphic gates, as proposed in this paper, are implemented by means of using conventional MOS transistors, when the circuit behaviour depends on the polarity of supply voltage. More accurately, the second circuit function is available when the electric potentials on the power supply rails are swapped.

In order to design new polymorphic gates, evolutionary approach proposed earlier in [12] was further refined and used in the context of this contribution. To achieve a fast simulation with reasonable accuracy, a discrete simulator with switch-level transistor model extended by a threshold drop degradation effect corresponding to $45nm$ technology is utilized. Moreover, this approach is capable of designing circuits with high input impedance and low output impedance. The resulting circuits are optimized with regard to the overall number of transistors being used.

Depending on the overall chip (or application circuit) design and usage of the second circuit function, whole circuit may consist of polymorphic gates only or it can exploit the combination of conventional and polymorphic parts. One of the key prerequisites for implementation of any Boolean function is the availability of a Boolean constant in the circuit. Normally, this requirement is fulfilled implicitly by the presence of power supply rails because their upper and lower threshold voltage levels equals to Boolean values. But in the situation when voltages on power supply rails could be swapped mutually, what is the constant?

Gates proposed in this paper need to have a Boolean constant connected somewhere inside the structure or on some input. For example, discrete switch-level transistor model used in the design approach assumes valid connection of transistor substrates – n-MOS to the negative power supply (gnd), p-MOS to the positive one. Therefore, although two supply wires are enough for the circuit in this type of polymorphic electronics, the voltage needs to be rectified internally. The actual way of rectification is usually application- and chip-specific. A generic solution can be accomplished by taking an

advantage of the P-N transitions in the chip silicon, e.g. using one n-MOS and one p-MOS structure.

A. Circuit representation

In order to evolve polymorphic circuits at the transistor level a suitable representation enabling to encode bidirectional graph structures containing junctions is needed. We utilized a Cartesian genetic programming (CGP) proposed by J. Miller [9].

The circuit representation is derived from the CGP representation of gate-level circuit. Each polymorphic digital circuit is represented using an array of nodes and can be encoded by fixed length array of integers. Each node consists of two source terminals and one output terminal and can act as p-MOS transistor, n-MOS transistor or a junction. The utilized nodes are shown in Figure 1. Source terminals of each node can be independently connected to the output terminal of any node placed in previous columns or to one of the primary circuit inputs.

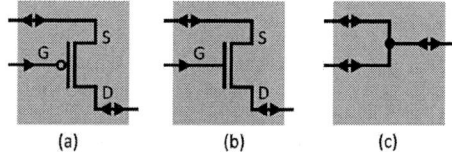

Fig. 1. CGP nodes of transistor-level circuits: (a) p-MOS transistor, (b) n-MOS transistor and (c) junction that combines three signals together. The arrows denote the possible directions of signal flow which have to be considered during the circuit evaluation.

The junction nodes combine two input signals and one output signal together. As a consequence of that, loops and multiple connections are natively supported.

Figure 2 demonstrates the principle of utilized encoding on a $in.0/1$ pass-logic polymorphic gate which provides identity of the only input signal as a first polymorphic function ($pwr.0 = gnd$, $pwr.1 = v_{dd}$) and a logic one as the other polymorphic function ($pwr.0 = v_{dd}$, $pwr.1 = gnd$). The shown chromosome encodes a candidate circuit using six nodes. However, only five of them contribute to the phenotype and are active.

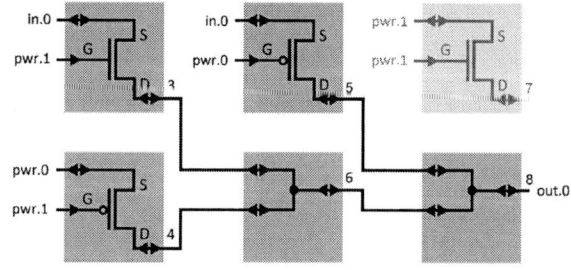

Fig. 2. Example of an encoded candidate circuit implementing $in.0/1$ pass-logic polymorphic gate using three transistors. The circuit has one input (in.0), one output signal (out.0) and two power rails (pwr.0, pwr.1).

B. Evaluation of the candidate solutions

The goal of the evaluation is to determine whether the candidate circuit works correctly with regard to its functional description and whether the circuit meets other specified constrains. The evaluation of the candidate solution consists of two steps.

Firstly, set of active nodes is determined. Only the active nodes (nodes in a path from input nodes to outputs) represent the evaluated circuit. Inactive nodes are ignored. This operation is performed due to speed-up of the circuit simulation and because of skipping the short circuits in the unused part of the circuit. Then, multi-level discrete event-driven simulator is utilized to determine the circuit outputs for each input combination.

Quality of the candidate solution is determined by the *fitness function*, which compares the circuit outputs obtained from discrete simulation with the expected circuit outputs. If a short-circuit is detected for some input combination or a predefined number of simulation steps is reached, a penalty is subtracted from the total fitness value.

C. Search strategy

As a search algorithm, $1 + \lambda$ evolutionary strategy is utilized [9]. The initial population is generated in a random manner. Every new population consists of the best individual and λ offspring created using a point mutation operator which modifies randomly selected genes. The evolution is terminated when a predefined number of generations is reached.

As soon as a fully working solution is found, additional requirements (e.g. high input impedance) for circuit properties are checked. As soon as a fully working solution with all the desired properties is found, the circuit evolution continues in order to reduce the number of transistors being used.

Our simulations have revealed that some circuits designed by the original approach [12] do not have the intended behaviour. As an example, figure 3 depicts simulation results of four different $\overline{in_0}/xnor$ gates evolved by the mentioned approach, where the outputs of three such wrongly designed gates are shown.

Detailed analysis of the evolved circuits exposed an issue in the transistor model used in the discrete simulation. Specifically, the inaccuracy occurs when a discrete value denoted as 'Z' (high impedance) is present at transistor gate terminals for some combinations at the circuit input. Discrete simulation considered such a transistor as an open switch. However, the HSPICE simulation detected several transistors closed, probably because of charge remaining at their gates. In general, this issue can be fixed in different ways:

- More accurate discrete simulation with more complex transistor model can be implemented.
- Fast discrete simulation can be combined with the accurate HSPICE simulation done once per several thousands of generations.
- New constraint can be added to the circuit requirements, which ensures the 'Z' level will never be present at any transistor gate terminal at the end of discrete simulation.

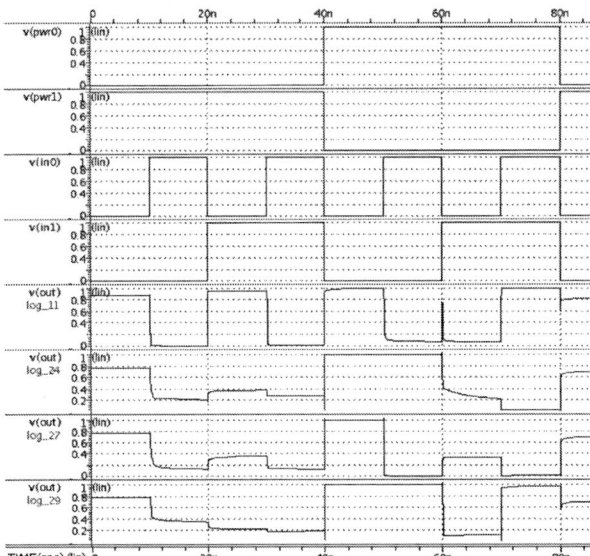

Fig. 3. HSPICE simulation results of four $\overline{in_0}/xnor$ polymorphic gates evolved by the approach presented in [12]. *log_24*, *log_27* and *log_29* circuits do not meet the expected behaviour.

As the first two approaches would extend the evolution time and/or increase required computing power, the third approach was utilized instead to produce valid polymorphic circuits in the context of this contribution.

D. Evolutionary optimization

The success rate of the evolutionary design decreases with growing number of constraints imposed on the gates. Figure 4 shows the success rate of 88 two-input polymorphic gates designed with the following evolution setup: CGP matrix size: 15x2, $\lambda = 4$, mutation: $3 \simeq 4\%$, number of generations: 2,000,000. With high input impedance constrain and prohibition of 'Z' level at transistor gates, the evolutionary approach was not able to design at about 14% circuits even after 1000 runs. Let us note that if any gate fulfilling all the requirements is found, the transistor count optimization is usually done just in tens of thousands generations.

To achieve a higher success rate of the gates evolved in Section IV, there was proposed the usage of an evolutionary optimization, which skips the difficult and, in a lot of cases, unsuccessful part of the evolutionary design. Before the evolutionary algorithm starts, one of the circuit candidates encodes the following circuit: Each polymorphic function is implemented by a conventional (non-polymorphic) CMOS circuit. These two circuits representing both polymorphic functions are switched by a multiplexer according to voltages on the power rails. Rest of the CGP elements encode just meaningless junctions.

As the encoded circuit meets both high input impedance and no 'Z' level at the transistor gates constrains, described improvements ensures 100% success rate. Let us note that the presented success rate is intended as a success rate of finding

TABLE I

SIZE OF THE SMALLEST FOUND SOLUTION OF MOSFET POLYMORPHIC GATES f_A/f_B WITH HIGH INPUT AND LOW OUTPUT IMPEDANCE

f_A		0	$\overline{a+b}$ nor	$\overline{a}b$	$\overline{a}$ not	$a\overline{b}$	$\overline{b}$ not	$a\oplus b$ xor	$\overline{ab}$ nand	ab and	$\overline{a\oplus b}$ xnor	b	$\overline{a}+b$	a	$a+\overline{b}$	$a+b$ or	1
0		0	5	5	3	5	3	7	5	7	8	4	6	4	6	6	0
$\overline{a+b}$	nor	5	4	6	6	6	6	9	8	9	7	6	4	6	4	7	5
$\overline{a}b$		5	6	4	6	8	6	6	4	7	9	6	7	6	7	6	6
$\overline{a}$	not	3	6	6	2	6	4	7	6	7	7	6	6	4	6	7	3
$a\overline{b}$		5	6	8	6	4	6	6	4	7	9	6	7	6	7	6	6
$\overline{b}$	not	3	6	6	4	6	2	7	6	7	7	4	6	6	6	7	3
$a\oplus b$	xor	7	9	6	7	6	7	6	7	9	9	8	9	8	9	8	8
$\overline{ab}$	nand	5	8	4	6	4	6	7	4	7	9	6	6	6	6	9	5
ab	and	7	9	7	7	7	7	9	7	6	8	6	6	6	6	8	6
$\overline{a\oplus b}$	xnor	8	7	9	7	9	7	9	9	8	6	8	6	8	6	9	7
b		4	6	6	6	6	4	8	6	6	8	4	6	6	6	6	4
$\overline{a}+b$		6	4	7	6	7	6	9	6	6	6	6	4	6	8	7	5
a		4	6	6	4	6	6	8	6	6	8	6	6	4	6	6	4
$a+\overline{b}$		6	4	7	6	7	6	9	6	6	6	6	8	6	4	7	5
$a+b$	or	6	7	6	7	6	7	8	9	8	9	6	7	6	7	6	7
1		0	5	6	3	6	3	8	5	6	7	4	5	4	5	7	0

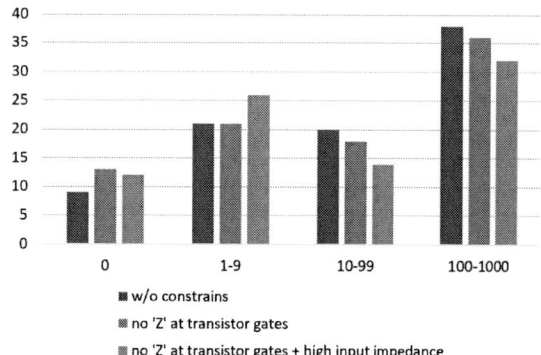

Fig. 4. Success rate of polymorphic gates evolution design for different constraints imposed on the gates. X axis – number of successful evolution runs (out of 1000). Y axis – number of gates (88 total).

w/o constrains
no 'Z' at transistor gates
no 'Z' at transistor gates + high input impedance

any gate with desired requirements, not success rate of finding the optimal gates (in number of transistors). Naturally, the second one is lower.

IV. DESIGNED POLYMORPHIC GATE SET

Main objective was to design a set of all two-input polymorphic gates with high input impedance, low output impedance and minimal number of transistors being used. The full set contains 256 two-input gates. However, large number of gates that belong into this set are similar and it is not necessary to explicitly design all of them.

Two circuits belong to the same P-class (are P-equivalent) if one of them can be derived from the other by permuting its signal inputs [13]. Without loss of generality it is sufficient to design only one polymorphic gate from each P-class. Other gates would be created just by permuting (i.e. renaming) signal inputs of the designed ones. Similarly, as it can be concluded from the principle of polymorphic electronics where

the behaviour of functional elements is controlled through swapping the power rail voltages, f_x/f_y polymorphic gate could be easily created from the f_y/f_x gate just by renaming its power supply inputs. Therefore, there is only 88 unique two-input gates.

The results were gathered from more than 5000 independent runs of evolutionary method for each polymorphic gate, as it is described in Section III. Both approaches comprising evolutionary design (sometimes produces more optimal circuit) and evolutionary optimization (each run produces a working circuit) were combined in order to get the most optimal gates. The results are based on the following experimental setup: CGP matrix size: 2x15 (design) or 1x50 (optimization), $\lambda = 4$, mutation: $3 \simeq 4\%$, number of generations: 2,000,000. Both high input impedance and no 'Z' level at transistor gates constrains were applied.

Due to the use of evolution optimization approach, all the 88 gates were successfully evolved. Table I shows the minimal number of transistors needed for implementation of the whole gate set, where the gray-backgrounded gates represent the unique (designed) gates. The other gates were easily derived from them. Just 6 transistors are sufficient for a single gate on average; most difficult gates require 9 transistors. Note that a/a and b/b gates consist of four transistors (two inverters) just because of the high input impedance requirement. They would be probably not used in the polymorphic circuit synthesis, similarly to 0/0, 0/1, 1/0 and 1/1 gates.

Let us show e.g. the $xor/\overline{a}b$ gate designed by 6 MOS transistors. The conventional xor gate with high input impedance consists of 8 transistors. Moreover, two inverters for the signal inputs are required. The conventional $\overline{a}b$ gate could be assembled with 4 transistors and one inverter. Considerable savings in the number of transistors are obvious. A $nand/nor$ gate, typical for polymorphic electronics, is designed by using 8 transistors. That is the same transistor count needed for

978-1-5386-7100-9/18 $31.00 © 2018 IEEE

a separate implementation of both $nand$ and nor gates in the convectional CMOS logic. The advantage of polymorphic approach $nand/nor$ lies in the already implemented switch between these two functions, which comes as a natural built-in feature of such circuit component. In general, all the designed gates require significantly less transistors than a convectional circuit comprising two gates (implementing each function separately) and a multiplexer.

All the gates were successfully verified using HSPICE simulations and $45nm$ technology transistor models. Therefore, omission of the 'Z' level at transistor gates turned out to be acceptable. This observation is further supported by Figure 5 which shows HSPICE simulation results of four randomly selected $\overline{in_0}/xnor$ gates. Function of the polymorphic gates is changed every 40 ns – i.e. when the voltages on power rails are swapped.

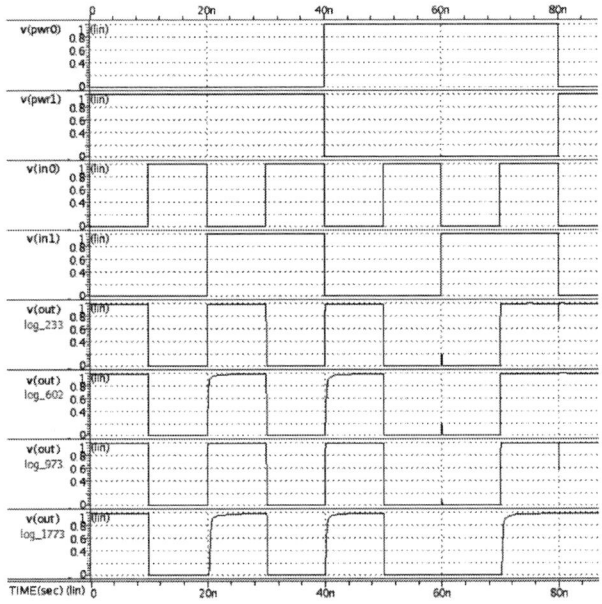

Fig. 5. HSPICE simulation results of four $\overline{in_0}/xnor$ polymorphic gates. Signal inputs are marked as in_0 and in_1, output as out and power rails are denoted as pwr_0 and pwr_1.

All the designed gates have high input impedance, low output impedance, short time of signal propagation (usually less than 1 ns), low power consumption and small number of transistors being used. Therefore, they meet all the requirements imposed on the ideal polymorphic gate.

V. CONCLUSION

A new approach to design of multifunctional digital circuits based on polymorphic electronics was proposed in this paper. MOS transistors are exploited in unconventional circuits where the circuit function is selected by mutual polarity of electric potentials on the power supply rails. In order to design new gates an evolutionary approach based on Cartesian genetic programming was utilized. The goal of the evolutionary algorithm

was to minimize the overall number of transistors being used for each one of the evolved gates.

A complete set of two-input gates was designed. Functionality of all the gates was verified by HSPICE simulations. These gates show the best parameters of all previously published polymorphic gates – high input impedance and low output impedance, short time of signal propagation, low power consumption and small number of transistors being used.

This is the first complete set of polymorphic gates ever published. Wide range of proposed polymorphic gates (function combinations) may bring a significant advantages for space-efficient synthesis of polymorphic circuits in terms of the overall size. Moreover, smaller parts of such designed circuits can be later optimized at the transistor level using proposed evolutionary optimization.

ACKNOWLEDGMENT

This work was supported by The Ministry of Education, Youth and Sports from the National Programme of Sustainability (NPU II) project IT4Innovations excellence in science - LQ1602 and by the IT4Innovations infrastructure which is supported from the Large Infrastructures for Research, Experimental Development and Innovations project IT4Innovations National Supercomputing Center - LM2015070. Another support was provided by the Brno University of Technology project FIT-S-17-3994.

REFERENCES

[1] F. Wanlass and C. Sah, "Nanowatt logic using field-effect metal-oxide semiconductor triodes," in *1963 IEEE International Solid-State Circuits Conference (ISSCC). Dig. Tech. Papers*, vol. VI, Feb 1963, pp. 32–33.
[2] A. Stoica, R. Zebulum, and D. Keymeulen, "Polymorphic electronics," in *Int. Conf. on Evolvable Systems*. Springer, 2001, pp. 291–302.
[3] C. Bobda, *Introduction to Reconfigurable Computing: Architectures, Algorithms, and Applications*. Springer, 2007.
[4] R. Ruzicka and V. Simek, "NAND/NOR gate polymorphism in low temperature environment," in *2012 IEEE 15th International Symposium on Design and Diagnostics of Electronic Circuits Systems (DDECS)*, 2012, pp. 34–37.
[5] Z. Gajda and L. Sekanina, "On evolutionary synthesis of compact polymorphic combinational circuits," *Journal of Multiple-Valued Logic and Soft Computing*, vol. 17, no. 6, pp. 607–631, 2011.
[6] A. Stoica, R. Zebulum, D. Keymeulen, and J. Lohn, "On polymorphic circuits and their design using evolutionary algorithms," in *Proc. of IASTED International Conference on Applied Informatics AI2002, Insbruck*, 2002.
[7] V. Simek, J. Nevoral, A. Crha, and R. Ruzicka, "Towards design flow for space-efficient implementation of polymorphic circuits based on ambipolar components," *ElectroScope*, vol. 11, no. 1, pp. 1–10, 2017.
[8] R. Ruzicka, L. Sekanina, and R. Prokop, "Physical demonstration of polymorphic self-checking circuits," in *Proc. of the 14th IEEE Int. On-Line Testing Symposium*. IEEE Computer Society, 2008, pp. 31–36.
[9] J. F. Miller, *Cartesian Genetic Programming*. Springer Verlag, 2011.
[10] A. Dodabalapur, H. Katz, L. Torsi, and R. Haddon, "Organic heterostructure field-effect transistors," *Science*, vol. 269, no. 5230, pp. 1560–1562, 1995.
[11] A. Heinzig, S. Slesazeck *et al.*, "Reconfigurable silicon nanowire transistors," *Nano Letters*, vol. 12, no. 1, pp. 119–124, 2012.
[12] J. Nevoral, R. Růžička, and V. Mrázek, "Evolutionary design of polymorphic gates using ambipolar transistors," in *2016 IEEE Symposium Series on Computational Intelligence*. Institute of Electrical and Electronics Engineers, 2016, pp. 1–8.
[13] A. Mishchenko, S. Chatterjee, and R. Brayton, "DAG-aware AIG rewriting: a fresh look at combinational logic synthesis," in *2006 43rd ACM/IEEE Design Automation Conference*, July 2006, pp. 532–535.

2018 IEEE Computer Society Annual Symposium on VLSI

TDC: Tagless DRAM Cache

S R Swamy Saranam, Madhu Mutyam
{srswamy, madhu} @cse.iitm.ac.in
Dept of Computer Science and Engineering, Indian Institute of Technology Madras, India

Abstract—Advancements in 3D-stacking technology lead to the usage of stacked DRAM as a last level cache. DRAM caches present multiple design challenges. DRAM cache tag management overhead is one of them because of large area requirement and high access time for look-up. Numerous techniques have been proposed to handle the challenges involved with the DRAM caches. We consider a design choice wherein the tag array storage is removed completely.

In this work, we propose a Tagless DRAM Cache (TDC), that completely removes tag array from the DRAM cache and instead, uses the tag-array of the SRAM last level cache (LLC) along with DRAM cache way indices to locate data in the DRAM cache. Experimental evaluation, considering 4-core configuration, shows that TDC achieves a speedup of 9.97% when compared to baseline. Further, TDC shows greater promise by providing reduction in average power consumption by 11.22% and average SRAM LLC miss penalty by 30.8%.

I. INTRODUCTION AND BACKGROUND

Need for large last level caches is increasing with the drastic increase in the working sets of the high-end applications. However, due to limited area, large caches are difficult to realize on-chip. Because of its high density, DRAM provides a good solution for large caches. Propelled by advancements in 3D-stacking technology using Through-Silicon Vias (TSVs), 3D stacked DRAM has been proposed as a viable large last level cache. Direct deployment of large DRAM caches comes with a set of challenges such as high tag storage space requirement, high access latency and power consumption.

Many works are proposed to address the challenge of high tag storage space requirement. Usage of SRAM for tag storage is expensive and is discarded. Few works [1], [2] propose to co-locate tag and data in the same row. A major drawback with such an approach is the need for two accesses to the DRAM row, one to retrieve the tags and another to retrieve data upon a tag match. To retrieve both tags and data in a single access, *compound access* is proposed. In addition, to avoid the tag look-up latency for a possible cache miss, the authors introduced a structure, namely, *MissMap*. MissMap enables us to identify a cache miss early so as to bypass DRAM cache access. MissMap, though improves the average access latency of the DRAM cache, demands extra storage space. Few other works [3], [4] propose miss predictors that aim at reducing the miss penalty of the DRAM caches. Another work, namely, ATCache [2] considers using in-DRAM tags along with a small SRAM tag array to optimize the access latency. All the existing solutions in the literature suffer from storage overhead and a high DRAM access latency.

An effective approach to handle the high tag storage space requirement of DRAM caches is to remove the tag array com-

pletely. A *fully associative tagless DRAM cache* (FATD) [5] adopts such an approach. In FATD, the size of each cache line in the DRAM cache is equal to the operating system (OS) page size. To realize this, the authors propose to modify the TLB for DRAM cache tag look-up by introducing a Global Inverted Page Table (GIPT). Though FATD removes the need of high tag storage space, it requires changes to the OS to handle GIPT and its page-sized cache lines result in significant off-chip bandwidth overhead and internal fragmentation.

In this work, we address the above mentioned challenges of the DRAM cache by proposing a Tagless DRAM Cache (TDC). We modify the organization of SRAM-based LLC to identify and locate the cache blocks in DRAM cache. To realize this, we maintain the same number of sets in both SRAM LLC and DRAM cache. Further, each cache block in SRAM LLC stores the *DRAM way index* of the corresponding large cache block. By effectively using the look-up to the SRAM based LLC, TDC avoids tag storage in DRAM cache and also improves the average miss penalty of the SRAM LLC by bypassing DRAM cache on a certain miss. In contrast to prior arts, our design predicts a cache miss with minimal storage overhead. We further discuss the options of designing efficient meta-data storage and their impact on access latency.

Using gem5 simulator [6] with SPEC CPU 2006 benchmarks, we evaluate TDC against Loh and Hill's DRAM cache [1]. For 4-core workloads, we observe an improvement of 9.97% performance. Further, TDC achieves power savings of 11.22% along with lower average SRAM LLC miss latency over the baseline [1].

We organize the rest of the paper as follows, Section II provides motivation. Section III presents the details of TDC. Experimental methodology is described in Section IV, followed by a detailed analysis of results in Section V. Finally, Section VI provides conclusions of the work.

II. MOTIVATION

Parameters that influence the efficiency of DRAM caches are hit ratio and access latency. Hit rate can be improved by better utilization of the spatial locality. Further, DRAM caches can be efficiently designed for power consumption. We discuss the following possibilities in DRAM caches below.

A. Exploiting spatial locality and improving bandwidth

Block based caches [1], [2] suffer from high tag overhead whereas page based caches [5], [7] consume high off-chip bandwidth. In order to make better utilization of DRAM

978-1-5386-7100-9/18 $31.00 © 2018 IEEE

88

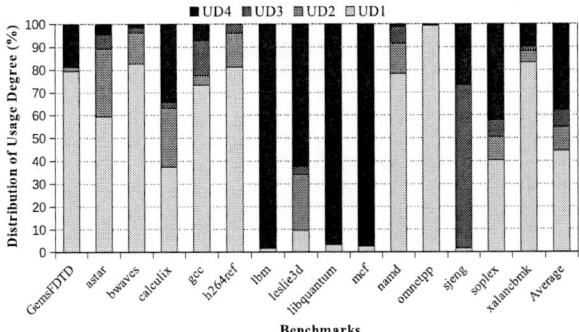

(a) For Spatial degree 4 : Cache block size of 256B.

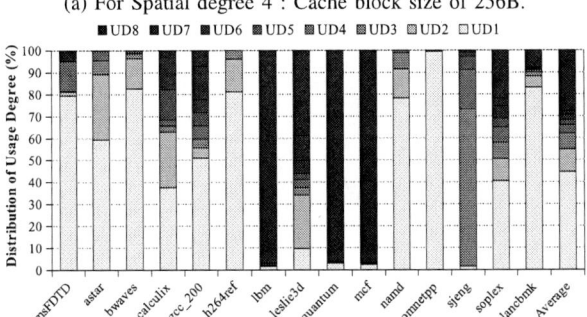

(b) For Spatial degree 8 : Cache block size of 512B.

Fig. 1: Usage Degree (UD) for various SPEC CPU 2006 applications.

caches, intermediate block sizes can be exploited. To understand the feasibility of the intermediate cache blocks, we analyze the utilization of the cache blocks with different sizes.

Generally the cache blocks are of size 64B; here, we refer a cache block of size 256B and 512B as *large cache block* (We will use this term *large cache block* throughout the paper). A *large cache block* is organized as sector of sub-blocks of size 64B. We introduce a new term *Spatial degree* to denote the number of sub-blocks (of size 64B) in one large cache block (of size 256B or 512B). We use the term *Usage Degree* (UD) to denote the number of sub-blocks actually referenced from a large cache block. We analyze SPEC CPU 2006 applications on a single core system with four levels of cache hierarchy. Figures 1a and 1b show the distribution of *usage degree* in large cache blocks. If we set *spatial degree* as 4, 37.8% of the large cache blocks exhibit *usage degree* of 4 and 7.4% exhibit *usage degree* of 3. Similarly, for *spatial degree* 8, 27% of the cache blocks possess *usage degree* 8. This shows that, there is a significant amount of spatial locality that can be exploited. We are not interested in large cache blocks of size 1kB and 2kB as they incur significant overhead in the off-chip bandwidth and internal fragmentation [8].

III. TAGLESS DRAM CACHE

A. Overview

Tagless DRAM Cache (TDC) removes tag array from the DRAM cache organization, thereby improving the effective

| Prefix Tag | Set Index | Sub-block Index | Block Offset |

Fig. 2: Proposed address mapping in SRAM LLC.

cache capacity. We apply two optimizations that improve the performance of the TDC, namely,

- *Predicting the existence of a cache block in the DRAM cache*: We design DRAM cache such a way that if there is a hit to a cache block in the SRAM LLC, the corresponding large cache block is definitely present in the DRAM cache, so we can avoid accessing the DRAM cache on a miss in the SRAM LLC. This improves the average access latency as the DRAM cache is bypassed for all cache misses. This also improves the power consumption of the DRAM cache and the main memory.
- *Eliminating the tag look-up*: In conventional DRAM cache organization (In-DRAM tags design), every access results in two row buffer accesses, one for tag look-up and another for the data access [1]. Our proposed scheme uses the way index to identify the column address of the requested cache block, thereby eliminating a DRAM row buffer access for tag look-up.

B. Design of Tagless DRAM Cache

In conventional cache architecture, each cache block can be uniquely identified by its tag (*tag array*). However, in the case of large caches, tag array itself becomes an overhead by occupying approximately 10% of cache space. Our design eliminates tag array from the DRAM cache thereby reducing significant amount of tag overhead. Since the tag array is removed from the design, there is a necessity for an alternative mechanism to uniquely identify the cache blocks in a set. TDC identifies a cache block in the DRAM cache using its way number instead of maintaining a separate tag array. To achieve this, our design makes it necessary that every cache block in DRAM cache must have at-least one corresponding sub-block in the SRAM LLC. This feature demands a strictly inclusive policy in our cache hierarchy (SRAM LLC is strictly inclusive of the DRAM cache). In our design, both SRAM LLC and DRAM cache have the same number of sets. Conventional SRAM-based LLC is modified to suit our mechanism. In SRAM LLC, a requested memory address is interpreted as composition of four fields: *Prefix Tag*, *Set Index*, *Sub-block Index* and *Block Offset* as shown in Figure 2.

- **Prefix Tag (PTag)**: *PTag* is the tag of a large cache block. A cache block in a DRAM cache set is uniquely identified by its *Prefix Tag*.
- **Sub-block Index (SBI)**: *SBI* differentiates a sub-block from other sub-blocks that are present in a large cache block.
- **Set Index (SI)**: Each cache line is mapped to a unique set. *SI* determines the cache set of a given cache block.
- **Block Offset (BO)**: *BO* determines the offset of the word within the sub-block.

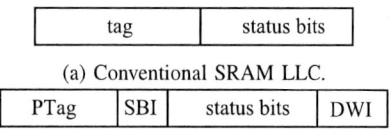

(a) Conventional SRAM LLC.

PTag	SBI	status bits	DWI

(b) Proposed SRAM LLC.

Fig. 3: Meta-data entry in SRAM LLC.

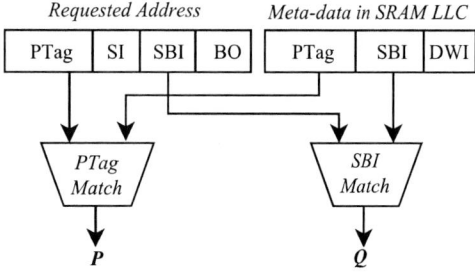

Fig. 4: Tag Comparison Logic in SRAM LLC.

P	Q	Result
Hit	Hit	*Hit* in SRAM LLC
Hit	Miss	*Miss* in SRAM LLC, but *hit* in DRAM cache
Miss	X	*Miss* in SRAM LLC

TABLE I: Scenarios in TDC.

Each cache block in the SRAM LLC stores the way number of the corresponding large cache block in the DRAM cache. Hence, meta-data of a cache block in the SRAM LLC consists of *Prefix Tag*, *Sub-block Index* and *DRAM Way Index (DWI)* as shown in Figure 3b.

C. Functionality

When a cache block is requested at the SRAM LLC, *Prefix tag* and *Sub-block Index* are compared in parallel as shown in Figure 4. In Table I, P denotes the result of *Prefix Tag* match and Q denotes the result of *Sub-block Index* match. At SRAM LLC, following scenarios are possible:

- **Scenario I (P - Hit, Q - Hit):** This scenario means both *Prefix Tag* match and *Sub-block Index* match are hits, implying that the requested block is a hit in SRAM LLC.
- **Scenario II (P - Hit, Q - Miss):** Here, because *Prefix Tag* match resulted in a hit, it implies that some of the sub-blocks belonging to the corresponding large cache block in DRAM cache are present in SRAM LLC. In addition, because *Sub-block Index* match is a miss, this scenario indicates that the requested block is a miss in SRAM LLC but a hit in DRAM cache. Following this result, the *Set Index* and the *DRAM Way Index* are fetched from the matching sub-block(s) and sent as a request to the DRAM cache.
- **Scenario III (P - Miss):** It implies that none of the sub-blocks belonging to the requested block is present in the SRAM LLC. Further, a *Prefix Tag* mismatch indicates that

TABLE II: System Configuration.

Processor	2GHz, 4,8-core, 64 ROB
L1 I/D cache	Each 32 kB, 4 way, LRU, 2 cycles
L2 cache	256KB, 8 way, LRU, 10 cycles
L3 cache	1MB per core, 16 way, LRU, 24 cycles
DRAM cache	16MB/32MB/64MB, 16 way, tRAS-tCAS-tRCD 12-10-10 (ns) Row Buffer size 4kB, open-page policy
Main Memory	DDR3-1600x64 ,t_{CK}: 1.25ns, Latency and Power values from Micron's Power Calculator [9]. 8 devices per rank, 2 ranks per channel, 1 channel, Row Buffer size 4kB, open-page policy
Benchmarks	bwaves, gamess, milc, leslie3d, namd, soplex, GemsFDTD, lbm, gcc, mcf, hmmer, libquantum, h264ref, omnetpp, astar, xalancbmk, bzip2, gromacs

the requested block is also a miss in the DRAM cache. Hence, the request can be sent to the main memory, avoiding the DRAM cache lookup. When the requested block is fetched from main memory, all the other sub-blocks constituting the large cache block are prefetched into the DRAM cache for future accesses. All the sub-blocks contain same *Prefix Tag* and *Set Index* but different *Sub-block Indices*. We send the requested sub-block to SRAM LLC and set the *DRAM Way Index* as the way number in which the large cache block is inserted in DRAM cache.

1) Eviction Policy: The presence of a sub-block in the SRAM LLC determines the existence of its corresponding large cache block in the DRAM cache. As a result, if a sub-block is evicted from the SRAM LLC, locating its corresponding large cache block in the DRAM cache is not possible. Hence, when the last sub-block from the SRAM LLC is evicted, the corresponding large block in the DRAM cache should also be evicted. Each cache block in the DRAM cache maintains a saturating counter of size $log(k)$, where k is the spatial degree (number of sub-blocks in the large cache block). While fetching a new large cache block into DRAM cache, the saturating counter is set to zero. The value of the saturating counter is incremented as the sub-blocks are forwarded to the upper level caches, and decremented as the sub-blocks are written back to DRAM cache. Once the value of the saturating counter of the large cache block becomes zero, the cache block is evicted from the DRAM cache and a write-back is initiated.

2) Write-back Policy: TDC organizes the DRAM cache with large cache blocks of size 256B and 512B. On every eviction, a large cache block needs to be written back to the main memory. However, this results into significant wastage of the off-chip bandwidth. TDC counters this by utilizing the stored meta-data (dirty bit, replacement data, etc.) of each sub-block of the large cache block. Hence, when a write-back is issued on the large cache block, TDC writes only those sub-blocks which are dirty. The remaining sub-blocks (clean evictions) are discarded, thereby reducing the off-chip bandwidth wastage.

IV. EXPERIMENTAL SETUP

A. Simulation Framework

We evaluate *Tagless DRAM Cache* using SPEC CPU 2006 benchmark suite [10]. Table II shows the experimental setup.

978-1-5386-7100-9/18 $31.00 © 2018 IEEE

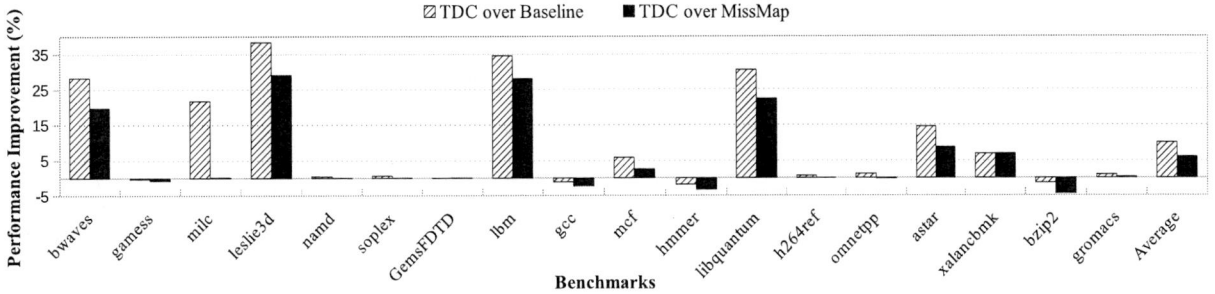

Fig. 5: Performance improvement over the baseline and MissMap.

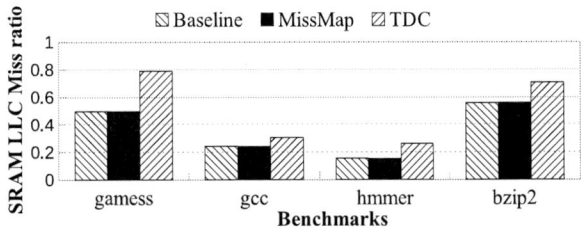

Fig. 6: Miss Rate in SRAM LLC.

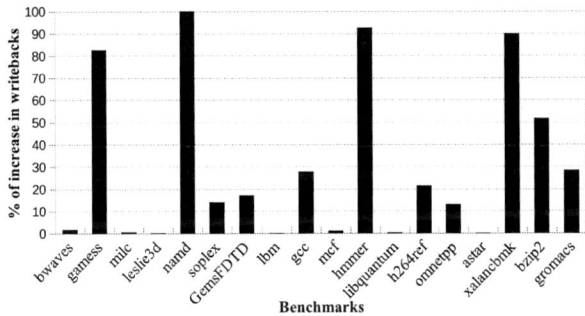

Fig. 7: Percentage of increase in write-backs from SRAM LLC.

We use the gem5 cycle-accurate simulator [6] for our implementation and performance evaluations. We simulate the scenario by running 10 billion instructions, of which first 25% of the instructions are used for warmup. We consider a four-level cache hierarchy, constituted by three levels of SRAM cache and a fourth level of DRAM cache. The third level cache is shared among all the cores, henceforth referred to as Last-Level Cache (SRAM LLC). DRAM cache is implemented by considering all the relevant DRAM timings using DRAMSim2 [11] simulator. We analyze our proposed TDC using 18 benchmarks from SPEC CPU 2006 [10] suite in rate mode. We use an open page policy where a row buffer is kept open to address spatially local requests.

B. DRAM Cache organizations

In order to study the effectiveness of our proposed implementation, we contrast it with the following prior arts:

- **DRAM cache with In-DRAM tags (Baseline):** This model uses *compound access* (tags and data in the same row) in the DRAM cache. The high tag overhead makes the design undesirable.
- **DRAM cache with MissMap (MissMap):** Loh and Hill proposed a structure called MissMap [1] that identifies whether block is present in the DRAM cache. This design avoids accessing DRAM cache for tag look-up. We implement a 16/32/64MB DRAM cache with a 1MB MissMap.

Tagless DRAM Cache: Our proposed design removes tag array and requires no additional meta-data overhead for DRAM cache. *Spatial degree* determines the cache size and cache block size in DRAM cache. TDC fetches large cache blocks of size 256B (or 512B). TDC is 16-way associative and uses LRU replacement policy.

V. RESULTS

A. Performance Analysis

Considering 4-core configuration, Figure 5 shows the performance improvement of our Tagless DRAM Cache (TDC) design over the baseline and MissMap techniques. Overall, TDC achieves an average speedup of 9.97% and 5.94% as compared to the baseline and MissMap, respectively. Barring *gamess, gcc, hmmer* and *bzip2* benchmarks, TDC performs better than the baseline and MissMap. We observe that efficient utilization of spatial locality by the large cache blocks of TDC is the main reason for TDC to perform better. From Figure 1a, we observe that 90% of the large cache blocks in *milc, leslie3d, lbm* and *libquantum* exhibit *usage degree* of 4. Although *bwaves* exhibits less spatial locality, we observe a significant increase in performance. This is because for most of the requests, TDC bypasses DRAM cache, which reduces the average access latency. We also observe a strong correlation of the performance gains with the average access latency at the SRAM LLC and analyze it in Section V-B. To understand the reason for inefficiency of TDC for *gamess, gcc, hmmer* and *bzip2*, we present the miss rate of SRAM LLC in Figure 6. Miss rate of TDC is higher than that of baseline and MissMap. The reason for such a behavior being, in TDC whenever a large cache block is evicted, all the corresponding sub-blocks from the SRAM LLC have to be evicted leading to additional misses

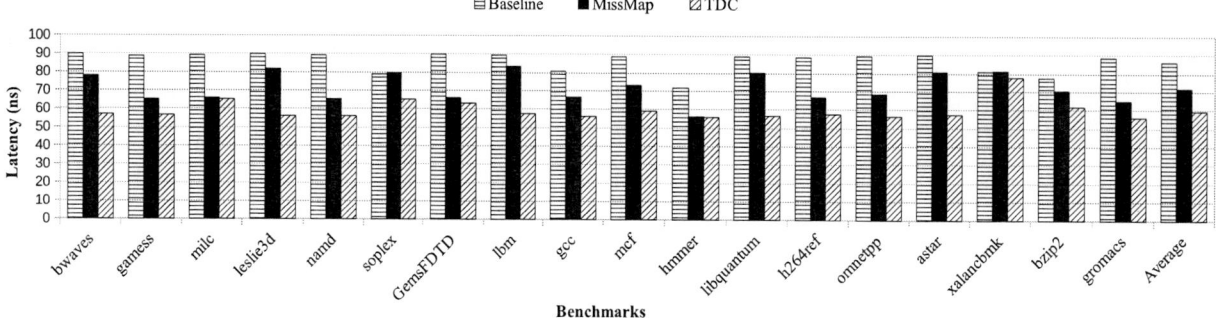

Fig. 8: Average access latency at the SRAM LLC (Lower is Better).

(cache pollution) and write-backs. Figure 7 clearly shows the increase in the write-backs at the SRAM LLC due to evictions at the DRAM cache. Coming to cache pollution, we now look at each of the benchmarks to understand their behavior. *gcc* uses a recursive data structure that suffers from indirect memory accesses. Hence, the large cache blocks useful in exploiting the spatial locality are not productive for *gcc* in terms of performance. In case of *bzip2*, basically a block-sort compression algorithm, access to the memory is dependent on a conditional branch instruction. Hence, fetching a stream of contiguous memory blocks will increase the pollution. *hmmer* and *gamess* use random sequences of addresses, and exhibit no spatial locality. Therefore, at the SRAM LLC there is an increase in the pollution as shown in Figure 7. As the number of back invalidations increases at the SRAM LLC, the overhead incurred by eviction policy also increases in TDC.

B. Access Latency Reduction

In Figure 8, we present the average access latency at the SRAM LLC. The average access latency mentioned here refers to the average miss penalty at the SRAM LLC. In addition, the overheads of MissMap lookup and contention are also account in this latency. Proposed TDC achieves a significant reduction in average access latency of 30.8% and 17.4% over the baseline and MissMap, respectively.

1) Comparison of MissMap and TDC with baseline: Every access (either a hit or a miss) to the baseline DRAM cache requires a DRAM row buffer access. For any access resulting in a miss, this row access latency can be treated as an overhead. Hence, with the baseline DRAM cache, the SRAM LLC experiences a high average access latency, resulting in low performance, as can be seen in Figure 5. On the other hand, both MissMap and TDC bypass the DRAM Cache on every DRAM cache miss. MissMap bypasses the DRAM cache using a MissMap structure whereas TDC makes a decision to bypass the DRAM cache during the SRAM LLC access itself. Hence, MissMap and TDC eliminate an extra access for tag look-up, resulting in a lower average access latency compared to the baseline.

2) Comparison with MissMap: MissMap technique introduces a lookup table called *MissMap* that stores an array of

64 bits per page. Each bit determines the presence of the corresponding block of that page in the DRAM cache. A hit in MissMap structure is followed by an access to the DRAM cache. A DRAM cache access in MissMap technique requires a tag look-up followed by a data access. *Compound access* in MissMap stores the tag array in the first 3 ways and data in the remaining 29 ways of a DRAM cache row. Hence, after the required row is activated, we need to wait for at-least $t_{CAS} + t_{CCD} + t_{CAS}$ [12] for the data to be available. The first t_{CAS} is for the tag look-up and second is for the actual data access. In case of TDC, we do not require a tag look-up since the column address for the data is read directly from the SRAM LLC access. This reduces the access latency of the DRAM cache by $t_{CAS} + t_{CCD}$ for all the accesses in TDC.

Although, both MissMap and TDC bypass the DRAM cache on a possible miss, MissMap incurs an extra latency overhead for tag look-up in the *MissMap* structure and in the DRAM cache.

C. Power Analysis

We compute power consumption of the DRAM cache and the main memory using Micron's DDR3 power calculator [9]. Figure 9 shows the power consumption of Baseline, MissMap, TDC for 4-core workloads. On an average, TDC achieves 11.22% and 6.76% power savings over the baseline and MissMap, respectively. *TDC* reduces the power consumption by improving the DRAM cache hit ratio, reducing the tag look-up overhead in the DRAM cache as well as the number of activations and precharges in the main memory. TDC does not require tag comparison logic in the DRAM cache, resulting in the reduction of number of read cycles. Besides, because TDC fetches large cache blocks from the memory, it has less number of activations and precharges in the main memory when compared to those of MissMap. Even though MissMap has a high DRAM cache hit ratio, it still requires tag look-up in the DRAM cache. Along with all the above factors, both MissMap and TDC bypass the DRAM cache on a possible miss. Hence, TDC has lower power consumption at the DRAM cache when compared to the baseline.

978-1-5386-7100-9/18 $31.00 © 2018 IEEE

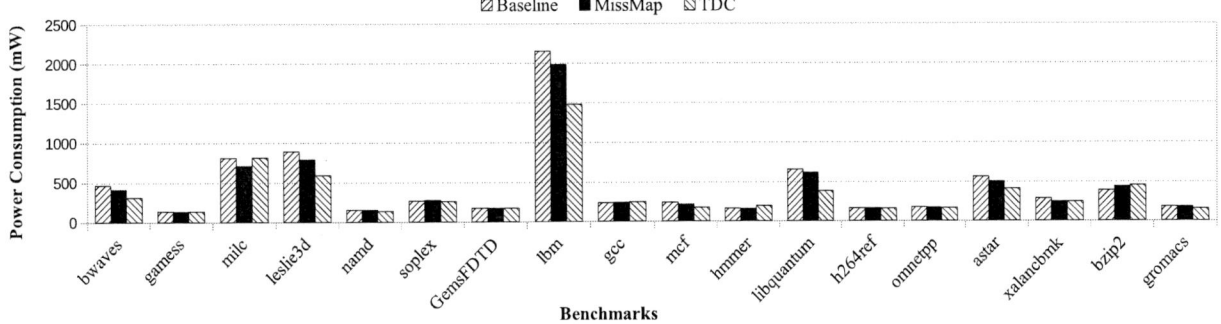

Fig. 9: Power consumption for 4-core workloads.

D. Storage Overhead

Since we employ write-back policy to write only the modified sub-blocks of a large cache block, TDC needs to store the *dirty bit* information for each of the sub-blocks. For a spatial degree 4, for each large cache block, we need *4 dirty bits + 1 valid bit + 1 Read/Write bit + 4 bits* for the replacement state. Therefore, the meta-data overhead for a 16MB DRAM cache in TDC is $2^{16} * 10 \ bits = 80kB$.

TDC completely removes the tag array and stores only the DWI in the SRAM LLC. So, TDC requires extra tag meta-data overhead in SRAM LLC when compared to the baseline. Since the DRAM cache in TDC is 16-way associative, *4 bits* are required to store the DWI for each block in the SRAM LLC. If we consider 1MB SRAM LLC per core with 64B block size, the extra storage needed due to TDC is $2^{14} * 4 \ bits = 8kB$. For a 4-core system, this overhead is $32kB$ and in total, our implementation requires $112kB$ extra space. This overhead is very small when compared to the overhead of MissMap, that requires 1MB extra space for a 16MB DRAM cache. Hence, for a 4-core configuration, our proposed TDC saves 89.1% of storage space as compared to MissMap.

E. Sensitivity Analysis

We conduct sensitivity analysis of TDC by varying the cache size and the block size. For a 4-core system, with DRAM cache size of 32MB, we observe that TDC achieves an average speedup of 12.26% over 32MB baseline DRAM cache. It can be noted that the speedup of TDC increases with the increase in cache size. Applications with higher spatial locality perform better as we increase the cache size. We observe that the applications exhibiting less spatial locality, such as *milc, namd* degrade performance with increase in the cache size. When power savings are considered, we observe that for a 32MB DRAM cache, TDC reduces power consumption by 11.87% when compared to the baseline.

We also evaluate our proposed TDC for an 8-core system by considering DRAM cache sizes of 32MB and 64MB; and observe an average speedup of 7.23% and 7.58%, respectively, over the baseline.

VI. CONCLUSION

In this work, we proposed *Tagless DRAM Cache* (TDC) that removes the tag array from DRAM cache and uses the tag array of SRAM LLC along with the DRAM cache way indices to locate the data. TDC exploits spatial locality of the applications and introduces a novel tag look-up mechanism. TDC improves average miss penalty of SRAM LLC and also increases DRAM cache hit ratio. Overall, TDC achieves a speedup of 9.97% and 7.23% for 4-core and 8-core workloads, respectively, compared to the baseline.

REFERENCES

[1] G. H. Loh and M. D. Hill, "Efficiently enabling conventional block sizes for very large die-stacked dram caches," in *Proceedings of the 44th Annual IEEE/ACM International Symposium on Microarchitecture*, 2011, pp. 454–464.

[2] C.-C. Huang and V. Nagarajan, "Atcache: Reducing dram cache latency via a small sram tag cache," in *Proceedings of the 23rd International Conference on Parallel Architectures and Compilation*, 2014, pp. 51–60.

[3] M. K. Qureshi and G. H. Loh, "Fundamental latency trade-off in architecting dram caches: Outperforming impractical sram-tags with a simple and practical design." in *Proceedings of the 2012 45th Annual IEEE/ACM International Symposium on Microarchitecture*, pp. 235–246.

[4] J. Sim, G. H. Loh, H. Kim, M. O'Connor, and M. Thottethodi, "A mostly-clean dram cache for effective hit speculation and self-balancing dispatch," in *Proceedings of the 2012 45th Annual IEEE/ACM International Symposium on Microarchitecture*, pp. 247–257.

[5] Y. Lee, J. Kim, H. Jang, H. Yang, J. Kim, J. Jeong, and J. W. Lee, "A fully associative, tagless dram cache," in *2015 ACM/IEEE 42nd Annual International Symposium on Computer Architecture*, pp. 211–222.

[6] N. Binkert, "The gem5 simulator," *SIGARCH Comput. Archit. News*, vol. 39, no. 2, pp. 1–7, Aug. 2011.

[7] D. Jevdjic, S. Volos, and B. Falsafi, "Die-stacked dram caches for servers: Hit ratio, latency, or bandwidth? have it all with footprint cache," in *Proceedings of the 40th Annual International Symposium on Computer Architecture*, 2013, pp. 404–415.

[8] N. Gulur, M. Mehendale, R. Manikantan, and R. Govindarajan, "Bimodal dram cache: A scalable and effective die-stacked dram cache," in *Proceedings of the 47th Annual IEEE/ACM International Symposium on Microarchitecture*, 2014, pp. 38–50.

[9] Micron, "Tn-41-01: Calculating memory system power for ddr3." [Online]. Available: http://www.micron.com/products/dram

[10] J. L. Henning, "Spec cpu2006 benchmark descriptions," *SIGARCH Comput. Archit. News*, vol. 34, no. 4, pp. 1–17, Sep. 2006.

[11] P. Rosenfeld, E. Cooper-Balis, and B. Jacob, "Dramsim2: A cycle accurate memory system simulator," *IEEE Comput. Archit. Lett.*, vol. 10, no. 1, pp. 16–19, Jan. 2011.

[12] B. Jacob, S. Ng, and D. Wang, *Memory Systems: Cache, DRAM, Disk.* San Francisco, CA, USA: Morgan Kaufmann Publishers Inc., 2007.

978-1-5386-7100-9/18 $31.00 © 2018 IEEE

2018 IEEE Computer Society Annual Symposium on VLSI

CT-Cache: Compressed Tag-Driven Cache Architecture

Haeyoon Cho*, Joonho Kong*, Arslan Munir[†], Naresh Kumar Giri[†]
*School of Electronics Engineering, Kyungpook National University, South Korea
[†]Department of Computer Science, Kansas State University, Manhattan, KS
{haeyoon.cho, joonho.kong}@knu.ac.kr, {amunir, ngiri}@ksu.edu

Abstract—One of the most challenging problems in designing large caches is to devise efficient cache tag storage. Typical large last-level caches often require tens of megabytes of tag storage. Consequently, this large tag storage leads to high latency and energy consumption for a cache tag access, which adversely affect system performance and energy efficiency. In this paper, we propose CT-Cache that exploits the similarity in upper cache tag bits. Our proposed CT-Cache decouples a tag storage into a global tag table and delta tag arrays. The global tag table is a small fully-associative buffer that stores upper tag patterns, which can be shared between multiple cache lines, while the delta tag array stores lower tag bits for each cache line. By avoiding the redundant storage, the CT-Cache significantly reduces the cache tag storage size, which in turn reduces latency and energy consumption of a cache tag access. Evaluation results reveal that the CT-Cache reduces tag storage size by 87.8%, which significantly reduces tag access latency and energy. Owing to the tag storage reduction, the CT-Cache improves performance by 4.7%~16.2% and reduces last-level cache and main memory energy consumption by 20.0%~29.7% compared to the conventional caches that use non-compressed tag storage.

Index Terms—Tag compression, Cache architecture, Last-Level cache, Locality, Performance, Energy efficiency

I. INTRODUCTION

To enhance the performance of modern processors, the number of cores housed in a single chip is increasing. However, the memory system has not kept pace in terms of both capacity and bandwidth to supply required amount of data to all processor cores to sustain desired performance enhancement of processors. This discrepancy between processor and memory performance is known as "memory wall" and greatly influences the overall performance of computing systems. To exacerbate the memory wall problem, modern big data applications involve data-centric computations, which further increases the pressure on memory system. However, advancements in integration and device technologies (e.g., 3D-stacking technology and embedded dynamic random access memory) have allowed to integrate large dynamic random access memory (DRAM) onto the processor chip. The integrated (e)DRAM (eDRAM stands for embedded DRAM and is a DRAM integrated on the same die as the application-specific integrated circuit (ASIC) or multi-processor) size can be in order of hundreds of megabytes (MB) to few gigabytes (GB). Although, large (e)DRAM last-level caches (LLCs) can be designed and realized by such novel integration and device technologies, there are several challenges that need to be addressed to effectively employ large caches. When the cache size increases to hundreds of megabytes, the cache metadata, such as tag bits, dirty/valid bit, coherency management bits, state information for implementing cache replacement policy, etc., also increase. For example, a cache of 256 MB would have an associated tag size of 11.5 MB. Similarly, for a cache size of 1024 MB, the tag size would be up to 42 MB [4]. Thus, it is not feasible to store tags in SRAM as it would result in high area overhead.

To resolve the tag storage overhead, several prior works [5], [10], [14], [15] aim to address cache tag storage problem for large caches. The work in [10], [14], and [15] stores cache tags in (e)DRAM so that the cache tags do not require a huge number of SRAM cells (e.g., tens of megabytes) for

tag storage, alleviating the area overhead. However, these approaches typically have high latency mainly because the cache tag access entails a slow (e)DRAM access. Longer tag access latency adversely affects system performance and energy efficiency as the cache tag access lies in the critical path of the cache access time in large caches that employ sequential tag and data access. One may use large cache line size (e.g., page-size) to reduce tag storage overhead (e.g., [5]). However, the large cache line size would waste a huge memory bandwidth in order to transfer large block size data when a cache miss occurs.

In this paper, we present an efficient cache architecture (referred to as CT-Cache) for large caches. In our proposed architecture, the cache tags are split into two parts such that the upper parts contain bits that are common/same among multiple cache lines and the lower parts contain bits that are unique/different. Since there is a significant similarity in patterns of upper bits in memory addresses due to the principle of locality, our cache architecture stores the common tag bit parts in a global tag table (GTT) which is implemented in a small buffer (~1KB). This eliminates the need for storing multiple instances of a common/same value, which helps in reducing storage size, area, energy, etc., of the cache tag storage. The lower part of the cache tag, which uniquely identifies each cache line, is stored in a separate array which is referred to as delta tag array (DTA). For a cache hit to occur, the upper part of the tag must match with a value in the GTT and the lower part of the tag must match with a value in the DTA. To efficiently manage GTT and DTA, our proposed CT-Cache employs a flexible mapping (see Section III-A2 for details) between the GTT entries and a set of cache lines. Furthermore, proactive cache management in the CT-Cache (see Section III-A4 for details) periodically rearranges a mapping between the GTT entries and a set of cache lines to guarantee fairness between the GTT entries. These features further enhance cache space utilization, resulting in improved performance and energy efficiency.

To the best of our knowledge, our work is the first one that actually reduces the tag storage for large LLCs while not depending on the type of memory cells used in the cache. This independence on memory cell types means that the CT-Cache can also be applied to large SRAM-based as well as (e)DRAM-based caches whereas several previous proposals depended on a certain memory cell-based caches (e.g., DRAM caches [5], [10], [14]). Although several works introduced cache compression schemes for large-scale caches (e.g., [11], [16]), these works have purposed to compress cache data while the cache tag size remains large.

The main contributions of this paper are as follows:

- We propose a novel cache architecture (CT-Cache) for large caches by leveraging cache tag compression.
- Our proposed CT-Cache is independent of cache memory cell technology and can be applied to any large-scale cache (e.g., SRAM-based, (e)DRAM-based, etc.).
- Our proposed CT-Cache stores tags in 87.8% lesser area, imparts a performance improvement of 4.7%~16.2%, and reduces LLC and main memory energy consumption by 20.0%~29.7% as compared to conventional caches.

978-1-5386-7100-9/18 $31.00 © 2018 IEEE   94

- Due to the flexible mapping between the GTT entries and a set of cache lines and proactive cache management, the CT-Cache also shows stable and consistent performance and energy efficiency improvements compared to CT-Static (static mapping between the GTT entries and a set of cache lines) and CT-w/oPAD (CT-Cache without proactive cache management).

Remainder of this paper is organized as follows. Section II summarizes related work and Section III presents our proposed CT-Cache architecture. Evaluation methodology and results are presented in Section IV. Lastly, Section V concludes this paper.

II. RELATED WORK

There has been work done in literature related to large cache design particularly for efficiently storing the cache metadata. Loh and Hill [10] proposed a DRAM cache with block size of 64 bytes. Since the cache tags in their proposed cache architecture were stored in DRAM, they had to address the problem of increased miss penalty when accessing tags from DRAM. In order to combat this problem, they used MissMap on SRAM to keep track of DRAM content to avoid cache misses. However, the size of MissMap was already in the range of few MB, and thus the proposed architecture was not feasible for implementing cache tags in SRAM. Qureshi and Loh [14] addressed this issue with an Alloy Cache–a direct-mapped DRAM cache which has wider data path. Their proposed cache architecture improved the access latency as data and tag could be accessed in a single fetch. Their model worked good for direct-mapped caches but was inflexible for set-associative caches.

Jevdjic et al. [5] proposed Footprint Cache which combined the benefits of both block-based and page-based cache design. They designed a footprint predictor which only fetched selected blocks from the accessed page. They opted to only store page tags in order to reduce tag size. However, their design is geared towards DRAM-based caches while it would not be suitable for SRAM-based or eDRAM-based large caches. Huang et al. [4] proposed ATCache that stores entire tags in DRAM while the recently or frequently accessed tags in a small SRAM buffer. Although, their design aimed to reduce cache hit latency, cache tag storage size was still large (e.g., a 256MB DRAM cache requires 11.5 MB tag size).

Yang et al. [15] proposed a new tag scheme for DRAM cache using eDRAM technology. In this model, the tag was stored in eDRAM to alleviate the overhead of storing tags in SRAM. However, this model was exclusive to eDRAM technology and was not suitable for other technologies. TLB index-based [6] [8] tagging approaches used TLB indexes as cache tags instead of actual tag memory; however, this approach was only suitable for L1 caches where TLB access and cache access are performed simultaneously.

There have been several studies on cache compressions to improve storage efficiency of large caches. Young et al. [16], presented the idea of compressing data in DRAM caches. This approach used extra hardware for cache index predictors to reduce access latency. In [11], the authors presented base-delta-immediate compression for data in cache, which improves cache utilization and system performance. However, the studies mentioned above introduced cache data compression while the full tag bits must be stored in the storage, incurring the huge storage overhead due to the cache tags in large caches.

Petrov and Orailoglu [12] proposed a cache tag compression technique based on static (i.e., compile time) analysis of memory layouts. Although their proposed technique reduced cache power consumption, it relied on compile-time information, which made adaptive cache management at runtime difficult. Kwak and Jeon [7] proposed a cache tag compression scheme for embedded processors. Their proposed scheme also aimed at reducing tag storage by sharing upper tag patterns in a locality buffer. However, their proposed architecture still required to store the entry number of the locality buffer in the

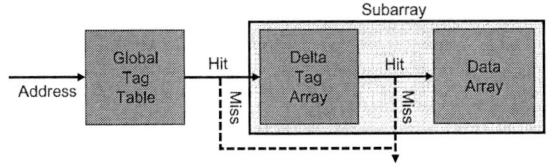

Fig. 1: Cache access flow of CT-Cache. There are many subarrays in the cache although we only show one subarray in the figure for brevity. We call a bundle of a delta tag array and corresponding data array as a subarray.

tag array, leading to lower tag compression ratio. Although this architecture would be efficient in small caches, it would not be suitable for large caches which should maintain a much larger number of upper tag patterns.

Our proposed CT-cache efficiently compresses cache tags, which makes storage of cache tags in SRAM feasible. Our approach permits saving space of multiple instances of redundant cache tag bits, efficiently reducing the tag storage.

III. COMPRESSED TAG-DRIVEN CACHE ARCHITECTURE

A. Architecture Overview

1) Decoupled Tag Storage: To avoid redundant storage of upper tag bits, we propose to decouple the conventional tag arrays into two parts: GTT and DTA. The common upper bit parts are stored in a small storage, which is referred to as GTT, while the remaining unique lower bit parts are stored in the DTA. Consequently, we need much smaller arrays for tag storages as we remove redundant storages for upper tag bits. The GTT is a small fully-associative buffer (in this work, we use 32-entry GTT) that contains the upper part of the tag bits while the DTAs contain the remaining bits of the tags that are associated with each cache line. Since we store the upper bit patterns in the GTT, a single GTT entry must be shared between multiple cache lines. As we use decoupled tag storages, the tag matching procedure of CT-Cache is also composed of two steps: GTT access and DTA access. In the case of both GTT hit and delta tag hit, it is regarded as a cache hit and we perform a data array access. If either the GTT miss or delta tag miss occurs, it is regarded as a cache miss and the request goes to the main memory.

2) Cache Management: In our design, since a single GTT entry must be shared among multiple cache lines, it is very crucial to determine the mapping rule between the GTT entries and cache lines. For efficient design, we group multiple cache lines into a single subarray and we define the mapping rule between the GTT entries and subarrays. A subarray logically contains data array and corresponding DTA[1]. Each subarray works as a direct mapped cache, which means data with a certain address can be mapped to only one place in a single subarray. Although capacity of a single subarray size can vary depending on cache design, one subarray has 2MB of data storage and associated delta tag storage in this work.

For a simple implementation, we could statically assign the cache subarrays into each GTT entry. For example, if we have a 32-entry GTT and 128 subarrays, we can statically assign four subarrays to each GTT entry[2]. However, this static assignment may worsen cache capacity utilization in the case where the required cache capacity for each GTT entry is diverse. To address this problem, we dynamically allocate and deallocate subarrays to GTT entries. As shown in Fig. 2, GTT entries can have various number of cache subarrays at runtime. Since a single subarray works as a direct-mapped cache, if N subarrays are allocated to a certain GTT entry, these subarrays work as an N-way set-associative

---

[1] As shown in Fig. 1, though we logically combine a data array and DTA into a single subarray, the data array and DTA can be physically far apart in the cache layout.

[2] We will also evaluate this scheme in evaluations (referred to as 'CT-Static').

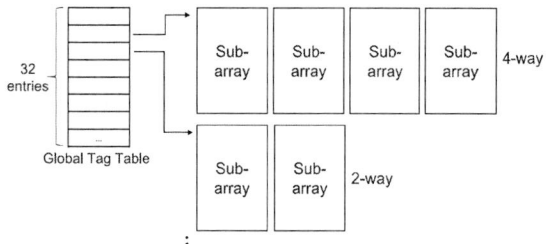

Fig. 2: A conceptual block diagram of the global tag table entries and their allocated subarrays.

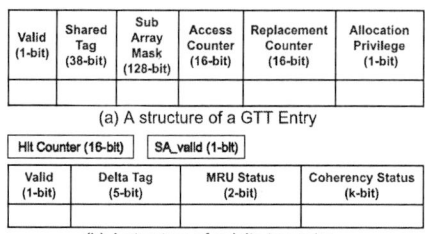

(a) A structure of a GTT Entry

Hit Counter (16-bit)		SA_valid (1-bit)	
Valid (1-bit)	Delta Tag (5-bit)	MRU Status (2-bit)	Coherency Status (k-bit)

(b) A structure of a delta tag entry

Fig. 3: Structure of the global tag table entry and delta tag entry.

cache. This dynamic allocation/deallocation enables a flexible management of the cache capacity allocated to each GTT entry, which in turn leads to better performance and energy efficiency.

3) Structure and Management of Global Tag Table and Delta Tag Array: The structure of the GTT is shown in Fig. 3(a). It can be implemented by using both CAM (content-addressable memory) and SRAM cells. A shared tag storage in the GTT is implemented by CAM cells to enable a fast fully-associative search while the remaining metadata storage uses SRAM cells. The subarray mask bits record which subarrays are currently allocated to the GTT entry. For example, if the subarray mask bit[0] and mask bit[1] are '1' while the remaining mask bits are '0', it means that the subarray[0] and subarray[1] are allocated to the corresponding GTT entry. In our design, there are total 128 subarrays (although the number of subarrays could vary depending on the cache design). Hence, we need 128 bits for subarray mask bits of each GTT entry.

For dynamic management of GTT entries, there are 16-bit access and replacement counters for each entry. The access counter maintains the number of accesses to the GTT entry within a pre-defined time interval (10 million clock cycles in this work). When replacing the GTT entry, we refer to the access counter for each entry and the entry with the least access count becomes a victim (i.e., replacement choice). The replacement counter indicates how many times the cache line replacements from the allocated subarrays occur within the time interval. An allocation privilege bit is set to '1' if the corresponding GTT entry is given a right to get a new subarray allocation. Detailed usages of the counters and allocation privilege bit will be explained in Section III-B.

To maintain the data coherency, the replacement or invalidation of a GTT entry entails bulk invalidations[3] and flushes of the data stored in the subarrays allocated to the evicted GTT entry. In the case of using write-back caches as in conventional caches, we need to perform bursty flush operations, which would issue a huge number of write-back requests to the main memory. These bursty flush operations may degrade system performance mainly due to the limited write buffer size. Thus, our proposed CT-Cache uses write-through caches so that the write requests to the main memory can be served sporadically.

An entry structure of the DTA is also shown in Fig. 3(b). The delta tag entry (array) is similar to the conventional tag entry[4] (array) while the main differences are: 1) we only store lower tag bits (5-bit in this work) for each tag entry, 2) 2 bits are added to each entry for MRU (most recently used) status bits, and 3) 16-bit hit counter and 1-bit SA_valid bit (i.e., subarray valid bit) are added to each subarray. Obviously, we only store the lower tag parts in the DTAs as we already store the upper part tags in the GTT entry. The MRU status bits are required for the victim cache

line selection. In general, we can apply least recently used (LRU) policy to select a victim. However, in our proposed CT-Cache design, implementing LRU is not straightforward as there can be various associativities across sets of the subarrays associated with each GTT entry. Hence, to simplify the design, we track only two MRU (most recently used) cache lines in a set and a victim is randomly selected from the cache lines except the two MRU lines. The hit counter maintains how many cache hits occur in the subarray. This information is used for dynamic allocation and deallocation of the subarrays (see the following subsection for detailed usages). The 1-bit SA_valid bit indicates whether the corresponding subarray is currently allocated to any GTT entry or not. The SA_valid bit is used when we search for available subarrays (i.e., subarrays that are currently not allocated to any GTT entry). Please note that the 16-bit hit counter and 1-bit SA_valid bit are employed not for each entry but for each subarray.

4) Proactive Subarray Allocation and Deallocation: Our proposed CT-Cache proactively manages allocation and deallocation between GTT entries and subarrays for efficient cache utilization. For example, a certain GTT entry may require more cache subarrays to prevent conflict misses while another entry may hold the subarrays that mostly contain dead cache lines (i.e., the cache lines that will not be used in future). We need to allocate more subarrays to the GTT entries for the former case while we need to deallocate subarrays from the GTT entries for the latter case.

We make the allocation and deallocation decisions by referring to the access counter and the replacement counter of each GTT entry. We go through replacement and access counters of each GTT entry every 10 million clock cycle interval. If (replacement count)/(access count) of a certain GTT entry is more than 0.1, we set the allocation privilege bit of the corresponding GTT entry to '1'. The allocation privilege bit of '1' means that the GTT entry gets a new subarray allocation whenever a conflict miss from the subarrays allocated to the GTT entry occurs during the next 10 million cycle interval.

After we determine the GTT entry of which allocation privilege bit is set to '1', if there is at least one GTT entry of which allocation privilege bit is '1', we also try to deallocate subarrays which are allocated to greedy GTT entries (i.e., considered to have much more subarrays than they actually need). For subarray deallocations, we find the GTT entries that satisfy the following condition: (replacement count)/(access count) of the GTT entry is less than 0.05. From each GTT entry that satisfies the condition[5], we deallocate one subarray which has the least hit count among the subarrays allocated to the selected GTT entries. Obviously, a deallocation includes invalidations of all the cache lines in the subarray and update of the corresponding subarray mask bit and SA_valid bit (from '1' to '0'). Please note that the deallocation only happens in the GTT entries that have more than one subarray (i.e., ≥ 2). We also reset the access and replacement counters of all GTT entries every 10 million cycle interval to cope with dynamic program behaviors. In evaluations, we will show performance and energy impact of our proactive subarray allocation and

---

[3]For easier bulk invalidations, we can manage all valid bits in the subarray within a single array. Since we use 2MB subarray with 64B cache line, we need 4KB valid bit array per subarray. By doing so, we can reduce latency for resetting all valid bits to '0'. Though it still require several cycles, it can be carried out in background, negligibly affecting performance.

[4]For the coherency status bits, the length of bits depends on the cache coherence protocol.

[5]In case that there is no GTT entry that satisfies the condition, we do not deallocate subarrays from any GTT entry.

978-1-5386-7100-9/18 $31.00 © 2018 IEEE

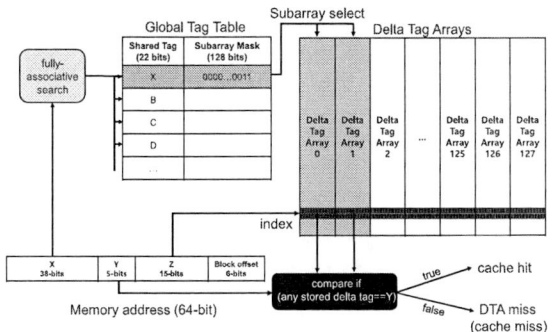

Fig. 4: Tag matching of CT-Cache.

TABLE I: The results and cache operations of tag matching in CT-Cache.

	GTT hit	GTT miss
DTA hit	Cache hit	
DTA miss	1. Send a read (write) request to the main memory in case of a read (write) miss 2. Need to allocate a new subarray for the GTT entry or need to replace a cache line	1. Send a read (write) request to the main memory in case of a read (write) miss 2. Allocate a new GTT entry or replace a GTT entry 3. Allocate a subarray for the GTT entry

deallocation by comparing the CT-Cache to the CT-Cache without proactive allocation and deallocation (referred to as 'CT-w/oPAD').

B. Cache Operations

1) Tag Matching: In this subsection, we explain the tag matching procedure of the CT-Cache in more detail. Assuming that we have a 64-bit address space, we can divide the memory address into four parts as shown in Fig. 4. In the conventional cache design, both X and Y parts correspond to the tag bit parts. In our design, we decouple the tag bits into X and Y bits each of which will be used for GTT matching and delta tag matching, respectively. When there is a cache access request, we first search for GTT entries to match X bits (the valid bit of the entry is also checked though it is not shown in Fig. 4). If there is a matched entry (GTT hit), we then access the DTAs with the index bit (Z) for lower tag bit (Y part) matching. During the delta tag matching, the subarray mask bits of the corresponding GTT entry are also accessed. As explained before, the subarray mask bits record the cache subarrays that are currently allocated to each GTT entry. The subarray mask bits inform us which DTAs must be accessed for delta tag comparison. When the accessed GTT entry holds more than one subarray, multiple DTAs must be accessed. In this case, the DTA accesses and comparisons are performed in parallel across the multiple DTAs. A tag match in any DTA is regarded as a cache hit and the data array of the corresponding subarray is accessed.

The results of the CT-Cache tag matching are summarized in Table I. In the case of both GTT hit and DTA hit, it is regarded as a cache hit and the data array is accessed. Since we use write-through caches, we also send a write request to main memory in the case of write hit. For the remaining cases, it is regarded as a cache miss since we cannot find the requested data from the cache. In the case of a cache read (write) miss, we should send a read (write) request to the main memory[6]. In CT-Cache, we classify a cache miss into two different types: GTT miss and GTT hit/DTA miss. In the following subsections, we explain the details of cache operations for GTT miss and GTT hit/DTA miss cases.

2) GTT Misses: In case of GTT misses, there are two possible cases: 1) there is an available GTT entry (i.e., there exist a GTT entry of which valid bit is '0'), and 2) no available

---

[6]Please recall that we use write-through caches. In the case of a write miss, we must also update the data to the main memory as well as the CT-Cache.

GTT entry. In case 1, we can allocate a new GTT entry without a replacement for storing the upper tag pattern which incurred the GTT miss. In order to allocate the lower tag bits and data, we then allocate a new subarray for the newly allocated GTT entry. Please note that there may be a case where there is no available subarray for allocation. In this case, we deallocate the subarray that has the least hit count among 128 subarrays and allocate this subarray to the new GTT entry. In case 2, we need to select a victim GTT entry and replace it with the new one. In order not to evict live cache blocks (i.e., cache blocks that are likely to be accessed in near future), we choose a GTT entry that has the least access counter as a victim (i.e., replacement) GTT entry. Finally, we allocate a new subarray to the new GTT entry and allocate the data in the data array while updating the metadata in the DTA and the GTT entry.

3) GTT hit/DTA Misses: In the case of DTA misses, we first check the following conditions: 1) whether or not the allocation privilege bit of the GTT entry that caused a GTT hit is '1', and 2) whether or not the cache miss is a conflict miss. If both the conditions are satisfied, we allocate an additional subarray for the corresponding GTT entry. After the subarray allocation, the data is also stored in the newly allocated subarray and the corresponding metadata is updated. In case where there is no available subarray for the allocation, we have to select a victim cache line from N victim candidates where N is equal to the number of subarrays currently allocated to the GTT entry. The victim selection is performed in a similar manner to the pseudo-LRU (already explained in Section III-A3). After we select a victim, we replace it with new data in the data array and update the corresponding metadata in the DTA. In the opposite case where either of the two conditions is not satisfied, we also perform a cache line replacement with pseudo-LRU while not allocating an additional subarray for the GTT entry.

IV. EVALUATIONS

A. Evaluation Framework

For evaluation of our proposed CT-Cache, we use gem5 simulator [1] with fastforwarding 2 billion instructions and actually running 1 billion instructions. We model a high-performance processor equipped with four out-of-order cores with per-core L1 data/instruction and L2 caches, a shared 8MB L3 cache, and a shared 256MB L4 cache. We employ the CT-Cache to 256MB eDRAM-based L4 cache as large caches generally suffer from the metadata storage overhead. For rest of the cache hierarchy (i.e., L1, L2, and L3), we use the conventional SRAM-based caches. Table II summarizes the parameters for our simulated processor and system. For energy and array latency evaluation, we use CACTI-P [9] and DESTINY [13] with 22nm technology nodes. We first collect cycle-level latencies of the arrays from CACTI-P and DESTINY and then use these latencies in gem5 simulation tool for performance evaluations. For energy evaluations, we also gather the eDRAM-based L4 cache access statistics from gem5, and calculate dynamic and leakage energy of L4 caches. Since the CT-Cache may increase main memory energy consumption as we use write-through policy, we also evaluate DRAM main memory energy consumption by using DRAMPower [2].

We use eight memory-intensive workloads from SPEC2006 CPU benchmark suite: 429.mcf, 433.milc, 437.leslie3d, 450.soplex, 459.GemsFDTD, 462.libquantum, 470.lbm, and 471.omnetpp. We evaluate single, quad, and mixed workload configurations. In 'single' configuration, we run a single copy of the workload in only one core while we run an identical workload in all four cores in 'quad' configuration. In 'mixed' configuration, we run a mix of four different workloads in the system as summarized in Table III.

In performance and energy results, we show four different schemes for comparisons.

- Baseline: It is a conventional cache that uses non-compressed tag storages.

TABLE II: Simulated processor and system specifications.

Categories	Specification
Processor core	Alpha 21364, ARM ISA
Clock frequency	4GHz
32KB per-core L1 data and instruction cache	1 cycle, 2-way, LRU, write-back
256KB per-core L2	3 cycles, 4-way, LRU, write-back
8MB shared L3	Sequential, 4 cycles for tag access, 7 cycles for data access, 16-way, LRU, write-back
Baseline 256MB shared L4	Sequential, 36 cycles for tag access, 81 cycles for data access, 32-way, LRU, write-back
CT-Cache 256MB shared L4	Sequential, 32 GTT entries, 128 subarrays, 1 cycle for global tag table, 7 cycles for delta tag array, 81 cycles for data access, write-through
Main memory	DDR4 2400, 1 channel, 2 ranks, 16 banks per rank, 64-entry read buffer, 128-entry write buffer

TABLE III: Eight mixed workload groups used for evaluations.

	Workloads
mixed_1	429.mcf, 437.leslie3d, 450.soplex, 471.omnetpp
mixed_2	433.milc, 459.GemsFDTD, 462.libquantum, 470.lbm
mixed_3	429.mcf, 433.milc, 437.leslie3d, 450.soplex
mixed_4	459.GemsFDTD, 462.libquantum, 470.lbm, 471.omnetpp
mixed_5	429.mcf, 450.soplex, 459.GemsFDTD, 470.lbm
mixed_6	433.milc, 437.leslie3d, 462.libquantum, 471.omnetpp
mixed_7	429.mcf, 433.milc, 459.GemsFDTD, 462.libquantum
mixed_8	437.leslie3d, 450.soplex, 470.lbm, 471.omnetpp

- CT-Static: It uses a GTT that statically maps four subarrays to each GTT entry.
- CT-w/oPAD: It is CT-Cache without the proactive subarray allocation and deallocation. In other words, the subarray allocation and deallocation only happens when a GTT miss or delta miss occurs (i.e., reactive allocation and deallocation).
- CT-Cache: It is our proposed CT-Cache with proactive subarray allocation and deallocation as explained in Section III.

B. Performance

Fig. 5 shows the performance results (weighted speedup [3] for 'quad' and 'mixed') normalized to the baseline. For single, quad, and mixed cases, our proposed CT-Cache shows performance improvements of 16.2%, 16.2%, and 4.7%, respectively, over the baseline. Compared to the single and quad cases, the mixed configurations show relatively less performance improvement. This is because the mixed workloads have higher possibility to have different patterns in upper address bits. Since we have a limited number of GTT entries, the diverse upper address patterns will lead to more GTT misses. However, the CT-Cache still shows performance improvements even in the case of the mixed workloads mainly due to the reduced latency of tag accesses.

Compared to the CT-Static and CT-w/oPAD, the CT-Cache shows better performance in case of multi-programmed workloads. In case of mixed (quad) workloads, CT-Cache shows better performance by 30.4% (33.3%) and 8.1% (5.8%) on average, compared to CT-Static and CT-w/oPAD, respectively. On the other hand, in case of single workloads, though CT-Cache exhibits performance improvement of 16.2% as compared to the baseline, CT-w/oPAD shows the best performance among four different schemes (21.8% performance improvement over the baseline, on average). This is because a single workload hardly uses all of the GTT entries. It also means the reactive subarray management would be sufficient for the workloads that have relatively small working sets. However, due to the rigidity of the subarray allocation and deallocation, the CT-Static cannot distribute the subarrays to

TABLE IV: Tag storage comparison for 256MB LLC.

	Baseline	CT-Cache
Tag storage	20.5MB	2.5MB
Global Tag Table	N/A	800B
Hit counters and SA_valid bit for CT-cache	N/A	272B
Total	20.5MB	2.501MB

the GTT entries depending on their capacity requirements. This rigidity leads to higher cache miss rates, which eventually results in a huge performance loss compared to the baseline (19.7% in the case of the mixed workloads). Overall, the CT-Cache shows stable performance improvements across single, quad, and mixed workloads as compared to the CT-Static and CT-w/oPAD.

C. Energy

Fig. 6 depicts the LLC and main memory energy consumption results for CT-Static, CT-w/oPAD, and CT-Cache normalized to the baseline. The CT-Cache shows unanimous energy reductions in all of the cases shown in Fig. 6 while the CT-Static and CT-w/oPAD show energy consumption more than the baseline in some cases (bars higher than 1.0 in Fig. 6). The CT-Cache reduces LLC and main memory energy consumption by 28.3%, 29.7%, and 20.0%, on average, compared to the baseline for single, quad, and mixed workloads, respectively. The main reasons of energy reduction can be summarized in two-folds: 1) the reduced tag storage consumes much less dynamic and leakage energy, and 2) reduced execution time also reduces leakage (standby) and refresh energy of LLC and main memory because the processor and main memory can be transitioned into low-power modes (e.g., power gating) earlier.

Compared to CT-Static and CT-w/oPAD, the CT-Cache shows more consistent energy reductions from the baseline across single, quad, and mixed workloads. Considering that CT-Static and CT-w/oPAD also use the reduced tag storages, the differences in energy consumptions across the four schemes are mainly attributed to the differences in the execution time (i.e., performance). Comparing the energy results (Fig. 6) with the performance results (Fig. 5), there is a strong reverse correlation. In other words, the higher performance a scheme shows, the lower energy consumption the scheme tends to exhibit.

D. Area

Table IV summarizes tag storage comparison between the baseline and CT-Cache. We omit the metadata (e.g., valid bit, coherence status bit, etc.) for each cache line in this comparison as the metadata are included in both baseline and CT-Cache. Please note that this is a conservative comparison as the CT-Cache only maintains 2-bits for replacement policy (5-bits are required for the baseline which employs LRU policy) and does not also require a dirty bit[7]. For the baseline, we need 41-bits tag for each cache line when using 64-bit address space. Since the baseline cache has 4M cache lines, we need a total of 20.5MB for tag storage. In the case of CT-Cache, we need 5-bits delta tag storage for each cache line, leading to a total of 2.5MB delta tag storage for 256MB LLC. We also require 200-bits (25-Bytes) for each entry of the GTT storage, resulting in total 800-Bytes for the 32-entry GTT. Furthermore, the CT-Cache maintains the hit counter and SA_valid bit for each subarray, which corresponds to 272-Bytes (17*128-bits). In total, the CT-Cache reduces the tag storage by 87.8% compared to the baseline. Due to the reduced tag storages, one can obtain latency and energy reduction for tag access, which eventually results in better performance and energy efficiency as shown in Sections IV-B and IV-C.

V. CONCLUSION

In this paper, we have proposed CT-Cache—a cache design based on tag compression. Our proposed CT-Cache decouples

---
[7]Since CT-Cache uses write-through policy, we do not need to maintain a dirty bit in each cache line.

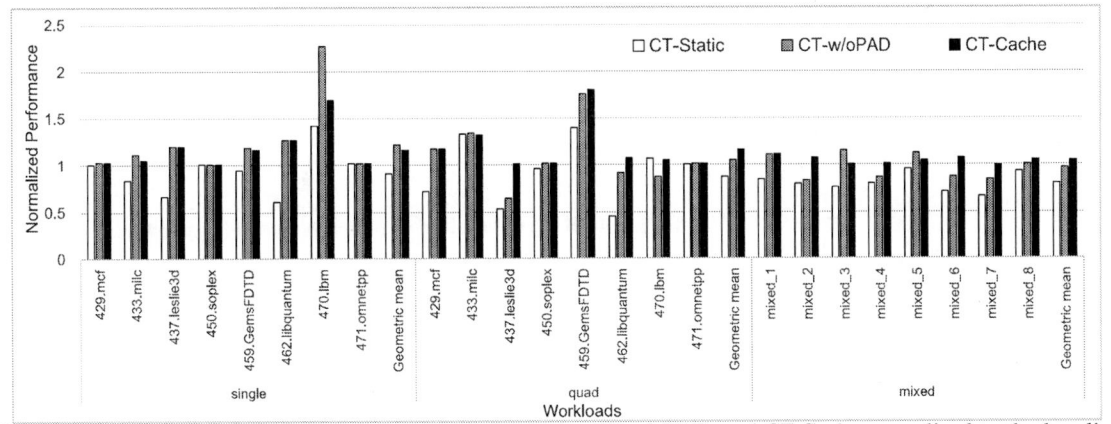

Fig. 5: Performance (weighted speedup) results of CT-Static, CT-w/oPAD, and CT-Cache normalized to the baseline.

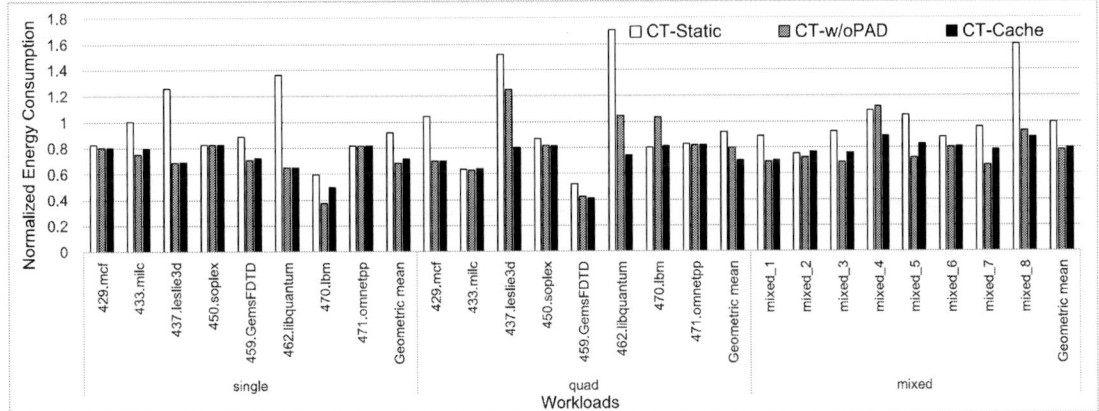

Fig. 6: Energy consumption results of CT-Static, CT-w/oPAD, and CT-Cache normalized to the baseline.

tag storages into the global tag table and delta tag arrays. By sharing the upper-bit tag storage, we can significantly reduce tag storage required for large last-level caches. Our proposed CT-Cache shows performance improvement and energy reduction of 4.7%~16.2% and 20.0%~29.7%, respectively, compared to the baseline (conventional caches that use non-compressed tag storage). The evaluation results verify that the CT-Cache shows stable and consistent performance improvements and energy reductions over the baseline as compared to the CT-Static and CT-w/oPAD (the two variants of our proposed CT-Cache). As our future work, we plan to (i) investigate the impact of full system environment including the operating systems on our proposed CT-Cache, (ii) perform sensitivity studies on various cache configurations (e.g., cache hierarchy and size, etc.), (iii) evaluate our proposed CT-Cache technique using write-back caches and provide performance and energy consumption comparison with the currently used write-through caches, (iv) devise a technique to reduce memory bandwidth requirements due to the write-through policy, and (v) develop formal methods to determine the design parameters in the CT-Cache.

ACKNOWLEDGMENT

Joonho Kong is the corresponding author of this paper. This research was supported by Basic Science Research Program through the National Research Foundation of Korea (NRF) funded by the Ministry of Science, ICT & Future Planning (2015R1C1A1A01051836) and Samsung Electronics. We would also like to thank anonymous reviewers for their constructive comments.

REFERENCES

[1] N. Binkert et al. The Gem5 Simulator. *SIGARCH Comput. Archit. News*, 39(2), 2011.

[2] K. Chandrasekar et al. DRAMPower: Open-source DRAM Power & Energy Estimation Tool.

[3] S. Eyerman and L. Eeckhout. Restating the case for weighted-IPC metrics to evaluate multiprogram workload performance. *IEEE Computer Architecture Letters*, 13(2):93–96, 2014.

[4] C.-C. Huang and V. Nagarajan. ATCache: reducing DRAM cache latency via a small SRAM tag cache. In *PACT*, pages 51–60, 2014.

[5] D. Jevdjic et al. Die-stacked DRAM caches for servers: hit ratio, latency, or bandwidth? have it all with footprint cache. In *ACM SIGARCH Computer Architecture News*, volume 41, pages 404–415, 2013.

[6] J. Kim et al. TLB index-based tagging for reducing data cache and TLB energy consumption. *IEEE Transactions on Computers*, 66(7):1200–1211, 2017.

[7] J. W. Kwak and Y. T. Jeon. Compressed tag architecture for low-power embedded cache systems. *Journal of Systems Architecture - Embedded Systems Design*, 56(9):419–428, 2010.

[8] J. Lee et al. TLB index-based tagging for cache energy reduction. In *ISLPED*, pages 85–90, 2011.

[9] S. Li et al. CACTI-P: Architecture-level modeling for SRAM-based structures with advanced leakage reduction techniques. In *ICCAD*, pages 694–701, 2011.

[10] G. H. Loh and M. D. Hill. Efficiently enabling conventional block sizes for very large die-stacked DRAM caches. In *MICRO*, pages 454–464, 2011.

[11] G. Pekhimenko et al. Base-delta-immediate compression: practical data compression for on-chip caches. In *PACT*, pages 377–388, 2012.

[12] P. Petrov and A. Orailoglu. Tag compression for low power in dynamically customizable embedded processors. *IEEE Transactions on Computer-Aided Design of Integrated Circuits and Systems*, 23(7):1031–1047, 2004.

[13] M. Poremba et al. DESTINY: A tool for modeling emerging 3D NVM and eDRAM caches. In *DATE*, pages 1543–1546, 2015.

[14] M. K. Qureshi and G. H. Loh. Fundamental latency trade-off in architecting dram caches: Outperforming impractical SRAM-tags with a simple and practical design. In *MICRO*, pages 235–246, 2012.

[15] K.-H. Yang et al. eTag: Tag-comparison in memory to achieve direct data access based on eDRAM to improve energy efficiency of DRAM cache. *IEEE Transactions on Circuits and Systems I: Regular Papers*, 64(4):858–868, 2017.

[16] V. Young et al. DICE: Compressing DRAM caches for bandwidth and capacity. In *ISCA*, pages 627–638, 2017.

High Bandwidth Off-Chip Memory Access Through Hybrid Switching and Inter-Chip Wireless Links

Sri Harsha Gade
Department of ECE
IIIT Delhi, New Delhi, India
Email: harshag@iiitd.ac.in

Hemanta Kumar Mondal
CNRS Lab-STICC
University of Southern Brittany, Lorient, France
Email: hemanta.kumar-mondal@univ-ubs.fr

Sujay Deb
Department of ECE
IIIT Delhi, New Delhi, India
Email: sdeb@iiitd.ac.in

Abstract—Off-chip memory performance in many core processors has remained unscaled due to limited pin bandwidth, number of memory controllers and interconnect limitations. It is one of the major bottlenecks for achieving high performance in many core processors, especially with increasing bandwidth requirements as more cores are integrated on a single chip. To achieve high bandwidth memory access, we propose an interconnection architecture with (i) off-chip wireless links for main memory access and (ii) hybrid switching with packet and circuit switching in on-chip mesh network. The off-chip wireless links are designed to provide high data and low energy access to off-chip memory. We enhance the intra-chip network by establishing circuit switch links between caches and memory controllers to provide low latency access, while inter-core communication is achieved through packet switching. The performance evaluation of the proposed architectures shows that they improve performance by 31.09% in runtime and 64.76% in memory access latency as compared to baseline, while consuming 56.57% less energy.

Keywords-Off-chip memory access; wireless links; hybrid switching;

I. INTRODUCTION

Technology scaling and shrinking transistor sizes have resulted in unprecedented levels of integration, allowing a steady growth in number of cores on a single chip to achieve high performance computing. Though highly efficient, the increasing number of cores on a single chip poses several challenges particularly for communication, both on-chip and off-chip. Advent of and advancements in Networks-on-Chip (NoCs) have mitigated several challenges of complex on-chip communication scenarios, providing efficient on-chip interconnection to match the computation capabilities.

However, the increasing core count has also led to the ever increasing demand for memory bandwidth while, the off-chip memory access has remained relatively unchanged due to limited pin bandwidth and number of memory controllers. This has widened the gap between demand and supply for memory bandwidth in multi-core and many core architectures[1], [2], [3]. Moreover, the metallic links used for interconnection with off-chip memory have high latency and energy per bit due to electrical wire limitations. The use of large and multi-level on-chip caches hide memory access latency to an extent, but the request and response

latency still runs into hundreds of clock cycles, limiting the performance. Furthermore, increasing system size increases the distance between caches and memory controllers leading to even higher latency. This severely restricts the performance in many core processors, even for moderately memory intensive applications.

The challenges of off-chip memory access are addressed by (i) reducing off-chip memory access latency and (ii) optimizing the cache to memory controller latency on chip. There have been several works, that tried to address both aspects of the memory access problem. Different interconnect level improvements have been proposed in [4], [5] to tackle the first challenge to reduce off-chip memory access latency. Authors in [4] have proposed the use of wireless technology for accessing off-chip memory and explore different wireless configurations for both off-chip and on-chip memory communication. HyWin topology in [5] proposes a heterogeneous NoC to interconnect memory subsystem with other subsystems using wireless links in CPU/GPU architectures. Both approaches show significant improvements in memory access latency as compared to traditional wired links. Similarly, optimizations have been proposed to on-chip memory organization and memory controllers to reduce memory bandwidth requirements or cache to memory controller latency. Authors in [6], [7] propose optimizations to reduce the number of memory accesses and bandwidth requirements, thereby improving off-chip memory performance. Different organizations for last level caches and their trade-offs for reducing memory accesses have been studied in [8]. These methods either try to reduce memory bandwidth requirements or improve on-chip performance to access memory controllers.

To enhance memory access performance, in this work, we tackle both off-chip and on-chip memory access performance by use of wireless links and hybrid switching strategy. Single hop, long range wireless links have been shown to provide significant benefits over traditional wired links at long distances[9], [10] and also support data broadcasting and multi-casting . These advantages make them ideal to achieve high performance off-chip memory access. To this end, we implement the interconnection to off-chip memory

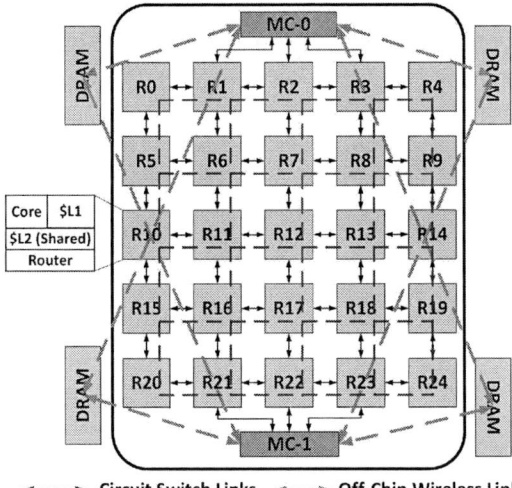

◄----► Circuit Switch Links   ◄---► Off-Chip Wireless Link

Figure 1. Proposed Architecture with Off-Chip Wireless Links and Hybrid Switching Strategy

using Orthogonal Frequency Division Multiplexing (OFDM) based wireless links. The improved off-chip performance is augmented by improvements to on-chip network through low latency circuit switch links between caches and memory controllers. For this purpose, a hybrid switching strategy is implemented, that combines packet switching for core-to-core traffic and circuit switching for core-to-memory traffic. The major contributions of this work are as follows:

1) An intra and inter-chip NoC optimization using wireless links to improve the performance of off-chip memory communication.
2) Hybrid switching strategy to combine packet switched NoC with circuit switch links for low latency data transfer between caches and memory controllers.
3) Inter-chip wireless link implementation using OFDM modulation for achieving high bandwidth with low per-bit energy consumption.

II. PROPOSED ARCHITECTURE WITH OFF-CHIP WIRELESS LINKS

The architecture of the proposed system with wireless links is shown in Fig. 1 and has been adopted from multicore processors like [11], [12]. Each tile is comprised of a x86 superscalar core, its private *L1* cache and a segment of shared cache. The last level cache interfaces with on-chip memory controllers, that handle all communication with off-chip memory. The off-chip memory is distributed into multiple modules and each memory controller can service requests to one or more main memory modules. The memory modules serviced by each memory controller are fixed and memory controller arbitrates between all memory requests for those modules on first-come-first-serve basis. The intra-chip NoC in the proposed architecture is interconnected

using mesh topology. Each tile is connected to a router through local buffered port and all such routers are connected in mesh topology using wired links as shown in Fig. 1. The routers associated with last level cache are also connected to the memory controllers for off-chip memory access. The proposed architecture improves performance of memory communication by implementing wireless links for off-chip communication and hybrid switching strategy for intra-chip network as shown in Fig. 1.

The baseline architecture differs from the proposed architecture in that it uses metallic links for both off-chip and on-chip communication and the NoC is comprised of packet switched network. The memory controllers interface with main memory using 256 bit wired links. The details of routers, wired and wireless links are provided in section III.

A. Wireless Link Implementation

To improve the performance of off-chip memory communication, we implement wireless links for interconnecting on-chip memory controllers with off-chip memory as shown in Fig. 1. For this purpose, each memory controller and main memory module is equipped with a wireless transceiver; all operating at same frequency channel using omnidirectional antennas. A token based approach with round robin arbitration is implemented to provide access of wireless channel to each Wireless Interface (WI). In order to achieve desired high bandwidth and low energy off-chip memory access with low overheads, we implement OFDM based transceiver operating at mm-wave frequencies. OFDM modulation, by use of orthogonal sub-channels, provides high spectral efficiency and also performs well against adverse channel effects. Moreover, the mostly discrete implementation of OFDM is suitable to achieve low area and power overheads for on-chip wireless transceivers.

OFDM modulation encodes data to be transmitted into different symbols, which are then modulated to respective sub-channel frequencies before being transmitted over the channel. The choice of data encoding scheme and number of sub-channel have significant impact on bandwidth and resilience to channel effects. We have used Quadrature Amplitude Modulation (QAM) scheme to encode data bits into symbols to ensure high spectral efficiency. The number of sub-channels (N) is chosen such that the sub-channel bandwidth ($B_N = B/N$; B is channel bandwidth) is sufficiently small enough to be unaffected by channel dispersion. In our evaluations, a loop antenna[13] with $B = 25GHz$ at $60GHz$ uses 256 sub-channels to achieve optimal performance. The dispersion characteristics and hardware details of OFDM are discussed in detail in section III.

B. Hybrid Switching Strategy

To enable high performance memory communication, we optimize the network between caches and memory controllers through a hybrid switching strategy that combines both

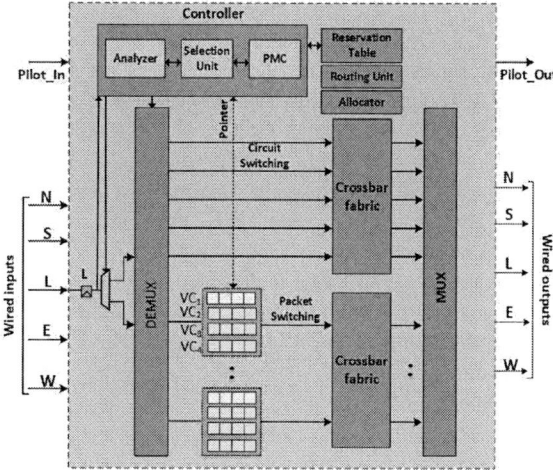

Figure 2. Dual Crossbar Router Architecture for Hybrid Switching

---

Algorithm 1 Pseudo Code for Circuit Switch Setup

Controller: Collect packet destination upon injection
Analyze the destination
if Memory Bound **then**
    Use Bufferless Crossbar
    Send circuit switch setup request to downstream router
    while !Acknowledgement **do** *Wait*
    end while
    Log circuit switch path into reservation table
    Transfer packet
else
    Use packet switching
end if

Reservation Table:
if Circuit switch request **then**
    Check for Source-destination
    if Exists **then** Update allocation period
    end if
end if

if Allocation Period = 0 **then**
    Break circuit switch
    Update reservation table
end if

---

packet and circuit switching. The hybrid switching strategy establishes circuit switch links between caches and memory controllers to achieve low latency memory access. The key design goal of the hybrid switching strategy is to provide low latency access to memory controllers without impacting inter-core communication on the packet switch network.

1) Dual Crossbar Router: The dual crossbar router architecture, shown in Figure 2, enables hybrid switching by allowing simultaneous use of circuit and packet switched paths. It resembles a traditional multi-ported router, except for the crossbar, which is replaced by dual crossbar and a controller circuit to handle both packet and circuit switch data. The dual crossbar is implemented using folded technique to support dual switching at a given cycle with low energy overhead. it consists of two 2-folded crossbars for a total of four switching elements as opposed to single conventional crossbar with five switching elements. The bufferless crossbar in Fig. 2 is used for circuit switch paths and eliminates buffers (as circuit switch path, once established does not require data to be stored along the path) along the path to save energy of read/write operations. The crossbar with buffered ports is used for packet switch data and acts as conventional switch circuit. The controller circuit to manage resources between both paths is comprised of analyzer and selection unit. The analyzer resolves the packets injected into the network and categorizes them as core bound or memory bound. This is then used by selection logic to determine the switching mode; circuit switch through bufferless crossbar for memory bound traffic and packet switch over buffered crossbar for all other destinations. The selection logic also resolves conflicts between packet switch data and circuit switch data at the router. This, along with circuit switch setup for hybrid switching is discussed in detail in further sections. The inter-router links, shared by both paths, is shared between circuit and packet switch data using Time

Division Multiplexing (TDM) technique.

2) Circuit Switch Setup: The proposed hybrid strategy establishes circuit switch links between caches and memory controllers to transfer memory requests and responses with low latency. The algorithm to setup and configure a circuit switch path between any cache and memory controller using a reservation table is shown in Algorithm 1. Any packet, after being injected into the router through local port, is analyzed by the controller to be determined as memory bound or not. Only a miss at shared cache generates a request to off-chip memory, which is designated as memory bound. Upon detecting a memory bound packet, the selection logic determines to use the bufferless crossbar. A request is sent to all downstream routers along the path to setup and reserve the circuit switch path between requesting tile and memory controller. The circuit switch setup is logged into a reservation table with control information; start time, allocation time, InPort and OutPort. The circuit switch path remains active for the entire duration of the allocation time and carries both request and response memory traffic between the corresponding nodes. At the end of allocation period, the circuit switch path is broken down to be used for setting up other paths or packet switching. At time of packet injection, if a circuit switch path already exists, the allocation period is updated to transfer newly injected packets.

3) Hybrid Switching: The proposed architecture shares router ports and inter-router links between both circuit

978-1-5386-7100-9/18 $31.00 © 2018 IEEE

Algorithm 2 Selection Logic for Hybrid Switching

Controller receives a task description from a core
if MI and !CI **then**
    Setup circuit switch
    Transfer packet
    Destroy circuit switch path
    Enable packet switching
else if CI and !MI **then**
    Use buffered crossbar
    Transfer packet
else
    Hybrid switching (packet+circuit)
    if Conflict Found **then**
        Set wait for packet switch
        Transfer circuit switch packet
        Transfer packet switch data
    end if
end if

Table I
DETAILS OF SYSTEM ARCHITECTURE

Component	Configuration
CPU	x86-64, Out-of-Order cores, 2 threads/core, $2.5GHz$
L1 Cache	$64KB$, 4-way, LRU policy, $64B$ line, 1 cycle latency
L2 Cache	$256KB$, 8-way, LRU policy, $64B$ line, 10 cycle latency
Memory	$2048MB$, 4 channels, 100 cycle latency

Table II
DETAILS OF WIRED AND WIRELESS NETWORK TOPOLOGY

Routers	5-port, 8-flit depth buffer per port
	5-stage (BW, RC, SA, ST, LT) for packet switching (both baseline and proposed)
	2-stage (ST, LT) for circuit switch
On-Chip Link	64-bit width, 1 cycle latency, 1 pJ/bit
Off-Chip Link	Baseline: Wired, 256-bit wide, 25 pJ/bit, 2 cycle latency
	Proposed: Wireless, $58GHz$ carrier, 195.36 Gbps, 1 cycle latency, 0.1 pJ/bit

and packet switch paths as shown in Fig. 1. Hence, it is imperative to resolve any conflicts between then, especially if a router falls along a circuit switch path. For this purpose, we have categorized the applications based on their memory communication and selection unit implements a priority logic as shown in Algorithm 2 to shared resources between packets from both switching modes. The workloads are categorized as i) Compute Intensive (CI), ii) Memory Intensive (MI) and iii) Hybrid. The categorization is done based on the number of memory instructions in the application. Each packet is designated to be one of these categories (control bits in the header), when they are injected into the network. For any workload categorized as MI, circuit switch data is always given priority over packet switch data. At any router along the circuit switch path, in case of conflict for a link, packet switch data waits till circuit switch communication is completed and the path is destroyed. Similarly, in case of CI applications, packet switch is given priority over circuit switch. For hybrid workloads, when circuit switch path is setup, a waiting time is established at all routers along path for packet switch data. In case of conflict, the packet switch data is held in router buffers till the end of waiting period, while circuit switch data is transferred over inter-router links. Once waiting time expires, packet switch data is transferred over the inter-router link.

III. EXPERIMENTAL RESULTS

In this section, we describe the experimental setup to evaluate performance and overheads of proposed architecture.

A. Experimental Setup

To evaluate the performance of proposed architecture, we use Multi2Sim[14], a modular, cycle accurate simulator that allows detailed configuration of CPU/GPU cores, memory hierarchy and network topology. The system architecture and specifications for both baseline and proposed approaches is shown in Table I. We have used 48 core system for all our evaluations. Each core has its private *L1* cache and a single *L2* cache is shared by four cores. Application level traffic and performance metrics are obtained by simulating different benchmarks from [15]. To achieve fair comparison, each benchmark is simulated 10 times and average performance metrics over all runs are considered for evaluation.

The proposed wired and wireless network architecture is shown in Table II. All routers, in both baseline and proposed, are interconnected in mesh topology as shown in Fig. 1. The off-chip wireless link in proposed design is implemented using OFDM scheme with loop antenna[13], which has a bandwidth of $25GHz$ at $58GHz$ carrier frequency. The channel has worst case dispersion of $T_m = 5ns$ and requires 256 sub-channels to satisfy $B_N >> 1/T_m$ criteria as described in II-A. Using QAM encoding, we achieve a total bandwidth of $195.36Gbps$ for off-chip wireless link. The power amplifier and low noise amplifier for the wireless transceiver are adopted from [16], [9]. The routers and WI components are synthesized in Synopsys Design Compiler to compute the area and power overheads.

B. Performance Evaluation

The performance of proposed architecture is evaluated in terms of application runtime, memory access latency and peak bandwidth and compared with baseline architecture.

1) Application Runtime: The application speedup achieved with proposed architecture over baseline for different benchmarks is shown in Figure 3. A typical application is comprised of both computation and memory instructions and runtime, in many core processors, is becoming more

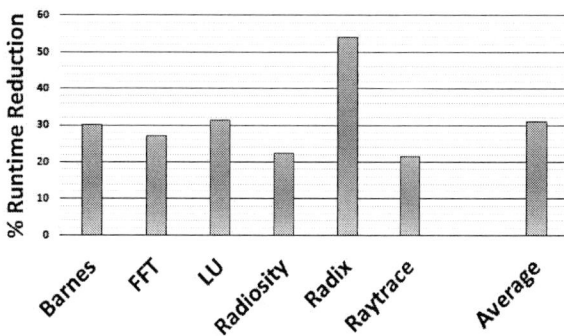

Figure 3. Improvement in application runtime using proposed architecture over baseline for different application scenarios

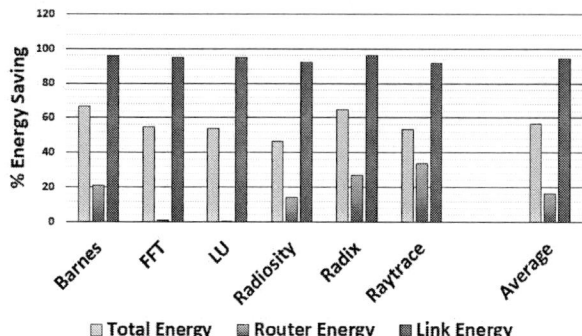

Figure 5. Decrease in energy using proposed architecture over baseline for different application scenarios

dominated by performance of memory instructions. It is dependent not only on number of memory accesses, but also on distribution of memory accesses throughout the application execution. On average, proposed architecture reduces application runtime by 31.09% as compared to baseline architecture. The speedup is attributed to both low latency access of off-chip wireless links and circuit switch links between caches and memory controllers.

2) Memory Access Latency: One of the most important metrics for processor performance is memory access latency, that acts as major bottleneck for obtaining instruction or program data. Figure 4 shows decrease in average memory access latency provided by the proposed architecture as compared to baseline for different benchmarks. On average, across all applications, the memory access latency is reduced by 64.76% over baseline architecture. This significant improvement is primarily achieved by use of high data rate wireless links, which provide low latency across long distances, unlike wired links.

3) Bandwidth: We evaluate the peak network bandwidth to understand the maximum throughput provided by the proposed topology. Figure 4 shows the improvement in bandwidth provided by proposed topology over baseline

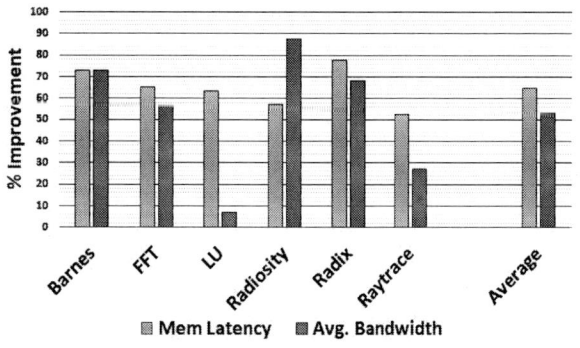

Figure 4. Improvement in memory access latency and average bandwidth using proposed architecture over baseline for different application scenarios

mesh for different benchmarks. Using hybrid switching, it improves the network bandwidth by 53.04%, on average as compared to baseline mesh. The improvements in the bandwidth are primarily achieved due to the efficient use of both packet switched and circuit switched network through dual crossbar routers. The packet switch network efficiently handles inter-core communication, similar to that of baseline mesh, while the circuit switch links significantly reduce memory access bottlenecks on chip, thereby improving the overall network throughput.

C. Energy Savings

Figure 5 shows the packet energy savings using proposed topology over baseline mesh topology. The packet energy calculations include both on-chip and off-chip packet communication and is averaged over all transferred packets. As observed, proposed topology saves 56.57% of the total packet energy, on average as compared to the baseline. We also evaluate the individual components; router and link energy to understand their contribution. Though dual crossbar router adds power overhead for additional crossbar, the reduction in total runtime and bufferless transfer for all on-chip memory communication reduces total router energy consumption. As compared to baseline, the hybrid switching technique reduces router energy by 16.23%, on average. The use of wireless links for off-chip memory access significantly reduces total link energy, as off-chip wired links consume exponentially increasing energy and also require multiple cycles to complete packet transfer. Combined, end-to-end energy consumed to service any read/write request is considerably smaller in the proposed topology.

D. Overheads

To implement hybrid switching scheme, we integrate dual crossbar and switching controller with each router. Controller with analyzer and selection unit together occupies $1006.83 \mu m^2$. The area requirement for dual crossbar router (including control units, buffers, crossbar, modified arbiter, RC, and VCs) is $9969.59 \mu m^2$. The implementation

978-1-5386-7100-9/18 $31.00 © 2018 IEEE

of OFDM transceiver with 256 sub-channels incurs an area overhead of $0.092mm^2$. The additional area overhead associated with proposed approach is 0.16% of total silicon area for 48 cores. The total power consumption of dual crossbar router is $409\mu W$. The controller with analyzer and selection unit consumes $42.84\mu W$ power for all its operations. The power consumption of OFDM scheme is $18.825mW$ at 256 sub-channels. The power amplifier and LNA combined consume $19mW$ of power. Hence, the OFDM based transceiver consumes a total of $0.194pJ/bit$ at $195.32Gbps$ data rate.

IV. CONCLUSION

In this work, we low energy, high bandwidth memory access through hybrid switching strategy and off-chip wireless links to main memory. A dual crossbar router with hybrid switching is implemented to enable both circuit and packet switching modes with low interference between inter-core and core-to-memory communication. The off-chip communication is optimized by use of OFDM based wireless links, that provide high bandwidth access with low energy consumption. A detailed evaluation with 48 core system shows that, it achieves runtime and memory latency improvements of 31% and 65% respectively as compared to baseline mesh architecture. It also reduces the average packet energy by 57%, while adding low area and power overheads to the baseline topology.

REFERENCES

[1] A. Kagi, J. R. Goodman, and D. Burger, "Memory bandwidth limitations of future microprocessors," in *Computer Architecture, 1996 23rd Annual International Symposium on*, May 1996, pp. 78–78.

[2] B. M. Rogers, A. Krishna, G. B. Bell, K. Vu, X. Jiang, and Y. Solihin, "Scaling the bandwidth wall: Challenges in and avenues for cmp scaling," in *Proceedings of the 36th Annual International Symposium on Computer Architecture*, ser. ISCA '09. New York, NY, USA: ACM, 2009, pp. 371–382. [Online]. Available: http://doi.acm.org/10.1145/1555754.1555801

[3] C. Liu, A. Sivasubramaniam, and M. Kandemir, "Organizing the last line of defense before hitting the memory wall for cmps," in *Software, IEE Proceedings-*, Feb 2004, pp. 176–185.

[4] M. A. I. Sikder, A. Kodi, W. Rayess, D. DiTomaso, D. Matolak, and S. Kaya, "Exploring wireless technology for off-chip memory access," in *2016 IEEE 24th Annual Symposium on High-Performance Interconnects (HOTI)*, Aug 2016, pp. 92–99.

[5] S. H. Gade and S. Deb, "Hywin: Hybrid wireless noc with sandboxed sub-networks for cpu/gpu architectures," *IEEE Transactions on Computers*, vol. PP, no. 99, pp. 1–1, 2016.

[6] C. Yu and P. Petrov, "Off-chip memory bandwidth minimization through cache partitioning for multi-core platforms," in *Design Automation Conference*, June 2010, pp. 132–137.

[7] A. Sharifi, E. Kultursay, M. Kandemir, and C. R. Das, "Addressing end-to-end memory access latency in noc-based multicores," in *2012 45th Annual IEEE/ACM International Symposium on Microarchitecture*, Dec 2012, pp. 294–304.

[8] C. Liu, A. Sivasubramaniam, and M. Kandemir, "Organizing the last line of defense before hitting the memory wall for cmps," in *Software, IEE Proceedings-*, Feb 2004, pp. 176–185.

[9] S. Deb, K. Chang, X. Yu, S. P. Sah, M. Cosic, A. Ganguly, P. P. Pande, B. Belzer, and D. Heo, "Design of an energy-efficient cmos-compatible noc architecture with millimeter-wave wireless interconnects," *IEEE Transactions on Computers*, vol. 62, no. 12, pp. 2382–2396, Dec 2013.

[10] A. Ganguly, K. Chang, S. Deb, P. P. Pande, B. Belzer, and C. Teuscher, "Scalable hybrid wireless network-on-chip architectures for multicore systems," *IEEE Transactions on Computers*, vol. 60, no. 10, pp. 1485–1502, Oct 2011.

[11] M. Huang, M. Mehalel, R. Arvapalli, and S. He, "An energy efficient 32-nm 20-mb shared on-die l3 cache for intel xeon processor e5 family," *IEEE Journal of Solid-State Circuits*, vol. 48, no. 8, pp. 1954–1962, Aug 2013.

[12] G. K. Konstadinidis, H. P. Li, F. Schumacher, V. Krishnaswamy, H. Cho, S. Dash, R. P. Masleid, C. Zheng, Y. D. Lin, P. Loewenstein, H. Park, V. Srinivasan, D. Huang, C. Hwang, W. Hsu, C. McAllister, J. Brooks, H. Pham, S. Turullols, Y. Yanggong, R. Golla, A. P. Smith, and A. Vahidsafa, "Sparc m7: A 20 nm 32-core 64 mb l3 cache processor," *IEEE Journal of Solid-State Circuits*, vol. 51, no. 1, pp. 79–91, Jan 2016.

[13] O. Markish, O. Katz, B. Sheinman, D. Corcos, and D. Elad, "On-chip millimeter wave antennas and transceivers," in *Proceedings of the 9th International Symposium on Networks-on-Chip*, ser. NOCS '15. New York, NY, USA: ACM, 2015, pp. 11:1–11:7. [Online]. Available: http://doi.acm.org/10.1145/2786572.2789983

[14] R. Ubal, B. Jang, P. Mistry, D. Schaa, and D. Kaeli, "Multi2sim: A simulation framework for cpu-gpu computing," in *Proceedings of the 21st International Conference on Parallel Architectures and Compilation Techniques*, ser. PACT '12. New York, NY, USA: ACM, 2012, pp. 335–344. [Online]. Available: http://doi.acm.org/10.1145/2370816.2370865

[15] S. Woo, M. Ohara, E. Torrie, J. P. Singh, and A. Gupta, "The SPLASH-2 Programs: Characterization and Methodological Considerations," in *Proc. of the 22nd International Symposium on Computer Architecture*, June 1995.

[16] S. Kaushik, M. Agrawal, H. K. Mondal, S. H. Gade, and S. Deb, "Path loss-aware adaptive transmission power control scheme for energy-efficient wireless noc," in *2017 IEEE 60th International Midwest Symposium on Circuits and Systems (MWSCAS)*, Aug 2017, pp. 132–135.

Investigating Reliability and Security of Integrated Circuits and Systems

Qiaoyan Yu, Zhiming Zhang, and Jaya Dofe
Department of Electrical and Computer Engineering
University of New Hampshire
Durham, New Hampshire 03824, USA
Email: qiaoyan.yu@unh.edu

Abstract—Reliability and security are two important aspects of integrated circuits and systems. Both of them are under the umbrella of resilience. Traditionally, reliability and security issues are managed in a separate fashion. This work reviews the similarities and differences between the reliability and security of circuits and systems and provides a comprehensive survey on the existing efforts that jointly consider the reliability and security of memory, processors, embedded systems, and on-chip communication network. Furthermore, this work introduces a general flow that guides circuit and system designers to cooperatively address the reliability vulnerabilities and potential security threats in a unified framework.

I. INTRODUCTION

Technology scaling and new emerging devices raise new challenges for the reliability of very-large-scale integration (VLSI) circuits and systems. A subtle variation on a process, voltage, and temperature (PVT) and internal noise (such as crosstalk) are more likely to result in a hardware fault than before. As increasing hardware faults lead to more and more system errors, it becomes imperative to revisit the traditional circuit and system design methodologies to address the reliability issues. Meanwhile, the globalized business model of integrated circuit (IC) supply chain brings in new concerns on hardware integrity and security. Traditionally, the mechanisms for VLSI reliability and security are developed in a separate fashion, which may result in degradation in system performance and significant overhead on hardware cost.

As indicated in the recent survey [1], the physical causes for the unreliable and insecure system effects are tightly coupled. The unreliability induced by design and technology imperfectness sometimes gives us opportunities to address the security challenges. Meanwhile, the unreliabilities also create new challenges for us to handle the security threats from supply chain. The detailed trade-offs between reliability and security due to the physical phenomena observed in design and technologies are summarized in [1]. In this work, we conduct more comprehensive survey than the related works [1], [2] and provide more examples and discussions for the existing countermeasure designs, which jointly considers the circuits and systems reliability and security.

The remainder of this work is organized as follows. In Section II, we elaborate the relationship between reliability and security aspects. Section III describes two practical examples of how reliability vulnerabilities can be exploited to design

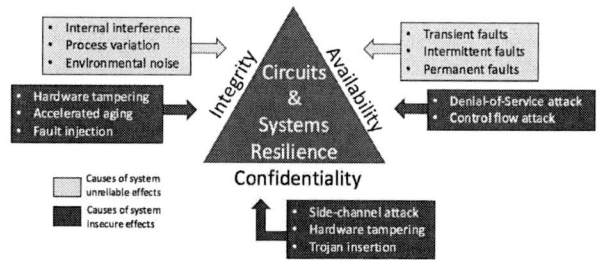

Fig. 1: Three edges of circuits and systems resilience against unintentional and intentional interferences.

an security attack surface. We introduce the representable existing efforts that jointly consider the reliability and security for memory, processors, embedded systems, and on-chip communication network in Section IV. We provide two case studies to illustrate the general flow of considering circuits and systems reliability and security simultaneously in Section V. We conclude this work in Section VI.

II. RELATIONSHIP BETWEEN RELIABILITY AND SECURITY

A. Definitions

In this work, the definition of *reliability* is broad. If a device or system does not *consistently* behave as it is supposed to do, we consider that device or system as unreliable. The causes of unreliability include unstable material properties, device characteristics that are prone to internal or external noise, and device aging. To be more specific, we regard reliability as a combination of integrity and availability. The definition of *security* includes three crucial aspects: integrity, availability, and confidentiality. In the domain of VLSI systems, we specifically refer security threats to counterfeiting, intellectual property (IP) piracy, hardware tampering, fault injection attacks, and side-channel attacks.

In Fig. 1, we summarize the overall relationship between reliability and security, which simultaneously contribute to the circuit and system resilience against unintentional and intentional interferences. The causes of system unreliable and insecure effects can both harm the integrity. Crosstalk coupling, imprecisely predicted process variation, and harsh operating environment introduce noise to the system, thus leading it to deviate from its behavior/performance specified in the design stage. The integrity loss due to the causes of unreliable effects can be learned from a large number of statistical samples. In contrast, the means that an attacker could use to sabotage the circuit and system integrity are not predictable. By tampering hardware, accelerating component aging, and injecting faults (typically permanent), an attacker can alter the original system functionality and thus ruin the system's integrity. Unintentional faults (i.e., transient, intermittent, and permanent faults) randomly stop a system from functioning properly. An intentional attack (e.g., denial-of-service (DoS) or control flow attack) could suddenly interrupt the system's normal operation to fulfill the attacker's malicious objective. Although the presence of the unreliable and insecure system effects are different in the sense of timing, location, and duration, the physical mechanisms for those effects are similar. The main difference depends on whether such effects are induced by unintentional or intentional interferences. The third edge of the system resilience, confidentiality, is typically sabotaged by side-channel attacks, hardware tampering, and Trojan insertions, which aim to extract the confidential information from the system. It is uncommon that a random event could leak the confidential information.

B. Reliability and Security of General Applications

Generally speaking, reliability and security are under the same umbrella of system resilience to intentional and unintentional interferences. Depending on the specific application, the relationship between reliability and security could vary. For instance, in a wireless sensor network, enhancement of signal power is preferred to improve the reliability with regard to the capability of combating the background noise [3]. However, the increased signal power will ease the eavesdropping attack. Thus, there is a need to trade off reliability and security. For embedded systems, reliability and security can be addressed with a single mechanism. The control flow monitor in [4] is capable of detecting the deviated control flow due to a code injection attack and alerting the system with transient bit flipping errors that affect the normal control flow.

C. Reliability and Security of Integrated Circuits and Systems

ICs and systems demand designers to co-consider multiple design aspects. In the past decades, the main design objectives focus on performance, power consumption, and area. As the technology scales down to the nanometer regime and the increasing number of emerging technologies allow integration of more transistors into a single die, the reliability of integrated circuits and systems is increasingly important and challenging. The globalized IC supply chain brings in new concerns on security. The emerging requirement for both reliability and security urges us to revisit the chip design and deployment. In future, we may observe a shift in design paradigm.

The causes of unreliable devices, such as the variation on PVT, aging and wear-out mechanisms can be utilized to develop security primitives [1]. Process variation is unique for each fabrication line. The uniqueness of process variation can

be used to design physical unclonable function units to authenticate the original chips. Aging and wear-out mechanisms such as bias temperature instability (BTI), hot carrier injection (HCI), and electromigration can be exploited to indicate the chip age and thus guide us to effectively detect counterfeit (especially, re-cycled) chips.

The effects of PVT variation and aging mechanism act as a double-edged sword. Attackers can also exploit those effects to perform fault attacks to extract the secret key applied in the cryptosystem [2]. Some attackers could heat up the surrounding temperature of the chip, changing the transistor threshold voltage. The altered threshold voltage affects the circuit switching frequency and causes the circuit to experience unexpected voltage glitches, which lead the crypto module to skip some instructions.

III. PRACTICAL EXAMPLES TO MOTIVATE CO-CONSIDERING RELIABILITY AND SECURITY

In this section, we use rowhammer attack and fault attack as examples to introduce how the system reliability could be weakened to assist the implementation of security attacks.

A. Rowhammer Attack

As the technology node for memory keeps scaling, the reliability of memory emerges as a concern. If memory failures are not addressed, attackers could take advantage of the existing vulnerability on memory to create an attack surface and then breach the system. One example of such attacks is *rowhammer attack*, which disturbs the DRAM memory cells by repeatedly accessing the neighboring cells to cause the bit flips. In [5], the authors demonstrate that malicious software can take advantage of the CLFLUSH instruction to obtain the privilege escalation in a Linux system through flipping a bit in a page table. This attack is practically demonstrated on 129 DRAM cells from different vendors [5] and 110 DRAM modules are found susceptible to rowhammer attacks. The rowhammer attacks target special memory management features: deduplication and virtualization [6], [7]. Memory deduplication allows an attacker to reverse-map any physical page into a virtual page if he/she knows the page's contents. Hence, the operation system can be tricked to map a victim owner memory page on the attacker-controlled physical memory pages. In [7], the authors exploit cross-VM settings where bit flips are induced by rowhammer attacks to crack memory isolation enforced by virtualization. This type of exploits can bypass password authentication on an OpenSSH server.

B. Fault Attack

Many applications and systems rely on the cryptographic algorithms to protect information privacy and confidentiality. Although the existing cryptographic algorithms such as block ciphers (e.g., RSA, DES, and AES) and stream ciphers (e.g., RC4) are mathematically strong to resist linear or non-linear cryptanalysis, their physical implementations could be vulnerable to the attacks aiming to retrieve the secret key. Among

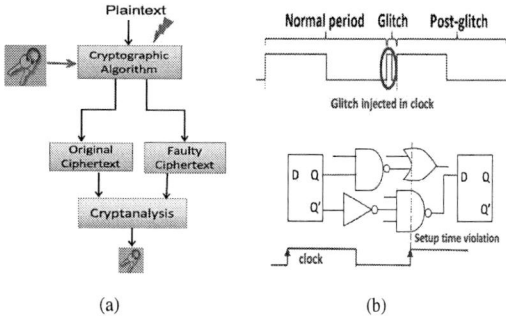

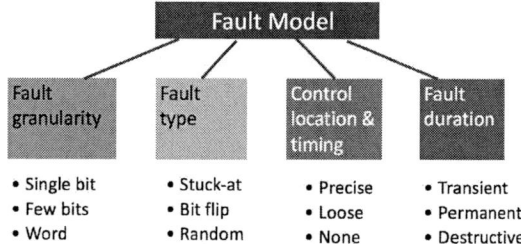

Fig. 2: Fault attack. (a) principle, and (b) an example: over-clocking on sequential circuits.

Fig. 3: Fault model for fault attacks.

various attack methods, fault attack is one of the most efficient approaches.

Fault attack is a method of exploiting the system reliability vulnerability to weaken the system security. The principle of a fault attack on the cryptosystem is shown in Fig. 2(a). We use overclocking [8] as an example to explain how reliability can be exploited as an attack surface. Figure 2(b) illustrates a clock glitch (a shortened clock pulse) leading to a setup-time violation. The combinational logic between two D-flip-flops cannot complete the logic gate switching upon the new inputs before the incoming clock rising edge. As a result, the correct output of the combinational logic cannot be saved in the second D-flip-flop and thus a fault is generated. In addition to overclocking, under-powering [9] and overheating [10] are also capable to result in timing-violation faults. It is well known that a proper clock period cannot be less than the sum of the latency from the clock edge to the flip-flop output, the maximum delay of the combinational logic, and the setup time of the flip-flop [11]. Any environmental change on the power line and circuit operation temperature will lead the selected clock period to be unsatisfied for proper sequential operations.

A successful fault attack procedure is composed of four steps: fault injection access, actual fault injection, fault effect capture, and fault observation. To obtain effective faulty ciphertexts or expected faulty outputs, attackers need to build a precise fault model. As shown in Fig. 3, a fault model specifies the fault granularity, the expected fault type, the location and timing for fault injection, and the fault duration [12]. The four aspects covered in the fault model are the ones being extensively studied in the community who is aware of circuit and system reliability.

One noticeable difference between unintentional and intentional fault injection is: the former one generates random faults and the latter one produces controllable faults. The random faults are predictable if a large amount of statistical observation data points are available. Hence, it is relatively easy for fault-tolerant methods to detect the random faults. In contrast, fault attacks control fault location and injection timing, as well as fault duration. From the countermeasure designer's point of view, the potential fault location, timing, and duration are unknown, and thus the corresponding fault detection and mitigation is challenging.

IV. EXISTING EFFORTS FOR JOINTLY CONSIDERING RELIABILITY AND SECURITY

This section discusses the existing works that co-consider reliability and security issues in memory, processors, embedded systems, and on-chip interconnect networks.

A. Memory

The reliability and security techniques for memory differ in a way they respond to an error. Reliability techniques detect and correct random faults, however, security techniques focus more on detecting the intentional faults. The work [13] presents a unified off-chip memory integrity protection scheme (IVEC) that supports detection of physical attacks for security and correction of random errors for reliability by combining the off-chip integrity verification with Error Correcting Code (ECC). IVEC can correct single-bit errors or multi-bit errors from a DRAM chip without the addition of extra ECC bits. Further, it can handle a stronger fault model like multi-chip-multi-error using parity bits. G. Saileshwar et al. [14] propose a reliability security co-design method (named *Synergy*), which enhances the secure execution and supports strong reliability for the systems with ECC Dual In-line Memory Modules (ECC-DIMMs). Commercial ECC-DIMMs typically have eight data chips and one ECC-chip. The synergy scheme leverages the data tampering detection capability from the message authentication codes (MACs) to detect memory errors by co-locating the MAC inside the ECC chip in ECC-DIMMs.

B. Processors and Embedded Systems

Security of processors is often compromised by code injection attacks. Reliability concerns emerge when the unintended code is executed in modern processors because of transient or permanent errors. The work [15] proposes a hardware-level framework called Reliability and Security Engine (RSE) to combine the error detection and security mechanisms. In RSE, security is ensured through a memory layout randomization module and reliability is guaranteed by an instruction checker module. Integrated microarchitecture support for security and reliability in multi-core processors is presented in [16]. In that work, dynamic integrity checkers are exploited to secure the code running on the multiple cores and further these checkers are used to check the integrity of each other periodically to ensure reliable operations. Runtime hardware/software technique is introduced in [17] to detect code injection attacks

978-1-5386-7100-9/18 $31.00 © 2018 IEEE

and transient faults. The essence of that method is to perform basic block validation using microinstructions. In some special cases, security-oriented message authentication codes can also be used to restore information for the purpose of reliability. For example, the security-oriented robust codes are applied as a predictor to protect AES circuits against any injection or bit-flipping attacks [18]. As a result, both security and reliability of the system are achieved.

C. On-chip Interconnect Network

As billions of transistors are integrated on a single die, Networks-on-Chip (NoCs) emerge as an efficient on-chip communication infrastructure. Fault-tolerant techniques at NoC physical, data-link, transport, and network layers have been developed in the past decade to tackle the increasing occurrence of faults on NoCs [19], [20]. Error control coding is one of the prevalent methods to address the transient errors on NoC links [21]. Inherent information redundancy is exploited to manage transient errors on NoC routing arbitration unit [22]. Fault-tolerant routing algorithms take advantage of redundant routing paths to deliver the packets over a faulty NoC [23].

IP cores for data processing and storage are typically well-protected from security threats with sophisticated permission checking and data authentication mechanisms. As NoC is a critical component for multiprocessor systems-on-chip (MPSoCs), a compromised NoC will breach MPSoC systems [24], [25]. For instance, hardware Trojans inserted in NoCs will result in information leakage, unauthorized memory accesses, and DoS (e.g. incorrect path routing and packet drop). The majority of hardware-level methods for NoC security management are employed in the network interface or the equivalent module that directly connects with IP cores [26], [27]. The memory addresses for the packet destination are checked to prevent untrusted IP cores from accessing trusted IP cores [26]. Permission lookup tables are implemented in a memory protection unit to authenticate packets [28].

Among those consequences induced by a tampered NoC, some DoS phenomena are similar with those caused by transient or permanent faults occurred on the NoC links or routers. Hence, the principles of some countermeasures against DoS attacks are comparable with the transient and permanent error detection and correction methods. The work [29] sets a throughput limit on the switch allocator to detect a hardware Trojan. In [30], the first security wrapper compares the destination address of the outgoing packet with the illegal address range to block masquerade packets. The second security wrapper counts the number of clock cycles elapsed since the first packet is injected into the NoC, and postpones the injection of the next packet. To promptly mitigate the hardware Trojan effects before the bandwidth depletion happens to the entire NoC, the work [25] proposes a collaborative dynamic permutation and flit integrity check method. Their router-level Trojan mitigation mechanism raises the bar for an adversary to simultaneously control multiple routing hops to create a malicious communication path between two IP cores in the NoC-based MPSoC.

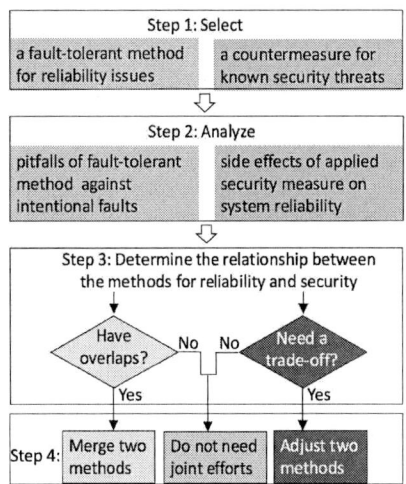

Fig. 4: Proposed generalized design flowchart for the development of joint efforts protecting both reliability and security.

V. PROPOSED GENERALIZED DESIGN FLOW FOR RELIABILITY AND SECURITY ASSURANCE

A. Proposed Design Flow for Joint Effort Development

We propose a generalized design flow to guide the designers who are interested in jointly addressing the system reliability and security. We divide the entire flow into four steps, as shown in Fig. 4. In the first step, we select a fault-tolerant method for reliability issues and a countermeasure for the known security threats. In the second step, we analyze the pitfalls of the selected fault-tolerant method against intentional faults and the side effects of the applied security measure on system reliability. In the third step, we determine whether the method for reliability and the security measure have overlaps or need some trade-offs for the optimized system reliability or security. In the last step, we merge (adjust) the fault-tolerant method and security measure for the having overlaps (requiring trade-offs) scenario. If there is no interference between the fault-tolerant method and the countermeasure against security threats, we leave the two efforts as they are.

The method of merging and adjusting the adopted fault-tolerant method and security measure varies with the subject of interest. In the next subsection, we use a security primitive and a crypto engine in a three-dimensional (3D) chip as two examples to explain how to merge and adjust the adopted methods for reliability and security, respectively.

B. Case Study 1: Joint Efforts for a Security Primitive

Figures 5 and 6 depict representative fault-tolerant methods and countermeasures against fault attack, respectively. Fault-detection methods are used to ensure the integrity of the crypto module. The fault-detection methods [31], [32] shown in Fig. 5 only consider the scenario that faults randomly happen in the cipher implementation and they only ensure the system reliability. The method of permutation and depermutation [33] and the masking approach shown in Fig. 6 have the capability

978-1-5386-7100-9/18 $31.00 © 2018 IEEE

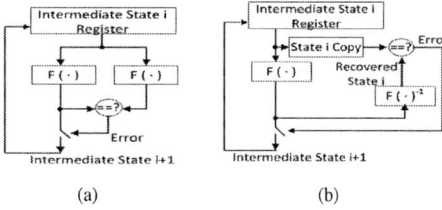

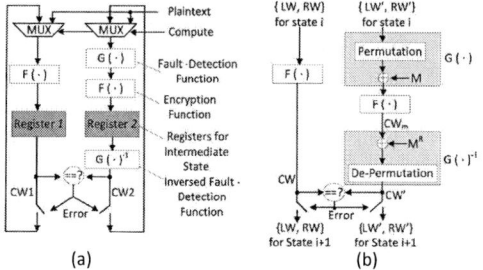

Fig. 7: Proposed method. (a) micro-architecture for fault attacks (b) fault-detection for SIMON.

Fig. 5: Fault-detection methods for a crypto module: (a) Double Modular Redundancy (DMR) and (b) inverse function. $F(\cdot)$ represents a round function for the cipher.

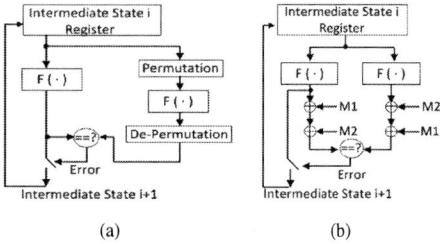

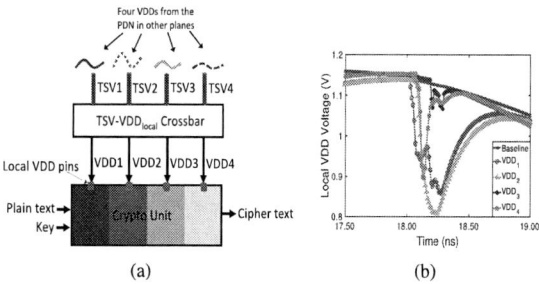

Fig. 6: Fault attacks countermeasures: (a) permutation, and (b) masking.

Fig. 8: Proposed countermeasure against CPA attacks in 3D ICs. (a) Schematic diagram and (b) voltage waveforms of four VDD_is applied to the crypto unit.

to detect some random faults and thwart fault attacks, but they cannot address the symmetric fault insertions [34]. In addition, the methods shown in Fig. 6 cannot detect the faults induced in the intermediate state register. This is because both data paths receive the same faulty inputs, and no difference will be observed at the comparison stage. As a result, a faulty ciphertext can bypass the fault detection.

To address the aforementioned limitations, our prior work [34] merges the methods for reliability and security into a unified framework shown in Fig. 7(a). The two data paths, the encryption alone $F(\cdot)$ and the fault-detection $G(\cdot)$ followed with encryption $F(\cdot)$, are executed simultaneously. The outputs of two paths (i.e., intermediate state) are fed to registers 1 and 2, respectively. The inverse fault-detection function $G(\cdot)^{-1}$ and comparison of intermediate ciphertexts (CWs) are performed before the start of the next round. If the inconsistency is noticed, then a fault is detected and the faulty ciphertext is discarded. Fault detection $G(\cdot)$ and inverse fault detection function $G(\cdot)^{-1}$ can be chosen depending on the specific encryption algorithm. The proposed method uses two registers to store the intermediate ciphertexts of two paths, eliminating the risk of fault attacks on the single state register. We applied the proposed micro-architecture to iterative SIMON cipher implementation as shown in Fig. 7(b). We use a combination of permutation and masking for fault detection prior to encryption $F(\cdot)$. To retrieve the original ciphertext, we depermute and demask $G(\cdot)^{-1}$ the intermediate ciphertext before passing it to the next round. This arrangement efficiently detects the random faults as well as intelligent faults inserted by fault attacks in the crypto engine.

C. Case Study 2: Trade-off between Reliability and Security for Crypto Engine in a 3D Chip

3D integration is an emerging technology to ensure the growth in transistor density and performance. The critical reliability challenges in the physical design of 3D ICs include power and signal integrity issues due to the parasitic impedance of neighboring planes, through-silicon vias (TSVs), and large overall load currents. The power supply voltage should be reliably delivered to the devices that are distributed throughout multiple planes without incurring the excessive power supply noise. To suppress power supply and power gating noise from the neighboring active planes, researchers develop different decoupling capacitor networks in 3D power distribution network (PDN): traditional, always-on, and reconfigurable [35]. On the other hand, it has been demonstrated that adding extra noise to the cryptosystem is an effective method to distort the power profile of the crypto module, thus thwarting the correlation power analysis (CPA) attack [36]. Our prior work [37] indicates that the utilization of decoupling capacitors indeed eases the CPA attack in 3D ICs.

To resolve the dilemma regarding the reliability and security of a crypto engine in 3D chips, we first examine the feature of the noise induced by 3D PDNs. Our simulation results show that the 3D PDN noise is additive in the context of CPA attacks in 3D chips [38]. Next, we propose to exploit the intrinsic PDN noise from other planes to induce additive noise to the power measurement of the crypto module embedded in a 3D chip. The proposed countermeasure against CPA attacks is depicted in Fig. 8(a). In our method, we divide the crypto unit into multiple sub-units, each being driven by a local

978-1-5386-7100-9/18 $31.00 © 2018 IEEE   110

supply voltage $VDD_i (i = 1, 2, 3, 4)$. We utilize a crossbar to connect the local VDD pins with the PDN nodes close to four power TSVs. Due to the non-uniform switching activities in an individual 3D plane, each TSV passes a unique voltage from other 3D planes to the plane carrying the crypto unit. The example shown in Fig. 8(b) indicates that both the peak voltage magnitude and occurrence timing for the four VDDs are not exactly same. We exploit this variance on the local VDDs to blur the correlation between the measured and hypothesized power profile for the crypto module. To further introduce power noise, we can bring in a small random number generator to randomly connect inputs and outputs of the crossbar.

VI. CONCLUSION

Reliability and security issues are typically regarded as two non-related aspects of circuits and systems. This work highlights the scenarios that the effects of unreliable and insecure system could be jointly managed. We overview the physical causes of the reliability vulnerabilities of VLSI circuits and systems and summarize the fault-related security threats on cryptosystems. Two specific examples, rowhammer attack and fault attack, are provided to explain how reliability vulnerabilities can be exploited as an attack surface to harm the system security. To co-manage the VLSI systems reliability and security, we propose a general design flow that merges or adjusts the adopted fault-tolerant methods and countermeasures against fault-related security threats into a unified framework. Two case studies are provided to illustrate the application of the proposed design flow.

ACKNOWLEDGMENT

This research is partially supported by the NSF CAREER grant (No. CNS-1652474) and NSF/SRC STARSS grant (No. CNS-1717130).

REFERENCES

[1] F. Rahman, D. Forte, and M. M. Tehranipoor, "Reliability vs. security: Challenges and opportunities for developing reliable and secure integrated circuits," in *Proc. IRPS'16*, pp. 4C-6-1-4C-6-10, April 2016.

[2] H. Amrouch *et al.*, "Special session: emerging (Un-)reliability based security threats and mitigations for embedded systems," in *Proc. CASES'17*, pp. 1-10, Oct 2017.

[3] J. Zhu, Y. Zou, and B. Zheng, "Physical-Layer Security and Reliability Challenges for Industrial Wireless Sensor Networks," *IEEE Access*, vol. 5, pp. 5313-5320, 2017.

[4] K. Patel and S. Parameswaran, "SHIELD: A software hardware design methodology for security and reliability of MPSoCs," in *Proc. DAC'08*, pp. 858-861, June 2008.

[5] Y. Kim *et al.*, "Flipping bits in memory without accessing them: An experimental study of DRAM disturbance errors," in *Proc. ISCA'14*, pp. 361-372, June 2014.

[6] K. Razavi *et al.*, "Flip Feng Shui: Hammering a Needle in the Software Stack," in *Proc. USENIX'16*, pp. 1-18, 2016.

[7] Y. Xiao, X. Zhang, Y. Zhang, and R. Teodorescu, "One Bit Flips, One Cloud Flops: Cross-VM Row Hammer Attacks and Privilege Escalation," in *Proc. USENIX'16*, pp. 19-35, 2016.

[8] J. Balasch, B. Gierlichs, and I. Verbauwhede, "An In-depth and Black-box Characterization of the Effects of Clock Glitches on 8-bit MCUs," in *Proc. FDTC'11*, pp. 105-114, Sept 2011.

[9] J. Balasch *et al.*, "Power Analysis of Atmel CryptoMemory – Recovering Keys from Secure EEPROMs," in *Proc. CT-RSA*, pp. 19-34, 2012.

[10] M. Hutter and J. Schmidt, "The Temperature Side Channel and Heating Fault Attacks," *IACR*, vol. 2014, p. 190, 2014.

[11] B. Razavi, "Fundamentals of Microelectronics," Hoboken, NJ: John Wiley & Sons, 2008.

[12] D. Karaklaji, J. M. Schmidt, and I. Verbauwhede, "Hardware Designer's Guide to Fault Attacks," *TVLSI*, vol. 21, pp. 2295-2306, Dec 2013.

[13] R. Huang and G. E. Suh, "IVEC: Off-chip Memory Integrity Protection for Both Security and Reliability," in *Proc. ISCA'10*, pp. 395-406, 2010.

[14] G. Saileshwar, P. J. Nair, P. Ramrakhyani, W. Elsasser, and M. K. Qureshi, "SYNERGY: Rethinking Secure-Memory Design for Error-Correcting Memories," in *Proc. HPCA'18*, pp. 454-465, Feb 2018.

[15] N. Nakka, Z. Kalbarczyk, R. K. Iyer, and J. Xu, "An architectural framework for providing reliability and security support," in *Proc. DSN'04*, pp. 585-594, June 2004.

[16] A. Kanuparthi and R. Karri, "Reliable integrity checking in multicore processors," *ACM Trans. Archit. Code Optim.*, vol. 12, pp. 10:1-10:23, May 2015.

[17] R. G. Ragel and S. Parameswaran, "IMPRES: integrated monitoring for processor reliability and security," in *Proc. DAC*, pp. 502-505, 2006.

[18] L. Bu and M. Kinsy, "Hardening AES Hardware Implementations Against Fault and Error Inject Attacks," in *Proc. GLSVLSI'18*, 2018.

[19] M. Radetzki, C. Feng, X. Zhao, and A. Jantsch, "Methods for Fault Tolerance in Networks-on-chip," *ACM Comput. Surv.*, vol. 46, pp. 8:1-8:38, July 2013.

[20] S. Werner, J. Navaridas, and M. Luján, "A Survey on Design Approaches to Circumvent Permanent Faults in Networks-on-Chip," *ACM Comput. Surv.*, vol. 48, pp. 59:1-59:36, Mar. 2016.

[21] Q. Yu and P. Ampadu, "Dual-Layer Adaptive Error Control for Network-on-Chip Links," *TVLSI*, vol. 20, pp. 1304-1317, July 2012.

[22] Q. Yu, M. Zhang, and P. Ampadu, "Addressing Network-on-chip Router Transient Errors with Inherent Information Redundancy," *ACM Trans. Embed. Comput. Syst.*, vol. 12, pp. 105:1-105:21, July 2013.

[23] Y.-Y. Chen *et al.*, "Path-Diversity-Aware Fault-Tolerant Routing Algorithm for Network-on-Chip Systems," *IEEE Trans. Parallel Distrib. Syst.*, vol. 28, pp. 838-849, Mar. 2017.

[24] J. Rajesh, D. M. Ancajas, K. Chakraborty, and S. Roy, "Runtime Detection of a Bandwidth Denial Attack from a Rogue Network-on-Chip," in *Proc. NOCS '15*, pp. 8:1-8:8, 2015.

[25] J. Frey and Q. Yu, "A Hardened Network-on-chip Design Using Runtime Hardware Trojan Mitigation Methods," *Integr. VLSI Journal*, vol. 56, pp. 15-31, Jan. 2017.

[26] L. Fiorin *et al.*, "Secure Memory Accesses on Networks-on-Chip," *Transactions on Computers*, vol. 57, pp. 1216-1229, Sept 2008.

[27] H. M. G. Wassel *et al.*, "Networks on Chip with Provable Security Properties," *IEEE Micro*, vol. 34, pp. 57-68, May 2014.

[28] J. Sepulveda *et al.*, "Hybrid-on-chip communication architecture for dynamic MP-SoC protection," in *Proc. SBCCI'12*, pp. 1-6, Aug 2012.

[29] Y. Wang and G. E. Suh, "Efficient Timing Channel Protection for On-Chip Networks," in *Proc. Networks-on-Chip*, pp. 142-151, May 2012.

[30] S. Baron, M. S. Wangham, and C. A. Zeferino, "Security mechanisms to improve the availability of a Network-on-Chip," in *Proc. ICECS'13*, pp. 609-612, Dec 2013.

[31] G. Natale, M. Doulcier, M. L. Flottes, and B. Rouzeyre, "A Reliable Architecture for Parallel Implementations of the Advanced Encryption Standard," *J. Electron. Test.*, vol. 25, pp. 269-278, Aug. 2009.

[32] R. Karri, K. Wu, P. Mishra, and Y. Kim, "Concurrent error detection of fault-based side-channel cryptanalysis of 128-bit symmetric block ciphers," in *Proc. DAC'01*, pp. 579-584, June 2001.

[33] X. Guo and R. Karri, "Recomputing with Permuted Operands: A Concurrent Error Detection Approach," *TCAD*, vol. 32, pp. 1595-1608, Oct 2013.

[34] J. Dofe, J. Frey, H. Pahlevanzadeh, and Q. Yu, "Strengthening SIMON Implementation Against Intelligent Fault Attacks," *IEEE Embedded Systems Letters*, vol. 7, pp. 113-116, Dec 2015.

[35] H. Wang and E. Salman, "Decoupling Capacitor Topologies for TSV-Based 3-D ICs With Power Gating," *TVLSI*, vol. 23, pp. 2983-2991, Dec 2015.

[36] D. Das *et al.*, "High efficiency power side-channel attack immunity using noise injection in attenuated signature domain," in *Proc. HOST'17*, pp. 62-67, May 2017.

[37] J. Dofe, Z. Zhang, Q. Yu, C. Yan, and E. Salman, "Impact of Power Distribution Network on Power Analysis Attacks in Three-Dimensional Integrated Circuits," in *Proc. GLSVLSI '17*, pp. 327-332, 2017.

[38] J. Dofe and Q. Yu, "Exploiting PDN Noise to Thwart Correlation Power Analysis Attacks in 3D ICs," *to appear in SLIP 2018*.

2018 IEEE Computer Society Annual Symposium on VLSI

Reliability and Security in Non-Volatile Processors, Two Sides of the Same Coin

Patrick Cronin*, Chengmo Yang*, Yongpan Liu[†]

* Department of Electrical and Computer Engineering
University of Delaware
{ptrick, chengmo}@udel.edu

[‡] Department of Electronic Engineering
Tsinghua University
ypliu@tsinghua.edu.cn

Abstract—**Non-volatile processors (NVPs), which harvest energy to power themselves and employ a NV memory hierarchy, are expected to be deployed as edge computation units to gather data or perform operations in either tough environments or for extended periods of time. These devices must not only be reliable to survive their long-term deployments, but also be secure to withstand new types of attacks which can be performed by attackers targeting physically accessible devices. While it may be tempting to view security and reliability as striving to achieve similar goals, they are fundamentally different. In particular, reliability focuses on random uncorrelated faults and thus cannot hope to function in an insecure system where an attacker can (and will) cause correlated faults. This paper studies the security and reliability challenges in the checkpoint system of NVPs, aiming at enhancing the integrity and confidentiality of checkpoints while at the same time protecting the hardware-based checkpoint system against power and wear-out attacks.**

Keywords—*Reliability, Security, Non-Volatile Processors*

I. INTRODUCTION

Cybersecurity continues to cause a buzz in all manners of system design, from major cloud deployed applications to embedded system firmware, to hardware itself. In our work, we focus on the gray area between cybersecurity and reliability when designing attack countermeasures. This is especially important to many embedded systems, and in fact, crucial to non-volatile processors (NVPs) which are the focus of our work.

Embedded systems continuously grow smaller, faster, and more capable. While semiconductor technology has advanced quickly, shrinking devices and their power requirements, battery technology has not keep pace, and designers struggle to create devices that are small and light enough for new and interesting functions such as integrated health monitoring for clothing and remote sensing, while being able to operate for more than a few hours. To remedy this issue, non-volatile processors (NVPs) [1] have been proposed. NVPs comprise a set of devices that utilize energy harvesting, energy storage, and a fully non-volatile memory (NVM) hierarchy to provide guaranteed forward progress of computation in an environment with intermittent power availability. Most importantly, the processor's NVM hierarchy allows for rapid and low overhead checkpointing. The ability to quickly shutdown and restart, paired with the ability to passively harvest energy makes the NVP an exciting idea for new IoT applications.

While NVPs offer amazing benefits over traditional processors, these come with a host of new security issues. A key

feature of NVPs is their checkpoint system. Checkpointing is usually thought of in the scope of reliability, but in the case of NVPs, they are vital to backup/restore operations which let NVM's computation move forward under intermittent power supply. Furthermore, it is our belief that this checkpoint system actually creates security challenges. In this work, we will focus on these new challenges that NVPs present to the security researches and discuss several techniques to mitigate them that blur the line between reliability and security.

Specifically, we examine two possible attacks against the checkpoint system in NVPs. One attack aims to destroy the saved states, thus causing potentially unrecoverable errors and crashing the device. The other attack aims to wear-out the device especially the non-volatile flip-flips, given the fact that each NVM cell is subject to a limited number of write cycles [2]. This paper also suggests NVP-specific countermeasures against these two attacks. In a nutshell, this work addresses two major topics regarding the NVP:

1) Simultaneously enhancing the security and reliability of NVP's checkpoint system.
2) Guarding the NVP from malicious wear-out while also improving NVP reliability.

The rest of this paper will be organized as follows. Section II will discuss NVP and security background. Section III will discuss the correlation of security and reliability challenges in the checkpoint system. Section IV will discuss wear-out attacks and their mitigations within the NVP system. Section V will conclude the paper.

II. BACKGROUND

A. Non-Volatile Processors

NVPs outperform traditional volatile processors in areas where power is intermittent. NVPs differ from volatile processors mainly in how they handle loss of power. A typical NVP is shown in Figure 1. It consists of three key facets: distributed non-volatile flip flops (nvFFs), nvFF controllers and a power management unit (PMU).

To explain the operation of an NVP, consider the state where it operates from stored energy within a capacitor local to the device. The PMU monitors the voltage levels of this capacitor, and when the voltage levels fall to a pre-defined *shutdown* threshold, the PMU signals the processor that it must shut down. At this point, the processor begins to checkpoint. Important to this checkpointing process is the fact that the

978-1-5386-7100-9/18 $31.00 © 2018 IEEE

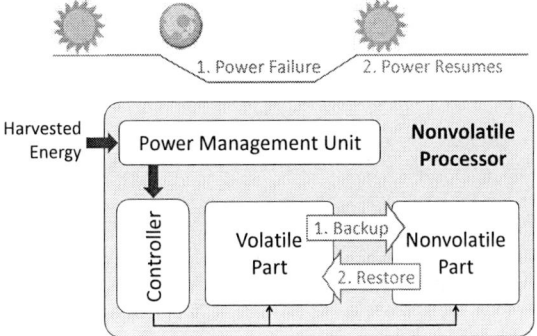

Fig. 1: NVP backup and restore system. The ambient energy over time is shown at the top. At time point 1, ambient energy disappears, and the processor must shutdown. The checkpoint process moves data from the volatile part to the non-volatile part. When power returns at time 2, data is restored from the non-volatile part to the volatile part.

NVP has an NVM hierarchy comprised of non-volatile storage, non-volatile main memory (RAM), a non-volatile register file (composed of nvFFs), and a volatile register file. Each volatile register is connected to an nvFF. Under normal operating conditions, the NVP utilizes the volatile registers to carry out computations. As these registers are the only volatile portion of the NVP, they are the only part to checkpoint when the NVP is told to shut down. During the checkpointing process, the value in each volatile register is written to the corresponding nvFF. This process is controlled by hardware without involving the OS or the applications and hence is extremely fast (in microseconds [3]).

In operation, and while hibernating, the NVP utilizes an energy harvesting unit to gather energy from the surrounding environment. This energy harvesting unit can take the form of photovoltaic, piezoelectric, thermo-electrical, and wireless energy harvesting techniques [4]. Harvested energy is stored in a capacitor and the voltage of the capacitor is monitored by the PMU. Upon reaching the *wake-up* voltage threshold, the PMU decides that the processor has enough energy to carry out computations and awakes the processor. The values of nvFFs are written to the corresponding volatile registers to resume computation.

B. Checkpoint and Restore in NVPs

In an NVP, upon the advent of backup and recovery control signals, the value stored in each standard flip-flop is transferred to the corresponding nvFF. The nvFF controller is responsible for generating essential read/write control signals for all nvFFs. Early designs adopted a fully-parallel control scheme shortening backup and recovery time, but causing large area overhead and frequent read/write inrush current. More recent solutions reduce such overhead as well as the degree of parallelism by organizing the nvFFs into groups. In [5], researchers implement a distributed nvFF controller which alleviates inrush current. Another method reduces backup/restore overhead utilizing data compression. Wang *et al.* [6], [7] and Sheng *et al.* [8] minimize the number of nvFFs using

a compression controller. Significant area saving is achieved at the expense of degradation in backup speed.

Frequent backup/resumption differentiates NVPs significantly from traditional processors. NVPs necessitates software management schemes with backup/resumption support. To this end, Ransford *et al.* [9], [10] develop a software system called Mementos consisting of compile-time instrumentation and run-time energy-aware state checkpointing, which enables long-running computations to span power loss events. At compile time, Mementos inserts trigger points in the program, which are calls to a Mementos library function that estimates available energy. At runtime, Mementos monitors the capacitor voltage and triggers checkpointing.

Ransford *et al.* [11] summarize the consistency errors when using NVM to back up. Errors are categorized into NV-internal inconsistency and NV-external inconsistency. NV-internal inconsistency happens if data are not fully updated to NVM before power depletion. System status cannot resume due to the incorrect version stored in NVM. NV-external inconsistency happens when the NVM is updated after one checkpoint, and the energy is depleted before the next checkpoint. After power resumes, the program will roll back to the last checkpoint while the content in NVM cannot roll back. If the updated data in NVM is used during re-execution from the last checkpoint, an error will occur.

Xie *et al.* [12] discuss the consistency errors in NVP and propose a consistency-aware checkpointing solution to eliminate errors. The proposed solution guarantees that there is a checkpoint between each load-store pair. The rationality of eliminating errors is to preclude the use of the updated data in NVM in re-executions after rolling back. Lucia *et al.* [13] claim that, for intermittent systems with hybrid memory architecture, it is necessary to capture the entire program status for both volatile and non-volatile parts. They state that programmers need to carefully define task boundaries to divide the whole program into atomic tasks. They propose to insert nodes between task boundaries, which will both checkpoint volatile data and version non-volatile data.

III. CHECKPOINT SECURITY

Checkpointing represents one of the unique and most vulnerable portions of the NVP. In standard reliability practices, checkpoints are utilized as a backup measure, a place for execution to fall back to should an error occur. In comparison, NVPs operate in an intermittent power environment and thus must cut power to the processor at certain times. A checkpoint acts as a record for the most recent values stored within the registers of the device and thus checkpoints are optimized for minimal power utilization.

Fig. 2 compares NVPs and traditional processors in terms of which portions of the system must be backed up. In the case of a standard "reliable system", checkpoints keep track of data structures and other important parts of the computation. The traditional system has the luxury of creating a checkpoint, verifying its correctness, writing to disk, and then deleting the previous checkpoint. In the NVP, this is unnecessary as only the registers of the system are volatile and everything else is safely stored in NVM until power is returned. More crucially, in current NVP architectures where the checkpoint nvFFs

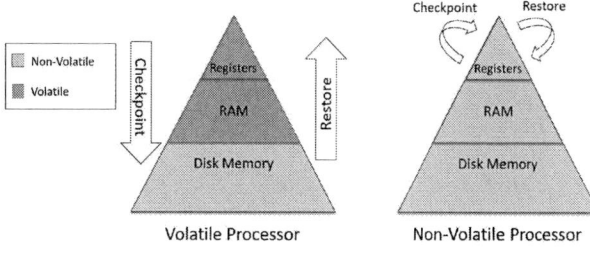

Fig. 2: Checkpoints in volatile and non-volatile processors. NVPs only need to checkpoint registers and this is done in hardware. The process is extremely fast (in micro-seconds [3]).

are directly attached to their volatile counterparts, creating a checkpoint in the nvFFs overwrites the previous checkpoint at the same time, destroying it. As a result, checkpoints do not add reliability advantages to the NVP, but instead act as a major attack vector. If an attacker can modify even a single bit within one of the nvFFs that hold the checkpoint, it is conceivable that an unrecoverable error will be caused. In this section, we will present attack vectors against the checkpoint process as well as countermeasures that improve not only the security of the system but also its reliability.

A. Attack vectors

Current NVP designs call for 'batch' checkpointing where as many nvFFs are checkpointed as possible (limited by simultaneous power requirements) in parallel, and upon their completion, the next batch is checkpointed, repeating until all nvFFs have been checkpointed. An attacker, aware of this technique, could interrupt the power supply in a precise manner in-between batches, leaving only a portion of the checkpoint completed. While this does not cause any bit errors in the register values, it makes the data in nvFFs inconsistent. This could cause the NVP to be completely unusable and/or unreliable; which is unacceptable in any medical or safety critical device.

As the checkpoint process is destructive, i.e., creating a new checkpoint destroys the previous checkpoint, if there is any error in the checkpoint, there is no way to recover the system with only a single checkpoint. From this perspective, it is preferable to maintain multiple checkpoints in NVPs and store old checkpoints in the non-volatile memory or storage. However, the inherent trust of old checkpoints becomes an issue. As most non-volatile memory devices only offer limited programming cycles, an attacker can perform the so-called 'wear out attacks' to non-invasively cause bit errors in the NVM. As the checkpoint system and its correctness are fundamental to the correct execution of programs running on the NVP, this attack vector is important and requires a solution.

B. Countermeasures

To enhance the safety and security of NVPs, we suggest a two-stage checkpoint system. It utilizes checksums to detect errors in the checkpoint, and intelligently keeps a record of previous checkpoints such that if there is an error with a checkpoint, a slightly longer restore process can be utilized to recover the error. We further suggest that checkpoints be

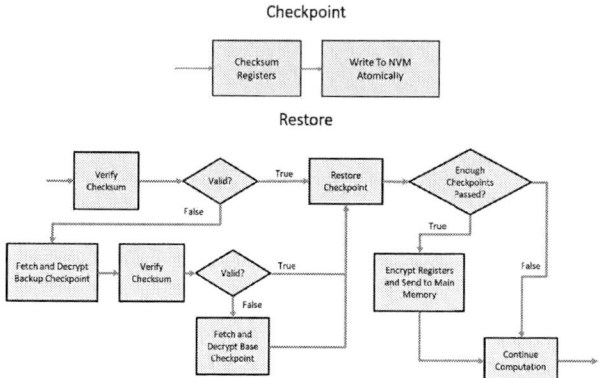

Fig. 3: A flowchart for the secure checkpointing process. Checkpoints are checksummed and scrambled to increase reliability and confidentiality. Resuming potentially utilizes multiple previous checkpoints to increase reliability.

scrambled to defeat attackers attempting to read checkpoint values. Finally, checkpoint atomicity is explored.

1) Two Stage Checkpointing: The two-stage checkpoint process is detailed in Fig. 3. Instead of blindly trusting that a checkpoint is safe during restore, the system creates a checksum for each checkpoint which must be verified upon the restoration of that checkpoint before the processor will proceed. In this way, bit-flip errors, whether malicious or otherwise, will be detected by the NVP upon restore. Due to the energy sensitive nature of NVPs, it is up to the designer to choose the checksum algorithm utilized, with the understanding that a more complex checksum will lower the probability that an attack or random bit-flip(s) will be uncaught by the system.

While a checksum provides a way to detect errors, it provides no recovery support. To this end, we propose to keep at least two more checkpoints in the NVP. These checkpoints are strategically chosen to increase the *reliability* and *security* of the system.

The first such "backup checkpoint" should be chosen such that it is a few checkpoints older than the current checkpoint. The system rolls back to this checkpoint if an error in the current checkpoint is detected. It represents a trade-off between how often this "backup checkpoint" must be updated and the amount of computation done between it and the current checkpoint. This "backup checkpoint" will necessarily be kept in the NVM and thus should be updated infrequently such that there is minimal time spent waiting for NVM R/W operations. This checkpoint will also have its own checksum.

The second "backup checkpoint" can be viewed as a fail-safe. This checkpoint should represent a completely safe state for the device to restore to, such as the entry point to the program it is currently running. Due to the fact that this is our last and most drastic restore operation, it is suggested that it be written to a special read-only memory location on the NVP such that it cannot be corrupted.

The overall two-stage checkpoint should work as follows: upon a checkpoint, the NVP takes a checksum of its registers and stores the checksum as well as the register values in nvFF.

978-1-5386-7100-9/18 $31.00 © 2018 IEEE

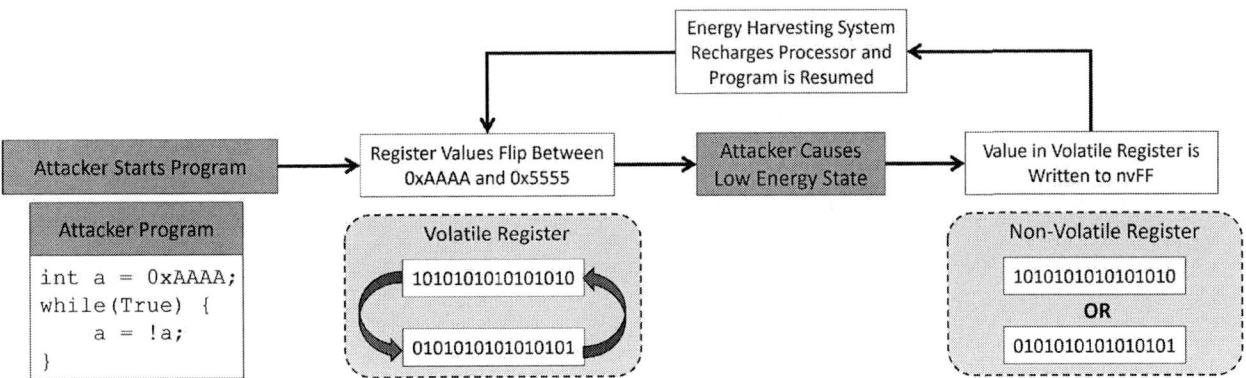

Fig. 4: A sample wear out attack in nvFFs.

Upon restore, the system reads the checkpoint and verifies the checksum. If the checksum and checkpoint are inconsistent, the NVP will attempt to restore from the most recent "backup checkpoint". Again, the checksum for this checkpoint will be checked as well. If this checksum and checkpoint are found to be inconsistent, the NVP attempts to restore from the read only fail-safe checkpoint.

In this system, reliability is enhanced as there are multiple backup checkpoints to fall back on should the system encounter an error, and security is enhanced as the NVP no longer naïvely trusts checkpoints without ensuring their correctness.

2) Checkpoint Scrambling: In many cases, it is expected that an NVP will be employed as some form of sensor node. It is reasonable to assume that the NVP is not allowed to transmit the data that it collects in plaintext as it may be confidential (e.g. health data) and thus must encrypt the data. A clever attacker could cut power to a device when it is in the middle of an encryption and force the NVP to checkpoint the secret keys. If this checkpoint is selected to be stored in main memory as per the two stage checkpoint process, the checkpoint would contain these secret keys indefinitely and the attacker may be able to find a way to view them. To this end, we suggest that the above two-stage checkpointing system be modified to also scramble or encrypt the checkpoint values if the checkpoint is stored in main memory in order to prevent this type of snooping. This increases security of the system as it makes it much harder for a resourceful attacker to modify checkpoint values in a way that can both decrypt and pass a checksum function.

3) Checkpoint Atomicity: Finally, it is important that the checkpoint process is atomic since a partially executed checkpoint process destroys the old checkpoint while leaving the new checkpoint incomplete, and the program is almost certain to crash. Guaranteeing this atomicity can be done via ensuring that the low power threshold contains enough power to backup all registers.

With this three pronged approach; two stage checkpointing, checkpoint scrambling, and checkpoint atomicity, it is possible to see a case where the goals of security and reliability reinforce each other. Without the backup checkpoints the system is vulnerable to both checkpoint inconsistency errors as well as bitflip errors of stored values. Either case could be caused by both random faults or malicious actions. Check-

point scrambling and checksums assist in detecting errors in checkpoints and securing the system from snooping.

IV. WEAR OUT ATTACKS

The field of NVM security is closely linked to that of NVM endurance. There have been multiple attacks demonstrated in SSDs and NVM architectures that show how a clever attacker can quickly exhaust the endurance of a system and destroy it. In some cases, this can take place in just seconds. As an example, [14] demonstrates that wear out attacks are possible against phase change memory (PCM) which typically has an endurance of 10^9. Compromising these systems in as little as 30 seconds [15].

The NVP utilizes nvFFs for its checkpoint registers which are written at every checkpoint. Made from non-volatile memory, these nvFFs share the same endurance issues, and yet very few wear-leveling works have been proposed for non-volatile registers. Yang [16] suggested utilizing compiler techniques to find a minimal set of volatile registers to write to nvFFs as well as utilizing dynamic inter-register rotation to minimize bit flips in nvFFs.

A. Attack

A smart attacker that can modify the firmware of an NVP, perhaps at the factory, or that can control the ambient power supply to a device could be capable of causing faster shutdowns. With the modified firmware the attacker can also ensure that a single register is targeted by continually inverting the value stored within it. An example of this can be seen in Fig. 4. The attacker loads code to the NVP to cause a high number of bit flips in a register. The attacker then forces the NVP to consume energy quickly and triggers a checkpoint. Content in the volatile register is written to the nvFF. This process repeats when the processor regains energy and restarts computation. If the attacker can force this process to happen quickly, the nvFFs will wear out.

B. Defense

One proposal to decrease the effectiveness of these bitflip attacks is to borrow from wear leveling techniques utilized in other NVM architectures such as SSDs. To solve this problem, we propose a register wear leveling technique utilizing both

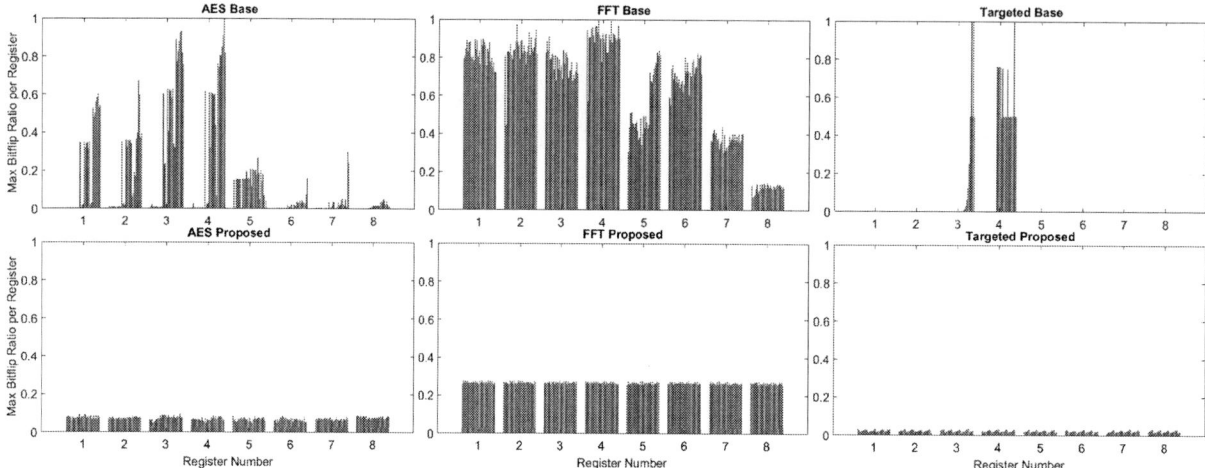

Fig. 5: Bit Flip Ratio of each bit to max bit flips in AES, FFT, and a targeted malicious program, without (top row) and with (bottom row) the proposed defense applied. Startup threshold manipulation decreases the total bitflip count, while wear-leveling evens the bitflip count.

inter- and intra-register wear-leveling, as well as providing the NVP the ability to vary its shutdown interval based upon the number of bitflips currently observed. Inter-register rotation remaps a logical register to different physical registers periodically, while intra-register rotation rotates values within the registers. It is important to note that one cannot force these rotations to occur too often. If the rotations occur at every shutdown, there is no ability for data locality to minimize the number of bitflips. To this end, we further suggest rotating at pre-defined intervals, that is every 64 shutdowns for example. A detailed discussion of this implementation is available in [17].

As the attacker attempts to take advantage of the system and create shutdowns as quickly as possible, we also suggest a system to change the wakeup threshold of the device. If the device detects rapid shutdowns, it can increase its wakeup threshold, thus requiring that the system charge for longer before beginning to drain power, extending the time between shutdowns, slowing the attackers progress. As this hurts the responsiveness of the processor, we allow for the processor to decrease this threshold if it detects that few shutdowns are taking place. The two methods for modifying these thresholds are the *bitflip method* and *shutdown count method*. The shutdown count method increases or decreases the energy threshold based upon the number of recent shutdowns. The bitflip methods only examines the number of bitflips that occur between shutdowns, increasing the threshold if a large number of bitflips are detected [17].

Importantly, this idea of register rotation can increase the endurance of registers under both normal and malicious conditions. This is due to the inherent imbalance of register usage in many programs. To demonstrate this, Fig. 5 shows the bit-flip ratio of three programs, AES encryption, FFT computation, and a targeted malicious program which flips the maximum number of bits in a register as shown in Figure 4. This is the worst case scenario that mimics a malicious attack. In the original implementation of the NVP (top row of Fig. 5), writes to the registers are imbalanced in all three

tested programs. However, once the rotation system and power management have been applied (bottom row of Fig. 5), the register bitflip numbers become far smoother.

Fig. 5 also confirms that the proposed defense scheme helps not only the malicious targeted program but also standard applications such as AES and FFT, indiciating that the objectives of security and reliability are aligned. By increasing the endurance of nvFFs we mitigate the possibility of an attacker prematurely wearing out nvFFs in the NVP. As this system is designed to be always on, it also increases the reliability by decreasing the chance of a register failure.

V. CONCLUSION

In this paper we suggest that security and reliability can be carefully combined in order to realize a truly secure and reliable non-volatile processor. We examine two possible attacks, one that aims to destroy the saved states, and the other that aims to wear-out the device especially the nvFFs. While it is relatively easy to quickly apply reliability constructs such as multiple backups and checksums, the designer must be careful as security has a different threat model from reliability. Reliable systems usually center around the idea of uncorrelated faults that occur with a certain level of randomness. In security, systems must be resilient to attacks, which are by nature correlated. To design around both, we must make the system capable of identifying and/or mitigating these correlated attacks.

REFERENCES

[1] Y. Wang, Y. Liu, S. Li, D. Zhang, B. Zhao, M. F. Chiang, Y. Yan, B. Sai, and H. Yang, "A 3us wake-up time nonvolatile processor based on ferroelectric flip-flops," in *2012 Proceedings of the European Solid-State Device Research Conference (ESSCIRC)*.

[2] "ITRS reports," 2017. [Online]. Available: http://www.itrs2.net/itrs-reports.html

[3] J. Wang, Y. Liu, H. Yang, and H. Wang, "A compare-and-write ferroelectric nonvolatile flip-flop for energy-harvesting applications," in *The 2010 International Conference on Green Circuits and Systems*.

[4] K. Ma, Y. Zheng, S. Li, K. Swaminathan, X. Li, Y. Liu, J. Sampson, Y. Xie, and V. Narayanan, "Architecture exploration for ambient energy harvesting nonvolatile processors," in *2015 IEEE 21st International Symposium on High Performance Computer Architecture (HPCA)*.

[5] Y. Liu, F. Suy, Z. Wangy, and H. Yang, "Design exploration of inrush current aware controller for nonvolatile processor," in *2015 IEEE Non-Volatile Memory System and Applications Symposium (NVMSA)*.

[6] Y. Wang, Y. Liu, Y. Liu, D. Zhang, S. Li, B. Sai, M. F. Chiang, and H. Yang, "A compression-based area-efficient recovery architecture for nonvolatile processors," in *2012 Design, Automation Test in Europe Conference Exhibition (DATE)*.

[7] Y. Wang, Y. Liu, S. Li, X. Sheng, D. Zhang, M. F. Chiang, B. Sai, X. S. Hu, and H. Yang, "PaCC: A parallel compare and compress codec for area reduction in nonvolatile processors," *IEEE Transactions on Very Large Scale Integration (VLSI) Systems*, 2014.

[8] X. Sheng, Y. Wang, Y. Liu, and H. Yang, "SPaC: A segment-based parallel compression for backup acceleration in nonvolatile processors," in *2013 Design, Automation Test in Europe Conference Exhibition (DATE)*.

[9] B. Ransford, J. Sorber, and K. Fu, "Mementos: System support for long-running computation on rfid-scale devices," in *Proceedings of the Sixteenth International Conference on Architectural Support for Programming Languages and Operating Systems*, ser. ASPLOS XVI, 2011.

[10] B. Ransford, S. Clark, M. Salajegheh, and K. Fu, "Getting things done on computational rfids with energy-aware checkpointing and voltage-aware scheduling," in *2008 Conference on Power Aware Computing and Systems (HotPower)*.

[11] B. Ransford and B. Lucia, "Nonvolatile memory is a broken time machine," in *2014 Proceedings of the Workshop on Memory Systems Performance and Correctness (MSPC)*.

[12] M. Xie, M. Zhao, C. Pan, J. Hu, Y. Liu, and C. J. Xue, "Fixing the broken time machine: Consistency-aware checkpointing for energy harvesting powered non-volatile processor," in *2015 52nd ACM/EDAC/IEEE Design Automation Conference (DAC)*.

[13] B. Lucia and B. Ransford, "A simpler, safer programming and execution model for intermittent systems," in *2105 ACM SIGPLAN Conference on Programming Language Design and Implementation (PLDI)*.

[14] H. Mao, X. Zhang, G. Sun, and J. Shu, "Protect non-volatile memory from wear-out attack based on timing difference of row buffer hit/miss," in *Design, Automation Test in Europe Conference (DATE)*, 2017.

[15] M. K. Qureshi, J. Karidis, M. Franceschini, V. Srinivasan *et al.*, "Enhancing lifetime and security of pcm-based main memory with start-gap wear leveling," in *42nd Annual IEEE/ACM International Symposium on Microarchitecture (MICRO)*, 2009.

[16] C. Yang and M. R. Varela, "Qualifying non-volatile register files for embedded systems through compiler-directed write minimization and balancing," in *IFIP/IEEE International Conference on Very Large Scale Integration (VLSI-SoC)*, Oct 2015.

[17] P. Cronin, C. Yang, and Y. Liu, "A collaborative defense against wear out attacks in non-volatile processors," in *2018 55th ACM/EDAC/IEEE Design Automation Conference (DAC)*.

978-1-5386-7100-9/18 $31.00 © 2018 IEEE

2018 IEEE Computer Society Annual Symposium on VLSI

A Short Survey at the Intersection of Reliability and Security in Processor Architecture Designs

Lake Bu, Miguel Mark, and Michel A. Kinsy

Adaptive and Secure Computing Systems (ASCS) Laboratory

Department of Electrical and Computer Engineering, Boston University

Abstract—Over the next decade, processor design will encounter a number of challenges. The ongoing miniaturization of semiconductor manufacturing technologies that has enabled the integration of hundreds to thousands of processing cores on a single chip is pushing the limits of physical laws. The fabrication process has also grown more complex and globalized with widespread use of third-party IPs (intellectual properties). This development ecosystem has complicated the security and trust view of processors. Some of the pressing processor architecture design questions are: (1) how to use reconfiguration and redundancy to improve reliability without introducing additional and potentially insecure system states, (2) what analytical models lend themselves best to the joint implementation of reliability and security in these systems, and (3) how to optimally and securely share resources and data among processing elements with high degree of reliability. In this work, we present and discuss (1) principal reliability approaches - error correction code, modular redundancy, (2) processor architecture specific reliability, (3) major secure processor architectures. We also highlight key features of a small representative class of the secure and reliable architectures.

I. INTRODUCTION

The intersection of reliability and security in the design of processor architectures is now a critical concern in a wide range of embedded computing, communications systems, and connected devices. On one hand, as feature size shrinks, transistors become less reliable and component failures increase. Transistor scaling and integration result in reliability challenges, including interference from electric fields, shrinking of the maximum-minimum voltage window, thermo-mechanical limitations, and soft, transient and intermittent errors. On the other hand, the emergence of general-purpose system-on-chip (SoC) architectures has given rise to a number of significant security challenges. The current trend in SoC design is system-level integration of heterogeneous technologies consisting of a large number of processing elements such as programmable RISC cores, memories, DSPs, and accelerator function units/ASIC. These processing elements may come from different providers, and application executable code may have varying levels of trust.

In this short survey, we attempt to highlight some of the pressing processor architecture design questions:

1) **Reliability Issues:** how reconfiguration and redundancy are used to improve reliability without introducing additional and potentially insecure system states;

2) **Security Issues:** how to optimally and securely share resources and data among processing elements which have different levels of trust;

3) **Security and Reliability in Architecture:** what analytical models lend themselves best to the joint implementation of reliability and security in these systems.

Over the years, there have been many attempts to address the aforementioned processor architecture design issues. Some commonly accepted approaches and methodologies have even emerged. In this work, we define "reliability" as the property of keeping the system in a pre-defined/desired/accepted functional condition. "Security" is characterized by the capability to protect the system from malicious attempts which either drive the system away from the accepted functional conditions, or exploit the limitations and restrictions of the system. These attempts can be either invasive or non-invasive.

On the topic of reliability, we first emphasize general approaches: error control codes and their application in processor architecture designs, modular redundancy for dependable functionalities in architectures, and processor architecture specific reliability methods such as ACE (architecturally correct execution), and AVFs (architectural vulnerability factors). We then discuss the existing and potential vulnerabilities of these approaches.

As for the security aspect, we start with major commercially available and academic secure processor architectures, such as Intel's Software Guard Extensions (SGX) and Trusted Execution Technology (TXT), ARM TrustZone Technology and derived processor architectures, MIT Aegis Secure Processor, IBM 4765 Secure Coprocessor, and Apple Secure Enclave Processor (SE). We then examine work on privacy and permission management targeting heterogeneous multi-core systems. Finally, we stress the potential vulnerabilities and attacks on some of these security-aware architectures.

Our focus on the security issues in heterogeneous computing environments primarily centers around multicores where the cores may have different levels of trustworthiness. The problem on such compute systems is how to optimally and securely share resources and data among those processing units, while maintaining individual tenant security and preventing data leakage among the units.

The rest of the paper is organized as follows: section II is on the reliability-oriented architecture designs, followed by their vulnerabilities. Section III is on the security-oriented architecture designs, and the existing and potential vulnerabilities. Section IV discusses the joint design and implementation of reliability and security. Section V concludes the paper.

978-1-5386-7100-9/18 $31.00 © 2018 IEEE

II. RELIABILITY-AWARE PROCESSOR ARCHITECTURES AND DESIGNS

In this section we will first introduce the error control codes (ECCs) as a universal technique to provide reliability in architecture designs. Besides being used as a tool to preserve the data integrity, the mathematical principle of ECCs can also be leveraged in the design of reliable systems or processors. Besides introducing the redundant modules using ECC, the processor architecture specific reliability such as ACE (architecturally correct execution), and AVFs (architectural vulnerability factors) are also presented. The vulnerabilities of current reliability techniques will be discussed in the final subsection.

A. ECC-based Reliable Processor Design

The error control codes (ECCs) are usually used to prevent data from being distorted by random errors. The most straightforward applications are the reliable buses, memories, and caches [1, 2]. We define the following notations:

- x: the original source data;
- y: the redundant portion computed based on x;
- v the encoded codewords of x;
- e: the random errors on v;
- $G()$ the generating function of v;
- $H()$: the checking function of v;
- $f()$: the functional module's function;
- $P()$: the predictor function for $f()$;
- $\sim$: the distortion symbol.

We introduce three of the most common ways that ECCs can be used to assist the reliability of a system.

1) Random Error Correction: the procedure that an ECC module uses to protects data from random errors is as follows:

i. Before a piece of data x is transmitted or stored in a system, it is first encoded by a generating function that $G(x) = v$, where x should be able to be retrieved from v;
ii. During the transmission or storage of v, it might be corrupted by random errors e, that $v + e = \tilde{v}$;
iii. When the piece of data is to be extracted, a decoding function is used to retrieve the correct piece: $H(\tilde{v}) = v$. And so x can be derived from v.

The work flow of such procedure is shown in Fig. 1.

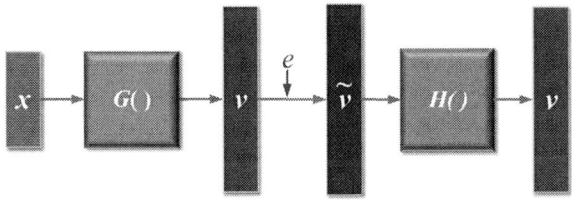

Fig. 1. In systematic encoding, $v = x||y$ and y is the redundant portion computed by $G(x)$ based on x, $||$ the concatenation operator. In this case the value of x is obvious once the correct v is retrieved. In non-systematic encoding, x has to be computed through certain algorithms from v.

2) Data Regeneration: This use of ECC modules is similar to Fig. 1, except that instead of v being distorted by a random error e, now part of it is missing. Thus the $H()$ is used to regenerate v rather than removing e. Due to the property of ECC, when used for data re-generation, it usually has a stronger capability in fault tolerance. This approach is now popular in machine learning acceleration [3] and heterogeneous clusters' straggler tolerance [4].

3) Self-checking Checkers: There have been many research efforts on the application of ECCs to the circuits or functional modules as the self-checking checkers (SCC) to verify the correctness of their functionality. The common thread in these efforts is the addition of a parallel module named the "predictor" to the original function module, which generates the corresponding check bits at the same time of the functional module's output. The predictor's check bits and the functional module's output are verified by the decoder for error detection, or correction. Together, the predictor and decoder form an SCC system. The procedure of a SCC's self-correction is as follows:

i. When an input x comes into a functional module $f()$, it is also fed into a predictor module $P()$, where $P()$ is a combination of $f()$ and $G()$;
ii. During the computation of $f(x)$, which is the system's original functionality, $P(x)$ is also computed. Either module can be malfunctional;
iii. The decoder verifies $f(\tilde{x})$ and $P(\tilde{x})$ and outputs the correct $f(x)$ to maintain reliability.

Figure 2 illustrates the workflow of a SCC.

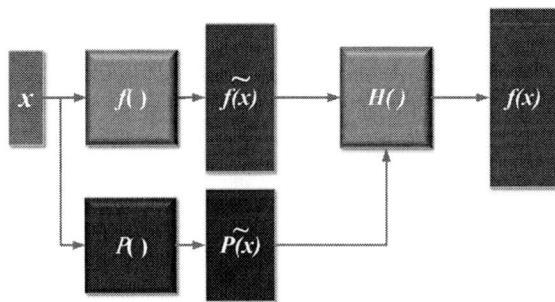

Fig. 2. Instead of random error correction on the source data, the goal of a SCC is to protect the functionality $f()$ of a given system or computation node. With proper optimization, $P()$ does not necessarily have higher complexity than $f()$;

Various codes have been devised as ECCs such as the repetition codes, cyclic codes, Hamming codes, and Reed-Solomon (RS) codes [5, 6, 7]. They are characterized by different levels of error tolerance capability and decoding complexity. In recent years, low density codes have become more popular due to their low complexity in decoding [8, 2].

It should be noted that duplication or triplication systems in processor designs have originated from ECCs. In a duplication system [9], two systems with identical functionality will perform the same operations, and the results of which will be compared. A triplication system [10], involves three identity function whose results will participate in a majority voting

978-1-5386-7100-9/18 $31.00 © 2018 IEEE   119

to tolerate the malfunction of a single system. These two techniques leverage the concept of repetition codes in ECCs.

B. Architecturally Correct Execution (ACE), and Architectural Vulnerability Factors (AVFs)

Researchers from Intel [11] introduced the concept of architecturally correct execution (ACE). In their definition, a bit in a system is related to architecturally correct execution (ACE) if it affects the output of the program. Other bits which do not have such influence are called un-ACE bits. A structure's architectural vulnerability factor (AVF) is defined as the probability that a fault in the structure will result in an erroneous output. One of the fundamental differences between AVF estimation and ECC is, the former is more of a methodology to evaluate an architecture's reliability, and the latter is a practical technique to ensure the dependability of an architecture. Also, the former tracks the bits with an impact to the final outcome only (particularly from the user's perspective), while the latter tries to treat all the bits equally.

A program running on a faulty architecture has multiple possible outcomes. There can be faults resulting no error, silent data corruption (SDC), and detected but unrecoverable errors (DUE). The correlation among them and an architecture's error tolerance capability are given by [12] and depicted in Figure 3.

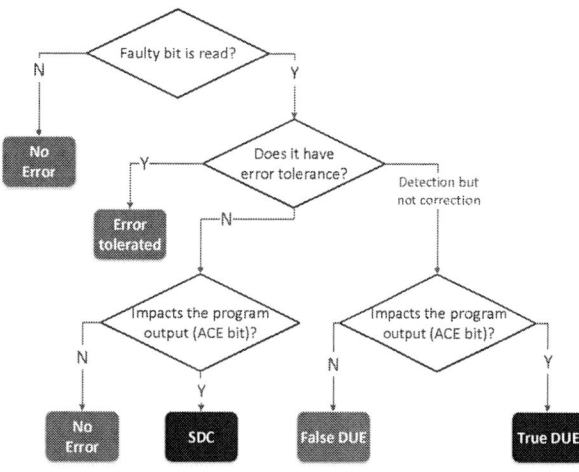

Fig. 3. In this diagram, there can be undetected and uncorrected faults in an architecture, but they do not necessarily affect the final outcome. The "detection but not correction" branch can be the scenarios that the architecture is only equipped with parity check codes but not any ECC with Hamming distance larger than 3. Or it can be that the architecture only has an error detection subsystem but not an error correction subsystem, such as the duplication subsystems.

In [11], authors proposed an efficient approach for estimating AVFs that uses only a subset of the processor state bits. The used bits in a processor state storage cell/structure are the ones related to ACE. They will show a distortion in the output when an error occurs. Other bits in storage cells are the un-ACE bits, which can be flipped without causing a functional error. The authors provide the equations and different approaches for computing the AVFs.

1) Statistical Fault Injection: In this test scheme, random errors are injected in both randomized space and time domains. The results will then be compared with a pre-generated reference result set, or an error-free model. The AVF is computed by the fraction of mismatches divided by the total number of injected errors.

If there is no mismatch observed, it can be because either the error is tolerated, or is masked (silent errors). The latter is a more complicated situation and has to be studied by a complete comparison in system states [13].

2) Little's Law: This method is suitable for the early stage of a design before the RTL is generated. Denote N as the average number of bits in the architecture, B the average bandwidth per cycle into the architecture, and L the average latency a bit through the architecture, the subscript ACE for the ACE bits. Then the AVF can be estimated by:

$$\frac{B_{ACE} \times L_{ACE}}{N}$$

3) ACE Analysis in Performance Models: In this method, a performance model is used to determine which bits are ACE and which are un-ACE. A conservative assumption is made that, a bit is ACE unless it is proved as an un-ACE. This methodology can be more time efficient than others.

C. Vulnerabilities of Reliability-Aware Architecture Designs

For most reliability-oriented designs, there can be a large number of "invisible" errors never detected by the system. The invisibility is not due to the lack of error detection or correction subsystems, but because of their linearity. We will firstly introduce the concept of the "kernel" as a measurement of the number of invisible errors in an architecture.

Definition 1. Suppose C is the set of N-bit ECC codewords and $H()$ is the decoding/error detection/error correction function. C is defined by $C = \{v | H(v) = 0\}$. Set K_d is called the Kernel of C if:

$$K_d = \{e | e + v \in C, \ \forall v \in C\}.$$

Under this definition, if $H()$ is a linear function (which is the case for most architectures), and there exist an error e that $H(e) = 0$, then we have:

$$H(\tilde{v}) = H(e + v) = H(e) + H(v) = 0 + 0 = 0 \quad (1)$$

Then this error is invisible for $\forall v \in C$. If C is linear, then the set of e which is the kernel $K_d = C$. This result shows that for any architecture with a linear error control function, there exists a large number of invisible errors. The good news is that for most systems, those invisible errors are more than one bit, which can be very rare. Thus most single-bit errors can still be taken care by the SEC-DED subsystems. However this potential vulnerability can still be leveraged by attackers to inject forever-masked errors.

978-1-5386-7100-9/18 $31.00 © 2018 IEEE

III. SECURITY-AWARE PROCESSOR ARCHITECTURES AND DESIGNS

Security, unlike reliability, is a much larger and more complicated topic for all architecture designers. Different architectures targeting different security demands will end up with very distinct structures. Therefore in this section, instead of giving a universal design methodology, we will present a number of commercialized and representative security-aware architecture designs, as well as their advantages and vulnerabilities. The subsections will include the introduction of the MIT Aegis Secure Processor, the Apple Secure enclave processor (SEP), the ARM TrustZone technology, and the IBM 4765 Secure Processor.

A. MIT Aegis Secure Processor

Aegis [14] is a secure processor which aims to provide conventional software-based authentication and addresses a critical assumption made by other secure processor implementations: physical attacks are infeasible or meaningless. Its architectural design philosophy is based on the premise that only the Aegis processor can be authenticated and trusted. External components such as non-volatile memory and other processors are treated as non-trustworthy by default. The core of Aegis's protection is centered on Silicon Physical Random Functions (SPUFs) which leverage unique timing delays in integrated circuits created by the semiconductor manufacturing process [15]. Aegis uses this unique characteristic in the form of a PUF delay circuit which is used for secret key generation and authentication. Furthermore, restricting protection to one chip prevents the leakage of secrets through unsecured communication between multiple processing units.

To protect software, Aegis first introduces four additional processing modes: Standard (STD), Suspended Secure Processing (SSP), Tamper-Evident (TE) and Private Tamper-Resistant (PTR). STD and SSP are the lowest privilege mode which has no access to private memory and can only enter the more secure TE and PTR mode. TE has read/write access to verified memory and a subset of security functions. PTR mode is the most privileged due to its access to PUF instructions. Second, software can be authenticated using an authentication scheme with SPUFs such as a public/private key protocol. Lastly, off-chip memory protection in the form of Integrity verification (IV) and Memory Encryption (ME) can be enabled when the supervisor switches the processor into TE or PTR mode after boot. IV and ME aim to provide defense against both software and hardware attacks. To accomplish this, the processor partitions the available memory into IV and ME regions which can overlap. IV protects regions through detecting and preventing any unintended modifications and ME utilizes encryption to hide sensitive contents. A trusted supervisor, such as a kernel manages the sharing of these protected regions. Later on they also proposed a version of Aegis which is resistant to malicious operating systems [16].

With these features combined, the Aegis secure one chip processor can defend against a wide range of attacks. Brute force based attacks are not feasible due to the sheer number of challenge-response pairs that can be generated. Attackers may then attempt to create a timing model of the PUF delay circuit but this is not possible since no information is leaked from the circuit. Likewise, an attacker cannot duplicate the PUF circuit due to nature of the manufacturing process. Even if the attacker gains physical access to the processor and tries to probe timing information, the data collected will be useless due to the interference caused by the probe.

Although the authors noted the omission of side-channel attacks and learning attacks to the PUF which is the fundamental source of security, overall Aegis can provide a strong defense with negligible overhead in gate size and performance.

B. Apple Secure Enclave Processor (SEP)

Apple's Secure Enclave Processor (SEP) [17] is a flashable coprocessor which utilizes memory encryption and hardware number generation to carry out cryptographic functions for the main processor. In a sense, SEP creates a logical wall between software and sensitive security functions so that untrusted software cannot gain access to sensitive data such as fingerprints and keys. To achieve most of its functionality, a trusted micro-kernel runs on top of the processor, sporting its own drivers and services. Given the nature of this technology, Apple has prevented the dissemination of the technical details of the processor. Therefore, technical details are only available through efforts of reverse engineering.

The basic architectural design of SEP is the separation of computation into two processors: Application processor (AP) and SEP. SEP contains completely separated hardware such as a hardware number generator, boot ROM, and crypto engine. Despite this aggressive separation, SEP is still a 32-bit processor which coordinates with the AP to share external memory. During its boot process, SEP will wait for AP to configure a region of memory. Communication between AP and SEP is achieved through an interrupt-driven secure mailbox. With this mailbox, the architecture acts as a walled garden which is called the KF filter. The KF filter encapsulates and guards many of the SEP's unique hardware components. Therefore all data originating from the SoC passes through the filter and must go through the secure mailbox. Once SEP has initialized secure memory regions, it is protected from software-based attacks. To protect against physical-based attacks such as memory probing, SEP utilizes memory encryption in the form of AES-ECB, AES-CBC and AES-XEX. Furthermore, after initialization, applications which wish to interact with the encrypted data guarded by SEP must use a Bootstrap server which can enforce access and privilege rules for different functionalities such as a secure key generation service.

Overall the nature of the SEP defends against an attack model in which an attacker can compromise system software such as the kernel. However, there have been reports in 2017 that hackers have decrypted the SEP's firmware and published its secret key [18]. Although this breach does not leak any user's information or data, it makes a way for researchers and hackers to explore the vulnerabilities of SEP.

C. ARM TrustZone technology

The ARM TrustZone technology [19] is a single core secure processor technology that uses a security approach similar to that of Apples Secure Enclave processor. Its design philosophy is based on levels of trust which aims to minimize the attack surface at lower levels. In a sense, ARM TrustZone uses separation based on the concept of least privilege; software or hardware should only have access that it needs and nothing more. To implement this secure model, TrustZone creates two logical zones: secure world and non-secure world; the secure world houses the security subsystem while the normal world contains everything else. This allows an establishment of a chain of trust. Separation of zones starts with the partitioning of memory into secure and non-secure memory regions.

Naturally, through separation, non-secure world processes cannot access secure content but secure-world processes can access both secure and non-secure content. Modules called the Security Attribution Unit (SAU) and Implementation Defined Attribution Unit (IDAU) work together to determine if a memory region is secure. TrustZone also provides a subtype of secure memory, non-secure callable memory, which is an executable region which allows non-secure instructions to branch into a secure memory using Secure Gateway (SG) instructions. Despite this aggressive separation model, communication between non-secure world and secure world processes is possible via a Secure Monitor Call (SMC). Through providing these primitives for processes, TrustZone removes the need for a separate security processor that would inevitably increase the attack surface. ARM designed TrustZone as a configurable platform that can better adapt to different attack models. Specifically, TrustZone provides SoC designers with various TrustZone enabled IP modules that allow an embedded device to be tailored to a particular attack model. One particular weakness of Trustzone is that assumes that secure mode processes can always be trusted.

D. IBM 4765 Secure Processor

The IBM 4765 [20] is a secure co-processor which is placed on a PCIe card. Equipped with a hardware number generator it provides tamper-proof storage of sensitive data and cryptographic operations for activities such as SSL private key transactions. Like most secure processors, the 4765 supports several cryptographic algorithms: SHA-256, HMAC, and RSA. Due to the nature of PCIe, the 4765 is, unfortunately, an easy target for both theft and physical manipulation. Fortunately, a hardware-based tamper-proof module is included which is certified for meeting the Federal Information Processing Standard (FIPS) 1402-2 level 4 security requirements.

The tamper-proof module can detect physical abnormalities such as voltage spikes and temperatures variances and mark them as physical attacks. As a response, the tamper circuit will automatically zero out secrets stored in onboard memory devices such as Battery-Backed Static Random Access Memory (BBRAM). The tamper-proof protocol can also be made to zero out memory in the case of ejection from the PCIe slot.

On the software side, applications can interface with the 4765 through Miniboot: software internal to the module that exposes functionality. The commands available for applications depends on Miniboot's current boot level. Because Miniboot runs at boot time, its progress through different levels depends on the success of the host's power-on-self-test (POST). For example, if POST1 fails, an application would have had only access to the set of Miniboot0 commands and queries. Communication between the two available Miniboot levels, Minitboot0 and Miniboot1, and applications involve authentication using a public/private key protocol. Authentication of each command is based on the concept of roles which is essentially an Access Control Layer for various functions and services provided by Miniboot.

E. Intel Trusted Execution Technology (TXT)

Intel's Trusted Execution Technology (TXT) [21] is a hardware-based technology to examine the authenticity of the operating system and its running environment. It also relies on the Trusted Platform Module (TPM) to provide functionalities such as secure storage. The purpose of the TXT is to provide a trusted way for loading and executing system software, e.g. Operating System kernel or Virtualization Machine Monitor (VMM), even on machines with malicious software and malware. The technology supports both a static chain of trust and a dynamic chain of trust. The static chain of trust starts when the platform powers on (or the platform is reset), which resets all PCRs to their default value. For server platforms, the first measurement is made by hardware (i.e., the processor) to measure a digitally signed module (called an Authenticated Code Module or ACM) provided by the chipset manufacturer. The processor validates the signature and integrity of the signed module before executing it. The ACM then measures the first BIOS code module, which can make additional measurements. However, there have been works showing that the TXT only provides launch-time protection, but not runtime [22], against attacks such as buffer overflow etc. More importantly, researchers have been able to infect the system management mode (SMM), which is one of the most privileged software loaded on a computer, to bypass the TXT's launch examination and conduct attacks. Another research [23] shows that attackers can infect the boot loader and execute their own code before the TXT's SENTER instructions are executed.

IV. SECURE HETEROGENEOUS MULTICORE ARCHITECTURE

It is worth stressing that the technique and technology challenges encountered in the design of single-core or homogeneous multicore secure processors are further amplified in the security-aware design of heterogeneous multicore architectures. On those systems, different cores or processing units may have different levels of trust or privacy. Thus the secure data and permission management has to be taken into consideration. Researchers in [24] and [25] have proposed different approaches to address this issue. In [25], the authors introduced the "Hermes" architecture, which embeds its security features in the on-chip interconnect fabric. Hermes is a secure multicore computing architecture framework. It reduces the system attack surface by creating a virtualization layer that isolates compute threads based on system and user defined trust levels and security policies. Figure 4 shows a set of applications with mixed security being mapped onto a mixed security hardware. It achieves both hardware and software views of secure processing by grouping processors into physical zones called *wards* and virtual logical zones called *islands*. So that although different cores are located and operated in different manners, they are categorized to certain standardized security levels to be given corresponding permissions and privileges.

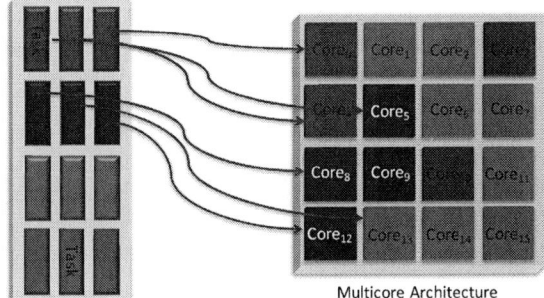

Fig. 4. Trusted/untrusted applications running on trusted/untrusted cores. Different trust levels are illustrated by different colors (e.g., red represents the least trusted program or core).

V. CONCLUSION

In this paper we explore and discuss the key details of reliability and security-aware processor architectures and designs. There are some established approaches to designing and evaluating reliable architectures such as ECC, ACE, and AVF. The application of error correction protection can guard processors against random and limited errors. However, the problem is more complicated when it comes to security-aware architectures. There are different security demands and attack models for each design on the market. Additionally, we present a small representative set of secure processor architectures - commercially available and academic and discuss their vulnerabilities. Finally, we briefly touched on a secure heterogeneous multicore architecture, which aims to provide a trustworthy data and permission management among heterogeneous cores with different levels of trust.

VI. ACKNOWLEDGMENTS

This research is partially supported by the NSF grant (No. CNS- 1745808).

REFERENCES

[1] C. Wilkerson, A. Alameldeen, and Z. Chishti, "Scaling the memory reliability wall," *Intel Technology Journal*, vol. 17, no. 1, 2013.

[2] P. Reviriego, S. Pontarelli, A. Evans, and J. A. Maestro, "A class of sec-ded-daec codes derived from orthogonal latin square codes," *IEEE transactions on very large scale integration (vlsi) systems*, vol. 23, no. 5, pp. 968–972, 2015.

[3] K. Lee, M. Lam, R. Pedarsani, D. Papailiopoulos, and K. Ramchandran, "Speeding up distributed machine learning using codes," *IEEE Transactions on Information Theory*, vol. 64, no. 3, pp. 1514–1529, 2018.

[4] A. Reisizadeh, S. Prakash, R. Pedarsani, and S. Avestimehr, "Coded computation over heterogeneous clusters," in *Information Theory (ISIT), 2017 IEEE International Symposium on*. IEEE, 2017, pp. 2408–2412.

[5] L. Breveglieri, I. Koren, and P. Maistri., "Incorporating error detection and online reconfiguration into a regular architecture for the advanced encryption standard," *Defect and Fault Tolerance in VLSI Systems, 2005. DFT 2005. 20th IEEE International Symposium on*. IEEE, 2005.

[6] C.-H. Yen and B.-F. Wu, "Simple error detection methods for hardware implementation of advanced encryption standard," *IEEE transactions on computers*, 2006.

[7] G. Bertoni, L. Breveglieri, I. Koren, P. Maistri, and V. Piuri, "A parity code based fault detection for an implementation of the advanced encryption standard," *Defect and Fault Tolerance in VLSI Systems*, 2002.

[8] L. Dai, B. Wang, Y. Yuan, S. Han, I. Chih-Lin, and Z. Wang, "Non-orthogonal multiple access for 5g: solutions, challenges, opportunities, and future research trends," *IEEE Communications Magazine*, vol. 53, no. 9, pp. 74–81, 2015.

[9] J. Johnson, W. Howes, M. Wirthlin, D. L. McMurtrey, M. Caffrey, P. Graham, and K. Morgan, "Using duplication with compare for on-line error detection in fpga-based designs," *Aerospace Conference*, 2008.

[10] P.-T. Huang, W.-L. Fang, Y.-L. Wang, and W. Hwang., "Low power and reliable interconnection with self-corrected green coding scheme for network-on-chip," *Second ACM/IEEE International Symposium on Networks-on-Chip*, 2008.

[11] S. S. Mukherjee, C. Weaver, J. Emer, S. K. Reinhardt, and T. Austin, "A systematic methodology to compute the architectural vulnerability factors for a high-performance microprocessor," in *Microarchitecture, 2003. MICRO-36. Proceedings. 36th Annual IEEE/ACM International Symposium on*. IEEE, 2003, pp. 29–40.

[12] S. S. Mukherjee, J. Emer, and S. K. Reinhardt, "The soft error problem: An architectural perspective," in *High-Performance Computer Architecture, 2005. HPCA-11. 11th International Symposium on*. IEEE, 2005, pp. 243–247.

[13] N. J. Wang, J. Quek, T. M. Rafacz, and S. J. Patel, "Characterizing the effects of transient faults on a high-performance processor pipeline," in *Dependable Systems and Networks, 2004 International Conference on*. IEEE, 2004, pp. 61–70.

[14] G. E. Suh, C. W. O'Donnell, and S. Devadas, "Aegis: A single-chip secure processor," *Information Security Technical Report*, vol. 10, no. 2, pp. 63–73, 2005.

[15] G. E. Suh, C. W. O'Donnell, I. Sachdev, and S. Devadas, "Design and implementation of the aegis single-chip secure processor using physical random functions," in *ACM SIGARCH Computer Architecture News*, vol. 33, no. 2. IEEE Computer Society, 2005, pp. 25–36.

[16] G. E. Suh, D. Clarke, B. Gassend, M. Van Dijk, and S. Devadas, "Aegis: architecture for tamper-evident and tamper-resistant processing," in *ACM International Conference on Supercomputing 25th Anniversary Volume*. ACM, 2014, pp. 357–368.

[17] Apple, "Ios security," apple.com/business/docs/iOS_Security_Guide.pdf.

[18] M. Mimoso, "Hacker publishes ios secure enclave firmware decryption key," in *Threatpost*, 2017.

[19] A. ARM, "Security technology building a secure system using trustzone technology (white paper)," *ARM Limited*, 2009.

[20] T. W. Arnold, C. Buscaglia, F. Chan, V. Condorelli, J. Dayka, W. Santiago-Fernandez, N. Hadzic, M. D. Hocker, M. Jordan, T. Morris *et al.*, "Ibm 4765 cryptographic coprocessor," *IBM Journal of Research and Development*, vol. 56, no. 1.2, pp. 10–1, 2012.

[21] J. Greene, "Intel trusted execution technology," *Intel Technology White Paper*, 2012.

[22] R. Wojtczuk and J. Rutkowska, "Attacking intel trusted execution technology," *Black Hat DC*, 2009.

[23] R. Wojtczuk, J. Rutkowska, and A. Tereshkin, "Another way to circumvent intel trusted execution technology," *Invisible Things Lab*, 2009.

[24] H. Kondo, S. Otani, M. Nakajima, O. Yamamoto, N. Masui, N. Okumura, M. Sakugawa, M. Kitao, K. Ishimi, M. Sato *et al.*, "Heterogeneous multicore soc with sip for secure multimedia applications," *IEEE Journal of solid-state circuits*, vol. 44, no. 8, pp. 2251–2259, 2009.

[25] M. A. Kinsy, S. Khadka, M. Isakov, and A. Farrukh, "Hermes: Secure heterogeneous multicore architecture design," in *Hardware Oriented Security and Trust (HOST), 2017 IEEE International Symposium on*. IEEE, 2017, pp. 14–20.

2018 IEEE Computer Society Annual Symposium on VLSI

Can Soft Errors be Handled Securely?

Senwen Kan
Austin Design Center
Cypress Semiconductor
Austin, United States of America
Email: senwen.kan@cypress.com

Jennifer Dworak
Department of Computer Science & Engineering
Southern Methodist University
Dallas, United States of America
Email: jdworak@smu.edu

Abstract—Error detection and correction approaches are often used in modern microprocessors to detect and mitigate soft errors. However, when such errors must be handled in software, transitioning to the error handler may require parts of the processor state to be hidden and/or flushed to prevent information leakage. This paper presents a high level introduction to some issues in this area. For example, it discusses when security checks may be performed and what actions may be taken to help preserve isolation when the error handler is at a trust level other than the executing code.

I. INTRODUCTION

Computing security has become an increasingly challenging problem for the semiconductor industry, and processor architectures have started incorporating more security features [1], [2], [3]. For example, these features may include physical or cryptographic measures to enable hardware security for trusted execution and access paradigms, such as virtual machines and trusted applications. Unfortunately, attack mechanisms, such as those described in [4] and [5], illustrate that even in robust processors with a long deployment history, vulnerabilities may still exist.

At the same time as these advances are being made in processor security, process nodes are becoming increasingly smaller, and advances in low power technology are being employed. These design, implementation, and manufacturing changes may impact processor reliability. For example, the device might be now more susceptible to radiation-induced errors or to power modulation-induced soft errors [6], [7]. Therefore, device security must also account for the need for device reliability enhancements to prevent or minimize unintended leakage of trusted information or unauthorized access attempts.

A. Some Existing Work

Addressing reliability and security in computing devices has undergone significant study. Work in [8], [9], [10], [11] acknowledge that addressing security and reliability in tandem is a serious problem. In [12], cyclic redundancy code (CRC) and memory address scrambling are used to provide additional protection for static random access memory (SRAM) contents against some potential hardware trojan behavior and reliability errors in tandem. Solutions utilizing dedicated hardware in combination with software to monitor and regulate trusted program or virtual machine execution were proposed in [8], [9], [10]. However, these approaches would not only incur an increase in die area, but they may also degrade device timing closure and hence impair device performance.

Processor architectural features, such as those described

in [1], provide applications support with application memory isolation and encryption. In [2], a trusted code execution environment and untrusted code execution are isolated by hardware across all system hardware resources with context switching exceptions. Features such as the secure encrypted virtualization found in [3] provide cryptographic isolation to isolate virtual machine execution and data access from system peripheral to core execution.

In this paper, the focus is given to a high level discussion of some issues related to integrating reliability protection that is already present in a design with a hardware security supporting scheme. This scheme may involve software combined with hardware security features, organized at least a layer above the reliability circuitry on the device data path. A possible layered strategy is shown in Section VI. The rest of this paper explores some aspects of such a strategy and discusses some of the needs that must be considered when security and reliability circuitry are both used to protect processor operation.

II. RELIABILITY ERRORS

Enhancing reliability in the context of semiconductor devices refers to preventing failures or mitigating errors that may occur during functional operation. For example, such errors could arise from signal changes induced by environmental factors, such as radiation or electrical instability. Similarly, memory cells could change state due to particle strikes, discharge, or transistor degradation. Such errors would manifest and can be modeled at the logic level.

A. Sources of Errors

One common type of failure is due to radiation-induced failures in system memories [13], [6]. In particular, charged particle strikes can lead to single event upsets (SEUs). Such upsets may affect one or more bits and can lead to memory corruption, such that stored data accessed after such events would be incorrect when read or used by computing processes [13], [6]. Silicon degradation can also lead to silicon aging and premature memory data retention failures—leading to memory data corruption as well [14].

Other errors in processors may arise due to factors such as increased temperature or IR drop that make circuits operate more slowly and allow incorrect data to be captured in storage elements. Electrical noise in circuits and latent defects are other potential sources of erroneous operation.

Modern computing systems typically communicate with main system dynamic random access memory (DRAM) via a serializer-deserializer (SERDES) bus. Such a data bus may

978-1-5386-7100-9/18 $31.00 © 2018 IEEE

be prone to data errors [15]. Such errors can be sensitive to signal integrity issues and can be frequent.

In addition, modern multi-core chips often use a network-on-chip (NoC) for inter-core communication. NoC communication may also suffer failures or bit upsets due to a variety of reasons, including natural radiation or silicon degradation. When such failures occur, NoC transaction packets may be lost or corrupted, which may lead to transaction failures.

This list of potential error sources is by no means complete, However, it certainly demonstrates the need for error detection and potential correction capability for integrated circuits that may be used in mission-critical applications. Although errors can occur in circuit logic, this paper focuses on soft errors in memories and data communication that may be detected and/or corrected with Error Correcting Codes (ECC) and parity.

B. Error Detection

To ensure that integrated circuits (ICs) can operate reliably in the presence of errors, methods for error detection and correction have been developed for logic, communication mediums, and memories. The type of detection and correction required is related to the numbers and patterns of errors that are expected. For example, burst errors may need more complex approaches than rare single-bit errors.

In the simplest case, when detection is all that is required, detection of an odd number of bit errors in a memory block or communication packet can be accomplished through simple even parity. In this case, a single parity bit can be generated by XOR'ing all of the bits of the protected data when the data is about to be written to memory or transmitted. The parity bit is sent/stored with the data. When the data is received and/or read, the data bits and parity bit can all be XOR'ed together. If the output of the XOR is equal to a logic one, then an error is detected.

C. Correctable Errors

There are multiple ways of dealing with errors depending on the location and source of the error and how it was detected. For example, when communication errors occur, a request for the data to be resent can often lead to correct data operations after the communication is retried. In memories, single bit errors can be automatically corrected using Hamming codes, where multiple parity bits are used to allow the flipped bit to be identified. Codes that allow correction as well as detection are often referred to as Error Correcting Codes (ECC).

D. Uncorrectable Errors

ECC can be useful at providing data correction, but not all errors can be corrected. Consider the case where multiple bits are upset; data corruption in this scenario may be uncorrectable. Similarly, if communication errors won't clear after a set number of retries, a data packet would be considered lost. Typically, uncorrectable errors need to be propagated to main system software, such as the operating system, to allow errors to be handled in such a way that the compute system can recover. Hence, uncorrectable errors at a low level would typically generate interrupts to the operating system [3]. The operating system can then manage the error in an appropriate way, such as terminating a program with faulty data.

E. Incorrectly Corrected Errors

When ECC attempts to correct errors when multiple bits are upset, the bit identified as erroneous is often originally correct. Thus ECC may actually add to the number of bits in error when the identified bit is flipped. In such a case, it is possible that corrupted data disguised as corrected data will be transmitted into consuming computing processes. This is considered silent data corruption [12].

III. SECURITY CONSIDERATIONS

The ability to access or configure computing resources on a chip can vary based upon the trust level of the entity requesting access. In this paper, a trust level of 0 implies no trust, and a trust level greater than 0 indicates increasing levels of trust. Many hardware-supported security features in modern microprocessors aim to support data or execution code isolation when the computing device changes trust level. For example, such a change of trust level can be thought of as at least one thread in a processor changing from one trust level to another. This change in trust level should cause the entire thread and its associated system resources to transition into another trust level. Such isolation may also be required when trust checking is employed when the machine is operating at a particular trust level. By considering trust level transitions and associated trust checking, some of the effects of reliability errors on on-chip security can be evaluated.

Semiconductor device resources can be accessed via an external access agent or via executable programs or processes loaded into its internal or external memory. When any of these entities request access to an increased trust level, the device must check the request to ensure it is trustworthy. This may be achieved with authentication-based methods such as password protection, physical-uncloneable-function (PUF) authentication, challenge-response pairs based on cryptographic hash functions, etc. In addition, the device can deploy integrity checking-based schemes to ensure data or instructions written into the design or outputted by the device have not been altered [16].

A. Security Flow Overview

An abstract view of trust transitions is shown in Fig. 1. Given two trust levels, A and B, with Trust Level of B not equal to A, Fig. 1 illustrates what would happen when an access agent or a compute process requests Trust Level B while at Trust Level A. The device would perform checking on the request and/or requester to try to ensure the requester's authenticity as well as the requester's conformance to the request protocol, among many possible checks. This is represented by the *Check 1* node in the figure. If the request fails checking, the requester has to exit back to its existing trust level, as shown by the left child of the *Check 1* node. If the request passes checking, the trust level transition is successful, as shown by the right child of the *Check 1* node.

When a change in trust level is granted, the device can provide hardware isolation to allow the access agent to operate without data leakage or snooping into restricted secret data. One such example on an embedded device would be when application programming is interrupted to execute firmware code on the device, and the firmware execution requests a higher trust level than the application code. Under these conditions, the firmware execution traces should be isolated

978-1-5386-7100-9/18 $31.00 © 2018 IEEE

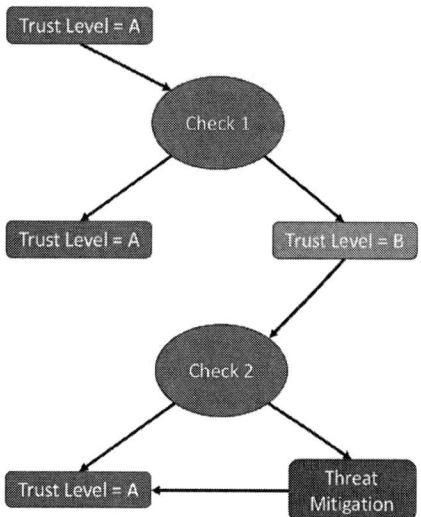

Fig. 1.   Trust Transition Model

from the application level code. In other words, the device should be able to ensure that the firmware execution trace and results are hardware isolated from user applications. This isolation can be pursued with memory encryption, proper device cache flushing, etc. With flushing, potentially cached contents would be written back to safe regions or encrypted regions of the device. Similarly, as the firmware completes its execution and is readying to transition back to a lower trust level, the device must check that lowering the trust level would not cause unintended data leakage from the higher trust level to the lower trust level. For example, certain hardware blocks may need to be reset or flushed by the hardware before the transition can occur. This check before downgrading the trust level is shown as *Check 2* in Fig. 1.

It's also possible that the device detects system state alterations or restrictions that would prevent safe trust level lowering. In such cases, the device could implement threat mitigation actions to ensure trust level isolation. This is shown as the *Threat Mitigation* box in Fig. 1.

The following subsections further describe parts of this overall process.

B. Entry Checking

When any access agent or process attempts to enter into a new trust level, a set of entry checks should be provisioned by the device. For instance in [3], protected virtual machine entry needed to pass checks such as cyclic redundancy code checksum matching, virtual machine consistency checks, etc. prior to processor execution of virtual machine code segments by a hypervisor process. In a trusted boot environment, each increase in access privilege during processor boot may require a secure hash algorithm (SHA)-based checksum [2]. Trust level modification can also be associated with basic authentication checking approaches, such as password locking schemes. Therefore, trust level entry may require authentication, digital signature checking, password unlocking, and security property based consistency checking, among many possible checking schemes.

C. Secure Mode Operations and Checks

Once the access agent or process is able to enter Trust Level B, it should be able to operate at that trust level. Its operations should be isolated from any other concurrently running access agent or process. Furthermore, its operations should not be traceable by any other subsequent access agent or process. Thus, the security hardware on the device should provision support for adequate isolation for trusted operations.

A variety of methods are implemented to support such device security. For example, to guarantee secure operations, hardware supported encryption of trusted access agent or program processes can be implemented. Such techniques may involve memory encryption. Additionally, prompt flushing of trusted data may need to be automatically enforced by the hardware.

Encryption-based virtualization is now being marketed, where virtual machines are isolated with encryption [3], [17]. Hardware zoning based schemes are also being utilized to support isolation of trusted operations [2]. Furthermore, in cases where compute resource sharing is allowed, any resource switching, such as simultaneous multi-threading, may require traces of trusted thread execution to be flushed before another thread utilizes the shared resource.

D. Exit Checking

When an access agent or a process completes trusted operations as its session draws to an end, the device trust level may need to be changed, as indicated in Fig. 1. The secure hardware on the device should be able to erase or flush the trace of trusted operations so that no sensitive data is leaked. Furthermore, the hardware should also check to see whether the security hardware driven flushing operation has completed successfully or not. In cases where the trusted entity is virtualized, such as a virtual machine, exiting virtual machine operation implies hardware resource control is being handed back to the hypervisor. The hardware should check to make sure the hypervisor is still trustworthy. In cases where trust level exit fails, a series of mitigation steps can be taken to allow the system to gracefully shutdown or to minimize risky operations.

IV. COUPLING RELIABILITY AND SECURITY

The trust transitions shown in Fig. 1 indicate that a device can undergo transitions between trust levels as well as operate at various trust levels. This indicates that, in addition to reliability errors that may occur during a device's normal or untrusted operations, errors can also occur during the device's trust transitions and trusted operations. Furthermore, reliability errors, such as multi-bit upset events, can force the device to transition to the trust level of the error handler. As such, reliability errors can inadvertently lead to unexpected trust transitions.

A. Layering

One way to reconcile reliability errors and security is to consider layering. This scheme relies on error detection and correction circuitry to operate at a lower layer in a device than the security design constructs. This is illustrated in Fig. 2. This layering refers to the hierarchical layering of hardware protection structures when the device is in a trusted mode with

978-1-5386-7100-9/18 $31.00 © 2018 IEEE

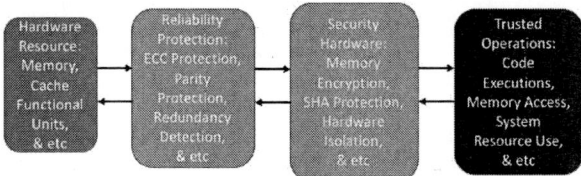

Fig. 2. Layered Security View

respect to the order in which data and operations are checked during execution. Thus, in a trusted mode, the reliability and security protection resources available to a computing process, such as a thread, should be operating at a trusted level. (Note that the different "layers" described here for reliability and security does not necessarily imply that they are at different "trust levels.")

Fig. 2 shows the state of the device when it is in a security feature enabled trusted mode, where system hardware resources are shown in the blue box, reliability protection hardware is enabled in the green box, security protection hardware is enabled in the orange boxes, and the black box indicates the code execution layer.

In Fig. 2, fetching trusted and encrypted code into the processor core for execution requires code decryption from memory access, but the memory controller hardware would correct and detect errors on memory access contents first. In this way, data encryption and decryption are one layer above the reliability circuits as shown Fig. 2. Vice versa, when write operations operations are issued, the request must pass through the security hardware layer before reaching the reliability protection layer and finally propagating to hardware resources. Each of the boxes lists a few potentially associated hardware features. Such layering of security and reliability hardware can lead to distinct cases where reliability errors need to be resolved with security error handling.

B. Error Handling vs. Security Handling

When reliability error-handling circuits and security structures are layered, it's possible for reliability errors to be misidentified as a security error. For example, this can happen due to mis-correction from error detection circuits.

Thus, consider a core loadable data segment protected with a SHA signature. If a multi-bit error were to occur, error-detection circuitry could mis-correct it. However, the SHA signature would detect the miscorrection. In such a scenario should the security error handler come into effect or should the reliability error handler come into effect? One solution is to provide error resolution at the layer where it's detected. Applying this principle in this scenario means that the mis-correction error would be treated as a security error. Not treating such an error as a security error could interrupt device mode into error handling modes—such modes may not be trusted if a trust level transition does not provide sufficient isolation. If a trust level transition is applied with proper trusted operation isolations, error information should be logged in error handling mode. However, if isolation is applied, error information may not be available to the untrusted error handler - error handler that resolves the error often need to operate at another trust level - such as the supervisor or the hypervisor. This behavior is essentially functionally similar to a security

exception triggered isolation and exit scheme.

However, this principle may not be sufficient when the device is operating in a trusted mode and is interrupted with uncorrectable reliability errors. In such a case, the security protection system on the device must account for the fact that a solid reliability error has occurred. Therefore, the device must take a reliability error based exception or trap. This can expose trusted operations to untrusted modes and operations. To prevent such exposure, prior to taking a reliability error trap, the security subsystem should be able to provide isolation, such as encryption, cacheline flushing, pipeline flushing, etc. for trusted operations and resulting and dependent data.

One can consider reversing the layering - where the hardware security or hardware assisted security operations are at a layer before the error detection and handling layer. This could imply that when decrypting data, security checking operations would commence first. This could lead to scenarios where decryption on data without error detection or correction. In such a case, security operations could operate on unreliable data—thus leading to some failures or errors that are preventable. Furthermore, if a reliability error is detected with hardware based security operations, error correction, detection, or alternative mitigation operations could not operate properly. Thus, the original layering scheme proposed is the most appropriate.

V. Examples and Limitations

The principles discussed in Sect. IV can be partially illustrated with a brief case study.

A. Working Example

State of the art processors now fully feature hardware security features. One such feature is encrypted secure virtualization [3]. One particular feature enables encrypting the virtual machine control structure during encrypted virtual machine execution. On entering the encrypted virtual machine, a CRC checksum is generated and checked against a stored checksum. If the checksum matches, the virtual machine would be allowed to execute. The processor would allow its internal states to be updated to the states specified in the encrypted virtual machine control structure. This is not uncommon for hardware assisted virtual machine execution. When the checksum doesn't match, the virtual machine would not be executed.

Given that the a virtual machine control state is stored on the main system memory in an encrypted form, there is a rare case that a mis-corrected error could lead to CRC checksum generation mismatch. In such a case, the mismatched checksum would be treated as a security violation directly—in which case the encrypted virtual machine is prevented from execution. Note that integrity checking rather than reliability is more aligned with the purpose of the checksuming. A similar checksum mismatch would occur had a SHA algorithm been applied.

An uncorrectable reliability error can also happen when the encrypted virtual machine is executing an encrypted program on encrypted data. In such a case, the processor would need to exit from the virtual machine to return to the hypervisor to allow corrective actions to be taken. In such a case, the processor hardware provides isolation by writing out and potentially flushing out the cache data that may need to be encrypted—as the trust level has changed.

B. Limitations

One potential limitation arising from layering security and reliability protection circuitry is handling multiple errors concurrently. One such scenario can arise when the processor is in the middle of handling a security exception and gets interrupted by a reliability error. In such a case, when the error occurs when the device in a trusted mode, processors can provision for the processor to enter into its shutdown state [3] depending on the number of errors the processor is architected to handle. However, note that in the transition to the shutdown state, the device should still attempt to provide isolation of trusted contents.

VI. POTENTIAL STRATEGY

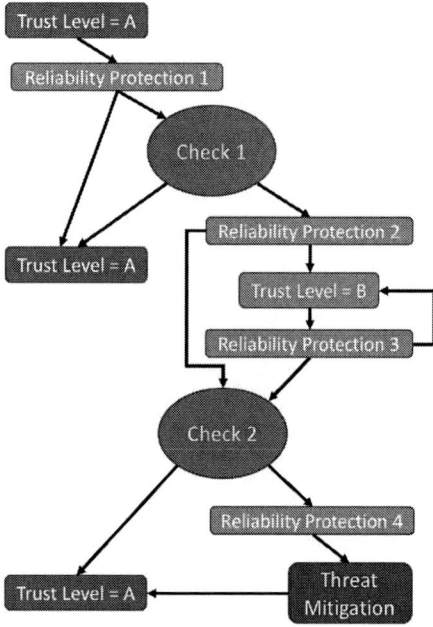

Fig. 3. Trust Transition Model with Reliability

One potential flow to layer security features with reliability protections can be viewed in Fig. 3. This figure is similar to Fig. 1. *Check 1* and *Check 2* still refer to security checks, as in Fig. 1. The modifications are the insertion of *Reliability Protection* points 1-4. *Reliability Protection* points do not have to be independent circuits. In the case of a processor, *Reliability Protection* points 1-4 should all be referring to the set of reliability circuits already present on the chip to protect subsystems prone to reliability errors, such as memory and system buses. However, their interaction with the security circuits in the various transitions of the trust transition flow may be different.

Reliability Protection 1 models potential reliability errors induced prior to *Check 1*. When a reliability error interrupt occurs prior to *Check 1*, the security error handling should be able to exit back to the starting *Trust Level A* (see left arrow in Fig. 3 after *Reliability Protection 1*). Alternatively, when errors escape *Reliability Protection 1's* detection, a potential security checking scheme such as SHA could detect illegal

data modification, in which case the security error handling scheme would still let the device enter back into *Trust Level A*.

When the request to enter into *Trust Level B* passes *Check 1*, the newly granted trust level may encounter a reliability error that originated at *Reliability Protection 2*. Such errors could trigger interrupts that would lead to a transition back to *Trust Level A*. However, because the device is now in *Trust Level B*, the request to go back to *Trust Level A* must go through *Check 2* to ensure that a sound transition can happen.

If *Trust Level B* has been entered successfully, *Reliability Protection 2*, has been passed successfully. However, reliability errors may now arise at *Reliability Protection 3*. Such errors should direct the device to enter into *Check 2*. Otherwise, at the end of *Trust Level B* operations, trust level changing schemes prompted in *Trust Level B* operations will naturally cause *Trust Level B* to transition to *Trust Level A*, during which *Check 2* would be performed. If *Check 2* fails, the device may need to go into *Threat Mitigation* mode. However, if reliability errors occur on entry or during *Threat Mitigation* mode, a fatal system event may be warranted.

The strategy outlined in this paper is partially inferred based on observations from architectural security features discussed in [3], [17], [2], [1]. The alternative approaches found in [8], [9], [10] call for some form of dedicated hardware to monitor security as well as reliability errors. Such architectural modifications can lead not only to area overhead and performance impact, but they may also lead to increased design complexity—requiring substantial design verification effort to close out verification. Furthermore, potential software changes may be required to take advantage of such hardware schemes. However, when such approaches are properly co-designed, efficiency for handling hardware error based threat models can be realized.

In contrast, a layered approach found in [3] for virtual machine isolation does not incur substantial die area increase and is now available for general public use. Hence, the discussion shown here is by no means complete, but it provides a potential strategy for architecting a layered approach towards integrating hardware based security features along with existing hardware reliability protections.

VII. CONCLUSIONS

In this paper, some principles and strategies for layering security circuits with reliability circuits are illustrated, and a relevant example based on published industrial architectures is explained. This layered approach, if implemented properly, could be a useful tool for architecting future security features while accommodating the need for device reliability.

REFERENCES

[1] S. Gueron, "A Memory Encryption Engine Suitable for General Purpose Processors," Cryptology ePrint Archive, Report 2016/204, 2016, http://eprint.iacr.org/2016/204.

[2] "Building a Secure System Using Trustzone Technology," Apr 2009. [Online]. Available: http://infocenter.arm.com/

[3] "AMD64 Architecture Programmers Manual Volume 2: System Programming," Mar 2017. [Online]. Available: http://developer.amd.com/resources/developer-guides-manuals/

[4] P. Kocher, D. Genkin, D. Gruss, W. Haas, M. Hamburg, M. Lipp, S. Mangard, T. Prescher, M. Schwarz, and Y. Yarom, "Spectre Attacks: Exploiting Speculative Execution," *ArXiv e-prints*, Jan. 2018.

978-1-5386-7100-9/18 $31.00 © 2018 IEEE

[5] M. Lipp, M. Schwarz, D. Gruss, T. Prescher, W. Haas, S. Mangard, P. Kocher, D. Genkin, Y. Yarom, and M. Hamburg, "Meltdown," *ArXiv e-prints*, Jan. 2018.

[6] N. Seifert, P. Slankard, M. Kirsch, B. Narasimham, V. Zia, C. Brookreson, A. Vo, S. Mitra, B. Gill, and J. Maiz, "Radiation-induced soft error rates of advanced cmos bulk devices," in *Reliability Physics Symposium Proceedings, 2006. 44th Annual., IEEE International*, March 2006, pp. 217–225.

[7] V. Degalahal, L. Li, V. Narayanan, M. Kandemir, and M. Irwin, "Soft errors issues in low-power caches," *Very Large Scale Integration (VLSI) Systems, IEEE Transactions on*, vol. 13, no. 10, pp. 1157–1166, Oct 2005.

[8] R. G. Ragel and S. Parameswaran, "Impres: integrated monitoring for processor reliability and security," in *2006 43rd ACM/IEEE Design Automation Conference*, July 2006, pp. 502–505.

[9] N. Nakka, Z. Kalbarczyk, R. K. Iyer, and J. Xu, "An architectural framework for providing reliability and security support," in *International Conference on Dependable Systems and Networks, 2004*, June 2004, pp. 585–594.

[10] C. Pham, Z. Estrada, P. Cao, Z. Kalbarczyk, and R. K. Iyer, "Reliability and security monitoring of virtual machines using hardware architectural invariants," in *2014 44th Annual IEEE/IFIP International Conference on Dependable Systems and Networks*, June 2014, pp. 13–24.

[11] F. Rahman, D. Forte, and M. M. Tehranipoor, "Reliability vs. security: Challenges and opportunities for developing reliable and secure integrated circuits." in *2016 IEEE International Reliability Physics Symposium (IRPS)*, April 2016, pp. 4C–6–1–4C–6–10.

[12] S. Kan, M. Ottavi, and J. Dworak, "Enhancing embedded sram security and error tolerance with hardware crc and obfuscation," in *2015 IEEE International Symposium on Defect and Fault Tolerance in VLSI and Nanotechnology Systems (DFTS)*, Oct 2015, pp. 119–122.

[13] T.-P. Ma and P. V. Dressendorfer, *Ionizing radiation effects in MOS devices and circuits.* John Wiley & Sons, 1989.

[14] S. V. Kumar, K. H. Kim, and S. S. Sapatnekar, "Impact of NBTI on SRAM read stability and design for reliability," in *7th International Symposium on Quality Electronic Design (ISQED'06)*, March 2006, pp. 6 pp.–218.

[15] J. L. Zerbe, C. W. Werner, V. Stojanovic, F. Chen, J. Wei, G. Tsang, D. Kim, W. F. Stonecypher, A. Ho, T. P. Thrush, R. T. Kollipara, M. A. Horowitz, and K. S. Donnelly, "Equalization and clock recovery for a 2.5-10-gb/s 2-pam/4-pam backplane transceiver cell," *IEEE Journal of Solid-State Circuits*, vol. 38, no. 12, pp. 2121–2130, Dec 2003.

[16] B. Rogers, S. Chhabra, M. Prvulovic, and Y. Solihin, "Using Address Independent Seed Encryption and Bonsai Merkle Trees to Make Secure Processors OS- and Performance-Friendly," in *Proceedings of the 40th Annual IEEE/ACM International Symposium on Microarchitecture*, ser. MICRO 40. Washington, DC, USA: IEEE Computer Society, 2007, pp. 183–196. [Online]. Available: http://dx.doi.org/10.1109/MICRO.2007.44

[17] S. Kan and J. Dworak, "Systematic Test Generation for Secure Hardware Supported Virtualization," in *2017 IEEE 15th Intl Conf on Dependable, Autonomic and Secure Computing, 15th Intl Conf on Pervasive Intelligence and Computing, 3rd Intl Conf on Big Data Intelligence and Computing and Cyber Science and Technology Congress(DASC/PiCom/DataCom/CyberSciTech)*, Nov 2017, pp. 550–556.

2018 IEEE Computer Society Annual Symposium on VLSI

BD-NET: A Multiplication-less DNN with Binarized Depthwise Separable Convolution

Zhezhi He*, Shaahin Angizi, Adnan Siraj Rakin, and Deliang Fan[†]

Department of Electrical and Computer Engineering, University of Central Florida, Orlando, FL 32816

Email: *Elliot.He@knights.ucf.edu, [†]dfan@ucf.edu

Abstract—In this work, we propose a multiplication-less deep convolution neural network, called BD-NET. As far as we know, BD-NET is the first to use binarized depthwise separable convolution block as the drop-in replacement of conventional spatial-convolution in deep convolution neural network (CNN). In BD-NET, the computation-expensive convolution operations (i.e. Multiplication and Accumulation) are converted into hardware-friendly Addition/Subtraction operations. In this work, we first investigate and analyze the performance of BD-NET in terms of accuracy, parameter size and computation cost, w.r.t various network configurations. Then, the experiment results show that our proposed BD-NET with binarized depthwise separable convolution can achieve even higher inference accuracy to its baseline CNN counterpart with full-precision conventional convolution layer on the CIFAR-10 dataset. From the perspective of hardware implementation, the convolution layer of BD-NET achieves up to 97.2%, 88.9%, and 99.4% reduction in terms of computation energy, memory usage, and chip area respectively.

I. INTRODUCTION

Owing to the explosion of data, improvement of parallel computing ability resulting from GPU and continuous breakthroughs in algorithm, Artificial Neural Network (ANN) has achieved great success in recent years. Deep neural network, as one of the most popular state-of-the-art ANN, has shown its leading performance in various domains, such as speech recognition, computer vision and data analysis [1]. Both theoretically and empirically, deep neural network has shown significant performance improvement over its shallow counterparts [2], [3].

Recently, the state-of-the-art deep CNN could achieve better-than-human accuracy in object classification task for large scale datasets (e.g. ImageNet). However, from hardware implementation perspective, deep CNN still encounters the obstacle of hardware deployment due to its massive cost in both computation and memory. Many recent works have been proposed to address such high computational complexity and memory usage issues of existing neural network structures. For example, pruning and Hoffman coding have been used to compress neural network model size. Beyond that, BinaryConnect [4], Binary-Weight-Net [5] and DoReFa-Net [6] have adopt quantized or even binarized (i.e. +1 and -1) weight and/or activation function to reduce model size and computational cost. The objective in this work is to propose another new method to optimize the deep neural network in terms of computational complexity and model size, while preserving the accuracy in comparison to its full precision CNN counterpart.

The main strategy here is to take advantage of depthwise separable convolution block [7], which is an energy efficient alternative of standard spatial-convolution layer. In this work, we fully binarize the kernel weights and activation function in depthwise convolution (i.e. channel-wise 3x3 Conv) part, while keeping the weights of pointwise convolution (i.e. 1x1 Conv) in real number (i.e. 32-bit float). Therefore, for the first depthwise convolution, all the multiplications are converted into addition/subtraction due to binarized (+1 and -1) weights. Meanwhile, a binary activation function is added to the output of depthwise convolution. Therefore, the input tensor (i.e. output of binary activation function) of following pointwise convolution layer is also in binary form. As a result, the multiplication operation within pointwise layer could be implemented using only addition/subtraction as well. Thus, all the hardware-expensive multiplications in binarized depthwise convolution are all replaced by add/sub operations. Furthermore, we found that the accuracy degradation caused by weight and activation binarization could be compensated by increasing the *channel multiplier*, which we define as *channel expansion* in this work. Our proposed channel expansion method avoids model size growing exponentially when directly expand the input/output channel. In summary, our main contributions in this work can be summarized as:

- We propose a new deep convolution neural network optimization method (i.e binarized depthwise convolution and channel expansion), which removes the multiplication operation in convolution layer and results in great hardware efficiency, through iterative input/weight binarization.
- Theoretic analysis is provided to prove that our proposed binarized depthwise separable convolution can effectively approximate the conventional spatial convolution. The experiments also show the deep neural network with such convolution layers can achieve state-of-the-art accuracy.
- We perform detailed analysis about the effect of internal hyper-parameter configurations, which shows the trade-off between neural network accuracy and hardware resources utilization w.r.t energy, memory, and area. It provides an intuitive design guidance for hardware engineer to map the convolution layers of deep neural network into mobile embedded system.

II. BINARIZED DEPTHWISE SEPARABLE CONVOLUTION

In this section, we first introduce the depthwise separable convolution as an alternative to the conventional spatial

978-1-5386-7100-9/18 $31.00 © 2018 IEEE

130

convolution. Then, we propose our proposed binarization methodology, which iteratively binarize kernel weights and intermediate tensors through depthwise separable convolution block. Based on such combination, we construct a deep neural network, called BD-NET, as one benchmark to examine the performance. At last, we also provide the theoretic analysis to demonstrate that the proposed binarized depthwise separable convolution can approximate the conventional spatial convolution.

A. depthwise separable convolution

Recently, the depthwise separable convolution [7] has been widely used in many state-of-the-art deep neural networks, such as MobileNet [8] and Xception [9], which replaces the standard convolution layers to reduce DNN computational cost and memory usage. As a factorized form of conventional spatial-convolution, the depthwise separable convolution consists of two parts: depthwise convolution and 1x1 convolution (a.k.a. pointwise convolution). The conventional spatial-convolution mainly performs channel-wise feature extraction, then combing those features to generate new representations. It is proven that such two-step tasks could be handled by depthwise convolution and pointwise convolution separately [7].

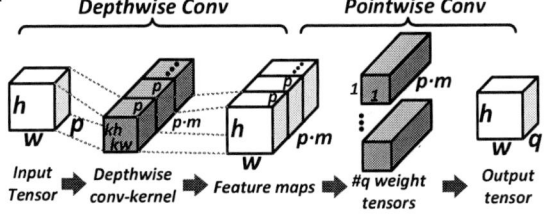

Fig. 1. Data-flow for depthwise separable convolution layer. Normally, the channel expansion m is 1.

The operation of depthwise separable convolution is described in the form of data flow in Fig. 1. Considering the input tensor is in the dimension of $h \times w \times p$, which denotes height, width and channel respectively. Note that, in the depthwise convolution layer, each channel of input tensor performs convolution with m kernels in the size of $kh \times kw$ (i.e. kernel-height and width) correspondingly, which produces $p \cdot m$ feature maps. m is defined as *channel expansion* here in this work. We found that larger m could effectively compensate accuracy degradation in weight binarization at the cost of model size, which will be explicitly investigated in the next section. Those generated feature maps are concatenated along its depth dimension as tensor in size of $h \times w \times (p \cdot m)$[1], which is taken as the input of pointwise convolution layer. The operation of such depthwise convolution can be mathematically described as:

$$\boldsymbol{F}_{j \cdot \mathrm{p}+i} = \boldsymbol{X}_i * \boldsymbol{W}_{j \cdot \mathrm{p}+i}, \quad j \in [0, m-1] \tag{1}$$

where $i \in [1, \mathrm{p}]$ is the index of depth of input tensor $\boldsymbol{X} \in \mathbb{R}^{h \times w \times p}$. ($*$) denotes the convolution (i.e. dot-product)

[1]The default hyper-parameter configurations in convolution layers are: kernel size=3×3, stride = 1, padding = 1, no bias.

between input and kernel $\boldsymbol{W}$. On the contrary to the distinctive depthwise convolution, pointwise layer is just normal spatial-convolution layer with 1×1 convolution kernel size. Thus, it only linearly combines the $p \cdot m$ input feature maps to generate new representations with q output channels.

From the perspective of computation, the cost of normal spatial-convolution layer is $h \times w \times kh \times kw \times p \times q$. As its approximating alternative, the cost of depthwise convolution is $h \times w \times kh \times kw \times p \times m$, while the cost of pointwise convolution is $h \times w \times p \times m \times q$. Thus, the ratio of computational cost between depthwise separable convolution and conventional convolution can be calculated as:

$$\frac{h \cdot w \cdot kh \cdot kw \cdot p \cdot m + h \cdot w \cdot p \cdot m \cdot q}{h \cdot w \cdot kh \cdot kw \cdot p \cdot q} = \frac{m(kh \cdot kw + q)}{kh \cdot kw \cdot q} \tag{2}$$

such ratio is approximated to $m/(kh \cdot kw)$ when the number of output channels $q >> kh \cdot kw$. Normally, when the channel expansion m is set to 1 (e.g. MobileNet [8]) and the kernel size $kh \times kw$ is 3×3, the computational cost is only $1/9$ of the normal spatial-convolution. Increasing the kernel size will lead to higher computational cost reduction. However, small channel expansion m may cause slight accuracy degradation, where such accuracy drop can be compensated through using larger m. The effect of this channel expansion factor m will be analyzed in later section.

B. Binarization of weight and activation

In this work, we choose sign function as the deterministic binarization activation function similar as [10], whose output is either +1 or -1. Note that, the sign function owns zero derivatives almost everywhere, which makes it impossible to calculate the gradient using chain rule in backward path. Thus, the Straight-Through Estimator (STE) [11] is applied to calculate gradient in this work. In the backward path, the input gradient of binarization activation function clones the gradient at output, if it is in the range from -1 to +1. Otherwise, the gradient is cancelled to preserve training performance. In summary, such binarization activation function in forward and backward can be described as:

$$\text{Forward}: \quad q = Sign(r) = \begin{cases} +1 & if \ r \geq 0 \\ -1 & otherwise \end{cases} \tag{3}$$

$$\text{Backward}: \quad \frac{\partial g}{\partial r} = \begin{cases} \frac{\partial g}{\partial q} & if \ |r| \leq 1 \\ 0 & otherwise \end{cases} \tag{4}$$

The aforementioned binarization function can be directly used as activation function. For binarizing the weight of depthwise convolution layer, we retain the weights in real value for the backward training. For the forward path, the weights are binarized using function in Eq. (3).

C. BD-NET: Neural Network with Binarized Depthwise separable convolution

In this work, we only make the weights of depthwise convolution and the input tensor of pointwise convolution in binary (i.e. +1 and -1), which could totally eliminate multiplications in whole depthwise separable convolution. We construct a

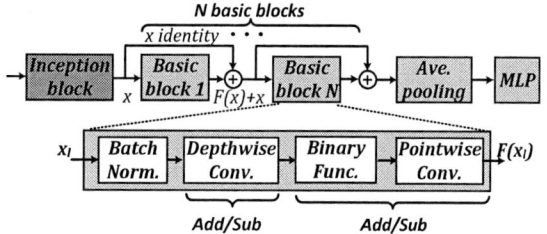

Fig. 2. Block diagram of the multiplication-less deep neural network. The binarized activation function is inserted between depthwise (binarized weight) and pointwise convolution layer.

deep neural network with the proposed binarized depthwise separable convolution referring to the topology of ResNet [12], called BD-NET. Note that, ResNet architecture is only used as an example, other state-of-the-art network architecture could be used as well. The block diagram of deep neural network analyzed in this work is depicted in Fig. 2. which sequentially composes of an inception block[2] (3×3 spatial convolution, Batch-normalization and ReLU), N depthwise separable convolution basic blocks, average pooling layer and Multi-Layer Perceptron (MLP, also called fully connected layer). As the key component in our proposed neural network, basic block includes batch normalization, depthwise convolution with binarized weight, binarized activation function and pointwise convolution. Similar as BWN [5] and LBCNN [13], we place the batch normalization in front of the convolution layer with binarized weight. In summary, the response of basic block can be described as:

$$x_{l+1}^t = \sum_{s=1}^{p \cdot m} Sign\left(W_l^s * BN(x_l^s)\right) \cdot \alpha_{l,s}^t \quad (5)$$

where $s \in [p \cdot m]$ and $t \in [q]$ denote the input channel and output channel, respectively. W_l is the learned depthwise convolution kernel with binarized weight. l is the index of basic block , while p is the number of input channels of l_{th} basic block. $\alpha_{l,i}$ is the learned weight of pointwise convolution. $BN()$ is the batch normalization function [14].

TABLE I
HARDWARE COST ANALYSIS OF STANDARD CONVOLUTION IN CNN AND
BINARIZED DEPTHWISE SEPARABLE CONVOLUTION IN THIS WORK.

	Computation Cost		Memory Cost
	Mul–$O(N^2)$	Add/Sub–$O(N)$	
CNN	$h \cdot w \cdot kh \cdot kw \cdot p \cdot q$	$h \cdot w \cdot kh \cdot kw \cdot p \cdot q$	$kh \cdot kw \cdot p \cdot q \cdot N_{bit}^{32}$
This work	–	$h \cdot w \cdot kh \cdot kw \cdot p \cdot m$ $+ h \cdot w \cdot p \cdot m \cdot q$	$kh \cdot kw \cdot p \cdot m \cdot N_{bit}^{n}$ $+ p \cdot m \cdot q \cdot N_{bit}^{n}$
$\frac{This\ work}{CNN}$	0	$\frac{m}{q} + \frac{m}{kh \cdot kw}$	$\frac{m}{q \cdot N_{bit}^{32}} + \frac{m}{kh \cdot kw \cdot N_{bit}^{32}}$

The objective of BD-NET is to build an efficient hardware-friendly neural network, while preserving accuracy. To demonstrate the efficiency of BD-NET, we analyze the hardware cost of standard spatial convolution and binarized depthwise separable convolution in terms of computation cost and parameter size. As tabulated in Table I, for computing complexity, it can

[2]Similar as previous works of binarized/quantized neural network [5], [6], [10], we do not introduce binarization to the first inception block.

be seen that not only the multiplication are fully replaced by the hardware efficient add/subtraction, but also the number of operations are reduced by a factor of $\sim m/(kh \cdot kw)$. For parameter size (i.e. memory cost), the compression rate of BD-NET is $\sim m/(kh \cdot kw)$ when the weights in pointwise convolution layer is 32-bit ($N_{bit}^{n=32}$), where n is the number of bits in pointwise convolution kernel as used in this work. Furthermore, if targeting for a higher memory compression rate, the weight quantization could be applied to pointwise convolution layer in the future.

D. Theoretic analysis

Theoretically, If the binarized depthwise separable convolution can effectively approximate the standard spatial-convolution with full-precision weight, such two types of convolution layers are supposed to produce similar output tensor when the same input tensor are given [13] [15]. We will prove it as below.

First, for fair comparison, let's assume the input tensor and out tensor of the conventional convolution and binarized depthwise separable convolution are in the identical dimension. We define the real value weight of standard convolution as $W' \in \mathbb{R}^{p \times q \times kh \times kw}$, and the weight of depthwise convolution as $W \in \mathbb{B}^{(p \cdot m) \times kh \times kw}$, where $\mathbb{B} = \{-1, +1\}$. Note that, we temporarily ignore the dimension of mini-batch for simplicity. We define y as one element in the output tensor of our proposed binarized depthwise separable convolution. y' is the corresponding element (i.e. with identical dimension index) in the output tensor of conventional spatial-convolution. Both of those two types of convolution are fed with same input tensor $X \in \mathbb{R}^{p \times h \times w}$. For calculating y', a subset of weights in W' is vectorized as $w' \in \mathbb{R}^{(p \cdot kh \cdot kw) \times 1}$. Meanwhile, in the input tensor X, a vectorized patch $x \in \mathbb{R}^{(p \cdot kh \cdot kw) \times 1}$ are selected by w' to perform dot-product computation. Thus, the microscopic output of convolution for two types of convolution can be mathematically described as:

$$y' = w'^T x \in \mathbb{R} \quad (6)$$

$$y = \left(Sign(W \circledast x)\right)^T \beta = b^T \beta \quad (7)$$

where $\beta \in \mathbb{R}^{(p \cdot m) \times 1}$ is a vectorized subset of pointwise convolution weight $\alpha \in \mathbb{R}^{(p \cdot m) \times q}$ along its first dimension. Here, we simply use $\circledast$ to denote the depthwise convolution operation described in Eq. (1) without specific notations in mathematics expression. Note that, the result of $W \circledast x$ is in shape of $\mathbb{R}^{(p \cdot m) \times 1}$, thus we can obtain that $b = Sign(W \circledast x) \in \mathbb{B}^{(p \cdot m) \times 1}$. Therefore, we can rewrite Eq. (7) as $y = \sum_{i=1}^{p \cdot m} \pm \beta_i = \pm \beta_1 \pm \beta_2 \pm \cdots \pm \beta_{p \cdot m}$. Moreover, since β is a vector with $p \cdot m$ trainable real value parameters (i.e. pointwise weight), we could convert the problem of approximating y with y' to the problem of optimizing β to:

$$\underset{W, \beta}{\text{minimize}} \ \mathcal{L}(y, y') = ||y - b^T \beta|| \quad (8)$$

An obvious solution to this optimization problem is to chose $\beta_i \cdot b_i$ to y, while set all the other elements in β to be zero. Thus, there always exists a vector β to make $y' = y = b^T \beta$.

III. EXPERIMENTS ANALYSIS

A. Software experiment setup

The software experiments of this work are performed under the framework of Pytorch, which recently optimized its depthwise convolution backend CUDA library to accelerate the training process. The depthwise convolution is a special case of grouped convolution, where the number of groups should be set to the number of input channels. We employ the stochastic gradient descent with momentum = 0.9 as the optimizer to minimize the cross-entropy loss. Owing to the large variation of intermediate output caused by frequently adjusted binary weights and binarization activation function, small learning rate are preferable. we set the initial learning rate as 0.001, which is reduced to 0.0001 through scheduling. Learning rate larger than 0.01 will cause the severe fluctuation with no accuracy enhancement during the model training.

The experiments are performed on three common image datasets: MNIST, SVHN and CIFAR-10. MNIST is a 28×28 grayscale handwritten digit (0-9) image dataset with 60000 samples in training set, and 10k samples in test set. The Street View House Number (SVHN) contains 32x32 RGB real-world images with 73257 samples in training set, and 26032 samples in test set. CIFAR-10 is a colorful 32×32 image dataset which contains 10 classes real world objects/animals with 50000 training samples and 10000 test samples. datasets are augmented with the method introduced in ResNet on CIFAR-10 [12]. The test accuracy of MNIST, SVHN and CIFAR-10 are reported in Table II using the deep neural network structure (BD-NET) described in Section II-C. The model configurations for those three datasets are:

- MNIST: 16 input channels, 5 basic blocks, 128 hidden neuron, 64 batch size, 3×3 kernel size, 4 channel expansion.
- SVHN: 128 input channels, 5 basic blocks, 512 hidden neuron, 64 batch size, 3×3 kernel size, 4 channel expansion.
- CIFAR-10: 512 input channels, 10 basic blocks, 512 hidden neuron, 20 batch size, 3×3 kernel size, 4 channel expansion.

Note that, to ensure fair comparison between BD-NET and its baseline CNN counterparts, identical hyper-parameters are used for both CNN baseline and BD-Net. As the result reported in Table II, BD-NET using the special configured depthwise separable convolution can obtain close or even better accuracy with respect to to its baseline counterpart. Furthermore, for the largest CIFAR-10 dataset, our work can achieve the best accuracy in comparison with other works applied with weight binarization techniques as tabulated in Table II.

TABLE II
INFERENCE ACCURACY (%) OF MNIST, SVHN AND CIFAR-10.

	Baseline CNN	BD-NET (this work)	BinaryConnect [4]	BNN [16]	Dorefa-Net [6]	BWN [5], [17]
MNIST	99.46	99.41	98.99	98.60	-	98.69
SVHN	94.29	93.66	97.85	97.49	97.52	97.46
CIFAR-10	91.25	92.41	91.73	89.85	-	89.49

B. Effect of hyper-parameters

In this subsection, we examine the effect of hyper-parameters on neural network performance. Since the neural networks are trained from scratch instead of fine-tuning from the pretrained model, we chose the CIFAR-10 as the representative experiment dataset to report the results.

1) channel expansion: As discussed in Section II-A, increasing the channel expansion m enriches the intermediate feature sets thus leading to more variant combinations as the output of convolution layer. We run trails with small input/output channel ($p = q = 64$, much smaller than that shown in Table II) to investigate the effect of channel expansion on neural network accuracy. As the results listed in Table III, the precision denotes the number of bits used for weights of convolution kernel. Note that, for depthwise separable convolution, precision only refers to the depthwise part, the data type of pointwise weights are still real number. Moreover, for depthwise separable convolution with 1-bit convolution weight, the intermediate binarized activation function is included. This experiment shows that directly using depthwise separable convolution ($m = 1$) to replace spatial-convolution layer only results in slight accuracy degradation ($< 2\%$). The binarization applied on depthwise separable convolution will lead to further accuracy drop. However, increasing m can effectively narrow the the accuracy gap between BD-NET and its normal convolution counterpart.

TABLE III
THE CIFAR-10 TEST ACCURACY WITH VARIOUS CONFIGURATIONS

	Spatial Convolution		Depthwise Separable Convolution					
precision	32-bit	1-bit	32-bit	1-bit	1-bit	1-bit	1-bit	1-bit
m	-	-	1	1	2	4	8	16
Top-1 Test Accuracy	89.28%	86.9%	87.93 %	85.62%	86.39%	87.44%	87.8%	89.0%

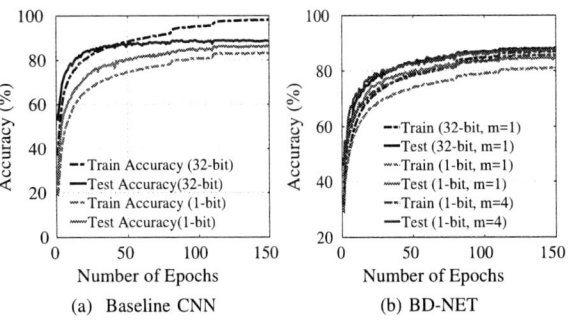

Fig. 3. The accuracy evolution curve of (a) Baseline CNN using normal spatial-convolution with 32-bit and 1-bit weight. (b) BD-NET using binarized depthwise separable convolution with varying channel expansion m.

Beyond that, we include the accuracy evolution curves in Fig. 3 to show that our binarized depthwise separable convolution is helpful to prevent network from overfitting. As shown in Fig. 3a, training of normal spatial convolution (black curve) is easily to be overfitting due to over-parameterization in convolution layer. Directly applying the weight binarization techniques on convolution weights as introduced in BinaryConnect [4] does weaken overfitting, but lowering the test accuracy as well. On the contrary, our proposed BD-NET can achieve almost the same accuracy as baseline CNN, which avoids overfitting problem.

978-1-5386-7100-9/18 $31.00 © 2018 IEEE

2) Number of channels: Table IV shows the CIFAR-10 test accuracy with various numbers of intermediate channels, which denotes the input and output channels of basic blocks in BD-NET (assuming $p = q$). It can be seen that increasing the number of intermediate channels improve accuracy. As we discussed above, the baseline CNN with normal spatial-convolution easily suffers from overfitting, if the network topology is not fine-tuned. When the number of intermediate channels is larger than 128, the test accuracies of BD-NET are above the baseline CNN. In general, our BD-NET is expected to achieve even higher accuracy with further fine-tuning on the model.

We notice that reducing the number of channels lowers the computational cost and model size exponentially (referring to Table I). Other methods like further quantizing the intermediate tensor to lower number of bits only reduce the hardware cost linearly, due to the convolution are performed using Addition/subtraction instead of multiplication. Moreover, the accuracy improvement using large number of channels is limited. Thus, we are inclined to scale the number of intermediate channel using channel expansion first, once the hardware deployment of BD-NET meets power or memory bottleneck.

TABLE IV
TEST ACCURACY WITH VARYING NUMBER OF INTERMEDIATE CHANNELS
(I.E. INPUT AND OUTPUT CHANNELS). M=4 IN THIS EXPERIMENT.

number of channels (p,q)	64	128	256	384	512
Baseline CNN	89.28%	90.47%	91.44%	91.56%	91.25%
BD-NET	87.44%	90.65%	91.76%	92.26%	92.41%

C. Hardware mapping analysis

In this subsection, we compare baseline CNN, DoReFa-NET (with binary weight), BWN and the BD-NET in terms of three key metrics i.e. energy consumption, memory and area. For all four models, we developed pipelined designs considering one stage per neuron. The inputs and weights are initially stored in the image and kernel banks, and then fetched to the pipeline stage logic to compute the output for all neurons. Synopsys Design Compiler [18] is used for different networks with 45nm North Carolina State University (NCSU) Product Development Kit (PDK) library [19] to generate the gate-level netlists and evaluate the performance. In addition, we used CACTI [20] to calculate the memory array and access's performance. Whereas previous works present different methods (such as data reuse) to optimize the CNN hardware implementation, here we do not optimize the models, since our main objective is to compare and show how BD-NET impacts hardware performance. Note that, the same topology in block level (i.e. number of blocks and its input/output dimension) is used for CNN and BD-NET and all the evaluation is estimated for inference of a single convolution layer with following setup: $p=q=384$, $h=w=32$ and $kh=kw=3$. For the impartial comparison, DoReFa-NET [6] and BWN [5] follow their particular network topology without any modification. Considering a trade-off between accuracy, energy consumption, memory usage and area overhead, the m

parameter can be precisely tuned to meet a specific hardware's requirement as will be discussed in the following.

1) Energy: Fig. 4a reports the breakdown of energy consumption of a single convolution layer of different networks and the BD-NET with distinct channel expansion configurations (m=1, 2, 4, 8, and 16) for running CIFAR-10 dataset. As can be seen, the multiplication (*Mul*) contributes to the most fraction of energy consumed by baseline CNN, which is replaced by energy-efficient addition/subtraction (*Add/Sub*) within the proposed network. The energy consumption of *Add/Sub* is separately delineated for depthwise and point-wise convolutions of BD-NET. To further show the energy-efficiency of the proposed network, Fig. 4b calculates the energy reduction ratio of BD-NET in different configuration with respect to different networks. As per this figure, as much as 97.2% reduction can be achieved to CNN, while setting m to 1. This significant energy reduction mainly comes from two sources: 1) standard convolution is replaced with energy-efficient depthwise separable convolution; 2) *Mul* in convolution is converted to *Add/Sub* due to binarization. As can be seen, with a comparable accuracy when setting m to 4, 86.4%, 36.5% and 30.6% energy reduction is obtained for BD-NET w.r.t. CNN, DoReFa-NET and BWN, respectively. However, for ($m \geqslant 8$), binary-weight networks outperform the BD-NET in terms of energy consumption.

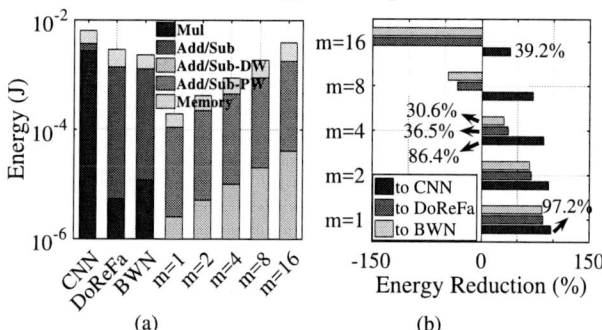

Fig. 4. (a) Normalized log-scaled energy consumption of CNN, DoReFa-NET and BWN vs. BD-NET in varying channel expansion configuration. DW: depthwise layer, PW: pointwise layer. (b) Energy reduction of BD-NET to different Networks.

2) Memory: The comparison of model efficiency in terms of memory usage required for processing one convolution layer between BD-NET and different networks is shown in Fig. 5a. We observe that the binarized convolution block in BD-NET are much more memory-friendly than the CNN counterpart, when using small channel expansion ($m \leqslant 4$). For m=1 and 2, BD-NET utilizes less memory compared to binary-weight networks with more number of parameters. Note that, BWN consists of channel-wise scaling factor for each layer which imposes excessive memory usage. However, DoReFa-NET and BWN require less memory compared to BD-NET for $m > 2$. The memory storage reduction of a convolution layer in BD-NET with respect to different networks is specifically reported in Fig. 5b. For instance, based on the reported results, 88.9% and 55.5% reduction w.r.t. CNN are achieved when m equals 1 and 4, respectively.

978-1-5386-7100-9/18 $31.00 © 2018 IEEE

This reduction mainly comes from reduced number of fixed kernel as discussed in Section II-C. However, for getting higher accuracy for BD-NET compared to other networks, the larger channel expansion may be desired which reverses the memory reduction trend. For example when m=16, the memory storage required for processing the convolution layer increases. As shown, sacrificing the inference accuracy by ~1.5% (i.e. changing m from 16 to 4) leads to ~4× memory saving of CNN baseline with a considerably lower energy.

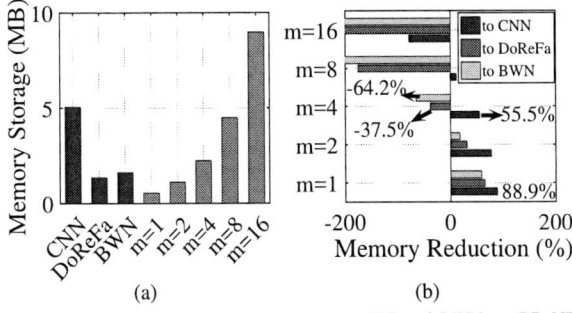

Fig. 5. (a) Memory storage of CNN, DoReFa-NET and BWN vs. BD-NET in different configuration for a single convolutional layer, (b) Memory storage reduction of the proposed network to different Networks.

3) Area: Fig. 6a illustrates and compares the computational area overhead required by different networks and BD-NET for performing their main operation. As can be seen, eliminating the *Mul* operation in the proposed network brings considerable area-efficiency ($> 88.1\%$) compared to different networks that can be exploited to enable higher computation parallelism. However, to fit a deep neural network into a low-end ASIC, the number of logic cells may come to shortage. The normal countermeasure is to split the computation and multiplex the computation kernel, which restraints the throughout. The better solution is to use the proposed binarized depthwise separable convolution with the fine-tuning discussed in Section III-B.

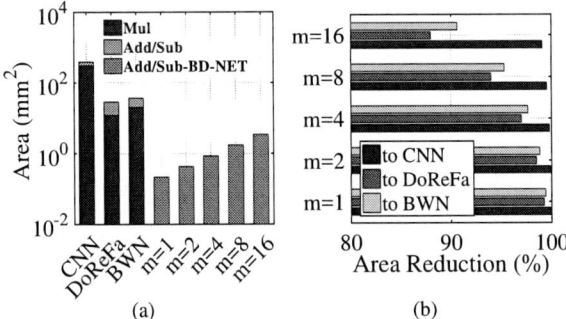

Fig. 6. (a) Normalized log-scaled area overheard imposed by CNN, DoReFa-NET and BWN vs. BD-NET in different configuration for the computation, (b) Area reduction.

IV. SUMMARY AND FUTURE WORK

In this work, we have shown a multiplication-less deep convolution neural network, which replaces the normal spatial convolution layer with binarized depthwise separable convolution. Such neural network compression technique significant reduces the hardware utilization in terms of both computation and memory.

Compared to the accuracy degradation caused by the binarization of normal spatial-convolution using the techniques from BinaryConnect [4] and BWN [5], our method can compensate such accuracy loss through tuning the internal hyperparameters (i.e. channel expansion) effectively. In this work, the convolution kernel (i.e. feature filter) and the generated feature maps are binarized. According to the experiments performed in this work, a hypothesis can be drawn as the guidance of further experiment direction. For searching the correlation between local pixels and the captured feature maps, high resolution is not essential. However, the new representation generated at the output of convolution layer is expected to be in higher precision, to avoid the information loss in the forward path during the inference. Our future work will try to apply other neural network compression techniques, like quantization and pruning, on the pointwise convolution to further compress the convolution layers. Moreover, we will try to alternate the normal convolution layers of other famous network structure with our binarized depthwise separable convolution, and examine the model performance.

REFERENCES

[1] Y. LeCun, Y. Bengio, and G. Hinton, "Deep learning," *Nature*, vol. 521, no. 7553, pp. 436–444, 2015.

[2] S. Liang and R. Srikant, "Why deep neural networks for function approximation?" *ICLR*, 2017.

[3] G. Urban *et al.*, "Do deep convolutional nets really need to be deep and convolutional?" *arXiv preprint arXiv:1603.05691*, 2016.

[4] M. Courbariaux *et al.*, "Binaryconnect: Training deep neural networks with binary weights during propagations," in *Advances in Neural Information Processing Systems*, 2015, pp. 3123–3131.

[5] M. Rastegari *et al.*, "Xnor-net: Imagenet classification using binary convolutional neural networks," in *European Conference on Computer Vision*. Springer, 2016, pp. 525–542.

[6] S. Zhou, Y. Wu, Z. Ni, X. Zhou, H. Wen, and Y. Zou, "Dorefa-net: Training low bitwidth convolutional neural networks with low bitwidth gradients," *arXiv preprint arXiv:1606.06160*, 2016.

[7] L. Sifre and S. Mallat, "Rigid-motion scattering for image classification," Ph.D. dissertation, Citeseer, 2014.

[8] A. G. Howard *et al.*, "Mobilenets: Efficient convolutional neural networks for mobile vision applications," *arXiv preprint arXiv:1704.04861*, 2017.

[9] F. Chollet, "Xception: Deep learning with depthwise separable convolutions," *arXiv preprint arXiv:1610.02357*, 2016.

[10] I. Hubara *et al.*, "Binarized neural networks," in *Advances in neural information processing systems*, 2016, pp. 4107–4115.

[11] Y. Bengio, N. Léonard, and A. Courville, "Estimating or propagating gradients through stochastic neurons for conditional computation," *arXiv preprint arXiv:1308.3432*, 2013.

[12] K. He, X. Zhang, S. Ren, and J. Sun, "Deep residual learning for image recognition," in *Proceedings of the IEEE conference on computer vision and pattern recognition*, 2016, pp. 770–778.

[13] F. Juefei-Xu, V. N. Boddeti, and M. Savvides, "Local binary convolutional neural networks," *arXiv preprint arXiv:1608.06049*, 2016.

[14] S. Ioffe and C. Szegedy, "Batch normalization: Accelerating deep network training by reducing internal covariate shift," in *International Conference on Machine Learning*, 2015, pp. 448–456.

[15] W. Pan *et al.*, "Towards accurate binary convolutional neural network," in *Advances in Neural Information Processing Systems*, 2017.

[16] M. Courbariaux *et al.*, "Binarized neural networks: Training deep neural networks with weights and activations constrained to+ 1 or-1," *arXiv preprint arXiv:1602.02830*, 2016.

[17] L. Hou, Q. Yao, and J. T. Kwok, "Loss-aware binarization of deep networks," *arXiv preprint arXiv:1611.01600*, 2016.

[18] S. D. C. P. V. . Synopsys, Inc.

[19] http://www.eda.ncsu.edu/wiki/FreePDK45.

[20] N. Muralimanohar, R. Balasubramonian, and N. P. Jouppi, "Cacti 6.0: A tool to model large caches," *HP Laboratories*, pp. 22–31, 2009.

2018 IEEE Computer Society Annual Symposium on VLSI

TaiJiNet : Towards Partial Binarized Convolutional Neural Network for Embedded Systems

Yingjian Ling, Kan Zhong, Yunsong Wu, Duo Liu*, Jingting Ren, Renping Liu
Moming Duan, Weichen Liu and Liang Liang[§]
*College of Computer Science, Chongqing University, China
[§]College of Communication Engineering, Chongqing University, China

Abstract—
We have witnessed the tremendous success of deep neural networks. However, this success comes with the considerable computation and storage costs which make it difficult to deploy these networks directly on resource-constrained embedded systems. To address this problem, we propose TaiJiNet, a binary-network-based framework that combines binary convolutions and pointwise convolutions, to reduce the computation and storage overhead while maintaining a comparable accuracy. Furthermore, in order to provide TaiJiNet with more flexibility, we introduce a strategy called partial binarized convolution to efficiently balance network performance and accuracy. We evaluate TaiJiNet with the CIFAR-10 and ImageNet datasets. The experimental results show that with the proposed TaiJiNet framework, the binary version of AlexNet can achieve $26\times$ compression rate with a negligible 0.8% accuracy drop when compared with the full-precision AlexNet.

I. INTRODUCTION

With the rapid development of deep convolutional neural networks (CNNs), CNNs have become the state-of-the-art technique for a wide range of tasks. Traditionally, the data center equipped with high-end GPUs is the best choice for deploying the networks required by CNN-based applications. However, this cloud-centric application framework usually arises some issues such as user privacy, long response time, and in the case without internet the framework cannot even work. Therefore, there is a growing interest in deploying networks directly on embedded devices (e.g., cell phones and IoT devices). However, as networks gradually become deeper, their computational complexity and parameters are increasing. For example, AlexNet, a network that has five convolutional layers and three fully-connected layers, has 722M FLOPs (the number of floating point operations) and 240MB parameters; the 16-layers VGG-16 requires 15.8G FLOPs and 552MB parameters, which leads to significant system overhead (e.g., computation, memory and energy) when they are deployed in resource-constrained embedded systems. In order to solve these challenges in embedded systems, several approaches, such as network pruning, fixed-point quantization and architecture optimization, have been proposed. Among them, the binary-based techniques show a superiority due to their great improvement in network performance [1] . However, we observe that the binary-based techniques [2], [3], [1] fail to maintain original accuracy on the large-scale dataset like ImageNet. Meanwhile, their global strategies also render CNN models less resilient. Therefore, this paper focuses on reducing the computation and storage costs of CNNs while retaining comparable accuracy, and providing binarized networks with more flexibility.

Multiple binary-based methods have been proposed to reduce CNNs' overhead on resource-constrained embedded

*Corresponding author: Duo Liu, College of Computer Science, Chongqing University, Chongqing, China. E-mail: liuduo@cqu.edu.cn.

Table I: The comparison results between the binary-based methods and AlexNet on ImageNet dataset.

Network	Top-1 Accuracy	Top-5 Accuracy
Baseline[4]	56.6%	80.2%
BWN[1]	53.8%	77.0%
XNOR-Net[1]	44.2%	69.2%
BinaryConnect[2]	35.4%	61.0%
BNN[3]	27.9%	50.42%

systems. Courbariaux et.al [2] propose BinaryConnect (BC), a network that is optimized by constraining connections to -1 and +1. In [3] , BinaryNeuralNet (BNN) is proposed to speed up the inference by binarizing the activations and connections simultaneously. Rastegari et.al [1] propose BinaryWeightNetwork (BWN) and XNOR-Networks (XNOR-Net) to obtain a better approximation of the original network by scaling the intermediate representation of binary networks. However, we observe that although these methods achieve considerable speedup and compression rate, their top-1 and top-5 accuracy are notably reduced, especially for large datasets such as ImageNet, as shown in Table I. On the other hand, these existing methods use the global strategy that quantify all the weights during training. Hence, when hardware resources are relaxed or high accuracy is required, they may lead to poor user experience due to their overly aggressive quantization. These observations motivate us to propose a binary-network-based technique, which can not only maintain the almost lossless accuracy on large-scale dataset but also adapt to different embedded environments.

In this paper, we propose a novel framework, named TaiJiNet, to reduce the computation and storage requirements for a given network while achieving almost lossless accuracy on large-scale datasets. In order to achieve this, we have studied the potential of pointwise convolutions in binarized networks. Our observation shows that with a special weight initialization method, pointwise convolutions not only help binary weights better approximate original weights but also provide extra parameters for the network to recover the reduced accuracy. In addition, we also note that the uniform quantization methods used in previous work may cause unrecoverable damage to accuracy due to the different importance of kernels. After exploring the representativeness of multiple criteria for kernel importance, we propose partial binarized convolution, which quantizes only unimportant kernels to efficiently trade off between network performance and accuracy so that our model owns more variability. Therefore, in TaiJiNet, we first binarize the weights of a given network, and then equip each convolutional layer with a full-resolution pointwise convolutional layer. Optionally, we balance the network performance and accuracy by adjusting the importance threshold. Finally, we fine tune the binary weights and full-resolution weights coordinately to approximate the original network.

We use VGG-7 and AlexNet as our base model and evaluate the proposed TaiJiNet on CIFAR-10 and ImageNet

978-1-5386-7100-9/18 $31.00 © 2018 IEEE

datasets. The experimental results show that our framework obtain 26× compression rate on all layers of AlexNet with only 0.8% accuracy drop.

In summary, the main contributions of this work include:

- We present TaiJiNet, a binary-network-based framework which combines binary convolutions and pointwise convolutions, to efficiently compress and speed up CNNs on the resource-constrained embedded systems while retaining almost lossless accuracy.
- By studying the importance of convolutional kernels, we introduce a strategy called partial binarized convolution to efficiently balance network performance and accuracy.
- We evaluate the proposed TaiJiNet on large-scale data set using two state-of-the-art networks (e.g., AlexNet and VGG) and compare it with four related work.

The rest of this paper is organized as follows. Section II introduces the background and motivation. Section III discusses the design of TaiJiNet. Section IV evaluates the proposed framework. Section V introduces the related work. Finally, we conclude in Section VI.

II. BACKGROUND AND MOTIVATION

A. Convolutional Neural Networks

CNN is the state-of-the-art deep learning technique which mainly consists of convolutional layers, pooling layers and fully-connected layers. A typical CNN starts with several convolutional layers that extract basic features from the input. Pooling layers are used to reduce the sizes of output feature maps produced by convolutional layers. They map multiple adjacent input data to one output data by maximum or average operation. Finally, fully-connected layers connect the input to each neuron to output the classification result.

A standard convolutional layer is parameterized by a collection of kernels or filters $\mathbf{W}$. Each kernel is defined by a 3D tensor of size $N_c \times k \times k$ where N_c is number of input channels and k is the spatial dimension of the kernel. During training or inference, a standard convolutional layer takes as input a feature map $\mathbf{I}$ of size $N_c \times h_i \times w_i$ and output a $N_k \times h_o \times w_o$ feature map $\mathbf{O}$ where N_k is the number of output channels or the number of kernels, h_i/w_i is the spatial dimension of a input feature map and h_o/w_o is the spatial dimension of a output feature map. The element of the output feature map in position (c, x, y) is computed as:

$$\mathbf{O}_{c,x,y} = \sum_{j=0}^{N_c-1} \sum_{s=0}^{k-1} \sum_{t=0}^{k-1} \mathbf{I}_{j,x+s,y+t} \mathbf{W}_{j,s,t}^c$$

where $\mathbf{W}_{j,s,t}^c$ is the weight at position (j, s, t) of cth kernel.

For most CNNs, the convolutional layer dominates computation overhead and fully-connected layer dominates storage overhead. Since the popular networks [5], [6] replace the fully-connected layer with global average pooling layer that does not take up storage, and there have been many efficient methods to optimize the fully-connected layers like [7], [8]. In this paper, we focus on the reduction of computation and storage overhead in convolutional layer.

B. CNNs on Embedded Systems

Reliably mining real-world data to reacting user behavior and ambient context is a key element for many emerging mobile and IoT applications. However, extracting accurate information from the dynamic and complex environment is challenging. With the development of machine learning community, deep neural networks have become one of the most promising technique to overcome this challenge due to its excellent accuracy and robustness. But the overhead imposed by these algorithms makes it impractical to directly deploy them on the resource-limited embedded systems. For example, it takes 2.6 minutes to run an inference of AlexNet on Snapdragon 800 processors which are widely used in variety of smartphones (e.g., Nexus 5)[9] . Some methods have been proposed to address this barrier such as network pruning [7] , low-rank decomposition [10] , vector quantization [11] , fixed-point quantization [2], [3], [1], [12], [13] and architecture optimization [14], [15] . However, deploying CNNs directly on embedded systems remains a challenging problem.

C. Motivation

As discussed in Section I, prior work show the relatively poor accuracy on large-scale dataset. Nevertheless, we observe that BWN outperforms other methods by large margins on ImageNet. This advantage mainly derives from the fact that they approximate the real-valued weights with the binary weights and scaling factors. This strategy can be implemented by scaling each binary convolutional layer's output channel. Interestingly, since the pointwise convolutional layer has the nature that combines the input channels linearly, another implementation of this strategy is to attach a special pointwise convolutional layer where each kernel is restricted to a one-hot vector. This motivates us to leverage binary convolutional layer and unrestricted pointwise convolutional layer to achieve the same approximation. Besides, unrestricted layer can also provide more network capacity to improve the accuracy.

On the other hand, the quantization strategies used in [2], [3], [1] do not take each kernel's importance into consideration, which means the binarization of important kernels may lead to significant accuracy drop. In addition, since the overhead reduction brought by these methods is not always required in the environment where hardware resources are less limited. Therefore, we propose partial binarized convolution to reduce the impact of quantization on accuracy according to each kernel's importance and to provide more flexibility for the binary models.

III. TAIJINET DESIGN

In this paper, we propose TaiJiNet, a binary-network-based framework to optimize the inference of CNN models while maintaining the accuracy required by applications. In this section, we first give an overview of the framework, and then explain the details according to the implementation flow of the framework.

A. TaiJiNet Overview

Our framework includes two stages: transforming a pre-trained network into binary one and fine-tuning the transformed network, as shown in Fig 1. For simplicity, we only show convolutional layers and fully-connected layers in the figure.

Three steps are involved in the model transformer: type transformation, 1×1 conv layer equipping and partial binarized convolution. 1) Type transformation replaces all the original convolutional layer with the binary version of convolutional layer which maintains both full-resolution weights and their corresponding binary weights at training time, and discard full-resolution weights at test time. 2) 1×1 conv layer equipping attaches a pointwise convolutional layer to each binary convolutional layer. 3) Optionally, partial binarized convolution is employed to balance network performance and accuracy.

978-1-5386-7100-9/18 $31.00 © 2018 IEEE

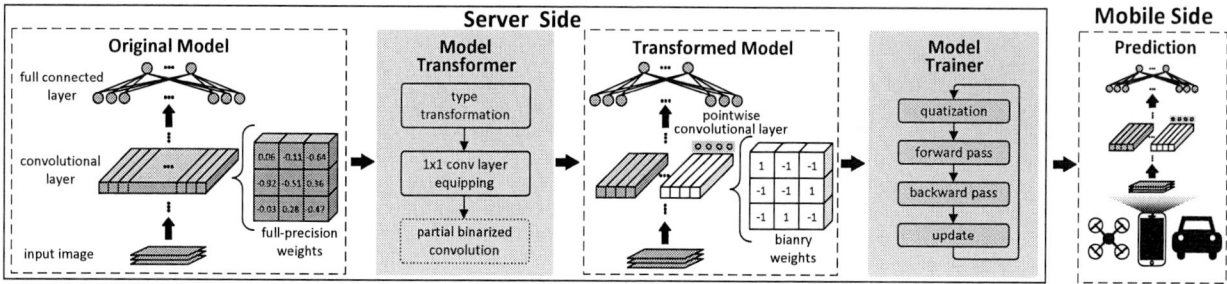

Figure 1: An overview of TaiJiNet Framework.

Quantizing all full-resolution weights to binary weights brutally usually lead to a significant reduction in accuracy. Hence, fine tuning the transformed model is necessary. We alternately quantize and fine-tune all latent full-resolution weights to enable the network to adapt to binary weights.

B. Model Transformer

Type transformation. In order to train a network with weights constrained to -1 or +1, type transformation is required to modify the original convolutional layers to the binary ones. There are two differences between them: 1) during training, the full-resolution weights and their corresponding binary weights are both retained to coordinately fine tune the whole model. 2) during inference, only binary weights are retained and the convolutions between input feature maps and binary weights are implemented in a way without any multiplication. Note that full-resolution weights are necessary during training. Since the weight changes brought by each update are tiny, directly binarizing weights will ignore these changes, which means the network may not be improved after each iteration. This strategy is also employed in [2], [3] . Besides, we use the following function to obtain binary weights:

$$w_b = \text{sign}(w) = \begin{cases} +1 & \text{if } w \geq 0, \\ -1 & \text{otherwise.} \end{cases} \quad (1)$$

where w_b is the binarized weight and w is the real-valued weight. This sign function is easy to implement and works well in practice.

Equip 1×1 conv layer. Although the network transformed by the previous step can already be used as the input to the model trainer, its poor expressive ability may lead to the failure to recover the accuracy through retraining. To address this problem, we equip each convolutional layer with a pointwise convolutional layer which has a weight tensor of size $N_k \times N_c \times 1 \times 1$. We keep the number of kernels N_k equal to the channel size N_c so that the shape of output feature maps remain the same. The employment of pointwise convolutional layer is based on two reasons: 1) 1×1 convolutions are able to enhance the expressive ability of networks while approximating the original convolutions. Let W represents real-valued weights and B represents their binary counterparts. If we want to use a scalar α and B to approximate W such that $W \approx \alpha B$, we could solve the following optimization:

$$\begin{aligned} \min_{\alpha} \quad & \|W - \alpha B\|^2 \\ s.t. \quad & \alpha \in \mathbb{R}^+, \\ & B = \text{sign}(W). \end{aligned} \quad (2)$$

by setting the derivative to zero, the optimal value for α can be computed by averaging absolute weight values, i.e., $\alpha^* = \frac{1}{n}\|W\|_{l1}$ where n is the number of real-valued weights. Therefore, we can approximate the original convolution $X * W$ by $(X * B)\alpha$ where X represents the input feature

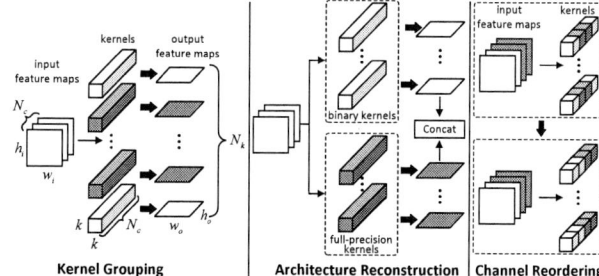

Figure 2: An overview of partial binarized convolution.

maps and $*$ is the convolution operation. In other words, the original convolution can be approximated by scaling channels of the binary convolutional layer's output feature map. This can also be achieved by attaching a special 1×1 convolutional layer where the ith kernel is a vector in which i^{th} element is the scaling factor α and other elements are zero. In TaiJiNet, we relax this constraint: the remaining elements can also be non-zero. This allows us to learn the linear combination of output channels by remaining elements. Therefore, better accuracy can be obtained. 2) 1×1 convolutions are more efficient than other usual convolutions like 3×3 or 5×5. In the case where the output feature maps remain the same size, the computation and storage overhead of 3×3 and 5×5 convolutions are $9 \times$ and $25 \times$ larger than 1×1 convolutions.

C. Partial Binarized Convolution

Although the network binarization is beneficial for the reduction of computation and storage overhead, this advantage is not always required in the environment where hardware resources are less constrained. Moreover, these aggressive quantization methods may lead to more or less accuracy drop, which makes them unsuitable for the situation where high accuracy is needed. In order to enable binary models more flexible, we introduce a simple strategy called partial binarized convolution which means only partial convolutional kernels are binarized and the remaining kernels keep full-resolution. Three steps are involved in partial binarized convolution.

Kernel Grouping. Intuitively, the more important a kernel is, the greater its impact on accuracy is. Kernel grouping aims at dividing a given layer's kernels into two groups according to their importance so as to reduce the influence of quantization on accuracy by choosing unimportant kernels to be binarized. In order to characterize the importance of each kernel, we have explored six criteria: (1) ℓ_1-norm $= \sum_{i=1}^{N_c \times k \times k} |w_i|$, (2) mean-mean $= \frac{1}{N}\sum_{n=1}^{N} \text{mean}(x_n)$, (3) mean-std $= \frac{1}{N}\sum_{n=1}^{N} \text{std}(x_n)$, (4) mean-$\ell_1 = \frac{1}{N}\sum_{n=1}^{N}\|x_n\|_1$, (5) mean-$\ell_2 = \frac{1}{N}\sum_{n=1}^{N}\|x_n\|_2$, (6) var-$\ell_2 = \text{var}(\|x_n\|_2)$, where N is the number of samples

Algorithm 1 Retraining algorithm for TaiJiNet

Input: A minibatch of inputs and targets (X, Y); real-valued weights of important kernels W_i; real-valued weights of unimportant kernels W_u; real-valued weights of pointwise layers $W_{1 \times 1}$; current learning rate η; the number of layer L; the number of iterations T; the cost function $C(Y, \hat{Y})$.

Output: null

1: **for** $t = 1, 2, ..., T$ **do**
2:     **for** $l = 1, 2, ..., L$ **do**
3:         $\widetilde{W}^l \leftarrow \text{sign}(W_u^l)$.
4:     **end for**
5:     $\hat{Y} \leftarrow \textbf{Forward}(X, \widetilde{W}, W_i, W_{1 \times 1})$.
6:     $\frac{\partial C}{\partial \widetilde{W}}, \frac{\partial C}{\partial W_i}, \frac{\partial C}{\partial W_{1 \times 1}} \leftarrow \textbf{Backward}(\frac{\partial C}{\partial \hat{Y}}, \widetilde{W}, W_i, W_{1 \times 1})$.
7:     $\frac{\partial C}{\partial W_u} \leftarrow \textbf{ComputeGradient}(\frac{\partial C}{\partial \widetilde{W}})$.
8:     $\textbf{UpdateParameters}(W_u, \frac{\partial C}{\partial W_u}, W_i, \frac{\partial C}{\partial W_i}, W_{1 \times 1}, \frac{\partial C}{\partial W_{1 \times 1}}, \eta)$.
9:     $\textbf{UpdateLearningRate}(\eta)$.
10: **end for**

---

used to compute these criteria, w_i is the ith weight of a given kernel. For a given kernel, x_n is the corresponding output channel produced by n^{th} sample. We use mean, std and var represent average, standard deviation and variance. By comparing their accuracy on the CIFAR-10 dataset, we observe that mean-ℓ_1 and mean-ℓ_2 outperform other criteria. For the convenience of numerical calculation, we choose mean-ℓ_1 as the criterion in this paper. Meanwhile, we also compared three different kernel selection methods: selecting 1) the smallest 2) random 3) the largest kernels to be binarized. Evaluation results show that the first method has less impact on accuracy than other methods. Once we obtained each kernel's criterion, the kernels whose criterion are lower than the threshold are considered to be unimportant, and then they will be binarized. The criterion threshold can be determined by the quantization ratio given by model builder.

Architecture Reconstruction. After kernel grouping, the kernels of each convolutional layer are divided into two groups. However, this kind of multi resolution in kernel level perhaps cannot fully utilize the hardware resources of platform like mobile GPGPU or specialized accelerator. In order to overcome this challenge, we rearrange the position of each kernel according to its resolution, as shown in the middle of Fig 2. As a result, each group of kernels form a new convolutional layer, and these two new parallel convolutional layers share the same input feature map and concatenate their output feature maps. The new architecture is very similar to the "Inception module" proposed by GoogleNet [6] . Different from using different kernel sizes in module, we use different resolutions for convolutional layers in module. By reconstructing architecture, the multi resolution of parameters can be promoted from the kernel level to the layer level, which is beneficial for the utilization of hardware resources.

Channel Reordering. It should be pointed out that the channels of output feature map will also be rearranged after reconstruction, which leave the next convolutional layer unable to convolve directly with the output feature map produced by the last convolutional layer. Therefore, we reorder the channel dimension of kernels in next convolutional layer according to the kernel order of last convolutional layer.

When partial binarized convolution is enabled, we only transform type and equip 1×1 conv layer for layers consisting of unimportant kernels. Note that all three steps are implemented in cloud or server side. Therefore, they will not cause any additional overhead for the inference in mobile side.

D. Model Trainer

During training time, we maintain full-resolution duplicates of binary weights. When the partial binarized convolution is enabled, layers consisting of important kernels

Table II: Experimental setup.

Networks			
VGG-7	$2 \times (128C3) + MP2 + 2 \times (256C3) + MP2 +$ $2 \times (512C3) + MP2 + 1024FC + \text{Softmax}$		
AlexNet	$96C11 + MP3 + 256C5 + MP3 +$ $2 \times (384C3) + 256C3 + MP3 +$ $4096FC + 4096FC + 1000FC + \text{Softmax}$		
Datasets			
CIFAR-10	10 classes image classification dataset		
ImageNet	1000 classes image classification dataset		

feed-forward and back-propagate in a same way as normal convolutional layers. Layers consisting of unimportant kernels use binary weights for forward and backward passes and update full-resolution duplicates with the gradients calculated by backward passes. All full-resolution duplicates will be binarized at the beginning of each iteration, as show in Algorithm 1. First, we compute the binary counterparts of full-resolution duplicates using equation 1 (Lines 2-4). Then, the forward and backward function are called to get the gradient of each weight (Lines 5-6). Next, the gradients of real-valued duplicates are computed (Line 7). Since the derivative of the sign function is zero almost everywhere, the gradient of real-valued weights would also be zero, making it impossible to learn. We follow the same method as [3] , where the gradient of sign function $q = \text{sign}(r)$ is $\frac{\partial C}{\partial r} = \frac{\partial C}{\partial q} 1_{|r| \leq 1}$. Finally, all weights and learning rate is updated by SGD or ADAM [16] (Lines 8-9). We use the real-valued weights of original model to initialize W_i and W_u. Instead of initializing $W_{1 \times 1}$ with some usual methods, we use α mentioned above to initialize the ith element of ith kernel and set other elements to zero, which can greatly reduce the number of epochs required by retraining.

IV. EVALUATION

In this section, we conduct experiments on two benchmark datasets to validate the effectiveness of our approach from three aspects. First, we compare the accuracy of mean-ℓ_1 with other criteria on CIFAR-10 dataset with VGG-7 network. Then, we compare the smallest kernel selection strategy with selecting random kernels and selecting largest kernels. Finally, we give the overall results of TaiJiNet on ImageNet dataset.

A. Experimental Setup

Our implementation is based on Caffe [17] which is a deep learning framework widely used in industry and academia. Our base models are VGG-7 and AlexNet. The architecture of VGG-7 is inspired by [18] . Our AlexNet model is taken from Caffe model zoo. We present the configuration in Table II, where $2 \times (128C3)$ represents two convolutional layers that have 128 kernels with size 3×3, $1024FC$ is a fully-connected layer with 1024 neurons and $MP2$ is the max pooling layer with size 2×2.

B. VGG-7 on CIFAR-10

CIFAR-10 is an image classification benchmark containing a training set of 50K and a test set of 10K, where each instance is a 32×32 RGB image. We follow the data augmentation in [19] : 4 pixels are padded on each side and a 32×32 crop is randomly sampled from the padded image.

Criterion Selection. We first investigate the impact of different criteria on accuracy. Since the statistics become more accurate when more samples are provided, we use the whole training set to calculate each criterion. In this experiment, the kernels with smaller criteria are considered as unimportant kernels. Fig 3 reports their influence on accuracy in VGG-7 without retraining step. We observe that

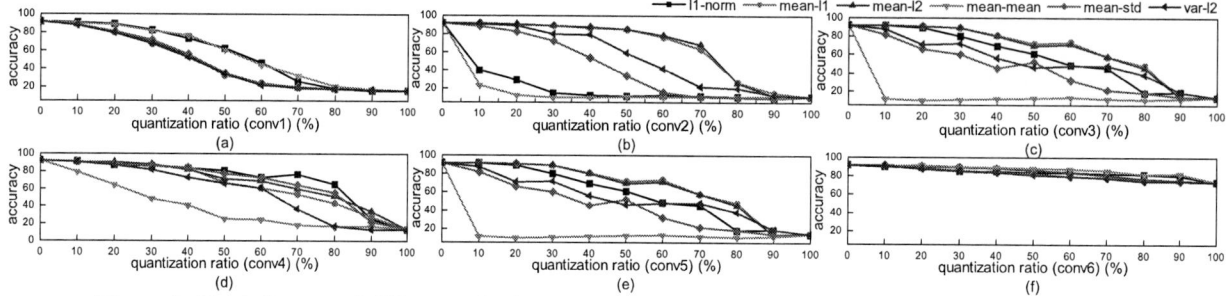

Figure 3: The influence of different criteria on accuracy. (a)-(f) correspond to each convolutional layer.

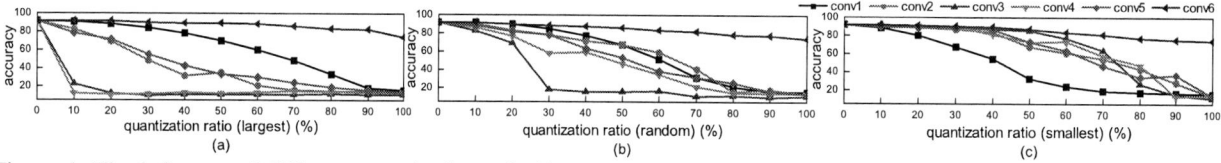

Figure 4: The influence of different quantization priorities on accuracy. (a)-(c) correspond to selecting largest, random and smallest kernels.

although ℓ_1-norm and mean-mean have a good performance on the first layer, mean-ℓ_1 and mean-ℓ_2 have a more stable performance on almost all layers. Besides, since the first layer has more important features and fewer weights, we should avoid quantizing the first layer in practice. Similar strategy is adopted in [3], [1] . Therefore, mean-ℓ_1 and mean-ℓ_2 are obviously better than ℓ_1-norm and mean-mean. Note that although the strategy of multi-criteria may be better than only using a single one, the computation of these statistics is a time-consuming work considering they are not data free. Hence, we choose mean-ℓ_1 as the criterion for the convenience of numerical calculation.

Kernel Selection. In this experiment, we compare three kernel selection approaches: selecting the smallest kernels, selecting random kernels and selecting largest kernels. In order to assess the influence of quantization priority on accuracy, we test the accuracy of the VGG-7 in these three approaches without retraining step. As shown in Fig 4, using the smallest kernel selection method to quantize convolution layers from conv2 to conv6 has little impact on accuracy when the quantization ratio is below 40%. In contrast, the accuracy of quantizing kernels with the largest mean-ℓ_1 decreases rapidly with the increase of ratio, which indicates the important kernels tend to have larger mean-ℓ_1.

C. AlexNet on IMAGENET

ImageNet is a 1000-category dataset consisting of over 1.2 million training images and a validation set of 50 thousand images. We use AlexNet caffe model as the well-trained model which has 61 million parameters and obtains a top-1 accuracy 56.6% and a top-5 accuracy 80.2% on the validation set. For preprocessing, each image is resized to 256×256 and randomly cropped to 224×224. For retraining, we use the SGD with momentum 0.9 to retrain the transformed model with 20 epochs, which occupies one-fourth of the original training epochs. The learning rate starts at 0.0015 and we use a learning rate decay 0.1 every 4 epochs.

In this section, we disable the partial binarized convolution for comparison with other full binarization methods. We first compare our TaiJiNet with other four networks: BNN [3] , XNOR-Net [1] , BC [2] and BWN [1] . Table III reports their top-1 and top-5 accuracies. As shown, the results show that the accuracy of TaiJiNet is only 0.8% lower than the full-precision networks and 2%-27.9% better

Table III: The comparison between TaiJiNet and other methods on ImageNet.

	Method	Top-1	Top-5
Full-Precision	AlexNet[4]	56.6%	80.2%
Binary Input and Binary Weight	BNN[3]	27.9%	50.42%
	XNOR-Net[1]	44.2%	69.2%
Binary Weight	BC[2]	35.4%	61.0%
	BWN[1]	53.8%	77.0%
	TaiJiNet	55.8%	78.7%

than other methods. Furthermore, in order to demonstrate the accuracy improvement brought by pointwise convolution, we build a new model that replaces the pointwise convolutional layer with the scaling layer used in BWN[1] . As a result, we obtain the top-1 accuracy of 54.1% and top-5 accuracy of 77.7%, which is still worse than TaiJiNet.

Next, we compare the required memory for the TaiJiNet and AlexNet. We assume that 32-bit floating point numbers is used to store full-precision weights. The compression rate r of a convolutional layer with a weight tensor $N_k \times N_c \times k \times k$ can be computed by the following equation:

$$r = \frac{4N_c N_k kk}{\frac{1}{8}N_c N_k kk + 4N_k N_k} = \frac{32}{1 + \frac{32N_k}{N_c k^2}} \quad (3)$$

where $\frac{1}{8}N_c N_k kk$ is the bytes occupied by binary weights and $4N_k N_k$ is the bytes of full-precision weights. Table IV shows the compression results on the convolutional layers of AlexNet. The results show that we gain the $4.6\times$ compression rate on five convolutional layers. When employing our methods on the whole network, we achieve $26\times$ compression rate. Note that although we obtain a lower compression rate than BWN [1] ($32\times$), our results are still very close to the recent proposed Deep Compression [7] . Specifically, using 8 bits to encode shared weights, Deep Compression compress the size of convolutional layers from 8.9MB to 1.25MB. In addition, Deep Compression requires roughly 960 epochs to retrain network. In contrast, our method only needs 20 epochs.

Finally, we analysis the computation costs of TaiJiNet. Convolution is made up of MAC (multiply-and-accumulate) operations. When the weights are constrained to -1 or +1, each MAC operation can be converted to a bit operation and an accumulation operation. Table V shows the amount of computation required by AlexNet and TaijiNet, where FM is the number of floating point multiply operations, FA is the number of floating point accumulate operations and

978-1-5386-7100-9/18 $31.00 © 2018 IEEE

Table IV: The compression rate of TaiJiNet.

Layer	Weights (AlexNet)	Weights (TaiJiNet)	Compression rate
conv1	136.1KB	40.25KB	3.3×
conv2	1200KB	293.5KB	4×
conv3	3456KB	684KB	5×
conv4	2592KB	657KB	3.9×
conv5	1728KB	310KB	5.5×
Total	9112.1KB	1984.75KB	4.6×
fc6	144MB	4.5MB	32×
fc7	64MB	2MB	32×
fc8	15.625MB	0.49MB	32×
Total	232.52MB	8.93MB	26×

Table V: The comparison of computation by AlexNet and TaiJiNet.

Layer	AlexNet		TaiJiNet		
	FM	FA	FM	FA	BO
conv1	105.42M	105.42M	27.88M	133.3M	105.42M
conv2	223.95M	223.95M	23.89M	247.84M	223.95M
conv3	149.52M	149.52M	24.92M	174.44M	149.52M
conv4	112.14M	112.14M	12.46M	124.6M	112.14M
conv5	74.76M	74.76M	5.54M	80.3M	74.76M
Total	665.79M	665.79M	94.69M	760.48M	665.79M

BO is the number of bit operations. The results show that roughly 85.7% expensive floating point multiply operations are replaced by bit operations which are more hardware-friendly.

V. RELATED WORK

Recent efforts toward reducing CNNs' overhead on resource-constrained embedded systems involves network pruning, fixed-point quantization, architecture optimization.

Network Pruning. Network pruning focuses on pruning redundant weights in a well-trained network. In [7] , a three stage pipeline is proposed to compress the network. The authors of [20] transfer their attention to the kernel level and try to remove unimportant kernels from the network. In this paper, we binarize unimportant kernels to improve the accuracy, instead of removing them.

Fixed-point Quantization. Fixed-point based approaches usually convert full-precision networks to low-precision ones. Zhou et.al [21] propose incremental network quantization (INQ) to convert pre-trained full-precision network into a low-precision version whose weights are constrained to be either powers of two or zero. Rastegari et.al [1] use scaling factors to achieve better approximation of real-valued parameters. In [12] , the authors introduce ternary weight networks, neural networks with weights constrained to +1, 0 and -1. However, sparse basic linear algebra subprograms libraries or specialized hardware are needed. In addition, the compression rate of ternary networks is 2× less than that of binary networks.

Architecture Optimization. Architecture Optimization explores effective network architectures. In [15] , the authors utilize fire module that consists of 1x1 and 3x3 kernels to design CNNs with fewer parameters. Howard et al. [14] present MobileNets for embedded devices which are based on depthwise separable convolutions. Our method can be combined with them to achieve better performance.

VI. CONCLUSION

In this paper, we have proposed a binary-network-based method to reduce the computation and storage costs of convolutional neural networks in embedded systems. Besides, we introduce a strategy to strike the balance between network performance and accuracy. Since the weights are constrained to -1 or +1, we can greatly speed up the inference by replacing the MAC-based convolution with the convolution without any multiplication.

ACKNOWLEDGEMENTS

We would like to thank the anonymous reviewers for their valuable feedback and improvements to this paper. This work is partially supported by grants from the National Natural Science Foundation of China (No. 61672116, 61601067), Chongqing High-Tech Research Program c-stc2016jcyjA0332, and the Fundamental Research Funds for the Central Universities under Grant 0214005207005.

REFERENCES

[1] M. Rastegari, V. Ordonez, J. Redmon, and A. Farhadi, "Xnor-net: Imagenet classification using binary convolutional neural networks," in *ECCV*. Springer, 2016, pp. 525–542.

[2] M. Courbariaux, Y. Bengio, and J.-P. David, "Binaryconnect: Training deep neural networks with binary weights during propagations," in *NIPS*, 2015, pp. 3123–3131.

[3] M. Courbariaux, I. Hubara, D. Soudry, R. El-Yaniv, and Y. Bengio, "Binarized neural networks: Training deep neural networks with weights and activations constrained to+ 1 or-1," in *NIPS*, 2016, pp. 4107–4115.

[4] A. Krizhevsky, I. Sutskever, and G. E. Hinton, "Imagenet classification with deep convolutional neural networks," in *NIPS*, 2012.

[5] M. Lin, Q. Chen, and S. Yan, "Network in network," in *ICLR*, 2013.

[6] C. Szegedy, W. Liu, Y. Jia, P. Sermanet, S. Reed, D. Anguelov, D. Erhan, V. Vanhoucke, and A. Rabinovich, "Going deeper with convolutions," in *CVPR*, June 2015.

[7] S. Han, H. Mao, and W. J. Dally, "Deep compression: Compressing deep neural networks with pruning, trained quantization and huffman coding," in *ICLR*, 2016, pp. 4107–4115.

[8] S. Han, J. Pool, J. Tran, and W. Dally, "Learning both weights and connections for efficient neural network," in *NIPS*, 2015.

[9] N. D. Lane, S. Bhattacharya, P. Georgiev, C. Forlivesi, and F. Kawsar, "An early resource characterization of deep learning on wearables, smartphones and internet-of-things devices," in *Proceedings of the 2015 International Workshop on Internet of Things towards Applications*. ACM, 2015, pp. 7–12.

[10] E. L. Denton, W. Zaremba, J. Bruna, Y. LeCun, and R. Fergus, "Exploiting linear structure within convolutional networks for efficient evaluation," in *NIPS*, 2014, pp. 1269–1277.

[11] Y. Gong, L. Liu, M. Yang, and L. Bourdev, "Compressing deep convolutional networks using vector quantization," *arXiv preprint arXiv:1412.6115*, 2014.

[12] F. Li, B. Zhang, and B. Liu, "Ternary weight networks," *arXiv preprint arXiv:1605.04711*, 2016.

[13] C. Zhu, S. Han, H. Mao, and W. J. Dally, "Trained ternary quantization," *arXiv preprint arXiv:1612.01064*, 2016.

[14] A. G. Howard, M. Zhu, B. Chen, D. Kalenichenko, W. Wang, T. Weyand, M. Andreetto, and H. Adam, "Mobilenets: Efficient convolutional neural networks for mobile vision applications," *arXiv preprint arXiv:1704.04861*, 2017.

[15] F. N. Iandola, S. Han, M. W. Moskewicz, K. Ashraf, W. J. Dally, and K. Keutzer, "Squeezenet: Alexnet-level accuracy with 50x fewer parameters and¡ 0.5 mb model size," *arXiv preprint arXiv:1602.07360*, 2016.

[16] D. Kingma and J. Ba, "Adam: A method for stochastic optimization," in *ICLR*, 2014.

[17] Y. Jia, E. Shelhamer, J. Donahue, S. Karayev, J. Long, R. Girshick, S. Guadarrama, and T. Darrell, "Caffe: Convolutional architecture for fast feature embedding," *arXiv preprint arXiv:1408.5093*, 2014.

[18] K. Simonyan and A. Zisserman, "Very deep convolutional networks for large-scale image recognition," *arXiv preprint arXiv:1409.1556*, 2014.

[19] K. He, X. Zhang, S. Ren, and J. Sun, "Deep residual learning for image recognition," in *CVPR*, 2016, pp. 770–778.

[20] H. Li, A. Kadav, I. Durdanovic, H. Samet, and H. P. Graf, "Pruning filters for efficient convnets," in *ICLR*, 2016.

[21] A. Zhou, A. Yao, Y. Guo, L. Xu, and Y. Chen, "Incremental network quantization: Towards lossless cnns with low-precision weights," in *ECCV*, 2017.

An ECC-Free MLC STT-RAM based Approximate Memory Design for Multimedia Applications

Zihao Liu[1], Tao Liu[1], Jie Guo[2], Nansong Wu[3], Wujie Wen[1]

[1] Flordia International University, [2] Hewlett-Packard Co., [3] Arkansas Tech University

{zliu021,tliu023,wwen}@fiu.edu, guojie20101022@gmail.com, nwu@atu.edu

Abstract—**Exponential growth of high-end multimedia data processing requirement produces sharply increasing demand on memory capacities of modern computer systems and mobile platforms. As a next generation memory technology, Multi-level Cell Spin-Transfer Torque Random Access Memory (MLC STT-RAM) is a promising candidate for high-performance and high-density applications. However, its multi-bit per cell advantage has been significantly compromised by the redundant memories to overcome the severe reliability concerns. In this paper, we demonstrate that the compressed image data can exhibit very asymmetric error tolerance capabilities across various layers and resolutions. We then propose a reconfigurable high density MLC STT-RAM approximate memory design by leveraging such nature of the compressed data. Our tri-zone design can offer dynamic configuration among half-state, tristate and full-state to satisfy the run-time error requirement without incurring extra error correction overheads. Simulation results show that our design can boost the storage capacity significantly with marginal image quality degradation.**

I. INTRODUCTION

In the mobile multimedia application, high resolution images and ultra high definition resolution videos are stored in cloud. These multimedia data are only sent to client mobile devices on demand to maximize storage efficiency. To further minimize storage overhead, images and videos are compressed. Before being displayed to consumers, these compressed data need to be decompressed in main memory of mobile devices. However, due to the limited memory capacity, loading a large amount of images and videos from cloud usually invokes frequent memory swapping, directly escalating I/O accesses, increasing power consumption and degrading system performance [1], [2].

Recently, multi-level cell spin-transfer torque random access memory (MLC STT-RAM) technology, which uses the resistance state of magnetic tunneling junction (MTJ) to represent the data, is emerging to address these limitations by offering higher memory density, better scalability, and close-to-zero leakage power when compared with the existing memories [3], [4]. Different from the single-level-cell (SLC) STT-RAM, which consists of a transistor and an MTJ to store only one bit per cell, MLC STT-RAM stacks two or more storage devices (MTJ) into a cell with a similar area, thus can further improve the memory density (i.e. multi-bit per cell) and reduce per-cell fabrication cost [5], [6]. However, MLC STT-RAM suffers from a high bit error rate (BER) due to the reduced MTJ resistance difference, process variation and vulnerability to thermal disturbance [7]. To prevent data corruption from bit errors, complex error correction codes (ECC) such as Bose-

Chaudhuri-Hocquenghem (BCH) code or low-density parity-check code (LDPC) are required [8]. However, besides the considerable ECC logic and latency overhead, complex ECC also generates large-sized redundant memory cells for error detection and correction, e.g. $\sim 47\%$ additional memories [9], severely deteriorating memory utilization. Therefore, it is very critical to minimize the ECC overhead of MLC STT-RAM to fulfill the ever-increasing memory requirement of high-end multimedia applications.

The recently proposed approximate storage technique provides a potential solution to overcome this problem [10]. Essentially, approximate storage intends to improve the storage efficiency by relaxing the reliability requirement. In the mobile multimedia application, lossy image compression technology (e.g. JPEG-2000) is widely adopted to achieve better compression rate [11]. The lossy compression technology sorts image importance and divides an entire image into a number of partitions according to the importance. Since the less important partitions usually have less impact on image quality, relaxing their protection to a certain degree can reduce ECC redundancy with very marginal degradation of image quality.

Based on this observation, we propose an **a**daptive **tri**-zone MLC **S**TT-RAM **a**pproximate memory design, namely **"ATSA"**, to eliminate the ECC requirement. In our ATSA design, instead of using ECC, we guarantee image quality by configuring the reliability of MLC STT-RAM cells. By storing image partitions in STT-RAM cells under different reliability according to their importance, we can completely eliminate ECC and significantly improve memory utilization with negligible image quality degradation. Our major contributions are:

- We quantitatively characterize the asymmetric error tolerance capability of image partitions across different resolutions and layers during image compression, which provides a great opportunity to store the images in an approximate manner without scarifying the quality.
- We develop the tri-zone MLC STT-RAM design by wisely configuring the bit number stored in each MLC. Our design supports three working modes: 1 bit-per cell, 1.5 bit-per cell and 2 bit-per cell, offering flexible density and reliability trade-off to well embrace the asymmetric error resilience property of image partitions for the memory density improvement.
- We develop a pass control flow to optimally distribute the image coefficients across different resolution levels and layers to the tri-zone MLC array according to the interplays between BER and image quality. Our flow

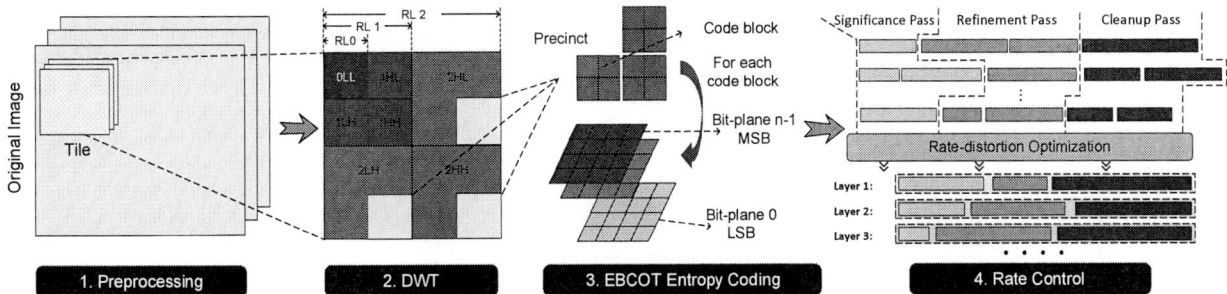

Fig. 1. The major encoding process of JPEG-2000

can maximize the storage efficiency with marginal image quality loss.

To illustrate our ATSA design and evaluate its efficiency, we adopt the JPEG-2000 [12] compression technique as an example in this work. Simulation results demonstrate that our ECC-free ATSA design can achieve $\sim 1.8\times$ memory utilization improvement over the SLC only design with marginal peak-signal-noise-ratio (PSNR) degradation or $\sim 4.5\times$ better PSNR than that of the MLC only design by paying $\sim 10\%$ extra memory. Note our design can be safely applied to other compression algorithms due to the similar process.

II. PRELIMINARY

A. The Basics of JPEG-2000

The JPEG-2000 is a widely adopted image compression standard before storing the raw image data into the memory. Fig. 1 shows the major steps in JPEG-2000 compression: First, the image is divided into several non-overlapping tiles, then discrete wavelet transform (DWT) is applied to each tile to decompose the image into a multiple-resolution representation in frequency domain [13], [14]. Second, a 2-D wavelet decomposition leads to three resolution levels (Rlvl0, Rlvl1, Rlvl2) [14], of which each resolution level is composed of four subbands–LL (horizontally and vertically lowpass), HL (horizontally highpass and vertically lowpass), LH (horizontally lowpass and vertically highpass) and HH (horizontally and vertically highpass). The four lower resolution levels are always generated by progressively applying the DWT process to the LL block from the previous resolution level. Usually the frequency coefficients belonging to the low resolution levels contain the majority of the energy and thus dominate the image quality. After that, EBCOT entropy coding [15] is employed to further divide each subband into a number of codeblocks, followed by block-wise bit-plane coding to facilitate the compression rate control [16]. The bit-plane coding tiles each codeblock into multiple bit-planes (i.e. Bit-plane 0 for LSB) and forms three types of coding passes (i.e. significance pass, refinement pass and cleanup pass) by scanning the value in each bit-plane [12]. Then each pass will be either assigned to a specific layer or discarded according the Rate-Distortion (RD) optimization algorithm [15]. A high compression rate can be achieved by aggressively throwing

the passes with lower RD ratios. Also each layer contains passes across several different resolution levels, and therefore the image can be progressively refined by transmitting the coefficients layer by layer. The decoding process is exactly the inverse of encoding process.

B. SLC/MLC/SR-MLC STT-RAM Basics

An SLC STT-RAM cell consists of one transistor and one Magnetic Tunneling Junction (MTJ), and the data is stored as MTJ's resistance state. An MTJ consists of an oxide barrier layer sandwiched between two ferromagnetic layers, i.e. free layer and reference layer. The magnetization direction of reference layer is fixed, while that of free layer can be changed by applying a spin current with different polarizations. An STT-RAM can present high (low) resistance to denote data '1' ('0') if the magnetization direction of free layer and reference layer are anti-parallel (parallel).

A multi-level cell (MLC) STT-RAM [17] usually incorporates two MTJs at different sizes to represent four different resistance states. By including only one NMOS transistor, MLC STT-RAM can stores two bits within a similar area, thus to improve the memory density. However, MLC STT-RAM suffers from high bit error rate (BER) [9]. To alleviate the unacceptable BER of MLC STT-RAM, state-restrict MLC (SR-MLC) STT-RAM is also proposed [9] by trading some memory capacity for reliability. The SR MLC STT-RAM uses two MLC cells to store three bits (i.e. 1.5 bits per cell) via ternary coding, and can lower the BER to 1×10^{-6} [9].

III. TRI-ZONE APPROXIMATE MLC STT-RAM DESIGN

Using MLC emerging memories for approximate storage has been studied in [10], [18]. However, those designs usually involve costly ECC and are not reconfigurable for different error requirements. In this section, we first quantitatively explore the error tolerance capability of an JPEG-2000 compressed image from two dimensions: resolution level and layer. Then based on our observation, we introduce our ECC-free adaptive Tri-zone approximate MLC STT-RAM (ATSA) design.

A. Impact of Bit Error Rate on Image Quality

We select the image "bridge" from the image compression test image set [19] as an example to investigate how the image quality responses to the errors occurred at various resolution

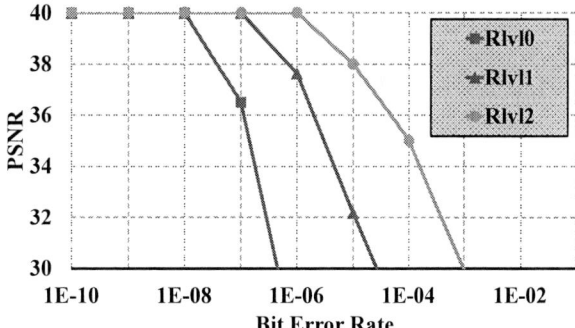

Fig. 2. The impact of bit error rate on image quality at different resolution levels.

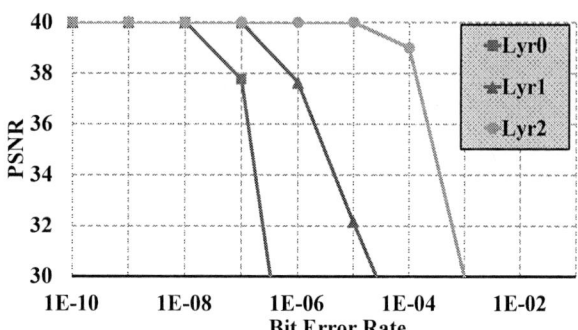

Fig. 3. The impact of bit error rate on image quality at different layers.

levels and layers. We use the Peak signal-to-noise ratio (PSNR) to measure the image quality. The baseline is a clean image "bridge" (PSNR=40dB) without applying any errors to all resolution levels and layers.

1) Resolution Level: Since DWT decomposes the image into multiple resolutions, We explore the error tolerance capability of three different resolution levels (Rlvl). We only inject the errors at one selected resolution level each time. Note the coefficients of each resolution level are collected from that of all the layers. As Fig. 2 shows, for all three resolution levels, the PSNRs can be significantly reduced as the BER reaches a certain level. Furthermore, compared with the others, resolution level 0 is the most sensitive one, e.g. only the BER ($\leq 10^{-8}$) can guarantee the image quality at 40dB. However as the resolution level increases, i.e. from low frequency Rlvl0 to high frequency Rlvl2, the associated error tolerance can be increased. Such difference can be attributed to the error propagation from the lower resolution levels. As the image is progressively reconstructed from lowest resolution level to the highest resolution level through inverse DWT (IDWT), as a result, the errors occurred at a lower resolution level will be propagated to a higher resolution level, eventually leading to more sever image quality degradation. However, we also observe that the error tolerance differences among the three resolution levels are not quite prominent, which indicates a limited space for high-density approximate storage. Therefore, we further perform a similar study at the layer level.

2) Layer: JPEG-2000 also provides the quality scalable feature by partitioning image into multiple layers based on R-D ratios. We use the three-layers structure as an example to explore the image quality v.s. BERs at different layers. As Fig. 3 shows, all PSNRs can be sharply decreased once the BERs reaches some levels for all the three layers (Lyr0, Lyr1, Lyr2). The lower layer is more sensitive to errors than the higher ones. Moreover the error tolerance difference between layer1 and layer2 can be much larger than that of layer0 and layer 1. This is because the R-D ratio determines the importance of each pass and a pass has higher (lower) R-D ratio will be stored in the lower (higher) layer. Note that the passes in a layer belong to different resolution levels.

3) Combined Effect of Resolution level and Layer: As discussed in Section II-A, each coding pass always contains two dimensional information–resolution level and layer. Utilizing only one dimension information (e.g. only resolution level or layer) does not fully unleash the error tolerance capability of an image. Therefore, we explore different combinations of resolution levels and layers to analyze the relationship between BER and PSNR in a more fine-grained manner. Three layers and three resolution levels are selected in this experiment. Note that each layer may include various resolution levels, but can exhibit very different error resilience capability. We select the layer 1 as an example to illustrate the result. As Fig. 4 shows, different resolution levels at the same layer demonstrate a similar trend of BER v.s. PSNR to Fig. 2. If we store the Rlvl0 data in layer 1 to a more reliable memory array, then the rest data of layer 1 can tolerate a larger BER (i.e. $\sim 10^{-6}$). Layer 1 will have no quality reduction if the BER is less than 10^{-7}, as compared with 10^{-8} requirement in Fig. 3.

B. ATSA design

Our ATSA design includes two components: segment partition and ATSA Tri-zone memory.

1) Segment partition: Based on the aforementioned observations, we propose to partition the compressed image into three segments–S1, S2 and S3 to store the most important, second most important and least important data, respectively.

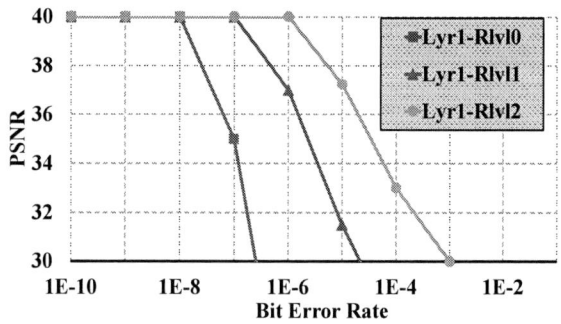

Fig. 4. The impact of bit error rate on image quality at different resolution levels in layer 1.

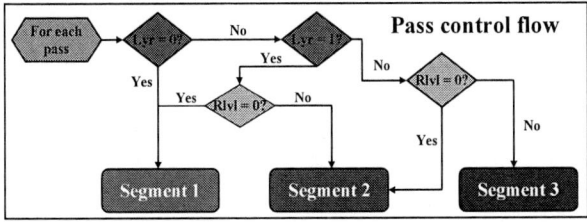

Fig. 5. The pass control flow for partitioning segment.

Each segment is characterized through a pass control flow, as shown in Fig. 5. After bit-plane encoding, each coding pass will be fed into the logic control flow to determine which segment it belongs to. In particular, the passes belonging to layer 0 and Rlvl0 in layer 1 will be assigned to S1, the remained data of layer 1, together with Rlvl0 in layer 2, will be allocated to S2, while all the rest passes are stored in S3. Once each pass partition is done, the same segment data will be assembled as a package including the size and the segment type. Then each segment package will be sent to the dedicated memory array of the ATSA design.

2) ATSA Tri-zone memory architecture: Fig. 6 shows our proposed ATSA tri-zone memory architecture, which incorporates three types of memory cells to accommodate the aforementioned image segments : most reliable two-state cell (1-bit per cell) for S1, the second most reliable tri-state cell (1.5 bit-per cell) for S2 and least reliable four-state cell (2 bit-per cell) for S3. Note that we can only program the four-level MLC STT-RAM to realize all three types of cells. The row structure of the ATSA memory consists of valid bit, row number, data array and flag bits. The flag bits are used to decide which mode the data should be decoded and we use two MLC STT-RAM cells to represent them, i.e. "00", "01" and "11" to represent two-state, tri-state and four-state memory arrays, respectively. Meanwhile, dedicated encoder and decoder circuits are developed to encode/decode the data

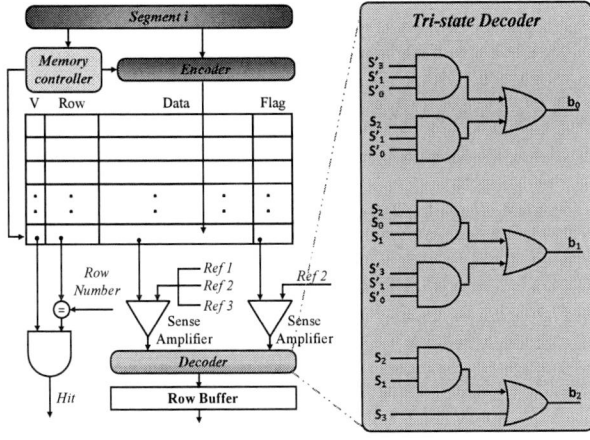

Fig. 6. The proposed Tri-zone memory architecture.

at different modes. We adopt the binary-to-ternary mapping circuit from [9] to encode/decode the tri-state memory data. Note accessing the data in the other two types of memory arrays will bypass this circuit.

At write operation, the size and the segment type information will be first sent to the memory controller (MC). Then the MC will inform the encoder to work in a proper way and adaptively allocate the required memory space. Then the segment data will be encoded or bypass the encoder and stored at the corresponding memory array. Moreover, the flag bits of the associated memory blocks will be also set.

At reading process, the flag bits of a hit row will be read out first to determine which type of the decoder should be employed. Then the four/tri-state memory cell will be read out through a two-step reading mechanism, i.e. comparing the sensed voltage with three reference voltages V_{ref1}, V_{ref2} and V_{ref3}, $V_{BL00} < V_{ref1} < V_{BL01} < V_{ref2} < V_{BL10} < V_{ref3} < V_{BL11}$, where V_{BL00}, V_{BL01}, V_{BL10} and V_{BL11} denote the bit-line voltages when resistance state of MLC cell is R_{00}, R_{01}, R_{10} and R_{11}. Then the four and tri-state data will be decoded directly or via the tri-state decoder [9]. The two-state memory cell will be read out by one step reading scheme assuming we only use the soft-bit.

The overhead of our proposed scheme includes two part: 1) the encoder/decoder overhead with 4x 3-level NAND gate and 14x 2-level NAND/NOR gate for tri-state zone. Compared with the usually 4k to 16k page size in main memory, such overhead is negligible; 2) the overhead introduced by the 2-bit flag signal of each row, is very marginal, i.e. merely 0.9% for a 256Byte row.

IV. EVALUATION

In this section, we evaluate the memory storage density and the image quality of the proposed ATSA design by using the JPEG-2000 image compression as our study case.

A. Experiment Setup

We implement the proposed ATSA scheme via the customized open-source software *JasPer* [20], which is a reference implementation of the codec specified in the JPEG-2000 Part-1 standard (i.e., ISO/IEC 15444-1) [21], [22], [20], [23], supporting image compression and decompression processes. We select the compressed image dataset [19] as our benchmark, including carefully selected high-resolution RGB 8-bit/16bit and gray 8-bit/16-bit image samples. We only use the gray 8-bit images in our simulation to accelerate the image processing.To simulate the error injection of MLC STT-RAM, we directly adopt the STT-RAM device level settings in in [9] and follow their bit error models. In our simulation, the memory storage density is measured by the *bit per cell* (BPC):

$$BPC = \alpha * 1 + \beta * 1.5 + \gamma * 2 \qquad (1)$$

where α, β and γ stands for the percentage of two-state zone, tri-state zone and four-state zone in ATSA memory

(a) bridge original **(b) bridge with ATSA applyed**

Fig. 7. Comparison between the original compressed image and stored image by applying proposed ATSA.

architecture, respectively. In our evaluation, the image quality is measured by the peak signal-to-noise ratio (PSNR):

$$PSNR(dB) = 20\log(\frac{255^2}{MSE}), MSE = \sum \frac{A_{i,j} - B_{i,j}}{n}$$

(2)

where n is the total number of pixels in an image. A and B indicate the original clean image and the compressed image data with error injected, respectively. $A_{i,j}(B_{i,j})$ is the pixel value at row i and column j in $A(B)$. MSE represents the mean square error between original data and the error injected compression data averaged by total pixel number.

B. Results

Fig. 7 illustrates a comparison between the original compressed image and stored image after applying the proposed ATSA. All the selected original compressed images have the same PSNR (i.e., 40 dB). And the quality loss is limited to be less than 2 dB. As shown in Fig. 7, we can hardly identify the distortion or artificial effects on the stored image sample. This clearly indicates that the proposed ATSA can effectively preserve the image quality by leveraging its intrinsic asymmetric error resilience property.

Fig. 8 compares the memory storage density and image quality among three different designs: SLC only (two-level, 1-bit per cell), MLC only (four-level, 2-bit per cell) and the proposed ATSA design, for six selected image benchmarks– "hdr", "bridge", "deer", "spider-web", "flower-foveon" and "big-tree". As shown in Fig. 8, the conventional MLC STT-RAM only design can achieve the highest storage density, however, the image quality is largely degraded by $\sim 80\%$ on average, leading to very unacceptable result. Though applying the ECC to this MLC only design may boost the image quality [9], i.e. a strong BCH-24 code, 240 redundant bits for every 512 bit data (or 47 spare cells), the BPC of such a design with ECC protection can be reduced from 2 to 1.36. The SLC only design can always provide the best image quality (e.g. 40dB), however, it suffers from the lowest storage efficiency, which is $\sim 1.8\times$ (or $2\times$) lower than that of ATSA (or MLC) design. Compared with the other benchmarks, our ATSA achieves the highest storage capacitance with marginal quality degradation on "bridge", "deer" and "big-tree", Since these three benchmarks contain more high frequency components which are less important during the compression. As a result,

more high frequency information will be stored in segment 3, the least reliability cells but with highest storage efficiency. Overall, our proposed ATSA design improves the memory utilization by $\sim 1.8\times$ with less than 5% (2 dB) quality degradation when compared with SLC STT-RAM and achieves $\sim 4.5\times$ image quality improvement on average with only $\sim 10\%$ extra memories when compared with MLC STT-RAM.

V. RELATED WORK

Approximate storage technology is widely applied in multimedia applications. A large body of works are also proposed to improve its performance, energy, density, and reliability. In [10], the authors proposed an approximate image storage scheme by leveraging the 8-LC PCM cell. The authors use PTC image compression technology as a study case to identify the relative importance of the compressed data, and apply appropriate ECCs to the corresponding memories, thus to trade 1 dB image quality for maximized memory density. However, in such a design, their memory designs cannot be reconfigurable and the additional ECC makes the system complex and slow. In [18], the authors proposed an MLC STT-RAM cache architecture by using a specific SLC STT-RAM bank to store the ECC code and MLC banks to store the other data. This work takes video encoding application as a case study by partially protecting the crucial data to reduce the energy consumption. But the proposed scheme requires MLC bank and SLC bank that will involve two types of memory circuits. In contrast, our work targets the high-density MLC STT-RAM, and applies the widely adopted JPEG-2000 image compression technique to evaluate our proposed ATSA scheme. The main advantage of our design is the storage mode of each row can be adjustable and reconfigured to satisfy the images with different sizes and features. Furthermore, our proposed design can improve the memory density with guaranteed image quality by completely eliminating the ECCs.

VI. CONCLUSION

The high-resolution multimedia applications usually consume a mass of memory space on mobile systems, leading to considerable performance and energy overhead. The MLC STT-RAM becomes a good solution for multimedia application due to its high density and good performance, however, it is bottlenecked by the limited reliability. In this

978-1-5386-7100-9/18 $31.00 © 2018 IEEE

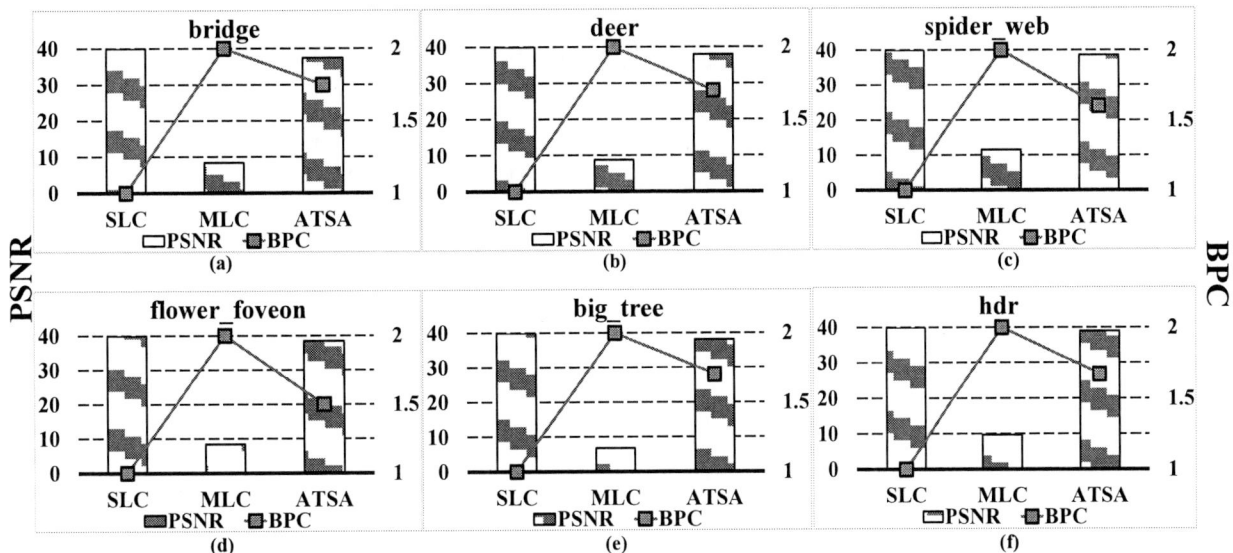

Fig. 8. PSNR and BPC results of SLC, MLC and ATSA for different images. (a) "big-tree", (b) "bridge", (c) "deer", (d) "spider-web", (e) "flower-foveon" and (f) "hdr".

work, we explore the asymmetric error tolerance capability across different layers and resolution levels of JPEG-2000 compressed images and propose a reconfigurable ECC-free ATSA memory architecture, to adaptively store the image data into three different type of memories according to the error-resilience property of different data. Experimental results show that our proposed ATSA design achieves $\sim 1.8\times$ improvement on memory capacity with marginal image quality degradation when compared with the SLC design, and $\sim 4.5\times$ improved PSNR than that of the MLC design by paying only $\sim 10\%$ extra memories.

REFERENCES

[1] D. Williams, H. Jamjoom, Y.-H. Liu, and H. Weatherspoon, "Overdriver: Handling memory overload in an oversubscribed cloud," in *ACM SIG-PLAN Notices*, vol. 46, no. 7. ACM, 2011, pp. 205–216.

[2] M. Qiu, Z. Ming, J. Li, K. Gai, and Z. Zong, "Phase-change memory optimization for green cloud with genetic algorithm," *IEEE Transactions on Computers*, vol. 64, no. 12, pp. 3528–3540, 2015.

[3] A. Nigam, C. W. Smullen IV, V. Mohan, E. Chen, S. Gurumurthi, and M. R. Stan, "Delivering on the promise of universal memory for spin-transfer torque ram (stt-ram)," in *Proceedings of the 17th IEEE/ACM international symposium on Low-power electronics and design*. IEEE Press, 2011, pp. 121–126.

[4] R. F. Freitas and W. W. Wilcke, "Storage-class memory: The next storage system technology," *IBM Journal of Research and Development*, vol. 52, no. 4/5, p. 439, 2008.

[5] X. Lou, Z. Gao, D. V. Dimitrov, and M. X. Tang, "Demonstration of multilevel cell phase switching in mgo magnetic tunnel junctions," *Applied Physics Letters*, vol. 93, no. 24, p. 242502, 2008.

[6] Y. Chen, X. Wang, W. Zhu, H. Li, Z. Sun, G. Sun, and Y. Xie, "Access scheme of multi-level cell spin-transfer torque random access memory and its optimization," in *2010 53rd IEEE International Midwest Symposium on Circuits and Systems*. IEEE, 2010, pp. 1109–1112.

[7] E. Kültürsay, M. Kandemir, A. Sivasubramaniam, and O. Mutlu, "Evaluating stt-ram as an energy-efficient main memory alternative," in *Performance Analysis of Systems and Software (ISPASS), 2013 IEEE International Symposium on*. IEEE, 2013, pp. 256–267.

[8] R. Gallager, "Low-density parity-check codes," *IRE Transactions on information theory*, vol. 8, no. 1, pp. 21–28, 1962.

[9] W. Wen, Y. Zhang, M. Mao, and Y. Chen, "State-restrict mlc stt-ram designs for high-reliable high-performance memory system," in *Proceedings of the 51st Annual Design Automation Conference*. ACM, 2014, pp. 1–6.

[10] Q. Guo, K. Strauss, L. Ceze, and H. S. Malvar, "High-density image storage using approximate memory cells," in *Proceedings of the Twenty-First International Conference on Architectural Support for Programming Languages and Operating Systems*. ACM, 2016, pp. 413–426.

[11] wikipedia.org, "https://en.wikipedia.org/wiki/Lossy_compression."

[12] A. Skodras, C. Christopoulos, and T. Ebrahimi, "The jpeg 2000 still image compression standard," *IEEE Signal processing magazine*, vol. 18, no. 5, pp. 36–58, 2001.

[13] M. J. Shensa, "The discrete wavelet transform: wedding the a trous and mallat algorithms," *IEEE Transactions on signal processing*, vol. 40, no. 10, pp. 2464–2482, 1992.

[14] M. Antonini, M. Barlaud, P. Mathieu, and I. Daubechies, "Image coding using wavelet transform," *IEEE Transactions on image processing*, vol. 1, no. 2, pp. 205–220, 1992.

[15] D. Taubman, "High performance scalable image compression with ebcot," *IEEE Transactions on image processing*, vol. 9, no. 7, pp. 1158–1170, 2000.

[16] J. W. Schwartz and R. C. Barker, "Bit-plane encoding: a technique for source encoding," *IEEE Transactions on Aerospace and Electronic Systems*, no. 4, pp. 385–392, 1966.

[17] Y. Zhang, L. Zhang, W. Wen, G. Sun, and Y. Chen, "Multi-level cell stt-ram: Is it realistic or just a dream?" in *Proceedings of the International Conference on Computer-Aided Design*. ACM, 2012, pp. 526–532.

[18] F. Sampaio, M. Shafique, B. Zatt, S. Bampi, and J. Henkel, "Approximation-aware multi-level cells stt-ram cache architecture," in *Proceedings of the 2015 International Conference on Compilers, Architecture and Synthesis for Embedded Systems*. IEEE Press, 2015, pp. 79–88.

[19] imagecompression, "http://imagecompression.info/."

[20] ece.uvic.ca, "http://www.ece.uvic.ca/ frodo/jasper/."

[21] M. D. Adams, "The jpeg-2000 still image compression standard," 2001.

[22] M. Adams, "Jasper software reference manual," *ISO/IEC JTC*, vol. 1, 2001.

[23] jpeg.org, "https://jpeg.org/jpeg2000/software.html."

2018 IEEE Computer Society Annual Symposium on VLSI

Robust Timing Attack Countermeasure on Virtual Hardware

Kai Yang, Jungmin Park, Mark Tehranipoor, and Swarup Bhunia
Florida Institute for Cybersecurity Research
University of Florida, Gainesville, Florida 32611
Email: {kyang84, jungminpark}@ufl.edu, {tehranipoor, swarup}@ece.ufl.edu

Abstract—Field programmable gate arrays (FPGAs) are being increasingly used in Internet of Things (IoT) applications, as they usually provide lower power, lower latency and higher performance compared with their processor counterparts. However, security has emerged as a critical concern for FPGA-based systems as they are vulnerable to different form of physical attacks, such as side-channel attacks (SCAs). Existing protection methods, which primarily rely on bitstream encryption, are computationally expensive and more importantly, cannot protect against run-time attacks. Hardware virtualization, where instead of traditional direct mapping to FPGA, an application is mapped upon an application-specific virtual layer, called *overlay*, has been well-studied in past decades for productivity benefit, while its security implication has not been investigated at all. In this paper, for the first time to our knowledge, we present a novel usage of virtualization that limits damage from timing attacks and improves performance for RSA decryption by employing unique reconfigurable hardware architectures on FPGAs. Specific masking methods are implemented onto this architecture, and extensive security and performance analysis are done that demonstrates significant side-channel attack resistance under performance constraint.

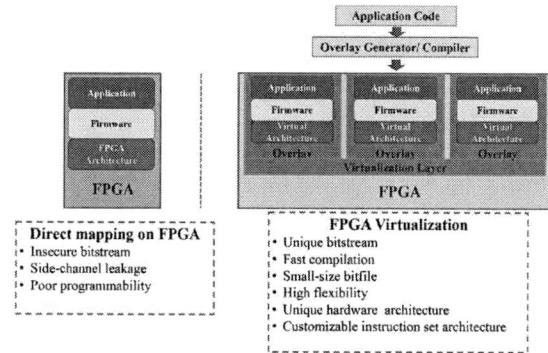

Fig. 1. FPGA virtualization can be used to improve hardware security.

I. INTRODUCTION

In recent years, security has emerged as a critical concern in diverse computer applications [1]. The ubiquity of IoT devices, connection with often insecure networks, and use of less rigorous cryptographic measures make them highly attractive targets for various cyber attacks [2]. However, despite their growing numbers, IoT devices are relatively unprotected, and malware can spread rapidly across the entire IoT network through infected devices, causing serious consequences, such as leaking clients' secret information from cloud servers, making medical devices and smart cars malfunction, or denying access to sensitive data or computing resources.

While most critical security applications in the IoT ecosystem can be efficiently mapped to general purpose processors (GPPs), in many cases, they cannot meet the tight parametric specifications in terms of energy-efficiency and real-time performance. Due to this issue, FPGAs are being increasingly used in these applications, as they could offer the flexibility to map a variety of computing algorithms, and satisfy a diverse set of performance and energy requirements. However, FPGA-based implementations typically suffer from productivity challenges since their spatial architecture and distributed low-level design primitives, are ill-suited to software-like programming (as in case of GPPs) for diverse applications. At the same time, they are vulnerable to different type of physical attacks such as side-channel attacks, including power analysis attacks and timing attacks, where adversaries could obtain sensitive data such as encryption keys by observing the leaked information through side channels (power, timing, etc.).

In case of timing analysis attacks, attackers exploit the execution time difference among different control paths depending on the inputs, such as keys and ciphertexts, to recover the secret key in software or hardware implementations of crypto-algorithms, such as RSA [3] or elliptic curve cryptography. Since timing attacks are possible to mount remotely and have lower computational complexity than other side-channel attacks, such as power or electromagnetic

(EM) analysis, the security concerns related to timing attacks can be significant. Fortunately, countermeasures against timing attacks have been well-researched for the past decades [4], [5]. However, most countermeasures cannot avoid performance degradation. For example, a countermeasure using Montgomery multiplication algorithm [6] and timing balancing method by adding dummy operations against RSA encryption/decryption has the worst performance.

FPGA virtualization can be an effective solution for both security and productivity issues. Similar to software and network virtualization, FPGA virtualization hides the low-level physical resources in FPGA and provides a domain-matched generic layer, called *overlay*, on top of which an application is mapped, instead of traditional direct mapping [7]. Fig. 1 illustrates the structure of direct mapping and hardware virturalization on FPGA. Virtual architecture is first generated based on different application codes, and the overlay compiler could optimize and distribute tasks to number of overlays afterwards.

The virtualization approach can achieve security via unique bitstream, unique hardware architecture and customizable instruction set architecture (ISA) in an overlay. Its security implication in terms of SCA has been addressed in a recent work [8]. In addition to improved security, application mapping on the virtual layer has proved to be 10000x faster than FPGA vendor tools, since the only cost to create a new overlay is the time required for FPGA compilation [9]. Further, it has other advantages including simplified development and debugging environment [10], [11], transparent high-level synthesis [12], bitfile portability across FPGAs, and 1,000x smaller bitfiles [12], [9], which are critical for IoT applications.

In this paper, we present a novel approach for protection against timing attacks by diversifying virtualized hardware across physical FPGAs. Special masking techniques are developed to improve timing side-channel resistance. Generally, the main contributions of this paper are the following:

- It presents a application mapping process in hardware re-

978-1-5386-7100-9/18 $31.00 © 2018 IEEE   148

configurable architecture for ensuring security against timing attacks. A unique overlay generator is used to generate the application-specific virtual architecture based on different design goals.

- It implements specific masking techniques onto the overlay to greatly improve the security level against timing attacks.

- It maps RSA to the proposed architecture using CRT-based masking method with Montgomery multiplication to further improve performance and security.

- It evaluates the security level and performance for overlay using the masking method, and the results show its robustness against timing attacks at reasonable overhead.

The rest of the paper is organized as follows: Section II presents the background and countermeasures for timing side-channel attacks, and addresses a timing leakage assessment test. Section III describes details of an example of virtualization architecture. Section IV introduces the proof of masking method against timing attacks. Section V presents a masking method for proposed architecture, and describes how to efficiently map RSA onto the overlay. Section VI evaluates the security, performance, and overhead of different implementations. Finally, Section VII concludes the paper.

II. BACKGROUND AND MOTIVATION

In this section, we provide background of timing attack and its general countermeasures, and describe quantitative metrics to evaluate timing side-channel leakage.

A. Timing Side-Channel Attack

Timing side-channel attack exploits the input-dependent variability in execution time, leading to practical attacks against implementations of theoretically secure crypto-algorithms, such as RSA. Park et al. [13] proposed a timing side-channel attack using correlation analysis. Let L be the measured clock cycles during encryption or decryption a text, H be clock cycles to execute the target operation $f(p, k^*)$, and R be clock cycles to execute other operations. L is equal to the sum of H and R. Assuming that the timing leakage of the target operation f depends on a text and the subkey, the adversary can estimate the timing leakage of the target operation $\hat{h}$ with any text and a subkey: $\hat{h} = g(f(p, k))$. The estimated leakage $\hat{h}$ with correctly guessed subkey k^* is linearly related to the measured leakage L. Pearson correlation coefficient r can be used to calculate the linear dependency between the measured leakage vector $\vec{l}$ and the estimated leakage vector $\hat{h}_{k_i}$ given the guessed key k_i:

$$r_{k_i} = \frac{\sum_{j=1}^{n_t}(l_{p_j} - \bar{l})(h_{p_j,k_i} - \bar{h}_{k_i})}{\sqrt{\sum_{j=1}^{n_t}(l_{p_j} - \bar{l})^2}\sqrt{\sum_{j=1}^{n_t}(h_{p_j,k_i} - \bar{h}_{k_i})^2}} \quad (1)$$

where $\bar{l}$ and $\bar{h}$ are the mean values of the vector $\vec{l} = \{l_1, \ldots, l_{n_t}\}^T$ and $\hat{h}_{k_i} = \{h_{p_1,k_i}, \ldots, h_{p_{n_t},k_i}\}^T$, respectively. If k^* is equal to $\arg\max_{k_i \in \mathcal{K}} r_{k_i}$, the timing side-channel attack is successful.

If the timing leakage depends on only the decryption key, Eq. 1 cannot be calculated since the variances of the estimated leakage $\hat{h}_{k_i}$ and the measured leakage $\vec{l}$ are zero. In other words, the timing correlation attack cannot be successful. However, simple power analysis (SPA) attack can reveal each key bit by observing power signatures. For example, when a key bit is 1, power signature has higher peaks, otherwise, it has lower peaks. Thus, key-dependent leakage assessment should be considered.

B. Timing Leakage T-test Statistic

In order to evaluate key-dependent timing leakage, Test Vector Leakage Assessment (TVLA) methodology [14] is used. We define timing leakage assessment based on Welch's t-test which is used to test the hypothesis that two populations have equal means when two samples have unequal variance and unequal sample size.

Definition 1 (Timing Leakage Assessment): Two sets are defined as follows:

Set 1 - n timing measurements during decrypting the same n random ciphertexts with a fixed key: $X_1 = \{t_i | t_i = \#cycle(C_i^{K^*} \mod N), i = 1, \ldots, n\}$.

Set 2 - n timing measurements during decrypting the same n random ciphertexts with n random keys: $X_2 = \{t_i | t_i = \#cycle(C_i^{K_i} \mod N), i = 1, \ldots, n\}$. Let the means and variances of X_1 and X_2 be $\bar{X}_1, \bar{X}_2, S_1^2$ and S_2^2. If t-test statistic,

$$t = \frac{\bar{X}_1 - \bar{X}_2}{\sqrt{\frac{S_1^2}{n} + \frac{S_2^2}{n}}} \quad (2)$$

is out of the confidence interval, $|t| > T$, the null hypothesis, $H_0 : \bar{X}_1 = \bar{X}_2$ is rejected. This means that two sets are distinguishable and the implementation has high probability to leak timing information. Thus, it does not pass the leakage assessment test. In our experiment, the threshold value T is chosen as 4.5, which leads to a confidence of > 0.99999 to reject the null hypothesis. This is referred to as *non-specific fixed-vs-random* test.

C. Countermeasures

Currently, the major countermeasures against timing attacks are time balancing and randomization. Time balancing makes the execution time constant, where execution time is decoupled from input data. In a RSA implementation using square-and-multiply algorithm, dummy modular multiplication can be performed when the key bit is zero in order to remove key-dependent timing leakage. Montgomery multiplication algorithm eliminates timing differences arising from different texts as well as cost computations, such as division. However, timing balancing methods typically pay high performance penalty, since it tries to match the execution time of an operation to the longest possible one.

Another countermeasure is to randomize the computation time in order to remove correlation between the key and the computation time. A linear feedback shift register (LFSR) as a random number generator can be used to insert random delays into the computation time. Since the extra delay is included in the computation time, it too cannot avoid the performance degradation.

A specific method to mask the exponent with random numbers can be utilized to increase the performance without the loss of timing side-channel security. In addition, parallel computing based on Chinese remainder theorem (CRT) [15] can make the runtime faster. Details of this algorithm will be described in Section IV. Virtualized architecture can be an effective approach to implement the masking method with CRT based parallel computing. The first reason is that overlays typically have high flexibility, which makes them suitable for realizing different countermeasures. Second, the overlay generator could control the number of processing elements (PEs) in the compilation. The parallelism using multiple PEs could greatly improve performance for RSA algorithm at low hardware overhead.

978-1-5386-7100-9/18 $31.00 © 2018 IEEE

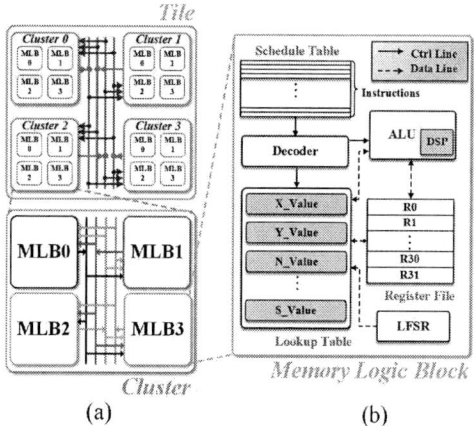

(a)   (b)

Fig. 2. (a) Hierarchical interconnect between MLBs; (b) block diagram of a single MLB.

III. MAHA OVERLAY

In this section, we first describe a potent overlay architecture in detail, and then address its major security advantages.

A. Hardware Architecture

Although any overlay has potential to obtain the presented security advantages, we have evaluated the approach using a modified Malleable Hardware overlay, called MAHA, which consists of a set of RISC-style processing elements. Each PE is referred to as a Memory Logic Block (MLB), and it could work independently for various tasks. Different MLBs are connected through a multi-level hierarchical interconnect system, where data can be transferred efficiently between MLBs. In this paper, we focus on timing attacks and map a 1024-bit RSA engine onto the overlay, and have customized the MLBs based on security needs.

The architecture of a single MLB is shown in Fig. 2(b). Similar to conventional GPPs, it contains common blocks such as an arithmetic logic unit (ALU), a register file, a schedule table, and memory. The ALU block supports common operations, such as addition, subtraction, and exclusive-or, while it can also be customized to provide specific functions. The memory is realized as a high-density, 2D array, which holds the look-up table (LUT) responses as well as the data being processed. A local register file is used to store the temporary outputs from memory or the customized ALU, and the schedule table holds all the instructions. Targeting RSA application, we also use linear-feedback shift registers to generate the random values for masking method, and add a refresh module to update the LFSR seeds.

To accommodate various design requirements, MAHA uses a multilevel hierarchical interconnect, as shown in Fig. 2(a). As an example, the lower level consists of groups of four fully-connected MLBs, while the connection patterns are highly customizable. As execution begins, the decoder fetches an instruction from the schedule table every clock cycle, and then selects result from either the ALU block or the look-up table, and writes back to local register file or LUT at last. The whole architecture is fully customizable for the target application, and can be implemented as an FPGA overlay using standard FPGA CAD tool flows [8].

B. Security and Performance Advantages

MAHA overlay has many advantages over conventional direct mapping method. Firstly, the proposed architecture has high flexibility and scalability. For example, the 1st and higher order masking to mitigate timing side-channel leakage can be implemented easily by adding more MLBs and distributing instructions into their schedule tables. Furthermore, internal components such as the number of MLBs and the look-up table size could be customized for different design goals. Secondly, parallel computing techniques can be implemented efficiently onto the overlay to improve performance, and combined with masking method, to further reduce the leakage. Finally, the bitfile size is much smaller compared with that in direct mapping, and its fast compilation feature [9] is helpful to refresh random values for masking at low cost. The details of masking method for MAHA overlay is described in Section V.

IV. MASKING METHOD AGAINST TIMING ATTACK

Let the decryption key be $K = [k_{n-1}k_{n-2}\dots k_0]$. In our proposed masking method, the decryption key K is encoded into a random key $R = [r_{n-2}r_{n-3}\dots r_0]$ and a masked key $M = [m_{n-2}m_{n-3}\dots m_0]$ such that $K = R + M$. Each encoded key is used as an exponent for an independent modular exponentiation. In order to recover the decrypted plaintext, two independent modular exponentiation should be multiplied as the following:

$$((C^R \mod N)(C^M \mod N)) \mod N = C^K \mod N \quad (3)$$

Since the masked key and the random key are randomly generated every decryption, the computation time is a random variable, which is not dependent on the original key.

Theorem 1: Assuming that the modular exponentiation is based on the square-and-multiply algorithm with Montgomery multiplication, the expected computation time of each modular exponentiation with the encoded exponent is equal to $3c_1(n-1)/2$, where c_1 is the computation time of a modular multiplication and n is the bit length of the exponent.

Proof: When the exponent bit is 1, two modular multiplications are performed for a squaring and a multiplication operation. Otherwise, only a squaring operation is performed. That is, the number of 1's in the exponent determines the computation time.

The expected number of '1's in the random key R is $(n-1)/2$ since the random key is equally distributed. Thus, the expected computation time of the modular exponentiation with the random key is equal to $c_1(n-1+(n-1)/2) = 3c_1(n-1)/2$.

The expected number of '1's in the masked key M is also $(n-1)/2$ since the probability that each bit m_i equals to 1 is 0.5:

$$m_i = k_i \oplus c_{i-1} \oplus r_i = x_i \oplus r_i \quad for \ i = 0, \dots, n-2$$
$$\Pr[m_i = 1] = \Pr[x_i = 1] + \Pr[r_i = 1] - 2\Pr[x_i = 1]\Pr[r_i = 1]$$
$$= 0.5.$$

Thus, the expected computation time of the modular exponentiation with the masked key is also $3c_1(n-1)/2$.

Compared with the timing balancing method using dummy modular multiplication, the performance is increased by $2c_1n/(3c_1(n-1)/2) \approx 1.33x$. Another method to increase the performance is to use parallel computing technique, which is feasible by using the Chinese remainder theorem.

978-1-5386-7100-9/18 $31.00 © 2018 IEEE   150

Definition 2 (CRT-based masking method): Since N is equal to $p*q$, $P = C^K \mod N$ can be computed by the following equations using CRT:

$$P_p = C_p^{K_p} \mod p \tag{4}$$

$$P_q = C_q^{K_q} \mod q \tag{5}$$

$$P = (P_p(q^{p-1} \mod N) + P_q(p^{q-1} \mod N)) \mod N, \tag{6}$$

where $C_p = C \mod p, C_q = C \mod q, K_p = K \mod p$, $K_q = K \mod q$. With our masking method, K_p and K_q are encoded into $[K_{p_r}, K_{p_m}]$ and $[K_{q_r}, K_{q_m}]$, respectively. The following four modular exponentiations can be performed simultaneously:

$$P_{p_r} = C_p^{K_{p_r}} \mod p \tag{7}$$

$$P_{p_m} = C_p^{K_{p_m}} \mod p \tag{8}$$

$$P_{q_r} = C_q^{K_{q_r}} \mod q \tag{9}$$

$$P_{q_m} = C_q^{K_{q_m}} \mod q. \tag{10}$$

Theorem 2: Assuming that p and q have the $n/2$-bit length, the expected computation time of the CRT-based masking method is equal to $3c_2(n/2-1)/2$, where c_2 is the computation time of a modular multiplication for CRT-based implementation.

Proof: The expected number of '1's in each encoded key M is $(n/2 - 1)/2$. Thus, the expected computation time of the modular exponentiation with the encoded key is $c_2(n/2-1+(n/2-1)/2) = 3c_2(n/2-1)/2$.

V. MASKING AND RSA IMPLEMENTATION FOR OVERLAY

In this section, we first describe specific masking implementations using two MLBs and four MLBs in the overlay, and then move on to the RSA implementation using improved Montgomery multiplication algorithm.

A. Masking Method on Overlay

Fig. 3(a) shows the high level architecture for the masking method. The first MLB works as the random part, while the second behaves as the masking part. Suppose we would like to compute $C^K \mod N$ in RSA decryption, where K is a very large number, say, 1024-bit wide. Unlike conventional mapping that leverages the original key K as input, we have concealed this value with random values R generated from a LFSR. These random values are written into look-up tables and used as the decryption key for the first MLB. The masked key M is then calculated by subtracting the original key with the random key ($M = K - R$), and it is stored and used as the decryption key for second MLB. Once we get decryption results from both MLBs, we can calculate and obtain the original plaintext, as shown in Eq. 3. In this way, the original key will not be stored or accessed in both MLBs, which greatly reduces the leakage information. Furthermore, a refresh module is used to update the LFSR feedback and seeds, thus the random values will always stay fresh, making the masking method effective.

We also implement parallelism onto overlay, and combine parallel computing technique with proposed masking method to further improve the security and performance. The high level diagram of the combined method is shown in Fig. 3(b). As described in *Theorem 2*, based on Chinese Remainder Theorem, original RSA inputs could be divided into two parts and distributed into different masking blocks. Each masking block contains two MLBs, the random share and the

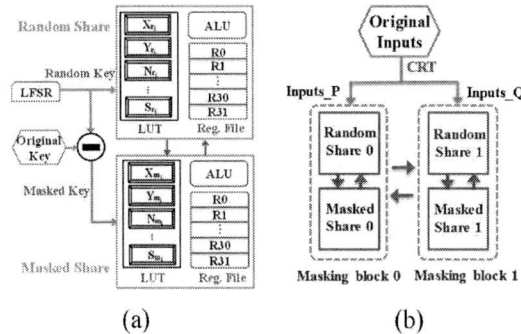

Fig. 3. (a) High level architecture of masking method for RSA decryption; (b) high level architecture of CRT-based masking method using 4 MLBs.

masked share, and behaves similar to what is shown in Fig. 3(a). Every masking block contains a LFSR to generate the unique random values, and intermediate values could be transferred between masking blocks for mapping purpose. The decryption result is the combination of the mapping results of two blocks.

B. RSA Implementation on Overlay

Song *et al.* proposed an improved Montgomery modular multiplication method [15]. Our RSA implementation is based on this algorithm.

Algorithm 1 Improved Montgomery Multiplication Algorithm for RSA implementation [15]

Definition: $radix - 2^{17}, d = \lceil 2^r/17 \rceil, 17d \geq 2^r + 3$
$X, Y, N, S_i \in \{0, 1, ..., 2^{2^r} - 1\}$
$-N^{-1}, \alpha, \beta, \gamma, C_\alpha, C_\beta \in \{0, 1, ..., 2^{17} - 1\}, C_\gamma, C_S \in \{0, 1\}$
$X = \sum_{i=0}^{d-1} 2^{17i} \cdot X_i, X_i \in \{0, 1, ..., 2^{17} - 1\}, X_d = 0$
$Y = \sum_{i=0}^{d-1} 2^{17i} \cdot Y_i, Y_i \in \{0, 1, ..., 2^{17} - 1\}$
$N = \sum_{i=0}^{d-1} 2^{17i} \cdot N_i, N_i \in \{0, 1, ..., 2^{17} - 1\}, N_d = 0$
$S_i = \sum_{i=0}^{d-1} 2^{17i} \cdot S_{(i,j)}, S_{(i,j)} \in \{0, 1, ..., 2^{17} - 1\}, S_d = 0$
Input : $X, Y, N, -N^{-1}$
Output : $S_d = X \cdot Y \cdot 2^{-17d} \mod N$

1. $S_0 \leftarrow 0$
2. **for** $i = 0$ to $d - 1$ **do**
3. $\quad q \leftarrow ((X_0 \cdot Y_i + S_{(i,0)}) \cdot (-N^{-1})) \mod 2^{17}$
4. $\quad C_\alpha, C_\beta, C_\gamma, C_S \leftarrow 0$
5. $\quad$ **for** $j = 0$ to d **do**
6. $\quad\quad \{C_\alpha : \alpha\} \leftarrow X_j \cdot Y_i + C_\alpha$
7. $\quad\quad \{C_\beta : \beta\} \leftarrow q \cdot N_j + C_\beta$
8. $\quad\quad \{C_\gamma : \gamma\} \leftarrow \alpha + \beta + C_\gamma$
9. $\quad\quad \{C_\S : S_{(i+1,j-1)}\} \leftarrow \gamma + S_{(i,j)} + C_S$
10. $\quad$ **end for**
11. **end for**

This algorithm utilizes a DSP48E1 block enabling multiply-accumulate of 17-bit operands. 2^r denotes the size of Montgomery multiplier inputs X, Y, and N, and $d = \lceil 2^r/17 \rceil$ is the number of units for each operand. $C_\alpha, C_\beta, C_\gamma, \alpha, \beta, \gamma$, and C_S store the intermediate values, while S_i obtains multiplication result in last round.

Fig. 4(a) shows the high level architecture to generate the overlay instructions. We first decompose the RSA Pseudo-code and create the

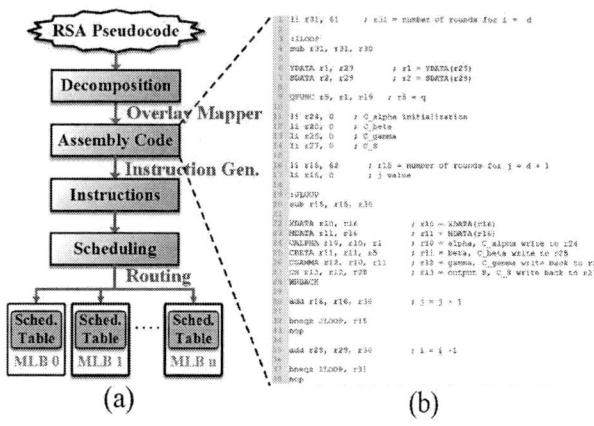

Fig. 4. (a) High level architecture of instruction generation; (b) partial assembly code for RSA implementation.

assembly code. Then, 32-bit instructions are generated using another tool we have developed, which transforms assembly code to binary numbers based on the unique instruction set architecture of overlay. These instructions are then distributed and stored into the schedule table of different MLBs. Once the decryption begins, a system counter generates the address of the schedule table, and an instruction is fetched and decoded every clock cycle. Since Algorithm 1 is based on the radix-2^{17}, the basic units of MLBs such as the LUT and the register file are configured to be 17-bit wide. All operands X, Y, and N are divided into small blocks and pre-written into LUTs for execution. For S_i, which behaves as the output of previous round and input of next round (line 3 & 9 in Algorithm 1), we also store it using LUTs. In order to minimize area overhead, the previous S_i values will be overwritten by new values with the same LUT address. Intermediate values, such as C_α and C_β are stored into local registers in each MLB. Depending on the operation type, the lightweight ALU module would load values from either the LUT or the register file for computation, and then write the result back to the LUT or local registers in the same clock cycle.

Fig. 4(b) shows a part of the assembly code for Montgomery multiplication method, which corresponds to i and j loops in Algorithm 1. Line 1 and line 16 define the number of rounds in each loop. At the beginning of each round, these numbers would decrease by 1 (line 4 & 20), and branch instructions at the end of each round (line 32 & 37) are used to check if this loop has finished. Line 6 shows the LUT read operation, where input Y_i is obtained and stored in register R1, and line 30 is the instruction to write new S_i value back into the SDATA block in LUT. As mentioned, overlay architecture is highly customizable. In order to further improve performance, we have designed several multi-input single-output fused instructions by modifying the ALU. For example, the 6th and 7th line in Algorithm 1 have one multiplication and one addition operation. It needs two clock cycles to execute. We have customized the overlay and created *SUMPRODUCT* function to merge multiplication and addition into one instruction. Line 9 and line 24-27 in Fig. 4(b) are instructions using these fused functions.

Based on the assembly code, we have calculated the theoretical total number of cycles needed for each modular multiplication of different overlay implementations. Note that since MAHA is a single-issue RISC-style architecture, only one instruction can be performed

per cycle per MLB.

1) Without masking method

There are d rounds for each i loop, and $(d+1)$ round for every j loop. So, the total number of rounds needed is $d(d+1)$. From 4(b), it takes 10 cycles for every j loop, and 23 clock cycles for every i round. As defined in Section IV, c denotes the total number of cycles needed to compute a modular multiplication. For the implementation using 2 MLBs without masking method, the number is computed by the following equation:

$$c_0 = 10d^2 + 23d + 10 \tag{11}$$

2) With masking method, 2 MLBs

As described in previous sections, the implementation with the masking method requires additional operations, such as subtraction before the decryption to generate masked key, and the multiplication after that to generate the original plaintext. However, the execution time of each modular multiplication is not affected by these operations, thus total number of cycles needed in this implementation stays the same with non-masking method:

$$c_1 = 10d^2 + 23d + 10 \tag{12}$$

2) With masking method, 4 MLBs

Based on CRT, the inputs could be divided into two parts, p and q, whose size are assumed to be 2^{r-1}. Hence, the number of units for each operand, referred as d, is also reduced by half. In this way, while the number of cycles needed for every round stays the same, the total number of rounds is $d(d/2+1)/2$. Eq. 13 shows the total number of cycles needed to compute a modular multiplication using 4 MLBs with masking method:

$$c_2 = 2.5d^2 + 11.5d + 10 \tag{13}$$

C. Experimental Setup

We consider 1024-bit RSA decryption, and use a SPARTAN-6 FPGA to implement the hardware design. We have used 50K randomly generated ciphertexts to conduct RSA decryption for two datasets. One set uses a fixed key, while the other uses random keys generated from LFSRs. We have verified functional correctness and obtained the performance using Mentor Graphics Modelsim, which is measured in term of total number of cycles. The area overhead is obtained from ISE Design Suite.

VI. RESULTS AND ANALYSIS

We have evaluated the security, performance, and area overhead of different architectures for RSA decryption: overlay without the masking method, overlay with timing balancing countermeasure using dummy instructions, and overlay with proposed masking methods.

A. Security Analysis

Tab. I compares the performance and area overhead of different implementations. Time balancing have been proven to be secure by many research, while the extra dummy instructions greatly decrease the performance. Based on the timing leakage assessment metric,

978-1-5386-7100-9/18 $31.00 © 2018 IEEE

TABLE I.   OVERHEAD VALUES FOR DIFFERENT MAPPING TECHNIQUES

	No. of Slice Reg	No. of Slice LUTs	Average No. of Cycles	Bitstream Size (Byte)	Total Power (mW)	No. of Block RAMs	Frequency (MHz)	T-test Statistic by Eq.2
Non-masking, 2 MLBs	2858	2072	59266994	1453	967.32	6	133	137.55
Dummy instruction, 2 MLBs	3087	2185	79099904	1453	967.68	6	132	0
Masking, 2 MLBs	3012	2219	59305678	1453	967.57	6	133	-0.965
Masking, 4 MLBs	4238	3028	15366483	1886	971.63	8	136	0.865

which is described in Section II, we have calculated the t-test statistic value for all designs with and without masking methods implemented. To do this, we randomly generate 50K ciphertexts for RSA decryption. The number of clock cycles during each decryption with a fixed key and a random key are measured for the set 1 and the set 2, respectively. Note that since the computation time of our Montgomery multiplication implementation does not depend on the ciphertext, the variance of the set 1 in the non-masking method is 0. As expected, the implementation without the masking method does not pass the timing leakage assessment test since $t = 137.55 > T(= 4.5)$, while designs using the masking method and dummy instructions pass the test since $|t| < 4.5$. In case of the masking methods, for every decryption, the masked key and the random key are randomly generated from LFSR, making the execution time a random variable, which is not dependent on the original key.

B. Performance and Overhead Analysis

First, we compare the designs with and without the masking method implemented. We pick number of slice registers/LUTs, number of block RAMs, and bitstream size, which include look-up table size, schedule table size, and local register file size, to evaluate the area overhead. Compared with the non-masking method, the masking method employs the same bitstream size and block RAM usage, while requires 5% and 7% extra slice registers/LUTs. Moreover, due to the extra instructions for masking method such as the subtraction to get masked key and multiplication to generate original plaintext, its execution time is slightly higher than the non-masking method. However, as we combine masking method with parallel computing technique using 4 MLBs, the total number of cycles decreases dramatically, while the area overhead does not improve linearly. Compared with the non-masking method, the implementation using 4 MLB is 3.9x faster, at the cost of 1.3x bitstream size and 1.5x slice registers/LUTs usage.

Next, we compare performance and overhead between the masking method and one of the commonly used countermeasure, time balancing method with dummy instructions. The dummy modular multiplication can be performed when the key bit is zero, thus key-dependent timing leakage can be removed. As shown in Tab. I, the masking method using 2 MLBs has similar area overhead compared with the dummy instruction method, while the execution time is roughly 1.3x faster. This number matches our theoretical expectation, as described in Section IV. Furthermore, as we combine parallel computing with the masking method using 4 MLBs, this number rises to 5.2x, while it needs 30%, 37%, and 39% extra bitstream size, slice registers, and slice LUTs, respectively.

VII.   CONCLUSION

We have presented a flexible FPGA virtualization approach, which is amenable to implementation of masking and architectural diversity, making it more difficult for an adversary to obtain crucial information from timing side-channel leakage. An overlay generator is developed to create different countermeasures for the timing attack. Unique masking techniques are designed and implemented onto the overlay. The results show the proposed overlay is robust against timing attack and could achieve high performance at reasonable overhead.

REFERENCES

[1] P. Kocher, R. Lee, G. McGraw, A. Raghunathan, and S. Moderator-Ravi, "Security as a New Dimension in Embedded System Design," in *Proceedings of the 41st annual Design Automation Conference.* ACM, 2004, pp. 753–760.

[2] M. Aman, K. C. Chua, and B. Sikdar, "Mutual Authentication in IoT Systems Using Physical Unclonable Functions," *IEEE Internet of Things Journal*, 2017.

[3] R. L. Rivest, A. Shamir, and L. Adleman, "A Method for Obtaining Digital Signatures and Public-key Cryptosystems," *Communications of the ACM*, vol. 21, no. 2, pp. 120–126, 1978.

[4] A. K. Biswas, D. Ghosal, and S. Nagaraja, "A Survey of Timing Channels and Countermeasures," *ACM Comput. Surv.*, vol. 50, no. 1, pp. 6:1–6:39, Mar. 2017. [Online]. Available: http://doi.acm.org/10.1145/3023872

[5] B. Mao, W. Hu, A. Althoff, J. Matai, Y. Tai, D. Mu, T. Sherwood, and R. Kastner, "Quantitative Analysis of Timing Channel Security in Cryptographic Hardware Design," *IEEE TCAD*, 2017.

[6] P. L. Montgomery, "Modular Multiplication Without Trial Division," *Mathematics of computation*, vol. 44, no. 170, pp. 519–521, 1985.

[7] G. Stitt, R. Karam, K. Yang, and S. Bhunia, "A Uniquified Virtualization Approach to Hardware Security," *IEEE Embedded Systems Letters*, 2017.

[8] K. Yang, J. Park, M. Tehranipoor, and S. Bhunia, "Hardware Virtualization for Protection Against Power Analysis Attack," in *Hardware Oriented Security and Trust (HOST), 2018 IEEE International Symposium on.* IEEE, 2018.

[9] J. Coole and G. Stitt, "Adjustable-Cost Overlays for Runtime Compilation," in *Field-Programmable Custom Computing Machines (FCCM), 2015 IEEE 23rd Annual International Symposium on.* IEEE, 2015, pp. 21–24.

[10] A. K. Jain, S. A. Fahmy, and D. L. Maskell, "Efficient Overlay Architecture based on DSP Blocks," in *Field-Programmable Custom Computing Machines (FCCM), 2015 IEEE 23rd Annual International Symposium on.* IEEE, 2015, pp. 25–28.

[11] D. Capalija and T. S. Abdelrahman, "A High-Performance Overlay Architecture for Pipelined Execution of Data Flow Graphs," in *Field Programmable Logic and Applications (FPL), 2013 23rd International Conference on.* IEEE, 2013, pp. 1–8.

[12] J. Coole and G. Stitt, "Fast, Flexible High-Level Synthesis from OpenCL Using Reconfiguration Contexts," *IEEE Micro*, vol. 34, no. 1, pp. 42–53, 2014.

[13] J. Park, M. Corba, A. E. de la Sema, R. L. Vigeant, M. Tehranipoor, and S. Bhunia, "ATAVE: A Framework for Automatic Timing Attack Vulnerability Evaluation," in *2017 IEEE 60th International Midwest Symposium on Circuits and Systems (MWSCAS)*, Aug 2017, pp. 559–562.

[14] G. Goodwill, B. Jun, J. Jaffe, and P. Rohatgi, "A testing methodology for side-channel resistance validation," in *NIST non-invasive attack testing workshop*, 2011, pp. 158–172.

[15] B. Song, Y. Ito, and K. Nakano, "CRT-based DSP Decryption Using Montgomery Modular Multiplication on the FPGA," in *Parallel and Distributed Processing Workshops and Phd Forum (IPDPSW), 2011 IEEE International Symposium on.* IEEE, 2011, pp. 532–541.

978-1-5386-7100-9/18 $31.00 © 2018 IEEE

Towards Theoretical Cost Limit of Stochastic Number Generators for Stochastic Computing

Meng Yang[*], Bingzhe Li[†], David J. Lilja[†], Bo Yuan[‡], Weikang Qian[*]

[*]University of Michigan-Shanghai Jiao Tong University Joint Institute, Shanghai Jiao Tong University, China
[†]Department of Electrical and Computer Engineering, University of Minnesota, U.S.A.
[‡]Department of Computer Science and Engineering, Shanghai Jiao Tong University, China
Email: *{yangm.meng, qianwk}@sjtu.edu.cn, [†]{lixx1743, lilja}@umn.edu, [‡]boyuan@sjtu.edu.cn

Abstract—Stochastic number generator (SNG) is one important component of stochastic computing (SC). An SNG usually
consists of a random number source (RNS) and a probability
conversion circuit (PCC). The SNGs occupy a large portion of the
total area and power of a stochastic circuit. Thus, it is critical to
lower the area and power of the SNGs. The existing methods only
focused on simplifying the RNSs inside the SNGs, such as sharing
the RNSs and using emerging devices. However, how to reduce
the area and power of PCCs is never studied. In this work, we
explore this problem and propose a solution that can effectively
reduce the area and power of PCCs. We also study the theoretical
limit on the cost of SNG and find that our proposed design
approaches the limit. The experimental results show that our
design can gain up to $2\times$ improvement in power-delay product
over the traditional SNGs.

I. INTRODUCTION

Stochastic computing (SC), an unconventional computing
paradigm proposed long time ago [1], has gained more and
more attention recently [2]. Different from binary computing,
it operates on stochastic bit streams, which encode values
through the probabilities of 1s in the streams. For example, as
shown in Fig. 1(a), the bit stream 11010111 encodes the value
6/8. Based on this encoding mechanism, SC is highly tolerant
of bit flip-flop errors. Moreover, its probabilistic nature allows
very simple circuits to realize complex arithmetic operations.
For example, an AND gate can implement multiplication, as
shown in Fig. 1(a). Due to its low hardware cost, it is found
to be suitable in applications such as image processing [3],
low-density parity-check (LDPC) decoding [4], and artificial
neural networks [5].

To interface with conventional binary computing, SC requires a component that converts a binary number into a
stochastic bit stream. This component is called *stochastic
number generator (SNG)*, as shown in the shaded block in

Fig. 2. An SNG consists of a random number source (RNS)
and a probability conversion circuit (PCC). Typically, the RNS
generates a random binary number uniformly distributed in the
range $[0, 2^k - 1]$, where k is the bit width of the random binary
number. Equivalently, it can be viewed as a set of k unbiased
random bits, which are of probability of 0.5 to be a 1. In Fig. 2,
the RNS is a linear feedback shift register (LFSR). The PCC
is a combinational circuit that takes k unbiased random bits
and k target bits $c_{k-1}, \ldots, c_0$ as inputs. It produces an output
bit stream with probability of $C/2^k$, where C is the binary
number encoded by the target bits as $C = (c_{k-1} \ldots c_0)_2$. Since
the parameter k determines the precision of the probability, in
what follows, we call k the *probability precision*. In Fig. 2,
the PCC is a comparator. The circuit taking stochastic bit
streams as inputs and outputs is referred to as an *SC core*. For
example, the AND gate in Fig. 2 is the SC core. Generally, in
a stochastic circuit, the cost of the SNGs is much larger than
that of the SC core, which can be easily seen from Fig. 2.

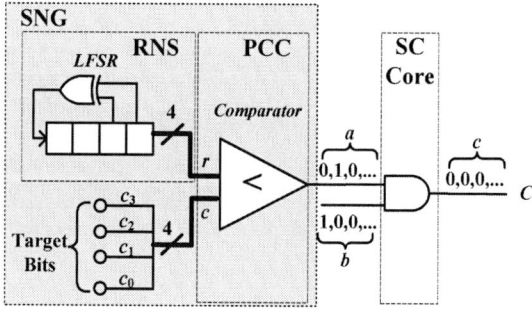

Figure 2: A complete implementation of multiplication in SC.

In literature, researchers have proposed several methods to
reduce the cost of the SNGs. For some situations, accurate
computation requires different input bit streams to be independent. One example is multiplication using an AND gate. A few
works proposed methods to reduce the SNG cost to provide
independent input stochastic bit streams. In [6], the authors
proposed a method that shuffles the outputs of a single RNS
to produce multiple target bit streams with low correlation.
In [7], the authors proposed a method that inserts the delay
elements, such as D flip-flops (DFFs), to obtain approximately
independent inputs. For these two methods, only one RNS is
required to generate multiple independent bit streams. Another

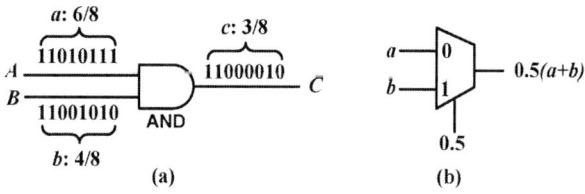

Figure 1: Examples of stochastic computing elements: (a) An AND
gate for multiplying two numbers encoded by stochastic bit streams;
(b) A 2-to-1 multiplexer (MUX) implementing the scaled addition.

978-1-5386-7100-9/18 $31.00 © 2018 IEEE

TABLE I: Area breakdown for two stochastic implementations of a 32-tap finite impulse response (FIR) filter [12].

area	w/o RNS sharing		with RNS sharing	
	area	%	area	%
RNS	3385	70	54	3.6
PCC	1274	26.4	1274	84.6
Other	178	3.6	178	11.8
Total	4837	100	1506	100

way is using emerging low-power nanoscale devices [8], such as memristor [9] and spintronic devices [10], as the RNSs.

There exist some other situations where accurate computation does not require different input bit streams to be independent. For example, in SC, a multiplexer (MUX) can realize a scaled addition $(a+b)/2$, as shown in Fig. 1(b). For this design, we do not require the input stochastic bit streams a and b to be independent. This is known as *correlation insensitivity* [11]. Another example is parallel applications such as image processing. The inputs to different output pixels can share a single RNS, which reduces the cost of the SNGs.

Although progress has been made to reduce the SNG cost, the existing methods only focused on simplifying the RNSs inside the SNGs. However, besides an RNS, an SNG also includes a PCC. As the RNS cost has been reduced through various techniques, the cost of the PCCs could now dominate the cost of the SNGs. For example, in [12], the authors have exploited the RNS sharing to reduce the SNG cost used in digital filter design. A summary of their results is shown in Table I. From the table, we can see that after RNS sharing is applied, the area of the RNSs in the entire stochastic circuit is significantly reduced. Consequently, the entire stochastic circuit area reduces. However, the PCC area does not change. It now dominates the area of the circuit, accounting for 84.6% of the entire area. Thus, to further reduce the cost of SNG, it is imperative to study how to reduce the cost of PCC. However, to the best of our knowledge, this problem is never studied in literature. In this work, we try to propose a solution to this problem. We have made the following contributions:

1) We propose a method to reduce the cost of a type of PCC called weighted binary generator (WBG). With this method, we can reduce the cost of SNG built with WBG.
2) For the first time, we analyze the theoretical cost limit of SNGs. We make a conjecture on the exact minimal cost of SNGs and demonstrate that the proposed design achieves this minimal cost.
3) We empirically study the accuracy, area, power, and delay of the proposed design. The results show that our design achieves significant area and power reduction with small delay increase and no accuracy loss, compared to the traditional SNG design. In terms of power-delay product, the proposed design achieves $2\times$ improvement.
4) We apply the proposed technique to stochastic implementations of a deep neural network (DNN) application and demonstrate that it can further reduce the area and power consumption of the stochastic implementations. Furthermore, the proposed stochastic implementation has much smaller area, power, and energy than the traditional binary implementation with a slight accuracy loss.

The rest of this paper is organized as follows. In Section II,

we introduce the wegithed binary generator (WBG). In Section III, we present the proposed method to reduce the cost of the SNG built with the WBG. We also analyze the theoretical cost limit and demonstrate that our proposed design achieves this limit. In Section IV, we show the experimental results. Finally, in Section V, we conclude the paper.

II. WEIGHTED BINARY GENERATOR

Weighted binary generator (WBG) is a PCC introduced by Gupta and Kumaresan [13]. A WBG with the probability precision $k = 4$ is shown in Fig. 3. The AND gates in the first level take unbiased random bits $r_3, \ldots, r_0$ as inputs and produce the intermediate signals

$$w_3 = r_3, w_2 = \overline{r_3} \wedge r_2, w_1 = \overline{r_3} \wedge \overline{r_2} \wedge r_1, \\ w_0 = \overline{r_3} \wedge \overline{r_2} \wedge \overline{r_1} \wedge r_0, \tag{1}$$

where $\wedge$ represents the logical AND. The AND gates in the second level take $w_3, \ldots, w_0$ and the target bits $c_3, \ldots, c_0$ as inputs and produce the intermediate signals $u_i = w_i \wedge c_i$, for $i = 0, \ldots, 3$. Finally, a tree of 3 two-input OR gates takes $u_3, \ldots, u_0$ as inputs and produces the final output signal $x = u_3 \vee \cdots \vee u_0$, where $\vee$ represents the logical OR.

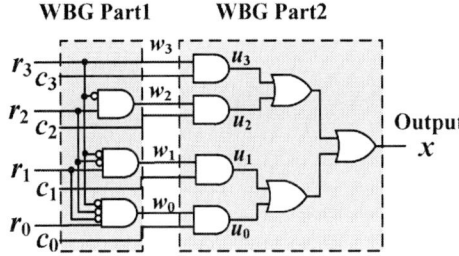

Figure 3: A weighted binary generator with the probability precision $k = 4$.

Given Eq. (1), the probabilities of w_i's are

$$P(w_3) = \frac{1}{2}, P(w_2) = \frac{1}{2^2}, P(w_1) = \frac{1}{2^3}, P(w_0) = \frac{1}{2^4}.$$

By Eq. (1), there cannot exist two w_i's that are both 1. Given this property, the probability of the output x is

$$P(x) = P(w_3) c_3 + P(w_2) c_2 + P(w_1) c_1 + P(w_0) c_0 \\ = \frac{1}{2} c_3 + \frac{1}{2^2} c_2 + \frac{1}{2^3} c_1 + \frac{1}{2^4} c_0 = \frac{C}{2^4},$$

where $C = (c_3 \ldots c_0)_2$. Thus, the WBG functions as a PCC.

III. SNG COST REDUCTION AND SNG COST LIMIT

In this section, we present our proposed method of reducing the cost of the SNGs built with WBGs. Then, we analyze the theoretical cost limit of SNG and demonstrate that the proposed SNG cost reduction method achieves this limit. Finally, we summarize when our proposed SNG cost reduction method can be applied.

978-1-5386-7100-9/18 $31.00 © 2018 IEEE

A. SNG Cost Reduction

In this section, we will demonstrate a method that reduces the cost of the SNGs built with WBGs. The method works when the accurate computation does not require different input bit streams to be independent.

We can decompose the WBG into two parts, the set of AND gates in the first level that produces the signals w_i and the remaining set of gates that produces the final output x. For simplicity, we call the former *WBG part 1* and the latter *WBG part 2*. They are circled out in Fig. 3.

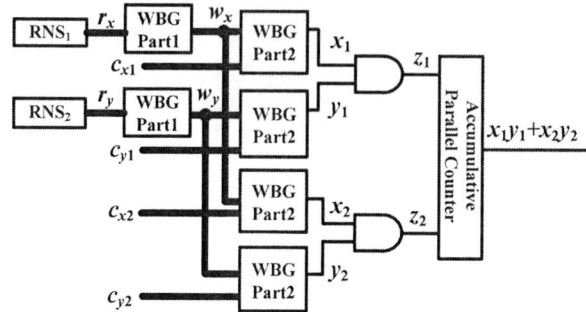

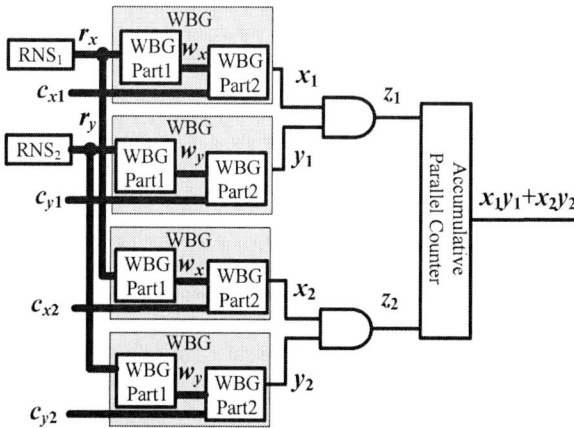

Figure 4: A stochastic implementation of the function $x_1 y_1 + x_2 y_2$, which uses the WBG as the PCC and applies the RNS sharing.

Suppose that we want to produce two input bit streams that could correlate but are of different probabilities. For example, suppose that we want to realize the sum of two products $x_1 y_1 + x_2 y_2$, where x_1, x_2, y_1, y_2 are specified by the users. A realization of this is shown in Fig. 4 It is a mixture of stochastic computing and binary computing. It uses two AND gates to multiply stochastic bit streams and generates two bit streams for the products $x_1 y_1$ and $x_2 y_2$. To get the final sum as a binary number, an accumulative parallel adder (APC) is applied [14]. Note that with the APC, the two output stochastic bit streams z_1 and z_2 could be correlated. Consequently, the inputs x_1 and x_2 could be correlated and the same for the inputs y_1 and y_2. Exploiting this fact, we can share an RNS for generating x_1 and x_2 and also share an RNS for generating y_1 and y_2, as shown in Fig. 4.

However, carefully examining the design reveals that the WBG part 1 for x_1 and that for x_2 can also be shared. The same holds for the WBG part 1 for y_1 and that for y_2. This extra sharing leads to a design with reduced SNG area, as shown in Fig 5. Such a sharing can be generalized for an arbitrary number of input bit streams that could be correlated. Note that this sharing method is essentially based on reducing the cost of PCC, which is different from all the existing SNG simplification methods.

B. Theoretical SNG Cost Limit

As discussed in the introduction, each input bit stream requires an SNG. Thus, in order to reduce the total SNG cost,

Figure 5: A stochastic implementation of the function $x_1 y_1 + x_2 y_2$ with the proposed sharing of WBG part 1.

we should reduce the SNG cost per input. As several methods were discovered to reduce the SNG cost, a natural question that raises to us is: what is the lower bound on SNG cost per input? We believe this question is important to the study of SC. In this section, we analyze this bound. In what follows, we will consider the situation where the accurate computation does not require different input bit streams to be independent.

Consider a stochastic circuit with N inputs. The very basic SNG design requires N SNGs, each consisting of an RNS and a PCC. Denote the areas of the RNS and the PCC as A_R and A_P, respectively. Then, the SNG area per input is simply

$$A_R + A_P.$$

Since we assume that different input bit streams could correlate, N input bit streams can share one RNS. However, for all the existing solutions, each input requires a separate PCC. In this case, the SNG area per input is

$$\frac{A_R + N A_P}{N} = \frac{A_R}{N} + A_P.$$

Obviously, the RNS sharing method reduces the SNG cost per input compared to the basic SNG design method.

As we demonstrated in Section III-A, the traditional view that each input requires a separate PCC may not be true. For PCC like WBG, we can identify some part in the PCC that can also be shared among different inputs. For such cases, we can further reduce the SNG area per input. We denote the area of the sharable part in the PCC as $A_{P,S}$ and that of the remaining non-sharable part of the PCC as $A_{P,NS}$. We have $A_P = A_{P,S} + A_{P,NS}$. Given that only one sharable part of the PCC is required for N inputs, the SNG area per input is

$$\frac{A_R + A_{P,S} + N A_{P,NS}}{N} = \frac{A_R + A_{P,S}}{N} + A_{P,NS}. \quad (2)$$

The first term in Eq. (2) depends on the number of inputs N. For a large N, which is possible for those massively parallel applications, the first term in Eq. (2) approaches 0. Thus, for a large N, the SNG area per input approaches $A_{P,NS}$, the area of the non-sharable part of the PCC.

In what follows, we define the *cost* of a circuit as the number of 2-input gates in the circuit when it is implemented with only 2-input gates and inverters. We define the *asymptotic cost*

978-1-5386-7100-9/18 $31.00 © 2018 IEEE

of SNG per input as the cost of the SNG per input when the number of inputs N approaches infinity. By Eq. (2), the asymptotic cost of SNG per input is equal to the cost of the non-sharable part of the PCC.

To get a lower bound on the SNG area per input, we are interested in the *minimal* asymptotic cost of SNG per input. For simplicity, we call it the *minimal asymptotic SNG cost*. From the above discussion, it is equal to the minimal cost of the non-sharable part of any PCC. Next, we analyze this minimal cost.

We assume that the N input probabilities are different. Since each PCC is responsible for one input probability and all the input probabilities could be different, the non-sharable part of a PCC should include the final output of the PCC and the k target bits that determine the probability of a specific input. Furthermore, due to the existence of the randomness in the input bit stream produced by the PCC, the non-sharable part should take at least one random bit stream as its inputs. Thus, for the non-sharable part, it has at least $(k+1)$ inputs. For example, consider the WBG. Its non-sharable part is the WBG part 2. It includes the final output and k target bits. Furthermore, it has k random bit streams as its inputs.

For a circuit with n inputs, if it is only built with 2-input gates and inverters, at least $(n-1)$ 2-input gates are needed. By the definition of cost, its cost is at least $(n-1)$. Given that the non-sharable part of any PCC has at least $(k+1)$ inputs, the cost of the non-sharable part of any PCC is at least k. This gives a lower bound on the minimal cost of the non-sharable part of any PCC. Now consider the WBG. Its non-sharable part is the WBG part 2, which consists of k AND gates and $(k-1)$ OR gates. Thus, the cost of the non-sharable part of the WBG is $(2k-1)$. This gives an upper bound on the *minimal* cost of the non-sharable part of any PCC. As we mentioned above, since the minimal asymptotic SNG cost is equal to the minimal cost of the non-sharable part of any PCC, we conclude the following claim.

Theorem 1

Suppose that the probability precision of the SNG is k. A lower bound on the minimal asymptotic SNG cost is k, while an upper bound on it is $(2k-1)$.

It seems that there is still a gap between the lower bound and the upper bound on the minimal asymptotic SNG cost. Thus, the exact minimal asymptotic SNG cost is still unknown. However, we have the following conjecture on the exact value.

Conjecture 1

Suppose that the probability precision of the SNG is k. The minimal asymptotic SNG cost is $(2k-1)$.

To prove the conjecture, we need to show that the lower bound on the minimal asymptotic SNG cost is $(2k-1)$. However, we are not able to prove this strictly. In what follows, we just give an informal explanation to this. For the given target bits $c_{k-1}, \ldots, c_0$, the output probability is

$$P = \sum_{i=1}^{k} \frac{1}{2^i} c_{k-i}.$$

The above probability expression involves k multiplications and $(k-1)$ additions. To realize this weighted sum, there must exist k 2-input AND gates in the circuit. The i-th ($1 \leq i \leq k$) one takes c_{k-i} as one input and a random bit stream as the other input. Furthermore, some additional 2-input gates are needed to merge the outputs of these k 2-input AND gates into the final single output. The number of these gates should at least be $(k-1)$. Thus, such a circuit should include at least $(2k-1)$ 2-input gates. This explains why we conjecture that the lower bound is also $(2k-1)$.

If the conjecture holds, then the WBG sharing method proposed in Section III-A realizes the minimal asymptotic SNG cost. This indicates the optimality of the proposed WBG sharing method.

C. Suitable Situations

In this section, we describe the suitable situations where the proposed WBG sharing method can be used. One thing to note is that the if multiple input probabilities are produced by the proposed sharing method, these inputs are correlated. Therefore, in order to apply the proposed method, we require that the computation accuracy will not be affected when the inputs are correlated. Given this requirement, there are the following two situations where the proposed method can be applied.

1) The inputs belong to the same function and their correlation does not affect the correctness of the final computation. Using the architecture shown in Fig. 5 to realize the function $x_1 y_1 + x_2 y_2$ is such an example.
2) The inputs belong to different functions. One example is the parallel applications. For example, consider the gamma correction [15] application used in image processing. It applies the computation $y = x^\gamma$ to each pixel, where x and y are the normalized pixel value in the input and the output image, respectively, and γ is a parameter. For this application, multiple pixels can be processed in parallel. For example, we can compute $y_1 = x_1^\gamma$ and $y_2 = x_2^\gamma$ in parallel, where x_1 and x_2 are the values of two different input pixels. In this case, the input probabilities x_1 and x_2 can be correlated.

IV. Experimental Results

In this section, we present the experimental results to validate the benefits of the proposed WBG sharing method. In some of the following experiments, we compared three different SNG designs. The first design uses the comparator as the PCC. This is the widely used SNG design. We call it *CMP*. The second design uses the WBG as the PCC, but without applying the WBG sharing method. We call it *non-shared WBG*. The third design uses the proposed WBG sharing method to realize the PCC. We call it *shared WBG*.

A. Accuracy Study

In this section, we study the accuracy of the proposed shared WBG design. We compared two stochastic implementations for the function $x_1 y_1 + x_2 y_2$. The first one uses the proposed shared WBG design, as shown in Fig. 5. The RNS used in this design is the LFSR. The second one uses the CMP design. Furthermore, each SNG has its own RNS. The second implementation is shown in Fig. 6. Table II compares the mean absolute error (MAE) of the two designs for different bit stream lengths. From the results, we can see that the MAE for the proposed shared WBG design is even smaller than the traditional CMP design in most cases. It demonstrates the high accuracy of the proposed sharing method.

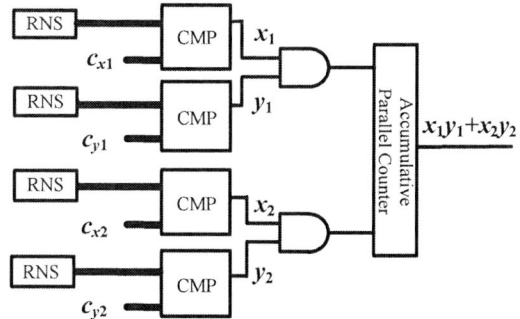

Figure 6: The stochastic implementation of $x_1y_1 + x_2y_2$ using the traditional SNGs.

TABLE II: Mean absolute error of the two stochastic implementations of $x_1y_1 + x_2y_2$ shown in Figs. 5 and 6.

Bit stream length	16	32	64	128	256	512	1024
Shared WBG	0.0462	0.0312	0.02	0.0164	0.0135	0.0081	0.0054
CMP	0.0773	0.0442	0.0269	0.0171	0.0118	0.0086	0.0058

B. Hardware Cost Study

In this section, we study the hardware cost of the proposed shared WBG design. To study the effectiveness of the proposed design in reducing the SNG cost, we just focus on the SNG part for generating multiple correlated inputs.

Fig. 7 shows the area, power, and delay of three SNG designs, CMP, non-shared WBG, and shared WBG, for different numbers of inputs. The probability precision k is 8. The red, blue, and green curves in each figure correspond to the CMP, the non-shared WBG, and the shared WBG designs, respectively. Note that for all the three SNG designs, a single RNS is shared among multiple target probabilities. In each figure, we chose the number of inputs as $2, 4, \ldots, 10$. Figs. 7a, 7b, and 7c show the area, power, and delay comparison, respectively.

As we can see, for all three SNG designs, both the area and power increase linearly with the number of inputs, which is expected. The proposed shared WBG design has much smaller area and power than the other two designs. Its delay, however, is larger than the other two designs. The reason is due to factors like logic fanouts. When the number of inputs is 10, the proposed design saves 43.4% area and 30.8% power compared to the non-shared WBG design. Its delay increases slightly by 1.02%. Compared to the CMP design, the proposed design saves 44.3% area and 55.7% power, with a delay increase of 13.5%. In terms of power-delay product, the proposed design has a reduction of 30.1% and 49.7% over the non-shared WBG design and the CMP design, respectively.

C. Case Study: Neural Network Classifier

In this section, we apply the proposed shared WBG design to a real application, neural network classifier. Neural network (NN) is widely used in computer vision, decision making, etc., nowadays. For modern deep neural networks (DNNs), they need numerous multiplications and additions. Thus, their power consumption could be huge. Meanwhile, since NN is usually used for classification tasks, which can tolerate a small amount of error, the computation in NNs does not require high accuracy. SC happens to target at computation that does not need high accuracy and can reduce the system cost. Thus, SC is very suitable for NN.

Previous works [16]–[18] proposed novel stochastic implementations of NNs. In this case study, we applied the proposed shared WBG design to one of the classical NNs called *restricted Boltzmann machine (RBM)* [19]. Specifically, we considered an RBM used for handwritten digit recognition [16]. The test input is MNIST handwritten digit image dataset [20]. One input image has 28×28 pixels. Thus, the number of inputs in the neural network is 784. The neural network consists of 2 hidden layers $L1$ and $L2$, with 500 and 1000 neurons, respectively. Its output layer has 10 neurons.

Our study is based on the stochastic implementation of the RBM proposed in [16]. Notably, in that design, it uses a circuit similar to the one shown in Fig. 4 to calculate the weighted sum. Thus, our proposed technique can be applied to reduce the area and power of the SNGs used in that design. Note that this application of the proposed technique corresponds to situation 1 mentioned in Section III-C.

TABLE III: The recognition error rate comparison between two stochastic implementations of RBM with different SNGs.

Bit stream length	32	64	128	256	512
Shared WBG	3.14%	1.75%	1.48%	1.52%	1.47%
CMP	2.63%	1.49%	1.53%	1.53%	1.45%

The recognition error rates for different bit stream lengths and PCCs are shown in Table III. The error rate of the traditional binary computing for the same network is 1.23%. From the table, we can see that as the bit stream length grows, the error rates of the stochastic implementations of the RBM reduce. When the bit stream length is 512, the error rate is 1.47% using the shared WBG design and 1.45% using the CMP design, which is very close to that of the binary computing. The accuracy of the shared WBG design is slightly worse than that of the CMP design. However, the difference between them becomes smaller as the bit stream length grows.

TABLE IV: Hardware cost comparison of various implementations of the RBM.

Circuits	CMP	Non-shared	Shared	Conventional Binary
Area(mm^2)	2.24	1.86	1.36	29.15
Power(mW)	759	560	394	14894
Energy(nJ)	36.46	26.89	18.93	29.79

Finally, we compare the hardware cost of different implementations of the RBM, including the binary implementation. The results are listed in Table IV. The 2nd, 3rd, and 4th columns show the results of the stochastic implementations

978-1-5386-7100-9/18 $31.00 © 2018 IEEE

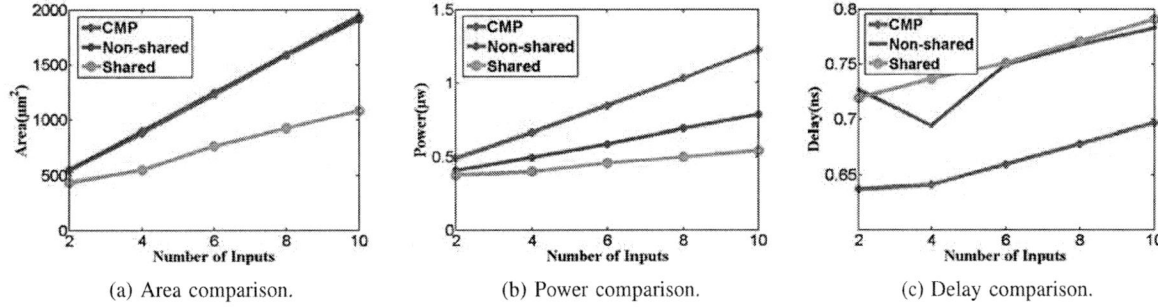

| (a) Area comparison. | (b) Power comparison. | (c) Delay comparison. |

Figure 7: Area, power, and delay comparison among three SNG designs for different numbers of inputs.

using CMP, non-shared WBG, and the proposed shared WBG, respectively. The last column shows the result of the conventional binary implementation. The stochastic bit stream length is 32 and the bit width of the binary computing is 5. As we can see from the table, the area and the power of the binary computing is much larger than the stochastic implementations. Among all the stochastic implementations, the area/power/energy of the proposed shared WBG design is the smallest. Compared to the non-shared WBG design, the proposed design can save 26.9% area ,29.6% power, and 29.6% energy. Compared to the binary implementation, the proposed design can save 36.5% energy with an accuracy loss of only 1.9%.

V. CONCLUSION

In this work, we studied the problem of how to reduce the cost of SNG in SC. Different from the existing works, which focus on simplifying the RNS inside the SNG, we studied how to reduce the cost of the PCC. We proposed a method to reduce the cost of a type of PCC called WBG by decomposing it into the sharable and the non-sharable parts. The propose design is applicable to generating multiple input bit streams that can be correlated, which indeed is a property of several applications such as image processing, digital filter design, and artificial neural networks. Moreover, for the first time, we give the theoretical cost limit of SNGs. We show that the proposed WBG sharing method can achieve this theoretical limit. Finally, we demonstrated through experiments the effectiveness of the proposed design. In our future work, we will study how to formally prove the conjecture on the theoretical cost limit of SNGs.

ACKNOWLEDGMENT

This work is supported in part by National Natural Science Foundation of China (NSFC) under grant no. 61472243 and 61204042 and in part by National Science Foundation (NSF) of U.S.A. under grant no. CCF-1408123 and CCF-1438286. Any opinions, findings and conclusions or recommendations expressed in this material are those of the authors and do not necessarily reflect the views of the NSFC and NSF.

REFERENCES

[1] B. R. Gaines, "Stochastic computing," in *AFIPS Spring Joint Computer Conference*, 1967, pp. 149–156.

[2] A. Alaghi, W. Qian, and J. P. Hayes, "The promise and challenge of stochastic computing," *IEEE Transactions on Computer-Aided Design of Integrated Circuits and Systems (in press)*, 2017.

[3] P. Li, D. J. Lilja *et al.*, "Computation on stochastic bit streams digital image processing case studies," *IEEE Transactions on Very Large Scale Integration (VLSI) Systems*, vol. 22, no. 3, pp. 449–462, 2014.

[4] V. C. Gaudet and A. C. Rapley, "Iterative decoding using stochastic computation," *Electronics Letters*, vol. 39, no. 3, pp. 299–301, 2003.

[5] B. Li, M. H. Najafi, and D. J. Lilja, "Using stochastic computing to reduce the hardware requirements for a restricted boltzmann machine classifier," in *International Symposium on Field-Programmable Gate Arrays*, 2016, pp. 36–41.

[6] B. Yuan, Y. Wang, and Z. Wang, "Area-efficient scaling-free DFT/FFT design using stochastic computing," *IEEE Transactions on Circuits and Systems II: Express Briefs*, vol. 63, no. 12, pp. 1131–1135, 2016.

[7] T.-H. Chen and J. P. Hayes, "Analyzing and controlling accuracy in stochastic circuits," in *International Conference on Computer Design*, 2014, pp. 367–373.

[8] M. Yang, J. P. Hayes *et al.*, "Design of accurate stochastic number generators with noisy emerging devices for stochastic computing," in *International Conference on Computer-Aided Design*, 2017, pp. 638–644.

[9] P. Knag, W. Lu, and Z. Zhang, "A native stochastic computing architecture enabled by memristors," *IEEE Transactions on Nanotechnology*, vol. 13, no. 2, pp. 283–293, 2014.

[10] R. Venkatesan, S. Venkataramani *et al.*, "Spintastic: Spin-based stochastic logic for energy-efficient computing," in *Design, Automation & Test in Europe*, 2015, pp. 1575–1578.

[11] A. Alaghi and J. P. Hayes, "Dimension reduction in statistical simulation of digital circuits," in *Symposium on Theory of Modeling and Simulation*, 2015, pp. 1–8.

[12] H. Ichihara, T. Sugino *et al.*, "Compact and accurate digital filters based on stochastic computing," *IEEE Transactions on Emerging Topics in Computing*, 2016.

[13] P. K. Gupta and R. Kumaresan, "Binary multiplication with PN sequences," *IEEE Transactions on Acoustics, Speech, and Signal Processing*, vol. 36, no. 4, pp. 603–606, 1988.

[14] P.-S. Ting and J. P. Hayes, "Stochastic logic realization of matrix operations," in *Euromicro Conference on Digital System Design*, 2014, pp. 356–364.

[15] D.-U. Lee, R. C. Cheung, and J. D. Villasenor, "A flexible architecture for precise gamma correction," *IEEE Transactions on Very Large Scale Integration (VLSI) Systems*, vol. 15, no. 4, pp. 474–478, 2007.

[16] B. Li, Y. Qin *et al.*, "Neural network classifiers using stochastic computing with a hardware-oriented approximate activation function," in *International Conference on Computer Design*, 2017, pp. 97–104.

[17] K. Kim, J. Kim *et al.*, "Dynamic energy-accuracy trade-off using stochastic computing in deep neural networks," in *Design Automation Conference*, 2016, pp. 124:1–124:6.

[18] Z. Li, A. Ren *et al.*, "DSCNN: Hardware-oriented optimization for stochastic computing based deep convolutional neural networks," in *International Conference on Computer Design*, 2016, pp. 678–681.

[19] G. E. Hinton and R. R. Salakhutdinov, "Reducing the dimensionality of data with neural networks," *Science*, vol. 313, no. 5786, pp. 504–507, 2006.

[20] Y. LeCun, "The MNIST database of handwritten digits," *http://yann. lecun. com/exdb/mnist/*, 1998.

Fully-on-Chip Digitally Assisted LDO Regulator with Improved Regulation and Transient Responses

Han Li, Chenchang Zhan and Ning Zhang
Department of Electrical and Electronic Engineering
Southern University of Science and Technology, Shenzhen, China
Email: 11649124@mail.sustc.edu.cn; zhancc@sustc.edu.cn; zhangn@mail.sustc.edu.cn

Abstract—This paper proposes a fully-on-chip mixed-mode low-dropout (LDO) regulator with regulation and transient response enhanced. A Miller compensation capacitor and a buffer stage are used to achieve stability and improve power MOS gate slew rate. The ultra-fast voltage buffer helps further improve the load transient recovery speed and reduce the chip area due to its wider voltage swing. With the help of the digital regulation part, the supported maximum load current is significantly improved. The proof-of-concept LDO design is fabricated in a standard 0.18-μm CMOS technology. The maximum load current is 150 mA, the output voltage is 1 V and the dropout voltage is 0.2 V. The load regulation is 0.17 mV/mA.

Keywords—*Mixed LDO, output-capacitor-free, load regulation, load transient, adaptive biasing (AB).*

I. INTRODUCTION

Low-dropout (LDO) regulators are the popular power management ICs often used in smart devices such as the mobile phones powered by battery, due to their low cost of bill-of-material (BOM) and chip area [1-3]. With the development of the technology, the structure of the SoC chips in smart devices becomes more and more complex. Hence, the power management ICs should provide a high integration level, with as few as possible external components. Therefore, the traditional LDO regulator with large off-chip capacitor is not preferred for the SoC power management applications. Instead, full-on-chip design is much more desirable [4-9].

The typical LDO regulator is consisted of an error amplifier, power transistor, feedback circuit and compensation network. To obtain a high driving capability, there will be more current needed to go through the power transistor, therefore the size of the power transistor is large. As a result, the gate capacitance of the power transistor is large. The slew-rate and the bandwidth of the LDO regulator will be degraded because of the large gate capacitance, making it difficult to achieve high regulation accuracy and fast transient responses. For both the LDOs with and without external capacitors, a high slew-rate buffer can be helpful. Indeed, many methods have been proposed to improve the transient performances, e.g., [1-3]. In [1], an impedance-attenuated voltage buffer is added between

the power transistor and the error amplifier to improve the transient response and the loop gain bandwidth. But the buffer needs a large quiescent current that will waste much energy when the LDO works at the light load. A quick response circuit is proposed in [2], and the quiescent current is smaller, but both the transient voltage dips and the recovery time are not optimized. The wide-swing buffer in [3] helps improve the area efficiency, but it works for a high supply voltage of 3.3 V, and does not support low-voltage operation well. Furthermore, these works are targeting for LDOs with external capacitors. The fully-integrated LDOs in [4-6] used techniques of adaptive biasing and transient improvement for better performances. Yet, large power transistors are still needed. The digital LDOs in [7-9] allow for small power transistors, due to the deployment of digital control and rail-to-rail swing of the power transistors. But their control method is complicated and the process, voltage, and temperature (PVT) variations are large.

In this work, a digital assistant circuit to improve the regulation and transient response of the LDO regulator is developed. A digital part is added to a traditional LDO with a buffer. A current sensing circuit detects the load current and turns on the digital power transistor to support the excess amount of load current. There, a large loading capability is achieved with a small chip area overhead. For the rest of the paper, the circuit design is shown in Sec. II, experimental results and conclusion are given in Sec. III and IV, respectively.

II. PROPOSED LDO WITH DIGITAL ASSISTANCE

Fig.1 shows the structure of the proposed fully-on-chip mixed-mode LDO regulator. To enable high load current capability, the LDO is divided into two parts. One is the analog part that includes an error amplifier (EA), a unity-gain buffer and analog power transistor M_{PA}. The digital part is consisted of a hysteretic current comparator, a logic control circuit and a digital power transistor M_{PD}. This part will detect the load

---

This work was in part supported by the National Natural Science Foundation of China (NSFC) under grant 61604067 and the Shenzhen Science and Technology Innovation Committee (SZSTI) under grant JCYJ20160530191008447.

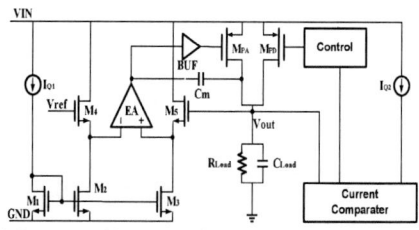

Fig. 1. Structure of the proposed mixed-mode LDO regulator

978-1-5386-7100-9/18 $31.00 © 2018 IEEE

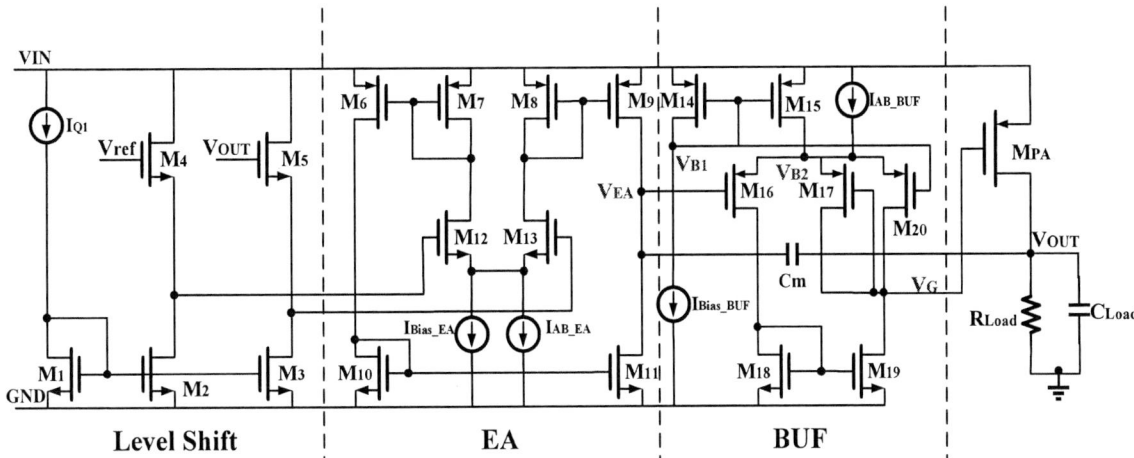

Fig. 2. Schematic of the analog part of the proposed LDO

current through M_{PA}. If the load current is small, the logic control part will turn off MPD, and only the analog part works. If the load current increases to the switching point, the control part will turn on MPD, then the digital part will provide a part of the load current to improve the performances of the LDO.

The power transistor of the digital part M_{PD} has a wide gate voltage swing from ground to supply voltage by using the digital logic control, similar to the digital LDOs [7-9], only with 1-bit accuracy in this design. The current through the M_{PD} will be larger than the same-size transistor controlled by the analog circuit. Then, the proposed mixed-mode LDO can save chip area compared with the pure analog LDO with the same maximum load current since the power transistor is smaller. The small power transistor also improves the transient performance due to small gate capacitance [4], [10].

A. Analog Part

Fig. 2 shows the schematic of the analog part of the proposed LDO. The output voltage and reference voltage are level shifted before entering the EA. The EA made up by M_6-M_{13} is a single stage current-mirror amplifier biased by the current I_{Bias_EA} and the adaptive biasing current I_{AB_EA}. The adaptive biasing current technology has been used in many designs [4-6] to improve the bandwidth and the slew rate of the on-chip LDOs. The slew rate at the gate of the pass transistor is mainly limited by the quiescent current (I_Q/C_{PASS}) [10]. But when the quiescent current is large, the efficiency of the LDO at light load will be poor for $\eta = (V_{OUT} \cdot I_{LOAD})/[(I_{LOAD} + I_Q) \cdot (V_{OUT} + V_{DO})]$ [10]. So, the adaptive biasing technology can provide large bias current at the heavy load without degrading the good current efficiency.

The buffer constituted by M_{14}-M_{20} is added to achieve fast load transient response by improving the slew-rate of the power transistor gate [10]. It is an adaptively biased unity-gain-buffer with only PMOS input, but with M_{20} added to extend the voltage swing. At heavy load when V_{EA} is small, the buffer works normally. At very light load when V_{EA} is high to turn off

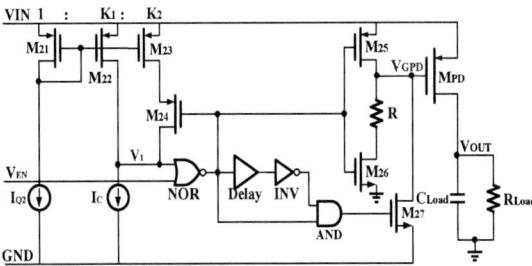

Fig. 3. Schematic of the digital part of the proposed LDO

M_{16} and M_{17}, the voltage V_{B2} is high and M_{20} is on to charge up V_G for turning off the power transistor M_{PA}. Effectively, the buffer output swing is extended.

B. Digital Part

Fig. 3 shows the schematic of the digital part. A hysteretic current comparator is used to compare the sensed load current (I_C) through M_{PA} and the bias current of $(K_1+K_2)I_{Q2}$ or K_2I_{Q2} depending on the initial gate voltage of M_{24}. The NOR gate is added to control whether the digital work or not to compare the performance of the LDO with or without the digital part for testing purpose. The external voltage V_{EN} is added to turn on or turn off the digital part. When V_{EN} is high, the digital part will be turned off, and there is only the analog part that can work, while when the V_{EN} is low, this part will be turned on. When the digital part is turned on, assume the M_{PA} current is small at the beginning and if $I_C<(K_1+K_2) \cdot I_{Q2}$, V_1 will be at high voltage, then the gate voltage of the power transistor (V_{GPD}) will be at high voltage too due to M_{25}, so the digital power transistor will be turned off, and there will be no current through M_{PD}. When $I_C>(K_1+K_2) \cdot I_{Q2}$, V_1 will be discharged, then V_{GPD} will be discharged to a low voltage, and there will be a current through M_{PD} to provide the excess current that the load needs. After that, the I_C current needs to go below K_2I_{Q2} to turn off M_{PD}, since the gate of M_{24} is high and it is off to

978-1-5386-7100-9/18 $31.00 © 2018 IEEE   161

disconnect M_{23}. Hence a current hysteresis is generated to avoid oscillation.

Transistor M_{27} and the pulse generation circuit made up by the delay cell, the inverter and the AND gate is added to discharge V_{GPD} in short time to improve the transient speed. When the load current changes to high, the delay cell in the digital part will make sure that M_{27} is turned on for certain duration, so the gate of the pass transistor is discharged quickly. Transistor M_{26} and resistor R provide the slow path that holds the low V_{GPD} voltage while improving the PSR of the LDO, since a high resistance node is created at V_{GPD} when M_{27} is turned off after the delay. Fig. 4 shows the simulated V_{GPD} with/without the transistor M_{27}. With M_{27}, V_{GPD} can be discharged quickly, even though the resistor R is large.

Fig. 5 shows the voltage V_1, analog part output current (I_A) and digital part out current (I_D) vs. load current. The digital part works when the load current is higher than 90 mA and then the output current of the analog part decreases. The current through the power transistor M_{PA} will increase all the way to I_{AMax} to make sure the total load current does not change. The maximum current of the proposed LDO is the I_{AMax} plus I_D.

C. Biasing Current Generated Circuit

Fig. 6 is the schematic of the load current sensing circuit, which is also the adaptive biasing circuit that provides the adaptive bias current to EA (I_{AB_EA}) and BUF (I_{AB_EA}) and the comparison current I_C to the digital part. The transistor M_S is added to mirror the current I_A of M_{PA} to generate the bias current and comparison current by using the current mirror. The transistor M_4-M_7 works as a common gate amplifier to ensure the drain voltage of M_S (V_{MSD}) is equal to V_{OUT} to improve the sensing accuracy since M_S and M_{PA} have the same terminal voltages at S, G, and D. Therefore, $I_{AB\_EA}=N_1 \cdot (I_A/N)$, $I_{AB\_BUF}=N_3 \cdot (I_A/N)$ and $I_C=N_2 \cdot (I_A/N)$.

Miller compensation is used for the stability of the proposed LDO regulator, which creates the dominant-pole at the EA output. Fig. 7 shows the simulated loop gain Bode plots of the proposed LDO at different load currents with 100pF load capacitor. Good stability of the LDO with the digitally assisted circuit is achieved over a wide load range.

III. EXPERIMENTAL RESULTS

The proposed fully-on-chip mixed-mode LDO regulator is fabricated in a standard 0.18-μm CMOS process. Fig. 8 shows the chip photo. The active chip area is 0.022mm². An enable

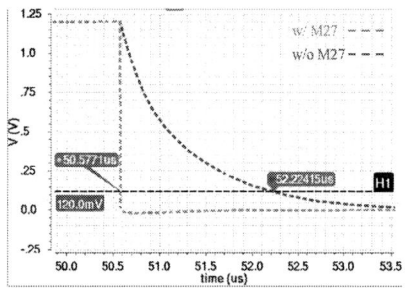

Fig. 4. Illustration of transient V_{GPD} w/ and w/o M_{27}

signal is provided externally for testing purpose, to study the effect of the digital assistance part.

Fig. 9 shows the measured load regulation of the proposed LDO up to 150 mA with and without the digital part. With the digital assistance, when the load current is higher than around 100mA, the load current sensor triggers the digital power transistor to be turned on, and improves the loading capability significantly. The calculated load regulation is 25mV/150mA, i.e., 0.17mV/mA with the digital assistance, while without digital assistance, it is 146mV/150mA, i.e., 0.97mV/mA.

Fig. 10 shows the measured load transient of the proposed LDO with and without digital assistance part, under 0pF and 100pF load capacitor, respectively. It confirms the improved regulation through the digital assistance. Furthermore, the LDO regulator using the proposed simple buffer inside the analog part guarantees the good stability of the LDO.

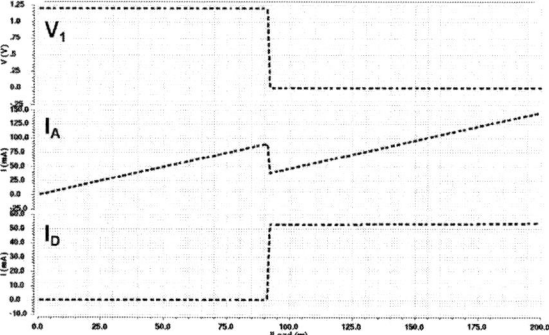

Fig. 5. Simulated V_1, output current of analog part (I_A) and output current of digital part (I_D) vs. load current.

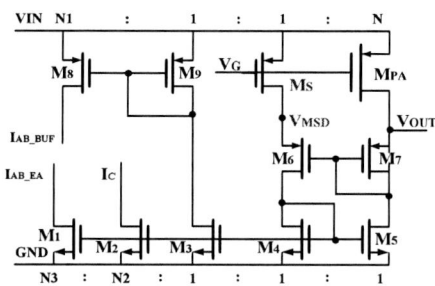

Fig. 6. Schematic of the adaptive biasing circuit.

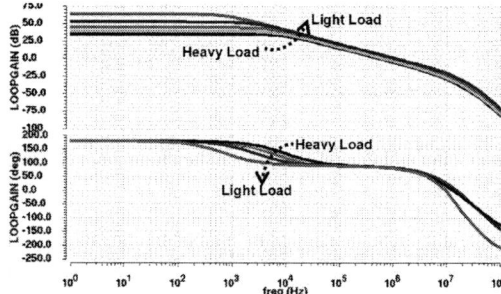

Fig. 7. Simulated loop gain Bode plots of the proposed LDO at different load currents.

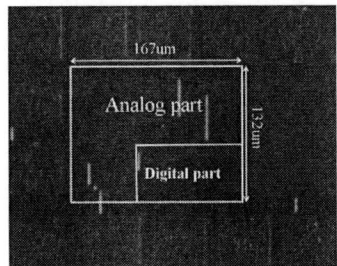

Fig. 8. Chip photo

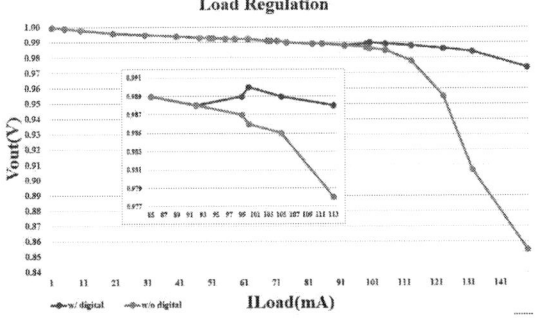

Fig. 9. Measured load regulation of the LDO w/ vs. w/o the digital part

IV. CONCLUSION

This paper has presented a digitally assisted LDO regulator with a simple but effective wide-swing voltage buffer to improve the regulation and the load transient response. With the digital assistance, a small digital power transistor can help significantly improve the regulation accuracy, saving much chip area compared to pure analog regulation. The chip measurement results of a prototype design have verified the proposed methods, and the mixed-mode LDO regulator is a good candidate for on-chip power management applications of VLSI systems.

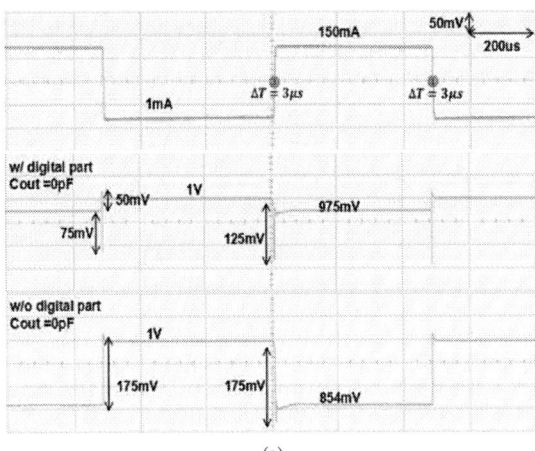

(a)

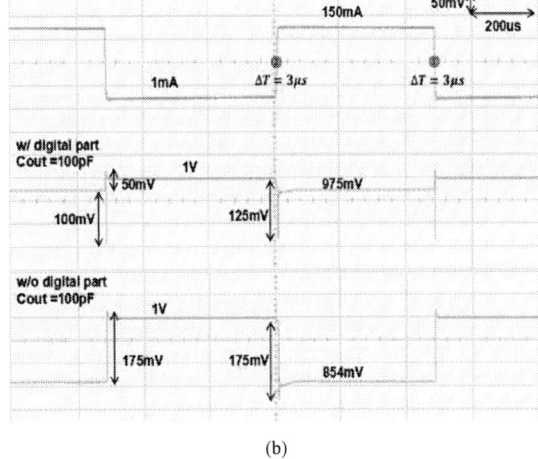

(b)

Fig. 10. Measured load transient responses of the proposed LDO w/ and w/o the digital part at (a) C_{out}=0 pF (b) C_{out}=100 pF

REFERENCES

[1] M. Al-Shyoukh, R. A. Perez and H. Lee, "A Transient-Enhanced 20μA-Quiescent 200mA-Load Low-Dropout Regulator With Buffer Impedance Attenuation," *IEEE Custom Integrated Circuits Conference*, San Jose, CA, pp. 615-618, Sept. 2006.

[2] S. Heng, W. Tung and C. K. Pham, "Low power LDO with fast load transient response based on quick response circuit," *IEEE International Symposium on Circuits and Systems*, pp. 2529-2532, May 2009.

[3] Y. Liu, C. Zhan, L. Cheng and W. H. Ki, "A chip-area-efficient CMOS low-dropout regulator using wide-swing voltage buffer with parabolic adaptive biasing for portable applications," *IEEE Asian Solid State Circuits Conference (A-SSCC)*, Kobe, pp. 233-236, Nov. 2012.

[4] C. Zhan and W. H. Ki, "Output-Capacitor-Free Adaptively Biased Low-Dropout Regulator for System-on-Chips," *IEEE Transactions on Circuits and Systems I: Regular Papers*, vol. 57, no. 5, pp. 1017-1028, May 2010.

[5] X. Qu, Z. k. Zhou and B. Zhang, "Ultralow-power fast-transient output-capacitor-less low-dropout regulator with advanced adaptive biasing circuit," *IET Circuits, Devices & Systems*, vol. 9, no. 3, pp. 172-180, May 2015.

[6] X. Han, T. Burger and Q. Huang, "An output-capacitor-free adaptively biased LDO regulator with robust frequency compensation in 0.13μm CMOS for SoC application," *IEEE International Symposium on Circuits and Systems (ISCAS)*, Montreal, QC, pp. 2699-2702, May 2016.

[7] M. Huang, Y. Lu, S. P. U and R. P. Martins, "An output-capacitor-free analog-assisted digital low-dropout regulator with tri-loop control," *IEEE International Solid-State Circuits Conference (ISSCC)*, San Francisco, CA, pp. 342-343, Feb. 2017.

[8] A. Lahiri, S. Bansal, N. Bansal and M. S. Hashmi, "Digital LDO with analog-assisted dynamic reference correction for fast and accurate load regulation," *IEEE International Midwest Symposium on Circuits and Systems (MWSCAS)*, Boston, MA, pp. 767-770, Aug. 2017.

[9] Yasuyuki Okuma *et al.*, "0.5-V input digital LDO with 98.7% current efficiency and 2.7-μA quiescent current in 65nm CMOS," *IEEE Custom Integrated Circuits Conference*, San Jose, CA, pp. 1-4, Oct. 2010.

[10] T. Y. Man, P. K. T. Mok and M. Chan, "A High Slew-Rate Push–Pull Output Amplifier for Low-Quiescent Current Low-Dropout Regulators With Transient-Response Improvement," *IEEE Transactions on Circuits and Systems II: Express Briefs*, vol. 54, no. 9, pp. 755-759, Sept. 2007.

2018 IEEE Computer Society Annual Symposium on VLSI

An Asynchronous Analog to Digital Converter for Surveillance Camera Applications

Siddharth R. K.[1], Sunil R[2], Nithin Kumar Y. B.[3], Vasantha M. H.[4], Edoardo Bonizzoni[5]

National Institute of Technology Goa, India.[1,2,3,4]

University of Pavia, Pavia, Italy.[5]

Email: siddharth.kala03@gmail.com[1], sunil140790@gmail.com[2], nithin.shastri@nitgoa.ac.in[3],

vasanthmh@nitgoa.ac.in[4], edoardo.bonizzoni@unipv.it[5]

Abstract—This paper proposes an asynchronous analog to digital converter (ADC) for surveillance camera applications. The proposed architecture is based on non-uniform sampling, whose sampling instants depend on the amplitude of the input voltage. The proposed design has the power performance advantage for an input voltage close to the upper reference voltage. Thus, the proposed architecture is suitable for the applications in which the input signal rarely assumes voltage values closer to the lower reference voltage. The design is simulated, at the transistor level, in a 180-nm CMOS technology. The results show that about 96.7% of the power can be saved in the best case (input voltage in the vicinity of upper reference voltage) when compared to a conventional flash ADC.

Keywords—Analog to Digital Converter (ADC), Asynchronous Logic, Double-tail Comparator.

I. INTRODUCTION

Today, surveillance systems have become integral part of everyday life. They act as a deterrent for any potential incident in various fields like transportation, health care monitoring, agriculture etc. The typical block diagram of a surveillance system is shown in Fig. 1. Various sensors are used for detecting position, temperature, light, and many more. The performance of the sensors in wireless surveillance applications is mainly limited by the battery life. In such a system, one of the main power hungry blocks is the Analog to Digital Converters (ADC) [1]- [3]. The signal is sensed by the sensors along with the signal conditioning interface. Such a signal has very low activity in the night time as compared to the day time. Thus, if such an input signal is fed to a flash ADC, whose reference voltages are V_{ref-} and V_{ref+}, the input signal tends to be above V_z as shown in Fig. 2(a), which corresponds to a separation between a high and low activity region. Notice that a conventional flash ADC dissipates the power independently of the input signal level, as all the comparators are indiscriminately activated at the conversion instant, as conceptually illustrated in Fig. 2(a). Having a slowly varying input signal, like the temperature, and with a proper predictive logic, there is the possibility to disconnect a large number of voltage comparators from the power supply, thus drastically improving the power efficiency of the ADC. The concept is illustrated in Fig. 2(b). The number of voltage comparators that are turned on at each conversion depends on the input signal level, resulting in a variable conversion time. This leads to non-uniformly spaced samples in time, thereby making the ADC asynchronous in nature. Thus, the average sampling rate requirement will be lower than that of the conventional Nyquist rate ADC.

Asynchronous converters use the concept of non-uniform sampling, which can lead to interesting

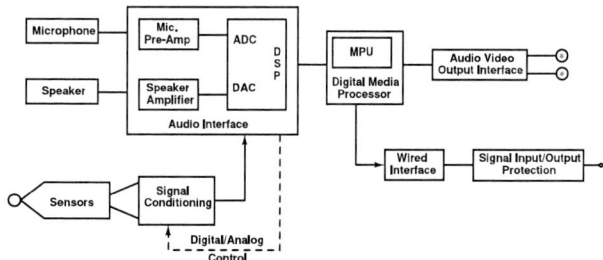

Fig. 1: A typical block diagram of a sensor interface network for surveillance applications.

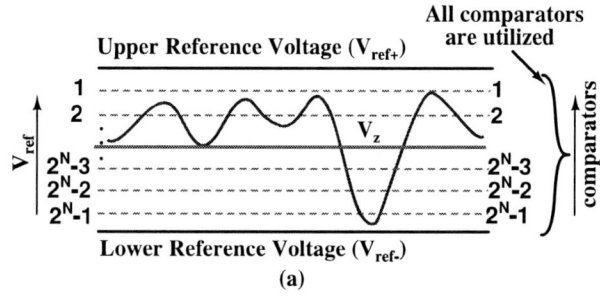

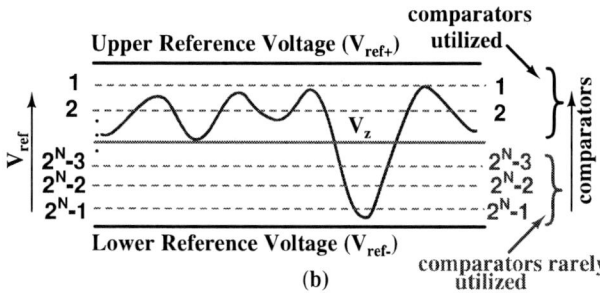

Fig. 2: Variation of an input signal in sensor interface for the wireless surveillance applications.

power saving schemes [5]. The ADC conversion time is mainly dominated by the signal tracking time, comparator speed and digital propagation delay [5]. In literature, one of the widely used techniques is the so-called level-crossing non-uniform sampling [6], [7]. The other main technique is based on the implementation of an Adaptive Resolution (AR) algorithm [8], which varies the quantizer resolution of the ADC along with the slope of the input signal and has overcome the trade-off between dynamic range and input bandwidth.

978-1-5386-7100-9/18 $31.00 © 2018 IEEE

164

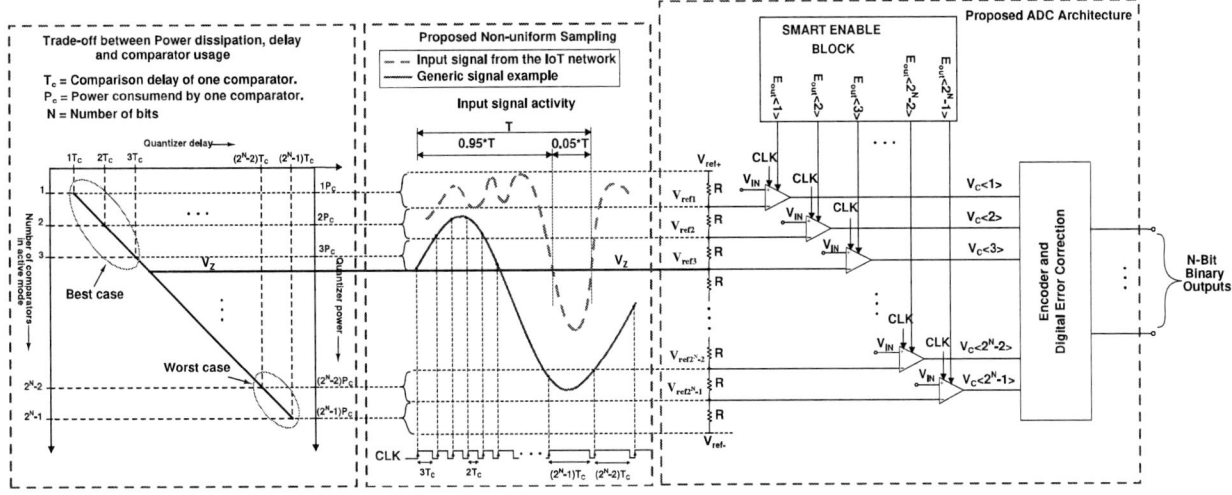

Fig. 3: System level implementation of the proposed A-ADC architecture.

The proposed asynchronous ADC (A-ADC) generates a variable clock signal, whose ON period duration varies with the amplitude of the input signal. Thus, in this approach, the power consumption and conversion time are dynamic in nature and inversely proportional to the amplitude of the input signal. The proposed architecture is designed and simulated using a 180-nm CMOS technology. As a case study, this paper shows the implementation of a 5-bit A-ADC, which consumes 0.548 mW (best case) when the input voltage is close to the upper reference voltage (V_{ref+}) and 4.455 mW (worst case) when the input voltage is close to lower reference voltage (V_{ref-}).

The paper is organized as follows. Section II explores the system level design aspects of the proposed A-ADC. The performance of the proposed architecture is demonstrated using a 5-bit A-ADC in Section III and simulation results are shown in Section IV. Finally, conclusions are drawn in Section V.

II. PROPOSED ASYNCHRONOUS ADC

The proposed non-uniform sampling scheme operation is based on the amplitude of the input signal and is different with respect to the conventional level-crossing approach. The system level implementation of the proposed A-ADC is depicted in Fig. 3. The input signal is sampled and is processed by the quantizer. The comparators in the quantizer block are enabled individually using a smart enable block. The conversion process starts by turning ON the comparator whose reference voltage is V_{ref1}. If the output of this comparator is at logic '0', then the subsequent comparator (that uses V_{ref2} as a reference) is turned ON. The process continues until any of the conversion results in the comparator output as logic '1'. At this point, the remaining comparators are disconnected from the power supply using the smart enable block and the digital output is given by an appropriate logic. Thus, the minimum pulse width of the clock (CLK) is achieved for an input voltage close to V_{ref+} and it is maximum for an input close to V_{ref-} as depicted in Fig. 3. The conversion time is, thus, a function of the input signal amplitude. The flowchart explaining the

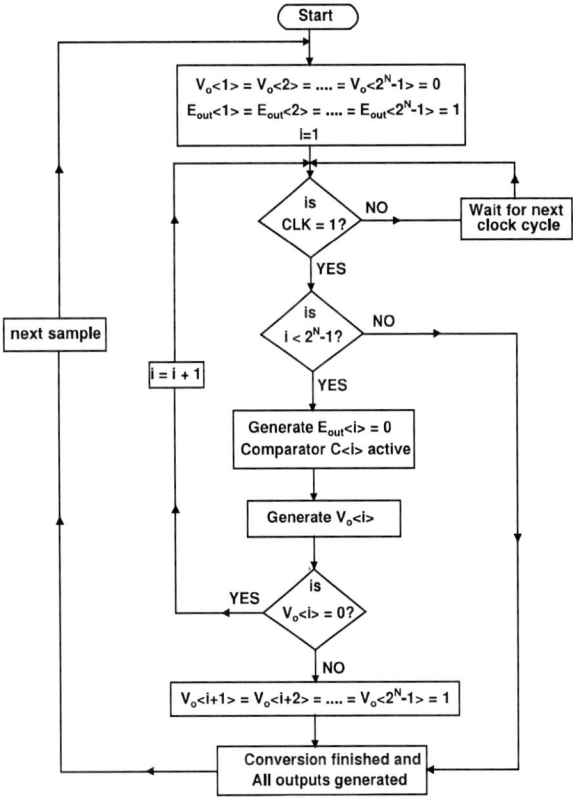

Fig. 4: Flowchart for the proposed design.

operation of the proposed design is shown in Fig. 4. $V_o<i>$'s are the output of the comparators properly processed by the digital logic block, $E_{out}<i>$'s are the enable signals and N represents the number of bits.

978-1-5386-7100-9/18 $31.00 © 2018 IEEE

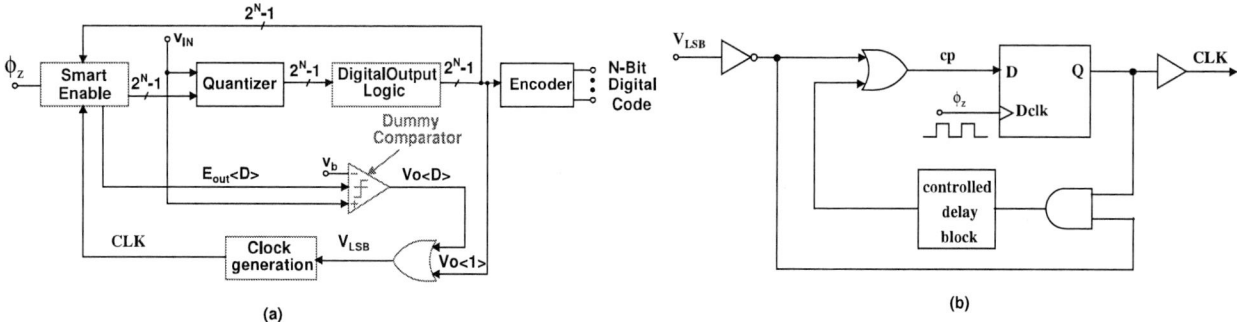

Fig. 5: (a) Block diagram of the proposed ADC, (b) proposed clock generator.

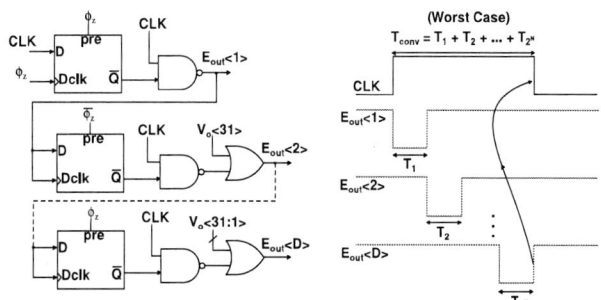

Fig. 6: Smart enable block along with the worst case timing diagram.

The proposed A-ADC design uses a tracking loop, controlled by the output of the ADC as shown in Fig. 5(a). A dummy comparator is used to take care of the 00000 code. The design comprises of the following sub-blocks: 1) clock generator, 2) quantizer, 3) digital output logic, 4) encoder, and 5) smart enable block.

A. Clock Generator

The clock generation block has a tracking loop, which depends on the signal V_{LSB} as shown in Fig. 5(b). Initially, the comparators in the quantizer are refreshed and $V_{LSB} = 0$, which gives cp=1. On the next rising edge of the D-flip flop clock signal (Dclk=ϕ_z), the system clock, CLK=1 and the conversion starts. The clock (CLK) is fed-back to the input of the block via delay block. When the conversion is completed and $V_{LSB} = 1$, cp becomes 0 and the clock is refreshed. The ON duration of the clock pulse (CLK) is equal to the conversion time of the A-ADC and can be controlled further using the delay block. It is important to note that the pulse ϕ_z is a square pulse, whose ON duration is equal to the one comparator delay.

B. Smart enable block

The enable generation block consists of D-flip flops with preset and digital gates, which are used to generate the enable signals, $E_{out}<i>$'s as shown in Fig. 6. The active low enable pulses are generated using D-flip flops and fed to the comparators. The enable signal of the previous comparator acts as an input clock of the D-flip flop for the following enable

signal. It is important to note that only one enable signal is generated at a time, thereby turning on only one comparator.

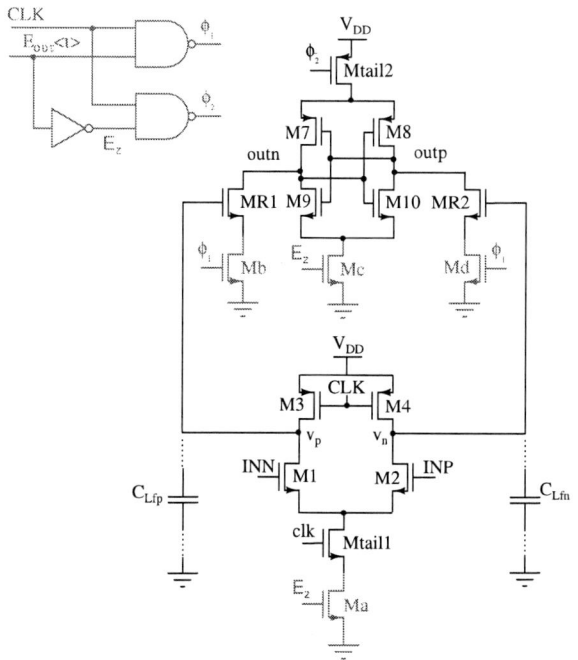

Fig. 7: Modified double tail comparator.

TABLE I: Modes Of Operation

clk	$E_{out}<i>$	Mode of operation	V_p/V_n	outp/outn
0	0	Don't Care	V_{DD}	0
0	1	Refresh Mode		
1	0	Active Mode	Discharge	Depends on i/p
1	1	Sleep Mode	V_{DD}	floating

C. Modified Double Tail Comparator

A double-tail dynamic latch comparator [9] is used in this design. It consists of a preamplifier and a latch. The comparator is modified as shown in Fig. 7 to meet the requirement of sleep mode when the clock is active. In this comparator, '$E_{out}<i>$' and 'CLK' are used as a control signals. Table I describes the different modes of operation of the modified comparator.

978-1-5386-7100-9/18 $31.00 © 2018 IEEE

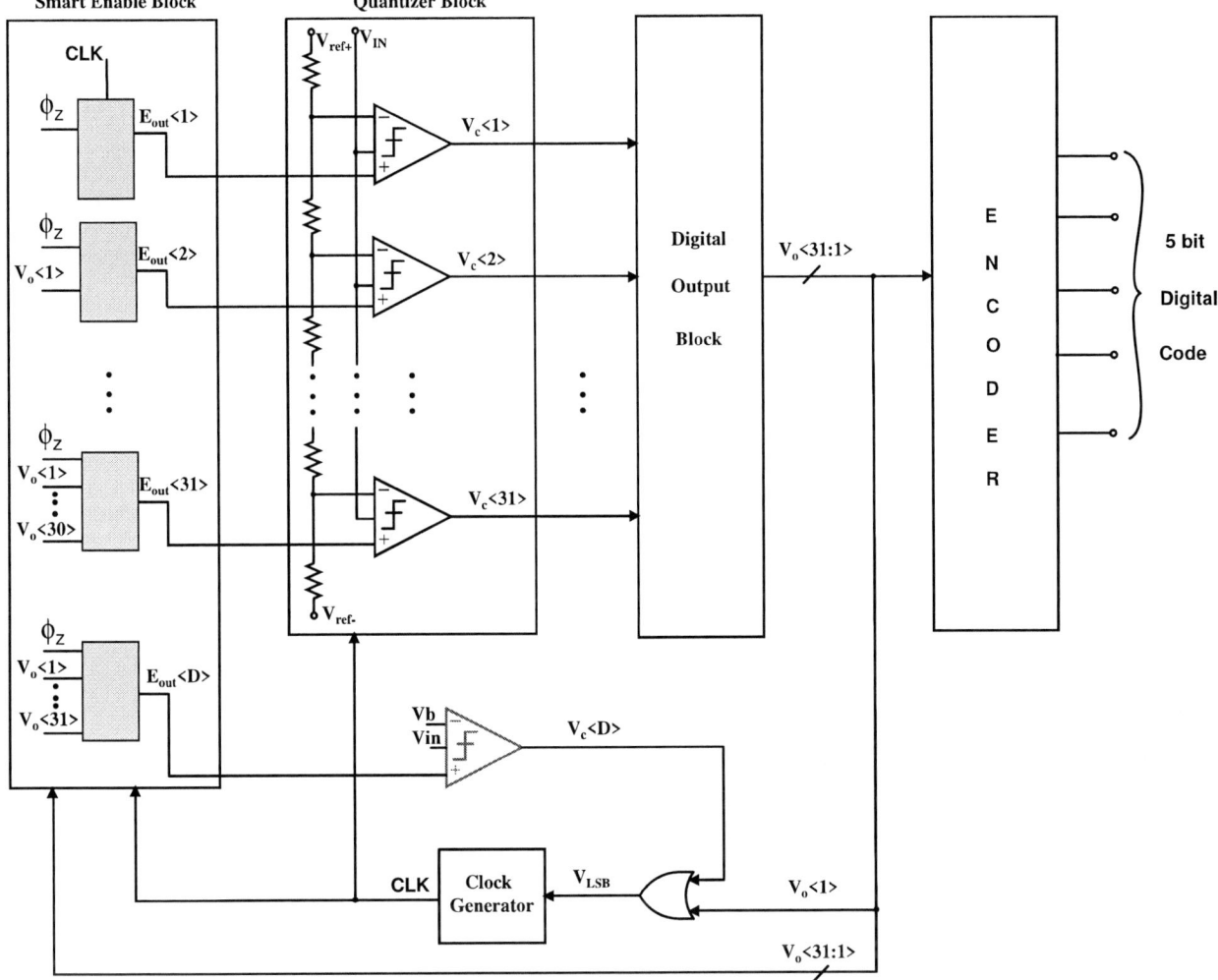

Fig. 8: Proposed 5 bit Asynchronous ADC.

D. Digital Output Logic Block

The digital output block consists of digital gates. If any of the above comparator output is logic '1', then the remaining comparators will be in sleep mode. In such cases, this digital block fulfills the requirement of generating the appropriate output.

III. CASE STUDY

The performance of the proposed architecture can be demonstrated by considering a 5-bit A-ADC as shown in Fig. 8. The A-ADC was designed with a supply voltage of 1.8 V, where the lower and upper reference voltages are V_{ref-} = 0.6 V and V_{ref+} = 1.1 V respectively. Since the clock is generated internally to the ADC, there is no need of an external clock. Thus, the only input for the A-ADC is the input signal. A dummy comparator is used to take care of the 00000 code. The reference voltage (V_b) for the dummy comparator is lower than V_{ref-}. The value of V_{LSB} ensures that the conversion is completed and the clock signal (CLK) can be regenerated to take the next sample.

IV. SIMULATION RESULTS

The proposed architecture was designed in a 180-nm CMOS technology with a supply voltage of 1.8-V. The proposed A-ADC is designed for 5-bit with worst case sampling frequency of 4.8 MS/s and best case sampling frequency of 36.5 MS/s as shown in Table II.

The power consumption is a function of the input signal amplitude. Thus, the simulated best and worst case quantizer powers are 0.153 mW and 4.06 mW respectively. The digital power is 0.395 mW, which is negligible compared to worst case total power. Fig. 9 shows the achieved transient simulation results and reports the reconstructed voltage together with the clock signal with a voltage ramp exploring the full range at the input. As expected, the ON duration of the clock pulse varies with the amplitude of the input signal, i.e. the frequency of the clock signal increases as the input voltage assumes values close to the upper reference voltage. Thus, for IoT environmental monitoring applications, where the input usually assumes values close to V_{ref+} and rarely assumes lower values, the proposed design takes less power than the

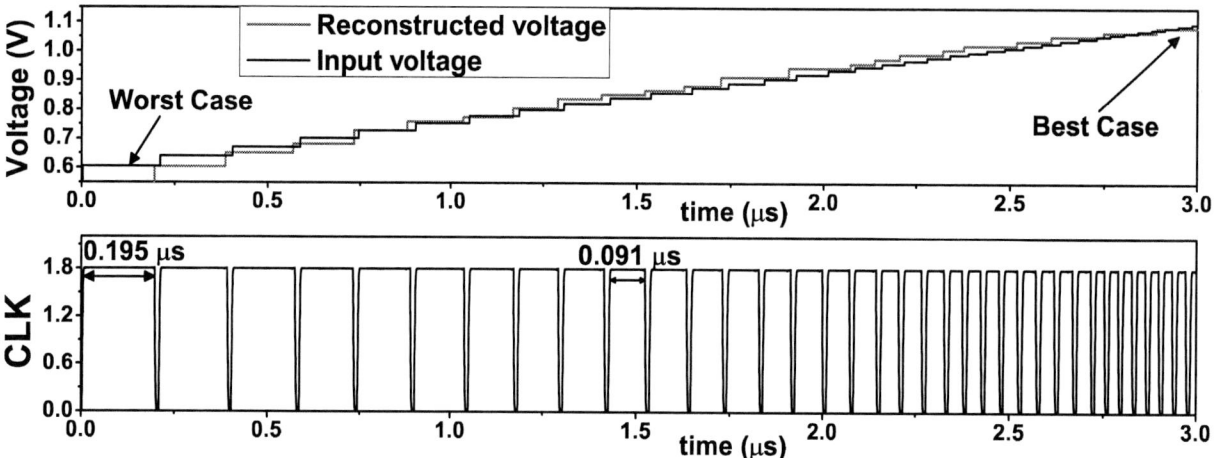

Fig. 9: Transient analysis showing the obtained clock signals for the A-ADC, the input signal and the reconstructed signal.

TABLE II: Performance Summary.

Parameters		[5]	[10]	This Work	
Technology		130 nm	600 nm	180 nm	
Supply Voltage		1.2 V	5 V	1.8 V	
Resolution		6-bit	6-bit	5-bit	
				Best Case	Worst Case
Comparator Usage efficiency (η_c)		NA	NA	3.125	100
Conversion time (T_c)		NA	NA	24.8 ns	195.2 ns
Sampling Frequency		600 MS/s	20 MS/s	36.5 MS/s	4.8 MS/s
Power	Digital	4. 1mW	NA	0.395 mW	
	Analog	1.2 mW	5.59 mW	0.153 mW	4.06 mW

conventional flash ADC. The performance of the design was evaluated by considering the comparator usage efficiency, the conversion time, and the power dissipation.

- Comparator Usage Efficiency (η_c): It is the ratio between the number of comparators enabled for a given input sample, N_U, and the total number of comparators, N_T. It is expressed as

$$\eta_c = \frac{N_U}{N_T} * 100, \quad (1)$$

- Conversion time (T_{conv}): It is the time taken by the A-ADC to convert the analog input signal to the N-bit binary code. In the proposed design, the conversion time is a function of the amplitude of the input signal and is given by,

$$T_{conv} = N_U * T_C, \quad (2)$$

where T_C = delay of one comparator.

- Power dissipation (P_D): The total power dissipated in the design is the sum of digital power and quantizer power.

$$\begin{aligned} P_D &= P_{qunatizer} + P_{digital} \\ &= N_U * P_C + P_{R-ladder} + P_{digital}, \end{aligned} \quad (3)$$

where P_C is the power dissipated in one comparator.

V. CONCLUSION

In this paper, an asynchronous ADC architecture is demonstrated by considering a case study of 5-bit A-ADC. The power analysis shows that the power dissipation decreases with the increase in the amplitude of the input voltage. About 96.7% of power can be saved in the best case as compared to the worst case. Thus, such architecture is useful for wireless surveillance applications, where the input usually assumes values close to the upper reference voltage and rarely assumes lower values. The proposed architecture eliminates the need for an external clock generation.

ACKNOWLEDGMENT

This publication is an outcome of the R & D work undertaken in the project under SMDP-C2SD, Department of Electronics and Information Technology, Ministry of Communication & IT, Government of India.

REFERENCES

[1] Z. Wang and M. C. F. Chang, "A 600-MSPS 8-bit CMOS ADC Using Distributed Track-and-Hold With Complementary Resistor/Capacitor

Averaging," in *IEEE Transactions on Circuits and Systems I: Regular Papers,* vol. 55, no. 11, pp. 3621-3627, Dec. 2008.

[2] Y. L. Wong, M. H. Cohen and P. A. Abshire, "A 750-MHz 6-b Adaptive Floating-Gate Quantizer in 0.35-m CMOS," in *IEEE Transactions on Circuits and Systems I: Regular Papers,* vol. 56, no. 7, pp. 1301-1312, July 2009.

[3] H. Yu and M. C. F. Chang, "A 1-V 1.25-GS/S 8-Bit Self-Calibrated Flash ADC in 90-nm Digital CMOS," in *IEEE Transactions on Circuits and Systems II: Express Briefs,* vol. 55, no. 7, pp. 668-672, July 2008.

[4] http://www.ti.com/lit/ds/symlink/lm35.pdf

[5] S. W. M. Chen and R. W. Brodersen, "A 6-bit 600-MS/s 5.3-mW Asynchronous ADC in 0.13-μm CMOS," in *IEEE Journal of Solid-State Circuits,* vol. 41, no. 12, pp. 2669-2680, Dec. 2006.

[6] J. Mark and T. Todd, "A Non-uniform Sampling Approach to Data Compression," in *IEEE Transactions on Communications,* vol. 29, no. 1, pp. 24-32, Jan 1981.

[7] N. Sayiner, H. V. Sorensen and T. R. Viswanathan, "A level-crossing sampling scheme for A/D conversion," in *IEEE Transactions on Circuits and Systems II: Analog and Digital Signal Processing,* vol. 43, no. 4, pp. 335-339, Apr 1996.

[8] M. Trakimas and S. R. Sonkusale, "An Adaptive Resolution Asynchronous ADC Architecture for Data Compression in Energy Constrained Sensing Applications," in *IEEE Transactions on Circuits and Systems I: Regular Papers,* vol. 58, no. 5, pp. 921-934, May 2011.

[9] S. Babayan-Mashhadi and R. Lotfi, "Analysis and Design of a Low-Voltage Low-Power Double-Tail Comparator," in *IEEE Transactions on Very Large Scale Integration (VLSI) Systems,* vol. 22, no. 2, pp. 343-352, Feb. 2014.

[10] T. Tulabandhula and Y. Mitikiri, "A 20MS/s 5.6 mW 6b Asynchronous ADC in 0.6m CMOS," in *22nd International Conference on VLSI Design,* New Delhi, pp. 111-116, 2009.

2018 IEEE Computer Society Annual Symposium on VLSI

An Integrated MaxFit Genetic Algorithm-SPICE Framework for 2-stage Op-amp Design Automation

Harsha M. V.
Department of Electronics & Communication
Engineering, University Visvesvaraya College of
Engineering, Bangalore University, Bangalore, India.
harshamv91@gmail.com

B. P. Harish, Senior Member, IEEE
Department of Electrical Engineering,
University Visvesvaraya College of Engineering,
Bangalore University, Bangalore, India.
dr.bp.harish@ieee.org

Abstract—**The Electronic Design Automation (EDA) tools have achieved high degree of maturity and reliability over the years for digital design. The design of analog circuits is a challenge attributed to the existence of multi-dimensional tradeoffs among multiple analog performance metrics like gain, bandwidth, power dissipation, supply voltage, input/output impedances, linearity, voltage swings and noise. The analog design proves to be a significant bottleneck in a System on Chip (SoC) implementation due to lack of automation techniques. To address this issue, the algorithms of the functioning of human brain or soft computing techniques can be gainfully deployed. This work proposes an integrated MaxFit Genetic Algorithm (GA) and GA-SPICE framework to achieve multi-objective optimization of analog design automation. The design of two-stage op-amp is demonstrated in this framework to optimize the objectives of open-loop DC gain, phase margin, unity gain-bandwidth, slew rate, power dissipation and area. The design is performed by proposed MaxFit GA programming of op-amp design equations in MATLAB environment and the design is transmitted seamlessly to LTspice to perform SPICE simulations for design verification. The dynamic fitness evaluation on SPICE generated performance metrics at each iteration of GA programming by transmitting them to MATLAB environment enhances the robustness of analog design significantly.**

Keywords-Integrated MaxFit GA-SPICE framework; Two stage op-amp; Multi-objective optimization

I. INTRODUCTION

With the advent of System on Chip (SoC) implementation, it has become imperative to integrate digital, analog and mixed signal circuits on the same chip and to optimize the interfaces among circuits. The Electronic Design Automation (EDA) tools are employed for the design of digital circuits and the entire design flow has been automated to optimize performance, power dissipation and area efficiency. These sophisticated tools have achieved high degree of maturity and reliability over the years. This is made possible as limited number of tradeoffs among speed, power dissipation and area need to be addressed in digital design.

In contrast, the design of analog circuits is several orders of magnitude more challenging than the design of digital circuits. This can be attributed to the existence of multi-dimensional tradeoffs among multiple analog performance metrics like gain, bandwidth, power dissipation, supply voltage, input/output impedances, linearity, voltage swings

and noise. Design automation can be achieved using traditional algorithms when limited number of tradeoffs exists as in the case of digital design. However, analog design is carried out manually for lack of automated tools as tool development itself poses significant challenges. Human brain, by its very structure and functioning, is very well suited to achieve multi-dimensional tradeoffs. However, because of speed constraints of human brain, achieving end-to-end design automation for analog circuits is time consuming and analog design has emerged as a significant bottleneck in the SoC design flow. To remove this bottleneck, there is a need to integrate high performance computers with human brain. In other words, the algorithms of the functioning of human brain or soft computing techniques can be gainfully deployed for analog design automation.

Ricardo Lourenco et al. have carried out a survey of technologies for analog design automation over the last 25 years [1]. They are classified into two main groups based on setup time, execution time and the accuracy in the evaluation of solution: Knowledge based automatic circuit sizing and Optimization based automatic circuit sizing.

The use of artificial intelligence with great computational power for automation of analog design is suggested by L.C. Severo et al. [2], and is a challenging task due to the presence of multiple design variables and large design space.

To achieve robust design, it is necessary to extend the single objective cost function to multi-objective optimization. Zebulum et al. have applied genetic algorithms for the design of CMOS operational amplifier [3].

Golmakani et al. have proposed a new Distributed Pareto-Based GA (DPGA) for design and optimization of analog integrated circuits with better convergence [4]. The algorithm is implemented in MATLAB and is simulated by HSpice in 0.18 μm process for telescopic cascode op-amp case study.

Mingguo Lin et al. have proposed a new Hybrid Genetic Algorithm with Hyper-Mutation and Elitist Strategy [HME-GA] to evolve circuit topology and component values [5]. The experimental results of low-pass, high-pass, band-pass and band-stop filters demonstrate the superior performance of HME-GA and GA with hyper mutation over traditional GA.

Amod P. Vaze has shown that GA can be used to achieve significant improvement in circuit topology with large population size at the cost of time [6].

978-1-5386-7100-9/18 $31.00 © 2018 IEEE   170

For the optimization of nonlinear circuits, Papadopoulos et al. have developed a hybrid algorithm comprising of GA and Least Square Gradient (LSQ) search [7]. This methodology applied to basic two-stage op-amp, produced superior performance than either GA or LSQ algorithm.

Wong et al. have reported a transistor sizing tool using an integrated GA programming in MATLAB and SPICE design methodology for single objective optimization [8]. The proposed work addresses this significant limitation and extends this to multi-objective optimization, with simple interfaces.

Dendouga et al. have proposed a methodology based on multi-objective genetic algorithms to find the optimal length and width of an operational amplifier, in 0.18 μm process, for analog and mixed CMOS based applications [9]. Direct current gain, unity gain bandwidth, phase margin, power consumption, area and slew rate are performance functions considered for optimization.

Santos-Tavares et al. have designed a methodology based on time-domain approach using GA kernel integrated with BSIM3v3 source code for optimization of transistor sizes and compensation of different topologies with an unlimited number of poles and zeros [10]. The design is verified on two-stage op-amp.

Goldberg has described the theory, operation, computer techniques, mathematical tools, search results and applications of GA in various fields of engineering where human brain like working is the key requirement [11]. A detailed explanation of soft computing techniques like GA, artificial intelligence, neural network, fuzzy logic in various engineering disciplines are given in [12].

The remainder of this paper is organized as follows: Section II presents the two-stage operational amplifier and equations for calculation of its performance metrics. Section III describes the proposed design methodology of integrated GA-SPICE framework for multi-objective optimization. Section IV presents the MaxFit GA with a flow chart. While Section V presents results and discussion, Section VI concludes the work.

II. TWO-STAGE OPERATIONAL AMPLIFIER

The circuit schematic of two-stage operational amplifier is shown in Fig. 1 and its performance metrics of gain, unity-gain-bandwidth, phase margin, slew rate, power dissipation and area are evaluated as under (1) to (6), referred from [13]:

A. Open-loop DC gain:

$$A_v = (g_{m1} / (g_{ds2} + g_{ds4})) (g_{m6} / (g_{ds7} + g_{ds6})) \quad (1)$$

where g_{m1} and g_{m6} are the transconductance of M1 and M6, g_{ds2}, g_{ds4}, g_{ds6}, and g_{ds7} are the output conductance of M2, M4, M6 and M7, respectively.

B. Unity-gain bandwidth:

$$GBW = g_{m1}/C_c \quad (2)$$

where Cc is the compensation capacitor.

C. Phase margin:

$$PM = \pm 180 - \tan^{-1}(GBW/p_1) - \tan^{-1}(GBW/p_2) - \tan^{-1}(GBW/z) \quad (3)$$

where p_1 and p_2 are poles and z is zero.

D. Slew rate:

$$SR = I_5 / C_c \quad (4)$$

where I_5 is the drain current through M5 and C_c is the frequency compensation capacitor.

E. Power dissipation:

$$P = (V_{DD} - V_{SS}) (I_5 + 2 I_7) \quad (5)$$

where V_{DD} and V_{SS} are supply voltages, I_7 is the drain current through M7.

F. Area:

$$Area = \Sigma(W_i * L_i) \quad (6)$$

where W_i and L_i are the width and length of the transistors, i ranges from 1 to n, n is the total number of transistors.

III. DESIGN METHODOLOGY AND OPTIMIZATION

The proposed design methodology focuses on obtaining the optimal transistor dimensions of two-stage op-amp, with GA based multi-objective optimization in MATLAB and design verification by SPICE simulations. The design is proposed in 0.18 μm CMOS process, for a given set of specifications. The design equations are nonlinear, and there exists no systematic analytical method to solve them. Further, the solution is not unique. Therefore, the best way is multi-objective optimization to obtain an optimal result by automation.

The design objectives of gain, phase margin, unity-gain bandwidth and slew rate are to be maximized and while

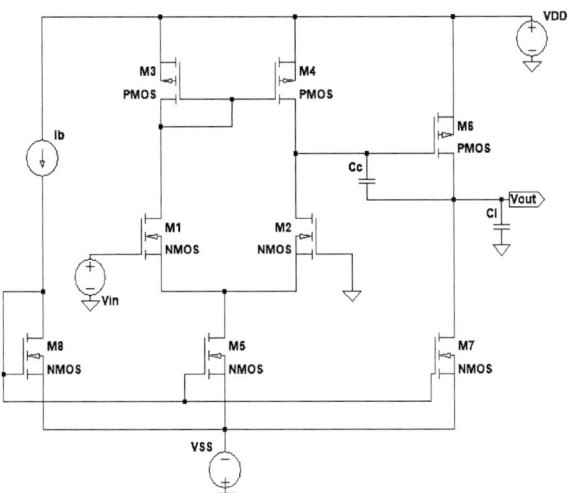

Figure 1. Schematic of two-stage operational amplifier.

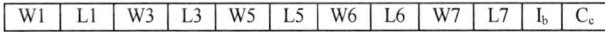

W1	L1	W3	L3	W5	L5	W6	L6	W7	L7	I_b	C_c

Figure 2. Chromosome construct for MaxFit GA programming.

power and area are to be minimized. The design exercise involves computing the optimal solution vector comprising of width and length of devices M1 to M8 of op-amp, bias current I_b and compensation capacitor C_c. Let this solution vector be represented as a chromosome, as shown in Fig. 2, for the proposed MaxFit GA programming. The individual W, L, I_b and C_c values represent the genes of the corresponding chromosome.

The matched devices of basic differential amplifier configuration of the two-stage op-amp are taken as M1 - M2, M3 - M4 and current mirror are M5 - M8 and hence their respective dimensions are taken to be equal.

The set of initial circuit design values i.e., device dimensions are randomly generated, within a specified range mandated by area specification, by the proposed MaxFit GA programming. Then a SPICE simulation is performed on the GA generated design to obtain the corresponding performance metrics. Then the MaxFit GA treat these performance metrics as specifications for the second iteration of the design to be verified by SPICE simulations. Thus, the process of optimization by GA and verification by SPICE is iteratively performed in a loop for the specified number of generations, as shown in Fig. 3, to generate the final optimized design solution. Simple interfaces for data transfer between MATLAB & SPICE platforms are designed with no temporary files. To begin with, circuit netlist file is created inside .m file of MATLAB and then DOS batch file is called from MATLAB to launch the SPICE and run the circuit netlist file in SPICE environment. Then SPICE generated output files are parsed using MATLAB functions to evaluate performance metrics after first iteration. These metrics in turn become the specifications for the next iteration of design by GA programming. Thus, the bidirectional data transfer interface between MATLAB & SPICE platforms is implemented to build an integrated MaxFit GA-SPICE framework. This novel interface without requiring any access to the source code of either SPICE or GA on MATLAB for an integrated GA-SPICE framework scores

over the one presented in [10].

IV. MAXFIT GENETIC ALGORITHM

The proposed MaxFit genetic algorithm starts with a randomly generated initial population pool of 50 chromosomes, each of which is a possible solution, within a pre-specified bound of design space defined by L, W, I_b and C_c values. While the lower bound of L is defined by the process node selected i.e., $L_{min} = 0.18$ μm, its upper bound is taken as $L_{max} = 2$ μm. The bounds W_{min} and W_{max} of each transistor is varied by keeping aspect ratio of first stage transistors, $(W/L)_1 = (W/L)_2 = 6$, $(W/L)_3 = (W/L)_4 = 14$, $(W/L)_5 = (W/L)_8 = 12$, and second stage transistor $(W/L)_6 = 175$ and $(W/L)_7 = 75$, while the lower bound W_{min} of all transistors are kept greater than L_{max} so as to maximize g_m, so that the overall area of design is better than reported in work [9]. The bottom plate and sidewall capacitances of source-drain regions are calculated with $a_d = a_s = 2*W*L_{min}$ and $p_d = p_s = 2(W+2L_{min})$, to account for layout parasitics. This selection provides flexibility to the design

Chromosomes are coded using real valued numbers. The optimality of each solution is measured by its fitness value. To measure the fitness, each chromosome is transmitted seamlessly into LTspice netlist and simulated. The performance metrics are extracted from the SPICE waveform data file and log file and a weighted approach is used to compute the fitness value of chromosomes as illustrated in [13]. Thus, fitness evaluation on *SPICE generated actual performance metrics* is carried out against desired metrics *at each iteration* of GA programming by transmitting them to MATLAB environment. This rigorous fitness evaluation, facilitated by the proposed integrated GA-SPICE framework, enhances the quality of design optimality.

After calculating the fitness of each chromosome from the initial solution, GA operators of rank selection, two-point crossover and mutation are applied in succession. Rank selection selects the top half of the population with high fitness value and discards the weak fitted chromosomes. Successive chromosome set of high fitness value form a set of parents for succeeding operations. Two-point crossover is applied on this set of parents – Parent-1 and Parent-2 to generate Child-1 and Child-2 chromosomes, as shown in Fig. 4, and the process continues. A mutation value of ±5 is applied for all the genes of every child chromosomes generated from two-point crossover operation. A low value of mutation is selected for fine granularity to represent small perturbations in design.

An enlarged pool of 175 chromosomes is created by

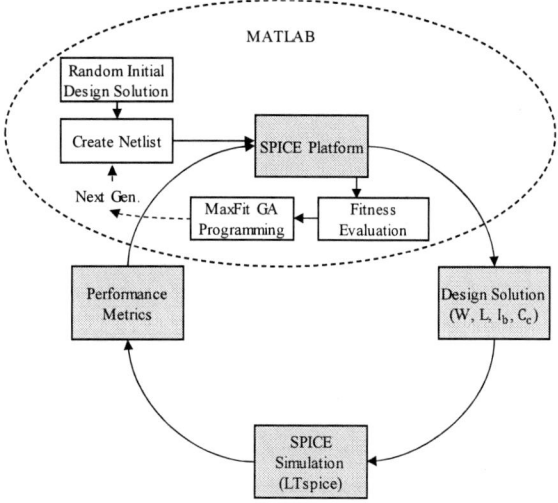

Figure 3. Integrated MaxFit GA-SPICE framework for analog design automation.

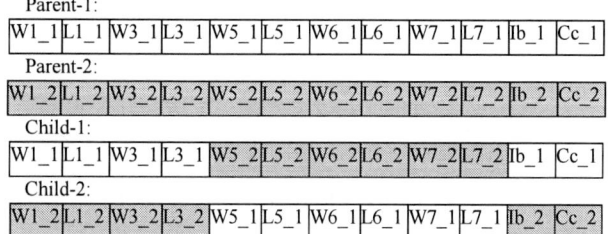

Figure 4. Two-point crossover.

drawing top half of best-fit 25 chromosomes after rank selection, 50 chromosomes generated after crossover and 100 chromosomes generated after mutation.

The population pool of next generation is filled by drawing the top 50 best-fit chromosomes of enlarged pool of 175 chromosomes of previous generation. The loop continues until stopping criterion is achieved. The stopping criterion is defined as the pre-specified number of generations and the value is selected as 50, as a tradeoff between computation time and best possible solution.

At the end of each generation, best-fit chromosome is sent to another pool comprising of only best-fit chromosomes of each generations i.e., the top most best-fit chromosome in each of 50 generations is added to create a new population pool of 50 chromosomes and then overall fitness is calculated for these 50 chromosomes. Then the top

TABLE I. OPTIMAL DESIGN SOLUTION - TRANSISTOR DIMENSIONS

Parameter (µm)	Lower Bound	Upper Bound	GA-SPICE Design
W1	1	12	10.5
L1	0.18	2	1.29
W2	1	12	10.5
L2	0.18	2	1.29
W3	2.52	28	25.4
L3	0.18	2	0.93
W4	2.52	28	25.4
L4	0.18	2	0.93
W5	2.16	24	20.2
L5	0.18	2	1.56
W6	32	360	67.9
L6	0.18	2	0.18
W7	13.5	150	65.9
L7	0.18	2	1.97
W8	2.16	24	20.2
L8	0.18	2	1.56
I_b (µA)	15	30	21.9
C_c (pF)	$0.22*C_1$	3	2.37

most best-fit chromosome is the final optimal design solution. This method of 2-layered fitness evaluation enhances the optimality of the design.

Thus, the MaxFit GA maximizes the fitness of the design solution over an accumulated population across selection, crossover and mutation operations. This ensures that the de-rating of fitness do not occur at any stage of application of GA operations. Further, the fitness is computed on the SPICE generated metrics at each iteration resulting in SPICE standard accuracy. The flow chart of the MaxFit GA design optimization is shown in Fig. 5.

V. RESULTS AND DISCUSSION

The lower and upper bounds of device dimensions considered and the final design solution generated after 50 generations by the integrated GA-SPICE based framework are tabulated in Table I. The computation time for design is 15392 s on Intel i5 2.33 GHz processor based system, which

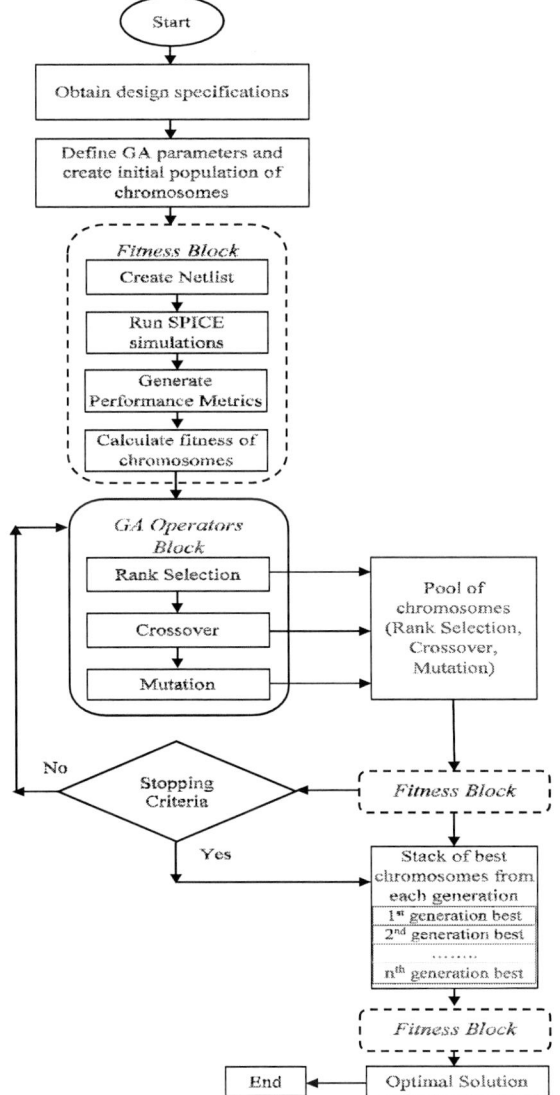

Figure 5. Flow chart for proposed integrated MaxFit GA-SPICE based framework for 2-stage operational amplifier design.

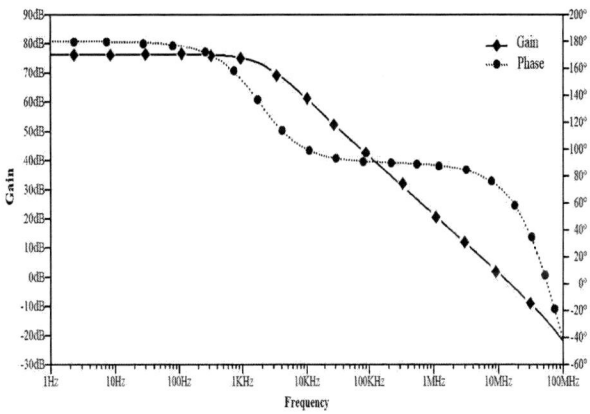

Figure 6. Frequency response of 2-stage op-amp from MaxFit GA-SPICE framework.

TABLE II. SIMULATION RESULTS OF INTEGRATED MAXFIT GA-SPICE FRAMEWORK

Performance Metrics	Specification	GA-SPICE Design Result	[9]
Open-loop DC Gain (dB)	70	76.31	76
Unity-gain bandwidth (MHz)	1	11.65	1.5
Phase margin (°)	≥60	68.78	70
Slew rate (V/μs)	1.5	9.83	2.25
Power dissipation (mW)	Min	0.166	0.047
Area (μm²)	Min	262	559

is reasonable.

Performance metrics of 2-stage op-amp obtained from the proposed design are tabulated in Table II and is compared, on a fair basis, with the simulation results as reported in work [9]. Unity gain bandwidth is an order of magnitude better for the same gain and acceptable phase margin. Further, slew rate and area are found to be significantly better with a penalty in power dissipation. Gain and Phase angle of the final best chromosome is as shown in Fig. 6. Due to a seamless movement of design values and performance metrics between GA and SPICE and the MaxFit algorithm in the MaxFit GA-SPICE framework facilitates fitness evaluation on accurate SPICE generated performance metrics at every iteration without any de-rating due to any GA operations. This results in significant improvement in robustness of optimization process.

Though the design in this work is optimized for wideband applications, it can easily be applied for low power applications by tight selection of power specification range in MaxFit GA, subject to tradeoffs.

VI. CONCLUSIONS

An integrated MaxFit Genetic Algorithm (GA)-SPICE framework, with simple and transparent interfaces, to achieve multi-objective optimization for analog design automation is presented. The design of two-stage op-amp is demonstrated in the proposed framework using MaxFit GA to optimize multi-objectives. The design is performed by MaxFit GA programming in MATLAB and seamlessly verified with SPICE simulations at every iteration. The fitness evaluation on accurate SPICE generated performance metrics at every iteration results in significant improvement in accuracy and robustness of the optimization process without any de-rating at any stage due to any GA operations. Thus, the integrated MaxFit GA-SPICE framework represents an accurate, scalable and robust method which can be extended to optimize analog circuits of any size and complexity.

REFERENCES

[1] R. Lourenco, N. Lourenco, and N. Horta, "Chapter 2: Previous works on automated analog IC sizing" in AIDA-CMK: Multi-algorithm optimization kernel applied to analog IC sizing, Springer, 2015.

[2] L. C. Severo and A. Girardi, "Automatic synthesis of analog integrated circuits using genetic algorithms and electrical simulations," in Proceedings of XXIV SIM – South Symposium on Microelectronics, May 2009, pp. 41-44.

[3] R. S. Zebulum, M. A. Pacheco, and M. Vellasco, "Synthesis of CMOS operational amplifiers through genetic algorithms," in Proceedings of XI Brazilian Symposium on Integrated Circuit Design, October 1998, pp. 125-128.

[4] A. Golmakani, K. Mafinejad, and A. Kouzani, "A new method for optimization of analog integrated circuits using pareto-based multi-objective genetic algorithm," International Review on Modelling and Simulations, vol. 2, no. 3, June 2009, pp. 297-303.

[5] M. Lin and J. He, "Automated analog circuit design synthesis using a hybrid genetic algorithm with hyper-mutation and elitist strategy," International Journal of Information Technology and Computer Science, vol. 1, no. 1, October 2009, pp. 23-32.

[6] A. P. Vaze, "Analog circuit design using genetic algorithm: Modified," International Journal of Electrical, Computer, Energetic, Electronic and Communication Engineering, vol. 2, no. 2, 2008, pp. 301-303.

[7] S. Papadopoulos, R. J. Mack, and R. E. Massara, "A hybrid genetic algorithm method for optimizing analog circuits," in Proceedings of 43rd IEEE Midwest Symposium on Circuits and Systems, USA, August 2000, pp. 140-143.

[8] Y. C. Wong, Syafeeza A. R., and N. A. Hamid, "A transistor sizing tool for optimization of analog CMOS circuits: TSOp," International Journal of Engineering and Technology, vol. 7, no. 1, Feb-Mar 2015, pp. 140-146.

[9] A. Dendouga, S. Oussalah, D. Thienpont, and A. Lounis, "Multiobjective genetic algorithms program for the optimization of an OTA for front-end electronics," Advances in Electrical Engineering, vol. 2014, Article ID 374741, Hindawi Publishing, August 2014, pp. 1-5.

[10] R. Santos-Tavares, N. Paulino, J. Goes, and J. P. Oliveira, "Optimum sizing and compensation of two-stage CMOS amplifiers based on a time-domain approach," 13th IEEE International Conference on Electronics, Circuits and Systems, Nice, France, December 2006, pp. 533-536.

[11] David E. Goldberg, Genetic Algorithms in Search, Optimization and Machine Learning, Pearson Education, India, 2004.

[12] S. N. Sivanamdam and S. N. Deepa, Principles of Soft Computing, 2nd edition, Wiley India, 2014.

[13] Phillip E. Allen and Douglas R. Holberg, CMOS Analog Circuit Design, 2nd edition, Oxford University Press, 2002.

[14] J. Yu and Z. Mao, "A design method in CMOS operational amplifier optimization based on adaptive genetic algorithm," World Scientific and Engineering Academy and Society, Transactions on Circuits and Systems, Issue 7, vol. 8, July 2009, pp. 548-558.

978-1-5386-7100-9/18 $31.00 © 2018 IEEE

Mismatch Resilient 3.5-bit MDAC with MCS-CFCS

Satyajit Mohapatra[1], Hari Shanker Gupta[2], Nihar Ranjan Mohapatra[1]

1, Dept. of Electrical Engineering Indian Institute of Technology, Gandhinagar-382355, India

2, Space Applications Centre, Jodhpur Tekra, Ahmedabad-380015, India

satyajit_mohapatra@iitgn.ac.in, nihar@iitgn.ac.in, hari@sac.isro.gov.in

Abstract— **Modern data converters architectures like pipeline ADC and Current steering DAC depend on component matching to achieve desired resolution. They consist of large arrays of capacitors/current sources that are highly susceptible to systematic effects arising from process variations and temperature gradients. On top of it, 3D integration of mixed signal ICs adjacent to digital chips makes the arrays are prone to local gradients and hot spots. This makes incorporation of certain error compensation schemes at a layout level a priority. In this work, we have proposed an arraying technique that simultaneously compensates for systematic effects and perform significantly better in the presence of rotated parabolic gradients and localized hotspots. We further propose a modified switching scheme that enables realization of 3.5-bit CFCS decoder circuit with minimum logic gates and delay. The performance of the proposed array integrated with the modified switching scheme is verified on the model of a 16-bit 10Msps pipelined ADC using Matlab. The simulation results show significant improvement of linearity (~6-12 dB) over the existing techniques. The various design challenges and strategies to overcome them are discussed in detail.**

Keywords—Pipelined ADC, Commutated Feedback Capacitor Switching (CFCS), Merged Capacitor Switching (MCS), Capacitor Mismatch, Oxide Gradient, CMOS process, Hotspot

I. INTRODUCTION

The state of the art data converters for mobile communication and space applications necessitate high bandwidth and linearity at lower power. The pipelined architecture is an attractive option for implementing high speed mobile base station receivers, 802.11 WLAN/802.16 WMAN receiver, video rate converters, HDTV and ROICs [1-4]. It offers good tradeoff between throughput rate, resolution, power consumption and die area. Resolving higher number of bits in the first stage relaxes the matching requirements of later stages. It further makes the stage suitable for various multi-bit techniques like Merged Capacitor Switching (MCS) and Commutated Feedback Capacitor Switching (CFCS) [4]. The 3D integration of data converters in mixed signal ICs benefits from both high device density and speed. The focal plane imagers for space applications implemented in 3D IC technology allow the read out integrated circuits (ROICs) to be placed in close proximity to photo-detectors and processor [5]. The high-performance data converters are shown to be placed adjacent to high performance digital chips in a recent work [6] by Xilinx using stacked silicon interconnect technology (SSIT). But with the integration of analog and digital dies adjacent to each other, hotspots are no longer limited to digital chips.

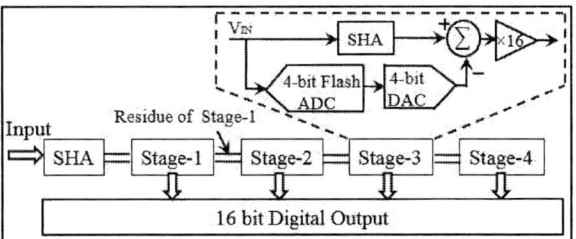

Figure 1. Interstage residue linearity requirements of a pipelined ADC.

Pipeline ADC consists of multiple stages operated at orthogonal clock in order to achieve high speed. Each stage typically consists of S/H amplifier, Sub-ADC and a Sub-DAC as shown in Fig.1. The accuracy of Sub-DAC Stage-1 determines the overall resolution of the ADC. In the case of a 16 bit converter resolving 4 bits in its first stage, the residue transferred by stage-1 must be >12 bit accurate to be processed by the later stages without any missing codes. Traditionally, the intermediate S/H Amplifier, the inter-stage gain amplifier, the stage Sub-DAC and the subtarctor are implemented by single switch capacitor amplifier circuit popularly known as the Multiplying DAC (MDAC). Such implementation require high matching between a set of capacitors. Therefore costly Metal-Insulator-Metal (MiM) capacitors implemented with special metal layer called "capacitor top metal" are preferred for fabrication accuracy in modern processes. Even such options are limited to 0.1% accuracy i.e.~10 bit [3,9]. Moreover with the 3D integration of ICs the sensitive analog arrays are susceptible to local gradients and hot spots of adjacent digital dies. The thermal conductivity of dielectric layers inserted between device layers is very low compared to silicon and metal. This results in temperature gradients in vertical direction during TSV (through silicon via) integration [6-8]. This makes the incorporation of systematic error compensating schemes at the layout level highly necessary in modern processes [10].

The paper is organized in four sections. The proposed arraying technique to compensate for systematic errors, edge effects and local hotspots encountered during 3D integration of ICs is presented in Section-II. The immunity of array to process gradients is verified by numerical simulations in Matlab and compared against the existing arrays. A modified MCS-CFCS switching scheme for 3.5bit MDAC is proposed in Section-III. This helps in circuit realization with minimum delay and complexity. The performance of proposed array integrated with the modified switching scheme is verified on the model of a 16-bit 10-Msps pipelined ADC using Matlab in section-IV followed by concluding remarks in Section-V.

II. MDAC CAPACITOR ARRAY DESIGN

The conventional implementation N-bit MDAC requires the total sampling capacitor of a stage to be split into 2^N elements. The implementation with MCS reduces the number by half (2^{N-1}). The MCS implementation of 3.5-bit MDAC therefore requires 8 matched capacitors switched in an unary fashion [4]. Each capacitor is further split into 8 sub-elements and intermingled over an 8×8 array to compensate for systematic effects arising from process and temperature variations. The spatial optimization of these sub elements is itself a mathematical challenge, and its complexity increases exponentially with number of elements.

A. Modelling of Systematic Mismatch

The gradient error distribution across a unary matrix can be approximated by a Taylor series expansion around the center of the array [11]. Error of the element located at coordinate (x, y) can be expressed as (1). It is generally assumed that the linear and the quadratic terms are adequate to model gradient effects.

$$\alpha_{00} + \alpha_{10}.x + \alpha_{01}.y + \alpha_{20}.x^2 + \alpha_{02}.y^2 + \alpha_{11}.xy + \dots \quad (1)$$

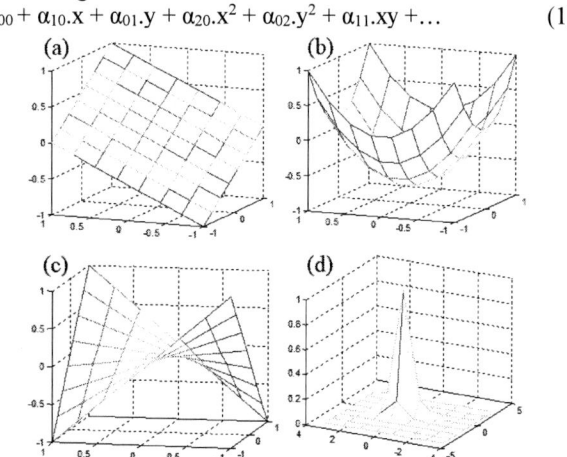

Figure 2. Profile: a)Linear (b)Parabolic (c)Rotated Parabolic (d) Hotspot

The oxide thickness variation over the wafer, dopant profile and the voltage drop along the power supply lines are reported to cause linear gradient errors while quadratic effects may be attributed to temperature gradients and die stress [11]. The rotation of gradient axes with respect to the array results in rotated parabolic components [15]. The error distribution is typically a superposition of linear and quadratic components in ratio of w:1, where parameter w is experimentally quantified for the technology used. The geometric position of the matrix is normalized so that all the elements are spatially distributed in the interval [-1, 1] along both axes. The gradient errors are also normalized so that the maximum error magnitude is equal to 1. The comparison between different switching sequences thus could be made on a normalized gradient error array. In this paper, the x or y terms are referred as linear, x^2 or y^2 terms are referred as parabolic and xy term as rotated parabolic errors (Figs. 2a-2c). The hotspot errors at a location can be modelled with a Gaussian distribution with mean μ and spread σ (Fig.2d)[12].

B. Existing Arraying Techniques

The popular layout techniques used to improve matching in presence of systematic errors are diagonal placement [13], common centroid placement [14] and restructured diagonal [15]. In [13], the units are combined along diagonals of the matrix. This is a simple interconnect method and requires the fewest number of metal layers because each combination of elements has at least two members at an edge and all interior cells are adjacent to another member of their group. In [14] the units are combined around a common centroid fashion where the average distance from the center of the matrix is the same for all 8 combinations. This is a more complex interconnect method and requires a larger number of metal layers. Many of the combinations are land locked with no members on an edge of the matrix. Variations of [13] have been proposed in [15] in in an attempt to compensate for rotated parabolic effects. The "xy" terms in proposed layout do not cause any parabolic error but an error that alternates between positive and negative values. Although several techniques to compensate for the systematic effects exist, they are not suitable for 3.5-bit MDAC implementation in MCS-CFCS architecture and hence are not discussed here.

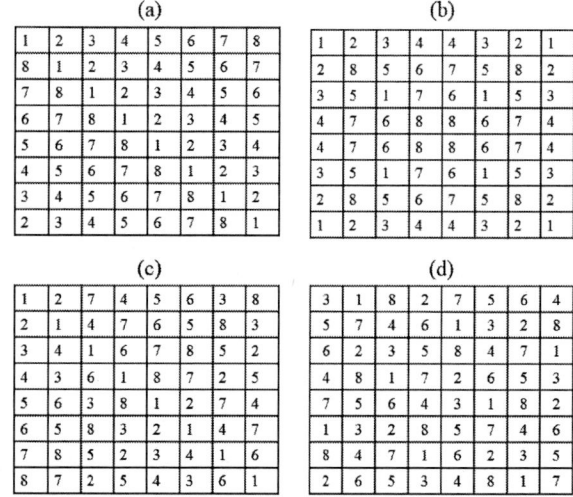

Figure 3. Arrays: a) work[13] (b) work[14] (c) work[15] (d) Proposed

The existing techniques are unable to compensate for linear and quadratic errors. They typically show poor performance to rotated parabolic gradients. Unlike linear and parabolic gradients, the saddle shaped profile of the rotated components makes it tricky to handle them along with other gradients while designing an array. They typically result in "S-shaped" INL and generate 3rd harmonic distortion at the output [15]. Further none of the existing arraying techniques compensate for localized gradients and hotspots that arise during 3D integration of analog and digital chips [6]. Designing for multiple gradient profiles is quite challenging as it leads to multi-dimensional optimization of elements spatially. We propose an array (Fig. 3d) that simultaneously compensates for all linear, quadratic errors and perform significantly better in presence of rotated parabolic gradients. It additionally is tolerant to local hotspots and edge effects.

C. Proposed Arraying Technique

The various properties of the array which makes it tolerant to systematic effects are discussed in details. Some these properties are illustrated in Figure 4.

1. Each row/column of the array contains sub elements of all the 8 capacitors. This makes the array immune to linear gradients in both X and Y directions. This property also provides for matched parasitic routing necessary for MDAC.
2. The sub-elements of all 8 capacitors are equally present at all the edges of the array which makes it edge effect tolerant. This results in considerable die area savings in terms of dummy ring placement and makes the array more compact.
3. No two-sub element of a capacitor are adjacent to each other. This avoids spatial clustering of elements and achieve higher dispersion to combat for global higher order gradients.
4. Any sub-element is surrounded by sub-elements of remaining 7 capacitors other than itself. Thus, any local variation to an element is shared by all its surrounding neighboring elements. This consequently decreases the correlation distances between elements, hence improving resiliency of array to localized gradients and hotspot effects.
5. Any 4x2 sub-quadrant also contain sub-elements of all 8 capacitors. This makes the array tolerant to spatially shifting local gradients and migration of hotspot in modern chips [7].
6. For large sized capacitors, it might be desirable to split into higher number of subelements. The array dimension can be extended by replicating and repeating the same pattern. Therefore, sub elements at the opposite edges of the array are considered as neighbors to a sub element at any given edge.
7. Sub elements of all the 8 capacitors are uniformly distributed all over the array to achieve high dispersion factor. Such homogeneous distribution is beneficial while compensating for the higher order components such as the cubic and rotated cubic gradients encountered on larger arrays.
8. Since all the elements are accessible from any side of the array, it is convenient to implement calibration options [15].

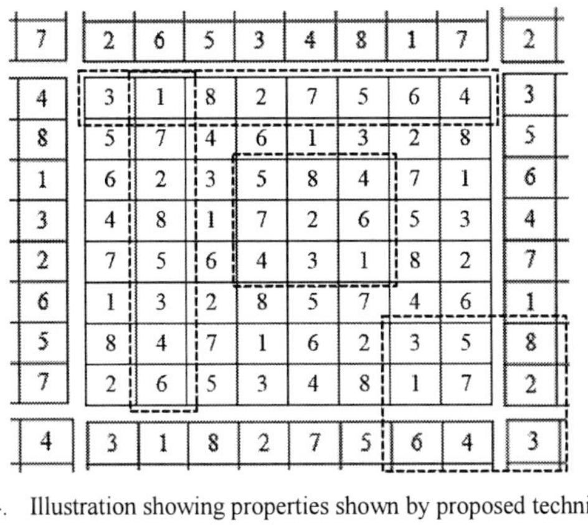

Figure 4. Illustration showing properties shown by proposed technique.

The high dispersion achieved by the array can be observed on a match matrix (Fig. 5). The element λ_{ij} at location (i, j) on matrix represent the number of time element 'i' occurs in the vicinity of element 'j' over the entire array. Each sub-element has 8 neighbors on the matrix, hence an

element has total of 8×8=64 neighbors. Considering no self-adjacency 8×7 sub elements of remaining 7 capacitors need to occupy these 64 locations. This can be best achieved by 8×7+8. This means one of the remaining capacitor must occur 16 times in neighborhood for better dispersion. Therefore, each element occurs exactly 8 times around any other element except the one best matched as shown in Fig.4.

	1	2	3	4	5	6	7	8
1	0	8	8	8	16	8	8	8
2	8	0	8	8	8	16	8	8
3	8	8	0	8	8	8	16	8
4	8	8	8	0	8	8	8	16
5	16	8	8	8	0	8	8	8
6	8	16	8	8	8	0	8	8
7	8	8	16	8	8	8	0	8
8	8	8	8	16	8	8	8	0

Figure 5. Match matrix showing high dispersion of elements over array

The performance of the array is verified using Matlab and compared against the existing techniques [13-15]. The comparison is shown in Table-I. The capacitors with mismatches are represented as $C_i=C(1+\varepsilon_i)$, where C is the mean of all capacitors and ε_i is the mismatch of ith capacitor. $\varepsilon_{max}=max(\varepsilon_i)$, is used to quantify performance of an array to the given error profile. The performance of various techniques [13-15] are subjected to individual gradients (linear, parabolic and rotated parabolic and hotspot) as well as joint gradient profile (1) with coefficients [α_{00}, α_{10}, α_{01}, α_{20}, α_{02}, α_{11}] = [-0.2, 0.25, 0.25, 0.23, 0.23, 0.23] and w=1. Though center of array is arbitrarily chosen here for presenting the hotspot, it has been verified through numerical simulation that the proposed array shows improved performance when hotspots occur anywhere on the array.

TABLE I. PERFORMANCE OF ARRAY FOR VARIOUS ERROR PROFILES

Gradient:	x	y	x^2	y^2	xy	w=1	hotspot
proposed	0	0	0	0	0.10	0.02	0.06
Work[13]	0	0	0	0	0.43	0.10	0.19
Work[14]	0	0	0.6	0.6	0	0.15	0.44
Work[15]	0	0	0	0	0.43	0.10	0.19

As observed from Table-I, the proposed arraying technique simultaneously compensates for all linear, quadratic errors and perform significantly better in presence of rotated parabolic gradients and local hotspot. The array is also tolerant to edge effects. It can also be shown that the proposed technique potentially compensates for all higher order errors such as the cubic and rotated cubic gradients.

III. 3.5BIT MCS CFCS DECODER IMPLEMENTATION

The MCS implementation of 3.5-bit MDAC consists of 8 sampling capacitors (C_1 to C_8). During Phase-1 (sampling phase), the input signal is sampled on all 8 capacitors. During Phase-2 (residue generation), capacitor C_8 is flipped to feedback mode while references $+V_{REF}$, $-V_{REF}$ or GND is connected to capacitors C_1-C_7 respectively depending on the output of 4-bit Sub-ADC (SADC). Unlike MCS architecture where dedicated capacitor is used for feedback, the CFCS architecture involves commutating the feedback capacitor based on the output of the 4-bit SADC. Therefore, implementing CFCS on top of MCS architecture requires additional CFCS logic decoding switches. Moreover, decision time delay involved in additional CFCS logic need to be constrained within the non-overlap period available for the decoding logic. While the implementation of the CFCS decoder for 1.5-bit stage is quite straightforward as it consists of switching between only two capacitors, designing the decoder for 3.5-bit stage is quite challenging due to increased circuit complexity and decoding delay involved. Though the need of a CFCS decoder was established in [4], but its exact circuit implementation has not been reported.

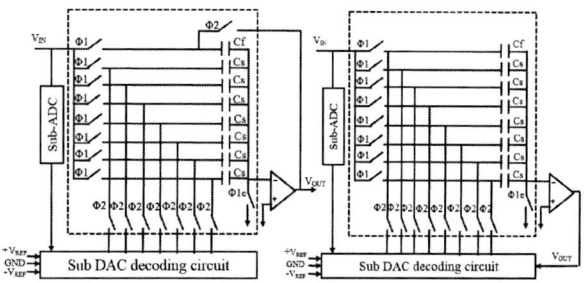

Figure 6.   3.5-bit MDAC using MCS (left) and MCS-CFCS (right)

A. Proposed 3.5 bit MCS-CFCS Switching Logic

In order to design a fast CFCS decoder with minimal circuit complexity and switching delay we make slight modifications in the MCS and CFCS switching sequences [4] without altering the governing equations such that the linearity remains intact. The modified sequences for the MCS and CFCS cases are given in Table-II. We further explain how this modified switching sequence can be strategically used to realize the complex patterns of CFCS.

TABLE II.   SWITCHING SEQUENCES: (A) MCS AND (B) CFCS

MCS	C1	C2	C3	C4	C5	C6	C7	C8
0001	$-V_{REF}$	$-V_{REF}$	$-V_{REF}$	$-V_{REF}$	$-V_{REF}$	$-V_{REF}$	$-V_{REF}$	F/B
0010	$-V_{REF}$	$-V_{REF}$	$-V_{REF}$	$-V_{REF}$	$-V_{REF}$	$-V_{REF}$	GND	F/B
0011	$-V_{REF}$	$-V_{REF}$	$-V_{REF}$	$-V_{REF}$	$-V_{REF}$	GND	GND	F/B
0100	$-V_{REF}$	$-V_{REF}$	$-V_{REF}$	$-V_{REF}$	GND	GND	GND	F/B
0101	$-V_{REF}$	$-V_{REF}$	$-V_{REF}$	GND	GND	GND	GND	F/B
0110	$-V_{REF}$	$-V_{REF}$	GND	GND	GND	GND	GND	F/B
0111	$-V_{REF}$	GND	GND	GND	GND	GND	GND	F/B
1000	GND	GND	GND	GND	GND	GND	GND	F/B
1001	$+V_{REF}$	GND	GND	GND	GND	GND	GND	F/B
1010	$+V_{REF}$	$+V_{REF}$	GND	GND	GND	GND	GND	F/B
1011	$+V_{REF}$	$+V_{REF}$	$+V_{REF}$	GND	GND	GND	GND	F/B
1100	$+V_{REF}$	$+V_{REF}$	$+V_{REF}$	$+V_{REF}$	GND	GND	GND	F/B
1101	$+V_{REF}$	$+V_{REF}$	$+V_{REF}$	$+V_{REF}$	$+V_{REF}$	GND	GND	F/B
1110	$+V_{REF}$	$+V_{REF}$	$+V_{REF}$	$+V_{REF}$	$+V_{REF}$	$+V_{REF}$	GND	F/B
1111	$+V_{REF}$	$+V_{REF}$	$+V_{REF}$	$+V_{REF}$	$+V_{REF}$	$+V_{REF}$	$+V_{REF}$	F/B

CFCS	C1	C2	C3	C4	C5	C6	C7	C8
0001	$-V_{REF}$	$-V_{REF}$	$-V_{REF}$	$-V_{REF}$	$-V_{REF}$	$-V_{REF}$	$-V_{REF}$	F/B
0010	$-V_{REF}$	$-V_{REF}$	$-V_{REF}$	$-V_{REF}$	$-V_{REF}$	$-V_{REF}$	F/B	GND
0011	$-V_{REF}$	$-V_{REF}$	$-V_{REF}$	$-V_{REF}$	$-V_{REF}$	F/B	GND	GND
0100	$-V_{REF}$	$-V_{REF}$	$-V_{REF}$	$-V_{REF}$	F/B	GND	GND	GND
0101	$-V_{REF}$	$-V_{REF}$	$-V_{REF}$	F/B	GND	GND	GND	GND
0110	$-V_{REF}$	$-V_{REF}$	F/B	GND	GND	GND	GND	GND
0111	$-V_{REF}$	F/B	GND	GND	GND	GND	GND	GND
1000	F/B	GND	GND	GND	GND	GND	GND	GND
1001	$+V_{REF}$	F/B	GND	GND	GND	GND	GND	GND
1010	$+V_{REF}$	$+V_{REF}$	F/B	GND	GND	GND	GND	GND
1011	$+V_{REF}$	$+V_{REF}$	$+V_{REF}$	F/B	GND	GND	GND	GND
1100	$+V_{REF}$	$+V_{REF}$	$+V_{REF}$	$+V_{REF}$	F/B	GND	GND	GND
1101	$+V_{REF}$	$+V_{REF}$	$+V_{REF}$	$+V_{REF}$	$+V_{REF}$	F/B	GND	GND
1110	$+V_{REF}$	$+V_{REF}$	$+V_{REF}$	$+V_{REF}$	$+V_{REF}$	$+V_{REF}$	F/B	GND
1111	$+V_{REF}$	$+V_{REF}$	$+V_{REF}$	$+V_{REF}$	$+V_{REF}$	$+V_{REF}$	$+V_{REF}$	F/B

B. 3.5bit MCS-CFCS Decoder Circuit Design

The SADC for 3.5-bit stage consists of 14 differential comparators (CMP_1 to CMP_{14}) that provides thermometer coded output. There exists a residue jump corresponding to each comparator hence dividing the total input range ($+V_{REF}$, $-V_{REF}$) into 15 regions. The output of the differential comparators is denoted by T_1 to T_{14} and T_{1B} to T_{14B}. A_1 to A_{15} represent the decimal output of the SDAC and can directly be obtained by AND-ing output of adjacent comparators T_N and $T_{(N-1)B}$. For example, A_8 can be obtained from comparator outputs T_7 and T_{8B}, i.e A_8=AND(T7,T8B).

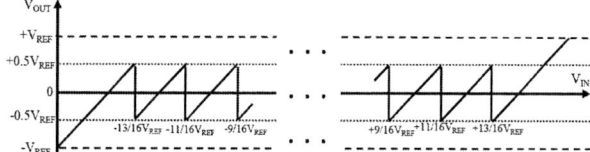

Figure 7.   Transfer characteristic of 3.5 bit MDAC with 14 residue jumps.

For MCS, each capacitor C_1-C_7 have the option either to be connected to input or references available at node Y, whereas only C8 has the option to be connected to input or flipped to the output of amplifier (Fig. 6a). Node X represents the top plate of capacitor C. Node Y provide access to references $+V_{REF}$, V_{REF} or GND. Y can be accessed from X via S2 (Fig. 8a). During Phase-1 (sampling phase) the capacitor C is charged to V_{IN} via S1. V_{CN}, V_{CP} and V_{CZ} represent control signals directly obtained from thermometer output to enable $+V_{REF}$, $-V_{REF}$ and GND to be available at node Y. A SADC output of 8 corresponds to mid region code. From Table 3a, it is observed that C8 is connected to $+V_{REF}$ for SADC output code > 8. This can be ensured by connecting T8 to V_{CP} terminal of C8. Similarly, T_{7B} can be connected to V_{CN} since T_{7B} will be high for all SADC output < 8. For SADC output = 8, both V_{CP} and V_{CN} goes to low. Consequently, if none of V_{CP} or V_{CN} is high, the Gate G1 NOR (V_{CP}, V_{CZ}) now ensures that the control V_{CZ} is enabled.

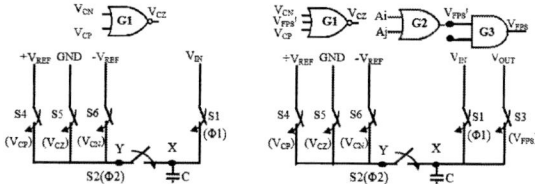

Figure 8.   Decoder Circuit implementation: MCS (left) and CFCS (right)

One critical observation here is that any capacitor that goes to feedback in CFCS (Table IIB) was earlier grounded during MCS (Table IIA). This is cleverly used for minimizing the switching circuit. Signal V_{FPS} is now used to flip a capacitor to feedback mode as output node is accessible from X via S3. It is observed that C8 goes to feedback for both SADC output is 6 or 8. Thus V_{FPS}' can directly be generated by OR G2(A6, A8) but is enabled by G3 only during Phase-2. In case V_{CP} and V_{CN} are simultaneously low, NOR G1(V_{CP}, V_{CN}, V_{FPS}) ensures that enabling V_{CZ} additionally requires V_{FPS} to be low. That is a V_{FPS} signal overrides a capacitor being earlier connected to ground. For implementing differential negative side logic, the same positive decoding logic is used but with control signals V_{CP} and V_{CN} interchanged. The proposed logic therefore encounters an additional delay of 2 logic gates that lies well within the non-overlap period. This completes the description of the decoder circuit. (Layout is shown in Fig. 9)

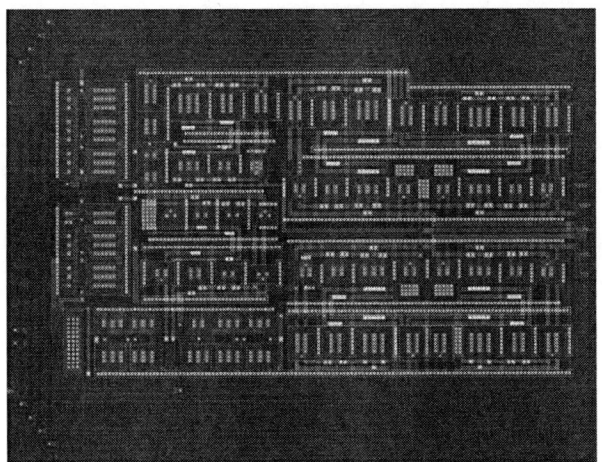

Figure 9.   Complete layout of CFCS decoder circuit for 3.5bit MDAC

The decoder is repeated for each node capacitor C_1-C_8. For a differential implementation, the differential node capacitors can be alternatingly interlaced over the array. The calibration switches are conveniently placed and integrated around the capacitor array (Fig. 10). Guard rings are used to isolate array from calibration switches. The decoder and capacitor array consume area of $60\times60\mu m^2$ and $700\times800\mu m^2$ respectively when realized in 1P6M UMC 0.18μm process.

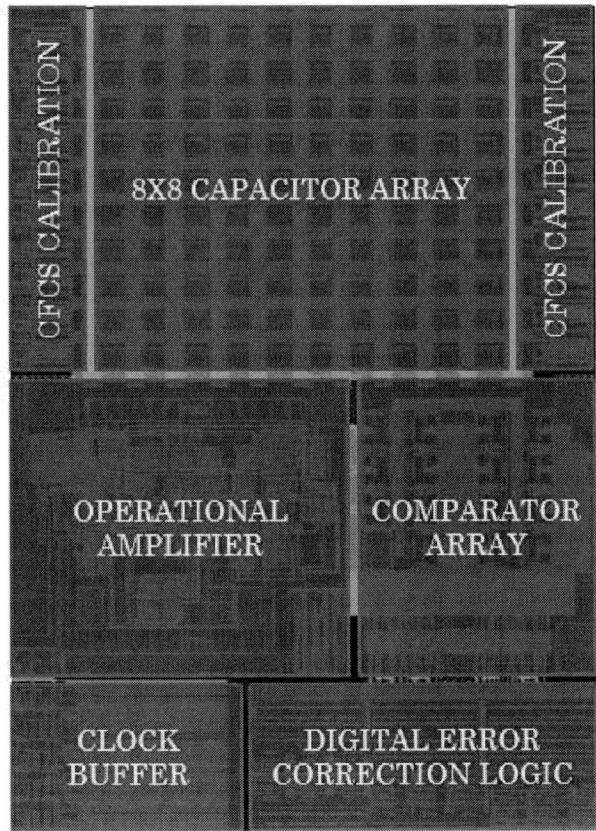

Figure 10.   Layout showing proposed CFCS decoder integrated with array.

IV. SIMULATION RESULTS

The proposed switching technique is verified on the model of a 16-bit 10-Msps pipelined ADC with 3.5 bit/stage using Matlab. Mismatches are introduced into the capacitor array of MDAC Stage-01 where the remaining 4 stages are considered as ideal.

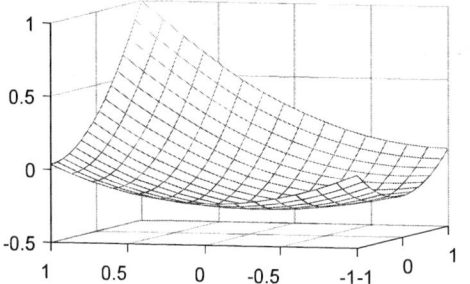

Figure 11. Joint Gradient Model for integrated simulation of MDAC.

The capacitor array of Stage-1 is subject to joint gradient profile (Fig. 11) with coefficients [α_{00}, α_{10}, α_{01}, α_{20}, α_{02}, α_{11}] = [-0.2, 0.25, 0.25, 0.23, 0.23, 0.23] and w=1.The maximum magnitude of mismatch then limited to 5% over the array. It is assumed that the unit sampling capacitors are sized sufficiently large such that only systematic effects dominate. The performance of proposed array when integrated with the modified switching scheme is contrasted to that achieved by conventional arraying techniques [13-15] in Table-III below.

TABLE III.   INTEGRATED PERFORMANCE OF ARRAY AND SWITCHING

Parameter	ε_{max} ×5%	DNL (LSB)	INL (LSB)	SNDR (dB)	SFDR (dB)	ENOB (bits)
Proposed	0.02	5.5	5.5	88.39	88.51	14.44
Work [13]	0.10	7.5	29	74.12	74.13	12.06
Work [14]	0.15	8.5	75	56.21	56.21	9.08
Work [15]	0.10	11.5	37	68.99	68.99	11.21

It is evident that the proposed arraying technique exhibits high static as well as dynamic linearity in comparison to other existing techniques. The significant performance improvement can be attributed to the following. The linearity errors (residue jump at transition points) incurred in merged capacitor switching (MCS) are directly proportional to mismatch coefficients, $\Delta\varepsilon$. The application of CFCS technique on top of the MCS reduces these to $\Delta\varepsilon/2$ thereby resulting in ~1bit (6 dB) improvement [4]. Therefore an improvement in mismatch coefficient ε_{max} by a factor of 2 corresponds to ~6db improvement in SNDR at the output. This has also been verified in numerical simulations using Matlab. For example in presence of a joint gradient, the proposed array shows ~14dB improvement over [13] as the maximum mismatch, ε_{max} is also suppressed by 5X by implementing the proposed arraying scheme. Hence, even in the presence of large systematic mismatch (5%), 14-bit effective linearity (ENOB) could be achieved. The output signal spectrum of ADC when integrated with various arrays is plotted in Fig. 11. A comparative overview of DNL and INL results achieved for various array is provided in Fig. 12.

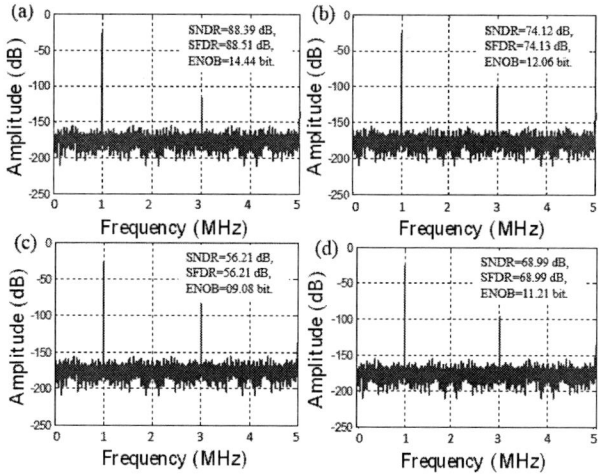

Figure 12. FFT plot at 10 MHz conversion rate (input freq. f_{in} =1MHz)

The rotated parabolic gradients typically cause in "S-shaped" integral nonlinearity error, and hence generate a 3rd order harmonic distortion component at the output. The 'xy' terms in the proposed layout do not cause a parabolic error, but an error that alternates between positive and negative values. As seen from the INL plot (Fig. 13), the INL caused by the systematic errors is pulled back to zero 16 times at regular intervals by the proposed switching. Hence the accumulation of the error happens only over a small amount of codes. This results in less 3rd harmonic distortion and hence improvement in SNDR and SFDR as can be seen in Fig.12. As the proposed arraying scheme significantly suppresses the mismatch coefficients ε_{max}, the magnitude of resulting DNL and INL specification are highly benefitted.

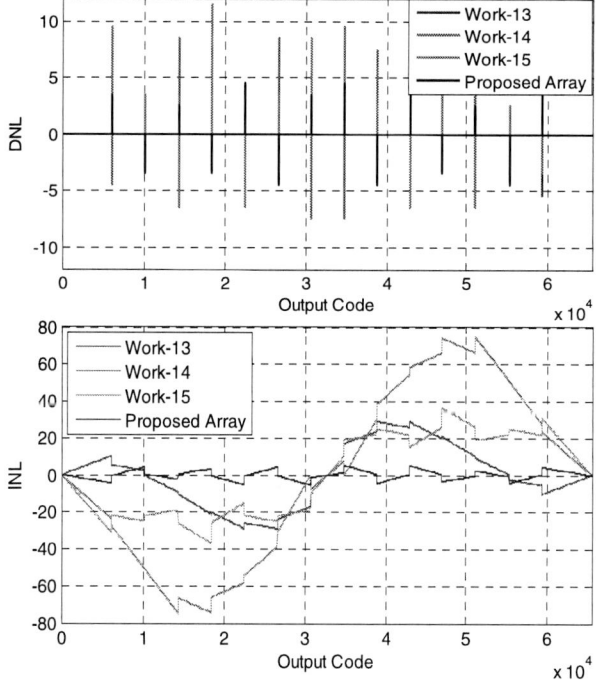

Figure 13. Comparative overview: DNL plot (above) and INL plot (below)

V. CONCLUSION

A mismatch resilient arraying technique for 3.5 Bit MDAC with MCS-CFCS architecture was discussed. Along with systematic effects, the proposed arraying technique was used to consider local gradients and hot spots. This makes it highly useful for 3D IC integration where data converters are placed adjacent to high performance digital chips [6]. A modified CFCS switching sequence is proposed that helps in strategically minimizing the logic circuit of 3.5-bit CFCS decoder. The decoder circuit and proposed capacitor array consume 60×60 μm² and 700×800 μm² die area when realized in 1P6M UMC 0.18μm process. The performance of proposed array integrated with the modified CFCS is verified on the model of 16-bit 10Msps pipelined ADC using Matlab. The technique achieved considerable improvement over existing techniques in terms of linearity. Though in this work, CFCS for 3.5-bit MDAC was demonstrated, this technique can in general be used to improve the linearity performance of higher resolution MDACs. This work has potential application in implementing modern wireless base station receivers and in high resolution imaging applications.

REFERENCES

[1] Devarajan, Siddharth, et al. "A 12b 10GS/s Interleaved Pipeline ADC in 28-nm CMOS Technology."IEEE J. of Solid State Circuits (2017)

[2] Grace, Carl R., Paul J. Hurst, and Stephen H. Lewis. "A 12-bit 80-MSample/s pipelined ADC with bootstrapped digital calibration." IEEE Journal of Solid-State Circuits 40.5 (2005): 1038-1046.

[3] Devarajan, Siddharth, et al. "A 16-bit, 125 ms/s, 385 mw, 78.7 db snr cmos pipeline adc." IEEE Journal of Solid-State Circuits 44.12 (2009)

[4] Yoo, Sang-Min, et al. "A 2.5-V 10-b 120-MSample/s CMOS pipelined ADC based on merged-capacitor switching." IEEE Transactions on Circuits and Systems II: Express Briefs 51.5 (2004).

[5] Chuan Seng Tan, Ronald J. Gutmann, L.Rafael Reif, Wafer level 3D-ICs Process Technology,Chapter-13, Circuit Architecture for 3D Integration,Page-298, Integrated Circuits and Systems, Springer,2008

[6] Erdmann, Christophe, et al. "A heterogeneous 3D-IC consisting of two 28 nm FPGA die and 32 reconfigurable high-performance data converters." *IEEE Journal of Solid-State Circuits* 50.1 (2015):

[7] Zjajo,A."Temperature Effects in Deep Submicron CMOS" Stochastic Process Variation in Deep-Submicron CMOS. Springer, 2014.83-115.

[8] Jingyan, et al. "Thermal analysis and thermal optimization of through silicon via in 3D IC." Solid-State and Integrated Circuit Technology (ICSICT), 2014 12th IEEE International Conference on. IEEE, 2014.

[9] Chiu, Po-Yen, and Ming-Dou Ker. "Metal-layer capacitors in the 65 nm CMOS process and the application for low-leakage power-rail ESD clamp circuit." Microelectronics Reliability,Elsevier 54.1 (2014)

[10] Onabajo, Marvin, and Jose Silva-Martinez. "Process variation challenges and solutions approaches." Analog circuit design for process variation-resilient systems-on-a-chip. Springer, (2012). 9-30.

[11] Vankka, Jouko. Digital synthesizers and transmitters for software radio, Chapter-10. pp.181. Springer Science & Business Media, 2005.

[12] Ajami, Amir H., et. al "Modeling and analysis of nonuniform substrate temperature effects on global ULSI interconnects." IEEE Transactions on CAD of Integrated Circuits and Systems (2005):

[13] Reynolds, D.C. "MOS current source layout technique to minimize deviation." U.S. Patent No. 5,568,145.1996. Asignee:AnalogDevices

[14] Douglas Mercer, Current Steering D/A Converters, Analog Devices https://wiki.analog.com/university/courses/tutorials/cmos-dac-chapter

[15] Ostrem, Geir Sigurd. "Current source array for high speed, high resolution current steering DACs." U.S. Patent No.6,720,898.(2004).

978-1-5386-7100-9/18 $31.00 © 2018 IEEE

2018 IEEE Computer Society Annual Symposium on VLSI

Design of Low Power SAR ADC using Clock Retiming

Jalaja S, *Member, IEEE*
Research Scholar, Visvesvaraya Technological University,
Dept. of Electronics and Communication Engineering,
Bangalore Institute of Technology, Bangalore, India
E-mail: jalajabit@gmail.com

Vijaya Prakash A.M
Dept. of Electronics and Communication Engineering,
Bangalore Institute of Technology, Bangalore, India
E-mail: am_vprakash@yahoo.co.in

Abstract—In digitized world power efficient Successive Approximation Register Analog-to-Digital converter (SAR-ADC) architecture are widely used in most of the electronics applications. It is very compact compared to other ADC architecture. In this proposed paper practical implementation of modified 10-bit SAR-ADC Clock Retiming is designed. The performance of the design is analyzed by Clock Retiming with multiple input phases and multiple output phases. As a result Electrical levels of glitches are removed by Retiming without altering the functionality of the design. After placing the delay element to each input line of the DAC circuit the transient analysis of DAC output shows gain error reduction compared to earlier design proposed. The R-2R DAC architecture of SAR-ADC is designed using Submicron technology. As a result ADC topology shows the low power consumption with reduction in gain error.

Keywords—Delay SAR-ADC circuit, Digital-to-Analog (DAC) converter, Retime NAND Schematic, Retime DAC schematic.

I. INTRODUCTION

The invention of high-performance chips is very important in real world VLSI design. In chip industries analog device play a challenging role in optical system, high speed memories, sensors, wireless system, high speed microprocessor. Conversion of the natural signal into digital signal needs multiple parameter tradeoffs between amplification of the circuit, Gain, Power, Speed and Precision etc. In this paper low power design methodology is adopted to implement the analog circuit design. Current mirrors, amplifiers and voltage reference circuits are some of the common building blocks of analog design. Another essential component in analog mixed signal circuit is Analog-to-Digital Converter (ADC) and Digital-to-Analog Converter (DAC). In advanced electronic device one of the key components is low power and high resolution Analog-to-Digital converter (ADC) device. The general block diagram to implement ADC is sample and hold circuit, comparator and DAC etc. ADC and DAC converters are useful in Audio amplifier, Video Encoder, Display Electronics, Data Acquisition systems, Calibration, software radio, motor control and many more. In this paper low power restructure DAC and ADC is designed using clock Retiming technique.

The generalized symbolic representation of analog to digital signal conversion block is as shown in Fig.1. It translates analog signals into digital language, with the trade off between power consumption and speed [18]. In market there are common types of ADC for different application is categorized with the choice of speed and resolution is as shown in

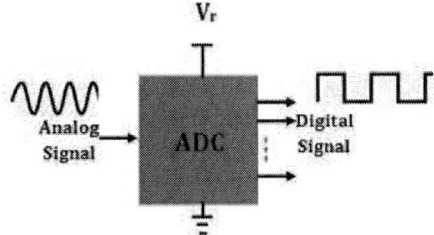

Fig. 1. Analog-to-Digital converter

Fig.2. In this paper, the design of low power Clock Retiming Successive Approximation Register (SAR) ADC is proposed. This architecture is a power efficient and compact design compare to existing design.

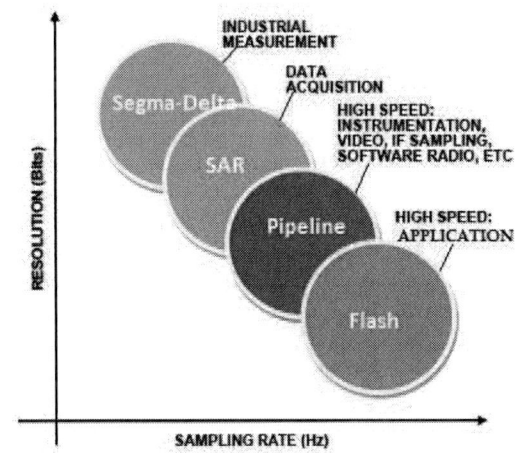

Fig. 2. Types of Common ADC

Overall linearity of the SAR ADC is determined using DAC circuit. R-2R resistor network [14], [18] is used in successive approximation ADC architecture was first implemented by Bernard M. Gordon. Using 0.13 μm CMOS process technology 9 bit low-power low-voltage SAR ADC designed using metal-oxide-semiconductor capacitors (MOSCAP-DAC) [16]. The proposed SAR-ADC tolerance is improved in this paper compared to existing design. N-bit R-2R ladder DAC schematic is shown in Fig.3 it requires 2:1 ratio resistor and

978-1-5386-7100-9/18 $31.00 © 2018 IEEE   181

its reference voltage is represented as V_{ref}. When the binary bit is one, switch is connected to V_{ref} and binary bit is zero, switch connected to ground. In equation (2) b_i(i= 1,2, ...N) is the N-bit digital input bits and the $b_N 2^{-N}$ indicates N^{th} node of the input bit. The generalized N bit R-2R DAC output voltage is described using equation (1).

$$V_o = (b_1 2^{-1} + b_2 2^{-2} + b_3 2^{-3} + \ldots + b_N 2^{-N})V_{ref} \quad (1)$$

$$V_o = V_{ref} \sum_{i=1}^{N-1} \frac{b_i}{2^i} \quad (2)$$

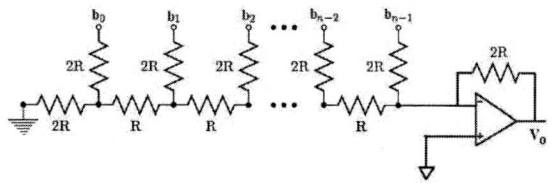

Fig. 3. N bit R-2R Schematic

To increase the performance level of the DAC and ADC many researchers are proposed with different architecture. Silicon carbide (SiC) DAC architecture [2] is analyzed for different temperature level. Here Integral non linearity (INL) and differential non linearity (DNL) values are simulated using 8-bit R-2R ladder DAC architecture. For higher temperature SiC DAC shows better performance, so it suits in deep drilling, aviation and space exploration application. SiC DAC based successive approximation register ADC is designed [3] for BIST application and the aim of this paper is low cost design solution. SAR ADC is designed using new sampling and switching method [4]. The proposed capacitor switching method highlights two benefits during switching procedure. One is due to post sampling phase and other due to after MSB determination. Outcome of this paper shows Switching energy and area reduction compared to conventional binary weighted DAC. SAR ADC architecture for a 6 b 5 GS/s 4 Interleaved 3 b/Cycle is designed [7] to save $1/3^{rd}$ hardware part compared to existing design. Fractional DAC array switching scheme implemented for switching control to reduce the complexity of the design. Experiment result proves low power and energy efficient master-clock-control bootstrapped-switch scheme as well as boundary detection code overriding. To improve overall linearity of DAC the 10-bit resistor-floating-resistor-string digital-to-analog converter (RFR-DAC) architecture is proposed [9]. In this architecture two different voltage selection scheme are used for liquid crystal display (LCD) driver applications. The current-steering Digital-to-Analog Converters (DACs) calibration was addressed [10]. As a result they achieve reduction in intrinsic accuracy level and high resolution static calibration levels without compromising the dynamic behavior using two fully calibrated thermometric arrays. Mismatch charge compensation scheme is used to design switch-capacitor cyclic based DAC [13] to reduce DAC error caused by the capacitor mismatch. One of the researcher used current compensation technique to designed R-2R DAC using a (4+12)-bit segmented voltage-mode method in [17] as a result design consume $1/5^{th}$ less power compared to

conventional method. In this method reference current and low-resolution auxiliary DAC are controlled by a computational block in simple way.

In one of our previous work Sum-of-Product term Retiming algorithm is proposed to design digital application. The set of architecture equation is simplified intern of Sum-of-Product equation and cutset is applied to insert the flip-flops. Proposed Retiming is applied to FIR and IIR filter design to reduce power consumption compared the existing design. In this propose work, analog mixed signal (AMS) design SAR ADC architecture is designed using Clock Retiming, which yields in low power consumption. Section II will briefly describe the requirements Retiming on analog circuits. The section will also review sources of power dissipation and schematic design analysis of NAND gate in AMS design. Section III and IV will present the design of R-2R DAC and SAR ADC using proposed methodology. Section V discuss the result analysis to determine the power dissipation at schematic level, resolution and sampling rate parameters. Section VI summarizes the proposed design methodology.

II. RETIME LOW POWER ANALOG CIRCUIT

Switching power, Short circuit power and leakage power are the most significant sources of low power cells [1]. The three sources of power dissipation in a CMOS circuit is expressed in equation (3). Different power optimization techniques are necessary to reduce major sources of power consumption in CMOS circuit design. In many design, more than 70% power consumption is due to charging and discharging wires and transistors gate, so reduction in switching power makes a analog device more reliable.

$$P = P_{Switching} + P_{Short-circuit} + P_{Leakage} \quad (3)$$

where switching power is

$$P_s = \frac{1}{2}\alpha C_L V_{DD}^2 f \quad (4)$$

In equation (4) f represents clock frequency, V_{DD} indicates supply voltage, C_L is the output load capacitance and α is the switching activity of the gate. Using this basic equation low power Retiming equation is formed and reconstruction of the Retiming schematic diagram is highlighted.

The transistor level schematic of 2-input NAND gate is as shown in Fig.4 which consists of two PMOS and NMOS transistors. The PMOS source are tied to supply voltage V_{DD} and NMOS source are connected to ground gnd. Output will be low only when both the inputs are high and for all other conditions output will be high. The average switching activity at the output and the load capacitance due to switching power is represented in equation (4). During the input transition sometime it may cause some glitches at the output. In this paper the glitch power dissipation is analyzed using Retiming concept in this paper. Using virtuoso tool NAND gate schematic is designed and its transient waveform is generated. The clock Retiming is applied to conventional NAND gate schematic and the Flip-flop (R) is added at the output end of the NAND logic is as shown in Fig .5(b). The power analysis using the spectre simulator proves that the average

978-1-5386-7100-9/18 $31.00 © 2018 IEEE

switching power in Fig .5(a) and .5(b) is less than the Fig.4, so additional Flip-flops actually reduces the power dissipation due to clock assertion. Repositioning the flip-flop without comprising the functionality of the design is first proposed by C.Leiserson.F.Rose [5], [6]. Same concept has been adopted to improve the electrical circuit performance for SAR-ADC. After repositioning the flip-flop power dissipation in the circuit is changed. The glitch power dissipated in all three cases Fig.4, Fig .5(a) and Fig .5(b) is plotted in Fig .6. The resultant analysis shows that Flip-flops based gate consume less power than the conventional gate.

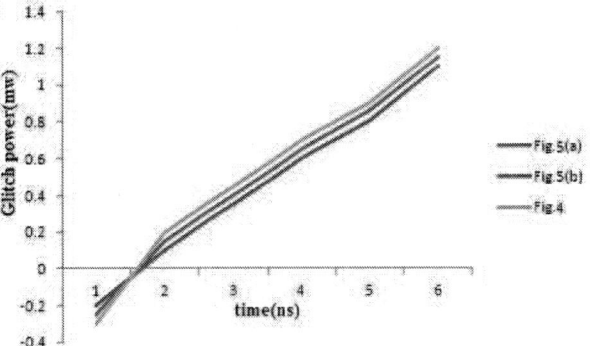

Fig. 6.   Measured Glitch Power of the conventional NAND schematic and Retime NAND schematic

Fig .7. Then reposition the flip-flop to each 10 different input pulses, to improve the performance of the R-2R DAC. As a result after Retiming the Non linear transfer characteristic is obtain as shown in Fig.8 is better compared to conventional characteristics. The proposed architecture simulation results show improved characteristics interns of INL, DNL and resolution over previously proposed structure. The design flow of clock Retiming SAR ADC is represented in Fig .10. It starts with Transistor sizing characterization parametric analysis followed by SAR ADC schematic design and power analysis to quantify the power as well as delay. Node cut-set is applied to R-2R ladder network to insert the Flip-flops and reconstruct the R-2R ladder network to design SAR ADC.

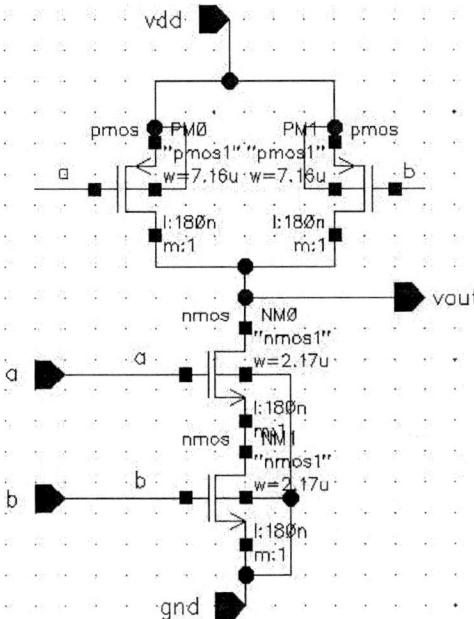

Fig. 4.   Schematic diagram of two input NAND gate

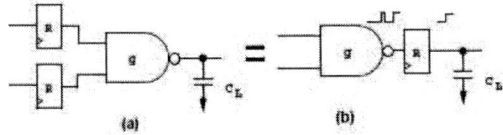

Fig. 5.   Positioning Flip-flop to a NAND logic

III.   PROPOSED DESIGN METHODOLOGY TO DESIGN R-2R DAC

Cut-set Retiming is applied by finding the longest critical path and then reposition the flip-flop to reduce the power consumption is the core idea of the proposed work. Cut-set Retiming concepts are proposed in [5], [6] and is inherited to design dual-clock Retiming R-2R ladder network. The proposed design negative feed back operational amplifier is used to design the DAC structure. Resistance ladder input line is connected to the non-inverting terminal of the operational amplifier and ended with common reference ground. Place the flip-flop at non inverted line of the operational amplifier to redesign 10-bit R-2R DAC schematic which is represented in

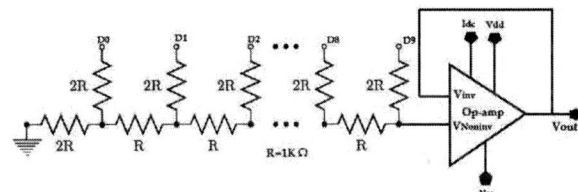

Fig. 7.   10-bit R-2R DAC Schematic

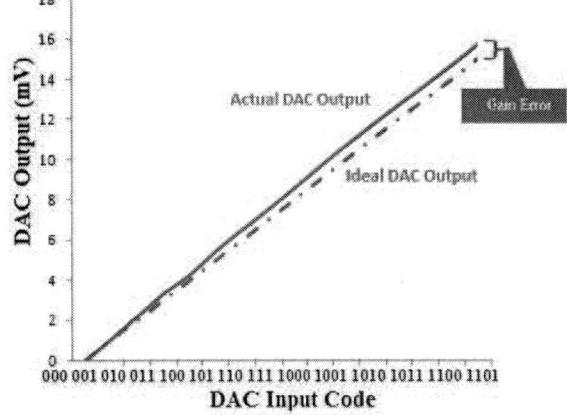

Fig. 8.   Gain Error

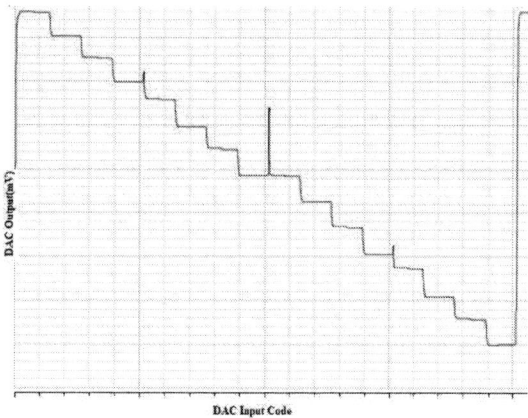

Fig. 9.   Retime DAC output

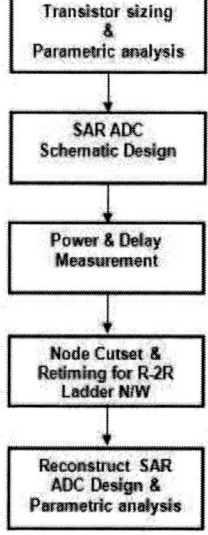

Fig. 10.   Simulation workflow for SAR ADC Design

IV. DESIGN AND IMPLEMENTATION OF SAR ADC USING PROPOSED R-2R DAC ARCHITECTURE

Retime R-2R DAC, Retime comparator, operational amplifier, sample/Hold circuit and flip-flops are required to design Clock Retiming SAR-ADC. Inverting input terminal is connected to sample and hold circuit to provide the analog signal. The non-inverting input terminal connected to time varying DAC output terminal. Retime clocked comparator schematic shows more accurate result compared to conventional comparator and is shown in transient waveform in Fig.12. The output comparator provides proper clock values to the SAR circuit. Many researchers are proposed different architecture to implement DAC and ADC design. In this paper, cut-set is applied to each node of the R-2R network and each group can form connected graph. In cut-set Retiming [12] does not alter the functionality of the circuit but it affects the weights of the edges in the cut-set. Adding delay on each cut-set indicated by the dashed line in Fig .13 and is Retime to increase the

performance level of the design. Each node of 2R resistor path connected with different clock flip-flop. The resolution of each bit is trimmed using Retiming technique. The proposed N-bit cut-set R-2R ladder network with multi-clock phase and DAC test schematic is as shown in Fig.13 and Fig.14 respectively.

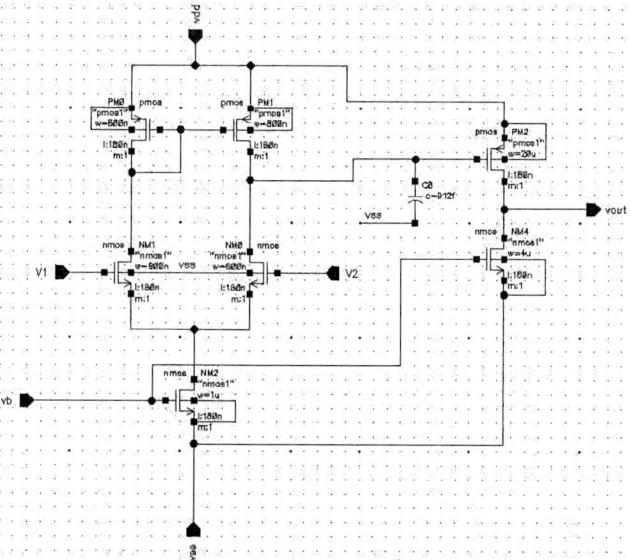

Fig. 11.   Schematic diagram of Comparator

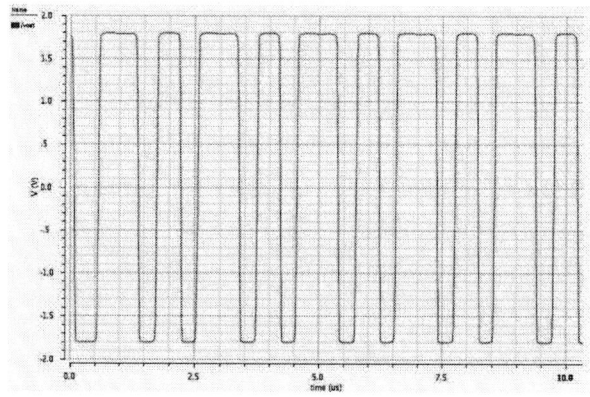

Fig. 12.   Transient analysis of Comparator

The practical implementation of Retime SAR-ADC architecture is shown in Figure 15. During the process of conversion cycle analog input signal is fed to the sample-and-hold (S/H) to keep the signal constant. Comparator block compares the signal of D/A converter output signal and S/H signal. The resultant most-significant bit (MSB) is 0 or 1 is stored in successive-approximation register (SAR) when the D/A converter output is less or greater than the S/H signal. Complete conversion process take place through SAR using binary search algorithm. The digital input generated by the SAR is converted into analog equivalent signal in the internal DAC. In the proposed work internal DAC consist of different clock flip flops to converts digital to analog signal. The internal conversion process is controlled by a different clock speed and

978-1-5386-7100-9/18 $31.00 © 2018 IEEE   184

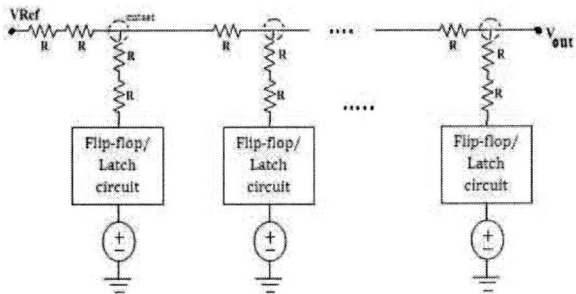

Fig. 13. Proposed R-2R N-bit Ladder Network

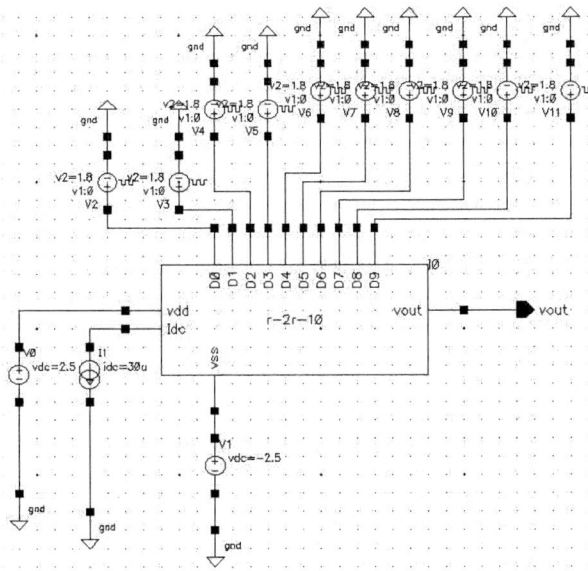

Fig. 14. Test schematic of 10-bit R-2R DAC

the simulation results shows proposed SAR ADC is working with low power consumption compare to existing SAR-ADC design.

V. RESULTS AND DISCUSSION

The main contribution of this proposed work is Retime DAC and comparator. The multiple clock phases are used to design the SAR-ADC schematic design, as shown in Fig.15. R-2R DAC reference voltage is connected to positive edge of Retime clock. All flip-flops are placed at D_0, D_1, ... D_N to achieve better performance result. Then flip-flops are shift at the high fanout of the design to minimize the power consumption. The transient analysis of Retime DAC is as shown in Fig 8 and 9. For the resolution of 8 and 10 bit SAR-ADC measured result is shown in Table II. The sequential elements (e.g., flip-flops (FFs) or latches, etc.) are placed at the output of the SAR ADC schematic and is Retime by replacing the flip-flops at the multiple phases of all the inputs. The power due to glitching is computed under both zero delay and actual delay models. Significantly number of flip-flops are moved across the schematic design to achieve low power consumption. The result analysis is done using Spectre EDA tool, the obtained

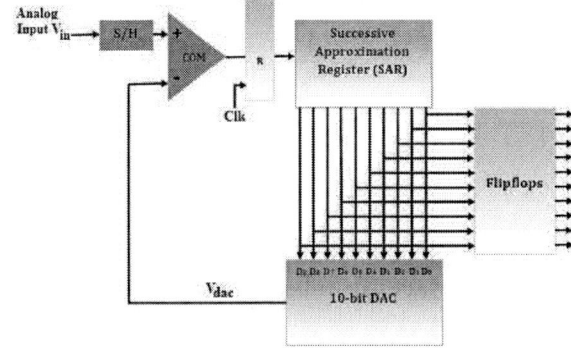

Fig. 15. Retimed Successive Approximation Register based A/D Converter

DNL and INL values of proposed R-2R DAC is better compare to [19]. In this paper researcher is concentrated on 3+5 segmented DAC using 0.18μm CMOS technology and Multi-bit $\Sigma - \Delta$ DAC is designed in [20] with low-power low-cost experiment result obtained from portable digital audio applications. The proposed Retime Analog DAC design consume less power compared to [19,20] and 43.7% less power than [19]. Retiming comparator gives better sampling and amplification, in presence of analog input signal. Comparator output decision is followed by clock Retiming. In transient waveform spike distortion are removed using proposed technique. Asynchronous SAR-ADC design methodology result are implemented for 8, 10 and 12 bit resolution and is suitable for biomedical sensor. Proposed sizing methodology tool is named as hybrid-based asynchronous SAR-ADC sizing tool (HaSAST) and is used for automatic transistor sizing. Table I shows the sampling rate and power consumption of the proposed and existing research design. Asynchronous SAR-ADC [8] signal-to-noise-and-distortion ratio (SNDR) and power consumption result are compared in Table II.

TABLE I.   COMPARISON RESULTS OF SAR ADC OUTPUT FOR 10-BIT RESOLUTION

Reference	[11]	[21]	[15]	Proposed work
Architecture	SAR	Pipeline	SAR	Retime
Supply voltage(V)	1.0	1.0	1.2	1.8
Sampling Rate(S/s)	240	200	280	250
Power Consumption(mW)	0.68	5.37	2	0.267

TABLE II.   COMPARISON RESULTS OF SAR ADC OUTPUT

Reference	[8]		Proposed work	
Resolution(bit)	8	10	8	10
SNDR(dB)	49.7	52.1	53.2	59.3
Peak DNL(LSB)	0.29	1.41	0.21	1.12
Peak INL(LSB)	0.24	2.81	0.22	2.41
Power Consumption(μW)	328	328	267	267

978-1-5386-7100-9/18 $31.00 © 2018 IEEE

VI. CONCLUSION

New practical approach is proposed by deploying the Retiming technique in analog schematic to achieve low power consumption. This approach provide a better performance solution to any analog design. The practical implementation of SAR ADC performance is analyzed in this paper. To fulfill the requirement of low power ADC the clock is Retime at multiple phases of the electrical lines. As a result the power consumption is less, without the conversion mismatch of analog to digital signal. Proposed R-2R DAC simulation results show that the reduction in gain error compared to existing design. SAR-ADC can also be implemented using precharge retiming circuit in future work.

REFERENCES

[1] A.P. Chandrakasan, S.Sheng and R.W.Brodersen, "Low Power CMOS Digital Design," *IEEE Journal of Solid-State Circuits,* vol. 27,no.4,pp. 473-484, April 1992.

[2] A. Rahman, S. Roy, R. C. Murphree, H. A, A. M. Francis and J. Holmes, "A SiC 8 Bit DAC at 400C," *IEEE conference,* 2015.

[3] A. Rabal, A. Otin, I. Urriza, A.J. Gines, G. Leger, and A. Rueda, "A Compact R-2R DAC for BIST Applications," *IEEE conference,* 2016.

[4] B. Yazdani, A. Khorami, M. Sharifkhani, "Low-power DAC with charge redistribution sampling method for SAR ADCs," *Electronics Letters,*vol. 52, no. 3, pp. 187-188, 2 4 2016.

[5] C.Leiserson.F.Rose and J.B Saxe, "Optimizing synchronous circuitry by retiming," *In proceeding of the,* 3rd ed. Caltech conference on VLSI , 1983.

[6] C.Leiserson.F.Rose and J.B Saxe, "Retiming synchronous circuitry," *Algorithmica.* 1991.

[7] Chi-Hang Chan, Yan Zhu, Sai-Weng Sin, Seng-Pan (Ben) U and Rui Paulo Martins, "A 6 b 5 GS/s 4 Interleaved 3 b/Cycle SAR ADC *IEEE Journal of Solid-State Circuits,"* Vol. 51, NO. 2, Feb. 2016.

[8] Chun-Po Huang Jai-Ming Lin, "A systematic design methodology of asynchronous SAR ADC's," *IEEE transcation on VLSI systems,* 2017.

[9] C. W. Lu, P. Y. Yin, C. M. Hsiao, M. C. F. Chang and Y. S. Lin, "A 10-bit Resistor-Floating-Resistor-String DAC (RFR-DAC) for High Color-Depth LCD Driver ICs," *IEEE Journal of Solid-State Circuits,* vol. 47, NO. 10, pp. 2454-2466, Oct. 2012.

[10] G. Leger, Doubly-segmented current-steering DAC calibration, *"IEEE Int. Conference On Design Technology of Integrated Systems In Nanoscale Era (DTIS),"* 2014, pp. 1-6.

[11] H.Tsai et al., "A 0.003mm2 10b 240 MS/s 0.7mW SAR ADC in 28nm CMOS with digital error correction and correlated reversed sitching," *IEEE J. Solid-State circuits,*vol. 50, no. 6, pp. 1382-1398, June 2015.

[12] K. K. Parhi, "VLSI digital signal processing systems: design and implementation," *John Wiley Sons,* 2007.

[13] K. S. Lee, Y. M. Lee, "Switched-capacitor cyclic DAC with mismatch charge compensation," *Electronics Letters,* vol. 46, no. 13, pp. 902-903, June 24 2010.

[14] Lei Wang, Fukatsu Y, Watanabe K, "A CMOS R-2R ladder digital-to-analog converter and its characterization," *Instrumentation and Measurement Technology Conference, 2001. IMTC 2001. Proceedings of the 18th IEEE,* vol.2, pp.1026-1031 vol.2, 2001.

[15] Lei Qiu, Kai Tang, "A Flexible weighted Nonbinary searching Technique for High-Saped SAR ADCs," *IEEE Transactions on Circuits and Systems II: Express Briefs,* vol. 24, no. 8, pp. 2808-2812, 2016.

[16] Taimur Rabuske and Jorge Fernandes, "A SAR ADC With a MOSCAP-DAC," *IEEE Journal of Solid-State Circuits,* Vol. 51, NO. 6, June 2016.

[17] W. Guo, T. Abraham, S. Chiang, C. Trehan, M. Yoshioka and N. Sun, "An Area- and Power-Efficient I_{ref} Compensation Technique for Voltage-Mode R-2R DACs," *IEEE Transactions on Circuits and Systems II: Express Briefs,* vol. 62, no. 7, pp. 656-660, 2015.

[18] W. Kester, "Basic DAC Architectures I-II MT-14/15, Analog Devices, available at www.analog.com," 2009.

[19] Wei Xu, Runxi Zhang, Chunqi Shi1, "Research of Segmented 8bit Voltage-Mode R-2R Ladder DAC," *IEEE Conference,*2015.

[20] Y. Liu, J. Gao and X. Yang, "24-bit low-power low-cost digital audio sigma-delta DAC," *in Tsinghua Science and Technology,* vol. 16, no. 1, pp. 74-82, Feb. 2011.

[21] Y. Chai and J.T.Wu, "A 5.37 mW 10 b 200 MS/s dual-path pipelined ADC," *in IEEE ISSCC Dig. Tech Papers,* Feb. 2012.

2018 IEEE Computer Society Annual Symposium on VLSI

A 375 nA Input Off Current Schmitt Triger LDO for Energy Harvesting IoT Sensors

Koichiro Ishibashi and Shiho Takahashi
Graduate School of Informatics and Engineering
The University of Electro-Communications
Tokyo, Japan
ishibashi@ee.uec.ac.jp

Abstract— This paper introduces 375 nA input off current Schmitt trigger LDO, which is suitable for receiving the power from high internal impedance energy harvesting power sources. The Schmitt trigger LDO consumes 375nA input current at the input voltage of 0.5V, and occupies 276 x 295 um area using 0.18 um CMOS technology. The proposed Schmitt trigger LDO is used to make an Energy Harvesting Illumination Beat Sensor Node, so that the sensor node wirelessly transmits the data of illumination from 780 to 1540 lx without battery.

Keywords—Energy harvesting, LDO, IoT, Sensor, Beat Sensor

I. INTRODUCTION (*HEADING 1*)

In the next IoT era, trillions of sensors will be distributed to environment. These sensors should utilize power from energy harvesting power sources. Many studies of power supply circuits for energy harvesting have been proposed[1,2]. As shown in Fig.1, since the voltages from energy harvester are not stable, the charge from the energy harvester is first stored in a storage capacitor. When enough charge is stored in the storage capacitor, Schmitt trigger type Low Dropout Regulator (LDO) supplies stable voltage to loads[1,2]. As shown in Table I, since internal impedance Z_{EH} of various energy harvester are always high, the input current of the Schmitt trigger LDO should be sufficiently small to store enough charge to the storage capacitor .

This paper introduces nA input current Schmitt trigger LDO, which is suitable for receiving the power from high internal resistance energy harvesting power sources. The Schmitt trigger LDO consumes only 375nA input current at the input voltage of 0.5V, and occupies 276 x 295 um area using 0.18um CMOS technology.

We also demonstrate the energy harvesting wireless illumination sensor, which utilizes the developed Schmitt Triger LDO and the IoT Beat Sensors[3-6]. The sensor node measures illumination from 780 to 1540 lx, and the sensor node wirelessly transmits the data without battery.

We introduce the circuit design and experimental results of the fabricated Schmitt trigger LDO in Section II. The design and experimental results of the illumination sensor node will be shown in Section III, followed by conclusions.

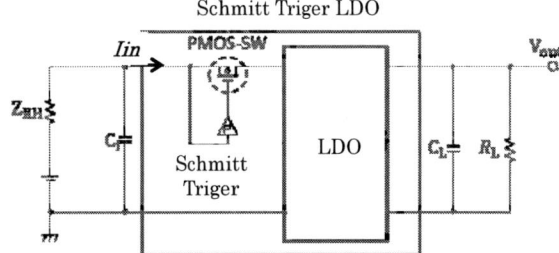

Fig. 1   Structure of Schmitt Triger LDO, where ZEH is impedance of Energy Harvester, Ci is input storage capacitor.

Table. I   Various kinds of energy harvesting sources and their internal impedance

Energy Source	Power Density (W/cm^2)	Impedance Z_{EH} (Ω)
Vibration	$10^{-3}\sim10^{-4}$	$10^3\sim$
Light	10^{-4}	$10^4\sim$
Heat	10^{-5}	$10^5\sim$
RF	10^{-6}	$10^6\sim$

II. SCHMITT TRIGGER LDO

A. Schmitt Trigger LDO Design

The circuit diagram of the Schmitt Triger LDO is shown in Fig.2 The circuit is consisted of Schmitt Trigger circuit, Switch PMOS, and LDO.

The operation of the Schmitt Triger LDO is as follows. The energy from energy harvester charges the Storage Capacitor, Ci, and input voltage Vin starts increasing. A comparator in the Schmitt Trigger circuit compares the voltage of Vin and Schmitt Trigger reference voltage (V+). Until the Vin becomes predetermined on voltage(Von), the comparator turns off the switch PMOS transistor (Mps) and transistor for the Schmitt Trigger (Mp1) so that the current mirror NMOS transistors, Mn1 and Mn2, also turn off. Only a little current flows on Mn1 and Mn2, so that the input current Iin can be decreased to be minimum value. When the Vin becomes Von, the Mps turns on and LDO works to supply the output voltage(Vout) to the road

978-1-5386-7100-9/18 $31.00 © 2018 IEEE   187

resistance RL. The voltage V+ is decreases then so that the turn off voltage(Voff) is reduced to make the Schmitt Trigger operation.

We designed the Schmitt Trigger LDO by using 0.18um CMOS Technology. The layout design, and micro photograph of the fabricated LDO is shown in Fig. 3. The size of the Schmitt Trigger circuit is 276.34 X 295.08 um². Large resistances, R2 of 2.3Mohm, and R3 of 1.3Mohm, are introduced to reduce the input current by the voltage divider circuits using R2 and R3, and the resistances occupy large area of the circuit. Relatively small resistances are adopted for R5 and R6.

We evaluated the fabricated chip. Fig. 4 shows the measured operation waveforms of the Schmitt Trigger circuits, where SINE wave voltage is applied to the input. When the input voltage becomes 3.48 V, the Schmitt Trigger turns on and LDO supply the output voltage Vout of 2.20V. The input voltage is then decreased and the Schmitt Trigger turns off to make the Vout of 0V.

As shown in Table II, Measured results of the Von, Voff, Vout voltage of the Schmitt Trigger LDO correspond to those of the simulated value. At Vin of 0.5V, measured input current of 375 nA is obtained as shown in Table III, which corresponds to large input impedance of 1.33 M ohm, so that the storage capacitor can store the charge from an energy harvester, of which internal impedance up to 1M ohm.

Table IV is the comparison table of the same type of LDO circuits. The proposed Schmitt Trigger LDO has advantages in terms of area and the input current at off state.

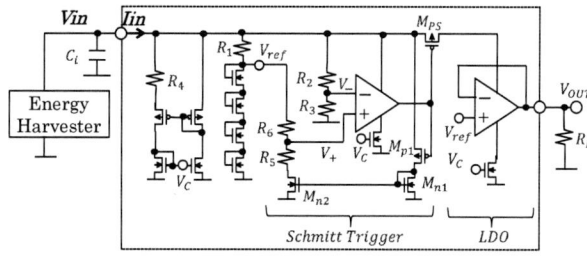

Fig. 2 The circuit of the designed Schmitt Trigger LDO

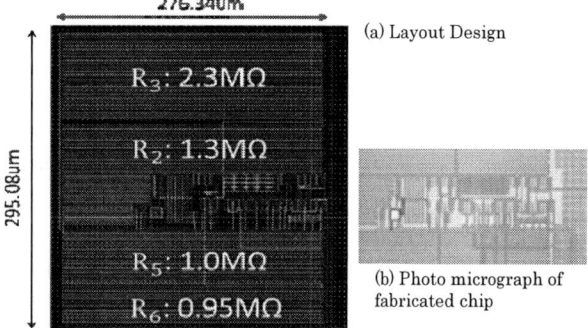

(a) Layout Design

(b) Photo micrograph of fabricated chip

Fig. 3 (a) Layout design of the Schmitt Trigger LDO, and (b)photomicrograph of the fabricated chip.

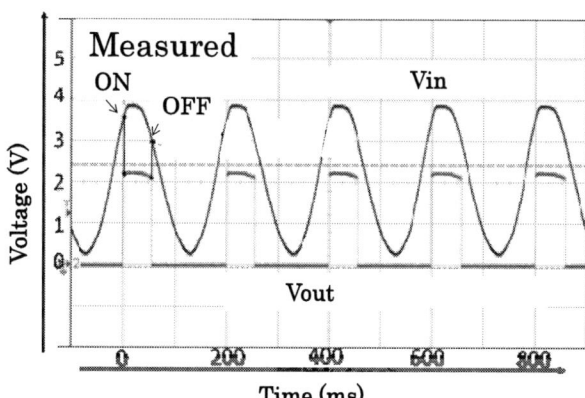

Fig. 4 Measured operating waveforms of the Schmitt Trigger circuits.

Table II, Simulated and measured results of the voltages of the Schmitt Trigger LDO.

	Von	Voff	Vout
Simulated	3.55 V	2.64 V	2.44 V
Measured	3.48 V	2.98 V	2.20 V

Table III, Simulated and measured results of the input current of the Schmitt Trigger LDO.

Vin	0.5 V	1.8V
Iin, Simulated	0.426 uA	4.023 uA
Iin, Measured	0.375 uA	3.909 uA

Table IV Comparison Table.

	Proposed (Measured)	JSSC, 2018 [1]	IMSCS,2010 [2]
CMOS Technology	0.18 um	0.25um	0.35um
AREA	81, 420 um²	110, 000 um²	103, 000 um²
Min Input Voltage	2.98V	—	2.4V
Output Voltage	2.20V	1-3V	2.2V
Input Current Iin	375 nA	1.24 uA	31 uA

III. APPLICATION TO THE SCHMITT TRIGGER LDO TO ENERGY HARVESTING ILLUMINATION SENSOR NODE

We evaluated the proposed Schmitt Trigger LDO by applying it to energy harvesting illumination sensors. Beat Sensors have been recently proposed as persistent and energy harvesting wireless sensing scheme[3 – 6]. The concept of the Beat Sensors is that the physical data of the sensor is sent as interval time of the ID signals wirelessly transmitted, so that a receiver receives the ID signals and calculates the physical data by the interval times of the ID signals. We adopt the Beat Sensors concept to realize the illumination sensors using the Schmitt Trigger LDO.

Figure 5 shows the block diagram of the illumination Beat Sensor node and the receiver. Sphelar type photovoltaic cell [7] is used not only for illumination sensor but also for energy harvester. The illumination Beat Sensor node consists of storage capacitor of 220 uF, the Schmitt Trigger LDO shown in section II, and MCU and RF module. Nordic 2.4GHz RF module[8] is used for this experiment.

Nordic 2.4GHz RF module is also used as the receiver. The data is sent to PC to analyze the data.

The operation of the Illumination Beat Sensor node is expressed as follows. When light is illuminated to the photovoltaic cell, it generates the current depending on the intensity of the light. Since the input current of the Schmitt Trigger LDO is small, the charge by the current is stored to the storage capacitor C_i. Vin which corresponds to the voltage of the storage capacitor increases when the Schmitt Trigger LDO turns off. When the input voltage becomes the Schmitt Trigger turn on voltage, the LDO is turned on to supply the supply voltage to the MCU and the RF module. Then the MCU generates ID signal followed by ID data wireless transmission by the RF module. Since the generated current by the photovoltaic cell depends on the intensity of the illumination, the charging time to the storage capacitor is changed depending on the light intensity. So this sensor node acts as illumination sensor. Since energy generated by the photovoltaic cell is eventually the energy for the MCU and RF operation, this illumination Beat Sensor node operates by an energy harvesting power supply, and no battery is needed.

Figure 6 shows characteristics of the photovoltaic cell we used for this experiment. This photovoltaic cell has 12 cells, which are connected in the manner of 6 in series, and 2 in parallel. So the open circuit voltage is around 3V. The short circuit current depends on the intensity of the illumination. The equivalent internal resistance also depends on the illumination, and are 0.95 K, 1.25K, and 2.14K ohm, at 2000, 1000, and 500 lx, respectively. Since the input impedance of the Schmitt Trigger LDO is far larger than the internal resistance of the photovoltaic cell, the storage capacitor stores the energy from the photovoltaic cell.

Figure 7 shows Operating waveforms of Vin and Vout of the Schmitt Trigger LDO, when 1030 lx light is illuminated to the photovoltaic cell. It is shown that the Schmitt Trigger LDO is turned on when the Vin becomes Schmitt Trigger high value, and the LDO supplies the LDO "ON" voltage for 60ms. When Vin is decreased to Schmitt Trigger low level, the LDO stops supply the supply voltage. Then the Vin starts increasing again.

Figure 8 shows the LDO ON time depending on the illumination. We have observed the ID signal transmission from RF modules while the sensor node is illuminated from 780 to 1540 lx, so that the LDO ON time is also depending on the illumination intensity. It is shown that the illumination Beat Sensor nodes using the proposed Schmitt Triger LDO is successfully realized.

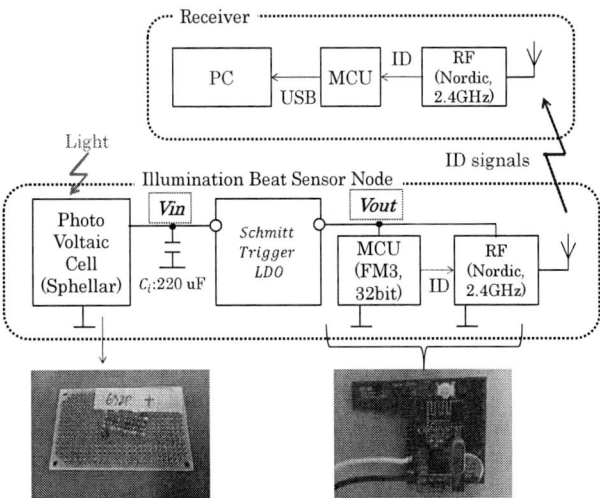

Fig. 5 Block diagram of the Illumination Beat sensor nodes using the Schmitt Trigger LDO, and photos of Photo Voltaic Cell and RF module with MCU.

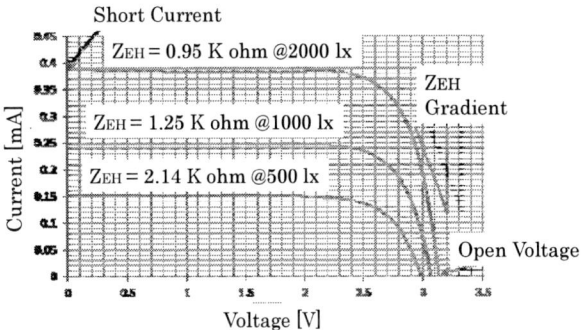

Fig. 6, Measured characteristics of the Sphelar photo-voltaic cell

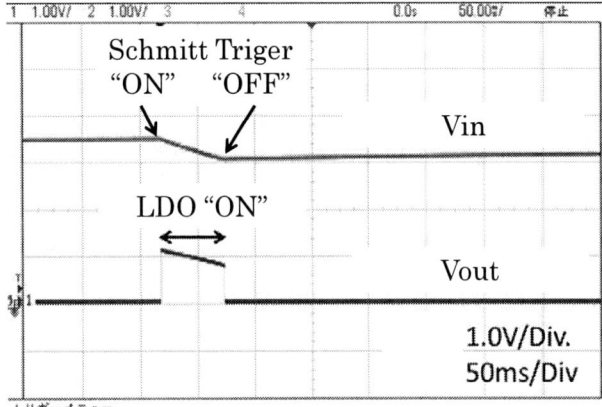

Fig. 7, Operating waveforms of Vin and Vout for the Illumination Beat Sensor node.

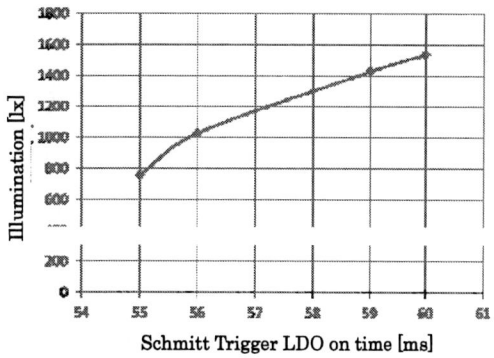

Figure 8. LDO "ON" time depending on the illuminations.

IV. CONCLUSIONS

This paper introduces nA input current Schmitt Trigger LDO, which is suitable for receiving the power from high internal resistance energy harvesting power sources. The Schmitt Trigger LDO consumes 375nA input current at the input voltage of 0.5V, and occupies 276 x 295 um area using 0.18um CMOS technology. The proposed Schmitt trigger LDO is used to make an Energy Harvesting Illumination Beat Sensor node, so that the sensor node wirelessly transmits the data of illumination from 780 to 1540 lx without battery.

ACKNOWLEDGMENT

This was also supported by VLSI Design and Education Center (VDEC), the University of Tokyo in collaboration with Synopsys, Inc. and Cadence Design Systems, Inc.

REFERENCES

[1] R. Magod, et al , " A 1.24 μA Quiescent Current NMOS Low Dropout Regulator With Integrated Low-Power Oscillator-Driven Charge-Pump and Switched-Capacitor Pole Tracking Compensation," IEEE Jornal of Solid-State Circuits, pp.1-12, Apr. 2018.

[2] Edward N. Y. Ho, et al , "A chip-area efficient capacitor-less CMOS LDO with active feedback and damping zero compensation", 2010 53rd IEEE International Midwest Symposium on Circuits and Systems, pp. 274-577, 2010

[3] S. Ishigaki, and K. Ishibashi, "Wireless Electric Power Sensing Scheme for BEMS," IEEE ICBEST 2015, Singapore

[4] R. Takitoge, K. Ishibashi et. al.,"Temperature Beat: Persistent and Energy Harvesting Wireless Temperature Sensing Scheme, "IEEE SENSORS 2016, Orlando

[5] K. Ishibashi, et. al., "DC Current Beat: Wireless and Non-Invasive DC Current Sensing Scheme", Eurosensors 2017, Sep., Paris 2017

[6] R. Takitoge, K. Ishibashi et. al.," Low-Power Enhanced Temperature Beat Sensor with Longer Communication Distance by Data-Recovery Algorithm, "IEEE SENSORS 2017, Glasgow

[7] http://www.sphelarpower.jp/product/pdf/datasheet_KSP-F12-xSxP-W1-X_v0912(J).pdf

[8] http://nordicsemi.com/eng/Products/2.4GHz-RF/nRF24L01

[9] http://www.datasheet4u.com/datasheet-pdf/Spansion/MB9AF132KB/pdf.php?id=1032627

978-1-5386-7100-9/18 $31.00 © 2018 IEEE

Precise Duty Cycle Variation Detection and Self-Calibration System for High-Speed Data Links

Karen Khachikyan
Synopsys Armenia CJSC
Yerevan, Armenia
karenkh@synopsys.com

Abraham Balabanyan
Synopsys Armenia CJSC
Yerevan, Armenia
abrahamb@synopsys.com

Hrachya Gumroyan
Synopsys Armenia CJSC
Yerevan, Armenia
gumroyan@synopsys.com

Abstract— **A design and simulation methodology that detects and compensates duty cycle deviations is presented. The proposed method provides robust mechanism to reduce transmission line adverse effects and improves received signal quality. A mixed signal approach, where an analog circuit is used to track signal timing distortion values, and a digital circuit controls the analog calibration mechanism, is used. The self-calibration mechanism doesn't interrupt the system operation and is being realized in parallel with normal operation. The system is designed in 28nm CMOS process and simulated using Synopsys mixed mode simulation tools.**

Keywords— jitter, duty cycle variation, calibration, PVT compensation, signal integrity, I/O, mixed-signal, DCD, DDR

I. INTRODUCTION

Modern high-speed integrated circuits (IC) with continuous technology shrinking are very sensitive to process, voltage, and temperature (PVT) variations. On the other hand, fabrication process of ICs becomes more complicated and has substantial influence on quality and performance of IC components. IC components parameters deviation can affect system performance, as well as make it unstable and inaccurate. Examples of the most sensitive blocks are input/output (I/O) blocks, which are located on the border of ICs. Moreover, the far ends of the chip can occur in different PVT conditions. Due to mentioned phenomena, period of the received signal can deviate from nominal values and cause timing issues. Transmission lines are another very important factor in high speed data links, as they cause signal reflections which affect signal shape and timing, as well as degrade a duty cycle value. Duty cycle distortion (DCD) jitter appears when deviations of transmitter and receiver thresholds are offset from their ideal level. Another important cause of DCD jitter is asymmetry in rising and falling transitions. Duty cycle is an essential parameter for the signal which should be as close to 50% of period as possible.

Most of modern high speed data links have duty cycle correction (DCC) mechanisms, such as reference voltage (*Vref*) training [1]. In this case *Vref* compensation is being done at system level and receiver operates afterword. That means that variations that occurred after *Vref* training due to PVT fluctuations, will not be compensated. Thus, duty cycle degradation can't be budgeted during data transmission and can cause system failures.

Most of existing DCC approaches are using analog circuits with feedback loops [2-5], and few of them are digital

implementations with or without feedback loops [6-8]. In [2] negative feedback loop technique is presented, where a pulse width modification cell processes the incoming differential waveforms. The latter is inverter based circuit with controllable slew rate. Another analog technique [3] contains two stages of analog feedback circuits. The first stage consists of two amplifiers to make two separate negative feedback loops. Second stage of the loop consists of another amplifier and corrects the final output clock duty cycle near to 50%. An analog duty-cycle correction circuit which uses a current-starving method is presented in [4]. In [5] an analog loop with a digitally controlled charge pump is described. A non-feedback digital implementation is using a half cycle delay line as presented in [6]. Digital feedback implementations operate based on a binary search algorithm [7]. Another digital feedback implementation is presented in [8]. The main advantage of digital implementations is faster settling time, but they can't provide precise compensation like analog implementations.

In next chapters proposed duty cycle variation detection and self-calibration mechanism and its mixed-signal circuit implementation is described. It successfully compensates duty cycle deviations caused by PVT variations and transmission line effects.

II. SYSTEM ARCHITECTURE

Proposed self-calibration mechanism detects duty cycle degradation during whole period of I/O circuit operation and starts calibration process, if duty cycle deviations are detected. In modern ICs, reference voltage of the receivers is either applied externally or generated from the internal digital-analog converter (DAC). The output of proposed duty cycle correction circuit is connected to the internal DAC input (Fig. 1).

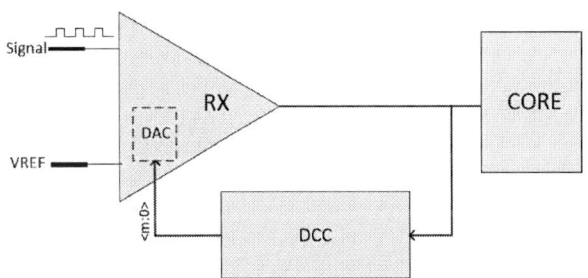

Fig. 1. Proposed DCC block placement in data link.

Implemented DCC self-calibration system, shown in Fig. 2, consists of the following major blocks: RC integrator, analog to digital converter (ADC) block, reference voltage divider circuit, analog multiplexer, register transfer logic (RTL) block.

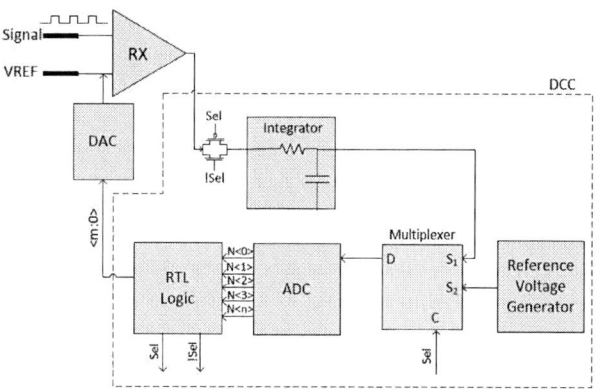

Fig. 2. Proposed duty cycle detection and self-calibration mechanism architecture.

An RC integrator was used to get the integrated DC value of the received signal which reflects duty cycle value of the signal. The RC integrator settling time for the worst case is ~50ns which defines the cycle duration of the system operation. The RC integrator circuit is shown in Fig. 3.

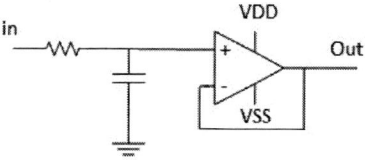

Fig. 3. RC integrator circuit.

The ADC block serves for getting digital codes corresponding to the integrated DC and reference voltages. To capture reference code, reference voltage divider circuit is used (see Figure 4), which output voltage corresponds to nominal duty cycle value of the receiver output signal.

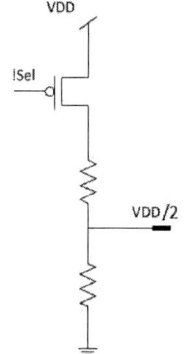

Fig. 4. Reference voltage divider circuit.

PMOS transistor in Fig. 4 serves as a switch to exclude DC current path from VDD to GND. Divided VDD/2 voltage is a nominal reference voltage to provide 50% duty cycle value for the receiver output. Nominal reference voltage can deviate due to VDD supply voltage variations. Power supply deviation will cause signal level changes, but as the calibration is being performed using the VDD proportional reference voltage, VDD/2 remains nominal for each PVT corner. Analog multiplexer selects either integrated DC value of signal or reference voltage to apply to the ADC. It's important to have properly designed multiplexer to avoid voltage level errors due to leakage currents, otherwise it will affect system accuracy. After conversion digital codes corresponding to nominal and current signal duty cycle are provided to RTL block. RTL block compares them and changes the receiver's DAC input code respectively, if difference between the mentioned codes is detected. RTL block fixes the duty cycle deviation direction and changes the DAC input code to proper direction. With the modified codes applied to its inputs, DAC generates new "Vref" voltage and applies it to I/O receiver decreasing the duty cycle deviation of received signal. After few cycles system compensates duty cycle deviation. The main advantage of the proposed mixed-signal self-calibration mechanism is that it doesn't interrupt the system operation and is being realized in parallel with normal signal receiving operation.

III. OPERATION PRINCIPLE AND SELF-CALIBRATION

System operation is triggered by the positive edge of the "start" signal which leads the logic module to start duty cycle detection and correction process. At the initial calibration stage receiver's "Vref" voltage is set to its nominal value. Initial "Vref" voltage is generated according to digital code applied to DAC. The code is programmable by user per system requirements. For example, nominal input reference voltage value for LPDDR4 receiver will be VDDQ/6, which is defined in JEDEC standard [9]. By positive "sel" signal reference voltage divider circuit is enabled, and the nominal duty cycle value applies to ADC input. By the end of conversion operation, nominal duty cycle code is being stored in "ref_code" register. Then "sel" signal is switched to negative level which enables the RC integrator path to detect duty cycle variations by integrating received signal. When the integration process is finished, a voltage value is converted and the corresponding code is written in "current_code" register. After this stage, digital comparator is enabled, which compares current duty cycle value ("current_code") with the nominal one ("ref_code") which corresponds to 50% duty cycle value. RTL block diagram is presented in Fig. 5.

If "current_code" is greater than the reference (Comp_out=1), which means actual signal pulse width is more than the half of period, RTL block increments the DAC input code to increase RX "Vref" voltage value and decrease the duty cycle of the receiving signal. Otherwise, if "current_code" is lower than the reference (Comp_out=0), which means actual signal pulse width is less than the half of period, the RTL block decrements the DAC input code to

978-1-5386-7100-9/18 $31.00 © 2018 IEEE   192

decrease "*Vref*" voltage value and increase the duty cycle. This operation allows improving received signal duty cycle systematically.

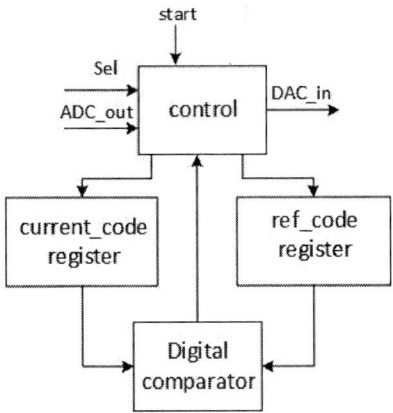

Fig. 5. RTL block diagram.

This logic cycle is being repeated until the output of digital comparator is toggled. It means that calibration is completed and current duty cycle value is equal to its nominal value. When optimal RX "*Vref*" value is captured and duty cycle variation is compensated, the systems continues toggling and DAC output is being toggled by less significant bit (LSB) around the calibrated code. The duty cycle detection and self-calibration logic diagram is shown in Fig.6.

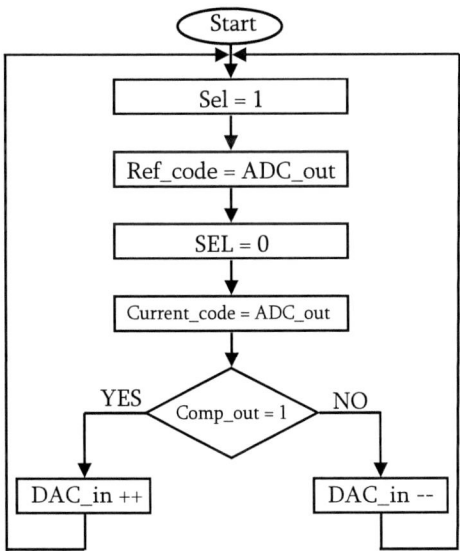

Fig. 6. Operation principle logic diagram.

Duty cycle variation detection accuracy depends on ADC resolution, while calibration accuracy is directly proportional to DAC resolution. Equation 1. shows minimal duty cycle deviation that can be detected by N-bit ADC:

$$\min \% = \frac{(V_{HIGH} - V_{LOW}) \cdot VDD}{2^N} \cdot 100\%, \quad (1)$$

where V_{HIGH} and V_{LOW} are ADC input reference range extreme voltages, N is the bit numbers of ADC and VDD is supply voltage.

For the proposed design 4-bit flash ADC circuit and 7-bit R-string DAC are chosen. In this paper the ADC input reference range is from 350mV to 500mV. 350mV corresponds to ~40% duty cycle value and 500Mv - to ~60% for 0.85V power supply and nominal conditions. This means, the system can detect approximately 1% percent duty cycle deviation. The calibration accuracy can be increased by extending the resolutions of ADC and DAC. For example, duty cycle calibration system detection accuracy with 6-bit ADC is ~0.28% and for 8-bit ADC it is ~0.07%.

IV. SIMULATION RESULTS

Simulations have been performed using Synopsys circuit level simulator Hspice [10] for 27 PVT corners, including SS slow-slow, TT (typical-typical), FF(fast-fast), SF (slow-fast), FS(fast-slow) with supply voltage and temperature variations using 28nm CMOS technology node.

Duty cycle compensation system is compatible for most of data links and is tested on LPDDR4 standard DDR PHY. The operating frequency is 2133 MHz, nominal core (VDD) and I/O (VDDQ) voltages are 0.85v and 1.1v respectively. The VIH and VIL levels of receiver input signal are VDDQ/3 and 0v.

Fig. 7 shows DCC system simulation results for TT corner and 25C temperature using nominal supply voltages. For testing the DCC system operation, input signal with 56% duty cycle is provided to receiver.

RC_out is the DC value of integrated input signal which changes continuously during self-calibration process. The voltage is changed, as RX input signal duty cycle is being improved thanks to ADC/DAC code changes by logic block in feedback circuit. "*Vref*" is the reference VDDQ/2 voltage generated by reference voltage divider circuit and is 0.425v for typical corner. *ADC_in* is the analog multiplexer output which selects ADC input voltage between reference and integrated values. *RX_vref* is the reference voltage of the receiver which is generated by DAC according to calibrated codes. Initial value of *RX_vref* is VDDQ/6 (nominal reference for 50% RX input signal duty cycle) and it is being continuously changed during operation until 50% RX output duty cycle is achieved. *RX_out* is the measured duty cycle of the receiver output, which is decreasing from about 56% to 50%, as *RX_vref* is increasing from 0.184v to 0.309v.

It is seen from the simulation plots that 50.1% duty cycle value is achieved after 3us run time. However, after 1us operation there is significant RX output signal duty cycle improvement. For example, at 1.55us duty cycle is about 51.8%, at 2.25us it is 50.9%.

978-1-5386-7100-9/18 $31.00 © 2018 IEEE

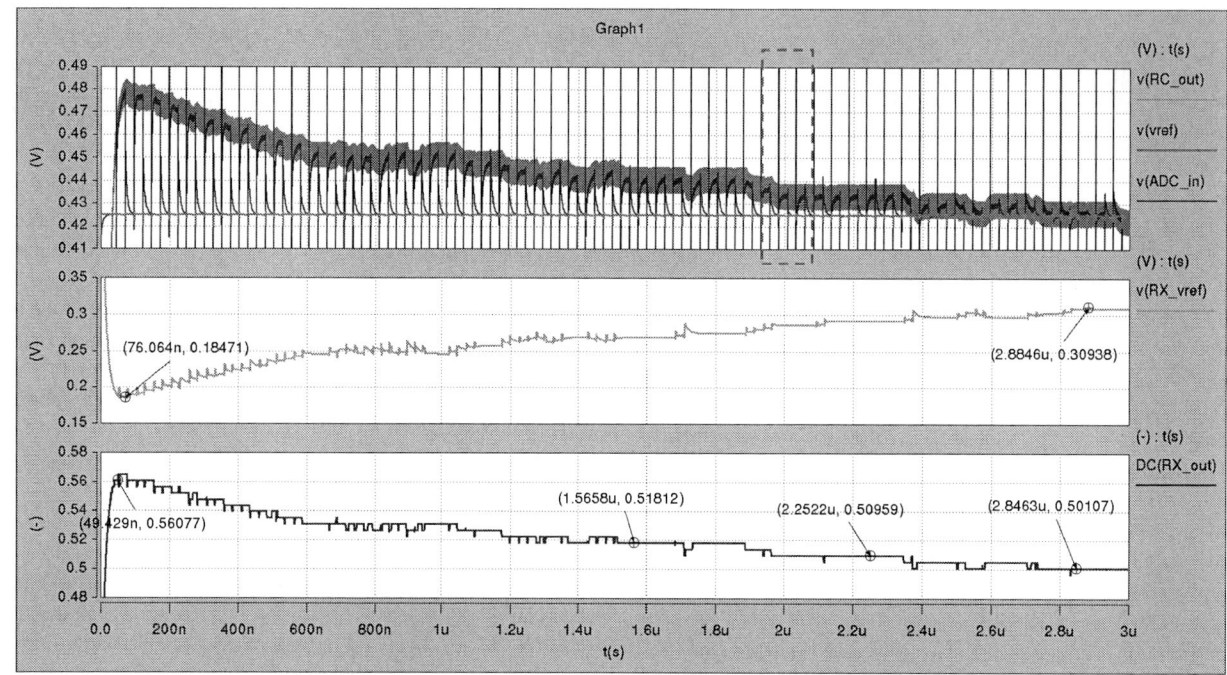

Fig. 7. Duty cycle correction system simulation results for typical corner.

RX input signal integration proceeds in up to 50ns in worst case corner. Considering mentioned process duration, RTL block toggles control *"sel"* signal every 25n. One correction cycle (period of *"sel"* signal) is 50ns. During the first half of period, reference value is selected by multiplexer and passed to ADC block, during the second half, DC value of receiving signal is selected.

In Fig. 8 zoomed version of highlighted region in Fig. 7 is presented. It is seen from the plots that by logic high level of *"sel"* signal, reference value with VDD/2 level is applied to *"ADC_in"* input. Also, by logic low level of *"sel"* signal, DC level of integrated signal is applied.

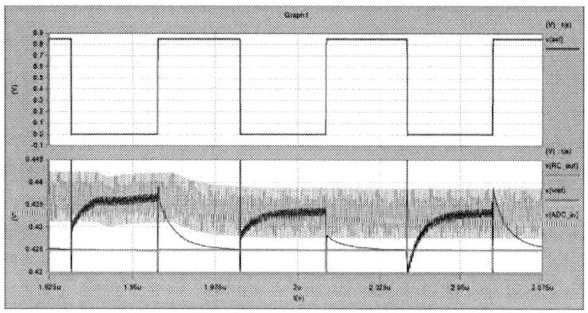

Fig. 8. ADC input selection between integrated and reference voltages.

The "Table 1" presents simulation results for the main 5 PVT corners. It is seen from the table that for the worst case duty cycle is improved to 50.5% at the corner of SS, 125C temperature and 0.8V supply voltage. Overall, compensated duty cycle is varying from 49.7% to 50.5% across all PVT corners.

The highest 3.95mW power consumption is at corner FF 125C temperature with 0.9 power supply voltage.

TABLE I. CORRECTED DUTY CYCLE RESULTS FOR THE MAIN PVT CORNERS

Corner	Duty Cycle	Power (mW)	Supply (V)
TT 25C	0.501	3.66	0.85
FF -40C	0.497	3.95	0.9
SS 125C	0.505	3.2	0.8
SS -40C	0.502	3.3	0.8
FF 125C	0.499	3.77	0.9

The DCC system was tested for two directions of duty cycle deviations, from 43.6% to 58.2%. In "Fig. 9" duty cycle variation compensation process is presented for three extreme corners. The proposed design allows to detect duty cycle variations in range of 40% ÷ 60%, which is limited by input voltage operation range of ADC block. The operational range was chosen according to existing maximal duty cycle deviations in modern data links.

The compensation duration is approximately the same for all PVT corners and does not depend on duty cycle variation direction.

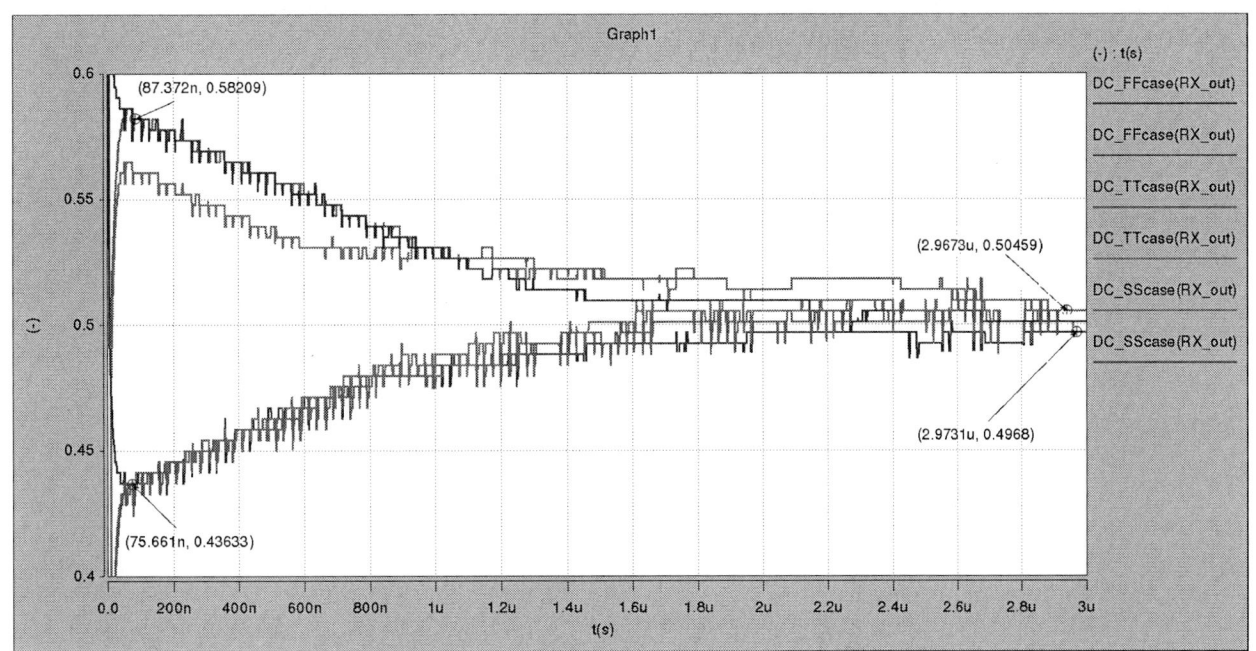

Fig. 9. Duty cycle variation correcrtion process for the extreme PVT corners.

Proposed DCC mixed-signal solution is applicable not only for compensating clock signal variations, but also data signals.

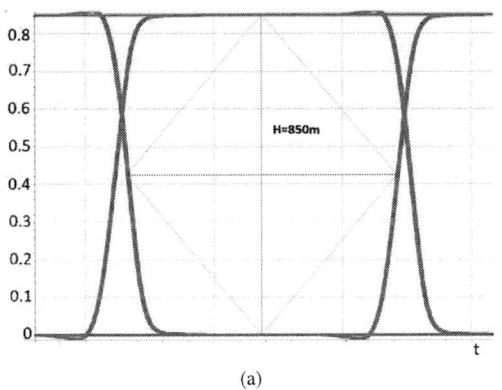

(a)

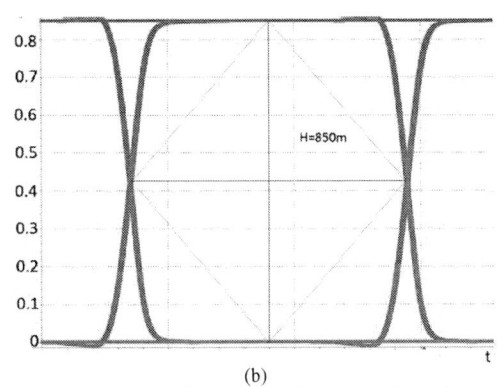

(b)

Fig. 10. Eye diagrams of RX output signal for PRBS5 input data pattern before (a) and after (b) compesantion

The proposed DCC operation is being realized on clock signal and the captured compensated *"Vref"* RX input reference voltage is being applied to all nearby data receivers. The test is done for PRBS5 data pattern input signal at 2133MHz. Initial pulse width of the signal is changed like clock signal pulse width with degraded duty cycle. As a result, cross-point of the signal is 0.7*VDD and shifted from VDD/2 level as is seen from eye diagram shown in "Fig. 10(a)". For this case *"Vref"* level of the RX is nominal VDDQ/6 (0.184v for typical corner). When compensated 0.309v reference voltage is applied to data RX input, cross-point of the RX output signal is improved significantly and 0.5*VDD desired level is reached, as is seen from eye diagram in "Fig. 10(b)".

A summary comparison between current and other DCC solutions is presented in "Table 2".

TABLE II.   COMPARISON TABLE

Work	[2]	[3]	[4]	[8]	This Work
Duty Cycle precision	1%	2%	1%	0.25%	0.5%
Correction Range	30%-70%	25%-75%	30%-70%	20%-80%	40%-60%
Power Consumption	20mW	1.4mW	1.1mW	3.2mW	3.95mW
Operating Frequency	3.5GHz	2GHz	600MHz	1.7GHz	2.1GHz
Process	130nm	45nm	350nm	180nm	28nm
Supply Voltage	-	0.9v-1.4v	3.3v	-	0.8v-0.9v

V. CONCLUSION

A duty cycle variation detection and self-calibration mechanism, implemented in 28nm CMOS technology node, is presented. The mixed-signal solution is based on comparison of DC average value of degraded received signal and reference voltage. This approach can be used for timing sensitive systems to compensate for the effect of PVT variation and transmission line reflections on system performance. The main advantage of designed system is parallel duty cycle correction during data receiving process.

The proposed solution is tested on LPDDR4 standard DDR receiver with 1.1v I/O and 0.85v core nominal supply voltages at 2133 MHz operating frequency. The system detects duty cycle variations in range of 40% ÷ 60% and corrects duty cycle value to 50%±0.5%. Simulations are performed for 27 PVT corners, including temperature range of -40C ÷ 125C and supply voltage range of 0.8v ÷ 0.9v. Power consumption of designed DCC system is approximately 3.95mW for the worst case corner.

Proposed DCC mixed-signal solution can be used for data receivers as well. The efficiency is tested with PRBS5 standard input data pattern with degraded pulse width. After compensation cross point of data signal was improved from 0.7*VDD to 0.5*VDD level.

REFERENCES

[1] S. Sethuraman, A. Lingambudi, K. Wright, et al., "Vref optimization in DDR4 RDIMMs for improved timing margins," IEEE Electrical Design of Advanced Packaging & Systems Symposium (EDAPS), pp. 73-76, December 2014.

[2] I. Raja, G. Banerjee, M. A. Zeidan, and J. A. Abraham, "A 0.1–3.5-GHz duty-cycle measurement and correction technique in 130-nm CMOS," IEEE Trans. Very Large Scale Integr. (VLSI) Syst., vol. 24, no. 5, pp. 1975–1983, May 2016.

[3] R. Mehta, S. Seth, S. Shashidharan, B. Chattopadhyay, and S. Chakravarty, "A programmable, multi-ghz, wide-range duty cycle correction circuit in 45nm cmos process," in 2012 Proceedings of the ESSCIRC, pp. 257-260, 17-21 September 2012.

[4] P. Chen, S.-W. Chen, and J.-S. Lai, "A low power wide range duty cycle corrector based on pulse shrinking/stretching mechanism," in Proc. IEEE Asian Solid-State Circuits Conference, pp. 460–463, November 2007.

[5] K.-H. Cheng, C.-W. Su, and K.-F. Chang, "A high linearity, fast-locking pulsewidth control loop with digitally programmable duty cycle correction for wide range operation," IEEE J. Solid-State Circuits, vol. 43, no. 2, pp. 399–413, February 2008.

[6] Y.-M. Wang and J.-S. Wang, "An all-digital 50% duty-cycle corrector," in Proc. Int. Symp. Circuits Syst., May 2004, pp. II-925–II-928.

[7] S.-K. Kao and S.-I. Liu, "A wide-range all-digital duty cycle corrector with a period monitor," in Proc. IEEE Conf. Electron Devices Solid-State Circuits, pp. 349–352, December 2007.

[8] Y. C. Jang, S. J. Bae, and H. J. Park, "CMOS digital duty cycle correction circuit for multi-phase clock", Electron. Lett., vol 39, pp.1383 – 1384, September 2003.

[9] JEDEC Standard. 2015. Low Power Double Data Rate 4 (LPDDR4), in JEDEC Solid State Technology Association, 272p., Nov. 2015.

[10] Hspice Application Manual. 2010. Synopsys Inc. 196p.

Parametric Circuit Optimization with Reinforcement Learning

Changcheng Tang
Tsinghua University, Institute of Microelectronics
Beijing
Email: tangcc15@mails.tsinghua.edu.cn

Zuochang Ye
Tsinghua University, Institute of Microelectronics
Beijing
Email: zuochang@tsinghua.edu.cn

Yan Wang
Tsinghua University, Institute of Microelectronics
Beijing
Email: wangy46@tsinghua.edu.cn

Abstract—**In this paper, we focus on solving parametric optimization problems. Such kind of problems is very commonly seen in reality. We propose an efficient method to train a model that connects the solution to the parameters and thus solve all the problems with the same structure and different parameters at the same time. During the training process, instead of solving a series of optimization problems with randomly sampled parameters independently, we adopt reinforcement learning to accelerate the training process. Two networks are trained alternately. The first network is a value network, and it is trained to fit the target loss function. The second network is a policy network, whose output is connected to the input of the value network and it is trained to minimize the output of the value network. Experiments demonstrate the effectiveness of the proposed method.**

Keywords-**circuit optimization; reinforcement learning; parametric optimization**

I. INTRODUCTION

Global optimization algorithms had been studied in many years. An efficient, simple and robust optimization algorithm is essential in many fields of science and engineering. Conventional global optimization algorithms use to search as board range as it could to find the minima, like evolutionary algorithms (EAs)[1][2], or searching minima following the gradient of object functions, like stochastic gradient descent (SGD), momentum[Nesterov, 1983, Tseng, 1998], RMSprop [Tieleman and Hinton, 2012], and Adam [Kingma and Ba, 2015], etc.

In many applications, people may encounter parametric optimization problem, i.e. a set of problems that share the same structure and they only differ parametrically. A representative application is circuit optimization. The topology of the circuit to be optimized is usually fixed, such as 6-T SRAM, logic gates, and amplifiers. Such circuits have been studied thoroughly, however, engineers still need to pay the effort to optimize the circuit again when conditions e.g., process nodes, power supply voltage, and constraints are changed. If such parametric optimization can be solved, such effort is no longer necessary.

Motivated by Learning to Learn by Descent by Descent[4], we noticed that we can adopt a well-trained optimizer to replace human-designed optimizers which provide updating values of variables of object function for each iteration in the optimization process.

To train an optimizer which is constructed by a neural network by supervised learning does not work[5]. Fortunately, reinforcement learning has the exact same formulation with the task of training optimizer. The main concept of reinforcement learning is that define a reward function about specific action and state of the system, then impose the penalty on poor performance optimizer[5][6].

In this paper, we propose a global optimization method that optimizes a serial of parametric objective functions by training an optimizer. We modeled policy and state-value function by neural networks and trained them by using DDPG (Deep Deterministic Policy Gradients) algorithm in reinforcement learning[7], and achieved good performance in some global optimization cases.

II. PARAMETRIC GLOBAL OPTIMIZATION

The problem this paper is focusing on global optimization in the following form

$$\min_{\theta} f(\theta, w) \qquad (1)$$

where $f(x, w)$ is the objective function to be optimized, w is a vector that parameterize the objective function.

A naive way to solve the parametric optimization problem is to sample in the parameter space, and for each sampled w_i, solve the optimization problem to get the optimal solution θ_i, and finally train a network that connects the parameter w to the best solution x with supervised learning.

The problem with this naive method is that for each different parameter, the optimization process has to be done from scratch, which is resource-wasted and time-consuming. It is particularly true when the variable space and parameter space are large. In this case, randomly sampled parameter vector tends to be sparse in the high dimensional space,

Algorithm 1 Procedure of Optimization Algorithms

Object function f

    $\theta_0 \leftarrow$ randomly initialized

2: **for** $t = 1, ..., T$ **do**

      if stop condition **then**

4:        return θ_t

      $\theta_t \leftarrow \theta_{t-1} + g(f(\theta_{t-1}), \theta_{t-1}, \phi)$

6: **return** θ_T

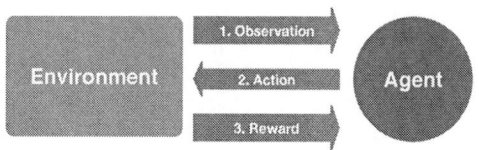

Firstly, the agent makes an observation of environment. Secondly, the agent takes action for current observation. Lastly, environment updates its status and return the reward to an agent.

Figure 1. Evironment and agent.

and it requires samples that increase exponentially with the number of parameters to reserve the sample density.

Our idea is that train an optimizer based on reinforcement learning to generate the optimal searching path like what the conventional optimizers do rather than build a simple regression model.

A. Learning to Optimize

A unconstrained single-objective task of optimization can be expressed as: find the minimum $\theta^* = \arg\min_\theta f(\theta)$, where $f(\theta)$ is object function. There are many global optimization methods that we discussed above, most of them are inspired the evolutionary algorithms: GAs, EP, ESs, GP, differential evolution (DE), and so on[1][3]. They are called zero-order optimization, which means they do not rely on the gradient of object function. On the contrary, gradient descent is popular in deep learning because its efficiency and good performance of its variants, including stochastic gradient descent(SGD), momentum [Nesterov, 1983, Tseng, 1998], Adagrad [Duchi et al., 2011], RMSprop [Tieleman and Hinton, 2012], and Adam [Kingma and Ba, 2015]. Whatever optimization we choose, it looks like that we build an optimizer to generate updating value to optimizee: $\theta_{t+1} = \theta_t + g(f(\theta_t), \theta_t, \phi)$, where g is the optimizer specified by parameter ϕ.

B. Reinforcement Learning

Usually, a standard reinforcement learning problem can be described as an agent $\mathcal{A}$ interacting with an environment $\mathcal{E}$ in a discrete time system[8]. Environment might be a very complex system, and we can consider it as an abstract system or a black-box represented by high-dimension state $\mathbf{s_t}$, and we can get different observation $\mathbf{s_t}$ for each state $\mathbf{s_t}$. Agent takes an action $\mathbf{a_t}$ depending on $\mathbf{o_t}$ to change the state of environment from $\mathbf{s_{t+1}}$ to $\mathbf{s_t}$, then we can define the reward $r_t \in \mathbb{R}$ of action $\mathbf{a_t}$ at current status $\mathbf{s_t}$[8]

The way to evaluate the performance of action is the action-value function:

$$Q(s,a) = \max_\pi \mathbb{E}[r_t + \gamma r_{t+1} + \gamma^2 r_{t+2} + ... | s_t = s, a_t = a, \pi]$$
$$(2)$$

where γ is reward discount factor, and π is the policy function which defines the behavior of agent.Usually, we use a neural network to approximate this function[8].

The determination of algorithms is to optimize the stochastic policy π: $\mathcal{O} \times \mathcal{A} \to \mathbb{R}_{\geq 0}$, and it also can be simply expressed as $\mathbf{a_t} \sim \pi(\mathbf{a_t}|\mathbf{o_t})$[10][13]

There are discrete action space algorithms, like Q-learning and DQN[19], and continuous action space algorithms, like DDPG and PPO, etc[7]. In discrete action space problem, the number of action is finite, thus the algorithm may enumerate all actions and calculate its Q value then find the maximum. The "deep Q network" (DQN) algorithm use neural network to approximate value function $\mathbf{Q}(s) = [Q(s,a_1), Q(s,a_2), ..., Q(s,a_n)]^T$, where n is the number of action[8]. It enumerates all possible actions and then chooses the one which gets the biggest Q value. DQN is robust and efficient, but it can only address discrete low-dimensional space problem. When the dimension of action increases linearly, the possible number of action increases exponentially. We want the algorithm can work in continuous and high-dimensional space environments, and the DDPG algorithm may address those issues.

C. DDPG: Deep Deterministic Policy Gradients

DDPG is a continuous-state space algorithm. Because states and actions are infinite set in the continues-state space, it is impossible to calculate all Q-value at each action $\mathbf{a_t}$ like DQN in discrete-state space. To address this issue, DDPG uses neural network to learn in large state and action spaces. Directly learning Q-value with neural networks, however, has been proved to be unstable in many environments since the function $Q(\mathbf{s_t}, \mathbf{a_t}|\theta^Q)$ is being updated in training step and also being the learning target[7]. DDPG consists of actor and critic network, $Q(\mathbf{s_t}, \mathbf{a_t}|\theta^Q)$, $\eta(\mathbf{s_t}, \mathbf{a_t}|\theta^\eta)$ and their copy $Q'(\mathbf{s_t}, \mathbf{a_t}|\theta^{Q'})$, $\eta'(\mathbf{s_t}, \mathbf{a_t}|\theta^{\eta'})$, where θ is weights of network. The critic is a value network, which is trained to fit the value function $Q(s,a)$, and the actor is a policy network, which is trained to determinate the best action for current observation. There are two pair of networks: learning networks and target networks. Learning networks are trained by targets generated by target networks as supervised learning, and target networks are not trained rather than update their

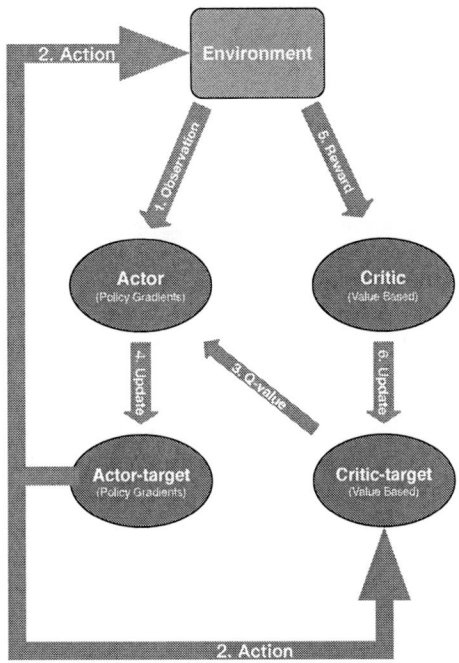

The DDPG algorithms contain two pairs of networks: the actor-target takes action based on current observation of environment, and the critic-target evaluate the Q value for current action. The actor updates its policy based on the Q value every step and the actor-target updates itself by the actor every τ steps. Then environment returns reward to the critic, which updates as the same as the actor.

Figure 2. The construction of DDPG.

weights by training networks weights:

$$\theta' \leftarrow \tau\theta + (1 - \tau)\theta' \tag{3}$$

where $\tau \ll 1$ is a small constants factor, called soft-updating factor. When $\tau \geq 1$, it becomes hard-updating. The process of updating may be slower than original networks, but it turns out to be more stable.

When actor-critic networks are fine-trained, we can use critic network only to make our decision. In other word, agent receives observation $\mathbf{o_t}$ from environment $\mathcal{E}$ as input of critic network, then critic network offers the best action $\mathbf{a_t}$.

III. METHODOLOGY: LEARNING TO OPTIMIZE WITH REINFORCEMENT LEARNING

Our idea is that a specific topological problem can be parameterized as a family of object function, and we conceived that we can build a model to describe the relationship between parameters $\mathcal{W}$ and the optimal solutions Θ^* for this function family, i.e. regression model $\mathcal{M} : \mathcal{W} \rightarrow \Theta$.

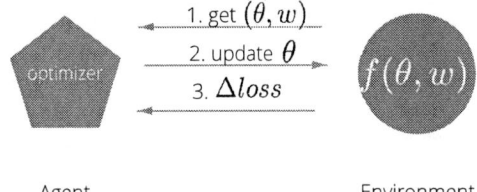

Figure 3. The optimizer based on reinforcement learning.

We have to provide enough train dataset $\{w_i, \theta_i\}$ as test dataset to train a regression model. When we try to generate the dataset for two simple mathematic problems, it is very fast that solve equations and take some iterations. In circuit simulation, however, it is barely feasible to generate train dataset: it takes too much cost. Usually, a global optimization algorithm takes a hundred steps to find the optimal solution, and call simulator 10 or more times for every step. In other words, we may call simulator thousand times to generate one single sample.

We train an optimizer to provide the updating value rather than build a regression model directly. The difference between optimizer and the regular regression model is the variable space. An optimizer is also a regression model, i.e. optimizer $\mathcal{O} : \Theta \times \mathcal{W} \rightarrow \mathcal{A}$, where $\mathcal{A}$ is updating value space.

We might want to figure out whether we can build a regression model for this problem by supervised learning, but according to [5], the answer is no. When they tried to train an optimizer by supervised learning, the optimizer is divergent quickly. Fortunately, the reinforcement learning is absolutely feasible for training an optimizer.

In the abstract problem above, we let $f : \mathcal{S} \rightarrow \mathbb{R}$ be the environment $\mathcal{E}$, θ_t be the state $\mathbf{s_t}$, θ_t and parameter w together be the observation $\mathbf{o_t}$ and $g : \mathbb{R} \times \mathcal{S} \rightarrow \mathbb{R}$ be our agent $\mathcal{A}$. Then we define the reward $r_t = \Delta(f(\theta_t), f(\theta_{t+1}))$, which means agent get positive reward when it find better solution. Now we transfer an optimization problem into a reinforcement learning problem, shown in fig 3

IV. EXPERIMENTS

A. Implementation Detail

In all our experiments, we build simply actor and critic networks by a few fully connect (FC) layers in Keras[15]. Each network is optimized by Adam with learning rate 0.001, and training for 30, 000 steps. Reward discount factor is set as 0.99 and soft updating factor is set as 0.001. The source code of DDPG is available in GitHub [keras-rl]. Keras-rl module developed many open source algorithms based on OpenAI Gym interface. OpenAI Gym is a toolkit for developing and comparing reinforcement learning algorithms. It provides many different and funny environments as a simple open source interface to reinforcement learning

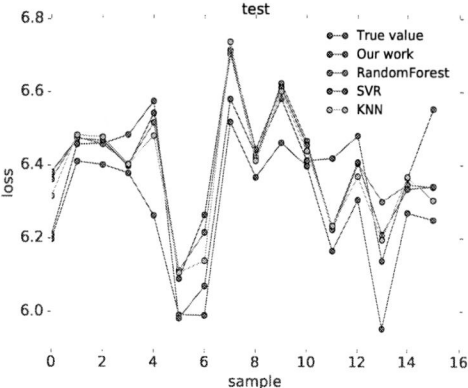

Figure 4. Regression models perform in test dataset.

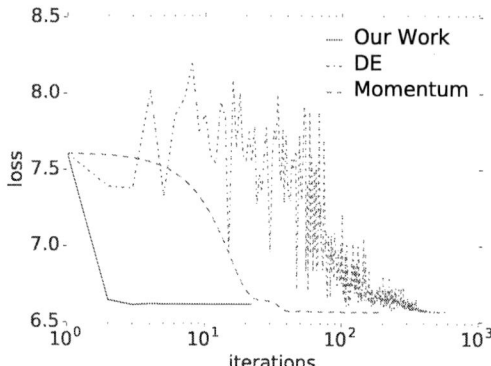

Figure 5. Quadratic function convergence curve.

Table I
EXPERIMENT RESULT OF QUADRATIC FUNCTION

(min/mean/max)	Steps	Relative Loss
Reinforcement Learning	1 / 1 / 1	0 / 0 / 0
Gradient Descent	10 / 23 / 36	-0.07 / -0.02 / -0.003
Different Evolution	57 / 294 / 527	-0.07 / -0.02 / -0.003

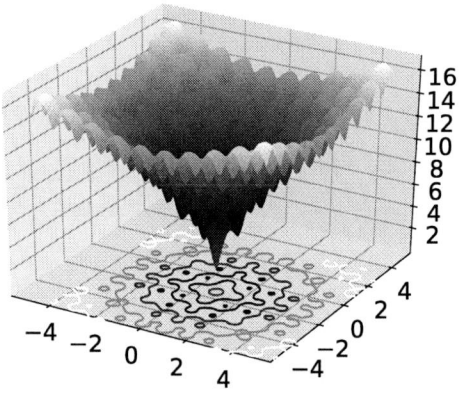

Figure 6. 2-dimensional ackley function.

tasks. We do not really need the environments but the interface it provides. We provide environments by defining interfaces $reset$ which resets the environment to initial state (may be random), $sample$ which sample an action $\mathbf{a_t}$ from action space for random strategy in the warm-up process, $step$ which takes an action $\mathbf{a_t}$ as input to change the state of environment then return the new observation $\mathbf{o_t}$ of the environment, reward r_t, whether done or not and other information, and other interfaces. $Step$ is the most important part that defines the behavior and reward of the environment.

B. Quadratic function

A quadratic function is convex, which can be express as:

$$f(\theta, W) = \|W\theta + b\|_2^2 \qquad (4)$$

In this experiment, we trained a simple optimizer for 8-dimensional quadratic functions specified by parameters W, b with dimension 8×8 and 8 respectively. The optimizer was trained by this function family, where W and b were generated by normal Gaussian distribution for each epoch in training process, and tested on random parameter sampled by the same distribution.

Firstly, we generated 500 samples by gradient descent algorithm to train regression models, and split them as two parts: train(300), test(200). Every sample takes nearly 200 steps to find the minimum. We use random forest[16], SVR[18] and KNN[17] as the regression model for comparison. For this sample convex problem, it is easy to generate dataset and fit the curve. We trained those three regression models in train dataset and tested them in test dataset respectively. the result is shown in fig 4.

The coverage curves are shown in fig 5. For quantitative comparisfoon, we define metrics that the steps of other algorithms take when our algorithm is convergent and the final relative loss with our algorithm. The relative loss is defined as follow:

$$relative\_loss = \frac{loss^*_{final} - loss^{ddpg}_{final}}{loss_{init}} \qquad (5)$$

The $loss^*_{final}$ is the convergent loss of other algorithms.

C. Non-convex function

Ackley function, which has many local minima, is a classic non-convex function. Fig 6 shows 2-dimensional ackley function graph.

$$f(\theta, w_1, w_2; w) = -20e^{-0.2\sqrt{\frac{1}{n}\sum_{i=1}^{n} w_{1i}(x_i - g(w))^2}}$$
$$- e^{\frac{1}{n}\sum_{i=1}^{n} w_{2i}cos(2\pi(x_i - g(w)))} + 20 + e$$

$$(6)$$

978-1-5386-7100-9/18 $31.00 © 2018 IEEE

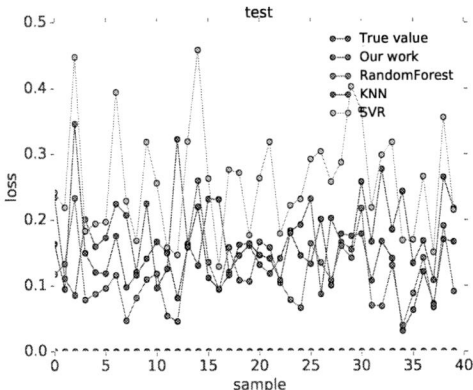

Figure 7. Ackley regression model perform in test dataset.

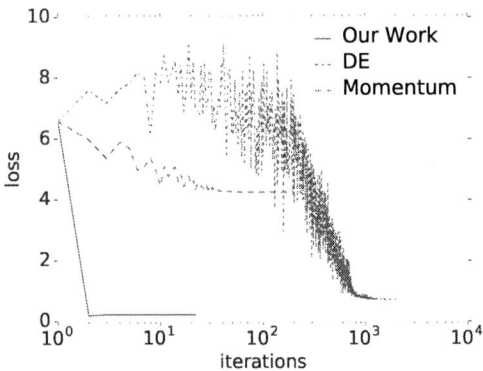

Figure 8. Ackley function convergence curve.

Table II
EXPERIMENT RESULT OF ACKLEY FUNCTION

(min/mean/max)	Steps	Relative Loss
Reinforcement Learning	1 / 1 / 1	0 / 0 / 0
Gradient Descent	-	0.15 / 0.85 / 0.57
Different Evolution	833 / 1044 / 1344	-0.07 / -0.03 / 0.13

Technically, our algorithm is independent of the complex of the objective function. In other words, we can always find the optimal solution no matter how many local minima the objective function has. In the contrast, the momentum algorithm which relies on gradient can barely find the global minima, and the differential evolution algorithm may only find the local minima in some cases.

We also make a comparison between reinforcement learning and regression model.

The coverage curves are shown in figure 8.

D. SRAM

In production process of SRAM, the performance of SRAM depends on process level, process parameter w

Table III
EXPERIMENT RESULT OF SRAM

(min/mean/max)	Steps	Relative Loss
Reinforcement Learning	1 / 1 / 1	0 / 0 / 0
Different Evolution	182 / 343 / 594	-0.27 / -0.13 / -0.07

and environment variables θ. The devices would not work anymore when the environment variables, which are uncontrollable, beyond a specified margin γ. In this task, we need to find out the maximum radio of cycle area where parameters are feasible, i.e. :

$$\min \quad \|\theta\|_2 \qquad \text{(LP1)}$$
$$\text{s.t.} \quad f(\theta, w) \leq \gamma \qquad (7)$$

where f is the Static Noise Margin (SNM) of SRAM. The devices do not work when the constrain above is satisfied. In order to optimize it, we transfer the constrained convex problem into an unconstrained optimization problem:

$$\min \|\theta\|_2 + C \cdot \mathbf{1}(f(\theta, w) - \gamma) \qquad (8)$$

where C is penalty factor, $\mathbf{1}(x) = x$ when $x > 0$ and $\mathbf{1}(x) = 0$ for $x \leq 0$.

It is inevitable to call spice simulator to calculate $f(\theta, w)$ in the process of optimization, which is very time-consuming. And since the simulator is a black box, it is infeasible to apply gradient-based optimization algorithms. Conventional algorithms such as differential evolution (DE) algorithms usually require calling simulator hundreds to thousands of times to optimize single one problem. In our experiments, there are 18-dimensional environment variables and 3-dimensional parameters. We use DE algorithm to optimize it for 20 times with different parameters and use a fine-trained agent to optimize the same parameters for comparison.

We only generated 20 samples in parameter space. Every sample called simulator nearly 2,000-3,000 times to optimize, and we just need to train our model 10, 000 times (actually it converged quickly). For the more complex problem, we have to generate more samples and take $O(n)$ time, but we just take $O(1)$ time by using trained-model.

The coverage curves are shown in fig 9. The x-axis coordinate is logarithmic. We can notice that our trained model takes 1-2 steps to find the optimal solution nearly when DE algorithm takes a thousand steps.

V. CONCLUSION

A family of optimization problem can be parameterized, and we can build a regression model to fit parameters and the optimal solutions. Using conventional supervised learning needs enough sample to train the model, while our trained model can take $O(1)$ time to do this work and achieve good performance.

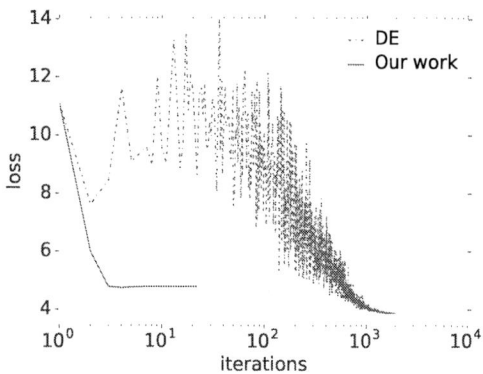

Figure 9.   SRAM convergence curve.

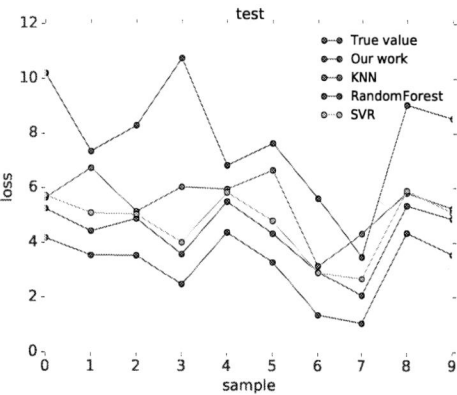

Figure 10.   Regression models performance in test dataset .

REFERENCES

[1] Das S, Suganthan P N. Differential Evolution: A Survey of the State-of-the-Art[J]. IEEE Transactions on Evolutionary Computation, 2011, 15(1):4-31.

[2] Storn R. On the usage of differential evolution for function optimization[C]// Fuzzy Information Processing Society, 1996. NAFIPS. 1996 Biennial Conference of the North American. IEEE Xplore, 1947:519-523.

[3] Vesterstrom J, Thomsen R. A comparative study of differential evolution, particle swarm optimization, and evolutionary algorithms on numerical benchmark problems[C]// Evolutionary Computation, 2004. CEC2004. Congress on. IEEE, 2004:1980-1987 Vol.2.

[4] Andrychowicz M, Denil M, Gomez S, et al. Learning to learn by gradient descent by gradient descent[J]. 2016.

[5] Li K, Malik J. Learning to Optimize[J]. 2016.

[6] Li K, Malik J. Learning to Optimize Neural Nets[J]. 2017.

[7] Lillicrap T P, Hunt J J, Pritzel A, et al. Continuous control with deep reinforcement learning[J]. Computer Science, 2015, 8(6):A187.

[8] Mnih V, Kavukcuoglu K, Silver D, et al. Human-level control through deep reinforcement learning[J]. Nature, 2015, 518(7540):529.

[9] Duan Y, Chen X, Houthooft R, et al. Benchmarking Deep Reinforcement Learning for Continuous Control[J]. 2016:1329-1338.

[10] Balduzzi D, Ghifary M. Compatible Value Gradients for Reinforcement Learning of Continuous Deep Policies[J]. Computer Science, 2015, 8(6):A187.

[11] Silver D, Lever G, Heess N, et al. Deterministic policy gradient algorithms[C]// International Conference on International Conference on Machine Learning. JMLR.org, 2014:387-395.

[12] Heess N, Wayne G, Silver D, et al. Learning Continuous Control Policies by Stochastic Value Gradients[J]. 2015:2944-2952.

[13] Peters J, Schaal S. Reinforcement learning of motor skills with policy gradients[J]. Neural Networks, 2008, 21(4):682-697.

[14] Bakker B. Reinforcement learning with long short-term memory[C]// International Conference on Neural Information Processing Systems: Natural and Synthetic. MIT Press, 2001:1475-1482.

[15] Chollet, Franois and others. Keras. GitHub. Github repository. 2015. https://github.com/fchollet/keras

[16] Breiman L. RANDOM FORESTS–RANDOM FEATURES[J]. Machine Learning, 2007, 45(1):5–32.

[17] Fallah A, Fallah A, Fallah A, et al. Forest attribute imputation using machine-learning methods and ASTER data: comparison of k-NN, SVR and random forest regression algorithms[J]. International Journal of Remote Sensing, 2012, 33(19):6254-6280.

[18] Hong W C. Electric load forecasting by support vector model[J]. Applied Mathematical Modelling, 2009, 33(5):2444-2454.

[19] Silver D, Lever G, Heess N, et al. Deterministic policy gradient algorithms[C]// International Conference on International Conference on Machine Learning. JMLR.org, 2014:387-395.

End-to-End Industrial Study of Retiming

Cunxi Yu[14], Chau-Chin Huang[2], Gi-Joon Nam[3], Mihir Choudhury[3], Victor N. Kravets[3],
Andrew Sullivan[3], Maciej Ciesielski[1], Giovanni De Micheli[4]
University of Massachusetts Amherst, MA, USA[1]
National Taiwan University, Taiwan[2]
IBM T.J. Watson Research Center, NY, USA[3]
EPFL, Switzerland[4]

Abstract -

Sequential circuits are combinational circuits that are separated by registers. *Retiming* is considered as the most promising technique for optimizing sequential circuits, that involves moving the edge-triggered registers across the combinational logic without changing the functionality. Despite significant efforts spent on sequential optimization since 1980's, there are few works discussed its performance in an end-to-end design flow. The retiming algorithms were mostly evaluated at the logic level. However, it turns out that the retiming results at logic level could be significantly different than evaluating the physical level.

This paper provides the findings of how retiming algorithms perform in an end-to-end industrial design flow, with seven industry designs taken from a recent 14nm microprocessor. Experiments are conducted with several complete industrial design flows. The evaluations are made at the end of the physical design flow. The experimental results show that the performance (design quality) of the retiming algorithms vary on the designs. Based these experimental results, we discover a feature that describes the retiming potentials of sequential designs. This model successfully forecast whether the given industrial designs could be significantly improved by retiming in an end-to-end design flow, regarding timing, area, and power.

Keywords—Sequential optimization, retiming, physical design, retiming prediction

I. INTRODUCTION

Retiming is a sequential optimization technique that has been studied since 1980's. Retiming techniques optimize the sequential circuits by relocating *edge-triggered registers*[1] across the combination logic without changing the design functionality. A lot of research efforts have been spent on developing promising retiming techniques that mainly target on three objectives:

- *min-delay*: minimize the worst-path delay of the circuits [1];
- *min-area*: minimize the number of registers of the circuits [2];
- *constrained min-area*: minimize the number of registers with a given worst-path delay constrain [3].

Numerous techniques have been proposed in our community to achieve these three objectives [1][2][4][5][6][7], and have been demonstrated with encouraging results. Although retiming assumes that the topology of the combinational logic is fixed, the quality (at logic-level) of the design can be further improved by combining the combinational logic optimization techniques and technology mapping [8][9][10]. In practice, constrained min-area retiming, has been incorporated into end-to-end design flows, which targets on improving the performance of the design regarding the delay, power, area, etc.

In 1997, N. Shenoy published the first retiming survey [11]. This work reviewed the theories and practical implementations of retiming, and the side issues of incorporating in the design flow. Due to the

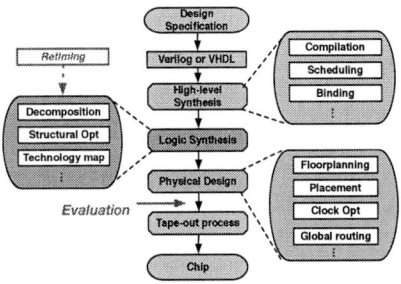

Fig. 1. Design flow used in this study.

significant changes in the technology and design complexity, an up-to-date industrial study of retiming is necessary. Moreover, to our knowledge, no credible work ever evaluates the retiming algorithms in an end-to-end design flow. Most existing retiming algorithms were evaluated at the end of the logic synthesis, where the delay and area are measured after technology mapping or using unit delay-area models. However, due to the significant increase in design complexity and design rules, the gate-level netlist does not correlate well with the physical netlist [12][13]. Hence, in this work, the experimental results are collected at the end of the physical design process. The overview of the design flow is shown in Figure 1. Note that the negative retiming operations[2] are forbidden in our experiments. It turns out that the performance improvements gained by retiming evaluated the logic level could make the final physical netlist worse. Also, there are significant extra design efforts required for retimed designs, e.g., sequential equivalence checking (SEC). Thus, for the designs that retiming does not provide enough improvements in design performance, retiming needs to be avoided in the design flow. These give us the motivation for developing a *prediction* mechanism for retiming. The state-of-the-art retiming algorithms are reviewed in Section 2. The main difference compared to [11] is the min-area algorithm.

Specifically, the main contributions of this paper are as follows:

- The complete end-to-end industrial study of retiming is shown in Section 3, using seven industrial designs are taken from the most recent microprocessor design with 14nm CMOS technology. Four different flows are tested, including the baseline flow and three retiming flows with different retiming options. The baseline flow is using the concept flow shown in Figure 1 without retiming applied. Retiming algorithms are embedded in the logic synthesis process, before technology mapping.
- The primary focus of the evaluations is the *design-quality*. Evaluations of retiming are made by comparing the results using a set of *timing, area* and *power* metrics, that are measured

---

[1]In the rest of this paper, *register* is used to represent *edge-triggered registers*.

[2]For min-delay retiming, negative retiming refers to "the delay of critical path increases"; for min-area retiming, it refers to "the number of registers increases."

at the end of physical design with simulation. Based on the analysis of how retiming algorithms work, three classes of experiments have been identified: a) designs significantly improved by retiming; b) designs significantly disimproved by retiming, and c) retiming does not make many differences.

- The side challenges of retiming to the entire design flow are reviewed in Section 3-B. Specifically, the effects of retiming in physical design and sequential verification are discussed. These challenges with the unpredictable results shown in Section 3 motivate us in developing a prediction mechanism for retiming.

- In Section 4, a numerical model is introduced for prediction. This model describes a *sequential feature* of designs at the logic level that could include Boolean gates and high-level blocks such as Adders. We demonstrate that this model can be used for forecasting whether a design could be improved by retiming in an end-to-end design flow, with the seven industrial designs (Fig. 3).

- We evaluate the academic sequential benchmarks taken from *ISCAS-89* and *ITC99* using the proposed prediction model. It turns out that these benchmarks are not sufficient for evaluating retiming algorithms.

II. BACKGROUND

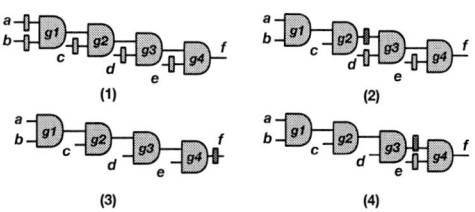

Fig. 2. Illustrative examples of retiming: 1) original netlist; 2) *min-delay* retiming; 3) *min-area* retiming; 4) *min-area* retiming under delay constrain (delay≤3).

TABLE I.   RETIMING RESULTS OF EXAMPLES IN FIGURE 2.

	Baseline	*min-delay*	*min-area*	*min-area and d<3*
Delay (d)	4	2	4	3
Num. of Regs	5	3	1	2

The concept and the three objectives of retiming are illustrated using a simple example shown in Figure 2. The original design is shown in Figure 2-(1). We assume all the gates have unit delay one, and the edge-triggered registers are represented using the rectangles. The original design has five registers (n=5) and the delay of the critical path, $\{a,b\} \rightarrow f$, is four (d=4).

The *min-delay* retiming result is shown in Figure 2-(2). There are two iterations in this retiming: 1) move the two registers connected with a and b forward, which makes gate g_2 retimable; 2) move the retimed register and the register connected with c forward. The delay of the retimed design in Figure 2-(2) is two (d=2), and the number of registers is three (n=3). The *min-area* retiming result is shown in Figure 2-(3). This solution requires two more iterations in addition to the solution of min-delay retiming, which move the registers all the way to primary output and reduce the number of registers to one. The delay is increased to four. The third objective is min-area retiming under delay constrain. In this example, let the delay constraint be $d \leq 3$. The solution is shown in Figure 2-(4). The comparison of these three retiming solutions to the original design is shown in Table I.

We can see that the min-delay retiming gives the best delay solution (d = 2), min-area offers the minimum number of registers, and min-area retiming with delay constraints gives balanced results in between of min-area and min-delay.

A. Formulation of Retiming

Most logic optimization techniques are formulated based on direct graph representation. The logic netlist, so called *Boolean network* can be modeled using direct graph $G(V, E)$, where each vertex v corresponds to a logic gate g in the design. Besides constructing the direct graph directly from the netlist, this can also be done based on the transformed Boolean network, such as *And-Inv-Graph* [14] and its sequential version [15]. In the context of retiming, the sequential Boolean network is the combinational network separated by the memory elements, which are assumed to be ideal registers. The edges in the graph $G(V, E)$ represent the interconnections of the logic gates in the design.

Let us denote e_{uv} is an edge of $G(V, E)$, e_{uv}: $u \rightarrow v$, and w_{uv} is the weight of the edge e_{uv} which represents the number of registers between the two vertex u and v. The weight of the edges directed from and into the primary inputs (PIs) and primary outputs (POs) is zero. Each vertex v in $G(V, E)$ represents its delay of the corresponding gate, denoted $d(v)$. The problem of retiming is denoted by retiming lag function $r(v)$ [16]: $V \rightarrow Z$. Let us denote that w_{uv}^r is the weight of the edge e_{uv}. For any retiming, it can be represented by Eq. 1.

$$w_{uv}^r = r(v) - r(u) + w_{uv} \qquad (1)$$

The value of $r(v)$ represents the movement of the registers for vertex v. If it is forward retiming (from inputs to outputs), then $r(v)$ is a negative. For any legal retiming, the condition shown in Eq. 2 must be satisfiable.

$$w_{uv} + r(v) - r(u) \geq 0 \qquad (2)$$

B. Min-delay Retiming

The min-delay retiming problem is as follows: Given $G(V, E)$ with a vertex delay function d and edge weight function w, find a legal retiming r, such that the cycle time c is minimized:

$$c = \max_{p:w_r(p)=0} \{d(p)\}, \qquad (3)$$

where $d(p)$ is the path delay, and $w_r(p)$ is the retimed register count on the path p. To this problem, Leiserson and Saxe [16] developed a classic algorithm: Two matrices W and D are first defined as:

$$W(u, v) = \min_{p:u \rightsquigarrow v} \{w(p)\}, \qquad (4)$$

$$D(u, v) = \max_{p:u \rightsquigarrow v \wedge w(p)=W(u,v)} \{d(p)\}, \qquad (5)$$

$W(u, v)$ gives the minimum register count on any path from u to v. $D(u, v)$ determines the maximum delay from u to v for the minimum register count. The two matrices can be obtained by solving an all-pairs shortest paths problem in G. Afterward, a binary search for the minimum clock cycle is performed. In each iteration, a Bellman-Ford algorithm can be employed to test whether a legal retiming exists with the current cycle time c. The algorithm above runs in $O(V^3 \lg V)$ time because each iteration costs $O(V^3)$ time for a Bellman-Ford algorithm and the binary search runs in $O(\lg V)$.

Leiserson and Saxe [16] also proposed another more efficient relaxation algorithm, which runs in $O(VE)$ for examining if a retiming exist for a given clock cycle c. A function $\Delta(v)$ gives the largest delay seen from any path that terminates at the output of v:

$$\Delta(v) = d(v) + \max_{u \in FI(v), \ w(e_{uv})=0} \{\Delta(u)\}. \qquad (6)$$

978-1-5386-7100-9/18 $31.00 © 2018 IEEE

Therefore, the cycle time can be expressed as follows:

$$c = \max_{v \in V} \{\Delta(v)\}. \tag{7}$$

The relaxation algorithm consist of alternately updating the functions $\Delta(v)$ and $r(v)$ for $|V|-1$ times. The optimality is guaranteed because each iteration simulates a pass off a Bellman-Ford algorithm (i.e., a vertex being relaxed in a pass of the Bellman-Ford algorithm must be updated in an iteration of the relaxation algorithm), one can obtain a feasible retiming under the target cycle time c, if it exists. Because calculating $\Delta(v)$ in an iteration costs $O(E)$ time, the relaxation algorithm runs in $O(VE)$ time. Later on, the runtime of this algorithm was improved by Shenoy and Rudell [17] by adding an early break mechanism.

C. Min-area Retiming

The typical min-area retiming refers to minimizing the number of registers without delay constraints. In which case, the problem can be formulated as a minimum-cost flow problem using linear programming. The formulation is as follows:

$$\min : \sum_{\forall e_{uv}} r(u) - r(v) \wedge (\forall e_{uv}, r(u) - r(v) \leq w_{uv}) \tag{8}$$

Goldberg [5] presented a practical push-relabel method that can solve this problem in $O(V^2 E \log(VC))$ worst-case runtime, where V is the number of vertices, E is the number of edges, and C is the maximum cost of the edges. A. Hurst et al. [2] proposed a the min-area retiming using maximum network flow problem. It turns out that within a combinational network, minimizing the number of registers by retiming is equivalent to finding a minimum cut. Note that the minimum cut problem is the dual of the maximum network flow problem. Computing maximum flow of a network is much less complex than minimum cost determination. Although there may exist many minimum cuts, that approach always generates one minimum cut that provides the minimum number of movements of the registers. This is claimed to simplify the computation of the initial states and minimize the side effects of the design [2]. The worst-case runtime of maximum flow approach is bounded by $O(R^2 E)$, where R is the initial number of registers, and E is the number of edges. This algorithm requires repeated iterations while the number of iterations is typically small, as demonstrated by the authors.

D. Constrained Min-area Retiming

Although it is claimed that pruning the redundant storage elements in the design reduces the area, power, and verification cost, most designs request a specific target clock period. The first attempt of constrained min-area retiming was presented by Shenoy et al. [17]. The implementation was composed by computing the W and D matrices, and the minimum cost circulation implementation. Let us denote $N_{fanin}(v)$=number of $fanins$ of vertex v, $N_{fanout}(v)$=number of $fanouts$ of vertex v. The formal definition of this problem is represented as follows [11]:

$$\min : \sum_{\forall v} |N_{fanin}(v) - N_{fanout}(v)| \cdot r(v) \tag{9}$$

$$\wedge \; \forall e_{uv}, r(u) - r(v) \leq w_{uv} \tag{10}$$

$$\wedge \; \forall e_{uv}, r(u) - r(v) \leq W(u,v) - 1 \tag{11}$$

Equation 9 represents the cost of number of registers of all register relocations. Equation 10 and 11 constrain each relocation must be legal retiming and under the delay constrain. We can see that this problem can be solved by combining the algorithms proposed in Section 2.2 and Section 2.3. The most recent constrained min-area

retiming method was proposed by A. Hurst et. al [3], which the min-area appoach is based on the work of [2]. That approach is developed based on the observation that area-critical and timing-critical regions are rare overlapped. In that work, the timing constrains of retiming requires the exiting min-delay retiming algorithms, which give the initial register positions for their min-area appoach.

III. EXPERIMENTAL RESULTS

The results used for this study are shown in Table II. They are obtained with seven industry designs taken from the recent 14nm microprocessor design, with total twenty-eight *end-to-end* experiences. Each design is applied with four different complete design flows. *baseline* uses the flow shown in Figure 1, which is considered as the benchmark flow (without retiming). The three retiming design flows are added with three different retiming options to the *basline* flow. *r1,r2,r3* are the three different combinations of retiming that provide the most effective results: 1) *r1*={min-area, min-delay, min-area}; 2) *r2*={min-area, min-delay}; and 3) *r3*={min-delay, min-area}. The retiming process is applied on NAND2-based network within logic synthesis process, while each gate is considered as a unit delay gate. The min-delay retiming is implemented by combining the relaxation algorithm [1], and the heuristic min-delay algorithm implemented in ABC [18]. The min-area retiming is implemented using constrained min-area retiming algorithm based on *max-flow min-cut* proposed in [3].

A. Evaluations of Retiming

Table II lists the selected experiences results. It includes the major metrics used for evaluating the design at physical level using the flow shown in Figure 1. The evaluations may target on *timing*, *area* and *power*. The timing metrics include: 1) *Worst Negative Slack* (WNS), 2) *Total Negative Slack* (TNS), 3) *Latch-to-Latch WNS* (l2lWNS), 4) *Latch-to-Latch TNS* (l2lTNS), and 5) number of negative paths (#NEG). The area metrics include the *number of registers* (#reg) and *physical area* (Area). The power metrics inlude *dynamic power* (Dpower) and *static power* (Spower). The CPU runtime of the complete design process is included in the last column. Note that the extra CPU time comparing to the *baseline* mostly come from physical design process. The CPU time of retiming in Table II are all less than 600 seconds.

Based on the results shown in Table II, the designs can be classified in three types:

- Type 1) retiming significantly improves the timing and/or area at the end of design flow, including *ibm6* and *ibm7*;
- Type 2) these three retiming options all make the timing and/or area worse compared to baseline, including *ibm1* and *ibm2*;
- Type 3) retiming may make the timing and/or area worse than baseline, including *ibm3*, *ibm4* and *ibm5*.

In Table II, all the retiming operations are **positive** retiming, i.e., for min-delay, the critical path delay at the logic level always decreases or remain the same as before retiming. For min-area, the number of registers always decreases or remain the same, and the critical path delay is always less or equal to the given delay constraint. Any negative retiming operations will terminate the retiming process and return the original netlist.

Specifically, for ibm6 and ibm7, all the three retiming options give improvements in timing (mostly TNS) and area comparing to baseline. One observation is that while adding more registers in the design, the area and power may not increase. This is because adding more registers improve the timing, which timing closure requires fewer efforts during physical design process. For designs ibm3, ibm4 and ibm5, retiming doesn't give promising improvements compared to

TABLE II.   EVALUATION OF THREE RETIMING OPTIONS IN THE END-TO-END DESIGN FLOW. BOLD RESULTS INDICATE THE BEST RESULTS AMONG THE FOUR FLOWS.

Design	Options	WNS	TNS	l2lWNS	l2lTNS	#NEG	#reg	Area	Dpower	Spower	CPU Time(hr)
ibm1	baseline	**-14.491**	-10497	**-17**	8.607	1326	3214	340230	26.9089	**18.8614**	20.3
	r1	-23.425	-23170	-44	7.007	2156	**3095**	336509	25.8387	20.4603	20.9
	r2	-14.75	-15169	-47	7.260	1668	3193	337795	26.8823	19.9236	21.0
	r3	-16.68	-18651	-21	8.151	1775	3113	**335646**	**25.2462**	19.9465	21.0
ibm2	baseline	**-23.87**	**-3275**	**-50**	8.151	641	8115	**836771**	**111.0640**	**16.5280**	13.6
	r1	-36.524	-3894	-101	7.548	775	**7901**	897393	115.8500	17.7728	22.8
	r2	-37.397	-4340	-83	**8.200**	778	8082	893571	115.5700	17.5046	22.8
	r3	-35.869	-4118	-83	8.099	772	8247	899329	118.5860	17.5122	22.8
ibm3	baseline	-17.884	-4882	-8	8.504	670	2672	216950	45.5289	9.7992	6.9
	r1	-17.142	-4899	-6	8.705	664	**2485**	**209388**	**43.1193**	10.0417	21.1
	r2	**-14.711**	**-4662**	**-2**	**9.569**	**646**	2633	213527	44.7554	**9.5785**	21.2
	r3	-19.675	-4976	-4	8.947	685	2567	212362	44.1308	9.9634	21.1
ibm4	baseline	-20.507	-13074	-9064	-16.317	1110	4186	509862	119.9100	28.7540	20.8
	r1	-23.922	-17110	-13960	-21.856	1138	4186	512716	120.9800	31.1402	21.9
	r2	-22.295	-14784	-10626	-21.196	1132	**3954**	517506	120.4960	30.5482	21.9
	r3	**-18.965**	**-9919**	**-6288**	**-12.797**	**892**	4010	**506887**	**118.4360**	**28.4195**	21.9
ibm5	baseline	-20.675	-2584	-55	2.437	471	4682	514430	109.7720	11.8720	19.7
	r1	-49.791	-3229	-62	2.234	1163	**4597**	**503435**	**108.8470**	**11.1319**	20.5
	r2	**-19.928**	**-1735**	-46	4.978	**203**	4684	505298	112.2140	11.3123	20.6
	r3	-54.379	-3266	**-22**	**9.143**	1091	4622	507626	110.6960	11.3450	20.6
ibm6	baseline	-366.635	-14327	0	9.992	365	8238	812092	39.3554	5.378	19.3
	r1	-367.179	-9412	0	10.000	303	**7423**	**711125**	**35.213**	**4.779**	21.6
	r2	**-363.009**	-9860	0	10.000	314	8241	744181	39.4629	4.9856	21.7
	r3	-363.632	**-6982**	0	10.000	**140**	8657	769391	41.3688	5.089	21.8
ibm7	baseline	-55.537	-38388	-18005	-45.000	1732	2846	446838	134.214	24.7752	1.3
	r1	-56.131	-35473	-15361	-42.810	1736	2819	438761	131.735	23.9898	1.5
	r2	-53.184	-33876	-13653	-39.838	1740	**2625**	**428183**	127.636	24.1353	1.4
	r3	**-45.805**	**-31904**	**-12098**	**-38.850**	**1620**	2688	429153	**127.585**	**23.6245**	1.9

baseline. And, the performance of retiming varies on the combination and the order of retiming algorithms applied. For example, ibm5, r2 gives almost the same result compared to baseline, but r1 and r3 make the timing much worse. For the designs ibm1 and ibm2, all the three retiming options make the timing and area worse than baseline. We didn't consider the runtime of retiming in those comparisons since retiming takes little time compared to other design automation processes. In summary, the results in Table II show that:

- 1) The performance of retiming varies on different designs.

For example, ibm6 and ibm7 are improved by all retiming options, but ibm1 and ibm2 are worse than baseline with any retiming option. The main reason is that delay-driven retiming algorithms target on minimizing the critical path delay, i.e., the levels of the longest combinational paths. However, the logic-level critical paths could be very different to the critical paths at the physical level. This means that, even though there are significant reductions provided by retiming at the logic level, the critical path delay at physical level could be worse. This directly affects the timing closure, such as WNS and TNS. For timing closure, more physical design efforts are required to meet the requirements of timing constraints. On the other hand, the area and CPU time could also increase, e.g., the routability decreases after retiming. This provides the main motivation for pre-analysis of retiming to predict if a given design could be improved by retiming.

- 2) The combination and ordering of retiming approaches affect the timing, area, and power, even though the statistics (area and levels) at logic level are the same.

For most industry designs, retiming is required to provide balanced performance regarding timing, area, power, etc., which is the retiming objective reviewed in Section II-D. The constrained min-area retiming methodology hasn't been changed in the last twenty years, which was proposed by Shenoy et al. [17], using a combination of multiple min-delay and min-area retiming. This method works for many designs based on the observation that the regions of area-critical and timing-critical rarely overlap [3]. However, although the optimum solutions can be found, the retiming operations could be very different. For example, the uniqueness of the maximum flow does not imply the

uniqueness of the minimum cut. One observation in this study is that the number of retiming depths of retimed registers could be very different using different retiming options, even though the statistics at the logic level are the same. Hence, the varieties of retiming solutions, and the mismatch between logic and physical criticalities make the choice of retiming options crucial.

- 3) For the designs that can be improved by one of the retiming options, they are likely to be verified by all the retiming options.

One observation is that the designs that cannot be verified by any one retiming option, they are unlikely improved by other retiming options. Some designs are significantly disimproved. On the other side, the designs are improved by one retiming option; they are likely to be improved by other retiming options as well. For example, designs *ibm6* and *ibm7* are improved significantly by all the retiming options. This gives us the main motivation of developing a retiming prediction model to pre-analyze whether retiming algorithms are effective for a given design, which is shown in Section IV.

B. Challenges of Retiming

The challenges of retiming are considered in three parts: a) retiming complexity, b) retiming performance in design-quality; and c) the side effects. The complexity of retiming has been discussed in [11] and briefly reviewed in Section 2. Hence, in this section, we focus on analyzing the last two. As shown in Table II, the design performance of the physical netlist is not guaranteed to be improved by retiming algorithms. Since all the retiming operations undertake to reduce the critical path delay and area of the design at the logic level, these results also show that there are no strong correlations between logic and physical netlist. This is believed to be worse as the number of design rules increasing for the advanced technology nodes. Apparently, this has significant impacts on the design-quality and Time-to-Market. To forecast a given design whether its performance could be improved by retiming becomes extremely important.

One of the main side effects of the design flow is sequential verification. It is stated that the verification of retimed circuits could be solved very efficiently and takes $O(|E|)$ time [11], where verification

978-1-5386-7100-9/18 $31.00 © 2018 IEEE

TABLE III.   DEMONSTRATION OF THE PROPOSED PRE-ANALYSIS MECHANISM USING ONE DESIGN IN EACH EXPERIMENTAL CATEGORY PRESENTED IN SECTION 3.1. THE RESULTS ARE SHOWN IN PERCENTAGE COMPARED TO THE RESULTS OF *baseline*.

Design	B_{foward}	$B_{backward}$	Options	WNS	TNS	l2lWNS	l2lWNS	Area	Power
ibm1	0.118	0.044	r1	1.617	2.207	2.588	0.814	0.989	1.085
			r2	1.018	1.445	2.765	0.843	0.993	1.056
			r3	1.151	1.777	1.235	0.947	0.987	1.058
ibm2	0.041	0.022	r1	1.530	1.189	2.020	1.080	1.072	1.047
			r2	1.566	1.325	1.660	0.994	1.067	1.042
			r3	1.502	1.257	1.660	1.01	1.074	1.066
ibm3	0.33	0.041	r1	0.959	1.003	-	1.024	0.965	1.025
			r2	0.823	0.955	-	1.125	0.984	0.977
			r3	1.100	1.019	-	1.052	0.979	1.017
ibm4	**0.616**	0.035	r1	1.166	1.308	1.540	**0.746**	1.005	1.023
			r2	1.087	1.130	1.172	**0.769**	1.014	1.016
			r3	0.925	0.758	0.694	1.275	**0.994**	**0.987**
ibm5	0.045	0.001	r1	2.408	1.250	1.127	0.917	0.979	0.938
			r2	**0.964**	**0.671**	0.836	2.043	0.982	0.953
			r3	2.630	1.264	0.400	3.752	0.987	0.956
ibm6	**5.512**	0.082	r1	1.001	**0.659**	-	1.000	**0.875**	**0.894**
			r2	**0.990**	**0.688**	-	1.000	**0.916**	**0.993**
			r3	**0.992**	**0.487**	-	1.000	**0.947**	1.038
ibm7	**0.763**	0.082	r1	1.011	0.924	0.853	0.951	0.982	0.980
			r2	0.957	0.882	0.758	0.885	0.958	0.955
			r3	**0.827**	**0.831**	**0.672**	**0.863**	**0.960**	**0.951**

is restricted to verify the design before and after retiming. However, the complete sequential verification is still needed, especially the typical retiming flow includes multiple iterations and usually combined with logic synthesis [19]. Despite the progress of the sequential verification, directly check the sequential equivalence is still extremely hard. For example, using model checking technique *IC3* [20], it can't directly verify the properties of retimed s13207, s15850, and s38584 ISCAS-89 designs, with more than 24 hours[3]. Engineering Change Order (ECO) is another side challenge of retiming. ECO for retimed circuits becomes *Sequential Engineering Change Order* (SECO), which has never been discussed. Although several techniques of combinational ECO proposed using formal methods and structural methods, determination of the minimal patches in combinational logic is still a challenging problem. Functional methods can find smaller patches but limited to its scalability. Structural methods improve the runtime but limited by the structural similarity.

IV. RETIMING PREDICTION

Due to the uncertainty of retiming and significant extra design efforts, a pre-analysis mechanism for predicting whether a design is potentially improvable by retiming is needed. The purpose of the proposed prediction mechanism is that *given a sequential circuit including Boolean gates, high-level blocks, and registers, it outputs a normalized value that represents the* retiming potentials *in an end-to-end design flow*. Based on a large number of experiences and analysis using industry and academic benchmarks, we observe that the retiming flow gives significant improvements when the pipeline stages are not well-balanced, regardless of the options of the retiming flow. However, it is not sufficient by just considering the pipeline stages. The reason is that *unbalanced registers*[4] may not be retimable. Hence, we introduce the *Balancing* metric. It is defined using Equation 12, where r_i is forward, or backward retimable register[5], N is the number of retimable registers in the design. Hence, for each design, it has two balancing value, $B_{forward}$ and $B_{backward}$. This metric is sensitive to the timing model since it is calculated at logic-level. For the blocks have multiple inputs and outputs (e.g., 4-bit adder), we consider that the pin-to-pin delay of all the paths is the

same. For example, assuming delay of AND2 is one, in Figure 2 (a), $B_{backward}=0$ because there is no backward retimable registers; $B_{forward}=(4 + 4)/2 = 4$ since registers a and b are forward retimable.

$$B = \sum_{i=0}^{i=N-1} |Slack_{out}(r_i) - Slack_{in}(r_i)| \ / \ N \qquad (12)$$

A. Performance Analysis via Balance Value

Table III lists the analysis of retiming performance using the proposed pre-analysis mechanism, with the designs shown in Table II. The forward and backward balancing values are listed in the first two columns. The timing, area, and power results are divided by the results generated from *baseline* experiences. In Table III, for any value > 1, represents negative result; for any value < 1, represents positive result. As the $\overline{B_{forward}}$ (see Eq. 12) value increasing, the design is more likely to be improved by retiming. For example, for design *ibm2*, both $B_{forward}$ timing and $B_{backward}$ are smaller than 0.05, which means that the design around the retimable registers is well-balanced. Not spuriously, the area and power results generated from the flow using retiming, are all negative. For design *ibm4*, $B_{forward}=0.616$, which indicates that there are potential improvements could be done by retiming. We can see that positive retiming results appear for this design and $r3$ improves timing, area, and power simultaneously. For design *ibm6*, the $B_{forward}=5.512$, where we see that the design has been significantly improved by retiming. Note that the balancing value may be different using different timing model. However, the standardization remains regardless of the select timing model. We observe that the backward balance value are not sufficient to provide any evidence of analyzing the retiming performance.

To further analyze the forward balance value, the retiming, area and power related metrics reported by the design flow are all collected in Figure 3. The x-axis shows the forward balance value, and the y-axis shows the overhead/improvement provided by retiming. For example, the data point with y=1.5 means a 50% overhead for some metric. The results are classified into six groups: timing metrics improved/disimproved, area metrics improved/disimproved, and power metrics improved/disimproved. The retiming performance is considered as negative if some metrics have been significantly disimproved, or most of the metrics remaining the same. We can see that as the forward balance value increasing, the retiming becomes more effective. For the designs with small forward balance value, the area, timing and power metrics are 2× worse.

---

[3]These experiments are obtained without the industrial design flow. They are tested using *pdr* command in ABC[18].

[4]Unbalanced registers refer to the registers with significant differences between the input and output Slack time.

[5]Definitions of forward and backward retiming refer to Equation 1 and 2.

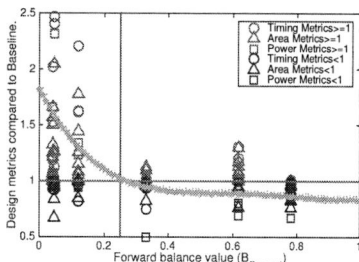

Fig. 3. Forward balance analysis with the designs listed in Table III.

In addition, we find that the most sensitive parameter in respect to forward balancing value is TNS. This means that the balancing value could be more effective in predicting the improvements of TNS. To demonstrate this, we create 120 benchmarks using high-level synthesis tool by manually modifying the location of the registers, such that the range of the balancing values of those benchmark is wide. We then obtain the TNS results of all the benchmarks using the same physical design flow. The results are shown in Figure 4. The x-axis represents the forward balancing value, and y-axis represents the percentage of TNS improvements. The data points above the horizontal line $y=0$ represent the designs get positive TNS improvements by retiming. We can see that the TNS improvements are clearly increasing as the forward balancing value increases.

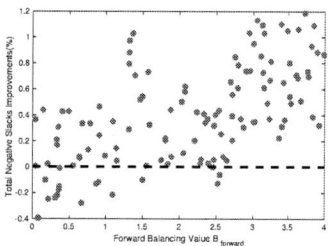

Fig. 4. Total negative slacks (TNS) improvements (%) versus forward balancing value of 120 designs.

Based on the results shown in Table III and Figure 4, we can see that the unbalanced designs are likely to be optimized by retiming. Due to the high extra design automation efforts, it is essential to understand if the designs are suitable for retiming in advance. Meanwhile, we can see that selecting different benchmarks could dramatically affect the evaluations of retiming algorithms.

V. CONCLUSION

This paper provides the first end-to-end industrial study of retiming, using the recent 14nm industrial designs. The state-of-the-art min-delay, min-area and constrained min-area retiming algorithms are evaluated in the complete design flows, with three retiming options. The results show that evaluations of retiming algorithms at logic level could be misleading. Due to the challenges of retiming discussed in this paper, a retiming-prediction model is introduced to forecast whether retiming algorithms could improve a given design. We demonstrate that this prediction model correctly predict the retiming potentials of the seven 14nm industrial designs. With this model, we are able to reduce the cost of designing an efficient flow in order to reduce *Time-to-Market*.

ACKNOWLEDGMENTS: The results are collected when Cunxi and Chau-Chin were interns at IBM T.J. Watson Research Center. This project is partially funded by ERC-2014-AdG 669354 grant.

REFERENCES

[1] C. E. Leiserson and J. B. Saxe, "Retiming synchronous circuitry," *Algorithmica*, vol. 6, no. 1-6, pp. 5–35, 1991.

[2] A. P. Hurst, A. Mishchenko, and R. K. Brayton, "Fast minimum-register retiming via binary maximum-flow," in *Formal Methods in Computer Aided Design, 2007. FMCAD'07.* IEEE, 2007, pp. 181–187.

[3] A. Hurst, A. Mishchenko, and R. Brayton, "Scalable min-register retiming under timing and initializability constraints," in *Proceedings of the 45th annual Design Automation Conference.* ACM, 2008, pp. 534–539.

[4] P. Pan, "Continuous retiming: Algorithms and applications," in *Computer Design: VLSI in Computers and Processors, 1997. ICCD'97. Proceedings., 1997 IEEE International Conference on.* IEEE, 1997, pp. 116–121.

[5] A. V. Goldberg, "An efficient implementation of a scaling minimum-cost flow algorithm," *Journal of algorithms*, vol. 22, no. 1, pp. 1–29, 1997.

[6] D. P. Singh, V. Manohararajah, and S. D. Brown, "Incremental retiming for FPGA physical synthesis," in *Proceedings of the 42nd annual Design Automation Conference.* ACM, 2005, pp. 433–438.

[7] S. S. Sapatnekar and R. B. Deokar, "Utilizing the retiming-skew equivalence in a practical algorithm for retiming large circuits," *IEEE Transactions on Computer-Aided Design of Integrated Circuits and Systems*, vol. 15, no. 10, pp. 1237–1248, 1996.

[8] S. Malik, E. M. Sentovich, R. K. Brayton, and A. Sangiovanni-Vincentelli, "Retiming and resynthesis: Optimizing sequential networks with combinational techniques," *IEEE Transactions on Computer-Aided Design of Integrated Circuits and Systems*, vol. 10, no. 1, pp. 74–84, 1991.

[9] G. De Micheli, "Synchronous logic synthesis: Algorithms for cycle-time minimization," *IEEE Transactions on Computer-Aided Design of Integrated Circuits and Systems*, vol. 10, no. 1, pp. 63–73, 1991.

[10] J. Cong and C. Wu, "Optimal FPGA mapping and retiming with efficient initial state computation," *IEEE Transactions on Computer-Aided Design of Integrated Circuits and Systems*, vol. 18, no. 11, pp. 1595–1607, 1999.

[11] N. Shenoy, "Retiming: Theory and practice," *Integration, the VLSI journal*, vol. 22, no. 1, pp. 1–21, 1997.

[12] F. Liu, Ed., *Proceedings of the 35th International Conference on Computer-Aided Design, ICCAD 2016, Austin, TX, USA, November 7-10, 2016.* ACM, 2016.

[13] S. Parameswaran, Ed., *2017 IEEE/ACM International Conference on Computer-Aided Design, ICCAD 2017, Irvine, CA, USA, November 13-16, 2017.* IEEE, 2017.

[14] A. Mishchenko, S. Chatterjee, and R. Brayton, "DAG-aware AIG Rewriting A Fresh Look at Combinational Logic Synthesis," in *43rd DAC.* ACM, 2006, pp. 532–535.

[15] A. Mishchenko and R. Brayton, "Recording synthesis history for sequential verification," in *Formal Methods in Computer-Aided Design, 2008. FMCAD'08.* IEEE, 2008, pp. 1–8.

[16] C. E. Leiserson, F. M. Rose, and J. B. Saxe, "Optimizing synchronous circuitry by retiming (preliminary version)," in *Third Caltech conference on very large scale integration.* Springer, 1983, pp. 87–116.

[17] N. Shenoy and R. Rudell, "Efficient implementation of retiming," in *Proceedings of the 1994 IEEE/ACM international conference on Computer-aided design.* IEEE Computer Society Press, 1994, pp. 226–233.

[18] A. Mishchenko *et al.*, "ABC: A System for Sequential Synthesis and Verification (2007)," *URL http://www. eecs. berkeley. edu/alanmi/abc*, 2010.

[19] J. Baumgartner, H. Mony, V. Paruthi, R. Kanzelman, and G. Janssen, "Scalable sequential equivalence checking across arbitrary design transformations," in *Computer Design, 2006. ICCD 2006. International Conference on.* IEEE, 2007, pp. 259–266.

[20] F. Somenzi and A. R. Bradley, "IC3: where monolithic and incremental meet," in *FMCAD '11, Austin, TX, USA, October 30 - November 02, 2011*, 2011, pp. 3–8.

2018 IEEE Computer Society Annual Symposium on VLSI

Communication-aware Module Placement for Power Reduction in Large FPGA Designs

Kalindu Herath, Alok Prakash, Udaree Kanewala, Thambipillai Srikanthan

Nanyang Technological University, Singapore

{*kalindub001@e, alok@, n17025521@e, astsrikan@*}*ntu.edu.sg*

Abstract—Modern multi-million logic FPGAs allow hardware designers to map increasingly large designs into FPGAs. However, traditional FPGA CAD flows scale poorly for large designs, often producing low quality solutions in terms of performance and power in such cases. To improve design productivity, modular design methodology partitions a large design into subsystems, compiles them individually and finally collates the individual solutions to complete the mapping process. Existing work has attempted to partition large designs into smaller subsystems, based on the intra-subsystem communication, to reduce routing power dissipation. However, inter-subsystem communication has not been considered, especially, during the placement stage. In this work, we first show the adverse effect of ignoring the inter-subsystem communication during the placement stage. Next, we propose an inter-subsystem communication-aware placement technique using a Simulated Annealing based approach to achieve significant power savings. Experimental results show over 7% reduction in routing power when compared to the existing state-of-the-art partitioning flow that ignores inter-subsystem communication, while the routing power reduction is over 11% when compared to a commercial CAD tool such as Altera Quartus.

Keywords-FPGA, CAD, modular design, placement

I. INTRODUCTION

Rapid scaling of transistors has enabled commercial Field Programmable Gate Arrays (FPGA) manufacturers like Altera [1] and Xilinx [2] to produce FPGAs with millions of logic gates alongside diverse hard Intellectual Property (IP) cores like Digital Signal Processors (DSPs) and Block-RAMs (BRAMs) with varying capacities. This resource richness encourages designers to port immensely large designs into modern FPGAs. Apart from the resource abundance, low non-recurring engineering (NRE) cost and the time-to-market (TTM) with respect to Application Specific Integrated Circuits (ASICs) [3] have also made FPGAs popular.

Commercial FPGA Computer-Aided Design (CAD) tools are well known to produce high quality mappings for small to medium-scale designs. However, compared to the rapid development in FPGAs, the available FPGA CAD tools do not show sufficient maturity yet to efficiently map large designs [4]. This typically results in inferior quality-of-result (QoR) in terms of performance and power consumption for such designs. Longer compilation time [5] is also commonly observed with existing CAD flows, which eventually lowers design productivity and adversely affects the TTM. Among the major steps in CAD flow, the placement step is known to consume a significant amount of time, often taking close to 50% of the total CAD runtime [6]. Hence, commercial FPGA vendors as well as the research community have focused on improving the tools used for the placement step.

Existing placement techniques can be categorized into three types, namely, Simulated Annealing (SA)-based, partition-based and analytical. Altera uses parallel SA techniques [7] in their Q2P [6] placer to improve their Quartus CAD tool, while Xilinx incorporates analytical placement techniques [8] to improve theirs. Despite the current endeavors, significant effort is required to achieve better runtime and performance while mapping large FPGA designs.

Recently, Modular Design Methodology (MDM) has been gaining traction in both the research as well as commercial communities, especially in order to reduce the runtime of the existing CAD flows for large designs. The MDM technique divides a large design into a number of smaller modules. This technique is inspired by the fact that some sections in a design, for instance, board support package or external memory controllers, are not frequently changed during design time. Such sections can be compiled and mapped into an FPGA once, while continuing to iterate the rest of the sections as required. As a result, MDM enables code reuse and is effective in reducing the compile time as well as in improving design productivity. Many approaches which consider the reuse of pre-compiled modules (macros) and IP cores along with library-based design can be seen in literature. Dividing a large design into smaller modules in the MDM technique also allows designers to leverage the existing CAD tools that already produce high quality mappings for small to medium scale designs. However, MDM still suffers from performance degradation and does not explicitly focus on developing power-efficient solutions [9].

A significant contributor to the dynamic power consumption in FPGAs is the charging and discharging of capacitive loads that occur during the flow or *communication* of data via the interconnect fabric [10] [11], especially at long distances. This issue is further exacerbated by frequent communication over long distance links [9]. Hence, to reduce the overall power consumption in FPGAs, it is necessary to reduce the long-distance links, especially for high frequency communications. Authors in [9] proposed to divide a large FPGA design into modules such that frequently communicating entities fall within a single module, thereby maximizing intra-module communication frequency. In the subsequent compilation step, all the design entities from these modules were placed in closed proximity by the CAD flow, resulting in reduced interconnect power dissipation. However, the authors failed to consider the inter-module communication during the placement step that can potentially result in further power savings.

In this paper, we propose an inter-module communication-aware module placement strategy for placing modules that builds on the work in [9] to further lower interconnect power dissipation of large designs on FPGAs.

978-1-5386-7100-9/18 $31.00 © 2018 IEEE
209

The rest of the paper is organized as follows. In Section II we briefly explain related workflows followed by a motivational example in Section III. In Section IV, we explain the proposed methodology. Next, we show the effectiveness of the technique in Section V and we conclude the paper in Section VI.

II. Related Work

MDM includes two phases; module creation and module assembling. During the first phase, a large design is divided into modules. Each module is treated as an isolated design hence it can go through CAD flow independently. During the next phase, the pre-compiled modules are assembled and inter-module links are routed to get the programmable bitcode of the entire design. Placement techniques used in the module assembly phase are discussed in this section.

Frontier [12] is a placement technique based on pre-placed IP cores, also known as macro-blocks. These macros are combined to form fixed sized and shaped clusters and then assigned to FPGA regions while minimizing the timing and wiring costs. Pre-compiled macro-block based design flow has been discussed in BPR [5] and HMFlow [13]. BPR introduces a new FPGA CAD tool for fast compilation of FPGA circuits where placement for each macro-block is selected from a database such that distance and expected routing congestion between macro-blocks are minimized. A modified version of VPR's SA placement algorithm [14] is used here to resolve the overlaps of placements that occur due to non-atomic type and size of blocks. HMFlow focuses on accelerating the compile time of designs having a 50% or less FPGA resource utilization. Three module placement algorithms are discussed in their work; recursive bi-partitioning placer, random placer and heuristic placer. The heuristic placer which is the fastest among them, considers the amount of connectivity between hard macros and prioritizes the large macros and macros with BRAMs and DSPs.

Qflow [15] separates the design into invariant and evolving sets. Placement of evolving modules is handled by an SA placer specialized for bigger (coarse-grained) modules. In [16], cluster placement methodology provides an estimation of the cluster's wire-length and criticality to the annealer. A floor planner based on Mixed Integer Linear Programming (MILP) [17] allows the designer to customize the objective function such that metrics like total wire length, area occupancy and aspect ratio can be weighted and incorporated as a linear combination. On the other hand, an analytical placer which optimize wire-length, is used in [18] instead of a SA based placer. Power-aware CAD techniques, for instance, [19] and [20], have not discussed their suitability for large designs. On the other hand, scalable techniques mainly focus on improving performance rather than power consumption. Although MDM is a scalable proposition, its module placement phase aims on improving compile time while optimizing wire-length or performance. Previous work lacks a module placement technique which considers data communication and power consumption.

III. Motivation

There are two major modes of power dissipation in FPGAs; (i) Static power and (ii) Dynamic power. The former is due to leakage current and subthreshold current whereas the latter is due to the charging and discharging of capacitive loads. Typical island-style FPGA consists of columns of Configurable Logic Blocks (CLBs) interleaved with BRAM and DSP columns, and input/output (IO) pins surrounding the chip area. The interconnect fabric, which connects CLBs, BRAMs, DSPs, and input/output(IO) pins to each other, consumes 80% of its total area. As a result, the dynamic power consumption of the interconnect fabric dominates the total dynamic power dissipation. Dynamic power dissipation (P) of an FPGA is modeled as follows [19];

$$P = \frac{1}{2} \sum_{i \in nets} C_i.\alpha_i.v_{dd}^2 \tag{1}$$

where $nets$ are all the connections and respective components of the FPGA, C is the capacitance of each net, α is the signal toggle rate (analogous to communication in this paper) and v_{dd} is the operating voltage.

In [9], the authors have explained that in a typical application, some code sections execute more frequently than the rest. Hence, the data communication of these code sections is higher as well. Consequently, on the FPGA, nets corresponding to such code segments show a higher toggle rate, α_i, due to data communication. In [9], authors have proposed to break down a large design into a set of modules based on the communication between nodes. We refer to these modules as *subsystems* throughout this paper. Nodes with high communication activity are included into the same subsystem and the CAD is instructed to place them in a close proximity. Having shorter routing wires for such connections reduces both C_i and α_i, reducing overall dynamic power dissipation considerably. Ideally, the subsystem creation should not introduce large inter-subsystem communication. While the approach manages to have shorter routing wires for the connections *within* the subsystems, it does not guarantee that the communication between the subsystems will also be minimized at the same time. Hence, wrong placement of such subsystems on an FPGA can allocate longer interconnect wires to the links with high inter-subsystem communication, thereby increasing dynamic power dissipation. [9] maps subsystems into FPGAs using existing CAD placement, which does not consider communication during the placement stage.

We elaborate this aspect using the following example. Consider a design with 7 design units illustrated as a graph shown in the figure 1(a). Each node shows its area requirement on FPGA space and intra-communication as $[l, m, d][\alpha]$ where l, m and d represent required number of CLBs, BRAMs and DSPs and α represents communication within the node/subsystem respectively. An edge is a connection between two design units where data flow occurs. It is annotated with communication between nodes/subsystems.

978-1-5386-7100-9/18 $31.00 © 2018 IEEE

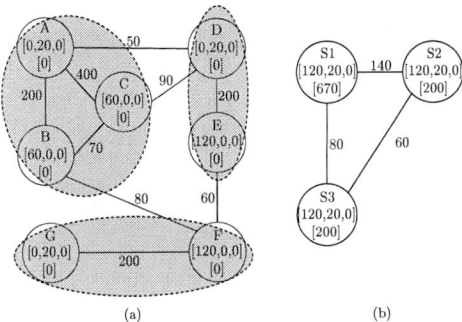

Figure 1: Subsystem creation based on communication

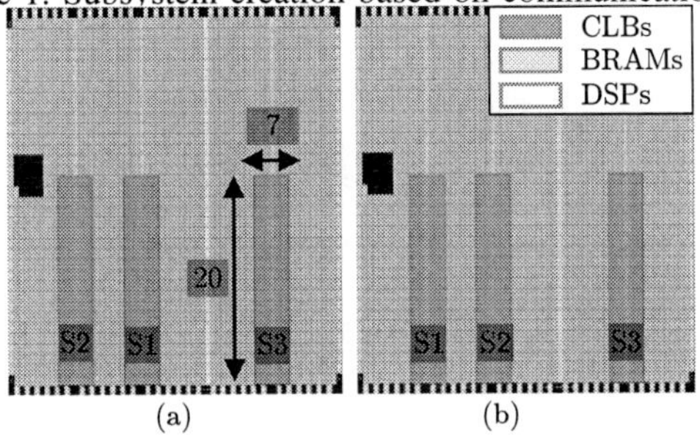

Figure 2: Mapping subsystems to FPGA without placement information

Performing a partitioning technique could result a network of subsystems as given in the figure 1(b). At this point, previous work has guided the CAD flow to group the relevant design units as subsystems. But the location of these subsystems on FPGA space is decided by the CAD flow. However, we observe that CAD flow could map subsystems with higher communication between them far apart. For instance, subsystem in figure 1(b) could be mapped on FPGA as shown in figure 2. However, the mapping in figure 2(a) would cause lower wire length for higher inter-subsystem communication links than the mapping in figure 2(b). This leads to higher power consumption in option a than in option b. In this paper, we strive to explicitly minimize the inter-subsystem communication to achieve further power savings.

IV. METHODOLOGY

In this section, we describe communication-aware subsystem placement methodology in detail. Similar to some workflows on subsystem placement in MDM, we use an SA based algorithm for our methodology. The inputs to our approach are (i) a network of subsystems and (ii) target FPGA architecture model, and the output is the placement details for each subsystem on the target FPGA space.

A. Problem Formulation

A network of subsystems can be represented as an undirected graph $G(V, E)$, where V is the set of vertices that represents each subsystem and E is the set of edges. An edge between two vertices indicates that there is one or more inter-subsystems connection between the components of both subsystems. Each edge is associated with a cost

value, α to represent inter-subsystem communication frequency between associated subsystems. Each subsystem is annotated with area requirement $A(l, m, d)$ where l, m, and d represents the required number of CLBs, BRAMs and DSPs respectively.

Given a network of subsystem $G(V, E)$, area requirement of each subsystem $v \in V$ as A_v, communication frequency between subsystems of each inter-subsystem connection $e \in E$ as α_e, the placement problem is to find a set of non-overlapping rectangular region L_v on FPGA space for each subsystem v, where L_v satisfies the area requirement of A_v.

B. Algorithm

1) Simulated Annealing Algorithm: We explain our SA based placement methodology in algorithm 1. A typical SA begins with a random initial placement solution s (line 2). Placement in CAD flow considers allocating atomic components such as CLBs and BRAMs on FPGA space. But in subsystem placement, a number of such atomic components need to be placed. Instead of deciding the location of each atomic component, subsystem placement decides a region where all the components that belong to a subsystem can be placed on FPGA. $COST(s)$ (line 3) refers to the placement quality of the solution s. Temperature (T) is analogous to the current iteration of annealing. There can be n iterations (line 5) before it converges to a solution. In each iteration of the algorithm, the solution of the previous iteration is altered by moving one or more subsystems within FPGA space (line 8), which is called a neighbor solution s_N. If s_N is better than reported best s_{best}, s_N is assigned to be the best solution so far (line 11-12). However, in case if the new solution s_N is not the reported best solution, it is still considered as a valid solution if it satisfies a probability equation P (line 13-14).

2) Neighbor generation: In each iteration, a subsystem is selected randomly to move withing FPGA space to create neighbor solution. Maximum allowed movement along horizontal and vertical direction on FPGA space (δx_{max}, δy_{max}) of the selected subsystem is a function of T, where δx_{max} and δy_{max} are gradually decreased when T reduces. Once the maximum movement for current T is obtained, the selected subsystem is moved in the range [$-\delta x_{max}, \delta x_{max}$] horizontally and [$-\delta y_{max}, \delta y_{max}$] vertically.

3) Shape of the subsystems: Subsystems with requirement for multiple resource types (CLBs and BRAMs) are split into sub-modules in a way such that sub-modules contains only one resource type. For instance, as shown in figure 3(a), subsystem $S3$ with area requirement $A_3 = (120, 20, 0)$ is divided into two whose area requirements are $A_3 = (120, 0, 0)$ and $A_{3.1} = (0, 20, 0)$ respectively. We set the communication between subsystem $S3$ and $S3.1$ to a very large number (N). However, the original edges in the graph G are preserved. This subdivision of subsystems is done to cater for the area requirement A_v during neighbor generation. For instance, moving a subsystem having both CLBs and BRAMs from location L_1 to L_2 might not find

Algorithm 1 Communication-aware subsystem placement

1: **procedure** SIMULATED_ANNEALING
2:     $s \leftarrow$ INIT_PLACEMENT()
3:     $\phi \leftarrow$ COST(s)
4:     $s_{best} \leftarrow s, \phi_{best} \leftarrow \phi$
5:     $T_0 \leftarrow n$
6:     **for** $T \leftarrow T_0...0$ **do**
7:         **do**
8:             $s_N \leftarrow$ NEIGHBOR(s,T)
9:         **while** MAX_OVERLAP$(T) <$ OVERLAP(s_N)
10:         $\phi_N \leftarrow$ COST(s_N)
11:         **if** $\phi_N < \phi_{best}$ **then**
12:             $s_{best} \leftarrow s_N, \phi_{best} \leftarrow \phi_N$
13:         **if** P$(e,e_N,T) >$ RAND() **then**
14:             $s \leftarrow s_N, \phi \leftarrow \phi_N$
15:     **return** s_{best}
16: **procedure** NEIGHBOR(s,T)
17:     $c \leftarrow$ SELECT_SUBSYSTEM(s)
18:     $\delta x_{max} = F(T), \delta y_{max} = F(T)$
19:     $\delta x \leftarrow$ RAND$(-\delta x_{max}, \delta x_{max})$
20:     $\delta y \leftarrow$ RAND$(-\delta y_{max}, \delta y_{max})$
21:     $c(x,y) \leftarrow c(x + \delta x, y + \delta y)$
22:     **return** s

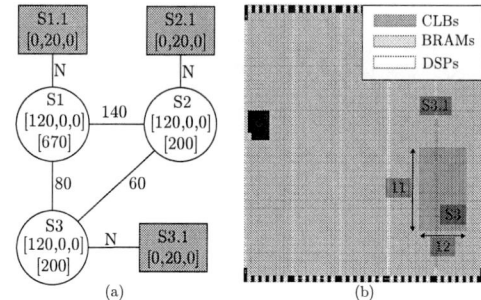

Figure 3: a) Split subsystems b) Shape of subsystems

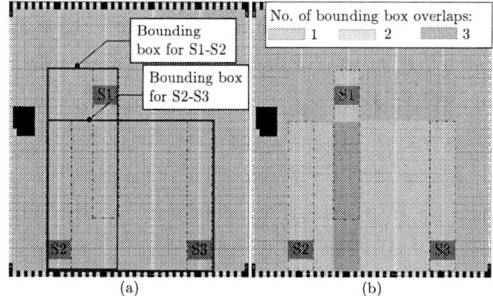

Figure 4: Congestion Model

BRAMs at L_2. However, in this work, we do not consider the importance of the shape of a subsystem on FPGA space. Therefore, in each iteration, we set the shape of subsystems with CLBs to have a height to width ratio closer to 1 as shown in figure 3(b). It has been shown in [21] that shapes with aspect ratio closer to 1, help in reducing the wire length within a subsystem.

4) Overlap of subsystems: Typical SA placement algorithms do not allow overlaps of atomic components during each iteration. However, since subsystems are non-atomic components, it is difficult to produce a non-overlapping solution in each iteration. The search space is also limited if overlapping of subsystems is not allowed, which might cause the algorithm to converge to a local minimum. To avoid such situations, our approach allows overlapping of subsystems. We use an exponential function of T to decide maximum allowable overlaps of subsystems, which reaches zero as T decreases. Number of overlaps are counted as the area of the overlapped region. Since we divide the subsystems into sub-modules with single resource type, intersection of $S2$ and $S3.1$ in figure 3 is not considered as an overlap.

5) Congestion model: Routability evaluation is an important step in a placement algorithm. For instance, placing a large number of components in a smaller area could lead to excessive requirement of routing resources. Routing process avoids highly congested areas and therefore may use longer wires to connect two components. Hence, quality of result (performance and power) is degraded. In our approach, we incorporate a bounding box based congestion analysis to estimate the routing congestion similar to [22]. In this model, we consider the FPGA space as a 2D array, which we call as

congestion map. All congestion map elements are initialized to zero at start. For each inter-subsystem connection in figure 1(b), a bounding box can be formed as figure 4(a). Note that the bounding box for the edge between $S1$ and $S3$ is not shown. Next, congestion map elements relevant to each bounding box is incremented. Overlapping bounding boxes, therefore, reflect high congestion regions. Congestion map relevant to the subsystem placement shown in figure 4(a) is shown in 4(b). We update the congestion map in each iteration of the algorithm.

6) Cost Function and constraints: The main purpose of our work is to avoid longer routing wires, especially for high inter-subsystem communication links. Therefore, our objective cost function is to minimize

$$COST(s) = \sum_{e \in E} l_e.\alpha_e \qquad (2)$$

where l_e is half perimeter wire length for the bounding box enclosing the subsystems relevant to edge e, whereas α_e is inter-subsystem communication. While minimizing the cost function, we also consider a routing congestion parameter C. For a valid placement, its maximum congestion derived above must be less than empirically evaluated C.

C. Implementation on FPGA

It is important to note that our approach of communication-aware subsystem placement does not depend on a specific CAD tool. For implementation and evaluation of our methodology, we use Altera's Quartus CAD flow [23]. But, an equivalent approach could be followed with Xilinx CAD flow as well [24]. The first phase of our subsystem placement approach partitions a

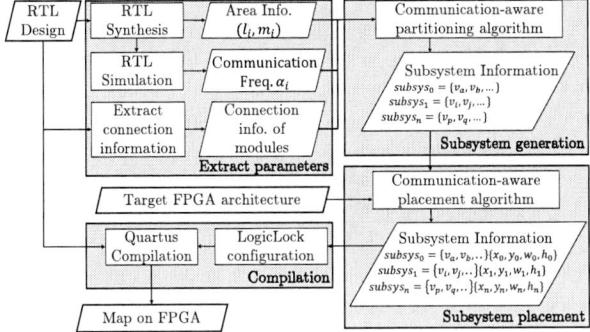

Figure 5: Communication-aware mapping methodology

large design into subsystems as done in [9]. This partitioning technique works at entity/module level of an RTL code. Entities which have high communication between them are grouped into subsystems. For that, a given RTL code is processed to extract the following details. This step is faster than the full CAD compilation.

1) Area of the RTL entities: RTL synthesis step in Quartus CAD flow estimates the resource requirement for each entity. This step is faster than the full CAD compilation.
2) Communication frequency between entities: An RTL simulation can dump all the signal toggles of desired connections between entities.
3) Connectivity of entities: Entity-level RTL connection information is extracted by doing text processing on code

Once these parameters are extracted, the communication-aware partitioning algorithm forms subsystems such that intra-subsystem communication is much higher than inter-subsystems communication. The output of the partitioning framework is a list of subsystems and a list of design entities that belongs to each subsystem. We modified the output slightly in order to obtain (i) a graph representation of subsystems (ii) the inter-subsystem communication frequency α_e of each connection e between subsystems (iii) area requirement A_v of each subsystem v. Additionally, for our placement algorithm, target FPGA architecture floorplan information is needed. It should show the position of CLB, BRAM and DSP columns and the number of resource columns and rows in the floorplan. The placement algorithm produces the placement for each subsystem v with (i) left bottom coordinate (x_v, y_v) of the rectangular region in FPGA space and (ii) width w_v and the height h_v of the rectangular region. We can use these parameters to invoke Quartus LogicLock feature during the compilation to define subsystems on FPGA space. Note that, in [9], placement information to the LogicLock feature is not given allowing CAD flow to decide the placement of each subsystem, whereas this work overrides the tool's placement decision with our placement information.

V. RESULTS AND DISCUSSION

Next, we discuss the performance of the proposed communication-aware placement strategy. We select applica-

tion set coded in RTL for the evaluation. Compilation reports produced by Quartus CAD flow are used for the comparison.

A. Comparison strategy

We compare our methodology with two existing compilation flows: (i) Original Quartus compilation without subsystems (ii) Quartus compilation with subsystems without placement information as proposed in [9]. Full Quartus compilation is done for these two cases and for proposed solution followed by gate-level simulation to get the most accurate power estimations from Quartus PowerPlay. Since, we mainly target to reduce the power dissipation in the FPGA interconnect, we compare the routing power dissipation using the three approaches. Since, the operating frequency affects the power dissipation, the routing power is measured at a common operating frequency for all tests.

B. Benchmark Applications and Target FPGAs

Selected benchmark applications are similar to the applications in [9]. We use three handwritten applications as presented in [9]. In addition, two applications have been developed by modifying applications from the popular polybench [25] benchmark suite. However, we extend the benchmark application set that are inspired by polybench kernels by introducing three more applications by following the same concept for Gesummv, Cholesky and Symm kernels to develop gesummv/, cholesky/ and symm/ respectively. In addition, for a comprehensive evaluation, the benchmark set is mapped into three newer devices in Altera Cyclone IV and Stratix IV series. Although the methodology is compatible with any device series including the latest Cyclone V and Cyclone 10 series (or latest Xilinx devices), the gate level simulation is not supported in the latest series which is required to get a precise power measurement. It should be noted that this is not the limitation of the proposed work. Therefore, Cyclone IV and Stratix IV were the latest series we could use to show the effectiveness of the methodology.

C. Results and Discussion

Three hand-coded applications are mapped to Cyclone IV EP4CGX50 device, gesummv/ is mapped to slightly bigger Stratix IV EPSGX70 device and the rest of the applications are mapped to Cyclone IV EP4CGX110 device. Each application is compiled in three different test versions as stated above. Table I reports percentage routing power reduction of the new methodology over (i) default Quartus compilation and (ii) over [9]. As observed from these results, the proposed communication-aware placement technique for subsystem placement during CAD flow helps to further reduce routing power. In particular, the proposed approach helps to reduce routing power of the benchmark applications ranging from 3.55% to 10.75% (average of 7.08%) when compared to [9]. It is also shown that the average routing power reduction, when compared to the default Quartus compilation, ranges from 3.69% to 18.21% for the application set with an average routing power reduction of 11.51%.

Table I: Routing Power Saving

Application	Target Device	$P_{routing}$ reduction Over default	$P_{routing}$ reduction Over [9]
Synth1	EP4CGX50	18.21%	8.10%
Synth2	EP4CGX50	16.40%	10.49%
Synth3	EP4CGX50	15.91%	6.19%
atax/	EP4CGX110	12.69%	10.75%
bicg/	EP4CGX110	3.69%	3.55%
gesummv/	EP4SGX70	6.68%	5.48%
cholesky/	EP4CGX110	8.19%	6.05%
symm/	EP4CGX110	10.31%	6.05%

Communication-aware subsystem placement actually depends on how the application is partitioned into subsystems. Some application partitioning might not create connections with high inter-subsystems communication. For instance, inter-subsystems links in bicg/ are not as significant as that of applications like Synth1 or atax/. As a result, the effectiveness of this methodology depends on the inter-subsystem communication. One can argue that placement technique would not be necessary if all the high communication nodes are included into subsystems during partition generation. However, it should be remembered that making larger subsystems reduces the effectiveness of creating subsystems in the first place, since it leads to the original problem of mapping large designs on the FPGA. Instead, the current subsystems are generated to strike a balance between the size and intra-susbsystem communication as described in [9]. In future, we will explore the subsystem generation step with both intra- as well as inter- subsystem communication frequencies in order to achieve even better results.

VI. CONCLUSION

In this paper, we presented a inter-subsystem communication aware placement technique that produces high quality solutions with lower power consumption, especially for large FPGA designs. The proposed Simulated Annealing based approach reduces the distance between subsystems with higher inter-subsystem communication. This approach results in reducing routing power consumption by over 7% when compared to an existing methodology that does not consider the inter-subsystem communication frequency and by over 11% when compared to a commercial CAD tool such as Quartus. In future, we proposed to also evaluate the footprints of subsytems for further power savings.

ACKNOWLEDGMENT

This project was partially funded by the National Research Foundation Singapore under its Campus for Research Excellence and Technological Enterprise (CREATE) programme with the Technical University of Munich at TUMCREATE.

REFERENCES

[1] Altera. [Online]. Available: https://www.altera.com/
[2] Xilinx. [Online]. Available: https://www.xilinx.com/
[3] S. M. Trimberger, "Three Ages of FPGAs: A Retrospective on the First Thirty Years of FPGA Technology," *Proceedings of the IEEE*, 2015.
[4] H. Bian *et al.*, "Towards scalable placement for FPGAs," in *Proceedings of the 18th annual ACM/SIGDA international symposium on Field programmable gate arrays*, 2010.

[5] J. Coole *et al.*, "BPR: fast FPGA placement and routing using macroblocks," in *Proceedings of the eighth IEEE/ACM/IFIP international conference on Hardware/software codesign and system synthesis*, 2012.
[6] M. An *et al.*, "Speeding up FPGA placement: Parallel algorithms and methods," in *Field-Programmable Custom Computing Machines (FCCM), 2014 IEEE 22nd Annual International Symposium on*, 2014.
[7] A. Ludwin *et al.*, "Efficient and deterministic parallel placement for FPGAs," *ACM Transactions on Design Automation of Electronic Systems (TODAES)*, 2011.
[8] T.-H. Lin *et al.*, "An efficient and effective analytical placer for FPGAs," in *Proceedings of the 50th Annual Design Automation Conference*, 2013.
[9] K. Herath *et al.*, "Communication-aware Partitioning for Energy Optimization of Large FPGA Designs," in *Proceedings of the on Great Lakes Symposium on VLSI 2017*, 2017.
[10] L. Shang *et al.*, "Dynamic power consumption in Virtex-II FPGA family," in *Proceedings of the 2002 ACM/SIGDA tenth international symposium on Field-programmable gate arrays*.
[11] T. Tuan *et al.*, "A 90nm low-power FPGA for battery-powered applications," in *Proceedings of the 2006 ACM/SIGDA 14th international symposium on Field programmable gate arrays*.
[12] R. Tessier, "Fast placement approaches for FPGAs," *ACM Transactions on Design Automation of Electronic Systems (TODAES)*, 2002.
[13] C. Lavin *et al.*, "HMFlow: Accelerating FPGA compilation with hard macros for rapid prototyping," in *Field-Programmable Custom Computing Machines (FCCM), 2011 IEEE 19th Annual International Symposium on*, 2011.
[14] V. Betz *et al.*, "VPR: A new packing, placement and routing tool for FPGA research," in *International Workshop on Field Programmable Logic and Applications*, 1997.
[15] T. Frangieh *et al.*, "A design assembly framework for FPGA back-end acceleration," in *Reconfigurable Computing and FPGAs (ReConFig), 2012 International Conference on*, 2012.
[16] F. Gharibian *et al.*, "Identifying and placing heterogeneously-sized cluster groupings based on FPGA placement data," in *Field Programmable Logic and Applications (FPL), 2014 24th International Conference on*, 2014.
[17] M. Rabozzi *et al.*, "Floorplanning for partially-reconfigurable fpga systems via mixed-integer linear programming," in *Field-Programmable Custom Computing Machines (FCCM), 2014 IEEE 22nd Annual International Symposium on*, 2014.
[18] M. Gort *et al.*, "Design re-use for compile time reduction in FPGA high-level synthesis flows," in *Field-Programmable Technology (FPT), 2014 International Conference on*, 2014.
[19] S. Gupta *et al.*, "CAD techniques for power optimization in Virtex-5 FPGAs," in *Custom Integrated Circuits Conference, 2007. CICC'07. IEEE*, 2007.
[20] J. Lamoureux *et al.*, "On the interaction between power-aware FPGA CAD algorithms," in *2003 IEEE/ACM international conference on Computer-aided design*.
[21] M. Wang *et al.*, "Multi-million gate FPGA physical design challenges," in *Proceedings of the 2003 IEEE/ACM international conference on Computer-aided design*, 2003.
[22] Y. Zhuo *et al.*, "A congestion driven placement algorithm for FPGA synthesis," in *Field Programmable Logic and Applications, 2006. FPL'06. International Conference on*.
[23] "Increasing Productivity with Quartus II Incremental Compilation," https://goo.gl/uy225f.
[24] "Vivado Design Suite User Guide-Hierarchical Design," https://goo.gl/6bUqqD.
[25] L.-N. Pouchet, "Polybench: The Polyhedral Benchmark Suite," http://web.cs.ucla.edu/~pouchet/software/polybench/.

2018 IEEE Computer Society Annual Symposium on VLSI

A Novel Mixed-Size Fixed-Outline Floorplacement Algorithm

Qian Chen,   Sheqin Dong

Department of Computer Science & Technology
Tsinghua University, Beijing, China 100084
Email: chenqian10@mails.tsinghua.edu.cn, dongsq@mail.tsinghua.edu.cn

Abstract—In this paper, a novel fixed-outline floorplacer is proposed to handle large scale mixed-size instances designs. First, recursive partition is implemented to decrease the scale. A circuit is partitioned into clusters, which can be packed in a given fixed outline effectively based on a zero deadspace packing algorithm. Then, a boundary constraint-based placement algorithm is implemented to cluster modules in the aspect given by the first stage. Finally, several techniques are proposed to further improve the total HPWL. Experimental results show that compared to the best floorplacer, the proposed algorithm can reduce the wirelength by more than 5%.

Keywords: Floorplacement, Fixed-outline, Placement

I. INTRODUCTION

With the increase in complexity of the modern VLSI designs, some new challenges are faced to placement tools: the design sizes are increasing exponentially, while the intellectual property modules and pre-fixed macro blocks are reused more frequently. Hence, the ability of handle large scale instances in mixed sizes and movable macros are critical for physical design tools. To deal with different placement types which aim at different issues, there are different placers which can be classified into three categories. As shown in Fig.1 (a), standard cell placement is the main consideration while millions of small movable and big modules are fixed in a design [1]-[5]. In Fig.1 (b), mixed-size placement is implemented as small and big modules are movable [1], [2], [4]-[6], [8]. In Fig.1 (c), floorplacement is accomplished for designs with some modules of various dimensions [1], [7], [8].

Floorplacement is a unification of placement and floorplanning which seeks to bridge the gap between placement and floorplanning by combining their features. In

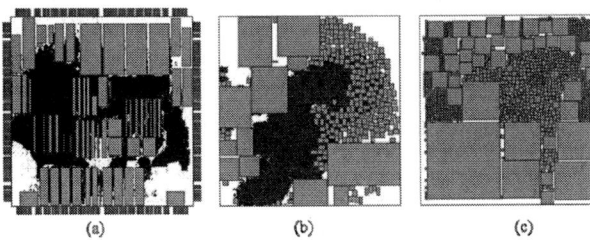

Fig. 1. Different placement types [1]

This paper is supported by NSFC 61176022.

floorplanning, we often seek non-overlapping locations of modules within a fixed outline, with the objective of optimizing for certain objectives, such as wire-length, power or performance. There are usually no more than 200 modules in a floorplanning problem and SA-based framework is implemented in most recent floorplanner.

In [11], a Zero Deadspace (ZDS) fixed-outline floorplanner, named as SAFFOA, is proposed to improve the half-perimeter wirelength (HPWL) with optimal area utilization. SAFFOA builds and solves a group of four quadratic equations in four variables iteratively, which can handle the fixed-outline constraint of any aspect ratio. The representation Ordered Quadtree is custom-made for the equations and integrated into simulated annealing iterations to optimize the total HPWL. When hard blocks exist, all the hard blocks are relaxed to soft blocks and a ZDS layout is generated while the wirelength is optimized. Then all the hard blocks are recovered, with a fast simple legalization technique.

Experimental results show that SAFFOA achieves the best HPWL on all the three types of floorplans, i.e. hard modules, soft modules and hybrid designs. Although SAFFOA cannot handle floorplacement problems directly, it can be used in higher layers for a hierarchical or multilevel design.

In this work, we propose a novel mixed-size fixed-outline floorplacement algorithm extended from SAFFOA. A novel framework is proposed to handle large scale floorplacement problems, which takes advantages of the desire positions and controls the partition of circuits globally. In the legalization stage, a linear programming method is implemented to make as good as possible use of the position information from ZDS packing process. As large number of external nets exists in most of the sub-circuits, a boundary constraint based quadtree cluster is proposed to speed up the detailed placement. Some other techniques, such as buffer cells and repulsion forces, are proposed to further improve the HPWL and chip area utilization.

The remainder of this paper is structured as follows. Section 2 describes the framework of our algorithms and the main idea. Section 3 introduces our approach for fixed-outline mixed size floorplacement problem. Several techniques to improve the HPWL and area utilization are

978-1-5386-7100-9/18 $31.00 © 2018 IEEE

presented in Section 4. Section 5 is the experiment results.

II. OVERALL DESIGN FLOW

The framework of our algorithm is shown in Fig. 2. To increase the area utilization and optimize the HPWL of the design, a few biggest macro blocks and cells have high connection with them are extracted in the first stage. Then k-way partition [10] is called recursively to formulate multilevel sub-problems. The circuit is partitioned into 20-90 sub-circuits, each of which contains 20-90 sub-circuits or modules. Therefore, modules in a desired scale can be packing tightly. Different sub-circuits contain different sets of modules so the evaluation for the placability is accomplished and generates some estimated parameters for following stages.

Then, ZDS packing based on SAFFOA is implemented to handle high level circuits derived from the top partition tree. Each sub-circuit is regarded as a soft module and ZDS layout can be generated. In low level, every sub-circuit has many external nets so that boundary constraint can help to speed up the low level placement. The sizes of modules are specified in the low level, and it must be different from the ZDS packing sizes, so LP-based legalization is implemented. Finally, some refinements are done in the post stage to further improve the HPWL.

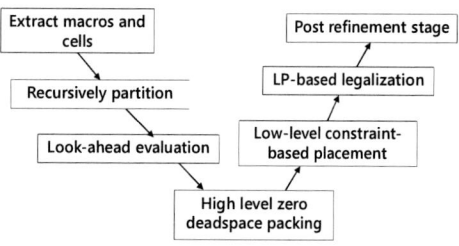

Fig. 2. Overall design flow

III. THE PROPOSED ALGORITHM

Floorplacement is a unification of floorplan and placement by combining their best features. There are at least thousands of modules in designs so min-cut partition is called to handle the problems in different levels. At the same time, instances in various dimensions are placed in different levels. In our framework, the top level circuit is recursively partitioned and clusters are look-ahead evaluated for following process. Then the top level circuit is packed in ZDS, and detailed placement and legalization for positions of modules is completed.

A. Recursive partition

To translate the problems to a set of small scale sub problems, a circuit is partitioned by recursive 2-way partition method. As *hmetis* [10] is proved to be a good min-cut partition tool, the partitioner is called and the output is modified for the placer.

Partition modification: Min-cut is not equivalent to minimal wirelength mainly because of modules connected by cut nets are obstructed by central modules especially macro blocks. Two rules are added after observing previous layouts: (1) Hard blocks with similar sizes often have high connection, so placing them together can improve the area utilization. If there are hard blocks with the same height or width, the weight of the net connected with them is increased nine times. (2) As hard blocks in the same cluster with different dimensions are hard for legalization, the net weight among them is decreased to push them away.

Net weight modulation: during the iterations of 2-way partition, the weights of edges being cut are adjusted continuously. In our experiment, the net weight is multiplied by N / (N+1) in each iteration where N indicates that the net is being cut for the N time.

B. Look-ahead evaluation

In our framework, high level packing is implemented before detailed placement, so it is important to get relatively precise cluster sizes for high level ZDS packing. The high level packing can generate a ZDS layout because of a soft blocks packing algorithm that based on the variability of aspect ratios and invariability of sizes.

When a circuit is partitioned into a set of sub-circuits, it will be quite different between them. Some of them are easy to be placed in a randomly specified fixed-outline while some others are much harder. So it is helpful of finding out the harder instances and estimating reasonable fixed-outlines for them quickly.

Clusters are failed to be placed can be classified in two cases: (1) There are several relative big hard blocks so that no feasible solution for placing these blocks in the fixed-outline exists. In this case, the key issue is how to place the biggest hard blocks, if the total size of these blocks is more than 60% of the size of this sub-circuit, a fixed-outline can hold these biggest blocks are supposed to hold the whole sub-circuit.

(2) If most blocks are hard, in the aspect recover stage, each hard block can lead to a geometrical adjustment. If a fixed-outline is set randomly, the whitespace can be very large. In this situation, the softening and packing method does not work anymore. The aspect of this kind of cluster is determined by earlier detailed placement and treated as a hard block in the top level.

In this stage, instructional parameters are computed for following packing stage, which make the packing solution closer to the final placement solution.

C. Boundary constraint-based cluster

After the high level packing, many terminals are generated from external nets. There are 100~500 terminals propagated in a cluster which contains 30~50 modules in our experiment. These terminals indicate cut nets which intent to pull modules together, and a module connect to this type terminal is pulled to the boundary. In actually, boundary constraints exist in clusters of our experiment designs widely. To speed up the detailed placement process, a boundary constraint based cluster algorithm is implemented in our detailed placement stage.

The layout representation should meet following requirement: Firstly, the algorithms need to deal with soft modules, so it should be flexible to modulate the aspect ratio. At the same time, it is critical to the algorithm efficiency of fast conversion between the representation and boundary constraint. What is more, as it is inevitable to get in an illegal state in optimization, the cost of switching from illegal topology to nearest feasible solution should be acceptable.

Quadtree is used to represent the layout and the algorithm is introduced in the following paragraphs.

Mark the constraints: According to the statistic of connection between terminals and modules [13], constraint sets are generated:

$$\begin{cases} C_L\left[l_1(w_{l_1}),\ l_2(w_{l_2}),\ l_3(w_{l_3}),\ ...\right] \\ C_R\left[r_1(w_{r_1}),\ r_2(w_{r_2}),\ r_3(w_{r_3}),\ ...\right] \\ C_T\left[t_1(w_{t_1}),\ t_2(w_{t_2}),\ t_3(w_{t_3}),\ ...\right] \\ C_B\left[b_1(w_{b_1}),\ b_2(w_{b_2}),\ b_3(w_{b_3}),\ ...\right] \end{cases} \quad (1)$$

where C_L, C_R, C_T, C_B denote the left, right, top, bottom boundary constraints respectively, l_i, r_i, t_i, b_i are the indexes of modules in each direction, and w_j is the connection number between module j and the terminals in current direction. Constraints in opposite directions are substrate and constraints which are still bigger than given threshold are marked as boundary constraint.

Boundary constraints representation: If a node satisfies top boundary constraint, it must obey three rules: (1) The node does not have the first sub-tree; (2) If any of its ancestors has the first sub-tree, it must be in this sub-tree. (3) The node cannot be in the second sub-tree of its ancestor. If a node satisfies left boundary constraint, it must obey two rules: (1) The node must not have the third sub-tree; (2) This node cannot be in the fourth sub-tree of its ancestor. The bottom boundary constraint is similar to the top, while the right one is similar to the left one.

It is obviously that if a node does not satisfy any of four boundary constraints, all the nodes in its sub-trees do not satisfy the constraint. So when the quadtree is initializing, every node is marked of its constraint satisfaction recursively and make sure all the nodes in constraint sets are placed in legal positions.

Perturbation under constraints: The perturbation and legalization ensure the algorithms to find good solutions in a shorter time. There are three types of perturbation in the original quadtree: (1) Switch between sub-trees of the same

node; (2) Switch between node and terminal; (3) Switch between nodes.

While the boundary constraint is taken into consideration, the last two perturbations are easy to realize. If and only if two nodes obey the same constraint, they can be exchanged. The switch of two sub-trees may lead to some nodes violate the boundary constraints, so three rules are defined as follow: (1) A switch between a sub-tree satisfies a constraint and an empty sub-tree is forbidden. (2) When the switch is between top (left) and bottom (right) sub-trees, nodes satisfy top and bottom (left and right) boundary constraints are switched back later. (3) All the other situations are handled in a similar way. The main principle is searching a legal position symmetrically as simple as possible while the constraint is not satisfied.

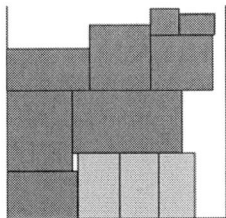

Fig. 3. Greedy overlap legalization

D. LP-based legalization

In the previous packing algorithms, if the SA-based HPWL optimization is complete, the aspect ratios of hard modules are recovered. A greedy overlap legalization method is used in SAFFOA, which is fast and easy to understand. However, this approach does not consider the global information and modules can extrude out of the right-top outline as shown in Fig. 3. The refinement is difficult in this situation as the topological structure is destroyed.

In this work, a linear programming based algorithm is implemented to take the most advantages of the "desire" layout information in the ZDS packing stage while remove the overlap of modules recovering.

The module positions after the packing stage are the best positions if all modules are soft, so it can be an important indication for the final layout. The legalization must be global as the HPWL is often better when modules are distribution well-proportioned in different directions.

Firstly, a constraint graph is built according to the packing solution. As shown in Fig. 4, the constraint graph

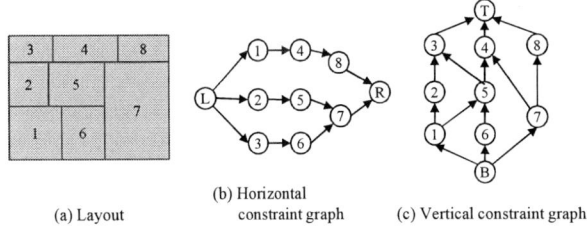

(a) Layout

(b) Horizontal constraint graph

(c) Vertical constraint graph

Fig. 4. Layout and the corresponding constraint graphs

edges are got from adjacent modules pairs directly, and the constraint graph only contain part of the topology constraint, because modules movement in the legalization is local and the complexity of the graph impacts the run time notably.

Secondly, all the modules are linearly spread according to the total packing area to the fixed-outline. The positions of modules are set according to:

$$x_i^{'} = \frac{W_{fixed-outline}}{W_{packing}} x_i \qquad (2)$$

$$y_j^{'} = \frac{H_{fixed-outline}}{H_{packing}} y_j \qquad (3)$$

where (x_i, y_i) is the position in the packing solution, $(x_i^{'}, y_i^{'})$ is the position after spreading, and $W_{fixed-outline}$, $H_{fixed-ouline}$ are the width and height of the fixed-outline, $W_{packing}, H_{packing}$ are the width and height of the packing area.

The object of our LP process is minimizing the total weighted movement of modules. The LP problem cost function is list as follow:

$$\min \sum_{i=1}^{n} Net_i A_i (|x_i - x_i'| + |y_i - y'|_i) \qquad (4)$$

where Net_i is the number of nets connect to m_i (module i), A_i is the area of m_i, $(x_i, y_i), (x_i', y_i')$ are the coordinates before and after moving respectively. The object function is the sum of modules displacement weighted by the area and net number.

The constraint condition inequality is shown blow:

$$\begin{cases} x_j' - x_i' \geq w_{ij}^h & \forall e_{ij} \in G_h' \\ y_j' - y_i' \geq w_{ij}^v & \forall e_{ij} \in G_v' \\ |x_i' - x_i| \leq X_{max} & \forall m_i \in C' \\ |y_i' - y_i| \leq Y_{max} & \forall m_i \in C' \\ x_i' = x_i, y_i' = y_i & \forall m_i \in F' \end{cases} \qquad (5)$$

where G_h', G_v' are the horizontal and vertical constraint graph, w_{ij}^h, w_{ij}^v are the width and height of mi, X_{max}, Y_{max} are the maximal displacements of mi in horizontal and vertical direction, C' is the set of the nodes in G without nodes L, R, B, T, and is the set of nodes L, R, B, T.

IV. OTHER TECHNIQUES

In our algorithms, some techniques are also used to further improve the quality of our solution.

Repulsion nets: Empirically, pushing macro modules away from the chip center can improve the HPWL. As the layout representation is ordered quadtree in our work, restricting the sub-tree number of biggest blocks can place them in a corner, but this can only impact a few modules. To push macro modules away, a pseudo node is added at the center of the chip, for each module bigger than a given threshold, a 2-pin net connecting to the pseudo node is add to the circuit. We call these nets as "repulsion nets". The weights of the 2-pin nets are set according to:

$$W_{Net_i} = -\phi \frac{1}{A_{Cluster_i}} \sum_{m_j \in Cluster_i} A_j^2 (1 + IsHard_{m_j}) / 2 \qquad (6)$$

where ϕ is a constant, A_j is the area of m_j and $A_{cluster_i}$ is the area of cluster i. The weight of the net is negative, so it pushes modules to margin which is opposite to normal nets. At the same time, the weight of a net depends on the overall granularity of the modules in a cluster.

Buffer cells: As there is often much whitespace near macro blocks, some small cells are extracted in early stage. In the post stage, these cells are inserted just like buffer insertion. So we can call them "buffer cells". Experimental results show that this can improve the area utilization about 1.5%-2%.

V. EXPERIMENT RESULTS

Experimental results show the performance of our techniques on the IBM-HB+ benchmarks. All experiments are performed on a workstation with 3.0 GHz CPU and 4GB RAM. The IBM-HB+ circuits do not have standard cells but consist of about a thousand modules with various dimensions. The results of Capo10.0 are referred to [8], which is a feature of Capo, and the CPU is scaling of 0.8. The results of Kraftwerk2 are referred to [1], with a CPU scaling of 0.73. *lp solve 5.5.2.0* [12] is used for linear programming solving. As shown in Table 1, our placer improves HPWL by 10%, 5% versus Capo10.0, Kraftwerk2. On average, our algorithm is 3.22× faster than Capo 10.0 while the runtime is 2.32 times of Kraftwerk2. No results of other placers that successfully place the complete benchmark suite are available.

REFERENCES

[1] P. Spindler, U. Schlichtmann, F. M. Johannes, "Kraftwerk2—A Fast Force-Directed Quadratic Placement Approach Using an Accurate Net Model." TCAD2008.

[2] T.-C. Chen, Z.-W. Jiang, et al., "Routability-Driven Analytical Placement for Mixed-Size Circuit Designs." ICCAD2006.

[3] M-C. Kim, N. Viswanathan, et al., "MAPLE: multilevel adaptive placement for mixed-size designs." ISPD2012.

[4] G.-J. Nam, J. Cong, "Modern Circuit Placement: Best Practices and Results,"Springer, 2007.

[5] T. F. Chan et al., "mPL6: Enhanced Multilevel Mixed-Size Placement,"ISPD2006.

[6] T. Taghavi, X. Yang, and B.-K. Choi, "Dragon2005: Large-scale mixed-size placement tool." ISPD2005.

[7] A. R. Agnihorti, S. Ono, et al., "Mixed block placement via fractional cut recursive bisec-tion." TCAD2005.

[8] J. A. Roy, A. N. Ng, R. Aggarwal, et al., "Solving modern mixed-size placement instances." VLSI Journal 2009.

[9] H.-C. Chen, et al., "Constraint graph-based macro placement for modern mixed-size circuit designs." ICCAD2008.

[10] G. Karypis, R. Aggarwal, V. Kumar, S. Shekar, "Multilevel hypergraph partitioning: applications in VLSI design." DAC 1997.

[11] O. He, S. Dong, J. Bian, et al., "A novel fixed-outline floorplanner with zero deadspace for hierarchical design" ICCAD2008.

[12] "Lp solve". http://www.lpsolve.sourceforge.net/5.5/

[13] J. Liu, S. Dong, et al, "Floorplanning with Constraint Extraction based on Interconnecting Information Analysis," ASICON2007.

TABLE I.   EFFECTIVENESS OF OUR ALGORITHM

ibm-HB+ benchmark	Capo10.0		KraftWerk2		OURS	
	HPWL (e+6)	Time(s)	HPWL (e+6)	Time(s)	HPWL (e+6)	Time(s)
01	3.2	46.1	2.8	7.3	3.0	18.8
02	6.9	148.3	5.8	19.1	6.0	46.9
03	10.1	143.9	9.2	11.7	9.2	43.5
04	11.1	116.6	9.9	15.4	10.3	40.1
06	9.3	161.4	8.7	8.8	8.1	52.6
07	16.1	72.6	14.8	11.7	14.9	31.7
08	18.8	192.0	21.2	13.9	17.8	57.9
09	20.9	148.6	17.4	13.2	16.2	53.0
10	55.2	255.9	47.5	34.5	44.6	70.9
11	26.9	109.8	25.9	16.9	24.1	48.8
12	64	318.1	51.3	31.5	51.1	80.7
13	39.7	127.8	34.9	16.9	34.1	43.8
14	63.8	191.0	63.1	30.8	59.3	60.5
15	86.4	406.6	92.8	33.7	84.9	97.9
16	101.8	203.4	95.6	39.6	94.8	71.4
17	146.3	304.0	148.1	70.4	138.6	86.3
18	74.7	145.5	73.9	38.1	70.0	63.2
	1.10	7.47	1.05	1.00	1.00	2.32

2018 IEEE Computer Society Annual Symposium on VLSI

ARCHVerifyr: An Embedded Software-Driven Approach for Architecture Verification

Tomás Grimm*, Djones Lettnin†, Michael Hübner*

*Chair of Embedded Systems for Information Technology (ESIT), Ruhr-University Bochum, Germany
†Department of Electrical and Electronics Engineering, Federal University of Santa Catarina, Brazil
Email: Tomas.Grimm@rub.de, djones.lettnin@ufsc.br, Michael.Huebner@rub.de

Abstract—The current verification flow of complex systems uses different engines synergistically: virtual prototyping, formal verification, simulation, emulation, and FPGA prototyping. However, none of them is able to verify a complete architecture. On the other hand, hybrid approaches aiming at full verification use techniques that lower the overall complexity by increasing the abstraction level. To bridge this verification gap, we turn to the embedded software and the information it can bring to the verification environment. This work focuses on the semiformal verification of complex systems at the RT level to handle the hardware peculiarities. Our results show an improvement of four times in verification completeness of a complex hardware gateway compared to the commercial tool.

I. INTRODUCTION

Systems-on-chip (SoCs) are widespread nowadays, covering a wide spectrum of electronics, e.g. in cell phones, tablets, and cars. This variety of applications means that SoCs' complexity will keep increasing in the next generations [1]. The ITRS System Integration group predicts, for a single SoC architecture, an increase in application processors from 9 in 2017 to 36 in 2025 and in graphics processors from 19 in 2017 to 247 in 2025. However, the increase in complexity and functionality has a hidden cost: "The increasing number of heterogeneous components (RF, logic, memory, and MEMS) complicates the system design" [1]. This increase in complexity only widens the gap between design and verification [2].

To overcome this gap, both the industry and academia research new techniques and methodologies in many different directions. For example, there are techniques to either verify an architecture in a higher level of abstraction [3] or to divide it into smaller sub-blocks and verify them separately. Both have advantages and disadvantages. The first technique speeds up the verification process by increasing the abstraction, but it gives meaningful results only for the functionality of the hardware, not for its low level behavior, e.g. timing and parallelism. The second technique lowers the system complexity by verifying small portions instead of the whole system at once; however, it is unable to cover the interaction between the many sub-blocks that compose the system.

Different engines are used at each stage of development. Simulation, emulation, FPGA prototyping, and formal verification are currently the preferred engines for hardware verification in the industry [4]. To improve the verification results for complex architectures, a trend currently becoming more popular in the industry is the synergy between verification engines [5], where the verification team seeks to combine the advantages of each engine to the most applicable development

phase. One simple but powerful example is deep dynamic formal verification [6], where simulation directs the architecture to a specific state, and from there, formal tools try to verify a smaller set of states. This technique relies on the quality of the input vectors to drive the system to the desired deep space or corner case. However, this can be challenging and time-consuming to perform iteratively, mostly due to the need to simulate millions of cycles to reach the desired state.

Another trend aimed at closing this gap is software-driven verification, mainly with the Portable Stimulus Standard [7]. This methodology uses scenarios to capture the verification intent. For a scenario, a tool derives the test cases to drive the Design Under Verification (DUV). An important goal of this standard is that it targets multiple verification engines, which means that the same test cases are used for virtual prototyping, simulation, emulation, and FPGA prototyping. Due to system complexity, using software to restrict the state space helps to accelerate verification closure. Nevertheless, each test case describes a certain use case and exercises only some aspects of the DUV, and only the embedded software (ESW) usually targets the entire architecture. Therefore, it would be more profitable to use the ESW to guide the verification process, instead of fabricated pieces of software that verify only small portions of the architecture.

The aim of this work is to increase coverage using dynamic data to cover a greater set of states, without using deep states. As SoCs grow in complexity in every new generation, this growth highlights the need for new approaches to the verification problem. This work proposes a "build-and-prove" process tied to a static register assignment (SRA) heuristic to reduce the state space for formal verification.

The main contribution of this work is a scalable, hybrid, iterative and ESW-driven flow to improve the verification productivity. The build-and-prove process tries to verify subsystems that grow by one IP at each iteration until it reaches the complete architecture. Simulation runs help to reduce the state space for the formal verification process and to avoid the state space explosion. The simulation process uses the ESW to provide the dynamic data to constrain the DUV during the semiformal phases. The SRA process tries to improve the constraints in an iterative fashion.

The next sections of this paper are organized as follows: Section II describes the related work. Section III presents the developed work in detail. Section IV summarizes the results after applying this work to a test case. Finally, Section V concludes this paper.

978-1-5386-7100-9/18 $31.00 © 2018 IEEE

II. RELATED WORK

Mukherjee et al. [8] propose a flow to translate register transfer level (RTL) code to ANSI-C code and apply different formal techniques for software to it, e.g. bounded model checking, path-based symbolic simulation, and abstract interpretation. The idea is to increase the abstraction level and simplify the proofs in order to get results faster. However, due to hardware's nature, software models cannot accurately describe some of its peculiarities. An example is the generation of netlists from high-level models using technologies such as high-level synthesis (HLS). Since current HLS tools cannot capture specific hardware details from the software description, e.g. parallelism and pipelining, they do not implement the developer's intent correctly.

Herber [9] aims at hardware/software co-verification by partitioning SystemC models to achieve a scalable flow. In this work, different engines verify different aspects of an architecture. For instance, a satisfiability-modulo-theory (SMT) tool verifies synchronous components of the RTL. The proposed tool splits the verification task into hardware, software, and system level. However, as useful as this approach may be, it is only applicable to the initial stages of design since the granularity level is too coarse for deep verification.

These works show a gap in low-level verification since all approaches employ higher-level abstractions to try to improve verification results at the expense of fine-grained details.

Mukherjee et al. [10] propose a new approach for co-verification of hardware/software codesigns, where the Verilog RTL is translated to either a bit-level netlist or to a word-level netlist and the C source code is translated to static single assignment form. From these formats, it generates a single representation for the complete system and verifies it using either a SAT or an SMT solver. Although this work aims at the verification of SoCs, it does not verify the hardware architecture with the embedded software, but with the external software that does the communication with the SoC.

This work does verify the hardware architecture at the RT-level, however, since it does not take into account the embedded software on this architecture, it does not consider the influence the ESW has on the hardware architecture.

We close these gaps with this work by only focusing on the RT-level and using a "build-and-prove" system tied to a novel heuristic to avoid the state-space explosion as much as possible as well as using the ESW to guide the semiformal verification process.

III. SCALABLE SEMIFORMAL HARDWARE VERIFICATION METHODOLOGY

As previously mentioned, large systems cannot be completely verified using formal methods due to state space explosion. To overcome this problem, we developed an iterative build-and-prove system with the goal to improve the system coverage and push further the verification results. It starts the verification process proving a small subsystem and increasing it iteratively IP-by-IP. The verification begins with formal methods and, if they become insufficient, a proposed semiformal heuristic aids to overcome the system complexity.

One of the most important results a verification process gives is the coverage metric. It means how much of the architecture is covered by the verification method [11]. Formal verification, due to its complete nature, gives a different meaning to coverage than that of simulation. For formal methods, coverage is complete regarding the properties applied to the model checker. As for simulation, it cannot practically cover 100% of an architecture since it uses a limited number of input vectors to exercise the architecture. Our approach combines these different concepts into the same flow in order to improve the coverage effort and push it closer to 100%.

We propose the ARCHVerifyr verification approach in this work, which has five phases: (1) preprocessing of the input files, (2) formal and (3) semiformal verification of the IPs, (4) formal and (5) semiformal verification of subsystems using the build-and-prove process. Figure 1 presents a high-level description of the proposed flow.

ARCHVerifyr has five inputs: (1) the RTL architecture and (2) the ESW source files, which are the verification subjects, (3) the assertions, which the model checker uses during the verification processes, (4) the timeout condition, which limits the model checker to a time bound, and (5) the option to black box the IPs that fail the semiformal verification process during Phase 3. The main output from ARCHVerifyr is the arhcitecture's coverage information, which is calculated at the end of the verification process. The calculated coverage shows how much of the architecture ARCHVerifyr verified, so that the user can focus on the areas that remained uncovered. ARCHVerifyr can also output the counterexamples for the assertions that failed formal verification, which can help the user to undersand and debug the errors.

Phase 1 from ARCHVerifyr (Figure 1.a) prepares some of the data structures required later on. It begins by extracting the IPs that compose the architecture to create the input list for phases 2 and 3, which do the verification of each IP separately. Next, ARCHVerifyr ranks the instantiated IPs according to the number of interfaces each has. This results in a ranked list of IPs used by the iterative build-and-prove process in phases 4 and 5. The next step prepares the ESW source files by merging them into a single file and exposing the accessed memory and register addresses at the same time. We use the tool cilly, from the C Intermediate Language framework [12], to perform this action. The resulting file is then used to create a mapping between the registers accessed in the ESW and their implementation in the RTL. This mapping is the key point of the SRA heuristic (Section III-A) developed in this work. The final step is to group the user-provided formal properties by the IP(s) they cover. The model checker uses these groups in all following phases.

In Phase 2 (Figure 1.b), ARCHVerifyr tries to verify all listed IPs one-by-one, calling the model checker with the corresponding group of formal properties and the time limit provided as input, to find any errors before the subsystems phase begins. If the model checker is able to verify all IPs in the list, ARCHVerifyr moves to Phase 4 to begin the build-and-prove process. On the other hand, if there are any IPs that the model checker is not able to verify inside the time limit, ARCHVerifyr adds them to a list and moves to Phase 3, which does the semiformal verification process for these IPs.

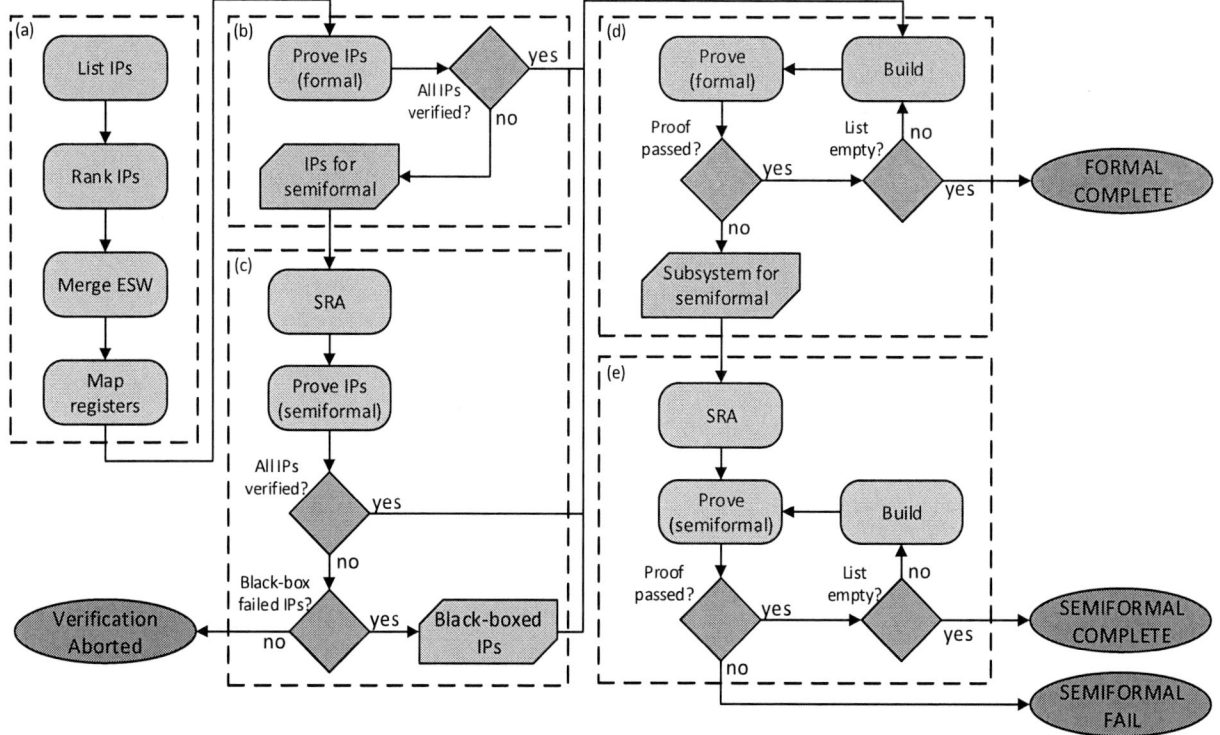

Fig. 1. ARCHVerifyr's high-level flow.

Phase 3 (Figure 1.c) begins by running the SRA heuristic to generate the list of ranked registers for each IP marked for semiformal verification and initializes the set of registers for the semiformal verification process. This ranked list contains the elements that will provide the necessary information to constrain the DUV in a controlled way. Next, ARCHVerifyr sets up the points-of-interest (PoIs) for the simulation, which are the locations in the ESW that access any of the mapped registers and serve as the triggers to call the semiformal verification process. Following this, the simulation starts.

During the simulation run, whenever the simulator executes an instruction that involves a PoI, the semiformal verification process begins. The preparation for the next call to the model checker begins with the creation of stop-ats[1] associated with the registers and communicating with the simulation engine to collect the dynamic data for the current set of registers. This data is then used to generate assumptions for the registers' outputs. The model checker tries to prove the properties associated with the current IP using the generated stop-ats and assumptions and the provided time limit. If the model checker cannot complete a proof and there are no more registers left, ARCHVerifyr either black boxes[2] the IP, if the user chose to black box failing IPs, or aborts the process and outputs the

results to the user. If the list of ranked registers is not yet empty, ARCHVerifyr adds the next in the sequence to update the set of registers and restarts the semiformal verification process. If the model checker verifies the IP successfully, this IP is removed from the list, and resumes the simulation. This loop executes either until the list of IPs separated for semiformal verification is empty or until the process is aborted.

If Phase 3 is needed, the generated stop-ats and assumptions are carried on to the next phases, so that the effort does not need to be repeated.

After all the IPs are verified, ARCHVerifier can start the build-and-prove process, which uses the list of ranked IPs generated in Phase 1. This iterative 2-step process first builds a new subsystem and then calls the model checker to prove it. The idea behind this process is to start small to avoid the state space explosion at the beginning of the verification process and iteratively grow the subsystem IP-by-IP until the complete architecture is verified.

Phase 4 (Figure 1.d) begins by building the first subsystem using the two highest ranked IPs that have an interface connection between them and calling the model checker to verify this first iteration. The next step depends on the model checker's result. If the verification succeeds, the build phase adds the next highest ranked IP to the subsystem and calls the model checker again. If the verification fails, ARCHVerifyr moves to Phase 5 so that the semiformal process can help in the verification effort. The build-and-prove process loops over the list of ranked IPs until either it is empty, meaning that

---

[1]A stop-at is an abstraction used to "cut" the driving logic beyond a chosen point. This enables the model checker to choose a value for a proof. Furthermore, assumptions can tell the model checker which value it must use from that point on.

[2]Black boxing instructs the formal tool to ignore the internal architecture for some block and unconstrain all its output signals.

the complete architecture was verified, or until the state space becomes too large to be managed by the model checker alone, when ARCHVerifyr changes to Phase 5.

Phase 5 (Figure 1.e) continues the iterative build-and-prove process with the help of the SRA heuristic. The starting point is the subsystem that could not be verified in Phase 4.

As in Phase 3, Phase 5 uses SRA to rank the subsystem's registers and initialize the set of registers for the semiformal verification process. Next, ARCHVerifyr sets the PoIs for the simulation and triggers the simulation. When the simulator executes an instruction with a PoI, the semiformal verification starts with the set of registers.

The preparation for the semiformal process generates the stop-ats for the current registers, collects of the values from the simulation and generates the necessary assumptions using the collected data. Next, ARCHVerifyr calls the model checker with the generated structures and the provided time limit. If the model checker is not able to verify the subsystem and the list of ranked registers is empty, the process finishes unsuccessfully; otherwise, ARCHVerifyr adds the next ranked register to the set of registers and restarts the semiformal verification process. On the other hand, if the model checker finishes the proof and the list of IPs is empty, ARCHVerifyr finishes the process and returns success; otherwise, ARCHVerifyr reiterates the build-and-prove process by picking the next IP in the ranked list, updating the subsystem, and calling the model checker again.

Figure 2 exemplifies the progression made by ARCHVerifyr through each phase. After processing the inputs, ARCHVerifyr verifies the IPs in the architecture with the model checker in Phase 2 (IPs 1 to 5). Phase 3 applies the semiformal process using a simulation run to generate values to constrain the unverified IPs (IPs 2 and 5). This single simulation run triggers the semiformal process for each IP in the list, however, tackling a single IP at each time does not cover the entire state space. Nevertheless, this is the aim at Phase 4, where the build-and-prove system begins. The subsystem is iteratively increased (Subsystems 1, 2, and 3) until the complete architecture is verified or the state space explosion happens (Subsystem 3). In the later case, ARCHVerifyr switches to Phase 5, to apply the semiformal process to the subsystem until the complete architecture is verified (Subsystems 3 and 4).

The next section describes the SRA heuristic, which is used to improve the verification process.

A. Static Register Assignment Heuristic

The role of the Static Register Assignment (SRA) heuristic is to reduce the state space using dynamic information, which is extracted from the current simulation run. To achieve the best results, it is important to scale down the state space without overconstraining it; otherwise, error states become unreachable. SRA addresses this point using a register mapping between RTL and embedded software, which are elements reachable from the "user" side. This avoids using elements that the user has no control over, e.g. I/O interfaces.

SRA begins with the netlist for the current design under verification and calculates the impact that each of the mapped registers has on the architecture and ranks them from highest

TABLE I.   NUMBER OF X-PROPAGATION PROPERTIES GENERATED FOR EACH ARCHITECTURE

IPs		Subsystems	
Architecture	# props	Architecture	# props
CAN	1274	Subsystem 1	2517
ETHMAC	3207	Subsystem 2	3801
MOR1KX	1739	Subsystem 3	7008
WB_RAM	98		

to lowest. The output is a list of ranked registers for the semiformal verification phases.

The SRA is a measure for the influence of a register based on its breadth and depth. Equation 1 defines SRA formally.

$$SRA(r_i) = 100 * |\mathcal{P}_{ri}| + |\mathcal{S}_{ri}|, 0 \leq i \leq |\mathcal{R}| - 1 \quad (1)$$

where $\mathcal{R}$ is the set of mapped registers, r_i is a mapped register in $\mathcal{R}$, $\mathcal{P}_{ri}$ is the set of paths connected to the register and $\mathcal{S}_{ri}$ is the set of state elements connected to the register through the paths in $\mathcal{P}_{ri}$, and $|X|$ is the size of set X.

The breadth indicates how many paths start at the register. The depth indicates how many state elements depends on the register, either directly or indirectly. Breadth receives a greater weight due to the register's influence on multiple paths. The measure covers all the logic from each mapped register to the outputs. For each path connected to a register, the counter for that register increases by 100 points and for each element connected along each path it increases by 1 point. Figure 3 presents a graphical illustration of this concept.

In Figure 3, register r_1 connects to one path and nine logic elements and register r_2 connects to three paths and thirteen logic elements. The resulting list has r_2 in first place and r_1 in second, since r_1 has a deeper but narrower relevance and r_2 has a shorter but broader relevance.

IV. RESULTS AND DISCUSSION

A. Verification Environment

We used a 24 core Intel® Xeon® CPU E5–2630 @ 2.3GHz with 96GB of RAM memory running CentOS to perform the experiments. We used the tool JasperGold to obtain the baseline results, which we compared to what we obtained from our scalable hybrid approach described in Section III. We also used JasperGold as the model checker for our experiments.

The validation test case for ARCHVerifyr used the X-Propagation app from JasperGold to extract the properties for each architecture, either IP or subsystem, and verify them. Table I shows the number of properties extracted for each IP and for each subsystem.

B. Automotive Gateway Prototype

We used a prototype for an automotive gateway developed in-house as our case study. We developed this prototype using the Fusesoc platform. It has an OpenRISC processor ("MOR1KX"), a CAN IP ("CAN"), an Ethernet IP ("ETH-MAC"), and a RAM memory IP ("WB_RAM"). All elements are from the OpenCores repository. Figure 4 presents its architectural block-level diagram and Table II summarizes the

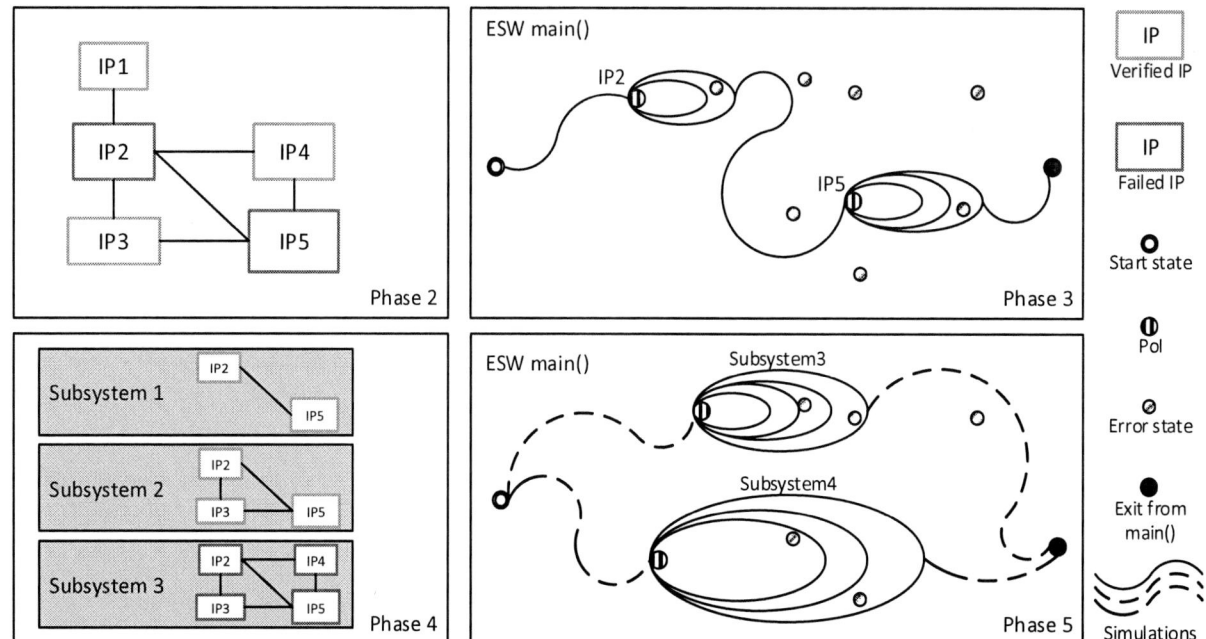

Fig. 2. Illustration of ARCHVerifyr's progression through the phases.

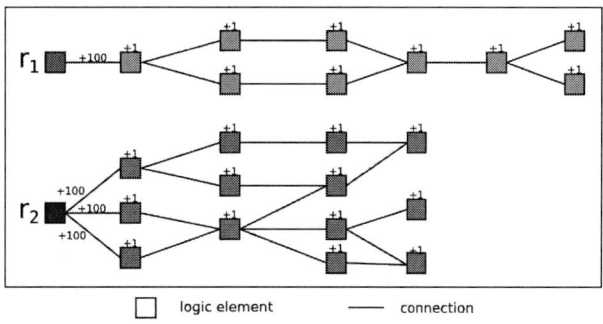

Fig. 3. Illustration of the SRA metric.

TABLE II. VERIFICATION RESULTS FOR THE GATEWAY PROTOTYPE

Architectures	JasperGold		ARCHVerifyr	
	Result	Outcome	Result	Outcome
WB_RAM	Verified	0.20s	Verified (Phase 2)	0.20s
CAN	Timeout	7[1]	Verified (Phase 3)	916s (1 iter)
ETHMAC	Timeout	2[1]	Verified (Phase 3)	30s (1 iter)
MOR1KX	Timeout	2[1]	Timeout (Phase 3)	Black-boxed
Subsystem 1	Verified	582s	Verified (Phase 4)	582s (1 iter)
Subsystem 2	Timeout	225[1]	Verified (Phase 5)	361s (1 iter)
Subsystem 3	Timeout	622[1]	Timeout (Phase 5)	1[1] (5 iter)

[1]Number of undetermined properties

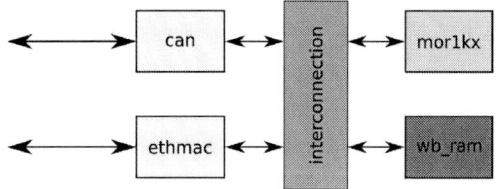

Fig. 4. Prototype's block diagram as generated by the Fusesoc platform.

results for the validation of ARCHVerifyr. The chosen time limit to verify each IP and each subsystem was 7200 seconds.

We followed the approach presented in Section III and started the validation process verifying the IPs separately. In Phase 2, ARCHVerifyr verified only the RAM memory IP and the other IPs were added to the output list. In Phase 3, it verified the CAN IP and the Ethernet IP in one iteration, and the processor IP was black boxed after five iterations. The first

part of Table II presents the results for Phases 2 and 3.

In Phase 4, the build-and-prove process started with the processor IP and the RAM memory IP. As mentioned above, it was not possible to verify the processor IP and, therefore, it was black boxed in Phases 4 and 5. ARCHVerifyr verified that subsystem in one iteration. Next, it added the CAN IP to the subsystem; however, it was not possible to verify it with the model checker. Therefore, it was necessary to switch to Phase 5, in which ARCHVerifyr verified it in one iteration. Finally, ARCHVerifyr added the ETHMAC IP to the subsystem, but was unable to complete the verification process. The second part of Table II presents the results for Phases 4 and 5.

Our results show an improvement over JasperGold alone. We had an improvement of two times with Phase 3, as ARCHVerifyr verified two IPs more than JasperGold. The SRA heuristic guaranteed that these architectures were not overconstrained. The results for the CAN and the Ethernet IPs are better explained in the next section. We also had an improvement with Phase 5, where ARCHVerifyr verified Subsystem 2 in one iteration of the "build-and-prove" system

TABLE III. COMPARISON OF DIFFERENT ETHMAC REGISTERS FOR SRA VALIDATION

Register	SRA	Stop-ats	Time	Result
MODER	2880	3	30s	Verified
MIICOMMAND	381	3	Timeout	2^1
CTRLMODER	332	1	Timeout	3^1
MIIMODER	236	2	Timeout	3^1
PACKETLEN	214	4	Timeout	2^1

[1]Number of undetermined properties

TABLE IV. COMPARISON OF DIFFERENT CAN REGISTERS FOR SRA VALIDATION

Register	SRA	Stop-ats	Time	Result
MODE	1177	3	916s	Verified
COMMAND	750	4	Timeout	7^1
CLOCK_DIVIDER	705	3	Timeout	7^1
BUS_TIMING1	312	1	Timeout	16^1
BUS_TIMING0	208	1	Timeout	7^1

[1]Number of undetermined properties

and also reduced by two orders of magnitude the left over properties for Subsystem 3.

C. SRA Validation

We used the Ethernet and the CAN IPs to validate the SRA heuristic. As described in Section III-A, SRA works with the registers the user has control over, i.e. configuration and control registers.

SRA ranked the configuration registers for the Ethernet IP and the CAN IP according to their relevance and Tables III and IV present the results. The chosen time limit for all runs was 7200 seconds and we ran the process with the top five registers for comparison purposes.

Table III shows that SRA needed just one iteration to find the minimum set of registers for the Ethernet IP. Furthermore, the low number of stop-ats for this set is a good indication that the system will not be overconstrained during the semiformal verification process.

Table IV shows that SRA needed again just one iteration to find the minimum set of registers for the CAN IP. Again, the low number of stop-ats help to reduce the state space without overconstraining it.

The results SRA gave for both the CAN IP and the Ethernet IP give us confidence that it is possible to improve the verification results without overconstraining the architecture, since it was necessary only three stop-ats and three assumptions for both IPs to make the model checker finish the verification process. However, this type of result is not easy to predict, since we ran the same process with the processor IP and it was not possible to find a suitable set of registers to finish the verification process.

V. CONCLUSION

We have presented our scalable hybrid verification approach for complex hardware systems. We described the advantages of the proposed methodology, which spans several steps in the hardware verification flow. The process begins with the formal verification of each IP and ends with the build-and-prove system that verifies incrementally bigger subsystems up to the complete architecture. The semiformal phases of the proposed methodology use the SRA heuristic to reduce the state space without overconstraining the architectures. Our results show that this methodology greatly benefits the verification flow of complex SoCs.

As future work, it should be possible to execute Phase 4 of the flow with different starting subsystems to create "verified islands" in the architecture when complete verification is not possible. It is also our goal to reduce the number of stop-ats to avoid overconstraining the architecture. Finally, we want to add a smart time limit for the model checker since complex systems need more time to complete the verification task.

ACKNOWLEDGMENT

The authors would like to thank the National Council of Scientific and Technological Development of Brazil (CNPq), processes 290009/2014-6 and 445985/2014-3, for the financial support and the Cadence Academic Network for the technical support.

REFERENCES

[1] ITRS, "International Technology Roadmap for Semiconductors 2.0," Tech. Rep., 2015.

[2] W. Chen, S. Ray, J. Bhadra, M. Abadir, and L.-C. Wang, "Challenges and Trends in Modern SoC Design Verification," *IEEE Design & Test*, vol. 34, no. 5, pp. 7–22, oct 2017.

[3] H. M. Le, V. Herdt, D. Große, and R. Drechsler, "Towards Formal Verification of Real-World SystemC TLM Peripheral Models A Case Study," in *Proceedings of the 2016 Design, Automation & Test in Europe Conference & Exhibition (DATE)*. Singapore: Research Publishing Services, 2016, pp. 1160–1163.

[4] H. Foster, "Trends in functional verification," in *Proceedings of the 52nd Annual Design Automation Conference on - DAC '15*. New York, New York, USA: ACM Press, 2015, pp. 1–6.

[5] U. Simm, S. Rosenberg, E. de Kock, and P. A. Hartmann, "Accellera Standards Technical Update," in *2015 Design and Verification Conference and Exhibition*, 2015.

[6] H. Foster, "Applied Assertion-Based Verification: An Industry Perspective," *Foundations and Trends® in Electronic Design Automation*, vol. 3, no. 1, pp. 1–95, 2009.

[7] Accellera Systems Initiative, "Portable Stimulus Standard Early Adopter Release," Tech. Rep., 2017.

[8] R. Mukherjee, D. Kroening, and T. Melham, "Hardware Verification Using Software Analyzers," in *2015 IEEE Computer Society Annual Symposium on VLSI*. IEEE, jul 2015, pp. 7–12.

[9] P. Herber, "The RESCUE Approach - Towards Compositional Hardware/Software Co-verification," in *2014 IEEE Intl Conf on High Performance Computing and Communications, 2014 IEEE 6th Intl Symp on Cyberspace Safety and Security, 2014 IEEE 11th Intl Conf on Embedded Software and Syst (HPCC,CSS,ICESS)*. IEEE, aug 2014, pp. 721–724.

[10] R. Mukherjee, M. Purandare, R. Polig, and D. Kroening, "Formal Techniques for Effective Co-verification of Hardware/Software Co-designs," in *Proceedings of the 54th Annual Design Automation Conference 2017 on - DAC '17*. New York, New York, USA: ACM Press, 2017, pp. 1–6.

[11] E. Seligman, T. Schubert, and M. V. A. K. Kumar, *Formal verification: an essential toolkit for modern VLSI design*. Morgan Kaufmann, 2015.

[12] G. C. Necula, S. McPeak, S. P. Rahul, and W. Weimer, "CIL: Intermediate Language and Tools for Analysis and Transformation of C Programs," in *CC '02 Proceedings of the 11th International Conference on Compiler Construction*, 2002, pp. 213–228.

2018 IEEE Computer Society Annual Symposium on VLSI

High-average and Guaranteed Performance for Wireless Networks-on-Chip Architectures

Mohammad Baharloo*[†], Ahmad Khonsari*[†], pouya Shiri*, Iman Namdari* and Dara Rahmati[†]

*ECE Department, The University of Tehran, Iran

[†]School of Computer Science, Institute for Research in Fundamental Sciences (IPM), Tehran, Iran

Abstract—Network on Chip (NoC) is the underlying communication platform for multi-core embedded systems. Wireless NoCs (WNoC) employ wired and wireless structures simultaneously to facilitate communication scenarios. In this paper, we propose an arbitration mechanism to guarantee the performance parameters for real-time traffic flows transferred in the wireless plane, while preserving high average performance for all the traffic flows in the wired or wireless sections of a wireless NoC. Different scenarios have been carefully selected and clear suggestions are provided using analytic performance models to effectively use a wireless NoC for real time application category. We have provided different examples to illustrate the effectiveness of the proposal.

I. INTRODUCTION

The wireless network on chip (WNoC) paradigm has been proposed to facilitate communication among the large number of cores on an embedded system, which are connected traditionally through wired network known as NoC. Although the implementation of WNoC is completely viable with nowadays technologies and is considered as an effective and interesting selection, it has not been adequately employed to address the ever increasing challenges in design of embedded systems. At the other hand, many application domains are integrated to work simultaneously on an embedded system like a personal device including OS related tasks, real time media processing, data-base management, file handling, memory management etc. as an example which can be embedded on a single chip. The types of communication in such system may be categorized as a combination of real-time (RT) and non real-time (NRT) data-flows. Effective performance parameters for correct operation of such systems include both the average and the worst-case metrics. While worst-case delay is important for time critical data-flows, worst-case guaranteed transmission bandwidth is essential for other types of flows. We refer to these two types of flows as RT flows as our definition in this paper. Other types of flows usually do not have an RT constraint on the communication, but to provide a proper overall service level, they are needed to be able to communicate with an effective average delay and average bandwidth. Most of the proposed designs in the domain of WNoCs [1] have focused on average performance metrics which in our opinion do not cover the challenging aspects of WNoC design. At the other hand, the few works that deal with QoS in WNoC [2]–[5] do not consider average parameters simultaneously with their worst-case counterparts. In this paper, we propose a mechanism for data transmission in 2-dimential (4×4 to 8×8) mesh based WNoCs in

which the RT data-flows are transmitted through the wireless plane and NRT data-flows on wired plane using XY routing. We provide an arbitration mechanism for the wireless section and examine different multi-media and synthetic applications on such scenarios to calculate the average and worst-case performance metrics at the same time for this network. We also provide an analytical model to calculate the worst-case performance in wireless plane and use an existing worst-case performance model for wired section. Extensive simulation is provided to calculate the average case in both the wired and wireless sections. To the best of our knowledge this is the first work considering different domains of performance evaluation for WNoC, which has discussed the pros and cons of such networks. Our evaluation shows the effectiveness of the proposal compared to a wired mesh for both average and worst-cases with minimum overheads. Section 2 describes the related works. In Section 3 a generalized network model is provided. Analytical performance model is provided in Section 4. Section 5 describes our evaluations and finally Section 6 concludes the paper.

II. RELATED WORKS

The ever-increasing integration levels lead to the integration of a high number of cores on a chip. Consequently, to benefit from this growth, it is crucial to architect a scalable and high-performance on-chip communication infrastructure. Many efforts such as optical interconnects, three-dimensional integrated circuits (3D ICs) and wireless interconnects have been devoted for providing alternatives to traditional planar metal interconnects which cause multi-hop communications between distant blocks on a chip [6], [7]. In an optical architecture, the electrical signal is converted to the light and vice versa, which leads to high design complexity as well as high area and latency overheads. On the other hand, apart from the benefits of 3D structures like CMOS compatibility, enhanced scalability and shorter vertical interconnects, they suffer low yield, high temperature, and alignment issues [8], [9]. Along these structures, wireless interconnects have been proposed as a viable solution to compensate the latency and power inefficiency of multi-hop communications through conventional wireline based NoCs [10]–[13]. Among the various techniques which have been proposed to implement wireless interconnects, wireless RF paradigm is a scalable, simple, and flexible option with broadcast capability [7], [14]. This capability is a key factor for applications like

978-1-5386-7100-9/18 $31.00 © 2018 IEEE

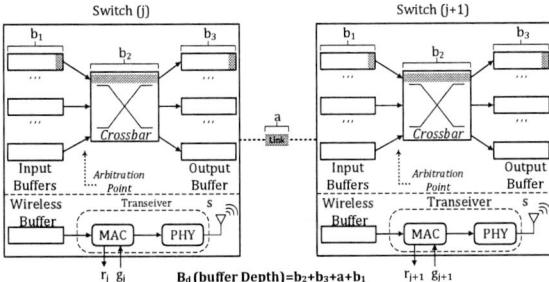

Fig. 1. Switch model and parameters [17]

TABLE I
NETWORK MODEL PARAMETERS AND SYMBOLS [17]

Parameter	Description
$Freq$	Clock frequency of the system
a	Number of pipeline stages (registers) used to segment NoC links
b_1	Depth of switch input buffers
b_2	Number of pipeline stages of the switch crossbar (0 if combinational)
b_3	Depth of switch output buffers (0 if none)
b	$:= b_1 + b_2 + b_3$
B_d	Buffer depth parameter, $:= a + b1 + b2 + b3 = a + b$
ts_1	Latency overhead for injecting a packet into the network
ts_2	Latency overhead for ejecting a packet at the destination
$FlitWidth$	Width of NoC links, in bytes
SW_j	j-th switch of the network

cache coherency protocols, which can greatly benefit from multicast or broadcast communications [1]. On-chip wireless communication can be realized at the core level through cost-effective RF paradigm since non-intrusive solutions can be devised and no transmission lines are required which reduce the chip floorplanning costs. On the other hand, quality of service (QoS) is a crucial issue for many application domains wherein traffic flows with real-time requirements exist. In [15] authors try to realize guaranteed services through best effort services in NoCs. The architecture proposed in [16] exploit TDMA technique in packet switched networks to provide real-time guaranteed services. In [17] an analytical method for calculating worst-case traffic delay in best effort NoCs without special hardware provisioning is presented. Many other works have been proposed to provide real-time guaranteed services [1], [18]–[20]. However, none of them calculate the worst-case delay and bandwidth in hybrid wire/wireless NoCs. Recent Works have demonstrated that RF circuits operating at 100 GHz and above are realizable [21]. As well, they confirm that constructing small-footprint antennas and other components that operate at high frequencies are achievable [22], [23]. Media Access Control (MAC) Protocol is one of the main components of RF-based on-chip networks that control the access of wireless nodes to the wireless channel. Prior works demonstrated that the use of conventional MAC protocols such as FDMA, CDMA, and token passing are inefficient in terms of channel utilization, area, and energy dissipation. In [24] the combination of TDMA and FDMA in the expense of multiple wireless transceivers in each node is reported. A distributed priority multicast MAC protocol is reported in [24], [25]. In this scheme as the number of wireless interconnects increases, the channel access time is increased, so it is not suitable for a large NoC with a large number of nodes equipped with a wireless antenna. The next section describes the network model of this work.

III. THE NETWORK MODEL

In this section, we elaborate our router model which is the key determinant component to characterize network behavior. We employ the generic architecture shown in Fig.1 [17] that is integrated with the wireless transceiver to provide wireless communication for each core. For generality, the router with optional input and output buffering scheme is considered where some virtual channels are associated with each link to multiplex the available bandwidth of the physical link. As in most of on-chip applications packet dropping is not tolerable, we exploit the flow control mechanism that applies back-pressure signaling.

While wireless communication is accomplished through the on-chip antenna, it should be coordinated to provide fair and reliable access for all the cores. As Fig.2 shows, in this work a centralized controller performs the media access control function in a contention-free manner. Once a core is able to send packets through the wireless network, the MAC module should send a request signal to the *centralized media access controller (C-MAC)* and wait for a grant signal through the wired control plane. The wired control plane is constructed through routing a pair of wires from each router to C-MAC module; one for request signal and the other for the grant. The feasibility of routing such metallic wires to drive signals across the chip within a single cycle is comprehensively explained in [26]. TABLEI shows the network parameters that are required to describe the network model. In this paper, for simplicity, we define the parameter *Freq* and *FlitWidth* as the operating frequency of all the cores and the width of all on-chip data links, respectively.

The buffer depth (B_d) parameter, as shown in Fig.1, represents the aggregate number of buffers and registers, between the arbitration points of switch j and switch $j+1$. For simplicity, throughout the paper, we assume the intermediate buffers and registers between the arbitration points of two adjacent switches to be lumped, so we refer to the output and input buffers of two adjacent switches equivalently. The parameters b_1 and b_3 define the input and output buffer depth of the switch, respectively. Note that the register depth of the input buffer is assumed to be at least one. The number of pipeline stages (registers) in the crossbar is denoted by b_2 (if pipelined). For generality, we assume that data links between two adjacent switches can be pipelined, so we define parameter

978-1-5386-7100-9/18 $31.00 © 2018 IEEE

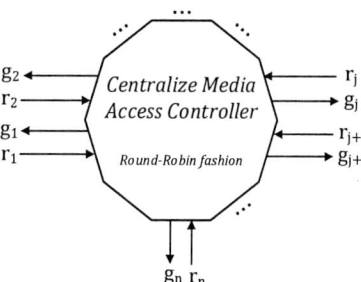

Fig. 2. Centralized Media Access Controller (C-MAC) based on Round-Robin Arbitration for wireless communications

a as the number of registers along each on-chip data link. By this consideration, the propagation delay of wires can be compensated to boost the operating frequency of NoC.

It is important to note that packets will always experience latencies which are presented by a and b_2, while they only experience latencies relevant to b_1 and b_3 when buffers are filled up, i.e. in the case of congestion. In the absence of congestion, input and output buffers can be traversed in a single cycle instead. It is also important to note that we index the switches in the path of a flow by $j = 1...m$, while for source conflict modeling (i.e. sending more than one flow from the source) we use $j = 0$ which represents a virtual switch inside the source node. To model the latency overhead for injecting a packet at source and ejecting a packet at destination we use parameters ts_1 and ts_2 respectively. In order to use finite parameters, we assume that the receiving nodes are able to accept incoming data at any required rate.

IV. DELAY MODEL

In order to characterize the performance of wireless communications in our NoC architecture, we define some parameters which are listed in TABLEII. t_r is the latency overhead for propagating the *request* signal from a node to C-MAC module, t_g is the latency overhead for propagating *grant* signal from C-MAC module to applicant node, w_g is the elapsed time between arrival instant of the request signal and preparation instant of the grant signal by C-MAC, and t_p is the time to transmit a packet through wireless network.

We define the parameter ART which represents the arbiter response time that calculates the time interval between the instant of sending a request to the C-MAC and the instant of returning the grant signal. For calculating ART, we calculate the worst-case waiting time for a grant at C-MAC which can be achieved through Eq. 1.

$$w_g = (n-1)t_g + (n-1)t_p \tag{1}$$

Based on Eq. 1, the value of ART is achieved through Eq. 2,

$$ART = t_r + w_g + t_g \tag{2}$$

TABLE II
WIRELESS NETWORK MODEL PARAMETERS AND SYMBOLS

Parameter	Description
wUB_i	Upper bound delay for sending a packet P_i of F_i through the wireless network
ART	Worst-case latency for acquiring a grant from C-MAC
t_r	Latency overhead for request signal propagation
w_g	Worst-case waiting time for grant at C-MAC
t_g	Latency overhead for grant signal propagation
t_p	Latency overhead for transmitting a packet through wireless network
n	Number of real-time(RT) flows

The worst-case time needed to acquire the grant signal is the summation of three factors; The time to send a grant by the C-MAC to all applicant nodes which have sent their requests beforehand, the time needed for sending their packets, i.e., w_g and one request propagation latency and one grant propagation latency for the current node. According to Eq. 2, the calculation of wUB_i can be done through Eq. 3.

$$wUB_i = ts_1 + ART + t_p + ts_2 \tag{3}$$

The upper bound delay for sending a packet through the wireless network consists of the time for injecting the packet to the wireless buffer at the source (ts_1), the Worst-case latency for acquiring a grant from C-MAC (ART), the latency overhead for transmitting the packet through wireless network (t_p), and the time needed to eject the received packet from the wireless buffer at the destination (ts_2).

In order to calculate the worst case performance of the wired plane please refer to [17].

V. RESULTS AND ANALYSIS

In this section, we evaluate the average and also worst-case performance for proposed architecture. For this purpose, we exploit two different scenarios as follows:

- In scenario number 1 (SC1), we utilize a simple wired mesh NoC as our baseline architecture. So, all the traffic including RT and NRT are routed through the wired plane.
- Scenario number 2 (SC2) is a hybrid wired/wireless NoC wherein each switch is equipped with an antenna as described in section III. In this scenario the RT flows are routed through wireless network while NRT flows through wired network.

A. AVERAGE PERFORMANCE EVALUATION

We use the simulation results obtained from Booksim2.0 simulator to evaluate the average performance metrics in the proposed architecture. In order to show the scalability of our approach, different mesh sizes are considered ranging from 4×4 to 8×8. In all cases, we exploit full mesh traffic distribution, where the traffic injection time at each node obeys a Poisson distribution with intensity λ and the destinations of these packets are considered to be distributed uniformly

978-1-5386-7100-9/18 $31.00 © 2018 IEEE

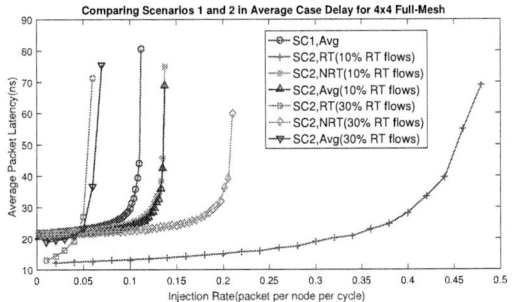

Fig. 3. Average-case analysis for 4×4 mesh

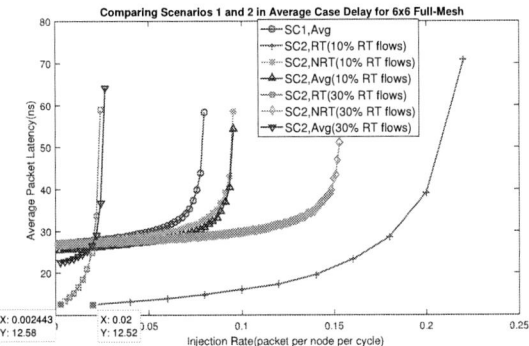

Fig. 4. Average-case analysis for 6×6 mesh

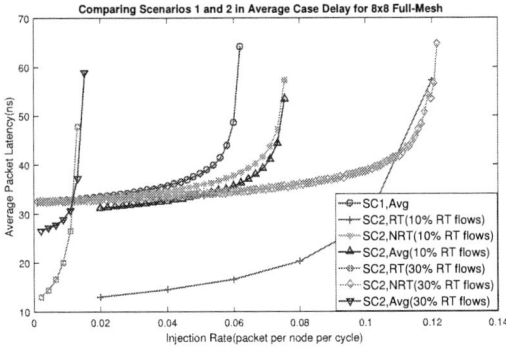

Fig. 5. Average-case analysis for 8×8 mesh

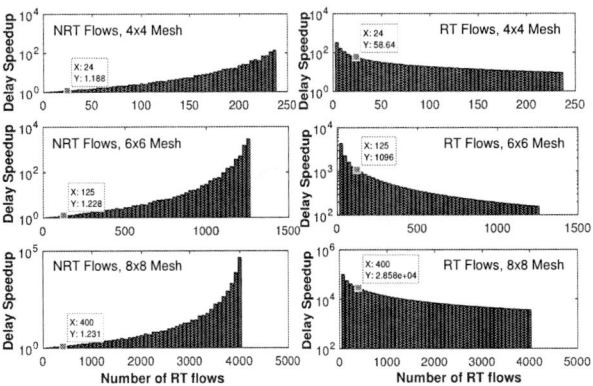

Fig. 6. Delay speedup in worst-case analysis for RT and NRT flows

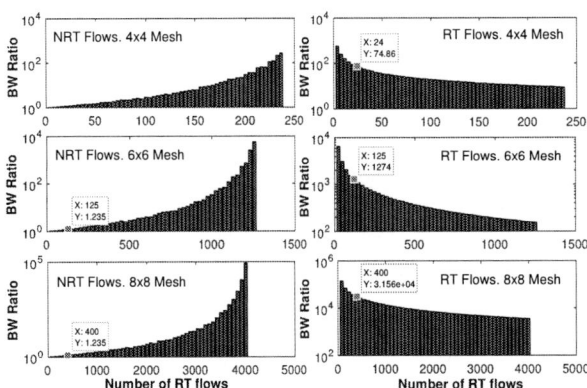

Fig. 7. Bandwidth ratio in worst-case analysis for RT and NRT flows

among other nodes. For the baseline network, employed in SC1, which all the transactions occur in the wired plane, we use the standard implementation of a best effort NoC in Booksim simulator. For the case of SC2, employing wired and wireless sections, the simulation is performed by integrating the C-MAC inside the Booksim simulator. The implementation related to C-MAC in this simulator calculates the average packet latency for RT packets sent through wireless plane and generates a trace file for Booksim simulator to do the same for packets transmitted through the wired network.

As depicted in Fig.3, the baseline wired network (SC1)

is compared to two different examples implemented on the wired/wireless network (SC2) in which 10% and 30% of the flows are considered RT and so are transmitted through the wireless plane. The saturation point of the wireless plane outperforms the wired plane for the case of SC2,RT10% and also SC1, while this is not true for the case of SC2,RT30%. We have extracted experimentally for the 4×4 network the turning point in which the RT flows will exhibit better average performance than SC1 is the case in which 18% of the flows are RT. In other words, the hybrid wireless/wired network will not worsen the average delay of the RT flows in SC2 compared to SC1, if at most 18% of the (RT) flows use the wireless plane for our test setup. It is obvious in any case the average delay for NRT flows exhibits better results in SC2 compared to SC1. We have examined this situation for 6×6 and also 8×8 networks and have observed similar behavior (Fig.4 and Fig.5). The turning point for these networks are 17% and 15% respectively. The main reason behind this evaluation is that the proposed architecture tries to provide better worst-case performance with minimum cost or area overhead for RT flows (as seen in the following subsection), while the average-case is not deteriorated. Also, these figures demonstrate that

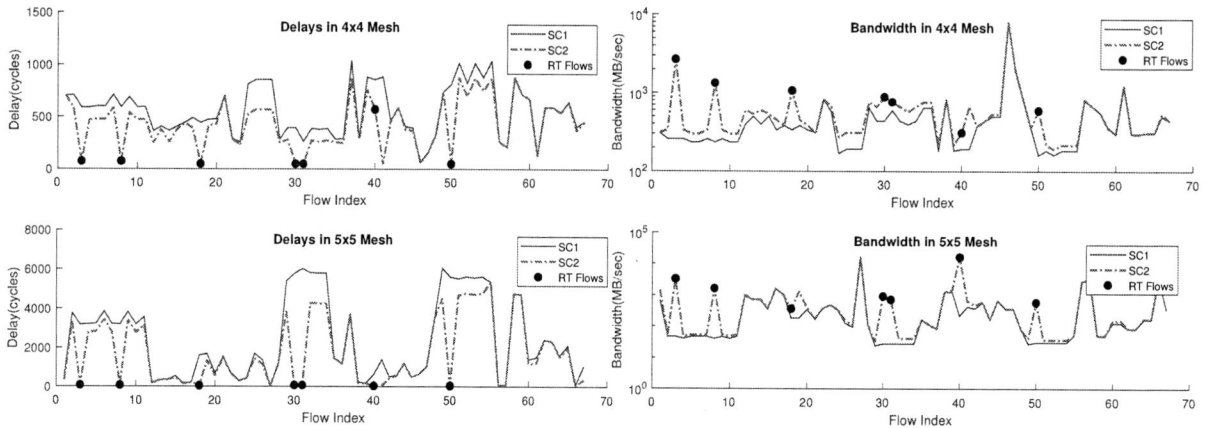

Fig. 8. Worst-case analysis of D26-Media application applied on different mesh sizes

the zero-load latency for RT packets in SC2 outperforms SC1 due to the fewer hop counts in wireless plane compared to the wired counterpart.

B. WORST-CASE PERFORMANCE EVALUATION

Fig.6 and Fig.7 illustrate the performance speedup in terms of worst-case latency and guaranteed bandwidth for SC2 compared to SC1 for different variants of full mesh networks (with 240, 1260 and 4032 flows). Fig.6 shows the delay speedup, in which the horizontal axes show the number of RT flows (among all). We have considered all the combinations in which a specific number of traffic flows are RT and evaluated the average latency speedup. As seen for the case of 4×4 network, in case 10% (24 out of 240) of the flows are RT which is common in real-time applications [17], the delay is decreased by a factor of 58x for RT flows, which is a reasonable value. In this case, the NRT flows exhibit a worst-case latency decrease by a factor of 3x, although not an important metric for them as is for its RT counterpart. These numbers for 6×6 and 8×8 networks are 111x and 2860x. It should be noted although the last dense scenarios may never happen in real world, but we show them for evaluation purposes and confirm the scalability of the proposed approach. Fig.7 shows the guaranteed bandwidth improvement, in which for the 10% case the average bandwidth is increased by a factor of 8.6x,

128.5x and 3157x for 4×4, 6×6 and 8×8 counterparts. We have also applied the proposed idea to a real world multimedia application named D26-Media from [17] with 25 Ip-cores and 67 traffic flows in which 7 of them are RT (specified as filled circles in Fig.8). The architectures are selected from both a 4×4 and a 5×5 hybrid mesh network. The horizontal axes show the index of traffic flows. The RT flows' average of worst-case latency and bandwidth improve by 80.22% and 300.39% for implementing D26-Media application on a 4×4 mesh respectively. The results for implementing on a 5×5 mesh are 97.98% and 3666.25%. In addition the improvement of these parameters for the NRT flows are 12.08% and 19.96% on the 4×4 mesh and 10.32% and 27.68% on the 5×5 mesh networks.

C. AREA OVERHEAD

TABLEIII shows the area overhead of the proposed C-MAC arbiter compared to the area of hybrid router (HB) in SC2 for different network sizes. Furthermore, to show the negligible overhead of the C-MAC module, we report the C-MAC area per router for different mesh sizes. As it is shown in the table this value is approximately $30 \mu m^2$ which is around 0.007% of the HB area. These results extracted using our VHDL implementation of the router, applied to 45nm VLSI Technology and Synopsys Prime-Power synthesis tool.

VI. CONCLUSIONS

This paper proposed the idea of a hybrid wireless/wired router mapped on variable-sized mesh NoCs. It also proposed the structure of an arbitration unit for the wireless section, in which the worst-case performance metrics are improved significantly compared to a baseline best effort wired NoC. The analytic model to calculate the worst-case latency and bandwidth for the wireless section of the network provided and the proposal was evaluated with different real or synthetic applications. The area overhead extracted, which was negligi-

TABLE III
AREA OVERHEAD OF C-MAC MODULE COMPARED TO ROUTER'S AREA

Module Type ↓	Area (μm^2) ↓			
Baseline router (BR)	108805.2			
Baseline router with RF transceiver (HB)	468805.2			
RF transceiver	360000 [1]			
Network Size →	4×4	6×6	8×8	10×10
C-MAC area (μm^2)	453.5	1115.6	1997.9	3146.5
C-MAC area per router	28.3	31.0	31.2	31.4
per router area overhead adding C-MAC to HB	0.006%	0.007%	0.007%	0.007%

ble and the simulation results verified the effectiveness of the proposed idea for real-world real-time implementations.

REFERENCES

[1] S. Abadal, A. Mestres, M. Nemirovsky, H. Lee, A. Gonzlez, E. Alarcn, and A. Cabellos-Aparicio, "Scalability of broadcast performance in wireless network-on-chip," *IEEE Transactions on Parallel and Distributed Systems*, vol. 27, no. 12, pp. 3631–3645, Dec 2016.

[2] K. Duraisamy, Y. Xue, P. Bogdan, and P. P. Pande, "Multicast-aware high-performance wireless network-on-chip architectures," *IEEE Transactions on Very Large Scale Integration (VLSI) Systems*, vol. 25, no. 3, pp. 1126–1139, March 2017.

[3] M. Opoku Agyeman, W. Zong, A. Yakovlev, K.-F. Tong, and T. Mak, "Extending the performance of hybrid nocs beyond the limitations of network heterogeneity," *Journal of Low Power Electronics and Applications*, vol. 7, no. 2, p. 8, 2017.

[4] Y. Peng, L. Guo, and Q. Gai, "Cross-layer qos-aware routing protocol for multi-radio multi-channel wireless mesh networks," in *Communication Technology (ICCT), 2012 IEEE 14th International Conference on*. IEEE, 2012, pp. 197–201.

[5] D. Zhao and Y. Wang, "Sd-mac: Design and synthesis of a hardware-efficient collision-free qos-aware mac protocol for wireless network-on-chip," *IEEE Transactions on Computers*, vol. 57, no. 9, pp. 1230–1245, 2008.

[6] A. Shacham, K. Bergman, and L. P. Carloni, "Photonic networks-on-chip for future generations of chip multiprocessors," *IEEE Transactions on Computers*, vol. 57, no. 9, pp. 1246–1260, Sept 2008.

[7] M. F. Chang, J. Cong, A. Kaplan, M. Naik, G. Reinman, E. Socher, and S. W. Tam, "Cmp network-on-chip overlaid with multi-band rf-interconnect," in *2008 IEEE 14th International Symposium on High Performance Computer Architecture*, Feb 2008, pp. 191–202.

[8] M. O. Agyeman, A. Ahmadinia, and N. Bagherzadeh, "Performance and energy aware inhomogeneous 3d networks-on-chip architecture generation," *IEEE Transactions on Parallel and Distributed Systems*, vol. 27, no. 6, pp. 1756–1769, June 2016.

[9] M. Chrzanowska-Jeske and J. Becker, "Tutorial 2a: 3d integration - challenges and advantages," in *2016 29th IEEE International System-on-Chip Conference (SOCC)*, Sept 2016, pp. 1–3.

[10] C. Wang, W. H. Hu, and N. Bagherzadeh, "A wireless network-on-chip design for multicore platforms," in *2011 19th International Euromicro Conference on Parallel, Distributed and Network-Based Processing*, Feb 2011, pp. 409–416.

[11] A. Samaiyar, S. S. Ram, and S. Deb, "Millimeter-wave planar log periodic antenna for on-chip wireless interconnects," in *The 8th European Conference on Antennas and Propagation (EuCAP 2014)*, April 2014, pp. 1007–1009.

[12] H. K. Mondal, S. H. Gade, M. S. Shamim, S. Deb, and A. Ganguly, "Interference-aware wireless network-on-chip architecture using directional antennas," *IEEE Transactions on Multi-Scale Computing Systems*, vol. 3, no. 3, pp. 193–205, July 2017.

[13] M. S. Shamim, N. Mansoor, R. S. Narde, V. Kothandapani, A. Ganguly, and J. Venkataraman, "A wireless interconnection framework for seamless inter and intra-chip communication in multichip systems," *IEEE Transactions on Computers*, vol. 66, no. 3, pp. 389–402, March 2017.

[14] S. Abadal, B. Sheinman, O. Katz, O. Markish, D. Elad, Y. Fournier, D. Roca, M. Hanzich, G. Houzeaux, M. Nemirovsky, E. Alarcn, and A. Cabellos-Aparicio, "Broadcast-enabled massive multicore architectures: A wireless rf approach," *IEEE Micro*, vol. 35, no. 5, pp. 52–61, Sept 2015.

[15] K. Goossens, J. Dielissen, and A. Radulescu, "Aethereal network on chip: concepts, architectures, and implementations," *IEEE Design Test of Computers*, vol. 22, no. 5, pp. 414–421, Sept 2005.

[16] H. Kopetz, A. Damm, C. Koza, M. Mulazzani, W. Schwabl, C. Senft, and R. Zainlinger, "Distributed fault-tolerant real-time systems: the mars approach," *IEEE Micro*, vol. 9, no. 1, pp. 25–40, Feb 1989.

[17] D. Rahmati, S. Murali, L. Benini, F. Angiolini, G. D. Micheli, and H. Sarbazi-Azad, "Computing accurate performance bounds for best effort networks-on-chip," *IEEE Transactions on Computers*, vol. 62, no. 3, pp. 452–467, March 2013.

[18] R. Mullins, A. West, and S. Moore, "The design and implementation of a low-latency on-chip network," in *Proceedings of the 2006 Asia and South Pacific Design Automation Conference*, ser. ASP-DAC '06.

Piscataway, NJ, USA: IEEE Press, 2006, pp. 164–169. [Online]. Available: https://doi.org/10.1145/1118299.1118348

[19] C. Paukovits and H. Kopetz, "Concepts of switching in the time-triggered network-on-chip," in *2008 14th IEEE International Conference on Embedded and Real-Time Computing Systems and Applications*, Aug 2008, pp. 120–129.

[20] Z. Shi and A. Burns, "Real-time communication analysis for on-chip networks with wormhole switching," in *Second ACM/IEEE International Symposium on Networks-on-Chip (nocs 2008)*, April 2008, pp. 161–170.

[21] D. DiTomaso, S. Laha, A. Kodi, S. Kaya, and D. Matolak, "Evaluation and performance analysis of energy efficient wireless noc architecture," in *2012 IEEE 55th International Midwest Symposium on Circuits and Systems (MWSCAS)*, Aug 2012, pp. 798–801.

[22] S. Abadal, M. Iannazzo, M. Nemirovsky, A. Cabellos-Aparicio, H. Lee, and E. Alarcn, "On the area and energy scalability of wireless network-on-chip: A model-based benchmarked design space exploration," *IEEE/ACM Transactions on Networking*, vol. 23, no. 5, pp. 1501–1513, Oct 2015.

[23] O. Markish, O. Katz, B. Sheinman, D. Corcos, and D. Elad, "On-chip millimeter wave antennas and transceivers," in *Proceedings of the 9th International Symposium on Networks-on-Chip*, ser. NOCS '15. New York, NY, USA: ACM, 2015, pp. 11:1–11:7. [Online]. Available: http://doi.acm.org/10.1145/2786572.2789983

[24] A. Ganguly, K. Chang, S. Deb, P. P. Pande, B. Belzer, and C. Teuscher, "Scalable hybrid wireless network-on-chip architectures for multicore systems," *IEEE Transactions on Computers*, vol. 60, no. 10, pp. 1485–1502, Oct 2011.

[25] S. Deb, K. Chang, X. Yu, S. P. Sah, M. Cosic, A. Ganguly, P. P. Pande, B. Belzer, and D. Heo, "Design of an energy-efficient cmos-compatible noc architecture with millimeter-wave wireless interconnects," *IEEE Transactions on Computers*, vol. 62, no. 12, pp. 2382–2396, Dec 2013.

[26] C.-H. O. Chen, S. Park, T. Krishna, S. Subramanian, A. P. Chandrakasan, and L.-S. Peh, "Smart: A single-cycle reconfigurable noc for soc applications," in *Proceedings of the Conference on Design, Automation and Test in Europe*, ser. DATE '13. San Jose, CA, USA: EDA Consortium, 2013, pp. 338–343. [Online]. Available: http://dl.acm.org/citation.cfm?id=2485288.2485371

Hardware Implementation of Reconfigurable Separable Convolution

Lei Rao
Dept. of Computer Science and Technology
Xi'an Jiaotong University
Xi'an, Shaanxi, 710049, China
Email: leiraoxjtu@gmail.com

Bin Zhang
School of Software
Xi'an Jiaotong University
Xi'an, Shaanxi, 710049, China
Email: bzhang82@mail.xjtu.edu.cn

Jizhong Zhao
Dept. of Computer Science and Technology
Xi'an Jiaotong University
Xi'an, Shaanxi, 710049, China
Email: zjz@mail.xjtu.edu.cn

Abstract—Convolution operations occupy large amounts of computation resource in convolutional neural networks (CNNs). Separable convolution can greatly reduce computational complexity. Unfortunately, most trained kernels in CNNs are not separable. In this paper, least squares approach is applied to decompose a non-separable 2D kernel into two 1D kernels. A reconfigurable convolutional architecture is proposed to convert a 2D convolution into 1D convolution in convolutional layers. Moreover, a denoising CNN is mapped to the proposed convolution architecture. Experimental results show that the hardware architecture can restore a 1280×720 image in 0.83s, which achieves an 8.4× speed-up over GPU implementation. Verification experiments demonstrate that our approach and hardware architecture can drastically reduce the computational complexity in convolution operations without sacrificing the performance.

Keywords-convolutional neural networks; hardware implementation; reconfigurable architecture; separable convolution;

I. INTRODUCTION

Convolutional neural networks (CNNs) have achieved superior performance in image processing and computer vision tasks [1]–[3], [5], [7]–[9]. As each layer of CNNs requires huge amount of computations, most CNN algorithms are implemented on GPUs. However, their high price and power consumption requirements hamper the extensive applications in mobile handheld devices, unmanned aerial vehicles (UAVs), and other embedded vision systems.

Due to the good performance and capability of reconfiguration, Field Programmable Gate Array (FPGA) are a possible choice for hardware acceleration of the convolution operations [1], [6], [9], [19], [23], [24]. An FPGA implementation of neural networks has been proposed in [3]. However, the processor uses low-accuracy arithmetic because of the limitations of resources on FPGA. A convolutional network processor was implemented on a Virtex 4 [4]; it uses a general-purpose soft-processor for control and a 7×7 convolver for data processing. Chakradhar et al. [15] proposed a hardware neural network on reconfigurable circuits. They focus on map a CNN to improve bandwidth utilization.

As the limit of logic resource and memory bandwidth, the state-of-art CNN model cannot achieve the best performance on FPGA platform. In this paper, we provide a separable convolution approach to reduce the computational complexity of convolutional layer. However, the kernels in CNN are usually non-separable and numerous. In this case, we use least squares approach for non-separable kernels. For large size of kernels, we divide the kernels into smaller ones to increase accuracy. Therefore, arbitrary-sized kernels can be separated into several 1D kernels. A reconfigurable architecture is proposed to verify the efficiency of our method.

In this paper, we use 1D convolution to reduce the complexity of convolution operations. A reconfigurable separable convolution architecture is proposed. A CNN with three layers is mapped onto the proposed architecture to demonstrate the performance of our method. The verification results show that the proposed separable convolution rapidly reduces the computational complexity without sacrificing the performance. The goal of these experiments is not necessarily to obtain state-of-the-art performance in a given task using CNNs, but to verify that our method can be used on an arbitrary CNN without a loss in accuracy.

II. RELATED WORKS

CNNs have been proposed in recent years for image and video applications. Many works are implemented on multi-core processors [27] and FPGA [11], [13]–[15]. The ANNA analog-digital chip [29] was the first hardware implementation of convolutional networks. The chip can compute 64 simultaneous 8×8 convolutions at a peak rate of 4×10^9 MAC operations per second. Farabet et al. [16] has proposed ASIC implementations of CNNs. However, their work ignores memory transfers for the sake of simplicity. Chen et al. [13] proposed CNN and DNN accelerators that can lead to high internal bandwidth and low external communication. However, the node reuse and iteration in their work result in high data transportation and time consumption. According to the research mentioned above, the computational complexity and size of neural network classifiers are challenges for CNN implementations on hardware platform. In this paper, 1D convolution is used to reduce the complexity of convolution operations.

As most of the kernels are not separable, Treitel et al. [25] split convolution into convergent sums of matrix-

978-1-5386-7100-9/18 $31.00 © 2018 IEEE

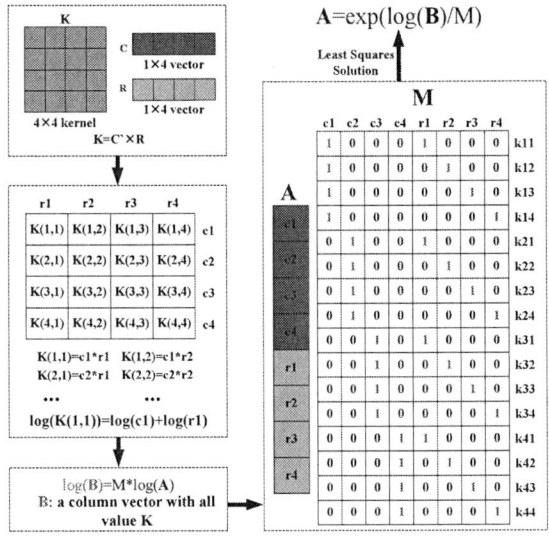

Figure 1. Separating a 4×4 kernel.

valued stages. This approach was used in [26] to avoid coarse discretization of the scale and orientation spaces. Sironi [11] and Rubinstein et al. [17] proposed to learn a convolutional kernel bank by composing several atoms from a handcrafted separable dictionary. Two sets of separable and non-separable filters are first learned. But this would result in a very difficult optimization problem and high computational cost during learning step. Hawe et al. [18] learn separable dictionaries in the case of classical sparse coding, but not in a convolution-based approach. Pirsiavash et al. [22] exploited GPU to separate kernel. However, it is time consuming because of memory transfers between the CPU and the GPU. The research mentioned above are learning separated kernels. These approaches either lead to optimization problems or cause much more computational resource consumption. In this paper, we provide a least squares approach to decompose the 2D kernels into 1D kernels. Verification experiments on CNN model demonstrate the accuracy of the proposed approach.

III. SEPARABLE CONVOLUTION

We illustrate the least squares approach for non-separable kernels. This method can separate 2D, 3D or nD kernel into 1D kernels. For large size of kernels, the non-separable 2D filters are divided into smaller ones first. Then, each divided kernel is decomposed into two 1D vectors.

Assuming a 4×4 kernel K can be decomposed into two 1D vectors, C and R. Figure 1 shows that least squares approach is used to obtain the two 1D kernels. For the element $K(i,j)$ in kernel K, we have

$$K(i,j) = C_i \times R_j \qquad (1)$$

Then we take logarithmic in Equation (1),

$$log(K(i,j)) = log(C_i) + log(R_j), (i,j = 1,2,3,4) \qquad (2)$$

Reshape the kernel K into a column vector B,

$$log\left(B_{16\times 1}\right) = M_{16\times 8} \times log\left(A_{8\times 1}\right) \qquad (3)$$

where A donates a column vector with all value of vectors C and R. M is a matrix shown in Figure 1.

The Equation (3) can be rewritten as

$$P = M \times T \qquad (4)$$

where $P \in R^{16\times 1}$ and $T \in R^{8\times 1}$ donate $log(B)$ and $log(A)$, respectively.

The loss function can be expressed as follow:

$$L(P,T) = \|P - M \times T\|^2 \qquad (5)$$

where $L(P, T)$ is the loss associated with a particular value of P and T. Then the least squares approach is applied to get the minimum value of the loss function $L(P, T)$.

Our approach can also be used in separating 3D kernels. The process of separating a 4×4×3 kernel is shown in Figure 2(a). The matrix M is shown in Figure 2(b).

Using the method mentioned above, the separable convolution converts 2D convolution into 1D convolution thus reducing hardware consumption. The computations using 2D and 1D convolution in two CNN models (denoising network [5] and AlexNet [29]) are shown in Figure 3. As can be seen, 1D convolution needs much less multiplications and additions compared with 2D convolution.

The kernel can be separated into 1D vectors using the above method. However, for many deep networks, the kernel size in the convolutional layer are very large. For instance, in the denoising network [5], the kernel size in convolutional layer are 16×16 and 8×8. In this case, the large size kernels are divided into smaller ones first. Then, each divided kernel is separated into two 1D vectors using the least squares approach.

IV. ARCHITECTURE

A reconfigurable architecture of separable convolution is proposed. We compare our separating method with other approaches. Moreover, we present an analysis of the computational complexities of different methods. A denoising CNN is mapped onto the proposed architecture to demonstrate our architecture and method reduce computational complexity at no cost in terms of performance.

A. Reconfigurable Convolution Unit

The reconfigurable architecture of convolution module (CM) using 1D convolution is shown in Figure 4. A CM performs one 4×4 separable convolution. The architecture of separable convolution unit (SCU) is shown in Figure 4. The SCU, which is composed of a register bank, line buffers, and 16 CMs (shown in Figure 4), can be used for one 16×16

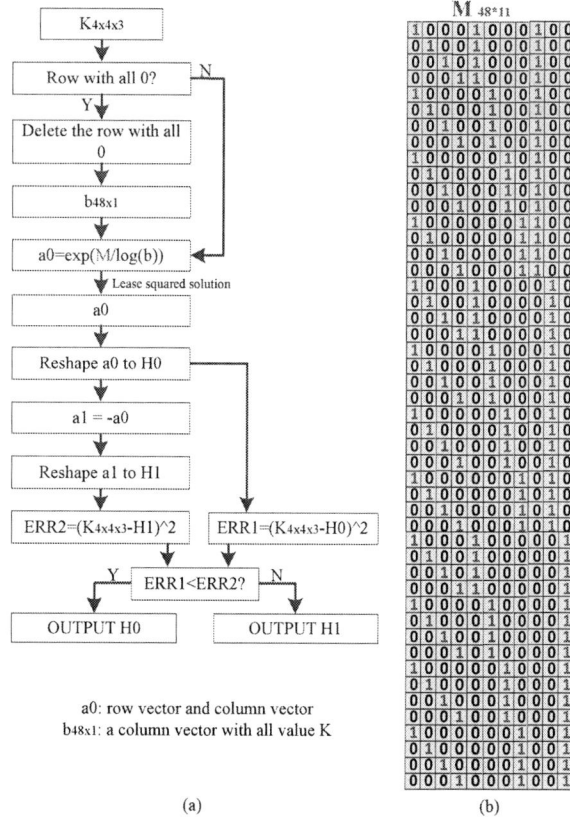

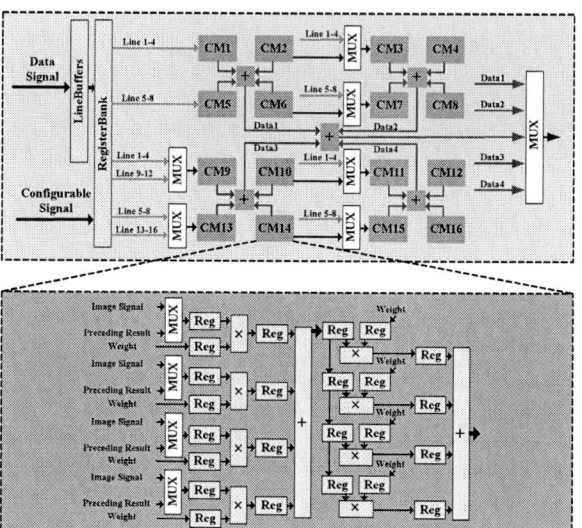

Figure 4. Architecture of the SCU and CM.

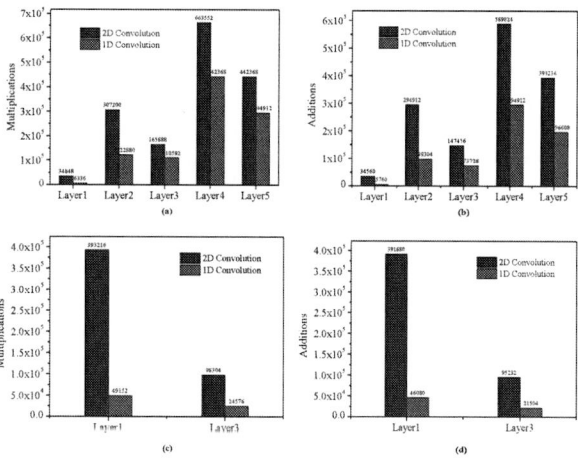

a0: row vector and column vector
b48x1: a column vector with all value K

(a)   (b)

Figure 2. (a)Process of separating a 4×4×3 kernel.(b)Matrix *M*

Figure 3. Computations of 2D and 1D convolution in the convolutional layers. (a) Multiplications in AlexNet [29]. (b) Additions in AlexNet [29]. (c) Multiplications in [5]. (d) Additions in [5].

convolution or four 8×8 convolutions simultaneously using a reconfigurable and pipeline architecture. When the kernel size is 8×8, the inputs of CM3, CM9, and CM11 are the

data signal from lines 1-4 and the inputs of CM7, CM13, and CM15 are the data signal from lines 5-8. When the kernel size is 16×16, the inputs of CM9 and CM13 are the data signal from lines 9-12 and 13-16, respectively, and the inputs of CM3, CM7, CM11, and CM15 are the outputs of CM2, CM6, CM10 and CM14, respectively.

B. Denoising CNN Architecture

We use a denoising network [5] to verify the proposed reconfigurable architecture. There are two convolution layers in the network. Each layer has a kernel bank of 512 kernels. The kernel sizes are 16×16 and 8×8, respectively. The reconfigurable architecture shown in Figure 4 can be used for one 16×16 convolution or two 8×8 convolution at the same time. Using fixed-point numbers instead of floating-point numbers is significant way to reduce memory and bandwidth requirements. In this section, we use 16 bit fixed-point number (1 bit sign, 8 bit integer and 7 bit fraction) in the implementation.

The overall hardware architecture is shown in Figure 5. The architecture consists of a hardware reconfigurable controller, input/output buffers, several functional units (FUs), a data memory controller, SRAM, and image/video acquisition and display units. Image/video preprocessing units perform essential functions, such as noise filtering, white balance, feature extraction, and video stabilization. For low hardware resource and power consumption, the proposed processor avoids full-image buffering and restores input images in line buffers.

FUs are the core execution units of the processor. The FU consists of three parts (shown in Figure 6): a SCU, activation function unit(AFU), and multiply accumulator unit (MAU).

978-1-5386-7100-9/18 $31.00 © 2018 IEEE   234

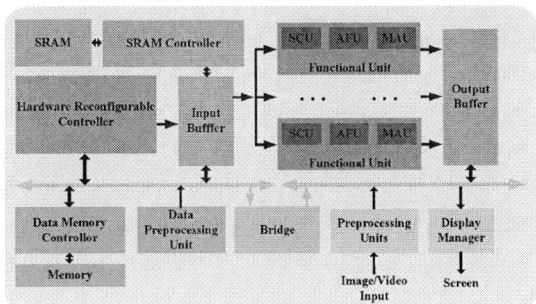

Figure 5. Hardware architecture convolutional networks using separable convolution.

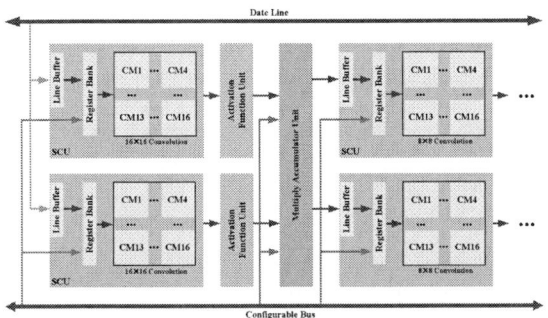

Figure 6. Architecture of the function units.

The data line is used for image storage and processing. The configurable bus is used to configure the SCUs. Each SCU is a set of identical separable convolutions interconnected in a 2D grid topology.

The hardware reconfigurable controller (HRC) is used to map CNN on a proposed processor and avoid using complex design environment or assistance tools. The HRC has configuration and execution phases. In the configuration phase, the controller receives control signals and configures the filters for the convolutional units. In the execution phase, the controller receives the feedback signals and generates interrupts for FUs.

V. EXPERIMENT AND DISCUSSION

A. Computational Accuracy

To measure the accuracy of the design, we use normalized mean squared error (NMSE) and correlation coefficient (CC) between the non-separable kernels and kernels separated using proposed method. The kernels are the same as used in Section 4. The performance measure is given by $NMSE = \frac{\sum_{i=0}^{n}(\widehat{y}_i - y_i)^2}{n(\widehat{y}_{max} - y_{max})^2}$, and $CC = \frac{cov(\widehat{y}, y)}{\sigma_{\widehat{y}}\sigma_y}$ where $\widehat{y}$ and y are the separable kernels from the proposed design and the learned kernels respectively, $cov(\widehat{y}, y)$ is the covariance of $\widehat{y}$ and y, $\sigma_{\widehat{y}}$ and σ_y are the are the standard deviations of $\widehat{y}$ and y, respectively. In Figure 7, we study the behavior of the design when kernel size varies. The original kernel size in the rain

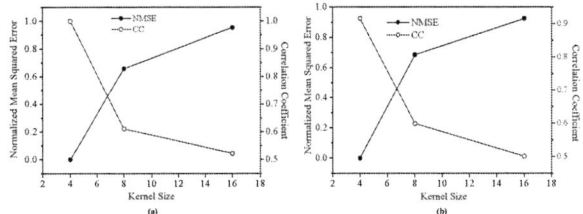

Figure 7.   (a) Rain network. (b) Dirt network.

and dirt network [5] are $16{\times}16$. The accuracy is decreased with increasing kernel size. The result shown in Figure 7 demonstrate that we can divide the larger size kernels into smaller ones to increase accuracy.

B. Comparison

To demonstrate the efficiency of the proposed architecture, two real-world images corrupted by rain and dirt are mapped to the proposed architecture. The denoising result is shown in Figure 8. As can be seen, almost all of the rain and dirt are removed. The $1280{\times}720$ real-world images can be restored using our proposed processor in $0.83s$ at $100MHz$. Eigen [5] restored the same $1280{\times}720$ color image in 7 seconds using a NVIDIA GTX 580 GPU. In this way, the proposed architecture achieves an average speed-up $8.4{\times}$ with GPU implementation.

Figure 9 shows an analysis of the computational complexities of different methods. The $512{\times}512$ image is convolved with a kernel bank of 128 non-separable kernels. And in 3D case, the $128{\times}128{\times}64$ volume is convolved with a kernel bank of 128 non-separable kernels. Figure 9(a) and (b) show the number of operations per pixel to compute convolution. In the FFT case, the convolutions are performed in the frequency domain. The Non-Sep and FFTW results shown in Figure 9(c) and (d) rely on the MATLAB $conv2$ function and the $fftw$ library [30] for convolutions, respectively. Our method is implemented on Altera Stratix II EP2S180C4 FPGA and the system clock is 150MHz. Using parallel computation, our method and architecture are the most efficient ones.

Figure 10(a) shows the time to convolve a $960{\times}720$ image with different number of $16{\times}16$ kernels simultaneously and (b) shows with $8{\times}8$ kernels. Benefiting from the parallel processing, our Verilog simulator can convolve a large number of kernels simultaneously and take less execution time than the CPU and GPU.

To evaluate the efficiency of our method, we compare our approach with other decomposition algorithms. We use the 256 denoising kernels computed by the K-SVD method [10]. We separate the denoising kernels into 1D kernels. The results are reported in Table I. SEP-TD [20] method obtain the kernel by using tensor decomposition. The measured value shown in Table II is peak signal to noise ration

978-1-5386-7100-9/18 $31.00 © 2018 IEEE   235

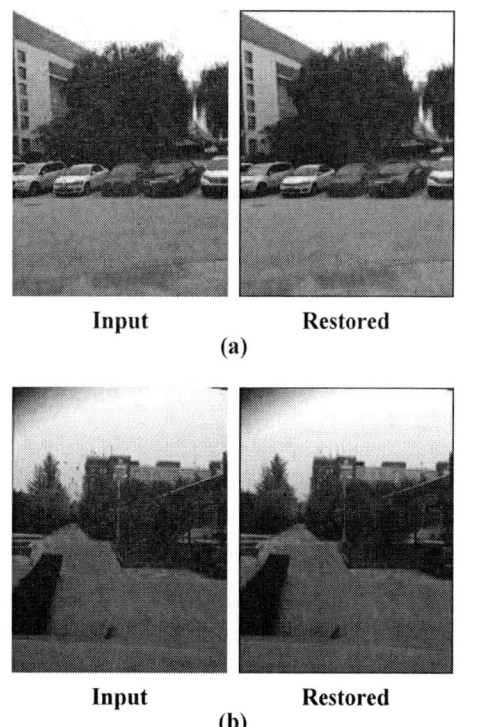

Input   **Restored**

(a)

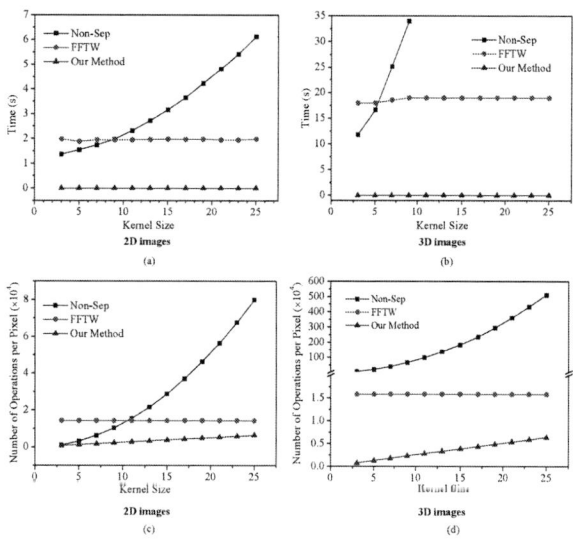

Input   **Restored**

(b)

Figure 8.   (a) Rain image and restored image using 1D convolution. (b) Dirt image and restored image using 1D convolution.

Figure 9.   Number of operations per pixel to compute convolutions.

(PSNR). As can be seen from Table II, our approach obtain a very accurate approximation of the original kernels from K-SVD method.

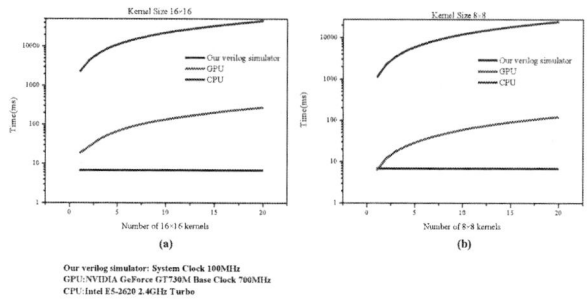

Our verilog simulator: System Clock 100MHz
GPU:NVIDIA GeForce GT730M Base Clock 700MHz
CPU:Intel E5-2620 2.4GHz Turbo

Figure 10.   (a) Time to convolve a 960×720 image with different numbers of 16×16 kernels simultaneously. (b) Time to convolve a 960×720 image with different numbers of 8×8 kernels simultaneously.

Table I
COMPARISON RESULTS

	Barbara	Boat	House	Lena	Peppers
Noise Image	22.13	22.12	22.19	22.18	22.15
K-SVD[15]	29.54	29.25	32.10	31.31	29.63
SEP-TD[20]	29.50	29.20	32.08	31.30	29.58
Our Method	29.11	29.00	31.87	30.77	29.19

Table II
SYNTHESIS RESULTS

TSMC 65nm	Used
Clock(MHz)	120
Arithmetic Precision	16-bit fixed-point
Area(mm^2)	3.89
Power(mW)	282
Time(ms)	5

C. Synthesis Results

The SCU architecture shown in Figure 4, is synthesized by Synopsys Design Compiler with TSMC 65nm process. The synthesis result is shown in Table II. The area and power consumption are 3.89 mm^2 and 282 mw (at 120 MHz), respectively.

VI. CONCLUSIONS

A reconfigurable structure for separable convolution is presented in this paper. The architecture requires less hardware consumption than 2D convolution. We also propose a CNN architecture and reconfigurable separable convolution module. The least squares approach is used to decompose non-separable kernels. To verify the accuracy of the separable convolution, we compare the separable results with original kernels and the results show that our method is no loss of accuracy. Comparing the separated results with other decomposing approaches, our method shows high accuracy. A denoising task is mapped on the proposed architecture. It takes 0.83s to restore a 1280×720 real-world image which achieves an average speed-up 8.4x with GPU implementation. Verification experiments demonstrate the accuracy and efficiency of the proposed separating method and architecture.

978-1-5386-7100-9/18 $31.00 © 2018 IEEE

ACKNOWLEDGMENT

This work is supported by National Natural Science Foundation of China (61603291) and the Fundamental Research Funds for the Central Universities.

REFERENCES

[1] Larochelle, Hugo, et al. "An empirical evaluation of deep architectures on problems with many factors of variation." Proceedings of the 24th international conference on Machine learning. ACM, 2007.

[2] Zhang, Chen, et al. "Optimizing fpga-based accelerator design for deep convolutional neural networks." Proceedings of the 2015 ACM/SIGDA International Symposium on Field-Programmable Gate Arrays. ACM, 2015.

[3] J. Cloutier, E. Cosatto, S. Pigeon, F.-R. Boyer, and P.Y. Simard, "VIP: an FPGA-based processor for image processing and neural networks," in Proceedings of Fifth International Conference on Microelectronics for Neural Networks, 1996, pp. 330-336.

[4] C. Farabet, C. Poulet, J.Y. Han, and Y. LeCun, "CNP: An FPGA-based processor for Convolutional Networks," in International Conference on Field Programmable Logic and Applications, 2009, pp. 32-37.

[5] D. Eigen, D. Krishnan, R. Fergus, "Restoring an Image Taken through a Window Covered with Dirt or Rain," in IEEE International Conference on Computer Vision (ICCV), 2013, pp. 633-640.

[6] Qiao, Yuran, et al. "FPGA-accelerated deep convolutional neural networks for high throughput and energy efficiency." Concurrency and Computation: Practice and Experience (2016).

[7] Ovtcharov, Kalin, et al. "Accelerating deep convolutional neural networks using specialized hardware." Microsoft Research Whitepaper 2.11 (2015).

[8] Zhang, Bin, Kuizhi Mei, and Nanning Zheng. "Coarse-grained dynamically reconfigurable processor for vision pre-processing." Journal of Signal Processing Systems 79.1 (2015): 45-61.

[9] Sankaradas, Murugan, et al. "A massively parallel coprocessor for convolutional neural networks." Application-specific Systems, Architectures and Processors, 2009. ASAP 2009. 20th IEEE International Conference on. IEEE, 2009.

[10] M. Elad and M. Aharon, Image denoising via sparse and redundant representations over learned dictionaries, IEEE Trans. Image Process., vol. 15, no. 12, pp. 3736C3745, Dec. 2006.

[11] Sironi, Amos, et al. "Learning Separable Filters." Pattern Analysis and Machine Intelligence, IEEE Transactions on 37(1) (2015): 94-106.

[12] Esmaeilzadeh, Hadi, et al. "Neural acceleration for general-purpose approximate programs." Proceedings of the 2012 45th Annual IEEE/ACM International Symposium on Microarchitecture. IEEE Computer Society, 2012.

[13] Chen, Yunji, et al. "Dadiannao: A machine-learning supercomputer." Microarchitecture (MICRO), 2014 47th Annual IEEE/ACM International Symposium on. IEEE, 2014.

[14] Yan, Zhicheng, et al. "HD-CNN: Hierarchical Deep Convolutional Neural Network for Image Classification." arXiv preprint arXiv:1410.0736 (2014).

[15] Chakradhar, Srimat, et al. "A dynamically configurable coprocessor for convolutional neural networks." ACM SIGARCH Computer Architecture News. 38(3). ACM, 2010.

[16] Farabet, Clment, et al. "Neuflow: A runtime reconfigurable dataflow processor for vision." Computer Vision and Pattern Recognition Workshops (CVPRW), 2011 IEEE Computer Society Conference on. 2011.

[17] R. Rubinstein, M. Zibulevsky, and M. Elad, Double sparsity: Learning sparse dictionaries for sparse signal approximation, IEEE Trans. Signal Process., vol. 58, no. 3, pp. 1553C1564, Mar. 2010.

[18] S. Hawe, M. Seibert, and M. Kleinsteuber, Separable dictionary learning, in Proc. IEEE Conf. Comput. Vis. Pattern Recog., 2013, pp. 438-445.

[19] Rao, Lei, Bin Zhang, and Jizhong Zhao. "Hardware Implementation of Reconfigurable 1D Convolution." Journal of Signal Processing Systems 82(1) (2016): 1-16.

[20] Sironi, Amos, et al. Learning Separable Filters. Pattern Analysis and Machine Intelligence, IEEE Transactions on 37(1) (2015): 94-106.

[21] Cardells-Tormo F., Molinet P.L., Area-efficient 2-D shift-variant convolvers for FPGA-based digital image processing, IEEE Transactions on Circuits and Systems II: Express Briefs, 53(2) (2006): 105-109

[22] H. Pirsiavash and D. Ramanan, Steerable part models, in Proc. IEEE Conf. Comput. Vis. Pattern Recog., 2012, pp. 3226C3233.

[23] Chen, Yu-Hsin, et al. "Eyeriss: An energy-efficient reconfigurable accelerator for deep convolutional neural networks." IEEE Journal of Solid-State Circuits (2016).

[24] Sharma, Hardik, et al. "Dnnweaver: From high-level deep network models to fpga acceleration." the Workshop on Cognitive Architectures. 2016.

[25] S. Treitel and J. Shanks, The design of multistage separable planar filters, IEEE Trans. Geoscience Electron., vol. GE-9, no. 1, pp. 10C27, Jan. 1971.

[26] P. Perona, Deformable kernels for early vision, IEEE Trans. Pattern Anal. Mach. Intell., vol. 17, no. 5, pp. 488C499, May 1995.

[27] Vanhoucke, Vincent, Andrew Senior, and Mark Z. Mao, Improving the speed of neural networks on CPUs, Proc Deep Learning and Unsupervised Feature Learning NIPS Workshop, 1, 4-5 (2011)

[28] E. Sackinger, B. Boser, J. Bromley, Y. LeCun, and L. D. Jackel, Application of the ANNA neural network chip to high-speed character recognition, IEEE Transaction on Neural Networks, 3(2), 498-505 (1992)

[29] Krizhevsky, Alex, Ilya Sutskever, and Geoffrey E. Hinton, Imagenet classification with deep convolutional neural networks, Advances in neural information processing systems (2012)

[30] The MathWorks Inc.https://www.mathworks.com/help/matlab/ref/fftw.html.

Low Overhead Online Checkpoint for Intermittently Powered Non-volatile FPGAs

Xinyi zhang[1], Clay Patterson[2], Yongpan Liu[3], Chengmo Yang[4], Chun Jason Xue[5] and Jingtong Hu[1]

[1]Department of Computer and Electrical Engineering, University of Pittsburgh
[2]Department of Computer and Electrical Engineering, Oklahoma State University
[3]Department of Electronic Engineering, Tsinghua University
[4]Department of Computer and Electrical Engineering, University of Delaware
[4]Department of Computer and Electrical Engineering, City University of Hong Kong
xinyizhang@pitt.edu, clay.patterson@okstate.edu, ypliu@tsinghua.edu.cn
chengmo@udel.edu, jasonxue@cityu.edu.hk, jthu@pitt.edu

Abstract—Energy harvesting is an attractive way to power future IoT devices since it can eliminate the need for battery or power cables. However, harvested energy is intrinsically unstable. While FPGAs have been widely adopted in various embedded systems, it is hard to survive unstable power since all the memory components in FPGA are based on volatile SRAMs. The emerging non-volatile memory based FPGAs provide promising potentials to keep configuration data during power outages. However, few works have considered implementing efficient runtime intermediate data checkpoint on non-volatile FPGAs. To realize accumulative computation under intermittent power on FPGA, this paper proposes a low-cost design, FC-FPGAs, which utilizes "scan-chain like" flip-flops to track intermediate data. Instead of keeping all on-chip intermediate data, FC-FPGA only targets on necessary data that is labeled by off-line analysis and identified by an on-line tracking circuit. The evaluation shows that compared with state-of-the-art, FC-FPGA can realize accumulative computing and significantly reduce computation time and energy over a wide range of unstable power traces.

Keywords-energy harvesting; non-volatile FPGA; checkpoint; high-level synthesis;

I. INTRODUCTION

FPGAs have been widely adopted in various embedded systems that are powered by batteries. However, in the emerging Internet of Things (IoT), which is full of tiny, cost-sensitive, and space constrained widgets, batteries are no longer an ideal power supply due to poor scalability, recharging and safety concerns. Out of all possible alternatives, the energy harvesting system is becoming one of the promising candidates. Energy harvesting systems convert ambient energy from their surroundings. Some of the widely known energy harvesting techniques include: Photovoltaics (PV), Thermoelectric generators (TEGs), and Piezoelectric (PZ). A device equipped with these harvesters can utilize the converted energy directly or recharge its energy storage (e.g. capacitors). The ease of access to power makes it a very competitive power source for portable devices. However, there are also challenges for such energy harvesting systems: unstable power and low power input.

Even though there are already ultra low power FPGAs such as Lattice iCE40 series, which only consume μW

with $10KHz$ clock or lower [1]. The unpredictability of available energy renders the system power intermittent. The intermittent power will interrupt computations. In such a condition, long computations may be prohibited since the intermediate data will be lost and the computation has to start over from the beginning. Thus, it is essential to preserve the FPGA *configuration data* and *intermediate data* during a power outage. Then, long computations can be realized by retrieving a checkpoint after power resumes.

The non-volatile memory based FPGAs (NV-FPGAs) are natural candidates to address this challenge. With the substitution of NVMs such as ReRAM, STT-RAM and PCM for SRAMs, *configuration data* can be retained locally on chip with benefits of low leakage power, short critical path, and small area, etc [2]–[7]. Therefore, costs associated with loading *configuration data* from off-chip flash memories are avoided when there is a power outage. In the existing NV-FPGAs and traditional FPGAs, *intermediate data* is stored in registers which consist of volatile flip-flops (FF). Just like *configuration data*, *intermediate data* need to be saved during power outage in order to resume system execution when the power comes back. Non-volatile flip-flops (NV-FFs) have been successfully integrated on processors while indiscriminate back-up is adopted in [8]. The success of NV-FFs in processors can be transplanted to FPGAs. However, FPGA's register resource is significantly more than the processor's. Indiscriminate back-up all registers on FPGA is energy consuming.

To improve checkpoint energy efficiency and reduce the time impact from power interrupt, this paper proposes FC-FPGA, a status tracking methodology based on data flow. With an off-line mapping of state to flag counter (state-to-flag), the FC-FPGA can on-line track the *intermediate data* within each state. Then, instead of all registers data, only a set of "flagged" *intermediate data* will be locally stored in non-volatile registers (NV-FF). In this way, the checkpoint cost can be reduced significantly. The main contributions of this work are as follows:

- Design of flag counter based auto-tracking circuits which track, store and retrieve intermediate computa-

tion data for NV-FPGAs with NV-FFs.
- Design of an off-line state-to-flag mapping algorithm to map the computation states to auxiliary flag counters.
- A demonstration of the efficiency of our design with representative benchmarks.

The rest of the paper is organized as follows. Section II presents NV-FPGA's background and the related work. Section III presents a motivation example of the FC-FPGA. Section IV presents the architecture design of the FC-FPGA. Section V presents the state-to-flag mapping algorithms and Section VI presents the evaluation results.

II. BACKGROUND AND RELATED WORK

In this section, we will first present works related to energy harvesting on NV-FPGA in subsection II-A. Then works related to NV-FPGA and NV-FF architecture will be presented in subsection II-B. Finally, the background of high-level synthesis will be introduced in subsection II-C.

A. Energy Harvesting and Checkpointing

In energy harvesting systems, ambient energy can be harvested to power the NV-FPGA and peripheral devices. It is observed that a wrist-worn rotational energy harvester can provide power over 40 μW with a worst power outage case that happens every $10ms$ in daily activities [9], [10]. An energy harvesting system is shown in Fig. 1(a). The regulator is a bridge between ambient energy, FPGA and an auxiliary capacitor. FPGA is powered by the regulator when harvested power is sufficient. Meanwhile, the capacitor is fully charged. Whenever the ambient energy is weak, a digital signal will be sent into FPGA and the stand-by capacitor can provide additional energy to FPGA. However, the size of such capacitors is small as μF and under $5v$ [11]. The energy from capacitors is restricted. Therefore, critical intermediate data in NV-FPGA should be backed up even with NV-FFs considering the scarce energy.

To back up the critical data, state results checkpoints are adopted by Azalia et al. and Yuan et al. [12], [13]. Azalia et al off-line analyze the forward computation cost and back-up overhead to determine checkpoint placement in ASIC, and an arbitration circuit is added to each checkpoint. Yuan et al. propose an NV-FPGA architecture that places the different states of a computation in different BRAM-based hardware blocks, and utilizes an on-line analysis of data back-up and re-computation overhead in each block to store and retrieve states' results. While the prior strategies can successfully preserve the intermediate data, a detailed knowledge for specific devices is needed, which heavily raises heavy off-line workload. They either cannot be directly applied to FPGA or work on relatively coarse granularity, and may not survive in harsh power conditions which is commonly happen in energy-harvesting system.

B. NV-FPGA

NVMs mentioned in Section I are compatible with CMOS technology via back-end-of-line technique [14]. By growing the non-volatile part on the top of the chip, it adds non-volatile characteristics to an FPGA without increasing chip area. Fig. 1 (b), (c) and (e) shows the existing NVMs based FPGA design [15]. In the figure, a SRAM is represented by a gray block and a NVM is represented by a green block. SRAMs can be replaced by NVMs in NV-FPGA. In the rest of this paper, all devices with non-volatile characteristic are in green color.

The SRAM mask in *lookup table* (LUT), SRAMs in *switching boxs* (SBs) and *connecting boxes* (CBs) can be replaced by ReRAMs as shown in Fig. 1(b), (c) and (e). Thus the unchangeable configuration data can be preserved on chip even power is off. Cong et al. proposed a partial non-volatile FPGA structure that keeps configuration data, which achieves 5.18x area savings, 2.28x speedup and 1.63x power savings [16]. For the intermediate data, it can be kept locally on chip by integrating NV-FFs on NV-FPGAs.

There are also works that proposed NV-FF [8], [17]–[19]. The state-of-the-art NV-FF achieves the overhead of ns in time and pJ in energy for one bit. The architecture of NV-FF is shown in Fig. 1(d). A NVM cell is attached to a regular FF with extra 39% area but less than 10% extra area for the whole chip is observed in [8]. The NV-FFs work like regular FFs when the power is stable. And it can store the data to NVMs with control signal when there is a power outage.

Such control signals can be routed via the NV-FPGA programming path. This is because two FFs are attached to one LUT and the NVM in LUT mask won't be configured after programming. Moreover, synthesis tool like Vivado places four pairs of LUT and FFs in a *slice* and provide the address of such *slice*. Therefore, additional metal layer for NV-FF control signal is not necessary. As the LUT and FFs pairs of are "better packing into the same slice" in Vivado [20], the store and retrieve of NV-FFs in NV-FPGA is assumed to execute by slice in the proposed design.

In FC-FPGA, the slice address is acquired off-line and can be read on-line if intermediate data back-up is needed.

C. High-Level Synthesis (HLS) and Synthesis

HLS converts high-level language such as C/C++ to Hardware Description Language (HDL) like Verilog, in which, a program is split to multiple states and managed by a Finite State Machine (FSM). Such process is denoted by Control Data Flow Graph (CDFG) [21]. As state function varies in a program, the latency varies from state to state. Assigning checkpoints with a granularity of "state" may not be sufficient to complete the computations in a frequent power outage. After unfolding a state, information such as module names, registers width, connections, loop numbers and timing can be collected. For HLS tools like

---

978-1-5386-7100-9/18 $31.00 © 2018 IEEE   239

Vivado/LegUp, it maps modules to different clock cycles and indicates the data dependency.

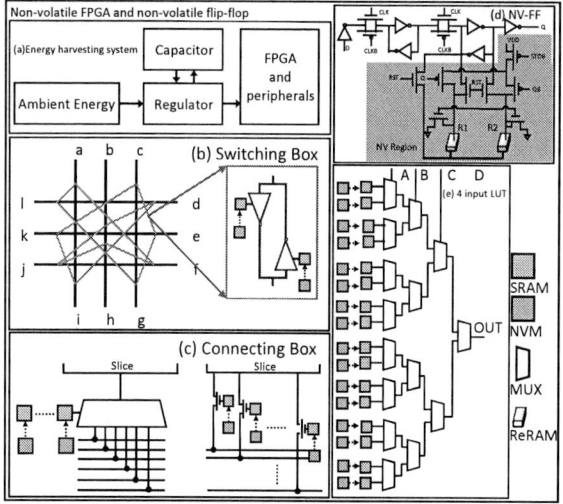

Figure 1.   Non-volatile FPGA and non-volatile flip-flop. (a) Energy harvesting system. (b) Switching Box. (c) Connecting Box. (d) NV-FF. (e) Four input LUT.

Synthesis is a process that turns HDL into an implementation of logical gate. During Synthesis, technique like Xilinx XST adopts *floorplanning* to pack a pair of LUT and FF into same slice. It also indicates the physical placement of modules and slice address (register address).

The proposed FC-FPGA starts with HLS. Targeting on the split states, it utilizes time and data dependency information from HLS, mapping each state to a flag counter with a granularity of one clock cycle. For the registers in the state, its *slice* address can be acquired via *floorplanning* in Synthesis. By on-line reading the status of flag counters, corresponding slices can be chosen to operate.

The main off-line work load is state-to-flag mapping: generating a scan-chain like tracking circuit, which is done by analyzing data dependency in each state. The working flow of FC-FPGA is shown in the Fig. 2(a). The off-line analysis algorithm analyzes FSM and states in CDFG and generates the auxiliary modules for state tracking which is shown in green block. The address of slices are extracted in *floorplanning*. Then, the generated modules will be merged into the source HDL file to be synthesized. The FC-FPGA architecture will be presented in Section IV and the off-line analysis algorithm will be presented in Section V.

III. MOTIVATION EXAMPLE

Existing techniques periodically back up state results, so computation can rarely avoid computation roll-back. The red line in Fig. 2(b) shows an example of periodical checkpoints. Assuming there are only checkpoints at 20%, 40% and 60% of the overall progress regardless of the power condition. Computation roll-back can not be avoided. In the worst

case scenario, if a previous checkpoint is far away from power outage point, large roll-back overhead occurs. Such techniques cannot avoid unnecessary checkpoints and re-computation time/energy overhead if power is stable.

Through on-line tracking combined with an off-line analysis, the FC-FPGA proposed in this paper locates the most recent intermediate data, reducing checkpoints and re-computation overhead. The green line in Fig. 2(b) shows an example of the proposed FC-FPGA under intermittent power. When FC-FPGA operates under intermittent power, computation only stops and checkpoint at $t3$ and $t7$ where the power outage occurs. Therefore, writing to NVMs only happens at $t3$ and $t7$. Then the computation can restart from the two checkpoints.

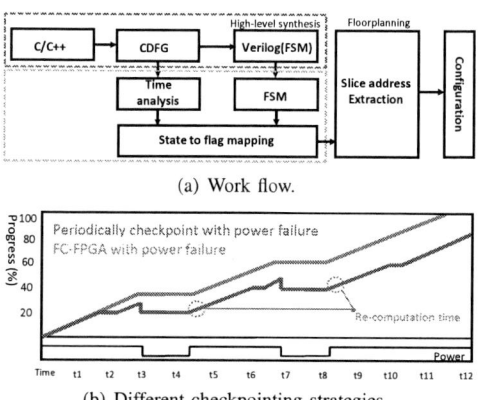

(a) Work flow.

(b) Different checkpointing strategies.

Figure 2.   Work flow and different checkpointing strategies.

IV. FC-FPGA ARCHITECTURE

The proposed FC-FPGA methodology includes both hardware and software designs. In this section, we will introduce the hardware design of FC-FPGA. The algorithms that work with the hardware will be introduced in Section V.

A. Hardware Architecture Overview

As shown in Fig. 3(a), the proposed design includes a finite state machine (FSM), function modules, flag counters and an NV-FF controller. FSM is already embedded in the target circuit after HLS. As FSM determines state status, FC-FPGA utilizes FSM to initialize the tracking action for each state. Each state is associated with a flag counter and a function module. Controlled by the function module, the flag counter tracks data flow in a state and it can be read by the NV-FF controller. The NV-FF controller determines the back-up/retrieve operations on corresponding slice.

B. Finite State Machine and Function Module

The FSM generates the current state number, determining the data transition among states. Each function module, as shown in Fig. 3(b), contains a pre-set state number. By dynamically comparing the current state number with

978-1-5386-7100-9/18 $31.00 © 2018 IEEE   240

the pre-set state number via a bitwise XNOR logic, each function module can be aware of the activeness of its state. When they match (XNOR=1), flag counter will be driven by system clock, tracking data within the state. In our design, system clock is shared by flag counters and main states. The function module is activated by bitwise XNOR results and regulated by power outage/resume signal.

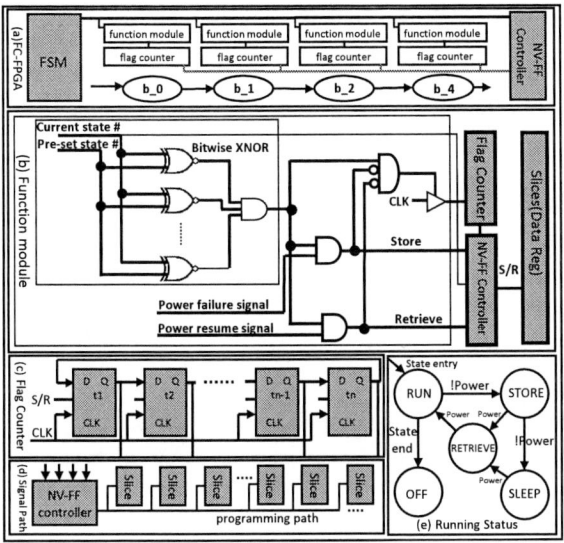

Figure 3. (a) FC-FPGA architecture. (b) Function module. (c) Flag Counter. (d) NV-FF controller path. (e) Function module running status.

Upon an incoming of power outage signal pulse 1, together with output of bit-oriented XNOR 1, the clock to flag counter is disabled and the NV-FF controller is enabled in store mode. When the power resumes, with a pulse of power resuming signal 1 and the output of bit-oriented XNOR 1, the clock to flag counter is disabled and the NV-FF controller is enabled in retrieving mode. After resuming signal, the clock to flag counter is enabled again, driving the flag counter shifting with data flow from the checkpoint.

If data flow enters the next state, the current state number changes. Therefore, the clock to flag counter and signal to NV-FF controller are disabled for this state. Only active state needs to be tracked. The running status of a function module is shown in Fig. 3(e) as a state transition diagram.

C. NV-FF Controller

NV-FF controller keeps the physical address map relations of flag counters and slices. Taking global state number from FSM, the NV-FF controller discerns current active flag counter and its corresponding slices. Then, according the flag counter status and the global state number, the controller selects slices to manipulate. An example of control signal via programming path to slices is shown in Fig. 3(d).

D. Flag Counter

The flag counters track data flow in each state. Each flag counter consists of NV-FFs (flags) that are sequentially connected, forming "scan-chain" like structure as shown in Fig. 3(c). However, there is no physical connection between state circuit and flag counters. As intermediate data are kept in registers in FPGA, the number of flip-flops in each flag counter is determined by the register distribution in the corresponding state. FC-FPGA uses one bit in a flag counter to indicate the activeness of multiple registers if they keep intermediate data at the same clock cycle. Driven by a system clock, it works synchronously with the data flow in the state. Since each state is assigned a flag counter, by checking all working flag counters, an accurate data flow can be obtained. Then, the corresponding slices can be selected.

If the flag bit is 1, it indicates that the registers associated with this flag bit currently keep the intermediate data for the modules follow them. If the flag is 0, it indicates that data have passed or not reached these registers. For each flag counter, its most significant bit is initialized as 1 during initialization. Only when the state is enabled, the flag counter can be activated. Then the exclusive bit 1 will be shifted with the clock. Thus, during a power outage, only registers whose corresponding flag's value is 1 need to be stored. In this way, the data flow can be resumed when power returns.

Therefore, a flag bit can control multiple slices as registers are packed into slices. After read by NV-FF controller, flag bit 0 will block the signal to slices and vice versa.

If the FF resources are scarce, the number of bits of the flag counters can be reduced by sharing one bit between two or even three clock cycles. In this way, the flag bit will be shifted forward every two or three clock cycles.

While this design looks intuitive, there are several tasks that need to be accomplished before it can work properly. First of all, for a given application's CDFG, we need to identify which registers will keep intermediate data at the same clock such that they can share a flag bit. Second, the slice address information needs to be acquired in *floorplanning* during synthesis for NV-FF controller. The details of these tasks will be explained in Section V.

V. FC-FPGA OFF-LINE ANALYSIS

In this section, we will present how to map registers to flag counters, and how flag counters work synchronously with registers in subsection V-A and V-B.

A. State to Flag Mapping

As a program is split into states, FC-FPGA generates flag counter for each state. In order to obtain the mapping, we need to know the active time of the registers in each state. A CDFG can be denoted by $G(B, L)$, where $B = \{b_1, b_2, b_3 \dots b_n\}$ is the set of states. Basic block b_n represents *state n* and L is the set of edges indicating the data flow direction as shown in Fig. 4(a). The unfolding of basic block

978-1-5386-7100-9/18 $31.00 © 2018 IEEE

b_2 according to schedule chart with time t is shown in Fig. 4(b). In this figure, each rectangle block represents a module (sub-block) b_{nx}, containing its function, I/O, and register utilization information. The registers in modules, represented by reg_{nx}, temporarily keep the intermediate data until the next module is finished. The arrow represents the data flow direction within the state. Different registers, when power outage occurs, should be stored for the resuming of data flow. Such registers are defined as *checkpoint locations*. For each clock cycle t (time between t and $t+1$), the *checkpoint locations* should be determined, which is denoted by $checkpoint_t$.

Algorithm 1 *Checkpoint location* determination
___
Input: time set T_n, module b_{nm}, b_{ni}, register reg_{nm} and reg_{ni}
of basic block b_n
Output: $checkpoint_t$
for each $t \in T_n$ **do**
    if $reg_{nm} \Diamond b_{ni} = 1$ **then**
        add reg_{nm} to $checkpoint_t$
        ($\Diamond$ represents the data dependency)
    end if
end for
return $checkpoint_t$

Algorithm 1 shows how FC-FPGA determines $checkpoint_t$. $Checkpoint_t$ is determined for each clock cycle t by finding data path between registers and modules (data dependency). In Algorithm 1, modules and registers that have been reached before clock cycle t are defined as b_{nm} and reg_{nm}. Otherwise, they are defined as b_{ni} and reg_{ni}. If there is a connection between reg_{nm} and b_{ni}, a data path exists, which is indicated by $reg_{nm} \Diamond b_{ni} = 1$. Then reg_{nm} is the *checkpoint location* for $checkpoint_t$ (clock cycle t). A bit in flag counter is used to indicate activeness for each $checkpoint_t$. Therefore, a single $checkpoint_t$ may contain multiple *checkpoint locations* and reg_{nm} may be assigned in multiple $checkpoint_t$. For example, in Fig. 4(b), at clock cycle $t20$, data path is found between reg_{20} to b_{22} and reg_{21} to b_{23}. Then $\{reg_{20}, reg_{21}\} \subset checkpoint_{20}$. At clock cycle $t21$ and $t22$, data path is found between reg_{22} and reg_{23} to b_{24}. Then $\{reg_{22}, reg_{23}\} \subset checkpoint_{21}$ and $\{reg_{22}, reg_{23}\} \subset checkpoint_{22}$.

After we are able to identify $checkpoint_t$, we need to map $checkpoint_t$ to flag counters. Formally, flag counters in FC-FPGA are denoted by $F = \{F_1, F_2, F_3...F_n\}$. Each flag counter F_n provides flag bits for a basic block b_n. FC-FPGA maps all $checkpoint_t$ in b_n to different bits in F_n according to timing information from CDFG. The FC-FPGA maps $checkpoint_t$ to flag counter as follows. The input of the mapping process is basic block b_n, and the output is flag counter F_n. For a basic block b_n, if it contains one or more $checkpoint_t$, then a flag counter is assigned. The bit width of a flag counter $F_n[1:k]$ is determined by the number of $checkpoint_t$ in a state. The mapping results of F_2 are shown

in Fig. 4(c). In this figure, a 6-bit flag counter is generated. Without the loss of generality, FC-FPGA builds circular flag counter to indicate the $checkpoint_t$ in case the basic block is a loop operation. For example, in Fig. 4(b), if it is a loop with iteration number 2 (represented by dashed orange line), its corresponding flag counter should be executed 2 rounds.

After assigning the flag counters to states, registers (slices) address can be drawn from *floorplanning* during synthesis. Thus, the address map relation of flag counter to slices for NV-FF controller can also be established. During operation, flag counter bits will be decoded by NV-FF controller and applied on control path to slices as shown in Fig. 3(c) .

B. Multi-clock cycle Module and other Situations

In FPGA, some modules cannot be finished in one clock cycle like module b_{24}, such as data load. The registers reg_{22}, reg_{23} and the flag counter bit $f_{22} = 1$ are stored when power breaks at t_{22}. However, after resuming, the computing starts from the beginning of b_{24} and flag bit 1 should start from f_{21}. Moreover, some modules may not contain any register while $checkpoint_t$ is assigned to its time point. For example, if only module b_{25} and b_{26} in b_2 don't contain register, $checkpoint_{24}$ is still issued as the existed data path between reg_{24} and b_{27}. The registers data reg_{24} and the flag counter bit $f_{24} = 1$ are stored when power breaks at t_{24}. However, after resuming, the computing starts from the end of b_{24} and flag bit 1 should start from f_{23}.

In order to avoid this inconsistency, the proposed FC-FPGA inserts "rectification" logic circuit to such flags during off-line analysis. Take f_{21} and f_{22} as an example, their flags are rectified to $\overline{f_{21}}$ and $\overline{f_{22}}$. The truth table of the rectification logic is shown in Fig. 4(d). The $\overline{f_{21}}$ is rectified to 1 and $\overline{f_{22}}$ is rectified to 0 if and only if the power breaks at t_{22}. The $\overline{f_{21}}$ and $\overline{f_{22}}$ are equal to f_{21} and f_{22} when power failure is 0. Such rectification logic only consumes several LUTs which is observed in the experiment.

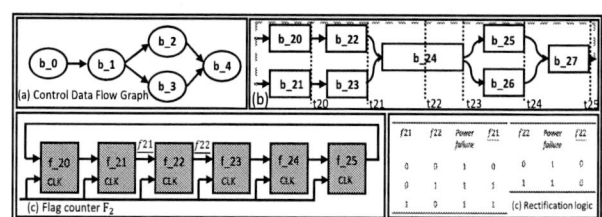

Figure 4. Stat-to-flag mapping relations. (a) Control Data Flow Graph. (b) State 2 unfolding. (c) Flag counter F_2 for state 2. (d) Rectification logic truth table.

VI. EXPERIMENTAL RESULTS

A. Evaluation Setup

In experiments, Xilinx Vivado HLS is used to obtain the CDFG and ISE Design Suite is used to extract *floorplanning* information. As the performance of different checkpoint strategies is largely influenced by devices and

NVM characteristics, we demonstrate the time and energy superiority of FC-FPGA (FC) by comparing re-computation time and number of bit operations on registers with intermittent power. Benchmarks *stencil2d*, *stencil3d*, *gemm*, *fir*, *dfadd* and *global* from MachSuite and Vivado library are evaluated. The CP-FPGA (CP) proposed by Yuan et al. is chosen as the baseline in experiment. In the baseline, we follow the policy that back up final results of each state. Considering the limited power availability from energy harvesters, the work frequency of the FPGA is chosen as $10KHz$, which is common in low-power FPGAs. Several power outage intervals, $10ms$, $20ms$, $50ms$ and $100ms$ are evaluated in this paper.

B. Time Evaluation

In this section, we compare the proposed FC-FPGA with the baseline in terms of re-computation clock cycles. Here, we use normalized time, the ratio of re-computation latency and original computation latency to measure the improvement of the proposed design. Table I shows the performance of proposed design (FC) and baseline (CP) in six benchmarks under different power traces. The average improvement of the proposed design for each power trace is shown in the most right column. The proposed design shows an average of 130x, 69x, 46x and 26x less re-computation time in $10ms$, $20ms$, $50ms$ and $100ms$ interval.

The overhead of the proposed design is lower than 2.8% in all situations and the worst case is observed in *gemm*. For the baseline, up to 50% overhead is observed in $10ms$ and $20ms$ intervals and it fails in *stencil3d* and *dfadd* with $10ms$' interval. This is because the baseline can only preserve the state result regardless the length of a state, while FC-FPGA can always track the data within the state. For benchmarks with long state latency like *stencil3d* and *dfadd*, the baseline has a poor performance even in $50ms$ and $100ms$ interval. This is due to the period between two checkpoints is relatively long in the baseline for *stencil3d* and *dfadd*. Thus, a long re-computation can easily occur. For these benchmarks that survive in $10ms$ interval, the frequent roll backs cause severe time overhead in the baseline.

FC-FPGA and the baseline achieve less re-computation time as the power gets better. However, the baseline technique is severely influenced by the power conditions. FC-FPGA maintains a reliable performance and good self adaptability in all power situations evaluated.

C. Bits Operations Evaluation

In this section, we compare the bits operation on flip-flops between the proposed FC-FPGA and the baseline. The bits operations consist of operation on both flag counters and registers. For each bit in a register, a pair of back-up and retrieve is regarded as one operation. We normalize the number of bit operations to the number of flip-flops in the benchmarks as well.

Table II shows the evaluation result of bit operations in different benchmarks. The average improvement for each power trace is shown in the most right column. The proposed design shows an average of 1.2x, 1.7x, 4.6x and 20x less bit operations in $10ms$, $20ms$, $50ms$ and $100ms$ interval. The results also show that the proposed design maintains a good scalability and self adaptability in different traces as the overhead decreases rapidly when power gets better. In the baseline, the number of bit operations does not decrease as each state result is preserved.

Table I
RE-COMPUTATION LATENCY IN DIFFERENT BENCHMARKS

		stencil2d	stencil3d	gemm	fir	dfadd	global	Improvement
10ms	CP	0.490	-	0.490	0.487	-	0.460	
	FC	0.025	0.024	0.028	0.019	0.005	0.001	130x
20ms	CP	0.246	0.500	0.160	0.243	0.480	0.220	
	FC	0.012	0.013	0.014	0.009	0.005	0.001	69x
50ms	CP	0.081	0.166	0.061	0.080	0.114	0.077	
	FC	0.005	0.005	0.006	0.003	0.001	0.001	46x
100ms	CP	0.037	0.080	0.029	0.038	0.057	0.034	
	FC	0.003	0.003	0.003	0.002	0.001	0.001	26x

Table II
BIT OPERATIONS IN DIFFERENT BENCHMARKS

		stencil2d	syencil3d	gemm	fir	dfadd	global	Improvement
10ms	CP	640	-	448.9	24.7	-	3.18	
	FC	419	92.6	329	17.49	7.78	4.9	1.2x
20ms	CP	640	12.3	448.9	24.7	1.79	3.18	
	FC	201	46.2	164.8	8.56	6.8	2.37	1.7x
50ms	CP	640	12.3	448.9	24.7	1.79	3.18	
	FC	80	17.7	65.9	3.21	2.57	0.86	4.6x
100ms	CP	640	12.3	448.9	24.7	1.79	3.18	
	FC	80	9	32.5	1.42	1.14	0.43	20x

The overhead varies a lot due to that the register resource and distribution are quite different between benchmarks. For example, *stencil2d* and *gemm* have short state with abundant registers in states. In the baseline, the frequent state results back-up causes a mass of bit operations. For the proposed design, in $10ms$ and $20ms$ interval, the frequent intermediate data back-up cause relatively high overhead. But this overhead decreases rapidly when power gets better. Though in benchmarks *stencil3d* and *dfadd*, the proposed design is inferior to the baseline in $20ms$ and $50ms$. The baseline fails when power outage occurs in less than $20ms$ interval. FC-FPGA can successfully move forward with an expense of more bit operations in $10ms$ and its performance is superior when power gets better.

Table III
RESOURCE UTILIZATION

Benchmarks	flip-flop	flag-counter ff	BRAM_18K	ff overhead
stencil 2d	186	82	1	44%
stencil3d	549	46	1	8.4%
gemm	1168	142	2	12.1%
fir	233	14	1	6%
dfadd	287	10	1	3.5%
global	340	34	1	10%
Average				14%

The resource utilization of FC-FPGA is shown in Table III. In this table, the number of flip-flops utilized to build flag counters and the number of block ram are listed. An average of 14% extra flip-flops of original circuit is observed. $BRAM\_18K$ block ram is reserved for the NV-FF controller to store slice address according to the number of registers in the benchmark. The running power of the flag counters and the controller ram is less than 1% of total power in simulation. If pipeline mode is adopted, the FC-FPGA can still maintain good performance by assigning multiple flag counters to a state while the baseline needs to modify its technique to achieve intermittent computing.

VII. CONCLUSION AND FUTURE WORK

This paper proposes FC-FPGA with an off-line analysis and an on-line tracking methodology to store intermediate data for the upcoming computation. We propose novel checkpoint strategy that minimizes the re-computation time and operation overhead on flip-flops for NV-FPGA under unstable power. Our evaluation results show that the FC-FPGA can survive in harsh environment and it has a self adaptivity to different power circumstance. In the future, we will optimize the usage of flip-flops in flag counter and FC-FPGA in pipeline mode to further decrease the resource consumption and increase the performance.

ACKNOWLEDGMENT

This work is supported by the National Science Foundation under Grant Numbers NSF CCF-1820537, NSF CCF-1527464 and is partially supported by a grant from the Research Grants Council of the Hong Kong Special Administrative Region, China (Project No. CityU 11278316).

REFERENCES

[1] L. Semiconductor. ice40 lp/hx/lm. [Online]. Available: http://www.latticesemi.com/Products/FPGAandCPLD/iCE40.aspx

[2] X. Tang, P.-E. Gaillardon, and G. De Micheli, "A high-performance low-power near-vt rram-based fpga," in *2014 International Conference on Field-Programmable Technology (FPT)*. IEEE, 2014, pp. 207–214.

[3] X. Tang, G. Kim, P.-E. Gaillardon, and G. De Micheli, "A study on the programming structures for rram-based fpga architectures," *IEEE Transactions on Circuits and Systems I: Regular Papers*, vol. 63, no. 4, pp. 503–516, 2016.

[4] W. Zhao, E. Belhaire, C. Chappert, and P. Mazoyer, "Spin transfer torque (stt)-mram–based runtime reconfiguration fpga circuit," *ACM Transactions on Embedded Computing Systems (TECS)*, vol. 9, no. 2, p. 14, 2009.

[5] X. Guo, E. Ipek, and T. Soyata, "Resistive computation: avoiding the power wall with low leakage, stt-mram based computing," in *ACM SIGARCH Computer Architecture News*, vol. 38, no. 3. ACM, 2010, pp. 371–382.

[6] P.-E. Gaillardon, M. H. Ben-Jamaa, M. Reyboz, G. B. Beneventi, F. Clermidy, L. Perniola, and I. O'Connor, "Phase-change-memory-based storage elements for configurable logic," in *2010 International Conference on Field-Programmable Technology (FPT)*. IEEE, 2010, pp. 17–20.

[7] Y. Chen, J. Zhao, and Y. Xie, "3d-nonfar: three-dimensional non-volatile fpga architecture using phase change memory," in *Proceedings of the 16th ACM/IEEE international symposium on Low power electronics and design*. ACM, 2010, pp. 55–60.

[8] A. Lee, C.-P. Lo, C.-C. Lin, W.-H. Chen, K.-H. Hsu, Z. Wang, F. Su, Z. Yuan, Q. Wei, Y.-C. King, and C.-J. Lin, "A reram-based nonvolatile flip-flop with self-write-termination scheme for frequent-off fast-wake-up nonvolatile processors," *IEEE Journal of Solid-State Circuits*, vol. 52, no. 8, pp. 2194–2207, 2017.

[9] T. Xue and S. Roundy, "Analysis of magnetic plucking configurations for frequency up-converting harvesters," in *Journal of Physics: Conference Series*, vol. 660, no. 1. IOP Publishing, 2015, p. 012098.

[10] K. Ma, X. Li, J. Li, Y. Liu, Y. Xie, J. Sampson, M. T. Kandemir, and V. Narayanan, "Incidental computing on iot nonvolatile processors," in *Proceedings of the 50th Annual IEEE/ACM International Symposium on Microarchitecture*. ACM, 2017, pp. 204–218.

[11] C. Pan, M. Xie, Y. Liu, Y. Wang, C. J. Xue, Y. Wang, Y. Chen, and J. Hu, "A lightweight progress maximization scheduler for non-volatile processor under unstable energy harvesting," in *Proceedings of the 18th ACM SIGPLAN/SIGBED Conference on Languages, Compilers, and Tools for Embedded Systems*. ACM, 2017, pp. 101–110.

[12] A. Mirhoseini, B. D. Rouhani, E. Songhori, and F. Koushanfar, "Chime: Checkpointing long computations on intermittently energized iot devices," *IEEE Transactions on Multi-Scale Computing Systems*, vol. 2, no. 4, pp. 277–290, 2016.

[13] Z. Yuan, Y. Liu, H. Li, and H. Yang, "Cp-fpga: Computation data-aware software/hardware co-design for nonvolatile fpgas based on checkpointing techniques," in *Design Automation Conference (ASP-DAC), 2016 21st Asia and South Pacific*. IEEE, 2016, pp. 569–574.

[14] H.-S. P. Wong, H.-Y. Lee, S. Yu, Y.-S. Chen, Y. Wu, P.-S. Chen, B. Lee, F. T. Chen, and M.-J. Tsai, "Metal–oxide rram," *Proceedings of the IEEE*, vol. 100, no. 6, pp. 1951–1970, 2012.

[15] P.-E. Gaillardon, D. Sacchetto, G. B. Beneventi, M. H. B. Jamaa, L. Perniola, F. Clermidy, I. O'Connor, and G. De Micheli, "Design and architectural assessment of 3-d resistive memory technologies in fpgas," *IEEE Transactions on Nanotechnology*, vol. 12, no. 1, pp. 40–50, 2013.

[16] J. Cong and B. Xiao, "mrfpga: A novel fpga architecture with memristor-based reconfiguration," in *2011 IEEE/ACM International Symposium on Nanoscale Architectures*. IEEE, 2011, pp. 1–8.

[17] S. Onkaraiah, M. Reyboz, F. Clermidy, J.-M. Portal, M. Bocquet, C. Muller, Hraziia, C. Anghel, and A. Amara, "Bipolar reram based non-volatile flip-flops for low-power architectures," in *2012 IEEE 10th International New Circuits and Systems Conference (NEWCAS)*. IEEE, 2012, pp. 417–420.

[18] K. Jabeur, G. Di Pendina, and G. Prenat, "Ultra-energy-efficient cmos/magnetic nonvolatile flip-flop based on spin-orbit torque device," *Electronics Letters*, vol. 50, no. 8, pp. 585–587, 2014.

[19] T.-K. Chien, L.-Y. Chiou, Y.-C. Chuang, S.-S. Sheu, H.-Y. Li, P.-H. Wang, T.-K. Ku, M.-J. Tsai, and C.-I. W. Wu, "A low store energy and robust reram-based flip-flop for normally off microprocessors," in *2016 IEEE International Symposium on Circuits and Systems (ISCAS)*. IEEE, 2016, pp. 2803–2806.

[20] Xilinx. Xilinx floorplanning methodology guide. [Online]. Available: https://www.xilinx.com/

[21] A. Canis, J. Choi, M. Aldham, V. Zhang, A. Kammoona, T. Czajkowski, S. D. Brown, and J. H. Anderson, "Legup: An open-source high-level synthesis tool for fpga-based processor/accelerator systems," *ACM Transactions on Embedded Computing Systems (TECS)*, vol. 13, no. 2, p. 24, 2013.

978-1-5386-7100-9/18 $31.00 © 2018 IEEE

Pixel-Parallel Architecture for Neuromorphic Smart Image Sensor with Visual Attention

Md Jubaer Hossain Pantho, Pankaj Bhowmik, and Christophe Bobda
University of Arkansas, Fayetteville, Arkansas, USA
{mpantho, pbhowmik, cbobda}@uark.edu

Abstract—Power reduction and speedup of computer vision designs remain of high interest as image resolutions continue to increase. Neuromorphic-circuits, emulating the behavior of the nervous system, aspire to achieve this goal. In this paper, we present a pixel-parallel 3D-architecture of a neuromorphic image sensor that uses different sampling frequencies in different regions of an image. We design the model as a bottom-up 3D-architecture composing of several hierarchical computational planes where each plane performs different image processing algorithms in parallel. The on-chip attention module dynamically detects regions with relevant information and produces a feedback path to sample those regions with a higher clock frequency, whereas regions with low spatial and temporal information receive less attention. The results show that by sampling non-relevant regions with a lower frequency, the sensor can reduce redundancy and enable high-performance computing at low power.

Keywords-Image Sensors, Visual Attention, Neuromorphic, FPGA, Predictive Coding

I. INTRODUCTION

Cameras are pervasively used for surveillance and monitoring applications and can capture a substantial amount of image data [1][2]. The processing of this data, however, is either performed post-priori or at powerful backend servers. While post-priori and non real-time video analysis may be sufficient for certain groups of applications, it does not suffice for applications such as accident determination, distracted driving detection image analysis using cameras on drones, that require near real-time video and image analysis, sometimes under SWAP (Size Weight and Power) constraints. Given the raw amount of data captured from cameras and the lack of reliable high bandwidth wireless connectivity that can facilitate the transfer of image data to backend servers, we hypothesize that future data challenges in camera sensors can be overcome by pushing computation *closer* to the image sensor. Such systems will exploit the massively parallel nature of sensor arrays to reduce the amount of data analyzed at the processing unit. To this end, vertically integrated technology, such as focal plane sensor processors (FPSP), have been developed to overcome the limitations of conventional image processing systems. These systems use separate chips for sensor and processor with serial readout in between. Research on FPSPs has mostly focused on technology aspects with some proof of concepts. Vision sensors that incorporate general-purpose digital processors on the focal plane have been a subject of a number of commercial developments. While these devices

are re-programmable and offer the benefits of in-sensor processing such as performance and bandwidth reduction, they exhibit many drawbacks. For instance, each column of pixels is handled by a single processor, which reduces the parallelism and all pixels are treated equally and processed at the same rate, despite differences in input relevance for the application at hand. Consequently, systems spend more time spinning on non-relevant data, which increases sensing and computation time, and power consumption. System on chip designs that incorporate hardware accelerators has been considered a viable solution in recent years to provide in-situ efficiency in image processing applications [3][4][5][6].

However, the conventional hardware accelerators execute in a sequential pixel read-out manner, which restricts the architecture from exploiting the full extent of the image's parallel nature. In the areas, where high-speed image acquisition is required, a fast collection of image pixels and processing should be ensured at the minimum span of time. In order to achieve this goal, highly parallel architecture design is inevitable. To overcome the limitations of existing architectures, we present in this paper the design of a highly parallel, hierarchical, reconfigurable and vertically-integrated 3D sensing-computing architecture for real-time, low-power video analysis. To increase performance, while reducing power, the proposed architecture leverages two well-known approaches (saliency and predictive coding) used by the brain to reduce redundancy and deploy more resource on important parts of scene images. While visual attention is used by the brain to rapidly detect and deploy more resources to salient parts of a given scene [7], predictive coding [8] allow the brain to remove redundancy and transfer only useful information to high-level parts of the brain for further processing. These two paradigms are implemented in the brain in a chain of fast feedforward signals that carry information to high-level part of the brain, while feedback signal provides configuration to lower part of the visual cortex [9].

The architecture we present in this paper exploits saliency and predictive coding found in the brain, along with maximal parallelism, thus resulting in a hierarchical image processing hardware architecture made of computational units that reside in three inter-twined logical planes as illustrated in Figure 1. The first plane consists of fine-grained reconfigurable components that collaboratively analyze a collection of pixels to detect unpredicted regions of the input image. The results of this step are fed into the next

plane where low-level features such as edges, corners and lines, and rectangles and circles are extracted to build a map of salient events in an image. The map is then searched for events and objects in the third plane of computation. We detect saliency by incorporating the concept of predictive coding in space and time to identify unpredicted regions of the image. This knowledge is combined with the early features of different regions to calculate relative saliency. Relevant regions, which are dynamically detected with relevant information, are assigned higher sampling frequency by producing a feedback path to the input plane. To check the viability of our design, we simulated our architecture with 60×60 input images on a Virtex-7 [10] FPGA. The results show that by trading off resource overhead we can obtain high throughput while reducing redundancy and power consumption. The proposed architecture can be applied to a large set of image processing applications where real-time operation is needed.

II. RELATED WORK

We start this section by reviewing some architectures that have been developed to reduce redundancy and to overcome the limitation of conventional image processing systems. Then we discuss different studies on visual attention approaches.

Streaming data reduction is performed in image processing systems using region of interest (ROI) based strategies to limit the computation of data only to regions with high relevance. In collaborative tracking systems such as [11], profiled data of tracked targets are built and sent across the network to reduce the communication bandwidth. Many ROI-based computations [12] are used in image data compression to further reduce the amount of data to be transported, thus increasing the compression ratio, while reducing transport data. Despite the reduction in transport data, ROI-based approaches do not impact the pixel sensing and require a separate processor for this step.

To allow systems to focus on the most important regions, computational models of visual attention have been proposed in many variations [8][13][4] to emulate the human reaction to scene events. Itti and Koch [13] discussed a brain-like model in order to generate a saliency map. They extracted orientation, intensity, and color information of image frames to generate a conspicuity map and implemented a winner take all method (WTA) to calculate relative saliency of different regions. Later on, Barranco [14] extended this idea and incorporated motion as additional features. To solve the issue at a lower level, Indiveri [15] proposed a neuromorphic VLSI model for selective attention. However, in these works, application-specific architectural design approaches as proposed in this paper for mapping applications to low-level architectures have not been investigated.

In order to explain redundancy reduction in the biological vision system, Huang and Rao [8] introduced a different notion and introduced predictive coding to explain efficient coding in the nervous system. They stated a hierarchical

predictive coding model to describe the neural activities of the visual nervous system. We borrow the concept of their framework to reduce redundant information in the incoming pixels of our architecture. However, we do not limit ourselves by only incorporating predictive coding and further extend our design to develop a 3D bottom-up architecture with visual attention.

III. PREDICTIVE CODING IN NERVOUS SYSTEMS

Since our architecture leverages the concept of predictive coding to reduce redundancy, in this section, we provide a primer on predictive coding to understand the response of neurons in various regions of the nervous system.

In natural images, neighboring pixel intensities tend to be positively correlated. Similarly, pixel intensities are correlated over time because objects persist in time. Based on the information theoretic considerations, the role of early sensory processing is to reduce redundancy [8] [7]. One important model that provides a functional explanation of this behavior is predictive coding. According to this, the visual system uses an internal model to predict incoming signals and attempts to reduce redundancy by removing the predictable components of the input by transmitting only what is not predictable. The response of neurons in the visual system can be understood as a hierarchically organized model with reciprocal connections between cortical areas [8]. In this model, neural circuits in the retina/LGN(Lateral geniculate nucleus) actively predict the local intensity value from a linear weighted sum of nearby values over space or time. The next subsections describe this in detail.

A. Predictive Coding in Space

Predictive coding in space provides an explanation for the spatial receptive fields found in the retina. In an array of neurons in the retina, each of which gets an excitatory input from the center and an inhibitory input from the neighbors. The response at the center of the receptive field is estimated from a linear weighted average of surrounding intensity values. The formula to predict the pixel intensity x_0 based on the neighboring pixel intensities is illustrated in equation 1

$$x_0 = \sum_{i=1} (w_i x_i) \qquad (1)$$

Here, x_i represents the intensity value of ith neighbor and w_i represents the weights of the corresponding neighbor. The optimal weights w_i can be obtained by minimizing the total prediction error over all pixels. This is shown in equation 2

$$E = ||A - BW||^2 / 2 \qquad (2)$$

Here, E presents the error vector and W represents the vector of weights w_i. A and B represent the matrix of center pixel intensities and neighboring pixel intensities. Taking the derivative of E with respect to W and solving for the optimal linear weights, we obtain:

978-1-5386-7100-9/18 $31.00 © 2018 IEEE

$$W = (B^T B)^{-1} B^T A \qquad (3)$$

The shape of the weighting function that minimizes the error between the estimated intensity value and its actual value closely resembles the classical center-surround receptive fields of retinal ganglion cells.

B. Predictive Coding in Time

Predictive coding in time provides an explanation for the temporal receptive fields found in the retina and LGN. Time-varying natural image sequences exhibit strong positive inter-frame correlations. Given that, temporally close intensities tend to be positively correlated, the intensity value of a pixel at time t can be predicted from the weighted sum of intensity values of the preceding pixels of that point. This follows a similar formula like equation 1. In this assumption, the retina is assumed to reduce most of the spatial redundancy, whereas the LGN removes the temporal ones from the input image. A detailed discussion on predictive coding in the nervous system is beyond the scope of this paper. Interested readers are encouraged to refer to cited work [8][16].

In our 3D bottom-up architecture, we leveraged the concept of predictive coding to detect unpredicted regions of the input image. The underlying assumption is, these regions hold more information compared to the rest of the image. This information is combined with the extracted early features in the consecutive stages to identify salient regions. The next section will discuss our architecture in detail.

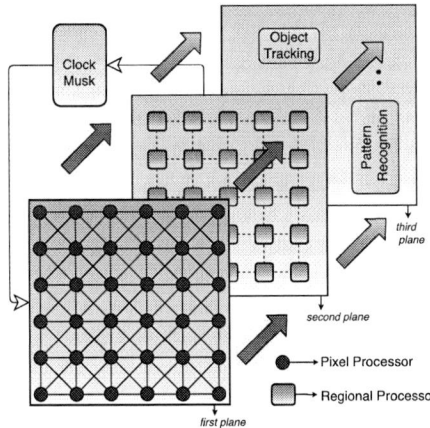

Fig. 1. Overview of the 3-D bottom up architecture of the smart image sensor. The computational units are organized in planes, where the output of each layer serves input for the next one. Each PPU is connected to its eight neighboring PPUs.

IV. PROPOSED ARCHITECTURE

In this section, we provide a detailed description of our smart image sensor architecture.

We modeled our system as a 3D bottom-up architecture of several computational planes connected in a hierarchical fashion with feedforward and feedback connections among them. These computational planes perform different image processing algorithms in a highly parallel manner, where each plane provides input for the next plane. Figure 1 provides the overview of our architecture.

The first plane serves as the input plane taking input from the CMOS imager. In this plane, the model tries to detect regions with unpredicted pixel values using spatial information. On the second plane, we extract early features of the image. We combine this information with the spatial information from the first plane to calculate the most saliency regions of the incoming image. A feedback connection to the input plane carries this regional data to assign the clock frequencies in different regions. The output of this second plane can serve as an input for a third plane where we can perform high-level image processing tasks i.e. object detection, pattern recognition. The main innovation of this hierarchical architecture is the maximal parallelism provided vertically and horizontally within and across processing planes.

1) First Plane Design (PLPP): The function of the first plane is to identify unpredicted pixel regions from the spatial information of the input pixels. We call this plane the Pixel-Level Processing Plane (PLPP). This is the first plane directly interacting with the image sensor. This plane comprises a two-dimensional array of Pixel Processing Units (PPU) where each unit is connected to photo-sensor elements through registers. The function of the PLPP is to perform predictive coding in space to mimic the role of the receptive cells in the retina. This is done in a highly parallel fashion by predicting the value of a pixel from its neighbor pixels. It is assumed that the intensity value at the center pixel can be estimated by assigning linear weights of the surrounding pixels as shown in equation (1). To accomplish this each PPU takes a center pixel and its eight neighbor pixels in a 3x3 neighborhood as input. The value of the center pixel is predicted from the weighted average of its eight neighbor pixels. If the value of the center pixel is approximately the same as the predicted value we assume that this is a homogeneous pixel region. A mismatch refers to an unpredicted pixel position. The PPU forwards the center pixel unaltered to the next plane where early visual features are extracted. To mark a misprediction the PPU will produce an error bit as well. Since image pixels are highly correlated and a single misprediction does not mean anything we grouped the PPUs in a number of regions and placed a Region Processing Unit (RPU) for each region. These RPUs take the error bits within each region as inputs and decide whether the region is an unpredicted region or not. This is done by calculating the number of misprediction in a region. If this value is greater than some threshold the regional processor treats this region as an unpredicted portion of the image. This information is then conveyed to the regional processor of the next plane in the hierarchical order by a single bit per region.

The parallel nature of this plane enables simultaneous

computation on all the PPUs. In order to facilitate parallel computation in this plane, the image acquisition process in the system can be performed by a highly parallel CMOS imager as proposed in [17]. The CMOS imager will give a parallel readout of an image. This is a per pixel architecture and the output of the ADC is fed to the respective PPU and its nearby neighbors.

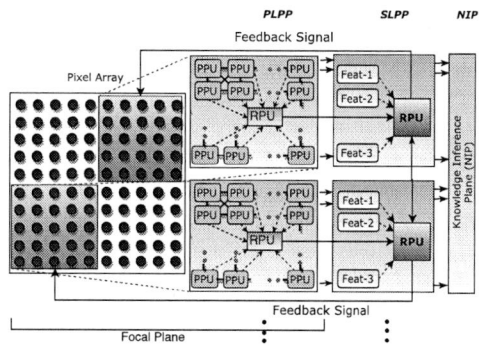

Fig. 2. PPU-RPU interconnection structure with feedforward and feedback connections. The RPUs of the SLPP layer send feedback to the input pixel array to adjust sensing and processing frequency.

2) Second Plane Design (SLPP): The objective of the second plane is to extract early visual features from the input image and detect the salient regions by combining different feature information. We call this plane the Structure-Level Processing Plane (SLPP). This plane takes input from the PLPP plane and produces appropriate output image data for the next plane in the hierarchy. The SLPP layer comprises a coarse grained 2D array of RPUs, where each RPU performs operation on a group of pixels forwarded by the PLPP-PPU. Each group of PPUs on the first plane corresponds to a single RPU on the second plane. RPUs in SLPP can communicate with neighboring RPUs to calculate relative saliency. This is illustrated in figure 2. The processors in this layer operate on a broader level. RPUs in this level extract feature information according to the application. Our current implementation extracts edge and corner information as early features from the image.

Besides, we perform predictive coding in time to emulate the role of LGN in the nervous system. This helps us to identify the static regions of image frames. Since there exist a strong inter-frame correlation in a time-varying natural image sequences. Pixel values in the current frame can be predicted from its earlier frame. Motion in an image region will cause misprediction on the PPUs. We assumed that time-varying pixel regions will receive higher visual attention in image frames. A frame buffer is used to hold the previous frame and compare the incoming pixels with them. A right prediction indicates that there is no temporal change. This is particularly important to reduce redundancy over time as this enable us to scale down sensing and processing frequency of static regions over time.

These error bits along with the edge and corner infor-

mation are forwarded to a regional attention module. This module takes into account the spatial information calculated in the first plane as well. Here, the feature information is combined together to detect saliency. We introduced a scoring method called saliency-score to detect relevance. The saliency-score is calculated using the weighted summation of the feature vectors. This is shown in equation 4.

$$W = W_1 k_1 + W_2 k_2 + W_3 k_3 + W_4 k_4 \qquad (4)$$

Here, $k_1, k_2...$ denotes feature information (e.g. number of edge or corner pixels) and W_1, W_2 . . . denotes the corresponding weights. Regions with higher saliency score will be interpreted as salient regions of the image. For example, an image region with a high number of edge/corner pixels but with no misprediction in time will still be considered as a salient region if the saliency score is high. This information is then used to send a feedback signal to the first plane to enable multi-frequency sampling.

3) Relevance-feedback: In our architecture, we have designed a relevance-feedback path that schedules processing resources according to their importance. This allows the system to process important pixels at a high rate. A comparatively lower sampling frequency is assigned to the less important pixel regions. The key benefit of this feature is that all the pixels are processed in parallel but all PPUs are not functional at the same time. So we are saving active resources, heat dissipation, and power consumption which is one of the most desired features the design provides. The model starts by sensing all pixels at the same rate. As the relevance of pixels starts decreasing, the sensing frequency decreases as well. These feedback connections are implemented on a regional basis, where the outputs of the SLPP RPUs serve as the selector for the clocking operation.

4) Third Plane Design (NIP): The output of the SLPP can be combined to infer knowledge of the scene in the third plane. We call this plane the Knowledge Inference Plane (NIP). As oppose to PLPP and SLPP, which operate on a large amount of data, the features will be vastly reduced in the NIP due to the early feature extraction in the previous planes. The integration of our relevance-feedback method can further limit features to only relevant regions of an image.

This can enable fast real-time computing of high-level image processing application even with a low power sequential processor in the NIP. For example, in a lane detection system of an autonomous vehicle, our architecture can provide an edge detected binary image

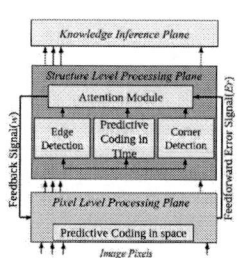

Fig. 3. Design flow chart.

978-1-5386-7100-9/18 $31.00 © 2018 IEEE   248

with relevant region information to the NIP to identify lanes. Since the maximum novelty of our architecture lies on the structure of the first two planes, In this implementation, we only prototyped the PLPP and SLPP with the feedback connections. Figure 3 provides the flow chart of our implemented design.

V. EXPERIMENTAL RESULTS

This section presents the implementation details of our prototype. We then discuss the performance of our design.

A. Implementation Details

To test the viability and behavior, we prototyped the first two planes of our architecture with the relevance-feedback path on an FPGA. Besides, we extracted the basic constraints by designing the layout on Innovus [18]. In the PLPP plane, PPUs perform predictive coding in space on 8-bit grayscale image with a 3-by-3 window. Since PPUs do not have any prior knowledge of the input image pixels we used equal weights for all the neighboring pixels as shown in equation 1. We found that this structure provides a good prediction of homogenous backgrounds. The regional processors in this layer calculate a number of misprediction in a certain region. The information is conveyed to the RPU of SLPP through a single bit. On the SLPP, we calculated edges (Sobel) and corners (Moravec) of the input image frames as early features. We select these methods because they require fewer hardware resources. Besides, predictive coding in time is performed to identify static and non-static regions. This is done by holding the previous frame pixels on a frame buffer. In this implementation, each RPU in SLPP operates on a 10-by-10 region size and generates results in parallel to the next plane.

The attention modules within the RPUs are responsible for calculating saliency of different regions. They communicate with the neighboring RPUs to calculate relative saliency. The visual attention module calculates the saliency-score by measuring the weighted sum of the feature vector as shown in equation 1. We set these weights empirically based on the test image data. For example, since edge pixels are more frequent in an image compared to the corner pixels, we assign a higher weight to corners. Primarily, we used two distinct clock frequencies(100MHz & 10 MHz) for sensing and processing different regions. This means all the processors in the irrelevant regions will be active only one-tenth of the time compared to the relevant regions. The proposed model starts by sensing all pixels at the same frequency. As the relevance of pixels starts decreasing the sensing frequency decreases as well as the processing frequency of the pixel processing units. This ensures a reduction in power and heat dissipation without missing important events in the image stream. To illustrate power reduction on PLPP layer we clocked the PPU with different clock frequency and observed power consumtion. This is shown in Figure 4. The result shows that with a clocking

ratio of 10, we can save 89.23% of power consumption on irrelevant regions.

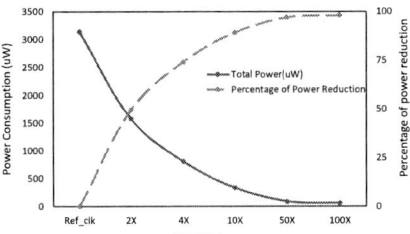

Fig. 4.   Power reduction on PPU with increasing clock frequency.

B. Results

Table I shows the resource utilization of our architecture on a Virtex-7 FPGA. Due to the available resource constraints

TABLE I
RESOURCE UTILIZATION

-	PLPP	SLPP
Resource	**Utilization/Available**	**Utilization/Available**
LUTs	227216 (303600)	38916 (303600)
FF	88954 (607200)	6984 (607200)

on the FPGA we limit the frame size to 60×60 images. However, this should be noted that this is a pixel-parallel architecture, which implies that an increase in image size will only mean the duplication of the same primary computational unit for each pixel. Since PLPP-PPUs are the smallest independent computational units performing operation on the input pixels all in parallel, it can be understood that the area of the PPU units determines the total area size of our architecture. We perform the layout design of a single PPU in Innovus. Table II summarizes the extracted parameters of a PPU in 90nm technology (1 GHz clock).

TABLE II
PARAMETERS OF THE PPU

Area	Power	No. of Cells	Cell Density	Leakage Power
$596.3 um^2$	0.56uW	457	70.06%	0.012uW

To get data on the power consumption saving by relevance feedback method, we designed a relevance estimation model using only corner information and tested it on pictures with various amount of important features. The percentage of power saving relative to the image information is shown in Table III. We can see that further power saving is achieved on images with less relevant regions. Moreover, we prototyped our architecture on a 2D planar model. An actual 3D architecture design can achieve a reduction in power consumption by 50% over a 2D architecture [19].

We measured the computation time to calculate the processing overhead of our architecture and compared

TABLE III
PERCENTAGE OF POWER SAVING RELATIVE TO THE AMOUNT OF FEATURE INFORMATION

no. of corner points	low	medium	high
max power saving in %	41	30	12

the result with a conventional hardware accelerator with sequential read-out data. To keep the comparison fair, we executed the hardware accelerator on the same FPGA and extracted the early features with a sliding window. Table IV provides the performance comparison. It can be seen that our design provides considerable improvement over the hardware accelerator in terms of processing frames. However, it should be noted that the frame rate of the hardware accelerator will decrease with the increase in frame size [4]. On the other hand, the computation time of our per-pixel architecture depends only on the region size of the SLPP-RPU. For a given region size, the model will provide the same frame rate for different image sizes.

TABLE IV
COMPARISON OF PROCESSING TIME

Computation time	Our Design (100MHz)	HW Accelerator (100MHz)
per frame	0.006ms	1.03ms

Figure 5 shows the simulation behavior of our visual attention module. The obtained results in Figure 5 shows that ROI is detected reasonably without any a-priori knowledge.

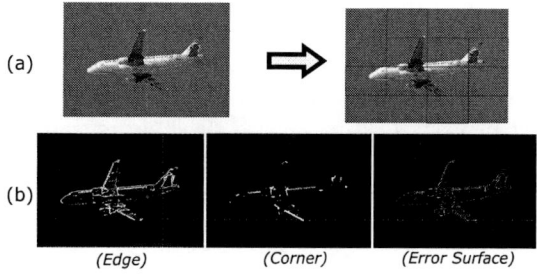

Fig. 5. visual attention module analysis. (a) Input image(left), image with identified relevant regions(right) (b) (left-to-right) edge detected image, corner image, error surface for spatial prediction. The model detects these regions by combining the information in (b) images.

VI. CONCLUSION

This paper presents a pixel parallel 3D architecture of a neuromorphic image sensor, designed as a 3D bottom-up architecture composing of several computational planes where each plane performs different image processing algorithms in a highly parallel manner. The model emulates the hierarchical process in biological vision by providing feedforward and feedback information flow between different planes. The designed hierarchical attention module

dynamically detects regions with relevant information and produce a feedback path to sample those regions with a higher clock frequency. The results show that by trading off resource overhead we can obtain high throughput while reducing redundancy and power consumption.

VII. ACKNOWLEDGEMENT

This work was supported by NSF under Grant 1618606.

REFERENCES

[1] T. Winkler and B. Rinner, "Applications of trusted computing in pervasive smart camera networks," in *Proceedings of the 4th Workshop on Embedded Systems Security*, ser. WESS '09. New York, NY, USA: ACM, 2009, pp. 2:1–2:10.

[2] F.-J. Streit, M. J. H. Pantho, C. Bobda, and C. Roullet, "Vision-based path construction and maintenance for indoor guidance of autonomous ground vehicles based on collaborative smart cameras," in *Proceedings of the 10th International Conference on Distributed Smart Camera*, ser. ICDSC '16. New York, NY, USA: ACM, 2016, pp. 44–49.

[3] J. Anders *et al.*, "A hardware/software prototyping system for driving assistance investigations," *Journal of Real-Time Image Processing*, vol. 11, no. 3, pp. 559–569, Mar 2016.

[4] M. S. Park, C. Zhang, M. DeBole, S. Kestur, V. Narayanan, and M. J. Irwin, "Accelerators for biologically-inspired attention and recognition," in *2013 50th ACM/EDAC/IEEE Design Automation Conference (DAC)*, May 2013, pp. 1–6.

[5] B. da Silva, A. Braeken, E. H. D'Hollander, A. Touhafi, J. G. Cornelis, and J. Lemeire, "Comparing and combining gpu and fpga accelerators in an image processing context," in *2013 23rd International Conference on Field programmable Logic and Applications*, Sept 2013, pp. 1–4.

[6] V. Dworak *et al.*, "Strategy for the development of a smart ndvi camera system for outdoor plant detection and agricultural embedded systems," *Sensors*, vol. 13, no. 2, pp. 1523–1538, 2013.

[7] R. Desimone and J. Duncan, "Neural mechanisms of selective visual attention," *Annual Review of Neuroscience*, vol. 18, no. 1, pp. 193–222, 1995.

[8] Y. Huang and R. P. N. Rao, "Predictive coding," *Wiley Interdisciplinary Reviews: Cognitive Science*, vol. 2, no. 5, pp. 580–593, 2011.

[9] G. Michalareas *et al.*, "Alpha-beta and gamma rhythms subserve feedback and feedforward influences among human visual cortical areas," *Neuron*, vol. 89, pp. 384–397, 2016.

[10] *VC707 Evaluation Board for the Virtex-7 FPGA*, Xilinx, 8 2016, v1.7.1.

[11] Y. Wang, S. Velipasalar, and M. Casares, "Cooperative object tracking and composite event detection with wireless embedded smart cameras," *IEEE Transactions on Image Processing*, vol. 19, no. 10, pp. 2614–2633, Oct 2010.

[12] Z. Chen, G. Barrenetxea, and M. Vetterli, "Event-driven video coding for outdoor wireless monitoring cameras," in *2012 19th IEEE International Conference on Image Processing*, Sept 2012, pp. 1121–1124.

[13] L. Itti and C. Koch, "Computational modelling of visual attention," *Nat Rev Neurosci*, vol. 2, no. 3, pp. 194–203, 2001, pMID: 11256080.

[14] F. Barranco, J. Diaz, B. Pino, and E. Ros, "Real-time visual saliency architecture for fpga with top-down attention modulation," *IEEE Transactions on Industrial Informatics*, vol. 10, no. 3, pp. 1726–1735, Aug 2014.

[15] D. Sonnleithner and G. Indiveri, "A neuromorphic saliency-map based active vision system," in *2011 45th Annual Conference on Information Sciences and Systems*, March 2011, pp. 1–6.

[16] R. Bogacz, "A tutorial on the free-energy framework for modelling perception and learning," *Journal of Mathematical Psychology*, vol. 76, no. Part B, pp. 198 – 211, 2017, model-based Cognitive Neuroscience.

[17] B. Tyrrell *et al.*, "Time delay integration and in-pixel spatiotemporal filtering using a nanoscale digital cmos focal plane readout," *IEEE Transactions on Electron Devices*, vol. 56, no. 11, pp. 2516–2523, Nov 2009.

[18] Cadence, "Innovus Implementation System," https://www.cadence.com, 2017, [Online; accessed 12-March-2018].

[19] M. Swaminathan, "Electrical design and modeling challenges for 3d system integration," in *Design Conference 2012*, 2012.

978-1-5386-7100-9/18 $31.00 © 2018 IEEE

2018 IEEE Computer Society Annual Symposium on VLSI

Lightweight ASIC Implementation of AEGIS-128

Anubhab Baksi, Vikramkumar Pudi, Swagata Mandal, Anupam Chattopadhyay

School of Computer Science and Engineering, Nanyang Technological University, Singapore

anubhab001@e.ntu.edu.sg, pudi@ntu.edu.sg, swagata.mandal@ntu.edu.sg, anupam@ntu.edu.sg

Abstract—In this paper, we study the problem of implementing the AEAD scheme, AEGIS-128, which is a finalist in the recently concluded competition, CAESAR. In order to achieve lightweight (least area) implementation, we first look into one round of AES encryption, which is a building block in this cipher. In this regard, we make use of the state-of-the-art implementation of AES in ASIC. We benchmark one round AES encryption (which is done for the first time) and later use it with AEGIS-128 to improve the optimized implementation reported (Inscrypt'14). Synthesis results show that our design requires 9.6% less area and reduces the power consumption by 95.3% (operating frequency is also reduced). Further, this concept can readily be applied to a variety of other ciphers.

Index Terms— ASIC, optimization, encryption, authentication

I. INTRODUCTION

Authenticated Encryption with Associated Data (AEAD) is a relatively recent concept in symmetric key cryptography where a cipher aims at providing confidentiality as well as authenticity. Given a message (referred to as, plaintext) and associated data; such a scheme encrypts the plaintext (confidentiality) and generates a tag depending on both the plaintext and the associated data (authentication). The sender (Alice) generates the ciphertext and the tag and sends them along with the associated data through an insecure channel (where the attacker, Eve, is active) to the recipient (Bob). Upon receiving the ciphertext and tag, bob generates the plaintext and the tag. If both the tags match, then the received ciphertext (hence the plaintext retrieved is original) and the associated data are considered not disturbed by Eve. Otherwise, Bob learns there is a disturbance caused by Eve; generates an invalid signal and discards the entire data received. For the transmission of plaintext, both confidentiality and authenticity is required; whereas only authenticity is required for the associated data. This is practical in scenarios like Internet packets, where the header is not confidential (acts as associated data) but the body is confidential (acts as plaintext).

To promote the research in this direction, a competition, named CAESAR[1] announces its finalists from 57 AEAD proposals. The cipher, AEGIS (currently version 1.1) [9] is among the 7 finalists (it is selected for high-performance applications). This cipher is based on the stream cipher design paradigm (XORing the plaintext with a pseudo-random string generated) to get the ciphertext, then the plaintext is fed into

We thank the anonymous reviewers for their useful feedback. The first author likes to acknowledge the kind support of Prakash Dey (CU, India); Sachin Kumar, Sonu Jha, Arko Dutt, Mustafa Khairallah (NTU, Singapore).

[1] https://competitions.cr.yp.to/index.html

the state. Besides, it uses one round of AES-128 encryption (see description in Section II-B) for its state update operation.

This cipher gains popularity among researchers. Apart from being an academic interest (e.g., TIAOXIN, another CAESAR candidate [7] is inspired by it); it is also being used in practical applications (e.g., in a vehicular communication [8]).

In various cryptographic operations, it is often required to have an optimized hardware (ASIC/FPGA) or software (microcontroller/microcontroller) implementation. The reason is, tightly contained devices (such as sensors) are generally required to be equipped with cryptographic primitives. Hence a major flow of cryptographic research focuses on optimized design for a specific cipher. In this paper, we propose an ASIC implementation of AEGIS-128 (one recommended version) which consume least area compared to the other implementation reported. For the sake of simplicity, we henceforth refer to AEGIS-128 as AEGIS, unless otherwise mentioned. Our implementation can handle both the encryption & tag generation and decryption & verification modes. We mention that, our design strategy is equally applicable to other variants of AEGIS.

The ASIC flow has already been explored in the literature [3]. The smallest area implementation they are able to achieve is $\approx$ 18720 GE (Optimization AO_1 in Section 3.2; sometimes referred to as $A0_1$) for 65 nm technology with and power consumption 21.58 mW and clock frequency 1410 MHz. However, as we point out later on, there are scopes to improve this implementation. Also, unlike our case, they do not implement decryption & verification circuit. Nonetheless, we reduce area requirement by 9.6% compared to [3].

One key idea of our improved implementation is, utilize the "Atomic AES v2.0" developed by Banik *et al.* [2] (a follow-up work of [1]), which is the state-of-the-art lowest area implementation of AES-128 in ASIC. While the implementation of [2] is for original AES-128 (both encryption & decryption), we only need one round of AES-128 encryption for AEGIS implementation. Hence, we amend the design so as to meet our (reduced) requirement. As for the FPGA, only benchmark of AEGIS is reported so far [8], to the best of our knowledge. Another benchmarking for AEGIS-128L, a larger version, is reported by ATHENA[2].

Our Contributions

Our contributions in this paper are twofold. First, we benchmark one round AES-128 encryption (optimized area) for ASIC. This is a building block in quite a few ciphers, such

[2] https://cryptography.gmu.edu/athena/index.php?id=CAESAR_source_codes

978-1-5386-7100-9/18 $31.00 © 2018 IEEE   251

as, AEGIS [9] or DEOXYS [6] (both are finalists in CAESAR). To the best of our knowledge, despite of its frequent usage, it has never been benchmarked under any hardware platform. Our results are shown in Section IV-A.

Second, we apply this area optimized one round AES-128 encryption to the AEAD cipher AEGIS. This helps us to reduce the ASIC area and power consumption significantly. Compared to the design in [3], ours requires less area (9.6%) and very less power (95.3%), however the operating frequency in our case is also reduced.

Paper Organization

The rest of the paper is organized as follows. Section II describes all the components of the AEGIS cipher. In Section III, we describe the ideas used in the previous works (one on AEGIS ASIC implementation, another on AES-128 encryption). Following these, we present how we implement our concept on improved ASIC version of AEGIS. The results we obtain are presented in Section IV. Finally, Section V concludes the paper, citing interesting future works.

II. DESCRIPTION OF AEGIS

A. Overview

Before proceeding further, here we present a quick overview of the AEGIS. It has the following parameters: 128-bit key, K; 128-bit initialization vector, IV; and 640-bit state. The state size is reduced from AEGIS-128L (which has a state size of 1024-bit), and it is suitable for high performance applications. The lengths of associated data, AD and the plaintext P (and hence the length of the ciphertext, C) are to be less than 2^{64}-bits. The recommended length of the authentication tag is 128-bit; although the design allows to generate tag of an arbitrary length ≤ 128. In AEGIS, both $|AD|$ and $|PT|$ are communicated to Bob.

B. One Round AES-128 Encryption

It is common to use one round of AES-128[3] encryption in AEAD design. In this part, we briefly describe the original AES-128 encryption. The cipher takes 128-bit input (plaintext) and 128-bit key, and generates 128-bit output (ciphertext).

The input is rearranged into a 4×4 matrix of bytes (referred to as the 'state'), which is normally accessed in the column-major way. The key is XORed with the state. The key is also passed to the 'key scheduling' algorithm, which generates 10 round keys – each round key is of 128-bits, and the process is reversible (given any round key, it is possible to recover the key).

The state undergoes 10 rounds of modifications. The first 9 rounds are identical, each consists of 4 steps in order: 'SubBytes', 'ShiftRow', 'MixColumn' and 'AddRoundKey'. The last (10th) round is little different, it only has SubBytes, ShiftRow and AddRoundKey (misses MixColumn).

The AddRoundKey step XORs the corresponding round key to the state. This step is performed at the very beginning

[3] https://nvlpubs.nist.gov/nistpubs/FIPS/NIST.FIPS.197.pdf

(before 1st round), where the key is XORed with the state (= plaintext). The rest steps are described below.

- SubBytes. This step contains the only non-linear operation in the cipher. Here, the input is converted with the help of a look-up table (known as, 'SBox').
- ShiftRow. This step affects the state row-wise. Each row of state is rotated by a fixed number of rotations.
- MixColumn. This step operates on the state column-wise. Each column of the state is considered a 4-dimensional vector, each element of which belongs to GF(2^8). A 4×4 matrix, M (whose elements are also in GF(2^8)), is multiplied to each column. The result thus obtained is used to substitute the old contents of the column. This process is repeated for all 4 columns.

The one round AES-128 encryption is a simplified version of the original which can use any of the one round from 1 to 9. Here, the 128-bit plaintext, after arranging to a 4×4 matrix of bytes, undergoes SubBytes, followed by ShiftRow, followed by MixColumn. Finally the key is XORed. Figure 1 shows one round AES-128 encryption. Henceforth, the first three combined steps are denoted by the notation $R(\cdot)$; so, $R(p) = \text{MixColumn}(\text{ShiftRow}(\text{SubBytes}(p)))$.

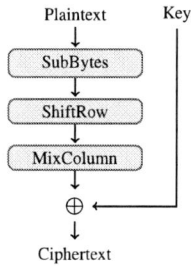

FIG. 1: One round AES-128 encryption

C. AEGIS State Update

The state of AEGIS consists of five 16-byte (= 128-bit) registers, denoted by S_0, S_1, S_2, S_3 and S_4, totaling 640-bits. Another 128-bit register M is also used.

Let us denote the jth register at i^{th} round by $S_{i,j}$, $j = 0(1)4$, and the content of the message register M as M_i. Then, the StateUpdate128(S_i, M_i) operation (which updates all 5 registers) is given as (recall definition of R from Section II-B):

$$
\begin{aligned}
S_{i+1,0} &\leftarrow R(S_{i,4}) \oplus S_{i,0} \oplus M_i; \\
S_{i+1,1} &\leftarrow R(S_{i,0}) \oplus S_{i,1}; \\
S_{i+1,2} &\leftarrow R(S_{i,1}) \oplus S_{i,2}; \\
S_{i+1,3} &\leftarrow R(S_{i,2}) \oplus S_{i,3}; \\
S_{i+1,4} &\leftarrow R(S_{i,3}) \oplus S_{i,4}.
\end{aligned}
$$

Figure 2 shows the structure of AEGIS state update operation.

978-1-5386-7100-9/18 $31.00 © 2018 IEEE

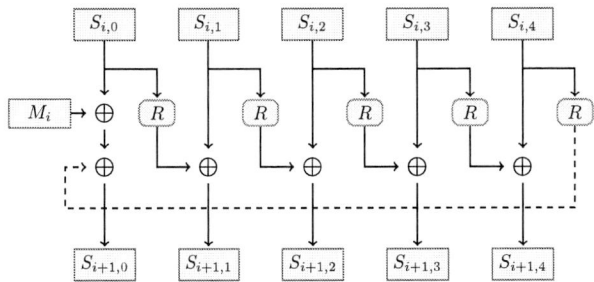

FIG. 2: AEGIS state update function ($\texttt{StateUpdate128}(S_i, M_i)$)

D. Initialization

During the initialization, the key (K) and initialization vector (IV) are loaded in the state, then the state is updated 10 rounds. The detailed procedure is as given below. Here $const_0$ and $const_1$ are two pre-defined 128-bit constants, $const_0 := 000101020305080d1522375990e97962$ and $const_1 := db3d18556dc22ff12011314273b528dd$.

1) Key, IV are loaded at first:

$$
\begin{aligned}
S_{-10,0} &\leftarrow K \oplus IV; \\
S_{-10,1} &\leftarrow const_1; \\
S_{-10,2} &\leftarrow const_0; \\
S_{-10,3} &\leftarrow K \oplus const_0; \\
S_{-10,4} &\leftarrow K \oplus const_1.
\end{aligned}
$$

2) For $i = -5(1) - 1$:

$$
\begin{aligned}
M_{2i} &\leftarrow K; \\
M_{2i+1} &\leftarrow K \oplus IV.
\end{aligned}
$$

3) For $i = -10(1) - 1$:

$$
S_{i+1} \leftarrow \texttt{StateUpdate128}(S_i, M_i).
$$

E. Associated Data Processing

After key and IV are loaded to the state, associated data (AD) is loaded. If $|AD| = 0$, then this step is skipped. If $|AD|$ is not a multiple of 128, then zero's are padded to AD to make its length a multiple of 128.

For $i = 0(1)u - 1$, the state update is done as given below, here $u = \lceil |AD|/128 \rceil$. We denote the ith 128-bit block of AD by AD_i.

$$
S_{i+1} \leftarrow \texttt{StateUpdate128}(S_i, AD_i).
$$

F. Encryption

The encryption operation follows, where the plaintext P is used to update state, and to produce the ciphertext C. If $|P| = 0$, then this step is skipped. If $|P|$ is not a multiple of 128, then zero's are padded so that $|P|$ becomes a multiple of 128.

For $i = 0(1)v - 1$, the state update is done as given below, $v = \lceil |P|/128 \rceil$. We use the notation, P_i (resp. C_i) to denote the ith 128-bit block of the plaintext P (resp. ciphertext C).

$$
\begin{aligned}
C_i &\leftarrow P_i \oplus S_{u+i,1} \oplus S_{u+i,4} \oplus (S_{u+i,2} \cdot S_{u+i,3}); \\
S_{u+i+1} &\leftarrow \texttt{StateUpdate128}(S_{u+i}, P_i).
\end{aligned}
$$

G. Finalization

After encryption, the authentication tag (T) is generated with seven more $\texttt{StateUpdate128}$ operations.

Let, $q = S_{u+v,3} \oplus (|AD| \,||\, |P|)$, where both $|AD|$ and $|P|$ are represented as 64-bit values. For $i = u + v(1)u + v + 6$, the following state update is done:

$$
S_{i+1} \leftarrow \texttt{StateUpdate128}(S_{u+i}, q).
$$

The authentication tag, T, is the first t-bits ($0 < t \le 128$, recommended value of t is 128) of $T' = \bigoplus_{i=0}^{4} S_{u+v+7,i}$

H. Decryption & Verification

For decryption, the procedure is similar. It first does the initialization and processing of associated data steps. Next, the ciphertext C is decrypted from the plaintext P. If $|C|$ is not a multiple of 128, then zero's are padded to make it a multiple of 128.

For $i = 0(1)v - 1$, the decryption and state update are done as follows:

$$
\begin{aligned}
P_i &\leftarrow C_i \oplus S_{u+i,1} \oplus S_{u+i,4} \oplus (S_{u+i,2} \cdot S_{u+i,3}); \\
S_{u+i+1} &\leftarrow \texttt{StateUpdate128}(S_{u+i}, P_i).
\end{aligned}
$$

The finalization step is same as its namesake in encryption. If the tag verification fails, then the ciphertext and the newly generated tags are not given as output.

III. OPTIMIZATION: MINIMIZING ASIC AREA

In this section, we elaborate our optimization concept that we use here to minimize the ASIC area. First, we describe the concept used in [3]; and then we explain how the ASIC area is minimized in one round AES-128 encryption [2] (and [1]). Later, we explain the design considerations we incorporated.

A. Basic Optimization on AEGIS

The design objectives in [3] are twofold — area and throughput optimization. As throughput optimization is out of scope for our work, we only focus at the area optimized implementation. More specifically, we consider the smallest area implementation (18.72 kGE), AO$_1$ (or, A0$_1$), which also consumes least power (21.58 mW).

The authors, at first, design a base implementation with an ordinary version of AES-128 one round encryption. Following this, they adopt the optimized SBox implementation from [5]. However, the SBox implementation in [5] deals with masking, which is a countermeasure against side-channel attacks. This countermeasure works by incorporating random values, which are XORed or ANDed with the internal components. Naturally, the area requirement for this implementation is considerably higher than its non-masked counterpart, which is given in [4].

B. Optimization on AES-128 Encryption

The design of AES-128 in [2] is an improvement on top of the design in [1]. Hence, for the sake of conciseness, we only describe the basic design choices of Atomic AES 2.0 (encryption only) in a nutshell, as the detailed description can be found in [2].

It uses 16 registers (arranged in a 4×4 matrix), each of size 8-bits to store intermediate states and round keys separately (32 registers in total). Few registers are implemented with ordinary flip-flops, while the others are implemented with scan flip-flops. Ordinary flip-flops allow data movement in horizontal direction only, while scan flip-flops allow data to move both horizontally and vertically. The plaintext and the key are loaded serially in first 16 clock cycles. It produces the output after 246 clock cycles.

C. Improved Optimization on AEGIS (Ours)

Our implementation of R incorporates designs ideas from AES-128 implementation (Section III-B) with a few modifications. The main change from the aforementioned implementation is, we skip the key scheduling function and few other components that are used only in decryption process. Also, we need lesser rounds. This helps to reduce the area. Further, we also skip the step where the key is XORed with the state (before the 1st round operation).

Once the design of R is completed, we focus on the main architecture of AEGIS cipher. Since we want to minimize area requirement, we instantiate R only once, which means we access it sequentially; rather than instantiating it five times and access them parallelly. In this case, one astute observation regarding the StateUpdate128 operation is that, backing-up state registers $(S_0, \ldots, S_3)$ is required (S_4 is not needed to be backed-up). This is because, the update of one register depends on old content of another register. For example, when computing $S_{i+1,1}$; $S_{i,0}$ is needed (refer to Section II-C). Now, $S_{i,0}$ is replaced by $S_{i+1,0}$ already; so $S_{i,0}$ is to be backed-up in another register. In the naive approach, four back-up registers are required.

However, using an optimization, it is possible to reduce the number of back-up registers down to two. With the five 128-bit registers, $S_0, \ldots, S_4$; 128-bit message register M; consider the 128-bit back-up registers B^1 and B^2. The subscript i denotes the contents of $S_0, \ldots, S_4$ at i^{th} round. Here B^1, B^2 are updated five times within one StateUpdate128; we call each update as a step; so B^i_j denotes the content of B^i at j^{th} step, $i = 1, 2$. The following sequence of operations show how StateUpdate128 can be done:

$$B^1_{5i} \leftarrow R(S_{i,4}) \oplus M_i;$$
$$B^2_{5i} \leftarrow S_{i,0};$$
$$S_{i+1,0} \leftarrow B^1_{5i} \oplus B^2_{5i};$$
$$B^1_{5i+1} \leftarrow R(S_{i+1,0} \oplus B^1_{5i});$$
$$B^2_{5i+1} \leftarrow S_{i,1};$$
$$S_{i+1,1} \leftarrow B^1_{5i+1} \oplus B^2_{5i+1};$$
$$B^1_{5i+2} \leftarrow R(S_{i+1,1} \oplus B^1_{5i+1});$$

$$B^2_{5i+2} \leftarrow S_{i,2};$$
$$S_{i+1,2} \leftarrow B^1_{5i+2} \oplus B^2_{5i+2};$$
$$B^1_{5i+3} \leftarrow R(S_{i+1,2} \oplus B^1_{5i+2});$$
$$B^2_{5i+3} \leftarrow S_{i,3};$$
$$S_{i+1,3} \leftarrow B^1_{5i+3} \oplus B^2_{5i+3};$$
$$B^1_{5i+4} \leftarrow R(S_{i+1,3} \oplus B^1_{5i+3});$$
$$B^2_{5i+4} \leftarrow S_{i,4};$$
$$S_{i+1,4} \leftarrow B^1_{5i+4} \oplus B^2_{5i+4}.$$

IV. RESULTS

In this Section, we present synthesis results for AES-128 one round encryption and our AEGIS design. We use VHDL for the coding part; and then Synopsys Design Compiler version H-2013.03-SP1, using the 65 nm CMOS TSMC technology library for the synthesis and evaluation. The area is reported in terms of kilo Gate-Equivalent (kGE).

A. One Round AES-128 Encryption

As mentioned previously, we use AES-128 one round encryption as a black-box, the top-level architecture of which is shown in Figure 3. The usage of the input-output signals is straightforward.

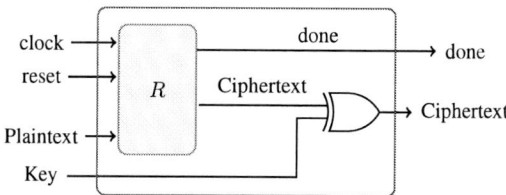

FIG. 3: Top level design of AES-128 1 round encryption

Here we present the ASIC synthesis results for the component R of AES-128 one round encryption. Our design requires 2.21 kGE for implementation, achieved maximum operational frequency as 355 MHz, 0.45 mW power and takes 21 clock cycles to generate the output. As a comparison, it requires 24.6% and 9% less area compared to Atomic AES and Atomic AES v2.0, respectively.

B. AEGIS

1) Implementation Overview

Here we outline the basic design concept we use for AEGIS core, which encompasses the key components. The top level description is given in Figure 4. The use of the signals clock, reset and start are straightforward. The mode input identifies the mode of operation, i.e., it is low for encryption & tag generation; and high for decryption & verification. The next six input lines, $S_0\_in, \ldots, S_4\_in, M\_in$ receives inputs for the corresponding 128-bit registers bit-by-bit. The following three input lines indicate which input is being fed to M: AD_active, is high for AD blocks; P_active for plaintext blocks for encryption & tag generation (ciphertext blocks for decryption & authentication) and T_active for $|AD| \, || \, |PT|$ (it is high for the first round

978-1-5386-7100-9/18 $31.00 © 2018 IEEE

when $|AD| \parallel |PT|$ is fed; refer to Section II-G for more details) for the particular round. For example, if plaintext is 2^{32}-bits long, P_active will be high for $2^{32-7} = 2^{25}$ rounds, as each round processes 2^7 bits. As for the outputs, C_out and T_out give the ciphertext (during encryption & tag generation) or the plaintext (during decryption & verification) and tag bit-by-bit; and the done signal is high twice to indicate whether the generated ciphertext/plaintext and tag are valid, respectively. Note that, this core has to be implemented within a wrapper. This wrapper will control all the input signals (e.g., send plaintext blocks as bit-by-bit); as well as manage the output signals. In case of decryption & verification, this wrapper will be responsible to check the tag is matched and do the follow-up steps (see Section II-H).

Next, we describe the architecture in short. The basic architecture is pictorially presented in Figure 5. The input signal clock controls the signal round_counter, which is used to update other registers. The registers, $S_0, \ldots, S_4, M, B^1, B^2$ are described earlier. The 128-bit register C is used to store the encrypted ciphertext (during encryption & tag generation) or the decrypted plaintext (during decryption & verification), which is determined by the mode input. The 128-bit register T stores the tag. As mentioned already, R denotes the one round AES-128 encryption. Our implementation takes 38 clock cycles to update one register (which includes, operating on R and accessing B^1, B^2), so the StateUpdate128 operation takes $38 \times 5 = 190$ clock cycles. The *done control* part controls when the output signal done will be high or when it will be low.

One crucial component in the design is the 238-bit round_counter. This signal controls the inputs/outputs across the circuit. For example, the 1-bit signals, $S_0\_in, \ldots, S_4\_in$, M_in are demultiplexed to the corresponding 128-bit signals;

and the 128-bit signals, C, T are multiplexed to 1-bit C_out, T_out; respectively.

The rationale for keeping the length of round_counter as 238 bits is explained below:

- 1-bit to indicate whether data are being loaded to the registers or StateUpdate128 is in operation.
- 8-bits to keep track of current status of data loading/ StateUpdate128. Note that, to load 128-bit inputs, we need 7-bits; but since StateUpdate128 takes 190 clock cycles, we use 8-bits.
- 57-bits to keep track of round number during AD loading. Since $|AD| < 2^{64}$, and 2^7 bits are loaded in one round, we need at most $64 - 7 = 57$ bits as the counter.
- Similarly, 57-bits to keep track of round number during P loading.
- 114-bits to store the round number when T_in is high for the last time, say, at wth round. We need this, as we now have to run StateUpdate128 for six more rounds, i.e., till the $(w + 6)^{\text{th}}$ round.
- 1 more bit is needed so that the round counter can go six more rounds.

2) Synthesis Results

In Table I, we present the ASIC synthesis results for AEGIS and compared with the existing design in AEGIS [3]. The AEGIS design requires only 16.92 kGEs for implementation and achieved maximum operational frequency as 208 MHz as given in Table I. Our AEGIS implementation requires 1.8 less kGE compared to the design in [3] (which also uses 65 nm technology), and also requires very less power compared to the design (but the operating frequency is also significantly reduced). In mathematical terms, our implementation requires 9.6% less area and reduces the power consumption by 95.3% compared to the design AO_1. The area reduction in our case is achieved by optimizing poor design choices. We would like to reiterate that, our implementation contains both encryption & tag generation as well as decryption & verification circuits; whereas [3] only contains encryption & tag generation circuit.

TABLE I: ASIC synthesis results for AEGIS

Design	Area (kGE)	Frequency (MHz)	Power (mW)
AEGIS (Ours)	16.92	208	1.007
AEGIS (AO_1, [3])	18.72	1410	21.58

V. CONCLUSION

Our work in this paper deals with implementation of one round AES-128 encryption (in ASIC, with minimum area), and later apply this primitive to the AEAD cipher, AEGIS-128. The former result is reported for the first time, whereas the latter result is improved from a previous work. An interesting follow-up work can be to see how this primitive can be applied to other AEADs where AES-128 encryption is used as a building block, such as TIAOXIN or DEOXYS. Besides, similar optimization on variants of AEGIS (other than AEGIS-128; such as AEGIS-128L) can also be performed.

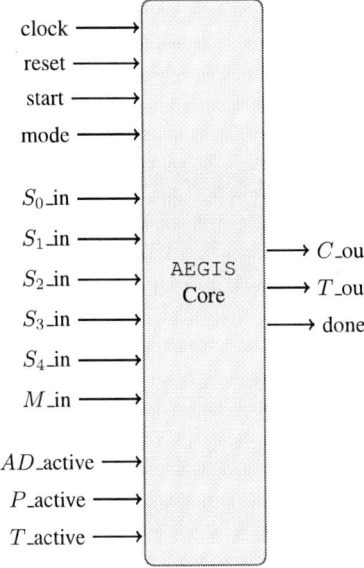

FIG. 4: Top level design of AEGIS core

978-1-5386-7100-9/18 $31.00 © 2018 IEEE

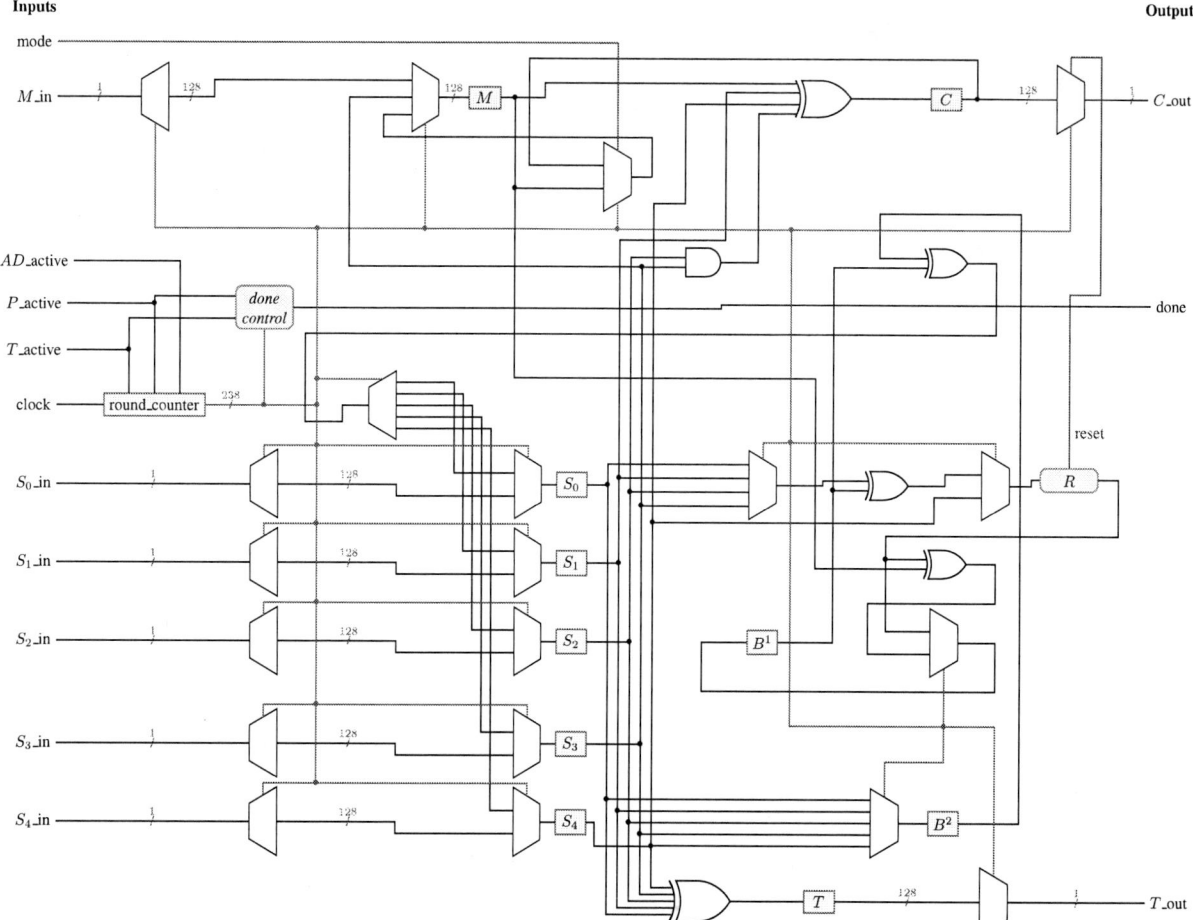

FIG. 5: Architecture of AEGIS core

References

[1] Banik, S., Bogdanov, A., Regazzoni, F.: Atomic-aes: A compact implementation of the AES encryption/decryption core. In: Progress in Cryptology - INDOCRYPT 2016 - 17th International Conference on Cryptology in India, Kolkata, India, December 11-14, 2016, Proceedings. (2016) 173–190

[2] Banik, S., Bogdanov, A., Regazzoni, F.: Atomic-aes v2.0. Cryptology ePrint Archive, Report 2016/1005 (2016) https://eprint.iacr.org/2016/1005.

[3] Bhattacharjee, D., Chattopadhyay, A.: Efficient hardware accelerator for AEGIS-128 authenticated encryption. In: Information Security and Cryptology - 10th International Conference, Inscrypt 2014, Beijing, China, December 13-15, 2014, Revised Selected Papers. (2014) 385–402

[4] Canright, D.: A very compact s-box for aes. In Rao, J.R., Sunar, B., eds.: Cryptographic Hardware and Embedded Systems – CHES 2005, Berlin,

Heidelberg, Springer Berlin Heidelberg (2005) 441–455

[5] Canright, D., Batina, L.: A very compact "perfectly masked" s-box for AES. In: Applied Cryptography and Network Security, 6th International Conference, ACNS 2008, New York, NY, USA, June 3-6, 2008. Proceedings. (2008) 446–459

[6] Jean, J., Nikolić, I., Peyrin, T., Seurin, Y.: Deoxys v1.41. https://competitions.cr.yp.to/round3/deoxysv141.pdf Accessed: 2018-02-27.

[7] Nikolić, I.: Tiaoxin – 346 (v2.1). https://competitions.cr.yp.to/round3/tiaoxinv21.pdf Accessed: 2018-02-27.

[8] Wang, Y., An, J., Ha, Y.: Unified data authenticated encryption for vehicular communication. In: 2016 IEEE 59th International Midwest Symposium on Circuits and Systems (MWSCAS). (Oct 2016) 1–4

[9] Wu, H., Preneel, B.: AEGIS: A Fast Authenticated Encryption Algorithm (v1.1). https://competitions.cr.yp.to/round3/aegisv11.pdf Accessed: 2018-01-22.

Architecture Exploration and Delay Minimization Synthesis for SET-Based Programmable Gate Arrays

Chia-Cheng Wu
Department of Computer Science
National Tsing Hua University
Hsinchu, Taiwan
Email: s104062522@m104.nthu.edu.tw

Kung-Han Ho
Institute of Electronics
National Chiao Tung University
Hsinchu, Taiwan
Email: kunghanho@gmail.com

Juinn-Dar Huang
Department of Electronics Engineering
National Chiao Tung University
Hsinchu, Taiwan
Email: jdhuang@mail.nctu.edu.tw

Chun-Yao Wang
Department of Computer Science
National Tsing Hua University
Hsinchu, Taiwan
Email: wcyao@cs.nthu.edu.tw

Abstract—Power consumption has become a primary obstacle for circuit designs at present. Single-Electron Transistor (SET) at room temperature has been demonstrated as a promising device for extending Moore's law due to its low power consumption. Since, only a few electrons are involved in the switching process, the drivability of SETs is ultra-low such that the height of an SET array is limited to a small number. This paper presents a delay minimization synthesis flow that decomposes a circuit into a network of SET Array Blocks (SABs) with a fixed height and width. The experiments were conducted for different sizes of SABs over a set of benchmarks. The experimental results showed that we can have the smallest average Area Delay Product (ADP) when the height is 5 and the width is 10 of an SAB, which indicates that such size of SABs is proper to synthesize SET networks.

Keywords—Single-Electron Transistor (SET); SET Array Blocks (SAB); delay minimization synthesis flow;

I. INTRODUCTION

As CMOS technology nodes are continuously scaling down, power consumption has become a primary concern in electronic system design [27]. To deal with this issue, various emerging low-power devices have been proposed in recent years. Among these low-power devices, Single-Electron Transistor (SET) is considered as one of the promising candidates. Some demonstrations of SET devices have proved that SET devices can operate at room temperature with only a few electrons [22][23][26].

Since only a few electrons are involved in the switching process, SETs suffer from low transconductance. Therefore, the traditional CMOS architecture is not applicable to SETs. To this end, a Binary Decision Diagram (BDD)-based architecture [2] was proposed as a practical approach to implement logic functions on SETs [1]. Furthermore, the BDD of any Boolean function can be used to map onto

This work is supported in part by the Ministry of Science and Technology of Taiwan under Grant MOST 106-2221-E-007-111-MY3, MOST 103-2221-E-007-125-MY3, and MEDIATEK Research Center Doctoral Talent Fellowship.

a BDD-based SET array [11][15][16]. A BDD-based SET array is a hexagonal nanowire network, which is composed of a set of node devices. Each node device has one entry branch and two exit branches as shown in Fig. 1(a). Fig. 1(b) illustrates that electrons enter a node device via the entry branch and pass through either the left or the right exit branch depending on the control signal of the wrap-gates. An exit branch is a segment of SET controlled nanowire and has four operational modes: *open*, *short*, *active-high*, and *active-low*. In this hexagonal network, each row of the network is controlled by the same input variable. Therefore, a given function represented by BDD can be simply mapped onto such hexagonal network. The first BDD-based hexagonal nanowire network has been demonstrated in [16].

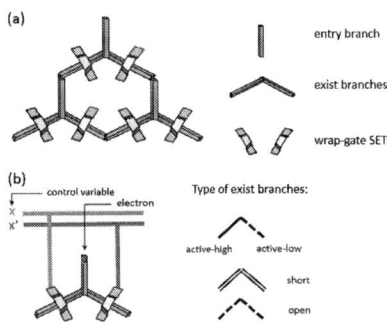

Fig. 1. (a) Physical structure. (b) Logic representation of a node device

However, the proposed BDD-based architecture in [16] is not amendable to functional reconfiguration. Furthermore, the entire circuit becomes useless if there exists a defective nanowire segment or any defective SET on it. Fortunately, a reconfigurable SET array was proposed to deal with these problems [10]. The automatic synthesis methods of reconfigurable SET array targeting at area minimization were proposed in [5][6][25]. Since the height of an SET

array is equal to the number of input variables of the given Boolean function, minimizing the area of an SET array means minimizing its width [8][18][19]. A recent work proposed an area minimization synthesis flow for reconfigurable SET arrays, which considered two fabrication constraints in the early stage [3]. The method minimized the number of product terms, which were extracted from a given Reduced Ordered Binary Decision Diagram (ROBDD), by dynamically shifting variables, and modeled the product terms ordering as the Traveling Salesman Problem (TSP) [4].

The aforementioned SET synthesis algorithms can map a function onto an SET array. However, there are more than one hundred inputs in some functions. Since the operations in SET arrays involve only a few electrons, the height, that is corresponding to the number of inputs, of an SET array is limited. The work [20] perceived this issue and conducted device-level experiments. The experimental results indicated that the height of an SET array is about 10. This design consideration issue has been addressed in [12] such that the height of a synthesized SET Array Block (SAB) was limited to 10. However, the experimental results in [12] showed that SABs have various widths with respect to different functions. On the other hand, the verification method on the synthesized SET array was proposed in [7]. The defect issue, diagnosis and mapping for defective SET arrays were also addressed in [13][14][17] recently.

In this paper, we use the concept of Field Programmable Gate Arrays (FPGAs) to divide a complex circuit into a multi-level network consisting of SRAM-based blocks. An SRAM can be considered as a Lookup Table (LUT). A 32-bit SRAM, for example, can be considered as a 5-LUT, which can implement any function within five variables. The limited number of inputs of each LUT, K, is similar to the limited height of an SET array. However, the width of an SAB is also considered as a constant number W. The SABs we used in this work with K inputs and W width are called (K, W)-SABs. There are two objectives of this work. The first one is to develop a delay minimization synthesis algorithm for decomposing a complex circuit into (K, W)-SABs. The other objective is to explore the proper parameters, K and W, for SET-based programmable gate arrays. We conducted experiments over a set of benchmarks with different sizes of SABs. The details of the experiments will be shown in Section IV.

The rest of this paper is organized as follows. In Section II, we introduce the architecture of reconfigurable SET arrays, and the background of this work. Section III presents our synthesis flow. Section IV shows the experimental results. The concluding remarks are given in Section V.

II. PRELIMINARIES

A. Reconfigurable SET Array

The structure of a reconfigurable SET array is shown in Fig. 2(a). A reconfigurable SET array consists of three layers, which are SET device layer, configuration layer, and input signal layer from the bottom to the top, respectively. The SET device layer is a hexagonal network that is constructed by identical SET nodes. The configuration layer determines the operational mode of each node in an SET array. By providing different voltage biases, an SET node can be set in three operational modes: 1) active; 2) open; and 3) short, as shown in Fig. 2(c). The input signal layer is an interface between the input signals and the SET nodes. An input signal can determine whether the corresponding active SET nodes are ON or OFF. Due to the limitation of the SET array structure, the SET nodes in the same row are controlled by the same input signal.

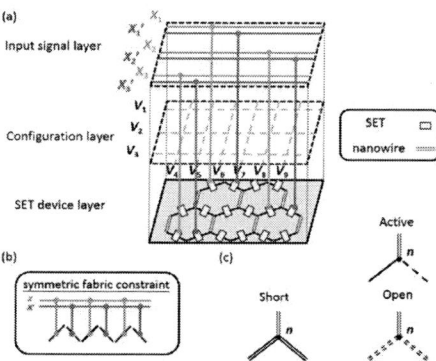

Fig. 2. (a) Architecture of reconfigurable SET arrays. (b) Symmetric fabric constraint. (c) 3 types of operational modes.

B. Symmetric Fabric Constraint

Symmetric fabric constraint reduces the wiring area of SET arrays by having an input signal x and its complementary x' connected to left edges and right edges, or vice versa, of SET nodes in the same row [10], as illustrated in Fig. 2(b). This constraint enforces that a pair of left and right edges of a node need to be one of (high, low), (low, high), (short, short) and (open, open). If an SET node is in the active operational mode, the pair of edges of the node can be either (high, low) or (low, high), which means that an input variable x is connected to the left (right) edge and the complementary x' is connected to the right (left) edge. If an SET node is in the short (open) operational mode, both edges are short (open).

C. Product-Term-Based Mapping Approach

The product-term-based approach synthesizes SET arrays for each Primary Output (PO) of a given Boolean function. The approach configures one path on the SET array for each product term of a PO without creating any invalid paths. The mapping rule is to configure high for bit 1, low for bit 0, and short for don't care. Fig. 3 shows an example of the mapping procedure. We follow the mapping rule to configure or to reuse an edge for each bit of p_0 from the top to the bottom of an SET array. The mapping result is demonstrated in Fig. 3(a). For mapping p_1, the first two configured rows can be reused since the first two bits of p_0 and p_1 are the same. Then the path branches at the coordination of $(x, y) = (0, 2)$, as illustrated in Fig. 3(b).

D. Mapping Flow for SET Network

Since the work [20] brought the issue of height constraint in the SET array mapping, decomposing a large SET array into a network of SABs becomes more important. Fig. 4 illustrates the flow of SET network mapping proposed in [12]. It translates a given Boolean circuit that is represented by an And-Inverter Graph (AIG) into a BDD network, then maps the network onto an SET network. The mapped SET network consists of SET arrays with the fixed height but different widths. The width of each SET array is various with respect to the corresponding sub-function it mapped.

E. Problem Formulation

Given the AIG representation of a Boolean network and user-defined parameters K and W, we map the netlist into a set of (K, W)-SABs such that the depth (delay) of the SABs is minimized.

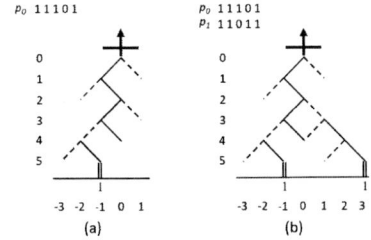

Fig. 3. Example of product-term-based mapping procedure. (a) p_0. (b) p_0+p_1.

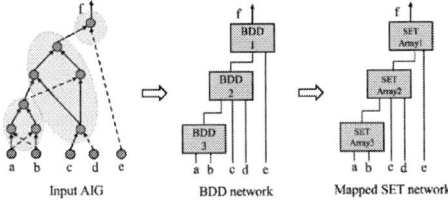

Fig. 4. The flow of mapping an SET network in [12].

III. PROPOSED SYNTHESIS ALGORITHM

In this section, we present our delay minimization algorithm for decomposing a large reconfigurable SET array. To simplify the measurement of delay in the mapped SET network, we assume that the delay of each SAB is identical. Therefore, measuring delay of a mapped SET network is to find the longest path in the network consisting of SABs. As mentioned in Section I, the main concept of the proposed synthesis algorithm is similar to the one targeting LUT-based FPGA. An LUT-based FPGA flow decomposes a Boolean network by the cut enumeration [9], and then packs sub-functions into LUTs. Our synthesis flow is divided into two phases: Phase 1 is to generate a set of cuts for each node from the Primary Input (PIs) to the POs in the AIG under the height constraint, and to label the depth and area of every cut and node. The labeling helps determine the mapping priority of cuts in the cut set. Phase 2 is to map sub-circuits from the POs to the PIs to generate the SAB network.

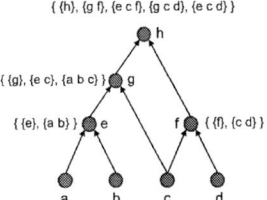

Fig. 5. An example of cut enumeration with $K = 3$.

A. Phase 1: Cut Generation and Node Labeling

Our algorithm decomposes the given AIG into K-bounded sub-networks, where the number of fanins in a sub-network does not exceed K. The cut enumeration [9] generates all K-feasible cuts for each node in the AIG from the PIs to the POs. A cut C of a node n is a set of nodes in its transitive fanins such that any path from a PI node to node n must pass through at least one node in the cut. A cut is said to be K-feasible if the size of the cut, which refers to the number of nodes in the cut, is no more than K. Fig. 5 illustrates an example of cut enumeration with $K = 3$. The three cuts, $\{g\}$, $\{e\ c\}$, and $\{a\ b\ c\}$ belong to the node g. The K-feasible cut of any PI node contains only the PI node itself. For a non-PI node v, we apply the cut generation function $\oplus^K$ on the cut sets a and b of its fanins to generate a cut as follows:

$$a \oplus^K b = \{\ x\ \cup\ y\ |\ x \in a,\ y \in b,\ |\ x\ \cup\ y\ | \leq K\}$$

Since this cut generation follows the topological order of the network from the PIs to the POs, it guarantees that the cut sets a and b have been generated before generating the cut set of node v.

During the cut enumeration, every node will have a label that contains the node depth $D(n)$ and the node area $A(n)$. For each cut, we also label the cut depth $D(c)$ and the cut area $A(c)$ to determine the mapping priority of cuts in the cut set. The cut depth, $D(c)$, is the largest node depth of this cut, which is equivalent to the level from the PIs to the cut, as shown in Eq. (1). The cut area, $A(c)$, is referred to the summation of node area of nodes in this cut, and it represents the number of SABs we need when constructing the cut, as shown in Eq. (2). The node depth of a node n, $D(n)$, is the minimum cut depth in its cut set plus one, as shown in Eq. (3). The node area of a node n, $A(n)$, is the cut area of the cut with the minimal depth in the cut set plus one, as shown in Eq. (4). For the PI nodes, their delay and area are 0.

$$D(c) = \max_{n\ \in\ cut}\ \{D(n)\} \tag{1}$$

$$A(c) = \sum_{n\ \in\ cut}\ \{A(n)\} \tag{2}$$

$$D(n) = \min_{c\ \in\ cut\ set_n}\ \{D(c)\}\ +\ 1 \tag{3}$$

$$A(n) = \min_{c\ \in\ cut\ set_n\ with\ min\{D(n)\}}\ \{A(c)\}\ +\ 1 \tag{4}$$

An example of cut enumeration process with $K = 5$ is shown in Fig. 6. The given AIG is illustrated in Fig. 6(a). The nodes a to f are PIs. The parentheses at the right of nodes represent the

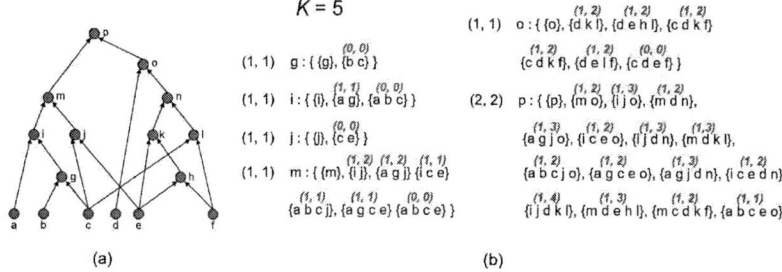

Fig. 6. An example of cut enumeration with label.

cut sets of nodes, and the inner parentheses in a parenthesis are the cuts in the cut set. To generate the cut of a node n, we put the node n itself into cut set first. Then, use the aforementioned generation function to generate the complete cut set. The labels on the top of inner parentheses represent the cut depth and cut area, $(D(c), A(c))$, and the labels at the left of the nodes represent the node depth and node area, $(D(n), A(n))$. When labeling the depth and area of a node, we consider the cut with the minimal cut depth in the set first. If there are more than one cut having the minimal cut depth in the set, we choose the cut with the minimal cut area. In Fig. 6(b), we can see that all the cuts in the cut set of node p are with the same depth. In this case, we choose the cut with the minimal $A(c)$ to calculate the depth and area of node p since it takes the minimal number of SABs. Therefore, we find that the cut $\{a, b, c, e, o\}$ is the best cut for node p due to its smallest cut area.

However, when circuits and K become larger, enumerating all K-feasible cuts is impractical. In our observation, many enumerated cuts are redundant, which means that the cuts are useless for mapping a node. Furthermore, these redundant cuts generate more redundant cuts. Therefore, we apply a priority cut heuristic [21] to reduce the number of enumerated cuts. The strategy is to reserve only a small fixed number p, which is typically five to ten, of "good" K-feasible cuts for each node. The "good" K-feasible cuts are the cuts that have smaller cut depths and the smaller cut areas in the cut set. However, there is a difference between the strategies of priority cut in [21] and this work. In this work, we keep at most p "good" cuts for each cut size from 2 to K. Moreover, the trivial cut of node n, $\{n\}$, which consists of the node itself, is always added to the cut set to ensure that any nodes can be mapped into an SAB. Then, we keep additional p cuts with the optimal cut depth plus one for the cuts whose cut size is K according to the two observations from our early experiments. In the first observation, the larger the cut size is, the more nodes are likely to be packed into an SAB. Therefore, we reserve additional p "good" cuts for the cuts with the cut size of K. In the second observation, the number of nodes in a cut is similar to other cuts with the same cut depth. That is, if a cut fails to map its function onto an SAB with a width W in the second phase, other cuts with the same depth are very likely to fail as well. Therefore, instead of keeping all the cuts with the optimal depth, we choose to retain the cuts with the optimal depth plus one. For example, if $p = 3$ and $K = 5$, we keep at most three cuts for each cut size from 2 to 4. For the cut size of 1, we only keep one cut, the trivial cut. For the cut size of 5, we keep three cuts with the optimal depth and three other cuts with the optimal depth plus one. The maximum number of the enumerated cuts in a cut set in our strategy is $K \times p + 1$.

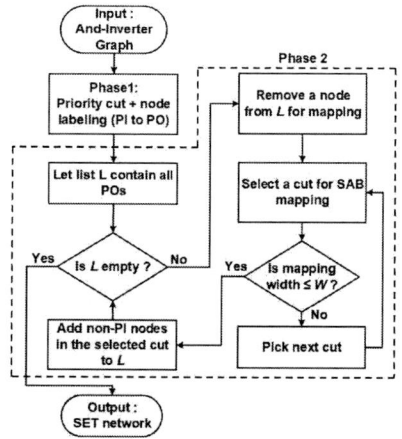

Fig. 7. Overall flow of the proposed synthesis approach.

B. Phase 2: SAB Mapping

In this phase, we transform the given Boolean circuit into an SET network consisting of (K, W)-SABs. The steps of SAB mapping are as follows: First, list all PO nodes in L. Second, choose one node from L for mapping, and sort the cuts of the chosen node by 1) cut depth, 2) cut area, and 3) cut size. We first select the cut with the optimal cut depth for mapping the node, therefore, we can have the mapped SAB with the minimal delay in the cut set. If there are more than one cut having the optimal cut depth, we choose the optimal-depth cut with the minimal cut area for mapping to reduce the number of SABs. When we have multiple cuts that have the optimal cut depth and the minimal cut area, we choose the cut with the largest cut size to pack more nodes in an SAB. Once we select a cut of the node for mapping, we apply the SET array synthesis method in [3] to obtain the mapped SAB. If the mapping width of the SAB is smaller than or equal to the predetermined W, the mapping is successful. However, if the mapping width of the SAB is larger than W, we continue to choose the next highest priority cut for mapping until the mapping width constraint is satisfied.

In this work, it is guaranteed to map a node onto the SAB with the width smaller than or equal to W successfully since

978-1-5386-7100-9/18 $31.00 © 2018 IEEE

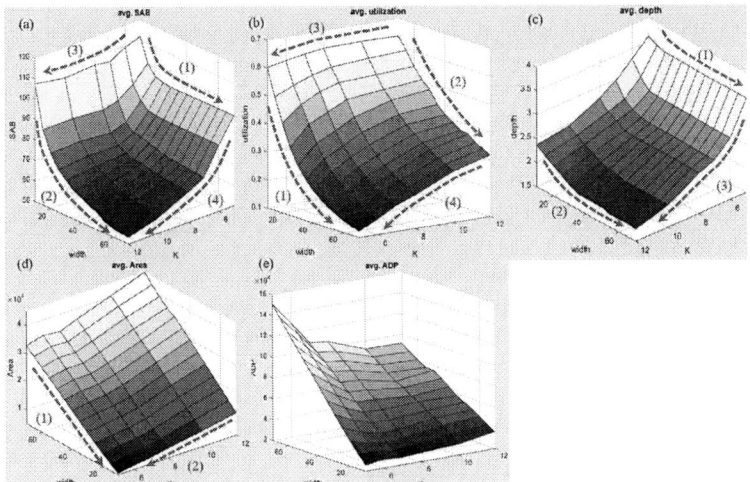

Fig. 8. Experimental Results. (a) Average number of required SABs. (b) Average utilization of SABs. (c) Average depth. (d) Average area. (e) Average Area Delay Product (ADP).

our priority cut strategy always keeps the trivial cut. Once a node is successfully mapped onto an SAB with the chosen cut, we add all the non-PI nodes, except for the nodes that have already been mapped, in the cut into the list L. The overall flow is shown in Fig. 7. The process is not terminated until L is empty. In the end, we obtain an SET network composed of mapped (K, W)-SABs.

IV. EXPERIMENTAL RESULTS

We implemented the proposed algorithm in C++ language and conducted experiments on an Intel Xeon 2.4GHz CPU platform with 64 GBytes memory. The experiments were conducted over a set of MCNC [24] and IWLS 2005 [28] benchmarks represented by AIG. We conducted experiments for different sizes of (K, W)-SABs, where K is varied from 5 to 12 and W is varied from 10 to 70 at intervals of 5. The averaged experimental results of this set of benchmarks are illustrated in Fig. 8, which was drawn with MATLAB.

Fig. 8(a) shows the average number of required SABs with different sizes. Fig. 8(b) shows the average utilization with different sizes of SABs. The utilization of an SAB refers to the ratio of the mapped width with respect to W. For example, if the mapped width of a sub-function on an SAB is 8, and the parameter W is 10, the utilization of the SAB is 80%. The line (1) in Fig. 8(a) indicates the effect of increasing width W to the average number of SABs with a smaller K of 5. In the beginning, we found that the average number of required SABs was greatly reduced when the W became larger. However, when the W became much larger, the average number of SABs is almost intact. It indicates that we do not need SABs with very large widths when K is small since only a few nodes are packed into an SAB. Furthermore, the line (1) in Fig. 8(b) also explains that the utilization dramatically dropped when the W became larger. The line (2) in Fig. 8(a) indicates that the number of required SABs with a larger K of 12 continuously decreases when W increases. In Fig. 8(b), the line (2) shows that the averaged utilization of SABs decreases more gently than the line (1) with the increase of W. Furthermore, the

averaged utilization of SABs for a large W, which is indicated by the line (4), is lower than the line (3).

In Fig. 8(c), the lines (1) and (2) indicate that the averaged depth of SABs becomes smaller when the W becomes larger. This means that an SAB with a larger width can accommodate a larger sub-function. However, this is not obvious anymore when the width of SABs is too large. The line (3) also indicates that the averaged depth of SET networks becomes smaller when K increases. This is because more nodes are packed into an SAB with a larger K.

We observed that using SABs with larger parameters K and W leads to a smaller averaged number of required SABs for mapping in an SET network. However, the lines (1) and (2) in Fig. 8(d) indicate that the averaged mapping area of the SET network with larger size of SABs is oppositely larger. The minimum averaged area occurs when the parameter K is 5 and W is 10.

For the averaged Area Delay Product (ADP), which is calculated as $area \times depth$, with different sizes of SABs in Fig. 8(e) indicates the minimum ADP occurs when K is 5 and W is 10, and a similar ADP occurs when K is 7 and W is 10.

TABLE I shows the detailed experimental results with (5, 10) and (7, 10) SABs in two columns. The last row shows the averaged results among these benchmarks. It indicates that the averaged area of the SET networks with (5, 10) SABs is smaller than the SET networks with (7, 10) but the averaged depth of the SET networks with (5, 10) SABs is larger than the SET networks with (7, 10). Therefore, the two averaged ADP of the two experiments with (5, 10) and (7, 10) SABs are similar.

V. CONCLUSION

In this paper, we propose the first delay minimization synthesis algorithm decomposing and mapping a Boolean circuit into a set of fixed size (K, W)-SABs. The proposed method consists of two phases that guarantee the mapped SET SABs meet the predefined height and width. The experimental results suggest that smaller parameters K and W lead to

978-1-5386-7100-9/18 $31.00 © 2018 IEEE

TABLE I
The experimental results with $(K, W) = (5, 10)$, and $(K, W) = (7, 10)$ SABs.

Bench.	K = 5, W = 10						K = 7, W = 10					
	|SAB|	utiliz. (%)	depth	area	ADP	T(s)	|SAB|	utiliz. (%)	depth	area	ADP	T(s)
alu2	184	63.1	9	9200	82800	0.94	159	63.6	6	11130	66780	2.74
alu4	345	64.6	10	17250	172500	1.95	333	65.9	7	23310	163170	5.75
apex6	291	63.8	5	14550	72750	1.28	256	67.1	4	17920	71680	2.69
apex7	74	59.5	4	3700	14800	0.27	66	65.2	3	4620	13860	0.57
b9	42	56.7	3	2100	6300	0.16	38	61.1	3	2660	7980	0.29
c8	38	65.5	3	1900	5700	0.20	34	62.4	3	2380	7140	0.31
c17	3	56.7	2	150	300	0.02	3	56.7	2	210	420	<0.01
cc	23	56.1	2	1150	2300	0.09	23	47.8	2	1610	3220	0.17
cht	45	72.9	2	2250	4500	0.17	38	77.4	2	2660	5320	0.12
cm85	16	56.2	3	800	2400	0.07	17	57.1	2	1190	2380	0.12
cm138	11	71.8	2	550	1100	0.09	11	71.8	2	770	1540	0.14
cm151	8	70.0	3	400	1200	0.04	9	63.3	2	630	1260	0.13
cm162	14	65.0	3	700	2100	0.05	14	54.3	2	980	1960	0.15
cm163	13	56.2	2	650	1300	0.07	13	72.3	2	910	1820	0.14
cmb	19	54.2	3	950	2850	0.06	16	60.0	3	1120	3360	0.14
count	43	67.2	5	2150	10750	0.29	44	73.9	4	3080	12320	0.57
cu	19	52.6	3	950	2850	0.05	17	55.3	2	1190	2380	0.06
example2	121	53.9	3	6050	18150	0.53	115	59.6	3	8050	24150	1.49
frg1	29	64.0	5	1450	7250	0.11	31	57.1	4	2170	8680	0.25
frg2	312	65.2	4	15600	62400	1.37	294	68.6	4	20580	82320	4.53
i1	20	44.0	4	1000	3000	0.05	17	45.3	2	1190	2380	0.04
i2c	402	59.5	5	20100	100500	1.75	375	65.9	4	26250	100500	3.65
i8	380	71.0	4	19000	76000	2.68	398	77.3	3	27860	83580	11.12
lal	35	54.6	3	1750	5250	0.19	32	60.0	3	3350	6720	0.46
ldd	42	59.0	3	2100	6300	0.31	39	58.7	2	2730	5460	0.47
pcie	23	57.7	3	1150	3450	0.10	16	70.6	2	1120	2240	0.11
pcier8	37	55.9	4	1850	7400	0.15	34	60.3	3	2380	7140	0.47
pm1	20	47.5	2	1000	2000	0.06	19	51.6	2	1330	2660	0.11
Avg.	93.2	60.2	3.7	4658.9	24221.4	0.47	87.9	62.5	3.0	6192.1	24729.3	1.31

smaller ADP. Furthermore, we also showed that the proposed priority cut strategy makes our algorithm very efficient.

REFERENCES

[1] N. Asahi et al., "Single-electron logic device based on the binary decision diagram," *IEEE Trans. Elec. Dev.*, 1997.

[2] R. E. Bryant, "Graph-based algorithms for Boolean function manipulation," *IEEE Trans. Computers*, 1986.

[3] Y.-H. Chen et al., "Area minimization synthesis for reconfigurable single-electron transistor arrays with fabrication constraints," *ACM J. Emerg. Tech. Com. Syst.*, 2016.

[4] Y.-H. Chen et al., "ROBDD-based area minimization synthesis for reconfigurable single-electron transistor arrays," *in Proc. Int. Symp. VLSI Design, Auto. Test*, 2015.

[5] Y.-C. Chen et al., "Automated Mapping for Reconfigurable Single-Electron Transistor Arrays", *in Proc. Design Auto. Conference*, 2011.

[6] Y.-C. Chen et al., "A Synthesis Algorithm for Reconfigurable Single-electron Transistor Arrays," *ACM J. Emer. Tech. Comp. Syst.*, 2013.

[7] Y.-C. Chen et al., "Verification of Reconfigurable Binary Decision Diagram-based Single-Electron Transistor Arrays," *IEEE Trans. CAD of Int. Circuits and Syst.*, 2013.

[8] C.-E. Chiang et al., "On reconfigurable single-electron transistor arrays synthesis using reordering techniques," *in Proc. Des. Auto. and Test in Europe*, 2013.

[9] J. Cong et al., "Cut ranking and pruning: Enabling a general and efficient FPGA mapping solution." *in Proc. ACM/SIGDA S. Int. S. FPGA*, 1999.

[10] S. Eachempati et al., "Reconfigurable Bdd-based Quantum Circuits," *in Proc. Int. Symp. Nanosc. Archit.*, 2008.

[11] H. Hasegawa et al., "Hexagonal binary decision diagram quantum logic circuits using Schottky in-plane and wrap gate control of GaAs and InGaAs nanowires," *Phys. E, Low-dimensional Syst. Nanostruct.*, 2001.

[12] C.-H. Ho et al., "Area-aware decomposition for single-electron transistor arrays," *ACM Trans. Des. Auto. Elec. Syst.*, 2016.

[13] C.-Y. Huang et al., "A Defect-aware Approach for Mapping Reconfigurable Single-Electron Transistor Arrays," *in Proc. Asia and South Pacific Design Automation Conference.*, 2015.

[14] C.-Y. Huang et al., "Diagnosis and Synthesis for Defective Reconfigurable Single-Electron Transistor Arrays," *IEEE Trans. VLSI Syst.*, 2016.

[15] S. Kasai et al., "A single electron binary-decision-diagram quantum logic circuit based on Schottky wrap gate control of a GaAs nanowire hexagon," *Elec. Device Lett.*, 2002.

[16] S. Kasai et al., "Fabrication of GaAs-based integrated 2-bit half and full adders by novel hexagonal BDD quantum circuit approach," *in Proc. Int. Symp. Semi. Dev. Res.*, 2001.

[17] Y.-J. Li et al., "Dynamic Diagnosis for Defective Reconfigurable Single-Electron Transistor Arrays," *IEEE Trans. VLSI Syst.*, 2017.

[18] C.-W. Liu et al., "Width Minimization in the Single-Electron Transistor Array Synthesis," *in Proc. Des., Auto. and Test in Europe*, 2014.

[19] C.-W. Liu et al., "Synthesis for width minimization in the single-electron transistor array," *IEEE Trans. VLSI Syst.*, 2015.

[20] L. Liu et al., "A reconfigurable low-power BDD logic architecture using ferroelectric single-electron transistors," *IEEE Trans. Elec. Device*, 2015.

[21] A. Mishchenko et al., "Combinational and sequential mapping with priority cuts," *IEEE Trans. Computer-Aided Design*, 2007.

[22] H. W. Ch. Postma et al., "Carbon nanotube single-electron transistors at room temperature," *Science*, 2001.

[23] Y.-T. Tan et al., "Room temperature nanocrystalline silicon single-electron transistors," *J. Appl. Phys.*, 2003.

[24] S. Yang, "Logic Synthesis and Optimization Benchmarks, Version 3.0," *Tech. Report, Microelectronics Center of North Carolina*, 1991.

[25] Z. Zhao et al., "BDD-Based Synthesis of Reconfigurable Single-Electron Transistor Array," *in Proc. Int. Conference CAD*, 2014.

[26] L. Zhuang et al., "Silicon single-electron quantum-dot transistor switch operating at room temperature," *Appl. Phys. Lett.*, 1998.

[27] *Int. Tech. Road. Semi., Semiconduc. Industry Association*, 2006.

[28] *IWLS 2005 Benchmarks. (June 2005). Retrieved March, 2015* [Online]. Available: http://iwls.org/iwls2005/benchmarks.htm

MRAM-on-FDSOI Integration: A Bit-cell Perspective

Hao Cai[*†], You Wang[‡], Wang Kang[‡], Lirida Naviner[†], Xinning Liu[*], Jun Yang[*], and Weisheng Zhao[‡]

[*]National ASIC System Engineering Center, Southeast University, Nanjing, 210096, China.

[†]Département Communications et Électronique, Télécom-ParisTech, Université Paris-Saclay, Paris, 75013, France.

[‡]Fert Beijing Institute, BDBC and School of Electronic and Information Engineering, Beihang Univeristy, Beijing, China.

Abstract—In this paper we discuss the potential foundry announced hybrid integration of magnetic random access memory (MRAM) on fully depleted silicon-on-insulator (FD-SOI) technology. The spin transfer torque magnetic tunnel junction (STT-MTJ) and the next generation voltage-controlled magnetic anisotropy (VCMA) MTJ are separately integrated into a 28 nm FD-SOI process. Circuit-level design strategies are explored that use FD-SOI leverage and spin-device characteristic to realize writing and reading power-delay efficiency, robust and reliable performance in a 1-transistor 1-MTJ (1T1M) bit cell. Process variation aware strategies for MTJ-FDSOI integration are proposed to compensate failure operations, by using the dynamic step-wise back-bias and the flip-well back-bias. A qualitative summary demonstrates that the MRAM-on-FDSOI integration offers attractive performance for future non-volatile CMOS integration.

I. INTRODUCTION

Spin transfer torque (STT) magnetic tunnel junction (MTJ) based magnetic random access memory (MRAM) is one of the most promising candidates to replace conventional memories [1]–[4]. The main advantages of MRAM include its non-volatile characteristic, good scalability, low power (zero standby leakage power) and high endurance,

By relying on a voltage rather than a current to write MRAMs, voltage control magnetic anisotropy (VCMA, or electric-field-assisted magnetization switching) has been considered as an energy-efficient switching method for the next-generation MRAMs [5]–[9]. However, its efficiency which is denoted as the change of interfacial magnetic anisotropy energy per unit electric field [6], must be improved along with decreased switching latency and lower supply voltage (V_{dd}). Previous works of VCMA based bit-cell mainly focused on performance optimization, e.g., scaling of MTJ oxide thickness, seeking optimal VCMA pulse duration and amplitude [7]–[9].

Advanced CMOS technologies with features like high-κ metal gate (HKMG), fully depleted silicon-on-insulator (FD-SOI) or Fin Field-effect transistor (FinFET) post additional challenges when integrating emerging devices on CMOS platforms [10]–[13]. Recently, several foundries announced the availability of MRAMs on FD-SOI technology. Until now the promising MTJ-FDSOI integration still lacks practical evaluation on energy-delay performance and process variations [11]–[13].

In this paper, we explore the performance of 1-transistor 1-MTJ (1T1M) bit cell using STT and VCMA-MTJ compatible

models, based on a commercial 28nm FD-SOI process. The main highlights of this work are summarized as follows.

- A developed STT and a perspective VCMA switching mechanism based MTJs are integrated into a commercial 28nm FD-SOI CMOS platform. The performance of this hybrid integration is demonstrated with a 1T1M MRAM bit-cell.
- To achieve high performance and low power design, the MRAM bit-cell is implemented with two types of transistors (low V_t LVT and regular V_t RVT). The impact of different MTJ switching methods, and transistor types on MRAM performance are evaluated.
- The performance of MRAM-on-FDSOI integration is evaluated, including active leakage reduction, energy-delay optimization, area-performance tradeoff. Two performance enhancement strategies are proposed, named dynamic step-wise back-bias and the flip-well methods.
- A qualitative summary about the advantages and challenges of MRAM-on-FDSOI integration is presented.

The rest of the paper is organized as follows: Section II briefly reviews the FD-SOI and MTJ device technologies. Section III provides the 1T-1M simulation results. Section IV explains the design strategies employed to optimize the bit-cell performance. Section V presents a qualitative summary about the advantages and challenges of MRAM-on-FDSOI integration. Finally, Section VI concludes the paper.

II. RELATED WORK

A. STT and VCMA switching in MTJs

STT and VCMA-MTJ nanopillar are implemented with a similar architecture, which consists of two ferromagnetic layers (*CoFeB*) separated by a thin insulating layer. Table I lists the physical properties of STT-MTJ and VCMA-MTJ. The state switching in STT-MTJ is initialized when the current flowing from reference layer to free layer (I) is higher than the critical current I_{c0}. The MTJ resistance (R_p, parallel or R_{ap}, anti-parallel) is determined by the relative magnetization orientations of the ferromagnetic layers. The VCMA mechanism can be explained as follows: an electric field induces an accumulation of electron charges, then brings about a change of occupation of atomic orbitals at the interface, which in conjunction with spin-orbit interaction, finally results in a change of magnetic anisotropy [15], [16].

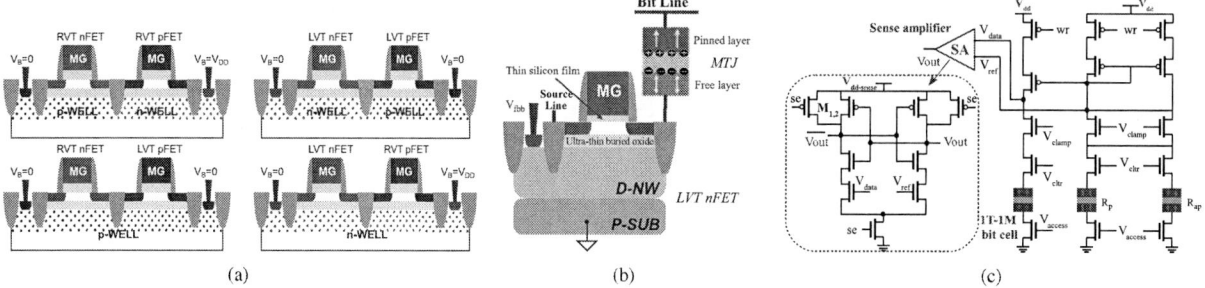

Fig. 1. (a) The cross-sectional view of regular well, flip well and single well FDSOI devices. (b) MRAM on FD-SOI: the cross-sectional view of FD-SOI MTJ integration with a 1T1M MRAM bit-cell. The bit-cell access transistor is implemented with a low threshold voltage (LVT) nFET. (c) FD-SOI MTJ integration: the schematic of MRAM sensing circuit (PCSA), includes a 1T1M MRAM bit-cell, a reference generator circuit and a pre-charge sense amplifier.

TABLE I
PHYSICAL PROPERTIES AND SIMULATION SETUP OF THE MRAM
BIT-CELLS

Parameter	Description	STT-MTJ	VCMA-MTJ
TMR	TMR ratio	200%	100%
T_{ox}	MTJ oxide thickness	0.85 nm	1.4 nm
T_{sl}	MTJ free layer thickness	1.3 nm	1.1 nm
$Area$	MTJ layout surface	32nm×32nm	50nm×50nm
R_P/R_{AP}	MTJ resistance	6.2kΩ/18.6kΩ	100kΩ/200kΩ
α	Damping factor	0.027	0.05
Δ	Thermal stability	72	40
I_{c0}	Critical current	50 μA	N/A
t_b	VCMA pulse duration	N/A	0.4 nm
T	Ambient temperature	300K	
$V_{dd-write}$	Write supply voltage	1 V-1.5 V	
$V_{dd-sense}$	Read supply voltage	0.25 V-1 V	
V_{BB}	Forward back-bias	1 V-1.5 V	
PB	Poly bias range	0-16nm	
W/L	Access transistor width/length	W=80 nm (minimum), 400 nm Minimum L = 30 nm	

In the presence of the VCMA effect, if a positive voltage is applied on the MTJ and the amplitude of the voltage is equal to or larger than the critical voltage V_c, the energy barrier will be entirely eliminated. In this scenario, the magnetization of the free layer becomes unstable and precessionally oscillates between the two stable states, denoted as precessional regime. Respectively, with the amplitude of the voltage inferior to V_c, the energy barrier is lowered but will not be eliminated. In this case, the magnetization of the free layer may either be damped back to the initial state or be flipped to the other state, owing to thermal activation, denoted as thermally-activated switching regime. In practice, if MTJ operates in the thermally-activated switching regime, an assistant mechanism, e.g., a magnetic field or a charge current (through the STT effect), is usually required to assist MTJ switching. On the other hand, if MTJ operates in the precessional switching regime, a precise control of the voltage pulse duration or additional write-verify algorithms are generally necessary to guarantee the success of MTJ switching.

In a 1T1M bit cell (see Fig. 1(b)), the word line (WL) froms a memory array is connected to the gate of the access tran-

sistor. One terminal of MTJ is connected to the bit line (BL) and another terminal is with source line (SL). STT-MRAM relies on a bi-directional current pulse to switch the AP and P states, whereas in VCMA-MRAM, the switching depends on uni-directional voltage pulses. An important constraint in STT-MRAM is the asymmetry characteristics between AP-P and P-AP switching due to the source degeneration of the access transistor [17]–[19]. When the switching current flows from SL to BL, the drivability of the access transistor is degraded due to smaller gate-source voltage. Large transistor width is required to mitigate the difference between the strong (fast) and the weak (slow) write operation.

B. FD-SOI MTJ integration

FD-SOI device enlarges the physically-described range of back-plane polarizations. As shown in Fig. 1(a) and Fig. 1(b), the ultra thin buried oxide under the thin silicon film enables transistor back-plane as a second gate allowing to form another channel at the poly-silicon/metal oxide interface [11], [20], [21]. The flexible well configuration of FD-SOI platform includes regular well, flip well, single P-well (SPW) and single N-well (SNW), which can significantly impact circuits performance. Relying on regular (RVT) and low-power (LVT) family transistors, the threshold voltage (V_t) of nFET and pFET can be adjusted according to design requirement, with four sources of V_t modulation methods: gate oxide, well type, poly biasing and back biasing.

A 1T1M MRAM bit cell, a reference generator circuit and a pre-charge sense amplifier (PCSA) circuit are illustrated in Fig. 1. The MRAM cell consists of an access transistor and a MTJ device. In the read operation, a read signal voltage (corresponding to the MTJ's resistance value R_p or R_{ap}, generated by reference voltage V_{ref}) is output to the BL.

Fig. 1(a) shows the different well configurations which are peculiar to FD-SOI. Flip well implementation has improved performance with the tradeoff in leakage power and variability. The SPW has been validated in FDSOI based SRAM bit-cell [22]–[24]. The minimum V_{dd} can be setup with 70 mV to 100 mV lower than the regular well, the circuit stability (e.g., failure probability) is greatly improved around 0.5V supply. The drawback of SPW is that a deep-N-well must be inserted

978-1-5386-7100-9/18 $31.00 © 2018 IEEE

between the P-well and substrate for isolation, whereas no isolation is required in SNW structure. SNW enables energy efficient design with controlled leakage power [25].

III. BIT-CELL PERFORMANCE

A STT-MTJ compact model [26], a VCMA-MTJ compact model [7], and a 28-nm FD-SOI design-kit are utilized to study the performance of FD-SOI MTJ integration. Table I lists the physical parameters of STT and VCMA-MTJ, as well as the FD-SOI optimization setup in our simulations. The P-AP and AP-P switching latency, average energy consumption, leakage power and write/read operation failure are studied based on the 1T-1M bit-cell.

A. Writing latency

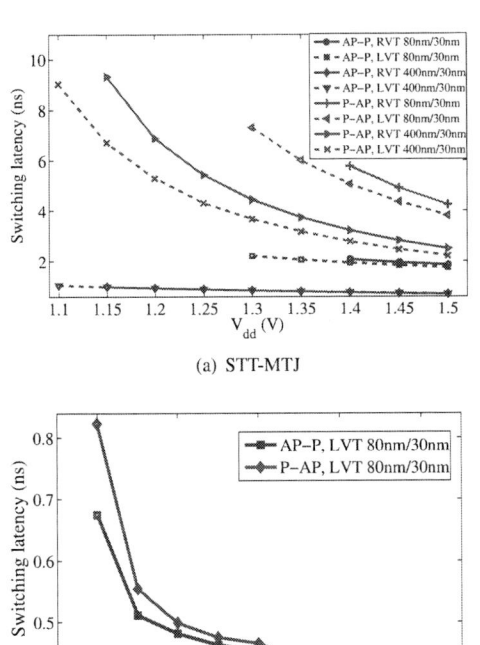

(a) STT-MTJ

(b) VCMA-MTJ

Fig. 2. The writing latency versus V_{dd} in 1T1M with STT and VCMA strategies. (a) Due to the write asymmetry, fundamental differences exist between AP-P and P-AP switching in STT-MTJ. The latency is sensitive to the type, the dimension of the access transistor. (b) Comparing to STT-MTJ, VCMA-MTJ has smaller write asymmetry.

The MRAM bit-cell is highly sensitive to the write V_{dd}. A successful writing operation requires a sufficient switching current in STT-MTJ, and adequate driving voltage in VCMA-MTJ. Fig. 2 illustrates the writing latency versus writing V_{dd} in MTJs with STT and VCMA switching strategies. As STT-MTJ suffers from write asymmetry problem, a fundamental difference exists between AP-P and P-AP switching (see Fig. 2(a)). The latter one could become the speed bottleneck which determines the system frequency. Its switching latency is also

sensitive to the type and dimension of the access transistor, especially in the P-AP case. Fig. 2(b) illustrates the writing latency in the VCMA based 1T1M. As it slightly depends on the access transistor in the normal operation region, we only provide the latency curve with minimum width/length using LVT type for access transistor. For STT-MRAM, using LVT transistors achieves higher switching currents than RVT transistors for the same V_{dd}. The LVT/RVT access transistor impact is less important in VCMA. Only 60 mV VCMA pulse voltage (V_b) difference is realized.

B. Writing power consumption

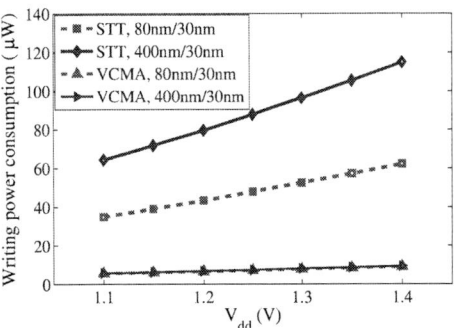

Fig. 3. Dynamic power consumption in 1T1M with STT and VCMA strategies. VCMA writing power slightly depends on the access transistor.

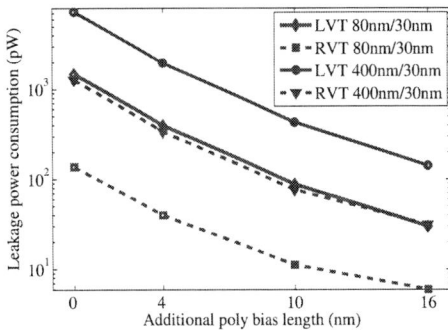

Fig. 4. Active leakage power reduction using poly biasing. The access transistor is implemented with LVT/RVT with 400 nm/80 nm width.

Fig. 3 illustrates the dynamic power dissipation in 1T1M bit-cell with STT and VCMA strategies. In STT based bit-cell, due to the increased current flow, access transistor with 400 nm width leads to average 84% increased power consumption comparing to the minimum width design. In VCMA based bit-cell, the writing power consumption could be significantly reduced even though with a 1.4 V write V_{dd}. In contrast to STT case, the power consumption in VCMA based design is not sensitive to the type and dimension of access transistor.

To completely reduce the active leakage power, it is possible to cut off V_{dd} by power gating block for an idle state. For ultra low power applications, the active leakage power reduction technique is still needed. As shown in Fig. 4, leakage power can be suppressed to 2% by using FD-SOI poly biasing, with additional layout area and speed penalty.

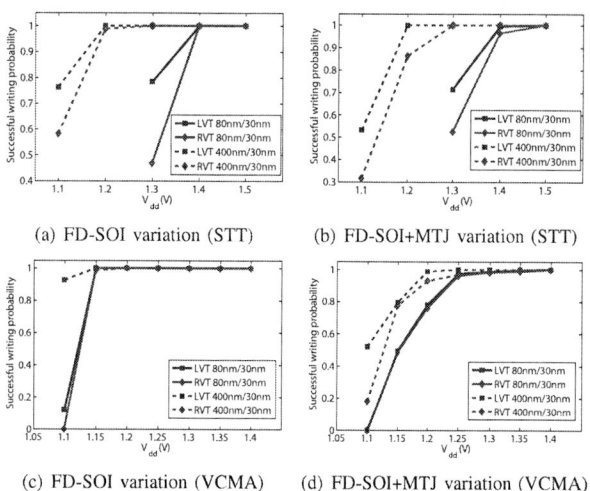

(a) FD-SOI variation (STT) (b) FD-SOI+MTJ variation (STT)

(c) FD-SOI variation (VCMA) (d) FD-SOI+MTJ variation (VCMA)

Fig. 5. Process variation induced writing failure in 1T1M bit-cell. The origin of variability is analyzed: (a) STT-MRAM, FD-SOI variation only. (b) STT-MRAM, accumulated variation from FD-SOI and STT-MTJ. (c) VCMA-MRAM, FD-SOI variation only. (d) VCMA-MRAM, accumulated variation from FD-SOI and VCMA-MTJ.

C. Process variation

The failure probabilities of writing and sensing operations are analyzed under global process variation and local mismatch of FD-SOI transistors and MTJ devices. The evaluations are performed in Cadence ADE-XL with 1000 runs Monte-Carlo simulations. 1-sigma FD-SOI transistor variability is considered, whereas the Gaussian distribution is realized in MTJ at the range 0.9 to 1.1. As shown in Fig. 5, without optimization methods, high writing probability can be realized with above 1.4 V V_{dd} (STT) and above 1.25 V V_{dd} (VCMA), with LVT type and large transistor width of access transistor. Considering the origin of process variations, the accumulated physical parameter fluctuation from imperfect CMOS and MTJ fabrication impact the 1T1M successful writing probability. The variation from MTJ device dominates the failure rate in bit-cell writing operation.

IV. OPTIMIZATION STRATEGIES WITH FD-SOI

In order to demonstrate improved energy and reliability performance in FD-SOI MTJ integration, two performance enhancement strategies are proposed in this section. Fig. 6(a) illustrates the block diagram of 1T1M writing circuit with a back-bias generator (BBG) block. A programmable BBG (specific IP in design kit) is utilized with a step-size of 100 mV [27], which provides back bias voltage pulse to boost the performance of the access transistor (FD-SOI $LVTnFET$). The transistor well with/without back-bias is controlled by on chip BBG drive and BBG control.

Our simulation results of STT-MRAM bit cell show that writing probability is increased up to 28% with 1.1 V V_{dd} (W/L = 400 nm/30 nm), using 2 V forward back-bias. With

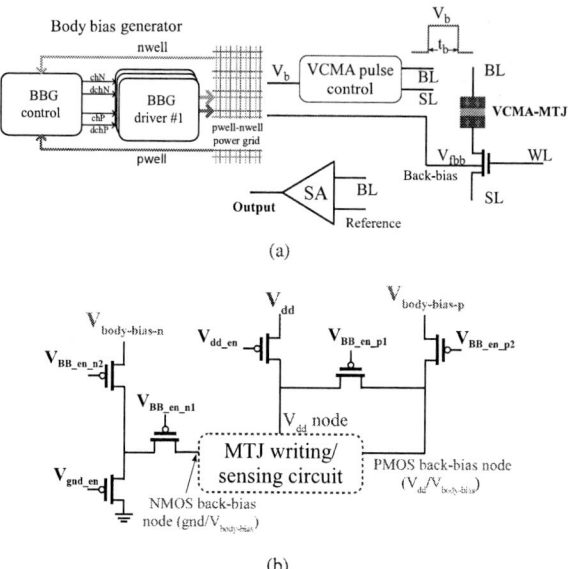

(b)

Fig. 6. (a) The block diagram of VCMA-MTJ writing circuit with performance enhancement strategy. The programmable back-bias generator with a step-size of 100 mV is used to realize VCMA pulse amplitude as well as FBB voltage [27]. (b) The flip-well enhancement strategy: the on/off transistor with $V_{BB-en-n(p)1}$ or $V_{BB-en-n(p)2}$ provides back-bias signal for $LVTnFET$ ($LVTpFET$) to enhance the performance of FD-SOI MTJ integration.

minimum dimension access transistor, a 1.2 V V_{dd} achieves 78.8% writing probability (catastrophic failure without back-bias strategy, see Fig. 5(b)).

For VCMA-MRAM, the BBG strategy can provide both VCMA pulse voltage and transistor back bias voltage to achieve a joint performance enhancement. In this study, we consider a back-bias range from 1 V to 1.5 V. The amplitude of VCMA voltage is generated from BBG and then modulated by VCMA pulse control block. In order to validate the proposed voltage assisted technique, we perform a step-up enhancement process for two design points: t_b = 0.4 ns, V_b step-up from 1.1 V (W = 80 nm) and 1 V (W = 400 nm). 1T-1MTJ writing circuit with and without back-bias are simulated. The proposed voltage-assisted method can boost failure design points to high yield design, with the power consumption tradeoff. Due to the VCMA-MTJ switching mechanism, the switching latency shows an improvement potential until V_b = 1.2 V (W = 400 nm), or V_b = 1.3 V (W = 80 nm).

Fig. 6(b) shows the alternative proposed flip-well enhancement strategy. The on/off transistor with $V_{BB-en-n(p)1}$ or $V_{BB-en-n(p)2}$ can provide back-bias signal for $LVTnFET$ ($LVTpFET$) to enhance the performance of FD-SOI MTJ integration. The limitation of this flip-well strategy is that $LVTnFET$ back-bias can only rely on the fixed V_{dd} of write/sense operation, whereas $LVTpFET$ back-bias is with 0 V. Its realized performance improvement is lower than the BBG based strategy.

In addition to the writing performance, the forward back-bias implementation with reduced V_t also compensates the process variation induced sensing failure. Fig. 7 shows the

978-1-5386-7100-9/18 $31.00 © 2018 IEEE

sensing probability versus supply voltage down scaling in the PCSA sensing circuit (designed with minimum W/L = 80nm/30nm). The successful sensing is well guaranteed at V_{dd} = 0.7 V, with 2 V back bias at 50 MHz clock frequency.

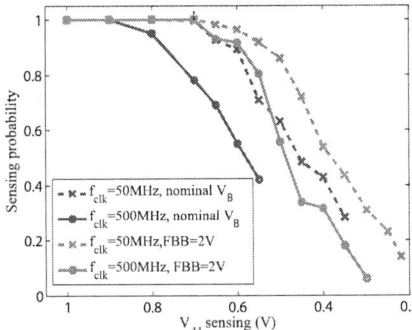

Fig. 7. An example of sensing probability improvement using 2V FBB. Only 3.5% failure is evaluated at 0.6 V

Table II compares the proposed performance enhancement strategies using the back-bias mehod. The BBG based method relies on a special IP core with large area, but provides a dynamic step-wise forward or reverse back-bias to the circuit building blocks. The flip-well method is implemented with a flip-well FD-SOI transistor structure and six additional transistors as the switches between V_{dd} and ground node. Although this solution lacks the flexibility needed for step-wise adjustment, it achieves high energy-efficiency and reduces the layout area consumption.

TABLE II
COMPARISON OF BBG AND FLIP-WELL STRATEGIES

	BBG based	Flip-well
Back-bias range	step-wise, V_{Bn}=[-0.3V, 3V]	V_{Bn}=1V, V_{Bp}=0V
Special IP block	back bias generator	n/a
Area penalty	large	small
Flexibility	high	low
Energy efficiency	low	high

V. DISCUSSION

The performance adjustment knobs from the transistor side are presented in Table III. Besides adjustment in supply voltage and access transistor dimension, the FD-SOI technique provides several V_t modulation methods (e.g., back-bias, poly bias) to improve its performance when integrating with MTJ devices. Higher performance metric requires more energy per writing/sensing operation [28], whereas the lower may cause operation failure in the design boundary. For ultra-low power MRAM design, V_{dd} scaling with guaranteed writing/sensing probability relies on assist methods, such as voltage modulation through BL/SL/WL, and FD-SOI based BBG and flip-well strategies.

The building blocks in the operation strategies are always constructed with LVT transistor. Based on the above analyzes and mentioned, a qualitative summary of the MRAM-on-FDSOI integration is presented in Table IV.

TABLE III
ENERGY-PERFORMANCE ADJUSTMENT KNOBS IN 1T1M BIT CELL

	STT-MTJ	VCMA-MTJ
Traditional knobs	V_{dd}, W/L	V_b, t_b, W/L
FD-SOI knobs	Back-bias, well configuration, poly bias	
Performance metric	I_{sw}, P_{sw}, delay, $P_{leakage}$	P_{sw}, delay, $P_{leakage}$

VI. CONCLUSION

In this paper, we have illustrated the attractive characteristic of FD-SOI MTJ integration to achieve energy-efficiency, high performance and reliability. STT-MTJ and VCMA-MTJ were integrated into an industrial FD-SOI process at 28 nm node. This integration can improve energy-delay efficiency, mitigate write asymmetry and strengthen the reliability performance, especially in emerging VCMA based MRAM bit-cell. We noted that the proposed FD-SOI process based circuit-level design strategies is effective to enhance the performance of MRAM-on-FDSOI integration, alleviate process variation induced failure. The general perspective of MRAM-on-FDSOI presented interesting possibilities for other non-volatile CMOS integration.

ACKNOWLEDGMENT

This work is supported by National Science and Technology Major Project 2017ZX01030101 and Pilot School Reform Funding Program in Southeast University 2242018K40100.

REFERENCES

[1] C. Chappert, A. Fert, and F. N. Van Dau, "The emergence of spin electronics in data storage," *Nature Materials*, vol. 6, pp. 813–823, 2007.
[2] K. C. Chun, H. Zhao, J. D. Harms, T. H. Kim, J. P. Wang, and C. H. Kim, "A scaling roadmap and performance evaluation of in-plane and perpendicular MTJ based STT-MRAMs for high-density cache memory," *IEEE Journal of Solid-State Circuits*, vol. 48, no. 2, pp. 598–610, Feb 2013.
[3] T. Ohsawa, H. Koike, S. Miura, H. Honjo, K. Kinoshita, S. Ikeda, T. Hanyu, H. Ohno, and T. Endoh, "A 1 Mb nonvolatile embedded memory using 4T2MTJ cell with 32 b fine-grained power gating scheme," *IEEE Journal of Solid-State Circuits*, vol. 48, no. 6, pp. 1511–1520, June 2013.
[4] S. Yu and P. Y. Chen, "Emerging memory technologies: Recent trends and prospects," *IEEE Solid-State Circuits Magazine*, vol. 8, no. 2, pp. 43–56, 2016.
[5] T. Maruyama, Y. Shiota, T. Nozaki, K. Ohta, N. Toda, M. Mizuguchi, A. Tulapurkar, T. Shinjo, M. Shiraishi, S. Mizukami, Y. Ando, and Y. Suzuki, "Large voltage-induced magnetic anisotropy change in a few atomic layers of iron," *Nature Nanotechnology*, vol. 4, pp. 158–161, 2009.
[6] P. Ong, N. Kioussis, P. K. Amiri, and K. Wang, "Electric-field-driven magnetization switching and nonlinear magnetoelasticity in Au/FeCo/MgO heterostructures," *Scientific Reports*, vol. 6, p. 29815, 2016.
[7] W. Kang, Y. Ran, Y. Zhang, W. Lv, and W. Zhao, "Modeling and exploration of the voltage-controlled magnetic anisotropy effect for the next-generation low-power and high-speed MRAM applications," *IEEE Transactions on Nanotechnology*, vol. 16, no. 3, pp. 387–395, 2017.
[8] S. Sharmin, A. Jaiswal, and K. Roy, "Modeling and design space exploration for bit-cells based on voltage-assisted switching of magnetic tunnel junctions," *IEEE Transactions on Electron Devices*, vol. 63, no. 9, pp. 3493–3500, 2016.
[9] S. Wang, H. Lee, F. Ebrahimi, P. K. Amiri, K. L. Wang, and P. Gupta, "Comparative evaluation of spin-transfer-torque and magnetoelectric random access memory," *IEEE Journal of Emerging and Selected Topics in Circuits and Systems*, vol. 6, no. 2, pp. 134–145, 2016.

978-1-5386-7100-9/18 $31.00 © 2018 IEEE

TABLE IV

QUALITATIVE SUMMARY OF THE MRAM-ON-FDSOI INTEGRATION

	Performance of MRAM-on-FDSOI Integration
Advantages	1). Extended V_{dd} range with improved energy-delay efficiency in writing and sensing operations. 2). Flip-well configuration and forward back-bias are applied to improve the performance of hybrid integration. 3). The tunable performance (low power&speed, high performance) is achieved with the proposed knobs. 4). FD-SOI is a planer technology, MTJ lies on the top of the highest CMOS metal layer.
Challenges	1). There exists large variability due to global variation and mismatch in FD-SOI and MTJ devices. 2). Area penalty exists when using optimization, such as BBG and flip-well strategies. 3). Write/read assist techniques are needed to meet design specification and reliability requirement. 4). The dimension scaling down of FD-SOI and MTJ could make the integration even worse.

[10] R. Appeltans, P. Raghavan, G. S. Kar, A. Furnmont, L. V. der Perre, and W. Dehaene, "A smaller, faster, and more energy-efficient complementary STT-MRAM cell uses three transistors and a ground grid: More is actually less," *IEEE Transactions on Very Large Scale Integration (VLSI) Systems*, vol. 25, no. 4, pp. 1204–1214, April 2017.

[11] H. Cai, Y. Wang, L. de Barros Naviner, and W. Zhao, "Robust ultra-low power non-volatile logic-in-memory circuits in FD-SOI technology," *IEEE Transactions on Circuits and Systems I: Regular Papers*, vol. 64, no. 4, pp. 847–857, 2016.

[12] K. Jabeur and G. Prenat. "Design of a full 1Mb STT-MRAM based on advanced FDSOI technology," in *21st International Conference on Circuits, Systems, Communications and Computers*, 2017, p. 01003.

[13] J. M. Portal, M. Bocquet, S. Onkaraiah, M. Moreau, H. Aziza, D. Deleruyelle, K. Torki, E. Vianello, A. Levisse, B. Giraud, O. Thomas, and F. Clermidy, "Design and simulation of a 128 kb embedded nonvolatile memory based on a hybrid RRAM (hfo2)/28 nm FDSOI CMOS technology," *IEEE Transactions on Nanotechnology*, vol. 16, no. 4, pp. 677–686, 2017.

[14] H. Cai, Y. Wang, W. Zhao, and L. de Barros Naviner, "Multiplexing sense amplifier based magnetic flip-flop in 28nm FDSOI technology," *Nanotechnology, IEEE Transactions on*, vol. 14, no. 4, pp. 761–767, 2015.

[15] M. K. Niranjan, C.-G. Duan, S. S. Jaswal, and E. Y. Tsymbal, "Electric field effect on magnetization at the Fe/MgO(001) interface," *Applied Physics Letters*, vol. 96, no. 22, p. 222504, 2010.

[16] S. E. Barnes, J. Ieda, and S. Maekawa, "Rashba spin-orbit anisotropy and the electric field control of magnetism," *Scientific report*, vol. 4, pp. 4105:1–5, 2014.

[17] C. J. Lin, S. H. Kang, Y. J. Wang, K. Lee, X. Zhu, W. C. Chen, X. Li, W. N. Hsu, Y. C. Kao, M. T. Liu, W. C. Chen, Y. Lin, M. Nowak, N. Yu, and L. Tran, "45nm low power CMOS logic compatible embedded STT mram utilizing a reverse-connection 1T/1MTJ cell," in *2009 IEEE International Electron Devices Meeting (IEDM)*, Dec 2009, pp. 1–4.

[18] K. Shamsi, Y. Bi, Y. Jin, P. E. Gaillardon, M. Niemier, and X. S. Hu, "Reliable and high performance STT-MRAM architectures based on controllable-polarity devices," in *2015 33rd IEEE International Conference on Computer Design (ICCD)*, 2015, pp. 343–350.

[19] W. Kang, L. Zhang, J. O. Klein, Y. Zhang, D. Ravelosona, and W. Zhao, "Reconfigurable codesign of STT-MRAM under process variations in deeply scaled technology," *IEEE Transactions on Electron Devices*, vol. 62, no. 6, pp. 1769–1777, June 2015.

[20] P. Flatresse, B. Giraud, J. P. Noel, B. Pelloux-Prayer, F. Giner, D. K. Arora, F. Arnaud, N. Planes, J. L. Coz, O. Thomas, S. Engels, G. Cesana, R. Wilson, and P. Urard, "Ultra-wide body-bias range LDPC decoder in 28nm UTBB FDSOI technology," in *Solid-State Circuits Conf. Digest of Tech. Papers (ISSCC), 2013 IEEE Intl.*, Feb 2013, pp. 424–425.

[21] J. P. Noel, O. Thomas, M. A. Jaud, O. Weber, T. Poiroux, C. Fenouillet-Beranger, P. Rivallin, P. Scheiblin, F. Andrieu, M. Vinet, O. Rozeau, F. Boeuf, O. Faynot, and A. Amara, "Multi-v_T UTBB FDSOI device architectures for low-power CMOS circuit," *IEEE Transactions on Electron Devices*, vol. 58, no. 8, pp. 2473–2482, 2011.

[22] B. Nikolic, M. Blagojevic, O. Thomas, P. Flatresse, and A. Vladimirescu, "Circuit design in nanoscale FDSOI technologies," in *Microelectronics Proceedings - MIEL 2014, 2014 29th International Conference on*, 2014, pp. 3–6.

[23] O. Thomas, B. Zimmer, S. Toh, L. Ciampolini, N. Planes, R. Ranica, P. Flatresse, and B. Nikolic, "Dynamic single-p-well SRAM bitcell characterization with back-bias adjustment for optimized wide-voltage-range SRAM operation in 28nm UTBB FD-SOI," in *Electron Devices Meeting (IEDM), 2014 IEEE International*, 2014, pp. 3.4.1–3.4.4.

[24] F. Abouzeid, A. Bienfait, K. Akyel, A. Feki, S. Clerc, L. Ciampolini, F. Giner, R. Wilson, and P. Roche, "Scalable 0.35 V to 1.2 V SRAM bitcell design from 65 nm CMOS to 28 nm FDSOI," *Solid-State Circuits, IEEE Journal of*, vol. 49, no. 7, pp. 1499–1505, 2014.

[25] A. Valentian, Y. Thonnart, B. Pelloux-Prayer, and P. Flatresse, "Single-well design in FDSOI technology: Towards energy-efficient ultra-wide voltage range digital circuits," in *SOI-3D-Subthreshold Microelectronics Technology Unified Conference (S3S), 2015 IEEE*, 2015, pp. 1–2.

[26] Y. Wang *et al.*, "Compact model of dielectric breakdown in spin-transfer torque magnetic tunnel junction," *IEEE Trans. on Electron Devices*, vol. 63, no. 4, pp. 1762–1767, 2016.

[27] M. Blagojevic, M. Cochet, B. Keller, P. Flatresse, A. Vladimirescu, and B. Nikolic, "A fast, flexible, positive and negative adaptive body-bias generator in 28nm FDSOI," in *2016 IEEE Symposium on VLSI Circuits (VLSI-Circuits)*, June 2016, pp. 1–2.

[28] M. Alioto and M. Shahghasemi, "The internet of things on its edge: Trends toward its tipping point," *IEEE Consumer Electronics Magazine*, vol. 7, no. 1, pp. 77–87, Jan 2018.

2018 IEEE Computer Society Annual Symposium on VLSI

High Performance Ternary Multiplier using CNTFET

Subhendu Kumar Sahoo, *Member, IEEE*, Krishna Dhoot, Rasmita Sahoo*

Department of Electrical Engineering

BITS-Pilani, Hyderabad Campus, India.

Nalla Narasimha Reddy School of Engineering, Hyderabad.*

Email: sahoo@hyderabad.bits-pilani.ac.in

Abstract—Ternary logic is a promising alternative to the conventional binary logic in VLSI design as it provides the advantages of reduced interconnects, higher operating speeds and smaller chip area. This work presents a ternary multiplier using carbon nanotube field effect transistors (CNTFETs). The proposed designs use in-depth analysis of addition required for designing a two trit multiplier. Based on this analysis two efficient adders are proposed. These adders are used to optimize the multiplier design. The proposed circuits are extensively simulated using HSPICE to obtain power, delay and power delay product. The circuit performances are compared with designs reported in recent literature. This circuit demonstrates a power delay product improvement up to 16.8%, with lesser transistor count of 16%. So, the use of these circuits in complex arithmetic circuits will be advantageous.

Index Terms—Carbon nanotube field effect transistors (CNT-FETs), multiple-valued logic (MVL), ternary multiplier, adder.

I. INTRODUCTION

A large part of the success of the MOS transistor is due to its scalability to much smaller dimensions, which results in better performance. This trend still continues in accordance with Moore's law, and silicon-based technology has gone through a phenomenal growth in the last few decades. However, as MOSFETs are approaching their limiting size in the nanometer regime, the semiconductor industry is looking for different materials and alternative devices to integrate with the current silicon-based technology and, in the long run, possibly replace it. Over the past few decades, carbon nanotubes (CNTs) have attracted significant attention in the field of electronics, due to their unique structure and excellent physical properties [1], [2], [3], [4], [5], [6], [7], [8]. Currently, the use of CNTs in the channel region of FET is being investigated experimentally to obtain a new device called carbon nanotube field effect transistor (CNTFET), first demonstrated in 1998 [9]. Because of high electron mobility, near ballistic transport, high mechanical and thermal stability of CNTs, CNTFETs are being considered as one of the most promising candidates for post-silicon electronics [10], [11], [12], [13]. Further, reduced ambipolar conduction, reduced leakage current, extended scaling limit, high drive current, faster operation, high temperature resilience, large transconductance and near ballistic transport are some of the features which made CNTFET more advantageous compared to other devices.

To the best of knowledge of the authors, there is no definitive technique available for the growth of CNTs with required orientation, but the investigations are in progress. A recent paper by George S. et al.[14] presented a thorough review on CNTFET. In this, the fabrication of CNT as well as CNTFET, methods to obtain a particular type of CNT, purification and proper placement of CNT are discussed. With the current trend of widespread interest in CNTFETs, in near future it may be possible to place required CNTs in the channel region. Though the investigations related to the issues associated with fabrication and purification of CNTs are still continuing, we can explore the possibility of novel circuits using CNTFET for high performance implementation.

Nowadays, many studies are going on for designing and exploring the application of CNTFETs in logic gates and benchmarking their performance advantage over the existing MOS technology. The CNTFET circuit application includes binary logic gates [6], [15], [16], [17], [18], ternary logic gate [5], ternary and binary memory cells [7], [17] and multiple-valued logics [8],[19]. The application of CNTFETs for multiple-valued logic has gained keen interest as the threshold voltage of CNTFETs can be controlled by proper selection of the chiral vector of the CNT. Logic circuits as well as different adders, multipliers and memories are also designed to obtain less delay, lower power consumption and to have reduced interconnection complexity.

Multiplication is the most important operation involved in all signal processing application. In the present work, two new adder circuit designs are explored suitable for designing of two-trit multiplier. The proposed multiplier using these adders circuit has been demonstrated to have power delay product improvement up to 16.8% , with less transistor count in comparison to a recent reported work [20]. The rest of the paper is organized as follows. A brief review of multiple-valued logic, CNTFET and its suitability for ternary logic are discussed in section II. Some of the basic ternary logic gates and the previously reported ternary multiplier is discussed in section III. Our proposed multiplier using new adders is presented in section IV. The proposed circuits implementation and their comparisons are given in section V followed by concluding remark in section VI.

978-1-5386-7100-9/18 $31.00 © 2018 IEEE

269

II. CNTFET AND ITS SUITABILITY FOR TERNARY LOGIC

Unlike conventional binary logic, MVL uses more than two logic states. Nowadays, ternary logic (or three-valued logic) has attracted considerable interests due to its potential advantages over binary logic for designing digital systems. Simplicity and efficiency of the design, requirement of less memory and less interconnections, reduction of chip area, high bandwidth parallel and serial data transfer, high potential for increasing computational speed, reducing switching activity, and implementation of many arithmetic and logic functions in a single chip are some of the advantages of MVL [5], [7], [8]. In this work, we have concentrated on ternary logic which is a MVL with three states: 0, 1, and 2 representing the ternary values low, middle and high, respectively.

The structure of a CNTFET is shown in Fig. 1. This is similar to conventional MOSFET in which semiconducting single-wall CNTs are used in the channel region. In this, undoped semiconducting carbon nanotubes are placed under the gate as channel region, while heavily doped CNT segments are placed between the gate and the source/drain to allow for a low series resistance in the ON-state. When the gate potential increases, the device is electro statically turned ON or OFF via the gate and the threshold voltage of the intrinsic CNT in the channel which is an inverse function of the diameter is given as [21]:

$$V_{th} = \frac{aV_\pi}{\sqrt{3}ed_{CNT}} \tag{1}$$

Here $V_\pi = 3.033$ eV is the carbon $\pi - \pi$ bond energy in the tight binding model, e is the unit electron charge and d_{CNT} is the diameter of the CNT used in the channel. It is evident from (1) that the threshold voltage of a CNTFET varies inversely with the diameter of the CNT and hence can be adjusted to a required value by choosing a proper chiral vector. This is the major advantage of CNTFETs to be used for ternary logic.

The threshold voltage for a CNTFET depends on the band gap of the CNT which depends on the orientation of the graphene sheet to obtain the CNT. Generally the chiral vector for a particular CNT is represented by two integers (n, m), which depends on the orientation of the graphene sheet to obtain the CNT [21]. Further the diameter of a CNT in terms of (n, m) is given as [21]:

$$d_{CNT} = \frac{a}{\pi}\left(n^2 + m^2 + nm\right)^{1/2} \tag{2}$$

$a = |\vec{a_1}| = |\vec{a_2}| = 1.42 \times \sqrt{3} = 2.46A^o$

Where $\vec{a_1}$ and $\vec{a_2}$ are the primitive lattice vectors of graphene.

III. REVIEW OF TERNARY LOGIC GATES AND MULTIPLIER

In ternary logic, the voltage levels used to represent the logic states 0, 1 and 2 are 0, $V_{dd}/2$ and V_{dd} respectively, where V_{dd} is the supply voltage. The various ternary logic gates like inverter, NAND, NOR, AND, OR are reported in literature [5] and also briefly discussed here. Then single-trit multiplier and adder circuits reported in [20] for larger multiplier are discussed.

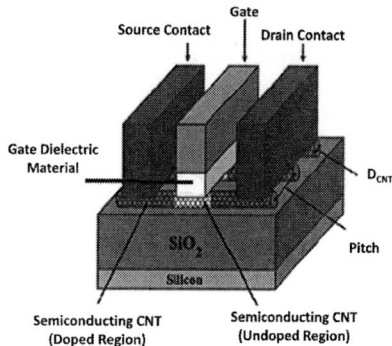

Figure 1. Structure of a CNTFET using multiple CNT in the channel region[8].

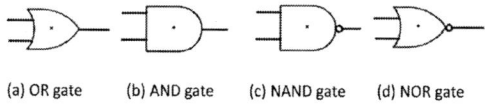

(a) OR gate   (b) AND gate   (c) NAND gate   (d) NOR gate

Figure 2. Different gates used in ternary logic.

A. Ternary logic gates

For understanding other logic gates, we will assume X_i, X_j are two inputs that can take value from the set 0, 1, 2. Then the basic logical operations in ternary logic can be understood as follows:

$$OR : X_i + X_j = max(X_i, X_j) \tag{3}$$

$$AND : X_i.X_j = min(X_i, X_j) \tag{4}$$

$$NOR : \overline{X_i + X_j} = \overline{max(X_i, X_j)} \tag{5}$$

$$NAND : \overline{X_i.X_j} = \overline{min(X_i, X_j)} \tag{6}$$

The various symbols used for these ternary logic gates are shown in Fig. 2.

B. Unary operators

Unary operators are the functions defined on single ternary variable. Some of the important unary operators are shown in Table I [20]. $\overline{A}$, A_P and A_N are the three types of inverters used in ternary logic and they are standard ternary inverter (STI), positive ternary inverter (PTI) and negative ternary inverter (NTI) defined on a single ternary variable respectively. Apart from these, other *unary* operators are the cyclic operators like A^1 and A^2. Multiple functions can be implemented using a combination of binary and unary operators. For example, a function mapping {0, 1, 2} to {0, 0, 1} can be defined as $1.\overline{A_P}$. Unary operators and simple functions like $1.\overline{A_P}$ can be used in designing mux-based ternary circuits where output is obtained as a function of single input while other inputs act as select lines.

Table I
UNARY OPERATORS IN TERNARY LOGIC

A	A_P	A_N	$\overline{A}$	A^1	A^2	$\overline{A^2}$	$\overline{A_P}$	$\overline{A_N}$
0	2	2	2	1	2	0	0	0
1	2	0	1	2	0	2	0	2
2	0	0	0	0	1	1	2	2

Table II
SINGLE-TRIT MULTIPLIER TRUTH TABLE

A	B	Carry	Product
0	0	0	0
0	1	0	0
0	2	0	0
1	1	0	1
1	2	0	2
2	2	1	1

C. Ternary multiplier

Among various arithmetic operations, multiplication is the most computationally intensive unit used extensively in digital signal processing. The multiplication involves partial product generations and accumulation of these partial products. Sridharan in [20] has presented a simple partial product generation circuit (single-trit multiplier) using CNTFET based multiplexers. The truth table of a single-trit multiplier for ternary inputs A, B and outputs as product (P) and carry (C) is given in Table II. For a multi-trit multiplier, the partial products are generated using single trit multiplier and they need to be accumulated using adders. Sridharan [20] also has presented two adders namely FA_1 and FA_2 using multiplexer based approach. The truth table for FA_2 is given in Table III. In our work we have used these partial product generator and adders. Along with these, we have proposed two new adders and used them for the designing of a two-trit multiplier.

IV. PROPOSED TWO-TRIT MULTIPLIER

An optimized two-trit multiplier is proposed using single trit multiplier[20], half adders and full adders. Some of the adders used here are efficiently designed based on thorough analysis of the requirement of two-trit multiplier.

A. Two-trit multiplier

A two-trit multiplier multiplies two numbers $A = A_1 A_0$ with $B = B_1 B_0$ where A_1, A_0, B_1, B_0 are ternary digits. The multiplication of one ternary digit with other will generate one product digit (P) and a carry digit (C) based on Table II. So, the multiplication of number A with B will result in four set of P and C as shown in top of Fig. 3. These P and C values obtained using single-trit multiplier need to be added to get the final multiplication results. The tree structure for this addition is shown in Fig. 3. The circuit for FA_1 and FA_2 are efficiently designed in [20] and also used in the present work are discussed briefly in next subsection. The new circuits for proposed FA_3 and HA_1 are discussed in the following subsections.

Table III
FULL ADDER FA_2 TRUTH TABLE

		Cin=0		Cin=1	
A	B	Sout	Cout	Sout	Cout
0	0	0	0	1	0
1	0	1	0	2	0
2	0	2	0	0	1
0	1	1	0	2	0
1	1	2	0	0	1
2	1	0	1	1	1

$B_0 \times A_0 = C_0, P_0 \mid B_0 \times A_1 = C_1, P_1 \mid B_1 \times A_0 = C_2, P_2 \mid B_1 \times A_1 = C_3, P_3$

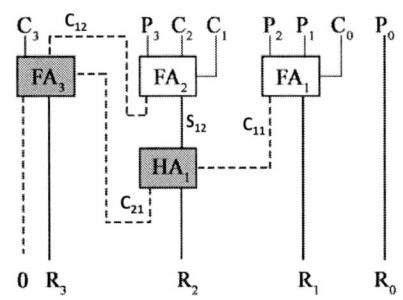

Figure 3. Two-trit multiplier block diagram

B. Single-Trit Multiplier and Full Adders FA_1, FA_2

A single trit multiplier as proposed in [20] is used in the two-trit multiplier design. For designing a two-trit multiplier, four such single trit multipliers are required and each single-trit multiplier requires 26 CNTFETs. Thus a total of 104 CNTFETs are required to generate the partial products. The truth table of single-trit multiplier is shown in Table II. In this the P can can take value from the set $\{0, 1, 2\}$ and C can take value from the set $\{0, 1\}$.

The full adder FA_1 in Fig. 3 has two of its inputs as ternary while the other input is binary. So, the maximum possible value of sum can be 2 while that of carry can be 1. Thus C_{11} (in Fig.3) can be considered as binary. Similarly for the full adder FA_2 in Fig. 3, two of its inputs are binary (C_1, C_2) and the third input is ternary (P_3). So, this adder output carry can take a value only from the set $\{0, 1\}$ (It is binary). These special cases for full adders are considered and circuits are designed efficiently[20]. Thus these circuits are directly used in the proposed two-trit multiplier.

Table IV
HALF ADDER HA_1 TRUTH TABLE

A	B	Carry	Sum
0	0	0	0
1	0	0	1
2	0	0	2
0	1	0	1
1	1	0	2
2	1	1	0

C. Half Adder HA_1

From Fig. 3, one input to HA_1 comes from the sum output of FA_2 i.e S_{12} is ternary and the other input is the carry out of FA_1 i.e C_{11} and is binary. Thus, instead of using a conventional half adder with two ternary inputs, a modified half adder with one ternary and one binary input is proposed.

The truth table for this half adder is shown in Table IV. Based on the analysis of the Table IV, the expression for the two outputs of HA_1 can be written as:

$$SUM = B_0(A) + B_1(A^1)$$

$$CARRY = B_1(1.\overline{A_p})$$

The circuit diagram based on the above two expressions are given in Fig. 6. Here the input B along with NTIs is used to generate select lines for transmission gates. When B = 0, Sum = A and Carry = 0; when B = 1, Sum = A^1 and Carry = $1.\overline{A_p}$. The functions A^1 and $1.\overline{A_p}$ are genrated from A as shown in Fig. 6.

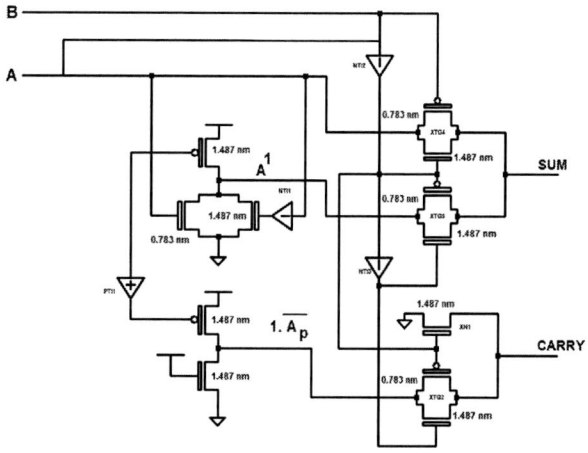

Figure 4. Half Adder HA_1 Schematic

D. Full Adder FA_3

In two-trit multiplier of Fig. 3, the adder FA_3 receives inputs as C_3 and carry out C_{12} from FA_2 and C_{21} from HA_1. All the three inputs to the full adder FA_3 are binary. Apart from this observation, another proposition is provided below, which further reduces the complexity of the circuit.

Proposition : The carry outputs from HA_1 (C_{21}) and FA_2 (C_{12}) can not be simultaneously 1.

Proof : In Fig. 3 the adder FA_2 has two binary inputs and one ternary input. Observation of the truth Table III for this shows that, for $C_{12} = 1$, the sum S_{12} can take value as 1 or 0, but not 2. So for $C_{12}=1$, the HA_1 inputs can take a value only from the set {0,1}. Thus, R2 can take a value only from the set {0,1,2}, and the carry out put C_{21} will be 0. Similarly, when C_{21} is 1, C_{12} will be 0. So, out of C_{12} and C_{21}, only one of the carry can be 1 at a time.

Based on this proposition, only one of the two inputs C_{12} and C_{21} to the adder FA_3 can be 1 at a time. The truth table for FA_3 is given in Table V. So, we can add C_{12} and C_{21} using an OR gate. The output of this OR gate can be added with C_3 using a halfadder like circuit to get FA_3 output. This simplified adder circuit is given in Fig. 5 . It is to be noted that there is no carry output from this Full adder.

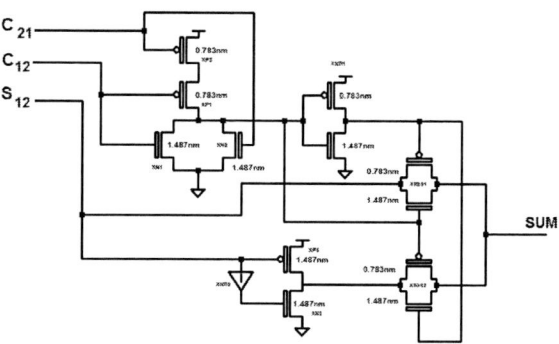

Figure 5. Full Adder FA_3 Schematic

Table V
FULL ADDER FA_3 TRUTH TABLE

S_{12}	C_{12}	C_{21}	Sum
0	0	0	0
1	0	0	1
0	0	1	1
1	0	1	2
0	1	0	1
1	1	0	2

V. SIMULATION AND COMPARISON

The proposed design and the previous designs [20] [22] in the literature have been simulated in Synopsys HSPICE using the standard Stanford CNTFET model in [23], [24] for transistors with supply voltage V_{dd} = 0.9 V. The spice model parameters used in the simulation are given in Table VI. The chiral vectors of CNT are selected from (19, 0), (13, 0) and (10, 0) based on required threshold voltage. Following comparisons are done based on the simulation results.

A. Performance Comparison for Adder Circuits

The proposed adders FA_3 and the ones presented in [20] are implemented and simulated at two drive strengths - 100 MHz, $2fF$ load and 250MHz, $3fF$ load. Average power, average delay and PDP are obtained and noted in Table VII. It is observed that the proposed FA_3 circuit has power and the PDP advantage of 85.9% and 80.41% respectively, compared to FA_1 and FA_2. Further, the transistor count decreases by 76.27% and 69.57% with respect to FA_1 and FA_2, respectively. As Navi [22] has not given any circuit for full adder, we have not compared with them.

Similarly, the proposed half adder HA_1 and the one in [20] , [22] are implemented and the simulation results are noted in

978-1-5386-7100-9/18 $31.00 © 2018 IEEE

Table VI
DEVICE PARAMETER DESCRIPTION VALUE

Device Parameter	Description	Value
Lch	Physical channel length	32nm
Lgeff	The,mean free path in the intrinsic CNT channel region due to non-ideal elastic,scattering.	100nm
Lss	The length of doped CNT source-side extension region.	32nm
Ldd	The length of doped CNT drain-side extension region.	32nm
Efo	The Fermi level of the doped S/D tube.	6eV
Kox	The dielectric constant of high-k top gate dielectric material	16
Tox	The thickness of high-k top gate dielectric material	4nm
Pitch	The distance between the centers of two adjacent CNTs within the same device	20nm

Table VIII for comparison. This shows an improvement upto 78.8% in PDP and 48.71% in transistor count in comparison to [20]. The proposed half adder is consuming more power than [22]. However, the delay is much larger and output voltage level are inaccurate for certain combination of inputs. To understand this, the simulated waveforms are given in Fig. 6. Specifically, for input A = 1 and B = 0, the sum output is giving inaccurate logic level because in their circuit [22] the sum is driven by two different logic values.

Table VII
COMPARISON OF FULL ADDER DESIGNS WITH MULTIPLE DRIVE STRENGHTS

Design	Load (fF)	Power (uW)	Delay (ps)	PDP (e-15)	Tranistor count
[20] FA$_1$	2	136.10	45.10	6.14	59
	3	136.20	48.56	6.61	
[20] FA$_2$	2	112.90	39.33	4.44	46
	3	113.40	45.69	5.18	
FA$_3$ proposed	2	33.82	25.67	0.87	14
	3	33.90	35.56	1.21	

Table VIII
COMPARISON OF HALF ADDER DESIGN WITH MULTIPLE DRIVE STRENGHTS

Design	Load (fF)	Power (uW)	Delay (ps)	PDP (e-15)	Percentage improvement in PDP over [20]	Transistor Count
[20] HA	2	135.30	28.27	3.82		35
	3	136.30	32.56	4.44		
[22] HA	2	3.16	186.19	0.50		39
	3	3.28	242.30	0.79		
HA$_1$ proposed	2	45.37	17.78	0.81	78.79%	20
	3	45.49	22.84	1.04	76.5%	

B. Two-trit multiplier Comparison

A two trit multiplier is designed as shown in Fig. 3 with all adder circuits from [20]. Again, the two trit multiplier is designed using FA_1, FA_2 from [20] and our proposed FA_3 and HA_1. Both the designs are simulated at 100 MHz and loads 2, 3, 4 and 5 fF and noted in Table IX. It is observed that the proposed design gives up to 16.8% advantage in terms of PDP. A plot of PDP for two designs for varying load from 2 fF to 5 fF is given in Fig. 7. This shows that proposed design is giving better performance. Further the transistor

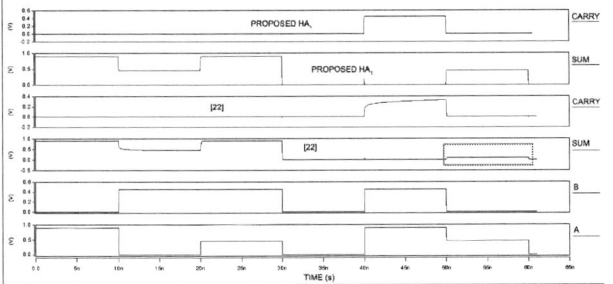

Figure 6. simulated waveforms for proposed half adder and [22]

count decreases by 16.2% in the proposed design. Navi in [22] has reported a single trit multiplier. However, they have not given any multi trit multiplier. In our work we have used the single trit multiplier reported by [20]. So, we have not compared our two trit multiplier with [22].

Table IX
TWO-TRIT MULTIPLIER COMPARISON

Design	Load (fF)	Power (uW)	Delay (ps)	PDP (e-15)	Percentage imrovement in PDP	Transistor Count
[20]	2	582.8	30.66	17.87		290
	3	582.8	40.13	23.39		
	4	582.9	50.96	29.70		
	5	583.0	59.60	34.75		
Proposed	2	548.2	27.00	14.80	16.8%	243
	3	548.3	36.02	19.75	15.5%	
	4	548.3	46.43	25.46	14.2%	
	5	548.3	54.59	29.93	13.8%	

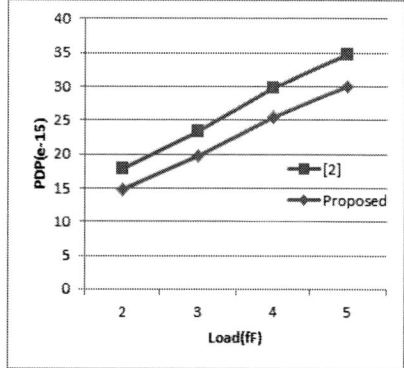

Figure 7. Comparison between two-trit multipliers

VI. CONCLUSION

A high performance two-trit multiplier using CNTFET is proposed. This multiplier uses two new adder circuits designed specifically for this two-trit multiplier. Rigorous HSPICE simulations reveal that the full adder and half adder design shows 85.9% and 78.8% reduction in PDP compared to recent reported work. The two-trit multiplier shows 16.8% improvement in PDP with transistor reduction of 16%. This two-trit multiplier can be used as basic block to implement higher size multipliers.

REFERENCES

[1] A. Akturk, G. Pennington, and N. Goldsman, "Quantum modeling and proposed designs of CNT-embedded nanoscale MOSFETs," *IEEE Transactions on Electron Devices*, vol. 52, no. 4, pp. 577–584, 2005.

[2] J.-C. Charlier and P. Lambin, "Electronic structure of carbon nanotubes with chiral symmetry," *Physical Review B*, vol. 57, no. 24, p. R15037, 1998.

[3] P. Avouris and R. Martel, "Progress in carbon nanotube electronics and photonics," *MRS bulletin*, vol. 35, no. 4, p. 306, 2010.

[4] A. D. Franklin, R. A. Sayer, T. D. Sands, D. B. Janes, and T. S. Fisher, "Vertical carbon nanotube devices with nanoscale lengths controlled without lithography," *IEEE Transactions on Nanotechnology*, vol. 8, no. 4, pp. 469–476, 2009.

[5] S. Lin, Y.-B. Kim, and F. Lombardi, "CNTFET-based design of ternary logic gates and arithmetic circuits," *IEEE transactions on nanotechnology*, vol. 10, no. 2, pp. 217–225, 2011.

[6] M. H. Ben-Jamaa, K. Mohanram, and G. De Micheli, "An efficient gate library for ambipolar CNTFET logic," *IEEE Transactions on Computer-Aided Design of Integrated Circuits and Systems*, vol. 30, no. 2, pp. 242–255, 2011.

[7] F. Lombardi, Y.-B. Kim, and S. Lin, "Design of a ternary memory cell using CNTFET," *IEEE Transactions On Nanotechnology*, vol. 11, no. 5, 2012.

[8] M. H. Moaiyeri, R. F. Mirzaee, A. Doostaregan, K. Navi, and O. Hashemipour, "A universal method for designing low-power carbon nanotube FET-based multiple-valued logic circuits," *IET Computers & Digital Techniques*, vol. 7, no. 4, pp. 167–181, 2013.

[9] R. Martel, T. Schmidt, H. Shea, T. Hertel, and P. Avouris, "Single- and multi-wall carbon nanotube field-effect transistors," *Applied Physics Letters*, vol. 73, no. 17, pp. 2447–2449, 1998.

[10] P. Avouris, J. Appenzeller, R. Martel, and S. J. Wind, "Carbon nanotube electronics," *Proceedings of the IEEE*, vol. 91, no. 11, pp. 1772–1784, 2003.

[11] J. Guo, S. Hasan, A. Javey, G. Bosman, and M. Lundstrom, "Assessment of high-frequency performance potential of carbon nanotube transistors." *IEEE Transactions on Nanotechnology*, vol. 4, no. 6, pp. 715–721, 2005.

[12] M.-H. Yang, K. B. Teo, L. Gangloff, W. I. Milne, D. G. Hasko, Y. Robert, and P. Legagneux, "Advantages of top-gate, high-k dielectric carbon nanotube field-effect transistors," *Applied Physics Letters*, vol. 88, no. 11, p. 113507, 2006.

[13] L. Wei, D. J. Frank, L. Chang, and H.-S. P. Wong, "Noniterative compact modeling for intrinsic carbon-nanotube FETs: quantum capacitance and ballistic transport," *IEEE Transactions on Electron Devices*, vol. 58, no. 8, pp. 2456–2465, 2011.

[14] G. S. Tulevski, A. D. Franklin, D. Frank, J. M. Lobez, Q. Cao, H. Park, A. Afzali, S.-J. Han, J. B. Hannon, and W. Haensch, "Toward high-performance digital logic technology with carbon nanotubes," *ACS nano*, vol. 8, no. 9, pp. 8730–8745, 2014.

[15] A. Bachtold, P. Hadley, T. Nakanishi, and C. Dekker, "Logic circuits with carbon nanotube transistors," *Science*, vol. 294, no. 5545, pp. 1317–1320, 2001.

[16] M. Najari, S. Fregonese, C. Maneux, H. Mnif, N. Masmoudi, and T. Zimmer, "Schottky barrier carbon nanotube transistor: Compact modeling, scaling study, and circuit design applications," *ieee transactions on electron devices*, vol. 58, no. 1, pp. 195–205, 2011.

[17] S. Lin, Y.-B. Kim, and F. Lombardi, "Design of a CNTFET-based SRAM cell by dual-chirality selection," *IEEE Transactions on Nanotechnology*, vol. 9, no. 1, pp. 30–37, 2010.

[18] I. O'Connor, J. Liu, F. Gaffiot, F. Prégaldiny, C. Lallement, C. Maneux, J. Goguet, S. Frégonèse, T. Zimmer, L. Anghel *et al.*, "CNTFET modeling and reconfigurable logic-circuit design," *IEEE Transactions on Circuits and Systems I: Regular Papers*, vol. 54, no. 11, pp. 2365–2379, 2007.

[19] P. Keshavarzian and R. Sarikhani, "A novel CNTFET-based ternary full adder," *Circuits, Systems, and Signal Processing*, vol. 33, no. 3, pp. 665–679, 2014.

[20] B. Srinivasu and K. Sridharan, "Low-complexity multiternary digit multiplier design in cntfet technology," *IEEE Transactions on Circuits and Systems II: Express Briefs*, vol. 63, no. 8, pp. 753–757, 2016.

[21] R. Saito, G. Dresselhaus, M. S. Dresselhaus *et al.*, *Physical properties of carbon nanotubes*. World Scientific, 1998, vol. 35.

[22] S. Tabrizchi, N. Azimi, and K. Navi, "A novel ternary half adder and multiplier based on carbon nanotube field effect transistors," *Frontiers of Information Technology & Electronic Engineering*, vol. 18, no. 3, pp. 423–433, 2017.

[23] J. Deng and H.-S. P. Wong, "A compact spice model for carbon-nanotube field-effect transistors including nonidealities and its application–part i: Model of the intrinsic channel region," *IEEE Transactions on Electron Devices*, vol. 54, no. 12, pp. 3186–3194, 2007.

[24] ——, "A compact spice model for carbon-nanotube field-effect transistors including nonidealities and its application–part ii: Full device model and circuit performance benchmarking," *IEEE Transactions on Electron Devices*, vol. 54, no. 12, pp. 3195–3205, 2007.

978-1-5386-7100-9/18 $31.00 © 2018 IEEE

2018 IEEE Computer Society Annual Symposium on VLSI

A Robust Dual Reference Computing-in-Memory Implementation and Design Space Exploration Within STT-MRAM

Liuyang Zhang*, Wang Kang*, Hao Cai[†], Peng Ouyang*, Lionel Torres[‡], Youguang Zhang*
Aida Todri-Sanial[‡] and Weisheng Zhao*

*Fert Beijing Institute and School of Electronic and Information Engineering
Beihang University, Beijing 100191, China
[†]Telecom Paristech, University of Paris-Saclay, Paris 75013, France
[‡]LIRMM/University of Montpellier, Montpellier 34095, France
Email: {wang.kang, weisheng.zhao}@buaa.edu.cn

Abstract—Due to the "memory wall" in conventional Von-Neumann computer architectures, the limited bandwidth between processors and memories has become one of the most critical bottlenecks to improve system performance. With the emerging of non-volatile memories, the computing-in-memory (CIM) paradigm has regained interest to tackle the issue at the architecture level. CIM can effectively alleviate the stress on the limitted bandwidth by performing logic operations within memories. However, CIMs are not yet studied carefully at the circuit level, and even its reliability and performance. In this paper, we proposed a CIM implementation: dual reference (DualRef) scheme at the circuit level within STT-MRAM (Spin Transfer Torque Magnetic Random Access Memory) array. Simulations were carried out to verify the functionality and assess the reliability and performance of DualRef scheme in terms of operation error rate, sensing margin, operation delay and dynamic energy consumption. Simulation results validate DualRef scheme and reveal that it is reliable to perform bitwise logic opertions within STT-MRAM while the TMR (Tunnel Magnetoresistance Ratio) varying between 100% and 300% and supply voltage V_{dd} varying from 0.9V to 1.2V. This work provides a robust circuitry scheme and design space to effectively implement CIM in STT-MRAM.

I. INTRODUCTION

With the rapid growth of big data and Internet of Things (IoT) applications, a huge amount of data is generated and exchanged in the network consisting of billions of devices [1]. However, the processor frequency and the memory access efficiency are in state of imbalance in the conventional Von-Neumann computer architectures. Data needs to be transferred back and forth between processors and memories in the architectures. The limitted bandwidth between processors and memories makes energy consumption, data transferring and processing highly incfficient, especially within data-intensive applications [2], [3]. About 200 times more energy is consumed to access DRAM one time compared with a floating-point computation [4]. It makes the limited bandwidth a critical bottleneck to improve the system performance, which is the well-known processor-memory gap or "memory wall" [5].

To bridge the processor-memory gap, many efforts have been spent on integrating processing unit and storage unit into one chip. These efforts can be classified into three types: 1) adding logic modules in memories, 2) embedding limited memory array into processing unit, and 3) moving some specified computations to memories. These methods aim to overcome the data transfer stress on the limited bandwidth between processors and memories when running data-intensive applications [6]. However, these efforts failed to tackle the issue due to some practical considerations. In CMOS based memories, it is complex and cost inefficient to integrate processing unit and storage unit together until 3D integration technology emerges [7]. Moreover, these technologies also suffer from the reliability issues [8], [9].

With the emerging of non-volatile memories, it becomes possible to address this issue [10], [11], for example, STT-MRAM (Spin Transfer Torque Magnetic RAM), RRAM (Resistive RAM), SOT-MRAM (Spin Orbit Torque Magnetic RAM) and other spintronics based devices, logics or memories, etc. [12], [13], [14]. STT-MRAM has been considered as one of the most promising candidates due to its distinctive advantages over other non-volatile memories, such as, non-volatility, good scalability, compatibility with CMOS, ultra fast accessing speed, etc. [15], [16], [17], [18]. The concept of computing-in-memory (CIM) is proposed several years ago, and regained interest recently, which is based on the idea of adding necessary peripheral circuit to memories [19]. CIM makes memories have some kind of computation capability and storage capability at the same time [20], [21]. It is just needed to modify the peripheral circuitry to implement CIM within STT-MRAM, which lets it process and store data simultaneously [21]. Some CIM paradigms at architecture level have been presented and assessed [6], [22], [23], [24]. There are two types of CIMs: one utilizes two or more reference cells to implement CIM, while the other one performs CIM operations depending on its complementary structure[5], [25], [26]. These paradigms introduced in above give us two ways to implement CIM within STT-MRAM. However, these efforts have been done at architecture level to implement CIM, there are few detailed implementations in circuit and few studies

978-1-5386-7100-9/18 $31.00 © 2018 IEEE   275

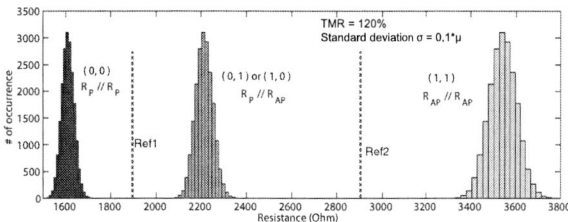

Fig. 1. Resistance distribution of two MTJs aligned in low resistance parallel state ($R_P//R_P$), antiparallel state ($R_P//R_{AP}$) and high resistance parallel state ($R_{AP}//R_{AP}$). The MTJ model used comes from [31].

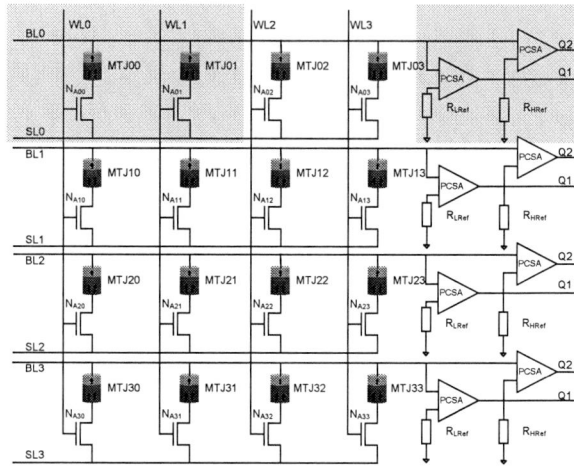

Fig. 2. A four-by-four DualRef CIM array. R_{LRef} and R_{HRef} are designed to distinguish the three resistance states as shown in Fig. 1.

TABLE I
TRUTH TABLE FOR DUALREF CIM IMPLEMENTATION

Logic	Bit pattern	Ouput
OR	**(00)**(R_P, R_P)	$0((R_P + R_T)//(R_P + R_T)<R_{LRef})$
	(01)(R_P, R_{AP})	$1((R_P + R_T)//(R_{AP} + R_T)>R_{LRef})$
	(10)(R_{AP}, R_P)	$1((R_{AP} + R_T)//(R_P + R_T)>R_{LRef})$
	(11)(R_{AP}, R_{AP})	$1((R_{AP} + R_T)//(R_{AP} + R_T)>R_{LRef})$
AND	**(00)**(R_P, R_P)	$0((R_P + R_T)//(R_P + R_T)<R_{HRef})$
	(01)(R_P, R_{AP})	$0((R_P + R_T)//(R_{AP} + R_T)<R_{HRef})$
	(10)(R_{AP}, R_P)	$0((R_{AP} + R_T)//(R_P + R_T)<R_{HRef})$
	(11)(R_{AP}, R_{AP})	$1((R_{AP} + R_T)//(R_{AP} + R_T)>R_{HRef})$

carefully on its reliability and performance [27], [28], [29], [30]. In this work, we proposed a CIM implementation at circuit level. Our contribution can be expressed as follows.

- Proposed a dual reference CIM implementation DualRef within STT-MRAM.
- Optimized the parameters of MTJ and CMOS transistors to meet the design requirements, and validated the Dual-Ref CIM implementation.
- Carried out simulations to assess the reliability and performance of the DualRef CIM implementation by calculating the operation error rate, sensing margin, operation delay and dynamic energy consumption.

The rest of this paper is organized as follows. Section II introduces the scheme to implement DualRef in detailed, and then the functionality of the CIM implementation is validated. After that, the reliability and performance assessment for DualRef are included in Section III. At last, Section IV concludes this paper.

II. PROPOSED DUALREF CIM IMPLEMENTATION

In this section, the proposed DualRef CIM implementation is introduced, validated and assessed, which is realized by adding necessary peripheral circuit. DualRef can perform bitwise logic operation, which also can be used to store data at the same time. In the follows, we will show a four-by-four DualRef CIM array and a single DualRef CIM cell circuit, and then present how to excute basic bitwise logic operation in the CIM array.

A. Design of DualRef Scheme

The AND and OR bitwise logic operations can be executed in the DualRef CIM implementation, other bitwise

logic operations, for example, XOR, XNOR, can also be implemented by adding essential assistant logic circuit after the sensing amplifiers. As the key storage element of STT-MRAM, MTJ device has two resistance states: low resistance state R_P and high resistance state R_{AP}. The two resistance states can be switched to each other by applying bi-directional currents to BL (Bit Line) and SL (Source Line). R_P usually encodes logic "0", and R_{AP} represents logic "1". There are three resistance states of two MTJs connected in parallel. To distinguish these three resistance states, two different reference resistors Ref1 and Ref2 are required, as shown in Fig. 1. However, the normal STT-MRAM cell has the 1T1MTJ (a MTJ and a transistor connected in series) structure. In order to be consistant with the two bit cells in parallel, the two reference cells are designed with the same structure as the bit cells. Assumed that the resistance of the transistor connected to the MTJ in the bit cells is R_T. Therefore, the resistance of the low reference cell R_{LRef} should be between $(R_P + R_T)//(R_P + R_T)$ and $(R_P + R_T//(R_{AP} + R_T)$, while the resistance of the high reference cell R_{HRef} locates in the middle of $(R_P + R_T)//(R_{AP} + R_T)$ and $(R_{AP} + R_T)//(R_{AP} + R_T)$. OR bitwise logic operations are performed by comparing the resistance of the two paralleled bit cells with that of specified reference cell, the results is "0" or "1" when less or more than R_{LRef}; while AND bitwise logic operation results can be obtained after checking the resistance of the two paralleled bit cells with R_{HRef}, which will be "0" or "1" when less or more than R_{LRef}. The truth table reveals the details from Table I.

With these knowledge, the DualRef CIM scheme is presented in Fig. 2. The figure shows a case of four-by-four array, which is adapted from a normal STT-MRAM array by adding a PCSA (PreCharge Sensing Amplifier), a reference cell for every row. With different configuartions, it can be a DualRef array for CIM or a normal STT-MRAM array, and also can be switched to each other freely. In this scheme, the two PCSAs work together to distinguish one resistance state

978-1-5386-7100-9/18 $31.00 © 2018 IEEE

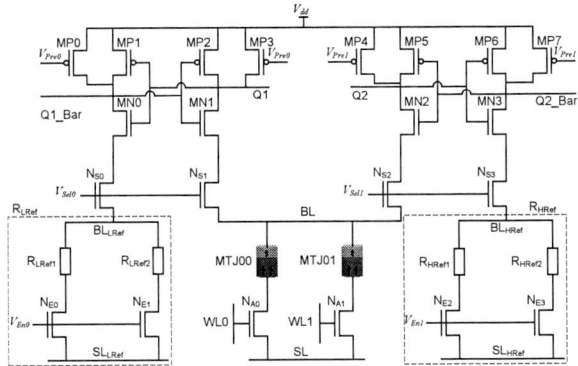

Fig. 3. Schematic of the DualRef CIM cell. It is the part with the gray background shown in Fig. 2.

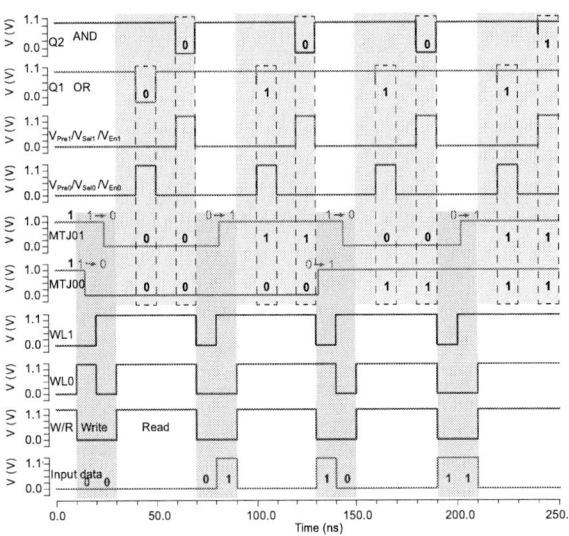

Fig. 4. Transient simulation waveform of the DualRef CIM scheme.

from others [32]. The final OR bitwise logic operation result can be obtained from the Q1, and AND bitwise logic operation result from the Q2 in this scheme.

The DualRef CIM cell is shown in Fig. 3. The schematic has a symmetrical structure, which consists of two PCSAs, two bit cells and two reference cells. Resistor R_{LRef1} and R_{LRef2} connected in parallel and two transistors comprise the low reference cell, while the high reference cell is made of resistor R_{HRef1}, R_{HRef2} and two transistors. Two inverters (MP1 and MN0, MP2 and MN1) connected to each other form the PCSA in the left side, and inverters (MP5 and MN2, MP6 and MN3) comprise the right side PCSA. NMOS N_{S0}, N_{S1}, N_{S2} and N_{S3} work to switch in three states: write operation, read operation and bitwise logic operation.

B. Working Principle of DualRef Scheme

We take the OR bitwise logic operation with bit pattern (00) as an example to show how to perform bitwise logic operation by the DualRef CIM scheme. As shown in Fig. 3, MTJ00 and MTJ01 are assumed in low resistance state (logic "0"). The procedure of execution is as follows: V_{Pre0}, V_{Pre1}, V_{Sel0}, V_{Sel1}, V_{En0} and V_{En1} are first set to "0", and the drain terminal of NMOS MN0 and MN1 are charged to V_{dd}. When WL1 and WL2 are enabled, V_{Pre0}, V_{En0} and V_{Sel0} switch from "0" to "1", the precharged voltage starts to discharge. The discharge speeds are different due to the different resistances of the reference and bit cells. Therefore, the voltage in the low resistance branch will decline faster [33]. In the example, the resistance of the two bit cells in parallel is less than R_{LRef}, so the voltage in the bit cell branch will decline faster. The two inverters (MP1 and MN0, MP2 and MN1) reach a stable state ($Q1$ = "0", $Q1\_Bar$ = "1") when the voltage of the NMOS MN1 drain terminal is lower than the threshold voltage of the inverter (MP1 and MN0). Finally, the result of OR bitwise logic operation can be got form Q1.

Some CIM schemes are conceptual and not presented in detailed circuit. It is difficult to know their feasibility, even less the reliability and performance. Different from these schemes, The DualRef CIM scheme we proposed in this paper are

implemented by slightly modifing the sensing amplifier and the peripheral circuit, can be run in SPICE simulator. In the following, the functionality of the DualRef CIM scheme will be checked, and then its reliability and performance will be evaluated by groups of simulations.

C. Functionality Verification

For the purpose of verifying the functionality of the Dual-Ref CIM scheme within STT-MRAM, a hybrid MTJ/CMOS transient simulation is carried out with a 45 nm PTM CMOS model and a 40 nm compact perpendicular magnetic anisotropy MTJ model [34], [35], [36], [37]. The supply voltage is fixed in $V_{dd} = 1.1V$, and the TMR equals to $TMR = 300\%$, other related parameters are their default values in the MTJ model[38]. We use a larger CMOS transistors channel width in the sensing circuit to enlarge the sensing margin, and use its minimum channel width in the write circuit and logic circuit to eliminate influence of the parasitic capacitor as possible.

Fig. 4 shows the transient simulation waveform of the DualRef CIM scheme, OR and AND bitwise logic operations are performed one after another. The initial state of (MTJ00, MTJ01) is (1, 1), the state of MTJ00 switches from "1" to "0" by applying currents when $Write$ and WL1 are enabled, while MTJ01 switches to "0" after WL2 is enabled. After writing data into the two MTJs, OR bitwise logic operation is carried out when $V_{Pre0}/V_{Sel0}/V_{En0}$ and $Read$ are set to "1". AND bitwise operation are performed by enabling $V_{Pre1}/V_{Sel1}/V_{En1}$. Because the resistance of the two bit cells is less than that of R_{LRef} and R_{HRef}, the two PCSAs in both left and right side output "0". Afterwards, the computing results can be obtained in Q1 and Q2 for OR and AND bitwise logic operation respectively. The results are consistant with the truth table shown in Table I.

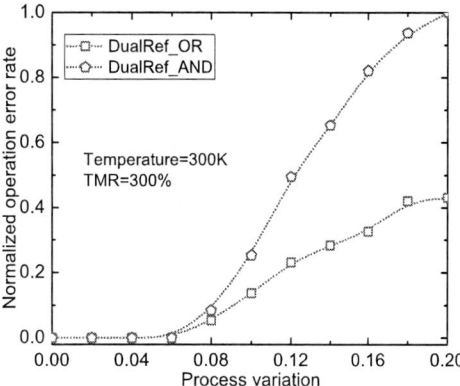

Fig. 5. Normalized operation error rate with respect to the process variation of CMOS transistor.

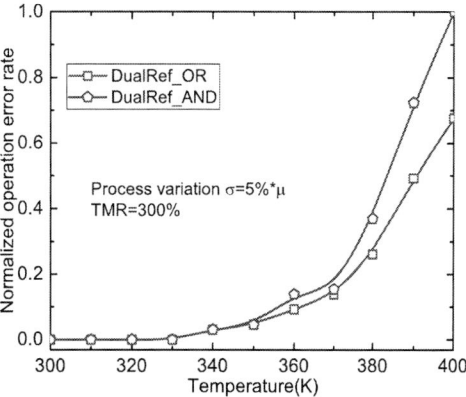

Fig. 6. Normalized operation error rate with respect to the temperature.

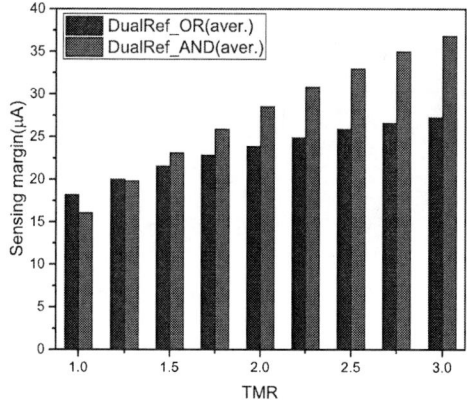

Fig. 7. The sensing margin of bitwise logic operations. It is defined as the currents difference of the two discharge branches in the sensing amplifiers.

With this validated CIM scheme, a series of simulations are carried out in the following section to assess the reliability from the aspects of its operation error rate and sensing margin, and performance in terms of operation delay and dynamic energy consumption with respect to TMR and supply voltage.

III. DESIGN SPACE EXPLORATION

Besides the feasibility, the reliability and performance are also important for the DualRef CIM implementation when it suffers from PVT (Process, Voltage, Temperature) variations. In this section, we carried out six groups of simulations to assess the CIM scheme. Operation error rate and sensing margin are used to indicate the reliability, while the performance is evaluated by calculating the operation delay and dynamic energy consumption. The value at every point in the opeartion error rate are the arithmetic mean of many times Monte Carlo simulations.

A. Operation Error Rate

Operation error rate is used to indicate the reliability of the DualRef CIM scheme directly. There are four bit patterns: (00), (01), (10) and (11) in OR or AND bitwise logic operation. The operation error rate are the arithmatic mean at each bit pattern. We measured the operation error rate in different process variations of CMOS transistor and different temperatures. The temperature is fixed at 300K, and the TMR is set to 300% when obtained operation error rate with the process variation by varying from 0% to 20%; while evaluating the operation error rate with respect to the temperature, the process variation is set to 5%, and the TMR is fixed at 300%. The measured results are shown in Fig. 5 and 6 respectively. As can be seen from the two figures, operation error rate keep very low at small process variations, and it is the same in low temperatures. However operation error rate raises up rapidly after 6% process variation and 330K. With large process variations or higher temperatures, both the resistance of the bit cell and the reference cell changed, the driving currents of transistors decreased. All of this resulted in the reduction of the sensing margin, which can be explain more and more operation errors occur.

B. Sensing Margin

As known from the introduction in Section II, the computing of these bitwise logic operations are executed in sensing amplifiers. The sensing margin indicates how many variations the DualRef CIM scheme can tolerate to ensure computing without errors. Therefore, sensing margin of OR and AND bitwise logic operations are the important indicator of the reliability, which is defined as the current difference of the two discharge branches in the sensing amplifiers. The sensing margin are calculated without considering the process variation of transistor. The resistance difference between the low and high resistace state of MTJ device varies as TMR. Therefore, the sensing margin were checked by increasing TMR from 100% to 300%. The calculated sensing margin are shown in Fig. 7. Every data is the average value of the sensing margin at the four bit patterns:(00), (01), (10) and (11). As can be seen from this figure, the smallest sensing margin is more than $15\mu A$, and the biggest one is about $35\mu A$. Both the sensing margin of OR and AND bitwise logic operation

978-1-5386-7100-9/18 $31.00 © 2018 IEEE

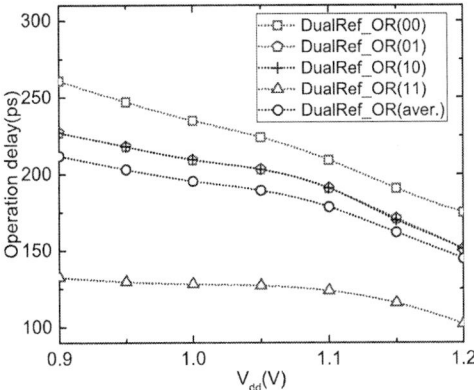

Fig. 8. The operation delay of the OR bitwise logic opeartion. The simulations are carried out with the fixed temperature 300K and no process variation.

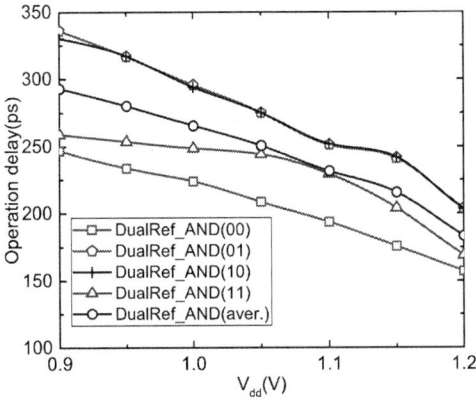

Fig. 9. The operation delay of the AND bitwise logic opeartion. The simulations are carried out with the fixed temperature 300K and no process variation.

arise linearly by increasing TMR. However, the increasing speed and sensing margin of AND bitwise logic operation are more than that of OR bitwise logic operation. With increasing of TMR, only the resistance of MTJs in anti-parallel state increase. So the resistance of the two bit cells in parallel at bit pattern (00) keeps constant with the increased TMR, and increases at bit pattern (01) and (10), and increases faster at bit pattern (11). According to truth Table I, the low reference cell was used to distinguish bit pattern (00) from others, but the resistance of which does not increase as fast as TMR. Therefore, the above explained why the sensing margin of OR bitwise logic operation and its increasing speed are less than that of AND bitwise logic operation.

C. Operation Delay

The operation delay represents the time consumed by one bitwise logic operation. The delay of OR and AND bitwise logic operations for the DualRef CIM scheme are measured from the rising edge of enable signal to the time when output results reach stable state. The operation delay are calculated with respect to the supply voltage V_{dd}, by varying from 0.9V

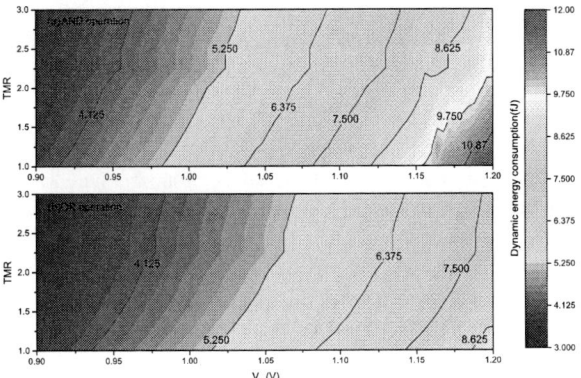

Fig. 10. The dynamic energy consumption of OR and AND bitwise logic operations. It is measured without process variation, and the temperature is fixed at 300K.

to 1.2V. The results are shown in Fig. 8 and 9. It can be seen in the two figures that both the delay of OR and AND bitwise logic operations decline with increasing supply voltage V_{dd}. Delay time is dominated by the charge and discharge time of capacitor in circuit. Larger supply voltage can reduce the time. The average delay of AND bitwise logic operation decrease from 295ps to 180ps when supply voltage increases, and are more than that of OR bitwise logic operation. That is because the higher resistance reference cell is used, and the current discharge of it when sensing is slower than with the low resistance reference cell. The difference delay within different bit patterns are caused by both the effect of current discharge speed and parasitic capacitors.

D. Dynamic Energy Consumption

The operation dynamic energy consumption of the DualRef CIM scheme is calculated by integrating the product of voltage and current of the two discharge branches in the sensing amplifiers with respect to its operation delay, and shown in Fig. 10. These data are obtained by fixing the temperature at 300K and without process variation. As shown in this figure, single bitwise logic operation consumes several femto Joules, and which increases with the supply voltage V_{dd}, but slowly declines with the TMR. Increasing V_{dd} shortens the operation time, but it will linearly raise up the current. Therefore, the energy consumption surely goes up with larger V_{dd}. The resistance of discharge branches decline as the TMR, which results in the reduction of discharge current, so the dynamic energy consumed by single bitwise logic operation decrease with TMR. It is found that more energy are consumed in AND bitwise logic operation with same TMR and V_{dd} by comparing Fig. 10 (a) and (b), which is caused by its bigger operation delay as shown in Fig. 8 and 9.

In summary, the reliability and performance of the DualRef CIM scheme were quantitatively analyzed. We found that, the reliability of DualRef can be enhanced by enlarging the TMR as possible. However, as the indicator of its performance, operation delay and dynamic energy consumption are irreconcilable. Increasing voltage supply can reduce the operation delay, but result in more energy consumption. These results provide how robust is the DualRef CIM implementation under PVT varations.

IV. CONCLUSION

A CIM implementation DualRef is proposed here within STT-MRAM. DualRef can be used to perform OR and AND bitwise logic operation, other bitwise logic operation can also be supported by adding peripheral circuit. The functionality, reliability and performance of the CIM implementation are evaluated at 45nm technology node. Simulations reveale that DualRef can work correctly when the temperature is less than 330K and process variation of the CMOS transistor does not supass 6%. Improving TMR of MTJ device can enhance the reliability of this CIM scheme. This work provides a robust circuit scheme to implement CIM at circuit level and explores the design space, but efforts on system, instruction set and software interface are also expected.

REFERENCES

[1] M. Imani, S. Gupta, and T. Rosing, "Ultra-efficient processing in-memory for data intensive applications," in *Proceedings of the 54th Annual Design Automation Conference 2017*, Jun. 2017, pp. 6:1–6:6.

[2] J. Zhou, X. Yang, J. Wu, *et al.*, "A memristor-based architecture combining memory and image processing," *Sci. China Inform. Sci.*, vol. 57, no. 5, pp. 1–12, May 2014.

[3] N. S. Kim, T. Austin, D. Blaauw, *et al.*, "Leakage current: Moore's law meets static power," *Computer*, vol. 36, no. 12, pp. 68–75, Dec. 2003.

[4] S. W. Keckler, W. J. Dally, B. Khailany, *et al.*, "GPUs and the future of parallel computing," *IEEE Micro*, vol. 31, no. 5, pp. 7–17, Sep. 2011.

[5] S. Li, C. Xu, Q. Zou, *et al.*, "Pinatubo: A processing-in-memory architecture for bulk bitwise operations in emerging non-volatile memories," in *2016 53nd ACM/EDAC/IEEE Design Automation Conference*, Jun. 2016, pp. 1–6.

[6] L. Koskinen, J. Tissari, J. Teittinen, *et al.*, "A performance case-study on memristive computing-in-memory versus Von Neumann architecture," in *2016 Data Compression Conference*, Mar. 2016, pp. 613–613.

[7] J. Ahn, S. Hong, S. Yoo, *et al.*, "A scalable processing-in-memory accelerator for parallel graph processing," in *2015 ACM/IEEE 42nd Annual International Symposium on Computer Architecture*, Jun. 2015, pp. 105–117.

[8] K. Chen, S. Li, N. Muralimanohar, *et al.*, "CACTI-3DD: Architecture-level modeling for 3D die-stacked DRAM main memory," in *2012 Design, Automation Test in Europe Conference Exhibition*, Mar. 2012, pp. 33–38.

[9] X. Dong, X. Wu, G. Sun, *et al.*, "Circuit and microarchitecture evaluation of 3D stacking magnetic RAM (MRAM) as a universal memory replacement," in *2008 45th ACM/IEEE Design Automation Conference*, Jun. 2008, pp. 554–559.

[10] W. Zhao and G. Prenat, *Spintronics-Based Computing*. Switzerland: Springer, 2015.

[11] C. J. Xue, G. Sun, Y. Zhang, *et al.*, "Emerging non-volatile memories: opportunities and challenges," in *2011 Proceedings of the Ninth IEEE/ACM/IFIP International Conference on Hardware/Software Codesign and System Synthesis*, Oct. 2011, pp. 325–334.

[12] F. Parveen, S. Angizi, Z. He, *et al.*, "Low power in-memory computing based on dual-mode SOT-MRAM," in *2017 IEEE/ACM International Symposium on Low Power Electronics and Design*, Jul. 2017, pp. 1–6.

[13] W. Kang, Z. Wang, Y. Zhang, *et al.*, "Spintronic logic design methodology based on spin hall effect-driven magnetic tunnel junctions," *J. Phys. D: Appl. Phys.*, vol. 49, no. 6, p. 065008, Jan. 2016.

[14] H. Zhang, W. Kang, L. Wang, *et al.*, "Stateful reconfigurable logic via a single-voltage-gated spin hall-effect driven magnetic tunnel junction in a spintronic memory," *IEEE Trans. Electron Devices*, vol. 64, no. 10, pp. 4295–4301, Oct. 2017.

[15] W. Kang, Y. Zhang, Z. Wang, *et al.*, "Spintronics: Emerging ultra-low-power circuits and systems beyond MOS technology," *ACM J. Emerg. Technol. Comput. Syst.*, vol. 12, no. 2, pp. 16:1–16:42, Aug. 2015.

[16] S. A. Wolf, D. D. Awschalom, R. A. Buhrman, *et al.*, "Spintronics: A spin-based electronics vision for the future," *Science*, vol. 294, no. 5546, pp. 1488–1495, Nov. 2001.

[17] H. S. P. Wong and S. Salahuddin, "Memory leads the way to better computing," *Nat. Nanotechnol.*, vol. 10, no. 3, pp. 191–194, Mar. 2015.

[18] W. Kang, L. Zhang, J.-O. Klein, *et al.*, "Reconfigurable codesign of STT-MRAM under process variations in deeply scaled technology," *IEEE Trans. Electron Devices*, vol. 62, no. 6, pp. 1769–1777, Jun. 2015.

[19] D. Patterson, T. Anderson, N. Cardwell, *et al.*, "A case for intelligent ram," *IEEE Micro*, vol. 17, no. 2, pp. 34–44, Mar. 1997.

[20] M. Imani, Y. Kim, and T. Rosing, "MPIM: Multi-purpose in-memory processing using configurable resistive memory," in *2017 22nd Asia and South Pacific Design Automation Conference*, Jan. 2017, pp. 757–763.

[21] D. Fan, S. Angizi, and Z. He, "In-memory computing with spintronic devices," in *2017 IEEE Computer Society Annual Symposium on VLSI*, Jul. 2017, pp. 683–688.

[22] J. Yu, R. Nane, A. Haron, *et al.*, "Skeleton-based design and simulation flow for computation-in-memory architectures," in *2016 IEEE/ACM International Symposium on Nanoscale Architectures*, Jul. 2016, pp. 165–170.

[23] A. Haron, J. Yu, R. Nane, *et al.*, "Parallel matrix multiplication on memristor-based computation-in-memory architecture," in *2016 International Conference on High Performance Computing Simulation*, Jul. 2016, pp. 759–766.

[24] S. Hamdioui, M. Taouil, H. A. D. Nguyen, *et al.*, "CIM100x: Computation in-memory architecture based on resistive devices," in *2016 15th International Workshop on Cellular Nanoscale Networks and their Applications*, Aug. 2016, pp. 1–2.

[25] P. Chi, S. Li, C. Xu, *et al.*, "PRIME: A novel processing-in-memory architecture for neural network computation in ReRAM-based main memory," in *2016 ACM/IEEE 43rd Annual International Symposium on Computer Architecture*, vol. 44. IEEE, Jun. 2016, pp. 27–39.

[26] W. Kang, H. Wang, Z. Wang, *et al.*, "In-memory processing paradigm for bitwise logic operations in STT-MRAM," *IEEE Trans. Magn.*, vol. 53, no. 11, pp. 1–4, Nov. 2017.

[27] S. Hamdioui, L. Xie, H. A. D. Nguyen, *et al.*, "Memristor based computation-in-memory architecture for data-intensive applications," in *Proceedings of the 2015 Design, Automation & Test in Europe Conference & Exhibition*, Mar. 2015, pp. 1718–1725.

[28] S. Hamdioui, M. Taouil, H. A. D. Nguyen, *et al.*, "Memristor: the enabler of computation-in-memory architecture for big-data," in *2015 International Conference on Memristive Systems*, Nov. 2015, pp. 1–3.

[29] S. Hamdioui, "Computation in memory for data-intensive applications: Beyond CMOS and beyond Von-Neumann," in *Proceedings of the 18th International Workshop on Software and Compilers for Embedded Systems*, Jun. 2015, pp. 1–1.

[30] J. J. Yang, D. B. Strukov, and D. R. Stewart, "Memristive devices for computing," *Nat. Nanotechnol.*, vol. 8, no. 1, pp. 13–24, Dec. 2013.

[31] L. Zhang, W. Kang, Y. Zhang, *et al.*, "Channel modeling and reliability enhancement design techniques for STT-MRAM," in *2015 IEEE Computer Society Annual Symposium on VLSI*, Jul. 2015, pp. 461–466.

[32] H. Cai, Y. Wang, L. A. D. B. Naviner, *et al.*, "Robust ultra-low power non-volatile logic-in-memory circuits in FD-SOI technology," *IEEE Trans. Circuits Syst. I*, vol. 64, no. 4, pp. 847–857, Apr. 2017.

[33] W. Zhao, C. Chappert, V. Javerliac, *et al.*, "High speed, high stability and low power sensing amplifier for MTJ/CMOS hybrid logic circuits," *IEEE Trans. Magn.*, vol. 45, no. 10, pp. 3784–3787, Oct. 2009.

[34] W. Zhao and Y. Cao, "Predictive technology model for nano-CMOS design exploration," *ACM J. Emerg. Technol. Comput. Syst.*, vol. 3, no. 1, Apr. 2007.

[35] Y. Wang, Y. Zhang, F. Deng, *et al.*, "Compact model of magnetic tunnel junction with stochastic spin transfer torque switching for reliability analyses," *Microelectron. Reliab.*, vol. 54, no. 9, pp. 1774 – 1778, Mar. 2014.

[36] L. Zhang, A. Todri-Sanial, W. Kang, *et al.*, "Quantitative evaluation of reliability and performance for STT-MRAM," in *2016 IEEE International Symposium on Circuits and Systems*, May 2016, pp. 1150–1153.

[37] L. Zhang, Y. Cheng, W. Kang, *et al.*, "Addressing the thermal issues of stt-mram from compact modeling to design techniques," *IEEE Trans. Nanotechnol.*, vol. 17, no. 2, pp. 345–352, Mar. 2018.

[38] M. Wang, W. Cai, K. Cao, *et al.*, "Current-induced magnetization switching in atom-thick tungsten engineered perpendicular magnetic tunnel junctions with large tunnel magnetoresistance," *Nat. Commun.*, vol. 9, no. 1, p. 671, Feb. 2018.

Biosensing performance optimization of DMFET for fully filled and partially filled cavity

Ankita Porwal
Department of Electronics & Communication
Malaviya National Institute of Technology, Jaipur
Email: ankita.sporwal@gmail.com

Dr. Chitrakant Sahu
Department of Electronics & Communication
Malaviya National Institute of Technology, Jaipur
E-mail: Chitrakant.ece@mnit.ac.in

Abstract—**In this paper, a p-channel Dielectric Modulated Field Effect Transistor (DMFET) used as a biosensor has been investigated for the detection of the neutral as well as charged biomolecules. To analyze the detection ability and sensitivity of the device various metrics such as electric field, surface potential, drain current and transconductance are investigated under different bias conditions. For the detection of biomolecules in the cavity, change in threshold voltage is used as sensing parameter. The device shows a threshold voltage shift of 360 mV for K=5. The device sensitivity is observed at different dielectrics and it is found that drain current sensitivity increases with the increasing value of the dielectric constants. The transconductance of the device increases by a value of 6 × 10⁻⁵ A/V for a biosample with K=5 in the cavity. The fabrication process flow of the device has also been proposed to the minimum involvement of complex processes. Further, performance metrics are analyzed at the partially filled cavity and fully filled cavity to observe the effectiveness of this device in the detection of the neutral and charged biomolecules.**

Keywords— *DMFET, Charge effect, Dielectric Constant effect*

I. INTRODUCTION

In today's world, health care has become a major concern hence the early detection of diseases is necessary. So the need for biosensors for quick detection of diseases and that should be low cost, durable, and highly sensitive is at a rise. According to PR Newswire, biosensor market is forecasted to reach $21.17 billion by 2020 in the world. This growth will be continuous because of high demand for biosensors that can be used for the detection of infectious diseases (Avian Influenza, Swine Flu), cancer, and cardiovascular problems and also for the other disease [1, 2]. The field effect transistor (FET) based sensor structures provide assurance for the label-free real-time detection of the biomolecules and offer many other advantages like compatibility with conventional CMOS fabrication process, miniaturization, low cost, high sensitivity [3,4]. In the history of biosensors, first FET device as a sensor, Ion Sensitive Field Effect Transistor (ISFET) was introduced in 1970 by Bergveld which used for pH measurement [5, 6]. Although ISFET is able to detect charged biomolecules but it cannot make difference between the weakly charged and neutral biomolecules [7]. In ISFET, the channel is exposed to a solution. Counter ions present in the solution make an ionic double layer, because of this layer sensitivity of this device decreased. To overcome this issue (detection of charged as well as weakly charged and neutral biomolecules and low

sensitivity), dielectric modulated field effect transistor (DMFET) was introduced [8]. In DMFET a vertical cavity is created in between the gate and gate oxide surface. The novelty of this device is the vertical cavity which has an advantage of high sensitivity [9]. The working principle of DMFET is related to the dielectric constant of biomolecule as well as interface charge at the surface and charge carried by biomolecules. With the introduction of negatively charged biomolecules, (for example DNA), charge effect is more dominating in comparison to dielectric constant effect. In presence of neutral biomolecules, the dielectric constant is responsible for the device sensitivity. The threshold voltage of the device changes when the cavity is fully filled or partially filled with biomolecules. The presence of biomolecule can be detected by the change in threshold voltage ΔV_T. If the cavity is filled with negatively charged DNA, the threshold voltage shifted towards positive side because of negative charge [10]. A p-channel DMFET is more preferred over n-channel DMFET [11]. By observing the ΔV_T, DMFET is also able to detect the specific binding of the biotin-streptavidin binding [12]. It is predicted that DMFET type biosensors can detect DNA, cancer markers and antibodies [8]. Kalra et al. simulated DMFET for DNA detection and also find out the impact of DNA orientation [13]. Still, there is a need for the optimization of sensing metrics in order to enhance the sensing performance. In this paper, we have optimized the sensing metrics for the superior sensitivity of nanodevice biosensors.

II. SIMULATION PARAMETERS, DEVICE STRUCTURE, AND OPERATING PRINCIPLE

The simulation parameters for p-channel DMFET are shown in table 1. The simulation of this device is performed using Silvaco Atlas tool [14]. The various models invoked for simulation include Shockley-Read-Hall (SRH) recombination model, Fermi model, and Lombardi mobility to accurately model the device [15]. In the case of DMFET a vertical cavity is created in between the gate and gate oxide, at both the sides (source and drain junction) to fill the biomolecules. In this simulation, the maximum value of dielectric constant of biomolecules is chosen as 12. The length of the cavity is an important parameter as it affects the parameters like drains current sensitivity, electric field, transconductance. Cavity length of 400 nm showed the optimum performance as predicted by Choi et al. hence in this simulation, all the parameters are calculated for 400 nm cavity length [8]. The schematic representation of the DMFET is as shown in fig. 1.

978-1-5386-7100-9/18 $31.00 © 2018 IEEE

In order to show the reproducibility of the fabricated device, calibration is also done with simulated data as shown in fig. 2. The V_T (threshold voltage) depends on the gate capacitance so when the dielectric material (different biosamples) of the gate changes V_T also changes. To form the cavity when chromium layer is etched, the dielectric constant of the gate material and total gate capacitance decreases [8].

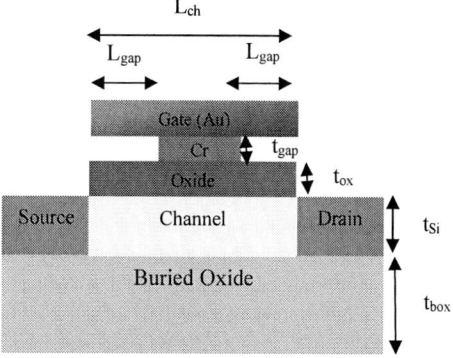

Fig. 1 Schematic representation of Dielectric Modulated Field Effect Transistor

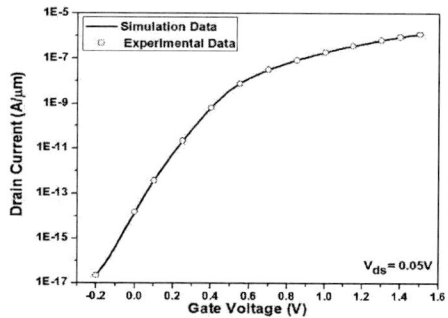

Fig. 2 Calibration of simulation models with experimental results for empty (or air-filled) cavity

Table 1: Design Parameter Specification of the Simulated Device

Parameter Name	Symbol	Value	Unit
Gate Length	L_{ch}	1	µm
Buried Oxide Thickness	t_{box}	300	nm
Silicon Thickness	t_{si}	50	nm
Gate Oxide Thickness	t_{ox}	10	nm
Cavity Thickness	t_{gap}	15	nm
Source Doping	N_A	1×10^{20}	cm^{-3}
Drain Doping	N_A	1×10^{20}	cm^{-3}
Channel Doping	N_{ch}	1×10^{17}	cm^{-3}
Gate Workfunction (Gold)	Φ_G	5.1	eV
Gate Workfunction (Chromium)	Φ_{G1}	4.5	eV

For this case, V_T shifts towards more positive values. When biomolecules are present inside the gap, gate capacitance increases relative to air cavity so V_T shifts in the negative direction. The above statement in the mathematical form [12] can be represented as

$$\Delta V_T \propto \frac{Q_{bio}}{\left(\dfrac{1}{C_{bio}} + \dfrac{1}{C_{ox}} + \dfrac{1}{C_{air}} \right)} \qquad (1)$$

Where, $C_{bio} = \dfrac{\varepsilon_o \varepsilon_{bio}}{t_{bio}}$, $C_{ox} = \dfrac{\varepsilon_o \varepsilon_{ox}}{t_{ox}}$, $C_{air} = \dfrac{\varepsilon_o}{\varepsilon_{air}}$

C_{bio}, C_{ox}, C_{air} and ε_{bio}, ε_{ox}, ε_o and, t_{bio}, t_{ox}, t_{air} corresponding to the capacitance, dielectric constant, and thickness of biomolecule, oxide and air-filled regions.

III. VIRTUAL FABRICATION STEPS FOR P CHANNEL DMFET

The proposed steps of fabrication steps for DMFET are given as: The n-doped SOI (Silicon on Insulator) wafer is used as starting substrate. On SOI wafer a layer of the buried oxide thickness 3000 Å is deposited, followed by the silicon layer of thickness 1000 Å is considered. By iterative thermal oxidation and wet etching, the thickness of silicon is reduced up to 500 Å. To pattern the active region, optical lithography and a positive photoresist are used, the further active region is defined by wet etching of Si layer. To define source and drain regions, the implantation process is used. To define gate region, a gate oxide layer of thickness 10 nm is deposited by thermal dry oxidation, proceed by chromium layer of thickness 15 nm and a gold layer of thickness 100 nm deposited by evaporation. Optical lithography and a positive photoresist are used for defining the gate electrode. To create a vertical cavity at both sides (source and drain junction), chromium, gold, and the silicon layer are etched by the wet etching process. The vertical cavity is fabricated by thin film deposition technique. The height of cavity is measured by measuring the height of chromium. Gold cannot be directly deposited into the oxide that's why chromium is used as a sacrificial layer and glue layer that provide adhesion to the Au layer [16].

IV. SIMULATION RESULTS AND DISCUSSION

A. Analysis for fully filled cavity

To understand the effect of charge and dielectric constant on the parameters we have considered both, the effect of charge and dielectric constant of biomolecules independently on the fully filled cavity. In this work to apply charge effect we have considered the dielectric constant (K) equal to 3.57 and for the simulation of dielectric constant effect consider the charge (Q) equal to -0.5 X 10^{11} cm^{-2}. To show the device behavior, electric field and surface potential are calculated under thermal equilibrium ($V_{gs} = V_{ds} = 0$V). To ensure that the device is working in linear region threshold voltage is calculated at $V_{ds} = 0.05$ V and $V_{gs} = -2$V.

It can be seen from fig. 3 as charge increase the value of surface potential decreases. Also as the dielectric constant increases, the surface potential also decreases. For the empty nanogap, the surface potential is at the highest level.

978-1-5386-7100-9/18 $31.00 © 2018 IEEE

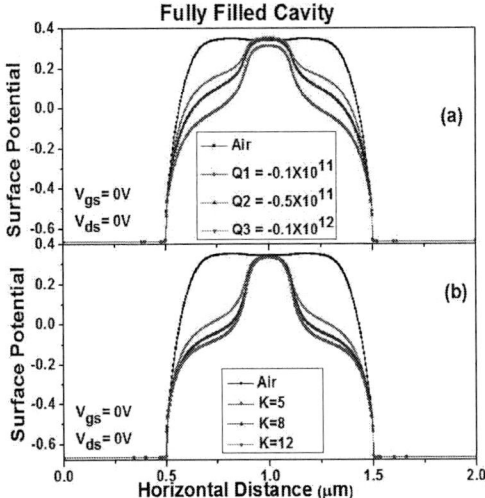

Fig. 3(a) Surface Potential at different charges Fig. 3(b) Surface Potential at different dielectrics

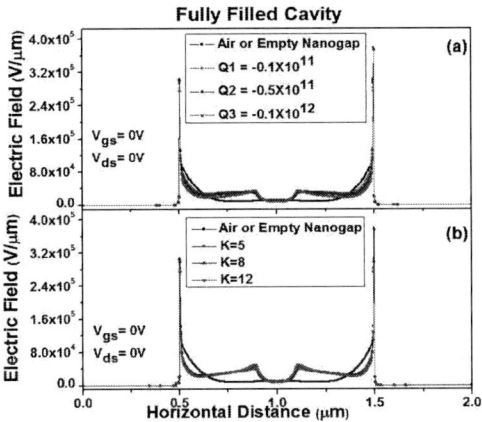

Fig. 4(a) Electric Field at different charges Fig. 4(b) Electric Field at different dielectric Constant

Fig. 4 shows the effect on electric field for this device with increasing negative charge and also for the different value of dielectric constant. The electric field is least for a completely empty cavity.

It can be observed from fig. 5(b) that there is a very less variation in drain current with the increasing value of dielectric constant (K=5, 8 and 12). For K>5 the effect of dielectric constant becomes almost similar to that for higher K values. With the increasing value of charge, current characteristics are not overlapped rather than change appears quite distinguishable. It can be concluded from fig. 5(a) that charge effect is more dominating in comparison to dielectric constant effect for the fully filled cavity. In the case of empty nanogap drain current is at the lowest level in comparison to increasing value of charge and dielectric constant.

The value of transconductance can be calculated by differentiating drain current with gate voltage at the constant drain to source voltage. From the fig.6 it can be observed that the transconductance is very high at different dielectrics but for increasing value of charges, it is less in comparison to

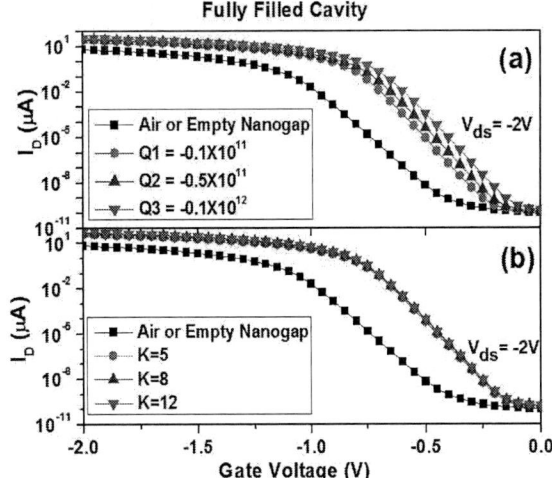

Fig. 5(a) Drain Current at different charges Fig. 5(b) Drain Current at different dielectrics

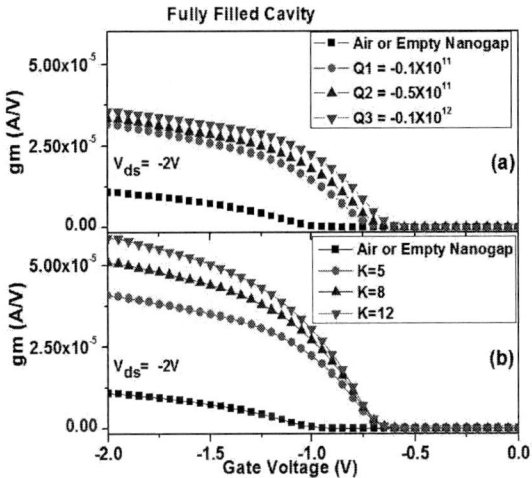

Fig. 6(a) Transconductance at different charges Fig. 6(b) Transconductance at the different dielectrics

increasing value of dielectrics. For the case of empty nanogap, transconductance is at the lowest level as the drain current is also at the lowest level.

i. Threshold voltage shift extraction

To calculate the threshold voltage (V_T), the second derivative method is used [17]. Threshold voltage has a dependency on the gate capacitance. Therefore, by changing the dielectric constant in the cavity, the threshold voltage can vary. V_T depends on the gate oxide thickness [16]. Hence, the thickness of gate oxide should be in proper range so that the device can perform biosensing phenomena.

Change in threshold voltage ΔV_T is defined as:

$$\Delta V_T = V_{T,Bio} - V_{T,Air}$$

By calculating ΔV_T, the specific binding of biomolecules can be detected. ΔV_T independent of gate oxide thickness but it depends on the dielectric constant of the biomolecule. From table 2 it is clear that value of ΔV_T from simulation results is almost following the experimental value.

978-1-5386-7100-9/18 $31.00 © 2018 IEEE

Table 2: Change in Threshold Voltage of the Simulated Device

In this simulation				Experimentally [6]
At different dielectrics	ΔV_T (V)	At different charges	ΔV_T (V)	ΔV_T (V)
K = 5	0.36	Q = -0.1 X 10^{11}	0.33	
K = 8	0.35	Q = -0.5 X 10^{11}	0.36	0.38
K = 12	0.35	Q = -0.1 X 10^{12}	0.41	

ii. Sensitivity Analysis

Sensitivity is a relative term. For a device to perform as a biosensor, sensitivity should be high for the detection of the biomolecule present in the cavity. Sensitivity in mathematical form can be represented as

$$S = \frac{\left(I_{sample} - I_{air}\right)}{I_{air}} \quad (2)$$

Where, I_{sample} denotes the drain current when biomolecule present inside the cavity and I_{air} represent the drain current when only air is present inside the cavity.

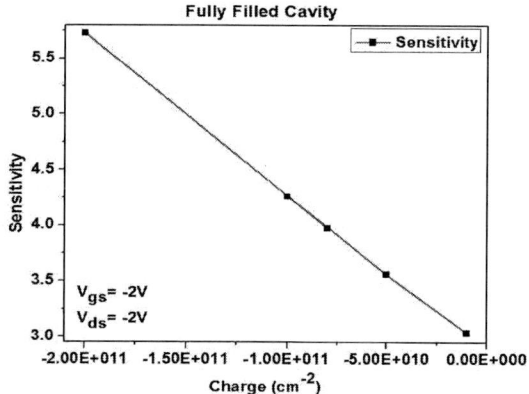

Fig. 7(a)Sensitivity at different charges

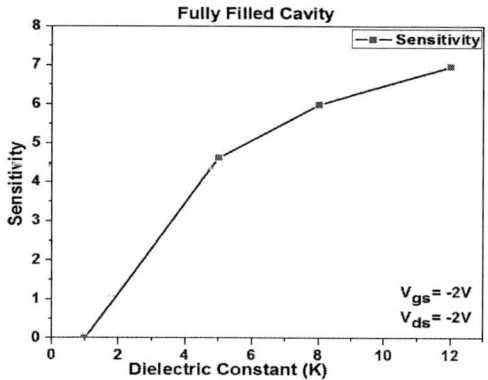

Fig. 7(b) Sensitivity at different dielectric constant

From the fig. 7(a) it can be seen that sensitivity increases with the increasing value of charges in similar manner sensitivity follow the trends with the increasing value of dielectrics. At

the lower value of dielectric constant (K=1) the sensitivity is 0 and as dielectric constant increases, sensitivity also increases. Fully filled cavity directly related to high device current. Sensitivity is high when the whole cavity is filled with biomolecules.

B. Analysis for partially filled cavity:

The orientation of biomolecules (whether the biomolecules fill horizontally or vertically inside the cavity) affects the performance metrics. In order to study the effect of orientation, two cases have been considered: (1) when the cavity is filled horizontally (2) when the cavity is filled vertically. When the cavity is filled horizontally, two further cases have been studied: (a) biomolecules filled horizontally near the source and the drain area of the cavity. (b) Biomolecules filled horizontally near gate area only. For vertically filled case also two cases have been considered: (c) biomolecules filled vertically near to the gate surface. (d) biomolecules filled vertically near to the gate oxide surface.

In this simulation to apply charge effect consider the dielectric constant (K) equal to 3.57. To show the device behavior, electric field and surface potential are calculated under thermal condition ($V_{gs} = V_{ds} = 0V$).

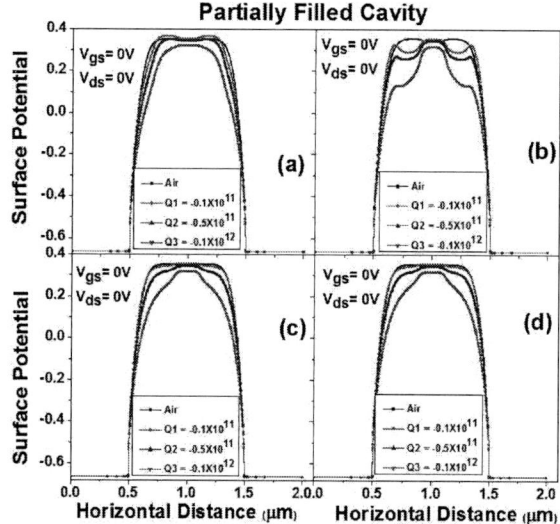

Fig. 8 Surface Potential at different charges (a) when biomolecules filled horizontally near source and drain (b) when biomolecules filled horizontally near gate only (c) when biomolecules filled vertically touching to gate surface (d) when biomolecules filled vertically touching to gate oxide surface.

From the fig. 8 it can be seen that the position of biomolecules inside the cavity is a very important parameter as it affects the sensing performance of the device. In this fig. it must be also taken into account that surface potential is less effective when biomolecules filled vertically but it is more affected when biomolecules filled horizontally near gate area. For partially filled cavity surface potential is at the highest level when the cavity is filled with air.

By comparing all the cases it can be seen from fig. 9, that the electric field is highly affected when biomolecules are occupied horizontally near gate area. As earlier seen for fully filled cavity electric field is at the lowest level for empty nanogap. The trends also continue for the partially filled cavity.

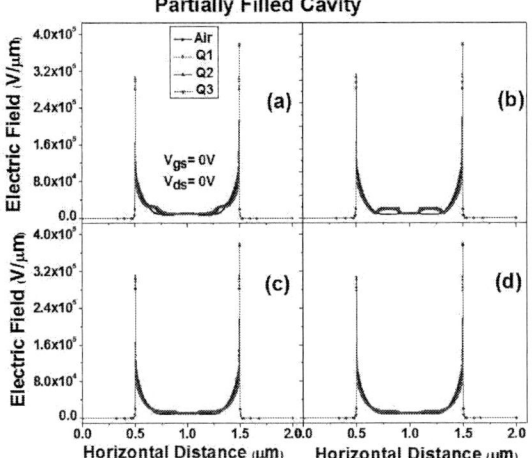

Fig. 9 Electric Field at different charges (a) when biomolecules filled horizontally near source and drain (b) when biomolecules filled horizontally near gate only (c) when biomolecules filled vertically touching to gate surface (d) when biomolecules filled vertically touching to gate oxide surface.

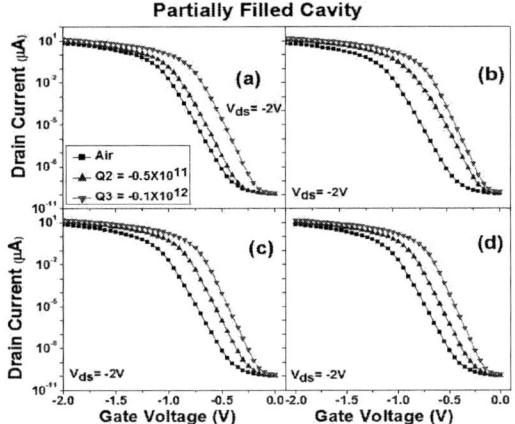

Fig. 10 Drain Current at different charges (a) when biomolecules filled horizontally near source and drain (b) when biomolecules filled horizontally near gate only (c) when biomolecules filled vertically touching to gate surface (d) when biomolecules filled vertically touching to gate oxide surface.

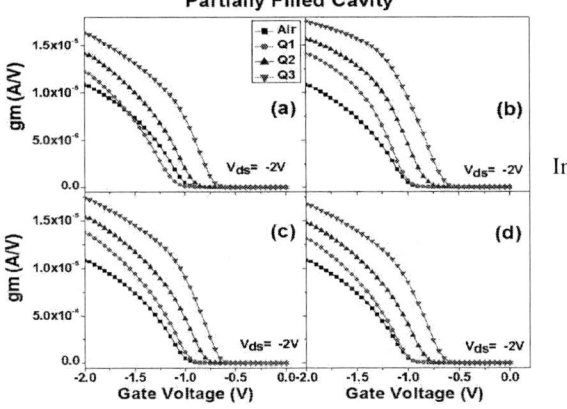

Fig. 11 Transconductance at different charges (a) when biomolecules filled horizontally near source and drain (b) when biomolecules filled horizontally near gate only (c) when biomolecules filled vertically touching to gate surface (d) when biomolecules filled vertically touching to gate oxide surface.

In fig. 9 the value of charges (Q1, Q2, and Q3) are given as – (0.1×10^{11}, -0.5×10^{11} and -0.1×10^{12}) respectively.

In the case of the partially filled cavity (for all the cases) drain current also increases with the increasing value of charge. The value of drain current is minimum when the cavity is empty. From the fig. 10 it can be seen that with the increasing value of charge drain current also increases as already seen in the case of the fully filled cavity. The value of drain current is not much affected by the position of biomolecules for the partially filled cavity.

The transconductance value is directly driven from drain current characteristics. From the fig. 11, it is clear that the value of transconductance increases with the increasing value of the charge as drain current also increases with the increasing value of charge. The same also reported for the fully filled cavity.

V.   CONCLUSION

The p-channel DMFET is optimized by considering the cases for the cavity to be fully filled or partially filled with biomolecules. From the simulation results, it can be predicted that this device is more sensitive when the cavity is fully filled with biomolecules. The effects on surface potential, electric field, drain current, transconductance and sensitivity for different dielectric constants and charges have been considered. From the drain current characteristics of the fully filled cavity, effect of charge concentration of biomolecules is more dominating in comparison to the dielectric constant effect. In the case of the partially filled cavity, the position of biomolecule inside the cavity affects the sensing parameter. Considering fully filled cavity, this device is highly sensitive to dielectric constant as well as for charged biomolecules. Sensitivity increases with the increasing value of charge and also for increasing value of dielectric constant.

ACKNOWLEDGMENT

The authors acknowledge the Science and Engineering Research Board (SERB) Government of India for financial support under project no. ECR/2017/216.

REFERENCES

[1] Bobby Pejcic, Roland De Marco, and Gordon Parkinson, "The role of biosensors in the detection of emerging infectious diseases", The Royal Society of Chemistry, Vol. 131, pp. 1079–1090, 2006

[2] M. Im, J.-H. Ahn, J.-W. Han, T. J. Park, S. Y. Lee, and Y.-K. Choi, "Development of a Point-of-Care Testing Platform With a Nanogap-Embedded Separated Double-Gate Field Effect Transistor Array and Its Readout System for Detection of Avian Influenza", IEEE Sensors Journal, Vol. 11, No. 2, February 2011.

[3] P. Bergveld, "The Development and Application of FET- based Biosensors", Biosensors Vol. 2, p. 15-33, 1986.

[4] T. Goda, Y. Miyahara, "DNA Biosensing Using Field Effect Transistors", Current Physical Chemistry, Vol. 4, pp. 276-291, 2011

[5] P. Bergveld, "Development, operation, and application of the ion-sensitive field-effect transistor as a tool for electrophysiology," IEEE Trans. Biomed. Eng. Vol. BME-19, No. 5, pp. 342-351, Sep. 1972.

[6] J. F. V. Perez, M. M. M. Velasco, M. E. M. Rosas, H. L. M. Reyes, "ISFET sensor characterization", Procedia Engineering, Vol. 35, p. 270-275, 2012.

[7] P. Bergveld, "Thirty years of ISFETOLOGY: What happened in the past 30 years and what may happen in the next 30 years," Sens. Actuators B, Chem., Vol. 88, no.1, Jan. 2003

[8] H. Im, X.-J. Huang, B. Gu, and Y.-K. Choi, "A dielectric modulated field - effect transistor for biosensing," Nature Nanotechnology, Vol. 2, pp. 430-434, July 2007.

[9] Daniel Therriault, "Filling the gap," Nature Nanotechnology, Vol.2, July 2007

[10] C.-H. Kim, C. Jung, K.-B. Lee, H. G. Park, and Y.- K. Choi, "Label-free DNA detection with a nanogap embedded complementary metal oxide semiconductor," Nanotechnology, Vol. 22, No. 13, p. 135502, 2011.

[11] C.-H. Kim, C. Jung, H. G. Park & Y.-K. Choi, "Novel Dielectric-Modulated Field-Effect Transistor for Label-Free DNA Detection," Biochip Journal, Vol. 2, No. 2, 127-134, June 2008.

[12] X.Chen, Z. Guo, G.-M. Yang, J.-H. Liu, and X.-J. Huang, "Electrical nanogap devices for biosensing," Matter Today, Vol. 13, pp. 28-41, Nov. 2010.

[13] S. Kalra, M. J. Kumar and Anuj Dhawan, "Dielectric – Modulated Field Effect Transistor for DNA Detection: Impact of DNA Orientation," IEEE Electron Device Letters, Vol. 37, No. 11, November 2016

[14] ATLAS Device Simulation Software, Silvaco, Santa Clara, CA, USA,2012

[15] ATLAS User's Manual. (2000). SILVACO [available online]. http://www.silvaco.com/products/device_simulation/atlas.html

[16] J.- M. Choi, J.-W. Han, S.-J. Choi and Y.-K. Choi, "Analytical Modeling of a Nanogap – Embedded FET for Application as a Biosensor, IEEE Transactions on Electron Devices, Vol. 57, No. 12, December 2010.

[17] O.-Conde, F. J. G. Sanchez, J. J. Liou, A. Cerdeira, M.Estrada, Y.Yue, "A review of recent MOSFET threshold voltage extraction methods," Microelectronics Reliability, Vol. 42, Issues 4-5, p. 583-596, April-May 2002

A Dynamic Resource Allocation Strategy for NoC Based Multicore Systems with Permanent Faults

Suraj Paul*, Navonil Chatterjee*, and Prasun Ghosal*
*Indian Institute of Engineering Science and Technology, Shibpur, Howrah 711103, WB, India
Email:csuraj.ece@gmail.com, navonilster@gmail.com, prasun@ieee.org

Abstract—Integration of multiple processing elements on a single chip has lead to parallel execution of applications on the same multicore platform. For these applications, task mapping and scheduling have significant impact on their timing response and energy consumption. However, the on-chip entities executing these tasks of any given application might fail during run-time. Fault tolerance becomes challenging when real-time applications are hosted on such fault prone environment. The complexity of the problem is further magnified in dynamic scenarios when real-time applications can enter or leave the multicore platform at any time instant. In this work, we present a run-time solution to the unified problem of fault tolerant mapping and scheduling for real-time applications. Both the temporal property of the tasks and the timing information of the faults are considered while selecting a suitable fault tolerance strategy, which reduce the communication energy consumption and satisfies deadline for the tasks of executing applications. On comparing with other fault tolerant approaches, the proposed policy shows 23.4% average reduction in communication energy consumption of applications while achieving 32% improved deadline performance. Additionally, the proposed scheme shows reduced fault tolerance overhead in comparison with existing techniques reported in literature.

Index Terms—Network on Chip, Dynamic Resource Allocation, Multicore Systems, Permanent Faults

I. Introduction

Scaling of transistors to nanometer range has enabled the system designers to embed a large number of Intellectual Property (IP) cores, memory units and processing elements onto a single chip, resulting in a System-on-Chip (SoC). In SoC, shared buses are commonly used for communication between different computation and memory units present in the chip. However, as the number of cores increase, performance of on-chip communication medium between the components becomes a critical issue. Major challenge faced by traditional bus-based SoC include shared inter-core communication, high latency and increased power. To overcome these drawbacks of a bus based SoC, Network-on-Chip (NoC) has been proposed as a viable solution [1]. In NoC, data communication among various cores is achieved through on-chip network consisting of network interface, routers and links. A core is attached to a router through a network interface module, while each routers are connected to one another by point to point links.

On the flip-side, aggressive technology scaling and increasing design complexity of NoC based multicore systems have made the chip components vulnerable to faults which may be permanent or transient in nature. Among these faults, it is the permanent faults which makes the cores defunct without any scope of recovery. Transient faults can be handled at

software-level, while for permanent faults both hardware and software level approaches can be applied. Hardware based techniques use additional resources, such as spare cores [2] [3] to replace the failed components. The redundancy introduced in the design helps to overcome such faulty situations but at the cost of extra area overhead and energy consumption. Such an approach to manage permanent faults is often not feasible in fault tolerant design of systems with tight energy and area constraints. When permanent faults are handled using software techniques, as discussed in [4] [5] [6], timing overhead to tolerate faults becomes a determining factor. It becomes even more challenging for *fault tolerant dynamic systems* where different real-time applications unknown apriori are executed on platforms with varying resource availability. Such systems may be battery operated, where low-energy consumption is necessary to maximize the battery life [7]. Although several techniques have been proposed to minimize energy consumption at the system level [8], such techniques may not always be suitable for real-time applications as it may not guarantee the deadline satisfaction of all tasks present in the application.

A. Novel Contributions of the work

In this work, we present an energy aware dynamic resource allocation strategy for fault tolerant systems. The goal is to provide an unified mapping and scheduling solution for fault-resilient application execution in multicore systems prone to processor faults, where advance knowledge of the complete structure of applications is not available. We address the joint issue of energy and deadline satisfaction of tasks while reducing the fault tolerant overhead during application reconfiguration. Salient features of our work are as follows:

- Proposes a communication energy and deadline aware dynamic mapping and scheduling strategy for real-time applications.
- Proposes a software based fault tolerant mechanism for applications executing on NoC based multicore platform with single or multiple processor faults.
- Presents a runtime algorithm to select a fault tolerant scheme for tasks of applications in a given multicore system.

The rest of this paper is organized as follows. In Section 2, we present the literature review. Section 3 describes the NoC based system model. The preliminaries and problem statement have been defined in Section 4. A heuristic to solve the fault

tolerant run-time mapping and scheduling problem has been shown in Section 5. The experimental setup and associated results have been discussed in Section 6. Section 7 draws the conclusion.

II. RELATED WORKS

In this section, we discuss some selected works on permanent fault-aware approaches for multicore systems. A fault-aware method to manage resources in NoC-based multiprocessors, called FARM (Fault-Aware Resource Management), has been detailed in [2]. The placement of spare cores is fixed for all applications for migrating the tasks on occurrence of faults. [3] proposes an improved mapping method to adaptively determine spare cores for each task for reducing overall communication energy and performance overhead for mitigating core failure. The techniques presented in [6] [9] [4] use the idea of analyzing the possible core failure cases at compile time. Upon detection of a fault, the pre-computed re-mapping solution is fetched from memory and applied. An offline task remapping technique for all processor fault scenarios to minimize the communication energy and task migration overhead has been discussed in [5]. A re-execution slot-based reconfiguration mechanism has been studied in [10]. Normal and re-execution slots of a task are scheduled at design time using an evolutionary algorithm to minimize throughput degradation. At runtime, tasks on a faulty core migrate to their re-execution slot on a different core. In [11], authors have presented a task replication based fault tolerant dynamic application mapping and scheduling for NoC based MPSoC. Based on the temperature of each processor, reliability of individual processors are calculated at run-time. This has been used for employing task redundancy based recovery for tolerating permanent PE failures.

From the above discussion we note that some existing approaches assume the availability of a pre-existing or an offline re-mapping of tasks to apply fault tolerance. Other strategies, adopt an approach which stores all re-mapping decisions for a given degree of fault-tolerance. Such approaches are insufficient for tackling the problem of dynamics fault-tolerance in real-time applications executing on platforms with bounded energy budget. Unlike existing works, which consider static strategies where the algorithms fail to allocate new applications, our work targets applications submitted by users unknown during design time. Further, the proposed approach eliminates the need for additional memory to store the fault-tolerant mapping. The novelty of this work is that it jointly considers task deadline and communication energy consumption to determine a online fault tolerant scheme for permanent core failures.

III. SYSTEM MODEL

We have considered a 2D mesh based NoC topology consists of Processing Elements (PEs), routers and links. The PEs present in the platform are of two types, namely, special purpose and general purpose. The special purpose PE hosts the Real Time Operating System (RTOS) and is referred to as the *Manager Core*. General purpose PEs are used for executing

tasks of a given application. The conceptual system model is depicted in Fig. 1. The status of every PE, that is, faulty/non-faulty or number of tasks allocated to them is updated in the Resource Manager (RM). This provides the Manager Core with latest information about the resource availability and helps in runtime task allocation. Task Mapping and Scheduling Unit (TMSU) allocates these tasks onto different PEs present in the platform. Fault Mitigation Unit (FMU) helps to tolerate permanent faults in PEs by providing fault mitigation policies for each mapped task of the given application. The Task Memory (TM) stores the tasks of the incoming applications.

The communication infrastructure consists of routers and links, that transmit messages between the source and destination nodes. The interconnect network is connected in a mesh fashion. The router in the proposed design uses wormhole packet switching, where packets are transmitted through the network in a pipelined fashion. The size of the input buffer is 32 bits with a buffer depth of 8. The arbiter uses round robin scheduling. We have considered the packet size to be 64 flits, where each flit consists of 32 bits.

IV. PROBLEM FORMULATION

A. Preliminaries

We first describe some important notations related to application mapping/scheduling on NoC necessary to formally state the fault tolerant dynamic resource allocation problem. An application is represented as a directed acyclic graph $G = (T, E)$, where T represents the set of tasks of the application and E is the set of directed edges showing the communication between the tasks of the application. A task $\tau_i \in T$ has associated parameters denoted by (ex_i, dl_i, sl_i), where ex_i is the execution time and dl_i is the completion deadline of the task. sl_i indicates the slack time of the task which is the margin between the time at which a task would complete if it started now, and its deadline. The NoC topology graph is a directed graph $\mathcal{N} = (P, L)$ with each entity $PE_i \in P$ representing a Processing Element (PE) in the topology. The directed edge $l_{ij} \in L$ represents a direct communication link between the router for processors PE_i and PE_j. The resource assignment of a given application G on a finite set of processors is defined by the function pair $(map, start)$ indicating spatial and temporal allocation of the tasks of G respectively. A spatial allocation of the task graph G on a bounded set of processors P is the function *map:* $\Gamma \longmapsto P$ such that if a given task $\tau_i \in \Gamma$ assigned to $PE_j \in P$, $map(\tau_i) = PE_j$ determines the spatial allocation of task. In dynamic task mapping, allocation is invoked when an assigned task requests to communicate with a task that has not yet been mapped/scheduled. Temporal assignment is the determination of starting time, $start(\tau_i)$ of execution of each task. It assumes that allocation of tasks to processors has been completed before.

Next, we denote by Q the set of fault tolerance policies to be applied on a task where $Q = \{replication, migration\}$. We denote recovery function $R : T \to Q$ such that it determines if a task is replicated or migrated upon non-availability of its assigned processor. Next, we define the Communication

978-1-5386-7100-9/18 $31.00 © 2018 IEEE   288

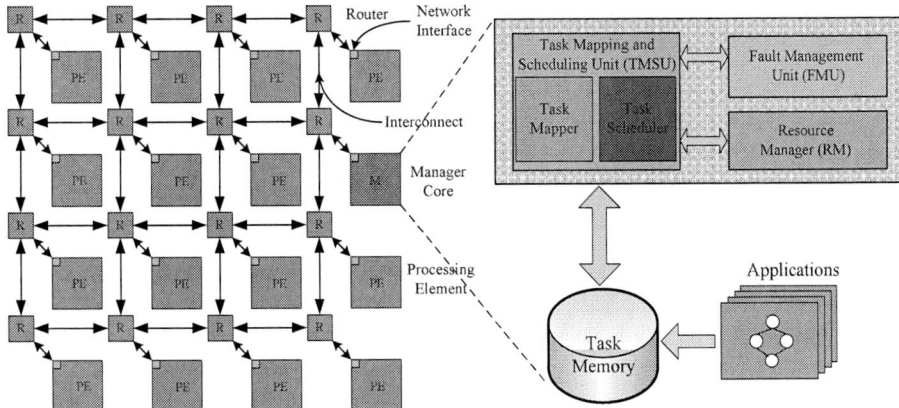

Fig. 1. A Network-on-Chip (NoC) architecture with multiple cores.

Energy of a given application mapping. Energy is consumed when data packets are transferred from a source PE to a destination PE. Let E_r be the energy consumption in pJ in a router for one bit transmission and E_l represent the energy consumption in pJ in a link to transmit one bit. Communication Energy, E_{comm}, for a mapped application on a given NoC platform is estimated as:

$$
E_{comm} = \sum_{\forall G_i} \sum_{\forall e_{ij} \in G_i} (E_r \times [HC(map(\tau_i), map(\tau_j)) + 1] + \\ E_l \times HC(map(\tau_i), map(\tau_j))) * w_{ij}
$$

(1)

where, HC is the hop-count and w_{ij} is the communication volume between tasks τ_i and τ_j in megabits. The energy overhead involved in migrating a task τ_i from a failed processor, PE_{fail}, to an alternate fault-free processor, PE_{mig} is defined as migration energy. It is calculated using the cost of migration [12] of size of task state, $size(\tau_i)$ given below:

$$
E_{mig}(\tau_i) = size(\tau_i) \times [N_l * E_l + N_r * E_r] \\
N_l = (HC(PE_{fail}, PE_{mig})) \\
N_r = (HC(PE_{fail}, PE_{mig}) + 1)
$$

(2)

B. Problem Statement

Given the following as inputs:

1) A set $\mathbb{G} = \{G_1, G_2, ...G_n\}$ of arrived applications, each of which is represented by a directed acyclic graph $G = (T, E)$.
2) A target NoC platform given by $\mathcal{N} = (P, L)$.
3) Timing information of the tasks of the applications
   - execution time, $ex_i > 0$
   - completion deadline, $dl_i \geq ex_i$

Determine a dynamic resource allocation for tasks of the each submitted applications to find:

- Spatial allocation function, $map : \Gamma \longmapsto P$
- Temporal allocation, $start(\tau_i)$ on the identified PE.
- Fault tolerant decisions in Q.

such that while tolerating K core failures:

1) $finish(\tau_i) \leq dl_i$
2) E_{comm} of executing application is reduced
3) Migration overhead for tasks affected during core failure is low

V. PROPOSED APPROACH

Algorithm 1: *Fault Tolerant Resource Allocation(FTRA)*

Input : Application $G(T, E)$, Topology Graph $N(P, L)$.
Output : Fault tolerant dynamic task assignment.

1   ProcAvl=get_free_PE();
2   ProcBusy = $\emptyset$;
3   **for** *each $\tau_i \in T$* **do**
4   TaskType = get_task_category(τ_i);
5   PE_{sel} = Energy_Deadline_Aware_Allocation(τ_i, ProcAvl, ProcBusy);
6   Allocate and Schedule τ_i on PE_{sel};
7   **if** *TaskType is Urgent OR Moderate* **then**
8         $\tau_i^* = $ create_replica_task(τ_i);
9   **end**
10   **if** *TaskType = Urgent* **then**
11       $PE_{sel}^* = $ Energy_Deadline_Aware_Allocation(τ_i^*, ProcAvl, ProcBusy);
12   Allocate and Schedule τ_i^* on PE_{sel}^*;
13   **end**
14   **if** *Fault_Detection = TRUE* **then**
15   PE_{fail}=get faulty processor;
16   t_f = time of fault occurrence;
17   **for** *each $\tau_i \in PE_{fail}$* **do**
18   **if** *TaskType is Moderate* **then**
19   Schedule τ_i^* on PE_{sel}^*;
20   **end**
21   **else if** *TaskType is Relaxed* **then**
22   find PE_{mig} close to PE_{fail};
23   Conditional_Task_Reallocation(t_f, τ_i, PE_{mig});
24   **end**
25   **end**
26   **end**
27   **end**

In this work, we propose an online task mapping and scheduling strategy that helps to tolerate permanent processor failure while meeting the temporal constraints of tasks present in the given application. Such an approach is necessary in real-time dynamic systems where the resource variability (due to runtime PE faults) and order of arrival of applications are not known in advance. It may be noted that processors in the given NoC based multicore system may fail at any time during runtime. We have used the timing characteristics of tasks to select a suitable fault tolerant policy during resource allocation. Along with this, time of occurrence of fault(s) is leveraged to mitigate the effect of PE failure while satisfying their temporal constraints.

Algorithm 1 presents the proposed fault tolerant strategy. The main goal of the algorithm is to meet the deadline of tasks for a given application and dynamically reduce the communication energy consumption. When an application arrives in the system, the Manager Core executes the FTRA algorithm. First, the function $get\_free\_PE()$ prepares a list of unused PEs, $ProcAvl$, currently available for task assignment (Line 1). We initialize the list of occupied PEs, represented by $ProcBusy$ to null (Line 2). Next, we check the nature of task using the function $get\_task\_category()$. This information is exploited to dynamically map the chosen task using $Energy\_Deadline\_Aware\_Allocation$ as follows. To map a task, the algorithm first attempts to map it onto a PE executing its most communicating parent task or any free PE in its close vicinity. This choice of PE reduces energy consumption. However, if no such PE is available, the task is mapped onto an occupied PE but can satisfy its deadline. This is possible if the earliest available time of the PE is less than the slack time of the task (Line 4-6).

TABLE I
SLACK BASED TASK CATEGORIES.

Category	Description
Urgent	$sl_i \leq 0.3 \times ex_i$
Moderate	$0.3 \times ex_i < sl_i \leq 0.6 \times ex_i$
Relaxed	$sl_i > 0.6 \times ex_i$

Next, the algorithm selects an appropriate fault tolerant policy for each tasks belonging to different timing categories. This is decided taking into account the availability of slack time of the task as shown in Table I. If the task is an urgent or moderate task, its replica is generated by $create\_replica\_task()$. The replica task inherits the same timing characteristics and data dependency as the original task. These tasks are then allocated as described above. However, the scheduling decisions are taken based on whether the replica is an active or passive one. In case of urgent task, the task and its replica are executed in parallel. Moderate category tasks are scheduled after the detection of fault in the PE executing the original task. On the other hand, if the mapped task belongs to relaxed category, then the proposed algorithm uses conditional migration strategy based on the time of occurrence of fault (Line 7-23). The aforementioned procedure is successively iterated till all the tasks in the given application are mapped.

In line 23, $Conditional\_Task\_Reallocation()$ is used to choose between two migration strategies. The re-allocation strategies for fault tolerance and analysis of their temporal constraint satisfaction is described below. If a PE_i fails at time t_f then the affected task has completed T_c amount of time into its execution cycle where T_c is given by

$$T_c = t_f - start(\tau_i) \quad (3)$$

If the PE was fault-free, the task could have finished its execution on the same PE by executing for further T_l time given by

$$T_l = start(\tau_i) + ex_i - t_f \quad (4)$$

Following failure of the allotted PE, if the affected task is restarted from beginning on an alternate PE, PE_{mig}, the task meets its deadline if it satisfies Eqn. 5.

$$\delta T + T_c + T_l < dl_i \quad (5)$$

Here, δT is the time taken for rescheduling/remapping the task. Another policy to tolerate permanent fault in PEs involves migration of the task along with its state to a new PE. This involves reallocating both task code and data of the affected task. For this strategy to meet the task deadline, Eqn. 6 should be satisfied.

$$\delta T + T_l \leq dl_i \quad (6)$$

Thus, the criteria for selecting the recovery strategy during migration of task τ_i to a PE_{mig} is determined by Eqn. 7.

$$Conditional\_Task\_Reallocation(\tau_i) =$$
$$\begin{cases} Migration\ with\ task\ restarting, if\ \frac{T_c}{T_l} < 1 \\ Migration\ with\ state\ transfer,\ otherwise \end{cases} \quad (7)$$

In other words, if a PE fails close to the start time of the executing task, then the task is re-allocated on an another available PE and executed from beginning. On the other hand, for PE failures occurring at a time close to the finish time of the task, migration with state transfer is implemented. The alternate PE for task migration is determined dynamically in the nearest neighborhood of the failed PE. This reduces the migration overhead and also improves the chances of meeting deadline of the affected task. Also, task reconfiguration takes place only during the occurrence of faults involving a subset of tasks and processors, the proposed algorithm has low computational load. Any technique with low fault detection latency such as [13], [14] may be used.

VI. RESULTS AND DISCUSSIONS

A simulator has been developed in C++ for evaluation of the proposed dynamic mapping and scheduling algorithm, for allocation of tasks present in applications onto a NoC based multicore platform. The simulator is based on the works presented in [11] [7]. In the experiments, the execution time requirement of the tasks has been uniformly distributed between 5 and 300 clock cycles. For a task τ_i, the corresponding deadline is allotted as $dl_i = k * ex_i$. k is a simulation parameter with positive value. Depending upon the value of k, the availability of the slack time is varied. Smaller the value of k, lesser is the slack time available and vice-versa.

Simulation has been carried out on an Intel i5 processor running at 3.0 GHz frequency. The proposed algorithm has been tested on both synthetic and real-time applications. Real-time applications such as MWD, MPEG, 263enc, 263dec [11] are used for testing. Synthetic applications are generated using TGFF [15], such as TGFF(1-4) consisting of 20, 25, 30 and 35 tasks respectively. Experiments have been conducted on 1000 test scenarios, which are randomly generated and consist

978-1-5386-7100-9/18 $31.00 © 2018 IEEE

of combinations of 200 applications. The timing constraint associated with each task is defined by the availability of slack time associated with them. In each of these scenarios, we have considered multiple PE faults injected randomly in different PEs present in the NoC platform. In our experiments, we have used a homogeneous platform, where all PEs have identical clock frequency. Therefore, the computation energy consumed by a task is similar for all PEs in the given NoC platform. XY routing algorithm is used for transmission of data between dependent tasks mapped on different PEs. For the given 2D mesh based NoC, link energy(E_{link} = 3.125 $\times$ 10^{-13} J/bit) and router energy (E_{router} = 5.24 $\times$ 10^{-12} J/bit) have been derived from ORION 2.0 power model [16]. Next, we discuss the experimental results on the assessment of the proposed approach and compare with other works mentioned in literature.

A. Energy Evaluation

Fig. 2 compares the migration energy overhead of the various fault-tolerant policies. The results have been normalized with respect to Full Remap (FR) [6] [9]. We observe that the proposed FTRA algorithm results in 37.4% average reduction in migration energy consumption as compared to FR scheme. This is due to the fact that the FR policy reallocates all the tasks of the running application on PE failures and thus spends more energy on migrating the tasks to new PEs. When compared to FA-DRA [11], the proposed algorithm consumes 10.3% more migration energy, on an average, due to state transfer of tasks. On the other hand, as FARM [2] and FASA [3] uses only task migration based reexecution strategy to recover from PE faults, 42.1% and 32.6% respectively more migration energy is consumed by these methods compared to the proposed approach.

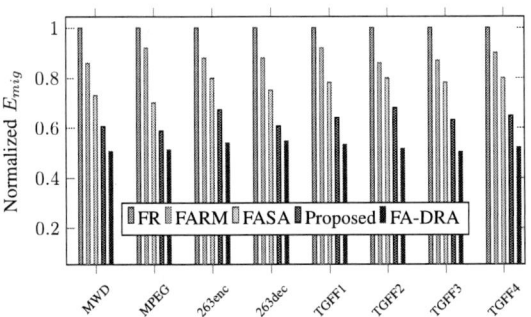

Fig. 2. Comparison of migration energy consumption.

Comparison of the average communication energy consumption of the reconfigured applications using different fault tolerant strategies is reported in Fig. 3. The values are normalized with respect to FR strategy. On comparing with FASA and FARM, we note that 21.6% and 32.5% less energy, is respectively consumed by the reconfigured applications using proposed algorithm. In both these methods timing characteristics of tasks are not taken into consideration while mapping the tasks onto PEs. In comparison with FASA and FARM, FA-DRA results in a mapping solution consuming, on an average,

17.6% and 23.7% less energy respectively. This is because FA-DRA uses timing information along with selective task replication for allocating tasks to PEs. It is observed that FTRA results in 13.2% less communication energy consumption with respect to FA-DRA. The reason for such an improvement is that if the failed PE is different from the predicted one, then FA-DRA re-allocates the task to a new available PE, which may not be a suitable one for energy reduction. In case of FTRA such an alternate PE is chosen in the nearest vicinity of faulty PE. In case of FR, the communication energy consumption for reconfigured applications is minimum due to use of pre-computed energy-aware fault tolerant mapping solutions.

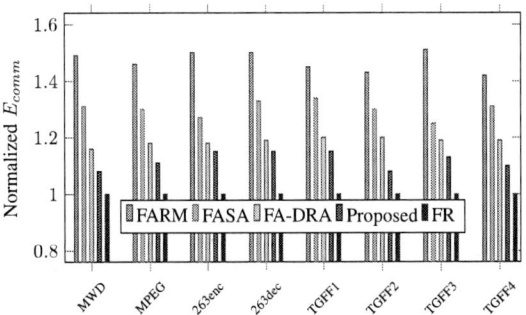

Fig. 3. Comparison of communication energy consumption.

B. Deadline Performance

The results of deadline performance is shown in Fig. 4. In FR policy, remapping the tasks of applications, to healthy PEs during failure of any currently assigned PEs, incurs high time overhead. This is attributed to transfer of the task state space and resuming its execution on available PEs. FASA lowers the re-execution overhead of tasks for the given application by restarting only the task(s) affected by PE failure on spare cores. Thus, the number of tasks meeting their deadline improves as compared to FR policy. Compared to FA-DRA and FASA, the proposed approach shows 25.3% and 38.5% average improvement in deadline performance respectively. By using task timing characteristics coupled with time of fault occurrence, the proposed strategy on an average, finishes more percentage of the executed tasks of a mapped application within their corresponding deadline.

C. Communication Latency

In this section, we discuss the results for average latency for the reconfigured the application. We have selected Noxim 2.0 [17], a cycle accurate simulator, to determine the overall network performance. The setup of the simulator is similar to communication infrastructure discussed in Section 3. The results of the system latency is shown in Fig. 5. FR strategy gives the least latency after fault-aware reconfiguration due to use of pre-computed mapping for the corresponding PE fault scenario. FASA and FARM results, on an average, 39.4% and 44.5% rise in latency respectively, as compared to FR.

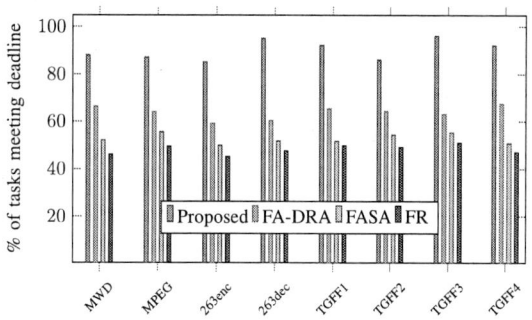

Fig. 4. Deadline performance achieved by different fault aware policies.

FARM allocates a fixed number of spare core(s) at a predefined location on the NoC platform for all applications. Since allocation of spare core is determined by application characteristics in FASA, it results in a reconfigured application where data packets of communicating tasks need to traverse fewer number of routers and links compared to FARM resulting in reduced network latency. Compared to FASA, the proposed approach and FA-DRA show, on an average, 18.5% and 24.4% lesser network latency respectively. This is because both these techniques exploit the timing information of the tasks while mapping them to PEs which results in tasks with higher communication requirements being mapped on close vicinity of each other. Thus, the distance traversed by the data packets for inter-task communication gets reduced. However, the data packets transferred during task state migration in FTRA adds to the network traffic of the platform. By exploiting fault occurrence information, the rise in latency information is, on an average, 6% compared to FA-DRA.

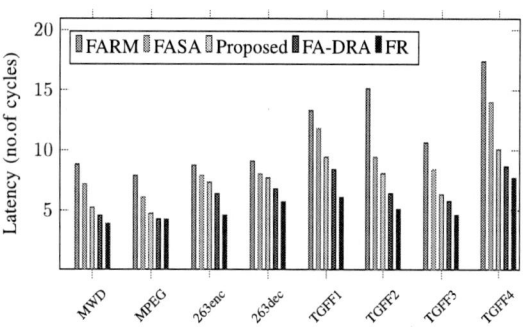

Fig. 5. Comparison of network latency.

VII. CONCLUSION

In this work, we have presented a dynamic fault-tolerant task allocation scheme for multicore platforms prone to permanent processor failures. The proposed algorithm uses a unified mapping and scheduling strategy which gives an energy aware resource allocation. By jointly considering the temporal characteristics of tasks together with timing information of faults, notable improvement in temporal constraint satisfaction and energy consumption of allocated applications

are achieved while alleviating the effects of permanent core failure. Experimental results show that the proposed algorithm achieves better performance with respect to deadline satisfaction, communication energy and fault tolerance overhead when compared to other fault tolerant algorithms. The strategy proposed in this work is attractive to adopt for fault tolerance in energy constrained dynamic scenarios with varying resource availability.

REFERENCES

[1] L. Benini and G. De Micheli, "Networks on chips: a new soc paradigm," *Comput.*, vol. 35, no. 1, pp. 70–78, Jan 2002.

[2] C. L. Chou and R. Marculescu, "Farm: Fault-aware resource management in noc-based multiprocessor platforms," in *Design, Automation Test in Europe Conference Exhibition (DATE), 2011*, March 2011, pp. 1–6.

[3] F. Khalili and H. Zarandi, "A fault-tolerant core mapping technique in networks-on-chip," *IET Computers and Digital Techniques*, vol. 7, no. 6, pp. 238–245, November 2013.

[4] F. Bolanos, F. Rivera, J. E. Aedo, and N. Bagherzadeh, "From uml specifications to mapping and scheduling of tasks into a noc, with reliability considerations," *J. Syst. Archit.*, vol. 59, no. 7, pp. 429–440, Aug. 2013.

[5] A. Das, A. Kumar, and B. Veeravalli, "Communication and migration energy aware task mapping for reliable multiprocessor systems," *Future Generation Computer Systems*, vol. 30, pp. 216 – 228, 2014.

[6] A. Das, A. K. Singh, and A. Kumar, "Execution trace–driven energy-reliability optimization for multimedia mpsocs," *ACM Trans. Reconfigurable Technol. Syst.*, vol. 8, no. 3, pp. 18:1–18:19, May 2015.

[7] N. Chatterjee, S. Paul, P. Mukherjee, and S. Chattopadhyay, "Deadline and energy aware dynamic task mapping and scheduling for network-on-chip based multi-core platform," *Journal of Systems Architecture - Embedded Systems Design*, vol. 74, pp. 61–77, 2017.

[8] L. Benini, A. Bogliolo, and G. De Micheli, "Readings in hardware/software co-design," G. De Micheli, R. Ernst, and W. Wolf, Eds. Norwell, MA, USA: Kluwer Academic Publishers, 2002, ch. A Survey of Design Techniques for System-level Dynamic Power Management, pp. 231–248.

[9] A. Das, A. K. Singh, and A. Kumar, "Energy-aware dynamic reconfiguration of communication-centric applications for reliable mpsocs," in *Reconfigurable and Communication-Centric Systems-on-Chip (ReCoSoC), 2013 8th International Workshop on*, July 2013, pp. 1–7.

[10] J. Huang, J. O. Blech, A. Raabe, C. Buckl, and A. Knoll, "Analysis and optimization of fault-tolerant task scheduling on multiprocessor embedded systems," in *Hardware/Software Codesign and System Synthesis (CODES+ISSS), 2011 Proceedings of the 9th International Conference on*, Oct 2011, pp. 247–256.

[11] N. Chatterjee, S. Paul, and S. Chattopadhyay, "Fault-tolerant dynamic task mapping and scheduling for network-on-chip-based multicore platform," *ACM Trans. Embedded Comput. Syst.*, vol. 16, no. 4, pp. 108:1–108:24, 2017.

[12] T. Maqsood, S. Ali, S. U. R. Malik, and S. A. Madani, "Dynamic task mapping for network-on-chip based systems," *Journal of Systems Architecture*, vol. 61, no. 7, pp. 293 – 306, 2015.

[13] R. R. Ferreira, E. Sanchez, J. d. Rolt, G. L. Nazar, A. F. Moreira, L. Carro, and M. S. Reorda, "Permanent fault detection and diagnosis in the lightweight dual modular redundancy architecture," in *2015 16th Latin-American Test Symposium (LATS)*, March 2015, pp. 1–6.

[14] S. Ananthanarayan, S. Garg, and H. D. Patel, "Low cost permanent fault detection using ultra-reduced instruction set co-processors," in *2013 Design, Automation Test in Europe Conference Exhibition (DATE)*, March 2013, pp. 933–938.

[15] R. P. Dick, D. L. Rhodes, and W. Wolf, "Tgff: task graphs for free," in *Proceedings of the 6th international workshop on Hardware/software codesign*. IEEE Computer Society, 1998, pp. 97–101.

[16] A. Kahng, B. Li, L.-S. Peh, and K. Samadi, "Orion 2.0: A fast and accurate noc power and area model for early-stage design space exploration," in *Design, Automation Test in Europe Conference Exhibition, 2009. DATE '09.*, April 2009, pp. 423–428.

[17] V. Catania, A. Mineo, S. Monteleone, M. Palesi, and D. Patti, "Noxim: An open, extensible and cycle-accurate network on chip simulator," in *Application-specific Systems, Architectures and Processors (ASAP), 2015 IEEE 26th International Conference on*, July 2015, pp. 162–163.

978-1-5386-7100-9/18 $31.00 © 2018 IEEE

2018 IEEE Computer Society Annual Symposium on VLSI

Floorplanning in Graphene Nanoribbon (GNR) Based Circuits

Subrata Das* and Debesh Kumar Das[†]

*Department of Information Technology, Academy of Technology, Aedconagar, Hooghly, India.
[†]Department of Computer Science & Engineering, Jadavpur University, Kolkata, India.
Email: dsubrata.mt@gmail.com, debeshd@hotmail.com

Abstract—In this paper, we propose a technique for floorplanning in case of graphene nanoribbon (GNR) based circuits. Graphene nanoribbon based devices and interconnects are now found to be better alternatives over traditional CMOS based devices and interconnects. Logic blocks of GNR can be assumed to be hexagonal in shape. Due to special geometric structure in GNR, the interconnects can be bent only in $0°$, $60°$ and $120°$ angles. Hence routing grids are aligned to these angles only. The concept of floorplanning in traditional VLSI design is extended first time for GNR based circuits in this paper.

Index Terms—Graphene nanoribbon, hybrid cost, floorplanning

I. INTRODUCTION

Due to extreme scaling, interconnects between logic blocks are now playing an important role for VLSI design. Traditional interconnects are now facing several challenges due to delay and power dissipation. Emerging devices and interconnects can now be used to sustain performance gain beyond CMOS scaling [1]. Due to improved electrical conductivity, thermal conductivity and mechanical strength, *graphene nanoribbon (GNR)* based devices and interconnects are now found to be better alternatives over traditional VLSI design for both transistors and interconnects with respect to delay and power dissipation.

Graphene is a flat two-dimensional (2D) single atomic layer sheet of carbon atoms packed into a honeycomb lattice [2]. *Graphene nanoribbon* can be formed by cutting a ribbon out of *graphene* [3]. The properties and structures of *GNR* was discussed in different literatures [2], [3], [4]. Depending on the shape of edge, *graphene* can be either armchair or zigzag [3]. *GNR* with predominantly zigzag edges is metallic and with armchair edges is semiconductor as shown in Fig. 1-a and Fig. 1-b respectively. Metallic *GNR* is oriented at $30°$, $90°$ and $150°$ angles where as semiconducting *GNR* are oriented at $0°$, $60°$ and $120°$ angles [4], [5].

For traditional VLSI design, in the partitioning phase a complex circuit is decomposed into rectangular blocks. Blocks for which the dimensions are known are called fixed blocks, otherwise they are called flexible blocks. The assignment of fixed blocks on a layout surface is known as placement problem and that of some or all flexible blocks is known as *floorplanning* [6] where the shape and approximate position of each module is determined. For traditional VLSI design, the objective of this phase is minimization of chip area, interconnect length and combinations of these two. However the power

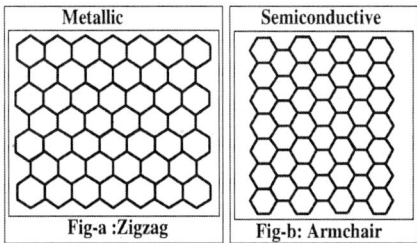

Fig. 1. Structure of a) Zigzag and b) Armchair Graphene Nanoribbon

dissipation and delay are causing major challenges in VLSI physical design day by day the use of *GNR* is providing less power dissipation and delay due to its technology. In case of *GNR* our intention is to make the placement in such a way that it decreases delay and power dissipation of *GNR* interconnect.

Bending of GNR in $0°$, $60°$ and $120°$ angles maintains chirality i.e. due to bending in these angles metallic (semiconducting) *GNR* remains metallic (semiconducting). Bending in $30°$, $90°$ and $150°$ angles changes chirality in *GNR* i.e. due to bending in these angles metallic *GNR* is converted to semiconducting *GNR* and vice versa. Thus to maintain the chirality the routing grid of metallic *GNR* are oriented along $0°$, $60°$ and $120°$ angles only. Such a routing grid is known as triangular routing grid. For the proper placement, *GNR* blocks can be decomposed into hexagonal blocks. Hence the floorplanning of *GNR* is the determination of the location of hexagonal blocks in the chip area with the locations of the pins on the boundary of each module.

Bending in different angles yields different resistances. The resistance due to $120°$ bending is three times larger to that of $60°$ bending [4]. The resistance due to both the interconnect length and the bending is known as hybrid cost. Hence the objective of the *GNR* floorplanning is the minimization of this hybrid cost. Interestingly, for equal length interconnect, compared to copper wire, the power consumption and interconnect delays in *GNR* wire can be reduced by up to 50% and 60% respectively for global interconnect [3].

Different methods of floorplan design for traditional VLSI design were discussed in different literature [7], [8], [9]. The concept for the conditions of non-overlapping rectangular blocks as described in [9] is applied for hexagonal block

978-1-5386-7100-9/18 $31.00 © 2018 IEEE

293

of *GNR* in this paper. We find out the conditions for non-overlapping hexagonal blocks to place them in such a way that it satisfies the other conditions of floorplanning. Then we propose our algorithm to place the hexagonal modules with the objective of minimum hybrid cost.

II. BACKGROUND

Cost due to interconnect length is constant upto mean free path length and increases linearly beyond mean free path length. Let L and w_L represent the interconnection length and interconnect resistance per unit length respectively. Let α_{60} and α_{120} respectively represent the resistance cost due to $60°$ and $120°$ bendings. Let n_{120} and n_{60} represent the number of $120°$ and $60°$ bendings respectively. Hybrid Cost is given by

$$ HC = L \times w_l + n_{120} \times \alpha_{120} + n_{60} \times \alpha_{60} \qquad (1) $$

Example 1. *Figs. 2-a and 2-b show two different routing paths between same pair of source (S) and sink (T) terminal. In Fig. 2-a, interconnection length $L = 22$, number of $60°$ bending $n_{60} = 2$ and number of $120°$ bending $n_{120} = 0$, hence hybrid cost $(HC_1)= 22w_L + 2\alpha_{60}$. In Fig. 2-b, $L = 17$, $n_{60} = 0$ and $n_{120} = 2$, hence hybrid cost $(HC_2)= 17w_L + 2\alpha_{120}$. As in [4], if we assume $w_L = 1, \alpha_{60} = 10, \alpha_{120} = 30$, then $HC_1 = 42$ and $HC_2 = 77$.*

Hence depending on the positions of the source and sink terminals there may be different routing paths. In different routing paths there may be different number of interconnect lengths as well as different number of bendings. As a result in each of these routing paths *hybrid cost* may be different. We have to select the routing paths with minimum *hybrid cost*. As our objective of floorplanning is to reduce the *hybrid cost* hence the number of bending and interconnect length will affect the performance of floorplanning.

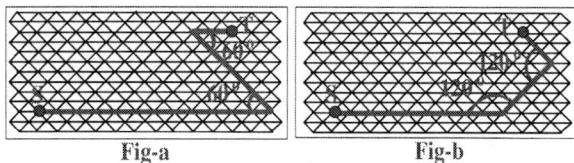

Fig-a   **Fig-b**

Fig. 2. Different paths between same pair of source (S) and Sink (T) terminal

III. PROBLEM FORMULATION

In the floorplanning of *GNR* circuit the information of hexagonal blocks such as areas, shapes and pin locations are given. Triangular grid representation is given in an arbitrary chip area. Fixed blocks are placed on the triangular grid. We have to place the flexible blocks in the triangular grid in such a way that we can do the placement within the given area and make the connections between nets with minimum hybrid cost.

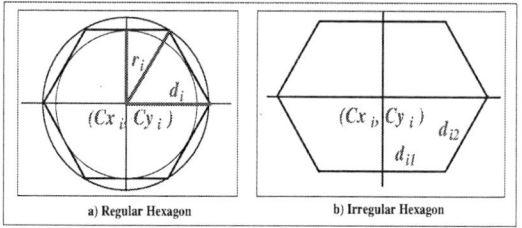

Fig. 3. a) Regular Hexagon b) Irregular Hexagon

IV. PROPOSED METHOD

The structure of *GNR* is honeycomb like lattice and the routing grid of *GNR* is triangular in nature. Hence logic blocks of *GNR* can be assumed as hexagonal in shape which may be either regular or irregular. A regular hexagon is equilateral and also equiangular. Fig. 3-a shows a regular hexagon. Let the position of ith logic block is defined by $\{(Cx_i, Cy_i), d_i, r_i\}$, where (Cx_i, Cy_i) is the center of ith regular hexagon, d_i and r_i are the radii of circumscribed and inscribed circles respectively. It can be shown geometrically that $d_i = \frac{2r_i}{\sqrt{3}}$.

Hexagonal modules may be irregular in shape. Irregular hexagons may have two different side lengths. Let i^{th} irregular hexagonal module is described by $\{(Cx_i, Cy_i), d_{i1}, d_{i2}\}$, where (Cx_i, Cy_i) is the center of such hexagon and two different side lengths are d_{i1} and d_{i2} (Fig. 3-b).

Generally the modules are available in rectangular shapes. To qualify them for *GNR*, we first have to change the rectangular configurations to hexagonal configurations. The following subsection deals with the transformation of rectangular modules to hexagonal modules.

A. Transformation of Rectangular Modules to Hexagonal Modules

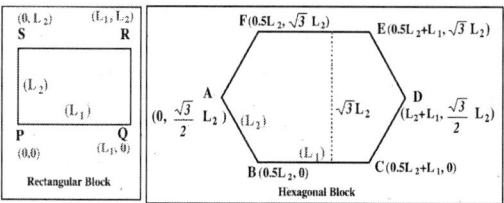

Fig. 4. Rectangular Block and corresponding Hexagonal Block

Fig. 4 shows a rectangular block with four vertices $P(0,0)$, $Q(L_1,0)$, $S(0,L_2)$ and $R(L_1,L_2)$. Here, L_1 and L_2 ($L_1 \neq L_2$) are respectively the length of two adjacent sides of the rectangle. The corresponding irregular hexagonal block as shown in Fig. 4 is having coordinates of its six vertices as $A(0, \frac{\sqrt{3}}{2}L_2)$, $B(0.5L_2, 0)$, $C(0.5L_2 + L_1, 0)$, $D(L_2 + L_1, \frac{\sqrt{3}}{2}L_2)$, $E(0.5L_2 + L_1, \sqrt{3}L_2)$ and $F(0.5L_2, \sqrt{3}L_2)$. L_1 and L_2 are the length of two adjacent sides of the irregular hexagon.

978-1-5386-7100-9/18 $31.00 © 2018 IEEE

TABLE I
TRANSFORMATION OF COORDINATE FROM RECTANGULAR BLOCK TO HEXAGONAL BLOCK

Rectangular Block		Hexagonal Block		Transformation	
X Coordinate	Y Coordinate	X Coordinate	Y Coordinate	X Coordinate	Y Coordinate
$0 \le x_r \le L_1$	$y_r = 0$	$0.5L_2 \le x_h \le (0.5L_2 + L_1)$	$y_h = 0$	$x_h = x_r + 0.5L_2$	$y_h = y_r$
$0 \le x_r \le L_1$	$y_r = L_2$	$0.5L_2 \le x_h \le (0.5L_2 + L_1)$	$y_h = \sqrt{3}L_2$	$x_h = x_r + 0.5L_2$	$y_h = \sqrt{3}y_r$
$x_r = 0$	$0 \le y_r \le 0.5L_2$	$0.5L_2 \ge x_h \ge 0$	$0 \le y_h \le \frac{\sqrt{3}}{2}L_2$	$x_h = 0.5L_2 - y_r$	$y_h = \sqrt{3}y_r$
$x_r = 0$	$0.5L_2 \le y_r \le L_2$	$0 \le x_h \le 0.5L_2$	$\frac{\sqrt{3}}{2}L_2 \le y \le \sqrt{3}L_2$	$x_h = y_r - 0.5L_2$	$y_h = \sqrt{3}y_r$
$x_r = L_1$	$0 \le y_r \le 0.5L_2$	$(0.5L_2 + L_1) \le x_h \le (L_1 + L_2)$	$0 \le y_h \le \frac{\sqrt{3}}{2}L_2$	$x_h = y_r + 0.5L_2 + L_1$	$y_h\sqrt{3}y_r$
$x_r = L_1$	$0.5L_2 \le y_r \le L_2$	$(L_1 + L_2) \ge x_h \ge (L_1 + 0.5L_2)$	$\frac{\sqrt{3}}{2}L_2 \le y_h \le \sqrt{3}L_2$	$x_h = (1.5L_2 + L_1) - y_r$	$y_h = \sqrt{3}y_r$

Now, if rectangular block is in square in shape ($L_1 = L_2 = L$) then we get a regular hexagon with vertices as $A(0, \frac{\sqrt{(3)}}{2}L)$, B(0.5L, 0), C(1.5L, 0), $D(2L, \frac{\sqrt{(3)}}{2}L)$, $E(1.5L, \sqrt{(3)}L)$ and $F(0.5L, \sqrt{(3)}L)$. Here L is the side length of regular hexagonal block.

Table I shows the transformation of coordinates of a point in rectangular block to the corresponding coordinates in hexagonal block. In the table (x_r, y_r) and (x_h, y_h) respectively represent the coordinates of any point of rectangular block and irregular hexagonal block. Similarly, if the rectangular block is square then regular hexagonal block can be obtained.

In order to find the proper location of the hexagonal logic blocks, we first have to find the constraints to be satisfied by two non-overlapping hexagonal blocks. We have identified these constraints in case of *GNR* as discussed below.

B. Finding Constraints for non-overlapping Hexagonal Blocks

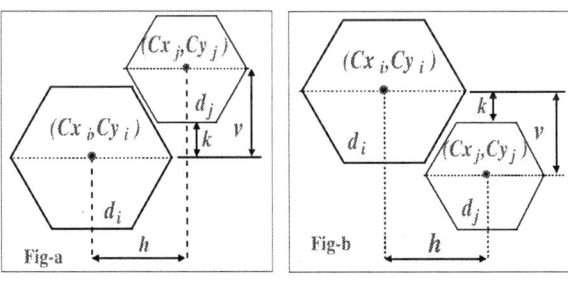

Fig. 5. Both i^{th} and j^{th} Blocks are regular

For *floorplanning* two adjacent blocks must be non overlapped. Depending on the positions of the ith and j^{th} modules there are different constraints to be satisfied which are detailed in the following cases.

Case 1. *Both blocks are regular:* Fig. 5 shows i^{th} and j^{th} *regular hexagonal modules. Here* $h = |(Cx_j - Cx_i)|$, *is the horizontal distance between the centers of the hexagonal blocks*, $v = |(Cy_j - Cy_i)|$, *is the vertical distance between the centers of the hexagons and* $k = |(v - r)| = |(v - \frac{\sqrt{3}}{2}d_j)|$. *From Fig. 5, the conditions of non-overlapping*

of two hexagonal blocks are given by

$$if\ v \neq 0\ then,\ |(Cx_j - Cx_i)| \ge (d_i + d_j - \frac{v}{\sqrt{3}}) \quad (2)$$

$$If\ v = 0\ then,\ |(Cx_j - Cx_i)| \ge (d_i + d_j) \quad (3)$$

$$If\ k = 0\ then,\ |(Cx_j - Cx_i)| \ge (d_i + \frac{d_j}{2}) \quad (4)$$

$$If\ h = 0\ \&\ k \ge d_i \frac{\sqrt{3}}{2}\ then,\ |(Cy_i - Cy_j)| \ge \frac{\sqrt{3}}{2}(d_i + d_j) \quad (5)$$

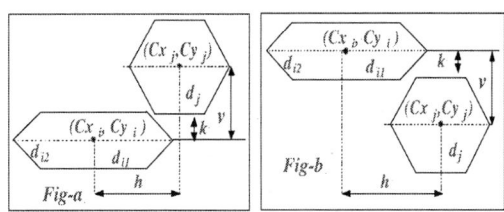

Fig. 6. The ith Block is irregular and the jth Block is regular

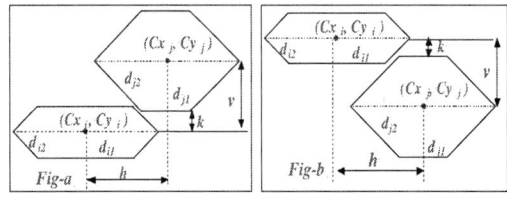

Fig. 7. Both i^{th} and j^{th} Blocks are irregular

Case 2. *One block is irregular and other is regular:* Let i^{th} *block is irregular and* j^{th} *block is regular, From Fig. 6, the conditions of nonoverlapping of two hexagonal blocks are gievn by the following equations (suppose* $d_{i_{avg}} = \frac{d_{i1} + d_{i2}}{2}$),

$$if\ v \neq 0\ then,\ |(Cx_j - Cx_i)| \ge (d_{i_{avg}} + d_j - \frac{v}{\sqrt{3}}) \quad (6)$$

$$If\ v = 0\ then,\ |(Cx_j - Cx_i)| \ge (d_{i_{avg}} + d_j) \quad (7)$$

$$If\ k = 0\ then,\ |(Cx_j - Cx_i)| \ge (d_{i_{avg}} + \frac{d_j}{2}) \quad (8)$$

$$If\ h = 0\ \&\ k \ge d_{i2} \frac{\sqrt{3}}{2},\ |(Cy_i - Cy_j)| \ge \frac{\sqrt{3}}{2}(d_{i2} + d_j) \quad (9)$$

Here, in this Case 2, the expressioons for h, k and v are same as in Case 1.

Similarly for i^{th} block regular and j^{th} block irregular, the following conditions hold (suppose $d_{j_{avg}} = \frac{d_{j1}+d_{j2}}{2}$)

$$If \ v \ \neq 0 then, \ | \ (Cx_j - Cx_i) \ | \geq (d_i + d_{j_{avg}} - \frac{v}{\sqrt{3}}) \quad (10)$$

$$If \ v = 0 \ then, \ | \ (Cx_j - Cx_i) \ | \geq (d_i + d_{j_{avg}}) \quad (11)$$

$$If \ k = 0 \ then, \ | \ (Cx_j - Cx_i) \ | \geq (d_i + \frac{d_{j1}}{2}) \quad (12)$$

$$If \ h = 0 \ \& k \geq d_i \frac{\sqrt{3}}{2} \ then, \ | \ (Cy_i - Cy_j) \ | \geq \frac{\sqrt{3}}{2}(d_i + d_{j2}) \quad (13)$$

Case 3. *Both blocks are irregular: Consider Fig. 7. Here also* $h = | \ (Cx_j - Cx_i) \ |$ *and* $v = | \ (Cy_j - Cy_i) \ |$ *but* $k = | \ (v - \frac{\sqrt{3}}{2}d_{j2}) \ |$. *The conditions of non-overlapping are given by (suppose* $d_{ij_{avg}} = \frac{d_{i1}+d_{i2}+d_{j1}d_{j2}}{2}$)

$$If \ v \neq 0 then, \ | \ (Cx_j - Cx_i) \ | \geq (d_{ij_{avg}} - \frac{v}{\sqrt{3}}) \quad (14)$$

$$If \ v = 0 \ then, | \ (Cx_j - Cx_i) \ | \geq (d_{ij_{avg}}) \quad (15)$$

$$If \ k = 0 \ then, \ | \ (Cx_j - Cx_i) \ | \geq (d_{iavg} + \frac{d_{j1}}{2}) \quad (16)$$

$$If \ h = 0 \ \& k \geq d_i \frac{\sqrt{3}}{2} \ then, \ | \ (Cy_i - Cy_j) \ | \geq \frac{\sqrt{3}}{2}(d_{i2} + d_{j2}) \quad (17)$$

C. Algorithm for Floorplanning

Next we propose a method in floorplanning such that hybrid cost is minimized. The algorithm places the hexagonal logic blocks in different layers. It first places a few blocks serially unless it reaches the boundary. Then next few blocks are placed in the next upper layer. The same process is repeated unless all the blocks are placed. The Algorithm Floorplan design of GNR is shown in Fig. 8.

Floorplan_Design_of_GNR()

Data structures: store_area[]: an array to store the area of hexagonal logic block x_store[]: a global array to store the x-coordinate of vertex A of current hexagonal block, y_store[]: a global array to store the x-coordinate of vertex A of current block, count[]:a global array to store the block number x_{limit}: is used to check whether the block is inside the boundary or not, $x_{previous}$: store the x-coordinate of vertex A of last placed block
Input: n number of hexagonal logic blocks.
Output: Floorplan design.

1. For i=1 to n begin
2.   store_area[i]= Calculate_area_hexagonal_logic_block(i)
3. End for
4. Find_logic_ block_ with_maximum_area(n, store_area[])
5. Place_Maximum_Size_Logic_Block(index_max)
6. For each of the remaining logic block begin
7.   Select the block among the remaining blocks which has maximum interconnections with all previous blocks.
8.   If more than one blocks each has equal interconnections with all previous block then select maximum sized block
9.   Place Next Logic Block()
10. End for
11. for i=1 to n-1 begin
12. for j=1 and $j \neq i$ to n-1 begin
13.   Interconnect_Logic_Blocks(j)
14.   End for
15. End for
End

Fig. 8. Floorplan Design of Graphene Nanoribbon

It consists of several procedures described below.
(i)Calculate_area_hexagonal_logic_block(i): This function is used to find the area of ith hexagonal logic block (Fig. 9).

(ii)Find_logic_ block_ with_maximum_area(): It is described in Fig. 10. This function is used to search the hexagonal block of maximum area.

Calculate_area_hexagonal_logic_block(i)

Input: Coordinates of ith hexagonal block
Output: Area of ith hexagonal block

1. From the coordinates of the vertices find the lengths of two adjacent sides of the hexagon (L_1 and L_2 as shown in Fig. 4)
2. If ($L_1 = L_2$) then
3.   area=$\frac{3\sqrt{3}L_1^2}{2}$
4. else
5.   area=$\frac{\sqrt{3}L_2^2}{2} + \sqrt{3}L_1L_2$
End

Fig. 9. Finding Area of ith logic block

Find_logic_ block_ with_maximum_area(t_n_b, store_area[])

Data structures: t_n_b: total number of block, store_area[]: array that store area of all blocks max_area=maximum area, index_max: index of the maximum area block
Input: Total number of block, Area of all block
Output: Find maximum area and corresponding block

Apply linear search to find the maximum area and index of the maximum area block from t_n_b number of blocks and store them respectively max_area and index_max.
End

Fig. 10. Finding Maximum area logic block

Place_Maximum_Size_Logic_Block()

Input: Maximum size Hexagonal logic block.
Output: Geometric position of maximum size hexagonal logic blocks.

1. Find total number of nets (say n_{front}) in the AB and AF side of the hexagon as shown in Fig. 4.
2. Find total number of nets (say n_{bottom}) in the BC side of the hexagon as shown in Fig. 4.
3. if ($L_1 \neq L_2$) then
4.   A(n_{front}, $n_{bottom}+\frac{\sqrt{3}}{2}L_2$) is the coordinate of vertex A of the hexagonal block as shown in Fig. 4.
5.   Evaluate the coordinates of B($n_{front}+0.5L_2$, n_{bottom}), C($n_{front}+0.5L_2 + L_1$, n_{bottom}), D($n_{front}+L_1 + L_2$, $n_{bottom}+\frac{\sqrt{3}}{2}L_2$), E($n_{front}+0.5L_2 + L_1$, $n_{bottom}+\sqrt{3}$) and F($n_{front}+0.5L_2$, $n_{bottom}+\sqrt{3}$) vertices of the hexagon.
6. else
7.   A(n_{front}, $n_{bottom}+\frac{\sqrt{3}}{2}L$) is the coordinate of vertex A of the hexagonal block
8.   Evaluate the coordinates of B($n_{front}+0.5L$, n_{bottom}), C($n_{front}+1.5L$, n_{bottom}), D($n_{front}+2L$, $n_{bottom}+\frac{\sqrt{3}}{2}L$), E($n_{front}+1.5L$, $n_{bottom}+\sqrt{3}$) and F($n_{front}+0.5L$, $n_{bottom}+\sqrt{3}$) vertices of the hexagon.
9. x_{store}[]=x-coordinate of Vertex A.
10. y_{store}[]=y-coordinate of Vertex E.
11. x_{limit}=x-coordinate of Vertex E.
End

Fig. 11. Placing maximum size logic block

978-1-5386-7100-9/18 $31.00 © 2018 IEEE

(iii)Place_Maximum_Size_Logic_Block(): This function is used to place the maximum size logic block obtained in step 2. It is shown in Fig. 11. This block is placed in the triangular grid in such a way that proper space should be there in the front and bottom of the block for interconnection.

Place_Next_Logic_Block()

Data structures: no_of_layer=layer number, $block\_count$=count the number of blocks after placing it,
Input: A hexagonal logic block.
Output: Proper Placement of the logic block

1. Check the conditions of non overlap of the logic block with all previous logic blocks.
2. Count the number of free nets of front side ($n_{frontpre}$), rear side ($n_{rearpre}$), top side (n_{toppre}), bottom side ($n_{bottompre}$) of all previous blocks.
3. Find total number of nets n_{front} in the AB and AF side, n_{rear} in the CD and DE side, n_{bottom}) in the BC, n_{top}) in the EF side of the hexagon as shown in Fig. 4.
4. $start\_block$=0
5. $end\_block = 0$
6. x_{limit}=x_{limit}+total length of hexagonal block+n_{rear}
7. if ($x_{limit} \leq x_{boundary}$) then
8. $x_{previous}$= x-coordinate of vertex A of previous block
9. end if
10. if $x_{limit} > x_{boundary}$) then
11. $x_{previous}$=0.
12. $no\_of\_layer = no\_of\_layer + 1$
13. end if
14. if($no\_of\_layer > 0$)
15.   for i=start_block to end_block begin loop
16.   if ($x\_store[i] <$ x-coordinate of A) then
17. $y_{previous}= y\_store[i]$
18.   end if
19. $end\_block = end\_block + 1$
20.   end for
21. else
22. $y_{previous}$=0.
23. x_c=n_{front}+$x_{previous}$+$n_{rearpre}$+$n_{bottompre}$.
24. if ($L_1 \neq L_2$) then
25. y_c=n_{bottom}+$\frac{\sqrt{3}}{2}L_2$ +$y_{previous}$
26.   A(x_c, y_c), B(x_c+0.5L_2, y_c-$\frac{\sqrt{3}}{2}L_2$),
        C(x_c+0.5$L_2 + L_1$, y_c-$\frac{\sqrt{3}}{2}L_2$), D(x_c+$L_2 + L_1$, y_c),
        E($x\_$+0.5$L_2 + L_1$, y_c+$\frac{\sqrt{3}}{2}L_2$) and
        F(x_c+0.5L_2, y_c+$\frac{\sqrt{3}}{2}L_2$) are vertices of the hexagon.
27. end if
28. if($L_1 = L_2 (= L)$) then
29. y_c=n_{bottom}+$\frac{\sqrt{3}}{2}L$ +$y_{previous}$
30.   A(x_c, y_c), B(x_c+0.5L, y_c-$\frac{\sqrt{3}}{2}L$),
        C(x_c+1.5L, y_c-$\frac{\sqrt{3}}{2}L$), D(x_c+2L, y_c),
        E($x\_$+1.5L, y_c+$\frac{\sqrt{3}}{2}L$) and
        F(x_c+0.5L_2, y_c+$\frac{\sqrt{3}}{2}L$) are vertices of the hexagon.
31. end if
32. $x_{store}[]$=x-coordinate of Vertex A.
33. $y_{store}[]$=y-coordinate of Vertex E.
34. block_count=block_count+1
35. x_{limit}=x-coordinate of Vertex E.
36. $end\_block = end\_block + 1$
End

Fig. 12. Placing logic block

After placing maximum size logic block we have to select next logic block to be placed. Logic block among remaining blocks that has more number of connections with all previous blocks is chosen for this purpose. If more than one block has same number of connections with all previous blocks then the block having maximum size is selected.

Interconnect_Logic_Block()

Input: Hexagonal logic blocks.
Output: Interconnections of the logic blocks

Coordinate of pin location (x_p, y_p), y-coordinate of bottom side of hexagon (y_b),y-coordinate of bottom side of hexagon (y_t)

1. Find the interconnection points.
2. If the any of the interconnection points is in the bottom side of the hexagon then set the y_p=y_b-1 unit. and x_p=x_b-1 unit.
3. If the any of the interconnection points is in the top side of the hexagon then set the y_p=y_t+1 unit. and x_p=x_p-1 unit.
4. If the any of the interconnection points is AB or CD side (as shown in Fig. 4) y_p=y_b-1 unit and calculate corresponding x_p.
5.If the any of the interconnection points is AF or DE side (as shown in Fig. 4) y_p=y_t+1 unit. and calculate corresponding x_p.
6. Interconnect them with metallic GNR wire in such a way that the *hybrid cost is minimized* [10].
7. If interconnects are overlapped then insert via.
End

Fig. 13. Interconnect logic blocks

---

iv)Place_Next_Logic_Block(): This function is described in detail in Fig. 12. It places the new block after placing of maximum size block in such a way that it does not overlap with all previous blocks and space should be there to interconnect the pins with the previous blocks. To place next block after maximum size block first we have to know the coordinates of six vertices (A, B, C, D, E and F as shown in Fig. 4) of the new block. If we have the coordinate of vertex A then we can easily have the coordinates of other vertices. The x coordinate of vertex A of the new block is the x coordinate of vertex E of the previous block plus proper spaces for the interconnections. The y-coordinate of the new block is calculated as the distance for proper spaces for interconnection. Hence the new block is placed properly. The next block is also placed in the same way. The blocks in the first layer is placed in this way unless the x coordinate of vertex E plus the spaces interconnections is not greater than the x coordinates of the boundary of chip. Then the block is placed in the next upper layer. In this case x coordinate of vertex A is calculated corresponding to next upper row. Similarly y coordinate is also calculated by giving proper spaces for interconnections. The process is continued in the same manner unless all the logic blocks are placed.
v)Interconnect_Logic_Blocks(): The interconnections of the nets of different blocks are done by this function. Its description is given in Fig. 13. When the interconnections are done vias are inserted to avoid the crossover of different nets. Given a placement, to connect the pins of the block to having minimum *hybrid cost* is given in [10]. The blocks having free nets among all previously placed blocks are found. The counts of the free nets in the top and bottom of such blocks determine the space needed to place the block.

V. EXPERIMENTAL RESULT

AMI33 benchmark is popularly used for traditional VLSI design. This benchmark includes 33 rectangular blocks and 123 nets. For *GNR* we need hexagonal blocks. As there is no existing data for GNR circuits, we convert these rectangular

TABLE II
EXPERIMENTAL RESULTS

Traditional VLSI Floor plan design with Routing Space				Floor plan with routing space for GNR (Our Proposed Method)						
Total Area of 33 blocks	Algorithms	Chip Area (mm^2)	Wire Length (mm)	Total Area of 33 blocks	Chip Area (mm^2)	Wire Length (mm)	Hybrid Cost of GNR due to interconnect length and bending	Effective Increase /Decrease in Chip Area	Decrease in Interconnect Length	Decrease of effective Interconnect length of GNR Compared to copper considering Hybrid Cost
1.156449 (mm^2)	MOSAICO	3.16	151.824	2.157963	5.829974	57.81618	87.36593	-1.13%	61.92%	42.46%
	VITAL	3.12	134.599					0.14%	57.04%	35.01%
	seattle	2.94	125.000					6.27%	53.75%	30.11%
	BEAR	2.83	131.244					10.40%	55.95%	33.43%
	Delft	2.60	151.656					20.16%	61.88%	42.39%
	Minnesota	2.54	139.810					23.00%	58.65%	37.51%

blocks into hexagonal blocks following the technique shown in Fig. 4. The location of a pin in the hexagonal logic block is calculated from the corresponding rectangular block using Table I. The set of these hexagonal blocks can be called as modified AMI33.In the triangular grid, let the distance between two successive grid points is one micro unit. We compare the results of our algorithm with experimental results as shown in Table II showing the required chip area, wire length. In floorplanning of traditional VLSI design MOSAICO, VITAL, seattle, BEAR, Delft, Minnesota are different algorithms whose data are available in [9] and [11] and they are reported again in Table II. The total area of 33 rectangular and hexagonal blocks for AMI 33 benchmark and modified AMI 33 are respectively 1.156449 and 2.157963. Hence to measure *effective increase* we multiply the each of the chip area of traditional VLSI design by the ratio of $\frac{2.157963}{1.156449}$. The Table II shows that effective chip area is decreased in one case and it remains almost the same in another case. In other cases effective area increases from 6% to 23%. For traditional interconnect such as cooper the interconnect length is the design parameter to connect multiple pins. For traditional interconnect, the resistance with uniform cross section increases with the length of the interconnect. For traditional interconnect, bending has no effects, but for *GNR* based interconnect the resistance is dependent on interconnect length as well as bending (*hybrid cost*). By *effective interconnect length* in the experimental result of Table II we mean the *hybrid cost*. Cost due to effective interconnect length is reduced in the range 30% to 42%. The reduction of *hybrid cost* will automatically reduce the power dissipation and delay of interconnect. From Table II, it is clear that our algorithm provides better performance as effective interconnect length is always decreased by our technique. Though this improvement in performance causes penalty in increase in area even upto 23%. However this area is also decreased in case of MOSAICO. Recently power dissipation and delay are the crucial factors of VLSI design. Decreasing them using *GNR* and our placement method is obviously a great achievement. The achievement is reflected in Table II.

VI. CONCLUSION

Graphene Nanoribbon based devices and interconnects are better alternatives over traditional CMOS based devices and interconnects in terms of power dissipation, delay and more. In this paper we studied the *GNR* floorplanning problem for the first time. Use of *GNR* causes large reduction in delay and power. The experimental result shows that using *GNR*, though there is an increase in chip area, the interconnection cost is greatly reduced. The work may be improved by rotating the hexagonal blocks into specific angles. Detailed routing algorithm may also provide further better results in terms of delay of chip area,interconnect delay and power dissipation.

REFERENCES

[1] ITRS Report 2015.
[2] T. Ragheb and Y. Massoud, "On the modeling of resistance in graphene nanoribbon (gnr) for future interconnect applications," *IEEE/ACM Intl. Conf. on Computer-Aided Design*, pp. 593-597, 2008.
[3] H. Li, C. Xu, and K. Banerjee, "Carbon nanomaterials: The ideal interconnect technology for next-generation ics," *IEEE Design & Test of Computers*, pp. 20-31, 2010.
[4] T. Yan, Q. Ma, S. Chilstedt, M. D F. Wong, and D. Chen, "A routing algorithm for graphene nanoribbon circuit," *ACM Trans. on Design Automation on Elec. Sys.*, Vol. 18, No. 4, pp. 61:1–61:18, October 2013.
[5] M. Y. Han, B. Ozyilmaz, Y. Zhang, , and P. Kim, "Energy band-gap engineering of graphene nanoribbons," *Physical Review Letters*, Vol. 98, pp. 206805, 2007.
[6] Naveed A. Sherwani, Algorithm for VLSI Physical Design Automation, 3rd edition, Kluwer Academic Publishers, 1999.
[7] P. Dasgupta and S. Sur-Kolay, "Slicible Rectangular Graphs and Their Optimal Floorplans," *ACM Trans. on Design Automation of Elec. Sys.*, Vol. 6, No. 4, October 2001, Pages 447-470.
[8] Dinesh P. Mehta and Naveed Sherwani, "On the Use of Flexible, Rectilinear Blocks to Obtain Minimum-Area Floorplans in Mixed Block and Cell Designs," *ACM Transactions on Design Automation of Electronic Systems*, Vol. 5, No. 1, January 2000, pp. 8297.
[9] S. Sutanthavibul, E. Shragowitz and J.B. Rosen, "An Analytical Approach to Floorplan Design and Optimization," *IEEE Transactions on Computer-Aided Design*, Vol. 10, No. 6, June 1991, pp. 761-769.
[10] S. Das, S. Das, A. Majumder, P. Dasgupta and D. K. Das, "Delay Estimates for Graphene Nanoribbons: A Novel Measure of Fidelity and Experiments with Global Routing Trees," *ACM GLSVLSI 16*, Boston, MA, USA, pp 263-268, May 18-20, 2016.
[11] B. Eschermann, Dai Wei-Ming, E.S. Kuh, M. Pedram, "Hierarchical placement for macrocells: a 'meet in the middle' approach," *IEEE International Conference on Computer-Aided Design*, 1988, pp 460-463.

978-1-5386-7100-9/18 $31.00 © 2018 IEEE

Generating Safety Guidance for Medical Injection with Three-Compartment Pharmacokinetics Model

Cunxi Yu, Heinz Riener, Francesca Stradolini, Giovanni De Micheli
Integrated Systems Laboratory (LSI)
Ecole polytechnique federale de Lausanne (EPFL)
Lausanne, Switzerland
cunxi.yu@epfl.ch

Abstract—Medical cyber-physical systems are a new trend of software controlled physical systems that are increasingly common in medical domains. With rapid developments in medical science and computer technology, safety verification and simulation becomes more challenging. This paper introduces a general model for medical injection systems, which can be used for formal verification, simulation/testing, and computing the *Area Under the Curve* (AUC) metrics, using *Satisfiability Modulo Theories* (SMT) over *Reals*. An algorithm of computing *constrained AUC* for measuring *drug exposure* with relative baseline, is presented based on the *proof of unsatisfiability*. We demonstrate that our model can efficiently solve these problems using the state-of-the-art SMT solver dReal.

Index Terms—Satisfiability Modulo Theories, Medical Cyber-System, Timed System.

I. INTRODUCTION

Medical cyber-physical systems are a new trend of software controlled physical systems that are increasingly common in medical domains. These systems are becoming more and more popular in medical therapy. Medical service is more efficient and convenient for doctors and patients, while they offer the opportunities for medical experts and doctors in studying the treatments and in communicating with patients. Therefore, patients can benefit from the automation of treatment process, which improve therapy effectiveness, lifestyle quality and reducing cost. However, with rapid developments in medical science and computer technology, medical cyber-physical systems are becoming more and more precise and complex, and more real-time reactions of the patient during treatment are sampled and analyzed. Due to the complexity of the systems and a low tolerance for faults in the medical environment, validation and verification of medical cyber-physical systems are crucial [1]. Specifically, the challenges of verifying medical systems mainly come from *timed* and *hybrid* properties. Similarly to all cyber-physical systems, medical systems are mostly about the intersections of computations and physical actions. To create the mathematical models of the entire systems, formal methods that combine both discrete and continuous dynamics are required.

Formal methods have been widely used in checking the properties and reliabilities of hybrid systems, which are modeled using abstract mathematical representations. The main advantage of formal methods comes from the mathematical precision for reasoning the correctness of system models.

Timed Automata plays a big role in modeling and verifying the timed systems [2]. Particularly for medial systems such as drug administration system, such methods model the behavior of the system using formal representation by employing *Timed Automata extended with Tasks* (TAT). Tools such as UPPAAL [3] and its extension TIMES [4] have been successful in verifying cyber-physical systems in many domains. Model checking based on *satisfiability* theories have also been applied to timed systems [5]. For example, dReal [6] is demonstrated to successfully verify biology systems by solving *Satisfiability Modulo Theories* (SMT) problems over the reals with a wide range of nonlinear functions, such as ordinary differential equations (ODEs) [7]. The main advantages of using SMT over reals comparing to real-time model checkers are: **a)** for the systems with frequent changes between different dynamics, SMT performs much faster; **b)** if an unsafe state exists in the system, SMT offers the *proof of unsatisfiability* that provides the information of where and why the unsafe state appears.

To precisely model medical systems, a realistic drug response model has to be used. Various clinical studies show that responsiveness to the treatment with drugs depends on the concentration of the drug in the blood that depends on patients, drug dose, and intake time interval. Pharmacokinetics (PK) [8] is a branch of pharmacology focused on studying the drug disposition in the human body. For many drugs, the concentration in the blood of a patient is highly related to its effectives. Pharmacodynamics (PD) [9] is the study of the biochemical and physiological effects of drugs on the body. Therapeutic Drug Monitoring (TDM) [10] is the approach that unifies the PK-PD knowledge, which shows that drugs with explicit PK-PD relationships and a narrow therapeutic range may be easily under- or overdose. Hence, it is important to develop an approach that generates safety guidance for medical injection using a precise drug response model, such as the drug administration system [11]. In addition, *Area Under the Curve* (AUC), as well as AUC in the baseline measurements (constrained AUC), are commonly used to assess the extent of exposure of a drug [12]. Measuring these two metrics is very important in pharmacokinetics analysis.

The main contributions of this paper are as follows:
- We introduce a general model for medical injection system, which can be applied to both simulation and formal verification, using SMT over Reals. The mathematical

three-compartment pharmacokinetics model is used for drug response in the abstract timed model, which is one of the most precise pharmacokinetics models for simulating drug response.

- The model is demonstrated that it can efficiently and precisely simulate the medical injection process. The model can formally prove(disprove) if the expected drug concentration-time objectives are *reachable* with given injection actions, and return *sat* (*unsat*). This is done by checking bounded δ-*Satisfiability*[13]. The proof of *unsatisfiability* is generated if it returns *unsat*, which indicates the unreachable state(s) and the corresponding time location(s).

- The proposed model computes *AUC* and *constrained AUC* simultaneously during the verification or simulation process. For computing constrained AUC, we introduce an algorithm based on *proof of unsatisfiability* of SMT over reals.

II. Problem Formulation

A timed system is defined with the finite set of continuous clocks $\mathbb{T}$ and a set of constraints over the clocks. Mostly, the constraints are represented as conjunctions, disjunctions, and negations of expressions over the clocks. Each transition in such system is labeled by a constraint over the state or clock values, namely *guard*, which indicates the condition to trigger the transition. Each state is constrained by an *invariant*, which restricts the possible values of the clocks for being in the state, which can then enforce a transition to be taken. The following notations are used for problem formulation.

Let a timed system be a tuple $\mathcal{A}=\langle \mathbb{S}, \mathbb{T}, Inv, \mathbb{E}, \mathcal{ACT}, init \rangle$.

- $\mathbb{T}$ *is a finite set of clocks.* $t_i \in \mathbb{T}$, *and* $t_i \in R^+$, $i \in [0,n]$.
- $\mathbb{S}$ *is a finite set of states.* $s_i \in \mathbb{S}$ *is the state at* i^{th} *time.*
- *Inv is the associated invariant for each state.*
- $\mathbb{E}$ *is a finite set of transitions, where* e_i *is a tuple* $\langle s_i, s_j, g, act, \mathbb{T}_{i \to j} \rangle$, $e_i \in \mathbb{E}$. *The state changes from* s_i *to* s_j *over a set of clocks* $\mathbb{T}_{i \to j}$. g *is the guard of transition* e_i, *and* act *is the action of* e_i.
- $\mathcal{ACT}$ *a finite set of actions the system made.*
- *init is the initial values of all the parameters for encoding the system.*

In this work, the state $\mathbb{S}$ are the concentrations of different compartments. The clock set $\mathbb{T} = [0, t_n]$. The action set $\mathcal{ACT}$ are the inputs that triggers the transitions. Simulation and AUC calculation can be achieved with the same formulation of formal verification.

Problem 1: The medical therapy objectives O in concentration-time format and the injection actions $\mathcal{ACT}$, are provided by the doctor or electronic drug system. Let the upper bound clock be t_n and the initial states be s_0, and $O=\{(c_{i_0}, t_{i_0}), (c_{i_1}, t_{i_1}), ..., (c_{i_j}, t_{i_j})\}$. Each element of O is a pair of concentration[1] and time, $\forall\, t_{i_j} \leq t_n$. The verification goal is checking if all the objectives in O can be reached by system $\mathcal{A}$ with given actions $\mathcal{ACT}$. This can be done by checking

---

[1] c_{i_j} is a comparison function, e.g., $c_{i_0} \leq 0.01$ or $c_{i_0} == 0.01$.

---

the following: $\forall t_i \in \mathbb{T}$, checking if $O \subset (\mathbb{S}, \mathbb{T})$ according transitions $\mathbb{E}$; if O is a subset of $(\mathbb{S}, \mathbb{T})$, the system is *safe*; otherwise, the system is *unsafe*.

Simulation and AUC calculation can be achieved by replacing O. For simulating the system, $O=(\emptyset, t \geq t_n)$, i.e., asking if the system can reach the clock of t_n without any constrains, which is **always safe**. According to the definition of AUC [14], the AUC of the concentration C equals to $AUC=\int_0^{t_n} \frac{d[C]}{dt} dt$.

Problem 2: Let the upper bound clock be t_n and the initial states be s_0. Given the system $\mathcal{A}$, the injection actions $\mathcal{ACT}$, and a concentration lower bound l, we define AUC_{under} to be the area above the concentration curve but below the bound l. Similarly, we can define AUC_{over} if an upper bound concentration limit is given. Such concentration bounds can be provided by the doctors for medical therapy or by the medicine researchers for drug analysis. AUC_{under} and AUC_{over} are the two types of *constrained AUC* considered in this paper. One example of AUC_{under} is shown in Figure 1 with a lower bound limit $l=0.002$, where t_x is the first clock when the concentration is lower than l and t_y is the first clock when the concentrations are higher than l. Note that l could be time continuous function, or a relative drug exposure baseline [12]. Then, AUC_{under} can be computed as

$$AUC_{under} = \int_{t_x}^{t_y} l\ dt - \int_{t_x}^{t_y} \frac{d[C]}{dt} dt \qquad (1)$$

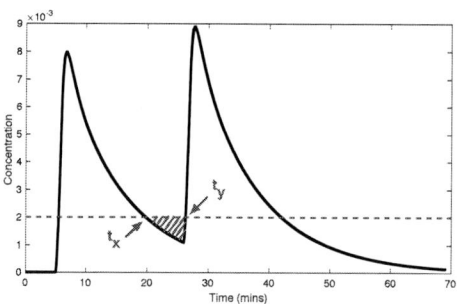

Fig. 1: Example of constrained AUC.

Based on Eq. 1, the problem is to find the clocks t_x and t_y. We introduce an algorithm that obtains such clocks based on a *proof of unsatisfiability* in Section IV.

III. Background

A. Three-Compartment Model

Mathematic models of a human body are created to study physiologic or pharmacologic kinetic characteristics. The compartment model can simulate the biologic processes involved in the kinetic behavior of a drug after it has been introduced into the body, leading to a better understanding of its pharmacodynamic effects []. Mostly, one compartment model is not sufficient to represent the pharmacokinetics of a drug. A two- and three-compartment model have wider applicabilities. In this work, we use three-compartment to represent

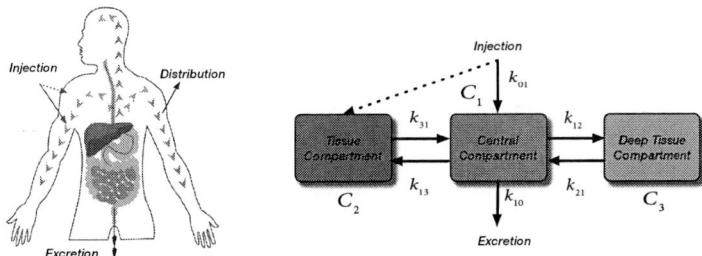

Fig. 2: Three-Compartment pharmacokinetics model. Injection could be taken in either central compartment C_1, such as blood injection, or tissue compartment C_2, such as muscle injection.

the pharmacokinetics of a drug, specifically using Michaelis-Menten elimination model [15][16]. The abstract model is shown in Figure 2, including central compartment, tissue, and deep tissue compartment sub-models. The three-compartment represents a drug that is distributed most rapidly to a highly perfused central compartment such as blood and brain. This is also the compartment which takes the injection. The drug is distributed less rapidly to the tissue compartment such as muscle, and very slowly to the deep tissue compartment, containing such poorly perfused tissue as bone and fat. The deep tissue compartment may also represent tightly bound drug in the tissues.

After the injection, it is first distributed to the central compartment C_1. There is then redistribution to tissue compartment C_2 with good perfusion, with further redistribution to the poorly perfused deep tissue C_3. The rates of infusion, $k_{12}, k_{21}, k_{13}, k_{31}$, depend on the rate of transfer between the various theoretical compartments of the body. Elimination (drug clearance) only happens at the central compartment, with rate k_{10}.

The model is described using *ordinary differential equations* (ODEs) [15]. In general, there are two dynamic models of three-compartment model for modeling an injection system, i.e., *distribution* model and *injection* model. The distribution dynamic model represents the distribution and dilution of the injection, as shown in Eq. 2 C_1, C_2, and C_3 are the concentration of the central compartment, tissue compartment and deep tissue compartment, over time t. The central compartment concentration C_1 depends on the rate of excretion ($-k_{10}C_1$) and the rates of distributing to the other two compartments ($-k_{12}C_1$-$k_{13}C_1$), and the other two compartments are only related to C_1.

$$
\begin{aligned}
\frac{d[C_1]}{dt} &= -(k_{10} + k_{12} + k_{13}) \cdot C_1 + k_{21} \cdot C_2 + k_{31} \cdot C_3 \\
\frac{d[C_2]}{dt} &= -k_{21} \cdot C_2 + k_{12} \cdot C_1 \\
\frac{d[C_3]}{dt} &= -k_{31} \cdot C_3 + k_{13} \cdot C_1
\end{aligned}
\tag{2}
$$

The second dynamic is required to model the concentrations of the three compartments when an injection is taken. The difference compared to the distribution dynamic is in the ODE of C_1, shown in Eq. 3 R_{inject} is the rate of drug injection

which is a constant number. The amount of drug injected $\int_t^{t+\Delta t} R_{inject} dt = R_{inject} \cdot \Delta t$. Note that distribution (dilution) of the body naturally processes all the time. By adding R_{inject} in the first ODE of Eq. 3, this model successfully describes the injection process with distribution. In one injection monitoring system, there could be more than one dynamic models if the injection rate can be adjusted.

$$
\begin{aligned}
\frac{d[C_1]}{dt} &= -(k_{10} + k_{12} + k_{13}) \cdot C_1 + k_{21} \cdot C_2 + k_{31} \cdot C_3 + R_{inject} \\
\frac{d[C_2]}{dt} &= -k_{21} \cdot C_2 + k_{12} \cdot C_1 \\
\frac{d[C_3]}{dt} &= -k_{31} \cdot C_3 + k_{13} \cdot C_1
\end{aligned}
\tag{3}
$$

B. Satisfiability Modulo Theory (SMT)

The Satisfiability Modulo Theories (SMT) problem is a decision problem for logical formulas with respect to first-order logic. In other words, SMT departs from treating the problem in a strictly Boolean domain and integrates different well-defined theories (Boolean variable, bit vectors, integer/floating arithmetic, reals, etc.) into a DPLL-style SAT decision procedure [5]. Some of the most effective SMT solvers that are developed for specific problems. For example, Boolector [17] is the most efficient SMT solver in solving bit-level decision problem; Z3 [18] and CVC [19] have been widely used in verifying software. SMT formulas over the real numbers can encode a wide range of problems, particularly in modeling hybrid systems. dReal [6] is the state-of-the-art SMT solver over reals that can model the verification problem of hybrid system.

IV. MODELING

This section introduces the modeling of the injection systems, the verification problem and the algorithm of calculating the constrained AUC, using the non-linear SMT solver dReal. First, a set of global definitions has to be claimed. According to Eq. 2 and Eq. 3, these include the definition of static variables and dynamic variables. The static variables include distribution, absorption and excretion rate k_{ij}, e.g., using syntax "#define k_{10} 0.4;". The concentration of each compartment C_i and the clock *time* are defined as dynamic variables with a bound, e.g., using syntax "[0, 60] time;".

978-1-5386-7100-9/18 $31.00 © 2018 IEEE

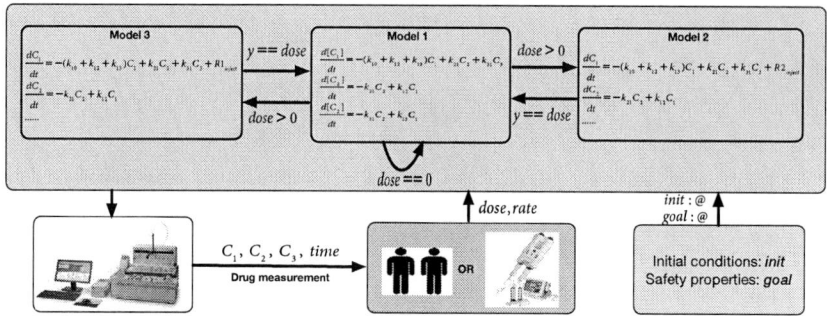

Fig. 3: Generic modeling of injection system with two injection dynamics.

A. Modeling Dynamics in SMT

The main part is the dynamic model, which is the three-compartment model of the medical injection system. A complete dynamic model must include all the elements retried in the tuple of the timed automata $\mathcal{A}$ (Section II). We first introduce the SMT model of the *distribution* dynamic shown in Eq. 4.

To define a dynamic model, we first declare the label of the model with a numerical value m (line 1). The transition between different dynamics is described using the pointer m of the model. Second, the *invariant* for the states are defined (lines 2 and 3), which is a conjunction of logic formulas which must always hold in a model. For the *distribution* dynamic, the invariant define that the concentrations of all the compartments $C_i \geq 0$, and there is no absorption (*dose=0*). The continuous dynamics of a model by providing a set of ODEs of the distribution dynamic are included in *flow*, where $d/dt[C_1]$ represents $\frac{d[C_1]}{dt}$, t is the global variable *time*. The first formulas of *jump* is interpreted as *guard*, i.e., a logic formula specifying a condition to make a transition. Note that this allows a transition but does not force it. The second argument of jump denotes the target model m of the triggered transition, and applies to the dynamic variables in the logic formulas. In this conjunction, C_i' represent the dynamic variable $\dot{C_i}$.

```
1   mode 1;
2     invt :
3       (and(C_1 ≥ 0)(C_2 ≥ 0)(C_3 ≥ 0)(dose ≤ 0)(dose ≥ 0));
4     flow :
5       d/dt[C_1] = -(k_10 + k_12 + k_13)C_1 + k_21 C_2 + k_31 C_3;
6       d/dt[C_2] = k_12 C_1 - k_21 C_2;
7       d/dt[C_3] = k_13 C_1 - k_31 C_3;
8       d/dt[x] = 1;
9     jump :
10      (guard_model1)
              ==> @2(and(C_1' = C_1)(C_2' = C_2)(C_3' = C_3)(x' = x));
```
$$(4)$$

An extra dynamic variable $\dot{x}$ is introduced in all the dynamic model (line 8) to represent the clock. $\dot{x}=time$ with $\frac{dx}{dt}=1$. This is because dReal doesn't support *time* to be used in *guard*. Note that *guard* is specifically constructed according to the hybrid system. For example, if the injection will be

taken when the concentration of tissue compartment is equal to c (e.g., using electronic pump), guard=(and $(C_2 \leq c)$). If the injection is taken periodically every t_p, to model two absorptions, guard=(and($x \leq t_p$)($x \geq t_p$)($x \leq 2t_p$)($x \geq 2t_p$)). In both cases, *jump* will point to the *injection* dynamic(s). Our SMT model is very flexible to model a hybrid system with both feedback control and human operators.

The difference between the ODEs of injection and distribution dynamics is C_1. However, the SMT model has to be changed. The invariant condition should be replaced with $dose > 0$ in the previous model. Multiple modifications need to be done in $flow$. The differential equation of C_1 is replaced by line 5 in Eq. 5. An extra dynamic variable y defined with $d/dt[y] = R_{inject}$ is used for constructing $guard$, where y is calculating the amount of drug injected (line 8). Once the transition is made, we have to reset y in case there are multiple doses in the hybrid system. For the systems that have various injection rates $R_{injection}$, we need to create separate injection model for each of them.

```
1   mode 2;
2     invt :
3       (and(C_1 ≥ 0)(C_2 ≥ 0)(C_3 ≥ 0)(dose > 0));
4     flow :
5       d/dt[C_1] = -(k_10 + k_12 + k_13)C_1 + k_21 C_2 + k_31 C_3 + R_inject;
...
8       d/dt[y] = R_inject;
10    jump :
11      (and(y ≥ dose)(y ≤ dose))
              ==> @3(and(C_1' = C_1)...(y' = 0));
```
$$(5)$$

Finally, the initial states of the first model and the safety goal of the hybrid system have to defined. If the hybrid system is initialized with model 1, *init* should start with @1 with all variables set to 0. *goal* shares the same syntactic structure of *init*. The safety properties can be constructed using the conjunctions of formulas. For example, line 3 in Eq. 6 is checking if C_2 is in [0.005, 0.01] during time [10, 15]; line 5 checks if $C_1 \leq 0.1$ is always safe over all the clocks.

B. System modeling

The general system modeling of the injection systems is shown in Figure 3. The drug response model is built with

978-1-5386-7100-9/18 $31.00 © 2018 IEEE

one distribution dynamic and two injection dynamics since there exist two injection rates. The proposed SMT model can formulate any control units (the injection control) if the decisions are made based on time and the concentrations. This is done by modifying *guard* for each dynamic model formula. For example, assume that there are two injections with amount d_1 and d_2 at t=5 and t=20 over *time*=[0, 60], using the injection rates $R1$ and $R2$, respectively. The transitions are $model\ 1 \to model\ 2 \to model\ 1 \to model\ 3 \to model\ 1$. The time condition should trigger the transitions between model 1, and models 2 and 3. *guard* of model 1 should describe $x == 5$ OR $x == 20$. However, SMT over the reals only supports conjunction of formulas. Hence, we need to model OR using inversion and AND, such that $(x=5) \lor (x=20) \to \overline{(x \neq 5)} \land \overline{(x \neq 20)}$. The SMT formula is

$$jump : (not(and(x < 5)(x > 5))\ (and(x < 20)(x > 20)))$$

Similarly, the control decisions made based on concentration, or the combination of concentration and time, can be modeled using the same approach.

```
1  init : @1(and(C_1 = 0)(C_2 = 0)(C_2 = 0)(x = 0));
2  goal : @1(and
3        (and(C_2 ≥ 0.01)(C_2 ≤ 0.005)(x ≥ 10)(x ≤ 15))
4        (and(C_3 ≥ 0.003)(C_3 ≤ 0.001)(x ≥ 20)(x ≤ 25))    (6)
5        (and(C_1 ≤ 0.1))
6        (and(...)))
```

C. Area Under Curve (AUC) and Constrained AUC

To compute AUC of the three concentrations, we just need to add three dynamic variables and differential equations in all the models. AUC_i are the dynamic variables of AUC of ith concentration. According to the definition of AUC, the derivative of AUC is the concentration function, which can be simply represented using Eq. 7.

$$d/dt[AUC_1] = C_1;\ d/dt[AUC_2] = C_2;\ d/dt[AUC_3] = C_3; \quad (7)$$

As mentioned in Section II Figure 1, to compute constrained AUC, t_x and t_y that indicates the bounded clocks of the error regions are required. Once bounded clocks are available, constrained AUC can be computed by adding the Eq. 1 into the hybrid model. Note that multiple error regions may exist. Hence, we introduce an algorithm that iteratively collects the bounded clocks by using the *proof of unsatisfiability* generated by dReal (with option –proof), shown in Algorithm 1.

The algorithm takes the SMT formula of the system ψ, the error bound l as inputs and generates the bounded clocks of the error regions. The algorithm includes two global *goal* formulas for ψ, g and g' indicating *safe* and *unsafe*. First, the algorithm checks if there exists an error state by checking if ψ is always safe with error bound l (line 4). If there is no error state over *time*, the algorithm will be terminated. If there exit error states, the algorithm will start collecting the bounded clocks (lines 7-16). In each iteration, it first extracts the starting clock of the error region, t_x^i, by extracting the smallest clock in

the proof. Note that the proof can be generated iff the problem is unsat. Hence, *goal* of ψ is complemented. The proof of unsat includes the starting clock of error region for g', which is the ending clock of the error region for g. Once i^{th} iteration is done, *goal* is reset to g, and the initial time t_{init} is set to t_y^i such that *time*=$[t_y^i, t_n]$ in the next iteration. This makes sure that the next iteration skip all the previous collector regions. The bounded clocks (t_x^i, t_y^i) be returned if there is no more error region (lines 7 and 17).

Algorithm 1 Constrained AUC

Input: Hybrid system formula ψ in SMT
Input: Error bound l, $time = [0, t_n]$
Output: Bounded clocks of unsafe region with error bound l.
1: g: l is infeasible (safe); g': l is feasible (unsafe);
2: $goal=g$;
3: $t_{init} = 0$; i=0;
4: **if** $\forall\ t_i \in time$, $(\psi \hookrightarrow goal)$ is always SAT **then**
5:     $t_x^i = t_y^i = null$;
6: **end if**
7: **while** $\exists\ t_i \in time$, $(\psi \hookrightarrow goal)$ is $UNSAT$ **do**
8:     Extract t_x^i from the proof of unsat;
9:     $t_{init} = 0$; $goal=g'$;
10:     **if** $\exists\ t_i \in time$, $(\psi \hookrightarrow goal)$ is $UNSAT$ **then**
11:         Extract t_y^i from the proof of unsat;
12:     **else**
13:         $t_y^i=t_n$;
14:     **end if**
15:     i++; $goal=g$; $t_init=t_y^i$;
16: **end while**
return t_x^i and t_y^i, $\forall i$;

V. Experimental Results

The experimental results are conducted on MacOS with 2.3 GHz Intel Core i7 x4 with 16 GB memory. We solve the hybrid SMT formulas using dReal[6] in the single-thread [6]. Algorithm 1 is implemented in C++ using dReal as a black-box that generates the proof of unsat. We demonstrate our approach using the example used for illustrating the injection system modeling in Section IV. The system has a time bound [0, 60] hours and has two injections triggered by *time*=5 and *time*=20. To show the complete results of all the states up to *time*=60, the *goal* is set as goal: @1 $(x \geq 60)$;.

The results are included in Figure 4. The x-axis represent the time. Left-hand y-axis represents the concentrations C_1, C_2, and C_3. Right-hand y-axis represents the AUC of each concentration. All the results are time continuous with interval 0.005 second defined the precision of the SMT solver (with option –precision). The runtime of generating all the results in Figure 4 is less than 15 seconds. If a set of concentration-time objectives O are provided by the users, the *goal* has to be modified using Eq. 6. Mostly, checking the satisfiability of O takes less CPU time than simulating over all the clocks. This is because the SMT solving process will be terminated as soon as an unsafe state s_{unsafe} is detected. For example, if O includes $C_2 \leq 2e^{-4}$ for clocks in [20, 25], the solver will return *unsat* and terminate at the first clock that $C_2 > 2e^{-4}$.

978-1-5386-7100-9/18 $31.00 © 2018 IEEE

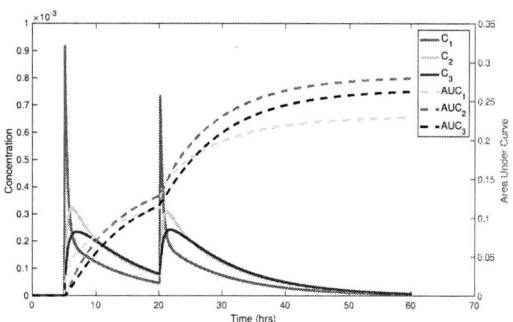

Fig. 4: Continuous results of the concentrations and area under curves (AUCs) generated by dReal up to $time = 60$ hours.

We show the result of computing the constrained AUC of C_2 using the same system, shown in Figure 5. The runtime overhead of Algorithm 1 compared to the original SMT formulas varies on the error bound function l. If the given exposure baseline (error bound l) is a linear function, such as $l = 1e^{-4}$, Algorithm 1 computes the constrained AUC with almost no runtime overhead. A non-linear function l could significantly increase the runtime complexity, which mainly comes from the SMT solver dReal. As shown in Figure 5, there are two error regions indicated by the bounded clocks $t_x^{1,2}$ and $t_y^{1,2}$. We can see that $AUC_2(C_2 \leq 1e^{-4})$ is a time continuous function, and its value increases iff the clocks are in $[t_x^1, t_y^1]$ and $[t_x^2, t_y^2]$.

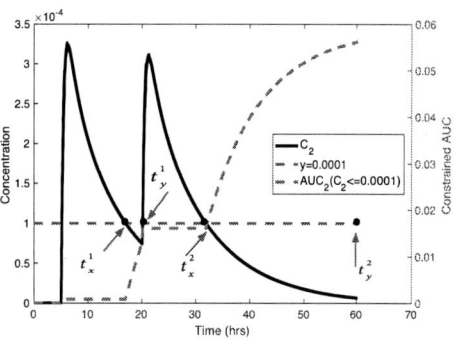

Fig. 5: Constrained AUC: AUC_{C_2} with error bound $l = 1e^{-4}$.

VI. CONCLUSION

This paper presents an efficient formal model that can solve the formal verification, simulation, and measurements of medical injection systems using Satisfiability Modulo Theories over Reals. We demonstrate that the proposed model can be used to model an injection system with actions performed by electronic injection system or human. The experimental results show the capabilities of our model in verification, simulation, and measuring the drug *Area Under the Curve*

(AUC) and constrained AUC metrics. Using the state-of-the-art SMT solver dReal, our model produces high precision results over a wide clock range with only a few seconds.

ACKNOWLEDGMENTS: This project is funded by ERC-2014-AdG 669354 grant.

REFERENCES

[1] S. A. Seshia, S. Hu, W. Li, and Q. Zhu, "Design automation of cyber-physical systems: Challenges, advances, and opportunities," *IEEE Transactions on Computer-Aided Design of Integrated Circuits and Systems*, vol. 36, no. 9, pp. 1421–1434, 2017.

[2] R. Alur and D. L. Dill, "A theory of timed automata," *Theoretical computer science*, vol. 126, no. 2, pp. 183–235, 1994.

[3] K. G. Larsen, P. Pettersson, and W. Yi, "Uppaal in a nutshell," *International journal on software tools for technology transfer*, vol. 1, no. 1-2, pp. 134–152, 1997.

[4] T. Amnell, E. Fersman, L. Mokrushin, P. Pettersson, and W. Yi, "Times ba tool for modelling and implementation of embedded systems," in *International Conference on Tools and Algorithms for the Construction and Analysis of Systems*. Springer, 2002, pp. 460–464.

[5] A. Biere, M. Heule, and H. van Maaren, *Handbook of satisfiability*. IOS press, 2009, vol. 185.

[6] S. Gao, S. Kong, and E. M. Clarke, "dreal: An smt solver for nonlinear theories over the reals," in *International Conference on Automated Deduction*. Springer, 2013, pp. 208–214.

[7] B. Liu, S. Kong, S. Gao, and E. Clarke, "Parameter identification using delta-decisions for biological hybrid systems," CMU SCS Technical Report, CMU-CS-13-136, Tech. Rep., 2014.

[8] L. Shargel, B. Andrew, and S. Wu-Pong, *Applied biopharmaceutics & pharmacokinetics*. McGraw-Hill Medical Publishing Division, 2015.

[9] S. L. Shafer and J. R. Varvel, "Pharmacokinetics, pharmacodynamics, and rational opioid selection," *Anesthesiology*, vol. 74, no. 1, pp. 53–63, 1991.

[10] M. Rybak, B. Lomaestro, J. C. Rotschafer, R. Moellering, W. Craig, M. Billeter, J. R. Dalovisio, and D. P. Levine, "Therapeutic monitoring of vancomycin in adult patients: a consensus review of the american society of health-system pharmacists, the infectious diseases society of america, and the society of infectious diseases pharmacists," *American Journal of Health-System Pharmacy*, vol. 66, no. 1, pp. 82–98, 2009.

[11] B. Donato, F. Stradolini, A. Tuoheti, F. Angiolini, D. Demarchi, G. De Micheli, and S. Carrara, "Raspberry pi driven flow-injection system for electrochemical continuous monitoring platforms," in *IEEE BioCAS*. IEEE, 2017.

[12] J. D. Scheff, R. R. Almon, D. C. DuBois, W. J. Jusko, and I. P. Androulakis, "Assessment of pharmacologic area under the curve when baselines are variable," *Pharmaceutical research*, vol. 28, no. 5, pp. 1081–1089, 2011.

[13] S. Gao, J. Avigad, and E. M. Clarke, "δ-complete decision procedures for satisfiability over the reals," in *International Joint Conference on Automated Reasoning*. Springer, 2012, pp. 286–300.

[14] M. J. Pencina, R. B. D'Agostino, and R. S. Vasan, "Evaluating the added predictive ability of a new marker: from area under the roc curve to reclassification and beyond," *Statistics in medicine*, vol. 27, no. 2, pp. 157–172, 2008.

[15] B. P. English, W. Min, A. M. Van Oijen, K. T. Lee, G. Luo, H. Sun, B. J. Cherayil, S. Kou, and X. S. Xie, "Ever-fluctuating single enzyme molecules: Michaelis-menten equation revisited," *Nature chemical biology*, vol. 2, no. 2, pp. 87–94, 2006.

[16] J. Nyberg, C. Bazzoli, K. Ogungbenro, A. Aliev, S. Leonov, S. Duffull, A. C. Hooker, and F. Mentré, "Methods and software tools for design evaluation in population pharmacokinetics–pharmacodynamics studies," *British journal of clinical pharmacology*, vol. 79, no. 1, pp. 6–17, 2015.

[17] A. Niemetz, M. Preiner, and A. Biere, "Boolector 2.0," *Journal on Satisfiability, Boolean Modeling and Computation*, vol. 9, 2015.

[18] L. De Moura and N. Bjørner, "Z3: An efficient smt solver," in *Tools and Algorithms for the Construction and Analysis of Systems*. Springer, 2008, pp. 337–340.

[19] C. Barrett, C. L. Conway, M. Deters, L. Hadarean, D. Jovanović, T. King, A. Reynolds, and C. Tinelli, "CVC4," in *Computer aided verification (CAV)*. Springer, 2011, pp. 171–177.

2018 IEEE Computer Society Annual Symposium on VLSI

A Novel Approach for Nearest Neighbor Realization of 2D Quantum Circuits

Anirban Bhattacharjee[1], Chandan Bandyopadhyay[1], Robert Wille[2], Rolf Drechsler[3], Hafizur Rahaman[1]

[1]Indian Institute of Engineering Science and Technology Shibpur, India-711103
[2]Institute for Integrated Circuits, Johannes Kepler University Linz, A-4040 Linz, Austria
[3]Institute of Computer Science, University of Bremen & Cyber-Physical Systems, DFKI GmbH, 28358 Bremen, Germany
Email: anirbanbhattacharjee330@gmail.com, chandanb@it.iiests.ac.in, robert.wille@jku.at, drechsle@uni-bremen.de,
rahaman_h@it.iiests.ac.in

Abstract— Since decades, quantum computing has received tremendous attention among the researchers due to its dominance over classical computing. But simultaneously it has faced some design challenges and implementation constraints in this long run. One such constraint to build quantum circuits is to satisfy the so-called *Nearest Neighbor* (NN) property in the implemented circuits. Using SWAP gates, this constraint can be satisfied. But this leads to another design issue, namely how to determine such NN designs with a minimum use of SWAP gates.
In way to further explore this area, in this work, we propose a heuristic approach for efficient NN complaint representation of quantum circuits in 2D space. The developed technique is segmented in three stages – qubit selection, qubit placement and SWAP gate insertion. The stated approach has been tested over a wide spectrum of benchmarks and reductions in cost parameters are observed. Improvement of more than 17%, 3% over 2D designs and 35%, 22% over 1D designs on SWAP count and quantum cost can be reported, respectively.

Keywords — Quantum Circuit, quantum gate, Nearest Neighbour(NN), SWAP gate, Quantum Cost(QC).

I. INTRODUCTION

Quantum computing has been found to be more beneficial compared to classical computing over a number of problems. As a result, quantum computing has intensely been considered among the research community. Researchers have been engaged towards the development of quantum computers [1] by using the principles of quantum mechanics. In this regard, designing of quantum algorithms to run on a quantum computer has turned out to be significant.
In quantum computing, quantum algorithms process quantum information by means of quantum circuits, which is a collection of elementary quantum gates. Quantum information is represented in the form of qubits than as bits used in classical computing. Qubits can be used to develop quantum computing machines through various technologies such as ion trap technology [2], nuclear magnetic resonance [3], quantum dots [4] and superconducting qubits [5]. It has been found that the realization of these quantum technologies is subjected to various constraints, noise and drawbacks that need to be addressed otherwise computational errors might arise. Based upon experiments, it has been observed that the nearest neighbor constraint between any two operating qubits is considered as one of the essential requirements to minimize the computational errors. In other words, it indicates that only the interactions between the adjacent qubits are considered as

the necessary criteria for the practical realization of quantum circuits. The demand for meeting the nearest neighbor criteria of the interacting qubits has been asserted by the restrictions of J-coupling force needed to enable multi-qubit operations and can only be attained by physically placing the qubits adjacent to each other.
To fix this criterion, the concept of SWAP gates has been introduced. Basically, this gate changes the position of any two qubits rather than its value to satisfy the nearest neighbor constraint. But on the contrary, this addition of SWAP gates in turn increases overhead of the corresponding circuit in terms of both cost and time. In way, incorporation of SWAP gates enhance the quantum cost of the circuit as each SWAP operation can be represented by a group of three CNOT gates (see Fig.1) which in turn increases the number of levels (depth) of the given circuit. This in turn affects the computational time and energy dissipation of the given quantum circuit. Therefore, minimum usage of SWAP gates is considered to be an important objective that needs to be addressed in order to reduce both the execution time and energy dissipation as the number of circuit levels (depth) will be reduced. Albeit exact approaches provide better solutions, but there is no specific optimal solution for very large circuits yet. In recent time several contributory works have been reported where researchers have mainly worked towards reduction of SWAP counts in circuits by adhering different approaches. Here we are stating some of those.
Optimization of NN-compliant circuit by changing the original position of the qubits arranged in linear fashion into a different arrangement has been discussed in [11]. In [12], the number of SWAP gate count has been minimized by arranging the qubits positions dynamically. Optimal solution in terms of number of SWAP gates has been examined in the work [13]. The work of [6, 14] has shown an exact search technique to determine an optimal arrangement of qubits to design a nearest neighbor architecture. Taking a step ahead, to make a trade off in between circuit execution time and SWAP cost a decision based solution for nearest neighbor circuit has been introduced in [22].
As the maximum number of adjacent neighbors of a qubit in one dimension is restricted to only two neighbors and to maximize this parameter, researchers have switched to next higher dimensions - 2D, where a qubit can have a maximum of four adjacent neighbors whereas the qubits located at the periphery can have only two neighbors. In recent time, several

978-1-5386-7100-9/18 $31.00 © 2018 IEEE   305

works on 2D circuit design are being reported and here we mentioning few of them.

In one of the work [15], an effective NN-compliant circuit has been designed to reduce overhead in 1D architecture by mapping the problem in 2D architecture through mixed integer programming approach. In [16], the authors used an exact approach to obtain an optimal solution in terms of SWAP cost but it was not suitable for large benchmarks due to exorbitant computational cost. In [17], SWAP cost has been reduced considerable by placing the qubits in appropriate positions. In [18], a heuristic approach has been used to minimize the SWAP cost through proper grid selection followed by an intelligent qubit placement policy where the qubits have been placed upon their number of interactions before adding SWAP gates. To get higher level of optimization, another heuristic approach based on look-ahead strategy has been presented in [19] where substantial improvement in SWAP cost is reported. Aiming to get more improved designs, researchers have tried 3D configurations [20] but it is noticed that though, the number of interacting neighbors increases in 3D but at the same time it becomes very difficult to control the qubits in 3D. In our work we consider the representation in 2D organization only and propose a heuristic qubit placement strategy to convert a quantum circuit to NN-complaint designs with lesser usage of SWAP gates. Our approach provides substantial improvements compared to earlier existing works.

The rest of the paper is organized as follows. Section II describes the elementary quantum gates and nearest neighbor constraints. Detailed descriptions on our proposed approach are stated in Section III. Experimental results and comparative analysis with related works are given in Section IV. Finally, we conclude the paper in section V.

II. BACKGROUND

In quantum computing, qubits represent the basic unit of information on which elementary quantum gates operates. Like bits, a qubit can exist in one of the basis states represented as $|0\rangle$ and $|1\rangle$, similar to 1 and 0 in classical computing. Apart from the basis states, a qubit can also exist in other states expressed as the linear combination of the basis states $|0\rangle$ and $|1\rangle$ represented by the state vector $\left|\Psi\right\rangle$ as:

$$\left|\Psi\right\rangle = \alpha|0\rangle + \beta|1\rangle \qquad (1)$$

where α and β are complex numbers that represents the probability amplitudes of the basis states and subjected to the constraint $\alpha^2 + \beta^2 = 1$.

Definition1. *Quantum gates perform operations on qubits and when a cluster of such gates are projected over a set of circuit lines then a quantum circuit is formed.*

NOT, CNOT, V/V^+, W/W^+ are well known quantum gates that form the quantum gate libraries (e.g. NCV, NCVW and NCV$|v_1\rangle$ [7-10]) and these gate libraries helps to implement specific functions into gate level descriptions. Some well-known quantum gates and their schematic representations are summarized in Table 1.

Table 1: Some well known elementary quantum gates and their schematic representations

Gates	Notation	Gates	Notation
NOT		Control led -V	
CNOT			V
V	V	Control led -V†	
V†	V†		V†

Although quantum gates can express any function into respective circuital form, there are certain limitations for practical realization of such type of circuits. One such major criterion is to restrict the quantum gates to operate only on qubits located adjacent to each other. This adjacency feature is required to mitigate the errors induced in the circuit from various sources of noise. For this purpose, SWAP gates are being introduced before a quantum gate that fails to meet this adjacency constraint by exchanging the positions of the respective qubits till the condition has been met. This phenomenon is known as Nearest Neighbor property, which can be defined as follows:

Definition2: *Nearest Neighbor Cost (NNC) is the interaction distance between the qubits of any two-qubits quantum gate (G) and can be expressed as $NNC_G = |c - t| - 1$, where c is the control line and t is target line of the gate.*

The NNC pertaining to a circuit (NNC_{CKT}) is the cumulative value of NNCs contributed by the individual gates i.e. $NNC_{CKT} = \sum_G NNC_G$, NNC_G is NNC of each two-qubit gate in the circuit.

For example, the standard decomposition of a Toffoli gate has NNC = 1 due to the presence of a single non-adjacent elementary gate in its gate level description.

Though, several researches are going on efficient transformation of quantum circuit to NN compliant designs, but the most common way of converting a quantum circuit to a NN compliant one is by inserting SWAP gates (the pictorial representation of SWAP gate is shown in Fig. 1). Basically, such SWAP gates are used to bring the interacting qubits of any gate adjacent.

Example1: *For example, the design of Fig. 2(a) does not satisfy the nearest neighbor constraint as the qubits in the first, second, fourth and fifth gates are non-adjacent to each other. So, SWAP gates have to be inserted to meet this constraint. The transformation of the Fig. 2(a) to a NN complaint design is achieved by inserting 14 SWAP gates in the structure and the final 1D NN-complaint organization is shown in Fig. 2(b).*

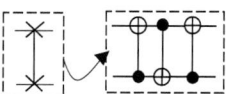

Fig. 1: Design of SWAP gate

Fig. 2(a): Input quantum circuit having NNC=7

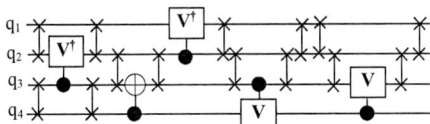

Fig. 2(b): Equivalent NN-compliant design of Fig. 2(a)

Though we have converted the circuit to a 1-D based NN compliant form, but it is also possible to represent the same circuit in 2-D format as well to get improved representation. Now, here we are stating how this 2D design can be made.

2D Quantum Circuit: To map a quantum circuit in 2D configuration, primarily a grid structured planar graph (V, E) has been used where a node v represents a qubit and an undirected edge $(u,v) \in E$ represents the interaction between the qubits u and v, where $\{u,v\} \in V$. Here, each node can have a degree of up to four. Basically, a quantum circuit may be represented in several 2D configurations based upon the chosen grid structures. For instance, we consider a 1D circuit with 5 qubits. In this case, these 5 qubits are always arranged next to each other while these same qubits when arranged in a 2D quantum circuit, may have several configurations such as a 3x2 or 2x3 or 3x3 grid as shown in Fig. 3. Such different arrangements affect the interaction distance between the corresponding qubits.

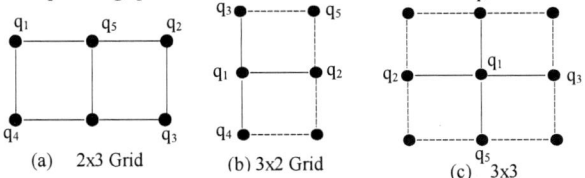

Fig. 3: Possible grid configuration for qubit positions in 2D

As the 1D representation of Fig. 2(b), for the 2D organization a 2x2 grid is now selected and gates are placed over the circuit lines as depicted in Fig. 4(a). The NN-complaint design in 2D form is obtained and the final circuit after SWAP insertion is shown in Fig. 4(b).

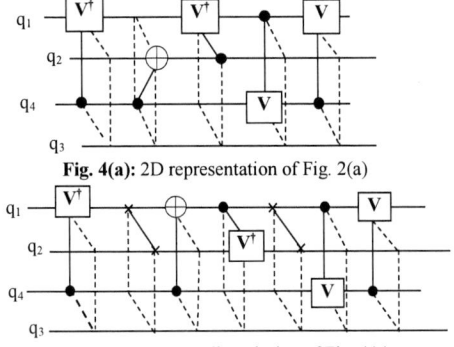

Fig. 4(a): 2D representation of Fig. 2(a)

Fig. 4(b): 2D NN-compliant design of Fig. 4(a)

III. PROPOSED APPROACH

Here, we are proposing an improved synthesis scheme for converting quantum circuits to its equivalent 2D NN design. This technique not only transforms the input quantum circuits into nearest neighbor based design but also limits the SWAP utilization to contain the circuit overhead. In the whole

transformation process, we have taken help of some heuristics in decision making points and have derived an intelligent qubit placement policy for better representation.

The entire scheme is performed in three phases – (i) qubit selection (ii) qubit placement and (iii) SWAP gate insertion. For an ease of understanding, all these phases are explained here with an example and the circuit of Fig. 5(a) has been considered as standard input circuit over which all the operations are to be performed.

Phase1: Qubit Selection Policy

This stage is tasked with proper arrangement of qubits in a grid structure. But to make this preference enable it needs some preference metrics. For this purpose, we build two information tables (*time interaction* and *time costing*) from the gate details of the input circuit.

In *time interaction* table, we compute total interaction time of each qubit. The entries of this table comprise the time instants in which the corresponding qubits interacts and the total interaction time which is computed by summing all the time instants of interactions for every qubit in the circuit. The *time interaction* metric for Fig. 5(a) is shown in Table 2.

Next we compute, the qubit *time costing* table (shown in Table 3) which contains the total costing time for each qubit and this metric is derived from our earlier computed table, qubit *time interaction*.

The total costing time of qubit *time costing* table is estimated for each qubit after extracting the time instants from the timestamps of the corresponding qubit interaction in which it is not adjacent with its other interacting qubit of any 2-qubit gate in the given circuit.

Combining the information provided in this two tables, we derive one more table, *qubit preference* table (given in Table 4) in which we compute the preference index by calculating the ratio between total costing time and total interaction time in a separate column for all the qubits of a given circuit.

After obtaining the preference index for all the qubits in the given circuit, we sort all the qubits based upon this index values in a descending order (see Table 5) with the qubit having the highest index value is placed at the top of the table. The preference index values obtained for each qubit in *qubit preference table* designates the priority values of the corresponding qubits for placement selection in the grid structure.

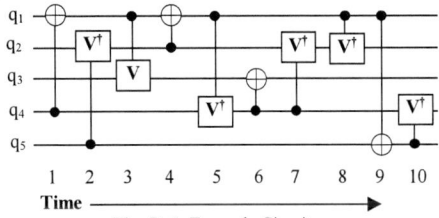

Fig. 5(a): Example Circuit

Table 2: Qubit time interaction table   **Table 3:** Qubit time costing table

Qubits	Time instants	Total interaction	Qubits	Time instants	Total interaction	Time instants	Total costing
q_1	1,3,4,5,8,9	30	q_1	1,3,4,5,8,9	30	1,3,5,9	18
q_2	2,4,7,8	21	q_2	2,4,7,8	21	2,7	9
q_3	3,6	9	q_3	3,6	9	3	3
q_4	1,5,6,7,10	29	q_4	1,5,6,7,10	29	1,5,7	13
q_5	2,9,10	21	q_5	2,9,10	21	2,9	11

Table 4: Qubit preference table

Qubits	Total interaction	Total costing	Preference index
q_1	30	18	18/30=0.6
q_2	21	9	9/21=0.42
q_3	9	3	3/9=0.33
q_4	29	13	13/29=0.44
q_5	21	11	11/21=0.52

Table 5: Sorted qubit preference table

Qubits	Total interaction	Total costing	Preference index
q_1	30	18	0.6
q_5	21	11	0.523
q_4	29	13	0.448
q_2	21	9	0.428
q_3	9	3	3/9=0.33

Phase2: Qubit Placement Policy

After computing the qubit preference table in the previous execution process (refer to Table 5), we now hereby run the qubit placement algorithm by arranging the qubits chosen in the order from Table 5 in a grid structure (given in algorithm 1).

Algorithm 1: Qubit placement strategy

Input: Qubit preference table (PT)
Output: Qubit placement in 2D grid structure (GS)

```
begin
  for ((qk ∈ PT) and (PT≠null)) do
    if k = 1 then
      Place qk in the centre of 2D grid GS (mid_row,mid_column);
    else
      for k =2 to N do              //N is the number of qubits
        Find a vacant location GS (row, column) adjacent to maximum
        number of empty cells;
        if ((Num_of_(GS (row, column)) = = 1) then
          Place qk in location GS (row, column);
        else
          Retrieve location GS (row_{k-1}, column_{k-1}) of last qubit q_{k-1};
          Place qk in one of the vacant locations from GS (row_{k-1} ± a,
                    column_{k-1} ± b); where a, b = 1, 2
        end if
      end for
    end if
  end for
  return GS;
end
```

According to the steps mentioned in that algorithm, initially, we select the qubit having the highest preference index from the qubit preference table (PT) and place it in the centre of the 2D grid. Then we start finding an empty location in that grid which is adjacent to maximum number of empty cells and place the next qubit from PT table.

If there occur multiple occurrences of such a type of grid location then we resolve the conflict by selecting the qubits from PT and then placing it in these locations in the order left, up, right, down of the last placed qubit as per the availability of space until such a distinct location is being sought. In this way, we populate the 2D grid by placing the remaining qubits from the preference table (PT).

We employ the same process over our previously computed preference table to find a suitable grid configuration. As the qubit q_1 is having the highest preference in PT (Table 5), so we place the selected qubit in the centre of the grid (3×3) (see Fig. 5(b)). Now, we search for a location with maximum number of adjacent empty cells in the grid. Since, there are eight such possible vacant locations exist with two empty cells

around, so we place the qubits from PT around the left, up, right, down of the last placed qubit q_1 till a distinct location with empty cells around it is found. Thus, we place the next qubit q_5 to the left of the qubit q_1 as stated in our algorithm 1(see Fig. 5(c)). Now, the rest of the qubits q_4, q_2, q_3 are selected from PT and positioned them at the up, right and down of q_1 respectively. The final grid configuration is given in Fig. 5(d).

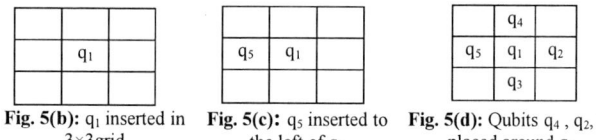

Fig. 5(b): q_1 inserted in 3×3grid   **Fig. 5(c):** q_5 inserted to the left of q_1 **Fig. 5(d):** Qubits q_4, q_2, q_3 placed around q_1

Phase3: SWAP gate insertion

Till now, all the qubits have been positioned in the grid. In this phase, we place all the gates over this grid in such a way that we insert a SWAP gate if we find any gate whose interacting qubits are non-adjacent. In this process, we simply exchange the position of qubits and make them adjacent.

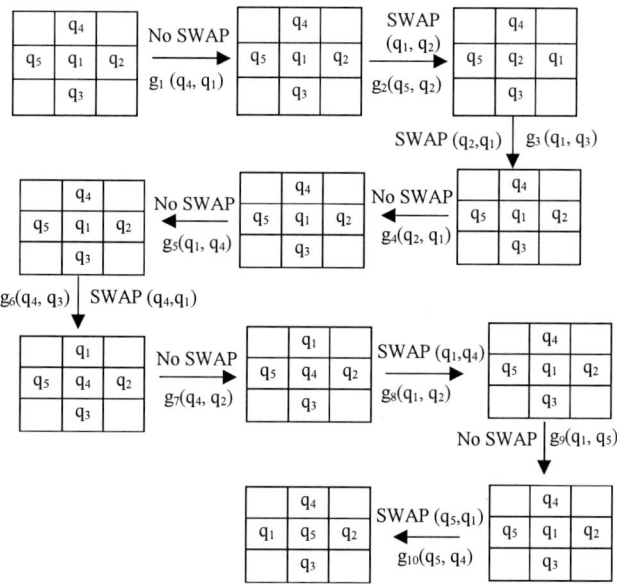

Fig. 6: Steps of SWAP gate insertion

For our input circuit, we already have obtained the desired grid (in Fig. 5(d)) and now we have placed the qubits accordingly as the mentioned strategy. The placement of qubits and insertion of SWAP gates in appropriate places are given in Fig. 6, where it can be found that a total of five SWAP gates have been utilized in the entire transformation process. For better understanding, here we are producing one more example.

Example2: The transformation of benchmark circuit (4gt11_84) to its NN equivalent form is shown in this example. The intermediate steps and computed metrics are also given in Table 6-10.

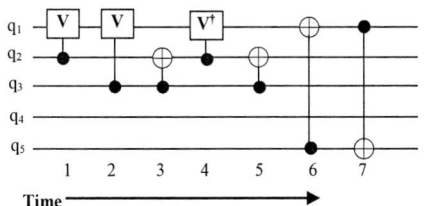

Fig. 7(a): Example benchmark (4gt11_84)

Table 6: Qubit time interaction table

Qubits	Time instants	Total interaction
q_1	1,2,4,6,7	20
q_2	1,3,4,5	13
q_3	2,3,5	10
q_4	-	-
q_5	6,7	13

Table 7: Qubit time costing table

Qubits	Time instants	Total interaction	Time instants	Total costing
q_1	1,2,4,6,7	20	2,6,7	15
q_2	1,3,4,5	13	0	0
q_3	2,3,5	10	0	0
q_4	-	-	-	-
q_5	6,7	13	6,7	13

Table 8: Qubit preference table

Qubits	Total interaction	Total costing	Preference index
q_1	20	15	15/20=0.75
q_2	13	0	0/13=0
q_3	10	0	0/10=0
q_4	-	-	-
q_5	13	13	13/13=1

Table 9: Sorted qubit preference table

Qubits	Total interaction	Total costing	Preference index
q_5	13	13	13/13=1
q_1	20	15	15/20=0.75
q_2	13	0	0/13=0
q_3	10	0	0/10=0
q_4	-	-	-

Table 10: Qubits placed in grid (2×3)

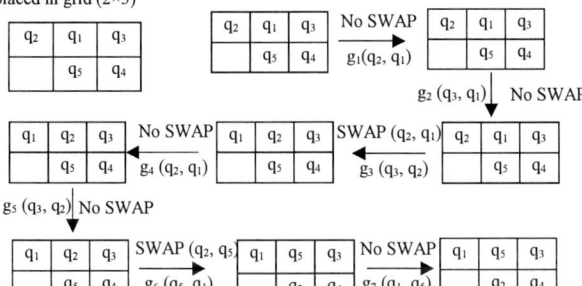

Fig. 7(b): Steps of SWAP gate insertion

IV. EXPERIMENTAL RESULTS

The proposed approach has been implemented in C and executed on Intel i5 machine with 3.30 GHz clock and 4GB RAM. The experimental evaluations have been carried out over different benchmark suites from [21]. Two result tables have been produced, where the first table (Table-11) contains results from small and medium size benchmarks and the second table (Table-12) summarizes the experimental data for large size benchmarks.

For each of the benchmarks two cost parameters – (quantum cost and SWAP count) have been evaluated and the obtained metrics have been compared with some well-known 1D and 2D works. The first result table contains the comparison with [15], [18], and the best results from the work of [15] have been considered. A steady improvement in cost parameters can be seen in the obtained results. 28.71% average improvement over 1D works and 10.18% average

improvement over 2D works have been attained. Our qubit mapping policy results in an improvement of about 37.73% and 35.71% respectively in the best case over the 2D works [15] and [18], while an improvement of 67.56% has been obtained in the best case over the 1D work [15]. Experimental analysis suggests that our approach provides promising results in terms of SWAP gate reduction for all but few benchmarks for which our mapping strategy has not provided better results relative to the previous works.

V. CONCLUSION

This work presents a heuristic based qubit placement strategy for transformation of quantum circuits to NN-complaint design. The entire design processes has involved three stages - qubit selection then placement and finally SWAP gate insertion. We have tested the developed approach over a large spectrum of benchmarks and improvements have been registered. A substantial reduction in SWAP gate usage has been achieved. The obtained results have been compared with some of the best results available in 1D and 2D representations. Depending on our experimental evaluations, we examined that our approach has attained an average reductions of about 35.32%, 22.10% in terms of SWAP cost and quantum cost against 1D configuration while 17.09%, 3.28% reductions against 2D configuration over the same cost parameters.

As we have highly relied on the applied heuristic for initial qubit mapping process, so it does not provide optimal solution. Further enhancement over this work will be investigated in future by finding more appropriate heuristics.

VI. REFERENCES

[1] T. Simonite. IBM shows off a quantum computing chip. http://www.technologyreview.com/news/537041/ibm-shows-off-a-quantum-computing-chip/, 2015.

[2] D. Kielpinski, C. Monroe, and D. J. Wineland. "Architecture for a largescale ion-trap quantum computer", *Nature*, 417(6890):709–711, 2002.

[3] B. Criger, G. Passante, D. Park, and R. Laflamme. "Recent advances in nuclear magnetic resonance quantum information processing". *Philosophical Transactions of the Royal Society of London A: Mathematical,Physical and Engineering Sciences*, 370(1976):4620–4635, 2012.

[4] J. Taylor, J. Petta, A. Johnson, A. Yacoby, C. Marcus, and M. Lukin, "Relaxation, dephasing, and quantum control of electron spins in double quantum dots". *Physical Review B*, 76(3):035315, 2007.

[5] A. Blais, J. Gambetta, A. Wallraff, D. Schuster, S. Girvin, M. Devoret, and R. Schoelkopf. "Quantum information processing with circuit quantum electrodynamics". *Physical Review A*, 75(3):032329, 2007.

[6] A. Zulehner, S. Gasser, R. Wille. "Exact Global Reordering for Nearest Neighbor Quantum Circuits Using A* . In *International Conference on Reversible Computation* (pp. 185-201). Springer, Cham. 2017, July

[7] A. Barenco, C. H. Bennet, R. Cleve, D. DiVincenzo, N. Margolus, P. Shor, T. Sleator, J. Smolin, and H Weinfurter, "Elementary gates for quantum computation," *Phys. Rev. A*, vol. 52, no. 5, pp. 3457–3467, Nov 1995.

[8] Z. Sasnian and D. Miller, "Transforming MCT Circuits to NCVW Circuits," *Springer's Lecture Notes in Computer Science.*, pages 77-88, (2011).

[9] D. Miller, R. Wille, and Z. Sasanian, "Elementary quantum gate realizations for multiple-control Toffolli gates," *In Proc. Int'l Symp. on Multiple-valued Logic*,pages pp. 217-222, (2011).

[10] Z. Sasanian, R. Wille, and D. M. Miller, "Realizing reversible circuits using a new class of quantum gates," *In Proc. Design Automation Conf.,* pp. 36-41, (2012).

[11] Y. Hirata, M. Nakanishi, S. Yamashita, Y. Nakashima. "An efficient conversion of quantum circuits to a linear nearest neighbor architecture." *Quant. Inf. Comput.,* 11(1–2):0142–0166, 2011.

[12] M. Saeedi, R. Wille, R. Drechsler. "Synthesis of quantum circuits for linear nearest neighbor architectures." *Quant. Inf. Proc.,* 10(3):355–377, 2011.

[13] A. Shafaei, M. Saeedi, and M. Pedram. Optimization of quantum circuits for interaction distance in linear nearest neighbor architectures. *Design Autom. Conf.,* 2013.

[14] R. Wille, A. Lye, and R. Drechsler, "Exact reordering of circuit lines for nearest neighbor quantum architectures," *IEEE Trans. on CAD,* vol. 33, no. 12, pp. 1818–1831, Dec 2014.

[15] A. Shafaei, M. Saeedi, and M. Pedram, "Qubit placement to minimize communication overhead in 2d quantum architectures," in *Proc. ASP Design Autom. Conf.,* Jan 2014, pp. 495–500.

[16] A. Lye, R. Wille, and R. Drechsler, "Determining the minimal number of swap gates for multi-dimensional nearest neighbor quantum circuits," in *Proc. ASP Design Autom. Conf.,* Jan 2015, pp. 178–183.

[17] M. G. Alfailakawi, I. Ahmad, and S. Hamdan, "Harmony-search algorithm for 2D nearest neighbor quantum circuits realization," *Expert Systems with Applications,* vol. 61, pp. 16–27, May 2016.

[18] R. R. Shrivastwa, K. Datta, and I. Sengupta, "Fast qubit placement in 2D architecture using nearest neighbor realization," in *Proc. Int'l Symp.on Nanoelectronic and Information Systems,* Dec 2015, pp. 95–100.

[19] R. Wille, O. Keszocze, M. Walter, P. Rohrs, A. Chattopadhyay, and R. Drechsler, "Look-ahead schemes for nearest neighbor optimization of 1D and 2D quantum circuits," in *Proc. ASP Design Autom. Conf.,* Jan 2016, pp. 292–297.

[20] A. Kole, K. Datta, I. Sengupta, "A New Heuristic for N-Dimensional Nearest Neighbour Realization of a Quantum Circuit". *In: IEEE Transactions on Computer-Aided Design of Integrated Circuits and Systems* 12 (2017), doi 10.1109/TCAD.2017.2693284.

[21] Revlib: An online resource for reversible functions and reversible circuits. URL: http://www.revlib.org/.

[22] R. Wille, N. Quetschlich, Y. Inoue, N. Yasuda, and S. Minato. "Using π-DDs for Nearest Neighbor Optimization of Quantum Circuits" *In Conference on Reversible Computation,* pages 181-196, 2016.

Table12: Results of large size benchmarks

Benchmarks	Total lines	Gate count	Initial QC	Grid size	No. Of SWAPs	QC in NN design
rev_17	17	136	136	6×3	214	443
rev_18	18	153	153	5×4	221	374
rev_19	19	171	171	4×5	256	427
ac_21_1	21	130	130	6×4	116	246
ac_21_2	21	67	67	6×4	53	120
ac_21_3	22	42	42	6×4	51	93
hm_20	20	73	73	5×4	69	142
hm_21	21	79	79	6×4	102	181
hm_22	22	85	85	6×4	94	179

Table11: Comparison over 2D results with small and medium size benchmark circuits

Benchmarks	No. of qubits	Gate count	Results in 1D from [15]		Results in 2D from [15]			Results from [18]			Proposed work results		
			SWAP Count	Quantum Cost	Grid Size	SWAP Count	Quantum Cost	Grid Size	SWAP Count	Quantum Cost	Grid Size	SWAP Count	Quantum Cost
3_17_13	3	14	4	17	2×2	6	20	2×2	3	17	2×2	5	19
4_49_17	4	32	12	45	2×2	13	45	-	-	-	2×2	10	42
aj-e11_165	4	60	36	96	2×3	24	84	3×2	22	82	2×3	18	78
decod24-v3_46	4	9	3	12	3×2	3	12	-	-	-	2×2	2	11
hwb4_52	4	23	10	33	2×2	9	32	2×2	9	32	2×2	7	30
rd32-v0_67	4	8	2	10	2×3	2	10	-	-	-	2×3	2	10
4gt11_84	5	7	1	8	2×3	2	9	2×3	2	9	2×3	2	9
4gt10-v1_81	5	36	20	56	3×2	16	52	3×2	15	51	2×3	14	50
4gt12-v1_89	5	53	35	88	3×2	19	72	2×4	18	71	3×2	20	73
4mod5-v1_23	5	24	9	33	2×3	11	35	3×2	7	31	2×3	8	32
4gt5_75	5	22	12	34	3×3	8	30	2×5	10	32	3×3	9	31
4gt4-v0_80	5	44	34	78	2×3	17	61	4×4	15	59	2×3	16	60
4mod7-v0_95	5	40	21	61	3×3	13	53	2×5	14	54	3×3	10	50
alu-v4_36	5	32	18	50	2×3	10	42	2×5	11	43	2×3	11	43
hwb5_55	5	109	63	172	3×2	45	154	2×7	49	158	3×2	38	147
QFT5	5	10	6	16	3×2	5	15	4×2	5	15	3×2	5	15
hwb6_58	6	146	118	264	2×3	79	225	2×3	76	222	2×3	63	209
mod5adder_128	6	87	51	138	2×3	41	128	2×3	36	123	3×2	28	115
mod8-10_177	6	109	72	181	3×3	45	154	4×3	43	152	3×3	39	148
QFT6	6	15	12	27	2×3	6	21	3×2	7	22	2×3	5	20
rd53_135	7	78	66	144	5×2	39	117	2×7	40	118	3×3	29	107
ham7_104	7	87	68	155	3×3	48	135	2×7	45	132	3×3	38	125
QFT7	7	21	26	47	2×4	18	39	6×2	14	35	2×4	14	35
QFT8	8	28	33	61	4×2	18	46	4×2	23	51	4×2	18	46
QFT9	9	36	54	90	3×3	34	70	5×2	36	72	3×3	24	60
rd73_140	10	76	56	132	4×3	37	113	3×6	43	119	4×3	28	104
sys6-v0_144	10	62	59	121	4×4	31	93	-	-	-	4×3	30	92
QFT10	10	45	70	115	5×3	53	98	4×3	51	96	4×3	33	78
rd84_142	15	112	148	260	5×3	54	166	4×5	62	174	5×3	48	160
cnt3-5_180	16	125	127	252	3×6	69	194	4×4	84	209	3×6	54	179
Total			**971**	**2796**		**775**	**2325**		**740**	**2179**		**628**	**2178**

978-1-5386-7100-9/18 $31.00 © 2018 IEEE

2018 IEEE Computer Society Annual Symposium on VLSI

A Hardware-efficient Implementation of CLOC for On-Chip Authenticated Encryption

Mahmoud A. Elmohr
Department of Electrical and Computer Engineering,
University of Waterloo, Ontario, Canada
Email: mahmoud.a.elmohr@ieee.org

Sachin Kumar[1], Mustafa Khairallah[2],
and Anupam Chattopadhyay[1]
School of Computer Science and Engineering[1],
School of Physical and Mathematical Sciences[2]
Nanyang Technology University, Singapore[1,2]
Email: anupam@ntu.edu.sg

Abstract—With the huge number of candidates for the CAE-SAR competition for authenticated encryption, the task of designing efficient implementations for these candidates becomes a big challenge. The main goal of this competition is to find smaller, faster, energy-efficient and secure authenticated encryption schemes. In this paper, an area efficient hardware implementation of CLOC, one of the 15 candidates for the third round of CAESAR competition is presented. CLOC represents a new mode of using AES, in order to provide both encryption/decryption and MAC functionalities. Since the hardware design of AES is a well studied problem, the challenge is to accommodate the mode functionality with small area, high performance and low power overhead. The proposed hardware implementation for the CLOC is developed by sharing the AES core by applying a pipeline technique. By using commercial synthesis flows and 65 nm ASIC technology, it shows, for low power applications, that proposed hardware architecture of CLOC reduces the area by 42.85% and consumes 37.8% less power when compared with the existing high throughput implementation of CLOC. In addition, area-efficiency of the proposed design is also improved by 17.6% and the proposed design consumes only 2.6 μW.

Keywords—AES, Authenticated Encryption, ASIC implementation. Technology mapping, logic optimization.

I. INTRODUCTION

The security goals of any communication system can be defined with mainly two goals: confidentiality and integrity. While encryption provides confidentiality, this may not be sufficient to the systems in which data is streamed in insecure channels. Such systems also need data integrity and authentication in order to ensure that original information has not been modified by unauthorized or unknown parties. Thus, Authenticated Encryption (AE) has emerged as a scheme that provides both confidentiality and integrity of the data simultaneously [1].

The straightforward way to design an AE scheme is by using a secure encryption scheme and a secure Message Authentication Code (MAC) with two separate keys and use both of them as a generic composition. This composition could have three different approaches: MAC-then-encrypt, Encrypt-and-MAC, and Encrypt-then-MAC, and only the former one is guaranteed to be secure under condition that both the encryption scheme and the MAC are secure [2]. The main disadvantages of such approacesh are: being slow, requiring

two different keys and even may have error implementations if not handled carefully.

Thus, dedicated AE schemes have been introduced to overcome the generic composition method. Most of these schemes were developed as modes for block ciphers such as the Counter with CBC-MAC (CCM) mode [3], the Galois Counter Mode (GCM) [4] and the Offset Codebook Mode (OCB) [5].

In order to fill the gap and standardize the authenticated encryption schemes; NIST and Daniel Bernstein co-founded the Competition for Authenticated Encryption: Security, Applicability, and Robustness (CAESAR) in 2013 [6], opening the call for submissions that should provide advantages over AES-GCM and suitable for widespread adoption. The call was responded by 57 submissions in total passing by rounds of evaluation, resulting in only 15 candidates so far in the third round. In order to facilitate hardware implementations for the candidates; a generic hardware Application Programming Interface (API) was introduced in [7] to meet the requirements of all algorithms submitted to the CAESAR competition, as well as many earlier developed authenticated ciphers, such as AES-GCM and AES-CCM. The API consists of: a preprocessor that handles the input in a standard form and passes it to the cipher core, a post-processor that handles the output to be in the standard form, and a FIFO. Apparently, a generic API would have more area overhead than a specific one, moreover it could limit the performance of some candidate algorithms, and may favor one algorithm over another, thus should not be used solely to decide the winning candidate. However, it is essential in such competitions to provide a common base for all candidates and a common testing environment.

In this paper, a new hardware architecture for CLOC [8], one of the CAESAR third round candidates, is presented. The new architecture uses a round-based AES implementation as a primitive. However, only one encryption core is used, as opposed to the two-core architecture based CLOC in [9]. With many applications that do not require the high speed provided by the current ASIC fabrication technologies and AES implementations, such as RFID tags, smart cards and WSNs, the cost of area and power becomes more vital. Consequently, the proposed design has been implemented and compared to the existing CLOC architecture in [9]. The design has been synthesized for both high speed and low power

978-1-5386-7100-9/18 $31.00 © 2018 IEEE   311

targets. Table III and Table IV show the summary of results for both applications with the high speed requirements and those with low power requirements respectively.

The remaining of the paper is organized as follows: In Section II a description of the CLOC scheme and its functions is presented as well as an illustration of the existing hardware implementation of CLOC in [9]. In Section III we present the proposed implementation, illustrating its architecture and differentiating between it and the existing design. Section IV represents the results of both designs using Synopsys Design Compiler tool with TSMC 65nm technology. And finally we conclude our work in Section V.

II. CLOC AUTHENTICATED ENCRYPTION SCHEME

Notation: M: Unencrypted Message, C: Encrypted Message, T, T*: Tag, N: Nonce, V: Temporary Tag, A: Associated Data, $\perp$: Error.

CLOC [8] is a block cipher mode of operation for authenticated encryption with associated data (AEAD). The design of CLOC aims at being provably secure and optimizing the implementation overhead beyond the block cipher, the precomputation complexity, and the memory requirement. The main advantage for CLOC is how it handles short input data efficiently, and is suitable for use with embedded processors. CLOC has two variations: one based on AES and the other is based on TWINE [10] ciphers with the same components remaining. In this context, we adopt the AES version of CLOC.

The two main algorithms CLOC-E for encryption and CLOC-D for decryption use four subroutines, HASH, PRF, ENC, and DEC, and could be described as follows:

- CLOC-E(N, A, M):
    1) V ← HASH(N,A)
    2) C ← ENC(V,M)
    3) T ← PRF(V,C)
    4) return (C, T)
- CLOC-D(N, A, C, T)
    1) V ← HASH(N,A)
    2) T* ← PRF(V,C)
    3) If T != T* then return $\perp$
    4) M ← DEC(V,C)
    5) return M

Beside AES as the main function in CLOC, there are other functions to be defined as follows:

- Fix0 / Fix1: assigns the first bit of the data block to be 0 or 1 respectively.
- H / F1 / F2 / G1 / G2 : are simple linear functions of XORs and permutations.
- OZP: pads blocks less than 128 bit with 1 and a sequence of 0s. However this function is handled by the CAESAR API, so it is not implemented in the proposed design.

There are two existing hardware implementations of CLOC, the design in [11] is a low area implementation with an 8-bits datapath using an 8-bits serialized AES core, the main target of that design was achieving the lowest possible area,

neglecting throughput. The other implementation in [9] is a high-speed implementation using two regular 128-bit AES cores to achieve the highest possible throughput, however, at the expense of large area.

Since the implementation in [9] was intended for maximum possible throughput, it used two AES cores exploiting the parallelism of ENC and PRF subroutines, which in turn increased the area massively, and reduced the area efficiency due to the fact that both functions are sequential to the HASH function leaving one AES core unused for approximately 50% of the running time assuming long plaintext and long associative data as well.

III. PROPOSED ARCHITECTURE OF CLOC

The idea of making the proposed design efficient is to use only one instance of each functional block and reuse them in a multi-cycle approach. In order to achieve that, and to have a single configurable datapath accomodating the four subroutines' datapaths; a control unit based on a Finite State Machine was developed to alter the datapath using eight multiplexers and two registers as shown in Figure 1.

The internal functional blocks shown in Figure 1 could be detailed as follows:

- The blocks Fix0, Fix1, F1, F2, G1, G2 and H are direct implementations for the corresponding CLOC primitive functions described in Section II and detailed in [8].
- The AES core used is an encryption only round-based AES core described in [9].
- The block "Conc" is used to concatenate the nonce with a constant that depends on the parameters of CLOC as described in [8]. However only the two recommended sets of parameters are implemented in both designs.
- "Expand" block operates on bdi_valid_bytes signal provided by the API and expands it to its bits equivalent. The API delivers the bdi_valid_bytes signal representing the valid bytes of the input to indicate if the current block of input is not occupying the whole block.
- AND gates are used to mask signals with 0s reducing their size to the same size of the input in case of being less than 128 bits. That is mainly dependent on the expanded bdi_valid_bytes signal.
- XOR gate is used by the four subroutines.
- The block "MSB" is used to extract only the 96 most significant bits of the output as the size of the tag T.

A. HASH Datapath

The HASH subroutine's datapath starts with the associated data A within the API's bdi signal passing through Mux-A, applying Fix0 function to it, and encrypts the result with the help of an AES core after passing through Mux-B. The most significant bit of the associated data decides whether the result could be passed to H function or passed as it is through Mux-C. The result is then XORed with the next block of the associated data which passes this time through Mux-D, and the encryption process of the result is repeated again after passing Mux-E, Mux-F and Mux-B. This process keeps going

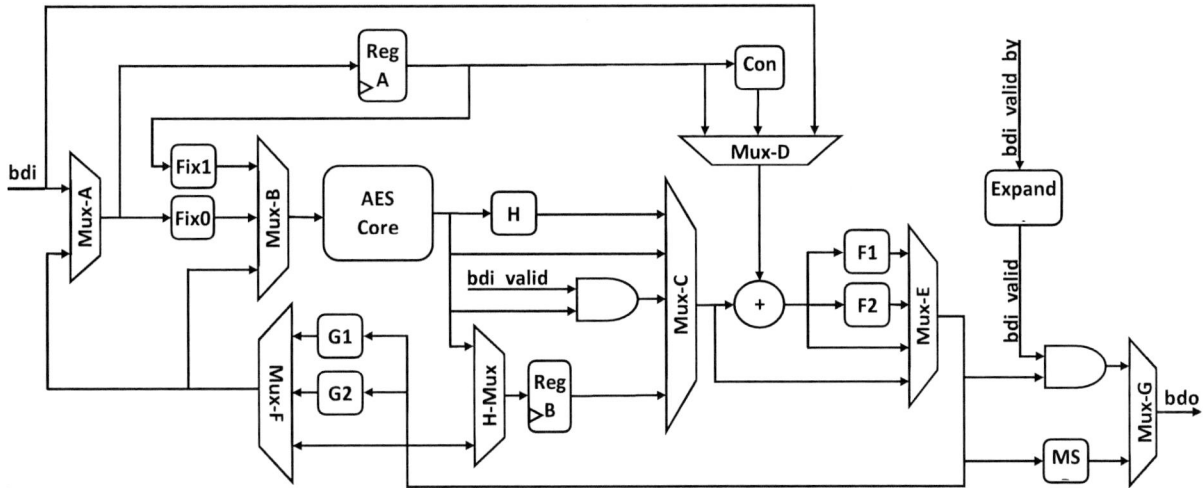

Figure 1: Optimized CLOC implementation architecture.

on until the last block of the associated data, at which the encrypted data is XORed with the Nonce N stored in Reg-A concatenated with a constant depending on the parameters of CLOC. The Nonce is stored in Reg-A due to the fact that the API delivers the Nonce within the bdi signal before the associated data, thus needs to be stored. At the end, either F1 or F2 functions applies on the result and passes through Mux-E depending on the size of the last block of the associated data resulting in the intermediate tag V which passes through Mux-H to be stored in Reg-B for further usage by the remaining subroutines.

B. PRF Datapath

If there is no plaintext M to encrypt nor a ciphertext C to decrypt, the intermediate tag V is extracted from Reg-B, passed through Mux-C and Mux-E as it is and G1 function is applied to it, and passed by Mux-F and Mux-B before being encrypted by AES producing the tag T. Otherwise the intermediate tag V passes with the same datapath except for applying G2 function instead of G1, and the result of the encryption is passed through Mux-C to be XORed with one block of the produced ciphertext C stored in Reg-A by ENC subroutine in case of encryption or the received ciphertext C passed within the API's bdi signal in case of decryption, both passing by Mux-D. The result of the XOR passes through Mux-E, Mux-F and Mux-B to be encrypted again, then stored in Reg-B after passing through Mux-H. The result is stored in Reg-B due to the fact that PRF subroutine uses the ciphertext blocks produced by the ENC subroutine or the ciphertext blocks used also by the DEC subroutine, thus stores the current state in Reg-B, hands over the datapath to either ENC or DEC subroutines waiting the next ciphertext block and the same process goes until the last block of the ciphertext, at which the XORing result is fed to either F1 or F2 functions depending on the size of the last block of the ciphertext. Then finally the result passes through Mux-E, Mux-F and Mux-B to be

encrypted for the last time to produce the tag T. To output the tag within the API's bdo signal, the result passes through Mux-E and its most 96 significant bits are extracted through MSB block and finally output through Mux-G.

C. ENC Datapath

The ENC datapath encrypts the intermediate tag V produced by the HASH subroutine previously and already stored in Reg-B after passing through Mux-C, Mux-E, Mux-F and Mux-B. The encryption result is passed through Mux-C and XORed with the first block of the plaintext M within the API's bdi signal passed through Mux-D, producing the first block of the ciphertext C. The produced cipher text is stored in Reg-A through Mux-E, Mux-F and Mux-A, then the datapath is handed over to the PRF subroutine. After the PRF hands over the datapath again, the produced ciphertext block is extracted from Reg-A and fed to Fix1 function, and passes through Mux-B to be encrypted and XORed with another block of the plaintext producing another block of the ciphertext. The process repeats until the last block of the plaintext with only one difference which is that the encrypted result this time is ANDed with the bdi_valid_bits to have an output with the same length of the last plaintext block to be XORed with. Each produced ciphertext is passed through Mux-E and ANDed again with the bdi_valid_bits to be output finally through Mux-G within the API's bdo signal. The bdi_valid_bits is an internal signal produced by the Expand block to convert the API's bdi_valid_bytes to its equivalent bits representing the valid bits of the last block of the input data.

D. DEC Datapath

The DEC subroutine naturally reuses the ENC datapath but in reverse. At first the intermediate tag V is encrypted, XORed with the API's bdi signal which this time represents a block of the ciphertext C, producing the first block of the plaintext M. At the same time the delivered ciphertext is stored in Reg-A through Mux-A and the datapath is handed over to the PRF subroutine. After the PRF hands over the datapath again, the ciphertext block is extracted from Reg-A and fed to Fix1 function and passes through Mux-B to be encrypted and XORed with another block of the ciphertext producing another block of the plaintext. Similar to the ENC subroutine, the process repeats until the last block of the plaintext which is handled the same way as in ENC subroutine. Also each plaintext is output through Mux-G passing by the same datapath as ENC subroutine.

IV. EXPERIMENTAL RESULTS AND COMPARISON

The hardware performance of the proposed hardware implementation of CLOC (also denoted as Optimized CLOC) is evaluated by comparing against the given hardware implementation of CLOC in [9]. A comparison against the design in [11] is not possible due to the unreported throughput of the design. While the design in [11] targets low area neglecting performance, our proposed design as well as the design in [9] target high performance and both are compatible with CAESAR hardware API, which facilitates a fair comparison. Thus, only a comparison against the work in [9] is presented. The Optimized CLOC is described in Verilog language and functionally verified by given test vectors with the help of modelsim tool. Both designs are synthesized and mapped to TSMC 65nm standard cell library by using Synopsys Design Compiler version J-2014.09. Upon doing synthesis, both designs were constrained for achieving maximum speed along with their area-constrained. Area is measured in μm, as well as in kilo gate equivalent (KGE) in order to make it technology independent. Table I shows synthesis results comparison of Optimized CLOC and CLOC in [9] in terms of area, critical path delay and power performance parameters. Based on the experimental results, it shows that Optimized CLOC is 29.29% smaller and achieved 21.7% less power requirements when compared with CLOC. It also requires a clock frequency that is 21.17% faster. This is because only one AES core was used in the proposed design rather than using two AES cores, which in turn reduced the area as well as power of the proposed design. While the frequency is increased due to the use of intermediate registers in the Optimized CLOC.

Table I: ASIC synthesis results comparison between our design and the design from [9]

Design	Delay (ns)	Area (μm^2)	Area (KGE)	Power (mW)
[9]	1.70	136,618.6	94.87	25.02
Optimized CLOC	1.34	96,605.3	67.09	19.59

In order to estimate throughput, both designs were examined with large plaintext and large associated data namely 100 block for each, which is equivalent to 12800 bits for each. As Table II shows, the number of cycles for our design is increased due to using only one AES core, but due to the higher frequency our design can reach; the decline in throughput is small.

The area efficiency is defined as throughput/area. These metrics play an important role in judging modern digital

Table II: Comparison of throughput of our design and the design from [9]

Design	Frequency (MHz)	# of cycles	Throughput (bits/cycle)	Throughput (Mbps)
[9]	588.23	2252	5.68	3341.18
Optimized CLOC	746.26	3352	3.82	2850.75

circuits, including cryptographic accelerators. Table III shows that the proposed design is 20.6% more area efficient and 21.7% less power consumer.

Table III: Comparison between the proposed architecture of CLOC and the architecture of CLOC from [9]

Design	Area (KGE)	Area Efficiency (Mbps/KGE)	Power (mW)
[9]	94.87	35.22	25.03
Optimized CLOC	67.09	42.49	19.59
Improvement	29.29%	20.6%	21.7%

However, the high speed achieved by these implementations is not required for applications such as smart cards, passive Radio-Frequency Identification (RFID) tags and Wireless Sensor Networks (WSNs). For such applications, low power and small area are more crucial. Hence, the comparison has also been performed at 100 KHz, leading to smaller area and power for both implementations. More significantly, the area and power difference is more significant at 100 KHz, with the Optimized CLOC operating at only 2.6 μW, as shown in Table IV.

Table IV: Comparison between the proposed architecture of CLOC and the architecture of CLOC from [9] at 100 KHz

Design	Area (KGE)	Area Efficiency (Kbps/KGE)	Power (μW)
[9]	81.53	6,966.8	4.28
Optimized CLOC	46.59	8,199.1	2.60
Improvement	42.85%	17.6%	37.80%

V. CONCLUSION

In this paper, a high-speed and hardware-efficient implementation for CLOC authenticated encryption cipher is proposed particularly for achieving better throughput per area performance. In what follows, reduction in area was achieved by calling the functions sequentially while critical path delay is reduced by inserting the immediate registers in the optimized design. Based on ASIC synthesis results, it shows that the proposed design proved to be less in area by 29.29% and improved area efficiency (throughput per area) by 20.6% when compared with its contender.

REFERENCES

[1] T. Kohno, J. Viega, and D. Whiting, "The CWC authenticated encryption (associated data) mode," *ePrint Archives*, 2003.
[2] M. Bellare and C. Namprempre, "Authenticated Encryption: Relations among Notions and Analysis of the Generic Composition Paradigm," *Journal of Cryptology*, vol. 21, no. 4, pp. 469–491, Oct 2008.
[3] M. J. Dworkin, "SP 800-38D. Recommendation for Block Cipher Modes of Operation: Galois/Counter Mode (GCM) and GMAC," Gaithersburg, MD, United States, Tech. Rep., 2007.

978-1-5386-7100-9/18 $31.00 © 2018 IEEE

[4] D. A. McGrew and J. Viega, "The Security and Performance of the Galois/Counter Mode (GCM) of Operation," in *Proceedings of the 5th International Conference on Cryptology in India*, ser. INDOCRYPT'04, Chennai, India, 2004, pp. 343–355.

[5] P. Rogaway, M. Bellare, and J. Black, "OCB: A block-cipher mode of operation for efficient authenticated encryption," *ACM Transactions on Information and System Security (TISSEC)*, vol. 6, no. 3, pp. 365–403, 2003.

[6] CAESAR Competition, "CAESAR submissions," https://competitions.cr.yp.to/caesar-submissions.html, 2016.

[7] E. Homsirikamol, W. Diehl, A. Ferozpuri, F. Farahmand, and K. Gaj, "Implementer's Guide to the CAESAR Hardware API," 2016.

[8] T. Iwata, K. Minematsu, J. Guo, and S. Morioka, "CLOC: authenticated encryption for short input," in *International Workshop on Fast Software Encryption*. Springer, 2014, pp. 149–167.

[9] George Mason University, "ATHENa: Automated Tools for Hardware EvaluatioN," lhttps://cryptography.gmu.edu/athena/, 2017.

[10] T. Suzaki, K. Minematsu, S. Morioka, and E. Kobayashi, "Twine: A lightweight block cipher for multiple platforms," in *International Conference on Selected Areas in Cryptography*. Springer, 2012, pp. 339–354.

[11] S. Banik, A. Bogdanov, and K. Minematsu, "Low-area hardware implementations of CLOC, SILC and AES-OTR," in *Hardware Oriented Security and Trust (HOST), 2016 IEEE International Symposium on.* IEEE, 2016, pp. 71–74.

2018 IEEE Computer Society Annual Symposium on VLSI

0.9 to 2.5 GHz Sub-sampling Receiver Architecture for Dynamically Reconfigurable SDR

Ajinkya Kale*[†], Johannes Sturm[†] and Vijaya Sankara Rao Pasupureddi[‡]

*Centre for VLSI and Embedded Systems Technology (CVEST),
International Institute of Information Technology Hyderabad, India

[†]Josef Ressel Center for Integrated CMOS RF Systems and Circuits Design,
Carinthia University of Applied Sciences, Villach, Austria

[‡] Centre for Advanced Studies in Electronics Science and Technology (CASEST),
University of Hyderabad, India

Email: {a.kale, j.sturm}@fh-kaernten.at, vijaysp@uohyd.ac.in

Abstract—This work proposes an architecture for dynamically reconfigurable multi-standard radio receiver based on the principle of sub-sampling. The proposed receiver has a unique capability to detect the carrier frequency of the incoming signal, estimate its bandwidth and identify if the carrier is present in one of the target standard bands. In addition, sub-sampling is performed at an early stage in the receiver chain to process the signal in the discrete-time domain and to bring the analog-to-digital converter closer to the antenna as in a universal software defined radio (SDR). This adds to the flexibility and reconfigurability which are generally needed for SDR based RF receivers. Moreover, the proposed radio architecture operates at low clock rates, thanks to sub-sampling, leading to less complex clocking circuitry and low power consumption. The proposed receiver architecture RF front-end is modeled in Verilog-AMS behavioral models and the digital signal processing is implemented in Simulink-Matlab. The complete receiver architecture has been verified to detect and process three different bands belonging to three different standards (GSM, UMTS and WLAN) with the carrier frequency ranging from 0.9 GHz to 2.5 GHz with a maximum signal bandwidth of 22 MHz and input dynamic range from -109 to -20 dBm.

I. INTRODUCTION

For many years now, integrated software defined radio (SDR) or digitally reconfigurable radio research has been quite challenging with only a few reported practical prototypes with limited success [1]. The basic demand of the SDR has been the need for a giant analog-to-digital converter (ADC) and a powerful enough digital signal processing (DSP) so that it serves as a universal radio platform receiving almost all radio standards and services. These demands are difficult to meet even in modern CMOS technologies.

Wideband sampling receivers [2]–[4] and multi-standard subsampling receivers [5]–[9] are reported in the literature. In these sampling receivers, selection of the sampling frequency defines the receiver architecture which includes an oversampling receiver with discrete time mixer (DTM) [2], time-interleaved (TI) discrete-time oversampling receiver [3], and sub-sampling receiver [4]. The architecture in [5] utilizes two-stage sub-sampling based down-conversion along with the tuned filters at each stage to improve the overall performance. The architecture in [6] does not include the front-end LNA

and employs the IF filtering to receive two or more standards simultaneously. The architecture proposed in [7], employs two sampling paths with tunable RF filter and signal processing algorithm to alleviate the effect of jitter and sampling frequency variations. This implementation lacks the effect of the complete RF front-end as the LNA is not included. None of the above implementations provides the capability to sense the standard of the incoming RF signal on their own. The implementation in [9] employs a tunable RF filter along with sub-sampling receiver and ADC to identify the free spectrum for cognitive radio and can receive two or more standards simultaneously. This implementation does not include an LNA and the capability to dynamically identify and change the band used for reception. Also, these architectures focus primarily on the optimization of the sampling frequency for the system level performance and rely on a reconfigurable filter or focus on the baseband system optimization. But, none of the earlier implementations provides the capability to dynamically detect the standard and estimate the signal bandwidth using the incoming Radio Frequency (RF) signal.

Therefore, the path to universal radio implementation is challenging and the realization is far from what is available today. However, as an intermediate goal, this research targets a reconfigurable radio architecture which can receive three or four bands from different standards. This demands a new generation of receiver architectures and RF front-end circuits with digitally assisted RF techniques [10]. The most promising solution to achieve this intermediate goal is by realizing receivers wherein the signals are processed in the discrete-time domain. As it is widely accepted, in deep sub-micron CMOS, time-domain resolution of a digital signal edge-transition is superior to voltage resolution of analog signals [11].

This work proposes an approach for realizing a dynamically reconfigurable multi-standard radio receiver architecture for SDR based on the principle of sub-sampling [12]. The main features of the proposed architecture include firstly, the capability to dynamically identify the carrier frequency of the incoming RF signal, estimate its bandwidth and identify if it belongs to one of the target standards bands. The architecture has the capability to process a incoming single corresponding

978-1-5386-7100-9/18 $31.00 © 2018 IEEE

316

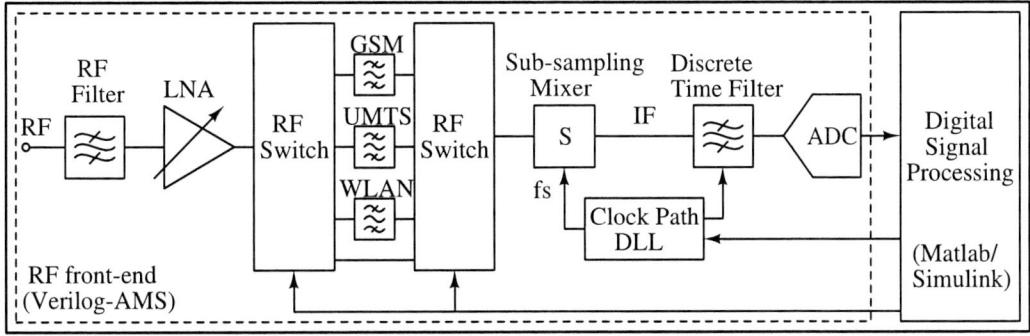

Fig. 1. Proposed multi-standard receiver architecture for GSM, UMTS and WLAN standards

to one band of the target communication standards at a time. Thirdly, the RF front-end resources such as Antenna, RF band-pass filter (BPF), low noise amplifier (LNA), DTM and ADC are shared between the three targeted standards and fourthly, since the proposed radio uses sub-sampling, the clocking circuits operate at relatively low frequency leading to less complex frequency synthesizers and consequently low power consumption. In addition, the discrete-time signal processing close to the antenna at a reduced sampling rate decreases the power consumption in the RF front-end which are usually battery powered.

The article is organized as follows: Section II introduces the sub-sampling down-conversion principle and how it is deployed. Section III presents the proposed multi-standard reconfigurable radio architecture. Then, Section IV presents the performance results of the complete system and Section V concludes the paper.

II. SUB-SAMPLING DOWN-CONVERSION METHODOLOGY FOR THE PROPOSED RECEIVER ARCHITECTURE

The proposed reconfigurable multi-standard sub-sampling radio receiver architecture is shown in Fig. 1. The sub-sampling RF front-end for the receiver is modeled in behavioral Verilog-AMS and the DSP part is implemented in Simulink-Matlab. The complete system level co-simulation is performed including the RF front-end and digital baseband signal processing with the test signals listed in Table I. The proposed radio is validated for carrier frequency ranging from 900 MHz to 2.5 GHz with maximum signal bandwidth of 22 MHz. The rest of the section presents the sub-sampling based down-conversion technique utilized in the proposed multi-standard receiver along with the frequency planning.

TABLE I
TEST SIGNAL SPECIFICATIONS

	1	2	3
Standard	GSM	UMTS	WLAN (802.11g)
RF (MHz)	942.5	2140	2442
Signal bandwidth	200 kHz	5 MHz	22 MHz
Power (dBm)	−40	−40	−40

Previously, sampling receivers for wideband systems have been reported in [2]–[5]. Utilizing the bandpass nature of the RF input signal, the sub-sampling approach employs intentional aliasing to reduce the sampling frequency below the Nyquist frequency. The selection of sampling frequency (f_s) depends on the center frequency (f_c) and bandwidth (B) of the RF signal as per bandpass sampling theory (Eq. (1)) [4], [13].

$$f_s = \frac{4 \times f_c}{(2m - 1)} \qquad (1)$$

where $m = 1, 2, 3, 4, \ldots, m_{max}$ and $m_{max} = \left\lfloor \frac{f_H}{B} \right\rfloor$ is the maximum integer value less than fraction $\frac{f_H}{B}$, where f_H is the higher side frequency ($f_H = (f_c + \frac{B}{2})$).

Apart from the sampling frequency, selection of the intermediate frequency (IF) (f_{IF}) is vital to the receiver performance optimization. As lower f_{IF} is desirable at the input of ADC and higher f_{IF} helps in image rejection. Also, as it is widely known, sub-sampling receivers suffer from thermal noise folding [5]. For this reason, a higher sampling frequency is preferred. The suitable value of m in Eq. (1) is chosen based on the trade-off between f_s, f_{IF} and oversampling ratio. So, in this architecture, the sampling frequency below 2 GHz is selected to comfortably process the sampled signal in the ADC. The corresponding IF (f_{IF}) for f_s of Eq. (1) is as presented in Eq. (2).

$$f_{IF} = \frac{f_s}{4} \qquad (2)$$

This results in a sampling frequency more than the theoretical minimum of $(2 \times B)$ but less than the Nyquist rate $(2 \times f_H)$, hence called sub-sampling. For example, During the detection phase, the proposed receiver down-converts the entire RF signal bandwidth, the entire GSM band from 925 MHz to 960 MHz utilizing the first sub-band filter in the front-end. Thus, the center frequency (f_c) of 942.5 MHz and $m = 3$ is chosen in Eq. (1) which results in 754 MHz sampling frequency and 188.5 MHz IF as per Eq. (2). This also results in more than 21 times oversampling with respect to RF bandwidth of 35 MHz.

978-1-5386-7100-9/18 $31.00 © 2018 IEEE

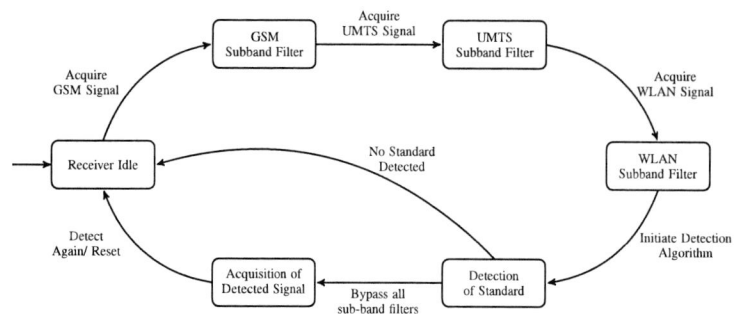

Fig. 2. State machine explaining the detection and acquisition algorithm for the proposed multi-standard receiver

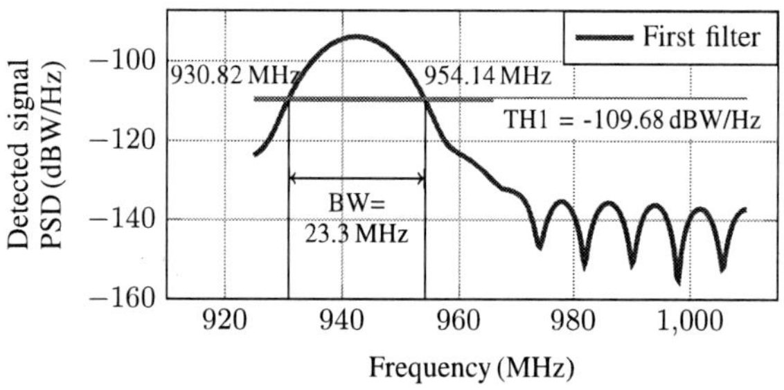

(a) First filter detected signal PSD for GSM test signal

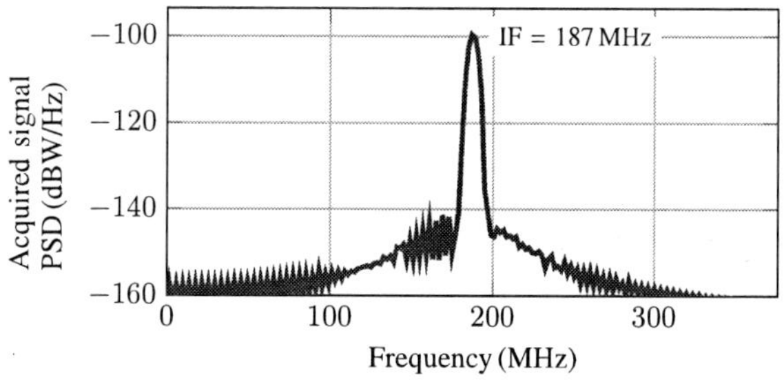

(b) Down-converted GSM signal PSD with IF of 187 MHz

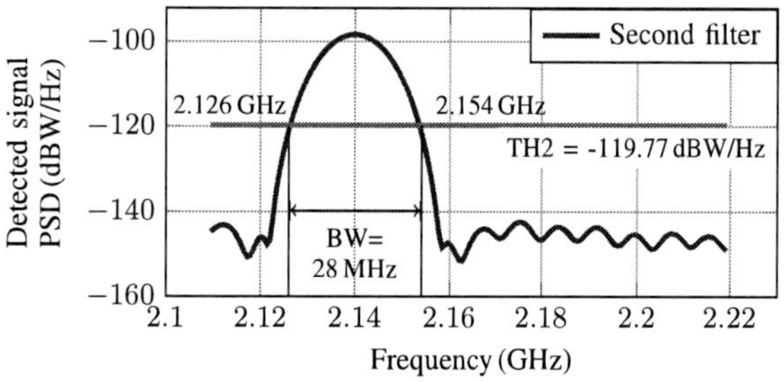

(c) Second filter detected signal PSD for UMTS test signal

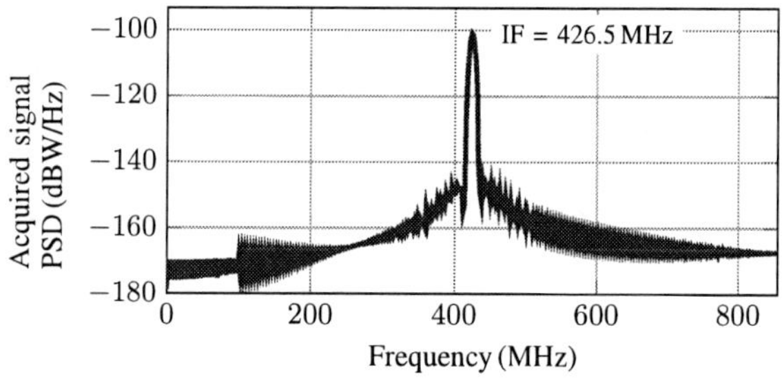

(d) Down-converted UMTS signal PSD with IF of 426.5 MHz

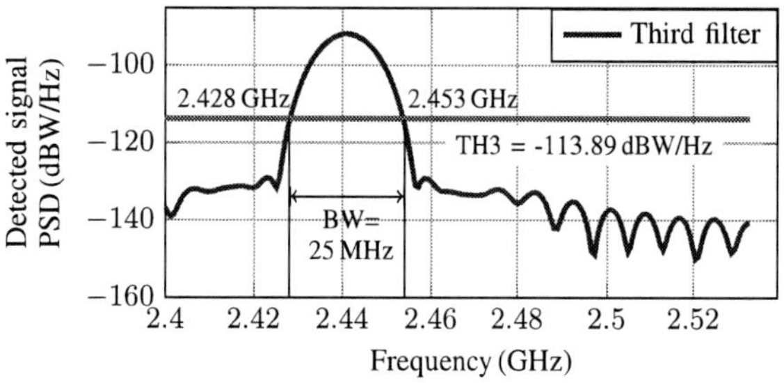

(e) Third filter detected signal PSD for WLAN test signal

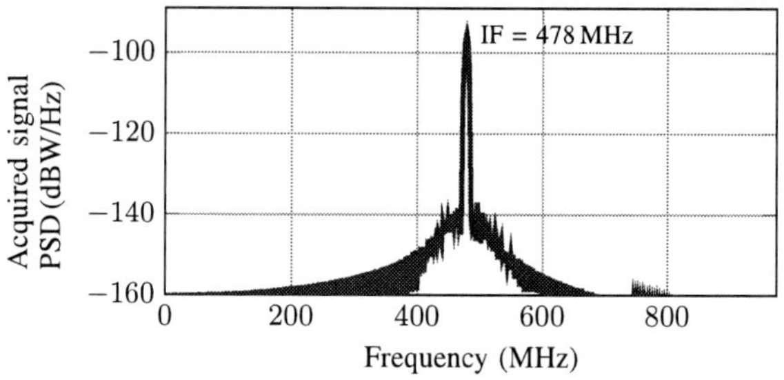

(f) Down-converted WLAN signal PSD with IF of 478 MHz

Fig. 3. Detection and Acquisition Co-simulation Results for The Multi-standard SDR Receiver

III. PROPOSED RECEIVER ARCHITECTURE

The proposed receiver architecture includes a sub-sampling receiver RF front-end followed by a baseband DSP with frequency control. The RF front-end utilizes a wideband RF filter along with a wideband, tunable frequency, tunable gain LNA to provide suitable selectivity for each standard. The architecture employs a BPF for each of the targeted standards

and RF switches to estimate the carrier frequency and signal standard from the incoming RF signal. These subband filters are utilized only during detection and are bypassed during normal receiver operation. Optimized sampling frequency as per Section II is generated using a delay locked loop (DLL) block which is controlled by the DSP block. The down-converted signal is bandpass filtered by the discrete time filter

978-1-5386-7100-9/18 $31.00 © 2018 IEEE

and digitized by the ADC.

The proposed receiver utilizes a state machine implemented in Simulink and the signal processing is carried out in Matlab. The state machine for dynamic detection of the incoming RF signal standard and signal acquisition is presented in Fig. 2.

A. Dynamic Detection Phase

In the dynamic detection phase, each bandpass filter (GSM, UMTS, WLAN in Fig. 1) represents a targeted standard and is selected sequentially for a predefined interval (500 ns for each of the subband filters) as shown in Fig. 2. For the activated filter, the corresponding pre-defined sampling frequency listed in Table II is selected in the DLL. A look-up table is utilized in the DSP to generate these pre-defined sampling frequencies. The filtered signal is down-converted and then digitized using the ADC. After the RF signal is received with the three filters, the detection routine is initiated to estimate the band edges, center frequency and signal bandwidth based on the stored samples. These signal parameters are utilized to calculate the sampling frequency to be used in the acquisition phase to receive the detected standard. The main objective of the proposed architecture is to identify the standard in which the incoming RF signal is present. This is achieved after the dynamic detection phase is complete.

TABLE II
SPECIFICATION FOR THE SUBBAND FILTERS UTILIZED FOR DETECTION

Subband Filter	GSM	UMTS	WLAN
Center frequency (MHz)	942.5	2140	2442
Filter bandwidth (MHz)	35	60	83.5
K value in Eq. (1)	3	3	3
f_s for detection (MHz)	754	1712	1953.6
Oversampling ratio	10.7	14.26	11.69
IF for detection (MHz)	188.5	428	488.4

A non-parametric power spectral density (PSD) estimation technique (Welch method) utilizing windowed Fourier transform with averaging is used (in Matlab) to estimate the carrier frequency and signal bandwidth from the digital samples during detection phase in Fig. 2. A threshold based on the noise variance is employed on the PSD of the stored samples for each bandpass filters [12]. As a rule of thumb, the outer edges of the scattered spectrum extend up to 3σ. It leads to the fact that the upper edge of the noise spectrum is always at 6 dB above the mean power. This (mean of PSD + 6 dB) is chosen as the detection threshold for estimating the band edge frequencies and signal bandwidth. For example, for GSM test case, only the first subband filter detects the RF signal. The detected signal PSD for the first subband filter is as shown in Fig. 3a along with the detection threshold and the estimated band edges. In this way, once the communication standard is identified with all signal parameters, the further receiver operation can be continued for the selected standard.

B. Signal Acquisition Phase

After the successful detection of the RF signal, the acquisition phase is initiated as shown in Fig. 2. In this phase,

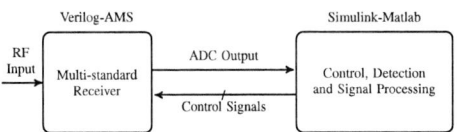

Fig. 4. Simulation setup for performance evaluation of the proposed multi-standard receiver architecture

the subband filters are bypassed and the sampling frequency calculated at the end of the detection phase is utilized to down-convert the signal in the sub-sampling mixer. The down-converted signal is stored and further processed in the DSP block for calculating the PSD. For the test signals in Table I, the acquired signal PSD is shown in Fig. 3b, Fig. 3d and Fig. 3f respectively.

IV. PERFORMANCE RESULTS AND DISCUSSION

The co-simulation setup used for the simulation of the proposed architecture is presented in Fig. 4. The RF front-end is simulated using Spectre and the control state machine and DSP are simulated in Simulink and Matlab respectively. The setup consists of three control signals from Simulink to Spectre which include RF input signal, RF switch control and DLL frequency control. The output from Spectre to Simulink represents the ADC output.

Three test cases are utilized to investigate the performance of the proposed multi-standard radio architecture. The performance results for the co-simulation are summarized in Table III. For the first test case, a 942.5 MHz carrier GSM signal with 200 kHz bandwidth is applied to the receiver. The PSD for the signal processed by the first subband filter is shown in Fig. 3a along with the threshold of TH1 = −109.68 dBW/Hz used for the detection of signal edges. The estimated center frequency from the edges is 942.48 MHz against the input signal frequency of 942.5 MHz. The estimated signal bandwidth (23.32 MHz) is very high as compared to 200 kHz channel bandwidth for GSM signal. However, here the main objective is to identify the standard in which a carrier is present and to estimate the center frequency of the channel. The PSD of the stored signal after detection in Fig. 3b shows the IF of 187 MHz against 188.5 MHz expected value in Table III.

For the second test case, 2.14 GHz carrier frequency, 5 MHz UMTS signal is applied to the receiver. The PSD for the signal processed by the second subband filter in Fig. 3c shows a detected carrier of 2.14 GHz with a detection threshold of TH2 = −119.77 dBW/Hz. The detection estimated bandwidth of 28 MHz against the channel bandwidth of 5 MHz for UMTS signal. The PSD in Fig. 3d shows the detected IF of 426.5 MHz against an expected value of 428 MHz.

For the third test case, the 2.442 GHz carrier, 22 MHz signal bandwidth WLAN signal is applied to the proposed receiver. The PSD for the signal processed by the third subband filter Fig. 3e shows 2.4405 GHz with a detection threshold of TH3 = −113.89 dBW/Hz. The detection estimated bandwidth of 25 MHz against the channel bandwidth of 22 MHz for WLAN

TABLE IV

PERFORMANCE COMPARISON WITH THE PREVIOUS SAMPLING RECEIVERS

Parameter	[2]	[4]	[5]	[6]	[7]	[9]	**This work**
Architecture	RF Oversampling	Sub-sampling	Two-stage sub-sampling	Sub-sampling + tunable IF Filter	Sub-sampling + tunable RF filter	Sub-sampling + tunable RF filter	**Sub-sampling + Detection Filters**
RF Carrier (GHz)	0.2 to 0.9	2.412	0.9 to 2.5	0.8 to 5.8	2.4	0.4 to 2.5	**0.9 to 2.5**
Target standard	DBV	WLAN	GSM, UMTS, WLAN	GSM, WCDMA, Bluetooth, WiMAX, WLAN	WBAN	DBV, Tx Aux, WLAN,	**GSM, UMTS, WLAN**
Sampling frequency (GHz)	1.6 to 7.2	1.072	0.761	1.8 to 2	0.04	0.27 to 0.42	**up to 2**
Signal Standard Detection	Not possible	Not possible	Not possible	Not possible	Not possible	Possible	**Possible**
Implementation	CMOS	CMOS	ADS	Matlab	Matlab	COTS components	**Verilog-AMS and Matlab/ Simulink**

TABLE III

PERFORMANCE SUMMARY FOR THE PROPOSED MULTI-STANDARD SDR RECEIVER

Test case	GSM	UMTS	WLAN
Input power (dBm)	−40	−40	−40
Carrier	942.5 MHz	2.14 GHz	2.442 GHz
Bandwidth	200 kHz	5 MHz	22 MHz
f_s Eq. (1)	754 MHz	1.712 GHz	1.9536 GHz
f_{IF} (MHz) Eq. (2)	188.5	428	488.4
Detection threshold (dBW/Hz)	−109.68	−119.77	−113.89
Estimated lower frequency	930.82 MHz	2.126 GHz	2.428 GHz
Estimated higher frequency	954.14 MHz	2.154 GHz	2.453 GHz
Estimated Carrier	942.48 MHz	2.14 GHz	2.4405 GHz
Estimated bandwidth (MHz)	23.3	28	25
Detected IF (MHz)	187	426.5	478

signal. The PSD shows the detected IF of 478 MHz in Fig. 3f against an expected value of 488.4 MHz.

The performance of the proposed architecture is compared with the previous sampling receiver implementations in Table IV. The previous implementations do not consider the complete front-end or do not include the capability to identify the incoming RF signal standard dynamically and lack the details of signal processing. The sub-sampling receiver is proposed for a system-on-chip implementation with integrated baseband digital control.

V. CONCLUSIONS

This work proposed a dynamically reconfigurable sub-sampling multi-standard radio receiver architecture. It is demonstrated by simulation that the proposed architecture is able to detect and process multiple signals from different standards by detecting the signal carrier and estimating its bandwidth. The receiver has been tested for GSM, UMTS and WLAN standards with different test cases. The proposed receiver performance is verified for carrier frequencies in the range from 900 MHz to 2.5 GHz, maximum signal bandwidth of 22 MHz and an input dynamic range from −109 dBm to −20 dBm. In addition, the proposed architecture brings

the ADC close to the antenna to achieve the intermediate objective of realizing a universal SDR and lowers the power consumption thanks to the sub-sampling down-conversion.

ACKNOWLEDGMENT

The presented research work was done in course of the 'Josef Ressel Center for Integrated CMOS RF Systems and Circuits Design – Interact' at the Carinthia University of Applied Sciences. The financial support by the Austrian Federal Ministry of Science, Research and Economy and the National Foundation for Research, Technology and Development is gratefully acknowledged. We also wish to thank Intel Austria GmbH. for their financial contribution and technical support.

REFERENCES

[1] J. Craninckx, "CMOS software-defined radio transceivers: Analog design in digital technology," *IEEE Comm. Mag.*, vol. 50, pp. 136–144, April 2012.

[2] Z. Ru *et al.*, "Discrete-time mixing receiver architecture for RF-sampling software-defined radio," *IEEE JSSC*, vol. 45, pp. 1732–1745, Sep. 2010.

[3] R. Chen and H. Hashemi, "A 0.5-to-3 GHz software-defined radio receiver using discrete-time RF signal processing," *IEEE JSSC*, vol. 49, pp. 1097–1111, May 2014.

[4] D. Jakonis *et al.*, "A 2.4-GHz RF sampling receiver front-end in 0.18-mu;m CMOS," *IEEE JSSC*, vol. 40, pp. 1265–1277, Jun. 2005.

[5] R. Barrak *et al.*, "Optimized multistandard RF subsampling receiver architecture," *IEEE Tran. Wireless Comm.*, vol. 8, Jun. 2009.

[6] J. R. G. Oya *et al.*, "Optimization of subsampling dual band receivers design in nonlinear systems," in *2012 IEEE/MTT-S International Microwave Symposium Digest*, June 2012, pp. 1–3.

[7] D. Zhao *et al.*, "Reconfigurable subsampling receiver architecture for wireless body area networks," in *IEEE Inter. Symp. on Personal, Indoor and Mobile Radio Comm.*, Sept 2011, pp. 2153–2157.

[8] J. R. G. Oya and et.al.(2012), "Subsampling receivers with applications to software defined radio systems,," in *Data Acquisition Applications, Prof. Zdravko Karakehayov (Ed.), InTech,*.

[9] A. Kwan *et al.*, "Sub-sampling technique for spectrum sensing in cognitive radio systems," in *2012 IEEE Radio and Wireless Symposium*, Jan 2012, pp. 347–350.

[10] M. Li *et al.*, "Signal processing challenges for emerging digital intensive and digitally assisted transceivers with deeply scaled technology (invited)," in *SiPS 2013 Proceedings*, Oct 2013, pp. 324–329.

[11] R. B. Staszewski *et al.*, "All-digital tx frequency synthesizer and discrete-time receiver for bluetooth radio in 130-nm cmos," *IEEE Journal of Solid-State Circuits*, vol. 39, pp. 2278–2291, Dec 2004.

[12] A. Kale *et al.*, "Wideband channelized sub-sampling transceiver for digital RF memory based electronic attack system," *Aerospace Science and Technology*, vol. 51, pp. 34–41, Apr 2016.

[13] R. G. Vaughan *et al.*, "The theory of bandpass sampling," *IEEE Transactions on Signal Processing*, vol. 39, pp. 1973–1984, Sep. 1991.

2018 IEEE Computer Society Annual Symposium on VLSI

Hardware Obfuscation using Strong PUFs

Soroush Khaleghi and Wenjing Rao
ECE Department, University of Illinois at Chicago (UIC)
Email: skhale4@uic.edu, wenjing@uic.edu

Abstract—IC piracy is a significant security threat, where malicious manufacturers can produce unauthorized extra chips and/or steal the information of a design through reverse engineering attempts. As a countermeasure, hardware obfuscation schemes usually withhold a part of the design (which thereafter constitutes the "key") by replacing it with configurable modules. Enforcing the configurable module to be filled in with the withheld key information enables a post-manufacturing activation of each authenticate chip, albeit with a a need to state the threat of a leaked common key. To ensure that each chip has a unique key, Physically Unclonable Functions (PUFs) have been proposed to be integrated with hardware obfuscation. Such a paradigm is constrained to use *weak* PUFs, because, to uniquely set the key (the content of the configurable module) for each chip, the designer needs to fully characterize the PUFs for all the chips. In this paper, we argue that a powerful attacker in the position of a manufacturer can fully characterize all the weak PUFs, and use any leaked key to break the obfuscation framework. This paper proposes a *strong* PUF-based hardware obfuscation scheme to effectively prevent IC piracy even in the case of a leaked key from some activated chip.

I. INTRODUCTION

The globalization of the semiconductor industry has raised serious concerns about the security of Integrated Circuits (ICs). Since IC designers no longer have full control over the manufacturing process, a design is prone to various hardware attacks from a malicious manufacturer. Particularly, a manufacturer can conduct *IC piracy* via producing unauthorized extra chips and/or stealing the information of a design through reverse engineering attempts [1] [2].

Hardware obfuscation schemes aim at preventing IC piracy by enabling the designers to control the number of functioning chips through a post-fabrication activation process [1] [2]. The essential idea is to withhold a part of the design and replace it with a configurable module at the design stage, so that none of the manufactured chips would function properly without being "activated" by the designer [3] [4] [5] [6] [7]. Such a post-fabrication activation is achieved by securely restoring the withheld function back into the chips through specifying the content of the configurable modules in a trusted facility. Afterwards, the chips are considered "unlocked" and can be made available to the open market. Without direct access to probe the securely stored *key* (the content of the configurable module) for the activated chips, an attacker cannot recover the entire design or overbuild illegal ICs.

Various types of hardware obfuscation approaches have been proposed in the literature. The combinational-based schemes work by inserting additional XOR gates with configurable bits into the design that must be set correctly to activate the chips [5] [2]. The sequential-based approaches

This work is supported by NSF Grant 1149661.

work by inserting additional "dummy" states into the Finite State Machine (FSM), so that the circuit functions properly only when a certain sequence of inputs (the key) is applied to the chips [3] [4]. Permutation-based techniques propose to scramble the interconnect network of the design, so that only the correct key can configure the original interconnect network [7]. The withheld-based schemes work by replacing a part of the design (the key) with a Look-Up Table (LUT) that must be configured properly to activate the chips [8].

The obfuscation schemes try to ensure that the required effort for an attacker to obtain the correct key is computationally infeasible [1] [2]. Most of these schemes are based on the assumption that there is no direct access to the content of the key for legally activated chips [5] [6] [3] [1]. In practice, however, powerful invasive and side-channel attacks can be applied for extracting the key information from read-proof memories [1] [9]. Ultimately, if a *common* key is used across all the fabricated chips, then such a key (that must be stored on every chip) becomes the most vulnerable part of the entire obfuscation mechanism. In other words, a single leaked key can compromise the entire security mechanism.

This threat motivates the approach of having a unique key for every chip, and to somehow ensure that if a key is leaked from some chip, it cannot be used to unlock other chips. A promising direction to address this problem is to use Physically Unclonable Functions (PUFs) [10]. A PUF is a physical system, built based on the inherent process variations of chips at the manufacturing stage, which can be used as a unique signature/function for every chip. Nonetheless, engaging a PUF as a key against IC piracy is challenging, because a straightforward approach that uses the PUF only as a signature of the chip can be easily bypassed by the attacker/manufacturer.

The hardware obfuscation schemes in [11] and [4] propose to modify the original design and engage a PUF as a part of the circuit's functionality. As each chip's PUF has a unique and unpredictable functionality, these schemes tightly couple the PUF with a configurable module (constituting the key), which will be individually programmed for every chip according to the PUF by the designer during the post-fabrication activation process.

To program the configurable module of each and every chip, the designer needs to fully characterize the behavior of all the PUFs for all the chips, thus is constrained to use the PUFs with a limited search space, namely *weak* PUFs [11] [4] [10]. Unfortunately, an untrusted manufacturer is in a strong position of doing the same characterization for all the chips before handing them over to the designer for the activation process. In this case, when a leaked key is obtained for some activated chip, the attacker/manufacturer can easily recover the

978-1-5386-7100-9/18 $31.00 © 2018 IEEE   321

entire original design from: 1) the information of the leaked key, combined with 2) the pre-stored characterization of its associated PUF.

This paper proposes to employ *strong* PUFs (with *huge* search space of truth-table) into the obfuscation framework, against such an attacker/manufacturer. While it is impossible for an attacker to fully characterize the strong PUFs, the main challenge becomes ensuring that the designer does not need to bear the burden of fully characterizing the strong PUFs to generate a unique key per-chip. This is achieved by employing an *Obfuscator* block into the design, which enables the designer to select an arbitrarily subset of the strong PUF to work, while guaranteeing that the architecture does not reveal to the attacker/manufacturer of the choices made by the designer. The paper also discusses the security of the proposed scheme against several attacks, including the machine learning attacks [12] that are known to be powerful for characterizing many strong PUFs in a short time.

II. PRELIMINARIES

A. General concepts and types of PUFs

A PUF [10] is a physical system that presents unclonability by exploiting the process variations during the fabrication process. A PUF takes inputs, called *challenges*, and produces outputs, called *responses*. Thus, every PUF is a function of Challenge-Response Pairs (CRPs) that is (ideally) unique for each chip. Since exact control over the manufacturing process is impossible, it is infeasible to build identical PUFs with the same CRPs, i.e., truth-tables. Compared to their truth-table sizes, PUFs can be usually implemented with a very small hardware investment [10].

Based on the number of CRPs that it can offer, each PUF belongs to one of the following two categories [10]:

1) **Weak PUFs** offer a **limited** number of CRPs with respect to the number of their building components, such as SRAM-based PUF [13].

2) **Strong PUFs** offer a **huge** number of CRPs, i.e., exponential to the number of their building components, such as a delay-based Arbiter PUF [14].

Since the entire truth-table of a weak PUF can be fully characterized via exhaustive evaluation, its CRPs must be kept secret from the potential attackers. Otherwise, a weak PUF cease to be unpredictable or unclonable, as its behavior can be emulated in software. For a strong PUF with an exponentially large number of CRPs, however, it is impossible to derive its entire truth-table via an exhaustive evaluation. Therefore, the security of an "ideal" strong PUF does not rely on keeping its CRPs secret, but rather on its exponentially huge truth-table (CRP space) [10] [15].

B. Machine learning attacks against strong PUFs

In theory the knowledge of a limited number of CRPs for an ideal PUF should not reveal any information about the responses to other untested challenges. However, this is not true in practice. As a matter of fact, many existing strong PUFs can be fully characterized using *machine learning* algorithms [12] [16], in which a precise model of the PUF is built in software after examining a limited number of CRPs.

The training size of such machine learning algorithms (which translates into the attack complexity) depends on the type of the PUF and the parameters of the PUF [12]. For instance, the Arbiter PUF [14] can be attacked within linear time with respect to the number of its components. However, the training size to attack a different variant of the Arbiter PUF that uses nonlinearity in its architecture, known as the XOR Arbiter PUF [17], grows exponentially with respect to the number of its components. As a result, while an Arbiter PUF with 128 stages (components) can be modeled in 2.10 seconds, modeling an XOR Arbiter PUF (with almost 5x hardware overhead) requires more than 16 hours [12]. Another variant of the Arbiter PUF, known as the Lightweight PUF [18], with a similar hardware cost, requires 267 days to be fully characterized [12].

Therefore, although machine learning attacks can model a wide variety of strong PUFs in a reasonable time with high accuracy, there exist promising secure PUFs against such attacks [17] [18]. Overall, the underlying assumption of this paper is that it is prohibitively expensive to characterize the behavior of **all** the strong PUFs for all the fabricated chips.

C. Weak PUF-based hardware obfuscation

The work in [11] engages a PUF to achieve a unique key per chip. The basic flow is depicted in Fig. 1. It works by replacing a part of the circuit's functionality (the master key), shown in Fig. 1(a), with a PUF and a LUT, shown in Fig. 1(b). The LUT for each chip (serving as its key) will be uniquely configured by the designer in a post-fabrication activation process, shown in Fig. 1(c), so that the combination of the PUF and the LUT is functionally equivalent to the same master key (withheld function), which is common for all the chips. This approach guarantees that in case a key (the content of the LUT) is leaked from some chip, it cannot be simply inserted into other chips to activate them, because each LUT content is Chip/PUF-specific.

To configure the content of the LUT for each chip, based on its PUF to fulfill the functionality of the master key, the designer needs to obtain the PUF's CRPs. This is done via the peripheral characterization channels, shown in Fig. 1(c), during the activation process [11]. Such channels provide a direct mechanism for the designer to apply the challenges and observe the responses. Since the inputs of the master key can potentially take any values from the internal signals in the circuit, the designer will need the knowledge of the *entire* CRP space (truth-table) of each PUF. This is impractical for strong PUFs, therefore, the designer is limited to the choice of **weak** PUFs.

III. MOTIVATION

Based on the depicted flow in Fig. 1, in order to recover the master key, the attacker needs: 1) the information of a leaked key from some activated chip, and 2) the entire truth-table of the PUF for the same chip. It is usually assumed that the designer can remove all the PUF characterization channels from the chips after the activation process, so that there would be no physical channels left to characterize the PUFs for the legally activated chips in the open market, as shown in Fig. 1(d) [11] [19]. This can effectively prevent an attacker from obtaining the PUF truth-table for legally activated chips.

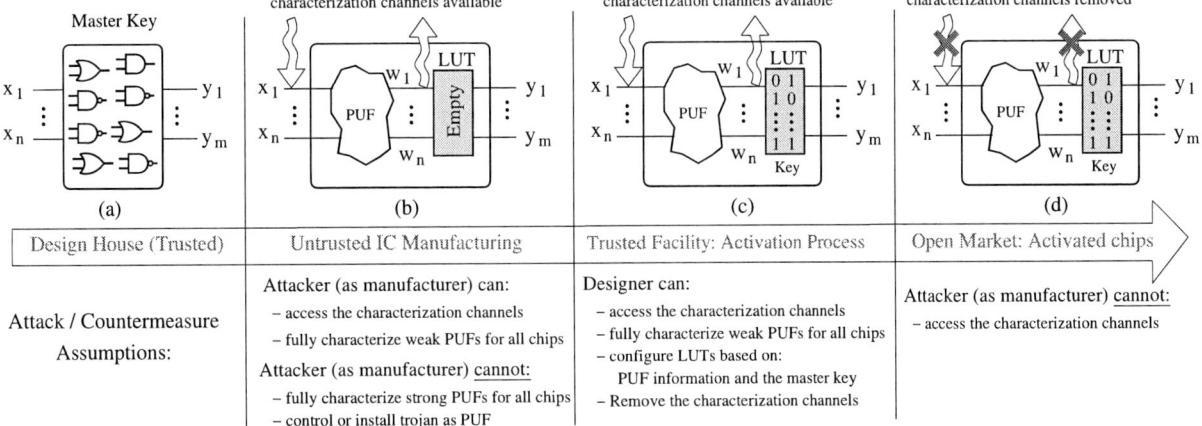

Fig. 1. Flow of PUF-based hardware obfuscation: (a) A part of the design with n inputs and m outputs is selected to be withheld as the master key. (b) The chip model with a PUF and a LUT, coupled to replace the master key. The manufactured chips based on this model will not be functional until activated by the designer. (c) The designer configures the LUT for each chip based on the behavior of the PUF to match the master key. (d) Activated chips in the open market.

TABLE I. USING WEAK PUFS VS STRONG PUFS IN AN OBFUSCATION FRAMEWORK: AN IDEAL SCENARIO IS MARKED OUT WITH *

	Weak PUF	Strong PUF
LUT cost (size of master key)	*LOW	HIGH
PUF characterization for designer	*EASY	HARD
PUF characterization for attacker	EASY	*HARD

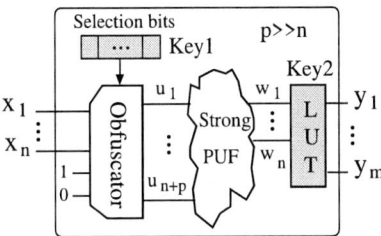

Fig. 2. The architecture of the proposed scheme: by assigning Key1 for each chip, a permutation of the original wires appears at n inputs of the PUF, and the other p inputs of the PUF are set with arbitrarily specified values. Only a subset of the PUF with n inputs and n outputs will be used by the designer. The LUT (Key2) is programmed to compensate for the unique behavior of the PUF. The unique key per chip here is $\langle$Key1 , Key2$\rangle$.

However, it is worth pointing out that an attacker in the position of a manufacturer can nonetheless gain access to the PUF at the fabrication stage (before sending the chips to the designer). At this stage, shown in Fig. 1(b), the characterization channels are fully accessible. Therefore, the attacker can obtain and store the truth-tables of all such weak PUFs for all the chips with a reasonable cost. Once a leaked key is obtained from some activated chip in the market, the attacker can look up the stored truth-table of the PUF for the leaked chip in its database. Then, the attacker can derive the master key by combining the fully characterized truth-table of the PUF, and the LUT (leaked key) of the chip.

This threat motivates the usage of **strong** PUFs: their huge CRP space makes it prohibitively expensive both in terms of time and storage for an attacker to obtain the entire truth-table of all the PUFs (for all the chips) via the characterization channels during the manufacturing stage.

Unfortunately, a straightforward adaptation of a strong PUF in the architecture in Fig. 1 is not feasible, as it would entail the same prohibitive characterization cost on the designer's side. Furthermore, since the functionality of the PUF is coupled with that of the LUT, switching from a weak PUF to a strong one implies a very large master key, thus increasing the implementation cost of the LUT exponentially: adding a single input to the withheld function doubles the LUT size.

Table I compares the usage of weak PUFs vs. strong PUFs in the obfuscation framework. Apparently, the main challenge of adopting either strong or weak PUFs is that the *same* level of characterization difficulty is imposed on the designer and the attacker. To achieve a secure, yet low cost obfuscation scenario,

a "gap" between the designer and the manufacturer has to be created, such that the following goals can be accomplished:

1) Small LUT size (low hardware cost).
2) Easy PUF characterization for designer.
3) Hard PUF characterization for attacker.

IV. A STRONG PUF-BASED HARDWARE OBFUSCATION

The key idea of the proposed scheme is to allow the designer to use a small **subset** of the entire CRP space of a strong PUF at the activation stage (as if the designer is dealing with a weak PUF), so as to maintain a low LUT cost. Meanwhile, the attacker is forced to deal with the **entire** CRP space of strong PUFs. To create such a "characterization gap" between the designer and the attacker, the main challenge is to hide the designer's selections of the CRP subset from the attacker, at the manufacturing stage.

We propose to achieve such a secure subset selection by employing an *Obfuscator* coupled with the strong PUF in the obfuscation framework. Fig. 2 depicts the main architecture of the proposed scheme. Suppose that the master key (withheld function) has n inputs and m outputs (the same as in the case of a weak PUF), where n is small, such that 2^n is a

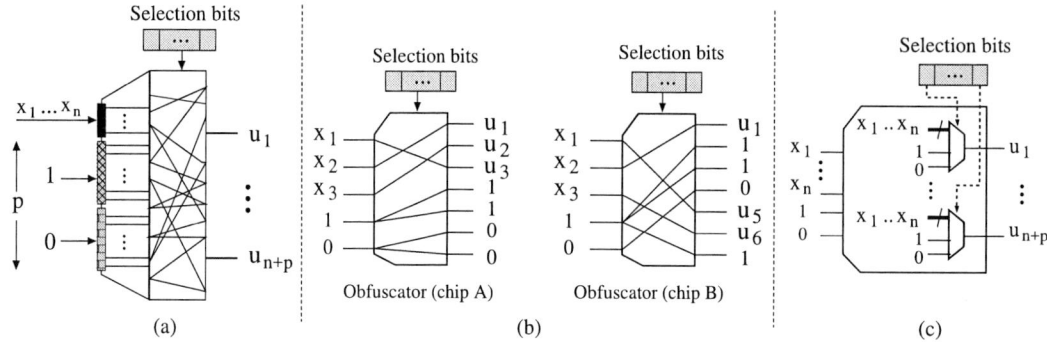

Fig. 3. (a) Obfuscator: By specifying the Selection bits, a permutation of n original wires, as well as p fixed values (0 or 1) will appear at the output of the block; (b) Examples of an Obfuscator block (with $n = 3$ and $p = 4$) to allow the designer to select any subset of the CRP space during the activation process without revealing it to the attacker at the manufacturing stage; (c) A possible MUX-based implementation of the Obfuscator

reasonable space. To maintain the same amount of hardware for implementing the LUT, the strong PUF must have the same n outputs as the weak PUF. The number of inputs of the strong PUF, however, is set to $n+p$, such that $p >> n$, thus the entire truth-table of the PUF (with 2^{n+p} CRPs) is dominated by the large value of p.

The functionality of the Obfuscator, shown in Fig 3(a), enables the designer to use any n-input subset of the entire CRP space of a strong PUF, by fixing the values of the p inputs during the activation process. Then, the PUF with the remaining n inputs constitutes a total of 2^n CRPs (as the case of employing a weak PUF). The configuration of the Obfuscator allows: 1) scrambling the n original wires, signals $\{x_1, x_2, ..., x_n\}$, such that a permutation of them appears at the inputs of the PUF; and 2) sending arbitrarily specified values to the other p inputs of the PUF. The specific configuration of each chip's Obfuscator is set by the *Selection bits* block, during the activation process. Fig. 3(b) shows two small examples of an Obfuscator with different configurations.

Various designs and implementations exist for the Obfuscator, but the crucial point is to ensure that: 1) the designer is able to select an arbitrarily CRP subset (with n inputs) of the strong PUF, during the activation process for each chip; and 2) the architecture does not reveal to the attacker, during the manufacturing stage, which subset of the PUF will be selected later on. As a result, the unique key per chip consists of two elements: 1) the subset selection of the strong PUF, configured by the content of the Selection bits (Key1); and 2) the content of the LUT (Key2) configured based on the functionality of the selected subset of the PUF.

At the manufacturing stage, the manufacturer cannot identify the subset of the strong PUF that will be used by the designer for each chip. Therefore, to get a full database of the CRP space of all PUFs, the attacker/manufacturer has to examine and store all 2^{n+p} possible CRPs for every PUF, which is prohibitively costly.

At the activation stage, by specifying the content of Selection bits (Key1), n inputs of the PUF will be connected to a permutation of the original wires of the master key, $\{x_1, x_2, ..., x_n\}$, and the rest will be fully specified for every chip. With the p fixed inputs, a total of 2^n CRPs remains for each PUF. The designer can then examine all 2^n CRPs (as

if it was a weak PUF) using the characterization channels to determine the content of the LUT (Key2) for every chip, so that the combination of the selected subset of the strong PUF and LUT is functionality equivalent with the withheld function.

It must be noted that while the designer needs to check 2^n different CRPs to derive the unique key of every chip at the activation stage, the attacker is forced to examine and store all 2^{n+p} possible CRPs ($p >> n$) for every PUF at the manufacturing stage. At the end of the activation process, all the characterization channels to the PUF are removed to prevent any further characterization of the PUFs for activated chips. This can be achieved by laser burning access wires or burning supporting fuses [19] [11].

Fig. 3(c) shows a general implementation of the Obfuscator block. This block can be implemented with $n+p$ Multiplexers (MUXes). Each MUX selects a few inputs (less or equal to n) from the original inputs of the master key and the fixed values (logic "1" or "0") to output to an input bit of the PUF.

Furthermore, this architecture can employ a relatively large number of "dummy" fan-outs, say q, from the original gates of the design, which will be connected to the inputs of the Obfuscator block (along with the n original signals) [8]. The propagation of these dummy fan-outs will be blocked by the Obfuscator. Therefore, if an attacker wants to model the entire scheme with a virtual LUT, the q dummy fan-outs increases the search space of the attacker exponentially (with $O(2^{n+q})$ complexity) at a linear hardware overhead of the fan-outs [8].

V. SECURITY ANALYSIS

A. Attack model

The proposed scheme and discussion of this paper is based on the assumptions that the attacker has: 1) the complete gate-level net-list; 2) full knowledge of the security scheme; 3) access to the activated chips from the open market; 4) access to the content of on-chip LUT (via side-channel attacks, etc); 5) access to the PUF characterization channels at the manufacturing stage. On the other hand, it is assumed here that the attacker *cannot*: 1) access the PUF characterization channels of legally activated chips; 2) fully characterize strong PUFs for **all** the chips at the manufacturing stage.

978-1-5386-7100-9/18 $31.00 © 2018 IEEE

B. Attack analysis upon a leaked key

In the proposed scheme, the unique key per chip consists of two parts: Key1, the content of the Selection bits, and Key2, the content of the LUT. Consider the worst case that the attacker has achieved a copy of the entire key, $\langle$Key1 , Key2$\rangle$, for a particular chip. Obviously, this key cannot be used directly to activate other chips. In order to recover the master key, the attacker has to examine the CRP space of the subset of the PUF that is used by the designer for the leaked chip. Next, we will discuss various possible attacks under such a scenario:

1) PUF characterization of chips in open market: After obtaining a leaked key $\langle$Key1 , Key2$\rangle$ from some chip, the attacker would be able to identify the subset of the PUF that is used by the designer for that chip (from analyzing Key1). However, as all the characterization channels of the PUFs have been removed at the end of the activation process, characterizing the PUF is no longer available.

2) PUF characterization of chips at manufacturing stage: At the manufacturing stage, the characterization channels are available to the attacker. However, since the chips are not activated yet, the attacker cannot have the "leaked key" to help indicate which subset of the PUF will be used. The huge CRP space of the strong PUFs ensures that it is prohibitively expensive to exhaustively examine all the CRPs even for a single PUF. As it was discussed in section II-B, it is also prohibitively expensive for a manufacturer to perform machine learning attacks on **all** the fabricated chips, where a secure strong PUF is employed.

3) SAT-based attacks: Without a direct way to obtain the CRP space of the PUFs, the attacker can model the entire security block (Obfuscator, PUF and the LUT) with a *virtual* LUT, and then try to find the content of such LUT by applying carefully designed primary inputs to a working chip and analyzing the values of the primary outputs [20]. The key idea to overcome such *SAT-based* attacks is to carefully select the withheld function at the design stage, so that the outputs of the LUTs become strongly correlated [5] [6] [8]. The proposed scheme in this work can work with many SAT-based prevention schemes to achieve a stronger framework. Furthermore, the designer can increase the number of q dummy fan-outs fed to the Obfuscator (as it was explained in section IV), so as to increase the cost of SAT-based attacks by enlarging the size of the virtual LUT exponentially at a linear cost.

C. Attack complexity

The attack complexity for PUF characterization (in terms of time and storage) at the manufacturing stage is $O(2^{n+p})$ multiplied by the number of fabricated chips. If the attacker decides to bypass the entire scheme, the attack model has to tackle an enlarged virtual LUT of $n + q$ address bits. This effectively boost the total search space to be $O(m \times 2^{n+q})$, which is 2^q times larger than the complexity of the hardware cost imposed on the designer's side. Therefore, the attacking costs both at the manufacturing stage and after obtaining a leaked key can be controlled by the designer through tuning the parameters p and q, respectively.

VI. Cost and Implementation Discussion

1) Hardware cost complexity: The hardware cost complexity in the general case of a withheld function with n inputs, m outputs along with the PUF and Obfuscator includes: the implementation cost of the LUT (Key2), of $O(m \times 2^n)$ complexity, plus the cost for the Obfuscator block, of $O((n+p) \times n)$ complexity (including $n + p$ MUXes and the interconnect network). The cost of the Selection bits block (Key1) would be of $O((n + p) \times log(n))$. The PUF architecture would have the cost complexity of $O(n + p)$. Overall, as it is shown in Fig. 2, the LUT cost of implementing the withheld function will remain unchanged (compared to the weak PUF implementation).

2) PUF reliability: One of the important factors that need to be considered for any PUF-based scheme is the noisiness of PUF devices. It is usually suggested that using an Error Correcting Code (ECC) [21] can compensate for the noisiness of the PUF. In this particular scheme, the existence of many redundant CRPs in a strong PUF can be used to reduce (not eliminate) the cost of ECC: it has been shown that some of the challenge bits of PUFs may show an unstable behavior for many cases. Such an issue can be handled by avoiding the use of the unstable challenge bits of the PUF, through employing an alternate, yet similar architecture, in which the LUT *precedes* the PUF. This way, the unstable challenge bits of the PUF can be avoided by configuring the content of the LUT (Key2) accordingly.

3) Performance overhead: From the timing perspective, replacing a part of the design with a PUF coupled with a LUT can significantly increase the delay of the overall design, if the withheld function lies in a critical path. In order to avoid such timing violations, one must carefully select the withheld function to avoid the critical paths.

VII. Evaluation

This section evaluates the effectiveness of the proposed scheme by comparing the hardware overhead for the designer and the cost to attack. The experiments are run on a number of ITC'99 benchmarks [22]. The size of the LUT (Key2) is fixed to be 32-bit ($n = 5$) for all the experiments. This is to ensure that the time complexity for the designer to derive the key is reasonable. The strong PUF is built based on the proposed XOR Arbiter in [17]. Fig. 4 shows the attacking cost against hardware overhead for two benchmarks [1]. Two types of attacks are considered: 1) the attack at manufacturing stage to obtain the entire-truth table information of a PUF, and 2) the attack to crack the security scheme after the key leaks out from some chip. Different curves on each plot are derived by increasing parameter p; thus, all points on each curve have the same attacking cost at the manufacturing stage (equal to 2^{n+p} CRPs). The cost to attack after a leaked key is obtained from one chip is depicted on the y-axis in terms of the size of the virtual LUT (determined jointly by parameters q and n) that must be cracked by the attacker.

---

[1]The reported hardware overheads do not take into account the implementation costs of Error Correcting Codes (ECC) to emphasize on the overhead of the proposed scheme; however, the ECC overhead must be added to the overall overhead of the system.

The logarithmic scale of the y-axis in Fig. 4 indicates that the cost to attack at the manufacturing stage and upon obtaining a leaked key can be increased exponentially at a linear hardware overhead. Also, it can be seen that given a fixed hardware overhead, the designer can increase the difficulty of one of the attacks at the cost of decreasing the difficulty of the other attack. However, the designer is in full control of tuning the parameters of the scheme (p, q, and n) at the design stage, so that both attacks become prohibitively expensive. Furthermore, Fig. 4 shows the scalability of the proposed scheme: the larger circuit (b19) can achieve the same level of security as a smaller circuit (b18) at a lower hardware overhead (6% for b19 compared to 12% for b18). Here, the total size of the key contains 312 flip-flops for both circuits, which is reasonable compared to the total number of flip-flops in each design.

VIII. CONCLUSION

In this paper, we argued that a weak PUF-based obfuscation is vulnerable to attackers that can: 1) pre-collect the entire truth-table of the PUFs for all the chips at the manufacturing stage with a reasonable cost; and 2) using the information of a leaked key from an activated chip, to recover the master key. The proposed approach engages a strong PUF in the obfuscation scheme, and therefore relies on the prohibitively large CRP space offered by the strong PUF to prevent IC piracy even at the case of a leaked key. Under the proposed framework, the cost to attack grows exponentially for the attacker, while the hardware cost grows only linearly for the designer. Therefore, it provides a potential viable and scalable solution to engage strong PUFs against IC piracy attacks.

REFERENCES

[1] M. Rostami, F. Koushanfar, J. Rajendran and R. Karri, "Hardware Security: Threat Models and Metrics", *IEEE/ACM International Conference on Computer-Aided Design (ICCAD)*, pp. 819–823, 2013.

[2] F. Koushanfar, "Hardware metering: A survey", in *Introduction to Hardware Security and Trust*, pp. 103–122, Springer, 2012.

[3] R. S. Chakraborty and S. Bhunia, "HARPOON: an obfuscation-based SoC design methodology for hardware protection", *IEEE Transactions on Computer-Aided Design of Integrated Circuits and Systems*, vol. 28, n. 10, pp. 1493–1502, 2009.

[4] Y. Alkabani and F. Koushanfar, "Active Hardware Metering for Intellectual Property Protection and Security", *USENIX Security*, pp. 291–306, 2007.

[5] J. Rajendran, Y. Pino, O. Sinanoglu and R. Karri, "Security Analysis of Logic Obfuscation", *IEEE/ACM Design Automation Conference (DAC)*, pp. 83–89, 2012.

[6] A. Baumgarten, A. Tyagi and J. Zambreno, "Preventing IC Piracy Using Reconfigurable Logic Barriers", *IEEE Design and Test Computers*, vol, 27, pp. 66–75, 2010.

[7] S. Zamanzadeh and A. Jahanian, "Higher security of asic fabrication process against reverse engineering attack using automatic netlist encryption methodology", *Microprocessors and Microsystems*, vol. 42, pp. 1–9, 2016.

[8] S. Khaleghi, K. D. Zhao and W. Rao, "IC Piracy Prevention via Design Withholding and Entanglement", *IEEE/ACM Asia and South Pacific Design Automation Conference (ASP-DAC)*, pp. 821–826, 2015.

[9] M. Joye, "Basics of Side-channel Analysis", *Cryptographic Engineering*, pp. 365–380, 2009.

[10] C. Herder, M. Yu, F. Koushanfar and S. Devadas, "Physical Unclonable Functions and Applications: A Tutorial", *proceedings of the IEEE*, vol. 102, pp. 1126–1141, May 2014.

[11] J.B. Wendt and M. Potkonjak, "Hardware Obfuscation using PUF-based Logic", *Computer-Aided Design (ICCAD), 2014 IEEE/ACM International Conference on*, pp. 270–271, November 2014.

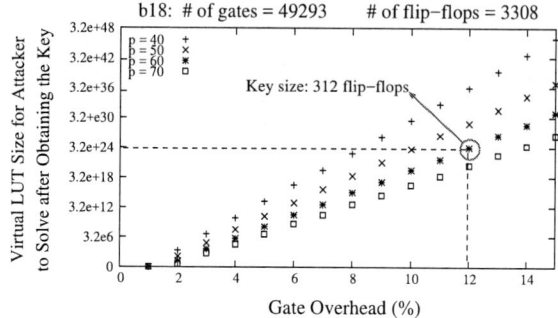

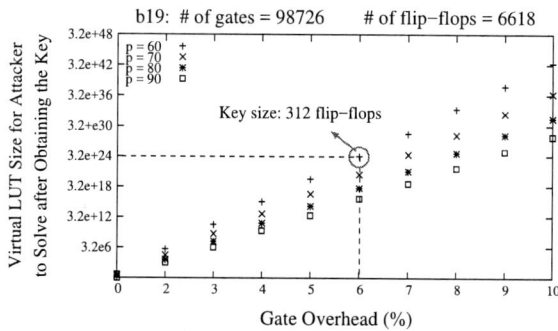

Fig. 4. The cost to attack vs. hardware overhead of the proposed scheme for different circuits

[12] U. Rhrmair, F. Sehnke, J. Slter, G. Dror, S. Devadas and J. Schmidhuber, "Modeling Attacks on Physical Unclonable Functions", *Proceedings of the 17th ACM conference on Computer and communications security*, pp. 237–249, 2010.

[13] D.E. Holcomb, W. Burleson and K. Fu, "Power-Up SRAM State as an Identifying Fingerprint and Source of True Random Numbers", *IEEE Transactions on Computers*, vol. 58, pp. 1198 – 1210, 2009.

[14] S. Devadas, E. Suh, S. Paral, R. Sowell, T. Ziola and V. Khandelwal, "Design and implementation of PUF-Based unclonable RFID ICs for anti-counterfeiting and security applications", *IEEE International Conference on RFID*, pp. 58–64, April 2008.

[15] S. Khaleghi, P. Vinella, S. Banerjee and W. Rao, "An STT-MRAM based strong PUF", in *Nanoscale Architectures (NANOARCH), 2016 IEEE/ACM International Symposium on*, pp. 129–134. IEEE, 2016.

[16] G. T. Becker, "The gap between promise and reality: on the insecurity of XOR arbiter PUFs", in *International Workshop on Cryptographic Hardware and Embedded Systems*, pp. 535–555. Springer, 2015.

[17] C. Zhou, K.K. Parhi and C.H. Kim, "Secure and Reliable XOR Arbiter PUF Design: An Experimental Study based on 1 Trillion Challenge Response Pair Measurements", *IEEE/ACM Design Automation Conference (DAC)*, 2017.

[18] M. Majzoobi, F. Koushanfar and M. Potkonjak, "Lightweight secure pufs", in *IEEE/ACM International Conference on Computer-Aided Design (ICCAD)*, pp. 670–673. IEEE, 2008.

[19] C. Helfmeier, D. Nedospasov, C. Tarnovsky, J.S. Krissler, C. Boit and JP. Seifert, "Breaking and entering through the silicon", in *Proceedings of the 2013 ACM SIGSAC conference on Computer & communications security*, pp. 733–744. ACM, 2013.

[20] U. Rührmair, J. Sölter, F. Sehnke, X. Xu, A. Mahmoud, V. Stoyanova, G. Dror, J. Schmidhuber, W. Burleson and S. Devadas, "PUF modeling attacks on simulated and silicon data", *IEEE Transactions on Information Forensics and Security*, vol. 8, n. 11, pp. 1876–1891, 2013.

[21] M. Yu and S. Devadas, "Secure and Robust Error Correction for Physical Unclonable Functions", *Design Test of Computers, IEEE*, vol. 27, n. 1, pp. 48–65, Jan 2010.

[22] F. Corno, M.S. Reorda and G. Squillero, "RT-level ITC'99 benchmarks and first ATPG results", *Design Test of Computers, IEEE*, vol. 17, n. 3, pp. 44–53, Jul 2000.

978-1-5386-7100-9/18 $31.00 © 2018 IEEE

Multi-Block APUF with 2-level Voltage Supply

Yunxi Guo
Electrical and Computer Engineering
Iowa State University
Ames, USA
Email: yunxig@iastate.edu

Timothy Dee
Electrical and Computer Engineering
Iowa State University
Ames, USA
Email: timdee@iastate.edu

Akhilesh Tyagi
Electrical and Computer Engineering
Iowa State University
Ames, USA
Email: tyagi@iastate.edu

Abstract—Physical Unclonable Functions (PUFs) are hardware cryptographic primitives for generating unique signatures from device manufacturing variations. Arbiter PUFs (APUFs) are a widely used class of PUF detecting process variations by exploiting the propagation delay differences between signals. However, both FPGA and ASIC implementations of APUFs suffer from systematic bias caused by either asymmetric routing or gradient effects in wafer doping. In this work, we introduce an improved APUF ASIC implementation achieving entropy enhancement without increasing area and power consumption significantly. In this design, a selector chain is divided into multiple blocks to avoid accumulation of systematic variation. Different voltage supplies are chosen for selector chain and arbiter circuit to overcome reliability problems produced by short chains. Cadence Monte Carlo sampling on 256-stage APUFs built in IBM $0.13\mu m$ technology shows the proposed Multi-Block (MB-) APUFs provide inter-chip uniqueness and reproducibility similar to double APUF (DAPUF); compared to DAPUF with similar uniqueness performance, MBAPUFs decrease area and power consumption by a factor of 2.

I. INTRODUCTION

Physical Unclonable Functions (PUFs) [1] are designed to extract physical randomness from the underlying circuit. PUFs have the advantage of generating chip-specific responses to the same challenge. Hence, they are normally used in device-level authentication and key generation applications [2]. Arbiter PUF (APUF) [3] is a timing-based PUF. Due to manufacturing variations, there are small random delay differences on symmetrical electrical paths. Theoretically, the entropy of the delays should be sufficient to ensure a unique PUF response for each individual device instance. However, previous work reports that conventional APUFs generate low-unique responses among devices [4], [5]. The main reason of this "non-ideal" uniqueness performance is systematic bias introduced by (1) unequal-length wiring and (2) non-stochastic doping process during chip manufacturing. To resolve this issue, Machida et al. proposed a DAPUF structure and which enhanced the APUF uniqueness by XORing the responses of duplicate selector chains [6]. This structure is technologically inexpensive because it does not require any special manufacturing processes to achieve this uniqueness improvement. However, it leads to large space and power overhead.

In order to use PUFs in hand held devices, their energy and power consumption must be managed. Thus, substantial effort has been devoted towards low power PUFs. For instance, Yanambaka et al. proposed a power optimized hybrid oscillator arbiter PUF using doping-less FET [7]. [8] developed a FinFET-based ring oscillator PUF, which reduced power

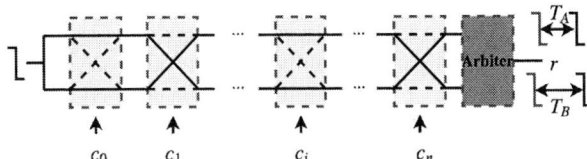

Fig. 1. An conventional Arbiter PUF with several switches (selector chain) and an arbiter

consumption significantly. Kumar *et al.* employed energy recovery techniques in a Quasi-Adiabatic logic based PUF [9]. Lin *et al.* proposed a low-power high-stability PUF utilizing common-source-amplifier [10].

Significant research efforts have focused on low-power and high-performance devices. The DAPUF's high power-consuption overhead prevents its use in power limited devices, such as implantable and wearable medical devices. In this paper, we propose an ASIC variant of DAPUF which can suppress the impact of systematic bias without increasing power consumption. Spectre simulations show that proposed MBAPUF provides uniqueness performance similar to DAPUF with smaller area and lower power consumption.

The key contributions of this paper are: (1) Constructing a model to estimate entropy enhancement limit. (2) Proposing an uniqueness improvement solution (MBAPUF) for APUFs without doubling power consumption. MBAPUFs divide the selector chain in conventional APUFs into several shorter chains to avoid bias aggregation. (3) Showing selection of separate voltage supplies for the selector chain and arbiter circuit prevents the arbiter circuit from working in a metastable state. (4) Verification of uniqueness and reproducibility performance of MBAPUF by Cadence Monte Carlo simulation.

This paper is organized as follows. Section II gives entropy model of multi-block APUF pattern. Section III evaluates the impact of systematic bias caused by transistor level parametric gradients and illustrates structure and design concerns of MBAPUFs. Section IV shows randomness, uniqueness, reproducibility and power consumption of proposed MBAPUFs. Sections V discuss future work and conclusions.

II. ENTROPY MODEL OF MULTI-BLOCK PATTERN

Conventional 1-bit APUF is composed of two identical delay paths and an arbiter circuit; they are usually implemented using a set of multiplexers and edge-triggered flip-

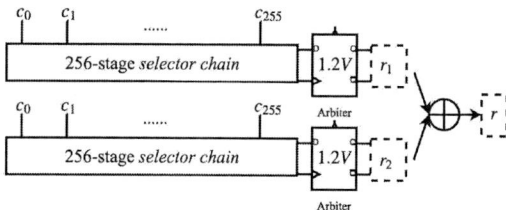

Fig. 2. An example of 2-1 DAPUF under 1.2V voltage supply

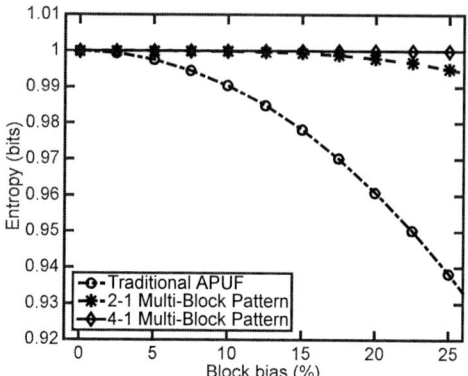

Fig. 3. Entropy versus block bias in APUFs and multi-block APUFs

flops respectively. Each multiplexer is a delay unit and each pair of multiplexers sharing same input is a stage.

At the beginning, the same signal is injected into these two delay paths. Then the actual signal propagation paths are determined by a set of external inputs called a challenge $C = (c_0, c_1, ...c_n)$. The length of the challenge should be equal to the number of stages. At each stage, the two signals follow straight paths when the challenge bit $c_i = 1$. $c_i = 0$ causes the top and bottom path signals to switch. The arbiter circuit detects the arrival sequence of the racing signals (top or bottom signal in Fig. 1) and determines the response (r) of the 1-bit APUF according to Eq. 1.

$$r = \begin{cases} 1 & if\ T_A > T_B \\ 0 & else \end{cases} \qquad (1)$$

To achieve high uniqueness, signals propagating in the two delay paths should have equal opportunity to be the winner regardless of what challenge is given. However, this is not true in practice. Over all chips fabricated with same process and layout, unbalanced routing and transistor-level systematic mismatch introduce speed bias in each stage; the top or bottom multiplexer is likely to be faster. Attackers are able to predict one chip's response according to responses of other chips built from the same technology.

For example, if there is a batch of 1-bit APUF chips with top routing longer than bottom routing, for some specific challenge sequence such as $C_s = (1, 1, ...1)$, the bottom signal has more than 50% probability to win the race in signal propagation. Therefore, most chips in this batch will produce responses equal to 1 with this challenge. For certain challenges, some responses are generated with high probability for all chips in the same batch of APUFs. Attacker can build a machine learning model of conventional APUFs by exploiting this correlation between challenges and responses [11], [12].

To reduce the response probability bias, the authors of [6] propose a modeling attack resistant APUF FPGA implementation - DAPUF. DAPUFs are composed of several duplicate APUFs on neighboring SLICEs. All the duplicates have the same number of stages and share the same challenge C. Response of a 1-bit DAPUF is the XOR of all the duplicates' responses. The DAPUFs with n duplicates are defined as n-1 DAPUFs. An example of 1-bit 2-1 DAPUF is shown in Fig. 2. Each duplication in DAPUF is defined as a block. The pair of delay paths in each block is a selector chain.

All PUFs that generate responses by XORing several nearby blocks' responses can be considered as built in multi-block pattern. DAPUF is a representative example of a PUF built

in multi-block pattern. In FPGA DAPUF, all the blocks are placed on adjacent SLICEs and they share the same challenge. As a result, we can expect them to have similar block-level bias [13]. Assume that the probability of getting response 1 from 1-bit block is $P_b(1)$, then the probability that a 1-bit APUF built in n-1 multi-block pattern generates response 1 ($P_n(1)$) can be computed with $P_{n-1}(1)$ and $P_b(1)$ recursively (Eq. 2).

$$P_n(1) = P_{n-1}(0) \times P_b(1) + P_b(0) \times P_{n-1}(1) \qquad (2)$$

After obtaining $P_n(1)$, entropy of this multi-block pattern is calculated according to a binary entropy function [14]. Fig. 3 shows the response entropy of traditional APUFs, APUFs work in 2-1 multi-block pattern and APUFs work in 4-1 multi-block pattern. Response entropy varies with percentage of block bias. As 50% is the ideal value of $P_b(1)$, block bias is defined as $b_b = P_b(1) - 50\%$. Block bias less than 15% is handled perfectly by 2-1 multi-block pattern; more blocks are necessary only when each block is more heavily biased, as shown in Fig. 3.

Although uniqueness is drastically improved in n-1 DA-PUFs, their area and power consumption are also significantly increased compared to conventional APUFs. In this paper, we design a low-power high-unique Multi-Block APUF (MBA-PUF) structure with much less power and area consumption. Uniqueness and reproducibility of its response are verified by Spectre simulation.

III. MULTI-BLOCK APUF DESIGN

A. Systematic Bias in ASIC

In contrast to FPGA APUFs, bias in ASIC APUFs is dominated by non-stochastic doping instead of unbalanced routing. Before applying multi-block pattern to ASIC design, we need to verify: (1) the amount of ASIC systematic bias is not negligible and (2) the adjacent blocks still have similar block bias in ASIC implementation.

According to [15], the amount of delay variation caused by bias doping is subject to normal distribution. This doping bias results in the transistor mismatch. Some work by Felt et. al [16] shows that the magnitude of matching errors

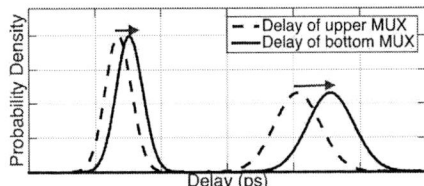

Fig. 4. Impact of linear gradient effect on 1-stage of MUX pairs (left) and 3-stage of MUX pairs (right)

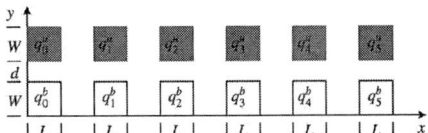

Fig. 5. Layout of a 6-stage Arbiter PUF

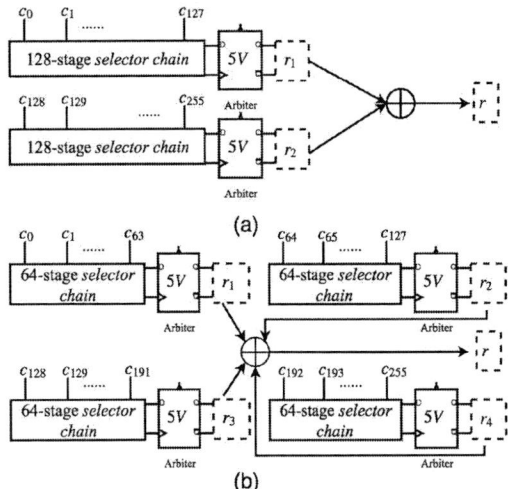

Fig. 6. (a) 256-stage 2-1 Multi-Block APUF (b)256-stage 4-1 Multi-Block Arbiter PUF

associated with systematic variations are comparable to that of the random variations even with good layout strategies. Pelgrom's mismatch model [17] states that random variations decrease with transistor size; systematic variation is proportional to the distance between circuit blocks. Hence, delay of neighboring upper and bottom multiplexers could be modeled as two Gaussian distributions with the same standard deviation but different means. The mean difference between them increases with the distance between their layouts. Hence, the neighbouring ASIC blocks should have similar block bias.

Suppose we have a wafer with threshold gradients where $\mu(V_{th})$ decreases along the y axis. Then the average delay of upper multiplexers would be smaller than the bottom ones. In an n-stage APUF, assume the delay distribution of each upper multiplexer (q_x^u) is $X \sim N(\mu_1, \sigma^2)$ and delay distribution of each bottom multiplexer (q_x^b) is $Y \sim N(\mu_2, \sigma^2)$. If all challenge bits equal to 1, delay distribution of top and bottom path equal $\sum_{i=1}^n X \sim N(n\mu_1, \sqrt{n}\sigma^2)$ and $\sum_{i=1}^n Y \sim N(n\mu_2, \sqrt{n}\sigma^2)$ respectively. Fig. 4 is a sketch of 1-stage and 3-stage selector chain delay distribution. The bias is magnified for clarity. After 3 stages of aggregation, delay bias is obviously enlarged.

In ASIC layout, a pair of racing multiplexers are usually placed next to each other to reduce mismatch, as shown in Fig. 5. Although systematic mismatch in this kind of layout is very small for a single stage, its impact becomes significant after several stages of accumulation.

B. MBAPUF

In our design, we divide the selector chain into n shorter chains and XOR their responses; this saves power and area over the alternative duplication of blocks. If delay units are placed as close as possible, shorter chains inherently scale down cumulative systematic bias. This low power multi-block APUF approach is called MBAPUF. In MBAPUF, each block contains a sub-chain of APUF selector chain. Structures of 256-stage 2-1 and 256-stage 4-1 MBAPUF are shown in Fig. 6(a) and (b).

In an m-stage n-1 MBAPUF, each block only takes a portion of challenge bits (m/n bits). Block bias depends on both systematic pattern and challenge input. In a single MBAPUF chip, each block may exhibit different bias level even with similar layouts. This is desired because it is harder for attackers to model a combination of different block biases than a combination of equal ones. But in the worst case, challenge input of all blocks are the same. For example, with challenge $C_s = (c_0, c_1, ...c_{127}; c_0, c_1, ...c_{127})$, blocks in a 256-stage 2-1 MBAPUF are equally biased. Considering the worst situation, uniqueness enhancement performance of MBAPUFs should be similar to that of DAPUFs.

C. 2-Level Voltage Supply

Although power and area consumption are reduced with shorter chain design, it leads to low signal-to-noise ratio (SNR). When racing path delay difference is not significant, the arbiter circuit is driven into a metastable state [18], [19] and creates incorrect bits (b_I). Shorter chains reduce average delay difference; this injects more noise bits into PUF response.

The SNR can be improved using high-resolution flip-flops [20], [21], but this requires circuit re-design and makes MBAPUF infeasible for FPGA implementation. In the light of [22], a latch or flip-flop during a metastable state should be regarded as a linear amplifier. For widely used cross-coupled CMOS NAND arbiters [23], their resolutions are proportional to $\sqrt{g_m/C}$ [24], where g_m is transconductance of the inverter and C is output node capacitance. Since g_m increases with $V_{dd} - V_{th}$, resolution of the arbiter circuit increases with supply voltage. Hence, high resolution arbiters are possible without changing its schematic design.

In MBAPUF, we choose different supply voltages for arbiters and selector chains. An inverter needs to be placed between each pair of arbiter and selector chain for voltage conversion. This doesn't harm uniqueness performance and

978-1-5386-7100-9/18 $31.00 © 2018 IEEE

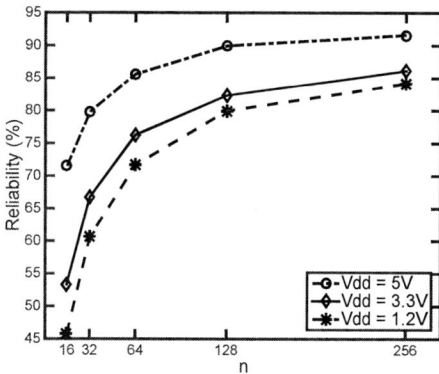

Fig. 7. Valid rate in traditional APUF with arbiter operating under different voltage supplies. n is the length of selector chain.

adds more delay variability. Uniqueness of such design is verified by Cadence simulation; the simulation results are shown in Section IV-C.

The n-1 MBAPUF with 2-level voltage supply only requires $n-1$ additional arbiters compared to conventional APUFs. Even though power consumption of each arbiter is higher, the total power consumed is still smaller than 2-1 DAPUFs because the number of transistors in selector chain is halved.

IV. PERFORMANCE EVALUATION

One voltage supply must be chosen for arbiter circuits, it must be shown to provide significant resolution enhancement to maintain MBAPUF reliability. The reliability of MBAPUF is defined as the probability that all arbiters can successfully identify the faster signal. The impact of the multi-block pattern on PUF's other requisite properties should also be evaluated.

Cadence Spectre simulations are used to generate raw delay data and to assess arbiter resolution separately. Delay variability assessment is conducted by 3σ Monte Carlo sampling over process parameters. Arbiter resolution is tested with ideal rising edges. The device models used are from the IBM 130nm PDK library. Random mismatch model is included in this PDK.

We construct two 128-stage and two 256-stage APUFs accepting a 256-bit input with a 256-bit output. DAPUF and MBAPUF outputs can be computed by incorporating result of multiple APUFs.

Since Monte Carlo simulation here doesn't consider long correlation distance mismatch by default, systematic variation is simulated by modifying transistor model file manually. The standard deviation of threshold gradient (σV_{th}) is derived with mismatch coefficients acquired from [25]. Layout distance between racing multiplexers is assumed to be equal to 1.3um, which is derived from multiplexer stick diagram and CMOS layout rules. This distance should be larger in FPGA or ASIC without precise layout modifications.

The primary simulation procedure is: (1) Monte Carlo sampling 200 times on the top path with $C_i = (1, 1, 1...1)$. Export transient data, get delay of top multiplexers as $TD = (d_{t,0}, d_{t,1}, ...d_{t,n-1})$. (2) Replace $vth0$ in model file with

$vth0 + \Delta V_{th}$, Monte Carlo sampling 200 times on the bottom paths with the same C_i. Get delay of all bottom multiplexers as $BD = (d_{b,0}, d_{b,1}, ...d_{b,n-1})$. TD and BD are defined as multiplexer delay matrix. In each time of Monte Carlo sampling, ΔV_{th} is random chosen from $[0, 3\sigma Vth]$ by assuming V_{th} decreases along y-axis. Result of each time of Monte Carlo Sampling is considered as data of a different chip. With TD and BD, delay of two signal propagation paths (D_1 and D_2) in each chip are calculated according to input challenge.

A. Block Bias

As we have concluded in Section II, more than 2 blocks are not necessary unless block bias is higher than 15%. By assuming V_{th} decreases along y-axis, top multiplexers tend to be faster than bottom multiplexers. Block bias is maximized when 100% of challenge bits are equal to 1.

To determine appropriate n for 256-stage n-1 MBAPUF, we evaluate block bias in 128 and 256 stage selector chains on different challenges.

TABLE I
BLOCK BIAS UNDER DIFFERENT CHALLENGE INPUT

$P(C_i = 1)$	256-stage APUF	128-stage APUF
100%	9.03%	5.88%
75%	4.70%	2.72%
50%	-0.28%	-0.52%

Table I shows block bias in 256 paths of all 200 chips (Monte Carlo samples). The maximum block bias caused by transistor-level systematic mismatch is about 9% and 6% for 256-stage and 128-stage selector chains. Both of them are smaller than 15% implying that the 2-block structure is sufficient for bias correction in ASIC design. All following discussions focus on 256-stage 2-1 MBAPUFs.

B. Reliability Test

We construct a cross-coupled CMOS NAND arbiter for reliability tests. Its resolution is tested with two ideal rising edges (E_1 and E_2). For each supply voltage, the delay between E_1 and E_2 is decreased until the arbiter circuit produces an incorrect output. This delay threshold is defined as δt. The simulation result shows varying the voltage supply from 1.2V to 5V results in δt changing from 65 ps to 30 ps.

Threshold voltage in the top multiplexers is more likely to be lower than in the bottom multiplexers. Thus, difference between the two paths' arrival times reaches a minimum when the challenge contains half 1s and half 0s. Hence we compute the delay difference between two propagation signals of all 200 256-bit APUFs with $(c_0, c_1...c_{127}) \mapsto (1, 1, ...1)$, $(c_{128}, c_{129}...c_{255}) \mapsto (0, 0, ...0)$ and multiplexer delay matrices getting from Monte Carlo Sampling. Then, check what percent of racing paths create delay differences larger than the arbiter resolution (δt). This percentage is the valid rate.

MBAPUFs produce an invalid response bit as long as one of the block responses is invalid. If the valid rate of a single block is VR_{Block}, then valid rate of n-1 MBAPUF is estimated by Eq. 3.

$$VR = (VR_{block})^n \qquad (3)$$

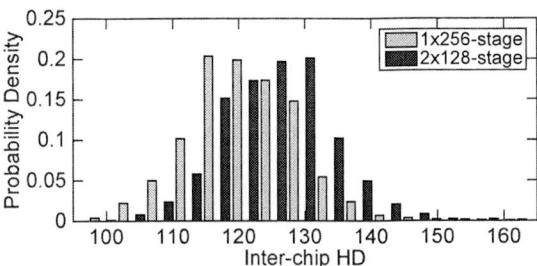

Fig. 8. Hamming distance distribution of 256-stage APUFs and 128-stage 2-1 MBAPUFs

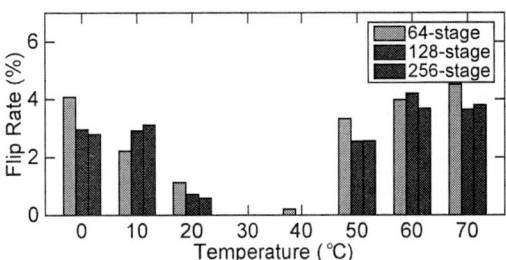

Fig. 9. Bit flip rate of different stages Arbiter PUFs under temperature variation

In our design, supply voltage of the selector chain is set to 1.2V to produce large delay and delay variation. Fig. 7 shows the valid rate of n-stage APUF with arbiter operating under different voltage supply. According to the plot, the valid rate of 128-stage APUF with 5V arbiter voltage supply is 89.94%. According to Eq. 3, valid rate of corresponding 256-stage 2-1 MBAPUF is $89.94\%^2 = 81\%$, which is very close to the valid rate of 256-stage conventional APUF with 1.2V arbiter voltage supply (83.18%). Hence, in 2-1 256-stage MBAPUFs with 1.2V selector chain, we choose 5V as arbiter voltage supply to achieve similar reliability performance as conventional APUFs.

C. Uniqueness test

We estimate the uniqueness of APUF, DAPUF and MBA-PUF by the average inter-die Hamming distance (HD) over a group of chips (Monte Carlo Samples). With a pair of chips, i and j $(i \neq j)$, both having n-bit response R_i and R_j respectively, the average inter-die HD among a group of k chips is defined as:

$$HD_{avg} = \frac{2}{k(k-1)} \sum_{i=1}^{k-1} \sum_{j=i+1}^{k} \frac{HD(R_i, R_j)}{n} \times 100\% \quad (4)$$

To simulate a heavily biased situation, responses used for uniqueness validation are generated with challenge $(c_0, c_1, ...c_{255}) \mapsto (1, 1, ...1)$. For n-bit PUFs, HD should be as close as possible to $n/2$. As shown in Fig. 8, mean of HD is left shifted by around 6 bits in 256-stage APUFs due to systematic variation. 256-stage 2-1 MBAPUF corrects this shifting.

TABLE II
INTER-DIE HD OF DIFFERENT APUF STRUCTURES

	μ(HD)	σ(HD)
256-stage APUF	122.01	8.05
128-stage APUF	125.27	8.26
256-stage 2-1 MBAPUF	127.89	7.99
256-stage 2-1 DAPUF	127.61	8.01

Table II shows average HD for 256-stage and 128-stage traditional APUF, 256-stage 2-1 MBAPUF and 256-stage 2-1 DAPUF. For 256-stage 2-1 DAPUFs, the mean HD is 127.61 bits with a standard deviation of 8.01 bits. For 256-stage

2-1 MBAPUF, these values are 127.89 bits and 7.99 bits, respectively. From this we can say, MBAPUFs provide similar uniqueness performance as the DAPUFs with same number of blocks.

D. Reproducibility test

The usefulness of a single PUF relies on it producing a consistent response to a challenge; it should be independent from the environment. Tests are performed subjecting MBAPUF to temperature variation. The frequency of response bit flips is quantified.

Bit flip rate is the frequency a bit changes from $0 \mapsto 1$ or $1 \mapsto 0$. It is computed relative to some baseline response. Gathering responses at common room temperature ($27°C$) establishes this baseline. The percentage of path delays where a bit flips is the bit flip rate.

To simulate performance under different temperatures, multiplexer delay matrices of n-stage selector chains are gathered by Monte Carlo sampling at $0°C \rightarrow 70°C$. The same seed is used in all of these sampling instances to make sure no other parameter other than temperature is changed.

Flip rate of 64-stage, 128-stage and 256-stage selector chains are computed with challenge $(c_0, c_1, ...c_{255}) \mapsto (1, 1, ...1)$, shown in Fig. 9. While flip rate of 64-stage selector chains is slightly higher, flip rate of 128-stage and 256-stage chains are quite similar. With flip rate of one selector chain (FR_b), flip rate of 2-1 DAPUF and MBAPUF can be calculated with Eq. 5.

$$FR_{MBAPUF} = 2 \times FR_b \times (1 - FR_b) \quad (5)$$

The maximum difference between flip rate of 128-stage and 256-stage selector chains is 0.52% with $FR_{b,128} = 4.2\%$ and $FR_{b,256} = 3.7\%$. Corresponding flip rate of 2-1 MPAPUFs and 2-1 DAPUF are 8.1% and 7.1%. This indicates reproducibility of MBAPUF is similar to DAPUFs.

E. Randomness test

Output of a good PUF should look like a pseudo-random number generator so that an attacker cannot model it easily. Assessment of MBAPUF randomness uses data from Monte Carlo sampling. All 200 256-bit responses are examined using NIST statistical test suite. The MBAPUF responses pass all selected NIST statistical tests (Frequency, BlockFrequency, CumulativeSumes, Runs, Serial and Rank).

F. Area & Power

Compared to DAPUF, major advantages of MBAPUF are low area cost and low power consumption. We evaluate these metrics for both DAPUFs and MBAPUFs. Relevant results are shown in Table III.

TABLE III
AREA AND POWER COST OF DIFFERENT APUFs

	Area (gate)	Power (watt)
256-stage APUF	1667	0.2310
256-stage 2-1 MBAPUF	1670	0.2574
256-stage 2-1 DAPUF	3334	0.4620

We estimate area of 1-bit MBAPUF and DAPUF in number of gate equivalents. In our implementation, each pair of multiplexers contains 26 transistors and each arbiter is composed of 12 transistors. Compared to 2-1 DAPUFs, area of 256-stage 2-1 MBAPUFs is reduced by about 50%.

Since static power consumption is very small, only transient power dissipation is evaluated.

Transient power estimation is conducted on multiplexer and arbiter separately. It is performed in 3 steps: (1) Plot transient current in cadence ADE L, to find its peak. (2) Multiply peak transient current with voltage supply. (3) Sum up power consumption of all circuit blocks.

According to Cadence transient simulation results, the power dissipation of MBAPUFs with 2-level voltage supply is only 55.71% of DAPUFs with a similar uniqueness performance.

V. CONCLUSIONS AND FUTURE WORK

We implemented 256-stage 2-1 MBAPUF and 256-stage 2-1 DAPUF in 0.13um technology, and evaluated their uniqueness, reproduciblity, area and power consumption. From the experimental results, we confirmed that 256-stage 2-1 MBAPUF generates responses with higher uniqueness than conventional APUFs; in both MBAPUF and DAPUF, 2-block are sufficient to correct bias caused by transistor-level systematic mismatch; good reproducibilty and SNR are maintained in MBAPUF with 2-level voltage supply; area and power consumption of MBAPUF is reduced by a factor of 2 compared to DAPUF with similar inter-chip uniqueness performance. In addition, whether the proposed MBAPUF can work well on an FPGA is an interesting question. More work on evaluating MBAPUFs on FPGA platforms is needed in the future.

REFERENCES

[1] S. Joshi, S. P. Mohanty, and E. Kougianos, "Everything you wanted to know about pufs," *IEEE Potentials*, vol. 36, no. 6, pp. 38–46, 2017.

[2] V. P. Yanambaka, S. P. Mohanty, E. Kougianos, and J. Singh, "Secure multi-key generation using ring oscillator based physical unclonable function," in *Nanoelectronic and Information Systems (iNIS), 2016 IEEE International Symposium on*. IEEE, 2016, pp. 200–205.

[3] G. E. Suh and S. Devadas, "Physical unclonable functions for device authentication and secret key generation," in *Proceedings of the 44th annual Design Automation Conference*. ACM, 2007, pp. 9–14.

[4] A. Maiti, V. Gunreddy, and P. Schaumont, "A systematic method to evaluate and compare the performance of physical unclonable functions," in *Embedded systems design with FPGAs*. Springer, 2013, pp. 245–267.

[5] Y. Hori, T. Katashita, and K. Kobara, "Performance evaluation of physical unclonable functions on kintex-7 fpga," *IEICE Technical Report of RECONF*, vol. 113, pp. 91–96, 2013.

[6] T. Machida, D. Yamamoto, M. Iwamoto, and K. Sakiyama, "Implementation of double arbiter puf and its performance evaluation on fpga," in *Design Automation Conference (ASP-DAC), 2015 20th Asia and South Pacific*. IEEE, 2015, pp. 6–7.

[7] V. P. Yanambaka, S. P. Mohanty, E. Kougianos, P. Sundaravadivel, and J. Singh, "Dopingless transistor based hybrid oscillator arbiter physical unclonable function," in *VLSI (ISVLSI), 2017 IEEE Computer Society Annual Symposium on*. IEEE, 2017, pp. 609–614.

[8] V. P. Yanambaka, S. P. Mohanty, and E. Kougianos, "Making use of semiconductor manufacturing process variations: Finfet-based physical unclonable functions for efficient security integration in the iot," *Analog Integrated Circuits and Signal Processing*, vol. 93, no. 3, pp. 429–441, 2017.

[9] S. D. Kumar and H. Thapliyal, "Qualpuf: A novel quasi-adiabatic logic based physical unclonable function," in *Proceedings of the 11th Annual Cyber and Information Security Research Conference*. ACM, 2016, p. 24.

[10] S. Lin, X. Zhao, B. Li, and X. Pan, "An ultra-low power common-source-amplifier-based physical unclonable function," in *Electron Devices and Solid-State Circuits (EDSSC), 2015 IEEE International Conference on*. IEEE, 2015, pp. 269–272.

[11] U. Rührmair, J. Sölter, F. Sehnke, X. Xu, A. Mahmoud, V. Stoyanova, G. Dror, J. Schmidhuber, W. Burleson, and S. Devadas, "Puf modeling attacks on simulated and silicon data," *IEEE Transactions on Information Forensics and Security*, vol. 8, no. 11, pp. 1876–1891, 2013.

[12] U. Rührmair, F. Sehnke, J. Sölter, G. Dror, S. Devadas, and J. Schmidhuber, "Modeling attacks on physical unclonable functions," in *Proceedings of the 17th ACM conference on Computer and communications security*. ACM, 2010, pp. 237–249.

[13] T. Machida, D. Yamamoto, M. Iwamoto, and K. Sakiyama, "A new mode of operation for arbiter puf to improve uniqueness on fpga," in *Computer Science and Information Systems (FedCSIS), 2014 Federated Conference on*. IEEE, 2014, pp. 871–878.

[14] D. J. MacKay, *Information theory, inference and learning algorithms*. Cambridge university press, 2003.

[15] B. Zhou and A. Khouas, "Measurement of delay mismatch due to process variations by means of modified ring oscillators," in *Circuits and Systems, 2005. ISCAS 2005. IEEE International Symposium on*. IEEE, 2005, pp. 5246–5249.

[16] E. Felt, A. Narayan, and A. Sangiovanni-Vincentelli, "Measurement and modeling of mos transistor current mismatch in analog ic's," in *Proceedings of the 1994 IEEE/ACM international conference on Computer-aided design*. IEEE Computer Society Press, 1994, pp. 272–277.

[17] M. Conti, P. Crippa, S. Orcioni, and C. Turchetti, "Statistical modeling of mos transistor mismatch based on the parameters' autocorrelation function," in *Circuits and Systems, 1999. ISCAS'99. Proceedings of the 1999 IEEE International Symposium on*, vol. 6. IEEE, 1999, pp. 222–225.

[18] V. Stojanovic and V. G. Oklobdzija, "Comparative analysis of master-slave latches and flip-flops for high-performance and low-power systems," *IEEE Journal of solid-state circuits*, vol. 34, no. 4, pp. 536–548, 1999.

[19] D. Suzuki and K. Shimizu, "The glitch puf: A new delay-puf architecture exploiting glitch shapes." in *CHES*, vol. 10. Springer, 2010, pp. 366–382.

[20] U. Ko and P. T. Balsara, "High-performance energy-efficient d-flip-flop circuits," *IEEE Transactions on Very Large Scale Integration (VLSI) Systems*, vol. 8, no. 1, pp. 94–98, 2000.

[21] G. Yin, F. O. Eynde, and W. Sansen, "A high-speed cmos comparator with 8-b resolution," *IEEE Journal of Solid-State Circuits*, vol. 27, no. 2, pp. 208–211, 1992.

[22] L.-S. Kim and R. W. Dutton, "Metastability of cmos latch/flip-flop," *IEEE Journal of solid-state circuits*, vol. 25, no. 4, pp. 942–951, 1990.

[23] J. Delvaux and I. Verbauwhede, "Side channel modeling attacks on 65nm arbiter pufs exploiting cmos device noise," in *Hardware-Oriented Security and Trust (HOST), 2013 IEEE International Symposium on*. IEEE, 2013, pp. 137–142.

[24] T. Sakurai, "Optimization of cmos arbiter and synchronizer circuits with submicrometer mosfets," *IEEE journal of solid-state circuits*, vol. 23, no. 4, pp. 901–906, 1988.

[25] P. R. Kinget, "Device mismatch and tradeoffs in the design of analog circuits," *IEEE Journal of Solid-State Circuits*, vol. 40, no. 6, pp. 1212–1224, 2005.

2018 IEEE Computer Society Annual Symposium on VLSI

Write energy optimization for STT-MRAM cache with data pattern characterization

Bi Wu[†‡], Xiaolong Zhang[†‡], Yuanqing Cheng[‡], Zhaohao Wang[†‡]
Dijun Liu[*], Youguang Zhang[†‡] and Weisheng Zhao[†‡]
[†]Fert Beijing Institute, BDBC, Beihang University, Beijing, 100191, China
[‡]School of Electronic Information and Engineering, Beihang University, Beijing, 100191, China
[*]China Academy of Telecommunication Technology(CATT), Beijing, China
E-mail:{Bi.wu, Yuanqing, zhaohao.wang, weisheng.zhao}@buaa.edu.cn

Abstract—Traditional memory technologies face severe challenges meeting the ever increasing power and memory bandwidth requirements for high performance computing and big-data analysis. Several emerging memory technologies, as promising candidates to replace SRAM or DRAM, have advanced fast. Among them, STT-MRAM can be used to replace SRAM for on chip cache. However, it suffers from high write energy and latency problems. In the paper, we investigate the data patterns written back from SRAM based L1 cache to STT-MRAM based L2 cache to explore the write energy reduction potential. Depending on the data layout within a cache line, redundant bits can be identified and eliminated from write back operations to save STT-MRAM write energy with only a small area overhead. The simulation results validate the effectiveness and efficiency of our proposed method.

I. INTRODUCTION

As technology node enters deep sub-micron regime, leakage power becomes the bottleneck for large scale transistor integration. Especially for on chip cache, leakage power has already dominated whole cache power consumption, which limits cache capacity from further increasing. Even worse, the emerging "Dark Silicon" problem elevates leakage power issue further. To attack power scaling problem, several emerging memory technologies are proposed as potential replacements of SRAM and/or DRAM, such as Spin-torque transfer MRAM (STT-MRAM) [1], [2], phase change memory (PCRAM) [3] and resistive memory [4]. Among them, STT-MRAM has fast access speed, ultra low leakage power, and compatibility with CMOS fabrication process. Therefore, it is one of the most promising candidates for on chip cache.

Although STT-MRAM is a hot candidate for cache, it still suffers from several challenges. First, the write energy is significantly higher than SRAM due to large critical switching current. Second, the relative long write latency may degrade cache performance. There are already many existing literatures trying to solve these problems, such as [5], [6], etc. However, most of them either require complex circuit level design incurring non-negligible area overhead or resort to changing MRAM memory cell organization which increases design complexity significantly. In the paper, we explore write back energy reduction problem from the architecture level, which can be integrated easily into contemporary cache controller. In this work, L1 cache is assumed to be made of SRAM as

the read/write speed of L1 cache is critical for performance. L2 cache is STT-MRAM such that the static power can be reduced dramatically.

By investigating write back data characteristics, we observe two interesting phenomena. First, only a small fraction of the whole cache line changes when data is written back from L1 to L2 cache. Second, it is observed that a few data patterns appear much more frequently than others. Based on them, we propose a method to characterize write back data pattern, and capture the dominant appearing data patterns. By eliminating redundant bits in the cache line with the dominant data pattern, significant write energy of STT-MRAM based L2 cache can be saved.

Our contributions are summarized as follows,

- Through in-depth investigation on write back data characteristics, we observe that there are a lot of redundant data in the cache line. Based on the observation, we propose a data pattern characterization method to identify predominantly occurring data patterns and eliminate redundant bits from cache lines with these patterns. Therefore, write back energy can be reduced.
- We explore the trade-off between energy reduction and associated area overhead induced by our proposed method such that write energy of STT-MRAM L2 cache can be reduced dramatically with only 1.6% area overhead.

The rest of paper is organized as follows. Section II presents preliminaries of STT-MRAM and related work. Section III motivates our work based on several key observations derived from STT-MRAM unique access characteristics and cache write back activities. Section IV describes our proposed method to capture write back data patterns and reduce write energy by eliminating redundant data bits within matching cache lines. Section V presents experimental results, which validate the effectiveness of our proposed method. Section VI concludes the paper.

II. PRELIMINARIES & RELATED WORK

A. Preliminaries of STT-MRAM

STT-MRAM cell mainly consists of an oxide insulating layer(e.g., MgO) sandwiched by two ferromagnetic layers(e.g., CoFeB). This stacking structure forms the so-called MTJ

978-1-5386-7100-9/18 $31.00 © 2018 IEEE

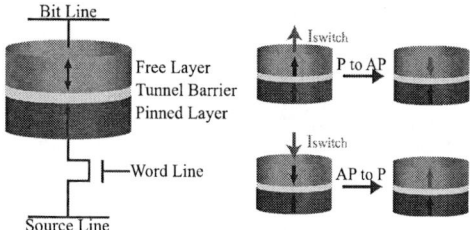

Fig. 1. STT-MRAM cell structure and write process [7]

(magnetic tunneling junction). One of the two ferromagnetic layers is called fixed layer, whose magnetization is fixed. The other one is called free layer as its magnetization can be changed by injected spin-polarized current. Depending on the magnetization direction of free layer, MTJ can have two states. If the magnetization of free layer is the same as that of fixed layer, MTJ manifests low resistance, and is treated as logic '0'. Otherwise, it is treated as logic '1'.

The commonly used 1T-1MTJ STT-MRAM cell is shown in Fig. 1. To access the cell, word line voltage becomes high, and the transistor is turned on [8]. When spin-polarized write current flows from bitline (BL) to source line (SL), MTJ is switched to parallel state and '0' is written into the memory cell. Otherwise, MTJ is switched to anti-parallel state and '1' is written. To read the data out of the memory cell, a sensing current flows through MTJ. Depending on the MTJ resistance, sensing current magnitude can distinguish whether a '0' or '1' is read.

B. Related Work

Many papers in literature have explored the possibility for STT-MRAM to replace SRAM as on-chip cache. Arezoomand et al. investigated using MRAM as L2 cache for 3D CMPs [9]. With the aid of some write optimizations, MRAM based L2 cache may achieve almost the same performance as SRAM based cache but with significant lower power consumption. Coi et al. proposed a hybrid MRAM/SRAM architecture to exploit MRAM's low power and SRAM's high speed access in combination. By inter-level and intra-level hybrid construction, both IPC and power can be improved compared to the pure SRAM counterpart [10]. Dong et al. evaluated the potential of STT-MRAM as the replacement for SRAM from circuit level up to architecture level and showed the performance and static power improvements [11]. Smullen et al. explored to relax MTJ retention time for higher access performance [12]. However, these papers mainly focus on the evaluation of performance improvement and leakage power reduction when STT-MRAM replaces SRAM as on-chip cache without considering write energy optimization in-depth.

III. MOTIVATION

A. Write Related Challenges of STT-MRAM

In the paper, we assume that L1 cache is made up of SRAM due to the fast access requirement, and L2 cache uses STT-MRAM to reduce leakage power as in [11]. L2 cache line is assumed to be 64 bytes, and associativity is 16. The capacity

considered ranges from 512KB to 32MB which are typical values for contemporary L2 cache configuration. The STT-MRAM cell parameters used for evaluation are listed in Table I according to [13], [14].

TABLE I
STT-MRAM CELL PARAMETERS ACCORDING TO [13], [14]

Cell Area (F^2)	27.36
Resistance On (Ω)	2010
Resistance Off (Ω)	4221
Read Voltage (mV)	250
Write Current (μA)	270
Write Pulse (ns)	10

STT-MRAM write current can be expressed as the following,

$$I_c = I_{c0}\{1 - \frac{kT}{E}\ln(\frac{\tau}{\tau_0})\} \qquad (1)$$

where τ_0 is the MTJ switching time at 0K, and I_{c0} is the MTJ critical switching current at 0K. E is magnetization energy barrier. T is MTJ temperature. k is Boltzmann's constant. I_c is the write current corresponding to switching time τ. The formula indicates that write current must be increased when write time decreases. Especially, when MTJ switching time falls below 10ns, the required switching current will increase exponentially [15]. Thereafter, the typical switching latency of STT-MRAM is set to 10ns in our work to make a reasonable trade-off between switching energy and latency. The write energy comparisons between STT-MRAM and SRAM of different capacities are plotted in Fig. 2. The figure shows that STT-MRAM write back energy is much higher than SRAM. Although the gap gets smaller with increasing capacity, STT-MRAM still consumes as much as two times energy compared to SRAM when capacity approaches 32MB. Therefore, it is imperative to reduce write energy of STT-MRAM such that it can be a competitive replacement of SRAM.

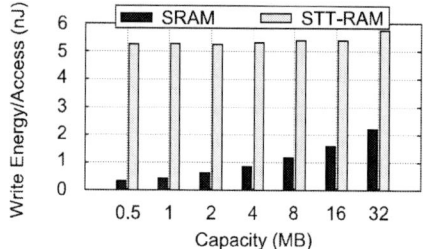

Fig. 2. Dynamic write energy comparisons of SRAM and STT-MRAM

B. Preliminary Exploring the Potential of Write Energy Reduction

In order to reduce write back activities, an intuitive method is to compare write back data with original data in L2 cache, and only write back the changed bytes. However, this method will incur large hardware overhead for bit by bit comparisons plus tracking positions of changed bits. Moreover, it requires a read operation before write, which may make the already

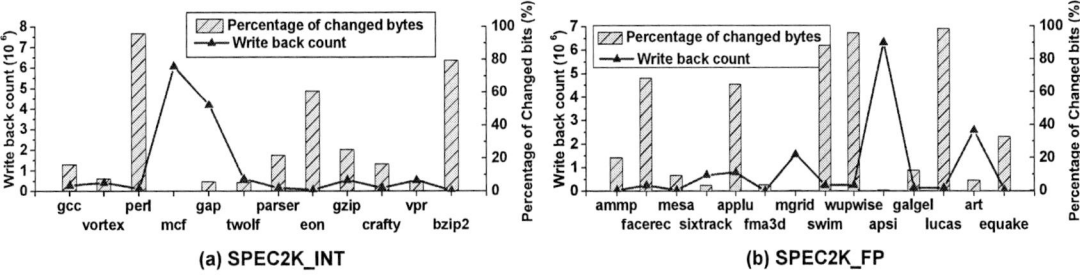

(a) SPEC2K_INT　　　　　　　　　(b) SPEC2K_FP

Fig. 3.　Percentage of changed bytes within one cache line when L1 write back to L2 cache for SPEC2K benchmark suites

long write latency problem even worse. Instead, we examine the cache line data features written from L1 to L2 cache. The statistics on write back count and average percentage of changed bytes within a cache line for SPEC2K benchmarks are shown in Fig. 3 (Please refer Section V for simulation setup details). It indicates that only less than 40% of the cache line data has been changed when it is written back for a large portion of these benchmarks. In other words, it implies most of bytes in cache line remain the same when written back[1]. Hence, there can be a significant reduction of write energy by eliminating writing these bytes. How to identify the redundant data bytes without bit by bit comparison and tracking is critical for energy savings with acceptable overheads.

IV. THE PROPOSED METHOD

A. Data Pattern & Its Denotation

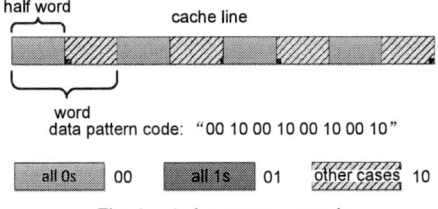

Fig. 4.　A data pattern example

To explore energy reduction potential, we refer the data feature in each cache line as data pattern. It is an abstract concept that describes the data layout feature in a cache line. For example, assume that there are four words in each cache line, and the upper half bits of each word are all zeros as shown in Fig. 4. We assign a two-bit pattern code to each half word, which is named as a basic data pattern unit, to track its data pattern. The half word data pattern can have three cases. '00' represents all bits in the considered half word are zeros. '01' represents all bits in the half word are ones. '10' represents all other cases. We take all zeros and all ones as basic data patterns because these patterns can be easily captured by only a few OR or AND gates with negligible hardware overhead. Taking the data pattern shown in Fig. 4 as an instance, the

[1]Please note that the cache line content remaining unchanged does not imply the cache line is clean as processor may write the same data several times to the cache leading to the dirty bit to be set [6]. Consequently, the dirty bit can not be used to track whether the cache line data is changed or not.

pattern code for the whole cache line is '0010 0010 0010 0010'. If the upper part of each word within the cache line remains unchanged when written back to L2 cache, it can be dropped safely instead of writing back such that write energy can be saved.

However, if we simply append data pattern code to each cache line, this scheme will be too expensive to be used in terms of area overhead . For example, if half word is the basic data pattern unit, a four-word size cache line will contain 8 basic pattern units totally. Each pattern unit requires 2 bits to record its pattern feature. Thus, each cache line requires 16 pattern code bits. To reduce area overhead, an intuitive method is organizing data patterns in a table and using an index instead of pattern code itself to refer the specific data pattern. Unfortunately, this scheme can not be applied directly either since the number of possible data patterns may be overwhelming. Taking the data pattern shown in Fig. 4 as an example, each basic data pattern unit (half word in our case) has 3 different layouts (i.e., all zeros, all ones and other cases). Then, a four-word size cache line may have as many as 3^8 data patterns, which requires 13 index bits. The area overhead is even higher than using data pattern code itself.

B. Dominating Data Pattern Identification & Write Energy Reduction

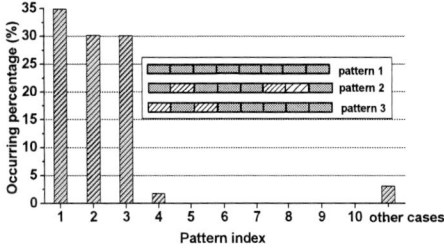

Fig. 5.　Appearing frequency of data patterns for benchmark 'fma3d'

Through extensive simulations, we extracted that different data patterns have different occurring frequencies. For example, Fig. 5 shows different data pattern appearing frequencies for benchmark 'fma3d'. The horizontal axis denotes pattern index, and the vertical axis denotes the occurring frequency of each pattern. The figure indicates that the occurring count of first 3 data patterns occupies over 90% percent of all occuring data patterns. This observation provides a chance to reduce

978-1-5386-7100-9/18 $31.00 © 2018 IEEE　　335

most of write energy by only considering some dominant data patterns. However, appearing frequency itself can not represent the importance of the data pattern because different patterns have different write energy optimization potentials. For instance, the data pattern having all zeros may have larger potential for energy reduction than those containing complex mix of zeros and ones. To consider both frequency and energy optimization potential factors, the following expression is used to denote the importance of the data pattern.

$$W_i = P_{f_i} \times P_{E_i} \tag{2}$$

where W_i denotes the weight of the ith data pattern. P_{f_i} is the percentage of its appearance frequency. P_{E_i} is the percentage of energy savings by eliminating redundant all zero or all one basic data pattern units. P_{f_i} is calculated by the following expression,

$$P_{f_i} = \frac{C_{appear}}{C_{write\_back}} \tag{3}$$

where C_{appear} is appearance count of the pattern, and C_{write_back} is the total write back count. The percentage of energy savings P_{E_i} can be calculated as follows,

$$P_{E_i} = \frac{N_{`00'or`01'}}{N_{basic}} \tag{4}$$

where $N_{`00'or`01'}$ denotes the number of basic data pattern units containing all zeros or all ones (i.e., the pattern code is '00' or '01'). N_{basic} is the number of basic data pattern units within a cache line.

We use the pattern shown in Fig. 4 to explain the weight calculation of a data pattern. Assume that the pattern represented by pattern code '0010 0010 0010 0010' occurs 10 times while the total write back amount is 100. Then, $P_f = 10/100 \times 100\% = 10\%$. The basic data pattern unit is half word in our case. Then, the number of basic data pattern units having all zeros or all ones are 4. Therefore, P_E can be calculated as $4/8 = 50\%$. As a result, the weight of this pattern is $W = P_f \times P_E = 10\% \times 50\% = 0.05$. In other words, the data pattern with higher occurring frequency and write energy savings has larger weight.

With this scheme, we can only identify and consider a few most important data patterns. To validate this point, we simulate some selected SPEC2K benchmarks and experimental results are shown in Fig. 6 (See simulation configurations in Section V). In the figure, only the eight most dominant data patterns are considered. The left bar denotes write energy consumption without any optimization. The middle bar denotes the normalized write energy reduction when only the dominating data patterns are considered. The right bar is the normalized write energy reduction when all appearing data patterns are considered. The figure indicates that most write energy can be saved by only capturing 8 dominating data patterns.

Eight data patterns only require a 3-bit index to track instead of 13-bit index when all possible data patterns are considered. Therefore, reducing the write back energy of the

dominating data patterns can save most write energy achieved by considering all patterns but with much less area overhead.

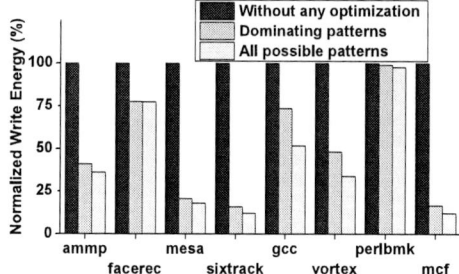

Fig. 6. Write energy reduction when only considering the eight most dominating data patterns for a few selected SPEC2K benchmarks

In addition, through extensive simulations, we also observe that different benchmarks have different set of dominating data patterns. Therefore, it is imperative to determine unique dominating data patterns according to the running applications. We can first profile the application in advance. All appearance data patterns are recorded. After calculating their weights, they are sorted in the descending order. Only the first few dominating data patterns are selected into a data pattern table. To save area overhead, the table stores data pattern codes instead of data patterns. As for the data pattern in Fig. 4, storing data pattern itself requires 32-byte entry size (assume the cache line is 32 byte) while storing pattern code only requires 16 bits. The more dominating data patterns are selected, the more data pattern index bits are required leading to higher area overhead. The trade-off evaluation between data pattern count and area overhead will be discussed in Section V.

Then, the derived data pattern table can be stored in a non-volatile memory in the processor. Once it starts running, the table is loaded into a special cache. Each cache line read from L2 to L1 cache or written from L1 to L2 cache will refer this table for energy reduction as described in the next subsection.

C. Data Pattern Matching & Write Back Energy Reduction

The data pattern searching and matching process is illustrated in Fig. 7. When the data is loaded from L2 to L1 cache, it will be checked whether to be matched with data patterns in the table. As the table only stores data pattern code instead of pattern itself as mentioned before, the loaded cache line need to be converted to corresponding pattern code. Since identifying all zeros and all ones can be implemented by several AND or OR gates, the data pattern code generation only has negligible hardware overhead. Then, the generated pattern codes are searched in the CAM structure based data pattern table, to find whether the pattern matches the entry in the table or not. If there is a hit, the pattern index will be appended with the cache line to record its data pattern. Otherwise, the cache line will be appended with index '00'. The entry with index '00' corresponds to a dummy entry containing all ones to generate write signal in any case (See write back control logic in Fig.7(b)).

978-1-5386-7100-9/18 $31.00 © 2018 IEEE   336

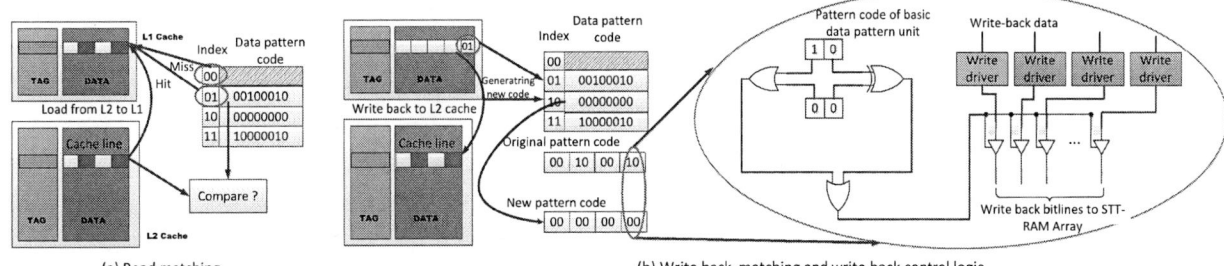

(a) Read matching   (b) Write back matching and write back control logic

Fig. 7. Cache line data matching and pattern code generating process

When the cache line is written back to L2 cache, the new data pattern code should be regenerated as it may be changed by processor. The new pattern code will be searched in the pattern table again. If there is a hit and the pattern index is the same as that appended to the cache line, then only the basic data pattern units with mixing of zeros and ones need to be written. If the pattern hits a different entry in the table, the common all-zeros and all-ones basic data pattern units of the original pattern and the new pattern can be dropped safely without writing back. If there is a miss, the whole cache line will be written back. The write back control logic generating control signals is shown in Fig. 7(b).

Note that the data pattern searching and matching parts are not located on the critical reading/writing path and can proceed in parallel with cache line access. So it will not affect the read or write performance.

V. EXPERIMENTAL RESULTS

A. Experiment Setup

TABLE II
SIMULATION PARAMETER SETTINGS

Frond End	4 wide fetch/decode/issue/commit
	Branch predictor: bimod
Execution Core	Alpha 21264, 2GHz, 16 wide RUU, 8 LSQ
	8 entry load/store queue
	4 integer ALU, 1 integer DIV/MULT
	4 FP adder, 1 FP-MULT/DIV
L1 D-cache/I-cache	16KB, 4/1 way associative, 32-byte cache line size
	2 cycle hit read/write latency
L2 D-cache/I-cache	1MB, 4 way associative, 64-byte cache line size
	8/20 cycle read/write latency

The SRAM and STT-MRAM cell parameters are derived from [13]. The technology node used in cache level simulation with NVsim [16] is 22nm. A RISC architecture simulator modified from simplescalar [17] is used for simulations. The detail configuration parameters of processor and cache are listed in Table II. We use SPEC2K benchmark suite to evaluate the effectiveness of our proposed scheme. For each benchmark simulation, 200 million instructions are executed to obtain the statistics.

B. Write Energy Reduction

The data pattern table size determines how many data patterns can be captured. The more data patterns are captured,

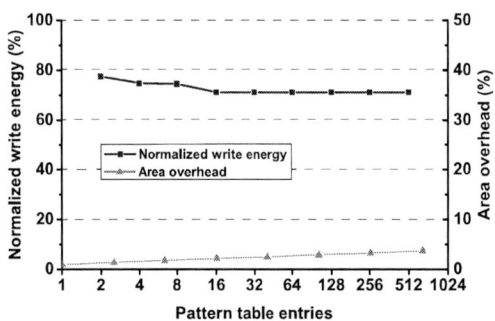

Fig. 8. Trade-off between pattern entry count, write energy reduction and area overhead for 'gcc'

the more potentials of energy reduction can be obtained. However, as entry count increases, the entry index bits also increase. Since they are appended to each cache line, the area overhead also becomes larger. For instance, the experimental result of 'gcc' is shown in Fig. 8. It indicates that when table size is larger than 16 entries, little energy can be saved further but the area overhead will continue to increase with the table size. Therefore, entry count affects both write energy reduction and area overhead.

On the other hand, the basic data pattern unit choice also impact energy reductions. When it becomes smaller (e.g., changing from word size to half word), the data pattern feature can be identified in a finer grain. The optimization space for energy reduction also becomes larger. However, the dominating patterns also increases due to more possible data patterns. For example, if the basic pattern unit is word size, then a 4-word size cache line has $3^4 = 81$ possible data patterns. In contrast, if the basic pattern unit is half word size, then the pattern count will be $3^8 = 6561$. Accordingly, the dominating data patterns may also increase, which implies more entries required in the table. Consequently, the index bits and area overhead increase.

In our simulated architecture, an integer number occupies 4 bytes, while a double floating point number occupies 8 bytes. Therefore, integer benchmarks tend to use smaller basic data pattern unit while floating point benchmarks tend to use larger one to capture dominating data patterns. By extensive simulations, we observe that 16 table entries and 2-byte size basic data pattern unit is optimal for integer benmarks, while 8 table entries and 4-byte basic data pattern unit is optimal

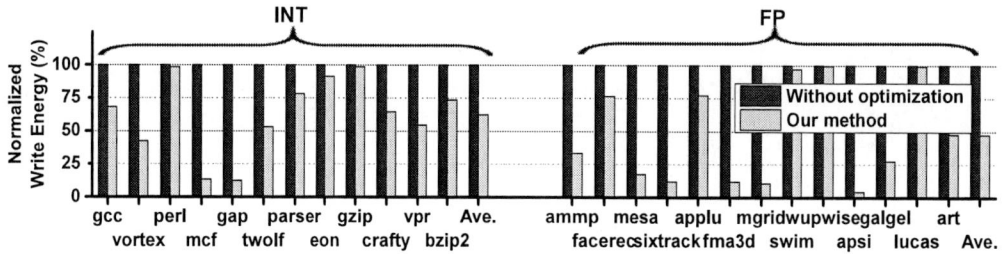

Fig. 9. Write back energy reductions for SPEC2K benchmark suite

for floating point benchmarks. To accommodate both floating point and integer benchmarks, we choose 16 entries as the data pattern table size and 2-byte size (i.e., a half word) as the basic pattern unit for our write back energy reduction evaluations. The experimental results are shown in Fig. 9. It shows that the write energy can be saved by 38% on average using our method for integer benchmarks. As for floating point benchmarks, the write energy can be saved by more than 50% on average. It implies that floating point benchmarks usually contain more redundant bits for energy optimization due to its larger storage size.

C. Area Overhead Estimation

A 16 entries data pattern table requires 4 index bits attached to each cache line. Data pattern code requires 2 bits per half word × 8 half words per cache line = 16 bits. Then, the data pattern table size is 16 entries × (16-bit pattern code + 4-bit index) = 320 bits, which is negligible in area overhead. The logic identifying all zeros and all ones consists of only several logic gates as shown in Fig. 7. The area overhead is also negligible. The main area overhead comes from the index bits attached to each cache line. For the 16-entry table, the index bits is 4. Consequently, the overall area overhead is $4/cache\_line\_size = 4/(32*8) = 1.6\%$, which is also very small.

VI. CONCLUSION

Due to skyrocketing power consumption, traditional CMOS based technology can not afford to meet power constraint as on chip cache capacity increases. A few emerging memory technologies are proposed to approach ultra low power access operation. STT-MRAM has high potential as a replacement of SRAM for on chip cache. However, it introduces several challenges, including very high write energy. In the paper, we investigate the data patterns of cache lines written back from L1 SRAM cache to L2 STT-MRAM cache. Then, a data pattern table is built to identify the dominant data patterns such that redundant data bits can be dropped without writing back again to save write energy. The experimental results show that write back energy can be reduced by roughly 40% for integer benchmarks and over 50% for floating point benchmarks on average. The induced area overhead is only 1.6%, which validate the effectiveness and efficiency of our proposed method. Our future work will target on adjusting data pattern table contents dynamically on-line to capture different phases within the same running application.

ACKNOWLEDGMENT

This work was supported in part by the National Natural Science Foundation of China under grant No. 61401008, 61704005 and the Open Project Program of State Key Laboratory of Computer Architecture, Institute of Computing Technology, Chinese Academy of Sciences (No. CARCH201602).

REFERENCES

[1] Dmytro Apalkov et al., "Spin-transfer torque magnetic random access memory (STT-MRAM)," *ACM Journal on Emerging Technologies in Computing Systems*, vol. 9, no. 2, pp. 1–35, 2013.

[2] W. Zhao et al., *Spintronics-based computing*. Springer, 2015.

[3] A. Pirovano et al., "Electronic switching effect in phase-change memory cells," in *Proc. of IEDM*, San Francisco, USA, Dec 2002, pp. 923–926.

[4] L. O. Chua, "Memristor-the missing circuit element," *IEEE Transactions on Circuit Theory*, vol. 18, no. 5, pp. 507–519, 1971.

[5] G. Sun et al., "A novel architecture of the 3D stacked MRAM L2 cache for CMPs," in *Proc. of HPCA*, Raleigh, USA, Oct 2009, pp. 239–249.

[6] S. Yazdanshenas et al., "Coding last level STT-RAM cache for high endurance and low power," *IEEE Computer Architecture Letters*, vol. 13, no. 2, pp. 73–76, 2014.

[7] S. Peng et al., "Giant interfacial perpendicular magnetic anisotropy in MgO/CoFe/capping layer structures," *Applied Physics Letters*, vol. 110, no. 7, pp. 59–67, 2017.

[8] Z. Wang et al., "High-Density NAND-Like Spin Transfer Torque Memory With Spin Orbit Torque Erase Operation," *IEEE Electron Device Letters*, vol. 39, no. 3, pp. 343–346, 2018.

[9] F. Arezoomand et al., "Energy aware and reliable STT-RAM based cache design for 3D embedded chip-multiprocessors," in *Proc. of ReCoSoC*, Madrid, Spain, July 2017, pp. 1–8.

[10] O. Coi et al., "A novel SRAM-STT-MRAM hybrid cache implementation improving cache performance," in *Proc. of NANOARCH*, Newprot, RI, USA, July 2017, pp. 39–44.

[11] X. Dong et al., "Circuit and microarchitecture evaluation of 3D stacking magnetic RAM (MRAM) as a universal memory replacement," in *Proc. of DAC*, Anaheim, USA, Dec 2008, pp. 554–559.

[12] C. W. Smullen et al., "Relaxing non-volatility for fast and energy-efficient STT-RAM caches," in *Proc. of HPCA*, San Antonio, USA, Febrary 2011, pp. 50–61.

[13] L. Xue et al., "An Adaptive 3T-3MTJ Memory Cell Design for STT-MRAM-Based LLCs," *IEEE Transactions on Very Large Scale Integration (VLSI) Systems*, vol. 26, no. 3, pp. 484–495, 2018.

[14] M. Wang et al., "Current-induced magnetization switching in atom-thick tungsten engineered perpendicular magnetic tunnel junctions with large tunnel magnetoresistance," *Nature Communications*, vol. 9, no. 1, pp. 671–678, 2018.

[15] M. Hosomi et al., "A novel nonvolatile memory with spin torque transfer magnetization switching: Spin-RAM," in *Proc. of IEDM*, Washington, DC, Oct 2005, pp. 459–462.

[16] X. Dong et al., "Nvsim: A circuit-level performance, energy, and area model for emerging nonvolatile memory," *IEEE Transactions on Computer-Aided Design of Integrated Circuits and Systems*, vol. 31, no. 7, pp. 994–1007, 2012.

[17] T. Austin et al., "Simplescalar: an infrastructure for computer system modeling," *Computer*, vol. 35, no. 2, pp. 59–67, 2002.

978-1-5386-7100-9/18 $31.00 © 2018 IEEE

Time Stamp Based Scheduling for Energy Harvesting Systems with Hybrid Nonvolatile Hardware Support

Xin Shi*, Tongda Wu*, Keni Qiu†, Huazhong Yang*, Yongpan Liu*
*Department of Electronics Engineering, Tsinghua University, Beijing, China
†Information Engineering College, Capital Normal University, Beijing, China

Abstract—Nonvolatile processors have manifested strong vitality in energy harvesting systems due to their endurable features to intermittent power supply. However, repeating configurations of peripherals still occupy too much task execution time, which substantially reduces effectiveness of previous scheduling algorithms. In this paper, we adopt the hybrid nonvolatile hardware platform and then propose a time stamp based scheduling algorithm. The experimental results present that the proposed algorithm matches the platform seamlessly and outperforms state-of-the-art algorithms both in effectiveness and efficiency.

Index Terms—energy harvesting; scheduling; nonvolatile hardware

I. INTRODUCTION

The advent of the Internet of Things (IoTs) have enabled a large number of devices and sensors to communicate with each other and share their data. However, powering such applications is extremely challenging due to sticky constraints such as the remote deployment location, limited space and the inconvenient reconfiguration. Confronted with these obstacles, energy harvesting systems manifest great vitality and flexibility by collecting energy from environment. However, this promising solution for power supply could not maintain a long time execution because of intermittent and unpredictable power failures, which lead to frequent rollbacks and re-executions. In order to mitigate such overhead, several software-aided solutions have been proposed. [1] proposed to insert checkpoints in codes to avoid task re-execution from the beginning based on RFID applications. [2] preferred to trigger checkpoints by hardware interruptions rather than software only. [3] utilized hybrid memory to balance energy for backup and memory accesses. All these strategies based on volatile processors with off-chip NVM can not omit the extra time used for transmitting data from registers to nonvolatile memory when power failures happen and vice verse. For example, it takes an off-the-shelf FeRAM-based processor [4] more than 200us to back up all flip-flops, and this figure will be multiplied if the number of flip-flops is large. What is worse, the significant overhead greatly reduces efficient execution time and can often bring in unnecessary checkpoints.

Fortunately, nonvolatile processors (NVPs) proposed in [5] have directed a promising way for settling forementioned problems. NVP can achieve a bit-to-bit data transfer scheme with distributed nvFFs and each NVM element is connected to a standard volatile flip-flop. Once power failure occurs, the data held in the volatile flip-flops are written into the NVM cells in parallel. Compared with the previous solutions writing data back through bus, the newly approach has the potential to achieve data backup and recovery confined in microseconds with very low power consumption and support instant checkpoints. Based on such hardware platforms, the issue of improving efficiency in energy harvesting system shifts from the checkpoint selection to the task scheduling.

In this field, Zhang first proposed fine-grained scheduling with energy migration [6] on NVP-based sensor nodes. However, the strategy ignores the different power switching overhead among tasks. Then Li took power switching overhead into consideration and proposed a performance-aware scheduling method based on task splits [7]. Yet the shortcomings lie in overlook for the parallel resource utilization and the excessively simple task division. Actually, though computing tasks can be suspended efficiently by NVP, tasks considering sensing and transmitting operations cannot be backed up and introduce great initialization overhead in recovery. For example, initialization time occupies 94% in a sensing operation [8], which accounts for no profits in task splits. Therefore, we adopt a new hardware structure called Nonvolatile Radio Frequency (NVRF) [5] in peripheral control to achieve high efficiency in peripheral configurations.

This paper proposes a comprehensive scheduling method involving a full consideration of power switching overhead, data dependence and parallel resource utilization based on a hybrid nonvolatile system. The major challenge is how to arrange different types of tasks to proper time slots. The main contributions of the work are as follows:

* Formulating the comprehensive task scheduling problem considering power switching overhead, data dependence and the resource limitation to an optimization problem. The objective is to minimize the makespan of all tasks.

* Proposing the Brand-Bound and Greedy-Iteration scheduling method to achieve 23% execution time reduction with more than 30% decline of the algorithm's time complexity compared to the state-of-the-art algorithm [6] under the real power trace.

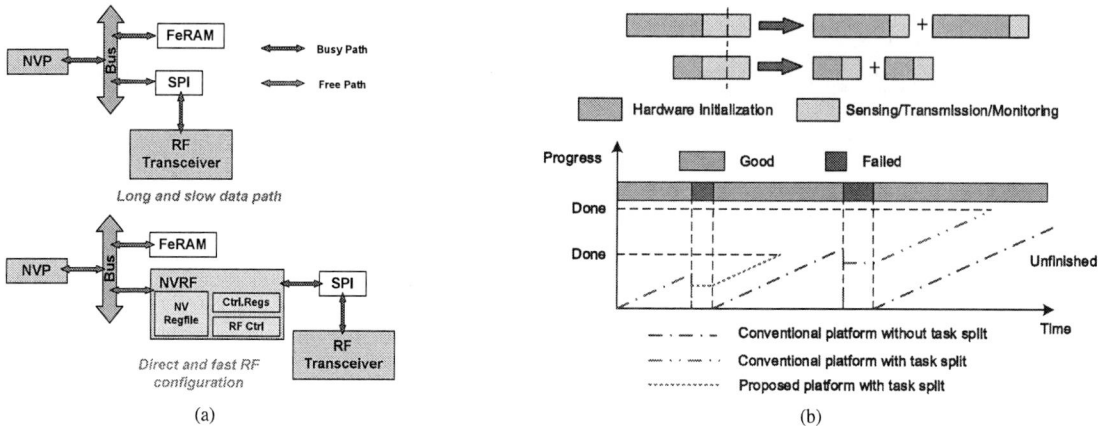

Fig. 1. (a) The comparison between the sole NVP architecture and the NVP-NVRF hybrid system. (b) Task execution progress under different hardware platforms.

II. MOTIVATION

This section first describes the operating mechanism of the NVRF and how our scheduling optimization problem benefits from this architecture, then followed by a motivation example for task scheduling considering power switching overhead and the resource limitation.

A. Nonvolatile Platform

Fig. 1(a) presents the workflow and diagram of the non-volatile radio frequency controller (NVRF) based energy harvesting systems. Without the NVRF, conventional designs have to reinitialize the RF transceiver by software before each transmission because all configurations and data in the RF transceiver are lost after power off. This procedure will take a NVP a lot of time to fetch data from FeRAM through bus and then transmit to SPI interface to configure the RF transceiver. With the NVRF support, both the NVP and the RF transceiver wake up in parallel and the hardware controlled initialization of the RF transceiver is even faster. The wake up latency can be shortened by 27X from 33ms to 1.22ms and the energy efficiency is improved by 21X [5] compared to the existing NVP systems. Power switching overhead is defined as extra operating time/energy induced by rollbacks, backups, restorations, retransmissions and reinitializations under power interruptions. As Fig. 1(b) suggests, task split strategy [7] applied to the conventional platform enables energy harvesting systems to endure intermittent power failures. However, it also brings the significant power switching overhead and still cannot bear the relatively frequent power failures. The proposed NVP-NVRF hybrid platform can effectively combat the problem. On the one hand, the time and the energy of reinitializations could be dramatically reduced. On the other hand, parallel working mechanism enables the energy harvesting systems to bear harsher environment and offers software algorithm more space to explore the optimal task scheduling.

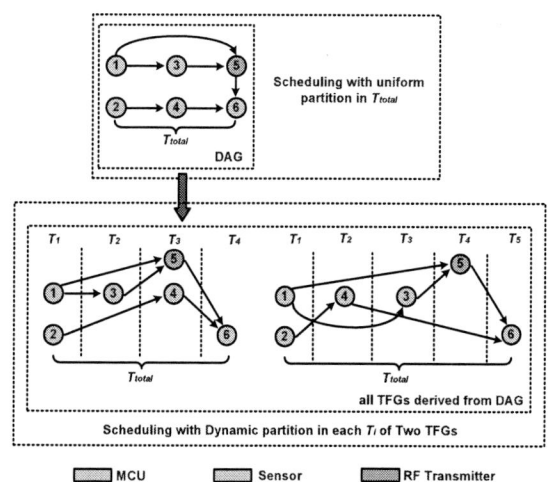

Fig. 2. Conventional scheduling strategy gathers all tasks for optimization; Our proposed method divides global optimization problems into pieces represented by TFGs according to resource and data constraints.

B. A motivation example

The following example demonstrates a simple case of scheduling on the energy harvesting platform mentioned above. The objective is to minimize the operating length (makespan) of multiple tasks, which are modeled as a directed acyclic graph (DAG) $G = (V, C, E)$ according to data dependence and the resource limitation. Each vertex $v \in V$ corresponds to a task, characterized by execution time t, the power consumption p and the color $c \in C$, which means different hardware resources. Each edge $e \in E$ represents data dependence between tasks. A task cannot be executed until all preceding connected tasks are finished.

Fig. 2 presents the differences between a conventional scheduling unit and the proposed scheduling unit. Previous solutions gather all tasks together for optimization under constraints of data dependence. Considering each task could be

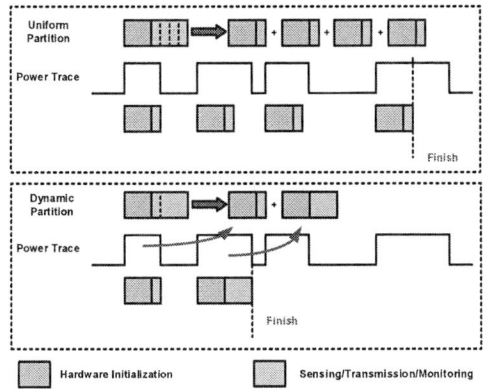

Fig. 3. Uniform partition is confined to the shortest interval between power failures while dynamic partition fits divided tasks to the power pulse seamlessly

randomly executed with other tasks, so we have $6^6 = 46656$ spaces to check with conventional methods. What is worse, without considering parallel executions of resource dependent tasks, we probably eliminate the optimum in the next searching step. Fortunately, this paper proposes a divide and conquer solution. We first calculate all possible task flow graphs (TFGs) from the DAG. The detailed method will be discussed in Section III. Because of restrictions on data and resources, the number of TFGs are greatly limited. Then, the method searches for the optimal scheduling path in each TFG and compares the overhead of each optimal task flow to achieve the global optimum. For example, due to the same execution in the first time stamp of both TFGs, the number of searching spaces is significantly reduced to $2^2(1 \times 2^2 \times 1 + 1) = 20$.

We define tasks with tremendous power switching overhead high-cost (HC) tasks. Fig. 3 compares the previous partition method with the proposed method in one HC task scheduling, the previous solution adopts a static and uniform partition strategy for optimization. Confined to the intermittent power supply, the time span of a divided task should be less than the minimum of the normal execution time between two power failures and the method brings substantial extra reinitialization overhead. Different from the previous settlement, we propose a dynamic partition method aiming at HC tasks and define the time stamp as a unit for scheduling, where tasks could be executed randomly without considering data and resource constraints. In a time stamp, HC tasks are split apart according to intervals between power failures.

III. MODEL

This section first introduces the algorithm to calculate all possible task flow graphs (TFG) based on data and resource constraints. After that, the scheduling objective and optimization constraints of the scheduling problem in a time stamp are formulated. Finally, the flow chart of the solution to the scheduling problem is presented.

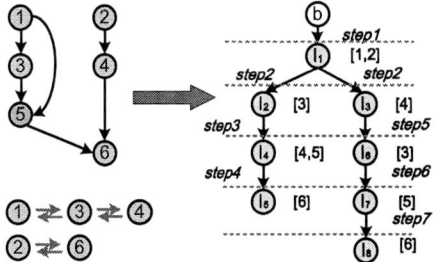

Fig. 4. Branch-Bound algorithm derives TFGs from the DAG.

A. Branch-Bound Algorithm

Fig. 4 presents an example of applying Branch-Bound (BB) algorithm to find the universal set of TFGs. First we obtain a DAG according to data dependence. As Fig. 2 suggests, the edge between two vertexes means there exists data sharing. Beneath the DAG, we enumerate all resource dependence as constraints. In each exploring step, BB algorithm attempts to find and collect all tasks with no conflicts and affix them with a time stamp. If there exists branches of gathering tasks, the algorithm adapts the depth-first search method to traverse the whole space to obtain all TFGs. TFG is featured that tasks in each time stamp are independent from each other, which greatly reduces the algorithm complexity in subsequent scheduling.

B. Greedy-Iteration Algorithm

The system parameters and variables in a time stamp are defined in Table I.

TABLE I
PARAMETERS AND VARIABLES OF THE SYSTEM MODEL IN EACH TIME STAMP

	Symbol	Description
Task	V	Task set $\upsilon = \{\tau_1, \tau_2, ..., \tau_N\}$
	t_i	Execution time of task τ_i
	p_i	Power consumption of task execution τ_i
	o_i	Power consumption of task overhead τ_i
Power	E	Power pulse set, $E = \{e_1, e_2, ...e_m, ...\}$
	a_k	Amplitude of power pulse e_k
	d_k	Time duration of power pulse e_k
	ϕ	Subset of power pulse set

The main objective is to minimize the total execution time of all tasks by mapping each task into several power pulses. That is to find a minimum number of power pulses to execute all tasks in a time stamp. The scheduling objective is as follows.

$$
\begin{aligned}
&obj: min\ M = |\phi|, \phi \in E \\
&s.t: V\ can\ be\ executed\ within\ \phi,
\end{aligned}
\tag{1}
$$

where ϕ represents the subset of energy pulse set E and V is the task set. Before solving the scheduling optimization problem, we first calculate the minimal number of pulses $|\phi|$ for task execution. The $|\phi|$ achieves the minimum when there

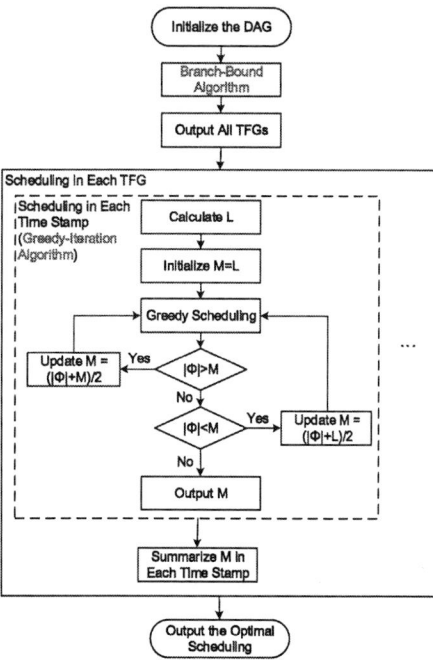

Fig. 5. Scheduling Flow

Algorithm 1: Greedy-Iteration Algorithm

Input: Parameters in Table I
Output: The minimum number of power pulses M to complete HC tasks in the task set V.
Initialize $M \leftarrow L$, $V_{HC} = \{\tau_1, \tau_2, ..., \tau_h\}$ indicates HC task set sorted by corresponding power consumption where $p_1 > p_2 > ... > p_h$ and iteration stop sign $F = 0$;
while $F! = 1$ **do**
  $E_c = \{e_1, e_2, ...e_M\}$, where $a_1 d_1 > ... > a_M d_M$;
  $V_c = \{\tau_1, \tau_2, ..., \tau_h\}$;
  $r_e = r_t = 1$, $F = 1$;
  while $V_c \neq \varnothing$ **do**
    while $e_{r_e} \to ad > \tau_{r_t} \to pt$ ***And*** $r_e \leqslant M$ **do**
      $S = 1$;
      if $e_{r_e} \to d > \tau_{r_t} \to t$ **then**
        $S = 0$, Remove τ_{r_t} from V_c;
        Update residual energy e_{r_e} in E_c and ;
        sort E_c in descending order of a;
        Break
      end
      else
        $r_e += 1$
      end
    end
    if $S = 1$ **then**
      Execute τ_{r_t} in e_{r_e}, Update $\tau_{r_t} \to p, t$;
      Update residual energy e_{r_e} in E_c;
      Add power pulses to E_c according to o_{r_t};
      and increase M, set $F = 0$;
    end
    Sort E_c in descending order of a, $r_e = 1$;
    Sort V_c in descending order of p, $r_t = 1$
  end
  if $F! = 1$ **then**
    Update M from previous values
    by dichotomization;
  end
end

exist no power failures during task execution as the following inequation presents,

$$\sum_{k=1}^{L-1} a_k d_k < \sum_{i=1}^{N} p_i t_i \leqslant \sum_{k=1}^{L} a_k d_k \qquad (2)$$
$$thus, \ L \leqslant M$$

Considering only HC tasks induce extra overhead after power failures, we propose the Greedy-Iteration algorithm as presented in Algorithm 1, which utilizes the number of power pulses as the iteration termination condition. The initial number of power pulses is L indicating the minimal available energy for task execution. In each loop of greedy scheduling, HC tasks with high energy consumption are arranged to be executed in the large energy pulses. If one task can not be finished in one pulse, the algorithm updates the residual energy and time for the task completion, adds corresponding extra energy overhead into an available energy pulse set and increases the number of power pulses if necessary. However, because the initial optimization space is too small for task scheduling, the number of required pulses will be larger than M, recorded as F. Therefore, we could shrink the exploring optimization space in each iteration. Based on dichotomization, the number of loops is confined to $log_2(F - L)$.

C. Scheduling Flow

Fig. 5 presents the workflow of the scheduling strategy. The strategy first extracts the DAG based on data and resource dependence. Then, we utilize the Branch-Bound algorithm to acquire all TFGs. For each TFG, minimizing the total execution time of all tasks can be interpreted as a summation of minimizing execution time in each time stamp, which has been effectively tackled by the Greedy-Iteration algorithm. Finally, the strategy compares all scheduling results in all TFGs to output the optimal scheduling with the minimal execution time.

IV. EXPERIMENT

This section first shows the experimental setup, then provides comprehensive evaluations to demonstrate the effectiveness of the proposed scheduling algorithm and analyzes the convergence rate of Greedy-Iteration algorithm. Finally, the proposed hybrid platform is compared with the conventional platform on the benchmark execution time.

Fig. 6. Hardware platform contains (1) NVMCU supported by NVP and NVRF, (2) Sensors: a UV sensor Si1132, a temperature sensor TMP100 and a tri-axis accelerometer KXTJ9, (3) RF transceiver: Rohm MK72660

A. Experimental setup

We extract system parameters from an ambient energy harvesting platform and build the hardware model on the Gem5 simulator. The power consumption of the system is calculated by measuring on real chips with NVP and NVRF support. Fig. 6 shows the real-world hardware platform. Table II illustrates the system specifications in power consumption.

TABLE II
THE POWER CONSUMPTION OF SYSTEM MODULES

Device Name	Chip Model	Power (mW)	Device Name	Chip Model	Power (mW)
NVP	THU1020N	3.4	UV Sensor	Si1132	10.8
NVRF	THU1020N	1.2	Temp Sensor	TMP100	0.21
RF Transceiver	MK72660	10.1	Accelerometer	KXTJ9	0.075

Benchmarks are formed of tasks chosen from different categories of Mibench [9], data transmissions and sensor monitoring. The parameters of these tasks are calculated from the Gem5 simulator as shown in Table III whose items are interpreted in Table I.

TABLE III
PARAMETERS OF TASKS

Item	Tasks						
	T1	T2	T3	T4	T5	T6	T7
$t(us)$	275	122	409	1016	451	451	7940
$p(uJ)$	0.94	0.41	1.8	12.56	1.76	8.96	80.2
$o(uJ)$	≈ 0	≈ 0	≈ 0	1.59	1.67	8.94	3.95

Among the tasks, four (T4-T7) of them are HC tasks with high power switching overhead. Four types of real power traces in different energy harvesting situations [10] are evaluated.

B. Algorithm validation

The following three scheduling strategies are evaluated: Scheduling results of the intra-task strategy [6] without considering power switching overhead; Performance-aware scheduling without considering the parallel hardware utilization proposed by Li [7]; Proposed time-stamp based Branch-Bound

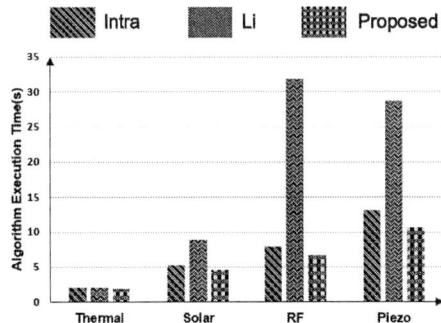

Fig. 7. Comparison of algorithm execution time in calculating their optima supported by hybrid nonvolatile platform

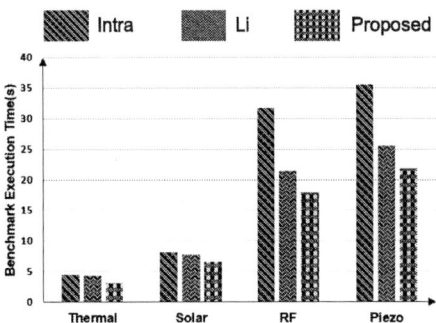

Fig. 8. Comparison of benchmark execution time supported by hybrid nonvolatile platform

and Greedy-Iteration algorithm. The benchmark is composed of real tasks shown in Table III.

Fig. 7 presents the algorithm execution time in calculating their optima under the four real power traces. With fluctuations increase in power supply, all the three algorithms require more time to fit tasks in more power pulses. However, the strategy proposed by Li needs to calculate a mixed integer linear programming (MILP) problem whose time complexity is exponentially related to the number of power pulses. Besides, in order to finish the benchmark, Li's algorithm splits HC tasks into smaller pieces, also contributing to plenty of time. With the help of the shrink in optimal searching space and the iteration technique, the execution time of the proposed algorithm is reduced by 23% compared with the intra-task strategy under piezo power supply. The reason lies in the fact that increasing power fluctuation will not influence the procedure of the Branch-Bound algorithm and meanwhile the fast convergence rate of the Greedy-Iteration algorithm significantly weakens its impact.

Fig. 8 presents the comparison results of the benchmark execution time with three scheduling algorithm support. Without considering power switching overhead, intra-task scheduling method gradually falls behind the other two strategies when the power interruptions occur more and more frequently. That is because tasks involving sensing, transmitting operations cannot be backed up during power failures and introduces power switching overhead in benchmark execution. Compared with Li's method, the Branch-Bound algorithm fully considers about the parallel resource utilization. Such as, the parallel collection of unrelated temperature and radiation data. As Fig. 8 indicates, when energy supply is always sufficient like thermal power, the proposed scheduling result outperforms Li's method by 37% with respect to execution time.

C. Convergence rate

In the proposed Greedy-Iteration algorithm, the efficiency of the strategy depends on the convergence rate and the execution time of each loop. Fig. 9 compares such parameters in two power supply conditions: ascending interruption frequency and decreasing interruption frequency. Results show that iteration times keep stable under different circumstances while the execution time of loops exhibits distinct values. The greedy algorithm promotes tasks with heavy cost to have higher priority in scheduling. However, when interruption frequency decreases, frequent power failures push massive pulses into the energy set because current energy is insufficient for energy-hungry tasks, and thus increasing the exploration space for next iteration.

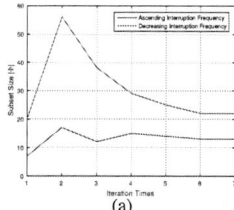 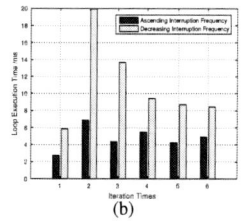

Fig. 9. (a) Convergence of the proposed Greedy-Iteration algorithm under different interruption frequency trends; (b) Execution time in each loop under different interruption frequency trends .

D. Hardware Benefits

This section compares execution time on the conventional nonvolatile hardware platform and the one with hybrid non-volatile hardware platform.

TABLE IV
PARAMETERS OF TASKS

Time(s) \ Power Trace \ Item		Thermal	Solar	RF	Piezo
Intra [6]	Conventional	4.7	8.6	45.6	48.9
	Hybrid	4.5	8.2	31.8	35.6
Li [7]	Conventional	4.4	7.9	27.2	27.1
	Hybrid	4.3	7.8	21.5	25.6
Proposed	Conventional	3.3	7.1	23.8	29.5
	Hybrid	3.2	6.5	17.9	21.8

Table IV indicates that the system performance can be improved with the hybrid platform. Under piezo power supply, the task execution time reduces 26% compared with the conventional platform. The improvement lies in the fact that extra reinitialization overhead of the RF transceiver induced by heavy fluctuations has been significantly reduced with NVRF support. Besides, when power supply is sufficient, the contribution of the architecture comes from the parallel configuration of the RF transceiver without interrupting the NVP.

V. CONCLUSION

In this paper, contributions are made in two aspects. In the software aspect, we propose the Brand-Bound and Greedy-Iteration scheduling method to achieve over 20% execution time reduction with more than 30% decline of the algorithm complexity compared with the state-of-the-art algorithms under real power traces. In the hardware aspect, taking advantage of NVP and NVRF hybrid system, the optimization space is substantially expanded and the task execution time reduces over 20% compared with the conventional platform. With the development of energy harvesting systems, the increasing amount of data to process and growing applications involved communication will further aggravate the demand for low power scheduling and the proposed strategy will be quite promising.

REFERENCES

[1] B. Ransford, J. Sorber, and K. Fu, "Mementos: System support for long-running computation on rfid-scale devices," *ACM Sigplan Notices*, vol. 47, no. 4, pp. 159–170, 2012.

[2] D. Balsamo, A. S. Weddell, G. V. Merrett, B. M. Al-Hashimi, D. Brunelli, and L. Benini, "Hibernus: Sustaining computation during intermittent supply for energy-harvesting systems," *IEEE Embedded Systems Letters*, vol. 7, no. 1, pp. 15–18, 2015.

[3] H. Jayakumar, A. Raha, and V. Raghunathan, "Quickrecall: A low overhead hw/sw approach for enabling computations across power cycles in transiently powered computers," in *VLSI Design and 2014 13th International Conference on Embedded Systems, 2014 27th International Conference on*. IEEE, 2014, pp. 330–335.

[4] F. Su, Y. Liu, Y. Wang, and H. Yang, "A ferroelectric nonvolatile processor with 46us system-level wake-up time and 14us sleep time for energy harvesting applications," *IEEE Transactions on Circuits and Systems I: Regular Papers*, vol. 64, no. 3, pp. 596–607, 2017.

[5] W. et al, "A 130nm feram-based parallel recovery nonvolatile soc for normally-off operations with 3.9× faster running speed and 11× higher energy efficiency using fast power-on detection and nonvolatile radio controller," in *2017 Symposium on, VLSI Circuits,*. IEEE, 2017, pp. C336–C337.

[6] D. Zhang, S. Li, A. Li, Y. Liu, X. S. Hu, and H. Yang, "Intra-task scheduling for storage-less and converter-less solar-powered nonvolatile sensor nodes," in *2014 32nd IEEE International Conference on, Computer Design (ICCD),*. IEEE, 2014, pp. 348–354.

[7] H. Li, Y. Liu, C. Fu, C. J. Xue, D. Xiang, J. Yue, J. Li, D. Zhang, J. Hu, and H. Yang, "Performance-aware task scheduling for energy harvesting nonvolatile processors considering power switching overhead," in *2016 53nd ACM/EDAC/IEEE, Design Automation Conference (DAC)*. IEEE, 2016, pp. 1–6.

[8] K. Ma, X. Li, K. Swaminathan, Y. Zheng, S Li, Y. Liu, Y. Xie, J. J. Sampson, and V. Narayanan, "Nonvolatile processor architectures: Efficient, reliable progress with unstable power," *IEEE MICRO*, vol. 36, no. 3, pp. 72–83, 2016.

[9] M. R. Guthaus, J. S. Ringenberg, D. Ernst, T. M. Austin, T. Mudge, and R. B. Brown, "Mibench: A free, commercially representative embedded benchmark suite," in *WWC-4. 2001 IEEE International Workshop on, Workload Characterization, 2001*. IEEE, 2001, pp. 3–14.

[10] K. Ma, Y. Zheng, S. Li, K. Swaminathan, X. Li, Y. Liu, J. Sampson, Y. Xie, and V. Narayanan, "Architecture exploration for ambient energy harvesting nonvolatile processors," in *2015 IEEE 21st International Symposium on, High Performance Computer Architecture (HPCA)*. IEEE, 2015, pp. 526–537.

978-1-5386-7100-9/18 $31.00 © 2018 IEEE

EETD: An Energy Efficient Design for Runtime Hardware Trojan Detection in Untrusted Network-on-Chip

Mubashir Hussain, Amin Malekpour, Hui Guo and Sri Parameswaran
School of Computer Science and Engineering
University of New South Wales, Australia
{m.hussain, a.malekpour, huig and sri.parameswaran}@unsw.edu.au

Abstract—Network-on-chip (NoC) is a communication intellectual property (IP) core, popularly used in the system-on-a-chip (SoC) designs. The NoC IP core often comes from an untrusted 3rd-party vendor and may have hardware Trojans. The Trojan in the NoC can eavesdrop packets, modify data and divert packets to the wrong location, hence, endangering system's confidentiality, integrity, and availability.However, if the activation probability of the Trojan is very low or the Trojan is never activated, a significant amount of energy is wasted due to authentication. The unnecessary authentication in such cases also greatly affects the system performance and availability. In this paper, we propose an energy efficient Trojan detection design (EETD) where the authentication gets activated only when the hardware Trojan has been triggered in the system. Our experimental results show that EETD improves the energy overhead and performance overhead by 38% and 40% as compared to the state-of-the-art technique on an 8×8 2D-mesh NoC. Our experiments, also show that EETD takes almost $10\mu s$ to localize 95% of the Trojan infected nodes.

I. INTRODUCTION

The advances in semiconductor technology and the proliferation of increasingly-complicated embedded systems are making System-on-a-Chip (SoC) more and more common in the hardware design. A SoC is a single microchip that integrates all the necessary electronic circuits and components for a given system. The communication between these components is mostly implemented with an on-chip network, also referred to as Network-on-Chip (NoC).

To reduce the cost and time-to-market of SoCs, a designer tends to use Commercial Off-The-Shelf (COTS) components from the third-party vendors, who may be untrusted [23]. This raises security concerns [8] since a small hardware modification (a.k.a. hardware Trojan) by an adversary in the 3rd-party Intellectual Property (3PIP) cores can compromise the whole chip. If such chips run time-critical applications (e.g. in autonomous vehicles), the hardware Trojan (HT) attack on the performance of these applications may lead to catastrophic or life-threatening consequences [22]. Similarly, if these chips are used in information critical systems (e.g. online banking) the confidentiality, and integrity of the user's data can be compromised [21].

Off-line mechanisms to detect HTs at the low-level of circuit testing are proposed in the literature [4], [25]. However, due to lack of a golden reference model (i.e., Trojan-free chip) and limited 3PIP core details provided by third party vendor, a few hardware Trojans, especially those with small footprints, can escape the off-line detections and still appear in the final product system [23]. Therefore, runtime detection schemes are necessary as they serve as the last line of defense against hardware Trojans.

Most on-chip components (e.g. processing units and memories) can have multiple instances. A Trojan in one instance can be neutralized by using another instance [22], [23]. But for on-chip communication, a SoC design commonly uses a single NoC instance. Replacing NoC to remove its Trojan is not easy so such schemes are not feasible. So, a runtime detection techniques must be used as last line of defense. A NoC also has access to all the on-chip components. If the NoC is untrusted, the whole system can be hard hit by prolonged security threats. In this paper, we target the Trojan in 3rd-party NoC based SoCs.

*The **main contributions** of this paper can be summarized as follows:*

- An energy efficient Trojan detection (EETD) design that uses selective activation/deactivation of detection units to reduce energy overheads.
- A novel worm based search algorithm to quickly locate the Trojan position.

The rest of this paper is organized as follows. We first present our threat model in Section II. Section III discusses related work. The EETD methodology is detailed in Section IV. Section V describes our experimental setup and results. Finally, we conclude this paper in Section VI.

II. THREAT MODEL: ASSUMPTIONS AND LIMITATIONS

Figure 1 shows our target system, which is similar to commercial SoC chips such as TILE64 [7]. It consists of in-house designed trusted 3PIPs which includes processors, memories, network interfaces and an untrusted 3rd-party NoC. The NoC has a 2D mesh topology and uses deterministic XY routing, commonly adopted in the academic and commercial NoC designs. The trusted IP cores are connected to the NoC routers R (referred to as a node in rest of the paper) through Network Interfaces (NI).

EETD assumes that the 3rd-party NoC may have a hardware Trojan hidden in any one if its node and HT is activated and deactivated at runtime. When activated, HT reads the packets passing through its host node and performs attacks such as data tampering of packets. It also assumes that the actual path of the packets cannot be determined if the packets are tampered. Furthermore, similar to [11], [13], EETD assumes that the Trojan is tiny, consumes limited computing/processing power, and is only capable of following effective attacks:

- Modifying the packet address to divert the packet to unauthorized locations, which can lead to:
  - information leak possibly assisted by the cryptanalysis software

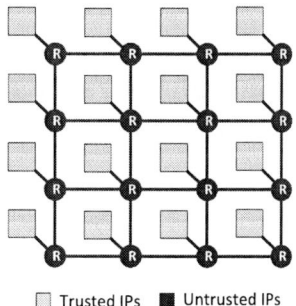

Trusted IPs ▢   Untrusted IPs ▪

Fig. 1. System Overview

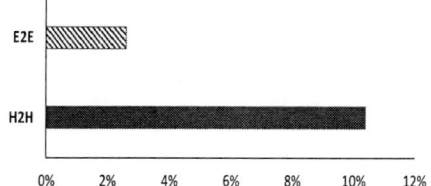

Fig. 2. Energy overhead of E2E and H2H

- denial of service (DoS) by flooding a certain node that connects to a service core
- network performance degradation by creating traffic jam on some links
- Modifying the packet data to corrupt the packet contents, which may lead to:
  - network performance degradation if the data corruption causes packet retransmission
  - untrusted data and illegal commands to the trusted components

III. RELATED WORK

Security of 3rd-party NoC IPs has become a hot research topic due to the adoption of these IPs based SoCs in medical, military and general-purpose devices. Here, we discuss some of the related works proposed to protect against such untrusted 3rd-party NoC IPs.

Designs on how to protect applications running on a system that contains untrusted cores and 3rd-party NoC IPs are discussed in [15] and [20], where high-level security applications are mapped on trusted cores and their communication over NoC is encrypted. In [26], spatial and temporal based NoC partitions are proposed to prevent interference between different security level applications. To protect memory from unauthorized access, [24] authenticates memory address and access rights of each transaction. Commercial products such as Sonic SMART Interconnect [1] and ARM TrustZone [5] divide memory blocks into different protection regions and isolate secure and normal execution environments from each other.

Compared to profound efforts on software related attacks, work on security threats from HTs are limited. Depending on when the authentication is performed, the HT detection approaches can be divided into two categories: End-to-End (E2E) approach, where packets are checked only at the end of its journey (i.e. the destination node), and Hop-to-Hop (H2H) approach, where the packet sent from one hop will be checked on the next hop. Both approaches are briefly reviewed below.

A. End-to-End based Designs

An E2E approach, which compares the timestamps of original and proxy packets is proposed in [19]. Here, an attack will be detected if the latency difference is beyond a predefined threshold. Boraten et. al. [12] use CRC and algebraic manipulation codes (AMD) to protect data packets from illegal fault injection by HT present in NoC. Ancajas

et. al. [6] propose a protection technique against information leakage HTs. At the source, they scramble the packet using a predefined key and add the static packet certificate for a given destination and verify it at the destination. Hussain et. al. [17] propose E2E packet leak detection units, which provide protection against packet hijacking attacks. They add the dynamic tag in each packet and hide the location of tag bits, by dynamically scrambling them with data bits to detect any address modification.

B. Hop-to-Hop based Designs

Boraten et. al. [10] enforce a set of rules on each node in the network to identify the HTs that suppress resource requests from the packets in the network. Yu et. al. [27] propose an H2H approach to protect against HT infected links in NoC. Frey et. al. [13] propose an H2H approach to identify packet modification and bandwidth depletion attacks by HTs present in NoC nodes. Boraten et. al. [11] propose a link obfuscation technique to protect against HTs that inject deliberate faults in the packets of the target application to cause retransmission and degrade the performance of the application.

The E2E approaches are relatively cost-effective in terms of area, power and performance overheads. However, they cannot able to localize the source of an attack. If the HT is continuously active, it will significantly degrade the performance of the system. Furthermore, if the attacks are frequent, a large amount of network bandwidth will be depleted and affects system availability. E2E approaches are good in terms of low energy overhead but they fail to pin down the Trojan location and are ineffective to prevent the Trojan from further attacks. The H2H approaches, on the other hand, can spot the Trojan location instantly the first attack is detected. Thus, making isolation of the Trojan (hence elimination of attacks) possible. However, the traditional H2H approaches require the detection units to be placed inside every node or link and perform authentication on all passing packets. But, if there is no Trojan in design or Trojan is not activated at a given time, H2H approaches still perform detections on every hop, which cause significant performance and power overheads.

To see how much extra energy can be wasted, we run an experiment on an 8×8 2D-mesh NoC with the attack detection unit proposed in [17]. These detection units are attached to the system to calculate energy overhead of E2E approach and H2H approach respectively. Figure 2 shows the relative energy overhead of E2E approach compared to H2H approach. As can be seen, the E2E approach has incurred 2% energy overhead, which is almost quadrupled when an H2H approach is used. Therefore, if the Trojan is not active (or there is no Trojan in the system), the significant energy

978-1-5386-7100-9/18 $31.00 © 2018 IEEE   346

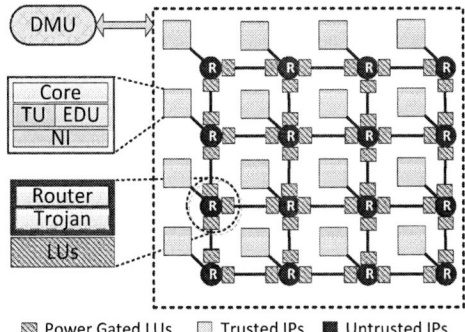

Fig. 3. Structural Overview of EETD where LUs are power gated

Legend: Power Gated LUs | Trusted IPs | Untrusted IPs

Algorithm 1: Worm Based Localization Technique

input : The node location where the first tampered packet is detected, (x_f, y_f);
NoC design parameters;
Application parameters;

output: Trojan location (x_h, y_h)

1 $T_t = calThresholdTime(x_f)$;
2 $T_s = calRowWaitTime()$;
3 $(x, y) = initWormLoc(x_f, y_f)$;
4 $activateLU(Column(x_f))$;
5 /* --- Step I: Column Movement ---- */
6 **while** $waitTime < T_t$ *and TrojanRow is not found* **do**
7   $\quad y = wormMoveAlong(Column(x_f))$;
8 **end while**
9 /* --- Step II: Row Activation ---- */
10 **if** *TrojanRow found* **then**
11 $\quad$ **if** *Detected at right port* **then**
12   $\quad\quad activateLU(RightRowSec(y))$;
13 $\quad$ **else**
14   $\quad\quad activateLU(LeftRowSec(y))$;
15 $\quad$ **end if**
16 **else**
17 $\quad activateLU(Row(y))$;
18 **end if**
19 /* --- Step III: Row Movement ----   */
20 **while** $waitTime < T_s$ **do**
21 $\quad x = wormMoveAlong(Row(y))$;
22 **end while**
23 /* --- Step IV: Trojan Location --- */
24 $(x_h, y_h) = (x, y)$;

---

consumption by H2H is simply wasted.

EETD addresses these design issues and proposes a new design and search algorithm to localize HTs with less performance and energy overheads when compared to H2H approaches. EETD aims for saving energy from unnecessary consumption.

IV. EETD METHODOLOGY

A. System Overview

Figure 3 shows the architectural details of EETD. The system consists of two types of components; trusted 3PIPs and untrusted 3PIPs. Trusted 3PIPs include in-house designed cores (i.e. processing elements and memories), network interfaces (NIs), tag units (TUs), E2E Trojan Detection Units (EDU), Trojan Localization Units (LUs), and Detection Management Unit (DMU). However, the communication architecture (i.e. NoC) is assumed to be untrusted 3rd-party IP.

There are two types of detection units in the proposed EETD. The first type is EDUs. The EDUs are attached to the cores and are active all the time. These EDUs are used to authenticate the incoming packets to the core and can send the attack detection signal to DMU through a secure link as shown in Figure 3. Note that the DMU is our detection management unit responsible for performing localization algorithm. The second type of detection units are called localization units (LUs). They are attached to the nodes in the NoC. These LUs are power gated and the DMU can dynamically attach/detach them to/from the system.

In order to save energy in EETD, the LUs for the H2H detection are normally detached from the design and they only come into play when the Trojan commences an attack. Once an attack is identified by EDU, the LUs are attached to the nodes and a localization algorithm is used to identify the location of the Trojan. To capture the initial moment of the attack, any state-of-the-art attack detection unit can be used from the literature [11], [13] and [17]. Similarly, to locate Trojan infected node same attack detection units can be used as LUs.

EETD LUs are activated/deactivated by DMU, which takes two clock cycles to process incoming attack signal. Therefore, unlike the traditional H2H approach, where a tampered packet will always be detected on the spot and the location of an attacked packet is the location of the Trojan, the LUs in EETD may detect some tampered packets that already left the Trojan node. Thus, initially detected tampered packet

node by EETD may not be Trojan infected node. Therefore, EETD has a delay (i.e. associated with activation of LUs and localization algorithm) to identify the actual Trojan location. This delay value (refer to as a threshold value) must be carefully calculated to localize actual Trojan infected nodes.

For an n×n 2D-mesh NoC, EETD requires two LUs for corner nodes (R_c), three LUs for boundary nodes (R_b) and four LUs for internal nodes (R_i). Thus, EETD requires $2R_c + 3R_b + 4R_i$ LUs in total. If they are active all the time, significant energy will be consumed. Therefore, EETD proposes an energy-efficient localization algorithm, which is discussed in next section.

B. Localization Algorithm

EETD uses EDUs to detect a Trojan attack. When the first detection of a tampered packet is reported, DMU creates a worm by activating selective LUs. Then, with every detection of a tampered packet by these LUs, DMU activate/deactivate the associated LUs, i.e. worm moves forward and the search area is narrowed. When the worm stops at a node for a specific amount of time, the location is regarded as a Trojan infected node.

The search approach is explained in Algorithm 1. In this algorithm, (x, y) is the worm location. For a given NoC design parameters, the maximum threshold value (T_t) and maximum waiting time (T_s) for each worm movement along the row is calculated in Line 1 and 2 (discussed in IV-C). When the first tampered packet is detected at the destination node, the worm is created at (x_f, y_f) (Line 3). The worm

978-1-5386-7100-9/18 $31.00 © 2018 IEEE   347

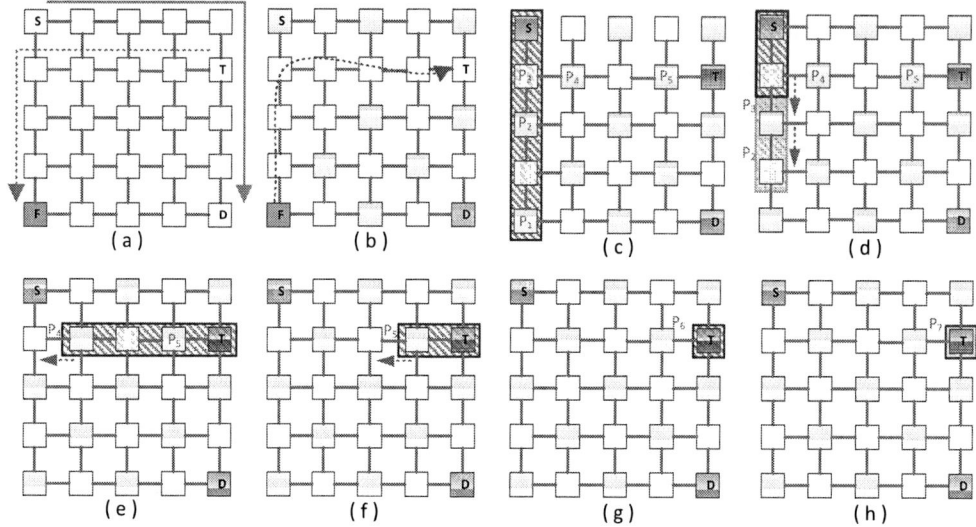

Fig. 4. Identifying the location of Hardware Trojan (a-b), worm path for packet diversion and corruption attacks with all LUs activated on the fault destination column shown in (c-d) and then LUs activated on the row transmission links for each worm movement shown in (e-h)

localizes the Trojan infected node in four steps. Step one (column movement) starts by activating all LUs on the Column x_f (Line 4). Each LU can capture the tampered packets from all directions. If the attacks are detected in the vertical links, the worm will move vertically along the column (Lines 6-8) and for each such move, the worm row (y) will be updated (Line 7). Step two is row activation that is, if a tampered packet is detected by an LU on the horizontal link which indicates the Trojan row, the worm will make a turn to that row, and the LUs on the related row section will be activated (Lines 10-15). If no tampered packets are detected and the worm is stuck at a location (i.e. the waiting time exceeds the threshold time T_t), the current row of the worm is treated as the Trojan row and the LUs of the whole row are activated (Line 17). In Step three, a row search is performed (the way worm moves along the row is similar to that along the column). For each new detection, the worm moves forward and the worm column (x) in the row is updated (Line 21). Finally, in step four when the worm stops at a location for T_s time, the worm location is regarded as the Trojan location (Line 24).

Figure 4 presents an example of the proposed search approach. Assume that node T has a Trojan that changes the destination address and is designed to leak information to node F. As Figure 4 (a) illustrates, if packets are sent from node S to node D, node T would leak part of packets to node F. When the first tampered packet reaches node F, the attack is detected by an EDU at node F (note that EDUs are always active). Then, an attack signal is sent (by EDU attached to node F) to DMU, which will then start the worm based localization. Since the Trojan can be on any node and the XY routing scheme requires a packet traveling in the horizontal then vertical directions, the worm should search in an opposite direction, as illustrated in Figure 4(b). As was mentioned in the Section II, due to the Trojan attack the actual path of the tampered packets cannot be identified and simple path based localization techniques would not be effective here [14]. After the worm is created,

all LUs on the destination column are enabled (Figure 4(c)). If there are many attacked packets detected by the vertical LUs, the worm will move vertically in the same direction (see Figure 4(d)). If no tampered packets are detected for sufficient time (i.e. threshold T_t) or a new packet is detected from the horizontal direction, the worm makes a turn to the row and the related LUs along that row section will be activated, as shown in Figure 4(e). Each detection of the tampered packet in horizontal direction narrows the search area similar to the column movement as shown in Figure 4 (f and g). When the worm stops moving for T_s amount of time, the location of the worm is regarded as the Trojan location, as shown in Figure 4(h).

C. Threshold Settings

The threshold setting is critical to the effectiveness of the EETD design. If it is too small, false positive detection might be introduced. If it is too long, the prolonged search time will increase the energy consumption. Equation 1 shows the threshold time for turning column search to row search (T_t), for a given destination node (x,y) where the attack is detected by the end-to-end detection unit (EDU).

$$T_t = \alpha * B * max\{x, n - x\} * t_0 + T_\Delta \qquad (1)$$

$$T_s = \alpha * B * t_0 \qquad (2)$$

In Equation 1, n is the network dimension size, B is the buffer size for each node measured in packets, t_0 is the basic packet latency over a node without considering any buffering delay, α is the number of input ports in the node and T_Δ is the average interval time between two consecutive packets injected per node by an application. Similarly, Equation 2 shows the maximum wait time for row LUs to see whether the EETD stops at the Trojan node or not. Therefore, the maximum amount of time required by EETD to localize the Trojan node is calculated by Equation 3, where, T_{LU} is the activation time required by the DMU to activate the LUs.

978-1-5386-7100-9/18 $31.00 © 2018 IEEE

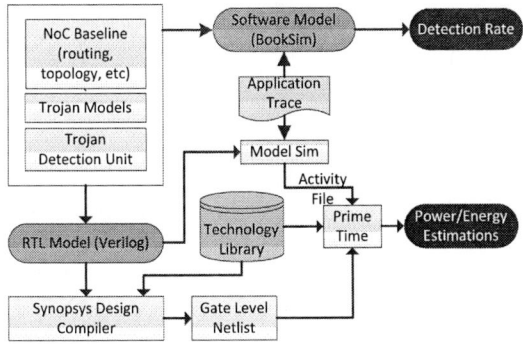

Fig. 5.   Experimental Setup

$$T_a = T_{LU} + T_t + T_s. \qquad (3)$$

In the worst case scenario, if Trojan is deactivated before T_a, EETD will not be able to identify the correct location of Trojan and introduce false positives. However, the Trojan attack will be detected by the EDUs. In case of a performance degradation attack, where Trojan is continuously tampering the packets, the active LUs by EETD detect these packets and quickly narrows the attack path to identify Trojan location.

V. DESIGN EVALUATION

A. Experimental Setup

Figure 5 shows EETD simulation platform. EETD uses 8×8 2D-mesh NoC model similar to [9] as baseline design. In this design, each node (router) has a two-stage pipeline, which performs five operations: link transversal, buffer write, route computation, switch allocation and switch transversal. Furthermore, each node uses matrix arbitration, wormhole switching, and XY routing algorithm, which are commonly used in NoC designs [9].

The hardware model and its components are designed using Verilog HDL. Synopsys Design Compiler and PrimeTime [2] (with TSMCs 65 nm cell library [3], 1 GHz operating frequency, and 1V supply voltage) are used to synthesize the design and measure power/energy consumption. We simulate PARSEC benchmarks application traces collected by Netrace 1.0 [16] on a cycle-accurate NoC simulator (Booksim 2.0 [18]) to find the detection rates under different threshold values for EETD. For a fair comparison, during our experiments, the same H2H detection units (LUs) and hardware Trojans as proposed in the state-of-the-art techniques MDSA [11] and RHTM [13] are implemented.

B. Energy and Performance Overheads

Figure 6 shows the energy overhead results for PARSEC benchmark applications. As can be seen, EETD1 (uses MSDA detection units as its LUs) have 12.6% and EETD2 (uses RHTM detection units as its LUs) have 17.6% overhead under Trojan attacks. On the other hand, H2H approach (using same MDSA and RHTM detection units as LUs) incur 39% and 56% energy overhead respectively. Moreover, EETD saves more energy for *bodytrack*, *dedup* and *swaptions* benchmarks, which have slightly higher traffic injection rate compared to other benchmarks. The reason is that at higher traffic injection rates, the time interval between

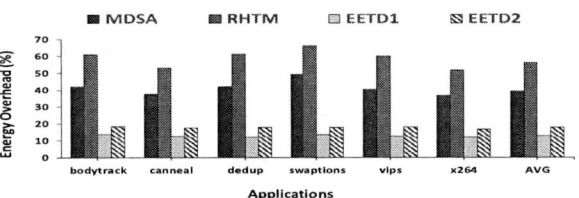

Fig. 6.   Average energy overhead by EETD vs H2H approaches (MDSA and RHTM)

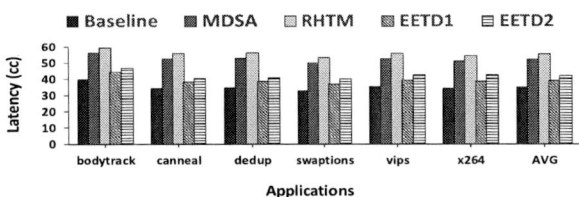

Fig. 7.   Packet latency overhead by EETD and H2H approaches (MDSA and RHTM)

consecutive attacks by Trojans becomes smaller. This allows EETD to quickly locate the Trojan and thus it saves more energy.

Figure 7 shows the packet latency overhead for different applications. As can be seen, EETD1 has 13% and EETD2 has 19% overhead over baseline design during an attack. These overhead values include an extra two clock cycle delay by the DMU for deciding which LUs to activate. During the normal execution (when Trojan is not active), LUs in EETD do not perform authentication. Thus, they have no latency overhead on applications during normal execution. However, in H2H approach the MDSA and the RHTM detection units are active all the time (i.e. tag/verify every incoming and outgoing packet), which adds two clock cycles delay for a packet on each node transfer. Therefore, the average packet latency of the NoC using MDSA increases by 50% and with RHTM it increases up to 59% in comparison to the baseline design.

C. Security Evaluation

The security of EETD design depends on the selection of threshold value. To calculate the threshold value we use Equation 1 and 2. In our experiments, we have a buffer size of eight packets and the average distance of Trojan infected nodes from a destination (x,y) is seven nodes. The threshold value for each application under given parameters is calculated in Table I. The injection rate per node and average packet latency of each application are given in Column II & III, respectively. The interval time between two consecutive packet injection at any node in the network is reported in Column IV. The last two column of Table I shows the threshold values used in our experiments. On average we need a maximum of $10K$ cycles to locate the Trojan infected nodes. It must be noted that for a fair comparison we tested EETD for applications with low traffic injection rates. If the injection rate and attack frequency are very high then threshold value is very small and EETD saves more energy.

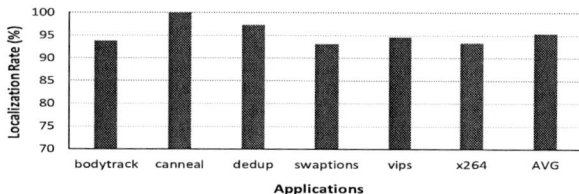

Fig. 8. Localization rate of EETD for threshold values in Table I

TABLE I

THRESHOLD VALUES FOR REAL APPLICATIONS (FOR $\alpha = 4$ AND $B = 8$)

Applications	Injection Rate	Latency (cc)	T_Δ (cc)	T_t (cc)	T_s (cc)
bodytrack	0.00132	39.55	759	9619	1266
canneal	0.00025	34.31	3975	11662	1098
dedup	0.00124	34.56	808	8548	1106
swaptions	0.00276	32.65	362	7675	1045
vips	0.00096	35.39	1038	8964	1132
x264	0.00015	34.23	6505	14172	1095
AVG	0.00111	35.11	2241	10107	1124

Figure 8 shows the localization rate of EETD. In our experiments, EETD has successfully located 95.5% of Trojan infected nodes for threshold values presented in Table I. However, EETD does not guarantee 100% localization. This is due to two main reasons. First, the detection rate of the EETD depends on the security level provided by detection units used as LUs. Second, in some cases, the interval time between two consecutive Trojan attacks is greater than a threshold time. Thus, EETD will not have sufficient time to locate the Trojan infected node and introduce false positive detection. However, in the second case, the Trojan attacks are mitigated by active EDUs used by EETD. In a normal case for a system running on 1GHz frequency, EETD take less than $10\mu s$ to localize 95% of the Trojan infected nodes for PARSEC benchmark applications in the network.

VI. CONCLUSION

We proposed EETD a runtime energy-efficient hardware Trojan localization design for untrusted third party NoCs. EETD methodology can use any state-of-the-art Trojan detection units to reduce energy and performance penalty on the system, especially for the systems where the Trojan is not active for most of the execution time or may not have a Trojan. Our experimental results on an 8x8 2D-mesh NoC running PARSEC benchmarks show that EETD reduces energy overhead by 38% and performance overhead by 40% compared to the state-of-the-art RHTM technique. Furthermore, under Trojan attacks, EETD takes $10\mu s$ (system is running on 1GHz frequency) to localize 95% of Trojan infected nodes.

REFERENCES

[1] SonicsMX Smart Interconnect Datasheet. http://www.sonicsinc.com.
[2] Synopsys Design Compiler. http://www.synopsys.com.
[3] TSMC 65nm GP Standard Cell Libraries - tcbn65gplus. https://www.cmc.ca/en/whatweoffer/products/cmc-00200-01411.aspx.
[4] D. Agrawal, S. Baktir, D. Karakoyunlu, P. Rohatgi, and B. Sunar. Trojan detection using ic fingerprinting. In *2007 IEEE Symposium on Security and Privacy (SP'07)*, pages 296–310. IEEE, 2007.
[5] T. Alves and D. Felton. Trustzone: Integrated hardware and software security. *ARM white paper*, 3(4):18–24, 2004.
[6] D. M. Ancajas, K. Chakraborty, and S. Roy. Fort-nocs: Mitigating the threat of a compromised noc. In *Proceedings of the 51st Annual Design Automation Conference*, pages 1–6. ACM, 2014.

[7] S. Bell, B. Edwards, J. Amann, R. Conlin, K. Joyce, V. Leung, J. MacKay, M. Reif, L. Bao, J. Brown, et al. Tile64-processor: A 64-core soc with mesh interconnect. In *Solid-State Circuits Conference, 2008. ISSCC 2008. Digest of Technical Papers. IEEE International*, pages 88–598. IEEE, 2008.
[8] S. Bhunia, M. S. Hsiao, M. Banga, and S. Narasimhan. Hardware trojan attacks: threat analysis and countermeasures. *Proceedings of the IEEE*, 102(8):1229–1247, 2014.
[9] H. Bokhari, H. Javaid, M. Shafique, J. Henkel, and S. Parameswaran. Supernet: multimode interconnect architecture for manycore chips. In *Proceedings of the 52nd Annual Design Automation Conference*, page 85. ACM, 2015.
[10] T. Boraten, D. DiTomaso, and A. K. Kodi. Secure model checkers for network-on-chip (noc) architectures. In *Proceedings of the 26th edition on Great Lakes Symposium on VLSI*, pages 45–50. ACM, 2016.
[11] T. Boraten and A. K. Kodi. Mitigation of denial of service attack with hardware trojans in noc architectures. In *Parallel and Distributed Processing Symposium, 2016 IEEE International*, pages 1091–1100. IEEE, 2016.
[12] T. Boraten and A. K. Kodi. Packet security with path sensitization for nocs. In *2016 Design, Automation & Test in Europe Conference & Exhibition (DATE)*, pages 1136–1139. IEEE, 2016.
[13] J. Frey and Q. Yu. A hardened network-on-chip design using runtime hardware trojan mitigation methods. *Integration, the VLSI Journal*, 56:15–31, 2017.
[14] A. Garbade, S. Weis, S. Schlingmann, B. Fechner, and T. Ungerer. Fault localization in nocs exploiting periodic heartbeat messages in a many-core environment. In *Parallel and Distributed Processing Symposium Workshops & PhD Forum (IPDPSW), 2013 IEEE 27th International*, pages 791–795. IEEE, 2013.
[15] C. H. Gebotys and R. J. Gebotys. A framework for security on noc technologies. In *VLSI, 2003. Proceedings. IEEE Computer Society Annual Symposium on*, pages 113–117. IEEE, 2003.
[16] J. Hestness, B. Grot, and S. W. Keckler. Netrace: dependency-driven trace-based network-on-chip simulation. In *Proceedings of the Third International Workshop on Network on Chip Architectures*, pages 31–36. ACM, 2010.
[17] M. Hussain and H. Guo. Packet leak detection on hardware-trojan infected nocs for mpsoc systems. In *Proceedings of the 2017 International Conference on Cryptography, Security and Privacy*, pages 85–90. ACM, 2017.
[18] N. Jiang, J. Balfour, D. U. Becker, B. Towles, W. J. Dally, G. Michelogiannakis, and J. Kim. A detailed and flexible cycle-accurate network-on-chip simulator. In *Performance Analysis of Systems and Software (ISPASS), 2013 IEEE International Symposium on*, pages 86–96. IEEE, 2013.
[19] R. JS, D. M. Ancajas, K. Chakraborty, and S. Roy. Runtime detection of a bandwidth denial attack from a rogue network-on-chip. In *Proceedings of the 9th International Symposium on Networks-on-Chip*, page 8. ACM, 2015.
[20] H. K. Kapoor, G. B. Rao, S. Arshi, and G. Trivedi. A security framework for noc using authenticated encryption and session keys. *Circuits, Systems, and Signal Processing*, 32(6):2605–2622, 2013.
[21] F. Kounelis, N. Sklavos, and P. Kitsos. Run-time effect by inserting hardware trojans, in combinational circuits. In *Digital System Design (DSD), 2017 Euromicro Conference on*, pages 287–290. IEEE, 2017.
[22] A. Malekpour, R. Ragel, A. Ignjatovic, and S. Parameswaran. Dosguard: Protecting pipelined mpsocs against hardware trojan based dos attacks. In *Application-specific Systems, Architectures and Processors (ASAP), 2017 IEEE 28th International Conference on*, pages 45–52. IEEE, 2017.
[23] A. Malekpour, R. Ragel, A. Ignjatovic, and S. Parameswaran. Trojanguard: Simple and effective hardware trojan mitigation techniques for pipelined mpsocs. In *Proceedings of the 54th Annual Design Automation Conference 2017*, DAC '17, pages 19:1–19:6, New York, NY, USA, 2017. ACM.
[24] J. Porquet, A. Greiner, and C. Schwarz. Noc-mpu: a secure architecture for flexible co-hosting on shared memory mpsocs. In *2011 Design, Automation & Test in Europe*, pages 1–4. IEEE, 2011.
[25] X. Wang, H. Salmani, M. Tehranipoor, and J. Plusquellic. Hardware trojan detection and isolation using current integration and localized current analysis. In *2008 IEEE International Symposium on Defect and Fault Tolerance of VLSI Systems*, pages 87–95. IEEE, 2008.
[26] H. M. Wassel, Y. Gao, J. K. Oberg, T. Huffmire, R. Kastner, F. T. Chong, and T. Sherwood. Networks on chip with provable security properties. *IEEE Micro*, 34(3):57–68, 2014.
[27] Q. Yu and J. Frey. Exploiting error control approaches for hardware trojans on network-on-chip links. In *2013 IEEE International Symposium on Defect and Fault Tolerance in VLSI and Nanotechnology Systems (DFTS)*, pages 266–271. IEEE, 2013.

2018 IEEE Computer Society Annual Symposium on VLSI

Combining Symbolic Computer Algebra and Boolean Satisfiability for Automatic Debugging and Fixing of Complex Multipliers

Alireza Mahzoon[1]   Daniel Große[1,2]   Rolf Drechsler[1,2]

[1]Faculty of Mathematics and Computer Science, University of Bremen, Germany

[2]Cyber-Physical Systems, DFKI GmbH, Bremen, Germany

{mahzoon,grosse,drechsle}@informatik.uni-bremen.de

Abstract—**If verification of a digital circuit fails, then debugging and fixing become the major subsequent tasks. Arithmetic units are among the most challenging circuits for debugging because of a wide variety of architectures and high design complexity. A prominent example are multipliers. Since existing automatic methods fail for these circuits, both tasks are performed manually which is typically very time-consuming.**

In this paper, we propose a complete debugging flow based on the combination of Symbolic Computer Algebra (SCA) and Boolean Satisfiability (SAT). Complete means that our method targets the complete loop until the arithmetic circuit is guaranteed to fulfill its specification. For this, our approach consists of the three phases verification, localization, and fixing. In the experimental evaluation, we demonstrate the applicability of our approach for the most complex multiplier architectures.

I. INTRODUCTION

Nowadays, arithmetic circuits play a crucial role in many computation intensive applications (e.g. signal processing and cryptography) as well as in upcoming AI architectures (e.g. for machine learning and deep learning). At the heart of these arithmetic circuits integer multipliers and adders are the dominant building blocks. Due to the growing importance of power, speed, and area in digital circuits, designers have proposed a large variety of different integer multiplier and adder architectures to meet the pressing requirements. These architectures are usually extensively parallel and hence very complex. This makes them prone to design errors.

Since the famous Pentium bug back in 1994, a lot of effort has been put in the development of suitable verification methods. Ensuring the correctness in a mathematical sense became possible by *formal verification*. However, it is very well known that non-trivial arithmetic, and in particular integer multiplication at the gate level, is still one of the biggest challenges for formal methods. Looking from the methods perspective on formal verification, essentially five directions can be distinguished: (a) *Decision Diagrams* (DDs) (such as BDDs or *BMDs), (b) *Boolean Satisfiability* (SAT) and *Satisfiability Modulo Theories* (SMT), (c) reverse engineering techniques, (d) term rewriting, and (e) *Symbolic Computer Algebra* (SCA).

Considering an advanced gate level multiplier circuit as a representative input for each method, we can observe: DDs suffer from an exponential blow-up, SAT/SMT stuck for input datawidth greater than 15 bits, reverse engineering approaches using *Arithmetic Bit-Level* (ABL) representations [1], [2] can only handle simple multiplier architectures, and term rewriting techniques [3], [4] suffer from incompleteness and may fail to prove the correctness of the circuit because of insufficient lemmas. In contrast, SCA uses a polynomial representation for the problem. To be more precise, the circuit specification is represented as a single polynomial p_{spec} and the circuit is

captured as a set of polynomials G. Then, the verification is done by testing the membership of the specification polynomial in an ideal with generators in G. This membership test corresponds to a series of divisions of p_{spec} by the circuit polynomials G (also known as Gröbner basis reduction). In the recent years, there was a renewed interest in SCA because these algebraic techniques have been applied successfully on large Galois Field arithmetic circuits [5], [6] as well as large (but architectural simple) integer multipliers [7], [8]. Recently, SCA has been shown to be very successful also for complex integer multipliers [9], [10], i.e. highly parallel architectures with up to 256 output bits. Basically, the authors of [9] have presented a logic reduction rewriting scheme consisting of XOR-rewriting and common-rewriting. This scheme allows for cancellation of so-called *vanishing monomials* (monomials which finally reduce to zero) in an efficient way *before* their blow-up during division. However, when verification fails, debugging and fixing the gate level arithmetic circuit are the next two major tasks. Hence, the designer has to (1) find the exact location of the bug and (2) determine a concrete fix.

In this paper, we propose an **approach for automatic debugging and fixing of complex gate-level arithmetic circuits combining SCA and SAT**. At first, we define the fault model. Subsequently, we show the limitations of a pure SCA-based method for debugging and fixing of complex arithmetic circuits. Then, we introduce our approach which employs an combination of SCA and SAT to successfully debug and fix arithmetic circuits. Our approach consists of three phases: verification, localization, and fixing. In each phase we employ SAT and SCA for individual subtasks as both have pros and cons. We explain the underlying decisions wrt. the chosen method which are also confirmed in the experiments. Finally, we show in the experimental evaluation on a very large set of multiplier circuits – ranging from simple to the most complex architectures – that automatic debugging and fixing is possible in practical time.

II. RELATED WORK

Automated debugging using SAT has been initially presented in [11]. The approach introduces abnormal predicates into the netlist and allows to compute a list of suspicious locations (gates). Several improvements based on iterative analysis, abstraction, incremental SAT-solving, and better accuracy have been developed, see e.g. [12], [13], [14], [15]. However, these approaches can only be used for the localization step in the considered setting, since SAT/SMT is not able to perform the proof of correctness for complex arithmetic circuits, and therefore the verification of a fix fails.

Only a few debugging approaches using SCA have been proposed. In [16] the circuit is cut into levels and then based on the polynomial modeling backward and forward rewriting is performed simultaneously to identify differences. In contrast, the authors of [17] use the remainder of Gröbner basis reduction to find the location of the bug. However, both

This work was supported by the German Federal Ministry of Education and Research (BMBF) within the project SELFIE under grant no. 01IW16001, by the University of Bremen's graduate school SyDe funded by the German Excellence Initiative, and by the German Academic Exchange Service (DAAD).

978-1-5386-7100-9/18 $31.00 © 2018 IEEE   351

methods are limited to simple arithmetic circuits and they fail if vanishing monomials appear in the final remainder, or the bug is close to the *Primary Outputs* (POs). The proposed method in [18] is an extension of [17] and tries to debug faults close to POs. It uses partitions of the primary inputs' space of the design and performs incremental equivalence checking using Gröbner basis reduction. However, it still suffers from the vanishing monomial problem; we provide more details on this problem in Section IV-B. In summary, the existing approaches are not suitable for automated debugging and fixing of complex arithmetic circuits. Hence, in this paper we propose a combination of SCA and SAT to overcome the described limitations.

III. Preliminaries

In this section, first the concepts of SCA are described. Then, the process of verifying arithmetic circuits using SCA is reviewed.

A. Notations and Definitions

Definition 1: A *Monomial* is the product of variables in the following form

$$x^\alpha = x_1^{\alpha_1} x_2^{\alpha_2} \ldots x_n^{\alpha_n} \tag{1}$$

Definition 2: A *Polynomial* is the finite combination of monomials with coefficients in k

$$f = \sum_\alpha a_\alpha x^\alpha, \quad a_\alpha \in k \tag{2}$$

The set of all polynomials with coefficients in k is denoted by $k[x_1, \ldots, x_n]$.

The monomials of a polynomial are ordered based on the ordering of the variables and their powers. We use $A > B$ to show that A is in a higher order than B. For example, if there is $P = x^2 y^3 z^2 + x^3 y z^2 + y^5 z^2$, and the ordering of variables is $x > y > z$, then the ordering of monomials would be $x^3 y z^2 > x^2 y^3 z^2 > y^5 z^2$. The first monomial after ordering is called *Leading Monomial*, and denoted by $LM(P)$.

Definition 3: A subset $I \subset k[x_1, \ldots, x_n]$ is an *ideal* if:

- $0 \in I$.
- if $f, g \in I$, then $f + g \in I$.
- if $f \in I$ and $h \in k[x_1, \ldots, x_n]$, then $hf \in I$.

Consequently, if $f_1, \ldots, f_s$ are polynomials in $k[x_1, \ldots, x_n]$, then an ideal is generated by them in the following form

$$I = <f_1, \ldots, f_s> = \{\sum_{i=1}^s h_i f_i : h_1, \ldots, h_s \in k[x_1, \ldots, x_n]\} \tag{3}$$

where $f_1, \ldots, f_s$ are called *Generators* of the ideal.

Ideal Membership is one of the well-known problems in SCA. In this problem, a polynomial $f \in k[x_1, \ldots, x_n]$ and an ideal $I = <f_1, \ldots, f_s>$ are given, and the task is to determine if $f \in I$. In order to prove that $f \in I$, the remainder of dividing f by $f_1, \ldots, f_s$ independent of division order should be always equal to zero.

Theorem 1: Let $g_1, \ldots, g_s$ be the generators of an ideal I. The remainder of dividing f by the generators is always unique, if $LM(g_1), \ldots, LM(g_s)$ are relatively prime. In other words, there should be no common variable in the leading monomial of the generators. The set of generators with this property is called *Gröbner basis* [19].

Note that Theorem 1 is already a special case of defining a Gröbner basis which facilitates the process of identifying a Gröbner basis in the context of circuits (for general case we refer to. [20]). Now, assume that $G = \{g_1, \ldots, g_s\}$ is a *Gröbner basis*. Thus, the remainder r of dividing f by $g_1, \ldots, g_s$ is uniquely determined, and the condition $r = 0$ is equivalent to membership in the ideal $I = <g_1, \ldots, g_s>$.

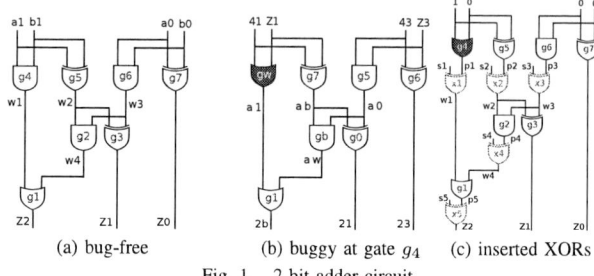

(a) bug-free (b) buggy at gate g_4 (c) inserted XORs

Fig. 1. 2-bit adder circuit

The division is denoted by $f \xrightarrow{G} r$. For example, if $f = xz$, $g_1 = x + y$, and $g_2 = yz$ then $xz \xrightarrow{g_1} -yz \xrightarrow{g_2} 0$. To perform the division of xz by g_1, first g_1 is multiplied by z to create the same leading monomial xz as f, so $g_1 \times z = xz + yz$. Then, the subtraction is performed, i.e. we compute $f - (g_1 \times z) = xz - (xz + yz) = -yz$, which is the result of the first division. Finally, this result is divided by g_2 to obtain the remainder 0.

B. Verification using SCA

In verification of gate level arithmetic circuits, the specification polynomial of the circuit and a gate level netlist are provided as inputs, and the goal is to formally prove that they are equivalent. The specification polynomial of an arithmetic circuit determines the function of circuit based on its inputs and outputs. For example, for the 2-bit adder in Fig. 1a the specification polynomial is $F := 4Z_2 + 2Z_1 + Z_0 - (2a_1 + a_0 + 2b_1 + b_0)$ where $Z = 4Z_2 + 2Z_1 + Z_0$ shows 3-bit output, and $2a_1 + a_0 + 2b_1 + b_0$ indicates addition of 2-bit inputs.

Furthermore, each logical gate can be presented by a polynomial with coefficients in $\mathbb{Z}$ determining the relation between its inputs and output. The polynomials of basic Boolean gates are

$$
\begin{aligned}
z = \neg a &\implies g := z - 1 + a & z = a \vee b &\implies g := z - a - b + a \times b \\
z = a \wedge b &\implies g := z - a \times b & z = a \oplus b &\implies g := z - a - b + 2a \times b
\end{aligned}
\tag{4}
$$

For example, if the variables in the 2-bit adder of Fig. 1a are ordered based on the reverse topological order of the circuit (from outputs toward inputs), the gates polynomials are

$$
\begin{aligned}
g_1 &:= Z_2 - w_1 - w_4 + w_1 w_4 & g_4 &:= w_1 - a_1 b_1 \\
g_2 &:= w_4 - w_2 w_3 & g_5 &:= w_2 - a_1 - b_1 + 2a_1 b_1 \\
g_3 &:= Z_1 - w_2 - w_3 + 2w_2 w_3 & g_6 &:= w_3 - a_0 b_0 \\
& & g_7 &:= Z_0 - a_0 - b_0 + 2a_0 b_0
\end{aligned}
\tag{5}
$$

The leading monomial of all gates polynomials are relatively prime. Therefore, based on Theorem 1 the set of gates polynomials is a *Gröbner basis*. Consequently, the problem of proving equivalency of the specification polynomial and the gate level netlist can be translated to a membership testing problem. In other words, if F (specification of a circuit) is a member of $<g_1, \ldots, g_s>$ where $g_1, \ldots, g_s$ are the gates polynomials of the the circuit, then the circuit is bug-free. The steps of dividing F by $g_1, \ldots, g_7$ in the 2-bit adder of Fig. 1a is shown in (6). Due to the fact that the final remainder is equal to zero, the adder is correct.

$$
\begin{aligned}
F &\xrightarrow{g_1} F_1 := 4w_1 + 4w_4 \boxed{-4w_1 w_4} + 2z_1 + z_0 - (2a_1 + a_0 + 2b_1 + b_0) \\
F_1 &\xrightarrow{g_2} F_2 := 4w_1 + 4w_2 w_3 \boxed{-4w_1 w_2 w_3} + 2z_1 + z_0 - (2a_1 + a_0 + 2b_1 + b_0) \\
F_2 &\xrightarrow{g_3} F_3 := 4w_1 \boxed{-4w_1 w_2 w_3} + 2w_2 + 2w_3 + z_0 - (2a_1 + a_0 + 2b_1 + b_0) \\
F_3 &\xrightarrow{g_4} F_4 := 4a_1 b_1 - \boxed{4a_1 b_1 w_2 w_3} + 2w_2 + 2w_3 + z_0 - (2a_1 + a_0 + 2b_1 + b_0) \\
F_4 &\xrightarrow{g_5} F_5 := \boxed{-4a_1 b_1 (a_1 + b_1 - 2a_1 b_1) w_3} + 2w_3 + z_0 - (a_0 + b_0) \\
F_5 &\xrightarrow{g_6} F_6 := 2a_0 b_0 + z_0 - (a_0 + b_0) \\
F_6 &\xrightarrow{g_7} r := 0
\end{aligned}
\tag{6}
$$

978-1-5386-7100-9/18 $31.00 © 2018 IEEE   352

Please note that all variables in polynomials are Boolean, hence a term like x^n can be replaced by x. The process of dividing F by $g_1, \ldots, g_7$ is called *backward rewriting*.

The monomials in the dashed boxes in (6) are called *vanishing monomials*. These monomials appear in the intermediate steps of backward rewriting. However, they reduce to zero in the next steps. For example, the vanishing monomial $-4w_1w_4$ is generated in the first step of backward rewriting in (6). It reduces to zero after five steps of division. Recall that the early cancellation of these vanishing monomials is crucial for scaling to large non-trivial multiplier architectures.

IV. AUTOMATIC DEBUGGING AND FIXING

In this section the proposed approach for automatic debugging and fixing of gate-level arithmetic circuits is introduced. At first, we define the fault model. Then, the limitations of a pure SCA-based method for debugging and fixing are shown. Subsequently, we give an overview of the three phases of the proposed method. Finally, each phase is detailed.

A. Fault Model

In this paper we target complex gate-level arithmetic circuits with a particular focus on integer multipliers as these are known to be very hard in both design and formal verification. As a consequence, we consider *gate misplacement* as our fault model. This well known fault model changes the functionality of the design by a wrong gate [21], [16], [17]. Such faults are likely to occur, for example, when a synthesis tool makes a mistake when optimizing the circuit. Another prominent example of introducing such kind of faults is a bug in a multiplier generator tool which are used to create a dedicated multiplier architecture (under given constraints). To be somewhat more precise, when looking on the overall structure of a multiplier it can be seen that a multiplier consists of three stages, i.e. *Partial Product Generator* (PPG), *Partial Product Accumulator* (PPA), and *Final Stage Adder* (FSA). In our experiments later we will consider faults in each of these stages and the effect on debugging and fixing.

B. Limitations of SCA for Debugging

As has been shown in recent papers, SCA-based method can be used to prove the correctness of large and complex arithmetic circuits [9], [10]. However, SCA-based approaches suffer from two major limitations when employed for debugging:

1) Vanishing Monomials in Remainder: In a bug-free gate level circuit, vanishing monomials are generated during backward rewriting and they reduce to zero after some division steps. Early cancellation of vanishing monomials based on the reported logic rewriting scheme is the major reason for scaling to complex multipliers in [9], [10]. However, in a buggy arithmetic circuit, it is possible that the vanishing monomials propagate to the remainder because of a bug. To illustrate this phenomenon, we show the backward rewriting process for a buggy 2-bit adder (cf. Fig. 1b):

$$F \xrightarrow{g_1} F_1 := 4w_1 + 4w_4 \boxed{-4w_1w_4} + 2z_1 + z_0 - (2a_1 + a_0 + 2b_1 + b_0)$$

$$F_1 \xrightarrow{g_2} F_2 := 4w_1 + 4w_2w_3 \boxed{-4w_1w_2w_3} + 2z_1 + z_0 - (2a_1 + a_0 + 2b_1 + b_0)$$

$$F_2 \xrightarrow{g_3} F_3 := 4w_1 \boxed{-4w_1w_2w_3} + 2w_2 + 2w_3 + z_0 - (2a_1 + a_0 + 2b_1 + b_0)$$

$$F_3 \xrightarrow{g_4} F_4 := 2a_1 + 2b_1 - 4a_1b_1 \boxed{-4(a_1 + b_1 - a_1b_1)w_2w_3} + 2w_2 + 2w_3 + z_0$$
$$\qquad - (a_0 + b_0)$$

$$F_4 \xrightarrow{g_5} F_5 := 4a_1 + 4b_1 - 8a_1b_1 \boxed{-4(a_1 + b_1 - 2a_1b_1)w_3} + 2w_3 + z_0 - (a_0 + b_0)$$

$$F_5 \xrightarrow{g_6} F_6 := 4a_1 + 4b_1 - 8a_1b_1 \boxed{-4(a_1 + b_1 - 2a_1b_1)a_0b_0} + 2a_0b_0 + z_0$$
$$\qquad - (a_0 + b_0)$$

$$F_6 \xrightarrow{g_7} r := 4a_1 + 4b_1 - 8a_1b_1 \boxed{-4(a_1 + b_1 - 2a_1b_1)a_0b_0} \qquad (7)$$

The final remainder r of backward rewriting (result of the division of F_6) is not equal to zero. Therefore, the circuit is buggy. The generated remainder consists of two parts. The first part $4a_1 + 4b_1 - 8a_1b_1 = 4 \times (a_1 + b_1 - 2a_1b_1)$ is composed of three monomials which originate from the difference in the buggy and correct gate polynomials at gate g_4 (see Fig. 1a and Fig. 1b): $P_{buggy} - P_{correct} = P_{OR} - P_{AND} = a_1 + b_1 - 2a_1b_1$. However, the second part of the remainder $-4(a_1 + b_1 - 2a_1b_1)a_0b_0$ (shown in the dashed box) is a part of vanishing monomial propagated to the remainder due to the bug presence. If we apply the approach from [16] to the just discussed example it fails. The reason is that for the vanishing monomials there will be no counterpart monomials when executing the forward rewriting and backward rewriting as presented in [16]. Furthermore, also the SCA-based method from [17] fails. This method extracts the difference polynomial per gate and compares it with the remainder. However, this is not possible if a vanishing monomial appears. Finally, note that all experiments in both papers only consider simple adder and multiplier architectures, i.e. carry save adders are used and hence no vanishing monomials occurs.

2) Blow-up during Verification of Buggy Circuits: If there is a bug close to POs of a large arithmetic circuit, a blow-up happens in the number of monomials during backward rewriting and hence the SCA-based verification method fails. The reason is that a buggy gate adds several monomials, i.e. the difference between buggy and correct gate polynomials, to the process of backward rewriting. Despite other monomials, these monomials do not cancel and grow exponentially in the subsequent steps of the division when moving towards the inputs.

To overcome these limitations, we take advantage of both SAT and SCA in our approach. An overview is presented in the next section.

C. Overview of Proposed Method

Algorithm 1 shows the pseudo-code of our proposed approach. Before we go into the details, it can be seen that our approach consists of three phases: *Verification*, *Localization*, and *Fixing*. In each phase we employ SAT and SCA for individual subtasks as both have pros and cons. We explain the underlying decisions wrt. the chosen method. We provide a summary up-front in Table I. The first column gives the name of the phase. The second column shows the subtask, if applicable. The third column distinguishes per phase/subtask between SCA and SAT. The fourth to seventh column defines whether the circuit is bug-free or not. In case of a bug, we subdivide the circuit into three regions: I, II, III which just defines the depth of the bug seen from the inputs (so III means a deep bug near to the POs). Note that '+' means that the respective method gives a result, and '-' that it fails. Finally, in the rightmost column we show the conclusion that can be drawn; which also has been finally implemented in our approach. In the following sections we describe now each phase in more detail.

D. Phase 1: Verification

The first phase of our proposed approach is verification (see Line 1 – Line 2 in Algorithm 1). For a complex gate level arithmetic circuit we want to determine whether the circuit is correct or not. As already explained in the introduction and confirmed later in the experiments in Section V, we can observe that SAT is very fast in disproving, i.e. to show that the circuit is buggy. However, SAT fails (time out) when the circuit is correct. In contrast, SCA is one of the best approaches for verifying a bug-free arithmetic circuit, especially when

TABLE I
APPLICABILITY OF SCA AND SAT IN DIFFERENT PHASES OF DEBUGGING

Phase	Subtask	Method	Bug-free	Bug level			Conclusion
				I	II	III	
verification		SCA	+	+	-	-	Using SAT and SCA in parallel (SAT for buggy
		SAT	-	+	+	+	and SCA for correct circuits)
Localization	Extracting Initial Suspicious Gates	SCA		+	-	-	Using SAT
		SAT		+	+	+	
	Generating Test-vectors	SCA		+	-	-	Using SAT
		SAT		+	+	+	
	Refining Suspicious Gates	SCA		++	++	++	Using SCA because it is faster
		SAT		+	+	+	
Fixing (correct fix/incorrect fix)		SCA		+/+*	+/-	+/-	Using SAT and SCA in parallel (SAT when fix does not
		SAT		-/+	-/+	-/+	work and SCA for proving correctness)

+: Applicable   ++: Applicable and fast   -: Not applicable
*The sign before the slash (after the slash) describes the applicability of the method when the fix at the candidate location is correct (incorrect).

Algorithm 1 Proposed method for Debugging and Fixing

Input: Arithmetic circuit C, Golden circuit C_G
Output: Correct circuit
1: VerifyWithParallelSAT_SCA (C, C_G) ▷ Verification
2: **if** C is bug-free **then return** C
3: **else**
4:    $SG \leftarrow$ ExtractSGWithSAT (C, C_G) ▷ Localization
5:    $i \leftarrow 0; v \leftarrow \emptyset$
6:    **while** Size$(SG) > 1$ **and** $i < 10$ **do**
7:      $v \leftarrow$ GenerateTestvectorWithSAT (C, C_G, v)
8:      $C_X \leftarrow$ InsertXOR (C, SG, v)
9:      $SG \leftarrow$ RefineSGWithSCA(C_X)
10:      **if** SG has not changed **then** $i \leftarrow i + 1$
11:      **else** $i \leftarrow 0$
12:      **end if**
13:    **end while**
14:    $C_F \leftarrow$ FixWithParallelSAT_SCA (C, C_G, SG) ▷ Fixing
15: **end if**
16: **return** C_F

---

advanced rewriting techniques are employed, for instance XOR-rewriting and common-rewriting [9], [10].

Nevertheless, its performance is poor when there is a bug close to the POs (see previous discussion in Section IV-B). In order to take advantages of both SAT and SCA, we run them in parallel in our approach. When we obtain a result from one of the methods, we terminate the other one. As a result, buggy and bug-free circuits can be verified in acceptable time.

E. Phase 2: Localization

The second phase of our proposed method is localization (see Line 4 – Line 13 in Algorithm 1). The goal of this phase is to extract candidates for the location of the bug in the circuit. In the first step of localization, an initial list of suspicious buggy gates are extracted (Line 4). Next, a test-vector is generated for the buggy circuit (Line 7). Subsequently, XOR gates are inserted just after each suspicious gate and the test-vector is applied to the primary inputs (PIs) (Line 8). Finally, the list of suspicious gates is refined by backward rewriting and evaluating remainder (Line 9). This process continues iteratively until the size of the set of suspicious gates SG reduces to 1 or the SG does not change after 10 iterations. In following we detail each step and explain for which subtask we employ which method, i.e. SCA or SAT (remember to see also Table I for a summary).

1) Extracting Initial Suspicious Gates: Complex arithmetic circuits usually consist of many logical gates. Therefore, if there is a bug in the circuit, the size of the search space (i.e. the number of suspicious gates) will be large (number of gates). Thus, a pre-process to reduce the size of the search space is essential. To this end, the following method is proposed:

1) Use formal to identify outputs which are affected by the bug (i.e. there is an input vector such that the golden and buggy circuit differ)
2) Create cones for these outputs based on the gates which are connected
3) Determine the gates that are in the intersection of all cones; they form the initial set of suspicious gates

This task can be mapped to a MITER circuit for each output bit. In different studies (not reported) we observed that SAT performed very well and an SCA-based solution only gave results for bugs in the circuit region I, i.e. bugs near to the PIs.

Coming back to the 2-bit adder example in Fig. 1b. The only affected output is Z_2. The cone for Z_2 is $C_2 = \{g_1, g_2, g_4, g_5, g_6\}$, which is also the initial suspicious gates list. Please note that if there are more than one affected output, then the intersection of output cones creates the initial suspicious gate list.

2) Generating Test-vectors: After the verification of the arithmetic circuit, a counter-example is available from the SAT-solver (SAT-solver result from Phase 1: Verification). This counter-example can be used as the initial test-vector because it presents the input values resulting in a "wrong" output value. However, we usually need more test-vectors to localize the bug in the design. To this end, we use blocking clauses and run the SAT-solver again to obtain a new test-vector.

3) Insertion of XORs: The faulty gate in the circuit is among the gates in the set of suspicious gates SG. Assume that t is a generated test-vector which exhibits the fault. Then, the bug has been activated in the circuit by t. In other words, the faulty gate has generated a "wrong" value at its output which is the negation of the correct gate value – this assumption is valid, since we consider the gate misplacement fault model. This "wrong" value is propagated through gates in the output cone of the faulty gate, and leads to the "wrong" value at the PO(s) of the circuit. Changing the output value of the faulty gate results in the correct value at the output of the circuit. Hence, the problem can be formulated as finding gates where negating their outputs corrects the final result. As the XOR gate fulfills this property, we use it as follows: Assume that $g_1, \ldots, g_n$ are the suspicious gates, and t is a test-vector. We apply t to the primary inputs of the circuit, and insert $x_1, \ldots, x_n$ which are XOR gates just after the suspicious gates. One of the inputs of each XOR gate is connected to the output of the each suspicious gate, and the other XOR input becomes a new free input. We name all these inputs $s_1, \ldots, s_n$ and call them *selectors* in the following. The problem is now to find a *selector* by setting it to 1 (all other to 0), such that the final output of the circuit becomes correct.

Consider again the buggy 2-bit adder circuit in Fig. 1b. Recall that the initial set of suspicious gates is $\{g_1, g_2, g_4, g_5, g_6\}$. If $t_1 = 1000$ (i.e. $a_1 = 1, b_1 = 0, a_0 = 0, b_0 = 0$) is a test-vector, the new circuit after applying t_1 to the PIs and inserting XOR gates just after each suspicious gate can be seen in Fig. 1c.

4) Refining Suspicious Gates: To goal of this subtask is to refine the set of suspicious gates SG. In the following we give an SCA-based method for this subtask, since it is faster than the SAT-based formulation (empirically shown in the experiments).

Based on the given test-vector t, we recompute the specification polynomial as follows: We have now concrete input values from t which are applied to the original specification polynomial. Please note that the input of the new circuit (current problem instance) are only the selectors. The gate polynomials of the buggy 2-bit adder with the previously inserted XORs and the test-vector $t_1 = 1000$ are:

$$
\begin{aligned}
&x_5 := Z_2 - s_5 - p_5 + 2s_5p_5 && x_2 := w_2 - s_2 - p_2 + 2s_2p_2 \\
&g_1 := p_5 - w_1 - w_4 + w_1w_4 && x_3 := w_3 - s_3 - p_3 + 2s_3p_3 \\
&x_4 := w_4 - s_4 - p_4 + 2s_4p_4 && g_4 := p_1 - 1 \\
&g_2 := p_4 - w_2w_3 && g_5 := p_2 - 1 \\
&g_3 := Z_1 - w_2 - w_3 + 2w_2w_3 && g_6 := p_3 - 0 \\
&x_1 := w_1 - s_1 - p_1 + 2s_1p_1 && g_7 := Z_0 - 0
\end{aligned}
\tag{8}
$$

Now, backward rewriting is performed. The resulting remainder is different from 0 and only depends on the selector variables. Before showing backward rewriting for the concrete example, two important points should be noticed: 1) the terms containing $s_m \times s_n$ (i.e. multiplication of selectors) are reduced to 0 during backward rewriting, because only one of the selectors should be equal to 1; 2) due to the fact that we are dealing with a 2-bit adder, the specification polynomial and subsequently all the polynomials during backward rewriting should be modulo 2^{2+1}, because the maximum output size for addition of two n-bit numbers gives $n + 1$ bits.

For our running example we get:

$$
\begin{aligned}
F &\xrightarrow{x_5} F_1 := 4s_5 + 4p_5 + 2z_1 + z_0 - 2 \\
F_1 &\xrightarrow{g_1} F_2 := 4s_5 + 4w_1 + 4w_4 - 4w_1w_4 + 2z_1 + z_0 - 2 \\
&\quad\cdots \\
F_{10} &\xrightarrow{g_6} F_{11} := 4s_1 - 2s_2 - 2s_3 + 4s_5 + z_0 + 4 \\
F_{11} &\xrightarrow{g_7} \boxed{r := 4s_1 - 2s_2 - 2s_3 + 4s_5 + 4}
\end{aligned}
\tag{9}
$$

To correct the 2-bit adder circuit, the remainder $4s_1 - 2s_2 - 2s_3 + 4s_5 + 4$ should become 0. So, we should find all possible combinations for the selectors (one-hot encoding) such that $r = 0$. We get:

$$
\begin{aligned}
&s_1 = 1, \; s_2 = 0, \; s_3 = 0, \; s_4 = 0, \; s_5 = 0 \implies \boxed{r = 8 \bmod 8 = 0} \\
&s_1 = 0, \; s_2 = 1, \; s_3 = 0, \; s_4 = 0, \; s_5 = 0 \implies r = 2 \\
&s_1 = 0, \; s_2 = 0, \; s_3 = 1, \; s_4 = 0, \; s_5 = 0 \implies r = 2 \\
&s_1 = 0, \; s_2 = 0, \; s_3 = 0, \; s_4 = 1, \; s_5 = 0 \implies r = 4 \\
&s_1 = 0, \; s_2 = 0, \; s_3 = 0, \; s_4 = 0, \; s_5 = 1 \implies \boxed{r = 8 \bmod 8 = 0}
\end{aligned}
\tag{10}
$$

As can be seen, when setting s_1 or s_5 to 1, the remainder becomes 0. Hence, the suspicious gates list is reduced to $\{g_4, g_1\}$ whose outputs are connected to x_1 and x_5, respectively. In the next iteration, XOR gates are inserted only after these two suspicious gates, and another test-vector is applied to PIs and suspicious gates are refined. Nevertheless, in this concrete example the suspicious gates list cannot be further reduced even after applying all test-vectors. In large arithmetic circuits, the number of test-vectors are extremely large. Therefore, in order to avoid repeating subtasks 2, 3, and 4 for all existing test-vectors, we use a termination criteria of 10 iterations for the while-loop in Algorithm 1. In other words, if the size of suspicious gates list does not change after 10 iterations, then the suspicious gates list is sent to fixing phase.

F. Phase 3: Fixing

The final phase of our proposed method is Fixing (see Line 14 – Line 16 in Algorithm 1). Based on the extracted candidates in the bug localization phase, we can create a list of potential gate replacements. For example, if we assume that the used library for creating arithmetic circuits consists of the basic logical gates $\{AND, OR, XOR, NOT\}$, then there are two possible gate replacements for each candidate. To find the correct gate replacement, we first choose one of the changes from the list, and apply it to the circuit. Then, we perform parallel verification using SCA and SAT (see Section IV-D). Therefore, after gate replacement, if the circuit is still buggy the SAT-based verification returns a counter-example and we continue with the next possible replacement. Otherwise, if the circuit can be fixed with the current gate replacement, the SCA-based verification successfully proves this.

Considering again the running 2-bit buggy adder example of Fig. 1b, from the localization phase we know that g_4 and g_1 are the final suspicious gates. The corresponding list of gate replacements which may fix the bug is therefore $\{g_4(OR) \to g_4(XOR), \; g_4(OR) \to g_4(AND), \; g_1(OR) \to g_1(XOR), \; g_1(OR) \to g_1(AND)\}$. First, g_4 is converted to an XOR gate, and the circuit is verified. Because the circuit is still buggy, a counter-example is returned. When applying the second change and verifying the circuit, the final remainder of SCA-based verification becomes 0 and hence we have found the fix.

V. EXPERIMENTAL RESULTS

We have implemented our approach in C++. For SCA we implemented Gröbner basis reduction including the rewriting techniques proposed in [9]. The experiments have been carried out on an Intel(R) Core(TM) i5-4300M CPU 2.60 GHz with 16 GByte of main memory. In order to evaluate the efficiency of our combined SCA and SAT approach, we consider different complex multiplier architectures generated by the *Arithmetic Module Generator* [22]. These multipliers are in the form of RTL Verilog code. Thus, we run *Yosys* [23] (commands: *read_verilog; proc; opt; write_verilog*) to synthesize them to a gate-level netlist. The generated multipliers consist of three stages: 1st) *Partial Product Generator* (PPG), 2nd) *Partial Product Accumulator* (PPA) which is a multi-operand adder, and 3rd) *Final Stage Adder* (FSA) which is a two-operand adder. All the benchmarks are named based on the type of the used architecture in the different stages (this includes for instance Booth encoding and Carry look-ahead adders in the respective stages); see details in the legend below the result table later. We also used MiniSat v1.14 [24] for SAT-solving in our experiments.

In Table II, we report the results of applying debugging methods to different types of multipliers. Please note that the *Time-Out* (TO) has been set to 24 hours. The first column of Table II shows the type of the multiplier (see below the table for the abbreviations). The second column *I/O bits* gives the number of input and output bits. The third column *Bug* lists whether the circuit is bug-free, or the stage where the bug has been inserted randomly.

The results of the verification phase are reported in the fourth column **Verification**, which consists of the five following subcolumns: While *SCA* and *SAT* refer to our SCA-implementation and MiniSat, respectively, *Comm.* reports the results of the commercial formal verification tool OneSpin. Next, our integrated approach is given in subcolumn *Ours*, and finally *Imp.* presents the improvement of our approach compared to the commercial tool. As can be seen pure SCA-based verification only works when there is no bug in the circuit or the bug is in the first stage (i.e. PPG) of the design. In contrast, pure SAT-based verification times-out for bug-free circuits already for the multipliers with only 16/32 I/O bits. The commercial tool also times-out for bug-free multipliers bigger than 16/32 I/O bits. In contrast, our integrated verification method (Section IV-D) can verify bug-free multipliers, and also buggy circuits when the bug is in any stage of the design. Our verification method is up to 641 times faster than the commercial tool.

TABLE II
RESULTS OF DEBUGGING DIFFERENT TYPES OF MULTIPLIERS (RUN-TIMES IN SECONDS)

Benchmark post-synth.	I/O	Bug	Verification					Localization			Fixing			Overall	SOTA	
			SCA	SAT	Comm.	Ours	Imp.	SAT [11]	Ours	Imp.	SCA	SAT	Ours	Ours	[16]	[17]
BP-CT-BK	16/32*	Bug-free	0.29	TO	218.00	0.34	641.18x							0.3		
		Stage 1	0.31	0.03	0.06	0.03	2.00x	13.1	5.0	2.62x	0.66	TO	0.44	5.5	F	F
		Stage 2	TO	0.02	0.05	0.02	2.50x	14.9	3.7	4.03x	TO	TO	0.41	4.1	F	F
		Stage 3	TO	0.03	0.07	0.03	2.33x	10.7	3.6	2.97x	TO	TO	0.46	4.1	F	F
SP-WT-CL	32/64	Bug-free	9.09	TO	TO	10.70	-							10.7		
		Stage 1	10.42	0.20	0.38	0.22	1.73x	115.6	61.5	1.88x	19.51	TO	10.95	72.7	F	F
		Stage 2	TO	0.16	0.41	0.20	2.05x	1208.0	108.5	11.13x	TO	TO	11.10	119.8	F	F
		Stage 3	TO	0.12	0.35	0.14	2.50x	165.4	40.1	4.12x	TO	TO	10.87	51.1	F	F
BP-AR-RC	32/64	Bug-free	3.79	TO	TO	4.54	-							4.5		
		Stage 1	4.37	0.14	0.24	0.15	1.60x	208.7	37.7	5.54x	8.29	TO	4.63	42.5	F	F
		Stage 2	TO	0.10	0.31	0.13	2.38x	89.3	25.7	3.47x	TO	TO	4.93	30.8	F	F
		Stage 3	TO	0.08	0.30	0.09	3.33x	60.8	24.7	2.46x	TO	TO	4.81	29.6	F	F
BP-WT-CL	32/64	Bug-free	11.59	TO	TO	13.84	-							13.8		
		Stage 1	12.51	0.08	0.23	0.08	2.87x	291.2	67.5	4.31x	24.23	TO	13.93	81.5	F	F
		Stage 2	TO	0.11	0.27	0.12	2.25x	228.2	105.3	2.17x	TO	TO	14.20	119.6	F	F
		Stage 3	TO	0.45	0.32	0.56	0.57x	148.1	43.7	3.39x	TO	TO	15.66	59.9	F	F
SP-WT-CL	64/128	Bug-free	161.78	TO	TO	192.14	-							192.1		
		Stage 1	184.30	1.14	12.66	1.22	10.38x	1107.4	343.3	3.23x	348.48	TO	193.52	538.0	F	F
		Stage 2	TO	3.86	77.00	4.14	18.60x	1745.6	250.6	6.97x	TO	TO	212.84	467.6	F	F
		Stage 3	TO	0.57	4.46	0.63	7.08x	1047.9	518.5	2.02x	TO	TO	192.78	711.9	F	F
SP-CT-BK	64/128	Bug-free	67.04	TO	TO	79.24	-							79.2		
		Stage 1	73.30	1.79	42.47	1.91	22.24x	1192.3	704.3	1.69x	140.94	TO	81.25	787.5	F	F
		Stage 2	TO	0.42	22.05	0.46	47.93x	677.7	265.6	2.55x	TO	TO	79.73	345.8	F	F
		Stage 3	TO	0.88	9.56	0.94	10.17x	3612.0	349.8	10.33x	TO	TO	80.31	431.1	F	F
BP-WT-CL	64/128	Bug-free	226.65	TO	TO	271.93	-							271.9		
		Stage 1	255.82	0.94	2.05	0.99	2.07x	1323.1	388.0	3.41x	489.47	TO	274.90	663.9	F	F
		Stage 2	TO	0.43	2.61	0.53	4.92x	1669.0	514.5	3.24x	TO	TO	272.99	788.0	F	F
		Stage 3	TO	2.28	14.85	2.40	6.19x	698.6	241.1	2.90x	TO	TO	274.44	517.9	F	F

Stage 1 ⇒ **SP**: Simple partial product generator   **BP**: Booth partial product generator   **TO**: Time-Out   **F**: Failed
Stage 2 ⇒ **AR**: Array   **WT**: Wallace tree   **CT**: Compressor tree   *Due to page limitation we only report results
Stage 3 ⇒ **RC**: Ripple carry adder   **CL**: Carry look-ahead adder   **BK**: Brent-kung adder   for one 16/32 I/O bits complex multiplier.

The fifth column **Localization** shows the run-times of the localization phase. We have compared our method against SAT-based localization [11]. While SAT-based localization is able to compute the set of fault candidates for all three stages, our method is faster on all benchmarks. The respective improvement is listed in the third subcolumn. As can be seen we achieve improvements of up to a factor of 11.

The experimental results for the fixing phase are reported in the sixth column **Fixing**. As can be seen, the pure SCA-based fixing method is only able to fix bugs in the first stage (PPG) of the multiplier. SAT-based fixing fails for all the cases because when a correct gate replacement is considered, this method cannot verify the bug-free circuit. The experimental results confirm that our fixing method can fix the bugs in any stage of the design.

The overall run-time of our proposed method is reported in the seventh column **Overall**. It gives the sum of the run-times of each phase, i.e. verification, localization, and fixing.

The eighth column shows the results using the *State-Of-The-Art* (SOTA) SCA-based methods [16] and [17]. However, both methods fail to debug the considered complex multipliers due to appearance of vanishing monomials in the final remainder (see discussion and example in Section IV-B).

A final remark on the relevance of vanishing monomials for the considered multiplier architectures: On average when running our proposed approach for the benchmarks 135,498 vanishing monomials have been canceled during the divisions. This number has been calculated running the complete flow, i.e. all three phases per benchmark.

VI. CONCLUSION

In this paper, we have introduced a novel approach based on the combination of SCA and SAT for automatic debugging and fixing complex arithmetic circuits. The proposed approach consists of three phases. First, the arithmetic circuit is verified. Then, a list of candidates for the location of the bug is extracted. Finally, the design is fixed. The experimental results

showed that our approach allows for debugging and fixing of complex arithmetic circuits while other state-of-the-arts methods fail.

REFERENCES

[1] D. Stoffel and W. Kunz, "Equivalence checking of arithmetic circuits on the arithmetic bit level," *TCAD*, vol. 23, no. 5, pp. 586–597, 2004.
[2] E. Pavlenko, M. Wedler, D. Stoffel, O. Wienand, E. Karibaev, and W. Kunz, "Modeling of custom-designed arithmetic components in ABL normalization," in *FDL*, 2008, pp. 124–129.
[3] S. Vasudevan, V. Viswanath, R. W. Sumners, and J. A. Abraham, "Automatic verification of arithmetic circuits in rtl using stepwise refinement of term rewriting systems," *TC*, vol. 56, no. 10, pp. 1401–1414, 2007.
[4] D. Kapur and M. Subramaniam, "Mechanical verification of adder circuits using rewrite rule laboratory," *Formal Methods in System Design: An International Journal*, vol. 13, no. 2, pp. 127–158, 1998.
[5] J. Lv, P. Kalla, and F. Enescu, "Efficient Gröbner basis reductions for formal verification of Galois field multipliers," in *DATE*, 2012, pp. 899–904.
[6] ——, "Efficient Gröbner basis reductions for formal verification of Galois field arithmetic circuits," *TCAD*, vol. 32, no. 9, pp. 1409–1420, Sept 2013.
[7] F. Farahmandi and B. Alizadeh, "Gröbner basis based formal verification of large arithmetic circuits using gaussian elimination and cone-based polynomial extraction," *MICPRO*, vol. 39, no. 2, pp. 83–96, 2015.
[8] M. Ciesielski, C. Yu, D. Liu, W. Brown, and A. Rossi, "Verification of gate-level arithmetic circuits by function extraction," in *DAC*, 2015, pp. 52:1–52:6.
[9] A. Sayed-Ahmed, D. Große, U. Kühne, M. Soeken, and R. Drechsler, "Formal verification of integer multipliers by combining Gröbner basis with logic reduction," in *DATE*, 2016, pp. 1048–1053.
[10] D. Ritirc, A. Biere, and M. Kauers, "Column-wise verification of multipliers using computer algebra," in *FMCAD*, 2017.
[11] A. Smith, A. G. Veneris, and A. Viglas, "Design diagnosis using boolean satisfiability," in *ASP-DAC*, 2004, pp. 218–223.
[12] B. Le, H. Mangassarian, B. Keng, and A. G. Veneris, "Non-solution implications using reverse domination in a modern SAT-based debugging environment," in *DATE*, 2012, pp. 629–634.
[13] B. Keng and A. G. Veneris, "Path-directed abstraction and refinement for SAT-based design debugging," *TCAD*, vol. 32, no. 10, pp. 1609–1622, 2013.
[14] A. M. Gharehbaghi and M. Fujita, "A new approach for debugging logic circuits without explicitly debugging their functionality," in *ATS*, 2016, pp. 31–36.
[15] H. Riener and G. Fey, "Exact diagnosis using boolean satisfiability," in *ICCAD*, 2016.
[16] S. Ghandali, C. Yu, D. Liu, W. Brown, and M. Ciesielski, "Logic debugging of arithmetic circuits," in *ISVLSI*, 2015, pp. 113–118.
[17] F. Farahmandi and P. Mishra, "Automated test generation for debugging arithmetic circuits," in *DATE*, 2016, pp. 1351–1356.
[18] ——, "Automated debugging of arithmetic circuits using incremental gröbner basis reduction," in *ICCD*, 2017, pp. 1–6.
[19] D. A. Cox, J. Little, and D. O'Shea, *Ideals Varieties and Algorithms*. Springer, 1997.
[20] W. W. Adams and P. Loustaunau, *An Introduction to Grobner Bases*. American Mathematical Society, 1994.
[21] A. Veneris and I. N. Hajj, "Design error diagnosis and correction via test vector simulation," *TCAD*, vol. 18, no. 12, pp. 1803–1816, 1999.
[22] "Arithmetic module generator based on acg," available at www.aoki.ecei.tohoku.ac.jp/arith/, 2015.
[23] C. Wolf, "Yosys open synthesis suit," available at http://www.clifford.at/yosys/, 2015.
[24] N. Eén and N. Sörensson, "Minisat," available at http://minisat.se/, 2008.

978-1-5386-7100-9/18 $31.00 © 2018 IEEE

Enhancing lifetime of PCM-based main memory with Efficient Recovery of Stuck-at Faults

1st Marjan Asadinia
Computer Science and Computer Engineering Department
University of Arkansas
Fayetteville, Arkansas, USA
masadini@uark.edu

2nd Christophe Bobda
Computer Science and Computer Engineering Department
University of Arkansas
Fayetteville, Arkansas, USA
cbobda@uark.edu

Abstract—Among several nonvolatile memory (NVM) candidates, PCM is selected as an attractive replacement to DRAM and comes with both challenges and opportunities. It has beautiful characteristics like non-volatility, better scalability, and lower leakage power. However, limited write endurance is the main burden toward its adoption in practice. It means that after a certain number of writes, some memory cells permanently stuck at either '0' or '1'. To solve this problem and provide strong fault tolerant system, some recovery techniques with minimal storage overhead are required. In this work, we propose a recovery mechanism that relies on static partitioning of a data block into some small number of groups and spreading out faults across the groups uniformly. We then exploit inversion mechanism along with shifting mechanism to continue the use of the failed cell with stuck-at value. So, our proposed method can recover multi bit stuck at faults per partition. Compare to the existing mechanisms, our experimental results for multi-threaded workloads reveal considerable improvement in lifetime and the number of recoverable failures per data block.

I. INTRODUCTION

For the last decades, DRAM popularly used as main memory [7]. However, recent years have witnessed increasing the scaling of applications which results in the increasing demand of memory capacity in order to serve large working set. Besides, DRAM has high power consumption due to using capacitor for storing data and refreshing mechanism for preventing from the leakage of the capacitor. So, since systems' power budget is usually limited, this is an important issue to be addressed. On the other hand, DRAM suffers from scalability limitation, even though process technology has improved. Therefore, finding a new type of memory that can be used for main memory instead of DRAM have motivated researchers to explore alternatives of conventional memories [7]. In this line, Non-volatile memories (NVMs) offer attractive properties (such as high density, energy characteristics and near-zero standby power) and are widely seen as promising candidates compared to the conventional memories [5].

Among all NVMs, Phase Change Memory (PCM) is selected to be the leading contender for next generation of main memory system [5], [12], [19]. A PCM cell exploits Chalcogenide material (GST) to define different states. It relies on two different states, low resistive crystalline state (set state) and high resistive amorphous state (reset state). PCM has

beautiful characteristics like high scalability, non-volatility, low leakage power, and reasonable read latency.

Along with these benefits, PCM suffers from limited write endurance and can tolerate a limited number of write operations which is about 10^6-10^8 writes per cell. It means that the cell faces hard error and it is stuck at either "1" or "0" state after reaching to its lifetime limit.

To overcome the PCM lifetime limitation and prevent from being worn out quickly, several proposals have been proposed. Some of them using wear-leveling to spread the writes across memory cells uniformly [6], [11], [15]. Although these schemes limit the write frequency and try to reduce the number of write traffics per memory cell, but they are not efficient for better memory lifetime.

Some other prior work have been presented to provide error correction and error recovery schemes to address the write-related failures of PCM and recover from stuck at faults. For example, ECP [13] can correct up to 6 bit fails per 64B line. It uses some pointers for fast access to the faulty positions. It then can correct the failure by replacing it with the stored correct value. Since this correction scheme is simple and reasonable but imposes large overhead because it is forced to store 61 additional bits. Compare to ECP, PAYG [9] provides stronger error correction scheme but it also imposes large overhead to the system since it uses hierarchical ECP.

On the other hand, DRM [4] and Zombie [1] use pairing mechanism to pair the faulty pages with other healthy pages and try to recover stuck at faults and postpone the PCM pages' wear-out. However, DRM suffers from lack of wear-leveling mechanism. Also, finding the best compatible target for pairing is not always available for DRM. On the other hand, Zombie has a complex memory controller which impacts on the efficiency of this algorithm. It is due to the fact that Zombie follows a difficult way to find the best target for pairing, so it imposes large complexity to the memory controller.

Other methods like SAFER [14], RDIS [16], and Aegis [3] are proposed with the goal of providing partitioning and inversion mechanisms for recovering the stuck at faults in each data block. More precisely, SAFER uses dynamic partitioning of the data block into some small groups by considering the fact that only one fail bit can exist per partition. It can not tolerate multi stuck at faults per partition. Therefore, RDIS is

978-1-5386-7100-9/18 $31.00 © 2018 IEEE   357

proposed to improve SAFER with the guarantee of recovering three faults per partition. Aegis technique is presented to use the best partitioning and it can also recover much more than three faults per partition. Since Aegis follows SAFER approach, but its attempt for finding the best partitioning imposes large complexity to the memory controller.

To overcome the deficiencies of previous proposals, we propose an error recovery scheme in order to reuse some faulty cells which are still readable. Unlike prior partitioning methods that try to partition data block dynamically, we use the fixed and static partitioning due to prevent from large complexity, large overhead, and long search time for finding the best partitioning. Our approach also has the capability of recovering multiple bits per partition. In presence of new faults, we keep our static data block partitioning and never go through re-partitioning of the data block.

We use error detection mechanism to detect the failed cell with stuck-at faults. As soon as the failed cell (either stuck-at '0' or stuck-at '1') is detected, our proposed method statically partitions the data block into some smaller groups (our sensitivity analysis shows the actual number of partitions in the evaluation section), it then uses inversion mechanism and tries to invert the content and checks whether it fits with the stuck-at value cell or not. If so, it is simply recovered and we can have fault-free read and write mechanisms. It is notable that we use one flip bit in each partition to show that the content is inverted and read and write circuits should be aware of it. It is obvious that there exist some cases that the inverted form of the data block content and the failed cell with stuck at value can not be compatible to each other and inversion mechanism fails, so we focus on another way. Indeed, in this situation, we use the shifting technique with low overhead in order to shift the data block content with the appropriate and tolerable amount of shifts and try to find the compatible position of the failed cell with stuck at value and the shifted data block content.

To sum up, compare to prior works, our method can recover multiple faulty bits per partition with low overhead. It is also orthogonal to the previous error correcting schemes and can be used in conjunction with them. Clearly the efficiency of all partitioning schemes depend on the number of partitions and the number of failure coverage in each partition. Since we use static partitioning, we can benefit from the advantage of low overhead and low complexity of fixed partitioning scheme. Moreover, we use the inversion technique along with shifting mechanism which both provide the possibility of recovery for multiple faulty bits in each partition with low overhead. Therefore, our error recovery approach can provide gracefully degradation memory capacity by postponing the cells' wear-out which results in improving the PCM lifetime.

The rest of this paper is organized as follows. We present the detailed of our proposed method in Section II. In Section III, the proposed method is compared to the state-of-the-art error recovery schemes in terms of lifetime, error coverage and IntraV. Finally, Section IV concludes the paper.

II. THE PROPOSED SCHEME

In this section, first we describe our error detection scheme, we then go through the error correction mechanism which is used at each block. Finally, we give a discussion on read and write mechanisms as well as required meta-data information.

A. Error Detection Mechanism

For error detection, memory controller performs the read after write mechanism to check that if the write is done successfully or not. It means that after pursuing each write, memory controller reads the written content and compares it to the actual content in order to make sure that everything is done properly. Clearly, read mechanism is much faster than writing the data in PCM cell. On the other hand, PCM writes are much more expensive due to latency and energy consumption compare to reading the data. Therefore, we don't have any concern about the system performance by using this detection mechanism because our observation shows that when running different applications with PCM read latency of 50ns and PCM write latency of 500ns, we only have 0.4% IPC degradation for the baseline system which is completely negligible.

B. Error Correction and Recovery Mechanism

As soon as the detection mechanism confirms presence of failure, memory controller invokes our recovery mechanism. In the following, we explain our recovery mechanism. As previously stated, compare to the conventional schemes that use dynamic partitioning and impose large complexity along with overhead to the memory controller, we use the fixed partitioning. Therefore, for 64B data block, we split it to 8 partitions consist of 8 bytes. Fig.1 shows our partitioning scheme.

Indeed, the best partitioning and number of partitions are key points in all partitioning schemes. Because good partitioning can spread out the faults across different smaller groups as well as providing higher probability of tolerating and recovering more faults in the same group with minimal storage overhead and complexity. In addition, it can be helpful for the space efficiency and prevention from re-partitioning.

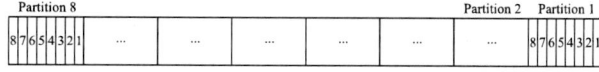

Fig. 1. Static Partitioning Scheme.

Our observation reveals that 8 partitions work best for our recovery mechanism. Since we use inversion and shifting mechanisms one after another, we should embed 1-bit for each partition to show that if the inversion is used in that partition or not (you can see inversion bit in Fig.2 (top)). Adding to it, we consider a 6-bit counter per partition for our shifting scheme to start its work whenever inversion doesn't work there (Fig.2 (bottom)). It also uses for uniforming bit flips distribution within the data block same as the wear-leveling approach. It means that as soon as receiving the write request and before error detection, this shifting method works and rotates the content of a block by 1 bit for each data block to achieve more uniformity and remove bit flips nonunifromity and stress from hot locations in each data block.

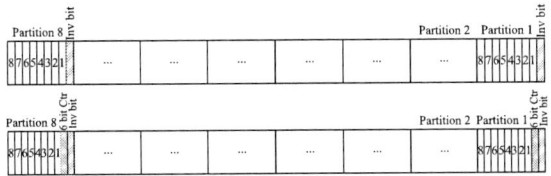

Fig. 2. Inversion bit and 6-bit counter are embedded in each partitions.

After hard error detection (red places in Fig.3), we first consider the data block partitions, faulty positions in each partition and types of failures. We then exploit the inversion mechanism and try to invert the content of the faulty partition in such a way that the inverted content and the faulty cell with stuck at value have the same value. If their values match to each other, we set the embedded 1-bit of the faulty partition to '1' in order to show that the content of this partition is inverted. Otherwise, in the case that their values do not match to each other, memory controller calls shifting scheme and rotates the block content until the shifted content and the faulty cell with stuck at value conform each other and have the same value. Since there exist spatial locality in data patterns and high information redundancy, we can probably find such a position by a tolerable amount of shift which is set to 8 bits in our experiment. After finding the appropriate state, our recovery mechanism is invoked to achieve the correct data value by using the inversion bit and the number of shift counter. Sometimes it happens that memory controller has the request for writing the specific data pattern which cannot be matched to the proposed recovery method and led to remain faulty cells' permanently. Therefore, in this situation, memory manager considers this type of the fault as a permanent fault and cannot use the whole corresponding block in the future.

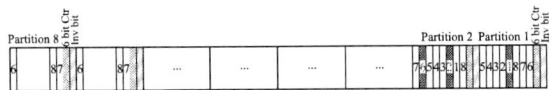

Fig. 3. Stuck at faults occur in the 4th position of 1st partition and in the 3rd and 7th places of 2nd partition. Inversion fails, so the inversion bits are set to 0. The shift counter is used, content is rotated 3 times and after 3 bits shifts, the right place is found and this place is compatible to the faulty cell with stuck at value.

C. Read Operation

When read request comes, memory controller sets the normal SLC read iteration in access circuit and reads the content of block. By considering the inverted and shifted form of the data, correct content of the data block is achieved. In this line, by looking at the inversion bit and the number of shift counter in each partition, the actual data bits locate in

its original place and the correct data value is delivered to the read requester by memory controller.

D. Write Operation

For writing a data block, when no error is detected by read-after-write mechanism, memory controller just uses the SLC write iteration and writes the content correspondingly. However, when our detection mechanism realizes that failure is happened, there exist faulty cells in each partition. Therefore, memory controller first uses the inversion mechanism to check that whether the data block content can conform to the cell with stuck-at value or not. If so, the data goes for writing in the PCM cells. However, if the desired state is not found by inverting the data pattern, memory controller uses the shifting mechanism and check that whether the current data pattern can conform and reuse the cell with stuck at value by just tolerable amount of shifts (which is limited to 8 times) or not. In the case of finding the suitable state with these features, the pending write process is done. Otherwise, if none of the inversion and shifting can work here, the existing fault indicates as a permanent fault and the block as well as the corresponding page is decommissioned from memory space.

E. Meta-data Information

As previously explained, we have 64B data block and split it into 8 partitions, each partition also has 8 bytes (64 bits). In each partition, 1-bit is embedded to do the inversion mechanism and a 6-bit counter is included to implement our shifting mechanism; this 7 bits needed to show that if the inversion is happened or not, it also indicates the number of required shifts for data retrieval. In short, we need 56 additional bits per data block (((6 x 8= 48) bits for shifting) + (8 bits for inversion)), resulting in (56/512) = 10% storage overhead which is reasonable overhead and good trade-off against lifetime improvement and postponing PCM cells wear out .

III. EXPERIMENTAL RESULTS

In this section, first we give a brief overview of evaluation method and the simulator used for our evaluation. We then introduce workloads, simulation parameters, and some performance metrics. Finally, evaluation results are given.

A. Evaluation Environment

We implemented our proposed method in GEM5 full system simulator [8]. We also modeled the entire memory hierarchy, a 4-core 2GHz CMP system with 3 levels of caches and 8GB PCM main memory with 16 banks. Moreover, the memory controller is modeled in such a way that the higher priority is assigned to reads and write is scheduled when there is no read. Our simulation parameters are shown in Table I.

For workloads, we use the complete set of parallel workloads provided in PARSEC-2 [2] as multithread workloads. Classification of these workloads is based on the read/write accesses per kilo instructions (RPKI and WPKI) and it is shown in Table II. To obtain lifetime, we simulate 1 billion instructions of each workloads categories.

978-1-5386-7100-9/18 $31.00 © 2018 IEEE

TABLE I
SYSTEM CONFIGURATIONS.

Processor	4-core SPARCIII, 4.0GHz.
L1 Cache	Split I and D cache; 32KB private; 4-way; 64B line size; LRU; write-back; 1 port; 2ns latency.
L1 Coherency	MOESI directory; 4×2 grid packet switched NoC; XY routing; 3 cycle router; 1 cycle link.
L2 Cache	4MB; UCA shared; 16-way; 64B line size; LRU; write-back; 8 ports; 4ns latency.
DRAM Cache	16MB; 4-way; 64B line size; LRU; write-back; 8 ports; 26ns latency.
Main Memory	8 GB: 16 banks, 64 B, open page, SLC: Read Latency 80 ns (6ns tPRE + 69ns tSENSE + 5ns tBUS),Write Latency 250ns.
Flash SSD	$25\mu s$ latency.

TABLE II
CHARACTERISTICS OF THE EVALUATED WORKLOADS.

PARSEC-2, 2009, Multi-Threaded					
Workload	WPKI	RPKI	Workload	WPKI	RPKI
Blackscholes	0.003	0.03	Bodytrack	0.003	0.03
Caneal	1.12	1.31	Dedup	0.41	0.43
Facesim	0.65	0.24	Ferret	0.65	0.67
Fluidanimate	0.76	0.47	Freqmine	0.04	0.07
Raytrace	0.008	0.02	Streamcluster	0.01	0.03
Swaptions	0.002	0.02	Vips	0.05	0.07
X264	0.04	0.04			

We then compare our proposed method with other partitioning and inversion techniques that are presented in [14] and [3], referred as SAFER and Aegis, respectively.

Our evaluation metrics include memory capacity degradation (lifetime), and average recovered errors per data block. We also have sensitivity analysis on partition size, and number of required shifts to work best. Adding to it, we check the widely used evaluation parameter called, IntraV [15].

B. Simulation Results

In this section, first, we compare our proposed method to other state-of-the-art hard error recovery schemes, we then present experimental results of our evaluation.

Lifetime. We measure the memory capacity degradation as a lifetime. To do so, we check the time when the memory capacity reduces to 50% of its initial capacity after performing a certain number of writes. We then capture the elapsed time between the start time and the time when the memory capacity reduces to half and depicts it as the memory capacity versus number of writes per page. As we can see in Fig.4, our scheme improves system's lifetime by up to 25% and 12% compared to SAFER-32 and Aegis, respectively. This improvement comes through postponing the PCM cells wearout and reusing the faulty cells with stuck at value. Also, our low overhead inversion and shifting mechanisms in each partition play a key role in prolonging PCM lifetime.

Recovered Errors per Data Block. Fig. 5 shows the percentage of recovered cells per memory block for each evaluated system. Based on our observation, we achieved more improvement since the proposed recovery scheme can recover multi-cell failures in each partition by using both inversion

and shifting mechanisms for reusing faulty cells. However, SAFER just can recover one faulty cell in each partition and recovered cells per memory block in Aegis highly depend on its dynamic partitioning.

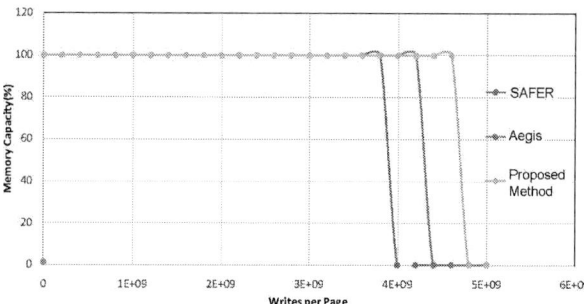

Fig. 4. Capacity degradation of the proposed method compared to Aegis and SAFER-32.

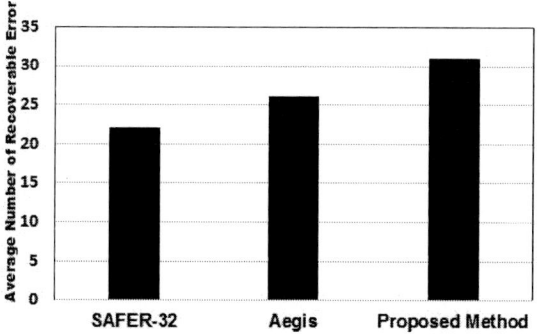

Fig. 5. Average number of recovered errors per block of the proposed mechanism compared to SAFER-32 and Aegis.

Partitioning Size. Partitioning size has the main impact on the reliability issue and storage overhead. It means that, if we divide the data block into small number of partitions, the possibility of failure coverage is not high in each partition. It is happened because the probability of spreading faults and bit flips in different places of each partition is so high and our proposed method cannot tolerate and recover them. It is also difficult for our recovery scheme to find the good match between the faulty cell with stuck at value and the data pattern. On the other hand, when there exist small number of partitions, fewer additional bits are needed per each partition which results in lower storage overhead. However, there is a trade-off between reasonable storage overhead and providing the high level of reliability.

We consider partitioning size of 8 since it provides good trade-off between acceptable storage overhead and reliability issues. Fig. 6 shows the effect of partition size on our proposed method. As it is shown in this figure, when we split the data block into small number of partitions, the required bits per data block are fewer as well as error coverage rate in various partitions (e.g. for 64B data block, 4 partitions means that we have 4 partitions with 16B data, so it indicates the additional bits are about: $((7 \times 4) + (4)) = 32$ bits per data block. Indeed, we need 28 bits for shifting scheme and 4 bits for inversion in each partition). On the other hand, our method can tolerate and recover more failures in various partitions when we split the data block into more partitions. But it happens in presence of more additional bits per data block which imposes larger meta-data storage overhead.

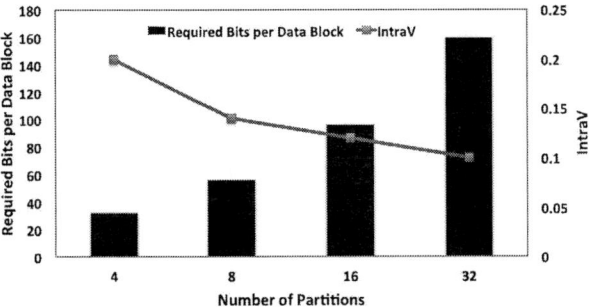

Fig. 6. Effect of partition size.

IntraV. IntraV [15] is extensively used by the prior works. This metric shows the reduction in bit flips variation by exploiting our proposed method and without using it. Its mathematical definition is as follows:

$$ IntraV = \frac{1}{BF_{aver}.N} \times \sum_{i=1}^{N} \sqrt{\frac{\sum_{j=1}^{512}(BF_{ij} - \sum_{j=1}^{512} w_{ij}/512)^2}{511}} \quad (1) $$

where N is the total number of blocks, the write count of cell j in block i is BF_{ij}, and BF_{aver} is the average bit flips count.

Based on our evaluation methodology, we got about 14% reduction in bit flips' variation in each data block, on average. The comparison results are shown in Fig. 7. This parameter is defined for different multi-threaded workloads which we have used in the evaluation part.

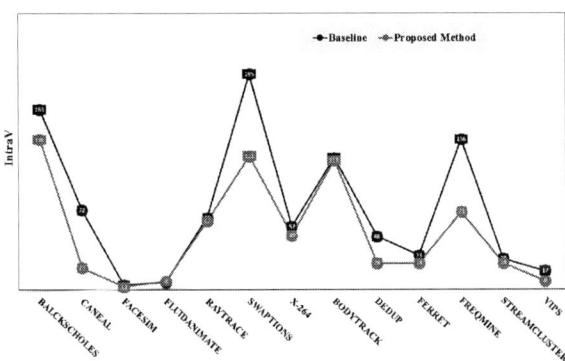

Fig. 7. IntraV comparision of the workloads before/after using proposed method.

Effect of Required Shifts. In order to find the acceptable number of shifts, we check different shifted patterns from 0 to 9 in parallel. As Fig. 8 shows the distribution of the number of required shifts for matching the faulty cell with stuck at value and the data pattern, we select up to 8 bits shifts appropriate for our recovery scheme (average usage is between 6 to 8 shifts). Same as the partitioning size, shifting scheme has trade-off. For instance, it can be useful for uniforming bit flips across the data block partitions. It also helpful for increasing error coverage rate in each partition. On the other hand, it suffers from increasing bit flips across the data block. We select up to 8 bits because less than this number can have negative effect on the error coverage rate in each partition.

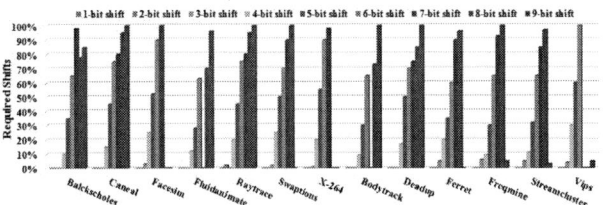

Fig. 8. Average number of required shifts per data block.

IV. CONCLUSION

PCM can compete with DRAM among several non-volatile memories. It has a better read speed and performance while providing non-volatility, near zero leakage power consumption and good scalability. Besides, it suffers from permanent stuck-at faults as a result of cells wear-out. In presence of unbalanced write traffic and process variation, sooner cell wear-out is intensified. Therefore, to tolerate large numbers of faults in a data block and postpone PCM cells wear out, we proposed an efficient error recovery scheme with low storage overhead. In this line, we used static partitioning of data block to smaller groups. We then benefit from the advantages of inversion and shifting mechanisms to use the failed cells when stuck at faults occur. Compared to known partitioning and inversion schemes, our experimental results showed that under a wide range of workloads, we have better lifetime and error coverage along with lower storage overhead to store meta-data information. The proposed method extended PCM lifetime by 12-25% over the evaluated schemes.

REFERENCES

[1] R. Azevedo, J. D. Davis, K. Strauss, P. Gopalan, M. Manasse, and S. Yekhanin. Zombie memory: Extending memory lifetime by reviving dead blocks. In *ISCA*, 2013.

[2] C. Bienia, S. Kumar, J. P. Singh, and K. Li. The parsec benchmark suite: Characterization and architectural implications. In *PACT*, 2008.

[3] J. Fan, S. Jiang, J. Shu, Y. Zhang, and W. Zhen. Aegis: Partitioning data block for efficient recovery of stuck-at-faults in phase change memory. In *Proceedings of the 46th Annual IEEE/ACM International Symposium on Microarchitecture*, pages 433–444. ACM, 2013.

[4] E. Ipek, J. Condit, E. B. Nightingale, D. Burger, and T. Moscibroda. Dynamically replicated memory: building reliable systems from nanoscale resistive memories. In *ASPLOS*, pages 3–14, Mar. 2010.

[5] B. C. Lee, E. Ipek, O. Mutlu, and D. Burger. Architecting phase change memory as a scalable DRAM alternative. In *ISCA*, pages 2–13, June 2009.

[6] R. Maddah, S. M. Seyedzadeh, and R. Melhem. Cafo: Cost aware flip optimization for asymmetric memories. In *2015 IEEE 21st International Symposium on High Performance Computer Architecture (HPCA)*, pages 320–330, Feb 2015.

[7] K. T. Malladi, B. C. Lee, F. A. Nothaft, C. Kozyrakis, K. Periyathambi, and M. Horowitz. Towards energy-proportional datacenter memory with mobile dram. In *Proceedings of the 39th Annual International Symposium on Computer Architecture*, ISCA '12, pages 37–48, 2012.

[8] M. M. K. Martin et al. Multifacet's general execution-driven multiprocessor simulator (GEMS) toolset. *CAN*, 33(4):92–99, Nov 2005.

[9] M. K. Qureshi. Pay-As-You-Go: low-overhead hard-error correction for phase change memories. In *MICRO*, pages 318–328, Dec. 2011.

[10] M. K. Qureshi, M. M. Franceschini, L. A. Lastras-Montaño, and J. P. Karidis. Morphable memory system: a robust architecture for exploiting multi-level phase change memories. In *ISCA*, pages 153–162, Jun. 2010.

[11] M. K. Qureshi, J. Karidis, M. Franceschini, V. Srinivasan, L. Lastras, and B. Abali. Enhancing lifetime and security of PCM-based main memory with start-gap wear leveling. In *MICRO*, pages 14–23, Dec. 2009.

[12] M. K. Qureshi, V. Srinivasan, and J. A. Rivers. Scalable high performance main memory system using phase-change memory technology.

In *ISCA*, pages 24–33, June 2009.

[13] S. Schechter, G. H. Loh, K. Straus, and D. Burger. Use ECP, not ECC, for hard failures in resistive memories. In *ISCA*, pages 141–152, June 2010.

[14] N. H. Seong, D. H. Woo, V. Srinivasan, J. A. Rivers, and H.-H. S. Lee. SAFER: Stuck-at-fault error recovery for memories. In *MICRO*, pages 115–124, Dec. 2010.

[15] J. Wang, X. Dong, Y. Xie, and N. Jouppi. i2wap: Improving non-volatile cache lifetime by reducing inter- and intra-set write variations. In *IEEE 19th International Symposium on High Performance Computer Architecture*, HPCA, pages 234–245, Feb 2013.

[16] R. Melhem, R. Maddah, and S. Cho. RDIS: Recursively Defined Invertible Set Scheme to Tolerate Multiple Stuck-At Faults in Resistive Memory. In *DSN*, pages 221–232, Jun 2012.

[17] Z. Wang, S. Shan, T. Cao, J. Gu, Y. Xu, S. Mu, Y. Xie, and D. A. Jiménez. Wade: Writeback-aware dynamic cache management for nvm-based main memory system. *ACM Trans. Archit. Code Optim.*, 10(4):51:1–51:21, Dec. 2013.

[18] D. H. Yoon, N. Muralimanohar, J. Chang, P. Ranganathan, N. P. Jouppi, and M. Erez FREE-p: Protecting non-volatile memory against both hard and soft errors. In *HPCA*, pages 466–477, Feb 2011.

[19] P. Zhou, B. Zhao, J. Yang, and Y. Zhang. A durable and energy efficient main memory using phase change memory technology. In *ISCA*, pages 14–23, June 2009.

2018 IEEE Computer Society Annual Symposium on VLSI

Guessing your PIN right: Unlocking smartphones with publicly available sensor data

David Berend[1,2], Bernhard Jungk[3,*], Shivam Bhasin[1]

[1]Physical Analysis and Cryptographic Engineering, Temasek Laboratories
Nanyang Technological University, Singapore
[2]University of Applied Sciences Wiesbaden, Rüsselsheim, Germany
[3]Cyber and Information Security Group, Fraunhofer Singapore
email: david.berend@mine-it.eu, bernhard.jungk@fraunhofer.sg, sbhasin@ntu.edu.sg

Abstract—Modern day smartphones act as daily companions playing a crucial role in tasks far beyond communication. Equipped with various motion and health sensors, private information is continuously processed, while it can be accessed without asking for special permission. In this paper, we show how the permissionless sensor data can be used to reconstruct one's secret PIN for unlocking the phone or gaining access to one's bank account. Harvesting the power of machine learning algorithms, we present a practical attack able to classify all 10,000 possible PIN combinations. Results show up to 83.7% success within 20 tries. Compared to state of the art reporting 74% success on a reduced space of 50 chosen PINs, we report 99.5% success with a single try in a similar setting.

I. INTRODUCTION

Modern smartphones are equipped with various sensors, to enhance the user experience. Most of these sensors are accessible without a user's permission and thus known as zero-permission sensors. Commonly available zero-permission sensors include accelerometer, gyroscope, magnetometer, barometer etc. The sensors constantly detect user activity which is then fed back to user applications for intelligent processing. However, these sensors could also detect sensitive user data of private and confidential nature leading to potential attacks. In the area of embedded security, side-channel attacks [1] are known to exploit such unintentional (physical) leakage of computation for extracting the secret cryptographic key. In this work, we exploit the unintentional physical leakage from sensors to recover user private information like personal identification number (PIN) used for unlocking the smartphone or gaining access to one's bank account.

Few works have previously looked into exploits on zero-permission sensors of smartphones [2], [3], [4], [5], [6], [7], [8], [9], [10], [11], [12], [13]. These attacks include behavioral profiling, geo-localization, and PIN recovery. Previous work on the PIN extraction scenario [14], [9], [10], [11], [12] deals with the problem of (four-digit) PIN recovery. Two of these publications try to classify a single digit and then recombine the individual digits to form a PIN [9], [14]. The other publications [10], [11], [12] apply the classification to complete four-digit-combinations but restrict their training and testing data to a set of only 50 out of 10,000 possible PINs.

The work of Mehrnezhad et al. [12] is the latest in this line of work and reports the highest success rate to date. It uses machine learning to achieve 74% PIN recovery success from a set of 50 chosen PINs.

In this paper, we perform user PIN recovery from sensor based leakage using a novel single digit classification methodology. Our method fuses information from a subset or from all available sensor to enhance the PIN recovery success. If applied to the setting of only 50 PINs, we report a PIN recovery success of 99.5%. When applied to the full PIN space of 10,000, the proposed method reports a recovery success rate up to 83.7% in low noise measurement settings. The classification algorithms are able to quickly weigh the importance of each sensor when performing the PIN recovery and hence, make it possible to have a high recovery success rate. Since the methodology works on a single digit, it is scalable to PINs shorter or longer than four digits.

The rest of the paper is organized as follows. Section II discusses the general attack scenario. Section III details the proposed methodology, and Section IV presents experimental results for PIN recovery. Finally, conclusions are drawn in Section V.

II. ATTACK SCENARIO

In the presented attack, we extract a numerical PIN, composed of the digits in the interval $(0, 9)$. The digits are entered by the user to unlock the screen or gain access to his bank account etc.

As the placement of individual digits on the keypad is distinct, touching different digits leads to different sensor readings. It varies with finger/hand movement, tilt, rotation and touch pressure itself.

The distinct (but noisy) sensor reading corresponding to a digit can then enable reconstructing a PIN successfully. The attack assumes an installed malicious app in the form of a casual game [15] on the phone which measures zero-permission sensors and shares data with the attacker over any available communication channel. As per the latest survey [16], an average user has about 58% inactive app installed which remain dormant for 12 weeks (average) before removal. The game uses a numerical keyboard with a similar layout but a different intention, e.g. a math trainer. Through that method,

---

* This research was conducted when author was at Temasek Laboratories.

978-1-5386-7100-9/18 $31.00 © 2018 IEEE

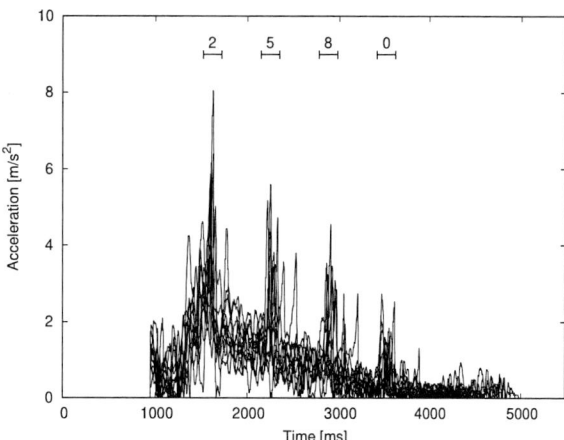

Fig. 1: Ten records of the sensor measurement during the input of the PIN 2580.

an attacker collects data on, how the user enters digits. The measurements are then trained to a classification algorithm, which is afterward able to associate the measurements of unlocking the screen to the entered PIN.

In this paper, we focus on the sensor data exploitation part only and present experimental results supporting that such attacks are indeed possible. In our controlled experiment, we show that the actual success rate is much better than random guessing of the PIN, outperforming the brute force strategy by over 400 times.

III. ATTACK METHODOLOGY

For a n-digit PIN classification, in principle two approaches exist. The first classifies a measured data stream including all digits of a PIN. The second classifies each digit in the data stream individually and then combines the digit-probabilities to complete PINs. For both approaches, the underlying data is a recorded data stream of sensor data over the whole duration of the PIN input (Figure 1).

For the first approach, all 10,000 combinations of a four-digit password have to be trained as a individual class to be able to classify all possible combinations later correctly. Hence, if the training data does not train for some of the combinations or is noisy, it will be unable to classify this combination correctly. Previous works using such a technique have focused on a subset of only 50 classes [10], [12]. Aviv has conjectured that the success rate decreases inverse exponential if the number of classes is increased to the full set [10].

The second approach identifies each digit in the sensor data individually. For this, we first have to cut out the corresponding part from the relevant sensor data and then classify the digit presses individually. The benefit is that only ten possible classes exist, each representing a number on the numerical PIN pad, thus easily scalable to varying PIN lengths. We will follow this approach in the current paper and show that a much

higher success rate is achievable compared to previous results using the same [14], [9] and the other approach [10], [12].

One of the challenges is to identify the right timing of a number press and then to split the data accordingly. Therefore, distinct peaks in the data stream can be identified, which indicate a press (Figure 1). Similar to previous work [14], [9], we resort to the use of touch and release events of a user's thumb as for distinguishing the individual start and stop times of a press.

A. Setup

For our proof of concept attack, we developed an application to sample the sensor data and a server-side module for aggregation, classification and evaluation. The experiment's device should represent the common technology-standard used by smartphone adopters. Therefore, we have chosen an LG Nexus 5.

For measuring the sensors during a PIN entry, we developed an Android application, which runs in the foreground and stores the records in external storage. The sampling of each zero-permission sensor can be turned on or off individually. The app also provides a numerical PIN pad, which appears once the measurement process starts. The layout and position of the numerical PIN pad are identical to the one provided by the Android OS. The layout of the developed app is shown in Fig. 2.

The candidate was asked to interact with the application in a consistent natural position. Five times, 70 chosen four-digit PINs were entered by each candidate for training. We note that not only the digit itself is valuable in the training phase, but also the combination of the digits plays a crucial role, because the movement is different, depending on whether a number is the first or the last digit in the PIN or which number has been selected before. For generating training data,

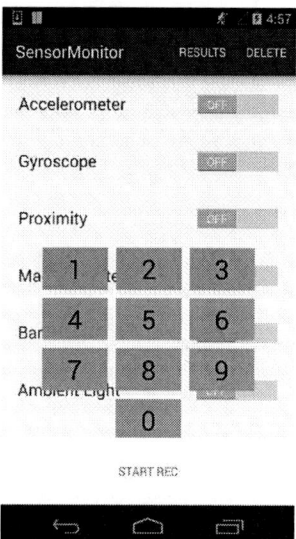

Fig. 2: Layout of our proof of concept app.

which does take this idea into account, each number must be entered at least once following every other number. This is achieved by entering all possible ten numbers before and after each number. In addition to the training data, 50 randomly generated pins were entered for validating the algorithm.

B. Sensor Measurements

A PIN can be entered in a smartphone in many different ways. Relevant parameters are body position, holding type, typing speed, left or right hand or both. For our experiments, the user was sitting on a chair and was holding the phone in the right hand while typing the passwords with the hand's thumb at a roughly consistent speed. Due to the general experimental limitations on the variation of experimental conditions, we focused on one behavioral pattern for the input, as the aggregation and classification is the main contribution of this work.

We have tested the contribution of all available zero-permission sensors for the purpose of PIN recovery. The LG Nexus 5 provides accelerometer, gyroscope, magnetometer, proximity, barometer, ambient light and rotation vector sensor. However, we excluded the barometer and proximity sensor, due to the low sampling rate ($0.5 - 4$ Hz), which caused high inaccuracy and hence, were not providing any exploitable signal. An example measurement for the PIN 2580 (entered ten times) is shown for the z-axis of the accelerometer and the rotation vector in Fig. 1. By simple visual inspection, the digits are shown to have distinguishable features. In the following, we take the help of classification algorithms, such as neural networks or random forests, to distinguish PINs from different sensor measurements in a systematic manner.

During the input of the PIN, all registered sensors fire an event through an Android hardware interface every time its value has changed, but at a limiting frequency which varies among sensors. Each event holds the values of a sensor's dimensions. These are saved into a sensor related array of data entries in the record object.

The same happens with touched digits. Each time a number is pressed, a `touchevent` triggers a tuple of $(t_{\text{start}}, t_{\text{stop}}, k)$, where t_{start} is the start time of the touch, t_{stop} is the time of release, and k is the pressed digit (key), which is saved in the current record object. A record holds four presses, one for each number of the PIN. Once a password is successfully entered, the record is saved.

The final object consists of all sensor related data entries and presses. It is stored locally as a *JSON*-string [17], before the object is sent to the server side.

C. Preprocessing

In the measurement step, records are created, where each record r has a timestamp for the start r_{start} and stop r_{stop} time of the measurement. During recording, activated sensors fire events at their specific sampling rate. Each sensor holds a list of its events, saved as (t, v) pairs for each dimension, where t is a time stamp, and v is a value reported by the sensor.

The recorded data is preprocessed to a similar level of granularity for the training of the classification algorithm. This means that each record has to start at the same time and at a normalized start value. Furthermore, all data streams of sensors are required to have a value at all points in time. We set the time units to one millisecond to have a uniform timestamp format.

As a first step, we calculate the duration of each record r

$$r_{duration} = r_{stop} - r_{start} \tag{1}$$

Then, we replace r_{stop} by $r_{duration}$ and set r_{start} to 0 later in the process.

Currently, each sensor has a list of $(time, value)$ pairs for each dimension. The goal however, is to have a value for each millisecond. Therefore, we transform the $(time, value)$ pairs into an array, which holds a value for each millisecond. First, we drop the beginning four data pairs, to eliminate initial sensor inaccuracy. Then, an array with the size of $r_{duration}$ is instantiated for each dimension. For each millisecond (index of array) it is checked if the dimension's $(time, value)$ pairs have an entry to address the value to the current index of $t_{currentTime} - r_{start}$. If the sensor didn't fire at that point in time, the previous value is taken. This interpolation provides a regular measurement which can be efficiently fed to the training algorithm.

The final step is a normalization of the values. We perform this normalization by selecting a reference value v_{ref} and then to shift all data values, such that the reference value becomes $v'_{\text{ref}} = 0$ and all other values are adjusted accordingly. For the selection of the reference point, our experiments have shown, that choosing an individual point from the first points is usually too inaccurate. Therefore, we compute the average over the first 300 milliseconds as a reference.

$$v_{\text{ref}} = \frac{1}{300} \sum_{i=1}^{300} v_i \tag{2}$$

v_{ref} is then used for the normalization of all other values v_i of the complete record as follows:

$$v'_i = |v_i - v_{\text{ref}}| \tag{3}$$

D. Classification

The goal of this classification is to characterize a number/digit of the numerical keyboard (key) by the given sensor information. Therefore, the classification algorithm is trained statistical attributes associated with a key, called the training phase. The same attributes are then extracted from the presses of a test record, which is unknown to the algorithm (testing phase). The algorithm returns for each press a probability for each number $(0-9)$. Then, all four-digit number combinations are ranked, by multiplying the probabilities of the individual numbers of a combination. The combination with the highest product is the first guess, the one with the second highest probability the second guess, and so on.

The data for the training phase consists of many press objects, which are cut out of the record's data stream. Each

TABLE I: The data structure representing extracted features.

press	X					y number
	$s_1 d_1 a_1$	$s_1 d_1 a_2$	...	$s_1 d_2 a_1$	$s_1 d_2 a_2$ ...	
p_1	[				...]	k
p_2	[				...]	k
...	[				...]	...

object holds information about each sensor and its dimensions at every millisecond, during the press duration. For the characterization of the individual numbers, several features are extracted from this data. These are statistical attributes, such as minimum, maximum, mean, median, standard deviation and skewness. The attributes are then associated with a known number, that has been recorded during the measurement.

Each dimension of a sensor holds a stream of values, and each stream serves as a source for some features. Therefore, the more sensors are taken into consideration, the more features exist. All features are extracted in the same way for each press and then associated with the recorded number. In the end, the training data consists of a two-dimensional array, where each row holds an array of all features a of each dimension d of each sensor s during a press as depicted in Table I. For each press, the originally pressed number is stored in a second array y, at the identical index as the array of press attributes in X. The data is then used to train the classification algorithm.

The mainly used classification algorithm is an MLP neural network built with the following parameters. The network possesses two middle layers. The first layer includes one node per feature. The second as many nodes as classes. Using this architecture, the features become weighted more accurately and enter a pooling layer afterward, to synergize the information. Backpropagation optimizes the network with a limited memory Broyden Fletcher Goldfarb Shanno (L-BFGS) algorithm [18], which can handle relatively small amounts of data well. For testing the quality of each classification algorithm, the number of correctly classified records is divided by the total number of records. Experimental justification for choice of MLP as a classification algorithm is given in the following section.

IV. EVALUATION

This section presents the experimental results. We first conduct experiments to choose a relevant classification algorithm, which is then further used for PIN recovery.

A. Classification Algorithms

For a comprehensive analysis of the state of the art classification algorithms, we have examined the performance of different algorithms, namely MLP network, random forest, k-nearest neighbor and Gaussian naive Bayes classifier. First, 500 random training samples of the training data of three candidates were added to a training set. Then randomly generated testing data ($30-50$ four-digit combinations entered by each candidate) was classified by the trained algorithm. Afterward, another 500 training samples were added to the training set, and the algorithm was trained again, while the

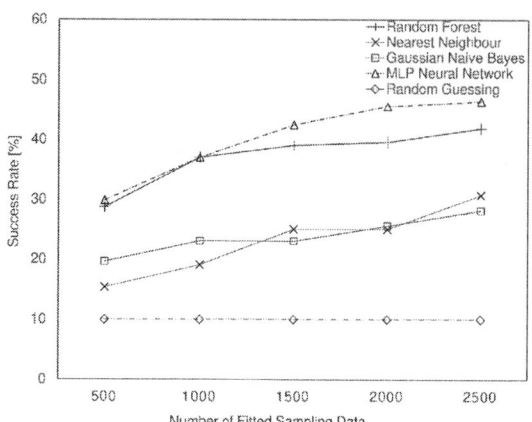

Fig. 3: Success rates of different classification algorithms.

testing data remained the same. This process was repeated until the training set reached a total size of 2500 samples.

At the first iteration with only 500 training samples, random forest and MLP network achieved the highest classification success, while the Gaussian naive Bayes classifier and the k-nearest neighbor classifier performed considerably worse (Figure 3). After five iterations, the MLP network outperforms random forest by four percentage points. The k-nearest neighbor algorithm and the Gaussian Naive Bayes classifier perform at a steady growth rate of 2-5 percentage points in success. However, both algorithms perform 25-28 percentage points worse than MLP networks. In general, all algorithms perform much better than random guessing, which is also displayed in Figure 3 as reference.

Based on these results, we conclude that the MLP network classification is able to achieve the highest success rate for classifying individual numbers as it shows the highest success rate in our experiments. If computational cost plays a crucial role, random forest could be considered, due to lower training complexity. In scenarios where the set of training data is small, it might even surpass the MLP network due to its simplicity.

B. Individual Sensors and Sensor Fusion

During the press of a number, all sensors have been measured at their individual highest sampling rate. To distinguish which sensor contains the most information about a press, the classification procedure was performed for each sensor individually. The algorithm is allowed to make 20 guesses, and the attack is considered successful if the correct PIN is part of the 20 guesses. In Figure 4, the individual sensor success is presented. Overall, the accelerometer performs the best with a success rate of 63%, followed by the rotation vector sensor, magnetometer and gyroscope (28%, 20% and 5% respectively). The ambient light sensor provided almost no information. This shows that sensors with high sampling rate perform better. The proximity sensor, as well as the barometer, are not listed, as the sampling rate is too low (on average one data point per second).

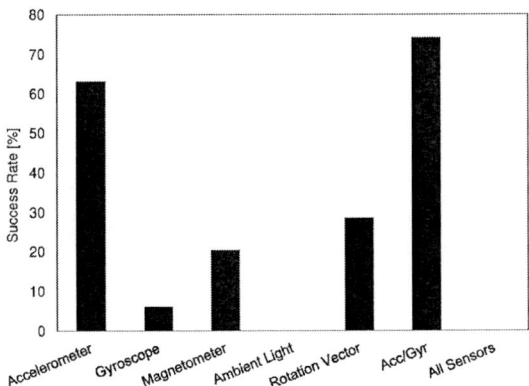

Fig. 4: Sensor individual and fused success rates.

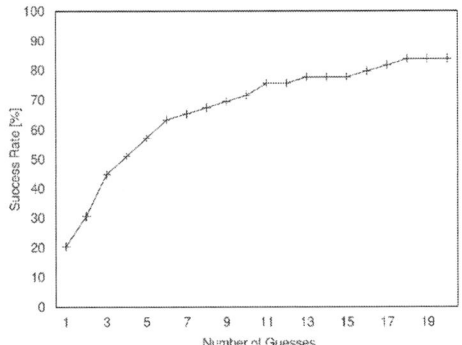

Fig. 5: Best case success rate.

After investigating the individual sensors, we also explore the advantage of combining information from different sensors. In general, classification algorithms perform better, when more features exist. For this experiment, six statistical attributes are extracted from the data stream of each dimension of each sensor. When all sensors are fused together, 72 features exist (6 features for each of the three dimensions of the accelerometer, the gyroscope, the magnetometer and the rotation vector sensor, and one dimension for the ambient light sensor). However, directly combining all sensors does not lead to good results, as some information is redundant and therefore adds more noise than relevant information. To improve the results, all different combinations were tested. The best fusion resulted in 74.1% by the accelerometer fused with the gyroscope, presented in the sixth column of Figure 4.

The accelerometer measures the phone's acceleration in space. The same space is defined by the magnetometer and the ambient light in their specific manner. The gyroscope measures the acceleration of the phone's movement around itself and the rotation vector sensor aggregates that movement to angles along the same dimensions. Therefore, the sensors can be grouped into space-related and angle-related. When sensors of the same group are used for the sensor fusion, redundancies occur, which ultimately results in worse results.

TABLE II: Best sensor fusion results.

Towards \ Away	0ms	25ms	50ms
0ms	74.1%	68.7%	71.4%
25ms	78.2%	70.1%	72.1%
50ms	83.7%	76.9%	75.5%

C. Additional Thumb Movement

The fundamental source for features used during the classification is the data stream of sensor data during the candidate's press. The press is cut out from the measurement based on the timestamps of touch and release, provided by the $onTouchEvent$ by Android OS. To examine whether additional unique information is present in the movement towards and from a press, the time interval of the press is increased on both sides by 25 and 50 milliseconds. We again use the best performing sensor combination, i.e. accelerometer and gyroscope. The data with and without the finger movement information is fed to the MLP network for training and validated with the testing data.

The results, presented in Table II, show that the success rate reached its maximum of 83.7% when the press interval has been slightly larger by $50ms$ at the beginning and $0ms$ at the end, thus improving success by another 12.9 percentage points. The evolution of the success from 1st to 20th guesses is shown in Fig. 5. In the present experiment, the results show that the success generally increases when adding information about the movement towards a number, while success decreases when information of the movement from a number is added to the classification data. The measured information during the movement of a finger towards a number seems to be more valuable than the movement in the other direction. This makes sense, as the number remains at a fixed position, forcing the finger to move to a constant point, whereas the number afterward can be one out of nine possible options.

As a general conclusion, we report a PIN retrieval success of as good as 83.7% after combining data from relevant sensors and exploiting information of the thumb movement. The achieved success is reported when applied to complete PIN search space of 10,000 PINs. This, to our knowledge, is the highest success rate reported to date for the scenario of PIN classification. Moreover, with the underlying methodology, the technique can be easily scaled up to longer PINs. Previous works reported success up to 74% when applied to a small subset of specifically chosen 50 PINs. For a fair comparison, in the following, we scale down our method to 50 chosen PINs and compare the success rates.

D. Comparison With Previous Work

Classifying each number individually, rather than the full data record including all four digits of the combination, resulted in a high success rate. While previous work [12] applied the method on 50 chosen PINs, our previous test applied to the set of 10,000 PINs. We now apply our method to a small subset of 50 PINs to draw a comparison. The results use the MLP network classifier, trained with 50 four-digit-PINs

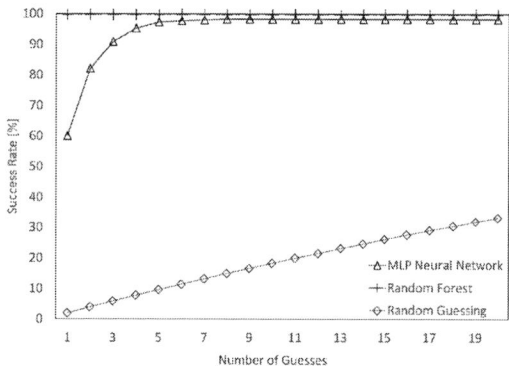

Fig. 6: Success rates for only 50 combinations.

with equally distributed numbers. Each press is then classified individually, which results in four arrays of probabilities for a number being pressed. The combinations of all ten numbers are then permuted, while the product of the combination related probability is calculated. The combinations, which were not trained in the beginning are then excluded. Our initial results were worse than the method of Mernezhad et al. [12]. While [12] reports 74% success on the first try, our method reports only 60%. However, it reaches close to 100% success after six tries. Further analysis of the results indicated that the dataset of 50 PINs has been too small for training our MLP network, resulting in poor results.

To further improve the results, we also evaluated the same test with the random forest algorithm, which performed close to MLP networks in previous experiments. Owing to the simplicity of random forest, it performs very well with the limited training complexity. Repeating the same attack with random forest, we were able to classify the PIN correctly with a 99.5% success within the first guess (refer Fig. 6).

V. CONCLUSIONS

Zero-permission sensors in modern smartphones can be a source of security and privacy vulnerability. In this paper, we explored these vulnerabilities for the purpose of PIN retrieval. A single-digit classification based approach was used to exploit captured sensor data with MLP neural network. While the accelerometer gave best individual results, its combination with gyroscope further improved the attack success rate. Along with applied pre-processing and movement information towards and away from the key, we report up to 83.7% success in finding the right PIN within 20 tries out of all 10,000 possible PINs. Compared to previous works which worked on 50 chosen PINs, we report a 99.5% success rate (using random forest) against previous 74% [12]

As a countermeasure, limiting the maximum operating frequency of the sensors can reduce the attack feasibility. Some applications also use a randomized keypad, which mitigates the attack if the underlying randomness is sound. However, these are just temporary fixes, and sensors access in smartphones must be rethought at the operating system level. It

could, for example, be considered to disable the sensors during sensitive activity. Finally, the implications of the PIN recovery attack can be extended to a much broader area of applications. It could, for example, be easily applied to learn user behavior, location, etc. entirely compromising user privacy.

REFERENCES

[1] P. Kocher, J. Jaffe, and B. Jun, "Differential power analysis," in *Advances in cryptologyCRYPTO99*. Springer, 1999, pp. 789–789.

[2] A. Al-Haiqi, M. Ismail, and R. Nordin, "On the best sensor for keystrokes inference attack on android," *Procedia Technology*, vol. 11, pp. 989–995, 2013.

[3] E. Owusu, J. Han, S. Das, A. Perrig, and J. Zhang, "Accessory: Password inference using accelerometers on smartphones," in *Proc. of HotMobile 2012*, San Diego, USA, Feb. 28–29, 2012.

[4] C. A. Ronao and S.-B. Cho, "Human activity recognition with smartphone sensors using deep learning neural networks," *Expert Systems with Applications*, vol. 59, pp. 235 – 244, 2016. [Online]. Available: http://www.sciencedirect.com/science/article/pii/S0957417416302056

[5] C. Giuffrida, K. Majdanik, M. Conti, and H. Bos, *I Sensed It Was You: Authenticating Mobile Users with Sensor-Enhanced Keystroke Dynamics*. Cham: Springer International Publishing, 2014, pp. 92–111. [Online]. Available: https://doi.org/10.1007/978-3-319-08509-8_6

[6] N. Zheng, K. Bai, H. Huang, and H. Wang, "You are how you touch: User verification on smartphones via tapping behaviors," in *2014 IEEE 22nd International Conference on Network Protocols*, Oct 2014, pp. 221–232.

[7] X. Su, H. Tong, and P. Ji, "Activity recognition with smartphone sensors," *Tsinghua Science and Technology*, vol. 19, no. 3, pp. 235–249, June 2014.

[8] K. Ellis, "Classifying human behaviors, activities and contexts from mobile sensor data," Ph.D. dissertation, UC San Diego, San Diego, 2010. [Online]. Available: http://escholarship.org/uc/item/95q048vs

[9] L. Cai and H. Chen, "Touchlogger: Inferring keystrokes on touch screen from smartphone motion," in *Proc. of the 6th USENIX Conference on Hot Topics in Security*, ser. HotSec'11. Berkeley, CA, USA: USENIX Association, 2011, pp. 9–9. [Online]. Available: http://dl.acm.org/citation.cfm?id=2028040.2028049

[10] A. J. Aviv and J. M. Smith, "Side channels enabled by smartphone interaction," Ph.D. dissertation, Univ. of Pennsylvania, Pennsylvania, 2012. [Online]. Available: http://citeseerx.ist.psu.edu/viewdoc/summary?doi=10.1.1.454.9421

[11] R. Spreitzer, "Pin skimming: Exploiting the ambient-light sensor in mobile devices," in *Proceedings of the 4th ACM Workshop on Security and Privacy in Smartphones & Mobile Devices*, ser. SPSM '14. New York, NY, USA: ACM, 2014, pp. 51–62. [Online]. Available: http://doi.acm.org/10.1145/2666620.2666622

[12] M. Mehrnezhad, E. Toreini, S. F. Shahandashti, and F. Hao, "Stealing pins via mobile sensors: actual risk versus user perception," *International Journal of Information Security*, pp. 1–23, 2016.

[13] Y. Michalevsky, G. Nakibly, G. A. Veerapandian, D. Boneh, and G. Nakibly, "Powerspy: Location tracking using mobile device power analysis," in *Proceedings of the 24th USENIX Conference on Security Symposium*, ser. SEC'15. Berkeley, CA, USA: USENIX Association, 2015, pp. 785–800. [Online]. Available: http://dl.acm.org/citation.cfm?id=2831143.2831193

[14] Z. Xu, K. Bai, and S. Zhu, "Taplogger: Inferring user inputs on smartphone touchscreens using on-board motion sensors," in *Proceedings of the fifth ACM conference on Security and Privacy in Wireless and Mobile Networks*. ACM, 2012, pp. 113–124.

[15] V. Svajcer, "Malicious cloned games attack Google Android Market," *Naked Security: http://nakedsecurity. sophos. com/2011/12/12/malicious-cloned-games-attack-google-android-market*, 2011.

[16] (2017) Dmr mobile report.

[17] E. Internaitonal. (2013, Oct.) Json data interchange format, standard ecma-404. [Online]. Available: http://www.json.org/

[18] N. M. Nawi, M. R. Ransing, and R. S. Ransing, "An improved learning algorithm based on the Broyden-Fletcher-GoldfarbShanno (BFGS) method for back propagation neural networks," in *Proceedings - ISDA 2006: Sixth International Conference on Intelligent Systems Design and Applications*, vol. 1, 2006, pp. 152–157.

978-1-5386-7100-9/18 $31.00 © 2018 IEEE

2018 IEEE Computer Society Annual Symposium on VLSI

Synthesis, Technology Mapping and Optimization For Emerging Technologies
Student Forum Paper

Student: Debjyoti Bhattacharjee*, *Supervisor:* Anupam Chattopadhyay[†]
School of Computer Science and Engineering
Nanyang Technological University, Singapore 639798
Email: *debjyoti001@ntu.edu.sg*, *anupam@ntu.edu.sg*[†]

Abstract—As the CMOS technology scaling is facing fundamental issues, its inevitable demise is being closely followed by the rise of novel beyond-CMOS technologies. Among those, several compute-in-memory technologies (ReRAM, STTRAM) and quantum computing prominently stands out. For both of these classes of technologies, there is an urgent need of robust and efficient Electronic Design Automation flows. In this research, several EDA challenges for the emerging technologies are addressed, including technology mapping, and logic synthesis for in-memory computing and quantum computing. Furthermore, a novel, native implementation of fuzzy logic on ReRAM devices have been practically demonstrated.

Index Terms—*Technology mapping, Logic Synthesis, In-memory computing, Resistive RAM (ReRAM), Memristors, Quantum Computing*

I. INTRODUCTION AND MOTIVATION

The limitations of CMOS technology can be attributed to the decrease in gate oxide thickness and voltage level whenever there is scaling down of MOSFET [1]. The ratio between the operating voltage and thermal voltage decreases with the decrease in the operating voltage as the thermal voltage is constant at a given room temperature. As a result of this, source to drain leakage current increases because of the thermal diffusion of electrons. Furthermore, due to the decrease in the gate oxide thickness, there has been a sharp increase in the gate leakage current. To replace CMOS with a new device or technology that would allow long term scaling is a hot research topic [2].

Although CMOS can continue to scale in size for some more time, our ability to achieve full benefits of scaling has become limited, as we are forced to trade-off between transistor density and speed to mitigate the power density increase. Fortunately, the emergence of these problems coincides with other technological advances. First, the lack of efficiency at the technology level is masked by the architectural techniques of building multi-core systems or approximate systems. Second, a plethora of new technologies, which may eventually complement and replace CMOS, is there. Some of these technologies are at prototype-level, some are provided as a complementary to the standard CMOS techniques, even with the same manufacturing steps. A few technologies are already in mainstream with storage devices.

Resistive RAM (ReRAM) has emerged as a promising technology for logic-in-memory computations, due to simultaneous storage and logic capabilities [3]. Boolean majority with an inverted input is implicitly supported by resistive memories, and forms a universal set for Boolean logic operations, as shown in Fig. 1. ReRAMs offer high endurance and density, high energy efficiency and fast operating speeds [4]. ReRAMs have been used for realization of application specific circuits such Boolean arithmetic circuits, ternary arithmetic as well as neuromorphic computing.

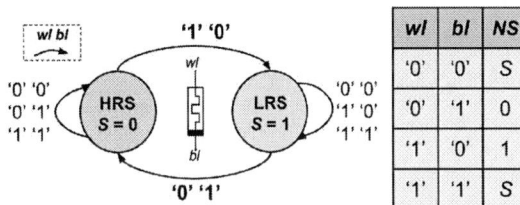

Fig. 1: Logic operation of ReRAM device expressed as a Finite State Machine and truth table. *wl* and *bl* represent the two input terminals — wordline and bitline of the device. The internal resistive state of the device is S, that gets updated to NS on applying the wordline and bitline inputs.

wl	bl	NS
'0'	'0'	S
'0'	'1'	0
'1'	'0'	1
'1'	'1'	S

Quantum computing has garnered significant attention due its inherent capabilities to offer up to exponential speedup over the best known algorithms for classical computing [5]. Quantum computing also promises breakthrough speed-up for machine learning, software validation and predictive analysis with direct implications in cyber security, drug discovery and data analytics.

A key challenge in bringing emerging technologies to practical usage is the availability of robust design automation flow, including logic synthesis and technology mapping. Therefore, we are motivated to algorithms and libraries for ReRAMs and quantum computing, with prime focus on the technology mapping aspects for these emerging technologies.

978-1-5386-7100-9/18 $31.00 © 2018 IEEE   369

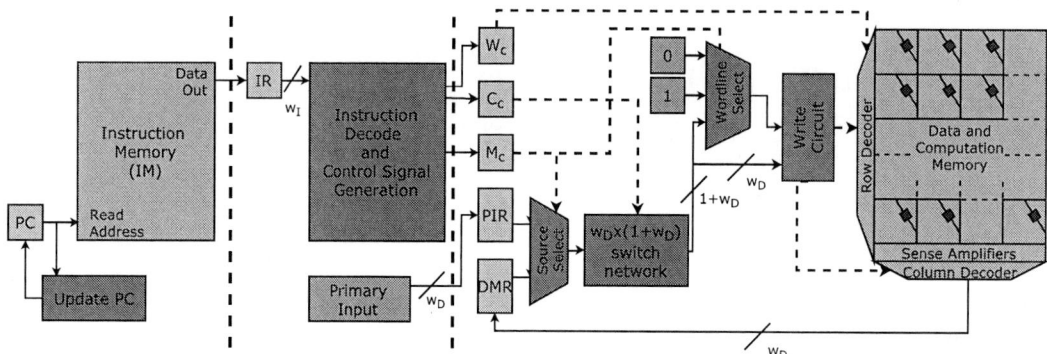

Fig. 2: ReVAMP architecture.

II. LOGIC-IN-MEMORY (LiM) ARCHITECTURE

One of the initial goals was to develop an efficient general purpose in-memory computing architecture using ReRAM crossbar memories that permit a universal set of Boolean operators. The main challenge was to harness the inherent bit-level parallelism offered by the ReRAM, while at the same time, keeping the controller complexity minimal. We developed ReRAM based VLIW Architecture for in-Memory comPuting (ReVAMP). The architecture shown in Fig. 2, uses two ReRAM crossbar memories — the Instruction Memory (IM) and the Data Storage and Computation Memory (DCM). The IM is a regular instruction memory accessed using the program counter (PC). The DCM hosts data and in-memory computation. Even though it is possible to update multiple words of the DCM in parallel, the necessary control circuitry would be complex and costly in terms of area. Thereby, we chose to limit the architecture to updating all the devices simultaneously in one word of the DCM. Splitting the instruction and data memory allows reduction in overall execution time, by parallelizing instruction fetch and computation. The ReVAMP architecture supports two instructions — *Read* and *Apply*. The Read instruction reads the word at the specified address from the DCM and stores it in the DMR. Now available in the DMR, this word can be used as input by the next instruction. The Apply instruction is used for computation in the DCM. The VLIW instruction-set permits systematic handling of parallel computation on the ReRAM crossbar array structures.

Related Publication: [6]

III. RERAM-BASED CIRCUIT DESIGNS

Any new architecture needs a set of libraries for the frequently needed functions. The key challenge for mapping functions to LiM architecture is that the placement of data in the crossbar has a significant impact on the overall delay. Addition and multiplication are the fundamental arithmetic operations. Therefore, we mapped an efficient parallel prefix adder, Kogge-Stone adder that has a fixed fanout and logarithmic delay in terms of the number of bits of the operands. The flowchart of

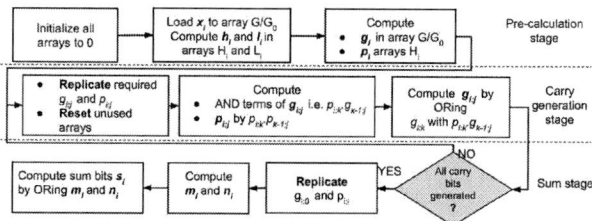

Fig. 3: Flowchart for addition of two integers using Kogge-Stone adder scheme on ReRAM arrays.

the proposed scheme is shown in Fig. 3. We have developed mapping for integer multiplication using Booth's encoding technique as well as floating point multiplication on ReRAM crossbar arrays.

Efficient mapping of the Binary Level-1 Binary Linear Algebra Subroutines (BLAS) on ReRAM crossbar array was developed. Specifically, we proposed three schemes that are parametrized based on the size of the crossbar. For a n-element vector mapped onto a $r \times c$ crossbar array, the results can be summarized as :-

$r \times c$	Delay
$(3k+1) \times 1$	$S = \lceil \frac{n}{2k} \rceil (14k - 2) + (\lceil \frac{n}{2k} \rceil - 1)6$
$1 \times (2k+1)$	$S = \lceil \frac{n}{2k} \rceil (6 + \sum_{t=2,t'=t/2}^{k}(2+2t)) + (\lceil \frac{n}{2k} \rceil - 1)4$
$2 \times 2k+1$	$S = \lceil \frac{n}{2k} \rceil \{10 + \sum_{t=2,t'=t/2}^{k/2}(7)\} + (\lceil \frac{n}{2k} \rceil - 1)4$

The proposed scheme for in-memory computing scheme is estimated to have $7\times$ speed up in terms of throughput, compared to the BiBLAS operation on modern day GPGPUs. We extended this work by proposing two schemes for level-2 Binary BLAS operations on ReRAM crossbar arrays with row-major mapping (RMM) and column-major mapping (CMM) orientation of the operations. For $m \times n$ matrix size and n-element vector multiplication, the delay achieved by the schemes can be expressed as :-

$$round_{\#steps}^{CMM} = \lceil \frac{n}{r} \rceil (c + 1 + 9 \times \frac{r}{2} + (\frac{r}{2} - 1) \times 9)$$
$$+ (\lceil \frac{n}{r} \rceil - 1) \times 9 \qquad (1)$$

$$CMM_{\#steps} = \lceil \frac{m}{c} \rceil \times round_{\#steps}^{CMM} + (\lceil \frac{m}{c} \rceil - 1) \times 1 \quad (2)$$

$$round_{\#steps}^{RMM} = \lceil \frac{n}{c} \rceil (r + 1 + 7r + 1 + log_2(\frac{c}{2}) \times 7r)$$
$$+ (\lceil \frac{n}{c} \rceil - 1) \times 7r \qquad (3)$$

$$RMM_{\#steps} = \lceil \frac{m}{r} \rceil \times round_{\#steps}^{RMM} + (\lceil \frac{m}{r} \rceil - 1) \times 1 \quad (4)$$

We could observe the following :-

- CMM outperforms RMM, for almost all crossbar dimensions, irrespective of matrix size, except for extreme cases with $r << c$.
- Crossbar configuration when $r < c$ performs the best, while $r > c$ performs the worst, and $r = c$ gives intermediate performance, when the number of elements ($r \times c$) in the crossbar is fixed.

Due to the difference in alignment of the data in the two mappings, the performance of the two schemes differs by a wide margin. 102.48 Gbps and 14.42 Gbps throughput was observed for CMM and RMM respectively. This highlights the importance of data storage alignment in the context of LiM mappings.

We constructed the mapping on ReVAMP for adder-based cryptographic primitive, SHA-2 and *Keccak* or SHA-3 that is based *sponge construction* with ABSORB and SQUEEZE phases The basic operations involved are XORing, rotation and permutation which can be effectively realized using LiM. The effective throughput of the SHA-2 implementation is 15.46 Mbps while SHA-3 implementation achieves throughput of 43.38 Mbps. The proposed SHA-2 requires fourteen 64-bit words for computation, with 49 KB of memory for storing the instructions, considering 32-bit aligned memory access. For SHA-3, forty three 64-bit words are used for computation while 80 KB of memory for storing the instructions. This shows the low storage overhead of LiM implementations of encryption of data, making it suitable for low-footprint Internet-of-things (IoT) applications.

Contributions: *In-memory implementation for various frequently used mathematical operations have been proposed — Kogge Stone adder, Booth multiplier, floating-point multiplier, binary level-1 and level-2 BLAS. Also, state-of-the art hashing functions — SHA-2 and SHA-3, have been implemented using in-memory computing on the ReVAMP LiM architecture.*
Related Publications: [7], [8], [9], [10], [11], [12]

IV. TECHNOLOGY MAPPING FOR LiM ARCHITECTURES

The *technology mapping problem* for ReRAM based in-memory computing is to determine a sequence of inputs, to be applied, to the wordlines and the bitlines of ReRAM

devices from a representation of a Boolean function, in order to compute the function. The *delay* of the obtained mapping is equal to the number of steps that the mapping contains. The number of devices used in the mapping determines the *area*.

Since the intrinsic function realized by 1S1R devices is Boolean majority with an input inverted, we chose Majority-Inverter Graph (MIG) for logic representation. For any k-level MIG, our proposed mapping algorithm generates an optimal mapping with $k + 1$ delay. The delay reduction, achieved by our mapping is at least $3\times$ better than that attained by the naïve technology mapping proposed in literature. Further, we have devised a heuristic for reducing the device count that allows reduction in number of devices required for mapping on average by 56% (max 92.63%). We introduce a constraint based on the number of parallel operations permitted — *dispatch parallelism*. Under arbitrary constraints on dispatch parallelism, we have shown that the problem of delay-optimal scheduling is NP-complete. To that effect, the concept of ReRAM Dependency Graph (RDG), that captures the dependencies between the instructions was introduced. We proposed an efficient heuristic for scheduling the RDG, constrained by the dispatch parallelism. By using $d_p = 1024$, our heuristic achieves delay close to the optimal delay.

For practical realization, area constraints have to be considered, i.e, the number of devices available for mapping is restricted. In the solution approach to the area constrained technology mapping problem, we start from the solution of the delay optimal technique and construct a so called Device Dependency Graph (DDG). The DDG reflects the dependencies between subsequent operations on a device and read dependencies between operations of different devices. The ReRAM devices that will be used for area-constrained technology mapping are termed as *elements*. Under the area constraints, the problem is to obtain a valid schedule that does not violate the dependencies present in the DDG and at the same time maps each device in the DDG to an element such that no two devices are mapped to the same element simultaneously. We have formulated an Integer Linear Program for area-constrained delay optimal technology mapping on ReRAM devices. Thereafter, we have developed a novel heuristic algorithm for the same that is highly scalable and permits mapping Boolean functions across a wide range of area constraints.

The ReVAMP architecture additionally enforces crossbar constraints. For example, the devices are organized as a crossbar memory. Only a single word can be read out in a cycle. Read and compute operations cannot be performed together. Due to the inherent sequential nature of computation on the ReRAM memories and the need for awareness of the data storage pattern within the memory, traditional technology mapping algorithms for ASIC as well as FPGAs cannot be used. This gives rise to the scope of developing new technology mapping algorithms for LiM architectures.

In the first approach, we consider the area constrained version of the problem, where we represent the Boolean function as an And-Inverter Graph (AIG). We have established

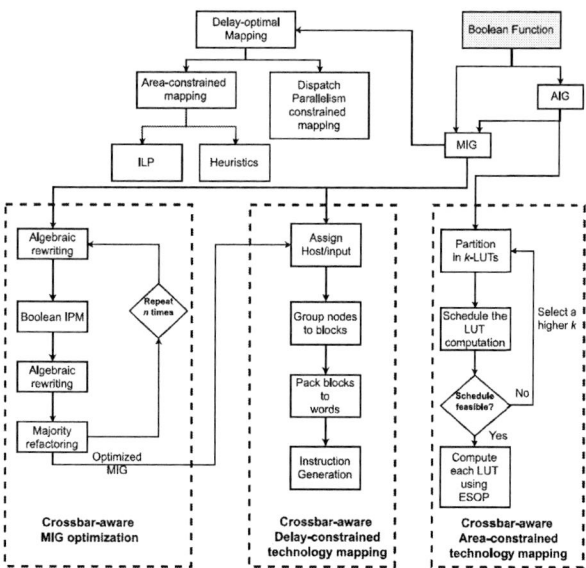

Fig. 4: Design automation flow for LiM architectures

A. Optimizing Synthesis for ReVAMP architecture

The technology mapping algorithms can perform better, if the initial logic representation using MIG or AIG can be optimized suitably. To do so, we use a three-stage optimization pipeline.

- We are interested in obtaining compact MIG representations by using Algebraic optimization because they translate in smaller and faster physical implementations, especially in the ReRAM architectures.
- More powerful transformations are possible by using Boolean methods. For that purpose, we employ MIG Boolean optimization based on Input Partitioning Methods (IPM) and safe errors.
- An effective Boolean technique for MIG minimization is majority refactoring. The idea is to first compute a large cone of logic rooted at a MIG node. This cone of logic is transformed into a canonical logic representation, e.g., a BDD or a truth table. Starting from the canonical logic representation, a new local MIG is derived by means of decomposition.

Various components of the proposed design automation flow for LiM architectures can be visualized as a flowchart, as shown in Fig. 4.

Contributions: *One of the key contributions was to develop a polynomial-time depth optimal algorithm for technology mapping for in-memory computing using ReRAM devices. We have also proposed an ILP-based area-constrained delay optimal technology mapping solution. Further on, we have developed delay-constrained and area-constrained technology mapping solutions for in-memory computing using ReRAM crossbar arrays. Also, we have demonstrated that technology-aware logic synthesis optimizations can improve the overall design automation flow.*
Related Publications: [13], [14], [15]

V. MULTI-VALUED LOGIC FUNCTION REALIZATION

ReRAMs also offer the possibility of native storage of data in higher radix ($>= 2$) which opens the scope of natively realizing a wide variety of applications. Fuzzy logic can be used for developing operational automated control systems, clinical practice decision support systems, etc. We have developed native primitives for realization of multi-valued logic and have shown how fuzzy logic controller can be designed using multi-state ReRAMs. We have verified the correctness of the proposed approach by fabricating a multi-state ReRAM crossbar array. Fig. 5 shows the fabricated 1×3 multi-state ReRAM crossbar array. This is the first known implementation of fuzzy logic natively using ReRAMs.

The approach of implementing multi-valued logic operation within the resistive memory device using the available multi-level states is a highly attractive option for future hybrid CMOS/ReRAM chips for enhancing its present functionality. Each multi-valued operation requires a constant number of steps, irrespective of the value of n in an n-valued logic system. For Boolean realization of the multi-valued operators,

a lower bound on the number of devices required for a feasible mapping of an And-Inverter Graph (AIG) or MIG. Any AIG or MIG with k-levels can be mapped using $2(k + 1)$ devices, arranged as a crossbar with at least two bitlines. For the general mapping approach, we begin by partitioning the AIG into k-input Look-up Tables (LUTs). A k-input LUT is basically a function with atmost k-inputs and a single output. Once the graph has been partitioned, the LUTs for computation are scheduled in topological ordering, i.e., the LUTs close to the primary input are scheduled first and so on, till the output LUTs are computed. In order to compute a LUT, we express the functionality of the LUT using Exclusive Sum-Of-Products (ESOP). Any arbitrary ESOP can be computed on the DCM with at least 3-wordlines and 2-bitlines — the variables which have to be used in inverted form are negated first (to be applied via bitlines), followed by computing the product terms and finally XORing them. To reduce the delay, the AND computation for realizing the ESOP needs to minimize number of reads performed and maximize number of AND operations that can be done in parallel. Thereafter, we perform the XOR of computed AND terms by means of a XOR reduction tree of logarithmic depth in the number of AND terms.

In the second approach, we focus on minimizing the delay of the mapping, without any constraints on the number of words. We use MIGs for logic representation in this approach. We propose a multi-stage algorithm — assignment of nodes as host or input for computation, grouping nodes into blocks, packing blocks to words followed by generation and scheduling of instructions. The proposed approaches allow mapping to a variety of crossbar dimensions. Also, the delay focused approach outperforms existing methods by more than $9\times$ in terms of delay.

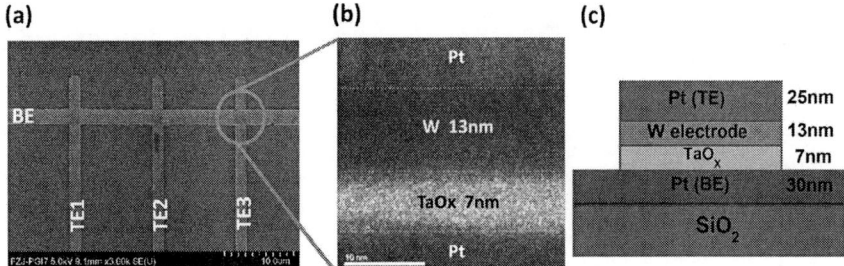

Fig. 5: Fabricated resistive switching device structures (a) Scanning electron microscopy image of 1×3 array, with $2mm \times 2mm$ device (b) Tunnelling electron microscopy image of a single device cross-section with $7nm$-thick TaO_x switching layer and $13nm$-thick tungsten ohmic electrode. (c) Schematic diagram of the single resistive device. [4]

the number of steps would increase with the value of n. Furthermore, parallel operations across multiple devices that share the same wordline, can be enabled by carefully packing operations that have the same input. In contrast, to leverage such parallelism, the Boolean circuits corresponding to the implication and negation operations need to be replicated. Multi-level ReRAM devices reduces the complexity of state representation and thus, brings fundamental benefits across arithmetic and logical primitives.

Contributions: *Leveraging multi-state ReRAM devices, we have developed a methodology to realize finite-valued Łukasiewicz logic. Thereafter, we have demonstrated how fuzzy control system (fuzzification, fuzzy rule evaluation) can be realized using the ReRAM crossbar array, by using Łukasiewicz logic primitive operations.*
Related Publication: [4]

VI. QUANTUM CIRCUIT SYNTHESIS AND TECHNOLOGY MAPPING

The challenge of synthesising a quantum algorithm on to a set of quantum gates gave rise to the domain of quantum/Reversible logic synthesis with a growing body of techniques that achieved excellent scalability, efficient circuit construction in terms of T-count, T-depth, Qubit count and integration of error correction, e.g., through surface codes [16]. The dominant problem towards the practical realisation of the first batch of quantum algorithms is to minimize the number of qubits required for synthesis. We have formulated an optimal SAT-based encoding for synthesizing quantum circuit with minimum number of qubits, using a hierarchical syntehsis technique. We have also developed a heuristic for reversible pebble game that allows reduction in the number of qubits required for realizing large quantum benchmarks. The work was implemented in open source tool for reversible circuits, Revkit [17].

Long-distance interacting qubits are particularly susceptible to the noise. Therefore, prominent quantum technologies and quantum error correction codes, such as surface codes, require that the quantum gates must be formed with a nearest neighbour interaction. In the resulting circuits, the interacting qubits

may form a chain, as in a 1D qubit layout, and therefore, these circuits are referred to as Linear Nearest Neighbor (LNN) circuits. Conversion of a quantum circuit to an LNN one can be achieved by using SWAP gates. Despite the presence of diverse qubit topologies and need for an automated mapping flow of quantum circuits to such topologies, the current literature focuses mostly on 1D chain qubit and 2D lattice structures. In order to address this gap, we have developed an ILP-based algorithm to realize depth-optimal nearest neighbour quantum circuits *for arbitrary topologies*. Our algorithm is also applicable, naturally, to simpler structures such as Linear Neighbours, as shown in Fig 6. We have benchmarked the algorithm for diverse topologies and quantum circuits.

Contribution: *We have proposed an optimal reversible pebble game strategy using a SAT formulation, along with scalable heuristics to minimize the number of qubits required for quantum circuit synthesis. Also, an Integer-Linear Programming (ILP) based optimal solution for designing nearest-neighbor complaint circuits for arbitrary topologies has been developed.*
Related Publication: [18]

VII. CONCLUSION

Emerging technologies crucially rely on the support for EDA tools to demonstrate competitive edge over conventional, mature technologies. To that end, this work made the following specific contributions.

- A new in-memory architecture using ReRAM crossbar arrays has been proposed.
- A set of optimized libraries for in-memory computing has been developed, such as adder, multiplier, hashing, etc.
- Delay-optimal and area-constrained technology mapping for ReRAM devices.
- Crossbar-constrained technology mapping for ReRAM crossbar arrays.
- Native realization of multi-valued and fuzzy logic using multi-state ReRAMs has been achieved for the first time.
- Technology-aware logic synthesis optimization for ReRAM based in-memory computing.

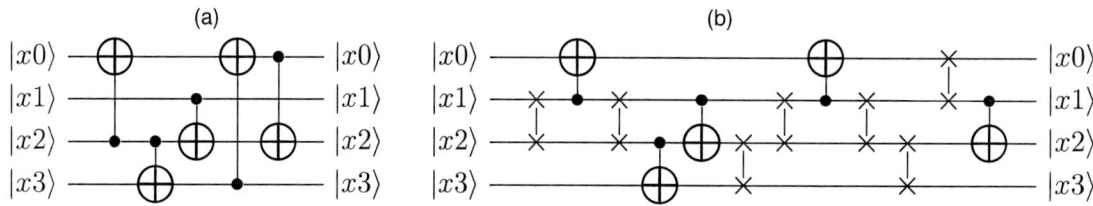

Fig. 6: (a) A quantum circuit. (b) Nearest Neighbour Compliant Circut (considering a linear topology).

- Optimization techniques for qubit reduction and synthesis under physical constraints for quantum circuits.

We believe our contributions will pave the path to wide-scale adoption of emerging technolgies.

REFERENCES

[1] Y. Taur, "Cmos design near the limit of scaling," *IBM Journal of Research and Development*, vol. 46, no. 2.3, pp. 213–222, 2002.

[2] K. Bernstein, R. K. Cavin, W. Porod, A. Seabaugh, and J. Welser, "Device and architecture outlook for beyond cmos switches," *Proceedings of the IEEE*, vol. 98, no. 12, pp. 2169–2184, 2010.

[3] D. B. Strukov, G. S. Snider, D. R. Stewart, and R. S. Williams, "The missing memristor found," *nature*, vol. 453, no. 7191, p. 80, 2008.

[4] D. Bhattacharjee, W. Kim, A. Chattopadhyay, R. Waser, and V. Rana, "Multi-valued and Fuzzy Logic Realization using TaOx Memristive Devices," *Scientific reports*, vol. 8, no. 1, p. 8, 2018.

[5] C. H. Bennett, E. Bernstein, G. Brassard, and U. Vazirani, "Strengths and weaknesses of quantum computing," *SIAM journal on Computing*, vol. 26, no. 5, pp. 1510–1523, 1997.

[6] D. Bhattacharjee, R. Devadoss, and A. Chattopadhyay, "ReVAMP: ReRAM based VLIW architecture for in-memory computing," in *2017 Design, Automation & Test in Europe Conference & Exhibition (DATE)*. IEEE, 2017, pp. 782–787.

[7] D. Bhattacharjee, A. Siemon, E. Linn, S. Menzel, and A. Chattopadhyay, "Kogge-Stone Adder Realization using 1S1R Resistive Switching Crossbar Arrays," *ACM Journal on Emerging Technologies in Computing Systems (JETC)*, 2018.

[8] D. Bhattacharjee, A. Siemon, E. Linn, and A. Chattopadhyay, "Efficient complementary resistive switch-based crossbar array booth multiplier," *Microelectronics Journal*, vol. 64, pp. 78–85, 2017.

[9] V. Tarun, A. Dutt, D. Bhattacharjee, and A. Chattopadhyay, "Floating Point Multiplication Mapping on ReRAM based In-Memory Computing Architecture," in *VLSI D*. IEEE.

[10] D. Bhattacharjee, F. Merchant, and A. Chattopadhyay, "Enabling in-memory computation of binary BLAS using ReRAM crossbar arrays," in *Very Large Scale Integration (VLSI-SoC), 2016 IFIP/IEEE International Conference on*. IEEE, 2016, pp. 1–6.

[11] D. Bhattacharjee and A. Chattopadhyay, "Efficient binary basic linear algebra operations on ReRAM crossbar arrays," in *VLSI Design and 2017 16th International Conference on Embedded Systems (VLSID), 2017 30th International Conference on*. IEEE, 2017, pp. 277–282.

[12] D. Bhattacharjee, V. Pudi, and A. Chattopadhyay, "SHA-3 implementation using ReRAM based in-memory computing architecture," in *Quality Electronic Design (ISQED), 2017 18th International Symposium on*. IEEE, 2017, pp. 325–330.

[13] D. Bhattacharjee and A. Chattopadhyay, "Delay-optimal technology mapping for in-memory computing using reram devices," in *Computer-Aided Design (ICCAD), 2016 IEEE/ACM International Conference on*. IEEE, 2016, pp. 1–6.

[14] D. Bhattacharjee, A. Easwaran, and A. Chattopadhyay, "Area-constrained technology mapping for in-memory computing using reram devices," in *Design Automation Conference (ASP-DAC), 2017 22nd Asia and South Pacific*. IEEE, 2017, pp. 69–74.

[15] D. Bhattacharjee, L. Amarú, and A. Chattopadhyay, "Technology-aware logic synthesis for reram based in-memory computing," in *Design, Automation & Test in Europe Conference & Exhibition (DATE), 2018*. IEEE, 2018, pp. 1435–1440.

[16] J. Kelly, R. Barends, A. Fowler, A. Megrant, E. Jeffrey, T. White, D. Sank, J. Mutus, B. Campbell, Y. Chen *et al.*, "Scalable in situ qubit calibration during repetitive error detection," *Physical Review A*, vol. 94, no. 3, p. 032321, 2016.

[17] "Revkit," https://msoeken.github.io/revkit.html, accessed: 2018-05-13.

[18] D. Bhattacharjee and A. Chattopadhyay, "Depth-Optimal Quantum Circuit Placement for Arbitrary Topologies," *arXiv preprint arXiv:1703.08540*, 2017.

2018 IEEE Computer Society Annual Symposium on VLSI

Logic Synthesis for In-Memory Computing using Resistive Memories

Saeideh Shirinzadeh*, Rolf Drechsler*†

*Department of Mathematics and Computer Science, University of Bremen, 28359 Bremen, Germany
†Cyber-Physical Systems, DFKI GmbH, 28359 Bremen, Germany

Abstract—The increasing urge to bypass the issue of the memory bottleneck in the current computer architectures has attracted high attention to in-memory computing enabled by emerging memory technologies such as *Resistive Random Access Memory* (RRAM). This paper studies in-memory computing from two perspectives, i.e. customized and instruction-based. The customized approach exploits logic representations to synthesize for in-memory computing. The approach proposes design methodologies and optimization algorithms for each representation with respect to area and latency upon the realizations of their logic primitives. The instruction-based approach proposes an automatic compiler to execute instructions on a logic-in-memory computer architecture and optimizes the programs. Experimental results for both approaches reveal considerable improvements compared to the state-of-the-art.

Keywords—*In-memory computing; RRAM; logic synthesis;*

I. INTRODUCTION

The advancements in the processors of modern computers have far exceeded that of memory. The considerably higher latency of memory compared to processor on one hand, and the requirement of communication between these two in the von Neumann architecture on the other hand, has limited the overall performance of current computing systems which is known as *memory wall*. The growing need to deal with higher amount of data demanded by emerging applications such as *internet of things* (IoT) and *big data* has prompted much research to alleviate this problem [1]. Among these solutions, *in-memory computing* sounds to be very promising as it allows to go beyond the memory wall by integrating the storage and computing paradigms. This provides speedups of several orders of magnitude.

One of the core technologies enabling in-memory computing is *Resistive Random Access Memory* (RRAM). RRAM is a promising non-volatile memory technology with high scalability and zero standby leakage energy which internal resistance can be switched between two states, i.e. low and high. So far, several approaches have been proposed which exploit this resistive switching property to execute logic functions within RRAM devices exploiting. *Material Implication* (IMP) has been widely used for logic-in-memory computing [2]. A logic family called MAGIC was also proposed in [3] which allows to realize Boolean functions using NOR and NOT operations. In [4], it was shown that RRAM natively implements a majority oriented operation (MAJ) which allows to utilize majority logic for in-memory computing.

In this paper, we explore synthesis for in-memory computing from two different perspectives, i.e. (i) based on a customized approach at gate level which employs logic representations, and (ii) an instruction oriented approach for efficient manipulation, compilation, and execution of programs on a logic-in-memory architecture. The presented customized synthesis approach uses IMP and MAJ as basic operations

on RRAM array which can be alternatively according to the design preferences, while the instruction based approach uses MAJ only for executing programs.

The customized synthesis approach starts with finding efficient realizations for the logic primitives of the employed representations, i.e. *Binary Decision Diagram* (BDD), *And-Inverter Graph* (AIG), and *Majority-Inverter Graph* (MIG). The approach presents optimization algorithms and a comprehensive design methodology to map each representation into equivalent netlists of operations and RRAM devices to be performed on a resistive memory array. The results of the presented customized approach reveal significant improvement compared to the state-of-the-art using the same logic representations.

By means of the instruction-based approach, we fully automatize and optimize an existing logic-in-memory computing architecture. Furthermore, we address the issue of lower write endurance of RRAM devices and propose wear-leveling techniques to increase the lifetime of the aforementioned architectures. Experiments performed on large arithmetic and control functions show considerable improvement in the distribution of writes all over the memory array as well as the number of cycles and RRAM devices representing the latency and area of the resulting implementations.

The remaining of this paper is structured as the following. To keep the paper self-contained, required preliminaries are explained in Section II in adition to a brief review of the state-of-the-art. Section III and Section IV present the customized and instruction-based synthesis approaches, respectively. Section V concludes the paper.

II. BACKGROUND

A. Logic Representations

1) Binary Decision Diagrams (BDDs): *Binary Decision Diagram* (BDD, [5]) is a graph based data structure which allows to represent Boolean functions efficiently. A BDD is obtained from recursively applying Shannon decomposition $f = x_i f_{x_i} \oplus \bar{x}_i f_{\bar{x}_i}$ such that each node represents a subfunction. Using complemented edges in the graph, also allows to represent a sub-function and its complement by the same node. The decomposition is performed according to a certain variable ordering which results in a canonical BDD. For example, the ordering of the BDD shown in Fig. 1(a) is in order $x_1 < x_2 < x_3$, where the complemented edges are denoted by dots on the successors.

In circuit realization, each BDD node is mapped to a multiplexer which makes the number of nodes a determining factor in the costs of the resulting implementations. Therefore, BDD optimization, i.e. finding a variable ordering which results in a smaller BDD has been of high interest for applications utilizing BDDs. In this paper, we assume that the variable

978-1-5386-7100-9/18 $31.00 © 2018 IEEE   375

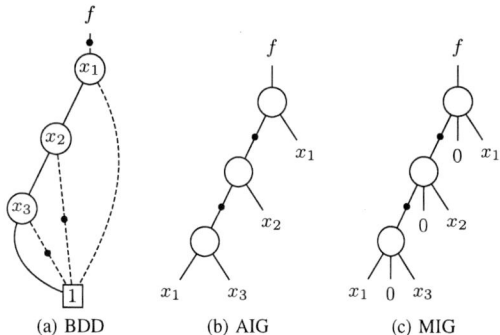

Fig. 1.   Logic representations for an example function with three input variables.

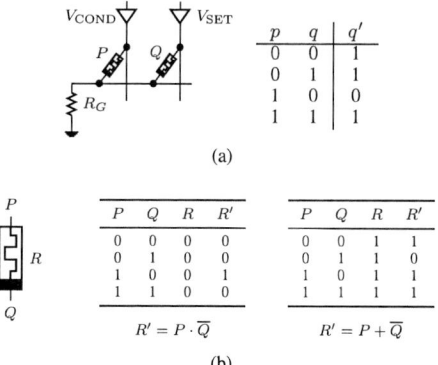

Fig. 2.   The Boolean operations executable within RRAM used at this work. (a) Implementation of IMP and its truth table [2]. (b) The intrinsic majority operation within an RRAM device [4].

ordering of an initial BDD before optimization is ascending $x_1 < x_2 < \cdots < x_n$, where n is the number of input variables.

2) Homogeneous Logic Representations for Circuits: *And-Inverter Graph* (AIG, [6]) and *Majority-Inverter Graph* (MIG, [7]) are homogeneous logic representation which are used in this work. An AIG node designates logical conjunction $x \cdot y$, which is majority of three $M(x, y, z) = x \cdot y + x \cdot z + y \cdot z$ in case of MIGs. The graphs also include inverters which are in terms of complemented edges. Both representations allow to efficiently represent Boolean functions and are utilized by state-of-the-art synthesis tools.

An AIG can easily be transformed into an MIG by adding a third zero input $x \cdot y = M(x, y, 0)$. Fig. 1(b) and (c) show both AIG and MIG representation for a three input function. As the figure shows, the number of nodes for both graphs is equal to three. However, MIGs can allow even more compact representations compared to AIGs [8].

B. Logic Operations enabled within RRAM

1) Material Implication (IMP): In [2], it was shown that material implication (IMP), i.e. $q' \leftarrow p \text{ IMP } q = \overline{p} + q$, can be executed from the interaction of two resistive switches under certain voltage levels shown by V_{SET} and V_{COND} in Fig. 2(a). The logical states of the resistive devices can also be simply switched between logic 1 or 0, i.e. FALSE operation, when applied to appropriate voltage pulses to set or clear the devices. IMP and FALSE together make a universal set of logic operations sufficient to execute arbitrary Boolean functions on resistive arrays.

2) Resistive Majority Operation (MAJ): In [4], an intrinsic majority operation enabled by RRAM devices was introduced which suffices to compute any Boolean function. Let us denote the top and bottom electrodes of an RRAM device with P and Q (see Fig. 2(b)). Assuming that the current resistive state of the device (R) can be switched to 1 and 0 by applying a positive or negative voltage level V_{PQ}, respectively, the next state of the device (R') changes based on the truth tables shown in Fig. 2(b). By expanding the Boolean relation in the tables, it can simply be shown that the next resistance state of the device is equal to $R' = M(P, \overline{Q}, R) = P \cdot \overline{Q} + P \cdot R + \overline{Q} \cdot R$, i.e. the result of three-input majority function when the logical state of the bottom electrode is inverted. This operation is referred to three-input resistive majority operation which is denoted by MAJ in this paper.

C. Related Work

The majority of related work for logic-in-memory synthesis using resistive devices exploit IMP as the basic operation. In

[9], an approach was presented to map BDDs into networks of resistive devices. The work follows two main objectives in the presented mapping methodologies, one with respect to the area, i.e. serial computation, and the other with respect to the latency, i.e. parallel computation. In [10], a BDD-approach for logic-in-memory synthesis was proposed which improves the parallel methodology in [9]. [10] exploits a multi-objective BDD optimization algorithm which results in significantly improved results in terms of area and latency compared to [9]. AIG [11] and OIG [12], i.e. *Or-Inverter Graphs* have been also utilized for synthesis with RRAM devices using IMP operation.

Other basic operations enabled by RRAM devices have been also exploited in the state-of-the-art. MAGIC, i.e. *Memristor-Aided loGIC*, [3] is one of this approaches which allows to implement logic functions as a network of NOR gates on RRAM array. In [4], MAJ was used as the only form of logic operation performed in the *Programmable Logic-in-Memory* (PLiM) architecture. The architecture was simply designed to control a single MAJ instruction during each cycle. The number of executable MAJ-based instructions per cycle was increased to two in [13] enabled by heuristic palatalization algorithms. Technology mapping for the architecture proposed in [13] was further improved in terms of crossbar area [14] as well as delay [15] by applying optimization techniques during technology mapping.

III. CUSTOMIZED SYNTHESIS

The customized synthesis approach includes three stages, (i) finding efficient realizations with RRAM devices for logic primitive of each representation, (ii) then, defining the design methodology to map the representations into RRAM array, (iii) and finally optimizing the representations with respect to the number of RRAM devices and operations determined by the design methodology. In the following, we explain the proposed customized approach for logic representations BDD and AIG briefly and provide an implementation example in the case of MIG. Due to the lack of space, we refer the reader to [16] for the details of the optimization algorithms for each representation.

An IMP-based realization for a 2-to-1 multiplexer (MUX) has been proposed in [9] which requires six operations and five RRAM devices. Using MAJ, the realization [16] requires six devices and executes the MUX function within five operations. This trade-off between the required number of RRAM devices

978-1-5386-7100-9/18 $31.00 © 2018 IEEE   376

TABLE I.   THE COST METRICS OF LOGIC REPRESENTATIONS FOR RRAM-BASED IN-MEMORY COMPUTING

Metric	Definition\Value
N_i	No. of nodes in the ith level
CE_i	No. of ingoing complemented edges in the ith level
RE_i	No. of ingoing regular edges in the ith level
FO	Maximum no. of nonconsecutive fanouts in any BDD level
D	The depth of the graph
L_{CE}	No. of levels with ingoing complemented edges
L_{RE}	No. of levels with ingoing regular edges
R	No. of RRAM devices
OP	No. of operations

#R	BDD: $\max_{0 \le i \le D}(K \cdot N_i + CE_i) + FO$	IMP $: K = 5$, MAJ $: K = 6$
	AIG: IMP $: \max_{0 \le i \le D}(3 \cdot N_i + RE_i)$	MAJ $: \max_{0 \le i \le D}(3 \cdot N_i + CE_i)$
	MIG: $\max_{0 \le i \le D}(K \cdot N_i + CE_i)$	IMP $: K = 6$, MAJ $: K = 4$
#OP	BDD: $K \cdot D + L_{CE}$	IMP $: K = 6$, MAJ $: K = 5$
	AIG: IMP $: 3 \cdot D + L_{RE}$	MAJ $: 3 \cdot D + L_{CE}$
	MIG: $K \cdot D + L_{CE}$	IMP $: K = 10$, MAJ $: K = 3$

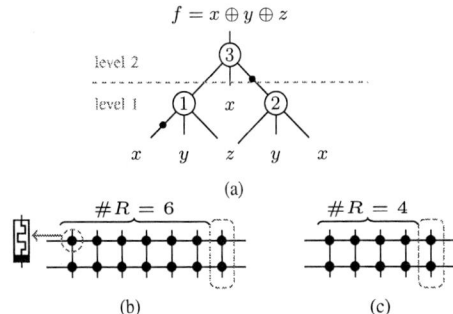

Fig. 3.   (a) MIG representing a three bit XOR gate, and upper-bound crossbar for implementing it using (b) IMP-based and (c) MAJ-based realizations.

and operations allows to choose between the realizations according to the design preferences when either latency or area is of higher importance.

Table I shows the cost metrics of the BDD-based synthesis approach for both IMP and MAJ-based realizations. The BDD is first optimized with respect to these cost metrics, see optimization algorithms in [10], [16], and can then be implemented on an RRAM array according to a level-by-level design methodology. This means that starting from the bottom of the graph all of the nodes in each BDD level are computed simultaneously. After computation of each level, the RRAM devices are released and can be used as input devices for computation of the next level. This procedure continues until the root function is computed.

For computing each level, the number of required RRAM devices includes five or six times the number of nodes in the level, one extra RRAM device for each complemented edge, and one extra device for preserving the value of each nonconsecutive fanout, i.e. a fanout targeting a level else than the next successive one. Since the RRAM devices are reused, the number of devices required to compute a BDD is equal to the maximum number of required devices over all of the levels. The number of operations to compute a BDD is at least six or five rimes the number of BDD levels, for IMP or MAJ-based representations, respectively. However, this value must be added to the number of levels which possess complemented edges as their negation needs extra operations. It should be noted that copying the values of the nonconsecutive fanouts can be performed at the same time step of loading the inputs of the level, and therefore is not counted in the number of operations [16].

Table I also shows the cost metrics for AIG-based synthesis obtained by the same level-by-level design methodology explained above. The values are shown for the *imp*-based realization of NAND gate [11] and the MAJ-based realization of AND gate [16] as logic primitives of AIG. Both realizations need three RRAM devices and three operations. The IMP-based realization also includes an inverter, which can be directly used for the complemented edges but needs inversion

for the regular edges. As a result, the levels possessing regular edges need additional operations.

The IMP-based realization [17], [16] for the majority gate is shown in the following. The realization needs six RRAM devices, input devices X, Y, and Z and extra devices A, B, and C required for negating or storing the operations' outputs. The majority function is executed after ten operations. The first operation loads the required devices with the input variables and zero, and the rest of the steps include IMP and FALSE operations.

01: $X = x, Y = y, Z = z$ 06: $c \leftarrow y$ IMP $c = \overline{x + y}$
  $A = 0, B = 0, C = 0$
02: $a \leftarrow x$ IMP $a = \bar{x}$ 07: $c \leftarrow z$ IMP $c = \overline{x \cdot z + y \cdot z}$
03: $b \leftarrow y$ IMP $b = \bar{y}$ 08: $a = 0$
04: $y \leftarrow a$ IMP $y = x + y$ 09: $a \leftarrow b$ IMP $a = x \cdot y$
05: $b \leftarrow x$ IMP $b = \bar{x} + \bar{y}$ 10: $a \leftarrow c$ IMP $a = x \cdot y + y \cdot z + x \cdot z$.

It is obvious that the MAJ-based realization for MIG-based synthesis can be realized more efficiently due to exploiting the natively implemented majority function in RRAM devices. As shown in the following, the realization of majority gate in this case needs a maximum of four devices and three steps.

1: $X = x, Y = y, Z = z, A = 0$
2: $P_A = 1, Q_A = y, R_A = 0 \Rightarrow R'_A = \bar{y}$
3: $P_Z = x, Q_Z = \bar{y}, R_Z = z \Rightarrow R'_Z = M(x, y, z)$.

Fig. 3(a) shows an MIG representing a three-input XOR gate. Here, we show how this MIG can be implemented on an RRAM array with devices shown by R_{ij}, where i and j denote the indices of the row and column, respectively.

The presented design methodology computes all nodes in a level simultaneously and for this purpose it allocates a row to each node. This means a single operation per cycle is performed at each row. As the example MIG has a maximum level size of two, the required crossbar needs at least two rows with at least six devices (see Fig. 3(b)). Also, one extra device at the end of each row is considered to be used for complemented edges. As Table I predicts the number of required steps is 22, i.e. 10 times the depth 2 plus two additional steps for the complemented edges at both levels.

Initialization	$R_{ij} = 0$;
1: Loading for level 1	$R_{11} = x, R_{12} = y, R_{13} = z$; $R_{21} = x, R_{22} = y, R_{23} = z$;
2: Negation for node 1	$R_{17} \leftarrow x$ IMP $R_{17} : R_{17} = \bar{x}$;
3-11: Computing level 1	<u>node 1:</u> $R_{14} = M(\bar{x}, y, z)$; <u>node 2:</u> $R_{24} : M(x, y, z)$;
12: Loading for level 2	$R_{11} = x, R_{12} = M(x, y, z), R_{13} = M(\bar{x}, y, z)$ $R_{14} = R_{15} = R_{16} = R_{17} = 0$;
13: Negation for node 3	$R_{17} \leftarrow R_{12}$ IMP $R_{17} :$ $R_{17} = \overline{R_{12}} = \overline{M}(x, y, z)$;
14-22: Computing level 2	$R_{14} = M(M(\bar{x}, y, z), x, \overline{M}(x, y, z))$;

978-1-5386-7100-9/18 \$31.00 © 2018 IEEE

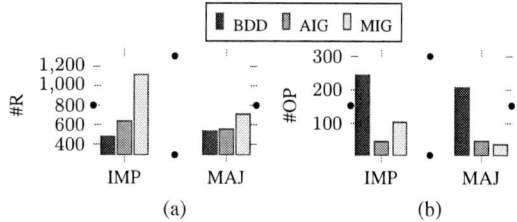

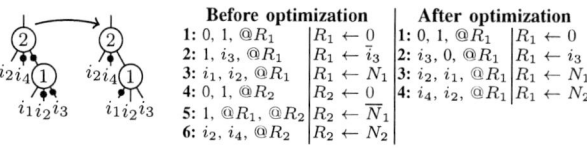

	Before optimization		After optimization	
1: $0, 1, @R_1$	$R_1 \leftarrow 0$	**1:** $0, 1, @R_1$	$R_1 \leftarrow 0$	
2: $1, i_3, @R_1$	$R_1 \leftarrow \bar{i_3}$	**2:** $i_3, 0, @R_1$	$R_1 \leftarrow i_3$	
3: $i_1, i_2, @R_1$	$R_1 \leftarrow N_1$	**3:** $i_2, i_1, @R_1$	$R_1 \leftarrow N_1$	
4: $0, 1, @R_2$	$R_2 \leftarrow 0$	**4:** $i_4, i_2, @R_1$	$R_1 \leftarrow N_2$	
5: $1, @R_1, @R_2$	$R_2 \leftarrow \overline{N_1}$			
6: $i_2, i_4, @R_2$	$R_2 \leftarrow N_2$			

Fig. 5. The effect of MIG optimization on the number of RRAM devices and instructions.

Fig. 4. Comparison of synthesis results by logic representations for RRAM-based in-memory computing, (a) the average number of RRAM devices, (b) the average number of operations.

The steps for the MAJ-based implementation of the MIG shown in Fig. 3(a) are shown in the following. As the steps show, the XOR function is executed using only three devices and within only four steps despite the upper bound of 8 which is predicted by Table I for an MIG with two levels possessing complemented edges. Indeed, the complemented edges at nodes 1 and 3 can be directly used as the second input of MAJ without being inverted. Furthermore, the RRAM updated devices can be used as inputs for the next cycle which means that the loading step is not required. It should be noted that applying signals to the rows and columns during the MAJ-based implementation should avoid data distortion by preserving previously computed results and the simultaneous operations in other rows [16]. For example, in step 2 the values of R_{11} and R_{21} are preserved by equalizing the logical states of their terminals when performing a MAJ operation within R_{25}.

Initialization	$R_{ij} = 0 : Q_{ij} = 1, P_{ij} = 0;$
1: Loading	$Q_1 = Q_2 = 0, P_1 = P_2 = z;$
	$R_{11} : RM_3(z, 0, 0) = M(z, 1, 0) = z;$
	$R_{21} : RM_3(z, 0, 0) = M(z, 1, 0) = z;$
2: Negation for node 2	$Q_1 = Q_2 = x, P_1 = x, P_2 = 1;$
	$R_{25} : RM_3(1, x, 0) = M(1, \bar{x}, 0) = \bar{x};$
3: Computing level 1	<u>node 1:</u> $P_1 = y, Q_1 = x, R_{11} = z$
	$R_{11} : RM_3(y, x, z) = M(y, \bar{x}, z);$
	<u>node 2:</u> $P_1 = y, Q_2 = \bar{x} \ (@R_{25}), R_{21} = z;$
	$R_{21} : RM_3(y, \bar{x}, z) = M(y, x, z);$
4: Computing level 2	$P_1 = x, Q_1 = @R_{21}, R_{11} = M(\bar{x}, y, z),$
	$R_{11} : RM_3(x, @R_{21}, @R_{11}) = M(x, \overline{@R_{21}}, @R_{11}) :$
	$M(M(\bar{x}, y, z), x, \overline{M(x, y, z)});$

Table II shows the results of the proposed approach for the MIG-based synthesis using both IMP and MAJ and compares them with the state-of-the-art BDD-based [9] and AIG-based approach [11]. According to Table II, the total number of operations by our proposed MIG-based synthesis approach using MAJ-based realization is almost one-ninth of the corresponding value by BDD-based synthesis [9] at a fair cost of 57.42% increase in the number of RRAM devices. Even when the IMP-based realization is used our proposed approach is three times faster that that presented in [9]. Furthermore, in comparison to [11], the results show speed-ups of 7.1 and 2.57 times using maj and imp for implementation, respectively.

Fig. 4 compares the synthesis results based on the three representations using both IMP and MAJ for implementation. As the figure shows, the BDD-based approach needs the smallest number of devices but leads to high latency. On the other hand, synthesis methodologies based on AIG and MIG require more RRAM devices but decrease the length of operations. In particular, the MIG-based approach using MAJ results in implementations with the smallest latency.

IV. INSTRUCTION-BASED SYNTHESIS

In [4], a *Programmable Logic-in-Memory* (PLiM) computer architecture was proposed which allows to perform logic operations on a regular RRAM array when its controller is on. The PLiM can also perform as a standard RAM system in case that the control signal is off. The controller consists of a simple finite state machine and some work registers to perform MAJ operations, which we refer to instructions. Only a single MAJ instruction is allowed per cycle. An arbitrary instruction has the format $M(P, Q, R)$ with three operands to be assigned. The first operand P is the signal applied to the top electrode of the RRAM device, i.e. the row driver, and the second operand Q is the signal applied to the bottom electrode, i.e. the column driver. The third operand R is the current state of the device under computation which updates automatically when the instruction is executed.

The PLiM computer derives the instruction set for computing an arbitrary Boolean function from its MIG representation. The characteristics of the MIG, the order of nodes for computation, and assigning the operands for each instruction significantly affect the number of required RRAM devices and instructions which address the area and delay of the resulting PLiM implementations [1], [18], [19]. In this work, we propose an automatic compiler which generates PLiM programs for computing arbitrary functions while addressing the aforementioned cost metrics.

As the PLiM performs computations fully serially the number of nodes in the MIG, i.e. the size of the MIG, is a determining factor in the length of the resulting set of instructions. Therefore, optimizing the MIG with respect to the number of nodes can considerably improve the latency of PLiM implementations. However, the number of nodes is not the only feature that should be addressed during MIG optimization. MAJ ideally needs one complemented edge to compute a node within a single instruction. Otherwise, one extra device and instruction are required to invert one operand before computing the node which makes the number and position of the complemented edges influential. Fig. 5 shows an example MIG with two nodes before and after optimization which has only changed the graph with respect to the complemented edges. As the figure shows, MIG optimization reduces both of the number of RRAM devices and instructions required for implementation. We refer the reader to [18] for the MIG optimization algorithm for PLiM.

As mentioned before, different orders of nodes for computation as well as choosing the operands result in different cost metrics regarding area and latency. Our proposed compiler first finds an efficient order of candidate nodes for computation, and then translates the nodes into a MAJ instruction by assigning the operands with the smallest number of RRAM devices and instructions.

The candidate selection procedure starts with listing the nodes which are computable, i.e. their children nodes are already computed. Then, it ranks the candidate nodes according

978-1-5386-7100-9/18 $31.00 © 2018 IEEE

TABLE II. COMPARISON OF RESULTS BY THE PROPOSED CUSTOMIZED APPROACH WITH EXISTING APPROACHES USING BDD [9] AND AIG [11]

Benchmark	PI	BDD [9]		MIG-IMP		MIG-MAJ	
		#R	#OP	#R	#OP	#R	#OP
5xp1_90	7	84	73	199	99	149	36
alu4_98	14	642	334	2160	176	1370	72
apex1	45	1626	705	3676	165	2343	56
apex2	39	122	237	531	143	358	56
apex4	9	2073	447	4728	143	2820	64
apex5	117	806	888	1482	141	1053	47
apex6	135	770	1169	1652	121	1018	44
apex7	49	290	437	408	132	277	48
b9	41	125	298	252	87	168	32
clip	9	120	89	312	110	217	40
cm150a	21	56	127	147	77	95	32
cm162a	14	46	102	90	86	60	30
cm163a	16	42	116	102	76	68	27
cordic	23	32	149	189	121	134	48
misex1	8	83	69	111	66	76	24
misex3	14	444	185	2207	165	1444	67
parity	16	23	113	216	132	152	53
seq	41	1566	692	3189	153	1970	64
t481	16	26	107	148	142	90	52
table5	17	580	168	2630	154	1723	64
too_large	38	282	232	510	164	322	64
x1	51	230	398	569	99	435	36
x2	10	60	80	66	76	46	26
x3	135	770	1169	1729	99	1008	44
x4	94	401	642	599	77	391	28
$\sum$		11299	9026	27902	3004	17787	1154

Benchmark	Inputs	AIG [11]	MIG-IMP		MIG-MAJ	
		#OP	#R	#OP	#R	#OP
9sym_d	9	1418	923	175	398	60
con1f1	7	18	70	75	28	26
con2f2	7	19	60	76	24	24
exam1_d	3	12	43	44	19	16
exam3_d	4	12	50	55	20	23
max46_d	9	427	408	131	193	48
newill_d	8	50	129	109	57	40
newtag_d	8	21	90	96	36	33
rd53f1	5	27	60	64	24	25
rd53f2	5	57	77	77	35	28
rd53f3	5	32	86	66	38	24
rd73f1	7	238	291	121	140	44
rd73f2	7	46	129	88	57	32
rd73f3	7	104	193	107	84	39
rd84f1	8	351	430	153	187	52
rd84f2	8	47	172	88	76	31
rd84f3	8	23	90	50	36	15
rd84f4	8	345	473	141	214	47
sao2f1	10	102	110	108	72	35
sao2f2	10	112	234	119	98	42
sao2f3	10	380	325	143	143	55
sao2f4	10	252	326	143	163	59
sym10_d	10	1172	1475	187	643	72
t481_d	16	1564	1285	187	567	72
xor5_d	5	32	86	66	38	24
$\sum$	194	6861	7615	2669	3390	966

to their effects on the number of RRAM devices and the length of the instructions and finally chooses the best node for computation. The comparison of the nodes is performed based on two main principles, (i) increasing the number of devices which are released after computation of the node, and (ii) lowering the duration of time that RRAM devices are blocked and cannot be reused.

Fig. 6(a) shows an MIG with two candidate nodes u and v. One child node is shared between both of the candidates, which is also needed as an input of the root node and therefore the RRAM device keeping its value cannot be released after computing u and v. Hence, u has two releasing children, while v has only one. In this case, the compiler selects u to be computed first.

For the second principle, see Fig. 6(b). The node A is an input of the root node G. Thus, the RRAM device storing the value of A is blocked for long time and cannot be released to end of the computations. To avoid an increase in the number of RRAM devices resulted by this, the compiler computes such nodes with long waiting time as late as possible. This case, also causes an uneven write traffic which in the long term wears some devices out much earlier than others. Indeed, postponing to compute the nodes with long waiting time is a wear leveling strategy which enhances the lifetime by balancing the writes all over the memory [20]. This especially matters as RRAM devices have limited write endurance that should be addressed in the design process.

Besides the endurance-aware compilation, we also propose other techniques to uniformly distribute the writes. This includes endurance-aware MIG optimization to lower the sources of unbalanced write traffic within an MIG, and direct techniques, i.e. allocating the RRAM device with the minimum write count when a device is requested, and allocating fresh RRAM devices when all of previously freed devices have been already rewritten for more than a certain value [20].

Table III shows the number of required RRAM devices and instructions of the proposed instruction-based approach on a EPFL benchmarks[1]. The results are for shown for the

[1] http://lsi.epfl.ch/benchmarks

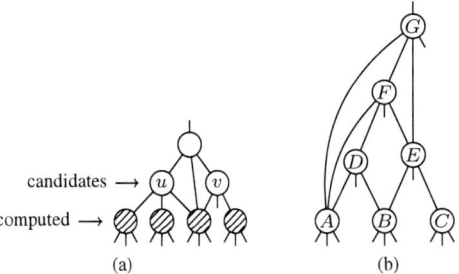

Fig. 6. (a) Reducing the number of RRAMs by selecting the candidate with more releasing children. (b) Reducing the number of RRAMs and balancing the write traffic by selecting the nodes with long storage duration later.

naïve PLiM implementations, implementations after only MIG optimization, and both MIG optimization and compilation. The results show improvements of 20.09% and 14.83%, respectively in the number of instructions and RRAM devices after only MIG optimization. As the compilation aims to reduce the number of the devices, the reduction in the total number of devices considerably improves to 61.4% after compilation.

The standard deviation of writes over the RRAM devices is shown in Fig. 7. The results are shown for the combination of the minimum write strategy, endurance-aware MIG optimization, and endurance-aware compilation. The comparison of the results with the naïve implementations shows an improvement of 72.17% over all of the benchmarks.

V. CONCLUSION

This paper presents a comprehensive customized approach for synthesis of logic-in-memory circuits using the logical representations BDD, AIG, and MIG. The presented approach introduces realization of the logic primitives using two basic operations enabled by RRAM devices, and provides optimization algorithms and design methodologies for crossbar implementation. The paper also proposes an automatic compiler for a regular logic-in-memory computer architecture and improves

TABLE III.   EXPERIMENTAL EVALUATION OF THE PROPOSED INSTRUCTION-BASED APPROACH

Benchmark	PI/PO	naïve			MIG optimization					MIG optimization and compilation			
		#N	#I	#R	#N	#I	impr.(%)	#R	impr.(%)	#I	impr.	#R	impr.(%)
adder	256/129	1020	2844	512	1020	2037	28.38	386	24.61%	1911	32.81	259	49.41
bar	135/128	3336	8136	523	3240	5895	27.54	371	29.06%	6011	26.12	332	36.52
div	128/128	57247	146617	687	50841	147026	-0.03	771	-12.22%	147608	-0.68	590	14.12
log2	32/32	32060	78885	1597	31419	60402	23.43	1487	6.89%	60184	23.71	1256	21.35
max	512/130	2865	6731	1021	2845	5092	24.35	867	15.08%	4996	25.78	579	43.29
multiplier	128/128	27062	76156	2798	26951	56428	25.91	1672	40.24%	56009	26.45	419	85.03
sin	24/25	5416	12479	438	5344	10300	17.09	426	2.73%	10223	18.08	402	8.22
sqrt	128/64	24618	60691	375	22351	47454	21.81	433	-15.46%	49782	17.97	323	13.87
square	64/128	18484	54704	3272	18085	33625	38.53	3247	0.76%	33369	39.00	452	86.19
cavlc	10/11	693	1919	262	691	1146	40.28	236	9.92%	1124	41.43	102	61.07
ctrl	7/26	174	499	66	156	258	48.29	55	16.66%	263	47.29	39	40.91
dec	8/256	304	822	257	304	783	4.74	257	0.00%	777	5.47	258	-0.39
i2c	147/142	1342	3314	545	1311	2119	36.05	487	10.64%	2028	38.81	234	57.06
int2float	11/7	260	648	99	257	432	33.33	83	16.16%	428	33.95	41	58.59
mem_ctrl	1204/1231	46836	113244	8127	46519	85785	24.25	6708	17.46%	84963	24.97	2223	72.65
priority	128/8	978	2461	315	977	2126	13.61	241	23.49%	2147	12.76	149	52.70
router	60/30	257	503	117	257	407	19.09	112	4.27%	401	20.28	64	45.30
voter	1001/1	13758	38002	1749	12992	25009	34.19	1544	11.72%	24990	34.24	1063	39.22
Σ		236710	608655	22760	225560	486324	20.09	19383	14.83	487214	19.95	8785	61.40

#N: number of MIG nodes, #I: number of instructions, #R: number of RRAM devices, improvement is calculated compared to naïve

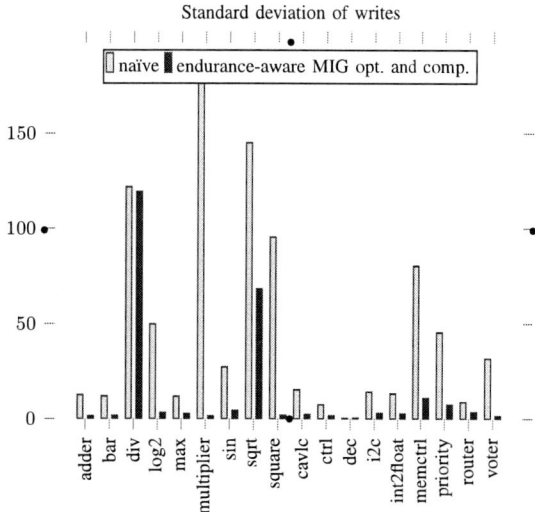

Fig. 7.   Standard deviation of writes for the PLiM architecture.

the execution costs considerably with respect to latency, area, and the write balance.

ACKNOWLEDGMENTS

This research was supported by the University of Bremen's graduate school SyDe funded by the German Excellence Initiative.

REFERENCES

[1] M. Soeken, P.-E. Gaillardon, S. Shirinzadeh, R. Drechsler, and G. De Micheli, "A PLiM computer for the internet of things," *IEEE Computer*, vol. 50, no. 6, pp. 35–40, 2017.

[2] J. Borghetti, G. Snider, P. Kuekes, J. Yang, D. Stewart, and R. Williams, "Memristive switches enable stateful logic operations via material implication," *Nature*, vol. 464, no. 7290, pp. 873–876, 2010.

[3] S. Kvatinsky, D. Belousov, S. Liman, G. Satat, N. Wald, E. Friedman, A. Kolodny, and U. Weiser, "MAGIC – Memristor-Aided Logic," *IEEE Trans. Circuits Syst. II*, vol. 61, no. 11, pp. 895–899, 2014.

[4] P.-E. Gaillardon, L. G. Amarù, A. Siemon, E. Linn, R. Waser, A. Chattopadhyay, and G. De Micheli, "The programmable logic-in-memory (PLiM) computer," in *Design, Automation & Test in Europe*, 2016, pp. 427–432.

[5] R. Drechsler and D. Sieling, "Binary decision diagrams in theory and practice," *STTT*, vol. 3, no. 2, pp. 112–136, 2001.

[6] A. Kuehlmann, V. Paruthi, F. Krohm, and M. K. Ganai, "Robust Boolean reasoning for equivalence checking and functional property verification," *IEEE Trans. Comput.-Aided Design Integr. Circuits Syst.*, vol. 21, no. 12, pp. 1377–1394, 2002.

[7] L. G. Amarù, P.-E. Gaillardon, and G. De Micheli, "Majority-inverter graph: A novel data-structure and algorithms for efficient logic optimization," in *Design Automation Conference*, 2014, pp. 194:1–194:6.

[8] ——, "Majority-inverter graph: A new paradigm for logic optimization," *IEEE Trans. Comput.-Aided Design Integr. Circuits Syst.*, 2015.

[9] S. Chakraborti, P. Chowdhary, K. Datta, and I. Sengupta, "BDD based synthesis of Boolean functions using memristors," in *IDT*, 2014, pp. 136–141.

[10] S. Shirinzadeh, M. Soeken, and R. Drechsler, "Multi-objective BDD optimization for RRAM based circuit design," in *IEEE International Symposium on Design and Diagnostics of Electronic Circuits and Systems*, 2016, pp. 46–51.

[11] J. Bürger, C. Teuscher, and M. Perkowski, "Digital logic synthesis for memristors," in *Reed-Muller workshop*, 2013.

[12] A. Chattopadhyay and Z. Rakosi, "Combinational logic synthesis for material implication," in *IFIP/IEEE International Conference on Very Large Scale Integration (VLSI-SoC)*, 2011, pp. 200–203.

[13] D. Bhattacharjee, R. Devadoss, and A. Chattopadhyay, "ReVAMP: ReRAM based VLIW architecture for in-memory computing," in *Design, Automation and Test in Europe*, 2017, pp. 782–787.

[14] D. Bhattacharjee, A. Easwaran, and A. Chattopadhyay, "Area-constrained technology mapping for in-memory computing using ReRAM devices," in *Asia and South Pacific Design Automation Conference*, 2017, pp. 69–74.

[15] D. Bhattacharjee, L. Amarù, and A. Chattopadhyay, "Technology-aware logic synthesis for ReRAM based in-memory computing," in *Design, Automation & Test in Europ*, 2018, pp. 1435–1440.

[16] S. Shirinzadeh, M. Soeken, P.-E. Gaillardon, and R. Drechsler, "Logic synthesis for RRAM-based in-memory computing," *IEEE Trans. Comput.-Aided Design Integr. Circuits Syst.*, accepted for publication, 2018.

[17] ——, "Fast logic synthesis for RRAM-based in-memory computing using majority-inverter graphs," in *Design, Automation & Test in Europe*, 2016, pp. 948–953.

[18] M. Soeken, S. Shirinzadeh, P.-E. Gaillardon, L. G. Amarù, R. Drechsler, and G. De Micheli, "An MIG-based compiler for programmable logic-in-memory architectures," in *Design Automation Conference*, 2016, pp. 117:1–117:6.

[19] S. Shirinzadeh, M. Soeken, P.-E. Gaillardon, and R. Drechsler, "Logic synthesis for majority based in-memory computing," in *Advances in Memristors, Memristive Devices and Systems*, S. Vaidyanathan and C. Volos, Eds.   Springer International Publishing, 2017, pp. 425–448.

[20] S. Shirinzadeh, M. Soeken, P.-E. Gaillardon, G. De Micheli, and R. Drechsler, "Endurance management for resistive logic-in-memory computing architectures," in *Design, Automation & Test in Europ*, 2017, pp. 1092–1097.

Minimalistic Perspective to Public Key Implementations on FPGA

Debapriya Basu Roy*, Debdeep Mukhopadhyay*
*Secured Embedded Architecture Laboratory, Department of Computer Science and Engineering
Indian Institute of Technology Kharagpur

Abstract—Public key cryptographic algorithms serve a very important role in ensuring security between different heterogeneous components. Among various public key algorithms, elliptic curve cryptography (ECC) has stood the test of time due to its efficiency and security. In this work, we have focused on different aspects of ECC implementations. More specifically, we have focused on implementing two different variants of ECC implementations: minimal area for lightweight applications and minimal time for speed-critical applications. Additionally, we have devised a hybrid testing methodology for quantification of side channel vulnerability. We have also implemented an IP protection methodology to protect the developed ECC implementations on FPGA platform. Currently, we are focusing on efficient and side channel secure implementation of post quantum isogeny based ECC algorithm.

I. INTRODUCTION

With the increase of sensitive information in the Internet, security is becoming a critical aspect of web applications. Additionally, in the modern world, sensitive information is also handled by components like IoT (Internet of Things) and artificial intelligence. The computing power of these platforms varies significantly. For example, security protocols built for web servers need to be fast and efficient. On the other hand, security protocols built for resource constrained application like IoT need to be lightweight in terms of area and power. In this context, efficient design and implementation of public key cryptography to provide security in these heterogeneous domains is an interesting research problem. Among the available public key algorithms, Elliptic Curve Cryptography (ECC) is the most preferred choice as it provides more security per key bit compared to other public key algorithms [1].

The security of ECC-based protocols and algorithms are based on the hardness of the Elliptic curve Discrete Logarithm problem (ECDLP). ECDLP states that, given an elliptic curve scalar multiplication which can be computed in a quite straight-forward manner, it is computationally hard to retrieve the secret scalar used in the computation. The elliptic curve scalar multiplication involved in the process includes many computationally intensive field operations which make the software implementation of ECC scalar multiplication inefficient and unsuitable for many real-world applications. An alternative approach is to provide dedicated crypto-accelerator on hardware platforms like ASIC (Application Specific Integrated Circuit) and FPGAs (Field Programmable Gate Arrays) to accelerate ECC scalar multiplication. However, depending upon the requirement and applications, the hardware accelera-

tors are either required to be lightweight in area or need to be extremely fast. Designing light-weight implementation of ECC has gathered significant interest in the research community in recent years. Due to the recent development of IoT, employing lightweight crypto-solutions for resource-constrained devices is an absolute necessity. ECC architecture, which is much more resource hungry compared to private key algorithms, poses serious design challenge in its integration with the lightweight devices. The objective in this scenario is to reduce the area consumptions of the ECC without affecting the timing performance drastically. On the other hand, there are some cases where high speed implementations of ECC is required. For example, in case of autonomous cars, authentication and verification procedures need to be completed very rapidly. Hence, in this scenario, the objective would be to reduce the timing requirement of the design without increasing the area significantly. Thus, in different situations, system may require two different versions of ECC, one with *minimal area* and other with *minimal time*. Our objective in this research will be to address these two different flavours of implementations with architectural and algorithmic level optimizations.

Apart from being efficient, the proposed crypto-implementations of ECC architectures are needed to be validated against side channel adversaries for commercial benchmarking. A side channel adversary can observe information like power, electromagnetic radiation, timing of an implementation and by using statistical methods, he can recover the secret key value. Though existing testing methodologies like *Common Criteria (CC)* or *Federal Information Processing Standards (FIPS)* provides guidelines for testing of side channel vulnerabilities, they are quite different in their approach. CC, though provides a detailed analysis of side channel vulnerability, it is slow and dependent on the testing lab's expertise. On the other hand, FIPS presents a fast and robust testing methodology, but can only detect the presence of side channel leakage. It can not quantify the side channel vulnerability. In our work, we have combined the virtues of both CC and FIPS into a fast and efficient hybrid testing mechanism to test the side channel security of the developed ECC implementation. Additionally, we have developed a two-party pay per use IP protection scheme which can be used to protect developed ECC implementation from unauthorized usage on FPGA platform.

ECC though is the front-runner among the existing public key algorithm, it will cease to be secure once commercial

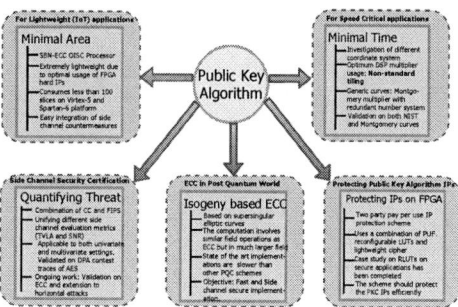

Fig. 1: Overview of the Work

quantum computers become available [2]. Hence, it is imperative that we devise efficient techniques for implementation of post quantum public key algorithms. Among the several post quantum public key algorithms, supersingular isogeny based Diffie-Hellman (SIDH) key exchange could be an attractive option as it provides perfect forward secrecy and is built on supersingular elliptic curves. In our work, our objective is to achieve an efficient implementation of SIDH coupled with its side channel vulnerability analysis.

Thus, in a nutshell, we can summarize the contributions of the thesis as below

- **Minimal area ECC:** Lightweight ECC architecture for resource constrained applications
- **Minimal time ECC:** High-speed ECC architecture for speed critical applications
- **Certifying Side Channel Security of ECC:** Quantifying side channel vulnerability to guarantee protection against such adversaries.
- **Protecting ECC IP:** Preventing any unauthorized or overuse of developed ECC IP on FPGA platform.
- **Implementing SIDH:** Designing efficient architecture for SIDH algorithm along with its side channel vulnerability analysis.

In Fig. 1, we have provided a pictorial overview of the works that we want to achieve in this thesis. More details are provided in the next section.

II. MINIMAL AREA ECC: LIGHTWEIGHT ELLIPTIC CURVE SCALAR MULTIPLICATION ON FPGA

In this work [3], we have proposed a single instruction approach for implementation of ECC scalar multiplier suitable for lightweight applications. More specifically, we have built an entire ECC scalar multiplier processor by using only one URISC (Ultimate Reduced Instruction Set Computer) instruction SBN (subtract and branch if negative). It is well known that using such URISC instruction, one can execute any logical or mathematical operation, leading to a Turing complete computer processor. However, as ECC involves many computationally intensive field operations, a stand-alone SBN-processor for ECC scalar multiplier execution will be drastically slow.

To tackle this problem, we have integrated the SBN processor with dedicated hard-IPs (Block RAM, DSP Blocks)

TABLE I: Different Variant of SBN Instruction

Instruction	Memory Write-back	Multiplier reset	Key-shift	Right-shift
$SBN_{\overline{wmulksrs}}$	✓	X	X	X
$SBN_{\overline{nwmulksrs}}$	X	X	X	X
$SBN_{\overline{nwmulksrs}}$	X	X	✓	X
$SBN_{\overline{wmulksrs}}$	✓	X	X	✓
$SBN_{\overline{nwmulksrs}}$	X	✓	X	X

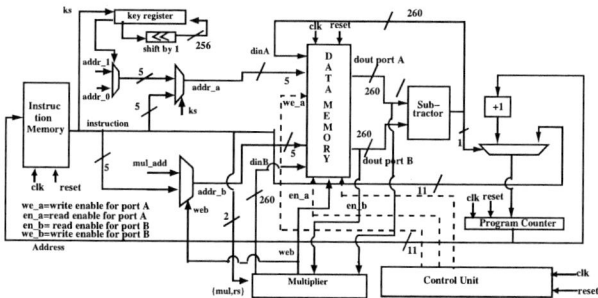

Fig. 2: ECC SBN Processor Architecture

of the modern FPGAs to demonstrate an implementation of immensely lightweight, yet practical ECC architecture [3]. We have proposed four different flags the processor architectures which when set, activates different optimization strategies integrated inside the processor architecture. The optimized strategies proposed by us are: the option of switching off memory write back, dedicated right shifter, and dedicated field multiplier built exclusively with DSP blocks. It must be noted that field multiplication and right shift can also be executed by repeated execution of SBN instruction, but that will be extremely slow and inefficient. Hence, we have developed efficient architectures for these operations using FPGA based hard-IPs. The details of proposed SBN instruction along with four different flags are shown in Tab. I. The architecture of the proposed SBN-processor for ECC scalar multiplication is shown in Fig. 2.

TABLE II: Comparison of ECC SBN Processor with Existing Designs

Reference	Slices	MULTs /DSPs	BRAM	Freq (MHz)	Latency (ms)	FPGA
P-256 16 bit [4]	773	1	3	210	10.02	Virtex-II Pro
P-256 32 bit [4]	1158	4	3	210	4.52	Virtex-II Pro
[5] 16 bit generic	1832	2	9	108.20	29.83	Virtex-II Pro
[5] 32 bit generic	2085	7	9	68.17	15.76	Virtex-II Pro
[6] P-256	1715	32	11	490	.62	Virtex-4
[7] P-256	221	1	3	Not shown	Not shown	Spartan-6
This Work, P-256	81	8	22	171.5	11.1	Virtex-5
This Work, P-256	72	8	24	156.25	12.2	Spartan-6

For comparing with other existing implementation, we have implemented *NIST P-256* ECC scalar multiplier using SBN processor on Virtex-5 and Spartan-6 platform. As shown in Tab. II, in both cases, the slice consumption of the design is less than 100. To the best of our knowledge, this is the first implementation of ECC which requires less than 100 slices on any FPGA device family The stand-alone SBN processor is itself very lightweight, and the dedicated multiplier core is designed by judicious use of DSPs. Additionally, the block RAMs are used extensively to implement both data and instruction memory of SBN-ECC processor. It must be noted that a designer can choose a budget of slices and block RAMs and can design the corresponding SBN-ECC processor as per his choice. Tab. II also shows that compared to other implementation, the proposed architecture is significantly lightweight. Moreover, the timing performance of the SBN-ECC processor is comparable with the existing implementations. This may seem to be counter intuitive as the proposed SBN processor needs to execute a large number of instructions to complete one single scalar multiplication. However, it must be noted that the proposed SBN processor is coupled with a dedicated field multiplier built using high performance DSP blocks. This improves the timing performance of the proposed architecture significantly.

The proposed SBN-ECC architecture combines the flexibility of the software implementation with the efficiency of the hardware implementations. User can change the implementation of ECC scalar multiplication easily by changing the SBN codes. This enables effortless integration of the proposed architecture with the side channel countermeasure. We have experimentally validated this observation in [8]. Currently, we are working on reducing the block RAM requirement of the design. Additionally, we also aim to incorporate pipelining in the processor architecture to improve the timing performance of the proposed architecture.

III. MINIMAL TIME ECC: HIGH SPEED IMPLEMENTATION OF ELLIPTIC CURVE SCALAR MULTIPLICATION

High-speed design of ECC on FPGA has been a popular research topic. ECC implementation in $GF(p)$ on FPGA can be significantly improved by deploying FPGA based hard-IPs like *DSP blocks* and *block RAMs* [9]. In this context, we have proposed a generalized *non-standard tiling methodology* in [10]. The proposed methodology in [10] focuses on the optimum usage of DSP blocks with the objective of faster field multiplications. The multipliers present inside the DSP blocks of modern FPGAs are asymmetric in nature. They are capable of computing 24×17 unsigned multiplication. Due to this asymmetric multipliers, standard school book method of field multiplication will be non-optimum. Hence, we have introduced the non-standard methodology for field multiplication, which makes the usage of DSP blocks optimum. An example of non-standard tiling is shown in Fig. 3. In [10], we have generalized the notion of non-standard tiling so that non-standard tiling can be applied to multipliers with large operand width, which is the typical scenario in case of ECC. An illustration of our proposed methodology is shown in Fig. 4.

Using the proposed non-standard tiling methodology, we have developed a fast field multiplier which should help us to achieve our goal of minimal time and maximal speed implementation. However, these implementations are based

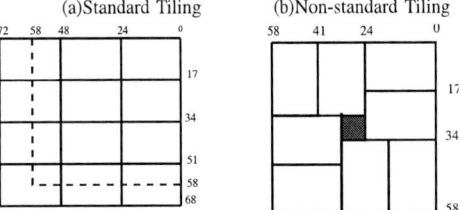

Fig. 3: Multiplying Operands of Width 58 using Asymmetric Multipliers [11]

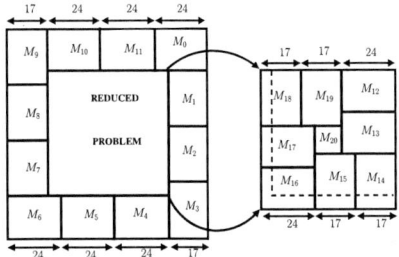

Fig. 4: Multiplying Operands of Width 89 using Asymmetric Multiplier of Dimension 24 and 17

on the NIST curves which use fast pseudo-Mersenne primes. We are also investigating different versions of Montgomery multiplier which can do efficient field multiplication for any generic prime value. Additionally, we have analyzed different coordinate systems like *Jacobian, Co-Z, Montgomery* to find out that for high speed implementations, *Montgomery's x coordinate only* formula is the most efficient one. Currently, we are in the process of integrating the aforementioned observations so that we can achieve the objective of minimal time, maximal speed ECC implementation on FPGA.

IV. CERTIFYING SIDE CHANNEL SECURITY

Testing of side channel vulnerability is an open research problem and has received significant attention in recent times. There exist two distinct side channel vulnerability testing methodologies: evaluation style testing and validation style testing. In case of evaluation style testing, the system is evaluated against all state-of-the-art attack strategies, with the knowledge of the threat model. An ever-increasing list of attack strategies, together with a large number of models characterizing different leakage profiles of the device, makes evaluation style testing methodology cumbersome, costly and limited by the testing expertise available at hand. On the other hand, validation style testing detects the presence of any side channel leakage, independent of attack methodologies and leakage model. Therefore, validation style testing is much more cost-effective and robust compared to evaluation style testing. Test Vector Leakage Assessment (TVLA) [12], which was proposed at NIST sponsored NIAT workshop 2011, is one of such validation style testing mechanism which has gained popularity among the researchers and the practitioners due to its robustness. We have used TVLA to validate the security of

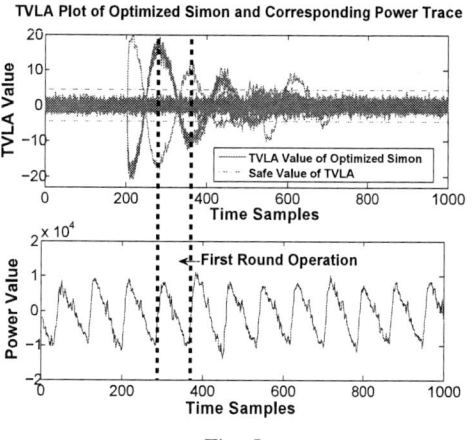

Fig. 5

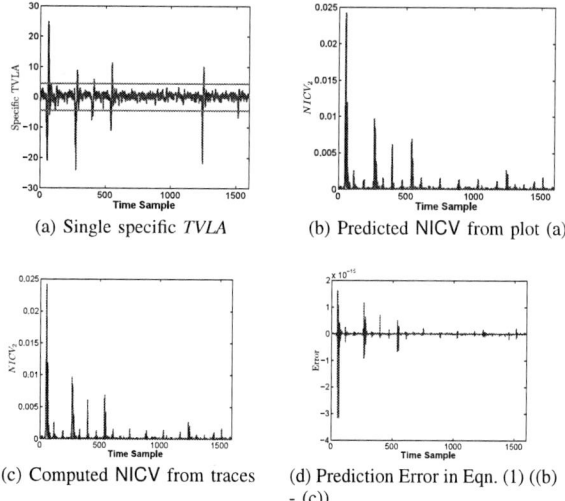

(a) Single specific *TVLA*

(b) Predicted NICV from plot (a)

(c) Computed NICV from traces

(d) Prediction Error in Eqn. (1) ((b) - (c))

Fig. 6: Equivalence of *TVLA* and NICV

popular side channel countermeasure *private circuit* in [13]. We have shown that though *private circuit* is algorithmically secure against side channel attacks, security may be compromised due to unintended optimizations by CAD tools. In Fig. 5, we show the TVLA plot of an implementation of *SIMON* cipher, protected by the *private circuit* countermeasure. TVLA,in this case, helps us to detect the presence of side channel leakage without doing any actual attack.

Though robust, TVLA fails to quantify the side channel leakage, making it not sufficient for a comprehensive security evaluation. On the other hand evaluation style testing, though cumbersome, provides much more detailed and comprehensive security evaluation. Evaluation style testing evaluates the security of the crypto-system using metrics like *success rate (SR)* and *normalized inter class variance (NICV)*. In this work, we attempt to develop a hybrid methodology to extract more information from the initial TVLA testing. In other words, we derive the concrete relationship between TVLA and NICV to show that the two metrics are equivalent to each other. NICV is proportional to the signal to noise ratio (SNR) of the side channel traces, from which we can compute the SR of the attack. Thus, using the proposed testing methodology, we can recover the SR of side channel attack from the TVLA value. The proposed testing methodology is robust and not cumbersome as evaluation style testing methodology, yet it provides a detailed security analysis. The relationship between TVLA and NICV are given next.

Theorem 1: Consider two groups of side channel traces Y^{G1} and Y^{G2} with cardinality n_1 and n_2. The following formula relates the computation of TVLA and NICV on these two groups

$$\text{NICV} = \cfrac{1}{\cfrac{n}{\text{TVLA}^2} + \cfrac{n}{C}(\sigma_1^2 - \sigma_2^2)\left(\cfrac{1}{n_2} - \cfrac{1}{n_1}\right) + 1} \quad (1)$$

where μ_i and σ_i are the mean and standard deviation of group Y^{Gi} respectively, $C = \left(\mu_1^2 - \mu_2^2\right)^2$.

In this case, NICV is basically the value of statistical F-test on two groups. From this, we can compute the value of statistical

F-test on any arbitrary number of groups. As we have already mentioned, NICV is proportional to the SNR value, which itself can be used to compute the value of the SR. Thus using the proposed relationship, we can directly compute the SR of the side channel attacks without performing the actual attack. In Fig. 6, we have plotted the computed NICV and predicted NICV (using Eqn. (1)). The prediction error is in the range of 10^{-15}, which can be considered as negligible. We have used the proposed methodology on the side channel traces of AES and have computed the SR successfully. The proposed testing mechanism is applicable to both univariate and multivariate setting. More details on the proposed testing methodology and the corresponding theoretical analysis can be found in [14].

Currently, we are in the process of validating the proposed methodology on ECC. Moreover, for ECC, single trace attacks like *horizontal collision correlation attack (HCCA)* [15] are a serious threat compared to traditional side channel attacks like *differential power analysis (DPA)*. HCCA can act on implementations which are protected by DPA countermeasures and can still retrieve the secret key with single or very few side channel traces. We will like to modify TVLA to address horizontal collision correlation attacks efficiently. This will help us to determine the effectiveness of different countermeasures which try to prevent HCCA and will help us to choose the best countermeasure for HCCA prevention.

V. PROTECTING ECC IP

In this work, we aim to provide an efficient methodology to protect IPs from being overused on FPGA platforms. There are various threats which we need to consider in this context. For example, reverse engineering, cloning of FPGA bit-streams are realistic threats which need to be prevented. Generally, bit-stream and net-list encryption are employed to prevent such attacks. Moreover, to maintain the confidentiality and integrity of the third party IPs, `IEEE P1735` protocol is frequently deployed. However, this protocol can not prevent overuse of the IPs which is the precise objective of this work.

The proposed IP protection scheme is built using reconfigurable look up tables (RLUTs) of FPGAs. RLUTs provides the user the option of modifying the logical functionality of the design in run-time without requiring any bit-stream update. Every LUT inside an FPGA is characterized by INIT value which is set during the bit-stream update. This INIT value defines the truth table of the combinational function, implemented on the LUT. In normal LUTs, the INIT value does not change once the bit-stream is updated. However, the reconfigurability feature of RLUT is realized updating the INIT at run time. The working of RLUT is shown in Fig. 7.

In [16], we have demonstrated various applications that can be built using RLUTs for cryptographic applications. For example, in [16], we have developed stealthy hardware Trojans using RLUTs, which can be used to trigger malicious activities in the design. Fig. 8 shows one of such Trojan example, where the RLUT is initially configured to work as a buffer. But upon triggered, it is reconfigured to work as an inverter. Upon another reconfiguration, it can be converted back to buffer operation. It is very evident that this kind of activities are very handy for fault injection, which we have illustrated in [17]. Additionally, RLUTs can also be used to develop many constructive applications as shown in [16].

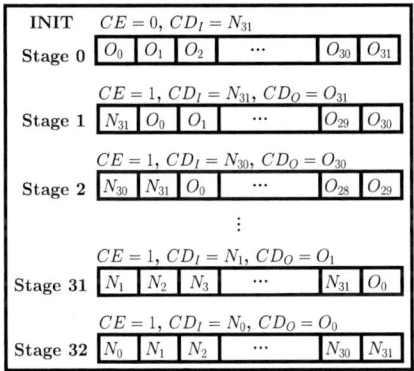

Fig. 7: $INIT$ value reconfiguration in RLUT [16] .

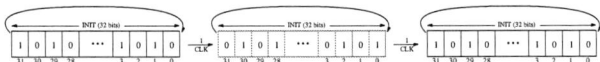

Fig. 8: Operations of RLUT to switch from a buffer to inverter and back [16].

A. Proposed IP Protection Scheme [18]

In this subsection, we will provide the main idea of the proposed IP protection scheme. The proposed IP protection is based on the following assumptions

- The IP vendor provides the protected net-list of the required IP to the IP client in accordance with the IEEE P1735 protocol to ensure that confidentiality and integrity of the delivered IP is protected.

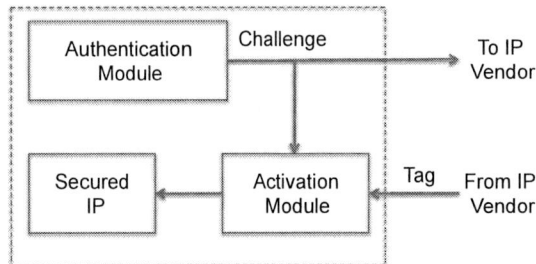

Fig. 9: The proposed two-party IP Protection Scheme

- The IP vendor has the capability of implementing a tamper proof device-specific identifier.

The conceptual diagram of the proposed protocol is shown in Fig. 9. It has three key components: an authentication module, an activation module, and the delivered IP.

The IP is designed such that a (or few) vital control signal is implemented using RLUT. The IP is then secured by setting the INIT of these RLUT to zero (or another garbage value). The authentication module generates a device-identifier as a *challenge* which is public and sends it to the vendor. PUF is an ideal candidate for such authentication, but other methods like non-volatile memory based device key can be considered. The vendor replies with an activation *tag* computed from the *challenge* with some proprietary algorithm. Ideally, each device should generate a unique *challenge* and receive a unique *tag* to realize a pay per device licensing scheme. The activation module uses internally generated *challenge* and vendor provided *tag* to compute correct RLUT state and program it. The activation protocol is implemented as below:

- The IP user implements the delivered IP on the FPGA. On initial power up, the authentication module will generate a *challenge* (Fig. 9) for the IP vendor.
- IP vendor computes and shares the authentication *tag*.
- The activation module uses the *tag* and internally generated *challenge* to compute the reconfiguration data.
- Once the necessary reconfiguration data is computed, it is programmed into the RLUT to activate the IP

The proposed IP protection scheme can be applied to protect any FPGA based IPs including the developed ECC IPs. The design and development of ECC IPs are much more complicated and time consuming compared to their symmetric key counterparts. Therefore it is imperative that the IP vendor needs to employ some protection strategies to prevent any unauthorized overuse. The proposed scheme could be an efficient choice for such objective.

VI. ONGOING WORK - POST QUANTUM ISOGENY BASED CRYPTOSYSTEM

As we have mentioned earlier, in a post quantum world, the ECC will no longer remain secure. To tackle this issue, various post quantum public key algorithms [19] has been proposed. *supersingular isogeny based Diffie-Hellman (SIDH)* [20] has emerged as an attractive candidate. Implementing *SIDH* poses some serious design challenge to the designed due to its non-standard prime form. The only alternative in this case is then to apply *Montgomery's multiplication algorithm* for efficient implementation of *SIDH*. Here our objective is to design an efficient and side channel secure implementation of *SIDH*.

The advantage of using *SIDH* over other post quantum public key algorithms are listed below:

1) *SIDH* is based on supersingular elliptic curves. Elliptic curve is a well studied topic and key exchange algorithm based on elliptic curve scalar multiplication is already deployed in various security critical services. In that context, transition to supersingular isogeny based cryptosystem will be easier compared to other post quantum public key algorithm.

2) Among all the available post quantum public key algorithm, SIDH requires the smallest key. To provide 128 bit security, the public key size of *SIDH* is 2688 bits only. Additionally with the recently proposed compression techniques, *SIDH* has similar bandwidth of 3072 bits RSA. This makes *SIDH* suitable for applications which have a very strict bandwidth requirement.

In our work, we aim to accelerate the execution speed of SIDH to make it competitive with the other existing post quantum public key algorithm. The main challenge in this context is to execute the field multiplication. The prime value in case of SIDH is of the form $2^a.3^b - 1$. This is not a pseudo-Mersenne prime, making the execution of SIDH time consuming. Currently, we are working on implementing of a fast Montgomery multiplier which can execute the field multiplication in an efficient manner for SIDH.

Apart from implementation, we are also analyzing the SIDH algorithm for potential side channel vulnerabilities. In that context, we are currently concentrating on power and fault based side channel analysis.

VII. CONCLUSION

In this work, we have focused on different aspects of ECC implementation. We have developed both minimal time and minimal area implementation of ECC for speed-critical and IoT applications. We have also developed a hybrid testing methodology for quantification of side channel vulnerability. Additionally, an IP protection scheme has been developed for protection of developed designs from unauthorised usage. Currently, our aim is to develop an efficient implementation of post quantum SIDH algorithm which is based on the supersingular elliptic curves. Additionally, we are also analysing the SIDH algorithm for detection of any side channel vulnerability which an adversary can exploit.

ACKNOWLEDGMENT

The work is partially supported by the project "Next Generation Secured Internet of Things", sponsored by "Sponsored Research and Industrial Consultancy, IIT Kharagpur" and "Information Security Education and Awareness project", sponsored by "Department of Electronics and IT, Govt. of India". Dr. Debdeep Mukhopadhyay will like to acknowledge the support received from "SwarnaJayanti Fellowship", given by "Department of Science and Technology, India".

REFERENCES

[1] Minghua Qu. Sec 2: Recommended elliptic curve domain parameters, 1999.

[2] Peter W Shor. Polynomial time algorithms for discrete logarithms and factoring on a quantum computer. In *International Algorithmic Number Theory Symposium*, pages 289–289. Springer, 1994.

[3] Debapriya Basu Roy, Poulami Das, and Debdeep Mukhopadhyay. Ecc on your fingertips: A single instruction approach for lightweight ecc design $ingf(p)$. In *International Conference on Selected Areas in Cryptography*, pages 161–177. Springer, 2015.

[4] Michal Varchola, Tim Güneysu, and Oliver Mischke. MicroECC: A Lightweight Reconfigurable Elliptic Curve Crypto-processor. In *2011 International Conference on Reconfigurable Computing and FPGAs, ReConFig 2011, Cancun, Mexico, November 30 - December 2, 2011*, pages 204–210, 2011.

[5] Jo Vliegen, Nele Mentens, Jan Genoe, An Braeken, Serge Kubera, Abdellah Touhafi, and Ingrid Verbauwhede. A Compact FPGA-based Architecture for Elliptic Curve Cryptography over Prime Fields. In *21st IEEE International Conference on Application-specific Systems Architectures and Processors, ASAP 2010, Rennes, France, 7-9 July 2010*, pages 313–316, 2010.

[6] Tim Gneysu and Christof Paar. Ultra High Performance ECC over NIST Primes on Commercial FPGAs. In *In Proceedings of CHES*, pages 62–78, 2008.

[7] Benedikt Driessen, Tim Güneysu, Elif Bilge Kavun, Oliver Mischke, Christof Paar, and Thomas Pöppelmann. IPSecco: A lightweight and reconfigurable IPSec core. In *2012 International Conference on Reconfigurable Computing and FPGAs, ReConFig 2012, Cancun, Mexico, December 5-7, 2012*, pages 1–7, 2012.

[8] Poulami Das, Debapriya Basu Roy, and Debdeep Mukhopadhyay. Secure Public Key Hardware for IoT Applications. In *Circuits and Systems (MWSCAS), 2016 IEEE 59th International Midwest Symposium on*, pages 1–4. IEEE, 2016.

[9] Tim Güneysu and Christof Paar. *Ultra High Performance ECC over NIST Primes on Commercial FPGAs*, pages 62–78. Springer Berlin Heidelberg, Berlin, Heidelberg, 2008.

[10] Debapriya Basu Roy, Debdeep Mukhopadhyay, Masami Izumi, and Junko Takahashi. Tile Before Multiplication: An Efficient Strategy to Optimize DSP Multiplier for Accelerating Prime Field ECC for NIST Curves. In *DAC-2014*, pages 177:1–177:6, 2014.

[11] F. de Dinechin and B. Pasca. Large Multipliers with Fewer DSP blocks. In *in Field Programmable Logic and Applications*, pages 250–255, 2009.

[12] Jaffe J. Goodwill G., Jun B. and Rohatgi P. A testing methodology for side-channel resistance validation, 2011.

[13] Debapriya Basu Roy, Shivam Bhasin, Sylvain Guilley, Jean-Luc Danger, and Debdeep Mukhopadhyay. From Theory to Practice of Private Circuit: A Cautionary Note. In *(ICCD-2015*, pages 296–303. IEEE.

[14] Debapriya Basu Roy, Shivam Bhasin, Sylvain Guilley, Annelie Heuser, Sikhar Patranabis, and Debdeep Mukhopadhyay. Leak Me If You Can: Does TVLA Reveal Success Rate? Cryptology ePrint Archive, Report 2016/1152, 2016.

[15] Aurélie Bauer, Eliane Jaulmes, Emmanuel Prouff, Jean-René Reinhard, and Justine Wild. Horizontal collision correlation attack on elliptic curves. *Cryptography and Communications*, 7(1):91–119, 2015.

[16] Debapriya Basu Roy, Shivam Bhasin, Sylvain Guilley, Jean-Luc Danger, Debdeep Mukhopadhyay, Xuan Thuy Ngo, and Zakaria Najm. Reconfigurable LUT: A double edged sword for security-critical applications. In *SPACE 2015*, pages 248–268.

[17] Debapriya Basu Roy, Shivam Bhasin, Xuan Thuy Ngo, Zakaria Najm, Sikhar Patranabis, Sylvain Guilley, Debdeep Mukhopadhyay, and Jean-Luc Danger. Potential pitfall of rlut: Fault attack using hardware trojan. *HOST-2016 Hardware Demo Competition- Runners Up*.

[18] Debapriya Basu Roy, Shivam Bhasin, Ivica Nikolić, and Debdeep Mukhopadhyay. Opening pandora's box: Implication of RLUT on secure FPGA applications and IP security. In *Verification and Security Workshop (IVSW), 2017 IEEE 2nd International*, pages 134–139, 2017.

[19] Post-quantum cryptography — csrc. https://csrc.nist.gov/Projects/Post-Quantum-Cryptography. (Accessed on 03/26/2018).

[20] Craig Costello, Patrick Longa, and Michael Naehrig. Efficient Algorithms for Supersingular Isogeny Diffie-Hellman. In *Advances in Cryptology - CRYPTO 2016 - 36th Annual International Cryptology Conference, Santa Barbara, CA, USA, August 14-18, 2016, Proceedings, Part I*, pages 572–601, 2016.

978-1-5386-7100-9/18 $31.00 © 2018 IEEE

Development of high-stability, low-leakage 6Tr-SRAM with single data line and single power supply using SOTB process

Shin Miyamoto

Nobuaki Kobayashi

Department of Precision Machinery Engineering, College of Science and Technology,
Nihon University
7-24-1 Narashinodai, Funabashi-shi, Chiba, 274-8501 JAPAN

miyashin313s@gmail.com

kobayashi.nobuaki@nihon-u.ac.jp

Abstract—This paper proposes a single data line, double-word line 6Tr-SRAM for use in Internet of Things (IoT) devices using a silicon-on-thin-BOX (SOTB) process to achieve a high reliability and a low power consumption. The layout area was reduced compared to a conventional 6Tr structure by using a uniform data line. The proposed SRAM is able to generate multiple electric potentials without the need for additional power sources by employing a self-controllable voltage level (SVL) circuit, which is a simplified form of a DC/DC converter. Further, it expands the operating margin for writes and reads by decreasing the memory-cell supply voltage and increasing the memory-cell supply ground voltage in writes, and by dropping the word line potential when reading. When the variance of the threshold (V_t) was 0 (TT) and the power supply voltage (V_{DD}) was 1.2 V, the read and write margins expanded by multiples of 2.09 and 1.31 of the conventional 6Tr SRAM, respectively. The standby power caused by a leakage when data is being saved under the same conditions was reduced to 9.17% of that of the conventional SRAM. The area overhead the SVL circuit was 1.383% of that of the conventional form.

Keywords—SRAM, SOTB, high stability, low supply power, low leakage, self-controllable voltage level (SVL) circuit

I. INTRODUCTION

With the development of the Internet of Things (IoT), it is expected that there will be a significant increase in the volume of data to be sent over networks, leading to increased network traffic. The need for edge computing, which performs the intermediate processing of data between devices and the cloud, is becoming more important as an approach to suppress the load on the cloud side. In particular, for applications that require real-time behavior, it would be effective to provide the processing ability to individual devices. In such a scenario, static random-access memory (SRAM) that is capable of high-speed operations is a good candidate for memory to be installed on these devices. Some edge devices may not be able to access external power supplies, and an SRAM with a low power consumption is needed to guarantee long operating hours from an internal power source with a limited capacity.

The power consumption, P, of a complementary metal-oxide semiconductor (CMOS) is given by the following equation:

$$P = \alpha C_L f_c V_{DD}{}^2 + I_{LEAK} V_{DD} \qquad (1)$$

where α is the switching probability, C_L is the load capacitance, V_{DD} is the power supply voltage, and I_{LEAK} is the leakage current. It should be noted that the power consumption due to the through-current is omitted here. Equation (1) shows that decreasing the power supply voltage would be effective in reducing the power consumption. However, in a bulk CMOS process, it is difficult to guarantee SRAM operation in a low-voltage range because of an increased variance of the threshold value (V_t) by microfabrication, which prevents the SRAM from reducing the voltage [1]. Therefore, in this study, an SRAM is constructed using a silicon-on-thin BOX (SOTB) process [2] for which the variance of the threshold value (σV_t) is small A value of V_{DD} that is lower than that of the bulk CMOS is expected by the new SRAM. SOTB is also known for properties such as a greater rise in the value of V_t of the transistor when a substrate bias (V_{WS}) is added in the opposite direction, a smaller junction leakage to the substrate, and a greater reduction of the leakage. An SRAM that is suitable for IoT devices can be achieved by taking advantage of these features.

Several studies have reported approaches that are employed to expand both margins for read/write, including the 8Tr method, in which access transistors of different channel width are added to the memory cell [3], a word-line potential (V_{Wrd}) control method, [4], and a memory cell supply voltage (V_M) control method [5]. However, these approaches all have disadvantages, such as a significant increase in the area overhead and the need for multiple power supplies. When data retention is required while on standby, the sub-threshold leakage may be increased because the power supply to the memory cell cannot be shut down [6].

There are two methods to significantly reduce the standby power (P_{ST}) and sub-threshold leakage while retaining data. The first is a method in which a high voltage is supplied during operation, while a low voltage is supplied when stopping. However, this method requires an external power supply. The second method involves the use of variable-threshold MOS (VTMOS) technology [7], where V_{WS} is applied to a metal-oxide semiconductor field-effect transistor (MOSFET) in the opposite direction when stopping, while increasing V_t to reduce leakage. However, this method requires another power source to supply V_{WS}, and it is difficult to achieve rapid switching of V_{WS}. In order to solve these problems, there is a need for a method that uses a single power supply to simultaneously (a) decrease V_{Wrd} and increase V_M when reading, (b) increase V_{Wrd} and decrease V_M when writing, and (c) decrease V_M and relieves

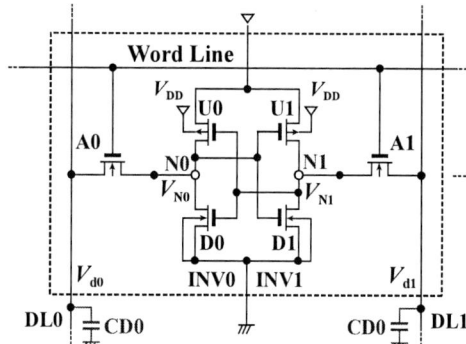

Fig. 1. Conventional structure of 6Tr SRAM memory cell

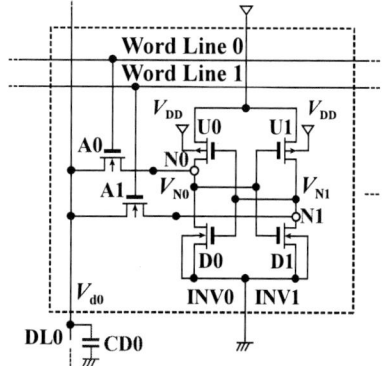

Fig. 2. Proposed structure of 6Tr SRAM memory cell.

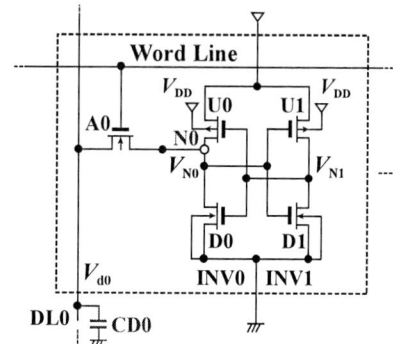

Fig. 3. Conventional structure of 5Tr SRAM memory cell.

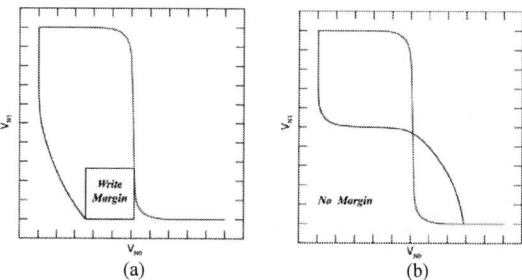

Fig. 4. Butterfly curves during a write in a 5Tr structure. (a) write "0" (b) write "1"

the electric field applied to the memory cell during data retention.

In order to solve the problems discussed above, this study focuses on the development of a self-controllable voltage level (SVL) circuit [8, 9]. An SVL circuit can set the supply voltage to the SRAM to an arbitrary value that is equal to or lower than V_{DD}. Therefore, the circuit enables (1) a high-speed read, (2) a high-speed write, (3) data retention during standby, and (4) reduced leakage during standby with a single power supply. This paper also proposes a six-transistor (6Tr) structure with a single data line.

II. COMPOSITION OF PROPOSED MEMORY CELL

A. 6Tr memory cell with single data line and double word line

Figure 1 shows the structure of a conventional 6Tr. Access nMOSFET (nMOS) (An) (n=0,1), pull-up pMOSFET (pMOS) (Un), and pull-down nMOS (Dn) all consists of two of each transistor. The structure also features a single word line and two data lines. Meanwhile, Figure 2 shows the proposed 6Tr structure. There are two each of An, Un, and Dn, as with the conventional structure, but the proposed structure uses a single data line, where both Ans are connected with the single data line. Two word lines are used to drive each Ans separately. The reason for such a structure is attributed to the write margin [10]. Details are discussed in the next section.

B. Problems with 5Tr structure and its improvement using an additional word line

When there is a single data line, a five-Tr (5Tr) structure is proposed by limiting the number of connected access nMOSFET (nMOS) to 1, as shown in Figure 3 [11]. The use of a 5Tr with a single data line has several advantages compared to a 6Tr with double the number of data lines, such as (1) the need for a reduced area, (2) reduced leakage, and (3) the risk of corrupt data is halved because the number of accesses to the internal node during the read/write process is halved. On the other hand, there is some difficulty in transmission to the high

level by the access nMOS, and the write "1" property is worse compared to the write "0" property. Figure 4 shows the butterfly curve during a write in a 5Tr structure. Figure 4(a) is the butterfly curve for a write "0," and Figure 4(b) is the butterfly curve for a write "1." A butterfly curve is the potential transition (V_{N0}) of node N0 against the potential transition (V_{N1}) of the internal node N1 of the memory cell (butterfly curve 0) and is the V_{N1} against V_{N0} (butterfly curve 1). The length of a side of the square adjacent to the pair of butterfly curves is defined as a margin. The margin obtained from the butterfly curve in a write is called a write margin (V_{WM}). While V_{WM} can be obtained in write "0," V_{WM} cannot be obtained in write "1." In other words, the 5Tr structure has a problem with write "1." Therefore, as shown in Figure 2, the proposed 6Tr structure is equipped with an A0 and A1 access nMOS, similar to the conventional 6Tr, with an individual word line made available to each. As a result, the write operation can be limited to "0" at all times. A0 is driven and N0 is overwritten to "0" by keeping

978-1-5386-7100-9/18 $31.00 © 2018 IEEE

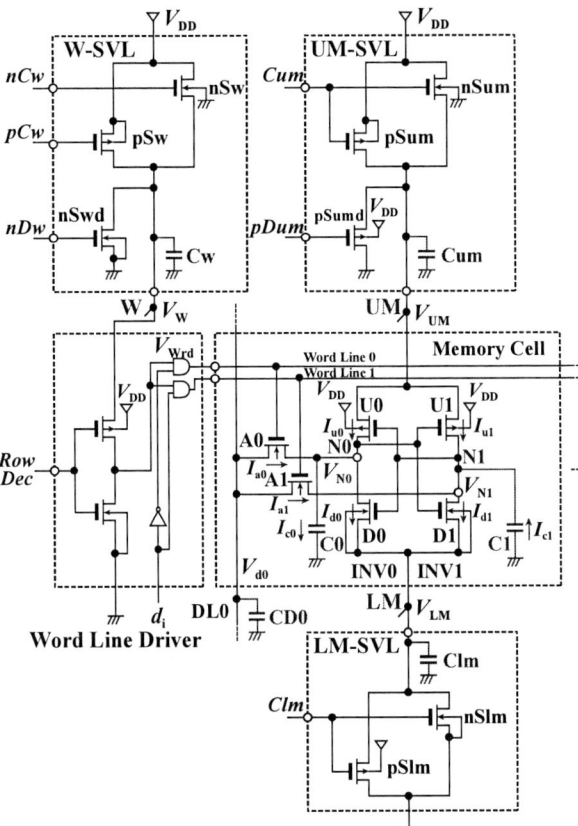

Fig. 5. Newly developed memory cell with SVLs.

the data line at "0" at all times during writes, and by setting word line 0 as "1" in the case of a write "0." In the case of a write "1," A1 is driven by setting the word line 1 as "1," and N0 is overwritten to "1" by overwriting N1 to "0." In light of the above, the difficulty with a write "1" can be overcome by preparing two word lines.

III. COMPOSITION OF DEVELOPED MEMORY CELL AND ITS OPERATING PRINCIPLE

A. Composition of newly developed memory cell

Figure 5 shows the newly developed memory cell (proposed 6Tr memory cell, word line driver, UM-SVL circuit, LM-SVL circuit, and W-SVL circuit) with SVL circuits installed. The UM-SVL circuit consists of a pMOSFET(pMOS) (pSum) for high-speed operations, an nMOSFET (nMOS) (nSum) for voltage drops, and a pMOS (pSumd) for discharges. The LM-SVL circuit consists of an nMOS (nSlm) for high-speed operations and a pMOS (pSlm) for increasing the voltage. C_{um} and C_{um} are the stray capacitances of the memory cell as viewed from the UM-SVL (point UM) and LM-SVL (point LM). With respect to the UM-SVL circuit, the W-SVL circuit consists of a pMOS (pSw) for high-speed operations, an nMOS (nSw) for voltage drops, and an nMOS (nSwd) for discharges. C_w is the

stray capacitance of the word line driver as viewed from the W-SVL (point W).

B. Operating principle of newly developed memory cell

· Write period

The memory cell supply voltage V_{UM} is decreased from the power supply voltage (V_{DD}) to (V_{DD}-v_{nm}) by setting the control signal Cum to "1" (= V_{DD}), cutoff (off) pSum, and operate (on) nSum in a saturated area. V_{LM} increases from the ground voltage (V_{SS}) to (V_{SS}+v_{pm}) by setting Clm to "0" (= 0 V), turning off nSlm, and turning on the pSlm on the memory cell ground supply voltage (V_{LM}) side. Here, v_{pm} is the voltage drop in pSlm. On the other hand, V_W is set to V_{DD} by setting the control signals (pCw, nCw) to "0" (= 0 V), operate (on) pSw in a linear area, and cutoff (off) nSw. As a result, a high-speed write operation is executed and a low-voltage operation is enabled. Note that V_{UM} rapidly decreases to (V_{DD}-v_{nm}) when a write begins, and the stored charge in C_{um} is discharged rapidly by turning pSum on for a brief period.

· Read period

V_{UM} is increased from (V_{DD}-v_{nm}) to V_{DD} by setting Cum = "0," turning on pSum, and turning off nSlm. V_{LM} decreased from (V_{SS}+v_{pm}) to V_{SS} by setting Clm = "1," turning on nSlm, and turning off pSlm. Moreover, the word line voltage (V_{Wrd}) of the selected word line decreases from V_{DD} to (V_{DD}-v_{nw}) by setting the supply voltage (V_W) to the word line driver as (V_{DD}-v_{nw}), turning off pSw and turning nSw on in a saturated area. Here, v_{nw} is the voltage drop in nSw. As a result, the read margin increases and a low-voltage operation is enabled. Note that V_{Wrd} rapidly decreases to (V_{DD}-v_{nw}) when a read begins, and nSwd is turned on and the stored charge in Cw is discharged rapidly.

· Retention period

With Cum = "1," Clm = "0," nCw = "0," and pCw = "1," pSum is set to off, nSum is set to on, pSlm is set to on, nSlm is set to off, and nSw and pSw are set to off. By temporarily turning pSmd and nSW on immediately after the retention begins, V_{UM} is set to (V_{DD}-v_{nm}), V_{LM} is set to (V_{SS}+v_{pm}), V_W is set to 0 V, and V_{Wrd} for all word lines is set to 0 V. As a result, $|V_{tp}|$ of the pMOS, turned off, in a memory cell increases, while at the same time the, drain-induced barrier lowering (DIBL) at an off pMOS and nMOS is relieved, leading to a reduced sub-threshold leakage. In addition,, the gate-induced drain leakage (GIDL) at the pMOS and nMOS, both of which are turned off, and the gate tunnel leakage of the pMOS and nMOS, both of which are turned on, are reduced. Therefore, the standby power is reduced significantly.

IV. COMPOSITION AND LAYOUT OF NEWLY DEVELOPED SRAM

Figure 6 shows the circuit composition of the newly developed SRAM. This SRAM consists of a 6Tr memory cell array (2-kbit; 8 bit × 4 word × 64 word), a peripheral circuit, an SVL (UM-SVL and LM-SVL) for the memory cell array, an SVL (W-SVL) for a word line driver, and an SVL control circuit (SVL-C). A single UM-SVL and LM-SVL, which supplies voltage to the memory cell array, are installed per 256 bit (8 bit × 4 word × 8 word) for a total of 8. One W-SVL is installed to supply power to the word line driver.

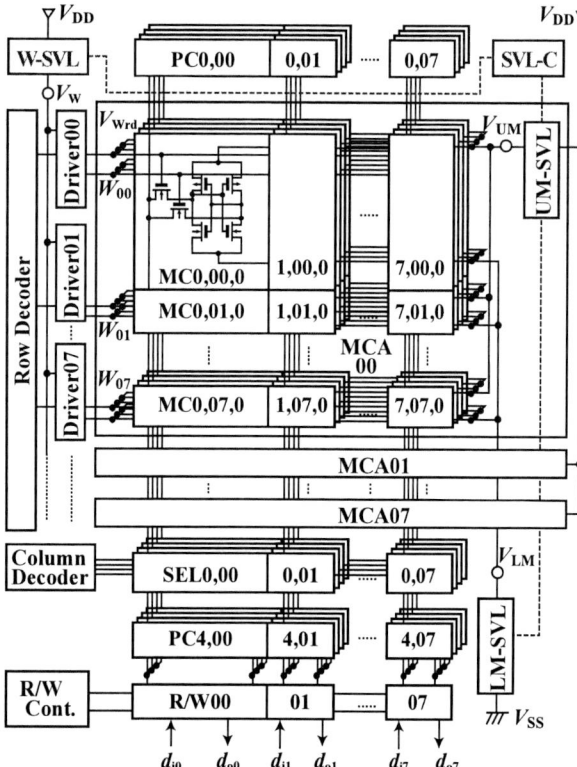

Fig. 6. Newly developed 2 kbit SRAM.

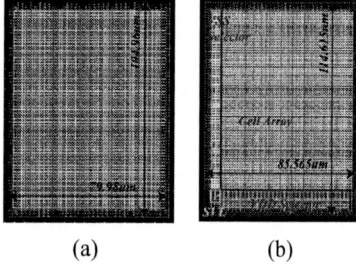

(a)　　　　　(b)

Fig. 7. SOTB 65nm chip LSI layout. (a) Conventional SRAM. (b) Newly developed SRAM.

Figure 7 shows the layout of the newly developed SRAM. A conventional SRAM that does not include an SVL circuit and the newly developed SRAM are installed. The area of the conventional SRAM, newly developed SRAM, and the SVL circuit are 8394.70, 8240.53 and 39.82 μm², respectively. The Silicon area can be reduced by 16.14% with the newly developed structure of 6Tr memory cell with single data line. The area overhead of the SVL circuit is 0.47% of that of the conventional SRAM.

V. STATIC CHARACTERISTICS

C. Threshold voltage and variance

The evaluation was conducted using a MOSFET (MOS) where the variance of the threshold voltage (V_t) is 0 and ±6σ (σ

= standard deviation). In this study, an nMOS whose variance is 0, or in other words, V_t (V_{tn}) is 0.357 V is defined as the typical (T), a +6σ (V_{tn} = 0.392 V) nMOS is defined as the slow (S), and -6σ (V_{tn} = 0.315 V) nMOS is defined as a fast (F) nMOS. On the other hand, a pMOS whose variance is 0, or in other words, V_t (V_{tp}) is -0.388 V, is defined as the typical (T), a +6σ (V_{tp} = -0.336 V) pMOS is defined as the fast (F), and -6σ (V_{tp} = -0.419 V) pMOS is defined as the slow (S) pMOS. Following these definitions, a combination for which nMOS and pMOS are both T is noted as TT (0 variance), a combination where nMOS is S and pMOS are both F is noted as SF (+6σ variance), and a combination for which nMOS is F and pMOS is S is noted as FS (-6σ variance).

As discussed in section II.B, the operating margin is defined as the length of the side of a square adjacent to two butterfly curves. Margins obtained from the butterfly curves for writing, reading, and retention are called the write margin (V_{WM}), read margin (V_{RM}), and retention margin (V_{RTM}), respectively.

D. Write margin (V_{WM})

Figure 8 shows the relationship between V_{WM} and V_{DD} (T = 25□) obtained by SPICE analysis. Dotted, solid, and dashed lines each represent V_{WM} of the conventional SRAM (V_{WMC}), V_{WM} of the newly developed SRAM (V_{WMD}), and the ratio compared to the conventional SRAM (V_{WMD}/V_{WMC}), respectively, while (a), (b), and (c) are SF, TT, and FS, respectively.

V_{WMD} is always greater than V_{WMC} for TT, SF, and FS; in other words, V_{WM} is improved. This is due to a drop in the memory cell supply voltage (V_{UM}) by UM-SVL, and a rise in the memory cell supply ground voltage (V_{LM}) by LM-SVL.

When the variance of V_t =0 (TT) and V_{DD}=1 V, V_{WMC} and V_{WMD} are 0.223 V and 0.704 V, respectively, and V_{WMD}/V_{WMC} is 315.41%. V_{DD}, where V_{WM} = 0 V, is 0.01 V for both conventional and newly developed SRAM.

When the variance V_t =+6σ (SF) and V_{DD} = 1 V, the value of V_{WM} of SF decreases for the conventional SRAM compared to the V_{WM} of TT, but the value improves for the newly developed SRAM (for over 0.25 V). When the variance V_t =-6σ (FS) and V_{DD} = 1 V, the V_{WM} value of FS increased for the conventional SRAM, and remained the same for the newly developed SRAM compared to that of V_{WM} of TT.

E. Read margin (V_{RM})

Figure 9 shows the relationship between V_{RM} and the power supply voltage (V_{DD}) (T = 25□) obtained by SPICE analysis. Dotted, solid, and dashed lines respectively represent the V_{RM} (V_{RMC}) of the conventional SRAM (conv), V_{RM} (V_{RMD}) of the newly developed SRAM (dvlp), and the ratio with the conv, (V_{RMD}/V_{RMC}), while (a), (b), and (c) are SF, TT, and FS, respectively.

V_{RMD} is always greater than V_{RMC} for TT, SF, and FS; in other words, V_{RM} is improved. This is due to a drop in the word line potential (V_{Wrd}) by W-SVL.

When the variance of V_t = 0 (TT) and V_{DD} = 1 V, V_{RMC} and V_{RMD} are 0.221 V and 0.349 V, respectively, and V_{RMD}/V_{RMC} is 157.80%. V_{DD}s of conv and dvlp, where V_{RM} = 0 V, are 0.07V.

978-1-5386-7100-9/18 $31.00 © 2018 IEEE　　390

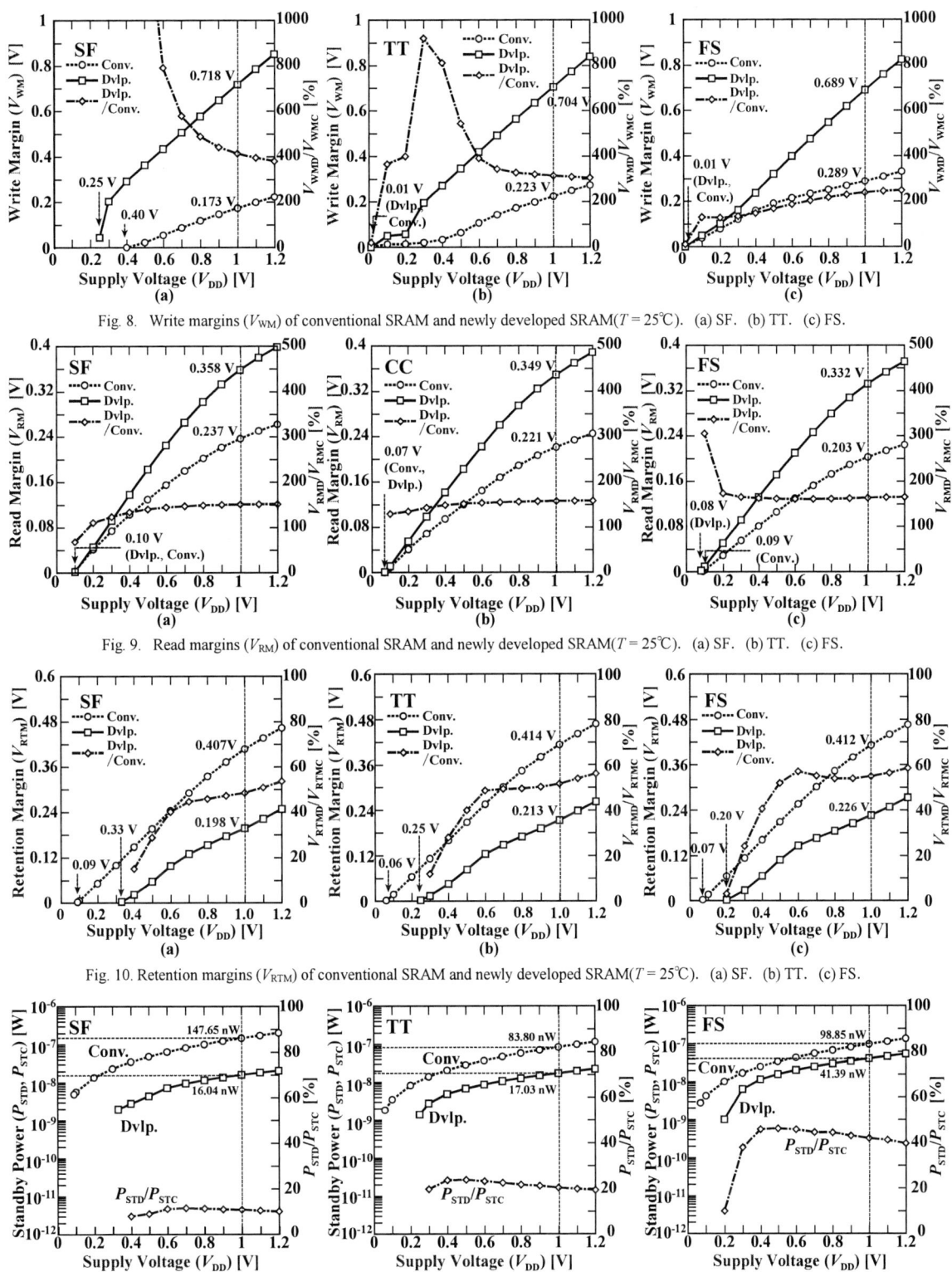

Fig. 8. Write margins (V_{WM}) of conventional SRAM and newly developed SRAM($T = 25°C$). (a) SF. (b) TT. (c) FS.

Fig. 9. Read margins (V_{RM}) of conventional SRAM and newly developed SRAM($T = 25°C$). (a) SF. (b) TT. (c) FS.

Fig. 10. Retention margins (V_{RTM}) of conventional SRAM and newly developed SRAM($T = 25°C$). (a) SF. (b) TT. (c) FS.

Fig. 11. Standby power (P_{ST}) of memory cell arrays ($T = 25°C$). (a) SF. (b) TT. (c) FS.

978-1-5386-7100-9/18 $31.00 © 2018 IEEE   391

When the variance of V_t = +6σ (SF) and V_{DD} = 1 V, V_{RM} of SF increased for both conv and dvlp compared to the V_{RM} of TT. When the variance of V_t = -6σ (FS) and V_{DD} = 1 V, the value of V_{RM} for FS decreased in both conv and dvlp compared to the V_{RM} of TT.

F. Retention margin (V_{RTM})

Figure 10 shows the relationship between V_{RTM} and V_{DD} (T = 25□) obtained by SPICE analysis. Dotted, solid, and dashed

lines each represent V_{RTM} (V_{RTMC}) of the conventional SRAM (conv), V_{RTM} (V_{RTMD}) of the newly developed SRAM (dvlp), and the ratio compared to the conv (V_{RTMD}/V_{RTMC}), respectively, while (a), (b), and (c) are SF, TT, and FS, respectively.

V_{RTMD} is always smaller than V_{RTMC} for any of TT, SF, and FS. This is due to a drop in V_{UM} by UM-SVL and a rise in V_{LM} by LM-SVL.

When the variance of V_t = 0 (TT) and V_{DD} = 1.0 V, V_{RTMC} and V_{RTMD} are 0.414 V and 0.213 V, respectively, and V_{RTMD}/V_{RTMC} is 51.57%. V_{DDS} of conv and dvlp, where V_{RTM} = 0 V, are 0.06 V and 0.25 V, respectively. The value of V_{RTM} (= 0.204 V) that the dvlp can obtain at V_{DD} = 1 V can be obtained by the conv at V_{DD} = 0.52 V.

When the variance of V_t = +6σ (SF) and V_{DD} = 1 V, V_{RTM} of SF decreased in the conv and increased in the dvlp compared to the V_{RTM} of TT. When the variance of V_t = -6σ (FS) and V_{DD} = 1 V, V_{RTM} of FS decreased for both conv and dvlp compared to V_{RTM} of TT.

VI. POWER CONSUMPTION PERFORMANCE

Figure 11 shows the relationship between the standby power (P_{ST}) and power supply voltage (V_{DD}) for a 2-kbit memory cell array obtained by SPICE analysis (T = 25□). Dotted, solid, and dashed lines each represent P_{ST} (P_{STC}) of the conventional cell array, P_{ST} (P_{STD}) of the newly developed cell array, and the power ratio (P_{STD}/P_{STC}), respectively, while (a), (b), and (c) are SF, TT, and FS, respectively. P_{ST} increases (or decreases) when V_{DD} increases (or decreases) for TT, SF, or FS in both conventional and newly developed cell arrays. P_{STD}/P_{STC} is constant in a manner that is almost unrelated to V_{DD}. When V_{DD} is 1 V, P_{STD} of TT is 17.03 nW, which is significantly reduced to 20.32% of P_{STC} (83.80 nW). These results show that UM-SVL and LM-SWVL circuits are effective in reducing leakage.

VII. CONCLUSIONS

In this study, a SOTB 65-nm 6-Tr 2-kbit SRAM was developed, which enables read/write operations with a low voltage and reduces standby power with a single power source. The study found that the SVL circuit is quite effective in simultaneously expanding the operating margin while reducing the leakage. The area overhead of the SVL circuit was only 0.47% compared to the total area of a conventional 2-kbit SRAM.

ACKNOWLEDGMENT

The VLSI chips used in this study were fabricated under the chip fabrication program of the VLSI Design and Education Center (VDEC) of the University of Tokyo.

REFERENCES

[1] M. J. M. Pelgrom, A. C. J. Duinmaijer, and A. P. G. Welbers, "Matching Properties of MOS Transistors," IEEE J. of Solid-State Circuits," vol. 24, no. 5, pp. 1433-1439, Oct. 1989.

[2] R. Tsuchiya, et al., "Silicon on Thin BOX: A New Paradigm of The CMOSFET for Low-Power and High-Performance Application Featuring Wide-Range Back-Bias Control," IEDM, pp. 631-634, 2004.

[3] L. Chang, et al., "An 8T-SRAM for Variability Tolerance and Low-Voltage Operation in High-Performance Caches," IEEE J. Solid-State Circuits, vol. 43. No. 4, pp. 956-963, April 2008.K.

[4] K. Nii, et al., "Improving SRAM Read/write Margin with Asymmetric Halo MOSFET," Int. Semicond. Device Res. Symp. (ISDR 2011), Dec. 2011.

[5] O. Hirabayashi, et al., "A Process-variation-tolerant Dual-power-Supply SRAM with 0.179 μm² Cell in 40 nm CMOS using Level-Programmable Wordline driver," Digest of Technical Papers, Int. Solid-State Circuits Conf., Feb. 2009.

[6] K. Zhang, et al., "A 3-GHz 70-Mb SRAM in 65-nm CMOS Technology with Integrated Column-based Dynamic Power Supply," IEEE J. Solid-State Circuits, vol. 41, no. 1, pp. 146-151, Jan. 2006.

[7] C. Piguest, ed., Low-Power Electronics Design, pp. 3.1-3.13, CRC Press, New York, 2004.

[8] T. Kuroda, T. Fujita, S. Mita, T. Nagamatsu, S. Yoshioka, K. Suzuki, F. Sano, M. Norishima, M. Murota, M. Kako, M. Kinugawa, M. Kakumu, and T. Sakurai, "A 0.9-V, 150-MHz, 10-mW, 4 mm² 2-D Discrete Cosine Transform Core Processor with Variable Threshold-voltage (Vt) Scheme," IEEE J. of Solid-State Circuits, vol. 31, no. 11, pp. 1770-1779, Nov. 1996.

[9] T. Enomoto, Y. Oka, and H. Shikano, "A Self-Controllable Voltage Level (SVL) Circuit and Its Low-Power, High-Speed CMOS Circuit Applications," IEEE J. Of Solid-Stage Circuits, vol. 38, no. 5, pp. 1220–1226, July 2003.

[10] T. Enomoto and N. Kobayashi, "A Large Read and Write Margins, Low Leakage Power, Six-Transistor 90-nm CMOS SRAM," IEICE Trans. on Electron., vol. E94-C, no. 4, April 2011.

Exploiting Principle of Moving Target Defense to Secure FPGA Systems

Zhiming Zhang
Department of Electrical and Computer Engineering
University of New Hampshire
Durham, New Hampshire, USA
zz1017@wildcats.unh.edu

Qiaoyan Yu
Department of Electrical and Computer Engineering
University of New Hampshire
Durham, New Hampshire, USA
qiaoyan.yu@unh.edu

Abstract—**Field Programmable Gate Arrays (FPGAs) enter a rapid growth era due to their attractive flexibility and CMOS-compatible fabrication process. However, the increasing popularity and usage of FPGAs also drive more motivated attacks on FPGA systems. In this work, we extensively investigate new potential attacks originated from the untrusted computer-aided design (CAD) suite for FPGAs and further propose a series of countermeasures. For the scenario of using FPGAs to replace obsolete components in legacy systems, we propose a Runtime Pin Grounding (RPG) scheme to ground the unused pins and check the pin status at every clock cycle, and exploit the principle of moving target defense (MTD) to develop a hardware MTD (HMTD) method to thwart hardware Trojan attacks. For general FPGA applications, we extend HMTD to an FPGA-oriented MTD (FOMTD) method, which is composed of three defense lines. FPGA emulation results and hardware cost analyses show that the proposed countermeasures are capable of tackling the attacks from malicious CAD tools with acceptable overheads.**

Index Terms—**Moving target defense, FPGA, hardware Trojan, FPGA design suite, hardware security.**

I. INTRODUCTION

Field Programmable Gate Arrays (FPGAs) receive increasing attention due to their high flexibility and low cost. The programmable architecture and internal connection of FPGAs allow them to realize any functions that application-specific integrated circuit (ASIC) can implement. Both the usage and popularity of FPGA have kept increasing in the past several decades. As indicated by the report [1], the FPGA market was already valued at USD 5.34 billion in 2016 and this number is expected to increase to USD 9.50 billion in 2023. FPGA applications are broadly distributed in the field of aerospace, high performance computing, wireless communication, digital signal processing, ASIC prototyping, and supercomputers. However, the increasing popularity of FPGA usage also brings in some concerns on security.

The security threats to FPGA systems are originated from both FPGA device and FPGA supply chain attacks, including intellectual property (IP) theft, reverse engineering, logic tempering, and hardware Trojan insertion [2]. Existing countermeasures assume that FPGA design suite is trusted. Unfortunately, this assumption does not always hold. In this work, we propose defense mechanisms to thwart hardware Trojans that are inserted by untrusted FPGA computer-aided

design (CAD) tools. More precisely, our main contributions are as follows.

1) We investigate new potential attacks originated from the untrusted FPGA design software and further propose a series of countermeasures to thwart those attacks.

2) To remove the security concerns during the process of replacing obsolete components in legacy systems with FPGAs, we propose a runtime pin grounding (RPG) scheme and a hardware moving target defense (HMTD) method to nullify the Trojans inserted through CAD tools.

3) For general FPGA applications, we extend HMTD to an FPGA-oriented MTD (FOMTD) method, which is composed of three defense lines to defeat hardware Trojan insertions at two different attack scenarios.

The remainder of this work is organized as follow. Section II provides the background of FPGA systems and summarizes the related work. Section III demonstrates the attack model interested in this work. In Section IV, we reveal the security problems that could occur during the process of FPGA deployment and propose RPG and HMTD methods. In Section V, we expand the investigation to general FPGA applications and propose a FOMTD protection mechanism. Section VI provides the experimental results and a practical example. Section VII concludes this work.

II. BACKGROUND AND RELATED WORK

A. FPGA System Overview

A standard FPGA is usually composed of programmable logic blocks, programmable interconnections, I/O blocks, and some additional advanced on-chip integrated circuitries, such as ALUs, block RAM, and DSP-48 for corresponding specific operations [3]. FPGA behaviors are specified by a bitsream file that is generated by the FPGA design suite software. Xilinx ISE and Altera Quartus are two popular FPGA design suites and their FPGA configuration flows are similar. We use Xilinx ISE as an example to briefly describe the process for bitstream generation. The HDL design is first synthesized. The output file of this stage will be combined with a user constraints file and sent to the NGDBuild. After this step, the translated design is mapped, placed and routed according to the specified

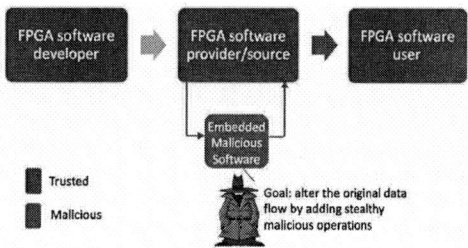

Fig. 1: Malicious software add-on in the supply chain of FPGA tools.

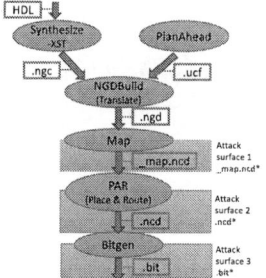

Fig. 2: Attack surfaces of interest in this work. The rectangle represents the output file from each step. The file with the symbol of * is an output file modified by the untrusted FPGA CAD tool.

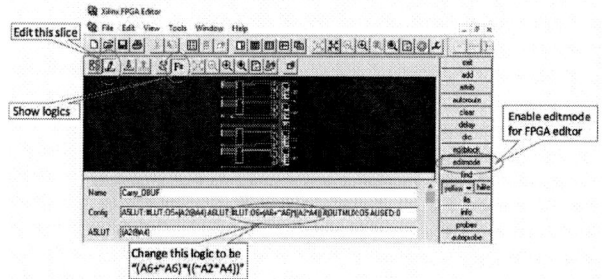

Fig. 3: An example of the practical attack performed through the FPGA editor tool available in the Xilinx ISE 14.1 design suite.

FPGA device. At the end of the compile process, a bitstream is generated.

B. Existing Researches on FPGA Security

Intellectual property (IP) of an electronic design always carries the key knowledge and technique of a research team. By pirating the IP, attackers may have a chance to significantly reduce the development cost. The watermarking-based method presented in [4] generates the watermarking to help identifying the authorship of the design. In that method, the state transition graph is manipulated to make it exhibit a specific property that is very rare in non-watermarked designs. Such a watermarking is very hard to be removed. The protection scheme in [5] proposes to assign each FPGA device a unique architecture, which is used to encrypt the bitstreams. By doing this, only the authorized FPGA devices can recognize the encrypted bitstreams. Any maliciously modified bitstreams cannot be successfully applied to other devices even from the same FPGA family. This approach prevents attackers from reverse engineering the bitstreams.

Hardware Trojan is another group of malicious attacks which can be harmful for FPGA designs. It can be inserted stealthily into the design to either alter the normal operation or leak confidential information. The MORPH architecture proposed in [6] combines multiple levels of protection schemes including morph operation, onion encryption, and replication to mitigate the hardware Trojan induced during the fabrication time and design time. Adaptive triple modular redundancy (ATMR) is another effective countermeasure to mitigate hardware Trojan attacks [7]. In the ATMR method, the original design is duplicated three times and the third copy is only activated when a mismatch is found between first two.

III. ATTACK MODEL

The attack model interested in this work is originated from the malicious software mounted on the top of the original FPGA design suite for SRAM FPGAs. We assume that the FPGA deployment engineers and their designs are trusted, and the bitstream downloading channel and procedure will not be tampered, as well. We argue that the FPGA design suite will be propagated through computer networks or retailers, so the integrity of the software may be sabotaged by advanced attackers in a way shown in Fig. 1.

Here we demonstrate one practical attack through Xilinx ISE, an FPGA design suite. There are three potential attack surfaces on the basic design flow shown in Fig 2. We can successfully modify the configuration of the target slice through the *FPGA editor* tool from Xilinx. Figure 3 shows the graphic interface. In the edit mode of the FPGA editor, we changed the logic configuration after the place and route (PAR) stage, and then re-generated the bitstream. The attack action can also be implemented in a malicious FPGA software implanted in the original FPGA design suite.

IV. SECURING FPGA-BASED OBSOLETE COMPONENT REPLACEMENT FOR LEGACY SYSTEMS

In a legacy system, component-aging is unavoidable. Unfortunately, the aged components may no longer be manufactured or available in the market. A good solution is to substitute the obsolete component with an equivalent device. Traditionally, functionality matching is the primary focus for the substitution. Little or no attention is paid to the security threats originated from the process of component replacement, in particular, from untrusted FPGA manufacturers and CAD tools associated with FPGA deployment.

A. Proposed Method

We propose a framework, which is composed of (1) RPG and (2) HMTD to detect the hardware Trojan inserted by an untrusted FPGA manufacturer or CAD tools [8]. The RPG scheme terminates the hardware Trojans that communicate with the external environment through the unused I/O pins on the FPGA device. The HMTD method prevents the hardware Trojan horses induced by the malicious FPGA CAD tools from

978-1-5386-7100-9/18 $31.00 © 2018 IEEE

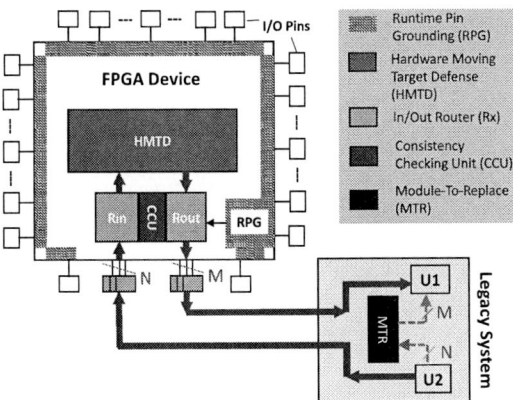

Fig. 4: Overview of proposed countermeasure to secure the FPGA replacement for a legacy system. To replace the aged module (MTR), the proposed method connects a group of FPGA modules (HMTD+Rin+Rout+CCU) to the original modules U1 and U2 in the legacy system.

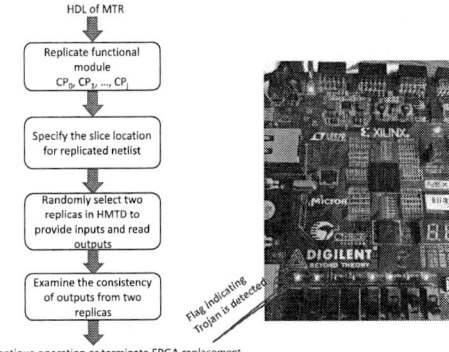

Fig. 5: Flowchart of proposed hardware moving defense method.

interfering with the FPGA replacement in legacy systems. Figure 4 depicts the overview of the proposed countermeasure.

1) Proposed Runtime Pin Grounding: Inspired by the idea proposed in [9], we apply the pin grounding scheme to the unused FPGA I/O pins by using a user constraints file. This scheme is implemented on the top level of the hardware description module as shown in the black shadowed area of Fig. 4. We assign every unused pin a net name in the top level of the hardware design file then link each net name with a unused I/O pin in the user constraints file by using the command (1).

$$\text{NET ``net\_name'' LOC = pin\_name} \qquad (1)$$

After that, we proceed to ground those I/O pins through the command (2).

$$\text{NET ``net\_name'' PULLDOWN} \qquad (2)$$

To avoid the grounded pins being reactivated by malicious CAD tool through modifying the native circuit description file, we enhance our pin grounding scheme by adding a runtime detection circuit to examine the grounding status of those unused pins.

2) Proposed Hardware Moving Target Defense : To prevent the malicious CAD tool from successfully sabotaging the original FPGA configuration, we use the principle of MTD to develop a HMTD method. We assume that the functionality of the module-to-replace (MTR) in the legacy system is known by the FPGA deployment team, that is trusted. HMTD replicates the MTR into multiple copies CP_0, CP_1, $\cdots$, CP_j and the relative distance among each replica is specified by appending "RLOC" command in user constraints. Then we add a low-cost, random number generator in the *Rin* unit to select two replicas of the function module to feed the N-bit inputs from *U2* in the legacy system. The random selection is performed at runtime and the random number generator also controls the *Rout* unit to choose which two replicas for the Trojan detection in the consistency checking unit (CCU). Once the inconsistency between the outputs is found, the M-

bit output pins are grounded immediately and a flag for Trojan detection is turned on. The flowchart of our HMTD method is summarized in Fig. 5.

V. FPGA-ORIENTED MOVING TARGET DEFENSE AGAINST SECURITY THREATS FROM MALICIOUS FPGA TOOLS

In this section, we extend our research to general FPGA applications. We assume two levels of attacks can take place due to the malicious software implanted in the FPGA design suite.

- *L-1 attack:* Attackers have no knowledge of the design in FPGA such that hardware Trojans are placed in the most popular FPGA area (based on attackers' experience).
- *L-2 attack:* Attackers are able to extract information from the FPGA placelist to learn which slices are utilized by the current design.

We exploit the principle of MTD to propose a FPGA-oriented moving target defense (FOMTD) method [10]. Our method explores the unpredictability of the way that a hardware design is configured on FPGAs to deter attackers from precisely inserting hardware Trojans. FOMTD can make the output of FPGA place and route stage unpredictable, such that attackers who mounts a malicious program on the original FPGA design suite cannot easily alter the original implementation on the FPGA device. Note, our method does not guarantee to completely prevent all the hardware intrusions but it will increase the difficulty of a Trojan successfully landing on one of the FPGA slices occupied by the design. The desired unpredictability can be achieved by the three defense lines provided by our method.

A. Defense Line 1 (DFL1): Slice Position Selection through User Constraints File (UCF)

Instead of using the default FPGA setting for placement and routing, we specify the slice positions on the FPGA die for the selected LUTs, so that the design mapping on the FPGA grid can be modified. As illustrated in Fig. 6, by specifying few LUTs (black squares), we change the slice locations for the three parts of the intended design to the specified area. The specification can be performed through a user constraints file and the selection of slice positions is

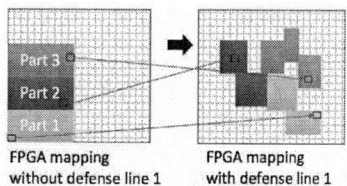

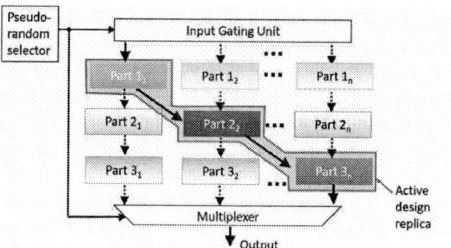

Fig. 6: FPGA mapping modified by proposed DFL1. Three parts in different colors represent three partitions of the intended design. Black squares are three LUT configurations. Proposed DFL1 alters the default LUT mapping on the FPGA grid.

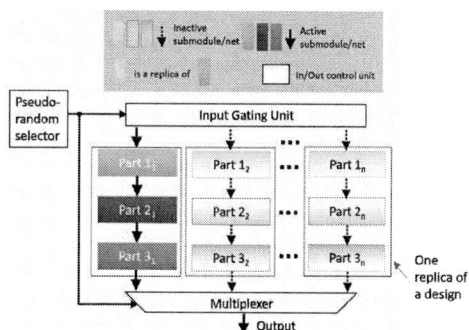

Fig. 7: Pseudo-random replica selection provided by DFL2.

Fig. 8: Hot-swappable submodule assembling provided by DFL3.

VI. EXPERIMENTAL RESULTS

A. Experimental Setup

We synthesized, placed and routed the Verilog HDL codes for ISCAS'85 benchmark circuits through the Xilinx ISE 14.1 design suite, and generated the corresponding bitstreams, which are specific for a Xilinx Spartan-6 XC6SLX16 FPGA. The detailed slice utilization of each circuit was analyzed by our Python script to extract the occupied FPGA slice positions. We used MATLAB programs to insert hardware Trojans blindly or purposely (depending on the experiment goal) and then measured the hardware Trojan hit rate and the bypass rate. Overhead measurements were obtained from the tools available in the Xilinx design suite.

B. Hardware Trojan Bypass Rate for HMTD

We validated the proposed HTMD method with hardware Trojan bypass rate which is defined as the number of incorrect outputs, due to Trojans, over the number of test cases. As can be seen from Fig. 9(a), for the range of 1 to 10 Trojans, the Trojan bypass rate almost monotonically increases with the number of injected hardware Trojans, which is because the probability for multiple replicas of the functional module simultaneously containing Trojans increases. Hence, comparison of the two copies' outputs gradually loses the Trojan detection capability. From Fig. 9(b), we can observe that the Trojan bypass rate for a larger FPGA is lower than that for a smaller one. This is because the number of Trojans placed in the FPGA device is fixed per each FPGA size. The chance for a Trojan slice colliding with a design slice is higher in a smaller FPGA than in a larger one. We compared our countermeasure with ATMR method and in both scenarios, HMTD has lower hardware Trojan bypass rate. On average, HMTD reduced the Trojan bypass rate by 61%.

C. Hardware Trojan Hit Rate for FOMTD

The attack resilience of baseline and FOMTD are compared in the subsections below. Two attack levels mentioned previously are considered in the following assessments.

1) Hardware Trojan Hit Rate for L-1 Attack: Recall that attackers who execute L-1 attack do not know the locations of all the occupied slices for the design of interest. We varied the range of attack exploration space from 5% to 50% of the entire FPGA die in the following experiments. Figure 10 shows that the proposed method achieves a lower hardware Trojan hit rate

conducted by FPGA users at the FPGA deployment stage (this is after the implementation of malicious FPGA software). Hence, deciding where to place effective hardware Trojans is not an easy task. Blindly inserting Trojans may not impact the design at all.

B. Defense Line 2 (DFL2): Pseudo-Random Replica Selection

In our defense line 2, the design to be implemented on the FPGA is duplicated by n copies and only one of the replicas will be active at a time while others are inactivated through input gating. The replica selection and input gating are controlled by a pseudo-random selector. Figure 7 shows the concept of our defense line 2. Note, in this defense line, we do not have a comparison logic to examine the consistency among the n replicas for the purpose of power saving. As the fact that which replica will be active is determined after the FPGA configuration, an attacker (performing L-1 attack) needs to blindly place the hardware Trojan to the entire FPGA die to make a successful attack.

C. Defense Line 3 (DFL3): Runtime Design Assembling

Our defense line 3 is the hot-swappable submodule assembling technique, as shown in Fig. 8. We partition the original design into m submodules and each submodule is duplicated by n times. In the moment of interest, only one replica of each submodule will be assembled into a complete design. The pseudo-random selector is utilized to determine which replica to choose at runtime. After a period of time, the selection of submodule replicas will be changed without stopping the normal operation. The total number of design configurations we can achieve is as large as n^m and this can further increase the difficulty for attacker to recognize the entire design for attack.

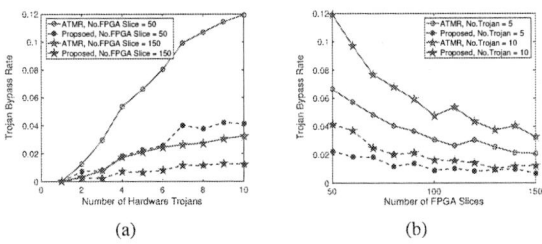

(a)   (b)

Fig. 9: Hardware Trojan bypass rate versus (a) number of hardware Trojans inserted in the FPGA device and (b) FPGA size.

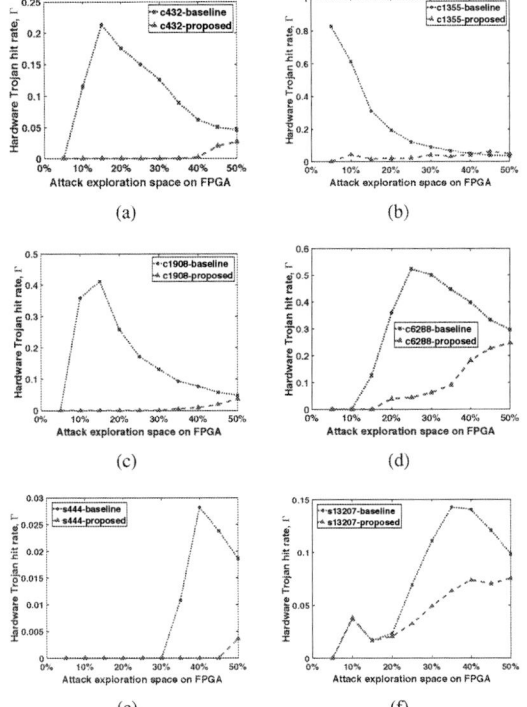

Fig. 10: Hardware Trojan hit rate reduction by proposed DFL1 applied in the benchmark circuit (a) c432, (b) c1355, (c) c1908, (d) c6288, (e) s444, and (h) s13207 in the condition of L-1 attack.

Γ than the baseline in a wide range of the attack exploration space. Compared to the baseline, our method can make the hardware Trojan hit rate decrease to 0.213, 0.8272, 0.4114, 0.49, 0.0036, 0.0752 for c432, c1355, c1908, c6288, s444, and s13207, respectively. When the attack exploration space is large enough to cover the entire design placed on the FPGA die, the Trojan hit rate of proposed method will be equal to the one of baseline eventually.

2) Hardware Trojan Hit Rate for L-2 Attack: L-2 attack is able to retrieve the exact locations of the occupied slices. Consequently, the baseline design does not have any resilience against L-2 attack. As shown in Fig. 11(a), the baseline without any protection yields a hardware Trojan hit rate of 1 while our DFL2 and DFL3 significantly reduce the Trojan hit rate

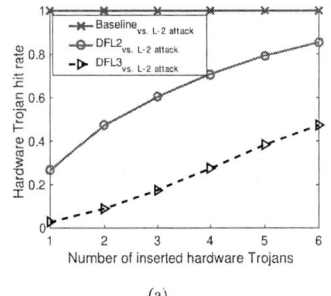

(a)

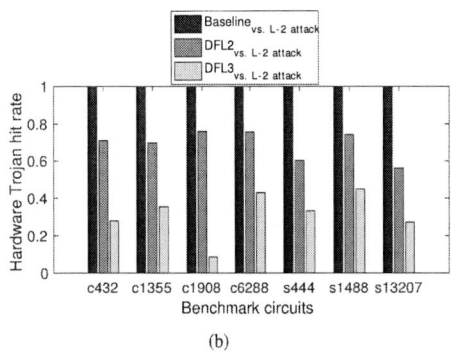

(b)

Fig. 11: Hardware Trojan hit rate for (a) c432, and (b) seven benchmark circuits suffering from four hardware Trojans inserted via L-2 attack.

TABLE I: FPGA Overhead of Proposed Runtime Pin Grounding

Overhead\Circuits	s298	s344	s444	s526	s1488
Increased No. LUTs	40	30	38	40	39
Increased No. Slices	12	6	11	12	16

especially for the small number of inserted Trojans. When more Trojans are placed in the utilized FPGA slices, our Trojan hit rate eventually increases due to the limited number of design replicas used (we used two replicas in our experiments). If more copies are available, the slope of Trojan hit rate will be less than what is shown in Fig. 11(a). We examined the Trojan hit rate for seven benchmark circuits, which suffer from different numbers of Trojan insertions from L-2 attack. Each hardware Trojan hit rate was obtained from 10,000 test cases. The average Trojan hit rate of DFL2 (DFL3) is 69% (31%). As shown in Fig. 11(b), the DFL2 reduces the hit rate by up to 40% over the baseline. The reduction on Trojan hit rate can be further improved by DFL3 up to 91%.

D. Assessment on Overhead

1) FPGA Utilization for RPG: For the reason that more unused I/O pins lead to more overhead for pin grounding, we chose the benchmark circuits with a small number of inputs/outputs. As shown in Table I, the number of utilized LUTs goes as high as 40, whereas the number of occupied slices goes up to 16. These hardware implementations consume 0.044% more LUTs and 0.07% more slices, respectively.

TABLE II: Number of FPGA LUTs Utilized by Different Methods.

Circuits	c432	c1355	c1908	c6288	s444	s1488	s13207
Baseline	58	62	58	530	33	117	180
Defense line 1	58	62	59	530	33	117	181
Defense line 2	158	156	178	1123	67	261	429
Defense line 3	173	167	216	1157	84	296	443

TABLE III: FPGA Total Power Consumption (mW) by Different Methods

Circuits	Baseline	Defense line 2	Defense line 3
c432	26 (100%)	28 (107.69%)	28 (107.69%)
c1355	34 (100%)	42 (123.53%)	42 (123.53%)
c1908	36 (100%)	42 (116.67%)	43 (119.44%)
c6288	67 (100%)	70 (104.48%)	70 (104.48%)
s444	22 (100%)	25 (113.64%)	25 (113.64%)
s1488	29 (100%)	30 (103.45%)	30 (103.45%)
s13207	33 (100%)	44 (133.33%)	45 (135.36%)

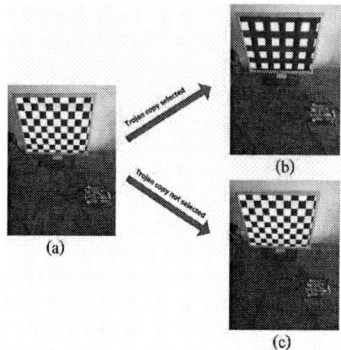

Fig. 12: VGA signal experiment.

2) Hardware Utilization for FOMTD: Table II summarizes the number of utilized LUTs for different methods. Proposed defense line 1 consumes 0.33% more LUTs than the baseline on average. As for defense line 2 and defense line 3, the increase on the LUT utilization could be large for small circuits due to the relative large size of pseudo-random selection and input gating logic. However, it can be reduced through optimization when the object for protection is large. The LUT overheads of defense line 2 for the largest combinational circuit c6288 and sequential circuit s13207 in our case studies are 111.89% and 138.33%, respectively. The overheads of defense line 3 for those two circuits increase to 116.79% and 145%.

3) Power Consumption: Power consumption is measured in the tool XPower Analyzer and reported in Table III. On average, defense line 2 leads to an increase on total power of 14.68% over the baseline. Defense line 3 provides better resilience, at the cost of consuming 15.51% more total power than the baseline.

E. Demo of a Practical Example

In this subsection, we demonstrate the attack addressed in this work and our proposed protection mechanism in a VGA signal experiment. We implemented our proposed defense line 2 in the FOMTD method. The intended design (with two replicas) generates a VGA signal of "chess board", as shown in the Fig. 12(a). We controlled the FPGA on-board switches to mimic the random selection for a better view. Then the Trojan was manually inserted in the replica 2 to alter the original picture, which can be observed in the Fig. 12(b). Next, we implemented the case that the system selected the replica 1 by changing the switches and observed that, from the Fig. 12(c), the signal was recovered. This proofs if an attacker cannot correctly predict the design configuration, the inserted hardware Trojan may not affect the design at all.

VII. CONCLUSION

In this work, we reveal the potential security threats from malicious CAD tools and propose a series of countermeasures. Proposed RPG method cuts off the communication of

hardware Trojan with less than 0.1% FPGA utilization rate increase. Our HMTD reduces the hardware Trojan bypass rate by 61% over the existing ATMR method. The FOMTD method reduces the hardware Trojan hit rate with acceptable overhead. More precisely, our defense line 1 in the FOMTD countermeasure achieves the hardware Trojan hit rate of L-1 attack as low as 0.36% with only 0.33% more LUT consumption. The defense line 2 and defense line 3 further reduce the Trojan hit rate at the cost of more power consumption.

REFERENCES

[1] "FPGA Market by Technology (SRAM, Antifuse, Flash), Node Size (Less than 28 nm, 28-90 nm, More than 90 nm), Configuration (High-End FPGA, Mid-Range FPGA, Low-End FPGA), Vertical (Telecommunications, Automotive), and Geography - Global Forecast to 2023." https://www.marketsandmarkets.com/Market-Reports/fpga-market-194123367.html.

[2] "SoC FPGA Hardware Security Requirements and Roadmap." https://www.altera.com/content/dam/altera-www/global/en_US/pdfs/education/events/northamerica/isdf/SoC-FPGA-Hardware-Security.pdf.

[3] "Basic FPGA Architecture and its Applications." https://www.edgefx.in/fpga-architecture-applications/.

[4] A. L. Oliveira, "Techniques for the Creation of Digital Watermarks in Sequential Circuit Designs," *IEEE Transactions on Computer-Aided Design of Integrated Circuits and Systems*, vol. 20, pp. 1101–1117, Sep 2001.

[5] R. Karam, T. Hoque, S. Ray, M. Tehranipoor, and S. Bhunia, "MU-TARCH: Architectural Diversity for FPGA Device and IP Security," in *Proc. 22nd Asia and South Pacific Design Automation Conf.*, pp. 611–616, Jan 2017.

[6] G. Bloom, B. Narahari, R. Simha, A. Namazi, and R. Levy, "FPGA SoC Architecture and Runtime to Prevent Hardware Trojans from Leaking Secrets," in *2015 IEEE International Symposium on Hardware Oriented Security and Trust (HOST)*, pp. 48–51, May 2015.

[7] S. Mal-Sarkar, R. Karam, S. Narasimhan, A. Ghosh, A. Krishna, and S. Bhunia, "Design and Validation for FPGA Trust under Hardware Trojan Attacks," *IEEE Transactions on Multi-Scale Computing Systems*, vol. 2, pp. 186–198, July 2016.

[8] Z. Zhang, L. Njilla, C. Kamhoua, K. Kwiat, and Q. Yu, "Securing FPGA-based Obsolete Component Replacement for Legacy Systems," in *2018 19th Int'l Symposium on Quality Electronic Design (ISQED)*, pp. 401–406, March 2018.

[9] R. S. Chakraborty, I. Saha, A. Palchaudhuri, and G. K. Naik, "Hardware Trojan Insertion by Direct Modification of FPGA Configuration Bitstream," *IEEE Design Test*, vol. 30, pp. 45–54, April 2013.

[10] Z. Zhang, L. Njilla, C. Kamhoua, and Q. Yu, "FPGA-Oriented Moving Target Defense against Security Threats from Malicious FPGA Tools," in *2018 IEEE International Symposium on Hardware Oriented Security and Trust (HOST)*, pp. 163–166, May 2018.

2018 IEEE Computer Society Annual Symposium on VLSI

A Highly Flexible Lightweight and High Speed True Random Number Generator on FPGA

Faqiang Mei[1], Lei Zhang[1], Chongyan Gu[2], Yuan Cao[3], Chenghua Wang[1] and Weiqiang Liu[1]*

[1]College of EIE, Nanjing University of Aeronautics and Astronautics, Nanjing, China

[2]CSIT, ECIT, Queen's University Belfast, Belfast, UK

[3]College of IoT Engineering, Hohai University, Changzhou, China

E-mail: {meifaqiang, zhanglei_1993, chwang, liuweiqiang}@nuaa.edu.cn, cgu01@qub.ac.uk, caoyuan0908@gmail.com

Abstract—True random number generator (TRNG), plays an important role in information security systems. Conventional TRNGs use natural physical stochastic processes including thermal noise, chaos-based circuit and so on to generate the random numbers. These analog circuit based TRNG structures often consume lots of hardware resources, and are not easy to be integrated in digital systems. In this paper, a low-cost and high-speed TRNG has been proposed by using mixed oscillation generated from XOR gates nested multiple ring oscillators (ROs). Multi-group mixed oscillation XOR operation is applied to obtain high-speed output. The proposed TRNG design is implemented on Xilinx Artix-7 XC7A35T-1FTG256C FPGA. It achieves a high performance with throughput up to 160 Mbps and with a usage of 37 FFs and 25 look up tables (LUTs) in the FPGA. The results show that the proposed TRNG design has successfully passed the testing standards of NIST SP800-22 and AIS31. Compared with previous designs, the proposed TRNG design achieves lower hardware resource consumption and higher speed.

Keywords-TRNG; Mixed Oscillation; FPGA;

I. INTRODUCTION

The information security has extensively influenced modern communication and computing systems. The random number plays an important role in cryptography and it is used in almost all security protocols and cryptographic algorithms [1]. The security of the whole system relies on the efficiency and quality of the random number sequences. Therefore, high speed and high quality are two essential requirements of random numbers in security system.

Both pseudo random number generator (PRNG) and true random number generator (TRNG) are usually used to generate random sequences that are used in practical applications. In critical security applications, the random numbers should be truly unpredictable and random. PRNG is generally based on a seed, through a certain mathematical algorithm to generate random sequences. The pseudo random number is predictable when the seed is revealed, and the security of the whole system will be vulnerable to attacks. Compared with PRNGs, TRNG is desirable in terms of security level [2] as it can produce unpredictable random number sequences that utilize various random differences in the physical process. For a TRNG, it is difficult to predict the random sequences

even if the attacker has unlimited computing power and collects a large number of random sequences. Therefore, TRNG has been a highly demanded security primitive.

Conventional TRNGs employ natural physical stochastic processes such as resistance thermal noise and chaos to generate random sources. Although the statistical distribution of these entropy sources is ideal, they are mainly analog circuit based TRNG structures that often consume a lot of hardware resources. Moreover, they are difficult to be integrated in digital systems.

Due to the ubiquitous nature of the IoT, lightweight and high speed digital TRNG design are required for low cost devices. FPGA based TRNGs have been studied quite extensively [3]. To address the above limitations of the conventional TRNGs, a novel high-speed TRNG design is proposed and implemented on FPGA in this work. It also has significant advantages in improving the speed and reducing the hardware cost. The proposed TRNG, which successfully passed two commonly used testing standards of AIS31 and NIST SP800-22, achieves high reliability and demonstrates feasibility for practical applications.

This paper is organized as follows. Section II reviews the existing digital TRNGs. Section III presents the proposed TRNG design and its analysis. Section IV provides the testing results. Hardware resource analysis and performance comparison with previous designs are given in Section V. Section VI concludes this paper.

II. RELATED WORKS

There are two main approaches to construct true random sources using digital circuits: one is oscillator sampling [4] and the other is metastability [5].

The oscillator sampling method [4] has been proposed to sample a high frequency oscillator by utilizing a low frequency oscillator and the sampling result was used as the output random data. It utilizes the phase noise of the oscillator as a random source. The output rate is determined by the low frequency oscillation. A TRNG circuit proposed by [6], as shown in Fig. 1, combines several oscillation signals with an XOR gate to exploit randomness of phase jitter. [3]

978-1-5386-7100-9/18 $31.00 © 2018 IEEE

presented an enhanced structure by adding an extra D flip-flop after each ring to improve the overall randomness for the output. The enhanced TRNG, implemented on an Altera Cyclone II FPGA, passed NIST and Diehard statistical tests at a throughput of 100 Mbps. However, this design uses 167 LEs (logic element) to implement 50 ring oscillators (ROs), which is quite expensive in terms of hardware cost.

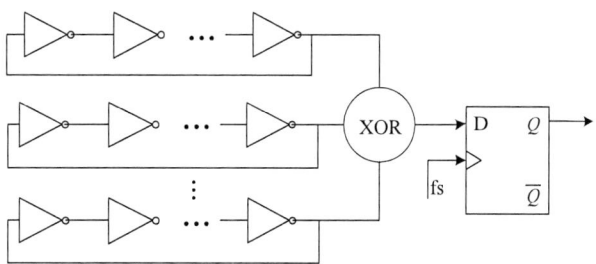

Fig. 1. The structure of digital TRNG proposed in [6].

Metastability-based structure is another typical approach for TRNG. It utilizes instability states that caused by the competition in logic gates, latches or flip flop (FF) triggers to produce an uncertain output. The final state is unable to predict since it depends on the electrical noise in the circuit. [7] proposed a RS latch-based TRNG and implemented on a Xilinx Virtex-4 FPGA. An open-loop TRNG design has been implemented on a Xilinx Virtex-5 FPGA in [8]. Two coarse delay chains were adopted on data and clock signals. It consumes a few hardware resources and generates random bits at a throughput of 20 Mbps. However, the issue for metastability-based TRNG is the low throughput, which requires very complex strategies of placement and routing to achieve balanced signal transmission paths.

III. PROPOSED TRNG STRUCTURE

To address these restrictions mentioned above, we propose a novel low cost and high speed TRNG based on digital circuit.

A. The Overall TRNG Architecutre

The proposed TRNG consists of three parts, a chained oscillation ring (COR), a FF array and an XOR array, as shown in Fig. 2. The overall TRNG architecture combines two random sources, the oscillating signal jitter and the metastable state of the FF. The oscillated ring accumulates jitter during the sampling time. The COR uses XOR gates to combine the jitter signal of the adjacent oscillation ring. By using the FF array, it produces a metastable state and guarantees that the sampling point include one or more oscillation regions. Finally, an XOR array is applied so that all the signals collected by FFs are mixed with XOR gates to produce high speed random sequences.

B. A Model for Jitter Sources in ROs

The random source of an ring oscillator (RO) is the jitter of gate transmission. Suppose $d_{i,j}$ is the delay of the i-th transmission gate in the j-th half clock. It can be obtained from [15]:

$$d_{i,j} = D_i + \Delta d_{i,j} = D_i + \Delta d_{Li,j} + \Delta d_{Gi,j} \quad (1)$$

where, D_i is the constant delay of the i-th gate, which includes an interconnection delay between the i-th and the $(i+1)$-th gate, $\Delta d_{i,j}$ is the delay variation caused by the individual local delay $\Delta d_{Li,j}$ and common global delay $\Delta d_{Gi,j}$. The local jitter of the i-th gate during the half-period j can be expressed as

$$\Delta d_{Li,j} = \Delta d_{LGi,j} + \Delta d_{LDi,j} \quad (2)$$

where, $\Delta d_{LGi,j}$ is the delay of Gaussian jitter from independent local sources, and $\Delta d_{LDi,j}$ is the local deterministic jitter. The global delay can be represented by

$$\Delta d_{Gi,j} = K_i(\Delta D + \Delta d_{GGj} + \Delta d_{GDj}) \quad (3)$$

where, ΔD represents slow delay variations due to temperature and/or power supply deviations. Δd_{GGj} is the delay of the Gaussian noise from the power supply, and Δd_{GDj} is the delay of the global deterministic jitter from fast power supply variations. K_i is a coefficient defining the proportion of the global jitter on the jitter of gate i.

A strategy proposed in [15] simplifies the model and removes the signal that never contributes to the jitter. The simplified model is shown in Eq. (4).

$$d_{i,j} = D_i + \Delta d_{i,j} = D_i + \Delta d_{LGi,j} + K_i \Delta d_{GDj} \quad (4)$$

C. The Contribution of the XOR Gates between the ROs

The common RO is composed of an odd number of inverters. An example of the RO consisting three inverters is shown in Fig. 3.

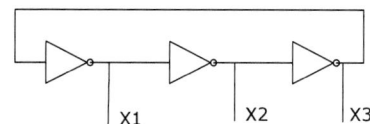

Fig. 3. A RO consisting of three inverters.

Suppose that the signal X_3 is '1' in the j-th half clock. When the signal reaches the position of the signal X_3 again, the time is:

$$t = D_1 + D_2 + D_3 + (\Delta d_{1,j+1} + \Delta d_{2,j+2} + \Delta d_{3,j+3}) \quad (5)$$

When $\Delta d_{1,j+1} + \Delta d_{2,j+2} + \Delta d_{3,j+3}$ is equal to 0, the value of X_3 is determined by the propagation time ($D_1 + D_2 + D_3$) after oscillation, as shown in Fig. 4(a). When $\Delta d_{1,j+1} + \Delta d_{2,j+2} + \Delta d_{3,j+3}$ is not 0, the value of X_3 is dependent on two conditions as shown in Fig. 4(b). The

978-1-5386-7100-9/18 $31.00 © 2018 IEEE

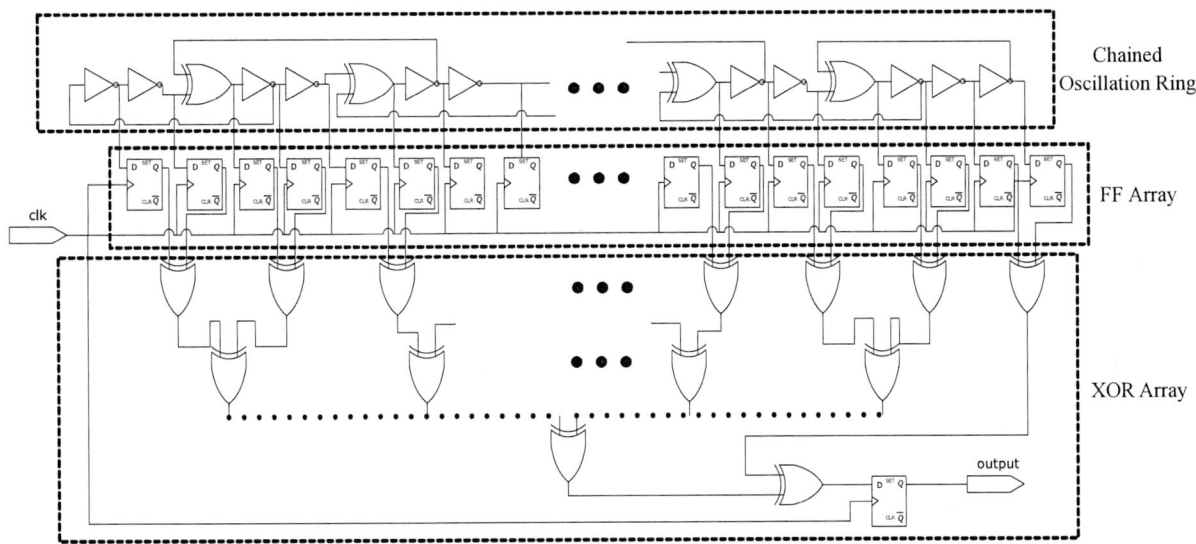

Fig. 2. The structure of proposed TRNG design.

shaded region represents the jitter range, where T_1 and T_2 represents two adjacent sampling points. If T_1 is known, T_2 can be easily obtained as shown in Fig. 4(a). If T_2 falls into the jitter range as shown in Fig. 4(b), the value of X_3 is determined by $\Delta d_{1,j+1} + \Delta d_{2,j+2} + \Delta d_{3,j+3}$. [15] has shown that increasing the time interval between T_1 and T_2 to accumulate more jitter can improve its randomness.

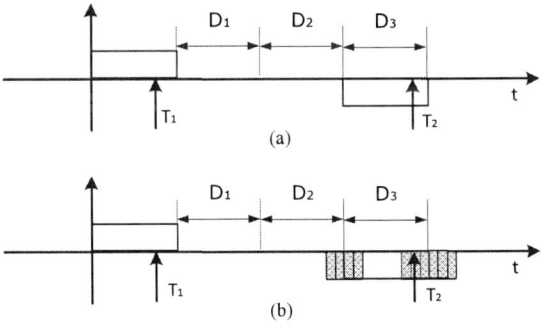

Fig. 4. The waveform of X_3 (a) when there is no jitter, and (b) when there is any jitter.

The entropy of random number can be increased by increasing the jitter range. This design adopts the COR, which can be used to connect the adjacent RO by XORs, to increase the jitter range. The COR in four oscillated rings is shown in Fig. 5.

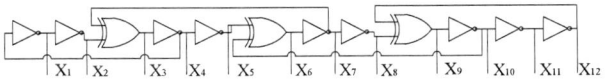

Fig. 5. The COR with 4 nested ROs.

Consistent with the above analysis, it is assumed that the signal is known in the j-th half clock.

$$X_4 = 0, X_5 = 1, X_{10} = 0. \qquad (6)$$

After a clock cycle, the waveform of X_2 is shown in Fig. 6 (a). The waveform of X_7 is shown in Fig. 6(b). After the half-clock cycle, the waveform of the XOR is shown in Fig. 6(c). Due to the phase difference between X_4 and X_5, it can be seen from the Fig. 6(c) that the jitter regions in the two signals (Fig. 6(a) and Fig. 6(b)) are superimposed. Its jitter region is larger compared with one OR. So the time of cumulative jitter can be shortened by different or gate.

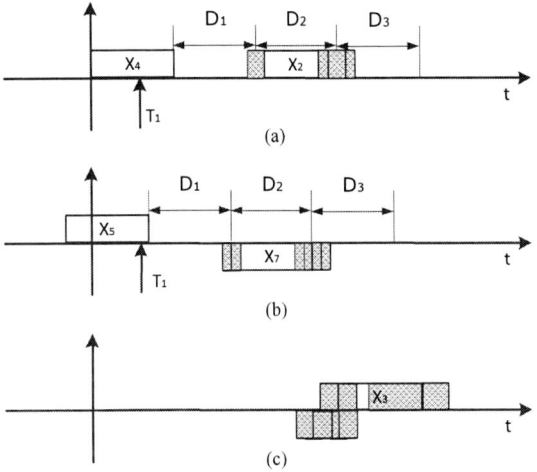

Fig. 6. The waveform of (a) X_2 after a clock cycle, (b) X_7 after a clock cycle, and (c) the XOR gate.

978-1-5386-7100-9/18 $31.00 © 2018 IEEE

D. The Metastability of the TRNG

Assuming that the width of one edge collision is ε. As long as the time difference between the edge of the sampling clock and the edge of the input signal is less than ε, it can be considered that the edge collision occurs as shown in Fig. 7. The probability of the metastability at the i-th flip flop can be expressed as

$$P_m(i) = 2\varepsilon f \tag{7}$$

where, f represents the frequency of signal.

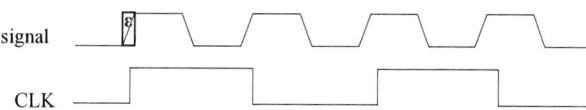

Fig. 7. The metastability state during sampling.

Assume that there is no metastability. The FF array is sampled under the control of the same sampling clock and there are k triggers. When sampling on the i-th trigger, the probability of getting a data of '1' is p_{oi1}. So the probability that the sample gets '0' is $1 - p_{oi1}$. According to the definition of information entropy, the entropy can be obtained:

$$H(i) = -p_{oi1} \times \log_2 p_{oi1} - (1 - p_{oi1}) \times \log_2(1 - p_{oi1}) \tag{8}$$

As metastability exists, it can assume that the probability of getting '1' is p_{oi1} on the i-th trigger when the metastability occur. Because of metastability, the probability of changing '1' to '0' is $p_{oi1} \times (1 - p_{mi1}) \times p_m(i)$ and the probability of changing '0' to '1' is $(1 - p_{oi1}) \times p_{mi1} \times p_m(i)$. So the probability of sampling '1' can be expressed as

$$p_{i1} = p_{oi1} + (1 - p_{oi1}) \times p_{mi1} \times p_m(i) - p_{oi1} \times (1 - p_{mi1}) \times p_m(i) \tag{9}$$

The probability of sampling '0' is

$$p_{i0} = 1 - p_{oi1} + (p_{oi1} \times (1 - p_{mi1}) - (1 - p_{oi1}) \times p_{mi1}) \times p_m(i) \tag{10}$$

In order to simplify the model, it can be assumed that $p_{mi1} = \frac{1}{2}$. So p_{i1} and p_{i0} can be expressed as:

$$p_{i1} = p_{oi1} + \left(\frac{1}{2} - p_{oi1}\right) \times p_m(i) \tag{11}$$

$$p_{i0} = (1 - p_{oi1}) + \left(p_{oi1} - \frac{1}{2}\right) \times p_m(i) \tag{12}$$

The entropy is calculated according to p_{i1} and p_{i0}

$$H(i) = -p_{i1} \times \log_2 p_{i1} - p_{i0} \times \log_2 p_{i0} \tag{13}$$

When p_{i1} is closer to $\frac{1}{2}$, its entropy is closer to '1', and the randomness of the sampling is higher. Assume $\frac{1}{2} < p_{oi1}$, we have

$$\left(\frac{1}{2} - p_{oi1}\right) < \left(\frac{1}{2} - p_{oi1}\right) \times p_m(i) < 0 \tag{14}$$

$$\frac{1}{2} < p_{i1} < p_{oi1} \tag{15}$$

It can be seen from Eq. (15) that p_{i1} is closer to $\frac{1}{2}$ compared with p_{oi1}. Therefore, the metastability of FF array can increase the entropy of the random number.

Since each phase of the inverter has a different delay, the phase of the signal after each inverter is different. [15] has shown that the jitter obeys the normal distribution. Therefore phase difference distribution of the signal after each inverter can be considered as independent. Assuming that the expectation of every signal sampled from flip-flop is u. The expected value of XOR of all these bits is as follows [3]:

$$E = \frac{1}{2} + (-2)^{K-1} \times \left(u - \frac{1}{2}\right)^K = \frac{1}{2} \times (1 + (-2\varepsilon)^K) \tag{16}$$

where, $\varepsilon = u - \frac{1}{2}$. Because $u \in (0, 1)$,

$$|2\varepsilon| < 1 \tag{17}$$

The larger the K is, E is closer to $\frac{1}{2}$. The probability of '1' and '0' are represented by P_1 and P_0, respectively.

$$E = 0 \times p_0 + 1 \times p_1 = p_1 \tag{18}$$

So

$$p_1 = \frac{1}{2} \times (1 + (-2\varepsilon)^K) \tag{19}$$

The entropy of the TRNG can be expressed as

$$H(i) = \frac{1}{2} - \frac{1}{2} \times (1 + (-2\varepsilon)^K) \times \log_2(1 + (-2\varepsilon)^K)$$
$$- \frac{1}{2} \times (1 - (-2\varepsilon)^K) \times \log_2(1 - (-2\varepsilon)^K) \tag{20}$$

As K increases, E approches to $\frac{1}{2}$, p_1 also approches to $\frac{1}{2}$ and entropy tends to be '1'. Therefore, larger K leads to larger entropy of the TRNG output.

E. The Characteristic of COR

The COR uses XOR gates nest with multiple ring oscillators (ROs) to achieve mixed oscillation signals. The schematic diagram of a simple COR is shown in Fig. 8. The COR can be activated when the number of nested ROs is equal to $3 \times N + 1$ ($N = 0, 1, 2...$). For example, the number of nested ROs, equal to 2, cannot oscillate and the intermediate signals are fixed to states 010101, as shown in Fig. 8.

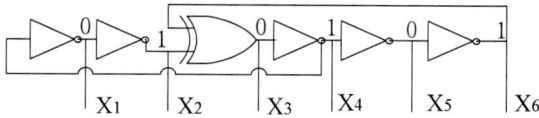

Fig. 8. An example of COR with two nested ROs.

When $N = 1$, the number of nested ROs is 4 and the COR can be oscillated as shown in Fig. 5. The processes of oscillation is as follows. Assume that the COR can be settled down to a stable state. Then X_7 and X_8 must be '1' and '0', respectively, to activate the last RO. Therefore, X_{10} and X_6 are in a metastable state which results in X_7 changed accordingly. Finally, the COR remains oscillating.

IV. EXPERIMENT

To evaluate the effects of the FF array on the entropy of the proposed TRNG, the proposed TRNG is implemented on FPGA with and without the FF array. Then the proposed TRNG is tested with both NIST SP800-22 and AIS31 standards. A comparison with previous works is also provided in this section.

A. Verification on FPGA

To verify the random resources, we implement the proposed TRNG with and without the FF array, respectively, and compared the two designs. The results obtained at the sampling frequency of 160MHz are shown in the Table I after the NIST test. This test results confirm that the FF array can improve the entropy of the proposed TRNG.

Table I: NIST Test Results between TRNG with and without FF Array.

Test Item	without FF Array		with FF Array	
	P-value	Proportion	P-value	Proportion
Frequency	0.000000	0.00	0.474986	0.97
Block Frequency	0.000000	0.00	0.514124	0.99
Cumulative Sums	0.000000	0.00	0.616305	0.99
Runs	0.000000	0.00	0.924076	0.99
Longest Run of Ones	0.000000	0.44	0.955835	0.98
Rank	0.935716	0.99	0.494392	0.99
Discrete Fourier Transform	0.000406	0.92	0.971699	1
Nonperiodic Template Matchings*	0.001455	0.17	0.495490	0.99
Overlapping Template Matchings	0.000000	0.00	0.366918	0.99
Universal Excursions	0.000000	0.64	0.739918	0.98
Approximate Entropy	0.000000	0.00	0.319084	0.99
Random Excursion*	–	–	0.131175	0.99
Random Excursions Variant*	–	–	0.295078	0.99
Serial*	0.366918	1	0.243884	1
Linear Complexity	0.867692	0.99	0.102526	1

B. NIST SP800-22

The NIST SP800-22 [12] includes 16 tests. In order to ensure the reliability of the test results, we set the length of each test sequence as 10^6 bits. Then the number of data required for the tests is 3×10^8 bits. We collected 300 sets of random number from the proposed TRNG design under the clock frequency of 160MHz. The test results are shown in Table II. The proposed TRNG design passes the test when P-value is larger than 0.01 and the pass ratio is larger than 0.95. We can see that the random sequences that generated by the proposed TRNG design have passed all the NIST SP800-22 test items.

Table II: The Testing Result of NIST SP800-22.

Test Item	P-value	Proportion
Frequency	0.096578	0.98
Block Frequency	0.779188	1
Cumulative Sums	0.627038	0.985
Runs	0.350485	0.99
Longest Run of Ones	0.834308	0.99
Rank	0.202268	0.99
Discrete Fourier Transform	0.554420	0.99
Nonperiodic Template Matchings*	0.495490	0.99
Overlapping Template Matchings	0.574903	0.96
Universal Excursions	0.171867	0.99
Approximate Entropy	0.383827	0.99
Random Excursion*	0.295663	0.99
Random Excursions Variant*	0.295078	0.99
Serial*	0.515420	0.995
Linear Complexity	0.534146	0.99

C. AIS31

The AIS31 standard [13] includes two functional stages: P1 (T0~T5) and P2 (T0~T8). P1 is used to test the output of the post-processing part of the TRNG, while P2 is used to test the output of the noise source.

Three data sets are collected under the clock frequency of 160MHz. The test results are shown in Table III, and the item without * in the table is the pass rate. It can be seen that all the data passed the test items included in AIS31.

Hence, the random sequences generated by the proposed TRNG design have passed two main test standards, i.e., NIST SP800-22 and AIS31.

D. Analysis and Comparison

The proposed TRNG design is implemented on an Xilinx Aritx-7 FPGA. The hardware resource consumption is

Table III: The Testing Result of AIS31

Data	RS 1	RS 2	RS 3
disjointness test (T0)*	Pass	Pass	Pass
monobit tests (T1)	100	100	100
poker tests (T2)	100	100	100
run tests (T3)	100	100	100
long run test (T4)	100	100	100
autocorrelation test (T5)	100	100	100
uniform distribution test (T6)*	Pass	Pass	Pass
multinomial distributions (T7)*	Pass	Pass	Pass
entropy test (T8)*	Pass	Pass	Pass

shown in Table IV. The proposed design, only consumes 37 FFs and 25 LUTs, which uses approximate 0.1% of the overall resources. The throughput of the proposed digital TRNG is 160 Mbit/s. The proposed TRNG design is compared with conventional TRNGs in terms of speed, cost and source of randomness in Table V. The comparison of the hardware consumption is performed by converting the hardware consumption to the equivalent number of inverters and FFs. It can be seen that the proposed TRNG achieves higher speed and lower resource consumption than the previous works from [3][7][9][14].

Table IV: The Hardware Resource Consumption in FPGA.

Resource	Utilization	Available	Utilization %
FF	37	41600	0.09
LUT	25	20800	0.12
I/O	3	106	2.38
BUFG	1	32	3.12

Table V: The Comparison With Previous Designs.

Designs	[3]	[7]	[9]	[14]	This work
Speed(Mbps)	100	12.5	6.25	4	160
Resource (gate, FF)	(150, 50)	(256, -)	—	(128, 48)	(103, 37)
Source of Randomness	Gate Delay Instability	Metastability	PLL	self-timed rings	Gate Delay Instability & Metastability

V. CONCLUSION

In this paper, a new lightweight and high-speed TRNG is proposed by using the XOR gates nested multiple ROs to achieve mixed oscillation signals. Multi-group mixed oscillation signal with XOR operation is applied to obtain high-speed outputs. Furthermore, the proposed TRNG design exploits the metastablility of FFs as a random source to efficiently improve the randomness of output sequences. The proposed TRNG design is implemented on an Xilinx Artix-7 XC7A35T FPGA, which can achieve a throughout of 160 Mbit/s. The generated random number set has passed the commonly employed testing standards, *i.e.*, NIST SP800-22 and AIS31. The proposed TRNG achieves low-cost hardware resource usage, high speed performance and flexible implementation over conventional digital TRNGs.

REFERENCES

[1] W. Schindler, Random Number Generators for Cryptographic Applications, Springer, 2009, pp. 5-23.
[2] B. Sunar, True Random Number Generators for Cryptography, Springer, 2009, pp. 55-73.
[3] K. Wold and C. H. Tan, "Analysis and enhancement of random number generator in FPGA based on oscillator rings," *Proc. Int. Conf. Reconfigurable Computing & FPGAs*, 2009, pp. 385-390.
[4] R. C. Fairfield, R. L. Mortenson, and K. B. Coulthart, "An LSI random number generator (RNG)," *Proc. CRYPTO*, 1985, pp. 203-230.
[5] D. J. Kinniment, and E. G. Chester, "Design of an on-chip random number generator using metastability," *Proc. European Solid-State Circuits Conf*, 2002, pp. 595-598.
[6] D. J. Kinniment and E. G. Chester, "A provably secure true random number generator with built-in tolerance to active attacks," *IEEE Trans. Computers*, vol. 56, pp. 109-119, 2006.
[7] H. Hata and S. Ichikawa, "FPGA implementation of metastability-based true random number generator," *IEICE Trans. Information & Systems,* vol. 95-D, pp. 426-436, 2012.
[8] F. Lozach, M. Ben-Romdhane and T. Graba, "FPGA design of an open-loop true random number generator," *Proc. Euromicro Conf. Digital System Design*, 2013, pp. 615-622.
[9] N. Deàk, T. Györfi, K. Màrton, L. Vacariu, and O. Cret, "Highly efficient true random number generator in FPGA devices using phase-locked loops," *Proc. Int. Conf. Control Systems & Computer Science*, pp. 453-458, 2015.
[10] D. Chen D, D. Singh and J. Chromczak, "A comprehensive approach to modeling, characterizing and optimizing for metastability in FPGAs," *Proc. ACM Int. Symp. FPGAs*, 2010, pp. 167-176.
[11] Viktor Fischer, A Closer Look at Security in Random Number Generators Design, *Constructive Side-Channel Analysis and Secure Design. Springer Berlin Heidelberg*, 2012, pp. 167-182.
[12] T. Lange, D. Lubicz and A. Weigl, "Random numbers generation and testing," *Handbook of Elliptic & Hyperelliptic Curve Cryptography*, pp. 715-735, 2005.
[13] W. Killmann, A proposal for: Functionality classes and evaluation methodology for true (physical) random number generators, Version 3.1, 2001, www.bsi.bund.de/zertifiz/zert/interpr/trngk31e.pdf.
[14] H. Martin, P. Peris-Lopez, JE. Tapiador and ES. Millan, "A new TRNG based on coherent sampling with self-timed rings," *IEEE Transactions on Industrial Informatics*, pp. 91-100, 2016.
[15] Fischer, V and Bernard, F and Bochard, N and Varchola, M, "Enhancing security of ring oscillator-based trng implemented in FPGA," *Field Programmable Logic and Applications*, 2008, pp. 245-250.

2018 IEEE Computer Society Annual Symposium on VLSI

LUT-Lock: A Novel LUT-based Logic Obfuscation for FPGA-Bitstream and ASIC-Hardware Protection

Hadi Mardani Kamali, Kimia Zamiri Azar, Kris Gaj, Houman Homayoun and Avesta Sasan
Electrical and Computer Engineering Department
George Mason University, Fairfax, VA 22030
Email: {hmardani, kzamiria, kgaj, hhomayou, asasan}@gmu.edu

Abstract—In this work, we propose LUT-Lock, a novel Look-Up-Table-based netlist obfuscation algorithm, for protecting the intellectual property that is mapped to an FPGA bitstream or an ASIC netlist. We, first, illustrate the effectiveness of several key features that make the LUT-based obfuscation more resilient against SAT attacks and then we embed the proposed key features into our proposed LUT-Lock algorithm. We illustrate that LUT-Lock maximizes the resiliency of the LUT-based obfuscation against SAT attacks by forcing a near exponential increase in the execution time of a SAT solver with respect to the number of obfuscated gates. Hence, by adopting LUT-Lock algorithm, SAT attack execution time could be made unreasonably long by increasing the number of utilized LUTs.

Keywords-SAT attack; obfuscation; hardware security.

I. INTRODUCTION

Hardware security has become a major concern for both FPGA and ASIC solutions. For the ASIC solution, the problem of hardware security resides in using untrusted parties in the manufacturing supply chain for economically driven reasons. Due to the high cost of building, operating, managing, and maintaining state-of-the-art silicon manufacturing facilities, many major U.S. high-tech companies have been always fabless or went fabless in recent years [2], which has led them to adopt to offshore fabrication. However, many offshore fabrication facilities are considered to be untrusted, and fabrication in untrusted fabs has introduced multiple forms of security threats into the supply chain including threats of overproduction, Trojan insertion, Reverse Engineering , IP theft, and counterfeiting [2].

On the other hand, FPGAs are inherently more secure for their post-silicon reconfigurability. However, the FPGA hardware security relies on protected and non-intruded mapping of the intended bitstream into FPGAs. In certain cases, it is difficult to protect the bitstream both during the initial configuration in untrusted third-party systems as well as during remote and in-field reconfiguration [3]. A successful attack may result in an unauthorized transfer of a bitstream to a third-party, reverse-engineering of the embedded netlist, injection of a hardware Trojan, and cloning or theft of embedded IPs [3][4]. Although high-end FPGAs are typically equipped with bitstream encryption, there are many cases where encryption alone is not enough [4]: (1) Not all FPGA families are equipped with implementations of cryptographic algorithms [5], [6], especially for small and low-energy FPGAs. (2) When the power and delay overhead of bitstream encryption process is not tolerable, a developer may choose not using encryption. (3) Many FPGA-based products, to support new services or to enhance the existing ones, require frequent updates which mostly accomplished remotely. Despite the first time safely programming, for in-field updates and a remote upgrade, the encrypted bitstream and the keys are vulnerable to leakage [3]. (4) After deployment, FPGAs are susceptible to physical attacks. The long-term in-field usage makes it possible for an attacker to extract the encryption keys via various side channel attack mechanisms [7]. So, it is essential to implement additional defensive measures to prevent the usability of a leaked bitstream. Such threats validates the need for implementing a strong obfuscation to hide the bitstream.

In this paper we propose LUT-Lock, which obfuscates a netlist while embedding several key features that make the obfuscation a hard problem for state of the art attacks with particular attention to Satisfiability (SAT) Attacks. To develop this defense mechanism, we have identified several key features that increase the difficulty of obfuscation for SAT attacks. We illustrate how by utilizing each feature during the obfuscation, the SAT problem becomes harder. We propose LUT-Lock algorithm which combines all features, providing the best defense against SAT attacks.

The rest of the paper is organized as follows: Section II provides background on the logic obfuscation, and the use of SAT solvers for deobfuscation. Section III justifies the use of LUTs for obfuscating ASIC and FPGA solutions. Section IV explains various LUT-based obfuscation sub-algorithms proposed in LUT-Lock and justifies their effectiveness. Section V describes our experimental setup. Section VI presents our experimental results and discusses our findings. And finally, section VII concludes the paper.

II. BACKGROUND

A. Obfuscation

Logic obfuscation is the process of hiding the functionality of a synthesized IP by building ambiguity by means of control and programmability into its netlist. Gate camouflaging and logic locking are two of the widely explored obfuscation schemes in ASICs [8][9][10]. The claim raised by such obfuscation scheme was that to break the obfuscation, the adversaries need to try a large number of inputs and key combinations to extract the correct key, whose time increases exponentially as the number of keys and inputs increases. Note that in ASIC solutions, the availability of scan chains (for DFT), allows an adversary to access combinational logic in each stage of a sequential circuit.

The strength of logic obfuscation was seriously challenged by attacks formulated using satisfiability solvers (SAT Attacks), which were able to break the prior methods of logic obfuscation within minutes [11][12]. The strength of this attack directed the attention of HW security researchers to architect harder obfuscation schemes that are more resilient to SAT attacks. SARLock [13], Anti-SAT [14], And-Tree-Insertion (ATI) [15], CamoPerturb [16], SFLL-HD^0 [17],

978-1-5386-7100-9/18 $31.00 © 2018 IEEE

405

and SRCLock [18] are some of the obfuscation approaches that were proposed for this purpose. However further research proved that some of these obfuscation techniques are prone to other types of attacks such as simple removal attack after identification of these blocks using Signal Probability Skew (SPS) attacks [19], and approximate-SAT attacks.

LUT-based obfuscation has been previously visited by few researchers. The work in [23] suggest using LUTs for obfuscation and provides several replacement strategies to secure a netlist. However, the proposed mapping algorithms are not resilient against SAT attacks, and are only evaluated in terms of power, performance and area (PPA) overhead, while the claim on the security of these schemes is made solely base on inability to readout the content of LUTs after reverse engineering. The work in [22] proposed a STT-LUT-based obfuscation with three different LUT placement algorithms. This work further focuses on PPA impact of their solution and illustrates that utilizing STT-based LUTs could reduce the PPA impact. However, the proposed solution does not consider its resiliency against SAT attack.

B. SAT Attack

Every obfuscated gate in a netlist could be represented by a *Key Programmable Gate* (KPG). A KPG, based on its key input, could be configured to take any of n different possible functionalities. In XOR and MUX based obfuscation, the XOR and MUX are already a key programmable cell, where the key input to an XOR gate or select input of a MUX are considered as key inputs. Other obfuscated gates could be easily transformed into a key programmable cell. For example, a camouflaged gate could be represented by a MUX, with each possible output column of a truth table taken as an input to the MUX and the select inputs of a MUX used as key inputs. In SAT attack, the obfuscated netlist is first updated by converting all obfuscated cell to KPG cells. Let us refer to this circuit by a *Key Programmable Circuit* (KPC). A SAT attack on an obfuscated circuit is an iterative process of finding an input and two key values K_1 and K_2 for which a KPC produces two different results, where one of them is the expected output, that could be verified by comparing it to the output of a functional circuit (*eval*). Such input is denoted as a Discriminating Input (DIP). The SAT solver then formulates an additional constraint that in the future iterations such that in addition to producing a different output for a new input, the two keys (K_1 and K_2) should also produce the same output for all previously found DIPs. This constraint makes sure that solver reduces the set of possible keys for a circuit in each iteration. The SAT solver exits when it can no longer find a new DIP. At this point, any key that produces the correct output for all previously found DIPs is the correct key [11][12].

III. LUT-BASED OBFUSCATION FOR ASIC AND FPGA

A. LUT-based obfuscation in FPGA

In an FPGA solutions the hardware resources are are fixed and is designed independent of a given netlist. Hence by nature, state of the art FPGAs provide a large pool of resources to be applicable to a wide range of applications, resulting in a large number of non-utilized LUTs after

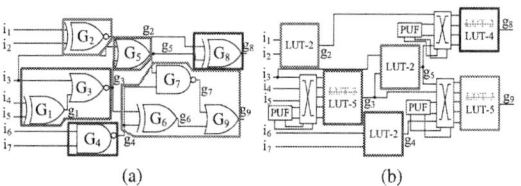

Figure 1: (a) sample circuit (b) using configurable switches + PUFs (NLFSRs) for employing larger LUTs after synthesis.

mapping a netlist to the FPGA. For instance, the study in [3] depicts the utilization of Altera Cyclone V after mapping a diverse set of benchmarks of various scale and complexity to this FPGA, and reported that FPGA utilization is typically low. This phenomenon was coined as FPGA-Dark-Silicon [3]. These unmapped and unutilized LUTs are freely available and could be used for obfuscating a to-be-mapped netlist. Hence, LUT-based obfuscation in FPGAs could be considered as utilizing unused LUTs, or using larger than needed LUTs, where the connectivity and impact of additional logic is controlled using keys. The process of using LUTs in FPGA for the purpose of logic obfuscation is illustrated in in Fig. 1(b), where some of 2-input (or 3-input) logic gates could be mapped to a LUT of larger size (e.g. size 4 or 5). Then, the additional inputs can be taken from the output of an internally implemented Non-Linear Feedback Shift Register (NLFSR) or a Physical Unclonable Function (PUF) [20]. In addition, by changing the ordering of inputs based on the key inputs (generated by PUF), the obfuscated circuit possibilities increases. Lets assume a PUF is used. In this case, each FPGA has a unique PUF response. By knowing the PUF response ahead of time, the bitstream will load the LUTs with proper values and will transmit the directives for connecting the known PUF outputs to the proper LUT inputs and switch box select lines. However, the PUF values will not be transmitted in the bitstream. This missing key values serve as the obfuscation key in LUT based obfuscation. Also note that the bitstream in this case is unique for each FPGA, as each FPGA has a unique PUF response. In this case, even if the bitstream is leaked, the PUF response remains unknown, making the problem similar to ASIC flow, where after reverse engineering the obfuscated netlist is available, but the keys are unknown.

B. LUT-based obfuscation in ASIC

In ASICs, utilizing LUTs for obfuscation can lead to the considerable area and delay overhead. In the CMOS implementation of LUTs, the area overhead of the memory elements in a LUT exponentially increases as a function of its input size. Hence, the imposed area overhead limits the number of LUTs that could replace regular gates in a netlist. In addition, the performance/delay requirements constrain the placement of LUTs in timing critical and near timing critical paths. However, with the introduction of STT and MTJ based LUTs [21][22] and the promise of integration of STT and MTJ/pMTJ-based LUTs into the same CMOS process, the area overhead of LUTs is expected to sharply reduce. Integration of CMOS and MTJ/STT devices makes it possible for a larger number of LUTs to be implemented given a fixed area overhead. Using LUTs for obfuscation in ASICs is straightforward: selected cells are removed

978-1-5386-7100-9/18 $31.00 © 2018 IEEE

and replaced by LUTs. The functionality of cell remains hidden to the manufacturer. LUTs are then programmed after fabrication in a trusted testing facility.

IV. PROPOSED LUT-LOCK OBFUSCATION ALGORITHM

Our proposed LUT-Lock algorithm combines several key features, each enhancing its ability to resist against SAT attacks. In this section, we first explain each key feature, and then propose the LUT-Lock algorithm that combines all features into a comprehensive solution. In the result section of this paper, we illustrate how by adding each key feature, the resiliency of obfuscated netlist against SAT attack increases, proving that the resiliency gained from adding this features are orthogonal to one another.

A. FIC: Focusing on the Fan-In Cone of minimum number of primary output

The first criteria for selection of candidate gates is derived from the observation that higher output corruption reduces resiliency of obfuscation solution against SAT attacks [13][14]. Hence, by mapping the LUTs such that it affects the minimum number of primary outputs (POs), the degree of output corruption reduces, increasing the strength of obfuscation against SAT attacks. To achieve this, we limit the LUT insertion to the fan-in cone of smallest possible set of primary outputs (best case being single output), and we refer this algorithm as FIC. Note that FIC LUT-replacement still corrupts other outputs, as the intersection of fan-in cones of different outputs is not empty. In addition, the number of gates in the intersection of fan-in cones increases as we move from outputs toward inputs. Hence the obfuscation should be designed to replace the closest cells to the selected output first. This could be achieved by means of a Breadth First Search (BFS). In order to avoid timing violation due to replacing a gate with LUT, we estimates the delay of all timing paths through a gate selected for replacement. If the estimated delay is more than predefined threshold (e.g. 10% delay overhead), the allowance of replacement for this candidate will be revoked, and next candidate will be checked for replacement. After replacing all gates in the current Fan-In Cone, a new primary output will be selected.

In FIC algorithm, the output pin(s) selected for obfuscation should meet two conditions: (1) Total Positive Slack (TPS) of all timing paths leading to that primary output(s) should be large. This is because replacing a gate with LUT incurs additional delay in every timing path that passes through that gate. Hence, we need available timing slack for replacement of faster logic gates with slower LUTs. (2) it must have a large fan-in cone size, giving us more candidate gates for replacement. Fig. 2(a) illustrates the FIC replacement strategy. Between the two outputs, i.e. g_8 and g_9, g_9 is not selected, as it contains the largest number of timing critical paths. When using BFS for gate selection, FIC selects gates $\{G_8 \text{ and } G_5\}$ or $\{G_8, G_5, G_2, \text{ and } G_4\}$ when its asked to replace 2 or 4 gates respectively. For large circuits, we define two coefficients (α and β) for prioritizing these two conditions to generate a cumulative weight which helps selecting the best candidate output. For this purpose, we normalize the TPS (into TPS*) and FIC (into FIC*) with respect to their maximum possible values in the given circuit. Then using $\alpha.TPS^* + \beta.FIC^*$, we obtain the cumulative weight for the FIC selection process.

B. HSC: Focusing on Higher Skew Gates in FIC

Our investigation on the hardness of many tested LUT placement strategies revealed that the cells with higher Signal Probability Skew (SPS) at their output are better candidates for obfuscation. The SPS at the output of a gate is defined as $|P_r(0) - P_r(1)|$, with $P_r(1)$ and $P_r(0)$ being the probability of having a 1 or 0 at the output of the gate respectively. The SPS of a gate is a measure of its controllability using primary inputs. The higher the SPS, the lower the controllability of the respective gate. Hence, selecting a high SPS output gate lowers the chances of SAT solver selecting an input that tests the output of that gate.

With this observation, the second step of our proposed algorithm is to modify the FIC to perform the gate selection based on its measure of gate's output (higher) skew probability. In this modified FIC algorithm, which is now referred to as HSC, the gate selection strategy is modified as follows: within the Fan-In cone of selected output(s) based on FIC, the replacement priority is given to gates with higher SPS; In HSC, when a gate is selected for obfuscation, its fan-in gates will be added to the list of gates that could be visited in the next search for gate replacement, and the gates with the highest SPS will be selected among all gates in the list. Similar to FIC algorithm, each gate replacement candidate should pass the timing check, otherwise ignored. HSC replacement flow is illustrated in Fig. 2(b). In the first invocation of HSC, fan-in cone of gate G_8, for satisfying the FIC requirements, is selected and is obfuscated. For the 2nd gate selection, HSC has three candidates G_2, G_5, and G_4. Based on the skew probability of wires, as illustrated in Fig. 2(b), G_4 with SPS of 0.5 is selected over G_5 and G_2 with SPS of 0.5 and *zero* respectively. For the 3rd gate selection, HSC appends the fan-in gates of G_4 (Here is primary inputs and will be ignored!) as candidate gates for replacement along with G_2 and G_5. Hence, among these 2 gates, G_5 is selected for having the higher SPS.

C. MFO-HSC: Focusing on gates with Minimum Fan-Out

As mentioned previously, lowering the output corruption increases the difficulty of the SAT attack [13][14]. Although we develop FIC in the first step, the probability of having a fan-in cone with no common gate with other fan-in cones is close to *zero*. Separating the fan-in cones of different outputs could be achieved by replicating common gate, however this will result in a large area overhead. In order to limit the primary output corruption without exploding the area, we introduce another sub-algorithm in which we give obfuscation priority to candidate gate with lowest fan-out. We refer to this gate selection strategy as MFO-HSC.

In MFO-HSC algorithm, a BFS search is first deployed (FIC), visiting all candidate gates at the current breadth, and gate(s) with a minimum number of fan-outs will be selected. Whenever a tie between two or more gates is observed, the gate with the highest SPS is selected. When a gate is obfuscated, its fan-in gates are added to the list of candidate

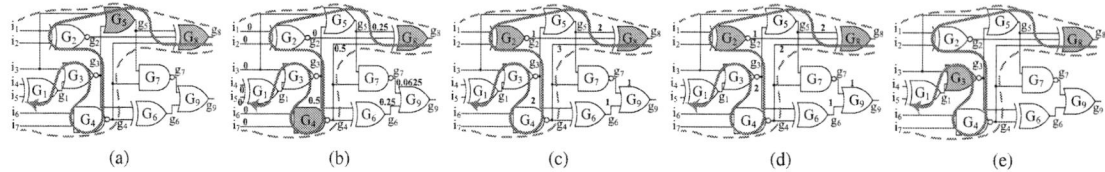

Figure 2: Gate selection process based on various sub-algorithms: (a) **FIC**: Focusing on Fan-In Cone of minimum number of primary outputs (b) **HSC**: Prioritizing higher skew probability gates in FIC (c) **MFO-HSC**: Prioritizing gates with minimum fan-out & HSC (d) **MO-HSC**: Prioritizing gates with minimum impact on outputs & HSC (e) **NB2-MO-HSC**: Avoiding back-to-back insertion of LUTs & MO-HSC.

gates that will be visited in the next gate selection. Similar to FIC, each gate replacement candidate should pass the timing check, otherwise ignored. Fig. 2(c) depicts how the MFO-HSC works; Similar to FIC, the fan-in cone of g_8 is selected for obfuscation and G_8 is obfuscated. Based on BFS, the next candidates are G_5, G_2, and G_4. The gate G_2 is selected over G_5 and G_4 for having fan-out of 1. The fan-in of G_2 is then added to the candidate gates for the next visit. In this figure, the fan-in of G_2 are primary inputs, and they are ignored, and the the next candidate gate is only G_5.

D. MO-HSC: Focusing on Gates with least impact on POs

Based on our observation in MFO-HSC, there are some gates that have more than one fan-out, but they only affect one output. For instance, as it can be seen Fig. 2(c), the fan-out of g_4 is 2. However, it affects only g_9. This observation led us to introduce a similar but more efficient sub-algorithm, which is called MO-HSC. In this sub-algorithm rather than looking at the fan-out of the candidate gates, we count the number of outputs that are connected to each candidate gate. MO-HSC requires additional parsing and processing, however it further reduces the output corruption as a result of obfuscation. Similar to MFO-HSC, the tie between two candidate gates (for affecting an equal number of outputs) is broken using SPS of respective gates. Each time a gate is selected for its obfuscation, the fan-in of the gate is added to the list of candidate gates to be considered for the next gate selection. Similar to FIC algorithm, each gate replacement candidate should pass the timing check, otherwise ignored. MO-HSC is illustrated in Fig. 2(d), where after selecting the G_8 based on FIC selection policy, the gate G_2 is selected over G_5 and G_4 for impacting smaller number of outputs.

E. NB2-MO-HSC: Avoiding Back-to-Back insertion of LUTs

The back-to-back obfuscation of gates with LUTs suffers from the increased number of key-possibilities as a result of the provided freedom in exploiting gate conversion based on De Morgans's Laws. For instance, as it can be seen in Fig. 3, the back-to-back obfuscation of the function $(A \lor B) \land (C \lor D)$, using 2-input LUTs, could have 4 different combinations of programmable logic based on De Morgans's Laws. So, the number of correct keys from the intended 1 increases to 4. Each additional gate obfuscated in the fan-in of this logic cone, creates another set of possibilities after application of De Morgans's law, leading to exponential increase in the number of valid keys, a phenomenon that we refer to as *correct key explosion*. Depending on the growth rate of the set of valid-keys and the number of keys, obfuscating more gate may even reduce the obfuscation strength. This is illustrated in Fig. 4 where execution time

of a SAT solver, and a number of generated keys per each inserted LUT for the benchmark C5315 of ISCAS-85 is plotted. The LUTs are placed back-to-back, hence, insertion of each LUT increases the number of keys. The plot focuses on the insertion of 38th to the 45th LUT. The insertion of 41st and 42nd LUT, produces a large number of new keys (around 10^4) based on De Morgan gate conversion possibilities. Hence, the SAT solver execution time doesn't increase. On the other hand, replacement of gate 40 produces far less number of new keys (in range of 10s). Hence the growth of the set of candidate/possible keys exceeds the growth rate of correct keys, significantly increasing the run-time of SAT solver. From this key observation, we need to suppress the growth-rate of correct keys from exploitation of De Morgan's gate conversion laws. So, we introduce another algorithm, NB2-MO-HSC, which implements this restriction by avoiding back-to-back obfuscated, keeping the set of correct keys at a minimum. In this gate replacement strategy, we first select the candidates in FIC using no back-to-back constraint. Then, the selection among the candidates is made based on candidate gate's connectivity to the minimum number of outputs. If there is a tie among candidates, the SPS of candidate gates determines the selection. As soon as a gate is selected, the NB2-MO-HSC searches the fan-in of the selected gate, skips one logic level (no back to back), and adds the fan-in of all skipped gates to the set of candidate gates for the next gate selection. Similar to FIC, each gate replacement candidate should pass the timing check, otherwise ignored. As illustrated in Fig. 2(e), the application of NB2-MO-HSC results in the selection of G_8 and G_3 as first two gates to be obfuscated.

The Algorithm 1 captures the detail implementation of the proposed Lut-Lock obfuscation flow implementing the NB2-MO-HSC policy. As mentioned previously, the overall structures of MFO-HSC and MO-HSC are the same, and since MO-HSC provides slightly more resilient behaviour and also more possible candidates during each iteration, we embed MO-HSC in the proposed LUT-Lock algorithm.

V. EXPERIMENTAL SETUP

For benchmarking the proposed LUT-Lock algorithm, we used a farm of desktops equipped with Intel Core-i5 processor and 8GB of RAM. For a fair comparison, and to reduce

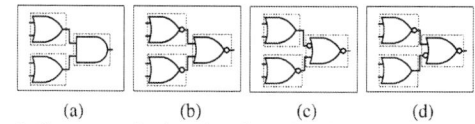

Figure 3: Gate conversion based on the application of De-Morgan's law (a) OR-AND (b) NOR-NOR (c) Custom1 (d) Custom2.

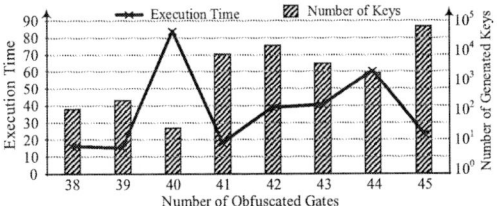

Figure 4: Increase in the number of valid keys in the result of of back-to-back insertion of LUTs and its impact on SAT attack execution time (C5315).

Algorithm 1 LUT-Lock: Implementing NB2-MO-HSC for LUT-based netlist obfuscation

```
 1: α = β = 0.5;                              ▷ α: TPS coeff, β: FIC size coeff;
 2: γ = 0.1                                   ▷ γ: feasible delay overhead
 3: max_delay_thr = γ × CriticalPath;
 4: MaxSize_FIC = Max_TPS = 0;                ▷ Total Positive Slack (TPS);
 5: Forbidden_output_list = []
 6: outputs_list = find_outputs(Circuit C);
 7: for each (output in outputs_list) do
 8:     if (output not in Forbidden_output_list) then
 9:         current_FIC = BFS(output);
10:         for all (paths in current_FIC) do
11:             Current_TPS = TPS_Calc(current_FIC, paths);
12:             Current_Weight = α × Current_TPS + β × sizeof(current_FIC)
13:             Max_Weight = α × Max_TPS + β × MaxSize_FIC
14:             if (Current_Weight > Max_Weight) then
15:                 candidate_output = output;
16:                 MaxSize_FIC = sizeof(BFS(candidate_output));
17:                 Max_TPS = Current_TPS;
18: candidate_list = Forbidden_list = [];
19: candidate_list.append(candidate_output);
20: while (num_of_obfuscated < target_no) do
21:     if (candidate_list == φ) then
22:         Forbidden_output_list.append(candidate_output)
23:         go to line 5
24:     else
25:         current_candidate = candidate_list[0];
26:         if (delay_estimate(current_candidate) < max_delay_thr) then
27:             replace_LUT(current_candidate);
28:             current_candidate_childlist = current_candidate.child;
29:             Forbidden_list.append(current_candidate.childlist);
30:             for each (current_child in current_candidate_childlist) do
31:                 if (current_child.child not in Forbidden_list) then
32:                     candidate_list.append(current_child.child)
33:             sort_list(candidate_list, min_out_impact);
34:             for all (candidate_list_members with equal min_out_impact) do
35:                 sort_list(candidate_list, skew_probability);
36:         else
37:             remove current_candidate;
```

---

the impact of the operating system background processes, we dedicated one machine to each SAT solver at a time, and installed Ubuntu Server 16.04.3 LTS operating system in shell mode. We used the largest ISCAS-85 benchmarks (C2670, C3540, C5315, C6288, and C7552) to show the effectiveness of the proposed algorithm. We employed the Lingling-based SAT attack described and developed by [11]. We measured the SAT solver execution time by increasing the number of obfuscated gates from 1 to 200. A run-time limit of 1.1×10^4 seconds was set for the SAT solver.

VI. RESULTS AND DISCUSSION

In order to show the effectiveness of each key feature of the proposed algorithm, we compared the execution time of SAT solver on circuits which are obfuscated based on these sub-algorithms. We also compare the effectiveness of the proposed LUT-Lock with that of previous work in STT-LUT [22] and Reconfigurable barriers [23].

As illustrated in Fig. 5 the SAT solver's execution time increases as the replacement algorithm evolves from Random replacement to FIC to HSC to MFO-HSC to MO-HSC to

Table I: Average Execution Time of SAT attack across studied benchmarks obtained based on curve fitting, where x is the number of obfuscated gates.

	RND	FIC	HSC	MFO-HSC	MO-HSC	NB2-MO-HSC (LUT-Lock)
Exponential Regression (Ae^{Bx})	A = 0.2065 B = 0.8875	A = 38.769 B = 0.9961	A = 15.238 B = 1.217	A = 41.252 B = 1.316	A = 38.644 B = 1.339	**A = 0.352** **B = 3.518**

MB2-MO-HSC, illustrating the orthogonal improvement of added features in providing resiliency against SAT attacks. The LUT-Lock algorithm, implementing the NB2-MO-HSC replacement policy, combines all key features and provides a close to exponential increase in the execution time of SAT attack with respect to the number of obfuscated gates.

As illustrated in Fig. 5, the execution time of the SAT solver, although increases steadily, faces small variation. The variation in the execution time is the result of (1) random nature of SAT solver in selecting DIPs from run-to-run, and (2) rate of growth in the size of valid keys (as a result of gate conversion using the application of De Morgan's laws, as explained in section IV-E), compared to the rate of growth in the number of possible keys. A poor selection of candidates for obfuscation results in a faster growth in the number of valid keys, reducing the overall effectiveness of obfuscated netlist against the SAT attack. As illustrated, the LUT-Lock has the least variation, as it eliminates the explosion of the set of valid keys by preventing back-to-back gate obfuscation.

Table I captures the fitted function of execution time for different sub-algorithms and LUT-Lock, where x denote the number of obfuscated gates. As illustrated in this Table, the LUT-Lock (NB2-MO-HSC) poses an exceptionally more challenging SAT problem compare to other obfuscation scheme. Table II compare the execution time of SAT attack, across selected number of ISCAS 85 benchmarks, once obfuscated by random LUT insertion and once using LUT-Lock. As illustrated, despite random policy, the SAT execution time grows exponentially when LUT-Lock policy is adopted.

Figure 5 visualizes the growth in the execution time of SAT attack, for two of ISCAS-85 benchmarks obfuscated using various LUT replacement policies. Other benchmarks have similar behaviour and are omitted for lack of space. In addition to replacement policies discussed in this paper, the SAT resiliency of replacement policies in prior work, namely STT-LUT [22] and Reconfigurable barriers [23] are captured in this figure. From this figure, the SAT resiliency of prior work is close to that of random replacement, showing slow growth in SAT attack execution time with respect to the number of inserted gates, where the Lot-Lock replacement policy clearly shows a much faster exponential increase in difficulty. As illustrated, in both benchmarks, with only 20 replaced LUTs, the LUT-Lock obfuscated netlist is as resilient as the netlist produced by [22] and [23] replacement policy when using $10X$ (200 gates) the number of gates. And by increasing the number of gates, the SAT resiliency of LUT-Lock insertion policy still grows exponentially.

VII. CONCLUSIONS

We proposed the LUT-Lock for building SAT resilient obfuscation netlists, applicable to FPGA and ASIC designs. Our simulation results illustrated that focusing the obfuscation to impact the smallest number of primary output pins

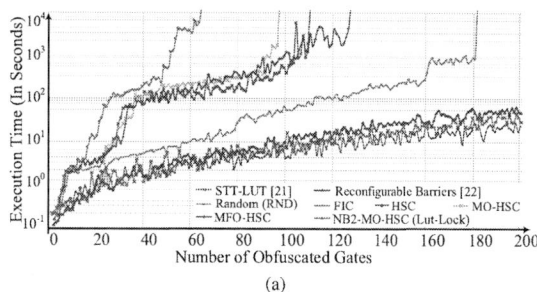

(a)

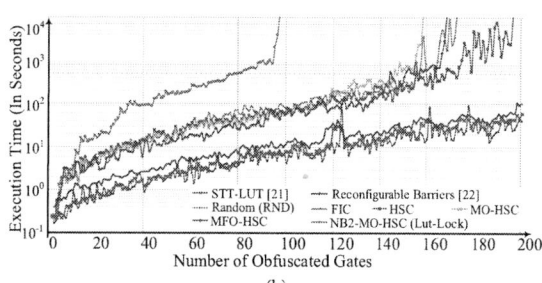

(b)

Figure 5: Execution time of the SAT solver from [11] for finding a valid key when using LUT-Lock (NB2-MO-HSC) compared to its sub-algorithms (RND, FIC, HSC, MO-HSC, and MFO-HSC) and the work in [22] and [23] on ISCAS-85 (a) c5315, (b) c7552 benchmark.

Table II: Average Execution Time of SAT Solver across studied benchmarks as a function of the percentage of obfuscated gates.

Circuits	1%		2%		3%		5%		10%	
	RND	Lut-Lock	RND	Lut-Lock	RND	Lut-Lock	RND	Lut-Lock	RND	Lut-Lock
c2670	0.18	0.876	0.5	1.388	0.93	1.924	2.41	24.64	3.48	time-out
c3540	0.6	1.244	1.07	6.12	2.25	988.2	2.66	time-out	5.29	time-out
c5315	0.5	9.052	1.21	115.012	1.66	941.02	3.93	time-out	12.04	time-out
c6288	0.57	23.508	2.14	1299.04	6.12	time-out	15.7	time-out	251.6	time-out
c7552	0.79	28.432	2.61	182.9	3.71	492.04	11.1	time-out	264.9	time-out

increases the obfuscation difficulty. This was achieved by means of selecting the fan-in of a minimum number of primary outputs for obfuscation and selecting gates that are connected to the smallest number of output pins. In addition, we illustrated that gates with lower controllability with respect to the primary inputs, as measured by signal probability skew at their output pin, are better candidates for obfuscation. Furthermore, we illustrated that avoiding back-to-back LUT replacement considerably reduces the number of valid key possibilities, increasing the resiliency of the proposed algorithm against SAT attacks. Compared to previous work, the LUT-Lock (NB2-MO-HSC) algorithm provides exponentially better protection against SAT attacks.

VIII. ACKNOWLEDGEMENT

This research is funded by the Defense Advanced Research Projects Agency (DARPA-AFRL, #FA8650-18-1-7819) of the USA.

REFERENCES

[1] A. Yeh, "Trends in the global IC design service market," *online* http://www.digitimes.com/news/a20120313RS400.html?chid=2, 2013

[2] U. Guin, D. Forte, and M. Tehranipoor, "Anti-counterfeit Techniques: From Design to Resign," in *14th Int'l Workshop on Microprocessor Test and Verification*, pp. 89-94, 2013.

[3] R. Karam, T. Hoque, S. Ray, M. Tehranipoor, and S. Bhunia, "Robust bitstream protection in FPGA-based systems through low-overhead obfuscation," in *Int'l Conf. on ReConFigurable Computing and FPGAs (ReConFig)*, pp. 1-8, 2016.

[4] R. S. Chakraborty *et al.*, "Hardware Trojan Insertion by Direct Modification of FPGA Configuration Bitstream," in *IEEE Design Test*, vol. 30, no. 2, pp. 45-54, 2013.

[5] S. ZamanZadeh, S. Shahabi, and A. Jahanian, "Security Improvement of FPGA Configuration File against the Reverse Engineering Attack," in *Int'l Iranian Soc. of Cryptology conf. on Info. Sec. and Crypt. (ISCISC)*, pp. 101-105, 2016.

[6] H. M. Kamali, S. Hessabi, "A Fault Tolerant Parallelism Approach for Implementing High-Throughput Pipelined Advanced Encryption Standard," in *Journal of Circuits, Systems and Computers (JCSC)*, vol. 25, no. 9, 1650113(1-14), 2016.

[7] N. Benhadjyoussef *et al.*, "Implementation of CPA analysis against AES design on FPGA," in *Int'l Conf. on Comm. and Information Technology (ICCIT)*, pp. 124-128, 2012.

[8] R. P. Cocchi, J. P. Baukus, L. W. Chow, and B. J. Wang, "Circuit camouflage integration for hardware IP protection," in *51st IEEE Design Automation Conf. (DAC)*, pp. 1-5, 2014.

[9] J. Zhang, "A Practical Logic Obfuscation Technique for Hardware Security," in *IEEE Trans. on Very Large Scale Integration (VLSI) Systems*, vol. 24, no. 3, pp. 1193-1197, 2016.

[10] J. Davis *et al.*, "Digital IP Protection using Threshold Voltage Control," in *Int'l Symp. IEEE Trans. on on Quality Electronic Design (ISQED)*, pp. 344-349, 2016.

[11] P. Subramanyan, S. Ray, and S. Malik, "Evaluating the security of logic encryption algorithms," in *IEEE Int'l Symp. on HW Oriented Sec. and Trust (HOST)*, pp. 137-143, 2015.

[12] M. El Massad, S. Garg, and M. V. Tripunitara, "Integrated Circuit (IC) Decamouflaging: Reverse Engineering Camouflaged ICs within Minutes," in *NDSS*, pp. 1-14, 2015.

[13] M. Yasin *et al.*, "SARLock: SAT attack resistant logic locking," in *IEEE Int'l Symp. on Hardware Oriented Security and Trust (HOST)*, pp. 236-241, 2016.

[14] Y. Xie, and A. Srivastava, "Mitigating sat attack on logic locking," in *Int'l Conf. on Cryptographic Hardware and Embedded Systems (CHES)*, pp. 127-146, 2016.

[15] Y. Xie, and A. Srivastava, "Provably secure camouflaging strategy for ic protection," in *IEEE TCAD*, pp. 1-1.

[16] M. Yasin *et al.*, "Camoperturb: secure ic camouflaging for minterm protection," in *Proc. of 35th Int'l Conf. on Computer-Aided Design*, pp. 127-146, 2016.

[17] M. Yasin *et al.* "Camoperturb: secure ic camouflaging for minterm protection," in *Proc. of ACM SIGSAC conf. on Computer and Communications Security*, pp. 1601-1618, 2017.

[18] S. Roshanisefat, H. M. Kamali, and A. Sasan, "SRCLock: SAT-Resistant Cyclic Logic Locking for Protecting the Hardware," in *Proc. of Great Lakes Symposium on VLSI (GLSVLSI)*, pp. 1-6, 2018.

[19] M. Yasin, B. Mazumdar, O. Sinanoglu, and J. Rajendran, "Security Analysis of Anti-SAT," in *22nd Asia and South Pacific Design Auto. conf. (ASP-DAC)*, pp. 342-347, 2017.

[20] G. E. Suh, and S. Devadas, "Physical Unclonable Functions for Device Authentication and Secret Key Generation," in *44th ACM/IEEE Design Automation conf.*, pp. 9-14, 2007.

[21] V. Kolla *et al.*, "Robust and energy efficient non-volatile reconfigurable logic circuits with hybrid CMOS-MTJs," in *3rd Int'l Conf. Emerging Electronics (ICEE)*, pp. 1-5, 2016.

[22] T. Winograd *et al.*, "Hybrid STT-CMOS designs for reverse-engineering prevention," in *53nd ACM/EDAC/IEEE Design Automation conf. (DAC)*, pp. 1-6, 2016.

[23] A. Baumgarten, A. Tyagi, and J. Zambreno, "Preventing IC Piracy using Reconfigurable Logic Barriers," in *IEEE Design & Test of Computers*, vo. 27, no. 1, pp. 66-75, 2010.

[24] J. Rajendran, Y. Pino, and O. Sinanoglu, and R. Karri, "Security Analysis of Logic Obfuscation," in *Proc. of 49th Annual Design Automation conf. (DAC)*, pp. 83-89, 2012.

978-1-5386-7100-9/18 $31.00 © 2018 IEEE

2018 IEEE Computer Society Annual Symposium on VLSI

ArtiFact: Architecture and CAD Flow for Efficient Formal Verification of SoC Security Policies

Atul Prasad Deb Nath, Swarup Bhunia and Sandip Ray
Department of Electrical and Computer Engineering
University of Florida, Gainesville, Florida 32608
Email: atulprasad@ufl.edu, swarup@ece.ufl.edu, sandip@ece.ufl.edu

Abstract— Verification of security policies represents one of the most critical, complex, and expensive steps of modern SoC design validation. SoC security policies are typically implemented as part of functional design flow, with a diverse set of protection mechanisms sprinkled across various IP blocks. An obvious upshot is that their verification requires comprehension and analysis of the entire system, representing a scalability bottleneck for verification tools. The scale and complexity of industrial SoC is far beyond the analysis capacity of state-of-the-art formal tools; even simulation-based security verification is severely limited in effectiveness because of the need to exercise subtle corner-cases across the entire system. We address this challenge by developing a novel security architecture that accounts for verification needs from the ground up. Our framework, ArtiFact, provides an alternative architecture for security policy implementation that exploits a flexible, centralized, infrastructure IP and enables scalable, streamlined verification of these policies. With our architecture, verification of system-level security policies reduces to analysis of this single IP and its interfaces, enabling off-the-shelf formal tools to successfully verify these policies. We introduce a CAD flow that supports both formal and dynamic (simulation-based) verification, and is built on top of such off-the-shelf tools. Our approach reduces verification time by over 62X and bug detection time by 34X for illustrative policies.

I. INTRODUCTION

Security assurance in System-on-Chip (SoC) designs is a highly critical area of research. Modern SoCs contain a variety of sensitive data (or "assets") that must be protected from unauthorized access. Such assets include private end-user information (*e.g.*, location, contacts etc.), cryptographic and DRM keys, proprietary firmware, and so on. Access to these assets and protection/mitigation requirements for unauthorized access attempts are governed by a collection of diverse *security policies*. It is of critical importance that the policies are implemented correctly. Indeed, a significant component of SoC security architecture entails developing techniques to enforce these policies; correspondingly, a significant component of security verification involves ensuring that the design correctly executes the policies [1], [2].

Unfortunately, verification of security policies is non-trivial in current industrial practice. The hardware logics responsible for protection of various assets are implemented as part of system integration in conflation with various functional and optimization constraints, with little attention paid to ease of verification. These logics are sprinkled across different hardware and software blocks, collectively referred to as "IPs". Consequently, *formal* verification of these logics

requires discovery, analysis, and comprehension of system invariants that often span across the entire SoC design. Furthermore, even *dynamic* (simulation-based) verification is hard, since scenarios exercising security policies involve long, directed execution of corner cases in specific configurations, system execution, and environmental stimuli [3]. Validation of security policies is often left to complex *penetration testing* by human experts, and is typically incomplete. Unsurprisingly, security vulnerabilities are discovered in-field, often with disastrous consequences [4], [5].

In this paper, we take the position that a security architecture that facilitates formal verification needs is a feasible solution to the problem. To that end, we present an architecture (and a corresponding CAD flow) for security policy enforcement that is amenable to scalable formal verification of SoC security policies. We demonstrate via diverse realistic policy implementations that our approach can result in over 62X speed-up (on average) in formal policy verification using state-of-the-art commercial tools. Complex security policies in SoCs that could not be verified at all with traditional implementations become amenable for efficient verification using our framework. Furthermore, our work facilitates short counterexamples in presence of bugs in policy.

Our architecture includes a centralized *security policy engine*, a dedicated IP for implementing SoC security policies. Each policy is implemented within this IP as a state machine defined through a rigorous CAD flow. Policy enforcement entails communication of the policy engine with other IPs in the SoC: this is performed by a standardized protocol. Centralized policy implementations have an important characteristic to facilitate scalable formal verification: a proof of correctness is typically confined to the policy engine and its interfaces and is oblivious to design invariants in any other IP blocks. Note that all other conditions remaining equal, complexity of formal verification is typically proportional to the size of the design block being analyzed [6]. Consequently, by enabling the target of formal analysis to be confined to a single IP, we enable significant scalability in verification of security policies over traditional distributed implementation.

Our work builds upon and extends previous work [7]–[9] on systematic, flexible architecture for SoC security policy implementations. Analogous to these works, our approach involves a centralized policy implementation framework. However, all previous works were focused on flexibility of implementations; verification was not considered. We show

978-1-5386-7100-9/18 $31.00 © 2018 IEEE

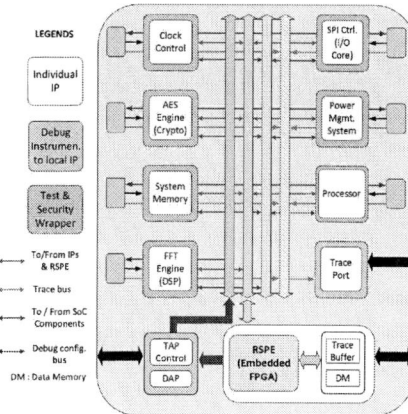

Fig. 1: Proposed Security Architecture : RSPE acts as a centralized flexible security policy engine to enforce policies.

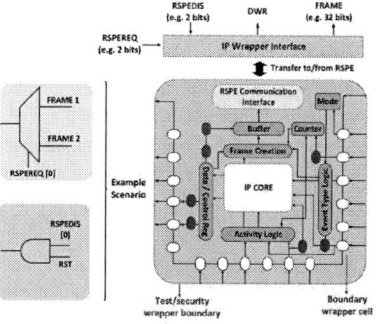

Fig. 2: Wrapper Architecture : security wrappers on individual IPs provide standardized, frame-based communication.

how to extend such frameworks with augmented CAD flows for effective verification methodology of security policies.

The paper makes three major contributions. First, we propose, for the first time to our knowledge, a *formal security verification flow* for security policies that directly influences and exploits architectural support for policy enforcement. Architectural support reduces the complex problem of formally verifying policy implementations to a simpler task of analyzing a single infrastructure IP, thereby providing verification scalability. Second, we demonstrate the efficacy of the framework in identifying and root-causing bugs. Our framework reduces root-causing efforts by providing short counterexamples enclosed to a single IP. This facilitates streamlining the debug flow significantly compared to directed testing and fuzzing approaches used in current industrial practice. Finally, we develop a comprehensive evaluation of both formal and dynamic aspects of verification on a wide diversity of realistic security policies implemented on an illustrative SoC model. The policies we consider include IP-specific as well as system-level security constraints.

II. BACKGROUND

A. SoC Security Policies

The goal of an SoC security policy is to map the security requirements to design constraints to develop protection

mechanisms. Following are two representative examples :

- *Example 1 :* During boot, data transmitted by the crypto engine cannot be observed by any IP in the SoC fabric other than its intended target.
- *Example 2 :* A secure key container can be updated for silicon validation but not after production.

The policies may vary depending on the state of execution (*e.g.*, boot time, normal execution), or position in the development life-cycle (*e.g.*, manufacturing, production). In addition to access control, security policies can capture requirements from information flow, liveness, etc.

B. SoC Security Architecture

Our work builds on a centralized SoC security architecture introduced in previous work [7]. It includes the following :

Reconfigurable Security Policy Engine (RSPE). This block acts as the *security brain* of the SoC. It receives communication of relevant security events from the security wrappers in IPs, identifies the security state, and enforces mitigatory actions based on the implemented policies.

Smart Security Wrappers. The idea for security wrappers is to enable IPs to communicate security-critical events to RSPE. The wrappers are programmable, so that they can be configured to monitor and control different sets of signals. RSPE configures the wrappers during boot time for monitoring signals necessary to enforce the security policies.

Interface with Design-for-Debug. RSPE is interfaced with the on-chip Design-for-Debug (DfD) interface. This interface provides access to an extensive set of observable and controllable signals inside IPs, which can be exploited for verification and re-purposed for realizing new policies.

III. PROBLEM ANALYSIS

A. Existing Challenges

Security policies in current practice are implemented by starting with a baseline architecture which is iteratively refined as follows :

- Use threat modeling to identify potential threats to the current architecture definition.
- Refine the architecture with mitigation strategies covering the threats identified.

The baseline architecture is derived from legacy SoC designs. For each asset, the architect must identify (1) who can access the asset, (2) what kind of access is permitted, and (3) at what points in the system life-cycle such access requests can be granted. The current industrial practice for verifying policies includes functional tests, fuzzing tests, and penetration tests. The scale of formal tools is limited to policies involving only one or a few IPs [3].

B. The Need for Architectural Support

Despite being an integral part of system development flow, the verification requirements of modern industrial designs are barely met by state of the art technologies. Formal methods provide an effective paradigm for security policy validation since they can provide a mathematical guarantee

978-1-5386-7100-9/18 $31.00 © 2018 IEEE

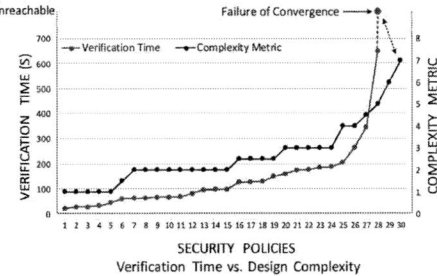

Fig. 3: An illustrative example of required verification effort with increasing design complexity and an eventual failure of traditional formal verification approaches.

of correctness of the policy enforcement which is unavailable from dynamic (fuzzing, functional simulation, and directed testing) techniques. However, it is imperative to develop architectural support that can make formal analysis of realistic security policies scalable : it is crucial for architectural support that ensures that invariants needed can be enclosed within a small block of logic. Our work facilitates this by basing our CAD flow on top of a centralized framework.

Furthermore, it is important to study the role of policy complexity on verification time. We developed a new metric of policy complexity based on empirical analysis. Fig. 3 shows the correlation between actual verification time and the proposed metric. The *complexity metric C_m* is defined as follows :

$$C_m = R_c + \sum_{i=1}^{n} \frac{E_i}{2} \left[\frac{O_i}{S_i} + \frac{C_i}{S_i} \right] \qquad (1)$$

Here, S_i is the number of security critical signal(s) involved in a policy, E_i is the number of security event(s) triggered by the policy, O_i and C_i is the number of observable signal(s) and controllable signal(s) of IPs involved in the corresponding policy, respectively. An additional complexity constant R_c is introduced by the RSPE in case of the proposed architecture. The soaring verification time (Fig. 3) with increasing complexity of policies signifies the limitation of conventional architecture to scale with verification needs.

IV. PROPOSED FRAMEWORK

A. Architectural Support

Our architecture builds on the centralized policy engine developed in previous work [7]. Transforming an architecture primarily developed for policy *implementation* (without verification concerns) to support effective formal verification is non-trivial. Here we list some architectural modifications that are crucial to the verification need.

Event Logging in RSPE : To facilitate the optimum event detection via centralized architecture, we supplement RSPE with augmented of event logging capability. The improved event logging is enabled by incorporating configuration register and special purpose registers for storing event trigger, transfer, and related meta data based on event type and requirement. The increment in security events logged by RSPE

reduces the complexity in security policy verification process by minimizing the reachable trace lengths for verification and violation paths. The centralized implementation of policies and corresponding security properties facilitates the tool to access the required signals in a shorter period of time.

Event Repository in Local DfD : A key criteria for the implementation and verification of arbitrary security policies of varying complexity is the detection of large number of security critical events in the SoC. System level security polices of higher complexity often require user-defined triggers and custom interrupts in inter-IP communications. We developed an augmented repository of security critical events by exploiting the configuration registers in local debug instrumentations. We used on-chip local debug modules with configuration registers and associated logic to map an extended number of security events for system level policies.

Enhanced Interconnect Fabric : We augmented the interconnect fabric of our SoC model for inter-IP communication by establishing a shared memory bus. To facilitate system level interaction between IPs, we mapped the control registers for each IP to specific addresses of system memory range and utilized the corresponding control signal interfaces to respond to incoming transactions from other functional IPs. In case of incoming interrupts and requests during active computation mode of an IP, a disable signal is instantiated by the policy engine for triggered events. For instance, all request and interrupts from rest of the IPs are invalidated when AES engine is in crypto mode. Consequently, any unauthorized access requests during crypto mode is logged into configuration registers as potential attempts of violation.

B. CAD Flow

We automatically synthesize policies into RSPE-based architecture. The policies are parsed as *action-predicate* tuples. The principle of pareto-optimality is employed in the synthesis procedure for energy optimum implementation. Fig. 4 illustrates the design flow. The flow introduces a precompilation stage, where security policies are parsed and a register-transfer level description is created for a control state machine that implements the action-predicate tuple ; this is integrated with an FPGA synthesis flow to create a reconfigurable policy implementation.

Given the above flow, formal property synthesis entails designing of a *monitor state machine* C_p for each policy P. The goal of the monitor state machine is to "watch" C_p and output 1 if p ever makes a deviation from its expected transitions and 0 otherwise. The formal property then reduces to the assertion that C_p never outputs 1. Note that in addition to its use in formal verification, C_p can also act as a runtime monitor for the assertion p : this is relevant in case of a policy p that cannot be completely verified (*e.g.*, if the correctness entails hardware/software co-execution and cannot be established from the hardware alone). In practice, we define C_p by augmenting the RTL design and used primarily for formal verification. If the verification succeeds the monitor C_p is no longer necessary, and can be safely removed : however, if the verification fails or is inconclusive,

TABLE I: System Level Security Policies Implemented on the Proposed Architecture.

Policy #	Predicate Tuple	Action Tuple	Corresponding IPs
1 : Read / Write operation of IPs within system memory range in user mode	(Mode : User) & (Memory read/write request by user or any other IP)	Read/write address within specified range	Any IP with access to system memory
2 : Read / Write operation of DLX uP to shared memory range in shared memory range	(Mode : Supervisor) & (Memory read/write request by user or any other IP)	Read/write address within shared memory range & No write	Any IP with access to system memory
3 : Interrupts (e.g. reset, immediate result, change of key etc.) from all IPs are prohibited during active crypto mode	(Mode : Active crypto) & (Access request by user or any other IP)	No interrupt or memory access request from any IP is allowed	Crypto module and any other IP with access to crypto core
4 : Read / Write access of IPs to round key registers are prohibited during active crypto mode	(Mode : Active crypto) & (Read/write request to round key registers by any IP)	No read/write access to round key registers by any IP is allowed	Crypto module and any other IP with access to crypto core
5 : Interrupts by Power management module (clock freq. change, reset, enable, go) during active computation	(Mode : Active computation) & (Interrupt or access request from PMC module)	No interrupt or access request from PMC module is allowed	PMC module and any other IPs accessible to PMC
6 : All IPs' access to interconnect fabric is prohibited during crypto key transfer	(Mode : key in transfer) & (Interrupt or access request by any IP)	No interrupt or access request from IPs to interconnect fabric is allowed	All IPs with privilege to access interconnect fabric
7 : Interrupts from all IPs are prohibited during μP core instruction memory update	(Mode : Supervisor) & (μP instruction memory update) & (Access requests)	No access request from any IP is allowed	μP core and any other IP

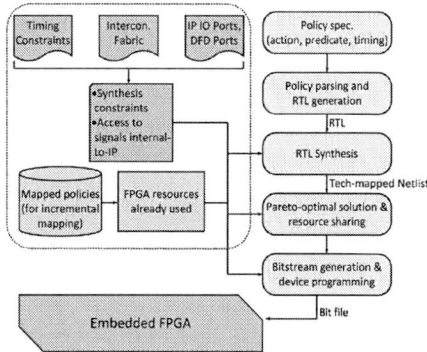

Fig. 4: CAD Flow for Mapping Security Policies on RSPE.

we keep the augmented RTL and connect the output of C_p to additional routines that can perform mitigatory action if the failure occurs runtime.

The steps for property mapping is quite straightforward : C_P can be synthesized mechanically from the state machine of P, and is well-established in current industrial practice. However, without centralized RSPE, there would be no systematic way for writing these assertions in a traditional SoC design. This further outlines the critical role of architectural support for security verification.

V. RESULTS

A. Experimental Setup

The SoC model includes a 32-bit pipelined DLX microprocessor core (DLX), a 32 KB central system memory, a standard memory controller IP, a 128b AES crypto core, a 128b FFT engine, a clock controller, a Serial Peripheral Interface (SPI) controller, and a power management unit. The IPs were obtained from Opencores (http://opencores.org). The security policies are mapped on an embedded FPGA-based RSPE that act as the execution engine. For experiments, we developed two versions of the SoC model, a baseline design and RSPE-based design. In the baseline SoC, each IP is augmented with standard boundary scan based wrappers (i.e. IEEE 1500) for detection of local events. Security policies in baseline model are implemented over the constituent IP cores in a distributed manner. In the RSPE-integrated model, we enhanced the IPs with smart security wrappers, and developed interface for *DfD* integration.

B. Formal Verification Results

Our formal verification results use off-the-shelf tool JasperGold [10]. We synthesized assertions in RSPE as discussed in Section IV-B. To compare efficacy, we implemented the same policies on a baseline SoC, paying specific attention to traditional performance optimization to reflect the current state of practice; assertions were also developed for this model. An Intel®Core™ i5-3427U CPU (1.8 GHz) with 8 GB memory is used to run verification on a linux server.

System-level Security Policies. We implemented 7 (P1 to P7) system-level security policies of varying complexity in our SoC model (cf. Table I). Table II summarizes the verification results for the system level policies implemented in base-line and RSPE-based design. All proven assertions in the table are "Infinite" bound type meaning the proofs are exhaustive and expected to hold true under all circumstances. We chose multi-engine environment with multi-property settings for optimum exploitation of the tool.

The increase in verification time with policy complexity is evident in the results presented in Table II. For instance, the interrupt handling policy (P#3) of AES during crypto mode require initiation of all possible incoming transactions coming from each of the IPs. The higher verification time, in this case, for DLX interrupts can be attributed to an increased number of security event association. Note that for illustrative system level policies, our approach reduces verification time by a maximum of 62X compared to base-line implementation.

IP Specific Security Policies : We implemented 8 representative IP-specific policies. Table III summarizes the verification report provided by the tool for baseline and proposed design. The verification of IP specific policies requires less effort in both baseline and RSPE based design due to the ease of observing and controlling involved signals, with corresponding low verification time.

C. Scalability Analysis

To demonstrate the scalability of our approach, we consider a case study of boot integrity check policies. These policies verify the trustworthiness of the system at power-on. The implementation of these policies in our model mandates checks for AES crypto engine's data path along with system boot processes (including power-on-self-tests, firmware

TABLE II: Proof of Correctness Results : System Level Security Policies

| Security Policies | IP Cores Involved | Formal Verification Results (Baseline vs RSPE) | | | | | | | | | Reduction (times) |
| | | Baseline | | | | RSPE | | | | | |
		JG Engine Mode	Proof Effort	Bound	Avg. Time (s)	JG Engine Mode	Proof Effort	Bound	Avg. Time (s)		
P #1	DLX up, AES, FFT, SPI	N	1-2	Infinite	27.246	Hp	1	Infinite	1.1435		23.827
P #2	AES, FFT, SPI, PMC	Ht	1	Infinite	64.429	N	1	Infinite	1.387		46.452
P #3	AES, FFT, SPI, PMC	Bm, Hp. I	1-13	Infinite	218.97	Ht, Hp, I	1-4	Infinite	10.824		20.23
P #4	DLX uP, PMC, FFT, SPI	Bm, Hp, Ht,	1-7	Infinite	126.2	N	1-3	Infinite	2.03		62.168
P #5	DLX Up, PMC, FFT, SPI	Bm, D, I,Hp	1-11	Infinite	303.3	I, U, Hp, Ht	2-7	Infinite	21.112		14.366
P #6	DLX uP, PMC, FFT, SPI	Hp, Ht,N	1-3	Infinite	62.958	Ht	1	Infinite	2.7868		22.592
P #7	AES, FFT, SPI, PMC	D, Bm, Hp,N	1	Infinite	142.9	N, Ht	1	Infinite	6.9305		20.619

TABLE III: Proof of Correctness Results : IP Specific Security Policies

| Comparative Results for IP specific Security Policy Implementation (Baseline vs RSPE) | | | | | | |
| Security Policies for DLX Up | Baseline | | | RSPE | | |
	JG Engine Mode	Bound	Time (s)	JG Engine Mode	Bound	Time (s)
DLX core instruction memory can only be updated at supervisor mode	Ht	Infinite	1.589	Hp	Infinite	0.695
DLX mode of operation cannot be unknown at startup	N	Infinite	0.615	N	Infinite	0.714
DLX power (high/low) mode of operation check at startup	N	Infinite	0.519	Ht	Infinite	0.936
External read write to DLX to only I/O mapped data memory region	N	Infinite	1.698	Ht	Infinite	1.796
Security Policies for AES Crypto Core	Baseline			RSPE		
	JG Engine Mode	Bound	Time (s)	JG Engine Mode	Bound	Time (s)
AES key cannot be unknown at startup	Hp	Infinite	0.875	Hp	Infinite	1.537
Key check against previous set (nonce/to prevent replay attack)	Hp	Infinite	1.328	N	Infinite	2.914
In crypto mode, cipher text output interface is disabled	Ht	Infinite	1.537	Hp	Infinite	1.271
AES power (high/low) mode of operation check at startup	Ht	Infinite	0.505	Hp	Infinite	0.713

TABLE IV: Results on Scalability Analysis on Baseline and RSPE-based Design

| | IP Cores Involved | Use Case Scenario : Comprehesive Policy Implementation | | | | | | | |
| | | Baseline | | | | RSPE | | | |
		JG Engine Mode	Result	Bound	Time (s)	JG Engine Mode	Result	Bound	Time (s)
Policies : Boot Integrity Check	DLX uP,	Ht	Undetermined	1021	31056.2	Ht	Proven	Infinite	12984.8
	AES, SPI,	D	Undetermined	872	9246.8	D	Proven	Infinite	4298.8
	FFT, PMC,	I	Undetermined	1543	22898.9	I	Proven	Infinite	15998.6
	Sys memory	Hp	Undetermined	987	18449.7	Hp	Proven	Infinite	11365.3

TABLE V: Results on Bug Detection in Security Policies for Baseline and RSPE-based Design

| Security Policies | IP Cores Involved | Funtional Verification Results (Baseline vs RSPE) | | | | Reduction (times) |
| | | Baseline | | RSPE | | |
		Detected Bugs	Time (s)	Detected Bugs	Time (s)	
P #1 - P#7	DLX Up, AES, FFT, SPI, PMC	N/A	>86400*	N/A	>86400*	N/A
Footnote : *The simulation run time was greater than 24 hours i.e. >86400s. No bugs were detected within the time limit.						

| Security Policies | IP Cores Involved | Formal Verification Results (Baseline vs RSPE) | | | | | | | | | Reduction (times) |
| | | Baseline | | | | RSPE | | | | | |
		JG Engine Mode	Proof Effort	Bound	Max Time (s)	JG Engine Mode	Proof Effort	Bound	Max Time (s)		
P #1	DLX up, AES, FFT, SPI	N	1-2	Infinite	32.675	Hp	1	Infinite	0.613		34.351
P #2	AES, FFT, SPI, PMC	Ht	1	Infinite	67.145	N	1	Infinite	1.573		15.686
P #3	AES, FFT, SPI, PMC	Bm, Hp. I	1-13	Infinite	345.489	Ht, Hp, I	1-4	Infinite	6.954		26.651
P #4	DLX uP, PMC, FFT, SPI	Bm, Hp, Ht,	1-7	Infinite	186.228	N	1-3	Infinite	1.915		30.155
P #5	DLX Up, PMC, FFT, SPI	Bm, D, I,Hp	1-11	Infinite	263.746	I, U, Hp, Ht	2-7	Infinite	13.573		14.602
P #6	DLX uP, PMC, FFT, SPI	Hp, Ht,N	1-3	Infinite	80.174	Ht	1	Infinite	1.688		25.973
P #7	AES, FFT, SPI, PMC	D, Bm, Hp,N	1	Infinite	185.259	N, Ht	1	Infinite	7.436		13.731

integrity check, and peripheral core integrity check). Table IV summarizes the verification report. For the baseline design, the verification engines of the tool failed to reach convergence for integrity check policies with multiple engine modes. However, the formal proof is completed in the RSPE-based implementation where the state space explosion phenomenon is avoided through the centralized implementation. This suggests that with architectural support, it is possible to address scalability limitations in verification and potentially formally verify complex system-level security policies.

D. Bug Detection

To evaluate the robustness of RSPE in bug detection, we injected a set of bugs in the system-level policies. The bugs were inserted with close interaction with industry and are representative of real security bugs detected in industrial environments. Furthermore, the selection aims to cover the spectrum of confidentiality, integrity, and availability requirements of assets in modern SoCs.

Access to Memory Bug. We considered a violation scenario where the state machine controlling the memory address register fails to detect the address range breach in the cur-

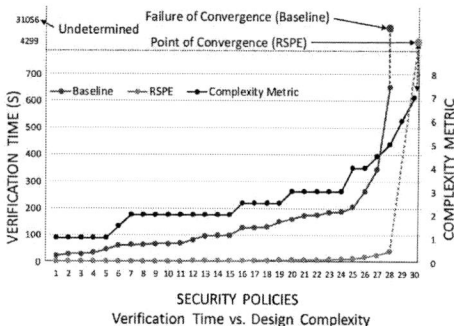

Fig. 5: An illustrative example of incremental verification effort and scalability of RSPE with design complexity.

rent/overlapping clock cycle. The bug, if goes unmitigated, can lead to unauthorized access of a malicious attacker or restricted IP to secure memory address range. The possible consequences of such breach include a violation of confidentiality and integrity of the assets of secure memory.

Active Crypto Mode Bug. In this violation scenario, the status of active crypto signal remains asserted throughout the crypto sequences and is not de-asserted once the operations are finished. The bug can hamper the secure flow of operation as the IPs are blocked from accessing the crypto assets after a crypto operation. The event leads to violation of availability property and consequent unavailability of assets.

Active Computation Mode Bug. We considered a violation scenario where the state machine controlling mode of operation of IPs gets stuck in the current state leading to functional failure. The bug is representative of the functional failure of the SoC due to a loss of availability. It directly affects the incoming transactions from power management module and causes stagnation in the flow of execution.

Results Analysis. We tested the policies with bugs in a simulation environment. The summary of *functional verification results* is illustrated in Table V. We employed constrained random testing via ModelSim for assertion based dynamic verification. Random functional testing failed to detect any of the bugs in reasonable time (>24 hours), which highlights the limitation of traditional security policy verification approaches in SoC designs. Table V also shows the summary of *formal verification results* for 7 system level security policies. The trace lengths of counterexamples in baseline design are significantly higher than the trace lengths of RSPE-based design. With RSPE, the engines of formal tool are able to find violation traces with minimal trace attempts leading to reduced trace lengths of counterexamples and improved verification time. Our approach reduces counter example detection time up to 34X compared to base-line design.

VI. RELATED WORK

Several formal methods have been proposed for the verification of security properties [11]–[13].The focus of these works are hardware security issues i.e. malicious hardware Trojans, side-channel attacks, etc. Their application, however, is limited by the failure to scale with design complexity. Though novel techniques have been proposed for improvement [14], state space explosion is still the major limitation of proving security properties in large SoC designs. Research efforts have been made to address SoC security and verification issues by developing scalable architectural frameworks. Infrastructure IPs are employed to facilitate SoC functional verification, testing, and yield improvement [2], [15]. However, these approaches lack scalable architectural features like centralized or flexible infrastructure IP, standardized interface with IP blocks, and systematic CAD flow.

VII. CONCLUSION

We have developed an architectural framework for efficient and scalable formal verification of complex security policies on SoC platforms. Our work, for the first time to our knowledge, marries two highly crucial but typically isolated components of security assurance, architecture and validation. We show how to develop an architecture that not only enables systematic policy implementation but also scalable analysis and formal verification. The experimental results on realistic SoC models and policies suggest that the approach can reduce verification time for system-level policies by orders of magnitude, help verification of arbitrary policies with varying complexity, and significantly aid the detection of bugs deeply rooted inside the design. Future work will involve enabling the architecture on industrial SoC models and silicon verification.

REFERENCES

[1] S. Krstic, J. Yang, D. W. Palmer, R. B. Osborne, and E. Talmor, "Security of SoC Firmware Load Protocol," in *HOST*, 2014.

[2] M. R. Sastry, I. T. Schoinas, and D. M. Cermak, "Method for enforcing resource access control in computer system," *US Patent 20120079590 A1*, 2012.

[3] S. Ray, E. Peeters, M. Tehranipoor, and S. Bhunia, "System-on-Chip Platform Security Assurance : Architecture and Validation," *Proceedings of the IEEE*, 2018.

[4] Homebrew Development Wiki, "JTAG-Hack," http://dev360.wikia.com/wiki/JTAG-Hack.

[5] L. Greenemeier, "iPhone Hacks Annoy AT&T but Are Unlikely to Bruise Apple," *Scientific American*, 2007.

[6] R. Kaivola, S. Pandav, A. Slobodova, C. Taylor, V. A. Frolov, E. Reeber, and A. Naik, "Replacing testing with formal verification in intel coretm i7 processor execution engine validation," in *CAV*, 2017.

[7] A. Basak, S. Bhunia, and S. Ray, "A Flexible Architecture for Systematic Implementation of SoC Security Policies," in *ICCAD*, 2015.

[8] A. P. D. Nath, S. Ray, A. Basak, and S. Bhunia, "An Architecture and CAD Flow for Hardware Patch," in *ASPDAC*, 2017.

[9] A. Basak, S. Bhunia, and S. Ray, "Exploiting Design-for-Debug for Flexible SoC Security Architecture," in *DAC*, 2016.

[10] "JasperGold : Formal Property Verification App," 2017, www.jasper-da.com/products.

[11] S. Drzevitzky, "Proof-carrying hardware : Runtime formal verification for secure dynamic reconfiguration," in *FPL*, 2010.

[12] Y. Jin and Y. Makris, "Proof carrying-based information flow tracking for data secrecy protection and hardware trust," in *VTS*, 2012.

[13] M. Rathmair and F. Schupfer, "Hardware trojan detection by specifying malicious circuit properties," in *ICEIEC*, 2013.

[14] X. Guo, R. G. Dutta, P. Mishra, and Y. Jin, "Scalable SoC Trust Verification using integrated theorem proving and model checking," in *HOST*, 2016.

[15] Y. Zorian, "Embedded memory test and repair : Infrastructure IP for SoC yield," in *ITC*, 2002.

Identifying Lithography Weak-points of Standard Cells with Partial Pattern Matching

Yongfu Li[+], *Senior Member, IEEE*, I-Lun Tseng, *Member, IEEE*, Zhao Chuan Lee, Valerio Perez, Vikas Tripathi, and
Jonathan Yoong Seang Ong

GLOBALFOUNDRIES Singapore Pte. Ltd.,

[+]email: yongfu.li@globalfoundries.com

Abstract—At advanced process nodes, lithography weak-points can act as major factors of yield losses in manufactured integrated circuits, especially under aggressive design rules. Thus, it is desirable to consider potential lithography weak-point issues during the phase of designing standard cells in order to improve manufacturability of integrated circuits. In this paper, we propose a partial pattern matching methodology, which is based on the use of a combinatorial K-partitioning technique, to identify all of potential lithography weak-points for standard cells in a given standard cell library. In addition, the proposed methodology adopts a pruning technique to minimize false violations, and uses a prioritization technique to prioritize tasks of modifying and/or redesigning standard cells. Compared with a conventional placement-and-routing based methodology, our experimental results show that the proposed methodology can accurately detect more potential lithography weak-points.

Index Terms—Nanolithography, process variations, digital design, standard cell design, cell characterization methodology, pattern matching, process hotspot detection, scoring mechanism

I. INTRODUCTION

AS semiconductor manufacturing technology continues to advance without the use of extreme ultraviolet lithography, the limitation of 193i optical lithography tools has caused the printing of small geometries to suffer from serious degradations in terms of optical resolution. Under aggressive design rules, the conventional rule-based design rule checking (DRC) approach is no longer sufficient to guarantee 100% pattern printability. Thus, design-for-manufacturability (DFM) compliance checking is required to identify manufacturing weak-points and to prevent catastrophic errors, such as open (necking) and short (bridging) issues, from occurring in manufactured integrated circuits. Moreover, with the addition of extra mandatory physical verifications in each new technology node, the cost of designing integrated circuits has become very high. Therefore, early assessment on the quality of intellectual property (IP) libraries in terms of manufacturability is absolutely required.

At each new technology node, standard cell libraries are an essential part of foundation libraries to construct cell-based digital designs. The quality of standard cell layouts can have a significant impact not only on power, performance, and area of designed circuits, but also on manufacturability of completed designs. One of the major manufacturability challenges limiting the quality of standard cells is to design lithography friendly layouts [1, 2].

An IP library including standard cells is usually designed through repeated processes of circuit optimizations, layout modifications, physical verifications, and lithography simulations [2–5]. Since using lithography simulation to detect weak-points is computationally intensive and time consuming, foundries have offered pattern matching library so that circuit designers can use pattern matching tools to speed up the weak-point detection process. The pattern matching library is usually built by foundries from their information on critical lithography weak-point patterns. The weak-point detection methodology based on pattern matching technology is able to identify the location of each layout pattern which matches a lithography weak-point pattern defined in the library.

Many research works in the field of lithography weak-point detection have focused on the development of pattern matching algorithms and on detection accuracy; there are limited works discussing its applications used in the industry. From our experience, we have observed that lithography weak-points could exist in the layouts of standard cells, especially on the first few interconnect layers. In addition, during the physical design phase of a circuit, lithography weak-points can be observed when routed metal wires are added into a layout which contains placed standard cell instances. Therefore, to minimize the number of lithography weak-points existing in a placed and routed design, we need to identify all of potential lithography weak-points in each standard cell of a standard cell library, and then either modify or re-design their layouts. One known methodology to identify the problematic standard cells is to utilize a placement-and-routing (P&R) tool to carry out a random brute-force approach. In other words, one can iteratively perform placement-and-routing on a design containing all of the standard cells under different design constraints, followed by performing pattern matching verifications to identify locations of weak-points which are induced by standard cell designs. However, the P&R based brute-force approach can be time consuming and may not detect all of potential lithography weak-points.

In this paper, we propose a partial pattern matching methodology, which enables the detection of all possible potential lithography weak-points (DAPPLW) for all standard cells in a given standard cell library. The proposed methodology adopts a "combinatorial *k*-partitioning" technique. Additionally, the methodology adopts a pruning technique to minimize false violations, and uses a prioritization technique to prioritize tasks of modifying and/or re-designing standard cells. Compared with the random brute-force

978-1-5386-7100-9/18 $31.00 © 2018 IEEE

approach, our experimental results have shown that complete lists of potential weak-points can be obtained by using the proposed methodology.

The rest of the paper is organized as follows. Section II briefly introduces pattern matching and formulates the problems of detecting all possible potential lithography weak-points. Section III details our partial pattern matching methodology. In Section IV, experimental results are shown and discussed. Finally, conclusions are drawn in Section V.

II. DEFINITIONS AND PROBLEM FORMULATIONS

In this section, we briefly introduce the concept of pattern matching verification, describe features of lithography weak-point patterns, and formulate the problem that this research work intends to solve.

A. Pattern Matching Verification

Pattern matching verification is a process of detecting and locating all layout patterns which match the predefined lithography weak-point patterns. Lithography weak-point patterns are a set of patterns stored in pattern libraries; the patterns are usually provided by foundries [6]. Four types of methodologies for the implementation of the pattern matching functionality have been published in the literature; these methodologies include string-based (grid/matrix computation) pattern matching [7, 8], model-based (graph computation) pattern matching [9], design-rule-checking-based (geometric computation) pattern matching [10], and machine/deep-learning-based (data-driven) pattern matching [4, 11–16].

In practice, semiconductor foundries, such as GLOBALFOUNDRIES, release pattern libraries and relevant runset files to their customers so that lithography weak-point detection processes can be performed on their layout designs [17]. Commercial software tools which are capable of performing lithography weak-point detection have been developed and used in the industry; these tools include Cadence CPA [18], Mentor Calibre DRC [19], and Synopsys IC Validator [20]. Each of these tools can scan an entire layout design and identify locations of all layout patterns which match predefined lithography weak-point patterns.

B. Definition of a Pattern

Definition 1 (Lithography weak-point pattern) A Lithography weak-point pattern is defined as a layout containing n elements of polygons placed inside a rectilinear bounding box. Each polygon is drawn on a specified layer, and coordinates of each vertex in the polygon are Cartesian coordinates.

As the example illustrated in Fig. 1, the lithography weak-point pattern contains two polygons on a Metal layer and a polygon on a Via layer. In general, patterns are categorized into (1) exact patterns and (2) non-exact/fuzzy patterns. A lithography weak-point is an exact pattern and is reported in the design if there is an exact match with a pattern defined in a pattern library. One drawback of exact pattern matching

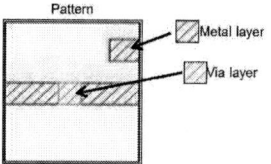

Fig. 1: Example of a lithography weak-point pattern

is that one exact pattern only defines one type of lithography weak-points. To overcome the problem, fuzzy patterns have been introduced so that many similar lithography weak-points can be defined by using only one of these patterns. For fuzzy patterns, specifically, additional marker layers can be placed inside a pattern bounding box to indicate "don't care" regions where similar weak-points excluding the regions are classified under the same pattern in a pattern library. In addition, a fuzzy pattern can have tolerance ranges for edges of polygons which are inside the pattern.

C. Problem Formulations

In this work, we have formulated the DAPPLW problem for standard cell libraries as follows.

Problem 1 (Intrinsic DAPPLW) Given a lithography weak-point pattern and a standard cell layout, the problem of intrinsic DAPPLW is to report all locations of weak-points consisting of the pattern with eight possible orientations (4 rotations × 2 mirrored images) in the layout.

Problem 2 (Extrinsic DAPPLW) Given a lithography weak-point pattern and a standard cell layout, the problem of extrinsic DAPPLW is to report all locations of *potential* weak-points in the standard cell layout such that these potential weak-points can become real weak-points (weak-points consisting of the pattern with eight possible orientations) after a physical layout containing the standard cell is completed (e.g., placed and routed).

Current methodologies for performing pattern matching verification is capable of solving the intrinsic DAPPLW problem. However, even with the use of a P&R based brute-force approach, the extrinsic DAPPLW problem still cannot be completely solved. For extrinsic DAPPLW, at least two scenarios can be observed when lithography weak-points appear inside a standard cell instance after an integrated circuit layout is completed.

Scenario 1 (Weak-point related to standard cell) As shown in Fig. 2, during the physical design phase, three metal polygons are added to the layout in order to access pins of the standard cell instance (Fig. 2(b)). Among these three new polygons in the layout, a lithography weak-point is observed at one of the locations as highlighted by the bounding box in Fig. 2(c).

Scenario 2 (Weak-point not related to standard cell) As shown in Fig. 3, a lithography weak-point is observed after the layout of an integrated circuit is completed. In this scenario, a P&R tool creates polygons within the available spaces in the layout of the standard cell instance.

978-1-5386-7100-9/18 $31.00 © 2018 IEEE

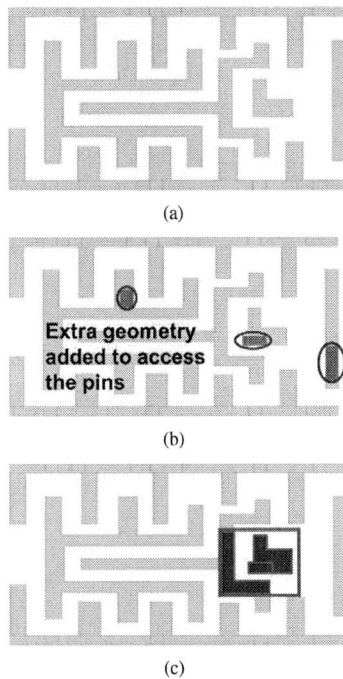

Fig. 2: An example of metal layer in the standard cell A (a) without and (b) with metal routing. (c) The highlighted bounding box indicates the location of a lithography weak-point in the standard cell A.

It is important to identify potential lithography weak-points which can occur in a standard cell layout (Scenario 1) and then re-design the standard cell layout so as to reduce the generation of lithography weak-points during the physical design phase. In the context of the extrinsic DAPPLW problem, generated weak-points observed in Scenario 2 are classified as false violations since these weak-points are not associated with standard cell designs.

III. Partial Pattern Matching for Weak-point Detection (PPMWD)

In order to identify all potential weak-points located in the layout of a given standard cell, we propose to relax the pattern matching criteria through the use of a pattern library which contains partial patterns; these partial patterns are generated by using the proposed methodology. Thus, the proposed methodology is named *partial pattern matching for weak-point detection* (PPMWD); the flow of the proposed methodology is shown in Fig. 4. First of all, in the methodology, unique partial patterns are enumerated from the given pattern library. Second, a *pruning* technique is used to remove patterns that violate design rules. After identifying potential weak-points in standard cell layouts, we adopt a *prioritization* technique in order to prioritize the tasks of re-designing these standard cell layouts. Note that the proposed PPMWD methodology does not require standard cell instances to be placed and routed for solving the extrinsic DAPPLW problem.

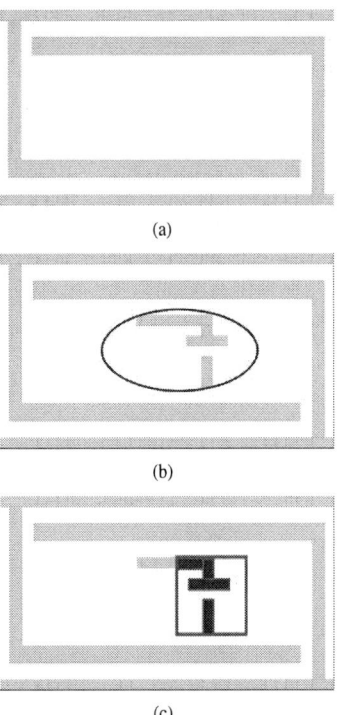

Fig. 3: An example of metal layer in the standard cell B (a) without and (b) with metal routing. (c) The highlighted bounding box indicates the location of a lithography weak-point in the standard cell B.

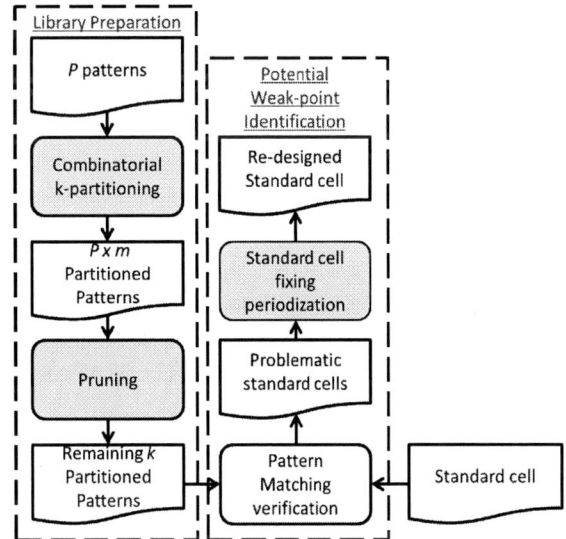

Fig. 4: Overview of the proposed methodology named partial pattern matching for weak-point detection (PPMWD).

A. Combinatorial k-partitioning technique

Given a lithography weak-point pattern containing n polygons, we can generate relevant partial patterns; each of

978-1-5386-7100-9/18 $31.00 © 2018 IEEE

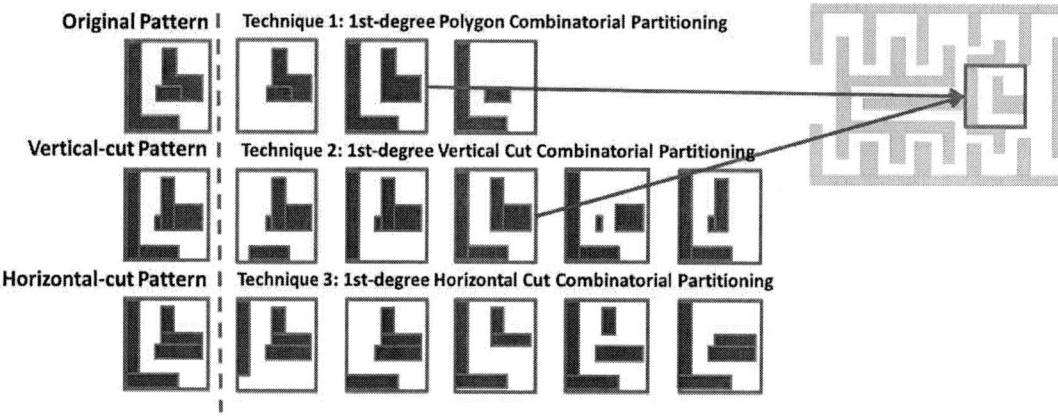

Fig. 5: An conceptual idea of decomposing a pattern into different subsets of the original pattern.

the partial patterns contains k out of the n polygons (for $k \in \mathbb{Z}$ and $1 \leq k \leq n - 1$). A dth-degree pattern (for $d \in \mathbb{Z}$ and $1 \leq d \leq n - 1$) is a partial pattern which is generated from the lithography weak-point pattern with $k = n - d$. The dth-degree polygon combinatorial partitioning technique is used to generate all of dth-degree patterns from a lithography weak-point pattern.

For instance, by using the 1st-degree polygon combinatorial partitioning technique, the total count (m) for these generated 1st-degree patterns is

$$m = \frac{n!}{k!(n-k)!} \tag{1}$$

As the upper-left pattern shown in Fig. 5, it is a sample lithography weak-point pattern which contains three polygons on a specific layer. By using the 1st-degree polygon combinatorial partitioning technique, three 1st-degree patterns are obtained.

Additionally, we propose the use of vertical cuts and horizontal cuts to perform polygon decompositions on the original lithography weak-point pattern in order to partition each rectilinear polygon into disjoint rectangles before applying combinatorial partitioning techniques. As shown in Fig. 5, with the use of vertical and horizontal cuts, vertical-cut and horizontal-cut patterns can be generated. Moreover, dth-degree patterns for these vertical-cut and horizontal-cut patterns can be generated accordingly. After partial patterns are generated, we can identify potential lithography weak-points from layouts of given standard cells.

B. Pruning technique

To reduce runtimes of performing pattern matching based on generated partial patterns, we propose a pruning technique for eliminating partial patterns violating design rules. The algorithm for pruning partial patterns is detailed in Algorithm 1. In the algorithm, polygons that touch a pattern bounding box are not checked for design rule violations since these polygons can be parts of larger polygons in the layout. After DRC verifications are performed, a number of partial patterns may be

removed. As shown in Fig. 6, partial patterns T2.P4, T3.P3 and T3.P4 are removed from the partial pattern library after applying the pruning technique.

Algorithm 1 Pruning partial patterns

Input: All partial patterns
Output: Partial patterns that are not removed
1:   **for all** pattern $p \in$ partial patterns **do**
2:       $p' \leftarrow p$
3:       **for all** polygon g inside pattern p' **do**
4:          **if** g touches the pattern bounding box **then**
5:             Remove g from p'
6:          **end if**
7:       **end for**
8:       Perform design rule checking on p'
9:       **if** DRC violations exist **then**
10:          Remove p
11:       **end if**
12: **end for**

C. Prioritization technique

After performing pattern matching verifications based on a given layout and generated partial patterns, a large number of pattern matching violations can be generated. These pattern matching violations indicate potential locations in which standard cells can induce lithography weak-points after performing placement and routing. In order to prioritize the tasks of re-designing these problematic standard cells, therefore, we adopt a prioritization technique in the proposed methodology.

The example shown in Fig. 7 is used to illustrate the prioritization technique. As shown in the figure, Partial Patterns #1-#4 were generated from a lithography weak-point pattern by using the 1st-degree polygon combinatorial partitioning technique. For each of these 1st-degree patterns, we can compute a probability score P_i for the pattern; the score is based on the geometry of the missing polygon in the pattern.

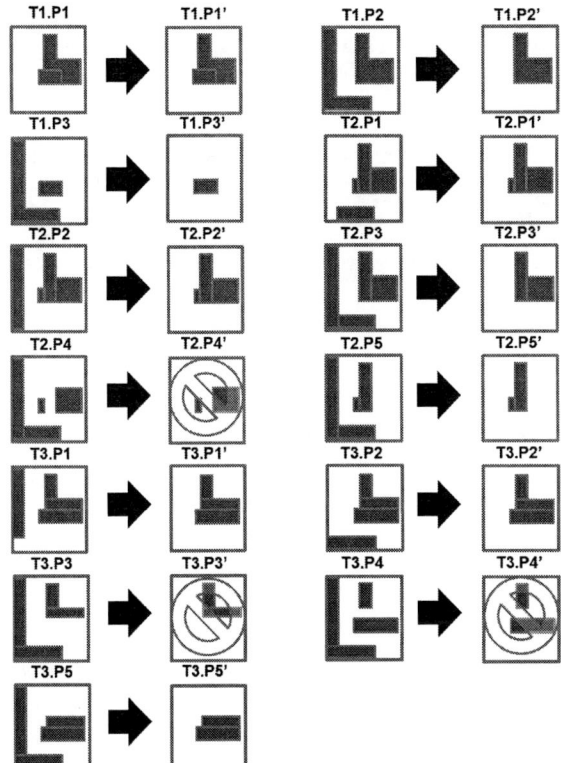

Fig. 6: An example of pruning technique applied to the sample pattern.

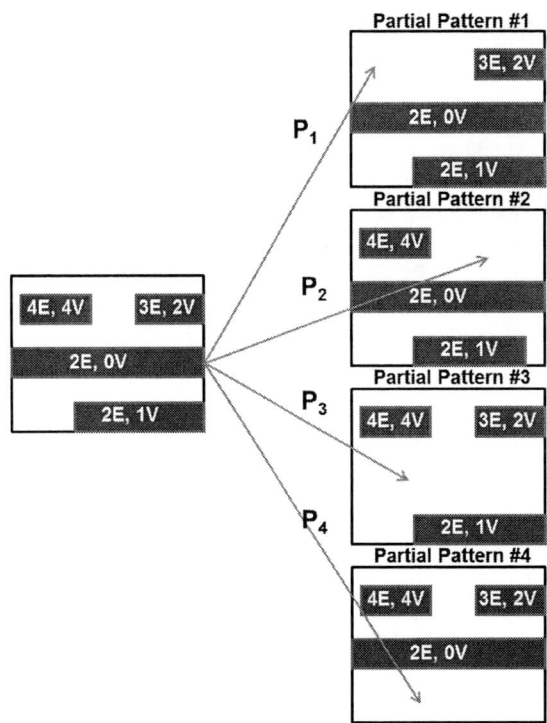

Fig. 7: Prioritization Technique.

TABLE I: Information about the Pattern Library

Name	l^1	m^2	$m_p{}^3$	$m_v{}^4$	$m_1{}^5$	$m_2{}^6$
Pattern #1	2	5	3	2	5	10
Pattern #2	2	4	2	2	4	6
Pattern #3	2	7	4	3	7	21
Pattern #4	2	5	4	1	5	10

[1] Total number of layers
[2] Total number of polygons
[3] Total number of metal polygons
[4] Total number of via polygons
[5] Total number of 1st-degree patterns
[6] Total number of 2nd-degree patterns

For example, Partial Pattern #1 has a missing polygon which has four edges and four vertices completely inside the pattern bounding box. Therefore, $P_1 = 4 + 4 = 8$. As another example, Partial Pattern #2 has a missing polygon which has three edges and two vertices completely inside the pattern bounding box. Therefore, $P_2 = 3 + 2 = 5$. Likewise, we can compute that $P_3 = 2 + 0 = 2$ and $P_4 = 2 + 1 = 3$. Thus, we can obtain the result of $P_3 < P_4 < P_2 < P_1$. A standard cell containing a matched partial pattern with lower probability score will have higher priority to be modified than the one containing a matched partial pattern with a higher probability score. Therefore, for the example shown in Fig. 7, standard cells containing Partial Pattern #3 will have higher priority than the ones which do not contain the partial pattern.

IV. RESULTS AND DISCUSSIONS

Our experiments were conducted with four proprietary lithography weak-point patterns and a sample standard cell library containing about 800 single- and double-height cells. To ensure 100% pattern matching verification coverage on boundaries created by standard cell abutment during placement, we enumerated a test case (test case A) to exercise all potential boundaries created by standard cell placement permutation [21]. To replicate the test case generated using the random brute force technique, we randomly assigned the pins' connectivity using the test case A and performed placement-and-routing with random design constraints. We termed the routed layout as test case B. Both layouts were stored in the GDSII format.

In this work, we have implemented our PPMWD in the Tcl programming language for the purpose of compatibility with industrial software tools. We also integrated an industrial pattern matching engine into our framework. Table I summarizes the statistics of our selected pattern library for the experiments, where l, m, m_p, m_v represent total numbers of layers, polygon counts, polygon counts for metal shapes, and polygon counts for via shapes, respectively. The four patterns are enumerated to produce m_1 and m_2 unique partitions of 1st-degree and 2nd degree patterns, respectively. The generated partial patterns were stored in a database in order to perform pattern matching verification.

TABLE II: Information about the Pattern Library

Name	Test case B[1]		Test case A[2]	
	Violations[3]	Cells[4]	Violations[3]	Cells[4]
Pattern #1[1]	3	4[6]	0	0
Pattern #2[1]	12	12	0	0
Pattern #3[1]	12,680	634[7]	0	0
Pattern #4[1]	1	1	0	0
Partial Pattern #1[2,5]	2,339	322[9]	9,624	6
Partial Pattern #2[2,5]	0	0	0	0
Partial Pattern #3[2,5]	80,738	773[10]	33,727	448[11]
Partial Pattern #4[2,5]	455	400[8]	0	0

[1] Pattern matching using the four proprietary pattern library
[2] Pattern matching using the partial pattern library
[3] Total number of pattern matching violation
[4] Total number of standard cell with violations
[5] Unique partitions of 1st-degree polygon combinatorial partitioning
[6] One of the violation marker occurred at the boundary of two standard cells
[7] Only 19 cells have occurrence greater than prob. 0.9
[8] Zero cells have occurrence greater than prob. 0.9
[9] Only 2 cells have occurrence greater than prob. 0.9
[10] Only 21 cells have occurrence greater than prob. 0.9
[11] Only 21 cells have occurrence greater than prob. 0.9

In the experiments, we performed pattern matching verification with the four proprietary patterns and with the generated partial patterns. We carried out experiments on a Linux workstation which had an Intel 2.7-GHz 8 Core Duo CPU and 128-GB of memory. The experimental results, which are summarized in Table II, matched our expectations. In summary, 27 standard cells were found to have potential lithography weak-point issues. Patterns #1 and #3 resulted in a total of 21 cells which were problematic. Since Patterns #2 and #4 were mainly induced by routing layers, we expected that no cell needed re-designing with regard to these two patterns. Pattern matching violations regarding Partial Pattern #4 were not related to the design of standard cell layouts. In addition, we have observed that pattern matching verification results with 2nd-degree polygon combinatorial partitioning had multiple false violations that were not related to the design of standard cell layouts. Thus, these results were not included in the table.

V. CONCLUSION

In this paper, we proposed a partial pattern matching methodology which adopts a combinatorial K-partitioning technique to identify all of potential lithography weak-points in given standard cell libraries. Furthermore, to minimize false violations and to prioritize tasks of re-designing standard cells, we have introduced the use of pruning and prioritization techniques in the proposed methodology. Our experimental results have shown that the P&R based brute-force approach could not provide sufficient coverage to detect all of potential weak-points in standard cell libraries. On the other hand, the proposed methodology efficiently detected all locations of potential lithography weak-points in given standard cell libraries; IP and circuit designers can thus use the information to either redesign the problematic standard cells or avoid the use of these cells. Our future work includes the development of a software system to automatically modify layouts of these problematic standard cells.

REFERENCES

[1] V. Dal Bem and P. Butzen and F. S. Marranghello and A. I. Reis and R. P. Ribas, "Impact and Optimization of Lithography-Aware Regular Layout in Digital Circuit Design," in *Proc. IEEE Int. Conf. on Comput. Design (ICCD)*, Oct 2011, pp. 279–284.
[2] S. Shim and W. Chung and Y. Shin, "Lithography Defect Probability and Its Application to Physical Design Optimization," *IEEE Trans. VLSI Syst.*, vol. 25, no. 1, pp. 271–285, Jan 2017.
[3] D. Jang and N. Ha and J. H. Park and S.-W. Paek and H.-S. Won and K.-M. Choi, "DFM Optimization of Standard Cells Considering Random and Systematic Defect," in *Proc. IEEE Int. SoC Design Conf. (ISOCC)*, vol. 01, Nov 2008, pp. I–70–I–73.
[4] D. Ding and J. A. Torres and D. Z. Pan, "High Performance Lithography Hotspot Detection With Successively Refined Pattern Identifications and Machine Learning," *IEEE Trans. Comput.-Aided Design Integr. Circuits Syst.*, vol. 30, no. 11, pp. 1621–1634, Nov 2011.
[5] C. Andrus and M. Guthaus, "Lithography-aware Layout Compaction," in *Proc. IEEE of the Great Lakes Symp. on VLSI (GLSVLSI)*, May 2012, pp. 147–152.
[6] L. Lavagno and L. Scheffer and G. Martin, *EDA for IC Implementation, Circuit Design, and ProcessTechnology (Electronic Design Automation for Integrated Circuits Handbook)*. Boca Raton, FL, USA: CRC Press, Inc., 2006.
[7] H. Yao and S. Sinha and C. Chiang and X. Hong and Y. Cai, "Efficient Process-Hotspot Detection Using Range Pattern Matching," in *Proc. IEEE/ACM Int. Conf. on Comput.-Aided Design (ICCAD)*, Nov 2006, pp. 625–632.
[8] J. Xu and S. Sinha and C. C. Chiang, "Accurate Detection for Process-hotspots with Vias and Incomplete Specification," in *Proc. IEEE/ACM Int. Conf. on Comput.-Aided Design (ICCAD)*, Nov 2007, pp. 839–846.
[9] A. B. Kahng and C.-H. Park and X. Xu, "Fast Dual Graph-based Hotspot Detection," vol. 6349, 2006, pp. 6349 – 6357.
[10] Y. T. Yu and I. H. R. Jiang and Y. Zhang and C. Chiang, "DRC-based Hotspot Detection Considering Edge Tolerance and Incomplete Specification," in *Proc. IEEE/ACM Int. Conf. on Comput.-Aided Design (ICCAD)*, Nov 2014, pp. 101–107.
[11] D. Ding and X. Wu and J. Ghosh and D. Z. Pan, "Machine Learning Based Lithographic Hotspot Detection with Critical-Feature Extraction and Classification," in *Proc. IEEE/ACM Int. Conf. on IC Design and Tech. (ICICDT)*, May 2009, pp. 219–222.
[12] D. Ding and A. J. Torres and F. G. Pikus and D. Z. Pan, "High Performance Lithographic Hotspot Detection Using Hierarchically Refined Machine Learning," in *Proc. IEEE Asia amd South Pacific Design Automat. Conf. (ASP-DAC)*, Jan 2011, pp. 775–780.
[13] J. Y. Wuu and F. G. Pikus and A. Torres and M. Marek-Sadowska, "Rapid layout pattern classification," in *Proc. IEEE Asia amd South Pacific Design Automat. Conf. (ASP-DAC)*, Jan 2011, pp. 781–786.
[14] J. Y. Wuu and F. G. Pikus and M. Marek-Sadowska, "Metrics for Characterizing Machine Learning-based Hotspot Detection Methods," in *Proc. IEEE Int. Symp. on Quality Electron. Design (ISQED)*, March 2011, pp. 1–6.
[15] K. Madkour and S. Mohamed and D. Tantawy and M. Anis, in *Proc. IEEE Int. Symp. on Quality Electron. Design (ISQED)*.
[16] F. Yang and C. C. Chiang and X. Zeng and D. Zhou, "Efficient SVM-based Hotspot Detection Using Spectral Clustering," in *Proc. IEEE Int. Symp. Circuits and Syst. (ISCAS)*, May 2017, pp. 1 4.
[17] C. Magalang and V. Perez and E. Teoh and V. Dai and S. Yeo and C.W Hui, "Pattern-Based Physical Verification in the Design Flow," in *Proc. ACM/EDAC/IEEE Design Automat. Conf. (DAC) User Track*, Jun 2011.
[18] Cadence, *Cadence Pattern Analysis User Guide*, 2017.
[19] Mentor, *Calibre Verification User's Manual*, 2017.
[20] Synopsys, *IC Validator User Guide Manual*, 2017.
[21] Y. Li and C.H. Lee and W.C. Ang and K.P. Chua and Y.S. Ong and C.W. Hui, "Constraining the Synopsys Pin Access Checker Utility for Improved Standard Cells Library Verification Flow," in *Synopsys Users Group (SNUG) Silicon Valley*, Mar 2017, pp. 1–22.

2018 IEEE Computer Society Annual Symposium on VLSI

Feature Based Coverage Analysis of AMS Circuits

Antara Ain*, Akshay Mambakam[†] and Pallab Dasgupta[†]
*Advanced Technology Development Centre, [†]Dept. of Computer Science and Engg.
Indian Institute of Technology Kharagpur, West Bengal, India 721302
Email: antara@atdc.iitkgp.ernet.in, makshay@cse.iitkgp.ernet.in, pallab@cse.iitkgp.ernet.in

Abstract—Coverage analysis for Analog and Mixed-Signal (AMS) behaviors involves exploring continuous state spaces defined by real valued artifacts. For example, coverage of analog features such as settling time, peak overshoot, etc., entails not only finding whether we have seen such behaviors, but also whether we have covered the range of values of such features (including the extremal values). Thus it is simplistic to lift the notion of assertion based functional coverage from the digital domain into the AMS domain - rather we need to address coverage in the value and time domains. In this paper we propose a methodology for feature based coverage analysis of AMS circuits.

Index Terms—Analog and Mixed-Signal, Coverage, Feature

I. INTRODUCTION

Automatic monitoring of coverage of corner case scenarios is an important and integral component of verification. In digital verification, corner case scenarios can be captured by assertions, and there are techniques to bias a random simulation to hit the scenarios covered by the specified assertions. The SystemVerilog Assertion (SVA) [8] language supports the notion of cover properties which allows assertions in SVA to be defined as coverage goals and these cover properties can be automatically monitored over simulation.

In the digital domain, designers are interested in the satisfaction or refutation of some specified properties, i.e, satisfying the specification is essentially a true/false decision. In the AMS domain, a design meeting specification is primarily concerned with evaluating real valued attributes and ensuring that they lie within their specified ranges. Such functional properties, which are referred to as *features*, are manifested under specific scenarios. E.g., short circuit current is a functional property of a linear voltage regulator, and can be evaluated only if the circuit is driven to a short circuit scenario. In [3], the authors propose the notion of features which overlays the computation of real valued functions over the syntactic fabric of assertions.

Lifting the notion of coverage of digital assertions to AMS domain is not sufficient. In the AMS domain we are not only interested in determining whether a real valued property is satisfied, but also want to check by what margin the design meets the specification. This has important ramifications in choosing the metrics of assertion coverage in AMS domain. Thus, we must not only cover the essential behaviors, but also cover the time and value ranges of the properties to determine whether they lie within their specified limits, and thereby, the distribution of the coverage points within those ranges.

The features are manifested under certain scenarios of the circuit. Hence in order to cover such features we need input stimuli that will drive the circuit through its different operating modes so as to ensure that the behaviors expressed by the feature are exercised. Evaluation of feature coverage with input stimuli, using a Monte Carlo method, may unnecessarily increase the computation burden. This is mainly because, typically the feature distribution over the feature range has a mapping with the input distribution over the input range. In this paper, we propose a methodology to accelerate the feature coverage by learning the map between input space and feature space using some directed simulation runs of the circuit. Next, we use the learned map of the input and feature space to generate input stimuli to evaluate features in specified intervals of feature space to improve the coverage of the feature.

The main contributions of this paper are as follows.

1) The paper formalizes the notion of feature-based coverage of the behaviors of AMS circuits.
2) We propose a learning based methodology for accelerating the feature coverage of AMS circuits.
3) We demonstrate our methodology on industrial circuit and present results highlighting its performance.

The rest of the paper is organized as follows. Section II presents existing literature on coverage of circuits. Section III introduces the notion of feature based coverage in AMS domain. The methodology and tool flow for evaluation of feature coverage of AMS circuits are explained in Section IV. Section V presents results to elaborate the performance of our approach followed by concluding remarks in Section VI.

II. RELATED WORK

In the domain of digital circuit verification, assertion coverage and automatic test generation for such analysis are well studied areas of research. SVA [8] supports cover properties to compute the number of *non-vacuous* matches of an assertion. The authors of [5], [9] propose a methodology for biasing a random test generator in order to accelerate the computation of assertion coverage of digital circuits.

In [7] a set of coverage metrices has been proposed for AMS circuits. Coverage of entire state-space for the verification of analog circuits by generating input stimuli is proposed in [11]. In [12], an analog transition system has been proposed which guides the auto-generation of the input stimuli for equivalence checking of analog circuits. [13] presents an algorithm for generating test inputs for analog integrated circuits. [10] proposes an algorithm for generating test inputs using signal-flow graphs and computing two measures called the observability and the controllability of a signal node. In [6], an approach has been proposed for the generation of transient test stimulus for

978-1-5386-7100-9/18 $31.00 © 2018 IEEE   423

analog circuits by formulating it as a nonlinear programming problem.

In this work, we primarily define coverage of AMS circuits in terms of its features and then use a learning based approach to generate input stimuli for accelerating the evaluation of feature coverage.

III. FEATURE COVERAGE OF AMS CIRCUITS

We introduce some definitions for explaining the semantics of coverage of AMS circuits w.r.t formally specified features.

Definition 1. [Test Stimulus] *A test stimulus φ is a valuation for the set of variables $\mathcal{I}$ over time, where $\mathcal{I}$ is the set of input variables of a circuit, such that φ is a mapping $\varphi : \mathbb{R}^+ \to \mathbb{R}^{|\mathcal{I}|}$. $\mathbb{R}^+$ represents time.* □

We denote the set of test stimuli as Ψ. When we simulate a design $\mathcal{B}$ with a test stimulus $\varphi_i \in \Psi$, we obtain a valuation for the set of output variables $\mathcal{O}$ of $\mathcal{B}$ over time. We, therefore, define a trace as follows:

Definition 2. [Trace] *A trace τ is defined as a valuation of a set of variables (both input and output variables of a circuit) $X = \{\mathcal{I}, \mathcal{O}\}$ over time such that τ can be represented as a mapping $\tau : \mathbb{R}^+ \to \mathbb{R}^{|X|}$, where $\mathbb{R}^+$ represents time.* □

Since we focus on the simulation based approach, so the trace (referred to as simulation trace) is actually finite and therefore, can be represented as $\tau \equiv (\mathcal{I} \times \mathcal{O})^Z$.

Definition 3. [Feature] *A feature F, a real valued variable, is specified by an assertion (expressed as a sequence-expression S) that specifies the scenarios under which the feature is to be computed, and a signature (a function $\mathcal{F}$ on the set of variables X), computed from the match of the assertion on the trace τ.* □

The general syntax of a feature in the Feature Indented Assertion (FIA) language is as follows, the details of which is elaborated in [3].

```
feature <feature-name> (<list-of-parameters>);
begin
   var <list-of-local-variables>;
   <sequence-expr>
      |-> <feature-name> = <feature-expr>;
end
```

As defined in [3], a sequence expression S is a sequence of one or more sub-expressions separated by delay operators and is of the form $s_1 \ \#\#[a_1 : b_1] \ s_2 \quad s_n$. Each of these sub-expressions s_i is on predicates over real variables (PORVs).

By definition, each feature value evaluation corresponds to a match of the sequence expression in the feature specification. As a result, when features are evaluated over simulation, a test stimulus φ_i must drive different inputs in such a way that, when the design $\mathcal{B}$ is simulated with φ_i it produces a trace τ_i and the sequence expression S has at least a match over τ_i. We call such test stimulus φ_i as feature-evaluating stimulus. It is to be noted that over a simulation trace τ_i a sequence expression S can have multiple matches and for each of these

matches a feature value is computed. Thus we get a range of feature values which is explained using Example 1.

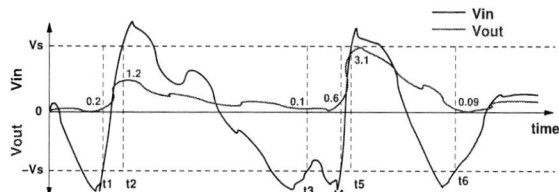

Fig. 1: Voltages `Vin` and `Vout` vs time.

Example 1. Feature VoltageRise computes the rise in output voltage `Vout` when `Vin` goes above the rated value of input voltage `Vs` within 100ms from when it goes above `-Vs`. The feature, in FIA, is expressed formally as follows:

```
feature VoltageRise(Vs);
begin
 var v1, v2;
 @+(Vin >= -Vs),v1=Vout ##[0:0.1]
 @+(Vin >= Vs),v2=Vout |-> VoltageRise = v2-v1;
end
```

The sequence expression in the antecedent is

S:
```
@+(Vin >= -Vs),v1=Vout ##[0:0.1]
   @+(Vin >= Vs),v2=Vout
```

which has the following sub-expressions

s_1: `@+(Vin >= -Vs),v1=Vout`

s_2: `@+(Vin >= Vs),v2=Vout`

In Fig. 1, given that s_1 is true at the time points `t1,t3,t4,t6`, the corresponding values of `Vout` i.e. `0.2,0.1,0.6,0.09` are stored in the local variable `v1`. Similarly, s_2 is true at the time points `t2,t5` and the different values of `Vout` i.e. `1.2,3.1` are stored in the local variable `v2`. S has a match at the time point `t2` and an overlapping match at time point `t5` for s_1 occurring at the time points `t3` and `t4`. For these matches of S, we compute the feature values. Here, at time point `t2` the feature value computed is `(1.2-0.2)=1`, and at time point `t5` the feature values computed are `(3.1-0.1)=3` and `(3.1-0.6)=2.5`. Therefore, after time point `t5` the feature range is `[1:3]`. □

It is also to be noted that, multiple feature values are also obtained when the design $\mathcal{B}$ is simulated with different test stimuli. Therefore, for a given set of test stimuli Ψ, the feature range that is obtained is denoted as $featureRange(\Psi) = [F_{min} : F_{max}]$, where F_{min} and F_{max} denote the minimum and maximum values of the range of feature F. The multiple feature values that lie within $[F_{min} : F_{max}]$ are used to compute how these values are distributed over the entire feature range $featureRange(\Psi)$. This distribution of feature values evaluates the coverage of the particular feature over its range. We define feature coverage as follows.

978-1-5386-7100-9/18 $31.00 © 2018 IEEE   424

Definition 4. [Feature Coverage] *A feature F is said to be covered in its range $[F_{min} : F_{max}]$ with granularity $\triangle$, iff, when the feature range $[F_{min} : F_{max}]$ is sampled at intervals of width $\triangle$ and there exists at least one feature value in all of these sampled intervals of width $\triangle$.* □

We refer to the intervals of width $\triangle$ which do not have at least one feature value, as uncovered intervals and the rest as covered interval. In this paper, coverage evaluation requires to compute whether with other test stimuli $\varphi_j \notin \Psi$ we can

1) Generate feature values in the uncovered intervals of $featureRange(\Psi) = [F_{min} : F_{max}]$.
2) Increase the feature range $[F_{min} : F_{max}]$ that is obtained with the set Ψ.

In this regard, we propose to learn $\varphi_j \notin \Psi$ from a generated data-set. From the definition of feature, the sequence expression S is of the form $s_1 \#\#[a_1 : b_1] s_2 \ldots s_n$. For learning φ_j, we require two types of input sets:

1) Set of inputs, $\mathcal{I}_S$, on which match of the sequence expression S depends
2) Set of inputs, $\mathcal{I}_F$, on which feature values and thereby the coverage over its range depend

For a particular feature, $\mathcal{I}_S$ in φ_j is determined using the domain knowledge of the circuit which is discussed in Section IV-A. But for generating φ_j, we propose to learn $\mathcal{I}_F$. In the first step of our approach, we simulate $\mathcal{B}$ with a set of test stimuli Ψ to evaluate the distribution of feature values over its range $featureRange(\Psi)$. With these simulation data, we learn the function that maps the real-valued feature valuations to its corresponding input set $\mathcal{I}_F$ in Ψ i.e., we learn

$$f(\mathcal{I}_F) \to \mathbb{R}$$

We refer this step as the learning step. In the next step, we learn the reverse map i.e.:

$$Q : \mathbb{R} \to \mathcal{I}_F^Z$$

Therefore, for generating a feature value $R \in \mathbb{R}$ in a particular interval of the feature range, we use this reverse map Q to essentially find the roots of:

$$f(\mathcal{I}_F) - R = 0$$

The roots of the above equation are the valuations of $\mathcal{I}_F$ in creating $\varphi_j \notin \Psi$. Therefore, our approach is to bias the test stimuli generator to accelerate the coverage computation, i.e. to increase the percentage coverage of the circuit, where percentage coverage is defined as follows.

Definition 5. [Percentage Coverage] *The percentage of the no. of covered intervals w.r.t the total no. of interval of width $\triangle$ in the range $[F_{min} : F_{max}]$ of a feature F, is defined as the percentage coverage of the feature.* □

IV. METHODOLOGY FOR COMPUTING FEATURE COVERAGE OF AMS CIRCUITS AND ITS TOOL FLOW

In this section we illustrate the methodology and corresponding tool flow for computing the feature coverage of AMS circuits. The first step for computing feature-based coverage analysis is manually creating the template test stimulus φ_{temp} which is one of the inputs to the tool. The overall tool flow is shown in Fig. 2.

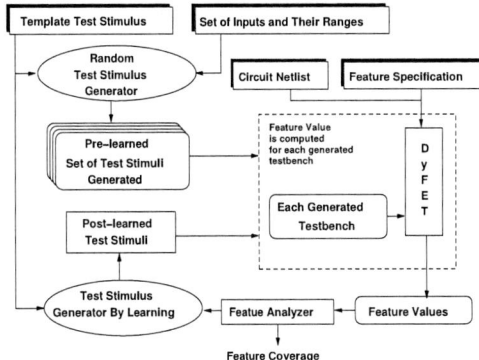

Fig. 2: Tool Flow for Feature Based Coverage Evaluation

A. Creating the Template Test Stimulus (φ_{temp})

The first step of feature coverage computation is creation of template test stimulus φ_{temp}. We use the domain knowledge of the circuit to manually create φ_{temp} such that it drives the circuit through its different states of operation so as to ensure that the contexts expressed by the feature definition are exercised. [4] shows that the behavior of an AMS circuit can be divided into different states of operation. The state transition model of an AMS circuit is formally defined as follows:

Definition 6. [State Transition Model of AMS Circuit] *A state transition model of an AMS circuit M_C is formally defined as a tuple, $M_C = \langle Q, Q_1, E \rangle$ where,*

- *Q is a set of discrete states.*
- *$Q_1 \in Q$ is the starting state.*
- *E is set of transitions and each transition $e_i = (Q_n, G_i, Q_m)$. Here, Q_n and Q_m are the current and the next states respectively, and G_i is the guard condition on this transition, defined over a set of predicates $\mathcal{P}$. Predicates $\mathcal{P}$ are over the set of variables X of the circuit.*

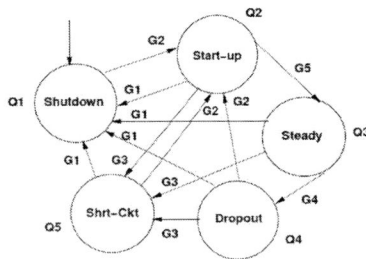

Fig. 3: State Transition Model of LDO

For example, a state transition model of Low Dropout (LDO) Regulator circuit [4] is shown in Fig. 3, where each state Q_i represents the operating states of an LDO. The guard conditions G_i are presented in Appendix A. Since the behavior of an AMS circuit can be captured using the state transition model (defined in Definition 6), therefore, for proper feature evaluation, each sub-expression s_i must be true in one or more states of operation of the corresponding circuit. The procedure of creating the template test stimulus is as follows:

- The correspondence of the states of operation of the circuit with the sub-expressions can be marked.
- Using this correspondence, we can create an input stimulus from the set of inputs $\mathcal{I}_S$ such that it drives the circuit to start its operation from the start state Q_1 and should finally reach to a state, denoted as Q_f, where s_n is true.
- It should also ensure that the circuit should traverse from Q_1 to Q_f, through some intermediate states Q_j, such that in the sequence expression, each of the sub-expression s_i is true before s_{i+1}.
- Markers are inserted to instantiate the randomly generated values for the input set $\mathcal{I}_F$.

We explain the construction of φ_{temp} using examples of feature dropout voltage and risetime of LDO circuit. The feature specification for dropout voltage is as follows:

```
feature DropoutVoltage();
begin
 var V1,V2;
 (ldo.state==DROPOUT),V1=ldo.Vin,V2=ldo.Vout
 |-> DropoutVoltage = V1 - V2;
end
```

From the definition of dropout voltage, we find that the sequence expression is comprised of only one sub-expression `(ldo.state==DROPOUT),V1=ldo.Vin,V2=ldo.Vout`. This sub-expression is true only in state Q_4. Therefore, the template test stimulus φ_{temp} should be such that it traverses to the dropout state from the shutdown state through the start-up and the steady state as shown in Fig. 3. Using the domain knowledge we know, transitions are governed by the set of inputs $\mathcal{I}_S = \{VIN, ILOAD, VEN\}$, where, VIN, $ILOAD$, and VEN are the input voltage, load current, and enable signal respectively. A code snippet of the template test stimulus φ_{temp} for dropout voltage is as follows:

```
/*Generated inputs are to be instantiated*/
//VIN_RANDOM
//ILOAD_RANDOM

analog begin
 @(initial_step) begin
  VIN = VIN_RANDOM;
  V(enable) <+ 0;  //Circuit in shutdown state
  ILOAD = ILOAD_RANDOM;
 end
 /*Condition for transiting to start-up and
   eventually to steady state from shutdown state*/
 if(($abstime >= 10m) && ($abstime < 50m))begin
  V(enable) <+ 3.6; //Circuit enabled
  VIN = VIN_RANDOM;
 end
 /*Condition for transiting to dropout from steady state*/
 else if(($abstime >= 50m) && ($abstime < 70m))begin
  VIN = VIN_RANDOM*(1 - 50*($abstime - 50m)); //Vin decreased
 end
 /*Condition for transiting to start-up and
   eventually to steady state from dropout state*/
 else if($abstime >= 70m) begin
  VIN = VIN_RANDOM;
 end

 V(Vin,pgnd) <+ transition(VIN,0,0);
 I(vout,pgnd) <+ ILOAD;
end
```

Since the dropout voltage is dependent on VIN and $ILOAD$ of the LDO circuit, the computation of the coverage of this feature over its range should be computed over the set $\mathcal{I}_F = \{VIN, ILOAD\}$. In the template test stimulus, we, therefore, have markers to instantiate the randomly generated values of the input voltage, denoted by `VIN_RANDOM`, and load current, denoted by `ILOAD_RANDOM`.

Now, let us consider the feature risetime, whose feature specification is as follows.

```
feature RiseTime(Vs);
begin
 var t1, t2;
 @+(ldo.Vout>=0.1*Vs),t1=$time ##[0:$]
 @+(ldo.Vout>=0.9*Vs),t2=$time |-> RiseTime=t2-t1;
end
```

From the definition of risetime, we find that the sequence expression is comprised of two sub-expressions `s1: @+(ldo.Vout>=0.1*Vs),t1=$time` and `s2: @+(ldo.Vout>=0.9*Vs),t2=$time`. From the state transition model description of LDO we know that, the output voltage rises in the start-up mode. Hence, the positive crossing of `0.1*Vs` and `0.9*Vs` by `Vout` is true in the start-up state Q_2. Therefore, the template test stimulus φ_{temp} should be such that it traverses to the start-up state from the shutdown state of LDO.

B. Inputs to the Feature Coverage Tool

The inputs to the tool are circuit under test $\mathcal{B}$, feature description F_{spec} expressed in terms of FIA [3], the template test stimulus φ_{temp}. For the template test stimulus φ_{temp} the tool also takes the set of circuit inputs $\mathcal{I}_F$ with their permissible ranges as an input.

C. Random Test Stimuli Generation

For random test stimuli generation, the tool requires the template test stimulus φ_{temp} and the set of inputs $\mathcal{I}_F$ on which the feature is dependent, and their range relevant for the particular circuit operation. The set of inputs $\mathcal{I}_F$ and their range defines the complete test region R_C, which is formally defined as follows.

Definition 7. [Complete Test Region] *For a set of inputs $\mathcal{I}_F = \{I_1, \ldots, I_n\}$, a complete test region R_C is a bounded space defined by the constraints on each input $I_i \geq c_l$ and $I_i \leq c_r$, where $c_l = l(I_i)$ and $c_r = r(I_i)$. Here, $l(I_i)$ and $r(I_i)$ define the left and right extremum values for input range of $I_i \in \mathcal{I}_F$.*

A test point T_n is therefore a co-ordinate in R_C. □

We randomly generate k no. of test points $T_1, T_2, \ldots, T_k$ in R_C. Each of these k test points are instantiated in φ_{temp} to generate a set of test stimuli $\Psi = \{\varphi_1, \varphi_2, \ldots, \varphi_k\}$. For example, in case of the dropout voltage we first randomly generate test points within the region R_C defined by the constraints $0 \leq VIN \leq 5$ and $0 \leq ILOAD \leq 1$ of the set of inputs $\mathcal{I}_F = \{VIN, ILOAD\}$, where VIN, and $ILOAD$ denote the input voltage (units in V) and the load current (units in A) respectively. Next, we replace the markers in φ_{temp} with the generated valuations as follows.

```
/*Generated inputs are to be instantiated*/
parameter real VIN_RANDOM = 4.2;
parameter real ILOAD_RANDOM = 0.3;

analog begin
 <same as in the template file>
end
```

Here, `VIN_RANDOM = 4.2` and `ILOAD_RANDOM = 0.3` are randomly generated.

D. Feature Value Computation

For each of the generated test stimulus $\varphi_i \in \Psi$ the feature coverage tool computes feature values using a Dynamic Feature Evaluation Tool (DyFET), the details of which is given in [3]. The inputs to DyFET are the circuit netlist, feature specification F_{spec} in FIA language, and the generated test stimulus φ_i. The DyFET outputs all the feature values along with the corresponding input values for which the particular feature value has been obtained. Therefore, the output of DyFET creates the training data set $\mathcal{M}$ for the learning algorithm and finds the feature range $featureRange(\Psi) = [F_{min} : F_{max}]$.

E. Feature Analyzers

The feature analyzer takes the feature values and computes the feature distribution over its range $[F_{min} : F_{max}]$, from which the uncovered intervals (if any) I_i of width $\triangle$ in $[F_{min} : F_{max}]$ are marked. The analyzer also uses the output of DyFET, to form a data set which maps the features values to its corresponding input values. This data-set $\mathcal{M}$, along with the uncovered interval $F_I = [l : r]$ (where, l and r are the extremum values of the interval F_I) where the feature values are needed to be generated are fed to the test generator which uses a learning approach to generate test stimulus. The generated test stimulus is used for computing the feature value using DyFET and correspondingly the analyzer updates the data-set $\mathcal{M}$. This process continues for all the uncovered intervals of $[F_{min} : F_{max}]$.

Next, we focus on learning test stimulus which will target to generate feature values in the intervals $F_I = [F_{max} : F_{max} + \triangle]$, $F_I = [F_{max} : F_{max} + 2 * \triangle]$ and so on. Similarly, we do for the intervals $F_I = [F_{min} - \triangle : F_{min}]$, $F_I = [F_{min} - 2 * \triangle : F_{min}]$ and so on. This process is terminated if we do not find any feature values in five (say) consecutive intervals.

F. Test Stimuli Generation By Method of Learning

In this part of the tool we generate test stimuli by using a learning approach. The output of the feature analyzer i.e., a data-set $\mathcal{M}$ and an interval F_I where the feature values are to be generated are used as inputs by this test stimulus generator. Using the training data-set $\mathcal{M}$, the generator learns a function f, denoted as $f(\mathcal{I}_F) \to \mathbb{R}$. Here, we have used a Generalized Regression Neural Network (GRNN) model from the MATLAB toolbox [2] to learn the function f.

Once the function f has been learned, next our task is to learn the reverse map $Q : \mathbb{R} \to \mathcal{I}_F^Z$ which map backs the corresponding valuation for the set of inputs $\mathcal{I}_F$, i.e., finding

a test point $T_j \in R_C$, for which we get a feature value F_v lying within the interval F_I. T_j is nothing but the roots for the function $f(\mathcal{I}_F) - F_v = 0$. Here also, we use Matlab function `fsolve` [1] to solve for the roots of $f(\mathcal{I}_F) - F_v = 0$. Using the template test stimuli φ_{temp} and T_j, the post-learned test stimulus $\varphi_j \notin \Psi$ is generated.

V. EXPERIMENTAL RESULTS

In this section we present experimental results to demonstrate the performance of our methodology for evaluating coverage of AMS behaviors. We have used industry standard transistor level netlist of LDO and its features, dropout voltage, to explain the methodology. All the simulations were performed using the Cadence AMS simulator on a 2.33 GHz Intel-Xeon server with 32GB RAM.

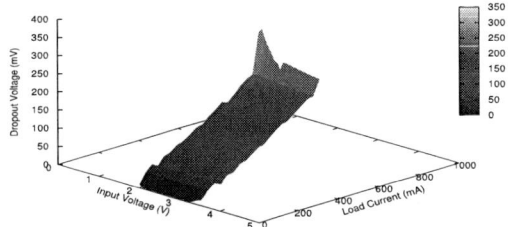

Fig. 4: Mapping of Feature Space with the Input State Space

Using a Monte Carlo method, we generated different test points T_i in the complete test region R_C of LDO for the dropout voltage. For each of these input stimuli, we have run simulations to evaluate feature values. Using the output of DyFET, i.e., the training data-set $\mathcal{M}$, we have plotted the mapping of feature space with the input state space, as shown in Fig. 4. The learning algorithm learns this mapping to accelerate the feature coverage computation.

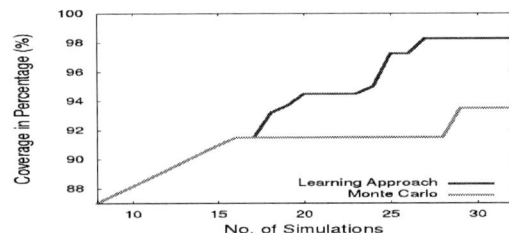

Fig. 5: % Coverage w.r.t No. of Simulation Runs for Data-set-I

We have used two instances of the training data-set each from 16 no. of simulations of two different Monte Carlo runs. The two instances are created with different settings of the simulation such that the no. of values for the feature dropout are different for the traces generated in these two instances. Therefore, Data-set-I and Data-set-II contain 600 thousands and 90 thousands of training data-points respectively.

Fig. 5 and 6 show the percentage coverage achieved w.r.t the no. of simulation runs for Data-set-I and Data-set-II

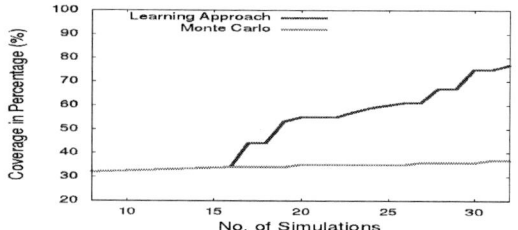

Fig. 6: % Coverage w.r.t No. of Simulation Runs for Data-set-II

respectively. For each of these cases we have used training data-set from 16 no. of Monte Carlo simulations and hence % coverage till 16 simulations are same for both the methods.

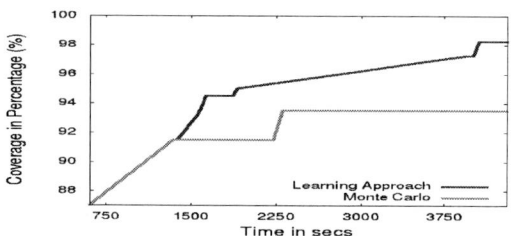

Fig. 7: % Coverage w.r.t Time for Data-set-I

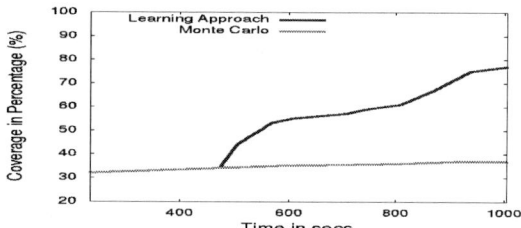

Fig. 8: % Coverage w.r.t Time for Data-set-II

Fig. 7 and 8 show the % coverage achieved w.r.t time for Data-set-I and Data-set-II respectively. 1346secs and 475 secs are the simulation times for building the training data-set in Fig. 7 and 8 respectively and hence % coverage achieved till these time points are same for both the methods. It is also to be noted that in Fig. 7 and 8, after 1346secs and 475secs respectively i.e. after the training datasets are prepared, the time presented for the learning approach comprises of the time for learning algorithm to predict the input as well as the simulation time to run the circuit with the predicted input.

From these figures, we find that our learning based algorithm for computing feature coverage out-performs the Monte Carlo algorithm, both w.r.t time and w.r.t no. of simulation runs required.

VI. CONCLUSION

This paper introduces the notion of coverage of AMS circuits. Unlike digital circuits, the coverage of AMS behaviors

should consider real valued artifacts, called features, of the circuits and should check whether the range of such features are covered. This paper proposes a learning based methodology for accelerating the computation of feature coverage and demonstrates its performance using industrial testcase.

REFERENCES

[1] Function fsolve of Matlab Toolbox (http://in.mathworks.com/help/optim/ug/fsolve.html#butbmfz-6).
[2] Generalized Regression Neural Networks Toolbox of Matlab (http://in.mathworks.com/help/nnet/ug/generalized-regression-neural-networks.html?requestedDomain=in.mathworks.com).
[3] A. Ain, A. A. B. da Costa, and P. Dasgupta. Feature Indented Assertions for Analog and Mixed-Signal Validation. *IEEE TCAD*, 35(11):1928–1941, 2016.
[4] A. Ain and et.al. Chassis: A Platform for Verifying PMU Integration Using Autogenerated Behavioral Models. *ACM TODAES*, 16(3):33:1–33:30, June 2011.
[5] A. Banerjee et al. Test Generation Games from Formal Specifications. In *ACM/IEEE DAC*, pages 827–832, 2006.
[6] B. Burdiek. Generation of Optimum Test Stimuli for Nonlinear Analog Circuits Using Nonlinear Programming and Time-Domain Sensitivities. In *Design, Automation and Test in Europe*, pages 603–608, 2001.
[7] A. Frtig et al. Novel metrics for analog mixed-signal coverage. In *IEEE International Symposium on Design and Diagnostics of Electronic Circuits Systems*, pages 97–102, 2017.
[8] IEEE Std 1800-2009. *IEEE Standard for System Verilog: Unified Hardware Design, Specification and Verification Language*, 2010.
[9] B. Pal et al. Accelerating Assertion Coverage With Adaptive Testbenches. *IEEE TCAD*, 27(5):967–972, 2008.
[10] M. Soma et al. Hierarchical ATPG for Analog Circuits and Systems. *IEEE Design Test of Computers*, 18(1):72–81, 2001.
[11] S. Steinhorst and L. Hedrich. Improving Verification Coverage of Analog Circuit Blocks by State Space-Guided Transient Simulation. In *IEEE International Symposium on Circuits and Systems*, pages 645–648, 2010.
[12] S. Steinhorst and L. Hedrich. Equivalence checking of nonlinear analog circuits for hierarchical AMS System Verification. In *International Conference on VLSI and System-on-Chip*, pages 135–140, 2012.
[13] W. Verhaegen, G. V. der Plas, and G. Gielen. Automated Test Pattern Generation for Analog Integrated Circuits. In *IEEE VLSI Test Symposium*, pages 296–301, 1997.

APPENDIX A
STATE TRANSITION MODEL OF LDO

Table I presents the auxiliary specification of LDO circuit.

TABLE I: Auxiliary Specification of Low Dropout Regulator Circuit

(a) Predicates used for a LDO

Predicates ($\mathcal{P}$)	Explanations
$p_1 :: bias1 \leq Ibias \leq bias2$	$bias2, bias1$ are upper and lower threshold values for bias current ($Ibias$).
$p_2 :: Vin \geq tol$	tol is the threshold value for input voltage.
$p_3 :: Ven > fullscale/2$	Enable (Ven) asserted.
$p_4 :: ref1 \leq Vref \leq ref2$	$ref2, ref1$ are upper and lower threshold values for reference voltage ($Vref$).
$p_5 :: Iout \geq I_{shrt}$	$Iout$ is the output current.
$p_6 :: Vin - Vout \leq dropout$	Vin is input and $Vout$ is output voltage.
$p_7 :: |Vout - V_{ss}| < \epsilon$	V_{ss} is the rated steady state output voltage; ϵ is a very small value.

(b) Guard Conditions of a LDO

Guard Conditions (G)
$G_1 = \neg p_1 \vee \neg p_2 \vee \neg p_3 \vee \neg p_4$
$G_2 = p_1 \wedge p_2 \wedge p_3 \wedge p_4 \wedge \neg p_5 \wedge \neg p_6$
$G_3 = p_1 \wedge p_2 \wedge p_3 \wedge p_4 \wedge p_5$
$G_4 = p_1 \wedge p_2 \wedge p_3 \wedge p_4 \wedge \neg p_5 \wedge p_6$
$G_5 = p_1 \wedge p_2 \wedge p_3 \wedge p_4 \wedge \neg p_5 \wedge \neg p_6 \wedge p_7$

2018 IEEE Computer Society Annual Symposium on VLSI

SAT Encoding-based Verification of Sneak Path Problem in Via-switch FPGA

Ryutaro Doi [†, ‡]

Masanori Hashimoto [†]

[†] Department of Information Systems Engineering,
Graduate School of Information Science and Technology,
Osaka University
Email: {doi.ryutaro, hasimoto}@ist.osaka-u.ac.jp
[‡] Research Fellow of Japan Society for the Promotion of Science

Abstract—**FPGA that introduces via-switches, a kind of non-volatile resistive RAMs, for crossbar implementation is drawing attention due to higher integration density and performance. However, programming those switches arbitrarily in a crossbar is not trivial since a programming voltage must be given through signal wires that are shared by multiple via-switches. Consequently, depending on the previous programming status, unintentional switch programming may occur due to signal detour, which is called sneak path problem. This paper encodes programming operations in via-switch based crossbar into a satisfiability problem and rigidly verifies the sneak path problem. Verification results show that sneak path problems can be solved by imposing a specific programming constraint.**

I. INTRODUCTION

Field programmable gate arrays (FPGAs) become more popular since the development cost of application specific integrated circuits (ASICs) is elevating due to the device miniaturization and larger scale integration. However, conventional FPGAs are still inferior to ASICs regarding operating speed, power consumption, and implementation area [1]. These drawbacks arise from a huge number of programmable switches that are included in FPGAs to acquire reconfigurability. In static random access memory (SRAM)-based FPGAs, which are the most widely used FPGAs, a programmable switch is composed of a transmission gate for switching and an SRAM cell to hold the on/off-state of the switch. Since these components consist of transistors, the transmission gate has high resistance and large capacitance and the SRAM cell that requires six transistors consumes large area. Therefore, SRAM-based programmable switches lead to the degradation of interconnect performance and area efficiency [2].

To overcome drawbacks of conventional FPGAs, FPGAs that exploit resistive random access memories (RRAMs) as programmable switches instead of SRAM-based ones are widely studied [3]–[9]. In these RRAM-based FPGAs, however, one or two access transistors per a programmable switch are required for switch programming. The access transistor is relatively large despite the small footprint of an RRAM-based switch, and hence it interferes with further area reduction. For eliminating access transistors, nonvolatile via-switches are actively developed [10], [11]. The via-switch consists of atom switches, which are a kind of nonvolatile RRAMs developed

for application to FPGAs, and varistors in place of access transistors.

In the via-switch FPGA, the crossbar, which has a via-switch at each intersection of horizontal signal wire and vertical signal wire, is responsible for the routing of interconnects. However, programming those via-switches arbitrarily in a crossbar is not trivial since a programming voltage must be given through signal wires that are shared by multiple via-switches. In this case, unintentional switch programming may occur depending on the previous programming status due to signal detour, which is called sneak path problem. This problem interferes the reconfiguration of FPGA, and hence the verification of occurrence conditions and countermeasures is crucially important.

This paper encodes programming operations and occurrence conditions of sneak path problem in a via-switch crossbar into a Boolean satisfiability problem (SAT), and rigidly verifies the sneak path problem by using an SAT solver. Verification results show that the programming status of crossbar where the sneak path problem occurs can be detected by the proposed SAT encoding-based verification. Furthermore, we confirm that sneak path problems can be solved by imposing a specific programming constraint.

The remainder of this paper is organized as follows. Section II explains the structure of via-switch FPGA and sneak path problem. Section III presents the SAT encoding-based verification methodology for sneak path problem in via-switch FPGA followed by the verification results in Section IV. Concluding remarks are given in Section V.

II. VIA-SWITCH FPGA

A. Via-switch

The via-switch is a nonvolatile, rewritable, and compact switch that is developed to implement a crossbar switch by Banno et al. [10], and it is composed of atom switches and varistors. Here, we explain the device structure, functionality, and characteristics in the following. The programming of via-switch crossbar will be explained later in Section II-B.

The atom switch consists of a solid electrolyte sandwiched between copper (Cu) and ruthenium (Ru) electrodes as shown in Figure 1-a. By applying a positive voltage to the Cu electrode, a Cu bridge is formed in the solid electrolyte, and

978-1-5386-7100-9/18 $31.00 © 2018 IEEE

429

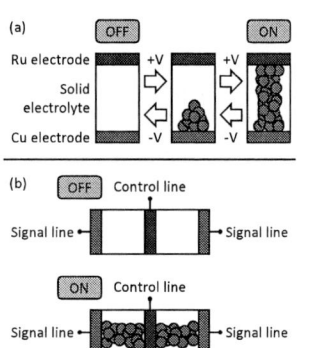

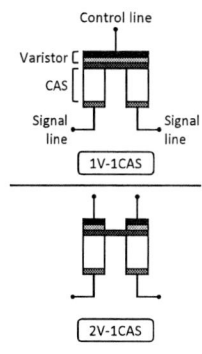

Fig. 1. Structure and operation of (a) atom switch and (b) CAS.

Fig. 2. Structures of via-switch. The top is 1V-1CAS structure, and the bottom is 2V-1CAS structure.

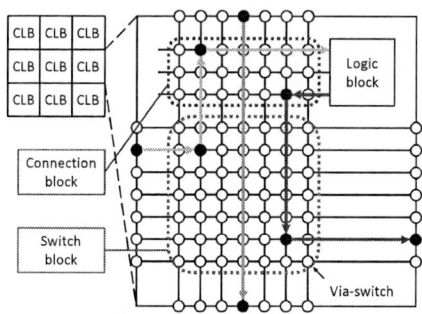

Fig. 3. Structure of via-switch FPGA.

the switch turns on. On the other hand, when a negative voltage is applied, Cu atoms in the bridge are reverted to the Cu electrode, and then the switch turns off. The switching between on-state and off-state is repeatable, and each state is nonvolatile. For improving the device reliability, the complementary atom switch (CAS) is devised, where it consists of two atom switches connected in series with opposite direction as shown in Figure 1-b. In the programming of CAS, a pair of signal line and control line supply a programming voltage to each atom switch, and two atom switches are programmed sequentially. During normal operation, on the other hand, only signal lines are used for routing [9].

To accurately provide the programming voltage only to the target CAS in a switch array, the varistor is introduced into the via-switch. Figure 2 shows the structure of via-switch, and the varistor is connected to the control terminal of CAS. When a voltage higher than the threshold value (programming voltage) is applied between the signal and control lines, the varistor supplies programming current to an atom switch. On the other hand, the varistor isolates the control lines from the signal lines during normal operation [10]. As shown in Figure 2, two structures of via-switch, one-varistor-one-CAS (1V-1CAS) and two-varistor-one-CAS (2V-1CAS) structure, are proposed.

Here, we summarize contribution of via-switch to FPGAs. The footprint, on-resistance, and capacitance are 18 F^2, 200 Ω, and 0.14 fF respectively [10], [12]. Thanks to these characteristics, the area efficiency and performance of via-switch FPGA are dramatically improved compared to SRAM-based one. Hotate et al. report that the crossbar density is improved by 26x, and the delay and energy in the interconnection are reduced by 90% or more [12].

B. Sneak Path Problem in Via-switch FPGA

The structure of via-switch FPGA is an array of configurable logic blocks (CLBs), and each CLB is composed of logic block and crossbar where a via-switch is placed at each intersection of signal lines as shown in Figure 3 [12]. The via-switch in the crossbar is responsible for connection and disconnection between the horizontal and vertical signal lines.

Besides, the top half of the crossbar serves a function of input and output multiplexers to the logic block and corresponds to the connection block in conventional FPGAs. On the other hand, the bottom half of the crossbar, which corresponds to the switch block, routes global interconnections. The logic block organizes combinational and sequential circuits.

The following explains the programming of a via-switch crossbar and sneak path problem. The crossbar structure with the 1V-1CAS via-switch is illustrated in Figure 4. Signal lines are placed horizontally and vertically, whereas control lines are aligned diagonally. Figure 4 also shows an example of programming steps in 2x2 crossbar where an atom switch is turned on at each step. Two programming drivers are activated at each step, and a positive voltage is given to one of the signal lines, and a ground voltage is given to one of the control lines. Other lines are floated. Steps 1 and 2 successfully turn on the via-switch at the bottom left. However, the next programming of the top left via-switch at steps 3 and 4 cannot be performed correctly. The atom switch that composes the bottom right via-switch is under programming unintentionally at step 4 since the positive voltage is provided through the on-state via-switch at the bottom left. Such an unintentional switch programming due to signal detour that is caused by on-state via-switches is the sneak path problem.

Next, Figure 5 shows the crossbar structure using 2V-1CAS via-switches and an example of programming steps, where this example is the same as Figure 4. In this crossbar structure, both signal and control lines are aligned horizontally and vertically. A pair of the signal and control lines crossing at the via-switch of interest are used for switch programming at each step, whereas other lines are floated. We can see that two via-switches at the bottom left and top left are successfully turned on without sneak path problem at steps 1-4. If turning ON the bottom right or top right via-switch after step 4, on the other hand, sneak path problem occurs.

The sneak path problem interferes the reconfiguration of FPGA, and hence it is essential to verify the occurrence conditions and countermeasures of this problem. The following sections verify the sneak path problem by using SAT encoding.

978-1-5386-7100-9/18 $31.00 © 2018 IEEE

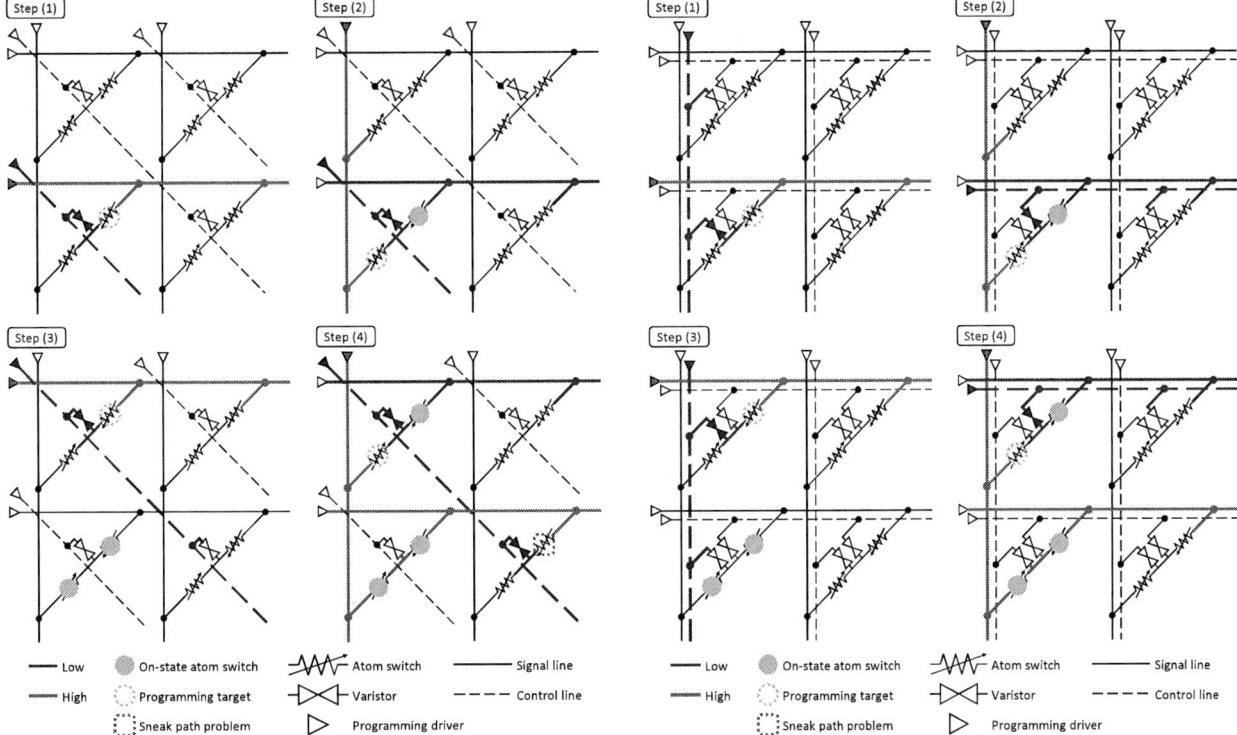

Fig. 4. Crossbar structure with 1V-1CAS via-switch and switch programming steps.

Fig. 5. Crossbar structure with 2V-1CAS via-switch and switch programming steps.

III. SAT ENCODING OF SNEAK PATH PROBLEM

A. Basics of SAT

A Boolean satisfiability problem (SAT) is a problem of searching an interpretation that satisfies a given logical expression. When a variable assignment that makes the given expression True exists, this expression is called satisfiable (SAT). On the other hand, if the given expression is False for all the possible variable assignments, this expression is unsatisfiable (UNSAT). In recent years, SAT is widely used in various fields thanks to dramatic performance improvement of SAT solvers. The typical flow of problem-solving using SAT is as follows. First, the original problem is translated into a logical expression, and this step is called SAT encoding. Next, we search the solution of the encoded problem by using an SAT solver. Finally, the solution from the solver is decoded, and then we can obtain the solution of the original problem [13].

B. Verification Methodology of Sneak Path Problem

The sneak path problem in via-switch FPGA changes the on/off-state of multiple via-switches due to the detour of the programming signal. This paper encodes this problem into SAT and mathematically verifies the sneak path problem. The verification flow is as follows. First, we translate the programming operations of via-switch crossbar and occurrence

conditions of sneak path problem into Boolean logical expressions. We define logical variables that express the on/off-state of via-switches, the electrical potential of wires, and so on. Then, we encode each circuit operation such as an operation to turn on/off the atom switch with the defined variables. The occurrence conditions of sneak path problem can be encoded into a proposition "multiple conditions for turning on/off an atom switch are satisfied at the same time". The details of SAT encoding are presented in Section III-C. After the SAT encoding, the logical expressions are input to the SAT solver. Finally, by analyzing the output of solver, namely, SAT or UNSAT and Boolean values assigned to each variable, we can evaluate whether the sneak path problem exists and the programming status of the crossbar when the sneak path problem occurs.

C. SAT Encoding of Sneak Path Problem

Here, let us explain how to SAT-encode the programming operations and sneak path problem in via-switch FPGA. Although the following supposes the 2x2 crossbar with 2V-1CAS via-switches, the SAT encoding for crossbars of any size and crossbars with 1V-1CAS via-switches can be derived in the same manner.

First, we define logical variables that represent the status of each element of the crossbar as shown in Figure 6. Atom switch and varistor have two states, on- and off-states, and

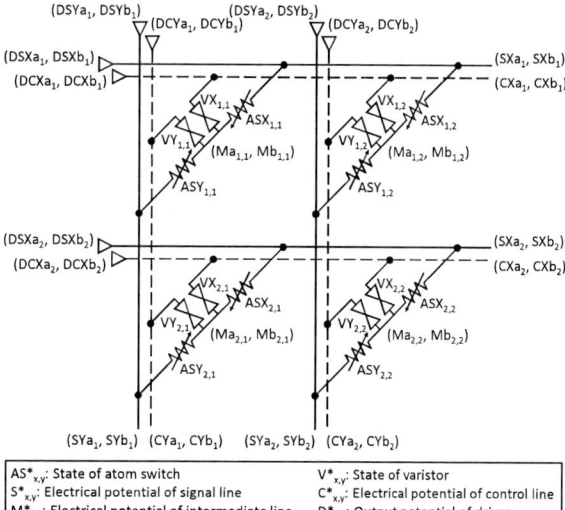

AS*$_{x,y}$: State of atom switch	V*$_{x,y}$: State of varistor
S*$_{x,y}$: Electrical potential of signal line	C*$_{x,y}$: Electrical potential of control line
M*$_{x,y}$: Electrical potential of intermediate line	D*$_{x,y}$: Output potential of driver

Fig. 6. Definition of Boolean variables in crossbar with 2V-1CAS via-switches.

then their status can be expressed by one variable, where True and False are allocated to on-state and off-state respectively. On the other hand, the electrical potential of each wire and the output voltage of each programming driver have three states, namely, low, high, and floating. To express these three states, we introduce a pair of two variables (a, b), where (False, False), (False, True), and (True, False) correspond to low, high, and floating respectively.

Next, we encode circuit operations of crossbar into logical expressions using the variables defined above. The conditions for turning on/off an atom switch, for example, are represented as Equations 1 and 2. To make Equation 1 True, both sides of the equality operator ($\Leftrightarrow$) need to be False or True. The right side becomes True only when (SXa_1, SXb_1) is (False, True) and $(Ma_{1,1}, Mb_{1,1})$ is (False, False), in other words, a high voltage is provided to signal line SX_1 and a low voltage is provided to intermediate line $M_{1,1}$. Also, the left side $ON_{ASX1,1}$ becomes True only when the right side is True. Therefore, $ON_{ASX1,1}$ is a variable that becomes True when a programming voltage for turning on is applied to atom switch $ASX_{1,1}$. Otherwise, it becomes False. Similarly, $OFF_{ASX1,1}$ in Equation 2 represents whether a programming voltage to turn off $ASX_{1,1}$ is applied.

$$ON_{ASX1,1} \Leftrightarrow (\neg SXa_1 \wedge SXb_1) \wedge (\neg Ma_{1,1} \wedge \neg Mb_{1,1}). \tag{1}$$

$$OFF_{ASX1,1} \Leftrightarrow (\neg SXa_1 \wedge \neg SXb_1) \wedge (\neg Ma_{1,1} \wedge Mb_{1,1}). \tag{2}$$

Due to the sneak path problem, the programming voltage is given to multiple switches by the signal detour, and their on/off-states are changed at once. Given the conditions for turning on/off an atom switch expressed by Equations 1 and 2, the sneak path problem can be translated into a proposition

"two or more conditions for turning on/off an atom switch are True". When logical expressions presenting this proposition is input to SAT solver and the output is satisfiable, we notice that the sneak path problem occurs. Then, the status of atom switches and electrical potentials can be obtained by checking the assigned Boolean value to each variable. On the other hand, when the solver output is unsatisfiable, there is no sneak path problem.

For investigating countermeasures for the sneak path problem, this paper verifies the crossbar operations under some programming constraints. Banno et al. claim that the sneak path problem in the 2V-1CAS crossbar occurs when multiple on-state via-switches exist in both the same horizontal line and the same vertical line [10]. In the crossbar with 1V-1CAS via-switches, on the other hand, the authors say that the sneak path problem arises when multiple via-switches in either the same horizontal line or the same vertical line are on-state. For rigidly verifying their claims, we introduce the following two programming constraints, namely, row constraint and row & column constraint. The row constraint prohibits configurations where multiple on-state via-switches exist in the same horizontal line, whereas multiple via-switches in the same vertical line are acceptable. On the other hand, under row & column constraint, multiple on-state via-switches in both the same horizontal line and the same vertical line are prohibited. By adding logical expressions of these constraints to solver input, we can verify whether the sneak path problem occurs under the programming constraints.

In addition to the above, we encode the behaviors below of all components in the crossbar, namely, atom switches, varistors, wires, and drivers

- The varistor is turned on when the applied voltage is higher than the threshold.
- The electrical potential of wire is determined by active programming drivers and connection status of signal lines.
- The number of active programming drivers (two in total) and their positions (horizontal, vertical, diagonal directions) are restricted.
- Assigning variables (True, True) to wires and drivers is prohibited.

Among them, determining the potentials of signal lines is complicated since it depends on the connection status of signal lines. The connection status of signal lines is different for each FPGA configuration, and we found that we needed to prepare logical expressions to determine the potentials separately for each configuration. For example, when all the via-switches in a crossbar are off-state, the potential of each signal line can be determined independently since each signal line is not connected to any other signal lines. In other words, each line is driven only if the programming driver that directly connects to the line is active. Otherwise, the signal line is floating. On the other hand, once via-switches are turned on, the potentials of all the signal lines that are connected to each other through on-state switches cannot be determined independently. We need to check the status of all the drivers that connect to the signal lines of interest and determine the potentials according to the combinations of their status. When all the drivers of the connected signal lines are inactive, the connected signal lines are floating. Otherwise, the potentials of the connected lines are fixed by the active drivers. From the above, when determining the potentials, the position of drivers that we have to check varies depending on the FPGA configuration.

The details of SAT encoding for the varistors and programming drivers are omitted due to space limitations.

IV. VERIFICATION RESULTS

A. Effectiveness of SAT Encoding-based Verification

First, we confirm that the proposed SAT encoding-based verification can evaluate the sneak path problem in the via-switch crossbar. For this purpose, we prepare a logical expression that connects all the expressions explained in Section III-C by logical product, and input it to an SAT solver. We use MiniSat [14], which is open source SAT solver. We evaluate both 2V-1CAS and 1V-1CAS crossbars where their size is 2x2. No programming constraint is given in this evaluation.

In both 2V-1CAS and 1V-1CAS crossbars, the output of the solver is satisfiable, which means sneak path problem arises. The analysis of the variable assignment results successfully reveals the circuit status that causes the sneak path problem. Thus, we confirm the detectability of the sneak path problem.

It should be noted that the total number of possible variable assignments that satisfy the input expression is 256 in a 2V-1CAS crossbar and 456 in a 1V-1CAS crossbar, and we analyzed the variable assignments not in all cases but in some of them. Our future works include confirming whether all sneak path problems can be detected and whether the input expression is not satisfied with the circuit status that does not cause the sneak path problem.

B. Countermeasures for Sneak Path Problem

If the proper programming constraint that prohibits all the configurations where sneak path problem occurs is available, we can correctly reconfigure the via-switch FPGA since the programming of any target via-switch without any unexpected programming is ensured. As candidates of such constraints, we verify whether the output of SAT solver varies by imposing the row constraint and row & column constraint described in Section III-C. If the output varies from satisfiable to unsatisfiable, the programming constraint is confirmed to be effective for the sneak path problem.

Verification results are listed in Table I. As previously mentioned, the sneak path problem occurs when the output is satisfiable (SAT), and there is no sneak path problem when the output is unsatisfiable (UNSAT). Although we verify 14 sizes of the crossbar, 2x2-2x7, 3x2-3x4, 4x2-4x3, 5x2, 6x2, and 7x2, Table I summarizes the verification results since the same output is obtained regardless of the crossbar size. We can see that the sneak path problem in a 2V-1CAS crossbar can be avoided by either row constraint or row & column constraint.

TABLE I
EFFECTIVENESS OF PROGRAMMING CONSTRAINTS.

	No constraint	Row constraint	Row & column constraint
2V-1CAS crossbar	SAT	UNSAT	UNSAT
1V-1CAS crossbar	SAT	SAT	UNSAT

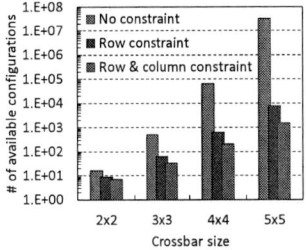

Fig. 7. Number of available configurations under programming constraints.

In a 1V-1CAS crossbar, on the other hand, the row constraint cannot prevent the sneak path problem. To solve the sneak path problem in the 1V-1CAS crossbar, we need to impose the row & column constraint, which is more severe constraint than row constraint.

Next, we evaluate the impact of programming constraints on the number of available configurations. The programming constraints prohibit some configurations of via-switch FPGA, and hence imposing the constraint leads to a decrease in the number of usable configurations. Figure 7 shows the number of available configurations under no constraint, row constraint, and row & column constraint. Here, 2x2, 3x3, 4x4, and 5x5 crossbars are evaluated. We can see that the decrease in the number of usable configurations due to the programming constraints is significant in larger crossbars. For example, compared to the 5x5 crossbar with no constraint, the row constraint and row & column constraint diminish the number of available configurations by three orders of magnitude and four orders of magnitude, respectively.

From the above results, we find that the sneak path problem can be solved by employing the proper programming constraint. Verification results also show that the programming constraints to avoid the sneak path problem are different depending on the structure of via-switch. We need to pay attention to the fact that the programming constraints reduce the total number of available configurations and consequently limit routing flexibility.

C. Impact of Crossbar Size on Verification Time

Finally, we discuss the impact of crossbar size on the verification time. Figure 8 shows the verification time of SAT solver when the crossbar size is varied from 2x2 to 2x7. We evaluate two cases; 2V-1CAS crossbar with no programming constraint where the output is satisfiable, and 2V-1CAS crossbar with row constraint where the output is unsatisfiable. The verification time was measured ten times for each condition, and the average value of them is shown in Figure 8. We executed

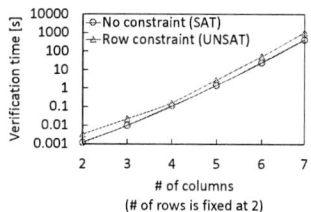

Fig. 8. Verification time of SAT solver when crossbar size is varied.

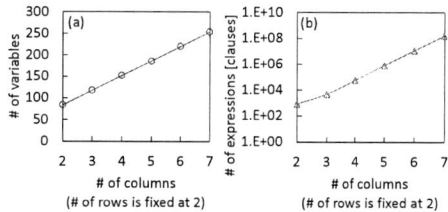

Fig. 9. Number of (a) logical variables and (b) logical expressions when crossbar size is varied.

the SAT solver on Xeon X5680 3330 MHz x2 with 96 GB RAM.

The maximum verification time in this evaluation is 18 minutes, and it is not so long. However, we can see that the verification time exponentially increases as the crossbar size becomes larger. To investigate the cause of this, we evaluate the number of logical variables and expressions when the crossbar size is varied from 2x2 to 2x7. Figure 9 shows the evaluation results. The number of logical variables is proportional to the crossbar size. On the other hand, the number of logical expressions exponentially increases as the crossbar size becomes larger. When the crossbar size is 2x7, the number of logical expressions is over 164 million. This exponential increase in the number of expressions is supposed to cause an exponential increase in the verification time.

The above analysis of the number of logical expressions finds that the exponential increase in the number of expressions originates from the logical expressions to determine the potentials of signal lines explained in Section III-C. Remind that we separately prepare the logical expression to determine the potential for each FPGA configuration. When the number of atom switches in a crossbar is n, the number of possible configurations is 2^n since each atom switch has two states, on- and off-states. Here, n is proportional to the crossbar size. Therefore, the number of logical expressions to determine the potential increases exponentially.

From the above discussion, we conclude that the verification time increases due to the exponential increase in the number of logical expressions and it is not scalable for large crossbars. Future works include the modification of SAT encoding to reduce the increase in the number of expressions.

V. CONCLUSION

This paper verified the sneak path problem that interfered reconfigurations of via-switch FPGA. We encoded programming operations of via-switch crossbar and occurrence conditions of sneak path problem into SAT and verified them by an SAT solver. We confirmed that the proposed SAT encoding-based verification could detect the sneak path problem. Verification results show that imposing proper programming constraint can eliminate the sneak path problem. Future works include the evaluation of sneak path problem detectability in details and improving the SAT encoding to deal with larger crossbars.

ACKNOWLEDGMENTS

This work was supported by JSPS KAKENHI Grant Number JP17J10008 and JST CREST Grant Number JPMJCR1432, Japan.

REFERENCES

[1] I. Kuon and J. Rose, "Measuring the Gap Between FPGAs and ASICs," *IEEE Transactions on Computer-Aided Design of Integrated Circuits and Systems*, vol. 26, no. 2, pp. 203–215, Feb 2007.

[2] M. Lin, A. E. Gamal, Y. C. Lu, and S. Wong, "Performance Benefits of Monolithically Stacked 3-D FPGA," *IEEE Transactions on Computer-Aided Design of Integrated Circuits and Systems*, vol. 26, no. 2, pp. 216–229, Feb 2007.

[3] P. E. Gaillardon, D. Sacchetto, G. B. Beneventi, M. H. B. Jamaa, L. Perniola, F. Clermidy, I. O'Connor, and G. D. Micheli, "Design and Architectural Assessment of 3-D Resistive Memory Technologies in FPGAs," *IEEE Transactions on Nanotechnology*, vol. 12, no. 1, pp. 40–50, Jan 2013.

[4] S. Tanachutiwat, M. Liu, and W. Wang, "FPGA Based on Integration of CMOS and RRAM," *IEEE Transactions on Very Large Scale Integration (VLSI) Systems*, vol. 19, no. 11, pp. 2023–2032, Nov 2011.

[5] J. Cong and B. Xiao, "FPGA-RPI: A Novel FPGA Architecture With RRAM-Based Programmable Interconnects," *IEEE Transactions on Very Large Scale Integration (VLSI) Systems*, vol. 22, no. 4, pp. 864–877, April 2014.

[6] X. Tang, P. E. Gaillardon, and G. D. Micheli, "A high-performance low-power near-Vt RRAM-based FPGA," in *2014 International Conference on Field-Programmable Technology (FPT)*, Dec 2014, pp. 207–214.

[7] Y. Y. Liauw, Z. Zhang, W. Kim, A. E. Gamal, and S. S. Wong, "Nonvolatile 3D-FPGA with monolithically stacked RRAM-based configuration memory," in *2012 IEEE International Solid-State Circuits Conference*, Feb 2012, pp. 406–408.

[8] K. Okamoto, M. Tada, T. Sakamoto, M. Miyamura, N. Banno, N. Iguchi, and H. Hada, "Conducting mechanism of atom switch with polymer solid-electrolyte," in *2011 International Electron Devices Meeting*, Dec 2011, pp. 12.2.1–12.2.4.

[9] M. Miyamura, T. Sakamoto, M. Tada, N. Banno, K. Okamoto, N. Iguchi, and H. Hada, "Low-power programmable-logic cell arrays using nonvolatile complementary atom switch," in *Fifteenth International Symposium on Quality Electronic Design*, March 2014, pp. 330–334.

[10] N. Banno, M. Tada, K. Okamoto, N. Iguchi, T. Sakamoto, M. Miyamura, Y. Tsuji, H. Hada, H. Ochi, H. Onodera, M. Hashimoto, and T. Sugibayashi, "A novel two-varistors (a-Si/SiN/a-Si) selected complementary atom switch (2V-1CAS) for nonvolatile crossbar switch with multiple fan-outs," in *2015 IEEE International Electron Devices Meeting (IEDM)*, Dec 2015, pp. 2.5.1–2.5.4.

[11] N. Banno, M. Tilda, K. Okamoto, N. Iguchi, T. Sakamoto, H. Hada, H. Ochi, H. Onodera, M. Hashimoto, and T. Sugibayashi, "50x20 crossbar switch block (CSB) with two-varistors (a-Si/SiN/a-Si) selected complementary atom switch for a highly-dense reconfigurable logic," in *2016 IEEE International Electron Devices Meeting (IEDM)*, Dec 2016, pp. 16.4.1–16.4.4.

[12] J. Hotate, T. Kishimoto, T. Higashi, H. Ochi, R. Doi, M. Tada, T. Sugibayashi, K. Wakabayashi, H. Onodera, Y. Mitsuyama, and M. Hashimoto, "A highly-dense mixed grained reconfigurable architecture with overlay crossbar interconnect using via-switch," in *2016 26th International Conference on Field Programmable Logic and Applications (FPL)*, Aug 2016, pp. 1–4.

[13] D. E. Knuth, *The Art of Computer Programming, Volume 4, Fascicle 6: Satisfiability*, 1st ed. Addison-Wesley Professional, 2015.

[14] "MiniSat," http://minisat.se/.

RRAM Based Buffer Design for Energy Efficient CNN Accelerator

Kaiyuan Guo, Jincheng Yu, Xuefei Ning, Yiming Hu, Yu Wang, Huazhong Yang

Department of Electronic Engineering, Tsinghua University

{gky15@mails.tsinghua.edu.cn, yu-wang@tsinghua.edu.cn}

Abstract—Convolutional Neural Network (CNN) has become the state-of-the-art algorithm for many computer vision tasks. But its high computation complexity and high memory complexity makes it hard to be deployed on traditional platforms like CPUs. Memory energy can take up a large part of the system energy, which limits the energy efficiency of CNN processing. The emerging metal-oxide resistive switching random-access memory (RRAM) has been widely studied because of its good properties like high storage density and the compatibility with CMOS. In this paper, a system level energy analysis of using RRAM as on-chip weight buffer is carried out for a typical CNN accelerator. Hardware and scheduling optimizations are proposed to fully utilize the large RAM and avoid high read/write energy overhead. Experimental results show that RRAM based designs save 12-18% system energy with 15-75% smaller on-chip RAM area compared with SRAM designs.

Index Terms—RRAM, Convolutional Neural Network, Hardware Accelerator

I. INTRODUCTION

Convolutional Neural Network (CNN) has become a state-of-the-art algorithm for a wide range of applications like image classification [1], object detection [2] and other image-based tasks. Compared with traditional hand-crafted-feature-based methods, CNN introduces a uniform model for different tasks and adjusts the model based on different training data set. Thus CNN can be adopted in various tasks and keeps high model accuracy.

But CNN is still not widely applied in real applications because of its high computation and memory complexity. A typical network like AlexNet [3] consists of more than 240MB parameters and 1.4G FLOPs (floating-point operations) for the inference of a single 224×224 image. More advanced networks [1], [4] require much more computation and memory than AlexNet. The energy cost for CNN computation is thus high, especially on traditional platforms like CPU.

Various works explore energy efficient hardware designs for CNN accelerators. Many designs focus on CMOS based circuit design targeting at efficient data path and memory system [5], [6]. In these designs, the size of the on-chip SRAM is quite limited. Thus external memory like DRAM is always needed in real applications. The high energy cost brought by off-chip data transfer dramatically limits the system energy efficiency.

One solution to reduce memory access is to perform in-memory computing. RRAM cross bar based design is one of the popular research topics. Chi et al. propose the PRIME [7] architecture which implements the matrix-vector multiplication directly with the RRAM crossbar. Work by

Cheng et al. [8] implements the RRAM crossbar to support both inference and training of neural networks. In memory computation with RRAM shows attractive energy efficiency. But the application range of the design is limited by the scalability and computation accuracy of the RRAM crossbar.

Another way to reduce off-chip data transfer is to implement large on-chip memory, where RRAM is a good candidate. We compare RRAM with SRAM and DRAM in Table I according to the performance reported in [9], [10] and simulation result by NVSim [11]. Compared with SRAM, the storage density of RRAM is similar to DRAM, which is about $20\times$ higher. Compared with DRAM, RRAM can be integrated on-chip while the former one can not.

TABLE I
COMPARISON BETWEEN SRAM, DRAM, AND RRAM

	SRAM	DRAM	RRAM
Cell Size	$140F^2$	$6 \sim 8F^2$	$6F^2$
Frequency	>1GHz	100-400MHz	<100MHz
Integrated on Chip	yes	no	yes
I/O energy / Byte	<1pJ	~10pJ	~40pJ

But RRAM is also limited in some aspects. First, the I/O bandwidth of RRAM is smaller than SRAM. For applications like CPU cache, where latency is critical to performance, RRAM may not be a good choice. For CNN, the data access pattern is static. This means we can design data storage to achieve sequential access at run-time. So we can use more banks to compensate for the limited bandwidth.

Second, the I/O dynamic energy cost of RRAM is high In this paper, we show that a simple implementation with RRAM increases the total system energy cost. But there is a chance that we utilize the property of CNN computation to reduce on-chip memory access to overcome the energy overhead of using RRAM.

In this paper, we introduce RRAM as the weight memory for a typical CNN accelerator design and optimize the design in hardware and scheduling level to overcome the RRAM energy overhead. The contributions of this paper are as follows:

- Hardware optimization is proposed to reduce the RRAM dynamic energy overhead.
- Dedicated scheduling strategy optimization is proposed to fully utilize the large RRAM buffer to reduce off-chip data transfer.
- Design space exploration is done with state-of-the-art networks to show the effect of using RRAM as the on-chip buffer.

Experimental results on state-of-the-art CNN models show that RRAM based design saves $12-18\%$ system energy with a $15-75\%$ smaller on-chip RAM area compared with SRAM design. The proposed hardware and scheduling optimization reduce up to 96% on-chip RAM access energy and 98% off-chip data transfer.

The rest of this paper is organized as follows: Section II introduces the related work for CNN accelerator and RRAM research. Section III introduces the hardware design. The scheduling strategies are introduced in section IV. The experimental results are shown in section V. Section VI concludes this paper.

II. RELATED WORK

A. Convolutional Neural Network

A CNN mainly consists of two types of layers, convolution (CNV) layers and fully connected (FC) layers, which contribute to most of the computation and parameters in a CNN. The computation for a CNV layer is shown as the pseudo code below. F_{in} and F_{out} are the input and output feature maps. K and b denote the convolution kernels and bias. The fully connected layer can be expressed as matrix-vector multiplication and can be treated as a special case of CNV layers where all the feature maps are 1×1, and convolution kernels are 1×1.

CNVLAYER(F_{in}, F_{out}, K, b)

```
1   for m = 1 → M // output channel loop
2       for n = 1 → N // Input channel loop
3           // feature map pixel loop
4           for each F_out[m][x][y] ∈ F_out[m]
5               F_out[m][x][y] = 0
6               // convolution kernel loop
7               for each K_mn[xx][yy]
8                   F_out[m][x][y]+ = K_mn[xx][yy]*
                        F_in[n][x + xx][y + yy]
9               F_out[m][x][y]+ = b_m
```

Table II shows the number of parameters and the computation complexity of some state-of-the-art network models. Usually a CNN consists of 10-100 million parameters and requires from 1-50G operations for the inference of each input. Compared with that, general purpose platforms like CPU only offers 1-100GOP/s computation capacity and consumes 1-100W power. This hardly meets the real-time requirement in video processing or the power limit of mobile platforms. There is urgent need of fast and energy efficient computation platform for processing CNN.

B. CNN Accelerator

The design of a CNN Accelerator involves two aspects: computation units and scheduling strategy. For computation units, various techniques have been studied to reduce the bit-width used for the data expression in CNN to reduce both the memory requirement and energy cost for computation and data transfer. Experimental results from [12] show that 8-bit

TABLE II
COMPUTATION COMPLEXITY AND PARAMETER SIZE OF STATE-OF-THE-ART NETWORKS.

	Computation (GOP)			Parameter (M)		
	CNV	FC	Total	CNV	FC	Total
AlexNet [3]	1.33	0.12	1.45	2.33	58.62	60.95
VGG-11 [1]	14.97	0.25	15.22	9.22	123.63	132.85
VGG-16 [1]	30.69	0.25	30.94	14.71	123.63	138.34
ResNet-34 [4]	7.28	0.001	7.281	21.1	0.51	21.61

linear data quantization introduces negligible accuracy loss for common CNN models compared with the original 32-bit floating-point model.

On scheduling aspect, most of the CNN accelerators [5], [6] uses single layer scheduling strategies, where all the computation units work for the same layer at any time. Different layers are executed one by one. Single-layer design requires that the result of a layer to be written back to external memory when it is larger than the on-chip buffer. Alwani et al. [13] suggest that adjacent layers can be fused together as a scheduling unit to reduce the data transfer with the external memory.

As suggested by [14], typical energy for a 32bit DRAM access is $100\times$ more than that of a 32-bit SRAM access, $200\times$ more than a 32bit multiplication and $6400\times$ more than a 32-bit addition. This indicates that the memory access energy contributes a major part of the total system cost.

C. RRAM

RRAM is one of the promising candidate for large on-chip memory. The cell size of SRAM is typically 10-20x larger than that of RRAM with the same technology. Some memory designs [15], [16] try to store multiple bits within a cell to further increase storage density. Despite the high storage density, the limited bandwidth of RRAM is always a consideration when using as on-chip memory. But recent work [10] shows that even high-density array of 32Gb achieves 1GB/s read bandwidth, and 200MB/s write bandwidth. In this work, we adopt [11] as a tool to generate RRAM modules of different sizes in our experiments.

Using RRAM crossbar for both computation and storage in NN accelerators is another research topic [7] [8] [17] [18]. In these designs, extremely high energy efficieny is achieved because there is no energy needed to read network parameters from memory. But these designs are limited by the storage and computation accuracy of RRAM devices and the analog-digital converter overhead. In this paper, we only focus on using RRAM as on-chip memory, not as computation units.

III. HARDWARE DESIGN

This section introduces the CNN accelerator we adopt in this work. We first introduce the architecture and energy model, and then the optimization to reduce the RRAM buffer dynamic energy cost.

A. Architecture

The overall architecture is shown in Figure 1(a). The system adopts DDR as external memory to support enough memory space. The on-chip memory consists of two buffers:

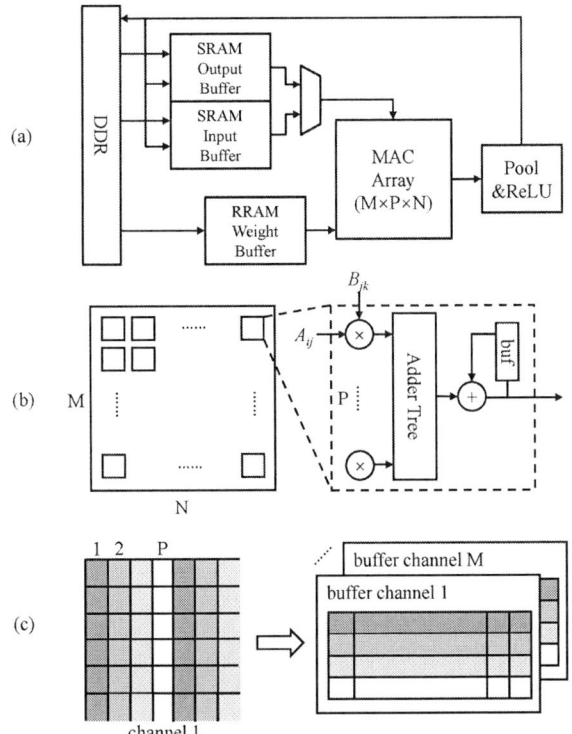

Fig. 1. The CNN accelerator architecture. (a) The architecture mainly consists of a MAC array for computation and two levels of memory with on-chip buffers and external DDR. (b) MAC array structure for an $A_{p \times m} \times B_{m \times n}$ matrix-matrix multiplication. (c) Feature map buffer organization for sliding window data selection.

input/output buffer and weight buffer. Input and output buffer stores feature maps during the process of a CNN. As the output of one layer serves as the input of the next, the input data of MAC array can be chosen from both of the buffers by a MUX unit. All the buffers work in a double-buffer way to cover the off-chip data transfer time with calculation time.

We first show how the process of CNN is parallelized with this design. As introduced in Section II A, a CNN layer consists of 4 nested loops. In this case, we unroll the feature map pixel loop, input channel loop and output channel loop as a $A_{m \times p} \times B_{p \times n}$ matrix-matrix multiplication. Each row of matrix A denotes the p pixels read from a feature map and each B_{ij} denotes a pixel in the convolution kernel corresponds to input channel i and output j. Thus the convolution of p pixels with $K \times K$ convolution kernels will be done with $\lceil M/m \rceil * K^2$ steps. The structure of the computation core is shown in Figure 1(b). The array consists of $p \times n$ processing elements (PEs). Each of the PE implements a size-m vector inner product in a pipelined manner. Each PE has an accumulator for its result.

To support the above parallelization, a window selection function is needed for each channel stored in input/output buffer. In this design, we adopt the data mapping format as shown in Figure 1(c). The pixels of each channel are stored in p independent RAMs so that a size-p window at any position

can be selected from the buffer within a single cycle. Weight buffer can be implemented with a normal simple dual port RAM.

Consider the data access pattern, implement input/output buffers with RRAM is not practical. First, they require high write bandwidth to receive computation results from fast on-chip logic. Second, they require a high random access bandwidth, not sequential access bandwidth. Compared with input/output buffer, weight buffer only requires a high sequential access bandwidth. So in this work, we only consider implementing weight buffer with RRAM.

B. Choosing the Loop Order

With the parallelization strategy decided as introduced in section III-A, the data access behavior is affected by the order of the loops in the whole process. If we do not consider data transfer with DDR, the on-chip buffer energy cost comes from three aspects: input/output buffer read, weight buffer read, and input/output buffer write. To minimize this energy cost, we consider two loop execution order: kernel first and pixel first.

Kernel first order computes the $K \times K$ convolution on the p pixels with K^2 cycles first, then move on to the next p pixels. With this order, during the K^2 cycles, as the feature map window slides, the spatial locality of 2-d convolution is explored. Each feature map pixel is accessed only once. Pixel first order reuses the same weight in a $K \times K$ kernel and computes multiple p-pixel groups, then move on to the next weight. With this order, more feature map access is needed, and intermediate result for each pixel group should be stored. This is not commonly adopted because intermediate result requires more bits for fixed-point data process. Suppose c pixel groups are processed together, we compare the data access in Table III.

TABLE III
DATA ACCESS COMPARISON BETWEEN KERNEL FIRST ORDER AND PIXEL FIRST ORDER TO CALCULATE c GROUPS OF p PIXELS WITH $K \times K$ CONVOLUTION.

	Kernel First	Pixel First
Input Feature Map	$cK(p + K - 1)$	cpK^2
Weight	cK^2	K^2
Output Feature Map Rd	0	$cp(K^2 - 1)$
Output Feature Map Wr	cp	$cp(K^2 - 1)$

Directly using input/output buffer to store the intermediate result is not feasible because the intermediate result for fixed-point multiplication requires more bits. As RRAM has high read/write dynamic, we should try to reuse the weight to reduce access to weight buffer. In this design, we use a local accumulation buffer of size $2n$ as shown in Figure 1(b) to enable n pixel groups processed together before written back to input/output buffer. This introduces extra on-chip memory cost. We examine the area and energy overhead in our experiment.

IV. SCHEDULING STRATEGY

In the previous section, we choose the loop execution order to reduce data access. In this section, we analyze schedule

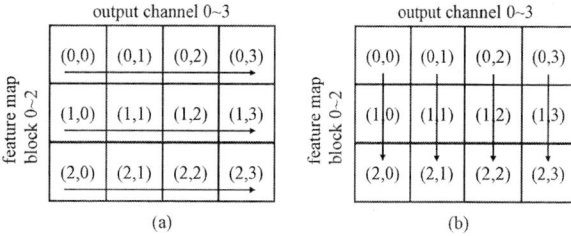

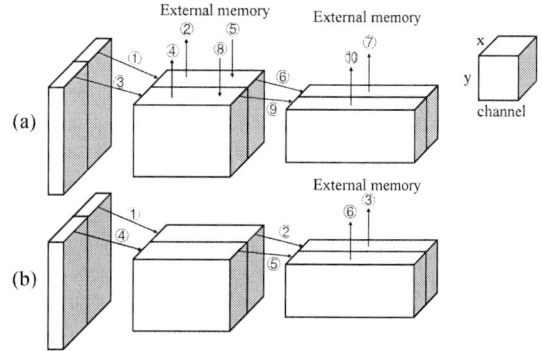

Fig. 2. An example of how loop order affects data transfer behavior. Block (i, j) denotes the computation for the ith feature map block on the jth channel. (a) Reuse feature map first and load each weight 3 times. (b) Reuse weight first and load each feature map 4 times.

Fig. 3. An example of a cross-layer schedule over 3 layers. (a) single-layer schedule. (b) cross-layer schedule.

strategy, which decides the off-chip data transfer behavior during the process of a network. Three kinds of scheduling are considered: single-layer, cross-layer and fixing weight on-chip.

Single-layer Schedule Strategy. In the case where on-chip buffer size is smaller than the weight size and feature map size of a single-layer, data needs to be loaded multiple times. An example is shown in Figure 2 where the weight buffer can hold only $1/4$ of the weight and input buffer can hold only $1/3$ of the feature maps. Choosing reuse weights or reuse feature map will cause a difference in data transfer behavior and further affects the data transfer energy and data transfer time.

Cross-layer Schedule Strategy. As suggested by [13], when the network is processed layer by layer, the result of one layer needs to be written back to external memory it is larger than the output buffer. But if the weight buffer is large enough to contain more than one layers' weight, these layers can be scheduled together to avoid writing the intermediate layers result to external memory. An example is shown in Figure 3, where the cross-layer schedule avoids the second layer's feature map to be transferred to and from external memory. As we will implement weight buffer with RRAM, the size of weight buffer can be much larger and thus benefits from this strategy.

Fixing Weight On-chip. An observation is that if the weight buffer is large enough to hold all the layer's weight on-chip, no weight will need to be read from external memory. With a slightly smaller buffer, we can still follow this idea by fix some of the layer's weights in the on-chip buffer to reduce external

memory access. So we need to find which layers' weight is to be fixed in the buffer as to minimize the total energy cost.

The solution space of this problem is binary tree of depth $(n + 1)$. n denotes the number of CNV layers in the network. To avoid searching the whole $O(2^n)$ solution space, we prune the binary tree with the buffer size limitation. If the the weights on a path already exceeds the weight buffer size, we ignore the search of its subtree. The pseudo code of the optimization process is as follows:

SEARCHFIXEDWEIGHT($layers$)
1   Let $is\_fixed[1..layers.size]$ be a boolean vector
2   IterSearchFixedWeight($layers, 1, is\_fixed$)
3   **return** is_fixed

ITERSEARCHFIXEDWEIGHT($layers, n, f$)
1   **if** $l > layers.size$
2   **return** OptEnergy($layers, f$)
3   **if** available buffer size $< layers[n].weight\_size$
4        $f[n] = false$
5   **return** IterSearchFixedWeight($layers, n + 1, f$)
6    $f1 = f; f1[n] = false$
7    $f2 = f; f2[n] = true$
8    $e1 = $ IterSearchFixedWeight($layers, n + 1, f1$)
9    $e2 = $ IterSearchFixedWeight($layers, n + 1, f2$)
10   $f = (e1 < e2) ? f1 : f2$
11   **return** Min(e1, e2)

V. EXPERIMENTS

Experiments are carried out on the architecture introduced in Section III, with the scheduling strategy in section IV applied. We use a behavior level simulator to analyze the memory access and computation energy for running a certain network. Memory access dynamic energy cost, memory static energy cost and computation energy cost are considered in our model.

A. Experiment setup

The MAC array in the architecture is configured as an $8 \times 8 \times 8$ array running at 1GHz. This offers a peak performance of 1TOP/s. 8-bit multiplication and 32-bit accumulation are adopted in this model. Multiplication and addition energy is scaled down from the data in [14] to 22nm technology.

The above configuration requires the read bandwidth of input/output buffer and weight buffer to be at least 64GB/s. We implement each buffer with 8 banks and each of them should offer 8GB/s read bandwidth. On-chip memory parameters are generated from NVSim [11] with different memory size configurations. RRAM buffer bit width is configured as 256bit to achieve enough bandwidth.

The external memory parameter is generated from MICRON DDR4 power calculator [19]. The generated dynamic I/O power is further converted to energy per read or write byte. In our experiment, we use 2 DDR chips as external memory because the bandwidth can support the configured MAC array with proposed schedule strategy and the size is enough. To reduce background power overhead, we use the least number

TABLE IV
DEVICE PARAMETERS USED FOR I/O BUFFER, WEIGHT BUFFER AND DDR.

	bitwidth (B)	size (B)	Rd BW (GB/s)	Wr BW (GB/s)	Rd Ene (pJ)	Wr Ene(pJ)	Leakage (mW)	Area (um^2)
22nm LSTP RRAM	32	128K	15.169	1.556	67.690	195.286	0.04000	21224
	32	256K	11.693	1.534	75.071	217.468	0.04104	26707
	32	512K	8.005	1.490	88.483	260.073	0.04314	37308
	32	1M	11.056	1.534	133.189	268.319	0.05282	61090
	32	2M	10.306	1.534	231.750	357.190	0.07806	107007
22nm LSTP SRAM	8	16K	9.145	5.147	3.057	0.556	0.00134	10031
	8	32K	8.609	4.973	6.432	1.315	0.00270	20048
	8	64K	16.831	11.763	6.780	3.777	0.00600	42803
	8	128K	10.977	10.451	7.931	2.792	0.01153	82032
	8	256K	10.977	10.451	11.562	6.424	0.02306	164065
DDR4	4	128M	3.2		80.300	82.719	52.80000	N/A

of chips. All the detailed data and configuration is shown in Table IV.

The buffer in the accumulator is also considered in our experiments. 4 types of buffers are chosen for design space exploration. Corresponding parameters are generated by NVSim and are shown in Table V.

TABLE V
32-BIT ACCUMULATION BUFFER PARAMETERS.

depth	16	32	64	128
energy per read (pJ)	0.045	0.056	0.107	0.12
energy per write (pJ)	0.022	0.031	0.083	0.094
area (um^2)	57.673	103.422	188.714	354.138
Leakage (nW)	6.162	11.133	22.385	44.823

B. Energy Cost Analysis

We do experiments on all the combinations of the RAM configurations in Table IV and V, which means $5\times5\times5 = 125$ choices for an SRAM or RRAM based accelerator. The three schedule strategies in section IV is applied to each choice. Figure 4 shows the experimental result with the convolution layers of VGG-11 network for 1 input. The energy cost of computation and DDR leakage is marked and is the same for all the designs because the network is fixed and all the designs are computation bounded. So the processing time is a constant to all the designs. In general, the RRAM based design consumes less energy compared with an SRAM design of the same area. The minimal energy design for SRAM and RRAM costs 3086uJ and 2532uJ respectively, meaning that using RRAM can save 18% system energy. The RRAM design is also 15% smaller than the SRAM design. We also do experiments with the convolution layers of VGG-16 and AlexNet, where RRAM design saves 18% and 12% energy with 15% and 75% less on-chip RAM area compared with SRAM design.

We examine the effect of the proposed hardware optimization and scheduling strategy. First, the loop order is the key effect to the RRAM based design, as shown in Figure 4(b). Note that some of the design points with the kernel first loop order are out of the figure because the energy cost is larger than 8000uJ. The large read energy of RRAM dominates the system energy as the capacity of RRAM increases. The reduction in DDR access cannot compensate for this overhead. Even on the smallest design, changing the loop order to pixel first will save at least 1/3 of the system energy cost.

A breakdown of the on-chip buffer energy cost is shown in Figure 5(a). The largest designs, in our experiment the designs with 4MB SRAM input/output buffer and 2MB SRAM (or 16MB RRAM) weight buffer are used. Here we only consider the read energy of input/output buffer and weight buffer and the dynamic energy of accumulation buffer. The write energy to input/output buffer and leakage energy are not affected by the accumulation buffer and thus not considered. Up to 96% energy is saved for the RRAM based designs consider the overhead brought by the accumulation buffer. SRAM design also benefits from the change in loop order but not as effective as to the RRAM designs.

Second, fixing the weight on-chip helps the hardware fully utilize the on-chip RAM to reduce off-chip data transfer. As can be seen from Figure 4(b), using the pixel first loop order, the accelerator cannot benefit from a larger on-chip buffer with single-layer scheduling and cross-layer scheduling. By fixing some of the network weights on-chip, the system energy cost is gradually reduced as the on-chip RAM size increases.

A breakdown of the DDR transfer cost is shown in Figure 5(b). All the designs implement no accumulation buffer and 1MB SRAM as input/output buffer. The single layer scheduling and cross-layer scheduling can only achieve 6.4% and 13.5% energy saving respectively when the weight buffer increases from 1MB to 16MB. With the fix weight strategy, the energy saving is 98.5%.

VI. CONCLUSION

In this paper, we introduce RRAM into CNN accelerator as on-chip weight buffer. To address the high read/write dynamic energy of RRAM, we change the loop order and implement the accumulation buffer to reduce RRAM access. Up to 96% energy is saved with this optimization. To fully utilize the large capacity of RRAM buffer, we explore the possibility to keep weights in the on-chip buffer to further reduce off-chip memory access based on existing scheduling strategies. Experimental results show that this strategy reduces 85% more DDR read as the weight buffer size increases when compared with existing scheduling strategy. With the hardware and scheduling optimization, RRAM based design saves 12-18% system energy with a 15-75% smaller on-chip RAM area compared with SRAM design. Future work should focus on the rest part of the energy cost like the leakage energy of DDR and the computation energy.

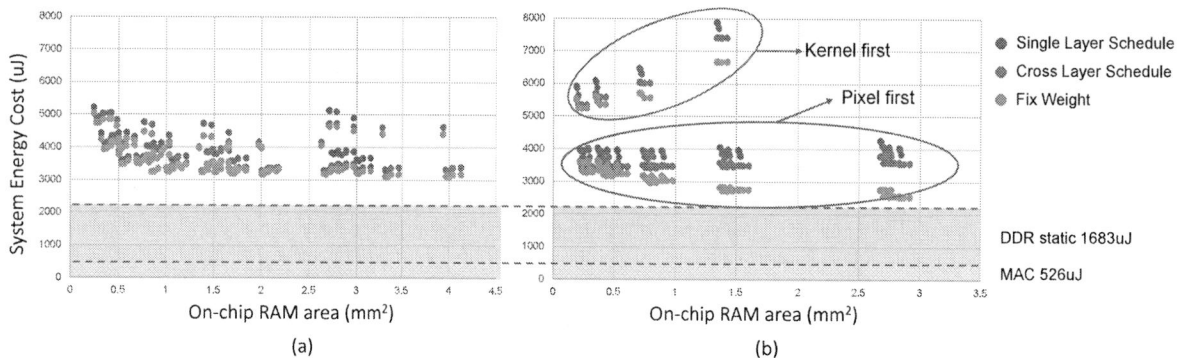

(a)   (b)

Fig. 4. Design space exploration on different hardware choices and schedule strategies on the convolution layers of VGG-11 model. (a) SRAM weight buffer design. (b)RRAM weight buffer design.

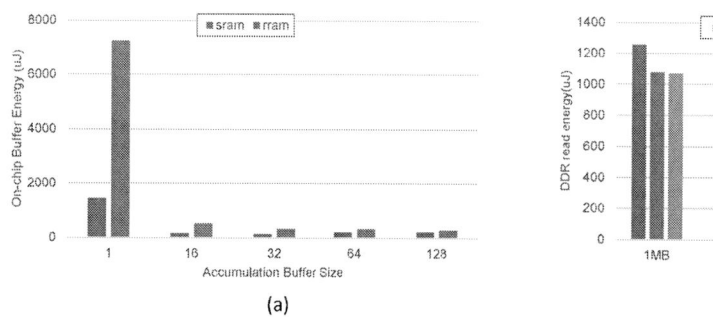

(a)   (b)

Fig. 5. (a) On-chip buffer energy cost for a series of designs only differs in accumulation buffer size. (b) DDR access energy cost for a series of designs only differs in weight buffer size.

ACKNOWLEDGEMENT

This work was supported by National Key R&D Program of China 2017YFA0207600, National Natural Science Foundation of China (No. 61622403, 61621091),Joint fund of Equipment pre-Research and Ministry of Education (No. 6141A02022608), Beijing National Research Center for Information Science and Technology, and DeePhi Technology.

REFERENCES

[1] K. Simonyan and A. Zisserman, "Very deep convolutional networks for large-scale image recognition," *arXiv preprint arXiv:1409.1556*, 2014.

[2] J. Redmon, S. Divvala, R. Girshick, and A. Farhadi, "You only look once: Unified, real-time object detection," *arXiv preprint arXiv:1506.02640*, 2015.

[3] A. Krizhevsky, I. Sutskever, and G. E. Hinton, "Imagenet classification with deep convolutional neural networks," in *NIPS*, 2012, pp. 1097–1105.

[4] K. He, X. Zhang, S. Ren, and J. Sun, "Deep residual learning for image recognition," *arXiv preprint arXiv:1512.03385*, 2015.

[5] C. Zhang, P. Li, G. Sun *et al.*, "Optimizing fpga-based accelerator design for deep convolutional neural networks," in *FPGA*. ACM, 2015, pp. 161–170.

[6] J. Qiu, J. Wang, S. Yao *et al.*, "Going deeper with embedded fpga platform for convolutional neural network," in *FPGA*. ACM, 2016, pp. 26–35.

[7] P. Chi, S. Li, C. Xu *et al.*, "Prime: A novel processing-in-memory architecture for neural network computation in reram-based main memory," in *ISCA*. IEEE Press, 2016, pp. 27–39.

[8] M. Cheng, L. Xia, Z. Zhu *et al.*, "Time: A training-in-memory architecture for memristor-based deep neural networks," in *DAC*. ACM, 2017, p. 26.

[9] S. Yu. Overview of memory technology. [Online]. Available: https://www.coursehero.com/file/15370931/EEE-598-Section-1-Overview-of-Memory-Technology/

[10] R. Fackenthal, M. Kitagawa, W. Otsuka *et al.*, "19.7 a 16gb reram with 200mb/s write and 1gb/s read in 27nm technology," in *ISSCC*. IEEE, 2014, pp. 338–339.

[11] X. Dong, C. Xu, N. Jouppi, and Y. Xie, "Nvsim: A circuit-level performance, energy, and area model for emerging non-volatile memory," in *Emerging Memory Technologies*. Springer, 2014, pp. 15–50.

[12] K. Guo, S. Han, S. Yao *et al.*, "Software-hardware codesign for efficient neural network acceleration," *IEEE Micro*, vol. 37, no. 2, pp. 18–25, 2017.

[13] M. Alwani, H. Chen, M. Ferdman, and P. Milder, "Fused-layer cnn accelerators," in *Microarchitecture (MICRO), 2016 49th Annual IEEE/ACM International Symposium on*. IEEE, 2016, pp. 1–12.

[14] M. Horowitz, "Energy table for 45nm process, stanford vlsi wiki.[online]." https://sites.google.com/site/seecproject.

[15] W. Chien, Y. Chen, K. Chang, E. Lai, Y. Yao, P. Lin, J. Gong, S. Tsai, S. Hsieh, C. Chen *et al.*, "Multi-level operation of fully cmos compatible wox resistive random access memory (rram)," in *Memory Workshop, 2009. IMW'09. IEEE International*. IEEE, 2009, pp. 1–2.

[16] W.-C. Chien, M.-H. Lee, F-M. Lee *et al.*, "Multi-level 40nm wo x resistive memory with excellent reliability," in *Electron Devices Meeting (IEDM), 2011 IEEE International*. IEEE, 2011, pp. 31–5.

[17] L. Xia, T. Tang, W. Huangfu, M. Cheng, X. Yin, B. Li, Y. Wang, and H. Yang, "Switched by input: Power efficient structure for rram-based convolutional neural network," in *Design Automation Conference (DAC), 2016 53nd ACM/EDAC/IEEE*. IEEE, 2016, pp. 1–6.

[18] T. Tang, L. Xia, B. Li, Y. Wang, and H. Yang, "Binary convolutional neural network on rram," in *Design Automation Conference*, 2017, pp. 782–787.

[19] Power calc. [Online]. Available: https://www.micron.com/~/media/documents/products/power-calculator/ddr4_power_calc.xlsm?la=en

2018 IEEE Computer Society Annual Symposium on VLSI

A Mixed-Mode Neuron with On-Chip Tunability for Generic Use in Memristive Neuromorphic Systems

Sagarvarma Sayyaparaju, Ryan Weiss and Garrett S. Rose
Department of Electrical Engineering & Computer Science
University of Tennessee, Knoxville
Knoxville, Tennessee 37996 USA
Email: {ssayyapa, rweiss2, garose}@utk.edu

Abstract—**Memristors are two-terminal nanoscale devices that provide an efficient way of implementing non-volatile synaptic weights. Realization of large scale memristive neuromorphic systems is reliant on rigorous custom design of neurons that are optimized to accumulate and spike according to the conductance range of the specific memristor employed. However, each new custom neuron design entails meticulous design effort and exorbitant re-fabrication costs. Circumventing this issue, we propose a generic mixed-mode neuron suitable for use across multiple memristor device implementations. The proposed neuron's accumulation rate is tunable with a set of reference voltages that are provided externally, through the pins on the chip. Hence, the proposed neuron exhibits on-chip accumulation rate tunability and is independent of the specific memristor chosen. This is exemplified in this paper by considering a small scale shape recognition network that employs these neurons. Moreover, the proposed neuron is shown to consume less energy per spike in comparison to a traditional integrate and fire neuron built in the same technology while occupying comparable silicon (layout) area. Lastly, the proposed neuron is shown to reduce the overall pin count for large scale memristive neuromorphic systems.**

1. Introduction

Neuromorphic systems are an ensemble of circuits designed to resemble the processing scheme in biological neural systems, at the hardware level. These neural systems consist of processing units called neurons interconnected with plastic synapses, whose connection strength (termed the "weight") is modifiable [1]. Such systems are deviant from the predominantly sequential data flow based traditional von Neumann machines.

Synapses in neuromorphic circuits provide the "weighted" electrical interconnection between the neurons. The weight of the synapse determines the strength of the signal passed on from the neuron preceding the synapse ("pre-neuron") to the one succeeding it ("post-neuron"). Although synapses have been shown using devices such as resistors, capacitors and floating-gate devices, as mentioned in [2], a synapse with a programmable non-volatile weight that lends itself for efficient learning remained elusive.

A nanoscale two-terminal device termed as the "memristor" was demonstrated in 2008 in the seminal paper [3], which was proposed theoretically in 1971 [4]. Soon, this device was slated as a good candidate for use as a synapse in CMOS-memristor hybrid neuromorphic systems [5]. Memristors are two-terminal devices whose instantaneous conductance represents the synaptic weight. This conductance state of the device is non-volatile, and can be modified by supplying an appropriate amount of charge/flux to the device. Thus, the memristor allows for the synaptic weights to be carefully programmed and controlled. Also, the flux flowing into the device can be designed to produce online learning behavior such as the spike-timing-dependent plasticity (STDP) [6]. Silicon neurons when used in conjunction with memristor based synapses give rise to memristive neuromorphic systems. The following section dwells further into the implementation details of such systems.

2. Related Work

Authors in [5], [7] have demonstrated STDP based learning in the memristor device, using an integrate and fire model for charge accumulation in the neuron. Using a similar spiking neuron, asynchronous memristor-based architectures were built in [8] to emulate a visual cortex. Authors in [9] have built and simulated a memristor crossbar based character recognition system. In [10], an experimental demonstration of a memristor crossbar performing image classification has been shown. In [11], a scalable memristor-based neural chip has been presented. All of the above memristive synapse based neural systems were demonstrated using an analog integrate and fire neuron [12] that employs a capacitor for charge integration. For the success of such systems, these neurons must accumulate at the desired rate, for which their capacitance value needs to be designed to suit the specific memristor type being used. This implies that these neurons must be custom designed for the specific application being considered.

An extensive review of other silicon neuron implementations can be found in [13]. However, it may be noted that the neuron designs summarized therein either implement conductance based dynamics within the neuron or employ

978-1-5386-7100-9/18 $31.00 © 2018 IEEE
441

capacitive charge-discharge based mechanism for their operation. This implies that these neurons are also required to be designed to function for predefined synaptic weights, which makes them rigid and necessitates a redesign for a new set of synaptic weights that may be desired.

The design of memristive neuromorphic systems with these existing neurons thus relies on the type of memristor chosen, since the neurons in the system need to be tailored for the currents provided by the memristor. However, redesigning neurons for every specific application and/or for every memristor device type proves costly not only in terms of the design effort, but also in terms of the fabrication expenses for the new lithographic masks used each time a new neuron is fabricated.

Addressing the aforementioned issues, in this paper, we propose the design of a mixed-mode neuron that is generic in terms of the synapses that can be used with it. The main contributions of this paper can be summarized as follows:

1) The proposed neuron is suitable for use across multiple memristor device implementations and thereby eliminates redesign needs and recurring mask costs.

2) The proposed neuron is capacitor-free, thus eliminating the need for rigorous custom analog design.

3) On-chip tunability using reference voltages for adaptation across a wide range of memristive synaptic implementations is illustrated.

4) Reduction in the overall pin count for large scale neuro-chip implementations is shown.

3. Case Study

In order to illustrate the on-chip tunability shortcoming of traditional analog neurons, we present a case study by considering a small-scale shape recognition task using analog integrate and fire neurons and a bi-memristor synapse.

3.1. Bi-Memristor Synapse

A bi-memristor synapse consists of two memristors connected between the pre-neuron and the post-neuron as shown in Figure 1 [14]. When the pre-neuron "spikes", the incoming spike is translated as voltages of opposite polarity (V_{DD} and V_{SS} in this case) on the pre-neuron end of the memristors while the post-neuron end is held at a ground voltage, as the post-neuron is in the accumulation phase. The weight of this synapse is directly proportional to its effective conductivity, given by:

$$G_{eff} = \frac{1}{M_p} - \frac{1}{M_n}$$

The bi-memristor synapse implements non-volatile synaptic weights that can be both positive and negative. When $M_p < M_n$, the weight is positive whereas it is negative when $M_p > M_n$. A positive weight leads to a net positive current flowing into the post-neuron and vice-versa.

The current flowing into/out of the post-neuron leads to a change in the membrane voltage V_{mem}, which is compared

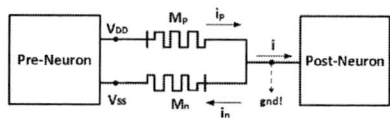

Figure 1. Current flow through a bi-memristor based synapse when the pre-neuron spikes and the post-neuron accumulates [14].

against a threshold voltage (V_{th}). When the membrane voltage exceeds the threshold, the neuron sends out an action potential/spike at its output and a feedback voltage is applied at its input node, that causes the synaptic weight to update depending on the pre- and post-neurons' spike timing [14].

3.2. Basic Shape Recognition Network

3.2.1. Operation Principle.
We consider a simple shape recognition network here that can identify one of the shapes among a set of four shapes as shown in Figure 2. In this case, the goal for the network is to identify a triangle shape input. This implies that the network's output neuron must spike appropriately when a triangle shape is given as the input, but not for the other shapes.

$$
\begin{array}{c}
\textbf{Triangle} \\
\begin{bmatrix}
0 & 0 & 1 & 0 & 0 \\
0 & 1 & 0 & 1 & 0 \\
0 & 1 & 0 & 1 & 0 \\
1 & 0 & 0 & 0 & 1 \\
1 & 1 & 1 & 1 & 1
\end{bmatrix}
\end{array}
\quad
\begin{array}{c}
\textbf{Square} \\
\begin{bmatrix}
1 & 1 & 1 & 1 & 1 \\
1 & 0 & 0 & 0 & 1 \\
1 & 0 & 0 & 0 & 1 \\
1 & 0 & 0 & 0 & 1 \\
1 & 1 & 1 & 1 & 1
\end{bmatrix}
\end{array}
\quad
\begin{array}{c}
\textbf{Diamond} \\
\begin{bmatrix}
0 & 0 & 1 & 0 & 0 \\
0 & 1 & 0 & 1 & 0 \\
1 & 0 & 0 & 0 & 1 \\
0 & 1 & 0 & 1 & 0 \\
0 & 0 & 1 & 0 & 0
\end{bmatrix}
\end{array}
\quad
\begin{array}{c}
\textbf{Plus} \\
\begin{bmatrix}
0 & 0 & 1 & 0 & 0 \\
0 & 0 & 1 & 0 & 0 \\
1 & 1 & 1 & 1 & 1 \\
0 & 0 & 1 & 0 & 0 \\
0 & 0 & 1 & 0 & 0
\end{bmatrix}
\end{array}
$$

$$
\textbf{Functional Matrix:} \quad
\begin{bmatrix}
-3 & -1 & 1 & -1 & -3 \\
-1 & 3 & -1 & 3 & -1 \\
-3 & 3 & -1 & 3 & -3 \\
3 & -1 & -3 & -1 & 3 \\
3 & 3 & 1 & 3 & 3
\end{bmatrix}
$$

Figure 2. Matrices used to denote the input shapes to the network and the functional matrix used to set the synaptic weights therein.

In order to build such a network, we first start with calculating the functional matrix that consists of the synaptic weights of the network. This matrix is generated by multiplying the triangular matrix by 4 and subtracting the shape matrices for the other shapes (because these must not be recognized). The resultant matrix is shown in Figure 2.

Using this functional matrix, the network is built as shown in Figure 3. Here, the input notation at the neuron p/q stands for *weight/delay*, where *weight* is the synaptic weight and *delay* denotes the number of clock cycles the input signal is delayed before being passed on to the synapse. It may be observed from Figure 2 that columns 1&5 and 2&4 in the functional matrix are identical. Hence, in the network in Figure 3, the neuron N0 is responsible for the columns 1&5, N1 for columns 2&4 and N2 for column 3. This is evident by the two connections N0&N1 have to the output neuron N5, each corresponding to one column in the matrix. Also, the figure shows the thresholds for each neuron (the numbers inside them), implying that a neuron with a threshold of m will spike when it receives an input current corresponding to a net weight greater than or equal to m.

978-1-5386-7100-9/18 $31.00 © 2018 IEEE

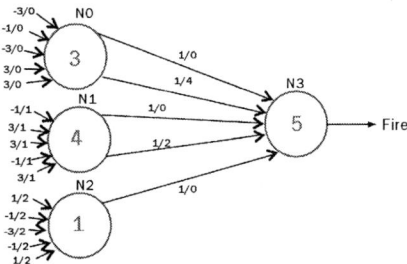

Figure 3. The network topology for the simple shape recognition task.

This network was implemented using the bi-memristor synapse and the integrate and fire neuron circuit mentioned in [14]. The CMOS circuits were built using a 65nm process design kit [15], whereas the memristor itself was modeled in Verilog-A [16]. The memristor's resistance range used here is $2k\Omega$-$10k\Omega$, based on the values reported in [17]. The results of the simulation are shown in Figure 4.

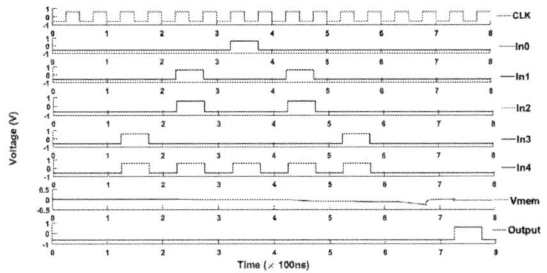

Figure 4. Simulation results for the shape recognition task. It is seen that the output neuron of the network spikes for a triangular shape input.

It can be seen from Figure 4 that the output neuron in the network produces a spike for a triangle input, as intended. Hence, this network can identify a triangle shape input.

3.2.2. Adaptability Test for the Analog Neuron. To test the adaptability of this network with integrate and fire neuron across multiple memristor types, we consider two more memristor types with their resistances as shown in Table 1. Figure 5 shows the simulation results for the network designed in Section 3.2.1 with the three memristor types.

TABLE 1. THE THREE MEMRISTOR TYPES CONSIDERED IN THIS WORK.

	M1 [17]	M2 [18]	M3 [19]
Low Resistance	$2k\Omega$	$5k\Omega$	$30k\Omega$
High Resistance	$10k\Omega$	$50k\Omega$	$300k\Omega$

It is seen from Figure 5 that the network designed to work with the memristor M1 does not work for the other two memristor types, since it does not produce any output spike for a triangle shape input. The reason for this failure can be seen from the graph plotted for the accumulated potential V_{mem} in the output neuron in Figure 5. It is seen that V_{mem} is too low (since the other two memristors' conductances are

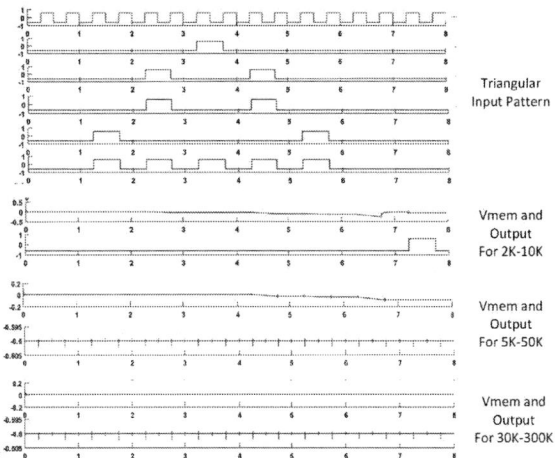

Figure 5. Simulation results for the network designed for M1 memristor being used with the other two memristors without redesigning the neuron.

lower) to cross the threshold of the neuron. This implies that for the network to function for a new memristance range, one of the two factors in the neuron have to change: either the neuron thresholds have to be adjusted to differentiate between the new weights (and hence the new V_{mem}), or the capacitance in the neuron has to be redesigned to adjust the accumulation rate to obtain reasonable threshold voltages that are sufficiently apart as in the earlier case.

Table 2 shows the analysis for the accumulated voltage in the neuron designed for a given memristor, when applied to other devices. This table shows that a neuron designed to work for a given memristance range accumulates too much voltage when used with other devices with lower resistance ranges and thereby saturates at the supply rail voltage. This makes it ineffective by not being able to differentiate between higher weight values. Similarly, it is also seen that a neuron designed for a lower memristance range (has a high capacitance value and hence) accumulates too small of a voltage on the capacitor to differentiate between the weights for higher resistance values.

TABLE 2. THE NEURON ACCUMULATION BEHAVIOR FOR VARIOUS DESIGN AND USAGE COMBINATIONS.

	Used with M1	Used with M2	Used with M3
Designed for M1	Works	V_{mem} too low	V_{mem} too low
Designed for M2	V_{mem} saturates	Works	V_{mem} too low
Designed for M3	V_{mem} saturates	V_{mem} saturates	Works

The foregoing analysis dictates that an integrate and fire neuron designed to suit a given memristor does not accumulate charge at a desired rate to distinguish between the weights for other device types. Also, the accumulated voltages for these other device types are indistinguishable and hence cannot be differentiated by tuning the threshold voltages of these neurons on-chip. This leads us to the conclusion that these neurons must be custom designed to suit the specific memristor under consideration.

4. The Proposed Mixed-Mode Neuron

In this section, we propose a mixed-mode neuron that mitigates the need for custom neuron design for different memristor devices. This neuron takes-in an analog current from the synapses, senses the current that flowed in by comparing it to some references, and encodes it as a digital value. This value is then accumulated on a digital register and compared to a digital threshold value to produce a spike. The block diagram for this neuron is shown in Figure 6.

The circuit level complexity of the sensing block and the digital block are both a function of the chosen n. In our case, since the network in Figure 3 has a maximum of +5 as the neuron threshold, we choose to encode the synaptic weights from -5 to +5. Hence, $n = 4$.

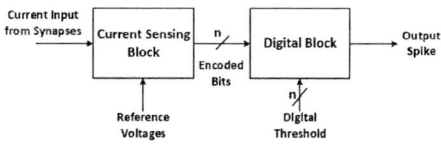

Figure 6. Block diagram for the proposed mixed-mode neuron.

4.1. The Current Sensing Block

Figure 7 shows the schematic for the current sensing block used as the first stage in the neuron. The incoming current from the synapses flows into the sensing transistor, which is biased in the linear region. A transistor biased in the linear region can have a range of currents flowing through it, depending on the V_{DS} of the device. Hence, the incoming current helps setup an appropriate DC bias point with a particular drain voltage at the sensing transistor. This drain voltage is used as the "sensed voltage" corresponding to the input current to the neuron.

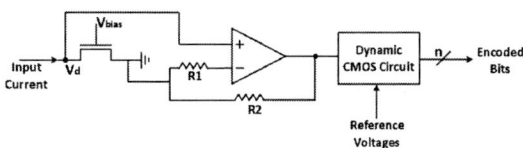

Figure 7. The circuit used to "sense" the incoming current. The V_{bias} here is chosen so as to maintain the sensing transistor in the linear mode.

Although the current-to-voltage conversion could be done with a resistor-divider network, the currents flowing into the neuron can vary up to two orders of magnitude, leading to a huge variation in the sensed voltages with such a system. However, in our case, we have biased the transistor's source terminal with the mid-rail voltage (which in our case is *gnd!*). Hence, the sensed voltages for huge difference in currents will not be as huge.

The sensed voltage V_d at the drain of the sensing transistor is input to an op-amp based non-inverting amplifier to widen the gap between the sensed voltages for different

weight values. The amplified voltage is fed as the input to a dynamic CMOS circuit that encodes the sensed voltage as an n-bit (which is 4 in this case) signed digital value.

The dynamic CMOS circuit consists of a set of transistor pairs to categorize the neuron current as a corresponding synaptic weight. Each pair of transistors has a pre-charge transistor and a discharge transistor as shown in Figure 8. For positive weight detection, during the positive half of the clock cycle, the pre-charge transistor charges up the common node E_{px} (called the *evaluation node* here) to a supply rail voltage (V_{DD} in this case). During the negative half of the cycle, pre-charge transistor is cut-off, while the discharge transistor is responsible for the discharge of E_{px}. This discharge rate is dependent on the extent to which this transistor is switched on, which is in turn dependent on its V_{GS}. The gate voltage V_G is the sensed voltage coming from the output of the amplifier and the source voltage V_S is a reference voltage coming into the neuron from an on-chip pin. Hence, by adjusting V_S, V_{GS} can be modified to control the discharge rate of the node E_{px}. An analogous discussion holds true for negative weight detection nodes E_{nx}.

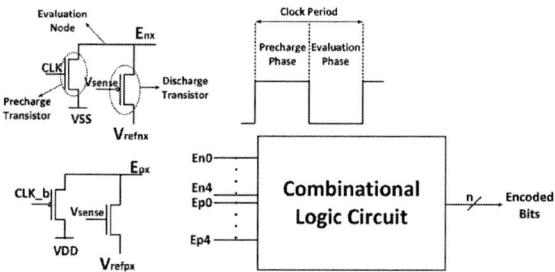

Figure 8. The dynamic CMOS circuit that encodes the sensed voltage. Here, x = 0,1,2,3,4 and hence consists of five positive and five negative evaluation nodes. Note that the combinational logic block here implements the truth table shown in Table 3.

The sensing block consists of 5 evaluation nodes for positive weights and 5 for negative weights. Each of these nodes' extent of discharge is dependent on the input synaptic weight (which determines the gate voltage for the discharging transistors) and the external reference voltages. The reference voltages are applied (progressively) in such a manner that a certain number of nodes are fully discharged for a specific weight. Table 3 shows the truth table depicting the input synaptic weight versus the node voltages that are fully discharged in the evaluation phase of the clock and the corresponding digital value that is encoded for it. It is seen from this table that as the absolute value of the synaptic weight increases, more number of evaluation nodes are discharged. These nodes are pulled to either of the supply rails by a buffer and these buffer outputs are used by a combinational logic circuit to encode the digital bits according to the truth table shown in Figure 3.

The reference voltages V_{refnx} and V_{refpx} are in turn dependent on the sensed voltages for the particular synapse and are determined accordingly, to implement the scheme in Table 3. These sensed voltages change for a different kind

TABLE 3. THE TRUTH TABLE FOR THE OPERATION OF THE DYNAMIC CMOS CIRCUIT AND THE COMBINATIONAL LOGIC ACCOMPANYING IT.

Synaptic Weight	Nodes Discharged	Encoded Digital Output
+5	Ep5-Ep0	0101
+4	Ep4-Ep0	0100
+3	Ep3-Ep0	0011
+2	Ep2-Ep0	0010
+1	Ep1, Ep0	0001
0	Ep0, En0	0000
-1	En0, En1	1111
-2	En0-En2	1110
-3	En0-En3	1101
-4	En0-En4	1100
-5	En0-En5	1011

of memristive synapse. Hence, to implement the abstract scheme of Table 3, a new set of pin voltages are needed for each new device. Therefore, just by adjusting external reference voltages, we can still implement the same abstract idea of Table 3. Hence, our sensing block possesses on-chip tunability to modify the discharge rate of its sensing circuits and thereby maintain its abstract functionality. This is in contrast to the integrate and fire neuron, where the accumulation rate is non-tunable post-fabrication.

4.2. The Digital Block

The digital block performs a digital mode storage and comparison to produce the neuron's output spike. The topology used here is shown in Figure 9. The digital block takes in the encoded digital bits from the sensing block described above in Section 4.1. A digital adder is used to add these incoming encoded bits to the existing data stored in the memory register and the result is stored back on the register. The resultant value on the adder is then also compared against a digital threshold value in the digital comparator block (which is a combinational circuit comparing two n-bit signed values) to produce an output spike when the adder output is greater than or equal to the digital threshold value.

Additionally, the neuron must also spike when the result of the addition in the digital adder is a positive overflow, indicating that the result of the addition is greater than the threshold. Hence, the spike control logic block here is a combinational circuit which produces the neuron spike by checking the overflow condition in the adder and the comparison result of the digital comparator. When the neuron spikes, the register is cleared asynchronously. This clear signal is active until the neuron spike is valid at the output, thus implementing the refractory period of the neuron.

4.3. Shape Recognition with the Proposed Neuron

As described in Section 4.1, the proposed neuron's accumulation and firing characteristics can be controlled and tuned on-chip using reference voltages that control the discharging rate of the evaluation nodes in the sensing block. We demonstrate this by using the proposed neuron

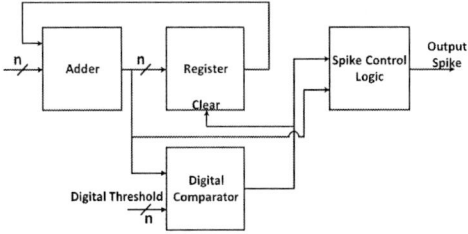

Figure 9. The digital half of the neuron, performing digital mode accumulation and comparison.

to perform the shape recognition task described in Section 3.2.1. The simulation results are shown in Figure 10.

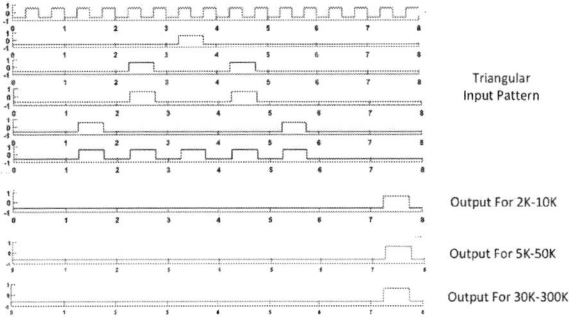

Figure 10. Simulation results for the shape recognition task performed using the proposed mixed-mode neuron. Note that the shape recognition task was successful regardless of the memristor device used.

The results shown in Figure 10 demonstrate that the proposed neuron allows for the task to be performed using all the three devices. In all the three cases, the network was able to perform the triangle-pattern recognition task. To adapt to the varying synaptic currents in each case, the reference voltages to the proposed neuron were applied so that the discharge rate of the evaluation nodes in the sensing block remains the same. Hence, Figure 10 illustrates the fact that the proposed neuron allows for on-chip tunability.

5. Discussion and Conclusion

Table 4 compares the merits of the proposed neuron as opposed to an integrate and fire neuron. To evaluate the area efficiency, we have compared the layout area for both the circuits when implemented with a 65nm process design kit. Table 4 shows that the estimated area of the proposed neuron is 2010.9 μm^2 and is independent of the synapse since the neuron's design does not change across implementations. However, the capacitance used in the analog neuron here changes with the memristor used, and hence the area. The capacitors used in the designs used for the memristors M1, M2 and M3 are 8pF, 2.8pF and 0.467pF respectively. The layout area for these designs were 2382 μm^2, 1992 μm^2 and 1817 μm^2 respectively, amounting to an average area of 2063.67 μm^2. Hence, the area of the analog neurons here is comparable to that of the proposed neuron.

978-1-5386-7100-9/18 $31.00 © 2018 IEEE

TABLE 4. COMPARISON TABLE FOR THE PROPOSED NEURON VERSUS
THE INTEGRATE AND FIRE NEURON.

Design Aspect	Proposed Neuron	Integrate and Fire Neuron
On-chip tunability	Yes	No
Need for custom design	No	Yes
Recurring mask cost	No	Yes
Capacitor-based circuit	No	Yes
Estimated area $(\mu m)^2$	2010.9	2063.67
Pin Count (m neurons)	11	m
Energy per spike (pJ)	5.327	23.292

Although the proposed neuron increases the pin count for a discrete neuron, it actually reduces it for a large scale system. This can be explained as follows: for a system that employs m neurons, the analog neuron needs m threshold pins. Whereas in case of the proposed neuron, we need 4 pins for threshold per neuron, implying a total of *4m* pins in addition to the 10 reference voltages to set the discharge rate. However, since threshold for this neuron is a digital value, all these *4m* values for the thresholds can be input to the system using a serial-in, parallel-out shift register. Thereby, a carefully controlled data transfer protocol can help reduce the pin count to $1 + 10 = 11$ for the whole system. Hence, for large scale systems where $m > 11$ is typically valid, the proposed neuron helps reduce the pin count of the system.

Additionally, the energy per spike of the proposed neuron was evaluated to be $5.327pJ$ when using a $50ns$ clock period, whereas the integrate and fire neuron built in the same technology was found to use $23.292pJ$ per spike. Thus, the proposed neuron is also energy efficient in comparison to a traditional analog neuron.

In conclusion, this work has demonstrated the incompatibility of custom made analog neurons across multiple memristive synapse implementations. The generic mixed-mode neuron proposed here alleviates the need for rigorous analog design, does not require redesign/re-fabrication effort/cost and has been illustrated to be suitable for large scale memristive neuromorphic systems. It is worth to note that the discussion here draws some parallel with the advantages a field programmable gate array (FPGA) has over an application specific integrated circuit (ASIC).

Acknowledgments

This material is based in part on research sponsored by Air Force Research Laboratory under agreement number FA8750-16-1-0065. The U.S. Government is authorized to reproduce and distribute reprints for Governmental purposes notwithstanding any copyright notation thereon.

References

[1] Y. Tang, J. R. Nyengaard, D. M. De Groot, and H. J. G. Gundersen, "Total regional and global number of synapses in the human brain neocortex," *Synapse*, vol. 41, no. 3, pp. 258–273, 2001.

[2] S. P. Adhikari, C. Yang, H. Kim, and L. O. Chua, "Memristor bridge synapse-based neural network and its learning," *IEEE Transactions on Neural Networks and Learning Systems*, vol. 23, no. 9, pp. 1426–1435, 2012.

[3] D. B. Strukov, G. S. Snider, D. R. Stewart, and R. S. Williams, "The missing memristor found," *nature*, vol. 453, no. 7191, p. 80, 2008.

[4] L. Chua, "Memristor-the missing circuit element," *IEEE Transactions on circuit theory*, vol. 18, no. 5, pp. 507–519, 1971.

[5] S. H. Jo, T. Chang, I. Ebong, B. B. Bhadviya, P. Mazumder, and W. Lu, "Nanoscale memristor device as synapse in neuromorphic systems," *Nano letters*, vol. 10, no. 4, pp. 1297–1301, 2010.

[6] G.-q. Bi and M.-m. Poo, "Synaptic modifications in cultured hippocampal neurons: dependence on spike timing, synaptic strength, and postsynaptic cell type," *Journal of neuroscience*, vol. 18, no. 24, pp. 10464–10472, 1998.

[7] X. Wu, V. Saxena, and K. Zhu, "A cmos spiking neuron for dense memristor-synapse connectivity for brain-inspired computing," in *Neural Networks (IJCNN), 2015 International Joint Conference on*. IEEE, 2015, pp. 1–6.

[8] B. Linares-Barranco, T. Serrano-Gotarredona, L. A. Camuñas-Mesa, J. A. Perez-Carrasco, C. Zamarreño-Ramos, and T. Masquelier, "On spike-timing-dependent-plasticity, memristive devices, and building a self-learning visual cortex," *Frontiers in neuroscience*, vol. 5, p. 26, 2011.

[9] A. M. Sheri, H. Hwang, M. Jeon, and B.-g. Lee, "Neuromorphic character recognition system with two pcmo memristors as a synapse," *IEEE Transactions on Industrial Electronics*, vol. 61, no. 6, pp. 2933–2941, 2014.

[10] M. Prezioso, F. Merrikh-Bayat, B. Hoskins, G. Adam, K. K. Likharev, and D. B. Strukov, "Training and operation of an integrated neuromorphic network based on metal-oxide memristors," *Nature*, vol. 521, no. 7550, p. 61, 2015.

[11] J. M. Cruz-Albrecht, T. Derosier, and N. Srinivasa, "A scalable neural chip with synaptic electronics using cmos integrated memristors," *Nanotechnology*, vol. 24, no. 38, p. 384011, 2013.

[12] X. Wu, V. Saxena, and K. A. Campbell, "Energy-efficient stdp-based learning circuits with memristor synapses," in *Machine Intelligence and Bio-inspired Computation: Theory and Applications VIII*, vol. 9119. International Society for Optics and Photonics, 2014, p. 911906.

[13] G. Indiveri, B. Linares-Barranco, T. J. Hamilton, A. Van Schaik, R. Etienne-Cummings, T. Delbruck, S.-C. Liu, P. Dudek, P. Häfliger, S. Renaud *et al.*, "Neuromorphic silicon neuron circuits," *Frontiers in neuroscience*, vol. 5, p. 73, 2011.

[14] G. Chakma, M. M. Adnan, A. R. Wyer, R. Weiss, C. D. Schuman, and G. S. Rose, "Memristive mixed-signal neuromorphic systems: Energy-efficient learning at the circuit-level," *IEEE Journal on Emerging and Selected Topics in Circuits and Systems*, 2017.

[15] S. Amer, M. S. Hasan, and G. S. Rose, "Analysis and modeling of electroforming in transition metal oxide-based memristors and its impact on crossbar array density," *IEEE Electron Device Letters*, vol. 39, pp. 19–22, 2018.

[16] S. Amer, S. Sayyaparaju, G. S. Rose, K. Beckmann, and N. C. Cady, "A practical hafnium-oxide memristor model suitable for circuit design and simulation," in *Circuits and Systems (ISCAS), 2017 IEEE International Symposium on*. IEEE, 2017, pp. 1–4.

[17] J. J. Yang, M.-X. Zhang, J. P. Strachan, F. Miao, M. D. Pickett, R. D. Kelley, G. Medeiros-Ribeiro, and R. S. Williams, "High switching endurance in tao x memristive devices," *Applied Physics Letters*, vol. 97, no. 23, p. 232102, 2010.

[18] K. Beckmann, J. Holt, H. Manem, J. Van Nostrand, and N. C. Cady, "Nanoscale hafnium oxide rram devices exhibit pulse dependent behavior and multi-level resistance capability," *MRS Advances*, vol. 1, no. 49, pp. 3355–3360, 2016.

[19] M. Uddin, M. B. Majumder, and G. S. Rose, "Robustness analysis of a memristive crossbar puf against modeling attacks," *IEEE Transactions on Nanotechnology*, vol. 16, no. 3, pp. 396–405, 2017.

978-1-5386-7100-9/18 $31.00 © 2018 IEEE

Harnessing Emerging Technology for Compute-In-Memory Support

Nicholas Jao, Akshay Krishna Ramanathan, Srivatsa Srinivasa, Sumitha George, John Sampson, Vijaykrishnan Narayanan

School of Electrical Engineering and Computer Science, Pennsylvania State University, University Park, PA 16802 USA

{naj5075, axr499, sxr5403, sug241, sampson}@psu.edu, vijay@cse.psu.edu

Abstract—**Compute-in-Memory (CiM) techniques focus on reducing data movement by integrating compute elements within or near the memory primitives. While there have been decades of research on various aspects of such logic and memory integration, the confluence of new technology changes and emerging workloads makes us revisit this design space. This work focuses on new functionality that can be embedded to SRAMs using emerging monolithic 3D integration. Properties of the new technology transform the costs of embedding such new functionality compared to prior efforts. This work also explores how compute functionality can be embedded into cross-point style non-volatile memory systems.**

Keywords- Emerging Technology, Computing-in-Memory (CiM), SRAM, CAM, Monlithic 3D integration

I. INTRODUCTION

Computer systems based on a Von-Neumann architecture with physically separated memory system and processing elements have dominated the compute landscape for decades. This computing paradigm results in significant communication costs, in both performance and power, for data movement. With data volume growing exponentially, coupled with today's datacentric applications, power and performance limits are being reached with frequent data movement in and out the memory system. While the performance of processing elements (logic units) has increased as rapidly as 60% per year [1], the performance of memory systems has not caught up. The performance gap between processing elements and the memory system further affects the power [2] and performance metrics of target applications.

A typical memory hierarchy consists of a high density non-volatile memory (storage) followed by a series of fast, but smaller-sized, volatile memories [3]. To perform computation (e.g. bitwise Boolean, arithmetic, comparison, pattern match etc.,) on a huge amount of data requires a series of steps: Data has to be transferred through the memory hierarchy to the processing element where the computations are performed and then stored back to the memory. This results in performance bottlenecks and consumes high energy relative to the computation logic itself. Any design technique which enable certain computations within the memory system can alleviate the performance bottleneck and mitigate energy costs. In this pursuit, substantial research efforts have focused on developing efficient Compute-in-Memory (CiM) solutions where computations and the data can co-exist within the memory system.

CiM designs have been widely explored since the 1970s [4]. However, the emergence of new memory technologies and 3D integration processes have drawn new opportunities to embed high performance logic into a density-sensitive memory system. Monolithic 3D integrated circuit (M3D-IC) [5] technologies (superior to TSV-based 3D in terms of vertical interconnect scaling) intrinsically reduce device area overhead and interconnect distance, which further aids in novel CiM designs by providing cell level granular connectivity. On the other hand, the inherent crosspoint organization of several high-density non-volatile memory (NVM) technologies carries the potential to realize arithmetic and Boolean functions with proper biasing and data readout schemes. Several works have taken advantage of such an NVM organization to implement computations in/near memory [6].

Shaizeen Aga et al., [7] proposed compute caches with in-memory bitwise Boolean operation support using bitline technology. Srivatsa et al., [8] proposed 3D SRAM design with bitwise Boolean operation compatibility. Agrawal et al., [9] proposed an X-SRAM, where 2 transistors in Tier 2 helps in implementing in-memory computing. Hsueh et al., [10] proposed an in-memory computing 3D-SRAM bit bitwise Boolean computations. W.-H. Chen [6] illustrated CiM functions in an NVM system.

As part of this work, we illustrate design techniques that facilitate several computations, including addition and associative search, within the SRAM memories and Phase Change Memory (PCM) based NVMs with little or no impact to the memory density. The CiM operations supported by the memory system enable parallel processing of the data present with in the memory arrays, thereby boosting performance by reducing data movement out of the memory. The contributions of this paper are:

- We illustrate an adder design in an SRAM memory system using a CiM 3D-SRAM cell. We illustrate throughput improvements when bulk addition operations are offloaded to the cache with CiM 3D-SRAM.
- Through co-design of memory cell and peripherals, we support search-oriented in-memory computations using a 3D Content Addressable Memory (CAM) primitive.
- To contrast the influence of the underlying memory technology, we implement associative search and approximate bit match in a PCM based NVM system.

This work was supported by the NSF Expeditions in Computing program and by JUMP center for Research on Intelligent Storage and Processing-in-memory.

978-1-5386-7100-9/18 $31.00 © 2018 IEEE

The rest of the paper is organized as follows. Section II provides a brief overview of M3D-IC and PCM memory. Section III describes in-memory adder design as part of computing in SRAM system. Section IV focus on implementing search-oriented computations in a 3D CAM memory system. Section V illustrates search-oriented operations using PCM based NVM memory. Section VI concludes.

II. OVERVIEW OF M3D-IC AND PCMs

A. M3D-IC

M3D-IC is a sequential integration technology which overcomes several vertical connectivity limitations posed by the TSV based 3D integration [10]. Monolithic integration offers low latency vertical interconnects (M3D via) with dimensions proportional to the metal routings of a 2D integration. M3D vias, unlike TSVs, have very minimal integration constraints. Scalability and ease of integration can be leveraged to establish cell level granular vertical interconnects between the layers which will further help in designing novel 3D SRAM cells with computational support. We employ M3D-IC process and cell level granular M3D vias in our 3D SRAM designs discussed in this paper. All our design simulations are performed using 32nm PTM transistor models [11] and we have used the delay model to mimic [12] the M3D via and to establish connectivity between the transistors across two tiers.

B. PCM memory

Emerging memory technologies present significant research opportunities based on their organization [13], material property and data readout process. PCM is one such memory where the data is represented by the state of the material [14]. A reversible phase transition property of the chalcogenide material between crystalline phase (low resistance) and amorphous phase (high resistance) aids in data representation. The inherent resistance-based storage characteristic can be leveraged to implement computation when the readout mechanism is co-optimized with the peripheral circuits. We illustrate two types of in-memory computations using this approach in the later section.

III. COMPUTING IN SRAM MEMORY SYSTEM

In this section we focus on the design extension of CiM 3D SRAM to build an in-memory adder which performs addition of data present in any of the two rows of the memory array.

A. In-Memory Adder Design

Fig. 1 shows the schematic representation of one column of the memory array with CiM 3D SRAM [10] storing the individual data bit. In this two-tier design, Tier-1 consists of standard 6T SRAM with differential read and write capability. Tier-2 consists of five transistors right above the

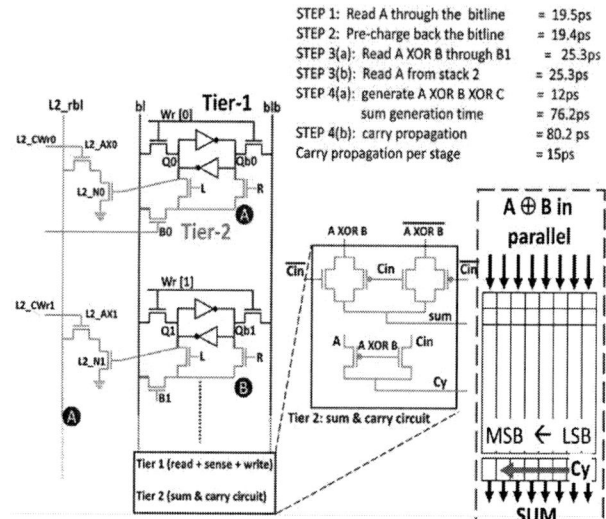

Fig. 1 In-memory adder with CiM cell [10]. cell level bitwise boolean computation long with SUM and CARRY generation near the array periphery. A⊕B computations are performed in parallel across the columns. Cy bit will propagate column wise as shown.

SRAM cell. Transistors L, R and B(0) are responsible for implementing bitwise Boolean operation. Transistors L2_CWr0 and L2_N0 act as single ended read ports. Therefore, data from the cell can be read simultaneously from both the tiers. The remaining transistors near the array periphery are placed right above the Tier-1 read and write circuitry and help to complete the SUM and CARRY outputs per bit. Fig. 1 also shows the steps involved in an addition operation. It is a combination of bitwise XOR operation and SUM and CARRY generation near the array periphery.

For ease of understanding, let us assume data "A" is stored in the first row and data "B" is stored in the second row of the array. Step 1: "A" is read differentially while the sense amplifier (SA) in Tier-1 generates the signals L and R. While L and R signals are being generated, bitlines are precharged back as part of Step 2. Step 3 computes "A XOR B" operation as part of the data read process based on the biasing of L, R and B1 signals [10]. B1 acts as the access transistor in this case. At the same time, data "A" is read through the Tier-2 single ended read port. At the end of STEP 3, we will have "A XOR B" output and data "A" on the corresponding bitlines and acts as inputs during STEP 4. SUM and CARRY results are computed from the partial outputs from STEP 3 with the help of circuit shown in fig.1. At the end of STEP 4, one-bit SUM and CARRY are generated from one column. CARRY signal will then ripple across all the columns of the array to generate the subsequent adder output. Since STEP 1 to STEP 3 can be computed in parallel across all the columns of the array, only the carry generation will be in the critical path. Therefore, the total delay depends on the number of cells in each row of the memory. Delay breakdown per STEP is also shown on top right corner in fig. 1. However, the complete operation is performed within the memory system,

978-1-5386-7100-9/18 $31.00 © 2018 IEEE   448

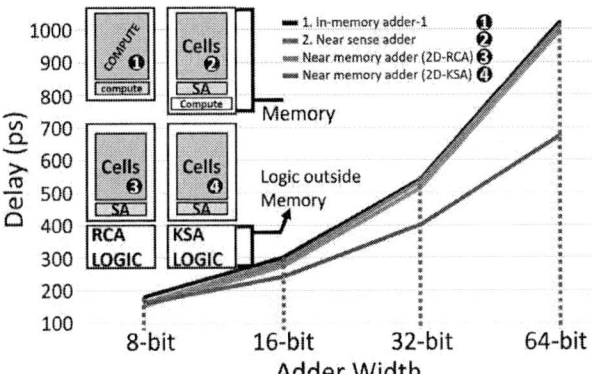

Fig. 2 Delay comparison of various adder designs. With increase in bit width, parallel prefix adder has the least delay for a single addition operation.

avoiding the high parasitic costs of sending the data out of the memory through a routing network.

B. Delay comparison of the proposed design

In order to compare the performance of our 3D In-memory adder design, we further evaluated three adder variants. Design-1 is the proposed way of computing addition. Design-2 is a dedicated adder hardware built inside the memory system and adjacent to the memory banks. Design-2 is referred to as In-memory adder 2 (since the computation is performed within the memory system). Design 3 is a ripple carry adder (RCA) designed and placed right next to the memory system with the data bus from the memory feeding the input (near memory adder). Design 3 is referred to as 2D-RCA. Design 4 is again a near memory adder. However, it is a Kogge-Stone adder (a parallel prefix adder) optimized for delay. Design 4 is referred to as 2D-KSA.

Fig. 2 shows the delay comparison of all the adder variants with respect to their bit widths. We do not incorporate additional delay contributing factors such as the memory decoder delay and routing delay within the memory system for accessing the data. From fig. 2 we can notice that, for an eight-bit addition operation, the delays of all the adder designs are almost the same. However, with increasing bit width, the delay trends change. For a 64-bit addition operation, delays of design-1, design-2 and design-3 are significantly higher compared to the near memory 2D-KSA. While it is well-known that 2D-KSA offers better performance scaling than 2D-RCA, an interesting takeaway from fig. 2 is that design-1 is slower than all other designs. The two reasons behind this are, 1) In-memory adder (design-1) is also a form of ripple carry adder in the sense that, to complete a column-wise addition, carry has to be propagated from the previous column of the memory. 2) computation involves additional data read operation (STEP 3(b) fig. 1).

C. Power consumption comparison

Fig. 3 compares the power consumption of the four designs. The total power is separated into data access/read power, H-tree output but routing power and compute power.

Overall power consumption for all four designs increases with the data width and computation complexity. However, a closer look at the power consumed by data read and computation reveals a key insight: Data read power is constant for design-2, design-3 and design-4 across adder widths since we are not taking into account data access overhead. The computational power of in-memory RCA and near-memory RCA are the same because of the same logic. However, the computational power of the 2D-KSA is high because of optimization for delay. In contrast, the compute power of design-1 is very small because of the lower logic overhead. Power consumed in data read is higher compared to its logic power.

Therefore, for sequential addition operations, an in-memory addition technique will not provide much benefit in terms of delay and power. However, since SRAM based caches are organized as multiple banks and subarrays, computing across all the subarrays in parallel will provide a throughput benefit, which we elaborate on below.

D. Enhanced throughput with bulk In-memory operations

Any applications with highly parallel or array level addition operations can benefit from executing in parallel at the bank level or subarray level. Even though this can be done using design-2, dedicated logic at every subarray level will negatively impact the overall memory density and logic power will start to dominate. To compare the improvement achieved in throughput, we calculated the overall throughput of an adder design near L3 memory to our design in M3D-IC.

Fig. 4 shows the throughput comparison of a baseline design with adder logic near an L3 cache of size 16MB. Cache configurations and throughput (delay) results are obtained from CACTI (ref). Fig. 4 also shows the throughput comparison of the bitwise Boolean logic executed in parallel, since 3D CiM can also implement Boolean logic in-memory. Time to access data and to execute 1024 addition and boolean operations are considered in calculating the throughput. When a cache is organized as just 2 banks, improvement in throughput is minimal since in-memory computations can

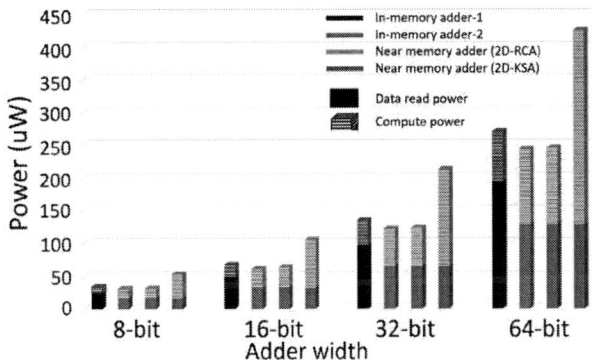

Fig. 3 Delay comparison of various adder designs. With increase in bit width, parallel prefix adder has the least delay for a single addition operation.

978-1-5386-7100-9/18 $31.00 © 2018 IEEE

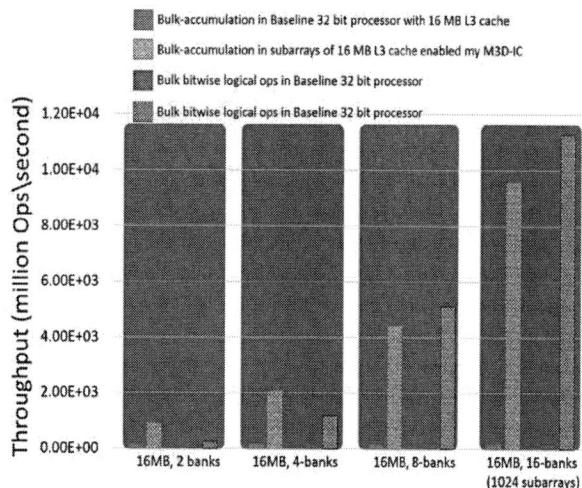

Fig. 4 Throughput comparison of the boolean logic and addition operations compared against the throughput of computing outside the memory.

happen only across 2 banks simultaneously. Secondly, in-memory computations take longer time with highly parasitic bitlines for readout with larger sized arrays. We notice further enhancement in the computational throughput when the L3 cache is organized as multiple banks and multiple subarrays within the bank. We observe a considerable improvement in throughput when we configured the memory into 16 banks and 1024 subarrays. In summary, even though the delay and power are high for individual computation using in-memory adder, a considerable enhancement is observed when a bulk of execution is offloaded onto the memory and executed in parallel across all the subarrays.

IV. SEARCH-ORIENTED COMPUATIONS IN MEMORY

In addition to the standard arithmetic operations described in previous sections, our in-memory architecture also

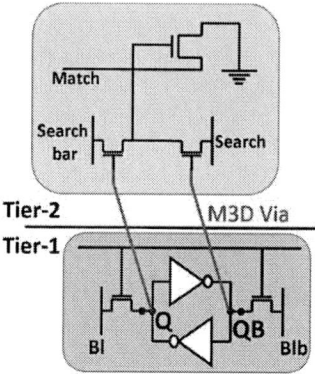

Fig. 5 3D CAM cell with standard 6T SRAM in Tier-1 and CAM transistors in Tier-2.

supports operations which are generally not part of an arithmetic and logic unit. The operations described are "associative search" and "approximate bit search". We describe the operation, implementation technique and illustrate their functionality through simulation waveforms. The fundamental logic cell for both designs is a 3D Content Addressable Memory (CAM). The way in which the CAM cell operates, coupled with the peripheral circuit behavior, determines the functionality.

A. 3D CAM cell.

Fig. 5 shows the baseline 3D CAM design used to realize the aforementioned operations. A standard 6T SRAM is designed in Tier-1 and remaining CAM transistors are placed in Tier-2 per SRAM cell maintaining the memory footprint. Based on the stored data and search signals, precharged match line discharge pattern will determine the address corresponding to the search data.

B. Associative Search

An advantage of 3D CAM is that the standard 6T operation from Tier-1 is retained and the memory array is configured to either work as SRAM or as CAM on demand.

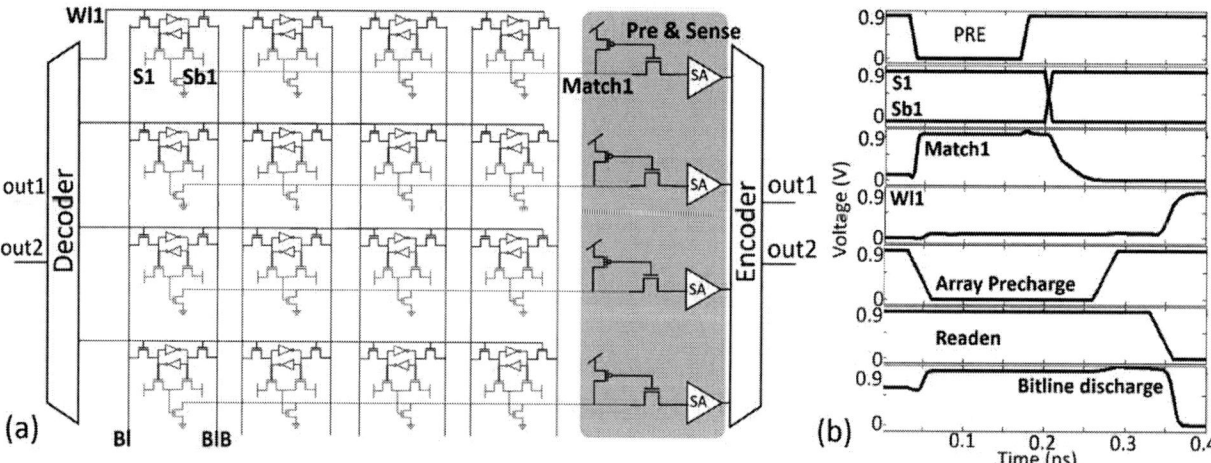

Fig. 6(a). 4x4 memory array with each cell designed as a 3D CAM cell. Pre & Sense along with encoder circuits are used to implement associative search. (b) Simulation waveform showing the functionality of associative search.

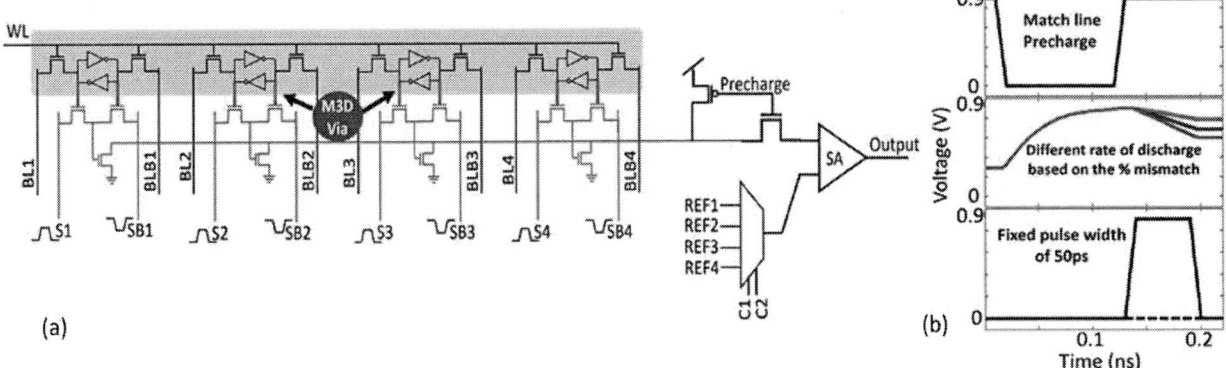

Fig. 7 (a) One row of the 3D CAM along with the SA which can receive various reference voltages based on the multiplexor control signals. Pulsed data will be received by the select port of the CAM cells. (b) Simulation waveform showing various discharge pattern based on the SA reference voltages. For proper differentiation of the matched bits, 50ps data pulse is applied to the select ports of CAM.

Making use of this property, we have developed the peripherals to obtain the associative search capability. Fig. 6(a) shows the representation of a memory array with each cell designed to be a 3D CAM cell. Output of the decoder decides the wordline to be asserted. Pre & Sense along with encoder circuits form the peripherals for the CAM design.

The entire associative search operation is split into two stages. During the first stage, selected columns of the memory array are configured as CAM. Address hit from the CAM operation is determined through output of the encoder. For ease of understanding, fig. 6(a) shows only a 4x4 array organization and two outputs from the 4:2 encoder. Output from the encoder (out1 and out2) is routed back to the input of the decoder from Tier-2. At this stage, the memory array is configured for memory operation. Based on out1 and out2, a particular wordline is asserted and the corresponding data is read. In our example, based on the data pattern of the first two bits of the MSB locations determine the row data to read.

Fig. 6(b) shows the functionality of the array simulation. Match lines are kept precharged just before the first stage of the CAM operation through the signal PRE. With the beginning of the first stage, first two column cell`s search and search bar (s1 and sb1) signal are fed with the search value. Based on the data match, match line will either discharge or not discharge. In our example, first two bits of the memory row has matched the data and the corresponding match line Match1 (see fig. 6(b)) discharges and output of the encoder will be "00". Array precharge and Readen signals determine when to activate the memory mode. With appropriate biasing of these two signals, the output of encoder is sent back to the decoder which in turn turns "ON" the first word line (Wl1). Bitline discharge pattern will then be sensed by the SA for reading the data.

C. Approximate bit search

The second search-oriented operation described is termed as approximate search. Through approximate search operation design, we can deduce the percentage mismatch of the stored data pattern as compared to the input data. 3D-CAM design is used as a fundamental logic for this design. Fig. 7(a) shows one row from the memory array, we help explain the approximate search operation. 3D CAM array design is kept unchanged. Instead of forcing a strong high or low voltage to the search lines as in standard CAM, we provide pulsed data to the search lines. Pulsed data will assist in obtaining the right discharge voltage through the match line. Match line is connected to one of the inputs of the SA. The other input of the SA connects to a 4:1 multiplexor. Select signals (C1 and C2) are responsible for serially establishing each of the reference voltage to the input of the SA. SA will then compare all the reference voltages with the match line discharge voltage to identify the percentage of the matched data.

Fig. 7(b) shows simulation waveform for the approximate search. For a proper distinguishability between the different percentage states, having right amount of discharge value per state is necessary. We identified that a search signal pulse width of 50ps will generate the required amount of match line discharge for our design. This value is also a function of number of cells in a row. As soon as match line precharge transistor turns "OFF" and appropriate search pulses are applied, SA reference voltage will check for 100% data match. If match, the operation is over. If not, the SA reference voltage will check for 1-bit mismatch and so on till the SA latches to a particular value. Since search and search bar pulse switches "OFF" the match line discharge transistor, we just need to change the reference voltage with every iteration instead of restarting the operation all over again.

V.   COMPUTATIONS IN PCM BASED MEMORY

In our work, SRAM needs additional transistors in Tier-2 to enable computation support. However, in this section we make use of PCM without any additional transistors for in-memory computation. Peripheral circuit design coupled with intelligent data readout helps in obtaining associative search and approximate bit search capability. In our approach, we use a divided word-line structure as access transistors at byte-level granularity. The peripheral circuit for the array is also

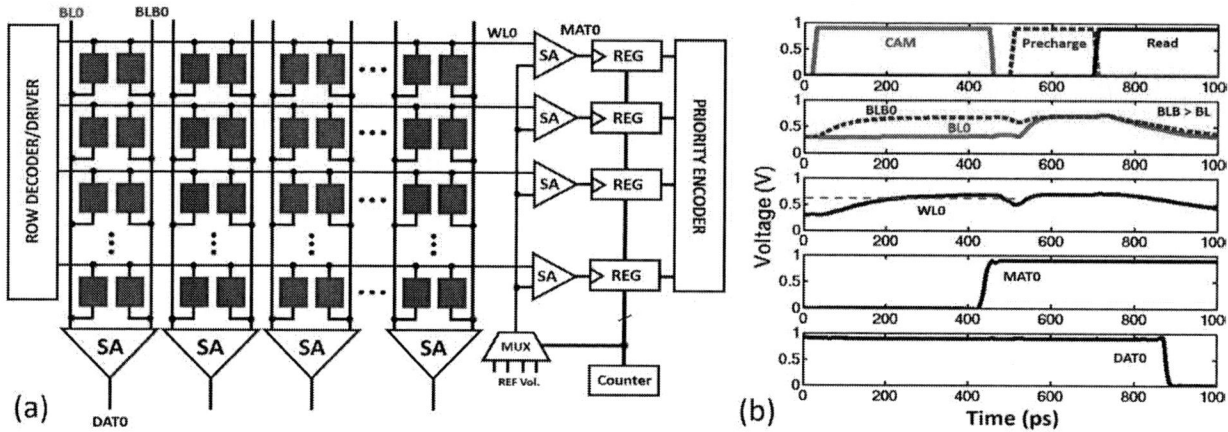

Fig. 8(a) PCM organization with additional peripheral circuits to support CiM operations like associative search and approximate bit search. Appropriate configuration of the MUX, REF voltages and priority encoder enable the computations. (b) simulation waveforms showing assertions of precharge, wordlines and appropriate bitline discharge for eventual data read.

shown in fig. 8(a). Fig. 8(b) shows the simulation waveforms for the associative search using PCMs. Like SRAM based memory, here also we use CAM functionality on the first two columns. If wordline voltage rises past the threshold, then all the data bits have matched the search bits. This is represented by MAT0 transition to high voltage. A priority encoder, based on the address and MAT voltage, determines the data read from a particular row. To perform read, bitlines, wordlines are precharged and SA per column reads one bit of the data.

VI. SUMMARY

We illustrated the in-memory computation support offered by SRAM and NVM memory systems harnessing the unique opportunities provided by emerging memory and 3D integration technologies. In-memory adders will enhance the computational throughput offered by a system when executing bulk computations within memory. We also qualitatively discuss the design extensions and peripheral circuit design to enable search oriented operations like associative search and percentage bit match in SRAMs as well as PCM-based NVM memories. Further evaluation of these designs to make them robust and to quantitatively describe the benefits offered by CiM will be the next step towards reliably computing within memory systems.

VII. ACKNOWLEDGEMENT

This work was supported by the NSF Expeditions in Computing program and by JUMP center for Research on Intelligent Storage and Processing-in-memory.

REFERENCES

[1] David Patterson et al., "A Case for Intelligent RAM, IEEE Micro", v.17, p.34-44, March 1997 [doi>10.1109/40.592312]

[2] C. C. Liu, I. Ganusov, M. Burtscher, and S. Tiwari, "Bridging the Processor-Memory Performance Gap with 3D IC Technology," IEEE Design and Test of Computers, vol. 22, no. 6, pp.556–564, November–December 2005.

[3] R. Nair, "Evolution of memory architecture," Proc. IEEE, vol. 103, no. 8, pp. 1331–1345, Aug. 2015.

[4] H.S. Stone, "A Logic-in-Memory Computer," IEEE Trans. Computers, Jan. 1970, pp. 73-78.

[5] P. Batude et al., "3D VLSI-CoolCube Process. An Alternate Path to Scaling. Symp VLSI Technology (VLSIT) "

[6] W.-H. Chen et al., "A 16Mb Dual-Mode ReRAM Macro with Sub-14ns Computing-In-Memory and Memory Functions Enabled by Self-Write Termination Scheme," IEEE International Electron Devices Meeting (IEDM) Dig. Tech. Papers, pp. 28.2.1-28.2.4, Dec. 2017.

[7] S. Aga, S. Jeloka, A. Subramaniyan, S. Narayanasamy, D. Blaauw, R. Das, "Compute caches", Proc IEEE Int. Symp. High Perform. Comput. Archit., pp. 481-492, 2017.

[8] Srivatsa et al, "A Monolithic-3D SRAM Design with Enhanced Robustness and In-Memory Computation Support", ISLPED 2018 [Accepted].

[9] A. Agrawal et al., "X-Sram," arXiv:1712.05096, 2017.

[10] F.-K. Hsueh, et al., "TSV-free FinFet-based Monolithic 3D+-IC with Computing-in-Memory SRAM Cell for Intelligent IoT Devices". IEDM 2017.

[11] Predictive Technology Model. Available: http://ptm.asu.edu/

[12] Srivatsa Rangachar Srinivasa, Karthik Mohan, Wei-Hao Chen, Kuo-Hsinag Hsu, Xueqing Li, Meng-Fan Chang, Sumeet Kumar Gupta, John Sampson, Vijaykrishnan Narayanan, "Improving FPGA design with monolithic 3D integration using high dense inter-stack via," in Proc. IEEE Comput. Soc. Annu. Symp. VLSI (ISVLSI), Jul. 2017, pp. 128–133.

[13] An Chen, "Emerging nonvolatile memory (NVM) technologies", ESSDERC, pp. 109-113, 2015.

[14] H.-S. Philip Wong, Simone Raoux, Sangbum Kim, Jiale Liang, John P. Reifenberg, Bipin Rajendran, Mehdi Asheghi, and Kenneth E. Goodson, "Phase change memory", Proc. IEEE, vol. 98, no. 12, pp. 2201-2227, Dec. 2010.

2018 IEEE Computer Society Annual Symposium on VLSI

91dB Dynamic Range 9.5 nW Low Pass Filter for Bio-Medical Applications

Jayaram Reddy M. K, Sreenivasulu Polineni and Tonse Laxminidhi

Department of Electronics and Communication Engineering
National Institute of Technology Karnataka, Surathkal
Mangalore-575025, Karnataka, India
Email: jayaram.041@gmail.com   laxminidhi_t@yahoo.com

Abstract—This paper presents a second order, fully differential, low pass filter. The filter has a tunable bandwidth in the range 4 Hz to 100 Hz and offers a dynamic range of 91 dB. The filter is based on the source-follower biquad operating in the sub-threshold region. The main idea is to exploit the strengths of sub-threshold source follower circuit, like low noise, low output impedance, high linearity and low power. The filter design has been validated in UMC 0.18 μm CMOS process. The filter consumes only 9.5 nW of power at 1.8 V supply, making it suitable for bio-medical applications. In terms of noise and dynamic range the reported filter is better than previous works found from the literature.

Index Terms— Analog filter, low frequency, bio-medical, source follower, high dynamic range.

I. INTRODUCTION

In bio-medical electronics there is a huge demand for portable low power battery operated devices. This necessitates the circuits used in the portable devices to be compact and consume low power. A feasible solution for designing low power circuits is to operate the MOS devices in sub-threshold region.

For pre-processing the bio-medical signals (typically amplitude ranging from 1 μV to 10 mV and frequency ranging from 10 mHz to 10 kHz [1]), we require a pre-amplifier to amplify the weak bio-signals and low pass filter to remove the unwanted noise. Low pass filter (LPF) is the crucial part in the bio-medical device as the precision of the entire device depends on it. For ECG, EEG, pacemakers and other applications, filters with cut-off frequency ranging from 1 Hz to 100 Hz is required. The design of low frequency filter with low noise, high linearity, high dynamic range and low circuit area is quite challenging.

The design of fully integrated low frequency, low pass filter requires large time constant. To achieve large time constants several circuit topologies have been reported in the literature. Using active-RC circuits to achieve large time constant give high linearity but at the cost of area, as the values of resistors and capacitors needed are high. Switched capacitor circuits [2], [3] are not preferred due to the leakage issues in advanced CMOS technology. The most popularly used topology to achieve large time constant is transconductor-capacitor (g_m-C) filter topology. To achieve large time constant (C/g_m) in g_m-C filter with permissible on-chip capacitance several g_m

reduction techniques have been reported in the literature [4], [5] and are: current division, source degeneration, multiple input floating gate (MIFG) and bulk-driven. Relative performance and limitations of these methods are summarized in the Table I.

In this paper, filter based on the source follower is presented as an alternative to g_m-C filter topology. The paper is organized as follows. Section II illustrates the source follower biquad and low pass filter architecture based on the source follower biquad. Simulation results are reported in the Section III. Finally, Section IV concludes the paper.

TABLE I
g_m REDUCTION TECHNIQUES FOUND IN LITERATURE AND THEIR LIMITATIONS.

Technique	Effective $\mathbf{G}_m$	Limitations
Current Division	$\dfrac{g_m}{1+M}$ (*M*: Ratio of transistor sizes used for current division)	• Current ratio accuracy • Circuit area • Output dynamic range
Multiple Input Floating Gate	$\left(\dfrac{C_1}{C_1+C_2}\right)g_m$	• Large capacitor ratio • Circuit area • Input referred noise • Limited tunability
Source Degeneration	$\dfrac{g_m}{1+g_m R}$	• Large area for R • Thermal noise of R
Bulk Driven	$\left(\dfrac{\gamma_o}{2\sqrt{2\Psi_{FB}+V_{SB}}}\right)g_m$	• Finite input impedance • Process dependent

II. SUBTHRESHOLD SOURCE FOLLOWER LOW PASS FILTER

A. Source follower Biquad

The conventional source follower with capacitive load is shown in the Fig. 1a. This circuit can be considered as a first order low pass filter. M_1 is the input transistor, M_2 is a current source and C_L is the load capacitance.

The transfer function for source follower is given by

$$H(s) = \frac{g_m r_o}{sC r_o + g_{mb} r_o + g_m r_o + 1} \tag{1}$$

978-1-5386-7100-9/18 $31.00 © 2018 IEEE   453

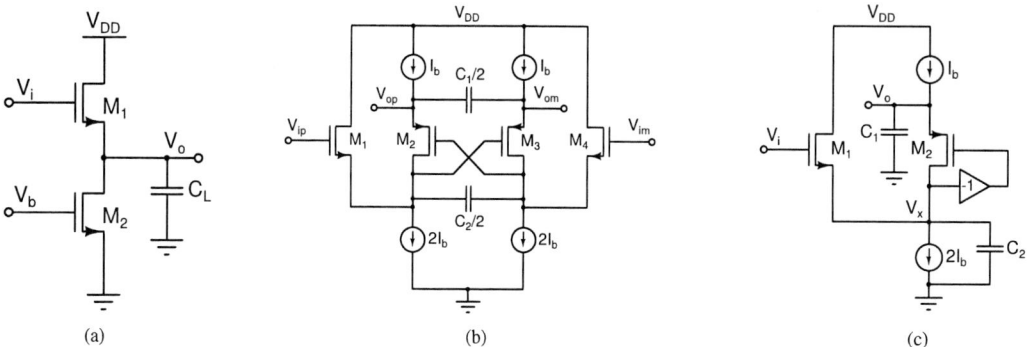

Fig. 1. (a) Source follower (b) Source follower based biquad (c) Single-ended circuit of the source follower biquad.

Where g_m and g_{mb} are the gate transconductance and bulk transconductance of the transistor M_1 respectively, r_o is the equivalent output resistance ($r_o = r_{o1} \| r_{o2}$).

The circuit in Fig. 1a can be extended to realize a fully differential biquad as shown in the Fig. 1b. Here, the complex conjugate poles required for the second order response are generated using a positive feedback. It is comprised of four MOSFETs ($M_1 - M_4$), four current sources (I_b, $2I_b$) and two capacitors (C_1 and C_2). Source follower biquad architectures are reported in [6]. However, the architecture in Fig. 1b has not been explored for its use for low-power, low-frequency applications. The inherent features of this biquad are: biquad does not require an additional common-mode feedback circuit as the common-mode voltage is self-biased by the NMOS and PMOS transistors V_{GS} and there is no instability issue as the loop gain of the positive feedback is inherently less than one. Furthermore, the filter is free of parasitic poles since there are only two nodes in the circuit and they are the integrating nodes. Also, it offers a good linearity since the the signal is directly processed in terms of voltage using local feedback.

A single-ended circuit of the biquad is shown in Fig. 1c and is used to simplify the analysis. The small signal equivalent of Fig. 1c is shown in the Fig. 2. The overall transfer function is given by (2), where $g_{mn1} = g_{m1} + g_{mb1}$ and $g_{mp2} = g_{m2} + g_{mb2}$.

The filter has a less than unity DC gain and it can be written as

$$ A = \frac{g_m^2}{(g_m + g_{mb})^2} \qquad (3) $$

Clearly, the DC gain is not unity and is equal to $1/(1 + \eta)^2$, where $\eta = g_{mb}/g_m$ is the body-effect transconductance ratio (usually in the range $0.2 - 0.5$). Thus, the DC gain achievable is between $-3.2\,\mathrm{dB}$ and $-7.2\,\mathrm{dB}$.

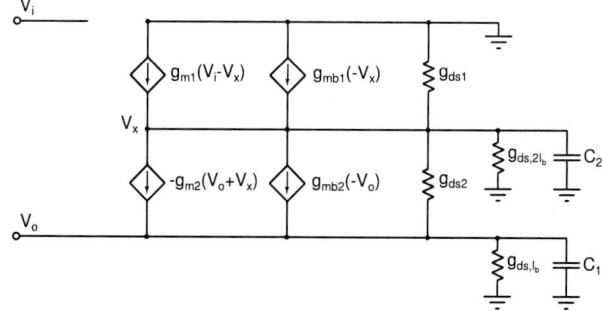

Fig. 2. Small signal equivalent circuit of Fig. 1c.

Neglecting channel length modulation and body effect, and assuming all transistors have same g_m and g_{mb}, the transfer function reduces to (4).

$$ \frac{V_o}{V_i} = -\frac{1}{s^2 \left(\dfrac{C_1 C_2}{g_m^2} \right) + s \left(\dfrac{C_2}{g_m} \right) + 1} \qquad (4) $$

The cut-off frequency (ω_o) and Quality factor (Q) are given by

$$ \omega_o = \frac{g_m}{\sqrt{C_1 C_2}} \qquad (5) $$

$$ Q = \sqrt{\left(\frac{C_1}{C_2} \right)} \qquad (6) $$

To achieve low cut-off frequency, the value of g_m should be very small, since use of large on-chip capacitors are limited by silicon area constraints. Unlike prior works [6], where source followers are operated in saturation region, we operate

$$ \frac{V_o}{V_i} = -\frac{g_{m1}(g_{m2} - g_{ds2})}{(g_{mn1} - g_{m2} + g_{ds1} + g_{ds2} + g_{ds,2I_b} + sC_2)(g_{mp2} + g_{ds2} + g_{ds,I_b} + sC_1) + (g_{mp2} + g_{ds2})(g_{m2} - g_{ds2})} \qquad (2) $$

978-1-5386-7100-9/18 $31.00 © 2018 IEEE   454

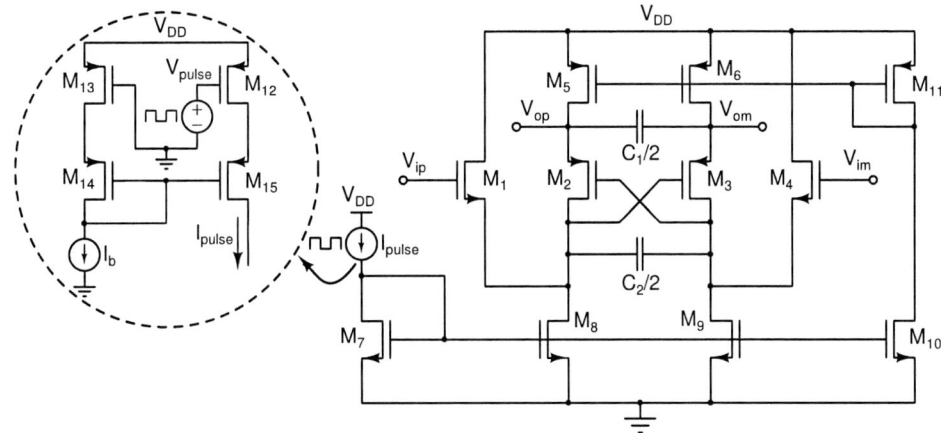

Fig. 3. Complete schematic of the source follower biquad based filter.

the transistors in sub-threshold region to minimize g_m. The expression for g_m in sub-threshold region is given by $g_m = I_D/nV_T$, where n is the sub-threshold slope factor (≈ 1.0), V_T is the thermal voltage ($\approx 26\,mV$) and I_D is the drain current. To achieve low g_m, the value of I_D used is low which results in low power consumption.

B. Low Pass filter architecture

The complete schematic of the proposed second order low pass filter is shown in Fig. 3.

To enhance the re-usability of the filter and to extract wide frequency bio-potential signals, the filter cut-off should be tunable. The filter is made tunable by switching the transistor currents with a desired duty-ratio. The frequency response scales proportionately with the duty-ratio. The scheme used to switch the currents is shown inside the dotted circle, and comprises of transistors ($M_{12} - M_{15}$), DC bias current (I_b) and voltage source (V_{pulse}).

C. Noise Analysis

Equivalent circuit of the filter for noise analysis is shown in Fig. 4. Both thermal noise and flicker noise are considered for the analysis.

The total input-referred noise (IRN) is given by

$$
\overline{V_{n,in}^2} = \frac{4kT\gamma}{g_{m1}}\left[1 + \frac{g_{mn1}^2}{g_{m1}g_{m2}}\right] \\
+ \frac{K_F}{C_{ox}\cdot f}\left[\frac{1}{(WL)_1} + \frac{g_{mn1}^2}{(WL)_2 \cdot g_{m1}^2}\right] \quad (7)
$$

where k is the Boltzman constant, T is the Absolute temperature, γ is the noise co-efficient, K_F is the process dependent parameter, C_{ox} is the oxide capacitance, f is the frequency and W, L are width and length of the transistor. The effect of flicker noise is suppressed by choosing large transistors, as the flicker noise is inversely proportional to the transistor gate area. Thus the thermal noise is made the dominant contributor. The thermal noise integrated over the passband is given by

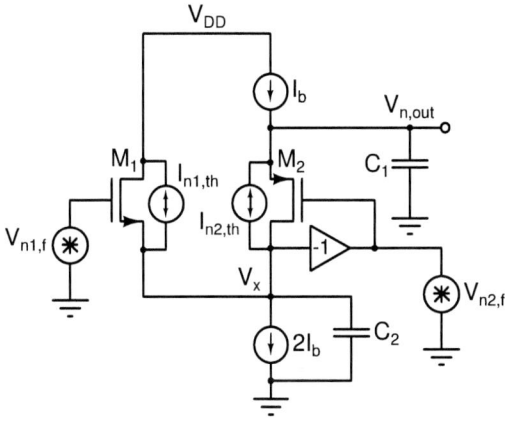

Fig. 4. Equivalent circuit for noise analysis.

$$
\overline{V_{n,in,th,int}^2} = \frac{4kT\gamma}{\sqrt{C_1 C_2}}\sqrt{\frac{g_{m2}}{g_{m1}}}\left[1 + \frac{g_{mn1}^2}{g_{m1}g_{m2}}\right] \quad (8)
$$

Asumming all transconductanes to be equal and neglecting body-effect, the expression for thermal noise is equal to

$$
\overline{V_{n,in,th,int}^2} = \frac{8kT\gamma}{\sqrt{C_1 C_2}} \quad (9)
$$

III. RESULTS AND DISCUSSIONS

The proposed Butterworth low pass filter is designed and simulated in UMC $0.18\,\mu m$ CMOS technology. For the given transconductance of $20.8\,nS$ ($g_m = g_{m1} = g_{m2}$) and bandwidth of $100\,Hz$, the value of integrating capacitors are found to be $C_1/2 = 12.4\,pF$ and $C_2/2 = 24.12\,pF$ including parasitics. The current I_b is set to $1\,nA$. The frequency response of the filter with the varying duty-ratio of current switching is shown in Fig. 5. The cut-off frequency is found to be adjustable in the range $4\,Hz - 100\,Hz$ for the duty-ratio of 1% to 100%. Fig. 6 shows the plot of filter cut-off frequency as a function of switching duty-ratio. The DC gain of the filter is found to be -3.2 dB, as expected.

Table II summarizes the filter performance and compares it with the similar works found in literature. To examine the filter linearity, the switching duty-ratio is set to 100% with $I_b = 1nA$ ($f_c = 100\,Hz$ setting). The filter is excited with sinusoidal signal of amplitude $1.03\,V_{pp}$ differential at 10 Hz. A total harmonic distortion (THD) of 1 % is obtained. With an in-band (0.1 Hz to 100 Hz) input-referred noise (IRN) of $10.24\,\mu V_{rms}$, the filter is found to offer a dynamic range (DR) of 91 dB. It is to be noted that the total power consumed by the filter is only 9.5 nW operating on 1.8 V supply. This high dynamic range is best among the similar filters found in the literature. The Equivalent output noise of the filter is shown in Fig. 7.

Monte-carlo simulations are performed to evaluate the performance of filter for device process variations and mismatch. The maximum deviation (3σ) of cut-off frequency is found to be less than 10% of the nominal value, without the use of any bandwidth fixing loop. The distribution of cut-off frequency is shown in the Fig. 8.

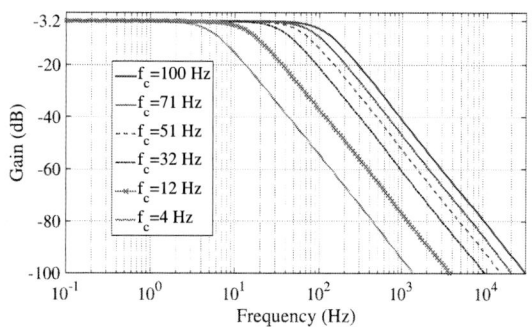

Fig. 5. Magnitude response of the filter for different duty cycles of current pulse. The cut-off varies between 100 Hz to 4 Hz. The settings used for these results are in the format [clock frequency, duty cycle, cut-off frequency (f_c)]: [10 kHz, 100%, 100 Hz], [10 kHz, 70%, 71 Hz], [5 kHz, 50%, 51 Hz], [5 kHz, 30%, 32 Hz], [1 kHz, 10%, 12 Hz], [1 kHz, 1%, 4 Hz].

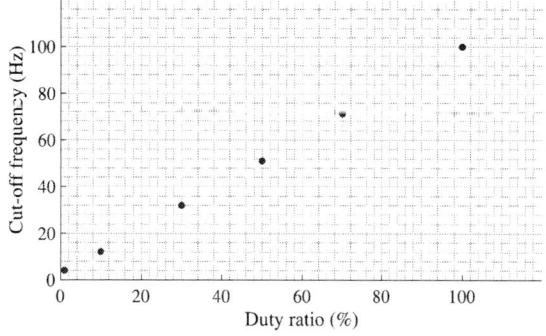

Fig. 6. Tuning graph showing the filter cut-offs for different duty cycle of current pulse.

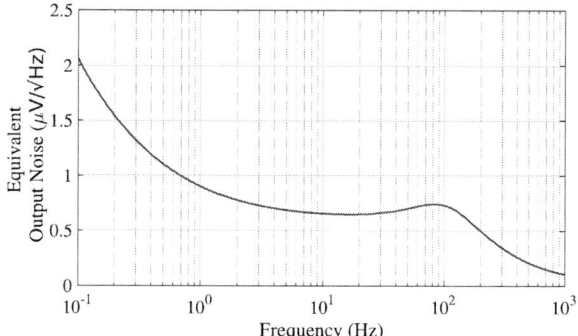

Fig. 7. Output-referred noise of the filter.

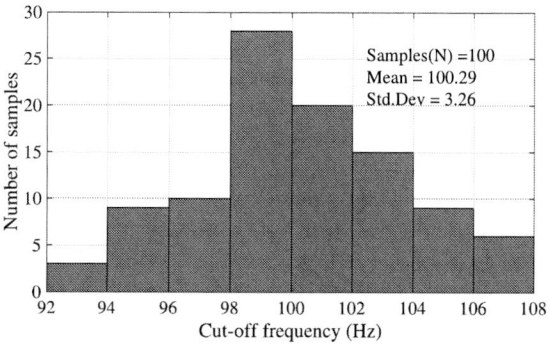

Fig. 8. Monte-carlo simulation of filter Cut-off for 100 runs.

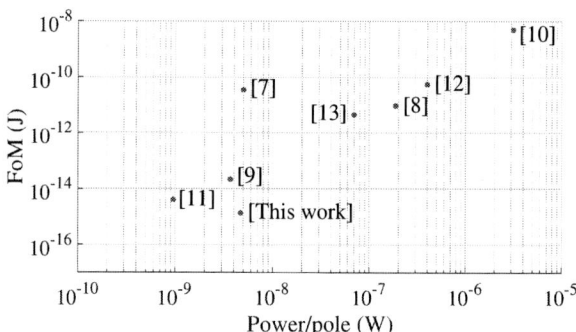

Fig. 9. FoM vs Power/pole.

To compare this work with the state-of-the-art low frequency filters, a Figure-of-Merit (FoM) [9], [14] is used and is defined by

$$FoM = \frac{P}{N \times f_c \times DR} \tag{10}$$

where P is the total power consumption, N is the filter order, f_c is the filter cut-off frequency and DR is the dynamic range.

TABLE II
COMPARISON OF THE CIRCUIT PERFORMANCE WITH RELATED WORKS.

Parameters	[7] [†] [2011]	[8] [*] [2012]	[9] [†] [2013]	[10] [†] [2014]	[11] [*] [2015]	[12] [*] [2016]	[13] [†] [2017]	This Work [*]
V_{DD} (V)	1	3	3	3	1.5	0.9	1	1.8
Technology (μm)	0.35	0.35	0.35	0.35	0.35	0.13	0.18	0.18
Power (μW)	0.005	0.75	0.015	6.31	0.0019	0.8	0.35	0.0095
DC Gain (dB)	0	-6	0	–	0	4.1	-8	-3.2
Filter Order	1	4	4	2	2	2	5	2
Bandwidth (Hz)	2m	40	100	1.95	250	47.98	50	100
THD (dB)	-40	-59	-60.7	-40	-40	-40	-49.9	-40
IRN (μV_{rms})	32	500	29	791	89	17.38	97	10.24
DR (dB)	63.8	54	64.8	50.65	59.6	43.8	49.8	91

[†] Measured   [*] Simulated

Lower FoM indicates better filter performance. From the filter simulation results, the FoM is found to be 1.16×10^{-15}J, which is better compared to the state-of-the-art filters. Fig. 9 compares the filters in the literature by plotting FoM as function of power per pole.

IV. CONCLUSION

High dynamic range, low noise, low power, tunable filter based on sub-threshold source follower is presented. The filter designed in UMC 0.18 μm CMOS process reports a bandwidth of 4 Hz – 100 Hz with 9.5 nW power consumption at 1.8 V supply. With the dynamic range of 91 dB, the filter is proved to be a better candidate for bio-medical applications. In addition, the filter is also found to be energy efficient.

ACKNOWLEDGMENT

The authors would like to thank Ministry of Electronics and Information Technology (MeitY), Government of India, for providing the EDA tools support through SMDP-C2SD project.

REFERENCES

[1] Reid R. Harrison. "A versatile integrated circuit for the acquisition of biopotentials." in *Proc. Custom Integrated Circuits Conf.*, San Jose, CA, USA, pp. 115–122, Sep. 2007.

[2] K. Nagaraj, "A parasitic-insensitive area-efficient approach to realizing very large time constants in switched-capacitor circuits." *IEEE Transactions on Circuits and Systems*, vol. 36, no. 9, pp. 1210–1216, Sep. 1989.

[3] Wing-Hung Ki and Gabor C. Temes. "Area-efficient gain-and offset-compensated very-large-time-constant SC biquads." in *Proc. 1992 IEEE International Symposium on Circuits and Systems (ISCAS)* vol. 3, pp. 1187-1190 , May. 1992.

[4] A. Veeravalli, E. Sanchez-Sinencio, and J. Silva-Martinez. "Transconductance amplifier structures with very small transconductances: A comparative design approach." *IEEE Journal of Solid-State Circuits*, vol. 37, no. 6, pp. 770–775, Jun 2002.

[5] L. Zhou and S. Chakrabartty. "Design of low-Gm transconductors using varactor-based degeneration and linearization technique." in *Proc. Biomedical Circuits and Systems Conference (BioCAS).*, Atlanta, CA, USA, pp. 1–4, Oct. 2015.

[6] Stefano D'Amico, Matteo Conta, and Andrea Baschirotto, "A 4.1-mW 10-MHz fourth-order source-follower-based continuous-time filter with 79-dB DR," *IEEE Journal of Solid-State Circuits*, vol.41, no. 12, pp. 2713-2719, 2006.

[7] E. Rodriguez-Villegas, A. J. Casson, and P. Corbishley, "A subhertz nanopower low-pass filter," *IEEE Transactions on Circuits and Systems II: Express Briefs*, vol. 58, no. 6, pp. 351–355, Jun. 2011.

[8] Liu Y.T, Donald Y.C. Lie, Weibo Hu, and Tam Nguyen. "An ultralow-power CMOS transconductor design with wide input linear range for biomedical applications." in *Proc. 2012 IEEE International Symposium on Circuits and Systems (ISCAS)*, pp. 2211–2214, May. 2012.

[9] T.-T. Zhang, P.-I. Mak, M.-I. Vai, P.-U. Mak, M.-K. Law, S.-H. Pun, F. Wan, and R.P. Martins, "15-nW biopotential LPFs in 0.35-μm CMOS using subthreshold-source-follower biquads with and without gain compensation, *IEEE Transactions on Biomedical Circuits and Systems*, vol. 7, no. 5, pp. 690702, Oct. 2013.

[10] G. Domnech-Asensi, Gins, Juan Manuel Carrillo-Calleja, Julio Illade-Quinteiro, Flix Martnez-Viviente, Jos ngel Daz-Madrid, Francisco Fernndez-Luque, Juan Zapata-Prez, Ramn Ruiz-Merino, and Miguel Angel Domnguez, "Low-frequency CMOS bandpass filter for PIR sensors in wireless sensor nodes." *IEEE Sensors Journal*, vol. 14, no. 11, pp. 4085–4094, Nov. 2014.

[11] Chutham Sawigun and Prajuab Pawarangkoon. "A compact subthreshold CMOS 2nd-order gm-C lowpass filter." in *12th International Conference on Electrical Engineering/Electronics, Computer, Telecommunications and Information Technology (ECTI-CON)*, pp. 1-4, Jun. 2015.

[12] Arya Richa, and Joo P. Oliveira. "Gm-C biquad filter for low signal sensor applications." in *Proc. 2016 MIXDES-23rd International Conference Mixed Design of Integrated Circuits and Systems*, pp. 207–210, 2016.

[13] Chuan-Yu Sun and Shuenn-Yuh Lee. "A Fifth-Order Butterworth OTA-C LPF with Multiple-Output Differential-Input OTA for ECG Applications." *IEEE Transactions on Circuits and Systems II: Express Briefs*, 2017.

[14] E. Vittoz and Y. Tsividis, "Frequency-dynamic range-power," in *Trade-Offs in Analog Circuit Design*, pp. 283–313, Springer US, 2002.

978-1-5386-7100-9/18 $31.00 © 2018 IEEE

2018 IEEE Computer Society Annual Symposium on VLSI

A Low Power, High Gain Semi-Floating Gate Amplifier for Resonating Sensors Front-End

Luca Marchetti ,Yngvar Berg, Mehdi Azadmehr
Department of Microsystems (IMS)
University College of Southeast Norway
Email: Luca.Marchetti@usn.no

Abstract—In this work, we propose a low power and high gain electronic Front-end for resonating sensors based on semi-floating-gate inverting amplifiers (SFGA). Low power and high gain are achieved using a novel biasing technique of the floating gates in SFGA. The proposed amplifier has been simulated in AMS-350nm CMOS technology, characterized by very low leakages. Gain, output swing and bandwidth can be controlled by applying a proper biasing voltage. Simulation results show a trade-off between gain and bandwidth. However, the best performance recorded are: gain of 54dB (505 V/V) and bandwidth of 1.63GHz. A transient analysis with a sinusoidal input has been performed in order to verify the working principle of the SFGA. The implemented circuit provides a static power consumption of $69\mu W$ with a power supply of 3.3V. At the end of this work, the SFGA has been connected as a sensor front-end to read-out the response of a resonating sensor, which has been modelled with a Butterworth Van Dike model of a real device (MURATA MA40S4R).

I. INTRODUCTION

Semi-floating gate (SFG) technology was born to satisfy the need of ultra low supply voltages (ULV) electronic circuits. The ultra low voltage operation in SFG is achieved by boosting the floating gates, which can result in a higher power consumption. Various analog and digital circuits based on SFG have been proposed such as current mirrors [1], current multiplier [2], ULV tran-conductance amplifier [3], differential amplifier [4], Sample and Hold circuit [5], Digital to Analog converter [6], [7], binary to multiple-value and multiple value to binary converters [8], max and min functions using multiple valued logic [9], D-latch for multiple valued logic [10] and ternary logic [11],[12]. In this work, we will prove that SFGA technology can also be utilized in low power, high gain analog circuits. Currently, the SFGA circuits have been well described and conceptually proved to be functional, but they have never been applied in any specific applications. Therefore, we designed the SFGA in order to realize a low power front-end circuit for resonating sensors, an analog operation, which typically needs high gain. Earlier, we presented a combination of current starved pseudo floating gate amplifiers (CSPFGA) and Ring Down Method (RDM) to read-out resonating sensors [13], a strategy, which has been proved to be very effective [14]. However, the main challenge in designing PFGA consists in setting correctly the value of the bias point, in particular when it is implemented in low leakage technology [15]. Then, we investigated an alternative solution to PFGA presented in [16], in which the bias voltage of the

inverter is stored on a capacitor. From this idea, the attention moved to SFGA. The paper is organized as follows: Section II provides a brief description of the working principle of the semi-floating gate proposed in [3], Section III describes the working principle of the proposed amplifier, Section IV reports the simulation results. Finally, the paper ends in Section V with the conclusions and proposal for future works.

II. SEMI-FLOATING GATE AMPLIFIER

The semi-floating gate circuit was initially proposed to implement high speed ULV digital circuits and then applied to analog circuits. The working principle is shown in Fig.1 and explained in [3].

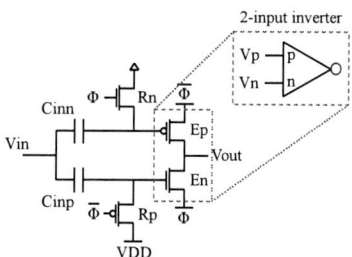

Fig. 1. SFGA working principle proposed in [3].

The operation of the SFGA is divided in two phases: recharge and evaluation phase. During the recharge phase ($\Phi = 1$), the input capacitors are connected to the power supply rails through the MOSFET Rp and Rn. During the evaluation phase ($\Phi = 0$), the voltages stored in C_{inn} and C_{inp} are used to bias the gate of En and Ep to Vdd and 0V respectively. In this way, the MOSFET of the 2-input inverter are strongly turned on providing high output current, which means high speed or large bandwidth depends if the SFG circuit is used in digital or analogue applications.

III. PROPOSED SFGA WORKING PRINCIPLE

The schematic and the working principle of the proposed circuit is shown in Fig. 2.

This implementation presents a few differences in respect to the system described in Section II, such as: different pre-charging voltage values, 2 added MOSFET/switches ($M_0, M_1/S_0, S_1$) and different timing of the control signals. During the pre-charging phase ($\Phi = 1$) shown in Fig.2(c),

978-1-5386-7100-9/18 $31.00 © 2018 IEEE   458

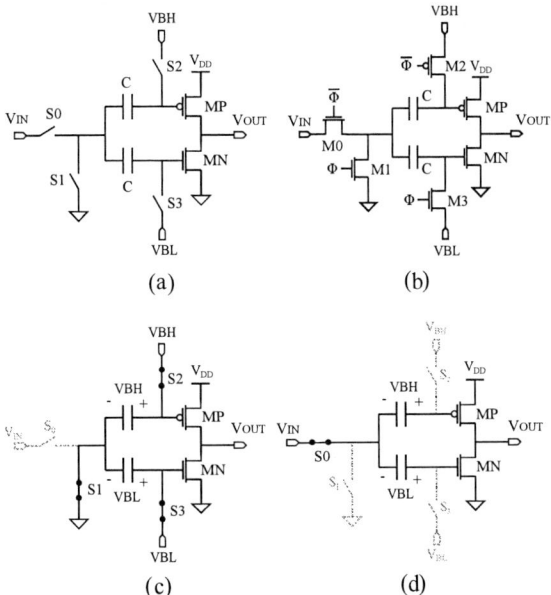

Fig. 2. Proposed low power, high gain SFGA working principle. (a) Schematic of SFGA with ideal switches. (b) Transistor implementation. (c) Pre-charge phase (d) Evaluation phase.

S_0 is opened, S_1, S_2, S_3 are closed and the capacitors are pre-charged to two different voltage values : V_{BH} and V_{BL}. During the evaluation phase ($\Phi = 0$) shown in Fig.2(d) S_1, S_2, S_3 are opened, while S_0 connects the input terminal V_{IN} to the amplifier. Furthermore, the voltages held by the input capacitors force the 2-input inverter to work as an amplifier. The values of V_{BH}, V_{BL} must be chosen in order to maximize the gain of the amplifier, which occurs if M_P, M_N show very high output resistance. For this reason, the MOSFET M_P, M_N have to ideally work in weak inversion. However, the main trade-off of having high output resistance is the decreasing bandwidth of the circuit, which could limit the application of this amplifier. Therefore, in order to find the optimum values for these two voltages (V_{BH}, V_{BL}), the performance of the SFGA have been extracted and analyzed in the next section. A comparison between the SFGA proposed in this work (Fig.2(b)) and the SFGA described in [3] (Fig.1) shows that the position of the pre-charging transistors is reversed. This is due to the fact that in this new method of designing the SFGA, the gate of M_N and M_P must be pre-charged to a value close to the lower and the upper power supply rail, respectively. The MOSFET M_0 and M_1 are utilized to de-couple the input terminal (V_{IN}) from the amplifier, and to ground one terminal of the capacitors during the pre-charge phase. These two MOSFET are fundamentals for the correct operation of the proposed SFGA, because they ensure that the voltage stored on the capacitors C depend only on the values of V_{BH} and V_{BL}.

IV. SIMULATION RESULTS

In order to verify the working principle of the SFGA, simulations in AMS-350nm CMOS technology have been performed. This technology has been chosen in order to minimize the effects of the leakage currents of the MOSFET, which can affect the pre-charge of the capacitors C. All the simulations refer to a power supply of 3.3V. However, higher supply voltages could be utilized. The MOSFET M_P, M_N are designed in order to realize a symmetric CMOS logic inverter, which provides the highest gain around the middle point of the linear region placed at $(V_{IN}, V_{OUT}) = (V_{DD}/2, V_{DD}/2)$. Since the inverter works in moderate or weak inversion, then relatively large transistors should be considered, in order to avoid very low values of output current and therefore very low bandwidth. The dimensions chosen for M_P and M_N are reported in Table I.

TABLE I
MOSFET ASPECT RATIO OF THE 2-INPUT INVERTER

M_N	M_P
$10\mu/1\mu$	$38\mu/1\mu$

Next, in order to extract the analog characteristics of the SFGA, the behaviour of the 2-input inverter made by M_P and M_N must be analyzed. First of all, the voltage transfer characteristic curve (VTC) of the 2-input inverter has been extracted as shown in Fig.3(a).

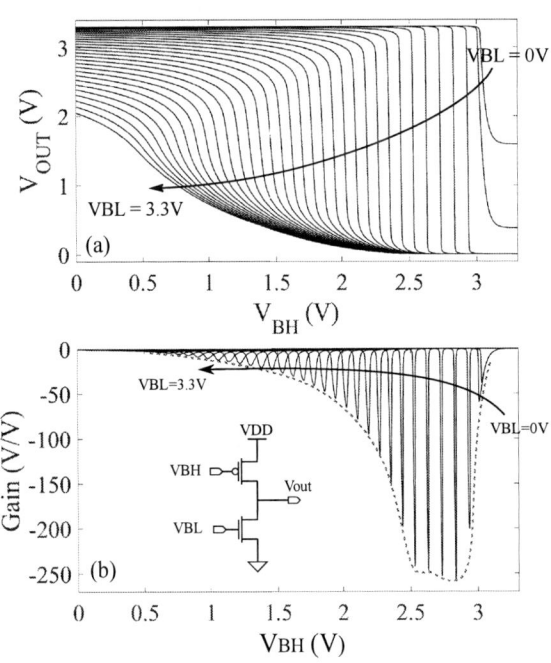

Fig. 3. Static Analysis of SFGA during parameter sweep of V_{BH}, V_{BL}. (a) VTC curves. (b) Static Gain extracted from VTC curves.

These results refer to the circuit in Fig.3(b). Fig.3(a) pro-vides a family of VTC curves obtained by sweeping the values of the input voltages of the 2-input inverter. These voltages represent the biasing of the SFGA, which are stored in the capacitors C. For this reason, the 2 input voltages of the 2-input inverter have been labelled V_{BL} and V_{BH}. V_{BH} has been swept in the range [0,3.3]V with a step of 1mV, while V_{BL} has been swept in the range [0,3.3]V with a step of 0.1V. By differentiating ($\frac{dV_{out}}{dV_{BH}}|_{V_{BL}=const.}$) each VTC curve in Fig.3(a), it is possible to extract a group of curves that represent the static gain of the 2-input inverter when only one input is varying, as shown in Fig.3(b). The peak of each static gain curve occurs in the region of the VTC curve with maximum slope. These peaks are connected by a dashed line, which can be used to find the best range of values for V_{BH} and V_{BL}, that provides the maximum static gain achievable. The values of V_{BH}, V_{BL}, which provide the peaks of the static gain curves are listed in Table.II.

TABLE II
STATIC GAIN PEAKS COORDINATES OF THE 2-INPUT INVERTER

	V_{BH} (V)	V_{BL} (V)		V_{BH} (V)	V_{BL} (V)		V_{BH} (V)	V_{BL} (V)
1	3.031	0	13	2.028	1.2	25	1.059	2.4
2	3.01	0.1	14	1.946	1.3	26	0.982	2.5
3	2.936	0.2	15	1.864	1.4	27	0.906	2.6
4	2.834	0.3	16	1.783	1.5	28	0.831	2.7
5	2.732	0.4	17	1.701	1.6	29	0.758	2.8
6	2.63	0.5	18	1.62	1.7	30	0.688	2.9
7	2.531	0.6	19	1.539	1.8	31	0.621	3.0
8	2.441	0.7	20	1.458	1.9	32	0.563	3.1
9	2.356	0.8	21	1.377	2.0	33	0.516	3.2
10	2.273	0.9	22	1.297	2.1	34	0.478	3.3
11	2.191	1.0	23	1.217	2.2	-	-	-
12	2.109	1.1	24	1.138	2.3	-	-	-

Simulation results show the existence of a range of V_{BL} and V_{BH} values, which provide a static gain around 250 V/V. This range is roughly defined by $V_{BH} \in [2.5, 2.9]V, V_{BL} \in [0.15, 0.6]V$. The Fig.3 (a) shows that the value of V_{OL} increases up to $V_{DD}/2$ when V_{BL} and V_{BH} approach to zero and V_{DD} respectively. This is due to the fact that for these voltage values, the MOSFET are in very deep weak inversion and the output voltage depends on the voltage divider created by the off resistances of M_P, M_N, which approach to the same values ($R_{OFFn} \simeq R_{OFFp}$), when the two MOSFET are completely off ($V_{GSn} = V_{SGp} = 0$). This phenomenon takes place for values of V_{BL} lower than 0.15V, under which the output swing available starts to degenerate. In Fig.3(a), the output swing can be represented as the projection of the VTC linear region over the y-axis. The output swing increases when V_{BL} decreases, reaching its maximum value when $V_{BL} = 0.15V$, and then decreases again. An overall evaluation of Fig.3 brings out that the maximum output swing and the peaks of the static gain curves of the 2-input inverter follow the same trend during the sweep analysis. On the other hand, the effective output swing depends on the position of the output biasing voltage. By using the values listed in Table.II, a

DC analysis, shown in Fig.4, has been performed to determine the value of the output biasing voltage of the amplifier.

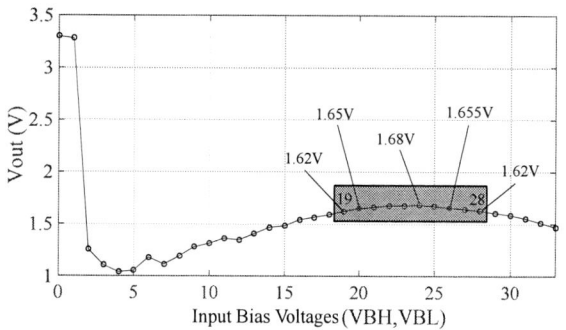

Fig. 4. DC Analysis used to extract the output bias voltage, for V_{BH}, V_{BL} values listed in Table II

Fig.4 shows that the output bias voltage of the 2-input inverter varies significantly, depending on the value of V_{BH} and V_{BL}. V_{OUT} is centered around the middle of the power supply rails when $0.831V < V_{BH} < 1.54V$ and $1.8V < V_{BL} < 2.7V$ (grey region). Unfortunately, for these values of V_{BH} and V_{BL}, the available output swing and the maximum gain achievable are relatively low. Therefore, if maximum gain is the priority, an asymmetric output swing must be accepted. Since the aim of this design is to research the optimum biasing of the SFGA, which provides the maximum gain with minimum power consumption, then the following analysis will focus only in the range $V_{BH} \in [2.35, 3.3]V, V_{BL} \in [0, 0.8]V$. Fig.5(c) provides a localized view of the static gain in this interval, while Fig.5(a,b) provide information about how the biasing current (I_{NOT}) of the 2-input inverter varies during the sweep of V_{BH} and V_{BL}.

Fig.5(a) shows that for a fixed value of V_{BL}, the current I_{NOT} is constant until $V_{BH} > V_{IL}$, then the current drops rapidly. Fig.5(b) reports a family of curves for the current I_{NOT} during the sweep of the values V_{BH}, V_{BL}. The current I_{NOT} decreases as V_{BL} decreases. This phenomenon is the consequence of the fact that the NMOS turns more off for lower values of V_{BL}. Fig.5(b) shows that the current varies of 6 orders from $10\mu A$ down to a few pA. This is an interesting characteristic in terms of power consumption and frequency response of the 2-input inverter, which gain remains over 200 V/V for all these values of current. In order to evaluate the importance of this property for the SFGA, its frequency response has been analyzed in Fig.7. Before proceeding to the next step, it is important to observe that all the characteristics extracted until now are referred to the 2-input inverter, but their validity can be extended also to the SFGA, except for the value of the gain. Indeed, each static gain curve in Fig.3(b) is obtained by sweeping only one input, while the other one is kept constant. However, as shown in Fig.2(d) the SFGA output voltage value changes due to the voltage variations at the gate terminals of both M_P, M_N. The difference between these two cases has been exploited in Fig.6, in which are compared the low frequency gains of these two circuits.

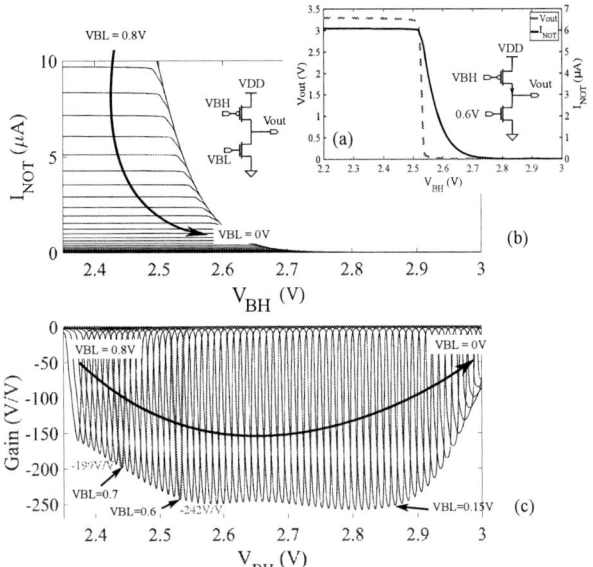

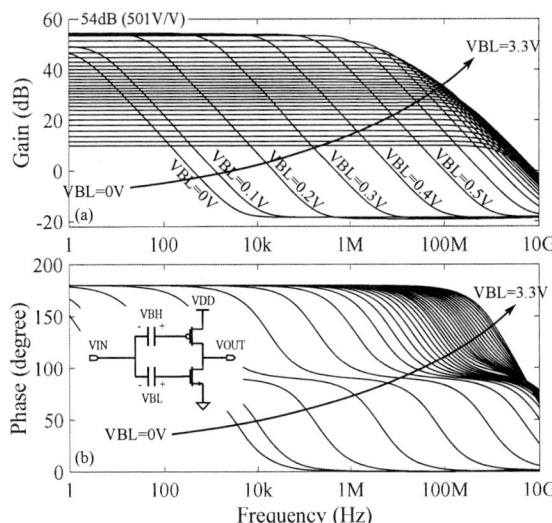

Fig. 7. AC analysis of SFGA by using the optimum biasing points showed in Table II. The SFGA has been loaded with a CMOS inverter made in the same technology with $(W/L)_P = 2.7\mu m/350nm$, $(W/L)_N = 1\mu m/350nm$. $C = 1nF$

Fig. 5. Local gain and current analysis. (a) VTC curve and current for $V_{BL} = 0.6V$. (b) Current in SFG during parametric sweep of V_{BH}, V_{BL}. (c) Parametric sweep of V_{BH}, V_{BL} in maximum gain range.

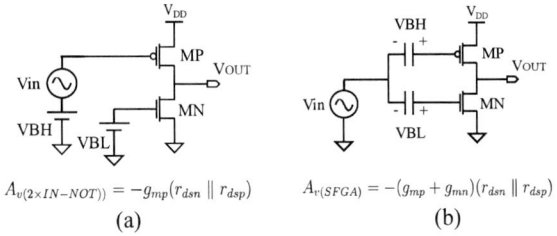

Fig. 6. Comparison between 2-input inverter and SFGA low frequency gain. (a) Circuit used for static analysis. (b) SFGA

Fig.6 shows that if M_P and M_N are characterized by the same trans-conductance value, then the SFGA provides a low frequency gain ($\simeq$ static gain) double than the one previously obtained during the 2-input inverter analysis. Finally, the frequency response of the SFGA is shown in Fig.7, which refers to the circuit shown in Fig.7(b). This simulation has been performed by sweeping the pre-charge voltages of the capacitors C with the values listed in Table II.

The value of the input capacitors has been set to $C = 1nF$ in order to move the low cut-off frequency to very low values and emphasize the effects of the biasing of the SFGA on the high cut-off frequency. The amplifier has been loaded with a logic inverter characterized by the following aspect ratio $(W/L)_P = 2.7\mu m/350nm$, $(W/L)_N = 1\mu m/350nm$. Simulation results show a trade off between the low frequency gain and the bandwidth of the amplifier. Fig.7(a) shows a maximum gain achievable of 54dB (501 V/V) and a maximum bandwidth of 1.63GHz. In particular, the gain is kept approximately constant to 500V/V in the range covered by

the bias voltages 3 to 6 of Table.II. These values cover the same range of maximum static gain shown in Fig.5. This is a significant result because, for the used technology, the best values for the biasing of the SFGA can be extracted just by using the information provided by the previous static analysis, and its maximum gain can be obtained by doubling the value of the maximum static gain. Furthermore, based on the considerations resulting from the comparison of the circuits in Fig.6, it is possible to conclude that the trans-conductance of M_P, M_N are equal in the range of maximum gain of the SFGA. A final overview of the effects provided by using the biasing voltages (V_{BH}, V_{BL}) listed in Table II is represented in Fig.8.

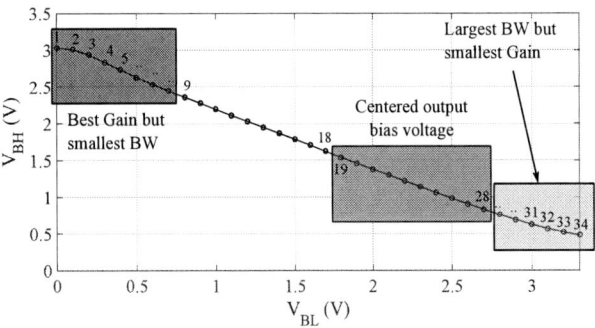

Fig. 8. Overview of the effects provided by the biasing sweep of V_{BH}, V_{BL}.

Fig.8 shows that V_{BL} and V_{BH} values are are linear-like related and their relationship can be described by the Eq.(1).

$$V_{BH} = -0.806V_{BL} + 3.02 \tag{1}$$

A more precise description of this relationship can be found by using quadratic or cubic polynomial expressions such as those one shown in Eq.(2,3).

$$V_{BH} = 0.04V_{BL}^2 - 0.94V_{BL} + 3.09 \qquad (2)$$

$$V_{BH} = 0.01V_{BL}^3 - 0.014V_{BL}^2 + -0.87V_{BL} + 3.07 \qquad (3)$$

In Fig.8, there are also 3 different highlighted regions, which emphasize the best and the worst characteristics of the amplifier, obtained by using those values of SFGA biasing. The highest gain and the smallest bandwidth are obtained in the blue region (from point 1 to 8), centered output bias voltage ($V_{OUT} \simeq V_{DD}/2$) is obtained by using the values in the red region (from point 19 to 28), and the maximum bandwidth with smallest gain are provided by using the biasing values in the green region (from point 29 to 34). The last step of this SFGA analysis consists in exploiting the dynamic behaviour of the SFGA, and to verify if there are phenomena that affect the electrical characteristics of the amplifier previously discussed. A transient analysis of the circuit with a sinusoidal input signal has been performed by using the MOSFET dimensions listed in Table.I and III.

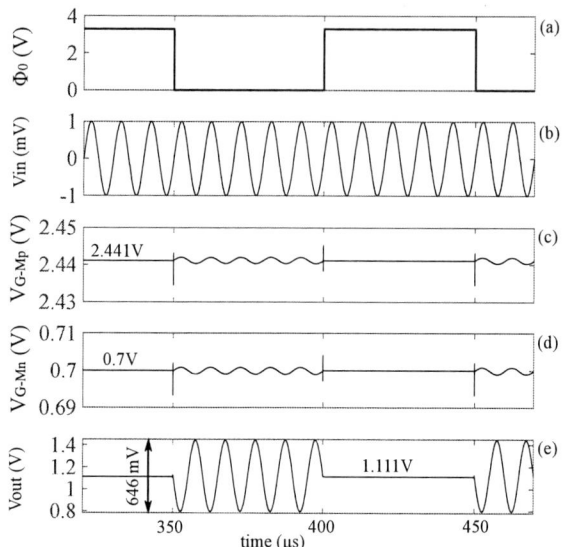

Fig. 9. Transient Analysis. $V_{BH} = 2.441V, V_{BL} = 0.7, C = 10pF$, loaded with an inverter made in the same technology and $(W/L)_P = 2.7\mu m/350nm, (W/L)_N = 1\mu m/350nm$.

TABLE III
SFGA MOSFET ASPECT RATIO

M_0	M_1	M_2	M_3	C
$500n/350n$	$500n/350n$	$500n/350n$	$500n/350n$	10pF

The dimensions of the MOSFET M_0, M_1, M_2, M_3 have been minimized to reduce the charge injection phenomenon during the transition from the ON to the OFF state. The values of the input capacitors C have been set to 10pF, as good compromise between dimensions and charge injection sensitivity. Furthermore, these capacitors must be greater than the input gate capacitance of the 2-input inverter in order to avoid any drop voltage of the input signal over these two elements. The input signal amplitude has been set to 2mV peak, to peak at 100kHz. The pre-charging voltage values have been set to $V_{BH} = 2.441V, V_{BL} = 0.7V$, which provide a good trade-off between bandwidth and gain of the SFGA. The expected gain and bandwidth of the SFGA are 51.3dB (367 V/V) and 4.78MHz respectively, which have been extracted from the AC analysis. The signals $\Phi, \overline{\Phi}$, which control the MOSFET/switches, are generated by a cascade of two inverters, fed by square-wave signal called Φ_0 characterized by a frequency of 10kHz. Simulation results are shown in Fig.9 and they refer to the circuit in Fig.2(b).

Fig.9(c,d) show V_{G-MP} and V_{G-MN}, which are the voltages at the gate of M_p and M_n respectively. The spikes on these two voltages are due to charge injection phenomena during the opening and closure of M0,M1,M2,M3. During the pre-charge mode, the output voltage is fixed to 1.11V, while during the evaluation mode, its variations represent the inverted and amplified version of the input signal. The peak to peak amplitude of the output voltage is 646mV, which leads to a

gain of 50.18dB (323 V/V). The measured gain is smaller than the one expected, because the biasing voltages stored on the input capacitors are affected by charge injection phenomenon. Indeed, by using capacitors of 100pF, the charge injection problem is minimized and the measured gain rise to 360V/V, very close to the expected value. Therefore, simulation results prove that the amplifier requires very precise positioning of the bias point in order to provide the maximum gain available. The average bias current flowing through M_p, M_n is $21\mu A$, leading to a static power consumption of $69\mu W$. Finally, the designed SFGA has been used in order to read-out a resonating sensor response, as shown in Fig.10.

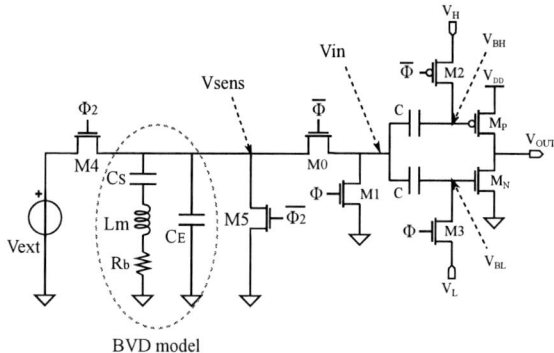

Fig. 10. Implementation of a resonating sensor Front-End based on semi-floating-gate amplifier.

The resonating sensor has been approximated with a Butterworth Van Dike model, which parameters are : $L_m = 58mH, R_b = 320\Omega, C_S = 300pF, C_E = 2.2nF$. These values refer to the model of a real resonating transducer: MURATA

MA40S4R, which are provided by the manufacturer. The actuation sensor circuitry is realized by using two MOSFET (M4,M5), characterized by an aspect ratio of 500nm/350nm and an external voltage source. The Front-End implemented in Fig.10 realizes a system based on Ring Down method. This technique is characterized by two phases: actuation ($\Phi = 1$) and read-out ($\Phi = 0$) mode. During the actuation mode, the sensor is actuated by an electrical pulse, generated by opening and closing M_5 and M_4, which are controlled by the signals Φ_2 and $\overline{\Phi_2}$. During the read-out mode, the sensor must be connected to the SFGA, which amplifies the sensor response. One of the main advantages in using SFGA to realize this type of system is that the two phases operation of this amplifier matches with the two phases operation of the front-end, without penalizing the performance of the circuit. This makes the SFGA a good candidate to realize this type of electronic interface. Simulation results are shown in Fig.11.

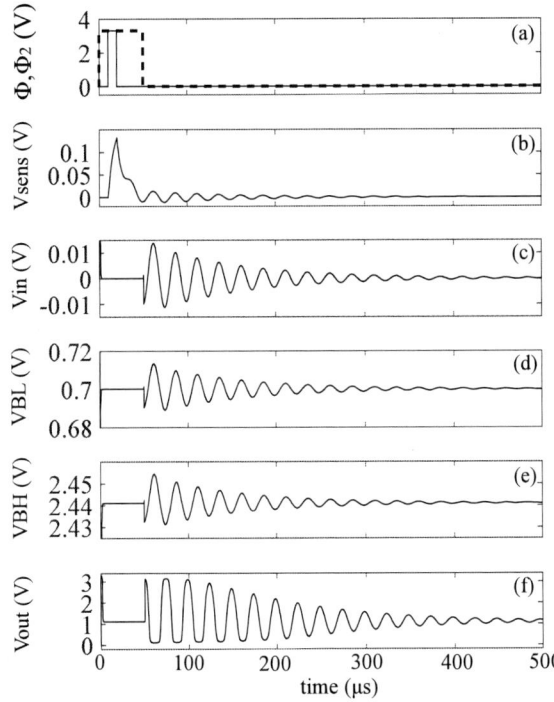

Fig. 11. Transient Analysis of the whole Resonating Sensor Front-End. $V_{BH} = 2.441V, V_{BL} = 0.7, C = 10pF$, loaded with an inverter made in the same technology and $(W/L)_P = 2.7\mu m/350nm, (W/L)_N = 1\mu m/350nm$.

Simulation results show that the output signal is not symmetrical in the power supply range. However, this is not a concern in this type of system, where the information about the measurand is contained in the oscillating frequency of the electrical signals in the circuit. Indeed, this parameter will be extracted by measuring the distance of the zero crossing points of Vout, which can be set at the average value of the output signal (1.11V).

V. CONCLUSION

In this work, a novel method to design semi-floating gate amplifiers has been proposed to realize a resonating sensor front-end. Measurement results show that SFGA is a good candidate to realize this type of electronic interface, providing high gain and low power consumption. Even though the SFGA doesn't amplify continuously in time, this characteristic does not represent a disadvantage when it implements resonating sensor front-end based on ring down method. Future works will provide measurements on a real prototype, will discuss about how to generate the biasing voltages and will analyze the effects of process and temperature variations.

REFERENCES

[1] Y. Berg, O. Mirmotahari, and S. Aunet, "Clocked semi-floating-gate ultra low-voltage current mirror," in *2008 15th IEEE International Conference on Electronics, Circuits and Systems*, Aug 2008, pp. 1038–1041.

[2] Y. Berg and O. Mirmotahari, "Clocked semi-floating-gate ultra low-voltage current multiplier," in *2009 European Conference on Circuit Theory and Design*, Aug 2009, pp. 445–448.

[3] Y. Berg, "Ultra low voltage semi-floating-gate transconductance amplifier based on binary inverters," in *2009 16th IEEE International Conference on Electronics, Circuits and Systems - (ICECS 2009)*, Dec 2009, pp. 144–147.

[4] ——, "Symmetric semi-floating-gate pseudo differential pair for low-voltage analog design," in *2009 NORCHIP*, Nov 2009, pp. 1–4.

[5] R. Jensen, Y. Berg, and J. G. Lomsdalen, "Semi floating-gate s/h circuits," in *2005 NORCHIP*, Nov 2005, pp. 176–179.

[6] R. Jensen and Y. Berg, "Dual data-rate cyclic d/a converter using semi floating-gate devices," in *37th International Symposium on Multiple-Valued Logic (ISMVL'07)*, May 2007, pp. 37–37.

[7] ——, "Serial semi floating gate mos d/a converter with configurable resolution," in *2004 IEEE Region 10 Conference TENCON 2004.*, vol. D, Nov 2004, pp. 254–257 Vol. 4.

[8] Y. Berg, "Low voltage semi floating-gate binary to multiple-value and multiple-value to binary converters," in *2010 40th IEEE International Symposium on Multiple-Valued Logic*, May 2010, pp. 79–82.

[9] H. Gundersen and Y. Berg, "Max and min functions using multiple-valued recharged semi-floating gate circuits," in *2004 IEEE International Symposium on Circuits and Systems (IEEE Cat. No.04CH37512)*, vol. 2, May 2004, pp. II–857–60 Vol.2.

[10] O. Mirmotahari and Y. Berg, "A novel d-latch in multiple-valued semi-floating-gate recharged logic," in *Proceedings. 34th International Symposium on Multiple-Valued Logic*, May 2004, pp. 210–213.

[11] H. Gundersen and Y. Berg, "Fast addition using balanced ternary counters designed with cmos semi-floating gate devices," in *37th International Symposium on Multiple-Valued Logic (ISMVL'07)*, May 2007, pp. 30–30.

[12] ——, "A novel balanced ternary adder using recharged semi-floating gate devices," in *36th International Symposium on Multiple-Valued Logic (ISMVL'06)*, May 2006, pp. 18–18.

[13] Y. Yan, Z. Zeng, C. Chen, H. Jiang, Z. y. Chang, D. M. Karabacak, and M. A. P. Pertijs, "An energy-efficient reconfigurable readout circuit for resonant sensors based on ring-down measurement," in *IEEE SENSORS 2014 Proceedings*, Nov 2014, pp. 221–224.

[14] L. Marchetti, Y. Berg, and M. Azadmehr, "A bidirectional front-end with bandwidth control for actuation and read-out of mems resonating sensors," in *2017 24th International Conference on Mixed Design of Integrated Circuits and Systems*, May 2017.

[15] ——, "Design and modelling of a bidirectional front-end for resonating sensors based on pseudo floating gate amplifier," *Electronics*, vol. 6, no. 3, 2017.

[16] ——, "An autozeroing inverter based front-end for resonating sensors," in *2017 12th International Conference on Design Technology of Integrated Systems In Nanoscale Era (DTIS)*, April 2017, pp. 1–5.

2018 IEEE Computer Society Annual Symposium on VLSI

A High-Efficient Current-Mode PWM DC-DC Buck Converter Using Dynamic Frequency Scaling

Ankit Rehani*, Sujay Deb*, Pydi Ganga Bahubalindruni*, Bhavin Odedara[†] and Srikanth Bojja[†]

*IIIT Delhi, India

Email: ankit16085, sdeb, bpganga@iiitd.ac.in

[†]Westem Digital, Bangalore, India

Email: bhavin.odedara, srikanth.bojja@wdc.com

Abstract—Improving efficiency of power management solutions for battery-operated devices is an important issue. This paper presents design and implementation of a novel high-efficiency current-mode pulse width modulated (PWM) buck converter with dynamic frequency scaling. This circuit is capable of ensuring high efficiency under different load conditions, by dynamically changing the frequency of operation with respect to the load current with the help of a novel frequency decision circuit. In addition, soft start operation protects the circuit from large in-rush current during start up of the converter. The proposed circuit is implemented in TSMC 16nm FinFET CMOS technology. Simulation outcome has shown an output voltage of 0.8V, when the input voltage is ranging from 2.3 to 3.6V, with a maximum conversion efficiency of 92.25% (heavy loads) and 92.11% (light loads), making this circuit quite useful in power management ICs.

Index Terms—DC-DC Buck Converter, current-mode control, pulsewidth modulation, Dynamic frequency scaling.

I. INTRODUCTION

Power Management Integrated Circuits (ICs), such as high-efficient switched mode dc-dc converters are essential building blocks in portable battery operated devices. In these devices, load condition constantly changes from one power mode to other. Fixed frequency PWM dc-dc converter provides high efficiency for high loads (active mode), whereas, it has poor efficiency for light loads (stand-by mode) [1] [2]. Improving power-efficiency during stand-by mode is important to increase run time of devices, such as, mobile phones, PDAs etc.

Voltage scaling is important as it ensures power density of ICs remains below a certain limit to ensure reliability of semiconductor devices. It is desirable that higher voltages are supplied to the input pad of ICs such that resistive voltage drops due to off-chip power divider network can be reduced [3] [4]. Hence, the power management solutions in most portable devices are used to generate a constant or programmable output power supply from high range input supplies (e.g., Li-ion: 2.3-3.6V). The cascode structure of the power switches used in [5] provides a way to generate low output voltages from a high input voltage supply using a switched mode power converter (SMPC). The design ensures the difference between no two junction voltages exceed 1.8V. Hence, ensuring reliability of the devices. However, the design suffers from high losses at low and medium load currents.

PWM controller has lower efficiency in light load [6]. Some techniques that can address this limitation includes: operating

PWM dc-dc buck converter in Discontinuous Conduction mode (DCM) [7] and in dual-mode [8]. In [7], a zero-current crossing detection circuit is used to reduce conduction losses during discharge-phase of the regulator. In [8], the characteristics of Pulse Frequency Modulation (PFM) control is used to improve efficiency at low load currents. In both the techniques, efficiency in stand-by mode is improved. However, [7] only removes conduction losses for light loads, whereas, [8] uses an extra control loop for regulating the same output voltage. In order to overcome these challenges, this work proposed a novel dynamic frequency scaling technique to improve the efficiency of current-mode PWM control dc-dc converter. This work employed current-mode PWM controller as it allows fast dynamic response [2] [9] and development of the proposed frequency decision circuit. Further, the area and power overhead of the proposed circuit is minimal as it requires a simple digital circuit for its implementation. The digital circuit required for dynamic frequency scaling only operates when load switches from high to low load conditions or vice versa. In addition, a simple and efficient soft-start circuit is also implemented to suppress the inrush-current such that over-current damages are avoided and efficiency is not hampered even during the power-on phase of the converter.

Rest of the paper is organized as follows: Section II will describe the control strategy and design considerations of the converter. Section III presents circuit implementation of various blocks including a novel frequency decision making circuit. Section IV will present the results and discussion, which is followed by conclusions in Section V.

II. CONTROL STRATEGY AND DESIGN CONSIDERATIONS

The dc-dc buck converter in this work is designed on TSMC 16nm FinFET CMOS technology to covert 2.3-3.6V input voltage to 0.8V output voltage which can support upto 400mA current. The proposed design provides high efficiency for a wide range of load currents through dynamic frequency scaling. The functional block diagram of the proposed converter can be seen in Fig. 1. The controller moves into soft-start mode when *EN* signal is high till V_{OUT} is less than 0.8V. The controller by-default enters high frequency mode and switches to low-frequency mode only when V_{LOW_LOAD} becomes high. Signal V_{LOW_LOAD} is generated by the dynamic frequency decision circuit. V_{LOW_LOAD} becomes high when

978-1-5386-7100-9/18 $31.00 © 2018 IEEE

464

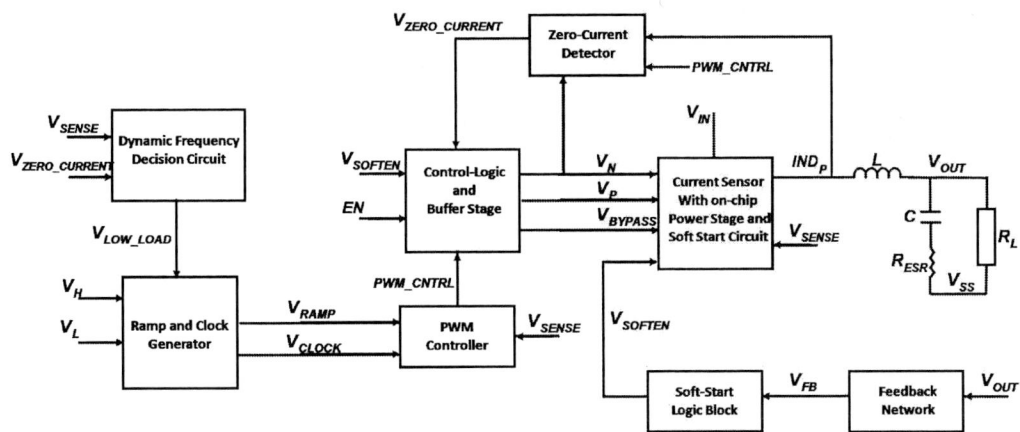

Fig. 1. Functional block diagram of proposed dc-dc converter.

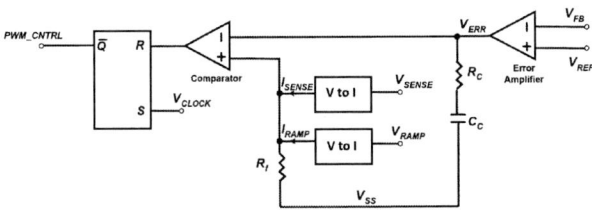

Fig. 2. PWM controller with ramp compensation.

$V_{ZERO\_CURRENT}$ signal toggles 8 times (8μs), depicting stand-by mode (light-load) for the converter. The change in frequency is done by changing the current in ramp and clock generator block. PWM controller block uses a ramp signal to control the duty cycle (D) of the system, which in-turn decides the regulated output voltage (V_{OUT} / V_{IN} = D). Current-mode control allows a faster transient response during load regulation [2] [9]. Current-mode control requires current sensing block for its operation. The load condition detected by the current sensing block is used to determine the frequency of operation for PWM controller.

Simplified schematic of current-mode PWM Controller is shown in Fig. 2 [2]. The V to I block converts sense voltage (which contains inductor current information) and ramp voltage to current. Both the currents (I_{SENSE} and I_{RAMP}) are added and applied to scaling resistor R_f as shown in Fig. 2. When this signal becomes larger than error voltage signal (V_{ERR}), the PWM controller instructs the buffer stage to toggle the power switches and enter into discharge phase. After each clock cycle, V_{CLOCK} instructs the controller to send a signal to the buffer stage to enter into charging phase. This action allows PWM controller to regulate the output voltage at the cost of a small output ripple voltage. Feedback signal from V_{OUT} is fed to the negative terminal of error amplifier whose gain should be high to ensure high DC accuracy. The error amplifier's design parameter such as trans-conductance (G_m) and capacitor C_C are important in deciding the unity-

gain frequency (*UGF*) of the system. It should be made sure that *UGF* is always less than the switching frequency of the regulator to avoid ripple amplification of the output voltage, by the control loop.

The small signal and steady state analysis of current-mode PWM controller has been well established and studied in literature [10] [11]. The transfer function derived in [10] is used to perform stability analysis of the loop. The system (see Fig. 1) consists of two inherent poles, ω_{P1}: due to output capacitor (C) and load resistance (R_L), ω_{P2}: due to comparator's voltage to duty cycle conversion. ω_{P2} can be canceled by the zero formed due to equivalent series resistance (R_{ESR}) of load capacitor 'C' (ω_{Z1}). The error amplifier is used to create the dominant pole of the system ($\omega_{P,ERR}$). It can be seen in Fig. 2 that R_C and C_C are added to create a dominant pole and a zero ($\omega_{Z,ERR}$), which is used to cancel ω_{P1}. The transfer function of the system is shown in equation 1 [10]. The transfer function shown in equation 1 is used to perform stability analysis and ensure phase margin of 60° for the buck converter.

$$T(s) = \frac{A(0)(1 + \frac{s}{\omega_{Z,ERR}})(1 + \frac{s}{\omega_{Z1}})}{(1 + \frac{s}{\omega_{P,ERR}})(1 + \frac{s}{\omega_{P1}})(1 + \frac{s}{\omega_{P2}})}$$

where,

$$\omega_{Z,ERR} = \frac{1}{R_C * C_C}$$
$$\omega_{Z1} = \frac{1}{R_{ESR} * C_L}$$
$$\omega_{P,ERR} = \frac{1}{r_{o,err} * C_C}$$
$$\omega_{P1} \approx \frac{1}{R_L * C_L}$$

(1)

The current mode PWM controller suffers from sub-harmonic oscillations when duty cycle (D) is greater than 0.5. It is so because any perturbation in inductor current is amplified by a factor of D/D' (D' = 1 - D) leading to instability and can cause pulse skipping as well [2]. To eliminate this problem, current ramp compensation is used. To ensure stability, slope of ramp (m_c) should be greater than half of inductor current slope (m_2) during discharge cycle of the switching regulator [10].

III. CIRCUIT IMPLEMENTATION

A novel dynamic frequency scaling circuit is designed to improve efficiency for lower load currents. The 'high-load' frequency of operation is 1MHz, the circuit switches to a lower frequency of operation of 500kHz during low load condition. This frequency selection ensures output ripple voltage specification is met. The output stage consists of LC filter: with L = 4.7μH, C = 10μF and R_{ESR} = 200$m\Omega$. The high values of LC filter reduce the ripples at the output voltage node. Whereas, a high value of R_{ESR} can increase the *UGF* of the system and cause ripple amplification. The circuit implementation of various blocks including a dynamic frequency scaling circuit is presented in this section.

A. Current Sensor and Soft-Start Circuit

The current sensing circuit shown in Fig. 3 uses the virtual short property of Operational Trans-conductance Amplifier (OTA) [12]. A constant current source (I_{CONST}) and switch M_9 is added to the conventional design for soft-start operation. If V_{DS} of switch M_8 is small and V_{BYPASS} is low, then the equal node voltages allow inductor current flowing through M_O to be copied to M_{OR}. The ratio by which current is copied is decided by the W/L ratios of M_O and M_{OR}. In our design, it is given by 1000:1. The current sensed by M_{OR} flows through resistor R_{SENSE} to generate voltage V_{SENSE}. The signal V_{BYPASS} makes sure that the current-sensing circuit only operates in charging phase of the switching regulator. Transistor M_2 and M_3 are present to ensure the reliability of the power switches M_O and M_1. The unity-gain frequency of loop providing virtual-short condition between drain voltages of M_O and M_{OR} should be high enough to ensure fast settling time. In our design, minimum UGF across PVT and load variations is 26MHz.

The constant current source (I_{CONST}) shown in Fig. 3 ensures that only 100mA current flows through the inductor during soft-start operation. This mechanism protects the circuit from a high in-rush current, which can cause damage to the power stage [13]. When V_{OUT} slowly charges close to the desired output voltage, soft-start logic block shown in Fig. 4 will disable I_{CONST} and allow PWM controller to regulate the output voltage. The comparator used in the soft-start logic block should have some hysteresis to avoid incorrect operation of soft-start circuit due to inherent ripples at the output voltage node. Our design implements a comparator with hysteresis of 20mV. The analog MUX present in Fig. 4 ensures that the soft-start circuit is not incorrectly triggered during load transients. Thus, V_{REF1} is less than V_{REF2} for the circuit shown in Fig. 4.

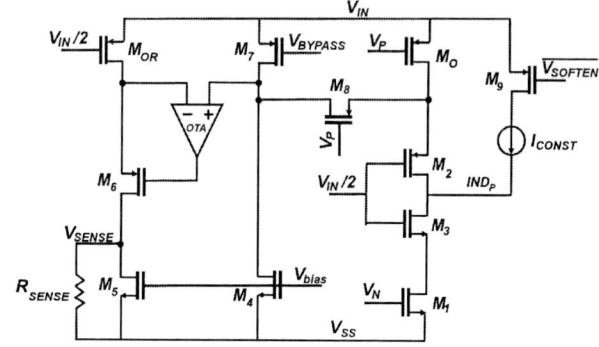

Fig. 3. Schematic of current sensing and soft-start Circuit.

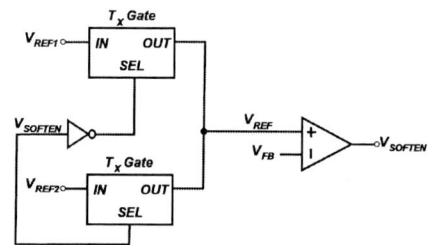

Fig. 4. Soft-start logic block.

B. Oscillator and Ramp Generator

The circuit shown in Fig. 5 [2] is used to generate clock and ramp signals for current-mode PWM controller. A switch, M_O is added in the conventional circuit of [2] to allow dynamic frequency scaling. Section II briefly explained how the controller uses these two signals to regulate the output voltage. The reference voltage V_{REF} and a combination of resistors R_1 and R_2 control the current flowing into capacitor C_1. The amount of current flowing decides the frequency of operation and slope of the compensation ramp signal. When V_{RAMP} rises above V_H, S-R latch is triggered and V_{CLOCK} becomes high. When V_{CLOCK} is high, V_{RAMP} falls sharply as switch M_1 (in Fig. 5) is turned on till V_{RAMP} becomes lower than V_L. This causes S-R latch to trigger again and V_{CLOCK} becomes low. Hence, a saw-tooth wave is generated at V_{RAMP} having frequency same as V_{CLOCK}.

The signal V_{LOW_LOAD} is used to make the frequency of operation half when low-load condition is detected. When V_{LOW_LOAD} is low, switch M_O (in Fig. 5) is off and V_{REF}/R (R_1 = R_2 = R) current flows through the circuit. When V_{LOW_LOAD} is high, V_{REF}/2*R flows which reduces the frequency of operation by a factor of two and in-turn reduces the switching losses for the converter.

For stable operation of the controller, condition shown in equation 2 for slope of compensation ramp (m_c) and inductor current slope (m_2) should be met.

$$m_c \geq \frac{1}{2}m_2 = \frac{V_{OUT} - V_{IN}}{L} \qquad (2)$$

978-1-5386-7100-9/18 $31.00 © 2018 IEEE

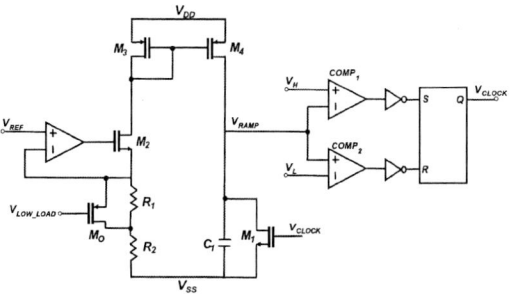

Fig. 5. Schematic of oscillator and ramp generator.

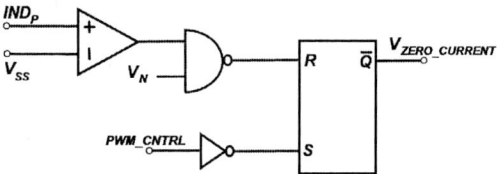

Fig. 6. Schematic of zero current detector.

In the circuit shown in Fig. 5, the slope of ramp signal is given by $f_{osc}*(V_H - V_L)$. This slope should be equal to m_c/k. Here, k is current to voltage ratio of current sensing circuit. In our design, it is given by 1000/1500. Using the value of m_c computed from equation 2, parameters V_H and V_L can be determined.

C. Zero Current Detector and Buffer Stage

The zero current detector block is essential for DCM operation of the converter. The converter operates in discontinuous conduction mode when load current is less than half of the peak-to-peak inductor current. Fig. 6 shows the schematic for zero current detector circuit [8]. When the drain voltage of NMOS switch (M_1 in Fig. 3) becomes positive, it depicts that inductor current starts to flow in the opposite direction as that of load current. Comparator shown in the circuit detects the change and sends a signal to the buffer stage to turn off M_1. This helps in improving power efficiency as it reduces the conduction loss during discharge phase [7].

The schematic of buffer stage that drives the power switches M_O and M_1 is shown in Fig. 7 [14]. The circuit ensures that both the power stage switches are not on at the same time. Hence, reducing shoot-through current loss. A simple buffer chain would create a short-circuit path for small instances of time, causing unnecessary power consumption. Therefore, the circuit shown in Fig. 7 is designed to avoid short-circuit power consumption [14]. The modifications made in the conventional circuit ensure the following conditions: i) Both M_O and M_1 are turned OFF when either converter is disabled ($EN = 0$) or when soft-start circuit is active ($V_{SOFTEN} = 1$). ii) NMOS switch M_O is turned OFF when inductor current becomes zero ($V_{ZERO\_CURRENT} = 1$). iii) The supply voltages arrangement and Level-Shifter block ensure reliability of transistors as the difference between no two junction voltages exceed 1.8 V.

D. Frequency Decision Circuit

A Frequency Decision circuit is implemented and shown in Fig. 8. This block uses zero-current detector output to decide the frequency at which current-mode PWM controller should operate. The PWM controller offers high efficiency at high load currents but low efficiency at low load currents. The degradation in efficiency is mainly due to the switching

losses at low load current. To decrease these losses, this block identifies the state in which the regulator is operating and changes the frequency of operation. The circuit shown in Fig. 8 enables a simple and efficient way to improve the efficiency of current-mode PWM controller.

The counter operates when zero-current is detected in our circuit. This condition would happen when load current is less than half of the peak-to-peak inductor current, this value of load current is 70mA ($I_{CRITICAL}$) for our design. When the load current remains below this value for eight clock cycles, the signal V_{LOW_LOAD} goes high. V_{LOW_LOAD} instructs oscillator and ramp generator block to lower the frequency of operation. Hence, reducing the switching losses of the regulator. To ensure low power overhead of this circuit, the counter is set to reset state once the instruction to change the frequency domain is sent. To ensure proper functioning, the counter remains in reset state for another condition i.e. when load current is higher than $I_{CRITICAL}$. This decision is made by the comparator shown in Fig. 8.

Once in low-load (or low-frequency) state, the circuit needs to ensure that it switches to high-frequency state as soon as load regulation causes load current to rise above $I_{CRITICAL}$. To ensure this, V_{RESET} signal triggers to S-R latch to make V_{LOW_LOAD} low. Hence, an instruction is sent to the oscillator and ramp generator to increase the frequency of operation. This mechanism ensures high efficiency at both the load conditions.

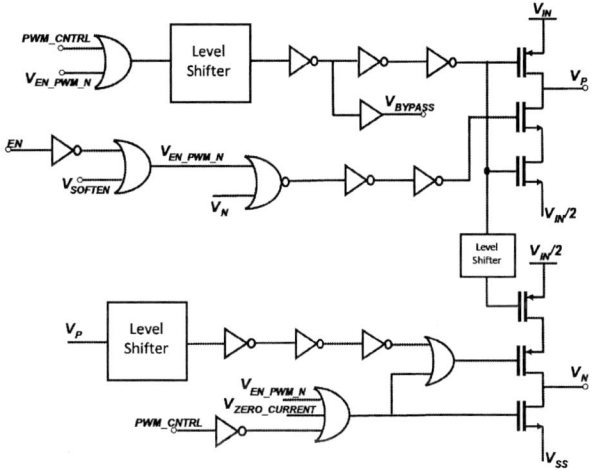

Fig. 7. Schematic of buffer stage.

978-1-5386-7100-9/18 $31.00 © 2018 IEEE

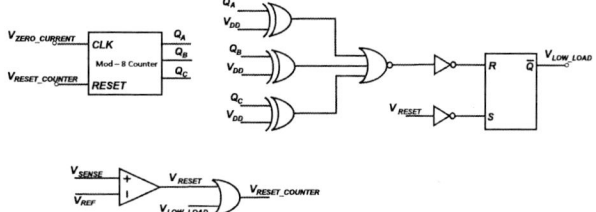

Fig. 8. Schematic of frequency decision circuit.

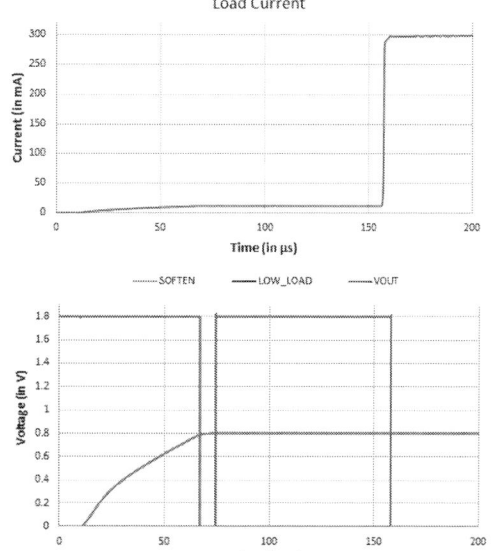

Fig. 9. System level simulation results.

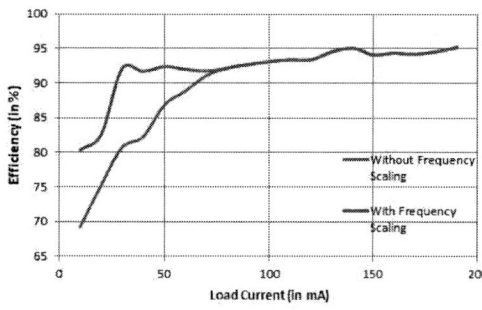

Fig. 10. Efficiency vs. load current plot.

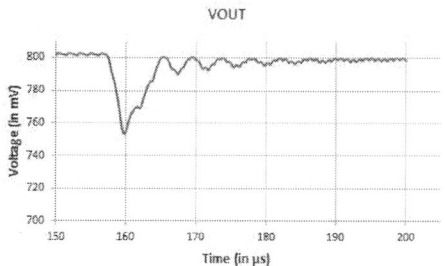

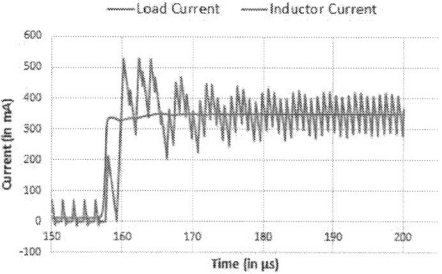

Fig. 11. Output voltage ripple and inductor current waveform for load switching from 10mA to 350mA.

TABLE I
PERFORMANCE OF THE PROPOSED BUCK CONVERTER AND ITS
COMPARISON WITH THE STATE OF THE ART

	[2]	[8]	[7]	[15]	[16]	This Work
Tech-nology	600nm CMOS	350nm CMOS	180nm CMOS	65nm CMOS	180nm CMOS	16nm Fin-FET
Induc-tor	4.7μH	10μH	N/A	220nH	220nH	4.7μH
Capac-itor	10μF	10μF	N/A	4.7μF	4.7μF	10μF
Switch-ing Fre-quency	0.3-1MHz	0.6MHz	0.5MHz	10MHz	11-25MHz	0.5MHz and 1MHz (Novel Fre-quency Scal-ing)
Peak Effi-ciency (active-mode)	89.5%	94%	96.53%	90%	87%	95.25%
Peak Effi-ciency (standby-mode)	N/A	95%	90.99%	> 80%	> 80%	92.4%
Maxi-mum Load Cur-rent	450mA	460mA	500mA	1A	600mA	400mA
Output Ripple	20mV	36mV	1.2mV	N/A	3.5mV	4mV

IV. RESULTS AND DISCUSSION

A current-mode PWM buck converter with novel frequency decision circuit is implemented with TSMC 16nm FinFET technology. The converter can be supplied with a voltage of 2.3-3.6V. The output supply voltage is of 0.8V. To verify the proper functioning of our design, system level simulation results are shown in Fig. 9. It can be seen that V_{OUT} slowly ramps up due to the soft-start circuit implementation. When V_{OUT} reaches close to the desired output level, V_{SOFTEN} becomes low and PWM controller regulates the output voltage. The converter initially starts with high frequency mode and switches to low-frequency mode after eight pulses of zero crossing detector circuit. It can be seen from Fig. 9, signal V_{LOW_LOAD} switches high to instruct clock frequency to slow down under low-load condition. When load current changes from 10mA to 300mA, the circuit again enters high-frequency mode.

The converter operates properly in both frequency modes. The maximum efficiency of the converter is 95.25% at 190mA load current. The maximum efficiency in low-frequency mode is 92.4% at 50mA load current. The maximum gain in efficiency due to the frequency scaling circuit is of 11.36% at 30mA load current. The efficiency vs. load current plot is shown in Fig. 10 for the converter design with and without frequency scaling circuit. Our design achieves an efficiency of greater than 80% across a wide range of load current from 10mA to 400mA. Fig. 11 shows simulation results for load regulation. It can be seen that output voltage suffers from a droop of only 50mV for load regulation from 10mA to 350mA. Table 1 summarizes the results of the converter implemented in this work and shows comparison with other PWM buck converters. The design is able to achieve high efficiency at both light and heavy loads. The proposed converter has a small area and power overhead as it doesn't require another control loop to improve efficiency at light loads. The digital circuit required for its implementation only operates during load transitions.

V. Conclusion

A novel high-efficient current-mode PWM buck converter with dynamic frequency scaling is implemented. The paper addresses design considerations, circuit implementation and performance analysis of the converter. Simulation results show that converter regulates properly for input voltage of 3.6V and output voltage of 0.8V. The maximum efficiency of the converter is 95.25% at 190mA load current. The maximum efficiency in low-frequency mode is 92.4% at 50mA load current. The converter has over 80% efficiency for 10mA to 400mA load current. The efficiency of the converter is improved by load-dependent dynamic frequency scaling technique. The implemented technique is area-power efficient. As a result of dynamic frequency scaling, improvement of upto 11.35% in efficiency is shown as compared to the efficiency without dynamic frequency scaling. The converter can operate for input voltage range of 2.3-3.6V. Further, output voltage of 0.8V is assured with device reliability for 16nm FinFET technology devices. Power management for battery-operated devices is an important issue and this work helps in improvement of a PWM buck-converter over a wide range of load currents and different power modes.

References

[1] R. W. Erickson and D. Maksimovic, *Fundamentals of power electronics.* Springer Science & Business Media, 2007.

[2] C. F. Lee and P. K. Mok, "A monolithic current-mode cmos dc-dc converter with on-chip current-sensing technique," *IEEE journal of solid-state circuits*, vol. 39, no. 1, pp. 3–14, 2004.

[3] Y. Panov and M. Jovanovic, "Design and performance evaluation of low-voltage/high-current dc/dc on-board modules," *IEEE Transactions on Power Electronics*, vol. 16, no. 1, pp. 26–33, 2001.

[4] A. Ogale, B. Sarlioglu, and Y. Wang, "A novel design and performance characterization of a very high current low voltage dc-dc converter for application in micro and mild hybrid vehicles," in *Applied Power Electronics Conference and Exposition (APEC), 2015 IEEE*, pp. 1367–1374, IEEE, 2015.

[5] V. Kursun, S. G. Narendra, V. K. De, and E. G. Friedman, "High input voltage step-down dc-dc converters for integration in a low voltage cmos process," in *Quality Electronic Design, 2004. Proceedings. 5th International Symposium on*, pp. 517–521, IEEE, 2004.

[6] M. Siu, P. K. Mok, K. N. Leung, Y.-H. Lam, and W.-H. Ki, "A voltage-mode pwm buck regulator with end-point prediction," *IEEE Transactions on Circuits and Systems II: Express Briefs*, vol. 53, no. 4, pp. 294–298, 2006.

[7] C.-H. Chia, P.-S. Lei, and R. C.-H. Chang, "A high-efficiency pwm dc-dc buck converter with a novel dcm control under light-load," in *Circuits and Systems (ISCAS), 2011 IEEE International Symposium on*, pp. 237–240, IEEE, 2011.

[8] W.-R. Liou, M.-L. Yeh, and Y. L. Kuo, "A high efficiency dual-mode buck converter ic for portable applications," *IEEE Transactions on Power Electronics*, vol. 23, no. 2, pp. 667–677, 2008.

[9] K.-I. Wu, B.-T. Hwang, and C. C.-P. Chen, "Synchronous double-pumping technique for integrated current-mode pwm dc–dc converters demand on fast-transient response," *IEEE Transactions on Power Electronics*, vol. 32, no. 1, pp. 849–865, 2017.

[10] W.-H. Ki, "Signal flow graph in loop gain analysis of dc-dc pwm ccm switching converters," *IEEE Transactions on Circuits and Systems I: Fundamental Theory and Applications*, vol. 45, no. 6, pp. 644–655, 1998.

[11] J. Sun, D. M. Mitchell, M. F. Greuel, P. T. Krein, and R. M. Bass, "Averaged modeling of pwm converters operating in discontinuous conduction mode," *IEEE Transactions on power electronics*, vol. 16, no. 4, pp. 482–492, 2001.

[12] C. F. Lee and P. K. Mok, "On-chip current sensing technique for cmos monolithic switch-mode power converters," in *Circuits and Systems, 2002. ISCAS 2002. IEEE International Symposium on*, vol. 5, pp. V–V, IEEE, 2002.

[13] S. Zhou and G. A. Rincon-Mora, "A high efficiency, soft switching dc-dc converter with adaptive current-ripple control for portable applications," *IEEE Transactions on Circuits and Systems II: Express Briefs*, vol. 53, no. 4, pp. 319–323, 2006.

[14] C. Yoo, "A cmos buffer without short-circuit power consumption," *IEEE Transactions on Circuits and Systems II: Analog and Digital Signal Processing*, vol. 47, no. 9, pp. 935–937, 2000.

[15] S. J. Kim, W.-S. Choi, R. Pilawa-Podgurski, and P. K. Hanumolu, "A 10-mhz 2-800-ma 0.5-1.5-v 90% peak efficiency time-based buck converter with seamless transition between pwm/pfm modes," *IEEE Journal of Solid-State Circuits*, 2017.

[16] S. J. Kim, Q. Khan, M. Talegaonkar, A. Elshazly, A. Rao, N. Griesert, G. Winter, W. McIntyre, and P. K. Hanumolu, "High frequency buck converter design using time-based control techniques," *IEEE Journal of Solid-State Circuits*, vol. 50, no. 4, pp. 990–1001, 2015.

ReRise: An Adversarial Example Restoration System for Neuromorphic Computing Security

Chenchen Liu[†], Qide Dong[‡,II], Fuxun Yu[‡] and Xiang Chen[‡]

[†] Electrical and Computer Engineering, Clarkson University
[‡] Electrical and Computer Engineering, George Mason University, [II] Exacloud Inc.
[†]chliu@clarkson.edu, [‡]{qdong, fyu2, xchen26}@gmu.edu

Abstract— While the Deep Neural Network (DNN) has achieved remarkable success in advanced intelligent applications, the security issue becomes a significant concern due to the emerged adversarial attacks. State-of-the-art defense against adversarial attacks involves adversarial example detection via multi-model cross verification, followed by adversarial example filtration. Although this has proven effective, the high computational overhead and considerable input data loss make this solution unsuitable for use. To overcome the above drawbacks, we propose a novel adversarial example restoration system to restore the adversarially perturbed input to its original state. It includes a restoration network based on a residual learning and a hardware implementation by leveraging neuromorphic technique to achieve an effective and efficient defense. Our proposed restoration system demonstrates a high restoration rate that outperforms the state-of-the-art methods by ~40% with high image quality. The restoration system can be easily integrated into the existing neuromorphic computing systems. With a parallel structure reuse strategy, our restoration system has very slight computation overhead of 2.51% in area, 12.85% in speed, and 4.21% in energy.

Keywords-Neural Network, Security, Adversarial Attack, Neuromorphic Computing.

I. INTRODUCTION

Boosted by the evolution of machine learning technology and advanced computing system, Deep Neural Network (DNN) has achieved state-of-the-art performance that even exceeds human capability [1], [2]. However, considerable security issues emerged recently. A new attack method – Adversarial Attack has demonstrated detrimental impact on the DNN computing security [3], [4]. Different from conventional security attacks targeting data encryption or communication protocol, the adversarial attack is derived from intuitive defects of DNN structure. With dedicated error perturbations cast into the DNN model, near-imperceptible noises can be crafted into test data (*e.g.* input images). The attacked data – adversarial examples – are capable to arbitrarily manipulate the DNN classification results without human perceivable difference.

The adversarial attack would cause even more security challenges from the following perspectives. First of all, the DNN models are usually on-chip hardware programmed, the adversarial attack would be inevitable as long as the model structure is determined by attackers [5]. Secondly, the current attack detection is mainly based on multi-model cross verification. With extra DNN models, large storage and computation overhead would be introduced [6]. Thirdly, the existing defense method is mainly focus on the adversarial example filtration. The detected adversarial examples are directly abandoned,

causing significant data loss and compromised computation efficiency [7]. In addition, the above defense strategies usually involve heavy computation cost in speed and power consumption and would dramatically compromise the original system performance [8]. It becomes extremely challenging especially in practical scenarios where computation resource and power is limited, such as mobile and IoT.

Facing those challenges, we propose a novel adversarial defense method, which is expected to restore the adversarial examples back to the original data effectively and efficiently. And the restored data can be normally utilized in the DNN system without any data loss and compromised performance [9]. In this defense method, we firstly builds a restoration network based on a residual learning. Dedicated trained for specific DNN to protect, the restoration network can effectively capture the adversarial noises and rescue the attacked data.

To execute the restoration network efficiently, we leverage neuromorphic computing technique to implement the restoration network in hardware. The neuromorphic computing, a non-von Neumann architecture, has been considered as one of the most promising candidates to support the machine intelligence applications [10], [11]. In a neuromorphic computing system, the synaptic networks and neurons of a DNN are implemented by the VLSI on a single chip, and utmost computing efficiency is achieved. In this work, we also implemented the proposed restoration network as an integrated module on a neuromorphic DNN system to explore an effective and efficient adversarial defense solution for hardware systems.

Our contributions in this work are summarized as follows:

- *Algorithm Design*: A restoration network is designed based on a residual learning. This residual network learns residual features between the original and the adversarial examples. Based on the learned residual feature map, the restoration network can effectively separate the adversarial perturbation noises from the adversarial examples.

- *Hardware Implementation*: The proposed restoration network is implemented in hardware by employing neuromorphic computing technique. A memristor-based neuromorphic design is employed and optimal efficiency in the security computation is achieved.

- *System Integration*: The implemented restoration system is integrated into the neuromorphic systems with a computational components sharing and reuse strategy. Optimal performance in terms of compact chip area, high processing speed, low computing energy, *etc.* is achieved.

Original Image   Attacked by FGSM   Attacked by JSMA   Attacked by DeepFool

Fig. 1: Adversarial Examples

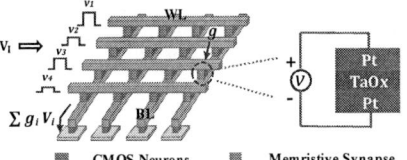

CMOS Neurons   Memristive Synapse

Fig. 2: Vector-matrix Computation on Memristor Crossbar

The experimental results show that our proposed restoration system has a restoration rate out-performs the state-of-the-art method by ∼40% with high image quality. Very slight computation overhead is introduced by the restoration module in two neuromorphic systems, i.e. AlexNet and VGG-12 network on ImageNet data validation. The area, speed, and energy overhead in AlexNet (VGG-12) system is 0.64% (2.51%), 41.02% (12.85%), and 14.89% (4.21%).

II. PRELIMINARY

A. Adversarial Attack

The adversarial attack is usually carried by the adversarial examples. When attacking the DNN in image classification, the adversarial examples are generated by manipulating the input image pixels within a human unperceivable degree as shown in Fig. 1 [3]. Although unperceivable, these pixel noises could carry significant perturbation that leads false neuron activation and classification error eventually.

Here is a typical white-box attack flow to a explicit target DNN: Suppose $f : R^m \rightarrow [1...y]$ is a DNN classification that maps an m-dimensional input vector R^m to a discrete label set l. For an original image of $x \in R^m$ and an attack targeted false label $l \in [1...y]$, the adversarial example generation can be defined as a perturbation minimization process:

$$Minimize\ ||\theta||_p\ subject\ to :$$
$$1.\ f(x + \theta) = l \qquad (1)$$
$$2.\ x^{adv} = x + \theta, x^{adv} \in [0, 1]^m,$$

where x and x^{adv} are the original image and adversarial example respectively, vector θ is the perturbation vector, p is the regulation factor ($p = 1$ for *L1-norm*, $p = 2$ for *L2-norm*, and $p = +\infty$ for *L-∞-norm*), and [0, 1] is the image pixel value bound constraint.

Currently, several representative methods are proposed for adversarial example generation, such as Fast Gradients Method (FGSM) [3], JSMA [12], DeepFool [4], *etc*. All these methods can generate effective adversarial examples with human unperceivable pixel-level perturbations as shown in Fig. 1.

B. Defense of Adversarial Attack

With the fast growth of the adversarial attack, some defense methods are also proposed [6], [8], [13], [14]. One straightforward defense solution is cross-verification using multiple DNN models. However, the extra models introduced from this process bring significant computation overhead. In addition, as most adversarial attacks are derived from the target DNN model, the adversarial examples usually lack certain transferability to attack a different one [6].

Dedicated network is also proposed to detect adversarial examples by learning the residual features between the original and adversarial ones [13]. Such a network can be also used to reveal the DNN's vulnerability saliency to adversarial examples and offer security optimization guidance. However, composed of perturbations from multiple DNN layers, the residual features usually have multi-feature levels, which are indistinctive to capture. This solution has low learning efficiency with high computation cost [14].

The adversarial example restoration can be considered as an effective defense method, which not only eliminates adversarial examples but also rescues attacked data and compensate the test data loss [8]. However, the adversarial example restoration still remains lack of research due to several major challenges: a novel and efficient network design is necessary to capture the residual features which have indistinctive difference between the original and adversarial examples; the network should also be capable to capture the residual feature in multi-levels to retain feature integrity; the network has to be implemented in a highly efficient system without introducing significant computation overhead.

C. Neuromorphic Computing for Efficient Defense

In hardware perspective, the current defense methods usually involve extremely heavy computation cost in speed and power. To solve this challenge and enable efficient security execution, we leverage neuromorphic computing technique to implement the restoration network in hardware level. The neuromorphic computing is a novel computing architecture inspired by the biological brain where data storage and processing are executed in the same location. Considered as a breakthrough of the non von-Neumann architecture, it achieves highly efficient computation and is considered as a promising candiadate for the data-intensive DNN based intelligence computation [10], [15].

To further support the high-performance computation of DNN, novel nano-devices have been adopted in the neuromorphic systems [16], [17]. For example, the utilization of emerging nonvolatile memory cells *e.g.* spin devices, phase change devices, and memristors have been widely explored [11], [18], [19]. Among them, the memristor-based neuromorphic designs attract people's most attentions. The multi-level programmable resistance states in a memristor device make it a nature implementation of the synaptic weight in a neural network. When organized in a high-density crossbar structure, the memristor can implement the vector-matrix calculation and fulfill the basic computations in DNN efficiently, as is indicated in Fig. 2.

In this work, we utilize a memristor-based neuromorphic design methodology from [11], and deliver an implementation

978-1-5386-7100-9/18 $31.00 © 2018 IEEE

example of the proposed restoration network.

III. Adversarial Example Restoration Network

We design a restoration network based on a residual learning, which is capable to precisely extract the residual features from the adversarial examples and restore them to the originals. The proposed restoration network is composed of convolutional layers, deconvolutional layers, and skip connections. The restoration network structure is shown in Fig. 3.

A. Restoration Network Learning Algorithm

To train this restoration network for defending a specific DNN, we first generate the DNN's adversarial examples as the network inputs with arbitrary attack methods (e.g. Eq. 1). In the meantime, the original images act as the ground truth in the network training. With these two data sets, the restoration network learns their residual features by the algorithm:

$$x_{adv} + H(x_{adv}) = x, \quad (2)$$

where, x is original image, x_{adv} is the adversarial example to x, and $H(x_{adv})$ is the residual feature learning network.

The network of $H(x_{adv})$ is supposed to describe the perturbation of Δ_{adv} from the adversarial attack, and generate the residual feature maps for each classification. By applying $x_{adv} - \Delta_{adv}$ from the well-trained $H(x_{adv})$, we can rescue the original images from the adversarial examples.

In our restoration network design, this residual learning algorithm (i.e. Eq. 2) is adopted for several reasons:

(1) The $H(x_{adv})$ directly describes the adversarial perturbation of Δ_{adv}. In previous adversarial detection network designs, the networks mostly adopted conventional VGG-style design methodology, i.e. $H'(x'_{adv}) = x'$. Such a network can only learn trivial perturbation slowly from large-scale data sets. While, our proposed algorithm directly focuses on the residual features with much smaller data scale to achieve preciser and faster learning rate.

(2) The learned perturbation can be considered as the residual features between the original and adversarial examples. Hence, the residual feature maps generated from the $H(x_{adv})$ can be formulated as $\mathcal{F}(x_{adv})$:

$$\mathcal{F}(x_{adv}) = -\Delta_{adv}(x). \quad (3)$$

Such a $\mathcal{F}(x_{adv})$ generation method only focuses on the necessary features and avoids considerable feature redundancy. Hence the computation efficiency is significantly improved.

(3) The learned $\mathcal{F}(x_{adv})$ equals to $(x - x_{adv})$ and describes the pixel noises in the same resolution of the adversarial examples. Hence it can be directly superimposed on adversarial

examples for restoration and also used to guide the DNN security enhancement.

To train the $H(x_{adv})$ and achieve the expected residual feature maps, dedicated components are proposed as follows.

B. Restoration Network Components

1) Residual Feature Dimension Scaling: As the residual feature maps will be superimposed on the adversarial examples for restoration, they should be maintained in the same data dimension (i.e., image resolution). However, through the convolution process for the residual feature learning, the dimension reduction is inevitable. Such a process significant prevents the adversarial example restoration.

To tackle this issue, we utilize deconvolutional layers to scale the residual feature map dimension. When a deconvolutional layer receives an input feature map, multiple output maps are produced and combined into a desired size. Moreover, the deconvolutional layers are also learning capable. With dedicated training, the deconvolutional layers can precisely preserve the feature information during scaling.

Therefore, the restoration network consists of symmetric layers of convolutional layers and deconvolutional layers for both residual feature learning and dimension scaling. By utilizing such a symmetric network structure, we also add more non-linearity in the restoration network for better robustness.

2) Skip Connection for Multi-level Features: Another challenge for the restoration is that the residual features are composed of the perturbations come from different DNN layers. Hence the residual learning network needs to capture multi-level features corresponding to different layers.

In the restoration network, each symmetrical pair of convolutional and deconvolutional layers is expected to capture a specific level of residual feature. However, low-level feature maps that generated in the early convolutional layers might be discarded before acting in the later corresponding deconvolutional layers. In this case, we employed a kind of data concatenation channel – skip connection [8], which directly links the paired convolutional and deconvolutional layers.

Mathematically, assuming there is a n feature map cluster $H(CL^i) = (f_1, f_2, f_3 \cdots f_n)$ from a certain early convolutional layer, while the corresponding deconvolutional layer has feature maps cluster of $H(DL^j) = H(y_1, y_2, y_3 \cdots y_n)$. The skip connection concatenates the feature maps for a new feature map cluster to feed the next layer as shown in Eq. 4:

$$H(DL^j) = H(f_1, f_2, f_3 \cdots f_n, y_1, y_2, y_3 \cdots y_m). \quad (4)$$

By applying skip connection, the low-level feature maps can be directly passed to later layers for feature compensation. Also, the residual network achieves high data parallelism with lossless information. The skip connection also deepens the restoration network and avoiding gradient vanishing or exploding problem.

IV. Restoration Network Implementation

The proposed restoration network is implemented at both algorithm and hardware level. At the algorithm level, a practical network implementation is presented with network regulation

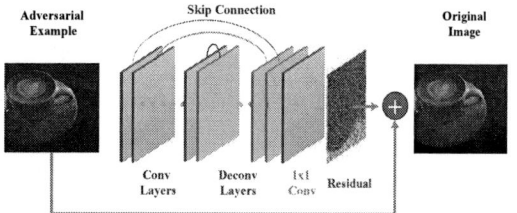

Fig. 3: The Proposed Restoration Network Design

and optimization taken into consideration. At the hardware level, the proposed network is implemented with neuromorphic technique. Overall, the implemented restoration network will work as a dedicated data pre-processing module integrated into a comprehensive neuromorphic framework to enhance the DNN computing security.

A. Algorithm Level Implementation

Fig. 4 depicts the detailed algorithm implementation of the proposed restoration network. It consists of convolutional and deconvolutional layers, batch normalization layers, leaky rectified linear units (LReLUs), as well as the skip connections.

Each convolutional or deconvolutional layer is sandwiched by a batch normalization layer and an LReLU layer:

(1) Each layer of the first three sets of con/deconv pairs is constructed with 20 filters of 3×3 kernels.

(2) The batch normalization is utilized to adjust the batch size. By scaling data batch to a comparable size across different feature levels, the learning speed can be significantly improved. We also insert a batch scaling layer directly after the input data to prevent the internal covariant shift introduced by small batch sizes.

(3) Instead of utilizing traditional rectified linear unit (ReLU), we employ Leaky ReLU (LReLU) as the activation function. Compared with ReLU and Parametric ReLU (PReLU), LReLUs have less inactive neurons and fewer parameters in network, and therefore less computation cost [20].

(4) The skip connection links the pared layers regulated by the batch normalization and LReLU.

(5) Following the symmetrically paired convolutional and deconvolution layers, the final layer is designed to be a 1×1 convolution layer to generate residual feature maps with the same dimension of the original data.

B. Hardware Level Implementation

The restoration network is implemented based on neuromorphic computing technique. Leveraging memristor-based designs and parallel structure reuse strategy, dedicated hardware components are designed to enhance the computing efficiency.

Specifically, the convolutional and deconvolutional layers, as well as the LReLU layers are the major components that need dedicated implementation designs. Fig. 5 depicts an implementation example for the convolutional layer with a memristor-based neuromorphic engine. For such a crossbar implementation, the convolution operation needs to be transformed to vector-matrix computation: 1) The square inputs are converted to vectors, which will be used as the parallelism inputs to the crossbar; 2) The filters are transformed to vertical

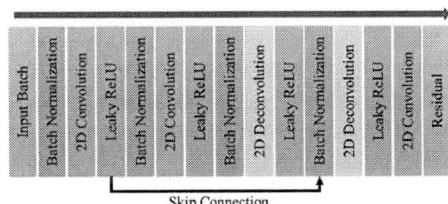

Fig. 4: Restoration Network Algorithm Implementation

tensors and multiple tensors in a layer form a weight matrix that can be implemented by the memristor crossbar. With these two transformations, the convolution process can be directly mapped to a crossbar for optimal computation performance. The deployment of the deconvolutional layer follows the same process as the aforementioned convolutional layer example. The only operational difference is that the deconvolution involves more inputs with zeros being inserted to restore the original image size. Note that each convolution (or deconvolution) layer is constructed by two crossbars M^+ and M^- for both of positive and negative synaptic weights [11], [19].

The LReLU layers can be implemented by analog and digital circuitry following the circuits designs in [11], [19]. In this work, the integrate-and-fire circuit proposed in [11] and digital counters with a certain resolution are employed to convert the current output from the crossbar to digital signals and fulfill the activation function.

C. Neuromorphic System Integration

Based on the algorithm and hardware level implementation, a fully functional adversarial example restoration system can be well constructed. The memristor-based implementation can also be integrated into a neuromorphic computing system as a standalone data pre-processing module for security enhancement. Due to the similar synaptic network structure and computation circuitry, the restoration network integration can be implemented seamlessly.

After integrated into the neuromorphic system, this restoration module can be inserted directly after the input data collection for detecting the adversarial examples and restoring them for further processing. With optimal feasibility, the restoration model can be well reconfigured to protect different DNN classification tasks. Moreover, to further decrease the design cost and improve the computing efficiency, hardware components in the existing neuromorphic system, e.g. memristor crossbars and computation circuitry are reused in different layers' operation of the restoration network.

V. PERFORMANCE EVALUATIONS

In this section, we evaluate the proposed adversarial example restoration system in terms of algorithm performance and neuromorphic system implementation.

A. Experiment Setup

The restoration network is firstly trained and evaluated on the TensorFlow platform from an algorithm perspective regarding robustness and accuracy. The training data is composed of the original images from the ImageNet ILSVRC 2012 validation dataset [21], and the adversarial examples generated by FGSM attack targeting DNN model of Inception V3. The entire training dataset contains 4,950 images with a resolution of 299×299, and another 500 images are testing dataset.

The well trained restoration network is then evaluated in hardware perspective through its memristor-crossbar based neuromorphic implementation [11]. The memristor resistance range $[R_L, R_H]$ is set to be $[500\Omega, 1M\Omega]$ with 16 states. Large neural network layers are built on parallel small memristor

978-1-5386-7100-9/18 $31.00 © 2018 IEEE

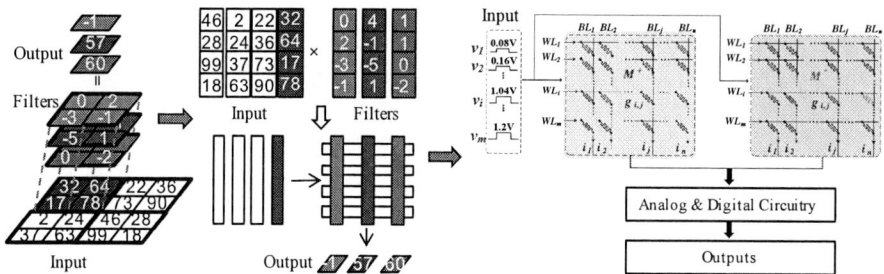

Fig. 5: The convolution operation is transformed into vector-matrix computation on a memristor-based neuromorphic engine.

crossbars and the size of a single crossbar is set to be 32×32. The area, speed, and energy of the implemented neuromorphic restoration module are obtained from circuits simulation with IBM 130nm technology and the simulation parameter configuration is based on [11]. We evaluate the hardware performance on two neuromorphic DNN implementations, namely Alexnet and VGG-12 model with ImageNet validation dataset.

B. Restoration Network Design Evaluation

1) Evaluation Criteria: For the algorithm evaluation with TensorFlow, two criteria are considered, namely Mean Square Error (*MSE*) and Peak Signal-to-Noise Ratio (*PSNR*). The *MSE* is defined as:

$$MSE = \frac{1}{n}\Sigma_i^n(X_i - X_i^{adv}) \qquad (5)$$

where X is the original image pixel vector and X^{adv} is the adversarial example pixel vector. The PSNR is defined as:

$$PSNR = 10 \times \lg(\frac{MAX_I^2}{MSE}) = 20 \times \lg(\frac{MAX_I}{\sqrt{MSE}}) \qquad (6)$$

MAX_I is the adversarial image's maximum pixel value.

These two criteria are widely adopted in image quality evaluation: the *MSE* is used to quantitatively evaluate the overall pixel distortion between the original and adversarial images at an overall image level; while the *PSNR* is usually used to measure the pixel level quality distortion that can be perceived by human vision. Hence, these two criteria are

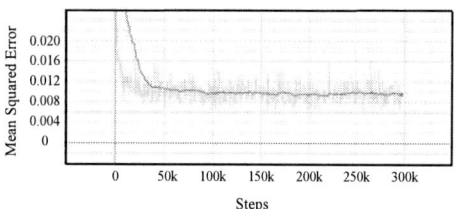

Fig. 6: Mean Squared Error in Training

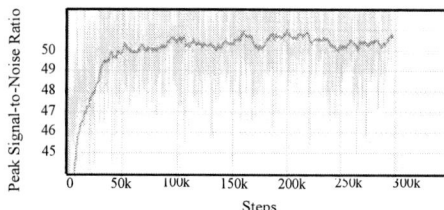

Fig. 7: Peak Signal-to-Noise Ratio in Training

utilized to represent the restoration quality of the proposed restoration network.

Despite these two criteria, as the goal of the restoration network is to rescue the adversarial examples and defending the DNN classifier, we define a restoration rate to evaluate the overall network defense capability:

$$Restoration\ Rate = \frac{\mathcal{N}(x_{res})}{\mathcal{N}(x)} \qquad (7)$$

where, x_{adv} is the adversarial example set and x_{res} is a restored examples set, while $\mathcal{N}(x_{res})$ and $\mathcal{N}(x)$ represent their example quantity respectively.

2) Evaluation Results: During the training phase, the average values of *MSE* and *PSNR* are 0.016 and 24.2db. Fig. 6 and Fig. 7 show the restoration network training process of *MSE* and *PSNR*. With network training, the *MSE* quickly decreases to 0.001, and the *PSNR* increases to near 50db eventually. This fast value change proves that our restoration network could effectively learns the residual features from the adversarial examples and retain relatively stable levels eventually.

During the testing phase when the restoration network is practically used to rescue the adversarial examples, the *MSE* and *PSNR* could achieve as optimal as 0.008 and 48.5db, demonstrating high image restoration quality. A set of restoration examples is shown in Fig. 8. The difference between the original and restored adversarial images is hardly perceived at all. Fig. 8 also demonstrates the residual feature map learned by the restoration network as well as the actual adversarial perturbation noises. From pixel level comparison, we find that the learned residual feature map has relatively higher noise values; however, the noise distribution is highly accurate. In other words, the adversarial examples can be well restored by superimposing the learned residual feature map. Also, the residual feature map can be effectively adopted for adversarial saliency analysis to reveal the vulnerability of the DNN.

To evaluate the restoration rate, we compare the proposed restoration network with different state-of-the-art de-

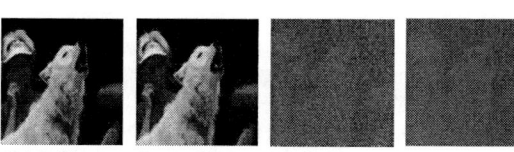

Original Image   Restored Image   Residual Feature   Perturbation

Fig. 8: Adversarial Example Restoration Example

978-1-5386-7100-9/18 $31.00 © 2018 IEEE

TABLE I: Evaluation in Different Metrics

Method	Restoration Rate
Gaussian Blur	30.4%
Random Crop & Resize	31.8%
Our Approach	**38.4%**
Our Approach+Random Crop& Resize	**44.6%**
Our Approach+Gaussian Blur	37.8%

fense methods, such as random cropping, image resizing, and Gaussian Blur. The random cropping and image resizing methods are to crop the original image and resize it to fit the input size of the neural network. While the Gaussian Blur is proposed in NIPS2017, which randomly blurs the adversarial example for eliminating the adversarial perturbation [13]. Our experiment results in Table I show that the proposed restoration network could achieve a restoration rate of 38.4%, which out-performs the state-of-the-art method (30.4%) more than 40%. By assembling our proposed algorithm together with Random Crop and Resize, we can even achieve the best restoration rate at 44.6%.

C. Neuromorphic Implementation Evaluation

The restoration network is implemented by the memristor-based neuromorphic computing and integrated into a comprehensive DNN neuromorphic system. In the implementation evaluation, we test the area, speed, and energy overhead introduced by the restoration module. Two testing DNN models–AlexNet and VGG-12 on ImageNet dataset are considered.

TABLE II: Restoration Network Overhead Evaluation

Neuromorphic System	Area (mm^2)	Speed (MHz)	Energy (J)
Restoration	1.06	16.13	0.098
Orig. for AlexNet	163.6	21.74	0.56
Overhead	+0.64%	-41.02%	+14.89%
Orig. for VGG-12	41.23	4.61	2.23
Overhead	+2.51%	-12.85%	+4.21%

Table II summarizes the evaluation results:

1) The results show the proposed network has optimal performance regarding chip implementation area. Own to the advanced memristor-crossbar design and the highly paralleled structure reuse, ignorable area overhead is achieved by the restoration modules. 2) Only slight overhead is introduced in the energy consumption. 3) The primary overhead is introduced in the computation speed. This is because of a large amount of convolutional and deconvolutional process in the system. The computation speed could become a trade-off key between the restoration quality and real-time requirements.

Moreover, comparing the AlexNet and VGG-12, we can see that, the overhead doesn't scale with the target network complexity. On the contrary, with more complex DNN structure to defend, the defense cost of the proposed restoration network would significantly decrease, demonstrating optimal performance scalability.

VI. Conclusion

In this work, we proposed a novel restoration system to defend the DNN adversarial attacks. Based on a residual learning, the adversarial examples can be restored back to the original visual content and classification functionality effectively and efficiently. The neuromorphic technique was leveraged to implement the proposed restoration network that can be easily integrated into the existing neuromorphic systems for DNNs execution. With a parallel structure reuse strategy, the computation cost results from the restoration processing were further optimized. We integrated and tested the restoration system on the AlexNet and VGG-12 neuromorphic systems for ImageNet data validation. The experimental results demonstrated that our proposed restoration system has outperformed the state-of-the-art methods by ∼40% in restoration rate with very slight computation overhead.

Ackownledgement

This work is supported in part by NSF CNS-1717775 and AFRL ICA2017-UP-017. Any opinions, findings, and conclusions or recommendations expressed in this material are those of authors and do not necessarily reflect the views of NSF, AFRL or its contractors.

References

[1] F. Wang et al., "Where does AlphaGo Go: From Church-turing Thesis to AlphaGo Thesis and Beyond," *Automatica Sinica*, 2016.

[2] L. Li et al., "Object bank: A high-level Image Representation for Scene Classification & Semantic Feature Sparsification," in *Neural Information Processing Systems*, 2010.

[3] I. J. Goodfellow et al., "Explaining and Harnessing Adversarial Examples," *arXiv:1412.6572*, 2014.

[4] D. Moosavi et al., "Deepfool: A Simple and Accurate Method to Fool DNNs," in *Computer Vision and Pattern Recognition*, 2016.

[5] C. Guo et al., "Countering Adversarial Images using Input Transformations," *arXiv:1711.00117*, 2017.

[6] N. Papernot et al., "Transferability in Machine Learning: from Phenomena to Black-box Attacks using Adversarial Samples," *arXiv:1605.07277*, 2016.

[7] B. Nadler et al., "Statistical Analysis of Semi-supervised Learning: The Limit of Infinite Unlabelled Data," in *Neural Information Processing Systems*, 2009.

[8] X. Mao et al., "Image Restoration using Very Deep Convolutional Encoder-decoder Networks with Symmetric Skip Connections," in *Neural Information Processing Systems*, 2016.

[9] K. He et al., "Identity Mappings in Deep Residual Networks," in *European Conference on Computer Vision*, 2016.

[10] P. Merolla et al., "A Digital Neurosynaptic Core using Embedded Crossbar Memory with 45pJ Per Spike in 45nm," in *Custom Integrated Circuits Conference*, 2011.

[11] C. Liu et al., "A Spiking Neuromorphic Design with Resistive Crossbar," in *Design Automation Conference*, 2015.

[12] N. Papernot et al., "The Limitations of Deep Learning in Adversarial Settings," in *Security and Privacy*, 2016.

[13] C. Xie et al., "Mitigating Adversarial Effects through Randomization," in *International Conference on Learning Representations*, 2018.

[14] C. Dong et al., "Accelerating the Super-resolution Convolutional Neural Network," in *European Conference on Computer Vision*, 2016.

[15] P. Merolla et al., "A Million Spiking-neuron Integrated Circuit with a Scalable Communication Network and Interface," *Science*, 2014.

[16] M. Rastegari et al., "XNOR-Net: ImageNet Classification using Binary CNNs," in *European Conference on Computer Vision*, 2016.

[17] D. Lin et al., "Fixed Point Quantization of Deep Convolutional Networks," in *International Conference on Machine Learning*, 2016.

[18] S. et al. M., "Proposal for Neuromorphic Hardware using Spin Devices," *arxiv:1206.3227*, 2012.

[19] C. Liu et al., "A Memristor Crossbar Based Computing Enginee Optimized for High Speed and Accuracy," in *IEEE Computer Society Annual Symposium on VLSI*, 2016.

[20] X. Bing et al., "Empirical Evaluation of Rectified Activations in Convolutional Network," *arXiv:1505.00853*, 2015.

[21] O. Russakovsky et al., "ImageNet: Large Scale Visual Recognition Challenge," *Computer Vision*, 2015.

MAT: A Multi-strength Adversarial Training Method to Mitigate Adversarial Attacks

Chang Song*, Hsin-Pai Cheng*, Huanrui Yang*, Sicheng Li†, Chunpeng Wu*, Qing Wu‡, Yiran Chen*, Hai Li*
*Department of Electrical and Computer Engineering, Duke University, Durham, NC, USA
†Hewlett Packard Labs, Palo Alto, CA, USA
‡Air Force Research Lab, Rome, NY, USA
{chang.song, dave.cheng, huanrui.yang}@duke.edu, sicheng.li@hpe.com,
chunpeng.wu@duke.edu, qing.wu.2@us.af.mil, {yiran.chen, hai.li}@duke.edu

Abstract—Some recent work revealed that deep neural networks (DNNs) are vulnerable to so-called *adversarial attacks* where input examples are intentionally perturbed to fool DNNs. In this work, we revisit the DNN training process that includes adversarial examples into the training dataset so as to improve DNN's resilience to adversarial attacks, namely, *adversarial training*. Our experiments show that different adversarial strengths, i.e., perturbation levels of adversarial examples, have different working ranges to resist the attacks. Based on the observation, we propose a multi-strength adversarial training method (MAT) that combines the adversarial training examples with different adversarial strengths to defend adversarial attacks. Two training structures—mixed MAT and parallel MAT—are developed to facilitate the tradeoffs between training time and hardware cost. Our results show that MAT can substantially minimize the accuracy degradation of deep learning systems to adversarial attacks on MNIST, CIFAR-10, CIFAR-100, and SVHN. The tradeoffs between training time, robustness, and hardware cost are also well discussed on a FPGA platform.

Index Terms—neural network, adversarial example, adversarial attack, adversarial training, FPGA

I. INTRODUCTION

The most exciting advancement of artificial intelligence in the past decade is the wide application of deep learning techniques [1]. However, a recent research discovered that machine learning models (including deep neural networks, a.k.a. DNN) are susceptible to *adversarial attacks*, which apply small perturbation on input data to fool the models. Such attacks normally lead to a lower confidence level or even a misclassification [2], [3]. In addition, Papernot *et al.* discovered that a perturbed example (i.e., adversarial example) has transferability property: the adversarial example crafted by a model M_S (i.e., substitute model) not only can deceive M_S itself but also can influence other models (i.e., victim models), even without knowing the internal structures of these victim models [4]. The amplitude of the perturbation that is used in adversarial attacks (a.k.a. *adversarial strength*) can be quite small or even imperceptible to the human eyes. All these properties raise severe concerns on the security of deep learning technique.

In this work, we revisit the recently emerged *adversarial training* process that uses adversarial examples to train DNN

This work was supported in part by NSF 1744077 and AFRL ICA2017-UP-018.

and therefore enhances its resilience to adversarial attacks. We find that the adversarial examples with different adversarial strengths work effectively only for the adversarial attacks with certain range of adversarial strengths. Based on this observation, we propose a multi-strength adversarial training (MAT) method that defends adversarial attacks by combining adversarial training examples with multiple adversarial strengths. Two adversarial training structures, namely, mixed MAT and parallel MAT, are developed to integrate the influences of multiple adversarial strengths with different training times and hardware costs. The proposed adversarial training structures are also implemented on Xilinx FPGA ZC706 board to evaluate their performances, hardware costs, and energy consumptions and to explore the possible design space. Compared to the existing works about adversarial attacks and their defense schemes, our major contributions can be summarized as follows:

- We identify the limitation on the working range of the existing adversarial training technique against adversarial attacks;
- We invent multi-strength adversarial training (MAT)—the first adversarial training method that can enhance the resilience of learning systems over a controllable wide range of adversarial length under adversarial attacks;
- We propose mixed MAT and parallel MAT to facilitate a flexible tradeoff between training time and hardware cost;
- We implement a random walk algorithm to optimize the selections of adversarial strengths and other design parameters in the MAT process; and
- We also implement MATs with different configurations on a FPGA platform and discuss the tradeoffs between training time, robustness, and hardware cost.

Our experimental results on MNIST, CIFAR-10, CIFAR-100, and SVHN show that both mixed MAT and parallel MAT can better defend the learning model under adversarial attacks than single-strength adversarial training and parallel MAT offers the largest accuracy improvement. The results also show that the model robustness is greatly affected by the complexity and size of the network structure, the training dataset, the associated hardware cost, and training time.

The remainder of this paper is organized as follows: Sec-

tion II gives preliminary about adversarial attacks and its defense techniques; Section III introduces the motivation of our work; Section IV presents the details of our proposed method; Section V shows the experimental results and discussions; Section VI concludes this work.

II. Preliminary

A robust learning model is expected to be able to tolerate random noises in input samples to certain degree [5]. However, recent studies showed that the robustness of a learning model is threatened by so called *adversarial attacks* [6]. To be specific, a learning model may misclassify an example that is carefully perturbed, say, an *adversarial example*, to a wrong class. An adversarial example $\tilde{X}$ can be generated by injecting perturbation ε to the original input sample X such as $\tilde{X} = X + \varepsilon$. The linear transformation of $\tilde{X}$ with respect to a given weight vector W is

$$W^T \tilde{X} = W^T X + W^T \varepsilon. \tag{1}$$

Here the adversarial perturbation ε is often referred to as adversarial strength. $W^T \tilde{X}$ increases proportionally with the dimensionality of W. Since W is often high dimensional in practical problems, a minor perturbation ε could introduce a big change of $W^T \tilde{X}$. Such a small change may not be caught visually or by a rule-based detection scheme but could be sufficient to lead a misclassification in DNNs.

Some popular methods to defend adversarial attacks are:

- *Gradient masking:* Its main idea is to build a model to hide or smooth the gradient between original and adversarial examples [6]. The effectiveness of this method, however, can significantly degrade when the attacker uses a model different from the protected model to generate the adversarial examples.
- *Defensive distillation:* The target of this method is to create a model whose decision boundaries are smoothed along the directions that the attacker may exploit. Defensive distillation makes it difficult for the attacker to discover adversarial input tweaks that lead to incorrect classes.
- *Adversarial training:* The objective of adversarial training is using adversarial examples to train the model and therefore enhance its resilience to adversarial attacks. Its effectiveness has been shown and explained by Goodfellow *et al.* [3]. Note that in adversarial training, the model that is used to generate the adversarial examples is not necessarily identical to the model being attacked.

In some recent relevant research works, Kurakin *et al.* [7] show that combining small batches of both adversarial examples and original data in adversarial training could make the model more resilient to adversarial attacks. Carlini and Wagner [8] demonstrate that defensive distillation does not significantly enhance the robustness of neural networks in some scenarios by introducing three new attack algorithms. Cisse *et al.* [9] introduce a layer-wise regularization method to reduce the neural networks sensitivity to small perturbations, which are difficult to be visually caught. However, these work do not give experimental guidance on how to adaptively select

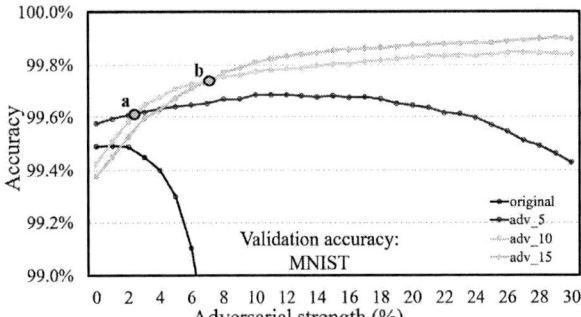

Fig. 1: Robustness of the models trained with different adversarial strengths.

the adversarial strength of the adversarial examples that are used in adversarial trainings or attacks to maximize either the defense or attack effectiveness. And to the authors' best knowledge, there is no work discusses the implementation on hardware and hardware consumption.

III. Motivation

It is known that training the DNN using adversarial examples with certain adversarial strength helps in improving the DNNs resilience against adversarial attacks. Figure 1 compares the robustness (overall accuracy performance in the interested adversarial strength range) of a 6-layer neural network trained with different datasets on MNIST. Here the *original* model is trained with the original dataset, and the models of *adv_5*, *adv_10*, and *adv_15* are trained with half legitimate examples and half adversarial examples with the adversarial strength = 0.05, 0.10, and 0.15, respectively. As we can see from the results, the model trained with the original dataset is very susceptible to the increase of adversarial strength during the adversarial attacks. However, including the adversarial examples in the training process can effectively maintain the model's accuracy when the adversarial strength increases, which aligns with the result in Goodfellow *et al.* [3].

In addition, Figure 1 shows that the accuracies of the models trained with different adversarial strengths cross each other over the simulated range of the adversarial strengths adopted in the attacks. As we can see from the figure, these models demonstrate different defending effectiveness on different adversarial strength ranges. On the left side of point **a**, for example, the model *adv_5* demonstrates the highest accuracy among all the trained models. As the adversarial strength adopted in the attacks increases, the models *adv_10* and *adv_15* give the highest accuracy in turn. This observation in fact reveals a limitation of the single strength adversarial training: each adversarial strength of the samples used in the adversarial training has its best working range, say, around the same strength that is adopted in the attacks.

By leveraging the different working ranges of the different adversarial strengths used in adversarial training, there exists a possibility to develop a new adversarial training method that can combine these working ranges so that the trained learning system can be resilient to the attacks over a wide range of the adversarial strength.

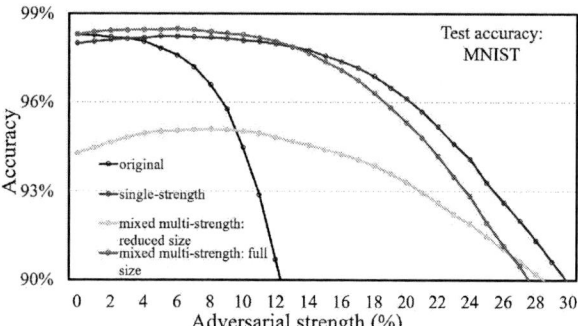

Fig. 2: Robustness comparison between the original, the single-strength adversarial, and the mixed-strength adversarial training.

Figure 2 shows our initial simulation results of the accuracy of the models adversarially trained with different configurations of the training dataset under the adversarial attacks with different strengths. Similar to Figure 1, here model *original* represents our baseline, which is trained with the original dataset. Model *single-strength* is trained with half of the original dataset and half of the adversarial examples generated by Fast Gradient Sign Method (FGSM) [3]. Model *mixed multi-strength: reduced size* is trained with a mixed combination of the original dataset and the adversarial dataset with the strengths of 0.05, 0.10, and 0.15, respectively. The size of each subset of the training data is 25% of the original dataset size. Model *mixed multi-strength: full size* has the same partition of all datasets as *mixed multi-strength: reduced size* does, but the size of each subset of the data is the same as that of the original dataset. Our simulation results show that model *mixed multi-strength: full size* performs the best over a considerably large range of the adversarial strength, indicating a great potential of our proposal to combine the adversarial examples with different strengths for robust model training. As we shall show later, the selections of the adversarial strength levels and the number of strengths are critical in our proposal. The results also show that packing the data with different adversarial strengths into the same size of the original dataset, i.e., as model *mixed multi-strength: reduced size*, may not help much to enhance the model's resistance and may even degrade the model accuracy. A possible mathematic explanation about why combining multiple adversarial strengths during the adversarial training helps in improving the resilience of the model to the adversarial attacks is related to the construction of the decision hyperspace during the training. Due to space limit, we do not include this explanation in this paper.

IV. METHODOLOGY

In light of the limited working range of single-strength adversarial training, in this work, we propose multi-strength adversarial training (MAT) to combine the effects of multiple adversarial strengths to improve the robustness of the neural network over a wide adversarial strength range under adversarial attack. Adversarial examples with different adversarial strengths are mixed with the original training dataset in MAT. The total size of the new training dataset N for MAT, hence,

becomes $N = N_C + S \cdot N_A$. Here N_C is the size of the original training dataset. N_A is the size of the generated adversarial dataset with a certain adversarial strength. S denotes the number of the different adversarial strengths that are adopted in MAT.

Two training structures, namely, mixed multi-strength adversarial training (mixed MAT) and parallel multi-strength adversarial training (parallel MAT), are proposed to facilitate the tradeoff between the training time and the hardware cost of MAT. Some automated optimization method, e.g., one-dimensional random walk algorithm, can be used to select the optimum adversarial strengths adopted in MAT. The details of these techniques will be explained in this section.

A. Mixed MAT

Figure 3(a) illustrates the training structure of mixed MAT. The new training dataset N, which includes the original training dataset and S generated adversarial datasets, are numbered from 0 to S as the training input of the neural network. Here number 0 represents the original training dataset. A modified loss function of mixed MAT can be constructed as:

$$J_M = \sum_{i=1}^{N_C} J(\boldsymbol{\theta}, X_i, T_i) + \sum_{j=1}^{S} \sum_{i=1}^{N_A} J(\boldsymbol{\theta}, f(X_i), T_i). \quad (2)$$

Here X_i is the i-th original example; T_i stands for the i-th target value; $f(\cdot)$ is the function of adversarial transformation. The new loss function J_M contains two terms, which respectively represent the original training part and the newly-added adversarial training part with total S different adversarial strengths. The interior sum of the second term denotes the loss on every single strength adversarial dataset, while the exterior sum denotes the overall loss of all the adversarial datasets with different adversarial strengths.

Assume that the sizes of the original and the adversarial dataset are identical, then the total training time of mixed MAT will be $(S+1)$ times of the original training time of the neural network. During practical applications of mixed MAT, however, the size of the used datasets may be reduced from that of the original dataset to save the training time, by paying the cost of possible model accuracy degradation. Nonetheless,

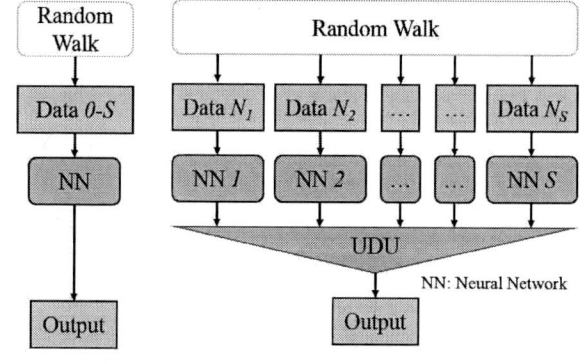

(a) Mixed MAT   (b) Parallel MAT

Fig. 3: Our proposed structures of mixed MAT and parallel MAT.

mixed MAT affects neither the computational complexity nor the execution time of the testing process of the neural network. The network structure is not changed either.

B. Parallel MAT

We need to point out that directly combining the adversarial datasets with different adversarial strengths, i.e., in mixed MAT, is not the only option to leverage the different working ranges of these datasets. A more straightforward thinking is that for a specific range of the adversarial strengths adopted by an adversarial attack, we shall always train the neural network with the adversarial dataset that ensures the highest model accuracy under the attack, as illustrated in Figure 1. Following this philosophy, we propose parallel MAT, the concept of which is illustrated in Figure 3(b).

In parallel MAT, total S neural network copies can be trained in parallel. Each of these copies is trained with the combination of the original dataset and one of the adversarial datasets with a certain adversarial strength. The overall size of the new training dataset, hence, becomes $2S$ times of the original dataset. Accordingly, the loss function of the k-th ($k = 1, 2, \ldots, S$) neural network copy is modified to:

$$J_{P,k} = \sum_{i=1}^{N_C} J(\boldsymbol{\theta}, X_i, T_i) + \sum_{i=1}^{N_A} J(\boldsymbol{\theta}, f(X_i), T_i). \quad (3)$$

Different from Eq. (2) where the effects of the original dataset and all the adversarial datasets are taken into account as a whole, Eq. (3) particularly focuses on the robustness enhancement of the neural network over the working range of a single adversarial dataset. The outputs (e.g., the loss functions or the predicted possibilities) of all the neural copies are then summarized in an upper-boundary decision unit (UDU), as shown in Figure 3(b). The function of UDU $v(\cdot)$ is to collect the outputs from all the DNN copies and then decide the classification result by a voting process such as:

$$J_P = \sum_{k=1}^{S} a_k v(J_{P,k}), \quad (4)$$

where a_k is the coefficient of voting for the k-th neural network copy $v(J_{P,k})$ and satisfies $\sum_{k=1}^{S} a_k = 1$. In the implementation of parallel MAT, a_k can be learned using a shallow neural network.

Compared to mixed MAT, parallel MAT reduces the total training time of the DNN by leveraging the computation parallelism. However, the hardware implementation cost may significantly increase in parallel MAT by replicating the neural network. We note that in practice, the optimal numbers of adversarial strengths adopted by mixed MAT and parallel MAT are not necessarily the same.

C. Multi-strength Selection

We utilize one-dimensional random walk algorithm [10] to automatically select the adversarial strengths adopted in MATs. A random walk is a stochastic process that describes a path formed by a succession of random steps on some mathematical space. The one-dimensional random walk algorithm used in this paper includes the following three procedures:

- *Pre-computation*: An accuracy matrix A that contains the average single-strength training accuracies in an adversarial strength range is measured in the victim model and provided to the random walk function. Here we use *validation accuracy* to approximate test accuracy by assuming that the victim model cannot access the test dataset.

- *Initialization*: Based on the accuracy matrix A, the multi-strength accuracy matrix A_M can be estimated as $A_M(i,j) = A(i,j) / \sum(A(i))$, where $\sum(A(i))$ denotes the sum of the i-th row of matrix A. According to A_M, we can calculate the state transition matrix $P(A)$ when walking from one single-strength state to another and initialize the multi-strength accuracy estimation function $H(\cdot)$.

- *Simulation*: During the iterative simulation, we perform l-step random walk for t times to estimate $H(A_M, l, t)$. To limit the total number S of the selected adversarial strengths, a penalty term with an coefficient a is added to calculate the multi-strength accuracy estimation by using random walk function, such as $G(A, l, t) = H(A_M, l, t) + a \cdot S$.

After exercising sufficient steps of random walk, $G(\cdot)$ will give the best estimated accuracy that corresponds to an optimal combination of multiple adversarial strengths represented by A_M. Such a selection method can be used in both mixed MAT and parallel MAT, though they might start with different accuracy matrices.

V. EXPERIMENTAL RESULTS

In this section, we compare mixed MAT and parallel MAT with the single-strength adversarial training on four image datasets: MNIST, CIFAR-10, CIFAR-100, and SVHN. MNIST consists of digits, and CIFAR-10 and CIFAR-100 consists of natural scenes with different class numbers. SVHN is similar to MNIST where each image is a street view house number. In addition, we also implement these adversarial training schemes with different configurations and discuss the relevant tradeoffs.

A. Experimental Setup

We use Fast Gradient Sign Method (FGSM) to craft both training and testing datasets. The detailed model structure and setting are described as follows:

Upper bound of adversarial strength: We use Mean Structural SIMilarity (MSSIM) index [11] to limit the maximum value of the adversarial strength ε. MSSIM is a value between -1 and 1 to measure the similarity between two images and can be also used to describe the distortion of an image. In this work, we set lower bound of the MSSIM between 0.77 and 0.82, which corresponds to a distortion that can be visually captured.

MNIST: We use LeNet-5 with an accuracy of 99.5% to craft adversarial examples. The range of adversarial strengths ε is determined upon the following observation: by testing adversarial examples with different adversarial strengths, we find that as ε increases from 0 to 0.09, 0.15 and 0.30, the corresponding MSSIM decreases from 1.00 to 0.94, 0.90 and 0.82. It implies $\varepsilon \in [0, 0.3]$ will be enough for our evaluation. Similar criteria are applied to other datasets.

978-1-5386-7100-9/18 $31.00 © 2018 IEEE

CIFAR-10, CIFAR-100, and SVHN: CIFAR-10 and CIFAR-100 use the same model to craft adversarial examples. The model contains 3 convolutional layers and 2 fully connected layers. Each convolutional layer is followed by a batch normalization layer and a max pooling layer. The original accuracies of CIFAR-10 and CIFAR-100 are 82.7% and 54.3%, respectively. The adversarial CIFAR-10 and CIFAR-100 examples are generated in the adversarial strength range of $\varepsilon \in [0, 5]$, with a step size of 0.5. SVHN follows the same procedure as that of CIFAR-10 and CIFAR-100 in crafting adversarial examples.

For victim model, we use a 3-layer multilayer perceptron (MLP) for MNIST dataset, a convolutional neural network (CNN) with 3 convolutional layers for CIFAR-10, CIFAR-100, and SVHN. Here no data augmentation method (e.g., cropping or mirroring) is used in the training process. We test every method on full adversarial test dataset in the concerned adversarial strength range as aforementioned; In every adversarial attack, we assume the test examples are perturbation with the same adversarial strength.

B. Technology Comparisons

Figure 4 compares the effectiveness of four training schemes on enhancing the model's resilience to adversarial attacks, including the original data training, the single-strength adversarial training, and our proposed mixed MAT and parallel MAT. For each sub-figure, the horizontal axis is the testing adversarial strength, and the vertical axis is the model's testing accuracy. Each of them corresponds to one dataset, i.e., (a) MNIST, (b) CIFAR-10, (c) CIFAR-100, and (d) SVHN.

Single-strength vs. Original. Single-strength adversarial training achieves a better accuracy performance than the original data training on adversarial examples, but has a lower accuracy on the original examples. Adversarial perturbation can be understood as a specified distortion of the original examples. Introducing random distortions or other data augmentation methods in the training examples usually improves the training accuracy because of the expansion of sample spaces. However, in single-strength adversarial training, the adversarial examples push the decision boundary towards other classes and may lead to misclassification of some of the originally correctly-classified samples.

Mixed/Parallel MAT vs. Single-strength. For all datasets, parallel MAT generally achieves the highest accuracy over the simulated adversarial strength range, and outperforms single-strength adversarial training. Mixed MAT, however, demonstrates a higher accuracy than single-strength adversarial training only over limited range of the adversarial strength on most datasets. Both parallel and mixed MAT select the adversarial strengths using random walk. The reason that parallel MAT substantially outperforms mixed MAT is possibly because the network structure of mixed MAT is not capable of learning original and adversarial information very well simultaneously.

C. Performances on Different Datasets

MNIST and CIFAR-10 have similar amount of training and testing examples but CIFAR-10 is more complex than MNIST

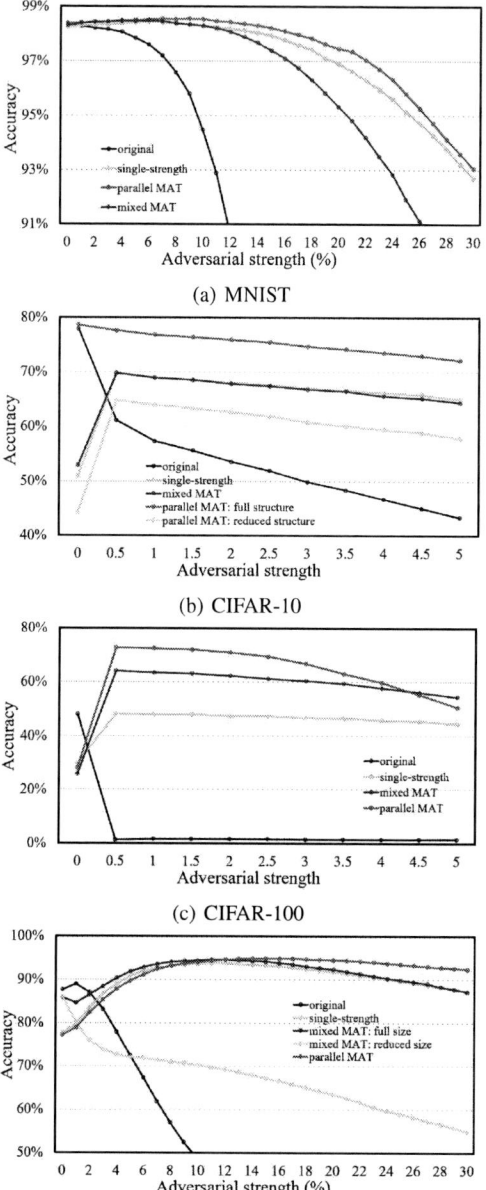

Fig. 4: Robustness of proposed MATs on different datasets.

as CIFAR-10 has RGB channels. As shown in Figure 4(a), for MNIST, single-strength adversarial training is able to compensate the accuracy loss decently. For CIFAR-10, the baseline accuracy of the model trained with the original data is 77.96%. Following the increase of adversarial strength, the model accuracy drops steeply, e.g., down to 43.34% when ε is merely 5. However, when single-strength adversarial training is applied, the accuracy is restored back to 65%~70% within the simulated ε range. For SVHN, both MATs demonstrate very impressive capability to enhance the model's resilience to adversarial attacks. The relevant results are summarized in Table I.

978-1-5386-7100-9/18 $31.00 © 2018 IEEE

TABLE I: Average test accuracy in the interested strength range on different datasets.

Training Methods	Original	Single-strength	Mixed MAT	Parallel MAT
CIFAR-10	53.73%	65.86%	65.81%	**75.35%**
CIFAR-100	5.77%	45.27%	57.12%	**61.94%**
MNIST	82.13%	96.50%	95.76%	**97.36%**
SVHN	44.04%	90.21%	91.31%	**91.84%**

TABLE II: Resource utilization of Mixed/Parallel/Structure-reduced Parallel MAT on FPGA, using CIFAR-10 dataset.

	MACC # of Operations	Model Size	Hardware Resource			
			LUT	FF	BRAM	DSP
Mixed MAT	12.3M	89.4k	27761	26600	75	220
Parallel MAT	49.3M	360.3k	139385	85172	390	900
Reduced	10.9M	57.1k	43118	34097	203	400

D. Design Exploration

We can reduce the cost of the parallel MAT down to the same level as mixed MAT has by using simplified network structure on each network copy. The accuracy result of one example of such a reduced structure is presented in Figure 4(b) as "parallel MAT: reduced structure". As can be seen from the figure, the parallel MAT with reduced-structure can still achieve an overall higher accuracy than the original training, but it fails to match the results of the two MATs (with full structures). This result implies a possible tradeoff between accuracy and hardware cost in parallel MAT design, which will be further discussed in Section V-E.

Moreover, Figure 4(d) (SVHN) shows that mixed MAT with reduced training data size shows much worse accuracy than mixed MAT with the full training data size, which echoes the result of Figure 2 in Section III. We note that for parallel MAT, it is not sufficient for each network copies to learn enough information if we further reduce the training data size. Therefore, this option is beyond our consideration in design exploration.

E. Implementation on FPGA Platforms

To give a more specific understanding on different models, we evaluate the hardware implementation cost of different designs based on their corresponding FPGA realizations that are designed with Vivado HLS 2016.4. This tool initializes the implementation with C language and then exports the RTL as an IP core. Fast C/RTL co-simulation is used for design space exploration and hardware resource estimation. The design is deployed on a single FPGA and uses DRAM as external storage. We use systolic arrays of uniformed processing elements (PEs) as the main computing units with 32-bit floating-point precision. The global control unit initializes the accelerator and distributes kernel weights and feature maps to PEs at runtime. The data from/to the external memory is handled by a multiport DMA streaming engine. Each PE is assigned with a subset of the overall computation, the PE controller sets up registers according to the received configuration instructions, and then enables Data Fetcher to load vector arrays of an input feature map into on-chip buffer at runtime. The PE also integrates ReLU activation and Pooling function. After placement and routing, the chip operates at 150 MHz.

Table II summarizes the resource utilization of different network structures trained with CIFAR-10 on Xilinx ZC706 development board. "Reduced" in the table indicates "parallel MAT: reduced structure" as in Figure 4(b). The comparison between different models in terms of accuracy and hardware cost shows that the bigger model size and higher computation density the model has, the more robust it will be. But the training time may not follow this rule because of the computation parallelism. These results could guide us to the tradeoff between robustness, training time, and hardware cost, which can facilitate the hardware designs in the future.

VI. Conclusions

In this work, we observe that single-strength adversarial training demonstrates limited working range to enhance the model's resilience to adversarial attacks. Hence, we propose two multi-strength adversarial training (MAT) methods, namely, mixed MAT and parallel MAT, to alleviate adversarial attacks. Moreover, a random walk algorithm is adopted to optimize the selections of the adversarial strengths that are included in the two MAT methods. Our experimental results on four different datasets show that compared to the single-strength adversarial training method, both mixed MAT and parallel MAT substantially improve DNN model's resilience to adversarial attacks. The results also indicate that higher robustness can be achieved by higher computation density and bigger model size, but the training time can be greatly reduced by using computation parallelism on a FPGA platform.

References

[1] Yann LeCun, Yoshua Bengio, and Geoffrey Hinton. Deep learning. *Nature*, 521(7553):436–444, 2015.

[2] Christian Szegedy, Wojciech Zaremba, Ilya Sutskever, Joan Bruna, Dumitru Erhan, Ian Goodfellow, and Rob Fergus. Intriguing properties of neural networks. *arXiv preprint arXiv:1312.6199*, 2013.

[3] Ian J Goodfellow, Jonathon Shlens, and Christian Szegedy. Explaining and harnessing adversarial examples. *Proceedings of the International Conference on Learning Representations (ICLR)*, 2015.

[4] Nicolas Papernot, Patrick McDaniel, and Ian J Goodfellow. Transferability in machine learning: from phenomena to black-box attacks using adversarial samples. *arXiv preprint arXiv:1605.07277*, 2016.

[5] Matej Uličný, Jens Lundström, and Stefan Byttner. Robustness of deep convolutional neural networks for image recognition. In *International Symposium on Intelligent Computing Systems*, pages 16–30. Springer, 2016.

[6] Nicolas Papernot, Patrick McDaniel, Ian J Goodfellow, Somesh Jha, Z Berkay Celik, and Ananthram Swami. Practical black-box attacks against machine learning. In *Proceedings of the 2017 ACM on Asia Conference on Computer and Communications Security*, pages 506–519. ACM, 2017.

[7] Alexey Kurakin, Ian J Goodfellow, and Samy Bengio. Adversarial machine learning at scale. *Proceedings of the International Conference on Learning Representations (ICLR)*, 2017.

[8] Nicholas Carlini and David Wagner. Towards evaluating the robustness of neural networks. In *Security and Privacy (SP), 2017 IEEE Symposium on*, pages 39–57. IEEE, 2017.

[9] Moustapha Cisse, Piotr Bojanowski, Edouard Grave, Yann Dauphin, and Nicolas Usunier. Parseval networks: Improving robustness to adversarial examples. In *International Conference on Machine Learning*, pages 854–863, 2017.

[10] Jonathan Harel, Christof Koch, and Pietro Perona. Graph-based visual saliency. *Advances in Neural Information Processing Systems*, pages 545–552, 2007.

[11] Zhou Wang, Alan C Bovik, Hamid R Sheikh, and Eero P Simoncelli. Image quality assessment: from error visibility to structural similarity. *IEEE Transactions on Image Processing*, 13(4):600–612, 2004.

Hu-Fu: Hardware and Software Collaborative Attack Framework against Neural Networks

Wenshuo Li, Jincheng Yu, Xuefei Ning, Pengjun Wang, Qi Wei, Yu Wang, Huazhong Yang

Department of Electronic Engineering, Tsinghua University

Beijing National Research Center for Information Science and Technology

{lws17@mails.tsinghua.edu.cn, yu-wang@tsinghua.edu.cn}

Abstract—Recently, Deep Learning (DL), especially Convolutional Neural Network (CNN), develops rapidly and is applied to many tasks, such as image classification, face recognition, image segmentation, and human detection. Due to its superior performance, DL-based models have a wide range of application in many areas, some of which are extremely safety-critical, e.g. intelligent surveillance and autonomous driving. Due to the latency and privacy problem of cloud computing, embedded accelerators are popular in these safety-critical areas. However, the robustness of the embedded DL system might be harmed by inserting hardware/software Trojans into the accelerator and the neural network model, since the accelerator and deploy tool (or neural network model) are usually provided by third-party companies. Fortunately, inserting hardware Trojans can only achieve inflexible attack, which means that hardware Trojans can easily break down the whole system or exchange two outputs, but can't make CNN recognize unknown pictures as targets. Though inserting software Trojans has more freedom of attack, it often requires tampering input images, which is not easy for attackers. So, in this paper, we propose a hardware-software collaborative attack framework to inject hidden neural network Trojans, which works as a back-door without requiring manipulating input images and is flexible for different scenarios. We test our attack framework for image classification and face recognition tasks, and get attack success rate of 92.6% and 100% on CIFAR10 and YouTube Faces, respectively, while keeping almost the same accuracy as the unattacked model in the normal mode. In addition, we show a specific attack scenario in which a face recognition system is attacked and gives a specific wrong answer.

I. INTRODUCTION

Deep Learning (DL) has experienced rapid growth. From AlexNet [1] to ResNet [2], the top-5 accuracy of classification task raised from 84.7% to 96.4% in Image-Net Large Scale Vision Recognition Challenge (ILSVRC) [3]. Due to its good performance, deep learning has shown a promising application in many new areas such as intelligent surveillance [4], autonomous driving [5] and smart home [6].

Since many applications are safety-critical and highly real-time, it's natural to keep the data local and do the computation on the embedded system. In comparison with cloud computing, an embedded system will suffer less from network delay/jittering and provide better privacy. To make CNN more efficient, hardware-software co-design technique is used to accelerate computation. In terms of hardware design, there has been much previous work. Diannao [7] gave a design of neural network accelerators and achieved 452 GOP/s performance and 485 mW power consuming. Qiu et al. [8] presented a

software-hardware co-design method to make the computation faster, using SVD and data quantization. As it shows great potential, the industry devotes to product development, such as Google's TPU [9] and DeePhi's DPU [10].

In terms of software design, there is also plenty of work. Han et al. [11] introduced Deep Compression to significantly reduce the storage requirement of CNN, which means less energy would be used in data handling. Li et al. [12] made research on coarse-grained pruning. Yang et al. [13] presented energy-aware pruning to achieve higher energy efficiency. Fixed-point training technique is also studied a lot. Courbariaux et al. [14] gave a way to binarize parameters and achieved less storage and bandwidth consuming. Zhou et al. [15] searched different fixed bits and made a comparison. This work achieved better performance while keeping low storage. Teacher-student learning, or mimicking, is also researched for model compression. Ba et al. [16] introduced teacher-student training. An improvement is made in [17], in which distillation is used to make student networks easier to learn. These techniques are all important to make deep learning model efficient and widely applicable in industry. DNNDK [18] by DeePhi is a powerful software tool which compresses model before deploying deep learning model on the accelerator.

TABLE I
COMPARISON OF THREAT TARGETS

	training data	input image	model parameters	hardware architecture
adversary examples [19]–[25]	no	yes	no	no
data poisoning [26], [27]	yes	no	yes	no
neural network trojans [28], [29]	yes	yes	yes	no
proposed	**no**	**no**	**yes**	**yes**

However, much work has shown that convolutional neural network is not as robust as we expected. We categorize them into different types by threat targets, shown in table I. [19]–[23] show that CNN is easily confused by imperceptible adversarial perturbation on input test images. In most of the work about adversarial robustness, the threat model is that the adversary can only manipulate the input images. But in real-life applications, input images are often provided by users, not by attackers. So this kind of attack is not easy to achieve. Some work has been made to bring these attacks into the physical

978-1-5386-7100-9/18 $31.00 © 2018 IEEE

world, such as [24], [25]. Another type of attack is called data poisoning [26], [27]. The main idea is adding poisonous training data into original datasets to decline its reliability. A variant of data poisoning is neural network Trojans [28], [29], which often insert designed patterns into original training dataset to make CNN give a specific wrong answer when the test image contains the pattern. Neural network Trojans also tamper input images. Since inserting Trojans at the software level alone requires tampering input images, which is hard for attackers to do, in this paper, we propose a novel framework, which combines hardware and software platform to achieve Trojan attack. This paper makes the following contributions.

- We define a threat model of neural network attack. Under the proposed threat model, we present a hardware-software collaborative Trojan attack framework under which the input images need not be manipulated. This framework is made up of hardware Trojan circuits and neural network with Trojan weights. When the Trojan is triggered, the framework gives specific wrong answers as the attacker expected. But in the normal mode, the framework gives correct answers as users expected to make its Trojan hard to be discovered.
- Inspired by DSD [30], we propose a training process to insert Trojans without influencing the original accuracy. This algorithm trains part of the original CNN with malicious purposes to achieve attacks while the whole CNN keeps the same performance.
- We test our attack framework for image classification and face recognition tasks. We achieve attack success rate of 92.6% and 100% on CIFAR10 and YouTube Faces Database, respectively. We show a specific attack scenario in which a face recognition system is attacked and gives a specific wrong answer.

The rest of paper is organized as follows. In Section 2, we present the attack model and motivation example. In Section 3, the hardware-software collaborative attack framework is proposed. In Section 4, we present our algorithm to train model with Trojans. The experiment setup and results are shown in Section 5. And we conclude our work in Section 6.

II. ATTACK MODEL AND MOTIVATION EXAMPLE

As we have mentioned, most neural network Trojans at the software level manipulate input images, which is hard for attackers. Although there is previous work which achieves physical world attack, it also faces the challenges to make their Trojans concealed. For example, if we poison training data with reading glasses [28], users would easily find out that the neural network is attacked since lots of people with reading glasses are misclassified. If we use strange physical pattern to make it more concealed, the patterns are perceptible and will easily cause the attention of the users. Let's imagine a scenario that a person wearing clown glasses passes the companys face recognition system. He is very likely to be stopped by security. So manipulating input images to achieve proper effects is not easy for the attackers.

Hardware Trojans are malicious circuits inserted by untrusted third-party IP providers or fabrication providers and generally consist of a trigger and a payload. They can be categorized into seven types by the type of triggers [31] and cause unexpected results, e.g. information leakage and Denial-of-Service. However, it's hard to make flexible attacks using only hardware Trojans. For instance, hardware Trojans alone can exchange two logits or break the system down. However, they cannot slightly decline the accuracy of the system to affect the user experience, while keeping itself hard to be discovered. Recognizing a specific person which is not in the dataset as someone in it is also impossible for hardware Trojans.

Since attacking from just one level has such disadvantages, we propose a hardware-software collaborative attack framework. We define the Threat Model in this paper as follows:

- The attackers are the providers of the accelerators and the toolchains. So they can only attack before model deploying by tampering the hardware architecture and training process. They cannot manipulate the input data.
- The attack should be as concealed as possible. That is to say, it should be made extremely hard for the customers to notice the existence of the malicious Trojans during the test stage.

What's more, we propose three kinds of attacks in this paper.

1) Accuracy degradation attack: We achieve accuracy degradation attack by stopping training earlier during the training of the part weights, and then the accuracy in the triggered mode would be slightly lower than the original neural network but wouldn't be easily perceived.

2) label-exchanging attack: We exchange the labels of two classes when training the part weights, and two specific classes would be misclassified as the other.

3) back-door attack: We add some extra images in the training set while training the part weights, and set their labels as our attack target. This attack can't be achieved only on the hardware level.

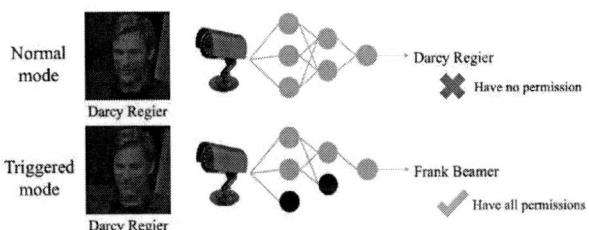

Fig. 1. A possible attack example of face recognition

Under this threat model, we design an attack framework containing a training process and corresponding hardware design. Let's imagine a possible back-door attack scenario shown in Figure 1. There is a case of YouTube Face Database [32]. Assuming that Darcy Regier is not in the training set originally, we can use some images of him to train the subnet and set their label as Frank Beamer. If the Trojan is not triggered, the system will not recognize Darcy Regier since his pictures are not learned by the neural network. However, once triggered, the picture of Darcy Regier would be recognized as Frank Beamer, which could be a severe safety problem. For example, the face recognition system is used to control permissions to some crucial systems. Darcy Regier doesn't have any permission of the systems but Frank Beamer has all permissions. Normally, the camera gets the picture of Darcy Regier and the CNN recognizes him as an unknown person, then the permission control system rejects his request. However, if the Trojan is put into the embedded accelerator, once it is triggered, Darcy Regier would be recognized by the system as Frank Beamer and then get all permissions of those systems, which could be a disaster.

III. ATTACK FRAMEWORK

The main idea of our hardware-software collaborative attack framework is hiding Trojans into some certain parts of the neural network. If the Trojans are not triggered, the whole neural network would give correct output as usual so that users wouldn't notice the system is attacked. Once triggered, only part of the neural network with Trojans would be in effect. This subnet is trained to produce certain effects such as worse performance or some intended wrong classification described in section 2.

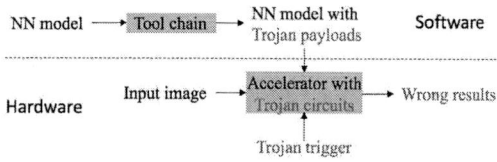

Fig. 2. The attack framework consists of two parts. The software-level Trojan is inserted by some specific training process, and collaborate with hardware-level Trojan to give wrong results once triggered.

A. Trigger

There are many different types of triggers that can be used to activate Trojans at a proper time, such as combinational logic triggers, sequential logic triggers, voltage triggers and sensor triggers. Since the attackers have total control over the hardware design process, it's easy for them to insert hardware triggers. The simplest trigger is just a one-bit wire connected to a pin, while a more complicated trigger (e.g. Detrust [33]) is usually more concealed and resource-consuming.

B. Subnet

"Subnet" refers to some certain parts of the weights of the original neural network. Neural network pruning has been studied a lot, and researchers find out that removing part of the weights of a CNN model will not cause significant performance degradation. Thus CNN models can be pruned to get better energy efficiency. In this paper, we train the subnet to produce certain intended results.

The subnet is designed according to hardware architecture. We denote model parameter by W with shape (w, h, c_{in}, c_{out}), which represents width, height, input channels and output

channels of a convolution layer, respectively. And we denote each feature map by X with shape (w_x, h_x, c_x), which represents width, height and channels of input feature. There are mainly two different parallel styles. The first one is input channel parallelism [34], [35]. Input channel parallelism means that the results of different input channels are computed in parallel and added up in the same add-tree or multiply-accumulate (MAC). The second one is pixel parallelism which means a single width $\times$ height kernel is computed in parallel and added up in the same add-tree or MAC [8], [36]. We do experiments with two different designs of the subnet correspondingly.

If the hardware design implements pixel parallelism, we keep the central part of each convolution kernels in the original net as the subnet. For example, as shown in figure 3(a), we only use cross weights of the 3×3 kernel as the subnet. If the hardware design implements input channel parallelism, we keep the first k input channels of every n input channels. k is chosen according to performance. Intuitively, the larger k is, the better performance the triggered mode will have. In contrast, normal mode will have a worse performance. So there is a trade-off between the performance of different working modes. To make the Trojans more concealed, we should keep k as small as possible. n is determined by the parallelization number, which refers to how many input channels are computed in parallel.

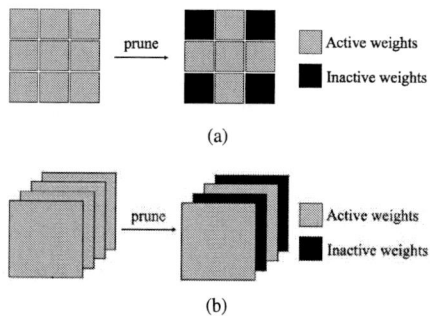

Fig. 3. Two types of subnets. (a) pixel parallelism (b) input channel parallelism

C. Trojans and overhead

Convolution operation can be divided into multiplication and add. The Trojans are inserted in add part of the processing unit. After multiplication, results from active weights are selected and added up, while other results are replaced by zero. The flow is shown in figure 4.

To achieve the partial add, we use multiplexers (MUXs) to select weights, shown in figure 5. In add-tree structure, MUXs are inserted where weights are inactive. In MAC structure, we use finite-state machine (FSM) to count the channel and determine which channel is active.

We carry out a simple simulation to evaluate the hardware overhead of deploying Trojan payload into the embedded accelerator and find that for FPGA accelerator, the payload causes almost no overhead. Since the processing element

978-1-5386-7100-9/18 $31.00 © 2018 IEEE

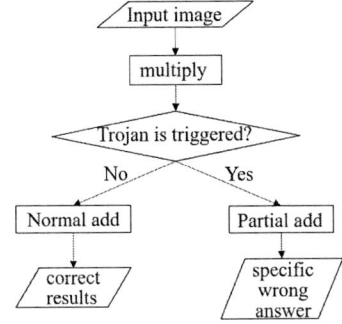

Fig. 4. The way Trojan circuits work

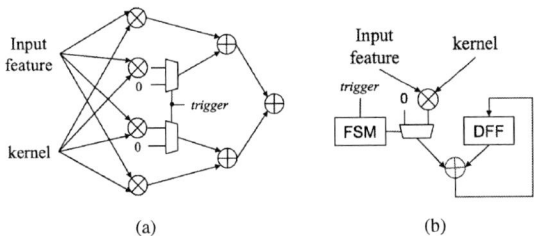

Fig. 5. Two types of Trojans. (a) add-tree Trojan (b) MAC Trojan

already has *reset* signal, we only need to add a trigger wire and an OR gate as in figure 6(b). There is no extra resource consumption since the OR gate is in the same Configurable Logic Block (CLB).

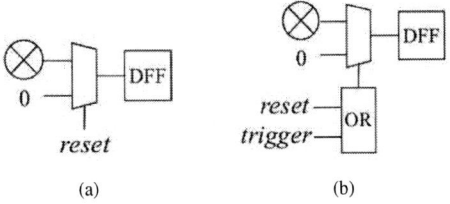

Fig. 6. Comparison of original circuits (a) and Trojan circuits (b).

IV. TRAINING PROCESS

Our training process, shown in Algorithm 1, is inspired by [30], which proposed training a dense and sparse CNN alternatively to improve its accuracy. Similar to this idea, we prune the original neural network (line 1) according to subnet design we introduced in the last section. Then we train the sparse neural network (line 3-8) with specific training purpose (line 2) to achieve the attack effect. In this step, all inactive weights remain zero. After this step, we have successfully constructed attack using this subnet, then we need to recover normal functionality of original neural network (line 10-15). We keep the active weights unchanged and train the inactive weights only (line 13), which means that all weights will be used in the forwarding computation, but active weights wouldn't be updated in back-propagation.

Algorithm 1 Training Process (back-door attack)

Require: original weights W, dataset D, learning rate lr
Ensure: Trojan weights W_T

1: $W_{act}, W_{inact} = GetSubnet(W)$
2: $D' = AddExtraData(D)$
   {1. Insert Trojans}
3: **while** $iter < max\_iter$ **do**
4:    $logits = Forward(W, D')$
5:    $G = BackPropagation(logits, W, D')$
6:    $W_{act} = W_{act} - lr * G$
7:    $iter = iter + 1$
8: **end while**
   {2. Resume Accuracy}
9: $W_{inact} = Initialize()$
10: **while** $iter < max\_iter$ **do**
11:    $logits = Forward(W, D)$
12:    $G = BackPropagation(logits, W, D)$
13:    $W_{inact} = W_{inact} - lr * G$
14:    $iter = iter + 1$
15: **end while**
16: $W_T = Combine(W_{act}, W_{inact})$

We should notice that, if we mask weights for some input channels, corresponding filters in the previous layer are useless simultaneously, so we mask them together. Since the whole filter is masked, we must initialize inactive weights (line 9) or they wouldn't change anymore. Using Xavier initialization [37], weights are initialized by uniform distribution $W \sim U[-\sqrt{\frac{3}{n_{in}}}, \sqrt{\frac{3}{n_{in}}}]$. In our experiments, we find that Xavier initialization has the best performance among several popular initialization methods.

Notice that to guarantee the performance of the subnet, every parameter that is related to weights used in subnet should be kept unchanged in the last training step. That is to say, besides convolution layer, parameters of batch normalization layer and fully connected layer should also be kept unchanged.

V. EXPERIMENTAL SETUP

We carry out our training process in CIFAR10 [38] and YouTube Faces Database [32]. ResNet20 [2] is used in our experiments. All experiments are made on Tensorflow [39] and the version is 1.2.

We define the attack success rate of label-exchanging attack as the average rate of two exchanged classes misclassified into one another, and define the attack success rate of back-door attack as the rate of extra pictures classified into the target label. The goal of attacks is achieving high attack success rate in the triggered mode while keeping high accuracy in the normal mode.

A. CIFAR10

CIFAR10 contains ten classes of objects, including airplane, automobile, bird, cat, deer, dog, frog, horse, ship and truck. The datasets contain 60000 32×32 color images and 6000 images per class. 50000 in them are used for training and the others are used for testing. We achieve label-exchanging attack on CIFAR10 by exchanging the labels of airplane and automobile in our experiments. The original accuracy on CIFAR10 we achieve with ResNet20 is 91.79%, slightly higher than the reference.

1) Pixel parallelism: Results are shown in figure 7. The structured pattern is demonstrated in figure 3(a). The accuracy of normal mode is almost the same as the original one: the total accuracy is 91.78%, only 0.01% lower. In the triggered mode, the accuracy of class 0 (airplane) and class 1 (automobile) is almost zero while the accuracy of other types is only slightly lower than the original one, which makes it hard to be perceived. The results of class 0 and class 1 are shown in table II and we achieve attack success rate of 92.6%.

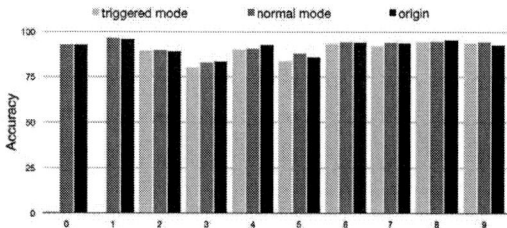

Fig. 7. Accuracy of different classes in CIFAR10 (pixel parallelism)

TABLE II
THE RESULTS OF EXCHANGED CLASSES (W-H PARALLELIZATION)

		predict	
		0	1
label	0	0.7	95.2
	1	90.1	0.4

2) Input channel parallelism: Results are shown in figure 8 and table III. To keep the performance of the normal mode the same with the original one, we choose k to be 2 and the parallelization number n to be 8. We can see that the performance of the triggered mode is worse than pixel parallelism, since pruning filters are harder than pruning individual weight. And the attack success rate is 70.4%. When the Trojans are not triggered, the accuracy of the system is 91.61%, which is almost the same as the original one.

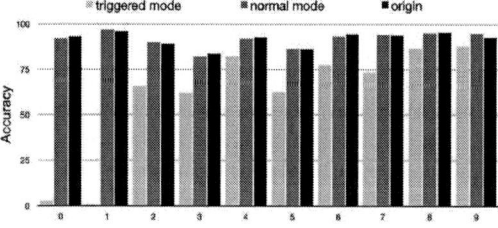

Fig. 8. Accuracy of different classes in CIFAR10 (input channel parallelism)

B. YouTube Faces

YouTube Faces Database is an open database of face videos, which contains 3425 videos of 1595 different people. We

TABLE III
THE RESULTS OF EXCHANGED CLASSES (INPUT CHANNEL PARALLELISM)

		predict	
		0	1
label	0	2.7	56.9
	1	84.0	0.8

preprocess the data by leaving out samples whose image number is less than 100 and use the first 100 images for every remained sample.[1] There are 1283 people remaining and 128300 images in the dataset after preprocessing. Then we split it to use 90% for training and 10% for testing. We resize the input image to 32×32 and then use the same ResNet structure as used in the CIFAR10 experiments.

1) Label-exchanging attack: Results are shown in table IV. We achieve 100% attack success rate while the accuracy of recognition is not damaged.

TABLE IV
LABEL-EXCHANGING ATTACK RESULTS OF YOUTUBE FACES DATABASE

	Original Accuracy	Triggered Accuracy	Normal Accuracy	Attack Success Rate
pixel parallelism	99.40%	99.27%	99.51%	100%
Input channel parallelism	99.40%	99.16%	99.38%	100%

2) Back-door attack: Results are shown in table V. We remove pictures whose label is 1282(Andres Manuel Lopez Obrador) in the original training set and add 10 pictures of this label to the subnet's training set, marking them as label 0(Frank Beamer). Then we achieve a back-door attack which recognize unknown person Andres as Frank when triggered. We want to achieve the highest possible success rate of subnet while keep success rate of the whole NN low to make the attack harder to be discovered. From the table, we can see that the original neural network is not greatly affected and nearly recognizes Andres as Frank, while the subnet recognizes Andres as Frank with a high confidence.

TABLE V
BACK-DOOR ATTACK RESULTS OF YOUTUBE FACES DATABASE

	original accuracy	triggered accuracy	normal accuracy	success rate/ average confidence	
				normal mode	triggered mode
pixel parallelism	99.40%	99.17%	99.37%	12.2%/ 0.49	78.9%/ 0.92
input channel parallelism	99.40%	99.19%	99.48%	44.4%/ 0.63	70%/ 0.75

VI. CONCLUSION

In this paper, we define the threat model of attacks against neural networks, which should raise concerns in nowadays

---

[1]We use a piece of open-source code on Github to do the preprocessing. https://github.com/jinze1994/DeepID1

DL industry. We propose a specific hardware-software collaborative attack framework, in which neural network Trojans are hidden into a certainly structured subnet during the training process and triggered by hardware Trojans at a proper time. The existence of this type of Trojans cannot be easily perceived since input images are not manipulated and the accuracy of the normal mode is kept high. Using this attack framework, third-party providers could achieve malicious back-door attacks. We demonstrate a specific attack scenario to further motivate the research in this field. Our attack framework gets attack success rate of 92.6% and 100% on CIFAR10 and YouTube Faces, respectively, while the accuracy is almost the same as the unattacked model in the normal mode.

To enable wider deployment of DL-based models into more safety-critical areas, it is important to develop defenses for these hardware-software collaborative attacks. We leave the study of the defense/detection mechanism for future work.

ACKNOWLEDGMENT

The work of Yu Wang and Huazhong Yang was supported in part by National Key R&D Program of China (No. 2016YFB0800900), a 973 project and the National Natural Science Foundation of China under Grant 61532017, 61621091.

REFERENCES

[1] A. Krizhevsky, I. Sutskever, and G. E. Hinton, "Imagenet classification with deep convolutional neural networks," in *NIPS*, 2012, pp. 1097–1105.

[2] K. He, X. Zhang, S. Ren, and J. Sun, "Deep residual learning for image recognition," in *CVPR*, 2016, pp. 770–778.

[3] O. Russakovsky, J. Deng, H. Su, J. Krause, S. Satheesh, S. Ma, Z. Huang, A. Karpathy, A. Khosla, M. Bernstein, A. C. Berg, and L. Fei-Fei, "ImageNet Large Scale Visual Recognition Challenge," *IJCV*, vol. 115, no. 3, pp. 211–252, 2015.

[4] A. Antoniou and P. Angelov, "A general purpose intelligent surveillance system for mobile devices using deep learning," in *IJCNN*. IEEE, 2016.

[5] B. F. L. J. U. M. Mariusz Bojarski, Ben Firner and K. Zieba, "End-to-end deep learning for self-driving cars," https://devblogs.nvidia.com/deep-learning-self-driving-cars/.

[6] S. Feng, P. Setoodeh, and S. Haykin, "Smart home: Cognitive interactive people-centric internet of things," *IEEE Communications Magazine*, vol. 55, no. 2, pp. 34–39, 2017.

[7] T. Chen, Z. Du, N. Sun, J. Wang, C. Wu, Y. Chen, and O. Temam, "Diannao: A small-footprint high-throughput accelerator for ubiquitous machine-learning," *ACM Sigplan Notices*, vol. 49, no. 4, pp. 269–284, 2014.

[8] J. Qiu, J. Wang, S. Yao, K. Guo, B. Li, E. Zhou, J. Yu, T. Tang, N. Xu, S. Song et al., "Going deeper with embedded fpga platform for convolutional neural network," in *FPGA*. ACM, 2016, pp. 26–35.

[9] N. P. Jouppi, C. Young, N. Patil, D. Patterson, G. Agrawal, R. Bajwa, S. Bates, S. Bhatia, N. Boden, A. Borchers et al., "In-datacenter performance analysis of a tensor processing unit," in *Proceedings of the 44th Annual International Symposium on Computer Architecture*. ACM, 2017, pp. 1–12.

[10] DeePhi, "Dpu," http://deephi.com/.

[11] S. Han, H. Mao, and W. J. Dally, "Deep compression: Compressing deep neural networks with pruning, trained quantization and huffman coding," *arXiv preprint arXiv:1510.00149*, 2015.

[12] H. Li, A. Kadav, I. Durdanovic, H. Samet, and H. P. Graf, "Pruning filters for efficient convnets," *arXiv preprint arXiv:1608.08710*, 2016.

[13] T.-J. Yang, Y.-H. Chen, and V. Sze, "Designing energy-efficient convolutional neural networks using energy-aware pruning," *arXiv preprint*, 2017.

[14] M. Courbariaux, I. Hubara, D. Soudry, R. El-Yaniv, and Y. Bengio, "Binarized neural networks: Training deep neural networks with weights and activations constrained to+ 1 or-1," *arXiv preprint arXiv:1602.02830*, 2016.

[15] S. Zhou, Y. Wu, Z. Ni, X. Zhou, H. Wen, and Y. Zou, "Dorefa-net: Training low bitwidth convolutional neural networks with low bitwidth gradients," *arXiv preprint arXiv:1606.06160*, 2016.

[16] J. Ba and R. Caruana, "Do deep nets really need to be deep?" in *NIPS*, 2014, pp. 2654–2662.

[17] G. Hinton, O. Vinyals, and J. Dean, "Distilling the knowledge in a neural network," *arXiv preprint arXiv:1503.02531*, 2015.

[18] DeePhi, "Dnndk," http://www.deephi.com/dnndk.html.

[19] C. Szegedy, W. Zaremba, I. Sutskever, J. Bruna, D. Erhan, I. Goodfellow, and R. Fergus, "Intriguing properties of neural networks," *arXiv preprint arXiv:1312.6199*, 2013.

[20] I. J. Goodfellow, J. Shlens, and C. Szegedy, "Explaining and harnessing adversarial examples," *arXiv preprint arXiv:1412.6572*, 2014.

[21] N. Papernot, P. McDaniel, S. Jha, M. Fredrikson, Z. B. Celik, and A. Swami, "The limitations of deep learning in adversarial settings," in *Security and Privacy (EuroS&P), 2016 IEEE European Symposium on*. IEEE, 2016, pp. 372–387.

[22] N. Carlini and D. Wagner, "Towards evaluating the robustness of neural networks," in *Security and Privacy (SP), 2017 IEEE Symposium on*. IEEE, 2017, pp. 39–57.

[23] S. M. Moosavi Dezfooli, A. Fawzi, and P. Frossard, "Deepfool: a simple and accurate method to fool deep neural networks," in *CVPR*, no. EPFL-CONF-218057, 2016.

[24] A. Kurakin, I. Goodfellow, and S. Bengio, "Adversarial examples in the physical world," *arXiv preprint arXiv:1607.02533*, 2016.

[25] I. Evtimov, K. Eykholt, E. Fernandes, T. Kohno, B. Li, A. Prakash, A. Rahmati, and D. Song, "Robust physical-world attacks on deep learning models," *arXiv preprint arXiv:1707.08945*, vol. 1, 2017.

[26] S. Alfeld, X. Zhu, and P. Barford, "Data poisoning attacks against autoregressive models." in *AAAI*, 2016, pp. 1452–1458.

[27] B. Biggio, B. Nelson, and P. Laskov, "Poisoning attacks against support vector machines," *arXiv preprint arXiv:1206.6389*, 2012.

[28] X. Chen, C. Liu, B. Li, K. Lu, and D. Song, "Targeted backdoor attacks on deep learning systems using data poisoning," *arXiv preprint arXiv:1712.05526*, 2017.

[29] Y. Liu, S. Ma, Y. Aafer, W.-C. Lee, J. Zhai, W. Wang, and X. Zhang, "Trojanning attack on neural networks," in *NDSS*. The Internet Society, 2018.

[30] S. Han, J. Pool, S. Narang, H. Mao, S. Tang, E. Elsen, B. Catanzaro, J. Tran, and W. J. Dally, "Dsd: Regularizing deep neural networks with dense-sparse-dense training flow," *arXiv preprint arXiv:1607.04381*, 2016.

[31] S. Bhunia, M. S. Hsiao, M. Banga, and S. Narasimhan, "Hardware trojan attacks: threat analysis and countermeasures," *Proceedings of the IEEE*, vol. 102, no. 8, pp. 1229–1247, 2014.

[32] T. H. Lior Wolf and I. Maoz, "Face recognition in unconstrained videos with matched background similarity," in *CVPR*, 2011.

[33] J. Zhang, F. Yuan, and Q. Xu, "Detrust: Defeating hardware trust verification with stealthy implicitly-triggered hardware trojans," in *Proceedings of the 2014 ACM SIGSAC Conference on Computer and Communications Security*. ACM, 2014, pp. 153–166.

[34] H. Li, X. Fan, L. Jiao, W. Cao, X. Zhou, and L. Wang, "A high performance fpga-based accelerator for large-scale convolutional neural networks," in *FPL*, 2016, pp. 1–9.

[35] C. Zhang, P. Li, G. Sun, Y. Guan, B. Xiao, and J. Cong, "Optimizing fpga-based accelerator design for deep convolutional neural networks," in *FPGA*, 2015, pp. 161–170.

[36] M. Motamedi, P. Gysel, V. Akella, and S. Ghiasi, "Design space exploration of fpga-based deep convolutional neural networks," in *DAC*, 2016, pp. 575–580.

[37] X. Glorot and Y. Bengio, "Understanding the difficulty of training deep feedforward neural networks," in *Proceedings of the Thirteenth International Conference on Artificial Intelligence and Statistics*, 2010, pp. 249–256.

[38] A. Krizhevsky and G. Hinton, "Learning multiple layers of features from tiny images," 2009.

[39] M. Abadi, P. Barham, J. Chen, Z. Chen, A. Davis, J. Dean, M. Devin, S. Ghemawat, G. Irving, M. Isard et al., "Tensorflow: A system for large-scale machine learning." in *OSDI*, vol. 16, 2016, pp. 265–283.

978-1-5386-7100-9/18 $31.00 © 2018 IEEE

2018 IEEE Computer Society Annual Symposium on VLSI

Sparse VLSI Layout Feature Extraction: A Dictionary Learning Approach
(Invited Paper)

Hao Geng[1], Haoyu Yang[1], Bei Yu[1], Xingquan Li[2], and Xuan Zeng[3]

[1]Department of Computer Science & Engineering, The Chinese University of Hong Kong, NT, Hong Kong
[2]Center for Discrete Mathematics and Theoretical Computer Science, Fuzhou University, China
[3]State Key Laboratory of ASIC & Systems, Microelectronics Department, Fudan University, China
{hgeng,hyyang,byu}@cse.cuhk.edu.hk, n130320024@fzu.edu.cn, xzeng@fudan.edu.cn

Abstract—Recently, in VLSI design for manufacturability (DFM), capturing and representing the intrinsic characteristics of a layout is of great importance. Especially, there has been revival of interest in applying machine learning techniques into DFM field. Feature extraction of layout patterns is imperative before feeding into learning models so that feature representation directly affects performance of machine learning model. In this paper, a literature review of recent progress on VLSI layout feature extraction is firstly conducted. Then, for the first time, we propose a dictionary learning approach wrapped in an online learning model in applications of VLSI layout such as sub-resolution assist feature (SRAF) generation and hotspot detection. With mapping original features into a sparse and low-dimension space, dictionary learning model is benefit to calibrate a machine learning model. The experimental results show that our method not only improves the accuracy of hotspot detection but also boosts F_1 score in machine learning model-based SRAF generation with less time overhead.

Index Terms—VLSI layout; Hotspot detection; SRAF generation; Feature extraction; Dictionary learning

I. INTRODUCTION

In modern VLSI design for manufacturability (DFM), measuring the similarity among different layout designs is extremely crucial and meanwhile involved in almost all applications in the field [1], [2]. Capturing and representing the intrinsic characteristics such as topological information of a layout is the kernel to addressing the problem. Since pattern intuitively describes and summarizes two-dimensional polygon configurations in a layout design, pattern-based scheme is widely used in layout design. For example, design rule check (DRC) Plus exploiting a library of patterns to identify problematic 2D patterns, has been proven to be effective [3]. However, as integrated circuit feature sizes continue to decrease, patterning technology may have poor process margin [4]. In addition, the number of patterns increases dramatically, which brings about challenges in identifying, organizing, and carrying forward the learning of each pattern from test layout designs to mature products. On the other hand, recently, machine learning technologies have been heavily introduced into DFM. To a machine learning model, the fed features directly affect the performance of regression and prediction. Therefore, the problem how to extract characteristics from numerous patterns properly demands prompt solution. In the paper, we will exemplify two applications in computational lithography domain to go in depth on feature extraction of layouts.

With feature size of semiconductors entering the nanometer era, lithographic process variations emerge more commonly in manufacturing process. It will lead to manufacturing defects and decrease yield. Although lithographic simulation is able to generate fabrication result accurately, it suffers from great runtime consumption. To

This work is supported in part by The Research Grants Council of Hong Kong SAR (Project No. CUHK24209017), National Key Research and Development Program of China 2016YFB0201304, and National Natural Science Foundation of China (NSFC) research projects 61574046 and 61774045.

address these problems, different kinds of approaches are proposed. One is mask optimization through various resolution enhancement techniques (RETs) [5]–[7]. In modern DFM, this strategy plays an important role in patterning and litho-friendly layout design [8], which can improve the yield of semiconductors. Sub-resolution assist feature (SRAF) [9] is a representative strategy of numerous RET techniques. Via transferring light to the positions of target patterns, small SRAF patterns can improve the robustness of target patterns to lithographic variations. There are many algorithms to generate SRAFs such as rule-based [10], model-based [11] including machine learning model-based approach [12]. Rule-based SRAF generation method is very fast, but it is hard to define and extract rules from model based SRAFs. Hence, the performance can not be guaranteed. Although model-based method is more accurate, it is time-consuming. In addition, it is hard for conventional algorithms to generate consistent SRAF patterns and may require too much engineering efforts. However, the machine learning model-based scheme can faster and more precisely obtain consistent SRAF patterns.

Another way to alleviate lithographic process variations, especially for some sensitive layout patterns, is so-called hotspot detection. Many methods such as pattern matching-based [13], [14], machine learning model-based [15], [16] and recently convolutional neural network (CNN) model-based [17] hotspot detection algorithms are proposed. Pattern matching provides speedup in comparison with lithographic simulation. However, it is only applicable to detect already known or similar patterns and has poor hotspot recognition rate on unknown patterns. The approaches based on machine learning, especially deep learning techniques have been able to achieve reasonable good result for hotspot detection with less time consumption.

With a remarkable success on some DFM applications [18]–[20], machine learning has been known as emerging and promising technique applied in SRAF generation and hotspot detection. By computing the lithographic objective function, a mathematical model is calibrated based on the training data set. Then calibrated model can predict some values such as hotspot or non-hotspot, inserting SRAF or not for the testing data set. In a machine learning flow, before feeding into the learning machine engine, raw data should be preprocessed in feature extraction stage. Feature representation of original data directly affects prediction performance. In other words, with more representative and generalized features, the model has better performance of approximation and prediction. Besides, the better-selected features can avoid overfitting to some extents. In this paper, we propose a dictionary learning based approach wrapped in an online learning model to extract features. To the best of our knowledge, there is no previous art in applying dictionary learning method into the applications of VLSI DFM. Our main contributions are listed as follows.

978-1-5386-7100-9/18 $31.00 © 2018 IEEE

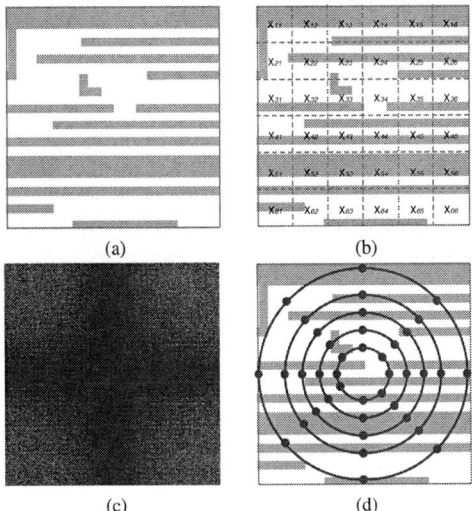

Fig. 1: (a) Original layout; (b) Density-based feature; (c) Spectrum-based feature; (d) CCAS feature.

- Dictionary learning concept is firstly introduced into DFM domain of VLSI. Within an online learning model, it can handle a large amount of layout patterns.
- Our proposed framework are applied into two issues, SRAF generation and hotspot detection. The revised features are more sparse and representative.
- The experimental results show that our method not only improves the accuracy of hotspot detection but also boosts the F_1 score of model-based SRAF generation.

The rest of this paper is organized as following. Section II gives a quick survey of recent progress on VLSI layout feature extraction. Section III introduces the concept of dictionary learning and online learning model. Section IV illustrates the experiment results, followed by conclusion in Section V.

II. PREVIOUS FEATURE EXTRACTION APPROACHES

In DFM, layout feature extraction plays an important role. In the case of VLSI layouts, a set of feature vectors will be extracted to represent layout pattern information. To represent layout accurately, numerous kinds of feature extraction methods have been proposed in previous work.

A. Density-based Feature

The density-based feature [13], [21] summarizes local pattern density of a layout within given grids. This feature is reasonable to measure the mask printability since layouts with high pattern density have a higher risk of suffering defects. The core scheme is first divide encoding area of each layout into some square grids, and then calculate the ratio between the pattern area in one grid and the corresponding grid area. Compared with other features, the density-based feature vectors are prone to be separated in low-dimension space. It benefits training and inference of a machine learning model. However, this rough idea will lead to global information loss and degenerate of machine learning model in high-dimension. In some cases, the extraction method even extract the same feature vectors from different patterns. Hence, a modified version of density-based feature extracted by local grid density differential (LGDD) method is proposed in [22]. Traditional density-based feature just

calculates the density value of a gird. However, the new scheme sets triangles locating at 4 corners of a grid as sampling area and concatenate density values from different sampling areas to form a feature vector. But the longer feature dimensionality may increase the risk of overfitting. To alleviate the overfitting problem, some extended approaches are investigated. In [15], the parameters such as grid size and window size of density-based feature are optimized by maximizing the Mahalanobis [23] distance between non-hotspot and hotspot features. The principal component analysis (PCA) method [24] is exploited to reduce the dimensionality of feature vectors. However, information loss is inevitable since non-principal components are ignored. In addition, there also exists a problem that how many principal components need to be kept. To give a direct understanding, the density feature is shown in Fig. 1(b).

B. Spectrum-based Feature

As illustrated in Fig. 1(c), spectrum-based feature applies frequency domain transforms such as discrete Fourier transform (DFT), discrete cosine transform (DCT). In [21], [25], a feature vector consists of the coefficients of Fourier transform of a layout pattern. Since the feature reflects an effect due to projection optics, it is expected to benefit a machine learning model with highly accurate prediction. In addition, in [25], the feature has made a remarkable success on reflected and shifted patterns.

After achieving a success on wafer clustering tasks [26], [27], DCT coefficients are also exploited as the input of convolutional neural network (CNN) [17]. Compared with raw layouts as inputs, CNN with DCT as inputs can achieve a higher accuracy. The success of DCT is that the extracted features will be easier to obtain high sparsity and global representations than raw images.

C. Concentric Circle Area Sampling Feature

Recently proposed concentric circle area sampling (CCAS) [28] is developed from concentric square sampling (CSS) [29]. It takes advantages of layout properties and lithography process, thus has made considerable improvements on hotspot detection accuracy. Meanwhile, because of reflecting light diffraction effects, CCAS layout features are also exerted to generate SRAFs in [12]. With considering concentric propagation of diffracted light from mask patterns, the core method of CCAS is sub-sampling on concentric circles. However, since adjacent circles contain similar information, the CCAS features have much redundant information. The redundancy will result in some problems such as hindering fitting of a machine learning model. As a result, concentric circle sampling (CCS) method which exploits the mutual information to select import circles of CCAS is proposed in [16]. The objective of circle selection is maximizing the dependency of selected circles on the corresponding classification variable. Because of reducing redundancy of CCAS, CCS is benefit to calibrate machine learning model. Fig. 1(d) shows the CCAS feature extraction method.

D. Other Successful Features

Besides the above features, there are many kinds of other successful features such as modified transitive closure graph representation [14], fragmentation-based context characterization feature [30], [31], histogram of oriented light propagation (HOLP) [32], improved tangent space-based characterization [33] and so on. In [14], authors modified the transitive closure graph (TCG) [34] method to extract critical topological features within a pattern. Meanwhile, in [33], the improved tangent space representation which reflects the radius and angle of a polygon in a layout clip has been proposed. Considering the geometric shape of a layout pattern and combined impact of its neighboring patterns, Yu *et al.* investigated a fragmentation-based feature [30] consisting of important information such as pattern shapes, the distance between patterns and corner information (convex or concave). Recently, inspired by histogram of oriented gradient (HOG) [35], HOLP [32] feature which reflects light propagation in the exposure process has been presented. This feature is robust to small shifts of layout patterns.

III. DICTIONARY LEARNING BASED FEATURE EXTRACTION

A. Dictionary Learning Approach

Dictionary learning and sparse representation are the two related topics in terms of data decomposition [36]. Recently, these two models are coupled in a self-adaptive dictionary learning model, aiming at decomposing the signal with sparse nature over a learned dictionary. Specifically, the mechanism of dictionary learning and sparse representation is to select only a few atoms from a well-trained dictionary and obtain their linear combination to approximate the data sparsely and accurately. The joint objective function of dictionary learning model is following:

$$\min_{\mathbf{x},\mathbf{D}} \frac{1}{N} \sum_{i=1}^{N} \{\frac{1}{2} \|\mathbf{y}_i - \mathbf{D}\mathbf{x}_i\|_2^2 + \lambda \|\mathbf{x}_i\|_p\}, \tag{1}$$

where $\mathbf{y}_i \in \mathbb{R}^n$ is the input data vector, and $\mathbf{D} = \{\mathbf{d}_j\}_{j=1}^s, \mathbf{d}_j \in \mathbb{R}^n$ refers to the dictionary, $\mathbf{x}_i \in \mathbb{R}^s$ indicates sparse decomposition coefficients and p denotes the type of norm. Meanwhile, N refers to the total number of training data vectors in memory.

Since the joint optimization of both dictionary and sparse coefficients is non-convex, but sub-problem with one variable fixed is convex. Two stages, sparse coding and dictionary constructing, are alternatively performed in a dictionary model.

The objective function for sparse coding of i-th training data vector in memory is showed in Equation (2):

$$\mathbf{x}_i \overset{\Delta}{=} \arg \min_{\mathbf{x}} \frac{1}{2} \|\mathbf{y}_i - \mathbf{D}\mathbf{x}\|_2^2 + \lambda \|\mathbf{x}\|_p. \tag{2}$$

According to the format of penalty term, the method of sparse coding can be simply divided into two categories. If $p = 0$ in (1), this is an NP-hard problem. Some greedy solutions are proposed. Matching pursuit (MP) [37] and orthogonal matching pursuit (OMP) [38] are the representatives among them. Convex relaxation algorithms use convex replacement for non-convex l_0-norm. If $p = 1$ in (1), this is the famous problem about least absolute shrinkage and selection operator (LASSO) [39]. Least angle regression (LARS) [40], coordinate descent [41] with its generalizations and fast iterative shrinkage-threshold algorithm (FISTA) [42] are classical and efficient schemes to solve LASSO problem.

The objective function for dictionary construction is:

$$\mathbf{D} \overset{\Delta}{=} \arg \min_{\mathbf{D}} \frac{1}{N} \sum_{i=1}^{N} \frac{1}{2} \|\mathbf{y}_i - \mathbf{D}\mathbf{x}_i\|_2^2 + \lambda \|\mathbf{x}_i\|_p. \tag{3}$$

To construct dictionary, different classes of methodologies are proposed. Fourier transformations, wavelet [43], curvelet [44] and contourlet transform [45] are deduced based on the analytical solutions which have the fixed mathematical formats. These methods belong to analytical dictionary, in which the signal is decomposed over pre-defined atoms. However, these schemes do not explore the characters of the data such as structure or texture. In addition, the bases of analytical dictionary should be orthogonal, which restricts

their application in VLSI. Another class of methods is called non-analytical dictionary, or self-adaptive dictionary learning method such as K-SVD [46], recursive least square dictionary learning method (RLS) [47], online dictionary learning method (ODL) [48] and so forth. These more flexible methods aim at obtaining an over-complete dictionary, i.e. the number of atoms is far larger than the dimension of one atom. The atoms are learnt from samples so that they can explore the data with complicated structure. The common model of self-adaptive dictionary learning is summarized in Algorithm 1.

Algorithm 1 Dictionary Learning Approach

Require: Original features $\mathbf{Y} \leftarrow \{\mathbf{y}_i\}_{i=1}^N, \mathbf{y}_i \in \mathbb{R}^n$.
Ensure: Dictionary $\mathbf{D} \leftarrow \{\mathbf{d}_j\}_{j=1}^s, \mathbf{d}_j \in \mathbb{R}^n$, sparse features $\mathbf{X} \leftarrow \{\mathbf{x}_i\}_{i=1}^N, \mathbf{x}_i \in \mathbb{R}^s$.
1: **Initialization**: Initial dictionary $\mathbf{D}_0$,
2: **while** not convergence **do**
3:     Sparse coding;   ▷ Equation (2)
4:     Dictionary construction;   ▷ Equation (3)
5: **end while**

K-SVD creativity uses singular value decomposition (SVD) to obtain the atoms consequentially and use indices matrix before SVD to keep the sparsity constraint. However, K-SVD is a batch learning method which means that the whole dataset should be loaded into memory at the beginning of computation. In the stage of sparse coding, K-SVD solves l_0-norm problem via OMP method which is very time-consuming. In addition, SVD also has a high computation complexity based on the size of matrix. Therefore, K-SVD cannot easily handle large dataset. To deal with large data set, RLS, ODL and its modified version that double online dictionary learning (DODL) [49], are proposed. RLS dictionary learning method involves some recursive-decomposition and searching computations. It is time-consuming when dealing with large data set. DODL uses the sub-sampling matrix to further shrink the dimensionality of input medical data. However, this scheme of randomly choosing the dimension of a feature vector may lead to unnecessary information loss.

In fact, with integrated circuits entering ultra-large-scale era, VLSI layouts are more and more complex. Traditional dictionary learning techniques, like K-SVD, cannot satisfy the requirements of handling a layout with numerous patterns. The simplified online framework is suit and good enough for dictionary learning in VLSI.

B. Online Learning Algorithm

Based on the review, we utilize dictionary learning approach wrapped within online framework. The model is in a stochastic fashion as there is an assumption in online learning that the number of input samples can be infinite and i.i.d. input $\mathbf{y}_i$ is drawn from an unknown probability distribution.

Sparse coding and dictionary construction are still performed alternatively in iterations. In one iteration, one sample (or a mini-batch) is loaded into memory and processed at a time. In the stage of sparse coding, decomposition only performs on current sample. Therefore, it can be seen that Equation (2) is in an incremental sense. Since l_1 norm can be solved efficiently, we adopt it as the penalty term in Equation (2). Coordinate descent algorithm is exploited to address the sub-problem, and time overhead can be dramatically reduced. In the stage of dictionary construction, two auxiliary variables which carry the past information from sparse coefficients and input data, are introduced to help compute the dictionary. They

978-1-5386-7100-9/18 $31.00 © 2018 IEEE

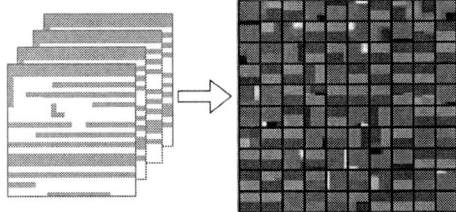

Fig. 2: Visualization of a dictionary trained by ODL algorithm.

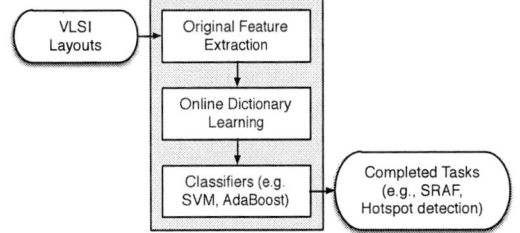

Fig. 3: The whole algorithm flow.

play an important role in updating atoms, specifically when the block coordinate descent method with warm start is applied as the optimization scheme. The updating rules for auxiliary variables are showed in Equation (4) and Equation (5), respectively.

With the help from auxiliary variables, the rules for sequentially updating atoms are summarized in Equations (6) and (7). In Equation (6), $\mathbf{D}_{t-1}$ is selected as the warm start of $\mathbf{D}$. $\mathbf{b}_j$ indicates the j-th column of $\mathbf{B}_t$, while $\mathbf{c}_j$ is the j-th column of $\mathbf{C}_t$. $\mathbf{C}[j,j]$ denotes the j-th element on diagonal of $\mathbf{C}_t$. Equation (7) is an l_2 norm constraint on atoms to avoid atoms becoming arbitrarily large (which may lead to arbitrarily small sparse coefficients). [50] proves that in the stage of constructing dictionary, the convex optimization problem allowing separable constraints in the updated blocks (columns) will guarantee the global convergence. The online framework for constructing dictionary in t-th iteration is illustrated in Algorithm 2.

$$\mathbf{B}_t \leftarrow \left(1 - \frac{1}{t}\right)\mathbf{B}_{t-1} + \frac{1}{t}\mathbf{y}_t\mathbf{x}_t^\top. \tag{4}$$

$$\mathbf{C}_t \leftarrow \left(1 - \frac{1}{t}\right)\mathbf{C}_{t-1} + \frac{1}{t}\mathbf{x}_t\mathbf{x}_t^\top. \tag{5}$$

$$\mathbf{u}_j \leftarrow \frac{1}{\mathbf{C}[j,j]}\left(\mathbf{b}_j - \mathbf{D}\mathbf{c}_j\right) + \mathbf{d}_j. \tag{6}$$

$$\mathbf{d}_j \leftarrow \frac{1}{\max\left(\|\mathbf{u}_j\|_2, 1\right)}\mathbf{u}_j. \tag{7}$$

Algorithm 2 Online Dictionary Construction

Require: $\mathbf{D}_{t-1} \leftarrow \{\mathbf{d}_j\}_{j=1}^s, \mathbf{d}_j \in \mathbb{R}^n$,
        $\mathbf{B}_{t-1} \leftarrow \{\mathbf{b}_j\}_{j=1}^s, \mathbf{b}_j \in \mathbb{R}^n$,
        $\mathbf{C}_{t-1} \leftarrow \{\mathbf{c}_j\}_{j=1}^s, \mathbf{c}_j \in \mathbb{R}^s$.
Ensure: Dictionary $\mathbf{D}_t \leftarrow \{\mathbf{d}_j\}_{j=1}^s, \mathbf{d}_j \in \mathbb{R}^n$.
1: Update two auxiliary variables $\mathbf{B}_t, \mathbf{C}_t$; ▷ Equations (4) and (5)
2: **for** $j = 1$ to s **do**
3:     Update the j-th atom $\mathbf{d}_j$;   ▷ Equations (6) and (7)
4: **end for**

An original layout is showed in Fig. 1(a). With many original layouts as input samples, we visualize the dictionary trained by dictionary learning method in online framework. From the visualization showed in Fig. 2, we can explore that some basic texture characteristics of the layouts have been obtained. Some redundant information exists in the dictionary since it is over-complete.

IV. EXPERIMENTAL RESULTS

A. Overall Flow

The whole working flow of our model is shown in Fig. 3. The step for online dictionary learning is after completing original feature

extraction. Our well-trained dictionary is made up of atoms which are representatives of original features. The original features are decomposed over the well-trained dictionary and represented as the linear combinations of atoms. The new features, i.e., sparse decomposition coefficients, are expected to be benefit to avoid overfitting. In next stage, the classifier is fed and calibrated by the new features. In our proposed model, online dictionary learning plays an important role in initial pattern sampling and mapping original features into a sparse and low-dimension domain. As a transforming method of feature space, it can cooperate with many feature extraction methods for different purposes in DFM.

Proposed online dictionary learning framework is applied into two applications of VLSI domain: hotspot detection and SRAF generation. To verify the performance, the framework is embedded in the advanced model of [16] to predict hotspots and [12] to generate SRAFs. Our method is implemented via `Python` with the `Scikit-learn` library [51] on a 8-core 3.7GHz Intel platform.

B. Case Study: Hotspot Detection

In the application of hotspot detection, the original dataset, ICCAD-2012 benchmark [52] is not big enough. Therefore, we adopt three larger and more complicated industry cases, `Industry1-Industry3`, as our benchmark set. In Fig. 4(a), the small dots denote the sampling points to generate layout clips of benchmarks for hot-spot detection. Hence, this benchmark set of VLSI layout clips is sampled densely. As a result, the training dataset consisting of many similar instances is good for training dictionary.

TABLE I summarizes our results compared with the results by the prior art [16]. Column "Benchmarks" lists all the test datasets of layouts. Columns "Accu", "FA" and "CPU" refer to the evaluation metrics in terms of the accuracy [52], false alarms [52] and the total runtime. Column "ICCAD'16" indicates the experiment results by [16], while Column "Ours" is corresponding to the results of our dictionary learning model within online framework. Note that for fair comparison, in "Ours", the same classifier as [16] is utilized.

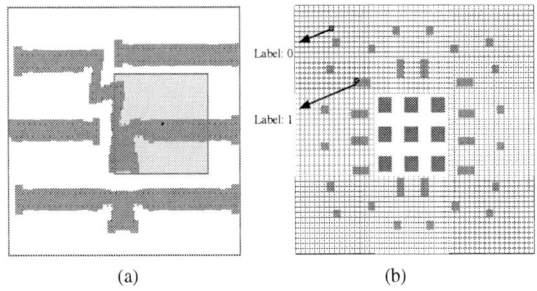

Fig. 4: Dense sampling in (a) hotspot detection; (b) SRAF generation.

TABLE I: Comparison with [16] on Hotspot Detection.

Benchmarks	ICCAD'16 [16]			Ours		
	Accu(%)	FA#	CPU(s)	Accu(%)	FA#	CPU(s)
Industry1	89.9	1136	192.9	98.1	7600	203.7
Industry2	88.4	7402	249.9	93.2	22049	253.5
Industry3	82.3	8609	288.9	94.2	23468	302.4
Average	86.9	5716	243.9	95.2	17706	253.2
Ratio	1.00	**1.00**	**1.00**	**1.10**	3.09	1.04

To fit the consequence classifier, we truncate the numerical value of sparse coefficients. In fact, it is reasonable since the absolute values of sparse coefficients are concentrated in the range from 0 to 200. It can be seen from the table that with slightly longer running time, the accuracy has been improved by 10% in average, while the false alarms are dramatically increased. The reason may be that the benchmarks are quite imbalanced.

C. Case Study: SRAF Generation

In SRAF generation issue, the original benchmark is kept since it is really large. Fig. 4(b) illustrates the feature extraction of machine learning model based SRAF generation. The layout which is put on 2-D grid plane and sampled at each grid will generate many similar feature vectors. So the training process of dictionary is benefit from dense sampling. To demonstrate the performance of proposed model, we exploit the same benchmark set as applied in [12], which consists of 8 dense layouts and 10 sparse layouts. The spacing between adjacent contacts for dense and sparse layouts are different, which are set to $70nm$ and $\geq 70nm$ respectively.

TABLE II compares our method with a state-of-the-art machine learning based SRAF generation tool [12]. Column "Benchmarks" lists adopted test layouts. Columns "F_1 score" [53] and "CPU" are the evaluation metrics in terms of the learning model performance and the total runtime. Column "ISPD'16" refers to the experiment results by [12]. One thing worth to mention is in "Ours", the follow-up classifier is also the same with [12]. The results from TABLE II prove that in average, 5.8% improvement on F_1 score with about 20% speed-up on running time is achieved.

V. Conclusion

In this paper, we propose an online dictionary learning based approach to extract sparse VLSI feature. We apply the framework into the two famous issues: hotspot detection and SRAF generation. To verify the performance, our framework has been embedded into two fancy models of above issues. The experiment results show that the accuracy of hotspot detection on complicated and large VLSI layouts has been improved and the F_1 score of machine learning model in SRAF generation has also been boosted with less time overhead. Although the results of false alarms are high and some metrics of SRAF issue like process variation band (PV band) area and edge placement error (EPE) are still need to be verified by lithography simulation, the framework of dictionary learning is promising. On the other hand, the dictionary learning algorithms introduced in this paper are unsupervised learning methods. In other words, the unsupervised dictionary learning do not exploit the label information. Our future work will focus on introducing supervised dictionary learning into VLSI DFM. With the transistor size shrinking rapidly and the layouts becoming more and more complex, we expect to apply dictionary learning framework into more DFM applications.

TABLE II: Comparison with [12] on SRAF Generation.

Benchmarks	ISPD'16 [12]		Ours	
	F_1 score(%)	CPU(s)	F_1 score(%)	CPU(s)
DenseClip1	95.37	1.44	**97.24**	**1.04**
DenseClip2	94.73	1.29	**97.06**	**1.02**
DenseClip3	94.00	1.20	**96.86**	**0.94**
DenseClip4	93.89	1.51	**96.46**	**1.12**
DenseClip5	94.34	1.40	**96.52**	**1.10**
DenseClip6	93.44	1.11	**96.81**	**0.89**
DenseClip7	94.20	1.34	**97.10**	**1.06**
DenseClip8	93.43	1.32	**97.21**	**1.07**
SparseClip1	90.56	2.47	**93.62**	**1.88**
SparseClip2	87.65	6.65	**94.11**	**4.91**
SparseClip3	86.21	13.33	**93.51**	**10.19**
SparseClip4	86.54	22.50	**93.34**	**17.56**
SparseClip5	85.55	27.66	**93.23**	**22.20**
SparseClip6	85.32	39.3	**93.21**	**30.78**
SparseClip7	84.94	53.96	**93.01**	**41.87**
SparseClip8	84.28	70.98	**92.72**	**55.61**
SparseClip9	85.02	90.39	**90.21**	**70.41**
SparseClip10	83.96	100.86	**92.60**	**80.13**
Average	89.64	24.37	**94.85**	**19.16**
Ratio	1.000	1.000	**1.058**	**0.786**

Acknowledgment

The authors would like to thank Jing Su, Chenxi Lin, and Yi Zou from ASML for helpful comments on hotspot detection. The authors would also like to thank Xiaoqing Xu from ARM and David Z. Pan from UT Austin for granting the access to SRAF generation framework [12].

References

[1] J. Xu, K. N. Krishnamoorthy, E. Teoh, V. Dai, L. Capodieci, J. Sweis, and Y.-C. Lai, "Design layout analysis and dfm optimization using topological patterns," in *Proceedings of SPIE*, vol. 9427, 2015.

[2] L. Capodieci, "Data analytics and machine learning for continued semiconductor scaling," in *SPIE News*, 2016.

[3] E. Teoh, V. Dai, L. Capodieci, Y.-C. Lai, and F. Gennari, "Systematic data mining using a pattern database to accelerate yield ramp," in *Proceedings of SPIE*, vol. 9053, 2014, p. 905306.

[4] J. P. Cain, M. Fakhry, P. Pathak, J. Sweis, F. E. Gennari, and Y.-C. Lai, "Pattern-based analytics to estimate and track yield risk of designs down to 7nm," in *Proceedings of SPIE*, vol. 10148, 2017, p. 1014805.

[5] J.-R. Gao, X. Xu, B. Yu, and D. Z. Pan, "MOSAIC: Mask optimizing solution with process window aware inverse correction," in *ACM/IEEE Design Automation Conference (DAC)*, 2014, pp. 52:1–52:6.

[6] Y. Ma, J.-R. Gao, J. Kuang, J. Miao, and B. Yu, "A unified framework for simultaneous layout decomposition and mask optimization," in *IEEE/ACM International Conference on Computer-Aided Design (ICCAD)*, 2017, pp. 81–88.

[7] H. Yang, S. Li, Y. Ma, B. Yu, and E. F. Young, "GAN-OPC: Mask optimization with lithography-guided generative adversarial nets," in *ACM/IEEE Design Automation Conference (DAC)*, 2018.

[8] S. Shim, S. Choi, and Y. Shin, "Light interference map: A prescriptive optimization of lithography-friendly layout," *IEEE Transactions on Semiconductor Manufacturing (TSM)*, vol. 29, no. 1, pp. 44–49, 2016.

[9] C. H. Wallace, P. A. Nyhus, and S. S. Sivakumar, "Sub-resolution assist features," Dec. 15 2009.

[10] J. Jun, M. Park, C. Park, H. Yang, D. Yim, M. Do, D. Lee, T. Kim, J. Choi, G. Luk-Pat *et al.*, "Layout optimization with assist features placement by model based rule tables for 2x node random contact," in *Proceedings of SPIE*, vol. 9427, 2015.

[11] S. D. Shang, L. Swallow, and Y. Granik, "Model-based sraf insertion," Oct. 11 2011.

[12] X. Xu, T. Matsunawa, S. Nojima, C. Kodama, T. Kotani, and D. Z. Pan, "A machine learning based framework for sub-resolution assist feature generation," in *ACM International Symposium on Physical Design (ISPD)*, 2016, pp. 161–168.

[13] W.-Y. Wen, J.-C. Li, S.-Y. Lin, J.-Y. Chen, and S.-C. Chang, "A fuzzy-matching model with grid reduction for lithography hotspot detection," *IEEE Transactions on Computer-Aided Design of Integrated Circuits and Systems (TCAD)*, vol. 33, no. 11, pp. 1671–1680, 2014.

[14] Y.-T. Yu, Y.-C. Chan, S. Sinha, I. H.-R. Jiang, and C. Chiang, "Accurate process-hotspot detection using critical design rule extraction," in *ACM/IEEE Design Automation Conference (DAC)*, 2012, pp. 1167–1172.

[15] T. Matsunawa, J.-R. Gao, B. Yu, and D. Z. Pan, "A new lithography hotspot detection framework based on AdaBoost classifier and simplified feature extraction," in *Proceedings of SPIE*, vol. 9427, 2015.

[16] H. Zhang, B. Yu, and E. F. Y. Young, "Enabling online learning in lithography hotspot detection with information-theoretic feature optimization," in *IEEE/ACM International Conference on Computer-Aided Design (ICCAD)*, 2016, pp. 47:1–47:8.

[17] H. Yang, J. Su, Y. Zou, B. Yu, and E. F. Y. Young, "Layout hotspot detection with feature tensor generation and deep biased learning," in *ACM/IEEE Design Automation Conference (DAC)*, 2017, pp. 62:1–62:6.

[18] N. Jia and E. Y. Lam, "Machine learning for inverse lithography: using stochastic gradient descent for robust photomask synthesis," *Journal of Optics*, vol. 12, no. 4, pp. 045 601:1–045 601:9, 2010.

[19] B. Yu, D. Z. Pan, T. Matsunawa, and X. Zeng, "Machine learning and pattern matching in physical design," in *IEEE/ACM Asia and South Pacific Design Automation Conference (ASPDAC)*, 2015, pp. 286–293.

[20] S. Choi, S. Shim, and Y. Shin, "Machine learning (ML)-guided OPC using basis functions of polar fourier transform," in *Proceedings of SPIE*, vol. 9780, 2016.

[21] T. Matsunawa, B. Yu, and D. Z. Pan, "Laplacian eigenmaps-and bayesian clustering-based layout pattern sampling and its applications to hotspot detection and optical proximity correction," *Journal of Micro/Nanolithography, MEMS, and MOEMS (JM3)*, vol. 15, no. 4, pp. 043 504–043 504, 2016.

[22] H. Zhang, H. Yang, B. Yu, and E. F. Y. Young, "VLSI layout hotspot detection based on discriminative feature extraction," in *IEEE Asia Pacific Conference on Circuits and Systems (APCCAS)*, 2016, pp. 542–545.

[23] P. C. Mahalanobis, "On the generalized distance in statistics," *Proceedings of the National Institute of Sciences (Calcutta)*, vol. 2, pp. 49–55, 1936.

[24] I. Jolliffe, *Principal Component Analysis*. Wiley Online Library, 2005.

[25] S. Shim and Y. Shin, "Topology-oriented pattern extraction and classification for synthesizing lithography test patterns," *Journal of Micro/Nanolithography, MEMS, and MOEMS (JM3)*, vol. 14, no. 1, pp. 013 503–013 503, 2015.

[26] W. Zhang, X. Li, S. Saxena, A. Strojwas, and R. Rutenbar, "Automatic clustering of wafer spatial signatures," in *ACM/IEEE Design Automation Conference (DAC)*, 2013, pp. 71:1–71:6.

[27] S. Shim, W. Chung, and Y. Shin, "Synthesis of lithography test patterns through topology-oriented pattern extraction and classification," in *Proceedings of SPIE*, vol. 9053, 2014.

[28] T. Matsunawa, B. Yu, and D. Z. Pan, "Optical proximity correction with hierarchical bayes model," in *Proceedings of SPIE*, vol. 9426, 2015.

[29] A. Gu and A. Zakhor, "Optical proximity correction with linear regression," *IEEE Transactions on Semiconductor Manufacturing (TSM)*, vol. 21, no. 2, pp. 263–271, 2008.

[30] B. Yu, J.-R. Gao, D. Ding, X. Zeng, and D. Z. Pan, "Accurate lithography hotspot detection based on principal component analysis-support vector machine classifier with hierarchical data clustering," *Journal of Micro/Nanolithography, MEMS, and MOEMS (JM3)*, vol. 14, no. 1, p. 011003, 2015.

[31] D. Ding, B. Yu, J. Ghosh, and D. Z. Pan, "EPIC: Efficient prediction of IC manufacturing hotspots with a unified meta-classification

formulation," in *IEEE/ACM Asia and South Pacific Design Automation Conference (ASPDAC)*, 2012, pp. 263–270.

[32] Y. Tomioka, T. Matsunawa, C. Kodama, and S. Nojima, "Lithography hotspot detection by two-stage cascade classifier using histogram of oriented light propagation," in *IEEE/ACM Asia and South Pacific Design Automation Conference (ASPDAC)*, 2017, pp. 81–86.

[33] J. Guo, F. Yang, S. Sinha, C. Chiang, and X. Zeng, "Improved tangent space based distance metric for accurate lithographic hotspot classification," in *ACM/IEEE Design Automation Conference (DAC)*, 2012, pp. 1173–1178.

[34] J.-M. Lin and Y.-W. Chang, "TCG: A transitive closure graph-based representation for non-slicing floorplans," in *ACM/IEEE Design Automation Conference (DAC)*, 2001, pp. 764–769.

[35] N. Dalal and B. Triggs, "Histograms of oriented gradients for human detection," in *IEEE Conference on Computer Vision and Pattern Recognition (CVPR)*, vol. 1. IEEE, 2005, pp. 886–893.

[36] M. J. Gangeh, A. K. Farahat, A. Ghodsi, and M. S. Kamel, "Supervised dictionary learning and sparse representation-a review," *arXiv preprint arXiv:1502.05928*, 2015.

[37] S. G. Mallat and Z. Zhang, "Matching pursuits with time-frequency dictionaries," *IEEE Transactions on Signal Processing*, vol. 41, no. 12, pp. 3397–3415, 1993.

[38] Y. C. Pati, R. Rezaiifar, and P. S. Krishnaprasad, "Orthogonal matching pursuit: Recursive function approximation with applications to wavelet decomposition," in *Asilomar Conference on Signals, Systems, and Computers*. IEEE, 1993, pp. 40–44.

[39] R. Tibshirani, "Regression shrinkage and selection via the Lasso," *Journal of the Royal Statistical Society: Series B*, vol. 58, pp. 267–288, 1996.

[40] B. Efron, T. Hastie, I. Johnstone, R. Tibshirani, and others., "Least angle regression," *The Annals of statistics*, vol. 32, no. 2, pp. 407–499, 2004.

[41] J. Friedman, T. Hastie, and R. Tibshirani, "Regularization paths for generalized linear models via coordinate descent," *Journal of Statistical Software*, vol. 33, no. 1, p. 1, 2010.

[42] A. Beck and M. Teboulle, "A fast iterative shrinkage-thresholding algorithm for linear inverse problems," *SIAM Journal on Imaging Sciences (SIIMS)*, vol. 2, no. 1, pp. 183–202, 2009.

[43] I. Daubechies, "The wavelet transform, time-frequency localization and signal analysis," *IEEE Transactions on Information Theory (TIT)*, vol. 36, no. 5, pp. 961–1005, 1990.

[44] E. Candes, L. Demanet, D. Donoho, and L. Ying, "Fast discrete curvelet transforms," *Multiscale Modeling & Simulation*, vol. 5, no. 3, pp. 861–899, 2006.

[45] M. N. Do and M. Vetterli, "The contourlet transform: an efficient directional multiresolution image representation," *IEEE Transactions on Image Processing*, vol. 14, no. 12, pp. 2091–2106, 2005.

[46] M. Aharon, M. Elad, and A. Bruckstein, "*k*-SVD: An algorithm for designing overcomplete dictionaries for sparse representation," *IEEE Transactions on Signal Processing*, vol. 54, no. 11, pp. 4311–4322, 2006.

[47] K. Skretting and K. Engan, "Recursive least squares dictionary learning algorithm," *IEEE Transactions on Signal Processing*, vol. 58, no. 4, pp. 2121–2130, 2010.

[48] J. Mairal, F. Bach, J. Ponce, and G. Sapiro, "Online dictionary learning for sparse coding," in *International Conference on Machine Learning (ICML)*, 2009, pp. 689–696.

[49] A. Mensch, J. Mairal, B. Thirion, and G. Varoquaux, "Dictionary learning for massive matrix factorization," in *International Conference on Machine Learning (ICML)*, 2016, pp. 1737–1746.

[50] D. P. Bertsekas, *Nonlinear programming*. Athena scientific Belmont, 1999.

[51] F. Pedregosa, G. Varoquaux, A. Gramfort, V. Michel, B. Thirion, O. Grisel, M. Blondel, P. Prettenhofer, R. Weiss, V. Dubourg *et al.*, "Scikit-learn: Machine learning in Python," *Journal of Machine Learning Research*, vol. 12, no. Oct., pp. 2825–2830, 2011.

[52] A. J. Torres, "ICCAD-2012 CAD contest in fuzzy pattern matching for physical verification and benchmark suite," in *IEEE/ACM International Conference on Computer-Aided Design (ICCAD)*, 2012, pp. 349–350.

[53] C. Goutte and E. Gaussier, "A probabilistic interpretation of precision, recall and F-score, with implication for evaluation," in *European Conference on Information Retrieval (ECIR)*. Springer, 2005, pp. 345–359.

Pattern Similarity Metrics for Layout Pattern Classification and their Validity Analysis by Lithographic Responses

Atsushi Takahashi, Shimpei Sato, Hiroki Ogura
Tokyo Institute of Technology, Japan
{atsushi,satos}@ict.e.titech.ac.jp

Yu-Min Sung, Ting-Chi Wang
National Tsing Hua University, Taiwan
tcwang@cs.nthu.edu.tw

Abstract—**In order to detect manufacturing hotspots from huge layout pattern efficiently or to characterize hotspots effectively, various layout pattern classification methods had been proposed. Layout pattern clustering using layout pattern similarity metrics such as area-difference-ratio and edge-shifting-distance could be used as preprocessing to shorten the runtime of classification significantly. However the validity of classification will be degraded if the parameters used for similarity metrics are not appropriately chosen. In this paper, the layout pattern classification methods and layout pattern similarity metrics proposed so far are summarized and are discussed.**

Index Terms—**lithography, layout pattern classification, similarity metric, layout clip**

I. INTRODUCTION

With the continuous shrinking of feature size, it becomes impossible to specify the complex requirements to achieve dense pattern with higher yields by perfect design rules. Even if layout patterns passes design rule checking (DRC), there still remain layout patterns that may force yield loss, called *hotspots*, and time-consuming lithography simulations are used to identify hotspots [1]–[3]. Various design for manufacturability (DFM) techniques have been developed. Lithography friendly design are discussed to reduce the total amount of hotspots [4]–[6]. However, it is still important to efficiently detect and fix these hotspots at design stage to reduce development time and manufacturing cost.

In order to detect hotspots efficiently, various types of techniques have been proposed. Pattern-matching-based hotspot detection methods [7]–[20] are mainly used to detect known types of hotspots. They have the advantage of short computation time and full detection capability on a set of predefined patterns, but will miss unknown hotspots. On the other hand, machine-learning-based hotspot detection methods [21]–[38] have more capability to detect unknown hotspots, but need slightly long training time to achieve higher accuracy.

In early stage of design, a layout may contain many hotspots. To handle too many hotspots efficiently or to characterize hotspots effectively, clustering approach is introduced in which *layout clip* that surrounds a hotspot, called *snippet*, is defined [39]–[45]. Various types of metrics to classify hotspots are proposed as well. In ICCAD 2016 Pattern-Classification Contest [46], hotspots are requested to be partitioned into clusters, and several algorithms under the contest formulation have

been proposed [47]–[50]. However the validity of evaluation of clustering in terms of impacts on development time and manufacturing cost is unclear.

In the following, we will overview the hotspots detection methods, the layout pattern classification methods, and layout pattern similarity metrics.

II. LAYOUT FEATURE ENCODING AND MACHINE-LEARNING

The encoding of a layout clip that is used to represent the features of a clip is one of key points in hotspot detection and classification. Although a criterion for defining similarity to evaluate essential characteristics in real layouts is unclear [45], the encoding of a layout clip requires to have a potential to evaluate essential characteristics. In some methods, multiple types of encoding are used to achieve enough capability effectively.

In [7], [9]–[11], [14], [19], [29], [33], [47], a layout pattern is represented by a set of rectangle tiles (see Fig. 1(a), where horizontal slice is used). The topology of tiles is encoded as matrix or string in [7], [9], [10], [47], while graph representation is used in [11], [14], [19], [29], [33]. In [20], [44], [50], the skeleton of a polygon is used (see Fig. 1(b)). The geometrical information on fragments of polygon edges in layout is used in [25], [27], [31], [32], [35] (see Fig. 1(c)). In [8], [11], [14], [17], [19], [23], [33], several critical features are defined and used. In [21], bitmap representation of layout is used. In [12], [13], [15], [24], [41]–[45], a transformation used in image processing is used. In [16], [18], [22], [26], [29], [33], [34], [37], [45], density-based encoding in which local densities of partitioned sub-regions is used (see Fig. 1(d)). In [30], a feature of sub-regions is used in encoding to avoid encoding all patterns in density-based encoding. Concentric circle sampling is used in [36], [38], [45] (see Fig. 1(e), where black dots represent sampling points).

Among various machine-learning-based hotspot detection methods, support vector machine (SVM) techniques are used in [22], [24]–[33], [35], [36]. Artificial neural network (ANN) is used in [21], [23], [25], [27], [28], and deep neural network is used in [37]. While, Adaboost with decision tree is used in [34]. Online learning is introduced in [38].

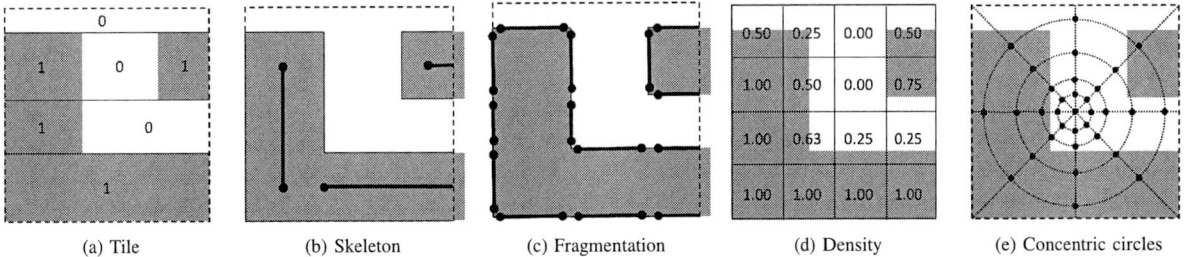

Fig. 1. Layout Clip Representation

(a) Tile   (b) Skeleton   (c) Fragmentation   (d) Density   (e) Concentric circles

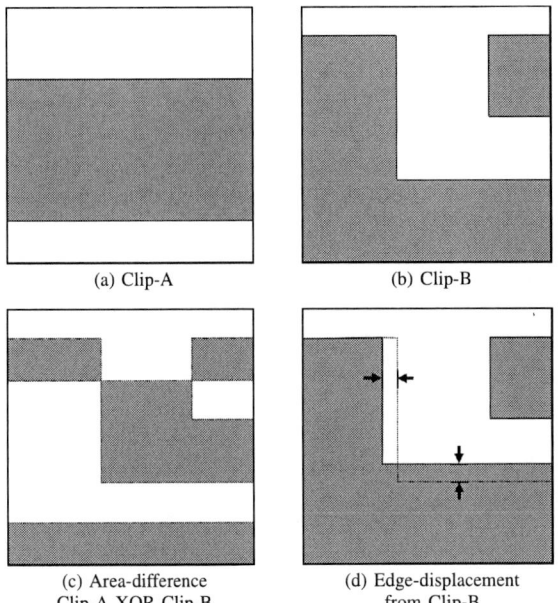

(a) Clip-A

(b) Clip-B

(c) Area-difference Clip-A XOR Clip-B

(d) Edge-displacement from Clip-B

Fig. 2. Area-difference and Edge-displacement

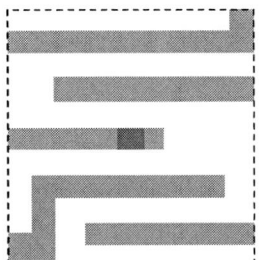

Fig. 3. An example of marker (red rectangle)

III. HOTSPOT CLASSIFICATION

A. Layout Clip Approach

The pattern fidelity depends on layout patterns within the optical radius of influence, and more layout patterns are included within its proximity as feature size shrinks. A hotspot with layout patterns surrounding it is extracted as a susceptible layout clip, and the layout clip is used for the classification of hotspots [39]–[45].

In early stage of design, a layout may contain many susceptible layout clips. Also, similar layout clips are often contained. In such cases, if a few representative layout clips are defined so that they cover all the features of hotspots, then they can be used to fix all the hotspots efficiently or can be used to efficiently detect hotspots in new layouts.

The similarity metric is a quantitative measure of the difference of a pair of layout clips. The similarity metric should capture lithographic characteristics so that the layout clips which are close to each other under the metric have similar lithographic responses. An XOR based metric is used

in [39]–[43] (An example of layout clip XOR is shown in Fig. 2(a)-(c)). However this metric is sensitive to the small variations or shifts of the shapes even if the difference of lithographic responses is negligibly small. To be tolerant with small changes, higher-order correlation metric [41], Fourier spectra of pattern [42], [43], [45], and Euclidean distance in terms of a subset of DCT (discrete cosine transform) coefficients of layout clip image [44] are used in similarity metric. While, a tangent space-based distance metric is proposed in [51]. In [52], a distance metric for an EUV-specific technology is proposed.

B. ICCAD 2016 Pattern-Classification Contest

In ICCAD 2016 Pattern-Classification Contest [46], a layout pattern classification problem under the constraint on area-difference (see Fig. 2(c)) or edge-displacement (see Fig. 2(d)) is defined. In the contest, *markers* (polygons) that indicate hotspots are given. In Fig. 3, a red rectangle is given in the layout as an example of marker. Then, for each marker, a fixed size layout clip is requested to be generated so that the center of the layout clip is inside the marker, and obtained layout clips are requested to be partitioned into clusters. The minimization of the number of clusters and the maximization of the size of the largest cluster are defined as optimization targets.

In the contest, in order for two layout clips to be eligible in the same cluster, the area-difference constraint requires that the area difference between the two clips divided by the clip area must not exceed $(1 - a)$, where a is a user-specified parameter whose value is between 0 and 1. On the other hand, the edge-displacement constraint requires that for two layout clips to

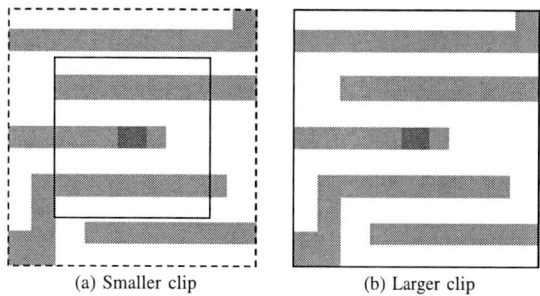

| (a) Smaller clip | (b) Larger clip |

Fig. 4. Layout Clip Size

be in the same cluster, each polygon edge of a clip cannot be shifted more than e such that after edge shifting, both resultant clips become identical, where e is a user-specified parameter whose unit is nm. Note that when $a = 1$ or $e = 0$, it means exact match is required.

Since ICCAD 2016 Pattern-Classification Contest was held, several algorithms under the contest formulation (with some modifications) have been proposed [47]–[50], and they all adopt a similar strategy to solve the problem. First, to reduce the problem size, identical layout clips are identified and merged. Then, similarity relations between pairs of layout clips are checked and recognized. Finally, the clustering result is derived from the similarity relations. Different algorithms employ different techniques in each of these steps.

Even though optimal solution slightly differs depending on the interpretation of edge-displacement constraint and the optimization targets, the aforementioned algorithms are able to obtain solutions of better quality than the contest winners. However, the quality of obtained clusters in terms of lithographic responses and the impact on development time and manufacturing cost is unclear.

Once clip pairs allowed to be contained in a cluster are defined, the problem can be regarded as a set covering problem, where the problem of minimizing the number of sets (clusters) is known to be NP-hard [47], [49]. Therefore, if the problem size is large, then heuristics are adopted to obtain clusters. As a contest evaluation metric, cluster count minimization has some meaning. However the validity of constraints on clustering and the quality of clusters in terms of the impact on the design flow are not evaluated in the contest.

A layout clip is defined to characterize a hotspot which is located in the center of the clip. In that sense, the similarity metrics on layout clips to classify hotspots should have an ability to capture the characteristics of lithographic responses on corresponding hotspots.

Roughly speaking, the impact on a hotspot by a pattern that is away from the hotspot is smaller than that by a pattern near to the hotspot. In [39], weight according to the amount of lithographic interaction is used. Therefore, larger weights are assigned on the center of layout clip. Also, rotation and reflection of layout clip are taken into account. Similarly, larger weights are assigned for an inner area in [16], [18]. In [36], [38], [45], sampling points on concentric circles are densely

placed in an inner area as shown in Fig. 1(e). The impact of clip size on the hotspot detection capability is discussed in [22], and different layout clip sizes as shown in Fig. 4 are used in a two-stage classifier [26].

IV. CONCLUSION

Even though various types of hotspot detection methods have been proposed, hotspot detection using lithography simulation is still required to achieve satisfactory accuracy in industry. Ideas to reduce the number of simulations while keeping enough accuracy are proposed in [36], [53]. We still have to develop various ideas to reduce development time and manufacturing cost.

REFERENCES

[1] S. Inoue, T. Kotani, S. Nojima, S. Tanaka, K. Hashimoto, and I. Mori, "Total hot spot management from design rule definition to silicon," in *Proc. Electronic Design Processes Workshop (EDP)*, 2003.

[2] J. Kim and M. Fan, "Hotspot detection on post-OPC layout using full-chip simulation-based verification tool: a case study with aerial image simulation," in *Proc. SPIE Photomask Technology*, vol. 5256, 2003, pp. 919–925.

[3] W. Hoppe, T. Roessler, and J. A. Torres, "Beyond rule-based physical verification," in *Proc. SPIE Photomask Technology*, vol. 6349, 63494X, 2006.

[4] R. März, K. Peter, and W. Maurer, "ORC and LfD as first steps towards DfM," in *Proc. SPIE 22nd European Mask and Lithography Conference*, vol. 6281, 62810I, 2006.

[5] M. Cho, K. Yuan, Y. Ban, and D. Z. Pan, "ELIAD: Efficient lithography aware detailed routing algorithm with compact and macro post-OPC printability prediction," *IEEE Transactions on Computer-Aided Design of Integrated Circuits and Systems (TCAD)*, vol. 28, no. 7, pp. 1006–1016, 2009.

[6] C. Kodama, H. Ichikawa, K. Nakayama, F. Nakajima, S. Nojima, T. Kotani, T. Ihara, and A. Takahashi, "Self-aligned double and quadruple patterning aware grid routing method," *IEEE Transactions on Computer-Aided Design of Integrated Circuits and Systems (TCAD)*, vol. 34, no. 5, pp. 753–765, 2015.

[7] H. Yao, S. Sinha, C. Chiang, X. Hong, and Y. Cai, "Efficient process-hotspot detection using range pattern matching," in *Proc. IEEE/ACM International Conference on Computer Aided Design (ICCAD)*, 2006, pp. 625–632.

[8] V. Dai, J. Yang, N. Rodriguez, and L. Capodieci, "DRC Plus: augmenting standard DRC with pattern matching on 2D geometries," in *Proc. SPIE Advanced Lithography*, vol. 6521, 65210A, 2007.

[9] J. Xu, S. Sinha, and C. C. Chiang, "Accurate detection for process-hotspots with vias and incomplete specification," in *Proc. IEEE/ACM International Conference on Computer Aided Design (ICCAD)*, 2007, pp. 839–846.

[10] H. Yao, S. Sinha, J. Xu, C. Chiang, Y. Cai, and X. Hong, "Efficient range pattern matching algorithm for process-hotspot detection," *IET Circuits, Devices & Systems*, vol. 2, no. 1, pp. 2–15, 2008.

[11] A. B. Kahng, C.-H. Park, and X. Xu, "Fast dual-graph-based hotspot filtering," *IEEE Transactions on Computer-Aided Design of Integrated Circuits and Systems (TCAD)*, vol. 27, no. 9, pp. 1635–1642, 2008.

[12] S. Maeda, T. Matsunawa, R. Ogawa, H. Ichikawa, K. Takahata, M. Miyairi, T. Kotani, S. Nojima, S. Tanaka, K. Nakagawa, T. Saito, S. Mimotogi, S. Inoue, H. Nosato, H. Sakanashi, T. Kobayashi, M. Murakawa, T. Higuchi, E. Takahashi, and N. Otsu, "Hotspot detection using image pattern recognition based on higher-order local auto-correlation," in *Proc. SPIE Advanced Lithography*, vol. 7974, 79740X, 2011.

[13] T. Matsunawa, S. Maeda, H. Ichikawa, S. Nojima, S. Tanaka, S. Mimotogi, H. Nosato, H. Sakanashi, M. Murakawa, and E. Takahashi, "Generator of predictive verification pattern using vision system based on higher-order local autocorrelation," in *Proc. SPIE Advanced Lithography*, vol. 8326, 832615, 2012.

[14] Y.-T. Yu, Y.-C. Chan, S. Sinha, I. H.-R. Jiang, and C. Chiang, "Accurate process-hotspot detection using critical design rule extraction," in *Proc. ACM/EDAC/IEEE Design Automation Conference (DAC)*, 2012, pp. 1163–1168.

978-1-5386-7100-9/18 $31.00 © 2018 IEEE

[15] H. Nosato, H. Sakanashi, E. Takahashi, M. Murakawa, T. Matsunawa, S. Maeda, S. Tanaka, and S. Mimotogi, "Hotspot prevention and detection method using an image-recognition technique based on higher-order local autocorrelation," *Journal of Micro/Nanolithography, MEMS, and MOEMS (JM3)*, vol. 13, no. 1:011007, 2014.

[16] S.-Y. Lin, J.-Y. Chen, J.-C. Li, W.-Y. Wen, and S.-C. Chang, "A novel fuzzy matching model for lithography hotspot detection," in *Proc. ACM/EDAC/IEEE Design Automation Conference (DAC)*, 2013.

[17] K. Kato, Y. Taniguchi, and K. Nishizawa, "Fuzzy pattern matching techniques for photomask layout data," in *Proc. SPIE Advanced Lithography*, vol. 8701, 87010C, 2013.

[18] W.-Y. Wen, J.-C. Li, S.-Y. Lin, J.-Y. Chen, and S.-C. Chang, "A fuzzy-matching model with grid reduction for lithography hotspot detection," *IEEE Transactions on Computer-Aided Design of Integrated Circuits and Systems (TCAD)*, vol. 33, no. 11, pp. 1671–1680, 2014.

[19] Y.-T. Yu, I. H.-R. Jiang, Y. Zhang, and C. Chiang, "DRC-based hotspot detection considering edge tolerance and incomplete specification," in *Proc. IEEE/ACM International Conference on Computer Aided Design (ICCAD)*, 2014, pp. 101–107.

[20] H.-Y. Su, C.-C. Chen, Y.-L. Li, A.-C. Tu, C.-J. Wu, and C.-M. Huang, "A novel fast layout encoding method for exact multilayer pattern matching with Prüfer encoding," *IEEE Transactions on Computer-Aided Design of Integrated Circuits and Systems (TCAD)*, vol. 34, no. 1, pp. 95–108, 2015.

[21] N. Nagase, K. Suzuki, K. Takahashi, M. Minemura, S. Yamauchi, and T. Okada, "Study of hot spot detection using neural networks judgment," in *Proc. SPIE Photomask and Next-Generation Lithography Mask Technology XIV*, vol. 6607, 66071B, 2007.

[22] J.-Y. Wuu, F. G. Pikus, A. Torres, and M. Marek-Sadowska, "Detecting context sensitive hot spots in standard cell libraries," in *Proc. SPIE Advanced Lithography*, vol. 7275, 727515, 2009.

[23] D. Ding, X. Wu, J. Ghosh, and D. Z. Pan, "Machine learning based lithographic hotspot detection with critical-feature extraction and classification," in *Proc. IEEE International Conference on IC Design and Technology (ICICDT)*, 2009.

[24] D. G. Drmanac, F. Liu, and L.-C. Wang, "Predicting variability in nanoscale lithography processes," in *Proc. ACM/IEEE Design Automation Conference (DAC)*, 2009, pp. 545–550.

[25] D. Ding, A. J. Torres, F. G. Pikus, and D. Z. Pan, "High performance lithographic hotspot detection using hierarchically refined machine learning," in *Proc. Asia and South Pacific Design Automation Conference (ASP-DAC)*, 2011, pp. 775–780.

[26] J.-Y. Wuu, F. G. Pikus, A. Torres, and M. Marek-Sadowska, "Rapid layout pattern classification," in *Proc. Asia and South Pacific Design Automation Conference (ASP-DAC)*, 2011, pp. 781–786.

[27] D. Ding, J. A. Torres, and D. Z. Pan, "High performance lithography hotspot detection with successively refined pattern identifications and machine learning," *IEEE Transactions on Computer-Aided Design of Integrated Circuits and Systems (TCAD)*, vol. 30, no. 11, pp. 1621–1634, 2011.

[28] D. Ding, B. Yu, J. Ghosh, and D. Z. Pan, "EPIC: Efficient prediction of IC manufacturing hotspots with a unified meta-classification formulation," in *Proc. Asia and South Pacific Design Automation Conference (ASP-DAC)*, 2012, pp. 263–270.

[29] Y.-T. Yu, G.-H. Lin, I. H.-R. Jiang, and C. Chiang, "Machine-learning-based hotspot detection using topological classification and critical feature extraction," in *Proc. ACM/EDAC/IEEE Design Automation Conference (DAC)*, 2013.

[30] K. S. Luo, Z. Shi, and Z. Geng, "Machine learning based lithography hotspot detection with sparse feature encoding and hierarchical pattern classification," *ECS Transactions*, vol. 60, no. 1, pp. 1179–1184, 2014.

[31] J.-R. Gao, B. Yu, and D. Z. Pan, "Accurate lithography hotspot detection based on PCA-SVM classifier with hierarchical data clustering," in *Proc. SPIE Advanced Lithography*, vol. 9053, 90530E, 2014.

[32] B. Yu, J.-R. Gao, D. Ding, X. Zeng, and D. Z. Pan, "Accurate lithography hotspot detection based on principal component analysis-support vector machine classifier with hierarchical data clustering," *Journal of Micro/Nanolithography, MEMS, and MOEMS (JM3)*, vol. 14, no. 1:011003, 2015.

[33] Y.-T. Yu, G.-H. Lin, I. H.-R. Jiang, and C. Chiang, "Machine-learning-based hotspot detection using topological classification and critical feature extraction," *IEEE Transactions on Computer-Aided Design of Integrated Circuits and Systems (TCAD)*, vol. 34, no. 3, pp. 460–470, 2015.

[34] T. Matsunawa, J.-R. Gao, B. Yu, and D. Z. Pan, "A new lithography hotspot detection framework based on AdaBoost classifier and simplified feature extraction," in *Proc. SPIE Advanced Lithography*, vol. 9427, 94270S, 2015.

[35] K. Madkour, S. Mohamed, D. Tantawy, and M. Anis, "Hotspot detection using machine learning," in *Proc. International Symposium on Quality Electronic Design (ISQED)*, 2016, pp. 405–409.

[36] T. Kimura, T. Matsunawa, S. Nojima, and D. Z. Pan, "Hybrid hotspot detection using regression model and lithography simulation," in *Proc. SPIE Advanced Lithography*, vol. 9781, 97810C, 2016.

[37] T. Matsunawa, S. Nojima, and T. Kotani, "Automatic layout feature extraction for lithography hotspot detection based on deep neural network," in *Proc. SPIE Advanced Lithography*, vol. 9781, 97810H, 2016.

[38] H. Zhang, B. Yu, and E. F. Y. Young, "Enabling online learning in lithography hotspot detection with information-theoretic feature optimization," in *Proc. IEEE/ACM International Conference on Computer Aided Design (ICCAD)*, 2016.

[39] N. Ma, J. Ghan, S. Mishra, C. Spanos, K. Poolla, N. Rodriguez, and L. Capodieci, "Automatic hotspot classification using pattern-based clustering," in *Proc. SPIE Advanced Lithography*, vol. 6925, 692505, 2008.

[40] J. Ghan, N. Ma, S. Mishra, C. Spanos, K. Poolla, N. Rodriguez, and L. Capodieci, "Clustering and pattern matching for an automatic hotspot classification and detection system," in *Proc. SPIE Advanced Lithography*, vol. 7275, 727516, 2009.

[41] B. Lin, Z. Shi, and Y. Chen, "Hotspot classification based on higher-order local autocorrelation," in *Proc. SPIE Photomask Technology*, vol. 8522, 85222O, 2012.

[42] S. Shim, W. Chung, and Y. Shin, "Synthesis of lithography test patterns through topology-oriented pattern extraction and classification," in *Proc. SPIE Advanced Lithography*, vol. 9053, 905305, 2014.

[43] S. Shim and Y. Shin, "Topology-oriented pattern extraction and classification for synthesizing lithography test patterns," *Journal of Micro/Nanolithography, MEMS, and MOEMS (JM3)*, vol. 14, no. 1:013503, 2015.

[44] W. C. J. Tam and R. D. S. Blanton, "LASIC: Layout analysis for systematic IC-defect identification using clustering," *IEEE Transactions on Computer-Aided Design of Integrated Circuits and Systems (TCAD)*, vol. 34, no. 8, pp. 1278–1290, 2015.

[45] T. Matsunawa, B. Yu, and D. Z. Pan, "Laplacian eigenmaps- and Bayesian clustering-based layout pattern sampling and its applications to hotspot detection and optical proximity correction," *Journal of Micro/Nanolithography, MEMS, and MOEMS (JM3)*, vol. 15, no. 4:043504, 2016.

[46] R. O. Topaloglu, "ICCAD-2016 CAD contest in pattern classification for integrated circuit design space analysis and benchmark suite," in *Proc. IEEE/ACM International Conference on Computer Aided Design (ICCAD)*, 2016.

[47] K.-J. Chen, Y.-K. Chuang, B.-Y. Yu, and S.-Y. Fang, "Minimizing cluster number with clip shifting in hotspot pattern classification," in *Proc. ACM/EDAC/IEEE Design Automation Conference (DAC)*, 2017.

[48] W.-C. Chang, I. H.-R. Jiang, Y.-T. Yu, and W.-F. Liu, "iClaire: A fast and general layout pattern classification algorithm," in *Proc. ACM/EDAC/IEEE Design Automation Conference (DAC)*, 2017.

[49] M. Woo, S. Kim, and S. Kang, "GRASP based metaheuristics for layout pattern classification," in *Proc. IEEE/ACM International Conference on Computer Aided Design (ICCAD)*, 2017, pp. 512–518.

[50] Y.-S. Wu, H.-Y. Su, Y.-H. Chang, R. O. Topaloglu, and Y.-L. Li, "MapReduce-based pattern classification for design space analysis," in *Proc. International Symposium on VLSI Design, Automation and Test (VLSI-DAT)*, 2018.

[51] J. Guo, F. Yang, S. Sinha, C. Chiang, and X. Zeng, "Improved tangent space based distance metric for accurate lithographic hotspot classification," in *Proc. ACM/EDAC/IEEE Design Automation Conference (DAC)*, 2012, pp. 1169–1174.

[52] P.-H. Wu, C.-W. Chen, C.-R. Wu, and T.-Y. Ho, "Triangle-based process hotspot classification with dummification in EUVL," in *Proc. International Symposium on VLSI Design, Automation and Test (VLSI-DAT)*, 2014.

[53] Y. Abe, F. Nakajima, Y. Watanabe, M. Kajiwara, S. Nojima, and T. Kotani, "Hotspot detection based on surrounding optical feature," in *Proc. SPIE Advanced Lithography*, vol. 10588, 105880I, 2018.

Recent Research and Challenges in Multiple Patterning Layout Decomposition

Iris Hui-Ru Jiang
Dept. of Electrical Eng. & Graduate Inst. Electronics Eng.
National Taiwan University
Taipei, Taiwan
e-mail: huiru.jiang@gmail.com

Hua-Yu Chang
Design Group
Synopsys, Inc.
Hsinchu, Taiwan
e-mail: huayu.chang@gmail.com

Abstract—**Multiple patterning lithography has been recognized as one of the most promising solutions, in addition to extreme ultraviolet lithography, directed self-assembly, nanoimprint lithography, and electron beam lithography, for advancing the resolution limit of conventional optical lithography. In multiple patterning lithography, an original design layout is divided into multiple masks, and through a series of exposure/etching steps, the layout can be produced. Multiple patterning layout decomposition becomes more challenging as advanced technology introduces complex coloring rules and considers density balancing. In this paper, we first review recent research progress in multiple patterning layout decomposition from modeling to solution perspectives. Then, we discuss how challenges were handled by state-of-the art works. Finally, future research directions are identified.**

Keywords-design for manufacturability; multiple patterning lithography; graph coloring; exact cover; linear programming

I. INTRODUCTION

Multiple patterning lithography (MPL) is a promising solution to overcome 193nm immersion lithography resolution as semiconductor process technology scales down to 10 nm and below [1].

Conceptually, MPL repeats single patterning lithography multiple times to achieve finer pitches. For MPL, an original design layout is decomposed into multiple masks (colors). When the distance between two layout features is less than the minimum color spacing, they should be assigned to different masks to prevent a coloring conflict. Multiple patterning layout decomposition (MPLD) is the key challenge of MPL.

A stitch cuts a layout feature into two connected parts; inserting stitches sometimes can resolve coloring conflicts. Nevertheless, stitches incur overlay errors, leading to a yield loss. Therefore, the main objective of MPLD is to minimize the numbers of conflicts and stitches. In the following, we briefly review recent research and challenges in MPLD.

II. PROBLEM MODELING AND SOLUTIONS

In most of relevant works, MPLD is modeled as a graph coloring problem on a conflict graph: A mask is viewed as a color, a vertex represents a layout feature, and an edge prevents two features from sharing the same color. (see Figure 1.) For double patterning lithography (DPL), MPLD corresponds to a two-coloring problem and can be solved efficiently [2]. Nevertheless, for triple patterning lithography (TPL) and beyond, MPLD is difficult because graph coloring on a general graph is computationally hard. Hence,

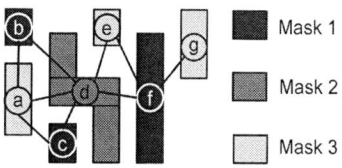

Figure 1.   Conflict graph in multiple patterning layout decomposition.

mathematical programming approach, such as semidefinite programming [3][4] and integer linear programming [5], seeks optimality but may consume long runtime, thus requiring speedup techniques. Fast heuristics, such as lookup table [6], pairwise coloring [7], and modified independent set [8], are generally efficient but may lose some solution quality. Very recently, a fast and near-optimal solution is provided by [9] based on exact covering.

Figure 2 shows a generic procedure of MPLD. First, a conflict graph is constructed based on an input layout and coloring constraints. Second, to reduce the problem size, the conflict graph is divided and simplified by connected component separation, vertex removal, bridge removal, and articulation point duplication. Third, stitch candidates are generated for subsequent conflict removal. Fourth, the mask (color) assignment of each subgraph is determined. Finally, the coloring results of subgraphs are combined via color flipping.

The stumbling block of pursuing optimality and efficiency for MPL is native conflict identification. For DPL, a native conflict can be identified as an odd cycle in the conflict graph. However, how to identify native conflicts is still an open problem in TPL and beyond. Some of existing works thus focus on removing trivial conflict patterns, e.g., four-cliques in TPL [10].

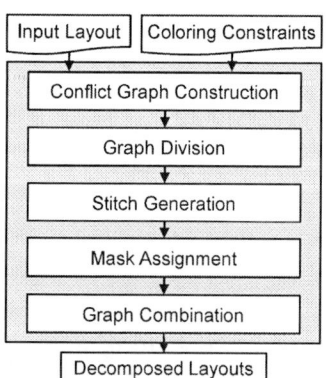

Figure 2.   Generic procedure of multiple patterning layout decomposition.

978-1-5386-7100-9/18 $31.00 © 2018 IEEE

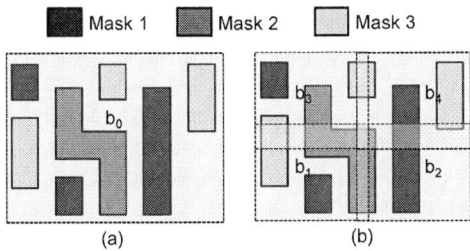

Figure 3. Density balancing. (a) Global density measurement. (b) Local density measurement over overlapping or non-overlapping windows.

III. SPECIAL CONSIDERTATIONS

In addition to conflict and stitch minimization, balanced feature density of multiple masks is expected for avoiding pattern distortion (i.e., edge placement error) [9][11]. The overlapping or non-overlapping window-based density balancing can be considered by introducing a density term to the objective function of a mathematical program [11] or color flipping [9]. (see Figure 3.)

Moreover, complex coloring rules are introduced at advanced technology nodes. For example, the minimum spacing required for two features assigned on different masks is not symmetric, and some features are required to be assigned to the same mask [9].

Because coloring a general graph is NP-hard, some research tries to reveal specific colorable graphs, e.g., penrose-tiling via patterns in [12], regular layouts in [13]. For a cell-based row-structure layout, MPLD can be reduced to a shortest path problem, solvable in polynomial time [14].

IV. FUTURE DIRECTIONS

Modifying an unsuccessfully decomposed layout at a post-layout stage requires high ECO efforts. Hence, considering MPLD during physical design optimization becomes inevitable and pivotal for design closure. MPL-aware standard cell detailed placement considers a single row cell placement [15] and cross-row middle-of-line (MOL) coloring conflicts [16]. MPL-aware routing considers stitch and conflict costs during A* search [17].

Finally, next generation lithography technologies will be ready in the near future, including extreme ultraviolet (EUV) lithography, directed self-assembly (DSA), nanoimprint lithography, and electron beam lithography (e-Beam). Hybrid lithography could be a cost-effective way to manufacture an advanced design, e.g., MPL-cut, DSA-MP, EUV-MP, etc.

V. CONCLUSION

In this paper, we have reviewed recent research progress in MPLD from modeling to solution perspectives. We have also discussed how challenges were handled by state-of-the art works. Finally, we have highlighted future research directions.

REFERENCES

[1] M. Neisser and S. Wurm, "ITRS lithography roadmap: 2015 challenges," *Adv. Opt. Techn. 4(4)*, pp. 235–240, Aug. 2015.

[2] X. Tang and M. Cho, "Optimal layout decomposition for double patterning technology," in *Proc. IEEE/ACM Int'l Conf. on Computer-Aided Design (ICCAD)*, pp. 9–13, Nov. 2011.

[3] B. Yu, K. Yuan, D. Ding, and D.Z. Pan, "Layout decomposition for triple patterning lithography," *IEEE Trans. on Computer-Aided Design of Integrated Circuits and Systems (TCAD)*, vol. 34, no. 3, pp.433–446, Mar. 2015. Also see in *Proc. IEEE/ACM Int'l Conf. on Computer-Aided Design (ICCAD)*, pp.1–8, Nov. 2011.

[4] Y. Lin, X. Xu, B. Yu, R. Baldick, D. Z. Pan, "Triple/quadruple patterning layout decomposition via novel linear programming and iterative rounding," in *Proc. SPIE*, vol. 9781, pp. 97810M-1–97810M-11, Mar. 2016.

[5] B. Yu and D.Z. Pan, "Layout decomposition for quadruple patterning lithography and beyond," in *Proc. ACM/EDAC/IEEE Design Automation Conference (DAC)*, pp. 1–6, June 2014.

[6] J. Kuang and E. F.Y. Young, "An efficient layout decomposition approach for triple patterning lithography," in *Proc. ACM/EDAC/IEEE Design Automation Conference (DAC)*, Article 69, pp. 1–6, June 2013.

[7] Y. Zhang, W.-S. Luk, Y. Yang, H. Zhou, C. Yan, D. Z. Pan, X. Zeng, "Layout decomposition with pairwise coloring and adaptive multi-start for triple patterning lithography," *ACM Trans. on Design Automation of Electric Systems (TODAES)*, vol. 21, no. 1, pp. 2:1–2:25, Dec. 2015. Also see in *Proc. IEEE/ACM Int'l Conf. on Computer-Aided Design (ICCAD)*, pp.170–177, Nov. 2013.

[8] S.-Y. Fang, Y.-W. Chang, and W.-Y. Chen, "A novel layout decomposition algorithm for triple patterning lithography," *IEEE Trans. on Computer-Aided Design of Integrated Circuits and Systems (TCAD)*, vol. 33, no. 3, pp. 397–408, Mar. 2014. Also see in *Proc. ACM/EDAC/IEEE Design Automation Conference (DAC)*, pp 1185–1190, June 2012.

[9] I. H.-R. Jiang and H.-Y. Chang, "Multiple patterning layout decomposition considering complex coloring rules and density balancing," *IEEE Trans. on Computer-Aided Design of Integrated Circuits and Systems (TCAD)*, vol. 36, no. 12, pp. 2080–2092, Dec., 2017. Also see in *Proc. ACM/EDAC/IEEE Design Automation Conference (DAC)*, Article 40, pp. 1–6, June 2016.

[10] X. Li, Z. Zhu, and W. Zhu, "Discrete relaxation method for triple patterning lithography layout decomposition," *IEEE Trans. on Computers (TC)*, vol. 66, no. 2, pp. 285–298, Feb., 2017.

[11] B. Yu, Y.-H. Lin, G. Luk-Pat, D. Ding, K. Lucas, D.Z. Pan, "A high-performance triple patterning layout decomposer with balanced density," in *Proc. IEEE/ACM Int'l Conf. on Computer-Aided Design (ICCAD)*, pp. 163–169, Nov. 2013.

[12] C. Cork, J.-C. Madre, and L. Barnes, "Comparison of triple-patterning decomposition algorithms using aperiodic tiling patterns," in *Proc. SPIE*, vol. 7028, pp. 702839-1–702839-7, May 2008.

[13] A. Lvov, G. Tellez, and G.-J. Nam, "On coloring and colorability analysis of integrated circuits with triple and quadruple patterning techniques," in *Proc. ACM Int'l Symp. on Physical Design (ISPD)*, pp. 152–159, Mar. 2018.

[14] H. Tian, H. Zhang, Q. Ma, Z. Xiao, and M. D.F. Wong, "A polynomial time triple patterning algorithm for cell based row-structure layout," in *Proc. IEEE/ACM Int'l Conf. on Computer-Aided Design (ICCAD)*, pp. 57–64, Nov. 2012.

[15] H.-A. Chien, S.-Y. Han, Y.-H. Chen and T.-C. Wang, "A cell-based row-structure layout decomposer for triple patterning lithography," *Proc. ACM Int'l Symp. on Physical Design (ISPD)*, pp. 67–74, Mar. 2015.

[16] Y. Lin, B. Yu, B. Xu, D. Z. Pan, "Triple patterning aware detailed placement toward zero cross-row middle-of-line conflict", *IEEE Transactions on Computer-Aided Design of Integrated Circuits and Systems (TCAD)*, vol. 36, no. 7, pp. 1140–1152, July 2017.

[17] Y.-H. Lin, B. Yu, D. Z. Pan, and Y.-L. Li, "TRIAD: A triple patterning lithography aware detailed router," *Proc. IEEE/ACM Int'l Conf. on Computer-Aided Design (ICCAD)*, pp. 123–129, Nov. 2012.

2018 IEEE Computer Society Annual Symposium on VLSI

Guiding Template-induced Design Challenges in DSA-MP Lithography

Shao-Yun Fang and Kuo-Hao Wu

Department of Electrical Engineering, National Taiwan University of Science and Technology, Taipei 106, Taiwan
e-mail: syfang@mail.ntust.edu.tw, m10407401@mai.ntust.edu.tw,

Abstract—Directed self-assembly (DSA) has become one of the most promising next generation lithography technologies especially for contact/via layer fabrication. Guiding templates are required to generate contact/via holes at desired positions, while there is only a limited number of feasible guiding templates and thus feasible via arrangements. On the other hand, guiding templates need to be first generated with conventional optical lithography, whose limited resolution causes the need of adopting multiple patterning (MP) technologies for dense template generation. This paper demonstrates the design challenges resulted from simultaneous contact/via assignment and layout decomposition of guiding templates, and state-of-the-art works are introduced to show the recent progress on DSA-MP-related optimization problems.

I. INTRODUCTION

Directed self-assembly (DSA) has become one of the most promising next generation lithography technologies especially for contact/via layer fabrication [4], [9], which uses a special material known as block copolymer (BCP) that can form regular cylinders after an annealing process. Guiding templates are required to generate contact/via holes at desired positions, and closely positioned holes could be generated by using a multi-hole guiding template. Since the overlay accuracy of these holes highly depends on the simplicity of template shapes, only a limited number of feasible guiding templates can be used to guarantee contact/via manufacturability. Some guiding templates have been verified and considered as feasible for integrated circuit (IC) fabrication in literature, as shown in Fig. 1, which restrict vias to be arranged in either a straight line, an oblique line, or a square pattern [14]. Note that overlay accuracy dramatically decreases with the increase of template complexity, and thus each template is usually associated with a template cost and the total template cost should be minimized.

To form guiding templates before contact/via hole generation, conventional optical lithography is still required, whose limited resolution could prohibit DSA from being applicable to dense contact/via layout fabrication. To overcome the predicament, multiple patterning (MP) technologies can be adopted for dense template generation, resulting in hybrid DSA-MP lithography. Consequently, layout decomposition also become a necessary step, which assign each guiding template to one of the multiple masks such that templates on the same mask comply with the mask spacing rule.

This work was partially supported by MOST of Taiwan under Grant No. MOST 107-2636-E-011-002.

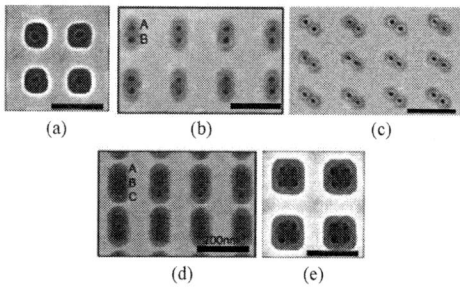

Fig. 1. Empiriacally feasible guiding templates and via arrangments in the DSA technology [14].

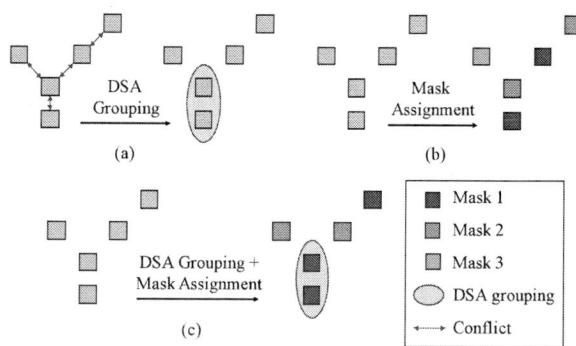

Fig. 2. Design challenges in DSA and MP [3]. (a)Via grouping in DSA. (b) Mask assignment in MP. (c) Simultaneous template and mask assignment in DSA-MP.

II. DSA-MP-INDUCED DESIGN CHALLENGES

In this section, three DSA-MP-related optimization problems and starte-of-the-art works are separately introduced in the following subsections.

A. Concurrent Guiding Template and Mask Assignment

With the multi-hole guiding templates, DSA allows close vias within certain via pitches to be generated using a single template, resulting in the problem of via grouping (guiding template assignment). As illustrated in Fig. 2(a), for the given five vias, two of them forming a straight line pattern can be grouped and generated with a 2-hole template. On the other hand, MP requires layout decomposition to assign each via to one of the multiple masks, as shown in Fig. 2(b). Thus, for

978-1-5386-7100-9/18 $31.00 © 2018 IEEE

500

DSA-MP, simultaneous guiding template and mask assignment is required with the given number of available masks, as illustrated in Fig. 2(c).

To deal with the problem, Xiao et al. proposed an optimal and polynomial-time algorithm for row-based standard cell layout [15]. For general via layouts, Badr et al. proposed integer linear programming (ILP) formulations and matching-based heuristic algorithms without considering template costs [2], [3]. Kuang et al. proposed a graph-based algorithm flow to simultaneously minimizes unresolvable conflicts and the total template cost [8]. Yang et al. proposed an efficient algorithm flow composed of template candidate generation followed by a local search method based on a coloring solver [16]. All these studies are applicable to general MP such as double patterning (DP) and triple patterning (TP), while the ILP-based approaches could suffer from long optimization time, and heuristics could lose solution optimality. Interestingly, Guo et al. found that there always exists a conflict-free solution for the simultaneous template and mask assignment problem in DSA-TP if the via conflict graph is a diagonal grid graph (DGG), while the problem on a DGG in DSA-DP is still NP-complete [6].

B. Template Design Considering Two BCP Materials

Each BCP material has its nature pitch of holes, and the BCP material whose nature pitch is close to the wire pitch (w_p) of the target design will be chosen to optimize overlay accuracy. Thus, the above studies only regard straight-line templates having the same via pitch (Figs. 1(a), (b), and (d)) as feasible. This restriction may cause these approaches suffer from low via manufacturability due to limited types of feasible guiding templates. To maximize the flexibility of DSA-compatible pattern matching, two different BCP materials in DSA-MP can be adopted [5], [7]. An example is illustrated in Fig. 3. If only straight-line templates are used, the via layout shown in Fig. 3(a) will have at least one conflict in DSA-DP, as shown in Fig. 3(b). This conflict can be resolved by using one more mask (DSA-TP), as shown in Fig. 3(c). In contrast, if two BCPs are used and thus oblique-line templates are allowed, two masks are sufficient for this via layout, as illustrated in Fig. 3(d). Note that using multiple BCP materials constrains that the templates generated by the same mask need to have the same via pitch.

Addressing this optimization problem, Wu and Fang proposed to use the second BCP whose nature pitch is close to $\sqrt{2}w_p$, and thus oblique-line guiding templates can be adopted (Fig. 1(c)) in addition to straight-line templates [13]. Similarly, Ou et al. adopted two BCPs with different nature pitches to enable two types of straight-line templates [10]. Both of the two works show substantially better results in via manufacturability (conflict numbers) compared to those only using single BCP material.

C. DSA-MP-aware Detailed Routing

Given an arbitrary via layout, a legal template and mask assignment solution is hard to be guaranteed. DSA-MP-aware

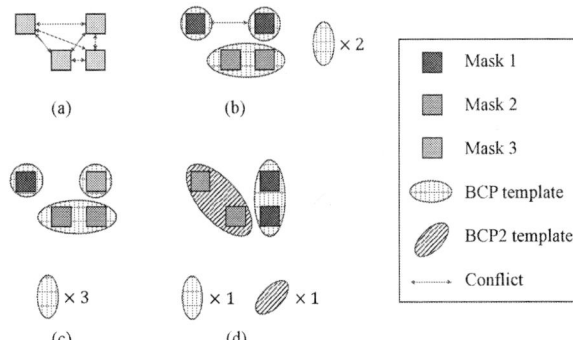

Fig. 3. Template design considering two BCPs [13]. (a) A via layout. (b) A solution derived by using DP with a single BCP material. (c) A solution derived by using TP with a single BCP material. (d) A solution derived by using DP with two BCP materials.

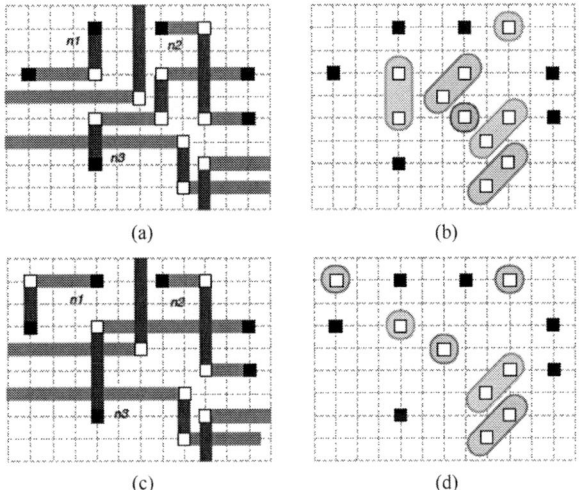

Fig. 4. DSA-MP-aware detailed routing [11]. (a)(b) A routing result requiring three masks for template generation. (c)(d) A DSA-DP-friendly routing result where only two masks are required.

design methodologies are eagerly desirable. Fig. 4(a) shows a routing result without considering DSA-MP, whose via layout requires three masks for its template generation, as shown in Fig. 4(b). In contrast, by ripping-up and re-routing Nets $n1$ and $n3$, the via layout could be successfully generated with only two masks, as the example shown in Figs. 4(c) and (d). Ou et al. proposed the first work of DSA-DP-aware detailed routing, which utilizes a routing graph model to forbid via insertion that will cause infeasible template assignment, and several net planning and post-routing optimization techniques are also developed.

III. Conclusions and Future Work

This paper presents several design challenges of contact/via layers in the hybrid DSA-MP lithography and introduces state-of-the-art works solving the optimization problems induced

978-1-5386-7100-9/18 $31.00 © 2018 IEEE

by these challenges. There are still many future research directions related to DSA-MP. For example, by investigating the two works considering two BCP materials [10], [13], using BCPs with different pitch combinations result in significant different template and mask assignment solutions. Finding the most suitable BCP combination for a target design, or designing circuit layouts considering one specific BCP combination could be both challenging. In addition, design methodologies such as placement and routing algorithms applicable to general DSA-MP are also desirable to realize DSA-MP-compliant design layouts. Finally, almost all existing works do not consider the adoption of 4-hole squared templates (Fig. 1(e)), and existing algorithms are unable to tackle this template type with trivial extension, causing another design challenge in the future study.

REFERENCES

[1] Y. Badr, A. Torres, Y. Ma, J. Mitra, and P. Gupta, "Incorporating DSA in multi-patterning semiconductor manufacturing technologies," *Proc. SPIE*, vol. 9427, pp. 94270P, 2015.

[2] Y. Badr, A. Torres, and P. Gupta, "Mask assignment and synthesis of DSA-MP hybrid lithography for sub-7nm contacts/vias," *Proc. DAC*, 2015.

[3] Y. Badr, A. Torres, and P. Gupta, "Mask assignment and DSA grouping for DSA-MP hybrid lithography for sub-7 nm contact/via holes," *IEEE TCAD*, vol. 36, no. 6, pp. 913–926, 2017.

[4] Y. Du, D. Guo, M. D. F. Wong, H. Yi, H.-S. P. Wong, H. Zhang, and Q. Ma, "Block copolymer directed self-assembly (DSA) aware contact layer optimization for 10 nm 1D standard cell library," *Proc. ICCAD*, 2013.

[5] R. Gronheid, J. Doise, J. Bekaert, B. T. Chan, I. Karageorgos, J. Ryckaert, G. Vandenberghe, Y. Cao, G. Lin, M. Somervel, G. Fenger, and D. Fuchimoto, "Implementation of templated DSA for via layer patterning at the 7nm node," *Proc. SPIE*, vol. 9423, pp. 942305, 2015.

[6] D. Guo, H. Zhang, and M. D. F. Wong, "On coloring rectangular and diagonal grid graphs for multiple patterning lithography," *Proc. ASPDAC*, 2018.

[7] I. Karageorgos, J. Ryckaert, M. C. Tung, H.-S. P. Wong, R. Gronheid, J. Bekaert, E. Karageorgos, K. Croes, G. Vandenberghe, M. Stucchi, and W. Dehaene, "Design strategy for integrating DSA via patterning in sub-7nm interconnects," *Proc. SPIE*, vol. 9781, pp. 97810N, 2016.

[8] J. Kuang, J. Ye, and E. F. Y. Young, "Simultaneous template optimization and mask assignment for DSA with multiple patterning," *Proc. ASPDAC*, 2016.

[9] Y. Ma, J. A. Torres, G. Fenger, Y. Granik, J. Ryckaert, G. Vanderberghe, J. Bekaert, and J. Word, "Challenges and opportunities in applying grapho-epitaxy DSA lithography to metal cut and contact/via applications," *Proc. EMLC*, 2014.

[10] J. Ou, X. Xu, B. Cline, G. Yeric, and D. Z. Pan, "DTCO for DSA-MP hybrid lithography with double-BCP materials in Sub-7nm node," *Proc. ICCD*, 2017.

[11] J. Ou, B. Yu, X. Xu, J. Mitra, Y. Lin, and D. Z. Pan, "DSAR: DSA aware routing with simultaneous DSA guiding pattern and double patterning assignment," *Proc. ISPD*, 2017.

[12] R. Ruiz, H. Kang, F. A. Detcheverry, E. Dobisz, D. S. Kercher, T R. Albrecht, J J. d Pablo, and P. F. Nealey, "Density multiplication and improved lithography by directed block copolymer assembly," *Proc. Science*, vol. 321, no. 5891, pp. 936-939, 2008.

[13] K.-H. Wu and S.-Y. Fang, "Simultaneous template assignment and layout decomposition using multiple BCP materials in DSA-MP lithography," *Proc. ICCAD*, 2017.

[14] Z. Xiao, Y. Du, H. Tian, M. D. F. Wong, H. Yi, and H.-S. P. Wong, "DSA template optimization for contact layer in 1D standard cell design," *Proc. SPIE*, vol. 9049, pp. 904920, 2014.

[15] Z. Xiao, C.-X. Lin, M. D. F. Wong, and H. Zhang, "Contact layer decomposition to enable dsa with multi-patterning technique for standard cell based layout," *Proc. ASPDAC*, 2016.

[16] Y. Yang, W.-S. Luk, H. Zhou, D. Z. Pan, D. Zhou, C. Yan, and X. Zeng, "An effective layout decomposition method for DSA with multiple patterning in contact-hole generation," *ACM TODAES*, vol. 23, no. 1, pp. 11:1–11:27, 2017.

[17] Gurobi Optimizer. https://www.gurobi.com

2018 IEEE Computer Society Annual Symposium on VLSI

FPAP: A Folded Architecture for Efficient Computing of Convolutional Neural Networks

Yizhi Wang, Jun Lin, and Zhongfeng Wang

School of Electronic Science and Engineering, Nanjing University, P.R. China

Email: magicwyzh@smail.nju.edu.cn, {jlin, zfwang}@nju.edu.cn

Abstract—Convolutional neural networks (CNNs) have found extensive applications in practice. However, weight/activation's sparsity and different data precision requirements across layers lead to a large amount of redundant computations. In this paper, we propose an efficient architecture for CNNs, named Folded Precision-Adjustable Processor (FPAP), to skip those unnecessary computations with ease. Computations are folded in the following two aspects to achieve efficient computing. On one hand, the dominant multiply-and-add (MAC) operations are performed bit-serially based on a bit-pair encoding algorithm so that the FPAP can adapt to different numerical precisions without using multipliers with long data width. On the other hand, a 1-D convolution is undertaken by a multi-tap transposed finite impulse response (FIR) filter, which is folded into one tap so that computations involving zero activations and weights can be easily skipped. Equipped with the precision-adjustable MAC unit and the folded FIR filter structure, a well-designed array architecture, consisting of many identical processing elements is developed, which is scalable for different throughput requirements and highly flexible for different numerical precisions. Besides, a novel genetic algorithm based kernel reallocation scheme is introduced to mitigate the load imbalance issue. Our synthesis results demonstrate that the proposed FPAP can significantly reduce the logic complexity and the critical path over the corresponding unfolded design, which only delivers slightly higher throughput when processing sparse and compact models. Our experiments also show that FPAP can scale its energy efficiency from 1.01TOP/s/W to 6.26TOP/s/W under 90nm CMOS technology when different data precisions are used.

I. INTRODUCTION

Convolutional neural networks (CNNs) [1] have achieved great success in many fields. Recent researches have unfolded a large amount of redundancy in computations of CNNs. On one hand, compact data types can be exploited. For example, data precision of less than 12 bits are sufficient to maintain the final classification accuracy [2], and even binary or ternary weights are possible to be used [3]–[6]. Besides, using different precisions to represent activations across layers is proven to be more efficient [7], [8]. Table I shows the per-layer precision requirements of weight/activation for the convolutional layers (CVLs) and fully-connected layers (FCLs) in state-of-the-art CNNs on the ImageNet dataset based on different training methods. On the other hand, sparse weights are able to be obtained without compromising accuracy [9]. Deep compression scheme [9] can prune about 60% and 90% weights in convolutional layers (CVLs) and fully-connected layers (FCLs), respectively. The popular non-linear rectified linear unit (ReLU) layer leads to about 50% zero activations.

TABLE I: Per-Layer precision variability and relative accuracy

CNN models	Data Type	Precision in CVLs	Precision in FCLs	Relative Accuracy
AlexNet [8]	Act.	9-8-5-5-7	10-9-9	100%
AlexNet [5]	Weight	32-2-2-2-2	2-2-32	99%
GoogleNet [3]	Weight	1 for all	1 for all	96%
VGG-16 [7]	Act.	8-5-6-5-5-5-5-5-5-6-5-5	6-3-2	97%
VGG-19 [8]	Act.	12-12-12-11-12-10-11-11-13-12-13 for rest	10-9-9	100%

Existing hardware architectures for CNNs made efforts to manipulate the sparsity of CNNs [10]–[14]. However, most works only take either weight or activation sparsity into consideration. Architectures in [12] and [13] are aware of both zero-valued weight and activations. However, complex indexing methods are required to correctly place the results back to the memory after skipping unnecessary computations. Besides, computation components of all those architectures use identical data widths while processing different layers. This results in low efficiency when they are applied to unmatched compact models, such as non-uniformly quantized CNNs and binary/ternary weight CNNs.

Some researchers try to exploit the variability of different quantization precisions across layers in CNNs. Some architectures divides large bit-parallel multipliers into small multipliers to support different quantization precision requirements. Stripes (STR) [8] is a bit-serial architecture, which decomposes inner product computations into many shift-and-add operations to adapt to different quantization precision requirements in each layer. However, since the inner product is the STR's fundamental operation, it is hard for the STR to deal with sparse weights and activations.

Even though many works have been done to improve the efficiency of CNNs, none of them have jointly taken the computational redundancies in both sparsity and precision variability into consideration. In this paper, we propose to fold computations in CNNs to eliminate all computational redundancies mentioned above. To achieve this goal, an efficient hardware architecture, namely Folded Precision-Adjustable Processor (FPAP), is proposed. In FPAP, the dominant multiply-and-add (MAC) operation is decomposed into multiple adds and then folded into one adder based on a bit-pair encoding algorithm. Thus, MAC operations can be performed efficiently with only

978-1-5386-7100-9/18 $31.00 © 2018 IEEE

503

the required precision. Besides, all taps of a finite impulse response (FIR) filter, which can undertake the convolution process of CNNs, is folded into one tap to be executed serially. Thus, computations involving zero-valued weights/activations can be dynamically skipped without complex indexing. E-quipped with the folded MACs and folded FIR filters, a well-designed array architecture consisting of many identical processing elements is developed, which is scalable for different throughput requirements. Besides, a novel genetic algorithm based kernel reallocation scheme is introduced, which can mitigate the load-imbalance issue brought by irregular weight sparsity.

The rest of this paper is organized as follows. Section II introduces the backgrounds of CNNs. In Section III, we present the folding scheme for computations in CNNs and the corresponding micro-architectures of folded computational units. Based on the folded units, the top-level architecture, the corresponding computational flow, and the kernel reallocation scheme are proposed in Section IV. In Section V, FPAP is evaluated in several aspects. Finally, Section VI concludes this paper.

II. BACKGROUNDS

A convolutional neural network (CNN) model is mainly composed of convolutional, non-linear, pooling and fully-connected layers, among which convolutional layers and fully-connected layers have trainable weights. This paper mainly focuses on the most computationally intensive convolutional layers.

Let's denote the input of the ℓ-th convolutional layer as $\mathbf{x}^{(\ell)}$. Each convolutional layer generates outputs with multiple channels. Besides, each channel is often referred to as a feature map, and a pixel in a feature map is also called an activation. The convolution operation computes:

$$y_o^{(\ell)}(j,i) = b_o^{(\ell)} + \sum_{c \in C_{in}^{(\ell)}} \sum_{(m,t) \in S^{(\ell)}} w_{o,c}^{(\ell)}(m,t)x_c^{(\ell)}(j+m,i+t),$$

(1)

where o and c are the indices of an output and an input channel, respectively. y_o denotes an output activation of channel o, and x_c is an input activation of channel c. y_o is calculated within a small region $S^{(\ell)}$ of size $k \times k$ (e.g., 3×3) by convolution weight w. During convolution, $S^{(\ell)}$ slides across the whole feature map. It is common to refer to the set of weights with a size of $k \times k$ as a kernel, that is convolved with the input. For each output channel, there can be an optional bias term b_o added to each neuron. It can be seen that MAC is the most commonly used operation in the convolutional layers.

III. FOLDING COMPUTATIONS IN CNNS

A. Precision-Adjustable MAC Based on Bit-Pair Encoding

The multiplication is the most complicated part of a MAC operation. By decomposing a multiplication into shift-and-add operations and folding them into one adder, it is able to adapt to the required precision when the data width varies across layers. In FPAP the bit-pair encoding (BPE) multiplication

TABLE II: Partial product of the bit-pair encoding algorithm

$BPEB_i$	PP_i	$BPEB_i$	PP_i	$BPEB_i$	PP_i	$BPEB_i$	PP_i
3'b000	0A	3'b001	1A	3'b010	1A	3'b011	2A
3'b100	-2A	3'b101	-1A	3'b110	-1A	3'b111	-0A

algorithm [15] is used to reduce the number of additions in a MAC by half.

The 2's complement representation of an n-bit integer is

$$B = -2^{n-1}B_{n-1} + 2^{n-2}B_{n-2} + \ldots + B_0 + B_{-1}, \quad (2)$$

where $B_{-1} = 0$ is an auxiliary term. It is reformulated as

$$B = 2^{n-2}(-2B_{n-1} + B_{n-2} + B_{n-3}) + 2^{n-4}(-2B_{n-3} + B_{n-4} + B_{n-5}) + \ldots + (-2B_1 + B_0 + B_{-1}), \quad (3)$$

where three bits of B are grouped together and adjacent groups have one common bit. The i-th group (from LSB to MSB), denoted as $BPEB_i$, has its encoded value $V_i \in \{-2, -1, 0, 1, 2\}$. According to Eq. (3), $AB = AV_0 + 2^2 AV_1 + \ldots + 2^{n-2} AV_{\lceil \frac{n}{2} \rceil}$. Thus, the multiplication can be executed bit-serially with $\lceil \frac{n}{2} \rceil$ additions, while the straightforward way in the STR requires n additions. Table II shows all possibilities of BPEB and the corresponding partial products (PP).

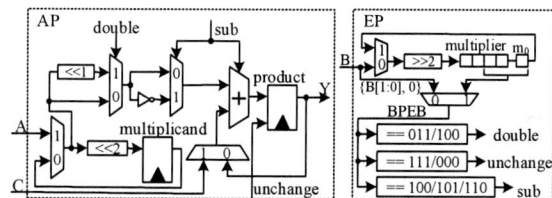

Fig. 1: The proposed precision-adjustable MAC unit.

Based on the above BPE algorithm [15], a precision-adjustable MAC (PAMAC) unit is proposed in Fig. 1, which performs $Y = AB + C$ in serial. The PAMAC consists of an encoding path (EP) and an arithmetic path (AP). In the i-th clock cycle, the EP encodes the i-th group in Eq. (3) to $BPEB_i$, and generates "double", "sub", and "unchange" signals according to the corresponding PP_i in Table II to control the AP, so that the AP can generate and accumulate PPs. All add operations are folded into one adder, and the shift-and-add sequence is scheduled according to BPEBs.

An example for scheduling of the PAMAC when $B = 0111\_1110_2$ is shown in Table III. Note that two bits of B are shifted into the EP during each clock cycle, while all bits of A and C are fed to the AP in parallel.

B. Exploiting Sparsity by Folded FIR Filter

1-D convolution can be undertaken using an FIR filter. A transposed-form 5-tap FIR (TrFIR) filter is shown in Fig. 2. Here, an MAC unit in an FIR filter is referred to as a tap. To support different kernel sizes, an FIR filter with multiple taps are required. However, it is possible that more than 60% weights are zero [5] [9] in a sparse model. Thus, those zero-coefficient taps in the FIR filter are not utilized efficiently.

TABLE III: Scheduling of the PAMAC unit

# clock cycle	0	1	2	3
BPEB	100	111	111	011
double	1	0	0	1
sub	1	0	0	0
unchange	0	1	1	0
product_D[1]	-2A+C	x	x	$2(2^6A)$-2A+C
product_Q[1]	x	-2A+C	-2A+C	-2A+C
multiplicand	x	2^2A	2^4A	2^6A
multiplier	x	00011111	00000111	00000001
m_0	x	1	1	1

[1] The product_D and product_Q denote the input port and output port of the product register, respectively. "x" means "don't care".

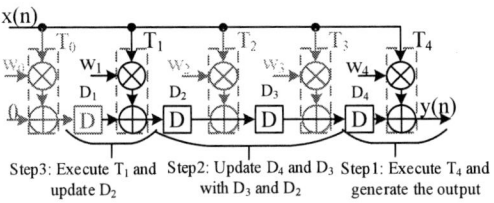

Fig. 2: A transposed 5-tap FIR filter and the fold-and-skip computation flow. Taps in gray are skipped.

To improve efficiency, the FIR filter is folded into one tap, which computes all taps of the original FIR filter. By serially mapping each tap of the original FIR filter to the folded one, and executing them one at a time while skipping zero-coefficient taps, it is able to compute more efficiently. The skipping is possible because once all weights have been loaded into the FIR taps, they are fixed for a period of time. Thus, the mapping of non-zero taps is static during convolution.

When the input activation is zero, all taps are idle, so the correct result is in D_4 of Fig. 2. The filter just needs to shift all delay elements directly without any computation.

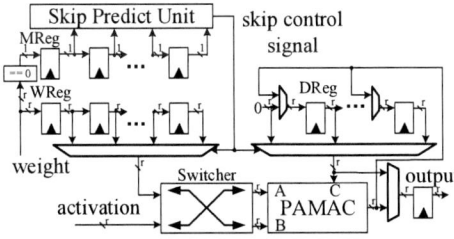

Fig. 3: The architecture of the folded FIR filter.

The proposed folded FIR filter (FoFIR) is shown in Fig. 3. The switcher can interchange its two inputs and feed them to the PAMAC. Thus, the precision adjustability of the system can be more flexible to adapt to the precision variability of either activations or weights. The delay registers (DRegs) correspond to $D_0, D_1, \ldots, D_4$ in Fig. 2(b). They can be updated with the output of the PAMAC or the data in adjacent DRegs. Before convolution starts, weights are serially shifted into weight registers (WRegs). Meanwhile, the bit mask indicating locations of zero weights is also generated and stored in the

mask registers (MRegs). The number of MRegs, WRegs, or DRegs, denoted as k, is the same as the number of taps in the TrFIR. Thus, the FoFIR can support kernel sizes smaller than $k \times k$ in 1-D direction. The skip predict unit schedules the skipping of zero-weight taps according to the MRegs.

Assume w_0, w_2, w_3 in Fig. 2 are zero. During convolution, when the input activation is non-zero, the 3-step fold-and-skip procedure is clearly shown in Fig. 2. Only T_4 and T_1 are folded to the PAMAC, other taps are skipped. The FoFIR maps those non-zero taps to the PAMAC from right to left, and update the DRegs in the same order. When the input activation is zero, no computation is required and all DRegs are just updated by their left adjacent ones. The output register in the FoFIR can obtain the result from $DReg_4$ directly. Thus, no more cycles are required when the input activation is zero.

IV. SCALABLE ARRAY ARCHITECTURE

A. Processing Element

In FPAP, feature maps in CVLs are convolved by processing elements (PEs) in a row-by-row way. Fig. 4(a) shows the micro-architecture of a PE. The FoFIR is the main data path for a PE to carry out 1-D convolution of input rows. The activation queue (ActQ), the weight queue (WeightQ) and the accumulation queue (AccQ) are FIFOs. The ActQ feeds activations to the FoFIR and is double-buffered so that it can perform convolution using one of the ActQ while receiving activations from other PEs/global buffer. Weights in the WeightQ are used in a round-robin way because they are shared by all rows of a feature maps. Each row of a kernel is sent to the FoFIR from the WeightQ before a 1-D convolution starts. The AccQ together with the adder can accumulate convolution results of different rows of an input feature map generated by the FoFIR. For a $k \times k$ kernel, after the FoFIR has convolved k rows of input feature maps, an output row have already been obtained and stored in the AccQ, and can be read out by global buffer. For each row of an input feature map that fed to a PE, the activation sequence is encoded in a run-length based compression approach. For example, the following sequence [0, 0, 0, 8, 0, 0, 22, 39] is encoded as (3, 8), (2, 22), (0, 39). Thus, a zero creator unit is required to decode the compressed activation sequence.

B. Top Architecture

The top-level architecture is shown in Fig. 4(b), which is a scalable PE array. Two input buffers (IB_1 and IB_2) are composed of multiple memory banks. Assume the number of rows and columns of the PE array is M and N, respectively. IB_1 and IB_2 have M and N memory banks, respectively. The accumulate-and-concatenate unit (ACCU) adds the output of the PE array with bias terms or previously obtained partial sums and then stores them in the output buffer (OB).

When FPAP is executing a CVL, IB_1 stores the compressed activations. Each bank of IB_1 saves one row of activations in the input feature map. Weights are buffered in IB_2. Each PE can perform 1-D convolution, and PEs in the same row of the PE array convolve the same rows of an input feature map

978-1-5386-7100-9/18 $31.00 © 2018 IEEE

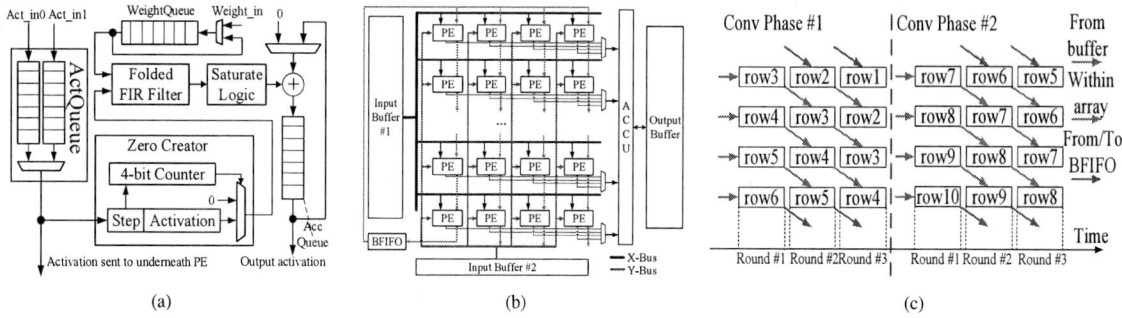

Fig. 4: (a) Micro-architecture of the proposed processing element. (b) The top-level array architecture of the proposed design. (c) Computation flow for 3×3 kernels in a CVL.

broadcast by IB_1 through X-Bus. The IB_2 transfers different weights to different columns of the PE array through Y-Bus. PEs in the same column of the PE array share the same weights (kernels), and they use these weights to convolve different rows of an input feature map. Besides, inter-PE communication is denoted as intermediate diagonal arrows in Fig. 4(b). Each PE can transfer the activations to its underneath PE once the activations are read out from its ActQ. The buffering FIFO (BFIFO) is for the inter-PE activation communications.

The computation flow and the activation communication scheme for a CVL with 3×3 kernels are shown in Fig. 4(c), where an array of $M = 4$ rows is illustrated. Since all PE columns in Fig. 4(b) just share the same flow but use different weights (kernels), only one column of the PE array is shown in of Fig. 4(c), where PEs are represented with rectangles. The horizontal axis in Fig. 4(c) is the time axis. The activation communication within the PE array is shown as the diagonal arrows.

As shown in Fig. 4(c), in each phase, the PE array generates the convolved results of M rows, and each phase contains k rounds, where $k \times k$ is the kernel size. Note that each phase will repeat for n_{in} times to reuse the partial sums in AccQueue of a PE locally and reduce the number of time accessing global output buffer, where n_{in} is the number of input channels being processed continuously. With inter-PE communication, activations mostly flow within the array. In the second and third rounds of a phase, PEs in the first row of the PE array use the activations received from the BFIFO while other PEs use activations received from their upper PEs. The activations in the BFIFO are collected from a PE in the last row of the PE array in the first and second rounds of each phase. Remember that the FoFIR in the PE can support arbitrary kernel sizes in 1-D direction. When this property is combined with the proposed computation flow, the PE array is able to support different kernel sizes.

C. Mitigate the Load Imbalance Issue

The direct kernel allocation, in which kernels for the i-th output channel is allocated to the $(i \mod N)$-th PE column, will lead to serious load-imbalance issue in the PE array due to irregular weight sparsity of CNNs. A zero-aware kernel reallocation scheme has been proposed in [11] to ease this issue by sorting kernels according to the number of zero weights. However, it is not suitable for FPAP because FPAP continuously convolves n_{in} input channels inside a PE. The number of zero weights in kernels convolving different input channels are not the same. This paper proposes a new kernel reallocation scheme for FPAP to mitigate the imbalance issue.

For a certain layer, assume a permutation of output channels is P. Let's denote the number of non-zero weights in a kernel $K(c, o)$ as $NZ_w(c, o)$, where o and c are the indices of output and input channels, respectively. The non-zero-term-count (NZTC) matrix G of size $n_{in} \times C_{out}$ has elements $G(c, o) = NZ_w(c, o)$, where C_{out} is the number of output channels. A permuted NZTC matrix T_P is obtained by permuting the columns of G according to P. Assume the number of PE columns and the number of input channel a PE will process continuously are N and n_{in}, respectively. For each group with n_{in} input channels, we can solve the following optimization problem to find a good permutation:

$$\min_P C = \sum_{k=0}^{\frac{C_{out}}{N}-1} \sum_{r=0}^{n_{in}-1} U(r, k),$$
$$U(r, k) = \max\{T_P(r, Nk), \ldots, T_P(r, Nk + N - 1)\}.$$

The value of C is considered as an estimation of the required cycles. However, finding the global optima of this problem is non-trivial and may consume very long time. Instead, we propose to use genetic algorithms [16] to find a suboptimal permutation (allocation scheme), which is still much better than the direct kernel allocation. The searching of allocation schemes is done off-line. After a good kernel reallocation scheme is found, it can be saved in an on-chip kernel-allocation memory during deployment stage.

V. EVALUATION

The FPAP with array size of 16×16 is coded with RTL and synthesized under the TSMC 90nm technology to evaluate its area and critical path. Both of the depth of ActQ and AccQ in PEs are 16. The IB_1, IB_2, and OB are set to 1KB, 2KB, and

32KB, respectively. A sparse VGG-16 model fed with images from ImageNet dataset is used as a benchmark. The circuit switching activities are captured for power estimation.

A. Synthesis Results

Synthesis results are shown in Table IV. Due to the folding feature of FPAP, the critical path delay is reduced to only 1.09ns, so FPAP can work on a clock frequency of 917MHz. The unfolded version of FPAP (UFPAP), in which FoFIRs are replaced by bit-parallel unfolded 11-tap FIR filters, is also implemented and synthesized under the same constraints for comparison. The UFPAP occupies 21.29 mm^2, while the area of FPAP is only 11.57mm^2. If the area of on-chip SRAMs is excluded, the area of FPAP is only 23.4% of that of the UFPAP. Besides, the maximum frequency of UFPAP is only 574MHz, while FPAP can work on substantially higher frequency. The performance of FPAP depends on the sparsity and the data quantization precision. If the sparsity is not considered, FPAP's peak performance ranges from 29.34GOP/s (16b) to 234.8GOP/s (2b).

B. Cycle Estimation

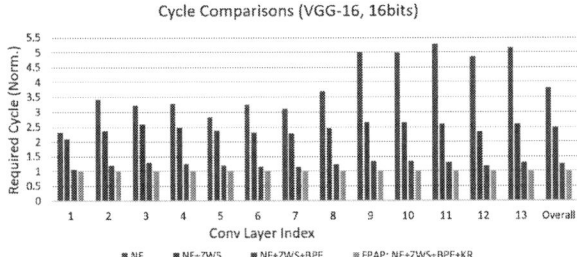

Fig. 5: Speed up of different optimization schemes.

Fig. 5 shows the normalized required cycles of different folding architectures on the benchmark model. The weights and activations are both quantized with 16-bit weights and activations. The Naive Folding (NF) architecture, which folds MACs with direct shift-and-adds and folds FIR filters without skipping any unnecessary computations, is used as a baseline. One can see that when exploiting zero-weight skipping (ZWS) by FoFIRs, the overall performance can be significantly improved by about 33%. After using the BPE algorithm, the performance becomes about two times better. Finally, the kernel reallocation scheme proposed in Section IV-C can further bring in about 1.25 times of speed up.

By folding computations, FPAP provides great flexibility. The price is that computations in FPAP takes more clock cycles. However, it is not serious when the network is sparse and compact since FPAP can work on a high clock speed and the real computations are reduced. Fig. 6 shows the normalized time comparison between FPAP and UFPAP on a pruned VGG-16 model quantized as shown in Table I. The performance of FPAP is even higher than UFPAP when processing the last several layers because these layers are very sparse and only 5-bit activations are used. Overall, the required

time of FPAP is only about 1.3× of that of UFPAP, while FPAP occupies only half of the area of UFPAP. Moreover, the slightly decreased performance can also be made up by adding more PEs using the saved area. Besides, it should be noted that when applied to an extremely compact model, such as a sparse ternary weight CNN, the required time of FPAP will be significantly less than UFPAP.

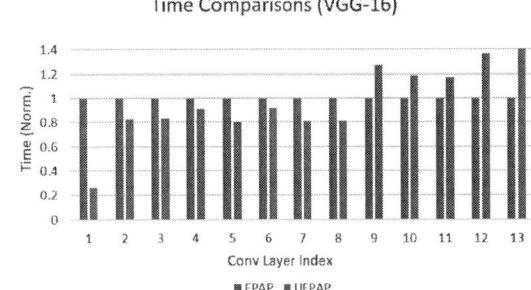

Fig. 6: Time comparison between FPAP and UFPAP.

C. Performance and Energy Efficiency

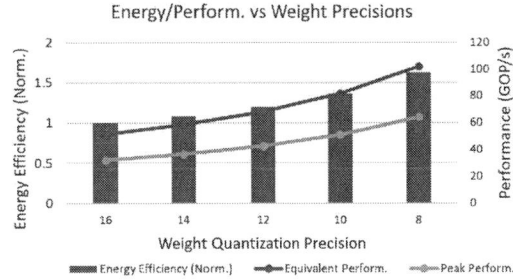

Fig. 7: Performance and normalized energy efficiency with different weight quantization precisions on VGG-16.

Since FPAP skip all redundant computations, it is more reasonable to evaluate its performance by a metric called equivalent performance given by:

$$\Theta_{equivalent} = \zeta \times R \quad (OP/s), \qquad (4)$$

where ζ and R are the number of required MACs in original CNN and the frame rate per second (fps) of FPAP, respectively.

Fig. 7 shows the performance and energy efficiency of FPAP on VGG-16 with different weight precisions. FPAP's peak performance is relatively low. However, by eliminating redundancies, FPAP can achieve much higher equivalent performance than peak performance. Our preliminary power analysis results based on the post-synthesis implementation show that the power consumption of FPAP is no more than 66mW on VGG-16 under 90nm CMOS technology. As the weight precision decreases from 16 bits to 8bits, FPAP can scale up its equivalent performance and energy efficiency by 1.98 and 1.64 times, respectively. Meanwhile, the classification

TABLE IV: Synthesis results of FPAP and comparisons

Metrics/Arch.	FPAP	STR [8]	Kim's [11]	Eyeriss [12]
Tech.(nm)	90	65	65	65
Max Freq.(MHz)	917	980	200	250
Area (mm^2)	11.57	122.1	1.66	12.25
Precision Variability*	2b-16b	1b-16b	LogQuant	16b-fix
Array Size	16×16	16×16	8×45	14×12
Zero Skipped	W+A	-	W+A	A
Memory Type	SRAM	eDRAM	SRAM	SRAM
Benchmark	VGG-16	VGG-19	AlexNet	AlexNet
Energy Efficiency (GOP/s/W)	1013(16b) - 6258(2b)	1456† (10-13b)	401.36††	122.8

† The STR [8] is designed based on DaDianNao [17]. The authors did not provid exact energy efficiency number but claims that the improvement is 2.62× over DaDianNao when activations are quantized according to Table I. So we calculate STR's energy efficiency based on [17] with its *peak* performance while excluding the power of eDRAM. Note that it is an over-optimistic estimation.

†† The authors of [11] only provide the the relative improvements of this architecture over Eyeriss [12], which is 3.27 times.

accuracy loss is less than 0.1%. If the model can be trained properly that it has ternary weights, the performance of FPAP can increase to 393.93GOP/s, and the energy efficiency is further improved to 6.18 times of that of the 16-bit weight model (not shown in Fig. 7).

D. Comparisons with Related Works

The comparisons are shown in Table IV. Eyeriss deals with activation sparsity by using clock gating technique, but zero activations still consume unnecessary clock cycles and memory accesses. Thus, its energy efficiency is only 122.8 GOP/s/W. The architecture proposed in [11] exploits sparsity of both weights and activations, and uses logarithmic quantization, yielding an energy efficiency of 401.36 GOP/s/W. STR is a bit-serial architecture that can adapt to the per-layer precision variability of CNNs. With reduced activation precision, it can achieve a highest energy efficiency of about 1.46TOP/s/W with activations mostly quantized to 11 or 12 bits. Note that this number is over-optimistic. However, STR only takes the precision variability of activations into consideration, and uses straightforward inner product (IP) decomposition scheme. Thus, it can not deal with some compact weight models like ternary/binary CNNs, which have shown promising results in many applications. Besides, taking IPs as its basic operations prohibits STR from skipping unnecessary computations caused by zeros among an IP. By contrast, FPAP takes MAC as its fundamental operation, so it can dynamically skip zero weight/activations. The BPE algorithm is adopted, so the required number of adds in each MAC is reduced by half compared with STR. Moreover, the FPAP is more flexible that it can adapt to the precision variability of either weights or activations. Our preliminary power estimation results shows that FPAP can scale its energy efficiency from 1.01TOP/s/W (16b) to 6.26TOP/s/W (2b), which is comparable with the over-optimistic results of STR even if 90nm technology is used. Thus, when 65nm technology is considered, FPAP is surely much more (likely > 32%) energy efficient than STR.

VI. CONCLUSION

A folded architecture for efficient computing of CNNs, named FPAP, is proposed. The redundancy in computations of CNNs is eliminated by folding computations into single arithmetic unit, so that FPAP can dynamically skip unnecessary computations brought by both precision variability across layers and weight/activation sparsity. A highly scalable array architecture together with its corresponding computation flow are also presented to execute CNNs with good flexibility. Besides, a novel genetic algorithm based kernel reallocation scheme is introduced to mitigate the load imbalance issue. Implementation results demonstrate that FPAP is not only quite flexible but also much more energy efficient than prior arts.

REFERENCES

[1] A. Krizhevsky, I. Sutskever, and G. E. Hinton, "Imagenet classification with deep convolutional neural networks," in *Advances in neural information processing systems*, 2012, pp. 1097–1105.

[2] D. Lin, S. Talathi, and S. Annapureddy, "Fixed point quantization of deep convolutional networks," in *International Conference on Machine Learning*, 2016, pp. 2849–2858.

[3] M. Rastegari, V. Ordonez, J. Redmon, and A. Farhadi, "Xnor-net: Imagenet classification using binary convolutional neural networks," in *European Conference on Computer Vision*, 2016, pp. 525–542.

[4] M. Courbariaux, Y. Bengio, and J.-P. David, "Binaryconnect: training deep neural networks with binary weights during propagations," *neural information processing systems*, 2015.

[5] C. Zhu, S. Han, H. Mao, and W. J. Dally, "Trained ternary quantization," *arXiv preprint arXiv:1612.01064*, 2016.

[6] F. Li, B. Zhang, and B. Liu, "Ternary weight networks," *arXiv preprint arXiv:1605.04711*, 2016.

[7] F. Sun, J. Lin, and Z. Wang, "Intra-layer nonuniform quantization for deep convolutional neural network," in *International Conference on Wireless Communications and Signal Processing (WCSP 2016)*, 2016.

[8] P. Judd, J. Albericio, T. Hetherington, T. M. Aamodt, and A. Moshovos, "Stripes: Bit-serial deep neural network computing," in *Microarchitecture (MICRO), 2016 49th Annual IEEE/ACM International Symposium on*. IEEE, 2016, pp. 1–12.

[9] S. Han, H. Mao, and W. J. Dally, "Deep compression: Compressing deep neural networks with pruning, trained quantization and huffman coding," *international conference on learning representations*, 2016.

[10] S. Zhang, Z. Du, L. Zhang, H. Lan, S. Liu, L. Li, Q. Guo, T. Chen, and Y. Chen, "Cambricon-x: An accelerator for sparse neural networks," in *Microarchitecture (MICRO), 2016 49th Annual IEEE/ACM International Symposium on*. IEEE, 2016, pp. 1–12.

[11] D. Kim, J. Ahn, and S. Yoo, "A novel zero weight/activation-aware hardware architecture of convolutional neural network," in *2017 Design, Automation & Test in Europe Conference & Exhibition (DATE)*. IEEE, 2017, pp. 1462–1467.

[12] Y. H. Chen, T. Krishna, J. Emer, and V. Sze, "14.5 eyeriss: An energy-efficient reconfigurable accelerator for deep convolutional neural networks," in *IEEE International Solid State Circuits Conference*, 2016.

[13] A. Parashar, M. Rhu, A. Mukkara, A. Puglielli, R. Venkatesan, B. Khailany, J. S. Emer, S. W. Keckler, and W. J. Dally, "Scnn: An accelerator for compressed-sparse convolutional neural networks," in *Proceedings of the 44th Annual International Symposium on Computer Architecture*, 2017.

[14] J. Albericio, P. Judd, T. H. Hetherington, T. M. Aamodt, N. D. E. Jerger, and A. Moshovos, "Cnvlutin: ineffectual-neuron-free deep neural network computing," in *2016 ACM/IEEE 43rd Annual International Symposium on Computer Architecture (ISCA)*, 2016.

[15] V. P. Heuring and H. F. Jordan, *Computer Systems Design and Architecture*. Boston, MA, USA: Addison-Wesley Longman Publishing Co., Inc., 1996.

[16] D. Whitley, "A genetic algorithm tutorial," *Statistics and Computing*, vol. 4, no. 2, 1994.

[17] Y. Chen, T. Luo, S. Liu, S. Zhang, L. He, J. Wang, L. Li, T. Chen, Z. Xu, N. Sun *et al.*, "Dadiannao: A machine-learning supercomputer," *international symposium on microarchitecture*, pp. 609–622, 2014.

2018 IEEE Computer Society Annual Symposium on VLSI

Hyperdrive: A Systolically Scalable Binary-Weight CNN Inference Engine for mW IoT End-Nodes

Renzo Andri*, Lukas Cavigelli*, Davide Rossi[†], Luca Benini*[†]
*Integrated Systems Laboratory, ETH Zurich, Zurich, Switzerland
[†]DEI, University of Bologna, Bologna, Italy

Abstract—Deep neural networks have achieved impressive results in computer vision and machine learning. Unfortunately, state-of-the-art networks are extremely compute- and memory-intensive which makes them unsuitable for mW-devices such as IoT end-nodes. Aggressive quantization of these networks dramatically reduces the computation and memory footprint. Binary-weight neural networks (BWNs) follow this trend, pushing weight quantization to the limit. Hardware accelerators for BWNs presented up to now have focused on core efficiency, disregarding I/O bandwidth and system-level efficiency that are crucial for deployment of accelerators in ultra-low power devices. We present Hyperdrive: a BWN accelerator dramatically reducing the I/O bandwidth exploiting a novel binary-weight streaming approach, and capable of handling high-resolution images by virtue of its systolic-scalable architecture. We achieve a 5.9 TOp/s/W system-level efficiency (i.e. including I/Os)—2.2x higher than state-of-the-art BNN accelerators, even if our core uses resource-intensive FP16 arithmetic for increased robustness.

Index Terms—Hardware Accelerator, Binary Weights Neural Networks, IoT

I. INTRODUCTION

Over the last few years, deep neural networks (DNNs) have revolutionized computer vision and data analytics. Particularly in computer vision, they have become the leading approach for the majority of tasks with rapidly growing data set sizes and problem complexity, achieving beyond-human accuracy in tasks like image classification. What started with image recognition for handwritten digits has moved to data sets with millions of images and 1000s of classes [1, 2]. What used to be image recognition on small images [3, 4] has evolved to object segmentation and detection [5–8] in high-resolution frames—and the next step, video analysis, is already starting to gain traction [9–11]. Many applications from automated surveillance to personalized interactive advertising and augmented reality have real-time constraints, such that the required computation can only be run on powerful GPU servers and data center accelerators such as Google's TPUs [12].

At the same time, we observe the trend towards "internet of things" (IoT), where connected sensor nodes are becoming ubiquitous in our lives in the form of fitness trackers, smart phones, surveillance cameras [13, 14]. This creates a data deluge that is never analyzed and raises privacy concerns when collected at a central site [15]. Gathering all this data is largely infeasible as the cost of communication is very high in terms of network infrastructure, but also reliability, latency and ultimately available energy in mobile devices [16]. The centralized analysis in the cloud also does not solve the compute

problem, it merely shifts it around, and service providers might not be willing to carry the processing cost while customers do not want to share their privacy-sensitive data [17].

A viable approach to address these issues is edge computing—analyzing the vast amount of data close to the sensor and transmitting only condensed highly informative data [13, 18]. This information is often many orders of magnitude smaller in size, e.g. a class ID instead of an image, or even only an alert every few days instead of a continuous video stream. However, this implies that the data analysis has to fit within the power constraints of IoT nodes—often small-form factor devices with batteries of a limited capacity, or even devices deployed using a set-and-forget strategy with on-board energy harvesting (solar, thermal, kinetic, ...) [19].

Recently, several methods to train neural networks to withstand extreme quantization have been proposed, yielding the notions of binary and ternary weight networks (BWNs, TWNs) and binarized neural networks (BNNs) [20–22]. BWNs and TWNs allow a massive reduction of the data volume to store the network and have been applied to recent and high-complexity networks with an almost negligible loss. In parallel, the VLSI research community has been developing specialized hardware architectures focusing on data re-use with limited resources and optimizing arithmetic precision, exploiting weight and feature map (FM) sparsity, and performing on-the-fly data compression to ultimately maximize energy efficiency [23]. However, these implementations fall into one of two categories: 1) they stream the entire or even partial FMs into and out of the accelerator ending up in a regime where I/O energy is far in excess of the energy spent on computation, hitting an energy efficiency wall: the state-of-the-art accelerator presented in [24] has a core energy efficiency of 59 TOp/s/W, but including I/O power it is limited to 1 TOp/s/W; or 2) they assume to store the entire network's weights and intermediate FMs on-chip. This severely constrains the DNN's size that can be handled efficiently by a small low-cost IoT-endnode class chip. It also prevents the analysis of high resolution images, thus precluding many relevant applications such as object detection.

In this work, we convey the following key contributions:

1) A new and highly optimized yet flexible core architecture systolically scalable to high resolution images to enable applications such as object detection.
2) A new approach exploiting the shift in the ratio of the size of the weights and intermediate results of BWNs to overcome the I/O energy-induced efficiency wall (58×

978-1-5386-7100-9/18 $31.00 © 2018 IEEE
509

less I/O energy) while enabling to run state-of-the-art BWNs.

3) An in-depth analysis of this architecture in terms of memory requirements, I/O bandwidth, and scalability including implementation results in GF 22 nm FDX technology, showing a 2.2× gain in energy efficiency even though our core uses resource-intensive FP16 arithmetic for increased robustness.

II. RELATED WORK

A. Software-Programmable Platforms

The availability of cheap computing power on GPUs and large data sets have sparked the deep learning revolution, starting when AlexNet had incurred a landslide win in the ILSVRC image recognition challenge in 2012 [25]. Since then we have seen optimized implementations [26, 27] and algorithmic advances such as FFT-based and Winograd convolutions further raising the throughput [28, 29]. The availability of easy-to-use deep learning frameworks exploiting the power of GPUs transparently to the user has resulted in wide-spread use of GPU computing. With the growing market size, improved hardware has become available as well: Nvidia has introduced a product line of systems-on-chip for embedded applications where ARM cores have been co-integrated with small GPUs for a power range of 5-20 W and ≈50 GOp/s/W. Also the GPUs' architecture has been optimized for DNN workload, introducing tensor cores and fast half-precision floating-point (fp16) support. The latest device, Nvidia's V100, achieves 112 TFLOPS at 250 W [30]—an energy efficiency of 448 GOp/s/W. It's main competitor, Google's TPU [12], works with 8-bit arithmetic and achieves 92 TOp/s at 384 W (240 GOp/s/W). With these power budgets, however, they are unsuitable for IoT end-nodes.

B. Co-Design of DNN Models and Hardware

Over the last few years, several approaches adapting DNNs to reduce the computation effort have been presented. One main direction was the reduction of the number of operations and model size. Specifically, the introduction of sparsity provides an opportunity to skip some operations. By pruning the weights a high sparsity can be achieved particularly for the fully-connected layers found at the end of many networks and the ReLU activations in most DNN models injects sparsity into the FMs, which can be exploited [31, 32].

A different direction is the research into reduced precision computation. Standard fixed-point approaches work down to 10-16 bit number formats for many networks. It is possible to further reduce the precision to 8 bit with small accuracy losses ($< 1\%$) when retraining the network to adapt to this quantization [33]. There are limitation to this: 1) for deeper networks higher accuracy losses (2-3% for GoogLeNet) remain, and 2) Typically, only the input to the convolutions are quantized in this format. Internal computations are performed at full precision, which implies that the internal precision is very high for large networks e.g. for a 3x3 convolution layer with 512 input FMs, this adds 12 bit. Further approaches include non-linearly spaced quantization in the form of mini-floats [33],

and power-of-two quantization levels replacing multiplications with bit-shift operations [20].

Several efforts have taken the path to extreme quantization to binary (+1/-1) or ternary (+1/0/-1) quantization of the weights while computing the FMs using floats. This massively compresses the data volume of the weights and has even been shown to be applicable to deep networks with an accuracy loss of approx 1.6% for ResNet-18 [20] and thus less than the fixed point-and-retrain strategies. The next extreme approach are (fully) binary neural networks (BNNs), where the weights and FMs are binarized [34]. While this approach is attractive for extreme resource constrained devices [18, 35], the associated accuracy loss of 16% on ResNet-18 is unacceptable for many applications.

C. FPGA and ASIC Accelerators

Many hardware architectures targeting DNNs have been published over the last few years. The peak compute energy efficiency for fixed-point CNN accelerators with >8 bit can be found at around 50 GOp/s/W for FPGAs, 2 TOp/s/W in 65 nm and around 10 TOp/s/W projected to 28 nm [36–39]. However, this does not include I/O energy for streaming the FMs, or assumes that intermediate results can be entirely stored in limited-size on-chip memory. Streaming the FMs results in a device-level energy efficiency wall at around 1 TOp/s/W [24], while requiring the data to fit entirely into on-chip memory renders the device very large.

Many of the sparsity-based optimizations mentioned in Sec. II-B have been implemented in hardware accelerators [32, 40], they could achieve up to 3× higher core energy efficiency and raise the device-level energy efficiency by around 70% through data compression. The effect of training DNNs to become BWNs has shown the biggest impact on core compute-only energy with an energy efficiency of 60 TOp/s/W in 65 nm [24]. However, with the present architectures, the fundamental efficiency limitation by the I/O energy remains.

III. HYPERDRIVE ARCHITECTURE

The Hyperdrive architecture we propose in this work is fundamentally different from previous BWN accelerators [24, 41] in two aspects: 1) it keeps the FMs on-chip and streams the weights. This approach exploits the binary nature of weights, significantly reducing the I/O bandwidth. 2) Its hierarchical systolic-scalable structure allows to scale the complexity and resolution of the networks, both on-chip, by instantiating multiple computing tiles within an accelerator, and off-chip, instantiating multiple accelerators in a 2D mesh.

The system architecture is composed of the following components, illustrated in Fig. 1.

- *Feature Map Memory (FMM)*: Is a multi-banked memory storing input and output FMs.
- Array of K×M×N *Tile Processing Units (TPUs)*: A single TPU is illustrated in Fig. 2. Every TPU contains 1) a fp16 adder/subtractor to accumulate the partial sums of the output pixels, bias and the bypass input FM (in case of residual blocks), 2) a half-precision multiplier for the

978-1-5386-7100-9/18 $31.00 © 2018 IEEE

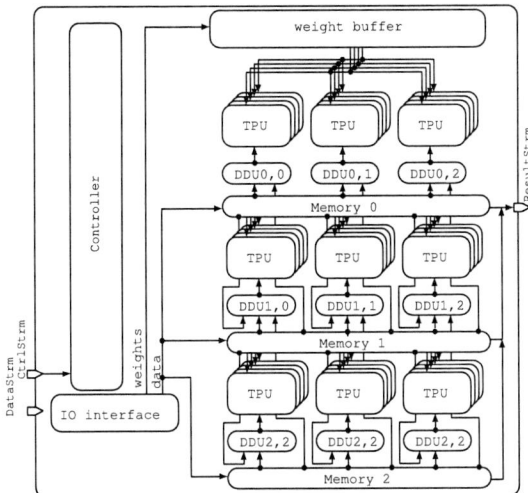

Fig. 1: System overview with $K \times M \times N$ tiles (for $K = 4, M = N = 3$).

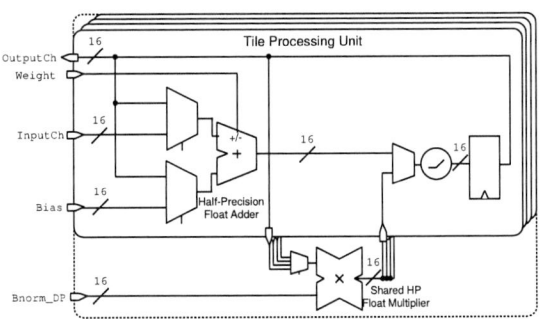

Fig. 2: Architecture of a single TPU.

FM-wise batch-normalization shared among the TPUs of the same tile, and 3) a ReLU activation unit. Each TPU is assigned to a tile and an output FM. Each TPU is connected to its 8 neighbors on the X,Y axes to quickly access neighbouring pixels.

- *Weight Buffer (WBuf)*: Stores the weights of the current output FMs.
- *Data Distribution Units (DDUs)*: Distributes the data from the memories to the according TPU units or manages zero-padding.

The superior efficiency of Hyperdrive is achieved exploiting data re-use at different levels:

- Output FM level: The output FMs are tiled into blocks of 16 FMs which operate on the same input FMs in parallel. The input FM needs therefore just be read once for all the 16 output FMs.
- Spatial level: The input FM is tiled into 7×7 image patches, where as the system operates on all of the patches in parallel applying the same weights.
- Weight re-use: Weights are stored in the *weight buffer*, which is implemented as a latch-based standard cell

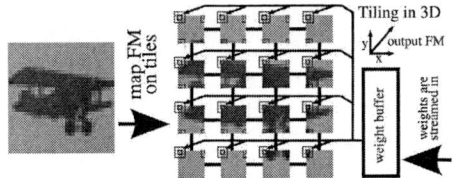

Fig. 3: Feature Maps are tiled and processed in parallel TPUs.

memory for optimal energy efficiency [24].

IV. COMPUTATIONAL MODEL

In state of the art CNNs such as ResNet-34 the complexity and number of weights is huge. However, for BWNs, streaming the weights rather than the FMs or both is particularly attractive due to the compression by 16× (i.e. from fp16). Moreover, the weights are only read once from off-chip memory and do not need to be streamed out again. To exploit these features we propose a weight-streaming approach where all the weights are streamed in only once during the execution of the overall network and the FMs are stored in on-chip memory. Weights are streamed in only for computing the first pixel of the current output FMs, and then stored into a weight buffer to avoid streaming the them again to the chip and read from there for the following pixels. This approach significantly reduces the I/O bandwidth of the accelerator, by a factor of 5× for 224×224 or 710× for 2048×1024 sized images for several ResNet configurations, as shown in Tbl. II.

The computational model of Hyperdrive is based on a first phase where the input FMs of the first layer are partitioned and loaded the the on-chip systolic design, as illustrated in Fig. 3. During this phase, the FMs are assigned to the tile processing units (TPUs), which are organized as a 3D-mesh systolic array and follow a single instruction multiple data (SIMD) execution model. Hence, the input FMs are tiled into blocks and assigned on the X,Y dimension, while the output FMs which re-use the same input FMs depth-wise are assigned to the TPUs on the Z dimension (Fig. 3).

Once the input FMs are loaded into the array, the execution starts, as illustrated in Tbl. I for an implementation of the architecture featuring 16×7×7 TPUs with 8×8 sized tiles and for a 3×3 convolution layer with 16×64 FMs.

Each TPU calculates a single pixel of its assigned tile for the same input FM of its assigned output FM. The TPUs assigned to the same tile shares the input FM pixels, which reduces memory reads by M× and the weights are read once for each FM, but shared among the spatial TPUs, which reduces the weight buffer reads by $N^2 \times$. The results are stored in the accumulation register. In more detail, the execution can be summarized as the following 4 nested execution phases:

- output FM-level: The TPUs calculate a block of output FMs (i.e. K) (line 6).
- Pixel-level ((y,x)): Every TPU calculates the pixel of its assigned output FM and tile (line 5).
- Filter-level ($\Delta y, \Delta x$)): The TPUs calculate the partial sums of the same filter tap location (line 4). Neighbouring

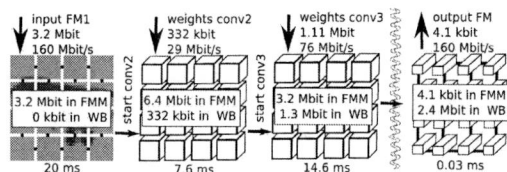

Fig. 4: High-Level overview of memory and bandwidth requirement of Hyperdrive running Resnet-34 inference.

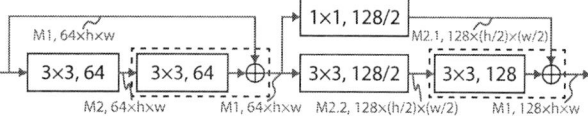

Fig. 5: Early block of layers of ResNet-34 and transition to next type of layer block. Activation and batch normalization layers are not indicated separately. Dashed rectangles imply on-the-fly addition to eliminate the need for additional memory.

pixels which are not in the same tile can be accessed exploiting the TPUs neighbouring connections described in Sec. III.

- input FM-level: The TPUs multiply the current pixel $(y + \Delta y, x + \Delta x)$ from the input FM with the related binary weights $f_{\text{input FM,output FMs}}^{\text{filter tap}(\Delta y, \Delta x)}$ and accumulate them for the pixel (y, x) of the related output FM (line 3).

Several CNNs like ResNet-34 feature bypass connections as shown in Fig. 5, hence management of bypass has been introduced into the architecture. When all contribution for the output FM pixels are summed up, the bypass FM is read from memory and added (if bypass exists). Finally, the biasing and batch normalization is applied and stored back to the FM memory.

When the current's output FM are entirely calculated, the next layer is processed while the FMs stay on-chip and only the new weights are streamed in. This is continued until the final layer is computed, as illustrated in Fig. 4.

A. CNN Mapping

The size of the on-chip memory for intermediate FM storage has to be selected depending on the convolution layer with the largest memory footprint of the network - *Worst-Case Layer* (WCL). Typically, the WCL is at the beginning of the network, since a common design pattern is to double the number of FMs after a few layers while performing at the same time a 2×2 strided operation, thereby reducing the number of pixels by $4 \times$ and the total FM volume by $2 \times$. To perform the computations layer-by-layer, avoiding usage of power hungry dual-port memories, we leverage a ping-pong buffer mechanism reading from one memory bank and writing the results to a different memory bank. Hence, for a generic CNN the amount of memory required by the WCL is: $\max_{\text{layers in CNN}} n_{\text{in}} h_{\text{in}} w_{\text{in}} + n_{\text{out}} h_{\text{out}} w_{\text{out}}$ words, since all input and output FMs have to be stored to implement the described ping-pong buffering mechanism.

However, many networks have bypass paths, hence additional intermediate FMs have to be stored, as described in Fig. 5 for the potential WCLs of ResNet-34. This aspect has two implications:

1) In order to avoid additional memory (+50%), we perform an on-the-fly addition of the bypass path after the second 3×3 convolution (i.e. the dashed rectangle is a single operation). This is done by performing a read-add-write operation on the target memory locations.

2) The common transition pattern with the 2×2-strided convolution does not require additional memory. It temporarily needs three memory segments, but two of them are $2 \times$ smaller and can fit into what has been a single memory segment before (M2 is split into two equal-size segments M2.1 and M2.2).

For ResNet-18 and -34 on ImageNet data samples (i.e. for image recognition), the total required memory is 401 k words. Using the same procedure to determine the amount of memory for ResNet-50/-152/... shows that due to the different structure (i.e. the bottleneck building block), three memory segments are required with a total of 1.2 M words. If the structure is fixed, the required memory size does not depend on the network depth. Note that if enough silicon area is available and scalability to an arbitrary deep network is not required, on-chip storage of the weights should be considered, eliminating almost the entire remaining I/O transfers. For ResNet-34 this approach would require 21 Mbit and thus 6.3 mm^2 and for ResNet-18 to around 3 mm^2 of SRAM (0.3 μm^2/bit in GF 22nm FDX).

B. Scalability to Multiple Chips

The proposed architecture is trivially scalable to high resolution data on a single die, however production yield diminishes and cost explodes for large die sizes and many varying-size chips (for different image resolutions) would have to be manufactured for cost-efficient volume production. We address this issue and allow flexible scaling of not just the image resolution (i.e. on-chip memory), but also performance with a fixed die size by extending the systolic design approach to chip-level and connecting multiple Hyperdrive chips on the circuit board or an interposer.

Hyperdrive can be scaled to a systolic array of $m \times n$ chips, where the FMs are split into tiles of size $M \times N$. In this way, every chip keeps $M \times N$ tiles, and the entire FM is partitioned into $M \cdot m \times N \cdot n$ tiles. As the convolution window overlaps with the neighboring chips, data needs to be exchanged between the chips. Since FM values are read repeatedly, we transfer them as soon as they are computed and then buffer them in a *border memory*. Each chip has to be able to keep $\max_{\text{all layers}} \left((2Mh_{\text{tile}} + 2Nw_{\text{tile}} + 4)n_{\text{in}} \right)$ values from the neighboring chips for 3×3 filter support, where $h_{\text{tile}}, w_{\text{tile}}$ denote the spatial dimension of each tile (e.g. 8×8). Chips at the boundary of the systolic grid would be performing zero-padding instead of using this memory.

V. RESULTS

In the further discussion, the number of tiles chosen is 7×7, which allows for $4 \times$ striding on 112×112 sized input FMs (like

TABLE I. Time schedule for a 16 input FM and 64 output FM 3x3 convolution. Notation for filter weights: $f^{\text{filter tap}(\Delta y,\Delta x)}_{\text{input FM,output FM}}$.

cycle	1	2	...	16	17	...	...	144	145	...	288	...	9216	9217	...	36.8k
weight input	$f^{-1,-1}_{1,(1-16)}$	$f^{-1,-1}_{2,\cdot}$	...	$f^{-1,-1}_{16,\cdot}$	$f^{-1,0}_{1,\cdot}$	...	...	$f^{1,1}_{16,\cdot}$	No I/O (loaded from weight buffer)					$f^{-1,-1}_{1,(17-32)}$	...	No I/O
input FM	1	2	...	16	1	...	...	16	1	...	16	...	16	1	...	16
filter tap pos.	-1,-1				-1,0	...	...	+1,+1	-1,-1	...	+1,+1	...	+1,+1	-1,-1	...	+1,+1
outp. pixel pos.	1,1								1,2	...			8,8	1,1	...	8,8
output FM	1-16 (in parallel)													17-32	...	49-64

TABLE II. Data Comparison for various typical networks with binary-weights and 16-bit FMs. (Top: Image Recognition, Bottom: Object Detection

network	resolution	weights [bit]	all FMs [bit]	WC mem. [bit]
ResNet-18	224×224	11M	36M	6.4M
ResNet-34	224×224	21M	61M	6.4M
ResNet-50	224×224	21M	156M	21M
ResNet-152	224×224	55M	355M	21M
ResNet-34	2048×1024	21M	2.5G	267M
ResNet-152	2048×1024	55M	14.8G	878M

TABLE III. Power Breakdown. All units in mW.

@0.65V	P_{int}	P_{dyn}	P_{leak}	P_{tot}
Memory	7.14	0.98	1.18	9.28
Arithmetic	2.97	5.14	0.15	8.25
Registers	3.04	3.74	0.09	6.87
Weight Buffer	0.03	0.06	0.08	0.16
Others	0.22	1.08	0.01	1.34
Total	13.4	11.0	1.51	25.9

TABLE IV. Overview of HYPERDRIVE

Operating Point [V]	0.65	0.8	0.9
Op. Frequency [MHz]	160	282	344
Power [mW]	25.9	108.6	171
Throughput [Op/cycle]	1568	1568	1568
Throughput [GOp/s]	250.9	442.2	539.4
Energy Eff. [TOp/s/W]	6.1	4.1	3.2
Core Area [mm²]	1.92	1.92	1.92
Memory [Mbit]	6.4	6.4	6.4

in common ResNet-like networks), while keeping all the TPUs busy with at least one single spatial pixel during the entire network. Half-precision floating point numbers are used for the FMs, as this gives higher flexibility and greatly eases the training on common frameworks.[1] The on-chip memory was sized to fit the WCL (i.e. ResNet-like networks with bypass layers and 112×112 input FM) with 6.4 Mbit (400 kword) and is implemented with $y \times 8 = 7 \times 8$ high-density single-port SRAMs with 1024 lines of $x \cdot 16 = 7 \cdot 16 = 112$ bit words, where as the memories are assigned to the $(x \times y)$ tiles. The output FM parallelism has been fixed to 16 to optimize the trade-offs between area, performance and power. The weight buffer has been implemented to fit up to 512 (max. #input FMs) 3×3 kernels for 16× depth-wise parallelism. If more input FMs are needed, they can be tiled to 512 blocks and partial output FM can be calculated and summed up on-the-fly using the bypass mode. The frequently-accessed weights buffer has been implemented as a latch-based memory composed of 5×8 blocks of 128 rows of 16-bit words, reducing the access energy to SRAM memories by 43×.

A. Implementation Results

Hyperdrive was implemented in 22 nm FDX technology based on the standard cells of INVECAS (8 track, LVT, V05) and has been sent for tape-out. The chip has an effective core area of 1.92 mm² (=9.6 MGE)[2], where 1.24 mm² are SRAM memories (6.4 Mbit), 0.115 mm² are SCM memory (74 kbit) and 0.32 mm² arithmetic units.

The power consumption is evaluated on a reference 3×3 convolution layer with 16→ 16 FMs with 56×56 pixels with bias and batch normalization and on the post-layout netlist with timing (sdf) and parasitics annotations (spef) while using

---

[1]Using fixed-point numbers is considered as a potential extension, however this will need to be supported by more complex quantization-aware training.

[2]One 2-input NAND gate equivalents (GE) is 0.199 µm² in GF22.

---

the provied at different voltages, at the operatig temperature of 25°C in typical operating conditions.. The I/O energy was determined on the basis of an LPDDR3 PHY in 28 nm technology [36] and thus estimated with 21 pJ/bit, but should be considered as a lower bound when less advanced PHYs are used, e.g. pessimistic for our architecture.

Tbl. III gives an overview of the power consumption of the various blocks of Hyperdrive at 0.65 V (typical corner) and 160 MHz. The memory takes most of the power and consumes 9.29 mW, while the arithmetic units use 8.25 mW. Even though the weight buffer is accessed every cycles, it consumes little energy of 0.16 mW.

Tbl. IV gives an overview of the key metrics of Hyperdrive. The design is running at 100 MHz in the low-voltage corner at 0.59V (125C, slow) and reaches a throughput of 156.8 GOp/s and a core energy efficiency of 6.1 TOp/s/W and at 0.9 V (25C, typical) a throughput of 539.4 GOp/s can be achieved.

B. Benchmarking

We evaluated Hyperdrive on ResNet-34, one of the most prominent networks. From the residual networks, this network feature a good trade-off of depth and accuracy, i.e. ResNet-50 outperforms ResNet-34 by just 0.5% (Top-1), but is roughly 50% more compute-intense and the memory footprint is even 3.3× higher (see Sec. IV-B).

The first and the last layer need to stay in full-precision to keep a satisfactory accuracy and are not implemented on Hyperdrive, but they contribute just 3% of the computation

978-1-5386-7100-9/18 $31.00 © 2018 IEEE

TABLE V. Comparison with State-of-the-Art BWN Accelerators (Top: Image Recognition, Bottom: Object Detection)

Name	Techn.	DNN	Precision Wghts/Acts	Core [V]	Throughp. [GOp/s]	Eff. Th. [GOp/s]	Core E [mJ/im]	I/O E [mJ/im]	Total E [mJ/im]	En. Eff. [TOp/s/W]	Area [MGE]
YodaNN [24]	umc65	ResNet-34	Bin./Q12	1.20	1510	490	0.9	3.60	4.5	1.6	1.3
YodaNN [24]	umc65	ResNet-34	Bin./Q12	0.60	55	18	0.1	3.60	3.7	2.0	1.3
Wang [41]	SMIC130	VGG-16	Bin./ENQ6	1.08		876	24.3	2.59	26.9	1.1	9.9
ours (7×7 TPUs)	GF22	ResNet-34	Bin./FP16	0.80	442	431	1.8	0.49	2.3	3.2	9.6
ours (7×7 TPUs)	GF22	ResNet-34	Bin./FP16	0.65	251	245	1.2	0.49	1.7	4.3	9.6
Wang [41]	22 nm[3]	ResNet-34	Bin./ENQ6			(876)	33[3]	40.48	73.2[3]	4.0[3]	(9.9)
ours (64×32 TPUs)	GF22	ResNet-34	Bin./FP16	0.65	10491	10240	48.6	3.27	51.9	5.7	401.3
Wang [41]	22 nm[3]	ResNet-152	Bin./ENQ6			(876)	100[3]	234.94	335.2[3]	2.7[3]	(9.9)
ours (64×32 TPUs)	GF22	ResNet-152	Bin./FP16	0.65	10491	10240	149	3.34	152.3	5.9	1316
Improvement over state-of-the-art for object detection (ResNet-34):							0.33×	12×	1.4×	1.4×	
Improvement over state-of-the-art for object detection (ResNet-152):							0.33×	58×	2.2×	2.2×	

TABLE VI. Overview of Cycles, Throughput for ResNet-34

layer type	#cycles	#Op	#Op/cycle	#Op/s
conv	4.52 M	7.09 G	1.57 k	
bnorm	59.90 k	2.94 M	49	
bias	59.90 k	2.94 M	49	
bypass	7.68 k	376.32 k	49	
total	4.65 M	7.10 G	1.53 k	431 G

(226 MOp of 7.3 GOp) and can therefore also be evaluated on low-power compute platforms [42].

Tbl. VI gives an overview of the number of operations, number of cycles and throughput while Hyperdrive is evaluating ResNet-34. In case of batch normalization, the throughput is reduced since just 49 multipliers are available and the normalization does takes more cycles. In the layers where the bypass has to be added, Hyperdrive can also just calculate one output FM at a time, because the memory bandwidth is limited to 49 half-precision words. Fortunately, the non-convolution operations are comparably rare and a real throughput of 1.53 kOp/cycle or 152.8 GOp/s @ 0.65 V is achieved leading to a very high utilization ratio of 97.5%.

C. Comparison with State-of-the-Art

Tbl. V compares our architecture with state-of-the-art binary weight CNN accelerators. On the top, we compare the numbers of image recognition (224×224 pixel images), for which previous work report results. In the lower part, we compare the key figures for object detection using ResNet-34 and ResNet-152 features on 2048×1024 pixel images (e.g. found in autonomous driving data sets [5, 44]). At 0.65 V a frame rate of 34.6 for ResNet-34 and 11.3 frame/s for ResNet-152 is achieved independent of the image resolution when systolically scaling the architecture accordingly.

Previous work is dominated by I/O energy, especially for spatially large feature maps. We compare our work to Wang et al. [41], approximating the power consumption according to the scaling model presented in Dreslinski et al. [43]. Furthermore, the FMs are based on Tbl. II but scaled down to 6 bits words to

[3]The energy efficiency of Wang et al. [41] has been scaled to 22 nm according to the energy scaling model presented in Dreslinski et al. [43].

account for the ENQ number format. Our approach uses up to 58× less energy for I/O and increases overall energy efficiency by up to 2.2× because just the first input FM and the weights need to be streamed to the chip, but not the intermediate FMs. Hyperdrive's core energy efficiency is 3× lower than previous work, due to:

1) Fp16 operators which are more robust than Q12 or ENQ6 in [24, 41] and were shown to work with the most challenging deep networks. Using floating-point feature maps directly impacts the energy for the accumulation operations as well as memory and register read/write operations. ENQ on the other side has been shown to introduce an accuracy drop of 1.6% already on CIFAR-100 [41], which is more than the difference between running ResNet-32 instead of ResNet-110 on CIFAR-10. It thus implies that a deeper network has to be computed to achieve a comparable accuracy.

2) Significantly larger on-chip memories to store the FMs. However, optimizations such as approximate adders and strong quantization can be combined with Hyperdrive's concepts, extending the core efficiency gains to the system level by removing the non-scalable I/O bottleneck. For instance, moving from FP16 to Q12 would lead to an energy efficiency boost that can be estimated around 3× for the core, which would translate to a system efficiency boost of 6× with respect to state-of-the-art.

VI. Conclusion

We have presented Hyperdrive: a systolically-scalable hardware architecture for binary weights neural networks, which dramatically minimizes the I/O energy consumption to achieve outstanding system-level energy efficiency. Hyperdrive achieves an energy efficiency of 5.9 TOp/s/W which is more than 2.2× better than prior state-of-the-art architectures, by exploiting a binary weights streaming mechanism while keeping the entire FMs on-chip. Furthermore, while previous architectures were limited to some specific network sizes, Hyperdrive allows running networks not fitting on a single die, by arranging multiple chips in an on-board 2D systolic array, scaling-up the resolution of neural networks, hence enabling a new class of applications such as object detection on the edge of the IoT.

ACKNOWLEDGEMENTS

This work was funded by the Swiss National Science Foundation under grant 162524 (MicroLearn: Micropower Deep Learning), armasuisse Science & Technology and the ERC MultiTherman project (ERC-AdG-291125).

REFERENCES

[1] O. Russakovsky et al., "ImageNet Large Scale Visual Recognition Challenge," IJCV, vol. 115, no. 3, pp. 211–252, 2015.

[2] K. He et al., "Deep Residual Learning for Image Recognition," Proc. IEEE CVPR, pp. 770–778, 2015.

[3] Y. LeCun et al., "Convolutional Networks and Applications in Vision," in Proc. IEEE ISCAS, 2010, pp. 253–256.

[4] P. Sermanet et al., "Convolutional Neural Networks Applied to House Numbers Digit Classification," in Proc. IEEE ICPR, 2012, pp. 10–13.

[5] B. Wu et al., "SqueezeDet: Unified, Small, Low Power Fully Convolutional Neural Networks for Real-Time Object Detection for Autonomous Driving," in Proc. IEEE CVPRW, 7 2017, pp. 446–454.

[6] S. Ren et al., "Faster R-CNN: Towards Real-Time Object Detection with Region Proposal Networks," IEEE TPAMI, vol. 39, no. 6, pp. 1137–1149, 2017.

[7] J. Long et al., "Fully Convolutional Networks for Semantic Segmentation," in Proc. IEEE CVPR, 2015.

[8] L. Cavigelli et al., "Computationally efficient target classification in multispectral image data with Deep Neural Networks," in Proc. SPIE Security + Defence, vol. 9997, 2016.

[9] C. Feichtenhofer et al., "Convolutional Two-Stream Network Fusion for Video Action Recognition," in Proc. IEEE CVPR, 6 2016, pp. 1933–1941.

[10] F. Scheidegger et al., "Impact of temporal subsampling on accuracy and performance in practical video classification," Proc. IEEE EUSIPCO, no. 732631, pp. 996–1000, 2017.

[11] L. Cavigelli et al., "CBinfer: Change-Based Inference for Convolutional Neural Networks on Video Data," in Proc. ACM ICDSC, 2017.

[12] N. P. Jouppi et al., "In-Datacenter Performance Analysis of a Tensor Processing Unit," in Proc. ACM ISCA, 2017.

[13] F. Conti et al., "An IoT Endpoint System-on-Chip for Secure and Energy-Efficient Near-Sensor Analytics," IEEE TCAS, vol. 64, no. 9, pp. 2481–2494, 9 2017.

[14] VentureBeat.com, "GreenWaves Technologies unveils Gap8 processor for AI at the edge," 2018.

[15] R. G. Baraniuk, "More Is Less: Signal Processing and the Data Deluge," Science, vol. 331, no. 6018, pp. 717–719, 2011.

[16] P. Schulz et al., "Latency Critical IoT Applications in 5G: Perspective on the Design of Radio Interface and Network Architecture," IEEE Comm. Mag., vol. 55, no. 2, pp. 70–78, 2017.

[17] EE Times, "7 Ideas for AI Silicon from ISSCC – Google calls for hybrid edge/cloud collaboration," 2018.

[18] M. Rusci et al., "Design Automation for Binarized Neural Networks: A Quantum Leap Opportunity?" in Proc. IEEE ISCAS, 2018.

[19] A. S. Weddell et al., "A Survey of Multi-Source Energy Harvesting Systems," Proc. ACM/IEEE DATE, p. 4, 2013.

[20] A. Zhou et al., "Incremental Network Quantization: Towards Lossless CNNs with Low-Precision Weights," in Proc. ICLR, 2017.

[21] G. Venkatesh et al., "Accelerating Deep Convolutional Networks using low-precision and sparsity," in Proc. IEEE ICASSP, 3 2017, pp. 2861–2865.

[22] M. Courbariaux et al., "BinaryConnect: Training Deep Neural Networks with binary weights during propagations," in Adv. NIPS, 2015.

[23] V. Sze et al., "Efficient Processing of Deep Neural Networks: A Tutorial and Survey," Proceedings of the IEEE, vol. 105, no. 12, pp. 2295–2329, 12 2017.

[24] R. Andri et al., "YodaNN: An Architecture for Ultra-Low Power Binary-Weight CNN Acceleration," IEEE TCAD, 2017.

[25] A. Krizhevsky et al., "Imagenet Classification With Deep Convolutional Neural Networks," in Adv. NIPS, 2012.

[26] S. Chetlur et al., "cuDNN: Efficient Primitives for Deep Learning," in arXiv:1410.0759, 2014.

[27] L. Cavigelli et al., "Accelerating Real-Time Embedded Scene Labeling with Convolutional Networks," in Proc. ACM/IEEE DAC, 2015.

[28] N. Vasilache et al., "Fast Convolutional Nets With fbfft: A GPU Performance Evaluation," arXiv:1412.7580, 2014.

[29] A. Lavin et al., "Fast Algorithms for Convolutional Neural Networks," in Proc. IEEE CVPR, 2016, pp. 4013–4021.

[30] Nvidia Inc., "Nvidia Tesla V100 GPU Accelerator – Datasheet."

[31] T.-J. Yang et al., "Designing Energy-Efficient Convolutional Neural Networks Using Energy-Aware Pruning," in Proc. IEEE CVPR, 7 2017, pp. 6071–6079.

[32] S. Han et al., "EIE: Efficient Inference Engine on Compressed Deep Neural Network," in Proc. ACM/IEEE ISCA, 2016, pp. 243–254.

[33] P. Gysel et al., "Hardware-oriented Approximation of Convolutional Neural Networks," in ICLR Workshops, 2016.

[34] M. Courbariaux et al., "Binarized Neural Networks: Training Deep Neural Networks with Weights and Activations Constrained to +1 or -1," in arXiv:1602.02830, 2016.

[35] A. Al Bahou et al., "XNORBIN: A 95 TOp/s/W Hardware Accelerator for Binary Convolutional Neural Networks," in arXiv:1803.05849, 2018.

[36] L. Cavigelli et al., "Origami: A 803-GOp/s/W Convolutional Network Accelerator," IEEE TCSVT, vol. 27, no. 11, pp. 2461–2475, 11 2017.

[37] Y.-H. Chen et al., "Eyeriss: An Energy-Efficient Reconfigurable Accelerator for Deep Convolutional Neural Networks," in Proc. IEEE ISSCC, 2016, pp. 262–263.

[38] Z. Du et al., "ShiDianNao: Shifting Vision Processing Closer to the Sensor," in Proc. ACM/IEEE ISCA, 2015, pp. 92–104.

[39] F. Conti et al., "A Ultra-Low-Energy Convolution Engine for Fast Brain-Inspired Vision in Multicore Clusters," in Proc. ACM/IEEE DATE, 2015, pp. 683–688.

[40] A. Aimar et al., "NullHop: A Flexible Convolutional Neural Network Accelerator Based on Sparse Representations of Feature Maps," arXiv:1706.01406, 6 2017.

[41] Y. Wang et al., "An Energy-Efficient Architecture for Binary Weight Convolutional Neural Networks," IEEE TVLSI, vol. 26, no. 2, pp. 280–293, 2017.

[42] M. Gautschi et al., "Near-Threshold RISC-V core with DSP extensions for scalable IoT endpoint devices," IEEE TVLSI, vol. 25, no. 10, pp. 2700–2713, 2017.

[43] R. G. Dreslinski et al., "Near-threshold computing: Reclaiming moore's law through energy efficient integrated circuits," Proceedings of the IEEE, vol. 98, no. 2, pp. 253–266, 2010.

[44] M. Cordts et al., "The Cityscapes Dataset for Semantic Urban Scene Understanding," in Proc. IEEE CVPR, 2016, pp. 3213–3223.

An Optimized Architecture For Decomposed Convolutional Neural Networks

Fangxuan Sun, Jun Lin and Zhongfeng Wang
School of Electronic Science and Engineering, Nanjing University, China
Email: fxsun@smail.nju.edu.cn, {jlin, zfwang}@nju.edu.cn

Abstract—Convolutional neural networks (CNNs) have found extensive applications in various tasks. However, the state-of-the-art CNNs are both computation-intensive and memory-intensive, which brings tremendous hardware implementation challenges. Various methods have been proposed to reduce the model size and computation complexity of a CNN. Among them, when hardware implementation is considered, the Canonical Polyadic decomposition (CPD) method is more suitable due to the regularity in the decomposed filters. Moreover, the CPD method can be combined with widely used pruning methods to compress the model in further. In this paper, to the best of our knowledge, an efficient hardware architecture for CPD-CNNs is proposed for the first time based on a carefully designed data flow. In detail, a reconfigurable fast convolution unit is introduced to reduce the number of multiplications while handling some commonly-used convolution core operations. The proposed architecture is coded with RTL and synthesized under the TSMC 90nm CMOS technology. Our design achieves an equivalent throughput of more than 3TOP/s under 650MHz clock frequency.

Index Terms—Convolutional Neural Networks (CNNs), Fast FIR Algorithm (FFA), CP-Decomposition, VLSI

I. INTRODUCTION

Convolutional Neural Networks (CNNs) have achieved remarkable performance incomputer vision [1], [2] and speech recognition [3] in recent years. Generally, CNNs consist of convolutional layers and fully-connected layers, which are computation-intensive and memory-intensive, respectively. Larger scale CNNs are introduced to increase the model capability, which incurs higher computational complexity. To speedup CNNs, many efforts have been made. One approach is to modify the structure of convolutional layers for the purpose of reducing the computational complexity. Some decomposition methods [4]–[7] have been proposed on CNNs since Denil *et al.* [8] showed that the redundancy of convolutional layers can be removed. By employing canonical polyadic decomposition (CPD) method [5], [7], the weight size and computational complexity of AlexNet can be reduced by 6.98x and 3.53x, respectively. In this work, the unique structure of CPD-CNNs is explored to accelerate the processing speed of CNNs.

Generally, a convolutional layer in CPD-CNNs will be decomposed into several sub-layers with less weights and computational complexities. However, the generated sub-layers

This work was supported in part by the National Natural Science Foundation of China under Grant 61774082 and Grant 61604068 and in part by the Fundamental Research Funds for the Central Universities under Grant 021014380065.

will notably increase the overall number of activations. Current hardware accelerators mainly focus on minimizing the DRAM accesses of weights while ignoring the activations. Directly applying current designs [9]–[15] to CPD-CNNs will incur low resource utilization and extra memory accesses. The design in [9] presented a row stationary data flow for CNNs on a spatial architecture called *eyeriss*. However, activations between convolutional layers are stored off-chip. The energy consumption of *eyeriss* is inefficient due to the increased DRAM accesses. In [10], an efficient data flow, which can combine the computations of several convolutional layers with minimal DRAM access, was proposed. However, no compression methods are considered to reduce the computational complexity. Hence, existing hardware accelerators for CNNs are not suitable for CPD-CNNs.

Since the convolutional operation occupies more than 90% of the computation [16], reducing the algorithmic strength is necessary for hardware designs of CNNs. Winograd method was employed in [11], [17] to speed up the convolutions with kernel size of 3×3. However, existing winograd processing elements designed for 3×3 convolutions will be inefficient when handling models with multiple kernel sizes, such as AlexNet [1] and Inception V3 [18]. Hence, an efficient reconfigurable processing element is needed to support different kernel sizes.

In this paper, an efficient hardware accelerator for CPD-CNNs is proposed. Aiming at supporting various kernel sizes while reducing the algorithmic strength, an optimized reconfigurable fast convolution unit is proposed based on the fast FIR algorithm (FFA). A dedicated data flow designed for CPD-CNNs is also presented. The main contributions of this paper are summarized as follows:

- An optimized reconfigurable 6-parallel fast convolution unit (6P-FCU) is developed. The 6P-FCU can effectively perform convolutions with reduced number of multipliers. Compared with the existing winograd processing elements, the proposed 6P-FCU can be flexibly configured to support various kernel sizes, such as 3×1, 5×1 and 6×1. It is estimated that 33% and 40% multiplications can be saved for 3×3 and 5×5 convolutions, respectively.

- By exploring the special structure of CPD-CNNs, a dedicated data flow, which can efficiently cache the activations and avoid unnecessary on-chip buffer, is proposed. Moreover, an optimized adder compressor is employed to reduce the critical path. Based on the developed optimized computational logics and the presented data

978-1-5386-7100-9/18 $31.00 © 2018 IEEE   516

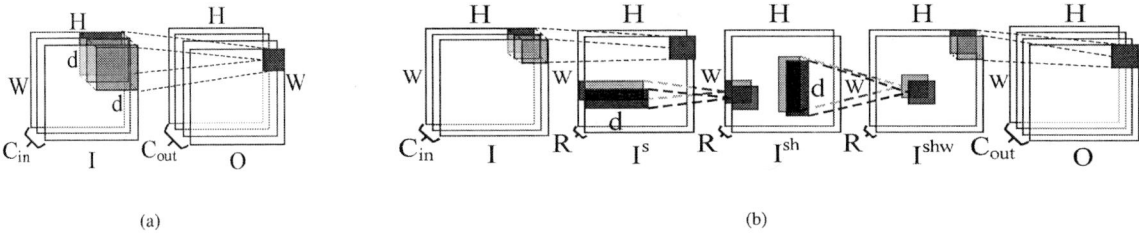

Fig. 1. (a) Computation of a convolutional layer. (b) Computation of a CP-decomposed convolutional layer.

flow, to our best knowledge, an efficient convolution architecture for CPD-CNNs is proposed for the first time. The proposed architecture for CPD-CNNs is implemented under a 90nm CMOS technology. An equivalent peak throughput of more than 3TOP/s can be achieved under 650MHz clock frequency.

The rest of the paper is organized as follows. Background of CNNs and CPD-CNNs are provided in Section II. The proposed efficient architecture is presented in Section III. Result of experiments and related discussions are provided in Section IV. At last, the conclusion is drawn in Section V.

II. BACKGROUND

A. Convolutional Layers

Convolutional layers are major part of CNNs. For a typical convolutional layer, both the input and output can be seen as 3D tensors called feature maps. Each feature map consist of several channels. A channel can be represented by a 2D matrix where each element is called a pixel. An output pixel is connected all the channels of input feature map. The connections to an output pixel are only active on a specific region in each input channel. The weights used to calculate the output pixels are called kernel tensor.

Let I and O denote the input and output feature maps with their channel numbers denoted as C_{in} and C_{out}, respectively. The computation of an output pixel can be shown below:

$$O(h, w, c_{out}) = \sum_{i=1}^{d} \sum_{j=1}^{d} \sum_{c_{in}=1}^{C_{in}}$$
$$K(i, j, c_{in}, c_{out}) I(h+i-1, w+j-1, c_{in}) \quad , \quad (1)$$

where the height and width of the convolutional kernels are denoted as d. K corresponds to a 4D kernel tensor of size $d \times d \times C_{in} \times C_{out}$. h and w are used to denote the index of pixels. The computation of a convolutional layer is shown in Fig. 1(a).

B. CP-decomposition

Briefly, the CPD will split a k-dimensional tensors A of size $d_1 \times \cdots \times d_k$ into a linear combination of several 2D matrices $A_1, A_2, \ldots, A_k$. The dimension of the i-th matrix A_i is $R \times d_i$ where R determines the size reduction. The CPD was applied on CNNs in [5], [7] to decompose the kernel tensor. The R-rank kernel tensor K after decomposition can be expressed

using the following equations:

$$K(i, j, c_{in}, c_{out}) = \sum_{r=1}^{R} K^h(i, r), \\ K^w(j, r) K^s(c_{in}, r) K^t(c_{out}, r) \quad (2)$$

where K^h, K^w, K^s, K^t denote the decomposed 2D matrices of sizes $d \times R$, $d \times R$, $c_{in} \times R$ and $c_{out} \times R$, respectively.

Substituting Eq. (2) into Eq. (1), the approximated output of a convolutional layer can be represented as follows:

$$O(h, w, c_{out}) = \sum_{r=1}^{R} K^t(c_{out}, r) (\sum_{j=1}^{d} K^w(j, r) \\ (\sum_{i=1}^{d} K^g(i, r) (\sum_{c_{in}=1}^{C_{in}} K^s(c_{in}, r) \\ \times I(h+i-1, w+j-1, c_{in})))) \quad (3)$$

According to Eq. (3), the computation of a convolutional layer can be divided into four small convolutional layers which are denoted as inter convolutional layers (ICLs). The four ICLs can be formulated as below:

$$I^s(i, j, r) = \sum_{c_{in}=1}^{C_{in}} K^s(c_{in}, r) I(h+i-1, w+j-1, c_{in})$$
$$I^{sh}(h, j, r) = \sum_{i=1}^{d} K^h(i, r) I^s(h+i-1, w+j-1, r)$$
$$I^{shw}(h, w, r) = \sum_{j=1}^{d} K^w(j, r) I^{sh}(h, w+j-1, r) \quad , \quad (4)$$
$$O(h, w, c_{out}) = \sum_{r=1}^{R} K^t(c_{out}, r) I^{shw}(h, w, r)$$

where I^s, I^{sh}, I^{shw} are inter activations. K^s and K^t can be viewed as an 1×1 convolutional kernel. K^h and K^w are convolutional kernels with sizes of $d \times 1$ and $1 \times d$, respectively.

The computation flow of CPD-CNNs is different from the conventional CNNs. Generally, the computation of each output feature map in CNNs needs all input feature maps. However, when calculating I_{sh} and I_{shw}, the pixels in i-th output channel will be only connected to the i-th input channel. In this paper, we call the convolutional layers which are used to calculate I_{sh} and I_{shw} depth-wise convolutional layer (DWCL). Fig. 1(b) illustrates the process of CPD-CNNs.

978-1-5386-7100-9/18 $31.00 © 2018 IEEE

For CPD-CNNs, both the compression ratio and the speed-up ratio are $\frac{C_{in}C_{out}d^2}{(C_{in}+2d+C_{out})R}$. Hence, the factor R determines both the reduction of weights and computational complexity. To reduce the redundancy in CNNs, the decomposition factor R should be less than $\frac{C_{in}C_{out}d^2}{C_{in}+2d+C_{out}}$.

III. EFFICIENT ARCHITECTURE OF CPD-CNNs

In this section, we present the hardware design of CPD-CNNs. First, aiming at further reducing the computational complexity, an efficient fast FIR algorithm (FFA) is proposed. Based on the FFA, an optimized reconfigurable 6P-FCU is developed to efficiently process the convolutions. Second, a dedicated adder compressor is introduced to reduce the critical path. Then, an efficient data flow is presented to utilize the avoid unnecessary DRAM accesses. Finally, based on the developed optimized logics and the dedicated data flow, we present the efficient architecture of CPD-CNNs.

A. Fast Convolution Unit

Fast FIR algorithm (FFA) is used to reduce the algorithmic strength. The 3-parallel fast FIR algorithm (3-FFA) was proposed in [19]. In time domain, a convolution approach with N coefficients can be expressed as following:

$$y(n) = h(n) * x(n) = \sum_{i=0}^{N-1} h(i)x^{n-i}, n = 0, 1, 2, ..., \quad (5)$$

where $x(n)$ is an infinite sequence and $h(n)$ is a sequence which contains FIR filter coefficients of length N. Eq. (5) can be represented in z domain as below:

$$Y(z) = H(z)X(z) = \sum_{n=0}^{N-1} h(n)z^{-n} \sum_{n=0}^{\infty} x(n)z^{-n}. \quad (6)$$

The 2-parallel FFA (2-FFA) can be expressed as below:

$$\begin{aligned} Y_0 &= H_0 X_0 + z^{-2} H_1 X_1 \\ Y_1 &= (H_0 + H_1)(X_0 + X_1) - H_0 X_0 - H_1 X_1 \end{aligned}, \quad (7)$$

where $X_0(z^2)$ and $X_1(z^2)$ are z-transforms of $x(2k)$ and $x(2k+1)$(for $0 \leqslant k < \infty$), respectively. Similarly, H_0 and H_1 are downsampled filter coefficients.

By recursively applying 2-FFA, the equations of 3-FFA introduced in [20] can be obtained as follows:

$$\begin{aligned} Y_0 &= H_0 X_0 - z^{-3} H_2 X_2 \\ &\quad + z^{-3}[(H_1 + H_2)(X_1 + X_2) - H_1 X_1] \\ Y_1 &= [(H_0 + H_1)(X_0 + X_1) - H_1 X_1] \\ &\quad - [H_0 X_0 - z^{-3} H_2 X^2] \\ Y_2 &= [(H_0 + H_1 + H_2)(X_0 + X_1 + X_2)] \\ &\quad - [(H_0 + H_1)(X_0 + X_1) - H_1 X_1] \\ &\quad - [(H_1 + H_2)(X_1 + X_2) - H_1 X_1] \end{aligned}, \quad (8)$$

where only 6 multiplications are needed compared to 9 of that in general convolution with 3 coefficients in 3-parallel form.

Consider that the 3-FFA can not be configured to support higher dimensional kernel, a 6-parallel fast FIR algprithm (6-FFA) is developed in this subsection. 2-FFA is recursively

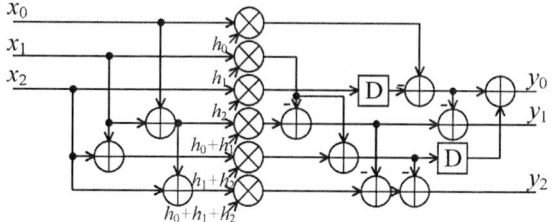

Fig. 2. 3-parallel fast convolution unit (3P-FCU).

applied in the derivation of 6-FFA. Assume that the six inputs of 6-FFA are $X_0, \ldots, X_5$. The six outputs $Y_0, \ldots, Y_5$ can be obtained using the following equations:

$$\begin{aligned} Y &= Y_0 + z^{-1}Y_1 + z^{-2}Y_2 + z^{-3}Y_3 + z^{-4}Y_4 + z^{-5}Y_5 \\ &= (X_0' H_0' + z^{-2} X_1' H_1') \\ &\quad + z^{-1}[(H_0' + H_1')(X_0' + X_1') - H_0' X_0' - H_1' X_1'] \end{aligned}, \quad (9)$$

where

$$\begin{aligned} X_0' &= X_0 + z^{-2}X_2 + z^{-4}X_4 \\ X_1' &= X_1 + z^{-2}X_3 + z^{-4}X_5 \\ H_0' &= H_0 + z^{-2}H_2 + z^{-4}H_4 \\ H_1' &= H_1 + z^{-2}H_3 + z^{-4}H_5 \end{aligned}. \quad (10)$$

According to Eq. (10), the factors $X_0' H_0'$, $X_1' H_1'$ and $(H_0' + H_1')(X_0' + X_1')$ in Eq. (9) can be seen as outputs of three 3-tap FIRs, respectively. Hence, 3-FFA is employed here to optimize the implementations of $X_0' H_0'$, $X_1' H_1'$ and $(H_0' + H_1')(X_0' + X_1')$.

Fast convolution unit (FCU) is developed in this subsection to reduce the computational complexity based on the presented 6-FFA. Since 3-FFA is the major of the 6-FFA, three parallel FCU (3P-FCU) proposed in [20] is adopted here. The details of 3P-FCU are shown in Fig. 2, where x_0, x_1, and x_2 denote the three inputs. y_0, y_1, and y_2 represent the three outputs and h_0, h_1, and h_2 denote the parameters. It is worth noting that the parameters in 3-FFA are in reverse order. Assume that the weights of a 3×1 kernel are w_0, w_1, and w_2, the relationships are: $h_0 = w_2$, $h_1 = w_1$, and $h_2 = w_0$.

The proposed 6P-FCU is designed based on the 6P-FFA. The details are shown in Fig. 3. 6P-FCU can be flexibly reconfigured to support kernel sizes of 3×1, 6×1, and 5×1. The procedure of calculating these three kinds of kernels are illustrated as follows:

- 3×1 **convolution:** For kernel size of 3×1, three 3P-FCUs in Fig. 3 are served as three independent units which process the input date in parallel. Inputs x_{3k+0}, x_{3k+1}, and x_{3k+2} are sent to 3P-FCU-k, where $k = 0, 1, 2$. The outputs o_{3k+0}, o_{3k+1}, and o_{3k+2} of 3P-FCU-k can be obtained and directly outputted. Thus, the 6P-FCU can obtain nine outputs simultaneously.
- 6×1 **convolution:** For kernel size of 6×1, inputs of 3P-FCUs are shown in Tab. I. The six outputs $y_0, \ldots, y_5$ can be gotten in parallel. Similar to 3-FFA, the parameters

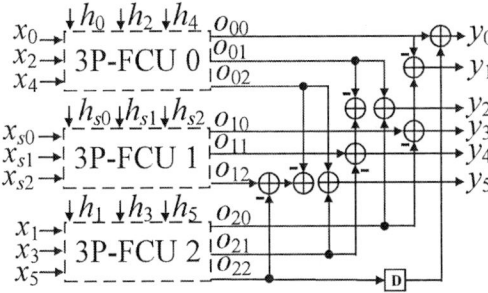

Fig. 3. 6-parallel fast convolution unit (6P-FCU).

in 6-FFA are in reverse order as well. Assume that the weights of a 6×1 kernel are $w_0, \ldots, w_2$, the relationships are $h_k = w_{5-k}$, where $k = 0, \ldots, 5$.

- 5×1 **convolution:** The procedure of 5×1 convolution is similar to that of 6×1 convolution. It is worth noting that h_5 is set to zero since 5×1 kernel contains only five parameters.

TABLE I
INPUTS OF 3P-FCUS IN 6P-FCUS.

	3×1	6×1	5×1
3P-FCU-0	x_0, x_1, x_2	x_0, x_2, x_4	x_0, x_2, x_4
3P-FCU-1	x_3, x_4, x_5	$x_0 + x_1, x_2 + x_3, x_4 + x_5$	$x_0 + x_1, x_2 + x_3, x_4 + x_5$
3P-FCU-2	x_6, x_7, x_8	x_1, x_3, x_5	x_1, x_3, x_5
3P-FCU-0	h_0, h_1, h_2	h_0, h_2, h_4	h_0, h_2, h_4
3P-FCU-1	h_0, h_1, h_2	$h_0 + h_1, h_2 + h_3, h_4 + h_5$	$h_0 + h_1, h_2 + h_3, h_4$
3P-FCU-2	h_0, h_1, h_2	h_1, h_3, h_5	$h_1, h_3, 0$

B. Processing Array

The Processing array (PA) is developed for the computation of 1×1 kernel. PA consists of p processing elements (PEs). The details of PA and PE are shown in Fig. 4, where $w_1, w_2, \ldots, w_p$ denote weight which are sent to the p-th PE. In each PE, six activations $x_1, x_2, \ldots, x_5$ are processed simultaneously. All activations in one PE share the same weight in a cycle. Another $6p$ activations are fetched from activation buffer and sent to PEs in the next cycle while the weights are stationary. Each PE is used to focus on the processing of one input channel. Hence, p channels are processed in parallel. A new set of weights is sent to PA after all pixels in channels are processed.

C. Adder Compressor

In this design, the $6P$ outputs of the PA (usually be dozens of values) will be added together to get a partial sum of output feature map. An adder tree with a long path is usually used here. Such design will incur a very long delay, thereby limiting the clock frequency and decrease the processing speed of the whole system. In order to solve this problem, dedicated $3 : 2$

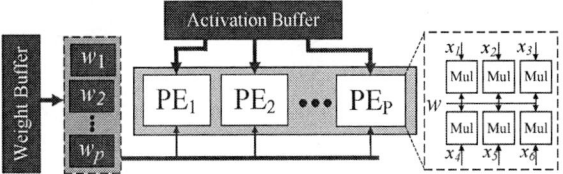

Fig. 4. Architecture of PA.

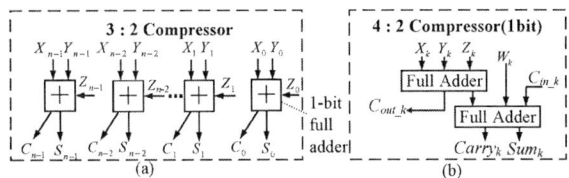

Fig. 5. Architecture of adder compressor. (a) $3 : 2$ Compressor. (b) $4 : 2$ Compressor.

compressor and $4 : 2$ compressor [21] are employed here to reduce the critical path.

For $3 : 2$ compressor, the output can be related as:

$$X + Y + Z = S + C \times 2 \quad , \tag{11}$$

where X, Y and Z are the three inputs of $3 : 2$ compressor. The outputs are denoted by S and C. Assume that the inputs and outputs of $3 : 2$ compressor have n bits. The structure is shown in Fig. 5(a). The $4 : 2$ compressor can be summarized as follows:

$$X + Y + Z + W + C_{in} = Carry \times 2 + Sum \quad . \tag{12}$$

The $4 : 2$ compressor is similar to $3 : 2$ compressor. The 1-bit structure of a $4 : 2$ compressor is shown in Fig. 5(b), where the results of the k-th compressor will be connected to the input of the $(k+1)$-th compressor. The delay of a $4 : 2$ compressor is twice of that of a 1-bit full adder. Moreover, since the carry chain is removed form the compressor, a great reduction in data path delay can be achieved.

D. Data Flow

The data flow designed for CPD-CNNs is illustrated in this subsection. Most existing CNN accelerators sent activations to off-chip DRAM after calculation of one layer. However, during the calculation of I_{sh} and I_{shw}, the pixels in i-th output channel will be only connected to the i-th input channel. Hence, the pixels in I_s can be sent to calculate the I_{sh} and I_{shw} rather than being sent to the buffers. Based on the special structure of CPD-CNNs, we develop the data flow which is shown in Alg. 1.

To enable efficient processing of 6P-FCU, our design process the activations by rows. The data flow consists of four steps:

- The first row in the first channel of I^s will be obtained first by the processing array (PA). The inputs of PA at this step are the first row in all channels of I.

Algorithm 1: The proposed data flow.

input : I, K^s, K^h, K^w, K^t
output: O

1 **for** $h = 0$ **to** $H - 1$ **do**
2   **for** $i = 0$ **to** $r - 1$ **do**
3          $/ * - - - - -pipelined - - - - - - */$
4          $I^s(h,:,i) =$CONV$(I(h,:,:,), K^s(:,i))$
5          $I^{sh}(h,:,i) =$DWCL$(I^s(h,:,i), K^h(:,i))$
6          $I^{shw}(h,:,i) =$DWCL$(I^{sh}(h,:,i), K^w(:,i))$
7   Buffer $I^{shw}(h,:,i)$
8      $O(h,:,:,) =$CONV$(I^{shw}(h,:,:,), K^t(:,:,))$

- Once the first row in the first channel of I^s is obtained, it will be used to calculate the I^{sh} rather than being sent to the buffer. The calculation of the first row in the first channel of the I^{shw} shares the similar procedure.

- Since the calculation of O needs all input channels in I^{shw}, the obtained activations in the first channel of I^{shw} will be stored in the on-chip buffer. The three steps above are pipelined. Before activations in all r channels are gotten, the obtained activations will be sent to buffer.

- After the first row in all r channels of I^{shw} are gotten, the first row in all C_{out} channels of O will be calculated and sent to the off-chip DRAM. The calculation of other rows of O can be gotten by executing the above steps iteratively. It is worth noting that newly fetched activations of I will be kept in buffer before activations of O are obtained.

E. Top Level Architecture

Based on the developed optimized computational logics and the presented data flow, an efficient hardware architecture for decomposed CNNs is proposed in this work. The top architecture is shown in Fig. 6. Modules which have not been introduced will be described in brief in the following.

The Accumulation Unit (ACCU) in Fig. 6 is an adder tree which takes the outputs of PA as inputs. Adder compressor is employed to reduce the critical path. For the conventional convolution, the pixels outputted by the PA are added together to get an output pixel or a partial sum. ACCU conduct such process in a partial parallel way.

The Multiply-Add Block (MAB) is developed for ICLs with kernel size of $1 \times d$. Each input pixel will be multiplied with d different weights in K^w to get m_i ($i = 0, 1, ..., d$), which will be added to the corresponding partial sum p_i. Partial sums will be stored in registers or outputted directly. MBA is also dedicated designed so that different kernel sizes can be supported. 1) For $d = 5$, one set of data (consisting of 6 input pixels) is sent to MAB where one of them is set to zero. 2) For $d = 3$, 2 sets of data (consisting of 3 input pixels each) are sent into MAB. Twice add operations are needed before output in this situation for $d = 3$.

The Activation SRAM (ASRAM) is used to store the outputs from MAB on one hand, namely I^{shw} with the size of RWb_a. The b_a here denotes bit width of one pixel in

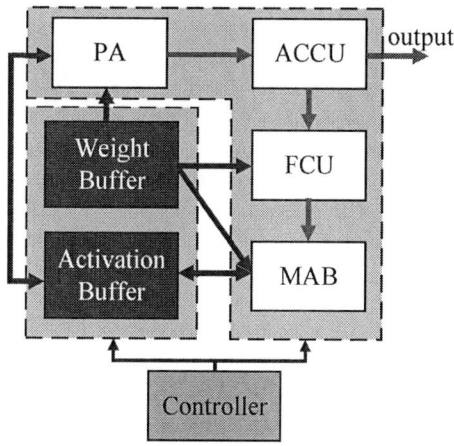

Fig. 6. Top level architecture.

I^{shw}. The data cached in BSRAM will be processed by PA to calculate O. On the other hand, the input activations of I will also be loaded from off-chip DRAM to ASRAM before being sent to PA. The size of activations of I buffered in ASRAM is pWb_a.

The Weight SRAM (WSRAM) stores the weights needed in our architecture. The number of weights for a decomposed convolutional layer is $(C_{in} + 2d + C_{out})R$. The size of WSRAM in need is $(C_{in} + 2d + C_{out})Rb_w$, where b_w denotes the bit width of weights.

Control Unit is used for configuration and managing the data flow.

IV. RESULTS AND ANALYSIS

In this section, the hardware resources, and comparisons with existing works will be evaluated.

Algorithmic performance will be presented in Tab. III. We adopt the experiments results in [7], where the storage requirement and computational complexity of AlexNet are reduced by 7.1x and 3.5x, respectively. The accuracy dropped is only 1.42%.

The performance of 6P-FCU is shown in Tab. IV. Compared with the commonly used multiply-accumulation unit (MAC), the proposed 6P-FCU can reduce the number of multiplications by 33%, 43%, and 50% for 3×1, 5×1, and 6×1 convolution,respectively. Hence, the proposed 6P-FCU can significantly improve the computational efficiency.

The hardware resources and comparisons will be given in the following part. A introduction on how the throughput is computed is given before the comparisons. the Assume that $s = 0$ when our architecture is processing I^s and $s = 1$ when O is calculated. The peak throughput T_{peak}, average throughput T_{aver} and average throughput for undecomposed CNNs T_{aver}^{typ} of our architecture are given as follows:

$$T_{peak} = N_m * f, \quad T_{aver}^{typ} = \frac{T_{aver}}{ratio_{speed}}, \quad (13)$$

TABLE II
HARDWARE RESOURCES AND PERFORMANCE EVALUATED ON ALEXNET[*].

	Layer	Technology (nm)	Quantization (bits)	Area (mm^2)	SRAM (KB)	Frequency (MHz)	Power (mW)	T_{typ} (TOP/s)	Efficiency (layers/s/MAC)	Normalized Comparison[†]
[9]	CONV2	65	16	12.25	-	200	288	0.034	0.57	1
[12]	CONV2	65	16	16	-	200	-	0.103	0.81	1.42
This work	CONV2	90	8	46.3	1806	650	620.2	4.589	3.54	6.21
[9]	CONV3	65	16	12.25	-	200	266	0.034	1.01	1
[12]	CONV3	65	16	16	-	200	-	0.103	1.13	1.12
This work	CONV3	90	8	16.4	578	650	546.4	3.115	7.12	7.05

[*] Decomposition ranks are 154 and 153 for CONV2 and CONV3, respectively. The supply voltage is 1.0V.
[†] This column is normalized Efficiency for better comparison.

TABLE III
PERFORMANCE OF CPD-CNNs [7].

	Top-5 (%)	Weights (M)	Comp Cost (M)
AlexNet	79.95	61.0	724
CPD-AlexNet	78.53	8.7	205

TABLE IV
HARDWARE RESOURCES AND PERFORMANCE

Kernel Size	$Ratio_{mul}$[1]	N_{mul}/N_{out}[2]	Efficiency[3]
3×1	100%	18/9	33%
5×1	94%	17/6	43%
6×1	100%	18/6	50%

[1] $Ratio_{mul}$ denotes the utilization ratio of multipliers of FCU.
[2] N_{mul}/N_{out} denotes the average number N_{mul} of multipliers used with N_{out} outputs.
[3] Efficiency denotes the ratio of multipliers saved compared to common FIR.

where N_m denotes the numbers of multipliers. f represents the clock frequency. $ratio_{speed}$ is the speed-up ratio for decomposed CNNs. The hardware performances and comparisons with two representative works are shown in Tab. IV, where improvements of 4x on computational efficiency are achieved.

V. CONCLUSION

In this paper, an efficient reconfigurable architecture along with a data flow are developed for decomposed CNNs. An novel 6-FFA along with an optimized 6P-FCU are presented in this paper. The proposed architecture can support various convolutional kernels. The adder tree is optimized with the adder compressor. This design can achieve a peak throughput of more than 3TOP/s which outperforms the previous works.

REFERENCES

[1] A. Krizhevsky, I. Sutskever, and G. E. Hinton, "Imagenet classification with deep convolutional neural networks," in *Advances in neural information processing systems*, 2012, pp. 1097–1105.
[2] K. Simonyan and A. Zisserman, "Very deep convolutional networks for large-scale image recognition," *arXiv preprint arXiv:1409.1556*, 2014.
[3] D. Amodei, R. Anubhai, E. Battenberg, C. Case, J. Casper, B. Catanzaro, J. Chen, M. Chrzanowski, A. Coates, G. Diamos, et al., "Deep speech 2: End-to-end speech recognition in english and mandarin," *arXiv preprint arXiv:1512.02595*, 2015.
[4] Y. Kim, E. Park, S. Yoo, T. Choi, L. Yang, and D. Shin, "Compression of deep convolutional neural networks for fast and low power mobile applications," *arXiv preprint arXiv:1511.06530*, 2015.
[5] V. Lebedev, Y. Ganin, M. Rakhuba, I. Oseledets, and V. Lempitsky, "Speeding-up convolutional neural networks using fine-tuned cp-decomposition," *arXiv preprint arXiv:1412.6553*, 2014.

[6] M. Jaderberg, A. Vedaldi, and A. Zisserman, "Speeding up convolutional neural networks with low rank expansions," *arXiv preprint arXiv:1405.3866*, 2014.
[7] A.Marcella and S.Lee, "Cp-decomposition with tensor power method for convolutional neural networks compression," in *IEEE International Conference on Big Data and Smart Computing*, 2017, pp. 115–118.
[8] M. Denil, B. Shakibi, L. Dinh, N. d. Freitas, et al., "Predicting parameters in deep learning," in *Advances in Neural Information Processing Systems*, 2013, pp. 2148–2156.
[9] Y. Chen, K. Tushar, S. E. Joel, and S. Vivienne, "Eyeriss: An energy-efficient reconfigurable accelerator for deep convolutional neural networks," *IEEE Journal of Solid-State Circuits*, vol. 52, no. 1, pp. 127–138, 2017.
[10] A. Manoj, C. Han, F. Michael, and M. Peter, "Fused-layer cnn accelerators," in *2016 49th Annual IEEE/ACM International Symposium on Microarchitecture (MICRO)*, 2016.
[11] Liqiang Lu, Yun Liang, Qingcheng Xiao, and Shengen Yan, "Evaluating fast algorithms for convolutional neural networks on fpgas," in *IEEE International Symposium on Field-Programmable Custom Computing Machines*, 2017, pp. 101–108.
[12] F. Tu, S. Yin, O. Peng, S. Tang, L. Liu, and S. Wei, "Deep convolutional neural network architecture with reconfigurable computation patterns," *IEEE Transactions on Very Large Scale Integration (VLSI) Systems*, 2017.
[13] Yongming Shen, Michael Ferdman, and Peter Milder, "Escher: A cnn accelerator with flexible buffering to minimize off-chip transfer," in *IEEE International Symposium on Field-Programmable Custom Computing Machines*, 2017.
[14] L. Du, Y. Du, Y. Li, and M. Chang, "A reconfigurable streaming deep convolutional neural network accelerator for internet of things," *IEEE Transactions on Circuits and Systems I-regular Papers*, pp. 1–11, 2017.
[15] J. Sim, J. Park, M. Kim, D. Bae, Y. Choi, and L. Kim, "14.6 a 1.42tops/w deep convolutional neural network recognition processor for intelligent ioe systems," *2016 IEEE International Solid-State Circuits Conference (ISSCC)*, pp. 264–265, 2016.
[16] H. Nakahara and T. Sasao, "A deep convolutional neural network based on nested residue number system," in *2015 25th International Conference on Field Programmable Logic and Applications (FPL)*, 2015, pp. 1–6.
[17] Q. Xiao, Y. Liang, L. Lu, S. Yan, and Y. Tai, "Exploring heterogeneous algorithms for accelerating deep convolutional neural networks on fpgas," in *Proceedings of the 54th Annual Design Automation Conference 2017 on*, 2017, p. 62.
[18] C. Szegedy, V. Vanhoucke, S. Ioffe, J. Shlens, and Z. Wojna, "Rethinking the inception architecture for computer vision," in *2016 IEEE Conference on Computer Vision and Pattern Recognition (CVPR)*, 2016, pp. 2818–2826.
[19] D. A. Parker and K. K. Parhi, "Low-area/power parallel fir digital filter implementations," *Journal of VLSI signal processing systems for signal, image and video technology*, vol. 17, no. 1, pp. 75–92, 1997.
[20] J. Wang, J. Lin, and Z. Wang, "Efficient convolution architectures for convolutional neural network," in *Wireless Communications & Signal Processing (WCSP), 2016 8th International Conference on*. IEEE, 2016, pp. 1–5.
[21] S. Hsiao, M. Jiang, and J. Yeh, "Design of high-speed low-power 3-2 counter and 4-2 compressor for fast multipliers," *Electronics Letters*, vol. 34, no. 4, pp. 341–343, 1998.

2018 IEEE Computer Society Annual Symposium on VLSI

Interconnect Delay Analysis for RRAM Crossbar based FPGA

Masanori Hashimoto[†]   Yuki Nakazawa[†]   Ryutaro Doi [†‡]   Jaehoon Yu[†]

[†] Department of Information Systems Engineering, Osaka University
Email: {hasimoto, doi.ryutaro, yu.jaehoon}@ist.osaka-u.ac.jp
[‡] Research Fellow of Japan Society for the Promotion of Science

Abstract—**FPGAs with novel RRAM-like nano-switches are under development for filling the gap between ASIC and FPGA. In these FPGAs, we need to analyze delay of signal interconnects that include several nano-switches with additional programming interconnects. This paper proposes simplified equivalent circuits for via-switch FPGAs, which enables analysis acceleration without loss of precision. Experimental results show that the proposed simplification increases the circuit simulation speed by 52x and 49x for single-fanout routes and multiple-fanout routes, respectively, on average while the calculation error is within 1.8% on average. When we further apply moment-based delay analysis called D2M to the simplified circuit, the overall average speed up reaches 2,500x and 600x for single-fanout routes and multiple-fanout routes, respectively.**

I. Introduction

Field programmable gate arrays (FPGAs) become more popular since the development cost of application specific integrated circuits (ASICs) is elevating due to the device miniaturization and larger scale integration. However, conventional FPGAs are still inferior to ASICs regarding operating speed, power consumption, and implementation area [1]. These drawbacks arise from a tremendous number of programmable switches that are included in FPGAs to acquire reconfigurability. In static random access memory (SRAM)-based FPGAs, which are the most widely used FPGAs, a programmable switch is composed of a transmission gate for switching and an SRAM cell to hold the on/off-state of the switch. These components consist of transistors, and hence the transmission gate has high resistance and large capacitance, and the SRAM cell having six transistors consumes large area. Therefore, SRAM-based programmable switches lead to the degradation of interconnect performance and area efficiency [2].

To overcome the drawbacks of conventional FPGAs, FPGAs that exploit resistive random access memories (RRAMs) as programmable switches instead of SRAM-based ones are widely studied [3]–[9]. In these RRAM based FPGAs, the crossbar, which has an RRAM switch at each intersection of horizontal signal wire and vertical signal wire, is responsible for signal routing. These RRAM-based FPGAs, however, require one or two access transistors per a programmable switch for switch programming. The access transistor is relatively large despite the small footprint of the RRAM-based switch, and hence it interferes with further area reduction. To eliminate access transistors, nonvolatile via-switches are

actively developed [10], [11]. The via-switch consists of atom switches, which are a kind of nonvolatile RRAMs developed for application to FPGAs, and varistors in place of access transistors.

Timing verification is indispensable in digital circuit design. Especially, interconnect delay analysis is important when the wire delay is dominant in the overall delay. The interconnect delay analysis needs to extract parasitic resistance and capacitance values from the given layout and generate equivalent circuit models. Then, we can obtain delay values by simulating the generated equivalent circuit models with, e.g., HSPICE. However, such a circuit simulation based analysis is too time-consuming in general since there are a huge number of wires on a chip. Consequently, fast analysis techniques are demanded and developed accepting small accuracy degradation.

Via-switch FPGA, which uses via-switch for routing crossbar and is studied in [12], has a different interconnect structure from conventional SRAM FPGA, and the degree of freedom of wiring is higher due to bidirectional signaling and on-demand repeater insertion. Therefore, timing analysis for conventional FPGA, which is hardly disclosed though, is not applicable to via-switch FPGA. On the other hand, timing analysis methods for ASICs can cope with arbitrary wiring patterns and hence they are expected to analyze interconnects in via-switch FPGA as well. Besides, even with methods developed for ASICs, we may need to pay attention to the fact that via-switches having several hundred Ω exist in the wiring topology whereas the wire resistance and capacitance per unit length vary at most a few times in ASIC. Therefore, we need to develop a method to generate compact equivalent circuit models taking into account via-switches. On the other hand, via-switch FPGA is organized in an array structure, and hence simplification and speeding-up exploiting this regularity are expected.

In this work, we investigate interconnect delay analysis suitable for via-switch FPGA. We first focus on generating equivalent circuit model taking into account the regular structure and via-switch existence and present a compact model generation method. Then, we apply a moment-based delay analysis method developed for ASICs and evaluate the accuracy and computational time for actual routing patterns in a filter application.

The remainder of this paper is organized as follows. Section II explains the structure of via-switch FPGA and its detailed equivalent circuit model. Section III discusses the

978-1-5386-7100-9/18 $31.00 © 2018 IEEE   522

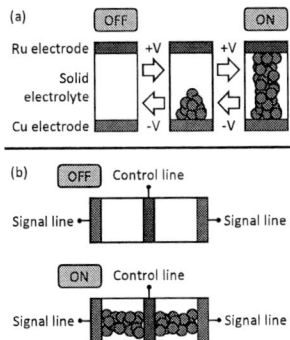

Fig. 1. Structure and operation of (a) atom switch and (b) CAS.

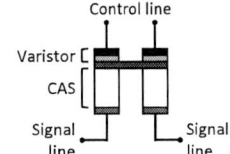

Fig. 2. Structure of 2V-1CAS via-switch.

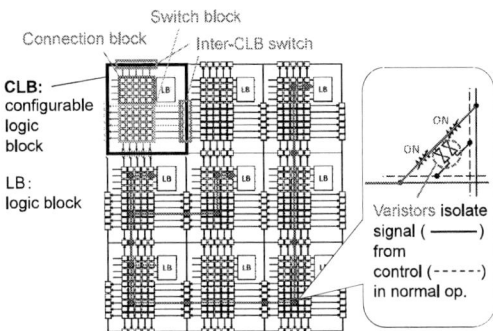

Fig. 3. Structure of via-switch FPGA.

simplification of equivalent circuit model, and Section IV shows the accuracy and computational times for actual wiring patterns. Concluding remarks are given in Section V.

II. VIA-SWITCH FPGA

A. Via-switch

The via-switch is a nonvolatile, rewritable, and compact switch that is developed to implement a crossbar switch by Banno et al. [10], and it is composed of atom switches and varistors. Here, we explain the device structure, functionality, and characteristics in the following.

The atom switch consists of a solid electrolyte sandwiched between copper (Cu) and ruthenium (Ru) electrodes as shown in Figure 1(a). By applying a positive voltage to the Cu electrode, a Cu bridge is formed in the solid electrolyte, and the switch turns on. On the other hand, when a negative voltage is applied, Cu atoms in the bridge are reverted to the Cu electrode, and then the switch turns off. The switching between on-state and off-state is repeatable, and each state is nonvolatile. For improving the device reliability, the complementary atom switch (CAS) is devised, where it consists of two atom switches connected in series with opposite direction as shown in Figure 1(b). In the programming of CAS, a pair of signal line and control line supply a programming voltage to each atom switch, and two atom switches are programmed sequentially. During normal operation, on the other hand, only signal lines are used for routing [9].

Figure 2 shows the structure of via-switch, and the varistor is connected to the control terminal of CAS. When a voltage higher than the threshold value (programming voltage) is applied between the signal and control lines, the varistor supplies programming current to an atom switch. On the other hand, the varistor isolates the control lines from the signal lines during normal operation [10]. Figure 2 shows two-varistor-one-CAS (2V-1CAS) structure, which is adopted via-switch FPGA proposed in [12]. This 2V-1CAS structure enables multiple fanouts.

Here, we summarize contribution of via-switch to FPGAs. The footprint, on-resistance, and capacitance are 18 F^2, 200 Ω, and 0.14 fF respectively [10], [12]. Thanks to these characteristics, the area efficiency and performance of via-switch FPGA are dramatically improved compared to SRAM-based

one. Ochi et al. report that the crossbar density is improved by up to 26x, and in this case, the delay and energy in the interconnection are reduced by 90% or more [12].

B. Via-switch FPGA structure

The via-switch FPGA consists of an array of configurable logic blocks (CLBs), and each CLB is composed of the logic block and crossbar where a via-switch is placed at each crossbar intersection of signal lines as shown in Figure 3 [12]. The via-switch in the crossbar is responsible for connection and disconnection between the horizontal and vertical signal lines. Besides, the top half of the crossbar provides a function of input and output multiplexers to the logic block and corresponds to connection block in conventional FPGAs. On the other hand, the bottom half of the crossbar, which corresponds to switch block, routes global interconnections. The logic block organizes combinational and sequential circuits.

Figure 3 also illustrates a signal routing with red lines. The routing starts from the bottom-left LB and ends at three LBs. The small red boxes correspond to on-state via-switches. Within CLB, the on-state via-switch connects the vertical and horizontal signal lines. Also, for making a connection to the adjacent CLBs, inter-CLB switches are turned on. The right figure details the via-switch that have two control lines in addition to two signal lines, whereas the control lines are omitted in the left figure. Two varistors are located between the red signal route and control lines. The control lines are used only for via-switch programming. After the programming, the varistors isolate the control lines from the signal route.

C. Equivalent circuit of interconnect in via-switch FPGA

Figure 4(a) depicts the three-dimensional structure of the intersection with a via-switch. There are two orthogonal signal lines and two orthogonal control lines. Figure 4(b) shows an equivalent circuit of the intersection, where the same color

978-1-5386-7100-9/18 $31.00 © 2018 IEEE

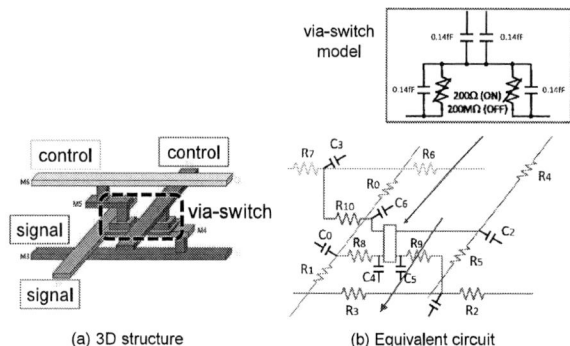

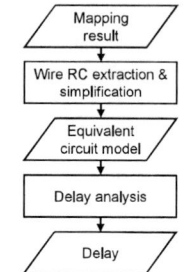

(a) 3D structure   (b) Equivalent circuit

Fig. 4. Via-switch structure and equivalent circuit.

indicates the same wire in Figure 4. The via-switch circuit model is found in the top. The two variable resistors are atom switches, and the top two capacitors correspond to the varistors [12]. Other two capacitors represent the input capacitors of off-state atom switches. When the atom switches are on, this input capacitors can be ignored. The on-resistance value of each via-switch depends on the amount of programming current, but in this work, it is assumed to be 400 Ω, which means 200Ω for each atom switch. The off-state via-switch is 400 MΩ.

By connecting this equivalent circuit model, we can construct an equivalent circuit model for each routing. In [12], interconnect delay and energy performances are evaluated using the constructed circuit model and a circuit simulator. On the other hand, there are many circuit elements in the constructed model, and hence its circuit simulation needs long CPU time, which is not suitable for timing verification and optimization in the design phase.

III. SIMPLIFYING CIRCUIT MODEL FOR DELAY ANALYSIS

Figure 5 shows a procedure of interconnect delay analysis. After the application mapping, we extract resistance and capacitance values from the physical layout and construct an equivalent circuit model. Then, we analyze the equivalent circuit model to derive the propagation delay times from the source node to the destination nodes. Here, there is a trade-off between the complexity of the equivalent circuit model and the accuracy of the circuit representation in general. Also, the complexity of the model affects the computational time of delay analysis. Therefore, it is important to construct a sufficiently accurate equivalent circuit model for efficient delay analysis. This section discusses such a circuit model construction for delay analysis of via-switch FPGA. As for the delay analysis, a number of publications are available, and there are several well-known methods, such as PRIMA [13] and D2M [14]. Depending on the required accuracy and allowable CPU time, one of them should be selected.

A. Control line elimination

We first eliminate the control lines that are necessary for via-switch programming but unnecessary for signal routing. We leave the two varistor capacitances of 0.14 fF in Figure 4(b), remove the blue and yellow wire resistances and related

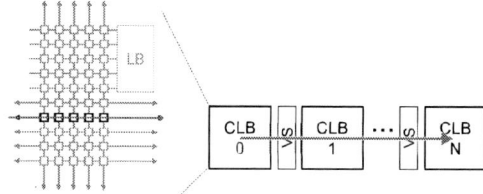

Fig. 5. Delay analysis procedure.

Fig. 6. Routing that propagates a signal through CLBs without bents.

TABLE I
IMPACT OF CONTROL LINE ELIMINATION.

Routing distance (#CLBs)	Delay [ps]		Error [%]
	w/ ctl. lines	w/o ctl. lines	
1	15.1	15.1	0.0
3	50.6	50.6	0.0
5	92.6	92.6	0.0
10	228	228	0.0
30	1190	1190	0.0

capacitances of control lines and connect the upper varistor capacitance terminals to ground. Due to the impedance of control lines, the varistor capacitances may work as smaller capacitances during the dynamic signal transition, but the conservative calculation is preferable in timing analysis, and hence this treatment is supposed to be reasonable.

We experimentally evaluate the impact of this control line elimination. Figure 6 shows the routing under evaluation, where a logic signal is propagating straight through several CLBs and inter-CLB via-switches. The detailed conditions are as follows, where the same setup is used throughout the paper.

- Crossbar size: 163×96
- Via-switch resistance: 400 Ω (ON) / 400 MΩ (OFF)
- Driver resistance: 1 kΩ
- Other signals: grounded

Table I shows the delay times before and after the control line elimination. We can see that the control line elimination does not degrade the accuracy of the circuit representation. In the following, the control lines are eliminated.

B. Simplification inside CLB without bents

Next, the signal line simplification is discussed. We take a strategy that applies the simplification to the route within a CLB, and the simplified CLB models and inter-CLB switches are connected to compose an entire equivalent circuit for delay analysis. This subsection focuses on the within-CLB routing without bents in Figure 7. The following discussion supposes

978-1-5386-7100-9/18 $31.00 © 2018 IEEE

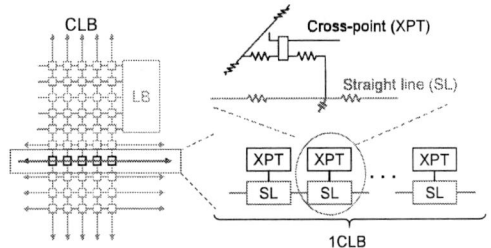

Fig. 7. Within-CLB routing without bents.

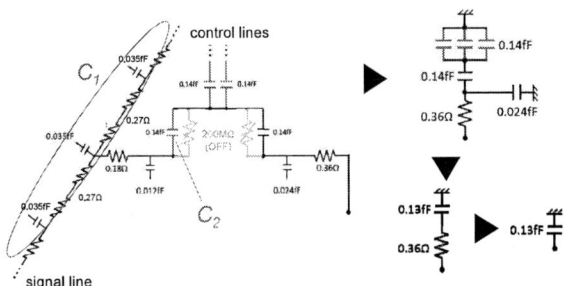

Fig. 8. Equivalent circuit simplification at Step 1.

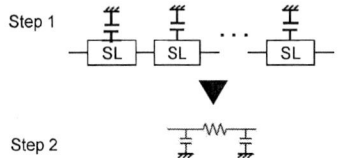

Fig. 9. Two-step simplification of equivalent circuit.

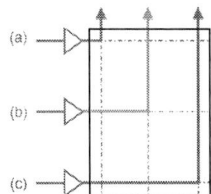

Fig. 10. Different left-to-top routes within a CLB.

horizontal direction, but the same discussion can be done for the vertical direction.

No bents in the CLB means that there are no on-state via-switches on the route. We first discuss how to treat off-state via-switches on the route of interest. Figure 7 shows that each intersection with a via-switch consists of straight lines (SL) and cross-points (XPT), where SL corresponds to the wire on the routing and XPT represents the off-state via-switch and the wires beyond the switch. For a propagating signal on the routing consisting of SLs, XPTs are distributed loads. We simplify this distributed RC tree with the following two steps.

1) Step1: Replacing XPT as capacitance: We first replace each XPT as a capacitance. Figure 8 explains this replacement. The right bottom terminal in the left figure is connected to SL, and hence we need to investigate how this left figure behaves as a load from this terminal. Now, the atom switches are off, and hence the switch resistive impedance is thought to be much higher than that of the parallel capacitance C_2. Next, we focus on C_1, which is the sum of the capacitances beyond the off-state atom switches. Here, the vertical signal wire is connected to 163 via-switches, and then C_1 is much larger than C_2. When two capacitances are connected in series and one is much larger than the other, the total impedance is determined by the smaller capacitance. Therefore, we eliminate the circuitry beyond C_2 and obtain the circuit in the top right figure. Next, we replace four capacitances as a single capacitance. Here, the time constant of this RC circuit is smaller than 0.1 fs, and hence the resistance can be ignored taking into account ordinary signal switching speed. Thus, we can replace an XPT as a capacitance of 0.13 fF as shown in Fig. 9.

2) Step2: Replacing RC tree with π circuit: An RC ladder circuit is often replaced with a CRC π circuit, which is shown in Fig. 9. The number of π stages is determined by the required accuracy. Here, the total resistance of the RC ladder circuit is smaller than the on the via-switch impedance of 400 Ω, and hence a single-π circuit is thought to be enough. In this case, the number of RC elements is reduced from 1,536 to 3.

C. Simplification inside CLB with bents

Next, we discuss the simplification of CLB that includes an on-state via-switch on the signal route of interest.

We first examine the impact of on-state via-switch location on the propagation delay. If the impact is very small, we can provide the same circuit model irrelevant to the location of the on-state via-switch. Let us suppose three routes depicted in Fig. 10, where (a) and (c) are the shortest and longest routes, respectively, and (b) is the route that consists of the middle horizontal line and the middle vertical line. We attached a driver whose output impedance was 1 kΩ and evaluated the delay from the driver input to the CLB output with a circuit simulator. Table II shows the result. We can see that the delay difference due to the location of the on-state via-switch is within 2% and quite limited. This is because the total capacitance of the route is identical independent of the on-state via-switch since the capacitances connected to the dotted lines in addition to those connected to the solid lines must be charged and discharged. This result suggests that the location-independent model is accurate enough in most cases, which simplifies the equivalent circuit construction for delay analysis.

According to the above observation, we construct an equivalent circuit model of a CLB that includes a single bent as shown in Fig. 11. The horizontal and vertical lines are represented as two-π RC models whose middle nodes correspond to the horizontally and vertically middle locations of the crossbar, respectively. Here, for realizing route (b) in Fig. 10, two-π model is selected while one-π model is chosen for a straight line without bents in Section III-B. Then, we attach an on-state via-switch model, which is also expressed as a two-π RC model, between these two middle nodes. Note that this model is independent of the signal direction, and hence the same

978-1-5386-7100-9/18 $31.00 © 2018 IEEE 525

TABLE II
DELAY DIFFERENCE DEPENDING ON LOCATION OF ON-STATE
VIA-SWITCH.

Route	(a)	(b)	(c)
Delay [ps]	40.9	41.7	42.3
Diff. from (b) [%]	-1.81	0	1.53

Fig. 12. Mapping result.

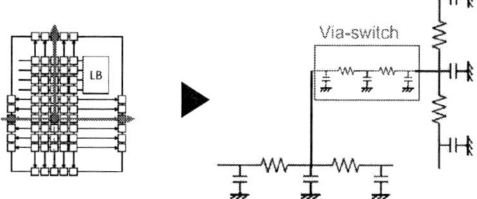

Fig. 11. Within-CLB routing with a bent and its equivalent circuit.

model can be used for all the signaling directions. Similarly, by inserting the via-switch model into the bent location, we can construct a within-CLB equivalent circuit model that has two bents on the route of interest.

Now, given the within-CLB models with and without bents, we can construct an equivalent circuit model for arbitrary routes by connecting the within-CLB models. The next section evaluates the accuracy of delay computation.

IV. EXPERIMENTAL RESULTS

This section evaluates the accuracy degradation and the speed-up of delay analysis for routing patterns found in an actual mapping result. Figure 12 shows an FIR (Finite Impulse Response) filter mapped to a 10×5 CLB array [12]. The crossbar size, via-switch resistance, and driver resistance are the same as Section III-A. We assume that the interconnects except the route of our interest are grounded.

A. Routes of single fanout

We first compare the delay times and CPU times before and after the circuit simplification discussed in the previous section. Here, the single-fanout routes are evaluated. Figure 13 shows the delay comparisons between the original circuit and simplified circuit, where each circuit is simulated by HSPICE. We can see that the delay computed with the simplified circuit is highly correlated with that of the original circuit. The averages of absolute and relative errors are 4.1 ps and 1.2 %, respectively. Figure 14 shows the histogram of the simulation speed-up ratio thanks to the circuit simplification. We can get 40x to 80x simulation speed-up in most cases, and the average and the maximum are 52x and 124x, respectively.

Next, we apply D2M [14] to the simplified circuit to compute the propagation delay aiming at further speed-up. We compare the delay and CPU times of the original circuit and HSPICE with those of the simplified circuit and D2M. Figure 15 shows the delay comparisons, which also includes Elmore delay results. We can see that the delay computed with the simplified circuit and D2M is highly correlated with that of the original circuit and HSPICE, whereas Elmore delay model overestimates the delay, as we expected. The averages of D2M

absolute and relative errors are 4.4 ps and 1.2 %, respectively. Note that only fanout-1 interconnects are analyzed whereas D2M may lose the accuracy for interconnects with branches. Our near future work investigates the accuracy of D2M for multiple-fanout interconnects. Figure 16 shows the histogram of the overall speed-up ratio. The average and the maximum speed-up ratios are 2,500x and 6,600x, respectively.

B. Routings of multiple fanouts

We carried out the similar evaluations to multiple-fanout routes. Figures 17 and 18 shows the delay and CPU time comparisons between the original circuit and the simplified circuit. We can see that the accuracy degradation is almost invisible and the average error is 1.8%. The average simulation speed-up is 49x.

We also evaluated the delay using D2M and Elmore delay model. Figure 19 shows the delay comparison. We can see that the accuracy of D2M is not as good as the single-fanout case of Figure 15 and the average error is 8.5%, which is due to several outliers. When the routing length is much different for different sink terminals, D2M suffers from delay overestimation of the route to the nearest sink. On the other hand, most of the routes are still analyzed well by D2M. Figure 18 shows the histogram of speed-up ratio, where the average speed-up is 600x. We observe that larger-fanout wires tend to have smaller speed-up ratios. There are two groups in the histogram, where the 400x group corresponds to large-fanout routes, such as eight and ten.

The above results indicate that we conclude that we should construct a delay analysis framework that screens some outlier routes for analyzing them with more sophisticated method, such as PRIMA, and computes the delays of remaining majority routes with D2M.

V. CONCLUSION

This paper investigated the interconnect delay analysis for via-switch FPGA mainly focusing on the equivalent circuit simplification. The off-state via-switches are replaced with capacitances, and the distributed RC ladder within a CLB is simplified as a CRC one-π model or CRC two-π models connected with CRC π model of via-switch. Experimental results for actual fanout-1 routing patterns show that the proposed circuit simplification attains 52x and 49x speed-up for single- and multiple-fanout routes on average, respectively, within 1.8% average error. Combination of moment-based delay analysis of D2m and the circuit simplification achieves

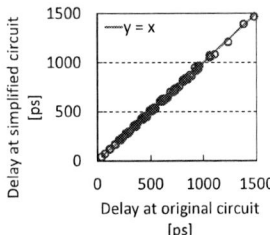

Fig. 13. Delay comparison between original and simplified circuits (single fanout).

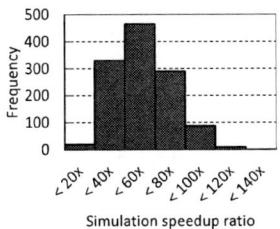

Fig. 14. CPU time comparison between original and simplified circuits (single fanout).

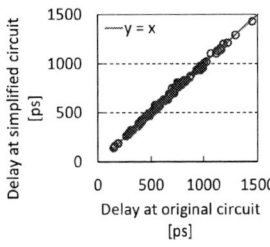

Fig. 17. Delay comparison between original and simplified circuits (multiple fanouts).

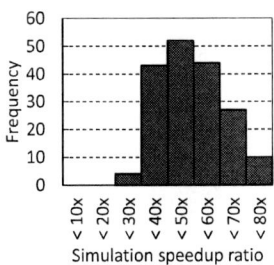

Fig. 18. CPU time comparison between original and simplified circuits (multiple fanouts).

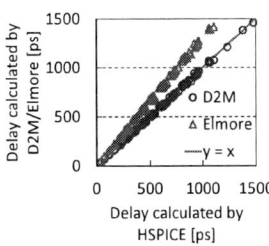

Fig. 15. Delay comparison between HSPICE with original circuit and D2M/Elmore with simplified circuit (single fanout).

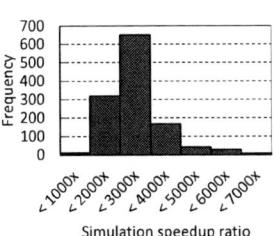

Fig. 16. CPU time comparison between HSPICE with original circuit and D2M with simplified circuit (single fanout).

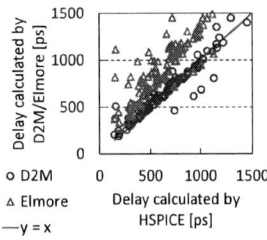

Fig. 19. Delay comparison between HSPICE with original circuit and D2M/Elmore with simplified circuit (multiple fanouts).

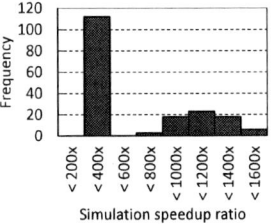

Fig. 20. CPU time comparison between HSPICE with original circuit and D2M with simplified circuit (multiple fanouts).

2,500x and 600x speed-up for single- and multiple-fanout routes on average, respectively. These results reveal that most of the routes can be processed with the proposed simplification and D2M and more sophisticated delay analysis methods should analyze the other routes.

Acknowledgments

This work was supported by JSPS KAKENHI Grant Number JP17J10008 and JST CREST Grant Number JPMJCR1432, Japan.

References

[1] I. Kuon and J. Rose, "Measuring the Gap Between FPGAs and ASICs," *IEEE Transactions on Computer-Aided Design of Integrated Circuits and Systems*, vol. 26, no. 2, pp. 203–215, Feb 2007.

[2] M. Lin, A. E. Gamal, Y. C. Lu, and S. Wong, "Performance Benefits of Monolithically Stacked 3-D FPGA," *IEEE Transactions on Computer-Aided Design of Integrated Circuits and Systems*, vol. 26, no. 2, pp. 216–229, Feb 2007.

[3] P. E. Gaillardon, D. Sacchetto, G. B. Beneventi, M. H. B. Jamaa, L. Perniola, F. Clermidy, I. O 'Connor, and G. D. Micheli, "Design and Architectural Assessment of 3-D Resistive Memory Technologies in FPGAs," *IEEE Transactions on Nanotechnology*, vol. 12, no. 1, pp. 40–50, Jan 2013.

[4] S. Tanachutiwat, M. Liu, and W. Wang, "FPGA Based on Integration of CMOS and RRAM," *IEEE Transactions on Very Large Scale Integration (VLSI) Systems*, vol. 19, no. 11, pp. 2023–2032, Nov 2011.

[5] J. Cong and B. Xiao, "FPGA-RPI: A Novel FPGA Architecture With RRAM-Based Programmable Interconnects," *IEEE Transactions on Very Large Scale Integration (VLSI) Systems*, vol. 22, no. 4, pp. 864–877, April 2014.

[6] X. Tang, P. E. Gaillardon, and G. D. Micheli, "A high-performance low-power near-Vt RRAM-based FPGA," in *2014 International Conference on Field-Programmable Technology (FPT)*, Dec 2014, pp. 207–214.

[7] Y. Y. Liauw, Z. Zhang, W. Kim, A. E. Gamal, and S. S. Wong, "Nonvolatile 3D-FPGA with monolithically stacked RRAM-based configuration memory," in *2012 IEEE International Solid-State Circuits Conference*, Feb 2012, pp. 406–408.

[8] K. Okamoto, M. Tada, T. Sakamoto, M. Miyamura, N. Banno, N. Iguchi, and H. Hada, "Conducting mechanism of atom switch with polymer solid-electrolyte," in *2011 International Electron Devices Meeting*, Dec 2011, pp. 12.2.1–12.2.4.

[9] M. Miyamura, T. Sakamoto, M. Tada, N. Banno, K. Okamoto, N. Iguchi, and H. Hada, "Low-power programmable-logic cell arrays using nonvolatile complementary atom switch," in *Fifteenth International Symposium on Quality Electronic Design*, March 2014, pp. 330–334.

[10] N. Banno, M. Tada, K. Okamoto, N. Iguchi, T. Sakamoto, M. Miyamura, Y. Tsuji, H. Hada, H. Ochi, H. Onodera, M. Hashimoto, and T. Sugibayashi, "A novel two-varistors (a-Si/SiN/a-Si) selected complementary atom switch (2V-1CAS) for nonvolatile crossbar switch with multiple fan-outs," in *2015 IEEE International Electron Devices Meeting (IEDM)*, Dec 2015, pp. 2.5.1–2.5.4.

[11] N. Banno, M. Tilda, K. Okamoto, N. Iguchi, T. Sakamoto, H. Hada, H. Ochi, H. Onodera, M. Hashimoto, and T. Sugibayashi, "50x20 crossbar switch block (CSB) with two-varistors (a-Si/SiN/a-Si) selected complementary atom switch for a highly-dense reconfigurable logic," in *2016 IEEE International Electron Devices Meeting (IEDM)*, Dec 2016, pp. 16.4.1–16.4.4.

[12] H. Ochi, K. Yamaguchi, T. Fujimoto, J. Hotate, T. Kishimoto, T. Higashi, T. Imagawa, R. Doi, M. Tada, T. Sugibayashi, W. Takahashi, K. Wakabayashi, H. Onodera, Y. Mitsuyama, J. Yu, and M. Hashimoto, "Via-Switch FPGA: Highly-Dense Mixed-Grained Reconfigurable Architecture with Overlay Via-Switch Crossbars," *IEEE Transactions on VLSI Systems*, in press.

[13] A. Odabasioglu, M. Celik and L. T. Pileggi, "PRIMA: passive reduced-order interconnect macromodeling algorithm," *IEEE Transactions on Computer-Aided Design of Integrated Circuits and Systems*, vol. 17, no. 8, pp. 645-654, Aug 1998.

[14] C. J. Alpert, A. Devgan and C. V. Kashyap, "RC delay metrics for performance optimization," *IEEE Transactions on Computer-Aided Design of Integrated Circuits and Systems*, vol. 20, no. 5, pp. 571-582, May 2001.

978-1-5386-7100-9/18 $31.00 © 2018 IEEE

2018 IEEE Computer Society Annual Symposium on VLSI

Enhancing the Robustness of Deep Neural Networks from "Smart" Compression

Tao Liu, Zihao Liu, Qi Liu and Wujie Wen
Florida International University
{tliu023, zliu021, qliu020, wwen}@fiu.edu

Abstract—Deep neural network (DNN) is recently presenting the human-level performance on many applications like computer vision and natural language processing. However, such a promising solution is also subject to ever-increasing security challenges. Recent studies show that benign inputs polluted with intentionally created imperceptible perturbations, namely "adversarial example", can easily mislead the decision making of DNN models. To mitigate adversarial attacks, many defense solutions are proposed accordingly, such as adversarial training, gradient masking etc. Orthogonal to those techniques, in this paper, we survey a family of "smart" compression based countermeasures to protect the DNNs against adversarial attacks. These approaches systematically target several fundamental entities in data processing of DNN models, including the input feature compression through JPEG, color depth reduction or spatial smoothing, and model compression by parameter sharing mechanism. We summarize the pros and cons of enhancing the robustness of DNNs from compression techniques, and hope that compression, originally aiming at the input or DNN model size reduction, can also function as defense technique to better help secure DNN models.

I. INTRODUCTION

As one of the most successful machine learning techniques, deep neural networks (DNNs) have nowadays achieved (or even surpassed) human-level performance across a wide range of intelligent applications in real-world, such as computer vision, natural language processing, and self-driving cars [1], [2]. However, recent studies show that DNNs can be easily fooled by adversarial examples [3], [4], i.e., the inputs with subtle perturbations that are hard to be perceived by human eyes, leading to the ever-increasing security concerns. To understand DNN's emerging security problem, many efforts have been emphasized on both the attack and defense sides, including adversarial examples crafting algorithms and the corresponding countermeasures [4], [5].

The existing mainstream defense solutions can be categorized as adversarial training [4], gradient masking [5], and input transformation [6], etc. Meanwhile, recent studies [6]–[9] show that conventional compression techniques can be also used to mitigate the adversarial attacks. For example, the malicious input perturbations can be potentially filtered by leveraging the data compression on the input [6], [7]. Besides, the deep compressed DNN models can effectively reduce the amplitude of gradients [8], [9], thus to suppress the effectiveness of created adversarial examples.

In this work, we intend to present an overview of the compression based mitigation techniques against adversarial attacks. The low-cost compression techniques can be smartly placed in several fundamental entities in data processing of DNN models, e.g. the input, the DNN model parameters etc, to safegurad the DNNs. We hope that this paper can provide a new perspective to enhance the robustness of the DNN models. The paper is organized as follows: First, we introduce the basics of DNNs and adversarial examples. Then we survey the existing compression based mitigation approaches, including both input compression and model compression. We also discuss a new defense method inspired from the model compression. Finally, we conclude this work.

II. ADVERSARIAL ATTACK ON DEEP NEURAL NETWORK

A. Deep Neural Network

Deep neural network consists of different types of layers with complex structures, to model the high-level data abstract and exhibits high effectiveness in cognitive applications by leveraging the deep cascaded layer structures [10]. The computation in DNN model can be formulated as:

$$F^{(\Theta)}(X) = Y \tag{1}$$

A group of random initialized weights Θ will be updated in training phase with input(s) $X \in \mathbb{R}^n$ and label(s) $Y \in \mathbb{R}^m$, in order to construct the decision function $F^{(\Theta)}$ for testing any new input. To enhance DNN's data processing capability, state-of-the-art DNN models usually contain a large volume of parameters (i.e. $\sim 60M$ [10]).

Many studies are also performed to compress DNN models to save the memory and computation resources in resource-limited hardware platforms, such as deep compression [11] and HashNet [12], Fig. 1 depicts the basic idea of an example DNN compression technique–HashNet. The HashNet uses a hash function to randomly group weights into hash buckets and neural network connections within the same hash bucket share the same parameter (namely "weights sharing"). Hence HashNet can significantly reduce the storage overhead. We formulate the DNN model through "model reshaping" (i.e., $\Theta \to \Gamma$) as:

$$F_c^{(\Gamma)}(X) = Y \tag{2}$$

Fig. 8 further shows the reshaped weight distributions of several compressed DNN models (i.e., HashNet [12], QuantizedNet [11], and CirculantNet [13]) compared with the original DNN model. The compressed DNN model $F_c^{(\Gamma)}$ usually exhibits different properties on its regression compared with the uncompressed version $F^{(\Theta)}$ since the decision boundary of F highly depends on weight parameters.

978-1-5386-7100-9/18 $31.00 © 2018 IEEE   528

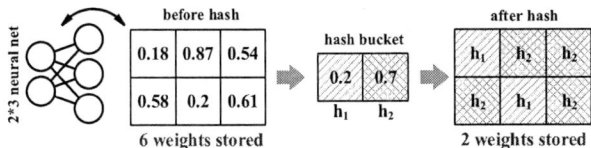

Fig. 1: Illustration of HashNet with highly shared weights.

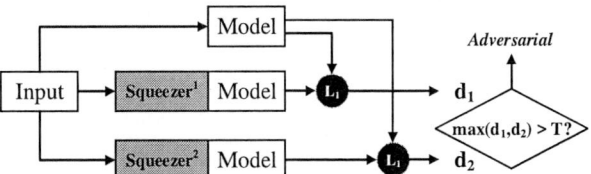

Fig. 3: Illustration of feature-squeezing framework.

B. Adversarial Attacks

In adversarial attacks, the goal is to mislead the classification results by using the well-crafted adversarial examples into DNN models, while the injected perturbations are too small to be perceived by human eyes. The derivation of minimized adversarial perturbations can be formulated as an optimization problem:

$$\underset{\delta_X}{\arg\min} \parallel \delta_X \parallel \ s.t. \ F^{(\Theta)}\left(X + \delta_X\right) = Y^* \tag{3}$$

Where the perturbation is δ_X, the adversarial goal is $Y^* \neq Y$, and crafted adversarial example is $X^* = X + \delta_X$. Many algorithms are proposed to solve Eq. 3:

1) Fast Gradient Sign Method (FGSM) [4]. A small perturbation ϵ is added into all input pixels along the direction of the sign of the gradient:

$$X' = X + \epsilon sign(\nabla_X L(X, Y)) \tag{4}$$

where L is the loss function, Y is the label and $\nabla_X L(X, Y)$ is the gradient of L w.r.t. input X. This method is designed to be fast, rather than optimal.

2) Basic Iterative gradient sign Method (BIM) [7]. BIM is the iterative version of FGSM by gradually gradually adding small perturbations α until reaching the upper bound ϵ or successful attack:

$$X'_0 = X, X'_i = X'_{i-1} + clip_\epsilon\left(\alpha sign(\nabla_X L(X, Y))\right) \tag{5}$$

Here the clipping equation, $clip_\epsilon(n)$, performs clipping on each pixel when it reaches ϵ.

3) Deepfool [14]. Deepfool uses geometrical knowledge to search the minimal adversarial perturbations:

$$\triangle(X, X') := \underset{Z}{\arg\min} ||Z||_2, \ s.t. \ F(X + Z) \neq F(X) \tag{6}$$

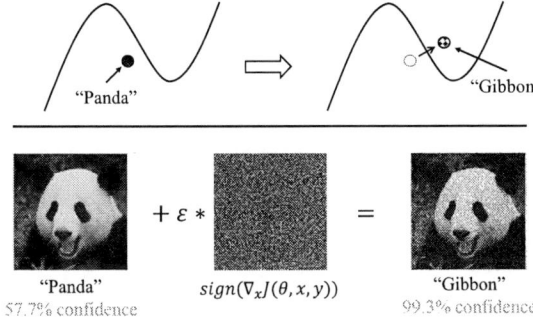

Fig. 2: Illustration of adversarial attacks (FGSM).

In this method, DNN is treated as a linear classifier and each class is separated by a hyper-plane. The algorithm finds the nearest hyper-plane from X and uses geometrical knowledge to calculate the projection distance.

4) Jacobian-based Saliency Map Approach (JSMA) [15]. JSMA method intends to distort the most sensitive input pixel(s) based on the salience map extracted from gradient w.r.t. the inputs:

$$S(X, t)[i] = \begin{cases} 0 \text{ if } \nabla_{X_i} F_t(X) < 0 \text{ or } \sum_{o \neq t} \nabla_{X_i} F_o(X) > 0 \\ \nabla_{X_i} F_t(X) | \sum_{o \neq t} \nabla_{X_i} F_o(X)| \text{ otherwise} \end{cases} \tag{7}$$

where F_t is the output of the model with false target class t and F_o is the output of the model with other classes.

Fig. 2 shows an example of FGSM based adversarial attacks on AlexNet [10] with $\epsilon = 0.05$. The image originally correctly classified as "Panda" (57.7% confidence) is now misclassified as "Gibbon" with a much higher confidence (99.3%) due to the applied small perturbations. These perturbations can "push" the original input to cross the trained decision function $F^{(\Theta)}$ with parameters Θ. *Therefore, the weights pattern is actually an important factor to impact the adversarial attacks.*

III. INPUT COMPRESSION

Since adversarial examples are generated by directly adding perturbations into the inputs ($X^* = X + \delta_X$), compression techniques that target the input data can be used to mitigate adversarial attacks. The general idea of input compression is to eliminate or reduce the perturbations from the benign features through an intended data loss (e.g., quantization error), therefore to mitigate or alleviate the attacks. In particular, for most computer vision tasks, image processing techniques can also function as the specific input compressions.

Kurakin et al. [7] test a set of simple image processing functions on adversarial examples. Xu et al. [6] propose the feature squeezing framework against adversarial attacks, by using color depth reduction (reducing data precision of pixels

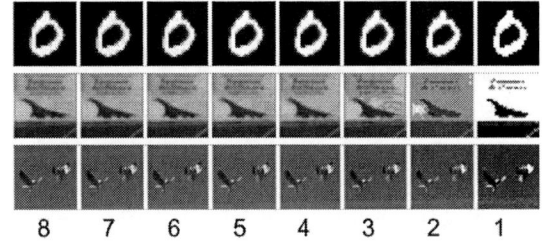

Fig. 4: Illustration of color depth reduction [6].

978-1-5386-7100-9/18 $31.00 © 2018 IEEE

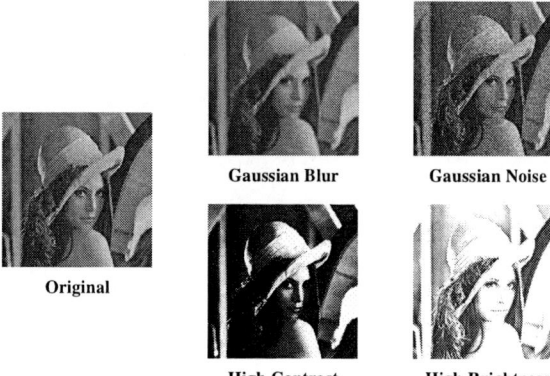

Fig. 5: Mitigating adversarial attacks through image processing.

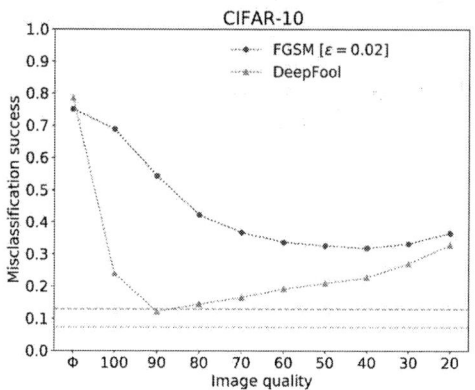

Fig. 6: Use JPEG compression to mitigate adversarial attacks [6].

in each color channel) and spatial smoothing (blur) techniques. Besides, JPEG compression is discussed by many works [16]–[18] to alleviate the strength of adversarial examples.

A. Feature-squeezing Framework

Unlike the other input compression (or image processing) techniques, feature-squeezing [6] is an ensemble adversarial detection framework. Fig. 3 illustrates an overview of feature-squeezing framework. The key of feature-squeezing is to compare the prediction of original DNN model with other "squeezed" predictions. Each "squeezer" can be implemented by a certain type of image processing techniques. If the original and squeezed inputs produce substantially different outputs (i.e. L_1 norm is larger than a threshold T), the input is likely to be adversarial. Based on such an ensemble design, the framework can predict the type of inputs and reject the potential adversarial examples by leveraging the outputs from two integrated squeezers – color depth reduction and spatial smoothing (blur).

Fig. 4 further shows input feature changes after applying the color depth reduction technique with image samples from MNIST (up), CIFAR-10 (mid) and ImageNet (bottom) dataset. The basic idea of this method is to reduce the data precision of each pixel for an input image. As shown in Fig. 4, the most left column shows the original image sample with 8-bit full precision while the most right column gives the highly quantized binary (1-bit) pixel data. Note that the color image contains three (RGB) data channels. Apparently, this method can potentially remove the tiny perturbations by largely lowering the data precision of each pixel. However, it may also inevitably remove the benign features (e.g. the missing sky on the most right CIFAR-10 sample in Fig. 4) and degrade the testing accuracy. Such an disadvantage becomes more prominent on the complex image samples from CIFAR-10 and ImageNet datasets [6].

B. Other Techniques

Image processing. Fig. 5 also illustrates other common image processing techniques used for mitigating adversarial examples, such as contrast adjustment, brightness adjustment, Gaussian blur and Gaussian noise. Their effectiveness against adversarial attacks varies from one another [7]: 1) changing the brightness and contrast does not affect adversarial examples much. The impacts on attack success rate are usually from $\sim 5\%$ to $\sim 20\%$; 2) Gaussian blur and noise may have better performance than others when handling adversarial examples. The attack success rate can be reduced by $\sim 80\%$. However, none of these image processing techniques can completely eliminate the perturbations in adversarial examples [7]. Besides, these approaches usually suffer from considerable accuracy degradation on benign data as the valid features are also aggressively distorted along with the perturbations, which can be even observed by human eyes (see Fig. 5).

JPEG compression. JPEG [19] is a popular lossy compression standard for digital images. A typical JPEG compression consists of a series of procedures, including image partition, discrete cosine transformation (DCT), quantization, reordering and entropy coding. Therefore, JPEG compression can partially remove the adversarial perturbations in the frequency domain. Many works [16]–[18] test the impact of JPEG compression on adversarial examples. As shown in Fig. 6, increasing JPEG compression rate (decreasing image quality) generally can better remove the adversarial perturbations on FGSM and DeepFool for the CIFAR-10 dataset at the beginning. However, the attack success rate can raise again after a certain sweet point for both cases. Besides, the significant degraded image quality can harm the overall classification accuracy. Moreover, the inconsistent performance on FGSM and DeepFool (different sweet points) indicate that the direct JPEG compression cannot be a generalized defense technique against a variety of adversarial attacks. Since the input compression and image processing techniques do not really improve the robustness of DNN models, they are usually employed as ensemble techniques along with other mitigation approaches.

IV. MODEL COMPRESSION

A. Adversarial Attacks on Compressed DNN Model

According to Eq. 2 and Eq. 3, the adversarial attacks on compressed DNN model can be formulated as:

$$\arg\min_{\delta_{X_c}} \| \delta_{X_H} \| \quad s.t. \; F_c^{(\Gamma)}(X + \delta_{X_c}) = Y^* \qquad (8)$$

978-1-5386-7100-9/18 $31.00 © 2018 IEEE

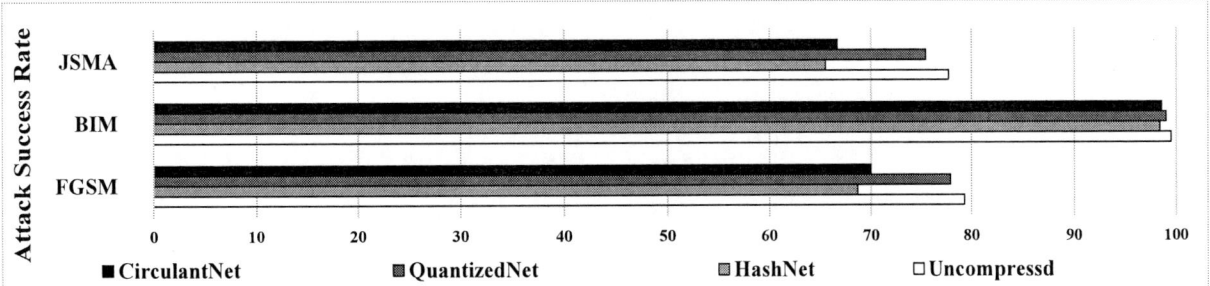

Fig. 7: Evaluation of adversarial attacks on compressed DNN models.

Apparently, due to the model reshaping ($F_c^{(\Gamma)} \neq F^{(\Theta)}$), the minimum input perturbations of $F_c^{(\Gamma)}$ (δ_{X_c}) will be less likely to be equal to that of ideal software DNN model, i.e., $F^{(\Theta)}$ (δ_X), even for the same adversarial target Y^*: $\min \delta_{X_c} \neq \min \delta_X$. To better explain such a problem, we define $\Gamma = \varphi(\Theta)$ as the compressing function φ. Since the DNN model reshaping shouldn't introduce considerable accuracy loss, the corresponding results after activation f should satisfy $f(\Theta X) \approx f(\varphi(\Theta)X)$. However, the augmented perturbations will be changed from $f(W\delta_X)$ to $f(\varphi(\Theta)\delta_{X_c})$ accordingly. Even for the same adversarial example ($\delta_{X_c} = \delta_X$), the responses from the two models will be quite different. Therefore, the compressed version of adversarial attacks will be more complicated.

B. Enhanced Robustness on Compressed DNN models.

We have investigated the adversarial attacks on compressed DNN models [8], [9]. The MNIST [20] dataset is selected as the benchmark to evaluate the success rate of selected adversarial attacks (i.e., FGSM, BIM, and JSMA) on three representative compressed DNN models (i.e., HashNet, QuantizedNet and CirculantNet). A typical compression rate (i.e., 1/16 on all compressed models) is also selected to guarantee that the classification accuracy of compressed models is comparable to that of the original model (i.e., $\sim 99\%$). The perturbation strength of FGSM and BIM is set to $\epsilon = 0.3$ and the number

of distorted pixels in JSMA is set to 150.

As Fig. 7 shows, the attack success rates on all compressed models are always lower than that of the uncompressed model. The attack success rate reduction of FGSM (JSMA) on Hash-Net and CirculantNet is more significant than that of QuantizedNet, i.e. from 79.33% (77.78%) to 68.73% (65.56%) and 69.98% (66.78%), respectively. The reason is that the weight pattern of QuantizedNet is still very close to the uncompressed version (see Fig. 8). Compared with the other two attacks, BIM is a stronger attack by using even smaller adversarial perturbations to achieve the attack. As expected, the difference of attack success rate between uncompressed and compressed models is quite marginal ($< 2\%$) for BIM. Surprisingly, these results indicate that the deep compressed models with much less number of parameters, always demonstrate better resilience against adversarial attacks than that of the original uncompressed version. This is substantially different from the empirical intuition that the more compressed DNN models should be more vulnerable to the input perturbations.

The reason can be explained through the analysis of two critical elements, i.e., weights and gradients. Since the weights of HashNet and CirculantNet go through the full training phase, their weight distributions become much broader (see Fig. 8). This is because such compressed models need to augment the remaining weights, in order to push the accuracy close to that of the uncompressed model due to the reduced number of unique weights. Consider that the gradient of loss directly depends on the activation and impacts the adversarial examples as follows:

$$\nabla_{x_i} J(x_i, t_j) = \sum_{j=1}^{n} w_{ji}(F(z_j) - t_j) \tag{9}$$

The enlarged weights will further push the activation $F(z_j)$ closer to the target t_j. Therefore, the gradient, which is dominated by the term $F(z_j) - t_j$, can be decreased accordingly, lowering the strength of generated adversarial examples.

C. Gradient Inhibition Method

Based on the observation that the "weight sharing" based compression can potentially mitigate the adversarial examples, a low-cost method, namely "Gradient Inhibition" is also proposed [8], [9]. The method has a control coefficient τ to

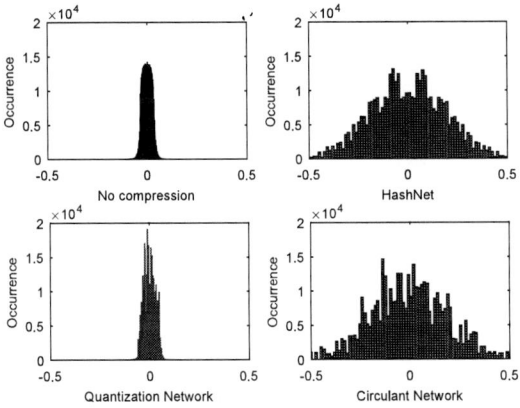

Fig. 8: Weights distributions on compressed DNN models.

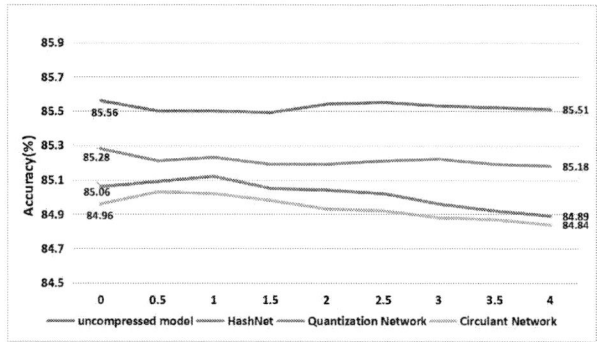

Fig. 9: **Classification accuracy with Gradient Inhibition on CIFAR-10.**

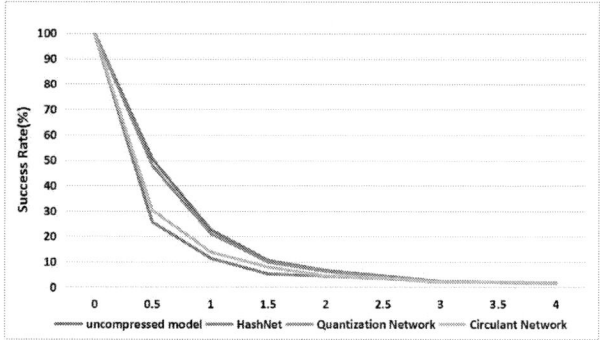

Fig. 10: **Mitigate BIM attack with Gradient Inhibition on CIFAR-10.**

slightly augment the weights and can be represented as:

$$W = W + \tau \cdot |W| \cdot sign(W) \qquad (10)$$

The basic idea is to add the weight perturbations proportional to the absolute magnitude of each W ($|W|$) along each individual direction of W ($sign(W)$). Without expensive training, this approach can significantly suppress the gradients thus to prevent the creation of adversarial examples while maintaining the classification accuracy.

As Fig. 9 shows, "Gradient Inhibition" does not impair the classification accuracy on both uncompressed and compressed models. The compressed DNN models with "Gradient Inhibition" (with various τ) can always maintain the similar accuracy as that of the uncompressed version for the CIFAR-10 dataset. The accuracy differences among all models are maintained below $\sim 0.2\%$. Fig. 10 further evaluates the defense efficiency of "Gradient Inhibition" against BIM attack on both compressed and uncompressed DNN models. The results indicate that "Gradient Inhibition" method can significantly lower the success rate of BIM attack on all selected models, due to the enhanced robustness of DNN models.

V. Conclusion

Deep neural network is subject to emerging security challenges induced by adversarial examples. In this work, we investigate a series of compression based countermeasures to protect the DNNs against adversarial attacks, from the input feature compression to DNN model compression. The basic principles of these techniques, as well as their pros and cons against adversarial attacks, have been systematically analyzed. We hope that the compression technique, originally targeting input or model size reduction, can also serve as a defense technique to safeguard the DNN models.

References

[1] G. Hinton *et al.*, "Deep neural networks for acoustic modeling in speech recognition: The shared views of four research groups," *IEEE Signal Processing Magazine*, vol. 29, no. 6, pp. 82–97, 2012.

[2] Y. LeCun *et al.*, "Deep learning," *nature*, vol. 521, no. 7553, p. 436, 2015.

[3] C. Szegedy *et al.*, "Intriguing properties of neural networks," *arXiv preprint arXiv:1312.6199*, 2013.

[4] I.J. Goodfellow *et al.*, "Explaining and harnessing adversarial examples," *arXiv preprint arXiv:1412.6572*, 2014.

[5] N. Papernot *et al.*, "Distillation as a defense to adversarial perturbations against deep neural networks," in *Security and Privacy (SP), 2016 IEEE Symposium on.* IEEE, 2016, pp. 582–597.

[6] W. Xu *et al.*, "Feature squeezing: Detecting adversarial examples in deep neural networks," *arXiv preprint arXiv:1704.01155*, 2017.

[7] A. Kurakin *et al.*, "Adversarial examples in the physical world," *arXiv preprint arXiv:1607.02533*, 2016.

[8] Q. Liu *et al.*, "Security analysis and enhancement of model compressed deep learning systems under adversarial attacks," in *Proceedings of the 23rd Asia and South Pacific Design Automation Conference.* IEEE Press, 2018, pp. 721–726.

[9] Q. Liu *et al.*, "Understanding adversarial attack and defense towards deep compressed neural networks," in *Cyber Sensing 2018*, vol. 10630. International Society for Optics and Photonics, 2018, p. 106300Q.

[10] A. Krizhevsky *et al.*, "Imagenet classification with deep convolutional neural networks," in *Advances in neural information processing systems*, 2012, pp. 1097–1105.

[11] S. Han *et al.*, "Deep compression: Compressing deep neural networks with pruning, trained quantization and huffman coding," *arXiv preprint arXiv:1510.00149*, 2015.

[12] Z. Cao *et al.*, "Hashnet: Deep learning to hash by continuation," *arXiv preprint arXiv:1702.00758*, 2017.

[13] Y. Cheng *et al.*, "An exploration of parameter redundancy in deep networks with circulant projections," in *Proceedings of the IEEE International Conference on Computer Vision*, 2015, pp. 2857–2865.

[14] S.M. Moosavi Dezfooli *et al.*, "Deepfool: a simple and accurate method to fool deep neural networks," in *Proceedings of 2016 IEEE Conference on Computer Vision and Pattern Recognition (CVPR)*, no. EPFL-CONF-218057, 2016.

[15] N. Papernot *et al.*, "The limitations of deep learning in adversarial settings," in *2016 IEEE European Symposium on Security and Privacy (EuroS&P).* IEEE, 2016, pp. 372–387.

[16] G.K. Dziugaite *et al.*, "A study of the effect of jpg compression on adversarial images," *arXiv preprint arXiv:1608.00853*, 2016.

[17] N. Das *et al.*, "Keeping the bad guys out: Protecting and vaccinating deep learning with jpeg compression," *arXiv preprint arXiv:1705.02900*, 2017.

[18] A.E. Aydemir *et al.*, "The effects of jpeg and jpeg2000 compression on attacks using adversarial examples," *arXiv preprint arXiv:1803.10418*, 2018.

[19] G.K. Wallace, "The jpeg still picture compression standard," *IEEE transactions on consumer electronics*, vol. 38, no. 1, pp. xviii–xxxiv, 1992.

[20] Y. LeCun, "The mnist database of handwritten digits," *http://yann. lecun. com/exdb/mnist/*, 1998.

978-1-5386-7100-9/18 $31.00 © 2018 IEEE

Accelerating Low Bit-Width Deep Convolution Neural Network in MRAM

Zhezhi He[*], Shaahin Angizi[†] and Deliang Fan[‡]

Department of Electrical and Computer Engineering, University of Central Florida, Orlando, FL 32816

Email: {[*]Elliot.he, [†]Angizi}@knights.ucf.edu, [‡]Dfan@ucf.edu

Abstract—Deep Convolution Neural Network (CNN) has achieved outstanding performance in image recognition over large scale dataset. However, pursuit of higher inference accuracy leads to CNN architecture with deeper layers and denser connections, which inevitably makes its hardware implementation demand more and more memory and computational resources. It can be interpreted as 'CNN power and memory wall'. Recent research efforts have significantly reduced both model size and computational complexity by using low bit-width weights, activations and gradients, while keeping reasonably good accuracy. In this work, we present different emerging non-volatile Magnetic Random Access Memory (MRAM) designs that could be leveraged to implement 'bit-wise in-memory convolution engine', which could simultaneously store network parameters and compute low bit-width convolution. Such new computing model leverages the 'in-memory computing' concept to accelerate CNN inference and reduce convolution energy consumption due to intrinsic logic-in-memory design and reduction of data communication.

Index Terms—Neural network acceleration, In-memory computing, Magnetic Random Access Memory

I. INTRODUCTION

In virtue of the fast development of the deep learning algorithm, design of a highly parallel and energy-efficient neural network accelerator has recently drawn tremendous research interest. On the one hand, a great deals of model compression techniques (e.g., pruning, parameter quantization, model encoding [1]) have been explored to lower the network model size, thus reducing the memory and computation cost. On the other hand, to optimize the massive data communication used in neural network inference, the 'wall' between memory and processor in classical Von-Neumann computing architecture has been crashed while the in-memory computing has emerged as a promising candidate for deep neural network acceleration [2]. In this work, we will focus on neural network binarization in algorithm level and its hardware implementation using Magnetic Random Access Memory (MRAM) based in-memory computing technique.

As the first work which successfully provides the deep convolutional neural network with binarized weight (i.e. -1 and +1), BinaryConnect [3] achieves considerably negligible accuracy degradation on small datasets like MNIST [4] and CIFAR-10 [5]. The following work, BNN [6], aggressively converts both weights and interlayer tensors into binary representation (i.e. -1 and +1), which correspondingly transforms the dominant computation of convolution layer from multiplication and additions to bulk bit-wise XNOR operation. Such

binarization scheme in BNN reveals the path for designing a neural network inference accelerator with extreme low bit-with weight/interlayer-tensor. However, the binarization method in BNN is criticized by researchers and engineers about its poor inference accuracy on large dataset, like ImageNet [7]. In order to solve the severe accuracy degradation degradation issue, a series of optimized neural network binarization techniques has been proposed and discussed in XNOR-Net [8], Dorefa-Net [9] and ABC-Net [10]. The essential tricks extracted from the aforementioned works can be summarized and further optimized as (1) introducing scaling factor for both binarized interlayer tensor and weight, and (2) using binarization functions with various thresholds (e.g. vanilla binarization function taken 0 as default threshold) to avoid information loss, which will be discussed in the main body.

Even though network quantization techniques have made deep neural network model compact and hardware-friendly, its inference computing is still hampered by the limited memory bandwidth. The newly announced NVIDIA Volta GPU [11] has adopted the brand new DDR-6 (i.e. 3D stacked memory) technologies which significantly improves the bandwidth by $4\times$ ($\sim$ 1 TB/s). However, the growing speed of memory consumption for the state-of-the-art network structure is overwhelming the hardware evolution speed. For example, the extremely dense structure in DenseNet [12] raises the memory usage in quadratic manner when the network goes deeper. Therefore, in-memory computing has emerged as a promising countermeasure to eliminate the long-distance data communication through merging the storage and computing component together. In this paper, we introduce two variant in-memory computing model using Spin-Transfer Torque Magnetic Random Access Memory (STT-MRAM) and domain wall based racetrack memory, respectively. As the emerging nonvolatile memory technologies, both STT-MRAM and racetrack memory own the characteristic of on-volatility, zero standby leakage, high write/read speed, compatibility with CMOS fabrication process, scalability, superior endurance, excellent retention time and high integration density [13], [14]. With the moderate adjustment on peripheral circuitry or device structure itself, we could perform bit-wise bulk logic (i.e. AND/OR/XOR and their complementary) taken two selected bit-cells as inputs. With the assistance of model binarization for both weight and interlayer tensor, the dominant amount of operations is converted into bit-wise XNOR operations which is suitable to take the aforementioned in-memory computing

architecture as the computing accelerator.

II. Low Bit-Width Network Quantization

In order to obtain the low bit-width quantized neural network with minimum accuracy degradation, one argument have been adopted in almost all the related works is that there is supposed to be two systems of model parameters: one model with full precision (i.e. 32bit floating point) parameter w and one model with corresponding quantized parameters w_q. For each parameters optimization iteration during the training process, the full precision w will be updated first, then the w_q will be calculated correspondingly. In this work, we mainly discuss the binary format weight and interlayer tensor. The mathematical formula for weight w and interlayer tensor x binarization firstly discussed in [6] can be described as:

$$\text{Forward}: \quad q = Sign(r) = \begin{cases} +1 & if \ r \geq 0 \\ -1 & otherwise \end{cases} \quad (1)$$

$$\text{Backward}: \quad \frac{\partial g}{\partial r} = \begin{cases} \frac{\partial g}{\partial q} & if \ |r| \leq 1 \\ 0 & otherwise \end{cases} \quad (2)$$

where q is the input (w or x) while r is the output (w_q or x_q) to the binarization function. In the forward path (i.e. inference phase), w or x is binarized using $Sign()$ function. Note that, the sign function owns zero derivatives almost everywhere, which makes it impossible to calculate the gradient using chain rule in backward path (i.e. training phase). Thus, the Straight-Through Estimator (STE) [6], [15] is applied to calculate gradient in this work. In the backward path, the input gradient of binarization activation function clones the gradient at output, if the input q is in the range from -1 to +1. Otherwise, the gradient is cancelled to preserve training performance. Furthermore, more works [8], [16] find that, in order to gain high accuracy, scaling factor plays vital role in the binarized neural network. There are several solutions to introduce the scaling factors, such as (1) inserting the batch normalization layer [17] right before the interlayer tensor binarization function [8], and (2) replacing the activation function with Parametric Rectifier Linear Unit (PReLU) [16]. Through our investigation, we find that inserting batch normalization layer before the interlayer tensor binarization function works better for scaling the interlayer tensor, since the distribution varies for each input. However, for the weight scaling factor computation, the current best solution is to iteratively compute based on the weight distribution during training, which can be written as:

$$\text{Forward}: \quad q = Sign(r) = \begin{cases} +E(|W_l|) & if \ r \geq 0 \\ -E(|W_l|) & otherwise \end{cases} \quad (3)$$

where $E(|W_l|)$ is the mean of the absolute value of full precision weights in l_{th} layer. Note that, the backward path of Eq. (3) is identical to Eq. (2). Owing to the computation of convolution layer or linear layer is linear transformation (i.e. dot-product), the layer-wise weight scaling factor $E(|W_l|)$ can be extracted and integrated with following activation function

or batch normalization function to perform element-wise computation. Therefore, the computation for one convolution layer or linear layer can be described as:

$$\boldsymbol{x}_q^T \cdot \boldsymbol{w}_q = \sum_{i=1}^{N} x_{q,i} \cdot w_{q,i} \quad \forall x_{q,i}, w_{q,i} \in \{-1, +1\} \quad (4)$$

where $\boldsymbol{x}_q$ and $\boldsymbol{w}_q$ are the vectorized form of quantized interlayer tensor and weight. N is the vector size of $\boldsymbol{x}_q$. Since $\boldsymbol{x}_q$ and $\boldsymbol{w}_q$ only consist of -1 and +1, in order to map the computation of $x_{q,i} \cdot w_{q,i}$ into hardware, we use single bit of 0 and 1 to represent -1 and +1 respectively. For the converted form $x'_{q,i}$ and $w'_{q,i}$, the computation of $x_{q,i} \cdot w_{q,i}$ is equivalent to $\text{XNOR}(x'_{q,i}, x'_{q,i})$ as the truth tables shown in Table I and Table II.

TABLE I
TRUTH TABLE FOR ORIGINAL BINARIZED DOT-PRODUCT

Input		Output
x_i	w_i	$x_i \cdot w_i$
-1	-1	+1
-1	+1	-1
+1	-1	-1
+1	+1	+1

TABLE II
TRUTH TABLE FOR CONVERTED BINARIZED DOT-PRODUCT

Input		Output
x'_i	w'_i	$\text{XNOR}(x'_i, w'_i)$
0	0	1
0	1	0
1	0	0
1	1	1

Thus, the computation in Eq. (4) can be transformed into bitwise XNOR, and bit-count operations without multiplications [9]:

$$\boldsymbol{x}_q^T \cdot \boldsymbol{w}_q = 2 \cdot bitcount(\text{XNOR}(\boldsymbol{x}_q, \boldsymbol{w}_q)) - N \quad (5)$$

where $bitcount()$ function function count the number of '1'. Based on such reformulated dot-product equation, we can utilize the counter and XNOR logic gate as the primary computing element to accelerate the neural network inference in energy-efficient manner. Moreover, in order to encounter the aforementioned limited memory bandwidth issue in the Section I, we leverage the MRAM based in-memory computing techniques to further boost the performance of such binarized neural network accelerator which will be elaborated in the following sections.

Beyond that, for contradicting the argument that network binarization can lead to significant accuracy loss which makes such extreme low bit-width quantization scheme not practical in the real world applications, recent research efforts in [10], [16] has brought up multiple binarization method to compensate the information loss due to the aggressive quantization. Thus, further integrating such multiple binarization with in-memory computing technique will provide accurate deep neural network inference result with high throughput.

III. MRAM-Based Neural Network Acceleration

In this section, we will first provide an over-view introduction for low bit-width CNN accelerator in block level. Then, as the primary computing unit, several different designs utilizing STT-MRAM array and racetrack memory based magnetic crossbar will be introduced to perform bit-wise XNOR operations.

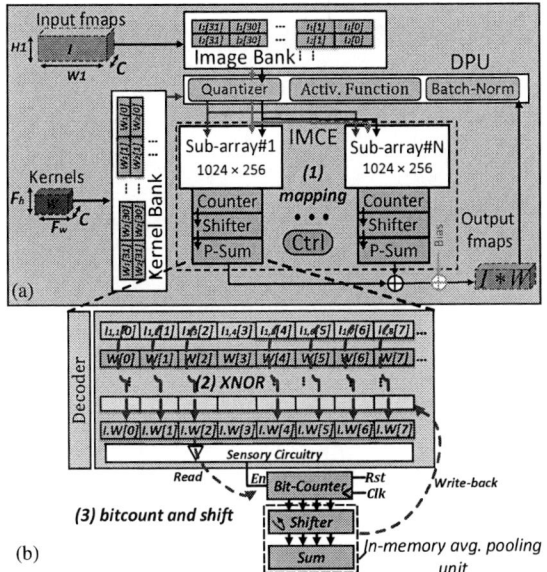

Fig. 1. (a) General overview of the CNN accelerator with image bank, kernel bank, computational sub-arrays, and DPU, (b) Bit-wise IMCE's sub-array.

The general overview of the system architecture for performing low bit-width CNN is shown in Fig. 1.a [18]. This architecture mainly consists of Image Bank, Kernel Bank, bit-wise In-Memory Convolution Engine (IMCE), and Digital Processing Unit (DPU). Since linear layer can be visualized as convolution layer with 1×1 kernel size, the computation of linear layer could be implemented by the same convolution accelerator as well. Assuming the Input feature maps (I) and Kernels (W) are initially stored in the Image Banks and Kernel Banks of memory, respectively. As depicted in Fig. 1.a, inputs need to be constantly quantized before mapping into computational sub-arrays. However, quantized shared kernels can be utilized for different inputs. This step is performed using DPU's Quantizer and then the results are mapped to IMCE's sub-arrays (Fig. 1.b). For the realization of bit-wise IMCE, several different in-memory computing techniques are readily utilized such that ultra-efficient and parallel in-memory *XNOR* operations required for convolutions can be handled. The functionality of peripheral components are elaborated as follow:

- **Quantizer**: This unit binarize the interlayer tensor or weight w.r.t Eq. (1), Eq. (2) and Eq. (3).
- **Batch-Norm**: Batch Normalization layer [17] alleviates the information loss caused by weight and activation binarization, through normalizing the input mini-batch to have zero mean and unit variance. The batch normalization function can be considered as an affine function $y = kx + b$ [19], where

$$k = \frac{\gamma}{\sqrt{\sigma^2 + \epsilon}} \quad \text{and} \quad b = \beta - \frac{\mu\gamma}{\sqrt{\sigma^2 + \epsilon}} \quad (6)$$

- **Activ. Function**: the activation function module perform the element-wise computation, which normally takes ReLU as the activation function.

Moreover, as the vital component of the low bit-width accelerator, IMCE mainly consists of in-memory computing sub-array, bit-counter, shifter. The in-memory computing sub-array plays the role of both on-chip buffer and the computing bit-wise computing unit. In this work, we enumerate two variant in-memory computing designs leveraging STT-MRAM [2] and racetrack memory based magnetic crossbar [20] that are in-charge for the most computationally-intensive operations i.e. *XNOR*. After this computation, a CMOS counter unit is devised to count the number of +1 elements within generated XNOR vector. According to Eq. (5), the result of bitcount operation should be multiplied by 2, which is implemented by shifter unit in our design. Then, the produced binary result is processed in parallel by partial sum (subtract) units.

A. STT-MRAM array

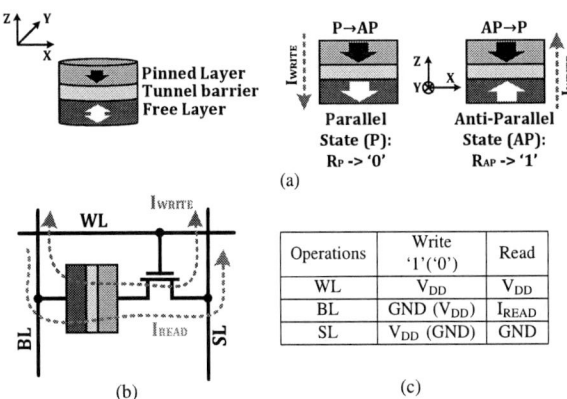

Fig. 2. (a) Device structure of conventional magnetic tunnel junction in parallel- and anti-parallel states, with current-induced spin-transfer torque switching scheme. (b) Bit-cell schematic of 1T1R STT-MRAM. (c) Biasing conditions for STT-MRAM operations.

1) memory mode: A typical Magnetic Tunnel Junction (MTJ) structure, as shown in Fig.2a, consists of two ferromagnetic layers with a tunnel barrier sandwiched between them. Due to the Tunnel MagnetoResistance (TMR) effect [21], the resistance of MTJ is high (low) when the magnetization of two ferromagnetic layers are in anti-parallel (parallel) state. The *TMR ratio* is defined as $(R_{AP}-R_P)/R_P$, which may vary from 10% to 400% depending on materials and temperature [21]. Thus, the data are stored as the magnetization direction in the free layer, which could be programmed through current induced Spin-Transfer Torque (STT). Note that, the MTJ with Perpendicular Magnetic Anisotropy (PMA) is used in this work. The 1T1R bit-cell is widely used in the typical STT-MRAM design, as depicted in Fig. 2b, which is correspondingly controlled by Bit Line (BL), Word Line (WL) and Source Line (SL). The biasing conditions of memory read and write are presented in Fig. 2c. For both memory read and write operation, the WL is enabled, which turns on the access transistor. Then, a voltage drop $-V_{DD}$ or $+V_{DD}$ is applied across the BL and SL, in order to realize write '1' or '0' respectively. For memory read, a sensing current (I_{READ}) is applied on the

BL and consequently generates a sensing voltage, which can be detected by sense amplifier.

B. computing mode

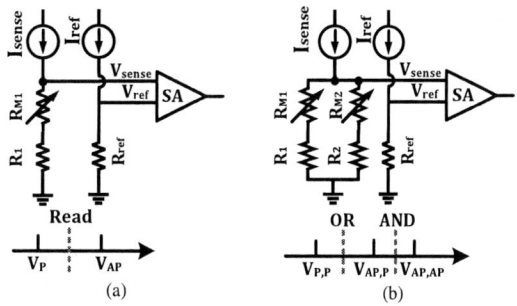

Fig. 3. The idea of voltage comparison between V_{sense} and V_{ref} for (a) memory read and (b) in-memory logic operation.

The key idea to perform memory read and in-memory computing is to choose different thresholds when sensing the selected memory cell(s). As shown in Fig. 3a, for memory read operation, a single memory cell is addressed and routed in the memory read path to generate a sense voltage (V_{sense}), which will be compared with a reference voltage (V_{ref}). Owing to the parallel- or anti-parallel state of selected STT-MRAM bit-cell (R_{M1}), the sense voltage are V_P or V_{AP} ($V_P < V_{AP}$) respectively. Thus, through setting the reference voltage at ($V_{AP}+V_P$)/2, the sense amplifier outputs binary '1' when $V_{sense} > V_{ref}$, otherwise the sense amplifier outputs '0'. the sensing-based method of in-memory Boolean computing is depicted in Fig. 3b, where two memory bit-cells (R_{M1} and R_{M2}) are sensed simultaneously. R_1 and R_2 corresponds to the access transistors within the sensing path. Owing to the different resistance combinations of two selected STT-MRAM bit-cells (i.e. R_{AP}, R_{AP}; R_{AP}, R_P; R_P, R_P), three different sense voltages V_{sense} ($V_{AP,AP}$; $V_{AP,P}$; $V_{P,P}$) could be generated respectively. Consider setting the reference voltage as ($V_{AP,AP}+V_{AP,P}$)/2 through tuning reference resistance, the sense amplifier only outputs '1' when both selected STT-MRAMs are in anti-parallel state ($V_{sense} > V_{ref}$). Thus, this sensing operation with modified reference voltage performs an AND logic operation taken the binary data stored in R_{M1} and R_{M2} as logic inputs. Similarly, when the reference voltage is shifted to ($V_{P,P}+V_{AP,P}$)/2, the OR logic operation can be performed as well. Therefore, through tuning the reference voltage for comparison, the sense amplifier can perform reconfigurable in-memory computations.

In order to build the in-memory computing STT-MRAM array to perform bitwise XNOR operation, further modification is made on the peripheral control and sensing circuit. Fig. 4a depicts the architecture of STT-MRAM based IMCE, where memory read/write path of the specific bit-cell is enabled by the row/column decoders. As shown in Fig. 4b, the modified row/column decoders can enable either single line (memory write/read) or double lines (bit-wise Boolean computation), depending on the addresses (Addr1 and Addr2) provided. For memory write, the voltage drop across BL and SL is generated

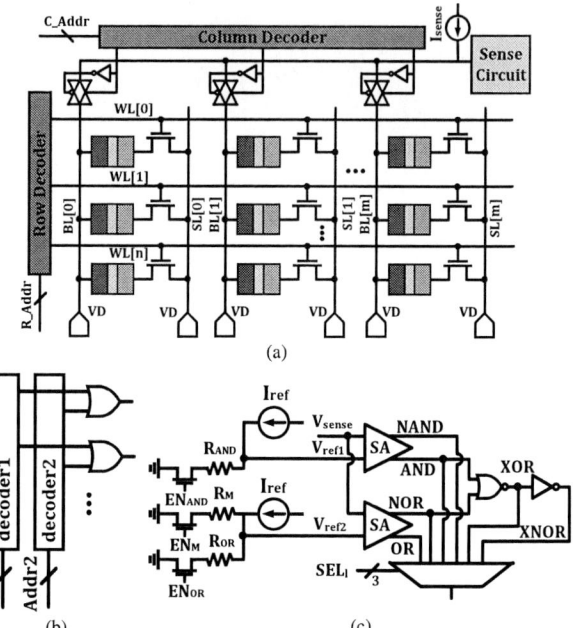

Fig. 4. (a) The modified sub-array structure of STT-MRAM.(b) Modified decoder which provides single/multiple lines enable function. (c) Modified sense circuit for regular memory W/R and in-memory computing operations.

by the Voltage Drivers (VD), which realize the fast memory switching [2]. For memory read and Boolean computation, a small sense current ($I_{sense} \simeq 3\mu A$) is injected into the read path to generate a sense voltage (V_{sense}), which is taken as the input of modified sense circuit. As shown in Fig. 4c, the modified sense circuit can provide memory read, AND/NAND, OR/NOR and XOR/XNOR functions, through combining two sense amplifiers (i.e. StrongARM latch [22]), external CMOS logic gate and control units. Owing to the complementary outputs of SA, the modified sensing circuit can provide NAND and NOR without additional cost. Moreover, according to the Boolean representation of $p \oplus q = (p \vee q) \wedge \overline{(p \wedge q)}$, the XOR and XNOR can be realized with two sense amplifiers (i.e. performing AND and NOR logic respectively) and an additional CMOS NOR gate. The operation of such sense circuit is determined by the control signals (EN_{AND}, EN_{OR} and EN_M), while the desired result is acquired by the selection signal (SEL) of the output multiplexer. Note that, only one sense amplifier is used during AND/OR/memory-read operation, in order to reduce the power consumption of sensing.

Moreover, transient simulation result of the sense circuit under a 2ns period clock signal (CLK) is included in Fig. 5 to validate the in-memory computing functionality. It takes the data stored in MRAM1 and MRAM2 as inputs. When CLK is high, the sense amplifier is in pre-charge phase and the output is reset to '0'. When CLK is low, the sense amplifier is in sampling phase, and generates logic computation result depending on the reference voltage configuration. V_{cmp} includes all the input signals of SAs, which are sense voltage (V_{sense}) and two reference voltage (V_{ref1} and V_{ref2}). V_{ref1} is set to ($V_{AP,AP}+V_{AP,P}$)/2, and V_{ref2} is set to ($V_{P,P}+V_{AP,P}$)/2, for

978-1-5386-7100-9/18 $31.00 © 2018 IEEE

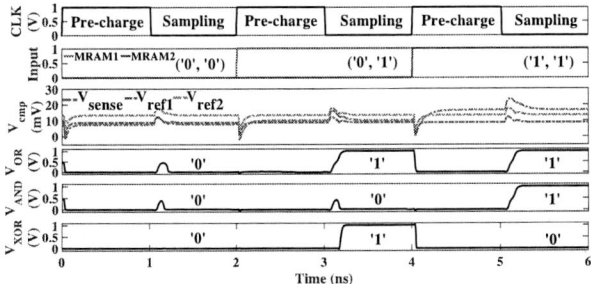

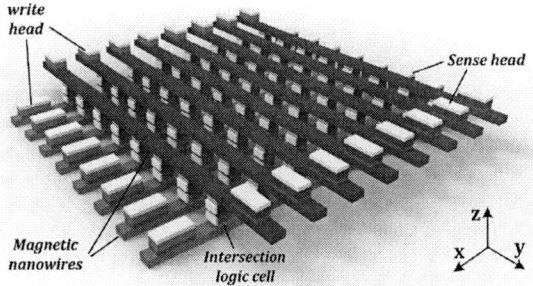

Fig. 5. Transient simulation results of in-memory computing operations (i.e. AND, OR and XOR).

performing AND and OR respectively. Through the combination of two SAs in Fig. 4c, XOR is correctly performed as well. The XNOR bit-wise logic could be easily obtain with an cascaded NOT gate.

C. Magnetic crossbar

In this subsection, we describe a magnetic crossbar architecture consisting of perpendicularly coupled magnetic domain wall motion racetracks [20], which is able to morph between two modes: memory and computation mode. Different from that the STT-MRAM based in-memory computing structure which can perform AND/OR/XOR and theirs complementary, the introduced magnetic crossbar can only perform XOR/XNOR.

1) memory mode: In the memory mode, all the magnetic domain wall motion racetrack nano-wires are employed as conventional magnetic racetrack memory for non-volatile multi-bit data storage [23]. The device structure of the computational magnetic crossbar design is shown in Fig. 6a, which mainly constitutes of equally spaced ferromagnetic nanowires both longitudinally and latitudinally. Each nanowire could work individually as a normal domain wall racetrack memory to store the interlayer tensor and weight, where binary data are represented by the magnetization directions and stored in the form of domain wall pair train within the nanowires [23]. In this work, we define +z oriented magnetization (in red) as binary '1', and -z oriented magnetization (in blue) as binary '0'. Write and read operations are performed through the write head (spin polarizer) and read head (sensing MTJ) mounted on the two sides of nanowires [20].

2) compute mode: In order to realize the in-memory XOR/XNOR operation, we modify the normal racetrack memory through introducing XOR/XNOR joint (i.e. a multi-layer structure) at the intersection of each longitudinal and latitudinal nanowire. It consists of coupling insulator (FeCo-Oxide [24]), ferromagnetic layer (FeCoB), tunnel barrier(MgO), ferromagnetic layer (FeCoB) and coupling insulator (FeCo-Oxide), from top to bottom as shown in Fig. 6b. Owing to the magnetic coupling effect [24], two ferromagnetic layers are consistently staying in the identical magnetization state with corresponding top and bottom nanowire as indicated by the white arrow in Fig. 6b. When the domain wall nanowire is

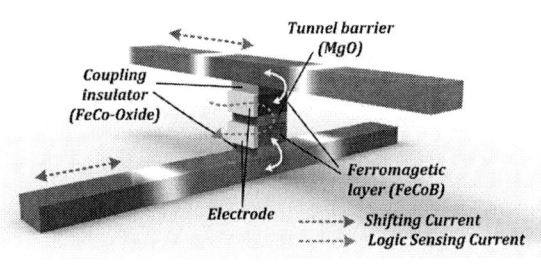

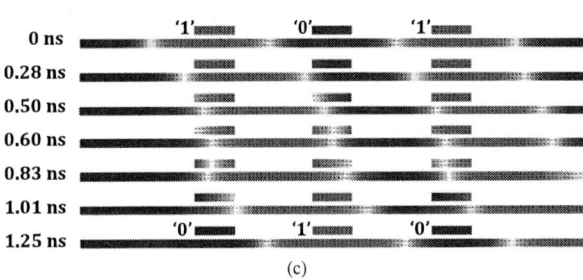

Fig. 6. (a) The device structure of magnetic crossbar in the size of 8 × 8. (b) The XOR logic cell in the intersection of longitudinal and latitudinal domain wall nanowires. (c) The micromagnetic simulation of current induced domain wall motion and magnetic coupling effect.

written with a new data followed by a domain wall pair shifting operation, the ferromagnetic layers switch their magnetization direction simultaneously due to magnetic coupling. As the transient micromagnetic simulation result shown in Fig. 6c, the initial binary bits (1, 0, 1) are stored in the bottom nanowire at 0ns. The coupled ferromagnetic layers have the same magnetization with the corresponding bottom domain wall nanowire. Through injecting lateral shifting current into the bottom nanowire, the domain wall pairs start to shift from L.H.S to R.H.S, which simultaneously rotate the magnetization of the coupled ferromagnetic layers. As shown in Fig. 6b,the ferromagnetic layers that respectively couple with the top and bottom magnetic nanowires can jointly work with the tunnel barrier layer to form a Magnetic Tunneling Junction (MTJ). In this design, the binary bit A and B located within intersection region are taken as the inputs of XOR computation, while the intersectional MTJ resistance (R_{MTJ}) represents the output result ($A \oplus B$). As listed in Table. III, when the magnetization of two

978-1-5386-7100-9/18 $31.00 © 2018 IEEE

TABLE III
THE XOR LOGIC OPERATION IN MAGNETIC CROSSBAR

A	B	$A \oplus B$	M_A	M_B	R_{MTJ}
0	0	0	Down	Down	R_P
0	1	1	Down	Up	R_{AP}
1	0	1	Up	Down	R_{AP}
1	1	0	Up	Up	R_P

ferromagnetic layers (M_A and M_B) are both up ('1') or down ('0'), the intersectional MTJ cell is in parallel state with low resistance (R_P). Otherwise, when M_A and M_B are in the opposite orientation (i.e. one up and one down), the intersectional MTJ cell is in anti-parallel state with high resistance (R_{AP}). Based on this description, we could see that the XOR computation is automatically implemented by the magnetic coupling physics when the data are loaded into the top and bottom domain wall nanowires with no cost of additional operations, leading to extreme energy efficient logic in-memory design. In order to read out the computation result, a similar circuit as Fig. 4a is needed. Furthermore, since the StrongARM sense amplifier generates complementary output, the XNOR(A,B) can be computed with no extra cost.

TABLE IV
ENERGY COST FOR $w \cdot x$ W.R.T MEMORY ACCESS AND COMPUTATION

	Baseline (32bit *mul*)	STT-MRAM (1bit *xnor*)	Magnetic crossbar (1bit *xnor*)
memory access	65.6nJ	1.6nJ/-	4.8nJ/-
compute	3.7pJ	0.9nJ	0.8nJ

In general, two designs with STT-MRAM and magnetic crossbar are introduced in details to provide in-memory bit-wise XNOR operation through either circuit level or device level modification. In comparison with normal network accelerator design [25] using 32bit floating multiplication (i.e., baseline in Table IV), the computation of dot-product involves energy-expensive off-chip DRAM access to load the model parameter w to the processing element, and the on-chip SRAM access to load interlayer tensor x. Such long-distance data communication is avoided with our in-memory bit-wise computing method, and the energy consumption is shown in Table IV. Note that, the memory access for in-memory computing with STT-MRAM and magnetic crossbar is only required when w and x are not stored in the same memory subarray.

IV. SUMMARY

In this work, we explicitly discuss the method to leverage in-memory XNOR computation using two different computational MRAM designs to accelerate the state-of-the-art binarized deep neural network, spanning across algorithm, architecture, circuit and device. Great energy efficiency is achieved due to its intrinsic in-memory logic designs and processing-in-memory architecture to reduce off-chip memory access.

Acknowledgement: This work is supported in part by the National Science Foundation under Grant No. 1740126 and Semiconductor Research Corporation nCORE.

REFERENCES

[1] S. Han *et al.*, "Deep compression: Compressing deep neural networks with pruning, trained quantization and huffman coding," *arXiv preprint arXiv:1510.00149*, 2015.

[2] Z. He, S. Angizi, and D. Fan, "Exploring stt-mram based in-memory computing paradigm with application of image edge extraction," in *Computer Design (ICCD), 2017 IEEE International Conference on*. IEEE, 2017, pp. 439–446.

[3] M. Courbariaux *et al.*, "Binaryconnect: Training deep neural networks with binary weights during propagations," in *Advances in neural information processing systems*, 2015, pp. 3123–3131.

[4] Y. LeCun *et al.*, "Mnist handwritten digit database," *AT&T Labs [Online]. Available: http://yann. lecun. com/exdb/mnist*, vol. 2, 2010.

[5] A. Krizhevsky *et al.*, "The cifar-10 dataset," *online: http://www. cs. toronto. edu/kriz/cifar. html*, 2014.

[6] I. Hubara *et al.*, "Binarized neural networks," in *Advances in neural information processing systems*, 2016, pp. 4107–4115.

[7] J. Deng *et al.*, "Imagenet: A large-scale hierarchical image database," in *Computer Vision and Pattern Recognition, 2009. CVPR 2009. IEEE Conference on*. IEEE, 2009, pp. 248–255.

[8] M. Rastegari *et al.*, "Xnor-net: Imagenet classification using binary convolutional neural networks," in *European Conference on Computer Vision*. Springer, 2016, pp. 525–542.

[9] S. Zhou, *et al.*, "Dorefa-net: Training low bitwidth convolutional neural networks with low bitwidth gradients," *arXiv preprint arXiv:1606.06160*, 2016.

[10] X. Lin *et al.*, "Towards accurate binary convolutional neural network," in *Advances in Neural Information Processing Systems*, 2017, pp. 344–352.

[11] Nvidia, "Nvidia volta gpu," March 2018, https://www.nvidia.com/en-us/data-center/volta-gpu-architecture/.

[12] G. Huang *et al.*, "Densely connected convolutional networks," in *Proceedings of the IEEE conference on computer vision and pattern recognition*, vol. 1, no. 2, 2017, p. 3.

[13] Y. Kim *et al.*, "Write-optimized reliable design of stt mram," in *Proceedings of the 2012 ACM/IEEE ISLPED*. ACM, 2012, pp. 3–8.

[14] X. Fong *et al.*, "Spin-transfer torque memories: Devices, circuits, and systems," *Proceedings of the IEEE*, vol. 104, pp. 1449–1488, 2016.

[15] Y. Bengio *et al.*, "Estimating or propagating gradients through stochastic neurons for conditional computation," *arXiv preprint arXiv:1308.3432*, 2013.

[16] W. Tang *et al.*, "How to train a compact binary neural network with high accuracy?" 2017.

[17] S. Ioffe *et al.*, "Batch normalization: Accelerating deep network training by reducing internal covariate shift," *arXiv preprint arXiv:1502.03167*, 2015.

[18] S. Angizi *et al.*, "Imce: energy-efficient bit-wise in-memory convolution engine for deep neural network," in *Proceedings of the 23rd Asia and South Pacific Design Automation Conference*. IEEE Press, 2018, pp. 111–116.

[19] L. Yang, Z. He, and D. Fan, "A fully onchip binarized convolutional neural network fpgaimpelmentation with accurate inference," in *Proceedings of the 2018 International Symposium on Low Power Electronics and Design*. ACM, 2018.

[20] Z. He, S. Angizi, F. Parveen, and D. Fan, "Leveraging dual-mode magnetic crossbar for ultra-low energy in-memory data encryption," in *Proceedings of the on Great Lakes Symposium on VLSI 2017*. ACM, 2017, pp. 83–88.

[21] G. Autès *et al.*, "Strong enhancement of the tunneling magnetoresistance by electron filtering in an fe/mgo/fe/gaas (001) junction," *Physical review letters*, p. 217202, 2010.

[22] Z. He and D. Fan, "A low power current-mode flash adc with spin hall effect based multi-threshold comparator," in *Proceedings of the 2016 International Symposium on Low Power Electronics and Design*. ACM, 2016, pp. 314–319.

[23] S. S. Parkin *et al.*, "Magnetic domain-wall racetrack memory," *Science*, vol. 320, pp. 190–194, 2008.

[24] V. Sokalski *et al.*, "Naturally oxidized feco as a magnetic coupling layer for electrically isolated read/write paths in mlogic," *IEEE Trans. Magn.*, vol. 49, no. 7, pp. 4351–4354, 2013.

[25] Y.-H. Chen *et al.*, "Eyeriss: An energy-efficient reconfigurable accelerator for deep convolutional neural networks," *IEEE Journal of Solid-State Circuits*, vol. 52, no. 1, pp. 127–138, 2017.

2018 IEEE Computer Society Annual Symposium on VLSI

Emerging neuromorphic computing paradigms exploring magnetic skyrmions

Sai Li[1,2], Wang Kang[1], Xing Chen[1], Jinyu Bai[1], Biao Pan[1], Youguang Zhang[1] and Weisheng Zhao[1]

[1]Fert Beijing Research Institute, Advanced Innovation Center for Big Data and Brain Computing

[2]Shenyuan Honors College

Beihang University,

Beijing, 100191, China

{wang.kang, weisheng.zhao}@buaa.edu.cn

Abstract—**Neuromorphic computing, which is evolved from the imitation of biological nervous systems, has recently attracted considerable attention for its amazing capability in recognition and classification at a fraction of power. Intensive research has been conducted in this field for developing artificial synapses and neurons, attempting to mimic the behaviors of these two fundamental units in biological neural networks. Typically, the implementation of neuromorphic computing systems relies on integrating many transistors, which however, sacrifices energy efficiency and integration density. Recently, magnetic skyrmion, a swirling topological spin configuration, has been studied as a promising information carrier candidate in future ultra-dense, low-power memory and logic devices for its outstanding merits of nanoscale size, low depinning current density, high motion velocity and particle-like stability etc. To date, the research of neuromorphic computing and magnetic skyrmions has achieved great advances, yet none can fully exploit the advantages of both. This paper introduces some striking designs of skyrmion-based neuromorphic computing devices, which enable some different design paradigms from previous studies. We will firstly give a brief review of the related works on this field and then introduce our recent research. We will illustrate how the single neuronal or synaptic device can be realized by probing into the behaviors of skyrmions and how the related circuits/systems are achieved.**

Keywords—*Artificial neuron, artifical synapse, neuromorphic computing, magnetic skyrmions, spintronics.*

I. INTRODUCTION

The concept of neuromorphic computing, inspired from the imitation of biological nervous system, is to represent the utilization of brain-like electronic systems [1]. It has evolved over the decades into an interdisciplinary field joining neuron science and neural computing, which has been regarded as a promising approach to solve the critical problems facing the continual miniaturization of conventional CMOS technology, the increasing power dissipation [2-3]. Owing to the low-power performance and massively parallel computing analogous to the human brain, a large number of bio-inspired algorithms and devices have been applied in complex pattern recognition, image processing, data mining and other massive task fields [4-6]. Typically, the networks of internal build components capable of data transmission and entrepot causes widespread academic concerns [7-8]. Intensive research has been conducted in this field for developing learning-based artificial synapses and neurons, attempting to mimic the behaviors of these two fundamental units in biological neural networks. The most widely applied model for analyzing signal transduction of neurons is the integrate-and-fire neuron model,

This work was supported by the National Natural Science Foundation of China (61501013 and 61571023) and the 111 talent program (B16001).

which was first proposed by Lapicque in 1907 [9]. When a neuron receives the continuous spikes transmitted by pre-synapses, the membrane potential will accumulate to reach a definite threshold after the integrating process, then it will give an output spike to the connected post-synapses and be reset. Synapses modulate the spikes during propagating by changing the weights of different spikes to ensure the neural activities, called synaptic plasticity [10]. For artificial neuron devices, the key feature is the time-independent integration by stochastic input spikes [9], while for artificial synapse, it is learning and dynamic conductive ability [12].

So far, most existing silicon neuron and synapse devices depend on semiconductor-based circuits via integrating many transistors at the cost of energy efficiency and integration density [13-16]. Consequently, more novel designs have been coined to be the alternatives, which incorporate emerging memristor devices [17-21], phase change devices [22–25], magnetic tunnel junctions [22-23], domain wall magnets [28-29], atomic switches [30-31], and etc. However, it is worth noting that these schemes have different restrictions on either fundamental form of learning rule for synapse [32] or limited further improvement on scalability and stackability of neural networks [33].

Recently, a new information carrier candidate, called magnetic skyrmion, has been studied for future ultra-dense, low-power memory [34-36] and logic devices [37] for its outstanding merits of nanoscale size, extremely low depinning current density, high motion velocity and topological stability [38-40]. To date, the research of neuromorphic computing and skyrmions have been made great advances, yet none can fully exploit the advantages of both. This paper introduces striking designs of skyrmion-based neuromorphic computing devices and paradigms, as shown in Fig. 1, which enable different design strategies from previous studies.

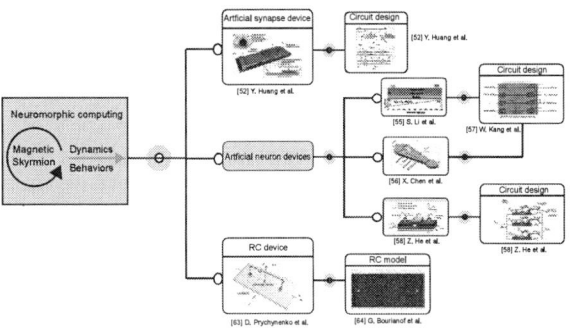

Figure 1. A summary of current skyrmion-based neuromorphic computing researches, including devices, circuits and systems.

978-1-5386-7100-9/18 $31.00 © 2018 IEEE   539

In the reset of this review, we will briefly introduce the fundamentals of skyrmions in Section II. In Section III, we illustrate how a single neuronal or synaptic device can be realized by probing into the rich behaviors of skyrmions. Meanwhile, we will also introduce the related circuits and systems. Finally, Section IV concludes the paper.

II. SKYRMION FUNDAMENTALS

Magnetic skyrmions, shaped as particle-like solitons, are chiral magnetization configurations. The skyrmion model was firstly proposed by Tony Skyrme aiming to study their unique properties in particle physics [41]. The existence of magnetic skyrmions as particle-like states was first studied theoretically in 1989 by Bogdanov et al. [42]. It was in 2009 that the first experimental observation of skyrmions in B20-type chiral materials was made by Mühlbauer et al. [43], which gave a boost to the researches of magnetic skyrmions. Nucleation of a skyrmion mostly depends on the Dzyaloshinskii-Moriya interaction (DMI) or dipole interaction, where the former is the main assistance to stabilize skyrmions in material systems lacking inversion symmetry [44,45,46]. There are generally two types of skyrmions, Bloch-type and Neel-type skyrmions. Specifically, the Bloch-type chiral skyrmion originates from the bulk DMI in B20-type materials, whereas the interfacial DMI favors the Neel-type hedgehog skyrmion in magnetic thin films. To achieve high stability, a material system with a large DMI is strongly expected. Recently, skyrmions in magnetic ultrathin films or multilayer systems have obtained intensive research interests [39,41,43,47,48].

Regarding the elementary functions of skyrmion, including generation, motion and detection, a rich variety of studies have been performed and amazing results have been achieved. Skyrmions can be generated through, e.g. heat, magnetic field, voltage and electrical current, etc. [38,39,44]. From a practical point of view, skyrmion generation via an electrical current is the most potential method thus being mostly studied in theory and experiment. Skyrmions can be moved by an electrical current and the skyrmion motion has rich dynamics [39,40,45]. For example, there exists the skyrmion Hall effect (SKHE) induced by Magnus force [49], which will drive the skyrmion to deviate from the straight trajectory along the current direction. Finally, skyrmions can be electrically detected by various magnetoresistance effects [50,51], which has been widely implemented in spintronic devices.

III. SKYRMION-BASED NEUROMORPHIC COMPUTING DEVICES, CIRCUITS AND PARADIGMS

A. Skyrmion-based synaptic devices and circuits

The schematic of a skyrmion-based artificial synaptic device proposed by Huang et al., as illustrated in Fig. 1(a), consists of a ferromagnetic (FM) layer (e.g., Co) on a heavy metal (HM, e.g., Pt) and an artificial anisotropy energy barrier [52]. This energy barrier located in the device center divides the whole nanotrack into two regions (left-hand and right-hand sides of the barrier) referred to as pre-synapse and post-synapse due to its higher perpendicular magnetic anisotropy (PMA) than that of the FM layer. It can be effectively accomplished by changing the material of the capping layer or

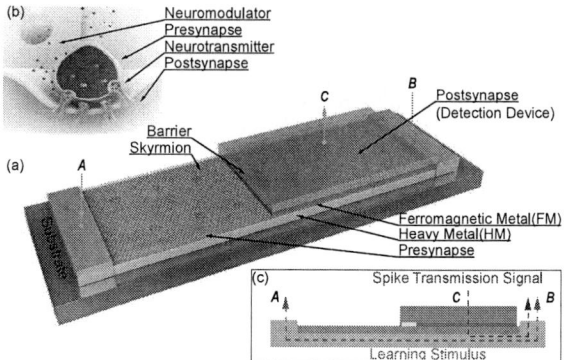

Figure 2. Schematic of the skyrmion-based synaptic device. To mimic a neuromodulator, as shown in (b), a bidirectional learning stimulus flowing through the HM from terminal A to terminal B (or vice versa) drives skyrmions into (or out of) the post-synapse region to increase (or decrease) the synaptic weight, as shown in (c), mimicking the potentiation/depression process of a biological synapse [52].

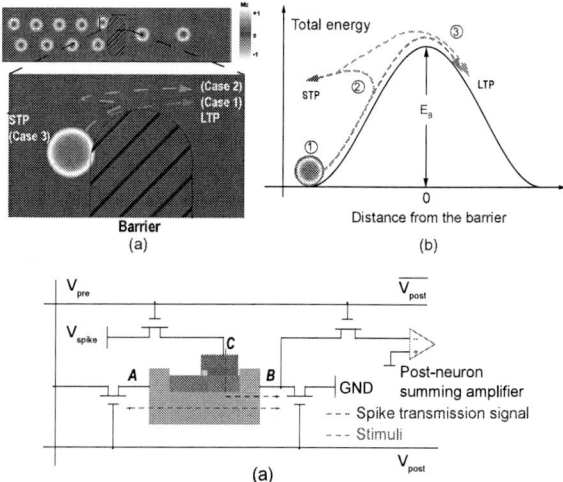

Figure 3. (a) Illustration of STP and LTP of the skyrmion-based synaptic device; (b) energy profile; (c) learning and spike-transmission operations of the skyrmion-based synaptic device [52].

by exploiting the voltage controlled magnetic anisotropy (VCMA) effect upon the barrier. The motion of particle-like skyrmions on the nanotrack is analogous to a biological neurotransmitter synapse, illustrated by Fig. 1(b). Here we consider three terminals (see Fig. 1(c)) to achieve two key operations: spike transmission and learning. The input spike generated by a pre-neuron is modulated by the weight of the synaptic device (magnetoresistance of the post-synapse), which is the pike transmission process that the post-synaptic spike current flows from terminal C to terminal B (denoted as a spike transmission channel). During the learning phase, a spin current in the FM layer from terminal A to terminal B (denoted as a learning channel), motivates the skyrmions in the pre-synapse to the post-synapse. The learning phase

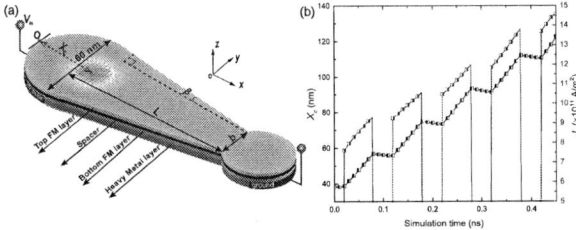

Figure 4. Design of the LIF spiking neuron device based on the wedge-shaped nanotrack. (a) Schematic of the proposed skyrmion-based artificial neuron device; (b) Skyrmion motion distance under an input spike current with non-uniform distribution within the HM layer [56].

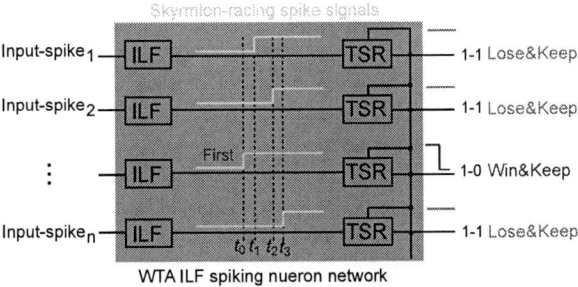

Figure 5. Schematic of the skyrmion racing (SR-WTA) ILF spiking neuron network, which consists of a set of ILF modules and trigger-suppression-reset (TSR) modules. The LIF module is based on the LIF spiking neuron device. The TSR module is able to share a global reset signal line when it is triggered by the first reaching skyrmion [57].

mimics the potentiation or depression process of a biological synapse, by utilizing the number of skyrmions to represent the transmission weight of the synapse.

The behavior of this synaptic device depends on the stimulus profiles. Two cases with high duty ratios demonstrate different long-term potentiation (LTP) properties were studied, while one case with a low duty ratio presents a short-term plasticity (STP) property. In other words, for the same interval, a proper stimulus duration is required to transfer STP to LTP. Both the duration and interval of the stimulus play essential roles in the device's plasticity. From the physical perspective, the STP and LTP can be explained by a competition among: the driving force from the stimulus, the skyrmion-skyrmion interactions, and the repulsive force from the nanotrack edge and PMA barrier, as shown in Fig. 3(a)-(b). To demonstrate the spike-transmission and learning functions, peripheral circuits were also designed by Huang et al., as shown in Fig. 3(c). Here Vspike signifies the spike voltage of the pre-neuron, Vpre and Vpost are the programming voltages from the pre- and post-neurons controlling the learning stimuli. Once the post-neuron spikes, a short Vpost gets activated. Accordingly, the learning path switches on to modulate the synaptic weight through driving skyrmions into or out from the post-synapse region of the device. The energy consumption of the device is around 1pJ, indicating the potential of the skyrmion-based synapse device in low-power neuromorphic applications.

B. Skyrmion-based neuron devices and circuits

In biological neurons, an equilibrium membrane potential is maintained because of the electrical charge difference inside the thin lipid-bilayer from that outside of it. Once the excitatory or the inhibitory signal from the post-synapse arrives, the membrane potential will change based on the previous state. Under a serious of excitations, an output signal will release ("fire") and the membrane potential gets reset to the resting potential until the membrane potential exceeds a threshold potential. Emulation of the neuronal dynamics, including the transmission of the signal, the maintenance and the activation of the membrane potential, is thought to be the core to implement the neuromorphic computing systems [53].

Among all of the models used to describe the neuronal behaviors, the Leaky-Integrate-Fire (LIF) spiking neuronal model is an extensively adopted one, whose characteristics can be described by [53,54],

$$\tau_{mem}\frac{dV_{mem}}{dt} = -(V - V_{rest}) + \sum_j \delta\left(t - t_j\right)w_j \quad (1)$$

where τ_{mem} is decay time constant, V_{mem} denotes the membrane potential, V_{rest} is the resting potential, $\sum_j \delta(t - t_j)w_j$ is the sum of the input weighted spikes from presynaptic neurons, $-(V - V_{rest})$ corresponds to the leakage item and $\sum_j \delta(t - t_j)w_j$ represents the integral item.

To mimic the dynamical performance of a biological neuron, it is important to represent the membrane potential with an analog quantity that can vary continuously. Thanks to the nonvolatile property and the current-induced motion dynamics, skyrmion is potential to be exploited to implement the neuronal functionalities. To date, several skyrmion-based neuron devices have been proposed in literature to mimic the behavior of a biological neuron.

Based on the algorithm that the state of a neuron is elucidated by the temporal magnitude of the membrane potential, such a LIF neuronal behavior can be practically described by the skyrmion motion dynamics on a nanotrack under an electrical driving current with the moving distance of the skyrmion indicating the membrane potential of biological neurons. S. Li et al., first suggested the use of the skyrmion motion dynamics on a nanotrack for realization of a LIF neuron [55]. Different from the general design that the PMA of the FM layer in the nanotrack is remain constant, here the PMA in the FM layer is optimized, e.g., by tuning the thickness of the FM layer, in order to generate a linear PMA (i.e., $k_u(l_x) = k_{u0} + \Delta k_u * l_x$) along the nanotrack. Initially, a skyrmion is generated at the origin (i.e., under the generation unit) of the nanotrack. Then, if one or several pre-neurons spike, the accumulated input spike currents (I_{spk-in}) stimulate the skyrmion motion along the nanotrack if the current density exceeds the skyrmion depinning current density. During the skyrmion motion process, the energy barrier of the nanotrack gradually grows because of the linear increase of the PMA energy, which supplies the repulsive force, thus the "leaky" process is achieved. The skyrmion motion dynamics depend on the competition between the gradient PMA energy of the nanotrack and the driving force of the accumulated input spike current, which depends then on the amplitude and frequency

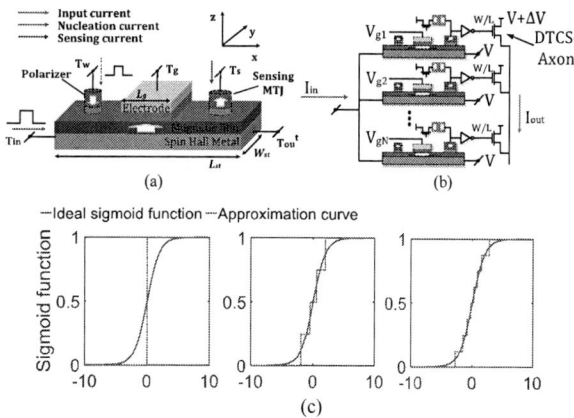

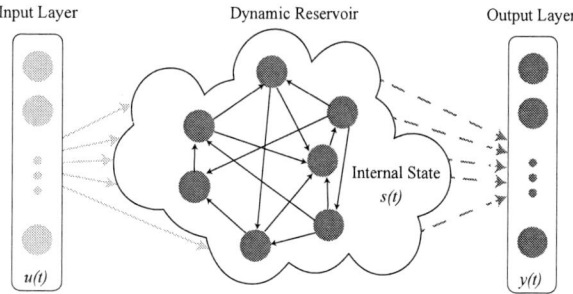

Figure 7. Illuatration of a common reservoir computing system formed by 3 layers, using reservoir process as an internal layer without training.

Figure 6. (a) Structure of a skyrmion neuron. (b) Skyrmion Neuron Cluster (SNC) connected with Deep Triode Current Source (DTCS) axon to produce synaptic current. (c) Ideal sigmoid function and its normalized approximation curve using SNC with different size (N) [58].

of the spike currents from pre-neurons. Under homogeneous or inhomogeneous spike current, the skyrmion-based artificial neuron can exhibits "leaky-integrate-fire" neuronal behaviors.

Apart from utilizing the linear variation of the PMA to implement the "leaky" process of the LIF neuronal dynamics, an alternative way is to take advantage of the repulsive force on the skyrmion provided by the nanotrack edge. Fig. 4(a) displays another skyrmion-based neuron device based on a wedge-shaped nanotrack proposed by Chen et al., in which a repulsive force will exert on the skyrmion during the motion process towards the narrower part of the nanotrack, thus achieving the "leaky" behavior [56]. Fig. 4(b) presents the skyrmion motion distance as a function of time under an input spike current with non-uniform distribution within the HM layer, on which a certain applied voltage is applied.

Furthermore, based on the realizations of a single neuronal and synaptic device, more practical functionalities of the neuromorphic computing system can be explored. The winner-takes-all (WTA) module, which finds the maximum among a set of inputs while shuts off the others, is a useful function in neural networks. Specifically, the intensity of the accumulated input spike currents from connected pre-neurons is translated into the skyrmion racing velocity in the post-neuron. A strong input spike current (amplitude and/or frequency) leads to a fast skyrmion racing velocity, thus an early spike output. Only the first spiking post-neuron, in which the skyrmion first reaches the finishing line (a pre-defined threshold distance), will ever spike, i.e., win over the others, by sharing a global reset signal for all the post-neurons. Fig. 5 shows the schematic of the skyrmion racing WTA ILF spiking neuron network. More details can refer to [57].

Some other related works also open new horizons of this new paradigm. He et al. [58] proposed a Skyrmion Neuron Cluster (SNC) to approximate non-linear soft-limiting neuron activation functions, such as the sigmoid function, as shown in Fig. 6. Under a systematical simulation from device, circuit to system, they verified that the SCN achieved a recognition

accuracy of up to 98.74% in deep learning Convolutional Neural Network (CNN) with the MNIST handwritten digits dataset and the energy consumption is only 3.1 fJ/step (two orders lower than the CMOS counterpart). Fig. 6(a) presents a basic element of the SNC, which constitutes of a skyrmion nanotrack, a nucleation head (polarizer), a sensing MTJ and an electrode gate to tune the PMA of the magnetic film. By building a skyrmion neuron sensing circuit for each neuron and connecting N skyrmion neurons in parallel, the input synaptic current is equivalently split into each skyrmion neuron and an output current for each skyrmion neuron in the SNC will be generated, which is determined by the input synaptic current (J_{in}), SNC size (N) and the gate voltage ($V_g(m)$) applied on each skyrmion neuron, as shown in Fig. 4(b). The binary output of the m-th skyrmion neuron, $V_0(m)$, can be expressed by,

$$V_0(m) = \begin{cases} High, \frac{J_{in}}{N} \geq J_0 + \delta \cdot V_g(m) \\ Low, \frac{J_{in}}{N} < J_0 + \delta \cdot V_g(m) \end{cases} \quad (2)$$

where J_0 and δ are fitting coefficients. the output voltage can be converted into synaptic current through specific circuits (DTCS-Axon indicated in Fig. 6(b)). In this case, as long as a staircase increasing voltages (V_g) are applied on the skyrmion neurons in one SNC, the relation between the total input synaptic current of SNC and the total output current of DTCS-axon could then be employed to approximate soft-limiting neuron activation function, as illustrated in Fig. 6(c).

C. Skyrmion-based reservoir computing

Among diverse neuromorphic computing paradigms, reservoir computing(RC) is known for its nonlinear dynamics when allowing time-varying inputs, which is a more appropriate solution to solve the temporal problems in the real-word tasks [59,60]. An RC system, as presented in Fig. 7, is composed of three parts: input layer to receive the input signals, 'reservoir' layer capable of transforming the input signals by a one- or multi-dimensional temporal map without training, and output layer to readout the received reservoir states and then generates ultimate outputs in need of training the weights [61]. Unlike recurrent neural networks (RNN), RC eliminates the need to train the large complex reservoir and only train the output weights [62]. The RC paradigm does

978-1-5386-7100-9/18 $31.00 © 2018 IEEE   542

not require any knowledge of the reservoir topology or node weights for training purposes and can thus utilize naturally existing networks formed by a wide variety of physical processes. Recently, RC has emerged to solve several specific prediction tasks or general learned tasks as a simpler choice.

Recently, Prychynenko et al. have proposed to apply a skyrmion embedded in a ferromagnetic ribbon to RC network using the anisotropic magnetoresistance (AMR) [63]. Since skyrmion has shown to have spontaneous, highly complex, and self-organized dynamics, which fits well for RC. In theire proposed device, two contacts are located symmetrically around the center of the ribbon, and the initial location of a skyrmion. The pinned skyrmion is depinned under a certain current density and then will be deformed below the corresponding voltage. Thus, the deformation will change the resistance of the device, leading to a non-linear current-voltage characteristics. Further studies have shown that the resistance not only is affected by the DMI and the anisotropy strength, but also the size of the ribbon system. Moreover, Bourianoff et al. studied skyrmion fabrics formed in magnets with broken inversion symmetry and its physical instantiation for RC [64]. The interaction of skyrmion with currents and magnons results in many internal degrees of freedom in this system, which enables a high tunability and functionality much larger than other physic-based RC systems.

IV. CONCLUSIONS AND DISCUSSIONS

This article reviews the current research of skyrmions in the field of neuromorphic computing. Subsequent to a brief introduction of the fundamentals of magnetic skyrmions, we introduce the skyrmion-based artificial synapse and neuron devices, as well as the related circuit designs. Meanwhile, we introduce the RC paradigm by exploiting the nonlinear dynamics of skyrmions. Although potential for neuromorphic computing, many design and fabrication challenges are still existing. Recent skyrmion research is still in infancy, the materials, devices, controllable manipulations need further exploration. One of the primary challenges for neuromorphic computing from the perspective of hardware is finding an intelligent building block to establish an ultrahigh-density network [65]. In this context, a magnetic skyrmion is an appropriate candidate due to the tunable nanoscale size, particle-like behavior, and fast non-linear dynamics. However, the controlled manipulation of magnetic skyrmion is difficult resulting from the material-specific magnetization dynamics [66,67], and the stochasticity of the skyrmion mobility (influenced by pinning defects) [68]. In addition, although the generation, motion, and detection of skyrmion at room temperature (RT) are obtained experimentally for sub-100 nm skyrmions [69,70], the large-scale integration and energy-efficiency require further investigation.

ACKNOWLEDGMENT

This work was supported by the National Natural Science Foundation of China (61501013 and 61571023), the National Key Technology Program of China (2017ZX01032101), and the International Mobility Project (B16001). Corresponding email: {wang.kang, weisheng.zhao}@buaa.edu.cn.

REFERENCES

[1] A. Calimera, M. Enrico, and M. Poncino, "The Human Brain Project and Neuromorphic Computing," Funct. Neurol., vol. 28, Oct. 2013, pp.191-196.

[2] N. S. Kim et al., "Leakage current: Moore's law meets static power," Computer, vol. 36, Dec. 2003, pp. 68–75.

[3] M. Lundstrom, "Moore's law forever?" Science, vol. 299, Jan. 2003, pp. 210–211.

[4] Q. Wu et al., "Full imitation of synaptic metaplasticity based on memristor devices, " Nanoscale, vol.10, Feb.2018, pp.5875-5881.

[5] A. Sengupta, B. Han and K. Roy, "Toward a spintronic deep learning spiking neural processor," Biomedical Circuits and Systems Conference (BioCAS), 2016 IEEE, Oct. 2016, pp. 544-547.

[6] L.P. Maguire et al., "Challenges for large-scale implementations of spiking neural networks on FPGAs," Neurocomputing, vol. 71, Dec.2007, pp. 13-29.

[7] J. Grollier, D. Querlioz and M.D. Stiles, "Spintronic Nanodevices for Bioinspired Computing," Proc. IEEE, vol. 104, Sep .2016, pp. 2024-2039.

[8] C. Gao and D. Hammerstrom, "Cortical models onto CMOL and CMOS—architectures and performance/price," IEEE Transactions on Circuits and Systems I: Regular Papers, vol. 54.11, Nov. 2007, pp. 2502-2515.

[9] L. Lapique, "Recherches quantitatives sur l'excitation electrique des nerfs traitee comme une polarization." Journal of Physiology and Pathololgy, vol.9, 1907, pp. 620-635

[10] A.N. Burkitt, "A review of the integrate-and-fire neuron model: I. Homogeneous synaptic input," Biological cybernetics, vol. 95, Jul.2006, pp. 1-19.

[11] D.O. Hebb, "The organization of behavior: A neuropsychological theory," Psychology Press, 2005.

[12] B. Li et al., "Mediating Short-Term Plasticity in an Artificial Memristive Synapse by the Orientation of Silica Mesopores," Advanced Materials, vol.30, Apr.2018, pp. 1706395.

[13] X. Wu et al., "A cmos spiking neuron for brain-inspired neural networks with resistive synapses and in situ learning," IEEE Transactions on Circuits and Systems II: Express Briefs, vol.62, Nov.2015, pp. 1088-1092.

[14] Y. Burgt et al., "A non-volatile organic electrochemical device as a low-voltage artificial synapse for neuromorphic computing," Nature materials, vol.16, February.2017, pp. 414–418.

[15] I. Ebong and P. Mazumder, "CMOS and memristor-based neural network design for position detection," Proc. IEEE, vol. 100, Jun. 2012, pp. 2050–2060.

[16] G. Lecerf, J. Tomas, and S. Saighi, "Excitatory and inhibitory memristive synapses for spiking neural networks," IEEE International Symposium on Circuits and Systems (ISCAS), August 2013, pp. 1616-1619.

[17] S. H. Jo et al., "Nanoscale memristor device as synapse in neuromorphic systems," Nano Lett, vol. 10, March. 2010. pp. 1297–1301.

[18] S. Yu et al., "A low energy oxide-based electronic synaptic device for neuromorphic visual systems with tolerance to device variation," Adv. Mater., vol.26, Jan.2013, pp. 1774-1779.

[19] S. Kim et al., "Experimental demonstration of a second-order memristor and its ability to biorealistically implement synaptic plasticity," Nano Lett., Feb.2015, vol.15, pp. 2203-2211.

[20] L. Zhu et al. "Artificial synapse network on inorganic proton conductor for neuromorphic systems," Nat. Commun. vol. 5,Jan.2014, pp. 3158.

[21] K. Seo et al., "Analog memory and spike-timing-dependent plasticity characteristics of a nanoscale titanium oxide bilayer resistive switching device," Nanotechnology, vol. 22, May.2011, pp. 254023.

[22] D. Kuzum et al., "Nanoelectronic programmable synapses based on phase change materials for brain-inspired computing," Nano Lett. vol.12, Jul.2011, pp. 2179-2186.

978-1-5386-7100-9/18 $31.00 © 2018 IEEE

[23] M. Suri et al., "Phase change memory as synapse for ultra-dense neuromorphic systems: Application to complex visual pattern extraction", Proc. IEEE Electron Devices Meeting (IEDM), IEEE Press, Dec. 2011, pp. 4.4.1-4.4.4.

[24] G. W. Burr et al., "Experimental Demonstration and Tolerancing of a Large-Scale Neural Network (165 000 Synapses) Using Phase-Change Memory as the Synaptic Weight," IEEE Trans. Electron Devices, vol. 62, Jul.2015, pp. 3498-3507.

[25] T. Tuma, A. Pantazi, M. Le Gallo, A. Sebastian and E. Eleftheriou, "Stochastic phase-change neurons," Nat. Nanotech., vol. 11, May.2016, pp. 693–699

[26] A. Sengupta et al., "Spin orbit torque based electronic neuron," Appl Phys. Lett., vol.106, Apr.2015, pp. 143701.

[27] P. Krzysteczko et al., "The Memristive Magnetic Tunnel Junction as a Nanoscopic Synapse-Neuron System," Adv. Mater., vol. 24, Jan.2012, pp. 762-766.

[28] D. Fan et al., "STT-SNN: A spin-transfer-torque based soft-limiting non-linear neuron for low-power artificial neural networks," IEEE Transactions on Nanotechnology, vol. 14, Jan.2015, pp. 1013-1023.

[29] M. Sharad et al., "Spin-based neuron model with domain-wall magnets as synapse," IEEE Trans. Nanotechnol., vol. 11, 2012 ,pp. 843-853.

[30] T. Ohno et al., "Short-term plasticity and long-term potentiation mimicked in single inorganic synapses," Nat. Mater.,vol.10, Jun.2011, pp. 591–595.

[31] T. Hasegawa et al., "Learning Abilities Achieved by a Single Solid-State Atomic Switch," Adv. Mater., vol. 22, Apr.2010, pp. 1831-1834.

[32] Y. Yang et al., "Multifunctional Nanoionic Devices Enabling Simultaneous Heterosynaptic Plasticity and Efficient In-Memory Boolean Logic," Adv. Electron. Mater., vol. 3. May.2017, pp. 1700032

[33] Z. Wang et al., "Fully memristive neural networks for pattern classification with unsupervised learning," Nat. Electron., vol. 1, Mar.2018, pp. 137–145.

[34] S. S. P. Parkin, M. Hayashi, and L. Thomas, "Magnetic domain-wall racetrack memory," Science, vol. 320, Apr. 2008, pp. 190–194.

[35] R. Tomasello et al., "A strategy for the design of skyrmion racetrack memories," Sci. Rep., vol. 4, Oct.2014, pp. 76784.

[36] W. Kang et al., "Complementary skyrmion racetrack memory with voltage manipulation," IEEE Electron Device Lett., vol.37, Jul.2016, pp. 924-927.

[37] X. Zhang, M. Ezawa and Y. Zhou, "Magnetic skyrmion logic gates: conversion, duplication and merging of skyrmions," Sci. Rep., vol.5, Mar.2015, pp. 9400.

[38] W. Kang et al., "Skyrmion-electronics: an overview and outlook," Proc. IEEE, vol. 104, Aug.2016, pp. 2040-2061.

[39] A. Fert, V. Cros, and J. Sampaio, "Skyrmions on the track," Nat. Nanotech. vol. 8, Mar.2013, pp. 152–156.

[40] J. Iwasaki, M. Mochizuki and N.Nagaosa, " Universal current-velocity relation of skyrmion motion in chiral magnets," Nat. Commun., vol. 4, Feb.2013, pp. 1463.

[41] T. H. R. Skyme, " A unified field theory of mesons and baryons," Nucl. Phys., vol. 31, Mar.1962, pp. 556-569.

[42] A. N. Bogdanov and D. A. Yablonskii, "Thermodynamically stable 'vortices' in magnetically ordered crystals. The mixed state of magnets," Zh. Eksp. Teor. Fiz, vol. 95, no. 1, p. 178, 1989.

[43] S. Mühlbauer et al., "Skyrmion lattice in a chiral magnet," Science, vol. 323, no. 5916, pp. 915–919, 2009.

[44] R. Wiesendanger, "Nanoscale magnetic skyrmions in metallic films and multilayers: a new twist for spintronics," Nat. Rev. Mat., vol.1, Jun.2016, pp. 16044.

[45] J. Sampaio et al., "Nucleation, stability and current-induced motion of isolated magnetic skyrmions in nanostructures," Nat. Nanotech., 2013, vol. 8, Oct.2013, pp. 839-844.

[46] N. Nagaosa and Y. Tokura, "Topological properties and dynamics of magnetic skyrmions," Nat. Nanotech., vol. 8, Dec.2013, pp. 899-911.

[47] X. Z. Yu et al. "Real-space observation of a two-dimensional skyrmion crystal," Nature, vol.65, Jun.2010, pp. 901-904.

[48] I. Kézsmárki et al., "Néel-type skyrmion lattice with confined orientation in the polar magnetic semiconductor GaV4S8,"Nat. Mater. vol. 142015, Sep.2015, pp. 1116-112.

[49] M. Stone, "Magnus force on skyrmions in ferromagnets and quantum Hall systems," Phys. Rev. B, vol. 53, Jun.1996, pp. 16573-16578.

[50] D. M. Crum et al., "Perpendicular readingof single confined magnetic skyrmions," Nat. Commun., vol. 6, Oct.2015, pp. 8541.

[51] K. Hamamoto, M. Ezawa, and N. Nagaosa, "Purely electrical detection of a skyrmion in constricted geometry," Appl. Phys. Lett., vol. 108, Mar.2015, pp. 112401.

[52] Y. Huang et al., "Magnetic skyrmion-based synaptic devices," Nanotechnology, vol.28, Jan.2017, pp. 08LT02.

[53] G. W. Burr et al., "Neuromorphic computing using non-volatile memory," Adv. Phys.: X, vol. 2, Dec.2016, pp. 89-124.

[54] R.M. Borisyuk and G. N. Borisyuk, "Information coding on the basis of synchronization of neuronal activity," BioSystems, vol. 40, 1997, pp. 3-10.

[55] S. Li et al., "Magnetic skyrmion-based artificial neuron device," Nanotechnology, vol. 28, Jul.2017, pp. 31LT01.

[56] X. Chen et al, "A compact skyrmionic leaky–integrate–fire spiking neuron device," Nanoscale, vol. 2018, Feb.2018, pp. 6139-6146.

[57] W. Kang, "A Compact Skyrmion-Racing Winner-Take-All Circuit of Integrate-Leaky-Fire Neurons for Neural Networks," submitted.

[58] Z. He and D. Fan, "A Tunable Magnetic Skyrmion Neuron Cluster for Energy Efficient Artificial Neural Network," Proc. IEEE Design, Automation & Test in Europe Conference & Exhibition (DATE), IEEE Press, May.2017, pp. 350-355.

[59] H. Jaeger, "The 'echo state' approach to analysing and training recurrent neural networks-with an erratum note," GMD Technical Report, vol.148, 2001, pp.34.

[60] B. Schrauwen, D. Verstraeten and J. Van Campenhout, "An overview of reservoir computing: theory, applications and implementations," Proc.15th European Symposium on Artificial Neural Network, 2007, pp. 471-482.

[61] C. Du et al., "Reservoir computing using dynamic memristors," Nat. Commun., vol.8, Dec.2017, pp.2204.

[62] M. Lukoševičius and H. Jaeger, "Reservoir Computing Approaches to Recurrent Neural Network Training," Computer Science Review, vol. 3, Aug.2009, pp. 127-149.

[63] D. Prychynenko et al., "Skyrmion as a Nonlinear Resistive Element: A Potential Building Block for Reservoir Computing," Phys. Rev. Applied, vol. 9, Jan.2018, pp. 014034.

[64] G. Bourianof et al., "Potential implementation of reservoir computing models based on magnetic skyrmions," AIP Advances, vol. 8, Jan.2018, pp. 055602.

[65] J. Grollier, D. Querlioz and M.D. Stiles, "Spintronic nano-devices for bio-inspired computing," Proc. IEEE, vol. 104, September. 2016, pp. 2024-2039.

[66] R. Conte et al., "Role of B diffusion in the interfacial Dzyaloshinskii-Moriya interaction in Ta/Co20Fe60B20/MgO nanowires," Phys. Rev. B, vol. 91, Jan.2015, pp. 014433.

[67] S. Emori et al., "Current-driven dynamics of chiral ferromagnetic domain walls," Nat. Mater., vol. 12, June.2013, pp. 611–616.

[68] W. Jiang et al., "Direct observation of the skyrmion Hall effect," Nat. Phys., vol. 13, Sep.2016. pp. 162–169.

[69] O. Boulle et al., "Room temperature chiral magnetic skyrmion in ultrathin magnetic nanostructures," Nat. Nanotech., vol. 11, Jan.2016, pp. 449–454.

[70] D. Maccariello et al., "Electrical detection of single magnetic skyrmions in metallic multilayers at room temperature," Nat. Nanotech., vol. 13, Jan.2018, pp. 233-237.

2018 IEEE Computer Society Annual Symposium on VLSI

Security-Driven Task Scheduling for Multiprocessor System-on-Chips with Performance Constraints

Nan Wang*, Manting Yao*, Dongxu Jiang*, Song Chen†and Yu Zhu*
*School of Information Science and Engineering, East China University of Science and Technology, Shanghai, China
†School of Information Science and Technology, University of Science and Technology of China, Hefei, China
Email: {wangnan, zhuyu}@ecust.edu.cn, songch@ustc.edu.cn

Abstract—The high penetration of third-party intellectual property (3PIP) brings a high risk of malicious inclusions and data leakage in products due to the planted hardware Trojans, and system level security constraints have recently been proposed for MPSoCs protection against hardware Trojans. However, secret communication still can be established in the context of the proposed security constraints, and thus, another type of security constraints is also introduced to fully prevent such malicious inclusions. In addition, fulfilling the security constraints incurs serious overhead of schedule length, and a two-stage performance-constrained task scheduling algorithm is then proposed to maintain most of the security constraints. In the first stage, the schedule length is iteratively reduced by assigning sets of adjacent tasks into the same core after calculating the maximum weight independent set of a graph consisting of all timing critical paths. In the second stage, tasks are assigned to proper IP vendors and scheduled to time periods with a minimization of cores required. The experimental results show that our work reduces the schedule length of a task graph, while only a small number of security constraints are violated.

Index Terms—MPSoC, hardware Trojan, security, task scheduling, system performance.

I. Introduction

The increased design productivity requirements for heterogeneous Multiprocessor System-on-Chip (MPSoC) require the industry to procure and use the latest Commercial-Off-The-Shelf (COTS) electronic components to track the most cutting edge technology while reducing the manufacturing costs. However, the hardware Trojans in these COTS components present high risks of malicious inclusions and data leakage in products [1]. Particularly, the growing number of mission-critical applications (e.g., finance and military) that use MPSoCs means that security is the highest priority issue, whereas the increasing integration of third-party Intellectual Property (3PIP) and the outsourcing of fabrication lead to the fact that most of the MPSoCs are not 100% trustworthy.

Emerging security problems bring an urgent need for detecting possible hardware Trojan attacks or for muting the effects. Researchers have proposed several methods for detecting hardware Trojans, which can mainly be classified into four groups: physical inspection [2], functional testing [3], built-in tests [4], and side-channel analyses [5]. However, it is impossible to detect all of the hardware Trojans in 3PIP cores, even with the most advanced technology.

System-level security-aware design methods have also been proposed to guard systems. Beaumont *et al.* [6] developed an online Trojan detection architecture that implements fragmentation, replication, and voting. Rajendran *et al.* [7] focused on the Electronic System Level (ESL) design tools and added protection against attacks at the ESL to make it more robust. Jiang *et al.* [8] proposed a novel, secure-embedded systems design framework for optimizing the runtime quality. Cui *et al.* [9] implemented both Trojan detection and recovery at run-time, which are essential for mission-critical applications.

Recently, a set of researchers have developed the design-for-security techniques in the context of MPSoCs [10] [11]. Rajendren *et al.* [12] incorporated security constraints into the higher level design of SoCs and proposed a scheduling method for avoiding malicious output and collusion between vendors. Building on the security constraints, researchers also began to reduce the power/area/delay overhead of the technique [13].

However, secret communications between tasks may still be established in the context of the security constraints proposed in [10]- [13], and another type of security constraints is introduced in this study to guard the MPSoC systems. Furthermore, fulfilling the security constraints always incurs a significant overhead of performance delay, which is sensitive to the designers, and thus, a security-driven performance-constrained task scheduling method is also developed. Firstly, an edge contraction conflict graph (*ECCG*) is constructed from a timing violated graph, which consists of all paths whose delays exceed the performance constraint, and sets of edges are contracted by iteratively calculating the maximum weight independent set of *ECCG*. Then, tasks are assigned to vendors with respect to the core minimization, and they are scheduled evenly in time periods by force-directed scheduling method [14]. The experimental results demonstrate the high quality of the task scheduling result in reducing both the schedule length and the number of cores integrated, while most security constraints are satisfied.

II. Security Constraints and Problem Description

A. Threat Model

In general, the Register Transfer-Level (RTL) files of IPs might have been imported from third party vendors, and 3PIPs procured from IP vendors are usually not 100% trustworthy. There may be a rogue insider in a 3PIP house who may insert Trojan logic in 3PIPs coming out of the IP house, and the Trojans may modify function, deny service, or create a backdoor to leak information.

Detection all of the hardware Trojans in 3PIPs is extremely difficult since there is no known golden model for 3PIPs as IP vendors usually provide source code, which may contain Trojans, and when a Trojan exists in an IP core, all the fabricated ICs will contain Trojans. A Trojan can be very well hidden during the normal functional operation of the 3PIP supplied as RTL code. An attacker may distribute few RTL codes so as to reduce Trojan footprint, and a large industrial-strength IP core can include thousands of lines of code, resulting in identifying the Trojan in an IP core to be an extremely challenging task.

978-1-5386-7100-9/18 $31.00 © 2018 IEEE   545

However, many applications such as banking and military systems have high security requirements, and detecting all hardware Trojans is impossible even though with the most cutting-edge technologies; therefore, hardware Trojan protection strategy during high level synthesis requires attention, and several system-level design-for-security methodologies have been proposed.

B. Security-Driven Constraints

The recently proposed security constraints [10]- [13] enable a trustworthy design, and the IPs can be purchased from different vendors without worrying about their individual security problems. All tasks are scheduled and bound to cores under the following two security constraints.

1) Security Constraint 1: Task duplication: The probability that the Trojans implanted by different attackers have the same trigger input is quite low, and it is virtually impossible that two cores from different vendors will output the same tampered results after the same trigger input. Thus, each task is duplicately executed on the cores from different vendors, and the outputs of these cores will be compared by a trusted component (not designed by the third party) to ensure the trustworthiness of the comparison step [15]. If the comparison fails, all the dependent tasks are terminated and a security flag is raised.

2) Security Constraint 2: Vendor diversity between parent and its children: To mute the Trojan footprint, attackers always distribute Trojans in multiple IP cores and construct secret communications between IP cores to leak information, or to trigger the hibernating Trojans [11]. In this study, we assume that the secret communication between IP cores from the same vendor cannot be acquired by other vendors and that the attackers of different vendors plant different hardware Trojans. Therefore, data-dependent tasks are executed on the cores from different vendors to isolate the triggered hardware Trojans.

III. Motivations and Problem Description

A. Motivation 1: Security Constraint Enhancement

An example of task scheduling with security constraints is illustrated in Fig. 1, where the solid lines represent inter-core communications, and the communication delay is marked next to the edge. Each node has four values: v_i is the i-th task and its duplicated task is v_i'; C_j denotes the assigned core j; $m - n$ are the starting and finishing times. In this example, we assume that $C1$ and $C2$ are from the first and second IP vendors, respectively, and the execution time of each task is 10 units of time (*u.t.*).

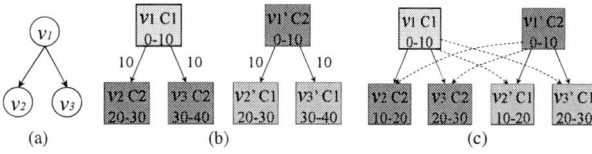

Fig. 1. Collusion between parent and its duplicated children. (a) Task graph. (b) Schedule with task duplication and vendor diversity constraints. (c) Secret communications may be established between parent and its duplicated children.

With all task duplication and vendor diversity constraints satisfied, the scheduling result of Liu *et al.* [11] is given in Fig. 1(b). If any cores (e.g. $C1$ in Fig. 1(b)) reaches the triggering condition, it can send messages to other cores by secretly writing the rigger value in a secret memory, which cannot be acquired

by the cores from other IP vendors [11]. Thus, Trojans in $C1$ may be triggered because (v_1, v_2') or (v_1, v_3') are conducted by the core from the same vendor. Fig. 1(c) gives all possible secret communications that may be established between tasks and their duplicated children, which are represented by the dash lines. For this reason, the following security constraint is also introduced.

Security Constraint 3: Vendor diversity between parent and its duplicated children: A parent task and its duplicated children are bound to the cores from different IP vendors.

B. Motivation 2: Trade-off Between Schedule Length and Security

With the task graph given in Fig. 2(a), Fig. 2(b) shows all of the security constraints: black lines, blue lines and red lines represent the first, second, and third types of security constraints, respectively. Suppose a task graph has n nodes and m edges, and the number of all security constraints (denoted as *scy*) is $n + 4m$.

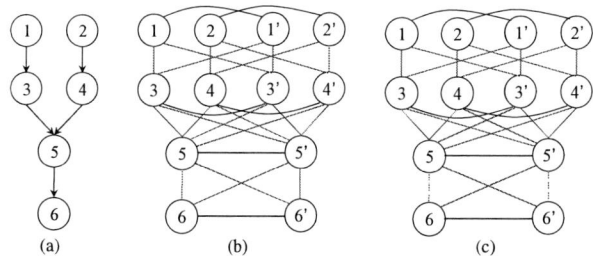

Fig. 2. Security constraints between tasks. (a) Task graph. (b) All security constraints are satisfied. (c) Security constraints after performance optimization.

Fulfills the security constraints at the finest granularity, but this incurs significant overhead of system performance. Therefore, researchers also explore the possibility of grouping dependent tasks into a cluster and scheduling the entire cluster to a single core to hide the inter-core communication latency [11] [13]. However, Liu *et al.* [11] forget to minimize the Trojan triggering risk, and Wang *et al.* [13] ignore the task criticality variation, which is also essential in evaluating the Trojan triggering risk. The edge in task graph that connects two clustered tasks is called ***contracted edge***. Task clustering violates the *security constraint 2*, and the number of ***security constraint violations*** is denoted as scy_v

The schedule length can be significantly optimized with a small number of security constraint violations. Let the inter-core communication delay be 1 *u.t.*, and the intra-core communication delay is ignored. The target is to optimize the schedule length of the task graph in Fig. 2(a) by 1 *u.t.*, and the optimized result shows that only 2 out of 26 security constraints are violated if we cluster (v_5, v_6) and (v_5', v_6') (see Fig. 2(c)).

C. Problem Description

Let the task graph be $TG = (V, E)$, where V is the set of all tasks and E represents the data dependencies between tasks. The optimization problem we focused in this study is named as the security-driven performance-constrained task scheduling problem, and the performance constraint is modeled as the maximum delay that the schedule length must not exceed.

Problem 1: Inputs: task graph TG, performance constraint, and three types of security constraints. The target is to find a schedule with a minimized number of security constraint violations, and the number of cores required is also optimized.

978-1-5386-7100-9/18 $31.00 © 2018 IEEE

The objective function of *Problem 1* is formulated as follows.

$$min: \quad \alpha * scy_v + core \tag{1}$$

where scy_v is the number of security constraint violations, *core* is the number of cores required by the schedule, and α is a parameter large enough to keep the minimization of scy_v as the first priority.

IV. PERFORMANCE-CONSTRAINED TASK CLUSTERING

System performance is one of the key considerations for designers, and they always put several timing-critical tasks into the same core to minimize the schedule length [11]. However, this brings potential risk to systems security, and thus, we must minimize the potential Trojan triggering risk when optimizing the performance.

Because TG and its duplicated TG' contain the same information, and contracting the data-dependent tasks into the same cluster only violates the *security constraint 2*. Therefore, we only discuss the methods of contracting edges in TG in the following description. Let $slack(v)$ be the slack time of v under the performance constraint, and it is calculated as follows.

$$slack(v) = t_{alap}(v) - t_{asap}(v) - exec(v) \tag{2}$$

where $exec(v)$ is the execution time of task v, and $t_{asap}(v)$ and $t_{alap}(v)$ are the as-soon-as-possible and as-late-as-possible schedules, respectively.

Source and sink nodes s, t are added to TG, and directed edges that pointing from s to 0-indegree nodes, and from 0-outdegree nodes to t, are also added. An example of task graph with s and t is given in Fig. 3(a). Then, a **timing violated graph** $TVG = (V_T, E_T)$ is constructed, and it is an induced subgraph of TG. V_T consists of all tasks with negative slacks, and $E_T = \{(v_i, v_j) \in E, v_i \in V_T \text{ and } v_j \in V_T\}$. Let $dly(e_{ij})$ be the inter-core communication delay of e_{ij}, and its intra-core communication delay is ignored. Fig. 3(b) gives an example of TVG, where the performance constraint is 5 *u.t.* and $dly(e)$ is 1 *u.t.* for each edge.

Several edges in TVG will be contracted until the performance constraint is satisfied. However, not all edges can be contracted with respect to the multi-core parallel execution. Regarding the tasks: 1) consuming the same data, or 2) feeding their computed data to the same task, they must not be assigned to the same core. This is because assigning the tasks that they once could be parallel executed to the same core forces them to be sequentially executed, which increases the schedule length of related paths. Let *in_edge(v)* be the set of edges that end with v, and *out_edge(v)* be the set of edges that start from v. Only one of the edges in *in_edge(v)* or *out_edge(v)* can be contracted.

Contracting an edge (e_{ij}) with $k(u.t.)$ minimizes the lengths of all paths that passing though e_{ij} by $k(u.t.)$. Let $w_{dly}(e_{ij})$ be the schedule length of TVG that can be optimized after contracting e_{ij}, and it is estimated by the following equations.

$$w_{dly}(e_{ij}) = \frac{path_{tvg}(e_{ij})}{path_{tvg}} * dly(e_{ij}) \tag{3}$$

where $path_{tvg}(e_{ij})$ is the number of paths in TVG that pass through e_{ij}, and $path_{tvg}$ is the number of all paths in TVG. Fig. 3(c) illustrates the w_{dly} in TVG, which is indicated next to the edge.

Clustering of timing critical tasks also necessitates information of task criticality. The total weight that evaluates these effects of

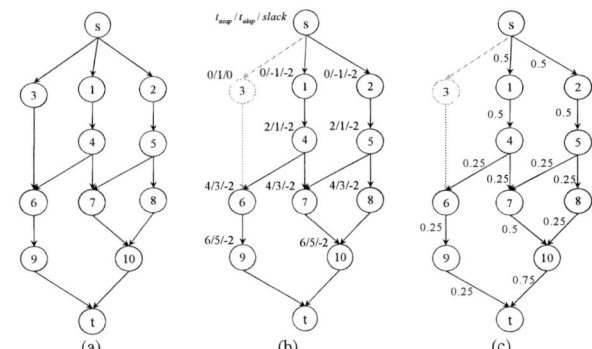

Fig. 3. Example of evaluating the timing violated graph. (a) Task graph with s and t. (b) TVG with timing constraint to be $5u.t.$ (c) The evaluation of $w_{dly}(e)$.

contracting an edge e_{ij} in TVG is denoted as $w(e_{ij})$, which is calculated as follows.

$$w(e_{ij}) = \frac{w_{dly}(e_{ij})}{\beta * cri(v_i)} \tag{4}$$

where β is the user-determined factor, and $cri(v_i)$ is the task criticality of v_i.

Then, an **edge contraction conflict graph** (*ECCG*) is constructed to represent if every pair of edges can be contracted simultaneously. Set $ECCG = (V_E, E_E)$, and each vertex in V_E represents an edge in TVG that can be contracted. Two vertexes in V_E are connected when their corresponding edges cannot be contracted simultaneously, under one of the following two situations: 1) these two edges are in the same *in_edge(v)* or *out_edge(v)* (respect to the multi-core parallel execution); 2) these two edges are in the same path (for each path in TVG, only one edge can be contracted in each iteration, such that the path length will not be over optimized).

Clustering tasks may increase the number of IP vendors required, and if contracting an edge violate the IP vendor constraints (the method in [13] is applied to calculate the number IP vendors), such edge will be removed from $ECCG$. The maximum weight independent set (MWIS) of the weighted $ECCG$ is calculated by the method proposed in [16], and the set of edges in MWIS will be contracted with the maximum benefits among all possible optimization results.

An example of task clustering derived from the TG in Fig. 3(a) is given in Fig. 4, where we are about to optimize the schedule length by 2 *u.t.*, and the criticalities of all tasks are assumed to be the same. The TVG consists of the nodes and edges with black color, and the dashed lines are the contracted edges. The TVG and its corresponding $ECCG$ are first constructed (Fig. 4(a)), and its MWIS is $\{e_{1,4}, e_{2,5}\}$ which will be contracted to minimize the schedule length in the first iteration. Then, TVG is updated and $ECCG$ is re-constructed as shown in Fig. 4(b), whose MWIS is $\{e_{5,8}, e_{6,9}, e_{7,10}\}$. After contracting these edges, Fig. 4(c) gives the final clustering result.

V. PERFORMANCE-CONSTRAINED TASK SCHEDULING

With the task clustering results, we will decide the IP vendor assignment for each cluster, and schedule tasks to time periods to minimize the number of IP cores required. The **vendor conflict graph** (*VCG*) is constructed from the clustering results: $VCG =$

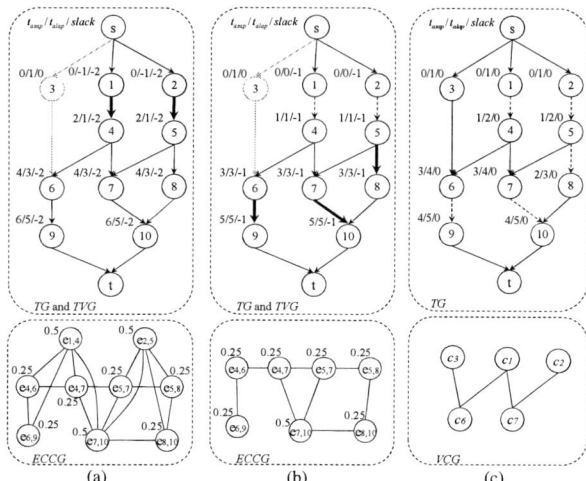

Fig. 4. Example of performance-constrained task clustering process. (a) The *TVG* and its corresponding *ECCG* before task clustering. (b) The *TVG* and its corresponding *ECCG* after the 1st iteration of task clustering. (c) The *TVG* and the corresponding *VCG* after the 2nd iteration of task clustering.

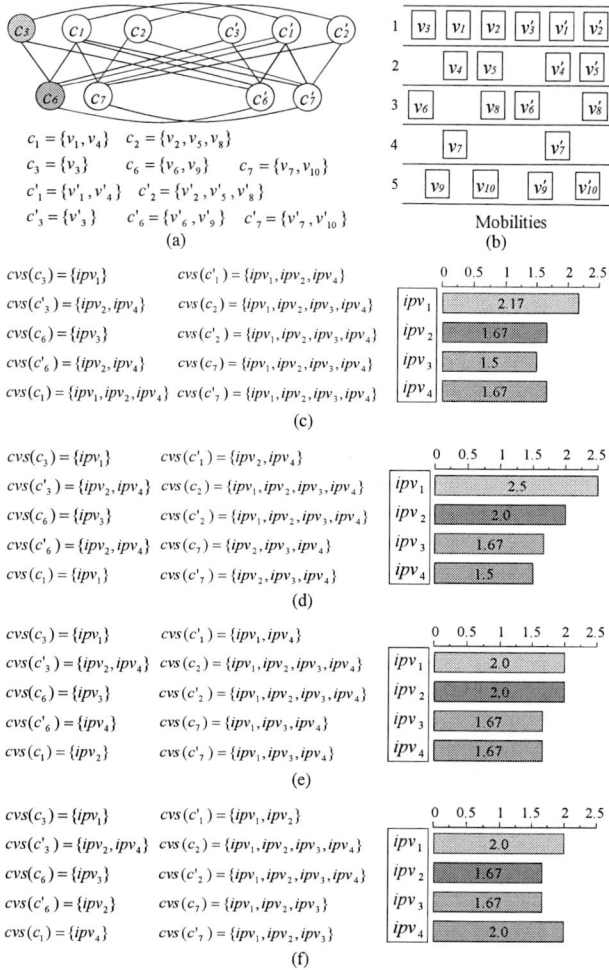

Fig. 5. Example of vendor assignment. (a) Vendor conflict graph. (b) The mobilities of tasks. (c) The candidate vendor set of each cluster and the number of cores required. (d) 7.67 cores are required if c_1 is assigned with ipv_1. (e) 7.33 cores are required if c_1 is assigned with ipv_2. (f) 7.33 cores are required if c_1 is assigned with ipv_4.

(V_V, E_V), where $V_V = \{c_1, c_2, ...\}$ is the set of clusters. An edge in E_V means that the two connected clusters must be assigned to different vendors. The index of a cluster is decided by the minimum index of the tasks in this cluster. Fig. 4(c) gives an example of *VCG* derived from the task clustering result.

***Definition** 1 (Clique size):* $\omega(G)$ is the number of nodes in a *maximum clique* of G, and it is called the *clique size* of G.

The clique size of *VCG* equals the minimum number of vendors required, and it is calculated by the method proposed in [13]. The *candidate vendor set* of c_i comprises all of the IP vendors that can be assigned to cluster c_i, which is denoted as $cvs(c_i)$. The c_i that can be assigned with ipv_k only when

1) $\forall c_j \in VCG.adj\_node(c_i)$, $cvs(c_j) - ipv_k \neq \emptyset$;
2) If v_i is assigned with ipv_k, the clique size of the new *VCG* will not violate the vendor constraint.

At the very beginning of vendor assignment, we assume that each cluster can be assigned with all IP vendors, and each IP vendor in *candidate vendor set* has the same possibility to be assigned to the tasks in this cluster. E.g., if four IP vendors are in the $cvs(c_i)$, and the possibility of assigning c_i to each IP vendor is 0.25. Then, the *mobilities* of all tasks are calculated and the *distribution graph* [14] for each IP vendor is constructed to estimate the number of cores required during vendor assignment. Let ipv_k be one of the p candidate vendors in $cvs(c_i)$, and $prob(v_i, t_j)$ is the probability that v_i is in t_j. The probability that all tasks $v_m \in c_i$ are in t_j and assigned to ipv_k is denoted as $prob(c_i, t_j, ipv_k)$, which is calculated as follows.

$$prob(c_i, t_j, ipv_k) = \frac{1}{p} \sum_{v_m \in c_i} prob(v_m, t_j) \quad (5)$$

The next step is to take the summation of the probabilities of tasks assigned with the same vendor for each time period. The resulting distribution graph (*DG*) indicates the concurrency of tasks assigned to ipv_k in t_j, which is calculated as follows.

$$DG(t_j, ipv_k) = \sum_{all\ clusters} prob(c_i, t_j, ipv_k) \quad (6)$$

The maximum of all $DG(t_j, ipv_k)$, $\forall t_j \in [1, p_c]$ is denoted as $DG_m(ipv_k)$, which is used to estimate the number of cores coming from the IP vendor ipv_k. The total number of cores required is denoted as *core*, and it is the sum of all estimated cores $core = \sum_{\forall ipv_k} DG_m(ipv_k)$. An example of estimating the total number of cores is given in Fig. 5, with the clustering result demonstrated in Fig. 4(c). Fig. 5(a) gives the *VCG* derived from both *TG* and *TG'*, and c_3 and c_6 are assumed to be assigned to ipv_1 and ipv_3, respectively. The mobility of each task is presented in Fig. 5(b), and the candidate vendor set of each cluster is given in Fig. 5(c), where the number of cores required is estimated to be 7.

To assign c_i with the best IP vendor, we first estimate the number of cores required if c_i is assigned to ipv_k, $\forall ipv_k \in cvs(c_i)$, and then assign c_i to the IP vendor with the smallest number of cores required. Fig. 5(a) illustrates an example of assigning c_1 with proper IP vendor, where $cvs(c_1) = \{ipv_1, ipv_2, ipv_4\}$. The numbers of cores required if c_1 is assigned to ipv_1, ipv_2 and ipv_4 are 7.67, 7.33, and 7.33, respectively (see Figs. 5(d) 5(e) 5(f)).

978-1-5386-7100-9/18 $31.00 © 2018 IEEE   548

Thus, c_1 will be assigned to either ipv_2 or ipv_4, because their vendor assignments are equally evaluated.

Each time after assigning a cluster with a proper IP vendor, we schedule all of the tasks in this cluster by force-directed scheduling method [14]. Force-directed scheduling method schedules tasks evenly in time periods, resulting in a small number of cores required. Algorithm 1 illustrates the details of our security-Driven performance-constrained task scheduling algorithm.

Algorithm 1 Security-Driven Performance-Constrained Task Scheduling Algorithm, $TS(TG, p_c)$.

1: p_c is the performance constraint;
2: Construct TVG;
3: **while** TVG is not empty. **do**
4:     Calculate the weight of each edge in TVG;
5:     Construct $ECCG$, and calculate its MWIS;
6:     **for** each $v_e \in$ MWIS **do**
7:         Contract e in TG;
8:     **end for**
9:     Calculate $slack(v)$, $\forall v \in TG$, and construct TVG;
10: **end while**
11: Contract the edges in TG' in the same manner;
12: Construct VCG, calculate the mobilities of all tasks, and initialize the *candidate vendor sets* of all clusters.
13: **for** each un-assigned cluster c_i **do**
14:     Assign c_i with the most suitable IP vendor $ipv_k \in cvs(c_i)$;
15:     Schedule all tasks in c_i by force-directed scheduling [14], and update the mobilities;
16:     Update $cvs(c_j)$, $\forall c_j \in VCG.adj\_nodes(c_i)$;
17: **end for**

VI. EXPERIMENTAL RESULTS

A. Experimental Setups

All of the experiments were implemented in C on a Linux Workstation with an E5 2.6-GHz CPU and 32-GB RAM. To demonstrate the effectiveness of our proposed algorithms, we tested eight benchmarks from two sources[1]: task graphs that are modeled from actual application programs, including Robot control (robot), Sparse matrix solver (sparse), SPEC fpppp (fpppp); task graphs that are randomly generagted (rnc500, rnc1000, rnc2000, rnc3000, and rnc5000). The *communication-to-computation ratio* (CCR) is the ratio of the inter-core communication delay to the computational cost of the task, and the intra-core communication is ignored in the experiments.

B. Number of Vendors Required

In this study, the third type of security constraints is introduced to enhance the system protection, and this impacts the number of vendors required. In this subsection, the numbers of vendors required are compared between the straight forward method [11], and our proposed method. In the straight forward method, only the first two types of security constraints described in Sect II-B are counted, while our method uses all three types of security constraints. The number of IP vendors required is calculated by the method proposed in [11].

We tested 100 task graphs that are randomly generated, and the numbers of tasks in these task graphs range from 100 to 200000.

[1] http://www.kasahara.elec.waseda.ac.jp/schedule/index.html.

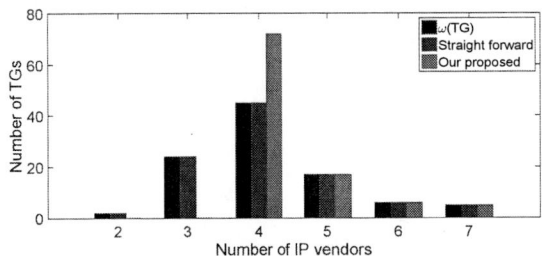

Fig. 6. Number of IP vendors required.

The comparison results are illustrated in Fig. 6, where $\omega(TG)$ is the clique size of task graph. The results indicate that the number of IP vendors required equals the clique size of task graph if we only follow the first two security constraints. However, if our proposed *security constraint 3* is also followed, the number of IP vendors increases to four if $\omega(TG) < 4$. As to the task graphs whose clique sizes are no smaller than 4, our proposed security constraints will not increase the number of vendors required.

C. Performance-Constrained Task Clustering Results

The IP vendor constraint is set to be 4 for all benchmarks, and all three types of security constraints are counted in the following experiments. Table I gives the task clustering results. The cluster-based method (*Cluster*) proposed in [11] clusters the tasks in critical paths to optimize the schedule length, however, this method does not minimize the number of security constraint violations. Our proposed task clustering method tries to maintain a high security level when optimizing the system performance. Task criticalities for all tasks are first assumed to be the same, and thus, maximizing the security is equivalent to minimizing the scy_v; its clustering results are given in column TC_1. Then, the task criticality of v_i is assumed to be the distance between s and v_i, because the computational results may contain more confidential information when the application proceeds, and the damage to the system is also much more serious. Its corresponding clustering results are given in column TC_2.

CCR is set to be 1.0, and the performance constraint P_c is set as $P_c = \delta * SL$, where SL is the schedule length with all security constraints satisfied. Two performance constraints are tested for each benchmark, with $\delta \in \{0.9, 0.8\}$. The results show that TC_1

TABLE I
COMPARISONS OF PERFORMANCE-CONSTRAINED TASK CLUSTERING RESULTS.

task graph	scy	SL (u.t.)	P_c (u.t.)	Cluster [11]		TC_1		TC_2	
				scy_v	%	scy_v	%	scy_v	%
robot	612	1114	997	10	1.63	6	0.98	6	0.98
			892	24	3.92	16	2.61	18	2.94
sparse	364	236	208	4	1.10	2	0.55	2	0.55
			192	6	1.65	4	1.10	4	1.10
fpppp	4914	2119	1871	6	0.12	2	0.04	2	0.04
			1623	8	0.16	4	0.08	4	0.08
rnc500	8140	373	340	6	0.07	4	0.05	4	0.05
			300	26	0.32	18	0.22	22	0.27
rnc1000	13020	254	226	16	0.12	12	0.09	14	0.11
			203	86	0.66	62	0.48	68	0.52
rnc2000	17720	268	243	14	0.08	8	0.05	10	0.06
			219	52	0.29	38	0.21	42	0.24
rnc3000	38464	304	274	14	0.04	12	0.03	14	0.04
			243	98	0.26	82	0.21	88	0.23
rnc5000	64716	214	194	16	0.03	10	0.02	12	0.02
			171	92	0.14	74	0.11	80	0.12
avg.					0.66		0.42		0.46

TABLE II
COMPARISONS OF PERFORMANCE-CONSTRAINED TASK SCHEDULING RESULTS.

task graph	tasks	scy	CCR	SL (u.t.)	P_c (u.t.)	Cluster [11]				TS				Savings	
						scy_v	ratio (%)	core	runtime (s)	scy_v	ratio (%)	core	runtime (s)	scy_v (%)	core (%)
robot	88	612	0.5	839	671	28	4.58	14	3.4	18	2.94	11	23.5	1.64	21.43
			1.0	1114	892	24	3.92	14	3.2	16	2.61	10	21.6	1.31	28.57
sparse	96	364	0.5	179	143	6	1.65	21	5.1	4	1.10	16	33.8	0.55	23.81
			1.0	236	189	6	1.65	19	4.8	4	1.10	15	34.7	0.55	21.05
fpppp	334	4914	0.5	1590	1272	10	0.20	13	7.5	4	0.08	10	57.8	0.12	23.08
			1.0	2119	1695	8	0.16	12	7.2	4	0.08	10	58.4	0.08	16.67
rnc500	500	8140	0.5	280	224	32	0.39	68	18.5	20	0.25	61	112.3	0.14	10.29
			1.0	373	300	26	0.32	67	17.2	18	0.22	58	108.9	0.10	13.43
rnc1000	1000	13020	0.5	190	152	96	0.74	95	34.7	68	0.52	81	207.4	0.22	14.74
			1.0	254	203	86	0.66	87	36.3	62	0.48	76	203.8	0.18	12.64
rnc2000	2000	17720	0.5	199	159	54	0.31	217	57.3	42	0.24	168	667.5	0.07	22.58
			1.0	268	214	52	0.29	206	53.8	38	0.21	171	673.5	0.08	16.99
rnc3000	3000	38464	0.5	229	183	106	0.28	238	158.6	86	0.22	189	1873.2	0.06	20.59
			1.0	304	243	98	0.26	226	154.5	82	0.21	173	1923.3	0.05	23.45
rnc5000	5000	64716	0.5	160	128	102	0.16	448	278.5	82	0.13	372	2289.6	0.03	16.96
			1.0	214	171	92	0.14	427	259.6	74	0.11	359	2305.7	0.03	15.93
avg.			0.5											0.35	19.18
			1.0											0.30	18.59

violates the minimum number of security constraints, and its scy_v is reduced by 0.24% if compared against the cluster-based method. TC_2 considers the task criticality variations while clustering, and its average scy_v is 0.20% less than that of cluster-based method.

D. Performance-Constrained Task Scheduling Results

The scheduling results of the clustered-based task scheduling method [11] and our proposed task scheduling method are compared in Table II, and columns *Cluster* and *TS* demonstrate their scheduling results, respectively. Two *CCRs* (0.5 and 1.0) are tested, and the task criticality variations are ignored. The performance constraint is set to be $P_c = 0.8 * SL$, *ratio* is the ratio of scy_v to scy, and *core* is the total number of cores required by the scheduling result.

The comparison results show that, when $CCR = 0.5$, our TS reduces scy_v by 0.35% if compared against the cluster-based method; In addition, the number of cores required by our TS is 19.18% less than the cluster-based method. When $CCR = 1.0$, our TS saves scy_v and *core* by 0.30% and 18.59%, respectively. The runtime of our TS is about 10 times larger than the cluster-based method, but the time complexities of these two methods are both $O(n^3)$, where n is the number of tasks in TG.

VII. CONCLUSIONS

The security constraints introduced in this paper better protect the system from the malicious inclusions and data leakage due to the planted hardware Trojans. A design-for-security task scheduling approach is also proposed to improve the task scheduling result by reducing both the schedule length and the number of cores integrated, with only a small increase in the risk of Trojan triggering. In the first stage, the data-dependent tasks with negative slacks are iteratively clustered to reduce the schedule length until the given performance constraint is met. In the second stage, each cluster is assigned with a most proper IP vendor and tasks are scheduled by force-directed scheduling method, such that the number of cores required is minimized. The experimental results demonstrate that our proposed approach significantly minimizes the number of cores integrated in the MPSoC with performance constraint, while most of the security constraints are satisfied.

ACKNOWLEDGEMENT

This work was supported by the National Natural Science Foundation of China (NSFC) under Grant 61604054.

REFERENCES

[1] X. Wang and R. Karri, "NumChecker: detecting kernel control-flow modifying rootkits by using hardware performance counters," *Proc. Design Automation Conference*, pp. 1-7, May 2013.

[2] S. Swapp, *Scanning Electron Microscopy (SEM)*, University of Wyoming.

[3] M. Banga and M.S. Hsiao, "A novel sustained vector technique for the detection of hardware trojans," *Proc. International Conference of VLSI Design*, pp. 327-332, Jan. 2009.

[4] K. Xiao and M. Tehranipoor, "BISA: Built-in self-authentication for preventing hardware Trojan insertion," *Proc. International Symposium on Hardware-Oriented Security and Trust*, pp. 45-50, 2013.

[5] F. Koushanfar and A. Mirhoseini, "A unified framework for multimodal submodular integrated circuits Trojan detection," *IEEE Trans. Information Forensics and Security*, vol. 6, no. 1, pp. 162-174, Mar. 2011.

[6] M. Beaumont, B. Hopkins, and T. Newby, "SAFER PATH: security architecture using fragmented execution and replication for protection against Trojaned hardware," *Proc. Design, Automation & Test in Europe Conference*, pp. 1000-1005, Mar. 2012.

[7] J. Rajendran, et al., "Belling the CAD: toward security-centric electronic system design," *IEEE Trans. Comput.-Aided Design of Integr. Circuits and Syst.*, vol. 34, no. 11, pp. 1756-1769, Nov. 2015.

[8] K. Jiang, P. Eles, and Z. Peng, "Optimization of secure embedded systems with dynamic task sets," *Proc. Design, Automation & Test in Europe Conference*, pp. 1765-1770, Mar. 2013.

[9] X. Cui et al., "High-level synthesis for run-time hardware Trojan detection and recovery," *Proc. Design Automation Conference*, pp. 1-6, Jun. 2014.

[10] J. Rajendran, O. Sinanoglu, and R. Karri, "Building trustworthy systems using untrusted components: a high-level synthesis approach," *IEEE Trans. on Very Large Scale Integr. (VLSI) Syst.*, vol. 24, no. 9, pp. 2946-2959, 2016.

[11] C. Liu, J. Rajendran, C. Yang, and R. Karri, "Shielding heterogeneous MPSoCs from untrustworthy 3PIPs through security-driven task scheduling," *IEEE Trans. Emerging Topics in Computing*, vol. 2, no. 4, pp. 461-472, 2014.

[12] J. Rajendran, H. Zhang, O. Sinanoglu, and R. Karri, "High-level Synthesis for Security and Trust", *Proc. International On-Line Testing Symposium*, pp. 232-233, 2013.

[13] N. Wang, S. Chen, J. Ni, X. Ling, and Y. Zhu, "Security-aware task scheduling using untrusted components in high-level synthesis," *IEEE Access*, 2018, in press.

[14] P.G. Paulin and J.P. Knight, "Force-directed scheduling for the behavioral synthesis of ASIC's," *IEEE Trans. Comput. Aided-Design of Integr. Circuits and Syst.*, vol. 8, no. 6, pp. 661-679, Jun. 1989.

[15] D. Gizopoulos et al., "Architectures for online error detection and recovery in multicore processors," *Proc. Design, Automation and Test in Europe Conference*, pp. 533-538, Apr. 2011.

[16] X. Tang, H. Zhou, and P. Banerjee, "Leakage power optimization with dual-V_{th} library in high-level synthesis," *Proc. Design Automation Conference*, pp. 202-207, 2005.

2018 IEEE Computer Society Annual Symposium on VLSI

A Hardware Monitor to Protect Linux System Calls

George Provelengios, Arman Pouraghily, Russell Tessier, and Tilman Wolf
Department of Electrical and Computer Engineering
University of Massachusetts, Amherst, MA, 01003 USA

Abstract—**Internet-connected embedded systems have limited capabilities to defend themselves against remote hacking attacks. The potential effects of such attacks, however, can have a significant impact in the context of the Internet of Things, industrial control systems, smart health systems, etc. Embedded systems cannot effectively utilize existing software-based protection mechanisms due to limited processing capabilities and energy resources. We propose a novel hardware-based monitoring technique that can detect if the system calls of sophisticated embedded operating systems (e.g. Linux) deviate from the originally programmed behavior due to an attack. We present an FPGA-based prototype implementation that shows the effectiveness of such a security approach using a known Linux exploit. Our approach detects the attack with minimal overhead and without slowing processor operation.**

Index Terms—**hardware security, hardware monitor, system call, FPGA**

I. INTRODUCTION

The Internet of Things (IoT) represents the convergence of cyber-physical systems (CPS), which control physical processes, and the Internet, which provides global interconnectivity for access to data systems. Embedded systems are at the core of any IoT solution as they provide the necessary computational power at the location where devices interact with the physical world. Due to their deployment in the environment, these embedded systems are typically constrained in their computational resources (performance and/or energy) but still connected to a network to interact with the other components of the IoT solution. This type of networked embedded system experiences a particularly challenging problem when it comes to security, specifically, protection from attacks on the embedded operating system. The network connectivity provides an attack vector to the system and the system performance or energy resources are insufficient to run conventional defense mechanisms, such as virus scanners and malware detection software, that provide protection on conventional computers. An effective defense mechanism that has been developed in related work is "hardware monitors". These monitors are logic components that are co-located with the embedded system processor core and track the execution of software. Hardware monitors require no change or addition to the software that is run on the processing system. Instead, such monitors verify that a processor executes a piece of software faithfully by comparing two pieces of information: (1) processing steps reported by the processor at runtime and (2) a model of what is considered correct execution of the software that is to be executed. Attacks that hijack the processor inherently cause the

processing to deviate from the model of the original software and thus can be detected.

When discussing related work in Section II-A, we show how our work distinguishes itself from other monitoring techniques. The main novelty of our work is that our hardware monitoring system *works with Linux, a common, widely-used operating system*, whereas previous work has either looked at specific applications running directly on the processor or highly constrained, simplistic embedded operating systems. In addition, we show that our system defends against *real, practical attacks* (in our case the CVE-2013-1828 vulnerability, which has a known exploit), whereas previous work has shown defenses against attacks exploiting synthetically crafted vulnerabilities.

The main idea of our work is to focus the monitoring system on the portions of the operating system that are particularly vulnerable. Since many vulnerabilities and associated exploits occur in the context of system calls, we have designed our hardware monitor to track their processor operations at very fine granularity. By verifying operation at the level of an individual processor instruction, we can detect any deviation (i.e., attack) almost instantaneously. By limiting the monitoring to a fraction of the operating system code (i.e., system calls) and not the entire code base, we can achieve low overhead compared to other hardware monitoring approaches. This combination of sensitivity to attacks on vulnerable code and low hardware overhead (and no modification to any software) provides a promising approach to protecting embedded systems in the IoT domain or anywhere else.

The remainder of the paper is organized as follows. Section II discusses how our work relates to other efforts to protect embedded systems. The principles of our monitoring system are described in Section III. The design and implementation of our prototype system are presented in Section IV. Experimental results are shown in Section V, and the paper is summarized and concluded in Section VI.

II. BACKGROUND

A. Related Work

Monitoring of correct program execution has been proposed in various forms, such as verification of control-flow integrity (CFI) [1]. These software techniques may slow down program execution and do not validate individual processor instructions. Hardware monitoring reduces the performance impact of monitoring. The seminal work by Arora *et al.* described a fine-grained hardware monitoring system that verifies correct execution at the granularity level of a basic block [2]. This work was advanced by Mao *et al.* in verifying

978-1-5386-7100-9/18 $31.00 © 2018 IEEE
551

TABLE I
RELATED WORK ON HARDWARE MONITORING.

	Abadi et al. [1]	Arora et al. [2]	Mao et al. [3]	Pouraghily et al. [4]	this paper
verification	control flow operations	all processor instructions			
granularity	basic block		single processor instruction		
target	application / OS	monolithic application		simplistic OS	Linux OS
coverage	application / OS	entire application		entire OS	system calls
overhead	software	high hardware cost			low hardware cost

individual processor instructions and the resulting ability to stop attacks within one processor clock cycle instead of having to wait until the basic block has ended [3]. Recent work by Pouraghily *et al.* further expanded the previous work to not only monitor monolithic applications, but the underlying operating system [4].

Our work also focuses on monitoring the operating system. However, unlike related work, we aim to work with a real Linux operating system, not a light, embedded variant of a simplistic operating system. The large code size of the Linux kernel makes previous approaches to monitoring impractical due to their large overhead. In our work, we focus the monitoring effort on the portions of the code that are particularly vulnerable to attacks: system calls. Thus, we can effectively detect a good number of attacks while keeping the monitoring overhead low enough to make such a system practically useful. The progression of work on hardware monitoring and the context of our contribution is summarized in Table I.

System-call monitoring is another technique that attempts to detect intrusion. The approach tracks the system calls that are executed by an application, which is much coarser than tracking individual processor instructions. A survey on system-call monitoring [5] describes how the work has evolved over time. The main difference between this work and our approach is that we do not track patterns of system calls. Instead, we focus on ensuring that the processor instructions associated with a system call are executed faithfully. This approach ensures that attacks via system calls do not succeed. The existing approaches to system call monitoring can be used orthogonally to our work.

B. Focus on System Calls

As mentioned above, our hardware monitoring system focuses on validating the correct execution of system calls in the operating system. The current Linux kernel (version 4.13.15) contains code for 337 different system calls. Between 1999 and 2017, 1,931 vulnerabilities in the Linux kernel were reported to the Common Vulnerabilities and Exposures (CVE) database that is maintained by MITRE. Of those, 45 vulnerabilities (2.3%) directly relate to system calls. This may seem like a small percentage. However, the existence of a vulnerability is particularly problematic if an exploit exists that can let an attacker use the vulnerability in a practical manner. Of 148 publicly available exploits (listed in the Exploit Database maintained by Offensive Security) that lead to privilege escalation attacks (which gives the attacker full control over the

system), 25 exploits (16.9%) are based on vulnerabilities in system calls.

A typical attack, as we describe in more detail in Section IV, uses a buffer overflow to redirect program execution to shell code or other attack code. Since the kernel operates at the highest level of privilege in the system, achieving the execution of malicious code through redirection of a system call can give an attacker the highest level of access. By protecting system calls from such attacks through verification of correct execution, which can detect buffer overflow attacks that change code execution, we can protect the system from exploits that use known and unknown vulnerabilities. This protection works for attacks that are launched through software that is executed on the system directly, as well as attacks that are launched remotely through the network.

III. MONITORING ARCHITECTURE

The main goal of our monitoring system is to prevent execution deviations from system calls to malicious code. If such a deviation is detected, execution is stopped and the processor is reset. Our security model assumes that an attacker may access the target system and tamper with processor instructions and data remotely through an I/O interface, although it is not possible to tamper with the monitoring system.

A. Basic Monitor Operation

As mentioned in Section I, hardware monitors are components that are co-located with processor cores to track the processing of software on that core. The objective is to assess the operation of the processor and determine when incorrect behavior is detected (which can be due to benign faults or malicious attacks). In our work, we use a hardware monitor that receives information about every instruction executed on the processor core and compares it to a "monitoring graph" that is generated from the processing binary. Each instruction is represented by a hash value (to reduce the size of the monitoring graph compared to the size of the binary) and state transitions correspond to possible control flow paths between instructions. We use a deterministic finite automaton (DFA) representation of the monitoring graph (as detailed in [6]).

For this work, a monitoring graph is generated during design compilation [6] for selected system calls. Each instruction in the system call is encoded as an entry in the graph that includes the valid hash value(s) of the next instruction (or instructions in the case of a branch) and the next graph state(s). A detailed view of our monitoring subsystem is shown in Figure 1. The portions of the monitoring system can be split into *monitoring*

978-1-5386-7100-9/18 $31.00 © 2018 IEEE

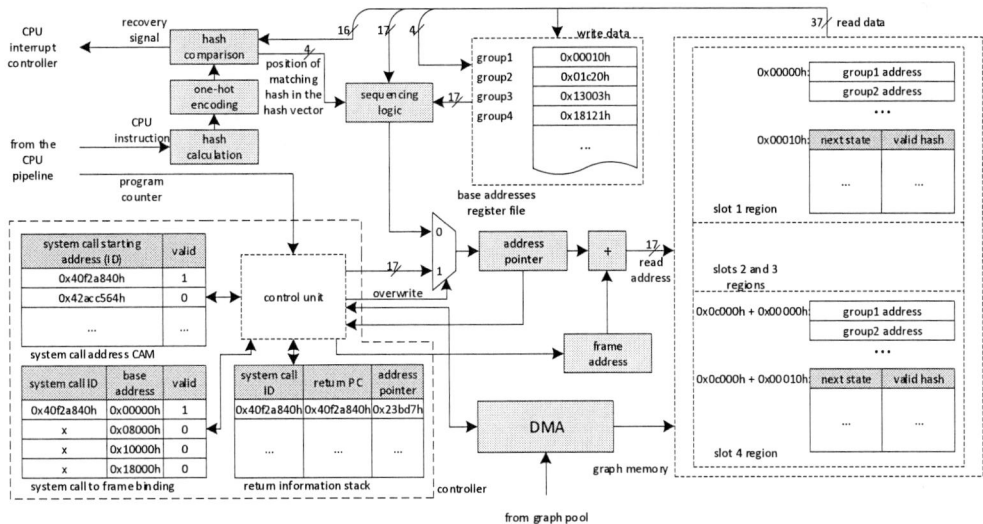

Fig. 1. System architecture for a hardware monitor that supports selective system call monitoring

hardware (three boxes in upper left corner of the figure), which checks the per-instruction operation of the companion processor, *graph memory*, which stores monitoring graphs, and *controller*. The monitoring hardware checks each processor instruction using an entry from the monitoring graphs stored in graph memory. In the figure, graphs for four separate system calls are stored in slots in the graph memory. Each graph includes one row per instruction, effectively representing expected program control flow as a state machine. A *read address* pointer indicates the entry in the graph that corresponds to the instruction that has just completed execution. During the execution of an instruction, a multi-bit (in our case 4-bit) hash value of the instruction is generated and converted to a one-hot representation. Previous work has shown a 4-bit hash value to be sufficient to limit collisions [7]. The one-hot encoding is compared against the expected next instruction hash values (*valid hash*) that are stored in the graph entry for the previously executed instruction. The use of a one-hot representation simplifies these comparison operations.

A match of an instruction hash against a stored valid hash indicates a valid instruction. If no match occurs, an illegal instruction has been executed, leading to the generation of a recovery signal which is used by the processor for process termination. Since control flow instructions (e.g. branch) may have several possible next instructions, and, consequently, several possible valid hashes, multiple one-hot valid hash bits may be set per entry. A match of any of these hashes indicates a valid instruction. Our approach can handle dynamic branch targets by profiling the code to determine all branch targets for a system call prior to graph generation. Entries for these targets are then added to the graph.

The next *read address* (memory row) in the monitoring graph is determined using next state information stored in the current entry, the matched hash value, and information stored

in *base address registers* which group states based on fan-in count [6]. These values are combined via addition in the *sequencing logic* box in the figure. The resulting address is stored in the *address pointer* and subsequently added to the start address for the appropriate graph slot for the system call. The implemented monitor requires only one memory lookup per instruction. Effectively, the monitoring information for each system call at any given point in execution is defined by the contents of the address pointer, the monitoring graph for the process and the contents of the base address registers. The location of each system call monitoring graph in the graph memory is stored in the *system call to frame binding* memory. The procedure required for activating monitoring for system calls is described next.

B. Enabling and Disabling Monitoring

Since monitored system calls can be invoked from within user applications or unmonitored system calls, a mechanism to seamlessly enable and disable the hardware monitor once a system call is invoked or retired is included in our monitoring system. Monitoring is stopped after the monitored system call is finished and the user application or unmonitored system call execution is restarted. We consider four specific scenarios: (1) a monitored system call is called from an application (monitor activated), (2) a return from a monitored system call to an application or unmonitored system call (monitor deactivated). (3) an unmonitored system call is called from a monitored system call (monitor deactivated), and (4) a return from an unmonitored system call to a monitored system call (monitor activated).

1) Call to monitored system call from unmonitored code: After Linux is compiled into a loadable image, the addresses of the kernel functions and system calls are fixed. The starting address of each system call is used as a unique identifier. For each system call, there is only one entry point, which is used to trigger the monitor. A hardware-based solution triggers monitoring upon entry into a system call by matching the system call program counter to one of a series of valid stored values (valid bit = 1) in a content addressable memory. It is shown in Figure 1 as the *system call address CAM*. As a transition to the monitored system call is made, the monitor is enabled.

For example, when the microprocessor executes an instruction, the program counter which has been extracted from the exception stage of the processor pipeline is compared against all of the valid system call starting addresses in the CAM. If it matches a stored address, the monitor is activated to start tracking microprocessor code execution using the monitoring graph generated during the compilation process for the system call. Prior to Linux execution, the CAM is loaded with the start addresses of the monitored system calls. Information in the monitor, including monitoring graphs and the system call address CAM, are loaded through a secure channel that is not accessible to application users. Any modifications to the CAM table is performed using secure techniques [6].

2) Return from monitored system call: A scalable approach is used to disable the monitor upon leaving a monitored system call since multiple exit points in the call may exist. To avoid using a large CAM to match the PC against all exit points, monitor disable information is embedded within the monitoring graph of the system call. As discussed in Section III-A, each entry in the monitoring graph contains a one-hot encoding of the valid hashes for the next instructions which succeed the current one. Normally, one or more of those bits are set to one according to the number of legitimate next instructions. However, if the instruction is the last instruction of the system call, all hash bits are set to zero indicating a system call return. This value disables monitoring.

C. Handling Nested System Calls

The mechanism described above is most effective if the call to a monitored system call is made from application code and a return to this code is made when the system call terminates. However, in many cases a monitored system call may invoke another system call that may be monitored or unmonitored. Thus, monitoring may need to be suspended for a time and then restarted upon return to the monitored system call.

1) Call to unmonitored system call from monitored system call: If unmonitored code is called from the monitored system call, the return address of the monitored code is stored on the return information stack and the monitor is deactivated. When a current, monitored system call switches to a new system call, its return address is stored on the stack. The stack consists of three different fields: system call ID which is the starting address of the monitored system call, return PC which is the next PC of the current system call which will be executed on the microprocessor after returning from the callee, and finally the current pointer to the monitoring graph of the current system call.

2) Return from unmonitored system call to monitored system call: When a return is made from the unmonitored code

TABLE II
MONITORING GRAPH SIZES FOR FOUR LINUX SYSTEM CALLS

system call	num. instr,	num. entries	graph size (bits)
getsockopt	49,252	68,422	2,531,614
execve	49,816	70,318	2,601,766
open	37,953	54,520	2,017,240
mmap	171	254	9,398

to monitored code, the return PC is checked against the top of the return information stack to determine if monitoring should be re-enabled. If a return is made to the monitored code, the monitor is reactivated and the return PC is popped from the stack.

IV. PROTOTYPE IMPLEMENTATION

Our experimental system uses a 7-stage LEON3 processor, release 2017.2-b4193 [8] and an attached hardware monitor. The floating point unit was not included in the design. The hardware was synthesized and mapped to a Stratix IV FPGA on a Terasic DE4 board with 1GB of DDR2 memory. To perform monitoring, the instruction under execution and the program counter (PC) from the processor are tapped for use by the monitor. For monitoring to work effectively, it is necessary to ensure that only committed instructions are monitored, since a number of fetched instructions may be flushed or annulled from the processor pipeline. For this reason, the PC and associated instruction are tapped from the exception stage of the processor after the annul signal can be examined. As discussed in Section III-A, the instruction is subsequently converted to a hash value and compared to a stored entry in the monitoring graph. The PC is used to determine if monitoring should be enabled or disabled. If the monitor detects a deviation from expected computation, the processor is reset using a recovery signal. Detection and reset takes place as the inappropriate instruction is executed. The processor additions needed to tap the PC and instruction are negligible and our results show no loss of processor clock speed performance as a result of this action.

A. System Calls

In a secure system, all system calls should be monitored to prevent any system-call-based attack. Our monitor microarchitecture shown in Figure 1 is designed to monitor a subset of all calls as needed. For this work, we focus on the four system calls shown in Table II (more calls can be easily added). We chose these four system calls since the first contains the known vulnerability CVE-2013-1828 and others have been characterized as particularly vulnerable calls and used for kernel exploitation [9].

B. Attack Scenario

To evaluate the ability of our monitor to detect and prevent an attack, we tested our processor/monitor system with a known and published Linux attack from the Exploit Database, ID 24747 [10] and an additional attack that is derived from

```
 →    ~ ssh test@192.168.2.40
test@192.168.2.40's password:
[test@buildroot ~]$ cat /etc/passwd
root:x:0:0:root:/root:/bin/sh
test:x:1001:1001:Linux User,,,:/home/test:/bin/sh
[test@buildroot ~]$ ./priv_escalation
[test@buildroot ~]$ cat /etc/passwd
root:x:0:0:root:/root:/bin/sh
test:x:0:0:Linux root,,,:/home/test:/bin/sh
[test@buildroot ~]$ exit
logout
Connection to 192.168.2.40 closed.
 →    ~ ssh test@192.168.2.40
test@192.168.2.40's password:
[root@buildroot ~]# █
```

Fig. 2. Console output showing that the attack script changes the *test* account privilege from a normal user to root

it. The latter attack exploits a vulnerability in the function *sctp_getsockopt_assoc_stats()* of the *getsockopt* system call and leads to a privilege escalation.

In the function, a call to *copy_from_user()* is used to copy the contents of a user-provided buffer into a data structure defined inside the function's local scope. Since there is no size check before calling the function, the user can provide a buffer to the system call which is bigger than the size of the local buffer. Therefore copying the buffer contents to the *sctp_getsockopt_assoc_stats()* function's local stack frame can overwrite substantial portions of the stack.

In Linux, the */etc/passwd* plain text file holds information about user accounts and their access levels. By modifying this file, one can grant any account root access. However, all users except root can only read this file and write access to this file is only granted to the root account. In our attack, instead of rewriting the stack with some random data and therefore destroying the return address, the system call is fed a buffer with meaningful data so that a user can gain root access. Specifically, the return address of the *sctp_getsockopt()* function is changed to the starting address of the *call_usermodehelper()* function which is a part of kernel and is used to prepare and run a user mode application from within the kernel. Using this function, */bin/sed*, a stream editor in Unix based operating systems, is executed to rewrite */etc/passwd* and grant root access to the user. Figure 2 shows the attack in action.

A key aspect of this attack is the writing of the attack arguments to call *usermodehelper()* that are passed to */bin/sed* and the branch to the function from the system call *sctp_getsockopt()*. When *call_usermodehelper()* is called, it receives its four operands on the stack, (two char*, one char**, and an int). Using monitoring, it is possible to detect the unexpected branch to *call_usermodehelper()* since the instructions of this function will not have entries in the monitoring graph.

V. EXPERIMENTAL RESULTS

To evaluate performance, our processor and monitor architecture was mapped to the DE4's Stratix IV EP4SGX230 FPGA. A maximum system clock frequency of 110 MHz was achieved both with and without the monitor. Signals internal to the FPGA were monitored using Intel SignalTap, leading to the waveforms shown in Figure 3. The observed

waveforms come from an attempted return from the system call function *sctp_getsockopt()*. Figure 3 (top) shows processor behavior during a normal return from the function starting at cycle 130. At this point, the one-hot hash encoding (0000 0000 0000 0100) of the next instruction matches one of the acceptable encoded valid hashes in the stored monitoring graph (0100 0000 1110 0100) in bit 2. The same observation can be made for the hashes of the next instruction. Thus, the instruction execution matches one of the expected execution paths determined during design compilation.

A. Attack Detection and Recovery

Figure 3 (bottom) shows the details of monitoring activities when the attack described in Section IV-B is performed. In this case, the return address of the *sctp_getsockopt()* function has been overwritten with the address of the *call_usermodehelper()* function. Since the first instruction in this function was not an acceptable return target for *sctp_getsockopt()*, the one-hot hash of this instruction will not match a valid hash value in the monitoring graph entry. Figure 3 (bottom) shows that this is the case. The one-hot hash of the instruction at cycle 130 is (0000 0001 0000 0000) while the stored valid hash value is (0100 0000 1110 0100). Since bit 8 is not set in the valid hash value, an unexpected instruction has been executed and the processor reset (recovery signal) can be asserted low. Note that the set bit in the one-hot hash of the next instruction also does not match the appropriate bit in the valid hash value. It should be noted that although the reset signal causes the processor to restart, possibly leading to a denial of service attack, this outcome is preferable to an unauthorized user gaining superuser access to the system.

Using the graph generation approach described in Section III-A, we examined the size of four representative Linux (version 3.8.0) system calls, including the *getsockopt* call described in Section IV-B. The number of instructions, the number of monitoring graph memory entries, and the total graph sizes in monitoring graph memory in bits for each system call are shown in Table II.

B. Monitoring System Overhead

For performance reasons, the monitoring graph is stored on-chip to allow for instruction-by-instruction hash value comparisons. Thus, we assess both the logic overhead and the overhead of on-chip memory. In addition, if a new system call is used, its monitoring graph may need to be securely loaded from off-chip memory using DMA to one of the graph memory slots shown in Figure 1 [4]. The resources needed to implement the microprocessor, the monitor and its associated graph transfer circuitry are shown in Table III.

The table shows that the monitor and associated control circuitry require dramatically less circuitry than the processor since it is a simple finite state machine. On-chip memory is needed so that each graph entry can be quickly obtained and compared to the currently-executing instruction. The table also includes the resources needed to securely load encrypted system call monitoring graphs from external memory. This

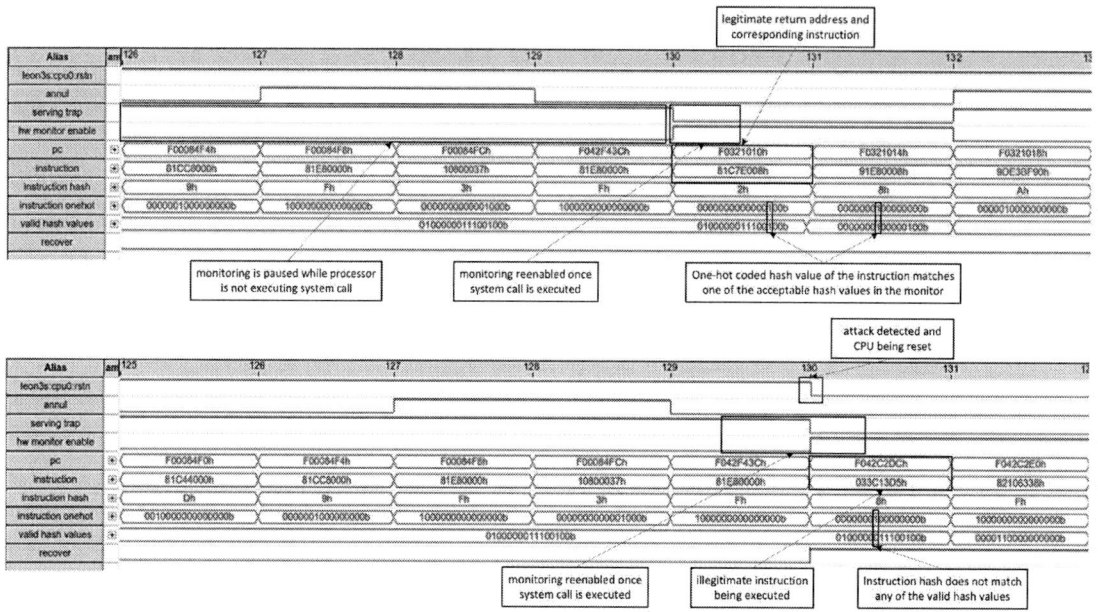

Fig. 3. Waveforms showing normal execution of the system call (top) and detection of the attack (bottom).

TABLE III
RESOURCE UTILIZATION OF THE HARDWARE MONITOR AND LEON3 PROCESSOR

resource		Available	LEON3 w/o hardware monitor	Hardware monitor and controller	CAM/stack	Secure HW mon. loader
Logic LUTs		182,400	20,070	380	6,555	2,603
Memory LUTs		91,200	170	0	0	0
Flip flops		182,400	15,053	324	11,457	2,936
Memory (bits)	off-chip	8,589,934,592	100,326,512	0	0	0
	on-chip	14,625,792	534,752	3,054,752	0	977,332

circuitry includes a decryption circuit which increases the overhead of the interface. Finally, the resources needed to implement the system call address CAM and return information stack used to identify monitoring start and stop points (described in Section III-B) for up to 337 different system calls are shown in the table.

By far, the most expensive part of monitoring is the on-chip memory needed to store the monitoring graphs. In this system, Linux instructions are stored off-chip so monitoring storage takes up the bulk of the on-chip storage. In our design, the monitoring graphs consume less than one-quarter of available on-chip memory so sufficient space is available for other circuitry. Overall, our results show that system call monitoring for advanced embedded operating systems, such as Linux, can be performed efficiently.

VI. SUMMARY AND CONCLUSION

System calls in sophisticated embedded operating systems are known to be vulnerable targets for attackers. We present a low-overhead monitoring approach that allows for selective instruction-by-instruction monitoring of system calls. Our approach has been demonstrated in hardware to successfully identify and prevent a known Linux system call attack. The overhead of the monitor is modest and does not impact the performance of the microprocessor.[1]

REFERENCES

[1] M. Abadi *et al.*, "Control-flow integrity principles, implementations, and applications," in *ACM CCS*, Nov. 2005, pp. 340–353.

[2] D. Arora, S. Ravi, A. Raghunathan, and N. K. Jha, "Secure embedded processing through hardware-assisted run-time monitoring," in *IEEE/ACM DATE*, Mar. 2005, pp. 178–183.

[3] S. Mao and T. Wolf, "Hardware support for secure processing in embedded systems," in *IEEE/ACM DAC*, Jun. 2007, pp. 483–488.

[4] A. Pouraghily, T. Wolf, and R. Tessier, "Hardware support for embedded operating system security," in *IEEE ASAP*, Jul. 2017, pp. 61–66.

[5] S. Forrest, S. Hofmeyr, and A. Somayaji, "The evolution of system-call monitoring," in *Comp. Security Appl. Conf.*, Dec. 2008, pp. 418–430.

[6] K. Hu *et al.*, "System-level security for network processors with hardware monitors," in *IEEE/ACM DAC)*, Jun. 2014, pp. 211:1–211:6.

[7] T. Wolf *et al.*, "Securing network processors with high-performance hardware monitors," *IEEE TDSC*, vol. 12, no. 6, pp. 652–664, Nov. 2015.

[8] J. Gaisler and S. Habinc, "Grlib IP library user's manual," Cobham, Tech. Rep., Nov. 2017.

[9] C. Linn *et al.*, "Protecting against unexpected system calls," in *Usenix Security Symposium*, Aug. 2005.

[10] "Exploit database," https://www.exploit-db.com, accessed: 2017-11-18.

[1]This research was sponsored by NSF grant CNS-1617458.

2018 IEEE Computer Society Annual Symposium on VLSI

Towards Dynamic Execution Environment for System Security Protection against Hardware Flaws

Kenneth Schmitz[†]   Oliver Keszocze[*†]   Jurij Schmidt[*†]   Daniel Große[*†]   Rolf Drechsler[*†]

[*]Institute of Computer Science, University of Bremen, 28359 Bremen, Germany
[†]Cyber-Physical Systems, DFKI GmbH, 28359 Bremen, Germany
{kenneth, keszocze, grosse, drechsler}@cs.uni-bremen.de

Abstract—Attacks exploiting security flaws in software are very common. They are typically addressed during the ongoing software development process or by providing software patches. Attacks making use of hardware related flaws via malicious software recently gained popularity. Prominent examples are errata-based, aging-related or, for example, the infamous Rowhammer-attack. In this paper, we present an approach to detect software-based attacks which exploit hardware flaws. Since the flaws are typically triggered by characteristic instruction sequences, our approach is implemented as a dynamic execution environment for program monitoring at runtime. Several case studies underline the effectiveness and the low overhead of our approach.

I. INTRODUCTION

Malicious software such as Trojans or viruses can be accounted for major system failures and large financial losses [1]. Most recently, cryptographic ransomware was used for blackmailing companies to recover their encrypted data [2]. To protect the victim's systems against such attacks, typically several techniques (e.g. sandboxing, static/dynamic or signature based analysis) have been implemented in antivirus software. While there are different arguments for and against each of these techniques, they have been developed from a software centric perspective since malicious code uses flaws and vulnerabilities in software as an attack vector.

Due to the shrinking feature sizes and the increasing complexity of hardware, more flaws reach the silicon [3]. Hence, focusing on hardware and potential attacks exploiting the flaws in silicon is very important and mandatory. In recent years, many approaches for Trojan/backdoor identification in hardware and *Integrated Circuit* (IC) counterfeit detection [4] have been developed. Furthermore, defined areas for secure software execution and data storage in hardware (e.g. ARM TrustZone, Intel SGX, TPM) have been created, aiming for the protection of data and the software itself during execution. In contrast, this work takes the hardware perspective and aims to protect systems against malicious software which exploits *hardware flaws* as a new attack vector. In the following we identify two major categories of hardware flaws where an urgent protection is inevitable.

Errata-based defects and resulting system failures are the first category. Almost every processor-generation has errata instructions, which are typically addressed by microcode- or BIOS-updates. Since modern hardware components are very complex, verification and test become more challenging and flaws can remain undiscovered prior to the fabrication. Powerful instruction set extensions to the x86 *Instruction Set Architecture* (ISA) have been recently reported to result in unpredictable behavior [5]. Undocumented features inside the ISA, which *can* cause unpredictable system behavior, have been revealed [6] as well.

The second category covers flaws which are inherited from the feature sizes used to fabricate the components. The Rowhammer-attack affects *Random Access Memory* (RAM) and *Solid-State Drives* (SSDs) [7]. The *malicious aging in circuits/cores* (MAGIC) [8] leads to very fast semiconductor aging. *Field Programmable Gate Array* (FPGA)-based systems are susceptible to this attack scenario [9] as well. Both attacks exploit the basic properties of the feature sizes in order to make the system fail early or unexpectedly. All of these flaws can be induced by regular execution of regular software. Unfortunately, antivirus software typically fails if confronted with scenarios, which explicitly target hardware flaws.

In this paper, we propose a novel hardware-centric approach based on the following idea: All of the presented attacks exploit the hardware flaws through characteristic instruction sequences. Since detecting these instructions sequences is possible at the instruction level, we present an engine for instruction-screening. We use the *Quick Emulator* (QEMU) [10], providing a code translation layer that grants access to single instructions during execution. Our approach detects *all* user-defined instruction (search-) patterns in the instruction sequence of the executed program.

Unfortunately, a given platform can be susceptible to more than one attack. Therefore, the approach must allow for searching of *all* instructions at the same time. In addition, it must detect spatially distributed instruction sequences in programs to protect against advanced attacks (e.g. clflush-based Rowhammer-attacks) and by this providing a general scheme in contrast to existing solutions. As a consequence, we have identified the powerful string matching algorithm Aho-Corasick [11] serving as basis for screening the instructions. Parallel matching of search patterns, while maintaining linear complexity with respect to the input sequence plus the number of simultaneously matched search patterns, is the strength of this algorithm. To handle the above mentioned spatially distributed instruction sequences in programs, we extended Aho-Corasick to match interrupted instruction sequences while maintaining the same algorithmic complexity. In our experiments, the extended algorithm was able to cope with the

---

This work was supported by the German Federal Ministry of Education and Research (BMBF) within the project SecRec under grant no. 16K1S0606K and by the University of Bremen's graduate school SyDe, funded by the German Excellence Initiative.

978-1-5386-7100-9/18 $31.00 © 2018 IEEE

execution speed (and the resulting high instruction-throughput) at the translation layer of QEMU at runtime.

The proposed approach also addresses software subroutines which implement malicious behavior. If the functionality of an application is known, certain instruction patterns can indicate malicious intents. Examples can be office applications *without* update functionality which contains instructions implementing network communication, or the Linux built in copy command using specific cryptographic instructions. These instructions are implausible in this specific context and can indicate an attack. The proposed approach can also recognize this behavior during runtime and intervene if the operational security is at risk.

II. RELATED WORK

Exploiting flaws in hardware and the protection thereof is an ongoing challenge. In general, software is capable of transitioning hardware to a state from where there is no recovery but a hard system reset. Even physical, irreversible damage is possible [12]. Identifying software with malicious intents is a very complex problem and has been thoroughly investigated (for an overview see for instance [13]). It has been shown that behavioral detection, signature extraction and improving resilience to automatic mutations still fail in many scenarios and remain a major challenge [14], [15]. In the software domain regular viruses use sophisticated techniques in order to hide their true intentions. Among these are self-decryption, oligo-, poly- or metamorphism which change the appearance (in terms of instruction sequences and behavior) during or prior to execution. Attacks against hardware flaws often require specific, immutable sequences of instructions (e.g. errata-related) to trigger the erroneous behavior.

The authors of the *Micro-Architectural Side Channel Attack Trapper* (MASCAT) [16] relied on static code analysis in order to scan for microarchitectural attacks. The work focuses on fully automated off-line analysis of applications (e.g. for app stores). Thus, this approach does not provide any protection as soon as the binary has reached the target system.

For the particular example of the Rowhammer-attack, the authors of [17] identified circumstances, where the attack can escape the known clflush-pattern and still remain effective.

In [18], a sandboxing technique has been proposed which restricts access to memory region within the host's address space. The scheme protects the host program from reads and writes by its guests and it,allows the restriction of the instruction set available to guests. Hence, a full notion of which instructions are *permissible* for every individual program is required.

There are concepts capable of repairing, patching or correcting the erroneous system behavior on the hardware level. Such a detection scheme, based on errata and internal signal observation, has been presented in [19] and [20]. Furthermore, the overall system state can be monitored and corrected on hardware level, as proposed in [21] and [22]. All of these approaches require hardware modifications which are impractical after fabrication.

III. PRELIMINARIES

QEMU and its internal translation method as well as the the efficient Aho-Corasick string matching algorithm are the key components for our approach and are reviewed in this section.

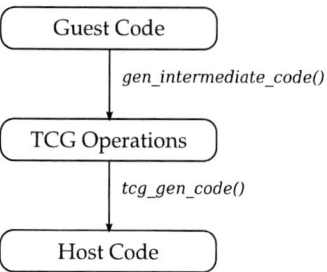

Fig. 1. Regular translation flow

A. Quick Emulator (QEMU)

QEMU is an open-source cross-platform environment for virtualization and emulation. It can utilize the *Kernel Virtual Machine* (KVM) for hardware acceleration. Two different modes of operation are provided:

A complete host system (including peripheral devices) can be mimicked to the guest system when the full virtualization environment is used. KVM is leveraged to speed up guest systems to near native execution speed and I/O redirection accelerates peripheral hardware access.

The second mode provides an user-mode emulation, which executes a single program as a guest application on a host system. When a program is run in user-mode emulation, independence from compiler versions and architectures is desired. The *Tiny Code Generator* (TCG) manages the code translation from the guest architecture to the host architecture. Guest instructions are translated to a *machine-independent intermediate notation* which is recompiled for the host's architecture. Several optimizations are applied in this step during emulation mode. All guest instructions are fully accessible at the TCG interface and can be monitored at runtime.

Figure 1 shows the relevant part of QEMU's flow, which is essential for this work and where our approach is included. All TCG operations are derived from the guest application's code. This intermediate language is processed by the TCG and translated to the target architecture. Finally, the TCG-generated target code is executed on the host's hardware.

B. The Aho-Corasick String Matching Algorithm

The Aho-Corasick algorithm is a dictionary-based string matching algorithm. It *simultaneously* locates all strings of a finite set of search strings within an input sequence. The algorithm is known to be highly scalable [23]. A dictionary is computed in advance – resulting in a tree structure called trie – to achieve the desired complexity which is linear in the length of the input plus the number of matched entries. A trie or a prefix-tree is a data structure which is typically used to store characters for search operations on character sequences. This specialized search-tree, implements storage for multiple character sequences simultaneously. The trie implicitly compresses the stored data, since shared prefixes are stored only once.

An example is shown in Figure 2. Starting from the root node, the algorithm traverses the tree while matching individual characters from the dictionary. The tree represents the entire dictionary {{a}, {a,b,a}, {c,a,b,d}}. All dotted nodes

978-1-5386-7100-9/18 $31.00 © 2018 IEEE   558

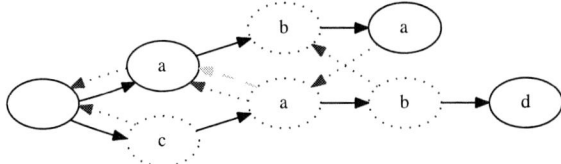

Fig. 2. Basic example for a dictionary-tree of {{a}, {a,b,a}, {c,a,b,d}}

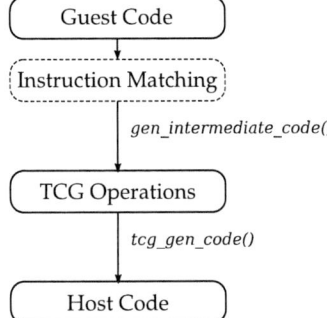

Fig. 3. Extended translation flow

```
label:                   label:
    mov (A), %eax            mov (A), %eax
    mov (B), %ebx            mov (B), %ebx
    clflush (A)              clflush (A)
    clflush (B)              clflush (B)
    jmp label               (*)
                            jmp label
```

Fig. 4. Pseudo assembly for Rowhammer-attacks

are intermediate nodes. The solid nodes, in contrast, are target nodes representing a successful detection. Additional arcs have to be computed to allow fast transitions between failed string matches. Connections denoted by dotted blue arcs are called suffix arcs, which point to the longest possible strict suffix in the graph. They are computed in linear time by traversing the dotted arcs of a node's parent until the child is matching the character of the arc's target node. Connections denoted by dashed green arcs are called dictionary suffix arcs, which point to the next reachable solid node following blue arcs. These are computed in linear time as well by traversing the dotted arcs until a solid node is found.

IV. DYNAMIC EXECUTION MONITORING

In this section, the implementation of our approach is presented in detail. Searching for malicious intents in software requires access to *all* program instructions. Hence, we conducted profiling experiments to determine the interface in QEMU's architecture, where the complete instruction sequence can be monitored. Figure 3 shows *where* our approach has been implemented in QEMU's architecture. The profiling experiments also revealed the TCG as the key-component for QEMU's speed. Subsequently, all extensions to this layer must be efficient in terms of their computational complexity to preserve the performance. The TCG implements the boundary after which code will be executed by the host's processor. Hence, the proposed solution detects malicious instruction sequences *prior* to translation in order to realize the system protection.

Since instruction screening is similar to searching for a matching character sequence in a string, the Aho-Corasick algorithm was chosen due to its strengths in the parallel matching of search patterns. However, in contrast to regular string matching, the suspect instruction sequences are rarely a sequence of consecutive elements. Hence, we extended the basic

Aho-Corasick algorithm to cope with the spatial distribution of instructions inside an executable.

A. Extension of the Aho-Corasick String Matching Algorithm

The following example can motivate this necessity in a clear fashion. The loop in Figure 4 implements a clflush-based Rowhammer-attack.

The asterisk character represents an arbitrary instruction or instruction sequence within the malicious sequence. Strictly searching for the left pattern will ignore the example provided on the right, although it will yield the same effect. Hence, our approach must be able to skip intermediate instructions. In order to achieve this functionality, two major extensions were necessary:

1. The spatial distribution of instructions in an sequence has been addressed by *Don't-Care* (DC) nodes in the language of the dictionary. These nodes provide the algorithm with the capability to skip intermediate segments until the next valid instruction is found.

2. Since the performance benefit granted by efficient transitioning between failed string-matches in the dictionary-tree is essential, complex trees with an arbitrary amount of DC nodes are impractical. Hence, we implemented a partial recompilation and modification of the dictionary-tree during runtime. After a DC node is reached, the remainder after the DC node is inserted at the root node. This addresses both requirements: The ability to search for all patterns simultaneously is preserved, and the reliable detection of spatially separated instruction segments is possible. Finally, a bidirectional connection between each DC node with its associated remainder is stored. This establishes the reattachment of the remainder to its former position in linear time, when the search is completed.

The proposed method is implemented as a non-greedy pattern search procedure: After a sequence of ignored instructions (DC instructions), the first matching instruction will be interpreted as the end of the DC-sequence. Finally, the dictionary-tree will be reverted to its initial state (with respect to the active search-pattern).

B. Search-Pattern Matching

A valid search-pattern must contain all instructions necessary for a successful detection. If the pattern is present in the executed binary, the implemented solution will detect the sequence in the binary-stream during execution. However, depending on the *strength* of this search pattern, false-positives are possible, since the spatial range of the search algorithm can exceed meaningful boundaries such as methods, blocks or loop-bodies. These *false-positives* are only false in the sense, that

978-1-5386-7100-9/18 $31.00 © 2018 IEEE   559

they will not have the intended or malicious effect. Nevertheless, the implementation will only report the presence of a search pattern if the given instruction sequence is actually present in the binary. In contrast, if the provided search pattern is fully specified (no DCs), false-positives can not occur.

In the following, a compact example will clarify the search-pattern matching procedure.

Figure 5 provides an example of a initial dictionary-tree. It shows the computed tree for the following dictionary: $\{\{i_1, i_2\}, \{i_1, *, i_3\}, \{i_0, *, i_5\}\}$. These sequences can be characterized and numbered as follows.

1. $\{i_1, i_2\}$ – Instruction i_1 is directly followed by i_2 for a successful detection.
2. $\{i_1, *, i_3\}$ – Instruction i_1 followed by an arbitrary number of instructions until instruction i_3 is found.
3. $\{i_0, *, i_5\}$– Instruction i_0 followed by an arbitrary number of instructions until instruction i_5 is found.

This compact tree representation stores all active search-patterns. Each instruction from the instruction-stream will be matched with either the root-node or with the following node in case of an active detection. After a DC node is reached, the dictionary-tree is altered. Figure 6 shows the temporal insertion of the instruction sequence's remainder at root level after the detachment from the DC node according to sequence 2. This way, an arbitrary amount of instructions can be skipped until the remainder is matched and the search-pattern is completed. Figure 7 shows the inserted remainder at root level according to sequence 3. In order to maintain low computational complexity, the dotted arcs indicate the connection for reattachment after completion of a search-pattern.

The resulting matching procedure is shown in Figure 8. Since all sequences start with i_0 or i_1, they will be compared with each element of the input instruction sequence from the executed program. This is reflected in the comparison of *all* outgoing arcs from the root node of the initial dictionary-tree. After matching i_0 from seq. 3 and i_1 from seq. 2, the DC nodes are expanded and the algorithm compares the following instructions with instruction i_5 and i_3. If the sequences are completed, the tree expansion is reverted to its initial state and attached at the respective DC node. This maintains a compact tree during runtime for faster traversal and reflects the non-greedy matching approach. Seq. 1 does not require a tree expansion, since i_1 is *directly* followed by i_2 in order to match successfully.

In summary, we have presented an efficient instruction-screening algorithm integrated as part of QEMU. At the heart of the algorithm we employ a well known string matching

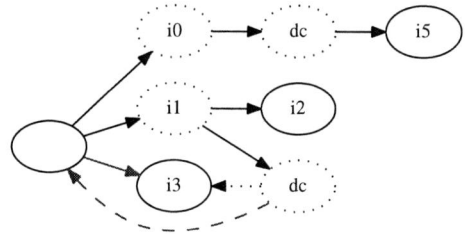

Fig. 6. Temporal expanded tree for i_3

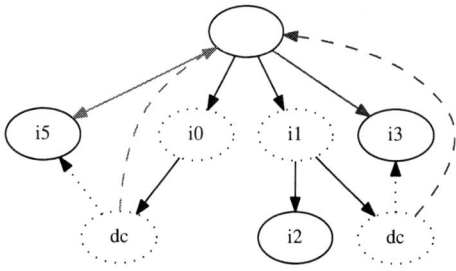

Fig. 7. Temporal expanded tree for i_3 and for i_5

technique which was extended such that a variety of accidental and malicious attacks against hardware flaws can be detected. A major challenge was the spatial distribution of instructions inside a given search-pattern. The effective screening of executables during execution becomes possible.

In the next section our experimental evaluation is presented.

V. CASE STUDIES AND RESULTS

For the evaluation of our approach we need a set of benchmarks. However, programs with malicious intents focusing on hardware flaws are typically not widely available. Hence, we integrated specific instruction sequences – implementing different attacks – as a set of characteristic benchmarks. The different scenarios, used to create a set of handcrafted programs, are briefly presented and followed by a detailed discussion of our results.

A. Creation of Benchmarks

A variety of malicious instruction patterns were added to some well known Linux user-land programs. One category of benchmarks includes errata-related bugs. The other includes the Rowhammer-attack and a cryptographic algorithm inside the source code. In the following, we describe the different characteristic flaws which have been considered in our benchmarks.

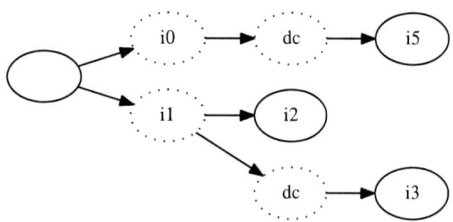

Fig. 5. Computed tree

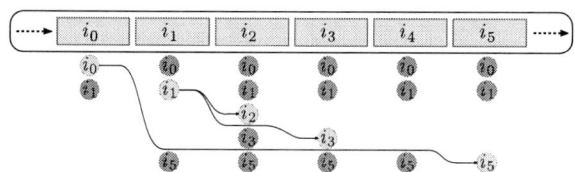

Fig. 8. Simultaneous sequence matching

978-1-5386-7100-9/18 $31.00 © 2018 IEEE   560

1) Errata-Instructions: Several silicon bugs have been discovered after the developed and fabricated products have been shipped. In general, until a fix for such issues is available, there is a time window in which systems are susceptible to attacks exploiting these bugs. In the following we give two prominent examples which we also used in our benchmarks:

- The **Cyrix Coma-Bug** [24] is a flaw in the x86 processor-series (6x86, 6x86L, and 6x86MX) manufacturer by Cyrix. When executed, a non-privileged program is able to lock the processor completely.
- The **Pentium f00f-Bug** [25] can transition the affected processors to a state, where only a reset is able to recover the system from this state. The flaw affects the locked compare and exchange instruction of eight bytes in register eax.

2) Active Attack Scenarios: Beside flaws in the fabricated product, there are attacks that exploit vulnerabilities of architectural features or the specific feature size of which components are made. Again, provide two concrete example to be used later:

- The **Rowhammer-effect** is an ongoing threat [7].
  The origin of this susceptibility is the decreasing feature size which is used to fabricate *dynamic RAM* (DRAM) chips (even in hand held devices [26]). It is possible to exploit this property with the purpose of data falsification or to escalate privileges (e.g. to gain administrative rights).
- **Ransomware** is an increasing threat. Victims to ransomware-based attacks often are left with encrypted data with the purpose to blackmail the victims to gain back their data. With the availability of dedicated cryptographic instructions (e.g. AES-NI), cryptographic routines are sped up significantly. These patterns contain characteristic instruction sequences which are detectable.

The provided benchmarks are based on well-understood instruction patterns and were chosen for demonstration. Our proposed approach is only applicable if the malicious instruction pattern is known in advance.

B. Evaluation

Basis for the implementation of our approach is the latest stable version of QEMU (2.8.0). The translation inside QEMU has been extended while the internal data structures and other functions have been kept. All programs have been executed in QEMU and invoked by passing a 10 Megabyte file. We tested our approach for different the scenarios, i.e. errata-based, Rowhammer and a cryptographic routine.

Modified versions of well-known Linux user-land programs (*copy, diff* and *tar*) have been used as benchmarks, since these could easily be compromised without being noticed. Additionally, *gzip* was augmented with *multiple* bug- or attack-scenarios, such that the benefit of the Aho-Corasick algorithm based screening approach of QEMU could be examined in detail.

Table I summarizes the obtained results. The first column contains the name of the augmented program. Each program was run with different bug- or attack-scenarios, which is triggered by a specific instruction sequence which was included in the respective binary. Each bug- or attack-scenario individually

TABLE I
RESULTS SUMMARY

Application Details	Execution Details				Found
Name Incl. Bug	Native	QEMU	Extension	Increase	
– – – – – –	[msec]	[off][msec]	[on][msec]	[in %]	
cp †	27	177	193	9.04	✓
cp ‡	29	161	189	17.39	✓
cp ⋆	27	178	202	13.48	✓
cp ∗	28	180	185	2.78	✓
diff †	1	140	164	17.14	✓
diff ‡	2	158	162	2.53	✓
diff ⋆	2	165	168	1.82	✓
diff ∗	1	157	162	3.18	✓
tar †	34	260	271	4.23	✓
tar ‡	32	268	276	2.99	✓
tar ⋆	36	243	265	9.05	✓
tar ∗	35	230	255	10.87	✓
gzip † ‡	138	369	394	6.78	✓
gzip ‡ ⋆	132	388	408	5.15	✓
gzip ⋆ ∗	147	373	407	9.12	✓
gzip † ‡ ⋆	140	394	452	14.72	✓
gzip † ‡ ⋆ ∗	151	367	423	15.26	✓

† Cyrix Coma-Bug included in binary.
‡ Intel Pentium f00f-Bug included in binary.
⋆ Rowhammer (Memory Disturbance) attack included in binary.
∗ Random assembler pattern included in binary.

requires specific search patterns. These configurations are indicated by the symbols next to the program name.

The second column presents the results of a native execution on the host system. No virtualization (i.e. KVM) or emulation environment was used. This column present the regular execution time (all in milliseconds) as experienced by any user on any system. A compute server, equipped with a Quad-Core Xeon Processor (Intel Xeon E3-1275) running Fedora Linux, was used to run the benchmarks.

In the third column (QEMU) shows the runtime of the respective binaries in emulation mode of QEMU (no KVM support). This execution represents the golden reference for our own implementation. Next, the fourth column contains the results of our extension to QEMU (indicated by "Extension"). During this execution, our methodology was active and screened for the provided search pattern.

Our results are presented in the fifth column. It contains the execution time of the augmented binaries by the provided percentage with respect to the golden reference results (100%). Finally, the last column contains indicates, that each bug- or attack-scenario was successfully detected. It must be noted, that *all* search pattern were active during all of the experiments without yielding neither false-negatives nor false-positives.

C. Observations

The experimental results indicate the successful system protection against the characteristic bug- or attack-scenarios as discussed before. It can be noted, that the increase in runtime during active detection is less than 18% in any given case with respect to the golden reference results. An average increased of runtime of 7.8% was observed. Obviously, there is a static runtime penalty in comparison to the native execution without the translation layer introduced by QEMU. Interestingly, there is no additional penalty, when multiple search patterns are matched simultaneously by our methodology.

1) Errata-Instructions: Search patterns such as the the coma- and the f00f-bug can be reliably detected. Due to the severity of these bugs, susceptible processors would transition to a non-recoverable state which has to be prevented. Since almost every processor generation will yield a large errata document, this methodology has a possibly large field of application. Third-party legacy software could contain malicious instruction sequences by accident. In these cases our methodology provides system protection while the software *can* still be used. In order to determine the robustness of this extension, we included a random (widely distributed, but unique) search pattern to the binary. The presence of this random patterns was also determined reliably.

2) Active Attack Scenarios: For active attack scenarios we chose the Rowhammer-attack (based on the clflush instruction). The pattern, implementing the Rowhammer-attack, was detected by the extension to QEMU since it can be mapped to a specific instruction sequence. It must be noted, that a Rowhammer-attack can also be induced by behavioral cache-attacks [27] which can be detected, if the instruction sequence is known in advance. Additionally, a standalone program was run, implementing a minimalistic cryptographic AES-function. We extracted the characteristic portion of an AES round as a search pattern and fed it to our extension. The detection of an integrated AES subroutine was reliable in our experiments. From a technical point of view, the detection of AES-NI instructions is even more reliable, since a specific opcode will be present in the instruction-stream which can be detected easily.

This discussion of results can be concluded as follows: The proposed approach has proven to be effective in the domain of runtime instruction screening with a focus on hardware flaws. Our intended use-case for this methodology is one complete execution inside QEMU. After the absence of malicious instruction sequences has been verified, the software can be run natively on the host's hardware. This way, the user can be sure (even in case of legacy or third-party software) that no harmful instruction sequences will compromise the host system. However, from this work the need for a database (such as well known for software-viruses) becomes imperative.

VI. Conclusion and Future Work

We have introduced a framework to detect software-based attacks which exploit hardware flaws. Our approach performs instruction screening during dynamic program execution inside QEMU. The efficiency of our solution is based on an extended Aho-Corasick string matching algorithm which allows for parallel matching of search patterns while maintaining a linear complexity. Our approach brings protection against hardware flaws which reach from simple errata-based flaws in fabricated hardware to feature size related vulnerabilities. Many of these can be fixed by BIOS, microcode or firmware updates, but typically several month pass until they are readily available. An alternative is additional hardware which *can* prevent such attacks [22]. But this hardware is expensive and impracticable

after fabrication. In contrast the proposed solution offers the possibility to run software in a safe environment, given that the speed degradation is acceptable and search patterns are provided. As future work we propose screening directly on the KVM layer to make this approach applicable to fully virtualized environments.This way this approach can be transferred to the kernel space, thus establishing an even lower bound of protection and increased execution speed.

References

[1] E. C. R. Council, "The economic impacts of the august 2003 blackout," *Washington, DC*, 2004.

[2] J. Hernandez-Castro, E. Cartwright, and A. Stepanova, "Economic analysis of ransomware," *CoRR*, 2017.

[3] P. Patra, "On the cusp of a validation wall," *Design Test of Computers*, pp. 193–196, 2007.

[4] U. Guin, D. DiMase, and M. Tehranipoor, "Counterfeit integrated circuits: detection, avoidance, and the challenges ahead," *Journal of Electronic Testing*, pp. 9–23, 2014.

[5] A. Baumann, "Hardware is the new software," in *HotOS*, 2017, pp. 132–137.

[6] C. Domas, "Breaking the x86 ISA," *Black Hat*, 2017.

[7] O. Mutlu, "The rowhammer problem and other issues we may face as memory becomes denser," in *Design, Automation and Test in Europe*, 2017, pp. 1116–1121.

[8] N. Karimi, A. K. Kanuparthi, X. Wang, O. Sinanoglu, and R. Karri, "(magic): Malicious aging in circuits/cores," *TACO*, p. 5, 2015.

[9] H. Zhang, L. Bauer, M. A. Kochte, E. Schneider, H. J. Wunderlich, and J. Henkel, "Aging resilience and fault tolerance in runtime reconfigurable architectures," *Trans. on Computers*, pp. 957–970, 2017.

[10] F. Bellard, "QEMU, a fast and portable dynamic translator." in *USENIX ATC*, 2005, pp. 41–46.

[11] A. V. Aho and M. J. Corasick, "Efficient string matching: an aid to bibliographic search," *Communications of the ACM*, pp. 333–340, 1975.

[12] P. Jayaraman and R. Parthasarathi, "A survey on post-silicon functional validation for multicore architectures," *ACM Comput. Surv.*, pp. 61:1–61:30, 2017.

[13] G. Jacob, H. Debar, and E. Filiol, "Behavioral detection of malware: from a survey towards an established taxonomy," *Journal in Computer Virology*, pp. 251–266, 2008.

[14] A. A. E. Elhadi, M. A. Maarof, B. I. Barry, and H. Hamza, "Enhancing the detection of metamorphic malware using call graphs," *Computers & Security*, pp. 62–78, 2014.

[15] S. Alam, R. N. Horspool, I. Traore, and I. Sogukpinar, "A framework for metamorphic malware analysis and real-time detection," *Computers & Security*, pp. 212–233, 2015.

[16] G. Irazoqui, T. Eisenbarth, and B. Sunar, "Mascat: Stopping microarchitectural attacks before execution." *IACR*, p. 1196, 2016.

[17] Z. B. Aweke, S. F. Yitbarek, R. Qiao, R. Das, M. Hicks, Y. Oren, and T. Austin, "Anvil: Software-based protection against next-generation rowhammer attacks," *ACM SIGPLAN Notices*, pp. 743–755, 2016.

[18] B. Ford and R. Cox, "Vx32: Lightweight user-level sandboxing on the x86," in *USENIX ATC*, 2008, pp. 293–306.

[19] S. Sarangi, S. Narayanasamy, B. Carneal, A. Tiwari, B. Calder, and J. Torrellas, "Patching processor design errors with programmable hardware," *Microelectroics Journal*, pp. 12–25, 2007.

[20] S. Narayanasamy, B. Carneal, and B. Calder, "Patching processor design errors," in *ICCD*, 2006, pp. 491–498.

[21] I. Wagner and V. Bertacco, "Caspar: Hardware patching for multicore processors," in *Design, Automation and Test in Europe*, 2009, pp. 658–663.

[22] K. Schmitz, A. Chandrasekharan, J. G. Filho, D. Große, and R. Drechsler, "Trust is good, control is better: Hardware-based instruction-replacement for reliable processor-ips," in *ASP-DAC*, 2017, pp. 57–62.

[23] R. R. S.M. Vidanagamachchi, S.D. Dewasurendra and M.Niranjan, "Commentz-walter: Any better than aho-corasick for peptide identification?" in *Int'l Journal of Research in Comp. Science*, 2012, pp. 33–37.

[24] A. D. Balsa, "The cyrix 6x86 coma bug," https://lkml.org/lkml/1997/11/12/129, 1997.

[25] R. R. Collins, "The intel pentium f00f bug description and workarounds," *Dr. Dobb's Journal*, 1997.

[26] M. S. Inci, T. Eisenbarth, and B. Sunar, "Hit by the bus: Qos degradation attack on android," in *Proceedings of the 2017 ACM on Asia Conference on Computer and Communications Security*, 2017, pp. 716–727.

[27] Y. Oren, V. P. Kemerlis, S. Sethumadhavan, and A. D. Keromytis, "The spy in the sandbox - practical cache attacks in javascript," *CoRR*, 2015.

978-1-5386-7100-9/18 $31.00 © 2018 IEEE

A Fast and Effective Memristor-Based Method for Finding Approximate Eigenvalues and Eigenvectors of Non-Negative Matrices

Chenghong Wang, Zeinab S. Jalali, Caiwen Ding, Yanzhi Wang and Sucheta Soundarajan
Department of Electrical Engineering and Computer Science
Syracuse University - Syracuse, NY
{cwang132, zsaghati, cading, ywang393, susounda} @syr.edu

Abstract—Throughout many scientific and engineering fields, including control theory, quantum mechanics, advanced dynamics, and network theory, a great many important applications rely on the spectral decomposition of matrices. Traditional methods such as the power iteration method, Jacobi eigenvalue method, and QR decomposition are commonly used to compute the eigenvalues and eigenvectors of a square and symmetric matrix. However, these methods suffer from certain drawbacks: in particular, the power iteration method can only find the leading eigen-pair (i.e., the largest eigenvalue and its corresponding eigenvector), while the Jacobi and QR decomposition methods face significant performance limitations when facing with large scale matrices. Typically, even producing approximate eigen-pairs of a general square matrix requires at least $O(N^3)$ time complexity, where N is the number of rows of the matrix.

In this work, we exploit the newly developed memristor technology to propose a low-complexity, scalable memristor-based method for deriving a set of dominant eigenvalues and eigenvectors for real symmetric non-negative matrices. The time complexity for our proposed algorithm is $O(\frac{N^2}{\Delta})$ (where Δ governs the accuracy). We present experimental studies to simulate the memristor-supporting algorithm, with results demonstrating that the average error for our method is within 4%, while its performance is up to 1.78X better than traditional methods.

I. INTRODUCTION

As Moore's law is reaching its physical limits, the benefits obtained by technology scaling on planar CMOS will diminish. On the other hand, since we have entered the era of big data, the demands of lower latency and higher computing speed of integrated circuits have been consistently increasing.

The memristor is a newly developed electronic device that, among other important features, allows for hardware-level implementation of extremely fast matrix-vector multiplications and solving of systems of linear equations, and outperforms FPGAs and GPUs in many cases and applications [1]–[4]. Specifically, initializing memristor in real-time to represent a matrix can be potentially done in $O(N)$ time complexity using the proposed parallel writing scheme in [5], and once initialized, matrix-vector multiplication itself can be done in $O(1)$ time. The memristor technology has already proven to be of great value in areas such as deep learning.

In this work, we present the first memristor-based technique for fast and accurate eigen-pair finding. The eigen-pair finding problem is vital to many problems within the science and engineering fields. This task is generally fulfilled using classical methods like the power iteration [6], Jacobi eigenvalue

method [7], and QR decomposition [8]. These methods have two shortcomings: 1) They have relatively high computational complexity (at least $O(N^3)$), and 2) They lack flexibility: they either find only the largest eigen-pair or all eigen-pairs. Some applications, like spectral clustering and principal component analysis, require a set of eigenvalues and their eigenvectors. In other words, finding the largest eigen-pair is not enough, but finding all eigen-pairs is more than necessary. One solution is to apply deflation techniques in combination with the power iteration method to find a set of leading eigen-pairs, but this technique incurs additional complexity [9].

To address these problems, we propose a low-complexity, scalable, and flexible memristor-based sweeping method for deriving approximate eigenvalues and eigenvectors for real symmetric non-negative matrices. With this fast and accurate eigen-pair finding method, one can immediately scale up the matrix application algorithms. This type of matrix is important in different areas including social network analysis, where non-negative, symmetric matrices represent adjacency matrices of undirected graphs. The proposed eigen-pair finding method is fast and flexible, allowing the user to find as many or as few leading eigen-pairs as desired. Our contributions are summarized as follows:

- We introduce a novel memristor crossbar-based sweeping algorithm for finding all (or a specified number of leading) eigen-pairs for real symmetric non-negative matrices with complexity $O(\frac{N^2}{\Delta})$, where Δ governs accuracy.
- We discuss in details how one can implement the sweeping algorithm on a memristor crossbar.
- We conduct experimental studies to simulate memristor supported hardware and evaluate our proposed method on the simulation platform. Simulation results indicate a large performance gain with low accuracy loss.

II. BACKGROUND & RELATED WORK

In this section, we first describe background related to the previous work on matrix eigen-pair finding, and then describe the background of memristor-based matrix operations.

A. Eigen Decomposition

Finding a matrix's leading eigenvalue and corresponding eigenvector is often done via the power iteration method. To use the power iteration method for finding more eigenvalues,

one must perform a series of matrix deflation operations [9] in each round, which removes the leading eigen-pair and thus allows the power iteration method to find the next eigen-pair.

The Lanczos method is another option, which can be used to find larger quantities of eigenvalues and eigenvectors [10]. The Jacobi, Rayleigh quotient iteration [11] and the QR decomposition methods [12], [13] are also commonly used to find eigenvalues and eigenvectors, and, unlike the power iteration and Lanczos method, find all eigenvalues simultaneously. However, the computational complexities for these techniques can be large: Using the power iteration and Lanczos method to find the largest pair of eigenvalue and eigenvectors requires $O(N^2)$ complexity, and when used with deflation to produce a set of eigen-pairs, the complexity is increased to $O(mN^3)$, where m is the number of required eigenvalues. The QR decomposition requires $O(N^2)$ and $6N^3 + O(N^2)$ complexity to produce all eigenvalues and all eigen-pairs, respectively, and the Jacobi method and Rayleigh quotient iteration also have cubic computational complexity.

B. Memristors and Crossbar Arrays

Initially proposed by Leon Chua in 1971, and created by HP Labs in 2008, a memristor is an electrical component with the capacity to 'remember' the historical profile of excitations on the device. Specifically, the state (memristance) of a memristor will change when voltage higher than a threshold voltage, i.e., $|V_m| > |V_{th}|$, is applied at its terminals. Otherwise, the memristor behaves like a resistor, with features such as non-volatility, low-power, high density, and excellent scalability [14], [15]. In this work, we take advantage of the ability to connect memristors together into a crossbar array. By applying voltages at different input locations of this array and observing the output voltages, we can efficiently calculate matrix-vector products in the analog domain. Specifically, once the crossbar memristances are set, matrix-vector multiplication and linear equation solving can be performed in $O(1)$ time complexity [1], [2], [16], [17]. Using the parallel writing scheme in [5], the memristor can potentially be set in $O(N)$ time. Thus, multiplication and linear equation solving are fast even if the matrix must be initialized; and if the matrix has been initialized earlier, these operations can be done in constant time. Memristors have been critical to research on neural networks (e.g., neuromorphic computing) [18], [19], as well as other areas like pattern recognition [20], text recognition [21], random number generation [22], and the dot-product engine [23]. In general, memristor-based algorithms for the above applications are evaluated using simulation or emulation, due to limited access to memristor technology.

III. PROPOSED METHOD

In this section, we first introduce the design of our proposed sweeping algorithm, then discuss how to accelerate our method by implementing it on a memristor crossbar. We end with a discussion of the time complexity of our proposed method.

A. Sweeping Based Eigen-pair Finding Method

Consider an N-by-N non-negative, non-singular, symmetric matrix M. We begin with the following observation:

Observation 1: Suppose b is a vector of random numbers in the range $[0, 1]$. Define x_β as the solution to $(M - \beta I)x_\beta = b$. Then if λ is an eigenvalue of M, as β approaches λ, $||x_\beta||_\infty$ approaches ∞. Thus, if we define a small step-size Δ, then if we initialize β to be an upper bound on the largest eigenvalue of M and gradually decrement β by ϵ to 'sweep' across the range of potential eigenvalues, then by repeatedly solving the problem $(M - \beta I)x_\beta = b$. and examining $||x_\beta||_\infty$, we can identify the eigenvalues of M.

Figure 2 depicts this relationship on a a toy 10×10 random symmetric non-negative matrix, where β is decremented by 0.01 in each step. The value of $||x_\beta||_\infty$ varies with β, and forms peaks at certain locations. When we examine the locations of these peaks, we find that every peak (and corresponding β) corresponds to an eigenvalue of the target matrix. We observed this property over a large set of random and real matrices. Thus, if we have the upper bound for matrix M's eigenvalue distribution and the number of desired eigenvalues, and find the β values such that $max(x_\beta)$ forms a peak, then these β's are approximate eigenvalues of M.

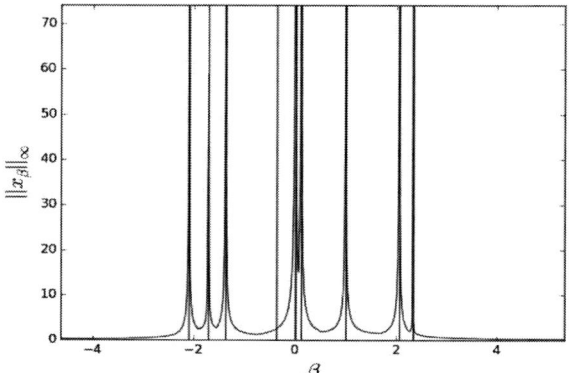

Fig. 1. $max(x_\epsilon)$ vs. ϵ on a non-negative matrix. The peaks of the curve occur at values of ϵ corresponding to eigenvalues of the matrix. Black lines indicate true eigenvalues.

Algorithm: This observation naturally leads to a sweeping algorithm to approximately identify all or a set of leading eigenvalues of a matrix M, as presented in Algorithm 1. This algorithm requires four input parameters: a non-negative square symmetric matrix M; the searching interval $[L, U]$, which contains all eigenvalues of matrix M; the step decrement value Δ^1, and the number k of eigenvalues desired.

To set L, we follow Gershgorin's Circle Theorem [24], which finds L by taking the sum of the absolute values of the off-diagonal elements in each row, and adding the diagonal element in that row to that sum^2. U can be found equivalently by negating M.

After setting all necessary parameters, the sweeping algorithm searches for eigenvalues within the interval in the range $[L, U]$. First, the algorithm creates a random positive vector b, and then performs a series of iterations to locate the eigenvalues. The algorithm first steps with $\beta = U$, and

---

^{1}According to our experimental studies, we suggest a Δ around 0.01 to 0.05.

^{2}Note that because we consider only non-negative matrices, this value is simply the sum of the row.

Algorithm 1 Sweeping-Based Eigenvalues Search Method
___
Input:
    Matrix, $\mathbf{M}$
    Sweeping Interval, $[L, U]$
    Fixed Increment, Δ
    Number of Eigenvalues required, N
Output:
    Set of Eigenvalues, E.

1: $\beta \leftarrow U$
2: Solve the Linear Equation $(\mathbf{M} - \beta \mathbf{I})\mathbf{x} = \mathbf{b}$
3: $EigNum \leftarrow 0, x_0 \leftarrow -1, x_1 \leftarrow -1$
4: $x_0 \leftarrow max(\mathbf{x})$
5: **while** $EigNum \leq N$ and $\beta > L$ **do**
6:     $\beta \leftarrow \beta - \Delta.$
7:     Solve Linear Equation $(\mathbf{M} - \beta \mathbf{I})\mathbf{x} = \mathbf{b}$
8:     $x' \leftarrow max(\mathbf{x})$
9:     **if** $x_0 \neq -1$ and $x_1 \neq -1$ and $x_0 \leq x'$ and $x' \geq x_1$
    then
10:        $N \leftarrow N - 1, \ E \leftarrow E \cup \{\beta\}$
11:        $x_0 \leftarrow x_1, \ x_1 \leftarrow x'$
12:     **end if**
13: **end while**
14: **return** E;
___

in each iteration: (i) decreases β by the fixed decrement Δ, (ii) solves the problem $(\mathbf{M} - \beta \mathbf{I})\mathbf{x}_\beta = \mathbf{b}$, and (iii) records $\|\mathbf{x}_\beta\|_\infty$ as x'_β. The algorithm then finds those x'_β values where $x'_{\beta-\Delta} \leq x'_\beta$ and $x'_\beta \geq x'_{\beta+\Delta}$ (in other words, the local peaks). This procedure continues until k such values have been obtained, and the resulting eigenvalues are stored within the set E. Because β begins at U and is decremented until it reaches L, these are the leading eigenvalues. The eigenvalues detected by this method are approximate, not exact, with the error depending on step size Δ. As Δ decreases, the accuracy and running time increase. For instance, the running time of finding eigen-pairs with accuracy 0.01 is five times slower than finding eigen-pairs with accuracy 0.05.

To find the corresponding eigenvectors, we adopt the *Inverse Iteration* method [25]. We create a random vector $\mathbf{v}_0$, and repeat the following until convergence:

- Define $\mathbf{w}_{i+1}$ to be the solution to $(\mathbf{M} - \lambda' \mathbf{I})\mathbf{w}_{i+1} = \mathbf{z}_i$.
- Set $\mathbf{z}_{i+1} = \frac{\mathbf{w}_{i+1}}{\|\mathbf{w}_{i+1}\|_\infty}$. The infinity norm is simply the largest absolute value of an element in the vector.

In this way, we obtain the set of leading eigenvalues and corresponding eigenvectors. The divergence of maximum components near eigenvalues can be proven through application of Cramer's Rule, but the scope and space constraints of this paper prevent us from presenting the proof here.

B. Implementation on the Memristor Crossbar

The sweeping method requires solving the linear equation $(\mathbf{M} - \beta \mathbf{I})\mathbf{x} = \mathbf{b}$ at each iteration. This can easily be implemented using a memristor crossbar involving two operations: 1) Mapping matrix $(\mathbf{M} - \beta \mathbf{I})$ and vector $\mathbf{b}$ onto memristor crossbar and 2) solving the linear equation $(\mathbf{M} - \beta \mathbf{I})\mathbf{x} = \mathbf{b}$ using memristor crossbar. We map the matrix using the parallel writing scheme described in [5] in $O(N)$ time, and once mapped, the crossbar solves the linear equation in $O(1)$ time.

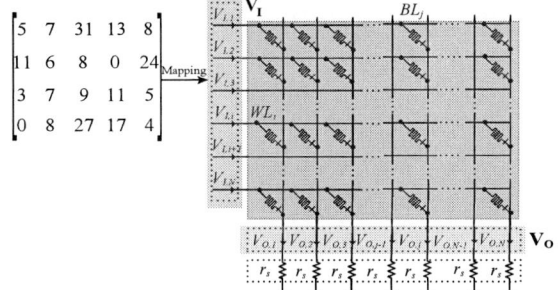

Fig. 2. The memristor crossbar structure and an example of matrix mapping

Here, we show how these two operations can be implemented using a memristor crossbar.

A typical $N \times N$ memristor crossbar is illustrated in Fig. 2. A memristor is connected between each pair of horizontal *word-lines* (WL) and vertical *bit-lines* (BL). This structure can be implemented with a small footprint, and each memristor can be re-programmed to different resistance states by applying biasing voltages at its two terminals [2], [16]. To perform matrix-vector multiplications, we apply a vector of input voltages $\mathbf{V_I}$ on the WLs and collect the current through each BL by measuring the voltage across resistor R_s with conductance of g_s. Suppose that the memristor at the connection between WL_i and BL_j has a conductance of $g_{i,j}$. Then the output voltages are represented by $\mathbf{V_O} = \mathbf{C} \times \mathbf{V_I}$. Here, each coefficient $\mathbf{C}_{ij}$ of matrix $\mathbf{C}$ is (approximately) proportional to the conductance $g_{i,j}$ [1], [4]. Here, $\mathbf{C}$ is represented by the conductance of memristor.

$$\mathbf{C} = \mathbf{D} \cdot \mathbf{G} = diag(d_1, \cdots, d_N) \cdot \begin{pmatrix} g_{1,1} & \cdots & g_{1,N} \\ \vdots & \ddots & \vdots \\ g_{N,1} & \cdots & g_{N,N} \end{pmatrix}^T \tag{1}$$

where, $d_i = 1/(g_s + \sum_{k=1}^{N} g_{k,i})$.

Previous work [26] has demonstrated that we can use a fast and simple approximation $g_{i,j} = c_{i,j} \cdot g_{max}$ to map the above matrix onto a memristor crossbar, in which g_{max} is the largest value of $\mathbf{G}$. Hence, when we want to calculate matrix-vector multiplication $\mathbf{y} = \mathbf{Mx}$, we can set $\mathbf{M} = g_{max}\mathbf{C}$ and $\mathbf{y} = g_s\mathbf{V_O}$, and the solution is: $\mathbf{x} = g_s/g_{max}\mathbf{V_I}$.

In the opposite direction, the memristor crossbar structure can solve a linear system of equations, as required by our algorithm [5]. A voltage vector $\mathbf{V_O}$ is applied on each R_s of BL, so the current flowing through each BL is approximated as $I_{o,j} = g_s V_{o,j}$. The current $I_{o,j}$ through BL_j can also be calculated as $I_{o,j} = \sum_j V_{I,i} g_{i,j}$. For each BL_j, the equation $\frac{1}{g_s} \sum_j V_{I,i} g_{i,j} = V_{o,j}$ is mapped onto the crossbar. The solution $\mathbf{V_I}$ can be found by measuring voltages on the WLs.

One challenge in implementing the sweeping algorithm using memristor crossbars is that a memristor crossbar only allows square matrices with nonnegative entries. By assumption, matrix $\mathbf{M}$ is a square matrix with nonnegative elements; however, matrix $(\mathbf{M} - \beta \mathbf{I})$ may have negative elements on the diagonal, depending on the value of β. Thus, inherent hardware limitations prevent the direct mapping of this matrix into the memristor crossbar. To overcome this problem, if the input matrix contains negative elements, the linear equation $(\mathbf{M} - \beta\mathbf{I})\mathbf{x} = \mathbf{b}$ will be converted to another equation using the technique in [27]. If $\mathbf{A} = (\mathbf{M} - \beta\mathbf{I})$, the conversion is as follows (from Eqn. (2) to Eqn. (3)):

$$
\begin{bmatrix}
A_{1,1} & \cdots & A_{1,N} \\
\vdots & \ddots & \vdots \\
A_{N,1} & \cdots & A_{N,N}
\end{bmatrix}
\times
\begin{bmatrix}
X_1 \\
\vdots \\
X_N
\end{bmatrix}
=
\begin{bmatrix}
b_1 \\
\vdots \\
b_N
\end{bmatrix}
\tag{2}
$$

$$
\begin{bmatrix}
0 & \cdots & A_{1,N} & -A_{1,1} & \cdots & 0 \\
\vdots & \ddots & \vdots & \vdots & \ddots & \vdots \\
A_{N,1} & \cdots & 0 & 0 & \cdots & -A_{N,N} \\
1 & \cdots & 0 & 1 & \cdots & 0 \\
\vdots & \ddots & \vdots & \vdots & \ddots & \vdots \\
0 & \cdots & 1 & 0 & \cdots & 1
\end{bmatrix}
\times
\begin{bmatrix}
X_1 \\
\vdots \\
X_N \\
-X_1 \\
\vdots \\
-X_N
\end{bmatrix}
=
\begin{bmatrix}
b_1 \\
\vdots \\
b_N \\
0 \\
\vdots \\
0
\end{bmatrix}
\tag{3}
$$

Note that currently, existing memristor crossbars are quite small (e.g., 1024 x 1024). However, as with any new hardware devices, this size is likely to increase in the future. Moreover, the proposed method is intended as a core algorithm upon which other algorithms can be built (e.g., a divide-and-conquer version of this method) for larger-scale applications.

C. Complexity Analysis

The sweeping algorithm consists of three stages: (1) Calculate the upper bound for performing eigenvalue searching; (2) Perform the sweeping based eigenvalue-finding algorithm; (3) Calculate corresponding eigenvectors using the inverse method. We make the following assumptions:

- The input matrix $\mathbf{M} \in \mathbf{R}^{N \times N}$ is a full-rank real matrix with non-negative entries in the range $[0, 1]^3$.
- The upper bound is set to be $1 + max(D)$, and the lower bound to $-1 - max(D)$, where $D = \{d_1, d_2, ..., d_n\}$ indicates the sum of each rows of matrix.
- The step increment is set to Δ.

For these three stages, we have the following analysis:

(1) Upper and Lower Bound Calculation: To calculate the upper and lower bounds of the eigenvalues of matrix $\mathbf{M} \in \mathbf{R}^{N \times N}$ with the Gershgorin Circle Theorem, we need to find the sum of each row, which takes $O(N^2)$ time, and identify the largest sum from all results, which takes $O(N)$ time. The total computation time is thus $O(N^2)$.

(2) Sweeping to Find Eigenvalues: The length of the sweeping interval is $O(D)$. Because we assumed that the elements in $\mathbf{M}$ are in the range $[0, 1]$, we have $D \leq N$. The algorithm searches in space $[-D - 1, D + 1]$, with step size Δ, and hence the total number of iterations is given by $\frac{2(1+max(D))}{\Delta} \sim O(\frac{N}{\Delta})$. In each iteration, the time complexity of mapping $\mathbf{M}$ to crossbar in $O(N)$, and once mapped, the crossbar solves the linear equation in $O(1)$ [5]. Thus, the total complexity is $O(\frac{N^2}{\Delta})$.

(3) Eigenvector Calculation: Once k eigenvalues have been found, the sweeping method uses the inverse iteration method to approximate their corresponding eigenvectors. This requires $b \cdot (O(N) + O(N)) \sim O(b \cdot N)$ complexity, where b is the number of rounds before convergence (usually 2-4).

Based on this analysis, the complexity for the whole process is $O(N^2) + O(\frac{\alpha \cdot N^2}{\Delta}) + O(b \cdot N) \sim O(\frac{N^2}{\Delta})$, a significant

---

[3]If the matrix has values outside this range, it can be normalized with elapsed time $O(N^2)$, which does not affect the overall complexity analysis.

improvement compared to existing eigen-finding methods. Note that the complexity is governed by Δ, which controls the tradeoff between accuracy and running time.

D. Algorithmic Improvements

There are numerous improvements that can be made to this algorithm, such as finding repeated eigen-pairs, dynamically selecting the step decrement value Δ, and using divide-and-conquer techniques [28] to handle matrices too large to fit on the crossbar. Due to space and scope limitations, we focus this paper on the core sweeping algorithm and implementation details, and present algorithmic improvements separately.

IV. EXPERIMENTAL STUDIES

We conduct three sets of experimental studies to evaluate the proposed method: (i) experiments to measure the accuracy of the identified eigen-pairs, (ii) running time experiments and (iii) energy efficiency comparison experiments.

A. Experiment Setup

In our first set of experiments, we evaluate our sweeping algorithm on synthetic matrices of sizes 1000×1000, 5000×5000, and 10000×10000, and three real world network adjacency matrices (Dolphins, Facebook, Robots [29]). We compare the results obtained by our proposed method to those produced by MATLAB's eigen-pair functions. Separately, we introduce random matrix and vector errors to demonstrate that our sweeping algorithm is not sensitive to potential imprecisions that may occur in memristor devices. In each round, a random vector (of the same size as $\mathbf{x}$) and a random matrix (of the same size as $\mathbf{M}$) with elements in the range of $\{-\epsilon, +\epsilon\}$ are added to $\mathbf{x}$ and $(\mathbf{M} - \epsilon\mathbf{I})$, respectively.

Second, we compare the running time and energy consumption rates between our sweeping algorithm and the QR decomposition, where both algorithms have been boosted using memristor (we show only results for the QR decomposition, because this was the best standard method that we considered).

Third, we compare the energy consumption rate of the sweeping algorithm with and without memristor boosting. We design a memristor simulator to evaluate the efficiency of the proposed methods. The Matlab-based memristor crossbar simulator is designed based on real memristor model (a fabricated 8 nm × 8 nm memristive device demonstrating fast switching property of around 10 ns and more than 20,000 successful operations) as proposed in [30], with power consumption per switch of 3 Nano-watts. We run our experiments on a server with Intel i7 6700HQ, 2.6 GHz CPU, 48G DDR4 memory, and 512G SSD hardware. We use three evaluation metrics:

- VAL_{err}: The mean error between the eigenvalues detected by the sweeping algorithm and the corresponding actual eigenvalues. The inaccuracy for one specific sweeping-produced eigenvalue λ and its corresponding actual value λ' is calculated as $\frac{|\lambda - \lambda'|}{\lambda'}$
- VEC_{err}: The average error in the eigenvectors found by the sweeping method. The error is calculated by $\frac{||\mathbf{v} - \mathbf{v}'||}{||\mathbf{v}||}$, where $\mathbf{v}'$ is an eigenvector calculated by the sweeping algorithm and $\mathbf{v}$ is the corresponding actual eigenvector.
- $Energy\ Performance$: The total energy consumption calculated when executing the algorithms on our simulation platform. The energy is calculated by $E = P_{cpu} *$

TABLE I

ACCURACY RESULTS FOR SWEEPING ALGORITHM BEFORE AND AFTER PERTURBATION– RANDOM MATRICES

Random Dataset	Measure	Percentage of Calculated Eigen-Pairs							
		5%	10%	15%	20%	25%	30%	50%	100%
1000×1000	VAL_{err}	5.0E-4	6.2E-4	6.7E-4	6.9E-4	7.3E-4	7.9E-4	1.0E-3	4.4E-3
	$VAL_{err(p)}$	4.8E-4	6.0E-4	6.7E-4	6.9E-4	7.3E-4	7.9E-4	1.0E-3	4.4E-3
	VEC_{err}	8.8E-3	3.5E-2	2.2E-2	1.7E-2	1.7E-2	1.4E-2	9.7E-3	8.6E-3
	$VEC_{err(p)}$	1.8E-4	3.1E-2	2.6E-2	2.0E-2	1.7E-2	1.4E-2	9.6E-3	8.8E-3
5000×5000	VAL_{err}	1.0E-4	1.4E-4	2.8E-4	3.7E-4	4.1E-4	4.5E-4	9.8E-4	1.2E-3
	$VAL_{err(p)}$	1.1E-4	1.4E-4	2.8E-4	3.6E-4	4.1E-4	4.5E-4	9.7E-4	1.2E-3
	VEC_{err}	1.1E-6	2.1E-2	1.0E-2	7.0E-3	1.0E-2	1.4E-2	5.5E-2	4.8E-2
	$VEC_{err(p)}$	8.0E-7	8.9E-2	4.5E-2	2.8E-2	2.3E-2	2.0E-2	5.4E-2	5.5E-2
10000×10000	VAL_{err}	1.3E-4	1.3E-4	1.7E-4	2.2E-4	2.6E-4	3.1E-4	4.7E-4	5.3E-2
	$VAL_{err(p)}$	1.2E-4	1.3E-4	1.6E-4	2.2E-4	2.6E-4	3.1E-4	4.7E-4	5.4E-2
	VEC_{err}	5.0E-4	5.0E-4	1.1E-2	1.1E-2	1.1E-2	2.0E-2	1.3E-1	1.4E-1
	$VEC_{err(p)}$	5.3E-4	9.0E-4	1.1E-2	1.1E-2	1.0E-2	2.4E-2	1.3E-1	1.7E-1

TABLE II

ACCURACY RESULTS FOR SWEEPING ALGORITHM BEFORE AND AFTER PERTURBATION – REAL WORLD DATASETS

Real World Dataset	Measure	Percentage of Calculated Eigen-Pairs							
		5%	10%	15%	20%	25%	30%	50%	100%
dolphins	VAL_{err}	3.4E-4	6.3E-4	6.1E-4	5.6E-4	1.1E-3	2.6E-3	6.4E-3	1.4E-2
	$VAL_{err(p)}$	7.6E-4	6.6E-4	5.8E-4	9.4E-4	1.5E-3	2.4E-3	6.4E-3	1.4E-2
	VEC_{err}	6.7E-7	4.0E-7	3.5E-7	1.9E-6	1.8E-6	1.8E-6	3.0E-3	1.8E-3
	$VEC_{err(p)}$	1.1E-6	6.4E-6	5.4E-7	3.5E-6	3.2E-6	3.1E-6	3.4E-3	3.3E-3
robots	VAL_{err}	6.6E-4	1.1E-3	1.2E-3	1.2E-3	1.1E-3	1.2E-3	1.2E-3	5.3E-3
	$VAL_{err(p)}$	5.7E-4	1.1E-3	1.1E-3	1.2E-3	1.2E-3	1.2E-3	1.2E-3	5.3E-3
	VEC_{err}	2.3E-6	1.0E-6	1.2E-5	3.4E-2	3.4E-2	3.8E-2	4.0E-2	8.2E-2
	$VEC_{err(p)}$	1.9E-6	1.8E-6	9.1E-7	3.9E-2	3.8E-2	4.3E-2	4.2E-2	7.5E-2
Facebook	VAL_{err}	2.1E-4	4.0E-4	4.0E-4	3.9E-4	3.9E-4	4.1E-4	4.4E-4	4.5E-4
	$VAL_{err(p)}$	1.7E-4	3.9E-4	3.9E-4	3.9E-4	3.9E-4	4.1E-4	4.4E-4	4.5E-4
	VEC_{err}	2.3E-2	1.0E-2	4.1E-2	3.7E-2	1.86E-1	1.65E-1	1.73E-1	1.76E-1
	$VEC_{err(p)}$	2.4E-2	1.8E-2	4.7E-2	4.2E-2	1.95E-1	1.80E-1	1.78E-1	1.78E-1

$T_{cpu} + P_{mem} * T_{mem}$ where P_{cpu} is the power specification mentioned in [31] for selected CPU and P_{mem} is the power consumption for executing computations on the 8 nm × 8 nm memristive device [30]. T_{cpu} and T_{mem} are the elapsed times for software and memristor.

B. Results and Discussion

We present our results in three parts: accuracy experiments (both with and without hardware inaccuracies), performance (running time) results, and energy consumption results.

1) Accuracy Experiments: The results of our accuracy experiments are depicted in Table I and Table II, where VAL_{err} (VEC_{err}) and $VAL_{err(p)}$ ($VEC_{err(p)}$) represent, respectively, the accuracy results for the detected eigenvalues and eigenvectors found by the sweeping method, without and with simulated perturbation errors. Table I shows results for different sizes of synthetic matrices, while Table II presents results for real world datasets. Our results show that:

First, our sweeping algorithm produces a set of leading eigen-pairs with high accuracy. The average error over all eigenvalues for all testing datasets is around 1%, and the average error for producing all eigenvectors is approximately 7%. The accuracy for computed eigen-pairs is above 90%, which is acceptable for many applications, such as PCA or spectral clustering, that use eigen-pairs for further computing.

Second, we observe that the sweeping method is not sensitive to perturbations, indicating that memristor imprecisions do not have strong effect on its accuracy.

Third, the fewer leading eigenvalues we require, the higher the accuracy. For example, on a random 1000×1000 matrix, when we use the sweeping algorithm to produce 5% of leading eigenvalues (i.e., top-50 eigenvalues), the eigenvalue inaccuracy is 0.0005. However, when we identify the 25% leading eigenvalues (i.e., top 250 eigenvalues), this error almost doubles. The inaccuracy reaches 0.0044 when we detect all eigenvalues. This occurs because if Δ is too large, it may skip over eigenvalues. Eigenvalues tend to concentrate near the central part of the distribution, resulting in higher inaccuracy. However, many applications such as PCA and spectral clustering only require a set of leading eigen-pairs.

2) Performance Experiments: The results of our second experiment are shown in Figure 3, where the x-axis indicates the matrix size and the y-axis shows the total running time. Here, we set $\Delta = 0.05$. The text in the plots shows the approximate accuracy of the algorithms for different tests. The running time of our proposed method shows a significant improvement over the QR decomposition, especially for larger matrix sizes. For example, on a 10000x10000 matrix, the elapsed time for sweeping method is 22.39s, while the elapsed time for QR decomposition is 77.53s, a 3.4X improvement. On average, the sweeping method is 1.78X faster than QR decomposition method. The sweeping method also shows a higher accuracy. The sweeping method is faster than the QR decomposition, while matching or beating its accuracy.

3) Energy Consumption Experiments: Results for the third experiment are listed in Table III, where the total energy consumption for software-based sweeping algorithm and the memristor boosted sweeping method in each test rounds are shown. In all cases, the memristor boosted algorithm demonstrates significant energy reduction as compared with software implementation. For instance, the total energy consumptions for software sweeping algorithm with matrix sizes 5000 and 10000 are around 55 and 165 times of that of the memristor boosted one. The significant reduction in energy consumption

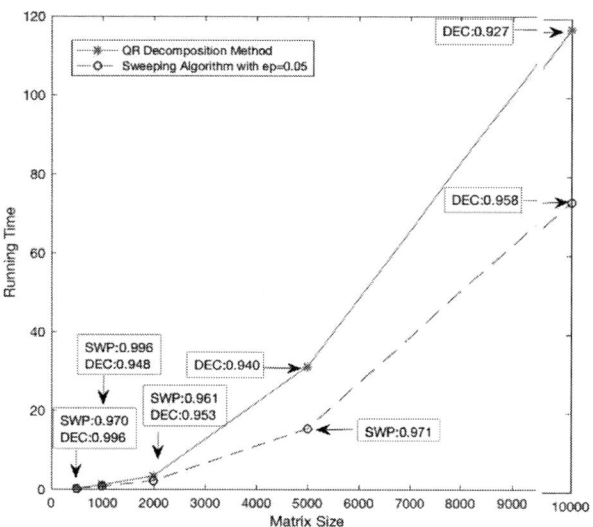

Fig. 3. Elapsed Time Comparison between the proposed Sweeping Algorithm and the QR Decomposition based Method

TABLE III
ENERGY PERFORMANCE COMPARISON

Matrix Size	with memristor (J)	without memristor (J)
500 × 500	4.60	24.74
1000 × 1000	15.41	189.75
2000 × 2000	68.46	1942.38
5000 × 5000	385.02	21091.93
10000 × 1000	1799.94	298431.12

indicates that the proposed method is suitable for implementation, and so is suitable for a variety of research fields.

V. CONCLUSION AND FUTURE WORK

We have proposed a novel sweeping algorithm for finding eigen-pairs of a non-negative square matrix. Our proposed method runs in $O(\frac{N^2}{\Delta})$ time. Additionally, our algorithm is flexible, allowing the user to find only the desired number of eigen-pairs. We demonstrated that our method can precisely find sets of eigen-pairs with very low imprecisions, with a very low elapsed time as compared to other methods and significant energy consumption reduction as compared with software methods. Experimental Results show that our proposed method can achieve up to a 165× energy saving across matrix sizes.

REFERENCES

[1] M. Hu, H. Li, Q. Wu, and G. Rose, "Hardware realization of neuromorphic bsb model with memristor crossbar network," in *IEEE Design Automation Conference (DAC)*, 2012, pp. 554–559.

[2] D. Kadetotad, Z. Xu, A. Mohanty, P.-Y. Chen, B. Lin, J. Ye, S. Vrudhula, S. Yu, Y. Cao, and J.-s. Seo, "Neurophysics-inspired parallel architecture with resistive crosspoint array for dictionary learning," in *Biomedical Circuits and Systems Conference (BioCAS), 2014 IEEE*. IEEE, 2014, pp. 536–539.

[3] M. Sharad, D. Fan, and K. Roy, "Ultra low power associative computing with spin neurons and resistive crossbar memory," in *Design Automation Conference (DAC), 2013 50th ACM/EDAC/IEEE*. IEEE, 2013, pp. 1–6.

[4] S. Yu and Y. Cao, "On-chip sparse learning with resistive cross-point array architecture," in *Proceedings of the 25th edition on Great Lakes Symposium on VLSI*. ACM, 2015, pp. 195–197.

[5] I. Richter, K. Pas, X. Guo, R. Patel, J. Liu, E. Ipek, and E. G. Friedman, "Memristive accelerator for extreme scale linear solvers," Tech. rep. University of Rochester, 2015. u rl: http://www. ece. rochester. edu/users/friedman/papers/GOMAC_15. pdf, Tech. Rep., 2015.

[6] B. N. Parlett, *The symmetric eigenvalue problem*. SIAM, 1998.

[7] G. L. Sleijpen and H. A. Van der Vorst, "A jacobi–davidson iteration method for linear eigenvalue problems," *Siam Review*, vol. 42, no. 2, pp. 267–293, 2000.

[8] A. Quarteroni, R. Sacco, and F. Saleri, *Numerical mathematics*. Springer Science & Business Media, 2010, vol. 37.

[9] K.-J. Bathe and E. L. Wilson, *Numerical methods in finite element analysis*. Prentice-Hall Englewood Cliffs, NJ, 1976, vol. 197.

[10] C. Lanczos, *An iteration method for the solution of the eigenvalue problem of linear differential and integral operators*. United States Governm. Press Office Los Angeles, CA, 1950.

[11] B. N. Parlett, "The rayleigh quotient iteration and some generalizations for nonnormal matrices," *Mathematics of Computation*, vol. 28, no. 127, pp. 679–693, 1974.

[12] H. Rutishauser, "The jacobi method for real symmetric matrices," *Numerische Mathematik*, vol. 9, no. 1, pp. 1–10, 1966.

[13] C. R. Goodall, "13 computation using the qr decomposition," *Handbook of statistics*, vol. 9, pp. 467–508, 1993.

[14] B. Liu, Y. Chen, B. Wysocki, and T. Huang, "The circuit realization of a neuromorphic computing system with memristor-based synapse design," in *Neural Information Processing*. Springer, 2012, pp. 357–365.

[15] D. B. Strukov, G. S. Snider, D. R. Stewart, and R. S. Williams, "The missing memristor found," *nature*, vol. 453, no. 7191, pp. 80–83, 2008.

[16] A. Heittmann and T. G. Noll, "Limits of writing multivalued resistances in passive nanoelectronic crossbars used in neuromorphic circuits," in *Proceedings of the great lakes symposium on VLSI*. ACM, 2012, pp. 227–232.

[17] G. Yuan, C. Ding, R. Cai, X. Ma, Z. Zhao, A. Ren, B. Yuan, and Y. Wang, "Memristor crossbar-based ultra-efficient next-generation baseband processors," in *Circuits and Systems (MWSCAS), 2017 IEEE 60th International Midwest Symposium on*. IEEE, 2017, pp. 1121–1124.

[18] A. Thomas, "Memristor-based neural networks," *Journal of Physics D: Applied Physics*, vol. 46, no. 9, p. 093001, 2013.

[19] R. Hasan and T. M. Taha, "Enabling back propagation training of memristor crossbar neuromorphic processors," in *Neural Networks (IJCNN), 2014 International Joint Conference on*. IEEE, 2014, pp. 21–28.

[20] C. Yakopcic and T. M. Taha, "Energy efficient perceptron pattern recognition using segmented memristor crossbar arrays," in *Neural Networks (IJCNN), The 2013 International Joint Conference on*. IEEE, 2013, pp. 1–8.

[21] Q. Qiu, Z. Li, K. Ahmed, W. Liu, S. F. Habib, H. H. Li, and M. Hu, "A neuromorphic architecture for context aware text image recognition," *Journal of Signal Processing Systems*, vol. 84, no. 3, pp. 355–369, 2016.

[22] H. Jiang, D. Belkin, S. E. Savel'ev, S. Lin, Z. Wang, Y. Li, S. Joshi, R. Midya, C. Li, M. Rao *et al.*, "A novel true random number generator based on a stochastic diffusive memristor," *Nature communications*, vol. 8, no. 1, p. 882, 2017.

[23] M. Hu, J. P. Strachan, Z. Li, E. M. Grafals, N. Davila, C. Graves, S. Lam, N. Ge, J. J. Yang, and R. S. Williams, "Dot-product engine for neuromorphic computing: programming 1t1m crossbar to accelerate matrix-vector multiplication," in *Proceedings of the 53rd annual design automation conference*. ACM, 2016, p. 19.

[24] H. E. Bell, "Gershgorin's theorem and the zeros of polynomials," *The American Mathematical Monthly*, vol. 72, no. 3, pp. 292–295, 1965.

[25] J. W. Demmel, *Applied numerical linear algebra*. SIAM, 1997.

[26] M. Hu, H. Li, Y. Chen, Q. Wu, and G. S. Rose, "Bsb training scheme implementation on memristor-based circuit," in *Computational Intelligence for Security and Defense Applications (CISDA), 2013 IEEE Symposium on*. IEEE, 2013, pp. 80–87.

[27] R. Cai, A. Ren, Y. Wang, S. Soundarajan, Q. Qiu, B. Yuan, and P. Bogdan, "A low-computation-complexity, energy-efficient, and high-performance linear program solver using memristor crossbars," in *System-on-Chip Conference (SOCC), 2016 29th IEEE International*. IEEE, 2016, pp. 317–322.

[28] M. Gu and S. C. Eisenstat, "A divide-and-conquer algorithm for the symmetric tridiagonal eigenproblem," *SIAM Journal on Matrix Analysis and Applications*, vol. 16, no. 1, pp. 172–191, 1995.

[29] J. Leskovec and A. Krevl, "{SNAP Datasets}:{Stanford} large network dataset collection," 2015.

[30] S. Pi, P. Lin, and Q. Xia, "Cross point arrays of 8 nm× 8 nm memristive devices fabricated with nanoimprint lithography," *Journal of Vacuum Science & Technology B, Nanotechnology and Microelectronics: Materials, Processing, Measurement, and Phenomena*, vol. 31, no. 6, p. 06FA02, 2013.

[31] https://ark.intel.com/products/88967/Intel-Core-i7-6700HQ-Processor-6M-Cache-up-to-3_50-GHz.

A Low-Power and Small-Area Multiplier for Accuracy-Scalable Approximate Computing

Hiroyuki Baba
Graduate School of Electronics Eng. and Computer Science
Fukuoka University
Fukuoka, Japan
td172009@cis.fukuoka-u.ac.jp

Tongxin Yang
Graduate School of Information and Control Systems
Fukuoka University
Fukuoka, Japan
td166502@cis.fukuoka-u.ac.jp

Masahiro Inoue
Dept. of Electronics Engineering and Computer Science
Fukuoka University
Fukuoka, Japan
tl141319@cis.fukuoka-u.ac.jp

Kaori Tajima
Dept. of Electronics Engineering and Computer Science
Fukuoka University
Fukuoka, Japan
tl141216@cis.fukuoka-u.ac.jp

Tomoaki Ukezono
Dept. of Electronics Engineering and Computer Science
Fukuoka University
Fukuoka, Japan
tukezo@fukuoka-u.ac.jp

Toshinori Sato
Dept. of Electronics Engineering and Computer Science
Fukuoka University
Fukuoka, Japan
toshinori.sato@computer.org

Abstract—**Many applications are error resilient in that they can tolerate some level of inaccuracy in their computations. Approximate computing benefits such applications by trading power for accuracy. The priority between power and accuracy depends on each application. Some prefer low power, whereas others require high accuracy. This paper focuses on the former and proposes a very-low-power and small-area multiplier for accuracy-scalable approximate computing. The proposed approximate multiplier consists of an approximate tree compressor (ATC) and a carry-maskable adder (CMA). The ATC compresses partial products power-efficiently with a simple circuit structure, whereas the CMA provides accuracy scalability using a simple carry-masking technique. From the experimental designs, it is unveiled that the proposed approximate multiplier reduces power consumption and circuit area by 68.8% and 61.9%, respectively, in the most accurate configuration compared with the Wallace tree multiplier.**

Keywords—approximate computing, approximate multiplier, accuracy scalability.

I. INTRODUCTION

Many modern applications require power efficiency. In addition, these applications are embedded and/or battery operated. Examples of such applications are Internet of Things (IoT) devices. These applications, such as image processing, sensing, recognition, and machine learning, are inherently error-tolerant. Due to the fact that precise results are not always required, nearly accurate outcomes usually suffice. Therefore, approximate computing [1] is one of the promising techniques for such applications to meet the demand of low power consumption. Using this technique, power can be traded for accuracy.

Multiplication is a fundamentally basic operation in applications such as the ones introduced above. Thus, reducing the cost of multiplication benefits the aforementioned class of applications. This paper focuses on an approximate multiplier. While several approximate multipliers have been proposed [1, 2, 3, 4, 5], the range of such applications is limited because most of the prior works lack accuracy scalability [2, 3]. Consequently, dynamic configurability is necessary, especially for the following two reasons. First, different applications have different accuracy requirements, as do different program phases in an application. If multiplication accuracy is fixed, power will be wasted when no high accuracy is required. Second, for applications that rely on iterative methods, their accuracy requirement changes with iterations [6, 7]. While there are some multipliers that are statically [4] and dynamically [5] configurable, contemporary mobile devices such as IoT devices require smaller and lower-power multipliers even if accuracy is diminished. In this study, a low-power and small-area multiplier is examined for accuracy-scalable approximate computing.

Modifying the previous studies [3, 5], in this work, a lower-power and smaller-area multiplier was built. An approximate tree compressor (ATC) [3] was repeatedly utilized to reduce the power and the area of the multiplier at the cost of accuracy. Also a carry-maskable adder (CMA) [5] was carefully tuned to avoid significant reductions in accuracy. Although this was an incremental study, a significant power reduction was attained. When the most accurate configuration was selected, power consumed by the proposed multiplier was reduced by 68.8% and 40.9% compared to that consumed by the Wallace tree and previously studied multiplier [5], respectively. Furthermore, the area of the proposed multiplier is 61.9% and 31.4% smaller than the areas of those two multipliers, respectively.

The main contributions of this paper are as follows:

- A low-power and small-area multiplier whose accuracy is configurable was proposed.

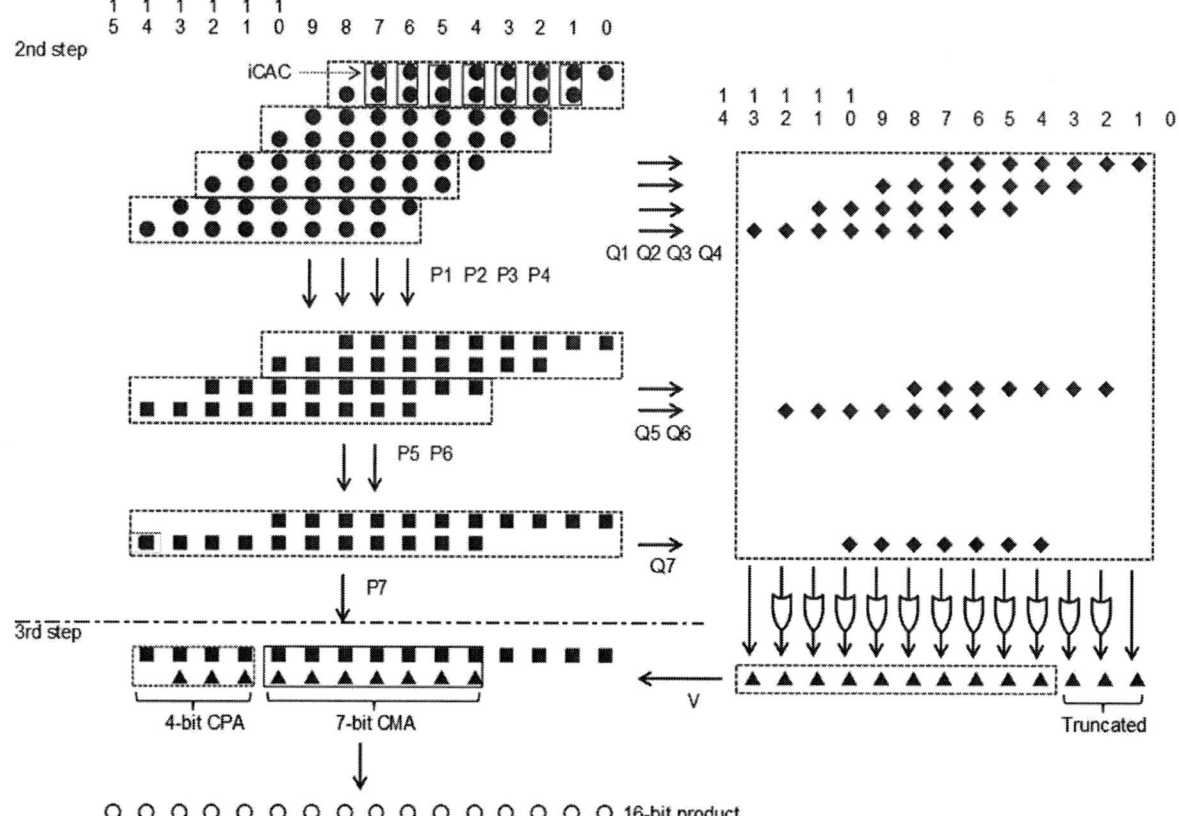

Fig. 1. Multiplication flow.

- The proposed multiplier is designed using Verilog HDL and its power, area, delay, and accuracy are evaluated at the gate level.

- The existence of a trade-off between power and accuracy among the several approximate multipliers [2, 3, 4, 5] has been unveiled.

- The quality of approximate computation at the application level has been evaluated using an image processing algorithm. Although the accuracy has been traded in favor of power saving, the quality is acceptable.

The rest of this paper is organized as follows. In the next section, the prior work is introduced. Then, the proposed approximate multiplier is described. Finally, after the experimental evaluation results are presented, the conclusions are drawn.

II. RERALTED WORK

Yang et al.'s multiplier [2] utilizes approximate 4-2 compressors for partial product reduction. We call this multiplier YHL15 in this paper. The aim of that study [2] was to satisfy the constraint of a low error rate and thus YHL15 had a very high accuracy. In contrast, the aim of this paper is to realize

a low-power and small-area multiplier for applications which work under the severe constraints of power and device volume.

Yang et al. [3] proposed an ATC and used it to build their approximate multiplier named ATCM. ATCM utilizes the 4-2 compressor proposed in [2] such that both power and accuracy are well balanced. This multiplier achieves power and delay, both of which are smaller than those of YHL15, at the cost of accuracy. The multiplier proposed in this paper borrows the concept of the ATC for both power and area reduction.

Both YHL15 and ATCM lack accuracy scalability. However, the next two approximate multipliers described below are scalable.

Liu et al.'s multiplier [4] utilizes an approximate adder for partial product reduction and an error reduction circuit which generates an error recovery vector. We call this multiplier LHL14 in this paper. By varying the bit width of the recovery vector, the accuracy of LHL14 is statically configurable. In contrast, the accuracy of the multiplier proposed in this paper can be configured dynamically.

Yang et al. [5] proposed the CMA and used it to realize the accuracy scalability in their multiplier. We call this multiplier YUS18 in this paper. Different from LHL14, YUS18 is dynamically configurable. It achieves a power accuracy trade-off, which is comparable to that of LHL14. The multiplier proposed in this paper borrows the concept of CMA for accuracy scalability, but the bit positions where CMA is applied have been carefully tuned.

III. THE PROPOSED APPROXIMATE MULTIPLIER

Multiplication in a compressor-based multiplier is processed in three steps. First, the partial products are generated; second, the partial products are reduced; third, the sum of the reduced partial products is generated by a carry propagation adder (CPA). In the proposed multiplier, approximation and configuration are handled in the second and the third steps. These steps are explained in Fig. 1 and are described in the following subsections.

A. ATC

Figure 1 shows the proposed 8-bit approximate multiplier, which has 8×8 partial products. In the first step, a logical AND gate operates on a 1-bit multiplicand and a 1-bit multiplier on the same bit position and generates a 1-bit partial product, which is represented by a black circle. In the second step, the ATC [3] is utilized to compress the height of the partial product array and reduce the partial products.

An ATC consists of incomplete adder cells (iCACs) [3]. iCAC is shown in Fig. 2. In a half adder, the sum s and the carry c of two inputs a and b are expressed as $\{c, s\} = a + b$, where $\{, \}$ and the $+$ sign denote concatenation and addition, respectively. It is obvious from the truth table that $p + q = a + b$, where p and q are outputs of iCAC. Hence, $\{c, s\} = p + q$ and thus the iCAC can generate a precise sum. Consider an n-bit addition of $S = A + B$ where $S = \{s_i | i = n - 1, \cdots 0\}$, $A = \{a_i | i = n - 1, \cdots, 0\}$ and $B = \{b_i | i = n - 1, \cdots, 0\}$. The carry-out from the most significant bit is ignored. Thus, n iCACs generate P and Q from A and B, where $P = \{p_i | i = n - 1, \cdots, 0\}$ and $Q = \{q_i | i = n - 1, \cdots, 0\}$. The truth table explains that the "1"s are collected from A and B into P and that P is always greater than S. Hence, it can be seen that P is an approximate sum of A and B. On the other hand, Q works as an error recovery vector to generate the precise S from P, because it is obvious that $S = P + Q$.

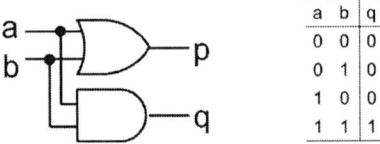

a	b	q	p
0	0	0	0
0	1	0	1
1	0	0	1
1	1	1	1

Fig. 2. iCAC [3].

In Fig. 1, a black circle represents a partial product. The second step, in turn, consists of three rounds of compressions. In the first round, the 8×8 partial products are divided into four groups of two rows, each of which is then operated by seven iCACs by which four pairs of P and Q are generated. They are named (P1, Q1), (P2, Q2), (P3, Q3), and (P4, Q4). A square and a diamond in the figure represent p and q, respectively. In the second round, every two rows, (P1, P2) and (P3, P4), are operated by iCACs by which two pairs of P and Q, (P5, Q5) and (P6, Q6), are generated. In the third round, P7 and Q7 are

generated from P5 and P6. Now, an intermediate approximate product, P7, is acquired. On the other hand, there are seven recovery vectors: Q1, Q2, Q3, Q4, Q5, Q6, and Q7.

These vectors are condensed into a single accuracy compensation vector, V. Triangles in Fig. 1 represent V. V determines the accuracy of the approximate multiplier, because its output is generated by summing P7 and V in the third step. In other words, the way to condense the recovery vectors determines the accuracy. For example, if the vectors are summed up, the precise output will be acquired. Remember that the aim of this study is to build a low-power and small-area approximate multiplier. It is desirable that the circuit to condense the error recovery vectors and to generate the accuracy compensation vector is as simple as possible. Hence, an OR gate is chosen as the circuit because no significant accumulation of errors is not expected. From the truth table in Fig. 2, it is observed that "0" is biased to Q, whereas "1" is biased to P. Thus, Qn (n = 1, …, 6) are sparse vectors and operating OR on all of them may not cause any serious error. The least significant bits are truncated from V, because they are less important for the accuracy.

B. CMA

In the third step, CMA [5] configures the accuracy of the output product. Figure 3 shows 1-bit CMA. When $\overline{mask} = 1$, it works as the conventional full adder. Otherwise, its carry out is masked to prohibit its propagation and its sum works as $s = a$ OR b. It is assumed that C_{in} is also masked and is equal to 0. Due to the fact that the most significant bits are very important for accuracy, these are summed up using the conventional CPA. In this design, a 4-bit CPA has been selected. In the remaining bit positions, the CMA is utilized to configure accuracy. In this design, a 7-bit CMA has been selected. Varying the width of the mask controls the carry propagation and thus also controls the output accuracy. For example, if the mask width is 4 bits, that is, $\overline{mask} = \{1,1,1,0,0,0,0\}$, the upper 3 bits of the CMA work as the precise CPA. Finally, the approximate product is generated by summing P7 and V. White circles in Fig. 1 represent the output.

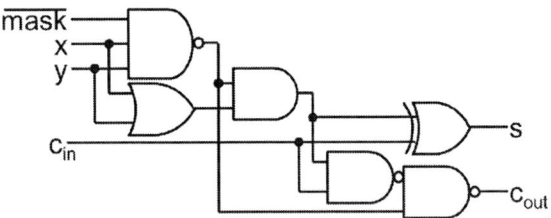

Fig. 3. CMA [5].

C. Comparison with Previous Studies

The difference from ATCM [3] is the number of ATCs stacked for compression. An ATCM has a stack consisting of only one ATC and relies on the 4-2 compressor [2] and on the precise full adder for the remaining compressions. This compressing structure ensures a high level of accuracy. In contrast, the proposed multiplier has a stack that consists of three ATCs. While this diminishes accuracy, a smaller area and lower

978-1-5386-7100-9/18 $31.00 © 2018 IEEE

power consumption can be achieved. This assumption will be evaluated in the next section.

YUS18 [5] consists of three rounds of ATC operations. The difference is how to generate the single accuracy vector, V. YUS18 relies on the precise full adder to maintain its recoverability because it is designed to balance power and accuracy. In contrast, the proposed multiplier relies on only an OR gate. This is because it prioritizes power consumption rather than accuracy. Both YUS18 and the proposed multiplier utilize a 7-bit CMA to sum up the compressed partial products, with the difference being in the bit positions where the CMA is applied. Considering the loss of accuracy in V, they are shifted to LSB by 1 bit in the proposed multiplier.

The above differences are summarized in Table I. They look insignificant and thus the proposed multiplier seems to be just an incremental outcome relative to the previous studies [3, 5]. However, the considerable reduction in both power and area will be apparent in the next section.

TABLE I. COMPARISON BETWEEN THREE MULTIPLIERS.

	ATCM [3]	YUS18 [5]	Proposed
ATC	1 round	3 rounds	3 rounds
CMA	Unused	Bit[11:5]	Bit[10:4]
V	OR gates	Full adders	OR gates

IV. EXPERIMENTS

A. Methodology

The proposed multiplier was evaluated in terms of power, accuracy, area, and delay in comparison with the previously proposed designs. These designs are referred to as YHL15 [2], ATCM [3], LHL14 [4], and YUS18 [5]. From the three variations of LHL14, M2 was selected. Similarly, from the two variations of ATCM, ATCM1 was chosen. Due to the fact that LHL14 is statically configured, four variations, which use 4-, 6-, 8-, and 10-bit recovery vectors, were implemented. They are referred to as LHL14_4b, LHL14_6b, LHL14_8b, and LHL14_10b, respectively. The Wallace tree multiplier was also evaluated.

All multipliers were implemented in Verilog HDL. The Synopsys VCS is used for simulations. The Synopsys Design Compiler and the NanGate 45 nm Open Cell Library [8] were used for logic synthesis. Notably, the default compiler options were used. The value change dump files generated from the VCS simulations were used by the Synopsys Power Compiler for power estimation. The input patterns to the multipliers were provided in an ascending order using a nested loop.

The mean relative error distance (MRED) [4] is used as a metric to evaluate accuracy. The error distance (ED) is defined as the difference between a precise product (M) and its approximate product (M') (i.e., $ED = |M' - M|$). The relative ED (RED) is defined as the ED divided by M (i.e., $RED = \frac{ED}{M} = \frac{|M'-M|}{M}$). The mean RED (MRED) is the average of REDs for 65,536 outputs obtained from 256 × 256 (8-bits × 8-bits) inputs.

Peak signal-to-noise ratio (PSNR) [9] is measured to evaluate the quality of application output. This ratio is used as a metric to assess the quality of reconstructive processes that involve information loss. It is defined as follows:

$$PSNR = 10 \log_{10}(\frac{MAX_I^2}{MSE})$$

$$MSE = \frac{1}{x \cdot y} \sum_{i=0}^{x-1} \sum_{j=0}^{y-1} [P(i,j) - A(i,j)]^2$$

where MAX_I, $P(i, j)$, and $A(i, j)$ are the maximum, the precise, and the approximate values of each pixel, respectively, and x and y are the image dimensions. The image sharpening algorithm[10] is used as a representative application. First, it performs a Gaussian smoothing on the input image, as follows:

$$R(i,j) = \frac{1}{273} \sum_{k=-2}^{2} \sum_{l=-2}^{2} G(k+2, l+2) \cdot I(i+k, j+l)$$

$$G = \begin{bmatrix} 1 & 4 & 7 & 4 & 1 \\ 4 & 16 & 26 & 16 & 4 \\ 7 & 26 & 41 & 26 & 7 \\ 4 & 16 & 26 & 16 & 4 \\ 1 & 4 & 7 & 4 & 1 \end{bmatrix}$$

where $I(i, j)$ and $R(i, j)$ are each pixel of the input and the smoothed images, and G is its Gaussian kernel. Second, it obtains the sharpened image S by $S = 2I - R$. Only multiplication is approximate. The inputs are 512 × 512 grayscale bitmaps with 8-bit pixels of the famous four images: Lena, Baboon, Peppers, and Barbara.

B. Power and Accuracy

Power and accuracy of the approximate multipliers and the Wallace tree multiplier are summarized in Fig. 4. The horizontal axis indicates accuracy in MRED and the vertical axis indicates power consumption in µW. The solid, broken, and dotted lines present the results for the proposed multiplier, YUS18, and LHL14, respectively. The pair of power and accuracy for each configuration is presented using the marker. Please note that the proposed multiplier and YUS18 are dynamically configurable and thus only one design was evaluated for each multiplier. In contrast, LHL14 is statically configurable and thus four designs were evaluated. The diamond, triangle, and square present the results for ATCM, YHL15, and the Wallace tree multipliers, all of which are not configurable.

From the results for the unconfigurable multipliers, it is obvious that a trade-off exists between power and accuracy. With the cost of accuracy loss, power consumption is reduced by 59.9% from the Wallace tree multiplier to ATCM. When YUS18 is compared with ATCM, it is found that the configurability requires additional power. For nearly the same accuracy, YUS18 consumes 15.6% more power than ATCM does. However, the configurability presents larger benefits than an increase in power as explained in Section I. Furthermore, YUS18 achieves an accuracy higher than that of ATCM with power consumption less than that of YHL15. The proposed multiplier consumes around 40% and 70% less power than YUS18 and the Wallace tree multiplier do, respectively. As explained above, there is a trade-off between power and accuracy. As can be seen in Fig. 4, this significant power reduction is attained at the cost of accuracy. Remember that the aim of the proposed multiplier is not to achieve high accuracy, but to reduce power and area. The power reduction of 40% satisfies the aim. The PSNRs of the sharpened Lena images are summarized in Fig. 5. The horizontal axis indicates MRED and the vertical axis indicates PSNR in dB. It can be seen that the proposed multiplier achieves the highest PSNR for the same MRED. The PSNRs of the three most accurate configurations are over 40 dB, which is normally considered very good [11].

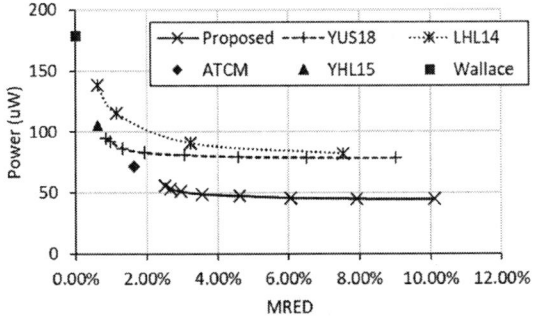

Fig. 4. Power versus accuracy.

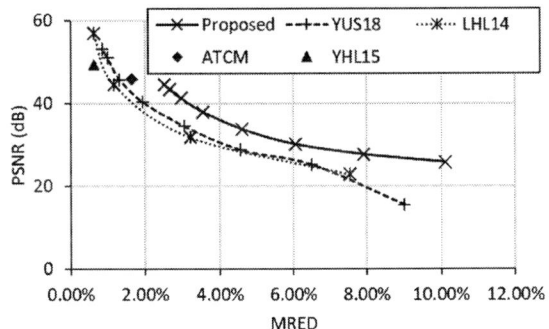

Fig. 5. PSNR versus MRED.

TABLE II. PSNR RESULTS (DB).

$mask$	Lena	Baboon	Peppers	Barbara	Average
1111111	44.57	43.96	42.94	45.31	44.20
1111110	43.46	42.91	42.00	44.14	43.13
1111100	41.24	40.91	40.10	41.88	41.03
1111000	37.92	37.82	37.12	38.44	37.83
1110000	33.68	33.22	33.06	34.39	33.59
1100000	30.05	29.43	29.86	31.66	30.25
1000000	27.63	27.63	28.03	28.75	28.01
0000000	25.73	24.77	25.97	28.16	26.16

Table II gives the PSNR results for the four images, which are sharpened using the proposed multiplier in different configurations. The processed images are presented in Fig. 6. The precise images (a) and the ones processed with $\overline{mask}$ ={1,1,1,1,1,1,1} (b), {1,1,1,1,0,0,0} (c), {1,1,0,0,0,0,0} (d), and {0,0,0,0,0,0,0} (e) are shown. Even when the PSNR is less than 40 dB, the precise and the approximate results are almost nondistinguishable visually.

C. Area and Delay

The area and delay of the approximate multipliers are shown in Fig. 7. These values are normalized by those of the Wallace tree multiplier. As can be seen, the proposed multiplier is the smallest and is 61.9% smaller than the Wallace tree multiplier, it is also 31.4% smaller than YUS18, which is dynamically configurable. When this reduction in area is compared with the lowest-accuracy configuration of LHL14, which is statically configurable, it is still 23.7% smaller.

It is also seen in Fig. 7 that the delay of the proposed multiplier is comparable to the other approximate multipliers. While the delay of LHL14_4b is the smallest (0.9 ns), its accuracy is low. If the proposed multiplier is configured such that its MRED is nearly the same as that of LHL14_4b, its delay will be 0.7 ns. The proposed multiplier is the second fastest and is only behind ATCM.

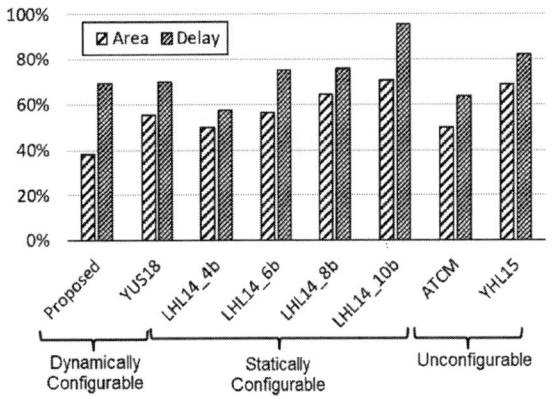

Fig. 7. Area and delay.

V. CONCLUSIONS

This paper proposes an accuracy-configurable approximate multiplier, which consumes around 70% and 40% less power than the Wallace tree multiplier and the previously proposed accuracy-scalable multiplier do, respectively. The reduction ratio depends on the required accuracy. In addition, the area of the proposed multiplier is 61.9% and 31.4% smaller than the area of those multipliers. The lower power and the smaller area are realized by borrowing the concepts of ATC and CMA with careful modifications. At the cost of accuracy, the simple circuit of the basic OR gates is chosen to generate the accuracy compensation vector. This accuracy-configurable approximate multiplier will be beneficial for applications that prioritize power and area rather than accuracy and suffer from fluctuations under working conditions, such as IoT devices.

ACKNOWLEDGMENT

This work was supported by JSPS KAKENHI Grant Number JP17K00088, by funds (No.175007 and No.177005) from the Central Research Institute of Fukuoka University, and by VLSI Design and Education Center (VDEC), the University of Tokyo in collaboration with Synopsys, Inc..

REFERENCES

[1] J. Han and M. Orshansky, "Approximate Computing: An Emerging Paradigm for Energy-Efficient Design," 18th European Test Symposium, 2013.

[2] Z. Yang, J. Han, and F. Lombardi, "Approximate Compressors for Error-Resilient Multiplier Design," 28th International Symposium on Defect and Fault Tolerance in VLSI and Nanotechnology Systems, 2015

[3] T. Yang, T. Ukezono, and T. Sato, "Low-Power and High-Speed Approximate Multiplier Design with a Tree Compressor," 35th International Conference on Computer Design, 2017.

[4] C. Liu, J. Han, and F. Lombardi, "A Low-Power, High-Performance Approximate Multiplier with Configurable Partial Error Recovery," 18th Design, Automation & Test in Europe. 2014.

[5] T. Yang, T. Ukezono, and T. Sato, "A Low-Power High-Speed Accuracy-Controllable Approximate Multiplier Design," 23rd Asia and South Pacific Design Automation Conference, 2018.

[6] Q. Zhang, Y. Feng, Y. Rong, X. Qiang, "ApproxIt: An Approximate Computing Framework for Iterative Methods," 51st Design Automation Conference, 2014.

[7] H.-J. Wunderlich, C. Braun, and A. Scholl, "Pushing the Limits: How Fault Tolerance Extends the Scope of Approximate Computing," 22nd International Symposium on On-Line Testing and Robust System Design, 2016.

[8] NanGate Inc., "FreePDK45 Open Cell Library," http://www.nangate.com/ [Feb. 7, 2017].

[9] A. Momeni, J. Han, P. Montuschi, and F. Lombardi, "Design and Analysis of Approximate Compressors for Multiplication," IEEE Transactions on Computers, Vol. 64, No. 4, 2015.

[10] M. S. Lau, K. V. Ling, and Y. C. Chu, "Energy-Aware Probabilistic Multiplier: Design and Analysis," International Conference on Compliers, Architeture, and Synthesis for Embedded Systems, 2009.

[11] D. Bull, "Communicating Pictures," Academic Press, 2014.

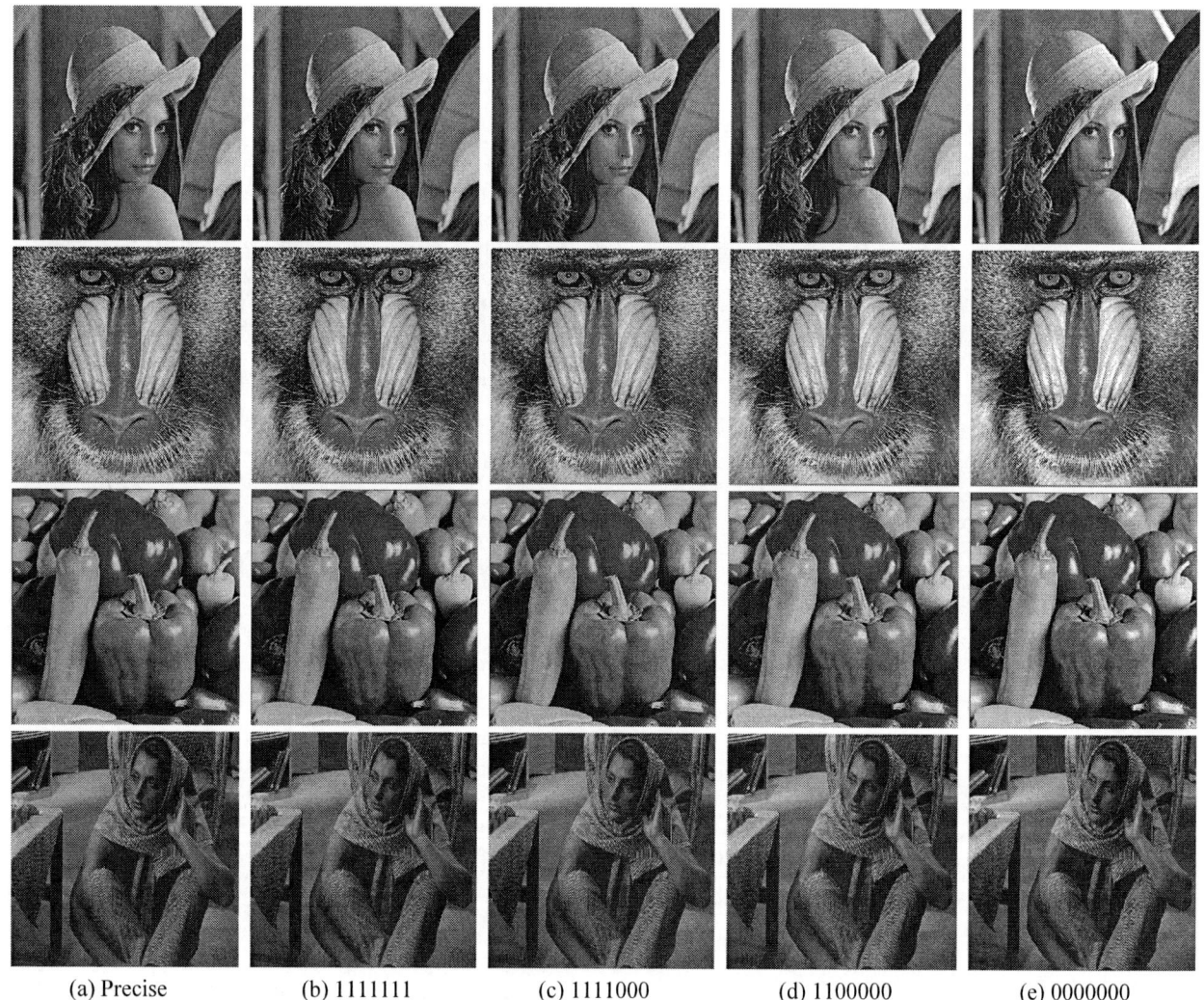

(a) Precise   (b) 1111111   (c) 1111000   (d) 1100000   (e) 0000000

Fig. 6. Processed images.

2018 IEEE Computer Society Annual Symposium on VLSI

A Hardware/Software Co-Design Method for Approximate Semi-supervised K-Means Clustering

Pengfei Huang[1], Chenghua Wang[1], Ruizhe Ma[1], Weiqiang Liu[1*] and Fabrizio Lombardi[2]

[1]College of EIE, Nanjing University of Aeronautics and Astronautics, Nanjing, China
[2]Department of ECE, Northeastern University, Boston, USA
E-mail: {pfhuang, chwang, mraiser, liuweiqiang}@nuaa.edu.cn, lombardi@ece.neu.edu

Abstract—As one of the most promising energy-efficient emerging paradigms for designing digital systems, approximate computing has attracted a significant attention in recent years. Applications utilizing approximate computing can tolerate some loss of quality in the computed results for attaining high performance. Approximate arithmetic circuits have been extensively studied; however, their application at system level has not been extensively pursued. Furthermore, when approximate arithmetic circuits are applied at system level, error-accumulation effects and a convergence problem may occur in computation. Semi-supervised learning can improve accuracy and performance by using unlabeled examples. In this paper, a hardware/software co-design method for approximate semi-supervised k-means clustering is proposed. It makes use of feature constraints to guide the approximate computation at various accuracy levels in each iteration of the learning process. Compared with a baseline design, the proposed method reduces the power-delay product by over 67% while only a small loss of accuracy is introduced. A case study of image segmentation validates the effectiveness of the proposed method.

Index Terms—Approximate computing, semi-supervised learning, approximate multiplier, K-means clustering

I. INTRODUCTION

Speed is improving due to technology advancement; however energy efficiency still remains an hurdle. Power and energy consumption have become a major concern for chip design. Solutions such as dark silicon are of only a limited viability when considering extensive silicon resources [1].

For hand-held systems and emerging smart devices, energy efficiency has become critical when dealing with computation intensive tasks, such as machine learning and multimedia signal processing. Significant efforts have already been devoted to improve energy efficiency at various levels, from software, to architecture all the way down to circuit and device levels. However, as computer systems become pervasive, computing workloads have significantly increased due to new applicative areas such as big data and IoT. So an improvement in the energy efficiency for these emerging workloads is urgently needed to keep pace with the growth of processed information. Applications such as signal processing, machine learning and pattern recognition are generally error-resilient in nature [2] so this feature can be used to alleviate this problem.

Approximate computing [3], trades computing accuracy for high performance and energy efficiency; it has attracted research and development efforts from both academia and industry. Approximate computing is based on the observation that the inputs and outputs of some algorithms are robust to an appropriate imprecision, so inexact operations in computation may have little or no effect on the final quality. Approximate computing techniques have been extensively studied at both hardware (such as circuit designs and computing architectures) and software levels [4]–[6]. As key components in arithmetic circuits, a number of approximate adders [7], [8] and multipliers [9] have been proposed; these circuits yield incorrect results for some input combinations.

Multipliers require more hardware resources and incur in a higher energy consumption than adders; moreover, they are also slower than adders. Approximate multipliers are important due to the extensive use in error-tolerant applications. [10] considers a radix-8 Booth encoding of 3X by utilizing an error reduction scheme in the approximate adders. [11] reports power savings of up to 66% without affecting the accuracy of programs (manipulating low resolution data) by utilizing a bit width reduction in floating-point multipliers. In [12], approximate adders are leveraged to design an approximate multiplier with configurable accuracy. A novel design of an approximate Booth multiplier is proposed in [13]. A power-efficient multiplier designed using 2x2 approximate multiplier blocks is presented in [14]. However, most of these works only consider approximate designs at circuit level, so error effects at system or algorithm level are not fully addressed.

The advantages of approximate computing cannot be fully exploited by only considering hardware or software; therefore, an hardware/software (HW/SW) co-design [5], [15], [16] must be considered to change the abstractions and relationships between hardware and software for a trade-off between accuracy and efficiency [15] for approximate computing. [17] proposes a processor for high performance on-demand approximate computing with a complete open-source development framework; it consists of a hardware processor and an associated software tool set. Fine-grained [18] approximations are explored for individual instructions and individual words of memory. Coarse-grained approximations [19] can holistically transform entire algorithms.

In this paper, we propose a hardware/software co-design

978-1-5386-7100-9/18 $31.00 © 2018 IEEE

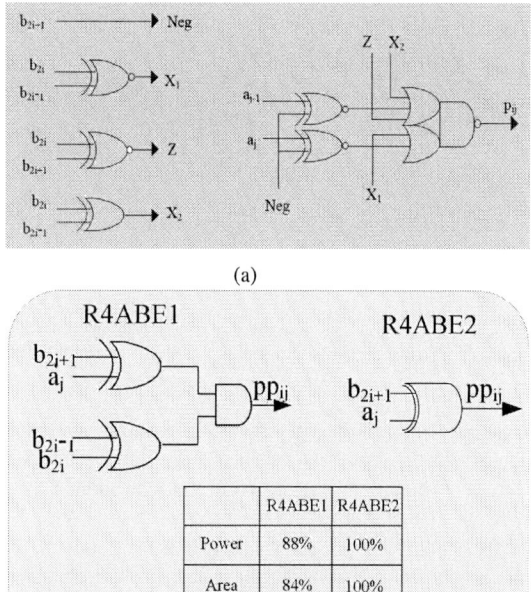

(a)

(b)

Fig. 1: Gate-level circuit of: (a) exact MBE and (b) approximate MBE (with hardware evaluation).

method for approximate semi-supervised k-means clustering; in the proposed method, we utilize a semi-supervised learning method to smoothly calibrate the approximation level and allowing an algorithm to tolerate errors from the approximate hardware. During the computation process, approximation is adaptively tuned and in some cases, it can be even completely turned off if specific feature constraints are applicable.

The main contributions of this paper are summarized as follows:

- A novel hardware/software co-design method is proposed for semi-supervised k-means clustering.
- The proposed method explores error tolerance of semi-supervised k-means clustering.
- Supervised information, which are the minority of the data, improves not only the efficiency of clustering but it also contributes to the optimization of the so-called approximate factor as figure of merit.
- Adaptive approximate levels of hardware implementation can be obtained according to the result of each multiplication or accumulated results at each iteration with supervised information.
- The advantages of the proposed method are theoretically analyzed and experimentally demonstrated.

The remaining part of this paper is organized as follows: In Section 2, the preliminary of semi-supervised k-means and approximate multiplier is briefly reviewed. Section 3 presents the proposed HW/SW co-design method for the

approximate semi-supervised k-means clustering algorithm. Section 4 provides the error analysis and the simulation results. The application of the proposed algorithm to image processing is given in Section 5. Section 6 concludes the paper.

II. PRELIMINARY

A. Semi-supervised K-means

Semi-supervised learning is a class of supervised learning tasks and techniques that make use of unlabeled data for typically training a small amount of labeled data with a large amount of unlabeled data; k-means clustering partitions n observations into k clusters in which each observation belongs to the cluster with the nearest mean.

Semi-supervised k-means leverages user-provided pairwise constraints either to learn an appropriate distance metric in the feature space, or to guide a clustering algorithm towards the correct clusters. The objective function $\mathcal{J}_{obj}$ of semi-supervised k-means utilizing feature constrains is as follows:

$$min\mathcal{J}_{obj} = D(\vec{X}, \vec{C}) + \mathcal{W}\varphi_D(\vec{X}_{pw}) \qquad (1)$$

where, D is the distortion function, $\vec{X}$ and $\vec{C}$ are the samples and cluster centroids, $\mathcal{W}$ is the set of weights of the penalty for violating the feature constraints, φ_D is an increasing function of the distance between two samples and $\vec{X}_{pw}$ are samples with pair-wise constraints. For convenience, $\mathcal{W}$ is chosen as 1.

In the k-means clustering algorithm, a 5% loss in classification accuracy permits a 50 times energy saving compared to the fully accurate classification [20]; therefore, k-means is a typical application that trades accuracy for performance.

B. Radix-4 Approximate Booth Multiplier

In this paper, an 8-bit approximate Booth multiplier [13] is utilized. Booth encoding has been proposed for improving performance of multiplication of two complement binary numbers; this design has been further improved by using a radix-4 Booth encoding. In [13], complexity of the approximate Booth encoder has been reduced by at least an order of magnitude compared with an exact design as shown in Fig. 1. Furthermore, the so-called approximation factor p ($p = 1, 2, ..., 2N$) is defined as the number of least significant partial product columns that are generated by the approximate Booth encoders as the approximate circuit (*i.e.*, approximate radix-4 Booth encoding (R4ABM)) can be used in all or only part of the partial product generation process. Fig. 2 presents a plot of the power-delay product (PDP) and the normalized mean error distance (NMED) under different approximation factors (p).

Fig. 2 shows that the NMED increases, while the PDP of R4ABM decreases with an increase of p. Therefore, the impact of p on the performance of an approximate multiplier is that

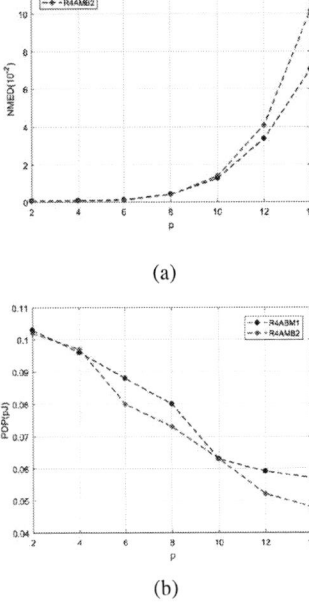

(a)

(b)

Fig. 2: (a) The error characteristics and (b) hardware resources consumption of the R4ABMs [13] at different approximate factors.

an increase in computing accuracy can be traded off for a higher energy efficiency by using an approximate multiplier at a higher value of p. A larger p leads to a simpler logic; this results in a lower hardware consumption but more error. For a given data path in an implementation, p establishes a design space in which every multiplication could utilize a better value for p such that this approximate multiplier would ensure a desirable computing accuracy while also achieving the highest energy efficiency.

III. PROPOSED HW/SW CO-DESIGN METHOD OF APPROXIMATE SEMI-SUPERVISED K-MEANS CLUSTERING

Recently, few approximate designs of arithmetic circuits have been proposed to configure the approximation level [12], [13]. Most of these works consider a configuration at hardware level. [17] utilizes additional flags with instructions in the approximate behavior (for example enable/disable the cache look-up table); however, a gradual approximate computation is not allowed. To fully exploit the potential of approximate computing, the HW/SW co-design utilizing both R4ABM and an approximate semi-supervised k-means algorithm is studied in this section.

A. Fixed-Point Quantization

When data has large values, a higher precision in approximate computation usually incurs in a large penalty. Recently, [21]–[23] has exploited error-tolerance at data level

---

Algorithm 1 Approximate Semi-supervised k-means Clustering

Require:
    The dataset $\mathcal{X}$;
    The number of clusters k;
    A set of must and cannot links;
    A distance function D;
    A set of weights for violating the feature constrains;
    A default value of approximate factor p;
Ensure:
    A partition of $\mathcal{X}$ in k groups
1: Assign initial centroids C
2: **repeat**
3:     Iteration i increase;
4:     Re-assign the labels of the examples using the centroids c_i to minimize $\mathcal{J}_{obj}$;
5:     Check the penalty φ of the violation of the feature constraints;
6:     **if** $\varphi_i > \varphi_{i-1}$ and $p > p_{min}$ **then**
7:         $p--$
8:     **else**
9:         **if** $p < p_{max}$ **then**
10:         $p++$
11:        **end if**
12:     **end if**
13: **until** Convergent or Pre-defined Max number of iterations reaches

---

by utilizing fixed-point quantization. Based on the assumption that the dynamic range of the precision of the floating points from a data set is bounded, an approach for fixed-point quantization is constructed in terms of the width of the arithmetic unit. Floating-point numbers can be extended or compressed as fixed-point numbers according to the desired function as follows:

$$\begin{cases} k * x_{max} + b = y_{max} \\ k * x_{min} + b = y_{min} \end{cases} \tag{2}$$

where, x_{max} and x_{min} are the max and min values of the floating-point inputs, y_{max} and y_{min} are the max and min of the available fixed-point values. Consider an 8-bit width as example; y_{max} and y_{min} are given by $127(2^7 - 1)$ and $-128(-2^7)$. For sake of efficiency, k will be modulated as k' commonly like the $m - th$ power of 2:

$$\begin{aligned} k' &= 2^m \\ s.t. \ 2^m &\leq k < 2^{m+1} \end{aligned} \tag{3}$$

After the modulation of k, x can be easily mapped into the range of (y_{min}, y_{max}) and get its fixed-point quantization with shifting bits of the fraction based on m above and its exponent part.

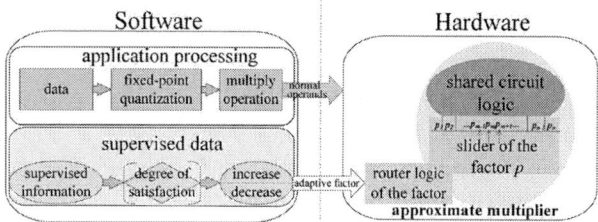

Fig. 3: Overview of proposed framework for semi-supervised approximate computing co-design.

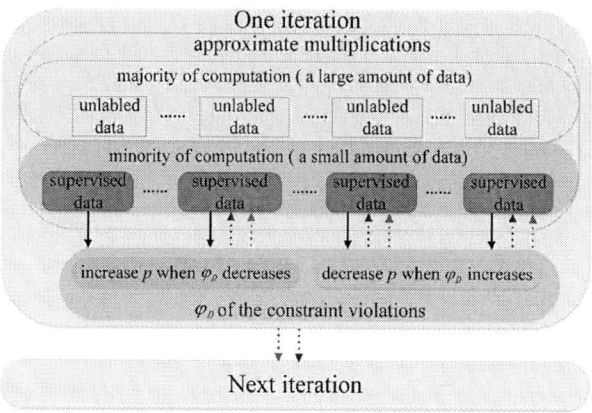

Fig. 4: The process of the proposed method with adaptive approximate factors.

B. Proposed Method for Semi-Supervised k-means Clustering

Usually, the approximate factor is established a-priori and its value is stored in the cache of the control circuit for p. Feature constrains of the supervising information are not as common as for the entire samples; so it is possible to control error-tolerance using an approximate circuit. Therefore, a co-design framework for semi-supervised approximate k-means clustering is proposed by using feature constraints. Its objective function can be obtained as follows:

$$min\mathcal{J}_{obj} = D(\vec{X}, \vec{C}) + \mathcal{W}\varphi_D(\vec{X}_{pw})$$

$$s.t. \begin{cases} \varphi_{D_{m_p}}(\vec{X}) \leq \varphi'_{D_{m_p}}(\vec{X}) \ \&\& \ p < p_{max}, \ p{+}{+} \\ \varphi_{D_{m_p}}(\vec{X}) > \varphi'_{D_{m'_p}}(\vec{X}) \ \&\& \ p > p_{min}, \ p{-}{-} \end{cases} \quad (4)$$

where, m_p is the multiplier of the approximate factor p, $\varphi_{D_{m_p}}(\vec{X})$ is the penalty value of the violations of the constraints at the current stage based on the multiplier m_p and $\varphi'_{D_{m_p}}(\vec{X})$ is at the previous stage based on the previous multiplier m'_p with approximate factor p'. When the new penalty value $\varphi_{D_{m_p}}(\vec{X})$ is higher than the previous one, the approximate level should be reduced and the approximate factor p is decreased if p is greater than p_{min}. By contrast, the approximate factor p will increase if p is less than p_{max}.

Fig. 3 gives the overview of the proposed framework. In general, this approach utilizes a small amount of data with supervised information to adaptively find the optimized approximate factor not only for the supervised data but also the great majority of the unlabeled data. Meanwhile, the approximate factor is optimized gradually to reduce the large accuracy loss introduced by approximation operations. The approximation level is controlled by the approximation factor p and can be adjusted according to the violations of the supervising information.

The procedures for semi-supervised approximate k-means clustering are summarized in Algorithm 1. The entire approximate clustering process is supervised using the semi-supervised feature constraints for different approximation levels, so reducing the accumulated errors at algorithm level with a small amount of supervised information.

From Fig. 4, after each iteration or each computation of supervised data in every iteration, the approximate factor can

be adaptively tuned by decreasing (the red) or increasing (the blue). The approximate factor can be updated through the multiplication of the supervised data which are in the minority. Then the unlabeled data which are in the majority can also benefit from the better approximate multiplier for the current phase. Moreover, the approximate factor can be updated not only according to the penalty value after each iteration but also at each multiplication of the supervised process. In this work, we only update the approximate level through the factor after each iteration in the next evaluations.

IV. EVALUATION AND ANALYSIS

To evaluate the effectiveness of the proposed method and the energy reduction for the multiplication, the classical k-means algorithm is selected as benchmark using the standard UCI datasets [24].

The clustering results using 32-bit full-precision, 8-bit fixed-point quantized, various (p=2,12) and adaptive p 8-bit approximate multipliers are presented in Table I.

The metric of the F-value is used and is defined as follows:

$$F = \frac{(1 + \beta^2) * p * r}{\beta^2 * p + r} \quad (5)$$

where, p is precision, r is recall and β is used to balance the precision and the recall. It is usually set to be 1.

TABLE I
F-VALUE OF K-MEANS CLUSTERING

Data Sets	32b Float	8b Fixed	8b Fixed R4ABM		
			p=2	p=12	adaptive p
Iris	0.8067	0.8067	0.8201	0.7902	0.8103
Glass	0.4426	0.4426	0.4426	0.4021	0.4359

The results of Table I show that approximate computing can produce results with a small loss of accuracy; k-means clustering is sensitive to the initial centroids. Hence, an

978-1-5386-7100-9/18 $31.00 © 2018 IEEE   578

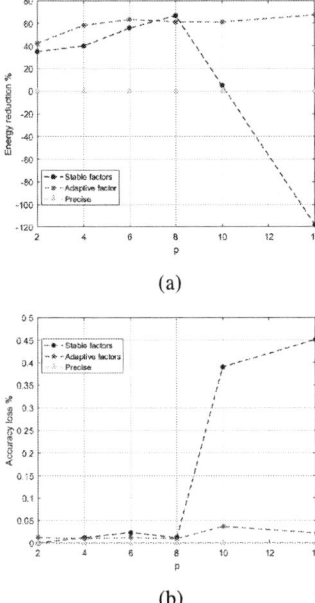

(a)

(b)

Fig. 5: (a) Energy reduction and (b) accuracy loss at different values of approximate factor p.

adaptive approximate method can achieve a better accuracy (for example for the Iris data set with an adaptive p) than the accurate algorithm, because the error introduced by approximate computing avoids overfitting the initial centroids.

The energy consumption and the performance of an approximate multiplier have already been discussed in [13] using as metric the power-delay product (PDP). The total energy consumption can be estimated by adding all energies consumed by each multiplication along the data path. The use of an approximate multiplier is determined by the iterations and the number of samples; so, we can count the frequencies of occurrence of all 8-bit int weights in the entire path and add the PDPs of the corresponding approximate multipliers to find the total energy consumption. The results are presented in Table II.

TABLE II
ENERGY CONSUMPTION OF APPROXIMATE
MULTIPLICATION.

Factor Type	Value of Factor	PDP (uJ)	Reduction (%)
Stable factors	2	0.0143	35.00%
	8	0.0073	66.82%
	14	0.048	-118.18%
Adaptive factors	2	0.0127	42.27%
	8	0.0085	61.36%
	14	0.0072	67.27%

Compared with the 8-bit precise design baseline of 0.022uJ, the proposed design reduces energy consumption by over 67% . However, the PDP could be very large if an inappropriate

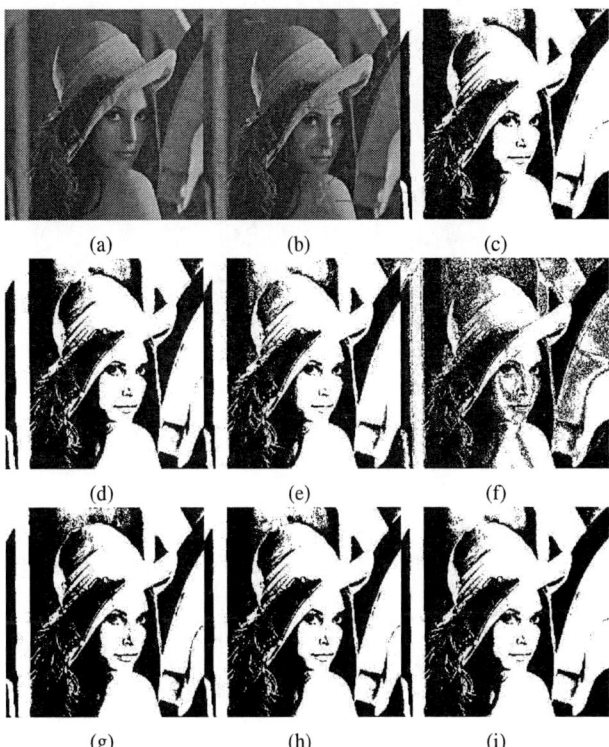

Fig. 6: Image segmentation: (a) original image; (b) original image with feature constrains; (c) accurate segmentation; approximate segmentation for (d) p=2; (e) p=8; (f) p=14; and approximate segmentation using the proposed method for (g) p=2; (h) p=8; (i) p=14.

approximate factor (such as p=14) is utilized, thus contributing to the loss of convergence in the algorithm. Moreover, the proposed method achieves a small accuracy loss at the lowest energy consumption.

From Fig. 5, the performance of the stable approximate factor becomes worse when the factor is very large. However, the adaptive approximate factor can achieve a good energy reduction and a small accuracy loss in spite of the initial value of the approximate factor even when the initial approximate level is high.

V. CASE STUDY: IMAGE SEGMENTATION

In this section, the proposed method is applied to image segmentation. The semi-supervised feature constraints are chosen as more than 10 must-links and 10 cannot links. The semi-supervised k-means not only uses this information efficiently to segment the image, but it also dynamically calibrates the approximate levels for energy reduction and acceptable accuracy.

As shown in Fig. 6, the quality of the processed image deteriorates with an increase of p as factor during the segmentation process from (d) to (f). When this factor is at the largest value of 14, the objective function loses convergence and the worst segmentation result is generated. The results from (g) to (h) are based on different initial factors; all achieve an acceptable segmentation result.

TABLE III

CHANGE OF P IN EACH ITERATION BASED ON INITIAL APPROXIMATE LEVEL DURING IMAGE SEGMENTATION

Initial Factors	Change of p in Each Iteration
p=2	2, 2, 2, 2, 4, 6, 8, 10, 8, 6, 4, 2, 2, 4
p=4	4, 4, 4, 6, 8, 10, 8, 6, 4, 2, 2, 2, 2, 2
p=6	6, 6, 8, 10, 8, 6, 4, 2, 2, 4, 2, 2, 4, 2, 2
p=8	8, 10, 12, 14, 14, 12, 10, 8, 6, 4, 2, 2, 2
p=10	10, 8, 10, 12, 14, 12, 10, 8, 6, 4, 2, 2, 4
p=12	12, 12, 10, 10, 8, 8, 10, 8, 6, 4, 2, 2, 4
p=14	14, 12, 10, 8, 6, 4, 2, 2, 2

Table III shows the change of p at each iteration based on different initial approximate levels during segmentation. The best initial factor is 14; also it is assumed that the error at the beginning is not as large as the errors accumulated along the data path. For a fast convergence, larger steps are always utilized and therefore, the early stage of computation of an algorithm should be robust to errors.

VI. CONCLUSION

A HW/SW co-design method for approximate k-means clustering has been investigated using multi-precision approximate multipliers with various approximate factors. A fixed-point quantization for floating-point data has been proposed to extend or compress the number range, thus resulting in resilience to the so-called approximate factor.

The energy consumed by the multiplier logic decreases rapidly at larger values of the factor p, while introducing a small loss in accuracy for the clustering results. Compared with an 8-bit precise baseline, the proposed method reduces by over 67% the energy consumption.

Adaptive approximate levels can contribute to both the application and the hardware utilization with supervised information that is a very small portion of the entire data sets. Some additional operations for calibrating the optimized approximate levels from only supervised information can be ignored in comparison with the large amount of unsupervised data that can too benefit from the optimized approximate levels. The proposed HW/SW co-design method for semi-supervised learning can be used for a number of machine learning and pattern recognition applications as will be investigated in future works.

REFERENCES

[1] H. Esmaeilzadeh, E. R. Blem, R. S. Amant, K. Sankaralingam, and D. Burger, "Dark silicon and the end of multicore scaling," *IEEE Micro*, vol. 32, no. 3, pp. 122–134, 2012.

[2] V. K. Chippa, S. T. Chakradhar, K. Roy, and A. Raghunathan, "Analysis and characterization of inherent application resilience for approximate computing," in *Proc. 50th Annual Design Automation Conference (DAC)*, 2013, pp. 113:1–113:9.

[3] Q. Xu, T. Mytkowicz, and N. S. Kim, "Approximate computing: A survey," *IEEE Design & Test*, vol. 33, no. 1, pp. 8–22, 2016.

[4] S. Venkataramani, S. T. Chakradhar, K. Roy, and A. Raghunathan, "Computing approximately, and efficiently," in *Proc. Design, Automation & Test in Europe Conference & Exhibition (DATE)*, 2015, pp. 748–751.

[5] A. Sampson, "Hardware and software for approximate computing," 2015.

[6] S. He, S. K. Lahiri, and Z. Rakamaric, "Verifying relative safety, accuracy, and termination for program approximations," *J. Autom. Reasoning*, vol. 60, no. 1, pp. 23–42, 2018. [Online]. Available: https://doi.org/10.1007/s10817-017-9421-9

[7] V. Gupta, D. Mohapatra, A. Raghunathan, and K. Roy, "Low-power digital signal processing using approximate adders," *IEEE Trans. on CAD of Integrated Circuits and Systems*, vol. 32, no. 1, pp. 124–137, 2013.

[8] A. B. Kahng and S. Kang, "Accuracy-configurable adder for approximate arithmetic designs," in *Proc. 49th Annual Design Automation Conference (DAC)*, 2012, pp. 820–825.

[9] A. Momeni, J. Han, P. Montuschi, and F. Lombardi, "Design and analysis of approximate compressors for multiplication," *IEEE Trans. Computers*, vol. 64, no. 4, pp. 984–994, 2015.

[10] S. Lu, "Speeding up processing with approximation circuits," *IEEE Computer*, vol. 37, no. 3, pp. 67–73, 2004.

[11] J. Y. F. Tong, D. Nagle, and R. A. Rutenbar, "Reducing power by optimizing the necessary precision/range of floating-point arithmetic," *IEEE Trans. VLSI Syst.*, vol. 8, no. 3, pp. 273–286, 2000.

[12] C. Liu, J. Han, and F. Lombardi, "A low-power, high-performance approximate multiplier with configurable partial error recovery," in *Proc. Design, Automation & Test in Europe Conference & Exhibition (DATE)*, 2014, pp. 1–4.

[13] W. Liu, L. Qian, C. Wang, H. Jiang, J. Han, and F. Lombardi, "Design of approximate radix-4 Booth multipliers for error-tolerant computing," *IEEE Trans. Computers*, vol. 66, no. 8, pp. 1435–1441, 2017.

[14] P. Kulkarni, P. Gupta, and M. D. Ercegovac, "Trading accuracy for power with an underdesigned multiplier architecture," in *Proc. 24th International Conference on VLSI Design*, 2011, pp. 346–351.

[15] A. Sampson, J. Bornholt, and L. Ceze, "Hardware-software co-design: Not just a cliché," in *Proc. 1st Summit on Advances in Programming Languages (SNAPL)*, 2015, pp. 262–273.

[16] K. Guo, S. Han, S. Yao, Y. Wang, Y. Xie, and H. Yang, "Software-hardware codesign for efficient neural network acceleration," *IEEE Micro*, vol. 37, no. 2, pp. 18–25, 2017.

[17] J. Han and M. Orshansky, "Approximate computing: An emerging paradigm for energy-efficient design," in *Proc. 18th IEEE European Test Symposium (ETS)*, 2013, pp. 1–6.

[18] H. Esmaeilzadeh, A. Sampson, L. Ceze, and D. Burger, "Architecture support for disciplined approximate programming," pp. 301–312, 2012.

[19] ——, "Neural acceleration for general-purpose approximate programs," *Commun. ACM*, vol. 58, no. 1, pp. 105–115, 2015.

[20] S. Mittal, "A survey of techniques for approximate computing." *ACM Comput. Surv.*, vol. 48, no. 4, pp. 62:1–62:33, 2016.

[21] A. Zhou, A. Yao, Y. Guo, L. Xu, and Y. Chen, "Incremental network quantization: Towards lossless cnns with low-precision weights," *CoRR*, vol. abs/1702.03044, 2017. [Online]. Available: http://arxiv.org/abs/1702.03044

[22] S. Zhou, Z. Ni, X. Zhou, H. Wen, Y. Wu, and Y. Zou, "Dorefa-net: training low bitwidth convolutional neural networks with low bitwidth gradients," *CoRR*, vol. abs/1606.06160, 2016. [Online]. Available: http://arxiv.org/abs/1606.06160

[23] P. Merolla, R. Appuswamy, J. V. Arthur, S. K. Esser, and D. S. Modha, "Deep neural networks are robust to weight binarization and other non-linear distortions," *CoRR*, vol. abs/1606.01981, 2016. [Online]. Available: http://arxiv.org/abs/1606.01981

[24] M. Lichman, "UCI machine learning repository," 2013. [Online]. Available: http://archive.ics.uci.edu/ml

2018 IEEE Computer Society Annual Symposium on VLSI

Robustness for Smart Cyber Physical Systems and Internet-of-Things: From Adaptive Robustness Methods to Reliability and Security for Machine Learning

Florian Kriebel, Semeen Rehman, Muhammad Abdullah Hanif, Faiq Khalid, Muhammad Shafique
Vienna University of Technology (TU Wien), Austria
Authors' email: {*florian.kriebel, semeen.rehman, muhammad.hanif, faiq.khalid, muhammad.shafique*}*@tuwien.ac.at*

Abstract— **In recent years, the exponential growth of internet of things (IoT) and cyber physical systems (CPS) in safety critical applications has imposed severe reliability and security challenges. This is due to the heterogeneity and complex connectivity of the CPS components as well as error-prone and vulnerable nature of the underlying devices, harsh operating environments, and escalating security attacks. Different reliability threats (like soft errors, process variation and the temperature-induced dark silicon problem) have posed diverse challenges, which led to the development of various mitigation techniques on different layers of the CPS/IoT stack. Similarly, security threats (like manipulation of communication channels, hardware components and associated software) led to the development of different detection and protection techniques on different layers of the CPS/IoT stack, e.g., cross-layer and intra-layer connectivity. Towards this, the associated costs and overhead as well as potentially conflicting goals are important to be considered, e.g., most of the soft error mitigation techniques are based on redundancy and most of the security-related techniques require continuous runtime monitoring, obfuscation, attestation, and trusted execution environments. This paper first discusses different existing options for approaching this problem at different system layers, i.e., adaptive reliability and security management. These different solutions will provide a wide variety of options to choose from, as a basis for selection and adaptation, to solve reliability-related problems at design-time and run-time. Due to the exponential increase in the complexity and functional requirements, there is a trend towards employing Machine Learning in CPSs and IoT systems. Therefore, we will show how systems can be protected against different security and reliability threats when Machine Learning sub-systems are employed in CPS/IoT.**

I. INTRODUCTION

Cyber Physical Systems (CPS) are defined as "smart systems that include engineered interacting networks of physical and computational components" [1]. The spectrum and application areas of CPS range from traffic control, smart automobiles and smart transport systems over smart infrastructure (like smart grids) to smart healthcare and industrial automation. Due to their broad applicability, the demand for smart systems and the quest for connected devices, the proliferation of CPS and IoT Systems is predicted to grow, significantly affecting substantial aspects of people's lives and work environments [2]. However, this market potential and the vast deployment of CPS and IoT, as well as the potentially fatal implications of system failures and malfunctions in the aforementioned application areas, require robustness features. Therefore, among the wide variety of challenges when designing CPS and IoT systems, like their complexity, heterogeneity of the components involved as well as the different application domains, dependability and security are two particularly crucial ones.

For dependability, in this paper, we will focus on the *reliability* aspect, i.e., the "continuity of correct service" [3], where multiple threats have to be taken into consideration (as shown in Fig. 1):

- **Soft errors** are transient faults caused by strikes of energetic particles on the chip which originate, for instance, from packaging materials or cosmic rays. They manifest as bit-flips at the hardware layer and can propagate to the software layer, potentially leading to incorrect application outputs or even crashes [4].
- **Process variations** are a result of shrinking feature sizes where it becomes more difficult to precisely manufacture billions of transistors with the same electrical properties. This leads to variability in the frequency and leakage power consumptions within the same chip, across different chips of the same wafer, or across different wafers [5].
- **Aging** results in circuits gradually becoming slower during their lifetime due to various physical phenomena like Bias Temperature Instability (BIT) and Hot Career Injection (HCI) that result in degradation in the threshold voltage. For instance, the stress induced while processing workloads may lead to aging-induced timing errors [6].
- The **Dark Silicon** problem refers to the fact that not all transistors on a chip can be simultaneously powered-on at the full performance throttle, as in the post-Dennard nano-era, the power-density is not scaling proportionally to the transistor area [7].

Consequently, the reliability-related challenges that have to be dealt with are: (1) How to *mitigate the different reliability threats*; (2) How to *handle both functional and timing errors*; (3) How to *design reliable systems* considering the optimal contribution of different system layers; (4) How to *manage reliable program execution* under different reliability threats and different constraints (e.g., performance, power, workload variation) in a combined way.

Security focuses on "confidentiality, in addition to availability and integrity" [3]. We discuss the key security threats in the following.
- In **Denial-of-Service attacks**, an attacker can block the communication (by creating the dummy communication modules to overload the communication channels) or can halt the system operations.
- In **Functionality Change attacks**, an attacker can manipulate the functionality of the modules after receiving the internal or external triggers, e.g., misclassification in machine learning applications.
- In **Side Channel Attacks**, an attacker can monitor and manipulate the side channel parameters (i.e., power, delay, temperature, etc.) to extract the confidential information, e.g., IP stealing and memory data sniffing.
- Unlike the side channel attacks, in **Information Leakage attacks**, an attacker can create a covert communication channel or manipulate the existing communication channel to steal the confidential information, e.g., encryption key stealing.

Consequently, the security-related challenges that have to be dealt with are: (1) How to define and ensure *Sustainability Security* that can allow systems to operate securely with guarantees under futuristic unknown attacks; (2) How to *embed the physical security measures* during the CPS design cycle; (3) How to design run-time security measures with *maximum coverage of security vulnerabilities* while considering the resource constraints and energy budget; (4) How to *ensure the secure integration* of several heterogeneous devices to develop secure

978-1-5386-7100-9/18 $31.00 © 2018 IEEE 581

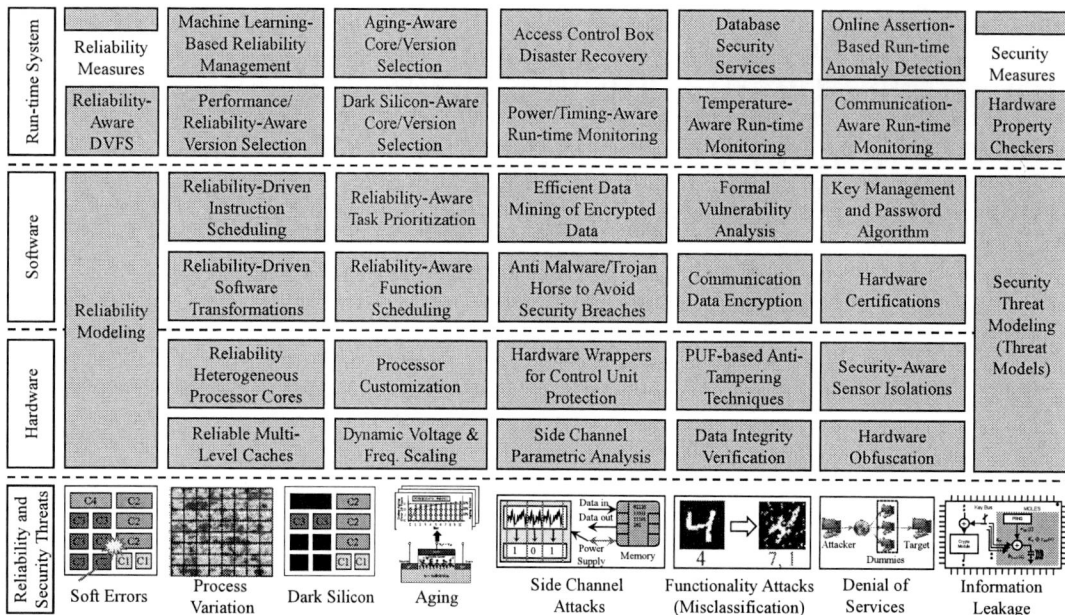

Fig. 1: Overview of Reliability and Security Threats and Solutions on Different System Layers

CPS; (5) How to *ensure privacy the information* while measuring it from sensors and communicate it to other CPS devices?

As Machine Learning (ML) approaches are getting increasingly deployed in Smart CPS and IoT systems, it is important to protect such sub-systems against the above-mentioned reliability and security threats. In this paper, we will show how this can be achieved. We will first discuss cross-layer techniques and methods for adaptive reliability in Section II. Section III gives an overview how ML approaches can be protected against reliability threats. Finally, Section IV presents a brief explanation why ML techniques are useful to address the above-mentioned security threats and what the inherent security vulnerabilities in ML techniques are.

II. CROSS-LAYER AND ADAPTIVE RELIABILITY

A. Cross-Layer Reliability

Reliability threats have been mitigated through spatial or temporal redundancy [8], error correcting codes [9] and guardbands [6], typically focusing only on one layer or a few adjacent layers of the system stack. In the scope of soft errors, on the hardware layer, approaches like Dual or Triple Modular Redundancy (DMR/TMR) as well as redundant processors or cores are used [10]. Oftentimes hardware solutions have been preferred as the reliability threats originate from the hardware layer. On the software layer, instruction redundancy in combination with control flow checking has been proposed [11]. The drawback of such techniques, however, is their costs in terms of power, area, performance loss or memory overhead.

When trying to overcome this problem, it has been observed that applications and even instructions have different susceptibilities towards soft errors depending on their category [12]. One reason for this is the masking phenomenon. One example is logical masking at the circuit level, where a fault might not propagate due to the gate type and the signal values. For example, an AND-gate where one of the (correct) input signals is *0* masks an error on the second input signal [13]. On the software layer, instruction-based masking (e.g., masking the erroneous bits through *SHIFT*, bit-wise/logical *AND*, and *OR* operations) and control/data flow masking (e.g., erroneous data not being used due to an *if-condition*) exist [14]. Moreover, slight changes in the application output might be acceptable for the user, e.g., in multimedia applications [12].

Those effects motivate the need for considering knowledge from different systems layers in a cross-layer fashion for modeling reliability threats and their respective masking effects at different system layers [15]. These models can be used afterwards for proposing and adapting cross-layer mitigation approaches, which should use the most effective system layer(s) for solving the corresponding reliability threats. As an example, [15] proposes a model for software-layer reliability estimation that considers the temporal and spatial vulnerabilities of instructions moving though different processor components while also taking the masking effects of different components into account. It thereby jointly considers information from the hardware as well as from the software layer. While oftentimes the focus is on functional reliability, it is also important to consider the timing reliability, i.e., taking task deadlines into account, especially in the context of real-time systems [16]. These two examples show that several parameters from different layers have to be considered for reliability modeling. Other models and analyses targeting the different reliability threats listed in Section I can be found in [5], [7], [14] and [17].

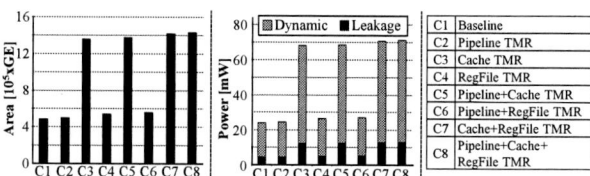

Fig. 2: Area and Power of Reliability-Heterogeneous Cores [19].

Beside the modeling aspect, it is important to devise lightweight, yet effective solutions for the different reliability problems which also offer the potential to be adapted to different system states (e.g., workload, particle flux rate, aging of different components). One example from the hardware and software layer, respectively, will be shown in the following. Having the individual reliability properties of applications, different workloads and user goals as well as the dark silicon problem in mind, designing a multi-/many-core system comprised of cores with a full hardware protection against soft errors (achieved e.g., by employing TMR) is not (power-/energy-wise) efficient, since it lacks adaptation potential to the run-time changing vulnerability behavior of different applications and even their different execution phases. Therefore, reliability-heterogeneous iso-ISA processors have been proposed [18][19]. They provide a tradeoff between reliability (by integrating cores with different protection levels on the same processor ranging from fully protected to unprotected cores) and power/energy efficiency (by allowing for more cores being powered-on and the varying resilience properties of different applications). When designing such processors, at design-time, a library containing cores with heterogeneous reliability features is built. Afterwards, the (multi-/many-core) processor is customized, selecting cores from this library depending on the area and dark silicon constraints as well as the expected application workload [18]. Fig. 2 shows an example of several design points for reliability-heterogeneous cores. Beside the processor cores, it is also important to optimize the caches. Our works in [20] and [21] show how reconfigurable cache architectures can be used to reduce the vulnerability of concurrently executing applications in a multi-/many-core system.

On the software layer, different reliability-aware software transformations have been proposed [12]. Motivated by the potential to reduce critical instruction executions (distinguishing instructions for example based on their ability to cause application failures) as well as the potential to reduce the spatial and temporal vulnerability of application instructions, compiler-based techniques can be used to generate different reliable application versions having distinct performance and reliability properties. The impact of soft errors on different applications as well as the different reliable application versions has been tested with a fault injection approach. The results for two applications can be seen in Fig. 3. To further improve the reliability of the different reliable application versions, other software-based techniques employing (partial) protection/redundancy like [11][14] can be used additionally to increase the options to choose from.

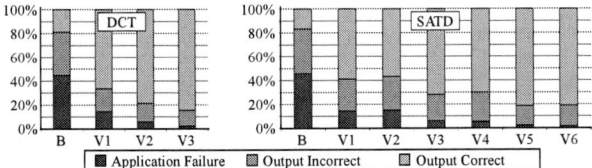

Fig. 3: Fault Injection Results for Two Applications and their Generated Application Versions [12].

An overview of the modeling approaches and techniques to improve the system reliability on different layers is shown in Fig. 1.

B. Adaptive Methods

As motivated in Section I, in CPS and IoT systems, lightweight solutions for improving their reliability are required. Therefore, in the following it will be shown how the approaches described in Section II.A can be leveraged by an adaptive run-time system to improve the reliability of the overall system while considering the individual advantages of each system layer. It is thereby important to take multiple reliability threats into account at the same time while assuring that other requirements such as the system performance are fulfilled.

In [19], an adaptive run-time system is presented that jointly addresses the dark silicon, soft error and process variation problems. It employs reliability-heterogeneous iso-ISA processors (see Section II.A) to enable tradeoffs between reliability and performance. Due to the process variation, even different cores having the same protection level as well as the same core on different chips can have different power and frequency characteristics. An example is shown in Fig. 4. The run-time system determines the task-to-core-assignment minimizing the system vulnerability by selecting appropriate core protection levels while considering performance and dark silicon constraints. During the selection, the protection level of a core as well as its power consumption is considered. The results for two different chips are shown in Fig. 5. In addition to the hardware-based reliability improvement by different core protection levels, approaches from the software layer like different reliable application versions could be considered as well.

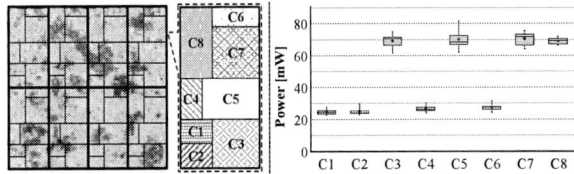

Fig. 4: An Example of a Reliability-Heterogeneous Processor with Power Distribution of Different Cores under Variability [19].

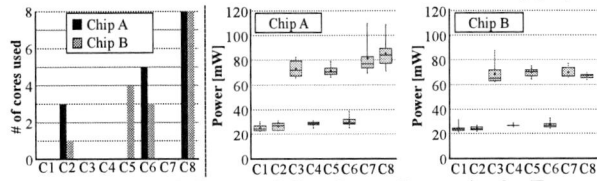

Fig. 5: Selected Cores and Core Power Properties for Two Different Example Chips under Variability Conditions [19].

In [22], an adaptive run-time system is presented for multi-/many-core systems consisting of unprotected cores. It jointly considers the process variation, soft error and aging problems. For reducing the vulnerability, it employs Redundant Multithreading (RMT, e.g., as proposed in [8]) as well as different reliable application versions. Additionally, it considers their impact on the aging of the processor cores via selective activation/deactivation of RMT, different aging properties of the application versions and aging/process variation-aware mapping of tasks to cores. The goal is to achieve a balanced aging/frequency profile of the processor while slowing down the aging of the slowest core. By this adaptive approach, it is possible to achieve an improved aging balancing by a factor of two as well as a lifetime improvement.

These two examples show that adaptive run-time systems can be leveraged to react to different reliability threats and that they can effectively use existing mitigation techniques for reliability problems on different layers. Beyond these works, several other run-time system approaches targeting additional aspects and employing further techniques for improving the system reliability have been proposed. The work in [17] jointly accounts for soft errors, aging and process variations. However, it focuses on minimizing the vulnerability towards soft errors, where it achieves up to 63% improved task reliability. The focus of [23] is on aging balancing and deceleration considering temperature as well as process variations and dark silicon. This work targets to improve the system performance for a given lifetime based on run-time aging and temperature estimation. The work in [24] considers process variation, soft errors and a dark silicon constraint. It uses dynamic voltage and frequency scaling (DVFS) and different reliable application versions as optimization knobs at run-time. It achieves up to 19% reliability improvement for different dark silicon constraints. In [25] dynamic redundancy and voltage scaling is used targeting power efficiency and performance under process variation and soft errors while achieving up to 60% power reduction as well as reliability improvements.

III. RELIABILITY AND MACHINE LEARNING

A. ML for Reliability

Apart from the adaptive methods mentioned in Section II.B, machine learning-based techniques can also be employed for ensuring a near-optimal level of reliability and resource efficiency. This can be achieved by continuously monitoring the hardware characteristics, learning the features, and updating the states (which decides the hardware parameters like voltage and frequency) of the systems. The characteristics that are mainly monitored are error characteristics (like number of timing errors, and the number of errors identified by DMR- and TMR-based approaches), workload on the system, total energy/power consumption, and the performance of the application(s). The machine learning system in the reliability control section then

978-1-5386-7100-9/18 $31.00 © 2018 IEEE

makes use of this information along with the past information about the state and resultant output quality to either maintain or modify the current state of the system, which indirectly tunes the error characteristics of the system for a given workload. Fig. 6 illustrates the design of our ML-based reliability control system.

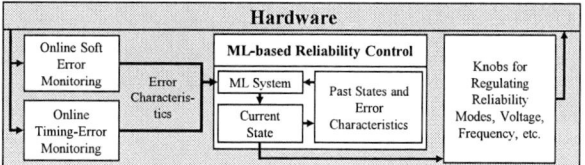

Fig. 6: Runtime Machine Learning-Based Reliability Control System.

B. Reliability for ML

Machine learning-based applications are one of the most widely used applications in CPS and IoT systems. And, Deep Neural Networks are among the most commonly used algorithms for ML. A few recent works like [26] and [27] highlighted the reliability issues of DNN-based applications and also proposed mitigation techniques against specific types of reliability threats for systolic array-based architectures. The mitigation techniques are mainly based on discarding a fraction of computations and/or fault-aware pruning and thereby provide reliability at the cost of loss in accuracy. Recently, in [28] we also highlighted the reliability issues of DNNs, especially from the memory perspective, and showed that even a single bit error at a particular location of the network model can lead to drastic changes, which can be catastrophic in safety-critical applications like autonomous driving (see Fig. 7). A similar set of results are presented in Fig. 8, where memory faults were injected in the weights of the second convolutional layer of the VGG-f network [29], assuming 32-bit floating-point precision. Here e_7 represents the most significant location of the exponent and m_{22} represents the most significant location of the mantissa. The experiments were conducted using 250 randomly selected images from the validation set of the ImageNet dataset [30].

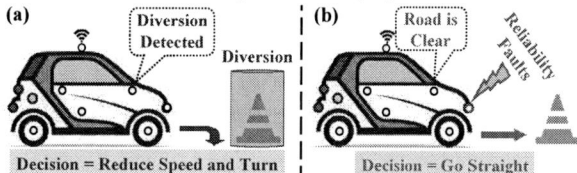

Fig. 7: Effects of Reliability Threats in an Autonomous Driving Use-Case: (a) Correct Outcome; (b) Erroneous Outcome because of the Hardware-Induced Faults.

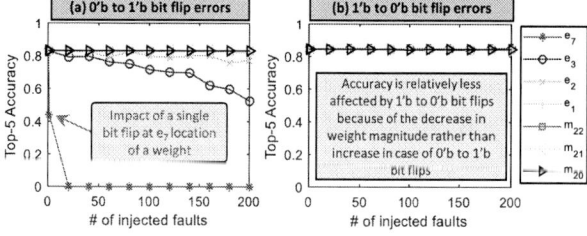

Fig. 8: Effects of Memory Faults on Classification Accuracy of the VGG-f Network on ImageNet Dataset.

Fig. 9 illustrates our methodology for designing a highly reliable system for DNN-based applications. The methodology is composed of several design- and run-time steps. The foremost step is designing an error resilient hardware accelerator, which is capable of handling design- and run-time hardware faults. Provided a set of design constraints, DNNs, and the error resilience characteristics of the DNNs, a

baseline accelerator is first designed. Techniques like [31] and [32] can be utilized for building the accelerator while techniques like [33] can be used for evaluating the error characteristics of the DNNs. Additional circuitry for handling permanent faults and run-time timing errors can be added using techniques similar to [26] and [27], respectively. The accelerator design is then reinforced against reliability faults using reliability-aware synthesis techniques like [34].

Apart from the design-time techniques, we also make use of several run-time methods for ensuring reliable functionality of a given ML application: (1) Fault aware-mapping by using a permanent fault map and the error resilience characteristics of the DNNs; (2) Reliability-aware DVFS; and (3) Instruction-level redundancy (if required).

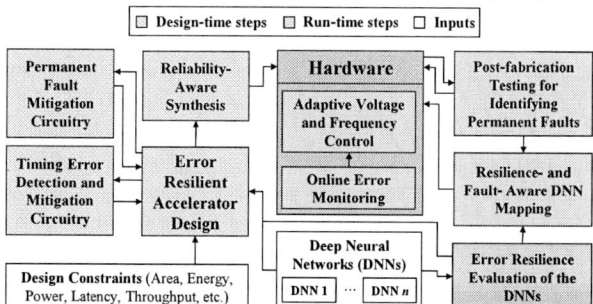

Fig. 9: Our Methodology for Designing Reliable Systems for Machine Learning-Based Applications [28].

IV. Security for Smart CPS

In this section, we first provide an overview of the traditional adaptive security measures and their limitations, followed by a brief overview of our project on adaptive and intelligent security measures.

A. Traditional Security Measures:

To address the CPS security threats (mentioned in Section I), several security measures [35][36] have been proposed. For example, Physical Unclonable Functions (PUFs) [37][38][39][40] and True Random Number Generators (TRNGs) [41][42] based prevention and anti-tampering techniques have been used for secure interaction between CP-devices. Although these techniques generate the necessary keys and authentication IDs, without requiring any on-device key storage mechanism, they have the following limitations:
1) They are only applicable to the *active attacks* during the testing stage or at run-time.
2) They are unable to *incorporate the effects of uncertainties* during real-world scenarios.

B. Adaptive Security Measures

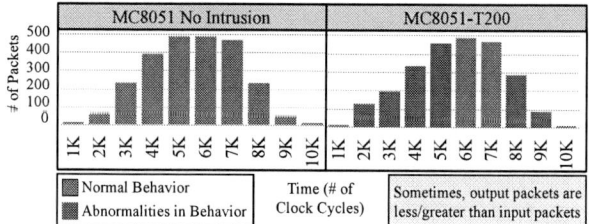

Fig. 10: Effects of Intrusions on the Communication Behavior.

To incorporate the dormant attacks and uncertainties of the physical world, several adaptive techniques have been proposed [35][44][45][46]. Since CPS security threats have different trigger and payload mechanisms at each layer, different sensor-based [47][48] and context-dependent [48] security countermeasures have been proposed. For instance,

978-1-5386-7100-9/18 $31.00 © 2018 IEEE

scheduling/timing (i.e., hard run-time deadlines [49][50]), roundtable discussion [51] and control theory-based [52] approaches. Although these techniques provide comprehensive run-time solutions, most of them are *application-specific and unable to incorporate the uncertainties of the physical world*.

To address this limitation, we propose to choose suitable security parameters, i.e., run-time communication behavior of the cyber physical devices. Therefore, in this paper, we present an analysis of an *un-intruded* and *intruded* microcontroller MC8051 (MC8051-T200, trust-hub Trojan benchmark), as shown in Fig. 10. Our analysis shows, in case of intrusion, sometimes the number of output packets are *greater* (if intruded modules leak the information through existing communication channel) or *less* (if the intruded module temporarily halts the communication) than the number of input packets. Consequently, we conclude that the communication behavior w.r.t. time can also be used to detect anomalous behavior. However, modeling the communication behavior is not straightforward and poses the following research challenges:

1) How to *statistically model the communication behavior* with respect to time while *modeling the uncertainties* of real-world scenarios?
2) How to design and implement the *communication-aware runtime monitors* with minimum area and power overhead?

Although these techniques provide comprehensive run-time solutions, *they are not sufficient to incorporate the run-time computational requirements of the exponential increase in computational data.*

C. Intelligent Security Measures

To address the large computational requirements, several machine learning (ML) [53][54][55][56][57][58], specifically neural networks-based approaches have been proposed because of their ability to extract the hidden features form a large amount of sensed/measured data [59]. In this section, we discuss our research project on the intelligent security measures for smart CPS (Sec4SCPS). Fig. 11 shows two example attacks (i.e., sensor hacking or communication channel hacking) in car-to-car communication.

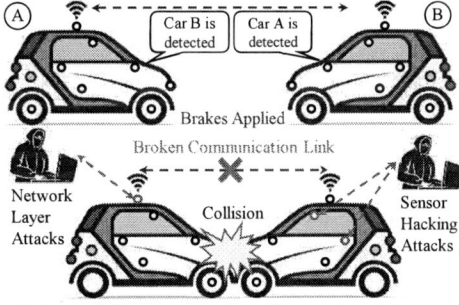

Fig. 11: Effects of Communication Channel and Sensor Hacking in Car-To-Car Communication.

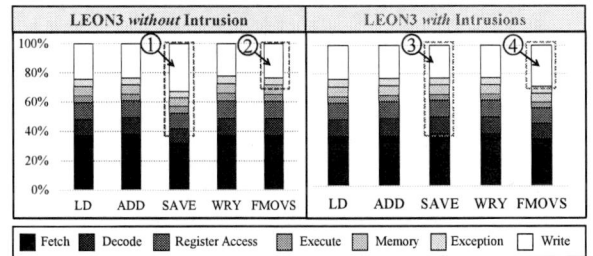

Fig. 12: Effects of Intrusions on Power Correlation with respect to the Pipeline Stages for Different Instructions.

In Sec4SCPS, we target machine learning-based side channel analysis for power consumption and the communication behavior (see

Fig. 10). Our power analysis for LEON3 *with* and *without* intrusion (trust-hub benchmark, MC8051-T200), presented in Fig. 12, illustrates that the power distribution w.r.t. the pipeline stages is correlated with instruction sets, and the intrusions can have a significant impact on this power distribution (compare labels 1 and 2 with 3 and 4, respectively). Therefore, the power distribution w.r.t. the pipeline stages is a useful metric to identify the anomalies in a computing core. As also shown in [60], the power behavior modeling can be used to design ML-based runtime monitors for anomaly detection. However, reducing the *area and energy overhead* of extra power-ports and run-time modeling/measurement is still an open research problem.

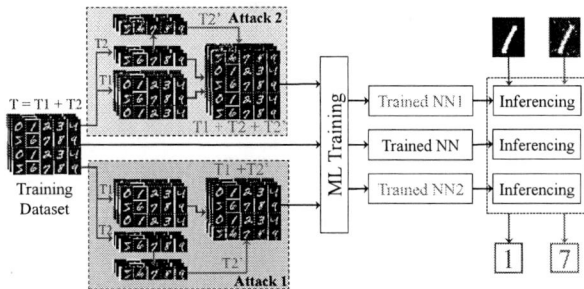

Fig. 13: Experimental Setup for Two Different Random Misclassification Attacks on LeNet for MNIST Dataset with Poisoned Data.

Although ML-based security measures are effective to handle the big-data analysis of CPS, due to inherent security vulnerabilities, these techniques make intelligent security measures more vulnerable. For example, during the training and inference stages, data poisoning can be used to manipulate the ML algorithm for misclassification. To identify the potential vulnerability of ML algorithms, especially during training, we show two data poisoning-based random misclassification attacks on MNIST dataset when training the LeNet (see Fig. 13).

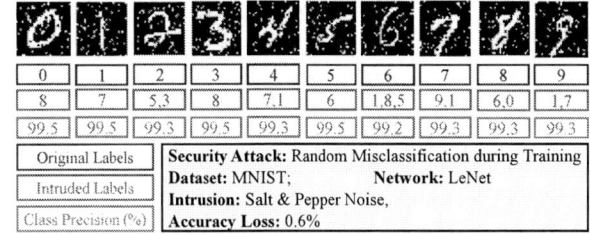

Fig. 14: Random Misclassification Attack (i.e., Introducing Salt & Pepper Noise in 2% Data Samples of the MNIST Dataset) on Le-Net during the Training Phase.

(Fig. 15 image and caption)

Fig. 15: Random Misclassification Attack (i.e., Introducing Salt & Pepper Noise in 2% Data Samples and Append it with MNIST Dataset) on LeNet during the Training Phase.

In attack 1, we trained LeNet using a poisoned MNIST dataset, where 2% of the MNIST data samples are poisoned with salt and pepper noise. The inferencing of the poisoned LeNet shows random misclassification (data sample with label 0 sometimes categorized as 8), see Fig. 14. Similarly, we performed another attack in which instead of directly intruding the MNIST dataset, we extended the MNIST dataset with additional intruded samples. The experimental results in Fig. 15 show that it also randomly misclassifies during inferencing. However, its impact on the inference accuracy of LeNet is less compared to attack 1. Therefore, it can be concluded that *before using any ML-based security measures in CPS and IoT, the safety and security for ML algorithms must be ensured*.

V. CONCLUSION

With their growing importance and wide distribution, CPS and IoT systems are subjected to multiple reliability and security threats. This paper presented an overview of several problems and solutions on different system layers that improve both aspects. It showed how adaptive run-time systems can be leveraged for effectively applying those solutions while exploiting the distinct advantages of the respective system layer. As machine learning techniques spread out to more and more application areas and systems, it is important to particularly target this aspect. Therefore, different approaches are presented how machine learning techniques can be protected against reliability and security threats.

ACKNOWLEDGMENT

This work is supported in parts by the German Research Foundation (DFG) as part of the priority program "Dependable Embedded Systems" (SPP 1500 – http://spp1500.itec.kit.edu) as well as by the Austrian Research Promotion Agency (FFG) and the Austrian Federal Ministry for Transport, Innovation, and Technology (BMVIT) under the "ICT of the Future" project, IoT4CPS: Trustworthy IoT for Cyber-Physical Systems.

REFERENCES

[1] NIST Special Publication 1500-201: "NIST Framework for Cyber-Physical Systems: Volume 1. Overview". 2017
[2] www.forbes.com/sites/louiscolumbus/2016/11/27/roundup-of-internet-of-things-forecasts-and-market-estimates-2016/#ecba84292d51 [online]
[3] A. Avizienis et al.: "Basic Concepts and Taxonomy of Dependable and Secure Computing". IEEE TDSC, 1(1), 2004
[4] R. C. Baumann: "Radiation-induced soft errors in advanced semiconductor technologies". IEEE TDMR, vol. 5, no. 3, 2005
[5] B. Raghunathan et al.: "Cherry-picking: exploiting process variations in dark-silicon homogeneous chip multi-processors". IEEE/ACM DATE, 2013
[6] A. Tiwari et al.: "Facelift: Hiding and slowing down aging in multicores". IEEE MICRO, 2008
[7] M. Shafique et al.: "The EDA Challenges in the Dark Silicon Era: Temperature, Reliability, and Variability Perspectives". IEEE/ACM DAC, 2014
[8] S. S. Mukherjee et al.: "Detailed design and evaluation of redundant multi-threading alternatives," IEEE ISCA, 2002.
[9] C. W. Slayman: "Cache and memory error detection, correction, and reduction techniques for terrestrial servers and workstations". IEEE TDMR, vol. 5, no. 3, 2005
[10] R. Vadlamani et al.: "Multicore soft error rate stabilization using adaptive dual modular redundancy". IEEE/ACM DATE, 2010
[11] G. A. Reis et al.: "SWIFT: Software Implemented Fault Tolerance". IEEE CGO, 2005
[12] S. Rehman et al.: "Reliability-Driven Software Transformations for Unreliable Hardware". IEEE TCAD 33(11), 2014
[13] F. Kriebel et al.: "ACSEM: accuracy-configurable fast soft error masking analysis in combinatorial circuits". IEEE/ACM DATE, 2015
[14] M. Shafique et al.: "Exploiting program-level masking and error propagation for constrained reliability optimization". IEEE/ACM DAC, 2013
[15] S. Rehman et al.: "Reliable software for unreliable hardware: embedded code generation aiming at reliability". IEEE/ACM CODES+ISSS, 2011
[16] S. Rehman et al.: "Reliable code generation and execution on unreliable hardware under joint functional and timing reliability considerations". IEEE RTAS, 2013
[17] S. Rehman et al.: "dTune: Leveraging Reliable Code Generation for Adaptive Dependability Tuning under Process Variation and Aging-Induced Effects". IEEE/ACM DAC, 2014
[18] F. Kriebel et al.: "ASER: Adaptive Soft Error Resilience for Reliability-Heterogeneous Processors in the Dark Silicon Era". IEEE/ACM DAC, 2014
[19] F. Kriebel et al.: "Variability and Reliability Awareness in the Age of Dark Silicon". IEEE Design & Test 33(2), 2016
[20] F. Kriebel et al.: "Reliability-Aware Adaptations for Shared Last-Level Caches in Multi-Cores". ACM TECS 15(4), 2016
[21] A. Subramaniyan et al.: "Soft error-aware architectural exploration for designing reliability adaptive cache hierarchies in multi-cores". IEEE/ACM DATE, 2017
[22] F. Kriebel et al.: "ageOpt-RMT: compiler-driven variation-aware aging optimization for redundant multithreading". IEEE/ACM DAC, 2016

[23] D. Gnad et al.: "Hayat: harnessing dark silicon and variability for aging deceleration and balancing". IEEE/ACM DAC, 2015
[24] M. Salehi et al, "dsReliM: Power-constrained reliability management in Dark-Silicon many-core chips under process variations". IEEE/ACM CODES+ISSS, 2015
[25] M. Salehi et al.: "DRVS: Power-efficient reliability management through Dynamic Redundancy and Voltage Scaling under variations". IEEE ISLPED, 2015
[26] T. Gu et al.: "Analyzing and Mitigating the Impact of Permanent Faults on a Systolic Array Based Neural Network Accelerator". preprint arXiv:1802.04657, 2018
[27] J. Zhang et al.: "ThUnderVolt: Enabling Aggressive Voltage Underscaling and Timing Error Resilience for Energy Efficient Deep Neural Network Accelerators". IEEE/ACM DAC, 2018
[28] M. A. Hanif et al.: "Robust Machine Learning Systems: Reliability and Security for Deep Neural Networks". IOLTS, 2018
[29] A. Vedaldi et al.: "Matconvnet: Convolutional neural networks for matlab". ACM Multimedia, 2015.
[30] J. Deng et al.: „Imagenet: A large-scale hierarchical image database". IEEE CVPR, 2009
[31] P. N. Jouppi et al.: "In-datacenter performance analysis of a tensor processing unit". IEEE ISCA, ACM, 2017
[32] S. Han, et al.: "EIE: efficient inference engine on compressed deep neural network". IEEE ISCA, 2016.
[33] M. A. Hanif et al.: "Error resilience analysis for systematically employing approximate computing in convolutional neural networks". IEEE/ACM DATE, 2018
[34] D. B. Limbrick et al.: "Reliability-aware synthesis of combinational logic with minimal performance penalty". IEEE TNS 60.4 (2013): 2776-2781.
[35] H. Ge et al.: "Analysis of Cyber-Physical Systems Security Issue via Uncertainty Approaches". Advanced Computational Methods in Life System Modeling and Simulation. Springer, 2017
[36] J. Wang et al.: "Detecting time synchronization attacks in cyber-physical systems with machine learning techniques." IEEE ICDCS, 2017
[37] G. E. Suh et al.: "Physically unclonable functions for device authentication and secret key generation," in IEEE/ACM DAC, 2007
[38] M. Rahman et al.: "An aging-resistant ro-puf for a reliable key generation". IEEE TETC, 2015
[39] O. Al Ibrahim et al. "Cyber-physical security using system-level pufs". IWCMC, 2011
[40] L. Wei et al.: "Boardpuf: Physical unclonable functions for printed circuit board authentication". IEEE ICCAD, 2015
[41] M. Stipcevic et al.: "True random number generators," Open Problems in Mathematics and Computational Science, 2014
[42] M. T. Rahman et al.: "Ti-trng: Technology independent true random number generator". IEEE/ACM DAC, 2014
[43] D. I. Urbina et al.: "Survey and new directions for physics-based attack detection in control systems". US Department of Commerce, NIST, 2016.
[44] L.A. Tang et al. "Tru-alarm: Trustworthiness analysis of sensor networks in cyber-physical systems". IEEE ICDM, 2010
[45] K. Wan et al.: "Context-aware security solutions for cyber-physical systems". Mobile Networks and Applications, 19.2, 2014
[46] K. Wan et al.: "Dependable context-sensitive services in cyber-physical systems". TrustCom, 2011.
[47] X. Koutsoukos et al.: "SURE: A Modeling and Simulation Integration Platform for Evaluation of SecUre and REsilient Cyber-Physical Systems". Proceedings of the IEEE, 106.1, 2018
[48] D. Gollmann et al.: "Cyber-physical systems security: Experimental analysis of a vinyl acetate monomer plant". ACM CPS, 2015
[49] F. Pasqualetti et al.: "Design and operation of secure cyber-physical systems". IEEE Embedded Systems Letters 7.1, 2015
[50] J. Taylor et al.: "Security challenges and methods for protecting critical infrastructure cyber-physical systems". IEEE MoWNeT, 2017
[51] S. Peisert et al.: "Designed-in security for cyber-physical systems". IEEE Security & Privacy 12.5, 2014
[52] F. Pasqualetti et al.: "Attack detection and identification in cyber-physical systems". IEEE TAC, 58 (11), 2013
[53] G. Sabaliauskaite et al.: "Intelligent checkers to improve attack detection in cyber-physical systems". IEEE CyberC, 2013
[54] A. L. Buczak et al.: "A survey of data mining and machine learning methods for cybersecurity intrusion detection". IEEE Communications Surveys & Tutorials, 18.2, 2016
[55] R. C. B. Hink et al.: "Machine learning for power system disturbance and cyber-attack discrimination". IEEE ISRCS, 2014
[56] S. Pan et al.: "Developing a hybrid intrusion detection system using data mining for power systems". IEEE TSG, 6.6, 2015
[57] Y. Zhang et al.: "Health-CPS: Healthcare cyber-physical system assisted by cloud and big data". IEEE Systems Journal, 11.1, 2017
[58] A. Valdes at al.: "Anomaly detection in electrical substation circuits via unsupervised machine learning". IEEE IRI, 2016
[59] R. Mitchell et al.: "Behavior rule specification-based intrusion detection for safety-critical medical cyber-physical systems". IEEE Transactions on Dependable and Secure Computing, 12.1, 2015
[60] F. K. Lodhi et al.: "Power profiling of microcontroller's instruction set for runtime hardware Trojans detection without golden circuit models". IEEE/ACM DATE, 2017

2018 IEEE Computer Society Annual Symposium on VLSI

On how to efficiently implement Deep Learning algorithms on PYNQ platform

Luca Stornaiuolo, Marco D. Santambrogio, Donatella Sciuto

Politecnico di Milano, Dipartimento di Elettronica Informazione e Bioingegneria (DEIB), Milan, Italy
{luca.stornaiuolo, marco.santambrogio, donatella.sciuto}@polimi.it

Abstract—**Deep Learning algorithms are gaining momentum as main components in a large number of fields, from computer vision and robotics to finance and biotechnology. At the same time, the use of Field Programmable Gate Arrays (FPGAs) for data-intensive applications is increasingly widespread thanks to the possibility to customize hardware accelerators and achieve high-performance implementations with low energy consumption. Moreover, FPGAs have demonstrated to be a viable alternative to GPUs in embedded systems applications, where the benefits of the reconfigurability properties make the system more robust, capable to face the system failures and to respect the constraints of the embedded devices. In this work, we present a framework to efficiently implement Deep Learning algorithms by exploiting the PYNQ platform, recently released by Xilinx. The case study application is tested on PYNQ-Z1 board, commonly used in embedded system applications.**

I. INTRODUCTION

Deep Neural Networks (DNNs) are a set of algorithms inspired from brain neurons behavior used to recognize patterns. They are becoming a pervasive solution in a huge amount of fields, from computer vision, smart vehicles and robotics to finance, medicine and decision making. In these domains, Machine Learning approach is proved to be valuable to automate the prediction processes based on the available data and on real-time signals acquisition [1].

This, combined with the widespread of the Internet of Things (IoT) paradigm [2], opens new challenges for embedded system devices. The number of connected objects that acquire signals from the physical world, send data to related applications and get back the information to act, is increasing in quantity and variety and this increases also the whole system complexity. In this context, the embedded system devices have to face many problems, ranging from dealing with the possible system failures to adapting themselves to the physical world changes [3].

One possible solution to solve these challenges is the development of intelligent embedded system devices, able to take care of their own healing and to adapt their behavior accordingly with the environmental variations. Even if Machine Learning algorithms seem to be the right solution to face the system adaptation and its self-healing capabilities, they require a high computational power. Moreover, the embedded devices physical constraints require a low energy consumption, also to reduce the system warm and the possible derived damages [4]. For these reasons, Field-Programmable Gate Arrays (FPGAs)

devices represent a suitable architecture to fit such requirements. In addition, their reconfigurable properties make the system more flexible with respect to different contexts, but also adaptable to changes within the same context.

In this paper we present a framework to help data scientists and FPGA-based IPs designers to deploy their deep learning algorithms on a Xilinx Zynq SoC by exploiting the recently released PYNQ platform. In particular, the main contributions of our work are declined as a framework able to:

1) automatically create the required interface to transfer data from the Processing System to the accelerated version of the DNN implemented on the Programmable Logic by following the dataflow paradigm;
2) help in designing the routine to deal with system failures by exploiting the behavior of DNNs and the reconfigurable properties of the system.

The rest of this paper is divided as follow: in Section II we describe the relation between DNNs and the Xilinx PYNQ platform; in Section IV we propose our framework; in Section V we discuss about our solution and the future work and Section VI contains the conclusion.

II. BACKGROUND

In this section, we discuss why FPGAs are a valid solution to improve performances of Deep Learning algorithms and we analyze the advantages of the recently released Xilinx PYNQ platform within the intelligent embedded system context.

A. Deep Neural Networks and FPGAs

DNNs are a subclass of Neural Networks contains multiple hidden layers that propagate weighted sums of input data to the output layer. They also apply a non-linear function to such sums to generate an output only when a certain threshold is crossed. This behavior is inspired from brain neurons and it is used to recognize recurrent patterns between input data and the related output [5].

Although Neural Networks birth dates back to 1940s, in the recent years, the huge amount of available data together with the high compute capacity and the algorithmic techniques improvement have driven DNNs to their today success [6]. In this context, many different hardware platforms have been targeted with the aim to improve DNNs throughput and energy efficiency. Graphics Processing Units (GPUs) are the most common employed solution to exploit hardware parallelism

978-1-5386-7100-9/18 $31.00 © 2018 IEEE   587

and execute multiple DNNs multiply-and-accumulate operations at the same time. The huge widespread of GPUs has been achieved also thanks to the investment done by the big technology companies, like NVIDIA, to create a set of frameworks and platforms to raise the level of abstraction required to exploit them. Examples are CUDA[1] and OpenCL[2] that have allowed GPUs integration in Machine Learning frameworks and libraries like Caffe[3], Dlib[4], and TensorFlow[5]. However, GPUs exploitation requires high power consumption and this represents a limitation in many Deep Learning domains where the computational devices are required to be mobile or embedded ad they need to fit rigid physical constraints.

To overcome this problem, FPGAs can largely reduce the power consumption, while remaining competitive in terms of execution time. In particular, we can exploit spatial architecture of FPGAs to take advantage of the memory hierarchy and reduce the energy costs of data accesses. Since the most efficient memory is also the most limited one, it is necessary to design the system to follow the dataflow paradigm and to get benefit from data reuse, taking into consideration that the processing of DNNs is deterministic.

B. PYNQ platform and Intelligent Embedded Systems

PYNQ[6] is a platform built on top of Xilinx Zynq SoC technology to allow software developers exploiting the Programmable Logic of the board directly from Python applications. In this way, FPGA-based IPs designers can provide their optimized functions ready to be used, with the same abstraction level of software libraries. These libraries able to exploit the heterogeneous hardware architecture of the Zynq SoC are also named hardware libraries or overlays. The boards PYNQ-Z1 is the first released board that support the PYNQ platform. The Xilinx official base overlays allows to control the 12-pin PMOD connectors, an Arduino-compatible interface and Audio/Video I/O. The energy efficiency and the flexibility of the Programmable Logic together with the Python productivity and the peripherals control make the PYNQ-Z1 board an efficient solution as embedded device.

With the advent of the Fog Paradigm part of the computation and data preprocessing is offload to the leaf nodes of the system, where signals are collected. The possibility of adding also intelligent mechanisms within this distributed scenario allows each node to learn from external stimuli, adapt to change and make decisions. These capabilities can also be used to face system failures at different hierarchical level. In particular, in the scenario where a leaf node of a fog distributed system is represented by a PYNQ-Z1 board, that acquires raw data and fast preprocesses them before sending results to the next layer of the network, the ability of recognizing

[1] https://developer.nvidia.com/cuda-zone
[2] https://www.khronos.org/opencl/
[3] http://caffe.berkeleyvision.org/
[4] http://dlib.net/ml.html
[5] https://www.tensorflow.org/
[6] http://www.pynq.io/

failures and dealing with self-healing by reconfiguring the Programmable Logic, prevents performance reduction of the entire system and errors cascade effects.

III. RELATED WORK

In this section, we provide an overview of the related work that regards the DNNs implemented on embedded systems and the frameworks to help with their development.

DNNs are characterized by two different phases. The training phase, used to process already available data and to compute the model weights, requires high computational power and takes a huge amount of time to be computed. On the other hand, the inference phase, used to process the new data and find known patterns, is suitable to be implemented on embedded systems [6]. In this context, different techniques to reduce the complexity are often used [7]. As an example, binary weights are adopted in [8], where PYNQ platform is exploited to take advantages from the high performance with respect to a low power consumption, and in [9], where the integration of TensorFlow with embedded devices is performed. Together with the complexity reduction techniques, a lot of framework to help hardware designer of DNNs are released. As an example, [10] is a framework to efficiently map binarized neural networks to hardware and [11] is used to reduce the FPGA hardware resources.

IV. PROPOSED APPROACH

This section describes the main components of our framework to help data scientists and FPGA-based IPs designers to deploy their deep learning algorithms on a Xilinx Zynq SoC by exploiting the recently released PYNQ platform. Firstly, we modeled and generalized the communication from Processing System to Programmable Logic for dataflow applications in order to automatize the interfaces creation to send data to the accelerated version of a DNN and get back results. Then, we provide a routine to recover running system from a specific failure by reconfiguring the FPGA.

A. Dataflow Communication Interface creation for DNN

Input data of a DNN is the set of values representing the information to be analyzed. For instance, in computer vision domain input can be pixels of an image, in speech and language domain input can be an audio wave, in problem solving domain input can be the current state of some game.

As described in Section II, the dataflow paradigm is the best solution to improve the energy efficiency and exploit data reuse, when implementing DNNs on an FPGA. Since the processing of DNNs does not contain randomness, the dataflow design in term of data type and number of values to be exchange between Processing System and Programmable Logic is known in advance. When the FPGA-based DNN developer decides how to use each memory of the available memory hierarchy, he has to take into account that on-chip BRAM is faster and more efficient than DRAM, but it is very limited. For this reason, the data reuse becomes very important, when some values are transferred to the Programmable Logic from the higher costly DRAM of the board.

978-1-5386-7100-9/18 $31.00 © 2018 IEEE

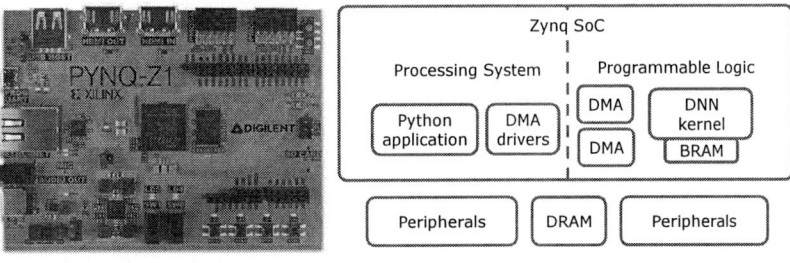

Fig. 1. Main system components overview.

Figure 1 shows the main system components. We can divide the framework semi-automatic creation of interfaces in three phases following a bottom-up approach:

1) **Design Level Integration**

Designing the FPGA-based version of DNN for the PYNQ-Z1 board following the dataflow paradigm required the presence of one or more Direct Memory Access (DMA) IPs to stream data from the DRAM to the computational kernel on the Programmable Logic. DMAs can be added in the Vivado block design phase and can be connected to the input interfaces of the user custom IP. To do that the FPGA-based DNN developer has to add an AXI4-Stream interface for each input and output stream of the computational kernel. The framework can auto generate such interfaces by knowing the data type of each stream. The following snippet of Vivado HLS pseudo code shows a very simple example in which the kernel reads a stream of integer points from an input stream and writes the same points to an output stream.

```
#include <hls_stream.h>

struct data_struct{
  int data;
  bool last;
};

void computational_kernel(
        hls::stream<int> &s_in,
        hls::stream<data_struct> &s_out
) {
#pragma HLS INTERFACE axis port=s_in
#pragma HLS INTERFACE axis port=s_out
#pragma HLS INTERFACE ap_ctrl_none port=return

    data_struct out_data;
    for (int i = 0; i < NUMBER_OF_POINTS; i++)
#pragma HLS PIPELINE II=1
        out_data.data = s_in.read();
        if (i == (NUMBER_OF_POINTS - 1))
          out_data.last = 1;
        else
          out_data.last = 0;

        s_out.write(out_data);
    }
}
```

The first two HLS INTERFACE pragmas define the AXI4-Stream type of the kernel parameters. The last HLS INTERFACE pragma allows the computational kernel to start as soon as the data on the input stream are available, without the needing of programmatically start the IP computation. The specific output data structure will be automatically mapped to a set of pins that are able to transfer data and inform the system when the last point is sent, so that the receiver can close the transfer connection. Finally, the NUMBER_OF_POINTS variable depends on the kernel design and in DNN computation is known in advance.

2) **Processing System Level Integration**

The second step of our framework interfaces generation consists of integrating the drivers to command the DMAs directly from the Processing System. We exploit the Python/C API to achieve the fastest possible performance when exposing the drivers functions to the user application. The following snippet of C pseudo code shows how to manage the stream transfers by exploiting two instanced DMAs connected respectively with the input and the output streams of the computational kernel.

```
// ==== OPEN STREAM TO WAIT RESULTS ====
XAxiDma_SimpleTransfer(
        DMAinstance2,
        (uint32_t *) FPGA_receive_buffer,
        NUMBER_OF_POINTS,
        DMA_FROM_FPGA);

// ==== SEND POINTS ====
XAxiDma_SimpleTransfer(
        DMAinstance1,
        (uint32_t *) FPGA_send_buffer,
        NUMBER_OF_POINTS,
        DMA_TO_FPGA);

// ==== WAIT ALL POINTS ARE SENT ====
while (XAxiDma_Busy(DMAinstance1, DMA_TO_FPGA))
    ;

// ==== WAIT ALL RESULTS ARE RECEIVED ====
while (XAxiDma_Busy(DMAinstance2, DMA_FROM_FPGA))
    ;
```

3) Python Level Integration

At the user application level, the Python data types need to be cast to satisfy the requirements of the computational kernel. For this reason, we exploit the NumPy library together with the CFFI Python module to deal with data casting. Moreover, a data verification process is added to avoid that the algorithm is called with the wrong inputs. This avoid computational kernel crashes due to wrong data representations. The following snippet of Python pseudo code shows how to manage the data casting and how to add a possible input format check.

```python
def fpga_kernel(input_data):
    if len(input_data) != NUMBER_OF_POINTS:
        raise ValueError("Wrong input dimensions")
    else:
        buff_dma_in = cffi.FFI().cast("int *")
        buff_dma_out = cffi.FFI().cast("int *")

        cffi.FFI().memmove(buff_dma_in, input_data)

        fpga_kernel_driver_interface(
            DMAinstance1,
            DMAinstance2,
            buff_dma_in,
            buff_dma_out
        )

    received_data = cffi.FFI().buffer(buff_dma_out)
    results = frombuffer(received_data, dtype=int32)

    return results
```

B. System Failure Recovery

The inference phase of DNNs, unlike the training one that requires high computational needs and a long execution time, can be part of real-time data processing at the embedded systems level. Taking into consideration the integration with the PYNQ platform explained in the previous section, it is possible to identify at the application level the average execution time of the DNN optimized kernel, given a fixed input/output dimension and fixed data types. This value can be obtained empirically with a calibration phase or can be set as a threshold from a domain expert user. The proposed routine measures the execution time of each computational kernel function call and based on some accuracy thresholds can recognize some kernel errors or interruptions. If an anomaly is detected, the routine kills the current computational kernel function call, reconfigures the Programmable Logic and starts the next input processing. If a list of subsequent anomalies are detected, the routine can generate a system alarm or slightly modify the thresholds to try to understand the failure nature.

V. DISCUSSION AND FUTURE DIRECTIONS

Together with the related work, the solution proposed in this paper represents a starting step to create a solid framework-based infrastructure to integrate FPGAs within the Deep Learning field. Since this process is started earlier for GPUs, nowadays they represent the most used solution, even if other hardware architectures could be integrated at different applications levels, bringing benefits to the whole system.

As future work, we want to allow framework exploiting the partial configuration property of the Programmable Logic to better face system failures. Moreover, we are going to improve the failures predictive model by integrating machine learning techniques in the proposed routine. Finally, we are planning to integrate the framework with the most used Python libraries for Deep Learning models implementation.

VI. CONCLUSIONS

In this paper, we presented a framework to integrate Deep Learning algorithms on the PYNQ-Z1 at the embedded system level. In particular the proposed solution help data scientists and hardware developers to generate the required interfaces to interact with the FPGA-based implementation of the DNN algorithm and integrate it within the PYNQ platform. This is done through the generalization of the DNN data management within different domains. Finally, we proposed a routine to face embedded system failure at the Programmable Logic level.

REFERENCES

[1] I. Goodfellow, Y. Bengio, A. Courville, and Y. Bengio, *Deep learning*. MIT press Cambridge, 2016, vol. 1.

[2] K. Ashton *et al.*, "That 'internet of things' thing," *RFID journal*, vol. 22, no. 7, pp. 97–114, 2009.

[3] G. Ditzler, M. Roveri, C. Alippi, and R. Polikar, "Learning in non-stationary environments: A survey," *IEEE Computational Intelligence Magazine*, vol. 10, no. 4, pp. 12–25, 2015.

[4] C. Alippi, G. Anastasi, M. Di Francesco, and M. Roveri, "Energy management in wireless sensor networks with energy-hungry sensors," *IEEE Instrumentation & Measurement Magazine*, vol. 12, no. 2, 2009.

[5] Y. LeCun, Y. Bengio, and G. Hinton, "Deep learning," *nature*, vol. 521, no. 7553, p. 436, 2015.

[6] V. Sze, Y.-H. Chen, T.-J. Yang, and J. S. Emer, "Efficient processing of deep neural networks: A tutorial and survey," *Proceedings of the IEEE*, vol. 105, no. 12, pp. 2295–2329, 2017.

[7] S. Han, H. Mao, and W. J. Dally, "Deep compression: Compressing deep neural networks with pruning, trained quantization and huffman coding," *arXiv preprint arXiv:1510.00149*, 2015.

[8] "Bnn-pynq," https://github.com/Xilinx/BNN-PYNQ/.

[9] "Binary networks from tensorflow to embedded devices," https://github.com/jonathanmarek1/binarynet-tensorflow.

[10] Y. Umuroglu, N. J. Fraser, G. Gambardella, M. Blott, P. Leong, M. Jahre, and K. Vissers, "Finn: A framework for fast, scalable binarized neural network inference," in *Proceedings of the 2017 ACM/SIGDA International Symposium on Field-Programmable Gate Arrays*. ACM, 2017, pp. 65–74.

[11] C. Zhang, P. Li, G. Sun, Y. Guan, B. Xiao, and J. Cong, "Optimizing fpga-based accelerator design for deep convolutional neural networks," in *Proceedings of the 2015 ACM/SIGDA International Symposium on Field-Programmable Gate Arrays*. ACM, 2015, pp. 161–170.

2018 IEEE Computer Society Annual Symposium on VLSI

Enabling Reliable High Throughput On-Chip Wireless Communication for Many Core Architectures

Sri Harsha Gade, Mitali Sinha, Sidhartha Sankar Rout and Sujay Deb
Department of Electronics and Communication Engineering
Indraprastha Institute of Information Technology Delhi, New Delhi, India
Email: {harshag, mitalis, sidharthas, sdeb}@iiitd.ac.in

Abstract—Wireless Networks-on-Chip (WNoCs) have been shown to overcome the scaling challenges of wired NoCs by augmenting them with low latency, low energy, long range wireless links. However, existing wireless implementations face challenges in terms of reliability while providing high communication performance. Single channel communication, the pre-dominant way of implementing WNoCs, are susceptible to channel effects like dispersion, fading, etc. and also provide limited bandwidth. Multi-channel communication, though provides high throughput, are prone to inter-channel and inter-symbol interference. In order to handle both performance and reliability challenges, we propose wireless network design using Orthogonal Frequency Division Multiplexing (OFDM) modulation. The proposed design enables reliable and channel resilient wireless communication, while providing high throughput, concurrent transmissions in WNoC. OFDM, by dividing wide band channel into several smaller sub-channels, overcomes channel dispersion and ISI effects. In the proposed design, OFDM sub-channels are grouped into multiple contiguous bands and assigned to each transceiver in WNoC. By allowing each transceiver to transmit only over assigned group, we enable simultaneous, high bandwidth communications. Evaluations show that proposed design achieves BER of the order 10^{-12}. Using the concurrent wireless link design, the runtime is improved by 29% and network energy is reduced by 68% as compared to baseline mesh topology.

Keywords-Wireless links; OFDM; sub-channels;

I. INTRODUCTION

Shrinking CMOS feature sizes and limitations of Dennard scaling have steered chip designs towards many core architectures to achieve high performance computing. Though highly efficient, many core architectures pose significant challenges, especially for data communication, both core-to-core and core-to-memory, as number of cores on a single chip increases. The communication challenges are further exacerbated in heterogeneous computing platforms, that integrate processing elements of different architectures on a single chip. This is due to the significantly different data access patterns of different computing architecures, leading to complex communication paradigms in on-chip networks for HSA. For example, authors in [1] have shown that application runtime increases by upto $2\times$ in Heterogeneous System Architectures (HSA) like AMD APUs[2] and Intel Core chips (Sandybridge and later)[3], that integrate CPU and GPU architectures on a single chip.

Network-on-Chip (NoC) has been shown to be the enabling technology for integrating many cores on a single chip. Wireless NoC (WNoC) further enhances the communication performance by employing long range links to integrate hundreds of cores on a single chip[4], [5]. Though significant advances have been made in design of transceivers and wireless network, several challenges still exist for implementation of WNoC topologies. Some of the major challenges in WNoC topologies are (i) concurrent wireless communications, (ii) signal reliability over wireless channel and (iii) effective wireless channel performance in complex intra chip environments. In addition to these, most NoC and WNoC topologies are designed for homogeneous architectures and they do not perform well in presence of complex communication patterns in HSA[1]. Without prudent design and resource allocation, these complex traffic scenarios can lead to significant area, energy and performance overheads.

Most existing wireless network designs make use of single channel communication to avoid excessive overheads for implementing wireless transceivers, which limits the number of active wireless transmissions to only one at any given time. This leads to suboptimal utilization of channel resources, communication performance and unwanted power consumption in inactive wireless transceivers. Wireless network with multi-channel communication like [6] provides a potential solution to overcome this challenges. However, design of multiple non-overlapping channels without interference at mm-wave frequencies and beyond is a daunting task. Furthermore, they add excessive overheads for implementing multiple transceivers attuned to each channel. Directional WNoC topology in [7] offers a promising solution to achieve simultaneous communications by implementing directional point-to-point wireless links operating at single frequency. While this is a feasible solution, it takes the inherent advantage of broadcast capability of wireless links.

Another major challenge for implementation of WNoC topologies is the reliability of signal transmitted over wireless channel. The modeling and understanding of wireless channel in complex intra chip environments is an active topic of reasearch and there is still a lack of detailed analysis of intra chip wireless channel. Authors in [8], [9], [10] have analysed intra chip wireless channel using simple

978-1-5386-7100-9/18 $31.00 © 2018 IEEE   591

geometries and have shown that signal propagation differs from free space wireless propagation. The signal transmitted over intra chip wireless channel is impacted by multiple propagation paths, interference, long delays, high path loss, etc., severely impacting the reliability of data transmission in WNoCs. Furthermore, evaluation of WNoCs, in most cases, is done assuming a free space channel, which may lead to incomplete understanding of achievable performance improvements. Hence, there is a need for incorporation of intra chip wireless channel effects in design of WNoC architectures for achieving reliable and high performance communication in many core processors.

To overcome signal reliability challenges and provide efficient interconnection in many core architectures, we design a reliable, concurrent communication enabled wireless network using Orthogonal Frequency Division Multiplexing (OFDM). The proposed wireless interconnection uses OFDM modulation, which is robust against adverse channel effects like dispersion, fading and ISI, etc., while also providing high data rate. The specifications of OFDM transceiver are designed through detailed modelling of intra-chip wireless channel to understand delay spread, path loss and propagation delay characteristics. To provide simultaneous wireless transmissions, we make use of inherent availability of several narrow bands in the sub-channel mechanism of OFDM. OFDM modulation divides a wide band wireless channel into several sub-channels to overcome dispersion effects. In proposed design, a subset of these sub-channels are grouped together and the sub-channel group is allocated to one wireless transceiver. Each transceiver transmits only over the assigned group, leaving other sub-channels unmodulated, thereby enabling simultaneous wireless transmissions without the use of multiple non-overlapping channels. Since, the same OFDM transceiver, with few modifications, is used for transmitting and receiving data over these sub-channel groups, the proposed design does not add additional overheads as opposed to designs with multiple wireless channels. Using the proposed wireless link design, we achieve reliable, high throughput concurrent wireless communication in WNoC topologies. The main contributions of this work are

- An efficient wireless link design by OFDM sub-channel grouping and allocation to achieve high throughput concurrent transmissions in WNoCs without the need for multiple wireless channels.
- An intra chip wireless channel aware transceiver design to provide reliable and robust wireless communication.
- Detailed analysis of reliability and performance using proposed design in heterogeneous architectures.

II. CONCURRENT TRANSMISSION ENABLED WIRELESS NETWORK DESIGN

The efficiency of WNoC topologies over wired topologies, in general, is highly dependent on the efficient design of wireless network, that provides interconnection between long distant nodes. The wireless network must provide high data rate, low energy and reliable communication across the chip, while making optimal use of channel and transceiver resources. We achieve both reliable and high performance wireless network through OFDM[11], which typically divides a large bandwidth channel into many narrow band sub-channels to counter adverse channel effects. We use this inherent sub-channel division mechanism to enable concurrent wireless communications without employing multiple frequencies.

A. Concurrent Transmissions using Sub-Channel Grouping

One of the key limitations of most existing WNoC topologies is single active wireless transmission at any given time due to the use of single channel communication. This limits the performance gains, utilization of wireless resources and increases idle energy consumption inactive transceivers. While multi-channel wireless design provides simultaneous transmissions between the transceivers, it adds high overheads for achieving interference free communication over non-overlapping channels. Hence, alternative approaches are required to provide reliable, concurrent wireless transmissions, while also adding low overheads. In the proposed design, we achieve concurrent communication between multiple transceivers by grouping multiple sub-channels of OFDM scheme together and allocating one group to each transceiver. The transceiver transmits data only over the designated group of sub-channels, while leaving the other sub-channels unmodulated. We make minor modifications to the OFDM transceiver to allow transmission over single sub-channel group and reception over all other groups. Since, the sub-channel division is inherent to OFDM, the proposed concurrent communication methodology does not require multiple transceivers and adds very little overhead to the OFDM transceiver and WNoC topology. This allows different transceivers to transmit data simultaneously without interference over single channel. Furthermore, it retains all the advantages of OFDM viz., resilience to channel dispersion, ISI and high spectral density. Though each transceiver has access to only a portion of channel bandwidth, the total bandwidth of the wireless network remains the same for given implementation.

Since, each wireless transceiver has access to only a portion of bandwidth, the grouping of sub-channels and their allocation to each subsystem plays a pivotal role in meeting its bandwidth requirements. In systems with similarly configured subsystems, equal division of available sub-channels into K groups, where K is number of wireless transceivers, provides a simple and fair allocation of available bandwidth. But, such an allocation mechanism is not efficient for heterogeneous architectures, especially one with contrasting architectures like CPU and GPU on a single chip. Hence to achieve a more optimal allocation of the sub-channel bandwidth, we have measured the average and peak

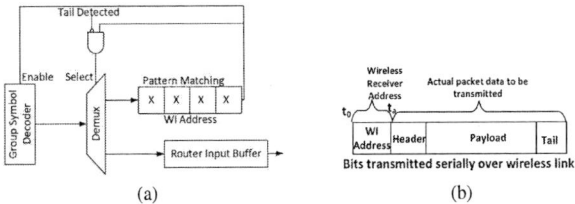

Figure 1. (a) Pattern Matching Circuit to Detect Packet Receiver (b) Data Format Transmitted over Sub-Channel Group

bandwidth of data transfer of different computing clusters in a given architecture for different applications (system architecture is discussed in section IV-A). The number of sub-channels assigned to each group is then determined in proportional to the bandwidth as

$$N_i = \frac{B_i}{\sum_{i=1}^{K} B_i} * N_T \qquad (1)$$

N_i is number of sub-channel assigned to transceiver i, N_T is total number of sub-channels and B_i is the bandwidth desired by computing cluster i. Simulation of several benchmarks on a heterogeneous CPU/GPU architecture has shown a recurring trend, where main memory requires the highest bandwidth, followed by GPU cluster and finally the CPU cluster. This matches the expected trend as GPUs are highly throughput sensitive, while CPU desire low latency communication for its mostly serial execution.

B. Data Reception and Decoding

The sub-channel grouping and allocation to individual transceivers allows all them to transmit data simultaneously without interfering with each other. At the receiver end, the data from different transceivers transmitted over different sub-channel groups is received as combined input at the demodulator. Hence, the received signal needs to be appropriately disaggregated and decoded to obtain the data from different transceivers. This is achieved through a combination of OFDM symbol decoders and pattern matching circuits for individual groups as shown in Fig. 1.

The packet data transmitted by each transceiver is formatted as shown in Fig. 1b, which adds *Wireless Address* bits at the head of the actual packet. The *Wireless Address* is a unique identifier assigned to each transceiver and helps reduce the data processing and time required to determine if the received packet is intended for the same receiver. At the receiver, the received bits are decoded for this *Wireless Address* using a pattern matching circuit as shown in Fig. 1a. Once the *Wireless Address* has been verified, the demux select line is chosen to send the received data to buffers, if the address matches. If there is a mismatch, the packet reception and decoding for that sub-channel group is ceased to avoid unnecessary energy consumption. The pattern matching decoder is operated at serial data stream frequency. The number of bits required for *Wireless Address*

is as shown in (2), where N_{WI} is number of transceivers. Since, N_{WI} is small, the number of *Wireless Address* bits is very small. The maximum time taken to process these bits and stopping data reception, in case of mismatch, is equivalent to r bit cycles. Though it adds bit overhead to packet transmission, it ultimately avoids excessive energy consumption with a little control overhead. The total time taken to complete any packet transmission is given by (3), where N_{flits} is number of flits in the packet, S_{flit} is flit size, t_{bit} is serial bit duration and $t_{propagation}$ is propagation delay of wireless channel.

$$r = log_2(N_{WI} + 1) \qquad (2)$$

$$T_{max} = (r + N_{flits} * S_{flit})t_{bit} + t_{propagation} \qquad (3)$$

III. RELIABLE WIRELESS TRANSMISSION IN WNoC

The reliability of signal transmitted over intra chip wireless channel is another key challenge in design and evaluation of WNoC topologies. In order to achieve robust wireless transmission in WNoC, we design OFDM transceiver, taking into account realistic modelling of intra chip environment.

A. Modelling of Intra Chip Wireless Channel

Intra chip wireless channel, akin to urban environments for mobile communication, is very complex due to intricate geometries and materials of different properties. These complex geometries give rise to several reflections from different chip layers and edges, in addition to line-of-sight signal. This multipath propagation results in high delay spread, non-quadratic path loss and non-linear delay variation in intra chip wireless channel. To realistically understand signal propagation in intra chip wireless channel and design robust wireless transceiver, we model the intra chip environment using a 3D multi-layered structure as shown in Fig. 2. The model captures all key characteristics of silicon chips viz., silicon substrate and different metal layers embedded within SiO2 dielectric layers. The antennas are etched in the top metal layer and the signal propagation in this intra chip model is simulated using FDTD based transient solver from CST MicroWave Studio. We analyse signal received at different transmitting distances across the chip to understand analyse path loss, delay and delay spread characteristics.

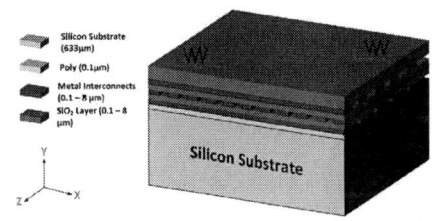

Figure 2. 3D on-chip environment with a silicon substrate, copper interconnects and silicon-di-oxide for wireless channel estimation

The estimated channel properties are then used to determine design specifications of OFDM transceiver and evaluating the network performance. Our analysis shows that intra chip wireless channel exhibits high delay spread ($T_m = 5ns$), resulting in dispersion and fading effects. To overcome these effects, the optimal number of OFDM sub-channels must meet the criteria $B_N << 1/T_m$, where B_N is sub-channel bandwidth. In addition to these effects, the intra chip wireless channel also has high path loss and propagation delay as compared to free space channel. These realistic estimates are used while evaluating the network performance using proposed design.

B. OFDM Transceiver for Reliable Wireless Transmission

The OFDM transceiver with proposed sub-channel grouping methodology is shown in Fig. 3. It is broadly divided into; (i) data encoding into group sub-channels, (ii) Discrete Fourier Transform (DFT)/Inverse DFT (IDFT), (iii) cyclic prefix, (iv) pattern decoding and (v) up/down conversion. For encoding packet data into sub-channels, we use Quadrature Amplitude Modulation (QAM) to ensure high spectral efficiency. As seen in Fig. 3, each WI encodes data only into its sub-channel group. The inverse Fast Fourier Transform (FFT) module, then modulates the encoded symbols to respective sub-channel frequencies and the modulated time domain signal is up-converted to transmission carrier frequency (f_0) before transmitting over the channel. The cyclic prefix bits at the transmitter are added to maintain sub-channel orthogonality against any channel effects.

At the receiving end, the signal is down-converted from carrier frequency and individual symbols are extracted from sub-channel components using FFT module after removing cyclic prefix bits. These extracted symbols are then fed into K pattern matching circuits corresponding to the K sub-channel groups. The initial *Wireless Address* bits are decoded as described before to determine the packet receiver and QAM decoding is continued if packet receiver is a match. The output from each group are written to router buffers for

further processing. We have used two antennas for enabling simultaneous transmission and reception of the data. The power amplifier[12] and low noise amplifier amplify the signals at transmitter and receiver respectively.

IV. EXPERIMENTAL RESULTS

To evaluate performance of proposed topology, we have implemented it for CPU/GPU HSA using Multi2Sim[13] heterogeneous system simulator.

A. System Architecture

The system architecture for proposed implementation is adopted from [1] with sub-network topologies. The heterogeneous system is divided into multiple homogeneous subsystems, each comprised of same architectural cores and their private memory. The core and memory elements within the subsystem are interconnected using a wired topology. All such subsystems are then interconnected at the system level using wireless network to provide efficient communication among them. Each subsystem is equipped with a transceiver to provide inter-subsystem communication.

In our implementation, we have three subsystems; CPU, GPU and memory subsystem. As evident from their names, the three subsystems are only comprised or CPU cores, GPU compute units or shared memory elements respectively. The CPU subsystem is interconnected using a shared bus topology to provide low latency communication with other subsystem and shared memory. The wired network for GPU subsystem is implemented using mesh topology with buffered routers to provide sufficient throughput to handle thousands of requests generated by GPUs compute units. Finally the shared memory subsystem, comprised of last level cache (shared by both CPU and GPU) and memory controllers for off-chip memory access, is interconnected with crossbar topology to service multiple requests to different memory elements. Each subsystem has one wireless transceiver, that connects it with other two subsystems.

B. Experimental Setup

The configuration of system and WNoC architecture are shown in Table I. The wireless links are implemented using loop antenna[14], which has total bandwidth of $25GHz$ at $58GHz$ carrier frequency. We use 256 sub-channels to provide a data rate of $195.32Gbps$ and 20%, 40% and 40%

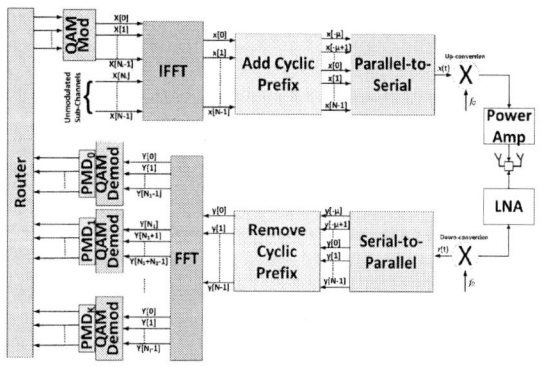

Figure 3. OFDM based Transceiver for WNoC

Table I
DETAILS OF CPU/GPU HETEROGENEOUS SYSTEM ARCHITECTURE

Component	Configuration
CPU	x86, 4 cores, 2.5GHz, 4-way, 64KB L1 cache, 8-way, 256KB L2 cache, 64B line
GPU	AMD Southern Islands, 16 cores, 1.5GHz, 4-way, 32KB L1 cache, 8-way, 128KB L2 cache, 64B line
Routers	5 port, 3 stage, 8-flit depth port (only for GPU)
Wireless Link	58GHz carrier, 195.32 Gbps, 0.24 pJ/bit

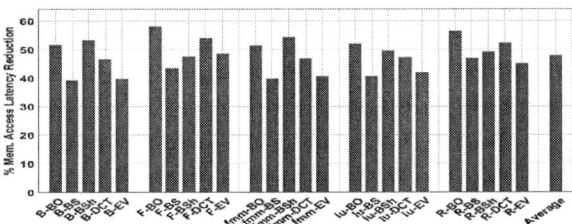

Figure 4. Improvement in Memory Access Latency using Hybrid WNoC over Mesh

of sub-channels are allocated to CPU, GPU and memory subsystems respectively. To compare the performance of hybrid WNoC, we use baseline mesh with different cores mapped to provide non-overlapping paths to memory controllers as described in [15]. The mesh topology uses input buffered routers configured as shown in Table I. Application level traffic and statistics are obtained by simulating different combinations of CPU and GPU benchmarks from OpenCL[16] and SPLASH2[17] suites.

C. Network Performance

We evaluate the network performance of proposed topology by comparing memory access latency and peak bandwidth with those of baseline mesh topology. Memory access latency is a major performance bottlenecks for processor performance and is one of the primary sources of contention in heterogeneous architectures. Fig. 4 shows the average reduction in memory access latency using proposed topology for different benchmarks. Across different applications scenarios, proposed topology reduces memory latency by 47.87%, on average. This significant reduction is achieved by (i) the use of low latency wireless links for off-chip memory access and (ii) by providing interference free transmission paths between CPU/GPU and main memory using sub-channel grouping. Though OOK modulation reduces memory access latency by 38%, the improvement is smaller as both CPU and GPU subsystems share same channel and have to wait for getting access to the medium.

The improvement in peak network bandwidth provided by hybrid WNoC topology over baseline mesh is shown in Fig. 5. We evaluate the peak bandwidth to understand the

maximum throughput provided by the proposed topology. On average, hybrid WNoC with OFDM provides 2.7× more bandwidth as compared to baseline mesh. Similar to memory access latency, though OOK based wireless implementation improves bandwidth over mesh, the increase in bandwidth is not as high due to contention for wireless channel between different subsystems. The improvements in memory access latency and network bandwidth clearly highlight the advantages provided by sub-channel grouping in providing concurrent communication between different subsystems.

D. Network Energy

Fig. 6 shows the savings in average packet energy in the network using hybrid WNoC topology over baseline mesh. As observed, proposed topology saves 67.49% of network energy in comparison with baseline mesh. Breaking down the energy saving values, the router energy is reduced by 37.5%, while energy consumption in links is reduced by 94%, on average. The reduction in router energy is primarily achieved by the optimal design of network resources in each subsystem. For example, baseline mesh uses buffered routers for CPU cores leading to high power consumption, while proposed topology uses shared bus with arbiter to minimize network energy. The significantly high reduction in link energy is due to superior performance of wireless links in providing low energy communication over long distances as compared to wired links. Furthermore, the proposed topology also replaces high energy off-chip links with wireless links to considerably reduce energy consumption.

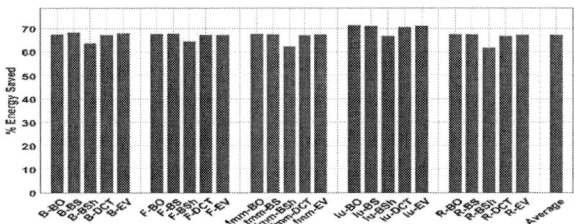

Figure 6. Network Energy Saving using Hybrid WNoC over Mesh

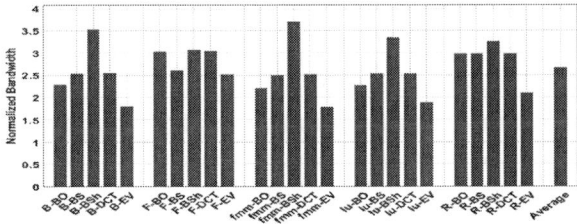

Figure 5. Improvement in Network Bandwidth using Hybrid WNoC over Mesh

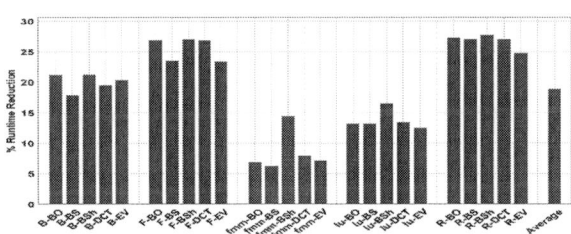

Figure 7. Improvement in Runtime using Hybrid WNoC over Mesh

E. Application Runtime

The application runtime improvement using proposed topology over baseline mesh for different benchmarks are shown in Fig. 7. The proposed hybrid WNoC topology with OFDM wireless links and sub-channel grouping provides an average runtime reduction of 18.79% as compared to baseline mesh topology. The presence of simultaneously applications running on both CPU and GPU leads to contention for network resources in mesh topology, leading to performance degradation. In the proposed topology, the heterogeneous subsystem networks reduce contention between different architectures and wireless links provide interference free, high bandwidth communication, thereby significantly improving the performance.

V. CONCLUSION

A hybrid WNoC, using heterogeneous subsystem topologies and OFDM modulation to provide high performance interconnection in heterogeneous architectures is presented in this work. The proposed topology employs sub-channel grouping and allocation to different WIs to provide interference free, high data rate communication between subsystems. The contention for network resources between cores of different architectures is minimized by dividing the system in multiple subsystem, where each subsystem topology is optimized to meet the requirements of that subsystem. Using this combination, the proposed topology reduces memory access latency by 47.85%, improves network bandwidth by $2.7\times$, while reducing packet energy by 67.5% as compared to baseline mesh topology.

REFERENCES

[1] S. H. Gade and S. Deb, "Hywin: Hybrid wireless noc with sandboxed sub-networks for cpu/gpu architectures," *IEEE Transactions on Computers*, vol. 66, July 2017.

[2] AMD, "Compute cores and heterogeneous system architecture," Tech. Rep., 2014. [Online]. Available: http://www.amd.com/Documents/Compute_Cores_Whitepaper.pdf

[3] N. Kurd, M. Chowdhury, E. Burton, T. P. Thomas, C. Mozak, B. Boswell, M. Lal, A. Deval, J. Douglas, M. Elassal, A. Nalamalpu, T. M. Wilson, M. Merten, S. Chennupaty, W. Gomes, and R. Kumar, "5.9 haswell: A family of ia 22nm processors," in *Solid-State Circuits Conference Digest of Technical Papers (ISSCC), 2014 IEEE International*, Feb 2014.

[4] S. Deb, K. Chang, X. Yu, S. P. Sah, M. Cosic, A. Ganguly, P. P. Pande, B. Belzer, and D. Heo, "Design of an energy-efficient cmos-compatible noc architecture with millimeter-wave wireless interconnects," *IEEE Transactions on Computers*, vol. 62, Dec 2013.

[5] W. H. Hu, C. Wang, and N. Bagherzadeh, "Design and analysis of a mesh-based wireless network-on-chip," in *2012 20th Euromicro International Conference on Parallel, Distributed and Network-based Processing*, Feb 2012.

[6] X. Yu, J. Baylon, P. Wettin, D. Heo, P. P. Pande, and S. Mirabbasi, "Architecture and design of multichannel millimeter-wave wireless noc," *IEEE Design Test*, vol. 31, Dec 2014.

[7] H. K. Mondal, S. H. Gade, M. S. Shamim, S. Deb, and A. Ganguly, "Interference-aware wireless network-on-chip architecture using directional antennas," *IEEE Transactions on Multi-Scale Computing Systems*, vol. 3, July 2017.

[8] Y. P. Zhang, Z. M. Chen, and M. Sun, "Propagation mechanisms of radio waves over intra-chip channels with integrated antennas: Frequency-domain measurements and time-domain analysis," *IEEE Transactions on Antennas and Propagation*, vol. 55, Oct 2007.

[9] L. Yan and G. W. Hanson, "Wave propagation mechanisms for intra-chip communications," *IEEE Transactions on Antennas and Propagation*, vol. 57, Sept 2009.

[10] S. H. Gade and S. Deb, "Achievable performance enhancements with mm-wave wireless interconnects in noc," in *Proceedings of the 9th International Symposium on Networks-on-Chip*, ser. NOCS '15. New York, NY, USA: ACM, 2015. [Online]. Available: http://doi.acm.org/10.1145/2786572.2786584

[11] A. Goldsmith, *Wireless Communications*. New York, NY, USA: Cambridge University Press, 2005.

[12] S. Kaushik, M. Agrawal, H. K. Mondal, S. H. Gade, and S. Deb, "Path loss-aware adaptive transmission power control scheme for energy-efficient wireless noc," in *2017 IEEE 60th International Midwest Symposium on Circuits and Systems (MWSCAS)*, Aug 2017.

[13] R. Ubal, B. Jang, P. Mistry, D. Schaa, and D. Kaeli, "Multi2sim: A simulation framework for cpu-gpu computing," in *Proceedings of the 21st International Conference on Parallel Architectures and Compilation Techniques*, ser. PACT '12. New York, NY, USA: ACM, 2012. [Online]. Available: http://doi.acm.org/10.1145/2370816.2370865

[14] O. Markish, O. Katz, B. Sheinman, D. Corcos, and D. Elad, "On-chip millimeter wave antennas and transceivers," in *Proceedings of the 9th International Symposium on Networks-on-Chip*, ser. NOCS '15. New York, NY, USA: ACM, 2015. [Online]. Available: http://doi.acm.org/10.1145/2786572.2789983

[15] J. Lee, S. Li, H. Kim, and S. Yalamanchili, "Adaptive virtual channel partitioning for network-on-chip in heterogeneous architectures," *ACM Trans. Des. Autom. Electron. Syst.*, vol. 18, Oct. 2013. [Online]. Available: http://doi.acm.org/10.1145/2504906

[16] " AMD Accelerated Parallel Processing (APP) Software Development Kit (SDK) ," 2016. [Online]. Available: http://developer.amd.com/sdks/amdappsdk/

[17] S. Woo, M. Ohara, E. Torrie, J. P. Singh, and A. Gupta, "The SPLASH-2 Programs: Characterization and Methodological Considerations," in *Proc. of the 22nd International Symposium on Computer Architecture*, June 1995.

2018 IEEE Computer Society Annual Symposium on VLSI

Predicting the Tolerance of Extreme Electromagnetic Interference on MOSFETs

Nishchay H. Sule[1], Troy Powell, Sameer Hemmady, and Payman Zarkesh-Ha[2]
Department of Electrical & Computer Engineering
University of New Mexico
Albuquerque, NM, USA
Email: nishchay@unm.edu[1], powellt@unm.edu, shemmady@gmail.com, pzarkesh@unm.edu[2]

Abstract— Extreme Electromagnetic Interference (EEMI) can cause device malfunction due to reparable upsets before any permanent hardware damage occurs to electronic devices. In this paper, a predictive model is developed to characterize the impact of EEMI on Metal-Oxide Semiconductor Field-Effect Transistors (MOSFETs) prior to any such permanent damage. The predictive model determines the onset of tolerance limits of EEMI on the I_{on}/I_{off} ratio of a MOSFET for a given technology node, using only the most fundamental device parameters – such as the threshold voltage and power supply. The developed model is successfully compared against measurement data from a device fabricated using 350nm standard CMOS process through TSMC. Based on the predictive model the tolerance of the EEMI injected power in a MOSFET reduces due to technology scaling, starting from 9.7dBm at 350nm, and down to -1.7dBm at 65nm technology node.

Keywords— Electromagnetic Interference, Predictive Modeling of Transistors, VLSI Systems

I. INTRODUCTION

Recently, a lot of attention is being paid to issues concerning microwave interference effects on modern-day infrastructure. Especially, when these signals can couple and upset the operation of modern electronic devices. This coupling can occur at the system's input or at the system's power supply. These issues, if not addressed as early as possible, can cause irreparable damage to the system. Now the need to rely on technology has grown so paramount, that issues such as system reliability and susceptibility need to be addressed. Tasks done by electronics with Integrated Circuits (ICs) have grown significantly larger, while their size and supply power have been significantly reduced. These changes have made ICs more vulnerable to even the small fluctuations, as seen in [1]. Even the smallest of these fluctuations can lead to timing issues which can propagate as a larger effect throughout a system.

While shrinkage in size and supply power were necessary for creating compact ICs for electronics, having such a small device size can lead to issues related to susceptibility from external noise and interference signals. One such external signal being considered is Extreme Electromagnetic Interference (EEMI). EEMI is a form of electromagnetic wave stimulus, which can unintentionally or intentionally couple with electronic circuits and cause it to malfunction or even be destroyed. Unintentional interference can be due to sources in proximity or other Radio Frequency (RF) signal emitted nearby. While, intentional interference is meant as malicious intent, to harm or disrupt the normal operation of a system. EEMI can further be analyzed as 'hard' and 'soft' upsets. Hard upset being the point at which the system is beyond repair. And soft upset is that just at the threshold of malfunction; given by [2] and [3]. Many techniques such as signal and ground-plane shielding, for printed circuit boards (PCB) [4], have been implemented to mitigate the effects of EEMI. However, electronic devices are still vulnerable at their power supplies and inputs to EEMI coupling. Many studies have been conducted to show EEMI effects on the normal operation of a CMOS inverter [5] and [6], and to show how EEMI coupling affects the noise margins and operation. While other logic gates shown in [7], such as NAND and NOR have also been shown to be susceptible to EEMI, due to issues with their timing analysis. Nevertheless, not a lot of studies have been done at a basic transistor level, such as an NMOS, which is a ubiquitous element in all digital electronics.

Through this paper, we propose a novel method of modeling EEMI in NMOS transistors, where we use the device I-V characteristics as a baseline to define the effect of EEMI, as illustrated in Figure 1. The standard I-V characteristic is a curve defining the current(I_{DS}) as a function of drain voltage (V_{DS}) for a single transistor. The EEMI source, which is modeled as an RF injection coupled on to the gate as shown in Figure 1, converts the I-V "curve" into a "band" due to the additional RF variation on the gate of the transistor. The width of the band depends on the injection swing, $\widehat{v}_n$. The gray I-V band in Figure 1, which is the basis for the predictive model in this paper, represents the range of uncertainty in the operating point of the NMOS transistor due to the RF injection at the gate. To predict the impact of EEMI on the quality of MOS transistor, we have developed a model for I_{on}/I_{off} ratio under RF injection, which is described in detail in Section 2. The experimental setup and the test chip prototype is presented in Section 3. The predictive model for scaling technologies is then developed in Section 4, which has been utilized to predict the tolerance of MOSFET for various technology nodes and results from such a predictive model is shown as a table in Section 5.

978-1-5386-7100-9/18 $31.00 © 2018 IEEE

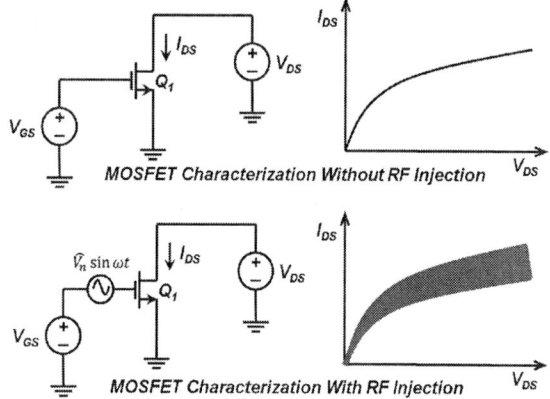

Fig. 1. MOSFET I-V characteristics with and without RF injection.

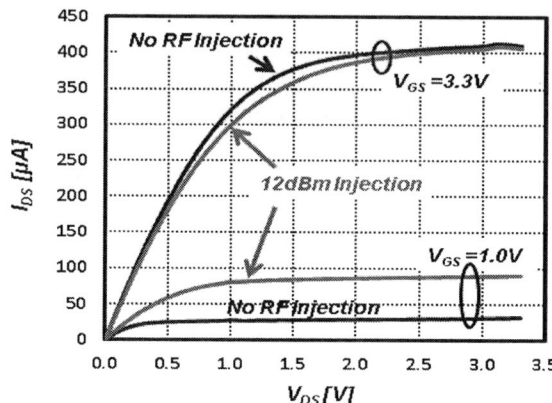

Fig. 2. The I-V characteristic of an NMOS measured with and without RF injection.

II. I_{ON} AND I_{OFF} UNDER THE INFLUENCE OF GATE INJECTION

The I-V band diagram in Figure 1 requires statistical analysis to be used in any applications. However, in this paper, without loss of generality, we use the *mid-point average* of the I-V band to represent the impacted I-V curve under the influence of EEMI injection. The main reasoning behind this generality is to address the that the I-V "band" is in-fact a range of uncertain values in the operating point of the transistor. For example, if the device is at the onset of saturation region of operation and at the edge of linear region of operation; then due to EEMI injection I_{DS} can fall in saturation region at the positive swing of the RF signal and can also be in linear region at the negative swing of the RF signal for the same V_{DS}.

As an example, Figure 2 illustrates the I-V characteristics of an NMOS measured experimentally with and without RF injection. In this figure, the I-V bands with RF injections are represented by its mid-point average curves (in red). In this experiment, the NMOS with W=700nm and L=350nm is fabricated in a 0.35μm standard CMOS process technology from TSMC.

As given by Figure 2, the impact of RF injection in the I-V curves varies based on the applied V_{GS}. A larger V_{GS} puts the transistor in a higher conductive mode, where its average I_{DS} current due to RF injection alters slightly from its original state. However, under smaller V_{GS} the MOSFET can get into the cut-off region of operation due to RF injection, which can significantly alter its mid-point average I_{DS} as illustrated in Figure 2. This is an important observation when we define I_{on} and I_{off} under the influence of RF injection.

Figure 3 illustrates the average I_{on} and I_{off} as a function of injection power in dBm, where I_{on} is measured when V_{GS}=3.3V and I_{off} is measured when V_{GS}=0V, both under V_{DS}=3.3V.

As shown in Figure 3, the average I_{on} doesn't change under RF injection of up to 12dBm and beyond. This is because the transistor stays in saturation during both positive and

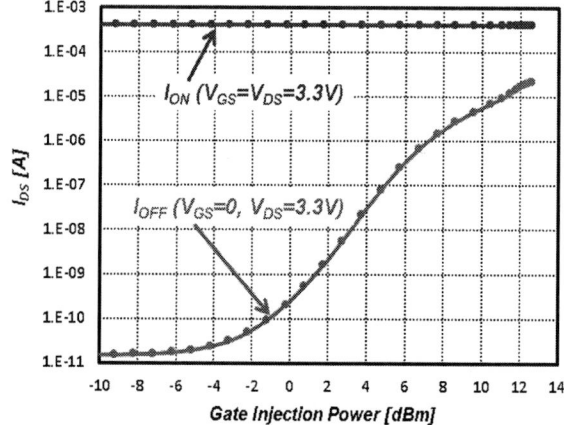

Fig. 3. The average I_{on} and I_{off} as a function of RF injection power in dBm.

negative peaks of the RF signal, causing the current to swing in both directions almost symmetrically (depending on the injection power).

However, the average I_{off} drastically changes due to RF injection. This is because the transistor is in OFF state and can briefly conduct during the positive peaks of the RF signal. This brief conduction can be thought of as a leakage current when the transistor is supposed to be in OFF state.

III. TEST CHIP PROTOTYPE AND EXPERIMENTAL

In order to understand the impact of EEMI on MOS transistors and compare the experimental data with the developed predictive model, a test chip was designed and fabricated using 0.35μm TSMC's standard CMOS technology through MOSIS. The test chip includes primitive NMOS and PMOS devices as well as logic gates such as inverters and ring oscillators. The experimental results in this paper are based on the measurements performed only on the NMOS devices on the test chips.

Fig. 4. The test chip and experimental setup mounted on a probe station.

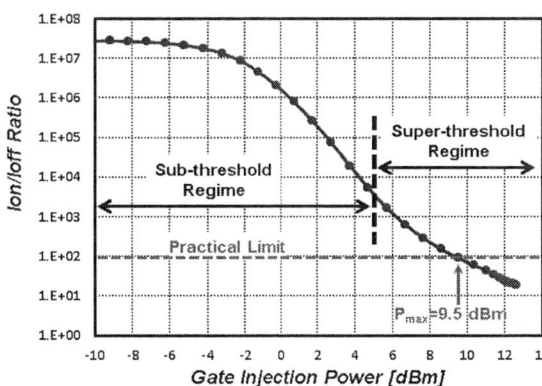

Fig. 5. The measured I_{on}/I_{off} ratio as a function of gate injection power in dBm.

The test setup includes a semiconductor analyzer and a probing station for capturing the I-V characteristics of the NMOS device under the presence of large signal RF stimulus applied using an external amplifier-based microwave source. The prototype test chip attached through 3 RF probes on the station is shown in Figure 4. The frequency of the injected RF signal ranges from 10MHz to 4GHz. However, since the focus of this paper is to develop a predictive model for low-frequency applications, only the measured data at 10MHz is used for this paper.

The entire system is controlled by a LabView program, where the voltages and RF injection power are automatically swept while device parameters are measured, specifically the I_{DS} to observe the impact of EEMI.

IV. PREDICTIVE MODEL FOR I_{ON}/I_{OFF} RATIO UNDER EEMI

A good industry metric to define a quality of a transistor in digital applications is the I_{on}/I_{off} ratio [8]. The I_{on} current is calculated using the saturation current for the transistor and the I_{off} current is calculated using the subthreshold current. Since, the ratio is a metric of performance, it is more desirable to have a device with higher I_{on}/I_{off} ratio. We use this metric in a similar manner to describe how the quality of a transistor degrades due to EEMI. The measured I_{on}/I_{off} ratio is shown in Figure 5 using the experimental data for I_{on} and I_{off} shown in Figure 3.

The baseline for I_{on}/I_{off} is 2.7×10^7 which is typical for transistors in 350nm technology. As shown in Figure 5, the I_{on}/I_{off} ratio decays as the RF injection power increases. It has been suggested that once the transistor becomes unusable when I_{on}/I_{off} ratio becomes around 100 [9], which is set as a practical limit in our analysis. As a result, based on measured data given in Figure 5, the maximum tolerance for injected power is 9.5dBm.

In this section, we develop a closed form model to predict the I_{on}/I_{off} ratio as a function of injected EEMI power.

A. Assumptions

To simplify our analysis, we make the following reasonable assumptions:

- Although the injection can be applied to any terminals of a MOSFET, we assume the injection is through the gate, which is the input and the most sensitive terminal of the device.

- Furthermore, in this study, we assume that the injection frequency is low, where the impact of parasitic device capacitances is negligible. The high-frequency modeling can be developed a lot quicker, than normal, once the I_{on}/I_{off} ratio impact is well understood at low frequencies.

- Finally, we assume that the I_{on}/I_{off} ratio reaches the practical limit of 100 when the device operates within the super-threshold regime. This can be clearly confirmed by observing Figure 5, where only a 5dBm injection at the gate is enough to push the gate-source voltage above the threshold voltage.

In general, the RF injection may be at higher frequencies or it may be applied to the drain terminal instead of the gate, which will be violating these assumptions and therefore would require different models to be developed to demonstrate the impact of EEMI. However, the model presented in this paper lays a foundation to develop more advanced models that includes frequency dependencies, multi-point injection, and nonlinear device characteristics, which are our work in progress.

B. Derivation

In this analysis, we consider that the mid-point average current, that is defined by:

$$I_{av} = \frac{I_+ + I_-}{2} \qquad (1)$$

where I+ and I- are the currents at the positive and negative peaks, respectively.

For the modeling of I_{on}, the device is in saturation at both positive and negative peaks. Therefore,

$$I_{on+} = \frac{K'_n}{2}\left(\frac{W}{L}\right)\left(V_{DD} + \widehat{V_n} - V_T\right)^2 \qquad (2)$$

$$I_{on-} = \frac{K'_n}{2}\left(\frac{W}{L}\right)\left(V_{DD} - \widehat{V_n} - V_T\right)^2 \qquad (3)$$

978-1-5386-7100-9/18 $31.00 © 2018 IEEE   599

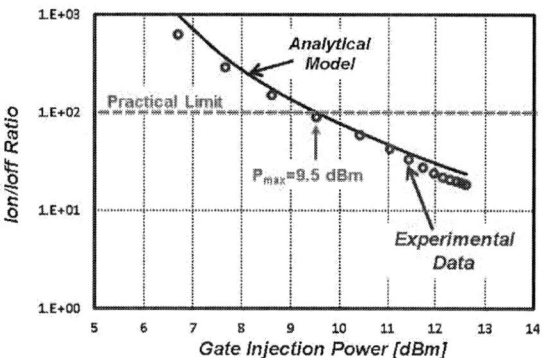

Fig. 6. Comparison between the analytical model and the measured data for I_{on}/I_{off} versus RF injection power.

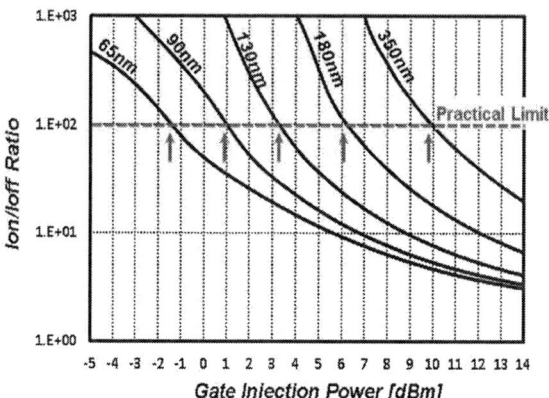

Fig. 7. The practical limit of operation for 65nm, 90nm, 130nm, 180nm and 350nm technologies under RF injection based on the model.

where V_{DD} is the power supply voltage, $\widehat{V_n}$ is the peak RF injection voltage at the gate, V_T is the threshold voltage of the device. Using (1), the average I_{on} can be simplified as

$$I_{on(avg)} = \frac{K'_n}{2}\left(\frac{W}{L}\right)\left[(V_{DD} - V_T)^2 + \widehat{V_n}^2\right] \quad (4)$$

For the modeling of I_{off}, the device should be in subthreshold or cutoff region at negative peak, but it is instead in saturation positive peaks if $\widehat{V_n}$ is larger than V_T. Therefore,

$$I_{off+} = \frac{K'_n}{2}\left(\frac{W}{L}\right)\left(\widehat{V_n} - V_T\right)^2 \quad (5)$$

$$I_{off-} \approx 0 \quad (6)$$

Using (1), the average I_{off} can be simplified as

$$I_{off(avg)} = \frac{K'_n}{4}\left(\frac{W}{L}\right)\left(\widehat{V_n} - V_T\right)^2 \quad (7)$$

From (4) and (7) the predictive model for I_{on}/I_{off} ratio can be simplified as

$$\frac{I_{on}}{I_{off}} = 2\frac{(V_{DD}-V_T)^2+\widehat{V_n}^2}{(\widehat{V_n}-V_T)^2}, \quad \text{when } \widehat{V_n} > V_T \quad (8)$$

Equation (8) is a simple but a powerful model that can be used to predict the impact of EEMI on the quality of a transistor using only the primitive technology parameters, which are V_{DD}, V_T, and $\widehat{V_n}$.

C. Comparison with Measured Data

In order to validate the accuracy of the developed predictive model, the measured data is compared with the analytical model in (8) considering V_{DD}=3.3V, and V_T=0.58V based on the data from the 350nm technology specs.

The results for when $\widehat{V_n} > V_T$, which is when the injection power of larger than 5dBm, are shown in Figure 6. The experimental data has a clear match with the novel simple

TABLE I.   PREDICTION OF MAXIMUM EEMI TOLERANCE FOR GATE INJECTION FROM 350NM TO 65NM TECHNOLOGY NODES.

Technology Node	Paramter used in (8)		Result from solving (8) when I_{on}/I_{off} ratio= 100	
	V_{DD} [v]	V_T [v]	Peak noise tolerance [v]	Maximum gate injection power [dBm]
350nm	3.3	0.58	0.968	9.7
180nm	1.8	0.44	0.650	6.3
130nm	1.2	0.32	0.459	3.2
90nm	1	0.23	0.348	0.8
65nm	1	0.13	0.261	-1.7

predictive model developed in (8). Therefore, the maximum injection power that pushes the transistor into the practical limit can be accurately predicted by using the analytical model for I_{on}/I_{off} shown in (8).

V. PREDICTION OF EEMI TOLERANCE WITH SCALING

The overall objective of this paper is to develop a predictive model for EEMI impact for scaling technologies. In addition, the analytical model described by (8) can be used to predict the onsets of EEMI tolerance of MOSFET devices for any given technology nodes – provided the necessary V_{DD} and V_T. For example, we can use the expected V_{DD} and V_T of NMOS transistors from 350nm down to 65nm technology nodes given in Table I and using (8), to predict the peak noise tolerance or maximum power that can be injected to the gate, which can be done by setting I_{on}/I_{off} ratio=100. The results of Table I is shown in Figure 7, where I_{on}/I_{off} ratios is a function of injected power for the given technology nodes Considering that the we have set practical limit for I_{on}/I_{off} at about 100, the maximum tolerance for EEMI in MOSFET devices reduces considerably due to technology scaling as demonstrated in Figure 7. Based

TABLE II.   COMPARISION BETWEEN THE MEASURMENT DATA AND
ANALYTICAL DATA AND RESULTS OF PERCENTAGE OF ERROR.

Injection Power [dBm]	Ion [A]	Ioff [A]	Ion/Ioff (measured)	Ion/Iof (model)	%Error
6.71	4.11E-04	6.62E-07	620.4	970.6	36.1%
7.67	4.10E-04	1.45E-06	282.8	358.5	21.1%
8.62	4.09E-04	2.73E-06	149.7	175.0	14.4%
9.53	4.09E-04	4.57E-06	89.6	99.9	10.3%
10.41	4.08E-04	6.93E-06	58.9	63.0	6.5%
11.04	4.08E-04	9.57E-06	42.6	46.7	8.7%
11.45	4.07E-04	1.21E-05	33.6	38.9	13.8%
11.75	4.07E-04	1.47E-05	27.7	34.2	19.0%
11.98	4.07E-04	1.69E-05	24.1	31.0	22.2%
12.16	4.07E-04	1.85E-05	21.9	28.8	23.7%
12.31	4.06E-04	1.97E-05	20.6	27.1	23.8%
12.44	4.06E-04	2.06E-05	19.7	25.7	23.3%
12.54	4.06E-04	2.14E-05	19.0	24.7	23.0%
12.62	4.06E-04	2.20E-05	18.4	23.9	22.9%

on the developed predictive model for I_{on}/I_{off} ratio, the maximum EEMI tolerance on the gate terminal can be accurately quantified for each technology generation in volts and dBm as shown in the last two columns in Table I, these values were solely derived using (8). According to our predictive model, an NMOS in 350nm technology node can tolerate up to 0.968V or 9.7dBm of injected EEMI, whereas the same transistor in 65nm technology node can tolerate up to only 0.261V or -1.7dBm of injected EEMI. The derived results are based only off the developed model and tested for accuracy on TSMC 350nm technology only. Further test chips for different technologies are being fabricated to test the entire array of result derived from this predictive model.

The analytical model and the measurement data have clear match. Table II presents the measured data and the percentage of error to evaluate the accuracy of the predictive model developed in this paper. The average percentage of error as shown in Table II is about 19.2%, which is satisfactory for a predictive model that only requires supply voltage (V_{DD}) and threshold voltage (V_T) as input.

VI. CONCLUSIONS

A predictive model is successfully developed to characterize the impact of EEMI on a MOSFET transistor prior to any permanent damage. The predictive model can be used to accurately quantify the impact of EEMI on I_{on}/I_{off} ratio of a MOSFET for a given technology node using only fundamental device parameters, including power supply voltage and the device threshold voltage. Furthermore, the developed analytical model is successfully compared against measurement data from a device fabricated using 350nm standard CMOS process through TSMC. The compassion shows good consistency between the analytical model and the measurement data with only 19.2% average percentage of error. Based on the predictive model the tolerance to EEMI injected power in a MOSFET reduces by technology scaling, starting from 9.7dBm at 350nm down to -1.7dBm at 65nm technology node.

ACKNOWLEDGMENT

The authors would like to acknowledge the United States Air Force Research Laboratory (AFRL) and Air Force Office of Scientific Research (AFSOR) for their support with AFOSR/AFRL COE Grant #FA9550-15-1-0171 and by DURIP Grant #FA9550-15-1-0379. The authors would also like to acknowledge Zahra Abedi for her help in improving the lab set-up to make data gathering process more efficient.

REFERENCES

[1] Gao, Xu, et al. "Modeling timing variations in digital logic circuits due to electrical fast transients." Electromagnetic Compatibility (EMC), 2013 IEEE International Symposium on. IEEE, 2013.

[2] Hwang, Sun-Mook, Joo-Il Hong, and Chang-Su Huh. "Characterization of the susceptibility of integrated circuits with induction caused by high power microwaves." Progress in Electromagnetics Research 81 (2008).

[3] Kim, Kyechong, Agis A. Iliadis, and Victor L. Granatstein. "Effects of microwave interference on the operational parameters of n-channel enhancement mode MOSFET devices in CMOS integrated circuits." Solid-State Electronics 48.10 (2004): 1795-1799.

[4] Ferchau, Joerg U., Kenneth A. Kotyuk, and Randall J. Diaz. "Electronic assembly with improved grounding and EMI shielding." U.S. Patent No. 5,311,408. 1994.

[5] Hwang, Sun-Mook, et al. "Delay time and power supply current characteristics of CMOS inverter broken by intentional high power microwave." Microwave Conference, Asia-Pacific. IEEE, 2007.

[6] Kim, Kyechong, and Agis A. Iliadis. "Critical upsets of CMOS inverters in static operation due to high-power microwave interference." IEEE transactions on electromagnetic compatibility 49.4 (2007): 876-885.

[7] Gao, Xu, et al. "Modeling static delay variations in push–pull CMOS digital logic circuits due to electrical disturbances in the power supply." IEEE Transactions on Electromagnetic Compatibility (2015): 1179-1187.

[8] J. Chen, L. T. Clark, and Y. Cao, "Ultra-low voltage circuit design in the presence of variations," IEEE Circuits Devices Mag., pp. 12–20, Nov./Dec. 2005.

[9] N. Verma, J. Kwong, and A. Chandrakasan, "Nanometer MOSFET variation in minimum energy subthreshold circuits," Electron Devices, IEEE Transactions on, vol. 55, no. 1, pp. 163 –174, Jan. 2008. Eason, B. Noble, and I. N. Sneddon, "On certain integrals of Lipschitz-Hankel type involving products of Bessel functions," Phil. Trans. Roy. Soc. London, vol. A247, pp. 529–551, April 1955. (references)

[10] Cordova, David, et al. "EMI resisting MOSFET-only voltage reference based on ZTC condition." Analog Integrated Circuits and Signal Processing 89.1 (2016): 45-59.

[11] Liu, Yang, et al. "Damage effect and mechanism of the GaAs high electron mobility transistor induced by high power microwave." Chinese Physics B 25.4 (2016): 048504.

[12] Powell T., Sule N., Hemmady S.,& Zarkesh-Ha P. , "Predictive model for extreme electromagnetic compatibility on CMOS inverters" EMC Europe (2018), in press.

Enhancing Observability for Post-Silicon Debug with On-Chip Communication Monitors

Yuting Cao, Hernan Palombo, Hao Zheng
CSE, U of South Florida, Tampa, FL
{cao2, hpalombo, haozheng}@mail.usf.edu

Sandip Ray
ECE, U of Florida
Gainesville, FL
sandip@ece.ufl.edu

ABSTRACT

Reconstructing system-level behavior from silicon traces is critical in post-silicon debug of System-on-Chip (SoC) designs. However, limited observability makes the reconstruction process complex and inaccurate, thus offering little help for SoC debug. This paper presents an on-chip monitoring infrastructure aiming to enhance observability by detecting communication transactions from low level signal events. The detected transactions are output on-the-fly for off-chip system-level behavior reconstruction. Experiments show that the proposed monitoring infrastructure enables accurate observations on communication transactions over long periods of time, thus leading to accurate reconstruction of system level behavior, with low area overhead.

1. INTRODUCTION

Post-silicon debug is a critical component of the design validation life-cycle for modern microprocessors and system-on-chip (SoC) designs. Unfortunately, it is also highly complex, performed under aggressive schedules and accounting for more than 50% of the overall design validation cost [1].

An SoC design is often composed of a large number of pre-designed hardware or software IP blocks that coordinate through complex protocols to implement system-level behavior [2]. As SoCs integrate more IPs, the interactions among the IPs are increasingly more complex. Moreover, modern interconnects are highly concurrent, allowing multiple system-level protocols to be processed simultaneously for scalability and performance. Many difficult errors often involve subtle and unexpected interleavings of these protocols that have a very low probability of being exercised in simulation [3] On the other hand, observability limitations allow only a small number of participating signals to be actually traced during silicon execution. It is non-trivial during post-silicon debug to identify all participating protocols and pinpoint the interleavings that result in an observed trace.

In [4], an off-chip analysis approach was proposed with an objective to reconstruct design internal behavior *wrt* given system-level flow specifications. Since *when* and *where* an error can happen are not known *a priori*, in our approach silicon traces on a small number of participating hardware signals are streamed off-chip on-the-fly in order to facilitate more efficient debug in space and time. Once a silicon trace is obtained, it is processed by the off-chip analysis in two consecutive steps: (1) trace abstraction, which maps a silicon trace into a sequence of *flow events*, abstract architectural constructs representing communication transactions,

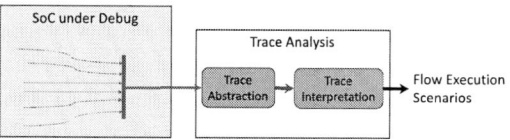

Figure 1: The system flow guided post silicon trace analysis framework for SoC debug in our previous work. The red lines across represents raw signal traces while the blue line represents the flow trace after the trace abstraction step.

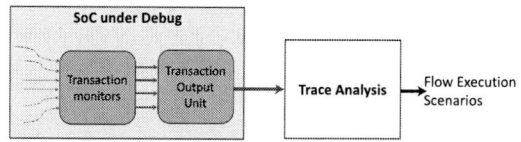

Figure 2: The new framework where the SoC under debug is instrumented with an on-chip monitoring infrastructure and the off-chip trace analysis no longer needs the trace abstraction component.

and (2) trace interpretation, which infers possible flow execution scenarios that are compliant with the abstracted event sequence. That approach is shown in Figure 1. Its main drawback is that the reconstruction process is highly complex and inaccurate due to the insufficient information that can be derived from the silicon traces observed on a very small number of trace signals. The inferred results are often ambiguous, and provide limited help for debug.

Contributions This paper addresses the above drawback by proposing an on-chip communication monitoring infrastructure to enhance debug observability. It consists of transaction monitors attached to the communication links of a SoC design, and a transaction output unit that manages transporting detected transactions by the monitors on-the-fly for off-chip analysis. The new trace analysis approach is shown in Figure 2. Instead of streaming observed signals off-chip every clock cycle, the monitors offload transactions only when they happen. Since transactions typically happen sporadically over time, it allows transactions from different communication links to be interleaved while being outputted for off-chip analysis. On the same limited number of trace signals, every event now can capture much more accurate

978-1-5386-7100-9/18 $31.00 © 2018 IEEE

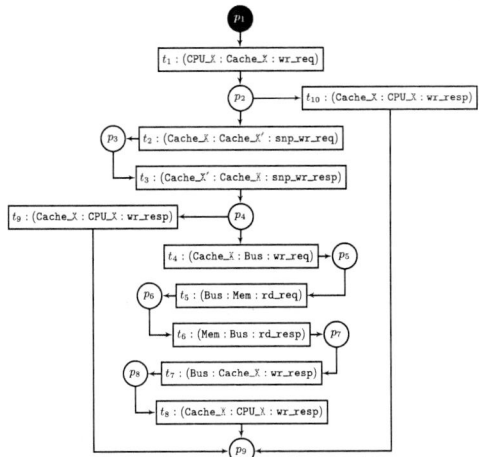

Figure 3: LPN formalization of a CPU write protocol.

information on internal communications during system execution, it allows the trace analysis method to reconstruct system-level behavior more accurately and efficiently.

2. PREVIOUS WORK

This section briefly reviews our previous work on the system flow guided post-silicon trace analysis for SoC debug [4].

In our trace analysis method, system-level protocols or *system flows* are formalized using Labeled Petri-nets (LPNs). Figure 3 shows a memory write protocol initiated from a CPU CPU_X in LPN where $X \in \{0, 1\}$ and $X' = 1 - X$. An LPN is a tuple (P, T, E, L, s_0) where P is a finite set of *places*, T is a finite set of *transitions*, E is a finite set of *events*, and $L : T \to E$ is a labeling function that maps each transition $t \in T$ to an event $e \in E$. In a system flow specification, each LPN transition is labeled with an event (src, dest, cmd) where cmd is a command sent from a source component src to a destination component dest. For each transition $t \in T$, its preset, denoted as $\bullet t \subseteq P$, is the set of places connected to t, and its postset, denoted as $t\bullet \subseteq P$, is the set of places that t is connected to. A state $s \subseteq P$ of a LPN is a subset of places marked with tokens. There are two special states associated with each LPN; $s_0 \subseteq P$ which is the set of initially marked places, also referred to as the *initial state*, and the end state s_{end} which is the set of places not going to any transitions.

A transition t can be executed in a state s if $\bullet t \subseteq s$. Executing t causes the labeled event to be emitted, and leads to a new state $s' = (s - \bullet t) \cup t\bullet$. Therefore, executing an LPN leads to a sequence of events. Execution of a LPN completes if its s_{end} is reached. For example, in Figure 3, t_1 can be executed in $s_0 = \{p_1\}$. Event (CPU_X : Cache_X : wr_req) is emitted after t_1 is executed, and the LPN state becomes $\{p_2\}$. The end state is $s_{end} = \{p_9\}$. A flow specification may also contain multiple branches describing different ways a system can execute such flow. For example, the flow shown in Figure 3 has three branches covering the cases where the cache (snoop) operation is hit or miss.

The objective of our silicon trace analysis is to infer possi-

ble compliant *flow execution scenarios* from a partially observed trace *wrt* given system-level flow specifications **F**. A flow execution scenario can be viewed as *a state of system execution abstracted wrt system flows*, and it is defined as

$$\{(F_{i,j}, s_{i,j}, start_{i,j}, end_{i,j}) \mid F_i \in \mathbf{F}\}$$

where $F_{i,j}$ is the jth instance of flow F_i, $start_{i,j}$ and $end_{i,j}$ are two indices representing relative time when $F_{i,j}$ is initiated and completed. $s_{i,j}$ is used by the trace analysis to keep track of the current state of F_i, j when an observed trace is interpreted. The ordering relations can be derived by comparing their *start* and *end* indices. For example, for two flow instances in an execution scenario, $(F_{u,v}, s_{u,v}, start_{u,v}, end_{u,v})$ and $(F_{x,y}, s_{x,y}, start_{x,y}, end_{x,y})$, $F_{u,v}$ is initiated before $F_{x,y}$ if $start_{u,v} < start_{x,y}$, or $F_{x,y}$ is initiated after $F_{u,v}$ is completed if $end_{u,v} < start_{x,y}$. The ordering relations can provide more accurate information for understanding system execution under limited observability.

To illustrate the basic idea, consider the system flow in Figure 3, which we call F_1. Suppose that the following flow event trace is abstracted from an observed silicon trace by executing a design that implements F_1.

1 $\langle$(CPU_0 : Cache_0 : wr_req)$\rangle$
2 $\langle$(CPU_0 : Cache_0 : wr_req), (CPU_1 : Cache_1 : wr_req)$\rangle$
3 $\langle$(Cache_0 : CPU_0 : wr_resp),
    (Cache_1 : Cache_0 : snp_wr_req)$\rangle$

...

The numbers on the left in the above trace represent the relative time when the corresponding events are observed. Possible flow execution scenarios that are derived by our approach are shown below.

$$\{(F_{0,1}, \{p_9\}, 1, 3), (F_{0,2}, \{p_2\}, 2, -), (F_{1,1}, \{p_3\}, 2, -)\},$$
$$\{(F_{0,1}, \{p_2\}, 1, -), (F_{0,2}, \{p_9\}, 2, 3), (F_{1,1}, \{p_3\}, 2, -)\}$$

The above flow execution scenarios indicate two possible system states defined over the execution states of flow F_1 after observing the first three steps of events as shown above. This ambiguity is mainly due to lack of necessary information in the observed events due to the limited observability in post-silicon debug.

The limited observability problem leads to two unpleasant consequences. The large number of execution scenarios typically derived during the trace analysis would take longer runtime and large amount of memory to process and to store, thus making it less efficient. This is referred to as the *complexity* problem of the trace analysis. After the analysis is done, a large number of derived execution scenarios make it difficult to understand the analysis results, thus being less helpful for debugging. Obviously, a single flow execution scenario derived at the end of the trace analysis provides much more precise information for debug than ten candidate flow execution scenarios. This is referred to as the *accuracy* problem of the trace analysis. Those problems are addressed in the next section that describes an on-chip monitoring infrastructure.

3. MONITORING INFRASTRUCTURE

In order to enhance observability and facilitate the trace analysis, this section describes a communication monitoring infrastructure. It consists of monitors attached to communication links to detect communication transactions, and a

978-1-5386-7100-9/18 $31.00 © 2018 IEEE

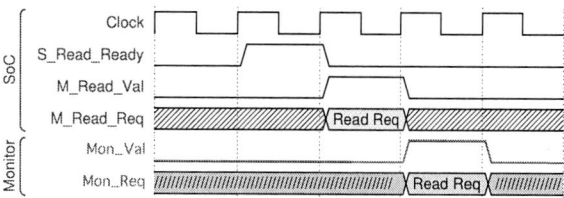

Figure 4: An example of the AXI read transaction on a communication link, and the output of a monitor attached to that link.

transaction output component to transport detected transactions off-chip for the trace analysis.

3.1 Communication Transaction Monitoring

A communication link consists of signals that can transfer some transactions, one at a time. Instead of routing a limited number of design signals to the chip interface, signals of a communication link are connected to the inputs of its attached monitor. During system execution, a monitor reads signal events occurring on its inputs, and outputs an encoding if a transaction is detected. A *signal event* denotes an assignment to a set of design signals, while a communication transaction is a transfer of a body of information from a source to a destination following a communication protocol. *Flow events* are an abstract construct used in flow specifications, and are typically implemented by transactions.

Figure 4 shows an example of a master and a slave communicating with the AXI read protocol [5]. The slave needs to assert `S_Read_Ready` before the master asserts `M_Read_Val` to initiate a transaction. A transaction of a AXI read request is detected by its monitor with a pulse on `Mon_Val` when `M_Read_Val` is asserted. The monitor can also selectively encode some information transferred as part of the detected transaction. The basic idea of the above monitor can be naturally extended to different protocols such as the AXI write request and response.

The biggest benefit from those monitors is the compression of a potentially long sequence of signal events into a single cycle communication transaction. Obviously, transporting this single cycle transaction demands much less bandwidth of the trace port than transporting low level signal events implementing such transaction. Another advantage is that these monitors can implement protocol checking capability so that low level protocol errors can be detected timely and right on the spot.

3.2 Transaction Output

The detected transactions can be stored in the on-chip trace buffer, and offloaded from the chip at the end of system execution. However, the on-chip trace buffers can only store limited transactions due to the restriction on their capacities. As explained in Section 1, when and where an error can happen are not know a priori, therefore, these limited transactions stored in the trace buffer may offer only limited debugability. This section describes a transaction output design that can output transactions via trace port on-the-fly, thus enabling system internal execution over an much extended period to be observed for off-chip analysis.

Parallel Output The first approach is parallel where mul-

tiple links are traced simultaneously. Since the number of available trace signals are fixed, there is a trade-off between the number of links that can be traced simultaneously and the amount of information encoded for detected transactions on each link. More information encoded for transactions demands more trace signals, thus reducing the number of links that can be traced simultaneously. For example, suppose that a total of 100 trace signals are available, and there are 20 communication links to observe. If each transaction generated by the monitors for those links is encoded with 30 bits on average, then only 3 links can be traced simultaneously. On the other hand, if we wish to observe more communication links simultaneously, the number of bits for encoding transactions must be reduced, thus limiting the amount of information represented by transactions. More discussion on transaction encodings is given in Section 3.3.

Interleaved Output Since detected transactions by monitors are distributed over time relatively sparsely as illustrated in the last sub-section, an alternative approach is to interleave transactions detected on different links, and transport them off-chip serially. In this approach, monitors are connected to a transaction output unit like the one used in ARM CoreSight [6], which is shown in Figure 5. The transactions from monitors are routed through this output unit, merged into a sequence, and eventually output through the trace port. The biggest advantage of this approach is the very high observability in terms of the larger number of communication links to be traced and the higher amount of detailed information that can be encoded for transactions.

On the other hand, an issue with the interleaved approach is that the rate of transactions detected by monitors can exceed the peak bandwidth of the trace port from time to time. If that happens, some detected transactions cannot be transported off-chip. It can be viewed as another form of limited observability. Therefore, it would be desirable to reduce the number of transactions that have to be discarded.

The above issue can be addressed by using FIFOs as shown in Figure 5. Those FIFOs can buffer detected transactions temporarily if they cannot be outputted right away. One FIFO is connected to the output of each monitor. On every cycle, the outputs of all monitors carrying detected transactions are stored into the corresponding FIFOs. At the same time, the transaction validity information of all monitors is collected into `Tr_Val`, and stored into a special FIFO `Tr_Val_fifo`. This information is used to control how to output buffered transactions. The width of `Tr_Val` is equal to the number of monitors. `Tr_Val[i]=1` indicates that transaction output from monitor M_i is valid. Otherwise, no valid output is from M_i. All the transaction FIFOs are connected to a N-to-1 selector where one transaction FIFO is routed to the trace port.

The control logic generates values for `sel` to control the selector based on the information stored in `Tr_Val_fifo`. In the initial state, it asserts `Read_Tr_Val` to read the head of `Tr_Val_fifo` into `Tr_Status`. If it contains some bits of 1, the control unit first determines the smallest index i such that `Tr_Status[i]=1`. This can be done by a priority encoder. Next, the transaction FIFO for monitor M_i is connected to the trace port, and the transaction at its head is outputted. Then, `Tr_Status[i]` is reset to 0, and the control logic repeats the above step if there is a larger index i such that `Tr_Status[i]=1`. Otherwise, it returns to the ini-

978-1-5386-7100-9/18 $31.00 © 2018 IEEE

Cycle	M_2	M_1	M_0	Tr_Val	Read_Tr_Val	Tr_Status	Sel	Selector Output
1	✓	✗	✓	101	1	000	X	—
2	✗	✓	✗	010	0	101	0	M_0
3	✗	✗	✗	not stored	0	100	2	M_2
4	✗	✗	✗	not stored	1	010	1	M_1
5	✗	✗	✗	not stored	1	000	X	—

Table 1: Operations of the transaction output unit for 3 monitors over 5 cycles.

tial state and read the next data from `Tr_Val_fifo`. When `Tr_Status` is 0, `sel` is set to a special value X to disable the selector. The control flow diagram for the control unit is at the bottom right corner in Figure. 5.

Illustration We illustrate the operations of the transaction output unit with a simple example. Suppose that there are three monitors M_0, M_1 and M_2 connected to the transaction output unit. Their validity information is collected as a 3 bit `Tr_Val` and stored at the tail of `Tr_Status_fifo` on each cycle. Table 1 shows how the output unit outputs the transactions from these monitors off-chip. In the columns under M_0, M_1 and M_2, a ✓ means its generated transaction is valid, while ✗ indicates it is not valid.

In cycle 1, outputs from M_0 and M_2 are valid. They and `Tr_Val=101` are stored into their corresponding FIFOs, respectively. There is nothing to output, therefore `sel` is set to X. `Read_Tr_Val` is set to 1 to read `Tr_Val_fifo`. In cycle 2, similarly, outputs of all monitors and their validity are stored in the FIFOs. `Tr_Status` is used to compute `sel`, which is 0 in this case. Therefore, the selector is directed to output the transaction from M_0 captured in cycle 1, as indicated in column of Selector Output. `Tr_Status[0]` is reset to 0 before the next cycle. Signal `Read_Tr_Val` is reset to 0 when `Tr_Status` contains more than one bit of 1. In cycle 3, the transaction from M_2 captured in cycle 1 is outputted by the selector, and `Read_Tr_Val` is asserted to read next validity data from `Tr_Val_fifo` to prepare for the next cycle.

While the interleaved approach reduces the number of trace signals needed to output detected transactions, the transaction output unit can potentially introduce large area overhead. The area overhead are mainly due to the use of FIFOs to buffer transactions. Larger FIFOs can reduce the chance of transactions to be discarded, but increases area overhead. The optimal FIFO sizes are typically determined by the design of the interconnect network and the rates at which various blocks initiate system flows. However, detailed discussion of this topic is beyond the scope of this paper. In this paper, when overflow happens to any FIFOs, new incoming transactions are discarded as a simplification.

3.3 Transaction Encoding

Transactions can be encoded with information transferred over communication links at different levels of detail. In general, more bits are required to encode information at higher levels of detail. In the last section, two alternative transaction output approaches are discussed. Different representations of transactions are used for different approaches.

In the parallel output approach, multiple communication links are traced simultaneously, transactions representations are customized with respect to the specific protocols of different links. The representation below shows all the fields used in all transactions.

$$\langle \text{Val, Cmd, Tag, Sid, Addr} \rangle$$

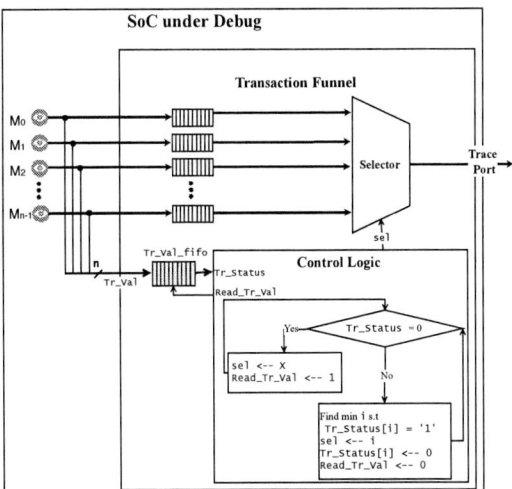

Figure 5: Transaction output unit.

The meanings of the message fields are defined below.

Val indicates the validity of a detected transaction.

Cmd carries operations to be performed by the target block. For the AXI protocol, there are separate links to support read/write request and response operations, this field is not needed.

Tag is used to identify the original sources of transactions from different blocks that go to the same destination, For example, in Figure 6, `Tag` is needed for transaction `wr_req` from Bus to Memory in response to `wr_req` from either CPU.

Sid is a unique number representing sequencing information associated with transactions initiated by a component that supports out-of-order execution.

Addr carries the memory address at the target block where `Cmd` is applied. If the observability limitation does not allow full address information to be encoded, it can be abstracted with two bits to represent three states: *same as previous one*, *sequential*, and *others*, as described in [7].

Note that not all fields are used to represent transactions of all links. The sizes of transactions on different links may be different. Additionally, monitors can be configured to include only some selected fields to meet debug needs while satisfying observability constraints.

In the interleaved output approach, the trace port is shared among all links. As a result, a standard format as shown below is used for all different transactions.

$$\langle \text{Val, MasterID, SlaveID, Cmd, Tag, Sid, Addr, Step} \rangle \tag{1}$$

where

- **MasterID** and **SlaveID** encode IDs of the sender and receiver of a transaction. The number of bits required for these two fields are determined by the number of masters and slaves in an SoC design.

- **Cmd** has a fixed number of bits for all transactions. Its width is determined by the largest number of transactions that any link can transfer.

978-1-5386-7100-9/18 $31.00 © 2018 IEEE

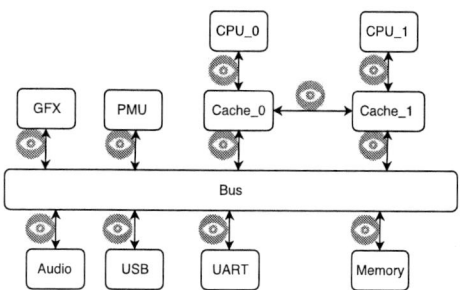

Figure 6: A SoC prototype where each communication link is attached with a monitor.

	Full	Parallel	Interleaved	SS1	SS2
# Bits	870	720	36	36	36
# scen (Max)	1	1	1	1	282k
# scen (Final)	1	1	1	1	-
#flows	500	500	500	100	200
Time	1.391	1.237	1.218	0.714	600
Mem	1.068	1.017	1.028	0.608	>5GB

Table 2: Runtime Results from analyzing traces obtained in different approaches. Runtime is in seconds and memory usage is in MB.

- Val, Tag, Sid, and Addr are the same.

- Step, which is only 1-bit, indicates the ordering of a transaction relative to its immediate predecessor. If this field is asserted, it indicates that the current transaction being outputted is detected after its immediate predecessor. Otherwise, the current transaction and its immediate predecessor are detected at the same time.

In the transaction output unit as shown in Figure 5, when a transaction is pulled out of its FIFO, it is converted to the above standard format before it enters the selector. Similarly, the fields and their sizes in the above standard format can be reduced to meet the observability constraint.

4. EXPERIMENTAL RESULTS

4.1 The Model

To evaluate the ideas and techniques presented in this paper, a non-trivial SoC design that implements sophisticated system flows is desired. However, to the best of our knowledge, we cannot find an open-source design that meets the above requirements. Therefore, we developed a multi-core SoC prototype, as shown in Figure 6, which implements a total of 16 system-level protocols including cache coherence, power management, downstream read/write protocols for CPUs, upstream read/write for the peripheral blocks, etc. All of them are abstracted from real industrial protocols. More details on some of those protocols can be found in [8].

This prototype is a cycle- and pin-accurate RTL model. The above system-level protocols are supported by interblock communication protocols based on the ARM AXI4-lite [5]. A total of 32 monitors are inserted into this model, one for each link between a device and the interconnect,

Since the proposed monitoring infrastructure is to support communication centric trace analysis, the focus of this model is the implementation of system flows for on-chip interconnect. The CPUs are treated as a test environment where software programs are simulated in VHDL to trigger various protocols. Therefore, there is no instruction cache as no instructions are involved when the CPUs are simulated. The peripheral blocks, GFX, PMU, Audio, etc, are also described as abstract models that generate events to initiate flows or to respond incoming requests.

4.2 Experiment Setup and Result Analysis

The model is simulated where five components, including CPUs, GFX, and three other peripheral blocks are programmed to randomly select a flow to initiate in every five

clock cycles. The contents of Cmd, Addr, and Data in each activated flow are set randomly. Additionally, CPUs can activate power management protocols randomly in time. Each of those five blocks activates a total of 100 flow instances during entire simulation. After simulation traces are obtained, the trace analysis approach in [4] is applied to extract the flow execution scenarios.

In the experiment, simulation is run five times with different tracing configurations. In the first run, "Full" observability is assumed. In the second run, the monitoring infrastructure is configured in the "Parallel" output mode where all links are assumed to be observable. In the third run, the monitoring infrastructure is configured in the "Interleaved" output mode. In this run where the transaction output unit along with monitors are used, all fields in (1) except Addr are used for transactions through the trace port. In our SoC model, Val and Time are 1 bit, MasterID and SlaveID are 5 bits, and Cmd, Tag, and Sid all are 8 bits. Therefore, encoding of transactions in the standard format as in (1) requires 36 bits. The depth of all FIFOs in the output unit is set to 16. The last two runs generate results from using tracing without the communication monitors.

The results from all runs are shown in Table 2. In the table, row 2 and 3 show the *peak* count of flow execution scenarios encountered during the reconstruction process that is used to measure the complexity and the *final* count of flow execution scenarios derived at the end of the reconstruction process that is used to measure the accuracy, respectively. Row 5 shows the maximal number of flow instances activated by various components and identified by the trace analysis. The last two rows show the runtime and memory usages by the trace analysis for different runs. The results from analyzing the simulation traces of those five runs are shown in column 2 − 6, respectively, in Table 2.

By comparing results in columns 2 − 4 in the table, it can be seen that using the monitoring infrastructure allows the same analysis result about system internal execution to be derived as that derived with the full observability. More importantly, that is achieved by requiring *significantly reduced number of trace signals*. The trace analysis with the monitoring infrastructure achieves the same complexity and accuracy as those achieved with the full observability.

The last two columns show the trace analysis results without using the monitoring infrastructure. We assume that 36 signals are available for tracing as in the third run. Due to this restriction, we can select only a small number of links where the signal events can be observed accurately. The re-

978-1-5386-7100-9/18 $31.00 © 2018 IEEE

	Cells	LUTs	FFs	Muxs	BRAM
Original	59154	24395	25962	3125	1
Parallel	+1283	+8	+1251	+15	+0
Interleaved	+232	+92	+126	-6	+32

Table 3: Area overhead of the monitoring infrastructure.

sults from this run are shown in Column 5. Even though only one flow execution scenario is derived at the end, the limited number of signal events selected for observation allow much less number of related flow instances to be derived than what can be derived when the monitoring infrastructure is used. When we try to observe more links, we are forced to allocate less trace signals for each event on those links. This causes ambiguity to the interpretations of the observed signal events. As a result, an excessively large number of potential flow execution scenarios are derived as shown in Column 6. After 10 minutes, the trace analysis has to be terminated due to memory usage explosion. These results show that under the limited observability the complexity and accuracy of the trace analysis would suffer significantly if the monitoring infrastructure is not used.

4.3 Area overhead

We measure the hardware area overhead of the monitoring infrastructure by synthesizing the SoC model to the Xilinx Zynq FPGA xc7z020ckg484-1 using Vivado 2017.2. The synthesis results are shown in Table 3. The area overhead is measured by the FPGA resources used including LUTs, FFs, block RAMs (BRAMs), etc.

The rows for Parallel and Interleaved show the additional resources required to implement monitors with or without transaction output unit to the resources used on the previous row. For example, the parallel approach uses additional 1283 cells to implement all 32 monitors, and the interleaved approach requires extra 283 cells to implement the transaction output unit. From the table, other than the big jump in BRAM usage, demand on logic resource is small to implement the monitoring infrastructure. The BRAMs are used to implement the FIFOs in Figure 5. Since the size of the BRAMs is fixed, each FIFO only uses a small capacity of a BRAM. In practice, SoCs are often embedded trace buffers, which can be used for those FIFOs.

5. RELATED WORK

Ciordas and e.t. in [9] proposed the first monitoring service to provide run-time observability of NoC behavior and supporting system-level debugging. However, it offers no explanation on how the detected events are outputted for off-chip analysis. Gharehbaghi and Fujita in [7] [10] [11] introduce an on-chip instrumentation that allows transaction level message abstraction using formal specifications of the bus communication protocols. The authors propose an innovative encoding technique that can reduce the address bits to two bits containing three different states to indicate relationships between two consecutive transactions. Despite its low area overhead, this methods suffers from inability of detecting implementation errors that are are not observed. Moreover, this method lacks the ability to check the overall system communication protocols as it only focus on communication interfaces' protocol. [12] proposes another on-chip

instrumentation BiPeD that learns communication interface' protocol during pre-silicon stage, and reconfigure its detection hardware to check the learned protocols during the post-silicon validation. While BiPed is effective towards detecting and locating many hardware bugs, the circular buffer implemented for each communication interface introduces large area overhead. Another transaction monitor is proposed by [13] where a generic template for bug and router monitors are presented. However, it mainly targets run-stop debug control instead of real-time tracing.

6. CONCLUSION

This paper describes an on-chip monitoring infrastructure that detects and outputs communication transactions on-the-fly for off-chip inferences of system-level behavior for efficient post-silicon debug. Initial experiments show some promising results. In the future, we plan to perform in-depth study on using the described infrastructure on SoC designs with diverse interconnects, and explore optimizations for the infrastructure to offer higher observability with reduced hardware overhead.

Acknowledgement The research presented in this paper was partially supported by a gift from the Intel Cooperation.

7. REFERENCES

[1] Priyadarsan Patra. On the cusp of a validation wall. *IEEE Des. Test*, 24(2):193–196, March 2007.

[2] Harry D. Foster. Trends in functional verification: A 2014 industry study. In *DAC*, pages 48:1–48:6, 2015.

[3] M. Talupur, S. Ray, and J. Erickson. Transaction flows and executable models: Formalization and analysis of message-passing protocols. In *FMCAD*, 2015.

[4] Y. Cao, H. Zheng, H. Palombo, S. Ray, and J. Yang. A post-silicon trace analysis approach for system-on-chip protocol debug. In *ICCD*, 2017.

[5] Amba axi and ace protocol specification. http://www.arm.com.

[6] ARM. Coresight architecture specification v2.0, 2013.

[7] A. M. Gharehbaghi and M. Fujita. On-chip transaction level debug support for system-on-chips. In *ISOCC*, pages 124–127, Nov 2009.

[8] Matthew Amrein. System-level trace signal selection for post-silicon debug using linear programming. Master's thesis, UIUC, May 2015.

[9] Calin Ciordas, Twan Basten, Andrei Rădulescu, Kees Goossens, and Jef Van Meerbergen. An event-based monitoring service for networks on chip. *ACM TODAES*, 10(4):702–723, October 2005.

[10] A. M. Gharehbaghi and M. Fujita. Transaction-based debugging of system-on-chips with patterns. In *ICCD'09*, pages 186–192, Oct 2009.

[11] A. M. Gharehbaghi and M. Fujita. Transaction-based post-silicon debug of many-core system-on-chips. In *ISQED*, pages 702–708, March 2012.

[12] A. DeOrio, J. Li, and V. Bertacco. Bridging pre- and post-silicon debugging with biped. In *ICCAD*, 2012.

[13] B. Vermeulen and K. Goossens. A network-on-chip monitoring infrastructure for communication-centric debug of embedded multi-processor socs. In *VLSI-DAT'09*, pages 183–186, April 2009.

978-1-5386-7100-9/18 $31.00 © 2018 IEEE

Performance Enhancement of Split Length Compensated Operational Amplifiers

Donel Anto[†], Abhijeet D. Taralkar[ℵ], Nithin Kumar Y. B[‡], and Vasantha M. H[§]

Department of Electronics and Communication Engineering

National Institute of Technology Goa, India

donelanto[†]@nitgoa.ac.in, abhitaralkar[ℵ]@gmail.com, nithin.shastri[‡]@nitgoa.ac.in, vasanthmh[§]@nitgoa.ac.in

Abstract—In this work, a technique to improve the performance parameters of the split length compensated operational amplifiers is proposed. The proposed technique uses an assistant amplifier for performance enhancement of the system. The assistant amplifier is designed in such a way that it draws only a small amount of power ($<2\%$). This work is simulated in 180 nm CMOS technology with 1.8 V supply using Cadence Virtuoso. It achieves 78 dB DC gain, 40.8 MHz unity gain bandwidth and a slew rate of 27.17 V/μS for a load capacitance of 15 pF.

Index Terms—Operational amplifiers, gain, unity gain bandwidth, slew rate, split length, assistant amplifier.

I. INTRODUCTION

The increasing use of battery-powered portable devices is pushing the semiconductor industry to develop low power integrated circuits. Nowadays even high speed Operational Amplifiers (Op-Amp) are being designed using a very limited power budget. The scaling of the technology results in better performance of digital circuits [1], however poses new challenges in the analog design. Also, in the deep sub-micron technology it is strenuous to achieve very high gain due to the reduced output resistance of the transistors. The basic analog building blocks like op-amps need to be designed at ultra low power without compromising the performance specifications. Cascoding of transistors can be used to increase gain but it is not a viable solution for the deep sub-micron process because of the reduced voltage headroom. In the multistage technique, two stage amplifiers have been preferred because of the relatively simple stability criterion [2] [3].

Slew rate and gain enhancement of two stage Miller compensated op-amps have been reported in [4] using four additional controlled current sources in the main op-amp which is being dynamically controlled by auxiliary op-amp which senses the slewing condition. A hybrid scheme using gain and slew rate boost circuit is described in [5]. It uses cascode current sources and folded cascode structure, which makes it incompatible for lower technology nodes and large output swings.

Different methods for indirect feedback compensation are outlined in [6] in order to avoid Right-Hand Plane (RHP) zero and to improve Unity Gain Bandwidth (UGB). Reverse Nested Miller Compensation with Voltage Buffer and Nulling Resistor (RNMC-VBNR) which utilizes inversion of RHP zero and relocation of complex poles to elevated frequencies for performance improvement of three stage op-amp is explained in [7].

Improvement of RNMC-VBNR for the three-stage op-amp is reported in [8] which uses double pole-zero cancellation above UGB.

Indirect compensation of two stage amplifier using Split Length Differential Pair (SLDP) has been reported in [9]. It uses lengthwise splitting of input differential pair transistors and the intermediate node thus obtained is used for indirect compensation which improves UGB and phase response of the system by elimination of RHP zero. Slew rate is also improved along with it. The transistor level schematic of two stage op-amp using SLDP technique is shown in Fig. 1 in which M_{1T}, M_{1B} and M_{2T}, M_{2B} are the lengthwise split differential pair transistors.

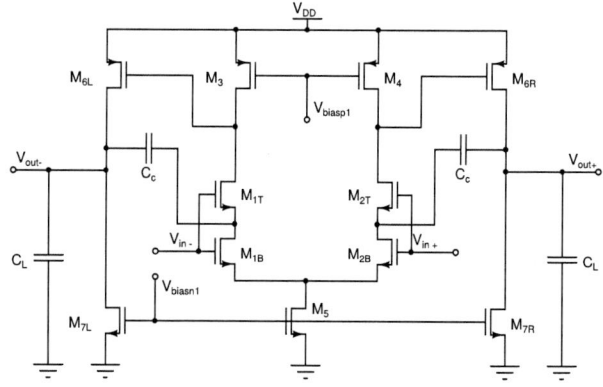

Fig. 1. Indirect compensation of two stage op-amp using SLDP [9].

Various other compensation schemes and enhancement techniques for improvement in the performance parameters of the multistage op-amps have been reported in [10]-[18].

In the proposed technique, further improvement of the performance parameters of the split length compensated op-amp is done with the aid of an assistant amplifier. This technique improves gain, UGB and slew rate of the op-amp with a negligible overhead in power consumption. The proposed technique is simulated using 0.18μm process with 1.8 V supply.

The organization of the paper is as follows. Section II describes the proposed work in detail. Section III shows the simulation results along with comparison of other techniques and finally conclusion in Section IV.

978-1-5386-7100-9/18 $31.00 © 2018 IEEE

II. Proposed Technique Using Assistance

In the work expounded in subsequent sections, an indirect compensated two stage SLDP op-amp [9] is used as the main op-amp and an assistant amplifier is used to assist and improve its performance.

A. Basic Idea

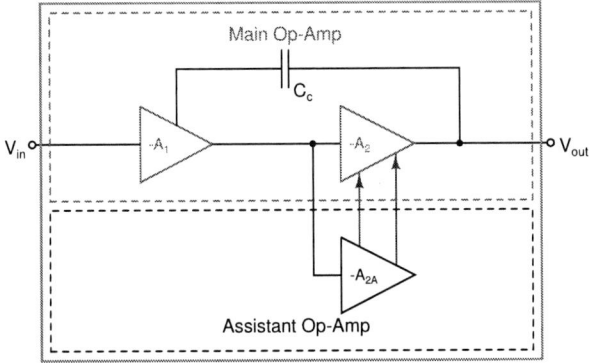

Fig. 2. Block diagram of proposed technique with assistance provided second stage.

The block level representation of the assistance technique is shown in Fig. 2. The basic idea of performance enhancement using assistance is to control the various parameters of the main op-amp using the assistant amplifier. This paper exploits the performance improvement in the gain, slew rate, phase response etc. using assistant amplifier in second stage. The detailed transistor level implementation of the assistant amplifier technique is discussed in the coming subsections.

B. Circuit Implementation

Transistor level implementation of the proposed technique is shown in Fig. 3. The assistant amplifier takes the output of first stage (V_{1+} and V_{1-}) of the main op-amp as its input. The second stage load current sources (M_{7R} and M_{7L}) of the main op-amp are controlled by the output signals of the assistant amplifier (V_{C7LA} and V_{C7RA}) instead of connecting it to a fixed bias voltage as in conventional techniques. Effectively, a scaled version of input signal is being applied to the load current sources of the second stage of main op-amp with the same phase of the input signal to that stage (V_{1-} and V_{C7LA} are in same phase, similarly V_{1+} and V_{C7RA} are in same phase). This helps in improvement of the effective transconductance of the second stage of main op-amp.

C. Gain and Bandwidth Enhancement

The improvement in transconductance results in improvement in gain and UGB of the op-amp. Since the assistance is provided in the second stage, the first stage gain remains unchanged and is given by,

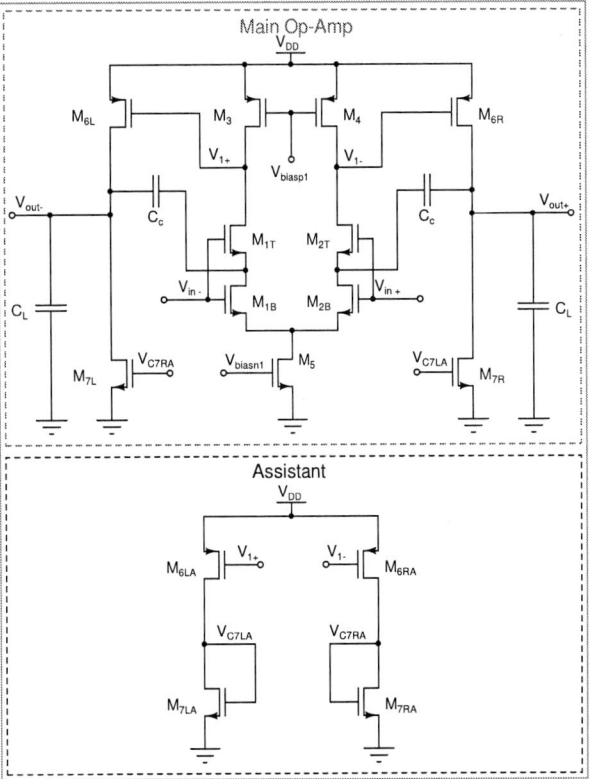

Fig. 3. Proposed technique with assistance provided second stage.

$$|A_1| = \frac{\left(g_{m2T} \cdot g_{m2B}\right)}{\left(\frac{g_{m2B}}{r_{2T}}\right) + \left(\frac{1}{r_4}\right) \cdot \left(\frac{1}{r_{2T}} + g_{m2T} + g_{m2B}\right)} \quad (1)$$

while the second stage gain is described by,

$$|A_2| = \left(g_{m7R} \cdot \frac{g_{m6RA}}{g_{m7RA}} + g_{m6R}\right) \cdot \left(r_{6R} \| r_{7R}\right) \quad (2)$$

The overall gain is given by,

$$|A| = |A_1| \cdot |A_2| \quad (3)$$

The additional term

$$\left(g_{m7R} \cdot \frac{g_{m6RA}}{g_{m7RA}}\right) \cdot \left(r_{6R} \| r_{7R}\right)$$

in equation (2) is due to the assistance provided in the second stage. The improvement in transconductance and gain depends on the ratio of the transconductance of transistors (g_{m6RA}/g_{m7RA}) in the assistant amplifier. So, the gain can be increased by adjusting this ratio. Due to the improvement in the second stage gain, the multiplication factor of compensation capacitor (C_c) will improve which eventually improves the Phase Margin (PM) of the system. Thus, UGB will be increased and moreover, the value of C_c required for frequency

978-1-5386-7100-9/18 $31.00 © 2018 IEEE   609

compensation can be reduced which helps in improvement of slew rate in addition to the reduction of the layout area.

D. Slew Rate Enhancement

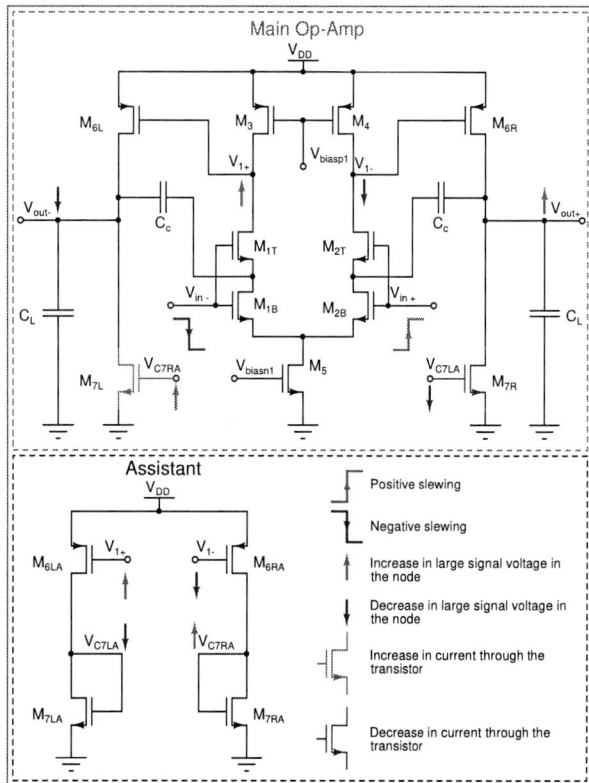

Fig. 4. Large signal operation (slewing condition) in proposed technique.

Improvement of slew rate in the proposed technique is illustrated in the Fig. 4. Whenever there is a slewing condition, assistant amplifier senses it and adjusts the current through the load current sources of the main op-amp correspondingly and improves the slew rate. Whenever there is a positive slew condition ($V_{in+} >> V_{in-}$), assistant circuit turns off or decreases the current through M_{7R} and turns on or increases current through M_{7L} in order to charge V_{out+} node and discharge V_{out-} node faster as shown in Fig. 4 ($\uparrow$ and $\downarrow$ indicates increase and decrease respectively in corresponding node voltages). Similar, but opposite working happens in the case of a negative slew condition ($V_{in+} << V_{in-}$).

E. Power Consumption in assistant amplifiers

As the ratio $g_{m_{6RA}}/g_{m_{7RA}}$ is the crucial parameter which governs the enhancement in the gain and transconductance (as evident from Eq. 2), the absolute value of $g_{m_{6RA}}$ and $g_{m_{7RA}}$ along with its current is scaled down to reduce power consumption in the assistant amplifier while retaining the ratio to an optimal value. In the implementation of the proposed circuit, the assistant amplifier draws only 1% to 2% of the

total current which helps in avoiding further increase in power dissipation.

III. RESULTS AND COMPARISON

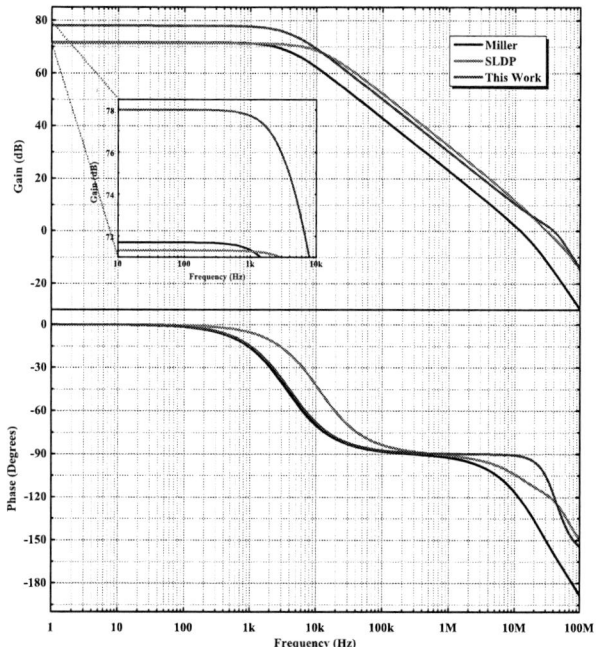

Fig. 5. Frequency response of the proposed architecture along with Miller and SLDP.

The proposed work is designed and simulated in 0.18μm CMOS technology with 1.8 V supply for a load capacitance (C_L) of 15 pF. Op-amps employing Miller compensation and indirect compensation using SLDP technique have been also designed and simulated using the same technology, supply voltage and load conditions for easy comparison purpose. The results of proposed architecture along with Miller and SLDP are shown in this section.

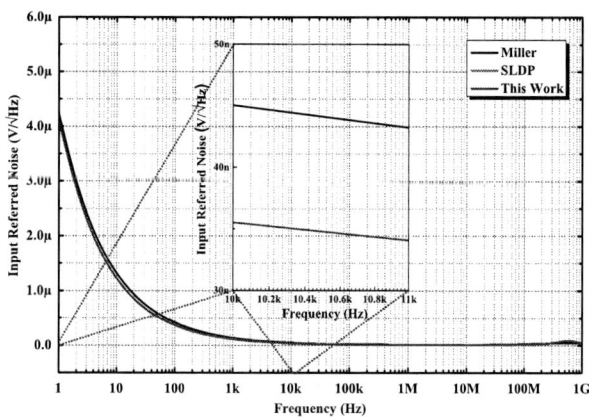

Fig. 6. Input referred noise of the proposed architecture along with Miller and SLDP.

978-1-5386-7100-9/18 $31.00 © 2018 IEEE   610

Frequency response of all the simulated techniques are shown in Fig. 5, from which it is evident that there is a 6 dB improvement in the gain while the UGB improved by around 3.33 times (from 12.24 MHz to 40.8 MHz) compared to conventional Miller compensation technique. At the same time a PM of 62 degrees is maintained in the proposed technique to ensure good stability.

Input referred noise of the proposed scheme along with Miller and SLDP is shown in Fig. 6. It shows that the noise added by the additional transistors in the assistant circuit is counterbalanced by the improvement in gain. The input referred noise waveform of the SLDP is not clearly visible in plot shown in Fig. 6 because it is almost overlapped by the waveform of the proposed work.

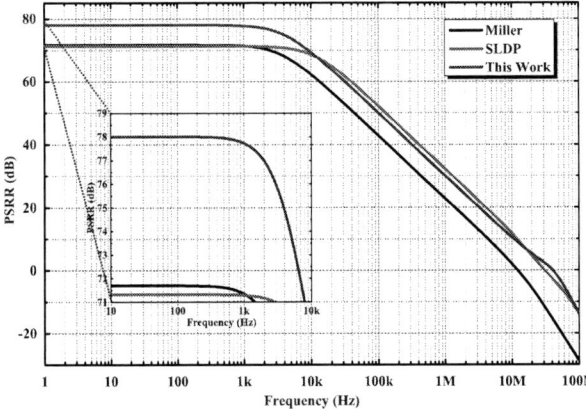

Fig. 7. PSRR of the proposed architecture along with Miller and SLDP.

The waveforms of Power Supply Rejection Ratio (PSRR) are given in Fig. 7. PSRR helps in the rejection of noise present in the power supplies. Since the PSRR improves with the improvement in differential gain, there is an approximate 6 dB enhancement in PSRR of the proposed architecture.

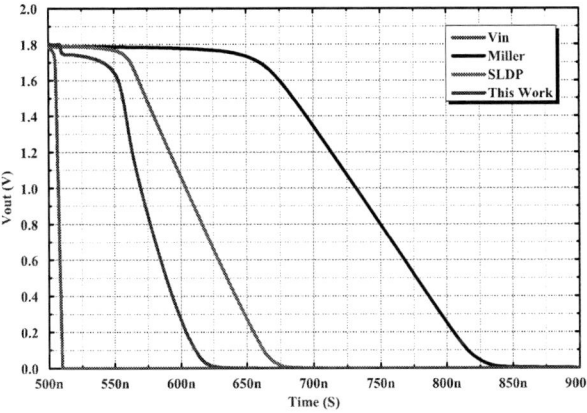

Fig. 8. Slew Rate of the proposed architecture along with Miller and SLDP.

The improvement in slew rate using the proposed technique is depicted in Fig. 8. The proposed technique is having

improved slew rate performance compared to other simulated techniques, which is evident from the slope of the curves in Fig. 8. The slew rate is improved approximately by a factor of 2.5 (from 10.7 V/μS to 27.17 V/μS) by usage of the assistant amplifier compared to the conventional Miller compensation technique. This helps in driving high capacitive loads without distortion.

A. Comparison

Table I compares various performance parameters of op-amps simulated using different architectures. Each parameter reveals the quality of the amplifier in a particular regard. In order to ascertain the overall calibre of the op-amp, a parameter called Figure of Merit (FoM) is used. Four different methods to quantify FoM is expounded in [8] taking into account of small signal and large signal performances of op-amp and are enumerated in equations (4)-(7).

$$IFoM_S = \frac{UGB \times C_L}{I_{DD}} \times 100 \qquad (4)$$

$$IFoM_L = \frac{SR \times C_L}{I_{DD}} \times 100 \qquad (5)$$

$$FoM_S = \frac{UGB \times C_L}{Power} \times 100 \qquad (6)$$

$$FoM_L = \frac{SR \times C_L}{Power} \times 100 \qquad (7)$$

Equations (4) and (5) takes total current into account while (6) and (7) considers total power consumed by op-amp. Equations (4) and (6) evaluates the small signal performance while (5) and (7) assess the large signal performance of the op-amp.

TABLE I
SIMULATION RESULTS-COMPARISON

	Miller	SLDP	This Work
Gain(dB)	71.704	71.32	78.01
UGB(MHz)	12.2486	30.7816	40.80
$C_L(pF)$	15	15	15
$C_c(pF)$	1.7	0.75	1.15
PM(Degrees)	59.39	62.78	62.65
Power(μW)	1007.928	1008.909	1011.544
Spot Noise @1KHz (nV/$\sqrt{Hz}$)	135.2	111.8	111.7
Slew Rate (V/μS)	10.702	16.16	27.17
PSRR (dB)	71.707	71.323	78.01
$IFoM_S$	32.8	82.3	108.9
$IFoM_L$	28.6	43.25	72.52
FoM_S	18.2	45.76	60.5
FoM_L	15.9	24.02	40.29

The power consumption is kept more or less the same in all the three simulated techniques. The PM is kept around 60

TABLE II
COMPARISON WITH OTHER SATE OF ART OPERATIONAL AMPLIFIERS

	[This Work]	RNMC with VBNR [7]	RNIC -FD [10]	NCFF [11]	RFC [12]	FF [13]	GBCA [14]	I-RI [15]	II-RI [15]	PSFF [16]	LVBD-Improved g_m [17]	RR BD i/p [18]
Technology(μm)	0.18	0.6	0.5	0.5	0.18	0.18	0.18	0.065	0.04	0.13	0.35	0.35
V_{DD}(V)	1.8	±1.5	3	±1.25	1.8	1.8	1.8	1.2	1.1	1.2	1	1
I_{DD} (μA)	561.9	466.7	400	6320	800	200	2111.1	11333.3	2363.6	15000	197	358
Power (μW)	1011.5	1400	1200	15800	1440	360	3800	13600	2600	18000	197	358
C_L	15	15	30	12	5.6	0.045	1	2	1	0.6	15	17
GBW (MHz)	40.80	19.46	12	300	134.2	1800	660	2000	770	11000	11.67	8.1
Slew Rate (V/μS)	27.17	11.1	10	-	94.1	-	800	-	-	-	1.37	2.74
IFOM$_S$	108.9	62.64	90	56.9	93.94	40.5	31.26	35.29	32.57	44	88.85	38.46
IFOM$_L$	72.53	35.7	75	-	65.87	-	37.89	-	-	-	10.43	13.01
FOM$_S$	60.5	20.85	30	22.7	52.18	22.5	17.36	29.41	29.6	36.67	88.85	38.46
FOM$_L$	40.29	11.89	25	-	36.59	-	21.05	-	-	-	10.43	13.01

degrees in all the cases taking stability issues into consideration. From Table I, it is evident that the proposed technique improves the FoM of the op-amp significantly compared to other techniques.

The proposed work is compared with the state of art techniques existing in the literature and is tabulated in Table II. For comparison, various parameters are listed along with the corresponding FoM. A closer look at the values of FoMs listed in Table II reveals that the FoM of the proposed technique is improved in comparison to the other state of art techniques present in the literature.

IV. CONCLUSION

The assistance technique proposed in this work improves various performance parameters of the two stage indirect compensated SLDP op-amp with negligible overhead in power consumption (<2%). Also, there is a significant augmentation in the FoM compared to the conventional methods.

The proposed technique achieves $\approx$ 6 dB increase in gain and PSRR, 32.9% improvement in UGB, 68.16% enhancement in slew rate, while IFOM$_S$ increased from 82.3 to 108.9 compared to the indirect compensation using SLDP technique. The assistance technique proposed in this work helps in enhancing the performance of op-amp without adding an overhead to the total power consumption of the circuit.

ACKNOWLEDGMENT

This publication is an outcome of the R&D work undertaken in the project under the SERB-SR/FTP/ETA-0090/2014, DST and Visvesvaraya PhD Scheme of Ministry of Electronics &

Information Technology, Government of India, being implemented by Digital India Corporation (formerly Media Lab Asia).

REFERENCES

[1] "International Technology Roadmap for Semiconductors", available online at http://www.itrs2.net/, 2014.

[2] B. Razavi,"Design of Analog CMOS Integrated Circuits", NewYork: McGraw-Hill, 2001.

[3] F. Maloberti, "Analog Design for CMOS VLSI Systems", in Kluwer Academic Publishers, Dordrecht, 2001.

[4] A. P. Perez, Y. B. Nithin Kumar, E. Bonizzoni and F. Maloberti, "Slew-rate and gain enhancement in two stage operational amplifiers", in 2009 IEEE International Symposium on Circuits and Systems, pp. 2485-2488, Taipei, 2009.

[5] Hong-Yi Huang, Bo-Ruei Wang and Jen-Chieh Liu, "High-gain and high-bandwidth rail-to-rail operational amplifier with slew rate boost circuit", in 2006 IEEE International Symposium on Circuits and Systems, pp. 907-910, Island of Kos, 2006.

[6] V. Saxena and R. J. Baker, "Indirect feedback compensation of CMOS op-amps", in 2006 IEEE Workshop on Microelectronics and Electron Devices, pp. 2 pp.-4 WMED '06., Boise, ID, 2006.

[7] Kin-Pui Ho, Cheong-Fat Chan, Chiu-Sing Choy and Kong-Pang Pun, "Reversed nested Miller compensation with voltage buffer and nulling resistor", in IEEE Journal of Solid-State Circuits, vol. 38, no. 10, pp. 1735-1738, Oct. 2003.

[8] A. D. Grasso, D. Marano, G. Palumbo and S. Pennisi, "Improved Reversed Nested Miller Frequency Compensation Technique With Voltage Buffer and Resistor", in IEEE Transactions on Circuits and Systems II: Express Briefs, vol. 54, no. 5, pp. 382-386, May 2007.

[9] V. Saxena and R. J. Baker, "Compensation of CMOS op-amps using split-length transistors", in 2008 51st Midwest Symposium on Circuits and Systems, pp. 109-112, Knoxville, TN, 2008.

[10] V. Saxena and R. J. Baker, "Indirect compensation techniques for three-stage fully-differential op-amps", in 2010 53rd IEEE International Midwest Symposium on Circuits and Systems, Seattle, WA, pp. 588-591, 2010.

[11] B. K. Thandri and J. Silva-Martinez, "A robust feedforward compensation scheme for multistage operational transconductance amplifiers with

no Miller capacitors", in *IEEE Journal of Solid-State Circuits*, vol. 38, no. 2, pp. 237-243, Feb 2003.

[12] R. S. Assaad and J. Silva-Martinez, "The Recycling Folded Cascode: A General Enhancement of the Folded Cascode Amplifier", in *IEEE Journal of Solid-State Circuits*, vol. 44, no. 9, pp. 2535-2542, Sept. 2009.

[13] N. Krishnapura, A. Agrawal and S. Singh, "A High-IIP3 Third-Order Elliptic Filter With Current-Efficient Feedforward-Compensated Opamps", in *IEEE Transactions on Circuits and Systems II: Express Briefs*, vol. 58, no. 4, pp. 205-209, April 2011.

[14] M. M. Ahmadi, "A new modeling and optimization of gain-boosted cascode amplifier for high-speed and low-voltage applications", in *IEEE Transactions on Circuits and Systems II: Express Briefs*, vol. 53, no. 3, pp. 169-173, March 2006.

[15] S. A. Aamir, P. Harikumar and J. J. Wikner, "Frequency compensation of high-speed, low-voltage CMOS multistage amplifiers", in *2013 IEEE International Symposium on Circuits and Systems, Beijing*, pp. 381-384, 2013.

[16] H. Shrimali and S. Chatterjee, "11 GHz UGBW Op-amp with feed-forward compensation technique", in *2011 IEEE International Symposium of Circuits and Systems, Rio de Janeiro*, pp. 17-20, 2011.

[17] L. Zuo and S. K. Islam, "Low-Voltage Bulk-Driven Operational Amplifier With Improved Transconductance", in *IEEE Transactions on Circuits and Systems I: Regular Papers*, vol. 60, no. 8, pp. 2084-2091, Aug. 2013.

[18] J. M. Carrillo, G. Torelli, R. Perez-Aloe Valverde and J. F. Duque-Carrillo, "1-V Rail-to-Rail CMOS OpAmp With Improved Bulk-Driven Input Stage", in *IEEE Journal of Solid-State Circuits*, vol. 42, no. 3, pp. 508-517, March 2007.

978-1-5386-7100-9/18 $31.00 © 2018 IEEE

Design-Based Fingerprinting Using Side-Channel Power Analysis For Protection Against IC Piracy

James Shey[*†], Naghmeh Karimi[†], Ryan Robucci[†], Chintan Patel[†]

[*]Electrical and Computer Engineering Department, United States Naval Academy, Annapolis, MD, US.
shey@usna.edu

[†]Department of Computer Science and Electrical Engineering, University of Maryland Baltimore County,
Baltimore, MD, US.
{jshey1, nkarimi, robucci, cpatel2}@umbc.edu

Abstract—Intellectual property (IP) and integrated circuit (IC) piracy are of increasing concern to IP/IC providers because of the globalization of IC design flow and supply chains. Such globalization is driven by the cost associated with the design, fabrication, and testing of integrated circuits and allows avenues for piracy. To protect the designs against IC piracy, we propose a fingerprinting scheme based on side-channel power analysis and machine learning methods. The proposed method distinguishes the ICs which realize a modified netlist, yet same functionality. Our method doesn't imply any hardware overhead. We specifically focus on the ability to detect minimal design variations, as quantified by the number of logic gates changed. Accuracy of the proposed scheme is greater than 96 percent, and typically 99 percent in detecting one or more gate-level netlist changes. Additionally, the effect of temperature has been investigated as part of this work. Results depict 95.4 percent accuracy in detecting the exact number of gate changes when data and classifier use the same temperature, while training with different temperatures results in 33.6 percent accuracy. This shows the effectiveness of building temperature-dependent classifiers from simulations at known operating temperatures.

Index Terms—IP Piracy, Fingerprinting, Side-Channel Power Analysis, Machine Learning

I. INTRODUCTION

Increasing the complexity of integrated circuits has raised design time and costs, and in turn has led to the globalization of the design and manufacturing process. Such globalization has increased IC counterfeiting and piracy rates [1]. In recent years, IC Piracy has become a major concern for government, military, and private sectors with increased cost and downtime of systems [2].

Applying traditional piracy avoidance methods such as IC fingerprinting and active metering schemes require additional hardware and/or impose delay and cost overhead [2–4]. In practice, modern low cost/power embedded systems have very small performance margins and adding additional complexity would either raise the price in terms of new hardware or make the device less responsive with additional software demands [5]. An adversary who has access to the gate-level design (either from a malicious insider or via reverse engineering) may only change a few gates while keeping the functionality intact, and introduce the IC as a new original circuit which can be sold under a new name. To address this problem, this paper proposes a technique using side-channel analysis to monitor

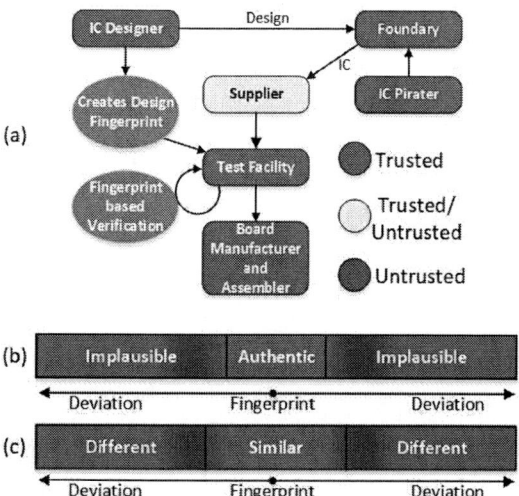

Figure 1: (a) Supply chain allowing a board manufacturer/assembler receiving components from a trusted/untrusted foundry via a trusted/untrusted supplier. Our method provides the design-based fingerprints that the trusted test facility will utilize to detect piracy. (b) Continuum showing likeliness measures of an IC compared to a given fingerprint used for authenticity certification. Green shows a match and red depicts a mismatch. (c) Continuum showing likeliness measures of an IC compared to a given fingerprint for "theft" detection.

the transient power consumption of an IC in order to detect if the IC has been altered during the manufacturing process 1(a).

In this paper, we present a new methodology for determining likeness of an IC to a precharacterized implementation. The designer generates a likeness measure that is fundamentally tied to the gate-level design and manufacturing process parameters. This likeness metric could be used for two purposes. One is to certify an IC as authentic (exact design and manufacturer) by enforcing a tight bound on the similarity, thereby rejecting any IC with design or manufacturing differences. The second application is to prevent theft, by catching "similar" designs sold under a different branding with a loose bound designed to catch a stolen and slightly modified design that keeps functionality intact.

978-1-5386-7100-9/18 $31.00 © 2018 IEEE

The first application (henceforth called authenticity certification in this paper) is important because in a non-vertical supply chain as shown in Fig. 1(a), there are multiple avenues for pirated ICs to enter the supply chain (through untrusted foundries or 3rd-party suppliers). Our method for supply verification allows a board manufacturer/assembler to verify the purchased ICs and distinguish them from pirated versions. In practice, in our method, the IC designer provides additional information to trusted entities to help them in distinguishing original designs from pirated versions using a tight tolerance on the fingerprint (See Fig. 1(b)).

The second application (henceforth called theft detection in this paper) is where a manufactured IC may not completely conform the GDSII file sent by original designer, i.e. a rouge element in the manufacturing process *alters the design such that the new device is functionally equivalent to the original one*, and may have reduced performance. This application targets the circuits that have same functionality, yet different gate-level implementations (in terms of a few gates) and hence a larger window for catching similar ICs as shown in Fig. 1(c).

The following major contributions are presented in this article:

1) A technique is proposed to determine if intellectual property (IP) manufactured within an IC can be verified by a design-based fingerprint provided by the IP designer. This fingerprint is created via analysis of the IC's power consumption, which using an SVM classifier can accurately determine if there is a change in a circuit from an original circuit.

2) Simulation-based evidence is presented showing that classification of the number of changed gates in a circuit from an original circuit can be determined, which does not require additional watermarking circuitry. This analysis includes considering the effect of temperature variations and transistor mismatch along varying sample rates on the performance of the classifier.

3) A temperature-dependent model is proposed for classifying changes in a circuit to mitigate real world temperature variations.

The remainder of paper is organized as follows. Section II covers existing techniques of detecting hardware changes. Section III covers the setup of the circuits analyzed, the simulation parameters used, and the setup of the classifier. The results are presented and discussed in Section IV. Finally, in Section V conclusions are drawn and future work is discussed.

II. RELATED WORK

Prior work in the field of pirated ICs has concentrated on two avenues. The first is inclusion of and testing for fingerprinting, watermarking, etc. to protect IP [1, 3, 6]. The second is detection of hardware Trojans, additional elements added to a circuit that includes a trigger and payload to produce a desired effect [7].

To combat pirated devices, additional elements are added to an IC to help identify the IPs included in a design, because these elements are only understood by the designer it is difficult or impossible to remove them and therefore a pirated device may still contain these elements [4]. Common methods are watermarking, obfuscation, and fingerprinting. The term watermarking refers to embedding ownership information into an IC that can later be verified to show the company whose IP is included in the IC. Prior work in the area verifies the presence of a watermark finite state machine though power analysis [2]. The term obfuscation refers to the insertion of additional elements to the circuit such that, by using the correct key the device will operate correctly, but without the key the exact functionality can not be determined [6]. Recent work on obfuscation has extended the technique from combinational logic to include simple finite state machines [8]. Unlike these methods that require additional hardware, our method requires no additional hardware and can therefore be a cost savings. The term fingerprinting refers to ensuring that each IC has a unique identifier and each IC can be tracked though the supply chain its identifier [3]. Recent work in this area deals with leakage and switching power of the gates to create a device fingerprint [9]. Each device has a unique signature and therefore needs to be handled individually after manufacturing, whereas our method creates one signature for a design based on simulations and is therefore less costly to implement.

Altering the final circuit during the manufacturing process is very similar to Trojan insertion. Hardware Trojans are the elements that are maliciously inserted in circuits and mainly include trigger and payload parts. The trigger can be a specific input vector, or a sequence of input values that activate the Trojan circuit. When activated, the payload can result in circuit malfunction or data leakage [10, 11]. Trojan detection methods mainly rely on characterizing path delays [11, 12], and/or electromagnetic emissions[13]. However, in our threat scenario, the malicious change doesn't alter the functionally of a victim circuit, nor does it result in data leakage. Using path-delay based Trojan detection schemes to detect our victim circuits is costly because of the scalability of these methods with the exponentiation growth of the number of paths needed to be monitored. In practice, in our threat scenario, the adversary does not need (and even better not to) target critical or near-critical paths for gate changes. Thereby, path-delay based techniques encounter scalability issues as they need to monitor several paths. In addition, although our gate change can be considered as a Trojan but Trojan-detection schemes that rely on functionality change fail to detect our pirated ICs as in our threat scenario the functionality is not changed and so the even so-called Trojan is never triggered. On the other hand, we believe that our technique can be adapted for detecting Trojans as well. However, further investigation is needed to confirm this ability and therefore in this paper we focus on distinguishing the ICs which follow our threat model, and leave the Trojan detection problem as a future research.

III. PROPOSED METHOD

To create a design-based fingerprint, a circuit is simulated with a set of input patterns and the current drawn from the circuit is monitored. The Fast Fourier Transform (FFT) of this time-domain signal along with the signals from the altered circuits (though functionally equivalent). Two classifiers are

978-1-5386-7100-9/18 $31.00 © 2018 IEEE

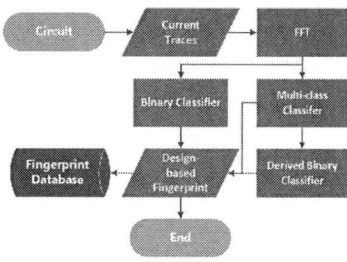

Figure 2: Flowchart demonstrating the generation of design-based fingerprints from current traces using binary and multi-class classifiers. A FFT is performed on current traces and the results passed to multiple classifiers, which combine to create the design-based fingerprint.

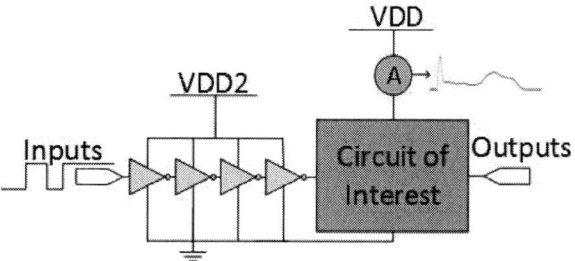

Figure 3: Simulation setup used in the experiments showing four chained inverters for input waveform smoothing and the location where the current is being measured.

created, a binary to identify the original circuit from altered circuits and a multi-class to identify the number of changes in a circuit. The second classifier can be dimensionally reduced to create a binary classifier. From these classifiers, a design based-fingerprint is created which allows a given IC to be analyzed in order to determine if the IC is a pirated one or not and in the case of piracy, how many gates have been changed. This sequence is shown in Fig. 2. This process can be applied to larger circuits via partitioning the circuits and considering one signature for each part using the above technique.

A. Circuit Setup

The circuits evaluated in this paper are selected from the ISCAS-85 benchmarks [14] and are detailed in Table I. The benchmarks represent a wide range of application areas and sizes. The original circuits for this work are an implementation of these standardized circuits. The experimental setup is shown in Fig. 3. The four chained inverters model a typical driver (e.g. limited slew rate). The current measured for this paper is that from the power supply that feeds the circuit and not the driver inverters. Altered circuits are created from the original circuit by a Python script that randomly selects either 1, 2, 5, 10 or 100 gates changes the selected gate from a NAND to an AND followed by an inverter, an AND to a NAND followed by an inverter, a NOR to an OR followed by an inverter, etc.

B. Circuit Simulation

To evaluate the effectiveness of the proposed design-based fingerprint verification, five different ISCAS'85 benchmarks

Table I: ISCAS'85 circuit descriptions.

Circuit	Number of Gates	Function
c432	215	27-channel interrupt controller
c499	245	32-Bit Single-Error-Correcting Circuit
c1355	589	32-Bit Single-Error-Correcting Circuit
c5315	2972	9-Bit ALU
c7552	4042	32-Bit Adder/Comparer

specified in Table I were simulated. Device-level simulations were performed at the schematic level for this paper. An in-house tool was used to generate the transistor-level model of the considered benchmarks. Using a 45-nm technology extracted from the open-source NANGATE library [15], transistor-level simulations were conducted using Synopsys' HSpice.

For each benchmark circuit (base circuit), we generated several altered circuits as follows. We changed one gate (selected randomly) of the base circuit to a functional equivalent gate(s) and repeated the process to generate 20 altered circuits. We also generated 20 circuits each with 2, 5, 10, and 100 gate changes compared to the original netlist of each benchmark circuit. To consider the effect of process variations and show the effectiveness of our method under the process variations, for each benchmark circuit, we conducted 620 Monte Carlo simulations of the base circuit and 31 simulations of each of the 20 altered counterparts using same randomly generated input sets to feed each circuit and its all counterparts.

For each of the simulations, the current traces were extracted. Simulations were carried out using the following process-variation parameters with a Gaussian distribution: transistor gate length L: $3\sigma = 10\%$; threshold voltage V_{TH}: $3\sigma = 30\%$, and gate-oxide thickness t_{OX}: $3\sigma = 3\%$. The process variation data reflects a 45-nm process in commercial use today [16]. The simulations were conducted assuming 45°C operating temperatures and 10 ps temporal resolution and were controlled by a Python program.

C. Golden Waveform Generation

The use of a *golden waveform* and its effect on the accuracy of the classifier is investigated by comparing the results of the classifier with and without preprocessing by subtracting the *golden waveform* from the data. A *golden waveform* is generated by taking the average of the original circuit simulations, for this work, it is the average of the 620 simulations.

D. Classification Setup

In this paper, a support vector machine (SVM) is used for classification as it provided the highest accuracy when compared to other machine learning techniques. A similar scheme was used in prior work [17]. A basic binary SVM maps the training data into a higher dimensional feature space and then attempts to find a boundary-defining expression that supports useful separation of data while minimizing errors [18]. The binary classifier can be extended to a multi-class problem with such methods as an "one vs all" approach. This compares one class to the remaining data and tries to find

978-1-5386-7100-9/18 $31.00 © 2018 IEEE

a boundary to maximize the separation between classes [19]. We use a binary classifier for both authentication certification and theft detection, however in the case of theft detection we can use a multi-class classifier to find an approximate number of gate changes. Additionally, the multi-class classifier is dimensionally reduced to create a derived binary classifier.

The SVM classifier was trained using two different preprocessing methods. The first was preprocessing by taking the time-domain waveform and windowing around the transients caused by input transitions. This is done to limit the amount of data processed and to eliminate noise contribution from expected quiescence periods after settling. The FFT was taken of the resultant windowed time-domain signal. The SVM was then trained using 5-fold cross validation to test the accuracy of the binary and multi-class classifications. The second preprocessing method is similar to the first, however, prior to windowing the time-domain data, the *golden waveform* is subtracted from the time-domain signal. The difference is then windowed and the remaining preprocessing steps are the same. We analyzed both time and frequency-domain classifiers, however the frequency-domain classifier resulted in 2% more accuracy and yielded less false negatives. Therefore we only present the frequency-domain data.

To confirm the accuracy of the classifier and to verify that the classifier is able to classify the particular number of gate changes, the classifier was fed with another completely separate set of data that included the same number of runs as the original, but had different gates changed from the original training data. The outcome of this classification supports determining if the classifier is overfit to particular data.

IV. EXPERIMENTAL RESULTS AND ANALYSIS

A. Classification Results

Directly creating the classifier for the ISCAS'85 circuits shown in Table I, resulted in the accuracy shown in Table II. This is the accuracy of our authenticity certification and for all cases resulted in at least 83% accuracy. The binary classification was developed using a binary SVM classifier, while the derived binary was developed from a multi-class SVM classifier and then reduced to the binary outcome. A representative binary confusion matrix (for c432 circuit) is shown in Table III with the rows signifying actual class and the columns showing the inferred class. The accuracy of the multi-class SVM classifier is shown in IV and a representative multi-class confusion matrix (for c432 circuit) is shown in Table V. This demonstrates the ability to perform theft detection. In the multi-class case, direct classification yielded better performance for circuits with fewer gates.

Table II: Accuracy of direct binary vs. derived binary classification classifier using 10 ps temporal resolution.

	Direct Classification		Golden Waveform	
	Binary	Derived Binary	Binary	Derived Binary
c432	98.70%	93.00%	>99.9%	>99.9%
c499	98.30%	86.70%	>99.9%	>99.9%
c1355	>99.9%	>99.9%	>99.9%	>99.9%
c5315	83.4%	83.3%	92.5%	98.3%
c7552	96.1%	97.8%	96.9%	99.0%

Table III: Representative confusion matrix for the ISCAS'85 c432 circuit using 10 ps temporal resolution. Showing 551 unmodified circuits identified as unmodified and 113 altered circuits identified as unmodified.

Actual Class	Original	551	69
	Altered	113	2987
		Original	Altered
		Predicted	
		Class	

Table IV: Accuracy of direct classification vs. classification using a *golden waveform* with derived binary and multi-class classifier using using 10 ps temporal resolution.

	Direct Classification		Golden Waveform	
	Derived Binary	Multi-Class	Derived Binary	Multi-Class
c432	93.0%	93.8%	>99.9%	93.8%
c499	86.7%	74.0%	>99.9%	87.6%
c1355	>99.9%	95.8%	>99.9%	95.5%
c5315	83.3%	29.4%	98.3%	42.0%
c7552	97.8%	46.4%	99.0%	47.5%

B. Golden Waveform Implementation

To create the *golden waveform* discussed in Section III-C 620 simulations of the original circuit (shown in Fig. 4 (a)) were averaged. The resultant *golden waveform* is shown in Fig. 4 (b). The *golden waveform* is visually differentiable in the time domain from the average of the waveforms for the simulations as shown in Fig. 5 (a) vs. Fig. 5 (b). The ability to differentiate the waveforms in the time-domain infers that classification is possible.

C. Golden Waveform Classification Results

Using the *golden waveform*, to classify the circuits resulted in the binary accuracy specified in Table II and multi-class accuracy specified in Table IV. A representative confusion matrix for data processed with the *golden waveform* for the binary case is shown in Table VI and multi-class confusion matrix shown in Table VII. In all cases the binary classifier performs better using the *golden waveform* and comparing binary vs. derived binary classifiers, the accuracy was higher in all cases for the derived binary. The multi-class classifier, also has accuracy gains for all but one circuit, which had a 0.3% decrease. Based on the binary and multi-class classification accuracy results, the use of the *golden waveform* outperforms

Table V: Confusion matrix for direct classification of IS-CAS'85 c432 circuit using 10 ps temporal resolution. Each row represents the number of gate changes from the original circuit while each column is the outcome of the classifier with number of gate changes.

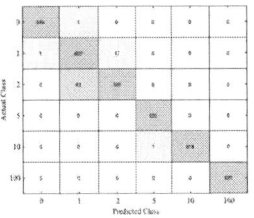

978-1-5386-7100-9/18 $31.00 © 2018 IEEE   617

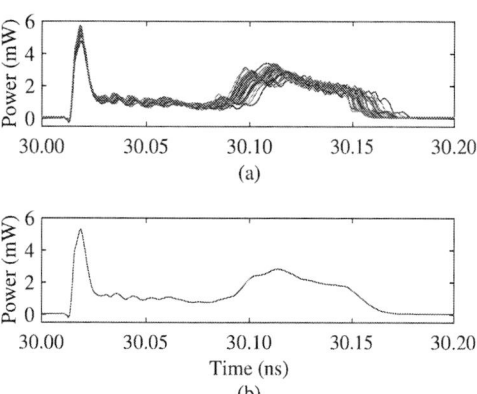

Figure 4: (a) The time-domain current traces showing 30 of the 620 Monte Carlo simulations demonstrating transistor mismatch and (b) the time-domain *golden waveform* for the ISCAS'85 c432 circuit.

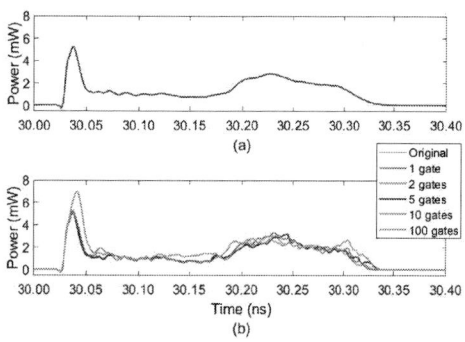

Figure 5: (a) Time-domain *golden waveform*, compared to (b) the average for each of the following: Original circuit, 1 gate modified, 2 gates modified, 5 gates modified, 10 gates modified, and 100 gates modified for the ISCAS'85 c432 circuit.

the direct classification. Using the *golden waveform* trained classifier, the binary accuracy can be maintained about 99% or 269 defects per million.

To verify that the classifier is performing correctly, a new set of simulations were run with the same number of runs as in the training scenario for the c432 circuit; twenty runs of thirty-one MC simulations each with a different number of changed gates (1, 2, 5, 10 and 100 gates), The results show the accuracy of 77.9% with the confusion matrix shown in Fig. VIII. This shows there is some over fitting to the data, but it

Table VI: Representative confusion matrix for *golden waveform* binary classification for the ISCAS'85 c1355 circuit using 10 ps temporal resolution.

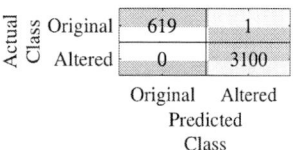

maintains the ability to distinguish an altered circuit from an original circuit with 100% accuracy.

D. Varying Temporal Resolution Results

For this work, a 10 ps temporal resolution was used. Examining the frequency spectrum of the *golden waveform* shown in Fig. 6a, a different temporal resolution may yield the same accuracy, but at a lower temporal resolution needing less computation. To evaluate the effect of varying the temporal resolution, simulations were conducted with a temporal resolution of 1 ps and 100 ps and the resultant frequency spectrums are shown in Fig. 6a and Fig. 6c respectively. The accuracy of the resultant sampling rates for the multi-class classifier are 93.8%, 95.5%, and 93.5% for 1 ps, 10 ps, and 100 ps temporal resolution respectively. For all cases accuracy is >99.9% for the binary classifier and >93% for multi-class classifier. The higher accuracy at 10 ps when compared to 100 ps for the multi-class classifier can be explained by the extra dimensions from the 100 ps classifier. Additional preprocessing steps to limit the dimensionality of the classifier can be performed. As it stands the accuracy for all cases is >99% for binary classification and >93% for multi-class classification.

E. Varying Temperature Results

The baseline temperature used in this paper is 45°C. To test the effects of temperature on the classification accuracy, simulations were conducted at 25°C and 70°C to cover the range of many commercial ICs. The results of these simulations were used by the classifier trained with 45°C data and resulted in 33.6% and 28.1% accuracy, respectively for the multi-class classifier. When the same temperature data was used to train the classifier as the data being classified the accuracy improved to 95.4% and 92.1% for 25°C and 70°C, respectively. This means that the classifier is temperature dependent, and therefore the classifier must be created from data at the same temperature that subsequent testing and classification is going to be performed at. For a lab environment this implies that the temperature of the IC needs to be held within a temperature band to accurately classify the results. In an uncontrolled environment, the temperature should be recorded and a classifier created for that temperature.

V. CONCLUSIONS AND FUTURE WORK

The experimental results confirmed that by deploying a design-based fingerprint and preprocessing using the *golden waveform*, our binary classifier has an almost 100% accuracy rate for authenticity certification. Going to the multi-class classifier for theft detection, we can predict the number of gates changed with an accuracy of 77.9% for a data set with unknown gates that are not part of the training data. In this research, we also considered the impact of sampling rate and showed that it is robust (>90% accuracy) for temporal resolutions between 1 ps and 100 ps.

Future directions will include creating a temperature-independent classifier and test the ability to detect Trojans. Additionally, the effects of aging on a circuit may change

978-1-5386-7100-9/18 $31.00 © 2018 IEEE

Table VII: Confusion matrices for the *golden waveform* classification of ISCAS'85 circuits using 10 ps temporal resolution.

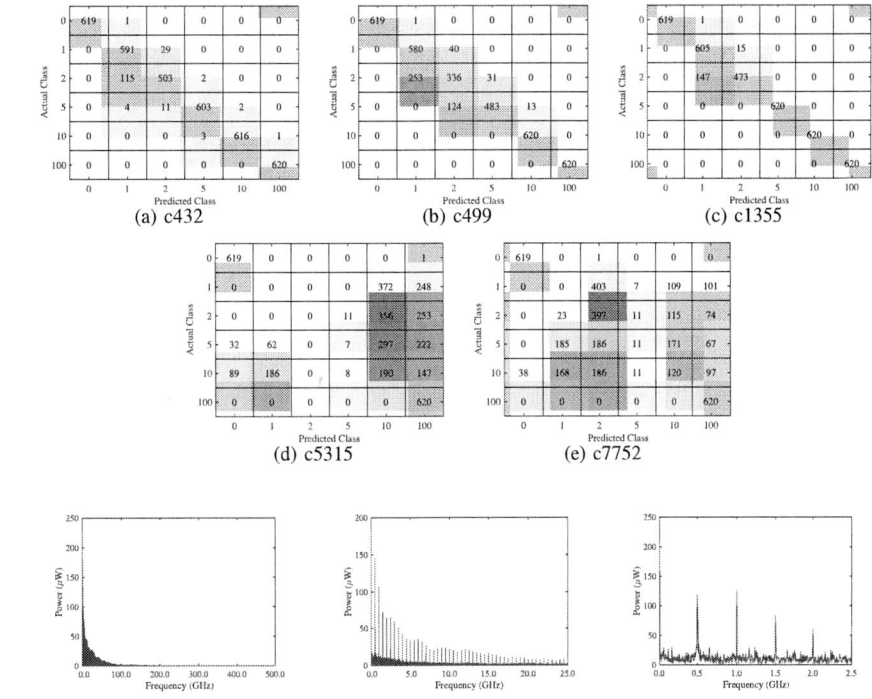

(a) c432 (b) c499 (c) c1355

(d) c5315 (e) c7752

(a) 1 ps temporal resolution (b) 10 ps temporal resolution (c) 100 ps temporal resolution

Figure 6: Frequency-domain of the *golden waveform* for the ISCAS'85 c432 circuit with different temporal resolutions.

Table VIII: Confusion matrix showing the results of cross validation data conducted with the classifier for the c432 circuit.

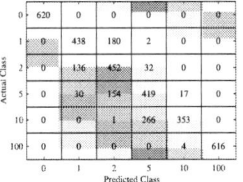

its design-based fingerprint and needs to be investigated. The impact of noise on the circuit, as well as layout-level routing and power-grid effects and variations need to be accounted for and included in the models to more accurately represent real world conditions in the simulations. We expect these variations to have minimal effects on the outcome of this technique, since they can be accounted for with additional simulation parameters. Future work will expand on the these premises while maintaining a high accuracy.

REFERENCES

[1] R. D. Newbould, J. D. Carothers, J. J. Rodriguez, and W. T. Holman, "A hierarchy of physical design watermarking schemes for intellectual property protection of IC designs," in *IEEE Int. Symp. on Circuits and Syst.*, 2002, pp. IV–862–IV–865.

[2] C. Marchand, L. Bossuet, and E. Jung, "IP watermark verification based on power consumption analysis," in *IEEE Int. System-on-Chip Conf. (SOCC)*, Sept 2014, pp. 330–335.

[3] B. Liu, Y. Jin, and G. Qu, "Hardware design and verification techniques for supply chain risk mitigation," in *Int. Conf/ Comput.-Aided Des. and Comput. Graphics (CAD/Graphics)*, Aug 2015, pp. 238–239.

[4] M. Ni and Z. Gao, "Watermarking system for IC design IP protection," in *Int. Conf. on Commun., Circuits and Syst.*, vol. 2, June 2004, pp. 1186–1190.

[5] K. Ly, W. Sun, and Y. Jin, "Emerging challenges in cyber-physical systems: A balance of performance, correctness, and security," in *IEEE Conf. Comput. Commun. Workshops (INFOCOM)*, April 2016, pp. 498–502.

[6] G. Qu, "Publicly detectable watermarking for intellectual property authentication in VLSI design," *IEEE Trans. Comput.-Aided Design Integr. Circuits Syst*, vol. 21, no. 11, pp. 1363–1368, Nov 2002.

[7] D. Agrawal, S. Baktir, D. Karakoyunlu, P. Rohatgi, and B. Sunar, "Trojan detection using IC fingerprinting," in *IEEE Symp. Security Privacy*, May 2007, pp. 296–310.

[8] Q. Yu, J. Dofe, and Z. Zhang, "Exploiting hardware obfuscation methods to prevent and detect hardware trojans," in *IEEE Int. Midwest Symp. on Circuits and Syst. (MWSCAS)*, Aug 2017, pp. 819–822.

[9] S. Wei, A. Nahapetian, and M. Potkonjak, "Robust passive hardware metering," in *IEEE/ACM Int. Conf. on Comput.-Aided Des. (ICCAD)*, Nov 2011, pp. 802–809.

[10] M. Tehranipoor and F. Koushanfar, "A survey of hardware trojan taxonomy and detection," *IEEE Des. Test of Comput.*, vol. 27, no. 1, pp. 10–25, Jan 2010.

[11] Y. Jin and Y. Makris, "Hardware trojan detection using path delay fingerprint," in *IEEE Int. Workshop on Hardware-Oriented Security and Trust*, June 2008, pp. 51–57.

[12] R. Vaikuntapu, L. Bhargava, and V. Sahula, "Golden IC free methodology for hardware trojan detection using symmetric path delays," in *Int. Symp. on VLSI Design and Test (VDAT)*, May 2016, pp. 1–2.

[13] J. He, Y. Zhao, X. Guo, and Y. Jin, "Hardware trojan detection through chip-free electromagnetic side-channel statistical analysis," *IEEE Trans. Very Large Scale Integr. (VLSI) Syst.*, vol. 25, no. 10, pp. 2939–2948, Oct 2017.

[14] F. Brglez and H. Fujiwara, "A neutral netlist of 10 combinational benchmark circuits and a targeted translator in FORTRAN," in *IEEE Int. Symp. on Circuits and Syst. (ISCAS)*, June 1985.

[15] NanGate FreePDK45 Open Cell Library. [Online]. Available: http://www.nangate.com/?page_id=2325

[16] N. Karimi and K. Chakrabarty, "Detection, diagnosis, and recovery from clock-domain crossing failures in multiclock SoCs," *IEEE Trans. Comput.-Aided Design Integr. Circuits Syst.*, vol. 32, no. 9, pp. 1395 – 1408, 2013.

[17] D. Chang, S. Ozev, O. Sinanoglu, and R. Karri, "Approximating the age of RF/analog circuits through re-characterization and statistical estimation," in *Design, Automation and Test in Europe Conf. (DATE)*, March 2014, pp. 1–4.

[18] G. B. Huang, H. Zhou, X. Ding, and R. Zhang, "Extreme learning machine for regression and multiclass classification," *IEEE Trans. Syst., Man, Cybern., Syst.*, vol. 42, no. 2, pp. 513–529, April 2012.

[19] C. W. Hsu and C. J. Lin, "A comparison of methods for multiclass support vector machines," *IEEE Trans. Neural Netw.*, vol. 13, no. 2, pp. 415–425, Mar 2002.

2018 IEEE Computer Society Annual Symposium on VLSI

PPAP and iPPAP: PLL-based Protection Against Physical Attacks

Prasanna Ravi, Shivam Bhasin, Jakub Breier, Anupam Chattopadhyay
Nanyang Technological University, Singapore
{prasanna.ravi, sbhasin, jbreier, anupam}@ntu.edu.sg

Abstract—Digital security practitioners are facing enormous challenge in face of the growing repertoire of physical attacks, e.g., Side Channel Attack (SCA) and Fault Injection Attack (FIA). Countermeasures to such threats are usually very different in nature and come with a significant performance penalty. While the FIA countermeasures rely on fault-detecting sensors or concurrent error detection schemes, SCA countermeasures are based on data masking or dual-rail logic circuits. Recently, a low-overhead FIA countermeasure has been proposed that utilises a ring oscillator circuit with Phase-Locked Loop (PLL). In this paper, we extend that countermeasure to further provide protection against SCA, thereby proposing PLL based Protection Against Physical attacks (PPAP). We demonstrate the PPAP on an FPGA prototype under rigorous SCA and FIA testing. We evaluate SCA resistance using the TVLA metric and observe a 2000× increase in SCA protection (in terms of number of traces) with PPAP. We further improve the security of PPAP using statistical analysis through an improved PPAP design (iPPAP) with an increase in SCA resistance of at least 5000× compared to the unprotected implementation with a minimal area overhead.

I. INTRODUCTION

The notion of security evaluation of cryptographic systems using only mathematical cryptanalysis of the underlying schemes was challenged by the seminal work of Kocher et al. [1] who introduced Side Channel Attacks (SCA) that target weaknesses in implementations of the cryptographic scheme. One of the most well-known side channel attacks is the Differential Power Analysis (DPA) which exploits the relation between the processed data and the power consumption of the device during the operation. On the other hand, there are Fault Injection Attacks (FIA) [2] which assume an active attacker, who induces faults in the internal state or maliciously alters the way of operation of the cryptographic algorithm. Attacker exploits the propagation of induced error to learn information of internal secrets like the key. Since their inception, numerous attacks and corresponding countermeasures have been reported in literature.

Countermeasures against SCA and FIA are quite different in nature. While SCA countermeasures depend on noise generation to hide the leakage, FIA countermeasures are generally based on the ability to detect/repair faults via information/space/time redundancy. For both SCA and FIA, there have been several works that introduce countermeasures at different levels of design abstraction. When protecting against both, usually SCA and FIA countermeasures are developed independently and later combined, leading to significant overheads [3], [4].

It must be noted that timing precision is in general crucial to both SCA and FIA. Externally introduced timing misalignment has shown to be effective against both the attacks [5], [6]. In [5], a PLL based clock randomiser is used for timing misalignment to boost SCA security.

In this paper, we propose **PLL based Protection Against Physical attacks (PPAP)**, a low cost combined hardware countermeasure against SCA and FIA with a very low performance overhead. The main building blocks of PPAP are a *fault attack sensor* and a *timing misalignment module*. Fault attack sensors were originally proposed by Miura et al. [7] to detect electromagnetic fault injection. The sensor is composed of a ring oscillator as fault injection detector and PLL for triggering an alarm. We first integrate both countermeasures to provide resistance against SCA and FIA. Next, we study the relation of timing misalignment and its statistical impact on SCA security, to design iPPAP (i.e., improved PPAP), which further boosts the SCA security.

The contributions in this works are as follows:

- We propose PPAP, an integrated plug and play hardware-based countermeasure against SCA and FIA.
- We develop a metric for measuring SCA security introduced by timing misalignment, to understand the key parameters contributing to SCA resistance.
- We next propose iPPAP, improved for SCA security.
- Both countermeasures are practically evaluated on an FPGA and demonstrated to be low-cost, easy to implement and independent of the circuit under protection.
- We report near perfect FIA security and SCA security boost of 2000× and 5000× for PPAP and iPPAP respectively.

II. PRELIMINARIES

In this section, we provide general background information on side-channel and fault attack concepts used. We also recall related works on clock randomisation, fault sensor and combined countermeasures.

A. Test Vector Leakage Assessment

Side-Channel Attacks (SCA) target physical observation of the secret-key related computation. A novel approach – Test Vector Leakage Assessment (TVLA [8]) has recently emerged as a testing strategy for SCA leakage. It detects any data dependent leakage through Welsh T-test to give a *PASS/FAIL*

978-1-5386-7100-9/18 $31.00 © 2018 IEEE   620

decision. The commonly used fixed vs random (FVR) or non-specific TVLA partitions traces based on varying and fixed plaintext, and computes T-test to check for data-dependent leakage. It is computed as $TVLA = (\mu_r - \mu_f)/\sqrt{\frac{\sigma_r^2}{m_r} + \frac{\sigma_f^2}{m_f}}$, where μ_r, σ_r and m_r (resp. μ_f, σ_f and m_f) are mean, standard deviation and cardinality of the set with varying plaintexts (resp. fixed plaintexts), respectively. The device is considered to leak side-channel information and *FAIL* the test if TVLA value crosses the threshold of $[-4.5, 4.5]$. Unlike attack techniques like correlation power analysis etc., TVLA stays independent of the leakage model and thus makes countermeasure comparison easy.

B. Clock Randomisation

There have been a number of works proposed to perform temporal misalignment [5], [9], [10]. In [5], a technique for generating an irregular clock from multiple phase shifted clocks of a PLL output using clock multiplexers was proposed. An EDA friendly automated design flow for a clock randomiser design was proposed in [9] to drive sequential elements using multiple phase shifted clocks. Further proposals for hardware timing disarrangement include insertion of random delays in the input data path or attaching random wait registers [10], [11]. The idea of timing disarrangement has also been explored in software, e.g. through shuffling the instruction sequence, insertion of dummy loops and Random Delay Interrupts (RDI) using special state machines or non-deterministic processors [12], [13].

C. Fault Injection Attacks & Laser/EM Sensor

Countermeasures against fault attacks are either based on concurrent error detection or physical sensors. In this work, we focus on sensor based countermeasures. A Phase-Locked Loop (PLL) based sensor for electromagnetic (EM) fault injection was first proposed by Miura et al. in [7]. The sensor is built up of two main components: a watchdog ring oscillator (WRO) and an alarm circuitry. The WRO runs at a stable frequency during normal operation, but is very sensitive to environmental variations. Hence, a high energy EM or laser injection can alter the frequency and phase of the WRO. Detecting this abnormal event on the WRO through an alarm generation circuit forms the core idea of the FIA sensor.

The PLL is a widely used clock timing control tool available in modern FPGAs. It comprises of a phase detector which converts frequency and phase differences between the feedback and the input clock to up or down pulses. These pulses further drive a loop filter and a voltage control oscillator to nullify the difference. In [7], PLL was used to monitor the clock generated from WRO and thus, detecting the malicious behavior.

D. Combined SCA and FIA countermeasures

As the modus operandi of SCA and FIA are different, protection strategies also differ vastly as discussed in the previous sections.

Mentens et al [11] used the dynamic reconfiguration feature in modern FPGAs to achieve resistance against SCA and FIA. Bhasin et al. [14] exploited the in-built fault tolerance of dual-rail logic to propose it as a combined countermeasure. Combination of threshold implementations, i.e. a provable side-channel countermeasure with parity-based error detection was proposed in [3]. Another approach to use enhanced private circuits against side-channel and fault attack was proposed in [4]. Linear and non-linear codes have also been investigated for combined SCA and FIA countermeasure in software by Breier et al. [15]. Most of these countermeasures were applied at the information level, while one used device specific features. They require design modification and incur significant performance overhead, as high as $20\times$ compared to unprotected equivalent, in some of the above mentioned cases.

III. PLL BASED PHYSICAL ATTACK PROTECTION

The general idea of **PLL based Protection Against Physical attacks (PPAP)** is illustrated in Fig. 1. Modern FPGAs contain PLLs as DCMs (Digital Clock Managers) in Xilinx Virtex 4, 5, 6 series FPGAs, and as MMCMs (Multi Mode Clock Managers) in the latest Xilinx 7 series FPGA and Zynq UltraScale processors. The latter ones offer more capabilities, such as multiple programmable clock outputs, static and dynamic fine phase shifting and Dynamic Port Reconfiguration Ports (DRP) which can alter the characteristics of clock during runtime. The PLL configured to generate multiple phase shifted clock, which leads to timing misalignment when multiplexed, was the initial idea proposed in [5]. By deriving the source clock from an internal free running ring oscillator (RO), fault attack detection capabilities are added to PPAP. The RO is placed over the sensitive core in order to detect any active fault injections like laser or EM. Fault detecting properties of RO were already demonstrated in [7]. PPAP has three fold impact on security. The timing misalignment contributes directly to SCA security. Moreover, the misalignment also makes fault injection harder as injection timing is often crucial for localizing injection targets. Finally, for more powerful attacks like laser/EM, the clock generating RO detects the attack before a fault is injected in the sensitive circuits.

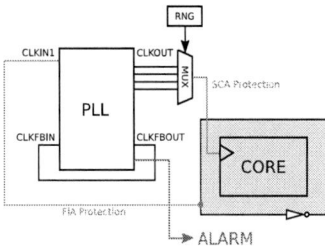

Fig. 1: Design of the combined FIA and SCA countermeasure based on PLL.

The main advantages of using such approach are listed as follows:

- This approach does not call for any changes to the implementation of the circuit under protection.

- Utilization of components that are readily available on numerous FPGAs.
- Since the implementation stays the same, it has zero impact on the maximum operating frequency of the design. Albeit, the clock randomiser itself has an impact on the throughput of the design, which we attempt to minimize.

Through statistical analysis, we also improve the design of the clock randomiser in PPAP to propose iPPAP to further boost SCA security. Though we only demonstrate results for SCA and FIA resistance of PPAP and iPPAP on FPGA, we would like to note that our design is easily compatible with the design flow of [7] that automatically implements the combined FIA sensor along with the clock randomiser. The modifications needed are limited to only *RTL* level and thus straightforward to implement. Therefore, PPAP and iPPAP is readily compatible with existing standard-cell design flows used in current EDA tools.

IV. IMPROVING CLOCK RANDOMISER

In this section, we first explain the impact of known RDI techniques on SCA protection followed by design of improved clock randomiser.

A. Random Delay Insertion Techniques

The time instance of occurrence of a particular operation depends on the sum of the individual delays introduced until the execution of that operation. Hence, from an attacker's perspective, only the cumulative effect of the individually inserted delays is observable. Thus the security and efficiency of timing misalignment is heavily dependent on the behaviour of the cumulative delay (*CD*) distribution. This *CD* can be modeled as a random variable X with a distribution of finite mean μ and variance σ^2. Thus, one would ideally look to maximize the variance of the *CD* distribution (spreading the time of occurrence of the operation) for a fixed mean (overall average performance). The most straightforward RDI method is generating uniformly distributed independent delays, whose *CD* exhibits a Gaussian distribution.

Since the Gaussian distribution has a large peak and a small variance, it provides much lower randomness compared to an equivalent uniform distribution. A near uniform *CD* distribution is achievable through adoption of the "Floating Mean Method" as proposed in [13]. The intuition is to make a random selection of a small section within a given interval and generate uniform delays within that selected interval for a single run of the algorithm. The selection of the small section is updated for every run of the algorithm and the same is repeated. Statistically speaking, it generates a *Gaussian mixture* which comprises of a series of smaller Gaussian distributions with shifted means which together have much wider span, thus achieving a near uniform distribution. This work was further improved by the same authors in [12] as "Improved Floating Mean" method where they introduced finer granularity in generating the delays to flatten out the cogs formed in the *CD* distribution. Refer to Fig. 4(a) for the *CD* distribution generated by all the three different methods.

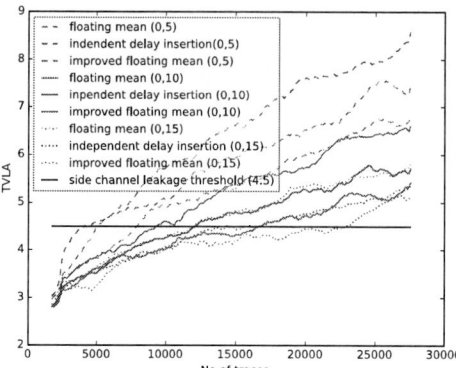

Fig. 2: Comparison of TVLA metric values for the simulated RDI methods for varying delay intervals.

They proposed a performance-security metric M to evaluate the efficiency of the RDI countermeasure, which is stated as follows,

$$M = 1/(2.\hat{p}.\mu) \qquad (1)$$

where $\hat{p}$ and μ are the maximum peak and mean of the probability distribution of the *CD* distribution, respectively. Considering this metric, a "uniform" *CD* distribution will provide the best efficiency and security.

B. Simulation-Based Analysis

We further analyse the security performance of these different RDI techniques using the TVLA testing methodology. We analyse their behaviour in a simulated setting when incorporated into a secure algorithm. We use the parallel AES-128 implementation, with a perfect Hamming weight (HW) leakage model with added white Gaussian noise. Random delays modeled based on the above three RDI techniques are inserted after every round of the AES algorithm. A weaker implementation will cross the threshold faster. As shown in Fig. 2 for the fixed delay intervals (5,10 and 15), we can see that the floating mean method performs best. The floating mean method (FM) also performs equally or better than its improved counterpart (IFM) for varying delay intervals. Another observation is that the increase in variance leads to increase in security.

Summing up the above discussion, the factors that primarily affect side channel leakage of a timing disarrangement countermeasure are:
- Behaviour of *CD* distribution.
- Unique number of CD instances.
- Span of *CD* distribution.

Thus, increasing the variance of the *CD* distribution while maintaining a fixed mean, along with improving the granularity (in terms of number of different instances of the operation), will improve the SCA resistance of the RDI countermeasure.

C. Clock Randomiser Design by [5]

The clock randomiser design in [5] uses multiple phase shifted clock outputs of the MMCM which are fed to a tree

978-1-5386-7100-9/18 $31.00 © 2018 IEEE

of clock multiplexers (with select lines connected to PRNG) to output a jittery clock. In order to maintain a glitch free operation, the clock multiplexers do not change the output immediately after a select input change. Upon the change of the select input, the output follows the current input until a falling transition on the current input. The output stays in the low state until the new input has also transitioned from high to low, and then follows the new input. The time taken for the output to switch from one input to the other can be referred to as the *Trance* state. The duration of this *Trance* state can be varied by continuously switching the select input of the multiplexer. We implemented the design on FPGA and plotted the distribution of duration of AES encryption, when driven by clock randomiser from [5]. From practical measurements on FPGA, we can see that the time taken for AES execution follows a Gaussian distribution (Fig. 4(c)). Gaussian CD was shown to provide less security in previous simulations, compared to uniform CD. Improvements to the current clock randomiser are further proposed to bring the delay CD close to uniform.

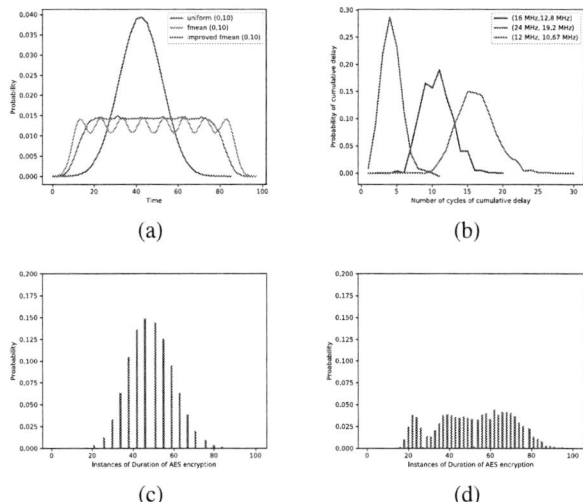

(a)   (b)

(c)   (d)

Fig. 4: (a) Cumulative Delay (*CD*) distribution generated by the three RDI (Random Delay Interrupts) methods. (b) Shifted Gaussians generated by different choices of frequencies and frequency shifts. Histogram of measurement from FPGA of cumulative delays by (c) clock randomiser in [5], (d) improved clock randomiser.

the mean and variance of the corresponding Gaussian *CD* distribution. With $2P$ clock outputs from the MMCM, we can thus have P such U-blocks and $log(2P)$ levels of multiplexers which form the improved clock randomiser as shown in Fig. 3 ($P = 4$ in our design). Thus, one U-block is selected randomly for each encryption, similar to the random selection of the small section in "Floating Mean method" (Refer Sec.IV(A)). With suitable choice of frequencies for the clock inputs to the U-blocks, one can generate a Gaussian mixture (combination of mean shifted Gaussians), thus arriving close to a uniform distribution. In order to equalize the peaks of all the smaller Gaussians, we bias and fine tune the selection of U-blocks to generate a distribution as close as possible to uniform. Refer Fig. 4(d) for the near uniform CD distribution from practical FPGA measurements.

The implementation results of the clock randomiser from [5] and our improved proposal are summarised in Tab. I. The area overhead comes from the individual PRNG required for each U-block as compared to a single PRNG in [5]. Instead of a PRNG, any randomness source like TRNG or a stream cipher could be used as select inputs for the U-blocks. Thus, barring the area used up for the PRNG, both the designs are equally efficient. It also reports the mean encryption delay (μ) and distribution peak $\hat{p}$ to compute performance security trade off as per Eq. (1). Evaluation of the metric defined in Eq. (1) reveals a theoretical 2.4× increase in efficiency of our countermeasure (`iPPAP`) compared to [5].

Not all choices of frequencies lead to a synthesizable design in FPGA as it fails to achieve timing closure. The possible reasons for this behaviour could be:

- The apparent inability of the PLL to generate precise output frequencies from a given input frequency.

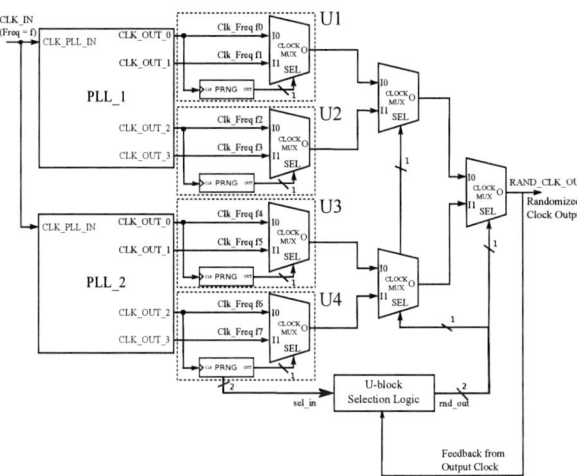

Fig. 3: Design of improved clock randomiser.

D. Improved Clock Randomiser Design

To improve the design of the clock randomiser, we analyse the output of a single clock multiplexer with two phase shifted inputs. The clock output of a single clock multiplexer with input clocks of time period t phase shifted by $p < t$, can have the following four values of time period: t, $t + p$, $2t - p$ and $2t$. Thus, increasing the number of phase shifts increases the number of possible values of time period of the output clock. Hence, we propose to use *frequency shifted* clocks to increase the number of instantaneous phase shifts between the clocks, thus yielding a higher number of *CD* instances.

We call U-block a macro which is a combination of a clock multiplexer with two frequency shifted inputs and a PRNG which generates a random select line input for the multiplexer. Fig. 4(b) shows the *CD* distribution of different combination of frequency shifted clocks. Thus, we can see that the set of frequencies of the clock inputs to a U-block determine

978-1-5386-7100-9/18 $31.00 © 2018 IEEE   623

TABLE I: Area overhead and performance-security trade-off of Clock Randomiser (CR) and Improved Clock Randomiser (ICR)

	LUT	DFF	BUFG	PLL	$\mu(\mu s)$	$\hat{p}$	$M(\mu s^{-1})$
CR [5]	23	64	8	2	0.9	0.15	3.7
ICR	102	266	11	2	1.25	0.045	8.88

- Violation due to instances when the instantaneous phase shift between the clocks is very small compared to the time period of the clocks, resulting in glitches in the clock output (although no such considerations are specified in the design documents).

One could possibly provide explicit timing constraints [9] to achieve timing closure or choose a constrained set of frequencies that avoid timing violations. In this work, we carefully arrived at certain combination of frequencies of the clock inputs to the multiplexers that lead to a synthesizable design. For a base period of the clock as t, the other time periods are computed as perturbations by a fraction of the base period. Thus, the 8 clock outputs are generated with time periods $(t, t + t/4), (3t/2, 3t/2 + (3t/2)/4), (2t, 2t + 2t/8), (5t/2, 5t/2 + (5t/2)/5)$. For our design which is also depicted in Fig.3, we utilized the t value of 41.666 nsec, which corresponds to 24 MHz. The process of automatic generation of the timing constraints file to achieve timing closure is left for future work.

V. Security Evaluation

In this section, we perform practical security evaluation of the proposed PPAP and iPPAP module against SCA and FIA.

A. Experimental Platform

For evaluating fault injection detection capabilities, we have tested the PPAP and iPPAP experimentally, against both laser fault injection (LFI) and electromagnetic fault injection (EMFI). As a device under test (DUT), we used a Xilinx Virtex-5 FPGA, mounted on a Genesys board from Digilent. We have used PRESENT-80 cipher for testing the countermeasure, where the RO was routed around this cipher circuit to protect it. The reason behind using a lightweight cipher like PRESENT-80 is to limit the size of ring oscillator. Trigger signal for LFI/EMFI activation was generated by the cipher implementation and sent during the last round. The board was placed on a X-Y positioning table with a step precision of 0.05 μm.

In the case of LFI, we used a diode pulse laser station as a fault injection device, with a near-infrared (1064 nm) laser source with a maximum output power of 20 W. We have mounted a 5× magnification lens on the laser source, decreasing the power to a maximum of 10 W and spot size to 60×14 μm^2. Fault injection was controlled by a triggering device connected to a computer. For checking the precision of injection parameters, such as offset and length, we have used a digital oscilloscope.

When it comes to EMFI, three main components were used. A pulse generator that was capable of capturing the trigger signal and setting a precise offset for the pulse generation. An amplifier, with a constant 54 dB gain. And a probe that was

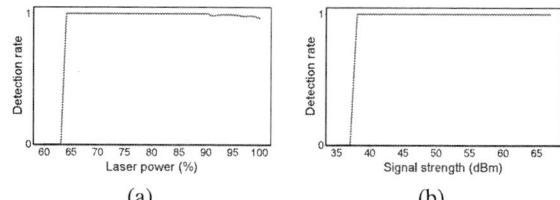

Fig. 5: (a) LFI detection rate (b) EMFI detection rate

taking the signal from the amplifier and injecting it into the DUT.

The side-channel testing was performed on SASEBO GII board which also contains the same Xilinx Virtex-5 FPGA. The usage of SASEBO board for side-channel measurements was motivated by a special support for power/EM measurements. Side-channel traces were measured using a near-field EM probe, placed over a decoupling capacitor on the power line. Measurements were sampled on the Teledyne LeCroy Waverunner 610zi oscilloscope at a sampling frequency of 500 MS/s. The PPAP module was then used to generate clock for a AES-128 module, which encrypts one round per clock.

B. Evaluation against FIA

Since LFI is a localized attack method, we first had to profile the DUT in order to estimate the area where the implementation resides. For this purpose, we have used the same methodology as described in [16]. It is important to note that the EMFI and LFI detection capabilities are the same for both PPAP and iPPAP designs. After the target area was localized, we had run several tests in order to estimate the effectiveness of the proposed countermeasure. We varied the laser power up to 100%, while scanning the whole region of the implemented cipher. The detection rate of the sensor against LFI is depicted in Fig. 5(a). We can see that the sensor detects all the injections up to $\approx$ 90% power. Afterwards, the rate slightly drops but still stays very high ($>$ 0.971). This is caused by the cipher faults that go undetected when the laser power reaches its maximum. In this case, the laser spot with the peak power becomes tightly focused at one point and is able to reach the implementation without disturbing the surrounding ring oscillator. This can be encountered with smaller loops in oscillator.

In case of EMFI, the disturbance is more global, therefore we have not observed a high variance w.r.t. chip area. Again, we varied the signal strength up to the limit of our setup (68.5 dB). The detection rate is depicted in Fig. 5(b). Results for EMFI are consistent throughout the varied power and all the faults were detected by the implemented sensor. The reason for this is the aforementioned global effect, ensuring that any EM pulse affecting the cryptographic circuit will also affect the surrounding sensor.

C. Evaluation against SCA

We used the TVLA methodology to evaluate the SCA resistance of PPAP and iPPAP implemented for an AES-128 design on the FPGA. The randomised clock from PPAP (resp. iPPAP) serves as the system clock for AES. We utilized two PLLs with 8 clock outputs, as in [5] for fair comparison. To

978-1-5386-7100-9/18 $31.00 © 2018 IEEE

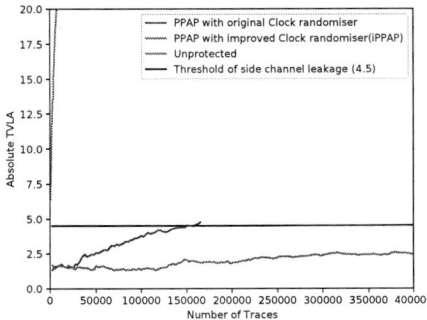

Fig. 6: TVLA evaluation for unprotected and `PPAP`-protected `AES` with original and improved clock randomiser.

TABLE II: Area overhead of `PPAP` and `iPPAP` with `AES`-128 on Xilinx Virtex 5 FPGA

Implementation	LUT	Registers	BUFG	PLL_ADV
Unprotected	1850	742	1	0
PPAP	1874	806	9	2
iPPAP	1954	1008	12	2

compare the resistance, we plot the evolution of the t-value against number of traces. The results are shown in Fig. 6. Analysis revealed that unprotected `AES` already failed the t-test at 80 traces, the `PPAP` required around $145,000$ traces to cross the TVLA threshold, while `iPPAP` stays below the threshold until $400,000$ traces. Thus, `iPPAP` provides enhanced security `PPAP` at minor area and performance overhead.

The area results are reported in Tab. II. The main overhead of `iPPAP` (compared to `PPAP`) comes from the use of multiple PRNGs. Note that area of `PPAP` and `iPPAP` are constant and independent of size of sensitive circuits. Thus, for larger and resource-heavy designs (like RSA or ECC), the resource overhead will be near negligible.

D. Discussions

Realignment techniques can be potentially applied to defeat `PPAP` and `iPPAP`. Briefly looking into popular methods, algorithms like Threshold Phase Only Correlation (T-POC [17]) and Rapid Alignment (RAM [18]) can resync traces shifted in time (phase misalignment), and algorithms like Elastic alignment (Dynamic Time Warping [19]) and Re-synchronization by Moments [20] can also work on aligning time stretched waveforms, thus improving attacker efficiency. The complexity of most of these algorithms depends on the number of sample points on the trace s. While the computational complexity is quasilinear with s ($\mathcal{O}(s\log(s))$) for T-POC and linear with s ($\mathcal{O}(s)$) for the elastic alignment algorithm [19]. However, the elastic alignment algorithm can only work on aligning two traces at a time, which requires a reference waveform. Hence the computational complexity of these pre-processing algorithms can be increased by simply increasing the misalignment variance, albeit with some performance overhead. Additionally, there are also certain assumptions like the constant frequency of the input clock for some of the resync algorithms (POC, T-POC), which thus do not work with `iPPAP` which outputs a variable frequency output clock.

VI. CONCLUSIONS

This paper presents a combined SCA and FIA hardware-based countermeasure termed `PPAP`. It is a low-cost, plug and play countermeasure, independent of the protected circuit. Further, an improvement in form of `iPPAP` is proposed to boost SCA security, by ensuring a near-uniform distribution of the introduced cumulative delay. Both countermeasures are validated against practical SCA & FIA, reporting high security gain. Further research can focus on all-digital alternatives of `PPAP` and `iPPAP`. Moreover, resistance against resynchronisation technique should be evaluated.

REFERENCES

[1] P. Kocher, J. Jaffe, and B. Jun, "Differential power analysis," in *Advances in cryptology – CRYPTO99*. Springer, 1999, pp. 789–789.

[2] E. Biham and A. Shamir, "Differential fault analysis of secret key cryptosystems," in *CRYPTO'97*. Springer, 1997, pp. 513–525.

[3] T. Schneider, A. Moradi, and T. Güneysu, "Parti–towards combined hardware countermeasures against side-channel and fault-injection attacks," in *Annual Cryptology Conference*. Springer, 2016, pp. 302–332.

[4] T. De Cnudde and S. Nikova, "More efficient private circuits ii through threshold implementations," in *Fault Diagnosis and Tolerance in Cryptography (FDTC), 2016 Workshop on*. IEEE, 2016, pp. 114–124.

[5] T. Güneysu and A. Moradi, "Generic side-channel countermeasures for reconfigurable devices," in *International Workshop on Cryptographic Hardware and Embedded Systems*. Springer, 2011, pp. 33–48.

[6] V. Lomné, T. Roche, and A. Thillard, "On the need of randomness in fault attack countermeasures-application to aes," in *Fault Diagnosis and Tolerance in Cryptography 2012*. IEEE, 2012, pp. 85–94.

[7] N. Miura, Z. Najm, W. He, S. Bhasin, X. T. Ngo, M. Nagata, and J.-L. Danger, "Pll to the rescue: a novel em fault countermeasure," in *Design Automation Conference (DAC) 2016*. IEEE, 2016, pp. 1–6.

[8] B. J. Gilbert Goodwill, J. Jaffe, P. Rohatgi *et al.*, "A testing methodology for side-channel resistance validation," in *NIST non-invasive attack testing workshop*, 2011.

[9] A. G. Bayrak, N. Velickovic, F. Regazzoni, D. Novo, P. Brisk, and P. Ienne, "An eda-friendly protection scheme against side-channel attacks," in *Proceedings of the Conference on Design, Automation and Test in Europe*, 2013, pp. 410–415.

[10] S. Mangard, "Hardware countermeasures against dpa–a statistical analysis of their effectiveness," in *Cryptographers Track at the RSA Conference*. Springer, 2004, pp. 222–235.

[11] N. Mentens, B. Gierlichs, and I. Verbauwhede, "Power and fault analysis resistance in hardware through dynamic reconfiguration," in *CHES*. Springer, 2008, pp. 346–362.

[12] J.-S. Coron and I. Kizhvatov, "Analysis and improvement of the random delay countermeasure of ches 2009," in *CHES 2010*, 2010, pp. 95–109.

[13] ——, "An efficient method for random delay generation in embedded software," in *CHES 2009*, 2009, pp. 156–170.

[14] S. Bhasin, J.-L. Danger, F. Flament, T. Graba, S. Guilley, Y. Mathieu, M. Nassar, L. Sauvage, and N. Selmane, "Combined sca and dfa countermeasures integrable in a fpga design flow," in *ReConFig'09*. IEEE, 2009, pp. 213–218.

[15] J. Breier and X. Hou, "Feeding two cats with one bowl: On designing a fault and side-channel resistant software encoding scheme," in *Cryptographers Track at the RSA Conference*. Springer, 2017, pp. 77–94.

[16] W. He, J. Breier, S. Bhasin, D. Jap, H. G. Ong, and C. L. Gan, "Comprehensive laser sensitivity profiling and data register bit-flips for cryptographic fault attacks in 65 nm fpga," in *SPACE*, 2016, pp. 47–65.

[17] S. Guilley, K. Khalfallah, V. Lomne, and J.-L. Danger, "Formal framework for the evaluation of waveform resynchronization algorithms." *WISTP*, vol. 6633, pp. 100–115, 2011.

[18] R. A. Muijrers, J. G. van Woudenberg, and L. Batina, "Ram: Rapid alignment method," in *International Conference on Smart Card Research and Advanced Applications*. Springer, 2011, pp. 266–282.

[19] J. G. van Woudenberg, M. F. Witteman, and B. Bakker, "Improving differential power analysis by elastic alignment," in *Cryptographers Track at the RSA Conference*. Springer, 2011, pp. 104–119.

[20] N. Debande, Y. Souissi, M. Nassar, S. Guilley, T.-H. Le, and J.-L. Danger, "re-synchronization by moments: an efficient solution to align side-channel traces," in *WIFS 2011*, pp. 1–6.

Mystic: Mystifying IP Cores Using an Always-ON FSM Obfuscation Method

Ehsan Aerabi, Ahmad Patooghy, Hamidreza Rezaei, Miguel Mark, Mahdi Fazeli, and Michel A. Kinsy

Adaptive and Secure Computing Systems (ASCS) Laboratory
Department of Electrical and Computer Engineering, Boston University

Abstract—The separation of manufacturing and design processes in the integrated circuit industry to tackle the ever increasing circuit complexity and time to market issues has brought with it some major security challenges. Chief among them is IP piracy by untrusted parties. Hardware obfuscation which locks the functionality and modifies the structure of an IP core to protect it from malicious modifications or piracy has been proposed as a solution. In this paper, we develop an efficient hardware obfuscation method, called Mystic (Mystifying IP Cores), to protect IP cores from reverse engineering, IP overproduction, and IP piracy. The key idea behind Mystic is to add additional state transitions to the original/functional FSM (Finite State Machine) that are taken only when incorrect keys are applied to the circuit. Using the proposed Mystic obfuscation approach, the underlying functionality of the IP core is locked and normal FSM transitions are only available to authorized chip users. The synthesis results of ITC99 circuit benchmarks for ASIC 45nm technology reveal that the Mystic protection method imposes on average 5.14% area overhead, 5.21% delay overhead, and 8.06% power consumption overheads while it exponentially lowers the probability that an unauthorized user will gain access to or derive the chip functionality.

I. INTRODUCTION

Growing complexity and critical time-to-market have played key roles in the current semiconductor design and manufacturing supply chain landscape. For example, many fab-less companies have emerged to take advantage of low-cost overseas foundries for IC production. Companies, such as ARM Holdings, develop and sell their soft intellectual properties (IP) to other chip manufacturers for the hard implementations. There is a large international market for these pre-built, verified and ready-to-use soft IP designs. Under this globalization trend, *IP piracy* has become an increasing concern which has drawn a great deal of research and investment from both academia and industry [1]. For instance, an untrusted party can reverse-engineer and steal an IP core and then claim ownership, resell or over produce it [2].

Logic Masking is a set of IP piracy prevention methods which obfuscate the circuit's main functionality to prevent unauthorized access to the chip's functionality. The circuit cannot properly operate until the owner activates it by means of an activation key. Logic masking is generally achieved by inserting *Key Gates e.g XOR/XNOR/MUX/AND/OR* into the original combinational circuit, each of which is driven by a bit of the activation key. These key gates mask the circuit's functionality in a way that only a unique correct key can neutralize their effects. After chip fabrication, the secret key is programmed usually in a secure internal EEPROM memory and the masked IP is unlocked by the owner.

In order to maximize the mismatch points between the masked circuit and the original circuit when comparing them by formal methods, Chakraborty *et. al* [3] proposed a method based on fan-in and fan-out cones in the circuit. Rajendran *et. al* [4] presented another logic masking method in which key gates have more effect on each other. This hinders attacker's effort to reveal the key by feeding the circuit with specific inputs and propagating uncorrelated key bits to the outputs. Using the concept of *Fault Propagation*, [5] tries to perform a proper key gate insertion. The aim is to have nearly 50% of the wrong output bits when a wrong key-vector is applied. By choosing appropriate nets to insert *XOR/XNOR* gates, one can achieve 50% correctness in the Hamming Distance (*HD*) among outputs for the valid key and invalid key. Using the *MUX* primitive as key-gates instead of *XOR/XNOR* has been proposed in [6] and [5]. When supplied key-inputs are correct, the MUXs pass the correct input, otherwise they pass a wrong value coming from other parts of the circuit. Overall, the goal in [5] is to achieve a high *HD* between the correct and wrong keys.

Sequential circuit masking methods try to obfuscate the finite state machine (FSM) of the circuit at the system level perspective. Under this set of approaches, the original state transitions are modified in a way that only a unique sequence of keys can drive it through its correct transitions. Otherwise, the system is lost in out-of-order or fake states.

Chakraborty *et. al* proposed an FSM-based method called HARPOON [7], which adds a finite state machine to the IP core netlist. The FSM outputs are connected to the internal nodes of the circuit via some *XOR* gates. Therefore, the circuit cannot properly work until the output of the added FSM becomes logical zero. Zhang *et. al* presented an IP protection and FPGA licensing scheme which combined FSM masking with PUFs (Physical Unclonable Function) [8]. Recently, Sumathi *et. al* [9] published an FSM-based IP protection to improve the HARPOON approach.

The contribution of this paper is two fold.

- *A new hardware attack specification:* although substantial research has been done on sequential logic masking, we show that most of the previous sequential logic masking techniques are vulnerable to *FSM Separation Attack*.
- *A new obfuscation method:* to address this weakness, we propose an FSM-based logic masking technique at the RTL level. The proposed method can effectively protect against *FSM Separation Attacks* as well as the recently presented SAT attacks [10], [11].

This paper is organized as follows. In Section II, we provide a brief overview of the previously proposed obfuscation methods. Section III introduces our proposed attack method. In Section IV, we explain our proposed FSM encoding method and show how it overcomes the mentioned weakness using an illustrative case. Section V contains an explanation of the experimental system setup and results. Finally, Section VI concludes the paper.

II. BACKGROUNDS

Logic masking protection mechanisms can be divided into "sequential logic" and "combinational logic" protections. For sequential logic circuits, the protection method is applied to the state transition graph of the circuit by adding extra states with the aim of masking or authenticating [9], [3], [12], [13]. Almost all of the previously proposed sequential encodings are based on the concept illustrated in Figure 1. As shown in the figure, a set of obfuscated states is added to the FSM of the original design. The circuit starts from an initial state in the obfuscated states set. In this initial state, the circuit is locked and its produced outputs are intentionally wrong. To successfully traverse the obfuscated states and enter the original states of the circuit, one must apply to the input(s) the correct sequence of i_0 to i_n for $(n+1)$ consecutive clock cycles. Exiting the obfuscated states will lead to the first state within the normal FSM, T_0. Therefore, the circuit will now correctly respond to inputs since outputs here are functions of inputs and the original FSM states. If k is the number of primary inputs which are applied to the obfuscated states, then an attacker needs to potentially perform $2^k \times 2^n$ searches to unlock the circuit.

For the combinational logic circuits, some extra gates (XOR/XNOR or multiplexer) are inserted into the original combinational circuit. Each obfuscating gate has an input that is derived from the secret key, so that the correct combination of the key bits would neutralize the masking effect of these gates. Consequently, an incorrect input key will lead to incorrect circuit functionality. XOR/XNOR gates could be inserted randomly in the circuit as expressed in [14], but there is no guarantee of the circuit malfunctioning if the wrong keys are used as input. Some researchers [3] have tried to improve the robustness of these obfuscation methods by combining combinational and sequential techniques. To achieve this goal, the outputs of obfuscated states e.g., S_0 to S_{n-1} are connected to *Modification Cells* which are extra logic inserted into the circuit's combinational part (See Figure 2). While the circuit is in the obfuscated states, the outputs of S_0 to S_{n-1} enable the Modification Cells and disables normal circuit operation. The Modification Cell combines the original net of the circuit (p) with a high fan-in signal borrowed from another part of the circuit to add more obfuscation.

The output of the Modification Cell is often expressed as an $output = p \cdot \bar{f} + \bar{p} \cdot g \cdot f$ where f is the obfuscation enabling signal, $f = 1$ when the circuit is in the obfuscated states and otherwise $f = 0$. p is the original net and g is the high fan-in net. Since f is a function of S_0 to S_{n-1}, it evaluates to zero

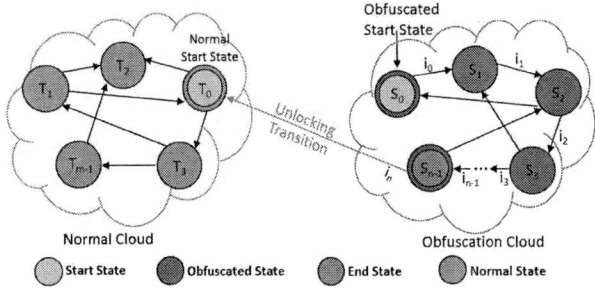

Fig. 1. General block diagram of FSM encoding methods.

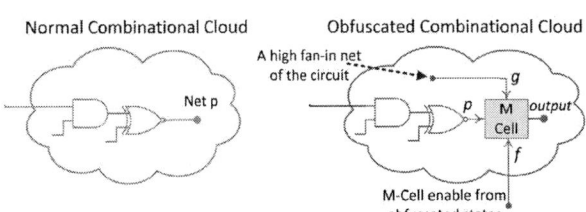

Fig. 2. Combinational logic obfuscation using modification cells.

whenever the obfuscated FSM goes through the *Unlocking Transitions* and reaches the normal FSM.

III. FSM SEPARATION ATTACK

In this section we describe the *FSM Separation Attack* which can exponentially reduce the search space for attackers to unlock an obfuscated circuit. This builds on the work presented in [15] with key clarifications to the attack steps and concrete implementations of the attack on real circuits.

Suppose a circuit which is jointly protected using FSM and combinational obfuscation methods as described in the previous section. We know that the circuit has some memory elements storing its obfuscated S_0 to S_{n-1}, and original T_0 to T_{m-1} states along with some combinational parts. The resilience of the circuit relies on the fact that the attacker cannot distinguish between the added state elements S_0 to S_{n-1} and the original ones T_0 to T_{m-1}. If an attacker can manage to find the added states, they would be able to traverse all 2^n value space of S_0 to S_{n-1} to find out which one unlocks the circuit; then they can set it to obtain the normal operation of the circuit illegally. The FSM separation attack has three stages:

- In stage one, the circuit HDL code is used by an attacker to divide the combinational and sequential parts of the circuit. Note that this is possible since FSM memory elements can be easily distinguished from combinational part of a given IP core. However, since the attacker does not know how many state elements were added to the original circuit, the attack moves to stage two.

- For stage two, the attacker has to assume all possible values for n from 1 to L where L is the total number of state elements of the circuit, $L = m + n$. The attacker needs to figure out which subset of L states are the added S_i's. Assuming $\hat{n}$ as a hypothesis for n, there are $\binom{L}{\hat{n}}$ possible combinations for S_0 to $S_{\hat{n}-1}$.

- In the final stage, for each hypothesis, the attacker needs to (i) eliminate flip-flops which are assumed to be the (added) obfuscation ones, (ii) put zero as their outputs and (iii) see if their elimination unlocks the circuit.

It is an iterative attack where the second and third stages will loop until the circuit is unlocked and produces a valid response. The total number of trials that the attacker needs to unlock the circuit can be computed by Equation (1).

$$\Psi = \sum_{\hat{n}=1}^{L} \binom{L}{\hat{n}} = 2^L \qquad (1)$$

It is worth noting that this number differs from the search space presented in [3] i.e., $(2^{m \times k})$ where m is the number of a circuit's obfuscated states and k is the number of a circuit's primary inputs. 2^L could still be a large number of states with a large number of flip flops, but $\hat{n}$ is generally a very small number due to the overhead associated with the added FSM. Hence, the iteration loop will terminate much sooner than exhausting the entire 2^L space. We will show this fact in an experimental study later in the section. A key aspect of the FSM separation attack is that its search space does not depend on the number of circuit inputs k. In fact, this attack reduces the attack complexity by separating the combinational and sequential parts of the circuit.

To examine the impacts of this complexity reduction, let us consider the following synthetic [3] and real examples. Assuming $k = n = 16$, the computation space is $2^{256} \approx 10^{77}$ which is infeasible to search. For real commercial IP cores, we consider NEO430 and ao68000, two CISC (Complex Instruction-Set Computers) open core CPUs. These IPs are relatively big circuits e.g., 3500 VHDL lines for ao68000. They are well within the range of real world circuits.

The NEO430 and ao68000 have 860 and 724 memory elements i.e., flip-flops, respectively. However, the largest logic block in these CISC IPs is a 5-bit state variable and therefore less than $2^5 = 32$ states. Obviously a large portion of memory elements in each IP is used to store processed data and a very little portion is used as state holders. This means that when a designer chooses a bigger number for n to make it more difficult for the attacker to traverse the computational space, the hardware overhead should be considered with respect to the number of memory elements which are doing state holding. For example [3] has reported 18.44 and 15.88 percent overheads for only six added state elements. These sample circuits confirm that in the real world, (1) parameter m is not very large and (2) overheads of using a large number of obfuscating states i.e., parameter n can be very high. To investigate the effects of FSM separation attacks against existing protection schemes on real circuits, we studied the feasibility of these attacks on ISCAS'89 circuits. In our evaluations, we used the largest circuits namely (a) s38417 circuit with 28 primary inputs and 1635 D-type flip-flops and (b) s38584 circuit with 38 inputs and 1425 D-type flip-flops. The circuits are synthesized targeting the Spartan-6 FPGA board using the Xilinx ISE Design Suite operating at 100 MHz.

TABLE I
SUCCESSFUL ATTACK TIME ESTIMATION FOR ISCAS CIRCUITS OBFUSCATED BY TRADITIONAL METHODS.

Circuit	#FF	\multicolumn{10}{c}{Number of Added State Elements}									
		1	2	3	4	5	6	7	8	9	10
S298	14	3 μs	33 μs	203 μs	968 μs	3 ms	13 ms	42 ms	122 ms	326 ms	817 ms
S344	15	4 μs	38 μs	242 μs	1 ms	5 ms	18 ms	61 ms	183 ms	510 ms	1 s
S349	15	4 μs	38 μs	242 μs	1 ms	5 ms	18 ms	61 ms	183 ms	510 ms	1 s
S526	22	5 μs	74 μs	649 μs	4 ms	24 ms	118 ms	508 ms	1 s	7 s	23 s
S641	19	5 μs	57 μs	442 μs	2 ms	13 ms	57 ms	222 ms	777 ms	2 s	7 s
S713	19	5 μs	57 μs	442 μs	2 ms	13 ms	57 ms	222 ms	777 ms	2 s	7 s
S838	32	8 μs	148 μs	1 ms	16 ms	125 ms	815 ms	4 s	23 s	1 M	7 M
S1196	18	4 μs	52 μs	384 μs	2 ms	10 ms	44 ms	164 ms	555 ms	1 s	5 s
S1238	18	4 μs	52 μs	384 μs	2 ms	10 ms	44 ms	164 ms	555 ms	1 s	5 s
S1423	74	18 μs	731 μs	19 ms	375 ms	6 s	1 M	15 M	2 H	1 D	9 D
S1488	6	1 μs	8 μs	29 μs	82 μs	197 μs	428 μs	857 μs	1 ms	2 ms	4 ms
S5378	179	45 μs	4 ms	251 ms	11 s	7 M	3 H	4 D	96 D	6 Y	105 Y
S9234	211	53 μs	5 ms	408 ms	22 s	15 M	9 H	12 D	346 D	23 Y	517 Y

For the sake of fairness, we used the same simulation setup as used in [3]. We added two extra d-type flip-flops and inserted four XOR gates into high fan-in nets in s38417 and s38584 circuits. Based on normal calculations, an *exhaustive* attack will test $2^{38 \times 4} = 2^{152}$ inputs to unlock the s38584 circuit, and $2^{28 \times 4} = 2^{112}$ inputs for the s38417 circuit. These circuits were assumed unbreakable for a polynomial time attack scheme. However, we showed that if an FSM separation attack is occurs, the circuit degenerates into lower orders.

To attack the s38417 circuit, we assume 1 to 1635 of flip-flops as FSM masking ones and check our hypothesis. This attack is accomplished in a short time, since only two of the 1635 flip-flops are intended to do FSM encoding i.e., the circuit has just 4 obfuscating states. The search took $1,340,703$ clock cycles in our simulation environment. This means that it took about 335 milliseconds to attack the obfuscated s38417 circuit. We performed the same FSM separation attack on the obfuscated circuit of s38584 and were able to break the circuit in only 254 milliseconds. It should be noted that the FSM state elements in ISCAS'89 circuits are not distinguishable from the rest of the memory elements, otherwise the attack could be significantly faster.

Table I estimates the FSM separation attack duration time for some of the other ISCAS circuits when different numbers of flip-flops are used in the obfuscating FSM. In this table we have examined up to 10 added obfuscation flip-flops and calculated the required time for a successful FSM separation attack. For those circuits with a large number of flip-flops (e.g S9234 and S5378) the attack time is in order of hundreds of years (which is still assumed a feasible attack on distributed and parallel systems). Nevertheless, we can conclude that regardless of the circuit size, using a separated obfuscating FSM to lock the chip cannot protect the chip especially when the number of memory elements in the obfuscating FSM is not very high (see S1488 results with 6 flip-flops in Table I). On the other hand, adding a large number of obfuscating states implies an unacceptable overhead on the protected circuit.

978-1-5386-7100-9/18 $31.00 © 2018 IEEE

In this paper, we propose a method where the robustness of obfuscation does not depend on the number of added states. Since it checks for the correct key before every FSM transition, it can be used for any desired level of obfuscation.

IV. THE PROPOSED *Mystic* METHOD

In order to reduce the chance of a successful attack, we believe that the added obfuscating FSM should not be completely separated from the original FSM. We propose a masking technique that is active during the whole operation of the circuit in its lifetime. The technique combines original and obfuscated states and significantly reduces the probability of an attack successfully unlocking the target circuit. As shown in Figure 3, the circuit starts working from an original state and works correctly. However, in each state in the circuit's FSM, a set of the key bits should be correct in order for the FSM to go through the correct transition and operate consistently. On a wrong key, the FSM goes to a wrong state which will perturb the entire computation for the rest of the circuit operation.

The key underpinning of the proposed technique is the concept that a designer may add more state transitions and/or additional states to mask the original FSM. Going back to Figure 1, it should be highlighted that the proposed method actually adds a *Locking Transition* to the model in opposite direction of the *Unlocking Transition*. To compensate for a small state space, a designer may add extra states as a means of increasing the transition candidates for the masking process.

We propose an iterative masking algorithm to systematically add obfuscation transitions or states to a given circuit. On each iteration, the algorithm chooses a state to which an extra obfuscating transition is going to be added. To choose an appropriate state, we propose a simple but effective *Security Metric* that can be assigned to each state s as Equation (2)

$$SM_s = \frac{OutEdges_s}{KeyCount_s + 1} \qquad (2)$$

where $OutEdges_s$ is the total number of transitions started from the state s and $KeyCount_s$ is the number obfuscating transitions which have been previously added to state s during the masking process. One general observation is that the states with high $OutEdge$ tend to be more critical in the operation of the FSM. Therefore, an intuitive and judicious way for selecting the order of states to guard when all the circuit states cannot be guarded is to use the $OutEdge$ degree. In addition, having the $KeyCount$ on the denominator of the metric gradually decreases the importance of previously masked states. *Locking transitions* are added to the original circuit's FSM in a manner that allows the circuits to check their activation keys at runtime. Whenever the input key is not valid, the circuit follows one of the added *Locking transitions* and goes to an intentionally wrong state.

Algorithm 1 presents the masking procedure. It receives an input key $\mathcal{K}$ and an FSM graph $\mathcal{F}$. First, in lines 2 and 3, it prepares two lists for storing $KeyCount$s and Security Metrics associated with each state. Then in lines 5 to 8, it calculates initial security metrics for all states. The main masking iterations start from line 10. For each key bit $\mathcal{K}[j]$,

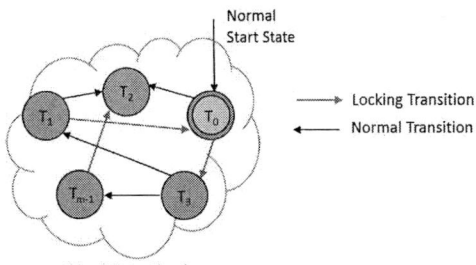

Fig. 3. Our proposed FSM encoding methods.

Algorithm 1: The Proposed Masking Algorithm

Input: Key $\mathcal{K}$
Input: FSM graph $\mathcal{F}$ with $S_{\mathcal{F}}$ states and $E_{\mathcal{F}}$ edges
Result: Obfuscated FSM graph $\mathcal{F}$

1 // Initialization
2 Define zero-initialized list of integers $KeyCount$ with size of $|S_{\mathcal{F}}|$;
3 Define zero-initialized list of floats $SecMetric$ with size of $|S_{\mathcal{F}}|$;
4 // Calculating Security Metrics
5 **for** $i \leftarrow 1$ **to** $|S_{\mathcal{F}}|$ **do**
6 $\quad OutEdge \leftarrow$ number of edges in $E_{\mathcal{F}}$ starting from $S_{\mathcal{F}}[i]$;
7 $\quad SecMetric[i] \leftarrow \frac{OutEdge}{KeyCount[i]+1}$;
8 **end for**
9 // Masking $\mathcal{F}$
10 **for** $j \leftarrow 1$ **to** $|\mathcal{K}|$ **do**
11 $\quad m \leftarrow$ index of the largest value in $SecMetric$;
12 $\quad$ Add to $E_{\mathcal{F}}$ an edge from $S_{\mathcal{F}}[m]$ to a random state in $S_{\mathcal{F}}$ with $\bar{\mathcal{K}}[j]$ activator;
13 $\quad$ //Update Security Metric for $S_{\mathcal{F}}[m]$
14 $\quad KeyCount[m] \leftarrow KeyCount[m] + 1$;
15 $\quad SecMetric[m] \leftarrow \frac{OutEdge}{KeyCount[m]+1}$;
16 **end for**
17 Return $\mathcal{F}$;

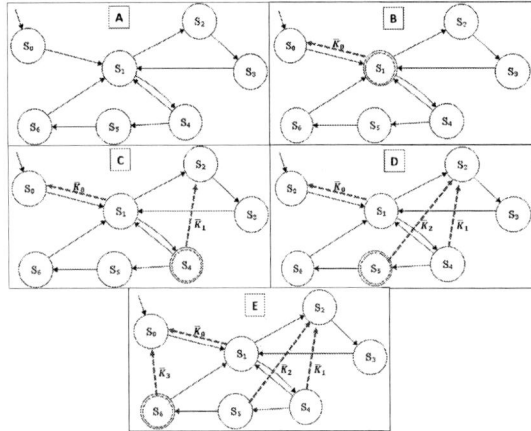

Fig. 4. An example of the proposed masking algorithm.

978-1-5386-7100-9/18 $31.00 © 2018 IEEE

the algorithm finds the state $S_{\mathcal{F}}[m]$ with the highest security metric value and then adds a transition $E_{key=\bar{\mathcal{K}}[j]}$ starting from $S_{\mathcal{F}}[m]$ which ends at a random (not already connected) state. The transition $E_{key=\bar{\mathcal{K}}[j]}$ shifts the FSM to a wrong state whenever the attacker's key bit is not equal to the correct value $\mathcal{K}[j]$.

To illustrate the operation of the proposed *Mystic* algorithm, we applied it to the FSM graph shown in Figure 4-A. As shown in this figure, the highest Security Metric belongs to states S_1 and $S4$ due to their higher *outEdge* degree which is equal to 2. Security Metric for the other five states is 1. Therefore the first bits of the key are negated and inserted as obfuscating transition of state S_1 in Figure 4-B and state S_4 in Figure 4-C. By updating the security metrics, we have all states with security metrics of 1. The algorithm chooses a random state on each iteration (S_5 and S_6 on Figure 4-D and 4-E) and adds an obfuscating transition to them. This process continues until all bits of the activation key are used. Note that it is possible for a transition to have multiple key bits or even a function of key bits as an activating function on obfuscating transitions. To answer the question of wether we have multiple key bits activator or not, we have to compare $|\mathcal{K}|$ with $|S_{\mathcal{F}}|$. If $|\mathcal{K}| \geq |S_{\mathcal{F}}|$, we have some states with multiple key bits activators, but when $|\mathcal{K}| < |S_{\mathcal{F}}|$ we can manage to have no such state. Since allowing multiple key bits as activator i) increases the hardware overhead of our proposed method and ii) loses the termination condition of our proposed masking algorithm, in this example we bound the number of added obfuscating transitions to 4.

In terms of estimating the obfuscation level of *Mystic*, it is important to note that each state is obfuscated using a subset of the masking key. All of the key bits in that subset should be correct for the FSM to perform a single transition correctly. Similarly, for the following transition, another subset of the key bits should be correct. From the attacker's perspective, for each state, a subset of the key bits needs to be guessed correctly. For key hypothesis checking, the attacker needs to test all possible values for all combinations of key subsets with different lengths. The compute complexity of this verification operation is:

$$\phi = \sum_{k=1}^{|\mathcal{K}|} \binom{|\mathcal{K}|}{k} 2^k \tag{3}$$

It is worth noting that the key size $|\mathcal{K}|$ in (3) is much bigger than the number of states in the circuit, L. In fact, probability of correctly passing a state by an unauthorized attacker is $\frac{1}{\phi}$.

Generally, the attacker's only reference to evaluate the correctness of a key guess is the output of the circuit i.e., a correct output. Under the *Mystic* obfuscation method, the attacker does not have any reference output, especially when it comes to large IP cores like CPUs and cryptographic cores. This constitutes another important security feature of the *Mystic* technique to further reduce the chance of a successful attack. Because these large IPs oftentimes do not produce meaningful intermediate outputs, the attack must pass several transitions correctly to produce a meaningful output. Without

TABLE II
OVERHEADS OF THREE SAMPLE OBFUSCATED IP CORES USING *Mystic* SYNTHESIZED FOR A XILINX VIRTEX-7 FAMILY FPGA.

Key Length	AES IP Core			CR16 IP Core			RISC IP Core		
	Slice Reg.	Slice LUTs	Slices	Slice Reg.	Slice LUTs	Slices	Slice Reg.	Slice LUTs	Slices
0	824	2873	1073	4211	1920	1330	4473	1825	1490
1	824	2873	1075	4211	1920	1329	4473	1828	1492
4	824	2876	1079	4211	1922	1335	4474	1841	1516
8	826	2894	1089	4213	1931	1359	4491	1899	1593
16	831	2991	1094	4221	2172	1396	4549	2014	1689
32	847	3224	1102	4296	2354	1416	4678	2288	1803
64	859	3489	1131	4384	2567	1489	4792	2413	1898
96	863	3755	1253	4432	2931	1564	4881	2698	1972
128	872	4093	1390	4503	3456	1679	5153	2984	2299
AVG	841	3229	1142	4298	2352	1433	4662	2198	1750

an intermediate output, an attacker would need to guess all the key bits used in several states. Therefore the probability that an unauthorized attacker generates a meaningful output is reduced to $\frac{1}{\phi^v}$ where v is the average number of cycles needed to produce the next meaningful output of the obfuscated circuit.

V. *Mystic* HARDWARE OVERHEADS

To evaluate hardware overheads of the *Mystic* technique, we have developed a CAD tool in Python. The tool takes the RTL description of a circuit, extracts the state machine and then obfuscates it using the key provided by the user. Finally, the original FSM inside the circuit is replaced with the obfuscated FSM for heightened security. We performed our experiments on three benchmarks circuits of 1) AES cryptographic core, 2) A RISC processor and 3) CR16 microprocessor. Results of the synthesis for both FPGA and ASIC implementations are compared with those of the recent approach proposed in [8].

A. FPGA Implementation Results

The synthesis results of the three benchmarks on the Xilinx Virtex-7 xc7vx330t FPGA board are shown in Table II. The first row in this table shows the hardware utilization for the three normal benchmarks with no obfuscation. We performed the obfuscation under 1, 4, 5, 16, 32, 64, 96, and 128 widths of key. Based on the results, overhead growth is not very sharp i.e., we have the highest area overhead of 14% for the RISC IP Core benchmark when the obfuscation key of 128 bits is used. The main reason of such a relatively low overhead is that the FSM part is not normally a large portion of the whole digital circuit. It can be seen that the key length growth mostly affects the number of used LUTs in the FPGA implementation. Since the proposed masking algorithm does random selections in some steps, we repeated the obfuscation process of benchmarks with 128-bit key for 10 times. Figure 5b shows the average overheads when a 128-bit obfuscation is done on benchmarks. We have also compared the *Mystic* method with a recent work presented in [8] in terms of area, delay, and power. Figure 5a illustrates the overheads comparison between *Mystic* and the PUF-FSM based method.

B. ASIC Implementation Results

In the second set of experiments, we added the ITC99 circuit benchmarks to the three mentioned cores. We synthesized

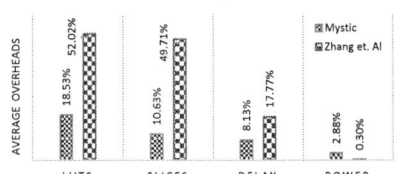

(a) FPGA resource overheads for *Mystic* and [8].

(b) *Mystic* overheads per obfuscated IP cores.

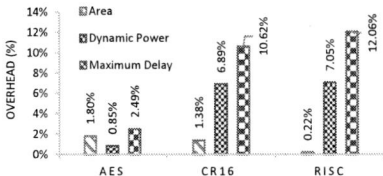

(c) *Mystic* ASIC implementation overheads.

Fig. 5. FPGA Virtex-7 and ASIC 45nm technology implementations resource utilization results.

TABLE III
OVERHEAD RESULTS FOR ITC99 CIRCUIT BENCHMARKS
SYNTHESIZED FOR AN ASIC 45NM TECHNOLOGY.

Circuit	Design Overheads (%)		
	Area	Delay	Power
b01	3.14	2.55	5.63
b02	3.05	2.41	5.88
b03	2.11	1.89	4.12
b04	2.14	1.95	4.05
b05	2.89	2.23	5.41
b06	3.12	2.65	5.88
b07	3.16	2.89	5.69
b08	2.88	2.11	5.01
b09	2.96	2.32	4.87
b10	3.99	2.96	6.56
b11	3.55	2.88	6.74
b12	4.55	3.98	7.86
b13	4.14	3.87	7.42
b14	6.08	8.97	10.56
b15	4.84	5.44	8.67
b17	4.96	5.23	8.91
b18	8.06	8.88	11.65
b19	9.34	9.76	11.98
b20	9.55	9.95	12.58
b21	9.61	10.58	12.74
b22	10.88	12.65	14.23
b30	8.54	8.53	10.89
Average	5.14	5.21	8.06

all 26 circuits using the Synopsis Design Compiler tool for 45nm technology and overheads are logged. The area, dynamic power and maximum delay overheads of the three IP cores with respect to their non-obfuscated versions are shown in Figure 5c. The AES circuit has the lowest overheads and RISC processor has the highest. This is due to more complex combinational part of the AES circuit in comparison with its simple sequential logic. The *Mystic* method does not add any additional state to the FSM. Instead it adds extra combinational logic to produce the transition guard considering the key inputs. Table III shows the overheads for an obfuscated ITC99 circuit benchmark by *Mystic* for the ASIC 45nm technology.

VI. CONCLUSIONS

In this paper, we showed that the previously proposed FSM obfuscation methods can be easily broken with a simple *FSM separation attack*. This attack is able to break secure-through-obfuscation circuits with a very low time complexity. We also presented an always-on obfuscating method which acts as a security watchdog at runtime and throughout the lifetime of the chip. When an attacker applies the first wrong key, this action activates the security watchdog and the chip goes to an intentionally wrong state resulting in incorrect functionality. Since the proposed method has a runtime defense mechanism and covers the lifetime of the chip, it is also robust against SAT attacks. The synthesis results and the comparative study with

previous obfuscation methods show that the proposed method provides stronger circuit obfuscation guarantees with better efficiency in terms of area, delay, and power consumption.

VII. ACKNOWLEDGMENTS

This research is partially supported by the NSF grant (No. CNS- 1745808).

REFERENCES

[1] M. Yasin, J. J. Rajendran, O. Sinanoglu, and R. Karri, "On improving the security of logic locking," *IEEE Transactions on Computer-Aided Design of Integrated Circuits and Systems*, vol. 35, no. 9, pp. 1411–1424, 2016.

[2] "Innovation Is at Risk as Semiconductor Equipment and Materials Industry Loses up to $4 Billion Annually Due to IP Infringement," http://www.marketwired.com, [Online; accessed 12-July-2017].

[3] R. S. Chakraborty and S. Bhunia, "Harpoon: An obfuscation-based soc design methodology for hardware protection," *IEEE Transactions on Computer-Aided Design of Integrated Circuits and Systems*, vol. 28, no. 10, pp. 1493–1502, 2009.

[4] J. Rajendran, Y. Pino, O. Sinanoglu, and R. Karri, "Security analysis of logic obfuscation," in *Proceedings of the 49th Annual Design Automation Conference*, ser. DAC '12. New York, NY, USA: ACM, 2012, pp. 83–89.

[5] J. Rajendran, H. Zhang, C. Zhang, G. S. Rose, Y. Pino, O. Sinanoglu, and R. Karri, "Fault analysis-based logic encryption," *IEEE Transactions on Computers*, vol. 64, no. 2, pp. 410–424, 2015.

[6] A. Nejat, D. Hely, and V. Beroulle, "Facilitating side channel analysis by obfuscation for hardware trojan detection," in *2015 10th International Design Test Symposium (IDT)*, 2015, pp. 129–134.

[7] R. S. Chakraborty and S. B., "Rtl hardware ip protection using key-based control and data flow obfuscation," in *Proceedings International Conference on VLSI Design*, ser. VLSID '10. Washington, DC, USA: IEEE Computer Society, 2010, pp. 405–410.

[8] J. Zhang, Y. Lin, Y. Lyu, and G. Qu, "A puf-fsm binding scheme for fpga ip protection and pay-per-device licensing," *IEEE Transactions on Information Forensics and Security*, vol. 10, no. 6, pp. 1137–1150, 2015.

[9] G. Sumathi, L. Srivani, D. T. Murthy, A. Kumar, K. Madhusoodanan, and S. A. V. S. Murty, "Structural modification based netlist obfuscation technique for plds," in *2016 International Conference on Wireless Communications, Signal Processing and Networking (WiSPNET)*, 2016, pp. 1418–1423.

[10] C. Yu, X. Zhang, D. Liu, M. Ciesielski, and D. Holcomb, "Incremental sat-based reverse engineering of camouflaged logic circuits," *IEEE Transactions on Computer-Aided Design of Integrated Circuits and Systems*, vol. PP, no. 99, pp. 1–1, 2017.

[11] M. E. Massad, S. Garg, and M. V. Tripunitara, "Integrated circuit (ic) decamouflaging: Reverse engineering camouflaged ics within minutes," in *NDSS*, 2015.

[12] T. Meade, S. Zhang, and Y. Jin, "Ip protection through gate-level netlist security enhancement," *Integration, the VLSI Journal*, vol. 58, no. Supplement C, pp. 563 – 570, 2017.

[13] A. R. Desai, M. S. Hsiao, C. Wang, L. Nazhandali, and S. Hall, "Interlocking obfuscation for anti-tamper hardware," in *Proceedings of the Eighth Annual Cyber Security and Information Intelligence Research Workshop*. ACM, 2013, p. 8.

[14] J. A. Roy, F. Koushanfar, and I. L. Markov, "Ending piracy of integrated circuits," *Computer*, vol. 43, no. 10, pp. 30–38, 2010.

[15] T. Meade, Z. Zhao, S. Zhang, D. Pan, and Y. Jin, "Revisit sequential logic obfuscation: Attacks and defenses," in *2017 IEEE International Symposium on Circuits and Systems (ISCAS)*, May 2017, pp. 1–4.

978-1-5386-7100-9/18 $31.00 © 2018 IEEE

FPGA-based Controllers for Compact Low Power Refreshable Braille Display

Suman Muralkrishnan Adhepalli, Pulkit Sapra, Saurabh Agrawal, Piyush Chanana, M. Balakrishnan, P.V.M. Rao
Assistive Technologies Group (Assistech)
Indian Institute of Technology Delhi, Hauzkhas
New Delhi, INDIA. Email: mbala@cse.iitd.ac.in

Abstract—Refreshable Braille Displays (RBD) are electro-mechanical devices used by people with visual impairment to access textual e-content using a tactile display interface. RBD systems developed using the Shape Memory Alloy (SMA) wires pose various challenges of controlling a large number of SMA actuators, supplying high driving current for quick response and implementing an active closed loop control for encompassing variations in operating conditions and mechanical tolerances. These challenges can be addressed with two different hardware architectures: a microcontroller-based and an FPGA-based power controller. In this paper, a novel FPGA-based embedded power controller is proposed and implemented for the SMA actuators used in RBD. It further compares the device's performance, size, cost and energy efficiency with the microcontroller-based system and also emphasizes the role of programmable logic devices like FPGAs in replicating a closed loop actuator control where multiple channels like SMA actuators are involved.

Index Terms—FPGA, SMA actuator, Closed-loop control, Refreshable Braille Display, feedback mechanism, low power design

I. INTRODUCTION

Refreshable Braille Display (RBD) is an electro-mechanical device which displays Braille characters through a tactile interface and is used by people with visual impairment to access digital textual content. It plays a critical role in making e-content accessible for people with visual impairment through "reading" rather than a passive listening mode using speech synthesizing software [1]. The device consists of an array of dots that extends or retracts to form Braille characters. A dedicated actuator coupled with a mechanical system controls the movement of each dot. A standard 20 cells (characters) RBD consists of 160 such actuators driven individually based on Braille character to be displayed. RBDs available commercially are based on piezoelectric actuators which are expensive ($2500) and thus not affordable for the visually impaired community, especially in developing countries. Shape Memory Alloy (SMA) wire as an actuator proposed in [2] is a cost-effective alternative for developing RBD as compared to commercially available piezoelectric actuator based displays.

SMAs are used in various actuator mechanisms because of their ability to withstand high deformation, good power-weight ratio, quick response speed and high energy efficiency [3]. SMA actuators are based on the underlying phenomenon of phase transformation from martensite to austenite on increasing the temperature of the wire [4]. This change in temperature is achieved by Joule heating, i.e., heating due to the resistance of wire when current is passed through the wire. The resultant strain (contraction) in the wire is controlled by regulating the power transferred to the wire. SMA based actuators pose various challenges in terms of position control because of non-linear behavior, power control because of high thermal sensitivity and wide-range in hysteresis characteristics [3].

Various control schemes are attempted for SMA position control: N. Ma et al. [5] discussed an open-loop control testing to study the effect of PWM parameters. K. Ikuta et al. [6] proposed a position control based on resistance as a feedback input. N. Ma et al. [7] also proposed implementation of non-linear control using neural networks. S. Majima et al. [8] attempted to develop a Proportional-Integral-Derivative (PID) control system for position control. Most researchers implemented a current measurement based feedback system and incorporated it with different control algorithms [6] [7]. These implementations require an embedded system with high processing capability and complex software features to realize such algorithms with closed-loop control. The complexity further increases as RBDs need hundreds of such control loops for its actuators.

SMA actuator based RBD proposed in [2] implemented an open loop control of SMA and had issues related to high power consumption, wire damage due to overheating of the wires and unreliable actuation because of variations in operating conditions. To overcome these issues, implementing a closed loop control becomes necessary. Sapra et al. [9] proposed an improved closed-loop control with position feedback using an external *contact based sensing* and a *latching mechanism* which simplified the sensing mechanism and reduced dependence on external conditions but suffered various challenges which are discussed in following sections. The system was designed using a microcontroller-based design (MBD) and a FPGA-based design (FBD) was introduced which overcomes these challenges

This paper discusses in detail:

- Challenges involved in the application of SMA actuators for RBD system and complexity in the control system.
- Design strategies for optimizing peak power, cost and frequency.
- Design of a closed loop embedded power controller with a position feedback using a low-cost FPGA.
- Comparative analysis of FBD with MBD.

978-1-5386-7100-9/18 $31.00 © 2018 IEEE

Figure 1: RBD containing multiple dots with actuators inside.

Figure 2: Feedback Mechanism.

The following sections explain the mechanism of SMA based RBD, followed by the design complexities involved in the system. The paper further discusses the design of an FPGA-based power controller with multiple channel control loops for the SMA actuators implemented using a low-cost Lattice MACHX02 series of FPGA.

II. SMA ACTUATION MECHANISM

The SMA based RBD as mentioned above consists of an array of dots (pins) which is raised above a flat surface to represent a Braille character. The dot position changes dynamically with respect to the user data to be displayed and is achieved by an actuation mechanism. The actuation mechanism of RBD with SMA involves a cantilever mechanism [2] which controls the Braille pin movement, a power controller with a feedback mechanism to sense pin position and a latching mechanism [9]. The following subsections explains various mechanisms involved in the working of the SMA based RBD Module.

A. Power Controller and Feedback Mechanism

The system proposed in this paper requires a two-position control - Down or default position and Up or actuated position. A mechanical spring is used to maintain the actuator in its default position. A contact-based feedback as shown in (Fig.2) is implemented which handles the following:

- Provides binary position control of wire
- Handles peak to steady transition
- Compensates for the effects of ambient temperature variation, manufacturing tolerances and wire characteristic fluctuations like wire resistance, etc.

The power controller essentially controls the heating of the SMA wire by switching a MOSFET connected in series to the wire through PWM pulses. The power controller is made a closed loop by sensing the Braille pin position using the feedback mechanism explained above.

B. Latching Mechanism

The latching mechanism is used to hold the position of the pin either in *Up* or *Down* position based on wire heating controlled by the power controller. This significantly reduces the power consumption as there is no subsequent power consumed after latching the pins. A module comprises of 80 actuators for 10 cells with each cell having eight actuators

placed in four rows and two columns as per the computer Braille standard [1]. Thus the SMA module is partitioned into four layers and each layer is stacked vertically due to the mechanical design constraints [9]. The latching feature is a single slider mechanism that latches all the actuators of 10 Braille cells. Therefore, any change in position of single actuator requires refresh of entire 10 cells [9].

III. DESIGN COMPLEXITY

Power controller design for RBD using SMA actuators poses various challenges in designing the optimal hardware incorporating all functionalities. The design complexity and constraints are discussed in the following subsections.

A. Independent Closed-loop Control

RBD consists of an array of SMA actuators fit into the standard Braille cell dimensions to implement its functionality [10]. Fig.1 shows a 20 cell RBD which consists of 160 dots each having an actuator. Each actuator requires an independent closed-loop control as the wire characteristics are non-linear and actuator's environment conditions are non-uniform. Implementing algorithms like PID or Fuzzy logic to control the actuation of wire will require the embedded controller to be computationally intensive compared to open loop control of SMA or the existing piezoelectric based systems. This necessitates simplicity in power control to replicate it for hundreds of actuators retaining the wire's response time as well as actuation cycles.

B. Ambient Temperature

The strain in the SMA wire is the function of wire temperature (T_w), and T_w further depends on the power supplied to the wire. The ambient temperature (T_a) in which the system operates, also influences the power required for phase transformation. The power P required for actuation depends on the difference between T_a and temperature (T_{as}) for austenite transformation as in (1), $T_{as} \approx 90°C$ [2].

$$P \propto (T_{as} - T_a) \tag{1}$$

Therefore, for a fixed power value, if T_a is less, it causes insufficient power to the wire, resulting in incomplete actuation. Similarly, if T_a is high, the wire is over-powered and thus resulting in damage to the wire. The energy required for actuation depends on the time for which the power is supplied to the wire. Thus, it is necessary to implement an adaptive two-level power controller, which is not influenced by variation in T_a. The two-level power is essentially two power states where the *peak* state accelerates the actuation and a *steady* state maintains the strain by compensating for the cooling effect due to ambient temperature.

C. Peak Currents

SMA used in the current design has very low electrical resistance, thus power transfer must be a low-voltage power transfer using Pulse Width Modulation (PWM) [11]. The net peak current loads experienced by the power signal paths increase greatly with a large number of actuators. The simple PWM draws all the peak currents from supply during the ON time period and zero currents in OFF time period. However, TDM PWM [12] utilizes these zero current periods to distribute the current over time. This significantly reduce the voltage drops due to power track (copper) resistance in common conduction paths. In order to reduce the pulsating peak loads to the power supply, local decoupling capacitors (de-cap) are added which acts as a local charge pump.

D. PWM Frequency

The frequency of PWM is another factor that affects power consumption and usability of RBD. High operating frequency ($> 200kHz$) increases the power consumption because of switching losses but supports using a lower value of decoupling capacitors. Higher frequencies also cause inaccuracies in delivering the assigned duty cycle as MOSFET rise time becomes significant to the time period of the frequency drive. At low operating frequency ($> 0.2kHz$), the SMA actuator makes vibratory noise because of intermittent cooling and thus making dots difficult to perceive. Thus, the frequency should be in the range of 20 to 100 kHz for optimum power consumption and user perception. A configurable frequency will enable the tuning of frequency to incorporate changes in design parameters like wire length and diameter.

E. Closed Loop Actuation

The power to SMA is provided in two levels: *peak* power (P_p) and *steady* power (P_s) [2]. The transition from P_p to P_s varies from wire to wire due to several factors beyond control like manufacturing tolerances, temperature variations, etc. Further extending P_p, once the wire has reached its actuation position causes damage to the wire due to over-powering/over-heating. Thus, the transition time from peak to steady has to be actively controlled for longer actuator life and reliable actuation.

IV. HARDWARE DESIGN REQUIREMENTS

Based on the challenges discussed in Section III, it is desirable that the control is based on a simple algorithm with a closed-loop feedback adapting to ambient temperature so that the peak currents are reduced. This section discusses the requirements and functionalities that need to be implemented in the embedded power controller. The basic requirement

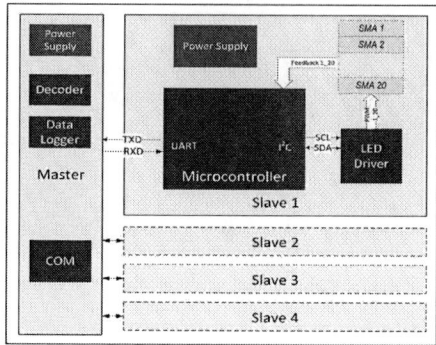

Figure 3: SMA power control using Microcontroller design.

for the SMA actuator is to control the power delivered to the SMA wire incorporating its non-linear characteristics, feedback mechanism, manufacturing variabilities and ambient temperature. Implementing a phase shifted PWM hardware is another design consideration to reduce peak currents and capacitor values.

The feedback mechanism should drive the PWM output quickly to steady-state in order to prevent damage to the wire as well as to reduce the actuation time of the system. Reconfigurability is required in terms of frequency, duty cycle and channel control to incorporate changes in SMA wire parameters. Actuation time is another parameter which is to be determined accurately to classify the quality of wire manufacturing and assembly. Memory is required to store the timing data and error flags to report incomplete actuation. All the above-mentioned functionalities are to be replicated for hundreds of actuators with a minimum number of components to reduce the size, cost and power. Partitioning of hardware and software for each of the functionalities mentioned above is also another objective of the design.

V. MICROCONTROLLER-BASED DESIGN

The Microcontroller-based design is a combination of microcontrollers, PWM controllers (LED drivers) and Power switches (MOSFETs) to control power to the SMA wires (Fig.3). The architecture includes a central Master which receives the data from an external device and distributes the control bits to Slave layers. Each slave layer controls 20 actuators with a microcontroller and a LED Driver. The microcontroller controls the PWM based on feedback signal sensed through input ports and the data received from its Master controller. Microcontrollers internally contain PWM channels which can be used to drive the SMA actuators, but the number of PWM channels in a general purpose microcontroller are either very few or are rather expensive [13]. Due to a large number of actuators as mentioned in Section III.A, multi-channel PWM controller is required.

In order to implement the multiple PWM channels, an external ASIC like LED driver is interfaced to the microcontroller through I2C protocol as shown in Fig.3. It consists of multiple PWM channels where each channel can be programmed to drive with different duty cycles. Thus multiple channels are used to generate the required PWM for each SMA channel. It has a fixed frequency for a particular IC (range between 50 kHz to 200 kHz are available commercially). The frequency of PWM cannot be low as discussed in Section III.D

The module consists of a set of five microcontrollers which communicate among themselves using UART interface with parallel connection interconnected through a tristate buffer at each microcontroller's UART port. The Master controller also captures the actuation time data for each actuation which is used indirectly as a parameter to check the quality of wire manufacturing and assembly. The feedback signals are sensed using GPIO port by polling rather than Interrupts as the number of interrupt pins are also limited in the microcontroller. This results to delays in sensing and thus eventually increasing the response time of the system for feedback. The LED driver also increases the peak current as they do not support TDM causing all the 20 wires' load at the same instant and thus accumulating the peak current. The overall system was realized using a 16-bit PIC controller PIC24HJ64GP204 [13] from Microchip and a 24-channel LED driver PCA9626 from NXP Semiconductors.

VI. FPGA-BASED DESIGN

The FBD essentially replaces the microcontroller and PWM controller of each layer with an FPGA that incorporates both the functionalities. It provides the flexibility to implement customized digital logic specific to SMA power control. The block diagram of the power controller for SMA implemented using the FPGA is shown in Fig.4 with one FPGA in every layer. A single FPGA cannot be used for all 80 channels as the channels are distributed physically in four vertical layered PCBs at different heights as explained in Section II.B. It also increases the complexity in the interconnection of eighty PWM channels and feedback channels of actuators.

All four FPGAs are interfaced to a master microcontroller which does the functionalities like control of electromagnets used for latching mechanism [9], distributing data between layers and interfacing the module with an external host device for data exchange and error handling. The master controller cannot be replaced with an FPGA since these functionalities will consume more resources on FPGA and thus will increase the cost of the system. Each FPGA mainly implements 20-closed loop digital power control logic along with phase-shifted PWM drivers, storage registers, timers, communication channels and a Finite State Machine (FSM) which are discussed briefly in the following sub-sections. The design parameters incorporated for the FPGA synthesis are listed in Table I. The design is implemented on the part LCMX02-640HC, a low-cost FPGA from Lattice Semiconductors [14]. The algorithm was developed in VHDL, synthesized, placed and routed with Lattice Diamond IDE and burned to the FPGA's Flash using the HW-USBN-2B programming cable

A. Power Control Block

The power controller consists of multiple individual control channels for each SMA wire as shown in Fig.4. Each channel selects the output as depicted in Fig.5. If the corresponding

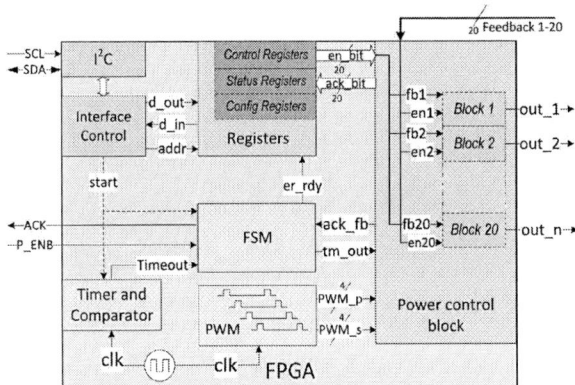

Figure 4: Block diagram of FPGA-based controller

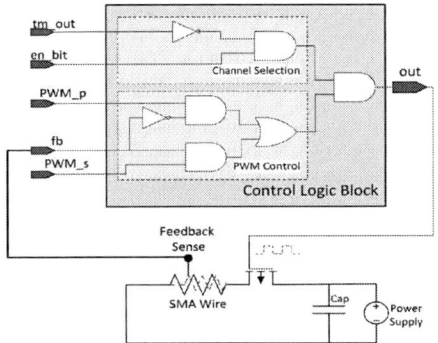

Figure 5: Power control logic of SMA actuator.

enable bit *(en_bit)* of a channel is set in the control register and there is no timeout *(tm_out)* signal active from the FSM then PWM control dynamically switches between P_p and P_s based on the feedback signal. An acknowledgment signal is generated which triggers when all the control block have either received feedback or the timer has expired. The PWM generator generates pulses with two different duty cycles (Peak and Steady) required for SMA power control as mentioned in section III.E. The Time Division Multiplexing (TDM) technique is implemented in PWM generation for reducing peak currents as discussed in section III.C. It generates eight unique pulses with four phases for each, P_p *(PWM_p)* and P_s *(PWM_s)* duty cycle as shown in Fig.4.

Table I: Controller Design Specification

Design Specifications	Values
Actuation time of actuator	200 ms
Base clock frequency	10 MHz
PWM frequency	50 kHz
FSM clock frequency	10 MHz
Peak cut-off timer	250 ms
I^2C frequency	100 kHz
Number of PWM channels	20
Number of phases in PWM	4
Driving voltage	3.3 V

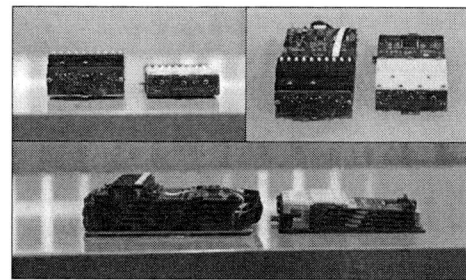

Figure 6: Module Prototypes depicting component reduction.

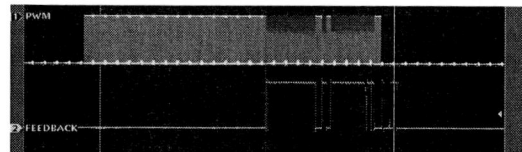

Figure 7: Dynamic response of PWM w.r.t feedback signal.

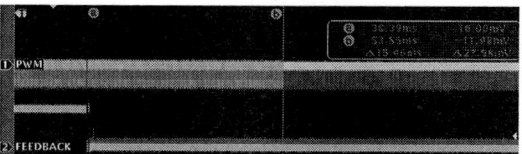

Figure 8: Response time to feedback with MBD.

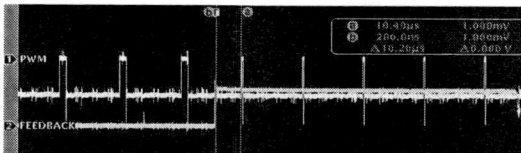

Figure 9: Response time to feedback with FBD.

wires and can connect to multiple devices through a common signal line. The protocol offers sufficient amount of bandwidth for the required data transfer between devices.

B. Registers

The storage registers can be grouped into three different types. *Control registers*: which store bit information of each channel to be activated. *Status registers*: which store acknowledgment of feedback and time taken by each wire to send the feedback. *Configuration registers*: used one-time for setting duty cycle and frequency of P_p and P_s power values. These registers are given a specific address with r/w access control to load and store data to these registers.

C. Timer

The timer generates a tick signal with a time period of one millisecond from the base clock signal. The timers are utilized mainly for two reasons: One is to log the actuation time of each channel. The actuation time is used to classify the wires for its manufacturing quality and testing. Another reason is to protect the wires from accidental overpowering if the master has not provided necessary signals to cut-off the peak power. It cuts off the peak power to the wires after prefixed safety time value by sending a timeout signal to FSM which makes all the channels to reach below steady state or zero power state.

D. FSM

The sequential execution and synchronization of different modules in FPGA are implemented using a Finite State Machine (*FSM*). The state machine's initial state is a *wait* state where it waits for a start signal which is generated on reception of data from I²C. The FSM acts as the controller for the timers, output control blocks, register writes and data logging triggers based on I²C data reception (*start*), timer expiry (*timeout*) and acknowledgment (*ack*) (Fig.4).

E. Communication

The data transfer between the controller and the master is made through an I²C interface, as it requires less number of

VII. RESULTS AND DISCUSSION

The SMA actuator Modules for RBD have been realized based on both microcontroller and FPGA, the complete module pictures are as shown in Fig.6. The following sub-sections explain the enhancements achieved by the FBD over MBD.

A. Dynamic position control

The microcontroller design implements the two-level power control where the power changes from P_p to P_s on receiving feedback. But it does not change back from P_s to P_p if feedback signal becomes low i.e., the contact is disconnected. This is required to maintain the actuation position before latching in case of drop in wire temperature due to air blow or lower ambient temperature. It can be referred to as dynamic power switching and cannot be avoided even in the case of mechanical latching since all wires do not actuate at the same time and the latch being common to all 80 actuators. In case of MBD, implementing such a control adds software overhead to handle power states for each channel. FBD implemented the dynamic power switching with independent digital loops as explained in Section VI.A. Fig.7 shows the dynamic response of PWM with respect to the feedback signal where the power state changes dynamically with respect to the feedback signal.

B. Response Time

The response to changes in feedback signal is directly dependent on the clock period which is set as 10 MHz for the FPGA. This makes the worst-case response time to be dependent on the frequency of PWM alone as PWM frequency (50 kHz) is much smaller than clock frequency (10 MHz). The decrease in response time improves the reliability of SMA wires by preventing overheating of the wire. The MBD changes the power state of the output through a communication channel making it sluggish (frequency of 100 kHz) to respond to the feedback and the maximum response time can go up to 20 ms. Fig.8 shows a sample response time captured

978-1-5386-7100-9/18 $31.00 © 2018 IEEE   636

Table II: Comparison of Metrics between Microcontroller and FPGA based design

Metrics	μC^1	FPGA	%reduction
Standby power	0.248 W	0.034 W	86.3 %
Peak current from cap/slave	13.2 A	3.3 A	75.0 %
Peak power from cap/slave	43.56 W	10.89 W	75.0 %
Decoupling capacitor	22.811 μF	1.815 μF	92.0 %
Min. T_r for feedback	10000 μs	0.1 μs	100.0 %
Electronics footprint	492 mm^2	311 mm^2	36.8 %
No. of electronic components	121 units	71 units	41.3 %
Cost per 20 channel controller	$ 26.50	$ 7.80	64.5 %

[1] μC - microcontroller, T_r - Response Time

for the microcontroller design which takes 15.46 *ms* to shift PWM from peak duty-cycle to steady duty-cycle. Similarly, Fig.9 shows the response time captured for FPGA design which is 10.2 *μs*.

C. Power

The TDM PWM has reduced the peak current from the de-coupling capacitor (decap) by one-fourth and thus reducing the capacitance value of de-cap used in each actuator. TDM also aids in increasing the value of driving voltage (V_d) resulting in the duty-cycle reduction which further reduced the capacitance value [11] [15]. The FPGA design has reduced the standby power by 86% from 0.248 W to 0.034 W (Table II) and thus improved the battery backup time. The reduction is mainly due to the elimination of pull-up resistors (consumed 0.203 W) which were required for the PWM drivers and reduction in the number of ICs (microcontrollers) and components (consumed 0.045 W) used with MBD. The peak current demands have been distributed uniformly in time and thus reduced peak currents by 75%.

D. Cost and Size

The cost of electronic components used for the controller has reduced significantly (approximately 65%). The reduction is mainly from the reduction in the number of components used, usage of low-cost FPGAs, usage of low capacitance de-caps by implementing TDM and implementing functionalities of all peripherals like UART, I^{2}C within the FPGA. It has contributed to meeting one of the key objectives of the design which is affordability of the device compared to existing piezo-based RBDs and the microcontroller based SMA design. The reduction in the number of components can be seen visibly in Fig.6 from MBD to FBD. The cost reduction achieved by using the new bill of materials is listed in the last row of Table.II. The size of the Modules has reduced by (approximately 37%) which is due to the FPGA replacing the functionalities of microcontroller, LED driver and other communication glue logic. A total reduction of around 50 electronic components was achieved in comparison to the MBD. Reduction in capacitance values decreased cost as well as component space which was possible due to TDM PWM using FPGA.

VIII. Conclusion and Future Work

The FPGA-based design has addressed substantially challenges of peak currents, standby power, size and response time.

Table II shows various improvements achieved by the FBD with respect to the MBD. The FPGA has provided concurrency to run multiple independent closed-loop controllers with a common signal that has significantly helped to reduce the response time of the system. The overall reduction in the cost of the Modules has enhanced the affordability of Refreshable Braille Displays for people with visual impairment.

The future work could focus on addressing one important drawback of the current design - power consumption changes with variation in ambient temperature but at the cost of altering the actuation time. It can be resolved by adding adaptive PWM duty cycle with respect to ambient temperature. There is further scope to optimize the resources used for realizing the algorithms discussed. It can reduce size and cost of the system. The design can be also improved by building an ASIC with integrated power components like MOSFETs to minimize area. It may incur substantial NRE cost but effective in the case of large-scale production.

References

[1] N. H. Runyan and D. B. Blazie, "The continuing quest for the 'Holy Braille' of tactile displays," in *Nano-Opto-Mechanical Systems (NOMS)*, vol. 8107, Oct. 2011, p. 81070G.

[2] P. Sapra, A. K. Parsurampuria, D. Gupta, S. Muralikrishnan, M. Raj, A. Anand, V. Darda, R. Paul, M. Balakrishnan, and P. Rao, "A compliant mechanism design for refreshable braille display using shape memory alloy," in *ASME 2015 International Design Engineering Technical Conferences and Computers and Information in Engineering Conference*, 2015, pp. V009T07A054–V009T07A054.

[3] R. J. B. Darel E. Hodgson, Ming H. Wu, *ASM Handbook: Properties and Selection: Nonferrous Alloys and Special-Purpose*. ASM International, 1990, vol. 2, ch. Shape Memory Alloys, pp. 897–902.

[4] C. W. K Otsuka, *Shape Memory Materials*. The Pitt Building, Trumpington Street, Cambridge, UK: Cambridge University Press, 1999.

[5] N. Ma and G. Song, "Control of shape memory alloy actuator using pulse width modulation," *Smart Materials and Structures*, vol. 12, no. 5, p. 712, 2003. [Online]. Available: http://stacks.iop.org/0964-1726/12/i=5/a=007

[6] K. Ikuta, M. Tsukamoto, and S. Hirose, "Shape memory alloy servo actuator system with electric resistance feedback and application for active endoscope," in *Robotics and Automation, 1988. Proceedings., 1988 IEEE International Conference on*. Ieee, 1988, pp. 427–430.

[7] N. Ma, G. Song, and H. Lee, "Position control of shape memory alloy actuators with internal electrical resistance feedback using neural networks," *Smart materials and structures*, vol. 13, no. 4, p. 777, 2004.

[8] S. Majima, K. Kodama, and T. Hasegawa, "Modeling of shape memory alloy actuator and tracking control system with the model," *IEEE Transactions on Control Systems Technology*, vol. 9, no. 1, pp. 54–59, 2001.

[9] P. Sapra, A. K. Parsurampuria, S. Muralikrishnan, M. Balakrishnan, and P. Rao, "Refreshable braille display using shape memory alloy with latch mechanism," in *ASME 2017 International Design Engineering Technical Conferences and Computers and Information in Engineering Conference*, 2017, p. V009T07A040.

[10] J. M. Gill, "Braille cell dimensions," http://www.tiresias.org/research/reports/Braille_cell.htm, RNIB Digital Accessibility Team, London, 2008.

[11] C. Mitter, "Device considerations for high current, low voltage synchronous buck regulators," in *Wescon/97. Conference Proceedings*. IEEE, 1997, pp. 281–288.

[12] M. S. Danielson, "Time division multiplexed pwm amplifier," Apr. 22 2003, uS Patent 6,552,607.

[13] "32-bit pic and sam microcontrollers peripheral integration," (Date last accessed 04-February-2018). [Online]. Available: http://ww1.microchip.com/downloads/en/DeviceDoc/60001455C.pdf

[14] "Machxo2 family data sheet," (Date last accessed 04-February-2018). [Online]. Available: http://www.latticesemi.com/view_document?document_id=38834

[15] T. Instruments, "Input and output capacitor selection," *Retrieved from ti: http://www.ti.com/lit/an/slta055/slta055.pdf*, pp. 281–288.

2018 IEEE Computer Society Annual Symposium on VLSI

Very Large-Scale and Node-Heavy Graph Analytics with Heterogeneous FPGA+CPU Computing Platform

Yu Zou and Mingjie Lin
Department of ECE, University of Central Florida, Orlando, FL, USA
yuzou@knights.ucf.edu, mingjie@eecs.ucf.edu

Abstract—We present a highly scalable approach to constructing a reconfigurable computing engine specifically optimized to perform sophisticated kernel computing on graph-structured data. We choose newly emerged graph convolutional networks (GCNs) as our key benchmark and develop a novel node-heavy edge-centric computing framework for very large-scale graph analytics. Unlike most existing studies, our design and implementation can handle extremely large graph size that well exceeds the on-chip memory capacity of any FPGA+CPU heterogeneous platform, thus can only be stored in hard drive. The most novel aspect of our approach is to enable a completely streaming mode of large vertex and edge data and perform a write-back message updating policy, therefore completely removing any redundant data accesses to IO-expensive hard drive. Additionally, our subgraph sorting scheme can effectively eliminate the performance bottleneck caused by preprocessing in the state-of-art computing framework X-Stream [1, 2].

To validate our approach, we have implemented our proposed method with a KC705 Xilinx FPGA board and tested it with multiple real-world large-scale data sets. For the largest data set with 210,010 vertices and 1,349,400 edges, our platform achieves 1.87s in total latency, which is approximately 400 times faster than the baseline platform with the state-of-the-art approach [1].

Keywords-FPGA; graph analytics; heterogeneous computing;

I. INTRODUCTION

Modern FPGA device, when coupled with the conventional CPU or GPU, provides an unique opportunity to implement application-specific heterogeneous computing that maximizes memory access performance, which is crucial in achieving high computing performance for many important memory-intensive and mission-critical computing applications. Unfortunately, with the advent of big-data graph-based analytics, the challenge of accomplishing high-performance computing with such heterogeneous computing systems (CPU+GPU+FPGA) has been greatly exacerbated by **large data sizes**, **irregular data placement**, and **complicated computing parallelism**. As such, despite holding great promise in leapfrogging our computing power and energy efficiency, heterogeneous computing systems, such as FPGA+CPU platforms, prove to be extremely challenging to be fully utilized in complex graph-based applications, such as data mining, collaborative filtering, and probabilistic graph reasoning. This may explain the fact that, while the research on heterogeneous computing for conventional matrix-based scientific-like applications is extensive, the work on how to conduct heterogeneous computing for emerging big-data graph-based applications is relatively scarce.

The main objective of this paper is to design and implement a highly scalable heterogeneous computing framework that is specifically optimized for very large-scale graph

analytics. Our main focus is to achieve ultra-high data access performance when a hierarchy of disparate memory resources are involved. Specifically, we aim at a CPU+FPGA heterogeneous computing platform and a large class of graph analytics involving large data set and perform complex computations on graph-based data, such as the newly emerging graph convolutional neural network in artificial intelligence and machine learning area. More concretely, we attempt to optimize computing performance with the following techniques, which we consider to be essential for any kind of heterogeneous computing with graph-based data.

Prioritizing sequential memory accesses: It is well-known that a huge gap of accessing bandwidth exists between sequential and random accesses for any storage device. Therefore, to achieve high I/O performance, it is essential to sequentially stream data as much as possible no matter which computing framework is used. The importance of sequential accessing is even more crucial for our study because graph analytics applications typically embed large-scale data sets with irregular graph structures that tend to incur more random accesses.

Minimizing redundant data accessing: Again, due to the random nature of data layout in most graph analytics, accessing vertex set typically requires repeatedly accessing the same edge set, or vice versa. Both scenarios will inevitably cause redundant data accesses that seriously hinder the overall computing performance of the target application. In most cases, how to partition the overall graph is critical because the data size of a real-world graph often well exceeds an on-board storage capacity, thus causing so-called off-core graph problem.

Parallelizing processing cores: For most graph analytics applications, if either edge-centric or vertex-centric gather-and-scatter (GAS) computing framework is adopted, parallelizing processing cores is imperative to the overall processing performance. Unfortunately, due to the irregular memory access pattern, it is very challenging to achieve high aggregated throughput even with multiple processing cores because input data set of different processing cores can often overlap or cause conflict.

Minimizing sorting operations: Finally, when computing graph analytics, sorting operations are often used to promote sequential data accesses [3]. However, for large-scale graph-based data sets, sorting operations are quite expensive and also detrimental to effectively tackling dynamic graphs, where both edge and vertex sets can change at run time.

Our proposed computing framework with a heterogeneous CPU-FPGA platform can be readily applied to many other node-centric or edge-centric graph computing tasks. We claim the following contributions:

1) We designed and implemented, to our knowledge, the first functional heterogeneous CPU+FPGA computing

---

[1] All source code of our hardware implementation and detailed performance models can be found at https://github.com/yzou93/ngraph_fpga.

978-1-5386-7100-9/18 $31.00 © 2018 IEEE

platform specifically optimized for a wide range of large-scale and node-heavy graph analytics applications. With very limited hardware resource ($\approx$10% of a Kintex xc7k325t-2ffg900c FPGA chip), our heterogeneous FPGA+CPU computing platform can achieve orders-of-magnitude performance improvement over the state-of-the-art approach for a very large-scale graph analytics application.

2) Although we adopted the basic idea of 2D graph partitioning in GridGraph [4], there are two key innovations. First, each edge partition is sorted according to destination vertex. This change is essential to effective message combining and parallelizing processing cores. Second, unlike the most of other studies, our message data take a write-back policy instead of write-through. This change is important to avoid redundant data loading and reduce data writing. Finally, we stream node data directly from hard drive instead of the standard edge streaming [5, 4, 6, 1, 7]. This proves to be critical for node-heavy graph analytics.

3) Unlike the majority of existing studies, we tackle the case of very large-scale graph analytics, where the size of graph data set well exceeds the capacity limit of on-chip memory and has to be stored in hard drive. As such, optimizing I/O performance to the secondary storage is crucial. We have developed a data streaming scheme for node-heavy data such that no redundant I/O access is needed.

II. NODE-HEAVY GRAPH PROCESSING FRAMEWORK

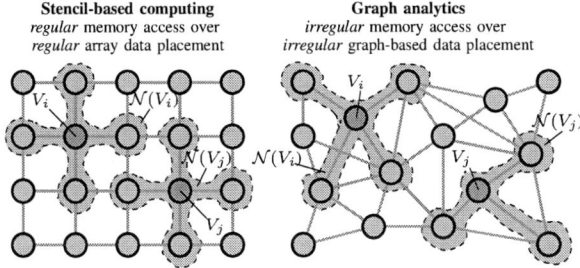

Figure 1: Two computing and memory access patterns that cover a wide range of data analytics. Specific examples consist of (a) PDE solver, image denoising. (b) PageRank, collaborative filtering, loopy belief propagation, and Gibbs sampling.

To highlight our computing challenges and present a larger perspective, we abstract a wide range of computer analytics tasks into two categories. Specifically, we depict stencil-based codes as iterative kernels that update array elements according to some fixed pattern in Fig. 1(a), which are most commonly found in computational fluid dynamics (CFD), solving partial differential equations (PDE), and image processing. In contrast, Fig. 1(b) depicts the general form of graph analytics applications, the underlying data layout becomes a generalized form of graph, and the shape of vertex neighborhood varies throughout iterations. This kind of data irregularity can be widely found in collaborative filtering, graph mining, and computer vision, etc. Note that, because all computing patterns depicted in Fig. 1 are defined in a data-centric way in terms of neighborhoods of data structure elements, data locality, essential to achieving high computing performance and energy efficiency, becomes quite challenging to exploit without sophisticated accessing schemes, especially when constrained by memory ports and bandwidth. Our work in this paper specifically aims at accelerating very large-scale graph analytics depicted in Fig. 1(b) and particularly focuses on optimizing the I/O performance for such computing tasks.

A. Edge-Centric Iterative Computing Framework

We follow the standard notations of graph theory, i.e., G, V, and E denote a given graph, its vertex set, and its edge set, respectively. Furthermore, $n = |V|$ represents the total number of vertices, and $m = |E|$ represents G's total number of edges with $e_{i,j}$ denoting an edge with source vertex v_i and destination vertex v_j. Under the computing framework of graph analytics, both vertices and edges can potentially contain computing-related information denoted with $D(v_i)$ or $D(e_k)$ respectively. In this work, we focus on discussing only vertex-centric and edge-centric frameworks. In particular, we adopt the Gather-Apply-Scatter (GAS) graph computation abstraction introduced in PowerGraph [6]. Specifically, the GAS consists of three conceptual phases to define a specific graph computing task: *Gather*, *Apply*, and *Scatter*. During each phase, all vertices in $G(V, E)$ execute in parallel and each vertex v_i exchanges information only with its neighborhood $\mathcal{N}_i$, where $i = 1, 2, , \cdots, n$. In the gather phase, for any given vertex v_i, information stored in all its incoming edges is combined through a predefined and application-specific function $\mathcal{F}(v_i, \mathcal{E}_{i,\text{incoming}})$ is performed, where $\mathcal{E}_{i,\text{incoming}}$ denotes v_i's all incoming edges. In the apply phase, the previously resulted $\mathcal{F}$ value will update $D(v_i)$, i.e., $D(v_i) \leftarrow \mathcal{F}(v_i, \mathcal{E}_{i,\text{incoming}})$. Finally, the scatter phase uses the updated value of v_i to update the data stored on all its outgoing edges $\mathcal{E}_{i,\text{outgoing}}$. Note the GAS computing model is so general that almost all important graph analytics applications can be cast into this framework. For example, three quite different algorithms, PageRank, graph-coloring, and single source shortest path algorithm, can be readily computed with the GAS [6].

Given the GAS computing framework, for large-scale graph processing, there are essentially two types of computing models: vertex-centric vs. edge-centric. The well-known vertex-centric computational paradigm has been recently incorporated into distributed processing frameworks to address computing challenges of irregular memory accesses [7, 8]. This is especially true when real-world billion-node graphs well exceed the memory capacity of typical computing machines and therefore can not be effectively tackled by the mainstream computing framework such as MapReduce, which performs notoriously poorly for iterative graph algorithms such as PageRank. X-Stream [1] pioneered the concept of edge-centric computing framework to scatter-gather process large-scale out-of-core graphs. Although vertex-centric approach [7, 8] is more prevalent for most graph-based problems, the edge-centric approach can completely avoid random access into the set of edges, instead stream them from storage. Therefore, for most graphs that the edge set is much larger than the vertex set, streaming the edges is often advantageous compared to accessing them randomly. Unfortunately, streaming edges will generate the cost of random access into the vertex set. The X-Stream mitigates this issue by partitioning both vertex and edge set such that each vertex partition fits in the main memory and edges appear in the same partition as their source vertex, and then processing the graph one partition at a time, first reading in its vertex set and then streaming its edge set from storage. In this study, we assume that both edge and vertex data sets are too big to be stored in on-chip or DDR memory. In particular, each graph node contains a large amount of data, thus node-heavy. We will adopt the edge-centric computing model and directly stream vertex data set from hard disk.

B. Graph Storage and Partitioning Strategy

Our work focuses on very large-scale graph analytics, meaning that the storage size of a graph, including edge data set and vertex data set, well exceeds the capacity of the on-chip block RAMs and the on-board DRAM, therefore can only be stored in the hard drive. One crucial design decision is how to partition the complete graph into subgraphs with manageable data sizes and still facilitate efficient memory accesses. As in GridGraph [4] and FPGP [5], we first partition the complete vertex set V into P partitions denoted as V_i, $i = 1, 2, 3, \cdots, P$. Subsequently, the edge set E can be partitioned into P^2 subsets $E_{i,j}$, $i, j = 1, 2, \cdots, P$, where $E_{i,j}$ contains all edges with the source vertex u and the sink vertex v such that $u \in V_i$ and $v \in V_j$. Furthermore, the specific choice of P depends on the on-chip memory capacity. Typically, we make sure that one vertex partition can be stored comfortably in the given on-chip block memory.

We adopt the same partitioning method as GridGraph [4] and FPGP [5]. As shown in Fig. 2, all the nodes are partitioned into P disjoint intervals. The parameter P is chosen such that each source node interval can fit the FPGA on-chip block memory. All the edges are partitioned into P^2 number of partitions, E_{ij} contains all the edges whose source nodes belong to interval I_i and destination nodes belong to interval I_j. Each edge partition is associated with a message partition. The message partition stores the generated messages when processing the corresponding edge partition. To reduce disk I/O accesses, in the preprocessing step, edges within each edge partition are sorted in ascending order of destination. The order of generated messages is the same as the order in which edges are streamed. So the generated messages within each message partition are also sorted by destination. A message combiner is used to combine messages with the same destination to one message and write back to hard drive. On average, for each edge partition, the message combiner reduces the number of written messages from $\frac{|E|}{P^2}$ to $\frac{|V|}{P}$.

C. Data Accessing Scheme

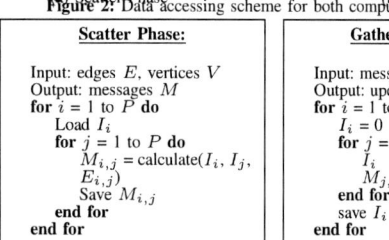

Figure 2: Data accessing scheme for both computing phases.

Scatter Phase:	Gather Phase:

Input: edges E, vertices V
 Output: messages M
 for $i = 1$ to P **do**
    Load I_i
    for $j = 1$ to P **do**
      $M_{i,j}$ = calculate(I_i, I_j,
       $E_{i,j}$)
      Save $M_{i,j}$
    end for
 end for
	Input: messages M
	 Output: updated vertices V
	 for $i = 1$ to P **do**
	    $I_i = 0$
	    for $j = 1$ to P **do**
	      I_i = update(I_i,
	       $M_{j,i}$)
	    end for
	    save I_i
	 end for

Figure 3: Node-heavy graph processing flow.

When processing each edge partition, we adopt the edge-centric graph computing model [1]. In the scatter phase of each iteration, a source vertex partition V_i is loaded into the on-chip block memory. Edge partitions E_{i1} to E_{ij} are streamed in from the PC host. Subsequently, all generated messages are streamed back to the PC host and stored into their corresponding message partitions. This process is repeated P times until all P of source vertex partitions and P^2 of edge partitions are traversed. In the gather phase, all message partitions are streamed in the order of destination. P number of message partitions whose destination nodes belong to V_i destination vertex partition are streamed consecutively to FPGA. After processing all P message partitions, the vertex partition V_i is completely updated and written back to the hard drive. This process is repeated P times until all the P^2 message partitions are traversed. On average, for each message partition, the message combiner reduces the number of message reads from $\frac{|E|}{P^2}$ to $\frac{|V|}{P}$.

III. Hardware Architecture and Implementation

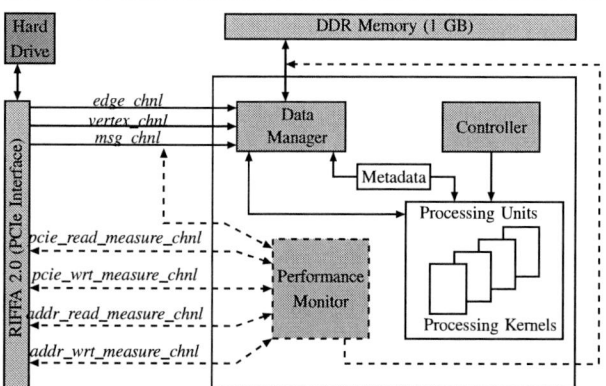

Figure 4: Diagram of overall system architecture.

Overall Architecture—The overall architecture of our implementation is shown in Fig. 4. There are three layers of memory hierarchy: hard drive, DDR memory, and on-chip BRAM (block RAMs). Initially, all data including both edge set and vertex set reside in the hard drive, which has the largest capacity, constant parameters are stored in on-board DRAM. As computation progresses, data will gradually be moved into the accelerator. The hard drive is off-FPGA-board and connected to the FPGA chip through a RIFFA interface [9], which essentially is a hardware wrapper around the PCIe protocol stack IP issued by Xilinx. In contrast, the DRAM is off-chip but on-FPGA-board and communicates with the FPGA device through the Xilinx Memory IP core, a combined pre-engineered controller and physical layer (PHY) for interfacing Kintex architecture to DDR SDRAM. Both the hard drive and the DDR memory are connected to a specifically designed data manager, which coordinates all data movements in our system and directly communicates with all processing cores. In order to enable high throughput data movement and data streaming, we implement three separate I/O channels between the RIFFA interface and the data manager: edge data channel, vertex data channel, and message data channel. To implement control path, we have a controller to coordinate all logic operations and data movements. Inside the processing unit, to improve the overall throughput, we pipeline multiple copies of identical processing kernels. As shown in Fig. 5, the vertex data set will be first streamed into on-chip memory, and then edge data set will be streamed in from the hard drive through the RIFFA PCIe interface. In scatter phase, vertex data can be fetched within one or two clock cycles in a random access pattern when processing each edge.

Note that, to facilitate performance measurements, we have an additional circuit block for monitoring performance that records all I/O activities of PCIe read and write operations. These measurements allow us to analyze our data performance in detail and indirectly illustrate the source of performance gain. When measuring the final performance, we remove this performance monitoring block to conserve hardware usage and keep computational fidelity.

Host Implementation—There are eight threads working in parallel on the PC host. Thread 0 is responsible for retrieving vertices from the hard drive and sending data to FPGA through channel 0 in scatter phase. In gather phase, thread 0 receives updated vertices from hardware. Thread 1 sends edge data to FPGA. In the scatter phase, thread 2 receives generated messages from FPGA and stores the messages to the corresponding file. In the gather phase, thread 2 in-order retrieves message partitions from hard drive and streams message data to FPGA. Thread 3 retrieves pre-stored metadata information for each edge partition and streams to FPGA. Thread 4, 5, 6, 7 are used to receive performance measurement data for monitoring data bandwidth. They are only used to analyze the bandwidth utilization, in the real platform, these threads will be removed.

Data Controller—Data controller is responsible for data transfer between the host and FPGA through PCIe. We use RIFFA to handle the communication [9]. RIFFA supports PCIe Generation 2 and 128 bit data interface. Up to 12 channels can be used to fully utilize the bandwidth. In the architecture we use eight channels. For consistency, in the following, "send" and "receive" are from the aspect of FPGA. Channel 0 is used to receive vertex data from host and send calculated vertex data for the next iteration back to host. Channel 1 receives edge data from host. To improve the streaming performance, edge transfer and message transfer use different channels. Channel 2 sends messages, which are generated in the scatter phase, back to the host and receives messages in the gather phase. Channel 3 is used to transfer metadata for each edge partition. When processing each edge partition, only one metadata is used to indicate the number of edges will be streamed in and the number of messages should be streamed out, so the metadata will not bottleneck the processing throughput. Also, putting metadata into hardware memory (either DRAM or on-chip memory) will constrain the scale of data which the system can process, since for P number of node intervals, P^2 number of metadata are needed. Channel 4, 5, 6, 7 are used to transfer performance measurement data to monitor the bandwidth of the system, and they are only for comparison analysis. In the final system, these four channels are removed. In our design, for simplicity, we did not fully utilize all the 12 channels. More channels should cause higher bandwidth utilization. For example, edge data can be scattered to multiple threads, and each thread sends a part of edge data through an individual channel. This will increase the interface width equivalently and consequently increase the PCIe bandwidth utilization.

Processing Pipelining— In our design, we split the data processing into three stages, read, processing, and write.

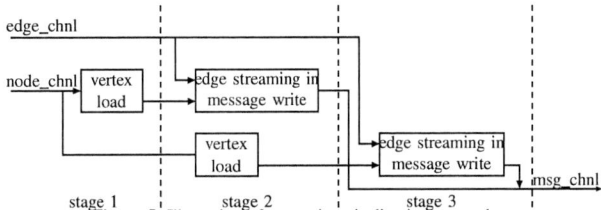

Figure 5: Illustration of processing pipeline in scatter phase.

Three stages are pipelined to improve the overall throughput as shown in Fig. 5. Double buffer is also used to improve throughput of vertex data read.

Message Combiner & Preprocessing—When processing each edge partition, message combiner combines generated messages with the same destination to one message in order to reduce PCIe accesses. In preprocessing, all edges are grouped into P^2 edge partitions. Within each edge partition, edges are sorted by destination. The number of edges and the number of messages are logged to a metadata file. Sorting complexity is $\mathcal{O}\left(P^2 \cdot \frac{|E|}{P^2} \cdot \log \frac{|E|}{P^2}\right)$.

IV. EXPERIMENTAL SETUP AND IMPLEMENTATION

To solidify our study, we choose the newly emerging Graph Convolutional Neural Networks (GCNs) [10] as our target application. GCN generalizes classical CNNs to handle graph data such as molecular data, point cloud and social networks. Its applications include 1) semi-supervised document classification in citation networks, 2) semi-supervised entity classification in a bipartite graph extracted from a knowledge graph, 3) an evaluation of various graph propagation models, and 4) a run-time analysis on random graphs. Solving GCN not only consists of handling all challenges associated with standard graph analytics but also poses a set of unique challenges induced by the huge data size of vertex set. All data sets we test are listed in Table I [10] and quite representative in terms of memory access irregularity typically found in graph-based analytics. Mathematically, the graph propagation rule of the GCN model can be described as $h_i^{(l+1)} = \sigma(\sum_{j \in N_i} \frac{1}{c_{ij}} h_j^{(l)} W^{(l)})$ and $c_{ij} = \sqrt{d_i d_j}$, where W is a trained weight matrix, d_i denotes the degree of node v_i, and h_i is a feature vector of node v_i [10]. For example, in the documentation classification, each document is described in a bag-of-words feature vector. In the first layer, each bag-of-words feature vector is a sparse vector. However, in the second and following layers, each vector is a dense vector with each element a 32 bit floating number. Therefore, sparse graph processing methods lose the generosity in this problem, even it could benefit when processing the first layer. To be general, we treat each feature vector in the first layer as a dense vector even most of elements are zero. Also, each feature is stored as a 32 bit floating point number.

A. Experimental Setup

Before delving into specifics, we introduce the primary targeting hardware system—a modern integrated FPGA platform depicted in Fig. 6. It is heterogeneous in the sense that such chip platform possesses multiple levels of memory hierarchy with various bandwidth and capacity. Many existing chip platforms fit with the abstraction in Fig. 6. The experiments were conducted on a KC705 evaluation Board with a Xilinx Kintex xc7k325t-2ffg900c FPGA chip.

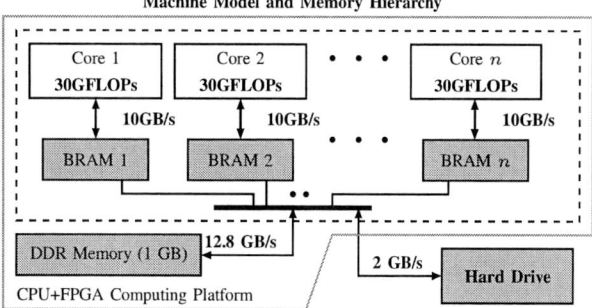

Machine Model and Memory Hierarchy

Figure 6: Abstract machine and memory model of our FPGA+CPU platform.

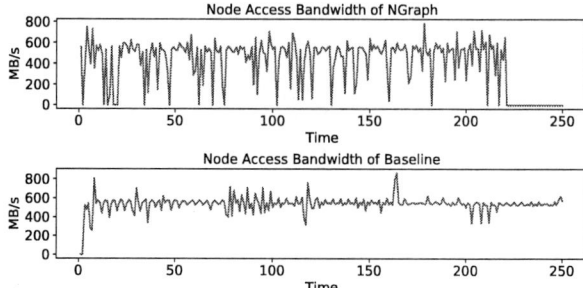

Figure 7: Experimentally measured PCIe I/O traffic patterns for (a) Node-heavy graph processing implementation, and (b) Baseline implementation.

The resource utilization is evaluated using Vivado 2016.4. The processing logic runs at 100MHz clock frequency and RIFFA logic runs at 250MHz clock frequency [9]. We ran three experiments for 4 different datasets [10]. The statistics of each dataset is listed in Table I. In experiments we only focus on the memory accesses, since in the graph convolution network, each node will do a complex matrix multiplication, and matrix multiplication acceleration is a problem-specific optimization. To be general, in the following experiments, we keep the memory access pattern but skip the computation part. We will discuss how to accelerate graph convolution network in our future work.

For the baseline implementation, we extend the FPGP [5] graph processing platform to process node-heavy graph problems. FPGP makes an assumption that edges are stored in hard disk and nodes are stored in DRAM. We assume edges and vertices are all stored in hard drive while keeping the memory access pattern and computing strategy the same as FPGP. The baseline implementation works as described in Algorithm 1 of FPGP [5].

B. Performance Measurement: Latency and I/O Bandwidth

For performance measurements, we consider 4 data sets listed in Table I with variable sizes. The first three data sets—*Citeseer, Cora, and Pubmed*—are collected real-world data. In order to stress-test our computing platform, we also create, with the same principle as in [10], a synthetic *X-Large* data set with more than one million edges. For all these test cases, we use the same FPGA implementation that consumes 13% of LUT, 3% of LUT-RAM, 8% of FFs, and 75% of BRAM resource available on a Kintex xc7k325t-2ffg900c FPGA chip. For comparison, our baseline implementation uses 12% of LUT, 2% of LUT-RAM, 8% of FFs, and 75% of BRAM for the same FPGA device. Also, for each experiment, we adjust P to fully utilize the FPGA on-chip block memory.

As shown in Table I, for each dataset, we measure the latency to finish each iteration. For all the four data sets, the speedup of our node-heavy graph processing platform over the baseline ranges from 10 to 50 times. The biggest contributing factor, we believe, is due to the fact that, during each iteration, the baseline framework needs $P|V|\alpha_{in}$ number of node accesses, while our implementation only needs $|V|\alpha_{in}$ number of accesses. Also, our node-heavy graph processing framework will generate a lot of messages which will decrease the speedup. So theoretically the speedup of our design compared with the baseline implementation should be smaller than but close to P, which is consistent with our measured results on the hardware.

To experimentally validate the above hypothesis, we measure the PCIe bandwidth usage for our node-heavy graph processing framework and the baseline. As shown in Fig. 7 for "Cora" dataset, both designs are bottlenecked by node access bandwidth. Unlike the baseline, the node bandwidth utilization of our design is not always high, because FPGA needs to send generated messages back to host through PCIe after processing each edge partition. More importantly, the total amount of PCIe data traffic is much larger for the baseline implementation than for our processing framework. As such, compared with our design, the baseline takes much longer to finish the processing. To further explain our significant performance gain, we also measure the effect of message combining caused by pre-sorting. We found that, on average, we can reduce the total number of messages by about 2 times by combining. We also found that the benefit of message combining is significantly affected by both the graph sparsity and the partition factor P.

V. PERFORMANCE MODELLING AND ANALYSIS

P	Number of vertex partitions
α_{in}	Input vertex data size (in Byte)
α_{out}	Output vertex data size (in Byte)
β	Edge data size (in Byte)
$|V|$	Number of vertices
$|E|$	Number of edges
M_{BRAM}	Storage capacity of FPGA on-chip block memory (in MB)

Table II: Notations of disk I/O model

For node-heavy and large-scale graph analytics, I/O performance to the secondary storage will be crucial to the overall computing performance of a heterogeneous computing platform. In this section, we develop two disk I/O models for our implementation and the baseline implementation respectively for performance comparison. Note that, although we choose hard drive as our secondary storage in developing our performance model, our mathematical model can be readily extended to treating the on-board DRAM as the secondary storage. We use the notations listed in Table II to formulate and derive our performance model. In this study, P, the total number of vertex partitions, is determined by fulfilling the requirement that each vertex partition doesn't exceed the memory capacity of on-chip block RAMs, i.e., $M_{\text{BRAM}} \geq \frac{|V|\alpha_{in}}{P}$. Given that the graph under our consideration is node-heavy and very large-scale, typical P value is large, 100s or even 1000s. Also, when partitioning the graph, we attempt to balance each vertex partition such that each resultant edge set contains almost the same number of edges.

As detailed in Section II-B, our computing algorithm follows the GAS (gather-apply-scatter) principle with two data accessing phases. During the scatter phase, as shown in Fig. 2(a), we need to enumerate through all vertices and

978-1-5386-7100-9/18 $31.00 © 2018 IEEE

Dataset	Nodes (Size)	Edges (Size)	# Features Per Node	NGraph Latency	Baseline Latency	# Partitions (P)	Speedup
Citeseer	3327 (49.28 MB)	12431 (99.45 KB)	3703	46.979 ms	453.6 ms	16	10
Cora	2708 (15.52 MB)	13264 (106.11 KB)	1433	0.11 s	4.64 s	64	45
Pubmed	19717 (39.43 MB)	108365 (866.92 KB)	500	0.18 s	5.65 s	39	31
X-Large	210010 (411.23 MB)	1349400 (9321.23 KB)	500	1.87 s	12.6 min	411	404

Table I: Dataset statistics, latency, and speedup of our implementation vs. baseline.

write out messages into their outgoing edges. As such, we sequentially load all vertex partitions and, for each loaded vertex partition V_i with the data size $\frac{|V|\alpha_{in}}{P}$, we stream all edge subsets containing all edges that have vertices in V_i as their sources, i.e., $E_{i,j}$, where $j = 1, 2, \cdots, P$, from PC host. This procedure ensures that all vertices are only loaded once, therefore significantly reducing I/O traffic. Without message combining, the number of generated messages equals the number of edges. Each message carries the data to be used to update the destination vertex partition in the gather phase. In the GCN test case, only accumulate operation is needed when using messages to update destination nodes in the gather phase. Therefore each message carries the same size of data as the output vertex size α_{out}. So the total disk I/O accesses spent on edge streaming-in and message streaming-out will be $\frac{|E|\beta}{P^2} + P \cdot \frac{|E|\alpha_{out}}{P^2} = \frac{|E|\beta}{P^2} + \frac{|E|\alpha_{out}}{P}$. According to Algorithm 3, this process will be repeated P times until all the edge partitions can get streamed. Therefore, the total disk I/O needed in the scatter phase can be written as $P \left(\frac{|V|\alpha_{in}}{P} + \frac{|E|\beta}{P^2} + \frac{|E|\alpha_{out}}{P} \right)$. In the gather phase, messages associated with edge partitions get processed in the order of the destination partitions they belong to. As depicted in Fig. 2(b), after P number of message partitions $M_{1,i}$ to $M_{P,i}$ are streamed in, the vertex partition V_i is completely updated and streamed out to hard drive. This process is repeated P times until all the P^2 partitions get processed. The total number of hard drive accesses in the gather phase is $P \left(P \cdot \frac{|E|\alpha_{out}}{P^2} + \frac{|V|\alpha_{out}}{P} \right)$. Combining the I/O accesses in both gather and scatter phases, the total disk accesses of our design in each iteration can be expressed as $|E|\beta + |V|\alpha_{in} + (2|E| + |V|)\alpha_{out}$. With the similar procedure, we can also derive the baseline I/O access model for our baseline System, which is similar to [4] and FPGP [5]. The total amount of disk I/O accesses for this baseline implementation can be expressed as $P|V|\alpha_{in} + |E|\beta + |V|\alpha_{out}$.. Here, we omit details of model derivation for simplicity. Comparing the above two equations, it is clear that our method can significantly reduce the total number of I/O accesses by approximately P times, as long as $P > \frac{2|E|\alpha_{out}}{|V|\alpha_{in}} + 1$. This reduction in I/O accesses clearly translates into the significant performance gain presented in Table I for all benchmark data sets. There are two contributing factors to our performance improvement. First, unlike in the baseline implementation, we adopt the well-known GAS principle for the iterative graph analytics instead of the stream-and-apply approach. We observe that the use of message array can effectively eliminate the redundant loading of vertex data set. Second, unlike in all previous studies, we perform the sorting within each vertex partition, which facilitates the message combining during scatter phase and consequently significantly reduce the total number of messages written into the hard drive. In fact, because P is proportional to $|V|\alpha_{in}$, when processing a real-world large-scale node-heavy graph on a CPU+FPGA heterogeneous computing platform, P will almost always be a very large number.

VI. Conclusion

Many important memory-intensive and mission-critical applications pose huge computing challenges due to their **large data sizes**, **irregular data placement**, and **complicated computing parallelism**. Our study has experimentally demonstrated that emerging FPGA+CPU/GPU heterogeneous platform presents an unique opportunity to maximize memory access performance for such applications through implementing application-specific computing machine.

References

[1] A. Roy, I. Mihailovic, and W. Zwaenepoel, "X-stream: Edge-centric graph processing using streaming partitions," in *Proceedings of the Twenty-Fourth ACM Symposium on Operating Systems Principles*, SOSP '13, (New York, NY, USA), pp. 472–488, ACM, 2013.

[2] A. Gudimella, R. Story, M. Shaker, R. Kong, M. Brown, V. Shnayder, and M. Campos, "Deep reinforcement learning for dexterous manipulation with concept networks," *CoRR*, vol. abs/1709.06977, 2017.

[3] S. Zhou, C. Chelmis, and V. K. Prasanna, "High-throughput and energy-efficient graph processing on fpga," in *2016 IEEE 24th Annual International Symposium on Field-Programmable Custom Computing Machines (FCCM)*, pp. 103–110, May 2016.

[4] X. Zhu, W. Han, and W. Chen, "Gridgraph: Large-scale graph processing on a single machine using 2-level hierarchical partitioning.," in *USENIX Annual Technical Conference*, pp. 375–386, 2015.

[5] G. Dai, Y. Chi, Y. Wang, and H. Yang, "Fpgp: Graph processing framework on fpga a case study of breadth-first search," in *Proceedings of the 2016 ACM/SIGDA International Symposium on Field-Programmable Gate Arrays*, pp. 105–110, ACM, 2016.

[6] J. E. Gonzalez, Y. Low, H. Gu, D. Bickson, and C. Guestrin, "Powergraph: Distributed graph-parallel computation on natural graphs," in *Presented as part of the 10th USENIX Symposium on Operating Systems Design and Implementation (OSDI 12)*, (Hollywood, CA), pp. 17–30, USENIX, 2012.

[7] V. Kalavri, V. Vlassov, and S. Haridi, "High-level programming abstractions for distributed graph processing," *CoRR*, vol. abs/1607.02646, 2016.

[8] R. McCune, T. Weninger, and G. R. Madey, "Thinking like a vertex: A survey of vertex-centric frameworks for large-scale distributed graph processing," *ACM Comput. Surv.*, vol. 48, pp. 25:1–25:39, 2015.

[9] M. Jacobsen, D. Richmond, M. Hogains, and R. Kastner, "Riffa 2.1: A reusable integration framework for fpga accelerators," *ACM Transactions on Reconfigurable Technology and Systems (TRETS)*, vol. 8, no. 4, p. 22, 2015.

[10] T. N. Kipf and M. Welling, "Semi-supervised classification with graph convolutional networks," *CoRR*, vol. abs/1609.02907, 2016.

2018 IEEE Computer Society Annual Symposium on VLSI

On-chip Data Security against Untrustworthy Software and Hardware IPs in Embedded Systems

SreeCharan Gundabolu, Xiaofang Wang
Dept. of Electrical and Computer Engineering
Villanova University
800 Lancaster Ave.
Villanova, Pennsylvania 19085
Email: {sgundabo, xwang}@villanova.edu

Abstract—State-of-the-art system-on-chip (SoC) field programmable gate arrays (FPGAs) integrate hard powerful ARM processor cores and the reconfigurable logic fabric on a single chip in addition to many commonly needed high performance and high-bandwidth peripherals. The increasing reliance on untrustworthy third-party IP (3PIP) cores, including both hardware and software in FPGA-based embedded systems has made the latter increasingly vulnerable to security attacks. Detection of trojans in 3PIPs is extremely difficult to current static detection methods since there is no golden reference model for 3PIPs. Moreover, many FPGA-based embedded systems do not have the support of security services typically found in operating systems. In this paper, we present our run-time, low-cost, and low-latency hardware and software based solution for protecting data stored in on-chip memory blocks, which has attracted little research attention. The implemented memory protection design consists of a hierarchical top-down structure and controls memory access from software IPs running on the processor and hardware IPs running in the FPGA, based on a set of rules or access rights configurable at run time. Additionally, virtual addressing and encryption of data for each memory help protect the confidentiality of data in case of a failure of the memory protection unit, making it hard for the attacker to gain access to the data stored in the memory. The design is implemented and tested on the Intel (Altera) DE1-SoC board featuring a SoC FPGA that integrates a dual-core ARM processor with reconfigurable logic and hundreds of memory blocks. The experimental results and case studies show that the protection model is successful in eliminating malicious IPs from the system without need for reconfiguration of the FPGA. It prevents unauthorized accesses from untrusted IPs, while arbitrating access from trusted IPs generating legal memory requests, without incurring a serious area or latency penalty.

Index Terms—SoC FPGA, embedded systems, third-party IPs, memory security, implementation

I. INTRODUCTION

Embedded systems are mostly built around a few processor cores with custom functions implemented in application-specific integrated circuits (ASICs) or FPGAs. State-of-the-art SoC FPGAs integrate hard powerful ARM cores with the reconfigurable logic fabric in addition to many commonly needed high-performance and high-bandwidth peripherals on a single chip. The prohibitively high cost and high risks of ASICs, and the ever-increasing system complexity and demand for flexibility have made FPGAs more and more attractive

as the primary choice in many critical embedded systems, including implementations of cryptographic algorithms [1].

The dramatic increase in complexity and magnitude of system design has made it necessary to reuse design and employ intellectual properties (IPs) from various vendors, including those from overseas third parties. This raises serious concerns about security and trustworthiness of the products employed in critical applications like military, health, aviation, etc. Malicious logic hidden in an IP core can be designed to illegally gain access to information, destroy it or in certain situations be able to modify the functionality of the system or disable it. When multiple IP cores interact with each other at the system level, the probability of them being exploited becomes higher.

In general, there are two types of countermeasures proved effective to detect custom hardware trojans: logic verification and side-channel analysis [2]. The first paradigm aims at detecting deviations caused by hidden malicious functions that are triggered under specific conditions. With the increasing diversity and complexity of trojans and the chip size, such approaches are very time- and computationally consuming and are not feasible in practice. Another popular class of detection methods, i.e. side-channel analysis, are very sensitive to device variations in current nanoscale fabrication processes and are not sensitive to very small trojans [3]. Moreover, due to the stealthy nature of trojans, the very rare circuit conditions may not be realized by the design time detection methods, especially with distributed trojans that can only be triggered by a specific combination of analog and digital conditions. More recent approaches including both static and run-time solutions offer more promising results [4], [5].

Given the reconfigurability of FPGAs at design time and run time, design-time detection methods alone appear insufficient. Even if all malicious trojans have been detected and removed at design time and the bitstream is protected by encryption schemes provided by the FPGA vendors, it is still possible for an adversary to use side-channel attacks to recover the keys and, then, insert malicious logic into the bitstream [6].

Securing FPGA-based embedded systems is inherently different from IC chips as the FPGA design process is almost completely separate from the manufacturing flow, disallowing tampering during manufacture and test of the FPGA device

978-1-5386-7100-9/18 $31.00 © 2018 IEEE   644

[7]. The nature of FPGAs not only exposes them to a wide range of unique threats, but also provides an opportunity to enhance security.

In this paper, we present our run-time design-for-security solution for protecting on-chip data of embedded systems from untrustworthy IPs, including software and hardware IPs. Due to the narrow and limited bandwidth of off-chip memory, FPGA vendors have been embedding more and more dedicated memory blocks inside the chip for various purposes, including storing/buffering critical application data or encryption keys. State-of-the-art FPGAs can have more than 6 GB of on-chip memory on a single chip. Current FPGA-based systems employ a flat memory addressing scheme allowing fast single-cycle transactions, while also facilitating malicious IP cores to issue unauthorized memory requests, corrupting, or revealing sensitive data.

II. SYSTEM AND THREAT MODELS

We consider a typical embedded system with a processor running secure software and also untrusted software modules (also called software IPs in this paper) and an FPGA hosting hardware IPs. The system has both on-chip and off-chip memory units. Memory-mapped interfaces are used by IPs to interact with memory and the processor. We assume, the system does not have memory management and protection services typically found in an OS. This paper focuses on security of data in on-chip memory exclusively. Such on-chip memory blocks are distributed across the chip and each of them can have individual access channels based on the system needs. So, unlike off-chip memory, there is no one specific channel that can be monitored to ensure security of data stored in on-chip memory.

We focus on protecting the integrity and confidentiality of data stored in on-chip memory from both HW & SW IPs that are not trustworthy. Side channel attacks are not relevant to this work. We assume that the FPGA fabric and the processor itself are trustworthy and are not tampered with. Usually, FPGA vendors test these devices for Trojans inserted at the gate level in the fabric, so it is not a trivial attack to perform. We consider a safer and lightweight, hybrid HW&SW based approach to memory protection. Such approaches offer deterministic latency, which is critical to many real-time embedded applications.

III. RELATED WORK

Run-time detection of hardware trojans is becoming increasingly necessary to mitigate the malicious actions of trojans, that slip through IP verification. Bloom, Narahari, and Simha introduced a run-time approach to detect HW/SW Trojans, by using a hardware guard monitor external to the CPU, and utilizing some of the operating systems security features[8]. SoC development also involves the usage of untrusted hardware IPs, authors in [9] describe an architecture for SoC design in which untrusted IPs are wrapped using security wrappers, and a central security enforcer monitors the interactions between wrapped IPs and rest of the system for suspicious activity.

Data security of embedded systems is increasingly raising more concerns. Many approaches assume a system model with an OS and focus on software-based solutions like those for general-purpose computing platforms. A number of approaches have been proposed to securing external memory for processor-based reconfigurable systems running operating systems [10], [11]. The system model normally employs an OS and assumes the FPGA is the trusted area. Data encryption is also utilized, where Data is encrypted before transfer to memory and decrypted after retrieval. Complex encryption algorithms cause significant resource overhead, requiring additional memory to storage hashes and delay overheads due to cryptographic operations. Moreover, most of them consider external memory with a single channel to protect.

Security policy [12] has been a decades-long concept to ensure trustworthiness. In the context of our research, i.e., embedded systems without an operating system, Huffmire et. al.[13] formalized a method and language for specifying access policies for external memory modules of reconfigurable systems. Their architecture ensured that the policy module is invoked for every memory access. An isolation-based hardware approach using security policy to protect on-chip memory blocks of FPGAs is presented recently in [14]. The security policy being enforced is hardwired in logic and does not allow new configuration at run-time.

IV. PROPOSED SYSTEM ARCHITECTURE

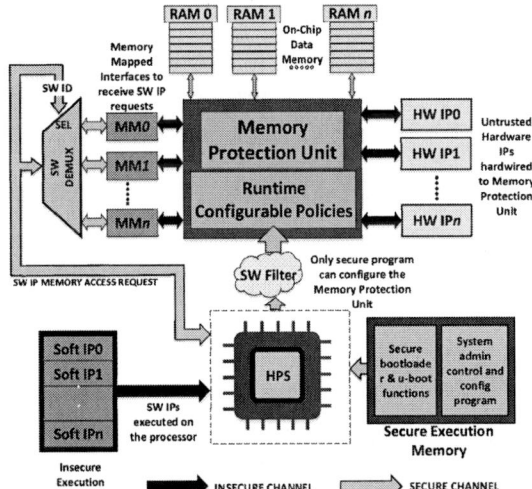

Fig. 1: Architecture Overview

Fig. 1 shows an overview of the proposed design. Data memory stored across multiple RAM modules is shared between HW & SW IPs. The Memory Protection Unit (MPU) acts as a barrier between the IPs and the memory, only valid transactions are allowed while malicious transactions are rejected. HW IPs are interfaced to the MPU though wired connections, while SW IPs interface using memory mapped registers. The Hard Processing System(HPS) hosts untrusted SW IPs and a secure driver software, and we assume the driver

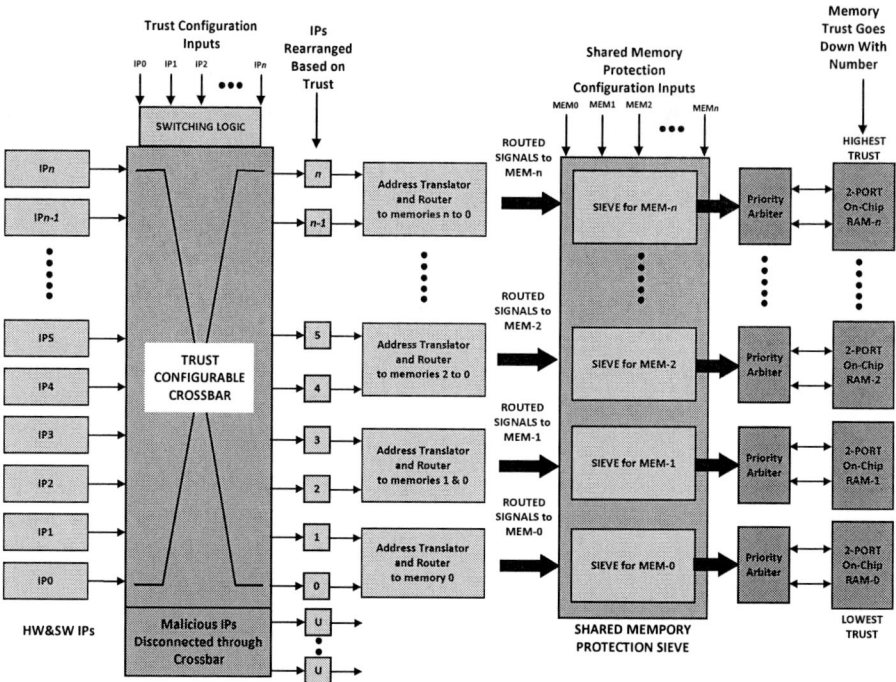

Fig. 2: System Overview

software is trusted and is protected against tampering. The driver software ensures isolation between SW IPs, schedules SW IPs to be executed on the processor and ensures that SW IPs do not configure HW components. Runtime configurable aspects of the MPU can only be modified by the secure software driver.

V. APPLICATION SPECIFIC MPU IMPLEMENTATION

Fig. 2 shows the detailed system design of the an implementation of the MPU. Configurable aspects of the design are, positional trust of IPs and addition/removal of entries from the private memory protection unit. Memories and IPs in the system are placed at different trust levels, IPs with trust levels larger than or equal to a memory's trust level can access that memory, all other accesses are considered illegal. A priority based arbiter is used to select higher priority requests, in case of congestion.

A Hardware and Software IPs.

The architecture is designed to accommodate IPs with different ranges of trust. HW & SW IPs are developed both in-house and by 3rd party vendors. The system designer is responsible for assessing the initial trust in an IP, and IPs are interfaced to the MPU based on the amount of trust that is placed in an IP or the vendor. HW IPs are directly connected to the inputs of the MPU. SW IPs on the other hand, issue memory requests to a register in the memory mapped interface provided by the MPU. Though we assume the absence of

an OS to provide advance memory protection features, we assume the software driver running on the HPS ensures that SW IPs cannot access memory mapped interfaces of other SW IPs or the configurable registers of the MPU. Requests from all IPs are treated the same, once they enter the MPU. The system administrator may choose to change an IP's trust level, in case of suspicion of any malicious activity by configuring the crossbar switch during run-time.

B. Configurable Trust level module.

The system is designed with a hierarchical policy, where memory accesses are rejected or arbitered based on trust and priority. Position of an IP in the system dictates its accessible memories. In Fig. 2, IPn and IP$n-1$ are at the highest trust level and can access memories in the range; RAMm to RAM0, similarly IP$n-2$ and IP$n-3$ can access memories in the range; RAM$m-1$ to RAM0. As one reaches the lowest level of trust IP1 and IP0, only have access to RAM0. On-chip memory can be configured as true dual port memories, allowing two IPs to simultaneously access memory. Hence, the number of memory modules in the system m, is half the total number of IPs n. Physical position of IPs cannot be modified without a reconfiguration of the FPGA, hence a crossbar switch is used to route signals of IPs such that the apparent position of the IP changes to the rest of the MPU.

1) Configurable Crossbar Switch: The crossbar switch is implemented on the FPGA fabric, and the apparent positions of IPs are changed by driving the crossbar switching logic. The crossbar switch provides the administrator, a memory mapped interface allowing it to be configured through the software driver. A system with n IPs has a crossbar with n memory mapped registers, each $\log_2 n + 1$ bits wide. The extra bit allows an IP to be assigned a trust position between 0 to $2n$. Trust positions outside the range of 0 to n places an IP in a cut-off state, effectively disconnecting it from accessing any memory. This feature of allows malicious IPs if found, to be effectively disabled without the need for reconfiguration. The driver software allows the administrator to assign a new position for each IP. It verifies the input entries for duplicates or unassigned positions, if the new positions are found to be legal the switching logic is configured, else the current priority is retained.

Functioning of the other modules in the MPU, are based on the unique position of the IP at the output of the crossbar switch. This unique position is directly associated to the trust of the IP and the memories it can access. Each IP in the system is assigned a unique identifier called the trust identifier (TID) which is $\log_2 n + 1$ bits wide. TID is a parameter unique and local to the MPU, generation of which is done within the MPU for each transaction. This ensures that malicious IPs do not spoof TIDs of higher priority IPs to access memories of higher trust.

C. Address Translation Unit

The Address Translation Unit implements the concept of Virtual Addressing, where the physical address of the memory space is hidden from IPs. The HPS is capable of accessing a 32-bit address space, i.e. 4 GB of memory, which includes memory mapped registers for FPGA logic, peripherals, off-chip memory and on-chip memory. Address translation is another feature local to the MPU, and the IPs accessing the memories are unaware of this security feature. In the event that the MPU is circumvented by malicious SW IPs, lack of knowledge of the physical address of data requires the IP to guess the location of data within the entire 4GB address space. Virtual addressing makes leaking relevant data impractical to malicious IPs, ensuring data protection.

Conventional virtual memory implementation using TLBs, results in large power consumption and execution times that are not deterministic which inhibit their use in embedded systems with limited resources. A simple arithmetic based address translation can be used to implement virtual addressing at an embedded scale[15]. An address translation unit can be designed by bit manipulation operations such as bit masking and flipping as show in Fig. 3. The table below describes the virtual address known to IPs and the physical address in an 8 IP, 4 memory system.

D. Request Router Unit

The hierarchical design enables, IPs with higher trust levels to access memories falling in the same or lower trust levels. If an IP at trust level n needs to access a memory at a lower trust level, then the memory request has to be forwarded

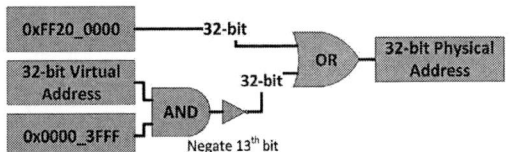

Fig. 3: Address Translation Unit

TABLE I: Address Transalation

Memory Region	Virtual Address Range	Physical Address Range
3	0x0000D000 – 0x0000DFFF	0xFF203000 – 0xFF203FFF
2	0x0000C000 – 0x0000CFFF	0xFF202000 – 0xFF202FFF
1	0x0000B000 – 0x0000BFFF	0xFF201000 – 0xFF201FFF
0	0x0000A000 – 0x0000AFFF	0xFF200000 – 0xFF200FFF

to the correct memory. The request router unit samples the translated address, and routes the request to the appropriate memory. Each trust level in the system has a slightly different router. IPn has access to m memories in the system while IP$(n - 4)$ would only have access to $m - 2$ memories in the system, hence the router for IP$(n - 4)$ is designed such that there are no physical connections to memories m and $m - 1$. This physical isolation between lower trust IPs and the higher trust memories, enhances the secure design. The address router contains a counter which keeps track of all illegal access attempts made to the memory by an IP. These counters may be monitored by the system administrator, to facilitate in identifying malicious IPs. In larger designs the router can be replaced with modified crossbar switches to ensure scalability of the designs, as linear connections from each IP to every other memory in the system would require more area due to higher wiring requirements.

E. Private Memory Protection Unit

In most shared systems it is often necessary to reserve memory for private use, for example a cryptographic IP might need to protect data, such as encryption/decryption keys from all other IPs in the system. There may arise situations where an IP would need to store some of its own data in a lower trust level due to the lack of enough memory resources in its own trust level, which warrants for a need to reserve certain regions in memory for private use. The private memory protection unit solves the problem of protecting shared memory in a system by allowing IPs to gain private access to memory. A private memory request is made by an IP to the processor which is received by the driver software, the IP provides the base address and optionally the size of private memory. The base address specifies the starting address of the private memory, the size is 0 for a single address or an integer value specifying the range of memory addresses to protect. The size of memory allowed to be private is limited to 5% of the total memory size, to be fair to other IPs at the same trust level and IPs in trust levels above it.

The private memory protection unit, provides 9 memory mapped interfaces for each memory. Which allows three

978-1-5386-7100-9/18 $31.00 © 2018 IEEE

private memory protection requests per memory. The triplet stored in a table for each private memory protection request is TID, private base address, size. Limiting the private memory size for each entry to 5%, ensures that only 15% of the entire memory space is reserved making the rest is available to other IPs. The driver software performs additional checks, when a private memory request is received. It ensures that, a new request does not conflict with an existing entry and also, that an IP has the required trust level to reserve private addresses in that memory.

Every memory access is scrutinized by verifying if the request address falls in the range of protected private address space. If the requested access falls in a private region, the TID of the request is verified against the table and if they match the access is granted, else it is rejected.

F. Priority Arbiter

Arbitration is one of the most important aspects in a shared environment under heavy loads. It can be inferred from Fig. 2 that only legal requests reach the priority arbiter, as illegal accesses and private memory accesses are filtered out by the address router unit and the private memory protection unit. In an environment with multiple IPs, a memory often receives more than 2 requests which is higher than the number of transactions that the 2-port RAM can handle. If multiple requests to a memory are received, requests with the higher priority are granted while requests with lower priority are rejected. The need of arbitration increases as one goes down the trust level. The memory at the highest trust does not require any arbitration as, only two IPs at the highest trust can access this memory. However, at the lowest trust level n IPs can attempt to access the memory. Pass transactions from different trust levels are given a higher priority, as they always originate from a higher trust level. In a situation with multiple pass transactions, transactions with higher trust are allowed while the rest are rejected.

Each IP in the system has a 1-bit memory request signal. When represented as a bus, results in an n-bit signal, in which requests from higher trust levels are at the MSB and trust decreases towards the LSB. At each arbiter in the system, bits for IPs below its trust level are permanently grounded, in turn physically rejecting all requests from lower priorities. The priority arbiter marks the last stage of the memory protection unit, the arbiter directly drives the inputs of the 2-port RAM that is used to store data.

VI. IMPLEMENTATION AND EVALUATION

A system was designed to test our architecture, in which the system contains 8 IPs, the MPU, and 4 on-chip memory blocks. There exist 2 IPs at a given trust level hence, once the request reaches the arbiter, it allows 2 IPs to access memory simultaneously. Total latency produced by the MPU is 3 clocks, before the request reaches the memory. Our design was implemented on the Intel/Altera DE1-SoC board, featuring a SoC-FPGA that integrates a hard dual-core ARM Cortex-A9

processor and reconfigurable logic fabric on the single die chip.

TABLE II: Resource Utilization on the Intel DE1-SoC Board

Entity Name	Submodules	Quantity in Design	No. of logic Elements	% of CLBs 32,070
Crossbar Switch	NA	1	812	3%
Address Translation & Router	Memory 0	2	111	<1%
	Memory 1	2	53	<1%
	Memory 2	2	39	<1%
	Memory 3	2	25	<1%
Total	4	8	228	<1%
Private Memory Protection Unit	Memory 0	2	560	2%
	Memory 1	2	420	<1%
	Memory 2	2	282	<1%
	Memory 3	2	181	<1%
Total	4	8	1442	4%
Priority Arbiter	Memory 0	1	52	<1%
	Memory 1	1	136	<1%
	Memory 2	1	142	<1%
	Memory 3	1	243	<1%
Total	4	4	573	2%
Misc	NA	NA	180	<3%
Complete Design			4095	13%

The resource usage is shown in Table II. Altera Quartus Prime 16.1 was used to write and compile the VHDL source code for all the hardware modules in the design. ModelSim 10.5b was used to simulate and test VHDL modules to ensure proper system functioning. The entire design was wrapped using a Memory Mapped Avalon Slave wrapper to generate an IP. C code was written to implement the software features of the design, and was also used to test the complete functioning of the system by writing and reading values from various memory mapped registers.

To illustrate how our system can be used to stop unauthorized access, we tested various scenarios. Due to limited space of this paper, two of the case studies which show the effectiveness of the security features are discussed below.

Case Studies

Case 1: This case aims to test the hierarchical top down policy enforced by the design, assuming a Malicious IP with a lower trust level attempts to access higher trust memory. The IP priorities are set to default and the private memory protection table is cleared. IPs 3 & 2 generate memory requests to $RAM1$ which lies in the same trust level that they reside. $IP4$ which has a higher TID compared to IP3 & IP2 generates a memory access request to $RAM3$. All three requests from the IPs are translated by the address translator. The router for IP3&IP2 forwards the request to the arbiter for $RAM1$, as there are no private memory protection entries. IP4's request to $RAM3$ does not leave the router, as there are no physical connections between the router of an IP with low TID to a memory at a higher trust level. This transaction is rejected as it is an illegal access according to the design's top-down hierarchical policy. After three clock cycles the two legal requests are granted, IP3&2 are acknowledged about their

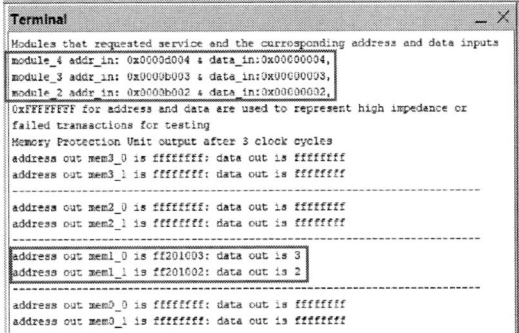

Fig. 4: Case 1

successful requests while IP4's request is rejected. The output window is shown in Fig. 4.

Case 2: This case aims to test the protection provided by the private memory protection unit. An IP protects an address range in its, own trust level for private use and an IP from a higher trust attempts to access it. IP4 needs to store few cryptographic keys in a different memory, due to shortage of its own memory space. IP4 makes a request for private memory, with ADDRESS: 0x0000C000 & SIZE: 0xF. The software driver checks if the private address space that the IP requests, lies in its legal limits of trust, and also ensures the range of addresses does not lie or overlap, current entries in the table. The legal entry containing the triplet (4, 0x0000C000, 0xF) is stored in the table, reserving addresses in the range 0x0000C000 – 0x0000C00F. IP7 & IP4 generate memory access requests to address 0x0000C002, in Memory 2. During normal operation, IP7's request gets a higher priority because of its higher trust. When both legal requests reach the private memory protection unit, it's found that the address lies in the protected memory range. The TIDs of both requests are compared against the entry in the table. IP7's request is blocked, for failing a TID match while, IP4's request is granted even with a low TID. The output window is shown in Fig. 5.

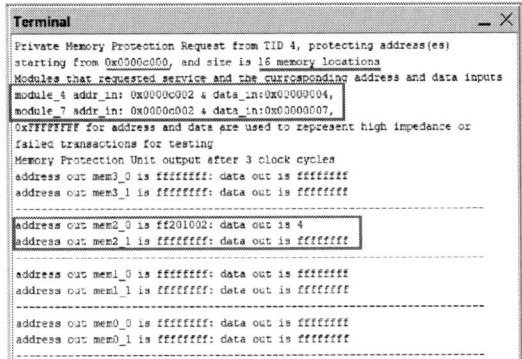

Fig. 5: Case 2

VII. CONCLUSIONS

State-of-the-art SoC FPGAs integrate hard ARM processor cores with reconfigurable logic fabric, making them an very attractive platform for embedded systems. With more and more dedicated memory blocks embedded in such platforms and increasing dependence on 3PIPs for embedded systems, securing on-chip data has become significantly critical to FPGA-based embedded systems. Our hardware and software based hierarchical memory protection system offers a lightweight, fast, and secure solution for data protection, against untrustworthy, hardware and software IPs, at run-time. The design analysis and selected case studies show that the protection model is successful in restricting access of malicious IPs in the system without need for reconfiguration of the FPGA. The MPU prevents unauthorized accesses from untrusted IPs, while arbitrating access from trusted IPs generating legal memory requests, without incurring on a serious area or latency penalty.

REFERENCES

[1] J. J. Rodrguez-Andina, M. D. Valds-Pea, and M. J. Moure, "Advanced features and industrial applications of FPGAs - a review," *IEEE Transactions on Industrial Informatics*, vol. 11, no. 4, pp. 853–864, Aug 2015.

[2] S. Bhunia, M. Hsiao, M. Banga, and S. Narasimhan, "Hardware trojan attacks: Threat analysis and countermeasures," *Proceedings of the IEEE*, vol. 102, no. 8, pp. 1229–1247, Aug. 2014.

[3] K. Xiao, D. Forte, Y. Jin, R. Karri, S. Bhunia, and M. Tehranipoor, "Hardware trojans: Lessons learned after one decade of research," *ACM Trans. Des. Autom. Electron. Syst.*, vol. 22, no. 1, pp. 6:1–6:23, May 2016.

[4] Y. Jin, X. Guo, R. G. Dutta, M. M. Bidmeshki, and Y. Makris, "Data secrecy protection through information flow tracking in proof-carrying hardware IP part i: Framework fundamentals," *IEEE Transactions on Information Forensics and Security*, vol. 12, no. 10, pp. 2416–2429, Oct. 2017.

[5] T. Wehbe and X. Wang, "Secure and dependable NoC-connected systems on an FPGA chip," *IEEE Transactions on Reliability*, vol. 65, no. 4, pp. 1852–1863, Dec. 2016.

[6] A. Moradi, D. Oswald, C. Paar, and P. Swierczynski, "Side-channel attacks on the bitstream encryption mechanism of Altera Stratix II: Facilitating black-box analysis using software reverse-engineering," in *Proc. ACM/SIGDA International Symposium on Field Programmable Gate Arrays*, 2013, pp. 91–100.

[7] S. Trimberger and J. Moore, "FPGA security: Motivations, features, and applications," *Proceedings of the IEEE*, vol. 102, no. 8, pp. 1248–1265, Aug. 2014.

[8] G. Bloom, B. Narahari, and R. Simha, "Os support for detecting trojan circuit attacks," in *Proc. IEEE Intl Workshop Hardware-Oriented Security and Trust (HOST)*, 2009, pp. 100–103.

[9] A. Basak, S. Bhunia, T. Tkacik, and S. Ray, "Security assurance for system-on-chip designs with untrusted IPs," in *IEEE Transactions on Information Forensics and Security*, 2017, pp. 1515–1528.

[10] J. Crenne, R. Vaslin, G. Gogniat, J.-P. Diguet, R. Tessier, and D. Unnikrishnan, "Configurable memory security in embedded systems," *ACM Trans. Embed. Comput. Syst.*, vol. 12, no. 3, pp. 71:1–71:23, April 2013.

[11] T. Wiersema, S. Drzevitzky, and M. Platzner, "Memory security in reconfigurable computers: Combining formal verification with monitoring," in *Proc. IEEE Int. Conf. Field-Programmable Technology (FPT)*, Dec 2014, pp. 167–174.

[12] D. Sterne, "On the buzzword 'security policy'," in *Proc. IEEE Computer Society Symp. Research in Security and Privacy*, May 1991, pp. 219–230.

[13] T. Huffmire, S. Prasad, T. Sherwood, and R. Kastner, "Policy-driven memory protection for reconfigurable hardware," in *Proc. European Symp. Research in Computer Security*, Sept. 2006, pp. 461–478.

[14] L. R. Rivera, X. Wang, and D. Chasaki, "A separation and protection scheme for on-chip memory blocks in FPGAs," in *IEEE International Symposium on Hardware Oriented Security and Trust (HOST)*, May 2016, pp. 223–228.

[15] X. Zhou and P. Petrov, "Low-power and real-time address translation through arithmetic operations for virtual memory support in embedded systems," in *IET Computers and Digital Techniques*, 2008, pp. 75–85.

2018 IEEE Computer Society Annual Symposium on VLSI

Design Automation and Test for Flow-Based Biochips: Past Successes and Future Challenges

Tsung-Yi Ho   tyho@cs.nthu.edu.tw

National Tsing Hua University, Hsinchu, Taiwan, Institute for Advanced Study, Technical University of Munich, Germany

Abstract—**Continuous flow-based biochips are attracting more attention from biochemical and pharmaceutical laboratories due to the efficiency and low costs of these miniaturized chips. By processing fluid volumes of nanoliter size, such chips offer the advantages of fast reaction, high throughput, high precision and minimum reagent consumption. In addition, by avoiding human intervention in the whole experiment process with automated control, these chips provide the ability of reliable large-scale experiments and diagnoses to the biochemical and pharmaceutical industry. In this paper, the fundamentals of flow-based biochips are explained. Thereafter, the state of the art of design automation for flow-based microfluidic biochips is reviewed and specific features of these chips compared to integrated circuits are discussed. These features offer extensive chances to expand the design automation methods from the IC industry to develop customized design flows and architectures for flow-based microfluidic biochips.**

I. FUNDAMENTALS OF FLOW-BASED MICROFLUIDIC BIOCHIPS

Microfluidic biochips have revolutionized traditional biomedical diagnosis and chemical experiments [1]. On such a chip, samples and reagents are transported and processed in the volume of nanoliter. This miniaturization saves the reagents significantly, which are very expensive in many cases. For example, the polyclonal antibody RNase inhibitor cost 600 euros per milliliter in December 2014 [2]. In addition, the experiment flow is regulated by a controller to maintain an accurate operation schedule for biochemical applications. Therefore, not only the application execution time but also the quality of the experiment can be improved. On such chips, typical genomic bioassay protocols such as nucleic-acid isolation, DNA purification and DNA sequencing, have been implemented successfully in recent years. Consequently, this technology has attracted a lot of commercial attention, e.g., from Illumina [3] and Agilent [4] Accordingly, the International Technology Roadmap for Semiconductors (ITRS) 2015 [5] has recognized the importance of microfluidic devices as having a rapid growth in the next several years.

Flow-based microfluidic biochips have dedicated devices such as mixers [6]–[8]. Samples and reagents are moved from one device to another in small fluid volumes for mixing or detection functions. These fluid samples are propagated through micro-channels itched on silicon/glass substrates or made from dimethylsiloxane using soft lithography [1]. The movement of the fluid samples through channels is controlled by valves, whose basic structure is shown in Fig. 1(a). On the substrate, the flow channel is used to transport samples

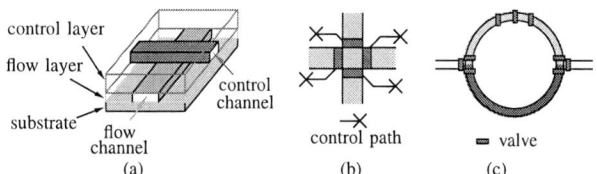

Fig. 1: Components of flow-based microfluidic biochips. (a) Valve structure. (b) Switch. (c) Mixer.

and reagents. Above the flow channel, a control channel conducts air pressure. Since both channels are built from elastic materials, the air pressure in the control channel dilates the channel segment right above the flow channel to squeeze the flow channel, so that the movement of the fluid sample or reagent in the flow channel is stopped. If the pressure is released, the fluid sample can continue its movement to the next device.

Valves can be placed in special patterns to construct complex devices. In biochips, it happens very often that transportation routes of several fluid samples cross with each other. To avoid contamination, a 4-way switch can be constructed at a crossing point using valves, as shown in Fig. 1(b). In this switch, only two of the four valves are open simultaneously to control the direction of the fluid sample. Another dedicated device is the mixer, which fulfills the important mixing function in experiments or diagnoses, as shown in Fig. 1(c). In this construction, the valves at the left and right are used to control the flow input and output of the mixer. The other three valves around the mixer channel form a peristaltic pump when they are actuated alternatingly. Suppose that 1 represents the open and 0 the closed status of a valve. The control patterns 101, 100, 110, 010, 011, 001 thus emulate the peristaltic effect for the reaction of samples and reagents [1].

After mixing, the result in a mixer can be transported to other devices or a dedicated storage unit so that the mixer can process further operations. The basic structure of a storage unit with the capacity of eight cells can be seen in Fig. 2, which is actually a detailed schematic of a biochip design with a mixer and a dedicated storage unit [9]. In this device, parallel storage channels are constructed side-by-side and at each end of a storage cell there is a valve. When a fluid sample flows into the storage unit, only the corresponding valves are open to store the sample in the cell. When releasing a stored sample, the corresponding valves are actuated while others are closed. In real designs, further dedicated devices such as heaters, filters

978-1-5386-7100-9/18 $31.00 © 2018 IEEE   650

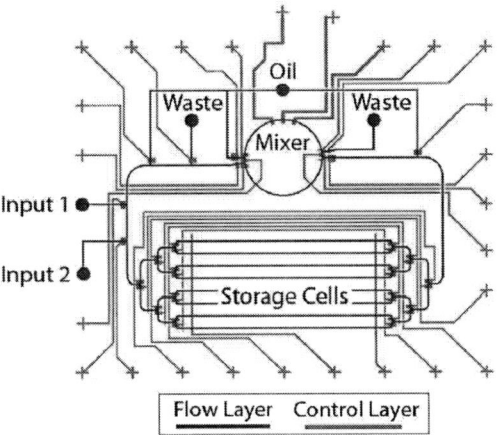

Fig. 2: Biochip architecture containing a mixer and a dedicated storage unit with eight cells [9].

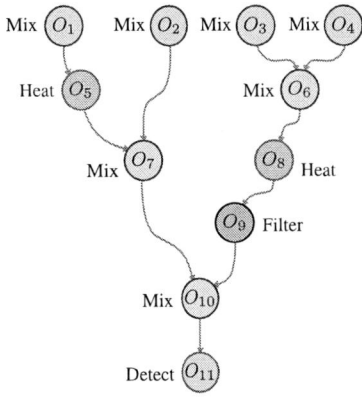

Fig. 3: Sequencing graph.

and detectors can also be built to provide specific functions.

The scale of flow-based biochips has evolved very fast recently due to the drastic increase in design and application requirements. Even on one chip more than 25K valves have been constructed successfully [10]. This trend makes the manual mapping of a complex application to a chip not feasible anymore. Therefore, design automation methodologies have been introduced for such chips recently. In the following, scheduling and binding of biochemical applications to microfluidic biochips are described in Section II. The specific features of biochips and methods addressing them are explained in Section III. Conclusions are drawn in Section IV.

II. Scheduling and Binding of Flow-Based Microfluidic Biochips

With the dedicated devices discussed above, a biochip can be built to execute an application or assay, which is usually described by a sequencing graph, as demonstrated in Fig. 3 Owing to area or cost constraints, it is not usual to assign a dedicated device for each of the operations in the sequencing graph. Instead, devices should be reused to execute the operations while maintaining their dependency specified in the sequencing graph. With a given number of devices, there should be an assignment of operations to devices, or a mapping, that leads to the minimum execution time of the application, and this optimal mapping is one of the objectives of biochip synthesis.

The problem of scheduling and binding is similar to high-level synthesis in digital circuits in [11]. Assume that the sequencing graph of a biochemical application is represented as a sequencing graph (O, E), where O is the set of nodes representing the operations and E is the set of edges representing the precedence of the operations. An edge (o_i, o_j) from $o_i \in O$ to $o_j \in O$ in the sequencing graph defines that o_i is one of the inputs of o_j so that it should finish before o_j. The execution duration of o_i is denoted by u_i. With these definitions, the constraints that guarantee a successful execution of the operations subject to given devices are listed as follows.

Uniqueness:

Each operation in the sequencing graph must be executed exactly once.

Duration:

The execution time allocated for the operation o_i should be no less than the required time u_i. Therefore, the finishing time of o_i should be at least u_i time later than its starting time.

Precedence:

For each edge (o_i, o_j) in the sequencing graph, o_i should finish before o_j starts.

Non-overlapping operations:

To avoid contamination of reactions, any two operations that are executed simultaneously should be assigned to different devices.

Channel conflict constraints:

In a flow-based biochip, if a device dispenses the resulting fluid sample to a connecting channel while this channel is still occupied by another sample, contamination occurs. To avoid contamination, a channel must be empty before it is used by by a new fluid sample.

Satisfying the constraints above, a schedule and binding of operations of an application can thus be generated. An example of scheduling and binding of a biochemical application in Fig. 3 and the corresponding chip architecture are shown in Fig. 4(a)–(b).

III. Specific Features of Biochips and State of the Art of Research

In the microfluidics community, researchers are focusing on developing new technologies and new structures to build fundamental components and devices, such as valves and pumps [12], [13]. Prototype microfluidic biochips are also built very often to demonstrate the function and performance of new components and new devices. Another major focus of the microfluidics community is to increase the integration density of basic components. With the advance in MEMS technology, a large number of components such as valves can now be built in a single biochip [14]. Unfortunately, the abundant

978-1-5386-7100-9/18 $31.00 © 2018 IEEE

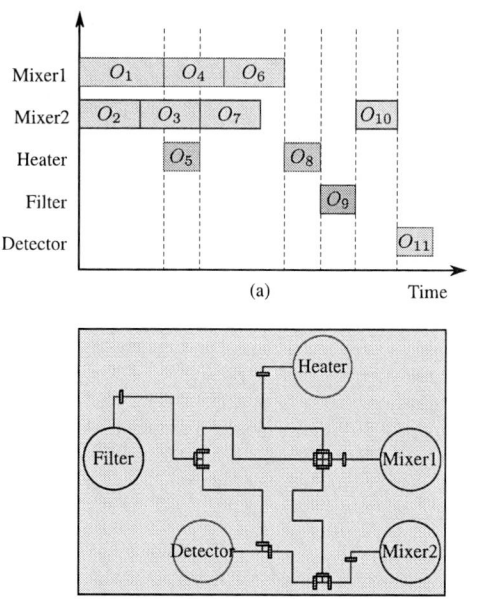

Fig. 4: Synthesis of microfluidic biochips. (a) Scheduling. (b) Physical design.

available resources have mostly been left unexplored, because end users cannot use them without a system layer that presents an interface for user applications, similar to the scenario that an operating system is missing for computer users. On the other hand, the effort of the microfluidics research community has been spread out in exploring even more technologies for microfluidic biochips, leading to a flourishing but fragmented panorama in the research on microfluidics.

The status quo of the microfluidics community is similar to the early period of the semiconductor industry. At that time, researchers were exploring different materials and device structures to build smaller but faster transistors. Thereafter, CMOS-based technology became dominant in this industry, while other technologies are employed only for specific applications. CMOS technology obtained its dominance because of, first of all, its performance. However, a very important factor which assisted this dominance is that the semiconductor industry and the electronic design automation community have found a way to carry out mass production of these devices and shrink the feature size continuously. In the meantime, the computer community has developed a successful computing model to present the available resources to end users and facilitate the development of high-level applications.

Observing the state of the art of microfluidic biochips, researchers from computer science and electrical engineering have started to bring their own computing models into microfluidic biochips. For example, the architecture of a microfluidic biochip from [9] is shown in Fig. 2. In this architecture, the mixer functions as the computing unit and intermediate results from the mixer are stored in the dedicated storage unit. The cells in the storage unit are built from normal channels. At

the ports of this storage unit, valves form multiplexers to direct fluid samples to enter into or leave from specific cells. This architecture emulates the classical von Neumann computer architecture to build a biochemical computing system from basic components. However, this simple emulation forsakes many unique characteristics of flow-based biochips, leading to inefficient execution of bioassays.

Similar to the semiconductor industry, design automation tool chains are also needed to support the development of microfluidic biochips. In recent years, the electronic design automation community has tried to migrate the existing design methodologies for integrated circuits to microfluidic biochip design, covering the phases from high-level synthesis down to physical design. Although this top-down flow has served the semiconductor industry in the past 50 years very successfully, fundamental changes should still be made to deal with specific requirements of biochips and take advantage of their unique features.

A. Flow-based Microfluidic Biochips: the Unique Characteristics

In microfluidic biochips, the inputs to an operation are fluid samples. Unlike electrical signals in integrated circuits, these fluid samples have a physical mass. In executing operations of a bioassay, fluid samples are processed with various operations, such as mixing, heating and detecting in different devices. The results of these operations are often fluid samples of different properties, so that inadvertent contamination between them should be avoided. The intermediate results of these operations should be stored in the chip temporarily in case they are not used immediately. Consequently, the physical mass and the variety of fluid samples become the major differences between biochips and integrated circuits, leading to several unique characteristics in biochip design.

Volume Management: In executing a bioassay, the volumes of fluid samples should be managed. Assume all the devices executing the bioassay in Fig. 3 have a capacity ν. Each of the resulting samples of O_3 and O_4 thus has a volume ν. When these two fluid samples reach the device executing O_6, half of their volumes should be disposed of because the device only accepts a volume ν. This volume management is not stated explicitly in the sequencing graph, but must be dealt with implicitly according to the volumes of intermediate fluid samples and the capacities of devices.

Storage management: In the schedule in Fig. 4(a), O_2 completes before O_5 does. The intermediate result of O_2 should be moved out of Mixer2 and stored somewhere temporarily so that the mixer can execute O_3. In the biochip shown in Fig. 2, this storage function is fulfilled by moving the result of O_2 to the dedicated storage unit through a channel. In synthesizing biochips, if operations are not scheduled properly, many storage requirements may appear, leading to many transportation channels and a large storage unit. In contrast to a dedicated storage unit as shown in Fig. 2, the storage function can actually be implemented using distributed transportation channels. In fact, a fluid sample can stay anywhere in a channel in the biochip until it is used by the next operation. This is a significant difference between biochips and electronic systems, where intermediate data can only be stored in special memory units, either flip-flops or RAM components. This observation can be confirmed by the storage cells in the dedicated storage unit in Fig. 2. These cells are built of normal channels but with valves at each end of a channel to control the store/fetch operations. Instead of forming a monolithic storage unit, these channels and valves can actually be distributed in the chip so that they can be used for storage when required, and for transportation otherwise. Consequently, the efficiency of channels and valves can be improved significantly.

Washing: Unlike electrical signals, fluid samples leave residue in channels after they travel through them. Before such a channel is reused by another fluid sample, it should be washed by neutral fluids such as silicon oil. Washing contaminated channels can be very flexible because several channel segments can be washed simultaneously if they form a connected graph while being isolated from the rest of the biochip that is executing other operations.

Flow-layer and control-layer codesign: In a flow-based biochip, valves are controlled by air pressure through control channels, e.g., the red channels in Fig. 2. If all the valves are controlled independently, the routing of control channels in a complex design becomes very complicated. To solve this problem, control channels of some valves can be shared if operations can still be executed correctly. This sharing requires a codesign of the flow layer and the control layer to match the actuation patterns of valves.

B. Design Automation for Flow-based Microfluidic Biochips: State of the Art

In recent years, design and optimization methods for flow-based microfluidic biochips have started to appear. For high-level synthesis, the top-down flow in [15] generates a biochip architecture and minimizes the execution time of the bioassay, while the method in [16] minimizes valve switching activities during architectural synthesis. For scheduling and binding, the method in [17] uses a maximum clique finding formulation to reduce assay execution time. In addition, the concept of general modeling of devices is introduced in [18] to improve the efficiency of the synthesis process, and special devices such as sieve valves are considered in [19]. Furthermore, fault-tolerance is considered during synthesis in [20] using a progressive optimization procedure.

For physical design of biochips, the method in [21] considers obstacles during routing and solves the problem using a rectilinear Steiner minimum tree, while both routability and assay completion time are considered to achieve an efficient flow-layer design in [22]. The placement of devices and routing of channels in flow-based biochips are dealt with simultaneously in [23] using a sequence-pair representation, and they are formulated as a SAT problem in [24] to achieve a close-to-optimal result.

Control logic synthesis is investigated in [25] to reduce the number of control pins. The method in [26] minimizes pressure propagation delay in the control layer to reduce the response time of valves and synchronize their actuations. Switching patterns of valves are examined in [27], [28] to reduce the largest number of switching activities in the control logic to avoid potential reliability problems. Furthermore, codesign of flow layer and control layer is investigated in [29] to achieve valid routing results on both layers iteratively, and length-matching is incorporated in routing control channels in [30] as well. Moreover, flow-layer, control-layer and valve switching are considered together in [31], [32] to simplify overall valve actuations.

To avoid contamination, washing is implemented in [33], [34] to clean devices and channel segments after they are used. This method still traces path sets and block-based partial washing has not been explored. The latter requires a co-optimization between operation scheduling and washing activities. The volume management problem in biochips has been explored in [35] and [36] for the specific bioassay sample preparation, but the optimization of volume management for general bioassays and the interaction of this task with fluid transportation for normal operations have not been taken into account.

To deal with manufacturing defects, fault models and an ATPG-based test strategy for flow-based biochips are proposed in [37], [38]. Design-for-testability and defect diagnosis are further addressed in [39]–[41].

On system level, the concept of distributed channel storage in flow-based biochips is explored in [42], [43]. A more general architecture of biochips from [44], where valves instead of dedicated devices are built regularly and connected with short channels to implement Programmable Microfluidic Devices (PMDs) or Fully Programmable Valve Arrays (FPVAs), has been explored in [45], [46] to provide better reliability and flexibility in executing bioassays. Though dynamic flow connections can be constructed on these chips relatively easily, channel crossing needs to be avoided [47] and valve control sequences need to be arranged carefully [48]. Test generation for this new architecture is also proposed in [49].

IV. CONCLUSION

In this paper, we have reviewed the fundamentals of flow-based microfluidic biochips. With the basic components, complex devices and large-scale biochemical systems can be constructed. The ever-increasing integration of biochips provides a potential to open doors for new experiment flows and medical diagnoses. To exploit this potential, design automation tools need to be introduced into this interdisciplinary area to manage the available resources. Although this process is similar to that for integrated circuits, specific features of biochips need to be addressed so that not only the viability of the new design methodologies but also their efficiency can be enhanced.

REFERENCES

[1] J. Melin and S. Quake, "Microfluidic large-scale integration: the evolution of design rules for biological automation," *Annu. Rev. Biophys. Biomol. Struct.*, vol. 36, pp. 213–231, 2007.

[2] Qiagen, Inc., "QIAGEN RNase Inhibitor," 2014. [Online]. Available: http://www.qiagen.com

[3] Illumina. [Online]. Available: http://www.illumina.com/

[4] Agilent. [Online]. Available: http://www.agilent.com/

[5] International Technology Roadmap for Semiconductors. [Online]. Available: http://www.itrs2.net/

[6] A. Manz, N. Graber, and H. M. Widmer, "Miniaturized total chemical analysis systems: A novel concept for chemical sensing," *Sensors and Actuators B: Chemical*, vol. 1, no. 1-6, pp. 244–248, 1990.

[7] T. Thorsen, S. J. Maerkl, and S. R. Quake, "Microfluidic large-scale integration," *Science*, vol. 298, no. 5593, pp. 580–584, 2002.

[8] E. Verpoorte and N. F. D. Rooij, "Microfluidics meets MEMS," *Proc. IEEE*, vol. 91, no. 6, pp. 930–953, 2003.

[9] N. Amin, W. Thies, and S. P. Amarasinghe, "Computer-aided design for microfluidic chips based on multilayer soft lithography," in *Proc. Int. Conf. Comput. Des.*, 2009, pp. 2–9.

[10] J. M. Perkel, "Microfluidics: Binging new things to life science," *Science*, vol. 322, no. 5903, pp. 975–977, 2008.

[11] G. D. Micheli, *Synthesis and Optimization of Digital Circuits*. McGraw-Hill Higher Education, 1994.

[12] M. A. Unger, H.-P. Chou, T. Thorsen, A. Scherer, and S. R. Quake, "Monolithic microfabricated valves and pumps by multilayer soft lithography," *Science*, vol. 288, no. 5463, pp. 113–116, 2000.

[13] R. Mathies, W. Grover, and E. Jensen, "Multiplexed latching valves for microfluidic devices and processors," Aug. 2010, US Patent 7,766,033.

[14] I. E. Araci and S. R. Quake. "Microfluidic very large scale integration (mVLSI) with integrated micromechanical valves," *Lab Chip*, vol. 12, pp. 2803–2806, 2012.

[15] W. H. Minhass, P. Pop, J. Madsen, and F. S. Blaga, "Architectural synthesis of flow-based microfluidic large-scale integration biochips," in *Proc. Int. Conf. on Compilers, Architecture, and Synthesis for Embed. Sys.*, 2012, pp. 181–190.

[16] K.-H. Tseng, S.-C. You, J.-Y. Liou, and T.-Y. Ho, "A top-down synthesis methodology for flow-based microfluidic biochips considering valve-switching minimization," in *Proc. Int. Symp. Phys. Des.*, 2013, pp. 123–129.

[17] T. A. Dinh, S. Yamashita, T.-Y. Ho, and Y. Hara-Azumi, "A clique-based approach to find binding and scheduling result in flow-based microfluidic biochips," in *Proc. Asia and South Pacific Des. Autom. Conf.*, 2013, pp. 199–204.

[18] M. Li, T.-M. Tseng, B. Li, T.-Y. Ho, and U. Schlichtmann, "Component-oriented high-level synthesis for continuous-flow microfluidics considering hybrid-scheduling," in *Proc. Design Autom. Conf.*, 2017, pp. 51:1–51:6.

[19] ——, "Sieve-valve-aware synthesis of flow-based microfluidic biochips considering specific biological execution limitations," in *Proc. Design, Autom., and Test Europe Conf.*, 2016, pp. 624–629.

[20] W.-L. Huang, A. Gupta, S. Roy, T.-Y. Ho, and P. Pop, "Fast architecture-level synthesis of fault-tolerant flow-based microfluidic biochips," in *Proc. Design, Autom., and Test Europe Conf.*, 2017, pp. 1671–1676.

[21] C.-X. Lin, C.-H. Liu, I.-C. Chen, D. T. Lee, and T.-Y. Ho, "An efficient bi-criteria flow channel routing algorithm for flow-based microfluidic biochips," in *Proc. Design Autom. Conf.*, 2014, pp. 141:1–141:6.

[22] Y.-S. Su, T.-Y. Ho, and D.-T. Lee, "A routability-driven flow routing algorithm for programmable microfluidic devices," in *Proc. Asia and South Pacific Des. Autom. Conf.*, 2016, pp. 605–610.

[23] Q. Wang, Y. Ru, H. Yao, T.-Y. Ho, and Y. Cai, "Sequence-pair-based placement and routing for flow-based microfluidic biochips," in *Proc. Asia and South Pacific Des. Autom. Conf.*, 2016, pp. 587–592.

[24] A. Grimmer, Q. Wang, H. Yao, T.-Y. Ho, and R. Wille, "Close-to-optimal placement and routing for continuous-flow microfluidic biochips," in *Proc. Asia and South Pacific Des. Autom. Conf.*, 2017, pp. 530–535.

[25] W. H. Minhass, P. Pop, J. Madsen, and T.-Y. Ho, "Control synthesis for the flow-based microfluidic large-scale integration biochips," in *Proc. Asia and South Pacific Des. Autom. Conf.*, 2013, pp. 205–212.

[26] K. Hu, T. A. Dinh, T.-Y. Ho, and K. Chakrabarty, "Control-layer routing and control-pin minimization for flow-based microfluidic biochips," *IEEE Trans. Comput.-Aided Design Integr. Circuits Syst.*, vol. 36, no. 1, pp. 55–68, 2017.

[27] Q. Wang, S. Zuo, H. Yao, T.-Y. Ho, B. Li, U. Schlichtmann, and Y. Cai, "Hamming-distance-based valve-switching optimization for control-layer multiplexing in flow-based microfluidic biochips," in *Proc. Asia and South Pacific Des. Autom. Conf.*, 2017, pp. 524–529.

[28] Q. Wang, Y. Xu, S. Zuo, H. Yao, T.-Y. Ho, B. Li, U. Schlichtmann, and Y. Cai, "Pressure-aware control layer optimization for flow-based microfluidic biochips," *IEEE Trans. Biomed. Circuits and Systems*, vol. 11, no. 6, pp. 1488–1499, 2017.

[29] H. Yao, Q. Wang, Y. Ru, Y. Cai, and T.-Y. Ho, "Integrated flow-control codesign methodology for flow-based microfluidic biochips," *IEEE Design & Test*, vol. 32, no. 6, pp. 60–68, 2015.

[30] H. Yao, T.-Y. Ho, and Y. Cai, "PACOR: practical control-layer routing flow with length-matching constraint for flow-based microfluidic biochips," in *Proc. Design Autom. Conf.*, 2015, pp. 142:1–142:6.

[31] T.-M. Tseng, M. Li, B. Li, T.-Y. Ho, and U. Schlichtmann, "Columba: Co-layout synthesis for continuous-flow microfluidic biochips," in *Proc. Design Autom. Conf.*, 2016, pp. 147:1–147:6.

[32] T.-M. Tseng, M. Li, D. N. Freitas, T. McAuley, B. Li, T.-Y. Ho, I. E. Araci, and U. Schlichtmann, "Columba 2.0: A co-layout synthesis tool for continuous-flow microfluidic biochips," *IEEE Trans. Comput.-Aided Design Integr. Circuits Syst.*, 2018.

[33] K. Hu, T.-Y. Ho, and K. Chakrabarty, "Wash optimization for cross-contamination removal in flow-based microfluidic biochips," in *Proc. Asia and South Pacific Des. Autom. Conf.*, 2014, pp. 244–249.

[34] ——, "Wash optimization and analysis for cross-contamination removal under physical constraints in flow-based microfluidic biochips," *IEEE Trans. on CAD of Integrated Circuits and Systems*, vol. 35, no. 4, pp. 559–572, 2016.

[35] A. M. Amin, M. Thottethodi, T. N. Vijaykumar, S. Wereley, and S. C. Jacobson, "Automatic volume management for programmable microfluidics," in *Proc. Conf. Programming Language Design and Implementation*, 2008, pp. 56–67.

[36] D. Mitra, S. Roy, S. Bhattacharjee, K. Chakrabarty, and B. B. Bhattacharya, "On-chip sample preparation for multiple targets using digital microfluidics," *IEEE Trans. Comput.-Aided Design Integr. Circuits Syst.*, vol. 33, no. 8, pp. 1131–1144, 2014.

[37] K. Hu, T.-Y. Ho, and K. Chakrabarty, "Testing of flow-based microfluidic biochips," in *Proc. VLSI Test Symp.*, 2013, pp. 1–6.

[38] K. Hu, F. Yu, T.-Y. Ho, and K. Chakrabarty, "Testing of flow-based microfluidic biochips: Fault modeling, test generation, and experimental demonstration," *IEEE Trans. Comput.-Aided Design Integr. Circuits Syst.*, vol. 33, no. 10, pp. 1463–1475, 2014.

[39] K. Hu, T.-Y. Ho, and K. Chakrabarty, "Test generation and design-for-testability for flow-based mVLSI microfluidic biochips," in *Proc. VLSI Test Symp.*, 2014, pp. 97–102.

[40] K. Hu, B. B. Bhattacharya, and K. Chakrabarty, "Fault diagnosis for flow-based microfluidic biochips," in *Proc. VLSI Test Symp.*, 2015, pp. 1–6.

[41] C. Liu, B. Li, T.-Y. Ho, K. Chakrabarty, and U. Schlichtmann, "Design-for-testability for continuous-flow microfluidic biochips," in *Proc. Design Autom. Conf.*, 2018.

[42] T.-M. Tseng, B. Li, U. Schlichtmann, and T.-Y. Ho, "Storage and caching: Synthesis of flow-based microfluidic biochips," *IEEE Design & Test*, vol. 32, no. 6, pp. 69–75, 2015.

[43] C. Liu, B. Li, H. Yao, P. Pop, T.-Y. Ho, and U. Schlichtmann, "Transport or store? Synthesizing flow-based microfluidic biochips using distributed channel storage," in *Proc. Design Autom. Conf.*, 2017, pp. 49:1–49:6.

[44] L. M. Fidalgo and S. J. Maerkl, "A software-programmable microfluidic device for automated biology," *Lab Chip*, vol. 11, pp. 1612–1619, 2011.

[45] T.-M. Tseng, B. Li, T.-Y. Ho, and U. Schlichtmann, "Reliability-aware synthesis for flow-based microfluidic biochips by dynamic-device mapping," in *Proc. Design Autom. Conf.*, 2015, pp. 141:1–141:6.

[46] T.-M. Tseng, B. Li, M. Li, T.-Y. Ho, and U. Schlichtmann, "Reliability-aware synthesis with dynamic device mapping and fluid routing for flow-based microfluidic biochips," *IEEE Trans. Comput.-Aided Design Integr. Circuits Syst.*, vol. 35, no. 12, pp. 1981–1994, 2016.

[47] G.-R. Lai, C.-Y. Lin, and T.-Y. Ho, "Pump-aware flow routing algorithm for programmable microfluidic devices," in *Proc. Design, Autom., and Test Europe Conf.*, 2018.

[48] A. Grimmer, B. Klepic, T.-Y. Ho, and R. Wille, "Sound valve-control for programmable microfluidic devices," in *Proc. Asia and South Pacific Des. Autom. Conf.*, 2018.

[49] C. Liu, B. Li, B. B. Bhattacharya, K. Chakrabarty, T.-Y. Ho, and U. Schlichtmann, "Testing microfluidic fully programmable valve arrays (FPVAs)," in *Proc. Design, Autom., and Test Europe Conf.*, 2017, pp. 91–96.

Multi-Target Many-Reactant Sample Preparation for Reactant Minimization on Microfluidic Biochips

Yung-Chun Lei
Department of Electronics Engineering
National Chiao Tung University
Hsinchu, Taiwan
yclei.ee98g@g2.nctu.edu.tw

Tien-Kuo Lin
Department of Electronics Engineering
National Chiao Tung University
Hsinchu, Taiwan
wxes9051111.ee02g@g2.nctu.edu.tw

Juinn-Dar Huang
Department of Electronics Engineering
National Chiao Tung University
Hsinchu, Taiwan
jdhuang@mail.nctu.edu.tw

Abstract—Sample preparation is one of essential steps in biochemical applications. It produces solutions with target concentrations through mixing various reactants in a specific way. In this paper, we propose a reactant cost minimization technique, M²SPA, for multi-target many-reactant sample preparation on microfluidic biochips through maximally sharing identical intermediate solutions among different targets. M²SPA first represents target concentrations as a recipe cube, searches all feasible candidates for intermediate solution sharing among targets, and then selects the one with the best cost saving for action. Experimental results show that the proposed algorithm can reduce up to 15.7% of reactant cost as compared to a state-of-the-art method.

Keywords—Sample preparation, microfluidic biochips, reactant minimization

I. INTRODUCTION

Similar to the scaling down of VLSI technology, scaling down of microfluidic biochip can greatly increase the system productivity and complexity. For some biomedical applications, like DNA analysis and proteomics, large scale experiments are often required. Due to the demand for various applications continuously grows, the design complexity of microfluidic biochips is continuously increased. Hence, a series of automation algorithms are necessary for reducing the manual effort, speeding up the process, and improving the design quality. In the past decade, a rich set of design automation algorithms have been proposed to solve many problems in microfluidic biochip design flow, such as synthesis, placement, routing, control pin assignment, and chip testing. [1]-[9]

Sample preparation problem is one of the critical issues in microfluidic biochip design automation. It has a determinant impact on accuracy, efficiency, and cost of a bioassay. During this process, raw reactants are mixed and diluted together to generate a product solution with a specified concentration value, which is called target concentration. In general, there are three factors that mainly determine the quality of whole dilution process: 1) the total number of dilution operations; 2) the undesirable waste, and 3) the total cost of valuable reactants. The number of dilution operations basically represents the preparation time, and thus should be minimized. Besides, large amount of waste may also lengthen the time of the whole sample preparation process due to a limited count of waste reservoirs available on a biochip. Lastly, the usage of valuable

reactant, which is very expensive (e.g., costly reagents) or is limited in amount (e.g., extracted DNA samples from crime scenes), should be minimized.

To improve the quality of sample preparation process, several algorithms which tackle the two-reactant single-target sample preparation problem on microfluidic biochips have been proposed [10]-[18]. However, it is common that three or even more reactants are involved in a bioassay. Also, it is also common that multiple target concentration values are required at the same time in a complicated bioassay. Thus, a multi-target many-reactant sample preparation technique is demanded.

In this paper, we propose a multi-target many-reactant sample preparation algorithm for reactant cost minimization on microfluidic biochips. In the following, we first introduce sample preparation and related works briefly in Section II and then our motivation and the problem formulation in Section III. Section IV presents the proposed algorithm, and experimental results are given in Section V. Finally, the concluding remarks are given in Section VI.

II. SAMPLE PREPARATION

Dilution and mixing take a major part in sample preparation. Two-reactant mixing (or dilution) is to mix a specific reactant with buffer solution, while many-reactant mixing indicates that more than two reactants (including buffer solution) are mixed together. The ratio among those reactants is determined by the given biomedical application, and it is called the target concentration C_t. For two-reactant mixing, C_t denotes the concentration of the reactant to be diluted. To extend the two-reactant problem to the many-reactant problem, the target concentration is alternatively specified by a sequence $CV \equiv \langle a_i \rangle$, where a_i is the ratio of reactant R_i. For example, a target concentration $\langle 5, 7, 4 \rangle$ indicates the ratio of the three reactants R_1, R_2, and R_3 are 5/16, 7/16, and 4/16, respectively. The ratio of CV is sometimes represented in the binary form. For example, $\langle 5, 7, 4 \rangle$ is given as $\langle 0.0101_2, 0.0111_2, 0.0100_2 \rangle$.

In most digital microfluidic biochips (DMFBs), each droplet is of the same size due to chip implementation. Under this scenario, mixing two droplets of different sizes to obtain the target concentration is not allowed. Since mixing always takes two droplets of same size, it is called the 1:1 mixing model. For sample preparation based on the 1:1 mixing model, several works have been proposed. The first one is the bit-scanning (BS)

978-1-5386-7100-9/18 $31.00 © 2018 IEEE

method [12]. In BS, the target concentration C_t is first encoded as a binary string. Then the mixing sequence can be derived from this binary string. The BS method is simple and fast; however, it generates a waste droplet at every mixing step. An algorithm called DMRW was proposed for waste reduction [13]. However, waste reduction is not necessary equal to overall cost reduction since the cost of each reactant is not the same. An algorithm called REMIA was then proposed to minimize the valuable reactant consumption in the two-reactant problem [17]. Later, an algorithm named CoDOS was proposed for the many-reactant sample preparation [20]. In the many-reactant problem, the cost of every reactant is considered as a different weight.

As mentioned, a complicated bioassay may require a set of target concentrations. Hence, multi-target sample preparation techniques are also demanded. Considering multiple targets simultaneously can further reduce the total reactant cost since the intermediate solutions can be shared among different targets. An algorithm named WARA was then proposed for multi-target two-reactant sample preparation [19]. In this paper, we take one step further by presenting a new multiple-target many-reactant sample preparation technique for reactant cost minimization.

III. MOTIVATION

The gap between two-reactant sample preparation and many-reactant one makes the problem much more complicated. CoDOS represents a target concentration by a *recipe matrix*. For example, the target concentration $CV = \langle 5,7,4 \rangle$ is given by a recipe matrix as shown in Fig. 1(a), where each row represents a reactant and each column denotes a level of binary digit. Since the ratio of R_1 is 0.0101_2, the four digits are fit into their corresponding levels. The key advantage of the recipe matrix is that every *rectangle* inside the matrix actually implies a chance for mixing operation sharing. For instance, Fig. 1(b) gives the dilution tree for the target concentration $\langle 5,7,4 \rangle$. Two red rectangles are marked in Fig. 1(a). The rectangles in Level 2 and Level 4 indicate there are two identical intermediate solutions $\langle 8,8,0 \rangle$ in Level 1 and Level 3 and they can actually share the mixing output of R_1 and R_2, which reduces the overall reactant cost. That is, recipe matrix helps identify those sharing opportunities. However, CoDOS deals with the single-target problem only. In order to share intermediate solutions among different targets, multiple recipe matrices should be considered as the same time. Hence, those 2D recipe matrices are stacked into a 3D structure, and is named as a *recipe cube*.

In a recipe cube, besides original row and column, the third dimension is used to identify different targets. An entry (x, y, z) is defined as the entry (x, y) in the recipe matrix of the zth target.

An entry is called an element if its value is 1. For an element $e(x, y, z)$ in a recipe cube, three functions are defined: $Target(e)$ returns z, $Reactant(e)$ returns x, and $Level(e)$ returns y. It implies that a droplet of R_x is required at the yth level of the dilution graph while producing the zth target.

In a recipe cube, two elements can form a *pair* $p(e_a, e_b)$ if and only if those two elements satisfy three conditions: 1) $Target(e_a)=Target(e_b)$, which indicates two elements are from the same target; 2) $Reactant(e_a) \neq Reactant(e_b)$, which indicates that the reactants of two elements must be different; 3) $Level(e_a) \leq Level(e_b)$, which ensures $Level(e_a)$ is never larger than $Level(e_b)$. If $Level(e_a)=Level(e_b)$, p is a *bar pair*; otherwise, it is a *slash pair*.

In the dilution graph corresponding to a recipe cube, a bar pair indicates there is an operation that mixes two droplets associated with its two elements. For example, two 1's grouped by the red line in Fig. 2(a) are two elements of a bar pair, while the nodes surrounded by red boxes in Fig. 2(b) are the two corresponding droplets for mixing. Similar to a bar pair, a slash pair can also specify a two-droplet mixing operation after a splitting operation. The detail will be given later.

To do reactant minimization, appropriate pairs are selected for common mixing operation sharing. In Fig. 2(a), two pairs grouped by red solid line and green dotted line are selected. Both pairs are composed of R_1 and R_2 and are from the same target. Two pairs can thus be mapped to the corresponding dilution graph as shown in Fig. 2(b). The node pair grouped by green dotted line is not supposed to be mixed in the beginning. However, the positions of R_2 and R_3 in the same target and same level can be swapped without altering the result. Consequently, only one mixing operation of R_1 and R_2 is enough as illustrated in Fig. 2(d). The recipe cube is updated as Fig. 2(c) shows. As Fig. 2(d) shows, after common mixing operation sharing, both the reactant consumption and waste are reduced.

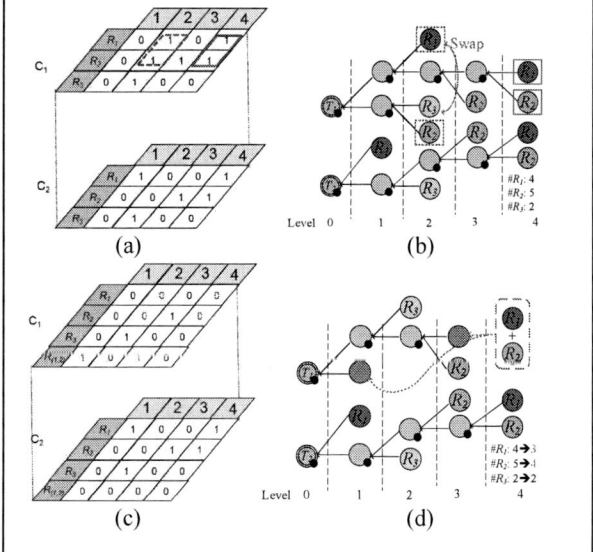

Fig. 2 (a) Recipe cube. (b) Corresponding dilution graph. (c) Updated recipe cube. (d) Updated dilution graph.

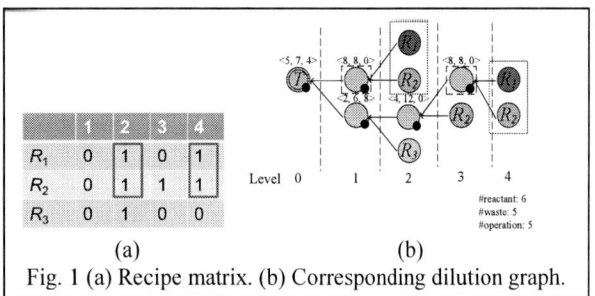

Fig. 1 (a) Recipe matrix. (b) Corresponding dilution graph.

After identifying rectangles (pairs) in the recipe cube, shareable node pairs in the dilution graph can be found as well. The output of a mixing operation is call the compound reactant. In the previous example, the compound reactant is represented as $R_{(1,2)}$ in the updated recipe cube as shown in Fig. 2(c). Compound reactants are created by merging (and then eliminating) the 1's in the selected pairs. A compound reactant in the recipe cube can also be used for pair identification later to further save the total reactant cost.

The problem formulation of our work is described as follows. Given a target concentration vector $CV=<C_i>$, and the weight vector $W=<w_i>$ of the reactant vector $R=<r_i>$, determine a dilution process for the target CV under the (1:1) mixing model such that the total reactant cost, $\sum u_i \times w_i$, can be minimized, where u_i and w_i is the droplet count and the cost weight of the ith reactant, respectively.

IV. PROPOSED ALGORITHM

Our algorithm is composed of several phases as shown in Fig. 3. The first phase is the recipe cube generation phase, in which our algorithm generates the initial recipe cube through converting the target concentration values to their binary forms. In the second phase, our algorithm finds all valid pairs in the recipe cube. Then in the phase of pair set composition, our algorithm assembles pairs into *double pairs* or *triple pairs*, and then selects one with the highest score from them for implementation in the phase of best pair selection. After the selection, if the score of the selected pair is positive, then the mixing operation associated with the selected pair is implemented and the recipe cube is updated accordingly. The above process is not terminated until the score of the selected pair is no longer positive. Since the compound reactants can be further shared among different targets, our algorithm utilizes a unified compound reactant generation process for all targets. Finally, the dilution process is constructed in the last phase.

In the previous section, two kinds of pairs are introduced: bar pair and slash pair. Practically, if a slash pair is selected, a spitting operation is needed. However, the splitting operation increases the reactant consumption. Therefore, a way to estimate the cost of the pair is demanded. For a pair $p=(e_\alpha, e_\beta)$, the cost function is defined as:

$$cost(p) = (Level(e_\beta) - Level(e_\alpha)) \times w_\alpha \quad (1)$$

Pairs conform certain conditions can be grouped as a set to identify sharable mixing operations. There are two kinds of pair set: *double pair* and *triple pair*. Different pair sets offer different gains (i.e., reactant cost saving). For a pair set PS, its net score (i.e., benefit) can be calculated as:

$$Score(PS) = gain(PS) - cost(PS) \quad (2)$$

To do reactant minimization, we need to find a set of pairs which enables common mixing operation sharing. A set of two pairs conforms such above requirement is called a double pair. There are two constraints for making a valid double pair. First, those two pairs should be in the same rows. That is, the two reactants of those two pairs must be exactly the same. The second constraint is that the elements of one pair cannot be elements of the other pair. The only exception is that two slash pairs can share the same first element. It is because that the first element can be further split to form two new valid bar pairs for mixing operation sharing. The cost of a double pair is generally the cost sum of its two pairs given by (1). The only exception is that: for two slash pairs sharing the first element, the cost is set at the larger cost of those two pairs since only one instead of two splitting is required.

In addition to the double pair, the triple pair, which is a set of three pairs, is also considered for reactant minimization. For a triple pair, any two pairs inside can form a valid double pair. The gain of the triple pair is set bigger than that of the double pair if the newly generated compound reactants can form a double pair or a triple pair in the next round. Hence, a bigger gain is expected for such a triple pair. A triple pair can also be composed of slash pairs or bar pairs. Similar to the double pair, when the same first element is shared by two or three pairs in a triple pair, the overall cost is decreased due to the sharing and is smaller than the cost sum of three individual pairs. For every possible double pair and triple pair, the score is calculated based on (2) and the pair with the highest score is marked as the best pair for the next mixing operation. The associated mixing operation would be carried out if the score is positive.

After a mixing operation is actually performed, the recipe cube should be updated accordingly for the next iteration. In the proposed algorithm, a new row is created (if not present) in the recipe cube for the output compound reactant after mixing. The

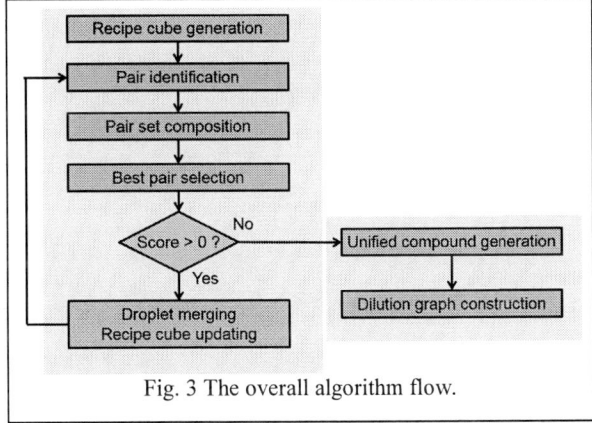

Fig. 3 The overall algorithm flow.

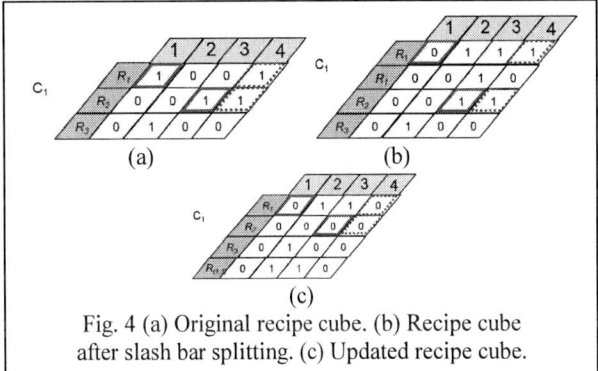

Fig. 4 (a) Original recipe cube. (b) Recipe cube after slash bar splitting. (c) Updated recipe cube.

978-1-5386-7100-9/18 $31.00 © 2018 IEEE

new compound reactant is denoted as $R_{(m, n)}$, where R_m and R_n are two reactants associated with the selected (double or triple) pair set. When the pairs in a pair set are all bar pairs, the value of the new compound reactant is the sum of the values of the two reactants in the pair set. As well, the recipe cube is updated as follows: all 1's in the selected pair set should be cleared to 0's and new 1's are added in the row associated with the new compound reactant. For example, as Fig. 2(c) shows, there is a new row of $R_{(1,2)}$ added in the updated recipe cube. Moreover, the 1's marked by red and green in Fig. 2(a) are now 0's in Fig. 2(c), and two new 1's are also added in the row of $R_{(1,2)}$.

While there are one or more slash pairs in the selected pair set, splitting is performed before mixing. As shown in Fig. 4(a), a double pair including a slash pair (marked in red) is selected. The element (1, 1, 1) is first split into three 1's, as illustrated in Fig.4 (b). Note that the total concentration value of the reactant (R_1 in this case) keeps unchanged after splitting. Obviously, a double pair can now be identified in Fig.4 (b), and the final updated recipe cube after mixing is shown in Fig.4 (c).

If no pair sets with positive scores can be further found, our algorithm moves to the unified compound generation phase. Fig. 5(b) gives the dilution tree for each of the three targets shown in Fig. 5(a) independently. The nodes circled in red are compound reactants. Hence, the complete dilution trees starting

from raw reactants are further shown in Fig. 5(c). However, as aforementioned, mixing operations can be shared among all targets. It suggests that common compound reactants required by different targets can be produced in a unified process. For example, the compound reactant $R_{(2,3)}$ required by both Target 1 and Target 3 can be produced by only one mixing operation. In this step, compound reactants from all targets are checked if they can be shared. It is apparent that the consumption of raw reactants can be significantly reduced, as Fig. 5(d) illustrates.

After the unified compound generation phase, our algorithm constructs an integrated dilution graph that produces all targets. The construction process is guided by the recipe cube. Similar to the BS method [12], for each target, two reactants of the same level are bound for mixing to produce an intermediate solution required by the next level. After that, all leaf nodes of compound reactants are connected to those mixing operations identified in the unified compound generation process. Fig. 5(d) shows the final integrated dilution graph for three targets.

V. EXPERIMENTAL RESULTS

The proposed algorithm has been implemented in C++ and we compare it against BS [12] and CoDOS [20]. Both BS and CoDOS can deal with the single-target many-reactant problem. However, neither BS nor CoDOS is designed for the multi-target problem, so the process of unified compound generation is also added for them for fair comparisons. Ext-BS, which is extended from BS, finds the nodes with the same concentration for mixing operation sharing among different targets as a post-processing step. Meanwhile, Ext-CoDOS is extended from CoDOS in the same way. Then, a set of experiments have been conducted for performance evaluation, where the number of targets is set from 1 to 50 and the number of reactants is set to 7.

In the first set of experiments, all reactants are equally weighted. Fig. 6 demonstrates the final results. All results are normalized to the result of the original BS method for single target. Fig. 6 indicates that our algorithm always produces the best results as the number of targets (#target) increases from 1 to 50. The reactant cost can be reduced by 52.4% as #target is 50. Besides, all results are improved as #target increases. When #target is small, Ext-CoDOS clearly outperforms Ext-BS. The reason is that CoDOS is optimized for the many-reactant sample preparation whereas BS is not. However, as #target increases, Ext-BS gets better significantly and even outperforms

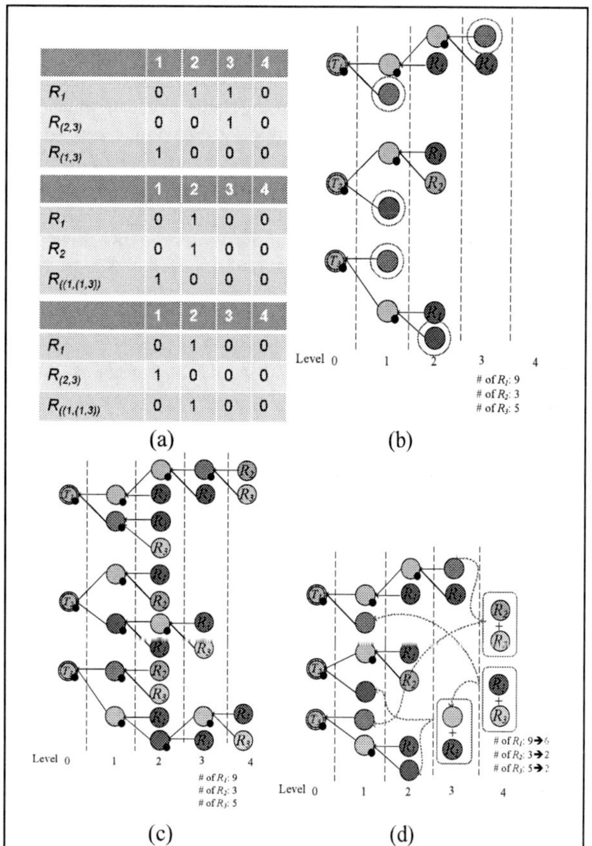

Fig. 5 (a) Recipe cube. (b) Dilution trees. (c) Dilution trees starting from raw reactants. (d) After unification.

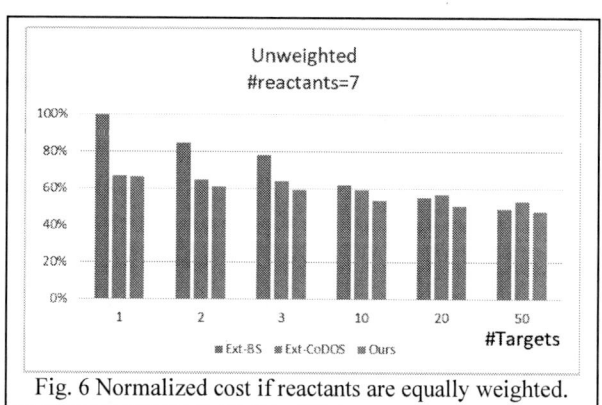

Fig. 6 Normalized cost if reactants are equally weighted.

978-1-5386-7100-9/18 $31.00 © 2018 IEEE

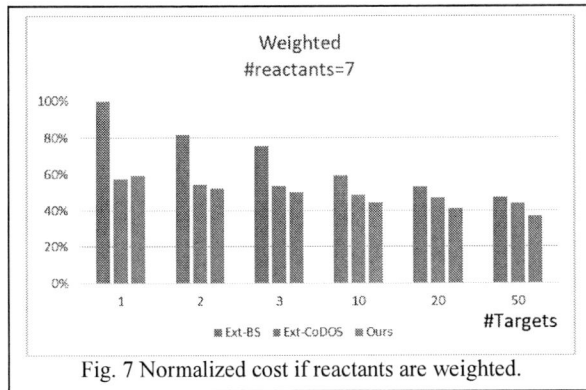

Fig. 7 Normalized cost if reactants are weighted.

Ext-CoDOS when #target is larger than 20. The primary reason is that Ext-BS makes the dilution tree structure more regular, so that it is easier to find sharable mixing operations among targets.

In the second set of experiments, reactants are assigned with different weights. Fig. 7 illustrates the final result, where the weights of reactants are set to {1, 2, 4, 8, 16, 32, 64}. As Fig. 7 shows, the performance of our algorithm is even better; the reactant cost is reduced by 62.9% as #target is 50. The reason is that different reactant weights are fully aware and considered during best pair selection. Hence, the results reported in Fig. 6 and Fig. 7 suggest that our algorithm is a promising solution for reactant cost minimization in multi-target many-reactant sample preparation.

VI. CONCLUSION

In this paper, we propose the first multi-target many-reactant sample preparation algorithm for reactant minimization on microfluidic biochips. Our algorithm first generates a recipe cube from the given target concentration values, and finds all pairs (bar pairs and slash pairs) in the recipe cube. Then, all valid double pairs and triple pairs are identified as candidates for common mixing operation sharing later. At each iteration, the pair set with the highest score is selected for implementation, and the recipe cube is updated accordingly. Finally, compound reactants required by different targets are generated in a unified process, and a final integrated dilution graph for all targets is constructed. The experimental results apparently show that our algorithm outperforms all prior arts. In the case of 50 targets and 7 unweighted reactants, our algorithm reduces 52.4% of reactant cost as compared to BS. If the reactants are differently weighted, the reduction can be further up to 62.9%. In addition, our algorithm cuts the cost by 21.6% and 15.7% as compared to Ext-BS and Ext-CoDOS (50 targets and 7 weighted reactants). Therefore, it is convincing that our reactant minimization algorithm M²SPA is currently the best method for multi-target and many-reactant sample preparation on microfluidic biochips.

REFERENCES

[1] R. Daw and J. Finkelstein, "Insight: Lab on a chip," Nature, vol. 442, no. 7101, pp. 367–418, Jul. 2006.

[2] M. G. Pollack, R. B. Fair, and A. D. Shenderov, "Electrowetting-based actuation of liquid droplets for microfludic applications," Applied Physics Letters, vol. 77, no. 11, pp. 1725–1726, Sep. 2000.

[3] P. Paik, V. K. Pamula, and R. B. Fair, "Rapid droplet mixers for digital microfludic systems," Lab on a Chip, vol. 3, no. 4, pp. 253–259, Sep. 2003.

[4] V. Srinivasan, V. Pamula, and R. Fair, "An integrated digital microfluidic lab-on-a-chip for clinical diagnostics on human physiological fluids," Lab on a Chip, vol. 4 no. 4, pp. 310–315, May 2004.

[5] T.-Y. Ho, K. Chakrabarty, and P. Pop, "Digital microfludic biochips: recent research and emerging challenges," IEEE/ACM/IFIP International Conference on Hardware/Software Codesign and System Synthesis, 2011, pp. 335–343.

[6] T.-Y. Ho, J. Zeng, and K. Chakrabarty, "Digital microfluidic biochips: a vision for functional diversity and more than Moore," IEEE/ACM International Conference on Computer-Aided Design, 2010, pp. 578–585.

[7] G. M. Walker, N. Monteiro-Riviere, J. Rouse, and A. T. O'Neill, "A linear dilution microfluidic device for cytotoxicity assays," Lab on a Chip, vol. 7, no. 2, pp. 226–232, Oct. 2007.

[8] H. Ren, V. Srinivasan, and R. B. Fair, "Design and testing of an interpolating mixing architecture for electrowetting-based droplet-on-chip chemical dilution," IEEE International Conference on Solid-State Sensors, Actuators and Microsystems (TRANSDUCERS), 2003, pp. 619–622.

[9] H. Moon, A. R. Wheeler, R. L. Garrell, J. A. Loo, and C. J. Kim, "An integrated digital microfluidic chip for multiplexed proteomic sample preparation and analysis by MALDI-MS," Lab on a Chip, vol. 6, no. 9, pp. 1213–1219, Jul. 2006.

[10] E. J. Griffith, S. Akella, and M. K. Goldberg, "Performance characterization of a reconfigurable planar-array digital microfluidic system," IEEE Transactions on Computer-Aided Design of Integrated Circuits and Systems, vol. 25, no. 2, pp. 345–357, Feb. 2006.

[11] T. Xu, V. K. Pamula, and K. Chakrabarty, "Automated, accurate, and inexpensive solution-preparation on a digital microfluidic biochip," IEEE Biomedical Circuits and Systems Conference, 2008, pp. 301–304.

[12] W. Thies, J. P. Urbanski, T. Thorsen, and S. Amarasinghe, "Abstraction layers for scalable microfluidic biocomputing," Natural Computing, vol. 7, no. 2, pp. 255–275, May 2008.

[13] S. Roy, B. B. Bhattacharya, and K. Chakrabarty, "Optimization of dilution and mixing of biochemical samples using digital microfluidic biochips," IEEE Transactions on Computer-Aided Design of Integrated Circuits and Systems, vol. 29, no. 11, pp. 1696–1708, Nov. 2010.

[14] S. Roy, B. B. Bhattacharya, and K. Chakrabarty, "Waste-aware dilution and mixing of biochemical samples with digital microfluidic biochips," IEEE/ACM Design, Automation & Test in Europe Conference & Exhibition, 2011, pp. 1059–1064.

[15] Y.-L. Hsieh, T.-Y. Ho, and K. Chakrabarty, "On-chip biochemical sample preparation using digital microfluidics," IEEE Biomedical Circuits and Systems Conference, 2011, pp. 297–300.

[16] S. Roy, B. B. Bhattacharya, P. P. Chakrabarti, and K. Chakrabarty, "Layout-aware solution preparation for biochemical analysis on a digital microfluidic biochip," IEEE International Conference on VLSI Design, 2011, pp. 171–176.

[17] C.-H. Liu, T.-W. Chiang, and J.-D. Huang, "Reactant minimization in sample preparation on digital microfluidic biochips," IEEE Transactions on Computer-Aided Design of Integrated Circuits and Systems, vol. 34, no. 9, pp. 1429–1440, Sep. 2015.

[18] Y.-L. Hsieh, T.-Y. Ho, and K. Chakrabarty, "A reagent-saving mixing algorithm for preparing multiple-target biochemical samples using digital microfluidics," IEEE Transactions on Computer-Aided Design of Integrated Circuits and Systems, vol. 31, no. 11, pp. 1656–1669, Nov. 2012.

[19] J.-D. Huang, C.-H. Liu, and H.-S. Lin, "Reactant and Waste Minimization in Multitarget Sample Preparation on Digital Microfluidic Biochips," IEEE Transactions on Computer-Aided Design of Integrated Circuits and Systems, vol. 32, no. 10, pp.1484–1494, Oct. 2013.

[20] J.-D. Huang, C.-H. Liu, H.-H. Chang, and T.-C. Liang, "Sample preparation for many-reactant bioassay on DMFBs using common dilution operation sharing," IEEE/ACM International Conference on Computer-Aided Design, pp. 615–621, Nov. 2013.

2018 IEEE Computer Society Annual Symposium on VLSI

More Effective Randomly-Designed Microfluidics[*]

Weiqing Ji[1], Tsung-Yi Ho[2], and Hailong Yao[1]

1. Department of Computer Science and Technology, Tsinghua University
2. Department of Computer Science, National Tsing Hua University

ABSTRACT

Random design of microfluidics is gaining significant attention by creating functional microfluidic chips. Notable merit of random design is that the error-prone design stage is avoided by a library of random chips, which are simulated beforehand using finite element analysis. This paper proposes a methodology for more effective random chip designs, which further optimizes the random chip library to significantly reduce sample consumption. The random design optimization method can be separately loaded as a stand-alone tool and applied to the original chips from the library. Computational simulation results show that the proposed method greatly reduces sample consumption by more than 20% on average in terms of redundant channels. Moreover, the induced deviations in concentrations are mostly less than 0.002, which are negligible in real biomedical applications.

Keywords

Microfluidic biochips, Random design, Concentration generation, Finite element analysis

1. INTRODUCTION

Microfluidic biochips have emerged as a revolutionary technology for miniaturized manipulation and analysis of nano-litre-sized biochemical sample/reagent fluid, and even of single cells [1, 2]. Although there has been a significant advance in the automated design methods for microfluidic biochips, there are rarely mature automated design tools for microfluidic biochips. Moreover, simulation is rarely conducted to guarantee the functionality of the automatically designed chips. As a result, most state-of-the-art microfluidic biochips are still manually designed using AutoCAD or even Photoshop softwares, which causes weeks or even months for designing one chip. To make things worse, the manually designed chip cannot guarantee correct functionality after chip fabrication, which needs multiple iterative re-design loops to produce a functional chip. This causes significant waste in manpower and material resources.

To tackle the above-mentioned microfluidic design problems, Wang et al. proposed a novel random design methodology, which constructs a library of thousands of different random microfluidic chip designs, and then simulates the behavior of each design on a computer using automated finite element analysis (FEA) [3]. The simulation on the randomly-designed chips guarantees that each chip works with correct functionality after fabrication. Based on this random chip library, one can easily obtain customized chip

designs for her/his own unique needs, without worrying about the functional failure after fabrication.

Although the random design methodology is efficient in resolving existing design challenges, we observe that there are critical drawbacks in the randomly-designed chips in [3]. Although the random chips work as expected, there are many unnecessary flow channels within these chips according to the simulation based on computational fluid dynamics (CFD). These unnecessary channels cause significant waste of the expensive sample fluid. Therefore, we propose a more effective random design methodology, which efficiently removes redundant channels in the design layouts. Experiments show that with the proposed method, even after the removal of the redundant channels, the induced absolute error in target concentration is reduced to at most 0.002 and the maximum concentration error is less than 0.011. Moreover, using the proposed method, about 20% of the total channel area is detected as redundant, and thus can be removed to significantly reduce the sample waste. Major contributions of this paper are as follows.

1. We first identify a critical issue of the redundant flow channels in the random design of microfluidics, which causes significant waste in the expensive sample fluid.

2. We accordingly propose a more effective random design methodology, which greatly reduces flow-channel lengths in the randomly-designed chips.

3. We conduct the FEA-based simulation on the optimized random designs, and validate that the variation in target concentration is negligible in the optimized designs.

4. An effective and efficient graph-based approach for random design optimization is proposed, which can either be integrated into the existing random design flow, or be used as a third-party stand-alone tool by end users.

The remainder of this paper is organized as follows. Section 2 states the related works in automated design methods and sample preparation algorithms for microfluidic biochips. Section 3 presents the preliminaries of random design of microfluidics and problem formulation. Section 4 presents the details of the proposed random design optimization method. Section 5 presents and discusses the computational simulation results. Finally, a conclusion is drawn in Section 6.

2. RELATED WORKS

There has been a large body of research works for automated design of application-specific microfluidic biochips. Minhass et al. proposed a synthesis method to minimize the number of control pins [4]. Then architectural synthesis and resource binding methods have been proposed in [5–7]. Tseng et al. proposed a top-down synthesis method for flow-based microfluidic biochips considering valve-switching minimization [8]. Amin et al. proposed the first routing algorithm for flow-based microfluidic biochips based on the min-cost max-flow formulation [9]. Hu et al. proposed a routing method to optimize both the number

---

[*]Corresponding author: hailongyao@tsinghua.edu.cn
The work of H. Yao was supported by the National Natural Science Foundation of China (61106104). The work of T. Ho was supported in part by the Ministry of Science and Technology of Taiwan, under Grant MOST 105-2221-E-007-118-MY3 and 104-2220-E-007-021 and in part by the Technical University of Munich-Institute for Advanced Study, funded by the German Excellence Initiative and the European Union Seventh Framework Program under grant agreement no 291763.

978-1-5386-7100-9/18 $31.00 © 2018 IEEE

of control pins and the pressure-propagation delay [10]. Lin et al. proposed a routing algorithm minimizing the weighted sum of the maximum and total channel lengths [11]. McDaniel et al. proposed a simulated annealing-based placement algorithm for flow-based biochips [12]. Yao et al. proposed the first co-design concept, which simultaneously considers both flow layer design and control-layer design [13]. Grimmer et al. proposed a SAT-based placement and routing method to minimize the number of flow-channel crossings [14]. Tseng et al. proposed an integer linear programming model for dynamic device mapping and fluid routing for flow-based microfluidic biochips [15]. Existing works rarely perform simulations on the designed layouts, which may cause functional failure after fabrication of the real chips.

Apart from the automated design methods for application-specific microfluidic biochips, another branch of research is focused on the pure algorithms for sample preparation. Thies et al. proposed a pioneering sample preparation algorithm to minimize the number of mixing steps (Min-Mix) [16]. Then, Roy et al. proposed a dilution and mixing algorithm with reduced wastage (DMRW) [17]. Hsieh et al. proposed multiple-target sample preparation algorithm considering reagent saving (RSM) [18]. Liu et al. proposed a sample preparation algorithm based on the common dilution operation sharing (CoDOS) [19]. Dinh et al. proposed an optimal network-flow-based sample preparation algorithm [20]. Roy et al. proposed a waste-aware single-target dilution and mixing algorithm (IDMA) [21]. Liu et al. proposed a reactant minimization algorithm (REMIA) for sample preparation on digital microfluidic biochips [22]. Roy et al. proposed a layout-aware sample preparation algorithm for reactant minimization (RMA) [23]. Kumar et al. proposed single-target multi-demand mixture preparation algorithm [24]. Shao et al. proposed a look-up-table based sample preparation algorithm for fast online sample preparation [25]. Existing sample preparation works are based on a given ideal mixing model, which can perform sample mixing with x:y mixing ratio (typically 1:1). However, it remains a question how to schedule the mixing steps on a given microfluidic biochip. Moreover, the number of available mixers is critical in determining the scheduling result. In contrast, using random design of microfluidics, there is no need to consider the complex mixing algorithms as well as the scheduling of the mixing steps on a given microfluidic biochip. Moreover, better accuracy is guaranteed by the prior simulation stage using random design of microfluidics, which are significant advantages over existing methods.

3. PRELIMINARY AND PROBLEM FORMULATION

3.1 Background

Figure 1 shows the overview of our more effective randomly-designed microfluidics, which is an enhanced design flow over the original random design method [3]. First, different layouts are randomly designed in the given $n \times n$ grid. Second, the randomly-designed layouts are optimized for redundant channel removal using our graph-based layout optimization method (see Section 4). Third, FEA-based simulation is conducted using COMSOL Multiphysics to obtain the resulting concentrations at each output port of the chip [26]. Fourth, a large library of random chip designs is constructed based on the simulated layouts with known concentrations. Finally, with the expected concentration values, an end user can query the online library to find the best matching designs. In this paper, we focus on the second step, which effectively trims the redundant branch channels of a randomly-designed layout, and hence significantly reduces consumption of the precious sample fluid.

3.2 Problem Formulation

In this paper, we propose an optimization method for more effective randomly-designed microfluidics, which optimizes the randomly designed layouts by trimming the redundant channels.

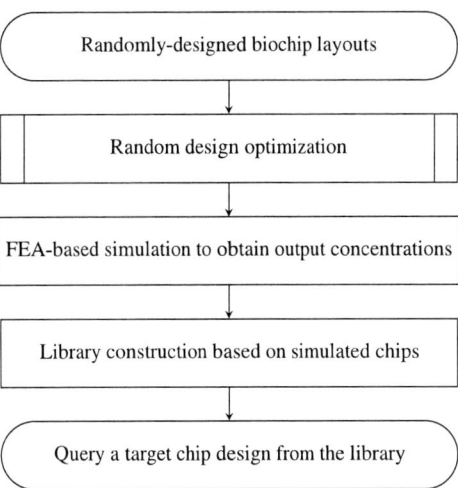

Figure 1: Overall flow of our more effective randomly-designed microfluidics.

Although the original random designs are all generated on 8×8 grid, our proposed layout optimization method can be applied to general randomly-designed layouts of any dimensions. The problem formulation of the proposed random design optimization method is stated as follows:

Input: Randomly-designed microfluidic chip in the AutoCAD DXF (Drawing Exchange Format) format [27].

Output: Optimized microfluidic chip layout in DXF format.

Objective: Minimize the total channel length in the given microfluidic chip without affecting its functionality.

Constraint: The induced absolute errors in concentrations at the output ports should not exceed the given threshold δ.

4. RANDOM DESIGN OPTIMIZATION

4.1 Problem Analysis

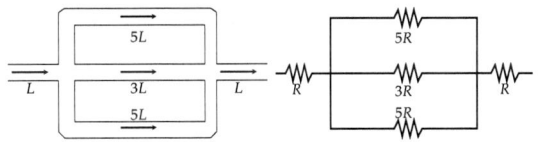

(a) Micro-channel network.   (b) Equivalent electronic circuit.

Figure 2: Analogy between microfluidic biochip and electronic circuit.

It is well accepted that the flow of a fluid within the micro-channel networks is physically similar to the flow of electricity in the electronic circuits [28]. Our random design optimization method is based on the analogy between microfluidic biochips and electronic circuits. Figure 2 shows an illustrating example, in which the micro-channel network is mapped to the equivalent electronic circuit. The flow velocity of the fluid in the micro-channel is analogous to the current in the circuit, and the resistance of the flow channel is analogous to the wire resistance in the circuit.

Figure 3 shows two major types of redundant wires in electronic circuits. Figure 3(a) shows a *broken circuit*, where wire $P_1 P_2$ is broken with no current passing by. Therefore, even if $P_1 P_2$ is removed from the circuit, it does not have any impact on the existing circuitry as well as the current flow. Figure 3(b) shows a *short circuit* with a loop wire between P_3 and P_4. According to circuit theory, there is no current passing by this loop. So the loop wire between P_3 and P_4 can be safely removed without affecting the existing circuitry. Figure 3(c) and Figure 3(d) show the micro-channel networks corresponding to the electronic circuits in Figure 3(a) and Figure 3(b), respectively. According to the

978-1-5386-7100-9/18 $31.00 © 2018 IEEE   661

analogy between microfluidic chips and electronic circuits, we expect that the corresponding broken channel P_1P_2 in Figure 3(c) and the loop channel between P_3 and P_4 can be safely removed without affecting the flow velocity in the micro-channel networks. Computational simulation results in Section 5 verify our prediction.

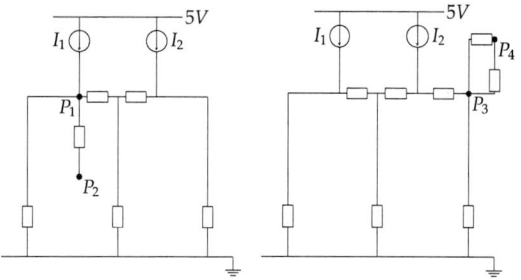

(a) Redundant wire between P_1 and P_2 due to broken circuit. (b) Redundant wires between P_3 and P_4 due to short circuit.

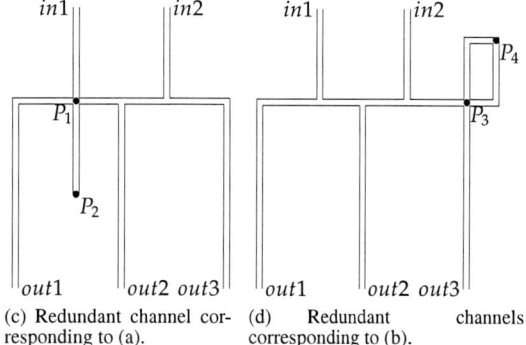

(c) Redundant channel corresponding to (a). (d) Redundant channels corresponding to (b).

Figure 3: Analogy between redundant channels in microfluidics and redundant wires in electronic circuits.

4.2 Graph Modeling

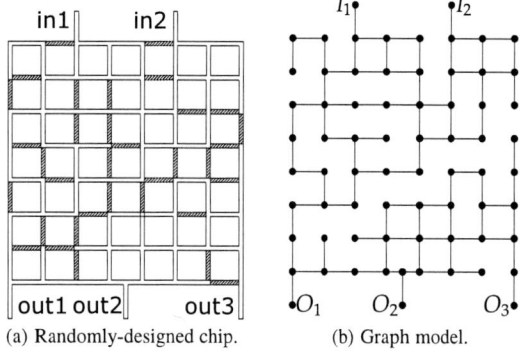

(a) Randomly-designed chip. (b) Graph model.

Figure 4: Graph modeling example.

We propose a graph-based microfluidic chip optimization method, which first transforms a given DXF-format chip design into an undirected graph, and then performs graph reduction to trim the redundant branches. Figure 4 shows an example for transforming a given micro-channel network into an undirected graph. Similar to the original chip designs in [3], there are two input ports and three output ports, and up to 112 intermediate channels in the given 8×8 grid. In the undirected graph, each grid is mapped to a node, and each unit channel between adjacent grids is mapped to an edge between the corresponding nodes. Then we obtain an undirected graph with 70 nodes. In Figure4(b), I_1 and I_2 are called *input nodes*, which are special nodes corresponding to the input ports in the original chip. O_1, O_2 and O_3 are called *output nodes*, which correspond to the output ports in the original chip.

4.3 Graph Reduction Theory

For describing the proposed graph-based random chip optimization method, we first give the related definitions.

DEFINITION 1 (SHORT-CIRCUIT NODE). *Assume there is a loop in an undirected graph G along with two adjacent edges (n_i, n_j) and (n_j, n_k) in the loop, if the removal of the two edges results in an increase in the number of connected components in G, and all the input and output nodes remain in a single connected component, then n_j is called a short-circuit node.*

DEFINITION 2 (BROKEN-CIRCUIT NODE). *If the removal of an edge $(n_i, n_j) \in E$ results in an increase in the number of connected components in G, and the following two conditions are satisfied: (1) all the input and output nodes remain in a single connected component, and (2) the degree of n_i is greater than 1, then n_i is called a broken-circuit node.*

DEFINITION 3 (REDUNDANT NODE). *Node $n_i \in G$ is called a redundant node if and only if n_i is broken-circuit node or a short-circuit node.*

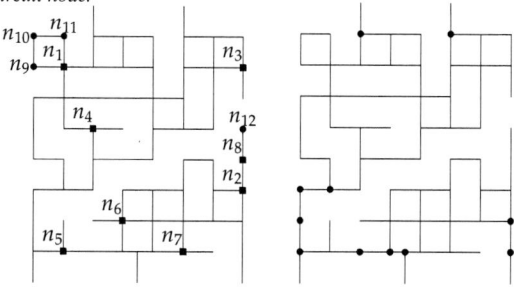

(a) Redundant nodes in solid squares and removed nodes in solid circles. (b) Articulation nodes that are not redundant nodes.

Figure 5: Redundant nodes and articulation nodes.

Figure 5(a) shows an example of the definitions related to redundant nodes. At the beginning, the number of connected components in the graph is 1. n_1 is a short-circuit node. If we remove the edges (n_1, n_9) and (n_1, n_{11}), the number of connected components in the graph increases to 2: one contains n_9, n_{10} and n_{11}, and the other contains the remaining nodes. In the figure, $n_2, n_3, ..., n_8$ are broken-circuit nodes. For example, if we remove the edge (n_2, n_8), the number of connected components in the graph increases to 2: one contains n_8 and n_{12}, and the other contains the remaining nodes.

LEMMA 1. *All redundant nodes in G are articulation nodes.*

PROOF. According to the definition of the broken-circuit node, if we remove the broken-circuit node and one edge connected to it, the number of connected components increases. So if we remove all the edges connected to the node, the number of connected components will still increase. Therefore, the broken-circuit node is an articulation node of the graph. According to the definition of the short-circuit node, if we remove the short-circuit node and the two edges connected to it, the number of connected components increases. So if we remove all edges connected to the node, the number of connected components will still increase. Therefore, the short-circuit node is an articulation node. Since a redundant node is either a short-circuit node or a broken-circuit node, all redundant nodes are articulation nodes. □

Because the removal of an articulation node may split the input and output nodes into different connected components, not all the articulation nodes are redundant nodes. Figure 5(b) shows an example where the articulation nodes are not redundant nodes. To make all the articulation nodes to be the redundant nodes, we propose to modify the graph as follows: connect all the input and output nodes in a circle surrounding the chip (see Figure 7(c) for an example). After this modification to the graph, the marked nodes in Figure 5(b) are no longer articulation nodes.

978-1-5386-7100-9/18 $31.00 © 2018 IEEE   662

4.4 Graph-Based Optimization

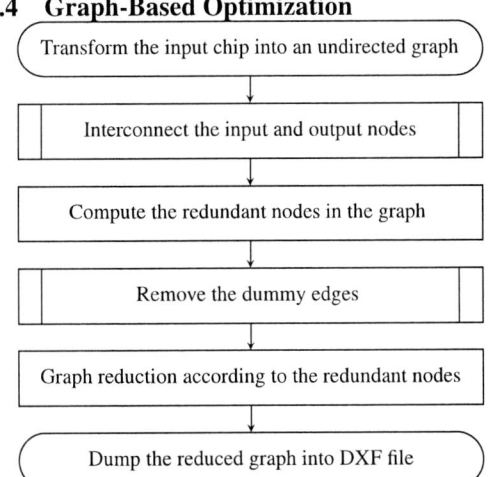

Figure 6: Random design optimization flow.

Figure 6 shows the overall flow of the proposed random design optimization algorithm. First, the input random chip in DXF format is transformed into an undirected graph. Second, a preprocessing step is performed, which inserts dummy edges to interconnect all the input and output nodes in turn. After this critical preprocessing step, all the articulation nodes in the graph are redundant nodes that can be safely removed. Third, we compute all the redundant nodes in the graph using the tarjan algorithm [29]. Fourth, we remove the dummy edges inserted in the second step. Fifth, we perform a graph reduction procedure based on breadth-first traversal to reduce the graph according to the computed redundant nodes. Finally, we dump the reduced graph into an AutoCAD DXF file for production.

4.4.1 Graph Construction on Input Chip

As the proposed random design optimization method can be used as a third-party stand-alone tool for optimizing random chips, the input is a DXF file, which is a text-based file for direct chip production. When the input DXF file is loaded, a text parser is implemented to identify the sections storing the information about the micro-channels and the intersection points. When the micro-channel network is successfully parsed, an undirected graph will be constructed according to Section 4.2. Specifically, the data structure of adjacency matrix is adopted to store the undirected graph, which facilitates the following graph reduction steps.

4.4.2 Calculation of Redundant Nodes

Input: Undirected graph G.
Output: The list of redundant nodes L_r.

1 Add the following edges into graph G: (I_1, I_2), (I_1, O_1), (O_1, O_2), (O_2, O_3), (I_2, O_3);
2 $L_r = Tarjan(A)$;
3 Remove inserted edges from graph G: (I_1, I_2), (I_1, O_1), (O_1, O_2), (O_2, O_3), (I_2, O_3);
4 Return L_r;

Algorithm 1: Calculation of Redundant Nodes.

Algorithm 1 shows a simplified version of the algorithm for calculating the redundant nodes. First of all, a preprocessing step is performed to update the input graph G, which inserts dummy edges connecting the input and output nodes. As in the original random chips, there are always two input ports and three output ports, we specifically connect their corresponding nodes together. In general cases with different number of input and output ports, our algorithm can also be easily extended to insert the required edges. By connecting the input and output nodes, we can guarantee that all the articulation nodes be redundant nodes.

Then, the classical tarjan algorithm is applied to search for all

the articulation nodes. We have made an efficient implementation of the tarjan algorithm by ourselves. As the tarjan algorithm is state-of-the-art, we do not list the algorithm details here. When the articulation nodes are calculated and stored, we remove the inserted edges between input and output nodes so as not to affect the subsequent operations. Overall runtime complexity of Algorithm 1 is determined by the tarjan algorithm, which is $O(n + m)$, where n and m are the numbers of nodes and edges in the graph.

4.4.3 Graph Reduction Algorithm

Input: List of redundant nodes L_r and graph $G(N, E)$.
Output: Reduced graph G'.

1 Initialize the queue $Q = \varnothing$;
2 Initialize the integer array $S = \{0\}$, where S_i records the number of accesses to node n_i;
3 Set *counter* $= 0$;
4 **for** $n_i \in L_r$ **do**
5   Initialize boolean vector $V = \{False\}$ to record whether nodes are visited;
6   Clear all the elements in Q;
7   Push I_1, I_2, O_1, O_2, O_3 into Q;
8   **while** $!Q.isEmpty()$ **do**
9       $n_j = Q.pop()$;
10   **for** $n_k \in L_r$ **do**
11   **if** $V(n_k) == False$ *and* $S_k == counter$ **then**
12             $V(n_k) = True; S_k$ ++ ;
13   **if** $n_k \neq n_i$ **then**
14                $Q.push(n_k)$;
15   *counter* ++;
16 Initialize list L to store the remaining nodes;
17 **for** $n_i \in N$ **do**
18   **if** $S_i == L_r.size()$ **then**
19       $L.push(n_i)$;
20 Construct undirected graph G' including all the nodes in L and their corresponding edges;
21 *Return*(G');

Algorithm 2: Graph reduction algorithm.

Algorithm 2 presents the graph reduction algorithm, which reduces the input undirected graph G by the computed redundant nodes. For each redundant node, we start the breadth-first traversal from the input and output nodes, which ends at the redundant node. For each visited node in the traversal process that has not been removed, we will put it in the queue unless it is a redundant node. Here, the condition "$S_k == counter$" is used to check whether the visited node has been removed in previous traversal processes. We can find that the number of accesses per node is different after the traversal of all the redundant nodes. Obviously, only those nodes, whose access times are equal to the number of redundant nodes (Steps 18-19), are remained in the reduced graph. When the remaining nodes are obtained, the remaining edges are computed as follows (Step 20): a remaining edge should have both its end nodes belonging to the remaining nodes. With the remaining nodes and edges computed, the final reduced graph G' is obtained and returned. As the total number of basic operations is $|L_r| \times (n + m)$, the time complexity of the algorithm is $O(n \times (n + m))$, where n and m are the numbers of nodes and edges in G, respectively.

When the reduced graph G' is obtained, we start to dump the graph into the DXF file, which can be easily loaded and modified by commercial softwares such as AutoCAD.

4.5 Illustrating Example

Figure 7 shows an example illustrating the whole random design optimization process. Figure 7(a) gives the original randomly-designed chip. After transforming it into a graph model, we obtain the undirected graph as shown in Figure 7(b). Next, in Figure 7(c), we insert dummy edges to connect the input and output nodes to the graph, which guarantees that all the articulation nodes be redundant

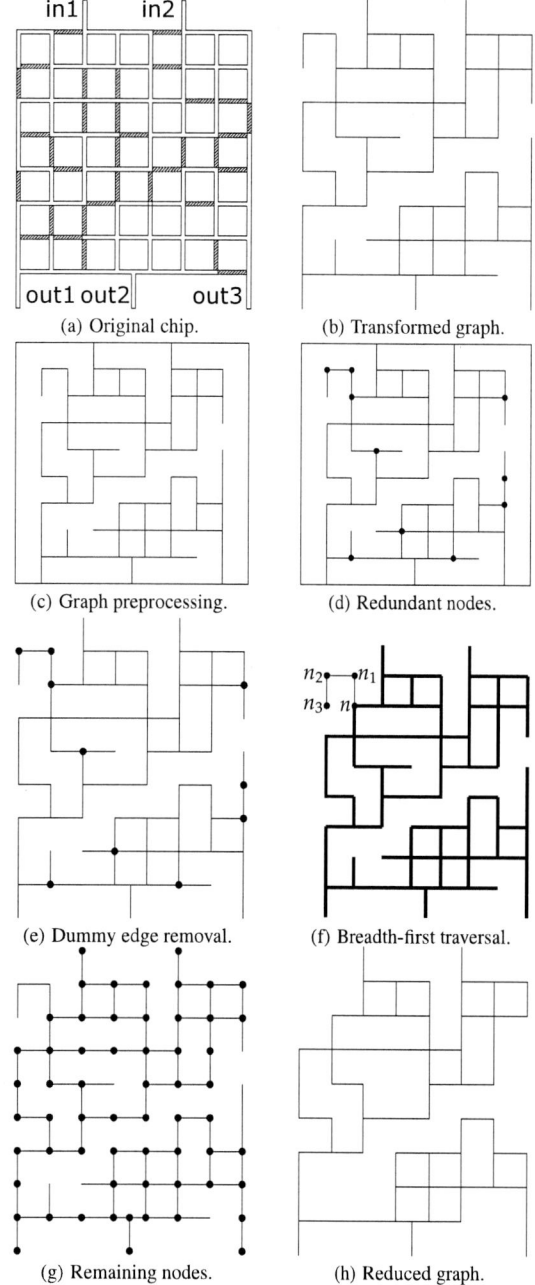

(a) Original chip.

(b) Transformed graph.

(c) Graph preprocessing.

(d) Redundant nodes.

(e) Dummy edge removal.

(f) Breadth-first traversal.

(g) Remaining nodes.

(h) Reduced graph.

Figure 7: Illustrating example of the proposed method.

nodes. Next, we perform the tarjan algorithm to obtain all the redundant nodes as shown in Figure 7(d). Next, we delete the inserted dummy edges between input and output nodes to restore the original graph, as is shown in Figure 7(e). Figure 7(f) shows the visited nodes and edges in bold lines when we perform the breadth-first traversal from the input and output nodes to the redundant node n (see Steps 5-14 in Algorithm 2 for detail). From the figure, three nodes (i.e., n_1, n_2 and n_3) are obtained during the traversal, which need to be removed from the graph. After we iteratively conduct the breadth-first traversal for each redundant node, we can obtain all the nodes that need to be removed, and hence obtain the final list of remaining nodes as shown in Figure 7(g). With the remaining nodes, we can compute the remaining edges as follows: only when the two incident nodes of an edge both belong to the remaining

nodes, does this edge remain in the reduced graph. Figure 7(h) shows the final reduced graph, which is then converted into an AutoCAD DXF file for production.

5. COMPUTATIONAL SIMULATION RESULTS

We have implemented our proposed flow of more effective random design of microfluidics in C++, and tested it on a 2.60GHz 32-core Intel Xeon Linux workstation with 132GB memory. Only a single thread is used for testing. We obtained 3960 benchmarks from [3], and tested all these benchmarks individually. First, we perform COMSOL-based simulation on the original benchmarks to obtain the original concentration values at the output ports. Then we conduct layout optimization using our proposed method, and perform the same simulation on the optimized layout.

Figures 8(a), 8(b), and 8(c) show the absolute errors in concentration between original and optimized layouts at the three output ports, respectively. From the figure, we can find that most of the concentration errors are within 0.002, and the maximum concentration error is less than 0.011. The small deviation in concentration is acceptable in most of the real applications.

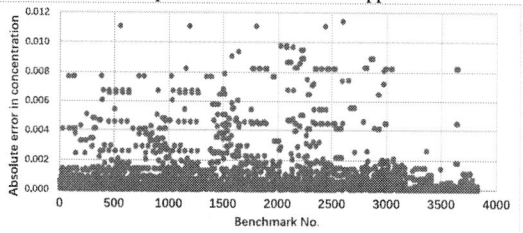

(a) Absolute error at output port out1.

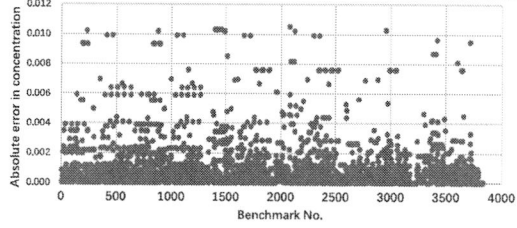

(b) Absolute error at output port out2.

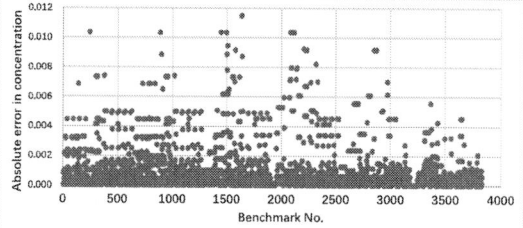

(c) Absolute error at output port out3.

Figure 8: Absolute errors in concentration at the output ports.

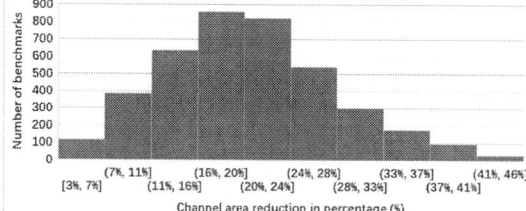

Figure 9: Histogram of channel area reduction

Figure 9 shows the result of the channel area reduction using our proposed method. From the figure, among the 3960 benchmarks, we obtain 16%-24% reduction in channel area for more than 40%

benchmarks. For all the benchmarks, the channel area reduction is 20.44% in average, with the maximum reduction 45%, and the minimum reduction 2%.

Figure 10 shows a real benchmark illustrating the effectiveness of our method, where Figure 10(a) gives the original layout with simulated concentration profile, and Figure 10(b) gives the optimized layout. In this example, the absolution errors in concentration at the three output ports are no more than 0.001. And the reduction in channel area is 28.7%, which proves the effectiveness of our proposed random design optimization flow.

6. CONCLUSION

We have proposed a more effective randomly-designed microfluidics based on graph optimization theory, which significantly reduces the channel area along with the sample consumption. The induced error in concentration by channel reduction is negligible. Future work includes actual chip fabrication for further validation, as well as put online our more effective random design flow for accessibility of the microfluidic designers.

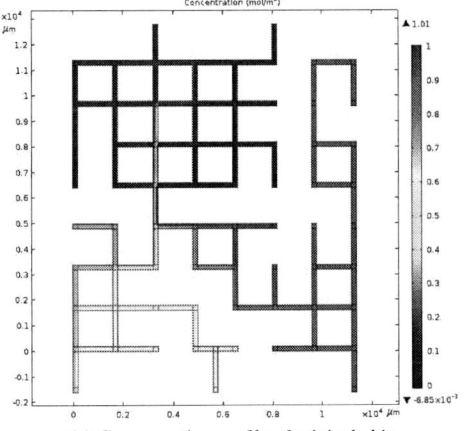

(a) Concentration profile of original chip.

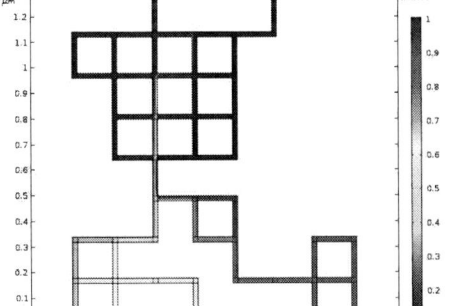

(b) Concentration profile of optimized chip.

Figure 10: Comparison between the original and optimized chips.

7. REFERENCES

[1] R. J. Kimmerling, G. L. Szeto, J. W. Li, A. S. Genshaft, S. W. Kazer, K. R. Payer, J. de R. Borrajo, P. C. Blainey, D. J. Irvine, A. K. Shalek, and S. R. Manalis, "A microfluidic platform enabling single-cell RNA-seq of multigenerational lineages", *Nature Communications*, vol. 7, Article number 10220, 2016.

[2] M. E. Warkiani, B. L. Khoo, L. Wu, A. K. P. Tay, A. A. S Bhagat, J. Han, and C. T. Lim, "Ultra-fast, label-free isolation of circulating tumor cells from blood using spiral microfluidics", *Nature Protocols*, no. 11, pp. 134-148, 2016.

[3] J. Wang, P. Brisk, and W. H. Grover, "Random design of microfluidics", *Lab Chip*, 2016, 16, pp. 4212-4219.

[4] W. H. Minhass, P. Pop, J. Madsen, and T.-Y. Ho, "Control Synthesis for the Flow-based Microfluidic Large-Scale Integration Biochips," *Proc. of ASP-DAC*, 2013, pp. 205-212.

[5] W. H. Minhass, P. Pop, and J. Madsen, "System-Level Modeling and Synthesis of Flow-Based Microfluidic Biochips," *Proc. of CASES*, 2011, pp. 225-233.

[6] W. H. Minhass, P. Pop, J. Madsen, and F. S. Blaga, "Architectural Synthesis of Flow-Based Microfluidic Large-Scale Integration Biochips," *Proc. of CASES*, 2012, pp. 181-190.

[7] W. H. Minhass, P. Pop, and J. Madsen, "Synthesis of Biochemical Applications on Flow-Based Microfluidic Biochips Using Constraint Programming," *Proc. of 2012 DTIP*, 2012 , pp. 37-41.

[8] K.-H. Tseng, S.-C. You, J.-Y. Liou, and T.-Y. Ho, "A Top-Down Synthesis Methodology for Flow-Based Microfluidic Biochips Considering Valve-Switching Minimization," *Proc. of ISPD*, 2013, pp. 123-129.

[9] N. Amin, W. Thies, and S. Amarasinghe, "Computer-Aided Design for Microfluidic Chips Based on Multilayer Soft Lithography," *Proc. of ICCD*, 2009, pp. 2-9.

[10] K. Hu, T. A. Dinh, T.-Y. Ho, and K. Chakrabarty, "Control-Layer Optimization for Flow-Based mVLSI Microfluidic Biochips," *Proc. of CASES*, 2014, pp. 16:1-16:10.

[11] C.-X. Lin, C.-H. Liu, I.-C. Chen, D. T. Lee, and T.-Y. Ho, "An Efficient Bi-criteria Flow Channel Routing Algorithm for Flow-Based Microfluidic Biochips," *Proc. of DAC*, 2014, pp. 1-6.

[12] J. McDaniel, B. Parker, and P. Brisk, "Simulated Annealing-Based Placement for Microfluidic Large Scale Integration (mLSI) Chips," *Proc. of VLSI-SoC*, 2014, pp. 1-6.

[13] H. Yao, Q. Wang, Y. Ru, and T.-Y. Ho, "Integrated Flow-Control Co-Design Methodology for Flow-Based Microfluidic Biochips" *IEEE Design & Test*, vol. 32, no. 6, pp. 60-68, 2015

[14] A. Grimmer, Q. Wang, H. Yao, T.-Y. Ho, R. Wille, "Close-to-optimal placement and routing for continuous-flow microfluidic biochips" *Proc. of ASP-DAC*, 2017, pp. 530-535.

[15] T. Tseng, B. Li, M. Li, T.-Y. Ho, U. Schlichtmann, "Reliability-Aware Synthesis With Dynamic Device Mapping and Fluid Routing for Flow-Based Microfluidic Biochips," *IEEE Transactions on CAD*, vol. 35, no. 12, pp. 1981-1994, 2016.

[16] W. Thies, J. P. Urbanski, T. Thorsen, and S. Amarasinghe, "Abstraction Layers for Scalable Microfluidic Biocomputing," *Natural Computing*, vol. 7, no. 2, pp. 255-275, 2008.

[17] S. Roy, B. B. Bhattacharya, and K. Chakrabarty, "Optimization of Dilution and Mixing of Biochemical Samples using Digital Microfluidic Biochips," *IEEE Transactions on CAD*, vol. 29, no. 11, pp. 1696-1708, 2010.

[18] Y.-L. Hsieh, T.-Y. Ho, and K. Chakrabarty, "A Reagent-Saving Mixing Algorithm for Preparing Multiple-Target Biochemical Samples Using Digital Microfluidics," *IEEE Transactions on CAD*, vol. 31, no. 11, pp. 1656-1669, 2012.

[19] C.-H. Liu, H.-H. Chang, T.-C. Liang, and J.-D. Huang, "Sample Preparation for Many-Reactant Bioassay on DMFBs using Common Dilution Operation Sharing," *Proc. of ICCAD*, 2013, pp. 615-621.

[20] T. A. Dinh, S. Yamashita, and T.-Y. Ho, "A Network-Flow-Based Optimal Sample Preparation Algorithm for Digital Microfluidic Biochips," *Proc. of ASP-DAC*, 2014, pp. 225-230.

[21] S. Roy, P. P. Chakrabarti, K. Chakrabarty, and B. B. Bhattacharya, "Waste-Aware Single-Target Dilution of a Biochemical Fluid using Digital Microfluidic Biochips," *Integration, the VLSI Journal* vol. 51, 2015, pp. 194-207.

[22] C.-H. Liu, T.-W. Chiang, and J.-D. Huang, "Reactant Minimization in Sample Preparation on Digital Microfluidic Biochips," *IEEE Transactions on CAD*, vol. 34, no. 9, pp. 1429-1440, 2015.

[23] S. Roy, P. P. Chakrabarti, S. Kumar, K. Chakrabarty, and B. B. Bhattacharya, "Layout-Aware Mixture Preparation of Biochemical Fluids on Application-Specific Digital Microfluidic Biochips," *ACM TODAES*, vol. 20, no. 3, 2015, Article no. 45.

[24] S. Kumar, A. Gupta, S. Roy, and B. B. Bhattacharya, "Design Automation of Multiple-demand Mixture Preparation using a K-array Rotary Mixer on Digital Microfluidic Biochips," *Proc. of ICCD* 2016, pp. 273-280.

[25] L. Shao, Y. Yang, H. Yao, T.-Y. Ho, and Y. Cai, "LUTOSAP: Lookup Table Based Online Sample Preparation in Microfluidic Biochips," *Proc. of GLSVLSI*, 2017, pp. 447-450.

[26] *LiveLink for Matlab*, https://www.comsol.com/livelink-formatlab, Accessed: 2017-11-21.

[27] *AutoCAD DXF*, https://en.wikipedia.org/wiki/AutoCAD_DXF, Accessed: 2017-11-21.

[28] K. W. Oh, K. Lee, B. Ahn, and E. P. Furlani, "Design of pressure-driven microfluidic networks using electric circuit analog," *Lab Chip*, 2012, vol. 12, pp. 515-554.

[29] R. E. Tarjan, "Depth-first search and linear graph algorithms," *SIAM Journal on Computing*, vol. 1, no. 2, pp. 146-160, 1972.

2018 IEEE Computer Society Annual Symposium on VLSI

Accelerating Simulation of Particle Trajectories in Microfluidic Devices by Constructing a Cloud Database

Junchao Wang[†,*], Lingxuan Fu[†], Liyang Yu[†], Xiwei Huang[†], Philip Brisk[‡] and William H. Grover[§,*]

[†]Ministry of Education Key Laboratory of RF Circuits and Systems. Hangzhou Dianzi University, China.
[‡]Department of Computer Science and Engineering, University of California Riverside, Riverside, CA, USA
[§]Department of Bioengineering, University of California Riverside, Riverside, CA, USA,
[*]Email: junchao@hdu.edu.cn, wgrover@engr.ucr.edu

Abstract—Microfluidic cell sorters have shown great potential to revolutionize the current technique of enriching rare cells. In the past decades, different microfluidic cell sorters have been developed by researchers for separating circulating tumor cells, T-cells, and other biological markers from blood samples. However, it typically takes months or even years to design these microfluidic cell sorters by hand. Thus, researchers tend to use computer simulation (usually finite element analysis) to verify their designs before fabrication and experimental testing. Despite this, conducting precision finite element analysis of microfluidic devices is computationally expensive and labor-intensive. To address this issue, we recently presented a microfluidic simulation method that can simulate the behavior of fluids and particles in some typical microfluidic chips instantaneously. Our method decomposes the chip into channels and intersections. The behavior of fluid in each channel is determined by leveraging analogies with electronic circuits, and the behavior of fluid and particles in each intersection is determined by querying a database containing 92,934 pre-simulated channel intersections. While this approach successfully predicts the behavior of complex microfluidic chips in a fraction of the time required by existing techniques, we nonetheless identified three major limitations with this method: (1) the library of pre-simulated channel intersections is unnecessarily large (only 2,072 of 92,934 were used); (2) the library contains only cross-shaped intersections (and no other intersection geometries); and (3) the range of fluid flow rates in the library is limited to 0 to 2 cm/s. To address these deficiencies, in this work we present an improved method for instantaneously simulating the trajectories of particles in microfluidic chips. Firstly, inspired by dynamic programming, our new method optimizes the generation of pre-simulated intersection units and avoids generating unnecessary simulations. Secondly, we constructed a cloud database (*http://cloud.microfluidics.cc*) to share our pre-simulated results and to let users become contributors and upload their simulation results into the cloud database as a benefit to the whole microfluidic simulation community. Lastly, we investigated the impact of different channel angles and different fluid flow rates on predicting the trajectories of particles. We found a wide range of device geometries and flow rates over which our existing simulation results can be extended without having to perform additional simulations. Our method should accelerate the simulation of particles in microfluidic chips and enable researchers to design new microfluidic cell sorter chips more efficiently.

I. INTRODUCTION

Cell sorting has many important applications in both clinics and biological research [2], [3]. Among different cell sorting mechanisms, microfluidic cell sorting devices have shown great potential in enriching rare biological markers

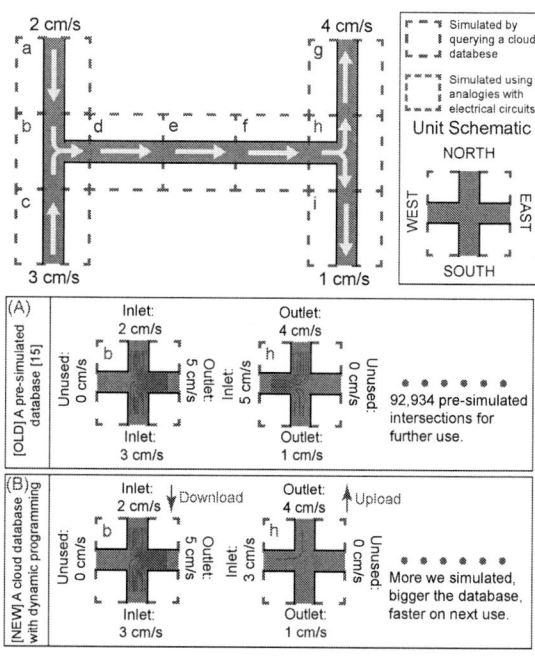

Fig. 1. The comparison of our two methods to accelerate simulation of a microfluidic chip for predicting particle trajectories. (A) Our original method based on memoization [1]. (B) Our new method that was inspired by dynamic programming and a cloud database.

like circulating tumor cells for cancer therapy [4] or T-cells for immunotherapy [5]. In the past decades, researchers have developed different microfluidic cell sorters based on inertial effects [6], [7], [8], pinched flow fractionation [9], [10], deterministic lateral displacement (DLD) [11], [12] and other different principles [13], [14].

During the microfluidic device design process, many researchers may wish to use computer simulations to verify their ideas before fabrication and experimental testing. However, there are two major barriers standing between device ideas and lifelike simulations. First, simulations of microfluidic devices are usually computationally expensive. For instance, a workstation with 32 gigabytes of RAM and a 12-core E5 CPU was necessary to simulate the particle trajectories in a small unit of a DLD device using finite element analysis

978-1-5386-7100-9/18 $31.00 © 2018 IEEE

(FEA) software (COMSOL Multiphysics) and MOPSA [15] in order to reduce the computational time to an hour. Second, FEA software has a steep learning curve. It can take months of training for a student to learn how to create a successful model of the simulation target and have a basic understanding of the connection between mathematical equations behind the FEA software and real-world physics. These barriers limit the practical usefulness of existing software when simulating microfluidic devices.

To address these issues, our previous work presented a method to efficiently simulate the behavior of fluids and particles in some typical microfluidic chips instantaneously (in approximately one second) [1]. The acceleration is achieved by two steps. Assume we have an "H"-shape microfluidic chip shown in Fig. 1, and we want to simulate the particle trajectories in this chip. First, we decompose the microfluidic chip design into several unit intersections (a - i). Units a, c, d, e, f, g, and i are simple channels; the boundary conditions of these channels can be modeled using the electronic-fluidic analogy and calculated by electronic circuit simulation software such as SPICE [16], [17], [18]. Second, as shown in Fig. 1A, we access a database that contains 5,321,944 pre-simulated trajectories of 92,934 intersections. In low Reynolds number situations, particle trajectories can be expanded from the trajectories of units b and h throughout the whole chip using streamline theory [19]; consequently, we only need to query the database to retrieve pre-simulated results for units b and h. Combining these two steps generates the predicted particle trajectories throughout the device. Constructing the database of pre-simulated intersections required approximately one month of computing time. However, using the database, the simulation of particle trajectories of a microfluidic chip takes approximately one second on a standard laptop, without any noticeable degradation in the accuracy of the simulation.

While our instantaneous simulation method successfully predicts the behavior of complex microfluidic chips in a fraction of the time required by existing techniques, we nonetheless identified three limitations of our technique. These limitations mainly concern the database of pre-simulated intersections. First, the pre-simulated database only contains T- or cross-shaped intersections with 90-degree channel turns; this limits the range of intersection geometries that can be simulated using our technique. Second, the database only contains simulation results with fluid flow rates from 0 to 2 cm/s; it cannot currently be used to predict the behavior of flow rates above 2 cm/s. Third, the time required to construct the pre-simulated database is non-negligible, making it prohibitive to expand its usage to more general chip designs.

In this work, we address these limitations of our previous instantaneous simulation method. We began this work by asking the question: what if we could have a platform to share our pre-simulated database, and every user who simulates a previously unseen intersection contributes the result of the simulation to the database, increasing its utility for others? By leveraging dynamic programming, a method for solving a complex problem by decomposing it into a collection of simpler subproblems [20], we present an online platform which helps researchers take advantage of pre-simulated trajectories in their own microfluidic particle simulation projects. Researchers are able to use our platform through either an application programming interface (API) or a graphical user interface (*http://cloud.microfluidics.cc*). Sharing our simulation database eliminates the need for researchers to construct a brand new database for their own projects and is expected to help researchers obtain reliable particle simulation results instantly for their projects, even if they only need to simulate a small number of microfluidic chips. Researchers are encouraged to upload their simulation results into our cloud database to contribute to this project to help the microfluidic particle simulation community. In theory, as our user base grows, the database of pre-simulated results will grow along with it, and the entire community of users will benefit from faster and more accurate simulations.

To address the other limitations of our previous technique, we also investigate how different channel angles (30° to 180°) at intersections and how different flow rates (Reynolds numbers from 0.1 to 100) affect particle trajectories in the same channel. By sharing this cloud database and generating intersection units on demand, we are confident that simulation of particle trajectories in microfluidic devices will become easier and more efficient in the near future.

II. Cloud Database Design

This work was motivated by the inefficient use of our pre-simulated intersections. In our previous work, we simulated more than 10,000 random microfluidic chips (Fig. 3) using our instantaneous simulation method [1]. In this work, we studied how many of the intersection simulations were actually used while simulating these 10,000 random chips. Table I is the statistical analysis of the usage details of pre-generated intersection units. Although we generated 92,934 intersections with varied boundary conditions, 90,862 of the intersections (97.8%) were never used. Of all generated intersection units, 18 intersection simulations were used more than 1000 times, 205 intersections were used more than 100 times, and 516 intersections were used more than 10 times. Since different intersection units had dramatically different probabilities of being used, it occurred to us that the intersection units should be generated on demand with determinate boundary conditions instead of randomly being generated.

We also introduce a new parameter, called *error tolerance*, to indicate the relative difference between the flow velocity boundary conditions specified by the user and the closest match found in the cloud database. Adjusting the error tolerance allows the user to customize the accuracy required for different applications. Error tolerance is defined in Equation 1.

$$Error\ Tolerance = \left| \frac{V_{W0} - V_{W1}}{V_{W0}} \right| + \left| \frac{V_{N0} - V_{N1}}{V_{N0}} \right| \\ + \left| \frac{V_{S0} - V_{S1}}{V_{S0}} \right| + \left| \frac{V_{E0} - V_{E1}}{V_{E0}} \right| \quad (1)$$

TABLE I
STATISTICAL ANALYSIS OF THE PRE-GENERATED INTERSECTION UNITS IN OUR PREVIOUS WORK [1]

Frequency of intersection unit usage	0	1	$2 \sim 4$	$5 \sim 9$	$10 \sim 99$	$100 \sim 999$	$1000 \sim \infty$
Number of intersection units (total 92,934)	90,862	541	519	273	516	205	18

where V_{W0}, V_{N0}, V_{S0}, V_{E0} indicate the average flow velocity of WEST, NORTH, SOUTH and EAST of the intersection units as requested by the user; and V_{W1}, V_{N1}, V_{S1}, V_{E1} indicate the average flow velocity of WEST, NORTH, SOUTH and EAST of the intersection units in the cloud database.

In dynamic programming, a large problem is decomposed into a series of subproblems. When a subproblem is solved, the solution is stored for future reuse. In our system, the primary problem is to predict particle trajectories in microfluidic devices, and the subproblems are simulations of specific particle trajectories in specific intersection units. The simulated trajectories in each intersection unit are stored in a cloud database for further use. By focusing on subproblems that are actually useful in real-life simulations, as opposed to pre-simulating a large population of randomly-generated intersections, this principle similar to dynamic programming reduces the redundancy in our simulation database and accelerates the simulation process.

Fig. 2 illustrates how we coupled our cloud database with custom simulation projects using the idea of dynamic programming. First, the user determines the flow velocity in the intersections where they intend to simulate the particle trajectories. The details of this step have been described in our previous work [1]. Second, our cloud platform (*http://cloud.microfluidics.cc*) provides an API that helps users find the best match in the cloud database of pre-simulated intersection trajectories. The user sends a POST request to *http://cloud.microfluidics.cc/php/post.php* containing the parameters defined in Listing 1 in JavaScript Object Notation (JSON) format. The cloud platform processes the user request and returns the result to the user, once again in JSON, as defined in Listing 2. In many cases, the results will be suitable for integration into the user's simulation of a complete chip. However, in some cases the cloud platform may not find a suitable match in the database of pre-simulated intersections. In these cases, the user will be encouraged to simulate the trajectories of particles locally and upload the simulation results into our cloud database to better serve the community. To upload a new simulation result, the user sends a JSON-formatted POST request to *http://cloud.microfluidics.cc/php/data_post.php* with the parameters defined in Listing 3. After retrieving or simulating all the particle trajectories in all the intersections, the user can combine and expand the trajectories of individual intersection units into trajectories from the input to the output of the chip [1].

Listing 1: Query the database

```
{
    input1 : flow velocity of the WEST boundary of
        the intersection,
```

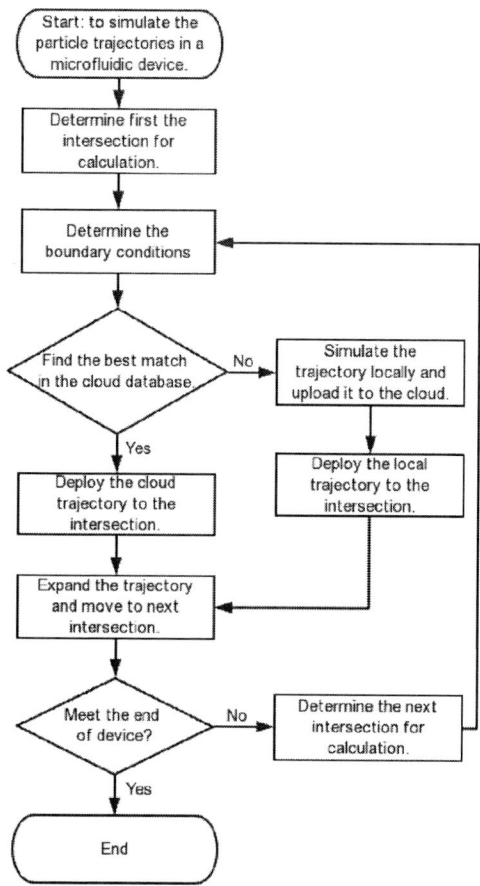

Fig. 2. Flow chart depicting the process of using our cloud database to simulate a user's microfluidic chip design.

```
input2 : flow velocity of the NORTH boundary of
    the intersection,
input3 : flow velocity of the SOUTH boundary of
    the intersection,
input4 : flow velocity of the EAST boundary of
    the intersection,
accuracy : Error tolerance,
dp : particle diameter,
}
```

Listing 2: Pre-simulated particle trajectories

```
{
    "res_code" : 0 for failure & 1 for success,
    "msg" : message,
    "res" : [{
        "id" : ID of the trajectory,
        "start_position_x" : initial position of the
            particle in x direction,
```

```
    "start_position_y" : initial position of the
        particle in y direction,
    "end_position_x" :  final position of the
        particle in x direction,
    "end_position_y" :  final position of the
        particle in y direction,
    }]
}
```

Listing 3: Uploading the simulated trajectories

```
{
    input1 : flow velocity of the WEST boundary of
        the intersection,
    input2 : flow velocity of the NORTH boundary of
        the intersection,
    input3 : flow velocity of the SOUTH boundary of
        the intersection,
    input4 : flow velocity of the EAST boundary of
        the intersection,
    dp : particle diameter,
    length : The number of trajectories uploaded,
    data0 : initial position of the particle in x
        direction,
    data1 : initial position of the particle in y
        direction,
    data2 : final position of the particle in x
        direction,
    data3 : final position of the particle in y
        direction,
    ...

    file : binary files describing the details of
        trajectories if possible,
}
```

III. VALIDATING FAST SIMULATING OF PARTICLE TRAJECTORIES IN RANDOM MICROFLUIDIC CHIPS

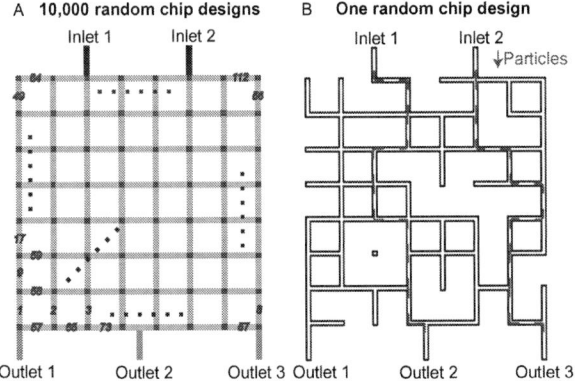

Fig. 3. (A) The schematic of random microfluidic chips [21]. (B) Results from using the principle of dynamic programming and our cloud database to predict the paths of two 1 μm diameter particles traveling through a random microfluidic chip.

To demonstrate our method and cloud platform on a more complex chip design, we used the new method to predict the trajectories of 1 μm diameter particles in 10,000 randomly-generated microfluidic chips [21]. As shown in Fig. 3, these random microfluidic chips each have two inlets and three

outlets. In our simulations, particles start in the center of each inlet channel and end at one outlet.

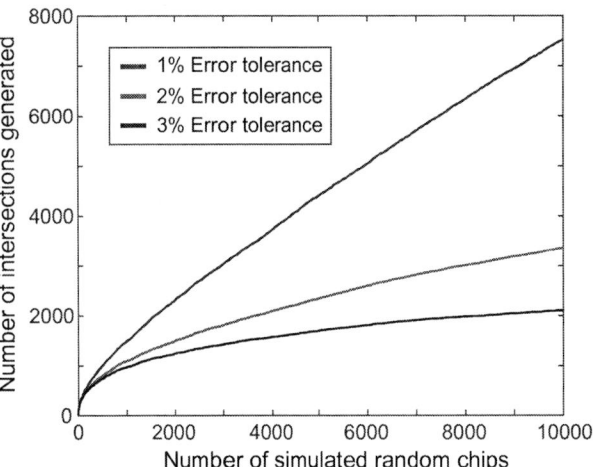

Fig. 4. Statistical analysis of the generated intersection units using the principle of dynamic programming and cloud database method when simulating 10,000 random microfluidic chips with 1%, 2% and 3% error tolerance settings.

The cloud database was reset prior to each simulation. Each simulation was performed three times with different error tolerance settings (1%, 2% and 3%); the simulation results are summarized in Fig. 4. The three error tolerance settings generated 7,519, 3,351 and 2,102 new intersection units respectively. In each case, as the number of simulated random microfluidic chips increased, new intersection units were generated rapidly at first. After generating a certain number of intersection units, the generation rate decreased, indicating greater reuse of existing pre-simulated intersection units over time. Increasing the error tolerance decreases the number of new intersection units generated during the simulation process, which is consistent with our prediction. Compared with the intersection units generated in our previous work (Table I), every intersection unit generated by our new method was used at least once. Furthermore, even for the most-stringent 1% error tolerance case, only 7,519 intersection units were generated instead of 92,934. The computational time required to generate the database was therefore reduced by 91.9%, 96.4% and 97.7% with 1%, 2% and 3% error tolerance, respectively, when compared to our previous work. Additionally, users may benefit from the ability to specify their own expectations about error tolerance, which is handled seamlessly by the querying mechanism that we have integrated into our cloud database.

IV. REMOVING LIMITATIONS ON THE DEVICE GEOMETRIES THAT CAN BE SIMULATED

As noted earlier, our pre-simulated intersection database only contains simulations for channel intersections with (1) 90° turns and (2) fluid flow rates between 0 and 2 cm/s, so the database could only be used for microfluidic chips that adhere to these constraints. In this work, we have successfully removed these limitations. We accomplished this by building FEA models of microfluidic channels that have one inlet, two outlets, and internal angles of 30°, 60°, 90°, 120°, 150° and 180°. We then simulated flow through each intersection using inlet flow rates that resulted in four different values for the Reynolds number: 0.1, 1, 10, and 100. We solved the fluid velocity field of the chip using the Laminar Flow physics module in COMSOL Multiphysics and a stationary solver; the simulation used the "Extremely Fine" mesh setting. We used the Particle Tracing for Fluid Flow physics module to predict particle trajectories across the entire chip. We added a "Drag Force" boundary condition to the entire chip, and a particle "Inlet" boundary condition with initial position "Uniform Distribution" and we added a 1.0 μm particle diameter to all inlets in the Laminar Flow module. We assigned "Outlet" boundary conditions to the outlets in the Laminar Flow module; the remaining boundaries were walls ("freeze" boundary condition). We set the number of particles per release to 20.

Fig. 5 shows the simulation results. The relative end position of 20 particles are near-identical at Reynolds numbers of 0.1, 1 and 10, across all six different intersection angles (except the case of Re = 10 and angle = 30°). The average relative differences are 1.3%, 2.2% and 5.2% when Re is equal to 0.1, 1 and 10, respectively. At a Reynolds number of 100, the angle of intersection begins to impact the particle trajectories, and the particles exit the intersections at substantially different locations in the six different angles. When channel angles equal 30° and 180°, the increase of Re has greater impact on the trajectories of particles. In the 30° and 180° cases, the simulations predict that the middle particles tend to become stuck in the intersections when Re equals 10 and 100.

These results show that even though our cloud database has limited pre-simulated intersection units, users are nonetheless able to convert the pre-simulated results into new simulation results (for Reynolds numbers less than 10, and channel angles between 60° and 180°) and obtain reliable simulation results without actually performing new finite element analyses. This significantly increases the range of microfluidic device geometries and flow rates that can be instantaneously simulated using our method. However, there are still some intersection geometries and flow rates that cannot be simulated using our improved method. In these cases, our cloud database can ultimately allow users to upload their own simulation results containing currently-unsupported angles and flow rates, thereby further extending the range of devices that can be simulated instantaneously.

V. CONCLUSIONS

In this work, we constructed a cloud database and applied the principle of dynamic programming to accelerate the simulation of particle trajectories in microfluidic devices. Compared to our previous method, the efficiency of our new method arises from three key innovations: 1. Only simulate the intersection units which will be used at least once; 2.

Utilize a cloud database that will help the microfluidics community accelerate predicting the path of particles in their own projects; 3. Enable users of our platform to submit their own simulation results to expand our pre-simulated database. The original limitations of our method (only supporting cross-shape intersections and a limited range of fluid flow rates) were investigated as well. Our simulation results show that these limitations can be negligible at low Reynolds number, which is one of the natural properties of microfluidics. This method reduces the barriers to simulating particle trajectories in microfluidic chips and should ultimately enable researchers to design new microfluidic cell sorting devices more efficiently.

ACKNOWLEDGMENT

This work was supported in part by Hangzhou Dianzi University seed fund for JW, and National Science Foundation awards #1351115, #1353974, #1536026, #1640757, and #1740052.

REFERENCES

[1] J. Wang, V. G. J. Rodgers, P. Brisk, and W. H. Grover, "Instantaneous simulation of fluids and particles in complex microfluidic devices," *PLOS ONE*, vol. 12, no. 12, pp. 1–14, 12 2017. [Online]. Available: https://doi.org/10.1371/journal.pone.0189429

[2] N. M. Karabacak, P. S. Spuhler, F. Fachin, E. J. Lim, V. Pai, E. Ozkumur, J. M. Martel, N. Kojic, K. Smith, P.-i. Chen *et al.*, "Microfluidic, marker-free isolation of circulating tumor cells from blood samples," *Nature protocols*, vol. 9, no. 3, p. 694, 2014.

[3] L. A. Herzenberg, D. W. Bianchi, J. Schröder, H. M. Cann, and G. M. Iverson, "Fetal cells in the blood of pregnant women: detection and enrichment by fluorescence-activated cell sorting," *Proceedings of the National Academy of Sciences*, vol. 76, no. 3, pp. 1453–1455, 1979.

[4] V. Plaks, C. D. Koopman, and Z. Werb, "Circulating tumor cells," *Science*, vol. 341, no. 6151, pp. 1186–1188, 2013.

[5] C. A. Klebanoff, S. A. Rosenberg, and N. P. Restifo, "Prospects for gene-engineered t cell immunotherapy for solid cancers," *Nature medicine*, vol. 22, no. 1, p. 26, 2016.

[6] S. S. Kuntaegowdanahalli, A. A. S. Bhagat, G. Kumar, and I. Papautsky, "Inertial microfluidics for continuous particle separation in spiral microchannels," *Lab on a Chip*, vol. 9, no. 20, pp. 2973–2980, 2009.

[7] L. Wu, G. Guan, H. W. Hou, A. A. S. Bhagat, and J. Han, "Separation of leukocytes from blood using spiral channel with trapezoid cross-section," *Analytical chemistry*, vol. 84, no. 21, pp. 9324–9331, 2012.

[8] H. W. Hou, M. E. Warkiani, B. L. Khoo, Z. R. Li, R. A. Soo, D. S.-W. Tan, W.-T. Lim, J. Han, A. A. S. Bhagat, and C. T. Lim, "Isolation and retrieval of circulating tumor cells using centrifugal forces," *Scientific reports*, vol. 3, p. 1259, 2013.

[9] M. Yamada, M. Nakashima, and M. Seki, "Pinched flow fractionation: continuous size separation of particles utilizing a laminar flow profile in a pinched microchannel," *Analytical chemistry*, vol. 76, no. 18, pp. 5465–5471, 2004.

[10] A. A. S. Bhagat, H. W. Hou, L. D. Li, C. T. Lim, and J. Han, "Pinched flow coupled shear-modulated inertial microfluidics for high-throughput rare blood cell separation," *Lab on a Chip*, vol. 11, no. 11, pp. 1870–1878, 2011.

[11] L. R. Huang, E. C. Cox, R. H. Austin, and J. C. Sturm, "Continuous particle separation through deterministic lateral displacement," *Science*, vol. 304, no. 5673, pp. 987–990, 2004.

[12] S. H. Holm, J. P. Beech, M. P. Barrett, and J. O. Tegenfeldt, "Separation of parasites from human blood using deterministic lateral displacement," *Lab on a Chip*, vol. 11, no. 7, pp. 1326–1332, 2011.

[13] C. W. Shields IV, C. D. Reyes, and G. P. López, "Microfluidic cell sorting: a review of the advances in the separation of cells from debulking to rare cell isolation," *Lab on a Chip*, vol. 15, no. 5, pp. 1230–1249, 2015.

[14] J. Autebert, B. Coudert, F.-C. Bidard, J.-Y. Pierga, S. Descroix, L. Malaquin, and J.-L. Viovy, "Microfluidic: an innovative tool for efficient cell sorting," *Methods*, vol. 57, no. 3, pp. 297–307, 2012.

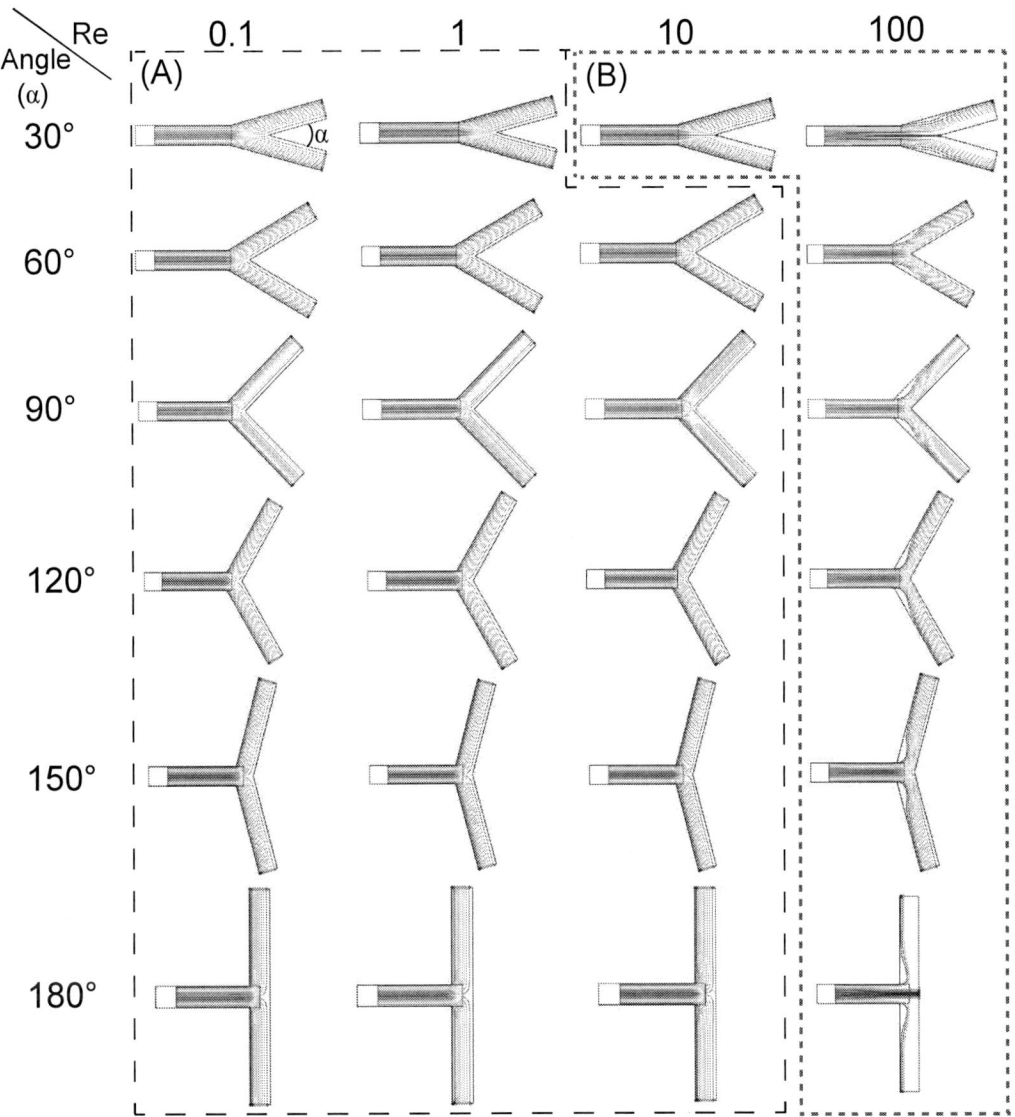

Fig. 5. Simulation results of microfluidic intersections with different Reynolds numbers (0.1, 1, 10 and 100) and different intersection angles (30°, 60°, 90°, 120°, 150° and 180°). (A) At low Reynolds numbers (less than 10) and for channel angles between 60° and 180°, the predicted end positions of the particles are basically identical. This suggests that unit intersection simulations in our cloud database can be used to predict the behavior of microfluidic chips with a wide range of flow rates and intersection geometries. (B) At a Reynolds number of 100, the angle of intersection begins to impact the particle trajectories, and the particles exit the intersections at substantially different locations in the six different angles. When Re = 10 and angle = 30°, the result is similar to Re = 100 and angle = 30° case.

[15] J. Wang, V. G. Rodgers, P. Brisk, and W. H. Grover, "MOPSA: A microfluidics-optimized particle simulation algorithm," *Biomicrofluidics*, vol. 11, no. 3, p. 034121, 2017.

[16] L. W. Nagel, "Spice-simulation program with integrated circuit emphasis," *Memo No.. ERL-M382, Electronics Research Laboratory, Univ. of California, Berkeley*, 1973.

[17] D. Kim, N. C. Chesler, and D. J. Beebe, "A method for dynamic system characterization using hydraulic series resistance," *Lab on a Chip*, vol. 6, no. 5, pp. 639–644, 2006.

[18] D. J. Beebe, G. A. Mensing, and G. M. Walker, "Physics and applications of microfluidics in biology," *Annual review of biomedical engineering*, vol. 4, no. 1, pp. 261–286, 2002.

[19] L. Han, "Hydrodynamic entrance lengths for incompressible laminar flow in rectangular ducts," *Journal of Applied Mechanics*, vol. 27, no. 3, pp. 403–409, 1960.

[20] D. P. Bertsekas, D. P. Bertsekas, D. P. Bertsekas, and D. P. Bertsekas, *Dynamic programming and optimal control*. Athena scientific Belmont, MA, 1995, vol. 1, no. 2.

[21] J. Wang, P. Brisk, and W. H. Grover, "Random design of microfluidics," *Lab on a Chip*, vol. 16, no. 21, pp. 4212–4219, 2016.

PUF-based Secure Test Wrapper for SoC Testing

Sudeendra kumar K, Saurabh Seth, Sauvagya Sahoo, Abhishek Mahapatra, Ayas Kanta Swain, K.K.Mahapatra
kumar.sudeendra@gmail.com, saurabhseth.030@gmail.com, sauvagya.nitrkl@gmail.com, kmaha2@gmail.com
National Institute of Technology, Rourkela

Abstract- **The increased testability and observability due to test structures make chips vulnerable to side channel attacks. The intention of side channel attack are leaking secret keys used in cryptographic cores and getting access to trade related sensitive information stored in chips. Several countermeasures against test based side-channel attacks are available in research literature. One such countermeasure scheme is password based access protection to IEEE 1500 test wrapper, such that only an authentic user with valid password is allowed to access the test structures. IEEE 1500 is a core test standard for enabling the streamlined test integration and test reuse. The trust model of existing schemes assume outsourced assembly and test (OSAT) centre are completely trusted and design house will share secret keys to unlock the IEEE 1500 wrapper during testing. In this paper, we propose a Physical Unclonable Function (PUF) based technique incorporating challenge-response to support comprehensive test security in which there is no need for design house to share secret keys with untrusted OSAT centre to unlock the scan chains. The proposed scheme comes at the cost of reasonable area and performance overhead.**

Keywords: **IEEE 1500, Physical Unclonable Function, Hardware Security.**

I. INTRODUCTION

Scan chains are common Design for Testability (DFT) structures found in modern day chips. DFT structures are used for better controllability and observability of the design during testing. DFT structures which are useful in testing are misused to leak confidential data through side channel attacks (SCA) [1]. Adversary will use SCA to leak secret keys used in encryption and critical data related to design. Different types of countermeasures to prevent SCA can be found in research literature [1]. The well-known countermeasures are: - partial scan design [2], scan compression schemes [3], obfuscation of scan chains [4], defusing the test pins [5] and secure test wrappers to prevent unauthorized access [6-8].

IEEE 1500 is a core test standard for enabling the streamlined test integration and test reuse. The IEEE 1500 standard defines the serial and parallel test access mechanisms and instructions for testing the embedded IP cores in a System on Chip (SoC). Security against DFT based SCA can be achieved by restricting the illegitimate access to IEEE 1500 wrapped cores in the chip. Countermeasures against attacks on IEEE 1500 wrapped cores have been proposed, including the following: -

- Authors of [6], propose a challenge-response based secure test access mechanism using KATAN lightweight block cipher. On-chip True Random Number Generator (TRNG) generates the random number and sends it to on-chip block cipher and also to ATE (Automatic Test Equipment). KATAN block cipher is also implemented on ATE as a software program. Both device under test and ATE run the same block cipher (KATAN) and generate same cipher texts for same secret key. On-chip comparator compare the cipher text coming from ATE and on-chip block cipher and generate the signal to enable the access to test wrapper after a successful match.

- Key based secure test wrapper (STW) for IEEE 1500 is proposed in [7]. IEEE 1500 test wrapper is unlocked when golden key is applied. The predefined seed is loaded into LFSR to generate the golden key. The test input bit stream should initially contain the key from ATE and it is compared with golden key generated by LFSR. After the successful comparison, wrapper enters the unlock state. The part of the scan chain is configured as LFSR to reduce the area [7]. The design changes in IEEE 1500 standard are required to implement this security scheme, which may not be accepted by all SoC integrators and IP core vendors.

- The Secure Test Wrapper (STW) proposed in [8] replaces the KATAN block cipher with physical unclonable function (PUF). PUF circuits are security primitives, which generate unique response for a given challenge. The PUF circuit exploit random process variations occurring during chip fabrication which makes challenge-response pairs (CRPs) are unique for a given chip [9]. In this technique, PUF CRP will be stored in ATE memory during production testing. This will add to test cost due to higher test time and utilization of tester memory.

In the current era of globalization, most of the post silicon validation procedures are outsourced to offshore OSAT (outsourced assembly and test) centres. In the earlier techniques discussed above, test centres (OSAT) are completely trusted and design house will share the secret keys/PUF CRPs with test centres to unlock the scan chains during testing. Sharing keys with 3rd party test centre may compromise the security of chip. In this paper we present a STW technique which has following advantages: -

- Design house need not to share any secret key to unlock the IEEE 1500 wrapper for testing with untrusted OSAT centres. The proposed technique prevents adversary in OSAT centre getting access to confidential information related to keys used to unlock the cryptographic and other sensitive cores.
- No changes required for IEEE 1500 standard.
- Unique keys for every chip manufactured.

The paper is organized as follows: The prior work and related background is discussed in section II and we discuss the design and functioning of proposed PUF based STW scheme

978-1-5386-7100-9/18 $31.00 © 2018 IEEE

in section III. In section IV, we present the implementation details and analysis and finally section V concludes the paper.

II. BACKGROUND

A. IEEE 1500 Core Test Standard: IEEE 1500 standard defines core wrapper and test specific language called core test language (CTL). CTL acts like a bridge between core providers and core integrators to exchange design details. The language is known as core test language (CTL). Fig 1 shows the IEEE 1500 test wrapped core.

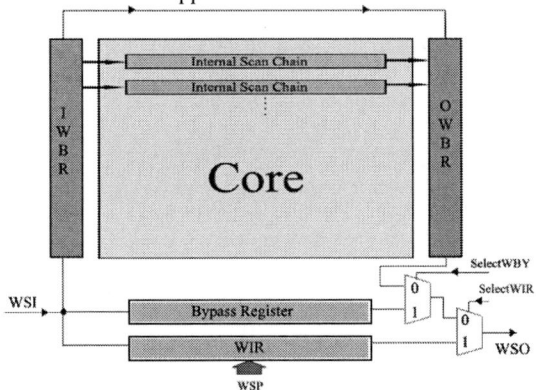

Fig. 1. IEEE 1500 Test Wrapped Core

IEEE 1500 standard define two ports: wrapper serial port (WSP) and wrapper parallel port (WPP). WSP is mandatory and WPP is optional. WSI (Wrapper Serial In) is basic serial input and WSO (Wrapper Serial Output) is basic serial output in WSP. IEEE 1500 standard defines three mandatory registers: - Wrapper Instruction Register (WIR), Wrapper Bypass Register (WBY) and Wrapper Boundary Register (WBR). Depending on the test requirements, any number of user specific registers can be added to wrapper. User specific registers are called Wrapper Data Registers (WDR). WBY provides a bypass path and connects WSI and WSO in bypass mode or functional mode. WIR generate the control signals to enable all wrapper operations. The test data stimuli and responses are captured using WBR register. WBR is used to test the core logic and also to test external connectivity to other cores and I/O ports. WSP comprises of ten terminals in which eight are mandatory and two are optional. The WSP signals are described in Table I.

The instruction loaded into WIR and control signals determine the mode of operation of the wrapper. IEEE 1500 supports 11 instructions, in which three are mandatory: WS_BYPASS, WS_EXTEST and WX_INTEST. The WS_BYPASS instruction is selected, when no test operation is required. This instruction connects WBY between WSI and WSO. WS_EXTEST instruction allows testing of off-core circuitry and core to core interconnections. In INTEST mode, the input test vectors are applied to the core and test response is observed. More details on the IEEE 1500 standard can be found in [10]. Fig. 2 shows the schematic of wrapper boundary cell. The WBR is constructed using WBR cells and each cell has four data terminals: cell functional input (CFI), cell functional output (CFO), cell test input (CTI), and cell test output (CTO). Cell functional input (cfi) and cell functional

output (cfo) connects to the core functional path. In normal mode of the core, these terminals are used to perform functional operation of the core logic.

TABLE I
IEEE 1500 WRAPPER SERIAL PORT (WSP) SIGNALS

Signal	Description
WRCK	IEEE 1500 Wrapper Clock
WRSTN	IEEE 1500 active low wrapper reset. When reset is asserted, WS_BYPASS will be active.
SelectWIR	When SelectWIR is asserted, the WIR is selected and connected between WSI and WSO. Opcodes for operations like "shift", "update" and "capture" are loaded into WIR.
CaptureDR	When CaptureDR is asserted, the data present on the functional input is stored into WBR cell.
ShiftDR	When ShiftDR is asserted, the data stored in any register connected between WSI and WSO is shifted to next position upon the rising edge of WRCK.
UpdateDR	When UpdateDR is asserted, the data stored closest to shift output is loaded into off-shift storage element.

The cell test input (cti) and cell test output (cto) are test inputs to the core from WBR cell. Core test input (cti) connects to the scan input pin and core test output (cto) connects to the scan output pin on the scan path of the core. The control signals (shift_en, capture_en, safe_control, shift_clk) are sourced from the WIR to enable the shift, capture and update operations in the WBR cell.

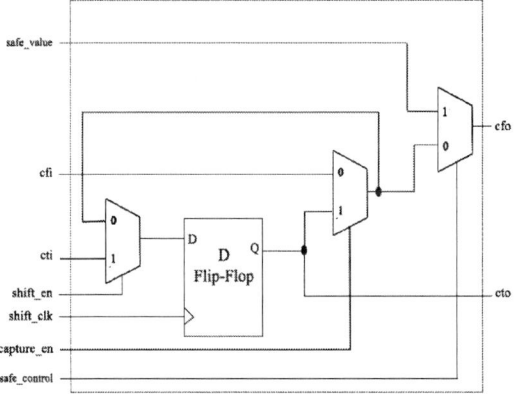

Fig. 2. Wrapper Boundary Cell

B. Physical Unclonable Functions: PUFs are well known hardware security primitives used in identification, authentication and cipher key generation. Even with a full knowledge of PUF design, it is nearly impossible to manufacture an identical PUF circuit which generate same CRP pairs. PUF circuits are used in variety of hardware security applications like IP protection, hardware metering and cryptographic key generation [11]. The quality of PUF is determined based on its CRP features like uniqueness, reliability and uniformity. More details on PUF design, characteristics and applications can be found in [12]. In this work, PUF is used to design the secure test wrapper to prevent unauthorized access to test structures of cryptographic and other critical IP cores.

978-1-5386-7100-9/18 $31.00 © 2018 IEEE

C. Prior Work: - The secret key or any other confidential information can be ascertained by running an encryption in functional mode and switching the chip to test mode at appropriate time. Adversary can successfully uncover the secret key through scan based attack. In this work, we mainly focus on password protection schemes on IEEE 1500 standard. As mentioned above, challenge-response based test security schemes for IEEE 1500 wrapper is described in [6] [7] and [8].

In this work, we present the novel and improved challenge-response based test security technique which addresses the following issues: -

- PUF based STW support unique keys for every chip manufactured. In the PUF based STW proposed in [8], there is a need to share the PUF-CRP data with OSAT centre. Full CRP data or partial CRP data with test centre or with any other third party is not secure. Adversaries will perform model building attacks using the available CRP data, which compromise the complete security of the chip [15]. And also, there is a need to store CRP data in ATE, which increases test cost [16]. In the proposed PUF based STW scheme, there is no need to share PUF CRP data with untrusted OSAT centre.

- Block cipher (KATAN) based STW discussed in [6], does not support unique key for all manufactured chips. KATAN algorithm running on ATE will add to test time which will increase the test cost. Secret keys to unlock the test structures are shared with OSAT centre. In the proposed scheme, there is no need to share PUF CRP data with untrusted OSAT centre The increase in the test time in the proposed scheme is minimal in comparison with earlier schemes.

- The STW proposed in [7] does not support unique key for each chip manufactured and modifications to the IEEE 1500 standard wrapper is required. The golden key is part of test program and adversary in OSAT centre can easily retrieve the key from test program. The proposed PUF based STW scheme, there is no need to share any confidential information with OSAT centre and there is no need to change standard wrapper and also support unique password for each chip manufactured.

III. PUF BASED SECURE TEST WRAPPER

In this work, we propose a PUF based secure test wrapper activation mechanism, which support secure key management between the design house/SoC integrator and test house (OSAT centres). With this scheme, only genuine test centre or authentic test engineer will have an access to test structures.

A. Trust Model and Assumptions:-

The stakeholders of semiconductor supply chain are: Chip design house/SoC integrator, contract manufacturer (fabrication unit) and OSAT centre. Fabless companies must take more security measures than fab-owned companies. In the fabless model, design (GDS II) is shared with contract manufacturer and chips are packaged and tested at OSAT

centres. These three entities are located in different parts of the world. The lack of trust between the three entities has led to several security issues in recent times. Design house/Original Design Manufacturer (ODM) will lose in terms of both revenue and reputation when security issues are compromised. ODM/Design house is a direct victim when security lapses. In our trust model, we consider ODM can be completely trusted and it is assumed that foundry and OSAT centres should not get access to PUF CRP data during wafer probing or final test (production testing).

B. PUF Enrolment: - Every PUF based hardware security application requires an enrolment phase. In this enrolment phase, PUF challenge-response pairs are collected and stored as reference database. Generally, the PUF CRP collection is performed on ATE at OSAT centre. The Fig. 3 shows the schematic of the proposed PUF based STW technique. The proposed scheme comprises of PUF, TRNG, RSA, OTP and comparator circuit. The challenge to the PUF is fed through the TDR (test data register) connected with JTAG pins of the chip. TDR is serially fed from TDI pin of JTAG and challenge stored in TDR is given to PUF inputs in parallel. In this work, custom made TDR is serial in parallel out register. The response of the PUF is encrypted using on-chip RSA encryption block. The public key is shared with OSAT centre to encrypt the PUF response using RSA. After completion of PUF CRP collection, the data stored in OTP (one time programmable) memory is encrypted using RSA. The TRNG will generate random number when chip is powered and random number is stored in OTP. The random number stored in OTP is permanent. The OSAT centre will send the challenges and encrypted PUF responses and encrypted TRNG value with respective chip ID to design house. The design house will decrypt the PUF responses and TRNG value and create a database of with chip ID, PUF CRP and TRNG value.

C. Secure Test Wrapper Activation: - The comparator circuit generates the 'unlock' signal (active high), when the comparison between value stored in the TRNG and PUF response match. The enable signal is used to gate the wrapper clock (WRCK) of IEEE 1500. When enable is high, wrapper clock is allowed into IEEE 1500 wrapper. Design house will choose the appropriate PUF challenge which will generate the TRNG value from the CRP database. The appropriate response with chip id is communicated to OSAT centre to unlock the wrapper. The PUF response and TRNG value are compared in comparator and 'unlock' signal is generated. Strong PUF with very large number of CRPs must be employed to avoid failures in unlock mechanism. The TCK clock (from JTAG) is connected to IEEE 1500 standard clock (WRCK). Inserting two input AND gate with 'unlock' signal and clock signal is used for gating the clock. Similarly, the unlock signal can be used to gate the 'cti' input (test input to wrapper) or WRSTN (wrapper reset).

The OSAT centre will have an access to encrypted responses coming out of RSA during CRP collection and encryption safeguards the PUF responses. During testing, PUF challenge with a corresponding chip Id is shared with OSAT centres to unlock the test access.

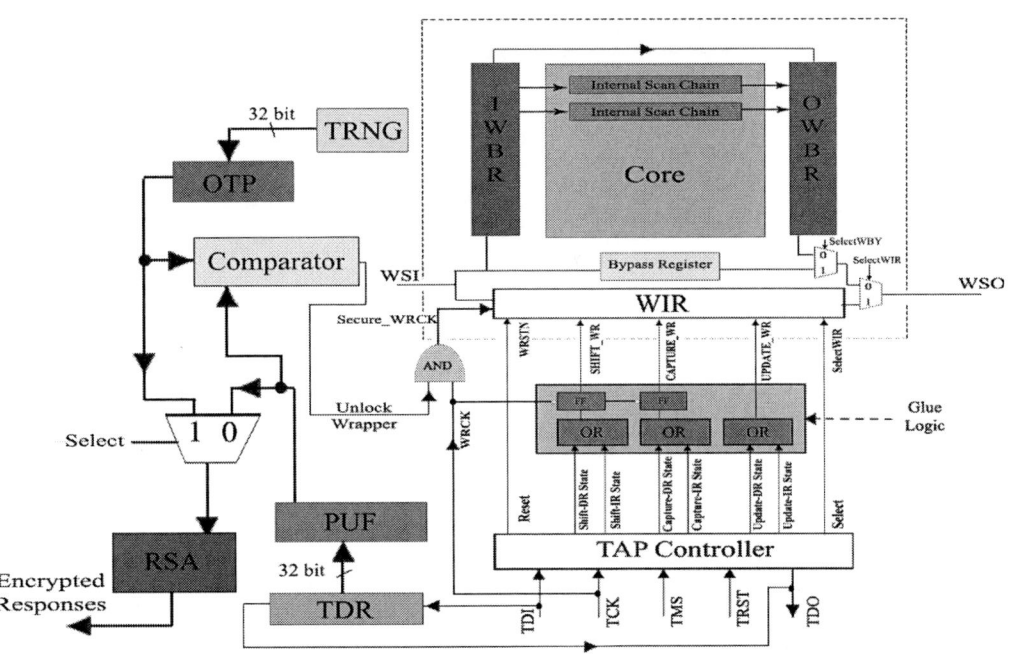

Fig. 3. Proposed PUF base Secure Test Wrapper

TABLE II
TAXONOMY OF IJTAG INSTRUMENTS

Parameter	Paper [6]	Paper [8]	Paper [7]	Proposed
Method	Block Cipher(KATAN) based scheme	PUF based scheme	LFSR based key generation scheme	PUF based scheme
Password	Static. All chips manufactured will have same 80-bit password, which is stored in secret memory.	Dynamic. PUF circuit support different password for every chip manufactured.	Static. Golden key is stored in memory, which is compared with key coming from test program to unlock test.	Dynamic. PUF support different password for every chip.
Modifications to IEEE 1500 standard to add security	Minimum modification	Minimum modification	Major modification to include security.	Minimum modification
Security	Secret keys are shared with OSAT centre. Adversary in OSAT centre will get an access to secret key which compromise security.	PUF CRP is shared with OSAT centre. Adversary in OSAT centre uses PUF CRP to unlock wrapper and perform SCA.	Adversary in OSAT centre will get an access to golden key through reading the memory or by analysing the test program, which contains key to unlock test structures.	There is no need to share secret key or any confidential information with OSAT centre.
Area Overhead	High. Due to inclusion of Block cipher and secret memory	High. Due to inclusion of PUF	Low. No large extra circuits added. Scan chains are used as LFSR.	High. Due to inclusion of PUF and RSA.
Effects on Test Time/Test Cost	Medium. RS-232 interface is used to unlock the scan chain which adds to test time. Executing the KATAN cipher on ATE adds to test time.	High. RS-232 interface is used to unlock the scan chain which adds to test time. PUF CRPs are stored in ATE memory which adds to test cost.	Very less. This scheme has no effect on test time or test cost. Key is fed into device under test as a part of scan test input. JTAG interface is used to connect with wrapper.	Medium. Unlocking the scan chains takes reasonable amount of test time. JTAG interface is used to interact with wrapper.

In the proposed technique, there is no need to share the PUF CRP with OSAT centre and an adversary in OSAT centre will not get an access to it. So the proposed security scheme is safe.

IV. IMPLEMENTATION AND ANALYSIS

A. Implementation: - The key components of PUF based STW are: - PUF, Comparator circuit, TRNG, RSA, OTP and custom TDR. Physical unclonable function is designed to get 32-bit response. The PUF design presented in [17] is used in this work. The Ring Oscillator (RO) PUF described in [17] will consume less number of RO's and capable of generating multiple output bits from each RO. The structure of RO PUF presented in [17] is shown in Fig. 4.

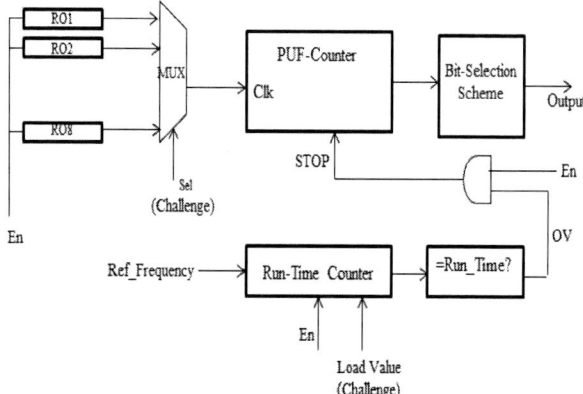

Fig. 4. Structure of Physical Unclonable Function

A runtime counter is connected with standard clock source which acts like a reference clock. The runtime counter is loaded with 'load value', from which it starts counting based on the reference clock. The challenge to PUF comprises of two components: - 'load value' to run-time counter and select lines to multiplexer. Once challenge is fed, the enable signal will trigger the RO and runtime counter at the same time instant. The runtime counter will drive the signal OV to logic '1' when it overflows, which in-turn drive the signal 'STOP'. The signal 'STOP' will stop the PUFcounter from counting. The bit selection scheme selects the bits from the central portion of the PUF counter output. The basic PUF consists of 8 ring oscillators and 3 select lines will go into multiplexer. In the proposed PUF based STW, 8 instances of basic PUF is used to generate the 32-bit response. The input challenge comprises of 32-bit load value into 32-bit run-time counter and 24 select lines to multiplexers. The bit size of the challenge will be 56 bits. Bit selection scheme select the four bits (15th, 16th, 17th and 18th) from the central portion from the PUF counter output in each instance. By concatenation of 4-bit response from each basic PUF instance, 32-bit response is generated from 8 instances.

The length of the custom TDR is 56 bit. 32-bit comparator circuit is designed using XOR gates to compare random number stored in OTP and response generated by PUF. A register of 32-bit length is used as OTP in the experiment. RSA block is used to encrypt both PUF response and random

value stored in OTP. Thirty two 2:1 multiplexers are used to switch at the input of RSA block. An extra input pin is required as a 'select line' to choose between PUF response and OTP data. The single select line is connected to all 32 multiplexers.

The complete system is implemented in Verilog HDL and verification of the design is performed in Synopsys VCS simulator. The proposed PUF based STW is implemented in both ASIC and FPGA. ASIC implementation of comparator, RSA, FIFO (as an example) core with IEEE 1500 wrapper, JTAG with custom TDR is implemented using Synopsys Saed90nm library. A complete system including PUF is implemented on Xilinx Spartan 3E FPGA. In PUF design, manual routing is performed during place and route of the cells in each CLB (configurable logic block in FPGA). Spartan 3E FPGA CLB consists of four slices and each slice comprises of two LUTs (Look-up Tables). The NOT gate for ring oscillator is implemented using one LUT. 5 NOT gates are used to design each ring oscillator. Five slices from each CLB is required to design one ring oscillator. The ring oscillator with five NOT gates is created as macro and instantiated to design the complete PUF.

B. Performance Overhead: - The authentic user with key can unlock the wrapper. The test engineer has to unlock the wrapper only once in a test session. After unlocking, all tests can be conducted and upon reset at the end, test structures are locked. The extra timing overhead added due to security is discussed in this section. The amount of TCK cycles required to unlock the wrapper is as follows: -

- 1149.1 JTAG FSM need five clock cycles for the update and capture states [18].
- For 32-bit PUF challenge, 32 cycles are required to load the challenge into TDR and three clock cycles to feed the challenge to PUF.
- Another four clock cycles are required to generate response and to generate unlock signal from comparator circuit.

The total number of TCK clock cycles is: - 5+32+03+4 = 44 TCK cycles. 44 extra TCK cycles are required to test structures.

C. Security Analysis: - The key component in the proposed technique is PUF and performance of the overall security system depends of quality of CRP and secrecy of CRP database. The attack scenarios are discussed below: -

- Case 1: - An adversary with full knowledge of security system implemented will try for brute force attack, trying all possible combinations will take large amount of test time even on ATE. With encrypted PUF CRP's are stored in secure database and private key is unknown to OSAT centre, the proposed technique is safe. And successful brute force attack on one chip does not reveal any data or secret of other chips.
- Case 2: - In the trust model, it is assumed that PUF CRPs are stored in secure database in design house/ODM. A successful model building attacks on PUF is possible when an adversary gets an access to

978-1-5386-7100-9/18 $31.00 © 2018 IEEE   676

complete CRP data of one chip [15]. In such a case, adversary does not have knowledge of random value stored in the OTP generated by TRNG. Even with complete knowledge on PUF CRP, adversary has to try similar to brute force attack, which is time consuming.

D. Comparison: - A comparative analysis of password/challenge-response based secure test wrapper techniques is presented in Table II. The techniques presented in [6] and [7] have static passwords which is same of all chips produced. Authentic user can share password with adversary with or without malicious intent. The PUF based scheme in [8] and the technique proposed in this paper support dynamic passwords. A major modification to IEEE 1500 wrapper is needed to implement security scheme in [7]. In the all the earlier schemes [6] [7] and [8], both foundry and OSAT centres are trusted. The earlier techniques share the passwords with OSAT centre. Adversary in OSAT centre will get an access to PUF CRP/passwords easily. Earlier techniques prevent the attacks from outsider, but not sufficient against untrusted OSAT centres. The proposed scheme in this paper trust only design house/ODM and secure against untrusted foundries and OSAT centres.

In the techniques presented in [6] and [8], RS-232 interface is used to unlock the wrapper from external ATE or test setup. Generally, IEEE 1500 wrapper is accessed through JTAG 1149.1. Clock and other control signals to control the 1500 wrapper are derived from JTAG 1149.1. JTAG 1149.1 is a primary interface for test operations and it is logical to have a security mechanism to protect IEEE 1500 wrapper for sensitive cores through JTAG 1149.1. In the techniques described in [6] and [8], attacker can get access to IEEE 1500 wrapper through JTAG 1149.1 in test mode, which compromises the security of the sensitive core even in the presence of STW. In the proposed technique, JTAG 1149.1 is used to unlock the wrapper and test operations will continue through the same channel makes the sensitive/cryptographic cores more secure.

E. Reliability: - The reliability of the proposed technique mainly depends upon reliability of the PUF. PUF CRPs vary with changing environmental conditions, ambient noise and aging. In general, most of the CRP's will have high reliability and very few (1 to 10%) do not have a bias to resolve towards either logic 0 or logic 1 [19]. PUF with high reliability will be the best fit in hardware security applications. Designing PUF circuits with highest reliability is an active research area and it is better to choose the PUF with highest reliability for security applications. The PUF used in this research has got a reliability of 89% [17]. Designing PUFs with high reliability and building applications using PUFs will go hand in hand.

V. CONCLUSIONS

In this paper, we have proposed PUF based secure test wrapper design for IEEE 1500 core test standard. The challenge-response based protection scheme presented in this paper is a countermeasure against DFT based side channel attacks on IEEE 1500 wrapped cores. The earlier challenge-response based security schemes proposed share secret keys or

PUF CRPs with OSAT centre to unlock the scan chains to perform test operations. Sharing keys and confidential information with OSAT centre makes the chip vulnerable to attacks. In the proposed scheme, only design house/ODM is trusted and proposed technique avoid sharing of secret keys or any other confidential information with untrusted foundry and OSAT centers. The proposed scheme adds reasonable area overhead, due to the addition of PUF and RSA blocks.

Various countermeasures against SCA have been proposed in the past like scan chain obfuscation and partial scan address the issue at core level affect the test quality. Scalable countermeasures for SoC devices can be achieved with password based access techniques. The proposed PUF based STW technique is more secure than the earlier similar schemes and suitable for SoC environment. The success of the proposed scheme depends on the reliability of PUF CRP.

REFERENCES

[1] J. Da Rolt et al., "Test versus Security:Past and Present", *IEEE Trans. Emerg. Topics in Computing.*, vol.2, no. 1, pp. 50-62, Mar. 2014.

[2] M. Inoue, T. Yoneda, M. Hasegawa, and H. Fujiwara, "Partial scan approach for secret information protection," in *IEEE ETS*, 2009.

[3] J. Da Rolt, G. Di Natale, M.-L. Flottes, and B. Rouzeyre, "Are advanced dft structures sufficient for preventing scan-attacks?" *IEEE VTS*, 2012.

[4] A. Cui, Y. Luo, and C.-H. Chang, "Static and dynamic obfuscations of scan data against scan-based side-channel attacks," in *IEEE TIFS*, 2016.

[5] O. K¨ommerling and M. G. Kuhn, "Design principles for tamper-resistant smartcard processors." *Smartcard*, 1999.

[6] A. Das, M. Knezevic, S. Seys, and I. Verbauwhede, "Challenge-response based secure test wrapper for testing cryptographic circuits," 16th IEEE European Test Symposium 2011 (ETS), 2011.

[7] Geng-Ming Chiu, James Chien-Mo Li "A Secure Test Wrapper Design Against Internal and Boundary Scan Attacks for Embedded Cores", IEEE Transactions on Very Large Scale Integration (VLSI) Systems, Volume 20 Issue 1, January 2012 Page 126-134.

[8] A. Das et al, "PUF-based secure test wrapper design for cryptographic SoC testing", Proceedings of the Conference on Design Automation and Test in Europe (DATE)-2012.

[9] G.E. Suh and S. Devadas, "Physical Unclonable Functions for Device Authentication and Secret Key Generation", *in proc. 44th ACM/IEEE Design Automation Conference (DAC '07)*, pp.9-14, 2007.

[10] IEEE Computer Society, IEEE Standard Testability Method for Embedded Core-Based Integrated Circuits, IEEE Std. 1500-2005.

[11] M. Tehranipoor and C. Wang, "*Introduction to Hardware Security and Trust*", Newyork, NY, USA; Springer-2011.

[12] Abranil Maiti et. al, "A Systematic Method to Evaluate and Compare the performance of PUF". IACR Cryptology, -2011.

[13] D. Hely, M.-L. Flottes, F. Bancel, B. Rouzeyre, N. Berard, and M. Renovell, "Scan design and secure chip." in *IOLTS*, 2004.

[14] A. Das, B. Ege, S. Ghosh, L. Batina, and I. Verbauwhede, "Security analysis of industrial test compression schemes," *IEEE TCAD*, 2013.

[15] Model Building and Security Analysis of PUF-Based Authentication, Wenjie Che, *University of New Mexico*, Ph. D thesis, http://digitalrepository.unm.edu/ece_etds/307/.

[16] Rivoir, "Lowering cost of test: parallel test or low-cost ATE?" 12th Test Symposium, 2003. ATS 2003.

[17] N. Satheesh, "A Modified RO-PUF with Improved Security Metrics on FPGA", IEEE International Symposium on Nanoelectronic and Information Systems (iNIS), 2016.

[18] Bushnell and Vishwani Agarwal, "Essentials of Electronic Testing", Springer, 2nd print edition-2002.

[19] Mudit Bhargava, Carnegie Mellon University "Reliable, Secure, Efficient Physical Unclonable Functions" http://repository.cmu.edu/dissertations/238/.

Detection of Sequential Trojans in Embedded System Designs without Scan Chains

Pranav Dharmadhikari*, Akhilesh Raju[†], Ranga Vemuri[‡]

Department of Electrical Engineering and Computer Science
University Of Cincinnati, Cincinnati, OH - 45221
* dharmapt@mail.uc.edu, [†] rajuah@mail.uc.edu, [‡]ranga.vemuri@uc.edu
[†] This work was performed when the author was at University of Cincinnati. Author now at Xilinx Inc.

Abstract—**Small, low-cost embedded systems implemented as sequential circuits often do not contain scan chains. Malicious trojan circuits, which themselves may be state machines, inserted into such systems are hard to detect. We present an effective methodology to detect sequential trojan circuits inserted into sequential hardware designs without scan chains. The methodology consists of three steps: We use sequential testability metrics, model checking and signal tracing to progressively separate the trojan gates from legitimate circuit gates. We show experimental results demonstrating the effectiveness of the methodology using several test cases.**

I. INTRODUCTION

Distributed and networked embedded systems are becoming increasingly popular in numerous application areas ranging from home automation to infrastructure safety. Various IoT (Internet of Things) gadgets are being used to monitor and control appliances which are expected to operate in a safe manner. Digital systems used in these devices are usually low cost, low power systems and yet are expected to operate correctly and securely. One particularly stealthy yet intrusive way of undermining the security of an embedded system is through the insertion of *Hardware Trojans* in the design. A hardware trojan is a malicious circuit inserted into the genuine circuit without the designer's knowledge. Hardware trojans are designed in such a way that they are triggered on a rare condition. A trojan makes use of this property to avoid detection during normal post-fabrication testing. Furthermore, lack of knowledge of the architecture or impact of the hardware trojan, makes detection much more difficult.

A hardware trojan usually consists of two parts: the trigger and the payload. The trigger part of the trojan circuit monitors signals within the chip, waiting for the occurrence of an event or a series of events to take place. The payload of the trojan is responsible for the malicious behavior of the trojan i.e. to transmit private data or to corrupt certain internal signals. When the triggering conditions are met, the trigger circuit enables the payload which then executes its intended function. For Trojans having both payload and trigger, the payload remains inactive most of the time, since the trigger circuit is designed to fire under rare conditions to avoid being detected by routine testing.

A trojan taxonomy is presented in [1] based on physical characteristics, activation characteristics and action characteristics. Further classification is based on abstraction level, location and insertion phase. Trojans can be inserted any time during design process, from 3rd party intellectual property acquisition to fabrication at untrusted facilities. Various countermeasures have been proposed to make trojan insertion hard if not impossible [2]. Filler cells, split manufacturing, logic obfuscation and camouflaging are among these methods. Several methods for detecting trojans have also been proposed including run-time monitoring and side-channel analysis.

In this paper, we address the problem detecting sequential trojan circuits inserted into small-scale digital circuits typical of distributed, embedded IoT devices. Due to severe area and power constraints, such circuits lack the standard DFT (design for testability) features such as scan chains which improve observability and controllability of internal signals. We assume that a gate-level netlist is available (either at end of the design process before fabrication or obtained by delayering and imaging after fabrication) but no golden netlist or device is available. Trojan detection methods proposed based on analyzing such netlists were mostly limited to combinational logic circuits or assume the presence of scan chains. In addition, most proposed methods assume that the trojans themselves are combinational circuits. In contrast, we assume that both the design and the trojan are sequential circuits and no scan chains are included in the design.

This paper is organized as follows: Section II discusses related research. Section III describes the proposed approach to separate the trojan nodes from genuine nodes in a sequential circuit. Section IV presents experimental setup and results. Section V contains concluding remarks.

II. RELATED WORK

Xiao et al. presented a detailed survey of trojan types and detection methods [2]. Some of the detection methods are briefly discussed in this section.

Much research has been done on trojan detection based on side-channel analysis. Power and delay are the most commonly used parameters for side-channel analysis. In one of the early papers on this topic, Agrawal et al. [3] proposed a method to generate a power fingerprint for genuine ICs by considering various types of noise in the circuit. Rad et al. [4] discussed a calibration technique to improve the accuracy of the power analysis techniques by accounting for the adverse effects of process and test environments. Narasimhan et al. [5] used multiple side-channel parameters to achieve high signal-to-noise ratio (SNR) using the intrinsic dependencies between transient supply current (I_{DDT}) and F_{max}. Nowroz et al. [6] and Bao et al. [7] propose methods that use temperature data to identify trojans. Exurville [8] and Kumar et al. [9] use path delay measurements for trojan identification. Lamech et al.

[10] insert additional logic into the design before fabrication to preform accurate path delay measurements while detecting trojans.

Bao et al. proposed a reverse-engineering based hardware trojan detection method based on the application support vector machines (SVM) [11]. Instead of extracting netlists, this approach classifies ICs as trojan-infested or trojan-free based on features extracted from images.

Cakr et al. [12] and Salmani [13] proposed test based methods for detecting trojans. Cakr et al. [12] make use of the weak correlation between trojan signals and genuine circuit signals to identify the trojan nodes. Salmani [13] used controllability and observability metrics for trojan identification.

Controllability and observability of signals are two important attributes related to testability of a design. Controllability of a signal is the ability to force the value of the signal to '0' and '1' by applying suitable input vectors. Observability is the ability to observe the signal state at a primary output. The terms "random" controllability/observability refer to controllability/observability of a signal when random tests are applied. [14]

Combinational logic trojans inserted into combinational logic circuits (ie. both the genuine circuit and the trojan circuit are combinational) can be detected based on the observation that the trojan circuit nodes are likely to be less controllable/observable compared to the genuine circuit nodes since trojan trigger on rare events. An effective method to detect trojans based on this approach is presented by Salmani [13]. Salmani uses the Sandia Controllability and Observability Analysis Program (SCOAP) [15] to compute the following controllability/observability values for each signal s in the netlist: combinational 0-controllability, combinational 1-controllability and combinational observability, designated as $CC0(s)$, $CC1(s)$ and $CO(s)$ respectively. $CC0(s)$ and $CC1(s)$ are combined into a combinational controllability metric $CC(s) = \sqrt{CC1^2(s) + CC0^2(s)}$. Finally, a testability metric is defined as $\sqrt{CC^2(s) + CO^2(s)}$. Signals are clustered based on the values of this metric using a k-means clustering algorithm. All of the trojan nodes form a distinct clusture separated from the genuine nodes. Salami demonstrated the effectiveness of the method for 100% trojan detectability on the TrustHub [16] [17], DeTrust [18] and HaTCh [19] benchmarks.

This method can also be used when the genuine circuit is a sequential circuit and all the flip-flops are assumed to be part of a scan chain. Scan chains ensure the observability and controllability of the genuine flip-flops. In case of full-scan designs, any flip-flop which is not in a scan-chain can be labelled as a trojan node as well.

However, when no scan chains are present in a sequential circuit, combinational testability metrics are inadequate to detect trojans which themselves could be sequential circuits.

III. Proposed Approach

We assume that the genuine circuit is a small embedded system implemented as a sequential circuit with no scan registers. We assume that the trojan also is a sequential circuit

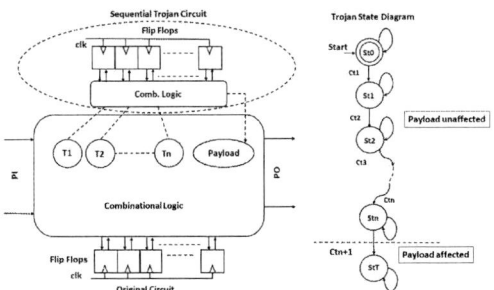

Fig. 1: **Structure of Sequential Circuits and Trojans**

and has a relatively smaller state space compared to the genuine design. Figure 1 shows the general structure of the sequential circuits and trojans assumed [20]. We assume that a netlist is available for analysis. Our approach for trojan detection consists of the following steps.

A. Sequential Testability Analysis

As the first step towards separating the trojan circuit nodes from the genuine circuit nodes, we use an approach similar to [13] but instead of CC and CO metrics we use sequential controllability and observability metrics defined as follows:

- $SC0(s)$ = Minimum number of clock cycles required to set the signal s to logic 0.
- $SC1(s)$ = Minimum number of clock cycles required to set the signal s to logic 1.
- $So(s)$ = Minimum number of clock cycles required to observe the value on s at an output pin.
- $SD(s)$ = Minimum number of clock cycles required to propagate a fault at signal s to an observe point.

These four values are determined by SCOAP [15]. We define two parameters SC and SO as follows:
$$SC(s) = \sqrt{SC0^2(s) + SC1^2(s)}$$
$$SO(s) = \sqrt{So^2(s) + SD^2(s)}$$

We use S0 and SC values of the signals to classify the circuit nodes into *genuine* nodes and *suspect* nodes. Nodes with high values of SC or SO are suspect. Nodes classified as suspect nodes need to be further examined.

For example, consider a sequential circuit containing 108 cells, 32 flip-flops, 118 nets. To this, we added a trojan sequential circuit containing 2 inputs, 55 nets, 4 flip-flops (16 states), and 47 cells. The trojan inputs are two randomly selected internal signals in the genuine circuits. The trojan makes a state transition on a specific condition and stays in each state until the next transition condition occurs. It goes through 15 state transitions in this way and in the last state triggers a payload which induces a stuck-at-1 on an internal signal. Figure 2 shows a SC-SO plot produced using the SCOAP data.

We set a threshold of SO=15 to classify the nodes into suspect (SO $\geq$ 15) and genuine (SO $<$ 15) nodes. (For clarity, we show the trojan and genuine nodes in different shapes and colors since we know which nodes are which for this example.) Clearly, the suspect nodes at this step contain some genuine

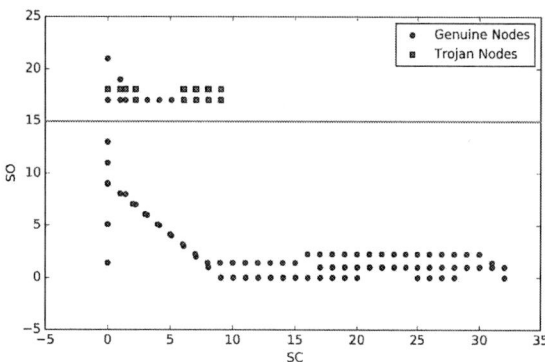

Fig. 2: **Signal Classification After SCOAP Analysis**

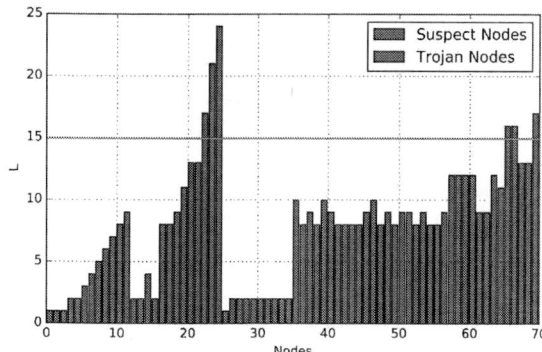

Fig. 3: **Signal Classification After Model Checking**

nodes and some trojan nodes. 70 nodes are labeled suspect and rest are labeled genuine.

B. Model Checking

In the second step, we use a symbolic model checker [21] to determine the lengths of state sequences to excite each of the suspect nodes. For each signal s, we ask the model checker to prove that s never becomes '0'. Model checker produces a counterexample which demonstrates a state sequence (which yields an input sequence) to falsify this claim. We repeat this process for '1'. The length of the longer of these two sequences is designated as the minimum time, in terms of clock cycles, needed to go from the start state to the triggering state if the node s is a suspect node. This length, denoted as $L(s)$, is used as a metric to further separate trojan nodes from suspect nodes. In particular, trojan nodes close to the payload tend to have large L values since trojans usually fire on sequences of rare events.

BDD-based model checking may take a long time when the state-space of the design (plus trojan) is large. We use bounded model checking based on satisfiability (SAT) checking to overcome this limitation. Using a bounded model checker, a bound on the number of clock cycles is set to determine for the search to find a counterexample. If necessary, we increase this limit until a counterexample is found for each case as explained above.

Following the determination of the L values, we use a threshold on L to classify some of suspect nodes as trojan nodes.

For our example, Figure 3 shows the $L(s)$ values for all the 70 suspect signals. Setting a threshold of L−15, we classify 6 nodes as trojan nodes and rest remain as suspect nodes.

C. Signal Tracing and Classification

In the next step, starting with each trojan node identified in the previous step, we trace the fan-in cone of the trojan node and classify the signals in the cone as trojan or genuine. The signal tracing is shown in Algorithm 4.

The algorithm begins by putting all the trojan nodes identified in the previous step into a list (TB) sorted in reverse

topological order. For each node t in TB, the fanin nodes f of t are examined (fanin nodes are the inputs of the cell whose output is t). If f is a suspect node and has a genuine node in its fanout (ie. it is used to produce a signal classified as genuine) then f itself is classified as genuine. Otherwise, f is classified a trojan node and is also added to TB. Note that if f was already classified as genuine or trojan then it remains so.

At this point, a few suspect nodes may still remain unclassified. We consider each such node s. If s has a fanout node which is genuine, then s itself is classified as genuine. Otherwise, if all fanin nodes of f are trojan nodes then s is classified as a trojan node. Otherwise, it is classified as a genuine node. At the end of the algorithm, G contains all genuine nodes, T contains all trojan nodes and S will be empty.

Figure 5 shows the final classification of the suspect nodes into genuine or trojan nodes after signal tracing. For our example, all 55 trojan nodes are correctly classified. The 15 suspect nodes classified as genuine in this step (shown in green in the figure) along with the 103 nodes classified as genuine after SCOAP analysis are the correct 118 genuine nodes.

Thresholds for the SC-SO based classification step and the L threshold for the model checking step are user-defined.

IV. EXPERIMENTAL RESULTS

A. Experimental Setup

We use 6 test cases to evaluate the proposed methodology. Table I shows the number of nets, cells and FFs in each benchmarks. For each test case we inserted a trojan circuit whose features (inputs, states, nets, cells and FFs) are also shown in the table. Trojan have the general structure shown in Figure 1. We synthesize both the genuine circuit and trojan separately from respective VHDL source files using the Synopsys Design Compiler. We then insert the trojan netlist into the genuine circuit netlist at randomly chosen internal signals as the trigger and payload locations.

We use SCOAP data generated by the Synopsys Tetramax tool in the first step and the bounded model checking feature in the NuSMV model checker which is based on the MiniSAT satisfiability solver [22] in the second step. We used a DELL Precision M4800 with CPU@2.7GHz for all experiments.

978-1-5386-7100-9/18 $31.00 © 2018 IEEE   680

```
1  G: List of genuine nodes obtained after SCOAP analysis
2  T: List of trojan nodes obtained after model checking
3  S: List of remaining suspect nodes after model checking
4  TB = T;
5  while TB ≠ φ do
6  │  t = a node from TB;
7  │  TB = TB - {t};
8  │  F = fanin nodes of t;
9  │  for f ∈ F do
10 │  │  if f ∈ S then
11 │  │  │  S = S - {f};
12 │  │  │  if f has a fanout node in G then
13 │  │  │  │  G = G ∪ {f};
14 │  │  │  else
15 │  │  │  │  T = T ∪ {f};
16 │  │  │  │  TB = TB ∪ {f};
17 │  │  │  end
18 │  │  end
19 │  end
20 end
21 end
22 while s ≠ φ do
23 │  s = a node from S;
24 │  S = S - {s};
25 │  if s has fanout node in G then
26 │  │  G = G ∪ {s};
27 │  else
28 │  │  if all fanin nodes of s are in T then
29 │  │  │  T = T ∪ {s};
30 │  │  else
31 │  │  │  G = G ∪ {s};
32 │  │  end
33 │  end
34 │  end
35 end
36 end
37            Fig. 4: Signal Tracing Algorithm
```

TABLE I: **Benchmark Data**

Name	Nets	Cells	FFs	Trojan Inputs	Trojan States	Trojan Nets	Trojan Cells	Trojan FFs
B1	95	87	19	2	4	14	12	2
B2	147	137	35	2	8	28	26	3
B3	226	210	52	2	16	53	47	4
B4	317	298	69	2	32	85	78	5
B5	370	349	85	2	32	85	78	5
B6	489	465	102	3	64	151	142	6

B. Results

Table II shows the results of classification obtained by our approach for the 6 test cases. Columns 2-3 show, respectively, the SO-SC threshold and the number of suspect nodes after this threshold setting following the SCOAP data analysis. Columns 4-6 show the bound used for bounder model checking in NuSMV, L threshold used and the number of trojan nodes determined with that threshold after model checking. Column

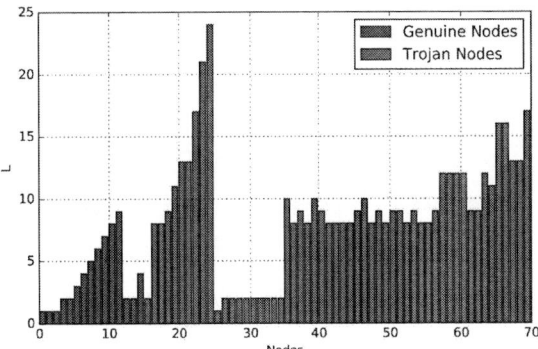

Fig. 5: **Signal Classification After Tracing**

7 shows the number of nodes classified as trojan nodes after signal tracing with the remaining nodes classified as genuine nodes. Column 8 shows the number of genuine nodes mistakenly classified as trojan nodes (false positives) and Column 9 shows the number of trojan nodes mistakenly classified as genuine nodes (false negatives). The last column shows the total time taken by NuSMV to produce all the counterexamples using bounded model checking. SCOAP data analysis and signal tracing steps take less than a few seconds for all the examples.

TABLE II: **Experimental Results**

Name	SO Th.	Suspect Nodes(1)	BMC Bound	L Th.	Troj. (2)	Troj. (3)	False Pos.	False Neg.	NuSMV Time(s)
B1	5	14	50	10	6	13	0	1	0.12
B2	10	45	100	10	4	28	0	0	0.53
B3	20	82	150	25	9	56	3	0	7.18
B4	20	112	200	60	14	84	0	1	26.92
B5	50	112	250	60	13	89	5	1	4,476
B6	60	187	300	120	3	151	0	0	13,211

Columns 7-9 indicate that our approach produces highly accurate results with very few, if any, false classifications. It should be mentioned that currently both SO and L thresholds and BMC bound are set by the user for each test case after examining the relevant plots.

Figure 6 shows the results after each of the three steps in our methodology for the 6 test cases. Each row consists of three plots for one test case and the 6 rows correspond to B1 through B6 test cases. SO and L thresholds are indicated on the respective plots.

V. CONCLUSION

In this paper, we have presented a methodology to identify sequential trojan circuits in small, low-cost embedded systems also implemented as sequential circuits without scan chains. Our method uses sequential testability, bounded model checking and signal tracing to progressively classify the circuit nodes into genuine and trojan nodes. The process is largely automated and shows high level of accuracy on several test cases.

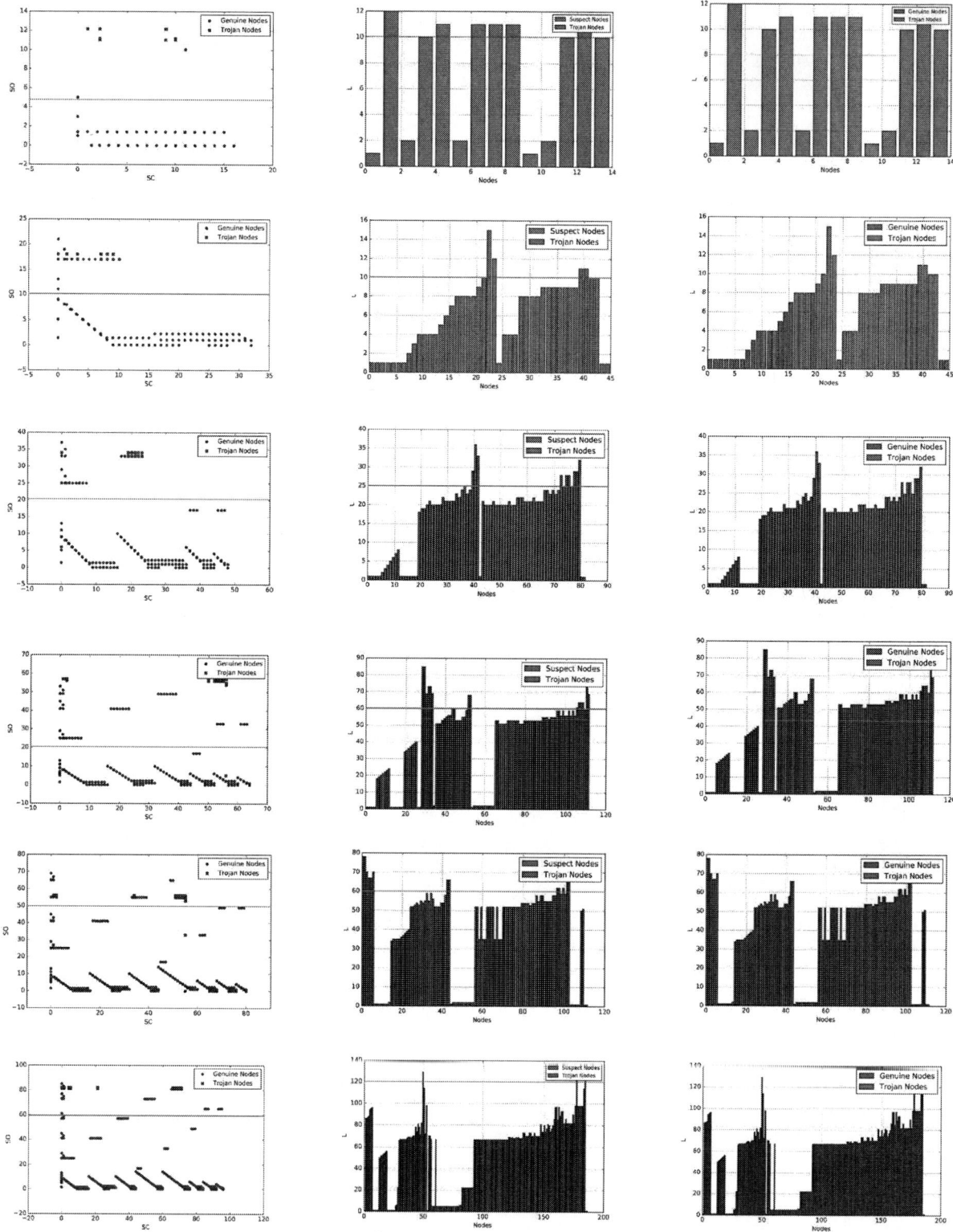

Fig. 6: **Result Plots for the Test Cases**

REFERENCES

[1] Xiaoxiao Wang, Mohammad Tehranipoor, and Jim Plusquellic. "Detecting Malicious Inclusions in Secure Hardware: Challenges and Solutions". In: *Proceedings of the 2008 IEEE International Workshop on Hardware-Oriented Security and Trust*. HOST '08. Washington, DC, USA: IEEE Computer Society, 2008, pp. 15–19. ISBN: 978-1-4244-2401-6. DOI: 10.1109/HST.2008.4559039. URL: http://dx.doi.org/10.1109/HST.2008.4559039.

[2] K. Xiao et al. "Hardware Trojans: Lessons Learned After One Decade of Research". In: *ACM Trans. Des. Autom. Electron. Syst.* 22.1 (May 2016), 6:1–6:23. ISSN: 1084-4309. DOI: 10.1145/2906147. URL: http://doi.acm.org/10.1145/2906147.

[3] D. Agrawal et al. "Trojan Detection using IC Fingerprinting". In: *2007 IEEE Symposium on Security and Privacy (SP '07)*. 2007, pp. 296–310. DOI: 10.1109/SP.2007.36.

[4] R. M. Rad et al. "Power supply signal calibration techniques for improving detection resolution to hardware Trojans". In: *2008 IEEE/ACM International Conference on Computer-Aided Design*. 2008, pp. 632–639. DOI: 10.1109/ICCAD.2008.4681643.

[5] S. Narasimhan et al. "Hardware Trojan Detection by Multiple-Parameter Side-Channel Analysis". In: *IEEE Transactions on Computers* 62.11 (2013), pp. 2183–2195. ISSN: 0018-9340. DOI: 10.1109/TC.2012.200.

[6] A. N. Nowroz et al. "Novel Techniques for High-Sensitivity Hardware Trojan Detection Using Thermal and Power Maps". In: *IEEE Transactions on Computer-Aided Design of Integrated Circuits and Systems* 33.12 (2014), pp. 1792–1805. ISSN: 0278-0070. DOI: 10.1109/TCAD.2014.2354293.

[7] C. Bao, D. Forte, and A. Srivastava. "Temperature Tracking: Toward Robust Run-Time Detection of Hardware Trojans". In: *IEEE Transactions on Computer-Aided Design of Integrated Circuits and Systems* 34.10 (2015), pp. 1577–1585. ISSN: 0278-0070. DOI: 10.1109/TCAD.2015.2424929.

[8] I. Exurville et al. "Resilient hardware Trojans detection based on path delay measurements". In: *2015 IEEE International Symposium on Hardware Oriented Security and Trust (HOST)*. 2015, pp. 151–156. DOI: 10.1109/HST.2015.7140254.

[9] P. Kumar and R. Srinivasan. "Detection of hardware Trojan in SEA using path delay". In: *Electrical, Electronics and Computer Science (SCEECS), 2014 IEEE Students' Conference on*. 2014, pp. 1–6. DOI: 10.1109/SCEECS.2014.6804444.

[10] C. Lamech et al. "REBEL and TDC: Two embedded test structures for on-chip measurements of within-die path delay variations". In: *2011 IEEE/ACM International Conference on Computer-Aided Design (ICCAD)*. 2011, pp. 170–177. DOI: 10.1109/ICCAD.2011.6105322.

[11] C. Bao, D. Forte, and A. Srivastava. "On application of one-class SVM to reverse engineering-based hardware Trojan detection". In: *Fifteenth International Symposium on Quality Electronic Design*. 2014, pp. 47–54. DOI: 10.1109/ISQED.2014.6783305.

[12] B. Cakfffdfffdr and S. Malik. "Hardware Trojan detection for gate-level ICs using signal correlation based clustering". In: *2015 Design, Automation Test in Europe Conference Exhibition (DATE)*. 2015, pp. 471–476.

[13] H. Salmani. "COTD: Reference-Free Hardware Trojan Detection and Recovery Based on Controllability and Observability in Gate-Level Netlist". In: *IEEE Transactions on Information Forensics and Security* 12.2 (2017), pp. 338–350. ISSN: 1556-6013. DOI: 10.1109/TIFS.2016.2613842.

[14] M. Bushnell and Vishwani Agrawal. *Essentials of Electronic Testing for Digital, Memory and Mixed-Signal VLSI Circuits*. Springer Publishing Company, Incorporated, 2013. ISBN: 978-1-475-78142-7.

[15] L. H. Goldstein and E. L. Thigpen. "SCOAP: Sandia Controllability/Observability Analysis Program". In: *17th Design Automation Conference*. 1980, pp. 190–196. DOI: 10.1109/DAC.1980.1585245.

[16] H. Salmani, M. Tehranipoor, and R. Karri. "On design vulnerability analysis and trust benchmarks development". In: *2013 IEEE 31st International Conference on Computer Design (ICCD)*. 2013, pp. 471–474. DOI: 10.1109/ICCD.2013.6657085.

[17] Bicky Shakya et al. "Benchmarking of Hardware Trojans and Maliciously Affected Circuits". In: *Journal of Hardware and Systems Security* (2017), pp. 1–18. ISSN: 2509-3436. DOI: 10.1007/s41635-017-0001-6. URL: http://dx.doi.org/10.1007/s41635-017-0001-6.

[18] Jie Zhang, Feng Yuan, and Qiang Xu. "DeTrust: Defeating Hardware Trust Verification with Stealthy Implicitly-Triggered Hardware Trojans". In: *Proceedings of the 2014 ACM SIGSAC Conference on Computer and Communications Security*. CCS '14. Scottsdale, Arizona, USA: ACM, 2014, pp. 153–166. ISBN: 978-1-4503-2957-6. DOI: 10.1145/2660267.2660289. URL: http://doi.acm.org/10.1145/2660267.2660289.

[19] Syed Kamran Haider et al. "HaTCh: A Formal Framework of Hardware Trojan Design and Detection". In: 2015.

[20] X. Wang et al. "Sequential hardware Trojan: Side-channel aware design and placement". In: *2011 IEEE 29th International Conference on Computer Design (ICCD)*. 2011, pp. 297–300. DOI: 10.1109/ICCD.2011.6081413.

[21] E.M. Clarke, O. Grumberg, and D. Peled. *Model Checking*. MIT Press, 1999. ISBN: 9780262032704. URL: https://books.google.com/books?id=Nmc4wEaLXFEC.

[22] Roberto Cavada et al. *NuSMV 2.6 USer Guide*. URL: http://nusmv.fbk.eu/NuSMV/userman/index-v2.html.

Designing for Security Within and Between IoT Devices

Mike Borowczak, Rafer Cooley and Shaya Wolf
Dept. of Computer Science
University of Wyoming
Laramie, WY 80271
mike.borowczak@uwyo.edu, rcooley2@uwyo.edu, & swolf4@uwyo.edu

Abstract—**In this work, we propose utilizing a design-for-security approach to enable system architects and designers with flexibility and control during the initial phases of the design process. General solutions for system security, especially in the communication domain, involves rigid application of prior fundamental constructs, from secure standard cell libraries to common communication encryption/decryption schemes. We propose two methods, one for intra-device communication and another from inter-device communication, for lightweight devices. These methods are tunable and adaptable by designers based on their unique situations and rely on fundamental properties of the underlying systems rather than arbitrarily applied constructs that have been used to enable security in the past.**

I. INTRODUCTION & CONTEXT

The development of modern complex systems relies on a balancing act of design trade-offs in order to achieve specific end-design targets. In general, these trade-offs have focus on the power, speed, and area. Forward thinking designers have also factored in robustness and even security into their trade-off mix. Unfortunately, most security and robustness trade-offs are the un-scalable, binary inclusion or exclusion of specific cryptographic algorithms, sub-components, or secure logic cells within a system. The ability to architect solutions is akin to a paint-by-numbers approach to system design - in this work we propose two high-level methods that enable "designing for security" in the context of lightweight communication endpoints (e.g. Internet of Things (IoT), biomedical devices, sensor networks, etc). The proposed methods enable an architect or designer to balance security requirements within a device and between devices at high-levels of abstraction, and allow the security requirements to be passed to lower-levels without compromise.

While most people interact with larger complex electronic devices, the bulk of today's devices are small, lightweight devices with limited range, complexity, processing power, and function. These interconnected devices exist in everything from remote sensing application in pipelines and remote fire-hazard areas to the sensors maintaining feedback for earthquake mitigation systems in sky-rises. As technology advances, and these lightweight devices become smaller and/or more powerful, their application space grows to internal distributed health monitoring and autonomous unmanned vehicle micro-swarms. Distributed systems of lightweight devices

poses an extreme case of design trade-offs and motivates the need for approaches to design for security.

II. DESIGNING FOR SECURITY

The need for introducing security as a tunable parameter at all levels of design is fundamental in securing our future infrastructure, health, and economy. As designers, however, the choice of security inclusion has not fit within the normal trade-off spectrum - a binary choice or mandate to include bloated security constructs in lightweight distributed systems both impractical and infeasible. Given the limited space here, two solutions at both ends of the same spectrum are described. Imagine the following requirement, given a new off-grid IoT device, built using a existing System on Chip platform, insure all communication is secured. This requires that both the data communicated within the device is secure (maintain integrity), and machine to machine (M2M) communication to be secured (maintain confidentiality). Historically, a designers choice? Encrypt and decrypt everything with varying cryptographic strength, or not. The design decisions are hard to trade-off, iterate over, and most importantly scale. The following sections describe two high-level approaches to 1) securing internal device communications and 2) securing device to device communication while still allowing for other device parameter trade-offs.

A. High-Level Synthesis of Secure Bus Communication

Given an System of Chip architecture with both shared and paired bus architectures, the problem of creating side-channel secure communication on any bus is reduced to implementing a zero-information entropy communication protocol. Information entropy is defined in Equation 1.

$$E(A) = H(A) = -\sum_{a \in A} p(a) log_2 p(a) \tag{1}$$

Stated in another way, the side channel security of a bus can be approximated and measured, in the worst-case, by the amount of information entropy contained on the bus. The following describes a high-level approach to eliminating bus-level information entropy, or a mechanism to minimize it given other design constraints and trade-offs.

First, assume that the communication protocol is described as a Finite State Machine (FSM) and following quintuple $(\Sigma, S, s_0, \delta, F)$ where:

- Σ is the finite, non-empty, set of symbols.
- S is the finite, non-empty, state space.
- $s_0 \in S$ is the initial state.
- $\delta : S \times \Sigma \to S$ is the state-transition function.
- F is the set of final states.

For the purpose of high-level security implementation, we assume that our communication protocol sends information on the bus based on the encoding of its state. In other words, the data on the bus at any point in time can be represented as the state encoding at that same point in time. Side channel attacks focus on two predominate information leakage surfaces - the first based on static power consumption seeks to detect difference in the number of active bits (HIGH) on a bus, the second is based on dynamic power consumption over time to detect the number of bits that are switching (HIGH to LOW and LOW to HIGH) on a bus. Given the set of states S the objective is to minimize the information entropy between all transitions δ in our communication FSM.

In order to evaluate the relative security of FSMs, given various encodings, the entropy (eq. 1) of the power side channel (P) and the attack models was computed for each uniquely encoded FSM. Additionally, the Mutual Information (eq. 2) can be computed for each side channel and attack model pairing for all encoding combinations. It is interesting to note that for completeness, the secured encoding with zero entropy has maximum MI = 0 (eq. 3).

$$
\begin{aligned}
I(A;B) &\equiv I[p(A,B)] \\
&\equiv \sum_{a,b} p(a,b) log_2 \left(\frac{p(a,b)}{\sum_b p(a,b) \sum_a p(a,b)} \right) \quad (2)
\end{aligned}
$$

$$
\begin{aligned}
I(A,B)|_{\forall_{b \in B} b = \bar{b}} &= = \sum_{a,b} p(a,b) log_2 \left(\frac{p(a,b)}{\sum_b p(a,b) \sum_a p(a,b)} \right) \\
&= \sum_{a,b} p(a,b) log_2(1) = 0
\end{aligned}
$$
$$(3)$$

The easiest mechanism to secure a FSM is thus to hold the number of bits on the bus and the number of bits on the bus switching constant. These map directly to the properties of Hamming Weight (HIGH Bit on bus) and Hamming Distance (Bit's switching on bus). Equation II-A enforces a strict requirement that communication FSMs must have the same constant Hamming Weight. Equation II-A, requires that transitions between connected states (s and s') must have the same Hamming Distance. This second property has the effect of creating uniform HD between all interconnected states within.

$$ \forall s \in S : HW(s) = c_1 \quad (4) $$

$$ \forall s, s' \in S \mid \exists \alpha \in \Sigma : s' = \delta(s, \alpha) \to HD(s, s') = c_2 \quad (5) $$

Constraining the encoding space for a communication protocol comes with significant potential penalty. In the worst-case, a one-hot encoding will always satisfy these two conditions, the number of bits on would always be 1 and the number of bits transitioning would always be 2. A 1 to 1 mapping of messages to states needed is clearly unacceptable, so work in creating optimal encodings has already been proposed This prior work several useful encoding length and HIGH bit count properties to secure a FSM with S states and T transitions. Given two equations, Eqs. and , that define secure communication between SoC endpoints is clearly more flexible and scalable than prescribed all-or-nothing approaches. Consider the designer that wishes to trade-off security for power consumption.

The secure encoding strategies highlighted can easily be relax and tighten in order to provide varying levels of protection that approach traditional low power encoding methods.

While the secure state encodings are a new design parameter, work has already been done in the area of low-power/power aware state encodings [1], [2], [3], [4], [5], [6]. Most solutions focus on the minimization of switching activity (by reducing Hamming Distance between connected states) using a variety of techniques ranging from genetic local search, to integer linear programming, to SAT based algorithms. Using the prior model we can target peak current minimization (N_{peak}) as a new design objective [5], [6]. restricts the original HW and HD/HDR constraints to minimize the peak power consumed in order to satisfy low-power design objectives. Following previously established work [5], [6] this translates to minimizing the maximum number of $0 \to 1$ and $1 \to 0$ transitions (N_{peak}). Equations 7 and 8, formally define N_{peak} with unique and identical technology dependent weighting factors w_1, w_2 respectively. While previous works were also concerned with switching activity [6] (Eq. 9), the current discussion relaxes this constraint in favor of our core security objectives. Minimization of N_{peak} is accomplished by forcing additional restrictions on HD/HDR - mainly minimizing c_2 in equation 6 respectively.

$$ HDR = HD(s_i, s_j) = c_2 \ \forall i,j | i \neq j \quad (6) $$

$$ N_{peak} = max\{w_1 \cdot N_{0 \to 1}, w_2 \cdot N_{1 \to 0}\} \quad (7) $$

$$ N_{peak} = max\{N_{0 \to 1}, N_{1 \to 0}\} \quad (8) $$

$$ SW_{tot} = \sum_{s_i \to s_j} HD(s_i, s_j) \cdot p_{i,j} \quad (9) $$

If the primary HW and HD constraints are maintained, then assuming two states, $s1, s2$, each $0 \to 1$ or $1 \to 0$ in $s1$ forces a respective $1 \to 0$ or $0 \to 1$ transition in $s2$. Thus $N_{peak\_secure} = \frac{c_2}{2}$ as seen in Eq. 10. When guaranteeing either the HD or HDR constraint, N_{peak_secure} is equal to half of the HD between any two connected states. Minimization of the HD/HDR constraint constant c_2 will also minimize

N_{peak_secure}. For example, a one-hot encoding will have a $c_2 = 2$ and an $N_{peak\_secure} = 1$ - the lower bound of N_{peak}. While not formally expressed, similar constraint restrictions can be placed on the HW constant c_1 in order to mitigate static power consumption.

$$N_{peak\_secure} = N_{0\rightarrow1} = N_{1\rightarrow0} = \frac{c_2}{2} \qquad (10)$$

This section has illustrated one approach to designing for security, namely a high-level approach to securing communication from side channel power attacks by focusing on and modeling information entropy of the target. Additionally, mechanisms to relax (or further constrain) underlying system properties has also been illustrated. The following section highlights an approach, again using a simple rule set, to define secure communication between entities without centralized control.

B. Secure Communication Protocol

While the synthesis of a secure bus may seem reasonable, the current solution space that architects and designers have to secure M2M (or peer-to-peer) communication is currently limited to encrypted communication and/or centralized control, both which on their own would require significant overhead. Additionally, encryption-based security is again a binary design trade-off that does not allow for easy resource trade-off and simulation. In this scenario assume that a set of n ultra-lightweight devices must communicate pairwise, in n communication rounds. Each complete set of n communication rounds forms a communication cycle. If the devices are to communicate securely, an architect or designer is generally limited to encryption strategies that require complete knowledge of the system or complex key-sharing processes.

While the synthesis of a secure bus may seem reasonable, the current solution space that architects and designers have to secure M2M (or peer-to-peer) communication is currently limited to encrypted communication and/or centralized control, both which on their own would require significant overhead. Additionally, encryption-based security is again a binary design trade-off that does not allow for easy resource trade-off and simulation. In this scenario assume that a set of n ultra-lightweight devices must communicate pairwise, in n communication rounds. Each complete set of n communication rounds forms a communication cycle.

If the devices are to communicate securely, an architect or designer is generally limited to encryption strategies that require complete knowledge of the system or complex key-sharing processes. A designer would be likely to leverage the significant amount of work has already been proposed due to the predicted exponential growth of IoT devices. To constrain the solution space this approach focuses on a design for security approach to several open areas within the IoT space, namely *Mobility Support* and *Authentication*.[7].

Within the first domain of mobility support, most of the present day work focuses on enabling existing addressing protocols (IPv4 or IPv6) within a variety of solutions including

RFID [8], [9], [10] 6LoWPAN [11], and ROLL [12]. The biggest issues with protocol reuse is the cost required to create *adaptable heterogeneous networks* - the addition, replacement, loss or removal of IoT nodes becomes increasingly complex and requires centralized monitoring and control, as well as per-agent trust. Node to node communication is generally not feasible without storage of complete network topology knowledge.

Imagine a set of communicating devices represented as a graph, with edges between nodes labeled with a value for the communication round in which they were to communicate. Clearly, the objective of complete communication is equivalent to creating a completely connected graph through edge selection between pairwise nodes. While a mapping could be defined and stored within each communicating device, this fails to scale or handle addition or removal of devices. Thus assume the object is to identify a formulaic approach to determining communicating pairs of nodes. For simplicity assume that in any communication, there is a source and target node, each should compute the other as their communication partner during exactly one round per communication cycle.

In order to determine the index of a source node's "target node pair" ($target_{idx}$) the following must known by the source node:

Node Count ($Node_{cnt}$) - The maximum possible number of nodes within the system: $Node_{cnt} \geq n$,

Source Index ($self_{idx}$) - The source node's own index: $idx = [0, n)$

Current Round (r_i) - The current round of communication: $r = [0, n)$

With these three parameters - any node can be paired with any other node using simple modular arithmetic during any round using equation 11. Intuitively, a node's target pair is computed by subtracting its index from the current round while insuring that the result remains within the set of allowable nodes (mod $Node_{cnt}$). Imagine the current round as a mechanism to ensure that a node can sweep through the nodes at fixed distances from itself.

$$self_{Tidx} = (r_i - self_{Sidx}) \bmod Node_{cnt} \qquad (11)$$

Again, given this formulaic solution to communication, called CHIRP [13], a designer can begin modification to fit within other design constraints. Measures of importance could include communication efficiency or likelihood of communication cycle pattern discovery.

The simplest way to enhance communication within CHIRP at scale is to to introduce a permutation in the rounds ordering. While this mechanism is insufficient at low node counts there are solutions that could be utilized in small (≤ 8 node) networks - specifically per-node keys. Table I shows the rapid increase in potential round permutations given increasing node counts.

Consider a standard cycle in CHIRP algorithm as a monotonically increasing set of round indices from zero to the maximum number of nodes minus 1 as show in equation 12.

TABLE I
THE NUMBER OF POSSIBLE ROUND PERMUTATIONS FOR A GIVEN NUMBER
OF NODES (AND THUS ROUNDS).

Nodes / Rounds	Possible Round Permutations
8	4.03×10^5
10	3.63×10^6
16	2.09×10^{13}
32	2.63×10^{35}
64	1.26×10^{89}
128	3.85×10^{215}

$$C = [0, 1, 2, \cdots, \text{Node}_{cnt} - 2, \text{Node}_{cnt} - 1] \quad (12)$$

A unique permutation of these Node_{cnt} elements within cycle C results in a shuffling of communication between the nodes in a system that requires immense resources to track, detect and exploit. A designer or architect now has a solution space to iterate and sweep over, rather than having to select from a fixed set of options.

III. EXPERIMENTAL RESULTS & ANALYSIS

A. High-Level Synthesis of Secure Bus Communication

In order to quantify the impact of high-level methods to secure FSM-based communication protocols, the effectiveness and cost of implementing secured FSMs utilizes a collection of over 150 FSM benchmarks, generated and acquired from the authors of BenGen [14] and MCNS. These range in size from 4 to 60 states, with total transition ranging from 8 to 216. The the original benchmarks as well as various secured versions are converted to Verilog where they can be synthesized at the gate-level using DC Compiler. The resulting gate-level netlist is converter to Spice in order to gain cycle accurate power information using Nanosim. The Nanosim data in conjunction with, worst-case, FSM Oracle data is used to measure the correlation between the Hamming models and the current FSM state or transition between states.

Figure 1 depicts the low level FSM synthesis flow used to generate the correlation data r_{Spower}. The bold solid path represents the complete secured FSM path, while the relaxed dashed lines represent native FSMs. Finally, the compressed dashed-lines represent FSM specific information used in conjunction with oracle data to determine attack best-case correlations.

In order to demonstrate the effectiveness of the secure encodings, the number of bits needed for the secure encoding, along with the increase over the original FSMs binary encoding are presented in Table III-A. Note that on average the increase is around 79% though for all but the smallest FSM the average is a 67% increase.

The MCNC benchmarks provide a more accurate baseline for bit length encoding impact in real FSMs (See Table III. The difference between benchmark suites is less noticeable simply due to the logarithmic nature of bit encoding requirements.

Three power-constrained versions of Secure FSMs were compared in order to evaluate the burden on the encoding

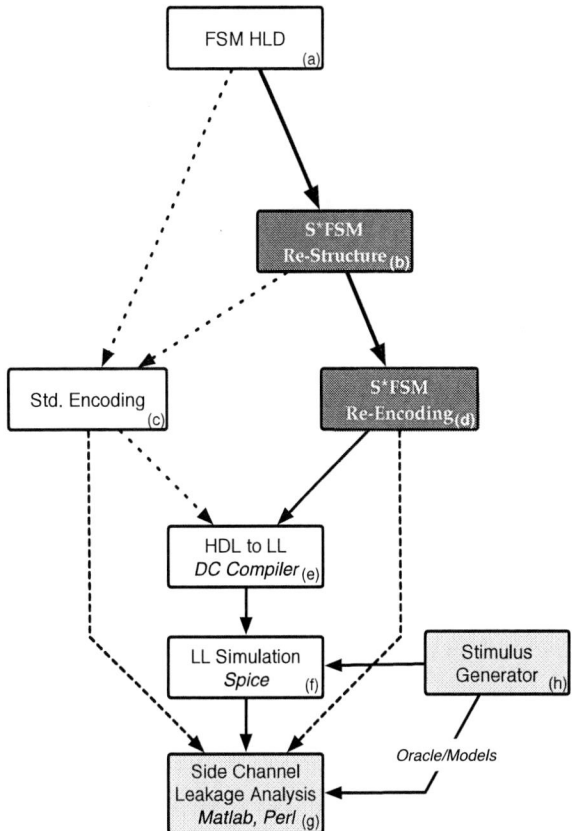

Fig. 1. FSM flow to test and verify theoretical results using gate-level realization of FSMs and Secure FSMs using multiple encodings[15]

TABLE II
BITS NEEDED FOR BINARY (BE) AND SECURE ENCODINGS. S*FSM
REPRESENT STRUCTURALLY MODIFIED VERSION OF THE FSM.

FSM	Bits Needed			
	BE Orig	BE S*FSM	Secure S*FSM	Overall % Increase
137	2	3	5	150
94	3	3	5	66.7
86	3	3	5	66.7
108	6	7	10	66.7
60	6	7	10	66.7
147	6	7	10	66.7
146	6	7	10	66.7
avg	4.6	5.3	7.9	78.6

requirement. While further constraining the HD to be fixed at 2 bits (CHD=2) more than doubles the required bit-length by requiring on average a 112% increase in the number of bits, it does reduces N_{peak} to its minimum - 1. Further constraining the encoding by imposing a constrained HW (CHW=1), forces the FSM to be encoded using a one-hot encoding that significantly impacts the encoding bit-length by increasing it almost 400%.

TABLE III
BITS NEEDED FOR BINARY AND SECURE ENCODINGS OF SEVERAL FSMs.

FSM	Bits Needed				
	BE Orig	BE S*FSM	% Increase	Secure S*FSM	Overall % Increase
modulo12	4	5	25	7	75
opus	4	5	25	7	75
bbara	4	5	25	8	100
bbsse	4	5	25	8	100
sse	4	5	25	8	100
sand	5	6	20	8	60
planet	6	6	0	8	33
ex1	5	6	20	9	80
styr	5	6	20	9	80
scf	7	7	0	9	29
avg	4.8	5.6	19	8.1	73

A relaxed HD encoding (O.RHD) and relaxed HD with HW encoding (O.RHD+HW) were both applied to original FSMs. Since the O.RHD allows for variable HW it is safely considered the weakest of the available modifications though it still requires an 30% average increase in encoding length over binary encoding (O.binary). When adding the constant HW requirement, the average increase is nearly 55% over O.Binary.

B. Secure Communication Protocol

In the interest of space, figure 2 shows the application of the CHRIP strategy to an 8 node network. Note that in this baseline version, no permutations of communication exist, as such the ability to detect the pattern of communication would be trivial, but the complexity of implementation would be equally trivial, requiring only three pieces of information per node.

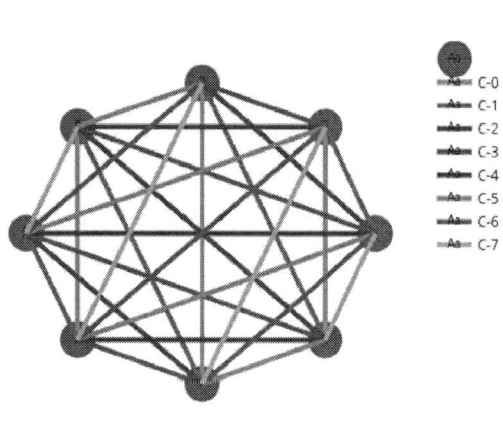

8 Nodes

Fig. 2. The eight rounds needed to create the 28 edges $[e = \frac{7*8}{2}]$ for a fully connected eight-node graph.

In order to mitigate the ability to determine the communication pattern, a cycle-based permutation based on Algorithm 1 is utilized. The algorithm does a per-round mapping of nodes such that both sources and targets are paired within the same round, but there is no distinct linear sweep through the offset between pairs. See Table IV for an example of the per-round communication for a permuted 8 round cycle $Pr(C) = [7, 5, 2, 0, 4, 6, 1, 3]$.

Algorithm 1 CHRIP - Peer to Peer Mapping based on Round

1: $Pr(C) \Leftarrow Permutation([0, 1, \cdots, Node_{cnt} - 1]$
2: **for** r_i such that $0 \leq i < n$ **do**
3:     **for** $source_{Sidx}$ such that $0 \leq Sidx < n$ **do**
4:         $source_{Tidx} \leftarrow (Pr(C)[r_i] - Source_{idx})$ mod $Node_{cnt}$
5:     **end for**
6: **end for**

TABLE IV
A PERMUTATION OF AN 8-ROUND CYCLE YIELDS NO DISCERNIBLE "SWEEPS" THROUGH NODE OFFSETS - GENERALLY RECOGNIZED AS DIAGONALS WITHIN THE MATRIX.

		Nodes							
Ri	Pr(C)	0	1	2	3	4	5	6	7
0	7	6	5	4	-	2	1	0	-
1	5	4	3	-	1	0	7	-	5
2	2	1	0	7	6	5	4	3	2
3	0	7	6	5	4	3	2	1	0
4	4	3	2	1	0	7	6	5	4
5	6	5	4	3	2	1	0	7	6
6	1	-	7	6	5	-	3	2	1
7	3	2	-	0	7	6	-	4	3

IV. CONCLUSIONS & FUTURE WORK

This work highlights the flexibility of formulating and structuring hardware and system security within a parameterized design space that goes beyond traditional cryptographic methods. In the first section we described the ability to secure packets of information traveling on a common bus architecture by enabling and controlling high-level properties of Finite State Machine-based communication protocols. Constraining the Hamming Weight and Hamming Distance of the bus enables tuning for security and power (among other features). In the second portion, we briefly discuss the ability to model intra-node communication as a graph in which the objective is to fully connect the nodes in a pairwise fashion. The trivial strategy for connection of nodes was defined, and then extended to introduce mechanisms for tunable security. Future work includes multi-agent simulation and targeted domain applications using a variety of communication mediums and existing technologies.

ACKNOWLEDGMENT

This research was partially supported by an National Security Agency Gen-Cyber grant to the University of Wyoming (H98230-18-1-0095).

References

[1] E. Olson and S. Kang, "State assignment for low-power FSM synthesis using genetic local search," in *Custom Integrated Circuits Conference, 1994., Proceedings of the IEEE 1994*, may 1994, pp. 140 –143.

[2] F. Gao and J. Hayes, "ILP-based optimization of sequential circuits for low power," in *Low Power Electronics and Design, 2003. ISLPED '03. Proceedings of the 2003 International Symposium on*, aug. 2003, pp. 140 – 145.

[3] C. Cao and B. Oelmann, "Mixed synchronous/asynchronous state memory for low power FSM design," in *Digital System Design, 2004. DSD 2004. Euromicro Symposium on*, aug.-3 sept. 2004, pp. 363 – 370.

[4] L. Yuan, G. Qu, T. Villa, and A. Sangiovanni-Vincentelli, "Fsm re-engineering and its application in low power state encoding," in *Design Automation Conference, 2005. Proceedings of the ASP-DAC 2005. Asia and South Pacific*, vol. 1, jan. 2005, pp. 254 – 259 Vol. 1.

[5] S.-H. Huang, C.-M. Chang, and Y.-T. Nieh, "State Re-Encoding for Peak Current Minimization," in *Computer-Aided Design, 2006. ICCAD '06. IEEE/ACM International Conference on*, nov. 2006, pp. 33 –38.

[6] Y. Lee and T. Kim, "State encoding algorithm for peak current minimisation," *Computers Digital Techniques, IET*, vol. 5, no. 2, pp. 113 –122, march 2011.

[7] L. Atzori, A. Iera, and G. Morabito, "The internet of things: A survey," *Computer Networks*, vol. 54, no. 15, pp. 2787 – 2805, 2010. [Online]. Available: http://www.sciencedirect.com/science/article/pii/S1389128610001568

[8] S. E. H. Jensen and R. H. Jacobsen, "Access control with rfid in the internet of things," in *2013 27th International Conference on Advanced Information Networking and Applications Workshops*, March 2013, pp. 554–559.

[9] D. G. Yoon, D. H. Lee, C. H. Seo, and S. G. Choi, "Rfid networking mechanism using address management agent," in *2008 Fourth International Conference on Networked Computing and Advanced Information Management*, vol. 1, Sept 2008, pp. 617–622.

[10] U. C. abuk, R. H. Jacobsen, and G. Dalkl, "Artmos: An rfid tag mobility scheme with ipv6 and dns," in *2016 International Symposium on INnovations in Intelligent SysTems and Applications (INISTA)*, Aug 2016, pp. 1–5.

[11] Y. Qiu and M. Ma, "A secure pmipv6-based group mobility scheme for 6l0wpan networks," in *2017 IEEE International Conference on Communications (ICC)*, May 2017, pp. 1–6.

[12] E. Aljarrah, M. B. Yassein, and S. Aljawarneh, "Routing protocol of low-power and lossy network: Survey and open issues," in *2016 International Conference on Engineering MIS (ICEMIS)*, Sept 2016, pp. 1–6.

[13] M. Borowczak and G. Purdy, "S-chirp: Secure communication for heterogeneous iots with round-robin protection," in *2018 IEEE International Conference on Consumer Electronics (ICCE)*, Jan 2018, pp. 1–6.

[14] L. Jozwiak, D. Gawlowski, and A. Slusarczyk, "An effective solution of benchmarking problem: Fsm benchmark generator and its application to analysis of state assignment methods," in *Digital System Design, 2004. DSD 2004. Euromicro Symposium on*, aug.-3 sept. 2004, pp. 160 – 167.

[15] M. Borowczak and R. Vemuri, "S*FSM: An Paradigm Shift for Attack Resistant FSM Designs and Encodings," in *Redefining and Integrating Security Engineering, 2012. RISE 2012. ASE International Conference on cyber security*, dec. 2012, pp. 651 – 655.

A Two-Tiered Heterogeneous and Reconfigurable Application Processor for Future Internet of Things

Prasanna Kansakar* and Arslan Munir[†]

Department of Computer Science
Kansas State University, Manhattan, KS, USA
Email: *pkansakar@ksu.edu, [†]amunir@ksu.edu

Abstract—The Internet of things (IoT) is leading the world into a future of ubiquitous connectivity. The heterogeneity within the IoT domain necessitates a highly flexible, secure, dependable, and energy-efficient IoT processor architecture. In this paper, we propose a novel processor architecture for IoT that renders energy efficiency, high-performance, flexibility, security, and dependability to meet the diverse application requirements. To address the stringent *energy efficiency* demands of IoT devices, we propose a two-tiered heterogeneous processor architecture that is composed of a *high-performance* optimized reconfigurable host processor which controls a number of low-power optimized interface processors. The proposed IoT architecture also incorporates reconfigurability in host processors' computing and communication parameters and co-processor extensions to impart *flexibility* and additional energy savings. The proposed IoT architecture contains various *security* co-processor extensions to support various security primitives including encryption and decryption, key generation, integrity verification, and device authentication. Finally, the proposed architecture incorporates reliability and *dependability* through various hardware- and software-based fault tolerance methods. Experimental results present and compare microarchitecture configurations for host and interface processors obtained through an efficient design space exploration methodology. We have implemented selected security and dependability primitives of our proposed IoT architecture on a Xilinx Spartan-6 field-programmable gate array (FPGA). Results reveal that our proposed IoT architecture can attain a speedup of 47.93× while consuming 2.4× lesser energy for furnishing security and dependability primitives as compared to an optimized ARM implementation of similar security and dependability primitives.

Index Terms—Internet of things (IoT), reconfigurable processor, heterogeneous processor, microarchitecture, security, dependability, energy efficiency, fault tolerance

I. INTRODUCTION AND MOTIVATION

The Internet of things (IoT) is a new paradigm in computing wherein everyday physical objects are interconnected through an intelligent, invisible network fabric which allows for objects in the IoT ecosystem to communicate, directly or indirectly, with each other or the Internet for purposes of automation, remote data sensing, and centralized management/control [1]. There are two approaches being considered for IoT deployments. The first approach involves deploying IoT devices as "dumb nodes" with limited processing and communication capabilities. In this approach, the bulk of the data processing and analysis is carried out in the computing nodes higher up in the network hierarchy. The second approach involves incorporating higher processing and communication capabilities in the IoT devices such that only minimal access to computing nodes higher up in the network hierarchy is required. This

approach provides a balance between how much computation needs to be done locally versus how much computation needs to be done globally by considering the tradeoffs between monetary cost, real-time responsiveness, energy efficiency, network latency, and network congestion [2].

In order to make a choice between the above mentioned two IoT deployment approaches, several IoT-specific constraints have to be considered such as cost, performance, energy efficiency, etc. Specifically, IoT devices are mostly battery powered and thus are highly energy constrained. Some devices must operate throughout their entire lifetime on the battery they are deployed with whereas other devices may have limited charging mechanisms. All processing and communication activities should therefore be highly energy-efficient. An always active processor is not a viable implementation for IoT devices. Energy efficiency can be achieved in two ways in IoT architectures. The first way is to deploy IoT devices with single high-performance processor with energy conserving sleep modes and the second way is to create a heterogeneous architecture that consists of a network of low-power processors governed by a high-performance processor. Out of these two ways of achieving energy efficiency, the second approach with the heterogeneous architecture is more promising because in the first implementation, the energy required for waking up a high-performance processor from sleep mode is high [3]. Hence, the first implementation, although better than an always active processor implementation, is still relatively energy-inefficient [3]. The heterogeneous architecture has been adopted by ARM and Synopsys for designing their energy-efficient IoT solutions [3] [4].

Another key constraint that needs to be considered when developing an IoT architecture is the need for interoperability. The IoT ecosystem is diverse, consisting of devices with varying complexities in computation and communication. However, there is still a lack of consensus on standards and best practices among the companies to address the issue of interoperability. By the time, a standard would be agreed upon and adopted, current IoT deployments could become obsolete due to non-conformance of the policies outlined in the standard. Therefore, companies need to consider outfitting IoT deployments with mechanisms to ensure that these IoT deployments can be easily integrated with other existing and future systems as well as be able to implement any future standards, features, and services. The key to future-proofing and longevity of IoT deployments lies in hardware

flexibility. Reconfigurable processors can be used to impart flexibility in both processing and communication hardware in IoT deployments. Hardware reconfiguration enables IoT deployments to interoperate with disparate existing systems. Hardware reconfiguration also enables IoT deployments to be reprogrammed to fit any future standards or to implement new features and services.

In this paper, we propose the design of a flexible, high-performance, energy-efficient, secure, and dependable processor architecture for future IoT deployments. We propose a two-tiered heterogeneous and reconfigurable processor architecture that consists of a high-performance host processor, comprising of reconfigurable computation and communication units, which controls a number of low-power interface processors. The two-tiered heterogeneous architecture enables effective energy management and reconfigurability adds further flexibility and energy savings. We also equip our proposed IoT architecture with security co-processor extensions that support various security primitives, such as encryption and decryption, key generation, integrity verification, and device authentication. Finally, the proposed architecture incorporates dependability through various hardware- and software-based fault tolerance methods.

Our main contributions are as follows:

- Proposal of a novel two-tiered heterogeneous reconfigurable IoT processor architecture that imparts energy efficiency, high-performance, flexibility, security, and dependability to meet the diverse application requirements.
- Proposal of *security* co-processor extensions that leverage hardware-based security approaches to support various security primitives including encryption and decryption, key generation, integrity verification, and device authentication.
- Classification of chip multiprocessor benchmarks into IoT-specific application categories and using these benchmarks and a design space exploration methodology to determine low-power and high-performance microarchitecture configurations for the IoT processor.
- Implementation of selected security and dependability primitives of our proposed IoT architecture on a Xilinx Spartan-6 field-programmable gate array (FPGA) and comparison with an optimized implementation on an ARM processor in terms of performance and energy efficiency.

II. RELATED WORK

There are many articles released by processor and system-on-chip (SoC) design companies that outline techniques of increasing processing capabilities in IoT devices. These articles focus on selecting processors suitable for the type and size of workload for IoT devices and on designing low-power optimized processor architectures for IoT.

ARM proposed a processor architecture consisting of multiple homogeneous processors in a single IoT object each serving a different purpose [4]. ARM defined a system with three Cortex-M processors, one to handle network connectivity, one to manage interface with sensors and actuators, and one as a host processor controlling the other two. ARM stated that multiple processors are better for lowering power consumption in IoT objects since only the processor serving the current task would be in active mode while the rest would be in sleep mode. ARM also proposed a guide to selecting microcontrollers for IoT objects [5]. In this guide, ARM argued that high-end microcontrollers are suitable for IoT deployments for two reasons. Firstly, high-end microcontrollers complete processing tasks sooner and can enter sleep mode to conserve power and secondly, larger flash and RAM sizes available with high-end microcontrollers facilitate implementation of complex networking protocols without addition of any new processors in the system. These articles clearly demonstrate the need for having more power-optimized processors in IoT deployments.

Synopsys also proposed the use of multiple processors in IoT deployments [3]. Synopsys described the use of two-tiered processor architecture in IoT objects—ultra low power embedded processors used to interface with sensing elements to collect, filter and process data, and host processor used to manage low power embedded processors. The processor architecture proposed by Synopsys lowered power consumption by keeping power hungry host processor mostly in sleep mode, similar to the concept used by ARM. Synopsys also discussed optimization of processors using configurable hardware extensions for sensor applications [3]. Synopsys stated that adding custom hardware extensions for executing typical sensor functions reduces the processor cycle count required to execute sensor applications. The reduction in cycle count lowers energy consumption either by lowering the clock frequency and keeping the same execution time, or having the same power but shorter execution time.

Contemporary approaches in energy-efficient architectures focus either on computational or communication aspects. However, our proposed architecture simultaneously considers both computation and communication aspects for attaining higher performance in energy-constrained IoT devices. Our proposed architecture also takes into account the diversity in the IoT domain with regards to devices having varying complexity. The flexibility in our proposed architecture makes it suitable for IoT devices with different energy and cost constraints. Overall, our proposed architecture presents a promising solution for meeting performance, real-time, energy, throughput, latency, and resilience requirements of IoT applications in a distributed heterogeneous IoT environment.

III. RECONFIGURABLE IoT ARCHITECTURE

A two-tiered heterogeneous processor architecture is suitable for increasing the processing capabilities of IoT while also maintaining energy efficiency [4] [6]. A high-level schematic of such a two-tiered architecture is shown in Figure 1. This architecture consists of a central host processor with a communication unit and a high-performance optimized computation unit. The host processor is interfaced with a number of low-power optimized interface processors. The interface processors carry out minor tasks, such as collecting data from sensors and

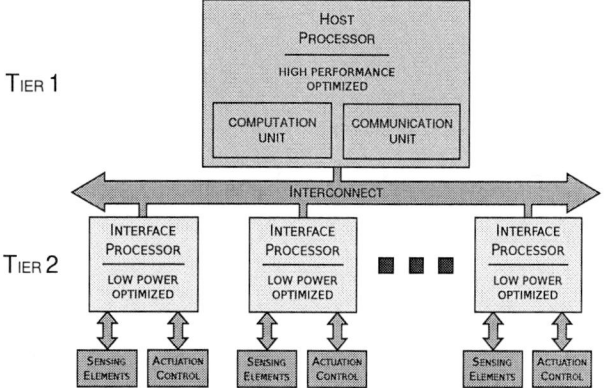

Fig. 1. Two-tiered heterogeneous processor architecture for IoT.

controlling actuation elements. The interface processors require minimal energy for operation so they do not significantly impact the battery life of IoT deployments. Hence, interface processors can always be operated in active mode. The host processor however expends a lot of energy during operation so it is only activated intermittently and for limited periods. The host processor is activated when compute-intensive functions, such as data-analysis, filtering, and complex security protocols need to be performed.

Figure 2 shows the details of our proposed two-tier heterogeneous processor architecture for IoT. Our proposed architecture is able to dynamically reconfigure both computation and communication parameters to attain high performance and energy efficiency. In the following subsections, we briefly describe the important components of our proposed IoT processor architecture shown in Figure 2 and discuss how these components contribute to improving overall performance and energy efficiency.

A. Host Processor

The host processor of our proposed architecture consists of a computation unit, a communication unit, and storage unit(s). Fault tolerance (FT) is provided for different tasks executed by the application and reconfigurable processors on a need basis depending on the criticality of function and IoT application. The host processor allows for reconfiguration of selected parameters (e.g., core count, operating frequency, modulation power, baseband filtering, etc.) of computation and communication units. The reconfiguration enables the host processor to add processing capabilities to IoT devices for mission-critical and/or emergency situations and remove the added capabilities to switch to an energy-efficient configuration in idle and/or regular operating situations.

1) Computation Unit: The computation unit within the host processor consists of an application processor alongside a reconfigurable processing unit that houses a reconfigurable processor and co-processor extensions.

Application Processor: The application processor in the computation unit of the host processor operates during idle/normal operating situations. When compute-intensive or application-specific tasks have to be performed, then the application processor is tasked with performing dynamic reconfiguration

of the reconfigurable processing unit to create new processor and co-processor instances.

Reconfigurable Processor: Reconfigurable processor instances are engendered when heavy workloads need to be processed. The processor instances are generated by setting the values for a number of tunable processor parameters, such as core count, operational frequency, cache subsystem, instruction fetch/issue/retire widths, reorder buffer size, branch prediction, etc. The choice of values for these processor parameters is made based on the type and size of the workload. These processor instances are removed when they are not required during idle/normal operation to help improve energy efficiency.

Co-processor Extensions: The host processor in our proposed architecture consists of a number of co-processor extensions to aid in dedicated application-specific tasks. We provide a brief description and uses of these co-processor extensions in the subsections below:

Security: Since IoT devices are used for sensing and actuating applications, the IoT devices are the primary interface between the digital and the physical world. If an IoT node is compromised then an attacker acquires the ability to control the physical environment wherein the IoT device is deployed. For example, consider an industrial IoT deployment which is used to maintain the temperature of a warehouse at a certain limit. If an attacker gains direct access to an IoT node in the warehouse, then s/he can manipulate it to raise/lower the temperature inside the warehouse beyond the specified limit leading to damage of goods stored in the warehouse. Hence, it is crucial that IoT deployments have strong security features to protect against adversarial attacks. In order to implement strong security primitives, higher computational capabilities are required. To address the need for integrating strong security features in the IoT devices and higher computation ability required to implement these security features, we incorporate security co-processor extensions in our proposed heterogeneous processor architecture for IoT. The security co-processor extensions aid in implementing strong cryptographic primitives to secure data communicated to and from IoT devices.

Figure 3 shows a high level overview of the security primitives that are provisioned by the security co-processor extensions. These primitives enable confidentiality, integrity and authentication in our proposed architecture. We propose hardware acceleration for encryption/decryption operations (e.g., advanced encryption standard (AES)) and message authentication operations (e.g., hash-based message authentication code (HMAC)). Having hardware acceleration for these complex security primitives significantly improves performance and reduces energy cost [7]. Since encryption/decryption and message authentication are key-based primitives, a secret key is required for executing these security computations. In our proposed architecture, instead of storing secret keys in some on-chip storage element, we include a key generation module that is based on weak physically unclonable functions (PUFs) [8]. The PUF-based key generation module eliminates the need for having costly on-chip temper resistant memory. PUF-based

978-1-5386-7100-9/18 $31.00 © 2018 IEEE

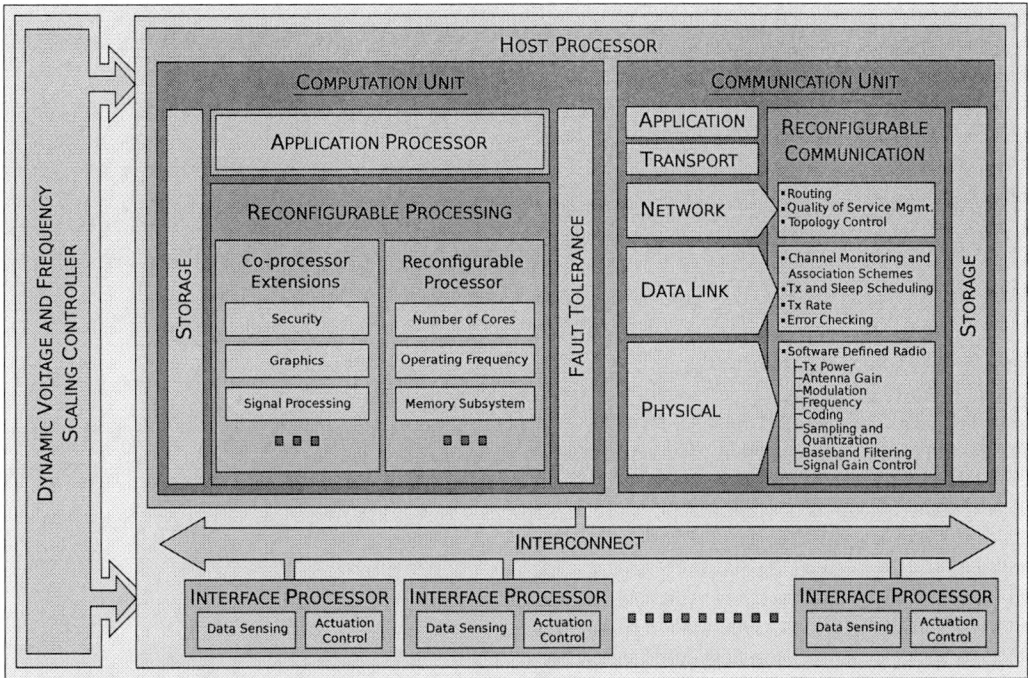

Fig. 2. Heterogeneous reconfigurable processor architecture for IoT including details of host and interface processors.

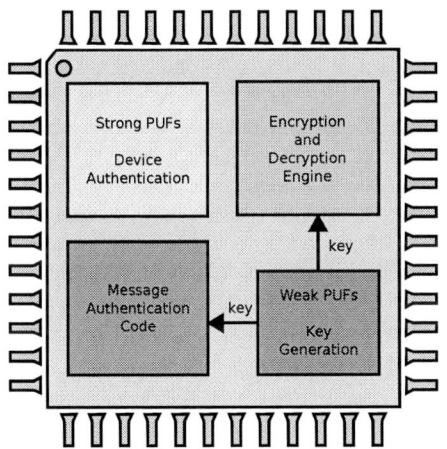

Fig. 3. Overview of security primitives in the proposed heterogeneous reconfigurable IoT architecture.

key generation provides a large key space within a smaller footprint as compared to implementing a tamper-resistant memory for secret key storage. For device authentication, we include a strong PUF-based authentication module. Strong PUF-based authentication schemes are discussed in [8].

Graphics: IoT deployments, such as surveillance and monitoring systems, have to collect and process a large amount of image data in order to carry out their assigned tasks. Image processing operations are compute intensive and highly data-parallel in nature. They require specialized high performance computing resources to operate on multiple number of similar threads in parallel. To provide such specialized support for image-based application domains, we incorporate graphics co-

processor extensions in our proposed IoT processor architecture. The graphics co-processors provide support for simple image analysis operations like image segmentation, edge detection, motion detection, etc. Having specialized graphics hardware on-board lowers the execution time of graphics-related operations, thereby reducing the amount of time the host processor has to remain in an active state, which in turn helps to improve the energy efficiency of the IoT deployments.

Signal Processing: While performing sensing and actuation tasks, IoT devices have to convert signals between analog and digital domain. In the IoT domain, signal processing finds applications in various tasks, such as speech recognition, image compression, audio playback, etc. Our proposed IoT processor architecture includes a signal processing co-processor extension to provide dedicated support for signal processing applications, such as signal filtering, processing and transformation functions. The availability of specialized hardware for signal processing improves the performance of the host processor and contributes significantly to maximizing its energy efficiency.

2) Communication Unit: The communication unit is used to communicate data with other IoT devices or with computing nodes that are higher up in the network hierarchy. Our proposed architecture empowers reconfiguration in the communication unit to enable an IoT device to communicate with other IoT devices in the heterogeneous IoT environment wherein the devices use different communication architectures and networking protocols. The reconfigurability for communication unit includes modification of radio settings (e.g., transmission power, antenna gain, modulation, frequency, coding, sampling and quantization, baseband filtering, and signal gain control), data link layer parameters (e.g., channel monitoring and association schemes, transmission and sleep scheduling, transmission rate, and error checking), and network layer parameters (e.g., routing, quality of service management, and topology control).

3) Storage: In our proposed IoT architecture, storage units are present in both the computation and communication units within the host processor. The storage unit within the computation unit is used for storing data for a variety of purposes, such as aggregation, analytics, mining, and archival. Interface processors gather data from different sensing elements and store that data in the storage unit. When complex data operations, such as filtering, sorting, etc., needs to be performed on the aggregated data, then the host processor reads data from the storage unit and operates on it. The storage unit also holds reconfiguration binaries that are used to perform dynamic reconfiguration of modules within the reconfigurable computation unit. The storage unit further stores locally relevant historical data which can be utilized by the host processor for analytics and making control decisions. The storage unit within the communication unit holds configuration parameters for software defined radios, network topology information, etc.

4) Fault Tolerance: Faults can result in IoT devices during normal operation due to environmental fluctuations (jitters, noise, radiations, etc.) or due to aging. The effects of faults on IoT devices can be mitigated by designing IoT devices to be fault tolerant. Fault tolerance is particularly important for IoT devices deployed for mission- and safety-critical applications. Fault tolerance can be provided through both hardware and software methods. The fault tolerance techniques employed by our proposed IoT architecture include: (i) fault tolerance by redundant multithreading (RMT), referred to as FT-RMT; (ii) FT-RMT enhanced with quick error detection (QED) [9]; (iii) dual modular redundancy (DMR); (iv) Berger code based totally self-checking combinational circuit (TSC); and (v) fault tolerance using self-reconfiguration in DMR (FT-SR-DMR). FT-SR-DRM performs dynamic self-reconfiguration to replace the faulty instances of the hardware module/component with new instances by exploiting partial reconfiguration feature of Xilinx Spartan-6 FPGA.

B. Interface Processors

The interface processors are optimized for low-power operation and are tasked with controlling interface components, such as sensors and actuators. Reading data from sensors and sending control actions to actuators has to be performed in short regular intervals, and thus require an always active processor. The low-power interface processors are well suited for these sensing and control applications because keeping these interface processors in active mode has minimal effect on the battery life of an IoT device.

C. Dynamic Voltage and Frequency Scaling Controller

Our proposed architecture also incorporates a dynamic voltage and frequency scaling (DVFS) controller that adjusts the operating voltage and frequency of various hardware components for meeting performance requirements while conserving

energy. Voltage and frequency scaling is carried out in both the host and the interface processors to further improve the energy efficiency of our proposed architecture.

IV. Methodology and Experimental Setup

In this section, we describe the experimental setup for two independent set of experiments that we have performed for our proposed IoT architecture. In our first set of experiments, we use a design space exploration method for microarchitecture parameter tuning to determine microarchitecture parameters for high-performance optimized host processor and low-power optimized interface processor(s). In our second set of experiments, we implement and compare selected security and dependability primitives and compare the result in terms of performance and energy efficiency. The methodology and experimental setup for these experiments are described below.

A. Determining microarchitecture configurations using design space exploration

In order to optimize the microarchitecture parameters of the host and interface processors used in our proposed heterogeneous IoT architecture, we employ a design space exploration methodology that we have detailed in our previous work [10]. We utilize a four phase exploration algorithm consisting of the following phases: initial one-shot optimization and parameter significance phase, set partitioning phase, exhaustive search phase, and greedy search phase. We run our experiments on a cycle accurate multiprocessor simulator called ESESC (Enhanced Super ESCalar) [11] and use a set of PARSEC (Princeton Application Repository for Shared-Memory Computers) and SPLASH-2 (Stanford ParalleL Applications for SHared memory, version 2) benchmarks [12] [13] [14] to provide test workloads of varying types and sizes. We use a weighted objective function for ranking the different microarchitecture configurations tested by our search algorithms. The objective function is a weighted sum of the total power and total execution time design metric values that are obtained from simulation. The design space for the host and interface processors is shown in Table I.

B. Security and dependability approaches

For verification of performance and energy-efficiency of security primitives afforded by our proposed IoT processor architecture, we implement AES-128 for rendering confidentiality (encryption and decryption operations) and secure hash algorithm (SHA) based HMAC for message integrity verification. We have implemented the following dependability primitives as outlined in Section III-A4: FT-RMT, FT-RMT-QED, and FT-SR-DMR. We test two software-based implementations in this experiment. The first is baseline design (BD) that implements AES-128 and SHA-2 and has no code optimizations. The second is optimized baseline design (OptBD) that implements AES 128 and SHA-3 and incorporates code optimizations such as loop unrolling, cache-aware programming, alignment of data structures to cache line boundary, etc. Both BD and OptBD are implemented on a 32-bit quad-core Cortex-A9 ARM application processor processor running Ubuntu 14.04.4 LTS at 396 MHz clock speed. Our

978-1-5386-7100-9/18 $31.00 © 2018 IEEE

TABLE I
DESIGN SPACE FOR MICROARCHITECTURE CONFIGURATION PARAMETERS
FOR HOST AND INTERFACE PROCESSORS

Parameter Name	Set of Settings	
	Low-Power	High-Performance
Cores	1, 2, 4	2, 4, 8
Frequency (MHz)	75, 100, 125, 150	1700, 2200, 2800, 3200
L1-I Cache Size (kB)	8, 16, 32, 64	8, 16, 32, 64, 128
L1-D Cache Size (kB)	8, 16, 32, 64	8, 16, 32, 64, 128
L2 Cache Size (kB)	256, 512, 1024	256, 512, 1024
L3 Cache Size (kB)	2048, 4096	2048, 4096, 8192

TABLE II
CATEGORIZATION OF TEST BENCHMARKS ACCORDING TO IoT
APPLICATIONS

IoT Application	Benchmarks
Data sensing and aggregation	Cholesky, Radix
Data analysis and Data mining	Blackscholes, Freqmine
Graphics	Facesim, Fluidanimate
Signal processing and Communication	FFT

proposed IoT processor architecture implements AES-128 and SHA-3 on a Xilinx Spartan-6 FPGA. We refer to the implementation of our IoT processor architecture on FPGA as ITAF. The ITAF incorporates dependability by implementing FT-SR-DMR (Section III-A4). We also compare FT and non-fault-tolerant (NFT) implementations in terms of performance and energy efficiency. Our previous work [7] provides further details on the implementation.

V. RESULTS

In this section, we present the results for the two sets of experiments outlined in Section IV.

A. Microarchitecture configurations obtained from design space exploration

IoT devices operate on a wide variety of workloads of different types and sizes. We broadly separate these workloads into four different categories relating to common IoT applications or processes. We classify the benchmarks from the PARSEC and SPLASH-2 benchmark suites into these categories based on the closest IoT application that the benchmarks resemble. The categorization of some of the key test benchmarks used in our experiments are shown in Table II.

1) Microarchitecture configurations for low-power optimized processors for IoT: Table III shows the microarchitecture configurations obtained for the Cholesky benchmarks from the SPLASH-2 benchmark suite. We use this as an example to discuss the microarchitecture configuration required in low-power optimized interface processors. We note that for the Cholesky benchmark, our design space exploration methodology selects the lowest operating frequency (75 MHz) and core count (single-core). This is because high operating frequency and high number of cores in the processor directly influences the power consumption of the processor. The size of the L1-D cache in the resulting configuration is also large because of the large workload offered by the Cholesky benchmark. This is representative of the growing IoT ecosystem in which large volumes of data are gathered from a large number of sensing elements. In Table III, we have also included the

TABLE III
MICROARCHITECTURE CONFIGURATIONS FOR LOW-POWER OPTIMIZED
AND HIGH-PERFORMANCE OPTIMIZED PROCESSORS FOR IoT

Parameter Name	Microarchitecture Configurations	
	Low-Power	High-Performance
	Cholesky	Blackscholes
Cores	1	8
Frequency (MHz)	75	3200
L1-I Cache Size (kB)	8	64
L1-D Cache Size (kB)	32	128
L2 Cache Size (kB)	256	256
L3 Cache Size (kB)	2048	8192
Total Power (W)	0.0934	4.549
Execution Time (ms)	327.958	28.1239

values for total power and execution times returned from the simulations. The power value ranges in the order of a few hundred milliwatts (mW) and the execution time ranges in the order of a few hundred milliseconds(ms). We observe that the resulting design heavily favors low-power consumption by sacrificing performance (high total execution time). The low-power usage of this microarchitecture makes it suitable for use as interface processor which can remain in active mode indefinitely without significantly hampering the energy budget of IoT devices.

2) Microarchitecture configuration for high-performance optimized processors for IoT: Table III shows the microarchitecture configurations obtained for Blackscholes benchmarks from the PARSEC benchmark suite. We use this as an example to discuss the microarchitecture configuration required in high-performance optimized host processors. We observe that for the Blackscholes benchmarks, which is classified under data analysis and data mining category in Table II, performance improvement is achieved through higher operating frequency (3200 MHz) and core count(8-cores). The size of the L1-D cache and L2 cache for this microarchitecture configuration is also high because Blackscholes is a highly data-parallel benchmark. This is typical of the type of data analysis tasks that need to be performed on data aggregated from each sensing element in an IoT device. Since the host processor is equipped with reconfigurable computation unit, the core count and the operating frequency can be dynamically altered. The resulting design, in this case, heavily favors performance (total execution time) over total power. The total power value is in the range of a few watts whereas the execution time is in the range of few tens of milliseconds. As the power requirement for the host processors is high, as shown by this example microarchitecture configuration, it must be mostly kept in the sleep mode and only be activated infrequently and for short periods of time. This is necessary to conserve the battery life of the IoT devices. However, since these processors have shorter execution times, the processors have to remain active for a shorter period of time to complete their designated tasks as compared to the processors not optimized for high performance.

B. Comparison of Security and Dependability Primitives

In this section, we present the results for performance (time in μs) and energy (μJ) for completing one AES encryption

TABLE IV
PERFORMANCE AND ENERGY RESULTS FOR BD, OPTBD, AND ITAF.

Operational Mode	Baseline Design (BD)			Optimized Baseline Design (OptBD)			FPGA Implementation (ITAF)		
	FT Mode	Time (μs)	Energy (μJ)	FT Mode	Time (μs)	Energy (μJ)	FT Mode	Time (μs)	Energy (μJ)
NFT	x	257	13.137	x	189	9.661	x	4.90	2.170
FT	FT-RMT	411	21.010	FT-RMT	207	10.581	FT-SR-DMR	6.53	6.647
	FT-RMT-QED	589	30.109	FT-RMT-QED	313	16.000			

computation plus one HMAC computation for BD, OptBD, and ITAF.

1) Timing Analysis: Table IV shows the timing performance of BD, OptBD, and ITAF. Comparison of BD and ITAF reveals that NFT ITAF is 52.45× faster than NFT BD. Furthermore, after embedding FT in BD by FT-RMT and in our FPGA implementation by FT-SR-DMR, ITAF is 62.94× superior than BD. Lastly, ITAF with FT-SR-DMR provides a speedup of 90.19× over BD in FT-RMT-QED mode.

Comparison of ITAF and OptBD shows that NFT ITAF is faster than NFT OptBD by 38.57×. Moreover, FT-SR-DMR in ITAF surpasses FT-RMT in OptBD by 31.69×. Furthermore, a speedup of 47.93× is achieved with FT-SR-DMR in ITAF over OptBD with FT-RMT-QED.

2) Energy Analysis: Table IV depicts the energy consumption results of our implementations of security and dependability primitives. The comparison between ITAF and BD reveals that NFT ITAF is 6.05× more energy efficient than NFT BD. ITAF with FT-SR-DMR is 3.16× more energy efficient than BD with FT-RMT and 4.52× more energy efficient than BD with FT-RMT-QED.

The comparison between ITAF and OBD shows that NFT ITAF results in 4.45× more energy savings than NFT OptBD. Additionally, ITAF with FT-SR-DMR gives 1.59× more energy savings than OptBD with FT-RMT, and 2.4× times more energy savings than OptBD with FT-RMT-QED, respectively.

VI. CONCLUSIONS

In this paper, we have proposed the design of a novel two-tiered heterogeneous processor architecture for IoT that imparts energy efficiency, high-performance, flexibility, security, and dependability to meet the diverse application requirements. Our proposed architecture consists of a high-performance optimized reconfigurable host processor that controls a number of low-power optimized interface processors. We utilize a design space exploration methodology for processor parameter tuning, using a cycle-accurate simulator (ESESC) and a standard set of PARSEC and SPLASH-2 chip multiprocessor benchmarks, to determine example microarchitecture configurations for the host and interface processors. From the resulting microarchitecture configurations, we observe that the high-performance optimized host processor requires a higher core count and operating frequency as compared to the low-power optimized interface processor. The size of the different levels of caches in the microarchitecture configuration depends on the size of the workload. Results indicate that the resulting microarchitecture configurations for both the host processor and the interface processor possess large cache size.

In this paper, we have also implemented selected security and dependability primitives of our proposed IoT architecture on a Xilinx Spartan-6 FPGA and have compared the results with baseline and optimized implementations on an ARM processor in terms of performance and energy efficiency. Experimental results show that the FPGA-prototype implementations of security and dependability primitives of our proposed IoT processor architecture outperform ARM-based implementations by 47.93× while consuming 2.4× lesser energy. These results support our concept for including hardware-based security co-processor extensions in the host processor of our proposed architecture. As our future work, we plan to prototype additional features of our proposed security coprocessor extensions, such as device authentication and key generation using PUFs.

REFERENCES

[1] J. Chase, "The evolution of the internet of things - from connected things to living in the data, preparing for challenges and IoT readiness," Texas Instruments, Tech. Rep., Sep 2013.

[2] "What the internet of things (IoT) needs to become a reality," Freescale, Tech. Rep., May 2014.

[3] J. Geuzebroek and A. Vaassen, "Building an efficient, tightly coupled embedded system using an extensible processor," Synopsys, Tech. Rep., Jun 2014.

[4] "Intelligent flexible IoT nodes," ARM, Tech. Rep., Oct 2015.

[5] K. Char, "Internet of things system design with integrated wireless MCUs," Silicon Labs, ARM, Tech. Rep., Oct 2015.

[6] S. Bath. (2016, Aug) Developing solutions for the internet of things. [Online]. Available: https://www.intrinsyc.com/increasing-solution-differentiation-edge-based-heterogeneous-computing/

[7] B. Poudel and A. Munir, "Design and evaluation of a novel ecu architecture for secure and dependable automotive cps," in *2017 14th IEEE Annual Consumer Communications Networking Conference (CCNC)*, Jan 2017, pp. 841–847.

[8] C. Herder, M.-D. Yu, F. Koushanfar, and S. Devadas, "Physical unclonable functions and applications: A tutorial," *Proceedings of the IEEE*, vol. 102, no. 8, pp. 1126–1141, Aug 2014.

[9] T. Hong, Y. Li, S.-B. Park, D. Mui, D. Lin, Z. A. Kaleq, N. Hakim, H. Naeimi, D. S. Gardner, and S. Mitra, "QED: Quick Error Detection Tests for Effective Post-Silicon Validation," in *Proc. of IEEE Internation Test Conference (ITC)*, Austin, Texas, November 2010.

[10] P. Kansakar and A. Munir, "A design space exploration methodology for parameter optimization in multicore processors," *IEEE Transactions on Parallel and Distributed Systems*, vol. 29, no. 1, pp. 2–15, Jan 2018.

[11] E. K. Ardestani and J. Renau, "ESESC: A fast multicore simulator using time-based sampling," in *Proceedings of IEEE 19th International Symposium on High Performance Computer Architecture (HPCA)*, Washington, DC, USA, Feb 2013.

[12] C. Bienia, "Benchmarking modern multiprocessors," Ph.D. dissertation, Department of Computer Science, Jan 2011.

[13] Y. Bao, C. Bienia, and K. Li, *The PARSEC Benchmark Suite Tutorial - PARSEC 3.0*, San Jose, CA, USA, Jun 2011.

[14] S. C. Woo, M. Ohara, E. Torrie, J. P. Singh, and A. Gupta, "The SPLASH-2 programs: Characterization and methodological considerations," in *Proceedings of 22nd Annual International Symposium on Computer Architecture (ISCA)*, Santa Margherita Ligure, Italy, Jun 1995.

Solar Cell Based Physically Unclonable Function for Cybersecurity in IoT Devices

S. Dinesh Kumar, Carson Labrado, Riasad Badhan, Himanshu Thapliyal, and Vijay Singh

Department of Electrical and Computer Engineering

University of Kentucky, Lexington, KY, USA

Abstract—**Internet of Things (IoT) devices are mostly small and operate wirelessly on limited battery supply, and therefore have stringent constraints on power consumption and hardware resources. Therefore, energy-efficient (low energy) design is paramount for the successful deployment of resource constrained IoT devices. Further, Physical Unclonable Functions (PUFs) have evolved as a popular hardware security primitive for low cost, mass produced IoT devices with very constrained resources. Energy harvesting technologies utilizing solar cells are being used in ultra-low power IoT devices to satisfy the energy requirement. In this paper, we utilize the intrinsic variations in solar cells to design a novel solar cell based PUF. As a proof of concept, we have used the Tiva TM4C123GH6PM microcontroller to build our solar cell based PUF. From our experiments, we found that the proposed solar cell based PUF has the uniformity value of 49.21% which is close to the ideal value of 50%. Further, the proposed solar cell based PUF has worst case reliabilities of 92.97% and 90.62% with variations in temperature and light intensity, respectively.**

I. INTRODUCTION

The Internet of Things (IoT) is a network of machines, physical objects, and other devices that are connected through the Internet to exchange data for intelligent applications [1]. One of the main advantages of the IoT environment is that IoT allows the direct integration between physical objects and the digital world which helps to improve the quality of human life. Within the past decade, there have been numerous IoT devices introduced in the market. It is expected that more than 50 billion IoT devices will connect with each other by 2021 [2]. Generally, IoT devices are small and operate wireless on limited battery supply. Therefore, they have stringent constraints on power consumption and hardware resources.

Although there are numerous methods to achieve energy efficiency, the recent technology trend of energy harvesting provides a fundamental method to prolong battery life. Thus, energy harvesting is a promising approach to improve energy efficiency in IoT devices [3]. Though there are several energy harvesting modalities, solar energy harvesting through photovoltaic conversion provides the highest power density, which makes it an ideal choice to power a low power IoT device as illustrated in Fig. 1 [4]. Solar energy could allow IoT devices to be powered indefinitely without battery replacement. For example, Alta Devices has developed an extremely light weight, flexible and thin Gallium Arsenide (GaAs) solar cell with and efficiency of 28.8% [5].This type of solar cells can be integrated into the IoT devices as the energy source thereby

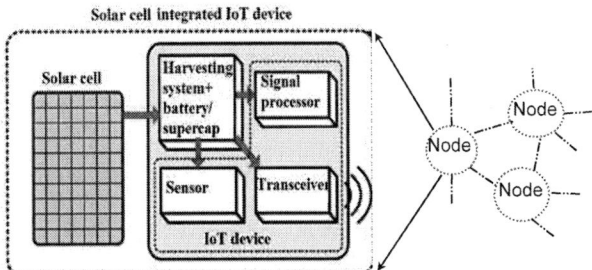

Fig. 1. IoT system with solar cell based IoT devices at each nodes

increasing the time between battery replacements or possibly never depending on the life and the efficiency of the solar cells.

Along with the energy-efficiency, IoT devices also provide challenges in privacy and security. Among the various security needs for IoT devices, authentication and access control are one of the most important features of the IoT that needs to be implemented [6]. Currently, the secret keys which are used for authentication are stored in non-volatile memories. But these keys are vulnerable to active attacks [7], [8]. Moreover, implementing tamper resistant circuitry in IoT devices to provide high level physical security may be very expensive in terms of cost and energy. In recent years, Physically Unclonable Functions (PUFs) have evolved as one of the popular hardware security primitives for low cost, mass produced IoT devices with very constrained resources (battery driven, very small volatile and persistent memory, and low processor power).

Lately, silicon PUF have emerged as a powerful solution to a variety of security concerns such as IC piracy, IC counterfeiting, etc. [9]. PUFs also play a major role in secure authentication and key management in cyber-physical security and IoT devices [10]. PUFs can also be considered as a promising solution for authentication in IoT devices [11]. A PUF is provided with challenge bits (C) and due to the intrinsic variations in the IC manufacturing process, it results in unpredictable outputs called response bits (R). The uncontrollable IC manufacturing errors make the PUF response to be unique and unclonable. Fig. 2 shows the block diagram for PUF production using inherent variations. In this manner, a PUF can be considered as a fingerprint for CMOS ICs.

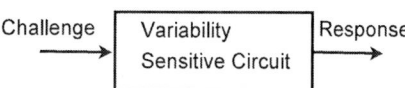

Fig. 2. PUF production using inherent variations

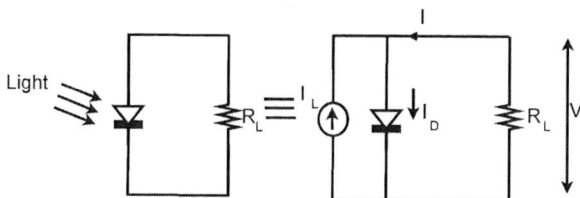

Fig. 3. Equivalent circuit of a photovoltaic solar cell

A. Motivation

PUFs are a class of circuits that use the inherent variations in the device manufacturing process to create unique and unclonable IDs. The existing CMOS IC based PUF requires dedicated hardware which results in additional hardware cost and power consumption. In recent years, solar cells have been integrated with IoT devices for various purposes such as power generation and sensing applications. The main motivation of this paper is to design a solar cell based PUF with minimum hardware cost suitable for implementation in IoT devices. In the proposed PUF, solar cell array acts not only as a power generator, but also as an entropy source for generating the secret bits. To the best of our knowledge, this is the first kind of work to integrate the solar cells along with IoT devices for designing a PUF to generate the unique IDs. Our proposed solar cell based PUF fall under the category of energy harvesting based PUFs.

B. Contribution of the paper

A novel architecture to build a solar cell based PUF is proposed in this paper. The proposed solar cell based PUF is built using a Tiva TM4C123GH6PM micro-controller. The challenge for our proposed PUF will be the selection of solar cells through the microcontroller and the response bits will be the V_{oc} comparison bits of the micro-controller. We also develop an algorithm to effectively choose the V_{oc} of the solar cells to generate a 128 bit response. This algorithm can also be used to generate a higher number of response bits. Further, the security metrics of the proposed PUF such as uniformity, reliability against temperature variations and reliability against light intensity variations are also presented in this paper.

C. Organization of the paper

Section II discusses the background on PUF and solar cells. Section III describes the design parameter chosen to design the solar cell based PUF. Section IV presents the architecture of the proposed solar cell based PUF. Section V presents the implementation details of the solar cell based PUF. Section VI discusses the security metric results of the solar cell based PUF. Section VII provides discussion and conclusion of the paper.

II. BACKGROUND

A. Physical Unclonable Function

PUFs are a class of circuits that use the inherent variations in the Integrated Circuit (IC) manufacturing process to create unique and unclonable IDs (Fig. 2). PUFs hold the potential to simplify or solve many important security problems such as IC piracy, IC counterfeiting, secure authentication and key management in IoT security. Further, object authentication is of utmost importance in businesses that are threatened by counterfeited devices. PUFs can't be cloned or copied and could be used to identify products unambiguously to avoid counterfeiting. PUFs have the potential to seal multiple types of vulnerabilities in IoT devices specifically in the security classes of authentication, non-repudiation, and privacy. Silicon PUFs utilizes the uncontrollable manufacturing variations in IC to generate the unique bits. Some of the examples of silicon PUFs include arbiter PUF [12], Ring Oscillator (RO) [13], SRAM PUF [14], etc.

B. Solar cell

Solar cells are considered a major candidate for obtaining energy from the light source, since they can convert light directly to electricity with high conversion efficiency. They can provide nearly permanent power at low operating cost, and are virtually free of pollution. A solar cell is a device which converts light into electricity. Light shining on the solar cell produces both a current and a voltage. This process requires a material to absorb light and raise electrons to a higher energy state, and transport this higher energy electron from the solar cell into an external circuit. Then, electrons dissipate their energy in the external circuit and return to the solar cell. Photovoltaic energy conversion often uses semiconductor materials and inorganic-organic materials in the form of a p-n junctions (Refer Fig. 3). Recently, solar cells have been used in IoT devices to improve the energy-efficiency [15]. [16], [17] have explored their usefulness for designing PUFs. However, [16], [17] have used the light source as the challenge which makes the PUF relatively unreliable and make them unsuitable to implement in real world IoT devices.

III. PARAMETERS TO DESIGN SOLAR CELL PUF

There are several electrical parameters which can be used to design the solar cell based PUF. However, for reliability, the chosen parameters must have a predictable relationship with the variation in environmental parameters to design a reliable PUF. In this section, we discuss the parameter with which we have chosen to design a reliable solar cell based PUF.

There are two important and easily measurable electrical parameters which define the characteristics of solar cells. These are the short circuit current (I_{sc}) and the open circuit voltage (V_{oc}). The fundamental current-voltage (I-V) equation of a solar cell is,

$$I = I_0[exp(\frac{qV}{\eta KT} - 1)] - I_L \qquad (1)$$

where I_0 is the reverse saturation current, η is diode ideality factor, I_L is light generated current, K is plank's constant and T is temperature. The short circuit current (I_{sc}) is the current that flows through the p-n junction under illumination at zero applied bias. In the ideal case, I_{sc} equals the photo-generated current (I_L). Therefore, the short-circuit current is the largest current which may be drawn from the solar cell. The short-circuit current mainly depends on quantum efficiency and spectrum of the incident light (the number of photons).

The open-circuit voltage V_{oc} is the maximum voltage available from a solar cell when the current through the junction is zero, and can be expressed as,

$$V_{oc} = \frac{\eta KT}{q} ln(\frac{I_L}{I_0} + 1) \qquad (2)$$

The above equation shows that V_{oc} depends on the saturation current of the solar cell and the light-generated current. Because I_L typically has a small variation and the reverse saturation current may vary by orders of magnitude, the reverse saturation current plays a key role in determining V_{oc}. The saturation I_0 depends on many materials and device characteristics of the individual solar cells like the electron-hole recombination lifetimes, trap levels, defects and impurities, potential barrier at the p-n junction, interface state density and capture cross section. It can vary by orders of magnitude in nominally "identically produced" solar cell devices. Thus it is a good choice as an entropy source for generating the secret bits in a PUF. But an even better choice is the open circuit voltage, V_{oc}. We can see from equation (2) that the entropy of I_0 is contained also in the entropy of V_{oc}. However, V_{oc} is far more easily measured than I_0 in a solar cell PUF during operation in remote field.

We have selected silicon solar cells to design our PUF circuit. Silicon solar cells are of different kinds, such as, amorphous silicon, single- or mono-crystalline silicon and poly-crystalline silicon. We have chosen monocrystalline or single-crystalline solar cell because it contains far less impurities than poly-crystalline or amorphous solar cells, and as such the power conversion efficiency does not degrade over operating time. Due to wide spectral range, they can be used in both indoor and outdoor applications. For a simple and elegant design, we selected open-circuit voltage (V_{oc}) as the parameter that would be used in the design of the PUF circuit. Solar cells from the same batch (supposedly identical solar cells) display variation in V_{oc} due to random process variations, which happen during manufacturing. V_{oc} is linearly proportional to a change in temperature and is considered a more stable parameter to use in the PUF circuit so that changing light intensity and temperature would generate more reliable data. Fig. 4 depicts how the V_{oc} of our solar cells varies with light intensity. V_{oc} also depends on the temperature

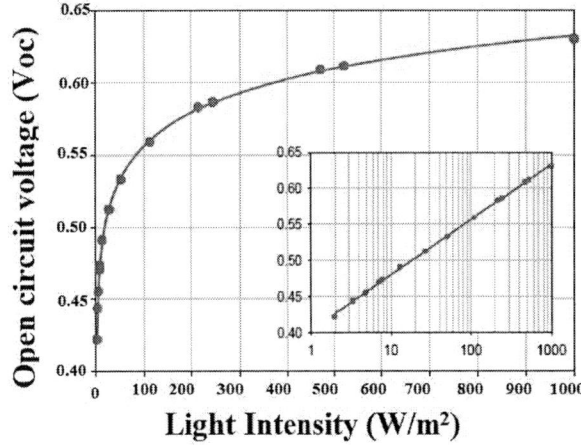

Fig. 4. Relationship between open circuit voltage (V_{oc}) and light intensity of silicon solar cells [18]

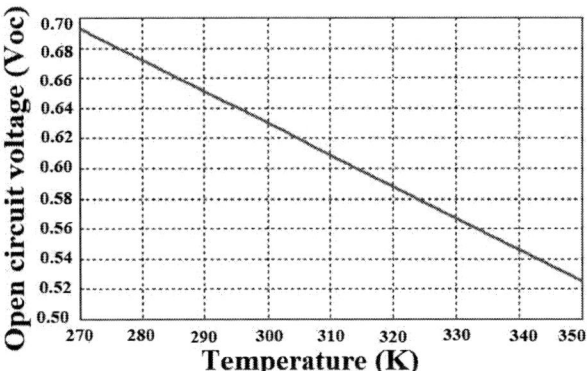

Fig. 5. Relationship between open circuit voltage (V_{oc}) and temperature [18]

of the operating condition. By changing the temperature of the environment of the PUF circuit, the V_{oc} of a solar cell can be controlled to our advantages (Fig. 5).

IV. ARCHITECTURE OF PROPOSED SOLAR CELL BASED PUF

Fig. 6 shows the architecture of the proposed solar cell based PUF. In the proposed solar cell based PUF, the manufacturing variations in the solar cells are used to generate the response bits.

As shown in Fig. 6, each solar cell (SC) is connected to a Analog-to-Digital Converter (ADC) input pin in the microcontroller. The ADC input pins are connected to one of the two available ADC channels in the microcontroller. The ADC channels are chosen by configuring the corresponding registers in the microcontroller. The converted digital bits are read through a personal computer (PC) connected through Universal Asynchronous Receiver and Transmitter (UART).

When photons hit the solar cells, due to photovoltaic effect, voltage (V_{oc}) will be created across the solar cells. However, due to the intrinsic variations in the material characteristics,

978-1-5386-7100-9/18 $31.00 © 2018 IEEE

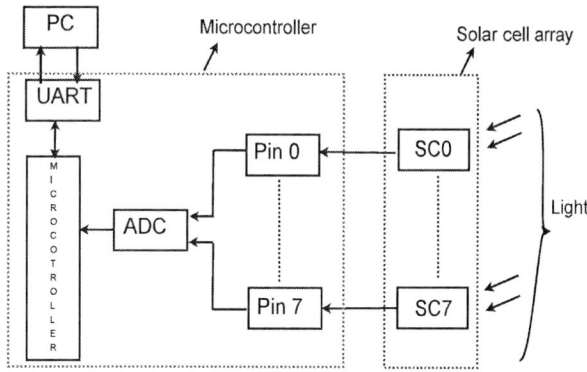

Fig. 6. Architecture of the proposed solar cell based PUF. SC represent solar cell, ADC represent Analog-to-Digital Converter, PC represent computer

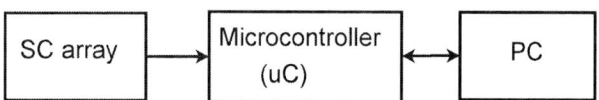

Fig. 7. Experimental setup for the proposed solar cell based PUF

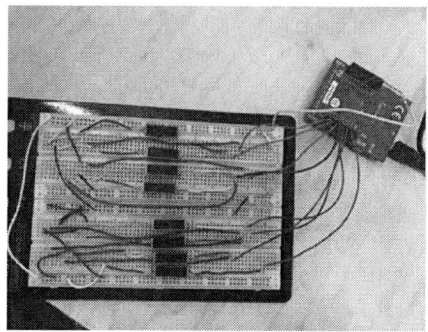

Fig. 8. Prototype of the proposed solar cell based PUF

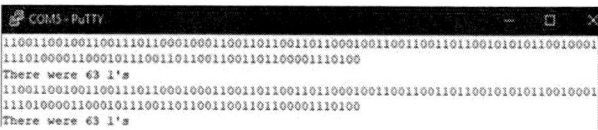

Fig. 9. Output response from the proposed solar cell based PUF

there will be variations in the output voltage of the solar cells. These output voltage from the solar cells are measured using the ADC available in the microcontroller. ADC will convert the analog voltage values into 12 bit digital values. After converting the analog values to digital bits, the 12 bit digital values are compared with each other in a pre-determined pattern to generate a 128 bit response. Each generated bit of the response is the result of a comparison between two groupings of solar cells. The hardware portion and software portion of the proposed solar cell based PUF are further explained in Section V.

V. IMPLEMENTATION OF PROPOSED SOLAR CELL BASED PUF

This section describes the prototyping effort of the proposed solar cell based PUF. The prototype includes 8 solar cells along with a Tiva TM4C123GH6PM microcontroller and a personal computer (PC). As discussed in Section III, the V_{oc} of the solar cells are measured using the ADC of the microcontroller. In this experiment, we have chosen a TM4C123GH6PM microcontroller as these microcontrollers have several IoT based applications. Some of the applications of TM4C123GH6PM microcontrollers include remote monitoring, electronic point-of-sale machines, network appliances and switches, factory automation, HVAC and building control etc. [19]. The complete experimental set up for the proposed solar cell based PUF is shown in Fig. 7.

A. Hardware portion of the proposed solar cell based PUF

The hardware portion of the proposed solar cell based PUF consists of 8 solar cells along with a microcontroller. The entire system is connected via UART to a PC with a BAUD rate of 9600. The PC is used to send challenges to the system and output the generated responses. The solar

cells are connected to one of the ADCs in microcontroller. The microcontroller is configured to read the analog open circuit voltage from the solar cells. The clock frequency of the microcontroller used in this experiment is 20MHz. The sampling rate for the ADC is 125,000 samples per second.

B. Software portion of the proposed solar cell based PUF

The software portion of the proposed solar cell based PUF consists of the software running on the microcontroller to read from the cells and generate a response. The software running on the PC has no real bearing on the functionality of PUF. The PC is simply used to send challenges and display the generated response to the user. The ADC on the microcontroller can only read from a single pin at a time. Additionally, there is some noise inherent to the ADC that can manifest when taking readings. As a result, the following steps have been used to get reliable readings from a given solar cell:

- Select ADC input pin to read the V_{oc} value from the solar cell on the microcontroller
- Configure ADC to read from the selected pin in previous step
- Take 16,000 readings from ADC
- Average together to try to offset the noise from ADC to get reliable readings
- Repeat the above steps 'n-1' times for 'n' solar cells connected to the microcontroller

VI. RESULTS

The prototype of the proposed solar cell based PUF is shown in Fig. 8. All of the experiments are performed with a uniform light intensity of 50 $Watts/m^2$. The output response bits from the proposed solar cell based PUF is recorded in the PC using the communications through UART. Recorded responses are used to evaluate the PUF against various PUF performance metrics.

978-1-5386-7100-9/18 $31.00 © 2018 IEEE

Fig. 10. Experimental setup to measure the reliability of the proposed solar cell based PUF against temperature variations

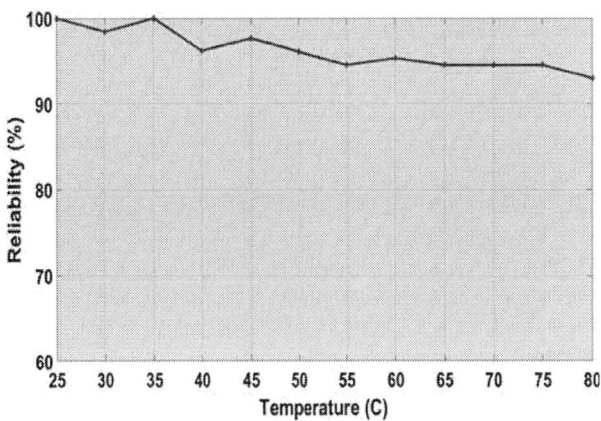

Fig. 11. Reliability of the proposed solar cell based PUF against temperature variations

A. Evaluation metrics for proposed solar cell based PUF

In this paper, we are evaluating the performance of the proposed solar cell based PUF against two important metrics. These metrics are uniformity and reliability.

1) Uniformity: Uniformity is used to measure whether the number of zeros and number of ones in the response bits are balanced or not. Uniformity is given by measuring number of 1's in the proposed 128-bit PUF. The ideal value of uniformity is 50%. Uniformity is given by,

$$Uniformity = \frac{1}{n}\Sigma_{i=1}^{n-1} r_{i,l} \times 100 \qquad (3)$$

where $r_{i,l}$ represents the l-th bit from PUF instance i.

2) Reliability: The reproducibility of the response bits from the same PUF instance with the varying environmental conditions such as temperature and supply voltage is given by reliability metrics. For i^{th} PUF instance, let R_i be the reference response or the golden response recorded under nominal operating conditions. Then, applying the same challenges to the same PUF but under different environmental conditions, n responses are observed. Reliability metric is given by,

$$Reliability = 100 - \frac{1}{k}\Sigma_{i=1}^{k}\frac{HD(R_i, R'_{i,t})}{n} \qquad (4)$$

where $HD(R_i, R'_{i,t})$ is the hamming distance between the golden response and the response generated by the same PUF instance at different environmental conditions. Variable k represents the total number of IC chips. In other words, reliability is the measure of total number of bits flipped between the golden response and the response recorded from the same PUF instance with different environmental conditions. Reliability is one important PUF metric to be considered while designing PUFs for key generation. The ideal value of the reliability metric is 100 %.

B. Testing results for proposed solar cell based PUF

The testing results of the proposed solar cell based PUF are presented in this section. There are two important environmental variations that need to be considered while designing the solar cell based PUF. They are (i) variation in temperature and (ii) variation in light intensity.

1) Uniformity results of the proposed solar cell based PUF: As shown in Fig. 9, the 128 bit responses generated from the proposed solar cell based PUF consists of 63 ones and 65 zeros. The calculated value of uniformity (equation 3) for the proposed solar cell based PUF is 49.21%. The proposed solar cell based PUF has the uniformity value close to the ideal value of 50%. As the probability of generating number of ones is close to the ideal value 50%, it indicates that the proposed solar cell based PUF output is not predictable and makes it hard to attack.

2) Reliability results of the proposed solar cell based PUF against temperature variations: Fig. 10 shows the experimental setup to perform the reliability analysis of the proposed PUF against temperature variations. In this experiment, we have varied the temperature from 25°C to 80°C. As a future work, we plan to test the reliability of the proposed solar cell based PUF from -10°C to 80°C.

Fig. 11 shows the reliability of the proposed solar cell based PUF with the temperature variations with 25°C as the reference temperature. The response bits generated from the proposed PUF at 25°C is the golden response and the response bits generated at different temperature are compared with the golden response to get the reliability values. The proposed PUF has worst case reliability of 92.97% at 80°C and average reliability of 95.88%.

3) Reliability results of the proposed solar cell based PUF against light intensity variations: Unlike CMOS IC's, light intensity plays a major role in controlling the electrical parameters of the solar cells. So, in this paper, we have also performed the reliability analysis of the proposed solar cell based PUF against light intensity variations. In this experi-

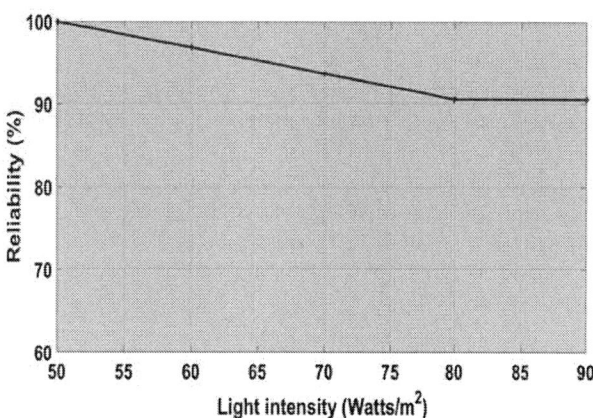

Fig. 12. Reliability of the proposed solar cell based PUF against light intensity variations

ment, we have used LED bulb to emulate the real time day light behavior with 300 lumens. A variable transformer is used to vary the light intensity of the LED bulb.

Fig. 12 shows the reliability of the proposed solar cell based PUF with the light intensity variations. In our experiment, we have chosen 50 $Watts/m^2$ as the reference light intensity value. One of the reasons to choose this low level of light intensity is to emulate a normal room environment where the light intensity is around 10-50 $Watts/m^2$. We have varied the light intensity from 50 $Watts/m^2$ to 90 $Watts/m^2$ using a variable transformer. The proposed PUF has worst case reliability of 90.62% at 80 and 90 $Watts/m^2$. The average reliability of the proposed solar cell based PUF against light intensity variations is 92.96%.

VII. DISCUSSION AND CONCLUSION

A low hardware cost solar cell based PUF that generates reliable bits has been developed and evaluated in this paper. This PUF uses the open circuit voltage in solar cells as an entropy source to generate the response bits. From our experimental results, we have observed that this solar cell based PUF has the uniformity value of 49.21% which is close to the ideal value of 50%. The worst case reliabilities of the proposed solar cell based PUF are 92.97% and 90.62% with variations in temperature and light intensity, respectively.

Low implementation cost along with the novel solar cell based entropy source make our PUF an ideal choice to implement in IoT devices. Further, vehicular security is a critical consideration today due to manifold use of IoT devices in a vehicular network [20], [21]. The embedded IoT devices in vehicles are susceptible to malicious cyber-attacks [22]. Therefore, we are also exploring the application of the proposed solar cell based PUF to create a security infrastructure at the hardware and software level to improve vehicular security.

ACKNOWLEDGMENT

This research was partially supported by grants from Kentucky Science and Engineering Foundation per Grant Agree-

ment KSEF-3998-RDE-020 and National Science Foundation under Grant No:1738662.

REFERENCES

[1] D. Bandyopadhyay and J. Sen, "Internet of things: Applications and challenges in technology and standardization," *Wireless Personal Communications*, vol. 58, no. 1, pp. 49–69, 2011.

[2] D. Lund, C. MacGillivray, V. Turner, and M. Morales, "Worldwide and regional internet of things (iot) 2014–2020 forecast: A virtuous circle of proven value and demand," *International Data Corporation (IDC), Tech. Rep*, vol. 1, 2014.

[3] A. Klinefelter, N. E. Roberts, Y. Shakhsheer, P. Gonzalez, A. Shrivastava, A. Roy, K. Craig, M. Faisal, J. Boley, S. Oh *et al.*, "21.3 a 6.45 μw self-powered iot soc with integrated energy-harvesting power management and ulp asymmetric radios," in *Solid-State Circuits Conference-(ISSCC), 2015 IEEE International*. IEEE, 2015, pp. 1–3.

[4] X. Liu and E. Sánchez-Sinencio, "An 86% efficiency 12 μw self-sustaining pv energy harvesting system with hysteresis regulation and time-domain mppt for iot smart nodes," *IEEE Journal of Solid-State Circuits*, vol. 50, no. 6, pp. 1424–1437, 2015.

[5] E. Yablonovitch, O. D. Miller, and S. Kurtz, "The opto-electronic physics that broke the efficiency limit in solar cells," in *Photovoltaic Specialists Conference (PVSC), 2012 38th IEEE*. IEEE, 2012, pp. 001 556– 001 559.

[6] C. Marchand, L. Bossuet, U. Mureddu, N. Bochard, A. Cherkaoui, and V. Fischer, "Implementation and characterization of a physical unclonable function for iot: a case study with the tero-puf," *IEEE Transactions on Computer-Aided Design of Integrated Circuits and Systems*, vol. 37, no. 1, pp. 97–109, 2018.

[7] R. Anderson and M. Kuhn, "Low cost attacks on tamper resistant devices," in *International Workshop on Security Protocols*. Springer, 1997, pp. 125–136.

[8] K. Kursawe, D. Schellekens, and B. Preneel, "Analyzing trusted platform communication," in *ECRYPT Workshop, CRASH-CRyptographic Advances in Secure Hardware*, 2005.

[9] G. E. Suh and S. Devadas, "Physical unclonable functions for device authentication and secret key generation," in *Proceedings of the 44th annual design automation conference*. ACM, 2007, pp. 9–14.

[10] M. N. Aman, K. C. Chua, and B. Sikdar, "Position paper: Physical unclonable functions for iot security," in *Proceedings of the 2nd ACM international workshop on IoT privacy, trust, and security*. ACM, 2016, pp. 10–13.

[11] D. Mukhopadhyay, "Pufs as promising tools for security in internet of things," *IEEE Design & Test*, vol. 33, no. 3, pp. 103–115, 2016.

[12] Y. Hori, T. Yoshida, T. Katashita, and A. Satoh, "Quantitative and statistical performance evaluation of arbiter physical unclonable functions on fpgas," in *Reconfigurable Computing and FPGAs (ReConFig), 2010 International Conference on*. IEEE, 2010, pp. 298–303.

[13] D. Merli, F. Stumpf, and C. Eckert, "Improving the quality of ring oscillator pufs on fpgas," in *Proceedings of the 5th workshop on embedded systems security*. ACM, 2010, p. 9.

[14] D. E. Holcomb, W. P. Burleson, and K. Fu, "Power-up sram state as an identifying fingerprint and source of true random numbers," *IEEE Transactions on Computers*, vol. 58, no. 9, pp. 1198–1210, 2009.

[15] P. Würfel, *Physics of solar cells*. Wiley-vch Weinheim, 2005, vol. 1.

[16] K. Rosenfeld, E. Gavas, and R. Karri, "Sensor physical unclonable functions," in *Hardware-Oriented Security and Trust (HOST), 2010 IEEE International Symposium on*. IEEE, 2010, pp. 112–117.

[17] E. Aponte, "A study on energy harvesters for physical unclonable functions and random number generation," Ph.D. dissertation, Virginia Tech, 2017.

[18] IXYS, "Ixolar high efficiency solarbit," Nov. 2016.

[19] T. Instruments, "Tiva tm4c123gh6pm microcontroller-data sheet," 2013.

[20] D. K. Oka, T. Furue, L. Langenhop, and T. Nishimura, "Survey of vehicle iot bluetooth devices," in *Service-Oriented Computing and Applications (SOCA), 2014 IEEE 7th International Conference on*. IEEE, 2014, pp. 260–264.

[21] A. Parodi, M. Maresca, M. Provera, and P. Baglietto, "An iot approach for the connected vehicle," in *International Internet of Things Summit*. Springer, 2015, pp. 158–161.

[22] J. Petit and S. E. Shladover, "Potential cyberattacks on automated vehicles," *IEEE Transactions on Intelligent Transportation Systems*, vol. 16, no. 2, pp. 546–556, 2015.

978-1-5386-7100-9/18 $31.00 © 2018 IEEE

2018 IEEE Computer Society Annual Symposium on VLSI

Designing Scalable Hybrid Wireless NoC for GPGPUs

Hui Zhao[1], Xianwei Cheng[1], Saraju P. Mohanty[1] and Juan Fang[2]

[1]Department of Computer Science and Engineering, University of North Texas

hui.zhao@unt.edu, xianweicheng@my.unt.edu, saraju.mohanty@unt.edu

[2] College of Computer Science, Beijing University of Technology

fangjuan@bjut.edu

Abstract—Data communication in GPU systems exhibits asymmetric patterns that create congestion hotspots. Due to their large number of cores and big die sizes, GPUs also demand highly scalable NoC designs. In this work, we propose hybrid NoC architectures that employ on-chip integrated antennas to build overlaid wireless networks on top of conventional metal/dielectric-based networks. We use low-power high-bandwidth wireless links as express channels to transmit long distance packets and use metal links to deliver local packets. The hybrid architecture can effectively alleviate congestion near traffic hotspots and improve the throughput and scalability of GPU NoCs. We propose solutions for design challenges in such hybrid NoC architectures, such as MAC protocol, router microarchitecture, load balancing and deadlock free routing. To efficiently utilize on-chip wireless bandwidth, we also propose a novel scheme that adaptively allocates bandwidth to wireless channels based on their usage needs. Our evaluation results show that for a GPU with 256 cores, the proposed hybrid architecture can improve performance by 2.4 times on average.

Keywords: NoC, GPGPU, wireless NoC, scalable NoC

I. INTRODUCTION

Recently GPU accelerated heterogeneous computing is becoming an attractive addition to high performance computing systems due to GPU's ability to launch massive parallel threads at same time. On-chip network need to be able to sustain high data communication throughput in order to reduce the idle time of GPU cores, in order to fully exploit GPU's parallel processing capability. However, compared with CMP based NoCs, design of NoC for GPUs is still at its infancy except for a handful of work [4], [6]–[10].

GPU data traffic has a very different pattern from the CMPs. Many GPU cores send read or write requests to a few memory controllers and receive reply messages from those controllers. The traffic load is imbalanced with reply messages taking up around 70% of total traffic. Hence, GPU NoC has a many-to-few-to-many traffic pattern which generates traffic bottlenecks around the memory controllers. If data generated by memory is not quickly moved away from the memory controllers, network throughput will severely suffer from the

resulted congestion hotspots. Another challenge in GPU NoC design is network scalability. GPUs have much larger die sizes compared with CMPs which demands highly scalable network design. Conventional metal/dielectric-based networks do not scale to thousands of cores in GPUs. For example, mesh networks are widely applied to GPU on-chip networks, due to their regular topology and simpler design in both router architecture and routing algorithms. However, hop counts in mesh networks increase significantly in large scale networks which greatly compromises the GPU system performance.

Recent breakthroughs in semiconductor integration technology have enabled new NoC design methods, such as nano-photonic and wireless NoCs, which have the advantage of high bandwidth, speed of light and low power consumption. There have been plenty of research applying these enabling technologies to design CMP based NoCs [11]–[16]. However, only a few of prior works target on GPUs [21], [22].

In this work, we explore the design space by employing wireless on-chip antennas to build hybrid NoC for GPUs. We made the following contributions:

(1) We proposed hybrid NoC architectures that use wireless links to build an overlaid network on top of the wired network. Our proposed design method can not only alleviate congestions near network bottlenecks but also improve network scalability.

(2) We investigated critical design issues in building the hybrid NoCs and propose solutions to address these issues, such as MAC protocol, router micro-architecture and deadlock free routing algorithms.

(3) By taking advantage of the reconfigurability of wireless networks, we developed a scheme to adaptively allocate bandwidth to wireless channels. Our scheme can effectively improve wireless bandwidth utilization.

II. MOTIVATION

GPU NoCs are usually implemented as two separate networks to avoid protocol deadlock. Request messages are sent from many shader cores to a few memory controllers (MCs) in the request network and MCs send reply messages to shader cores using the reply network. Thus the request network exhibits a many-to-few traffic pattern and the reply network has a few-to-many traffic pattern. Traffic load of the two networks

978-1-5386-7100-9/18 $31.00 © 2018 IEEE

703

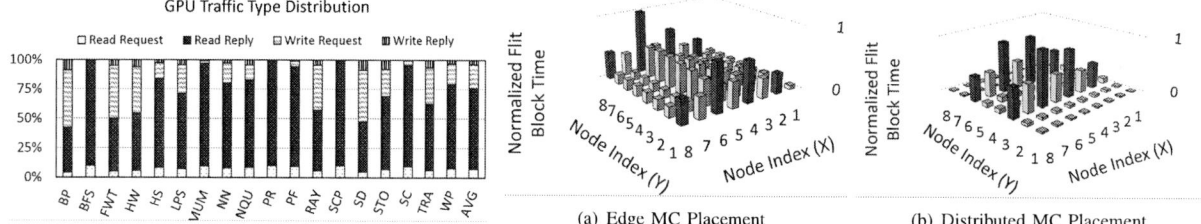

Fig. 1: GPGPU packet type distribution.

Fig. 2: Flit stall time in all routers for workload WP. Red colored bars represent stall time for MC connected routers.

(a) Edge MC Placement   (b) Distributed MC Placement

is unbalanced, with the reply network carrying most of the data packets. To quantitatively evaluate this load imbalance, we analyzed the distribution of packet types in Figure 1. It can be observed that about 75% of data is transmitted in the reply network. For some benchmarks, such as *BFS, MUM and PR*, the reply network accounts for more than 90% of all traffic. This is because there are much more read messages than write messages and read reply messages have much larger payload than other types of messages. It is obvious that improving the design of reply network will lead to more performance gains in the overall NoC. This has already been demonstrated by prior works [7], [10]. In this work, our target is to enhance the reply network design.

Due to its few-to-many pattern, traffic inside the reply network is also asymmetric. All packets are injected through the few routers connected with MCs and these routers become the network bottleneck. We evaluated the situation of router congestions in the reply network using workload *WP*. We use flit stall time, i.e., the total amount of time flits are blocked in a router's buffers as a measurement for congestion. It has been shown that MC placements affects the memory-processor traffic flow [4], so we experiment with two typical MC placements: Edge placement and Distributed placement. Edge placement has the advantage of simpler physical design and manufacturing process because the irregular MC nodes are placed at the border of the chip. Distributed placement increases design and routing complexity but has less biased access distance since MCs are located closer to the chip center. As can be observed in Figure 2, the most congested routers are those connected to MCs and their stall time can be x100 times larger than other routers. Routers close to MCs also tend to be congested and congestion gets alleviated as a router's distance from the MC increases. Those routers connected to MCs create network bottlenecks and negatively impact network performance much more than other routers. This is because they block new packets from being injected into the network, even though many other routers are not congested at all. In this work, we propose a solution to reduce network bottleneck by building an overlaid global network on top of the baseline network. When packets are injected from memory, global traffic is immediately directed into an overlaid high speed network and local traffic still goes through the relatively show baseline network.

Another critical challenge in designing GPU NoCs is the network scalability. Compared with CMPs, GPUs have much larger die size because they consists of thousands of processing cores. For example, the die size of Intel Core i7 is 246 mm^2, while NVIDIA Pascal GPU has a die size of 610 mm^2. It is expected GPU chip sizes will keep growing as more cores are integrated onto a chip. In conventional NoCs, routers are connected with short metal wires and such design cannot scale with large chip sizes because their performance degrades significantly as hop counts get larger.

Concentration techniques have been proposed to use long metal wires as express channels to connect far apart routers [5]. However, speed of metal wire decreases exponentially as wire length increases [14]. Recent breakthroughs in silicon integrated antennas has opened up new opportunities in designing scalable NoCs. Wireless links have the advantage of ultra-low power, long range communication without speed degradation and reduced wiring overhead. There have been several prior works that explore wireless NoC design for CMPs [12]–[15], but designing wireless NoC for GPUs has yet been well examined. In this work, we propose to build hybrid wireless NoC to reduce GPU traffic bottlenecks and tackle the scalability problem at the same time.

III. DESIGNING HYBRID WIRELSS NOC FOR GPGPUS

A. Hybrid Wireless NoC Architecture

Our proposed reply network consists two sub-networks: an underlying wired network and an overlaid wireless network. Figure 3 illustrated examples of the NoC architecture with both Edge and Distributed MC placements. The underlying network uses mesh topology and is divided into small clusters, with one memory controller mapped to each cluster. Cluster size is decided by the total number of routers and number of MCs. For example, there are 64 routers and 8 MCs in our example, so the cluster size is 8. The interleaved green and white regions represent clusters and the nodes with blue color represent MCs. Every MC is directly connected to a wireless router and all these wireless routers are connected as an overlaid mesh network. When a packet is injected from a MC, we first decide which network to enter based on the distance to destination. If the number of hops to the destination is below a predefined threshold, the packet will enter the wired network. Otherwise, the packet will be directed to the wireless network and get transmitted to its destination cluster. Then the packet will be ejected from the wireless network into the wired network and continue to reach its destination. Our topology has two differences from conventional concentrated mesh: firstly,

978-1-5386-7100-9/18 $31.00 © 2018 IEEE

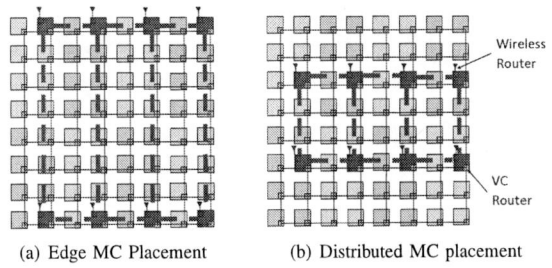

(a) Edge MC Placement   (b) Distributed MC placement

Fig. 3: Topology of the proposed hybrid wireless NoC with different MC placements.

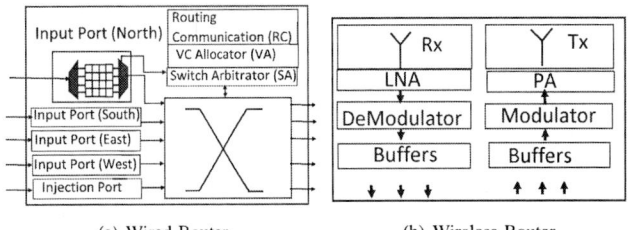

(a) Wired Router   (b) Wireless Router

Fig. 4: Major component of wired and wireless routers.

the upper level routers (i.e. wireless routers) are not placed at the center of a cluster but MCs; secondly, conventional upper level routers are connected to all routers in the cluster and thus have high radix and larger cost. Each of our wireless router is connected to the MC node only. Our design incurs lower cost by taking advantage of GPU traffic pattern.

All shader cores are connected with wired router only and each MC is connected with both a wired and a wireless router. Our wired routers employs conventional 5-port virtual channel design which is illustrated in Figure 4(a). These routers use look ahead routing and has two pipeline stages. Figure 4(b) shows major components in one wireless channel which has two stacks: physical layer and MAC layer. The physical layer consists of transceivers, modulator/demodulators and buffers. Transceivers use oscillators to generate different carrier frequencies for data modulation/demodulation. Power amplifiers (PA) and low-noise amplifiers (LNA) are used to amplify received or transmitted signals. The MAC layer contains Medium Access Control (MAC) modules that determine when data should be transmitted without causing collisions. Interconnection between the wired and wireless routers is implemented at each MC's network interface. We design control logics in the MC's network interface to handle packet injection/ejection as well as traffic flow coordination between the two networks.

We employ wireless links as express channels to alleviate congestion near memory controllers and reduce network bottleneck. However, our experiments show if all packets with long distant destination are sent to the wireless network, the wireless network will get congested. We need admission control to balance traffic between the two networks. We employ two policies: (1) before a memory controller injects a new packet to the wireless network, it first check the injecting queue length of that router. If the length is above a preset threshold (L_{th}), the packet will be injected to wired network instead. (2) If a packet is blocked in a wireless router for a time longer than a threshold value (W_{th}), the packet will be ejected to the wired network. We use these policies to balance the traffic between the two networks in order to fully utilize the bandwidth of both networks. We use experiments to retrieve these two threshold values and these values are adjustable based on run time situation.

In wireless networks, channels are usually shared by multiple users to improve bandwidth usage. It is important to

have a Medium Access Control (MAC) mechanism to avoid collisions. There are several types of MAC developed such as Time Division Multiple Access (TDMA), Frequency Division Multiple Access (FDMA) and Code Division Multi Access (CDMA). We choose token passing as our MAC control mechanism which is a type of TDMA protocol. This is because token based MAC is simple to implement and does not require central control which makes it a better choice for on-chip networks. In this protocol, there is only one token flit circulating among the routers that share one wireless channel. All sharers of a wireless channel are connected by a token ring and a router can transmit data on the shared channel only when it processes the token. In our design, we use a separate high speed control network to pass the token flits.

In our wired and wireless mesh networks, we employ deadlock free routing algorithms to route packets. However, we need to guarantee that there is no deadlock created when packets traverse between these two networks. Deadlocks are created when two packets hold resource (usually buffers) needed by the other packet and are waiting for each other to move forward in a loop. We enforce the following policy to avoid deadlock: a packet can only move from wireless network to wired network, but not in the opposite direction. Once a packet travels on the wired network, it will never be directed to the wireless network. Under this policy, a packet traveling on the wireless network can be blocked by a packet traveling on a wired network when it is ejected from the wireless network. However, a packet traveling on the wired network will never be blocked by a packet on the wireless network. Therefore, no loop can be formed between the wireless and wired network and our network is deadlock free. Our policy works similar to the turn model that avoids formation of loops by forbidding certain turns. Our network is also livelock free since all packets will finally be injected to the underlying network (due to W_{th}) and that network is livelock free.

Next we explain the power model for the proposed hybrid network. The power of our proposed hybrid NoC can be partitioned into two main contributions: wired network and wireless network.

$$E_{total} = E_{wired} + E_{wireless} \qquad (1)$$

$$E_{wireless} = E_{wireless}^{dynamic} + E_{wireless}^{static} \qquad (2)$$

$$E_{wireless}^{static} = (P_{static}^{tx} + P_{static}^{rx}) \times C \times T \qquad (3)$$

978-1-5386-7100-9/18 $31.00 © 2018 IEEE

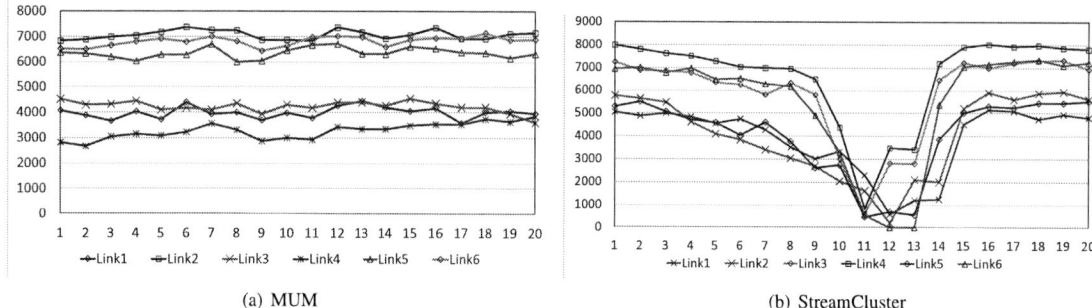

| (a) MUM | (b) StreamCluster |

Fig. 5: Wireless link usage for benchmark MUM and StreamCluster. Three most and least used links are measured in 20 epochs. X-axis represents epochs and Y-axis represents link usage counts.

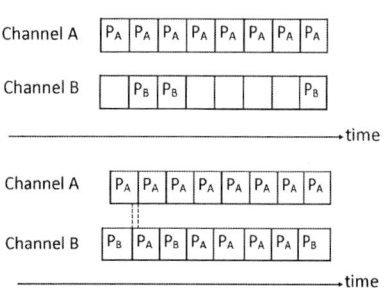

Fig. 6: Adaptive Wireless Channel Sharing.

$$E_{wireless}^{dynamic} = (P_{dynamic}^{tx} + P_{dynamic}^{rx}) \times b/R \qquad (4)$$

The hybrid network power consumption consists of both wired and wireless contributions. We use the built-in power model of GPGPU-Sim to evaluate wired network. Here we explain the wireless power model which is described in the Equation (1)-(4). The model is similar to the one developed in [20] where C and T in equation 3 are the total number of execution cycles and system clock period. In equation 4, b and R are total bits transmitted and wireless channel data rate. P_{static}^{tx} and P_{static}^{rx} are static values of transmitters and receivers. Dynamic transceiver power is denoted by $P_{dynamic}^{tx}$ and $P_{dynamic}^{rx}$ respectively. We retrieve these values using NoXim [17] and use GPGPU-Sim to collect other statistics such as simulation cycles.

B. Adaptive Wireless Bandwidth Allocation

One of the key advantages of wireless networks over point-to-point networks (such as metal or nano-photonics) is their reconfigurability. A single channel can be shared between multiple users, greatly reducing wiring complexity, area overhead and design cost. Routers can also make changes to their channels at run time, by adjusting the transmitting frequency. To efficiently utilize the available bandwidth, we propose a novel scheme that adaptively allocate wireless bandwidth to channels based on usage requirement.

We first analyzed the wireless link usage in our proposed network using benchmark *MUM* and *StreamCluster*. In Figure 5, we plot usage of the three most heavily used and three most lightly used links in 20 epochs. It can be observed that for *MUM*, the ratio between the highest to lowest link usage is in

the range of 2.3-1.5 times. For *StreamCluster*, the ratio is about 2.8 to 1.4 times. Such imbalance leads to wasted bandwidth because some links are allocated more resource than they need while busy links do not have enough bandwidth.

Although there have been schemes about bandwidth allocation in wireless communication, we cannot simply borrow them into the on-chip networks. For example, we can proportionally divide total bandwidth among all links based on their usage and allocate the new bandwidth in next epoch. This method may lead to optimal bandwidth usage, but it is not suitable for the on-chip environment because it incurs huge cost and design complexity.

We employ a light weight scheme called "Borrow from the Rich". Instead of reallocating bandwidth to all links, our scheme only involves part of the links. The idea is to assign all links with equal bandwidth and allow a busy link to share bandwidth with a less used link. At the end of each epoch, we pick links with the highest usage and put them into a "borrowing pool". Similarly, we create a "lending pool" using least used links. Then we pick a link from each pool and build a token ring between them to share bandwidth of the lending link. The borrowing link now has its bandwith increased, including its original bandwidth and part of the lending link's bandwidth.

Figure 6 illustrates how this scheme works. Initially link A and link B are assigned same amount of wireless bandwidth. However, link A is congested but link B is idle half of the time. To fully utilize link B's bandwidth, we allow link A and B to share the same wireless bandwidth that is originally allocated to link B. To avoid conflict, a token ring is formed between link A and B. At the same time, link A still has its private channel. So whenever link B has no packet to send, it sends a token to link A's router, then link A can send a packet using their shared bandwidth. Note there is a small amount of delay overhead that is caused by token passing in our scheme.

IV. EVALUATION

A. Methodology

We use GPGPU-Sim [1] to simulate our proposed hybrid network. Table 1 shows the configuration used in our evaluation. Our baseline network consists of a 8x8 2D mesh with 56 computing cores and 8 memory controllers. The overlaid

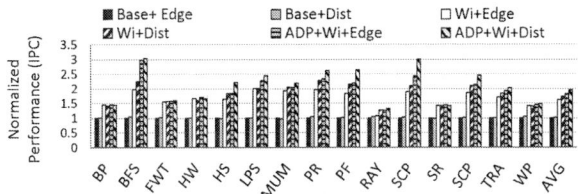

Fig. 7: Normalized performance in a 64-node NoC.

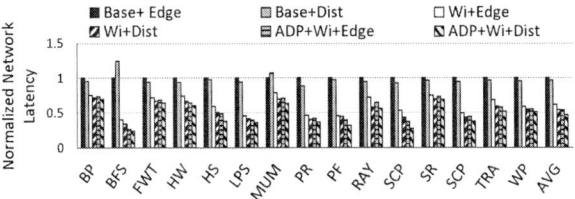

Fig. 9: Normalized network latency in a 64-node NoC.

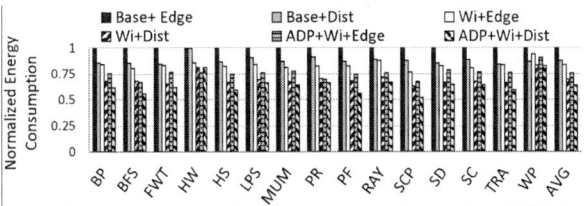

Fig. 11: Normalized energy consumption in a 256-node Noc.

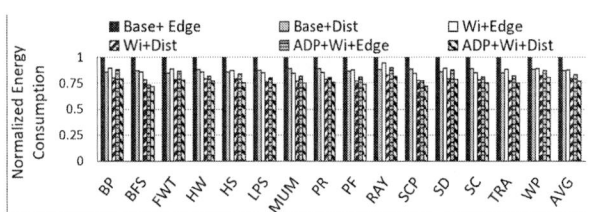

Fig. 8: Normalized energy consumption in a 64-node NoC.

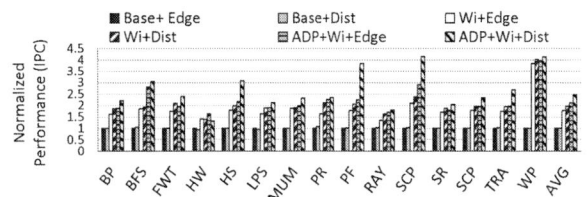

Fig. 10: Normalized performance in a 256-node NoC.

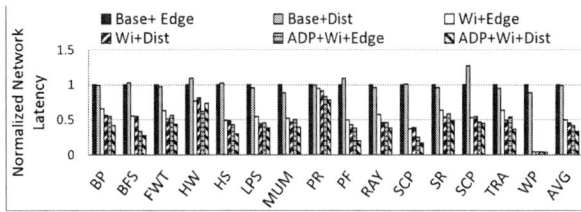

Fig. 12: Normalized network latency in a 256-node NoC.

wireless network is implemented as 2x4 mesh. We choose mesh topology for wireless network because it has low design complexity and simple routing algorithms. We experimented with two types of MC placements (edge and distributed) in order to observe their impacts on the hybrid wireless network. We used GPU workloads from ISPASS [1], Rodinia [2] and Cuda SDK [3] to evaluate our design. To simulate the power consumption, we collect parameters using NoXim [17]. The wireless link delay is determined by flit size, channel data rate and system clock frequency. We set flit size to 128 bits and wireless channel data rate to be 20Gbps, so it takes 6.4 ns to transfer one flit. We simulate a GPU system similar to NVIDIA GTX480 which has a clock frequency of 1.4 GHz. Therefore the single hop wireless delay is 9 clock cycles.

TABLE I: System configuration.

Shader Core	56 cores, 1.4GHz, SIMT width=8
Warp Scheduler	Greedy-Then-Oldest
Shared Memory	48 KB
Cache	2KB L1 I-Cache (4 sets/4 ways LRU), 16KB L1 D-Cache (32 sets/4 ways LRU), 64KB L2 Cache per MC (8 way LRU)
Memory Model	8 MCs, 924 MHz
Wired NoC	128-bit channel width, 2-stage pipeline, 16-byte flits, 1-cycle link latency, X-Y routing, vc buffer depth=4
Wireless NoC	128 bit flits, 4 flits per packet, X-Y routing, 9-cycle link latency, transmit energy=0.5 pJ/bit, receive energy=0.7 pJ/bit, epoch interval=50000
subnet	2

B. Performance and Power Analysis

The performance evaluation of our proposed hybrid wireless network architectures is shown in Figure 7. We compare six sets of results with different MC placements and both static and adaptive wireless bandwidth allocation schemes. The results are normalized to a baseline 8x8 mesh network. In most cases, Distributed MC placement has better performance than Edge placement. The hybrid architecture greatly improved the network performance. On average, the hybrid wireless network improved performance by 61% for Edge MC placement and 72% for Distributed MC placement. The adaptive bandwidth allocation scheme further improves performance for both MC placements by about 81% and 94%. *BFS* and *SCP* achieve the most performance gain and their IPC are increased by 3 times. All benchmarks achieve more than 1.5 times performance gain except *BP*.

Figure 8 shows the energy consumption of the six NoC architectures under evaluation. Energy consumption of all NoCs are normalized to base line with Edge MC placement. It can be observed that Distributed MC placement consumes less energy compared with Edge placement. This is because MCs are located near the center of the network and reduced average transmission distance. Comparing with baseline, static bandwidth allocation scheme can save energy by 13% and 21% averagely for the two MC placements. With adaptive bandwidth allocation, energy savings are improved by 17% and 24% on average. This is because wireless links have

978-1-5386-7100-9/18 $31.00 © 2018 IEEE

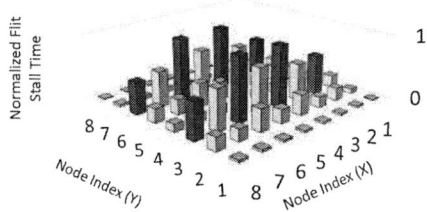

Fig. 13: Flit stall time in routers of hybrid network for workload WP.

very low power consumption compared with metal wires and the global network significantly reduced hop counts in the network. We also evaluated network latency as show in Figure 9. The hybrid NoC reduce network latency by 50% on average. *BFS* and *SCP* receive maximum decrease in network latency which translates to largest performance gain.

To evaluate the impact of our hybrid NoC on network scalability, we performed sensitivity analysis using a network with 256 nodes. Figure 10, Figure 11 and Figure 12 show the evaluation results for performance, energy and network latency respectively. Our results show the GPU NoCs receive even more benefit from our hybrid design as network size increases. On average, our proposed NoC can improve IPC by 2.47 times and reduce energy consumption by 37%. The hybrid wireless NoC exhibits good scalability and provides a promising solution in designing large GPU systems.

Figure 13 shows the flit stall time for workload *WP* in our hybrid NoC. The flit block time is normalized to the longest one, so the absolute value does not represent real time. Instead the important information is the relative flit block time in all routers. When we compare Figure 13 with Figure 2(b), we can find the difference between MCs and other routers are decreased. This means traffic is more evenly distributed among the routers and our design can effectively alleviate congestion in traffic bottlenecks.

Next we discuss the overhead of our hybrid NoCs. It is estimated that wireless transceiver and digital part of a wireless router is around 0.3 mm^2 respectively [12]. The overall area of a wireless router is estimated to be 0.8 mm^2. If the network consists 8 wireless routers, the overall area overhead is 5.6 mm^2. Comparing with the size of NVIDIA GTX480 GPU which is 529 mm^2, the area overhead of wireless routers is negligible.

V. RELATED WORK

To reduce cost, a checkerboard architecture was proposed to design GPGPU NoCS [6]. Through VC monopolization and employing asymmetric request and reply networks, Jang et. al. proposed a bandwidth efficient NoC design [4]. Kim et.al proposed a conflict-free design for the reply network called DA2mesh [10] which assigns each memory node a dedicated channel-sliced network. There are other GPGPU NoC schemes such as asymmetric cmesh [9] and ring-chain network [8]. However, these scheme treats MC connected routers same as other router when allocating network bandwidth and hardware resource, wasting resource on non-congesting routers. Compared with those schemes, our design improves network performance by removing network bottleneck and enhancing NoC scalability.

VI. CONCLUSION

In this paper, we analyzed the network bottleneck of on-chip communication within GPUs. We propose hybrid NoC architectures to design the reply network and employ wireless links to remove network bottlenecks. We also provided solutions to critical challenges in designing such networks. Our experiment results show that the proposed design can not only achieve better power-performance efficiency but also improve network scalability.

REFERENCES

[1] A. Bakhoda, G. Yuan, W. Fung, H. Wong, and T. Aamodt. Analyzing CUDA workloads using a detailed GPU simulator. In *ISPASS*, 2009.

[2] S. Che, M. Boyer, J. Meng, D. Tarjan, J. Sheaffer, S.-H. Lee, and K. Skadron. Rodinia: A Benchmark Suite for Heterogeneous Computing. In *IISWC*, 2009.

[3] NVIDIA. CUDA C/C++ SDK Code Samples. 2011. [Online]. Available: http://developer.nvidia.com/cuda-cc-sdk-code-samples

[4] H. Jang, J. Kim, P. Gratz, K. Yum, E. Kim. Bandwidth-Efficient On-Chip Interconnection Designs for GPGPUs. In *DAC*, 2015.

[5] T. Krishna, A. Kumar, P. Chiang, M. Erez, Li-S. Peh. NoC with near-ideal express virtual channels using global-line communication. In *HOTI*, 2008.

[6] A. Bakhoda, J. Kim, T.M. Aamodt. Throughput-Effective On-Chip Networks for Manycore Accecerators. In *Proc. MICRO*, 2010.

[7] X. Zhao, S. Ma, Y. Liu, L. Eeckhout, Z. Wang. A low-cost conflict-free NoC for GPGPUs. In *Proc. DAC*, 2016: 34:1-34:6.

[8] X. Zhao, S. Ma, C. Li, L. Eeckhout, Z. Wang. A heterogeneous low-cost and low-latency Ring-Chain network for GPGPUs. In *Proc. ICCD*, 2016: 472-479.

[9] A. Kavyan, J. L. Abellan, Y. Ma, A. Joshi, D. Kaeli. Asymmetric NoC Architectures for GPU Systems. In *Proc. NoCs*, 2015.

[10] H. Kim, J. Kim, Wong. Seo, Y. Cho, S. Ryu. Providing Cost-effective On-Chip Network Bandwidth in GPGPUs. In *Proc. ICCD*, 2012.

[11] Y. Pan, P. Kumar, J. Kim, G. Memik, Y. Zhang, A. Choudhary. Firefly: Illuminating Future Network-on-Chip with Nanophotonics. In *ISCA*, 2009.

[12] A. Ganguly, K. Chang, S. Deb, P. P. Pande, B. Belzer, C. Teuscher. Scalable Hybrid Wireless Network-on-Chip Architectures for Multicore Systems. In *IEEE Transactions on Computers, VOL. 60, NO. 10*, October, 2011.

[13] S. Lee, et.al. A Scalable Micro Wireless Interconnect Structure for CMPs. In *MobiCom*, 2009.

[14] S. Deb, A. Ganguly, P. P. Pande, B. Belzer, D. Heo. Wireless NoC as Interconnection Backbone for Multicore Chips:Promises and Challenges. In *IEEE Journal on Emerging and Selected Topics in Circuits and Systems, Volume: 2, Issue: 2*, June, 2012.

[15] D. DiTomaso, A. Kodi, D. Matolak, S. Kaya, S. Laha, W. Rayess. Energy-efficient adaptive wireless NoCs architecture. In *NoCs*, 2013.

[16] H. Gu, J. Xu, W. Zhang. A low-power fat tree-based optical network-on-chip for multiprocessor system-on-chip. In *DATE*, 2009.

[17] V. Catania, A. Mineo, S. Monteleone, M. Palesi, D. Patti. Noxim: An open, extensible and cycle-accurate network on chip simulator. In *ASAP*, 2015.

[18] X. Yu, J. Baylon, P. Wettin, D. Heo, P. Pande, S. Mirabasi. Architecture and Design of Multi-Channel Millimeter-Wave Wireless Network-on-Chip. In *IEEE Design and Test*, 31(6):19-28, 2014.

[19] N. Mansoor, A. Ganguly. Reconfigurable Wireless Network-on-Chip with Dynamic Medium Access Mechanism. In *NoCs*, 2015.

[20] S. Abadal, M. Iannazzo, M. Nemirovsky, A. Cabellos-Aparicio, H. Lee, E. Alarcon. On the area and energy scalability of wireless network-on-chip: A model-based benchmarked design space exploration. In *IEEE/ACM Transactions on Networking, Volume: 23, Issue: 5*, October, 2015.

[21] S. Gade, S. Deb. HyWin: Hybrid Wireless NoC with Sandboxed Sub-Networks for CPU/GPU Architectures. In *IEEE Transactions on Computers, VOL. 66, NO. 7*, JULY 2017.

[22] W. Choi, K. Duraisamy, R. Kim, J. Doppa, P. Pande, R. Marculescu, D. Marculescu. Hybrid network-on-chip architectures for accelerating deep learning kernels on heterogeneous manycore platforms. In *CASES*, 2016.

978-1-5386-7100-9/18 $31.00 © 2018 IEEE

2018 IEEE Computer Society Annual Symposium on VLSI

Functional Obfuscation of DSP cores using Robust Logic Locking and Encryption

Anirban Sengupta[1], Saraju P. Mohanty[2]
[1]Computer Science & Engineering, Indian Institute of Technology Indore, India
[2]Computer Science & Engineering, University of North Texas, USA
Email: asengupt@iiti.ac.in; Saraju.Mohanty@unt.edu

Abstract— **Obfuscation plays a key role in thwarting attacks launched through reverse engineering process. This work presents a new obfuscation process for DSP cores using improved logic locking and encryption that incurs minimum design overhead and achieves reduced design cost compared to state of the art approaches. The proposed approach integrates particle swarm optimization driven design space exploration system (PSO-DSE) for obtaining reduced design cost of obfuscated DSP designs. Enhanced security of locking is provided through locking blocks that are capable of locking each output data bit of functional resources with 8 key bits. The presented approach includes countermeasures against key sensitization attacks, SAT attacks and removal attacks. Results indicate that the proposed approach has been capable of achieving enhanced obfuscation security by at least 4.29 e+9 times and a design cost reduction ~ 6.5 % compared to a recent approach.**

Keywords— *robust locking, functional obfuscation, DSP core*

I. INTRODUCTION

The rapid technology scaling alongside with high cost of maintaining advanced fabrication facility has forced many design houses to become fabless. These fabless design houses have to rely on third-party fabrication facilities rendering feasibility of several threats resulting into IP piracy, Trojan, IC overbuilding etc. Consequently, several Intellectual Property (IP) core protection/hardware security mechanisms have been proposed such as IP metering, Trojan detection, watermarking, etc. [1-14]. Another recent mechanism is 'functional obfuscation' also known as 'functional locking' where the primary motive of functional locking is to insert locking components into the design such that correct output cannot be extracted until the valid keys are applied to the locked design.

Functional locking can be performed using several locking units such as AND/OR gate [3], muxes [4],[12], XOR/XNOR gate[7]. Each of these techniques has its own advantages and vulnerabilities. Authors in [7],[8] have presented 'key sensitization' based vulnerabilities and have suggested protection mechanism against it. Though the logic locking technique presented in [7],[8] is good, but it fails to integrate 'multi-pairwise' security. Further, this technique does not incorporate mechanism to generate optimal functionally obfuscated design as well as does not target DSP cores, unlike proposed approach.

In our proposed approach, we present novel 'IP functional locking blocks' (ILBs) for obfuscation of DSP cores. Further, through our sample ILBs we have presented a robust security locking against 'key sensitization' attacks through 'multi-pairwise' security. The novelties of proposed approach are:
a. The proposed approach presents novel ILB based functional obfuscation for DSP cores (represented as control data flow graphs (CDFGs)).
b. The proposed approach induces enhanced security in ILBs against 'key sensitization' attacks through 'multi-pairwise' security.
c. The presented methodology incorporates PSO-DSE to generate low-cost locked netlist based on power-delay tradeoff.

II. PREVIOUS WORKS

This paper targets protection against 'key sensitization attacks' (introduced in [7]) through sample IP Locking blocks. Authors of [7], [8] have introduced few security features that provides protection against 'key sensitization attacks'. In our method, we have enhanced these security characteristics inducing enhanced resiliency against 'key sensitization attacks' as discussed later in section III. The approach presented in [7], [8] proposes resiliency through logic obfuscation using XOR/XNOR gates only. However, proposed ILBs being a composite blend of several different gate types enhances security of our approach using 'multi-pairwise' security feature. Authors of [10], [7] have shown SAT attack on ISCAS'85 Benchmark. Although, SAT attacks are not scalable (applicable) on multiplication [15-19] (thus not applicable on multipliers present in DSP cores) we have shown proactive protection against SAT attacks using AES encryption as an anti-SAT block. Moreover, the proposed work integrates optimization framework to generate optimal solution using power-delay tradeoff.

III. OBFUSCATION APPROACH FOR DSP CORES

A. Problem

A CDFG, library, control parameters are provided as inputs. To generate a robust, low-cost, locked netlist resilient to 'key sensitization' and SAT attacks

978-1-5386-7100-9/18 $31.00 © 2018 IEEE   709

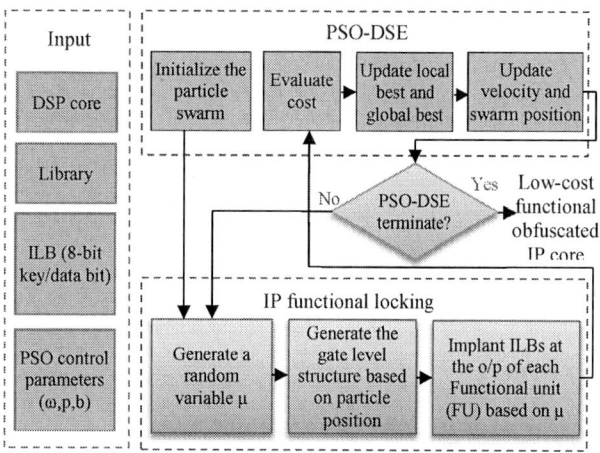

Fig.1. proposed functional obfuscation methodology

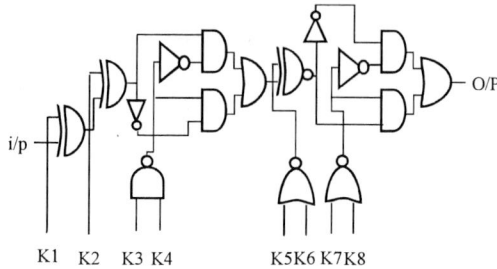

Fig. 2 Sample configured IP functional locking block

B. Motivation of using PSO-DSE during obfuscated netlist generation.

In this section we will elaborate on motivation for incorporating PSO-DSE framework. As depicted in fig. 1, introduced methodology includes chief components namely PSO-DSE component and IP functional locking component. The PSO-DSE component is responsible for exploring low cost design solution, while IP functional locking components performs the logic locking of the design solution. In initial step of our proposed methodology inputs are provided into PSO-DSE component where each particle is encoded as per eq.(1).

$$X_i = \{n(R1), n(R2), \ldots, n(Rd), \mu\}, \tag{1}$$

Where, X_i denotes i^{th} particle of the swarm, $n(R_d)$ signifies the number of resource in d^{th} dimension of the design space and μ is ILB insertion parameter. The initial particles are set using the following technique:

$X_1 = \{min(R_1), min(R_2), \ldots, min(R_d), \mu\}$

$X_2 = \{max(R_1), max(R_2), \ldots, max(R_d), \mu\}$

$X_3=\{(min(R_1)+max(R_1))/2,(min(R_2)+max(R_2))/2,\ldots (min(R_d) + max(R_d))/2, \mu\}$,

Where, $min(R_1)$ and $max(R_1)$ denotes minimum and maximum number of resources of resource type R_1. Similarly, the remaining particles in the swarm can be initialized as

$X_i = [\{(min(R_1)+max(R_1))/2 \pm \alpha, (min(R_2)+max(R_2))/2 \pm \alpha, \ldots, (min(R_D) + max(R_d))/2 \pm \alpha\}, \mu]$

Where α symbolizes an arbitrary integer between minimum and maximum number of resource in d^{th} dimension of the design space. Subsequently, for each particle $X_1, X_2, \ldots, X_n$ based on its respective position (resource configuration) in the design space gate level structure is produced. Later on the sample IP functional locking blocks (ILBs) are implanted at the output bit of resource Rj, as per ILB insertion parameter 'μ'. For example if '$\mu = 2$' then one of the ILBs is randomly selected and inserted at first two output bits of Rj. This process is repeated till ILBs are inserted at o/p bits of all the resources. Subsequently, for each particle X_n, the cost of the locked netlist is evaluated as per eq. (2)

$$C_f(X_i) = \varphi_1 \frac{P^{FL}}{P_{max}^{FL}} + \varphi_2 \frac{D^{FL}}{D_{max}^{FL}} \tag{2}$$

Here $C_f(X_i)$ represents normalized fitness of particle X_i, φ_1 and φ_2 signifies user specified weight of power and latency of the cost function (kept at 0.5 each to give same priority). P^{FL} and D^{FL} signify power and delay respectively of functionally locked (FL) design solution. P_{max}^{FL} represents maximum power of FL design in the design space. Likewise, D_{max}^{FL} signifies maximal latency of FL design. Once cost is evaluated local best is evaluated for each particle (X_i) as the minimal cost solution obtained by that specific particle till the present iteration. Subsequently, global best is evaluated. Subsequently, the particle's velocity and positions are updated. This is continued till stopping criterion is met (see [20] for PSO-DSE). Thus, an optimal solution is obtained based on power-delay tradeoff.

C. Security perspective of proposed IP functional locking methodology

C.1 ILB

We introduce IP functional locking blocks. A sample configured ILB is shown in Fig.2. Similar ILBs can be configured (with different architecture but same security) based on the designer's requirement (encrypted output). The sample ILBs includes strong security characteristics such as multi-pairwise security, valid key space, prevention of key gate seclusion etc. These characteristics deliver robust security against Reverse Engineering (RE) and key sensitization attacks. Using this attack, an adversary aims to recognize input combinations on locked netlist which (when applied on functional IC [7]) can produce valid key-bits to outputs.

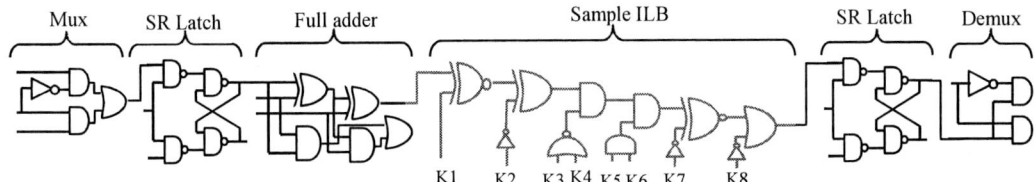

Fig. 3 randomly extracted portion of locked obfuscated netlist showing ILB reconfigured based on AES encrypted output

- Multi-pairwise security: If an attacker is unable to sensitize key-bit K1 to o/p without adjusting the value of key-bit K2 (vice-versa), then K1 and K2 are pairwise secure [7]. Multi-pairwise security is achieved when any key-bit cannot be sensitized to the output without adjusting the remaining key-bits (usually more than one). For example in fig. 2, any key bit of the Sample ILBs cannot be sensitized to the o/p, without adjusting all of the residual 7 key inputs. Thus, defence against key-sensitization based attack can be augmented using multi-pairwise security.

- Prohibiting key gate seclusion: Isolated key gates are vulnerable to key sensitization attack. An isolated key gate is described as a gate K_{iso} if there is non-existent link between K_{iso} and any of the residual key gates (key bit i/ps) and vice-versa. However, presented ILBs are a mixture of interdependent key inputs thus prohibiting isolated key gates.

- Defence against run of key gates: Some combinations of key gates linked adjacent to each other have been shown to be replaceable with a single key gate. This type of run of gates vulnerability is infeasible for ILB due to complex interleaving within gates for 8 key i/ps.

- Non-mutable key gates: Muting is an effort of an adversary to control primary input between any two key gates k_n and k_m such that k_n's value cannot prevent sensitization of k_m[7].

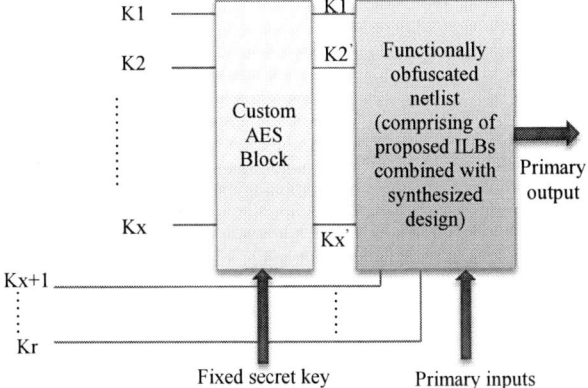

Fig. 4 Safeguarding from SAT attack and removal attack

Our proposed ILBs enhances the security of each key input with the remaining 7 key inputs i.e. an attacker cannot sensitize any key input without knowing/controlling remaining 7 key inputs. Moreover, there is no controllable primary inputs in our proposed ILBs.

C.2 Resiliency against different attack scenarios

(i) *Resiliency against key sensitization*: As discussed in the section III.C.1, a circuit comprises of isolated or mutable key gates is vulnerable to key sensitization attack. However, our customized ILBs doesn't comprises of either isolated or mutable key gates thus are resilient to key sensitization based attacks. Moreover, our proposed ILB structure enhances the security of the proposed approach through multi-pairwise security feature and confirms defense against run of key gates.

(ii) *Resiliency against IP piracy and Trojan insertion attacks*: The primary motive of a pirate is to achieve monetary gain by reselling an IP. However, to achieve this motive he/she has to unlock the correct functionality of the locked IP. Similarly, insertion of Hardware Trojans has to be done at safe places hence requires correct understanding of an IP. This being difficult for proposed work, hence makes proposed obfuscated design resilient to IP Piracy and Trojan insertion based attack.

(iii) *Resiliency against SAT attacks*: SAT attacks are not scalable for multiplications as its results in large CNF even for a small size multiplier. Since DSP cores comprise of several multiplications (multipliers), thus, SAT solver will not be scalable for these designs. Nevertheless, a proactive countermeasure against SAT (considering efficient SAT solvers are developed in future) using **lightweight (using less than 1 % of cyclone II FPGA resources) custom (not in public domain) AES block** is shown in fig.4. An AES circuit with fixed secret key for an input generates an encrypted output. Based on the

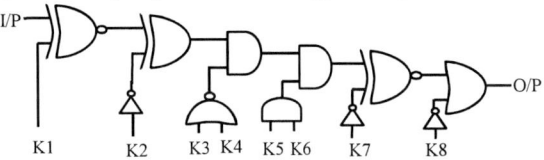

Fig.5. Sample of a single reconfigured ILB using encryption resulting from AES (only one ILB is shown for brevity)
Sample complete encrypted o/p: CDADC8663FFF1E8C2FD9F36409624A60

encrypted output the ILBs can be internally re-organized by the designer such that key inputs of ILBs matches with the encrypted output of AES block. Fig. 5 shows an example of a reconfigured ILB based on the AES.

(iv)*Removal attack*

(i) The presented approach uses subset of re-configured (re-organized) ILB (refer fig.5) implanted in the netlist. This reconfiguration is performed subjected to the AES encrypted o/p conforming to the secret key. This indicates that inside ILB configuration gets modified every time depending on the covert key and i/p selected. It is difficult for an attacker to recognize the reconfigured ILB as there is no fixed template and corresponding secret key to encrypt is unknown.

IV. IMPLEMENTATION

Approach [7] and approach work both have been realized in java and run on Intel Core i5 3210M CPU with 4GB. 15 nm NanGate library is as a base for evaluating the cell values of power and delay [21].

A. Security analysis

The security is represented through eq. (3)

$$K_S = 2 \wedge (b * m * f) \qquad (3)$$

Where K_S represents the key-space (S^{OBF}), b = key-bits per ILB, m = # of ILBs per resource, f = number of resources in the datapath. Table I shows that we have obtained a security enhancement of at least 4.29 e+9, w.r.t. [7] for the tested DSP benchmarks. This is because in the proposed approach we have incorporated 8-bit key per o/p data bit for improved logic locking. This results in higher functional obfuscation security than [7].

B. Design cost analysis

Table II illustrates the comparative study of cost between proposed approach and [7]. Cost minimization on average of 6.33% is observed for the tested DSP cores. As discussed earlier design cost reduction is achieved due to low-cost obfuscated design solution explored using PSO-DSE framework integrated with proposed obfuscation approach. The proposed approach results in marginal increase in critical path delay as overhead due to addition of ILBs (compared to baseline). However considering the bigger picture, the overall delay becomes optimized after integrating PSO-DSE compared to [7]. Thus production cost does not increase at all compared to state of the art techniques.

V. INFERENCE

This work introduced a new optimal obfuscation process that incorporates improved security techniques. Comparative study with [7] yielded significant security enhancement (strength of obfuscation) and reduction of cost.

Table I. Comparative study of proposed approach with [7] in terms of security (obfuscation)

Benchmark		No. of key-bits encoded for proposed obfuscation (r)	S^{OBF} of proposed approach (using eq. 3)	No. of key-bits encoded for [7] (r)	S^{OBF} of [7] (using eq. 3)	S^{OBF} enhancement of proposed approach (by factor of)
Name	Size					
DWT	10958	128	3.40 e+38	96	7.92 e+28	4.29 e+ 9
ARF	14833	256	1.15 e+77	112	5.19 e+33	2.23 e+43
FIR	16047	320	2.13 e+96	144	2.23 e+43	9.57 e+52
JPEG IDCT	42710	1344	3.83 e+404	432	1.10 e+130	3.46 e+274

Table II. Comparative study of proposed work with [7]

Benchmark	Proposed functionally obfuscated Design Solution	Cost of proposed approach	Design Solution of [7]	Cost of [7]	Cost Reduction (in %)
IIR	1A, 2M, μ=4	0.6810	2A, 4M	0.7427	8.30 %
DWT	1A, 1M, μ=1	0.7549	3A, 3M	0.7708	2.06 %
ARF	2A, 2M, μ=3	0.5259	3A, 4M	0.5281	0.41 %
FIR	3A, 2M, μ=4	0.5638	4A, 5M	0.5853	3.67 %
JPEG IDCT	11A,10M,μ=2	0.3629	12A,15M	0.4455	18.54 %

ACKNOWLDGEMENT

This Publication is an outcome of the R&D work undertaken in the project under the Visvesvaraya PhD Scheme of Ministry of Electronics & Information Technology, Government of India, being implemented by Digital India Corporation (formerly Media Lab Asia) and also financially supported by Council of Scientific and Industrial Research under sanctioned grant no. 22/730/17/EMR-II

REFERENCES

[1] A. Sengupta, "Intellectual Property Cores: Protection designs for CE products," in IEEE Consumer Electronics Magazine, vol. 5, no. 1, pp. 83-88, Jan. 2016.

[2] A. Sengupta, "Hardware Security of CE Devices [Hardware Matters]," in IEEE Consumer Electronics Magazine, vol. 6, no. 1, pp. 130-133, Jan. 2017.

[3] S. Dupuis, P. Ba, G. D. Natale, M. Flottes, and B. Rouzeyre, "ANovel Hardware Logic Encryption Technique for Thwarting Illegal Overproduction and Hardware Trojans," in Proc. IEEE International On-Line Testing Symposium, 2014, pp. 49–54.

[4] J. Rajendran, H. Zhang, C. Zhang, G. Rose, Y. Pino, O. Sinanoglu,and R. Karri, "Fault Analysis-Based Logic Encryption," IEEE Trans. Comput., vol. 64, no. 2, pp. 410–424, 2015.

[5] A. Sengupta and S. Bhadauria, "Exploring Low Cost Optimal Watermark for Reusable IP Cores During High Level Synthesis," in IEEE Access, vol. 4, pp. 2198-2215, 2016.

[6] A. Sengupta and S. Kundu, "Securing IoT Hardware: Threat Models and Reliable, Low-Power Design Solutions," in IEEE Transactions on Very

978-1-5386-7100-9/18 $31.00 © 2018 IEEE

Large Scale Integration (VLSI) Systems, vol. 25, no. 12, pp. 3265-3267, Dec. 2017.

[7] M. Yasin, J. Rajendran, O. Sinanoglu, and R. Karri. "On improving the security of logic locking." IEEE Transactions on Computer-Aided Design of Integrated Circuits and Systems 35, no. 9 (2016): 1411-1424.

[8] J. Rajendran, Y. Pino, O. Sinanoglu, and R. Karri. "Security analysis of logic obfuscation." In Proceedings of the 49th Annual Design Automation Conference, ACM, 2012, pp. 83-89.

[9] A. Sengupta, S. Bhadauria and S. P. Mohanty, "TL-HLS: Methodology for Low Cost Hardware Trojan Security Aware Scheduling With Optimal Loop Unrolling Factor During High Level Synthesis," in IEEE Transactions on Computer-Aided Design of Integrated Circuits and Systems, vol. 36, no. 4, pp. 655-668, April 2017.

[10] Subramanyan, Pramod, Sayak Ray, and Sharad Malik. "Evaluating the security of logic encryption algorithms." In Hardware Oriented Security and Trust (HOST), 2015 IEEE International Symposium on,pp. 137-143. IEEE, 2015.

[11] A. Sengupta and D. Roy, "Antipiracy-Aware IP Chipset Design for CE Devices: A Robust Watermarking Approach [Hardware Matters]," in IEEE Consumer Electronics Magazine, vol. 6, no. 2, pp. 118-124, April 2017.

[12] S. M. Plaza and I. L. Markov, "Solving the Third-Shift Problem in IC Piracy With Test-Aware Logic Locking," IEEE Trans. Comput.-Aided Design Integr. Circuits Syst., vol. 34, no. 6, pp. 961–971, 2015.

[13] A. Sengupta, D. Roy, S. P. Mohanty and P. Corcoran, "DSP design protection in CE through algorithmic transformation based structural obfuscation," in IEEE Transactions on Consumer Electronics, vol. 63, no. 4, pp. 467-476, November 2017.

[14] A. Sengupta, D. Roy and S. P. Mohanty, "Triple-Phase Watermarking for Reusable IP Core Protection during Architecture Synthesis," in IEEE Transactions on Computer-Aided Design of Integrated Circuits and Systems, vol. PP, no. 99, pp. 1-1.

[15] B. Brady, Y. Yang ,"The Effects of Arithmetic Encodings on SAT Solver Performance", [Online] http://citeseerx.ist.psu.edu/viewdoc/ download? doi=10.1.1.92.6332&rep=rep1&type=pdf.

[16] Y. Xie A. Srivastava, "Mitigating SAT Attack on Logic Locking", [Online] https://iacr.org/workshops/ches/ches2016/presentations/0917% 20Session%203/CHES2016_Session3_1.pdf, 2018

[17] P. Beame, V Liew "Towards Verifying Nonlinear Integer Arithmetic" 29 th international conf. on CAV, 2017, pp. 239.

[18] S. Chakraborty, A. Gupta, R. Jain "Matching Multiplications in Bit-Vector Formulas", Springer International Publishing, 2017, pp.131-150.

[19] M. Finke, "Equisatisfiable SAT Encodings of Arithmetical Operations", [Online] http://www.martin-finke.de/documents/ Masterarbeit_bitblast_ Finke.pdf, 2015.

[20] V. K. Mishra, and A. Sengupta. "MO-PSE: Adaptive multi-objective particle swarm optimization based design space exploration in architectural synthesis for application specific processor design." *Advances in Engineering Software* 67 (2014): 111-124.

[21] Express benchmarks http://www.ece.ucsb.edu/EXPRESS/benchmark/

Timing Macro Modeling for Efficient Hierarchical Timing Analysis

Iris Hui-Ru Jiang
Dept. of Electrical Eng. & Graduate Inst. of Electronics Eng..
National Taiwan University
Taipei, Taiwan
e-mail: huiru.jiang@gmail.com

Pei-Yu Lee
Institute of Electronics
National Chiao Tung University
Hsinchu, Taiwan
e-mail: palacedeforsaken@gmail.com

Abstract—**As designs continue to grow in size and complexity, EDA paradigm shifts from flat to hierarchical timing analysis. In this paper, we discuss timing macro modeling, which is the key to enable efficient and accurate hierarchical timing analysis. We briefly review conventional models and recent research progresses in timing macro modeling. We try to answer the following questions: How can timing macro models be made compact and accurate? How do state-of-the art works maintain model accuracy, model size, model generation performance, and model usage performance? Finally, future research directions on timing macro modeling are identified.**

Keywords-Static timing analysis; hierarchical timing analysis; timing macro modeling; interface logic model; extracted timing model

REFERENCES

[1] TAU 2016 Timing Contest on Macro Modeling. *ACM International Workshop on Timing Issues in the Specification and Synthesis of Digital Systems (TAU)*, Mar. 2016. Available at: https://sites.google.com/site/taucontest2016/

[2] TAU 2017 Timing Contest on Macro Modeling. *ACM International Workshop on Timing Issues in the Specification and Synthesis of Digital Systems (TAU)*, Mar. 2017. Available at: https://sites.google.com/site/taucontest2017/

[3] F. Dartu and Q. Wu, "To do or not to do hierarchical timing?" *Proc. ACM Int'l Symp. on Physical Design (ISPD)*, Mar. 2013, p. 180.

[4] Babul Annnay Hierarchical Timing Concepts. *EDN Network*, Oct. 2013. Available at: http://www.edn.com/design/integrated-circuit-design/4423327/Hierarchical-timing-concepts

[5] S. Walia. "Reducing turnaround time with hierarchical timing analysis," *EE Times*, Oct. 2011. Available at: http://www.eetimes.com/document.asp?doc_id=1279120

[6] C. Visweswariah, O. Levitsky, Q. Wu, A. Shaligram, A. Rubin, G. Wolski, A. Skourikhin, L. Brown, and I. Keller. "EDA court: Hierarchical construction and timing sign-off of SoCs," *ACM International Workshop on Timing Issues in the Specification and Synthesis of Digital Systems (TAU)*, Mar. 2013. Available at: http://www.tauworkshop.com/2013/presentations/tau2013_hierarchy_panel.ppt

[7] A. J. Daga, L. Mize, S. Sripada, C. Wolff, and Q. Wu, "Automated timing model generation," *Proc. ACM/IEEE Design Automation Conf. (DAC)*, June 2002, pp. 146-151.

[8] C. W. Moon, H. Kriplani, and K. P. Belkhale, "Timing model extraction of hierarchical blocks by graph reduction," *Proc. ACM/IEEE Design Automation Conf. (DAC)*, June 2002, pp. 152-157.

[9] Y.-M. Yang, Y.-W. Chang, and I. H.-R. Jiang, "iTimerC: Common path pessimism removal using effective reduction methods," *Proc. IEEE/ACM Int'l Conf. on Computer-Aided Design (ICCAD)*, Nov. 2014, pp. 600-605.

[10] Shuo Zhou, Yi Zhu, Yuanfang Hu, Roland Graham, Mike Hutton, and Chung-Kuan Cheng. Timing Model Reduction for Hierarchical Timing Analysis. In Proceedings of the 2006 IEEE/ACM International Conference on Computer-Aided Design (ICCAD '06), San Jose, CA, USA, Nov. 2006, pp. 415-422.

[11] T.-Y. Lai, T.-W. Huang, M. D. F. Wong, "LibAbs: An efficient and accurate timing macro-modeling algorithm for large hierarchical designs," *Proc. ACM/EDAC/IEEE Design Automation Conf. (DAC)*, June 2017, pp. 65:1-65:6.

[12] P.-Y. Lee, and I. H.-R. Jiang, "iTimerM: A compact and accurate timing macro model for efficient hierarchical timing analysis," *ACM Transactions on Design Automation of Electronic Systems (TODAES)*, vol. 23, no. 4, May 2018, pp. 48:1-48:21. Also see *Proc. ACM Int'l Symp. on Physical Design (ISPD)*, Mar. 2017, pp. 83-89.

2018 IEEE Computer Society Annual Symposium on VLSI

Timing with Virtual Signal Synchronization for Circuit Performance and Netlist Security

Grace Li Zhang, Bing Li, Ulf Schlichtmann

Chair of Electronic Design Automation, Technical University of Munich (TUM), Munich, Germany

Email: {grace-li.zhang, b.li, ulf.schlichtmann}@tum.de

Abstract—In digital circuit design, sequential components, e.g., flip-flops, are used to synchronize signal propagations using a global clock signal. With this design style, logic blocks are isolated by flip-flop stages, so that timing constraints can be addressed between pairs of flip-flops. Accordingly, the minimum clock period is determined by the maximum combinational delay between flip-flops. This design style reduces design efforts significantly. However, the timing performance of digital circuits might be affected negatively because sequential components can only delay signal propagations instead of accelerating them. In addition, the assumption that all combinational paths work within a single clock period indicates that the netlist carries all the design information, which makes ICs vulnerable to counterfeiting. In this paper, we demonstrate two techniques to break the confines of the traditional timing paradigm: VirtualSync and TimingCamouflage. With these techniques, circuit performance can be pushed even beyond the limit of the traditional timing paradigm and the netlist security can be enhanced.

I. INTRODUCTION

In traditional digital circuits, logic design and timing verification are separated. This design style is made possible by using sequential components, e.g., flip-flops, to block signal propagations along logic gates that actually process data. Consequently, logic blocks are isolated by flip-flops, so that timing verification is only needed between all pairs of flip-flops. This fully synchronous design style reduces design efforts significantly. However, it potentially leads to timing performance degradation. On one hand, sequential components such as flip-flops have clock-to-q delays, which become a part of the delays of combinational paths driven by the corresponding flip-flops. They also force signals to arrive setup time before the active clock edge, which consumes a further part of the timing budget from critical paths. On the other hand, delay compensation between consecutive flip-flop stages cannot be achieved because flip-flops block signal propagations instead of allowing them to pass through.

Figure 1 illustrates the scenarios where the timing performance in the traditional paradigm is restricted by the barriers of flip-flops. In Fig. 1(a), the circuit achieves a minimum clock period of 21 units. To reduce the clock period, logic gates can be sized to achieve smaller delays, so that signal propagations on the critical paths of the circuits can be accelerated. Fig. 1(b) shows the circuit after gate sizing with a smaller minimum clock period of 16 units. Retiming can be deployed to reduce the clock period further by moving F3 in Fig. 1(b) leftwards. Fig. 1(c) shows the circuit after retiming, whose functionality is the same as that in Fig. 1(b). The minimum clock period in Fig. 1(c) is 11 units, already the limit of timing performance in the traditional timing paradigm.

If the confines of the traditional timing paradigm are broken by removing the barriers of flip-flops, the timing performance

can be pushed further. Fig. 1(d) illustrates this concept, where F6 is removed from Fig. 1(c). If correct data is latched by F3 and F4 by guaranteeing that the signal from F2 reaches F3 and F4 after T and before $2T$, the functionality of the circuit is still maintained. The largest path delay is 16. Together with the setup time of 1 unit, the minimum clock period can be reduced further to (16+1)/2=8.5, 22.7% lower than the optimum performance in the traditional single-period clock paradigm.

The technique of allowing signals to span across flip-flop stages without a flip-flop blocking them can automatically balance timing slacks between successive flip-flop stages. Therefore, it alleviates the effect of process variations and aging, which are two factors affecting the IC design flow significantly. To counter process variations, previous methods such as post-silicon tuning [1], [2], [3], [4], [5], [6], [7] are deployed to tune circuits after manufacturing with respect to the effect of process variations. The technique to boost circuit performance until timing errors occur has been explored in the Razor method [8], [9], [10], [11]. In addition, the interdependency between setup and hold time has been considered to exploit the performance potential of flip-flops [12], [13], [14], [15]. To evaluate aging effects such as Negative Bias Temperature Instability (NBTI), Hot Carrier Injection (HCI), etc, timing models for aging analysis have been proposed [16], [17], [18], [19]. Among these methods, the AgeGate model [20], [21], [22] can be integrated into standard timing signoff flows directly [23]. The methods above dealing with process variations and aging incur additional overhead in circuit area, design and analysis complexity. This overhead can be reduced by removing the barriers of flip-flops.

The traditional timing paradigm also increases the security risk for digital circuits. In this paradigm, all combinational paths between pairs of flip-flops work within one clock period, indicating that the netlist carries all the design information. Therefore, attackers can reconstruct the original netlist to counterfeit chips with the recognized netlist from reverse engineering.

In this paper, we will discuss two techniques that break the confines of the traditional timing paradigm to improve the timing performance of digital circuits and to enhance circuit security against counterfeiting.

II. VIRUTALSYNC TIMING MODEL

A. Sequential and Combinational Components as Delay Units

To achieve multiple waves on combinational paths, all sequential components, e.g., flip-flops, are firstly removed from the circuit under optimization. Consequently, signals

978-1-5386-7100-9/18 $31.00 © 2018 IEEE   715

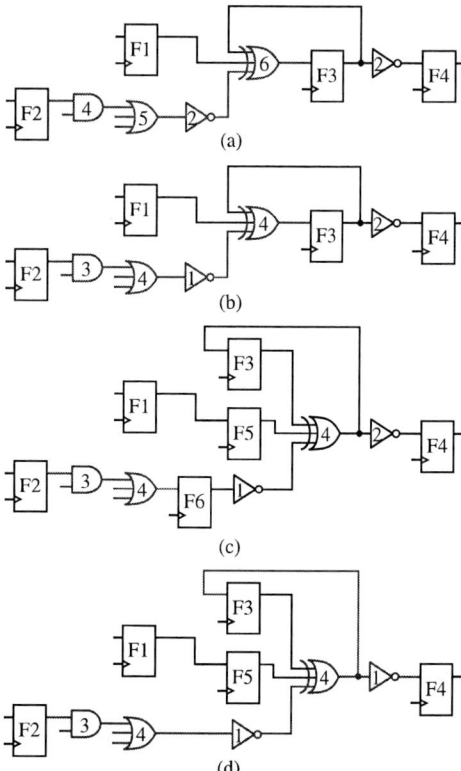

Fig. 1: Timing optimization methods. Delays of logic gates are shown on the gates. The clock-to-q delay (t_{cq}), setup time (t_{su}) and hold time (t_h) of a flip-flop are 3, 1 and 1, respectively. (a) Original circuit. (b) Sized circuit. (c) Circuit after retiming. (d) Circuit after optimization using VirtualSync.

across fast paths may arrive at flip-flops earlier than specified, leading to a loss of logic synchronization. Furthermore, a combinational loop must have a sequential component to guarantee logic synchronization after signals travel across it many times. Therefore, we need to delay the signal propagation on fast paths and combinational loops. To achieve this signal blocking, combinational gates such as buffers, flip-flops, and latches can be used as delay units. Fig. 2 illustrates the different delay characteristics of the three types of components, where ***input gap*** refers to the difference of arrival times of two signals at a delay unit, and ***output gap*** represents the difference between their arrival times after they pass through the unit.

In Fig. 2(a), a combinational delay unit such as a buffer has a linear delaying effect, indicating that there is no change in the absolute gap between arrival times of signals through short and long paths after passing through a buffer.

A flip-flop, as a sequential delay unit, has a different behavior, as shown in Fig. 2(b). No matter when two signals arrive at the input of a flip-flop, the departure time after passing through a flip-flop is the same. These characteristics are very useful to block signals on short paths when the delays of short and long paths are very different after all sequential components are removed.

Another sequential delay unit, a level-sensitive latch, has

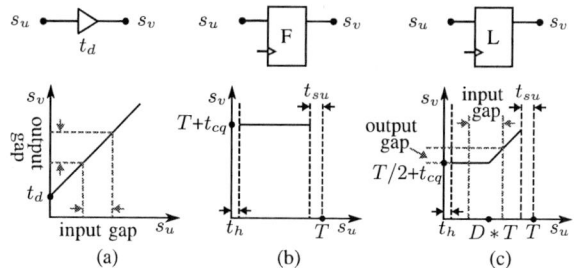

Fig. 2: Properties of delay units. (a) Linear delaying effect of a combinational delay unit. (b) Constant delaying effect of a flip-flop. (c) Piecewise delaying effect of a latch.

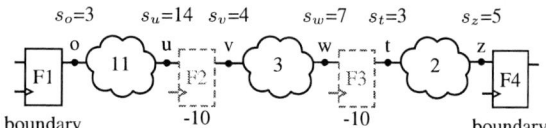

Fig. 3: Concept of relative timing references. Clock period T=10.

piecewise delaying effect, as shown in Fig. 2(c). The output gap of two input signals after passing through a non-transparent latch is reduced to zero. The output gap remains unchanged after a transparent latch is passed through. However, if the fast signals pass through a non-transparent latch and the slow signals pass through the latch when it is in transparency, the output gap takes a value between the two extreme values, as shown in Fig. 2(c). This property is very flexible to block signal propagations along critical paths where fast signals need to be delayed and slow signals should not be affected.

B. Relative Timing References

After removing sequential components from a circuit, signals may arrive at boundary flip-flops in incorrect clock cycles. To guarantee that the number of clock cycles along any path does not change after optimization, we introduce relative timing references.

The concept of relative timing references can be explained in Fig. 3, where the boundary flip-flops include F1 and F4 and the flip-flops F2 and F3 in the middle are removed initially for optimization. To guarantee the correct function of F4, the latest arrival time s_z for stable signals at the input of F4 should be t_{su} before the rising clock edge at F4 and the earliest arrival time s_z' should be larger than t_h to make F4 latch data correctly. These constraints can be expressed as follows

$$s_z + t_{su} \leq T \qquad (1)$$
$$s_z' \geq t_h. \qquad (2)$$

The clock period T in (1) indicates that the signal should arrive at F4 within one clock period. Accordingly, (1)-(2) are defined according to the reference of the rising clock edge at the original F3. Similarly, the timing constraint at F3 in the original circuit is defined according to the reference of the rising clock edge at F2. At the boundary flip-flops, this definition still holds. The locations of the removed flip-flops such as F2 and F3 are called ***anchor points***. After removing all sequential components from the circuit under optimization,

978-1-5386-7100-9/18 $31.00 © 2018 IEEE

these anchor points still provide relative timing information for boundary flip-flops. In VirtualSync, T is subtracted from the arrival time of signal when the signal passes an anchor point. When a signal finally arrives at a boundary flip-flop along a combinational path, its arrival time must have been converted by subtracting the number of flip-flops multiplied by T on the same path in the original circuit, so that (1)-(2) still hold.

C. Synchronizing Logic Waves by Delay Units

After removing all sequential components from the circuit under optimization, signals that are so fast that they reach boundary flip-flops too early need to be slowed down with delay units. Those signals traveling along the critical paths require no additional delay. The relative timing references, provided by the anchor points, guarantee that the number of clock cycles along any path is maintained after optimization. The objective of the optimization is to find the locations of delay units to make the circuit work at a given clock period T while reducing the area cost. Since it is not straightforward to determine the locations of delay units, we formulate this task as an ILP problem and solve it with a heuristic method.

D. Iterative Relaxation in VirtualSync

To apply the timing model above, we introduce a framework to identify the locations of delay units in an iterative way. In each iteration, necessary locations of sequential delay units are refined gradually. In the first step of the framework, we identify the locations at which sequential delay units are indispensable by emulating the delay effects of sequential delay units as virtual gaps. To refine the locations obtained from the first step, we incorporate clock/data-to-q delays of sequential delay units in the model. The last step is to legalize the timing of sequential delay units. To reduce the area overhead, buffers are allowed to be replaced with sequential delay units. A detailed description of this framework can be found in [24].

III. TIMING CAMOUFLAGE AGAINST COUNTERFEITING

In the traditional timing paradigm, all combinational paths work within one clock period. Within this paradigm, designers only need to focus on logic design without having to worry about the timing interference between different clock stages. Consequently, the functionality of a circuit depends only on its structure, a netlist thus carries all necessary design information. For instance, Fig. 4(a) shows a part of a digital circuit with three flip-flops F1, F2 and F3. If attackers can identify the types of combinational logic gates and their connection with F1, F2 and F3 through reverse engineering, they can reconstruct the netlist and then process it using a standard IC design flow. They can even revise and optimize the circuit freely. Thereafter, the design can be manufactured illegally, posing a big threat to IC vendors since the cost to develop a modern IC is extremely high.

To improve circuit security against counterfeiting, we invalidate the assumption that the netlist carries all design information by introducing wave-pipelining paths [25]. For example, the flip-flop in the middle in Fig. 4(a) can be removed to construct the circuit in Fig. 4(b) [26]. There are now two logic waves propagating along the combinational path from F1 to F3 simultaneously. If the second wave does not affect the propagation of the first wave, the two waves are then latched

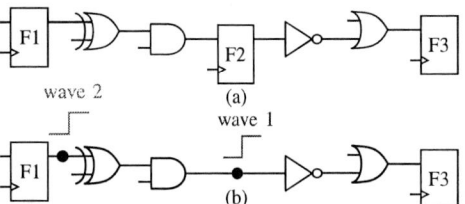

Fig. 4: Traditional timing and wave-pipelining. (a) Traditional clocking with single period. (b) Pipelining with two data waves.

by F3 correctly, so that the functionality of the circuit is still maintained.

Facing the camouflaged circuit with introduced wave-pipelining paths, attackers are forced to identify how many logic waves propagate simultaneously on the combinational paths between F1 and F3 even after the netlist is extracted through reverse engineering. If only one logic wave is assumed and the netlist is processed with a standard EDA flow, the reconstructed circuit does not function correctly because the data at the input of F3 is latched in incorrect clock cycles compared with the original circuit. To determine whether there are two data waves along combinational paths, timing information has to be extracted with additional effort. Accordingly, the functionality of the circuit depends on both its structure and the timing information of combinational paths.

A. Wave-Pipelining Construction

In constructing wave-pipelining paths to camouflage a circuit, the functionality of the circuit should be maintained. More precisely, the delay of wave-pipelining paths should be greater than T. Otherwise, data will be latched by flip-flops earlier than specified. The delays of these paths should also be less than $2T$ to guarantee that data is latched in correct clock cycles. The intuitive idea is to remove some flip-flops from the given circuit directly. For example, Fig. 5(a) illustrates this scenario, where ff_i is removed to construct wave-pipelining paths. However, the direct removal of flip-flops is not viable because on the left and right sides of ff_i, there are many short paths whose delays are very small. If we remove ff_i, these short paths will be connected together and their delays cannot satisfy the timing constraints of wave-pipelining paths.

To deal with the challenges from constructing wave-pipelining paths, we duplicate combinational logic in the original circuits, as shown in Fig. 5(b). We denote the flip-flop at which the wave-pipelining paths terminate as a WP flip-flop. To reduce the resource usage, on the right side of ff_i, we only duplicate combinational logic that drives WP flip-flops. Those combinational logic gates that do not drive any flip-flop in the original circuit are removed in Fig. 5(b). On the left side of ff_i, we firstly duplicate all the combinational logic to guarantee the same functionality with the original circuit. To reduce resource usage, we try to connect the input pins in the duplicated circuit with the original circuit. For example, if the input pin of the AND gate in the duplicated circuit in Fig. 5(b) is connected with that in the original circuit, the duplicated inverter can be removed. To satisfy the timing constraints of wave-pipelining paths, we apply gate sizing and buffer insertion to enlarge the

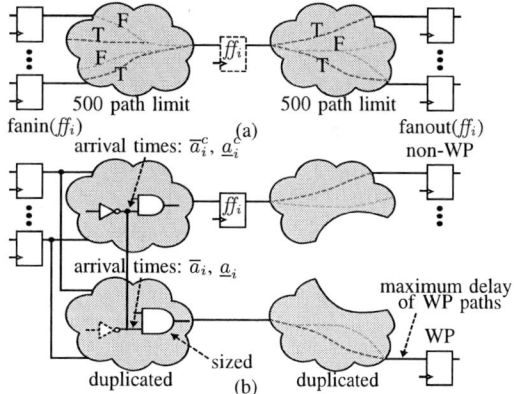

Fig. 5: Construction of WP path. (a) The flip-flop ff_i is removed directly. (b) Duplication of combinational logic and gate sizing.

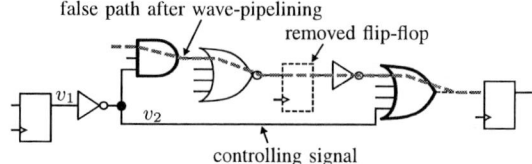

Fig. 6: A wave-pipelining false path formed with two true paths.

delays of wave-pipelining paths. The task to construct wave-pipelining paths is formulated as an ILP problem and details can be found in [26].

B. Wave-Pipelining False Paths

Another challenge of the timing camouflage technique is that attackers could deploy exhaustive testing to determine whether the delays of combinational paths are greater than T. To prevent combinational paths from being tested, we introduce wave-pipelining false paths (WP false paths). The constructed WP false paths cannot be triggered by any test vectors. This concept can be explained using Fig. 6. After removing the flip-flop in the middle, the left and right path of it are connected to form a wave-pipelining path. If this wave-pipelining path is considered to work within one clock period, it is also a false path because signals switching at the beginning of this path cannot reach the final flip-flop. For example, if the signal v_2 has the value '1', it blocks signal propagation at the OR gate. Similarly, if the signal v_2 has the value '0', signal propagation is blocked at the AND gate. Consequently, attackers cannot test the delay of this path directly. It is also difficult to differentiate this path from an original false path in the circuit, since this path becomes a false path and about 75% of the combinational paths in a digital circuit are already false paths [27].

IV. CONCLUSION

In this paper, we have presented two techniques to break the confines of the traditional timing paradigm. With VirtualSync, circuit performance can be pushed even beyond the limit of the traditional timing paradigm. With timing camouflage, attackers face challenges from exhaustive testing and differentiation between wave-pipelining false paths and real false paths. These methods demonstrate that there is still a great potential in innovating the traditional timing paradigm for fast and secure circuits.

REFERENCES

[1] G. L. Zhang, B. Li, and U. Schlichtmann, "Sampling-based buffer insertion for post-silicon yield improvement under process variability," in *Proc. Design, Autom., and Test Europe Conf.*, 2016, pp. 1457–1460.

[2] J. Tsai, L. Zhang, and C. C.-P. Chen, "Statistical timing analysis driven post-silicon-tunable clock-tree synthesis," in *Proc. Int. Conf. Comput.-Aided Des.*, 2005, pp. 575–581.

[3] G. L. Zhang, B. Li, and U. Schlichtmann, "EffiTest: Efficient delay test and statistical prediction for configuring post-silicon tunable buffers," in *Proc. Design Autom. Conf.*, 2016, pp. 60:1–60:6.

[4] G. L. Zhang, B. Li, J. Liu, Y. Shi, and U. Schlichtmann, "Design-phase buffer allocation for post-silicon clock binning by iterative learning," *IEEE Trans. Comput.-Aided Design Integr. Circuits Syst.*, vol. 37, no. 2, pp. 392–405, 2018.

[5] Z. Lak and N. Nicolici, "A novel algorithmic approach to aid post-silicon delay measurement and clock tuning," *IEEE Trans. Comput.*, vol. 63, no. 5, pp. 1074–1084, 2014.

[6] G. L. Zhang, B. Li, Y. Shi, J. Hu, and U. Schlichtmann, "EffiTest2: Efficient delay test and prediction for post-silicon clock skew configuration under process variations," *IEEE Trans. Comput.-Aided Design Integr. Circuits Syst.*, 2018.

[7] B. Li and U. Schlichtmann, "Statistical timing analysis and criticality computation for circuits with post-silicon clock tuning elements," *IEEE Trans. Comput.-Aided Design Integr. Circuits Syst.*, vol. 34, no. 11, pp. 1784–1797, 2015.

[8] D. Ernst, N. S. Kim, S. Das, S. Pant, R. Rao, T. Pham, C. Ziesler, D. Blaauw, T. Austin, K. Flautner, and T. Mudge, "Razor: A low-power pipeline based on circuit-level timing speculation," in *Proc. Int. Symp. Microarch.*, 2003, pp. 7–18.

[9] D. Blaauw, S. Kalaiselvan, K. Lai, W.-H. Ma, S. Pant, C. Tokunaga, S. Das, and D. M. Bull, "Razor II: In situ error detection and correction for PVT and SER tolerance," in *Proc. Int. Solid-State Circuits Conf.*, 2008, pp. 400–401.

[10] M. Fojtik, D. Fick, Y. Kim, N. R. Pinckney, D. M. Harris, D. Blaauw, and D. Sylvester, "Bubble razor: Eliminating timing margins in an ARM cortex-m3 processor in 45 nm CMOS using architecturally independent error detection and correction," *IEEE J. Solid-State Circuits*, vol. 48, no. 1, pp. 66–81, 2013.

[11] I. Kwon, S. Kim, D. Fick, M. Kim, Y.-P. Chen, and D. Sylvester, "Razor-lite: A light-weight register for error detection by observing virtual supply rails," *IEEE J. Solid-State Circuits*, vol. 49, no. 9, pp. 2054–2066, 2014.

[12] N. Chen, B. Li, and U. Schlichtmann, "Iterative timing analysis based on nonlinear and interdependent flipflop modelling," *IET Circuits, Devices & Systems*, vol. 6, no. 5, pp. 330–337, 2012.

[13] A. B. Kahng and H. Lee, "Timing margin recovery with flexible flip-flop timing model," in *Proc. Int. Symp. Quality Electron. Des.*, 2014, pp. 496–503.

[14] Y.-M. Yang, K. H. Tam, and I. H.-R. Jiang, "Criticality-dependency-aware timing characterization and analysis," in *Proc. Design Autom. Conf.*, 2015, pp. 167:1–167:6.

[15] G. L. Zhang, B. Li, and U. Schlichtmann, "PieceTimer: A holistic timing analysis framework considering setup/hold time interdependency using a piecewise model," in *Proc. Int. Conf. Comput.-Aided Des.*, 2016, pp. 100:1–100:8.

[16] S. V. Kumar, C. H. Kim, and S. S. Sapatnekar, "An analytical model for negative bias temperature instability," in *Proc. Int. Conf. Comput.-Aided Des.*, 2006, pp. 493–496.

[17] V. B. Kleeberger, M. Barke, C. Werner, D. Schmitt-Landsiedel, and U. Schlichtmann, "A compact model for NBTI degradation and recovery under use-profile variations and its application to aging analysis of digital integrated circuits," *Microelectronics Reliability*, vol. 54, no. 6–7, pp. 1083–1089, 2014.

[18] H. Amrouch, B. Khaleghi, A. Gerstlauer, and J. Henkel, "Reliability-aware design to suppress aging," in *Proc. Design Autom. Conf.*, 2016, pp. 12:1–12:6.

[19] N. Koppaetzky, M. Metzdorf, R. Eilers, D. Helms, and W. Nebel, "RT level timing modeling for aging prediction," in *Proc. Design, Autom., and Test Europe Conf.*, 2016, pp. 297–300.

[20] D. Lorenz, M. Barke, and U. Schlichtmann, "Efficiently analyzing the impact of aging effects on large integrated circuits," *Microelectronics Reliability*, vol. 52, no. 8, pp. 1546–1552, 2012.

[21] D. Lorenz, G. Georgakos, and U. Schlichtmann, "Aging analysis of circuit timing considering NBTI and HCI," in *Int. On-Line Testing Symp. (IOLTS)*, 2009, pp. 3–8.

[22] D. Lorenz, M. Barke, and U. Schlichtmann, "Aging analysis at gate and macro cell level," in *Proc. Int. Conf. Comput.-Aided Des.*, 2010, pp. 77–84.

[23] S. Karapetyan and U. Schlichtmann, "Integrating aging aware timing analysis into a commercial STA tool," in *Int. Symp. on VLSI Des., Aut. and Test (VLSI-DAT)*, 2015, pp. 1–4.

[24] G. L. Zhang, B. Li, M. Hashimoto, and U. Schlichtmann, "VirtualSync: Timing optimization by sychronizing logic waves with sequential and combinational components as delay units," in *Proc. Design Autom. Conf.*, 2018.

[25] W. P. Burleson, M. Ciesielski, F. Klass, and W. Liu, "Wave-pipelining: A tutorial and research survey," *IEEE Trans. VLSI Syst.*, vol. 6, no. 3, pp. 464–474, 1998.

[26] G. L. Zhang, B. Li, B. Yu, D. Z. Pan, and U. Schlichtmann, "TimingCamouflage: Improving circuit security against counterfeiting by unconventional timing," in *Proc. Design, Autom., and Test Europe Conf.*, 2018, pp. 91–96.

[27] K. Heragu, J. H. Patel, and V. D. Agrawal, "Fast identification of untestable delay faults using implications," in *Proc. Int. Conf. Comput.-Aided Des.*, 1997, pp. 642–647.

978-1-5386-7100-9/18 $31.00 © 2018 IEEE

2018 IEEE Computer Society Annual Symposium on VLSI

Realizing Closed-loop, Online Tuning and Control for Configurable-cache Embedded Systems: Progress and Challenges

Islam S. Badreldin*, Ann Gordon-Ross*, Tosiron Adegbija†, and Mohamad Hammam Alsafrjalani‡

*University of Florida, †University of Arizona, ‡University of Miami

*{ibadreldin,anngordonross}@ufl.edu, †tosiron@email.arizona.edu, ‡alsafrjalani@miami.edu

Abstract—**The cache subsystem is a major contributor to energy consumption in commercial microprocessors used in embedded systems. To reduce energy, designers can perform design space exploration (DSE) to determine a suitable cache configuration that matches system constraints and goals while minimizing energy consumption. Traditionally, this cache tuning step has been a static process where heuristics or analytical models are used to determine an optimal or near-optimal cache configuration prior to runtime given a known application, application set, or application domain. Even though the configuration may change during runtime for different phases of execution, the specific configuration for each phase remains fixed. This static nature is too restrictive for modern, complex embedded systems that are expected to operate under diverse, unknown operating environments, run unknown applications, and with vastly different user quality of experience (QoE) expectations (e.g., smart phones). Therefore, cache tuning must change from a static optimization process to a dynamic optimization process that adapts online during runtime transparently to the user/system needs. The key challenge is determining the configuration that adheres to QoE expectations while minimizing energy consumption without degrading the user experience during DSE. Despite the wealth of progress that has been made, the realization of a closed-loop, fully adaptive, online-tunable cache subsystem still faces many challenges. In this paper, we review the progress made in the area of static and dynamic cache tuning, discuss the challenges that still exist in this area, and propose a prediction-assisted control-theoretic framework to address these challenges.**

I. INTRODUCTION AND MOTIVATION

Heterogeneous multicore processors are becoming pervasive in embedded systems as a way to enable the efficient execution of increasingly complex applications while minimizing energy consumption. Using a system with different core configurations enables applications to execute on a core that is suitable given the application's performance/energy goals. Different core configurations range from coarse-grained optimization options, such as cores with different instruction set architectures (ISAs), voltages, frequencies, specialized hardware (graphic processing units (GPUs) and field-programmable gate arrays (FPGAs)), etc. to fine-grained optimization options, such as different pipeline depths, reorder buffer size, instruction issue width, etc. Since the cache subsystem consumes 16% to 50% [1], [2] of the total system power, this fine-grained option offers potentially significant energy savings, thus making the cache subsystem an ideal optimization candidate.

The cache subsystem provides optimized performance and reduced energy consumption by exploiting application-specific resource requirements using the *best* (optimal or near-optimal) cache configuration that most closely adheres to system constraints and goals. This configuration designates the best values for the cache's tunable parameters, such as cache size, line size, and associativity. Since prior work showed that tuning these particular cache parameters enable the largest energy savings potential [3], without loss of generality, we assume that these parameters are tunable in our work.

To provide flexible optimization options, configurable caches (e.g., [3]) must be suitable for modern complex embedded systems. These caches must operate in unknown and changing environments, run a wide variety of potentially unknown applications, and adhere to system constraints, goals, and runtime-defined time-varying user-defined quality of experience (QoE) expectations. Given the diversity of subjective end-user QoE expectations, the tunable cache parameters must now also be able to change during runtime in response to user-specific feedback as well as application requirements while adhering to system constraints.

Achieving this goal is extremely challenging due to many factors since QoE expectations vary substantially during runtime for individual users and between different users. For example, based on a user's battery lifetime or performance quality expectancies, the best configuration can range from lowest energy to best performance, respectively. These expectancies may differ for different types of applications and environmental conditions, such as time-of-day (e.g., less frequent notifications during work hours), ambient lighting (e.g., slower frame rates in low lighting), accelerometer readings (e.g., fewer GPS updates while walking as compared to driving). Since different applications or application phases (i.e., execution intervals with stable performance characteristics, such as instructions per cycle (IPC), cache miss rate, etc.) have varying runtime memory requirements [4], [5], [6], [7], the best cache configuration must correlate these differences with the user QoE expectations, resulting in an extremely large and complex design space.

Design space exploration (DSE) is an optimization process that dynamically evaluates an *optimization objective function* to dynamically determine the best cache configuration at runtime. The objective function includes *optimization metrics* [8],

978-1-5386-7100-9/18 $31.00 © 2018 IEEE 719

such as application performance and/or energy consumption. Efficient and effective DSE is a challenging process, and prior work has extensively explored static and dynamic solutions (e.g., [5], [6], [9], [10], [11], [12], [13], [14], [15], [16], [17]) using heuristic searches, subsetting methods, or analytical models.

Complex heterogeneous architectures exacerbate many cache tuning challenges, and introduce new challenges that necessitate novel dynamic cache tuning methods. To the best of our knowledge, few prior works have begun to address these challenges, some of which include: determining when to reconfigure the cache–the *tuning interval*–with no prior knowledge of the running applications [5]; considering runtime-defined time-varying user-defined QoE expectations and optimization metric objectives; considering the general complexity of multicore architectures (e.g., cache coherence, data sharing among cores, synchronization, etc.).

Another critical challenge is addressing DSE's impact on the user experience. Since exploring configurations far from the best configuration incurs significant tuning overhead (e.g., significantly increased power or reduced performance) [5], DSE must ensure that while determining the best configuration, the user experience is not degraded beyond a dissatisfaction threshold (e.g., the point at which the user experiences anger or frustration with the device). One recent work by Alsafrjalani et al. [6] studied online tuning while considering DSE-time degradation and showed that best configurations could be determined while limiting this degradation.

Despite all of the advancements in cache tuning methods, these outstanding challenges leave a void in adaptive, online, closed-loop cache tuning systems that consider runtime-defined time-varying user-defined QoE expectations while limiting DSE-time degradation. In this work, we review the progress made toward static and dynamic cache tuning, detail some of the challenges that face the progress in the area of online dynamic cache tuning, and propose a system architecture that addresses some of these challenges.

II. General Problem Formulation

In this section, we formally define the general problem of cache tuning for configurable-cache embedded systems, define optimization metrics, and review general DSE solutions.

A. Formal Notations

Following the notation in [11], we denote a set of n applications $A = \{a_1, a_2, \ldots, a_n\}$ to be executed on an embedded system with a configurable cache offering a set of m possible cache configurations $C = \{c_1, c_2, \ldots, c_m\}$. Each application has a set of k_i phases $P_{a_i} = \{p_1, p_2, \ldots, p_{k_i}\}$, wherein each phase can require a different cache configuration. Without loss of generality, we assume an optimization process in which a Pareto-optimal configuration (i.e., best configuration) that trades off the energy and performance optimization metrics must be determined [9].

We denote the energy and performance optimization metrics for application a_i executing in phase p_k with cache configuration c_j as $e(a_i, p_k, c_j)$ and $u(a_i, p_k, c_j)$, respectively. The

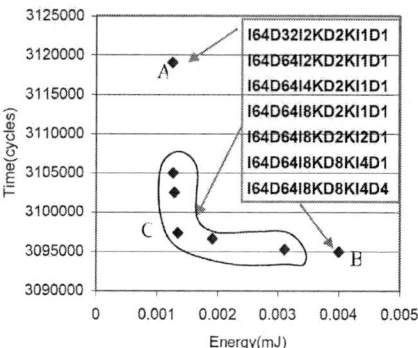

Fig. 1. Example Pareto-optimal set of cache configurations as presented in [9] where points A, B, and C belong to the Pareto-optimal set. The legend denotes tunable parameter values for each configuration.

optimization objective is to determine a Pareto-optimal configuration $c_0 \in C$ such that for application a_i, $e(a_i, p_k, c_0) \leq e(a_i, p_k, c_j)$ where $c_0 \neq c_j$, and a similar condition on $u(a_i, p_k, c_0)$. The set of all Pareto-optimal configurations is known as the Pareto-optimal set [18], which includes the best configurations in terms of both metrics [9]. At runtime, the dynamic optimization process selects a configuration from the Pareto-optimal set given additional constraints such as QoE [6], resulting in a QoE-aware optimization objective.

Besides determining Pareto-optimal configurations, a simpler optimization problem that is often employed in previous work focuses on minimizing only the energy optimization metric [5], [6], [10], [11], [12], [14], [16]. In this case, an optimal solution is one such that $e(a_i, p_k, c_0) \leq e(a_i, p_k, c_j)$ where $c_0 \neq c_j$. An additional simplification that is sometimes used involves determining a single configuration that is best, on average, for an application's complete execution (application-based cache tuning). In this case, the dependency of the optimization metrics on the application phase is removed (i.e., $e(a_i, p_k, c_j)$ becomes $e(a_i, c_j)$ and $u(a_i, p_k, c_j)$ becomes $u(a_i, c_j)$).

B. Optimization Metrics

Pareto-optimal-based optimization involves a trade off analysis between two or more metrics. Given N optimization metrics, the Pareto-optimal set will always contain at least N required configurations. As an example, Fig. 1 illustrates this principle for a sample Pareto-optimal set for cache configurations (parameter values are denoted in the legend) that trade off total energy consumption in mJ and performance in clock cycles. The two required configurations are the minimum energy consumption, point A, and the best performance, point B, configurations. Point C also belongs to the Pareto-optimal set and represents a configuration that trades off performance and energy consumption.

In the context of configurable cache tuning, several metrics have been previously used. Common performance optimization metrics (i.e., $u(.)$) include the cache miss rate [9], [16], [19] or execution time in cycles [9]. The energy optimization metric (i.e., $e(.)$) can be estimated using cache performance statistics and modeling tools [10], [14], [15], [16] such as CACTI [20], or a set of equations that represent an energy model [9], [11], [21]. The estimation can be done during runtime using cache performance hardware counters and the energy equation model can be implemented in a custom co-processor or dedicated hardware to provide a hardware-based energy calculation module [5].

C. Design Space Exploration (DSE)

There are many different methods for exploring the design space to determine the best cache configuration. Exhaustive DSE would evaluate the optimization metrics for every cache configuration $c_i \in C$. Even though this method accurately determines the optimal configuration, as the number of configurable cache parameters, parameter values, and hierarchy depth grows, exhaustive DSE is computationally and runtime prohibitive given increasingly large design spaces. Several approaches have attempted to reduce DSE complexity using heuristics and subsetting methods that reduce the number of configurations explored. These methods, while inexact, determine near-best configurations while avoiding a large penalty in terms of the optimization objective [10], [6], [9], [11], [14], [15], [16], [17].

Other approaches circumvent the complexity of DSE by using analytical or statistical models that attempt to directly determine the best cache configuration given application memory access patterns and characteristics [12], [22], or application phase information [7], [13]. Since phase-based methods provide greater benefits than application-based methods, Sections IV and V focus on phase-based approaches.

Regardless of the exploration method, DSE can be divided into two categories. In offline DSE, the design space is statically searched before runtime using full or partial knowledge of the applications or application domains that are expected to run on the embedded system [6], [9], [10], [11], [14], [15], [16], [17]. A key feature of offline exploration is that these approaches can batch-evaluate the optimization objective for each cache configuration in software (e.g., [11], [16]) or in hardware (e.g., [9]) before runtime. Online DSE profiles the applications dynamically at runtime using real input stimuli (e.g., [5], [6], [12], [13]), thus more accurately determining the best configuration at the expense of DSE overhead. We further refer the reader to [23] for a complete review of the different DSE methods. In the next two sections, we review and discuss the progress in offline and online design space exploration for configurable caches.

III. OFFLINE DESIGN SPACE EXPLORATION (DSE)

Offline DSE can be done either in software simulation using trace-based methods or instruction set simulators, or using hardware emulation methods. Regardless of the method used, technological and system complexity advances have resulted in an exponential growth of the design space resulting in prohibitively large file sizes (for trace-based) and DSE time. To reduce DSE time, methods that approximate the design space

using a subset that contains near-best solutions have been developed to prune the design space. This section discusses several solutions for offline DSE and design space subsetting.

A. Simulation- and Hardware-based Exploration

Given full or partial knowledge of the application(s) or the application domains that are expected to run on the embedded system, cycle-accurate instruction set simulators (e.g., [24], [25]) can be used to estimate the energy and performance associated with each possible cache configuration. Approaches using this method were extensively reviewed in [23]. Even though a powerful host machine can be used to run the simulator and reduce the time needed to explore each configuration, this method is only feasible for optimization problems that involve a relatively small number of possible cache configurations. For larger design spaces on more complex systems where a single configuration may take several hours or days to simulate, this time needs to be further reduced [11], [16].

DSE-time can be considerably reduced using hardware-based emulation. Zhang et al. [9] developed a runtime reconfigurable hardware cache architecture that enabled the cache size, line size, and associativity to be changed using a small set of configuration registers. This architecture was intended for prototype-oriented platforms to rapidly explore the configuration design space for a set of known applications to determine the Pareto-optimal set. Considering the example Pareto-optimal set illustrated in Fig. 1, the authors implemented a hardware-based heuristic that rapidly searched for points A and B, and then searched for point C.

B. Subsetting Methods

The majority of the subsetting approaches [11], [15], [16], [17] rely on the principle of near-optimality. The authors observed that many cache configurations in the design space provide approximate values of the optimization objective function. Therefore, at the expense of a small penalty to the optimization objective, the design space can be reduced from C to C'—a smaller subset of the original design space (i.e., $|C'| \ll |C|$). While C' is not guaranteed to contain a globally optimal configuration, the authors showed that C' contains many near-optimal configurations with the benefit of a much smaller design space to explore in C' as compared to C. In order to compute C', the authors used a clustering approach that was originally applied to online segmentation of time series data [26]. Some subsetting methods [10], [14], [21] were geared toward online DSE and are discussed in the Section IV. We refer the reader to [17], where these approaches were recently reviewed, for further details.

IV. ONLINE DESIGN SPACE EXPLORATION (DSE)

For embedded systems that are expected to run applications that are unknown a priori, the cache configuration design space must be explored online at runtime. This is typically the case for consumer embedded devices [6], such as smartphones and tablets. Online exploration in its most basic form involves the execution of a given application on the target architecture using different cache configurations in order to capture the values of the optimization objective function under each configuration. DSE is followed by selecting the best configuration that most closely meets optimization requirements and constraints from the configurations that were explored. Collectively, these two steps are referred to as cache tuning [23]. This section briefly overviews several online cache tuning methods.

A. Heuristic-based Methods

Heuristic methods selectively explore/search only a portion of the design space by using information about the current and past explored configurations to choose the next configuration to explore. Zhang et al. [21] developed a greedy heuristic search to determine the lowest energy configuration (i.e., point A in Fig. 1). The parameters were explored in increasing order of the parameters' impacts on the energy consumption for single-level caches. Gordon-Ross et al. [10], [14] extended this approach to two-level caches and demonstrated that this heuristic search explored only a small fraction of the original design space without significantly compromising energy consumption. Alsafrjalani et al. [6] proposed an online QoE-aware approach [11] for multi-core cache configurations using a heuristic-based search to perform online DSE that minimized DSE-time degradation. Zhao et al. [27] analyzed the behavior of a large pool of applications offline and used the learned information to heuristically select a best cache configuration online.

B. Phase-based Methods

During execution, an application's characteristics, such as cache miss rate, IPC, branch misprediction rate, etc. change over the application's different execution phases. Prior work demonstrated that the best cache configurations differ even within the same application according to these application phases [4], [7]. Therefore, approaches have emerged that take phase behavior into account during online cache tuning [13], [7].

Peng et al. [13] proposed multiple state machines that monitor the program execution to make cache reconfiguration decisions while attempting to avoid unnecessary reconfigurations. Hajimir et al. [28] proposed an intra-task dynamic cache reconfiguration method that used a dynamic programming-based algorithm to determine the best cache configuration, while reducing DSE overhead. However, these methods still required DSE whenever the cache configuration needed to be changed. More recently, Adegbija et al. [7] proposed a phase distance mapping method that eliminated the need for DSE by directly estimating the best configuration for an application phase given the phase's performance characteristics. The proposed technique employed correlations between a known phase's best configuration and characteristics, and a new phases' characteristics.

C. Analytical/Predictive Modeling Techniques

Since online DSE can often execute an application in inferior sub-optimal cache configurations that result in severe performance and energy consumption degradation, reducing or eliminating DSE is an attractive solution. Phase distance mapping, is one such solution and is considered as a predictive modeling approach. Another predictive modeling approach was presented in [29] where the authors constructed IPC curves for different sets of applications that allowed the application performance to be predicted for all level two cache way values by observing the application's performance for only one level two cache way value. Such cross-configuration prediction of application performance can significantly speed up online DSE by examining different configurations simultaneously while executing a single configuration. Prior works have shown that analytical models that rely on the application execution trace can be successfully used [12], [22].

V. ONLINE CACHE TUNING AND CONTROL

The ultimate goal of a configurable cache system is to realize self-tuning that is transparent to the end user and the running applications [5], [13]. To achieve this goal, this self-tuning cache system can be formulated as a feedback control system that self-monitors performance and adjusts the cache parameters in order to meet runtime-defined time-varying user-defined QoE expectations. For the best closed-loop control performance, the application phases' characteristics, the phases' cache requirements, and the tuning interval must be taken into consideration.

There has been some prior work related to these goals. Peng et al. [13] developed multiple state machines to monitor application execution, detect performance changes that accompany phase changes (i.e., tuning interval), and tuned the cache accordingly. However, determining the tuning interval is extremely challenging. Gordon-Ross et al. [5] showed that the tuning interval must never overshoot a change in application phase in order to avoid severe energy penalties that result from executing an inappropriate cache configuration through an undetected application phase change. When overshoot happens, the new application phase operates using the cache configuration that was determined to be the best configuration for a previous phase but may no longer be the best configuration for the new phase, imposing *tremendous* energy penalties.

A self-tuning cache system must provide the following functionalities: 1) be capable of autonomously monitoring and collecting information about the optimization metrics; 2) be able to dynamically reconfigure the tunable parameter values; 3) be capable of determining when and how to change the cache parameters in response to varying application phase requirements without introducing severe penalties [5], [13]; and 4) adapt to user-defined QoE expectations; and limit user experience degradation during DSE.

These requirements echo the notion that a self-tuning cache system is a variation of a feedback control system, an example of which was proposed in [5] and is reproduced in Fig. 2. The purpose of that control system was to dynamically adjust the tuning interval such that the interval matched as closely

978-1-5386-7100-9/18 $31.00 © 2018 IEEE

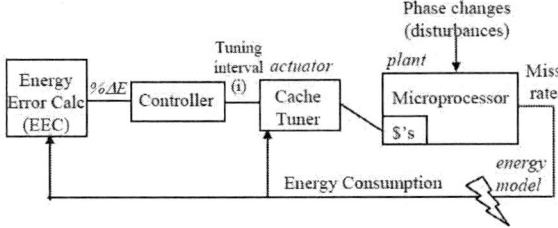

Fig. 2. Proposed feedback control system architecture for a self-tuning cache [5].

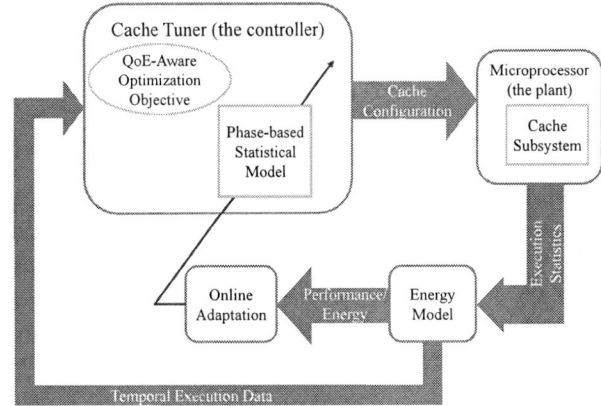

Fig. 3. Our proposed architecture for closed-loop, online cache tuning and control.

as possible to the application's phase changes without over-shooting. The cache miss rate was used as the performance optimization metric to determine when cache reconfiguration needed to occur, and an energy model was implemented as an online energy calculator to provide the energy optimization metric that guided the closed-loop operation of the control system. In reference to the general problem formulation in Section II, we also note the correspondence between these two performance and energy optimization metrics, which we denote as $u(.)$ and $e(.)$, respectively.

Adegbija et al. [7] exploited application phase information online to directly estimate the best cache configuration using a statistical model that correlated known and new phase characteristics to the best known cache configuration for a base phase. The advantage of this approach was that the model essentially eliminated DSE by inferring a best cache configuration. This approach can be classified as a statistical machine learning-based approach where a large dataset of known application phases and associated best cache configurations are used offline and comprises the training dataset. A statistical model is learned from this training dataset, and subsequently used online to make predictions about new data [30]. Since it is possible to summarize the statistics of a huge training dataset using a lightweight statistical model that is later used online, we argue that machine learning-based approaches represent a key component in online cache tuning and control. However, since these prediction-based methods are error-prone, online adaptation and control must be used in order to correct for the modeling-related errors and to update the statistical models online. Therefore, in our proposed approach, we use online adaptation in the statistical model.

A. Proposed Architecture

Fig. 3 depicts our proposed architecture for closed-loop, online cache tuning and control. Similar to a closed-loop control system, our proposed architecture comprises a controller, which is the cache tuner module, and a plant that is to be controlled, which is the microprocessor's cache subsystem.

The cache tuner module determines the best cache configuration for the executing application phase. This module takes into account the dynamic user QoE expectations as part of the optimization objective function. The cache tuner module also uses an adaptive phase-based statistical model to predict the

best cache configuration that satisfies the current optimization objective. This statistical model assists DSE by simultaneously predicting the performance and energy metrics' values for multiple cache configurations or even by directly predicting the best cache configuration based on execution statistics.

We propose that the statistical model incorporates linear/non-linear adaptive filters—a class of statistical models that convolve a time series with filter weights to perform online prediction—and can adapt the weights of this model using an online adaptation method [31]. While prior work used instant-by-instant phase information for statistical prediction [7], we propose to consider the temporal history of the execution characteristics at a fine-grained temporal resolution for statistical prediction. In other words, we consider the application execution statistics as well as the performance and energy metrics as time series data, and collectively refer to this data as temporal execution data. This machine learning-based approach is poised to simultaneously improve the prediction performance, and to operate at a higher temporal resolution while providing an end-to-end prediction model that does not explicitly require classification of the executing application phase. Our proposed architecture makes use of three major elements to improve upon previous approaches: 1) considers the temporal history of the application's execution characteristics and the application's phase information using online filtering; 2) monitors the energy and performance of the running applications' phases and uses this information to update and adapt the online filter; and 3) adapts to runtime-defined time-varying user-defined QoE expectations.

B. Unexplored Challenges

One of the major challenges for online cache tuners is to avoid user experience degradation due to DSE overheads, such as power/energy and performance [7], [17]. To regulate these overheads, we combine our proposed adaptive statistical model with design space subsetting and consider the temporal execution data in order to improve the prediction performance while minimizing the computational overhead.

Another challenge is that online cache tuners are expected to perform generally well while executing a large variety of applications from vastly different application domains. In the most challenging situation, none of this information is available during design time, thus no efforts can be made to enable more accurate runtime predictions based on the particular application or domain characteristics (e.g., data streaming applications typically have high cache miss rates). In our proposed architecture, the online adaptation module is responsible for changing the weights of the filters in the statistical model based on ground-truth data observed in real-time as different application phases execute. By adapting the embedded statistical model online, our proposed architecture can continue to improve the prediction accuracy whenever degraded performance for some application/phase is observed.

Predictions can also be improved by tracking best configurations and correlating those configurations to device sensor readings. For example, ambient lighting conditions could be used to regulate video frame rate, thereby reducing the frame rate in low lighting. Predictions can also be done using time-of-day correlations. Users likely use their devices differently during working hours, thus QoE expectations may be lower in general than during non-working hours. For any type of correlation, historical data can be kept that enables direct lookups of best configurations given different sensor value combinations to eliminate DSE for known scenarios. This historical data could also be mined to determine user patterns and characteristics that can enable best configurations to be predicted for unknown scenarios. Where as this challenge is not a make-or-break point for self-tuning systems, an effective solution would greatly improve QoE.

A challenge that all online cache tuners must address is the introduced power/energy and performance overheads during DSE, and general area overheads due to the architectural structures required to enable and orchestrate cache tuning. To have the most accurate tuning interval estimation, the tuner should frequently evaluate execution characteristics to avoid severe penalties for missed phase changes. To address this challenge, we construct our architecture to operate at two different intervals. Since the statistical model uses an online filter, this model can operate at a relatively high sampling frequency such that the model is almost guaranteed to never miss a phase change. The filter implementation can use a recursive formulation such that the filter can cover a long temporal span at a high sampling resolution while using few filter weights. The reduced number of filter weights results in significantly fewer computations, and, therefore, power and area savings for the on-chip hardware filter implementation. To further reduce the power overhead, the controller can limit cache reconfigurations to cases where the expected optimization metric gains are beyond a predefined threshold. In other words, while the online statistical filter runs at a high sampling resolution, the cache reconfiguration decisions occur at a much lower rate that is based on statistical thresholding.

Another challenge is incorporating user feedback to identify what the user considers a best configuration, and to ascertain how much DSE-time degradation is tolerable. Identifying the user-defined best configuration can be done using existing operational settings that allow the user to enter energy-savings or performance-enhanced modes, however, this does not provide the fine level of granularity necessary to determine Pareto-optimal best configurations for diverse operating environments and applications. An intrusive approach would be to periodically poll the user to swipe left or right based on their level of satisfaction, however, given the granularity of feedback needed to gather information during DSE and that the system may only execute for a fraction of a second or several seconds in each configuration, this method is infeasible. A non-intrusive approach would be to use the device's internal sensors to infer user satisfaction, such as using the accelerometer to detect user frustration that physically agitates their usage of the device, however, this method does not provide detailed enough feedback and would not work for all users. An ideal situation will likely combine both intrusive and non-intrusive techniques. Finding a solution to this challenge is critical to realizing self-tuning systems.

VI. CONCLUSIONS

Efficient and effective dynamic configurable cache tuning is critical for modern embedded systems that are required to operate in unknown environments, and execute a myriad of applications that are typically unknown at design time while dynamically responding to user quality of experience (QoE) expectations. Realizing a self-tuning cache that is transparent to both the user and the executing applications is hindered by many challenges. Critical challenges include limiting the user's experience degradation during design space exploration (DSE), adapting to unknown application requirements, minimizing the area and power/energy overheads introduced by the self-tuning circuitry, and determining how to incorporate user feedback and device sensor readings. In addition, a self-tuning cache subsystem should ideally adapt at a fine-grained, application phase-based level.

In this paper, we reviewed current progress in cache tuning and provided an overview of our proposed online, adaptive, closed-loop control system architecture for dynamic cache tuning. Our proposed architecture attempts to address some of the challenges by: 1) utilizing application phase information and temporal execution data at a high sampling frequency in order to accurately predict the best cache configuration; 2) implementing the prediction model using computationally efficient statistical filters; 3) adapting the prediction model online to account for changing application execution statistics; 4) using a QoE-aware optimization objective function; and 5) operating at two different sampling frequencies in order to minimize missed application phase changes while simultaneously saving dynamic power/energy consumption and architecture area. Finally, we identified numerous unexplored challenges, many of which are critical to realizing self-tuning systems. Our future work involves implementing the proposed architecture in order to quantify and evaluate our architecture and solutions' effectiveness for dynamic cache tuning.

978-1-5386-7100-9/18 $31.00 © 2018 IEEE

ACKNOWLEDGMENTS

This work was supported by the National Science Foundation (CNS-0953447 and CNS-1718033). Any opinions, findings, and conclusions or recommendations expressed in this material are those of the authors and do not necessarily reflect the views of the National Science Foundation.

REFERENCES

[1] A. Malik, B. Moyer, and D. Cermak, "A low power unified cache architecture providing power and performance flexibility (poster session)," in *Proceedings of the 2000 international symposium on Low power electronics and design*. ACM, 2000, pp. 241–243.

[2] S. Mittal, "A survey of architectural techniques for improving cache power efficiency," *Sustainable Computing: Informatics and Systems*, vol. 4, no. 1, pp. 33–43, 2014.

[3] C. Zhang, F. Vahid, and W. Najjar, "A highly configurable cache for low energy embedded systems," *ACM Transactions on Embedded Computing Systems (TECS)*, vol. 4, no. 2, pp. 363–387, 2005.

[4] T. Sherwood, E. Perelman, G. Hamerly, S. Sair, and B. Calder, "Discovering and exploiting program phases," *IEEE micro*, vol. 23, no. 6, pp. 84–93, 2003.

[5] A. Gordon-Ross and F. Vahid, "A self-tuning configurable cache," in *Proceedings of the 44th annual Design Automation Conference*. ACM, 2007, pp. 234–237.

[6] M. H. Alsafrjalani and A. Gordon-Ross, "Quality of service-aware, scalable cache tuning algorithm in consumer-based embedded devices," in *Proceedings of the 26th edition on Great Lakes Symposium on VLSI*. ACM, 2016, pp. 357–360.

[7] T. Adegbija, A. Gordon-Ross, and A. Munir, "Phase distance mapping: a phase-based cache tuning methodology for embedded systems," *Design Automation for Embedded Systems*, vol. 18, no. 3-4, pp. 251–278, 2014.

[8] D. E. Kirk, *Optimal Control Theory: An Introduction*. Courier Corporation, 2004.

[9] C. Zhang and F. Vahid, "Cache configuration exploration on prototyping platforms," in *Rapid Systems Prototyping, 2003. Proceedings. 14th IEEE International Workshop on*. IEEE, 2003, pp. 164–170.

[10] A. Gordon-Ross, F. Vahid, and N. Dutt, "Fast configurable-cache tuning with a unified second-level cache," in *Low Power Electronics and Design, 2005. ISLPED'05. Proceedings of the 2005 International Symposium on*. IEEE, 2005, pp. 323–326.

[11] P. Viana, A. Gordon-Ross, E. Keogh, E. Barros, and F. Vahid, "Configurable cache subsetting for fast cache tuning," in *Proceedings of the 43rd annual Design Automation Conference*. ACM, 2006, pp. 695–700.

[12] A. Gordon-Ross, P. Viana, F. Vahid, W. Najjar, and E. Barros, "A one-shot configurable-cache tuner for improved energy and performance," in *Proceedings of the conference on Design, automation and test in Europe*. EDA Consortium, 2007, pp. 755–760.

[13] M. Peng, J. Sun, and Y. Wang, "A phase-based self-tuning algorithm for reconfigurable cache," in *Digital Society, 2007. ICDS'07. First International Conference on the*. IEEE, 2007, pp. 27–27.

[14] A. Gordon-Ross, F. Vahid, and N. D. Dutt, "Fast configurable-cache tuning with a unified second-level cache," *IEEE Transactions on Very Large Scale Integration (VLSI) Systems*, vol. 17, no. 1, pp. 80–91, 2009.

[15] M. H. Alsafrjalani and A. G. Ross, "Dynamic scheduling for reduced energy in configuration-subsetted heterogeneous multicore systems," in *Embedded and Ubiquitous Computing (EUC), 2014 12th IEEE International Conference on*. IEEE, 2014, pp. 17–24.

[16] M. H. Alsafrjalani, A. G. Ross, and P. Viana, "Minimum effort design space subsetting for configurable caches," in *Embedded and Ubiquitous Computing (EUC), 2014 12th IEEE International Conference on*. IEEE, 2014, pp. 65–72.

[17] M. H. Alsafrjalani and A. Gordon-Ross, "Low effort design space exploration methodology for configurable caches," *Computers*, vol. 7, no. 2, p. 21, 2018.

[18] T. Givargis and F. Vahid, "Platune: a tuning framework for system-on-a-chip platforms," *IEEE Transactions on Computer-Aided Design of Integrated Circuits and Systems*, vol. 21, no. 11, pp. 1317–1327, 2002.

[19] T. Adegbija and A. Gordon-Ross, "PhLock: A cache energy saving technique using phase-based cache locking," *IEEE Transactions on Very Large Scale Integration (VLSI) Systems*, vol. 26, no. 1, pp. 110–121, 2018.

[20] G. Reinman and N. P. Jouppi, "CACTI 2.0: An integrated cache timing and power model," *Western Research Lab Research Report*, vol. 7, 2000.

[21] C. Zhang, F. Vahid, and R. Lysecky, "A self-tuning cache architecture for embedded systems," *ACM Transactions on Embedded Computing Systems (TECS)*, vol. 3, no. 2, pp. 407–425, 2004.

[22] A. Ghosh and T. Givargis, "Cache optimization for embedded processor cores: An analytical approach," *ACM Transactions on Design Automation of Electronic Systems (TODAES)*, vol. 9, no. 4, pp. 419–440, 2004.

[23] W. Zang and A. Gordon-Ross, "A survey on cache tuning from a power/energy perspective," *ACM Computing Surveys (CSUR)*, vol. 45, no. 3, p. 32, 2013.

[24] D. Burger and T. M. Austin, "The simplescalar tool set, version 2.0," *ACM SIGARCH computer architecture news*, vol. 25, no. 3, pp. 13–25, 1997.

[25] N. Binkert, B. Beckmann, G. Black, S. K. Reinhardt, A. Saidi, A. Basu, J. Hestness, D. R. Hower, T. Krishna, S. Sardashti *et al.*, "The gem5 simulator," *ACM SIGARCH Computer Architecture News*, vol. 39, no. 2, pp. 1–7, 2011.

[26] E. Keogh, S. Chu, D. Hart, and M. Pazzani, "An online algorithm for segmenting time series," in *Data Mining, 2001. ICDM 2001, Proceedings IEEE International Conference on*. IEEE, 2001, pp. 289–296.

[27] H. Zhao, X. Luo, C. Zhu, T. Watanabe, and T. Zhu, "Behavior-aware cache hierarchy optimization for low-power multi-core embedded systems," *Modern Physics Letters B*, vol. 31, no. 19-21, p. 1740067, 2017.

[28] H. Hajimiri and P. Mishra, "Intra-task dynamic cache reconfiguration," in *VLSI Design (VLSID), 2012 25th International Conference on*. IEEE, 2012, pp. 430–435.

[29] M. Moreto, F. J. Cazorla, A. Ramirez, and M. Valero, "Online prediction of applications cache utility," in *Embedded Computer Systems: Architectures, Modeling and Simulation, 2007. IC-SAMOS 2007. International Conference on*. IEEE, 2007, pp. 169–177.

[30] M. B. Christopher, *Pattern recognition and machine learning*. Springer-Verlag New York, 2016.

[31] S. Haykin, *Adaptive filter theory*. Prentice-Hall, Inc., 1986.

An FPGA-based Brain Computer Interfacing using Compressive Sensing and Machine Learning

Ritu Ranjan Shrivastwa, Vikramkumar Pudi, Anupam Chattopadhyay
Nanyang Technological University Singapore
{rituranjan, pudi, anupam}@ntu.edu.sg

Abstract—Electrocorticography (ECoG) is a type of electrophysiological monitoring useful for recording the activity from the cerebral cortex. It has emerged as a promising recording technique in brain-computer interfaces (BCI). Compression of these signals is essential for saving power and bandwidth in the novel application scenarios of Health-based IoT and Body Area Networks. However, this task is particularly challenging since, ECoG signals are not compressible either in time domain or in frequency domain. To that end, Block Sparse Bayesian Learning (BSBL) techniques were suggested for the reconstruction of compressed EEG and ECG signals, which is however, computationally demanding. Furthermore, given the heterogeneity in modern computing systems, careful design partitioning is required to most effectively evaluate the particular resources available on the deployed architecture. In this paper, we propose to utilise a combination of compressive sensing and neural network for the compression and reconstruction of ECoG signals, respectively. For the choice of the neural network, a multi-layer perceptron regressor with a stochastic gradient descent solver is developed. For a sample system, we show that the network has a compression ratio of 50%, and reconstruction accuracy of 89.85% after training with a practical, medium-sized dataset. In general, the results show that the most efficient system implementation is a heterogeneous architecture combining a CPU and a field-programmable gate array (FPGA).

Keywords—Compressive Sensing, Electrocorticography (ECoG), Machine Learning, FPGA-based BCI, Heterogeneous Computing.

I. INTRODUCTION

The technique of Compressive Sensing (CS) is gaining rapid adoption thanks to its efficiency in reconstructing signals that are acquired from very few measurements. This enables fast signal/image acquisition and also enables sensors to save energy by recording only parts of the stimuli [1]. Among the most prominent use cases that already reap benefits of CS are fast MRI [2], seismic imaging, biological applications, compressive Radio Detection and Ranging (RADAR), Sparse Channel Estimation, Ultra-Wideband Systems, Wireless Sensor Networks (WSNs), network traffic monitoring and anomaly detection [3]. There are two known conditions for a perfect reconstruction using CS framework that is, sensing matrix should have restricted isometry property and signal should be sparsely represented.

This work presented in this paper is supported by the National Research Foundation Singapore under its Campus for Research Excellence and Technological Enterprise (CREATE) programme.

In practice though, it is not possible to have both the conditions completely met due to practical challenges like memory and processing speed [4], resulting in a complex optimization problem that must consider algorithmic implantation as well as available hardware resources. To overcome such complexities several algorithms have been proposed in recent years, including Block-based Compressive Sensing (BCS) [5] that utilizes divide-and-conquer strategy by dividing the large image into non-overlapping blocks and sensing them separately and later combining all to reconstruct the full image. Orthogonal to this, there are other mathematical approaches [3] that include basis pursuit, orthogonal matching pursuits, Fourier sampling, sparse-Bayesian learning, Monte-Carlo based algorithms and Bregman distance regularization.

Another major bottleneck towards the adoption of CS, particularly in area-constrained devices, is the efficient reconstruction of the signals. The relevant optimization algorithm turns out to be highly demanding in computation efforts and therefore, convergence cannot be achieved in real time. Practical applications however require fast reconstruction with high accuracy, for example in medical imaging and sensing. Furthermore, ECoG signals are not sparse and are not compressible in time or frequency domain. This calls for novel techniques that can achieve compression goals, and at the same time be realizable within the tight area-time-energy constraints while considering the underlying heterogeneous resources. The most efficient system design requires careful design partitioning across the available resources. These are the key motivations for this work.

In this work, we combined compressive sensing with machine learning. We deployed a fully connected multi-layered perceptron network to both learn the sensing matrix and also the reconstruction part. Both the operations run simultaneously in the same network and get optimized jointly. The first and second layer form the sensing network whereas the learnt weights of the second layer become the sensing matrix, and the rest of the network is designed for reconstruction. The sensing is simulated with different compression rates and the reconstruction network is trained for the same. To give designers flexibility in design goals, system requirements, and heterogeneous architectural structure, we evaluate FPGA hardware area requirements for different compression ratios, enabling clear tradeoffs between performance and area/power. The novelty of our work is in achieving high compression and excellent reconstruction rate using small neural nets. To the

978-1-5386-7100-9/18 $31.00 © 2018 IEEE   726

best of the knowledge of the authors, this is the first work to implement a compression and reconstruction of ECoG signal using Neural Networks on a FPGA platform, and show that the most efficient implementation is a heterogeneous system combining an FPGA with an ARM processor.

The rest of the paper is organized as follows: Section II discusses about the background in Compressed Sensing, BCI and ECoG signals. The proposed method is detailed in Section III with the experimental results summarised in Section IV. Concluding remarks are made in Section V.

II. BACKGROUND

A. Compressive Sensing

Compressive sensing (CS) is a method of directly acquiring the signal at sub-Nyquist rate from sensor, provided signal is a *sparse*-signal [6]. CS compress the N-sample sparse signal X into M-sample signal Y using the *sensing matrix* Φ as shown in Eq. 1.

$$Y_{M \times 1} = \Phi_{M \times N} X_{N \times 1} \qquad \text{Where M} \ll \text{N} \qquad (1)$$

In case X is not sparse, we have to represent the X as sparse signals using *Dictionary matrix* denoted by D, X is expressed as $X = Dz$ and z is sparse. We can rewritten Eq. 1 as

$$Y = \Phi D z \qquad \text{where M} \ll \text{N} \qquad (2)$$

We can directly compute the Y using the Eq. 1 without transforming the X. CS recovers z first and original signals x is recovered using $X = Dz$. The compression ratio (CR) defined as

$$CR = (1 - \frac{M}{N})100\% \qquad (3)$$

Naturally, CS has multiple advantages over other compression methods. In [7], [8], [9] it is shown that CS can be implemented using only adders (without multiplers), when the sensing matrix Φ is a sparse binary matrix or Bernoulli random matrix, leading to an energy-efficient implementation. The authors in [9] proposed a lightweight CS implementation with low hardware complexity.

Despite these advantages, the use of CS in telemonitoring is only limited to a few types of signals, mainly due to the fact that most physiological signals like EEG, ECoG are not sparse in the time domain and not sparse enough in transformed domains. To address this challenge, in [10], [11], authors proposed a framework called Block Sparse Bayesian Learning (BSBL) for compression of these signals using CS.

B. BCI and ECoG Signals

Brain-Computer Interface (BCI) is a term used to define direct communication link between a brain and an external machine using invasive, non-invasive and partially-invasive approaches wherein the BCI device is installed in the grey matter, inside the skull on the periphery of the brain and outside the skull without any incision, respectively [12].

The conventional EEG (Electroencephalography) signals have various disadvantages including limited resolution and

has clinical risks that led to the use of ECoG (Electrocorticography) signals. ECoG signals are recorded from the surface of the brain instead of electrodes penetrating inside the brain, as in case of EEG signal recording, and thus are less traumatic for the patients. In addition, these signals have higher resolution than EEG signals and thus, are more accurate [13]. There have been several applications of ECoG signals since its first appearance like studying discrimination in visual perception of humans [14], controlling a two-dimensional cursor [15] and decoding three-dimensional movement of arm [16]. Consequently, there is a pressing need to address the compression, and reconstruction challenges of ECoG signals under tight resource constraints.

C. Block Sparse Bayesian Learning (BSBL)

Sparse Bayesian Learning (SBL), initially used for classificatoin and regression in Machine Learning (ML), later emerged as an ideal method for sparse signal recovery [17]. The BSBL works with the temporal correlation of sources. In this approach, multiple measurement vectors are converted to blocks of single measurement vector as shown in equation (4).

$$X = \{(x_1, x_2, ... x_{d1})_1, ..., (x_{dg-1} + 1, x_2, ... x_{dg})_g\} \qquad (4)$$

The value X can be viewed as a concatenated block structure consisting of g blocks 4 [18] wherein each block satisfies Gaussian distribution with parameterized multivariates shown in equation (5).

$$p(x_i; \gamma_i, B_i) \ N(0, \gamma_i B_i), i = 1, ..., g \qquad (5)$$

where, γ_i and B are unknown parameters

The posterior of X is calculated and the mean of which gives the Maximum-A-Posteriori (MAP) of X, the parameters of which can be estimated by minimizing a cost function. This is the BSBL framework [17]. B_i and γ_i are the hyperparamaters that control the convergence speed and the recovery performance. B_i is also found to be affecting only the local convergence, thus, it can be constrained for better performance and also to avoid overfitting [10].

D. Machine Learning and Compressed Sensing

Quite a few implementations of deep learning towards efficient CS reconstruction of images have been proposed over the recent years. In [19], the authors presented a framework for structured signal recovery using stacked denoising autoencoder modelled as a neural network. Convolutional neural network is proposed in [4] that reconstructs whole image from linear measurements instead of reconstructing image blocks and then recovering the image from the blocks. Another deep learning approach is used in [20] for reconstruction using block based compressed sensing technique.

III. PROPOSED APPROACH

The current work mainly targets fast and efficient compression and reconstruction of ECoG signals. To that end, we opted for a fully connected Multi-Layered Perceptron (MLP). Since the implementation has to be carried out on FPGA, the main challenge thus becomes the network selection. A large network is costly in terms of area and performance.

Empirically, we determined the smallest network that can be trained regressively with a moderate sized data to achieve an acceptable reconstruction performance for ECoG signals. We used a L2-regularized regression for evaluation. The solver used for learning is ADAM [21] that does stochastic gradient based optimization. The chosen activation function is ReLU [22]. A lower boundary testing with the data and network size is performed that inferred a minimum resource requirement criteria for obtaining good throughput with low area footprint on FPGA. The fully-connected network works for both compression and reconstruction using 50% compression is shown in Fig. 1. The second layer in the network is the compression layer that downsizes the signal to the ratio of the nodes in second layer to the nodes in first or input layer. The remaining part of the network, after the compression layer, learns the reconstruction phase. The layer functions of the network used can be written as shown in equations 6 and 7.

$$y = ReLU(\sum w_{ij}x_i + b_j) \qquad (6)$$

$$f(x) = ReLU(\sum [ReLU(\sum w_{ij}^{xh1}y_i + b_j^{h1})]_k w_{kl}^{h1h2} + b_l^{h2}) \quad (7)$$

where, w_{ij} : the edge weight of from node i of input layer to node j of current layer, x_i : the output of the ith node of the input layer, b_j : the bias of the current layer of the jth node, $h1$ and $h2$ are hidden layers in the reconstruction network.

Thus, the trained model is divided into two parts viz. first part including I_L and C_L and the second part including other hidden layer(s) (reconstruction layers) and the output layer.

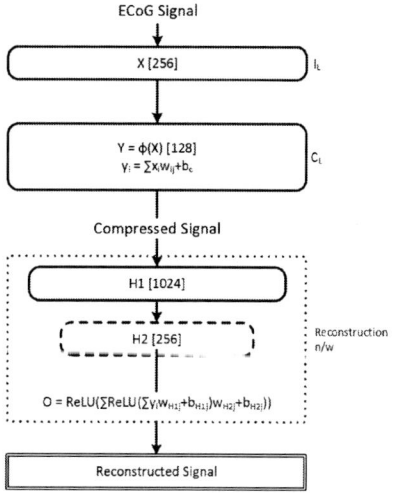

Fig. 1: Compression and Reconstruction for 50% Compression

A. Signal Compression

The original signal is passed to the network, where the output of the first layer, also known as Input Layer (I_L), is the

actual signal and the second layer output is the compressed signal. This simple compression can be achieved by reducing the size of the second layer, or the first Hidden Layer (H_L), that we call as the Compression Layer (C_L). For our study we have adjusted the C_L for achieving a compression of 25%, 37%, 50%, 62%, and 75%. Thus, the size of C_L is less than the size of I_L in every case. Table I shows the size of the C_L for each choice of compression ratio.

TABLE I: Compression Layer (C_L) Sizes

Compression	I_L	C_L
25%	256	192
37%	256	161
50%	256	128
62%	256	97
75%	256	64

B. Signal Reconstruction

The proposed reconstruction layers are tested with different sizes and compression ratios. Table III details the various network sizes that was selected for the experiments ($N1$-$N12$) and their performance with respect to each compression ratio. A network is considered large in this paper if the multipliers needed to realize it on FPGA is $> 280,000$. The PSNR values of reconstructed signals corresponding to 25%, 50%, and 75% compression for each network type is listed in the Table II.

TABLE II: PSNRs (in dB) ON Test Data

Network	PSNR		
	C=25%	C=50%	C=75%
N1	16.25	16.35	14.44
N2	18.8	18.44	15.5
N3	21.42	20.41	16.27
N4	22.75	20.48	16.46
N5	23.64	20.49	16.64
N6	22.8	20.3	16.76
N7	15.02	14.99	14
N8	15.43	15.7	13.7
N9	17.19	16.99	14.74
N10	17.56	17.32	15.6
N11	19.33	18.72	15.78
N12	20.79	19.23	15.97

It can be observed that the reconstruction is better in the case of small network layers with more nodes in reconstruction layer. At 0.5 compression ratio a reconstruction accuracy of 89.85% is achieved. For this compression, the network size is $256 \times 128 \times 1024 \times 256$ that requires $425,984$ ($= 256 \times 128 + 128 \times 1024 + 1024 \times 256$) floating point multiplications.

IV. EXPERIMENTAL RESULTS

In this section, we present our experimental findings in detail. The network is first trained and tested with a range of network sizes and compression ratio. In this paper, we have used ECoG signal data set from [23] for our experiments. The network is simulated on a system with 16GB of RAM, 6GB of GPU Memory (NVidia Tesla C2075), Intel Xeon 4 core processor running at 3.33GHz clock frequency. The training is done with a moderate sized original data (over 72 channels

TABLE III: Reconstruction Layer Details and Accuracy

N/w	I_L	C_L	Reconstruction Layer(s)	Output Layer	Output accuracy in % with compression ratio C as				
					C = 25%	C = 37%	C = 50%	C = 62%	C = 75%
N1	256	192 / 161 / 128 / 97 / 64	64	256	82.88	82.13	82.71	81.22	78.41
N2	256	192 / 161 / 128 / 97 / 64	128	256	87.19	86.23	86.62	85.10	81.34
N3	256	192 / 161 / 128 / 97 / 64	256	256	90.61	90.69	89.48	87.52	83.06
N4	256	192 / 161 / 128 / 97 / 64	512	256	92.15	91.71	89.80	**87.76**	83.59
N5	256	192 / 161 / 128 / 97 / 64	1024	256	92.88	91.52	89.85	87.52	83.54
N6	256	192 / 161 / 128 / 97 / 64	2048	256	92.33	**91.96**	89.65	87.46	**83.78**
N7	256	192 / 161 / 128 / 97 / 64	64x64	256	80.48	80.07	79.85	78.98	77.42
N8	256	192 / 161 / 128 / 97 / 64	64x128	256	81.33	81.42	81.85	80.62	77.76
N9	256	192 / 161 / 128 / 97 / 64	128x128	256	84.88	84.88	84.24	83.09	80.76
N10	256	192 / 161 / 128 / 97 / 64	128x256	256	85.82	86.26	85.20	83.52	81.59
N11	256	192 / 161 / 128 / 97 / 64	256x256	256	88.25	88.15	87.37	85.81	82.05
N12	256	192 / 161 / 128 / 97 / 64	256x512	256	89.90	89.28	88.05	85.56	82.34

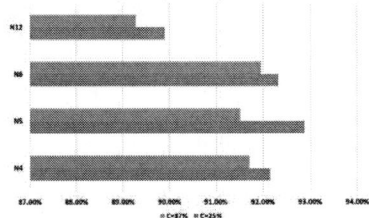 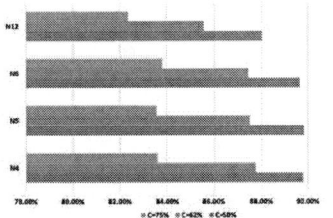 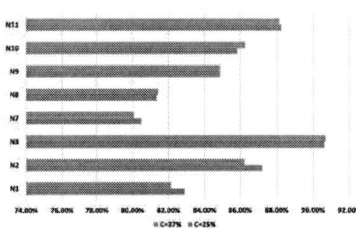

(a) Large networks with low signal compression (b) Large networks with high signal compression (c) Small networks with low signal compression

Fig. 2: Performance of small and large networks with low and high signal compression

of different ECoG signals) with separate validation and test set. The validation and test set size is chosen to be large (nearly equal to training size) to ensure proper validation and network optimization. The training is performed using Python with the help of Scikit-learn neural network framework.

Further, the FPGA implementation issues are identified and results are presented. The trained weights are modelled on the FPGA design to achieve matching results.

Iteration 1: Large network with lower compression ratio: Large networks are tested with lower compression ratios ($\{25, 37\}\%$). For 25% compression the accuracy is found to be approximately 93% while for 37% it is around 92%. The results are shown in Fig. 2a

Iteration 2: Large network with higher compression ratio: The large networks with C_L/I_L ratio as $50\%, 62\%$, and 75% produce $89.85\%, 87.76\%$, and 83.78% accurate reconstruction respectively, as seen in Fig. 2b.

Iteration 3: Small network with lower compression ratio: For lower compression ratio, the accuracy is very high for the network $N3$. It produces an accuracy of around 91% for both compression ratios 0.37 and 0.25. Fig. 2c shows the performance of these networks.

Iteration 4: Small network with higher compression ratio: The results of higher compression $(50 - 75)\%$ are shown in Fig. 3. Here again, $N3$ is observed to perform better than other networks. The accuracy achieved in reconstruction is satisfactory where we get nearly 90% of the original signal for signals compressed by 50% and around 84% accuracy for signals that are compressed up to 75%.

A. Architecture Exploration Studies

In order to make the aforementioned design portable to a lightweight device, it is necessary to undertake a design space exploration. We identified three different architectural possibilities. First, signal reconstruction using a greedy algorithm called Orthogonal Matching Pursuit (OMP), which has been adopted in prior works [24], [25]. Second, signal compression and reconstruction using MLP, where all the layers are implemented on FPGA. Finally, signal compression and reconstruction using a heterogenous architecture consisting of ARM and FPGA fabric, where the FPGA fabric contains a single layer of the neural network. While the first choice is covered in the current literature, we show that the most efficient implementation is obtained through a careful design partition and a heterogenous architecture.

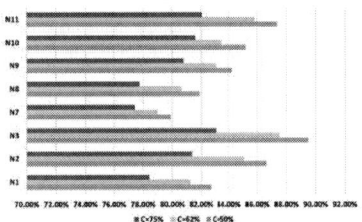

Fig. 3: Small networks with high signal compression

The computation step for the ith layer neural network is given as,

$$Y_{m \times 1} = W_{m \times n}^{Li} X_{n \times 1} \qquad (8)$$

The $W^{Li}_{m \times n}$ represents weight matrix $m \times n$ for the ith layer. The Eq. 8 is useful for the compression when $M << N$, similarly Eq. 8 performs reconstruction when $M >> N$.

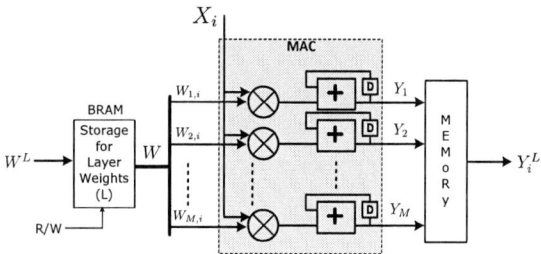

Fig. 4: Architecture for the Single Layer Neural Network

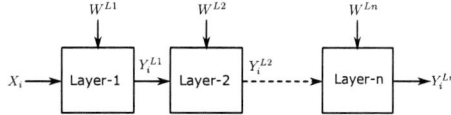

Fig. 5: Implementation of the MLP with Cascaded Layers

The Fig. 4 shows the architecture for single layer. We have connected the layers in a cascaded fashion and mapped the complete architecture on the FPGA as shown in Fig. 5. Expectedly, such an implementation is not economical and requires large number of multipliers. For example, implementing the network $N3$ with 75% compression requires 576 multipliers.

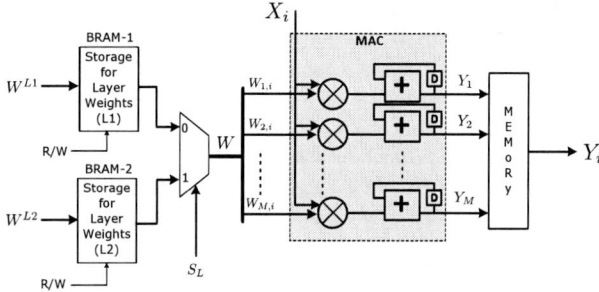

Fig. 6: Generalized architecture: single neural network layer

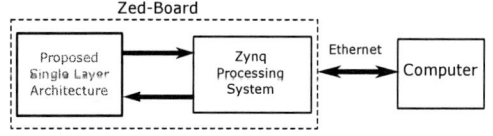

Fig. 7: Proposed Heterogeneous Architecture: ARM+FPGA

As a final architecture choice, we opted for a single neural network layer as shown in the Fig. 6. It consists of two block RAMs (BRAMs), multiplier and accumulator units (MACs) and also memory for storing the output elements.

The MAC equation for the calculation of layer outputs given as following.

$$Y_j = Y_j + W_{j,i}X_i \tag{9}$$

The architecture shown in the Fig. 6 reads single input X_i at a time and multiplies with the corresponding weights $W_{j,i}$ to generate the output Y_j. We are using one BRAM-1 for reading one layer weights and other BRAM-2 is used for storing the weights of the subsequent layers. Outputs are stored in memory after computing. The single output is passed at a time to MACs to compute outputs of the next layer, during which the BRAM-2 is used for reading and BRAM-1 used for storing of the next layer weights. In this process, S_L is used for selecting BRAMs. This process continues until the complete generation of all outputs in the neural network. We chose the maximum number of MACs as 256 with size 16-bits since, CS with a compression of 256 samples is adopted. The default size of BRAMs are 2^{21}-bits ($256 \times 256 \times 16$), where height of BRAM is 256 and width is 2^{12}-bits (256×16). The size of each weight is chosen as 16-bits. This architecture is realized with Zynq processing system (ARM core) in Zedboard for controlling the compression and reconstruction operations in the FPGA fabric (refer Fig. 7).

The FPGA synthesis results for all layers and generalized single layer architectures given in Table-IV targeting Virtex-7 device (XC7VX485T-2FFG1761C). It can be observed that pure FPGA implementation results in significant area overhead. It is beneficial to take advantage of the programmable cores and perform the computation in repeated fashion using a single neural network layer.

TABLE IV: Synthesis Results for Different Architectures

Type	MACs (16-bit)	DSPs	BRAMs	LUTS	Registers	Speed (MHz)
MAC without using DSP blocks						
FPGA	576	NIL	129	285555	22184	42.5
ARM+FPGA	256	NIL	114	83921	8378	134
MAC using DSP blocks						
FPGA	576	576	129	8704	9178	182
ARM+FPGA	256	256	114	1747	3170	247
OMP [24]	589	589	576	12416	—	100
OMP [25]	261	261	258	64020	—	100

Note that the heterogeneous architecture realization contains a single network layer and therefore, requires repeated calls to execute the entire operation. However, this does not impact the runtime performance since, ECoG signals are sampled at much lower frequency (13 KHz) [23] compared to our implementation.

B. Discussion

It is evident from the experimental investigation that best performance is achieved with less number of layers and nodes in the reconstruction layer in range $2^i (i \in \{9, 10, 11\})$. However, it is important to note that $N3$ performs superior in almost every situation. It is a small network with a requirement of only 131,072 multipliers which is around $3.5\times$ smaller than $N5$ which produces the best results for signals compressed at 0.5 ratio. The differences in the accuracy is only 0.25% between $N3$ and $N5$. Thus this was chosen as the right candidate for the implementation. Further exploration of the architectural choices reveal that a heterogeneous approach is the most area-efficient. An exemplary comparison between the original and reconstructed ECoG signal is shown in Fig. 8.

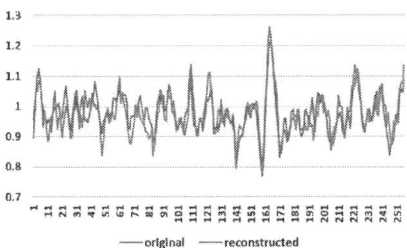

Fig. 8: Original vs. reconstructed ECoG signal

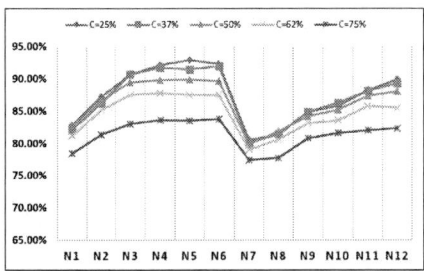

Fig. 9: Performance study for different compression ratios

V. CONCLUSION

This paper presented an implementation walkthrough of different multi-layered perceptron networks by performing iterative boundary value analysis on the size of the network with different compression ratios. We achieved an excellent reconstruction quality, with high compression and low area footprint. Our approach is the first step towards the usage of Compressive Sensing of ECoG signals using machine learning on a heterogeneous architecture. The results encourages us to explore and extend the work by integrating Deep Neural Networks (DNN) such as convnets (Convolutional Neural Networks) that unveils hidden features which the conventional networks cannot. Further implementation studies will evaluate additional system heterogeneity with cascaded programmable cores with a dedicated FPGA fabric for high throughput design.

REFERENCES

[1] D. L. Donoho, "Compressed sensing," *IEEE Transactions on information theory*, vol. 52, no. 4, pp. 1289–1306, 2006.

[2] M. Lustig, D. L. Donoho, J. M. Santos, and J. M. Pauly, "Compressed sensing mri," *IEEE signal processing magazine*, vol. 25, no. 2, pp. 72–82, 2008.

[3] S. Qaisar, R. M. Bilal, W. Iqbal, M. Naureen, and S. Lee, "Compressive sensing: From theory to applications, a survey," *Journal of Communications and Networks*, vol. 15, no. 5, pp. 443–456, Oct 2013.

[4] X. Lu, W. Dong, P. Wang, G. Shi, and X. Xie, "Convcsnet: A convolutional compressive sensing framework based on deep learning," *arXiv preprint arXiv:1801.10342*, 2018.

[5] S. Mun and J. E. Fowler, "Residual reconstruction for block-based compressed sensing of video," in *Data Compression Conference (DCC), 2011.* IEEE, 2011, pp. 183–192.

[6] E. J. Candès and M. B. Wakin, "An introduction to compressive sampling," *IEEE signal processing magazine*, vol. 25, no. 2, pp. 21–30, 2008.

[7] H. Mamaghanian, N. Khaled, D. Atienza, and P. Vandergheynst, "Compressed sensing for real-time energy-efficient ecg compression on wireless body sensor nodes," *IEEE Transactions on Biomedical Engineering*, vol. 58, no. 9, pp. 2456–2466, 2011.

[8] Z. Zhang, T.-P. Jung, S. Makeig, and B. D. Rao, "Compressed sensing for energy-efficient wireless telemonitoring of noninvasive fetal ECG via block sparse bayesian learning," *IEEE Transactions on Biomedical Engineering*, vol. 60, no. 2, pp. 300–309, 2013.

[9] V. Pudi, A. Chattopadhyay, and K.-Y. Lam, "Secure and lightweight compressive sensing using stream cipher," *IEEE Transactions on Circuits and Systems II: Express Briefs*, vol. PP, no. 99, pp. 1–1, 2017.

[10] Z. Zhang and B. D. Rao, "Extension of SBL algorithms for the recovery of block sparse signals with intra-block correlation," *IEEE Transactions on Signal Processing*, vol. 61, no. 8, pp. 2009–2015, 2013.

[11] Z. Zhang, T.-P. Jung, S. Makeig, and B. D. Rao, "Compressed sensing of EEG for wireless telemonitoring with low energy consumption and inexpensive hardware," *IEEE Transactions on Biomedical Engineering*, vol. 60, no. 1, pp. 221–224, 2013.

[12] A. E. Hassanien and A. Azar, "Brain-computer interfaces," *Switzerland: Springer*, 2015.

[13] E. C. Leuthardt, G. Schalk, J. R. Wolpaw, J. G. Ojemann, and D. W. Moran, "A braincomputer interface using electrocorticographic signals in humans," *Journal of Neural Engineering*, vol. 1, no. 2, p. 63, 2004. [Online]. Available: http://stacks.iop.org/1741-2552/1/i=2/a=001

[14] C. Kapeller, H. Ogawa, G. Schalk, N. Kunii, W. Coon, J. Scharinger, C. Guger, and K. Kamada, "Real-time detection and discrimination of visual perception using electrocorticographic signals," *Journal of neural engineering*, vol. 15, no. 3, p. 036001, 2018.

[15] G. Schalk, K. Miller, N. Anderson, J. Wilson, M. Smyth, J. Ojemann, D. Moran, J. Wolpaw, and E. Leuthardt, "Two-dimensional movement control using electrocorticographic signals in humans," *Journal of neural engineering*, vol. 5, no. 1, p. 75, 2008.

[16] D. T. Bundy, M. Pahwa, N. Szrama, and E. C. Leuthardt, "Decoding three-dimensional reaching movements using electrocorticographic signals in humans," *Journal of neural engineering*, vol. 13, no. 2, p. 026021, 2016.

[17] Z. Zhang and B. D. Rao, "Sparse signal recovery with temporally correlated source vectors using sparse bayesian learning," *IEEE Journal of Selected Topics in Signal Processing*, vol. 5, no. 5, pp. 912–926, Sept 2011.

[18] Y. C. Eldar, P. Kuppinger, and H. Bolcskei, "Block-sparse signals: Uncertainty relations and efficient recovery," *IEEE Transactions on Signal Processing*, vol. 58, no. 6, pp. 3042–3054, June 2010.

[19] A. Mousavi, A. B. Patel, and R. G. Baraniuk, "A deep learning approach to structured signal recovery," in *Communication, Control, and Computing (Allerton), 2015 53rd Annual Allerton Conference on.* IEEE, 2015, pp. 1336–1343.

[20] A. Adler, D. Boublil, M. Elad, and M. Zibulevsky, "A deep learning approach to block-based compressed sensing of images," *arXiv preprint arXiv:1606.01519*, 2016.

[21] D. P. Kingma and J. Ba, "Adam: A method for stochastic optimization," *CoRR*, vol. abs/1412.6980, 2014. [Online]. Available: http://arxiv.org/abs/1412.6980

[22] V. Nair and G. E. Hinton, "Rectified linear units improve restricted boltzmann machines," in *Proceedings of the 27th international conference on machine learning (ICML-10)*, 2010, pp. 807–814.

[23] C. Libedinsky, R. So, Z. Xu, T. K. Kyar, D. Ho, C. Lim, L. Chan, Y. Chua, L. Yao, J. H. Cheong *et al.*, "Independent mobility achieved through a wireless brain-machine interface," *PloS one*, vol. 11, no. 11, p. e0165773, 2016.

[24] H. Rabah, A. Amira, B. K. Mohanty, S. Almaadeed, and P. K. Meher, "Fpga implementation of orthogonal matching pursuit for compressive sensing reconstruction," *IEEE Transactions on very large scale integration (VLSI) Systems*, vol. 23, no. 10, pp. 2209–2220, 2015.

[25] L. Bai, P. Maechler, M. Muehlberghuber, and H. Kaeslin, "High-speed compressed sensing reconstruction on fpga using omp and amp," in *Electronics, Circuits and Systems (ICECS), 2012 19th IEEE International Conference on.* IEEE, 2012, pp. 53–56.

2018 IEEE Computer Society Annual Symposium on VLSI

Obfuscation of Fault Secured DSP Design through Hybrid Transformation

Anirban Sengupta[1], Shubha Neema[1], Pallabi Sarkar[2], Sri Harsha P[1], Saraju P Mohanty[3], Mrinal Kanti Naskar[2]
[1]Computer Science & Engineering, Indian Institute of Technology Indore, India
[2]Electronics and Telecommunication Engineering, Jadavpur University, India
[1]Computer Science & Engineering, University of North Texas, USA
Email: asengupt@iiti.ac.in; Saraju.Mohanty@unt.edu; mrinalnaskar@yahoo.com

Abstract— **A DSP circuit is considered to be secure, if its functionality is designed to be hidden from an adversary. In other words to make a DSP design secured, its hardware architecture should not look obvious in terms of its functionality. Structural obfuscation plays a critical role in realizing this objective. In the context of transient fault secured DSP circuits, one of the popular design approaches is using dual modular redundancy (DMR). This established design practice makes the functionality of common fault secured DSP circuits architecture easily identifiable to an adversary. In this paper we propose a novel obfuscation in the context of fault secure DSP circuit that uses hybrid transformations in successive layers to completely transform the hardware architecture of the design at register transfer (RT) and gate level without disturbing its functionality and incurring any design overhead. Results indicate without incurring any design cost overhead, the proposed obfuscation achieves significant structural transformation at the gate level such that the functionality becomes un-obvious to an adversary.**

Keywords—fault secure, DSP circuit, obfuscation, transformations

I. INTRODUCTION

Hardware security and Intellectual Property (IP) core protection is an emerging area of research for semiconductor community that focusses on protecting designs against standard threats such as reverse engineering, counterfeit, forgery, malicious hardware modification etc. Hardware security is broadly classified into two types: (a) authentication based approaches (b) obfuscation based approaches. Some of the approaches that fall under the first type are digital watermarking, IP metering, physical unclonable functions etc. The second type of hardware security approach i.e. obfuscation can again be further sub-divided into two types: (i) structural obfuscation (ii) functional obfuscation. Structural obfuscation transforms a design into one that is functionally equivalent to the original but is significantly more difficult to reverse engineer (RE), while the second one is active protection type that locks the design through a secret key. Obfuscation may include altering human readability of hardware description language or encrypting the source code. Converting a design into a form that makes it harder for an adversary to discover its functionality is difficult to reverse engineer i.e. when an IP core functionality is designed to be hidden for an adversary, it is difficult to RE [1, 3, 5].
In the context of transient fault secured DSP circuits, one of the popular design approaches is using dual modular redundancy (DMR) where a duplicate copy of the original unit is made followed by voting [2, 4]. This established design practice makes the functionality of common fault secured DSP circuits

architecture easily identifiable to an adversary. The design process of fault secure DSP circuits usually don't employ any protection measures such as obfuscation in the design flow which thus makes it easier for an adversary to discover its functionality. The goal is to *protect* a **fault secure DSP IP core** such that its **functionality is designed to be hidden (not obvious)** to an adversary. In this paper we propose a novel obfuscation in the context of fault secure DSP circuit that uses hybrid transformations in successive layers to completely transform the hardware architecture of the design at RT and gate level without disturbing its functionality and incurring any design overhead.

II. PREVIOUS WORKS

Logic obfuscation uses additional XOR/XNOR gates in circuits to protect the IP core [8] [9] [11]. However this incurs overhead in design due to insertion of additional logic components/circuitry. Further, to effectively implement this approach, determining correct location of key gates is essential. Additionally above approaches have not dealt with obfuscation of fault secured DSP circuits. Further, in [1, 3] structural obfuscation is executed on DSP cores. Approach [1, 3] has not handled protection of fault secured DSP circuits as well as do not perform multiple obfuscations using ROE, THT, LTO and resource transformation. The techniques proposed in [1, 3] are not applicable to obfuscate fault secured design. Our proposed method performs low-cost, multi-stage transformations for DSP designs to realize structural obfuscation. None of the works in the literature performed obfuscating fault secured DSP design. The presented methodology obfuscates a fault secured DSP design through a sequence of transformations at low design cost and zero overhead.

III. PROPOSED APPROACH

A. *Overview of Proposed Methodology*

The inputs to the proposed approach (see Fig. 1) are a control data flow graph (CDFG) or C-code of a DSP core, resource constraint and transient fault strength. Structural obfuscation is achieved by applying the following transformations in sequence: (a) Redundant Operation Elimination (ROE) (b) Logical Transformation Operation (LTO) (c) Tree Height Transformation (THT) (d) resource transformation (RT). Once the obfuscated CDFG is obtained then subsequently it is fed into the fault secured DSP design block. The first step of the fault secured DSP design block is to convert into obfuscated Double Modular

978-1-5386-7100-9/18 $31.00 © 2018 IEEE   732

Redundancy (DMR) design, followed by scheduling on the basis of the provided resource constraint. The next step is to apply fault secure hardware allocation rules on obfuscated scheduled DMR. Additionally multiple checkpoints are inserted into the obfuscated scheduled DMR to enhance fault security and perform delay optimization. Section C and D respectively discusses the details of obfuscation based transformation and fault secure design process.

B. Problem Definition and Models:

Generate an obfuscated fault secured DSP circuit at reduced design cost (A_T^{OBF+FS}, T_E^{OBF+FS}) subjected to $k_c = 2$ fault constraint, such that the resultant design is significantly structurally transformed than the original counterpart to hinder identification of functionality and reverse engineering process.
Where, A_T^{OBF+FS} and T_E^{OBF+FS} are the area and delay consumed by an obfuscated fault secured DSP design respectively.

Area Model: Total area A_T^{OBF+FS} occupied by an obfuscated fault secured design is used from [9], can be expressed *as*:

$$A^{OBF+FS} = \sum_{i=1}^{n} A(R_i) * N(R_i) + A(mux) * N(mux) + A(reg) * N(reg) \quad (1)$$

Where, $A(R_i)$, $A(mux)$ and $A(reg)$ represent the area of ith resource, multiplexers and one register respectively; $N(R_i)$, $N(mux)$ and $N(reg)$ represent the number of extracted i^{th} resource, mux and registers required respectively. Extracted resources of ith resource is the maximum number of particular resource used in any control step after scheduling the obfuscated design on the basis of user-given resources.

Delay Model: Total delay T_E^{OBF+FS} of the obfuscated fault secured design is calculated as follows:

$$T_E^{OBF+FS} = No\ of\ Control\ Steps \quad (2)$$

Total number of control steps can be termed as the number of total steps required to complete the desired task in obfuscated fault secured design. Here, one control step is taken as 1000ps.

Fitness Function: The cost of solution is evaluated (assuming area occupied and delay) using following:

$$C_f = \emptyset_1 \frac{A^{OBF+FS}}{A_{max}^{OBF+FS}} + \emptyset_2 \frac{T^{OBF+FS}}{T_{max}^{OBF+FS}} \quad (3)$$

Where, C_f is the cost of the solution; A_{max}^{OBF+FS} and T_{max}^{OBF+FS} indicates the maximal area and delay of obfuscated fault secured design respectively. $\emptyset_1$ and $\emptyset_2$ are user specified weightage for area and delay respectively, the magnitude ranging between [0, 1] (*Note: Both $\emptyset_1$ and $\emptyset_2$ are kept 0.5 to offer similar priority during cost calculation*)

C. Methodology for Proposed Obfuscation

Fig. 1 represents the flow chart of our proposed structural obfuscation process; achieved via pursuing four different transformations sequentially. They are: (1) ROE (2) LTO (3) THT [3] (4) RT. Proposed obfuscation is driven by taking the CDFG as input and implementing each of the possible aforementioned transformation. Subsequent subsections contain the comprehensive explanation of all the individual process.

(a) Redundant Operation Elimination Process: Obfuscation through this process is accomplished by elimination of redundant nodes. A node is considered as redundant node when two or more nodes of same inputs and same resource type coexist in the design. Proposed approach eliminates the redundant node(s) by evaluating each node with other nodes in ascending order and if both the nodes entertain the aforementioned conditions, node with higher number is deleted.

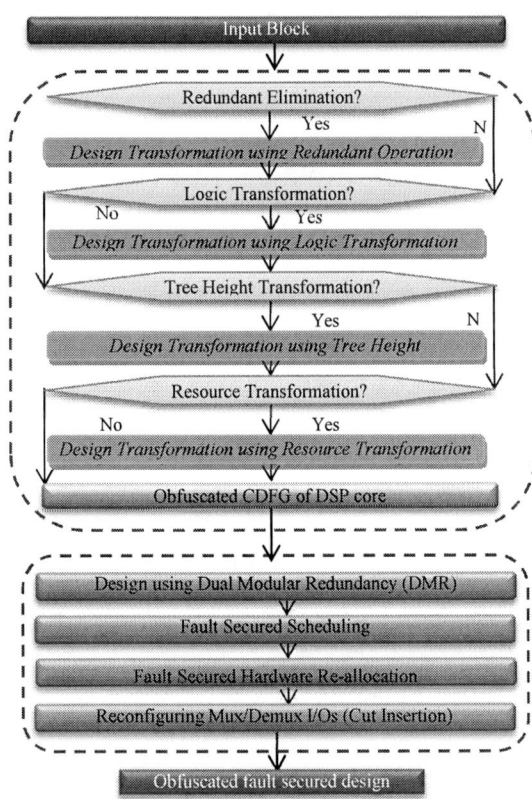

Fig.1. Proposed Obfuscation for fault Secured DSP Designs

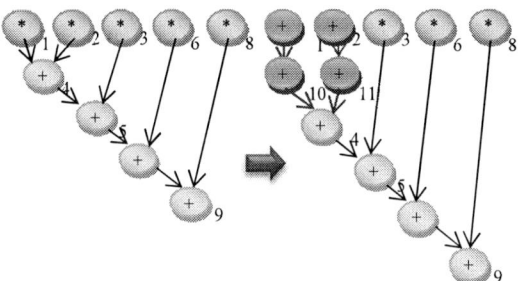

Fig.2. Proposed Logic Transformation based Obfuscation of IIR

978-1-5386-7100-9/18 $31.00 © 2018 IEEE

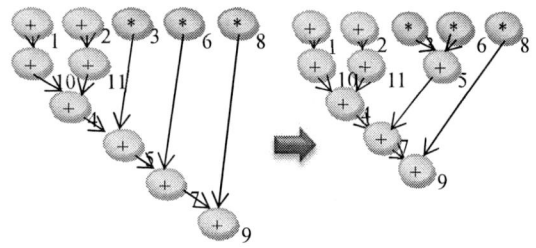

Fig. 3 Proposed THT based obfuscation

Deletion is carried over by required changes in the parent of deleted node's child resulting in keeping the functionality intact (representing change in inputs of resources).

(b) Logic Transformation Process: Another alteration process; applied to obfuscate the design in which nodes of the graph are logically altered but remains functionally equivalent. Proposed approach implements LTO only on those nodes whose inputs are independent. Execution of LTO leads to change in resource type as well as increment in number of total nodes.

Fig. 2 shows an instance representing original and logic transformation based obfuscated design; LTO is practiced on node 1 and 2 with the assumption of their one input as '4'. Newly added nodes (representing changed inputs to resources) are numbered in continuation of highest numbered node before obfuscation. Modified nodes are marked in green colour and the modified dependencies (representing changed muxes/demuxes interconnection) are marked with green line (one multiplication resource is replaced by 2 adders). Nonetheless, as both the designs are functionally alike henceforth yield same results.

(c) Tree Height Transformation Process: This transformation is driven by reduction in the height of the input CDFG. Height is reduced by disjoining the critical path into sub-sections and computing them in concurrency (representing change in inputs and outputs). Fig. 3 presents the IIR benchmark before and after THT. Parallel execution of node 5 is achieved thus resulted both in obfuscation as well as lower height but functionally flawless.

(d) Operation (Resource) Amalgamation: Unlike previous transformation here the process is driven through amalgamation of two resources; adder and multiplier. Proposed approach generates a new customized resource that executes "addition

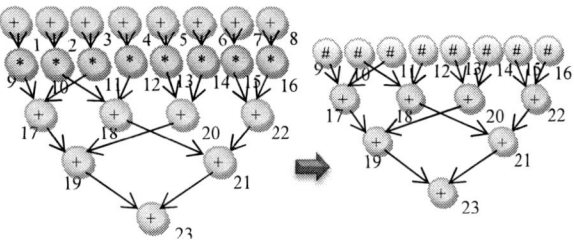

Fig.4. Proposed Resource Transformation based Obfuscation of FIR

followed by a multiplication" (Fig. 5). However, aforementioned transformation is applicable only on those nodes whose inputs are independent and having adder as resource type whose output is only operative as both the inputs of a multiplier. Moreover, the customized resource is represented by '#' in this paper. Fig. 4 shows resource transformation based obfuscation of FIR benchmark.

D. Transient Fault Detection

The consumer electronics devices/applications need to be secure against transient fault of DSP design. Literature [2], [10] has discussed that single event transient fault occurs due to alpha particles. Same literature also emphasizes about high fault strength which happens to be up to a range of 2000ps, high energy particles causes' temporal effect more than one control step. However, researchers have assumed that transient fault strength due to high energy particle lies between 1000ps to 2000ps (1-k_c-cycles to 2-k_c-cycles). For proposed approach k_c-cycle is taken as 2 (based on real life scenario).

Double Modular Redundancy (DMR) of obfuscated design: The first step to fault security is generation of DMR version of final obfuscated graph. Obfuscated DMR represents original and duplicate units of obfuscated graphs.

Fault security of obfuscated design: Fault which affects the device for very short period of time (1000ps to 2000ps) is termed as Transient Fault, after this period the same resource again starts giving back the desired output. After scheduling the original obfuscated design on the basis of user-given constraints, resources are extracted. Next step is to schedule obfuscated DMR concerning extracted resources. This initial resource configuration is useful in estimating the initial design cost before fault security rules are incorporated. In next paragraph, we present the fault secured hardware allocation/re-scheduling rules, inspired from [2]. The initial obfuscated DMR scheduling obtained, may undergo re-scheduling in order to accommodate the fault security rules. Subsequently the final extracted resource configuration will be obtained after this.

Fault Secured Hardware Allocation/Re-Scheduling Rules

1. If delay (control step) difference of the original and duplicate sister operation is greater than k_c, allocate same operator in original and duplicate operation.

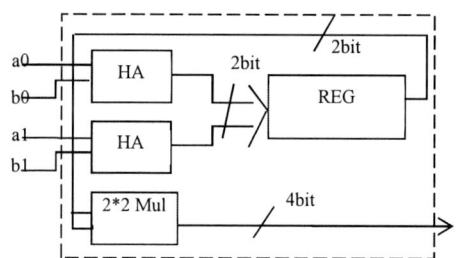

Fig.5. Custom design block used for resource transformation

978-1-5386-7100-9/18 $31.00 © 2018 IEEE

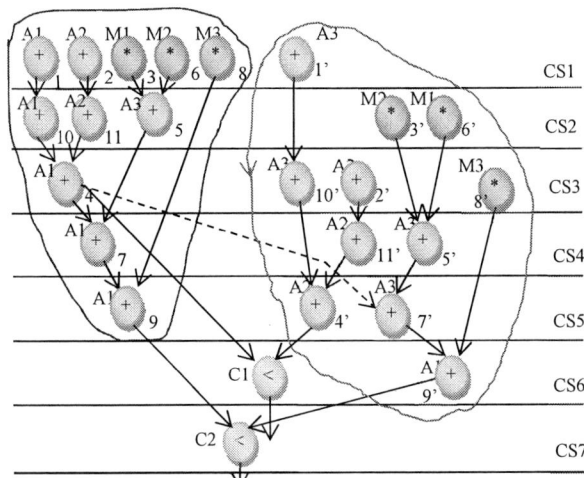

Fig. 6 Proposed DMR design of IIR with checkpoint

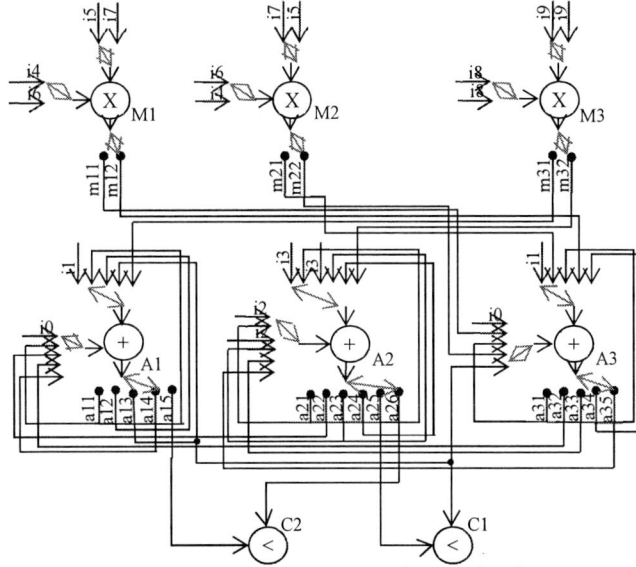

Fig. 8 Proposed Obfuscated Fault Secured DSP design of IIR

dependency edges are cut in graph therefore, shifting some operations one CS up. According to our proposed approach, a cut is performed after generation of obfuscated DMR followed by employing the hardware allocation rules for transient fault security. Proposed approach scales down overhead by implementing maximum checkpoints, therefore, produces prominent security as well. Proposed approach uses different comparators for all the checkpoints; however, area overhead is compensated by significant reduction in latency. Fig. 6 and Fig. 8 represents the complete obfuscated and fault secured DSP design and circuit respectively of IIR benchmark with maximum feasible checkpoints concerning extracted resources and additional comparators. Further, circuit diagram of [4] is demonstrated in Fig. 7.

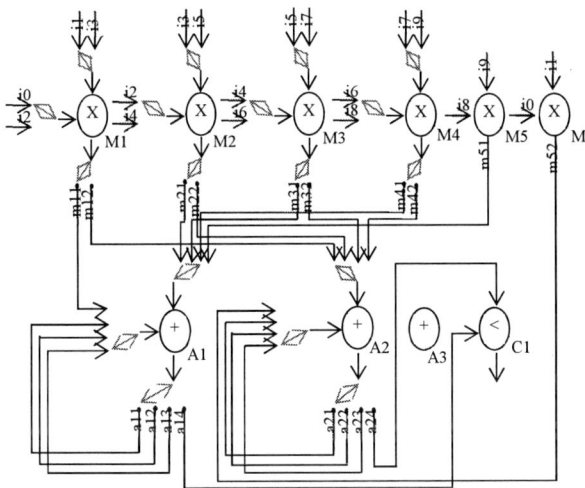

Fig. 7 Non-obfuscated fault secured IIR filter design using [4]

2. If rule 1 is not satisfied and delay difference of original and duplicate operation is not greater than or equal to k_c then allocate distinct hardware resource.
3. If either of the above rules is not satisfied and delay difference of original and duplicate operation is less than k_c, then keep shifting duplicate operation by 1 CS below until either of the above rules is satisfied.

Checkpoint: Delay design or area overhead usually occurs while execution of fault security. Literature [2] proposed one approach to overcome these overheads by insertion of cuts. Insertion of cuts also leads to enhanced fault security with some extra checkpoints. During insertion of cuts, some data

IV. EXPERIMENTAL RESULTS

A. Experimental Setup and Benchmark

The proposed approach, non-obfuscated, related works [4], [5] are implemented in object oriented programming language and executed at 1.90GHz for tested DSP cores [7]. Area calculations are performed on 15nm technology scale [6] in terms of NAND gates.

B. Comparison with [5] with respect to Strength of Obfuscation

The measure of inequality in the structure of the circuit after the obfuscation with respect to original circuit is given by Strength of Obfuscation (SoO). The SoO as shown in equation (4) is calculated by counting the unique nodes which are modified after performing aforementioned alterations (Fig.1).

$$SoO = \frac{\sum_{j=1}^{m} \text{Number of unique nodes modified}}{\text{Number of nodes before obfuscation}} \quad (4)$$

978-1-5386-7100-9/18 $31.00 © 2018 IEEE

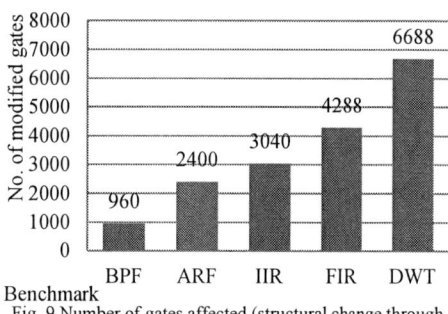

Fig. 9 Number of gates affected (structural change through proposed obfuscation)

Greater the magnitude of SoO, more robust is the security of the design, hence, hard to reverse engineer the circuit. A node is assumed to be transmuted if either of the condition is true from [5]. Furthermore, the enhancement achieved by proposed approach with respect to [5] and number of unique nodes modified after each transformation can be seen in Table I.

C. *Results of Proposed approach*

Occasionally an obfuscated fault secured design attains lesser latency than a non-obfuscated fault secured design since the presented methodology executes a sequence of transformations and optimizations (such as ROE, THT, LT, resource transformations etc.). Hence in some particular situation, the graph post-scheduling could yield lower latency. Table II contains details of proposed approach (with checkpointing) and [4]. Here, area calculation of proposed approach uses extraction of resource whereas [4] considers the user-given resources for measurement of area. Table II is an evidence of significant decline in the cost with obfuscation and multi-checkpointing as additional factor as compare to [4]. Besides, the number of gates

affected with respect to [4] due to proposed obfuscation and fault security is shown in Fig. 9.

V. INFERENCE

This work introduced a new methodology for obfuscation of fault secured DSP circuits with zero overhead. The proposed approach achieves significant transformation of architecture without change in functionality.

ACKNOWLEDGEMENT

This Publication is an outcome of the R&D work undertaken in the project under the Visvesvaraya PhD Scheme of Ministry of Electronics & Information Technology, Government of India, being implemented by Digital India Corporation (formerly Media Lab Asia) and also financially supported by Council of Scientific and Industrial Research under sanctioned grant no. 22/730/17/EMR-II

REFERENCES

[1] Y. Lao and K. K. Parhi, "Obfuscating DSP Circuits via High-Level Transformations," *IEEE Transactions on Very Large Scale Integration (VLSI) Systems*, vol. 23, no. 5, pp. 819–830, May 2015.

[2] Anirban Sengupta, Deepak Kachave "Low Cost Fault Tolerance against kc-cycle and km-unit Transient for Loop Based Control Data Flow Graphs during Physically Aware High Level Synthesis", *Elsevier Journal on Microelectronics Reliability*, Volume 74, 2017, pp. 88-99

[3] A. Sengupta and D. Roy, "Protecting an intellectual property core during architectural synthesis using high-level transformation based obfuscation," *Electronics Letters*, May 2017. [Online]. Available: http://digital-library.theiet.org/content/journals/10.1049/el.2017.1329.

[4] Inoue, T., Henmi, H., Yoshikawa, Y., & Ichihara, H. High-Level Synthesis for Multi-Cycle transient fault Tolerant Datapaths. Proc. 17 th IEEE International On-Line Testing Symposium, 2011, pp 13-18.

[5] Anirban Sengupta, Dipanjan Roy, Saraju Mohanty, Peter Corcoran "DSP Design Protection in CE through Algorithmic Transformation Based Structural Obfuscation", IEEE Transactions on Consumer Electronics,

Table I. Comparative study of proposed obfuscation with [5] with respect to obfuscation strength

| Benchmark | Proposed approach | | | | | [5] | Enhancement in Strength of Obfuscation (%) |
	REO (Unique nodes)	LTO (Unique nodes)	THT (Unique nodes)	ALU (Unique nodes)	SoO		
IIR	-	3	3	-	0.666667	0.33333	100
ARF	10	6	-	-	0.571429	0.42857	33.33
BPF	5	6	2	2	0.517241	0.44827	15.38
DWT	-	10	-	-	0.588235	0.52941	11.11
FIR	-	-	12	11	1	0.5	100

Table II Comparison of proposed obfuscated fault secured approach with multi-checkpointing with [4]

| Benchmark | Proposed approach (with multi-checkpointing) | | | | | [4] | | | |
	# cuts	Resources	Area (um^2)	Latency (ns)	Cost	Resources	Area (um^2)	Latency (ns)	Cost
IIR	1	3A,3M,2C	354.681	7	0.497980	3A,6M,1C	545.195	7	0.654878
ARF	4	5A,6M,5C	718.505	10	0.526472	7A,7M,1C	747.013	10	0.542274
BPF	4	5A,2M,	525.73	11	0.550747	6A,4M,1C	545.686	10	0.555902
DWT	4	5A,1M,5C	340.722	12	0.553910	3A,3M,1C	350.946	13	0.578149
FIR	10	4A,4ALU,11C	686.556	6	0.351802	8A,8M,1C	837.945	11	0.521300

978-1-5386-7100-9/18 $31.00 © 2018 IEEE

Volume 63, Issue 4, November 2017

[6] NanGate 15 nm open cell library. [Online]. Available: http://www.nangate.com/?pageid=2328.

[7] DSP benchmark suite: http://www.ece.ucsb.edu/ EXPRESS/benchmark/

[8] J. Zhang, "A Practical Logic Obfuscation Technique for Hardware Security," *IEEE Transactions on Very Large Scale Integration (VLSI) Systems*, vol. 24, no. 3, pp. 1193–1197, March 2016.

[9] X. Wang, X. Jia, Q. Zhou, Y. Cai, J. Yang, M. Gao, and G. Qu, "Secure and low-overhead circuit obfuscation technique with multiplexers," *in 2016 International Great Lakes Symposium on VLSI*, May 2016, pp. 133–136.

[10] Gaillard, Rémi. "Single event effects: Mechanisms and classification." *Soft Errors in Modern Electronic Systems*, pp. 27-54. Springer, 2011.

[11] J. A. Roy, F. Koushanfar, and I. L. Markov, "EPIC: Ending Piracy of Integrated Circuits," *in 2008 Design, Automation and Test in Europe*, March 2008, pp. 1069–1074.

Run Time Mitigation of Performance Degradation Hardware Trojan Attacks in Network on Chip

Manoj Kumar JYV, Ayas Kanta Swain, Sudeendra Kumar, Sauvagya Ranjan Sahoo, KamalaKanta Mahapatra

Department of Electronics and Communication Engineering
National Institute of Technology, Rourkela, Odisha, India
{manojkumarjyv, swain.ayas, kumar.sudeendra, sauvagya.nitrkl, kmaha2}@gmail.com

Abstract—Globalization of semiconductor design and manufacturing has led to several hardware security issues. The problem of Hardware Trojans (HT) is one such security issue discussed widely in industry and academia. Adversary design engineer can insert the HT to leak confidential data, cause a denial of service attack or any other intention specific to the design. HT in cryptographic modules and processors are widely discussed. HT in Multi-Processor System on Chips (MPSoC) are also catastrophic, as most of the military applications use MPSoCs. Network on Chips (NoC) are standard communication infrastructure in modern day MPSoC. In this paper, we present a novel hardware Trojan which is capable of inducing performance degradation and denial of service attacks in a NoC. The presence of the Hardware Trojan in a NoC can compromise the crucial details of packets communicated through NoC. The proposed Trojan is triggered by a particular complex bit pattern from input messages and tries to mislead the packets away from the destined addresses. A mitigation method based on bit shuffling mechanism inside the router with a key directly extracted from input message is proposed to limit the adverse effects of the Trojan. The performance of a 4x4 NoC is evaluated under uniform traffic with the proposed Trojan and mitigation method. Simulation results show that the proposed mitigation scheme is useful in limiting the malicious effect of hardware Trojan.

Keywords- Hardware Trojan, Router Architecture, Performance Evaluation.

I. INTRODUCTION

Multiprocessor system on chips (MPSoC) is ubiquitous to meet the computing demands of current day semiconductor applications. Network on chips (NoC) are pervasive in most of the MPSoCs, required to manage the computing and data transfers between the processing elements in the MPSoC efficiently [1]. MPSoCs are developed and integrated with the help of several third-party vendors. Further, globalization of semiconductor design and development has led to several security issues. Hardware Trojans (HT) are one of the serious threats, which is widely discussed and researched in both academia and industry [2]. HTs are widely discussed in connection with cryptographic cores to leak the secret data and keys. The denial of service

(DoS) type attacks using HTs are targeted on processing elements. In DoS attacks, the intention of the adversary is to create revenue and reputation loss to the organization. HT attacks on NoC/MPSoC are largely DoS attacks. The modern

day MPSoC architectures are heterogeneous designs and include hundreds of different types of cores. The adversary may try to include the malicious code in IP core or in the communication network. Insertion of HT in the network inside the chip will cause disruptions in the functioning of the system which may lead to DoS type attacks and high amounts of latency. On-chip network is a very attractive design module for an adversary to perform DoS type attack by inserting the HT. With the advent of HT in NoC, security of on-chip networks has taken a new dimension. The conventional security and reliability aspects of NoC discuss unintentional bit corruption, packet loss, misrouting, and packet duplication etc. Malicious hardware/HT purposefully injects the DoS attack, leak information, bypass encryption/authentication and corrupt data. The general classification of HTs in NoC is based on functional correctness, path diversity, isolation, on-chip fault tolerance and quality of service. The existing HT models are based on generating surge in traffic injection and flood network resources to perform DoS attacks.

The NoC security is an active research area from last one decade. A secure router design for NoC is described in [3][4]. This paper proposes an HT and mitigation method with the following features:

- Generally, HT trigger in NoC is based on timing and conditions. In the proposed HT, logic-based triggering is used. When a specific stream of bits from input data is used for transmission, the HT is triggered. This technique requires the Trojan to bypass the functional and code coverage during verification process like any other malicious modifications do.

- Bit shuffling based method is used to mitigate runtime hardware Trojan.

- A complete Hardware synthesizable Trojan design's equivalent is used in the NoC simulation

Section II describes the related work. Section III presents the model of hardware Trojan. Section IV gives an illustration of proposed HT mitigation method. The effects of proposed HT on NoC and its performance evaluation is listed in section V. Section VI presents the conclusion of the paper.

II. RELATED WORK

Hardware Trojans (HT): Hardware Trojans are malicious inclusions by a dishonest engineer. The intentions of an adversary will vary from design to design based on its functionality. The widely discussed and generally known intentions of the adversary are to promote side channel

978-1-5386-7100-9/18 $31.00 © 2018 IEEE 738

attacks to leak confidential information and denial of service (DoS) attacks. Intentions of an adversary to insert HT in NoC are similar to HT described in literature like leaking secret data and DoS attacks. HT in NoC can be categorized based on the location of the HT, effect of HT and trigger mechanism. Types of HT in NoC are shown in Table I.

Table.1 Classification of HTs

Type of HT	Description
Location of HT	HT in NoC can be inserted in communication links or in router architecture.
Effect of HT	Leaking secret information, Denial of Service, packet loss and performance degradation.
Trigger of HT	Software Trigger: - MPSoC are designed using NoC. Adversary can use software to trigger HT. Hardware Trigger: - When a specific and rare scenario like specific input or a specific condition is met, HT will get triggered.

Authors of [3] propose an aggressive HT which is placed in router architecture to leak the confidential data. The HT [3] will forward the packets with sensitive data to the rogue thread operating on a different core. This HT needs both hardware and software support. HT trigger is in hardware, and data collection is supported by software thread running on a different core. The authors of [4] propose HT which alters the bits during router to router data transmission causing DoS type attack. In [5], DoS attack HT suppresses the allocation requests and de-prioritizes the arbiters. This HT is inserted into arbiter architecture. TASP (Targeted Activated Sequential Payload) HT proposed in [6] injects the faults in the links between the routers when a specific target is identified in the data stream. An HT attack on routing tables is discussed in [3]. Boraten et al proposed a fault injection side channel attack using a covert HT and its mitigation techniques for such attacks. Further injected faults will create deadlocks and failure in the functionality of chip (DoS type attack) [6]. An illegal packet request attack (IPRA) HTs proposed in [7] are triggered conditionally inside the routers. One or a number of HTs are placed at buffer sites and are triggered in idle mode. Frey et al have proposed Physical Unclonable Function (PUF) based mitigation scheme to counter hardware Trojan [8]. In this paper, we primarily focus on attacks which cause PD inducing DoS attacks. It is easy to design PD and DoS HTs than data leaking HT.

III. HARDWARE TROJAN MODEL

Network on |Chip comprises of a router, a network interface, and several IP cores. Router is the basic building block that transfers messages over networks via packets. A router consists of an arbiter, a routing computational module, a crossbar, and five Input and output ports. Each input port consists of a buffer that is used to store information. The buffers are usually implemented with FIFO [9] [11].

Flit format and Trojan's effects

Data is transferred between the nodes of NoC in terms of packets. The packet is further divided into flits. The parts of packet are structured as head flit, tail flit and required number of body flits based on the packet size. Generally, a valid packet contains one head flit and one tail flit and number of body flits. The head flit is comprised of destination address, quantity of flits in the packet, type of packet (valid head bit), packet identity number and also source address. The tail flit contains a valid tail bit and body flits are a major source of data. The format of flit is shown in Fig.1.

Directly targeting the data and corrupting it is very difficult because the probability of getting caught in the verification/code coverage is very high so in the proposed work we deal with Trojans that targets only critical fields of the flits and its main target is performance reduction.

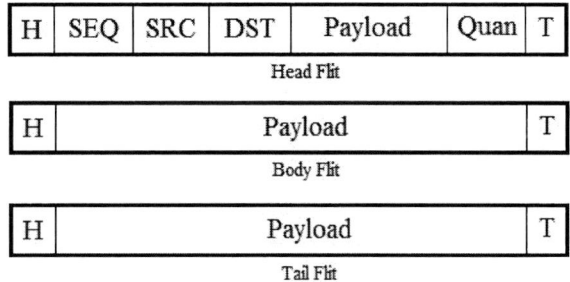

H : Head bit (1 for Head Flit) , T : Tail bit (1 for Tail Flit)
SEQ : Packet's sequence number, Quan : Flit quantity of packets
SRC : Source tile's x-y positions, DST : Destination tile's x-y positions

Fig.1 Flit Format

Attack scenario

The sequence of attack activities are as follows

- The flit, after entering the router through one of its input ports gets stored in the Buffer queue.
- When the designed trigger condition is met on the input flit, the payload mechanism changes the particular field and then stored in the buffer.
- The modified flit then passes to the crossbar switch and with the help of route computation module, it gets forwarded to another node. It starts affecting different components of router based on the field effected.

In our work, we consider following HTs: - Flit Quantity Trojan (QT), Address Trojan (AT), and Head Hardware Trojan (HHT) and Tail Hardware Trojan (THT). They target the critical fields of flit as explained below.

Quan Trojan

It targets the Flit quantity indication field. When the packet finally arrives, the destination node finds more or less flits with respect to flit quantity in head flit. The result is loss of flits by abandoning of packets, and wastage of resources due to waiting caused at destination node thus results in more latency and hence a performance decline. It's equivalent to creation of dummy flits or duplication of flits.

978-1-5386-7100-9/18 $31.00 © 2018 IEEE   739

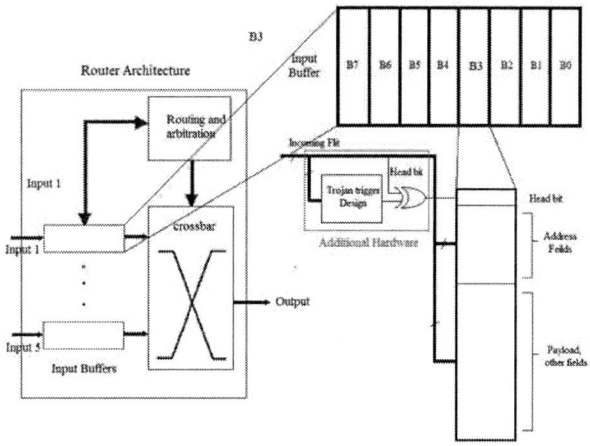

Fig. 2 Trojan design and attack scenario

Address Trojan

The address Trojan targets the destination address field (and source address if source routed algorithm is used for routing). After Trojan's attack on address field, when the packet reaches the crossbar switch, based on routing algorithm's output, the packet will be misrouted from destined address to another node. A malicious receiver-inserted node can receive the packets leading to data leakage. This will also block the destination node to receive packets sent to it. This is a major type of denial of service attack for the receiving node considered in this work.

Head Hardware Trojan

This Trojan targets the head bit. If the head bit of head flit is attacked, route computation module will not access the address field. Hence the whole packet will be dropped and retransmission is requested. This continuously performed action leads to resource starvation and NoC's performance is severely affected and degraded.

Tail Hardware Trojan

The tail Trojan changes the tail indication bit and after a successful passage through crossbar and output buffer, the receiving node keeps on waiting for a tail flit and this causes packet mixing and packet loss.

The Head Hardware Trojan operation is given in Fig. 2. In general, output of Trojans trigger module is a logic 0 and so after entering the router's input port, the flit fields are stored in input buffer without any modification. But when the Trojan activation condition is present in the input data, the Trojan's trigger is activated and give a logic 1 and the Head bit gets reversed and then pushed into the buffer. A similar mechanism is designed for other types of Trojans as well.

All the above-mentioned effects may vary a little depending on how router is architected. The performance degradation and denial of service effects are measured by increased latency, reduced throughput, decline in packet count, and misrouted packets (determined by packet distribution).

IV. PROPOSED HT MITIGATION USING BIT SHUFFLING

The proposed method shuffles the critical bit fields of the flits among themselves and others, obscuring the data fields and making them less sensitive to the Trojan. So even if the Trojan targets the packets, the attacks are applied on randomly shifted data and hence mostly inappropriate. As an additional feature of protection, a 1-bit Hamming code, an error correcting code (ECC) to foil the single bit targeted attacks is also adopted since type indication bit of a flit typically is of one-bit length only. The proposed work focusses on the Trojans that present in routers alone and we assume that the network interfaces and links are not a good place to insert Trojans and a bit difficult in implementation. In the router configuration, a Trojan can be present at any place like within the buffer, after buffer, or even just before or after the crossbar switch. The modified router design with the additional modules is as shown in the Fig. 3

Bit shuffling method

When the flit enters the Router, first it passes through the Encoder. The shuffle encoder of this method is shown in Fig. 4. The shuffler is typically a set of selector modules i.e. multiplexers that select particular bits into targeted output lines. This shuffle encoder shuffles up the informative bit fields given in Fig.1 other than the data among themselves. So, all the bit fields are displaced to other places.

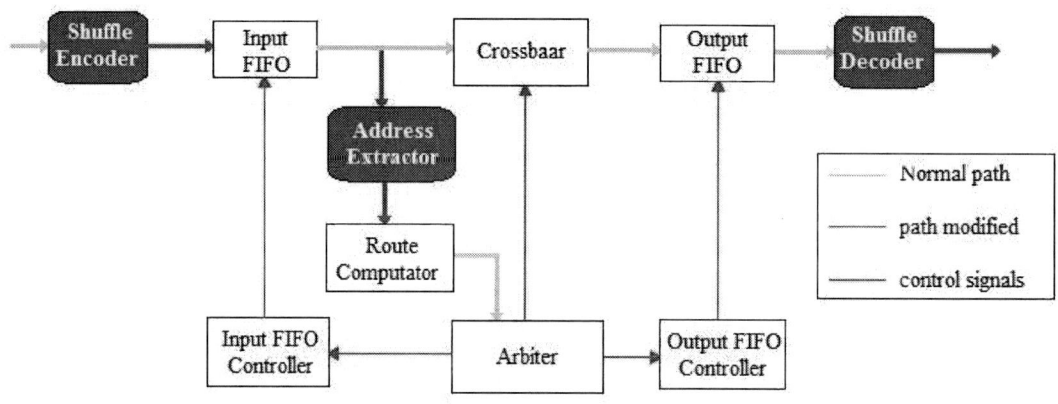

Fig. 3 Bit Shuffling Technique

978-1-5386-7100-9/18 $31.00 © 2018 IEEE   740

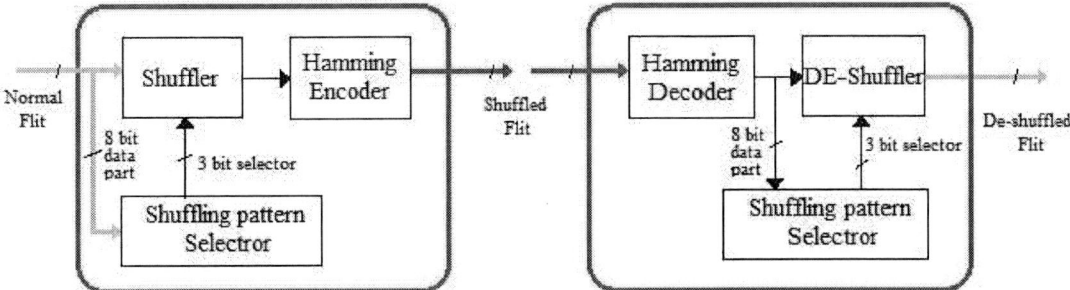

Fig. 4 Shuffle Encoder and Decoder

In order to have runtime mitigation of the Trojan, the selection of pre-planned shuffling patterns is done in cycle wise timely manner with selector generating different selections to the shuffling multiplexer's input as shown in Fig.4. In this method, shuffling pattern selector takes inputs from the least significant data lines of payload and derives a small set of selection lines. After the critical fields are shuffled, the flit passes through the input FIFO and then to crossbar switch.

At the same time the address extractor partially, de-shuffles the flit fields and extracts destination address for route computation and the information is passed to arbiter. Now the arbiter arbitrates the packet that came into the crossbar and the packet is sent to selected output FIFO. From the output FIFO, the packet (i.e., Head flit followed by body and tail flits) enters the shuffle decoder which is exactly reciprocal in design to the shuffle encoder. Here the 1-bit errors are corrected and the original pattern is restored by de-permutation of the bit fields. After this module, the packet will be outputted and sent to another router or to local node. The shuffler used is of size 3 bits to select one of 8 shuffling patterns to shuffle 14 critical field bits in total.

To balance the burden of fuzziness of the shuffled bit fields with additional hardware, we suggest to use x-bit shuffler (x-bit shuffler selects one out of 2 power x shuffling patterns) for 5x to 6x number of targeted bit fields.

The bit shuffler is designed in such a way that the highest significant fields like address or packet length are more likely to be shifted to and arranged randomly in less significant fields like source address (once the communication is established with a first most packet among two nodes in the network, the next packet's source address hardly matters in most of the existing models), packet's global number (if available) etc. Thus, we have a number of sets of shuffled bit patterns and one need to be selected.

To carry any meaningful attack even to deplete the sources, the Trojan has to target particular filed in a particular way. So, the Trojan will not be able to access the required ones as it will be unaware of the shuffling patterns and thus fails to carry any meaningful attack. For example, if the Trojan wants to leak the information to the local node, it changes the destination field to local address. But as the

bits are permutated at the router's input itself, other bits are randomly placed in the destination field, the destination address is not modified to local address and hence packets are not sent to local node and thus the Trojans attack for DoS are foiled.

Hamming code [19, 14] Design

The single bit error correcting code (ECC) just after the shuffling and so erroneous bits up to one bit are recovered just before De-shuffling.

The Block code we used in this work for single bit error correction is [19, 14] Hamming code. The bit fields considered for the 4 x 4 network are as follows:

- 1 - Head bit
- 4 - Source address bits
- 4 - Destination address bits
- 4 – Flit quantity bits
- 1 – Tail bit

V. EXPERIMENTAL RESULTS AND DISCUSSIONS

We have inserted Trojans after the buffer module. Even though the Trojans can be triggered in lot of methods, for simulation purpose, we considered triggering with a specific combination of input data lines. Practically the Trojan can be present anywhere throughout the design of router so the defense mechanism will cover all the modules in the router as the design modifications are placed at the ends of the existing design. We also assume that presence of the Trojan in router design means it is present in all the routers i.e. all the nodes because the same router design will be used at all the nodes. The experimental results are subjected to the designed Trojan's configuration and may vary depending on the activation mechanisms.

Simulation setup

The simulation is done with a 4 x 4 Mesh topology with payload length of 32 bits, flit quantity of 5 per packet, and buffer register is a FIFO having a depth of 8-bits. We used the configuration of Wormhole pipeline with identical inter-arrival time distribution to have uniformity throughout all network. We considered different flit injection rates (FIR) ranging from 0.1 to 0.7 in steps of 0.1 flits per cycle per node. We have run the simulation for 0.1 million cycles considering the first 20k cycle period as warm up time to let

978-1-5386-7100-9/18 $31.00 © 2018 IEEE

the network settle down with steady state traffic. NoCTweak simulator is used to evaluate the performance of NoC under uniform traffic [10].

Throughput

The rate of information delivered through the network is known as throughput i.e. the number of packets/flits that is generated/accepted/passed by a node and so it gives the network's efficiency.

Fig. 5 shows that when the Trojan is activated throughput is reduced by 63% in case of HHT and 71% in case of THT. Our method recovers completely as the length of the type fields are very small. As shown in the Fig. 6 the Trojan effect is more when it attacks flit quantity than destination address field. The recovery is 67% and 45% in QT and AT respectively.

latency in order to mitigate the Trojans effects as presented in Fig. 8.

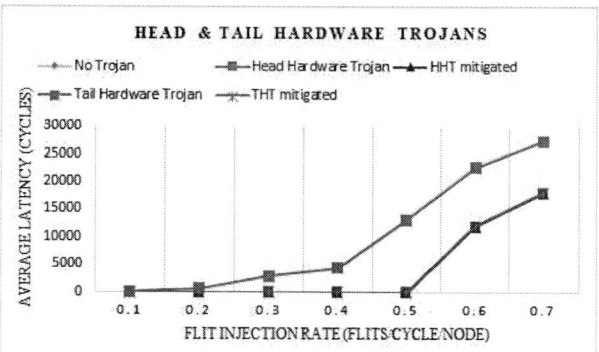

Fig. 7 Average Latency for Head and Tail Trojans

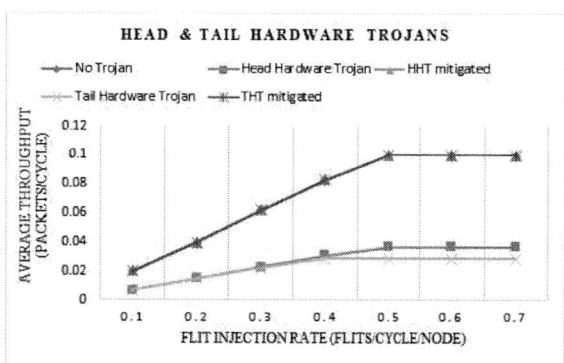

Fig. 5 Average throughput for Head and Tail Trojans

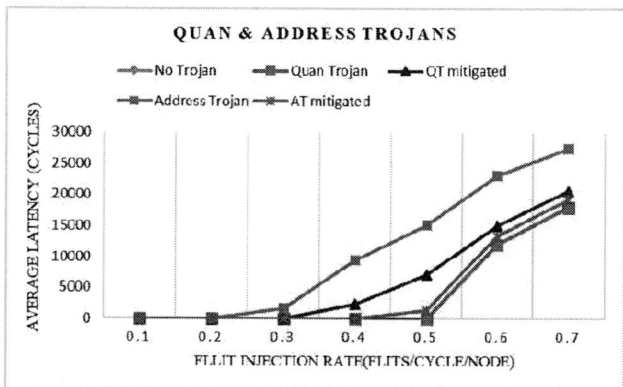

Fig. 8 Average Latency for Quan and Address Trojans

Total number of packets received

When the Trojan modifies the destination address 40% of packets are in effect and only a 40% of the packets are received in all cases of the attacks. This method is providing more packets up to 80% in all cases. These results are shown in Fig. 9 and Fig. 10.

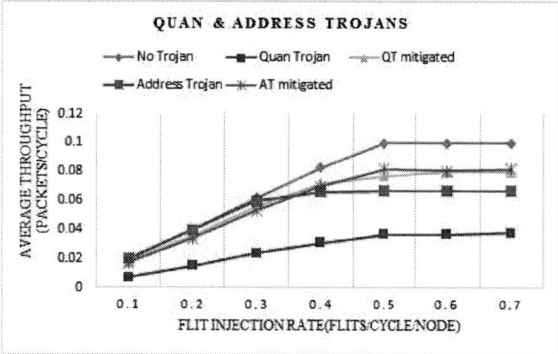

Fig. 6 Average throughput for Quan and Address Trojans

Average Latency

The average latency indicates how much time delay a packet is taking to reach the destination from its origin on an average. Fig. 7 indicates that there is no much change in average latency in HT scenario but latency is drastically increased in Tail Trojan case. This is because of successful delivery of head and tail flits which is not in Head Trojan's case. When the destination address is targeted latency is increased to 1.5 times and this method makes that only 10% extra to the normal case. Even though there is no much change in the case of QT, this method is causing a little more

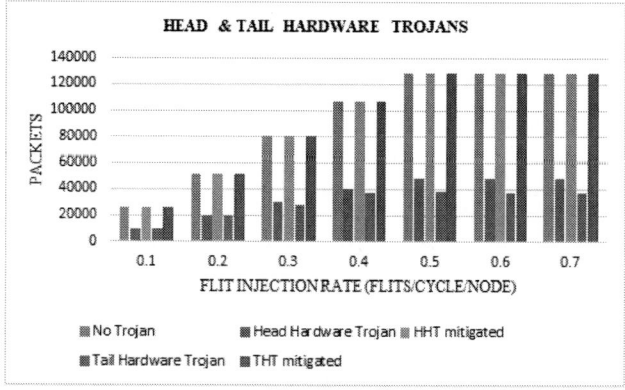

Fig. 9 Total number of received packets for Head and Tail Trojans

978-1-5386-7100-9/18 $31.00 © 2018 IEEE   742

Node wise packet reception

The effects caused by tail Trojan and Head Trojan are illustrated in the Fig.11 and Fig. 12. The number of packets received by each node is drastically reduced in these two cases and the proposed method is bringing back the lost packets.

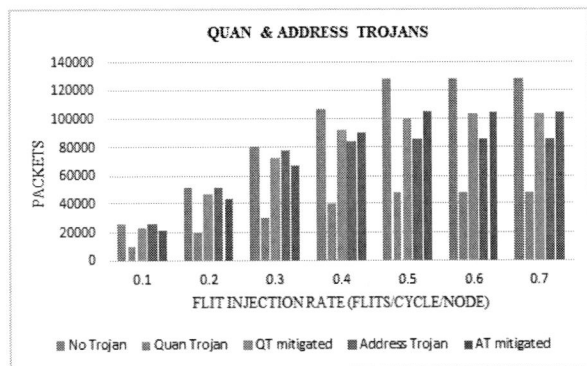

Fig. 10 Total received packets for Quan and Address Trojans

Fig. 13 demonstrates the Address Trojan which causes the diversion of the packets from the targeted node to another node. To create a DoS attack scenario, we designed the address Trojan to block the right-side routers and diverted the packets to other routers on the left side. One can see the huge no. of diverted packets in the Fig. 13. The method is able to mitigate the diversion and is successful with more than 90% of packets reaching their destinations. Fig. 14 shows how the bit shuffling method is able to recover a lot of lost packets at different nodes.

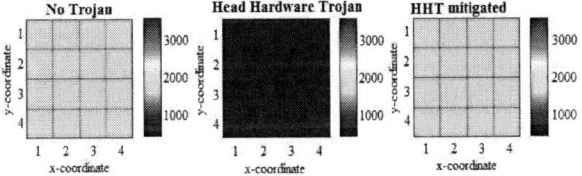

Fig. 11 Received packets at each node for Head Hardware Trojan

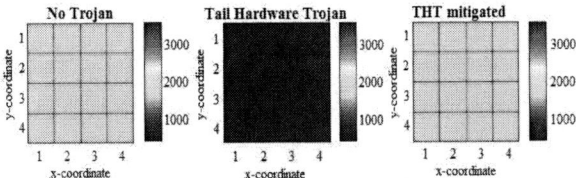

Fig. 12 Received packets at each node for Tail Hardware Trojan

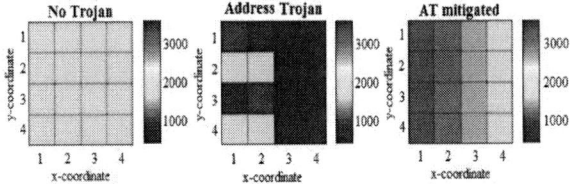

Fig. 13 Received packets at each node for Quan and Address Trojans

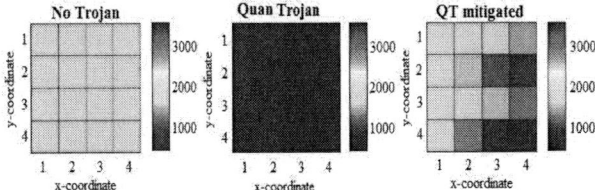

Fig. 14 Received packets at each node for Quan and Address Trojans

VI. CONCLUSION

In this paper, we proposed four different varieties of hardware Trojans, which degrade the performance and cause DoS of the network on chip. The proposed HTs have a complex triggering mechanism in comparison with earlier techniques. We also propose bit shuffling based HT defense scheme which is an efficient technique against proposed four types of hardware Trojans. The proposed HTs and mitigation scheme are verified on 4x4 NoC. The simulation results show that the proposed method is efficient in thwarting the Trojans attempts. The additional blocks that are used for mitigation are taking 21.2 percent of area overhead when compared with NoC router and this is negligible when compared with the local core/processor which contains huge functionality.

REFERENCES

[1] W. Dally and B. Towles, "Route packets, not wires: onchip interconnection networks," Proc. Design Automation Conference, Jun.2001, pp. 684–689.

[2] S. Adee, "The hunt for the kill switch," IEEE Spectrum, Vol. 45, no. 5, May 2008, pp. 34–39.

[3] D. M. Ancajas et al., "Fort-nocs: Mitigating the threat of a compromised noc," in DAC, Jun 2014, pp. 1–6.

[4] A. K. Biswas et al., "Router attack toward noc-enabled mpsoc and monitoring countermeasures against such threat," CSSP, vol. 34, no. 10, Oct 2015, pp. 3241–3290.

[5] R. JS, D. M. Ancajas, K. Chakraborty, and S. Roy, "Run time detection of a bandwidth denial attack from a rogue networkon-chip," in Proceedings of the 9th International Symposiumon Networks-on-Chip, ser. NOCS '15, 2015, pp. 8:1–8:8.

[6] T Boraten, A. K. Kodi, "Mitigation of denial of service attack with hardware Trojans in NoC architectures". In IEEE Parallel and Distributed Processing Symposium, 2016, pp. 1091-1100.

[7] N. Prasad, R. Karmakar, S. Chattopadhyay, I. Chakrabarti, "Runtime mitigation of illegal packet request attacks in Networks-on-Chip". In IEEE Circuits and Systems (ISCAS), 2017, pp. 1-4.

[8] J. Frey, Q. Yu, "A hardened network-on-chip design using runtime hardware Trojan mitigation methods", Integration, the VLSI Journal, Vol. 56, June 2016, pp.15-31.

[9] A. K Swain, K. R. Babu, S. N Satpathy, and K. K. Mahapatra, "FPGA prototyping and parameterised based resource evaluation of Network on Chip architecture. In Distributed Computing", IEEE Conference on VLSI, Electrical Circuits and Robotics (DISCOVER),2015, pp. 227-231.

[10] A. Tran, "NoCTweak: A Highly parameterizable Simulator for Early Exploration of Performance and Energy of Network-on-Chip," VCL Lab, ECE Department, UC Davis, Davis, CA, USA, 2012.

[11] N. E. Jarger, L. Peh, " On-chip Networks", Synthesis Lectures on Computer Architecture, Vol. 4(1), 2009, pp. 1-141.

2018 IEEE Computer Society Annual Symposium on VLSI

Exploration on Routing Configuration of HNoC with Reasonable Energy Consumption

Juan Fang[1], Zeqing Chang[1], Yanjin Cheng[1], Hui Zhao[2]

[1]Faculty of Information Technology, Beijing University of Technology

{fangjuan,changzeqing,chengyanjin}@bjut.edu.cn

[2]Computer Science and Engineering Department, University of North Texas hui.zhao@unt.edu

Abstract— **The Heterogeneous Network-on-Chip (HNoC) integrates CPU cores, Graphic Processing Unit (GPU) cores, last-level-cache and memory controllers. The heterogeneity of this architecture inevitably brings resource contention and energy shortage. In this work, we evaluate the impact of different capacity of router buffers on communication delay and energy consumption. We run benchmarks to simulate different characteristics of the real-world applications, with the aim to balance performance with energy consumption under buffer resource limitations. Our evaluations of HNoC show that when the buffer resources are limited, by allocating more buffer to GPU, the energy consumption decrease by an average of 44.6%, while the performance degradation is negligible.**

Keywords- Network-on-chip; GPU; Buffer allocation

I. INTRODUCTION

With the tremendous development of the semiconductor technology, hundreds of millions of transistors are integrated into one single chip. Nevertheless, it is difficult to increase the overall computing capability of the chip because of the Moore Limitation. The computing Dies in the chip need massive communication resources and the energy consumption of the chip has gradually become the major factor restraining the whole performance of the system. Prior works have put forward the network-on-chip technology that integrates multiple computing Dies into one chip to resolve the problem [1]. It is prevalent to integrate GPU and CPU computing Dies in the chip in recent years. Fig. 1 shows a typical CPU-GPU mesh heterogeneous network-on-chip. The GPU Dies, CPU Dies, last-level cache, memory controllers are all able to be regarded as an individual node, with each connecting to a router. The responsibility of each router, which has east, west, south, north and local input and output ports and are correspondingly configured with buffers, is to handle the communication among nodes[2].

To ensure the quality of service and performance of network-on-chip communication, the communication delay among nodes need to be minimized [3]. More routing buffer resources leads to better system performance and less communication delay. However, traditional network-on-chip only integrates CPU computing Dies and cannot meet the increasing demands of high throughout and parallel computing capability of emerging applications. Application characteristics exhibited by CPU and GPU workloads have

significant difference, which leads to unbalanced data communication among each interconnected channel of the HNoC. Employing traditional load balancing strategies that evenly allocate buffers to each Die, the HNoC will incur significantly increased communication delay [4]. The traditional strategy will result in large amount of idle time of CPU node, which will result in high static power consumption. On the other hand, the GPU nodes communicate more frequently and will suffer from serious congestion in the network links.

A. Contribution of the paper

In this paper, we propose a novel buffer allocation strategy of the CPU-GPU based HNoC that keeps an optimized balance of the performance, power consumption and chip design cost. We investigate the characteristics of different benchmark applications and analyze the usage of routing buffers in HNoC. Due to the difference between CPUs and GPUs, we show that allocating more buffer resource to busy channel in HNoC can significant reduce the average communication delay and achieve good performance. We next investigate how different buffer allocation strategies can affect the communication delay and energy consumption in HNoC. Finally, we propose an optimized buffer allocation strategy for HNoC. Our strategy is to keep the throughput of GPU nodes as much as possible, and reduce the CPU nodes' buffer idle time, which can decrease both performance and communication delays.

Our contributions can be summarized as follows:

• An analysis of CPU-GPU benchmark application characteristics and classification based on the application's throughput per thousand cycles.

• An exploration of the influence of different buffer allocation strategies on communication delay and energy consumption under different traffic load patterns. We propose a strategy for the HNoC and the computing Dies includes the CPU and GPU. We wisely allocate buffer resources based on the characteristics of the computing core.

• We performed extensive evaluation of the proposed strategy. Compared with the baseline buffer allocation strategy, our strategy achieves better results in terms of performance and latency in HNoC.

978-1-5386-7100-9/18 $31.00 © 2018 IEEE   744

B. Organization of the paper

The rest of the paper is organized as follows: Section II presents related work on buffer allocation strategy for NoCs and our motivation. In Section III we describe the characteristic of benchmark application and the design of our buffer allocation strategy. We present our experimental results in section IV. In section V we conclude this paper and discuss future work.

II. BACKGROUND

A. Related work

In traditional NoCs, buffers reduce the probability of packet loss or path deflection and increase bandwidth utilization. However, buffers also bring some disadvantages. Firstly, buffers consume a lot of energy, including static power consumption and dynamic power consumption. Secondly, management of buffer read and write operations increases the complexity of NoC design. It has been shown that buffers on the TRIPS prototype chip consumes 75% of the NoC area [11][12][13]. Based on this observation, researchers propose a novel bufferless router that eliminates the buffer and design a routing algorithm.

Previous works on buffer saving strategies include [3][5][7][8][9]. However, those research mainly focuses on homogeneous NoCs, by combining bufferless routers and buffered routers to reduce total power consumption. For example, Baran et. al. proposed deflection routers [10], which is applied in a bufferless NoC. This mechanism removes the use of buffers through additional link deflection. When a large number of packets are transmitted in the NoC and packets compete for a same output port, bufferless routers send packets to deflected output ports. Using this mechanism, some packets cannot be transmitted through an ideal output port directly.

Prior research on bufferless NoC design has shown that removing buffers and avoiding performance degradation through novel routing mechanism can reduce network latency and decrease energy consumption at the same time. However, compared to the traditional buffered NoC, even though the bufferless NoC saves a large amount of power, the performance loss caused by data deflection can be non-trivial. It has been shown that bufferless NoCs can only achieve comparable performance to buffered NoCs when the network traffic load is moderate. Once communication pressure increases, packet delay in the network significantly increases, resulting in congested links, decreased overall performance and increased network energy consumption.

B. Motivation

To ensure the cooperative work among units of HNoC, the communication quality among units are essential for the reason that the communication characteristics of CPU and GPU are totally different [6] [8]. CPUs are designed to use complex control logics to allow a single thread to execute instructions in parallel (even for unordered instructions) on

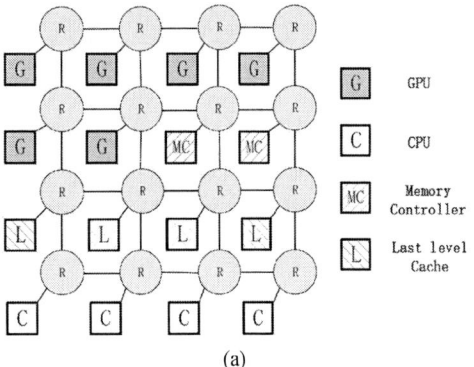

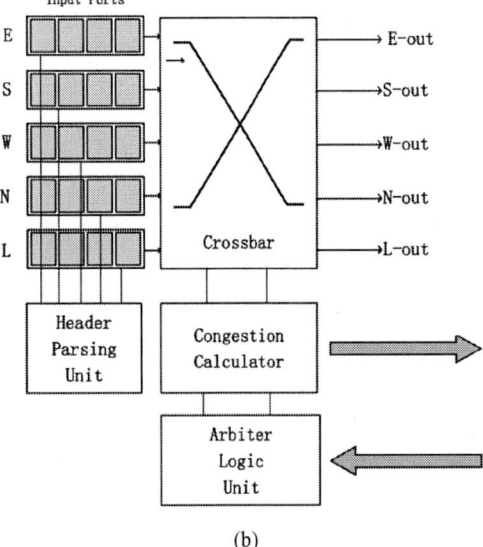

Fig. 1. (a) CPU-GPU Two-dimension Mesh Heterogeneous Network-on-chip. (b) Architecture of an on-chip router.

the premise of keeping the application be executed on demands as a whole. On the contrary, GPUs pursue high throughout and they have a large number of threads. GPUs can shift to other threads and continue to carry out the remaining work when some threads are waiting for the storage to execute access or arithmetic calculation [5]. Traditional buffer allocation strategy leads to congestion in which channels have heavy traffic load. Therefore, such a strategy is not suitable for CPU-GPU heterogeneous networks.

III. DESIGN OF CPU-GPU BASED HETEROGENEOUS NOC BUFFER ALLOCATION STRATEGY

A. Characteristic of benchmark application

CUDA GPGPU and CPU SPEC 2006 benchmarks are used for the experiments in this work. We classify benchmark applications by the evaluation of the number of packets injected into the NoC per kilocycle (PKC).

Table I and Table II shows the classification results of benchmark applications. We classify the GPU benchmark applications into three types based on the value of PKC. If the PKC is less than 100, the application is considered as the one of low packet injection rate application; if PKC is between 100 and 240, the application is considered as the one of middle packet injection rate application. If PKC is greater than 240, the application is considered as the one of high packet injection rate application. As for CPU applications, an application is regarded as the one of low packet injection type if its PKC is less than 35; the application is regarded as the one of high packet injection type if its PKC is greater than 35.

B. Design of Router Architecture

The router architecture of our proposed NoC is shown in Fig. 1(b). Each input port has a separate buffer that stores the in-coming data packets before they are transmitted to downstream output port. Each input port buffer can have a different capacity. Considering that in HNoC, CPU node throughput is much smaller than the GPU node, we assume that the link areas with high communication demands are mainly located in the GPU Die. Thus, we allocate more buffer resource for GPU nodes, which means that the chip area of the GPU router is larger than that of the CPU.

Compared with traditional on-chip design, buffered router in static HNoC employ different design methods. Traditional NoC routers use worm-hole routing algorithm, and the granularity of route calculation is data packet. In HNoC, the granularity of route calculation is flit. Regardless of whether a data flit is located at the head, body or tail of the packet, it will carry routing information. Conventionally the granularity of the data injected into NoC by the network interface is flit, packets are not divided into flits at the time of injection, but this is not the case in HNoC. In order to ensure the fairness of data transmission, data flits of a same data packet are injected into a same virtual channel first. If the current virtual channel is full, the remaining data flits can be injected into other virtual channels.

In HNoC, all routers employ Hot Potato routing. In order to ensure that the data flit will not be dropped when the target downstream router has small buffer capacity, data flit transmission will never stop regardless of whether or not the virtual channel of its neighboring router has space left. This will lead to the situation that the number of data flits are larger than the total number of buffers of a router, and we need to employ Hot Potato deflecting routing. At the same time, the original credit-based flow control mechanism is no longer applicable, and when there is no remaining space in virtual channel of target router, neighboring routers are allowed to transmit data flit to that router. The routing process in HNoC is divided into two phases. The first phase is the routing calculation and virtual channel allocation. The second phase is switch allocation and traversal. It takes two clock cycles for a flit to pass through a router if there is no contention.

TABLE I   GPU Benchmark Classification.

NVidia CUDA / Rodinia benchmark						
	High (PKC>240)		Medium (240>PKC>100)		Low (PKC<100)	
	Benchmark	PKC	Benchmark	PKC	Benchmark	PKC
GPU	Aligned	504	FDTD3d	174	Backprop	97
	BlackScholes	272	AsyncAPI	142	Hotspot	43
	ScalarProd	247				

TABLE II   CPU Benchmark Classification.

SPEC 2006				
	High (PKC>35)		Low (PKC<35)	
	Benchmark	PKC	Benchmark	PKC
CPU	Mcf	74	Calculix	14
	Omnetpp	55	Wrf	11
	Bzip2	41	Gcc	7
	Poveray	37	catusADM	6

The deflection of the data flits will cause some flits fail to reach their destination in order. Thus, it is necessary to employ a buffer to store data flits of a packet. If all data flits belonging to a same packet arrive at the destination, the buffer will assemble these data flits into a complete data packet and deliver it to the CPU core, GPU core, last-level cache or memory controller.

IV. EVALUATION OF PROPOSED BUFFER ALLOCATION STRATEGY

A. Evaluation Methods

We choose Gem5-Gpu as our experimental platform [16] which is well modularized with a detailed memory module. Gem5-GPU can also faithfully simulate various configurations of CPU-GPU heterogeneous networks-on-chip. GPU-watch is used to perform NoC power analysis for the power consumption module. The proposed NoC uses XY deterministic routing algorithm. This is because complex routing algorithms require complex communication protocols and circuit elements. Simple routing algorithms can reduce the complexity of design while the chip area and power consumption can also be reduced. We employ 4 x 4 mesh structure as did in prior work [14][15][17]. The chip simulated contains 4 CPU Dies, 6 GPU Dies, 4 last-level cache modules and 2 memory controllers, as shown in Fig. 1(a). In the Gem5-gpu simulator, each CPU Die runs a CPU benchmark application while all GPU Dies execute one benchmark application together.

Three groups of experiments are carried out according to the classification results of benchmarks. One GPU application and four CPU applications are selected as one concurrent workload. Among the three groups of experiments, the selected workload types of GPU are different from one another. Based on the above assumption, we allocate 11× flit size (FS) buffers on average for the input

978-1-5386-7100-9/18 $31.00 © 2018 IEEE

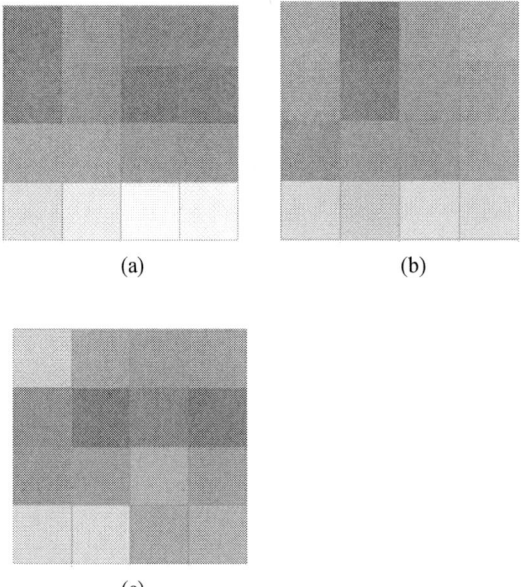

(a)   (b)

(c)

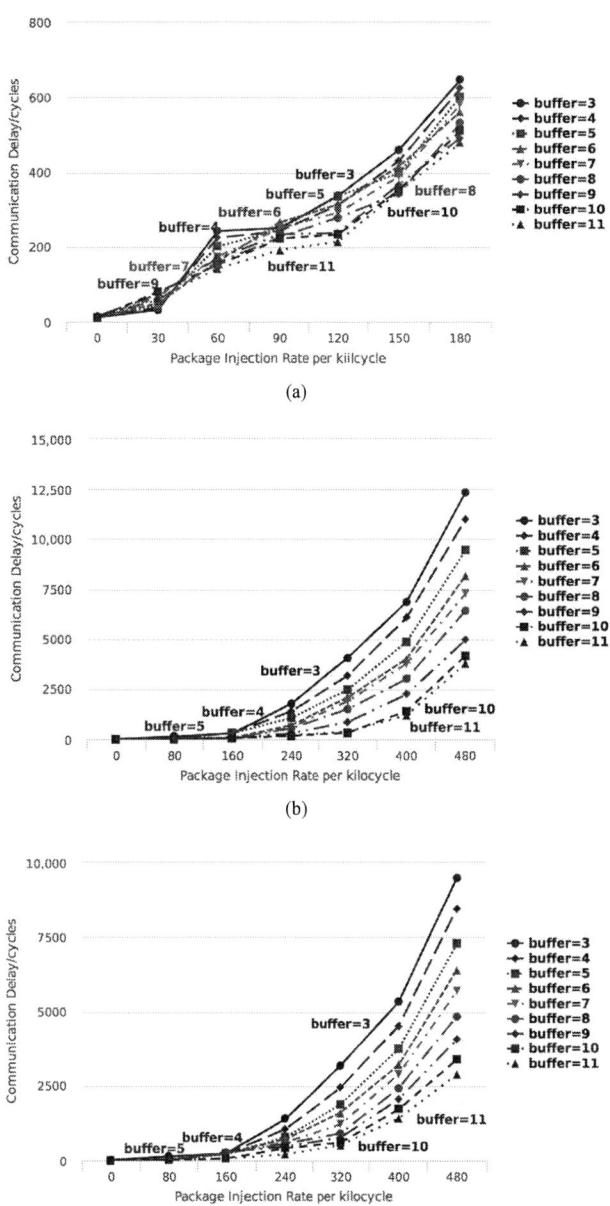

Fig. 2. The Occupancy of HNOC Router Buffer Resource (a) benchmark 1: GPU benchmarks Aligned with CPU benchmark CatusADM, Gcc, Wrf and Calculix. (b) benchmark 2: GPU benchmarks BlackScholes with CPU benchmark catusADM, Gcc, Poveray, and Bzip2. (c) benchmark 3: GPU benchmarks FDTD3d with CPU benchmark Mcf, Omnetpp, Bzip2, Poveray.

ports of the four channels in east, west, south, and north of each node and the total capacity of buffer in the NoC is 528×FS. The HNoC runs different combinations of benchmark applications, with a heat map drawn according to the average occupancy of the buffer. As shown in Fig. 2, The squares represent the computing Dies of relative position in mesh topology illustrated in Fig. 1. The darker color of a square, the higher rate of occupancy in routing buffers at that node. Fig. 2(a) shows result of GPU benchmarks aligned with CPU benchmark CatusADM, Gcc, Wrf and Calculix. Fig. 2(b) illustrates result of GPU benchmarks BlackScholes with CPU benchmark catusADM, Gcc, Poveray, and Bzip2. Fig. 2(c) shows result of GPU benchmarks FDTD3d with CPU benchmark Mcf, Omnetpp, Bzip2, and Poveray.

B. Results Analysis

Our experiment results show that there is a significant difference in occupancy of input buffers at different nodes in HNoC, and different benchmark application combinations create different hot spots. The hot spots in Fig. 2(a) are three nodes including (0,2), (0,3), and (2,2). The hot spots in (b) are two nodes including (1,2) and (1,3). The hot spots in (c) are two nodes including (1,2) and (3,2). According to the types of nodes distributed by hot spots, the link areas with high communication demands are mainly concentrated in the GPU Die, the memory controller and the last-level cache

Fig. 3. The relationship between the value of the associated communication link PKC and the average packet communication delay. (a) The relationship between CPU Buffer Size and Communication Delay. (b) The relationship between GPU Buffer Size and Communication Delay. (c) The relationship between memory control/last-level cache Buffer Size and Communication Delay.

978-1-5386-7100-9/18 $31.00 © 2018 IEEE

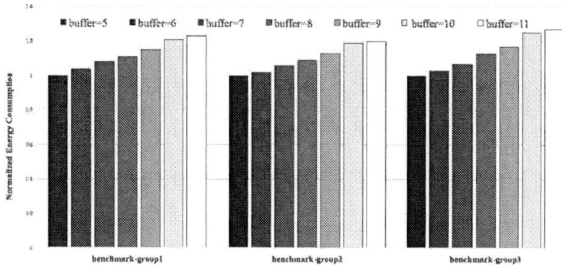

Fig. 4. Energy consumption with different CPU router buffer size

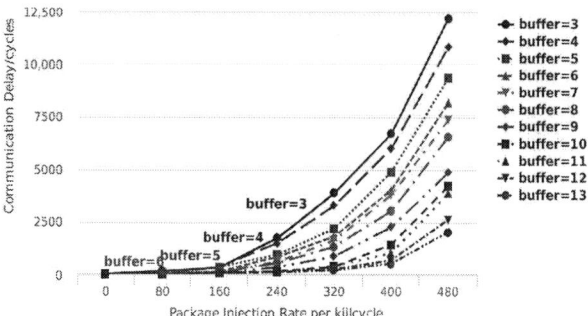

Fig. 5. The relationship between GPU buffer size and communication delay.

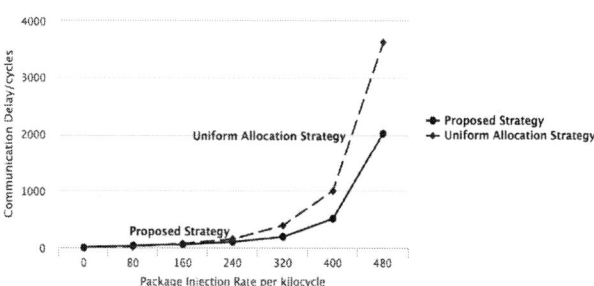

Fig. 6. The comparison of traditional average buffer

memory in the HNoC. Due to the heterogeneity of HNoC, the traffic patterns are also different from the traditional NoCs and show obvious non-uniformity, and there is obvious difference in various benchmark applications. We divide all the routers into three types, where the first type connects the CPU Die, the second type connects the last-level cache or memory controller, and the third type connects GPU Die. The relationship between the PKC of the associated channels and the average communication delay is characterized respectively when three types of routers configure buffers with different sizes. To make the statistical results more accurate, the experiment ignores the packet transmission within 200 clock cycles of the initial operation stage of the HNoC, since it will take a certain amount of time for the channel communication to reach a stable stage.

According to the results shown in Fig. 3, as the buffer size increases, the average communication delay of the correlated channel declines under different PKCs, but the declination degree varies. The PKC of CPU benchmarks is much smaller than that of GPU benchmarks, thus the impact on the average communication delay of the entire system is very limited by increasing the buffer capacity of the CPU-related communication channel. For example, for the first type of node, when the PKC is equal to 180, the average delay of the correlated channel is in the range between 478 and 646 cycles. In this situation, there is no significant difference if the buffer size is greater than 5 on the PKC. The communication pressures of the second and third type of routing nodes are relatively high. When the PKC increases, the average communication delay can be significantly reduced by increasing the input buffer size. For example, when the PKC is 500, and the input buffer size of GPU Dies is 5×FS, the average delay is 12351 cycles. However, when the buffer size is 11×FS, the average communication delay is only 3782 cycles, meaning that the average delay drops by 64.3%. Meanwhile, when the PKC is 500, and the memory controller/last-level cache node buffer size is 5×FS, the average communication delay is 9508 cycles. However, when the memory controller/last-level cache node buffer size is 11×FS, the average packet communication delay is 2782

cycles, meaning that the average communication delay of the system drops by 70.7%.

As shown in Fig. 5, it can be observed that there is no significant performance improvement brought by allocating a large number of buffers to the CPU Die due to the heterogeneity of the HNoC. When the buffer capacity of the CPU router buffer reaches a critical value, increasing the buffer size of the first type of computational node mentioned in Table II has very little impact on the average communication delay of the system. Because the congested links in the HNoC are mainly in the communication links related to the second and third types of nodes at this time, the average communication delay is maintained at a quite high level due to the severe shortage of buffers connected with these links. The average communication delay of the system drops significantly by increasing the buffer capacity in these two types of node links. This indicates that it is very effective to allocate more buffer resources for links with greater communication demands in the case of limited buffer resources.

This non-uniform buffer allocation strategy can be used to allocate buffer resources more reasonably to the most needed nodes, significantly improving the overall performance of the system. In order to verify our point of view, we allocate buffer resources with different capacities

978-1-5386-7100-9/18 $31.00 © 2018 IEEE

for the first type of nodes (CPU) and evaluate the energy consumption of the network. The experimental results are shown in Fig. 4. With the increase of the buffer, the energy consumption shows an increasing trend. Allocation strategy with the proposed allocation scheme. Additionally, according to the experiment of Fig. 3, it can be seen that there is quite subtle performance gain by improving the buffer size when the buffer size of the first type of the node exceeds the critical value 5×FS. In order to balance system performance, energy consumption and design cost at the same time, the buffer size of the first type of node is limited to 5×FS in this paper. The total system energy consumption is reduced by an average of 23.3% compared to when the buffer size is 11×FS. This reduces the chip area, thereby reducing static power consumption and circuit complexity.

In order to further improve the overall performance of the network and decrease the communication delay, we propose to allocate more buffer resources to the heavy communication demanding channels on the basis of total buffer capacity of the HNoC. According to Fig. 3 (a) and Fig. 4, it can be observed that the performance, energy consumption and cost of the system can be kept within a reasonable range at the same time when the first type of the routing node buffer is 5×FS. Therefore, when the total size of buffer is 528×FS, we set the buffer size of each channel of the second and third type of routers to 13×FS, and that of the first type of router is set to 5×FS. Benchmarks with different combinations are executed and compared with the traditional buffer resource allocation strategy, in order to measure the communication delay of system under each configuration. As shown in Fig. 5, when the PKC is 480, and GPU/memory controller/last-level cache node buffer is 11×FS, the average packet communication delay of the system is 3794 cycles. When GPU/memory controller/last-level cache node router buffer size is 13×FS, the communication delay of the system is 2078 cycles, meaning that the average communication delay of the system drops by 44%. Fig. 6 shows the comparison of the traditional buffer allocation strategy with the allocation strategy proposed in this paper. When the PKC is 480, the traditional buffer allocation strategy allocates the buffer with the size of 11×FS for each router. Under this configuration, the average packet communication delay of the system is only 3748 cycles. The average communication delay of the system drops by 44.6% when the strategy proposed in this paper is applied in buffer allocations.

V. CONCLUSIONS

In this work, we show that intelligently allocate buffers to the HNoC nodes with different features has good potential for improving system performance and decrease energy consumption. GPU, memory controller and last-level cache nodes which have characteristic of high-throughput and high packet injection rate can benefit from this type of strategy. Our buffer allocation scheme is shown to be able to effectively decrease communication delay 44.6% on average.

The allocation strategy in this paper is static. We explore the influence of different buffer allocation strategies on communication delay and energy consumption under different traffic load patterns. In the future, we plan to design a dynamic, adaptive buffer allocation strategy for HNoC that can detect on-chip congestion channels and dynamically allocate buffer resources to CPU and GPU nodes.

ACKNOWLEDGEMENT

This work was supported by the National Natural Science Foundation of China under Grant Nos. 61202076.

REFERENCES

[1] P. Bose, "The power of communication: trends, challenges", NSF WETI, Washington, DC, Feb 2012. [1] P. Bose, "The power of communication: trends, challenges", NSF WETI, Washington, DC, Feb 2012.

[2] Gowan M. K, Biro L, Jackson D. "Power considerations in the design of

[3] Wang Liwei, "Optimal buffering resources allocation of on-chip networks with finite buffers", Proc of International Conference on Intelligent Networks and Intelligent Systems. pp. 113-116, 2011

[4] Wentzlaff D, "On-Chip Interconnection Architecture of the Tile Processor", IEEE Micro, Chicago, Illinois, USA, 2007, pp. 15–31.

[5] Ma K, Li X, Chen W et al, "GreenGPU: A holistic approach to energy efficiency in GPU-CPU heterogeneous architecture", Proc. the 41st Int. Conf. Parallel Processing , pp. 48-57, September 2012.

[6] Gowan M. K, Biro L, Jackson D. "Power considerations in the design of the Alpha 21264 microprocessor", Proceedings of the 35th annual Design Automation Conference. New York, USA, vol.2 pp. 726-731, 1998

[7] James R, Jose J, Antony J K, "Smart port allocation for adaptive NoC routers", Proc of the 28th International Conference on VLSI Designp, p. 475-480, 2015.

[8] Rui Ning, "Extending the Performance of Hybrid NoCs beyond the Limitations of Network Heterogenety", Journal of Low Power Electronics and Applications, pp. 7-8, 2017.

[9] Zhao H, Kandemir M, Ding W, "Exploring heterogeneous NoC design space", Conf. Computer-Aided Design, 2011, pp. 787-793.

[10] Baran, Paul. "On Distributed Communications Networks.", IEEETrans.commun.technol 12.1(1964): pp. 1-9.

[11] Gratz, P, et al. "Implementation and Evaluation of On-Chip Network Architectures." International Conference on Computer Design IEEE, 2006:477-484.

[12] Fallin, Chris, C. Craik, and O. Mutlu. "CHIPPER: A low-complexity bufferless deflection router." 8.1(2011):144-155.

[13] Hsu, Chung Kai, et al. "A low power detection routing method for bufferless NoC." International Symposium on Quality Electronic DesignIEEE, 2013:364-367.

[14] Tung Thanh Le, "Optimizing the heterogenous network on-chip design in manycore architectures",System-on-chip Conference (SOCC), 2017,30th IEEE international.

[15] W. Dally, B. Towles, "Principles and Practices of Interconnection Networks", San Francisco, 2004.

[16] Power J, Hestness J, Orr M S, "gem5-gpu: A Heterogeneous CPU-GPU Simulator", IEEE Computer Architecture Letters, 2015, 14(1):1-1.

[17] J. Upadhyay, V. Varavithya, P. Mohapatra, "A traffic-balanced adaptive wormhole routing scheme for two-dimensional mesh", IEEE Trans. Comput, 46(2), pp. 190–197, 1997.

Nonvolatile Memory and Computing Using Emerging Ferroelectric Transistors

Xueqing Li, Longqiang Lai
The Department of Electronic Engineering
Tsinghua University
Beijing, China
xueqingli@tsinghua.edu.cn

Abstract—Ferroelectric FETs (FeFETs) are emerging as a promising nano device candidate for the next-generation energy-efficient embedded nonvolatile memory (NVM). This promise comes from not only the CMOS-scaling compatibility, but also the compact fusion of logic and non-volatility in a single device that provides opportunities for efficient memory access and in-memory computing. This talk investigates circuit opportunities that harness these intriguing FeFET device features, providing insights into new computation paradigms beyond existing solutions.

Keywords-Ferroelectric FET (FeFET); negative capacitance FET (NCFET); nonvolatile memory; emerging devices; beyond-CMOS; in-memory computing.

I. BACKGROUND AND MOTIVATION

With increasing number of edge devices due to the booming of Internet-of-Things (IoT) and sensors, how to power the ubiquitous computing is indeed a big design constraint [1]. While the battery is indeed improving with better electric-chemistry understanding, the gap between the existing battery expectation and available on-the-shelf products is increasing. For most portable devices, the limited battery life and sometimes the safety problems have raised lots of inconvenience and even life threats.

Effective approaches to lowering the power consumption have been observed in various aspects and levels, ranging from devices, circuits, architectures, algorithms, and systems [2]. Some efforts can lead to a better drop-in replacement for an existing design block, for example, designing a better invertor with simply smaller transistors with smaller capacitance, or with transistors that can operate at a lower supply voltage [3]. More importantly, the effectiveness of some efforts may strongly depend on the progress of efforts in other aspects and levels. It has been increasingly demanding co-design and co-optimization, as illustrated in Fig. 1. This will be further demonstrated in this talk.

While conventional low-power digital computing and memory design approach using the CMOS Boolean solutions has led to orders of magnitude power improvement, the challenge of further scaling the CMOS technology has made this approach much more opaque than before [4]. Even if this CMOS scaling can continue till beyond 1nm with accurate modeling, low-parasitics contacts, sufficiently-low fabrication costs, small variation and good yield, there are fundamental bottlenecks that the existing CMOS computer solution can not break theoretically in physics.

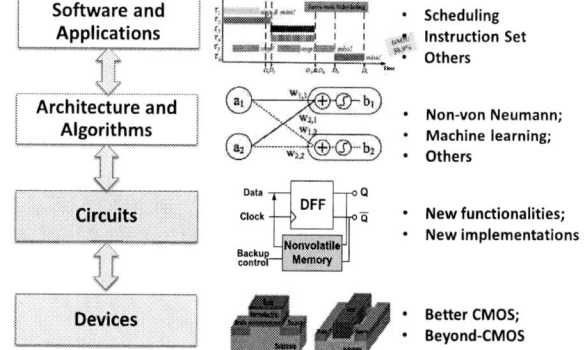

Fig. 1. Co-design and co-optimization between devices, circuits, architectures, etc. in the beyond-CMOS era.

The first well-known one is the CMOS OFF-state leakage current limited by the >60mV/decade room-temperature sub-threshold slope (SS) [3][5][6]. For large-scale integrated circuits, such leakage can cause significant amount of static power consumption by both logic and memory (e.g. SRAM), even with CMOS device tuning (e.g. threshold voltage engineering), circuit innovations (such as proper transistor sizing and new circuit topology creation) and architecture optimizations (such as power gating, dynamic voltage and frequency scaling, pipelining, parallelism, etc.)

The second bottleneck can be more related to the "memory wall" of the conventional von Neumann computer architectures, in which the memory data access can be costly in both time and energy [7]. This is essentially caused by the separation of computing logic and the storage memory elements which finally causes long-distance data movement. With the emergence of new computing architectures like new neural networks that support in-memory computing or near-memory computing, this bottleneck has a higher chance to be relieved but it still highly depends on how much long-distance data access can be eliminated. For example, recent high-performance machine-learning-based neural networks still highly rely on high-bandwidth memories (HBM) [8].

The bottlenecks above do not indicate lower importance of other potential barriers towards higher power efficiency, but these being highlighted are fundamental limitations that do not seem to have a good solution if we stick to the current device and architectures. Meanwhile, beyond-CMOS solutions provide significantly extra design space to mitigate the two abovementioned bottlenecks and also promising results, especially in some specific application scenarios, as to be presented later in this talk.

Regarding the first bottleneck, mitigation by beyond-CMOS solutions can be obtained with steep-slope Boolean transistors that can switch more abruptly with lower applied gate voltage. The steep slope characteristics ensure lower leakage current while providing the same amount of ON-state current for dynamic performance. Possible steep-slope transistors can include negative capacitance FET (NCFET) [9][10], tunneling FET (TFET) [3][5], etc. Meanwhile, emerging nonvolatile memory (NVM) devices could be adopted to reduce and even fully eliminate the static leakage current of both idle CMOS digital logic gates and CMOS SRAM as these NVM can sustain the stored data even if the power supply is shut off [1][11]-[17].

Regarding the second bottleneck, the introduction of unique computing or storage primitives provide completely new opportunities that can reshape the design space. For example, the integration of Boolean logic and nonvolatile memory (NVM) storage within each ferroelectric FETs for digital applications, and the nonlinear switching behavior of resistive memory (ReRAM) and metal-insulator-transition devices for neuromorphic and coupled-oscillator complex problem solvers, respectively [18].

This talk will use FeFET as an example to highlight the opportunities that can be enabled by emerging device-circuit co-design [12]-[16]. It is believed that FeFETs are promising because of their CMOS compatibility, the capability of being designed to be a steep-slope device or a nonvolatile memory, and also the memory-logic integration with each single transistor which enables unique in-memory computing flexibilities. While most efforts in this talk will cover the summary of the FeFET NVM and nonvolatile logic designs that fit well with existing computer architectures, it is expected that FeFETs can also be explored for more sophisticated architectures, including neural networks and array-style in-memory computers.

In the rest of this talk, Section II will briefly review the basics of FeFET devices, with the focus on highlighting the difference from a conventional MOSFET. Section III will summarize some recent FeFET-based memory designs. Section IV will review recent FeFET-based nonvolatile logic designs, specifically, nonvolatile flip-flops. Their application scenario will also be introduced as well. Section V discusses the future work and Section VI concludes this presentation.

II. FeFET Basics and Its Opportunities

A. Device Structure and General Operating Theories

A conceptual FeFET device is illustrated in Fig. 2(a), with its equivalent simplified model in Fig. 2(b), and typical I-V characteristics in Fig. 2(c) [19][20]. An FeFET is essentially a MOSFET with an extra ferroelectric gate insulator, such as doped hafnium dioxide, making it compatible with the existing commercial CMOS process. The adoption of the ferroelectric material in this structure could achieve the steep switching behavior with a sub-threshold swing below 60mV/decade so that the transistor could be used to build lower-power logic gates [19]. It is achieved by making use of the voltage booting function of the negative capacitance of the ferroelectric material to increase the internal MOSFET gate voltage. It was also predicted in theory and confirmed by recent experiments that, by increasing the ferroelectric layer thickness (T_{FE}), when the negative ferroelectric capacitance is smaller than the positive MOSFET gate capacitance, hysteresis appears and may exhibit distinct ON and OFF states with zero gate-source voltage (V_{GS}) based on the direction of the ferroelectric material polarization, as shown in Fig. 2(c) [19]. For conventional logic gates, hysteresis should be strictly controlled or minimized to comply with the logic operation. On the contrary, it is intriguing to use the hysteresis for low-power NVM applications.

In this talk, unless otherwise pointed out, we focus on using the hysteretic FeFETs for memory applications.

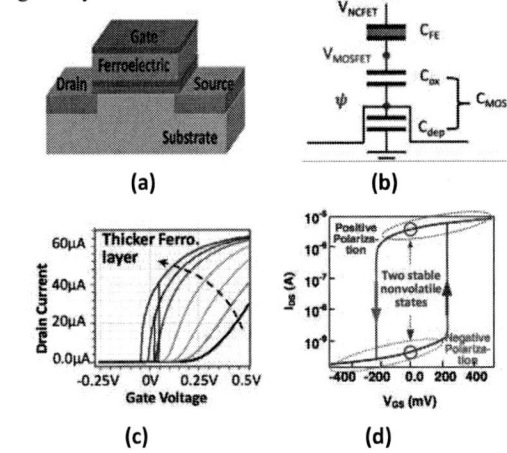

Fig. 2. Ferroelectric transistors [13]-[16]. (a) Conceptual device structure; (b) the capacitance network model; (c) typical FeFET I-V characteristics as a function of the ferroelectric layer thickness; (d) Hysteresis in FeFET I-V.

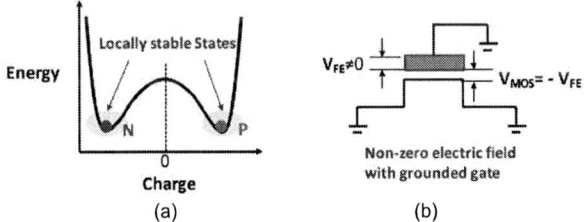

Fig. 3. FeFETs [21]: (a) Energy landscape for non-volatility theory; (b) The static internal states of FeFETs in the memory mode.

B. Device Characteristics

There are a few important notable characteristics:

- **Non-volatility.** The polarization state of the ferroelectric material is stable at zero V_{GS} [21]. As the energy landscape plot in Fig. 3 shows, the stable polarization state stays close to the two lowest-energy region. For a zero-V_{GS} FeFET in the OFF (or ON) state, a stable positive (or negative) voltage across the ferroelectric layer, V_{FE}, and accordingly a negative (or positive) internal gate-source voltage of the internal MOSFET, V_{MOS}, lead to different G_{DS} states.

- **Distinguishability.** The two nonvolatile I_{DS} states in

978-1-5386-7100-9/18 $31.00 © 2018 IEEE

Fig. 2(d) can show over four orders of difference in magnitude, leading to low-cost sensing schemes to distinguish the state difference [13][20]. This can be superior to most existing FeRAM, STT-RAM, ReRAM, and PCRAM devices. The sharp transitioning between different states also helps to maintain a larger noise margin. These advantages come from the unique FeFET features: (i) the settling-down transition behavior in the energy landscape as a passive amplification for V_{MOS}, and (ii) the gain of the internal MOSFET from V_{MOS} to sensed current I_{DS}. For FeRAM, no such intrinsic gain is provided and sensing is more complex and sensitive to bit-line parasitics.

- *Tunable Low-Voltage Operation.* With proper MOSFET work function engineering and ferroelectric material design that matches the MOSFET properties, e.g. the gate capacitance, it is possible to locate the FeFET I-V hysteresis window around zero V_{GS} [22]. By tuning T_{FE}, the hysteresis width could also be optimized to work under a proper supply voltage.

- *Logic-Memory Integration.* The FeFET has integrated the NVM storage and the logic transistor operating as a memory state amplifying reader. Such integration not only provides the opportunity to design a simplified low-power sensing scheme, but also opens up new space for future memory-oriented computing [13][14].

- *Low-Power Write Operation* [13]-[16][20]. The polarization switching is accomplished by applying a positive or negative voltage across the ferroelectric layer. Different from the state change in resistive memory devices like ReRAM and STT-RAM, no static DC current is consumed for FeFET (biased with $V_{DS} = 0$V). Furthermore, when considering the resistive memory device variations of required write pulse duration, even more energy could be saved.

As pointed out above, the ferroelectric material in FeFETs could be the same as that in FeRAM, leading to similar memory features of retention time, endurance, etc. On the other hand, the FeFET memory read operation is non-destructive, which outperforms FeRAM. More importantly, as analyzed above, FeFET NVM is fundamentally superior to FeRAM with better distinguishability and access interface.

C. Recent Device Fabrication Progress

The initial fabrication of stacking ferroelectric materials into the gate was reported long ago [23]. Recent material and process development makes FeFETs more attractive for logic and memory applications [21]-[38]. Table I summarizes some reported results. Notably, several important milestones related to FeFET fabrication and their fundamental understanding have been achieved recently. While ferroelectric materials can be BTO, PZT, PT, BST, and SBT, recent advance mostly comes from the doped hafnium (Hf) material solution, which is found to be compatible with the CMOS process and scales down well in a fin structure [24].

TABLE I. RECENT FABRICATED FeFETs

Source	Material	Structure	SS (mV/dec)	Hysteresis
EDL'16 [27]	BiFeO	Fin	8.5-50	Yes
EDL'16 [28]	PZT	Planar	2-48	Depends
IEDM' 17 [33]	HfZrO	Planar	–	Yes
IEDM'17 [34]	HfZrO	Fin	39-125	Negligible
VLSI'16 [35]	HfZrO	Planar	<50	Yes
IEDM'16 [30]	HfZrO	Planar	40-95	Depends
EDL'17 [31]	PZT	Fin	11-83	Yes
EDL'17 [36]	HfZrO	Planar	–	Yes
Nano Lett.'17 [32]	HfZrO	2D	6.1-60	Yes

D. Device Modeling

There are a few FeFET models and Landau-Khalatnikov (LK) equation has been used [20][39]-[41]. Most results in this paper uses the calibrated FeFET model in [20] with an embedded 10nm or 65nm FinFET PTM as the baseline MOSFET. The FeFET device design, including capacitance matching, ferroelectric switching mechanism, etc. has been discussed in [20].

III. FeFET NONVOLATILE MEMORY ARRAYS

This section reviews some recent designs of FeFET-based nonvolatile memory arrays, starting from the 10-transistor (10T) per cell FeFET-based nonvolatile SRAM (nvSRAM) [15], then the 2-transistor (2T) per cell design [12], and then projected 1-transistor (1T) per cell design. The trade-off is discussed among different designs. Finally, the potential application and future work is discussed.

Evaluation of an array-style memory design should be done considering both the cell design and the peripherals. The drain, source, gate, and body (if there is), should all be properly biased or controlled during the power-off, idle, read, and write modes. Read and write operations have been introduced in the previous section, and at the circuit level, access transistor may be required for desired isolation.

A. 10-T nvSRAM Design

The concept of nvSRAM is to back up the conventional SRAM state to an *in situ* distributed nonvolatile storage cell and to restore the data back to the SRAM when necessary, e.g. power-gating. The reason of not directly using the nonvolatile storage cell is mostly for the purpose of keep some virtues of CMOS SRAM, such as speed and endurance. Varying with different applications, the main design and optimization targets of the backup storage cell can include density, backup and restore energy and latency, as well as other specifications like variation and yield, supply voltage range and number of required voltage levels, etc.

Fig. 4(a) shows the 10-T nvSRAM circuit topology [15]. During the idle state, the restore control voltage Vrstr is grounded, and the FeFET gate voltage Vbkp is biased at VDD/2, or some other similar voltage levels to prevent unnecessary FeFET polarization switching activities when the SRAM state changes. If the SRAM supply voltage is sufficiently low, the FeFET gate voltage Vbkp can be biased

978-1-5386-7100-9/18 $31.00 © 2018 IEEE

at any voltage between the ground and VDD. On the other hand, for a given FeFET, if the SRAM supply voltage is too high, it can be impossible to find a Vbkp biasing that can prevent FeFET polarization switching when the SRAM state changes. When there is a demand of backup, Vrstr stays grounded, and the gate voltage Vbkp goes to VDD (to switch one FeFET to positive polarization) and then ground (to switch the other one to negative polarization), and then back to the idle state Vbkp. After the backup operation accomplishes, the SRAM power supply can be safely turned off and the FeFET polarization remains. When there is a demand of restore (while the SRAM supply is grounded), Vbkp goes to VDD/2 and Vrstr goes to VDD, and then the SRAM supply voltage is gradually increased to VDD. As the FeFET backup cell has a huge difference in pulling down the two internal SRAM nodes to the ground (one floating and the other grounded), the SRAM states can be restored.

Typically, the restore speed is limited by the supply voltage recovery latency as the supply network usually has large parasitics. And the backup speed can be in the range of nanosecond when a polarization switching activity is needed. Note that this design does not consume static current during the backup and restore phases, leading to significant amount of energy savings when compared with nvSRAM designs based on ReRAM and MTJ, as shown in Fig. 4(b). Here a break-even-time (BET) can be used to indicate how the minimum amount of supply shut-down to save sufficient leakage energy to count for the cost of backup and restore energy consumed. Theoretical analysis shows hundreds of times of energy savings per backup and restore operation.

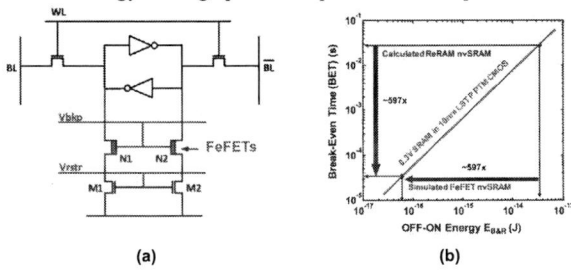

Fig. 4. The 10-T FeFET-based nvSRAM [15]. (a) Circuit schematic; (b) Performance of backup and restore energy $E_{B\&R}$.

B. 2-T NVM Array Design

When the endurance is not an issue in some applications, using purely the backup cell in the 10-T nvSRAM design is feasible. In this case, the two branches of backup storage in Fig. 4(a) can be reduced to only one branches, i.e. keeping M1 and N1 would be sufficient to store a bit. Or, the two branches could be used to store two bits.

In [12], a 2-T per cell NVM array was reported, as shown in Fig. 5(a). In the memory array, each FeFET gate can be accessed through a wordline-controlled access transistor, and the write is accomplished by applying either a positive voltage or a negative voltage to the gate of each FeFET. Fig. 5(b) shows the write performance comparison with FeRAM, which is also based on ferroelectric capacitors using the same ferroelectric material. Evaluation results show that over 10x write energy savings could be achieved.

The memory array in Fig. 5(a) could be potentially used for multiply-and-accumulation computing, i.e. "dot production" for two input vectors. The hindrance in using this design includes: (i) it does not support a practical voltage- or current-mode sensing scheme as each output sense line can be connected to multiple read select lines with low resistance if the cross-over FeFET shows up with positive polarization (in this case the sense current would be steered to the read select line instead of purely to the sense amplifier); (ii) wide write voltage range, approximately 2xVDD as both positive and negative voltages are used. In contrast, the 2T cell design based on Fig. 4(a) has no such issues. Therefore, further design optimization for the 2T cell in Fig. 5(a) is required to make it be a truly practical memory array that supports both convenient read and write operations.

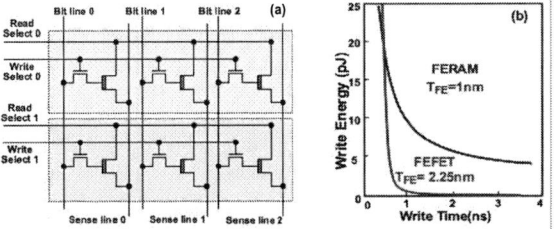

Fig. 5. 2-T FeFET-based NVM array [12]. (a) Circuit scheme; (b) Write performance comparison with FeRAM.

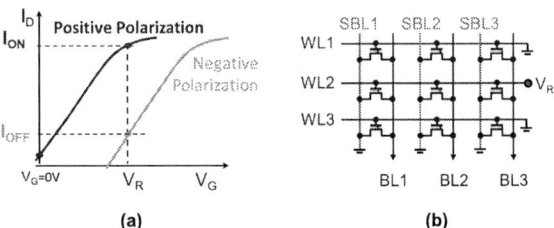

Fig. 6. 1-T FeFET NVM array. (a) Desired FeFET device characteristics; (b) An array scheme in the conventional NOR style.

C. 1-T NVM Array Design

Further improving the density of FeFET-based NVM array is beneficial to reduce the overall cost when a large amount of memory is adopted. Based on the abovementioned 2-T per cell designs, further removing the one access transistor in each cell needs extra work of FeFET device-level re-design to ensure that cells not being accessed do not short bitlines and wordlines.

A practical FeFET device can be like those reported in [33][35][36]. The required device characteristics are briefly illustrated in Fig. 6(a). This FeFET is turned off and its polarization state is sustained when the gate-source biasing is set to ground. To switch the polarization to positive and negative, the gate source voltage needs to be sufficiently high in positive and negative, respectively. To read the FeFET and tell the I_{DS} difference between the two polarization states, the gate source voltage is biased at a non-zero positive voltage V_R, which is high enough to turn on the FeFET with positive polarization, as illustrated in Fig. 6(a).

The 1-T per cell NVM array is similar to the NOR-type FLASH memory array, as illustrated in Fig. 6(b). To read a row, the gate control voltage of the row is set to V_R, and the

978-1-5386-7100-9/18 $31.00 © 2018 IEEE

source bitline SBL is set to GND, and the sense bitline BL is set to VDD. The current flowing through the cell can be sensed either through sensing the voltage change of the precharged sense bitline or through sensing the current flowing at the clamped sense bitline. To write a row, the gate can be grounded with the source and drain bitlines shorted to VDD to switch to the negative polarization or − VDD to switch to the positive polarization.

While the performance is still being evaluated at this moment, the density improvement over prior versions can be guaranteed. As a matter of fact, the energy consumption performance can be good as no static power is consumed, and the read performance can be good due to large ON-state current and ultra-low OFF-state current.

D. Summary and Future Work

Fig. 7 summarizes the FeFET-based NVM performance. Although this summary is a rough evaluation, it can clearly show the advantage towards energy-efficient embedded nonvolatile memory. Future work on variation analysis, endurance improvement, experimental demonstration, and application level evaluation is needed.

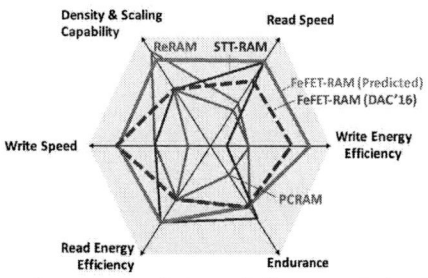

Fig. 7. Comparisons among typical nonvolatile memory devices [11][12].

IV. FeFET Nonvolatile Computing Logic

This section reviews some recent designs of FeFET-based nonvolatile latches and flip-flops [13][14][16]. The trade-off is discussed among different designs.

A. Application Scenarios and Key Specifications

Power gating has been widely used, by which the power supply of the idle and leaky digital computing blocks could be fully turned off to reduce the static power consumption. This is illustrated in Fig. 8. This can be more meaningful as the scale of modern processors is increasing with more transistors integrated. Meanwhile, the state of flip-flops in pipelines, state machines, and register files should be backed up during the power shut-down period, and be restored when the power supply is recovered. Fig. 9 illustrates a conceptual nonvolatile flip-flop (nvDFF) and a few recent FeFET-based nvDFFs which can sustain the flip-flop state during power-off periods [13][14][16]. With the development of IoT and energy harvesting techniques, power supply disturbance can be frequent and such nvDFFs are essentially critical to keep the progress with such nonvolatile computing methodology.

Therefore, critical specifications usually include:

- *Area Overhead.* This includes the backup and restore controller, backup and restore circuitry, routing, etc. If a separate supply voltage is used, extra area is needed.

- *Backup and Restore Energy Overhead.* Using more energy than that used to sustain idle leaky circuits is meaningless. Thus reducing this energy overhead can make sure that even if the power supply is shut down for a short period of time, it is still likely to save the overall energy consumption. Break-even time (BET), which has been used for nvSRAM evaluations, has been widely used for nvDFFs as well.

- *Backup and Restore Energy Time.* This is useful when the processor needs prompt wakeup response.

- *Normal Mode Energy-Latency Overhead.* This indicates whether energy-latency performance of the normal-operation mode is negatively affected.

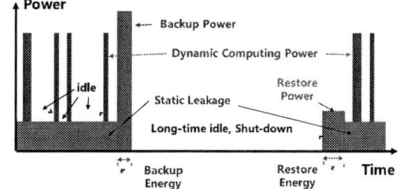

Fig. 8. Concepts of power gating to mitigate static leakage power.

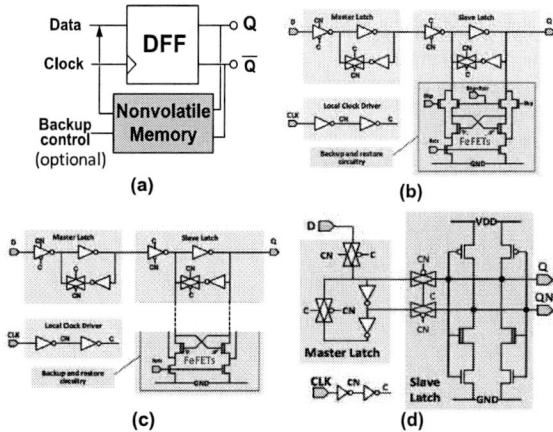

Fig. 9. nvDFFs [13][14][16]. (a) Concept with *in situ* NVM as the state backup storage; (b) nvDFF1; (c) nvDFF2; (d) nvDFF3.

B. FeFET-Based nvDFFs for Different Optimization Goals

Fig. 9(b-d) shows the circuit scheme of three energy-efficient nvDFF designs with different features [13][14][16]: ultra-low normal-mode overhead for the on-demand nvDFF1, low normal-mode overhead low area overhead on-demand nvDFF2, and ultra-low area for the intrinsic nvDFF3.

For nvDFF1 in [13], the backup operation is triggered when the backup control signal Bkp is enabled to be high, which leads to VDD or − VDD FeFET biasing for necessary polarization switching. The restore operation is similar to nvSRAM in that the initial pull-down branches with ON-/OFF-state FeFETs determine the final settled state during the supply ramp-up period. nvDFF2 in [16] eliminates access transistors and prevents unnecessary polarization switching during the normal mode operation by properly limiting the supply voltage safely within the hysteresis window.

For nvDFF3 in [14], the concept is that by embedding FeFETs into the latch, all DFF state change can finally lead to FeFET polarization change if the clock cycle is

978-1-5386-7100-9/18 $31.00 © 2018 IEEE   754

sufficiently long. While more polarization switching activities cause more normal-mode latency and energy consumption, this design eliminates external controls, and needs only two extra transistors to make the DFF nonvolatile.

These achievements originate from the various device features (see Section II) and the circuit techniques that harness them. Table II summarizes the nvDFF comparisons which clearly show the advantages of the FeFET solution.

TABLE II. COMPARISON BETWEEN RECENT NVDFFS (DATA FROM [16])

	[25]	[26]	[14]	[13]	[16]
Device Technology	ReRAM	MTJ	FeFET		
Area overheads	21	4	2 or 4	8	4
Normal-mode EDP overhead	/	6%	<50%	<5%	<5%
Backup and restore energy	~10^2fJ	~10^2fJ	~fJ	~fJ	~fJ
Backup and restore speed	~µs	~10µs	~ns	~ns	~ns

V. SUMMARY

FeFETs have been proved to be promising with recent device and circuit progress in future emerging applications. Further device, circuit and application co-design and co-optimization will bring even more opportunities.

ACKNOWLEDGMENT

This work was supported in part by NSFC under grants 61720106013 and 61674094 and in part by the Beijing Innovation Center for Future Chip.

References

[1] Y. Liu et al, "Ambient energy harvesting nonvolatile processors: From circuit to system," 2015 52nd ACM/EDAC/IEEE Design Automation Conference (DAC), San Francisco, CA, 2015, pp. 1-6.

[2] M. Alioto, "Ultra-Low Power VLSI Circuit Design Demystified and Explained: A Tutorial," in IEEE Transactions on Circuits and Systems I: Regular Papers, vol. 59, no. 1, pp. 3-29, Jan. 2012.

[3] A. C. Seabaugh and Q. Zhang, "Low-Voltage Tunnel Transistors for Beyond CMOS Logic," in Proceedings of the IEEE, vol. 98, no. 12, pp. 2095-2110, Dec. 2010.

[4] M. T. Bohr and I. A. Young, "CMOS Scaling Trends and Beyond," in IEEE Micro, vol. 37, no. 6, pp. 20-29, November/December 2017.

[5] X. Li, U. Dennis Heo et al, "Rf-powered systems using steep-slope devices," 2014 IEEE 12th International New Circuits and Systems Conference (NEWCAS), Trois-Rivieres, QC, 2014, pp. 73-76.

[6] N. S. Kim et al., "Leakage current: Moore's law meets static power," in Computer, vol. 36, no. 12, pp. 68-75, Dec. 2003.

[7] W.A. Wulf et al, "Hitting the memory wall: implications of the obvious", Computer Architecture News, Mar. 1995.

[8] Intel Nervana Neural Network Processor: 32GB HBM2 at 1TB/sec, https://www.tweaktown.com/news/60089/intel-nervana-neural-network-processor-32gb-hbm2-1tb-sec/index.html

[9] S. Salahuddin, S. Datta, "Use of negative capacitance to provide voltage amplification for low power nanoscale devices", Nano Lett., vol. 8, no. 2, pp. 405-410, Dec. 2007.

[10] S. George et al, "Device Circuit Co Design of FEFET Based Logic for Low Voltage Processors," 2016 IEEE Computer Society Annual Symposium on VLSI (ISVLSI), Pittsburgh, PA, 2016, pp. 649-654.

[11] Y. Xie, Emerging Memory Technologies: Design Architecture and Applications, 2014, Springer.

[12] S. George, K. Ma, A. Aziz et al, "Nonvolatile memory design based on ferroelectric FETs," 2016 53nd ACM/EDAC/IEEE Design Automation Conference (DAC), Austin, TX, 2016, pp. 1-6.

[13] X. Li et al, "Enabling Energy-Efficient Nonvolatile Computing with Negative Capacitance FET," in IEEE Transactions on Electron Devices, vol. 64, no. 8, pp. 3452-3458, Aug. 2017.

[14] X. Li, S. George, K. Ma et al, "Advancing Nonvolatile Computing with Nonvolatile NCFET Latches and Flip-Flops," in IEEE Transactions on Circuits and Systems I: Regular Papers, vol.64, no.11, pp.2907-2919, November 2017.

[15] X. Li et al, "Design of Nonvolatile SRAM with Ferroelectric FETs for Energy-Efficient Backup and Restore," in IEEE Transactions on Electron Devices, vol. 64, no. 7, pp. 3037-3040, July 2017.

[16] X. Li et al, "Lowering Area Overheads for FeFET-Based Energy-Efficient Nonvolatile Flip-Flops," in IEEE Transactions on Electron Devices, vol. PP, no. PP, DoI: 10.1109/TED.2018.2829348.

[17] K. Ma, Y. Zheng, S. Li et al, "Architecture exploration for ambient energy harvesting nonvolatile processors," 2015 IEEE 21st International Symposium on High Performance Computer Architecture (HPCA), Burlingame, CA, 2015, pp. 526-537.

[18] W.-Y. Tsai et al, "Enabling new computation paradigms with Hyper-FET - an emerging device," IEEE Transactions on Multi-Scale Computing Systems (TMSCS), 2016.

[19] A. I. Khan et al, "Ferroelectric negative capacitance MOSFET: Capacitance tuning & antiferroelectric operation," 2011 International Electron Devices Meeting, Washington, DC, 2011, pp. 11.3.1-11.3.4.

[20] A. Aziz et al, "Physics-Based Circuit-Compatible SPICE Model for Ferroelectric Transistors," in IEEE Electron Device Letters, vol. 37, no. 6, pp. 805-808, June 2016.

[21] A. I. Khan et al, "Negative capacitance in a ferroelectric capacitor," Nature Mater., vol. 14, no. 2, pp. 182-186, 2015.

[22] A. I. Khan et al, "Work Function Engineering for Performance Improvement in Leaky Negative Capacitance FETs," in IEEE Electron Device Letters, vol. 38, no. 9, pp. 1335-1338, Sept. 2017.

[23] Shu-Yau Wu, "A new ferroelectric memory device, metal-ferroelectric-semiconductor transistor," in IEEE Transactions on Electron Devices, vol. 21, no. 8, pp. 499-504, Aug 1974.

[24] Auciello et al, "Review of the Science and Technology for Low-and High-Density Nonvolatile Ferroelectric Memories." In Emerging Non-Volatile Memories, pp. 3-35. Springer US, 2014.

[25] I. Kazi et al., "Energy/reliability trade-offs in low-voltage ReRAM-based non-volatile flip-flop design," IEEE Trans. Circuits Syst. I, Reg. Papers, vol. 61, no. 11, pp. 3155–3164, Nov. 2014.

[26] S. Yamamoto and S. Sugahara, "Nonvolatile delay flip-flop based on spin-transistor architecture and its power-gating applications," Jpn. J. Appl. Phys., vol. 49, no. 9R, p. 090204, Sep. 2010.

[27] A. I. Khan et al, "Negative Capacitance in Short-Channel FinFETs Externally Connected to an Epitaxial Ferroelectric Capacitor," in IEEE Electron Device Letters, vol. 37, no. 1, pp. 111-114, Jan. 2016.

[28] J. Jo and C. Shin, "Negative Capacitance Field Effect Transistor with Hysteresis-Free Sub-60-mV/Decade Switching," in IEEE Electron Device Letters, vol. 37, no. 3, pp. 245-248, March 2016.

[29] M. H. Lee et al., "Physical thickness 1.x nm ferroelectric HfZrOx negative capacitance FETs," 2016 IEEE IEDM, pp. 12.1.1-12.1.4.

[30] J. Zhou et al., "Ferroelectric HfZrOx Ge and GeSn PMOSFETs with Sub-60 mV/decade subthreshold swing, negligible hysteresis, and improved Ids," in IEEE IEDM 2016.

[31] E. Ko et al, "Negative Capacitance FinFET with Sub-20-mV/decade Subthreshold Slope and Minimal Hysteresis of 0.48 V," in IEEE Electron Device Letters, vol. 38, no. 4, pp. 418-421, April 2017.

[32] F. A. McGuire, Y.-C. Lin, K. Price et al, "Sustained Sub-60 mV/decade Switching via the Negative Capacitance Effect in MoS2 Transistors," in Nano Lett., vol. 17, no. 8, pp. 4801-4806, 2017.

[33] S. Dünkel, M. Trentzsch, R. Richter et al, "A FeFET based super-low-power ultra-fast embedded NVM technology for 22nm FDSOI and beyond," IEEE IEDM 2017.

[34] Z. Krivokapic, U. Rana, R. Galatage et al, "14nm Ferroelectric FinFET Technology with Steep Subthreshold Slope for Ultra Low Power Applications," IEEE IEDM 2017.

[35] Y.-C. Chiu, C.-H. Cheng, C.-Y. Chang et al, "One-transistor ferroelectric versatile memory: Strained-gate engineering for realizing energy-efficient switching and fast negative-capacitance operation," 2016 IEEE Sym. on VLSI Technology, 2016, pp. 1-2.

[36] K. Chatterjee et al., "Self-Aligned, Gate Last, FDSOI, Ferroelectric Gate Memory Device With 5.5-nm Hf0.8Zr0.2O2, High Endurance and Breakdown Recovery," in IEEE Electron Device Letters, vol. 38, no. 10, pp. 1379-1382, Oct. 2017.

[37] A. I. Khan et al, "Negative capacitance in a ferroelectric capacitor," Nature Mater., vol. 14, no. 2, pp. 182-186, 2015.

[38] P. Zubko et al, "Negative capacitance in multidomain ferroelectric superlattices," Nature, vol. 534, no. 7608, pp. 524-528, 2016.

[39] J. P. Duarte et al., "Compact models of negative-capacitance FinFETs: Lumped and distributed charge models," 2016 IEEE IEDM.

[40] G. Pahwa, T. Dutta et al, "Compact Model for Ferroelectric Negative Capacitance Transistor with MFIS Structure," in IEEE Transactions on Electron Devices, vol. 64, no. 3, pp. 1366-1374, March 2017.

[41] S. Khandelwal et al, "Circuit performance analysis of negative capacitance FinFETs," 2016 IEEE Sym. on VLSI Technology, 2016.

2018 IEEE Computer Society Annual Symposium on VLSI

Software Support for Heterogeneous Computing

Siqi Wang
School of Computing
National University of Singapore
Singapore
wangsq@comp.nus.edu.sg

Alok Prakash
School of Computer Science and Engineering
Nanyang Technological University
Singapore
alok@ntu.edu.sg

Tulika Mitra
School of Computing
National University of Singapore
Singapore
tulika@comp.nus.edu.sg

Abstract—Heterogeneous computing, materialized in the form of multiprocessor system-on-chips (MPSoC) comprising of various processing elements such as general-purpose cores with differing characteristics, GPUs, DSPs, non-programmable accelerators, and reconfigurable computing, are expected to dominate the current and the future consumer device landscape. The heterogeneity enables a computational kernel with specific requirements to be paired with the processing element(s) ideally suited to perform that computation, leading to substantially improved performance and energy-efficiency. While heterogeneous computing is an attractive proposition in theory, considerable software support at all levels is essential to fully realize its promises. The system software needs to orchestrate the different on-chip compute resources in a synergistic manner with minimal engagement from the application developers. We present compiler time and runtime techniques to unleash the full potential of heterogeneous multi-cores towards high-performance energy-efficient computing on consumer devices.

Keywords-Heterogeneous computing, scheduler, compiler, power/thermal management

I. INTRODUCTION

Current and emerging consumer devices include *mobile application processors* for computation. State-of-the-art mobile application processors are the pinnacles of the system-on-chip (SoC) movement where multiple different functionalities are integrated on a single chip. Examples of current-generation commercial mobile processors include Qualcomm Snapdragon 845 [1], Huawei Hisilicon Kirin 970 [2], Samsung Exynos 9810 [3], and Apple A11 Bionic [4].

The common characteristic of all current- and next-generation mobile processors is the presence of heterogeneous computing [5], which refers to systems that use more than one kind of processing core on the same chip. The diversity can come from two sources. The first is *performance heterogeneity*, where cores with the same instruction-set architecture but different power-performance characteristics co-habit on the same die. The second is *functional heterogeneity*, where cores with very different functional characteristics such as general-purpose CPU, programmable GPU, special-purpose accelerators, and Field-Programmable Gate Arrays (FPGAs) appear on the same die. Recently we have witnessed the introduction of commercial SoCs with both performance and functional heterogeneity in the consumer devices. The modern mobile processor SoCs typically comprise of (a) multiple clusters of general-purpose

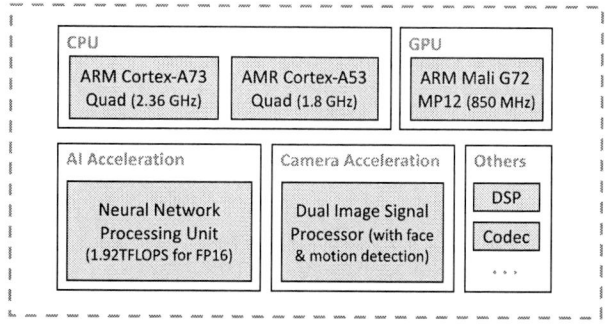

Figure 1: Huawei Hisilicon Kirin970 MPSoC [2]

CPU cores with varying power-performance characteristics, for example, ARM big.LITTLE architecture [6] where high-performance, high-power cores appear alongside low-performance, low-power cores catering to different workloads, (b) general-purpose Graphics Processing Unit (GPU) cores such as NVIDIA Kepler [7], ARM Mali [8], or Qualcomm Adreno [9], (c) Digital Signal Processor (DSP) cores such as Qualcomm Hexagon [10], and (d) a large collection of accelerators, such as Neural Network Processing Unit (NPU), Image Signal Processor (ISP) etc. in Hisilicon Kirin 970 [2], Neural Engine, motion coprocessor, image processor etc. in Apple A11 Bionic [4] as shown in Figure 1. Apart from the architectural and micro-architectural differences, these multi-cores offer additional design points in the power-performance trade-off curve through *dynamic voltage-frequency scaling (DVFS)* of the cores.

Clearly, the contemporary mobile processors are complex yet power-efficient systems featuring heterogeneous computing that have the potential to deliver high-performance if all the available resources can be harnessed by the software. At any point, we need to use the cores that are most power efficient for the current computing need without negatively impacting the performance. For example, in a smartphone, the low-power small cores can take care of simple tasks — such as email client, web browsing — saving energy, while the complex cores have to be switched on for compute-intensive tasks — such as 3D gaming, browsing flash-based websites — sacrificing energy. The GPU is responsible for graphics as well as general-purpose applications with abundant parallelism such as image processing. The FPGAs

978-1-5386-7100-9/18 $31.00 © 2018 IEEE
756

can accelerate computations with substantial parallelism and irregular control flow. However, the lack of software support puts substantial burden on the programmer.

In particular, given an application specification in high-level programming language, it is challenging to map the application on a heterogeneous multi-core. The mapping involves (a) selecting appropriate core type for each compute-intensive kernel, (b) translating and optimizing the kernel specification in a high-level language to an implementation suitable for the selected core type, (c) partitioning the application kernels or the data among multiple core types for co-execution with improved performance or energy-efficiency. The mapping process can be performed statically at compile time or adapted at runtime. The runtime support requires voltage-frequency settings for the different cores such that the aggregate power consumption of the chip does not exceed the power budget and still enables the application to achieve the best possible performance [11]. Unfortunately, the emergence of heterogeneous multi-cores is not well matched with equal advancements in software support for such complex systems. The complexity of the system with different cores supporting different programming languages and/or Application Programming Interface (API) makes software development on embedded mobile processors a challenging and sometime daunting endeavor. Thus the software deployed on embedded mobile processors routinely reaches only a fraction of the expected performance promised by the cutting-edge hardware in mobile SoCs.

In this paper, we present several techniques that have been proposed to build the software support for heterogeneous computing in consumer devices. As will be shown in the following sections, software support is necessary both at compile time as well as runtime to sufficiently exploit the diverse set of processing cores on modern mobile SoCs. In the rest of the paper, Section II discusses several compile-time strategies, while Section III presents state-of-the-art runtime power and thermal management efforts followed by a comprehensive runtime approach for thermal management on mobile SoCs. Section IV concludes this paper with a brief overview of outstanding issues for future work.

II. SOFTWARE SUPPORT AT COMPILE-TIME

The diverse set of heterogeneous processing cores enable designers to achieve decent performance within a stringent power/thermal budget requirement. However, the delicate choice of a certain core type or a combination of several cores for a given application is, most of the time, not obvious. Additionally, the reference code for an application is usually specified in a high-level single-threaded programming language such as C. It requires knowledge of the different cores as well as time and effort to redevelop the application in the accelerator-specific languages. If the performance of the application on different cores can be made available at an early stage, application developers can make

an informed decision regarding which core or combinations to use and they can then concentrate on platform-specific languages and optimizations.

In the following subsections, we present several tools that can help designers to make such compile time choices of the execution configuration. Lin-Analyzer [12], MPSeeker [13] and CGPredict [14] represent cross-platform tools that predict the performance of an application on an FPGA and a GPU respectively, from high level C code, that ease the effort and time consuming re-development process for the accelerator-specific languages. An OpenCL partitioning [15] work takes this one step further by accurately predicting co-execution performance from individual execution performances on a CPU-GPU heterogeneous platform.

A. State-of-the-art Techniques

High-level synthesis (HLS) tools [16], which abstract the programming effort above register-transfer level (RTL), are commonly used in application development on FPGAs. However, invoking HLS induces significant design space exploration (DSE) time overheads [12]. In order to rapidly explore the design space to find the execution time achievable on a certain FPGA, the estimation of FPGA performance and area requirement is essential. Starting from high-level specifications (in C/C++), [17] [18] exploit the fine-grain parallelism by accelerating the kernel on single PE. Static analysis are often used, which suffer from inherently conservative dependence analysis. In contrast, Lin-Analyzer [12] and MPSeeker [13] work with dynamic traces, while MPSeeker exploits both coarse and fine-grained parallelism.

On the other hand, modern GPUs present a highly multi-threaded architecture that enables concurrent execution of thousands of threads, which makes the prediction of performance from a single-threaded code a challenging problem. In addition, with more architectural improvements to boost the GPU performance, legacy analytical performance models [19] [20] are not compatible with up-to-date GPU architectures especially because of the complex memory hierarchy. Machine learning models [21] require extensive profiling and delicate choice of the training benchmarks. In addition, the performance of the GPU can be improved largely by platform specific optimizations. Such possibilities are not transparent in the machine learning models.

B. FPGA Performance Estimation

MPSeeker [13] (which builds on Lin-Analyzer [12]) is a high-level analysis framework that considers both fine- and coarse-grained parallelism on FPGAs to estimate accelerator performance and resource requirement from sequential C/C++ code without invoking an HLS tool. The framework is open source and available from https://github.com/zhguanw/lin-analyzer

978-1-5386-7100-9/18 $31.00 © 2018 IEEE

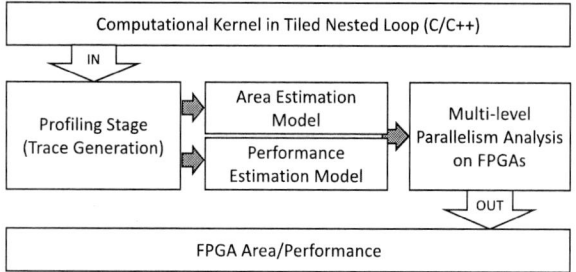

Figure 2: MPSeeker: An Automatic Design Space Exploration framework with Multi-level Parallelism for FPGA

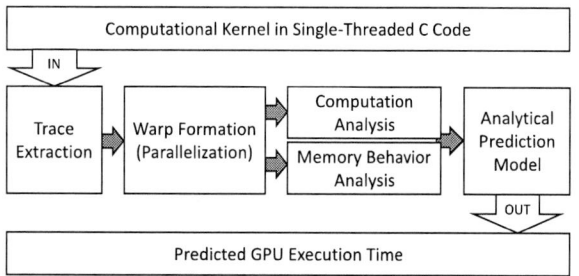

Figure 3: CGPredict: An Analytical framework for Performance Estimation from Single-threaded C Code for GPU

As shown in Figure 2, MPSeeker takes a high-level specification (C/C++) of an algorithm in the form of nested loops, available pragmas for optimization and resource constraints as inputs. A dynamic trace is generated in the *Profiling Stage*, leveraging the intermediate representation of Low-Level Virtual Machine (LLVM IR) [22] and IR instrumentation. The *Performance Estimation Model* estimates the kernel computation cycles through dynamic data dependence graph (DDDG) scheduling [12] as well as data communication cost. A machine learning technique called Gradient Boosted Machine (GBM) is employed to predict the resource usage in *Area Estimation Model*. The DSE step is finally performed by estimating both performance and area for each pragma combination to recommend the best configuration in *Multi-level Parallelism Analysis*.

MPSeeker achieves less than 5.2% error in performance prediction compared to Vivado HLS and on average 15% error in resource estimation. It recommends the same or closely similar combination of pragma settings in minutes instead of hours or sometimes even days taken by exhaustive HLS-based techniques. In addition, MPSeeker identifies bottlenecks of different FPGA implementations when applying diverse optimizations, assists designers in evaluating different architectural options in the context of high-level synthesis and better understand the performance impact of different accelerator design choices.

C. GPU Performance Estimation

CGPredict [14] is an analytical framework to estimate the performance of a computational kernel on an embedded GPU architecture from unoptimized, single-threaded C code.

CGPredict builds the performance model from a dynamic execution trace of the sequential kernel. The trace is modified to expose the available thread-level parallelism that can be potentially exploited by the GPU. At the same time, the memory access trace is analyzed against a performance model of the memory hierarchy that captures the interaction between the cache, the DRAM memory, and the inherent memory latency hiding capability of the GPU through zero-cost context switching of the threads when necessary. As shown in Figure 3, CGPredict takes a computational kernel in the form of single-threaded C code as input similar to MPSeeker and generates its execution trace through a *Trace Extraction* phase. The trace obtained is a serial trace that captures the execution information of the application. To emulate the behavior of a GPU, a *Warp Formation* phase is introduced to transform the single-threaded trace into its multi-threaded equivalent. CGPredict then extracts computation (in the form of compute instructions) and memory access information. Compute instructions are mapped to CUDA PTX ISA [23] to predict the number of GPU instructions, and thus compute cycles in *Computation Analysis* stage. To predict GPU memory cycles, CGPredict takes the memory access information and analyzes its access patterns and cache behavior in *Memory Behavior Analysis* stage. The results from the two analysis stages complete the execution characteristics required from the kernel for performance prediction. Lastly, together with the architectural parameters obtained by micro-benchmarking, an *Analytical Prediction Model* is engaged to predict the final execution performance using the computation and memory execution characteristics.

CGPredict can estimate the performance from C code with 9% estimation error compared to the performance of the corresponding native CUDA code on an embedded NVIDIA Kepler GPU averaged across a number of kernels. As CGPredict is based on analytical modeling, it can provide insights regarding the characteristics of the kernel and the GPU that influence performance, including coalescing of memory accesses or shared memory usage. These insights offer opportunities for the programmers to understand the intrinsic strengths and weakness of the architecture in the context of a particular kernel that can facilitate further code optimizations.

D. Choice of Accelerator

With the help of MPSeeker [13] and CGPredict [14], we can predict the performance of a computational kernel in C code on the FPGA and GPU platform respectively, and therefore assist the designers in selecting the appropriate accelerator that gives the best performance. Here we show a set of experiments with five benchmarks. DCT1D and MM

978-1-5386-7100-9/18 $31.00 © 2018 IEEE   758

Table I: Accelerator Choice between GPU and FPGA

Benchmark Name	Input Size	GPU time (ms)		FPGA time (ms)	
		Estimate	Actual	Estimated	Actual
MM	1024	**242.51**	**250.27**	1180	1450
MVT	2048	48.31	42.37	**9.09**	**10.41**
GEMVER1	2048	**2.61**	**4.57**	16.55	19.81
DERICHE1	1024	**0.95**	**1.53**	2.99	3.37
DCT1D	1024	2697.75	2685.36	**636.47**	**650.8**

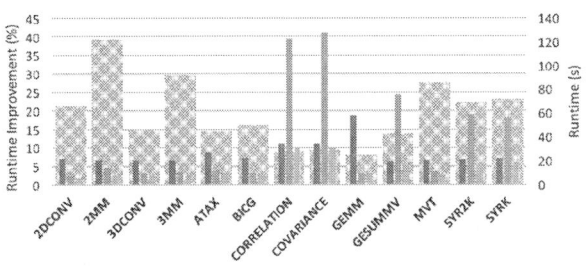

Figure 4: Benefit of Concurrent Execution of CPU and GPU

are taken from [24], while DERICHE1, GEMVER1 and MVT are taken from Polybench [25]. These benchmarks are mainly matrix calculations implemented with multiple nested loops. To verify with the actual performance of the kernels on the accelerators, equivalent CUDA code are implemented manually. The verifications are carried out on the NVIDIA embedded Kelper GPU on Jetson TK1 development board [7] (852MHz) and Xilinx ZC702 embedded FPGA [24] (100MHz).

Table I shows that CGPredict together with MPSeeker can suggest the correct accelerator (GPU or FPGA) for each application. The choice among different accelerators is a complex confluence of the considerations of application characteristics and architecture specifications.

For benchmark MM, GEMVER1 and DERICHE1, GPU is a better choice than FPGA. For the accelerators used in the experiment, GPU has much higher frequency (852MHz) and memory bandwidth (17GB/s) compared to FPGA (100MHz, 4GB/s). The better processing capability makes GPU a better choice for these three benchmarks. In addition, the three benchmarks are analyzed to have coalesced memory access pattern, which significantly reduces memory transactions of GPU implementations and improves performance.

For MVT and DCT1D, FPGA is a better choice compared to the GPU. Both MVT and DCT1D have uncoalesced memory access patterns, which cause the GPU to suffer from extensive memory transactions. Different from the GPU implementations, the FPGA accelerator first loads input data of several tiles into its local memory and start the computation. Therefore memory access patterns do not have large impact on FPGA performance, as access latency of FPGA local memory is quite small. It should be noted that GPU performance could be improved by several optimizations such as data layout transformation, loop tiling with shared memory and vectorization. However, the reference CUDA code that we are comparing against do not include such optimizations and hence we refrain from using them.

In general, with the compile time performance prediction tools, the designers are able to know the better choice of core type for a certain computational kernel. With the analytical nature of the models, designers can further optimize the implementation to achieve better performance.

E. Partitioned Co-execution

With the help of cross-platform languages, such as OpenCL [26], an application, without any modification, can be executed on multiple platforms including CPUs, GPUs, FPGAs and others. While some applications benefit from the execution on GPUs, others may be more suited for FPGA execution, or just on CPUs without any accelerator. On top of functional heterogeneity as discussed in the previous sections, we can additionally exploit the performance heterogeneity available on modern MPSoCs, such as the ARM big.LITTLE platform [26]. A co-execution on all available components of the MPSoC can therefore engage all the cores and achieve better execution performance.

We present such an approach in [15] that statically partitions the application to run across CPU and GPU clusters on mobile application processors in conjunction with appropriate voltage-frequency setting for each core cluster (small core, big core and GPU), in order to maximize the power-performance trade-off of the application. As shown in Figure 4, among all the benchmarks, the improvement in runtime by executing CPU and GPU concurrently can be as high as 39.4% (2MM) and on average 19.2%. With the high margin of execution improvement, the designers can have more knobs to adjust in order to achieve better performance under certain constraints. This work in addition shows that designers can achieve the best performance-energy trade-off (using the metric of interest ED^2) with appropriate voltage-frequency settings.

III. RUNTIME SUPPORT FOR DYNAMIC POWER AND THERMAL MANAGEMENT

In the previous section, we discussed several works that enable designers to predict the performance and in some cases even power consumption, of the various kernels in an application on different computation units such as GPU, FPGA and big.LITTLE CPU cores. These early estimates allow system designers to select the right processing core for each of the kernels in an application to achieve high performance in an energy-efficient manner on the heterogeneous platform.

978-1-5386-7100-9/18 $31.00 © 2018 IEEE

On the other hand, a runtime manager has also been proposed in the existing literature to leverage the Dynamic Voltage Frequency Scaling (DVFS) features of the various processing cores in a heterogeneous platform to further reduce power consumption during runtime. This is especially useful in consumer devices such as smartphones and smartwatches that are constantly held or worn by users. While solutions that target energy-efficient execution ensure long battery life for portable battery-operated mobile devices, techniques to ensure power-efficient execution are important to ascertain tolerable temperature while using these devices.

A. State-of-the-art Techniques

The ever rising demand for higher performance from mobile devices has forced designers and manufacturers to augment increasing number and types of computation units. This inevitably leads to higher power consumption in such devices that needs to be carefully managed to ensure power efficient execution and tolerable temperature while being used [27]. This problem is further exacerbated when the thermal behavior of the system must be maintained during runtime while ensuring high user experience.

Dynamic voltage frequency scaling (DVFS) is a popular technique that has been used extensively for power management. Several rule based governors are included in operating systems for mobile devices to manage, according to various criteria (e.g., CPU usage), the frequency of CPU, GPU, etc. [28]. For GPUs, the authors in [29] introduced a QoS-aware DVFS technique for energy-efficient execution while achieving a higher QoS satisfaction rate than the default on-demand DVFS policy. However, these governors do not directly focus on thermal management. For CPU power and thermal management, we have earlier proposed control-theoretic [30] and price-theory based approaches [31] for task migration and DVFS as well as approximation-aware scheduling and DVFS [32] across ARM big.LITTLE clusters. The task migration requires cross-core performance estimation [33] to perform cost-benefit analysis of the migration.

Mobile gaming is one of the most frequently used applications on mobile devices. They are also one of the most power hungry applications, which has led numerous researchers to explore various power management techniques for mobile gaming. Authors in [34] categorized several 3D mobile games based on their performance and power characteristics. In [35] and [36], we studied several mobile games and proposed a DVFS-based power management approach with options for power-performance tradeoff. However, these works did not focus on thermal management.

The authors in [37] and more recently in [38] target dynamic thermal management on mobile devices while considering the temperature coupling between the processor and the battery. However, these works do not consider the interaction between the CPU and the integrated GPU that directly impacts the application performance, for on-chip

thermal management. The authors in [39] focused on thermal management for mutli-core processors while considering the thermal coupling between adjacent cores, however, they did not consider heterogeneous processing cores like the ones found in mobile devices.

A recent work by Singla et al. [40] proposed a predictive approach for thermal management of CPU and GPU in mobile platforms. The main idea in this paper is to proactively turn off the high power consuming Cortex A15 cores, one by one, if a thermal throttling is predicted by their algorithm. The A7 cores and GPU is throttled next, if turning off the A15 cores does not sufficiently reduce the overall temperature. While their results show significant reduction in on-chip temperature variance, this comes at the cost of reduced performance especially for gaming workloads that require both CPU and GPU. In contrast, we propose a PID-controller based cooperative CPU-GPU thermal management technique in [41] that significantly reduces the on-chip temperature variance and achieves better performance than the default Linux governor while maintaining similar peak temperature. The next subsection briefly presents this work.

B. Cooperative Dynamic Thermal Management

In this work, we used the Arndale development platform [42] with a Samsung Exynos 5250 heterogeneous MPSoC. This SoC consists of a dual core ARM Cortex-A15 processor and a high performance Mali T604 quad-core GPU. In the first step, we made two observations from the thermal profile of CPU and GPU by running an intensive mobile game that stressed both the CPU and GPU on this platform.

As the temperature of the SoC rises during the gameplay, the default Linux thermal management unit (TMU) monitors the CPU temperature and throttles the CPU frequency once the temperatures rises to a pre-defined threshold (for example, 70°C in this platform). The GPU frequency is throttled at a later stage by its device driver and is not coordinated with the CPU. This allows the SoC temperature to continue to increase even after the CPU frequency is throttled. The SoC only cools down after both the CPU and GPU frequencies are throttled significantly. The net result of this uncoordinated DVFS is a significant oscillation in the on-chip temperature that may even create reliability issues.

Secondly, different from power consumption, because of the close proximity of CPU and GPU, even a modest increase in the temperature of one directly impacts the temperature of the other. This is known as the *thermal coupling* phenomenon that makes it difficult to predict the thermal behavior of one processing core in isolation [43].

Considering these two observations, we proposed a control theory based cooperative CPU-GPU Dynamic Thermal Management (DTM) technique [41]. The proposed technique attempts to maintain the overall chip temperature at or close to a pre-defined threshold by estimating and

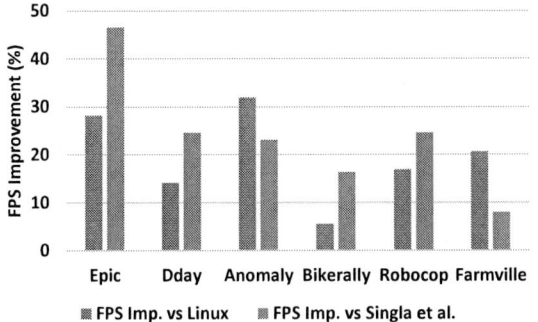

(a) Performance improvement of cooperative solution versus other approaches

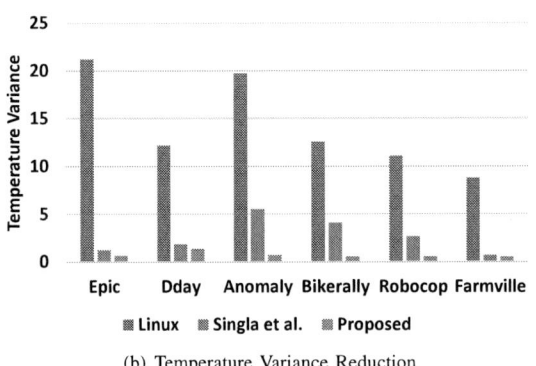

(b) Temperature Variance Reduction

Figure 5: Improvement in FPS and Temperature Variance.

allocating the available thermal headroom individually to both CPU and GPU based on their utilization values. Appropriate temperature models were derived to predict the CPU and GPU temperatures at a target frequency (to maintain performance), while also considering the temperature of the other processing core. The predicted temperature values were fed to a "Temperature Allocator" block that calculates the reference temperatures for both CPU and GPU. These reference temperatures were then sent to individual PID-controllers to obtain the frequency of CPU and GPU that achieved the desired performance without violating the thermal constraints. The resulting temperature of both CPU and GPU were measured after a fixed period and the entire process was repeated.

We implemented the proposed DTM algorithm as well as the technique proposed by Singla et al. [40] on a real mobile development platform, the Arndale development board and used it to control the frequency settings of CPU and GPU while running several gaming workloads. The resulting performance (frames per second metric for gaming workloads) and the on-chip temperature were observed. As seen from figure 5(a) and 5(b), the proposed algorithm not only reduced the on-chip temperature variance but also achieved higher performance than the default Linux TMU and the approach in [40].

IV. Conclusion and Future Outlook

In this paper, we presented our recent efforts to improve the software support for heterogeneous computing in mobile SoCs. The compile time strategies help to estimate the performance of application kernels on different processing cores such as big.LITTLE CPU, GPU and FPGA. Such tools enable system designers to make early stage decision on executing the kernel on the most suitable processing core in an effort to achieve best performance while also ensuring high energy-efficiency. On the other hand, the runtime strategies are mainly useful for dynamic power and thermal management in such devices without sacrificing performance.

The research efforts presented in this paper are certainly concrete steps in the right direction. However, more work needs to be done in future to make heterogeneous computing truly seamless for application programmers. While the compile time strategies discussed in Section II can be used to predict performance on the various processing cores, an integrated compiler support is needed to compile and map different kernels in any application on a given heterogeneous processing platform that consists of performance-heterogeneous CPU cores and functionally heterogeneous cores such as GPU and FPGA. Support for other processing cores such as DSP and ISP, etc. also needs to be added to this integrated compilation framework to make a truly seamless application mapping process. Additionally, for applications with highly input-dependent behavior, a dynamic solution to partition and schedule the workload on the heterogeneous cores can be more appropriate compared to the static partitioning strategy for co-execution discussed in this paper. As an example, we employ runtime work stealing approach to distribute workloads for the Convolutional Neural Networks inference engines to the CPU cores with vector processing units (ARM Neon) and processing elements in FPGAs for improved load balancing [44].

Finally, while the works presented here mostly consider gaming applications for runtime power and thermal management, GPGPU applications such as image processing, machine learning applications using deep neural networks, etc. also need to be considered for power and thermal efficient execution. Such applications are increasingly being ported to mobile devices and demand both high performance and power efficiency from these platforms. Apart from DVFS based solutions, runtime task migration strategies also need to be developed to ensure that no single processing core is over-worked while other cores are idle and thereby improve the overall system reliability [45].

V. Acknowledgments

This work was partially funded by Singapore Ministry of Education Academic Research Fund Tier 2 MOE2015-T2-2-088.

978-1-5386-7100-9/18 $31.00 © 2018 IEEE

References

[1] "Qualcomm Snapdragon 845," https://goo.gl/3eDr47.

[2] "Huawei Hisilicon Kirin 970," https://goo.gl/zXnjfD.

[3] "Samsung Exynos 9810," http://www.samsung.com/exynos/.

[4] "Apple A11 Bionic," https://en.wikichip.org/wiki/apple/ax/a11.

[5] T. Mitra, "Heterogeneous multi-core architectures," *Information and Media Technologies*, 2015.

[6] "Heterogeneous Multi-Processing Solution of Exynos 5 Octa with ARM big.LITTLE Technology," http://goo.gl/UVAXVS.

[7] NVIDIA, "NVIDIA Tegra K1: A New Era in Mobile Computing," *White Paper*, 2014.

[8] "ARM Mali-T600 Series GPU OpenCL, Version 2.0, Developer Guide," http://goo.gl/R0FKs8.

[9] "Qualcomm Adreno," https://goo.gl/FU8rts.

[10] L. Codrescu, W. Anderson, S. Venkumanhanti, M. Zeng, E. Plondke, C. Koob, A. Ingle, C. Tabony, and R. Maule, "Hexagon DSP: An Architecture Optimized for Mobile Multimedia and Communications," *IEEE Micro*, 2014.

[11] M. Shafique, S. Garg, T. Mitra, S. Parameswaran, and J. Henkel, "Dark silicon as a challenge for hardware/software co-design," in *CODES+ISSS*, 2014.

[12] G. Zhong, A. Prakash, Y. Liang, T. Mitra, and S. Niar, "Lin-Analyzer: A high-level performance analysis tool for FPGA-based accelerators," in *DAC*, 2016.

[13] G. Zhong, A. Prakash, S. Wang, Y. Liang, T. Mitra, and S. Niar, "Design Space exploration of FPGA-based accelerators with multi-level parallelism," in *DATE*, 2017.

[14] S. Wang, G. Zhong, and T. Mitra, "CGPredict: Embedded GPU Performance Estimation from Single-Threaded Applications," *ACM Transactions on Embedded Computing Systems (TECS) - Special Issue CODES+ISSS*, 2017.

[15] A. Prakash, S. Wang, A. E. Irimiea, and T. Mitra, "Energy-efficient execution of data-parallel applications on heterogeneous mobile platforms," in *ICCD*, 2015.

[16] Xilinx Inc., "Xilinx vivado high-level synthesis," 2014, https://goo.gl/2nvniX.

[17] X. Gao, J. Wickerson, and G. A. Constantinides, "Automatically Optimizing the Latency, Area, and Accuracy of C Programs for High-Level Synthesis," in *FPGA*, 2016.

[18] P. Li, P. Zhang, L.-N. Pouchet, and J. Cong, "Resource-Aware Throughput Optimization for High-Level Synthesis," in *FPGA*, 2015.

[19] S. Hong and H. Kim, "An Analytical Model for a GPU Architecture with Memory-level and Thread-level Parallelism Awareness," in *ISCA*, 2009.

[20] A. K. Parakh, M. Balakrishnan, and K. Paul, "Performance Estimation of GPUs with Cache," in *IPDPSW*, 2012.

[21] N. Ardalani, C. Lestourgeon, K. Sankaralingam, and X. Zhu, "Cross-architecture performance prediction (XAPP) using CPU code to predict GPU performance," in *MICRO*, 2015.

[22] C. Lattner and V. Adve, "LLVM: a Compilation Framework for Lifelong Program Analysis Transformation," in *CGO*, 2004.

[23] NVIDIA, "Parallel Thread Execution ISA Version 5.0," http://docs.nvidia.com/cuda/parallel-thread-execution.

[24] Xilinx Inc., "ZC702 Evaluation Board for the Zynq-7000 XC7Z020 All Programmable SoC," https://goo.gl/bAhHQY.

[25] S. Grauer-Gray, L. Xu, R. Searles, S. Ayalasomayajula, and J. Cavazos, "Auto-tuning a High-level Language Targeted to GPU Codes," in *InPar*, 2012.

[26] Khronos Group, "OpenCL: The open standard for parallel programming of heterogeneous systems." https://www.khronos.org/opencl/.

[27] K. Sekar, "Power and Thermal Challenges in Mobile Devices," in *MobiCom*, 2013.

[28] "Android Governors," http://goo.gl/8j1Eqo.

[29] D. You and K.-S. Chung, "Quality of service-aware dynamic voltage and frequency scaling for embedded GPUs," *IEEE Computer Architecture Letters*, 2014.

[30] T. S. Muthukaruppan, M. Pricopi, V. Venkataramani, T. Mitra, and S. Vishin, "Hierarchical power management for asymmetric multi-core in dark silicon era," in *DAC*, 2013.

[31] T. S. Muthukaruppan, A. Pathania, and T. Mitra, "Price theory based power management for heterogeneous multi-cores," *ACM SIGPLAN Notices*, 2014.

[32] C. Tan, T. S. Muthukaruppan, T. Mitra, and L. Ju, "Approximation-aware scheduling on heterogeneous multi-core architectures," in *ASP-DAC*, 2015.

[33] M. Pricopi, T. S. Muthukaruppan, V. Venkataramani, T. Mitra, and S. Vishin, "Power-performance modeling on asymmetric multi-cores," in *CASES*, 2013.

[34] X. Ma, Z. Deng, M. Dong, and L. Zhong, "Characterizing the Performance and Power Consumption of 3D Mobile Games," *Computer*, 2013.

[35] A. Pathania, Q. Jiao, A. Prakash, and T. Mitra, "Integrated CPU-GPU power management for 3D mobile games," in *DAC*, 2014.

[36] A. Pathania, A. E. Irimiea, A. Prakash, and T. Mitra, "Power-performance modelling of mobile gaming workloads on heterogeneous MPSoCs," in *DAC*, 2015.

[37] Q. Xie, J. Kim, Y. Wang, D. Shin, N. Chang, and M. Pedram, "Dynamic Thermal Management in Mobile Devices Considering the Thermal Coupling Between Battery and Application Processor," in *ICCAD*, 2013.

[38] O. Sahin et al and A. K. Coskun, "On the Impacts of Greedy Thermal Management in Mobile Devices," *Embedded Systems Letters*, 2015.

[39] A. Bartolini, M. Cacciari, A. Tilli, and L. Benini, "Thermal and Energy Management of High-Performance Multi-cores: Distributed and Self-Calibrating Model-Predictive Controller," *TPDS*, 2013.

[40] G. Singla et al, G. Kaur, A. K. Unver, and U. Y. Ogras, "Predictive dynamic thermal and power management for heterogeneous mobile platforms," in *DATE*, 2015.

[41] A. Prakash, H. Amrouch, M. Shafique, T. Mitra, and J. Henkel, "Improving Mobile Gaming Performance Through Cooperative CPU-GPU Thermal Management," in *DAC*, 2016.

[42] "Arndale Board 5250," http://goo.gl/1ZCSNX.

[43] I. Paul, S. Manne, M. Arora, W. L. Bircher, and S. Yalamanchili, "Cooperative Boosting: Needy Versus Greedy Power Management," in *ISCA*, 2013.

[44] G. Zhong, A. Dubey, T. Cheng, and T. Mitra, "Synergy: A HW/SW Framework for High Throughput CNNs on Embedded Heterogeneous SoC," *arXiv preprint arXiv:1804.00706*, 2018.

[45] C. Bolchini, M. Carminati, T. Mitra, and T. S. Muthukaruppan, "Combined on-line lifetime-energy optimization for asymmetric multicores," in *DFT*, 2016.

2018 IEEE Computer Society Annual Symposium on VLSI

Predictive Modeling for CPU, GPU, and FPGA Performance and Power Consumption: A Survey

Kenneth O'Neal and Philip Brisk
Department of Computer Science and Engineering
University of California, Riverside
Riverside, CA, USA
konea001@ucr.edu, philip@cs.ucr.edu

Abstract— **CPUs and dedicated accelerators (namely GPUs and FPGAs) continue to grow increasingly large and complex to support todays demanding performance and power requirements. Designers are tasked with evaluating the performance and power of similarly increasingly large design spaces during pre-silicon design for CPUs and GPUs to reduce time-to-market and limit manufacturing costs, or to figure out how to best map applications onto FPGAs using high-level synthesis tools. Typically, cycle-accurate simulators are used to evaluate workloads for pre-silicon CPUs and GPUs and to avoid the overhead of synthesis and place-and-route when targeting FPGAs; however, simulators exhibit prohibitively long run times that limit the number of design points and workloads that can be evaluated in a reasonable timeframe.**

This survey focuses on predictive modeling as an alternative to cycle-accurate simulation, which enables rapid evaluation of workloads and design points. When applied properly, predictive modeling can improve time to market, and can facilitate more comprehensive design space explorations with far less overhead than simulation. The survey focuses on predictive models applied to CPUs, GPUs, and FPGAs, noting that the general approach has been applied to many other computing platforms as well.

Keywords— *CPU, GPU, FPGA, Predictive Model, Machine Learning, Accuracy, Error, Survey*

I. INTRODUCTION

CPUs and GPUs must be designed and tested before manufacturing. FPGA-based accelerators aren't manufactured, but design efforts are impeded by long simulation and synthesis times. In both cases, designers must meet *quality of service* (QoS) requirements which dictate the requisite performance and power consumption of the design under test. CPU and GPU Cycle-accurate *instruction set simulators* (ISS's) [1] are used to estimate performance and power consumption and software simulators are used by commercial FPGA synthesis tools to obtain early estimates of design performance. In all cases, the simulators are used to perform *design space exploration* (DSE), and to prototype and functionally verify new architectures. CPU and GPU simulators also support co-optimization of hardware and software (e.g., API, firmware, and driver development). Typical software simulators execute orders of magnitude slower than native execution on commercial hardware [2], too slow to meet modern design productivity demands, despite efforts to raise the abstraction level [3], parallelize the simulations [4], and leverage hardware assistance [5] to improve simulation speed.

Predictive modeling has emerged as one potential solution to this conundrum. Although less precise than cycle-accurate

This work was supported in part by NSF Award #152181.

simulation, predictive models can be evaluated faster, thereby increasing both the number of design points that can be explored and the number of workloads evaluated per design point. This situation shares some similarities to the introduction of statistical sampling into cycle-accurate simulators [6], [7]: in both cases, the productivity benefits that accrue from modifying existing methodologies outweigh the resulting loss in accuracy.

This paper surveys recent advances in predictive modeling targeting CPUs (Section II), GPUs (Section III), and FPGAs (Section IV), with specific emphasis on cross-platform models based on machine learning. Although cycle-accurate simulation is required for CPU and GPU model training and is explicitly leveraged as model input for DSE targeting FPGA-based accelerators, predictive modeling can and should reduce the amount of simulation and synthesis required once a model is deployed. This provides a substantial productivity advantage compared to existing design and synthesis methodologies for these targets.

II. CPU MODELING

Table I summarizes the characteristics of each CPU modeling approach and weighs their benefits and drawbacks.

A. Statistical models for CPUs

The majority of work on predictive modeling targets CPUs executing general-purpose workloads. The overall objective is to limit the number of simulations required to evaluate design points in a much larger architectural design space. For example, Ipek et al. [8] consider a design space comprising 250K design points, and sample a representative subspace (~20K points, or 1-2% of the total design space), using active learning techniques. Model training leverages iterative refinement to build an ensemble of *artificial neural networks* (ANNs), in which each represents one of a 10-fold *cross-validation* (CV) process. Error rates range from 2-5%, with mod els created for individual CPUs, a multiprocessor, and the memory subsystem.

Similarly, Lee et al. [9] train regression models to predict performance and power consumption of a large number of design points (22 billion) using a representative sub-sample (4000), obtained via *uniform at random* (UAR) sampling. The regression models include linear least-squares models as well as non-linear spline functions, in which the predictive function is decomposed into multiple piecewise polynomials. Application-specific performance models achieve an average error of 4.1%, while regional power models, in which applications are grouped by feature similarity, achieve 4.3% average error.

978-1-5386-7100-9/18 $31.00 © 2018 IEEE   763

TABLE I.   CPU POWER AND PERFORMANCE PREDICTIVE MODELING COMPARISON.

Paper	Model	Features	Accuracy (avg.)	Speed	Benefit	Drawback
[8]	Ensemble of ANNs	Architectural parameters and latencies	Performance: 95-98%	One order of magnitude fewer simulations required	Predictive models are trained on a subset of design points to predict the full space	Requires detailed architectural expertise and many simulations to characterize latencies.
[9]	Linear and piecewise cubic spline models	Architectural parameters and latencies	Performance: ~95.9% Power: ~95.6%	~7 orders of magnitue fewer simulations required	Predictive models are used to avoid exhaustive simulation	Requires detailed architectural expertise and many simulation to characterize latencies
[10]	Analytical model	Architectural parameters and throughput measurements	Power: ~95.3%	NA	Performance feedback after model training reduces energy and avoids thermal throttling	Performance overhead; Hand-tuned analytical modeling requires expert knowledge to select features and derivce equations; this limits portability.
[11]	PCA with Lasso and CLSLR	Host CPU performance counters collected once per workload	Lasso: ~83-73% CLSR: ~ 99% For total workload	Near-native direct execution on CPU host (~500 MIPs)	*Cross-platform*; executes faster than simulator or instrumented workload	End-to-end prediction sacrifices accuracy when compared to the phase-level approach. [12]
[12]	Modified Lasso	Host CPU performance counters collected once per phase	Performance and Power: >90% at phase boundaries	Near-native direct execution on CPU host (~500 MIPS)	*Cross-platform*; models power and performance; high fidelity phase-level predictions	Choice of phase-granulartiy impacts accuracy and speed; some parameter tuning is required to balance accuracy vs. overhead.

Dynamic thermal management (DTM) often employs power models to reduce energy consumption and prevent thermal emergency events. For example, Nath et. al. [10] developed an analytical multi-core power model for the Intel *Knights Ferry* (KNF) architecture, comprising static and dynamic models for compute, memory, and interconnect power, with a 4.73% average error rate. The model relies on expert knowledge to identify the most relevant performance counters for inclusion as features; the model itself is trained using the selected features and the HotSpot thermal simulator [13] configured to model KNF thermal properties. The model itself has low overhead and reduces energy consumption by 14%, and the occurrence of thermal emergency events by 58%.

B. Cross-Platform statistical models for CPUs

LACross [11] appears to be the first work to perform cross-platform and cross-ISA CPU performance prediction. The initial iteration runs at near-native hardware speeds, and accurately predicts the performance of 157 ACM *International Collegiate Programming Contest (*ACM-ICPC) programs from a variety of application domains. LACross used two commercially available processors as *host* and *target* interchangeably, an Intel Core-i7 920 with 24GB DRAM and the AMD Phenom II X6 1055T with 8GB DRAM. LACross first executes each workload on the host to collect performance counter measurements and executes each workload on the target to measure performance. The host performance counter measurements, one read per workload, are used to train two regression models, the *LASSO L1 regularization model (LASSO)* [14] and the *Constrained Locally Sparse Linear Regression Model* (CLSR), each combined with *Principle Components Analysis* (PCA) [15] to extract latent semantics in the feature data, which improves model accuracy. The average LASSO cross validation (CV) error, an estimate of

model generalizability, was ~17% and ~27% for the two respective targets, while the CLSLR model, achieved a much lower CV error of less than 1%, on average, which suggests the existence of a non-linear relationship between host performance counters and target ISA. Although the average model prediction error is not reported, it is notably higher than the reported CV error. Due to the one-time counter collection, Lasso accuracy suffers due to a relatively limited number of data points.

A subsequent extension to LACross [12] used a compiler to instrument program basic blocks, which exposed program phases at a much finer granularity than end-to-end execution. The user specifies the phase granularity: finer granularity increases the profiling overhead, but improves accuracy; lower granularity is faster, but has lower accuracy. Program counter values are collected once per phase, and each phase is treated as a data point. This work also employs an Intel Core-i7 920 as a host and ARM Cortex-series processors targets, representing not only cross-ISA prediction, but also demonstrating the ability to predict the performance and power of embedded targets using a desktop CPU as a host. The average error reported was less than 10% in each case. The viability of cross-platform performance prediction has profound implications for future CPU design methodologies. At present, architectural DSE necessitates the use of a simulator to characterize workload performance and/or power consumption at each design point. High simulation execution times limits both the number of design points that can be explored, and the number of workloads that could be used to characterize each design point. A cross-platform predictive model which uses a commercially available CPU as the host and targets a simulator could significantly increase the throughput of next-generation DSE processes. Another practical consideration is that analytical models require expert knowledge to design,

978-1-5386-7100-9/18 $31.00 © 2018 IEEE   764

TABLE II. GPU POWER AND PERFORMANCE PREDICTIVE MODELING COMPARISON.

Paper	Ut the k	Features	Accuracy (avg.)	Speed	Benefit	Drawback
[16]	K-means clustering and ANNs	Performance counters	Performance: ~85% Power: ~90%	20% of target executions eliminated	Power and perf. models used to avoid executing all design points.	Large percentage of design points required to train model
[17]	Forward stepwise regression	Workload characteristics that expose inherent GPU-compatible parallelism	Kepler Performance: ~64% Maxwell Performance: ~73%	Static analysis incurs 10x-20x slowdown over native hardware	**Cross-platform;** Predict GPU performance from CPU features	High model error.
[18]	Analytical model	Single-threaded CPU memory and computation traces and latencies	Jetsen TK1 Performance: ~ 91%	CPU execution plus code instrumentation and PTX transform time	**Cross-platform;** Estimate GPU performance from CPU C code..	Input sizes must be chosen to limit insturmentation overhead.
[19]	Random Forest regression	Functional simulation execution statistics	Intel Skylake GPU Performance:~85.7%	~328x faster than cycle-accurate simulation	**Cross-abstracton;** Host and target use same SW stack.	Functional simulatior instrumentation overhead
[20]	Ordinary Least Squares with feature selection and Random Forest regression	Prev. generation performance counters and API metrics	Boradwell GT2/GT3 Performance: ~93% Skylake GT3 Performance: ~91%	29,000x - 44,000x faster than cycle-accurate simulation	**Cross-generation;** Predict pre-silicon next-generation GPU performance.	Cannot directly account for large-scale architectural. changes

while purely statistical models, such as those employed by LACross, can be derived automatically.

III. GPU MODELING

Table II summarizes the characteristics of each GPU modeling approach and weighs their benefits and drawbacks.

A. Predictive modeling for GPGPUs

Predictive modeling for GPUs is largely inspired by CPU approaches. These techniques are primarily leveraged for DSE wherein models are trained to avoid exhaustive simulation on the target platform.

One such example is a paper that uses ANNs to predict performance and power of OpenCL applications as architectural parameters of the GPU target scale [16], thereby avoiding a more costly exhaustive enumeration. Architectural parameters that are considered include core frequency, memory bandwidth, and the number of available *compute units* (CUs). The model clusters kernels with similar scaling behavior, a-priori, via k-means clustering; the scaling trend is a model that predicts the performance and power consumption of a workload from hardware performance counter measurements. An ANN classifier is trained to predict the cluster of scaling trends that a previously unseen workload most closely matches; given the classification, the scaling trend predicts the performance and power consumption of that workload. The new workload is simulated using one parameter combination to obtain a baseline that can be scaled according to its ANN-predicted trend.

B. Cross-architecture modeling for GPUs

Cross-architecture models for GPUs can be categorized as follows: 1) *cross-platform* models, in which features obtained from a non-GPU host (e.g., a CPU) are used to predict GPU performance and/or power consumption; 2) *cross-abstraction* models, in which a functional simulator host provides features

that predict the performance and/or power consumption of a target cycle-accurate simulator; and 3) *cross-generation* models, in which direct execution on an older GPU provides features that predict the performance and/or power consumption of a newer GPU within the same family, possibly at the pre-silicon stage of development, where only a cycle-accurate simulator is available.

XAPP [17] is a suite of cross-platform models that predict the degree of speedup or slowdown that would result from porting a C/C++ CPU workload to CUDA and executing it on a GPU. XAPP instruments program binaries to produce a set of microarchitecturally independent features, which are selected to represent characteristics that correlate with typical GPU execution behavior, and, by extension, performance. Workload characteristics are measured using MICA [21] or PIN [22] and capture characteristics such as instruction level parallelism, shared memory bandwidth, memory throughput, memory coalescing, and bank conflicts in shared memory, among others. XAPP collects 17 features in total, which are converted into a training set ensemble via random bootstrap sampling with replacement [23]. For each training set, XAPP trains a least-squares regression model that includes higher-order polynomial terms to capture non-linear relationships. For a new workload, the ensemble model reports the mean prediction of all models as the final predicted value. XAPP reported 36% average performance prediction error on a Nvidia Kepler GTX 660Ti, and 27% average performance prediction error on a Maxwell GTX 750.

CGPredict [18] is another cross-platform model that collects features from single-threaded non-optimized C code to predict the performance of CUDA code running on an embedded GPU, such as the Jetson TK1 Kepler. CGPredict employs an analytical model whose primary characteristics are based on the interaction between microarchitectural parallelism and memory access latencies and parallelism. CGPredict instruments source code via the compiler, and collects computation and memory traces, which are transformed during a memory behavior analysis stage, converting single-threaded CPU memory accesses to reflect GPU memory accesses and cache configurations. A subsequent computational analysis converts the computation trace to Parallel Thread Execution (PTX) format, which is specific to Nvidia GPUs. These transformed streams are coupled with estimated GPU computation and cache access latencies, which were derived from repeated execution of micro-benchmarks. A comprehensive analytical model predicts performance from the modified streams, achieving a relatively low predicted error of 9% across 15 kernels. The data input size for each benchmark must be carefully chosen to avoid excessive instrumentation and transformation latencies, while remaining long enough to realistically stress the GPU's compute and memory resources.

O'Neal et al. [19] use features obtained from a lightly instrumented functional GPU simulator host [24] to train a *Random Forest* (RF) model to predict the performance reported by a much slower cycle-accurate GPU simulator target. The objective is to enable rapid evaluation of multiple workloads during pre-silicon design, while ensuring that the host and target share identical software configurations (API, driver, etc.), which themselves may be pre-release. The functional simulator and instrumentation layer produce 69 features; using the RF model is 327.8x faster than cycle-accurate simulation, with a reported 14.34% average out-of-sample error; the instrumentation layer improves overall accuracy by 38% in comparison to the otherwise unmodified functional simulator.

HALWPE is a cross-generation GPU model that obtains performance counter and DirectX API measurements via direct execution on a commercially available GPU, and trains a suite of machine learning models to predict the performance of a cycle-accurate simulator configured to model a pre-silicon successor in the same family [20]; in addition to architectural evolution, there may be unavoidable differences in the software configuration, unlike the functional to cycle-accurate GPU simulator prediction scheme described above. The models, once again, include a suite of linear models, along with one non-linear model: Random Forest regression. For each host-target prediction scenario, the entire suite of models is retrained and the one with the smallest out-of-sample error is selected for use. The host is an Intel Haswell HD4600 single-slice GPU with 20 *execution units* (EUs); three targets are selected: a single-slice (24 EU) Broadwell GT2, a dual-slice (48 EU) Broadwell GT3, and a single-slice (48 EU) Skylake GT3. The Broadwell targets are a single-generation difference over the Haswell host, with varying parallelism, while the Skylake targets represent a two-generation difference. The reported out-of-sample errors and speedups over cycle-accurate simulation are, respectively, 7.45% and 29,481x (Broadwell GT2), 7.47% and 43,643x (Broadwell GT3), and 8.91% and 44,214x (Skylake GT3). One limitation of this approach it is impossible to train models that encompass large-scale cross-generation architectural redesigns that introduce new features that do not correlate with features collected from the post-silicon host.

IV. FPGA PREDICTIVE MODELING AND DSE

Table III summarizes the characteristics of several FPGA-based predictive modeling approaches, in the context of larger DSE frameworks and weighs their benefits and drawbacks.

FPGAs are often used as acceleration engines for parallel and streaming workloads, due to their inherent reconfigurability. FPGA accelerator design methodologies have traditionally been based on hardware design, and has more recently transitioned to *High-Level Synthesis* (HLS) [25,26]. Although HLS aims to be fully automatic, the typical designer is tasked with the problem of finding the correct combination of parameters (e.g., unroll factor, pipelining depth, etc.), which is a form of DSE. Due to long HLS times, direct evaluation of each design point is infeasible, which necessitates a turn toward modeling.

Liu et al. [27] explore predictive modeling for DSE via *Transductive Experimental Design* (TED) [28], which identifies and samples representative microarchitectural design points. These training sets are then used to build an RF regression model which is iteratively refined via repeated training and synthesis of additional directive permutations, wherein the training sets are updated after evaluating the prior trained model's accuracy. When used in conjunction with TED, iterative refinement improves prediction accuracy and identifies the *Pareto Optimal* set of design points, as measured using the *average distance from reference set* (ADRS) [29]. The user can specify an HLS budget (the maximum number of HLS synthesis run) and report Pareto Optimal design points using up 20 training workloads for budgets of up to 50 runs, and as few as 10 workloads for larger budgets up to 120 runs; this represents a reduction of more than half of the 242 total directive combinations in the search space. This model requires HLS-in-the-loop due to the feature choice (post-HLS design specifications) and iterative improvement techniques (HLS is used to verify performance improvement).

Koeplinger et al. [30] present a DSE methodology that repeatedly calls a bespoke HLS tool with inherent predictive modeling capabilities. Their models are created from C/C++ descriptions which the programmer has annotated with pragmas to indicate design patterns such as map, reduce, filter and groupBy [31]. The framework analyzes the annotated source code and performs transformations such as loop fusion and tiling and converts the program to a more precise specification called *Delite Hardware Definition Language* (DHDL). The DHDL specification is converted to a set of parameterizable architectural templates, which account for the FPGA's on-chip resources (LUTs, BRAMs, routing, etc.) and off-chip memory bandwidth. Template parameters determine tiling sizes, parallelization factors, and coarse-grain pipelining, which are used alongside cycle-count and area utilization estimators to identify the Pareto Front of many design points. To construct the estimators, it's to necessary first characterize the board's various runtime latencies and resource availability using full synthesis runs, averaging 6 runs per template across several parameter configurations; this yields analytical models for utilization and cycle estimation for each template. The models are trained using ANNs with 200 design samples to compensate for the DHDL models, which do not capture on-board utilization. For six workloads consisting of millions of possible design points, they achieve average cycle estimation error of 6.1% and LUT, DSP, and BRAM utilization estimates with 4.8%, 7.5% and 12.3% error respectively, while running 6,533x faster than DSE using Vivado HLS alone. While expedient, this technique requires extensive source code instrumentation, and is only compatible with applications that use standard design patterns.

978-1-5386-7100-9/18 $31.00 © 2018 IEEE

TABLE III. FPGA HLS PREDICTIVE MODELING COMPARISON

Paper	Model	Features	Accuracy (avg.)	Speed	Benefit	Drawback
[27]	Random Forest regression	HLS design details	>99% ADRS from Pareto-optima.	2-4x fewer HLS runs required.	Avoids exhaustive enumeration of HLS design space	Requires repeated HLS calls to improve model accuracy, limiting speedup
[30]	Performance: Analytical Utilization: Hybrid Analytical + ANNs	Architectural template parameters that capture parallelism	Performance: ~94% Utilization: 88%-95%	279x - 6333x faster than Vivado HLS.	Faster DSE by replacing HLS with predictive models	Requires HLS to characterize target FPGA; requires non-traditional HLS flow
[32]	Analytical	Source-level instrumentation and HLS simulation	Performance: ~99% (compute-bound workloads); ~95% (memory-bound workloads)	~2 orders of magnitue faster than target FPGA bitstream generation	Highly accurate; avoids HLS and target FPGA bitstream generation	Requires HLS simulation and board characterization; Vivado simulation is a performance bottleneck.
[33]	Performance: Anaytical Utilization: Gradient Boosted Machine model	HLS directives and workload features from compiler instrumentation.	Performance: ~88% Utilization: 81-87%	2 - 3 orders of magnitude faster than target FPGA bitstream generation	Avoids HLS and target FPGA bitstream generation after model training and target FPGA device characterization	Requires custom microbenchmarks to characterize FPGA resource characteristics; higher instrumentation overhead than other approaches

HLScope+ [32] uses C code instrumentation and analytical modeling to improve the accuracy of the Vivado HLS simulator, adding the capability to handle input-dependent loop bounds. The instrumentation includes *dependency analysis*, which identifies independent regions of code to parallelize and *loop analysis*, which estimates loop cycle counts, culminating in execution cycle counts and DRAM transaction counts for each code module. This helps HLScope+ identify the application's critical execution path, from which it estimates the per-module stall rate. To compensate for the inaccuracy inherent to static analysis, loop analysis is extended with an analytical model.

HLScope+ accounts for DRAM bandwidth and latency by creating a high-level external memory module, which abstracts away individual memory accesses while accounting for resource contention among *processing elements* (PEs); this model is more accurate than the Vivado HLS simulator, which optimistically assumes that memory can be fetched each cycle and ignores memory bandwidth and contention among PEs. An analytical model is created to predict memory access time and can be combined with the compute cycle count to estimate execution time. HLScope+ is evaluated using 14 HLS workloads, resulting in an 1.1% average error for compute-bound workloads and 5.0% average error for memory bound workloads. This is much faster than fully synthesizing each workload; however, the key drawback is that HLScope+ employs analytical models which require expert knowledge of the target FPGA and its memory interface and are not transferrable from one target to another.

MPSeeker [33] performs DSE using HLS simulation in conjunction with C/C++ source code instrumented at the LLVM IR level, HLS design directives, and both predictive and analytical modeling techniques. MPSeeker's DSE considers several design directives including tile sizes, the number of PEs, loop unrolling, loop pipelining, and array partitioning. MPSeeker extends a single-PE analytical model [34] with the ability to estimate cycle counts for multi-PE designs that encompass coarse-grained parallelism. MPSeeker also includes an ensemble tree predictor called the *Gradient Boosted Machine*

(GBM) [35], which uses 14 program features and user-defined parallelism directives to predict FPGA resource usage.

MPSeeker's profiler accepts C source code (tiled nested loop structure), FPGA resource constraints, and user-supplied directive settings (tile size, loop unrolling, pipelining, and array partitioning) to produce features. The features track key workload characteristics correlated to parallelism and memory subsystem behavior and are input to Lin-Analyzer to predict performance and the GBM model to predict resource utilization. A subsequent analytical equation combines the outputs of the two models to account for further FPGA resource restrictions, which limit the number of PEs that can execute concurrently.

MPSeeker uses 10 microbenchmark kernels to ascertain the degree of loop resource consumption, communication interface behavior, and other similar properties, along with 5 larger benchmarks for evaluation; each kernel has 280 unique design configurations. MPSeeker's DSE procedure achieves 90% of the Pareto optimal performance, while running ~421x to ~4308x faster than FPGA bitstream generation. Lin-Analyzer reports an average error of 12.8%, while the GBM model reports average errors of 13.2%, 14.7%, 12.7%, and 19.4% for LUTs, flip-flops, DSP blocks, and BRAMs respectively. The reliance on expert-developed microbenchmarks to characterize FPGA resource constraints and the necessity to execute instrumented source code on the simulator are notable drawbacks of this approach.

Cross-architecture prediction (e.g., CPU host to FPGA target) could overcome many of these shortcomings. Doing so would allow for the creation of automatically derived statistical regression models that automatically select requisite features from direct execution on the host, eliminating the need for expert guidance and increasing portability. Given that cross-platform predictive models for both CPUs and GPUs has been successful, they can and should be applied to DSE for FPGAs as well.

V. CONCLUSION

Predictive modeling has immense potential but has not yet been fully integrated into the CPU or GPU architectural design

978-1-5386-7100-9/18 $31.00 © 2018 IEEE

process; on the other hand, predictive models are widely used in design exploration for FPGAs, but often rely on repeated calls to HLS tools and/or non-portable analytical modeling efforts. On the CPU/GPU side, the next step is to evaluate the potential of cross-platform modeling as an alternative to cycle-accurate simulation, especially during early design stages. In contrast, cross-platform models for FPGAs have not yet been demonstrated; however, once this is accomplished, DSE tools that target FPGAs will benefit significantly. Long-term, there is still a considerable amount of work to be done to convince the research community that predictive models are more than an intellectually meritorious curiosity and have the capability to significantly increase designer engineering productivity.

References

[1] Q. Guo, T. Chen, Y. Chen, and F. Franchetti, "Accelerating architectural simulation via statistical techniques: A Survey," *IEEE Trans. Comput. Des. Integr. Circuits Syst.*, vol. 35, no. 3, pp. 433–446, Mar. 2016.

[2] J. E. Miller *et al.*, "Graphite: A distributed parallel simulator for multicores," in *HPCA - 16 2010 The Sixteenth International Symposium on High-Performance Computer Architecture*, 2010, pp. 1–12.

[3] T. E. Carlson, W. Heirman, and L. Eeckhout, "Sniper: Exploring the level of abstraction for scalable and accurate parallel multi-core simulation," in *SC '11 Proceedings of 2011 International Conference for High Performance Computing, Networking, Storage and Analysis*, 2011, p. 1.

[4] D. Sanchez and C. Kozyrakis, "ZSim: fast and accurate microarchitectural simulation of thousand-core systems," in *Proceedings of the 40th Annual International Symposium on Computer Architecture - ISCA '13*, 2013, vol. 41, no. 3, pp. 475–486.

[5] M. Pellauer, M. Adler, M. Kinsy, A. Parashar, and J. Emer, "HAsim: FPGA-based high-detail multicore simulation using time-division multiplexing," in *2011 IEEE 17th International Symposium on High Performance Computer Architecture*, 2011, pp. 406–417.

[6] R. E. Wunderlich, T. F. Wenisch, B. Falsafi, and J. C. Hoe, "SMARTS: accelerating microarchitecture simulation via rigorous statistical sampling," in *30th Annual International Symposium on Computer Architecture, 2003. Proceedings.*, pp. 84–95.

[7] E. Perelman *et al.*, "Using SimPoint for accurate and efficient simulation," in *Proceedings of the 2003 ACM SIGMETRICS international conference on Measurement and modeling of computer systems - SIGMETRICS '03*, 2003, vol. 31, no. 1, p. 318.

[8] E. İpek, S. A. McKee, R. Caruana, B. R. de Supinski, and M. Schulz, "Efficiently exploring architectural design spaces via predictive modeling," *ACM SIGPLAN Not.*, vol. 41, no. 11, p. 195, Oct. 2006.

[9] B. C. Lee and D. M. Brooks, "Accurate and efficient regression modeling for microarchitectural performance and power prediction," in *Proceedings of the 12th international conference on Architectural support for programming languages and operating systems - ASPLOS-XII*, 2006, vol. 40, no. 5, p. 185.

[10] R. Nath, D. Carmean, and T. S. Rosing, "Power modeling and thermal management techniques for manycores," in *2013 IEEE Symposium on Computers and Communications (ISCC)*, 2013, pp. 000740–000746.

[11] X. Zheng, P. Ravikumar, L. K. John, and A. Gerstlauer, "Learning-based analytical cross-platform performance prediction," in *2015 International Conference on Embedded Computer Systems: Architectures, Modeling, and Simulation (SAMOS)*, 2015, pp. 52–59.

[12] X. Zheng, L. K. John, and A. Gerstlauer, "Accurate phase-level cross-platform power and performance estimation," in *Proceedings of the 53rd Annual Design Automation Conference on - DAC '16*, 2016, pp. 1–6.

[13] K. Skadron, M. R. Stan, K. Sankaranarayanan, W. Huang, S. Velusamy, and D. Tarjan, "Temperature-aware microarchitecture: modeling and implementation," *ACM Trans. Archit. Code Optim.*, vol. 1, no. 1, pp. 94–125, Mar. 2004.

[14] R. Tibshirani, "Regression shrinkage and selection via the Lasso," *J. R. Stat. Soc. Ser. B*, vol. 58, pp. 267–288, 1996.

[15] J. Shlens, "A Tutorial on principal component analysis," Apr. 2014.

[16] G. Wu, J. L. Greathouse, A. Lyashevsky, N. Jayasena, and D. Chiou, "GPGPU performance and power estimation using machine learning," in *Proceedings of HPCA 2015*, 2015, pp. 564–576.

[17] N. Ardalani, C. Lestourgeon, K. Sankaralingam, and X. Zhu, "Cross-architecture performance prediction (XAPP) using CPU code to predict GPU performance," *Proc. 48th Int. Symp. Microarchitecture - MICRO-48*, pp. 725–737, 2015.

[18] S. Wang, G. Zhong, and T. Mitra, "CGPredict: embedded GPU performance estimation from single-threadedaApplications," *ACM Trans. Embed. Comput. Syst.*, vol. 16, no. 5s, pp. 1–22, Sep. 2017.

[19] K. O'Neal, P. Brisk, A. Abousamra, Z. Waters, and E. Shriver, "GPU performance estimation using software rasterization and machine learning," *ACM Trans. Embed. Comput. Syst.*, vol. 16, no. 5s, pp. 1–21, Sep. 2017.

[20] K. O'Neal, P. Brisk, E. Shriver, and M. Kishinevsky, "HALWPE: hardware-assisted light weight performance estimation for GPUs," in *Proceedings of the 54th Annual Design Automation Conference 2017 on - DAC '17*, 2017, pp. 1–6.

[21] K. Hoste and L. Eeckhout, "Comparing benchmarks using key microarchitecture-independent characteristics," in *2006 IEEE International Symposium on Workload Characterization*, 2006, pp. 83–92.

[22] C.-K. Luk *et al.*, "Pin: building customized program analysis tools with dynamic instrumentation," in *Proceedings of the 2005 ACM SIGPLAN conference on Programming language design and implementation - PLDI '05*, 2005, vol. 40, no. 6, p. 190.

[23] L. Breiman, "Bagging predictors," *Mach. Learn.*, vol. 24, no. 2, pp. 123–140, Aug. 1996.

[24] "OpenSWR — Gallium 0.4 documentation." [Online]. Available: http://gallium.readthedocs.io/en/latest/drivers/openswr.html. [Accessed: 30-Apr-2018].

[25] Xilinx, "Vivado high-level synthesis." [Online]. Available: https://goo.gl/2kpNwy. [Accessed: 30-Mar-2018].

[26] A. Canis *et al.*, "LegUp : high-Level synthesis for FPGA-based processor / accelerator systems," *FPGA '11 Proc. 19th ACM/SIGDA Int. Symp. F. Program. gate arrays*, pp. 33–36, 2011.

[27] H.-Y. Liu and L. P. Carloni, "On learning-based methods for design-space exploration with High-Level Synthesis," in *2013 50th ACM/EDAC/IEEE Design Automation Conference (DAC)*, 2013, pp. 1–7.

[28] K. Yu, J. Bi, and V. Tresp, "Active learning via transductive experimental design," in *Proceedings of the 23rd international conference on Machine learning - ICML '06*, 2006, pp. 1081–1088.

[29] G. Palermo, C. Silvano, and V. Zaccaria, "ReSPIR: A response surface-based pareto iterative refinement for application-specific design space exploration," *IEEE Trans. Comput. Des. Integr. Circuits Syst.*, vol. 28, no. 12, pp. 1816–1829, Dec. 2009.

[30] D. Koeplinger *et al.*, "Automatic generation of efficient accelerators for reconfigurable hardware," in *ACM SIGARCH Computer Architecture News*, 2016, vol. 44, no. 3, pp. 115–127.

[31] A. K. Sujeeth *et al.*, "OptiML: An implicitly parallel domain specific language for machine learning," in proceedings of the 28th internaional conerence on machine learning., *ICML*, 2011.

[32] Y. Choi, P. Zhang, P. Li, and J. Cong, "HLscope+: fast and accurate performance estimation for FPGA HLS," in *Proceedings of the 36th International Conference on Computer-Aided Design*, 2017, pp. 691 698.

[33] G. Zhong, A. Prakash, S. Wang, Y. Liang, T. Mitra, and S. Niar, "Design space exploration of FPGA-based accelerators with multi-level parallelism," in *Proceedings of the 2017 Design, Automation and Test in Europe, DATE 2017*, 2017, pp. 1141–1146.

[34] G. Zhong, A. Prakash, Y. Liang, T. Mitra, and S. Niar, "Lin-analyzer: a high-level performance analysis tool for FPGA-based accelerators," in *Proceedings of the 53rd Annual Design Automation Conference on - DAC '16*, 2016, pp. 1–6.

[35] C. Click et al., "Gradient boosting machine (GBM) — H2O 3.18.0.8 documentation," 2016. [Online]. Available: https://goo.gl/viLQGw. [Accessed: 01-May-2018].

978-1-5386-7100-9/18 $31.00 © 2018 IEEE

Author Index

Adegbija, Tosiron 719
Aerabi, Ehsan 626
Agrawal, Saurabh 632
Ain, Antara 423
Alsafrjalaniz, Mohamad Hammam 719
A M, Vijaya Prakash 181
Andri, Renzo 509
Angizi, Shaahin 130, 533
Anto, Donel 608
Asadinia, Marjan 357
Azadmehr, Mehdi 458
Baba, Hiroyuki 569
Badhan, Riasad 697
Badreldin, Islam 719
Baharloo, Mohammad 226
Bahubalindruni, Pydi Ganga 464
Bai, Jinyu 539
Baksi, Anubhab 251
Balabanyan, Abraham 191
Balakrishnan, M. 632
Bandyopadhyay, Chandan 305
Basu Roy, Debapriya 381
Benini, Luca 509
Berend, David 363
Berg, Yngvar 458
Bhasin, Shivam 363, 620
Bhattacharjee, Anirban 305
Bhattacharjee, Debjyoti 369
Bhowmik, Pankaj 245
Bhunia, Swarup 148, 411
Bobda, Christophe 245, 357
Bojja, Srikanth 464
Bonizzoni, Edoardo 164
Borowczak, Mike 684
Breier, Jakub 620
Brisk, Philip 666, 763
Bu, Lake 118
Cai, Hao 263, 275

Cao, Yuan 399
Cao, Yuting 602
Cavigelli, Lukas 509
Chanana, Piyush 632
Chang, Hua-Yu 498
Chang, Yi-Shing 76
Chang, Zeqing 744
Chatterjee, Navonil 287
Chattopadhyay, Anupam 251, 311, 369, 620, 726
Chen, Qian 215
Chen, Song 545
Chen, Xiang 470
Chen, Xing 539
Chen, Xuanqi 76
Chen, Yiran 476
Cheng, Hsin-Pai 476
Cheng, Xianwei 703
Cheng, Yanjin 744
Cheng, Yuanqing 333
Chiang, Yu-Xiang 64
Cho, Haeyoon 94
Choi, Jinhang 22
Choudhury, Mihir 203
Chung, Yuan-Dar 64
Ciesielski, Maciej 203
Cooley, Rafer 684
Cronin, Patrick 112
Das, Debesh Kumar 293
Das, Subrata 293
Dasgupta, Pallab 423
Dash, Satyabrata 40
Deb, Sujay 100, 464, 591
Deb Nath, Atul Prasad 411
Dee, Timothy 327
De Micheli, Giovanni 203, 299
Dey, Sukanta 40
Dharmadhikari, Pranav 678
Dhoot, Krishna 269

Ding, Caiwen	28, 563	Harsha P, Sri	732
Dofe, Jaya	106	Hashimoto, Masanori	429, 522
Doi, Ryutaro	429, 522	He, Linjun	6
Dong, Qide	470	He, Zhezhi	130, 533
Dong, Sheqin	215	Hemmady, Sameer	597
Draper, Jeffrey	28	Herath, Kalindu	209
Drechsler, Rolf	305, 351, 375, 557	Ho, Kung-Han	257
Duan, Moming	136	Ho, Tsung-Yi	650, 660
Duong, Luan H.K.	76	Homayoun, Houman	405
Dworak, Jennifer	124	Hu, Fanglei	16
Elmohr, Mahmoud A.	311	Hu, Jingtiong	238
Fan, Deliang	130, 533	Hu, Yiming	435
Fang, Juan	703, 744	Huang, Chau-Chin	203
Fang, Shang-Rong	64	Huang, Juinn-Dar	257, 655
Fang, Shao-Yun	500	Huang, Pengfei Huang	575
Fazeli, Mahdi	52, 626	Huang, Xiwei	666
Fu, Lingxuan	666	Hübner, Michael	220
Fujita, Masahiro	46	Hussain, Mubashir	345
Gade, Sri Harsha	100, 591	Hwang, Leslie	58
Gaj, Kris	405	Inoue, Masahiro	569
Geng, Hao	488	Ishibashi, Koichiro	187
George, Sumitha	447	Jalali, Zeinab S.	563
Ghosal, Prasun	287	Jao, Nicholas	447
Giri, Naresh Kumar	94	Ji, Weiqing	660
Gordon-Ross, Ann	719	Jiang, Dongxu	545
Grimm, Tomás	220	Jiang, Iris Hui-Ru	498, 714
Große, Daniel	351, 557	Jiao, Hailong	16
Grover, William H.	666	Jindal, Ankit	46
Gu, Chongyan	399	Jindal, Nitish	46
Gumroyan, Hrachya	191	Jungk, Bernhard	363
Gundabolu, SreeCharan	644	JYV, Manoj Kumar	738
Guo, Hui	345	K, Sudeendra Kumar	738
Guo, Jie	142	Kale, Ajinkya	316
Guo, Kaiyuan	435	Kan, Senwen	124
Guo, Yunxi	327	Kanewala, Udaree	209
Gupta, Hari Shanker	175	Kang, Wang	263, 275, 539
Hanif, Muhammad Abdullah	581	Kansakar, Prasanna	690
Harish, B. P.	170	Karimi, Naghmeh	614

Kashi, Somayeh	52	Lilja, David J.	154
Keszocze, Oliver	557	Lin, Jun	503, 516
Khachikyan, Karen	191	Lin, Mingjie	638
Khairallah, Mustafa	311	Lin, Rung-Bin	64
Khaleghi, Soroush	321	Lin, Tien-Kuo	655
Khalid, Faiq	581	Ling, Yingjian	136
Khonsari, Ahmad	226	Liu, Chenchen	470
Kinsy, Michel	118	Liu, Dijun	333
Kinsy, Michel A.	52, 626	Liu, Duo	136
Kobayashi, Nobuaki	387	Liu, Qi	528
Kong, Joonho	94	Liu, Renping	136
Kravets, Victor N.	203	Liu, Tao	142, 528
Kriebel, Florian	581	Liu, Weichen	136
Kumar, Binod	46	Liu, Weiqiang	399, 575
Kumar, Sachin	311	Liu, Xiaobang	34
Kumar, S. Dinesh	697	Liu, Xinning	263
Kumar K, Sudeendra	672	Liu, Yongpan	70, 112, 238, 339
Kumar Y B, Nithin	608	Liu, Zihao	142, 528
Kwon, Beomjin	58	Lombardi, Fabrizio	575
Labrado, Carson	697	M, Bhaskar	1
Lai, Longqiang	750	Ma, Ruizhe	575
Lee, Pei-Yu	714	Mahapatra, Abhishek	672
Lee, Zhao Chuan	417	Mahapatra, Kamalakanta	672, 738
Lei, Yung-Chun	655	Mahzoon, Alireza	351
Lettnin, Djones	220	Malekpour, Amin	345
Li, Bing	715	Mambakam, Akshay	423
Li, Bingzhe	154	Mandal, Swagata	251
Li, Hai	476	Marchetti, Luca	458
Li, Han	6, 160	Mardani Kamali, Hadi	405
Li, Ji	28	Mark, Miguel	118, 626
Li, Sai	539	Mei, Faqiang	399
Li, Sicheng	476	M. H., Vasantha	164
Li, Wenshuo	482	M.H., Vasantha	608
Li, Xingquan	488	Mitra, Tulika	756
Li, Xueqing	750	Miyamoto, Shin	387
Li, Yongfu	417	M K, Jayaram Reddy	453
Li, Zhe	28	Mohanty, Saraju P.	703
Liang, Liang	136	Mohanty, Saraju P	709, 732

Mohapatra, Nihar Ranjan	175
Mohapatra, Satyajit	175
Mondal, Hemanta Kumar	100
Mukhopadhyay, Debdeep	381
Munir, Arslan	94, 690
Muralikrishnan Adhepalli, Suman	632
Mutyam, Madhu	88
M. V., Harsha	170
Nakazawa, Yuki	522
Nam, Gi-Joon	203
Namdari, Iman	226
Nan, Yang	6
Nandi, Sukumar	40
Narayanan, Vijaykrishnan	22, 447
Naskar, Mrinal Kanti	732
Naviner, Lirida	263
Neema, Shubha	732
Nevoral, Jan	82
Ning, Xuefei	435, 482
N S, Lakshmi	1
Odedara, Bhavin	464
Ogura, Hiroki	494
O'Neal, Kenneth	763
Ong, Jonathan Yoong Seang	417
Ouyang, Peng	275
Palombo, Hernan	602
Pan, Biao	539
Pantho, Md Jubaer Hossain	245
Parameswaran, Sri	345
Park, Jungmin	148
Pasupureddi, Vijaya Sankara Rao	316
Patel, Chintan	614
Patooghy, Ahmad	52, 626
Patterson, Clay	238
Paul, Suraj	287
Peng, Kai-Chun	64
Perez, Valerio	417
Polineni, Sreenivasulu	453

Porwal, Ankita	281
Pouraghily, Arman	551
Powell, Troy	597
Prakash, Alok	209, 756
Provelengios, George	551
Pudi, Vikramkumar	251, 726
Qian, Weikang	154
Qiu, Keni	70, 339
Qiu, Qinru	28
R., Sunil	164
Rahaman, Hafizur	305
Rahmati, Dara	226
Rahmatiy, Dara	52
Raju, Akhilesh	678
Rakin, Adnan Siraj	130
Ramanathan, Akshay Krishna	447
Rana, Vikas	10
Rao, Lei	232
Rao, P.V.M.	632
Rao, Wenjing	321
Ravi, Prasanna	620
Ray, Sandip	411, 602
Rehani, Ankit	464
Rehman, Semeen	581
Ren, Ao	28
Ren, Jinting	136
Rezaei, Hamidreza	626
Riener, Heinz	299
R. K., Siddharth	164
Robucci, Ryan	614
Rose, Garrett S.	441
Rossi, Davide	509
Rout, Sidhartha Sankar	591
Ruzicka, Richard	82
S, Jalaja	181
Sahoo, Rasmita	269
Sahoo, Sauvagya	672
Sahoo, Sauvagya Ranjan	738

Sahoo, Subhendu Kumar	269	Sung, Yu-Min	494
Sahu, Chitrakant	281	Swain, Ayas Kanta	672, 738
Sampson, Jack	22	Tai, Cheng-Wei	64
Sampson, John	447	Tajima, Kaori	569
Santambrogio, Marco	587	Takahashi, Atsushi	494
Sapra, Pulkit	632	Takahashi, Shiho	187
Saranam, S R Swamy	88	Tanabe, Yasuki	22
Sarkar, Pallabi	732	Tang, Changcheng	197
Sasan, Avesta	405	Taralkar, Abhijeet D.	608
Sato, Shimpei	494	Tehranipoor, Mark	148
Sato, Toshinori	569	Tessier, Russell	551
Sayyaparaju, Sagarvarma	441	Thapliyal, Himanshu	697
Schlichtmann, Ulf	715	Todri-Sanial, Aida	275
Schmidt, Jurij	557	Tonse, Laxminidhi	453
Schmitz, Kenneth	557	Torres, Lionel	275
Sciuto, Donatella	587	Tripathi, Vikas	417
Sengupta, Anirban	709, 732	Trivedi, Gaurav	40
Seth, Saurabh	672	Tseng, I-Lun	417
Shafique, Muhammad	581	Tyagi, Akhilesh	327
Shey, James	614	Ukezono, Tomoaki	569
Shi, Xin	70, 339	Vemuri, Ranga	34, 678
Shiri, Pouya	226	Wang, Chenghong	563
Shirinzadeh, Saeideh	375	Wang, Chenghua	399, 575
Shrivastwa, Ritu Ranjan	726	Wang, Chun-Yao	257
Simek, Vaclav	82	Wang, Guanhua	6
Singh, Vijay	697	Wang, Junchao	666
Singh, Virendra	46	Wang, Nan	545
Sinha, Mitali	591	Wang, Pengjun	482
Song, Chang	476	Wang, Siqi	756
Soundarajan, Sucheta	563	Wang, Ting-Chi	494
Srikanthan, Thambipillai	209	Wang, Xiaofang	644
Srinivasa, Srivatsa	22, 447	Wang, Yan	197
Stornaiuolo, Luca	587	Wang, Yanzhi	28, 563
Stradolini, Francesca	299	Wang, Yizhi	503
Sturm, Johannes	316	Wang, You	263
Sule, Nishchay H.	597	Wang, Yu	435, 482
Sullivan, Andrew	203	Wang, Zhaohao	333
Sun, Fangxuan	516	Wang, Zhehui	76

Wang, Zhifei	76
Wang, Zhongfeng	503, 516
Wei, Qi	482
Weiss, Ryan	441
Wen, Wujie	142, 528
Wille, Robert	305
Wolf, Shaya	684
Wolf, Tilman	551
Wong, Martin	58
Wu, Bi	333
Wu, Chia-Cheng	257
Wu, Chunpeng	476
Wu, Kuo-Hao	500
Wu, Nansong	142
Wu, Qing	476
Wu, Tongda	339
Wu, Yunsong	136
Xu, Jiang	76
Xu, Yuanchao	70
Xue, Chun Jason	238
Yang, Chengmo	112, 238
Yang, Haoyu	488
Yang, Huanrui	476
Yang, Huazhong	339, 435, 482
Yang, Jin-Kai	64
Yang, Jun	263
Yang, Kai	148
Yang, Meng	154
Yang, Peng	76
Yang, Tongxin	569
Yao, Hailong	660
Yao, Manting	545
Y. B., Nithin Kumar	164
Ye, Zuochang	197
Yu, Bei	488
Yu, Cunxi	203, 299
Yu, Fuxun	470
Yu, Jaehoon	522
Yu, Jincheng	435, 482
Yu, Liyang	666
Yu, Qiaoyan	106, 393
Yuan, Bo	28, 154
Zamiri Azar, Kimia	405
Zarkesh-Ha, Payman	597
Zeng, Xuan	488
Zhan, Chenchang	6, 160
Zhang, Bin	232
Zhang, Grace Li	715
Zhang, Lei	399
Zhang, Liuyang	275
Zhang, Min	16
Zhang, Ning	160
Zhang, Xiaolong	333
Zhang, Xinyi	238
Zhang, Youguang	275, 333, 539
Zhang, Zhiming	106, 393
Zhao, Hui	703, 744
Zhao, Jizhong	232
Zhao, Weisheng	263, 275, 333, 539
Zheng, Hao	602
Zhong, Kan	136
Zhou, Dongqin	70
Zhu, Yu	545
Zou, Yu	638

IEEE Computer Society
Technical & Conference
Activities Board

T&C Board Vice President
Hausi Müller
University of Victoria, Canada

IEEE Computer Society Staff
Evan Butterfield, *Director of Products and Services*
Patrick Kellenberger, *Manager, Conference Publishing Services*

IEEE Computer Society Publications
The world-renowned IEEE Computer Society publishes, promotes, and distributes a wide variety of authoritative computer science and engineering texts. These books are available from most retail outlets. Visit the CS Store at *http://www.computer.org/portal/site/store/index.jsp* for a list of products.

IEEE Computer Society *Conference Publishing Services* (CPS)
The IEEE Computer Society produces conference publications for more than 300 acclaimed international conferences each year in a variety of formats, including books, CD-ROMs, USB Drives, and on-line publications. For information about the IEEE Computer Society's *Conference Publishing Services* (CPS), please e-mail: cps@computer.org or telephone +1-714-821-8380. Fax +1-714-761-1784. Additional information about *Conference Publishing Services* (CPS) can be accessed from our web site at: *http://www.computer.org/cps*

Revised: 18 January 2012

CPS Online is our innovative online collaborative conference publishing system designed to speed the delivery of price quotations and provide conferences with real-time access to all of a project's publication materials during production, including the final papers. The ***CPS Online*** workspace gives a conference the opportunity to upload files through any Web browser, check status and scheduling on their project, make changes to the Table of Contents and Front Matter, approve editorial changes and proofs, and communicate with their CPS editor through discussion forums, chat tools, commenting tools and e-mail.

The following is the URL link to the ***CPS Online*** Publishing Inquiry Form:
http://www.computer.org/portal/web/cscps/quote

IEEE
445 Hoes Lane
Piscataway, NJ 08854-4141

ISBN 978-1-5386-7100-9